ENCYCLOPEDIA OF ANALYTICAL CHEMISTRY

ENCYCLOPEDIA OF ANALYTICAL CHEMISTRY

Applications, Theory and Instrumentation

Edited by R. A. Meyers

RAMTECH Ltd, Tarzana, CA, USA

JOHN WILEY & SONS, LTD
Chichester · New York · Weinheim · Brisbane · Singapore · Toronto

Baffins Lane, Chichester
West Sussex PO19 1UD, UK

National: 01243 779777
International: (+44) 1243 779777
e-mail (for orders and customer service enquiries): cs-books@wiley.co.uk
Visit our Home Page on http://www.wiley.co.uk
or http://www.wiley.com

Other Wiley Editorial Offices

John Wiley & Sons Inc., 605 Third Avenue,
New York, NY 10158-0012, USA

Wiley-VCH Verlag GmbH, Pappelallee 3,
D-69469 Weinheim, Germany

Jacaranda Wiley Ltd, 33 Park Road, Milton,
Queensland 4064, Australia

John Wiley & Sons (Asia) Pte Ltd, 2 Clementi Loop #02-01,
Jin Xing Distripark, Singapore 129809

John Wiley & Sons (Canada) Ltd, 22 Worcester Road,
Rexdale, Ontario M9W 1L1, Canada

Library of Congress Cataloguing-in-Publication Data

Encyclopedia of analytical chemistry: applications, theory and instrumentation/
editor-in-chief, Robert A. Meyers
p. cm.
Includes bibliographical references and indexes.
ISBN 0-471-97670-9 (set : alk. Paper)
1. Chemistry, Analytic–Encyclopedias. I. Meyers, Robert A. (Robert Allen), 1936–

QD71.5. E52 2000-06-15
543′.003 – dc21 00-042282

British Library Cataloguing in Publication Data

A catalogue record for this book is available from the British Library.

ISBN: 0 471 97670 9

Typeset in 10/12 pt NewTimes by Laser Words, Madras, India
Printed and bound in the UK by Cambridge University Press
Cover design by Jim Wilkie, J & R Press, Hants, UK.

This book is printed on acid-free paper responsibly manufactured from sustainable forestation, for which at least two trees are planted for each one used for paper production.

EDITORIAL BOARD

Preface

By the Editor-in-Chief

The members of our team of over 800 authors, more than 600 peer reviewers, 42 Section Editors and 13 distinguished Board Members and Advisors have been working on the publication of the *Encyclopedia of Analytical Chemistry* for more than three years. Our objective was to prepare the largest, most comprehensive compendium of analytical chemistry in existence. Indeed, this work has twice the depth of any other such compendium, such that a chemist (organic, polymer, inorganic, biochemist, molecular biologist as well as, of course, an analytical chemist), or a physicist or engineer (environmental, industrial or materials) can find all the essential information required to analyze any analyte in any matrix for any purpose or application, interpret the results and also to gain a thorough knowledge of the theory and instrumentation utilized.

We define analytical chemistry as the measurement, characterization and mapping of chemical species or systems varying from (i) the components of life, e.g. proteins, carbohydrates, nucleic acids, clinical samples, biomedical spectroscopy, forensics, (ii) threats and safeguards to life as in chemical weapons agents, pesticides, environment, industrial hygiene and forensics, (iii) life critical or enhancing analyses, e.g. food and pharmaceuticals, and (iv) analyses required by industry, e.g. coatings, particle size, polymers, rubbers, metals, pulp and paper, process, petroleum and surfaces. The techniques utilized span the *in-situ* analysis of soil, water, waste, air and the human body, to laboratory analyses, remote sensing and stellar spectroscopy.

We chose to organize our Encyclopedia mainly according to the recent-literature summation sections comprising the annual *Applications* literature review and *Fundamentals* review of the most cited journal of chemical analysis, *Analytical Chemistry*. In fact, almost half of our Section Editors have served as corresponding Editors for *Analytical Chemistry*. We added some key sections not covered in the review issues: these are Biomedical Spectroscopy, Biomolecules Analysis, Chemical Warfare Chemicals Analysis, Nucleic Acids Analysis and Remote Sensing. Thus, the articles comprising our Encyclopedia are organized into twenty four *Applications* sections presented in alphabetical order from Biomedical Spectroscopy through to Surfaces – these make up the first ten volumes. Then there are sixteen *Theory and Instrumentation* sections presented in the next five volumes, beginning with Atomic Spectroscopy and concluding with X-ray Spectrometry. The final section contains cross-cutting articles (e.g. **Analytical Problem Solving: Selection of Analytical Methods**, **Quality Assurance in Analytical Chemistry**, **Literature Searching Methodology**) and this together with the Appendices and Indexes completes the fifteenth volume.

The Encyclopedia has been designed to assist readers in finding the appropriate information as simply as possible. For example, the user knows whether the analyte in question is considered a protein or a nucleic acid, or an environmental contaminant or food or petroleum or a pharmaceutical etc., and goes directly to the alphabetized section of that name in one of the clearly identified *Applications* volumes. One or more, generally instrumental, methods are given in sufficient detail to perform the needed analysis. If the reader desires theory, design or operational information on a chosen instrumental method (e.g. NMR, IR, MS, AA), the reader goes to the appropriate alphabetized section in one of the five clearly marked *Theory and Instrumentation* volumes to select the appropriate articles. Optionally, the reader can rely on the detailed subject index for the whole Encyclopedia. In addition, the *Theory and Instrumentation* articles include key example spectra which are cross-referenced within the *Applications* articles for easy access.

Our team methodology was as follows: the Section Editors prepared the lists of article titles and nominated authors. Appropriate Board Members then reviewed these lists, e.g. Professors Bard, Zewail, Siegbahn and Haddad overseeing theory and instrumentation; Professors Huber, Ganem and Hester overseeing biological applications analyses; Drs Slavin and Sparkman reviewing various applications especially environmental, and Drs Demers and Turner reviewing the clinical analysis areas. Our Advisors, Drs Roth (Library Science) and Maddalone (Environment) supported this effort. Professors Bard, Huber and Siegbahn and Dr Roth also contributed articles. Each article has been peer reviewed by scientists nominated by the Section Editors and Board. The Editor-in-Chief coordinated all of the above.

Robert A. Meyers

October 2000

CONTENTS

VOLUME 1

VOLUME 2

Clinical Chemistry *1085*

Coatings *1725*

VOLUME 3

Environment: Trace Gas Monitoring *1885*

Environment: Water and Waste *2247*

VOLUME 4

Environment: Water and Waste (cont'd) *2803*

Forensic Science *4333*

Industrial Hygiene *4577*

VOLUME 6

Industrial Hygiene (cont'd) *4695*

Nucleic Acids Structure and Mapping *4845*

Particle Size Analysis *5299*

VOLUME 7

Peptides and Proteins *5611*

Pesticides *6109*

VOLUME 8

Pesticides (cont'd) *6563*

Petroleum and Liquid Fossil Fuels Analysis *6607*

Pharmaceuticals and Drugs *6987*

VOLUME 10

Pulp and Paper (cont'd) *8441*

Remote Sensing *8499*

Steel and Related Materials *8849*

Surfaces *9029*

VOLUME 11

Atomic Spectroscopy *9355*

Chemometrics *9669*

Electroanalytical Methods *9839*

VOLUME 12

Electronic Absorption and Luminescence *10257*

Gas Chromatography *10627*

Nuclear Magnetic Resonance and Electron Spin Resonance Spectroscopy *11985*

VOLUME 14

Nuclear Magnetic Resonance and Electron Spin Resonance Spectroscopy (cont'd) *12087*

Nuclear Methods *12421*

Radiochemical Methods *12841*

VOLUME 15

Raman Spectroscopy *13017*

Thermal Analysis *13143*

X-ray Photoelectron Spectroscopy and Auger Electron Spectroscopy *13227*

X-ray Spectrometry *13267*

General Articles *13445*

Appendices *13671*

Lists and Index *13685*

Atomic Spectroscopy

SECTION CONTENTS

G.M. Hieftje
Indiana University,
Bloomington,
USA

Atomic Spectroscopy: Introduction

Gary M. Hieftje
Indiana University, Bloomington, USA

Atomic spectroscopy (also termed atomic spectrometry) is one of the oldest and most well established of the analytical methods. Its roots can be traced to very early observations in which the presence of specific salts in a chemical sample imparted characteristic colors to a luminous flame. Atomic spectrometry is perhaps the most prominent and widely used of the family of methods employed for elemental analysis.

All methods for elemental analysis, including atomic spectrometry, exploit quantized transitions characteristic of each individual element. For example, X-ray fluorescence spectrometry utilizes inner-shell electron transitions. Because inner-shell electrons are not involved in bonding, the transitions they undergo are perturbed very little by the presence of nearby atoms, by chemical bonds, or by the external environment of an atom. As a consequence, X-ray fluorescence spectra consist of narrow spectral lines that are indicative of each element. Moreover, the sample can be in virtually any state – solid, liquid, or gas. However, X-ray fluorescence spectrometry has its shortcomings. Detection limits are seldom below the level of one part per million in either a solid or liquid, it is difficult to measure elements of low atomic number, and X-radiation emitted by one element can be absorbed by certain others present in the same sample. A consequence of this last factor is rather severe matrix (interelement) interferences. That is, the signal that is observed from one element depends not only on its concentration, but also on the concentration of other elements in the sample.

Neutron activation analysis (NAA) is another approach for elemental analysis that relies upon discrete transitions typical of each element. In this case, however, the transitions are in the atomic nucleus. Thus, they too are unaffected by chemical bonding or by the presence of other nearby atoms. Unfortunately, most samples do not lend themselves immediately to analysis by means of nuclear transitions, since they do not contain appreciable amounts of naturally radioactive nuclei. As a consequence, the samples must be "activated" by placing them in an intense flux of slow neutrons. The resulting activated nuclei can then be employed for chemical analysis. However, the requirement of a slow neutron source limits the widespread applicability of the method.

Unlike these alternative methods, atomic spectrometry utilizes quantized transitions in the conveniently accessible ultraviolet, visible, and near-infrared regions of the electromagnetic spectrum. As a result, the spectroscopic instrumentation is relatively inexpensive, readily available, and easy to use. Of course, it is valence electronic transitions that occur in these spectral regions. For narrow-band, characteristic spectrum to be generated, each atom must be isolated from all others, so the transitions are not perturbed by neighboring atoms or by bonding effects. If this requirement is not met, the resulting spectra are representative more of molecules or molecular fragments than of atoms themselves.

Accordingly, the underlying requirement for all atomic methods of analysis is that a sample be decomposed to the greatest extent possible into its constituent atoms. Ideally, this atomization step should be quantitative; there should be no residual bonding in the gas-phase atomic cloud. Anything less than complete atomization will understandably yield lower sensitivities in any atomic method. Even more importantly, changes in the fraction of atomization from sample to sample or from sample to standard will cause errors in calibration. Thus, it is less important that complete atomization be achieved than that the fractional atomization be extremely consistent. To the extent that this condition is not met, interelement (matrix) interferences in atomic spectrometry can be extremely troublesome.

It is not surprising, then, that a consistent theme in the history of atomic spectrometric analysis is a search for improved methods of atomization for samples in solid, solution, or gaseous form. It is therefore appropriate that we consider such systems in some detail here.

1 SYSTEMS FOR ATOM FORMATION IN ATOMIC SPECTROSCOPY

Schemes for the atomization of various samples have followed several traditional paths. The most common first step is to dissolve the sample if it is not already in solution form. Although this step is an inconvenient and often time-consuming one, it also offers a number of important benefits. First, after a sample is dissolved, the principal constituent in the sample solution is the solvent. Consequently, most sample solutions look more or

For references see page 9361

less the same and similarly resemble standard solutions that are prepared. Secondly, samples in solution form are relatively easy to handle and lend themselves readily to automation. Third, sample solutions permit relatively simple and straightforward background correction, simply by use of a solvent or reagent blank. Lastly, other constituents can readily be added to sample and standard solutions to simplify such procedures as standard additions (spiking) and internal standardization.

Among the methods widely employed for automation of the solution-introduction step in atomic spectrometry is flow injection analysis (FIA). Unlike many other automation schemes, FIA can be employed not only to introduce solution samples directly, but to dilute them in an automatic and predictable fashion, to mix the sample solution with other reagents, to exploit standard-additions and internal-standardization approaches, and to speed up the sample introduction process. Through use of FIA, it becomes possible routinely to employ sample solution volumes as low as 10 µl, to reduce build-up on sprayers and other equipment in an atomic spectrometric system, and to improve precision. Some of the benefits and capabilities of FIA are described in **Flow Injection Analysis Techniques in Atomic Spectroscopy**.

Once a sample solution is prepared, it can be converted to free atoms in several alternative ways. One is to pipette a small aliquot of the sample solution into an electrothermal atomizer (ETA) such as a graphite furnace. This method is discussed in greater detail in **Graphite Furnace Atomic Absorption Spectrometry**. In the graphite furnace, the solvent is first evaporated from the sample at a moderate temperature. The furnace temperature is then raised so organic material is ashed, and the temperature next increased rapidly to the point where the sample is vaporized and ultimately atomized. Not surprisingly, the furnace temperature and the sample composition are extremely important in ensuring the efficiency and consistency of this atomization.

A second path to the atomization of solution samples is through a nebulization (spraying) process. In simple terms, nebulization serves to increase the surface area of the solution sample, so solvent evaporation (desolvation) can proceed more rapidly and so the resulted dried solute particles can be volatilized better. This scheme, which is probably the most common in analytical atomic spectrometry, is employed in flame atomic absorption, flame emission, and plasma emission spectrometry and is detailed in **Flame and Vapor Generation Atomic Absorption Spectrometry** and **Inductively Coupled Plasma/Optical Emission Spectrometry**. Once formed, droplets in the nebulized spray are sent into a high-temperature environment such as a chemical flame or flowing rare-gas plasma. There, desolvation and solute-particle vaporization take place, and the resulting vapor converted more or less efficiently into free atoms. Indeed, the environment in these discharges is often hot enough that many of the atoms that are formed wind up as positively charged ions. Also, the environment in these atomization sources is often energetic enough to yield strong emission from either the freed atoms or their ionic counterparts. Correspondingly, this general scheme lends itself well to a number of different detection approaches that will be discussed shortly: atomic emission spectrometry (AES), atomic absorption spectrometry (AAS), atomic fluorescence spectrometry (AFS), and atomic mass spectrometry (AMS).

For some elements, it is possible to employ a more straightforward means of generating free atoms. In the case of mercury, for example, free atoms can be formed simply by the chemical reduction of inorganic mercury in solution to the free atomic form. The neutral mercury atoms can then be driven from solution merely by passing an appropriate carrier gas (for example, air or argon) through it. The liberated atoms, present in a relatively cool environment, can then be measured alternatively by such techniques as atomic absorption or AFS.

For other elements, a chemical reaction will yield not free atoms but rather other volatile species that can be dissociated into free atoms at moderate temperatures. The most common example here is the use of a chemical reduction to form a stable hydride of such elements as tin, antimony, and arsenic. As in the case of mercury atoms, these volatile hydrides can be driven from solution by bubbles of an appropriate gas, which will carry them into a moderate-temperature flame or furnace for atomization. Once formed, those atoms can then be measured by atomic absorption, atomic fluorescence, or certain other spectrometric methods. These schemes are discussed in greater detail in **Flame and Vapor Generation Atomic Absorption Spectrometry**.

Naturally, it would be desirable in many cases to be able to analyze solid samples directly. Even more beneficial in special cases would be the possibility of measuring sample concentrations in a solid on a three-dimensional spatial basis. Some methods employed for sample volatilization and atomization have been aimed at exactly that goal.

Of such methods, the best established employs a glow discharge, detailed in **Glow Discharge Optical Spectroscopy and Mass Spectrometry**. In a glow discharge, ordinarily operated at pressures in the range of 1 Torr, the sample surface is bombarded by energetic rare-gas ions, usually of argon. Because of this steady bombardment, the surface is eroded on a layer-by-layer basis. Thus, atoms freed from the surface as a function of time are taken from successively deeper layers within the sample. Recording the spectrometric signal as a function of time then permits a "depth profile" of the sample to be obtained. In early work, glow-discharge devices were applicable only to conductive solid samples because of the need to impart a negative charge to the sample surface, in order to attract argon ions to it. More recent developments, however, have shown it possible to utilize radiofrequency-sustained discharges for the same purpose; such discharges permit the depth-resolved analysis of nonconductive samples as well.

In recent years, intense laser beams also have been exploited for depth resolution. Unlike the glow discharge, however, a laser beam can be focused to discrete, extremely small spots on the sample surface, so that not only depth resolved but also laterally resolved information can be obtained. In its most straightforward version, the laser is employed not as an atomization source but rather as a sampling device, as is described in greater detail in **Laser Ablation in Atomic Spectroscopy**. In recent years, the importance of the laser's wavelength and power density on the ablation process has been characterized more fully and has led to a burgeoning array of applications. Significantly, sample material ablated by a laser can be fed into any of several sources for further atomization. The most attractive combinations have been between laser ablation and either inductively coupled plasma (ICP) AES or ICP mass spectrometry.

A more direct approach to solid-sample analysis, described in **Laser-induced Breakdown Spectroscopy**, is to exploit the laser beam not only for sample vaporization but also to generate a plasma in which the sample vapor can be atomized, excited, and ionized. This technique, termed laser-induced breakdown spectroscopy (LIBS), is applicable to samples of many types and can also be applied to surfaces that are relatively remote from both the laser and the spectrometric measurement equipment.

Some sources employed in atomic spectrometry are not as good at generating atoms as they are at exciting or ionizing them. The microwave-induced plasma (MIP), discussed in **Microwave-induced Plasma Systems in Atomic Spectroscopy**, typifies such sources. Interestingly, the MIP (like some other spectroscopic sources) is well removed from local thermodynamic equilibrium. That is, the source appears to have several temperatures that are linked not by equilibrium processes but rather by kinetically controlled events. Thus, the temperature that would define the energy distribution of electrons moving in a MIP is generally much higher than the temperature that would pertain to the motion of heavier particles such as rare-gas atoms or ions. An important consequence of this dis-equilibrium is that features driven by the electrons (most importantly from our standpoint, ionization and excitation of atoms) are extremely efficient whereas other events that are controlled by gas-kinetic (thermal) operations are less efficient. Because solvent evaporation and solute-particle vaporization occur mostly through these latter thermal events, they are not efficient in a MIP.

The result of this situation is that the MIP is exceptionally useful for exciting and ionizing atoms, but that it cannot as readily effect vaporization. In fact, many MIP systems tolerate only a small loading of solvent vapor before they are visibly perturbed. The MIP therefore finds an important role in atomic spectrometry in the dissociation of gas-phase samples and in the production of atomic emission and mass spectra. It is therefore particularly attractive as a source for detection of gas chromatographic effluents, of gas-phase species generated by a chemical-based hydride-formation apparatus, or for atoms volatilized from an auxiliary source such as a carbon furnace.

2 DETECTION METHODS IN ATOMIC SPECTROSCOPY

Once atoms are in the gas phase, they can be probed by any of several spectrometric techniques, including AES, AAS, AFS, AMS, coherent forward scattering spectrometry, photothermal deflection spectrometry, atomic magneto-optic rotation spectrometry, and several others. Of these alternatives, the most common have become atomic absorption, atomic emission, and AMS.

AAS dates from the earliest observations by Fraunhofer of dark lines in the sun's spectrum and, as an analytical method, from the pioneering work by Walsh and by Alkemade and Milatz in 1955.[1,2] It has become a workhorse in atomic spectrometric analysis, in large part because of the simplicity of the instrumentation that it requires. To perform atomic absorption requires only a primary light source, usually a hollow cathode lamp, an atomization cell such as any of those described in the preceding paragraphs, a spectral isolation device such as a monochromator, an appropriate photodetector, and associated electronics. Working curves are usually constructed in the form of a Beer's-law plot familiar from molecular spectrophotometry. However, for Beer's law to be followed closely (that is, for linear calibration curves to be obtained), the band of detected radiation must be narrower than the atomic absorption line that is being measured. Because atomic absorption lines are extraordinarily narrow, the primary light source (hollow cathode lamp) must emit a band of light that is at least as narrow. It is for this reason that the hollow cathode lamp was chosen. Significantly, this choice also simplifies the measurement, since spectral lines emitted by the hollow cathode lamp are typical of the element being measured and are naturally locked onto the narrow atomic absorption lines of interest. Thus, spectral mismatch is almost impossible and spectral interferences are relatively uncommon because of the narrowness of the spectral lines that are being probed.

For references see page 9361

Unfortunately, it is not particularly convenient to measure more than one element at a time by means of AAS. There is a necessary straight-line geometry among the primary light source (hollow cathode lamp), the atom cell (flame or furnace) and the detection equipment. This alignment makes it difficult to incorporate more than a single source into the system; because each hollow cathode lamp emits efficiently the spectrum of only one, two, or three elements at a time, measuring additional elements requires substituting a new hollow cathode lamp. Although recent advances have been made in continuum-source atomic absorption, even those arrangements are somewhat limited: continuum sources that extend well into the important ultraviolet region of the spectrum are not widely available. Moreover, when furnace atomization is employed, it is often necessary to employ a different temperature program for each element.

One of the problems encountered in AAS is the generation of a broad-band spectral background. This background arises from absorption caused by residual molecules or molecular fragments in the atom source and by scattering from smoke or other airborne particulate matter. Because a hollow-cathode lamp is utilized as a light source, this molecular absorption or scattering cannot be distinguished from atomic absorption, since it simply attenuates the beam. If left uncorrected, this "nonspecific absorption" then makes element concentrations appear to be higher than they really are.

Not surprisingly, nonspecific absorption is more of a problem when furnace atomizers are employed than when a chemical flame is utilized. In a furnace, a dense smoke cloud often arises during atomization of the sample, whereas in a chemical flame the particulate matter is more thoroughly volatilized. Nevertheless, it is considered to be prudent to employ some form of background correction with either a furnace or flame atomizer.

Several alternative schemes for background correction have been developed for AAS and are covered in some detail in **Background Correction Methods in Atomic Absorption Spectroscopy**. The most common of these schemes, termed "continuum-source background correction" is particularly ingenious and is the first of such methods to be developed. It exploits the spectral difference between narrow-band atomic absorption and broad-band molecular absorption or scattering. However, two sources are required, a hollow cathode lamp and an auxiliary continuum source, usually a deuterium arc lamp. Still, it remains the most widely used method.

Another technique for background correction in atomic absorption utilizes the Zeeman effect. The Zeeman effect is a splitting of atomic lines that occurs when the atoms are present in a magnetic field. Accompanying this spectral splitting is a change in polarization of the spectral lines that are generated. By use of appropriate instrumentation, it is possible to distinguish atomic absorption from the background features. Although no auxiliary source is required, a magnetic field and polarizers usually are.

Another approach employed for background correction in atomic absorption utilizes a pulsed hollow cathode and is similar in concept to the continuum-source procedure. In essence, the same hollow cathode lamp is used as both a narrow-band and broad-band source. The broad-band spectrum arises from spectral line broadening that occurs at unusually high hollow cathode operating currents.

In contrast to AAS, AES is inherently a multielement procedure. In a high-temperature flame or plasma (such as the ICP), atoms are not only formed extremely efficiently, but also they generate intense atomic emission. The emission is isotropic and also occurs from all elements at the same time. It is therefore possible to perform multielement determinations simply by means of a multichannel detection system. Multichannel devices are routinely being introduced that employ two-dimensional spectral dispersion coupled with two-dimensional arrays of detector elements of unprecedented sensitivity and low noise.

Like AES, AMS is inherently a multielement approach. Moreover, AMS can provide information about isotopes present in a sample. Although a small spectral shift occurs in atomic emission spectra because of the "isotope effect", for most elements in the periodic table that shift is too small to be detected with a conventional spectrometer. In contrast, AMS can readily distinguish one isotope from another; one of the important applications of AMS has therefore become the measurement of isotope ratios in various kinds of samples.

AMS offers also the benefit of extraordinarily low background levels. In the simplest sense, this low background results from the fact that ions in a mass spectrometer can possess energies that are considerably greater than energies corresponding to photons in the ultraviolet or visible region of the spectrum. For a photon of such low energy to be detected with high sensitivity requires naturally that the detector be very sensitive and accordingly sensitive to thermal effects. Thermal noise generated in such detectors therefore makes it more difficult to observe the atomic transitions of interest. In contrast, detectors employed in mass spectrometry can be considerably less sensitive to thermal noise because they can employ detection surfaces with a high "work function". The result is background count rates in AMS that can be below one count per second, whereas background count rates in AES are of the order of 10^4 cps. The result is that typical detection limits in AMS are lower than those in AES by a factor of 100 or so. Recent manufacturers' literature suggests that it is now routinely possible to measure solution concentrations in ICP mass spectrometry at the level of one part per quadrillion or so.

For the ultimate in sensitivity, however, none of the above-mentioned techniques can better AFS. Indeed, a number of AFS studies over the last decade or so have demonstrated the detection of single atoms of a desired element. The reason for this extraordinary sensitivity is quite straightforward: in AFS, each atom can be detected many times, since it can be excited over and over again. Each time the atom emits a photon, it can be re-excited by an incident beam of photons, probably from a laser. In contrast, detection in AMS is usually destructive, so there is only a single chance to observe each atomic ion. Although AES also offers the opportunity to collect several photons from each atom, the background levels are usually high enough that it is difficult to distinguish the few photons an atom emits from others that a detector registers.

Several properties of a laser make it useful not only as a source for AFS, but for other measurements in atomic spectrometry as well. These features are documented in greater detail in **Laser Spectrometric Techniques in Analytical Atomic Spectrometry**. In short, the laser can deliver light to a sample very effectively, so it can be used not only as a source for laser sampling (laser ablation) and LIBS, as indicated earlier, but also as a source for atomic magneto-optic rotation spectrometry, for AAS, and for other more exotic measurements. To date, the widespread acceptance of these alternative methods has been limited by the cost and complexity of current laser systems. However, recent advances and engineering developments suggest that such impediments might be overcome in the future.

From this brief narrative, it should be clear that atomic spectrometric analysis is not only an extremely important area of chemical analysis, but also one that continues to evolve rapidly. Over the past decade, the nature of instrumentation for atomic spectrometry has changed dramatically. Benchtop units are now available that measure virtually the entire atomic emission spectrum of a complex chemical sample at once. Similarly, alternative kinds of mass spectrometric instrumentation (quadrupole mass filters, sector-field instruments, time-of-flight mass spectrometers, ion traps, and Fourier transform mass spectrometers) have been applied to AMS. These alternative approaches enable samples to be analyzed with unprecedented speed, accuracy, and precision. They permit isotope ratios to be determined on samples having elemental concentrations below 1 ppt and at a speed that would have been considered impossible only a few years ago. Because of this rapid development, the reader should consider the treatments in this section to be an excellent introduction to the field, but that continuing acquaintance will be necessary to keep abreast of ongoing developments.

ABBREVIATIONS AND ACRONYMS

AAS	Atomic Absorption Spectrometry
AES	Atomic Emission Spectrometry
AFS	Atomic Fluorescence Spectrometry
AMS	Atomic Mass Spectrometry
ETA	Electrothermal Atomizer
FIA	Flow Injection Analysis
ICP	Inductively Coupled Plasma
LIBS	Laser-induced Breakdown Spectroscopy
MIP	Microwave-induced Plasma
NAA	Neutron Activation Analysis

REFERENCES

1. A. Walsh, 'The Application of Atomic Absorption Spectra to Chemical Analysis', *Spectrochim. Acta*, **7**, 108–117 (1955).
2. C.T.J. Alkemade, J.M.W. Milatz, 'Double-beam Method of Spectral Selection with Flames', *J. Opt. Soc. Am.*, **45**, 583–584 (1955).

Background Correction Methods in Atomic Absorption Spectroscopy

Margaretha T.C. de Loos-Vollebregt
Delft University of Technology, Delft, The Netherlands

For references see page 9378

Atomic absorption spectroscopy (AAS) is based on the absorption of element-specific primary source radiation by analyte atoms. If part of the radiation is absorbed by molecules or lost due to scattering, a higher gross absorbance is measured. The difference between the net absorption of the analyte atoms and the measured gross absorbance is called background absorbance. Background absorption and scattering effects have much more serious effects on the results produced in electrothermal atomization atomic absorption spectroscopy (ETAAS) than flame atomic absorption spectroscopy (FAAS).

The amount of incident light deflected or absorbed by nonatomic species must be measured to obtain the correct, net absorbance of the analyte atoms only. Perfect background correction can be obtained only when the background absorbance measurement corresponds exactly in space, time and wavelength with the atomic absorbance measurement. Since exact coincidence of all three parameters is obviously impossible, it is customary to give priority to the equality in space and to make the difference in time and/or wavelength as small as possible. Although molecular absorption and radiation scattering are both broad-band phenomena, there is no spectral range where constant background attenuation can be guaranteed.

In all background correction systems two measurements are made. The total or gross absorbance is measured at the wavelength of the resonance line emitted by the hollow cathode lamp (HCL). The background attenuation is then subtracted from the measured analyte absorbance to obtain the analyte absorbance. There are several ways in which the nonatomic absorption at the resonance wavelength can be estimated, including two-line background correction, continuum-source or deuterium lamp background correction, Zeeman background correction, pulsed lamp or Smith–Hieftje background correction and wavelength-modulation correction methods. The principle of each method, the instrumentation and applications are discussed. Continuum-source background correction is widely used in FAAS and ETAAS, whereas Zeeman background correction and pulsed lamp background correction are often preferred in ETAAS.

1 INTRODUCTION

AAS is based on the absorption of element-specific primary source radiation by analyte atoms. If part of the radiation is absorbed by molecules or lost due to scattering, a higher gross absorbance is measured. The difference between the net absorption of the analyte atoms and the measured gross absorbance is called background absorbance. In fact, we do not know which part of the background absorbance arises from scattering and which part is due to molecular absorption.

Scattering of primary source radiation from nonvolatilized particles much smaller than the wavelength (λ) follows Rayleigh's law.The intensity of the scattering depends on the wavelength as λ^{-4}, so that short wavelengths are scattered more than than long wavelengths. Scattering is directly proportional to the number of scattering particles per unit volume and to the square of the particle volume. Consequently, radiation scattering occurs more strongly with increasing particle size. In FAAS, well-designed nebulizers and spray chambers are used and therefore virtually no unvolatilized particles pass into the optical beam. In graphite furnace AAS, considerable numbers of particles can be observed. They may arise from organic sample materials during the pyrolysis step or from volatilized inorganic materials that condense in the cooler parts of the atomizer. Small solid particles can also appear owing to boiling, sputtering and incomplete vaporization. Broad-band molecular absorption of radiation is caused by molecular species formed or vaporized in the atomizer during the atomization step, particularly alkali and alkaline earth metal halides. Sharp fine structure of the vibrational bands in electronic spectra of the molecules can be superimposed on broad-band background absorption. The effects of molecular absorption in FAAS have long been recognized. The $CaOH^+$ band has a maximum near the barium resonance line at 553.6 nm. Molecular absorption spectra of a number of halides, nitrates and sulfates have been reported. As

List of selected abbreviations appears in Volume 15

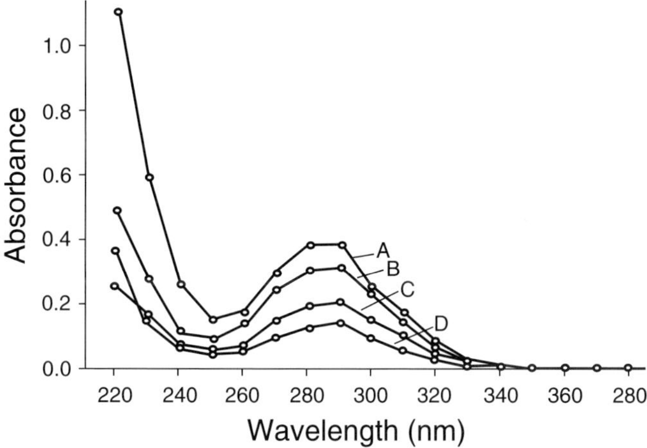

Figure 1 Continuous dissociation spectra of SO_3 (260–330 nm) and SO ($\lambda < 245$ nm) resulting from volatilization of sulfates in a graphite tube furnace. A, 0.01 g mL^{-1} Mg; B, 0.01 g mL^{-1} Fe; C, 0.01 g mL^{-1} Ni; D, 0.01 g mL^{-1}; Zn. Sample volume, 20 mL. (Reprinted from R.A. Newstead, W.J. Price, P.J. Whiteside, *Prog. Anal. At. Spectrosc.*, **1**, 267–298 (1978), Copyright 1978, with permission from Elsevier Science.)

an example, Figure 1 shows the continuous dissociation spectra of SO_3 and of SO resulting from volatilization of sulfates in a graphite furnace, reported by Massmann and Güçer[1] and Newstead et al.[2] Massmann and Güçer[1] also measured emission and absorption spectra resulting from the CN system with band heads at 386.2, 387.1 and 388.3 nm using nitrogen as the inert gas in ETAAS. Background absorption and scattering effects have much more serious effects on the results produced in ETAAS than in FAAS.

The amount of incident light deflected or absorbed by nonatomic species must be measured to obtain the correct, net absorbance of the analyte atoms only. In electrothermal atomization the spatial and temporal distribution of the atoms and the background-generating species may be entirely different.[3] Perfect background correction can be obtained only when the background absorbance measurement corresponds exactly in space, time and wavelength with the atomic absorbance measurement. Since exact coincidence of all three parameters is obviously impossible, it is customary to give priority to the equality in space and to make the difference in time and/or wavelength as small as possible. Although molecular absorption and radiation scattering are both broad-band phenomena, there is no spectral range where constant background attenuation can be guaranteed.

In all background correction systems two measurements are made. The total or gross absorbance ($A_{measured}$) of the atoms and the nonatomic species is measured at the wavelength of the resonance line emitted by the HCL. The background attenuation (A_{bg}) is then subtracted from the measured absorbance to obtain the analyte absorbance ($A_{analyte}$) (Equation 1):

$$A_{analyte} = A_{measured} - A_{bg} \qquad (1)$$

There are several ways in which the nonatomic absorption at the resonance wavelength can be estimated.

2 TWO-LINE BACKGROUND CORRECTION

In the two-line background absorption approach, the total absorbance is measured at the resonance line emitted by the HCL and the nonspecific absorbance is measured at a nearby line of the lamp, where the analyte atoms do not absorb.

Two-line background correction was first proposed by Willis,[4] using a nonresonant analyte line or a line from an element that is emitted by the source but not present in the sample. Argon or neon lines can also be used. Alternatively, another source, of an element which is not found in the sample, can be inserted for the background measurement. The background spectral character must be known to be the same at the second line as it is at the primary line and therefore the closer the second line is to the primary line, the better will be the correction. These conditions cannot be met for all elements and all matrices. In some wavelength regions, the steep slope of molecular absorption bands makes it difficult to find a nonabsorbing line close enough to the analyte line to ensure an accurate background correction. The atomization conditions for sequential measurements should be as similar as possible, even if an appropriate nonabsorbing line is available from the primary source itself. Dual-channel AAS instruments offer the possibility of simultaneous measurement at two lines. The monochromator of the first channel is set to the chosen absorption line for the analyte element and the second channel monochromator is set to the chosen line for the background correction. The signal-handling system of the instrument subtracts the absorbance measured in the second channel from the absorbance signal measured in the first channel. Two-line background correction went into practice earlier than the other methods. However, it has not been widely used. Lines have been recommended for background correction for the most commonly determined elements.[2] It is important that the chosen line should be close to the line of the analyte element. Because the absorption signal from background scatter varies with wavelength and increases with decreasing wavelength, the background correction error may become substantial at wavelengths below 220 nm if the line chosen is more than 1 nm from the analyte line. The use of an automatic system, chopping rapidly between both lines, is essential if the system is used

For references see page 9378

with an electrothermal atomizer because measurements made on two successive furnace cycles are unlikely to show the same background signal.

3 CONTINUUM-SOURCE BACKGROUND CORRECTION

3.1 History

To improve the sensitivity of AAS, Koirtyohann and Picket[5] used a 40 cm long, 10 mm i.d. tube that was heated in a hydrogen–oxygen flame, providing a much longer effective absorption path than was previously used in the more conventional burners. They observed that, at the higher sensitivities, absorption by matrix salts at the wavelength of an elemental resonance line can cause significant errors and proposed to use a hydrogen lamp to correct for matrix absorption. In fact, this background correction system is still widely used to correct for background absorption in FAAS and ETAAS.

3.2 Principle

The gross absorbance $A_{measured}$ is the sum of the atomic absorbance of the analyte atoms $A_{analyte}$ and the background attenuation A_{bg}. The gross absorbance is measured with the narrow HCL emission line of the element at the selected specific wavelength of the resonance line. An estimate of the background absorbance is obtained as an average over the spectral band-pass of the monochromator using a continuum-source of radiation, e.g. a hydrogen lamp or deuterium lamp in the ultraviolet (UV) wavelength region or a tungsten lamp in the visible part of the spectrum. The width of the atomic absorption line is about 100 times smaller than the band-pass of the monochromator, so that the analyte absorption of the continuum-source radiation can be neglected.

3.3 Instrumentation

For the correction of background attenuation, Koirtyohann and Picket[5] used a dual-channel system in which the radiation from an HCL was passed alternately with the radiation of a continuum hydrogen lamp through the flame. Nowadays, a deuterium lamp is most frequently used as the continuum-source. Figure 2 shows the basic schematic configuration of a continuum-source background correction system for FAAS. By using lamp pulsation or by means of a rotating chopper with a sector mirror, the radiation from the HCL with intensity I^0_{HCL} and the radiation from the deuterium lamp with intensity $I^0_{D_2}$ are passed alternately through the atomizer. After passing

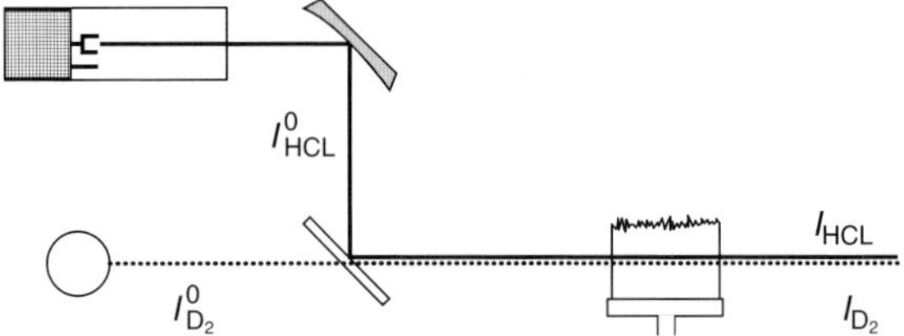

Figure 2 Schematic configuration of a continuum-source background correction system. Radiation from the HCL (I_{HCL}) is passed alternately with radiation from the deuterium lamp (I_{D_2}).

the monochromator, both radiation beams reach the same detector and the net atomic absorbance is derived.

The operation of the continuum-source background correction system is explained in Figure 3(a–f). The monochromator separates the resonance line from the emission spectrum of the element-specific HCL. The width of this line is a few picometers. From the deuterium lamp emission spectrum the monochromator isolates a band equivalent to the spectral band-pass, usually 0.2–1 nm. The primary intensities of the two beams of radiation are equalized to obtain $I^0_{HCL} = I^0_{D_2}$. When analyte atoms absorb HCL radiation, the intensity I^0_{HCL} decreases with an amount corresponding to the atom concentration. Naturally, $I^0_{D_2}$ is also attenuated at the resonance wavelength, but since the half-width of the atomic absorption line is about 5 pm whereas the continuum has a width equal to the spectral band-pass of 0.2–1 nm, the absorption of continuum radiation by the analyte atoms can be neglected. If nonspecific attenuation of the radiation occurs, whether through scattering or molecular absorption, both radiation beams will be attenuated to the same extent since background attenuation is usually a broad-band phenomenon. In continuum-source background correction it is assumed that the background is constant over the observed spectral range, i.e. the attenuation of $I^0_{D_2}$ is measured as the average over the spectral band-pass and used for background correction at the wavelength position of the analyte absorption line. If the background attenuation is not continuous within the observed spectral range, positive or negative background correction errors will occur.

The two lamps are observed by the detector alternately in time. Instrument electronics separate the signals and compare the absorbances from both sources. A net absorbance will be displayed only when the absorbance of the two lamps differs. Since the background attenuates both sources equally, it is ignored. Analyte atoms absorb the HCL radiation and negligibly absorb the broad-band continuum-source emission and it is therefore still measured and displayed as usual without background correction. Most modern AAS instruments permit almost

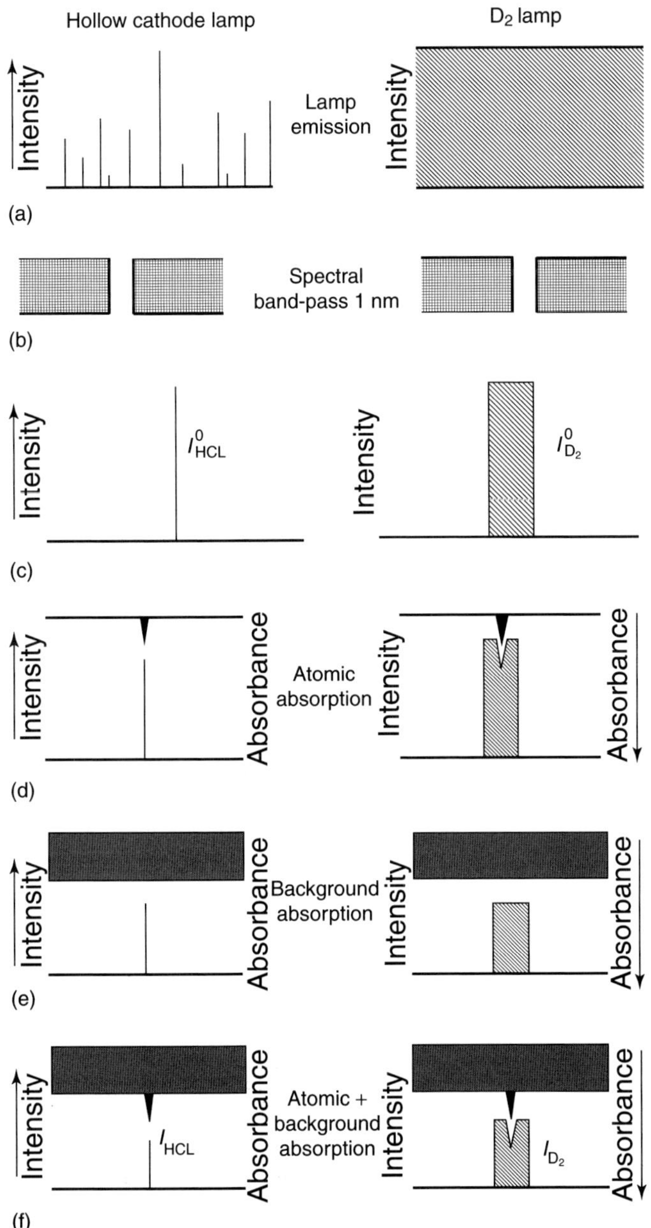

Figure 3 Mode of operation of a continuum-source background correction system. (a) The line emission spectrum of the HCL and the continuum emitted by the deuterium lamp; (b) the monochromator isolates the resonance line from the spectrum of the HCL and passes a band of radiation from the deuterium lamp corresponding with the spectral band-pass; (c) the intensities I^0_{HCL} and $I^0_{D_2}$ are equalized; (d) for atomic absorption by the analyte element I^0_{HCL} is attenuated by an amount corresponding to its concentration whereas $I^0_{D_2}$ is not significantly attenuated; (e) broad-band background attenuates the intensity of both sources to the same degree; (f) atomic absorption by the analyte in addition to the background attenuates the HCL intensity further as in (d) whereas the deuterium lamp intensity is not further attenuated. (Reproduced by permission of Wiley-VCH from B. Welz, *Atomic Absorption Spectrometry*, 2nd edition, VCH, Weinheim, 1985.)

simultaneous measurement of the uncorrected signal or background signal and the corrected signal. The speed of the background correction system should be high enough to follow the fast transient signals in ETAAS (see section 7.2).

Background correction with a continuum-source has some disadvantages. The alignment and superposition of the HCL or electrodeless discharge lamp and the continuum-source are critical. Both lamps are located at different positions in the instrument, the shape of the beams is different and the intensity distribution in the beam is not the same. Even if it is possible to align the beams exactly, different absorption volumes may be irradiated. The application of two light beams requires a more complicated optical system and in many instruments double-beam operation is available to compensate for intensity changes of the lamps. The deuterium lamp emits sufficient radiation in the short wavelength range from 190 nm to 330 nm, whereas a halogen lamp is used for background corrections above this range. In general, the signal-to-noise ratio (S/N) is worsened through the use of two radiation sources and the more complicated optical system that is used to measure their intensities. If a continuum-source is used outside its optimum range, the radiant intensity of the line source must be reduced, leading to further detoriation of the S/N. Finally, it should be noted that the principle of operation of continuum-source background correction limits successful use to continuous background absorption only.

3.4 Applications

Many commercial AAS systems are based on deuterium lamp background correction. In FAAS and hydride generation AAS it is the most widely applied background correction system. It is fully adequate for all applications of FAAS, except in some very unusual circumstances. During methods development, background correction is considered. In many applications, there is no need to use background correction at all. In general, in the lower UV wavelength region (<220 nm) background correction is required.

Background correction systems employing continuum-sources are incapable of correcting the background attenuation of electronic excitation spectra because these comprise many narrow lines. The actual background correction that is required depends on the degree of overlap between the elemental spectral line and the individual molecular rotational lines. An example is the absorption line of gold at 267.6 nm that lies exactly in the middle between two rotational lines of indium chloride so that the actual background attenuation is much lower than the mean background absorbance found over the spectral range. Similar interferences have been reported for the

For references see page 9378

determination of Bi at 306.8 nm and Mg at 285.2 nm through absorption bands of OH in an air–acetylene flame and also for Fe at 247.3 nm, Pd at 244.8 and 247.6 nm and Yb at 246.4 nm due to PO molecular absorption.[6] Another possible source of error with continuum-source background correction may arise from absorption by concomitant atoms from the matrix. This type of spectral interference has been reported by Manning[7] for the determination of Se in the presence of Fe. Problems with continuum-source background correction in FAAS seldom occur. The background correction system is an extremely useful device for FAAS that frequently leads to the correct results.

However, the continuum-source background correction system is not free from pitfalls. The limited energy of the deuterium lamp lowers the S/N and the correction becomes inaccurate when the background is structured. Most users distrust their results when the background level exceeds 0.5 absorbance unit. The limitations are probably related to the different geometries and possible misalignment of the optical beams of the HCL and the deuterium lamp. The increasing interest in ETAAS as a selective and highly sensitive analytical method has stimulated the development of improved background correction systems.

4 ZEEMAN BACKGROUND CORRECTION

4.1 Zeeman Effect

In 1897, the Dutch physicist Pieter Zeeman[8] (Figure 4) reported the observation that if a source of light is placed between the poles of a magnet the lines appear to broaden and, if the magnetic field is strong enough and the resolving power of the spectroscope is sufficiently high, to spit into a number of components. The first splitting patterns studied by Zeeman follow from lines of a singlet series. A single line was split into three polarized components.

Figure 4 Dr Pieter Zeeman (1865–1943). Dutch physicist who discovered the influence of a magnetic field on radiation in 1896.

Both splitting and polarization were explained by Lorentz with the classical electromagnetic theory. A few years later it was demonstrated that the lines of other series do not split into the simple triplet but to a group of four or more components. The classical theory was unable to account for this. Therefore, the two types of splitting were labeled the normal and the anomalous Zeeman effects. An example of the normal Zeeman effect is presented in Figure 5(a) and Figure 5(b) shows an example of the anomalous Zeeman effect. The splitting pattern is always symmetrical about the position of the nondisplaced line. The components polarized parallel to the magnetic field are referred to as π-components and the components polarized perpendicular to the magnetic field are referred to as σ-components (σ deriving from the German senkrecht). In the case of a normal Zeeman triplet, the π-component is located at the position of the original line at frequency ν and two σ-components are found at frequencies $\nu \pm \Delta\nu$. Such a Zeeman splitting is observed from the direction perpendicular to the magnetic field (transverse Zeeman effect). In the direction of the magnetic field, the central component is not observed whereas the two displaced components are circularly polarized and rotate in opposite directions (longitudinal Zeeman effect). Lines that do not belong to a singlet series split into more than three components. Their pattern is always symmetrical with respect to the position of the original line, with a central group of π-components and groups of σ-components on either side as illustrated in Figure 5(b) for a $^2S_{1/2}$–$^2P_{3/2}$ transition. All Zeeman splitting patterns are observed in both absorption and emission.

The sum of the intensities of the π-components is equal to the sum of the intensities of the σ-components and if the various components are combined, the resulting

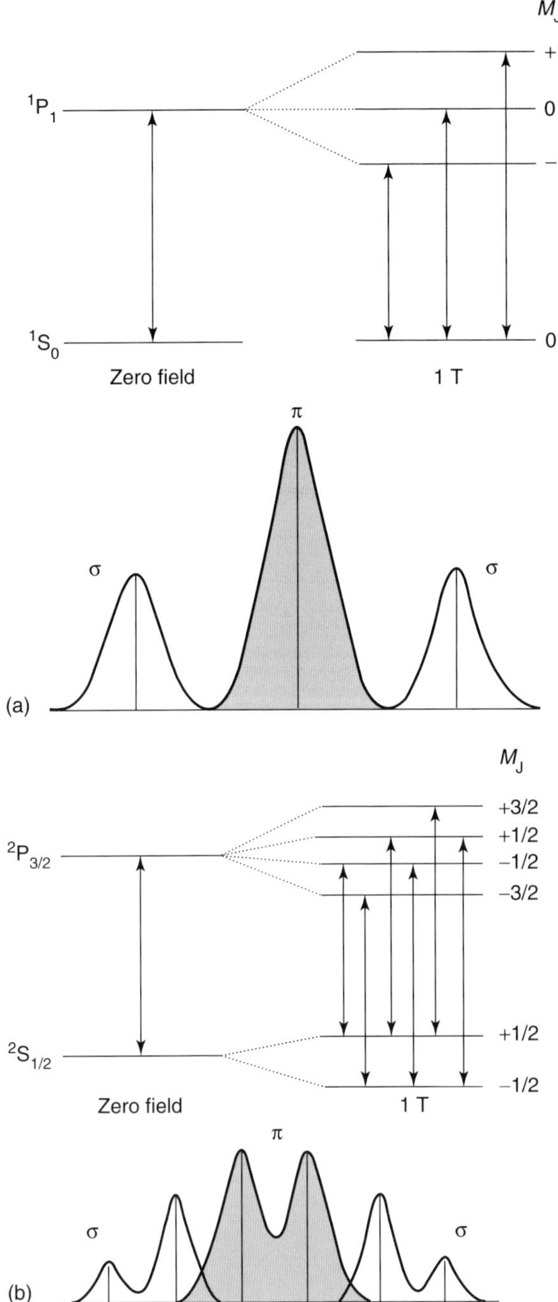

Figure 5 Magnetic splitting of atomic energy levels and the corresponding Zeeman splitting patterns. (a) $^1S_0-^1P_1$ transition, normal Zeeman effect; (b) $^2S_{1/2}-^2P_{3/2}$ transition, anomalous Zeeman effect. (Reprinted from M.T.C. de Loos-Vollebregt, L. de Galan, *Prog. Anal. At. Spectrosc.*, **8**, 47–81 (1985), Copyright 1985, with permission from Elsevier Science.)

beam is nonpolarized and the intensity is equal to the intensity observed without the magnetic field. The energy states generated around an original energy level E^0 are described by Equation (2):

$$E = E^0 + \mu_B B M_J g \quad (2)$$

where μ_B is the Bohr magneton ($9.274 \times 10^{-24}\,J\,T^{-1}$), B is the magnetic flux density (tesla), M_J is the magnetic quantum number and g is the Landé factor. Both emission and absorption transitions between energy levels are governed by the selection rule $\Delta M_J = 0, \pm1$, where $\Delta M_J = 0$ represents the π-components and $\Delta M_J = \pm1$ the σ-components. The displacement of each Zeeman component is proportional to the strength of the magnetic field. In a magnetic field of 1 T the displacement of the σ-components in the normal Zeeman effect is $0.467\,cm^{-1}$, which is equal to 4 pm at a wavelength of 300 nm. The normal Zeeman effect is shown by Be 234.9 nm, Mg 285.2 nm, Ca 422.7 nm, Sr 460.7 nm, Ba 553.6 nm, Cd 228.8 nm and Zn 213.9 nm. Several lines that belong to a triplet series also split into one π-component and two σ-components. However, the shift of their σ-components varies from half to twice the shift in the normal Zeeman effect, e.g. Ge 265.2 nm (1.5), Hg 253.7 nm (1.5), Hg 184.9 nm (1.0), Pb 217.0 nm (0.5), Pb 283.3 nm (1.5), Pd 244.8 nm (1.0), Si 251.6 nm (1.5) and Sn 286.3 nm (1.5). All other transitions used in AAS show anomalous Zeeman splitting patterns that consist of at least two π-components and at least two σ-components. The Zeeman splitting pattern in Figure 5(b) is observed for Ag 328.1 nm, Cu 324.8 nm, Au 242.8 nm, Na 589.0 nm and K 766.5 nm. Zeeman splitting patterns of the absorption lines used in AAS of various elements were classified in several groups by Koizumi and Yasuda.[9] Figure 6 shows the Zeeman splitting patterns for various transitions that are used in AAS.

4.2 Principle of Zeeman Background Correction

In AAS, the non- or slightly shifted π-components are used to measure the total absorbance of the analyte atoms and the background whereas the shift of the (groups of) σ-components from the original line position is used to measure the background absorbance. Figure 7(a) shows the resonance line emitted by the HCL and the absorption line of the atoms in the graphite furnace at zero magnetic field strength. Note that the width of the absorption line is about twice the width of the lamp emission line owing to the higher temperature and higher pressure in the atomizer compared to the HCL.

Figure 7(b) shows the situation where the graphite furnace is in a transverse magnetic field of 1 T. Absorption takes place within the π-components of the absorption line profile k_π^a whereas the σ-components of the absorption line profile k_σ^a are shifted away to wavelengths λ_1 and λ_2. When a π-component rejecting polarizer is positioned in the optical beam and when the magnetic field is switched on and off at high enough frequency, e.g. a 50 Hz alternating current (AC) magnetic field, the gross absorbance and the background absorbance can

For references see page 9378

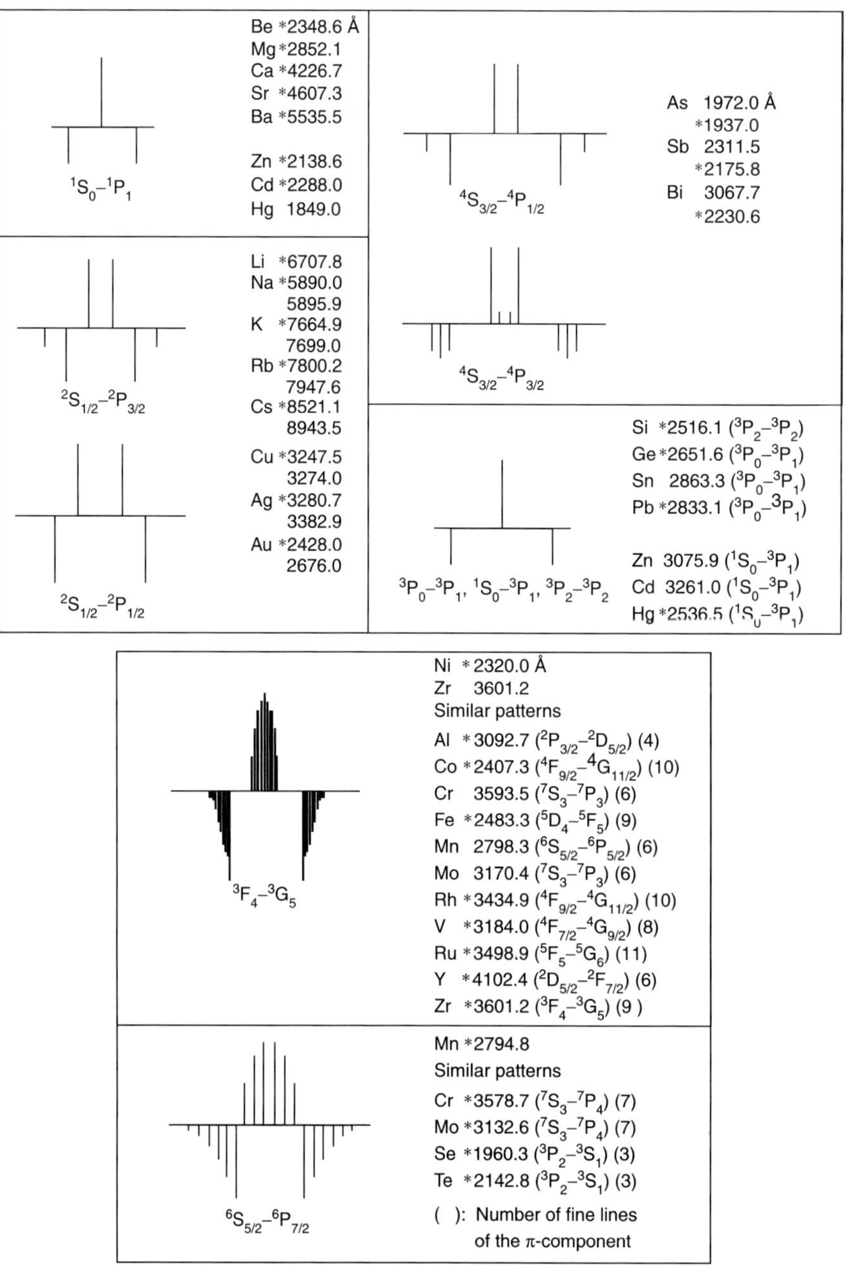

Figure 6 Classification of Zeeman patterns on various elements. $*$ indicates lines most frequently used in AAS. 1 Å = 10^{-10} m. (Reprinted from H. Koizumi, K. Yasuda, *Spectrochim. Acta, Part B*, **31**(10–12), 523–535 (1976), Copyright 1976, with permission from Elsevier Science.)

be measured alternately. The net analyte absorbance follows then from Equation (1). The background absorbance k^b is measured exactly at the wavelength of the analyte resonance line λ_a and within the width of the HCL emission line profile. Consequently, the background absorbance is measured correctly for any type of background, even if it is highly structured. The alternative approach is to use a constant magnetic field from a permanent magnet or a direct current (DC) magnet and to rotate the polarizer to permit alternate measurements with π-polarized radiation and σ-polarized radiation. The π-polarized radiation is then absorbed by the π-component of the analyte absorption line profile k_π^a and the background k^b providing the gross absorbance. The σ-polarized radiation is absorbed by the background only because the σ-components of the analyte absorption line profile are shifted to wavelengths λ_1 and λ_2 and they are too far away to absorb the lamp emission line. Combination of the two measurements again provides the net analyte absorption.

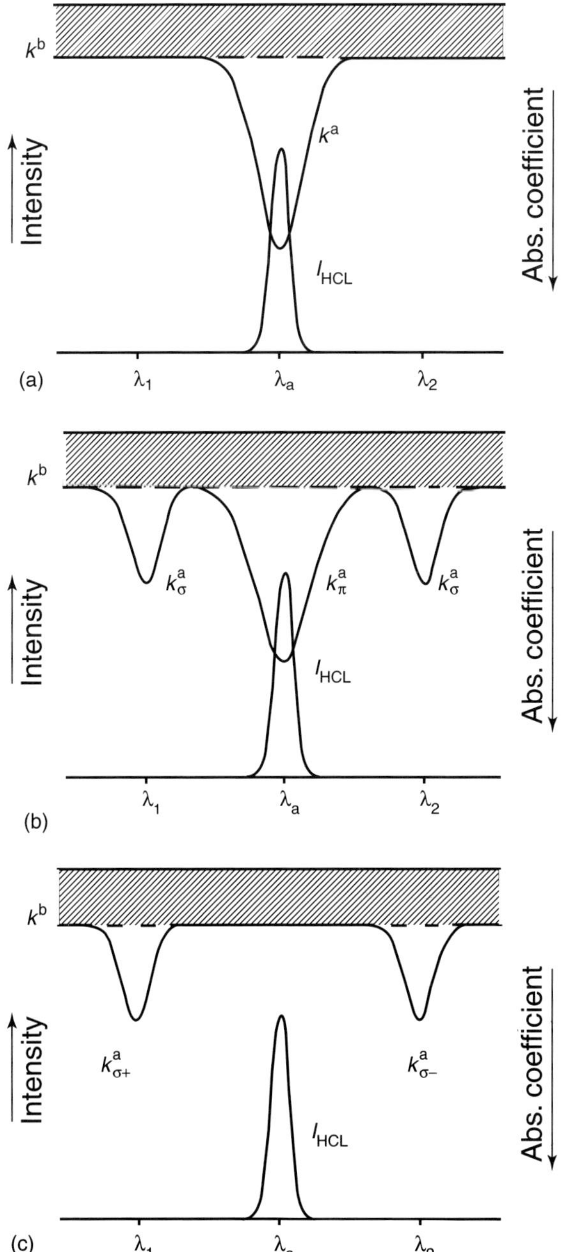

Figure 7 Profiles of the HCL analyte emission line (I_{HCL}) and the absorption coefficient of the analyte (k^a) at wavelength λ_a in the presence of background absorbance (k^b) for the situation of normal Zeeman effect. (a) Conventional AAS, no magnetic field; (b) in the presence of a transverse magnetic field; (c) in the presence of a longitudinal magnetic field.

Another possibility is to use a longitudinal magnetic field rather than a transverse magnetic field. Figure 7(c) shows that in the direction of the magnetic field only the circularly polarized σ-components of the analyte absorption line profile $k_{\sigma+}^a$ and $k_{\sigma-}^a$ are observed. In a longitudinal AC modulated magnetic field the gross absorbance is measured at zero-field (see Figure 7a) and the absorbance of the background is measured at exactly the wavelength position of the lamp emission line during the magnet on period. A longitudinal AC Zeeman background correction system does not require a polarizer and therefore the full intensity of the HCL resonance emission line can be used in both measurements.

In all AAS instruments using Zeeman background correction, two alternate intensity measurements are performed. One refers to the unshifted zero-field analyte absorption line (in an AC field) or the slightly shifted and broadened π-components (in a constant field). The intensity is given by Equation (3):

$$I_1 = I_1^0 \exp(-k_1^a) \exp(-k_1^b) \tag{3}$$

where I_1^0 is the incident intensity, k_1^b is the absorption coefficient of the background and k_1^a is the analyte absorption coefficient. The other measurement refers to the situation when the σ-components are shifted away from the original line position (AC field) or when σ-polarized radiation is used (constant field). The intensity is given by Equation (4):

$$I_2 = I_2^0 \exp(-k_2^a) \exp(-k_2^b) \tag{4}$$

where I_2^0 is the incident intensity, k_2^b is the absorption coefficient of the background and k_2^a is now the remaining analyte absorption coefficient of the shifted σ-components, which should be close to zero. A simple subtraction of the absorbances measured in both situations corresponds to taking the log ratio of the intensities (Equation 5):

$$\ln\left(\frac{I_2}{I_1}\right) = \ln\left(\frac{I_2^0}{I_1^0}\right) + (k_1^a - k_2^a) + (k_1^b - k_2^b) \tag{5}$$

This expression forms the basis for all Zeeman background correction systems in AAS. When the incident intensities I_1^0 and I_2^0 are equal and when the same background attenuation is measured, the net analyte absorption follows from Equation (6):

$$A = \log\left(\frac{I_2}{I_1}\right) = 0.43(k_1^a - k_2^a) \propto N_A \tag{6}$$

It is clear that a linear analytical response proportional to the number of analyte atoms N_A is obtained. The assumption that I_1^0 is equal to I_2^0 is reasonable because it is the same beam of radiation in the case of the AC magnetic field, or the π-polarized and the σ-polarized parts of the same beam in the case of the constant magnetic field. The background absorbance is measured within the analyte emission line profile in all Zeeman background correction systems with the magnetic field applied to the atomizer. The only requirement is that

Table 1 Comparison of relative sensitivities in different Zeeman background correction systems[9] (conventional AAS sensitivity is equal to 1)

Element	Wavelength (nm)	Zeeman pattern[a]	Constant field 1 T	Constant field optimized[b]	AC field 0.8 T
Ag	328.1	A	0.28	0.61	0.91
Al	309.3	A	0.97	0.97	0.90
As	193.7	A	0.43	0.66	0.89
Cd	228.8	N	0.94	0.94	0.98
Cr	357.9	A	0.42	0.68	0.88
Cu	324.8	A	0.49	0.49	0.53
Fe	248.3	A	0.73	0.76	0.92
Mn	279.5	A	0.61	0.65	0.91
Pb	283.3	N	0.86	0.86	0.83
Sb	217.6	A	0.68	0.81	0.95
Se	196.0	A	0.45	0.51	0.88
Sn	286.3	N	0.97	0.97	0.94
Zn	213.8	N	0.90	0.90	0.88

[a] Zeeman splitting pattern: N = normal, A = anomalous.
[b] Variable constant magnetic field up to 1 T.

the background is not influenced by the magnetic field when an AC magnet is used. From Equation (6) it can be concluded that the σ-components in the Zeeman splitting pattern should be shifted as far away as possible from the original line position to reduce k_2^a to a minimum, i.e. a strong magnetic field should be applied, especially when the analyte absorption line is broad by nature or when its Zeeman splitting pattern is relatively narrow. In contrast, maximum sensitivity also requires that k_1^a is large, i.e. close to the conventional AAS absorption coefficient. Optimum sensitivity is obtained when a strong AC magnetic field is used. In a constant magnetic field, the optimum magnetic field strength compromises a large shift for the σ-components simultaneous with minimum shift in the π-components. Table 1 shows a comparison of relative sensitivities for several elements obtained in a constant magnetic field of 1 T, the best sensitivity that can be obtained for the element at optimized constant magnetic field strength and, for comparison, also the corresponding sensitivity obtained in an AC magnetic field of 1 T.[10] Even at optimum field strength, the sensitivity is reduced by about a factor of two for many of the analyte absorption lines that show the anomalous Zeeman effect. This explains why the AC modulated magnetic field is preferred in many of the commercially available AAS instruments with Zeeman background correction.

In principle, Zeeman background correction systems can have the magnetic field either around the atomizer or around the primary source of radiation. When the primary source of radiation is in the magnetic field, σ-components of the lamp emission line are located at the wavelength positions λ_1 and λ_2. These lines are used to measure the background absorbance at both sides of the analyte atomic absorption line. Equations (3–6) apply also to this situation.

4.3 Instrumentation

The application of the Zeeman effect for background correction in AAS was proposed by Prügger and Torge[11] in 1969. The authors claim that a shift of the primary source resonance line of about 10 pm is required to allow successive measurements of the gross AAS signal and the background absorbance and propose a system based on magnetic splitting of the source line in a DC or AC magnetic field. The first commercial instrument became available in 1975, incorporating a 1.1 T permanent magnet around the atomizer.[12] The plane of polarization is rotated at 100 Hz to provide for background correction whereas the intensity of the HCL radiation is modulated at 1.5 kHz to eliminate the signal caused by the emission of radiation from the atomizer. Since that time, many other companies have marketed Zeeman AAS instruments. Only a few instruments have the magnetic field around the primary source of radiation. The main reason is that normal HCLs and electrodeless discharge lamps that are widely used in AAS do not fit into the gap of a magnet that has to provide a magnetic field strength close to 1 T at reasonable size, power requirements and price of the magnet. In addition, the direction of the discharge in the lamp should be parallel to the direction of the magnetic field to provide a stable beam of radiation. Special lamps have been designed for stable operation in a magnetic field. For nonvolatile elements, instruments were operated with glow discharge lamps in a permanent magnet, whereas low-frequency, high-voltage gas discharge lamps were used for volatile elements.

The more useful alternative is to apply the magnetic field to the atomizer, as discussed in the previous section. Only one manufacturer decided to apply Zeeman background correction to FAAS and therefore constructed a magnet with a gap as long as the conventional

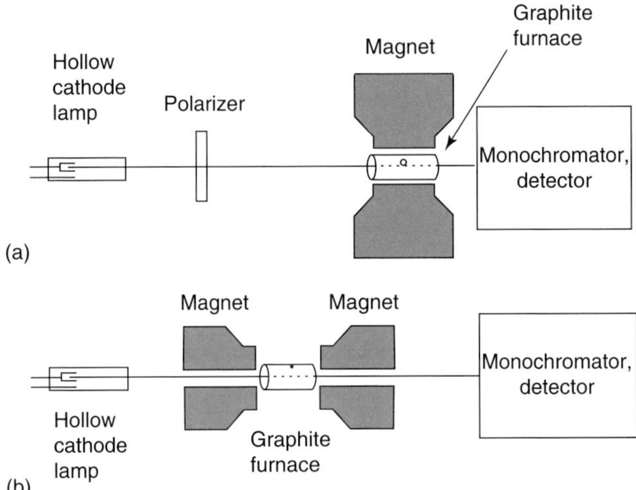

Figure 8 Schematic diagrams of different Zeeman background correction systems for AAS. (a) Transverse magnetic field, i.e. the magnetic field is perpendicular to the optical axis of the spectrometer. The polarizer provides alternately π- and σ-polarized radiation if the magnetic field is constant or transmits only σ-polarized radiation if an AC modulated magnetic field is applied; (b) longitudinal magnetic field oriented parallel to the optical axis of the spectrometer. No polarizer is required in this configuration.

10-cm burner that is routinely used in FAAS. In general, there is no need for Zeeman background correction in FAAS. The continuum-source background correction system works satisfactorily. In graphite furnace AAS, the transverse Zeeman background correction system in Figure 8(a) is marketed by various manufacturers of AAS instruments and most of them use an AC modulated magnetic field of about 0.8 T[10,13] in combination with a fixed polarizer to reject π-polarized radiation and transmit only σ-polarized radiation. The longitudinal Zeeman background correction system presented in Figure 8(b) was introduced by De Loos-Vollebregt et al.[14] and is also commercially available.[15] In this system, a transverse heated atomizer is placed within a longitudinal magnetic field and there is no polarizer in the optical system. This gives a considerable improvement in light throughput and simplifies the optical system so that detection limits are lower.

4.4 Applications

Zeeman background correction is nowadays widely used in ETAAS. The background correction is performed exactly at the wavelength of the atomic absorption line. High nonspecific absorbances (up to about 2) can be corrected and in the AC Zeeman system sensitivities are similar to those for conventional AAS. Nevertheless, there are a few elements that show somewhat reduced sensitivity because their lines are relatively broad and/or the Zeeman splitting pattern is relatively narrow. An overview of the early applications[16] of Zeeman background correction shows that a wide variety of elements have been measured in different samples and certified reference materials and in the recent years many applications followed. The relatively volatile elements, e.g. Pb and Cd, have been measured by many analytical chemists using ETAAS with Zeeman background correction. This is not surprising because in the determination of volatile elements the pyrolysis temperature is not high enough to remove most of the matrix compounds that are responsible for the background absorption during atomization. Pb and Cd are often determined in a rather difficult matrix such as blood and urine and the low concentrations of the analyte elements do not permit dilution of the samples. The results obtained with Zeeman background correction have been compared with those obtained with continuum-source background correction, looking at the baseline of the net AAS signals. Zeeman background correction seems to be superior, although a fair comparison would require that the same atomizer and atomization conditions are used. There are an increasing number of solid sampling applications where the Zeeman background correction system is the key to success.

In general, the dynamic range is slightly limited when Zeeman background correction is used. In a strong magnetic field the σ-components are shifted almost completely away from the absorption line profile so that, according to Equation (6), the shape of the AC Zeeman AAS analytical curve is completely determined by the zero-field absorbance. The analytical curve will therefore be very similar to the corresponding conventional AAS analytical curve measured without background correction. The relatively small contribution of the σ-components to the measurement of the background has only a minor influence on the slope and the curvature of the AC Zeeman AAS analytical curve. This is equally valid for the normal Zeeman effect and the anomalous Zeeman effect. The same is essentially true when a constant magnetic field is used as long as the spectral line displays a normal Zeeman effect. For the majority of the transitions that show an anomalous Zeeman effect, the splitting in the π-components broadens the absorption line profile and therefore the linearity of the Zeeman AAS analytical curve is somewhat improved, although the slope is lower.

At very high analyte concentrations, all the ETAAS signals show a dip (see Figure 9) when Zeeman background correction is applied, which can be explained from the phenomenon that is known as roll-over. From the very early days when Zeeman background correction was used, rumors were around that the analytical curve first increases, then reaches the maximum absorption level at relatively high analyte concentration and beyond the

For references see page 9378

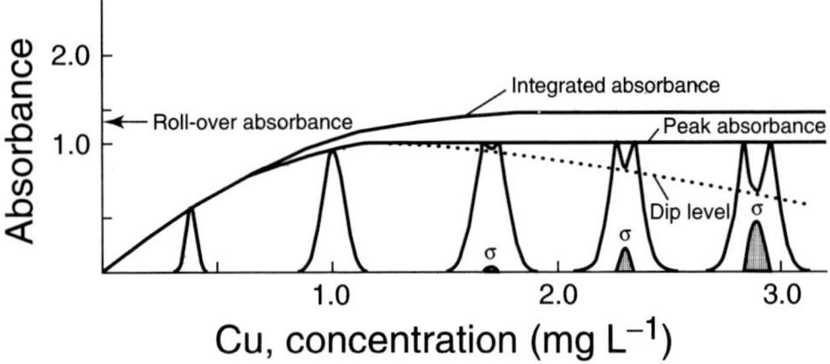

Figure 9 Analytical curve for Cu 324.8 nm measured in ETAAS with Zeeman background correction at magnetic field strength 0.8 T. The peak absorbance reaches a maximum at high analyte concentration whereas the integrated absorbance increases slowly beyond the roll-over absorbance.

maximum decreases again towards the concentration axis. Such a double-valued behavior has indeed been reported for FAAS systems with Zeeman background correction. From a theoretical analysis of the phenomenon[(16)] it became clear that the nonabsorbed radiation that is emitted by the primary source and reaches the detector is responsible for this. With increasing analyte concentration, the zero-field absorbance or π-component absorbance, k_1^a in Equation (6), increases much more rapidly than the σ-component absorbance, k_2^a. With increasing analyte concentration, k_1^a levels off. The σ-component absorbance is subject to the same curvature, but at much higher concentration. At a certain concentration a situation is reached where k_1^a and k_2^a show the same rate of increase, or equal derivative with respect to concentration. At that concentration, the difference signal according to Equation (6) shows zero slope, i.e. the analytical curve reaches a maximum. Thereafter, a FAAS analytical curve would bend back to the concentration axis. The absorbance and concentration of roll-over are indicated in Figure 9, where the dotted line illustrates the decreasing absorbance values. The height and the position of the maximum depend on the strength of the magnetic field and the amount of nonabsorbed source radiation (so-called stray light). A list of typical roll-over absorbances for ETAAS published by Slavin and Carnrick[(17)] is given in Table 2. The characteristic masses are also presented.

Figure 9 shows the net ETAAS analyte absorbance signals. At the lower analyte concentrations, the signals are similar to those in conventional AAS. However, beyond the concentration where the maximum peak absorbance is reached, the signals show a dip. The dip becomes deeper with increasing concentrations. In peak absorbance, the analytical curve levels off when the absorbance of roll-over is reached whereas in integrated absorbance the analytical curve continues to increase but the slope becomes much smaller. It is clear that there is no risk of misunderstanding the measured peak absorbances or integrated absorbances in ETAAS. At high analyte concentrations, the signals are analytically useless but there is no double-valued behavior. In FAAS, a warning system for roll-over is used that is based on the detection of the short periods of high absorbance at the start of nebulization and at the end where the atom density is increasing beyond the value that corresponds with the roll-over concentration. Detection of these peaks alerts the analyst that sample dilution is necessary to obtain the correct concentration values.

Table 2 Roll-over absorbance and characteristic mass data[(16)] measured in an AC magnetic field of 0.9 T

Element	Wavelength (nm)	Absorbance of roll-over	Characteristic mass[a] (pg)
Ag	328.1	>2	1.2
Al	309.3	0.8	10
As	193.7	1.3	17
Cd	228.8	0.8	0.4
Cr	357.9	1.6	1.8
Cu	324.8	0.7	3.0
Fe	248.3	0.8	2.8
Mn	279.5	1.1	1.7
Pb	283.3	1.5	7.0
Sb	217.6	1.4	38
Se	196.0	1.4	19
Sn	286.3	>2	20
Zn	213.8	1.0	0.4

[a] Mass of analyte in picograms that produces an integrated absorbance signal of 0.0044 s.

5 PULSED LAMP BACKGROUND CORRECTION

5.1 Principle

Ling[(18)] first advanced the idea of background correction based on the subtraction of absorbances measured alternately with a normal lamp emission line and a broadened lamp emission profile showing a dip in the center. He described a simple method to correct for nonatomic absorption in a portable mercury photometer that included two mercury vapor lamps emitting the mercury resonance line broadened to different extents. Smith and Hieftje[(19)] proposed a method for background correction in AAS based on the broadening which occurs in an HCL spectral line when the lamp is operated at very high currents. Under such conditions, the absorbance measured for a narrow atomic line is low, whereas the apparent absorbance caused by a broad-band background contributor remains the same as when the lamp is operated at conventional current levels. Background correction can therefore be effected by taking the difference in absorbances measured with the lamp operated at high and low currents. The so-called Smith–Hieftje background correction system applies its correction very near the atomic line of interest.

5.2 Instrumentation

The HCL power supply and drive circuitry[(19)] generate a 9-ms low-current pulse of 5–10 mA, followed by a 0.3-ms high-current pulse of 200–300 mA, as presented in Figure 10(a). The atom cloud in front of the hollow cathode dissipates rapidly for some elements whereas other elements (e.g. Ag, Cd, Cu, Pb) exhibit a persistence of the vapor cloud for more than 6 ms. The total pulse cycle repetition time of 50 ms allows the atomic cloud generated in the HCL during the high-current mode to clear before measurement is made again in the low-current mode for a broad range of elements. Figure 10(b) presents the conventional lamp emission line profile observed at low lamp current and the broadened and self-reversed lamp emission line during the high-current operation of the lamp. It is clear that the narrow line profile emitted under low-current operation is affected equally by atomic absorption or a broad-band spectral feature. In contrast, the broadened and self-reversed profile obtained at high current is affected by a broad-band absorber or scatterer but there is not so much overlap with the atomic absorption line profile. Absorbances calculated under low- and high-current operation are subtracted to yield a difference value which is free of any broad-band contribution.

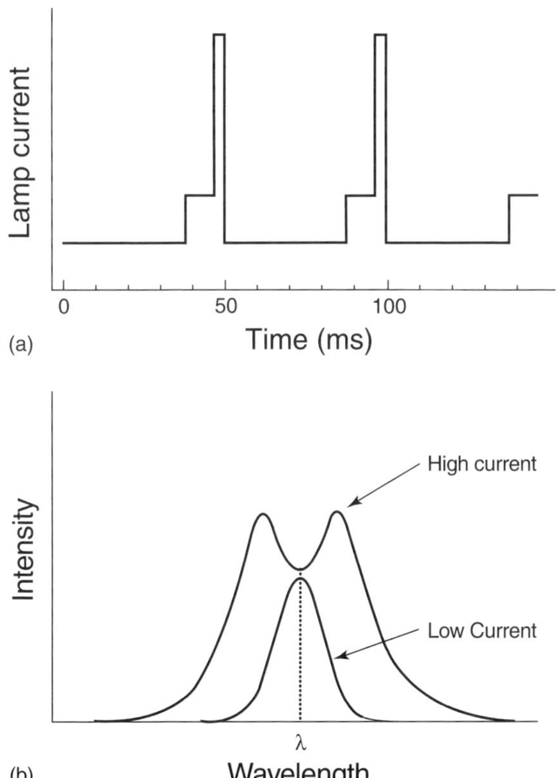

Figure 10 Pulsed HCL background correction. (a) Modulation of the lamp current that drives the HCL; (b) corresponding lamp emission line profiles generated under low- and high-current operation.

5.3 Applications

The pulsed lamp background correction method is equally applicable to FAAS, hydride generation AAS and graphite furnace AAS. For effective operation it is necessary to broaden appreciably the emission line of each element under investigation. However, the degree of broadening is different for each element and also depends upon the peak current at which the HCL is driven. To achieve the desired degree of linc broadening and line reversal while maintaining HCL reliability and acceptable lifetime, a special design of the lamp modulation circuitry and optimum high lamp current adjustment for each element are required. The slope of the analytical curves is decreased to some extent when pulsed HCL background correction is applied. The measurement at high lamp current is intended to measure the background. Unfortunately, the broadening of the lamp emission line profile is limited and self-reversal is not complete. The remaining radiation in the central part of the lamp emission line profile is therefore absorbed by analyte atoms and subtracted from the gross absorbance together with the background. For 30 elements most commonly determined by AAS the loss of sensitivity is 47%, ranging from 13% for Cd to 84% for Hg. Similarly to the Zeeman background correction system, the ETAAS signals show a dip at very high analyte concentrations when pulsed HCL background correction is applied. The ability of the pulsed HCL background correction system to overcome background interferences in real samples has been demonstrated. Like the Zeeman approach, the pulsed lamp background correction method requires only a single primary source of radiation and single beam optics so that alignment is simplified.

6 WAVELENGTH-MODULATION CORRECTION METHODS

6.1 Principle

The broad-band spectral distribution of a continuum-source conveniently permits off-line background measurements to be performed in AAS similar to the procedures employed traditionally in atomic emission spectroscopy. Background with a flat spectral distribution requires simply a two-point background correction procedure in which the background absorption at the analytical

For references see page 9378

absorption line is accounted for by a second measurement with the monochromator tuned to an arbitrary off-line wavelength position. The difference between the two measurements provides the net analytical signal. For an absorption line superimposed on a linearly sloping background, two background measurements, one on either side of the absorption line, would be required. The background absorption at the analytical absorption line in this case would be equivalent to the interpolated value of the two off-line measurements. Clearly, with increasing complexity of the background in the immediate vicinity of the analyte absorption line, manual off-line correction procedures would become exceedingly tedious and time-consuming. The use of wavelength-modulation in AAS in conjunction with a continuum primary source of radiation was first described by Snelleman.(20)

6.2 Instrumentation

A vibrating mirror, mounted on the axis of a milliammeter, is positioned in the optical path between the focusing mirror and the exit slit and driven by an AC current which vibrates the mirror at a fixed frequency. The analytical absorption line, when scanned across the exit slit, generates an AC component in the photosignal with a frequency twice that of the mirror vibration frequency. Phase-sensitive signal detection discriminates against DC and first-harmonic AC components of the photosignal due to flat and linearly sloping background distributions, respectively, over the narrow wavelength interval.

Zander et al.(21) combined a high-intensity continuum-source and a wavelength-modulated high-resolution échelle monochromator. A quartz refractor plate modulator was mounted vertically in front of the exit slit. Sinusoidal modulation of the quartz refractor plate scanned the image of the entrance slit rapidly back and forth across the exit slit. In later studies,(22) the refractor plate was positioned immediately behind the entrance slit in order to wavelength modulate the incident nondispersed radiation, permitting wavelength modulation of multiple predetermined wavelengths in the use of a multichannel échelle polychromator. Figure 11 shows a block diagram of a single-channel continuum-source AAS instrument based on a wavelength-modulated high-resolution échelle spectrometer. In this system the exit slit is replaced by a multichannel exit slit unit consisting of 20 prealigned exit slits for different atomic lines and a photomultiplier is positioned behind each slit. Later, more sophisticated detection systems as a linear photodiode array and a segmented array charge-coupled device detector(23) have been used in combination with the échelle optics.

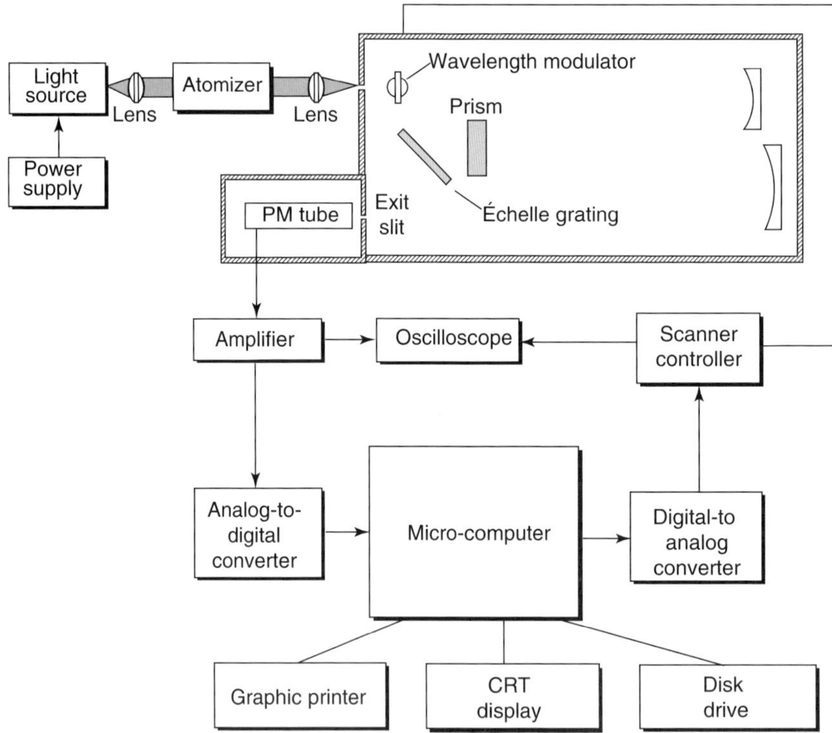

Figure 11 Block diagram of a single-channel continuum-source AAS spectrometer based on a wavelength-modulated, high-resolution échelle spectrometer. (Reprinted from T.C. O'Haver, J.D. Messman, *Prog. Anal. Spectrosc.*, **9**, 483–503 (1986), Copyright 1986, with permission from Elsevier Science.)

6.3 Applications in Continuum-source Atomic Absorption Spectroscopy

Simultaneous multielement AAS based on wavelength-modulation background correction provides the basic raw data to cover the wide range of concentrations likely to be encountered in different samples because the entire absorption profile can be observed. The intensities at several points across the modulation interval are measured. Using the points at the extreme ends of the modulation interval as I^0 and those near the center and along the sides of the line as I, several absorbances with different sensitivities are calculated. This set of absorbances is measured and stored for every sample, blank and standard solution. From the set of absorbances measured for a series of standard solutions, a set of calibration graphs with different slopes is obtained. The set of absorbances measured for each sample and blank are converted to concentrations by reference to the corresponding calibration graphs. This procedure yields a set of concentration estimates for each element in each sample. The final concentration estimate is a weighted average based on the concentration S/N of each individual estimate. The line-center absorbance is utilized at low concentrations and, as the concentration increases, the lower sensitivity absorbances are progressively utilized. The individual calibration graphs are not linear over the entire range. Therefore, a certain number of standard solutions are required to characterize each calibration curve adequately. The wavelength-modulation background correction approach shares with the Zeeman and pulsed lamp methods the advantage of using only a single light source, so that lamp alignment is not so much of a problem as in the continuum-source method, and of making the background measurement at wavelengths very close to the analytical wavelength. Consequently, errors due to background structure and matrix atomic absorption are less likely to occur. The background correction performance of the continuum-source wavelength-modulation method has been evaluated experimentally for a variety of samples.

7 COMPARISON OF METHODS

7.1 Sensitivity and Dynamic Range

From the principle of continuum-source background correction (section 3.1), it is clear that the conventional AAS sensitivity is not influenced by the background correction system. The additional measurement with the deuterium lamp provides the background absorbance and the analyte atoms do not contribute significantly to the background signal because of their narrow absorption line profile.

In all Zeeman background correction systems, the measurement that is intended to provide the background absorbance is slightly influenced by the absorption of the analyte atoms owing to the incomplete shift of the Zeeman σ-components away from the lamp emission line profile. The contribution of the remaining σ-components decreases the net analyte absorbance by about 10% for most elements and by up to 45% for a few elements (Cu, Be, Bi). A further decrease in sensitivity is observed in Zeeman background correction systems based on a constant magnetic field from a permanent magnet or a DC magnet for all elements that show anomalous Zeeman splitting patterns. The sensitivity reduction is more severe (about 50%) when a central π-component is missing. The best Zeeman AAS sensitivity for all elements is obtained in a strong AC magnetic field. The best result in a constant magnetic field of optimum strength is about half the conventional AAS sensitivity, except for those elements that show normal Zeeman splitting patterns.

Pulsed lamp background correction provides sensitivity close to conventional AAS for a few elements only. It is hard to achieve sufficient broadening and self-reversal in the lamp emission line profile during the high-current pulse. Similarly to Zeeman AAS, the contribution of the remaining part of the lamp line profile that is insufficiently shifted is subtracted from the gross absorbance together with the estimated background. The loss of sensitivity varies from 13% for Cd to 84% for Hg.

In wavelength-modulation background correction, analytical curves of different sensitivity are obtained simultaneously. The sensitivity can be selected between the normal value for conventional AAS and much lower values when the analytical information comes from the wings of the line profile. The various calibration graphs become available simultaneously. In comparison with atomic emission spectroscopy, the dynamic range of AAS is limited. Starting from the detection limit, absorbances can be measured over about three orders of magnitude, i.e. from 0.001 up to 1 absorbance unit. This range is not influenced by the continuum-source background correction system. In Zeeman and pulsed lamp background correction systems the dynamic range is further limited owing to the roll-over phenomenon. The analytical curves in peak absorbance level off at somewhat lower absorbances than in conventional AAS whereas the analytical curves in integrated absorbance continue to increase, slowly, up to very high analyte concentrations. In contrast, the wavelength-modulation background correction system offers various analytical curves of different slopes and using them all together the dynamic range is significantly increased in comparison with conventional AAS.

For references see page 9378

7.2 Frequency of the Background Correction System

All background correction systems involve sequential measurements of the gross absorbance and the background according to Equation (1). In the continuum-source background correction system this requires switching, at a certain frequency, between the HCL and the continuum source. The frequency should be high enough to obtain a good estimate of the background at the time of the measurement of the gross absorbance. In ETAAS, the atomic and background absorption signals may rise to a maximum in times ranging from 0.1 s or less to several seconds, depending on the furnace design, the heating rate and the composition of the sample. The beam switching frequency should be high enough to enable the fastest rates of change of the atomic and background absorption signals to be followed. The upper limit to the chopping frequency follows from practical considerations whereas HCLs can be pulsed at very high frequencies. A high enough background correction frequency of 200–400 Hz can easily be achieved.

In Zeeman background correction systems the frequency is related to the modulation frequency of the AC magnetic field or the frequency of rotating the polarizer in the case of a constant magnetic field. In most of the instruments the background correction frequency is 50–100 Hz. It is complicated to rotate a polarizer at high speed and it is expensive to modulate an AC magnetic field at a frequency higher than the mains frequency. The pulsed lamp background correction system also is not very fast (20 Hz) because the lamp needs time to recover after the high lamp current pulse. In all background correction approaches, an increase in the frequency decreases the time available to measure the signals and consequently decreases the S/N.

Figure 12 shows that for a short interval during the atomization step in ETAAS, each background correction system makes repeated, rapid, sequential measurements of the gross sample absorbance and background absorbance.[24] These measurements are usually spaced evenly in Zeeman background correction and in wavelength-modulation background correction whereas an uneven spacing of the measurements of gross sample absorbance and the background is found in pulsed lamp background correction systems. For accurate background corrections, the background absorbance must be known at the same time as the gross sample absorbance. Ideally, a mathematical function with the same shape as the background absorbance function would be fit to the background measurements and the background absorbance could be predicted at any time. Various interpolation procedures are used in ETAAS instruments to avoid or reduce background correction errors when the background is changing rapidly.

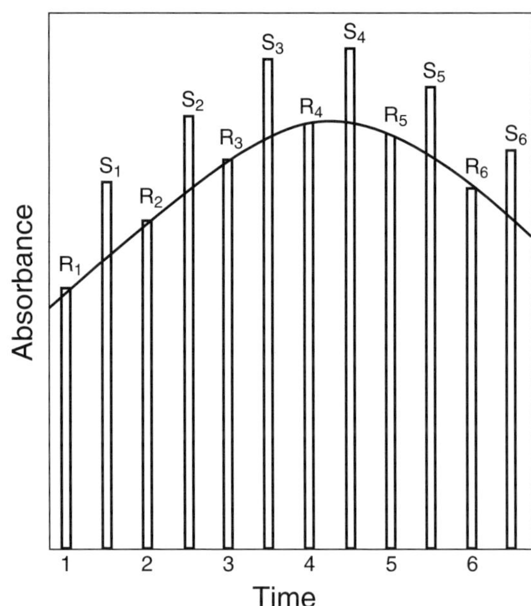

Figure 12 Illustration of repetitive, sequential data aquisition of the gross absorbances of the sample (S_1-S_6) and background absorbances (R_1-R_6) as a function of time for a short interval during the atomization of a sample in ETAAS with analyte and background present. (Reproduced by permission of The Royal Society of Chemistry from J.M. Harnly, J.A. Holcombe, *J. Anal. At. Spectrom.*, **2**, 105–113 (1987).)

7.3 Background Correction Performance

The accuracy of the correction for background absorbance must be perfect up to high levels of background absorption. In the absence of analyte, the baseline should be a perfectly straight line even at high background absorbance levels. Three conditions should be fulfilled to accomplish this. In the previous section we have seen that the transient ETAAS signal requires a good approximation of the background at the time of measurement of the gross absorbance signal. In addition, the background must be measured at the same position in the atomizer. This is easily achieved if only one source of radiation is used as in the Zeeman background correction system, the pulsed lamp background correction system and the wavelength-modulation system. In continuum-source background correction, lamp alignment is always difficult because of the different positions of the lamps in the instrument, the different optical paths, the different geometries of the beams and the different intensity distributions. Since the atomic vapor in both the flame and the graphite furnace is not truly homogeneous, nor is the radiant cross-section of either source, perfect matching is difficult. Background absorbance in FAAS is usually low and it can be corrected easily with the continuum-source background correction system.

In ETAAS, the continuum-source background correction system is often also successful. Background correction problems have been observed when the background is higher than about 0.5–1 absorbance unit, often owing to misalignment of the radiation beams. Many instruments are fitted with both a deuterium lamp for the UV and a tungsten lamp for the visible wavelength region. The intensity of the deuterium lamp is quite low when a low spectral band-pass is selected, which may be a limiting factor for the HCL intensity also because most instruments require more or less equal intensities of the two radiation beams.

Continuum-source background correction systems are not able to correct for structured background. The background is measured as an average background absorbance level over the spectral band-pass of the monochromator, i.e. within 0.2–2 nm. Numerous over- and undercorrections have been reported. Several of them are well known and can be found in text books on AAS, others are only observed under very special experimental conditions. Zeeman-, pulsed lamp- and wavelength-modulation-based background correction systems measure the background at exactly the wavelength position of the analyte absorption line or very close to line, i.e. within an interval of 10 pm. Although perfect background correction is not guaranteed, there is a better chance that the right background absorbance is measured at a wavelength position close to the absorption line in comparison with the average value over the full spectral band-pass. The Zeeman background correction system that applies the magnetic field to the atomizer is the only system that provides correction for the background exactly at the wavelength of the analyte absorption line. Even here, perfect background correction is not guaranteed because, in a very few cases, the background may change with the strength of the magnetic field in AC Zeeman systems. Most of the background correction errors reported for continuum-source background correction are not found when Zeeman, pulsed lamp or wavelength-modulation background correction is applied.

8 BACKGROUND CORRECTION PROBLEMS AND HOW TO AVOID THEM

The most frequent spectral interference in continuum-source background correction results from atomic absorption of the background radiation by massive amounts of metals in the matrix, usually at secondary absorption lines of these elements. However, some of the structured absorption errors result from molecular absorption in the vapor phase. Probably the first identified example of background correction error was the influence of large amounts of Fe in the determination of Se at the 196-nm line. A list of potential interferences of this type[17] shows errors that may occur, based on the wavelengths of the various lines. Some of the problems are very unlikely to be serious in FAAS. In the graphite furnace the amount of matrix is typically much larger relative to the analyte and therefore interferences can be more severe. However, during method development much time and effort are usually spent in obtaining an acceptable program for heating of the graphite furnace and the application of chemical modifiers is also considered. Owing to the improved graphite furnace design, the temperature program used and the application of chemical modifiers, potential interferences and background correction errors are not always found in real analysis.

Background correction errors are hard to identify when the background absorbance takes place simultaneously with the atomic absorption. If there is some time delay between the two processes, a positive or negative disruption of the baseline may be observed just before the atomic absorption starts or just after the atomic absorption is finished. It should be noted that the method of standard additions does not control background correction errors.

Several problems of background overcorrection have been noted that have not yet been fully explained whereas others are well understood. For example, the effect of phosphate on Se and As is related to molecular P_2 that is generated in the vapor phase of the furnace and the molecular band absorbs the continuum radiation.

With Zeeman background correction, the pulsed lamp background correction system or wavelength-modulation, most of the interferences reported for continuum-source background correction are not observed. Nevertheless, a few interferences have been reported.[16] In most of the commercial Zeeman instruments, background correction is achieved by using an AC magnetic field transverse to the optical axis. If matrix absorption lines or molecular bands that are very close to the analyte wavelength exhibit Zeeman splitting, correction errors may occur. A matrix absorption line that does not overlap with the lamp emission line at zero-field may do so at maximum field strength, or the other way around. An example of such problems is the PO molecular absorption band that causes an overcorrection for lead in the presence of phosphate. The overcorrection error is influenced by the strength of the magnetic field. It is diminished by decreasing the amount of phosphate or by replacing the standard transverse-heated graphite atomizer with an end-capped tube. Accurate results for the determination of Pb in bone show that such background correction errors can be avoided by optimization of the experimental conditions during method development.

For references see page 9378

ABBREVIATIONS AND ACRONYMS

AAS	Atomic Absorption Spectroscopy
AC	Alternating Current
DC	Direct Current
ETAAS	Electrothermal Atomization Atomic Absorption Spectroscopy
FAAS	Flame Atomic Absorption Spectroscopy
HCL	Hollow Cathode Lamp
S/N	Signal-to-noise Ratio
UV	Ultraviolet

RELATED ARTICLES

Environment: Water and Waste **(Volume 3)**
Flame and Graphite Furnace Atomic Absorption Spectrometry in Environmental Analysis

Environment: Water and Waste cont'd **(Volume 4)**
Slurry Sampling Graphite Furnace Atomic Absorption Spectrometry in Environmental Analyses

Steel and Related Materials **(Volume 10)**
Atomic Absorption and Emission Spectrometry, Solution-based in Iron and Steel Analysis

Atomic Spectroscopy **(Volume 11)**
Atomic Spectroscopy: Introduction • Flame and Vapor Generation Atomic Absorption Spectrometry • Graphite Furnace Atomic Absorption Spectrometry

REFERENCES

1. H. Massmann, S. Güçer, 'Physikalische und chemische Prozesse bei der Atomabsorptionsanalyse mit Grafitrohröfen – I. Messbedingungen, Versuchsanordnung und Temperaturmessungen', *Spectrochim. Acta, Part B*, **29**(11/12), 283–300 (1974).
2. R.A. Newstead, W.J. Price, P.J. Whiteside, 'Background Correction in Atomic Absorption Analysis', *Prog. Anal. At. Spectrosc.*, **1**, 267–298 (1978).
3. A.K. Gilmutdinov, T.M. Abdullina, S.F. Gorbachev, V.L. Makarov, 'Concentration Curves in Atomic Absorption Spectroscopy', *Spectrochim. Acta, Part B*, **47**(9), 1075–1095 (1992).
4. J.B. Willis, in *Methods of Biochemical Analysis*, ed. D. Flick, Wiley-Interscience, New York, 1–67, Vol. 11, 1963.
5. S.R. Koirtyohann, E.E. Picket, 'Background Corrections in Long Path Atomic Absorption Spectrometry', *Anal. Chem.*, **37**, 601–603 (1965).

List of selected abbreviations appears in Volume 15

6. H. Massmann, Z. El Gohary, 'Analysestörungen durch strukturierten Untergrund in der Atomabsorptionsspektrometrie', *Spectrochim. Acta, Part B*, **31**(7), 399–409 (1976).
7. D.C. Manning, 'Spectral Interferences in Graphite Furnace Atomic Absorption Spectroscopy. I. The Determination of Selenium in an Iron Matrix', *At. Absorpt. Newsl.*, **17**(5), 107–108 (1978).
8. P. Zeeman, 'On the Influence of Magnetism on the Nature of the Light Emitted by a Substance', *Philos. Mag.*, **43**(5), 226–237 (1897).
9. H. Koizumi, K. Yasuda, 'A Novel Method for Atomic Absorption Spectroscopy Based on the Analyte–Zeeman Effect', *Spectrochim. Acta, Part B*, **31**(10–12), 523–535 (1976).
10. F.J. Fernandez, W. Bohler, M.M. Beaty, W.B. Barnett, 'Correction for High Background Levels Using the Zeeman Effect', *At. Spectrosc.*, **2**(3), 73–80 (1981).
11. M. Prügger, R. Torge, 'Vorrichtung zur Spektrochemischen Feststellung der Konzentration eines Elementes in einer Probe', *Ger. Pat.*, 1964469, filed 23 December, 1969.
12. H. Koizumi, K. Yasuda, 'New Zeeman Method for Atomic Absorption Spectrophotometry', *Anal. Chem.*, **47**, 1679–1682 (1975).
13. P.R. Liddell, K.G. Brodie, 'Application of a Modulated Magnetic Field to a Graphite Furnace in Zeeman Effect Atomic Absorption Spectrometry', *Anal. Chem.*, **52**, 1256–1260 (1980).
14. M.T.C. de Loos-Vollebregt, L. de Galan, J.W.M. van Uffelen, 'Longitudinal a.c. Zeeman AAS with a Transverse Heated Graphite Furnace', *Spectrochim. Acta, Part B*, **43**(9–11), 1147–1156 (1988).
15. I. Shuttler, 'The Application of a Transversely Heated Electrothermal Atomizer with Longitudinal Zeeman-effect Background Correction to the Determination of Vanadium in Urine', *At. Spectrosc.*, **13**(5), 174–177 (1992).
16. M.T.C. de Loos-Vollebregt, L. de Galan, 'Zeeman Atomic Absorption Spectrometry', *Prog. Anal. At. Spectrosc.*, **8**, 47–81 (1985).
17. W. Slavin, G.R. Carnrick, 'Background Correction in Atomic Absorption Spectrometry', *CRC Crit. Rev. Anal. Chem.*, **19**(2), 95–134 (1988).
18. C. Ling, 'Sensitive Simple Mercury Photometer Using Mercury Resonance Lamp as a Monochromatic Source', *Anal.Chem.*, **39**, 798 (1967).
19. S.B. Smith, Jr, G.M. Hieftje, 'New Background-correction Method for Atomic Absorption Spectrometry', *Appl. Spectrosc.*, **37**, 419–424 (1983).
20. W. Snelleman, 'An a.c. Scanning Method with Increased Sensitivity in Atomic Absorption Analysis Using a Continuum Primary Source', *Spectrochim. Acta, Part B*, **23**(6), 403–411 (1968).
21. A.T. Zander, T.C. O'Haver, P.N. Keliher, 'Continuum-source Atomic Absorption Spectrometry with High-Resolution and Wavelength-Modulation', *Anal. Chem.*, **48**, 1166–1175 (1976).

22. T.C. O'Haver, J.D. Messman, 'Continuum-source Atomic Absorption Spectrometry', *Prog. Anal. Spectrosc.*, **9**, 483–503 (1986).
23. J.M. Harnly, C.M.M. Smith, D.N. Wichems, J.C. Ivaldi, P.L. Lundberg, B. Radziuk, 'Use of a Segmented Array Charge Coupled Device Detector for Continuum-source Atomic Absorption Spectrometry with Graphite Furnace Atomization', *J. Anal. At. Spectrom.*, **12**, 617–627 (1997).
24. J.M. Harnly, J.A. Holcombe, 'Mathematical Correction of Systematic Temporal Background-correction Errors for Graphite Furnace Atomic Absorption Spectrometry', *J. Anal. At. Spectrom.*, **2**, 105–113 (1987).

Flame and Vapor Generation Atomic Absorption Spectrometry

M.D. Amos
Emerald, Vic., Australia

Atomic absorption spectrometry (AAS) is a means of specific analysis for most of the metallic and metalloid elements. It achieves its specificity from the highly selective absorption, by free atoms of the analyte, of atomic emission of the same element at characteristic wavelengths. In its simplest form of application the free atoms of the analyte are created by aspiration of a solution of the sample to be analysed into a suitable flame, flame atomic absorption spectrometry (FAAS). The other principal means employed for the production of atoms (atomization) are graphite furnace atomization (GFAAS) and vapor generation atomization (VGAAS). Sample volumes of a

For references see page 9401

few milliliters are normally needed for FAAS and VGAAS, but smaller volumes are needed for GFAAS. For most of the approximately 65 elements amenable to measurement by AAS, detection limits are from parts per billion to parts per million by flame methods and from parts per trillion to parts per billion levels by GFAAS and VGAAS. FAAS is inherently simple and rapid to use, whereas GFAAS and VGAAS are less rapid. The instrumentation used for AAS is relatively inexpensive Since all of these methods generally require the sample to be in solution form, the need for dissolution may be seen as a limitation in some cases. In the most common instrumentation employed, elements are only measured one at a time. It follows that for the determination of many elements in a large group of samples the method must be seen to be slow by comparison with multielement techniques such as inductively coupled plasma mass spectrometry (ICPMS) and inductively coupled plasma atomic emission spectrometry (ICPAES). However, the equipment costs involved for these techniques are generally higher and they are more complex to employ.

1 INTRODUCTION

1.1 The Origins of Atomic Spectra

Although astronomical observations led to the discovery of the 'Fraunhofer (atomic absorption) lines' in the solar spectrum, it was not until about 1850 that the origin of these dark lines in an otherwise continuous bright spectrum started to be properly deduced. It appears that several workers, among them Foucault, Balfour Stewart and Stokes started to realize that these lines were the result of absorption of the continuous emission from the sun and other astronomical bodies by the atomic vapor of some elements occurring in their atmospheres. It was not until 1860 that Bunsen and Kirchhoff first comprehensively demonstrated experimentally, and then elucidated, the origin of such atomic emission and absorption spectra and drew attention to their interrelationship.

In what are now seen as classical experiments they demonstrated that atomic vapors of sodium and lithium, produced in a flame, would each not only emit radiation at specific wavelengths but would also absorb radiation at precisely the same wavelengths.[1,2] Kirchhoff's original paper[1] is given only for completeness but the paper, and of course the title, are in German. However an English translation of this paper is also available.[2] From these results Kirchhoff elucidated that atomic transitions between different energy levels might be observed either using absorption or emission.

He found that if, for example, a solution of a sodium salt was heated in a flame it would then spontaneously emit or absorb radiation at wavelengths characteristic *for sodium only*, the famous 'D lines'. Likewise a solution of a lithium salt would also, under the same circumstances, emit or absorb radiation at a wavelength different from those of sodium but equally characteristic *for lithium only*. By doing this he essentially drew attention to the potential usefulness of such observations to identify the presence or absence of sodium and lithium (qualitative analysis). This work was later extended, by Bunsen and Kirchhoff, to the qualitative determination of the presence of other elements (potassium, calcium, strontium and barium).[3]

As an interesting example of the powerful usefulness of atomic spectra, these workers also first discovered the existence of the elements rubidium and cesium by observing the presence of other characteristic (and unexpected) lines when employing solutions of natural salt for their experiments.[4] The salt they used was not produced from the evaporation of seawater but rather was natural salt from underground deposits.

(Note that the Kirchhoff articles[3,4] are also translations of articles originally published in *Annalen der Physik und Chemie*.)

Bunsen and Kirchhoff only employed these observations in a qualitative mode and seem to have drawn no attention to the possible quantitative use of the phenomena of atomic emission and absorption.

This does not seem too surprising given that in 1860 means of quantitative measurement of emission or absorption signals were not readily available.

1.2 First Analytical Applications of Atomic Spectra

Later, starting with the further development of the photographic process, the means of recording intensities for at least semiquantitative purposes was to lead to the development of emission spectrography using (initially) electric arcs as the means of excitation.

Flames were also seen to be a potential means of excitation for observing emission and by 1928 Lundegardh developed suitable burners (using acetylene as the fuel gas and air as the support gas) and methods for sample introduction.[5] From this time on, and for the period to the 1950s, developments continued in the above methods but the use of methods utilizing atomic absorption spectra were almost totally ignored. As will be described in section 2, this was in major part because the means of measuring atomic absorption with good sensitivity was not yet properly developed.

1.3 Suggestions for the Analytical Use of Atomic Absorption Spectra

After a significant period of research into atomic emission spectrometry (AES), Walsh became convinced that this neglect of the use of atomic absorption had gone on for too long. He recognized that there would be potential advantages in using the measurement of atomic absorption instead of atomic emission as a means of measuring the concentration of elements in samples.

The major advantages he postulated for the use of atomic absorption methods were as follows:

- Since ground-state atoms would be used for the measurement, the method should be independent of the excitation potential of the element to be measured.
- The method should be much less influenced by the temperature of the atomizing device used.
- There should be much less likelihood of spectral interferences impairing the accuracy of the method.

In 1955, Walsh published a landmark paper explaining, in major detail, these theories Additionally, however, he also defined a practical approach to the necessary key elements required for instrumentation to measure atomic absorption.[(6)]

Completely independently, Alkemade and Milatz also published an article suggesting the use of atomic absorption spectra.[(7)] They, however, did not develop the practical means of performing this measurement to the same extent as Walsh and this limited the importance of their paper in a practical sense.

In the following section, the theoretical basis of atomic absorption and its measurement will be dealt with in more detail.

2 THEORETICAL BASIS OF ATOMIC ABSORPTION SPECTROMETRY

2.1 General Comments

At this juncture, any comparisons made will be solely between flame emission and flame atomic absorption and no further reference will be made to emission measurements made following excitation by electric arc or spark or by inductively coupled plasma. These are all established methods in widespread use, although arc and spark spectrometry is currently in much less use than in previous years.

In order to measure atomic absorption, it is first necessary to produce free atoms of the element to be measured. However, it should be realized that it is also necessary to produce free atoms of any element to be measured by atomic emission. Atoms cannot be excited unless they are first converted to free atoms from any compound associations.

2.2 Atomization and Excitation

In most atomic absorption, or flame emission, methods atoms are produced from a sample in solution form by the application of heat. If a solution sample is introduced, for instance, into a flame then atomization will often occur. Depending on the excitation potential of the element in question, some of these atoms produced may also then be excited. It is this atomization, and subsequent excitation, which forms the basis of flame emission spectrometry. However, as Walsh pointed out, there are some important factors which should be considered. These are best expressed in some figures taken from a table in his first paper, shown here in Table 1.

The figures in Table 1 show that

- the fraction of excited atoms at any time is always a very small fraction of the total atoms present;
- the fraction of excited atoms changes by a huge amount with any major temperature change;
- since the fraction of ground-state atoms is equal to one minus the fraction of excited atoms, this remains sensibly constant at very close to 1, regardless of temperature change.

Since atomic emission signals originate from excited atoms and atomic absorption signals from ground-state atoms, the consequences of the above are the following:

- Atomic emission signals will be dramatically affected by temperature changes but absorption signals will be affected to a much lesser extent.
- For many harder to excite elements (such as Zn in Table 1), atomic absorption will be much more sensitive than atomic emission at such temperatures.
- All excited states will, at any given time, have a very small population of atoms. As a result, sensibly only those transitions which originate in the ground state will be able to be measured by atomic absorption. It

Table 1 Values of N_j/N_0 for various resonance lines[a]

Element	Wavelength of resonance line (nm)	N_j/N_0	
		2000 K	3000 K
Cs	852.1	4×10^{-4}	7×10^{-3}
Na	589.0	1×10^{-5}	6×10^{-4}
Ca	422.7	1×10^{-7}	4×10^{-5}
Zn	213.9	7×10^{-15}	6×10^{-10}

[a] N_j is the number of atoms in the excited state at any given time and N_0 is the number of atoms in the ground (unexcited) state at any given time, and consequently the ratio N_j/N_0 is the fraction of atoms in the excited state at any given time. The figures quoted are taken from a table which appears in Walsh.[(6)] They represent only some of the entries in that table and have been rounded off for convenience.

For references see page 9401

follows, then, that since there are only a very limited number of such transitions for each element, the atomic absorption spectrum of all elements will be very much simpler than the atomic emission spectrum of the same element. The major consequence of this is that the likelihood of spectral interference will be much lower for absorption measurements than for emission measurements.

Properly, the resonance line for any element is the line associated with the transition between the ground state and the first (lowest energy) excited state. There is therefore commonly only one true resonance line per element. In the years since atomic absorption analysis was introduced, the convention has arisen that all lines which are associated with transitions involving the ground state are called resonance lines. This atomic absorption convention will generally be followed throughout this article.

2.3 Absorption and Emission Lines

A purely diagrammatic representation of the energy levels involved in the production of an emission and absorption line is shown in Figure 1.

The relationship between the difference of the two energy levels involved in the transition and the frequency of the radiation emitted or absorbed is given by Equation (1):

$$E_1 - E_0 = h\nu \qquad (1)$$

where E_1 and E_0 are the two energy states, h is Planck's constant and ν is the frequency of the radiation emitted or absorbed. In spectroscopy in the ultraviolet and visible regions of the spectrum wavelengths are more commonly used rather than frequency, so Equation (1) may be rewritten as Equation (2):

$$E_1 - E_0 = \frac{hc}{\lambda} \qquad (2)$$

where c is the velocity of light and λ is the wavelength of emission or absorption. Rearranging, we obtain Equation (3):

$$\lambda = \frac{hc}{E_1 - E_0} \qquad (3)$$

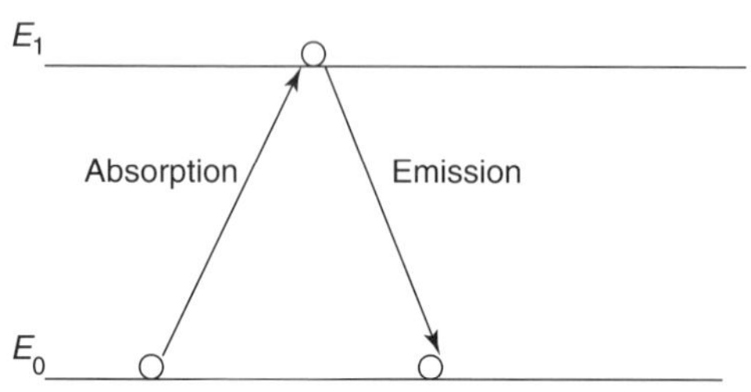

Figure 1 Relationship of emission and absorption lines.

Since h and c are both constants, it follows that the wavelength corresponding to the change of energy of the atom, from one energy level to another, is inversely proportional to the difference between the two energy levels. Since the same two energy levels are involved, it follows that the emission and the absorption wavelengths will be identical since the calculation is the same in both cases.

2.4 The Shape of Spectral Lines

The emission and absorption wavelengths are generally quoted in texts as being, for example, Cu 324.754 nm, and it might often be assumed that this means a line of virtually no shape, i.e. an infinitely narrow line. Such is not the case, however, and lines do have a finite width and shape, both of which will vary under certain conditions. The wavelength designated might be seen as the central wavelength, but there will be present under all conditions a distribution of wavelengths around that central wavelength (called the 'natural' line width). Various factors may increase the line width further ('broaden' the line).

In a short survey such as this article, only the two principle causes of line broadening will be considered. The first of these is Doppler broadening, caused by the fact that in a vapor cloud the atoms will be moving in all directions. Observations of emission from an atom coming towards the point of observation at the time of emission will record a wavelength shorter than that recorded from an atom going away from the point of observation at the time it emits. Obviously, as the temperature is increased the velocity of motion of the atoms is increased and this effect will therefore increase with increasing temperature, producing an increased line width.

The second cause is pressure broadening due to the concentration of atoms and molecules in the vapor in which the emission is occurring. The higher these concentrations are, the greater is the broadening which will occur.

The combined effects of these two (and some other minor) causes of broadening is to produce an emission line with an approximately Gaussian profile, as shown in Figure 2(a). The usual manner of describing the width of this line is the 'half-width', that is, the width at half the maximum peak intensity. The use of this manner of describing line width is necessary because the maximum width which exists, as zero intensity is approached, becomes progressively harder to measure and is dependent on the sensitivity of measurement available and the concentration of the atoms responsible for the absorption or emission.

Under the conditions prevailing in a flame, burning at atmospheric pressure at a temperature in the 2000–3000 K range, the half-width of the absorption lines

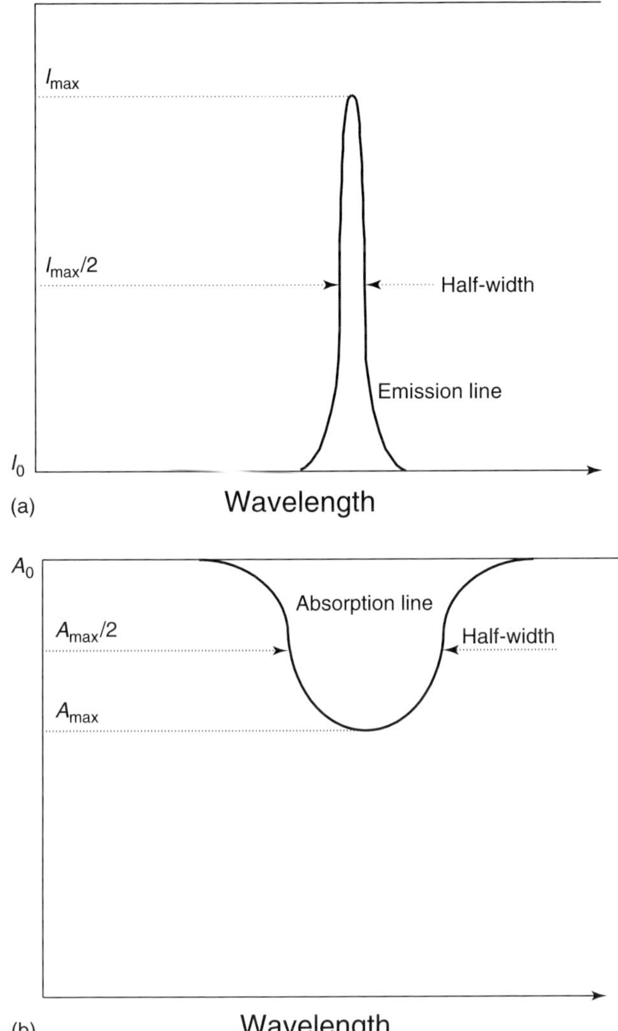

Figure 2 Representation of (a) an emission line and (b) an absorption line.

will be of the order of 0.002–0.01 nm. An absorption line may be drawn, as shown in Figure 2(b), following the convention of drawing absorption lines downwards and emission lines upwards. For the understanding of this diagram, and for descriptions which will follow later, it is necessary to describe just what it illustrates, as far as atomic absorption measurement is concerned.

For any absorption spectrometric method, the absorption signal is always designated in terms of "absorbance" (A) and this is defined as $\log_{10}(I_0/I_t)$ where I_0 is the intensity of the original source radiation and I_t the intensity of the same radiation after it has passed through the absorbing medium.

The absorbance (A) which will occur follows Beer's law, which may be expressed as Equation (4):

$$A = kcl \tag{4}$$

where k is the absorption coefficient, c the concentration of the absorbing species and l the absorption pathlength.

Returning to Figure 2(b), any point on the curve shown may be seen to represent the absorption coefficient at the particular wavelength. As a result, it may easily be understood that at the two wavelengths represented at the arrow points, where the half-width is defined, the absorption coefficient will be exactly half that at the peak of the absorption line, designated A_{max}. From the same reasoning, the absorption coefficient at the extreme left and right of the absorption line will be approaching zero.

2.5 Measuring Atomic Absorption

2.5.1 Using Continuum Source Emission

Early concepts of measuring atomic absorption were based on the idea of measuring the absorption of light from a continuum source, but this approach has limited value for the following reason. The spectral band-pass of a monochromator of modest resolution, say 0.2 nm, can be represented as shown in Figure 3 with a representation of an absorption line with a half-width of approximately 0.01 nm superimposed on it. It can be seen that the sensitivity of this measurement will be limited, since even if the peak absorption at the center of the absorption line is, for example, 40%, perhaps less than 1% of the total energy reaching the detector will be absorbed. In addition, it is obvious that changes of absorption with concentration of the atoms of the analyte will give rise to a nonlinear function.

Naturally, the sensitivity may be improved by using a higher resolution monochromator as represented, for example, as 0.05 nm in Figure 3. However, the absorption

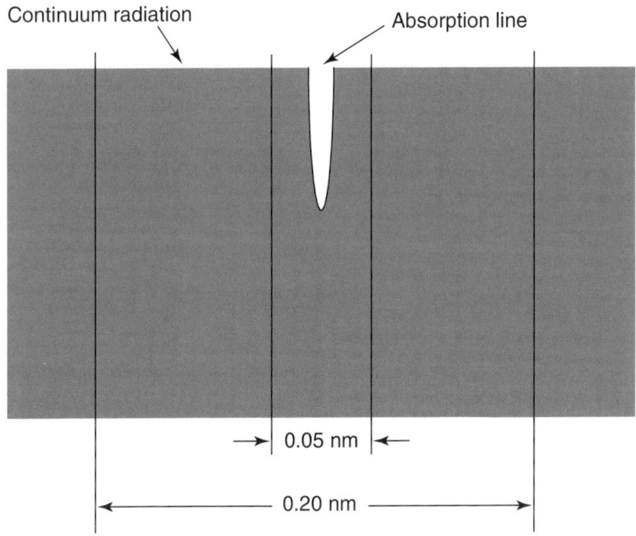

Figure 3 Measurement of atomic absorption using a continuum source.

For references see page 9401

will still probably be less than 2–3% of the total radiation. Also, and equally important, with such a narrow slit opening, the total intensity of radiation from a typical continuum source would be very low and the precision of the measurement function would therefore be limited by this low intensity.

2.5.2 *Using a Sharp Line Emission Source*

One of Walsh's major contributions to the subject was to suggest that peak absorption could be measured if use was made of a line source, providing emission at precisely the same wavelength as the absorption, and having a line width much less than that of the absorption line. This situation is represented in Figure 4.

The means of providing such a sharp line emission, at precisely the same wavelength as the absorption, was to use an individual hollow-cathode lamp (HCL) to provide the emission spectrum for each element to be determined. The construction and operation of such lamps will be discussed more in the next section, but at this stage it is useful to indicate why this sharp line emission occurs.

An HCL operates at reduced pressure, typically 1–2 kPa, so the concentration of atoms and molecules within the tube will be much less than in a flame at atmospheric pressure and this will reduce pressure broadening. Additionally, the temperature of the discharge will be only perhaps 500–600 K, compared with flame temperatures of 2000 K or more, and this will reduce Doppler broadening. Under these conditions, a HCL produces emission lines with half-widths of the order of 0.0004–0.002 nm.

As shown in Figure 4, it should be understood that the major part of the emission signal will be capable

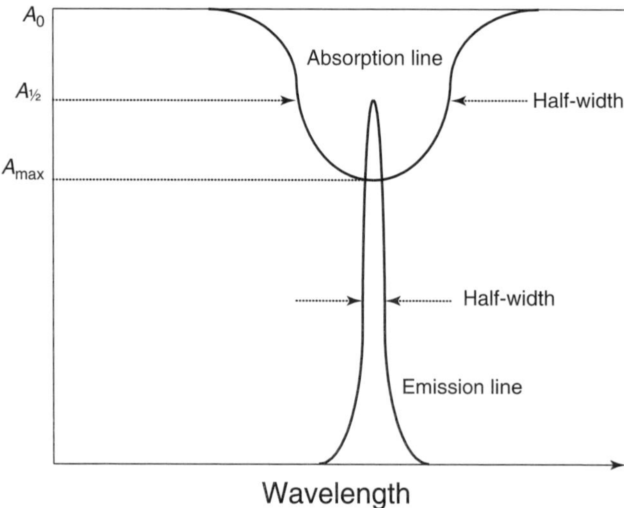

Figure 4 Measurement of atomic absorption using a sharp-line source.

of absorption at or near the center of the absorption line, that is, at the wavelength where the absorption coefficient is at its maximum. This will result in maximum sensitivity of measurement. Quantitatively, the amount of energy at the maximum and minimum wavelengths of the emission line (the left- and right-hand flanks of the line as shown) will be very small. As a result, the fact that at these wavelengths the absorption coefficient will be slightly smaller will have only a small effect on the overall sensitivity achieved.

2.6 Spectral Interferences in Atomic Absorption Spectrometry

True spectral interferences may be defined as spectral interferences which, if undetected and uncontrolled, may lead to an error in the results obtained. There is another phenomenon, here termed "pseudo-spectral interference", which can affect measurements and this will be described first. It is not really a spectral interference since it only contributes to the quality of the measurement function and does not cause incorrect results. This minor problem will be dealt with first.

2.6.1 *Pseudo-spectral Interferences*

As illustrated in Figure 4, it may be seen that the emitted line from the light source used to make the measurement may be absorbed by the atomic absorption process, the amount of absorption being dependent on the concentration of the analyte present. If for whatever reason there is an emission produced by the light source which results in a line within the spectral band-pass of the monochromator, but which is not overlapped by the absorption line, as shown in Figure 5, no interference will be produced.

However, since the extraneous line will not be absorbed, regardless of how great the concentration of the analyte is, some loss of sensitivity of the measurement will occur and the calibration function will be nonlinear. Two means are available which may successfully overcome this problem:

- employ a narrower spectral band-pass from the monochromator by altering the slit opening;
- remove the element which gives rise to the extraneous line.

With regard to the first of these solutions, it should be recognized that in Figure 5 the spectral band-pass is roughly 0.04 nm (if we assume that the half-width of the absorption line is about 0.01 nm). To eliminate the effect of the "interfering" line the slit opening would need to be closed to a spectral band-pass of about 0.02 nm and this would require a relatively high resolution monochromator.

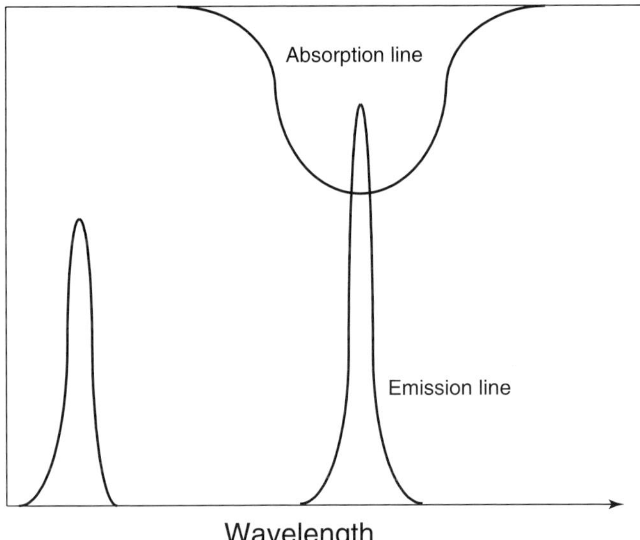

Figure 5 Atomic absorption measurement with a nearby emission line.

The second solution will be possible, for example, if the foreign line is due to the fill gas of the HCL by replacing it with a different fill gas. If the foreign line is due to an element other than the analyte, and which is introduced in the construction of the cathode, then the construction of the cathode might be modified to remove this element.

Should the foreign line be a nonresonance line of the analyte itself, then the second solution is not possible and only the first solution can be used either to remove or at least to minimize this curvature of calibration. It must be emphasized again, however, that curvature of calibration is only a minor operational problem and does not in itself cause errors in the analysis.

2.6.2 True Spectral Interferences

By comparison, again with Figure 4, if another element has an absorption line within the spectral band-pass of the monochromator, as illustrated in Figure 6, then no interference will occur. This is because there is no overlap of the other absorption line with the emission from the sharp line source at the resonance wavelength.

Should there be, however, a much closer absorption line, such as shown in Figure 7, then interference is possible since the left-hand flank of the absorption line of B does, as represented here, slightly overlap the right-hand flank of the emission line of the analyte (A).

It should be appreciated, however, that, in addition to the separation of absorption lines, the relative concentrations of the two elements A and B will affect this situation. In the absorption line for element B, at low concentration (shown as the full line) there is a small overlap but as represented it would cause only a slight absorption of a very small part of the total emission signal. However, the absorption due to a higher concentration (represented by the dotted line) is at a greater absorption coefficient, and the overlap is greater. This means that the potential interference will be more significant. Note that the absorption line, as represented, is actually effectively wider. This is because the concentration of atoms absorbing at these flank wavelengths is greater and so the effect will be more observable. Remembering that $A = kcl$, this may be seen as the effect that even though k may be very low as zero absorption coefficient is approached, the effect of multiplying a low k by a high c will result in a larger signal.

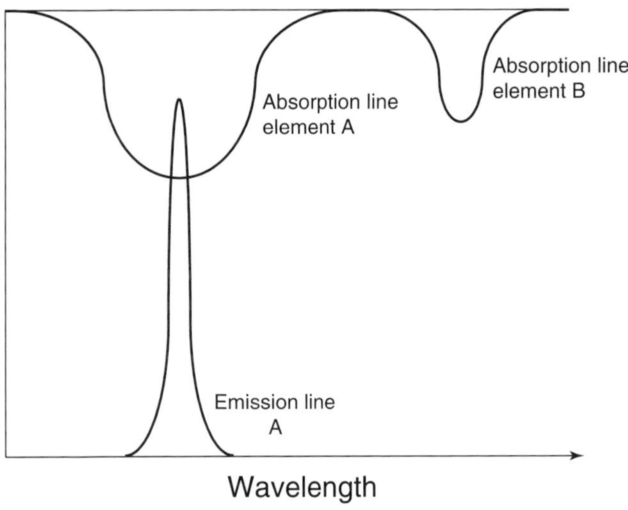

Figure 6 Atomic absorption measurement with a nearby absorption line.

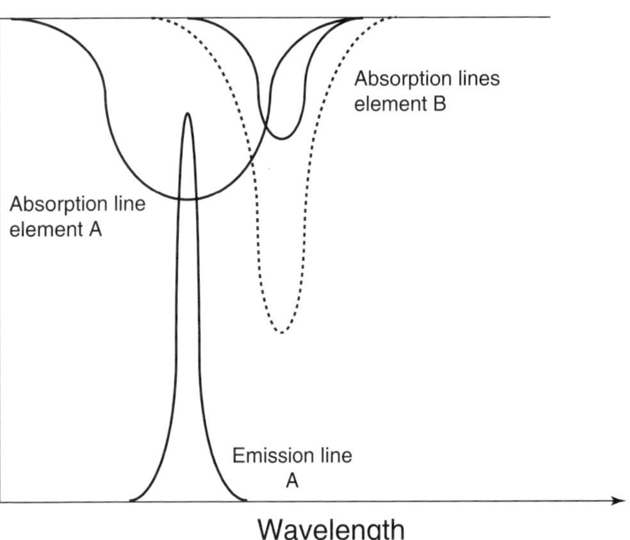

Figure 7 Interfering atomic absorption line.

For references see page 9401

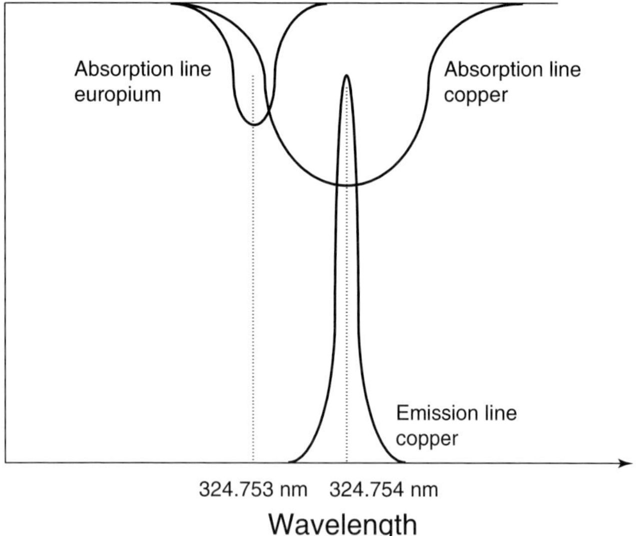

Figure 8 Atomic absorption measurement of copper in the presence of europium.

Nothing can be done about this type of interference except to use a different resonance line of A for the measurement, since clearly limiting the spectral band-pass of the monochromator will not remove the problem.

When the method was first proposed, it was suggested that since the total atomic absorption spectra of all elements were relatively simple, such interferences might be expected to be rare. This has proven to be the case and no more than 10–12 such interferences have been reported, often rather exotic and unlikely to be encountered in practice. For example, such an interference does occur between copper and europium, as illustrated in Figure 8. The most sensitive line of copper at 324.754 nm is very close to an absorption line of europium at 324.753 nm. Referring to Figure 8, if we assume that the half-width of the emission line is 0.0002 nm, then the separation of the central wavelengths of the two absorption lines shown is about 0.001 nm. As may be seen, the copper emission line from the HCL is slightly overlapped by the tail of the europium absorption line. This line is not the most sensitive line for europium so its absorption coefficient would be very low, and additionally the overlap is very small.

In fact, significant interference will only be observed if the Eu–Cu ratio is greater than about 500 : 1. However, if copper needed to be determined, in a sample which contained a very high concentration of europium an error would occur. In what is, after all, a rare analytical situation, the solution to the problem would be to measure copper at the alternative resonance wavelength of 327.4 nm where no such interference occurs and where the sensitivity for copper measurement is reduced by a factor of only about two.

Another case of true spectral interference is the possible absorption of the resonance line of the analyte by molecular absorption. Some molecular species, which might be produced in the atomizer from certain sample types, might have absorption bands which overly the resonance wavelength. Such nonspecific absorption can occur in the presence of high salt concentrations, particularly at wavelengths in the low-ultraviolet region of the spectrum (below 300 nm).

This type of interference is seldom very major in flame atomization but may be needed for the most precise analyses, particularly for measurements at very short wavelengths. It is, however, very important in furnace atomization methods and for this reason background correctors have been developed and will be discussed briefly in the next section.

3 INSTRUMENTATION FOR ATOMIC ABSORPTION SPECTROMETRY

Following the publication of his first paper,[6] Walsh, with some colleagues, described a suitable instrument for the practical measurement of atomic absorption spectra.[8]

The essential component sections of an atomic absorption spectrometer are shown in Figure 9. As illustrated these are as follows:

- a light source (usually an HCL) which provides a sharp line emission at the resonance wavelength for the element to be measured (analyte);
- an atomizer (shown here as a flame) which produces free atoms of the analyte;
- a wavelength selection device (usually a monochromator) which allows the isolation of this resonance line from other lines produced in the spectrum of the lamp;
- a detector (usually a photomultiplier tube) which measures the intensity of the signal which is passed by the wavelength selection device;
- an amplifier–read-out device which processes the signal from the photomultiplier and converts it into a suitable form from which the result is obtained.

Each of these component sections will now be examined in some detail.

3.1 Light Sources

3.1.1 Hollow-cathode Lamp

The principal light sources which were first used in the development of a practical atomic absorption instrument,

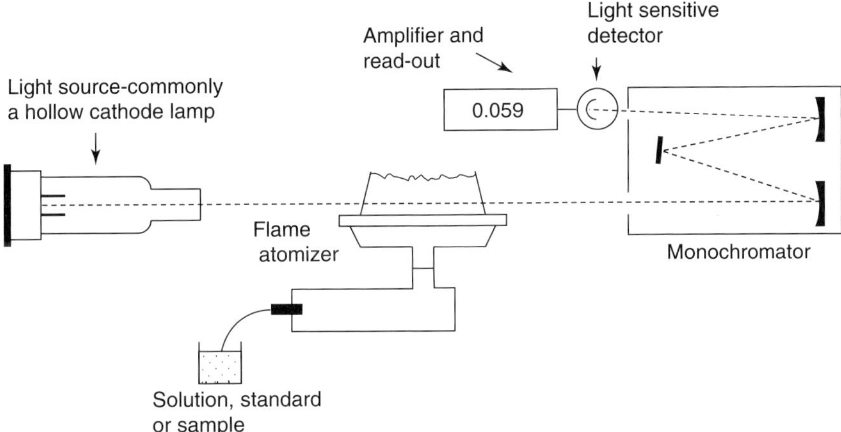

Figure 9 Diagrammatic representation of an atomic absorption spectrometer. (Reproduced by permission of Varian Australia Pty Ltd. from *Introducing Atomic Absorption*, 1983.)

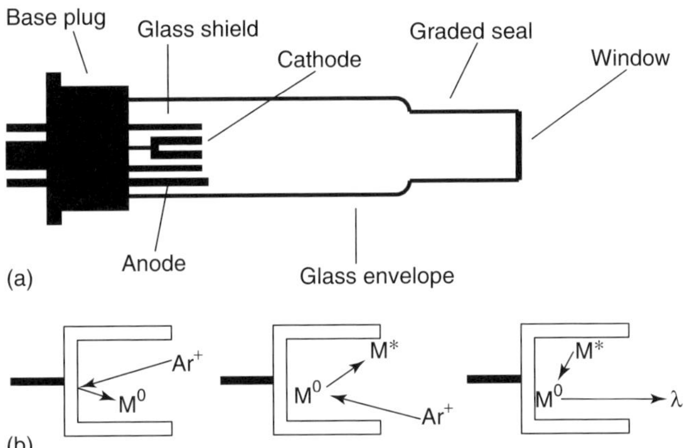

Figure 10 (a) An HCL and (b) its operation. (Reproduced by permission of Varian Australia Pty Ltd. from *Introducing Atomic Absorption*, 1983.)

in the early 1950s, were HCLs. Such lamps were, at the time, relatively uncommon and an initial task of Walsh and his colleagues was to perfect the means of making such lamps which were both stable and exhibited adequately long lives.(9) Possibly surprisingly, nearly 50 years later, such lamps still remain the basis of most atomic absorption instruments. A sketch of such a lamp and an illustration of its operation appear in Figure 10(a) and (b), respectively.

The lamp consists of a glass envelope into which are sealed the contacts leading to the anode and cathode. The cathode is either made from the element whose spectrum is sought, an alloy containing this element, or an insertion of the element or a compacted powder mixture containing the element.

A fused-silica window is sealed to the end of the lamp with a 'graded seal' joining the window to the envelope. This is necessary as the coefficient of expansion of the glass in the envelope is much greater than that of the silica and cracking would result following the heating and cooling which take place during lamp operation.

The lamp is filled with a monatomic inert gas such as argon, helium or neon, at a low pressure (1–2 kPa). Early lamps generally used argon as the fill gas but for some time neon has been used as the 'standard' fill gas. Argon is used for a few lamps, principally for lamps of elements where the presence of neon causes the emission of lines which are closely adjacent to the resonance lines commonly used for the measurement of that element. In operation, a potential difference of 400–500 V is applied between the anode and the cathode. Initially this ionizes some of the neon (or argon) and once current flow commences the potential difference across the electrodes will fall to about 250 V. The positively charged ions of the inert gas are attracted to the (negatively charged) cathode by electrostatic forces, causing the ions to be rapidly accelerated towards the cathode.

The rapidly moving ions of the inert gas, on striking the interior of the cathode, dislodge atoms from the cathode material ("cathodic sputtering"), creating a cloud of atoms at the mouth of the cathode. Many of these atoms are excited by collision with more of the rapidly moving ions of the inert gas and then, on spontaneously returning to the ground state, emit the emission spectrum of the element.

The total spectrum produced by this process will consist of atomic and ionic emission lines of the inert gas and atomic and ionic emission lines of the element(s) contained in the cathode. Of course, for any element the strongest atomic emission lines will be the resonance lines, since the resonance transitions will be the most likely to occur.

Such HCLs can be made for all of the elements which may be determined by atomic absorption, and are the

For references see page 9401

standard light sources used for most analytical atomic absorption measurements. They do, however, have finite lifetimes due to a process which is called "clean-up".

As described earlier, HCLs operate by some of the material of the cathode being removed, as atoms, by a sputtering process. After these atoms have been excited and given off their emission signals, they eventually condense on a cooler part of the lamp as a thin film of the metal. In general, most of this thin film of metal appears on the portion of the inner wall of the lamp which is immediately adjacent to the cathode itself as a mirror like layer. Since this film is literally deposited atom by atom, it has a very large surface area and adsorbs some small amount of the fill gas used in the lamp. Eventually this adsorption process reduces the fill gas pressure sufficiently that the lamp will no longer sustain a hollow-cathode discharge and the signal of the resonance line is no longer strongly produced. At this stage the lamp must be replaced.

Operating lifetimes of lamps were initially a problem, particularly for some volatile elements, but, with progressive development, today are typically from 1000 to as much as 10 000 h or longer.

Light sources other than HCLs have been used through the period of development of the technique but only two of these are commercially available and in serious use today. These are the boosted discharge HCL (initially called the "high-intensity lamp") and the so-called electrodeless discharge lamp.

3.1.2 Boosted-discharge Hollow-cathode Lamp

This has the essential construction of an HCL, but with additional auxiliary electrode(s) sealed into the lamp beside the open end of the hollow cathode, as shown in Figure 11.[10] Similar designs have been used by various manufacturers, variously called "high-intensity lamps" and "super lamps", but the principle remains essentially the same.

For probability reasons, such as described earlier for flame excitation, most of the atoms in a hollow-cathode discharge are in the ground state at any given time. Many of these atoms may be excited by the passage of a low-voltage discharge from the auxiliary electrode(s) through the atom cloud which exists in front of the cathode. The result is a more intense emission signal and, because the auxiliary discharge is of low energy (voltage), a greater proportion of the extra emission signal occurs at the resonance lines.

3.1.3 Electrodeless Discharge Lamps

A typical commercially available electrodeless discharge lamp is illustrated in Figure 12.

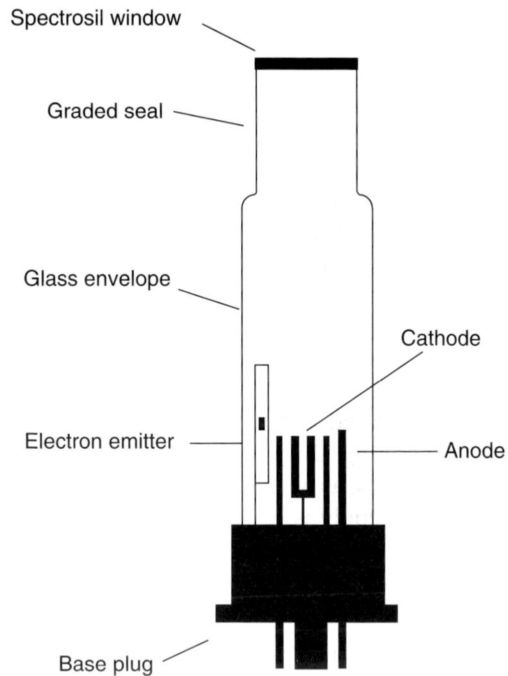

Figure 11 A boosted discharge HCL. (Reproduced by permission of Photron Pty Ltd. from *Photron Lamps*.)

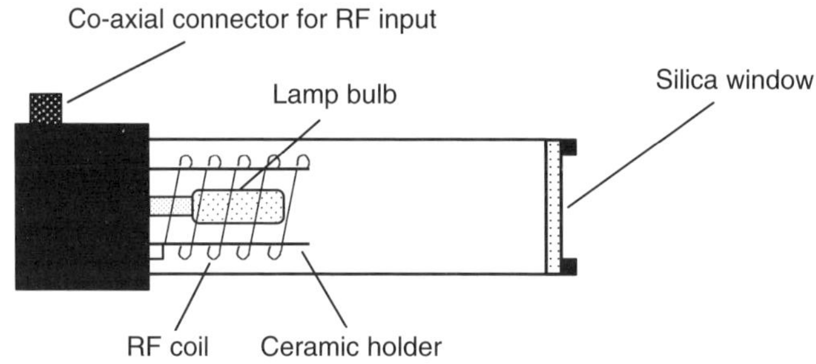

Figure 12 An electrodeless discharge lamp. (Reproduced by permission of Perkin-Elmer Corporation from *Concepts, Instrumentation and Techniques in Atomic Absorption Spectrophotometry*, 1978.)

An electrodeless discharge lamp consists of a small, sealed-off silica bulb containing a very small mass of the element whose emission spectrum is required and a low pressure of inert fill gas. Excitation is by a radiofrequency signal coupled into the contents of the bulb via an induction coil surrounding the bulb. Such lamps are only readily produced for relatively volatile elements such as As, Se, Na and K, but for these elements more intense signals are produced compared with HCLs. Additionally, since HCLs for these volatile elements have shorter lifetimes, as described above, the electrodeless discharge lamps often also have the advantage of longer operating lifetimes.

3.1.4 Operation of Light Sources

Since the most sensitive wavelengths for atomic absorption are generally also the most sensitive wavelengths for atomic emission, some interferences may occur for elements which are significantly excited in the flame atomizer. This is because the flame atomic emission signal would add to the total signal incident on the detector. To overcome this, the signal from the light source is modulated, and an alternating current (AC) amplifier is then able to discriminate this signal from the (essentially) direct current (DC) signal produced by flame atomic emission. Such a modulated signal may be produced by applying a modulated current to the light source or by mechanical chopping. Both techniques are in use in commercial instruments, as will be shown later in this section. A stable power supply is necessary, regardless of which of these modulation procedures are used, in order to produce a stable output from the light source.

3.2 Atomizers for Atomic Absorption Spectrometry

Without the production of free atoms, atomic absorption cannot be measured. It is in this section of the instrumentation that the operator of an atomic absorption spectrometer has the most control over the actual analysis and its veracity. Since the choice of the means of atomization employed and its detailed operation are of such fundamental importance to the technique, more detailed discussion will be postponed until later.

3.3 Wavelength Selection

As has already been discussed, the real resolution of atomic absorption measurement is provided by the absorption process itself. The function of the monochromator in the technique is therefore only to 'tidy up' the emission spectra from the light source. This is illustrated diagrammatically in Figure 13.

Here it may be seen that the light source is producing a spectrum, in the vicinity of the desired resonance wavelength R, containing a number of other lines. These lines may consist of other lines of the analyte, lines produced by other elements in the cathode material and lines of the fill gas of the lamp. Separating the resonance line (R) from all other lines in the vicinity is desirable in order to maximize sensitivity and limit calibration curvature.

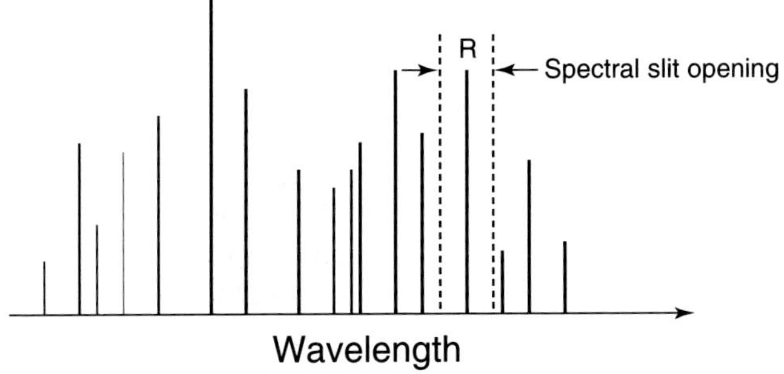

Figure 13 Isolation of a resonance line.

Generally a modest monochromator will achieve this fairly readily, with a spectral band-pass of 0.7–1 nm usually being adequate. In a few cases where the analyte itself has another line near the resonance line (and this occurs with Fe, Co and Ni), a spectral band-pass of 0.2 nm will be advantageous in the interests of achieving a calibration graph which is as linear as possible.

Initially monochromators which employed prisms for dispersion were commonly employed. However, for several years now most atomic absorption spectrometers have employed a monochromator with a diffraction grating as the means of wavelength dispersion. Such monochromators are more desirable than those employing prisms for dispersion as the wavelength settings are less prone to drift with temperature change. Obviously this is an important property when hot flames are being used as the means of atomization.

3.4 Detectors

The universal detectors of atomic absorption spectrometers for many years have been photomultiplier tubes. In the formative years of the technique it was often considered necessary, for the best results, to employ two different types. The range of AAS, as normally used, is from 193.7 nm for arsenic to 852 nm for cesium. In the 1950s there were no photomultipliers that covered this entire range with good sensitivity. Users who wished to measure arsenic and selenium as well as potassium needed two different types (rubidium and cesium, if needed, made this situation even more acute, but very few analysts measured these elements).

Two factors have altered this situation since that time. First, more intense light sources, including electrodeless discharge lamps, made the problem less acute. Second, the technology of photomultiplier tubes has improved and much better tubes with a wider wavelength range have been developed. Both of these improvements, of course, had a compound effect and today a single photomultiplier tube is all that is needed.

3.5 Amplifier and Read-out

As indicated before, the amplifier should respond only to a modulated AC signal and ignore any essentially DC signal resulting from flame emission. For the best performance this is accomplished by employing an amplifier tuned as closely as possible to the frequency at which the light source signal is modulated. Preferably

For references see page 9401

it should be a synchronously demodulated amplifier which is "locked in" to the frequency of the lamp modulation.

The relationship between the concentration of the analyte and its absorption is logarithmic so that the amplifier will incorporate a logarithmic function to achieve a linear, or close to linear, read-out. Modern amplifier systems also incorporate mathematical processing of curved calibration functions to achieve direct concentration read-out of the result, regardless of whether the calibration function is linear or otherwise.

4 ATOMIZATION SYSTEMS AND THEIR USE

4.1 Historical

Historically the first actual use of atomic absorption measurements for quantitative analysis occurred in the late 1930s, long before Walsh's first suggestions. The British company Adam Hilger (later Hilger and Watts) developed an instrument to measure the concentration of mercury in air samples.[11] This was a specialized approach to serve a particular need and was, of necessity confined to mercury. Of all the elements which may be measured by AAS, mercury is the only one which can exist at room temperature and pressure in the stable form of free atoms. Later discussions will concentrate on mercury individually but, for the moment, more general approaches need to be considered.

4.2 Flame Atomization

4.2.1 Early History of Flame Atomic Emission

Flame atomization methods had been in use for many years when atomic absorption methods started to be developed in the 1950s. They had been used in flame emission analysis since the beginning of the century although probably seriously only after the important work of Lundegardh in 1928.[5]

The fact that these were indeed means of flame atomization was obscured by the fact that they were also means of excitation and were viewed more in this vein. However, as stated earlier, in flame emission methods the process is first of all the production of atoms and only then the excitation of those atoms produced.

From what has been said before, it should be clear that since the wavelength of radiation resulting from atomic emission is inversely proportional to the energy level of the excited state, then low-energy excitation sources can only produce emission for elements which emit radiation at long wavelengths.

Although Lundegardh's development of the use of the air–acetylene flame did extend the range of elements which could be determined by atomic flame emission, the temperature of this flame (ca. 2500 K) was not high enough to excite very many elements significantly. Since inadequate temperature was obviously the major limitation, investigators in the 1940s and later were drawn to the use of gas mixtures which would give rise to hotter flames, notably the readily available oxygen–acetylene mixture, which produced a temperature of about 3300 K. However it was found that the use of this gas mixture in a premix system could easily lead to serious explosion hazards and so alternative systems were developed. Particularly this led to the development of the so-called "total consumption burner", in which the gases were only mixed with each other near the burner mouth. As will be mentioned later, such a burner system, although providing higher temperatures, introduced many problems to the technique.

4.2.2 Initial Developments of Flame Atomizers of Atomic Absorption

When investigations into the use of atomic absorption spectra for chemical analysis first commenced, the emphasis was initially on those elements which, although easy to atomize, were difficult to excite.[8] These include Zn (213.8 nm), Cd (228.8 nm), Ni (232.0 nm) and Pb (283.3 nm). Because of the short wavelengths associated with these resonance lines, the energy required to excite the atoms is relatively high.

Initial use was made of burner systems from existing (emission) flame photometers which employed premix systems and employed relatively cold flames such as air–coal gas (ca. 2100 K). It was immediately found that it was possible to measure elements such as Zn and Cd with excellent sensitivity, comparable, for example, to the sensitivity obtainable for Na and K. This immediately proved one of the major benefits, initially claimed for atomic absorption, that it would be independent of the excitation energy required to populate the first excited state. It was also realized that since the method did measure absorption, a longer absorption pathlength increased the sensitivity, and suitable elongated slot burners were soon developed. Such burners had to employ slot dimensions which were matched, as closely as possible, to the geometry of the beam of light from the HCL.

4.2.3 The Use of the Air–Acetylene Flame

It was soon realized that although high energy was not needed for excitation purposes, it was, for some elements, needed to achieve atomization. The development of elongated slot burners suitable for use with the air–acetylene flame (in the early 1960s) extended the technique to a number of additional elements so that by this period about 35 elements could be determined by AAS using this flame.

(The scientist most instrumental in the first development of air–acetylene burners, suitable for atomic absorption, was Allan[(12)]).

Relatively soon this flame became virtually the standard flame in use,[(13)] and this remains the situation today. Several of the additional elements, however, either had modest sensitivities or suffered from a number of matrix interferences, or both.

4.2.4 Development of Hotter Flames

It had become clear that a number of additional elements, although potentially amenable to measurement by atomic absorption, were not readily available because they were very difficult to atomize. These included important elements such as aluminum, titanium and silicon. These, and many other, elements readily formed very stable monoxides in the flame and at the temperature of the air–acetylene flame produced virtually no free atoms. Attempts were made to use total consumption burners, with various modifications, in order to use the oxygen–acetylene flame, but these proved to be of limited success.

The design problem lay in the very high burning velocity of the oxygen–acetylene flame compared with the air–acetylene flame.

In a premixed system, where the fuel and the oxidant gases are mixed in a chamber before passing through the burner slot, it is of vital importance that the velocity of the mixed gases through the burner slot is never exceeded by the burning velocity of the mixture of gases. Should it do so then a "flashback" of the flame into the chamber will result in a severe explosion.

In designing a suitable burner for safe operation with an air–acetylene flame, the slot opening had to be reduced compared with the opening in use with air–coal gas. However, examination of the burning velocities for the acetylene flames, supported by air and oxygen, respectively, made it clear that a huge (and impractical) reduction in slot dimensions would be required to burn oxygen–acetylene safely in a premixed system.

Some developments were first made using physical mixtures of air and nitrogen, as the support gas, and successful determinations of aluminum, titanium, silicon and many other elements were achieved with this approach. Mixtures containing as much as 85% oxygen were investigated, albeit with some hazards.[(14)]

The use of nitrous oxide as a support gas, however, was a major breakthrough as it permitted the safe burning of nitrous oxide–acetylene using a burner with a sensibly reduced slot opening.[(15)] As may be seen in Table 2, this flame provided a high temperature with a moderate flame velocity.

Compared with the potentially variable composition of a physical mixture of nitrogen and oxygen, nitrous

Table 2 Characteristics of various types of flames

Gas mixture	Burning velocity ($cm\,s^{-1}$)	Maximum temperature (K)	Burner slot (mm)
Air–coal gas	55	2100	100 × 1.5
Air–propane	85	2200	100 × 1.5
Air–acetylene	160	2500	100 × 0.5
Nitrous oxide–acetylene	280	3100	50 × 0.5
Oxygen–acetylene	1130	3300	–

oxide, on thermal decomposition in the flame, provides a mixture containing 67% (v/v) nitrogen and 33% (v/v) oxygen. However, it produces a flame more equivalent in temperature to that achieved with 50% N_2–50% O_2 owing to the exothermic decomposition of nitrous oxide.

With the additional use of this flame, a total of about 65 elements could, from 1965, be determined by AAS using flame atomization. More detailed examination of the flame atomization process will be presented later in this article.

4.2.5 Nebulizer–Burner Systems

Many different types of flame nebulizer–burner systems have been employed for atomic absorption. Some of these have been very useful for certain applications but for the purposes of this article the description will be confined to those commonly employed in available commercial instrumentation. A typical system is illustrated in Figure 14.

This system operates as follows. A flow of support gas (air or nitrous oxide), under pressure, passes into the nebulizer. The gas is expanded through a venturi throat and develops a pressure difference which draws a flow of

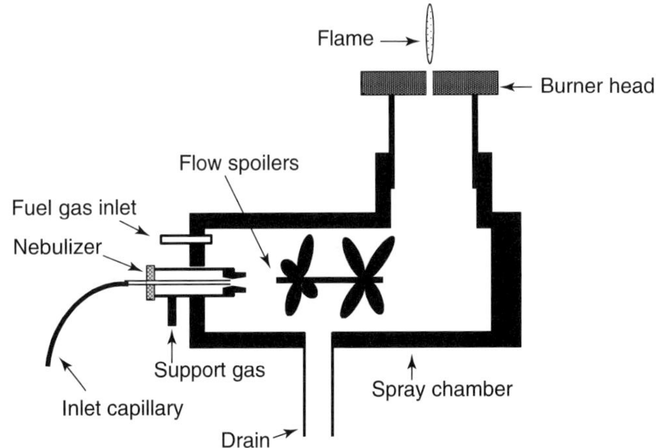

Figure 14 A nebulizer–burner system with flow spoilers. (Reproduced by permission of Perkin-Elmer Corporation from *Concepts, Instrumentation and Techniques in Atomic Absorption Spectrophotometry*, 1978.)

For references see page 9401

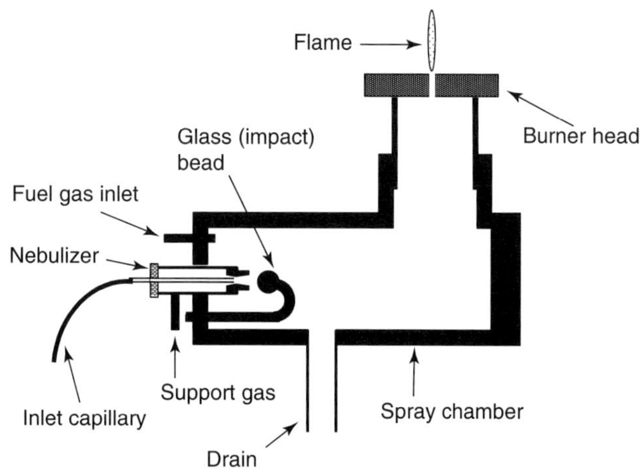

Figure 15 A nebulizer–burner system with a glass impact bead. (Reproduced by permission of Varian Australia Pty Ltd. from *Introducing Atomic Absorption*, 1983.)

sample solution through the inlet capillary. This solution, on emerging from the mouth of the venturi throat at a very high velocity, is shattered into droplets ranging in size from as small as 1–2 µm up to 100 µm or more, and this spray of droplets passes into the spray chamber. In this chamber, the oxidant gas and the droplets of sample solution are mixed with the fuel gas (acetylene). The flow spoiler, in this case a set of baffles, encourages the precipitation of the larger and heavier droplets of solution and these are drained from the chamber to waste. Most droplets larger than 10 µm in diameter will precipitate in the chamber. The mixture of oxidant, fuel and fine spray (aerosol) then passes into the burner head and through the burner slot to the flame, burning above the slot.

A minor variation of the system illustrated in Figure 14 is shown in Figure 15. In this system, an impact bead, adjustable in position, is set in front of the nebulizer throat. When this is adjusted to its optimum position (which is critical), larger signals are produced, indicating that a greater amount of fine aerosol is being produced, presumably owing to the bead causing a more efficient shattering of the larger droplets into smaller ones. Both of these systems are in use in commercial instruments currently available.

Overall, the total nebulizer–burner system has changed little in principle since the 1950s when Walsh and his colleagues first used such a system to demonstrate the method.

Nebulizers were developed which were more corrosion resistant when it was realized that the solutions required to keep some elements in stable solution were very corrosive.

Early burner heads were of relatively light construction, but the development of the nitrous oxide–acetylene flame made additional demands. In particular, since the flame was much hotter, more robust construction was needed to avoid distortion after protracted use. Also, a more massive construction was needed to provide a better "heat sink" to avoid overheating, which could detonate the combustible mixture of gases in the spray chamber. In general practice today the above system is used with two different types of burner head. First, for use with an air–acetylene flame, a burner with a slot of dimensions 100 × 0.5 mm is used. For burning a nitrous oxide–acetylene flame, a burner with slot dimensions of 50 × 0.5 mm is suitable.

4.2.6 The Atomization Process in the Flame

As the aerosol produced from the sample solution passes into the flame, the first process will be "desolvation", that is, the boiling off of the solvent (which is generally water). This will result in the production of extremely minute solid particles of the constituents of the dissolved material in the solution.

The effect of the higher temperature, as these particles move into the hotter part of the flame, will then be, in many cases, to dissociate the salts present and produce atoms of the constituent elements. However this is not always as simple as this might sound. Simple compounds such as NaCl and $ZnCl_2$ will very readily dissociate in the simplest manner possible, e.g. as shown in Equation (5):

$$NaCl + heat \longrightarrow Na^0 + Cl^0 \quad (5)$$

where Na^0 and Cl^0 represent free atoms of sodium and chlorine. For many elements this simple process will apply in the air–acetylene flame. However, for some elements the process will be more complex and this may be represented for aluminum as Equation (6):

$$Al(NO_3)_3 + heat \longrightarrow AlO + (\text{various gases}) \quad (6)$$

This is due to the great affinity for aluminum with oxygen and will occur regardless of which compound of aluminum is present. At the temperature of the air–acetylene flame, AlO is so stable it is virtually completely undissociated, so that no Al atoms are formed, and no atomic absorption signal can therefore be measured. However, at the temperature of the nitrous oxide–acetylene flame the next stage (Equation 7):

$$AlO + heat \longrightarrow Al^0 + (O) \quad (7)$$

will occur. Most of those elements which demand the nitrous oxide–acetylene flame for atomization will react in this manner as most of them form very stable monoxide compounds.

Some elements occupy an intermediate position between these two extremes, and this may be typified with calcium. In this case it is appropriate to represent the reactions as Equation (8):

$$Ca(NO_3)_2 + \text{heat} \longrightarrow (\text{various gases}) + CaO \rightleftharpoons Ca^0 + (O) \quad (8)$$

When air–acetylene is employed as the flame, the equilibrium represented is very much an incomplete reaction and only a portion of the calcium is converted to atoms. At the higher temperature of the nitrous oxide–acetylene flame, however, the equilibrium will be very much pushed in the direction of almost complete production of atoms and greater sensitivity of measurement will result.

It is worth adding some comments here about the above description as it relates to the so-called "total consumption" burner. Such burners were developed to provide safe oxygen–acetylene flames for flame atomic emission measurement. In such burners, constructed as shown in Figure 16, all of the solution droplets pass into the flame, since there is no intermediate spray chamber.

Referring to the description above it follows that, since very many large droplets will pass to the flame, relatively large solid particles are formed following desolvation. These very large solid particles are poorly dissociated in the obviously very limited residence time available. Additionally, such burners provide a much lower temperature than the theoretical oxygen–acetylene flame temperature owing both to the lack of adequate mixing of the two gases and to the cooling effect of the larger solvent volume. All of these factors make such burners very limited as atomizers, whether used for atomic emission or atomic absorption, compared with premix systems such as those described previously.

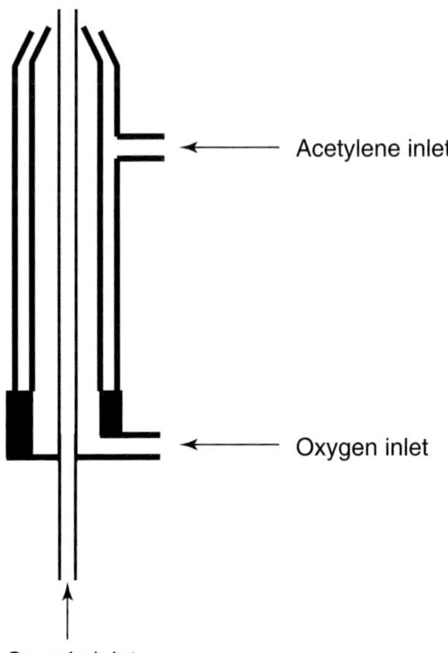

Figure 16 A total consumption burner. (Reproduced by permission of Wiley-VCH from B. Welz, *Atomic Absorption Spectrometry*, VCH, Weinheim, 1985.)

It is worth pointing out that in the discussions above the accent has almost solely on the temperature of the flame atomizer. This is almost certainly the major factor, but it is not the only factor. The chemical environment of the flame is extraordinarily complex, and of course the chemistry of every single element which may be measured by atomic absorption is different from that of almost every other element. Flames used for atomization may, for simplicity, be classified as stoichiometric, fuel lean or fuel rich. A stoichiometric flame may be defined as a flame where the amount of oxygen present is exactly the amount required to totally burn the hydrocarbon of the fuel in conformity to the equation of the combustion process. For example (Equation 9):

$$C_2H_2 + 5N_2O \longrightarrow 2CO_2 + H_2O + 5N_2 \quad (9)$$

For any given fuel and oxidant combination the highest temperature will be provided by such a stoichiometric flame.

A flame which has an excess of the fuel gas present, compared with the amount of oxidant present, is classified as fuel rich, is sometimes also called a reducing flame and will be somewhat cooler than the stoichiometric flame.

However, for many elements, better atomization, as indicated by larger atomic absorption signals, will occur for mildly or even strongly reducing flames. Chromium in the air–acetylene flame is a good example of this, as is silicon in the nitrous oxide–acetylene flame. The full understanding of all of the processes involved probably still evades the scientific investigators of flame chemistry, but as will be discussed later, a practical means of achieving optimum adjustments is still attainable.

4.3 Vapor Generation Techniques

4.3.1 Mercury Cold Vapor

It has been pointed out earlier that mercury was measured by atomic absorption in the 1930s. Using flame atomic absorption, the sensitivity obtained for this element is very poor, particularly when viewed against the background that since mercury is a very toxic element it often needs to be determined at very low concentrations. However, it was found that mercury can be released from any compound associations in solution, and converted to elemental mercury, by adding a reducing agent (such as stannous chloride) to the solution.[16] By bubbling a stream of a gas, such as nitrogen, through the solution the mercury atomic vapor is carried from the solution in this stream. The flow of gas is then passed through a flow cell placed in the light path of the atomic absorption apparatus, which then records the absorption signal.

For references see page 9401

Because the reaction is fairly rapid, a very high transient concentration of atomic vapor of mercury is achieved, giving rise to a large absorption signal peak. Whereas flame methods only allow the detection of about 1 $mg L^{-1}$ of mercury, the "cold vapor" method permits the detection of small fractions of 1 $\mu g L^{-1}$.

In general, the equipment commonly used for this type of measurement is the same as that used for hydride methods employed for some other elements, and which will be described below.

In the case of stannous chloride the reduction reaction, in solution, is as shown in Equation (10):

$$Hg^{2+} + Sn^{2+} \longrightarrow Hg^0 + Sn^{4+} \qquad (10)$$

It is also possible to use sodium borohydride ($NaBH_4$), as a means of generating hydrogen, as a reductant as described below for hydride methods.

4.3.2 Hydride Methods

Another group of metals, which exhibit poor sensitivities by flame atomization methods, may also be determined by a different vapor generation method. First suggestions of such an approach were made in 1969.[17] As now employed, the addition of a sodium borohydride solution to an acid solution of the analyte causes the production of hydrogen, and this in turn induces volatilization of hydrides of these elements. Such hydrides may be typified by arsine (AsH_3), which has a very low boiling point and so is readily evolved from solution once formed. These hydrides are readily decomposed, in a quartz flow cell, at modest temperatures (800–1000 °C) to produce free atoms of the analyte. A diagrammatic representation of the principle is shown in Figure 17.

The quartz cell may be heated either by mounting it above a flame produced using a normal burner system or by electrically heating the cell. (Initially the method used was to add metallic zinc powder to the acidic solution, but sodium borohydride gives a faster and more reproducible reduction of the metal to the hydride. In both cases the reduction depends upon the production of nascent hydrogen.) This method is only applicable to those elements which readily produce hydrides of very low boiling point, viz. arsenic, antimony, bismuth, selenium, tellurium and tin.

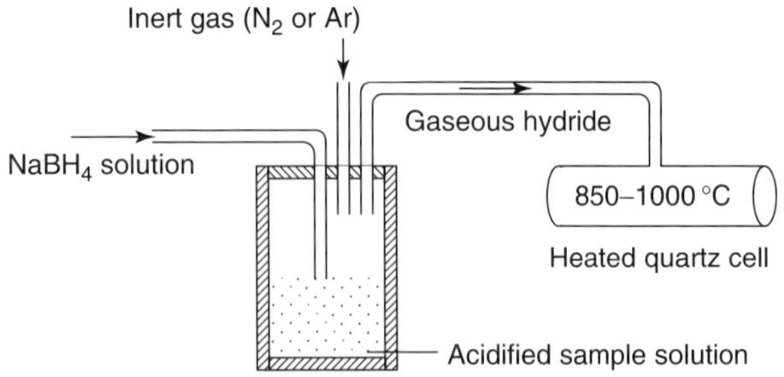

Figure 17 Diagrammatic representation of a vapor generation system. (Reproduced by permission of Wiley-VCH from B. Welz, *Atomic Absorption Spectrometry*, VCH, Weinheim, 1985.)

List of selected abbreviations appears in Volume 15

Using this method, most of these elements have detection limits of <0.1 $\mu g L^{-1}$, whereas in the flame their detection limits are at best around 1000 times poorer at ca. 0.1 $mg L^{-1}$.

Lead has been determined by this technique but with some operational difficulty, which limits the appeal of the method for this element.

4.4 Graphite Furnace Atomization

This method was first described in 1961[18] and has been in use with commercial equipment since about 1970. It produces detection limits for many elements which are several hundred times better than flame detection limits of the same elements. It also has the advantage that only very small sizes of sample are needed (a few microliters compared with a few milliliters typically required for flame methods). It has the disadvantages of requiring a moderately expensive accessory apparatus and of being a very slow method. However, it is readily automated, which overcomes at least some of the disadvantages of the slowness. This method is outside the scope of this article and will be dealt with in detail elsewhere.

5 OPTICAL DESIGN OF INSTRUMENTS FOR ATOMIC ABSORPTION SPECTROMETRY

5.1 Atomic Absorption Spectrometer Design

Two basic types of optical system have been employed since the first atomic absorption instruments were produced commercially. These are the single-beam system and the double-beam system and simplified diagrams of such types are illustrated here.

In Figure 18 is shown a very simple single-beam system. This contains an absolute minimum of optical components and is therefore optically very efficient, and also relatively inexpensive. The major disadvantage of such a system is that any instability in light source output will result in a variable base line reading.

In Figure 19 is shown a simple double-beam system where the beam is split and two paths of light, one through the atomizer (sample beam) and the other around it (reference beam), are produced. These are subsequently recombined and both signals measured and compared with one another.

The obvious advantage is that since comparison of the two signals is continually and frequently being made, variations in source intensity are cancelled out.

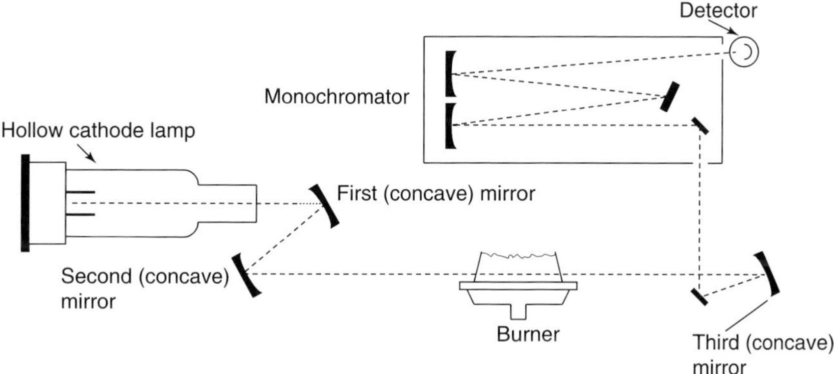

Figure 18 An optical system for a single-beam atomic absorption spectrometer. (Reproduced by permission of Varian Australia Pty Ltd. from *Introducing Atomic Absorption*, 1983.)

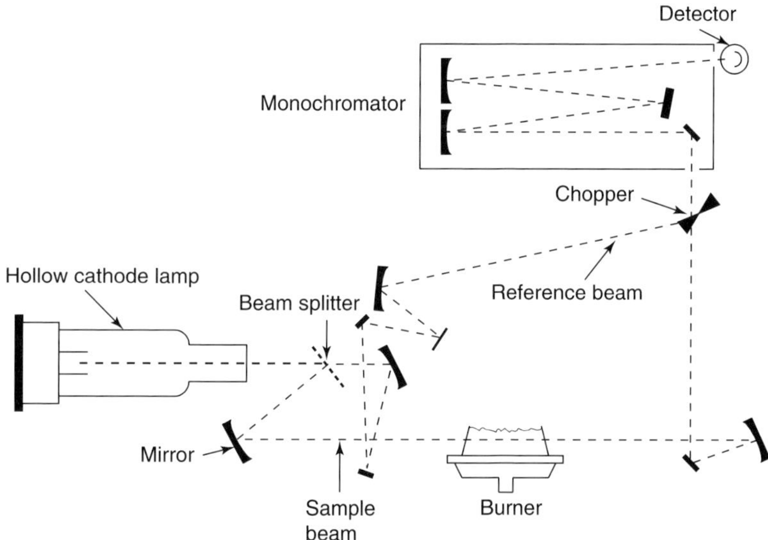

Figure 19 An optical system for a double-beam atomic absorption spectrometer with beam splitters. (Reproduced by permission of Varian Australia Pty Ltd. from *Introducing Atomic Absorption*, 1983.)

Two means of achieving this are in use in commercial instruments. That illustrated in Figure 19 employs a beam splitter and recombiner. The same result can be achieved by employing a system of chopping with a rotating mirror, as shown in Figure 20. In both cases the systems are not as simple as the single-beam system and consequently the cost is greater.

5.2 Optical Design for Background Correction

As mentioned previously, attenuation of the signal at the resonance wavelength by molecular absorption can cause errors in some cases, particularly when GFAAS is used. However, even in flame atomic absorption if the concentration of the analyte is low (giving rise to a very small atomic absorption signal), and the concentration of species producing molecular absorption at the analyte wavelength is high, errors will be introduced if correction is not made.

Figure 21 shows how a simple means of correction using a continuum light source may overcome this problem, a method first described in 1965.[19] This is illustrated by comparison with the double-beam instrument shown in Figure 19, but such an approach can be similarly used with the systems illustrated in Figures 18 and 20.

Alternative measurements with the sharp line source and the continuum source will measure as indicated in Equations (11) and (12):

$$\text{sharp line source} = AA + MA \tag{11}$$

$$\text{continuum source} = MA \tag{12}$$

where AA = atomic absorption and MA = molecular absorption. Subtracting the second reading from the first

For references see page 9401

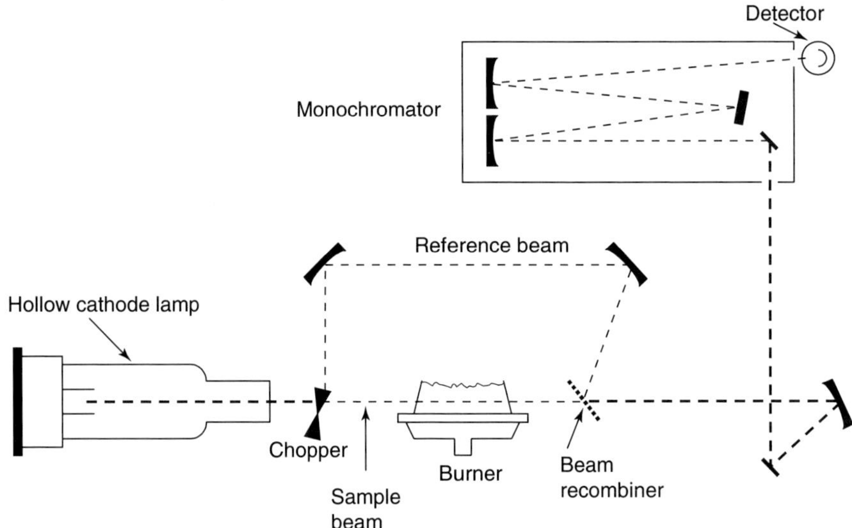

Figure 20 An optical system for a double-beam atomic absorption spectrometer with a reflecting chopper. (Reproduced by permission of Perkin-Elmer Corporation from *Concepts, Instrumentation and Techniques in Atomic Absorption Spectrophotometry*, 1978.)

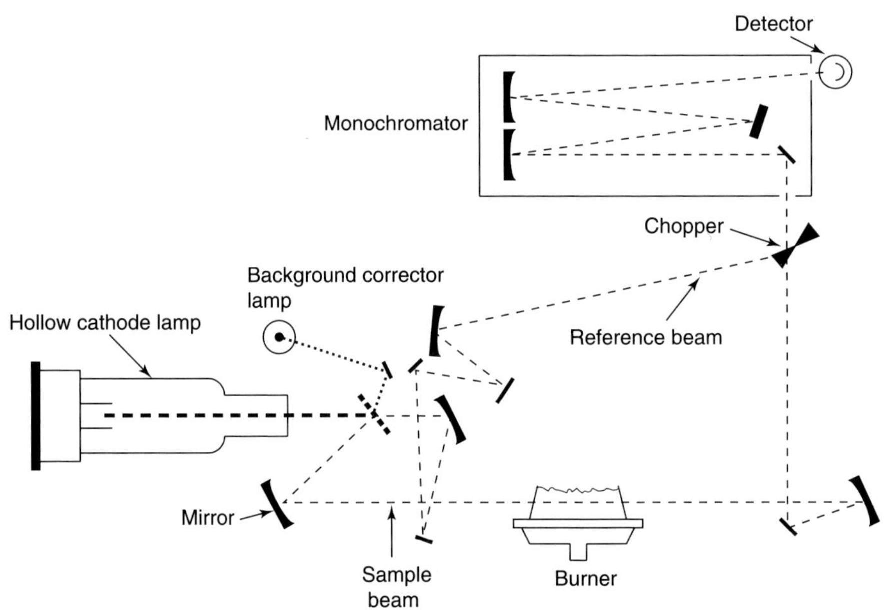

Figure 21 An optical system for a double-beam atomic absorption spectrometer with background correction. (Reproduced by permission of Varian Australia Pty Ltd. from *Introducing Atomic Absorption*, 1983.)

then gives (Equation 13):

$$(AA + MA) - MA = AA \tag{13}$$

Some of the radiation from the continuum source will be absorbed by atomic absorption. However, the instrument will be looking at the radiation from the continuum across a spectral band-pass of, say, 0.7 nm. Since the atomic absorption will only be over an absorption profile maybe 0.01 nm wide, less than 1% of this radiation will be absorbed and as a first approximation this may be ignored with very little resultant error.

The situation is completely different for GFAAS methods. Here it may be commonplace for the background absorption to be of a magnitude which is greater than that of the atomic absorption signal. Very careful design of the background corrector and its very careful adjustment are then most important. For this reason, a much better means of background correction is the use of the Zeeman effect background correction system. This is outside the scope of this article since it is applied only to the graphite furnace method.

6 PRACTICES OF ATOMIC ABSORPTION SPECTROMETRY

6.1 General

Like most spectroscopic methods, AAS is a comparative method, that is, it requires the calibration of the instrument with standards of known concentration before the analytical measurement of the samples may be made. However, as will be discussed later, it is generally not necessary to match the composition of the calibration standards to the samples with regard to elements other than the analyte. The calibration standards may generally be simple aqueous solutions containing nothing but accurately controlled concentrations of the analyte. Like almost all instrumental methods, the performance is limited by the signal-to-noise ratio, with regard to both the detection limit and the precision achieved. In the following sections these factors will be discussed more and attention given to means of optimization in order to obtain the best possible detection limits and precision.

6.2 Optimization of Instrumental Parameters

Because the optimization process involves two factors, signal (size) and noise, it often involves some sensible compromise. This will be briefly discussed in relation to the various parameters which may be set by the operator of an atomic absorption spectrometer.

6.2.1 Hollow-cathode Lamp Currents

Referring to earlier discussion on line shapes, the selection of lamp currents is very much a compromise. As the HCL current is increased, two things occur. A stronger signal is produced but inevitably the line width of the emission signal of the resonance line is also increased. The first of these changes will allow the use of a lower gain in the photomultiplier and amplifier system and this will decrease the "noise" produced from the photomultiplier and amplifier system. However, the broader emission line will decrease the sensitivity and this will reduce the absorbance for a given concentration and may also increase the curvature of calibration. In method development, the best compromise should be experimentally determined with reference to obtaining the best detection limit and/or the best precision.

6.2.2 Wavelength Selection

In general, for a majority of elements, the most sensitive line will be used for the measurement. However, this is not always the case and some brief examples will be mentioned.

In the measurement of lead, the most sensitive line is at 217.0 nm and this will produce a signal, for a given concentration, which is about twice as large as that obtained if the 283.3-nm line is used. However, the signal at 217.0 nm has a much greater noise component because of two factors:

1. The photomultiplier response at 217.0 nm is much poorer than at 283.3 nm so the photomultiplier will have to be operated at higher dynode volts, giving rise to more noise.
2. At 217.0 nm the flame gases absorb more radiation than they do at 283.3 nm and so minor fluctuations in the flame will generate some noise.

As a result, the best signal-to-noise ratio and hence the best detection limits and precision will be achieved using the 283.3-nm line even though it has the lesser sensitivity.

For the measurement of low concentrations of nickel, the 232.0-nm line (which is the most sensitive line for nickel) is preferable. However, owing to a very closely adjacent (and virtually non-absorbing) line of nickel, curved calibration graphs are produced for this element, this effect becoming very pronounced at high concentrations. Obviously, at the extreme of curvature when the calibration graph becomes almost flat, measurement is no longer possible. If samples containing high concentrations of nickel are to be measured, the less sensitive line at 341.5 nm will provide almost linear calibration graphs over a wide range of concentration.

Many similar cases may be cited – the rule should be that for optimum results each element and each application need to be investigated.

6.2.3 Monochromator Settings

The slit opening selected is generally a matter of recommendation by the equipment manufacturer as a function of the element to be measured and the wavelength to be employed. For example, considering the case of nickel discussed above, if this is to be measured at the 232.0-nm line a slit opening of, say, 0.2 nm should be chosen, rather than a wider slit to maximize the exclusion of the adjacent nonabsorbing line of nickel. However, if the 341.5-nm line is used, a larger slit opening should be chosen to increase the light throughput as this will improve the signal strength. Since there are no lines closely adjacent to this line, this may be done with impunity.

For references see page 9401

6.2.4 Flame Selection

For those elements which are easy to atomize, the air–acetylene flame should be used. It is less expensive to operate and will give rise to better sensitivity than if the nitrous oxide–acetylene flame is used (unnecessarily). In general, nitrous oxide will be used for those elements which are either atomized poorly, or not at all, with the air–acetylene flame.

There are some elements which occupy a central ground in this case. Elements such as calcium and strontium are only modestly atomized in air–acetylene and are subject to some interferences. By employing the nitrous oxide–acetylene flame, not only is the sensitivity improved but also the interferences are overcome. There are a number of such cases and again for each element the recommendations of the instrument manufacturers need to be accepted or the actual element and sample type need to be studied for each analysis by the analytical chemist.

6.2.5 Flame Optimization

It should be understood that each element has its own optimum atomization conditions, although for some elements the differences between them may be relatively small. This applies specifically to the composition of the flame used and the position of the burner in the light path.

Some elements give a more efficient atomization, resulting in a larger signal, if the flame is very "lean", that is, oxidizing or deficient in fuel. On the other hand, certain elements, such as chromium, require very 'fuel-rich' flames, that is, reducing and having an excess of fuel. Additionally for each element, and for each flame composition, the portion of the flame in which the maximum population of atoms occurs will be different. Since the maximum absorption signal will result when the beam from the HCL passes through this region of maximum atom population, efforts must be made to adjust the burner position to achieve this situation.

The best means of achieving both of the above two optimizations is, while aspirating a solution of the analyte, first to adjust the gas mixture to achieve maximum absorption and then to adjust the burner position. Sometimes, with elements for which the atomization conditions are critical, more than one iteration of these two adjustments might be necessary.

7 INTERFERENCES IN ATOMIC ABSORPTION SPECTROMETRY

7.1 Spectral Interferences

This subject has been already covered in some detail. True spectral interferences will only occur when an absorbing line of another element overlaps the emission line of the analyte from the light source and the interfering element is also present in the sample. This is, as previously stated, a rare occurrence and probably all such cases have, by now, been documented in the literature.

7.2 "Chemical" Interferences

Such interferences are sometimes also called "matrix interferences", and arise when an element, or compound, present in the sample (but not in the calibration standards) interferes with the production of atoms. A classical case has already been mentioned in passing.

If a simple solution of calcium of a given concentration is sprayed into the flame a certain reading will result. If, instead of the simple calcium solution, one containing the same calcium concentration, but also a significant concentration of phosphorus, is sprayed into the flame a much lower reading will be obtained. This is caused by the formation in the flame of some calcium phosphate type of compound which is harder to convert to atoms than the CaO usually formed. Obviously one could match the standards with the same P concentration but this would require the prior determination of P in each sample. Similar interferences will occur for all of the alkaline earth group – Ca, Mg, Sr, Ba and Mg – in the presence of P, Al and Si.

Such interference may be overcome in two ways:

1. By adding a large concentration of strontium or lanthanum a competing mechanism is set-up. Because the Sr or La is present in much greater concentration than Ca, and because Sr and La are chemically similar to Ca, the P will preferentially associate with the Sr or La rather than the Ca. In practice, the Sr or La is added both to the samples and to the calibration standards.
2. If the nitrous oxide–acetylene flame is used, the interference disappears because the flame is hot enough to dissociate readily the compounds formed.

These are probably the most severe cases but minor interferences of a similar nature occur with some other element combinations and the literature should always be consulted before embarking on any totally new application.

7.3 Ionization Interferences

At the elevated temperatures occurring in flame atomizers, the atoms of some elements will be ionized, that is, the equilibrium shown in Equation (14) will occur:

$$M^0 \rightleftharpoons M^+ + e \tag{14}$$

where M^0 is the ground-state atom of element M, M^+ is the ion of the same element and e is an electron. Since this particular atom is now an ion and not an atom it will not absorb radiation at the same wavelength.

This effect will occur with elements that have ionization potentials which are low relative to the energy of the flame used for atomization. With air–acetylene flames significant ionization will occur only for the alkali metals, Cs, Rb, K and Na, all of which have very low ionization potentials.

In the much hotter nitrous oxide–acetylene flame, however, many elements will be significantly ionized. The most extreme are the alkaline earths, Ca, Sr and Ba, but ionization will also occur significantly for many elements such as Al and the rare earth elements.

The ionization reaction shown above, being an equilibrium, is subject to the law of mass action (Equation 15):

$$K = \frac{[M^+][e]}{[M^0]} \tag{15}$$

where K is an equilibrium constant, $[M^+]$ is the concentration of the ion of the atom in the flame, $[M^0]$ is the concentration of the atom of the element in the flame and [e] is the concentration of electrons in the flame. If there is present in a sample, apart from the analyte, any element which readily ionizes (such as Na), then the additional electrons contributed to the flame by the ionization process will cause the equilibrium to shift back to the left in the equation and favor the production of more ground-state atoms. This will increase the atomic absorption reading. If the calibration standards have no sodium present then the readings obtained for the standards will be relatively lower. The net result will be a high result for the sample. In the case of the measurement of a number of samples with varying Na content, this will produce errors which will also vary.

To overcome this problem all solutions, calibration standards and samples should have added to them a large excess of a salt of an easily ionized element as an ionization suppressor. The most common such suppressor used is cesium, generally added as the chloride, which has the lowest ionization potential of all elements. This suppressor will ionize very substantially and in doing so will generate a high concentration of electrons. As described before, this high concentration of electrons will push the equilibrium back very strongly in the direction of simple (nonionized) atoms.

7.4 "Physical" Interferences

Some errors may also occur due to the physical composition of the sample solutions, by comparison with the calibration standard solutions. For example, if solutions of a solid sample are prepared by using high concentrations of sulfuric and/or phosphoric acid, the viscosity of the solutions will be high compared with simple aqueous solutions. This will diminish the uptake rate of the solution by the nebulizer and also will diminish the yield of fine spray to the burner. The error caused by this effect can only be prevented either by avoiding the use of such viscous acids or, if this is not possible, preparing the standards with the same concentrations of these acids. Similar viscosity effects may occur with very high concentrations of dissolved solids and again such problems can only be avoided in the same way.

The presence of a miscible organic compound may have the effect of reducing the viscosity of a solution and thus enhance the efficiency of nebulization. Ethanol, for example, will cause this effect. If an element in a sample of wine is measured by comparison with aqueous standards of the same element, high results will be obtained. Again, the solution to the problem will be to match, or at least approximate, the alcohol concentration of the calibration standards to that of the wine.

8 VALIDATION OF RESULTS OBTAINED BY ATOMIC ABSORPTION SPECTROMETRY

As may be seen from much of the preceding discussion, many factors may influence the veracity of results obtained using atomic absorption. In order to avoid errors due to an unexpected influence, it is wise to accompany any batches of samples analyzed with one or more standard reference materials, as close as possible in type to the samples. Control charts should be kept of the results obtained and remedial action taken if any deterioration in the quality of results is observed.

9 AUTOMATION IN ANALYSIS USING ATOMIC ABSORPTION SPECTROMETRY

Many automatic or semi-automatic atomic absorption spectrometers are now available. It is possible with such instruments, generally computer controlled, to store various parameters such as lamp currents, wavelengths, monochromator slit openings and flame gas flows. Although clearly such instruments are more costly, they do permit analyses to be performed by semiskilled operators once the parameters have been determined and stored by an expert analyst.

Automatic sample presentation to the instrument permits unattended operation, once a suitable type of automatic sampler has been loaded with samples and the analysis programmed. With instruments such as those described above, it is possible to measure sequentially a

For references see page 9401

large batch of samples for more than one element as part of an automatic process.

Caution is suggested, however, with the totally unattended use of flame atomic absorption instruments for obvious safety reasons.

One of the most important developments for automated analysis by AAS is the use of flow injection analysis (FIA). This method represents a means of faster automated analysis and also permits minimization of the sample volume. Suitable FIA methods and equipment have been developed for use with flame and vapor generation methods of atomization. Details of such methodology properly belongs in an article on FIA and so are not considered further here.

10 COMPARISON WITH OTHER ANALYTICAL TECHNIQUES

10.1 Flame Emission

At the time when atomic absorption methods were first introduced, the obvious comparison was with flame atomic emission. Today it can safely be said that the atomic absorption method has all but replaced flame atomic emission in practical analytical chemistry.

FAAS provides a method which is rapid and easy to use, combined with simplicity of operation, and is not very prone to unexpected errors. It has a good sensitivity for most elements, adequate for many purposes. However, it measures only one element at a time and so is not particularly rapid for multielement analysis.

10.2 Inductively Coupled Plasma Atomic Emission Spectrometry

By comparison with FAAS, ICPAES is more sensitive for the more difficult to atomize elements such as B, W and Si, but is not as sensitive as some of the easy to atomize elements such as Zn and Cd. ICPAES methods, however, are ideally suited for simultaneous, or rapid sequential, multielement analysis and are thus faster than FAAS for this purpose.

In general FAAS instruments will cost less than ICPAES equipment and are easier to use. For the determination of low concentrations of elements such as As, Se and Hg, vapor generation AAS is more sensitive than ICPAES methods.

10.3 Inductively Coupled Plasma Mass Spectrometry

In more recent times, ICPMS has offered a much faster means of determining extremely small traces of most elements, and provides real competition for GFAAS. However ICPMS requires very expensive equipment and very skilled operation.

10.4 Choice of Methods

As discussed immediately above, there is major competition for AAS as a means of analysis. Certainly, for the determination of very low concentrations of several elements in large numbers of samples, ICPMS is probably the method of choice by comparison with graphite furnace AAS. The very high cost of the equipment and the high operating costs, plus the skills requirement, however, will keep the former technique out of the reach of most small laboratories.

In a similar manner, but to a lesser extent, ICPAES instruments offer some advantages for some users but remain less attractive to the smaller laboratory. It seems unlikely that major developments will occur in AAS to alter this scene. Rather, the inevitable lowering of costs for ICPAES and ICPMS will make it increasingly hard for expensive types of AAS to survive. A versatile and major analytical service laboratory may certainly justify the installation of ICPMS, ICPAES and AAS with flame, vapor generation and furnace modes.

In such a situation, the desirability of having individual AAS instruments for flame and furnace operation should be considered for the sake of convenience and the avoidance of contamination of the clean environment needed for graphite furnace methods. With such a versatile combination, the advisability of having a Zeeman background-corrected furnace instrument might also be considered.

11 FUTURE DIRECTIONS

Because of the lower costs of equipment and the simplicity of operation, AAS appears to be likely to continue to be attractive to most analytical laboratories, particularly the smaller ones. It also seems, to the author, that low-cost atomic absorption modules, serving as specialized detectors, in association with sample preparation and handling techniques, such as those associated with flow injection methods, are one of the important directions in which development might occur. Such an approach would continue to use the highly specific capabilities of the atomic absorption method, which have made it so popular since its inception.

ABBREVIATIONS AND ACRONYMS

AAS	Atomic Absorption Spectrometry
AC	Alternating Current

AES	Atomic Emission Spectrometry
DC	Direct Current
FAAS	Flame Atomic Absorption Spectrometry
FIA	Flow Injection Analysis
GFAAS	Graphite Furnace Atomization
HCL	Hollow-cathode Lamp
ICPAES	Inductively Coupled Plasma Atomic Emission Spectrometry
ICPMS	Inductively Coupled Plasma Mass Spectrometry
VGAAS	Vapor Generation Atomization

RELATED ARTICLES

Clinical Chemistry **(Volume 2)**
Atomic Spectrometry in Clinical Chemistry

Environment: Water and Waste **(Volume 3)**
Flame and Graphite Furnace Atomic Absorption Spectrometry in Environmental Analysis • Flow-injection Techniques in Environmental Analysis • Hydride Generation Sample Introduction for Spectroscopic Analysis in Environmental Samples

Environment: Water and Waste cont'd **(Volume 4)**
Mercury Analysis in Environmental Samples by Cold Vapor Techniques • Optical Emission Inductively Coupled Plasma in Environmental Analysis

Steel and Related Materials **(Volume 10)**
Atomic Absorption and Emission Spectrometry, Solution-based in Iron and Steel Analysis

Atomic Spectroscopy **(Volume 11)**
Atomic Spectroscopy: Introduction • Background Correction Methods in Atomic Absorption Spectroscopy • Flow Injection Analysis Techniques in Atomic Spectroscopy • Graphite Furnace Atomic Absorption Spectrometry • Inductively Coupled Plasma/Optical Emission Spectrometry

NOTE

The number of references available in AAS is, by now, huge and the following represent only a judicious selection. There are many detailed text books available and those readers requiring a more detailed treatment of particular aspects are advised to consult any of these. One particular text book is, however, recommended as being very detailed in its blend of theoretical and practical details.[20]

REFERENCES

1. G. Kirchhoff, 'Über das Verhältniss zwischen dem Emissionsvermögen und dem Absorptionsvermögen der Körper für Wärme und Licht', *Annalen der Physik und Chemie Band CIX*, 275–301 (1860).
2. G. Kirchhoff, 'On the Relation between the Radiating and Absorbing Powers of Different Bodies for Light and Heat', *Phil. Mag.*, S4, No. 130, **20**, 1–21 (1860).
3. G. Kirchhoff, R. Bunsen, IX, 'Chemical Analysis by Spectrum-observations', *Phil. Mag.*, S4, No. 131, **20**, 89–109 (1860).
4. G. Kirchhoff, R. Bunsen, XLII, 'Chemical Analysis by Spectrum-observations – Second Memoir', *Phil. Mag.*, S4, No. 148, **22**, 329–349 (1861).
5. H. Lundegardh, *Die quantitative Spektralanalyse der Elemente*, Fischer, Jena; Part I, 1929 and Part II, 1934.
6. A. Walsh, 'The Application of Atomic Absorption Spectra to Chemical Analysis', *Spectrochim. Acta*, **7**, 108–117 (1955).
7. C.T.J. Alkemade, J.M.W. Milatz, 'A Double Beam Method of Spectral Selection with Flames', *Appl. Sci. Res.*, **B4**, 289–299 (1955).
8. B.J. Russell, J.P. Shelton, A. Walsh, 'An Atomic Absorption Spectrophotometer and its Application to the Analysis of Solutions', *Spectrochim. Acta*, **8**, 317–328 (1957).
9. W.G. Jones, A. Walsh, 'Hollow Cathode Discharges – the Construction and Characteristics of Sealed-off Tubes for Use as Spectroscopic Light Sources', *Spectrochim. Acta*, **16**, 249–254 (1960).
10. J.V. Sullivan, A. Walsh, 'High-intensity Hollow Cathode Lamps', *Spectrochim. Acta*, **21**, 721–726 (1965).
11. T.T. Woodson, 'A New Mercury Vapour Detector', *Rev. Sci. Instr.*, **10**, 308–311 (1939).
12. J.E. Allan, 'Atomic Absorption Spectrophotometry with Special Reference to the Determination of Magnesium', *Analyst*, **83**, 466–471 (1958).
13. B.M. Gatehouse, J.B. Willis, 'Performance of a Simple Atomic Absorption Spectrophotometer', *Spectrochim. Acta*, **17**, 710–718 (1961).
14. M.D. Amos, P.E. Thomas, 'The Determination of Aluminum in Aqueous Solution by Atomic Absorption Spectrometry', *Anal. Chim. Acta*, **32**, 139–147 (1965).
15. M.D. Amos, J.B. Willis, 'Use of High-temperature Premixed Flames in Atomic Absorption Spectroscopy', *Spectrochim. Acta*, **22**, 1325–1343 (1966).
16. N.S. Poluektov, R.A. Vitkun, Yu.V. Zelyukova, 'Determination of Milligamma Amounts of Mercury by Atomic Absorption in the Gaseous Phase', *Zh. Anal. Khim.*, **19**, 937–942 (1964).
17. W. Holak, 'Gas Sampling Technique for Arsenic Determination by Atomic Absorption Spectrophotometry', *Anal. Chem.*, **41**, 1712–1713 (1969).
18. B.V. L'vov, 'The Analytical Use of Atomic Absorption Spectra', *Spectrochim. Acta*, **17**, 761–770 (1961).

19. S.R. Koirtyohann, E.E. Pickett, 'Background Corrections in Long Path Atomic Absorption Spectrometry', *Anal. Chem.*, **37**, 601–603 (1965).
20. B. Welz, M. Sperling, *Atomic Absorption Spectrometry*, 3rd edition, Wiley-VCH, Weinheim, 1999.

Flow Injection Analysis Techniques in Atomic Spectroscopy

Rosario Pereiro and Alfredo Sanz-Medel
University of Oviedo, 33006 Oviedo, Spain

Flow injection analysis (FIA), developed originally for the automation of serial assays, has become a powerful tool most adequate for performing on-line any sample preparations before final measurement (e.g. sample dissolution, dilutions, matrix removal, preconcentration, etc.). It is not surprising that the combination of FIA with atomic spectrometric techniques has enlarged the analytical potential of atomic methods and expanded their wide field of applications. The collection of sample manipulation processes which can be covered today by flow operation procedures is amazing and therefore the general instrumentation required is reviewed in this contribution.

The description of the different flow strategies is carried out according to a hierarchy going from simple dilutions (including isotopic dilution of particular interest when using mass spectrometric detection), reagent mixing or standard additions, to more sophisticated flow manifolds such as those based on the use of two phases (e.g. gas–liquid, liquid–liquid or solid–liquid) for separation/preconcentration purposes. Modern approaches allowing for on-line decomposition/dissolution of solid samples (e.g. on-line chemical oxidation, photo-oxidation and microwave heating) are also described. Finally, manifolds for the in situ uptake of the sample and its complete on-line pretreatment are discussed.

List of selected abbreviations appears in Volume 15

The coupling of the above flow methodologies to a variety of atomization/excitation/ionization sources (flames, quartz tubes, graphite furnaces, inductively coupled plasmas (ICPs), microwave coupled plasmas, glow discharges (GDs), etc.) is detailed, in order to understand the usefulness of this combination both in atomic techniques based on photon measurements (absorption, emission and fluorescence spectrometry techniques) and on ions measurement mass spectrometry (MS).

1 INTRODUCTION

The progress and implementation of new techniques for the automation of sample preparation has improved the performance of sample pretreatment methods in the everyday work of trace element analytical laboratories. The processing of a larger number of samples per unit time, the improvement of the precision by minimizing sample contamination and human errors, the small sample volumes required and the decreasing costs (attributed to the reduction of both human participation and consumption of reagents), are among the goals which can be accomplished by automating the sample pretreatment step.[1]

Automation can be achieved by one of the following three approaches, or by combination of two of them:[2] using robots which imitate the way that samples are manipulated by human operators, by discrete systems (batch analyzers) in which each sample preserves its integrity in a vessel mechanically transported (by a belt) to various zones of the analyzer (where the different sample pretreatment steps are carried out), and by flow manifolds in which samples are transported in flowing streams.

Flow systems based on the use of a continuous stream of liquid or, much less often, a gas have reached a special interest. There are two types of flow analyzers: (a) gas segmented flow analyzers (originally developed in the 1950s and first commercialized by Technicon under the name of "AutoAnalyzer"), in which the flow is segmented by air bubbles aiming at preserving the integrity of samples, and (b) unsegmented flow analyzers, which can be classified according to whether samples are discretely injected FIA or continuously introduced into the system. Nowadays, the unsegmented flow analyzers are much more popular than the gas segmented flow ones.

The name of FIA was coined by Ruzicka in 1975[3] and he is generally credited as father of this technique, although Sarbeck et al.[4] published, back to 1972, a sample introduction approach for flame atomic absorption spectrometry (FAAS) which would be called today flow injection (FI) in its strict sense. The first reports on techniques explicitly defined as FI for atomic absorption spectrometry (AAS) were published in 1979.[5,6] Since then, the connection of FI to atomic spectrometric detectors has been extremely fruitful in terms of instrumentation versatility and variety of applications, giving rise to a great number of publications.[7–10]

FIA can be defined[11] as "information-gathering from a concentration gradient formed from an injected, well-defined zone of fluid, dispersed into a continuous unsegmented stream of a carrier". Since this definition does not facilitate visualization of the instrumentation required and the analytical interest of this versatile sample manipulation strategy, both aspects will be emphasized in the following paragraphs.

The basic equipment needed in FIA is relatively low cost, mechanically simple, robust and easily incorporated in automated systems. Figure 1 shows a general diagram of the most basic FI manifold, including the most commonly used instrumental units. The carrier solution and reagents are pumped continuously through a long, narrow tube, while a well-defined sample volume is introduced into the carrier stream (e.g. with the help of an injection valve). After the reactions occurring in the FIA system, the product of the reaction, which is monitored at the detector, gives rise to a transient response. Precise pumping and injection allows for identical treatment of every sample and standard in the FIA manifold, and this has the important consequence that there is no need to be restricted to reactions in which the reactants are stable. Also, precise timing is not necessary to obtain stable products nor to form them instantly. The use of FIA systems brings about several important advantages as compared to their manual batch analogs, both in terms of enhanced sample manipulation and of analytical performance (see Figure 2).

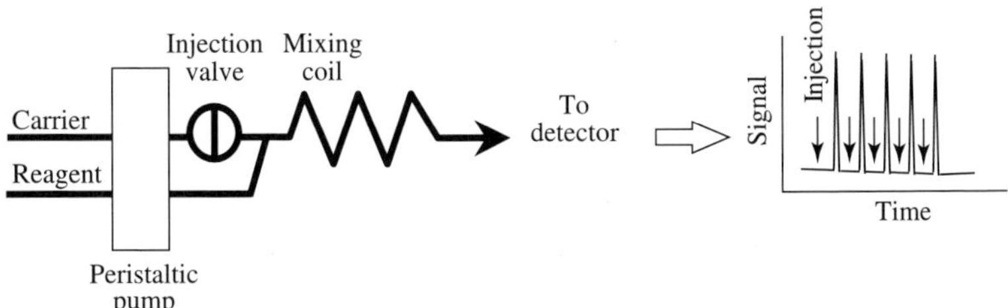

Figure 1 Schematic of a basic two-channel FI manifold including the most commonly used components.

For references see page 9422

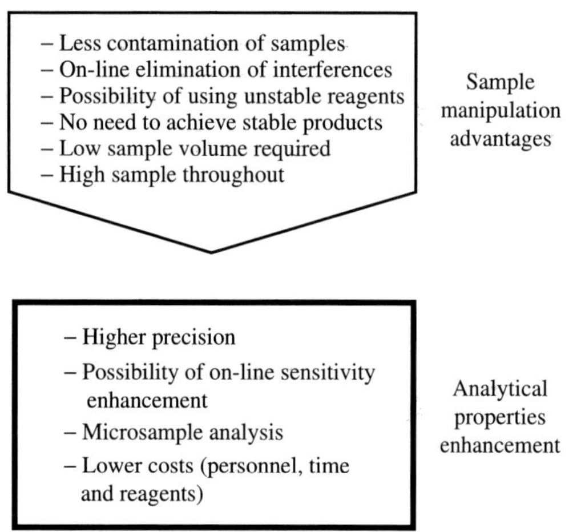

Figure 2 Analytical advantages of FIA systems in comparison with batch sample pretreatment procedures.

The sample manipulation processes collected in Figure 3, covered today by flow operation procedures extends over a wide scope and the instrumentation required for most of them will be reviewed in this contribution. Such flow strategies can be classified according to a hierarchy going from simple sample dilutions before final measurement to the in situ uptake of the sample and its complete on-line pretreatment.

Although processes like dilution (including isotopic dilution of particular interest when using mass spectrometric detection) and reagents mixing or standard additions, can be considered as the simpler pretreatment operations which can be carried out in flowing systems, however, they are none the less important and nowadays robust on-line flow manifolds are routinely used to face these sample pretreatment stages.[12,13]

A most interesting advantage of flow systems derives from their ability to integrate non chromatographic separation/preconcentration techniques based on the use of two phases (e.g. gas–liquid, liquid–liquid or solid–liquid) with a continuous detector in a straightforward manner. The general aim of these on-line flow separation/preconcentration techniques is to provide two sample fractions: one would contain the analyte(s) enriched and free from potential matrix interferences, while the other fraction would contain the matrix. In some particular cases this fractionation can offer a simple means of speciation;[14,15] however, general speciation problems demanding information on several compounds or species of a given element require more powerful separation techniques (e.g. chromatographic or electrophoretic in nature). Finally, it should be highlighted that the use of simple two-phase flow separation systems allowed a dramatic increase of analytical reproducibility achievable by atomic spectrometry indirect determinations of organic pharmaceutical products, as compared to batch procedures.[16]

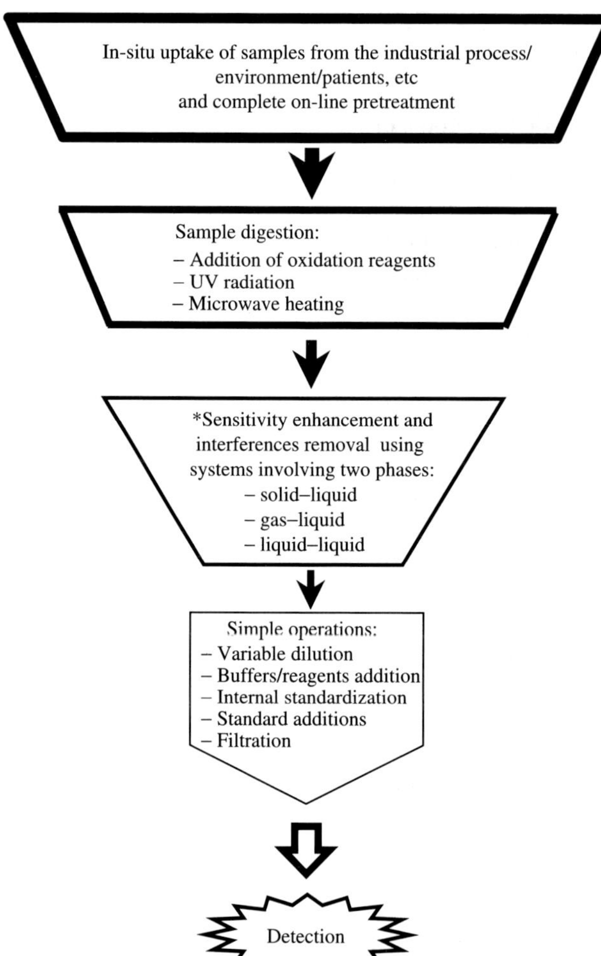

Figure 3 Scope of the sample pretreatment processes covered by flow procedures.

In a further extension of sample pretreatment processes enhanced by flow manifolds and keeping in mind that most atomic detectors are designed for the introduction of liquid samples, it is not surprising that a great deal of interest is attracted at present by the development of on-line flow systems allowing for the decomposition/dissolution of solid samples. Different approaches including on-line chemical oxidation,[17] photo-oxidation[18] and microwave heating[19] have been successfully exploited for these purposes.

Moreover, the in situ uptake of samples straightforward from their source and their on-line pretreament before atomic detection can be considered as the ideal to be pursued in terms of automation (see Figure 3). This goal has already been addressed and interesting manifolds (some already available commercially) have been proposed to carry out in situ environmental analysis, to wholly automate analytical measurements in the clinical laboratories (e.g. in vivo sample uptake of blood samples[20]) and for industrial process control.

2 BASIC CONCEPTS AND OPERATION PRINCIPLES

2.1 Basic Principles of Dispersion

The basis of FIA analyses relies on three aspects which should be kept controlled and reproducible: (i) volume of sample injected into the flow manifold, (ii) dispersion of the injected sample throughout the conduits of the system (precise pumping is required), and (iii) timing for each sequence carried out in the flow system (i.e. mixing with other streams, time to reach the detector, etc.).

Precise pumping and injection leads to a precise flow movement in a given manifold, allowing for reproducible chemical and physical manipulations. This is the fundamental feature that makes FIA so useful as an analytical technique.[11] In FIA, every sample and standard is subjected to identical treatment and measured after the same time interval, therefore, there is no need to wait to complete reactions.

In FIA analysis the profile "concentration versus time" observed at the detector, for an injected sample with an analyte concentration C_o, is a peak. This peak is the result of the dispersion processes occurring to the sample in the flow. The ratio of the injected analyte concentration, C_o, to the instantaneous concentration corresponding to any point of the sample volume reaching the detector, C_g, is known as the dispersion coefficient, D_g. In many cases, the value of the dispersion coefficient at the peak maximum, D, is used as the parameter to characterize the extent of mixing in a given FIA system.

The main experimental variables increasing the D value include: increases of the length and diameter of the tube, increases of the average flow rate and detector volume, and increases of the corresponding molecular diffusion coefficient. On the other hand, the most common way to obtain decreased D values is to increase the sample volume injected (other methods include the use of packed tubes or the modification of the viscosity). In most FIA systems D values are higher than unity. However, whenever a preconcentration step has taken place, the resulting D values are lower than unity.

Laminar flows are obtained under the experimental conditions used in FIA systems, i.e. on average, the molecules on the flow follow stream lines parallel to the walls of the narrow bore tubing. The walls of the tubing produces a drag on the flow. Thus, a gradient between the center of the stream and the wall is formed (the plot of the velocity profile from wall to the tube center is a parabola) giving rise to concentration gradients.

Besides, the concentration gradients will produce the diffusion of analyte molecules during the movement inside the tube. This diffusion will be axial (along the length of the tube) and radial (across the tube). It is important to note that, in the case of radial diffusion, the sample molecules on the leading boundary which diffuse towards the walls enter slower moving stream lines while molecules of sample close to the walls diffuse towards the middle of the tube and thus moving faster (see Figure 4). As a consequence of these processes, an interdiffusion between sample and carrier takes place which dilutes the sample or, in the case that appropriate reagents are introduced, produces chemical reactions between components of sample and carrier.

Figure 4 Effect of laminar flow and diffusion on the flow boundaries between sample and carrier after a few ms. Similar effects occur for the carrier and the sample.

In those cases requiring efficient mixing, other mixing mechanisms are added to this inherent interdiffusion process, such as the addition of other channels which merge with the sample at confluence points, the inclusion of mixing chambers, etc. Also, in some instances additional flow patterns to aid mixing are included, like bending or knotting the tube, packing a tube with glass beads, etc. Mixing of carrier streams in FIA procedures can be aimed at achieving on-line sample dilution, internal standardization, standard additions, chemical reactions, etc.

2.2 Chemical Reactions in a Flow

In many FIA procedures, a reaction between a component of the carrier stream and the analyte is needed to form a detectable product or when interference eliminations by chemical complexation are required. In both cases, a compromise between residence time and percentage of product formation should be chosen for maximum sensitivity: lower residence times would allow D decreases, however, product formation can be too low at short reaction times (in any case, it should be stressed again that, in many cases, chemical reactions do not need to be completed in FIA procedures).

In atomic spectrometry, on-line chemical reactions are increasingly exploited: (i) to form a volatile derivative of the analyte (e.g. hydride generation (HG)), (ii) to produce a hydrophobic analyte derivative able to be extracted on an organic solvent or on a hydrophobic solid surface, (iii) to fix and later release an analyte from an active solid support with chemical bonding groups, and (iv) to digest samples on-line (e.g. by chemical oxidation, with the aid of ultraviolet (UV) radiation, microwaves, etc.). There are also situations where the addition of a given reagent is required to promote a desirable

For references see page 9422

reaction in the atomizer (e.g. the addition of lanthanum as releasing agent in FAAS).

2.3 Flow Injection and Atomic Spectrometry Combination

There are two basic modes for detection in atomic spectrometry: photons measurement: AAS, atomic emission spectrometry (AES) or atomic fluorescence spectrometry (AFS) techniques, and ions measurement: MS techniques. A great variety of atomization/excitation/ionization sources is available to the analytical chemist in order to transform the analyte adequately for further measurement as ions or photons, including: flames, graphite furnaces and plasmas [e.g. ICP, microwave induced plasma (MIP), direct current plasma (DCP), stabilized capacitive plasma (SCP), GD, etc.].

Basic concepts and instrumentation of atomic spectrometry are beyond the scope of this contribution, but such concepts and general applications can be found in general text books as well as in monographs.(21–23) The most widely used instruments in atomic spectrometry are the FAAS, the graphite furnace atomic absorption spectrometry (GFAAS), also known as electrothermal atomic absorption spectrometry (ETAAS), and the inductively coupled plasma atomic emission spectrometry (ICPAES) while flame atomic fluorescence spectrometry (FAFS) is used to a lesser extent. Also, the continuously growing technique of inductively coupled plasma mass spectrometry (ICPMS) should be highlighted. Developments of all aspects of atomic spectrometry are still taking place exploiting FIA combinations with atomic spectroscopy and may be easily followed via the Atomic Spectrometry Updates review articles appearing in the *Journal of Analytical Atomic Spectrometry* of the Royal Society of Chemistry.

The combination of FIA strategies to atomic spectrometry has improved the analytical performance of such atomic instruments in terms of: (i) lower detection limits (through on-line preconcentration processes and matrix separations), (ii) greater accuracy (sample pretreatment is carried out in a highly reproducible fashion and contamination is reduced), (iii) high variety of samples can be directly processed (e.g. by on-line digestion) and, of course, (iv) higher sample throughput.

3 INSTRUMENTATION

3.1 Elemental Components of the Flow System

The three basic components of a flow manifold are the sample introduction system, the propulsion unit and the connecting tubes. Moreover, as was commented in the Introduction, special devices to allow for on-line separation/preconcentration/sample digestion, are becoming more and more popular inserted or integrated in the basic manifold.

3.1.1 Sample Introduction Systems

Two basic categories of sample introduction modes in FIA can be distinguished, namely volume-based and time-based. In the first case, samples are introduced in the FI manifold through an injection port, which consists of a septum or, most frequently, a valve. When time-based injection systems are used, the injected sample volume depends on the time period during which the sample is allowed to enter into the flow manifold.

In the early stages of development of FIA, the liquid samples were introduced with a syringe through a rubber septum.(24) However, important disadvantages of this sample introduction mode were distortion of the stream movement, lack of reproducibility in the manual injection and leakage after repeated injections on the same spot. The number of designs described to improve the performance of the septum has been considerable and different volume-based systems were proposed such as double proportional slider valves,(25) six(26) or eight(27) ports rotary injection valves, and hydrodynamic injectors (valveless systems),(28) as well as time-based injection systems including approaches based on two variable-speed computer-controlled peristaltic pumps and a commutator at the merging point(29) or timed solenoid injectors.(30)

In Figure 5(a) the schematics of a commercially available six-way rotary valve are shown. The sliding rotary valve has an external loop determining the sample volume to be injected. To introduce a different sample volume the external loop has to be changed. This type of valve is the most frequently used.

Figure 5(b) shows the principle of hydrodynamic injection:(28) the system is based on the introduction of the sample by a second propulsion system into a well defined volume of conduit of the main flow manifold and Figure 5(c) shows the schematics for a time-based/sample dilution system(29) allowing for the injection of a sample plug having an exactly predetermined concentration profile for the measured analyte.

Common insertion ports for the introduction of gaseous samples are gas-tight rotary valves and also the so-called 'exponential dilutors'(31,32) consisting of a magnetically stirred chamber of known volume, where the sample is diluted in the carrier gas. Injections are made through a septum by means of a gas-tight syringe. A set of two valves (or a three-way stopcock) permits the chamber to be isolated from the carrier gas flow during injections (see Figure 6). After the sample is stirred for some seconds

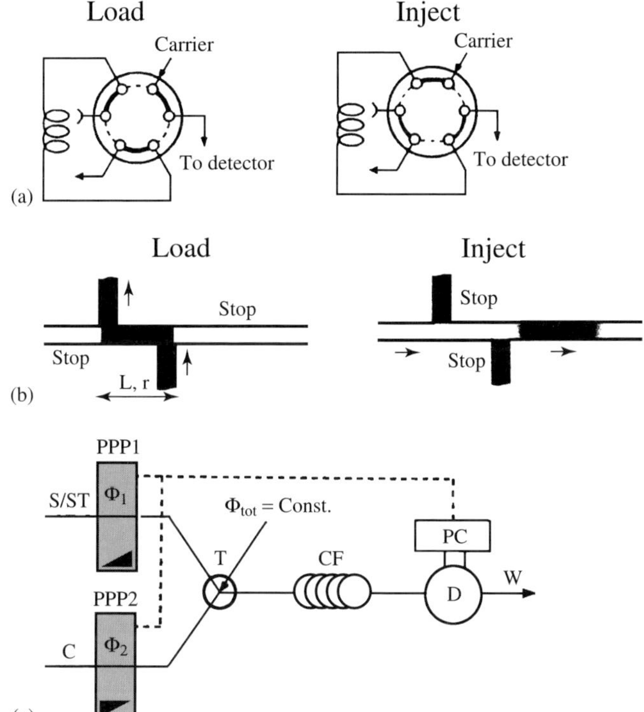

Figure 5 Examples of liquid sample introduction devices in FIA. (a) Six-way rotary valve. In the "load" position the loop is filled with the sample while in the 'injection' position the loop content is dragged through the manifold by the carrier solution. (b) Principle of hydrodynamic injection. A fixed volume of sample solution is metered into the conduit of length L and internal radius r (Load), and this volume is subsequently propelled downstream by the carrier (Injection). During the sampling cycle, the carrier stream circuit is stopped, and vice versa. (Reproduced with permission of Elsevier Science Ltd. from Ruzicka and Hansen[28] © 1983.) (c) Time-based/sample dilution system. S/ST: sample/standard; C: carrier; PPP1, PPP2: computer controllable peristaltic pumps; T: mixing point; CF: chemifold; D: detector; PC: personal computer; W: waste. (Reproduced with permission of Elsevier Science Ltd. from Novic et al.,[29] © 1999.)

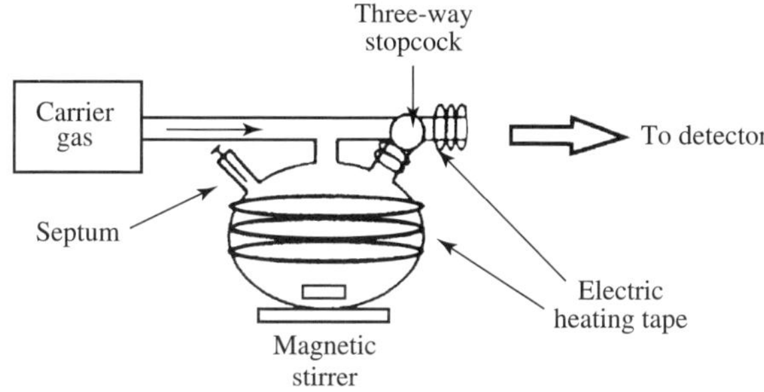

Figure 6 Schematic of an exponential dilutor for gas sample introduction.

in the isolated chamber, the valves are open giving rise to a signal peak with an exponential decay as a function of time.

3.1.2 Propulsion Systems

The propulsion system has to provide a continuous and reproducible flow rate of the solutions passing through the manifold. The ideal propulsion system should give a pulsed-free flow with an easily selected rate. Besides, in some cases, more than one solution has to be propelled and, therefore, multichannel capabilities are advisable.

Three main types of propulsion systems have been described:

1. Propulsion-less manifolds, which rely in the negative pressure generated by the nebulizer of an atomic detector to draw the liquid solution to the instrument.[33]
2. Gas-pressurized carrier reservoirs.[34] In this case, to keep a constant flow rate the pressure has to be maintained by means of a regulator. Although this system gives rise to a pulsed-free flow, drawbacks such as the consumption of gases or the difficulties to achieve constant flow rates have limited its use.
3. Pumps. Positive displacement pumps have been described as propulsion systems in flow manifolds; however, they introduce pulses in the analytical system and when two or more reagents are delivered in parallel these pressure pulses cause a non-uniform mixing of reagents reducing the accuracy and precision of the analysis. Although several depulsing systems have been described,[35] this type of pump does not enjoy great popularity. The peristaltic pump is by far the most used type of propulsion system. These pumps allow to use more than one channel and to achieve a constant and accurate flow rate, which can be easily modified according to the analyst needs. The pumping rate provided depends on the rotation rate and the inner diameter of the pumping tubes. Peristaltic pumps tend to give slight pulses which can be almost eliminated by an adequate adjustment of the clamps pressing the peristaltic tubes or by the use of coils located in the manifold right next to the propulsion unit.

3.1.3 Connecting Tubes

Inert and flexible tubing with inner diameters in the range from 0.3 to 1.5 mm allows the connection of the different units of the flow manifold. In some parts of the system, the tubing could be bent to allow for an efficient mixture of reagents and sample. Also, multiway connectors are used to mix flows from different channels.

3.1.4 Two-phase Separation Units

In this section, only a brief introduction to the basic instrumentation needed for gas–liquid, liquid–liquid and

For references see page 9422

solid–liquid separations in a flowing system will be presented. More details will be given in sections 5, 6 and 7 of this chapter.

3.1.4.1 Gas–Liquid Separation The co-occurrence of a gas and a liquid phase in a hydrodynamic system to achieve a two-phase separation gives rise to problems which call for ingenious technical solutions.(36–38) Gas liquid separators (GLSs) to be used in continuous separation systems should work smoothly and regularly to avoid lack of reproducibility and signal fluctuations and they should induce minimal dispersion or dilution of the gas phase (vaporized analyte). Many different designs have been evaluated as GLSs. The most common ones consist of a chamber, typically made of glass, in which a smooth separation between gases and liquids is aided by employing an inert purge gas. The use of gas permeable membranes, either flat(37) or concentric hollow cylinders(38) has been also described to separate both phases in flow systems. Examples of different GLSs used in flow systems are depicted in Figure 7.

3.1.4.2 Liquid–Liquid Separation A flow liquid–liquid extractor consists of a device able to merge two non miscible liquid phases to produce a single segmented flow, within which mass transfer takes place; the segmented flow passes through a phase separator, emerging from it as two separate streams, containing one phase each. The liquid–liquid phase separator must work efficiently and rapidly. There is a large variety of phase separators, based on two main principles, namely: density differences between the phases and selective wetting of appropriate inner surfaces. The most popular designs are based on the use of hydrophobic membranes which are only permeable to the organic solvent.

3.1.4.3 Solid–Liquid Separation Three main approaches have been described: sorption on a solid phase, generally consisting of a minicolumn, inserted in the flowing system,(26,39) precipitation and co-precipitation, based on the combination in a flow of three processes: precipitation of the analyte, its filtration and final dissolution,(40) and anodic stripping, consisting of the use of flow electrolytic cells to eliminate nondepositing sample matrix components and to preconcentrate analytes which are deposited at the working electrode.(15)

3.2 Assembled Flow Systems

Several companies are marketing sample introduction flow systems allowing for automated preparation of the most varied sample matrices. Representative examples of manifolds allowing for several sample pretreatment operations are reviewed below and in the next sections. However, it is important to emphasize that the

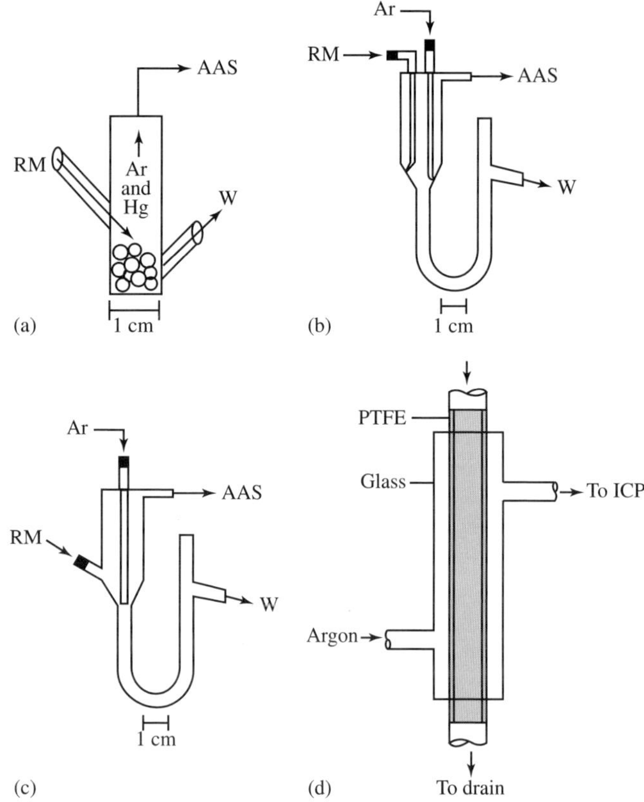

Figure 7 Examples of gas–liquid separators used in flow systems (not drawn at the same scale). RM: reaction mixture, W: waste. (a) Perkin-Elmer GLS; (b) PS Analytical GLS for sodium tetrahydroborate reductions; (c) PS Analytical GLS for tin chloride reductions; (d) Tubular membrane GLS. (Parts (a), (b) and (c) reproduced from Hanna et al.,(36) © 1993; part (d) reproduced from Barnes and Wang,(38) © 1988, with permission from the Royal Society of Chemistry.)

modular character of these manifolds allows to adapt them to almost every particular sample pretreatment needed.

Figure 8 presents several diagrams of different configurations of basic flow manifolds. The addition of a second channel to the system depicted in Figure 1 allows some manipulations or pretreatment of the sample (Figure 8a) such as dilution and addition of reagents or masking agents, while the set-up depicted in Figure 8(b), called "merging zones", permits important savings in reagent consumption(41) and is also of interest for on-line dilution or standard additions.

Figure 8(c) shows a manifold to inject simultaneously several plugs of a given solution into the same carrier stream to achieve overlapped zones as a consequence of dispersion and thus permitting the analysis of a wide range of analyte concentrations(42) and Figure 8(d) collects the schematics of a system to obtain a variety of dilution factors for a single injection without the need for controlled timing of any operation.(43) The manifold

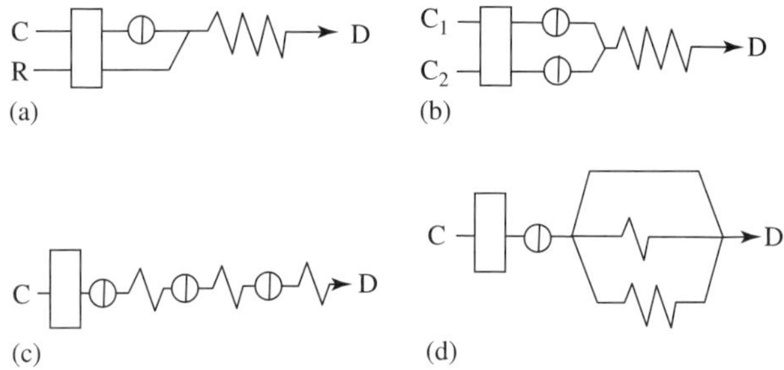

Figure 8 Examples of configurations of basic systems. C: carrier; R: reagent; D: detector. (a) Two line; (b) merging zones; (c) sequential injection flow system; (d) three-branch on-line sample dilution system.

consists of a three-branch network giving three partially overlapping peaks and five measurement points, three maxima and two minima.

Detailed instrumentation for separation/preconcentration manifolds including a solid–liquid, liquid–liquid or gas–liquid phases separation units as well as tandem on-line combinations of more than one separation strategy is thoroughly reviewed in sections 5 to 7 and the reader is addressed to those sections for specific information.

The possibility of carrying out decomposition/dissolution of solid samples on-line with the detector is attracting also a great deal of interest. Besides, in order to achieve volatile species generation from different analyte compounds, some organospecies need first to be decomposed, being this approach particularly interesting in metal speciation for species decomposition at the interface between the exit of a liquid chromatographic column and the atomic detector. Different approaches for on-line digestion of samples will be illustrated in section 8 in more detail.

3.3 Interfaces between Flow Systems and Atomic Detectors

The advantages of using flow manifolds as sample introduction systems for atomic detectors have been demonstrated for a variety of techniques including different atomic sources and/or detectors characterized by continuous operation (e.g. FAAS; ICPAES; MIP, SCP, and GD combined with AES; ICPMS, etc.) and even interesting approaches have been proposed for GFAAS, notwithstanding its discontinuous operation.

3.3.1 On-line Coupling to Continuous Atomic Spectrometric Detectors

The interfacing of flow manifolds to continuous atomic spectrometric detectors for the introduction of liquid samples requires usually a nebulizer and a spray chamber to produce a well-defined reproducible aerosol whose small droplets are sent to the high temperature atomic cell.

In FAAS, the concentric pneumatic nebulizer is the most frequently used.[44,45] Appleton and Tyson have given a theoretical modeling approach for nebulizer behavior in FIA coupled with FAAS.[44] The suitability of the response kinetics of the FAAS as well as the contribution of individual components to the overall dispersion of an injected sample has been investigated by Fang et al.,[45] being observed that the dispersion effects in the nebulizer – burner system generally showed a very limited contribution to the dispersion of the injected samples.

The following guidelines for optimum performance in terms of sensitivity and precision when using a barrel nebulizer and FAAS detection were given by Brown and Ruzicka:[46] (a) the flow rate of the carrier pumped into the nebulizer, should always be greater than the natural aspiration rate of the nebulizer; (b) the sensitivity is a function of the flow-rate of the carrier stream entering the nebulizer and the aspiration rate of the latter; (c) to achieve an optimum flow through the nebulizer, an additional stream may be added to augment the flow of the carrier without a significant decrease in sensitivity due to dilution.

Concerning plasma-based atomic sources, the more commonly employed is, by far, the ICP which was first used for AES and, more recently, for MS detection. A variety of nebulizers have been described for ICP work[47] including cross-flow pneumatic nebulizers, pneumatic concentric nebulizers, Babington pneumatic nebulizers, ultrasonic nebulizers, microwave thermospray nebulizers, jet impact nebulizers, electrospray, microconcentric nebulizers, oscillating capillary nebulizers, direct injection nebulizers, etc.

In the 1980s much reasearch work was devoted to the search of the "ideal" nebulizer, most of these designs also being tested as FIA interfaces to the plasma. It should be perhaps highlighted the direct injection nebulizer, ideally 100% efficient at transporting the introduced analyte into the ICP[48] (the relative detection limits found for 30 μL sample injection volumes were generally comparable to those obtained for continuous sample introduction into a conventional cross-flow nebulizer), the microconcentric nebulizer, inserted directly into the tip of a conventional sample introduction tube on an ICP torch to allow for a low dead volume interface,[49] or the thermospray interface allowing for 10 fold better FIA detection limits as compared to FIA cross flow.[50]

Probably the most frequently used spray chamber in ICP is the Scott type. However, a new type of spray chamber combining gravitational, centrifugal, turbulent and impact loss mechanisms in one apparatus to remove large droplets, increase transport efficiency and reduce memory effects was designed and evaluated by Wu and Hieftje:[51] this chamber

For references see page 9422

has at least 30% higher sample utilization efficiency, 2–3 times shorter sample clean-out time, half the cost, and simpler construction than the Scott type; moreover, it offers better detection limits and precision and when it was investigated for FI combined with plasma AES, the same sensitivity of detection was achieved by continuous nebulization than injecting volumes of 200 µL.

Considering alternative plasma sources to the ICP, the low-power MIP should be highlighted. This excitation source is inexpensive and simple to operate, but its major limitation is the low tolerance to liquid samples and susceptibility to interferences caused by easily ionized elements. However, a MIP source called microwave plasma torch (MPT), unlike more conventional MIP supporting torch/cavities structures, tolerates aqueous aerosols and molecular gases introduced into the discharge. A study exploring the potential of the FI mode for sample presentation to the MPT/AES with an ultrasonic nebulizer(52) allowed the reduction of memory effects of this nebulizer without loss of sensitivity or precision. Furthermore, by appropriate choice of sample dispersion, a significant reduction of the Na and K interferences were observed.

The introduction of analytes as gaseous derivatives offers special advantages in terms of sensitivity in atomic detection since 100% of the volatilized analyte goes into the atomic detector as compared to 1–5% of analyte introduced by liquid nebulization. The earliest flow volatile analyte generation systems connected online to ICPAES and AAS detector systems for the analysis of some metals and metalloids were reported in 1978(53) and 1982,(54) respectively, and they have been also coupled to other atomic excitation sources such as the MIP(55) and GD(56) for AES detection. Besides, AAS and AES, AFS(57) and MS(58) detection has been successfully employed as detectors for these manifolds. Using quadrupole-MS detectors, the elimination of matrix interferences in some particular applications is required and flow systems based on the formation of a gaseous analyte derivative can overcome these problems: for example, selenium is one of the most difficult elements to be determined sensitively by ICP combined with quadrupole MS because of isobaric interferences on its most abundant isotopes. However, using on-line flow HG coupled to ICPMS the sensitive analysis of Se is easily carried out.(58)

Plasmas such as MIPs, SCPs and GDs using He as plasma gas offer good sensitivity for non-metals (e.g. halogens) provided that the samples are introduced as a gas phase. The determination of halides with these plasmas and AES detection, by on-line previous chemical oxidation to the corresponding volatile haline, has been successfully investigated.(59,60)

3.3.2 *Semi On-line Coupling to Graphite Furnace Atomic Absorption Spectrometry*

The intrinsic discontinuous nature of the GFAAS technique has limited the interest of interfacing basic continuous flow manifolds to this detector. However, several flow approaches which will be further detailed in the next sections, offer some capabilities favoring their exploitation as important aids in GFAAS work. These approaches include:

- Separation and preconcentration by on-line column sorption, coprecipitation and solvent extraction, as a consequence of the compatibility of organic solvents with GFAAS.(39,40)
- Formation of volatile derivatives of the analyte and their preconcentration on a graphite tube ("in situ trapping").(61,62)
- Slurry sampling: although solid samples may be placed directly in the graphite furnace, the coupling of slurry sample introduction systems to GFAAS is proving to be very promising and versatile.(63)

4 FLOW INJECTION ANALYSIS STRATEGIES INVOLVING ONE-PHASE LIQUID FLOW SYSTEMS

4.1 Automated Variable Dilution

Most FI procedures give rise to an inherent dilution of the sample, i.e. dispersion coefficient higher than one. This dilution can be particularly notorious if additional channels (and the corresponding mixing points) or strategies such as mixing chambers are introduced into the manifold.

There are a large number of possibilities for the use of FI techniques to control dilution of stock solutions or samples. According to Fang,(64) FIA dilution systems are based on two basic mechanisms: sample dispersion (the measurement of the dispersion coefficient at different points of the peak profile is equivalent to the measurement of different dilutions of the injected sample) and flow manipulations (e.g. merging-flow, split-flow, etc.). Examples of strategies proposed in atomic spectrometry for on-line dilution, very useful also for standardization using a single standard, are summarized below:(65)

- Variation of sample volume: changing the sample loop, a linear dilution of the standard solution injected can be obtained.
- Variation of the mixing coil dimensions: higher dispersion of the sample bolus, and therefore higher

dilution of the sample, can be achieved by increasing the length and the inner diameter of the mixing coils.

- Merging flows: merging the sample carrier with another carrier provides a linear dilution of the sample which depends on the sample injected volume, the manifold volume and the ratio between the two mixing flow-rates.
- Time-based electronic dilution: consists of timing the readout delay following sample dispersion. This procedure is the basis of gradient techniques for standardization with a single standard.
- "Partial overlapping zones" strategy: based on multiple injections by using multiple loop valves or network manifolds, to obtain partial overlaps of sequentially injected sample zones which penetrate in each other. It offers a series of minima between neighboring zones where the injected samples are diluted to different degrees by the carrier.
- Mixing chambers: the exponential concentration gradient obtained by the dispersion of a discrete volume of a concentrated standard through a well-stirred mixing chamber provides a high dilution capacity.
- "Zone sampling" procedure: a portion of the dispersed sample is re-sampled from a section of the gradient and dispersed again.

4.2 Automated Addition of Reagents to Improve Analytical Performance

Addition of special reagents to the sample to improve the analytical performance characteristics of atomic spectrometry procedures could be classified in two groups:

1. Strategies for one phase flow manifolds, such as the use of releasing agents or ionization suppressors to overcome interferences in flames or plasmas, the method of isotopic dilution in ICPMS to improve precision and accuracy, the addition of stabilizers or the formation of slurries and the on-line digestion of samples.
2. Strategies involving two phase flow manifolds, needed for on-line chemical reactions to form a gaseous or a non water soluble analyte derivative or to retain/elute the analyte from a solid phase.

The measurement of isotope ratios in ICPMS can compensate for matrix effects and instrumental drift. In the isotopic dilution method, the isotopic ratio in a sample aliquot is modified by adding a spike enriched in one of the isotopes of the element to be measured. From the measurement of the relative abundance, R, of each one of the two considered isotopes, A and B, in the unspiked and the spiked sample, the concentration of the analyte can be calculated according to the formula:

$$C = \frac{(A_{sp} - R\,B_{sp})W_{sp}}{(R\,B_s - A_s)W_s}$$

being C the analyte concentration, A_s and B_s the mass percentage of A and B in the sample, A_{sp} and B_{sp} the mass percentage of A and B in the spike, W_s the weight of sample and W_{sp} the weight of the spike.

The on-line isotope dilution[66,67] can be carried out by merging two carrier streams or by the simultaneous injection of two solutions into the carrier stream. Advantages of the use of the isotopic dilution procedures using an on-line flow mode include: (i) only the sample volume consumed for measurement is mixed with the spike (the rest remaining intact), (ii) ease of selection of the proper sample to spike ratio in unknown sample concentrations, and (iii) possibility to achieve simultaneously on-line sample dilution and on-line isotope dilution.

The on-line flow addition of components such as anticoagulants in whole blood analysis, acidic solutions to form slurries, chemical modifiers for ETAAS analyses such as the mixture $0.1\,mol\,L^{-1}\,Mg(NO_3)_2 + 0.01\,mol\,L^{-1}\,Pd(NO_3)_2$[20] has also been described. Samples, in acidic media, can be on-line digested using microwaves. The degree of dissolution in the microwave cavity is mainly governed by three variables: microwave power, length and diameter of the tubing inside the microwave oven, and the acid slurry strength. This methodology has been already applied to a great variety of samples of clinical, biological and environmental interest, using FAAS, GFAAS, FAFS, ICPAES or ICPMS detection modes. Details about on-line microwave assisted digestion are given in section 8.3.

4.3 Standardization and Implementation of the Standard Additions Method

Besides the traditional use of a series of standards with increasing concentrations of the analyte, in FIA systems external calibration can be also carried out employing a single standard by following any of the on-line dilution strategies already described under the "automated variable dilution" section.

The standard additions method to correct for matrix interferences and the internal standard method to alleviate instrumental drifts can be also incorporated via FIA systems, giving rise to automated methods where the amount of solutions employed and the time and cost of the analyses are reduced. Procedures based on merging zones, partial overlapping zones, zone sampling, etc. can be readily implemented for these purposes.[65]

For references see page 9422

5 FLOW INJECTION SYSTEMS FOR ON-LINE SOLID–LIQUID SEPARATION AND PRECONCENTRATION

Nowadays on-line solid–liquid extraction procedures, and in particular, methods based on sorption of the analyte in columns and its subsequent release, are experiencing a great popularity for sample pretreatment because they can be coupled in a most straightforward manner to most atomic detectors. Simplified handling, reduced solvent consumption and increased preconcentration factors are among the favorable features of solid phase extraction procedures. Of course, the elimination of matrix interferences (e.g. alkaline and earth-alkaline elements in high salt content samples), is also another advantage of these couplings, which is important for ICP analysis (for example seawater, dialysis fluids, etc.) or in the frequent cases of eliminating polyatomic interferences using the popular quadrupole ICPMS.[58]

FI on-line solid liquid separation and preconcentration methods for atomic spectrometry may be classified into three main groups, according to the separation principle used for retention of the analyte, namely: sorption, precipitation/coprecipitation and anodic stripping voltammetry (ASV).

5.1 Sorption

Many sorbents have been developed so far and some of them are marketed as prepacked minicolumns. The characteristic features of the on-line preconcentration flow systems demand some properties of the packing material which may be only of minor importance in batch or traditional column procedures, since for flowing systems the column material has to be reusable, the kinetic processes or reactions have to be rapid and the flow uniform. Therefore, the special requirements may at least include: high mechanical resistance (in order to withstand high flow rates through the column and to maintain long column lifetimes), good kinetic properties (analytes have to be rapidly retained and readily eluted) and low degree of swelling and shrinking when being transformed from one form to another or with change of solvent conditions (the swelling/shrinking of some resins will give rise to backpressure and non uniform flow patterns).

Sorption preconcentration methods can be divided into two general groups: (a) analyte ions are collected directly by immobilized counter ions (ion-exchange) or immobilized chelating functions (sorption of ions), and (b) metal ions are adsorbed on suitable phases (e.g. activated carbon, octadecyl functional groups bonded silica gel, etc.) as hydrophobic metal chelates, previously formed by reactions in solution (sorption of metal chelates). The eluents more commonly used in the first case are diluted acids, bases or complexing agents, while for the second, the chelates are most frequently eluted with hydroorganic solvents.

The sorption of ions involves the use of "solid active phases" containing a suitable ion exchanger, a chemical reagent or even bioorganisms, immobilized on the solid support packed in a minicolumn. The solid active phases can be purchased (e.g. Chelex 100, Muromac A1, EDTrA-cellulose™, 8-hydroxyquinoline (8Q) bound to controlled pore glass (CPG)-8Q, 8Q bound to cellulose, etc.) or prepared at the working laboratory. The most common chemical mechanisms for reagent immobilization on the solid support to prepare the solid active phase are: adsorption (frequently used for hydrophobic reagents such as dodecanylquinolin-8-ol which is effectively fixed to nonionic supports), ion-exchange (the reagent is bound to the support by ionic groups which are not involved in the reaction with the analyte, such as the binding of 5-sulfonic oxine to resins with ammonium quaternary groups) and covalent binding (the reagent is covalently bound to solid supports such as controlled pore glass, silica gel, cellulose, etc.).

Chelex 100 was the first packing material used for on-line column preconcentration[68] and it has been widely used since then. This resin contains paired iminodiacetate ions coupled to a styrene divinylbenzene support. Despite its success, Chelex 100 does not fully meet the requirements mentioned at the beginning of the section for an ideal packing since the resin is subjected to some troublesome swelling properties: the particle size depends very much upon the counterion. These problems have been successfully overcome by the use of countercurrent flow elution. The Japanese product Muromac A1 which has also iminodiacetic acid functional groups was reported to be free from the troublesome swelling properties of the latter and has been used in a number of FAAS and ICPAES applications;[69,70] however, the material does not seem to be widely available and it was used mostly by Japanese workers.

The controlled pore glass-oxine is other frequently column packing material used in on-line preconcentration.[71] Its popularity can be probably attributed to its good mechanical and good kinetic properties derived by the immobilization of the oxine functional groups on easily accessible porous glass surface. Fibrous cellulose with EDTA groups (EDTrA-cellulose™)[72] and oxine immobilized on CM-cellulose (cellulose-8Q),[73] are also commercially available. Both materials show high capacity, high exchange kinetics and good mechanical and chemical stability.

Also, it is interesting that a minicolumn made of sulfhydryl cotton has been proposed for the speciation of inorganic mercury and methylmercury:[74] at pH 3–4 only the organomercury species are claimed to be retained and not the inorganic mercury. On the other hand, several applications of activated alumina in continuous flow-systems have been described both in the acidic form for the preconcentration of oxyanions and analysis by ICPAES[75] or in the basic form. In the latter case, the alumina has been applied, for example, for the separation of Cr(III) from Cr(VI).[76]

The ability of bioorganisms such as algae to accumulate trace metals by biosorption has been known for some years. This effect was used in water treatment systems, however, only recently it has been exploited for analytical measurement methods. In 1991 the first application was described of columns with immobilized algae on-line coupled to atomic detectors for trace metal preconcentration being analytes eluted with diluted acids.[77] Later, other applications such as the speciation of Cr(III) and Cr(VI) using an on-line column with immobilized algae has been also described.[78]

The sorption of metal chelates opens also interesting possibilities.[39,79–81] In this case the chelate is on-line formed in the flow system and then retained on a hydrophobic adsorbent, being later eluted with a hydroorganic solvent. The most commonly used reagents are dithio-carbamate derivatives, possessing very active sulfur binding sites. The most popular detectors for these applications seem to be FAAS and GFAAS (more compatible with hydroorganic solvents than plasmas).

Whenever the difference in the sorbent selectivity between the free chelating reagent and the formed metal chelate is small, the sorbent will absorb also free ligand (in addition to the metal chelate) during preconcentration. Once the capacity of the sorbent has been overcome, any remaining metal chelate still passing through the column will not be adsorbed. To avoid this problem, it has been proposed as strategy sorbing first a small amount of the chelating agent and then the analyte is preconcentrated by direct complexation with the solid sorbent agent.[82] In this case, since both the chelate and the reagent are eluted, the column has to be fed with the reagent for each analysis. A comparative view of general advantages and disadvantages of metal chelates sorption, as compared to the sorption of ions, is presented in Figure 9.

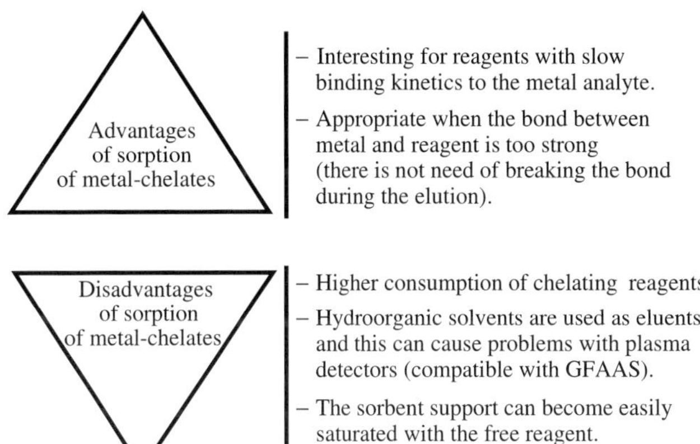

Figure 9 Comparison of "sorption of ions" and "sorption of metal chelates" procedures.

5.2 Precipitation and Co-precipitation

Continuous precipitation – filtration – dissolution processes can be described as continuous liquid–solid systems in which the second phase is first generated and then disappears in situ. The earliest flow on-line precipitation system with FAAS detection was reported by Valcarcel's group[83] employing on-line stainless filters to retain the precipitate. Usually, the precipitate is formed in the reaction coil, retained in the filter and subsequently dissolved by a suitable dissolution agent.

In 1991 Fang et al. presented an interesting approach based in analyte coprecipitation in the flow system. The analyte was coprecipitated quantitatively with the iron(II) – hexamethyleneammonium hexamethylenedithiocarbamate complex and collected in a "knotted reactor" made of ethyl vinyl acetate tubing. The precipitate was dissolved in isobutyl methyl ketone (IBMK) and introduced directly in the nebulizer of an FAAS.[84] Similar approaches have been exploited successfully using this or other collectors in combination with FAAS and GFAAS.[40,85] In comparison with on-line precipitation methods, coprecipitation procedures are less demanding on the solubility of the precipitate formed.

It is important to highlight how chemistry in precipitation and coprecipitation systems has some resemblance to metal–chelate sorption (the filter replacing the sorbent minicolumn), except that for precipitation/coprecipitation no column capacity limitations exist. Conversely, practical problems associated with continuous precipitation/coprecipitation systems can be encountered during the running of such manifolds.

5.3 Anodic Stripping Voltammetry

The use of flow-cells based on the principle of ASV coupled on-line to atomic detectors has also been investigated to obtain improved analysis of selected species (those which can be deposited at a working electrode and later released for atomic specific detection). These cells eliminate sample matrix components that are not electroactive and do not deposit during passage of the sample through the cell. Moreover, signal enhancement of analytes, as they may be preconcentrated at the electrode[86] are obtained. Besides, in some cases ASV can allow for oxidation state speciation of elements: for example As(III) and Se(IV) are electrochemically

For references see page 9422

responsive in most electrolytes while As(V) and Se(VI) are not.[15] Thus, on-line ASV has allowed the speciation of both analytes in urine using ICPMS detection. Similar approaches have been also proposed for the speciation of Cr(VI) and V(V).[87]

6 FLOW SYSTEMS FOR ON-LINE GAS–LIQUID SEPARATION AND PRECONCENTRATION

For some elements it is possible to increase the sensitivity of their atomic spectrometric determinations by generating a volatile derivative of the analyte outside the instrument and transporting it most efficiently to the atomizer.[88] Strategies for such chemical vapor generation (CVG) include mainly: HG, cold vapor generation (CV), formation of alkyl derivatives and formation of halines.

Despite their advantages in atomic spectrometry, sample introduction methods based on classical batch analyte vapor formation procedures have a number of pitfalls, resulting in their replacement by flow-based methods. Their relatively modest throughputs, their bad precision (e.g. poor control over reaction conditions such as time), their lack of applicability to non stable analyte vapor derivatives, the use of quite large volumes of sample, and the generation of by-products, particularly hydrogen (when released as a sudden burst, it can affect negatively the flame or plasma) are detrimental to their analytical performance. Most of these problems can be eliminated, or at least reduced, by the use of flow procedures allowing for on-line vapor generation and gas–liquid separation. It is worth noting that such flow systems require reactions exhibiting high analyte volatilization kinetics (e.g. fast chemical reactions) and efficient mass transfer between both phases.

6.1 Flow Systems for the Formation of Volatile Hydrides of Metals and Metalloids

Chemical formation on a FIA system of covalent volatile hydrides of the analyte, has been widely applied to the atomic spectroscopic determinations of elements such as As, Bi, Ge, Pb, Sb, Se, Sn and Te.[89] The earliest continuous HG systems connected on-line to ICPAES and AAS were reported by Thompson et al.[53] and by Astrom,[54] respectively. In flow hydride systems, the sample (introduced either continuously or in the FIA mode) is on-line mixed with a reducing agent. In some cases additional channels are included for the introduction of other reagents (e.g. for on-line prereductions). The volatile species formed are separated from the liquid in a GLS with the aid of a continuous flow

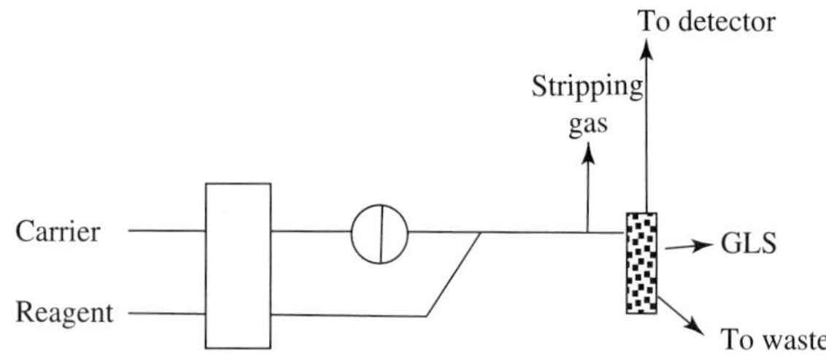

Figure 10 Basic FIA manifold to achieve gas–liquid separation.

of a stripping gas (see Figure 10). The stripping gas and the volatile species (analyte derivatives plus by-products continuously produced such as H_2, CO_2 or water vapor) are continuously swept to the detector, while the liquid is driven to waste. Argon, at constant flow rate, is most frequently used as the stripping gas (examples of designs of GLSs are shown in Figure 7).

On-line flow hydride procedures have been coupled to AAS,[54] AES,[53,55,56] AFS[57] and MS[58] specific detectors. For AES, the coupling of these manifolds has even been described to MIP[55] or GD;[56] however, the most commonly used excitation source is, by far, the ICP.[53] The use of this source in combination with mass spectrometric detection gives rise to extremely sensitive procedures due to the MS detector intrinsic sensitivity as compared to AES. However, on a routine basis, the more common detection systems used for these applications are AAS and AFS, where the analyte is swept to a heated quartz cell located on the optical path. Of course, it should be stressed that AAS and AFS have much less running costs than the ICP and they offer frequently excellent detection limits.

The most frequently used chemical reduction agent is tetrahydroborate. For As and Sb determinations both trivalent and pentavalent forms of analytes can yield hydrides but prereduction is highly desirable since the inorganic species As(III) and Sb(III) provide better sensitivity. Reagents such as L-cysteine or iodide have been used for on-line prereduction. For selenium, the best results are obtained by on-line compounds decomposition transforming the analyte species in its selenite state, and in this case it is frequently resorted to the use of 'strong' prereduction steps, which better correspond to section 8, such as the use of microwaves in $KBrO_3$/HBr medium. Unlike the situation for the hydride-forming elements of groups 14 and 15 of the periodic table, lead hydride is formed from a precursor having the metal in its highest oxidation state, Pb(IV). Among the different reaction conditions investigated for the best generation of plumbane, the greatest sensitivity, using FI with HG and AAS with quartz tube or graphite furnace atomization, was obtained in a hexacyanoferrate (III)/hydrochloric acid medium.[27,28]

Generation of volatile species for atomic spectrometry can also be facilitated by resorting to surfactant based "organized media", such as micelles and vesicles.[90] Analytical sensitivity and selectivity achieved with HG/atomic detection systems could be improved by the use of surfactants for the generation of the volatile species. In fact, surfactant aggregates can concentrate reactants at a molecular level and, therefore, change thermodynamic and kinetic reaction constants. Moreover, surfactants can solubilize, in a selective manner, analytes and reactants in the self-assembled "aggregates". As a result, the special microenvironment existing in or on these aggregates may change the reactions (and so the observed interferences) in aqueous media.

6.2 Alternative Strategies: Cold Vapors, Ethylderivatives and Halines

The capability of mercury to form a monoatomic vapor at room temperature has been widely exploited in atomic spectrometry for its sensitive determination in the most varied samples. The reduction of mercury ions to Hg^o CV was first utilized in 1968[91] to develop a CV with AAS "batch" analytical method for mercury determination. In recent years, most of the Hg analyses by CV technique are carried out by flow systems. Usually AAS detection,[36,37] but also AES[92] or MS[93] are being employed for detection. Tetrahydroborate or tin(II) chloride are common reduction agents. The use of tin(II) chloride avoids the massive production of molecular hydrogen, which often deteriorates the performance of the spectrochemical plasma source. Conversely, it is important to point out that $SnCl_2$ is not effective enough to reduce organomercury species. Therefore such organomercury species should be first oxidized to the inorganic form, preferably by resorting to a continuous on-line flow system.[17,19]

Mercury seemed to be the only metal proven to be able to form monoatomic vapor at room temperature. This property has been widely used to develop CV-atomic spectrometry techniques for very low levels of mercury as shown before. In 1995 evidence was shown[94] that cadmium is also able to produce 'cold atomic vapor' in appropriate conditions of reduction (e.g. surfactant-based vesicles). It appears that volatile cadmium hydride can be produced in the presence of several media and catalysts.[90] In this form, the metal can be transported to an absorption measurement cell where it dissociates spontaneously to atomic vapor of cadmium. More than 20 times improvements in Cd detectability have been demonstrated in this way by CV using flameless quartz tube with AAS,[94] in situ preconcentration and GFAAS,[62] CV with ICPAES[95] and recently by CV with ICPMS.[96]

Ethylation as a means of CVG is a promising (yet not fully explored) alternative to common HG and CV techniques.[88,97] The formation of volatile derivatives with sodium tetraethylborate is particularly interesting to enhance the detectability of some nonhydride generating elements (e.g. Zn, Cu, etc.) able to form ethylderivatives. Moreover, this type of derivatization is very useful to volatilize organospecies of Hg, Pg, Se, Sn, etc. (e.g. for speciation purposes), and is also of interest in the determinations of some hydride[98] forming elements which can be considered as 'problematic' such as Sn (high reagent blanks), Pb (low yield of HG generation) or Cd (unstable volatile derivatives). Recent work has also proposed the use of ethylation reactions in the presence of "organized media" (micelles and vesicles) in flow systems.[99]

Low power helium plasmas such as the He-MIP, the He-SCP or the He-GD possess high electronic temperatures allowing for good sensitivity by atomic emission determination of elements with high lying excitation levels in the visible region (e.g. most of the non-metals). Unfortunately, their rotational (kinetic) temperatures are so low that only gases or vapors are usually allowed in such plasmas. Thus, the chemical generation of chlorine, bromine, iodine, as a means of analyte introduction to these plasmas in flow mode has been successfully evaluated.[59,60] Different oxidizing mixtures were assayed such as H_2O_2/H_2SO_4, $K_2Cr_2O_7/H_2SO_4$, $K_2S_2O_8/H_2SO_4$ and $KMnO_4/H_2SO_4$, the latter being the most promising for the simultaneous generation of the three above mentioned halines. Halide detection limits achieved using such flow generation mode lie in the low $ng\,mL^{-1}$ level using MIP,[59] SCP[59] or GD/AES[60] detection.

6.3 Approaches for Preconcentration in the Gas Phase

Further increases of analytical sensitivity can be obtained by in situ trapping of the volatile form of the analyte transported to an adequate trap of the vapor. The most frequently used approaches for the in situ trapping/concentration of analytes after chemical vaporization are: (i) amalgamation of Hg in gold and release by heating, (ii) retention of mercury or volatile hydrides on a slightly heated surface (normally graphite coated with noble metals) and release by further heating, (iii) cryogenic trapping of hydrides and subsequent release by removal of the cryotrap.

The amalgamation of mercury CV with noble metals, and its posterior release by heating, is a well established method (commercial equipment is available) to further concentrate mercury vapor. The trap usually consists of gold-coated silica powder, gold coated sand, or amalgams such as gold-platinum gauze.[36]

For references see page 9422

Other approaches to trap the vaporized analytes involve the use of cryogenic traps[100] or the retention of volatile hydrides on a slightly heated surface (normally graphite coated with noble metals) followed in both cases by metal release by further heating. This latter approach has found an important range of applications in connection with final atomization by GFAAS.[61,89] In this case, the graphite furnace is used as both the hydride trapping medium and the atomization cell. The hydride purged from the generator is trapped in the preheated furnace, usually in the range 300–600 °C, until the evolution of hydride from the sample is completed. The trapped analyte is subsequently atomized at temperatures generally over 2000 °C. The scope of this technique has been steadily expanding in the last decade comprising, for example, most of the volatile hydride forming elements, CV of mercury and cadmium, ethylated derivatives of Pb, Se, Cd, alkyl derivatives of As, Pb, Sn, etc.

The technique FI using CVG with GFAAS, which is now fully automated and is commercialized, enhances the sensitivity significantly and eliminates effectively the possible influence of the HG kinetics on the signal shape. The nature of the graphite tube is expected to affect greatly the efficiency of hydride adsorption. It has been shown that the coating of the graphite tube with Pd, Ir Zr, Ag or mixtures, improves the sensitivity and precision significantly.

The sensitivities achieved with the above mentioned gas phase preconcentration methods are very high and in most cases they are limited by the blank values, being a major contribution to the contamination by the analyte in the reagents. In this sense, approaches such as the use of immobilized tetrahydroborate could open a way for further improvements in detection limits.[27]

7 FLOW SYSTEMS FOR ON-LINE LIQUID–LIQUID SEPARATION AND PRECONCENTRATION AND TANDEM ON-LINE COMBINATIONS

7.1 Flow Systems for Liquid–Liquid Extraction

Traditional set-ups (separation funnels) for liquid–liquid extraction usually require large vessels, large volume of reagents and laborious handling, being very time-consuming methodologies. To overcome most of these problems, the use of flow liquid–liquid extraction procedures has been investigated. A continuous liquid–liquid extractor includes a solvent segmentor, an extraction coil and a phase separator. The segmentor receives the streams of two non miscible phases and gives rise to a segmented flow; analyte mass transfer takes place through the multiple interfaces established throughout an extraction coil; finally, two separate streams (one with each phase) emerge from the liquid–liquid phase separator.[101–104]

The liquid–liquid phase separator is probably the most important unit in the system. This unit must separate efficiently and rapidly the aqueous phase from the organic solvent in a continuous manner and its operation should be smooth and constant throughout.

Making use of differences in density of the two liquid phases and the selective wetting of inner surfaces, a variety of phase separators has been described. Figure 11 collects four representative examples of such devices. The separator depicted in Figure 11(a) is exclusively based on density differences between the phases: it consists of a conical device into which the segmented flow penetrates to form a single interface in the center. The flow of the lighter phase leaves the separator from its top while the heavier phase goes through the bottom of the device. To avoid large dead volumes in FIA systems, the volume of the inner cavity is of the order of tens of microliters. Figure 11(b) shows an improved gravitational phase separator.[101] The separation unit was composed of two half blocks. The upper block was made from PTFE (poly(tetrafluoroethylene)), with a conical cavity and a single outlet for the separated phase, while the lower was of stainless steel, and furnished with an inlet and outlet for the segmented phase and waste flows, respectively. The dead volume of the conical cavity was about 45 μL. Other improved models, all-glass made based on density differences have been also proposed.[102]

The T-type phase separator depicted in Figure 11(c) relies on both gravitational and selective wetting effects (as can be seen, it contains a whisker of fluoroplastics to assist the separation of the organic flow). The basic design of the membrane-based separators are shown in Figure 11(d). They are constructed with membranes which are permeable to most organic solvents but which repel aqueous solutions. The membrane is placed between two blocks. Once the segmented aquo/organic phase enters the lower part of the minichamber, the organic phase (lighter) passes across the membrane while the aqueous phase continues; thus, the two phases are separated. The membrane separator provides perhaps the best performance in terms of separation efficiency, reproducibility and applicability to a great variety of organic solvents; however, the short lifetime of the membrane is one of its serious practical limitations.

Important pitfalls of the liquid–liquid approach include: (i) the limited compatibility of organic effluents with plasma atomic sources, (ii) comparatively low enrichment factors for preconcentration purposes as compared to solid–liquid system (the actual enrichment factors depend on the flow-rate ratio for the two liquids involved and operational impediments preclude the

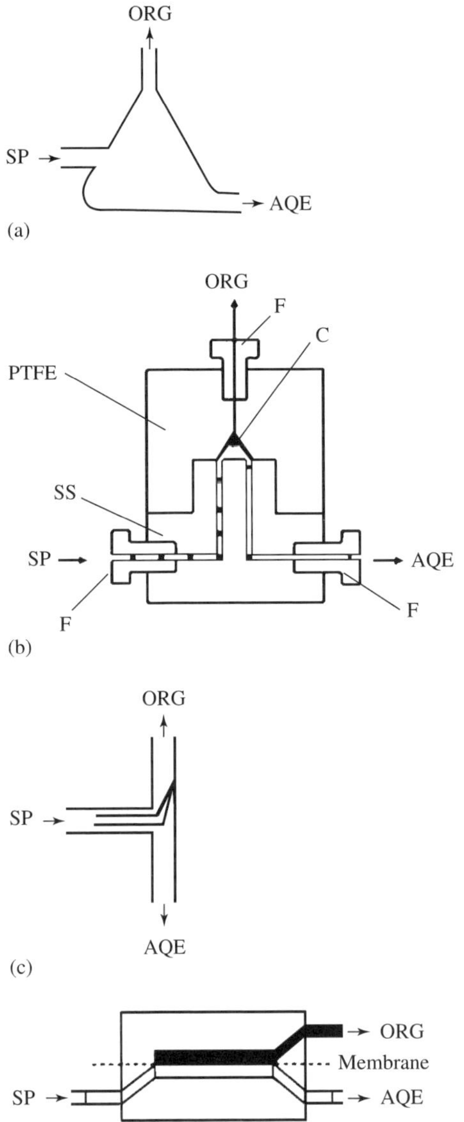

Figure 11 Diagrams of examples of liquid–liquid separation units (not drawn at the same scale). sp: segmented phase; aqe: aqueous phase; org: organic phase. (a) Gravity-based separator; (b) PTFE-stainless steel gravitational phase separator, C: conical cavity, F: threading fittings, SS: stainless steel; (c) T-type separator; (d) membrane type separator. (Parts (a), (c) and (d) reproduced from Lin et al., *Spectrochim. Acta*, Part B **51**, 1769–1775 © 1996; part (b) reproduced from Tao and Fang,[101] © 1995, with permission from Elsevier Science Ltd.)

use of highly different flow rates of the two phases), (iii) frequent problems associated to poor long term stability and robustness of the system requiring special operator skills.

Notwithstanding those limitations, flow liquid–liquid extraction systems have found important fields of application in connection with atomic detectors which withstand organic solvent introduction, such as FAAS and GFAAS. Finally, the use of these continuous separation systems for indirect analysis of organic pharmaceutical products should be highlighted here. Using the flow liquid–liquid separation strategy, it is possible to increase the analysis reproducibility as compared with the batch analogs, and some analyses would be only feasible when using these flow systems. For example, the determination of alkaloids has been described by formation of ion pairs between the sought drug and the inorganic complexes, BiI_4^- or $Co(SCN)_4^{2-}$, followed by liquid–liquid separation and measurement of bismuth or cobalt in the organic phase[103] or the determination of amphetamines based on their reaction with carbon disulfide to yield dithiocarbamic acids; these acids reacted with Cu(II), Ni(II) and Zn(II) and were extracted with IBMK being the salts or complexes on-line detected with FAAS.[104]

7.2 Integration of Two Continuous Separation Units: The Tandem On-line Concept

Occasionally, the complexity of the sample matrix or the sensitivity requirements may demand the use of more than one separation/preconcentration strategy to obtain appropriate analytical performance. Coupling two flow separation techniques in a tandem mode for sample preparation may be an effective solution to deal with such samples.

For example, we have seen that the sensitivity of analytical determinations by atomic spectroscopic methods can be enhanced by generation of volatile derivatives of the analyte. However, the chemical generation of these derivatives can be subjected to interferences (e.g. some metal transition ions in HG). Such interferences could be eliminated, for example, by using liquid–liquid or liquid–solid strategies. In this direction, a high degree of automation can be achieved by resorting to the 'tandem on-line' concept, i.e. the combination in a single on-line configuration of two continuous separation units: first, either a flow liquid–liquid extraction or a solid–liquid extraction step of the analyte; second, a volatile species generation from the second phase continuously fed to a gas–liquid separation unit; the first part allows improvements in both, selectivity and sensitivity, while the second permits the introduction of the analyte in gas phase to the detector, giving rise to further increases of the analytical sensitivity achievable.

This idea of "tandem on-line" separations for sample introduction in atomic spectrometry, was first applied to arsenic determinations in different samples by coupling in an on-line mode, a liquid–liquid and a gas–liquid separation device to an ICPAES.[105] A limitation of the technique could be the degree of compatibility of the plasma source with the organic solvent used for the liquid–liquid extraction (a certain amount of volatile

For references see page 9422

organic solvents can reach the plasma in the gas separation step). However, this problem is eliminated by using xylene to extract the analyte in the liquid–liquid separation process and by employing $NaBH_4$ in dimethylformamide for the CVG process and final gas–liquid separation step. A similar "tandem on-line" concept has been extended to indirect determinations of organic compounds[16] and to metal speciation, such as the analysis of inorganic Hg(II) and methylmercury based on the selective liquid–liquid extraction of methylmercury, into xylene, as bromide and CV generation of atomic mercury from the organic phase.[106]

Other on-line two-step approaches involve the elimination of interferences or the preconcentration of the analyte (prior to the HG step), using an ion exchange column (solid–liquid system), such as the manifold described for the determination of As, Sb and Se in a cobalt matrix.[107]

8 FLOW SYSTEMS FOR FIELD SAMPLING AND ON-LINE DIGESTION OF SAMPLES

8.1 Minicolumn Field Sampling and Flow Injection Analysis

The problem of securing during sample storage, an unchanged concentration of low levels of analyte or preserving the integrity of the species has received scarce attention from the analytical community both, for total metal and for metal speciation analyses. In recent years progress has been made in the development of true "field sampling" methods for the analysis of waters (seawater, river, lakes, etc.), which are mainly based on the isolation of the desired species in a stable form on appropriate minicolumns from which the analyte is later eluted for direct determination in the laboratory. A lateral advantage of this strategy is the easiness to transport, from the sampling source to the laboratory, small minicolumns containing the analyte (instead of transporting large containers with the liquid samples). This approach has been reported for the field sampling of Cr(III)/Cr(VI),[76] inorganic and organomercury species,[74] the determination of the 'fast reactive' aluminum fraction in waters,[108] etc.

Minicolumn field sampling is readily implemented using basic FI equipment. The sample delivery system can be a syringe, in the simplest case, or a peristaltic pump (battery-operated pumps are available). Although different approaches are possible, the minicolumns are generally transported back to the laboratory after sampling and on-line eluted to the detector. On-line elution reduces the risk of sample contamination or loss; on the other hand, since small elution volumes are used, high preconcentration factors can be achieved.

Different solid active phases can be used as column packing materials for field sampling (reviewed in section 5.1). However, it is important to point out that for minicolumn field sampling, additional parameters such as column to column variability and stability of immobilized species adsorbed in the column are of particular relevance and should be studied carefully.

8.2 On-line Photo-oxidation Flow Systems

The use of UV radiation to decompose organic matter is widely described in the literature; in particular, the oxidation reaction that combines a strong oxidizing agent with UV radiation has been described as an effective mode to destroy organic matter.

Concerning flow systems for analytical purposes, several photoreactor designs have been proposed to obtain the highest efficiency in such on-line decompositions. The favorable properties of PTFE for constructing tubular photoreactors is one of the most interesting features of this development, since in addition to its low UV absorption this material shows significant advantages over quartz coils, such as lower cost, easier handling and lower fragility. In atomic spectroscopy, most on-line photoreactors have been post-column coupled for metal speciation in hyphenated techniques using liquid chromatography for separation.[109]

FI systems involving on-line photo-oxidation schemes have been also proposed. Figure 12 shows the schematics of one such manifold used for the determination of organoarsenic compounds by on-line arsine generation.[108] A mercury lamp (OD 1.5 cm, length 20 cm) is used as UV source. The lamp is wrapped with a coil of 5 m of 0.56 mm ID. PTFE tubing through which the sample flows. To increase the light intensity reaching the coil and to prevent eye exposure to UV radiation, the unit was enclosed in aluminum foil.

8.3 On-line Microwave-assisted Digestion

A priori, continuous microwave digestion in a flow was expected to be associated with serious problems derived from the vigorous chemical conditions (e.g. excess of acids), resulting elevated temperature, high pressure, and long digestion times, frequently required to obtain complete decomposition. On the other hand, an additional problem to be faced is the evacuation of the gases produced during the digestion step. However, interesting instrumental designs have been developed trying to solve creatively these problems.

The microwave oven can be incorporated into flow manifolds which may be off-line or on-line connected to the atomic detector. For example, in order to analyze samples requiring long digestion times, the flow can be interrupted for a period of time, while the sample is in

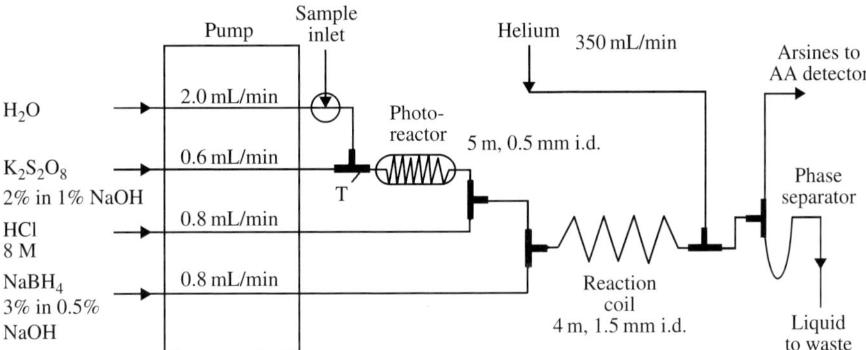

Figure 12 On-line flow sample digestion with UV radiation from a high intensity mercury lamp. (Reproduced with permission of Elsevier Science Ltd. from Atallah and Calman,[18] © 1991.)

the oven, resulting in stopped-flow digestion systems. A prototype of this type was developed by Karanassios et al. In their system, a sample plug was pumped into the center of the coil located inside the oven, but leaving about 50 cm of air at both ends of the tubing.[110] During digestion, the sample slowly rotates inside the coiled tube; this rotation served as a stirring mechanism and also helped to reduce the effects of non-uniform microwave heating, due to the formation of 'hot' spots. The oven was modified by placing an electric fan to vent hot air during operation and to help cooling the tube at the end of the digestion.

Figure 13 illustrates the instrumental set-up for an on-line FI-microwave digestion system described by Haswell and Barclay for the analysis of biological tissues using FAAS detection.[111] Here, samples were prepared as slurries in 5% v/v nitric acid and sample volumes of 1 mL were introduced into the flow manifold. In the microwave oven, a PTFE coil of 20 m length and 0.8 mm i.d. was used as digestion loop, being the flow rate around 5 mL min^{-1}. Various types of animal and botanical tissues were analyzed and element recoveries for Ca, Fe, Mg and Zn were typically found to be in the interval 94–107% with sample throughputs of 1–2 min per sample. As can be seen in Figure 13, a backpressure regulator was used to remove the fumes produced during acid decomposition of organic materials. Other alternatives described for this purpose were the use of an ice-bath or a diffusion cell connected to a vacuum pump.

As already mentioned in section 6.1, prior to HG the destruction of the organic matter and the transformation of the analytes into appropriate oxidation steps are frequently needed. On-line continuous flow microwave digestions are ideally suited for FI using CVG sample introduction systems and have been extensively used to shorten the pre-reduction steps and to decrease the problems associated with the volatility and adsorption losses of analytes such as As, Sb, Sn, Sn, Pb or Hg during traditional extensive pretreatment procedures: microwave energy allows the rapid conversion of the analyte from

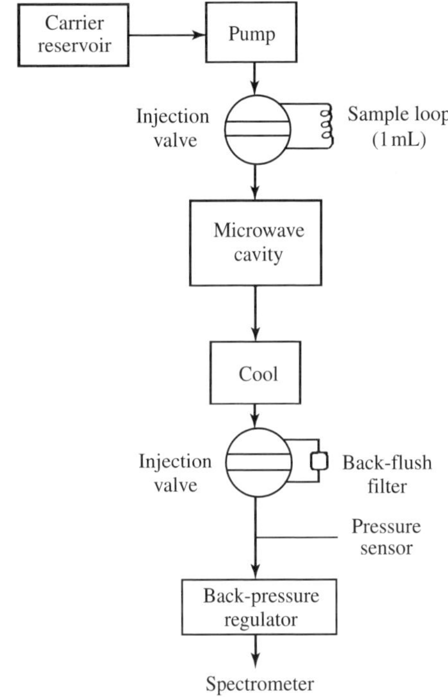

Figure 13 Experimental set-up for an on-line microwave digestion FI system for FAAS detection. (Reproduced with permission of the Royal Society of Chemistry from Haswell and Barclay,[111] © 1992.)

its form bound to the organic matrix to a free inorganic appropriate oxidation step. The general procedure consists of first mixing the samples with an appropriate reagent, being the mixture transported in the flow to the microwave oven; the resulting digest is on-line mixed with sodium borohydride to form the corresponding volatile species which are swept into the detector. A bromination mixture (bromate/bromide) and persulfate in acidic media proved effective for the on-line speciation of selenium and determination of arsenic, antimony, tin and lead.[112] Table 1 collects several examples of

For references see page 9422

Table 1 Selected examples of FI digestion procedures with microwave heating and atomic detection

Sample	Analyte	Comments	Detector	Ref.
Powdered botanical and biological reference materials	Al, Ba, Ca, Cu Fe, Mg, Mn, Zn	Off-line stopped-flow digestion system	ICPAES	110
Animal and botanical tissues	Ca, Fe, Mg, Zn	Samples are introduced as slurries in 5% HNO_3	FAAS	111
Urine and waters	Hg, As, Bi, Pb, Sn	Bromate–bromide and peroxidisulfate as oxidants	HG/AAS	112
Sediments and biological tissues	Hg	Organomercury is oxidized to inorganic Hg(II)	CV/AFS	19
Waters	Se(IV) and total inorganic Se	Microwaves were used only for total inorganic Se analyses	HG/AFS	113
Waters	Inorganic arsenic and organoarsenic	First report on FI/HG interface to ICPMS allowing for instantaneous reduction of both inorganic and organometallic As species	HG/ICPMS	114
Whole blood	Co	In-vivo uptake of specimens and on-line treatment of the samples	GFAAS	20

applications of on-line FI with microwave digestion coupled to a variety of atomic detection systems, which have been proposed by different authors.[19,20,110–114]

Perhaps the more advanced degree of automation using flow systems has been demonstrated with an on-line microwave assisted mineralization manifold designed for in vivo sample uptake of whole blood samples (Figure 14). The samples were drawn and pumped directly from the patient's forearm to a timed injector. This injector was automatically controlled to inject into the carrier stream a mixture on-line formed between the sample, an acidic solution and an anticoagulant.[20] Using this type of "closed-circuit" manifold, sample contamination problems from the environment should be reduced; besides, from the clinical point of view, it has also to be stressed that this type of manifold should eliminate risks of infection to nurses and analysts during the sample manipulation steps.

Before concluding this section it is important to point out that, although important progress has taken place in this field of on-line flow microwave digestion, the available instrumentation so far, does not yet seem to provide reliable performance. On a routine basis this technique needs further technical development to solve frequent malfunctions which will be probably overcome in the near future.

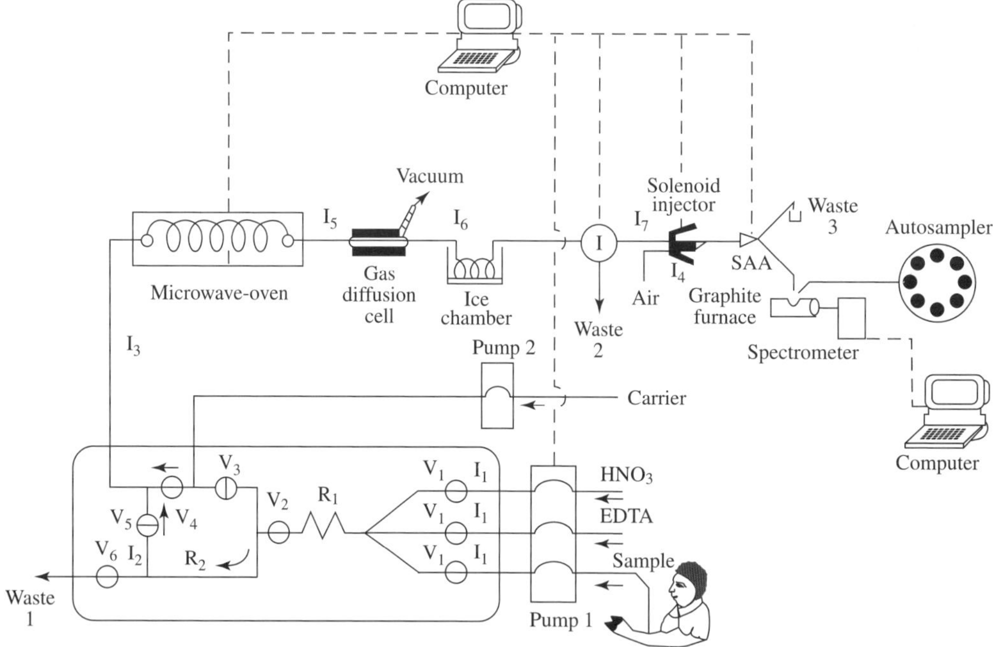

Figure 14 Diagram of a FI microwave oven ETAAS manifold for in vivo sample uptake of blood samples. I_1–I_7: tubing length; V_1–V_6: valves of the time-based solenoid injector; R_1: mixing tube; R_2: sample/reagent entrapment tubing; I: valve injector; SAA: sampling arm assembly. (Reproduced with permission of the Royal Society of Chemistry from Burguera et al.,[20] © 1995.)

9 POTENTIAL APPLICATIONS PREVIEW AND FUTURE DEVELOPMENTS

As a consequence of the ever-growing number of chemical analyses routinely carried out, a major demand in the area of research and development in chemical analysis is the need to make methods more cost-effective. On the other hand, the concept of quality is being fortunately installed in almost any aspect of our lives, analytical chemistry being a metrological discipline where the quality of results is of paramount importance. Hopefully, throughout this chapter it has become clear how both aspects could be improved by FIA methodologies.

There is a vast literature on the use of FIA techniques coupled to atomic detectors for a great variety of applications in different fields such as clinical, biological, environmental, etc. and several FIA methodologies, like volatile HG, are well implemented in routine analyses. However, the implementation of FIA strategies developed in research laboratories, in routine analysis is taking place at a slower pace, particularly for flow systems with higher degrees of complexity, even if they solve important problems. Since routine analyses demand robust instrumentation and procedures, this gap between routine practice and research laboratory work is being gradually reduced as reliable FI analyzers become commercially available.

From the upcoming research point of view, the great potential of the minicolumn 'field sampling' strategy should probably be highlighted; its merits have yet to be fully appreciated and exploited. Prospects for new developments and routine analysis in this area would be stimulated if custom sampling kits and minicolumn chemistries were commercially available. Also, the capability of flow systems to develop 'uncontaminating' methodologies, allowing sample contamination to decrease for modern ICPMS determinations and speciation cannot be overemphasized. The extreme sensitivity provided by this multielemental technique offers special attraction for its use in the analysis of very low level concentrations of any element and is unavoidable today for trace element speciation studies, particularly in biological systems. Besides, the ability of the ICPMS to accurately measure different isotopes opens new avenues in different fields such as in nutrition (e.g. to investigate the uptake of trace elements or their target organs and for identification of the source of environmental exposure to toxic elements). The simplification of the sample preparation steps and the reduction of contamination risks offered by FI methodologies, suggests that FI strategies will become the "natural" sample introduction mode in the ICPMS as in other detectors.

The field of analytical atomic spectrometry has undergone profound changes during the last decade. A sort of "metamorphosis" is taking place at the sunset of this millennium: instead of looking for information on the atomic/elemental composition of matter (for many decades the goal of atomic spectrometry), we are witnessing a return of many atomic spectroscopists to look for molecular information (as aimed at for present element speciating problems). Of course, as atomic methods are intrinsically non-speciating (i.e. they destroy molecular information) the most popular approach to overcome the problem consists of coupling a separating unit (for compounds separation) on-line with an atomic detector (for final element-specific detection). Particularly the use of MS analyzers has revolutionized this field by offering unprecedented sensitivity and isotope measurement capability.

It is clear that atomic spectrometry will expand its "natural" scope in the next millenium into molecular information gathering as well. To do so, flow analysis techniques (FIA conventional strategies, chromatography and also capillary electrophoresis) will be synergically combined/integrated with the rich array of atomic/specific detectors at our disposal nowadays.

ABBREVIATIONS AND ACRONYMS

AAS	Atomic Absorption Spectrometry
AES	Atomic Emission Spectrometry
AFS	Atomic Fluorescence Spectrometry
ASV	Anodic Stripping Voltammetry
CV	Cold Vapor
CVG	Chemical Vapor Generation
DCP	Direct Current Plasma
ETAAS	Electrothermal Atomic Absorption Spectrometry
FAAS	Flame Atomic Absorption Spectrometry
FAFS	Flame Atomic Fluorescence Spectrometry
FI	Flow Injection
FIA	Flow Injection Analysis
GD	Glow Discharge
GFAAS	Graphite Furnace Atomic Absorption Spectrometry
GLS	Gas Liquid Separator
HG	Hydride Generation
IBMK	Isobutyl Methyl Ketone
ICP	Inductively Coupled Plasma
ICPAES	Inductively Coupled Plasma Atomic Emission Spectrometry
ICPMS	Inductively Coupled Plasma Mass Spectrometry
MIP	Microwave Induced Plasma
MPT	Microwave Plasma Torch
MS	Mass Spectrometry

For references see page 9422

PTFE Poly(tetrafluoroethylene)
SCP Stabilized Capacitive Plasma
UV Ultraviolet

RELATED ARTICLES

Clinical Chemistry **(Volume 2)**
Atomic Spectrometry in Clinical Chemistry • Automation in the Clinical Laboratory

Environment: Water and Waste **(Volume 3)**
Flame and Graphite Furnace Atomic Absorption Spectrometry in Environmental Analysis • Flow-injection Techniques in Environmental Analysis • Heavy Metals Analysis in Seawater and Brines • Hydride Generation Sample Introduction for Spectroscopic Analysis in Environmental Samples • Inductively Coupled Plasma Mass Spectrometry in Environmental Analysis

Environment: Water and Waste cont'd **(Volume 4)**
Mercury Analysis in Environmental Samples by Cold Vapor Techniques • Sample Preparation for Elemental Analysis of Biological Samples in the Environment • Sample Preparation Techniques for Elemental Analysis in Aqueous Matrices

Food **(Volume 5)**
Atomic Spectroscopy in Food Analysis

Industrial Hygiene **(Volume 6)**
Metals in Blood and Urine: Biological Monitoring for Worker Exposure

Process Instrumental Methods **(Volume 9)**
Flow and Sequential Injection Analysis Techniques in Process Analysis

Atomic Spectroscopy **(Volume 11)**
Flame and Vapor Generation Atomic Absorption Spectrometry • Graphite Furnace Atomic Absorption Spectrometry • Inductively Coupled Plasma/Optical Emission Spectrometry • Microwave-induced Plasma Systems in Atomic Spectroscopy

REFERENCES

1. M. Valcárcel, M.D. Luque de Castro, M.T. Tena, 'Preliminary Operations: A Pending Goal of Today's Analytical Chemistry', *Anal. Proc.*, **30**, 276–279 (1993).
2. M. Valcarcel, M.D. Luque de Castro, *Automatic Methods of Analysis*, Elsevier, Amsterdam, 1988.
3. J. Ruzicka, E.H. Hansen, 'Flow Injection Analyses, Part I. A New Concept of Fast Continuous Analyses', *Anal. Chim. Acta*, **78**, 145–157 (1975).
4. J.R. Sarbeck, P.A. St. John, J.D. Winefordner, 'Measurement of Microsamples in Atomic Emission and Atomic Fluorescence Flame Spectrometry', *Mikrochim. Acta*, **149**, 55–60 (1972).
5. W.R. Wolf, K.K. Stewart, 'Automated Multiple Flow Injection Analysis for Flame Atomic Absorption Spectrometry', *Anal. Chem.*, **51**, 1201–1205 (1979).
6. N. Yoza, Y. Aoyagi, S. Ohashi, A. Tateda, 'Flow Injection System for Atomic Absorption Spectrometry', *Anal. Chim. Acta*, **111**, 163–167 (1979).
7. Z. Fang, S. Xu, G. Tao, 'Developments and Trends in Flow Injection Atomic Spectrometry', *J. Anal. At. Spectrom.*, **11**, 1–24 (1996).
8. Z. Fang, *Flow Injection Atomic Absorption Spectrometry*, Wiley, Chichester, 1995.
9. *Flow Injection Atomic Spectroscopy*, ed. J.L. Burguera, Marcel Dekker, New York, 1989.
10. *Flow Analysis with Atomic Spectrometric Detectors*, ed. A. Sanz-Medel, Elsevier, Amsterdam, 1999.
11. J. Ruzicka, E.H. Hansen, *Flow Injection Analysis*, Wiley, New York, 2nd edition, 1988.
12. I. Lavilla, B. Perez-Cid, C. Bendicho, 'Use of Flow-injection Sample-to Standard Addition Methods for Quantification of Metals Leached by Selective Chemical Extraction from Sewage Sludge', *Anal. Chim. Acta*, **381**, 297–305 (1999).
13. H. El Azouzi, M.Y. Pérez-Jordán, A. Salvador, M. de la Guardia, 'Extension of the Dynamic Range of Flame Atomic Absorption Spectrometry using Flow Injection Analysis with Variable-volume Dilution Chambers', *Spectrochim. Acta*, **51B**, 1747–1752 (1996).
14. L. Campanella, K. Pyrzynska, M. Trojanowicz, 'Chemical Speciation by Flow-injection Analysis. A Review', *Talanta*, **43**, 825–838 (1996).
15. J.R. Pretty, E.A. Blubaugh, J.A. Caruso, 'Determination of As(III) and Se(IV) Using an On-line Anodic Stripping Voltammetry Flow Cell with Detection by Inductively Coupled Plasma Atomic Emission Spectrometry and Inductively Coupled Plasma Mass Spectrometry', *Anal. Chem.*, **65**, 3396–3403 (1993).
16. A. Menéndez García, E. Sánchez-Uría, A. Sanz-Medel, 'Rapid Indirect Determination of Very Low Levels of Cocaine by Tandem On-line Continuous Separation and Inductively Coupled Plasma Atomic Emission Spectrometry Detection', *J. Anal. At. Spectrom.*, **11**, 561–565 (1996).
17. C.P. Hanna, J.F. Tyson, S. McInthosh, 'Determination of Total Mercury in Waters and Urine by Flow Injection Atomic Absorption Spectrometry Procedures Involving On and Off-line Oxidation of Organomercury Species', *Anal. Chem.*, **65**, 653–656 (1993).
18. R.J. Atallah, D.A. Kalman, 'On-line Photo-oxidation for the Determination of Organoarsenic Compounds

List of selected abbreviations appears in Volume 15

by Atomic-absorption Spectrometry with Continuous Arsine Generation', *Talanta*, **38**, 167–173 (1991).

19. K.J. Lamble, S.J. Hill, 'Determination of Mercury in Slurried Samples by Both Batch and On-line Microwave Digestion-Cold Vapor Atomic Fluorescence Spectrometry', *J. Anal. At. Spectrom.*, **11**, 1099–1103 (1996).
20. M. Burguera, J.L. Burguera, C. Rondón, C. Rivas, P. Carrero, M. Gallignani, M.R. Brunetto, 'In Vivo Sample Uptake and On-line Measurements of Cobalt in Whole Blood by Microwave-assisted Mineralization and Flow Injection Electrothermal Atomic Absorption Spectrometry', *J. Anal. At. Spectrom.*, **10**, 343–347 (1995).
21. *Atomic Absorption Spectrometry. Theory, Design and Applications*, ed. S.J. Haswell, Elsevier, Amsterdam, 1991.
22. A. Montaser, D.W. Golightly, *Inductively Coupled Plasmas in Atomic Analytical Atomic Spectrometry*, VCH, New York, 2nd edition, 1992.
23. *Plasma Source Mass Spectrometry. New Developments and Applications*, eds. G. Holland, S.D. Tanner, The Royal Society of Chemistry, Cambridge, 1999.
24. P.W. Alexander, R.J. Finlayson, L.E. Smythe, A. Thalib, 'Rapid Flow Analyses with Inductively Coupled Plasma Atomic Emission Spectroscopy Using a Micro-Injection Technique', *Analyst*, **107**, 1335–1342 (1982).
25. F.J. Krug, H. Bergamin, E.A. Zagatto, 'Commutation in Flow Injection Analysis', *Anal. Chim. Acta*, **170**, 103–118 (1986).
26. M.R. Pereiro García, A. López García, M.E. Diaz García, A. Sanz-Medel, 'On-line Aluminium Pre-concentration and its Application to the Determination of the Metal in Dialysis Concentrates by Atomic Spectrometric Methods', *J. Anal. At. Spectrom.*, **5**, 15–20 (1990).
27. J.F. Tyson, 'High-performance, Flow-based, Sample Pretreatment and Introduction Procedures for Analytical Atomic Spectrometry', *J. Anal. At. Spectrom.*, **14**, 169–178 (1999).
28. J. Ruzicka, E.H. Hansen, 'Recent Developments in Flow Injection Analysis: Gradient Techniques and Hydrodynamic Injection', *Anal. Chim. Acta*, **145**, 1–15 (1983).
29. M. Novic, I. Berregi, A. Rios, M. Valcárcel, 'A New Sample-injection/Sample-dilution System for the Flow-injection Analytical Technique', *Anal. Chim. Acta*, **381**, 287–297 (1999).
30. J.L. Burguera, M. Burguera, C. Rivas, M. de la Guardia, A. Salvador, 'Simple Variable-volume Injector for Flow-injection Analysis', *Anal. Chim. Acta*, **234**, 253–257 (1990).
31. T.K. Starn, R. Pereiro, G.M. Hieftje, 'Gas-sampling Glow Discharge for Optical Emission Spectrometry. Part I: Design and Operating Characteristics', *Appl. Spectrosc.*, **47**, 1555–1561 (1993).
32. J.F. Camuña-Aguilar, R. Pereiro-Garcia, J.E. Sánchez-Uría, A. Sanz-Medel, 'A Comparative Study of Three Microwave-induced Plasma Sources for Atomic Emission Spectrometry-II. Evaluation of their Atomization/Excitation Capabilities for Chlorinated Hydrocarbons', *Spectrochim. Acta*, **49B**, 545–554 (1994).
33. A.S. Attiyat, G.D. Christian, 'Nonaqueous Solvents as Carrier on Sample Solvent in Flow Injection Analysis Atomic Absorption Spectrometry', *Anal. Chem.*, **56**, 439–442 (1984).
34. S. Olsen, L.C.R. Pessenda, J. Ruzicka, E.H. Hansen, 'Combination of Flow Injection Analysis with Flame Atomic-absorption Spectrophotometry: Determination of Trace Amounts of Heavy Metals in Polluted Seawater', *The Analyst*, **108**, 905–917 (1983).
35. K.K. Stewart, 'Depulsing System for Positive Displacement Pumps', *Anal. Chem.*, **49**, 2125–2126 (1977).
36. C.P. Hanna, P.E. Haigh, J.F. Tyson, S. McIntosh, 'Examination of Separation Efficiencies of Mercury Vapor for Different Gas–Liquid Separators in Flow Injection Cold Vapor Atomic Absorption Spectrometry with Amalgam Preconcentration', *J. Anal. At. Spectrom.*, **8**, 585–590 (1993).
37. J.C. de Andrade, C. Pasquini, N. Baccan, J.C. Van Loon, 'Cold Vapor Atomic Absorption Determination of Mercury by Flow Injection Analysis Using a Teflon Membrane Phase Separator Coupled to the Absorption Cell', *Spectrochim. Acta*, **38B**, 1329–1338 (1983).
38. R.M. Barnes, X. Wang, 'Microporous Polytetrafluoroethylene Tubing Gas Liquid Separator for Hydride Generation and Inductively Coupled Plasma Atomic Emission Spectrometry', *J. Anal. At. Spectrom.*, **3**, 1083–1089 (1988).
39. R. Ma, W.V. Mol, F. Adams, 'Selective Flow Injection Sorbent Extraction for Determination of Cadmium, Copper and Lead in Biological and Environmental Samples by Graphite Furnace Atomic Absorption Spectrometry', *Anal. Chim. Acta*, **293**, 251–260 (1994).
40. Z. Fang, L. Dong, 'Flow Injection On-line Coprecipitation Preconcentration for Electrothermal Atomic Absorption Spectrometry', *J. Anal. At. Spectrom.*, **7**, 439–445 (1992).
41. E.A. Zagatto, F.J. Krug, H. Bergamin, S.S. Jorgensen, B.F. Reis, 'Merging Zones in Flow Injection Analysis. Part 2. Determination of Calcium, Magnesium and Potassium in Plant Material by Continuous Flow Injection Atomic Absorption and Flame Emission Spectrometry', *Anal. Chim. Acta*, **104**, 279–284 (1979).
42. E.A. Zagatto, M.F. Giné, E.A.N. Fernandes, B.F. Reis, F.J. Krug, 'Sequential Injections in Flow Systems as an Alternative to Gradient Exploitation', *Anal. Chim. Acta*, **173**, 289–297 (1985).
43. J.F. Tyson, S.R. Bysouth, 'Network Flow Injection Manifolds for Sample Dilution and Calibration in Flame Atomic Absorption Spectrometry', *J. Anal. At. Spectrom.*, **3**, 211–215 (1988).
44. J.M.H. Appleton, J.F. Tyson, 'Flow Injection Atomic Absorption Spectrometry: The Kinetic of Instrument Response', *J. Anal. At. Spectrom.*, **1**, 63–74 (1986).

List of selected abbreviations appears in Volume 15

45. Z. Fang, B. Welz, M. Sperling, 'Contribution of System Components to Dispersion in the Analysis of Microvolume Samples by Flow Injection Flame Atomic Absorption Spectrometry', *J. Anal. At. Spectrom.*, **6**, 179–189 (1991).
46. M.W. Brown, J. Ruzicka, 'Parameters Affecting Sensitivity and Precision in the Combination of Flow Injection Analysis with Flame Atomic-absorption Spectrophotometry', *The Analyst*, **109**, 1091–1094 (1984).
47. 'Sistemas de introducción de muestras líquidas en espectrometría atómica', J.L. Todoli, J. Mori, V. Hernandis, A. Canals, Secretariado de Publicaciones, Alicante University, Spain, 1996.
48. K.E. LaFreniere, G.W. Rice, V.A. Fassel, 'Flow Injection Analysis with Inductively Coupled Plasma-atomic Emission Spectroscopy: Critical Comparison of Conventional Pneumatic, Ultrasonic and Direct Injection Nebulization', *Spectrochim. Acta*, **40B**, 1495–1503 (1985).
49. K.E. Lawrence, G.W. Rice, V.A. Fassel, 'Direct Liquid Sample Introduction for Flow Injection Analysis and Liquid Chromatography with Inductively Coupled Argon Plasma Spectrometric Detection', *Anal. Chem.*, **56**, 289–292 (1984).
50. J.A. Koropchak, D.H. Winn, 'Thermospray Interfacing for Flow Injection Analysis with Inductively Coupled Plasma Atomic Emission Spectrometry', *Anal. Chem.*, **58**, 2558–2561 (1986).
51. M. Wu, Y. Madrid, J.A. Auxier, G.M. Hieftje, 'New Spray Chamber for Use in Flow-injection Plasma Emission Spectrometry', *Anal. Chim. Acta*, **286**, 155–167 (1994).
52. Y. Madrid, M. Wu, Q. Jin, G.M. Hieftje, 'Evaluation of Flow Injection Techniques for Microwave Plasma Torch Atomic Absorption Spectrometry', *Anal. Chim. Acta*, **277**, 1–8 (1993).
53. M. Thompson, B. Pahlavanpour, S.J. Walton, G.F. Kirkbright, 'Simultaneous Determination of Trace Concentrations of As, Sb, Bi, Se and Te in Aqueous Solution by Introduction of the Gaseous Hydrides into an Inductively Coupled Plasma Source for Emission Spectrometry', *Analyst*, **103**, 568–573 (1978).
54. O. Astrom, 'Flow Injection Analysis for the Determination of Bismuth by Atomic Absorption Spectrometry with Hydride Generation', *Anal. Chem.*, **54**, 190–193 (1982).
55. F. Lunzer, R. Pereiro-García, N. Bordel-García, A. Sanz-Medel, 'Continuous Hydride Generation Low-pressure Microwave Induced Plasma Atomic Emission Spectrometry for the Determination of Arsenic, Antimony and Selenium', *J. Anal. At. Spectrom.*, **10**, 311–315 (1995).
56. J.A.C. Broekaert, R. Pereiro, T.K. Starn, G.M. Hieftje, 'A Gas-sampling Glow Discharge Coupled to Hydride Generation for the Atomic Spectrometric Determination of Arsenic', *Spectrochim. Acta*, **48B**, 1207–1220 (1993).
57. W.T. Corns, P.B. Stockwell, L. Ebdon, S.J. Hill, 'Development of an Atomic Fluorescence Spectrometer for the Hydride-forming Elements', *J. Anal. At. Spectrom.*, **8**, 71–77 (1993).
58. H. Tao, J.W.H. Lam, J.W. McLaren, 'Determination of Se in Marine Certified Reference Materials by Hydride Generation Inductively Coupled Plasma – Mass Spectrometry', *J. Anal. At. Spectrom.*, **8**, 1067–1073 (1993).
59. J.F. Camuña, M. Montes, R. Pereiro, A. Sanz-Medel, C. Katschthaler, R. Gross, G. Knapp, 'Determination of Halides by Microwave Induced Plasma and Stabilised Capacitive Plasma Atomic Emission Spectrometry after On-line Continuous Halogen Generation', *Talanta*, **44**, 535–544 (1997).
60. J. Rodriguez, R. Pereiro, A. Sanz-Medel, 'Glow Discharge Atomic Emission Spectrometry for the Determination of Chlorides and Total Organochlorine in Water Samples via On-line Continuous Generation of Chlorine', *J. Anal. At. Spectrom.*, **13**, 911–915 (1998).
61. H. Matusiewicz, R.E. Sturgeon, 'Review. Atomic Spectrometric Detection of Hydride Forming Elements Following in situ Trapping Within a Graphite Furnace', *Spectrochim. Acta*, **51B**, 377–397 (1996).
62. H. Goenaga Infante, M.L. Fernández Sánchez, A. Sanz-Medel, 'Ultratrace Determination of Cadmium in Atomic Absorption Spectrometry Using Hydride Generation With In Situ Preconcentration in a Palladium-coated Graphite Atomizer', *J. Anal. At. Spectrom.*, **11**, 571–575 (1996).
63. M.A.Z. Arruda, M. Gallego, M. Valcárcel, 'Automatic Preparation of Milk Dessert Slurries for the Determination of Trace Amounts of Aluminium by Electrothermal Atomic Absorption Spectrometry', *J. Anal. At. Spectrom.*, **10**, 55–59 (1995).
64. Z. Fang, in *Flow Injection Atomic Spectroscopy*, ed. J.L. Burguera, Marcel Dekker, New York, 1989.
65. M. De la Guardia, in *Flow Analysis with Atomic Spectrometric Detectors*, ed. A. Sanz-Medel, Elsevier Science, Amsterdam, The Netherlands, Chapter 4, 1999.
66. M. Viczián, A. Lásztity, S. Wang, R.M. Barnes, 'On-line Isotope Dilution by Flow Injection and Inductively Coupled Plasma Mass Spectrometry', *J. Anal. At. Spectrom.*, **5**, 125–133 (1990).
67. J. Goossens, L. Moens, R. Dams, 'Determination of Lead by Flow-injection Inductively Coupled Plasma Mass Spectrometry Comparing Several Calibration Techniques', *Anal. Chim. Acta*, **293**, 171–181 (1994).
68. S. Olsen, L.C.R. Pessenda, J. Ruzicka, E.H. Hansen, 'Combination of Flow Injection Analysis with Flame Atomic-absorption Spectrometry: Determination of Trace Amounts of Heavy Metals in Polluted Seawater', *The Analyst*, **108**, 905–917 (1983).
69. S. Hirata, K. Honda, T. Kumamaru, 'Trace Metal Enrichment by Automated On-line Column Preconcentration for Flow-injection Atomic Absorption Spectrometry', *Anal. Chim. Acta*, **221**, 65–76 (1989).

70. S. Hirata, Y. Umezaki, M. Ikeda, 'Determination of Cr(III), Ti, Va, Fe(III), and Al by Inductively Coupled Plasma Atomic Emission Spectrometry with On-line Preconcentrating Ion-exchange Column', *Anal. Chem.*, **58**, 2602–2606 (1986).
71. B. Mohammad, A.M. Ure, D. Littlejohn, 'On-line Preconcentration of Aluminium with Immobilized 8-Hydroxyquinoline for Determination by Atomic Absorption Spectrometry', *J. Anal. At. Spectrom.*, **7**, 695–699 (1992).
72. P. Schramel, L.Q. Xu, G. Knapp, M. Michaelis, 'Application of an On-line Preconcentration System in Simultaneous Inductively Coupled Plasma – Atomic Emission Spectrometry', *Mikrochim. Acta*, **106**, 191–201 (1992).
73. G. Knapp, K. Müller, M. Strunz, W. Wegscheider, 'Automation in Element Pre-concentration with Chelating Ion Exchangers. Plenary Lecture', *J. Anal. At. Spectrom.*, **2**, 611–614 (1987).
74. W. Jian, C.W. McLeod, 'Field Sampling Technique for Mercury Speciation', *Anal. Proc.*, **28**, 293–294 (1991).
75. C.W. McLeod, I.G. Cook, P.J. Worsfold, J.E. Davies, J. Queay, 'Analyte Enrichment and Matrix Removal in Flow Injection Analysis – Inductively Coupled Plasma – Atomic Emission Spectrometry: Determination of Phosphorus in Steels', *Spectrochim. Acta*, **40B**, 57–62 (1985).
76. A.G. Cox, C.W. McLeod, 'Preconcentration and Determination of Trace Chromium (III) by Flow Injection – Inductively Coupled Plasma-atomic Emission Spectrometry', *Anal. Chim. Acta*, **179**, 487–490 (1986).
77. H.A.M. Elmahadi, G.M. Greenway, 'Immobilized Algae as a Reagent for Preconcentration in Trace Element Atomic Absorption Spectrometry', *J. Anal. At. Spectrom.*, **6**, 643–646 (1991).
78. H.A.M. Elmahadi, G.M. Greenway, 'Speciation and Preconcentration of Trace Elements with Immobilized Algae for Atomic Absorption Spectrophotometric Detection', *J. Anal. At. Spectrom.*, **9**, 547–551 (1994).
79. J. Ruzicka, A. Arndal, 'Sorbent Extraction in Flow Injection Analysis and its Application to Enhancement of Atomic Spectrometry', *Anal. Chim. Acta*, **216**, 243–255 (1989).
80. M. Fernández García, R. Pereiro García, N. Bordel García, A. Sanz-Medel, 'On-line Preconcentration of Inorganic Mercury and Methylmercury in Seawater by Sorbent-extraction and Total Mercury Determination by Cold Vapor Atomic Absorption Spectrometry', *Talanta*, **41**, 1833–1839 (1994).
81. R.E. Santelli, M. Gallego, M. Valcarcel, 'Preconcentration and Atomic Absorption Determination of Copper Traces in Waters by On-line Adsorption-elution on an Activated Carbon Minicolumn', *Talanta*, **41**, 817–823 (1994).
82. H.L. Lancaster, G.D. Marshall, E.R. Gonzalo, J. Ruzicka, G.D. Christian, 'Trace Metal Atomic Absorption Spectrometric Analysis Utilizing Sorbent Extraction on Polymeric-based Supports and Renewable Reagents', *Analyst*, **119**, 1459–1465 (1994).
83. R. Martinez-Jiménez, M. Gallego, M. Valcarcel, 'Preconcentration and Determination of Trace Amounts of Lead in Water by Continuous Precipitation in an Unsegmented Flow Atomic Absorption Spectrometric System', *Analyst*, **112**, 1233–1236 (1987).
84. Z. Fang, M. Sperling, B. Welz, 'Flame Atomic Absorption Spectrometric Determination of Lead in Biological Samples Using a Flow Injection System with On-line Preconcentration by Coprecipitation Without Filtration', *J. Anal. At. Spectrom.*, **6**, 301–306 (1991).
85. H. Chen, S. Xu, Z. Fang, 'Determination of Copper in Water and Rice Samples by Flame Atomic Absorption Spectrometry With Flow Injection On-line Adsorption Preconcentration Using a Knotted Reactor', *Anal. Chim. Acta*, **298**, 167–173 (1994).
86. J.R. Pretty, E.A. Blubaugh, E.H. Evans, J.A. Caruso, T.M. Davidson, 'Determination of Copper and Cadmium using an On-line Anodic Stripping Voltammetry Flow Cell With Detection by Inductively Coupled Plasma Mass Spectrometry', *J. Anal. At. Spectrom.*, **7**, 1131–1137 (1992).
87. J.R. Pretty, E.A. Blubaugh, J.A. Caruso, T.M. Davidson, 'Determination of Cr(VI) and V(V) Using an On-Line Anodic Stripping Voltammetry Flow Cell with Detection by Inductively Coupled Plasma Mass Spectrometry', *Anal. Chem.*, **66**, 1540–1547 (1994).
88. J. Dedina, D.L. Tsalev, 'Hydride Generation Atomic Absorption Spectrometry', Wiley, Chichester, 1995.
89. D.L. Tsalev, 'Hyphenated Vapor Generation Atomic Absorption Spectrometric Techniques. Invited Lecture', *J. Anal. At. Spectrom.*, **14**, 147–162 (1999).
90. A. Sanz-Medel, M.R. Fernández de la Campa, E. Blanco González, M.L. Fernández-Sánchez, 'Review. Organised Surfactant Assemblies in Analytical Atomic Spectometry', *Spectrochim. Acta*, **54B**, 251–287 (1999).
91. W. Ronald, W.L. Ott, 'Determination of Sub-Microgram Quantities of Mercury by Atomic Absorption Spectrophotometry', *Anal. Chem.*, **40**, 2085–2087 (1968).
92. J.M. Costa-Fernández, R. Pereiro-García, A. Sanz-Medel, N. Bordel-García, 'Effect of Plasma Pressure on the Determination of Mercury by Microwave-induced Plasma Atomic Emission Spectrometry', *J. Anal. At. Spectrom.*, **10**, 649–653 (1995).
93. A. Stroh, V. Völlkopf, 'Optimization and Use of Flow Injection Vapor Generation Inductively Coupled Plasma Mass Spectrometry for the Determination of Arsenic, Antimony and Mercury in Water and Seawater at Ultratrace Levels', *J. Anal. At. Spectrom.*, **8**, 35–40 (1993).
94. A. Sanz-Medel, M.C. Valdés-Hevia y Temprano, N. Bordel Garcia, M.R. Fernández de la Campa, 'Generation of Cadmium Atoms at Room Temperature Using Vesicles and Its Application to Cadmium Determination by Cold Vapor Atomic Spectrometry', *Anal. Chem.*, **67**, 2216–2223 (1995).

List of selected abbreviations appears in Volume 15

95. M.C. Valdés-Hevia y Temprano, M.R. Fernández de la Campa, A. Sanz-Medel, 'Generation of Volatile Cadmium Species With Sodium Tetrahydroborate From Organized Media: Application to the Cadmium Determination by Inductively Coupled Plasma Atomic Emission Spectrometry', *J. Anal. At. Spectrom.*, **8**, 847–852 (1993).
96. H. Goenaga Infante, M.L. Fernández Sánchez, A. Sanz-Medel, 'Vesicle-assisted Determination of Ultratrace Amounts of Cadmium in Urine by Electrothermal Atomic Absorption Spectrometry and Inductively Coupled Plasma Mass Spectrometry', *J. Anal. At. Spectrom.*, **13**, 899–903 (1998).
97. S. Rapsomanikis, 'Derivatization by Ethylation With Sodium Tetraethylborate for the Speciation of Metals and Organometallics in Environmental Samples. A Review', *Analyst*, **119**, 1429–1439 (1994).
98. J. Ashby, S. Clark, P.J. Craig, 'Methods for the Production of Volatile Organometallic Derivatives for Application to the Analysis of Environmental Samples', *J. Anal. At. Spectrom.*, **3**, 735–736 (1988).
99. M.C. Valdés-Hevia y Temprano, M.R. Fernández de la Campa, A. Sanz-Medel, 'Sensitive Inductively Coupled Plasma Atomic Emission Spectrometry Determination of Cadmium by Continuous Alkylation with Sodium Tetraethylborate', *J. Anal. At. Spectrom.*, **9**, 231–236 (1994).
100. M. Burguera, J.L. Burguera, M.R. Brunetto, M. De la Guardia, A. Salvador, 'Flow-injection Atomic Spectrometric Determination of Inorganic Arsenic(III) and Arsenic(V) Species by Use of an Aluminium-column Arsine Generator and Cold-trapping Arsine Collection', *Anal. Chim. Acta*, **261**, 105–113 (1991).
101. G. Tao, Z. Fang, 'On-line Flow Injection Solvent Extraction for Electrothermal Atomic Absorption Spectrometry Determination of Nickel in Biological Samples', *Spectrochim. Acta*, **50B**, 1747–1755 (1995).
102. S. Lin, Q. Shuai, H. Qiu, Z. Tang, 'Design and Application of a New Phase Separator for Flow Injection Liquid–Liquid Extraction', *Spectrochim. Acta*, **51B**, 1769–1775 (1996).
103. M. Eisman, M. Gallego, M. Valcárcel, 'Coupling of a Continuous Liquid–Liquid Extractor to a Flame Atomic Absorption Spectrometer for the Determination of Alkaloids', *J. Anal. At. Spectrom.*, **7**, 1295–1298 (1992).
104. R. Montero, M. Gallego, M. Valcárcel, 'Determination of Amphetamines by Use of Liquid–Liquid Extractor Coupled On-line to an Atomic Absorption Spectrometer', *Anal. Chim. Acta*, **252**, 83–88 (1991).
105. A. Menéndez García, E. Sánchez Uría, A. Sanz-Medel, 'Tandem on-line Continuous Separation and Determination of Arsenic by Inductively Coupled Plasma Atomic Emission Spectrometry', *Anal. Chim. Acta*, **234**, 133–139 (1990).
106. A. Menéndez García, M.L. Fernández Sánchez, J.E. Sánchez Uría, A. Sanz-Medel, 'Speciation of Mercury by Continuous Flow Liquid–Liquid Extraction and Inductively Coupled Plasma Atomic Emission Spectrometry Detection', *Mikrochim. Acta*, **122**, 157–166 (1996).
107. H. Wang, Y. Chen, J. Wang, 'Determination of Arsenic, Antimony and Selenium in Cobalt by On-line Injection, Cation Exchange, Hydride-generation, Inductively Coupled Plasma Atomic Emission Spectrometry', *Anal. Proc.*, **31**, 357–359 (1994).
108. B. Fairman, A. Sanz-Medel, P. Jones, 'Field Sampling Technique for the 'Fast Reactive' Aluminium Fraction in Waters Using a Flow Injection Mini-column System with Inductively Coupled Plasma Atomic Emission Spectrometric and Inductively Coupled Plasma Mass Spectrometric Detection', *J. Anal. At. Spectrom.*, **10**, 281–285 (1995).
109. R. Rubio, J. Albertí, A. Padró, G. Rauret, 'On-line Photolytic Decomposition for the Determination of Organoarsenic Compounds', *Trends Anal. Chem.*, **14**, 274–279 (1995).
110. V. Karanassios, F.H. Li, B. Liu, E.D. Salin, 'Rapid Stopped-flow Microwave Digestion System', *J. Anal. At. Spectrom.*, **6**, 457–463 (1991).
111. S.J. Haswell, D. Barclay, 'On-line Microwave Digestion of Slurry Samples With Direct Flame Atomic Absorption Spectrometric Elemental Detection', *Analyst*, **117**, 117–120 (1992).
112. D.L. Tsalev, M. Sperling, B. Welz, 'On-line Microwave Sample Pretreatment for Hydride Generation and Cold Vapor Atomic Absorption Spectrometry', *Analyst*, **117**, 1735–1741 (1992).
113. Y. He, H. El Azouzi, M.L. Cervera, M. De la Guardia, 'Completely Integrated On-line Determination of Dissolved Selenium (IV) and Total Inorganic Selenium in Seawater by Flow Injection Hydride Generation Atomic Fluorescence Spectrometry', *J. Anal. At. Spectrom.*, **13**, 1291–1296 (1998).
114. D. Beauchemin, 'Hydride Generation Interface for the Determination of Inorganic Arsenic and Organoarsenic by Inductively Coupled Plasma Mass Spectrometry Using Open-focused Microwave Digestion to Enhance the Pre-reduction Process', *J. Anal. At. Spectrom.*, **13**, 1–5 (1998).

Glow Discharge Optical Spectroscopy and Mass Spectrometry

Robert E. Steiner
Los Alamos National Laboratory, Los Alamos, USA

Christopher M. Barshick
Oak Ridge National Laboratory, Oak Ridge, USA

Glow discharge (GD) sources provide analytically useful gas-phase species from solid samples. These sources can be interfaced with a variety of spectroscopic and spectrometric instruments for both quantitative and qualitative analysis. Optical (atomic absorption spectroscopy, AAS; atomic emission spectroscopy, AES; atomic fluorescence spectroscopy, AFS; and optogalvanic spectroscopy) and mass spectrometric (magnetic sector, quadrupole mass analyzer, QMA; quadrupole ion trap, QIT; Fourier transform ion cyclotron resonance, FTICR; and time-of-flight, TOF) instrumentation are well suited for coupling to the GD.

The GD is a relatively simple device. A potential gradient (500–1500 V) is applied between an anode and cathode. In most cases the sample is also the cathode. A noble gas (e.g. Ar, Ne, Xe) is introduced into the discharge region prior to power initiation. When a potential is applied, electrons are accelerated toward the anode. As these electrons accelerate they collide with gas atoms. A fraction of these collisions are of sufficient energy to remove an electron from a support gas atom, forming an ion. These ions are, in turn, accelerated toward the cathode. These ions impinge on the surface of the cathode, sputtering sample atoms from the surface. Sputtered atoms that do not redeposit on the surface diffuse into the excitation/ionization regions of the plasma where they can undergo excitation via a number of collisional processes.

GD sources offer a number of distinct advantages that make them well suited for specific types of analyses. These sources afford direct analysis of solid samples, thus minimizing the sample preparation required for analysis. The nature of the plasma also provides mutually exclusive atomization and excitation processes that help to minimize the matrix effects that plague so many other elemental techniques. Unfortunately, the GD source functions optimally in a dry environment, making analysis of solutions difficult. These sources also suffer from difficulties associated with analyzing nonconducting samples.

In this article, the principles of operation of the GD plasma will first be reviewed, with an emphasis on how those principles relate to optical spectroscopy and mass spectrometry. Basic applications of the GD techniques will next be considered. These include bulk analysis, surface analysis, and the analysis of solution samples. The requirements necessary to obtain optical information will be addressed following the analytical applications. This section will focus on the instrumentation needed to make optical measurements using the GD as an atomization/excitation source. Finally, mass spectrometric instrumentation and interfaces will be addressed as they pertain to the use of a GD plasma as an ion source.

1 INTRODUCTION

We have always questioned the physical make-up of our universe, but only recently has science advanced to the point that some of the most fundamental questions can be answered. While all fields of science play a role in information gathering, this is the specific charge of the analytical chemist. The field of analytical chemistry is broad, and although no one area can claim to be more valuable than another, the area of elemental analysis has always been critical to our understanding of the chemical and physical properties of materials.

Many years ago chemists characterized the world in terms of the basic elements of fire, earth, air, and water. Today, we classify our world in a similar manner,

For references see page 9448

but in terms of carbon, nitrogen, oxygen, hydrogen, etc.; all that has really changed are the techniques we use – techniques that allow us to measure elemental and isotopic composition, elemental speciation, and spatial distribution of atoms and molecules.

This article focuses on two techniques used in elemental analysis: GD optical and mass spectrometries. We begin by describing the fundamental operation of a GD, including the sputtering and ionization processes. Next, we examine optical techniques that take advantage of the GD as an atomization source. Finally, we conclude by describing GD mass spectrometry, a powerful tool for multielement ultratrace analysis.

2 THE GLOW DISCHARGE

Although not always apparent, aspects of our everyday lives are permeated with technologies originally developed to solve scientific problems. An excellent example is the GD. Developed in the early 20th century[1–3] as a spectrochemical ionization source used for fundamental studies of atomic structure, a variation of the same GD now illuminates storefronts in the guise of the common neon sign.

The GD is a simple device. Before delving into the fundamental processes that characterize the glow, however, it is useful to define a few terms. A GD is one example of a general class of excitation/ionization sources known as "plasmas". This term refers to a partially ionized gas with equal numbers of positive and negative ions, and a larger number of neutral species.[4] Often the terms GD and gas discharge are interchangeable. A gas discharge is formed by passing an electric current through a gaseous medium.[5] For current to flow through this medium, a fraction of the gas must be ionized. In practice one applies a potential gradient between a cathode and an anode; in doing so, electrons are accelerated toward the anode. As the electrons move through the gas, they collide with gas phase atoms. A fraction of these collisions will be of sufficient energy to remove an electron from an atom, thus producing an ion and a secondary electron. The positively charged gas ions will, in turn, be accelerated toward and impinge upon the negatively biased sample cathode. Upon impact a variety of species will be liberated from the surface. Gas phase sample atoms are then free to diffuse into the plasma where they can undergo excitation and ionization. These dynamic processes are depicted in Figure 1.

A GD can be formed in virtually any vacuum cell that can be equipped with an inlet for a support gas. A variety of geometries have been investigated in the past. Table 1 lists some of the most common. The most popular source

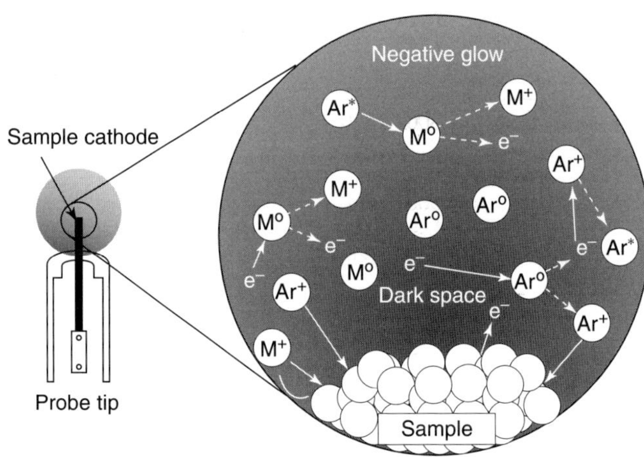

Figure 1 Illustration of various steady-state plasma processes. (Reproduced by permission from F.L. King, J. Teng, R.E. Steiner, *Glow Discharge Mass Spectrometry: Trace Elemental Determinations in Solid Samples*. Copyright (1995) John Wiley & Sons Ltd.)

is based on a cross-shaped vacuum housing that provides ports for spectroscopic viewing, sample introduction, gas introduction, pressure monitoring, and ion extraction.

2.1 Fundamental Glow Discharge Processes

The fundamental processes occurring in the discharge define a number of discrete regions. For the current discussion, only three regions – the cathode dark space, the negative glow, and the Faraday dark space – will be defined, because the other regions, while important, are often not visible due to limited separation of the cathode and anode. For a complete treatise on this subject the reader is directed to several excellent articles.[6–8] The dark space regions are characterized by low light intensities relative to other regions of the plasma. This lack of luminosity arises from an absence of collisions and consequently the absence of excitation and the radiative relaxation events that produce photons. The cathode dark space is located between the cathode surface and the negative glow region. The presence of a large positive space charge in the negative glow causes the development of a potential gradient. The bulk discharge potential decreases rapidly through this region leading to its common name, the cathode fall. This large potential gradient affects the acceleration of electrons that can ionize the discharge gas species and liberate secondary electrons in the negative glow[4] which help to sustain the plasma. Radiative relaxation of species excited in the negative glow region yields the characteristic emission for which it is named. Because the major charge carriers in this region are electrons, the net space charge is negative. As electrons collisionally cool, they slow, decreasing their cross-section for excitational collisions with atoms. This

Table 1 GD ion sources and their characteristics

Source type	Voltage (V)	Current (mA)	Pressure (Torr)	Advantages	Disadvantages
Coaxial cathode	800–1500	1–5	0.1–10	Can conform to various sample shapes and sizes Penning ionization dominated	Powders must be pressed
Grimm	500–1000	25–100	1–5	Depth profiling Compacted powder samples	Flat samples only High gas flow rates
Hollow cathode	200–500	10–100	0.1–10	High sputter rate Intense ion beams for mass spectometry Large localized atom populations for optical spectroscopy	Complicated geometry Charge exchange mechanism is important
Jet-enhanced	800–1000	25–30	1–5	High sputter rate Compacted powder samples	Flat samples only High gas flow rates

smaller excitation cross-section results in the Faraday dark space close to the anode. Within this region the net space charge is zero and the potential gradient approaches a constant.

2.1.1 The Sputtering Process

The GD is of particular utility when analyzing solid conducting samples. Most competing techniques used for elemental applications (i.e. inductively coupled plasma mass spectrometry (ICPMS), flame spectroscopies, and graphite furnace techniques) require a sample in solution form to facilitate aspiration or introduction into the ionization source. The GD source possesses the inherent characteristic of producing gas phase analyte atoms directly from the solid conducting sample material. This phenomenon, known simply as cathodic sputtering, can be most easily described using a basic billiard ball analogy. Positively charged discharge gas ions are accelerated toward the negatively biased sample cathode. Prior to impact, these high-energy ions recombine with Auger electrons released from the cathode surface. The resulting high energy neutral species impact the surface of the cathode, transforming their kinetic energy (KE) into the lattice of the sample, thus causing a cascade of collisional events, much like the breaking of a racked set of billiard balls. If the resulting energy transfer is sufficient to overcome a surface atom's binding energy, the atom will be released into the gas phase. The sputtering process liberates not only individual cathodic atoms, but also electrons, ions, and clusters of atoms and molecules. This process is illustrated in Figure 2.

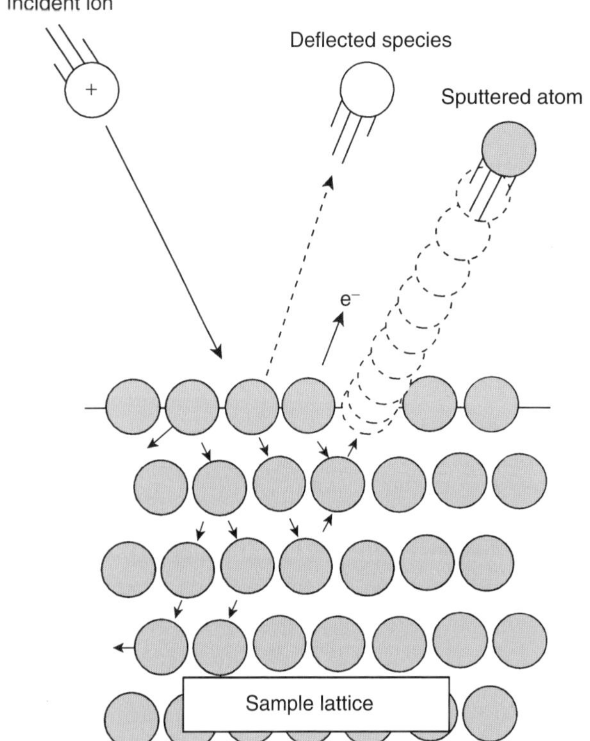

Figure 2 Illustration of cathodic sputtering process.

Sputtered electrons are accelerated across the cathode dark space into the negative glow where they can contribute to excitation and ionization of gas phase atoms. The ions formed by the sputtering event do not travel far from the cathode but are returned to the surface by the effects of the electric field. Once in the gas phase, the analyte atoms and neutral clusters are free to undergo collisions that may dissociate clusters and redeposit atoms at the surface. A fraction of these neutrals diffuse into the negative glow where they undergo excitation/ionization.

The effect of an ion's impact on the sample lattice can be measured by the sputter yield, S, as shown in Equation (1),[9]

$$S = (9.6 \times 10^4)\left(\frac{W}{M} i^+ t\right) \tag{1}$$

where W is the measured weight loss of the sample in grams, M is the atomic weight of the sample, i^+ is the ion current in amperes, and t is the sputtering time in

For references see page 9448

seconds. The ion current is related to the total current, i, by Equation (2),

$$i^+ = \frac{i}{1+\gamma} \tag{2}$$

where γ is the number of secondary electrons released, on average, by a single ion.

Much of the previous research involving sputtering has used secondary ion mass spectrometry (SIMS). Typically, ion beams generated from these types of sources are more tightly focused and of much higher energy than in GDs. This is important to note since there are a number of characteristics that affect the sputter yield for a particular system including the nature of the target, the nature of the incident species, the energy of the incident ions, and the angle of the incident ion beam.

2.1.2 *Excitation/Ionization Processes*

Although the atomization or sputtering process creates species essential for atomic absorption and fluorescence spectroscopies, it does not supply the excited species (such as ions) needed for atomic emission and mass spectrometric analyses. These excitation/ionization processes occur in the collision-rich environment of the negative glow. Collisions that occur within this region not only provide analytically useful species, but are also integral in maintaining the stability of the plasma. Figure 3 illustrates the two principal types of collisions that occur within the negative glow involving electrons, excited atoms, and ions. Excitation is dominated by electron ionization (Figure 3a), while ionization is governed by both electron ionization and Penning ionization (Figure 3b). These two processes account for roughly 90% of ionization occurring within the plasma.(10) Electron ionization involves collision of an energetic electron with a gas phase atom. If the electron is of sufficient energy, it can interact with an electron in the valence shell of the atom, transfer enough energy to eject the electron from the atom's electron

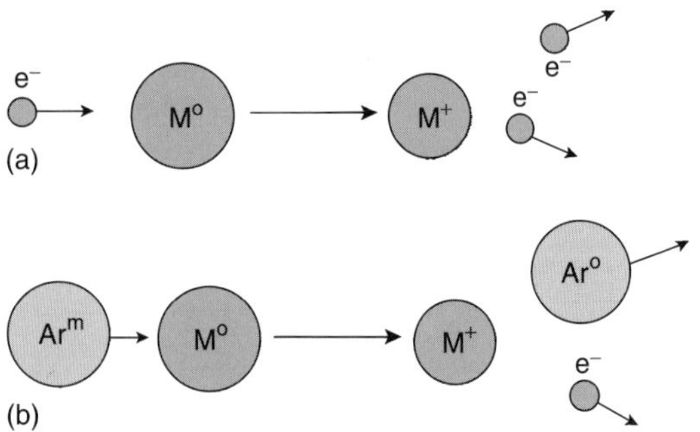

Figure 3 Illustration of the two major ionization pathways available in the GD plasma: (a) electron ionization; (b) Penning ionization.

List of selected abbreviations appears in Volume 15

cloud, and form an ion and a secondary electron.(11,12) There is only a small probability that such a collision will result in ionization; this is zero below a threshold level, increasing at a rate of $C^{1.127}$, where C is the excitational cross-section, as the electron energy increases.(13) This can be explained using classical collision theory. At the threshold, only the complete transfer of energy will result in ionization. As the electrons increase in energy, collisions with only partial transfer of energy will result in ionization and the cross-section will increase. At high electron energies the time and wavefunction overlap is too short for ionization to occur, and the cross-section begins to decrease.(5,13) The average energy of electrons found within the GD is not great enough to ionize most elements or the discharge gas.(4,14,15) However, a Boltzmann distribution of energies predicts the presence of a small percentage of electrons with sufficient energy to ionize all elements. Chapman(4) has performed calculations of the Maxwell–Boltzmann distribution of electrons to determine the percentage of electrons with energies above 15.76 eV, the first ionization potential of argon, the most common discharge gas. At an average energy of 2 eV, 0.13% of the electrons present are of sufficient energy to ionize argon. At an average electron energy of 4 eV, the percentage increases to 5.1%. A more thorough presentation of these calculations is given by Chapman.(4)

Penning ionization, named after F.M. Penning who discovered the effect in 1925,(16,17) involves the transfer of potential energy from a metastable discharge gas atom to another atom or molecule. If the first ionization potential of an atom or molecule is lower than the energy of the metastable atom, ionization will occur when they collide. Ionization cross-sections for most elements are similar for the Penning process, resulting in somewhat uniform ionization efficiencies. Metastable states are reached through either the activation of a discharge gas atom to an excited state from which radiative decay is forbidden or by the radiative recombination of discharge gas ions with thermal electrons. For argon, the most common discharge gas, the metastable levels are the $^3P_2$ and $^3P_0$ states with energy levels of 11.55 eV and 11.72 eV, respectively. Table 2 lists the metastable levels and energies of the discharge gases most commonly used to support GD plasmas. Metastable species are relatively long lived, existing for milliseconds under normal plasma conditions.(18) Their longevity within the plasma, along with a relatively large ionization cross-section and energy sufficient to ionize most elements, make this process a major contributor to ion production. Investigations of ionization processes in steady-state DC-powered discharges have indicated that 40–80% of ionization occurring within these plasmas can be attributed to the Penning ionization process.(19,20) Penning ionization also affords the unique advantage of discriminating against the

Table 2 Metastable spectroscopic notations, energies and first ionization potentials for common discharge support gases

GD support gas	Spectroscopic notation	Metastable energy (eV)	First ionization potential (eV)
Helium (He)	2^1S	20.6	24.5
	2^3S	19.8	
Neon (Ne)	$^1P_0$	16.7	21.6
	$^3P_2$	16.6	
Argon (Ar)	$^3P_0$	11.7	15.8
	$^3P_2$	11.5	
Krypton (Kr)	$^3P_0$	10.5	14.0
	$^3P_2$	9.9	
Xenon (Xe)	$^3P_0$	9.4	12.1
	$^3P_2$	8.3	

ionization of discharge contaminants whose ionization potentials are greater than the metastable energy of the discharge gas atoms.

There are a number of other processes that play minor roles in ionization within the GD plasma. These processes involve charge exchange or associative ionization. Resonant charge exchange involves transfer of charge from one ion to an atom of the same species, while nonresonant charge exchange involves the transfer of charge from an ion to an atom of a different species. Associative ionization involves the combination of a metastable species with a gas phase atom to form a molecular ion with the liberation of a secondary electron. These processes contribute only marginally to ionization within the plasma but are the principal mechanisms by which interfering metal argide ions are formed.[4]

2.2 Radiofrequency-powered Glow Discharge Operation

Traditionally, most GD devices have used a steady-state DC-powered source; recently however, the utility of radiofrequency (RF)-powered discharges as sources for atomic spectroscopies and mass spectrometry has been investigated.[21,22] A major advantage offered by the RF-powered plasma is the ability to analyze directly nonconducting samples such as ceramics and glasses.[23] In the past, these types of samples were powdered, mixed with a conductive matrix, and pressed into a pin or disk to allow analysis by the DC GD.[24–26] Utilization of an RF plasma allows the sputtering and subsequent analysis of these sample types without this added sample preparation step.

Current cannot propagate through an insulating material; when subjected to an applied DC potential, therefore, an insulator behaves in a fashion similar to a capacitor and begins to accumulate charge. With the application of a negative voltage, the surface potential of the insulating material decays to a more positive potential with time. This decay can be attributed to charge neutralization at the cathode surface. The DC discharge will sustain itself until its threshold voltage is reached and the plasma is extinguished. Application of a high frequency potential to a conductive material adjacent to the insulator will allow positive charges that accumulate at the nonconductor surface to be neutralized by electrons during the positively biased portion of the cycle. Therefore, application of a high frequency potential will allow a negative potential to be maintained on the nonconductor's surface.

Figure 4 illustrates the phenomenon known as cathode self-biasing that allows the discharge to be maintained for lengthy periods of time.[4] Self-biasing is based upon the mobility difference between ions and electrons. As a negative potential is initially applied, the surface charges quickly, reaches a maximum, and begins to decay (Figure 4b). When the potential is switched, electrons are accelerated and bombard the surface much like the positive ions during the negatively biased portion of the cycle. The electrons, however, have a greater mobility than the more massive positive ions, therefore the surface potential decays more quickly. After a number of cycles, the waveform will reach a steady DC offset. This DC offset potential is approximately one half of the applied peak-to-peak voltage and sustains the sputtering

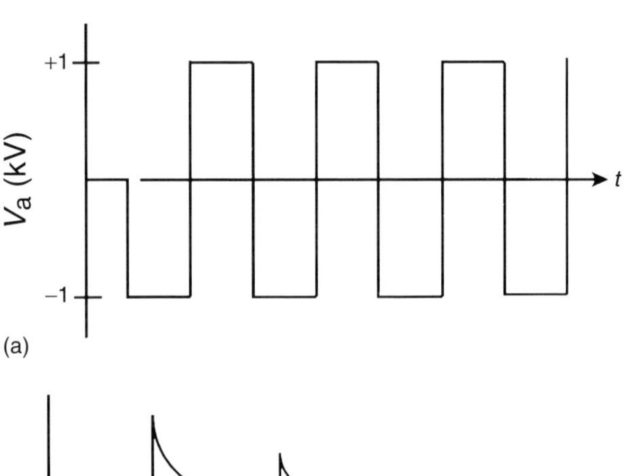

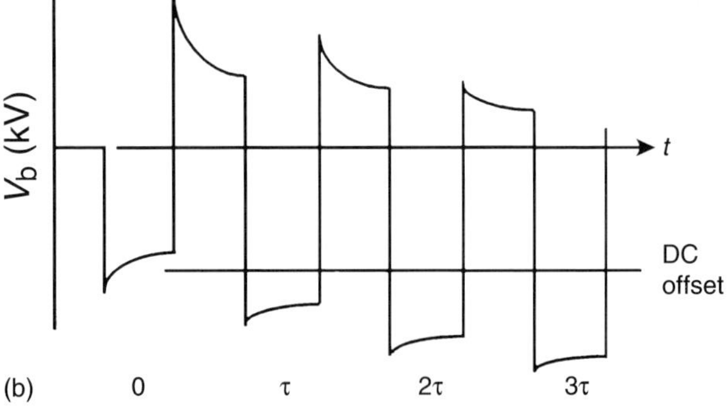

Figure 4 Electrode response to an applied square wave potential: (a) V_a, applied voltage; (b) V_b, response voltage. (Reproduced by permission from B. Chapman, *Glow Discharge Processes, Sputtering and Plasma Etching*. Copyright (1980) John Wiley & Sons Ltd.)

For references see page 9448

ion current. The nonconducting cathode is alternately bombarded by high-energy positive ions and low-energy electrons that support the sputtering process. Operation in this mode has proven useful for the analysis of materials such as nonconducting alloys, oxide powders, and glass samples.[27]

2.3 Pulsed Operation of the Glow Discharge

The GD plasma can be operated in a modulated power mode as well as the steady-state mode described previously. Pulsed power operation offers some distinct analytical advantages that make it attractive to the analyst.[28–32] Steady-state discharges are power limited because increased power application results in resistive heating of the cathode, eventually jeopardizing sample integrity.[33] Modulation of the applied discharge power permits operation at higher instantaneous power while keeping the average power at an acceptable level. This operation mode serves to increase the sputter yield by increasing the average energy and/or the number of incident ions while allowing the sample to cool during the off portion of the discharge cycle. Modulated operation also provides temporal segregation of discharge processes.

Modulated GD operation relies on a microsecond to millisecond square wave power pulse, with a duty cycle of 10–50%. These parameters allow sufficient time for cooling and for the removal of species from one pulsed event before the next one is initiated. Figure 5 depicts a typical pulse sequence showing both discharge gas (Ar) and analyte ion signal profiles. It is apparent that signal behaviors for discharge gas and analyte species differ dramatically. Upon power initiation, the discharge gas ion signal exhibits a sharp rise in intensity to a maximum[30,34] (Figure 5a). This "prepeak" results from the electrical breakdown of the discharge gas species upon power application. The short delay occurs because the acceleration of electrons and subsequent electron ionization of the discharge gas are not instantaneous processes. In contrast, the analyte ion profile behaves differently (Figure 5c). A much longer delay occurs before a more gradual signal increase is observed. This delay arises because sample atoms must first be sputtered from the cathode surface and diffuse into the negative glow region before they can undergo ionization. The temporal correlation between the observation of discharge gas ions and the appearance of analyte signal can be seen in the figure. Both signals reach equilibrium conditions about half way through the applied power pulse during the "plateau" region. During this time regime the plasma most closely approximates the behavior of a steady-state plasma.

Upon applied power termination, the two ion profiles again show markedly different behavior. The discharge gas ion profile decreases quickly ($\approx$500 µs) upon power termination. Previous studies have explained this signal decay as the rapid recombination of discharge gas ions with thermal electrons to form metastable discharge gas atoms.[35,36] The analyte ion profile quickly increases during the "afterpeak" region, reaching a maximum shortly after pulse power termination. This signal then decreases gradually to the baseline. Figure 5(b) represents the first derivative of the discharge gas ion signal profile. The maximum and minima observed in this portion of this figure represent the temporal location of the respective maximum signal intensity increases and decreases for the discharge gas signal. Interestingly, the maximum decrease in the discharge gas ion signal corresponds to the afterpeak maximum for the analyte species.[33] This correlation supports the theory that at power termination electrons are collisionally cooled and recombined with argon ions, forming metastable species. This increase in metastable argon atoms increases the probability of ionization via the Penning process, thus providing enhancements in the analyte ion signal.

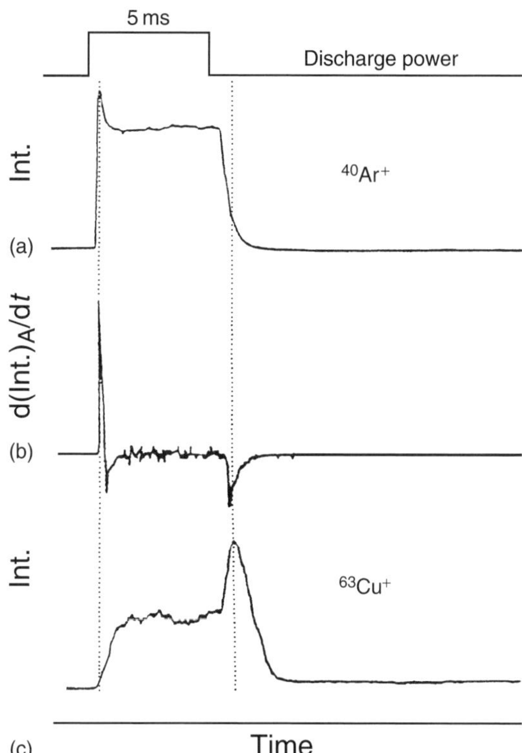

Figure 5 Temporal ion signal profiles: (a) discharge gas ions, $^{40}Ar^+$; (b) the first derivative of the $^{40}Ar^+$ profile; (c) sputtered analyte ions, $^{63}Cu^+$. (Reproduced by permission from King and Pan.[28] Copyright (1993) American Chemical Society.)

Temporally gated separation and detection of species found in these distinct plasma regions increases the utility of the GD devices.[33,37] The most analytically useful region is the afterpeak. Data acquired within this time regime offer two advantageous characteristics. As illustrated in Figure 5(c), the signal intensity for the analyte species increases substantially, thus potentially enhancing the sensitivity for analyte species in this region. The other advantage arises from the suppression of electron-ionized interfering species.(29,30) The first ionization energies of the discharge gas and molecular contaminants (e.g. H_2O, N_2, O_2) are too high to allow ionization through the Penning process. Upon power termination species that are ionized via electron impact are no longer excited. Thus, if the acquisition gate is moved far enough into the afterpeak, contributions from these species will be minimized.

2.4 Applications

In the early 1970s, as GD instrumentation and techniques were being transitioned from research applications to routine sample analysis, spark source techniques reigned as the analytical tool of choice for trace analysis of solids. Spark source spectrometry was limited, however, by its expense, complexity, and unreliability. These disadvantages associated with spark source spectrometry facilitated the acceptance of the GD for the analysis of solid samples.

2.4.1 Bulk Analysis

Historically, GD mass spectrometry and optical spectroscopies have proven most valuable for the analysis of bulk conductive solid samples. As explained earlier, GD methods provide a representative gas phase analyte population directly from samples in the solid form, thus dramatically simplifying sample preparation. GD techniques can also be used for the analysis of nonconducting samples such as glasses, polymers, and ceramics.(38–40) The first analyses of nonconducting materials involved mixing a powdered sample with an easily sputtered, conducting matrix, such as high purity copper or silver powder.(41–44) The mixture was pressed into a sample cathode using a die and hydraulic press system. Unfortunately, this complicated sample preparation, especially for samples that were not in powdered form. Homogeneity of the mixed sample also became of concern for the analyst. Two other methods for analysis of nonconducting samples have also be used. These methods do not require sample mixing but rather use an RF-powered or a secondary cathode GD source to sample directly solid nonconducting materials. Section 2.2 describes plasmas powered by RF sources. Secondary, or sacrificial, cathode GD systems utilize a monoisotopic conductive mask (e.g. Ta, Pt) that is placed on top of the sample to be analyzed. A potential is applied to the mask, which thus assumes the role of the cathode. As material is sputtered from the mask, a large portion of it is redeposited on the surface of both the secondary cathode and the nonconducting sample. At this point, the layer of cathode material deposited onto the sample becomes conductive and thus assumes the applied potential and attracts impinging ions. As ions impinge upon the cathodic layer, they ablate both the cathodic material and the underlying nonconductor, introducing both species into the gas phase for subsequent analysis. It is clear that the mask material must not contain species that are of interest because this will contaminate the sample and preclude accurate measurements. A number of references are available describing in detail the use of secondary cathodes.(45–47)

2.4.2 Surface Analysis

Recently, surface analysis by GD spectrometry has aroused great interest in the analytical community. In a sense, GD is always a surface analysis technique, acting as an atomic mill to erode the sample surface via the sputtering process. Atoms sputtered from the surface are subsequently measured using either optical or mass spectrometric techniques. GD sputtering consumes relatively large quantities of sample in a relatively short time period (up to milligrams per minute), making analysis of thin films (<500 Å) difficult to virtually impossible. Plasma conditions, along with discharge gas choice, can be adjusted to slow the ablation process, thus allowing the analysis of films with micrometer thicknesses. Figure 6 is a graphical illustration of a typical spectral intensity–time profile of a multilayer coating prepared by chemical vapor deposition of carbon steel. The outer layer consists of vanadium carbide and the inner layer chromium carbide. It is obvious from Figure 6 that signal responses do not follow an ideal square wave pattern for appearance and disappearance. Clearly, the metal

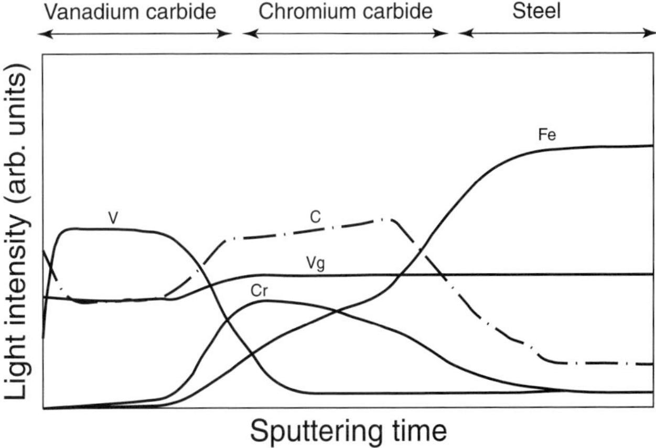

Figure 6 Qualitative analysis of multilayer coating. Vg, voltage. (Reproduced by permission from Hocquaux.(59))

For references see page 9448

sample layer is not all eroded at a well-defined time, but is scattered across a diffuse region that is gradually removed. This signal tailing is a manifestation of the redeposition that occurs from the plasma. It has been estimated[48] that up to 67% of the sputtered atom population is returned to the surface by collisions with argon atoms, to be resputtered before eventually escaping permanently from the surface. Redeposition, along with a relatively high sputter rate, prevents GD from being used to profile very thin films. By calibrating erosion rates using standard layered samples, the thickness of layers can be determined by analyzing the signal–time profiles, allowing the full characterization of sample layers.

2.4.3 *Analysis of Solution Samples*

Analysis of solutions has been performed using GD techniques even though their major advantage lies in the ability to directly analyze solid-state samples. It is important to note that aspiration of a solution directly into the GD will cause quenching of the plasma, making removal of any solvent from the sample preferable. The simplest and most direct method of doing this is to evaporate a solution onto a conducting cathode, leaving a dried residue as the sample. A more elegant method for depositing solution samples onto a cathode involves electrodeposition. This approach also permits analytes in large sample volumes to be preconcentrated before analysis. Solutions containing samples have also been mixed with a conductive powder and dried for subsequent analysis. This approach is similar to that used for the analysis of nonconductive powders.

Although limited by its difficulty, direct analysis of solution samples by GD techniques has been performed for specialized applications. Strange and Marcus[49] have used a particle injection system to introduce a solution sample into the GD. Steiner et al.[50] have also utilized a pulsed plasma TOF system to obtain concurrent elemental and molecular information for high vapor pressure liquid samples. For the foreseeable future however, GD techniques will likely remain focused on the analysis of solid-state samples.

3 SPECTROCHEMICAL METHODS OF ANALYSIS

The goal of the analytical chemist is often to identify a particular chemical species or to quantify the amount of that species in a sample. In a spectrochemical analysis, the chemist uses the intensity of radiation emitted, absorbed, or scattered by a particular species versus a quantity related to photon energy, such as wavelength or frequency, to make such measurements.[51]

In this section, the basic requirements necessary to obtain a spectrochemical analysis are reviewed, along with three spectrochemical methods that are used to effect the measurement: AAS, AES, and luminescence spectroscopy. Although there are many variations of each of these methods, we will review only the classical approaches and the variations that have been used with GD devices. Table 3 summarizes the quantity measured and gives examples associated with each measurement technique.[51]

3.1 Basic Requirements Necessary to Obtain Optical Information

For thousands of years scientists have been performing qualitative analyses based on color, smell, taste, size, and shape. Although first year college chemistry students are still taught to use their senses to help identify substances in qualitative laboratory, for the rest of us these less precise approaches have been replaced with chemical and instrumental methods that can measure not only pure materials but trace components in complex mixtures. Most materials are not willing to give up this information spontaneously, however. Instead, to obtain chemical information about a sample, it is necessary to perturb the sample through the application of energy in the form of heat, radiation, electrical energy, particles, or a chemical reaction.[51] The application of this energy often causes electrons in analytes to be excited from their lowest energy or ground state to a higher energy or excited state. Several spectroscopic phenomena depend on these transitions between electronic energy states.[51]

Table 3 Classification of spectrochemical methods[51]

Class	Quantity measured	Examples
Emission	Radiant power of emission, Φ_E	Flame emission, DC arc emission, spark emission, inductively coupled plasma and direct current plasma emission, GD emission
Absorption	Absorbance or ratio of radiant power transmitted to that incident, $A = \log(\Phi/\Phi_0)$	UV/visible molecular absorption, IR absorption, atomic absorption
Luminescence	Radiant power of luminescence, Φ_L	Molecular fluorescence and phosphorescence, atomic fluorescence, chemiluminescence

UV, ultraviolet; IR, infrared.

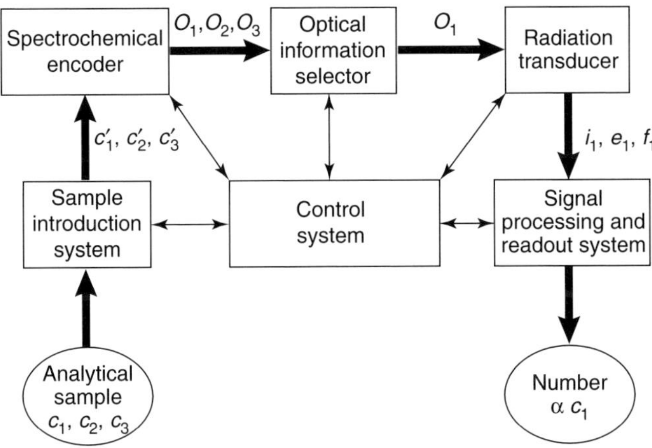

Figure 7 Spectrochemical measurement process. (Reproduced by permission from Ingle and Crouch.[51])

Information can be obtained by measuring the electromagnetic radiation emitted as the electron returns to its ground state from an excited state (emission), by measuring the amount of radiation absorbed in the excitation process (absorption), or by measuring the changes in the optical properties of the electromagnetic radiation that occur when it interacts with the analyte (e.g. ionization or photochemical reactions).[51] Qualitative information is extracted by observing a particular element-specific transition, while quantitative information is obtained by quantifying the amount of radiation emitted or absorbed in that transition. Figure 7 is an illustration used by Ingle and Crouch[51] to depict the many processes involved in converting concentration information into a number – the analytical chemist's goal in making a measurement. A sample introduction system presents the sample to the encoding system, which converts the concentrations c_1, c_2, c_3 into optical signals O_1, O_2, O_3. The GD is somewhat unusual (compared with other atomic sources like the flame or the inductively coupled plasma) because it serves as both the sample introduction system (through sputter atomization) and the spectrochemical encoder. The information selection system (often a monochromator) selects the desired optical signal O_1 for presentation to the radiation transducer or photodetector. This device converts the optical signal into an electrical signal that is processed and read out as a number. All spectrochemical techniques that operate in the UV/visible and IR regions of the spectrum employ similar instrumentation;[51] the only differences lie in the arrangement and type of sample introduction system, encoding systems, and information selection system.

3.1.1 Spectrometers

The optical information selector in Figure 7 sorts the desired optical signal from the many signals produced in the encoding process. Although it is possible to discriminate against background signals on the basis of time and position, most often discrimination is based on optical frequency (wavelength). The most widely used wavelength selection system is the monochromator, although there are a variety of other systems, including polychromators and spectrographs, and nondispersive systems like the Fabry-Perot, Michelson, Mach-Zender, and Sagnac interferometers. Our discussion will be limited to monochromators; for a thorough discussion of other types of wavelength selection systems, the reader is directed elsewhere.[51]

Monochromators isolate one wavelength from the countless number of wavelengths found in polychromatic sources. A monochromator consists of two principal components: a dispersive element and an image transfer system. Light is transferred from an entrance slit to an exit slit by a series of mirrors and lenses; along the way it is dispersed into its various wavelengths by a grating (or sometimes a prism). To change wavelengths, one rotates the dispersive element; the result is that different wavelength bands are brought through the exit slit in succession. One can easily calculate the angular dispersion (D_a) (i.e. the angular separation ($d\beta$) corresponding to the wavelength separation ($d\lambda$)) of a grating by knowing the angle of incidence, the angle of diffraction, the order of diffraction, and the groove spacing of the grating. For practical purposes, however, it is more important to calculate the linear dispersion, $D_l = dx/d\lambda$ (a value that defines how far apart in distance two wavelengths are separated in the focal plane), or the reciprocal linear dispersion, R_d (the number of wavelength intervals contained in each interval distance along the focal plane).

A monochromator's resolution is closely related to its dispersion in that dispersion determines how far apart two wavelengths are separated linearly while an instrument's resolution determines whether the two wavelengths can be distinguished. In many cases the resolution is determined by the monochromator's spectral bandpass, s, defined as the half-width of the wavelength distribution passed by the exit slit. If the slit width is large enough to ignore aberrations and diffraction, a scan of two closely spaced monochromatic lines of peak wavelengths λ_1 and λ_2 will be just separated (baseline resolution) if $\lambda_2 - \lambda_1 = 2s$. Therefore, the slit-width-limited resolution $\Delta\lambda_s$ is given by Equation (3):

$$\Delta\lambda_s = 2R_d W \qquad (3)$$

where W is the slit width.[51] By adjusting the monochromator so that $\Delta\lambda_0 = \Delta\lambda_1$ and $W = \Delta\lambda/R_d$, the image of λ_1 will be completely passed, while that of λ_2 will be at one side of the exit slit.[51]

For references see page 9448

3.1.2 Detectors

Two detectors are used most often in atomic spectroscopy: photomultipliers (the most common) and multichannel detectors. A photomultiplier is a more sophisticated version of a vacuum phototube. A cascade of electron collisions with dynodes of increasing potential and the subsequent ejection of electrons from each dynode's surface leads to the formation of an electrical current proportional to the number of photons striking the detector. The process begins with a photon striking a cathode made of a photoemissive material (e.g. alkali metal oxides, AgOCs, CS_3Sb). If the energy of the photon is above some threshold value, an electron is ejected from the cathode. Only a certain fraction of the photons with energy greater than threshold produce photoelectrons with sufficient KE to escape the photocathode. This fraction is called the quantum efficiency and is the ratio of the number of photoelectrons ejected to the number of incident photons. After leaving the photocathode, a photoelectron strikes the first dynode of the multiplier; this causes the subsequent ejection of two to five secondary electrons which are in turn accelerated by an electric potential to a second dynode where they cause the release of two to five more electrons. This multiplication process continues until the electrons reach the last dynode and impinge on the anode. A modern photomultiplier tube might have 5–15 dynodes (made of a secondary emission material like MgO or GaP) in a cascade. The result of this photomultiplication is a large charge packet a few nanoseconds in width produced at the anode for each photoelectron collected by the first dynode. Photomultipliers can be operated in either analog mode, where the average current that results from the arrival of many anodic pulse is measured, or in photon counting mode, where the number of anodic pulses, and not photons, is counted per unit time.

A wide variety of photomultipliers is available with both end-on and side-on viewing for adaptation to a wide variety of monochromators. Care must be given to the wavelength range over which one is working to ensure a uniform response. Other concerns for the spectroscopist, all of which are beyond the scope of this article, include the quantum efficiency of the photomultiplier, the multiplication factor of each dynode, the operating (accelerating) voltages applied to the dynodes, and the dark current generated when a potential is applied between the anode and cathode, with no photons hitting the photocathode.

The second type of detector that is widely used in atomic spectroscopy is the multichannel detector. These devices include early photographic detectors like photographic film or plates, as well as modern detectors such as photodiode arrays and charge-coupled and charge-injection devices. The idea behind the multichannel detector is simultaneous detection of dispersed radiation. Modern multichannel detectors usually take the form of some sort of solid state *pn*-junction diode device packaged in integrated-circuit form with a large number (e.g. 256, 512, or 1024) of elements arranged in a linear fashion. These devices often have linear dynamic ranges of two to four orders of magnitude. Limitations at the low end result from the noise associated with readout of a given dynode. Limitations at the upper end are the result of saturation; this is determined by the number of electron–hole pairs that can be created. A typical saturation charge is 1–10 pC. The reader is referred to several excellent references for a more thorough description of multichannel detectors.[52–54]

3.2 Atomic Emission Spectroscopy

AES is the simplest spectroscopic method for determining the elemental composition of a sample, and is the logical place to start talking about atomic spectrometry. Optical emission results from electron transitions occurring within the outer electron shells of atoms. These transitions give rise to line spectra where the wavelength of the lines relates to the energy difference of the levels according to Equation (4):

$$\Delta E = \frac{hc}{\lambda} \tag{4}$$

Spectroscopists often categorize spectral transitions according to term symbols. For a complete discussion of term symbols, the reader is referred elsewhere;[55] a brief discussion follows.

Each electronic state of an atom has five quantum numbers that define its electronic configuration. These include the principal quantum number, n, the orbital angular momentum quantum number, l, the orbital magnetic quantum number, m_l, the electron spin quantum number, s, and the spin magnetic quantum number, m_s. According to Ingle and Crouch,[51] for many-electron atoms, the hydrogen quantum numbers can be thought of as describing the individual electrons, but they are not "good" quantum numbers for the entire atom. Good quantum numbers are associated with operators that commute with the total atom Hamiltonian. These include the resultant orbital angular momentum quantum number, L, produced by coupling the orbital angular momenta of each electron, and the resultant spin quantum number, S. For atoms with weak spin–orbit interactions, L and S couple to produce a total angular momentum quantum number, J. A multiplet of closely spaced states with the same L and S values but different J values is called a spectroscopic term, and is designated as $n^{2S+1}\{L\}_J$, where n is the principal quantum number for the valence electrons, $2S+1$ defines the multiplicity, and J is the total angular momentum quantum number. Of all the possible transitions between states, only a fraction of them are observed. From quantum mechanical principles, it is possible to derive selection rules that tell which transitions are allowed (i.e. those that occur with high probability and give reasonably intense lines) and which are forbidden (i.e. those that occur with low probability and give weak lines); this is beyond the scope of this article. For this discussion it is sufficient to note that term symbols for almost all practical configurations have been tabulated(55) and tables of spectral line intensities have been assembled for nearly all the elements.(56)

In AES, the information relevant to an analysis can be found in the radiation emitted by excited analyte atoms decaying from a nonradiational activation event. The radiant power of this emission is a function of several factors, including the population density of the excited atoms, the number of photons emitted per second by each atom, the energy of each photon, and the volume of the emitting system. The reader is directed elsewhere for a more complete discussion of each of these factors.(51)

AES has the power to provide rapid, qualitative, and quantitative multi-element analyses. Although a qualitative survey of the elements (i.e. a plot of the analytical signal versus wavelength) in a sample may be useful, more often the desired information is the concentration of an analyte. Unfortunately, this is almost never obtained directly as the result of an absolute measurement of an optical signal because amplitude and elemental concentration are seldom related in a simple way. Obtaining the desired concentration from an optical measurement usually involves calibration, subtraction of blanks, comparison with standards, and other similar procedures.(51) Quantitative analysis of a single element is most easily accomplished in AES by monitoring the emission intensity as a function of the analyte's concentration under a given set of conditions (e.g. constant discharge gas pressure, voltage, and current at a given wavelength). Standards, often provided by the National Institute of Standards and Technology, provide a range of concentrations over which a calibration curve can be developed. The signal intensity of an unknown concentration is then compared with the intensity of the standards, thus providing the concentration of the analyte in question.

Although simple in principle, quantification is complicated by a number of factors, including spectral background, incomplete wavelength separation, self-absorption, peak broadening, etc., most of which are beyond the scope of this article. When one considers these complicating issues, it is clear why it is important to control conditions precisely and use standards for the most accurate quantification.

Instrumentation used for emission spectroscopy includes an excitation source, a sample container, a wavelength selector, and a radiant power monitor. Depending upon the spectrochemical method, the excitation source and sample container may be separate components or they may be combined, as is the case with the GD. Figure 8 illustrates one instrumental configuration used in our laboratory at the Oak Ridge National Laboratory for AES. A 0.5-m monochromator serves as the wavelength selector and the combination of a photomultiplier tube, preamplifier, and readout photometer comprise the radiant power monitor. Figure 8 also illustrates three common sources used with GD emission spectroscopy – a planar cathode discharge and two versions of the hollow cathode discharge.

The planar cathode discharge is thus termed because the portion of the sample exposed to the discharge is flat. It is often contrasted with the coaxial or pin-type cathode used more commonly with mass spectrometry and described elsewhere in this chapter. With the development of the Grimm lamp in 1968(57) and its eventual commercialization, the planar cathode discharge gained widespread use for emission spectroscopy. Although other planar cathode discharges have been developed in the past 30 years, the Grimm source still finds the greatest application today. One interesting feature of the Grimm source is that it is an obstructed discharge (i.e. the discharge is confined to the sample by the extension of the anode into the cathode dark space). Moreover, the vacuum in the anode–cathode inner space is lower than in the discharge region itself, necessitating a dual outlet pump with a larger throughput for the inner electrode space.(58) Another interesting feature of the Grimm source is that the cathode is located outside of the source itself; this provides for easy sample interchange, and means that the Grimm source is particularly amenable to the analysis of any flat conducting surface that can be brought up to the source opening, such as metal sheets or disks. Typical operating conditions for the Grimm source are 500–1000 V, 25–100 mA, and 1–5 Torr. The relatively high power produced by the source (12.5–100 W) means that the cathode is often water-cooled; this usually is not a problem in emission spectroscopy, but makes interfacing the Grimm source with a mass spectrometer (with its high vacuum requirements) more difficult.

Planar cathode discharges have been interfaced to a variety of commercial emission instruments.(7) Grimm-type sources find their greatest use in trace elemental analysis of solids and in depth profiling of layered metal samples.(58) Detection limits by emission spectroscopy are of the order of 0.1 ppm.(7) Precision of the order of 0.5–5% has been obtained for concentrations in the 0.01–10 mg g^{-1} range.(58) Ablation rates range from 0.1 to 3 mg min^{-1} depending upon the element, discharge area, current, and voltage.(58) At these rates, the Grimm source cannot be used for thin film analysis, but it is

For references see page 9448

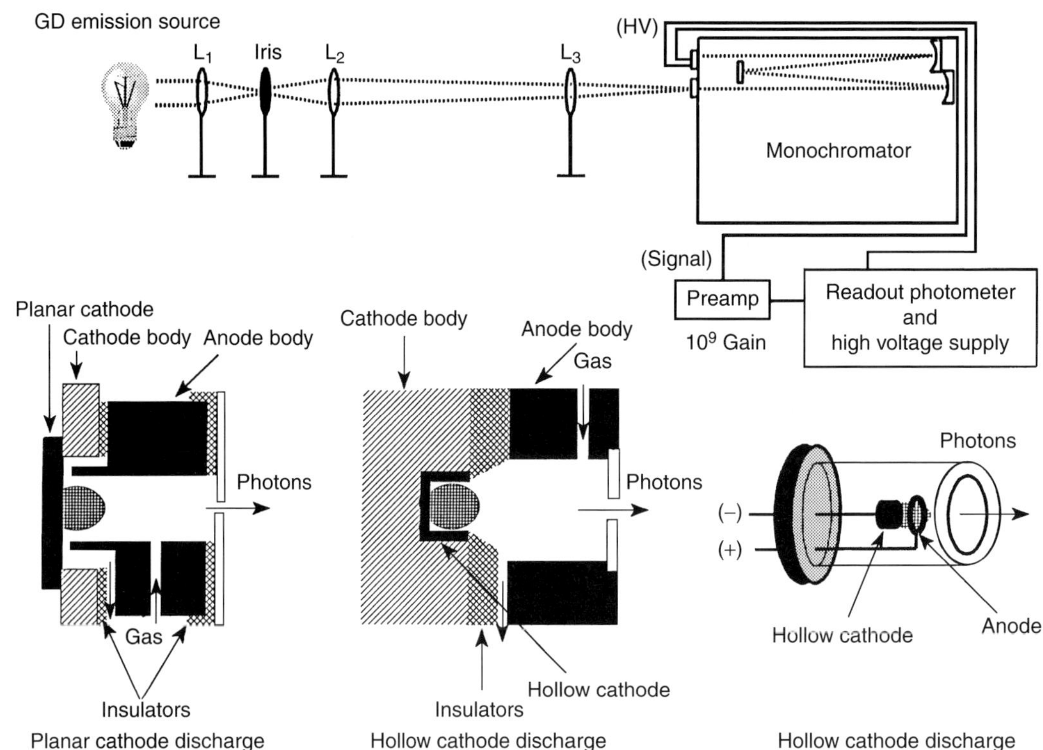

Figure 8 Schematic of an instrumental arrangement used at the Oak Ridge National Laboratory for AES. (Reproduced by permission from Harrison et al.[7] Copyright (1990) American Chemical Society.)

ideal for thin layer analysis or in-depth profiling where it may be necessary to profile from a few nanometers to several tens of micrometers in a relatively short time.[59] Recently, Hocquaux[59] wrote an excellent chapter on thin film analysis by GD emission spectroscopy.

The other two sources shown in Figure 8 are hollow cathode discharges. Although the hollow cathode discharge appears similar to the planar cathode physically, the shape of the cathode cavity provides some properties that make it appealing for atomic spectroscopy. This discharge derives its name from a cathode that has been drilled out to form a cylindrical cavity closed at one end.[60] The so-called hollow cathode effect can be visualized as a GD with two parallel cathode plates being brought sufficiently close to each other until the two cathode glow regions coalesce.[61] The result of this coalescence is an increase in current density that can be several orders of magnitude larger than a single planar cathode at the same cathode fall potential.[62] Coupling this increase in current density with the longer residence time that the analyte experiences in the negative glow region (due to the cathode's shape) results in a dramatic increase in the intensity of radiation emitted compared to a planar cathode. In addition, background intensities are low because electron number densities are low, resulting in a very high signal-to-background ratio.[63] To perform an analysis using a hollow cathode discharge, it is necessary to machine the sample into the shape of a cylinder, or to place powder or metal chips into a hollow cathode made of some inert material such as graphite. One can also analyze solutions by drying a residue on the hollow cathode surface. Operating conditions vary widely, but typically range from 200 to 500 V, 10 to 100 mA, and 0.1 to 10.0 Torr.[7] Detection limits have been reported in the picogram range,[58] but more typical results are in the nanogram range. Although hollow cathode discharges are used widely in atomic spectroscopy, the majority of these devices are light sources for AAS (see below). A typical hollow cathode lamp (HCL) is depicted at the extreme bottom right of Figure 8.

3.3 Atomic Absorption Spectroscopy

In absorption spectroscopy, spectrochemical information can be found in the magnitude of the radiant power from an external light source that is absorbed by an analyte.[51] To obtain information relevant to measuring an element's concentration, however, it is necessary for the frequency of the incident radiation to correspond to the energy difference between two electronic states of the analyte atoms being measured. Often, but not always, the atoms start in their electronic ground state and are excited to a higher lying electronic state by the incident radiation. The adsorption of this radiation usually follows Equation (5):

$$A = -\log T = \frac{-\log \Phi}{\Phi_0} = abc \qquad (5)$$

where A is the absorbance, T is the transmittance, a is the absorptivity, b is the path length of absorption, and c is the concentration of the absorbing species. This equation is commonly referred to as Beer's law. To calculate the concentration, one measures the incident radiation (M_0) and the transmitted radiation (M), calculates the absorbance, and relates the absorbance to concentration using a series of standards and calibration curves, similar to emission spectroscopy.

A typical instrumental configuration for an atomic absorption spectrometer is shown in Figure 9. A hollow cathode, fabricated from the elements of interest, is often used as the source of incident radiation, although an electrodeless discharge lamp may be used for some elements such as As, Se, or Te, where the emission from an HCL may be low. The HCL is focused to a point inside the discharge and then refocused into the entrance aperture of the monochromator. To obtain the background signal, one can use a shutter, or alternatively modulate the HCL and measure the background during the off period. In the arrangement shown in Figure 9, a mechanical chopper is used to facilitate background subtraction by providing a reference signal to a lock-in amplifier. Transmission is measured first with the discharge off and then with the discharge on, often for a range of currents and voltages. Using Beer's law, absorbance is calculated for a series of standards to produce a calibration curve; the absorbance of an unknown is then correlated with its concentration.

Two different discharge configurations are shown in Figure 9 for atomic absorption. The one on the left is an atomic absorption sputtering chamber developed by Gough.[64] A planar cathode is mounted near the top by pressing the sample against an O-ring that provides the vacuum seal. The gas flow of the cell was designed to provide transport of sputtered atoms into the observation zone, 1–2 cm from the sample. In 1987 Bernhard[65] took the idea of gas-assisted transport of atoms one step further, reporting on a design that used gas jets aimed at the sputtering surface to increase significantly the sampling rate as well as the absorption signal in a sputtering chamber. A commercial atomic absorption cell was designed based on this principle (Atomsource,[65] Analyte Corporation, Medford, OR), renewing the interest in AAS that began with Walsh[66] more than four

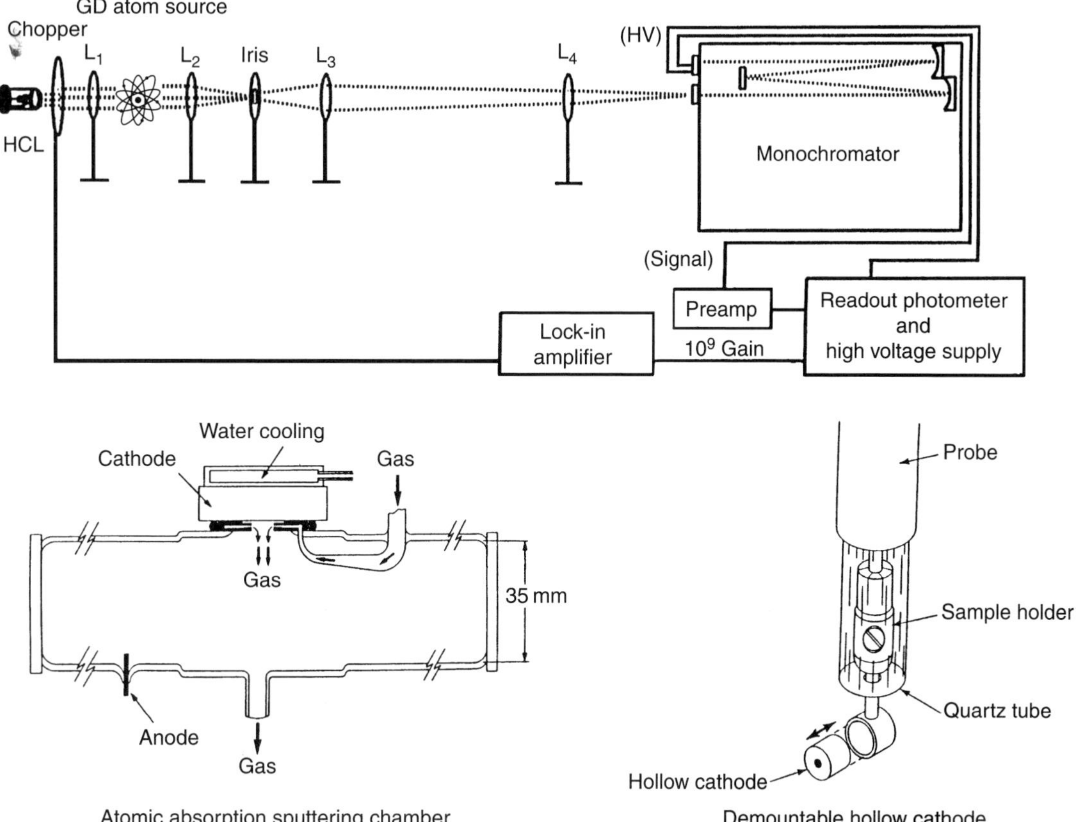

Figure 9 Schematic of an instrumental arrangement used at the Oak Ridge National Laboratory for AAS. (Reproduced by permission from Gough.[64] Copyright (1976) American Chemical Society.)

For references see page 9448

decades ago. The source on the right is a much simpler atom generator. It is based on the direct insertion probe (DIP) design of King.[67] The coaxial cathode in King's original design has been replaced by a stainless steel ring that accommodates a demountable hollow cathode (4.82 mm in diameter × 2.54 mm in length with a 3.18-mm hole in the center). The DIP facilitates alignment of the HCL emission, which is focused through the orifice (i.e. the region of highest atom density) and collected after it passes through a window in a six-way vacuum cross. Typical operating conditions for this source are 1.0–3.0 Torr, 500–2000 V, and 2–15 mA. Detection limits for GD atomic absorption are in the low parts per million range. Although GD atomic absorption is not as widely used as flame or graphite furnace atomic absorption, it has found its niche in applications where analysis by other atomic absorption methods (primarily solution-based) are difficult (e.g. the analysis of materials that are difficult to dissolve).

3.4 Atomic Fluorescence Spectroscopy

Atomic fluorescence is similar to AAS in that both rely on an external light source to produce an analytical signal from an atomic vapor. In fluorescence spectroscopy, the signal is contained in the emission of photons from the atom population after absorption of the incident energy. There are five basic types of fluorescence: resonance, direct-line, step-wise, sensitized, and multiphoton fluorescence.[51] For this discussion, it is not important to define these five types, only to say that the differences lie in the excitation and relaxation pathways that each follow to produce fluorescence. Resonance fluorescence (where the same upper and lower levels are involved in the excitation–de-excitation process so absorption and emission wavelengths are the same) finds the most widespread use in analytical spectroscopy because the transition probabilities and the source radiances are the greatest for resonance fluorescence when conventional line sources are used.

Figure 10 shows a conventional instrumental arrangement for a single-beam atomic fluorescence spectrometer. Radiation from the source is focused into the GD. Fluorescence photons are imaged onto the entrance aperture of a monochromator that isolates the analyte fluorescence from background emission and fluorescence from other species. Fluorescence is usually viewed at an angle of 90° with respect to the excitation source to minimize the collection of scattered source radiation.

One critical component of an atomic fluorescence spectrometer is the excitation source. HCLs, electrodeless discharge lamps, and metal vapor discharge lamps have all been used successfully, although today most fluorescence experiments use a laser. Lasers are superior sources

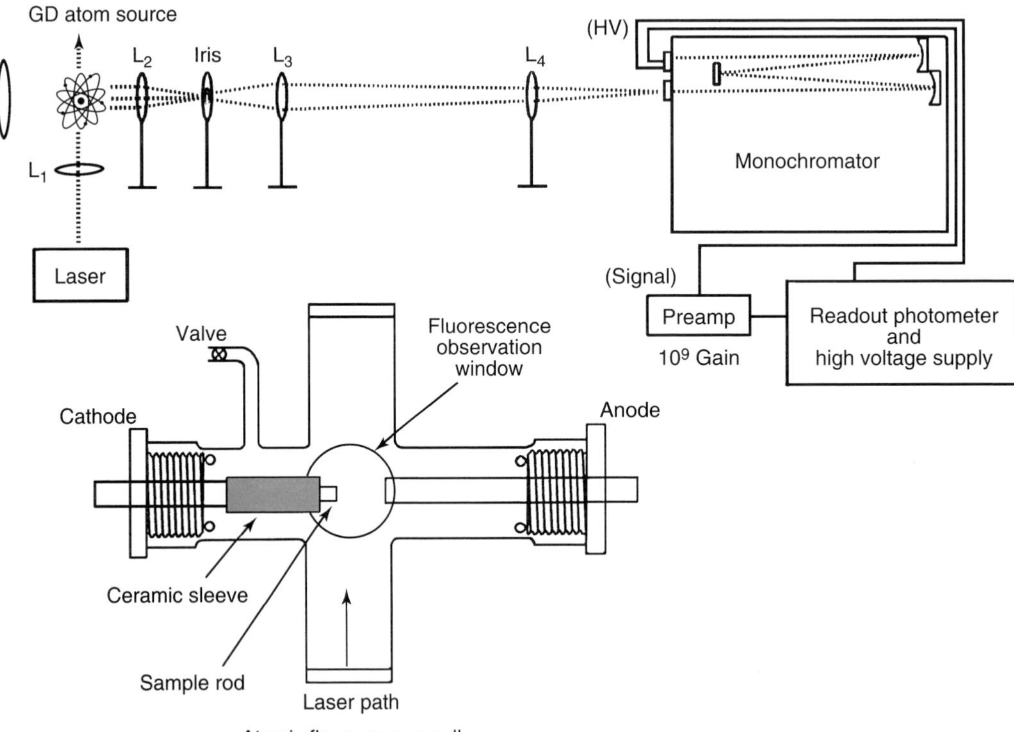

Figure 10 Schematic of a conventional instrumental arrangement for a single-beam atomic fluorescence spectrometer.

for atomic fluorescence because they provide a fluence several orders of magnitude greater than other sources, are tunable over a wide wavelength range, have spectral bandwidths much narrower than absorption line widths, and can be focused to very small spot sizes. Both continuous wave (with chopping) and pulsed lasers have been used, with dye lasers finding the most use because they can be tuned over a large number of wavelengths. An inductively coupled plasma has also been used as an excitation source; here, the excitation wavelength is governed by the analyte that is aspirated into the torch (usually at high concentrations).

Like AES and AAS, the ideal atomizer for fluorescence would produce a stable population of atoms of sufficient number density to make quantification of small concentrations practical. A fluorescence atomizer should also produce minimum thermal excitation to limit analyte and interference emission, a potential source of background. When this stipulation is met, Rayleigh scattering from atoms or molecules determines the fundamental limitation for background noise.

Most atomic fluorescence measurements have been made with flame atomizers,[(51)] but recently inductively coupled plasma has been used. Plasmas generally provide better atomization efficiency and a larger population of free atoms than flames. When analyzing a solid sample, however, an atomizer, like the one shown in Figure 10, is more practical.[(68)] This simple design consists of a Pyrex glass housing into which a 6.35-mm diameter sample rod is inserted through a ceramic sleeve and sealed to the cell with O-rings. Quartz windows are glued into the cell to allow a laser beam to pass. A fill port and an evacuation port are also provided, and the entire cell is pumped by a single rotary vacuum pump. A DC power supply (Electronic Measurements, Eatontown, NJ) provides voltages up to 600 V and currents up to 200 mA. A typical operating voltage is 570 V at 25 mA. The cell is pressurized with argon to between 900 and 1000 Pa (7–8 Torr).

More recent applications of atomic fluorescence by this group[(69)] have produced GD cells constructed from high vacuum components like ConFlat® crosses and flanges with sapphire windows, but the principles of operation remain the same – fluorescence is detected 90° to the laser beam path by a photomultiplier tube. The fluorescence signal is amplified by a wideband amplifier and processed by a gated integrator and boxcar averager; the result is a fluorescence spectrum as a function of wavelength which is indicative of those elements in a GD cathode for which a fluorescence transition is allowed.

3.5 Optogalvanic Spectroscopy

The final optical technique to be discussed is optogalvanic spectroscopy. The optogalvanic effect was first observed using a weak, incoherent light source in 1928;[(70)] light from one neon discharge affected the electrical characteristics of a nearby discharge. The process is quite simple; in a discharge there is an equilibrium established between the neutral species and the corresponding ions. If some means of energy is added to the discharge, the equilibrium position can be displaced and the fraction of ions altered; this is the case when a photon is absorbed by the gaseous atom or molecule. This permits one to measure optical absorption by an all-electronic means, that is, without the use of photodetectors. The electrical circuit employed to monitor the optogalvanic effect commonly includes a ballast resistor in series with the discharge resistance; the discharge impedance change is usually monitored as a change in the discharge voltage. The light source is modulated, and a lock-in amplifier is employed to measure the alternating current component of the discharge voltage induced by absorption of light as the source wavelength is scanned. In theory, the atomization source could be any of the discharges discussed thus far; in practice, however, we have found that the demountable hollow cathode operating in the same fashion as it does for atomic absorption provides the greatest flexibility for optogalvanic spectroscopy.[(71)]

4 MASS SPECTROMETRIC METHODS OF ANALYSIS

Much like the spectrochemical techniques described above, mass spectrometry offers the analyst a method for determining the identity and quantity of a particular species in a sample. This technique however, provides analytical information through the separation and subsequent detection of charged species associated with the sample. Ions are generated in the source region and selected by mass-to-charge ratio (m/z) usually using electrostatic or magnetic fields.

In this section, the basic requirements necessary to obtain mass abundance information are described, along with five types of mass spectrometers: quadrupole mass filters, magnetic sector mass analyzers, QITs, FTICR devices, and TOF mass spectrometers. These reviews will focus only on variants that have been coupled to the GD source. Table 4 provides basic information and characteristics that are unique to each of the mass spectrometric systems.

4.1 Basic Requirements Necessary to Obtain Mass Abundance Information

Mass spectrometric systems in general require a number of fundamental components to create, transport, separate, and detect ions, manipulate the resulting signal to account

For references see page 9448

Table 4 Mass analyzers and considerations for elemental analysis

Mass analyzer type	Advantages	Disadvantages
Magnetic sector mass spectrometer	Commercially available Reasonable resolution	Scan speed Complex Cost
Quadrupole mass filter	Commercially available Robustness Scan speed Peak hopping mode	Limited resolution
QIT mass spectrometer	Cost CID to remove interferences	Complicated interface to GD source
FTICR mass spectrometer	High resolution CID capability	Cost Complex Complicated interface to GD source Space charge limited
TOF mass spectrometer	Simplicity Cost Speed Simultaneous data acquisition Good resolution when operated in reflectron mode	Some ion extraction biases Poor isotope ratio measurements

CID, collision-induced dissociation.

for system inconsistencies, and provide useful information to the analyst (see Figure 11). A number of these components will be considered in the following section.

4.1.1 Ion Source

There are a number of GD source geometries that can be implemented for a variety of analyses (see Table 1). Some types were designed, and are well suited, for optical applications as outlined in the previous section. Many of the advantages afforded by the optically applicable sources stem from an the enhancement of the atom population, often a result of increased sputtering rate or a confined viewing region that facilitates optical viewing. Mass spectrometry requires a different set of criteria, however. Sources designed for mass spectrometric applications must provide a reasonable population of analyte ions (not atoms). These ions then must be extracted from the source region into the mass analyzer. For these reasons, one appropriate source for mass spectrometry is the coaxial cathode. Ions generated by the coaxial cathode are extracted through an ion exit orifice in the anode. Ionization is dominated by the Penning process that leads to an ion population with a narrow KE spread relative to other GD source geometries.(72) This geometry also facilitates sample introduction via a DIP, making trivial the appropriate adjustment of the plasma relative to the ion exit orifice.

4.1.2 Vacuum Systems

Mass spectrometric techniques impose stringent requirements on a vacuum system. One advantage of interfacing any of the currently available mass analyzers to the GD as compared to an atmospheric pressure source arises from the operating pressure of the ion source itself. Most mass spectrometers operate at pressures from 10^{-5} to 10^{-9} Torr. The GD source also operates at a reduced pressure (0.1–10 Torr with support gas), although not nearly as low as those required for mass analysis. The pressure differences associated with GD mass spectrometric implementation are overcome using differential pumping schemes. These types of systems employ a series of pumping regions to reduce the effect of the required pressure drop. A typical scheme would involve pumping the ion source with a rotary-vane roughing pump. This would evacuate the discharge cell to a pressure of approximately 10^{-3} Torr in the absence of discharge gas. The next differentially pumped region facilitates extraction of ions from the ion source. This region is often

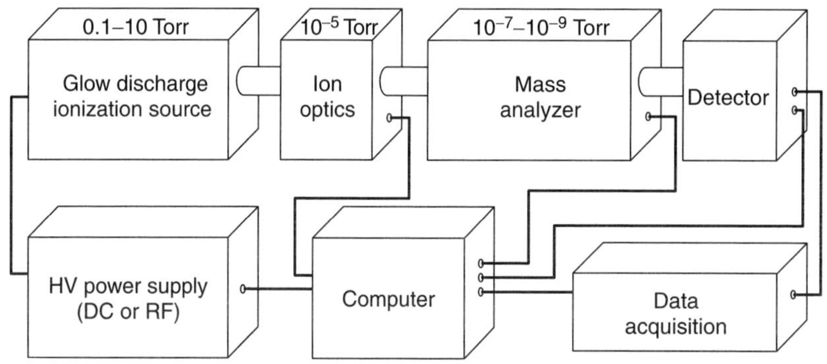

Figure 11 Block diagram of typical GD mass spectrometer system components.

evacuated using a turbo molecular pump or oil diffusion pump and maintains pressures of 10^{-4}–10^{-5} Torr during discharge operation. The analyzer region, which includes the mass separation components as well as the detector, is evacuated using a turbo molecular, oil diffusion, cryo, or ion pump and maintains pressures of 10^{-7}–10^{-9} Torr.

4.1.3 Ion Optics

Ions generated in the GD plasma must be transported efficiently and indiscriminantly to the mass separation region to facilitate accurate and precise analytical measurements. The transfer of ions from the source to the mass analyzer is usually accomplished using an optics system. One type of system used for ion transport is the Einzel lens.[12,73,74] This type of lens is based on three conducting tubular lenses of similar dimension mounted in series. Typically, similar potentials are placed on the first and third lenses while that of the middle lens is adjustable. Many times the middle lens is held at ground potential. As an ion beam passes through this lens system it is focused in a manner analogous to an optical beam, reaching a focal point on the opposite side. It is important to note that cylindrical lens systems like the Einzel lenses produce a cylindrical ion beam rather than a planar one. This characteristic may limit their use for some applications such as ion guides for magnetic sector instruments which require a ribbon-shaped ion beam.

A second type of ion optic is the Bessel box.[75,76] This technology is used in conjunction with QMAs to remove photons and neutral species from the ion beam while simultaneously limiting the KE spread of ions entering the quadrupole lens region. If the energy spread is too great quadrupole performance deteriorates, resulting in degradation of mass resolution, peak splitting, and asymmetrical peak shapes.[77,78] A Bessel box is constructed from a square entrance and exit electrode surrounded by sets of electrodes on each of the other four sides. A center plate or cone is located within the box parallel to the entrance and exit electrodes. Potentials are applied to each electrode. These potentials can be varied to permit the transmission of ions with a discrete KE. Photons and neutral species will not be affected by these potentials and will proceed into the box linearly and collide with the center stop, removing them from the beam. Ions entering the box with too little KE will be repelled toward the entrance plate. If the energy of the ion is too great it will not be steered around the center stop and will collide with the side electrode or exit plate. Only ions with the selected KE will travel around the center stop, through the exit aperture and into the quadrupole region In most cases, one or two lenses are located behind the exit aperture to focus and transport the selected ions.

Quadrupole lenses can also be used as ion guides when operated in an RF-only mode.[79,80] It will be shown later in this chapter that the application of both DC and RF potentials provides a notch filter that can be adjusted and scanned to provide mass unit resolution. When only an RF potential is applied, the quadrupole acts as an ion guide, focusing all ions through quadrupole lenses.

4.1.4 Detection Systems

Three types of detection systems are routinely used for GD mass spectrometric measurements: Faraday cups, electron multipliers, and microchannel plates (MCPs). Detector selection is often independent of the mass analyzer in use.

The detector used most often for applications with high ion abundance is the Faraday cup. Modern Faraday detectors are extremely quiet. When operated using high grade resistors and amplification components, these detection systems offer state-of-the-art measurements with respect to signal-to-noise ratio. Although not currently available on commercial GD mass spectrometric systems, recent developments in multi-Faraday array detection systems offer increasingly precise measurements for scanning instruments by negating the effects of source fluctuations on measurements. Each Faraday cup in the array is dedicated and positioned to measure a single isotope at a given dispersion setting. This allows simultaneous collection of the selected ions during the acquisition sequence without scanning the mass dispersion device, virtually removing any dependence on fluctuations in ion beam intensity arising in the source. Minimization of source fluctuation effects afforded by multicollector arrays is most important for applications involving the measurement of isotope ratios.

For applications requiring optimum sensitivity, discrete dynode electron multipliers operated in a pulse counting mode are required. Operation in this mode registers a signal pulse for every ion impinging on the first dynode. Each impinging ion generates a number of secondary electrons that are successively amplified by each dynode. Overall gains of 10^6–10^8 are common when using multipliers with 14–20 dynodes. After the pulse of electrons leaves the multiplier, it is amplified and proceeds to a discriminator that is set to remove pulses arising from dark noise. The signal is then sent to a universal counter that records each pulse, stores it for a given time, and passes it to a computer-based data acquisition system that presents the data in a usable form. It is important to keep the count rate low enough ($<10^6$ counts s^{-1}) to maintain the integrity of the pulse-counting system and to ensure that any pulses, not measured because of time lag in the electronics, will be statistically insignificant.

For references see page 9448

Daly detection systems behave in a manner similar to electron multipliers.[81] In the Daly system, the ion beam is accelerated to 10–20 kV and directed onto a highly polished aluminized steel electrode positioned directly behind the collector slit. Approximately eight electrons are liberated from the aluminized surface for every impinging ion. These electrons are repelled by the high negative potential applied to the electrode and directed onto a scintillator that produces a photon for each electron. These photons are counted by a photomultiplier located outside of the vacuum system. The resulting pulses are treated in a fashion analogous to that of the electron multiplier. Both Daly and electron multiplier-based detection systems can be operated in a mode that integrates the current of the impinging ion beam. This mode of operation is used for applications that do not impose such stringent sensitivity requirements.

A third detection system is used most often with TOF instruments and is built around an MCP detector. The MCP is characterized by a large, flat active area, high gain, and excellent time response. These operating parameters make it an ideal detector for TOF mass spectrometry. These detectors are fabricated from very thin glass wafers or plates perforated with microscopic channels oriented parallel to the impinging ion beam. The nature of the detector material is such that, in the presence of a potential bias (up to 1200 V), an ion impinging on the entrance to one of the channels will liberate one or more electrons which will, in turn, cascade through the channel, liberating further electrons with each wall collision. Amplification of the order of 10^4 is routinely realized for a single MCP. After emerging from the MCP, these electrons are collected by a positively charged electrode positioned parallel to the MCP. The resulting signal is further amplified and manipulated using a fast digitizing oscilloscope or any number of computer-based flash analog-to-digital converter (ADC) computer boards. Two or more MCPs can be arranged in a stack orientation to amplify further the signal thus increasing system performance.

4.2 Magnetic Sector Mass Analyzers

Initially used by Aston[82] for his studies of gaseous discharges, the magnetic sector mass analyzer, or mass spectrograph as Aston called it, is based on the spatial dispersion of ions with different m/z that is effected when they traverse an electromagnetic field. The magnetic field acts as a prism dispersing monoenergetic ions of differing m/z values across a focal plane; see Figure 12(a).[83] The radius of the curved flight path of an ion through a magnetic field is given by Equation (6):

$$r_{\mathrm{m}} = \left(\frac{144}{B}\right)\left(\frac{mV}{z}\right)^{1/2} \tag{6}$$

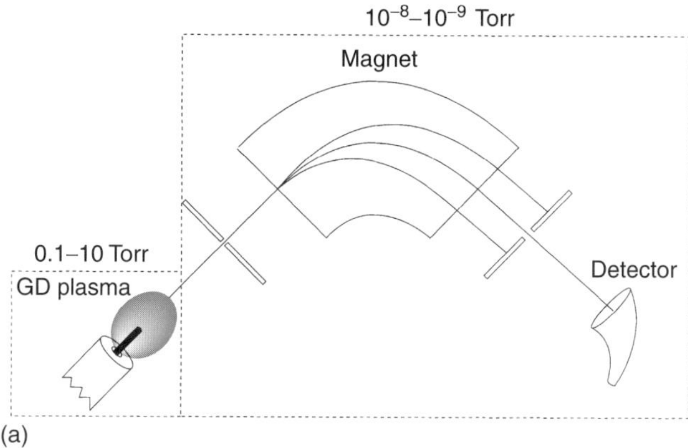

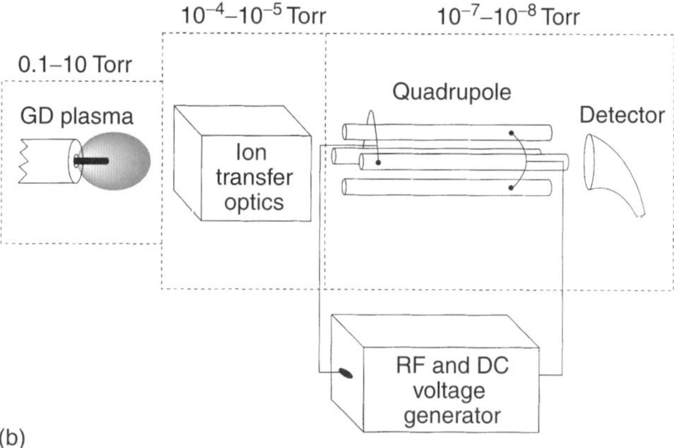

Figure 12 Schematic of (a) a magnetic sector mass spectrometer and (b) a QMA.

where B is the magnetic field strength in gauss, m is the atomic mass of the ions in amu, V is the acceleration voltage of the ion prior to entrance into the magnetic field in volts, z is the charge of the ion, and 144 is a constant prescribed by the units.[83] Because a sector instrument can be made to focus ions onto a plane, it can be designed as either a single or multi-collector instrument. A single electronic detector is used when operating in the sequential acquisition mode. Different m/z ions are brought into focus on the detector by varying either B or V. Simultaneous acquisition instruments utilize either a photographic plate or a detector array oriented in the focal plane. These systems allow the detection of a suite of different m/z ions at a given B and V setting. Although more expensive, the array-based systems offer shorter analysis times and the potential for more precise measurements because source fluctuation effects are minimized.

Mass resolution is another parameter that influences the ability of a system to solve an analytical problem. Mass resolution is a measure of the instrument's ability to separate ions having small mass differences. Mass resolution in magnetic sector instruments is defined by ion beam focusing in the focal plane. Single-focusing instruments rely on direction focusing to increase resolution. This is accomplished by narrowing the entrance and exit slit widths, thus reducing the width of the ion beam.[83] Mass resolution becomes more important when ions of the same nominal mass must be separated.

Multi-sector instruments are quite complex and expensive to build and maintain. They provide adequate to excellent resolving power, especially when an electrostatic sector is coupled to a magnetic sector to provide double (momentum and energy) focusing. When operated in the sequential detection mode, sector instruments are hindered by relatively slow scanning speeds which adversely affect analysis time and sample throughput. In the past, a number of commercial instruments were available from a variety of vendors based on the sector design. Currently, however, their availability has become somewhat limited.

4.3 Quadrupole Mass Filters

Since its development in the 1950s and early 1960s, the quadrupole mass spectrometer has become a powerful tool for the analysis of a variety of materials. Much of its popularity stems from the time of its development when it was viewed as a more rugged, more compact, and more cost-effective alternative to magnetic sector mass spectrometry systems, albeit with compromised performance. The quadrupole mass filter is a variable bandpass filtering ion optic, analogous to an optical bandpass filter. The quadrupole system is capable of transmission of all ions when operated in the RF-only mode (as described earlier), or of measuring only one m/z at a time as a sequential mass analyzer. The quadrupole offers the ability to scan the entire mass range very rapidly or to "peak-hop" among a series of selected isotopes.[84]

A quadrupole mass filter consists of four high precision, cylindrical, conducting rods or poles arranged in a square configuration, as shown in Figure 12(b). Mass filtering is accomplished by applying steady state DC and pulsed RF potenials to these poles. The application of these voltages results in the formation of hyperbolic electric fields, with the ideal quadrupole defined by Equation (7):

$$\frac{r}{r_0} = 1.148 \tag{7}$$

where r is the radius of each rod and r_0 is the radius enclosed by the electrodes.[85] The effects that the applied potentials and resulting electric fields have on a charged particle are best described by the ion trajectory in the $x-z$ and $y-z$ planes. The set of poles in the $x-z$ plane have a positive, time-independent DC voltage and a time-dependent RF voltage applied to them. The poles in the $y-z$ plane have a negative, time-independent DC voltage and a time-dependent RF voltage applied to them. The RF potential applied in the $y-z$ plane is 180° out of phase with the RF voltage in the $x-z$ plane. Ions enter and travel between the poles along the z-axis.

In the $x-z$ plane, larger mass ions are focused along the z-axis by the positively biased DC field, while the smaller mass ions are destabilized by the RF field. In the $y-z$ plane, larger mass ions are deflected away from the z-axis by the negatively biased potential while ions of smaller masses are stabilized by the RF field. The net result is a high-pass filter in the $x-z$ plane and a low-pass filter in the $y-z$ plane, allowing the stabilization and transmission of ions above a selected m/z, and the stabilization and transmission of ions below a selected m/z, respectively. When these two types of filters coexist, a narrow bandpass mass filter results. The magnitude of the DC potential and the frequency of the RF potential can be varied to allow transmission of different m/z ions through the quadrupole lenses, thus providing a means of m/z selection.

Quadrupole-based GD systems have been used extensively.[28,30] The relatively low cost and robustness of these instruments have made them an excellent choice for both the researcher and routine sample analyst. A major limitation of the quadrupole system is its relatively low resolving power, significantly increasing the deleterious effects of overlapping polyatomic interferences. Appropriate selection of operating conditions and discharge gas can minimize some of these concerns, but performance of quadrupoles does not match that of sector-based instruments.

4.4 Ion Trap and Fourier Transform Ion Cyclotron Resonance Devices

While sector and QMAs were being developed for elemental and isotopic applications, the QIT technology was being driven by needs in the organic community. FTICR mass spectrometry, another form of ion trapping technology, has filled a niche in the biological mass spectrometry community because of its unrivaled mass resolution.[86] Recently, these devices have been characterized and used by the inorganic mass spectrometric community. A number of external ionization sources have been implemented, including the GD.[86,87] This section will describe the operating principles of both the QIT and FTICR systems and cover some of their unique properties.

The operation of the QIT is very similar to that of the quadrupole mass filter. Two of the opposing rods in the quadrupole mass filter are connected to form a ring, and each of the remaining pair of rods is replaced by a hyperbolic endcap, as shown in Figure 13(a). The result is a three-dimensional quadrupole field that is symmetric with respect to rotation about the center. The end caps

For references see page 9448

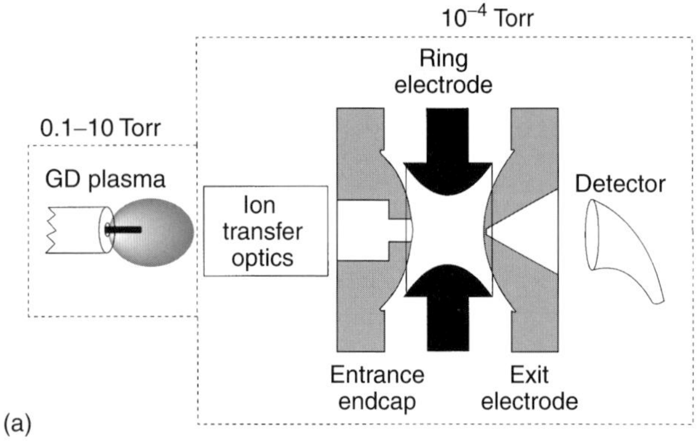

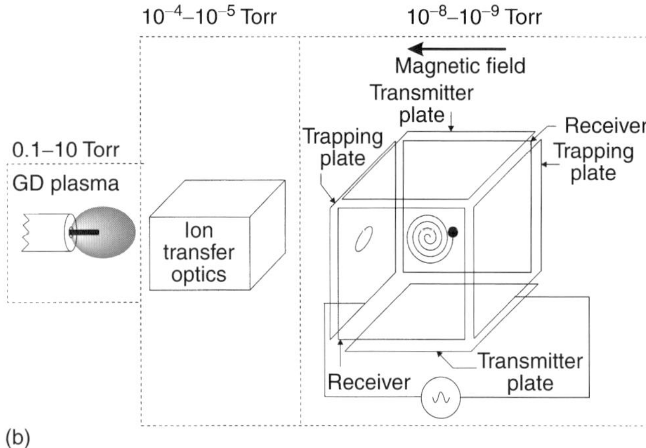

Figure 13 Schematic of (a) a QIT mass spectrometer and (b) an FTICR mass spectrometer.

are oriented along the former z-axis, and the x- and y-axes become a plane, r, symmetric about z. In the most common inorganic applications, externally generated ions are collected and stored within the trap. Typically the internal pressure of the QIT is 10^{-3}–10^{-4} Torr He. Helium is introduced to cool collisionally the precessing ions, thus allowing them to relax toward the center of the trap and increase trapping efficiency.

Stored ions precess with a frequency that is m/z dependent. Ions that are stored in the trapping fields have a fundamental secular frequency along the axial or z-axis. When a supplemental AC signal is applied to the end-cap electrodes, ions whose secular frequencies are in resonance with the applied frequency are excited to higher translational energies. The magnitude of this resonance excitation is directly proportional to the amplitude of the applied signal. At appropriately high amplitudes, the ions can either be lost in collisions with an electrode or ejected through apertures in the exit end-cap electrode. Mass analysis can thus be performed by scanning the frequency of the AC excitation signal. During a scan, ions become destabilized through the excitation process and are selectively ejected through the exit end-cap electrode and are then detected using an electron multiplier.

The relatively high operating pressure of the QIT makes it well suited for coupling to ion sources that operate at higher pressures, such as the GD. Ions from the GD can be directly introduced into the QIT without the need for an elaborate differential pumping scheme. The high pressure and trapping nature of the QIT also facilitate the use of ion–molecule interactions to provide a number of advantageous results, most notably the suppression of unwanted contributions from isobaric interferences.

The FTICR mass spectrometer was initially considered by the elemental mass spectrometry community for its high resolving power. This characteristic allows the physical separation of the analyte signal from interferences without actually removing the interfering species from the cell. High resolution does not come without a cost, however: FTICR systems are one of the most expensive types of mass spectrometers available. The basic operation of the FTICR system is similar to that of the QIT. Precessing ions are constrained spatially within a cubic cell using both electric and magnetic fields. A homogeneous magnetic field confines ions radially, while electrostatic potentials are applied to the end-caps of the cell to trap the ions axially; see Figure 13(b). Ions trapped by these fields are characterized by three motions: one that confines the ions between the two end-cap electrodes, magnetron motion, and cyclotron motion.[88] The ions orbit perpendicular to the applied magnetic field at the ion characteristic cyclotron frequencies, T_c, that are inversely proportional to their m/z values and proportional to the magnetic field strength, B,[89] as shown in Equation (8)

$$w_c = \frac{zB}{m} \tag{8}$$

Ions are excited to larger cyclotron orbits by the application of a resonant RF potential to transmitter plates. An image current is generated as the coherent ion packets of a given m/z come into close proximity to the receiver electrodes. This current is converted to a voltage, amplified, digitized, and stored as a transient signal. The Fourier transform converts these transient signals into their frequency components, T_c, which are, in turn, related to the m/z as shown in Equation (8). Extremely high mass resolution can be achieved because the basis for mass measurement lies in measuring frequency, which can be done with great precision.[90] It should also be noted that mass resolving power is inversely proportional to the ion m/z,[91] a fortunate circumstance for elemental analysis because all masses of interest are below 250 Da.

4.5 Time-of-flight Mass Spectrometers

Perhaps the simplest type of mass spectrometer, the TOF mass spectrometer, is depicted in Figure 14. Its operation

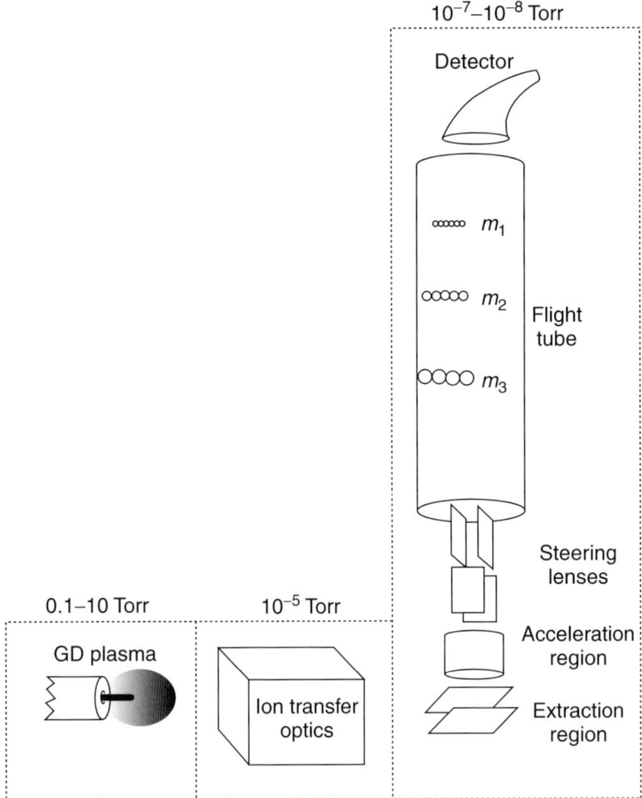

Figure 14 Schematic of a TOF mass spectrometer.

is based on the KE equation, where KE is a function of mass, m, and velocity, v; alternatively it can be expressed in terms of charge, z, and accelerating potential, V, as shown in Equation (9):

$$\mathrm{KE} = \tfrac{1}{2}mv^2 = zV \tag{9}$$

From this equation one can deduce that ions of different m/z accelerated to a common KE will have different velocities. The TOF instrument operates on the principle that if all ions leave the extraction grid at the same point in space and time with equal KEs, they will travel with different velocities, v_y, that are inversely proportional to their respective masses, m_y, as shown in Equation (10):

$$v_y = \left(\frac{2\mathrm{KE}}{m_y}\right)^{1/2} \tag{10}$$

The time needed to traverse the flight path distance D and to arrive at the detector is related to the m_y/z ion by Equation (11):

$$t_y = D\left(\frac{m_y}{2zV}\right)^{1/2} \tag{11}$$

Monitoring the current at the detector (often an MCP) yields a time-dependent signal that can be correlated to the m/z using Equation (11). Mass resolution is determined directly from the temporal resolution $\Delta t/t$, which is determined by the initial KE spread of the ions and the speed of the detection electronics.

The effects of this spread in initial energy can be minimized by using a reflectron TOF mass spectrometric instrument. This instrument geometry has a series of electrostatic lenses located at the end of the flight tube. Positive potentials applied to these lenses increase toward the detection end of the flight tube. Ions traveling down the flight tube enter this potential gradient and penetrate, slowing until they reach a point in the gradient equal to their initial KE. They are then accelerated in the opposite direction, traversing the flight tube a second time. They are then detected by an MCP located at the base of the flight tube. Ions of higher KE penetrate further into the reflectron field, increasing their flight distance (time) and effectively minimizing the impact on mass resolution.

TOF instruments are ideally suited for pulsed ion sources such as lasers or pulsed ion beam sputtering. SIMS using TOF technology has gained wide acceptance as a technique for the characterization of a wide variety of materials. TOF mass spectrometry is also well suited for use with the pulsed GD source.(37) The TOF mass spectrometer provides simultaneous detection of all ions with each injection pulse. This quality affords an excellent diagnostic tool for plasmas and is essential for the temporal characterization of millisecond and microsecond GD pulses.(37,92) Although no commercial GD TOF mass spectrometric systems are currently available, a number are being used for research activities.(37,38)

5 CONCLUSIONS

Optical and mass spectrometric techniques using GD ion sources are now established as routine methods for the direct analysis of solid samples of widely different origin and composition. This article has focused on the fundamental operation of the GD source in a number of operating modes as well as the optical and mass spectrometric instrumentation used for elemental determination. These source–instrument combinations will undoubtedly continue to offer advantages essential for specific applications in the future.

DISCLAIMER

Los Alamos National Laboratory, an affirmative action/equal opportunity employer, is operated by the University of California for the US Department of Energy under

For references see page 9448

contract W-7405-ENG-36. By acceptance of this article, the publisher recognizes that the US Government retains a nonexclusive, royalty-free license to publish or reproduce the published form of this contribution, or to allow others to do so, for US Government purposes. Los Alamos National Laboratory requests that the publisher identify this article as work performed under the auspices of the US Department of Energy. Los Alamos National Laboratory strongly supports academic freedom and a researcher's right to publish; as an institution, however, the Laboratory does not endorse the viewpoint of a publication or guarantee its technical correctness.

ABBREVIATIONS AND ACRONYMS

AAS	Atomic Absorption Spectroscopy
ADC	Analog-to-digital Converter
AES	Atomic Emission Spectroscopy
AFS	Atomic Fluorescence Spectroscopy
CID	Collision-induced Dissociation
DIP	Direct Insertion Probe
FTICR	Fourier Transform Ion Cyclotron Resonance
GD	Glow Discharge
HCL	Hollow Cathode Lamp
ICPMS	Inductively Coupled Plasma Mass Spectrometry
IR	Infrared
KE	Kinetic Energy
MCP	Microchannel Plate
QIT	Quadrupole Ion Trap
QMA	Quadrupole Mass Analyzer
RF	Radiofrequency
SIMS	Secondary Ion Mass Spectrometry
TOF	Time-of-flight
UV	Ultraviolet

RELATED ARTICLES

Coatings **(Volume 2)**
Atomic Spectroscopy in Coatings Analysis

Environment: Trace Gas Monitoring **(Volume 3)**
Laser Mass Spectrometry in Trace Analysis

Environment: Water and Waste **(Volume 3)**
Atomic Fluorescence in Environmental Analysis • Inductively Coupled Plasma Mass Spectrometry in Environmental Analysis

List of selected abbreviations appears in Volume 15

Environment: Water and Waste cont'd **(Volume 4)**
Laser Ablation Inductively Coupled Plasma Spectrometry in Environmental Analysis • Luminescence in Environmental Analysis • Optical Emission Inductively Coupled Plasma in Environmental Analysis

Forensic Science **(Volume 5)**
Atomic Spectroscopy for Forensic Applications • Mass Spectrometry for Forensic Applications

Atomic Spectroscopy **(Volume 11)**
Inductively Coupled Plasma/Optical Emission Spectrometry • Laser Ablation in Atomic Spectroscopy

Mass Spectrometry **(Volume 13)**
Mass Spectrometry: Overview and History • High-resolution Mass Spectrometry and its Applications • Inorganic Substances, Mass Spectrometric in the Analysis of • Isotope Ratio Mass Spectrometry • Literature of Mass Spectrometry • Quadrupole Ion Trap Mass Spectrometer • Tandem Mass Spectrometry: Fundamentals and Instrumentation • Time-of-flight Mass Spectrometry

REFERENCES

1. F. Paschen, 'Bohrs Heliumlinien', *Ann. Phys.*, **50**, 901–940 (1916).
2. H. Schuler, 'Uber die Anregung von Spektren zur Untersuchung von Hyperfeinstrukturen', *Z. Phys.*, **59**, 149–153 (1929).
3. H. Schuler, J.E. Keystone, 'Hyperfeinstrukturen und Kernmomente des Quecksilbers', *Z. Phys.*, **72**, 423–441 (1931).
4. B. Chapman, *Glow Discharge Processes, Sputtering and Plasma Etching*, John Wiley & Sons, New York, 1980.
5. A.M. Howaston, *An Introduction to Gas Discharges*, 2nd edition, Pergamon Press, New York, 1976.
6. R.K. Marcus (ed.), *Glow Discharge Spectroscopies*, Plenum Press, New York, 1993.
7. W.W. Harrison, C.M. Barshick, J.A. Klingler, P.H. Ratliff, Y. Mei, 'Glow Discharge Techniques in Analytical Chemistry', *Anal. Chem.*, **62**, 943A–949A (1990).
8. R.K. Marcus, T.R. Harville, Y. Mei, C.R. Shick, 'RF-powered Glow-discharges: Elemental Analysis Across the Solids Spectrum', *Anal. Chem.*, **66**, 902A–911A (1994).
9. P.W.J.M. Boumans, 'Studies of Sputtering in a Glow Discharge for Spectrochemical Analysis', *Anal. Chem.*, **44**, 1219–1228 (1972).
10. F.L. King, J. Teng, R.E. Steiner, 'Glow Discharge Mass Spectrometry: Trace Elemental Determinations in Solid Samples', *J. Mass Spectrom.*, **30**, 1061–1075 (1995).
11. T.D. Mark, G.H. Nunn, *Electron Impact Ionization*, Springer-Verlag, New York, 1985.

12. L. Valyi, *Atom and Ion Sources*, John Wiley & Sons, New York, 1977.
13. F.H. Field, J.L. Franklin, *Electron Impact Phenomena*, Academic Press, New York, 1970.
14. J. Delcroix, C. Ferreira, A. Richard, 'Metastable Atoms and Molecules in Ionized Gases', in *Principles of Laser Plasmas*, ed. G. Bekefi, John Wiley & Sons, New York, 159–233, 1976.
15. V.S. Bordin, Y.M. Kagan, 'Excitation of Helium in a Hollow-cathode Discharge', *Opt. Spectrosc.*, **23**, 108–110 (1967).
16. F.M. Penning, 'Uber die Erhehung der Zundspannung von Neon–Argon Gemischen durch Bestrahlung', *Z. Phys.*, **57**, 723–738 (1929).
17. F.M. Penning, 'Uber den Einfluss Sehr Gerginger Beimischungen auf die Zundspannung der Edelgase', *Z. Phys.*, **46**, 335–348 (1925).
18. D.H. Stedman, D.W. Stetser, 'Chemical Applications of Metastable Rare Gas Atoms', *Prog. React. Kinet.*, **6**, 193–238 (1971).
19. R.L. Smith, D. Serxner, K.R. Hess, 'Assessment of the Relative Role of Penning Ionization in Low-pressure Glow Discharges', *Anal. Chem.*, **61**, 1103–1108 (1989).
20. M. Saito, 'Relative Sensitivity Factors in Direct Current Glow Discharge Mass Spectrometry Using Kr and Xe Gas – Estimation of the Role of Penning Ionization', *Fresenius' J. Anal. Chem.*, **351**, 148–153 (1995).
21. S. Degent, W.W. Harrison, 'Parameter Evaluation for the Analysis of Oxide-based Samples with Radio-frequency Glow Discharge Mass Spectrometry', *Anal. Chem.*, **67**, 1026–1033 (1995).
22. C. Lazik, R.K. Marcus, 'Effect of Excitation-frequency on Source Characteristics in Radio-frequency Glow-discharge Atomic-emission Spectrometry, 1.2–30 MHz', *Spectrochim. Acta*, **49B**, 649–663 (1994).
23. D.C. Duckworth, D.L. Donohue, D.H. Smith, T.A. Lewis, R.K. Marcus, 'Design and Characterization for a Radio-frequency-powered Glow-discharge Source for Double-focusing Mass Spectrometers', *Anal. Chem.*, **65**, 2478–2484 (1993).
24. J. Teng, C.M. Barshick, D.C. Duckworth, S.J. Norton, D.H. Smith, F.L. King, 'Factors Influencing the Quantitative Determination of Trace Elements in Soils by Glow-discharge Mass Spectrometry', *Appl. Spectrosc.*, **49**, 1361–1366 (1995).
25. D.C. Duckworth, C.M. Barshick, D.A. Bostick, D.H. Smith, 'Direct Measurement of Uranium Isotopic-ratios in Soils by Glow-discharge Mass Spectrometry', *Appl. Spectrosc.*, **47**, 243–245 (1993).
26. D.C. Duckworth, C.M. Barshick, D.H. Smith, 'Analysis of Soils by Glow-discharge Mass Spectrometry', *J. Anal. Atom. Spectrom.*, **8**, 875–879 (1993).
27. D.C. Duckworth, R.K. Marcus, 'Radio-frequency Powered Glow Discharge Atomization Ionization Source for Solids Mass Spectrometry', *Anal. Chem.*, **61**, 1879–1886 (1989).
28. F.L. King, C. Pan, 'Time-resolved Studies of Ionized Sputtered Atoms in Pulsed Radio-frequency Powered Glow-discharge Mass Spectrometry', *Anal. Chem.*, 3187–3193 (1993).
29. J.A. Klingler, P.J. Savickas, W.W. Harrison, 'The Pulsed Glow-discharge as an Elemental Ion-source', *J. Am. Soc. Mass Spectrom.*, **1**, 138–143 (1990).
30. J.A. Klingler, C.M. Barshick, W.W. Harrison, 'Factors Influencing Ion Signal Profiles in Pulsed Glow-discharge Mass-spectrometry', *Anal. Chem.*, **63**, 2571–2576 (1991).
31. M.R. Winchester, R.K. Marcus, 'Emission Characteristics of a Pulsed, Radio-frequency Glow Discharge Atomic Emission Device', *Anal. Chem.*, **64**, 2067–2074 (1992).
32. M. Glick, B.W. Smith, J.D. Winefordner, 'Laser-excited Atomic Fluorescence in a Pulsed Hollow-cathode Glow Discharge', *Anal. Chem.*, **62**, 157–161 (1990).
33. C. Pan, F.L. King, 'Ion Formation Processes in the Afterpeak Time Regime of Pulsed Glow Discharge Plasmas', *J. Am. Soc. Mass Spectrom.*, **4**, 727–732 (1993).
34. W.W. Harrison, K.R. Hess, R.K. Marcus, F.L. King, 'Glow-discharge Mass-spectrometry', *Anal. Chem.*, **58**, 341A–356A (1986).
35. M.A. Biondi, 'Studies of the Mechanism of Electron–Ion Recombination', *Phys. Rev.*, **129**, 1181–1188 (1963).
36. M.A. Biondi, 'Diffusion, De-excitation and Ionization Cross-sections for Metastable Atoms', *Phys. Rev.*, **88**, 660–665 (1952).
37. R.E. Steiner, C.L. Lewis, F.L. King, 'Time-of-flight Mass Spectrometry with a Pulsed Glow Discharge Ionization Source', *Anal. Chem.*, **69**, 1715–1721 (1997).
38. C.R. Shick, P.A. DePalma, R.K. Marcus, 'Radio-frequency Glow-discharge Mass-spectrometry for the Characterization of Bulk Polymers', *Anal. Chem.*, **68**, 2113–2121 (1996).
39. X.H. Pan, R.K. Marcus, 'Direct Analysis of Glass Powder Samples by Radio-frequency Glow-discharge Atomic Emission Spectrometry (RF-GD-AES)', *Mikrochim. Acta*, **129**, 239–250 (1998).
40. J.A.C. Broekaert, T. Graule, H. Jenett, G. Tölg, P. Tschopel, 'Analysis of Advanced Ceramics and their Basic Products', *Fresenius Z. Anal. Chem.*, **332**, 825–838 (1989).
41. M. Dogar, K. Laqua, H. Massmann, 'Spektrochemische Analysen mit einer Glimmentladungslampe als Lichtquelle-II Analytische Anwengdungen', *Spectrochim. Acta*, **27B**, 65–88 (1972).
42. G.S. Lomdahl, R. McPherson, J.V. Sullivan, 'The Atomic Emission Spectrometric Determination of Non-conducting Materials with a Boosted Output Glow-discharge Source', *Anal. Chim. Acta*, **148**, 171–180 (1983).
43. T.J. Loving, W.W. Harrison, 'Dual-pin Cathode Geometry for Glow-discharge Mass-spectrometry', *Anal. Chem.*, **55**, 1526–1530 (1983).
44. S. Caroli, A. Almonti, K. Zimmer, 'Applicability of a Hollow-cathode Emission Source for Determining Trace-elements in Electrically Non-conducting Powders', *Spectrochim. Acta*, **38B**, 625–631 (1983).

45. W. Schelles, R.E. Van Grieken, 'Direct Current Glow Discharge Mass Spectrometric Analysis of Macor Ceramic Using a Secondary Cathode', *Anal. Chem.*, **68**, 3570–3574 (1996).
46. D.M. Wayne, R.K. Schulze, C. Maggiore, D.W. Cooke, G. Havrilla, 'Characterization of Tantalum Films on Analytical Surfaces: Insights into Sputtering of Nonconductors in a Direct-current Glow Discharge Using Secondary Cathodes', *Appl. Spectrosc.*, **53**, 266–277 (1999).
47. W. Schelles, R.E. Van Grieken, 'Quantitative Analysis of Zirconium-oxide by Direct Current Glow-discharge Mass-spectrometry Using a Secondary Cathode', *J. Anal. At. Spectrom.*, **12**, 49–52 (1997).
48. C.M. Barshick, W.W. Harrison, 'The Laser as an Analytical Probe in Glow-discharge Mass-spectrometry', *Mikrochim. Acta*, **III**, 169–177 (1989).
49. C.M. Strange, R.K. Marcus, 'Aqueous Sample Introduction into a Glow-discharge Device via a Particle Beam Interface', *Spectrochim. Acta*, **46B**, 517–526 (1991).
50. R.E. Steiner, C.L. Lewis, V. Majidi, 'Consideration of a Millisecond Pulsed Glow Discharge Time-of-flight Mass Spectrometer for Concurrent Elemental and Molecular Analysis', *J. Anal. At. Spectrom.*, **14**, 1537–1541 (1999).
51. J.D. Ingle, Jr, S.R. Crouch, *Spectrochemical Analysis*, Prentice Hall, Englewood Cliffs, 1988.
52. Y. Talmi, *Multichannel Image Detectors*, ACS Symposium Series 236, American Chemical Society, Washington, DC, Vol. 2, 1983.
53. Q.S. Hanley, C.W. Earle, F.M. Pennebaker, S.P. Madden, M.B. Denton, 'Charge Transfer Devices in Analytical Instrumentation', *Anal. Chem.*, **68**, 661A–667A (1996).
54. J.H. Giles, T.D. Ridder, R.H. Williams, D.A. Jones, M.B. Denton, 'Selecting a CCD Camera', *Anal. Chem.*, **70**, 663A–668A (1998).
55. E.U. Condon, G.H. Shortley, *The Theory of Atomic Spectra*, Cambridge University Press, Cambridge, 1963.
56. W.F. Meggers, C.H. Corliss, B.F. Scribner, *Tables of Spectral-line Intensities: Part II – Arranged by Wavelengths*, US Government Printing Office, Washington, 1975.
57. W. Grimm, 'Eine neue Glimmentladungslampe fur die Optische Emissionsspektralanalyse', *Spectrochim. Acta*, **23B**, 443–454 (1968).
58. J.A.C. Broekaert, 'Atomic Emission Spectroscopy', in *Glow Discharge Spectroscopies*, ed. R. K. Marcus, Plenum Press, New York, 113–174, 1993.
59. H. Hocquaux, 'Thin Film Analysis', in *Glow Discharge Spectroscopies*, ed. R.K. Marcus, Plenum Press, New York, 329–372, 1993.
60. E.H. Daughtrey, D.L. Donohue, P.J. Slevin, W.W. Harrison, 'Surface Sputter Effects in a Hollow-cathode Discharge', *Anal. Chem.*, **47**, 683–688 (1975).
61. W.W. Harrison, B.L. Bentz, 'Glow-discharge Mass-spectrometry', *Prog. Anal. Spectrosc.*, **11**, 53–110 (1988).
62. P.F. Little, A. von Engel, 'The Hollow-cathode Effect and the Theory of Glow Discharges', *Proc. R. Soc. London*, **224A**, 209–227 (1954).
63. S.L. Mandelstam, V.V. Nedler, 'On the Sensitivity of Emission Spectrochemical Analysis', *Spectrochim. Acta*, **17**, 885–894 (1961).
64. D.S. Gough, 'Direct Analysis of Metals and Alloys by Atomic-absorption Spectrometry', *Anal. Chem.*, **48**, 1926–1931 (1976).
65. A.E. Bernhard, 'Atomic Absorption Spectrometry Using Sputtering Atomization of Solid Samples', *Spectroscopy*, **2**, 24–27 (1987).
66. A. Walsh, 'The Application of Atomic Absorption Spectra to Chemical Analysis', *Spectrochim. Acta*, **7**, 108–117 (1955).
67. F.L. King, PhD Dissertation, University of Virginia, 1989.
68. B.M. Patel, B. Smith, J.D. Winefordner, 'Laser-excited Fluorescence of Diatomic Lead in a Glow-discharge Source', *Spectrochim. Acta*, **40B**, 1195–1204 (1985).
69. J.B. Womack, E.M. Gessler, J.D. Winefordner, 'Atomic Fluorescence in a Pulsed Hollow-cathode Discharge with a Copper Vapor Pumped Dye Laser', *Spectrochim. Acta*, **46B**, 301–308 (1991).
70. F.M. Penning, 'Demonstration of a New Photoelectric Effect', *Physica*, **8**, 137–140 (1928).
71. C.M. Barshick, R.W. Shaw, J.P. Young, J.M. Ramsey, 'Isotopic Analysis of Uranium Using Glow-discharge Optogalvanic Spectroscopy and Diode Lasers', *Anal. Chem.*, **66**, 4154–4158 (1994).
72. W.A. Mattson, B.L. Bentz, W.W. Harrison, 'Coaxial Cathode Ion-source for Solids Mass Spectrometry', *Anal. Chem.*, **48**, 489–491 (1976).
73. O. Klemperer, *Electron Optics*, Cambridge University Press, Cambridge, 1971.
74. A. Septier, *Focusing of Charge Particles*, Academic Press, New York, 1967.
75. B.L. Bentz, PhD Dissertation, University of Virginia, 1980.
76. *Electrostatic Energy Analyzer Model 616-1 Instruction Manual*, Etranuclear Laboratories, Inc., Pittsburgh, 1977.
77. S.S. Medley, 'Energetic Ion Mass Analysis Using a Radiofrequency Quadrupole Filter', *Rev. Sci. Instrum.*, **49**, 698–706 (1978).
78. W. Paul, H.P. Reinhard, U. VanZahn, 'Das Elektrische Massenfilter als Massenspektrometer und Isotopentrenner', *Z. Phys.*, **152**, 143–182 (1958).
79. M. Szilagyi, *Electron and Ion Optics*, Plenum Press, New York, 1988.
80. H. Wollnik, *Optics of Charged Particles*, Academic Press, Inc., Orlando, 1987.
81. N.R. Daly, 'Scintillation Type Mass Spectrometer Ion Detector', *Rev. Sci. Instrum.*, **31**, 264–267 (1960).
82. F.W. Aston, *Isotopes*, 2nd edition, Longman, New York, 1924.
83. J. Roboz, *Introduction to Mass Spectrometry: Instrumentation and Techniques*, Interscience, New York, 1968.

84. R.S. Houk, 'Mass-spectrometry of Inductively Coupled Plasmas', *Anal. Chem.*, **58**, 97A–105A (1986).
85. P.H. Dawson, *Quadrupole Mass Spectrometry and its Applications*, Elsevier Scientific, New York, 1976.
86. M.V. Buchanan, R.L. Hettich, 'Fourier-transform Mass-spectrometry of High-mass Biomolecules', *Anal. Chem.*, **65**, 245A–259A (1993).
87. S.A. McLuckey, G.J. Van Berkel, D.E. Goeringer, G.L. Glish, 'Ion-trap Mass-spectrometry of Externally Generated Ions', *Anal. Chem.*, **66**, 689A–696A (1994).
88. A.G. Marshall, P.B. Grosshans, 'Fourier-transform Ion-cyclotron Resonance Mass-spectrometry: the Teenage Years', *Anal. Chem.*, **63**, 215A–229A (1991).
89. N.M.M. Nibbering, 'Gas-phase Ion Molecule Reactions as Studied by Fourier-transform Ion-cyclotron Resonance', *Acc. Chem. Res.*, **23**, 279–285 (1990).
90. D.C. Duckworth, C.M. Barshick, 'Ion Traps: What do they Hold for Elemental Analysis', *Anal. Chem.*, **70**, 709A–717A (1998).
91. M.B. Comisarow, A.G. Marshall, 'Theory of Fourier-transform Ion-cyclotron Resonance Mass-spectrometry 1. Fundamental Equations and Low Pressure Line-shape', *J. Chem. Phys.*, **64**, 110–119 (1976).
92. W.W. Harrison, W. Hang, 'Pulsed Glow Discharge Time-of-flight Mass Spectrometry', *J. Anal. At. Spectrom.*, **11**, 835–840 (1996).

Graphite Furnace Atomic Absorption Spectrometry

James A. Holcombe
University of Texas at Austin, Austin, USA

Graphite furnace atomic absorption spectrometry (GFAAS) is an atomic spectroscopic technique in which a small sample is placed inside a graphite tube that is then resistively heated to accomplish sample desolvation (for liquid samples), ashing or charring (to decompose the sample and volatilize some of the matrix) and finally atomization. The light from a line source characteristic of the element being determined is passed longitudinally through the tube and the absorbance resulting from the presence of free analyte atoms in the gas phase is measured. The signal is transient in character, lasting approximately 1–5 s, and the area under this transient peak is generally used in the construction of a calibration curve. Modern instrumentation provides high levels of automation with capabilities of background correction as well as routine methods of sample analysis, for example quality assurance/quality control (QA/QC), standard additions, matrix modification, etc. Since the technique was first introduced in 1969, a lot of progress has been made in understanding the processes occurring within the graphite tube atomizer that ultimately produces the absorbance signal. An understanding of the free atoms formation process has facilitated the application of the technique to the analysis of a variety of complex samples.

Graphite furnace (also known as an electrothermal atomizer or ETA) atomic absorption (AA) is generally considered an ultratrace and microtrace analytical technique with limits of detection (LODs) in the low picogram range, precision of a few percent (relative standard deviation) and a dynamic range of about three orders of magnitude. In addition to its excellent sensitivity, it is

For references see page 9467

unique in its ability to handle microsamples including aqueous solutions, viscous liquids, slurries and even solids. In general, there is considerable literature detailing methods and procedures for the determination of a variety of analytes in complex matrices that can be used by the analyst to apply the approach to new, complex analytical needs. When used correctly, this analytical tool can provide precise, accurate analysis for a wide range of sample types.

1 INTRODUCTION AND OVERVIEW

What is commonly referred to as GFAAS has been referred to as flameless AA in the past and is officially designated by IUPAC as electrothermal atomic absorption spectrometry (ETAAS). Typically, a commercial ETA consists of a graphite tube approximately 2.5–3 cm in length and 4–6 mm inside diameter. There is a small (1–2 mm diameter) hole in the center of the tube for introducing the sample, which is generally a solution aliquot of up to ca. 50 µL and is more typically 20–25 µL. Figure 1 shows a simplified schematic of the furnace and some of its typical operating features. Commercial tube furnaces are ca. 2.5 cm long and 0.6 cm in diameter and are composed of graphite with a relatively impervious pyrographite coating. The tube is resistively heated by an external supply and is protected from oxidation by a flow of sheath gas (typically Ar) around the furnace (not shown in Figure 1). The light source passes through the furnace and Beer–Lambert's law relates the absorbance to the concentration. Shown in the graphical inserts are the temperature (heating) program of the furnace and the absorbance signal that results once the "atomization cycle" is reached. As can be seen, the signal is transient in character and can last up to several seconds.

2 BRIEF HISTORY

The ability of free atoms in the gas phase to absorb characteristic wavelengths of light is not a new concept.

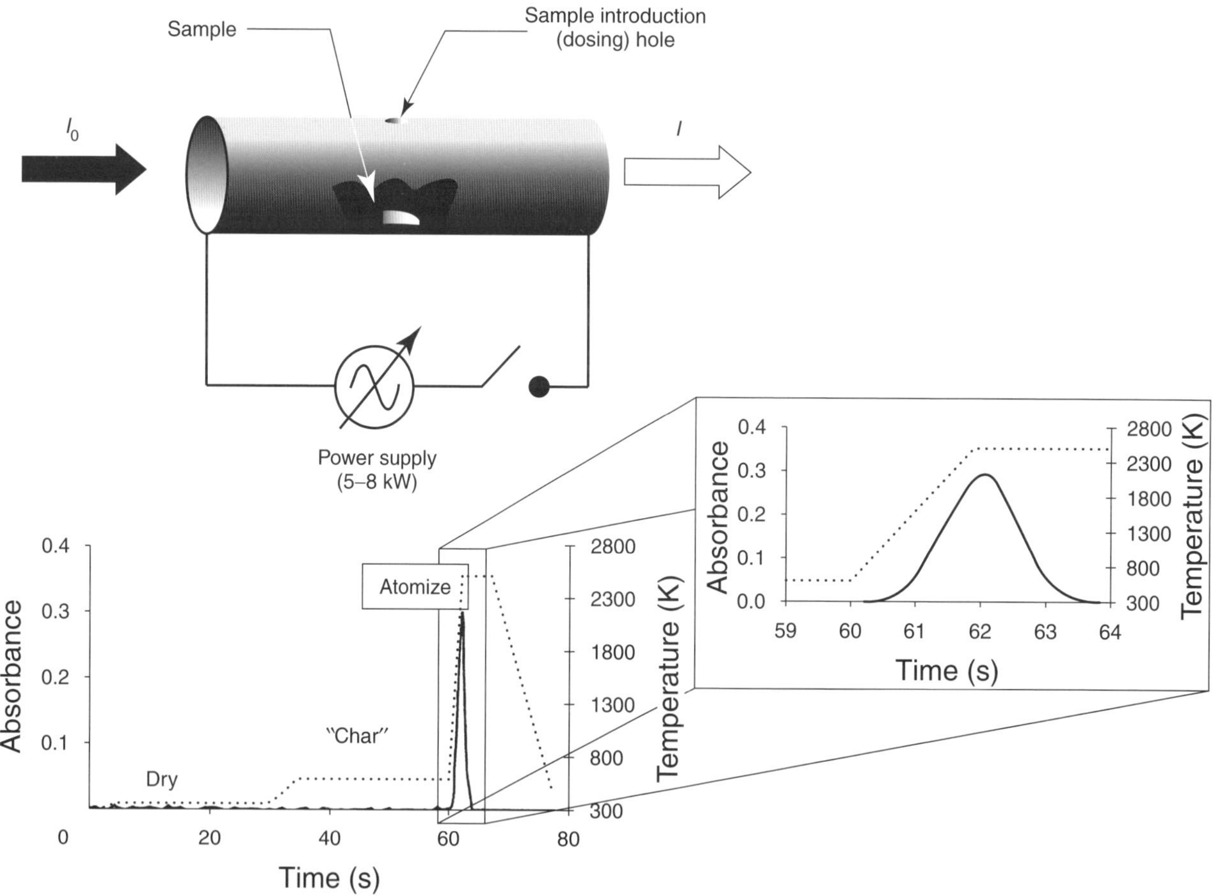

Figure 1 Conceptual diagram of a tubular ETA or graphite furnace atomizer. The atomizer is connected to a variable voltage power source and is resistively heated by current passing through the electrothermal vaporizer. Shown are the three basic heating regimes: dry, char and atomization cycles. The light source is passed through the tube and during the high temperature "atomization cycle" the analyte is vaporized and atomized. As a consequence an absorbance signal is registered.

List of selected abbreviations appears in Volume 15

In fact, the absence of discrete wavelengths in the yellow region of the solar spectrum was attributed to the absorption of these wavelengths by sodium as early as 1860 by Kirchoff. Walsh[1] is credited with the first practical employment of AA spectrometry as a quantitative analytical tool using a flame as the "atomic cell" where free gaseous atoms are produced. Light from a forerunner of today's hollow cathode lamp (HCL) was passed through the flame, the correct wavelength of radiation was isolated and detected. The attenuation of the light was then used for determining the concentration of analyte in the sample solution.

Prior to Wash's publication, a large, heated graphite tube ("King furnace") had been used for basic studies of the absorption characteristics of free atoms and molecules which were produced continuously from a large amount of material located within the furnace.[2] However, there was no attempt to use this for quantitative or trace analysis, and there was no suggestion that this might even be possible.

L'vov[3] was the first to combine the concepts of Walsh with a significantly altered version of the King furnace to demonstrate that quantitative analysis by AA could be conducted on discrete sample amounts by pulsed vaporization of the sample. Figure 2 shows a picture of this first graphite furnace atomizer for spectrochemical analysis by AA. This approach not only provided quantitative results but gave significantly improved sensitivities with LODs in the picogram range for many elements because of the significantly longer residence time within the furnace in comparison to a flame. This paper and the promising results spurred a great deal of activity in "flameless" atomizers.

Much of the initial work employed graphite-based devices that were resistively heated, although many

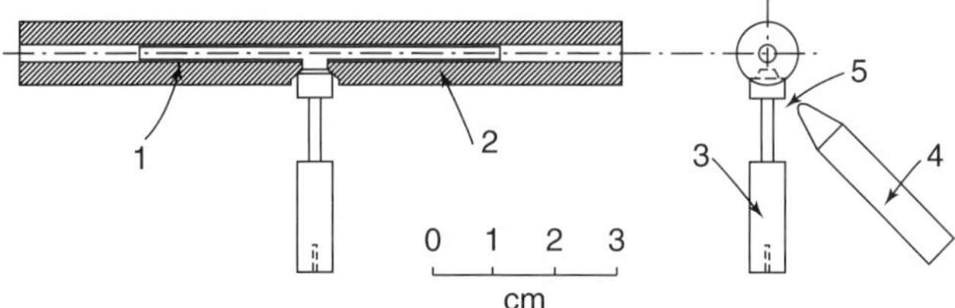

Figure 2 L'vov's original "graphite furnace atomizer. The sample was deposited on the "plug" (3) which was inserted into the preheated graphite tube (1). An arc (5) was then struck with an auxiliary electrode (4) to rapidly heat the plug and release the sample into the tube and optical path. A tantalum liner (2) was used in this original furnace to minimize analyte loss from vapor diffusion through the graphite (Modern furnaces use pyrolytically coated graphite through which minimal vapor diffusion occurs). (Reprinted from B.V. L'vov, 'The Analytical Use of Atomic Absorption Spectra', *Spectrochim. Acta*, **17**, 761–770, Copyright (1961), with permission from Elsevier Science.)

were in the form of rods rather than the original tubular design of L'vov. This was probably a result of the ease of in-house manufacturing of making low resistance contacts to carry the currents needed to heat the rods. Much activity persisted in the 1970s in trying different atomizer designs that ranged from rods, to braids to tubular designs. The tubular design ultimately dominated because of its ability to provide a higher temperature, semi-enclosed environment. The elevated temperature minimized compound formation and condensation while the semi-enclosed nature of the tubular design enhanced the residence time of the analyte within the optical path to improve sensitivity.

The atomizer material was not limited to graphite, and a number of high melting metals (e.g. W, Ta, Pt, etc.) were evaluated. The metals often provided the advantage of higher resistance and lower mass, thereby permitting lower powers and the use of higher voltages (with lower currents) to bring about rapid heating. For analytes that tended to form refractory carbides on graphite, these metal atomizers permitted complete vaporization at lower temperatures. However, there were also a large number of metals that dissolved in the metal atomizer or formed solid solutions that produced memory effects as they were slowly and continuously distilled from the atomizer with each subsequent firing. Graphite probably remained the material of choice for a number of reasons, although the most readily apparent were its advantage of being available at high purity, having a very high (ca. 3400 K) sublimation temperature, and oxidizing to form a surface "oxide" ($C-O_{(ads)}$) that was readily removed as $CO_{(g)}$ and $CO_{2(g)}$ from the surface with simple heating. The size of the tubular atomizers ranged from very small devices that heated very rapidly and could only accommodate a few microliters of sample, to the "Woodriff furnace"[4,5] which approached 25 cm in length and was preheated before the sample was introduced.

Significant among the developments was Massmann's[6] early design of a directly heated tube into which the sample was introduced. This design precluded the striking of an auxiliary arc as was required with the L'vov design. Most modern furnaces are smaller versions of this design with tube lengths typically ca. 2.5–3 cm with a 4–7 mm inner diameter (Figure 3). As shown in Figure 3, the furnace is heated from the ends (i.e. "end-heated") by direct passage of current through the furnace.

With the tube furnace, it became readily apparent that analyte elements were often "stuck" near the ends of the tube since the tube did not heat uniformly. Greater heat loss was experienced where the ends of the furnace contacted the water-cooled endcones and could result in a temperature gradient as large as 1000 K between tube center and end.[7] The heating of the furnace was more uniform during a rapid deposition of energy into it (e.g.

For references see page 9467

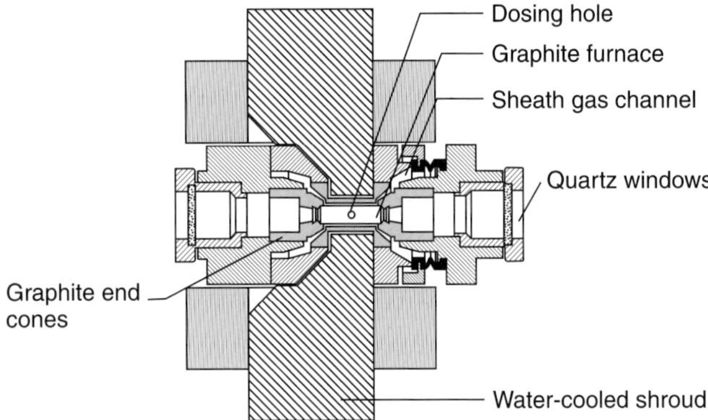

Figure 3 This is a typical, modern end-heated graphite furnace and workhead. The graphite furnace atomizer is in the center of the workhead that includes the electrical, gas and cooling connections. The furnace itself is ca. 2.5 cm in length with a 0.5 cm inner diameter. (Reproduced by permission of Varian, Inc.)

during a rapid thermal ramp) and the ends tended to cool as heating rates decreased or once a steady state temperature in the tube center was achieved. This often resulted in material vaporizing from the initial deposition site and recondensing on the now cooler ends of the furnace. As a result, the next heating of the atomizer produced a secondary release of these condensed metals, giving rise to memory effects. This was not uncommon, especially for the more refractory analyte elements. To minimize this problem, it was common to use higher atomization temperatures and "clean cycles" (i.e. high-temperature heating of the furnace after the analytical measurement was complete). Another alternative was a redesign of the tube so that the electrical contacts did not also serve as a heat sink that produced this longitudinal temperature gradient. Isothermal cuvettes (or "transversely heated graphite atomizers") were first introduced by Frech et al.[8] to minimize thermal gradients along the tube. Because of the more isothermal environment and encouraging preliminary results, this concept has made its way into the commercial market and is another option to the end-heated furnace concept.

Another major breakthrough in use of furnaces was initially suggested by L'vov et al.[9] and involves the use of a platform inserted into the furnace (Figure 4). Since the formation of free atoms is favored at elevated temperatures, it becomes problematic if a volatile analyte (or analyte-containing compound) vaporizes at a low temperature, which permits molecular species to form (or remain). A small graphite flat in the furnace onto which the sample is placed has minimal thermal contact with the heated tube walls. As a result of poor conductive heating of the platform by the wall, the temperature of the platform lags that of the furnace walls. Depending on the heating rate of the tube wall and the heat transfer rate

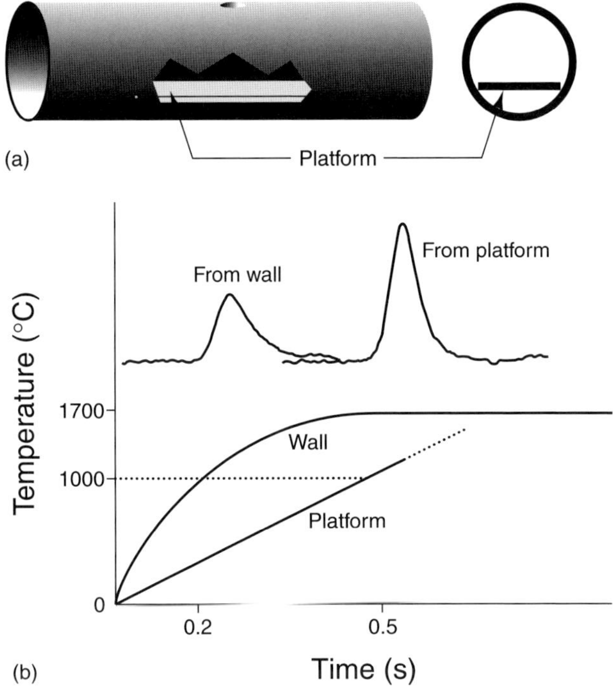

Figure 4 (a) Schematic diagram of a platform located inside a graphite furnace atomizer. The sample is initially deposited on the platform. (b) Representation of the temperature of the furnace wall and platform temperature during the pulse heating of the atomizer, that is "atomization cycle". (Reproduced by permission from K.W. Jackson (ed.), *Electrothermal Atomization for Analytical Atomic Spectrometry*, Wiley, New York, 13, 1999.) The horizontal dashed line at 1000 °C represents the appearance temperature. It should be noted that vaporization from the platform presents the analyte to the gas phase when the wall temperature is several hundred degrees hotter, thus promoting free atom formation.

between wall and platform, this lag can be as large as several hundred degrees. As a consequence, vaporization of the sample from the platform into the furnace takes place when the gas phase is at a higher temperature. This is illustrated in Figure 4. Generally, the higher temperature will sponsor molecular dissociation (atomization) and both improve atomization efficiency as well as minimize many interference effects. Because most platforms never reach the wall temperature, the simple design depicted in the figure is often not useable for the more refractory elements such as V, Mo or even Ni and Cr. Alternative designs of the platform where better thermal contact with the wall exists are starting to appear on the market and even determination of refractory metals with this new design is feasible.

In the early days of flameless AA, carbon rod atomizers, etc. the technique was plagued by extreme interference effects. Some of the modifications noted above (i.e. tube-type construction, isothermal heating and platform use) ameliorated many of these problems. In spite of these unique challenges, it was readily recognized that vapor formation using an ETA was not limited by factors that often plagued other spectroscopic techniques where nebulizers were used for sample introduction. For example, samples that were viscous, contained high concentrations of total dissolved solids, or contained suspended solids were often unusable with conventional nebulization but could be dealt with using ETA. As a result, expectations of high accuracy and precision were appearing for samples containing extremely complex matrices. These complex matrices placed heavy burdens on the ability of the ETA to "synthesize" atoms within the furnace consistently and with nearly 100% efficiency.

The use of chemicals added to the sample as "matrix modifiers" became common place as a means of minimizing matrix effects by permitting, for example, the removal of the offensive matrix during a thermal pretreatment step prior to the high temperature atomization heating cycle. Ediger[10] is often recognized as one of the first to introduce such a concept with the suggestion that Ni could be added to a sample that was being analyzed for Se. His suggestion was that the Ni bound the Se and "stabilized" it on the graphite surface to a higher temperature while offending matrix concomitants could be vaporized during the ash or thermal pretreatment cycle. The delayed vaporization of the selenium until the ETA was at a higher temperature also improved the atomization efficiency. The search for and employment of matrix modifiers continues to dominate the literature as an important means of achieving reproducible and accurate analytical results in ETAAS. Many of the features alluded to above will be discussed in more detail in subsequent sections.

It is important to stress at this point that ETAAS is a powerful analytical technique that is widely used for ultratrace and microtrace analysis. It has the advantages of a high degree of elemental selectivity, and once a method is developed, analysis is relatively simple. Modern instrumentation has empowered the user with a high degree of automation, which further simplifies the use of ETAAS for a large number of samples. Inherent to AA, however, is its traditional single elemental analytical capabilities, that is, one element at a time can be determined. However, recent work in the area of using multiple sources and continuum sources may arm the approach with simultaneous multielement capabilities in future instruments.

The first sections discuss the theory behind the technique in an attempt to explain what is taking place within the atomizer at the molecular level. In this chapter there is also a brief discussion of "interferences" that can occur and why they are occurring, again to provide a complete picture of what is taking place fundamentally. However, the practical analytical utility of the approach should not be covered up by these discussions since, it will be shown, most of these potential problems can be circumvented with proper choice of instrument operating conditions and, in many cases, correct selection of matrix modifiers.

3 THEORY

3.1 Absorption Fundamentals

The basics of AA are also covered in **Atomic Spectroscopy: Introduction**. In brief, the absorption of radiation by the analyte within the furnace volume for the purpose of conducting analysis assumes the existence of free, gas-phase atoms. Analyte atoms present as molecular species (e.g. PbO, MnCl, etc.) or as ions (e.g. Na^+) will not absorb at wavelengths used to detect the neutral atomic species. The preferred analytical lines for conducting AA in the ETA are identical to those used with flame atomic absorption (FAA) and are generally absorption transitions which originate in the unexcited atomic ground state and terminate in a higher lying electronic state, that is, resonance lines. For many atoms, there exists more than a single excited state that can be populated by an absorptive transition from the ground state. In these instances, the strength of the absorption is determined by the oscillator strength (or Einstein B coefficient for stimulated absorption). A larger value implies a higher probability of photon absorption and, hence, greater sensitivity. In some instances, the wavelength of the source wavelength may be noisier at this more strongly absorbing line and, as a consequence, the weaker line is used since it may provide an improved signal-to-noise ratio. For example, the 217-nm line for Pb is ca. 2.4 times more sensitive than the 283-nm resonance line, but the shorter wavelength is often noisier and more prone to interferences by scatter and molecular absorption. Hence, it is not uncommon to see the 283-nm line used in preference.

Lines which originate above the ground state generally show weaker signals since the strength of the absorption is directly dependent on the population of the low lying state. The magnitude of the absorption is governed by both the number of atoms within the optical path and the population of the absorbing state for those atoms. Generally 99+% of the atoms will exist in the lowest excited electronic state (i.e. ground state) even at temperatures of 3000 K. Any of the higher lying states will be less populated and vary with temperature as dictated by the Boltzmann distribution. Hence, if absorbance measurements are being taken as the temperature within the furnace is changing, the atom density is not strictly proportional to the absorbance since the population of the originating state is changing.

For references see page 9467

It is possible that the absorbance measurement can be too large, that is it lies outside the dynamic range. When a given analyte concentration produces too large an absorbance signal, the sample can be diluted or an alternate absorbing line can be used. An alternate line that originates in the ground state but terminates in an upper state with a smaller oscillator strength is preferred since it will show minimal temperature dependence in comparison to an absorption that originates in a state above the ground state.

In general, the absorbance signal is proportional to the gaseous analyte mass within the furnace volume at any given instant in time. If the rate of analyte release into the gas phase is constant from standard-to-sample and from sample-to-sample, then the height of the absorbance signal ("peak height absorbance") could be used to construct a calibration curve. However, the peak shapes may change because of changes in vaporization rates caused by altered ETA heating rates, changes in vaporization mechanisms, etc. Therefore, the peak area is generally used in constructing the calibration curve and results in improved analytical accuracy. The peak area is calculated by integrating the absorbance signal over the lifetime of the peak and is expressed as peak area and has units of "per second" (since absorbance is dimensionless).

As with all atomic absorbance techniques, the linearity of the calibration curve is limited and covers only two to three orders of magnitude in concentration. The cause of the nonlinearity and apparent violation of Beer's law, can be a consequence of stray light,[11,12] the finite width of the source line relative to the absorbing line[13] or other more subtle spectral line shape alterations[14] and spatial distributions of source radiation and analyte vapor.[15] Lasers (e.g. diode lasers[16,17]) can be more intense with less noise than conventional HCLs or electrodeless discharge lamps (EDLs). As a result, the lower concentration end of calibration curves may be extended, which results in improved detection limits and an enhancement in the dynamic range of ETAAS.

Like any AA technique, the likelihood of atomic spectral line interferences is low. This is a result of the narrow line widths of both the source and the absorber. Additionally, unlike emission spectroscopy, the number of atomic spectral lines that are capable of producing a significant absorbance signal is quite small. This is primarily a result of the need for a significant population of the originating state.

For any given radiation source (e.g. HCL), ETA is considered 100 to 1000 times more sensitive than FAA when comparing characteristic masses, that is the mass deposited in the furnace that is needed to produce a peak area of 0.0044 s. While some gain in sensitivity is garnered by the narrower absorbing line profile for atoms present in the furnace, most of the improvement results from the longer residence times of the atoms within the radiation beam with ETAAS. In a typical flame with a linear flame velocity of ca. $100\,cm\,s^{-1}$, the analyte atoms reside in the optical path for less than 10 ms. Atom residence times within the furnace are several tenths of a second to several seconds!

3.2 Production of the Atomic Absorption Signal

The typical sequence of events in the practical analysis of a sample involves the introduction of a sample (typically 10–25 µL of a solution) into the furnace through the dosing hole. A gentle dry cycle is initiated by heating the furnace to ca. 80–150 °C for 30–40 s to desolvate the sample. The proper choice of time and temperature is generally selected by viewing the sample using an inspection mirror (e.g. a small "dentist's mirror") through the furnace while drying to ensure that the droplet disappears completely and slowly, that is, no "splattering" from heating too rapidly. One or more thermal pretreatment cycles (sometimes referred to as an "ash", "char", or "pyrolysis") are used to decompose some of the matrix and possibly to even remove the matrix if it is more volatile than the analyte. The temperature and times for this step vary considerably depending on the analyte, sample and matrix. For example, a simple sample containing a volatile element such as Cd, Zn or Pb may be heated to no more than 300–400 °C while a more refractory element such as Ni might be heated to 1200 °C or higher without fear of analyte loss. Finally, the atomization cycle is initiated where the furnace is rapidly heated (e.g. $1000–2000\,°C\,s^{-1}$) to a final temperature that is sufficient to remove all of the analyte from the furnace (e.g. 2000–2700 °C). It is during this cycle that the absorbance is recorded. Following the ramp, the furnace is often held for a few seconds at this elevated temperature. A clean cycle (2800–2900 °C) is sometimes initiated before cooling down to ensure complete vaporization and removal of any residuals from the sample.

In the following sections are discussions of the processes occurring during these heating cycles that ultimately lead to the generation of the transient absorbance signal from which quantitative analysis will be conducted. The discussion will begin with a look at the last cycle first (namely, the atomization cycle) and discuss the key steps leading to the production of gaseous atoms within the furnace ETA.

3.3 Basic Vaporization (Desorption) Process

The basic analytical signal is shown in Figure 5. As one might expect it is transient in character because the vapor generated by vaporization is not "trapped" in the furnace but is lost through the dosing hole and the ends of the furnace. There are some characteristics of the peak

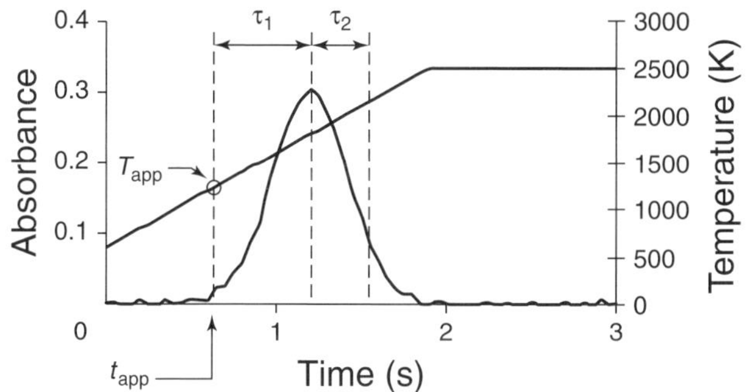

Figure 5 Typical analytical absorbance signal. The width of the peak can vary depending on the analyte, heating rate and furnace geometry. The appearance temperature is determined as the temperature or time where the analytical signal reaches a value three times the standard deviation of the baseline noise level. The time periods designated as τ_1 and τ_2 are times required to reach the peak absorbance signal and for the signal to decay to 1/e of its peak maximum, respectively.

that are often alluded to and these are labeled in the diagram. The appearance time and temperature of the signal are t_{app} and T_{app} respectively. This is typically the time or temperature when the signal level has reached a magnitude that is three times the baseline noise level. The values are not thermodynamic in their origin, but do give an indication of when a signal will first appear and provide guidance regarding the atomization temperature that should be used. The time designations of τ_1 and τ_2 provide some indication of the shape of the signal and have their origins in the original papers of L'vov who made the first attempts to explain the atom production process. τ_1 is the "atomization time" and is measured from t_{app} to the peak time. τ_2 is the time required for the peak absorbance to decay to 1/e of its value. In the initial treatment, L'vov suggested that minimizing τ_1 and maximizing τ_2 would place the maximum number of analyte atoms in the furnace volume and in the analytical volume. He proposed that rapid heating combined with a longer furnace would favor the total containment of the sample vapor in the furnace during the measurement period. Since this earliest work, we find now that the processes are somewhat more complex in many instances.

The mathematics describing the production of the analytical signal is a convolution of a time-dependent supply function, $S(t)$, and a removal (or dissipation) function, $R(t)$, as shown in Equation (1):

$$N(t) = \int_0^\infty S(t) - R(t)\,\partial t \qquad (1)$$

where $N(t)$ is the number of atoms in the analytical volume, that is, the furnace. $R(t)$ must incorporate loss by: diffusion out of the ends of the furnace and the dosing hole; reactions in the gas phase that form analyte-containing molecules; readsorption, condensation, or sticking of the analyte to the furnace wall; nucleation in the gas phase to form condensed phase particles; and any other process which depletes the free atom density within the volume where the absorption measurement is made. $S(t)$ includes the time-dependent production of gaseous analyte atoms by: vaporization (or desorption) from the surface; dissociation of analyte-containing molecules in the gas phase or on the surface (followed by metal vaporization); and any other source responsible for adding more atomic analyte to the gas phase.

3.3.1 Generation

Since the furnace is generally being heated while the transient absorption signal is produced, generation and loss functions must also account for the time-dependent temperature change, that is, nonisothermality of the furnace in time. Additionally, the furnace and the gas phase within the furnace may not be spatially isothermal, which must also be considered for an exact description of the shape of the signal. Thus, it is easy to see that even if the exact processes that contributed to $N(t)$ were known, the solution to Equation (1) could be quite complex. Using overly simplified assumptions, however, does produce an $N(t)$ versus t plot that is a reasonable approximation to what is observed. For example, one can assume that the number of gaseous atoms is proportional to the partial pressure of the analyte and assume that this pressure is dictated by equilibrium vapor pressures of the heated metal solid or liquid. Additionally, one can consider loss as a simple diffusive transport of material from the center of the furnace out the ends and ignore the dosing hole and any other loss pathway.

Since the number of analyte atoms on the surface is quite small, in many instances the analyte may not take on "bulk" thermodynamic characteristics. For example, if 20 pg of copper were evenly dispersed over a 2-mm-diameter area where the sample was initially dried, only about 5% of the surface would be covered. Thus, the energy required to produce copper vapor may not depend on the heat of vaporization (ΔH_v) of $Cu_{(l)}$ but instead the energy needed to desorb adsorbed copper from graphite. In some instances where the interatomic forces are strong, it is possible that the metals on the surface diffuse and form small droplets of the "pure metal" in which case ΔH_v is applicable to describe vapor production.

The mechanism of vapor production in an ETA has been an intense area of research interest for a number of years and is still not totally understood. One of the more interesting diagnostics that is often used attempts to extract the energy associated with the vaporization process and, by inference, to deduce the pathway by

For references see page 9467

comparing the experimental value to literature values. The commonly cited approach was proposed by Sturgeon et al.[18] and the resultant graph is often referred to as an "Arrhenius plot", because its derivation is similar to that used in deducing reaction kinetics. In brief, it may be given by Equation (2):

$$\ln(A) = \frac{-E_a}{kT} + C \qquad (2)$$

where A is the absorbance signal, E_a is the activation energy of release (often taken as ΔH), k is the Boltzmann's constant and T is the furnace temperature in Kelvin at the time the absorbance measurement is made. C is often assumed to be a constant and incorporates a number of terms including spectroscopic constants that relate an absorbance measurement to atom density (or pressure) as well as thermodynamic (or kinetic) values that are assumed to be temperature independent. In particular entropy change (or the pre-exponential factor when viewing the process as a kinetic process) is involved in this term.

Since the derivation of Equation (2) assumes that the analyte surface coverage (or condensed state activity) is not changing, the equation is most applicable (and most linear) at the very beginning of the absorbance signal. Smets[19] added a modification to this basic equation that attempted to account for changing surface coverage and extends the linear portion of this plot later in time. The temperature is usually measured using optical pyrometry by viewing the inside of the furnace wall through the dosing hole. This approach permits coverage starting at ca. 1100 K. Once a value for E_a (or ΔH) is obtained, interpretation to extract the vaporization step can often be tenuous.

It is also possible that gaseous atoms originate directly from molecular species. This includes thermal decomposition of metal-containing compounds in the gas phase or on the surface. In some instances, the surface reaction can be thought of as a reduction or "thermoreduction" process in which reactive carbon of the furnace participates in the free atom formation as the temperature increases, for example, reaction of a metal oxide with C to form free gaseous metal atoms and CO or CO_2.

3.3.2 Dissipation

Analyte loss from the analytical volume of the ETA can take a number of forms, but in all cases the loss must be irreversible. For example, metal gas colliding with the wall of the ETA and sticking but revaporizing at a later time during the measurement cycle would not be considered "irreversible loss". The simplest loss or dissipation is diffusion of analyte from the ends of the furnace or out of the dosing hole.

A simple equation, Equation (3), that considers the continuous diffusive loss from the furnace ends was proposed by L'vov

$$\tau_2 = \frac{L^2}{8D} \qquad (3)$$

where τ_2 is the time required for the absorbance signal to decrease to 1/e of its peak value, L is the furnace length and D is the temperature-dependent diffusion coefficient. This provides a reasonable approximation to the loss rate by diffusion, but does not include loss from the dosing hole. It also ignores the finite time at the onset of vaporization for material to diffuse from the center to the ends of the furnace to set-up a pseudo-steady-state concentration. It also does not take into account the more subtle loss mechanisms of recombination of the analyte metal to form nonabsorbing molecular species.

3.3.3 Putting it Together

Use of the more complex geometry of the furnace and the presence of a small, finite amount of sample makes the solution to Equation (1) more complicated. One approach to modeling the more realistic system has employed the Monte Carlo technique.[20] This is a stochastic approach that permits complex geometries and boundary conditions to be used in the construction of a model. Figure 6 shows the results using Monte Carlo simulation techniques to illustrate the time-dependent location and loss of analyte during a furnace heating cycle. This simulation is for the production of copper vapor. As can be seen, the leading edge of the gas phase signal rises

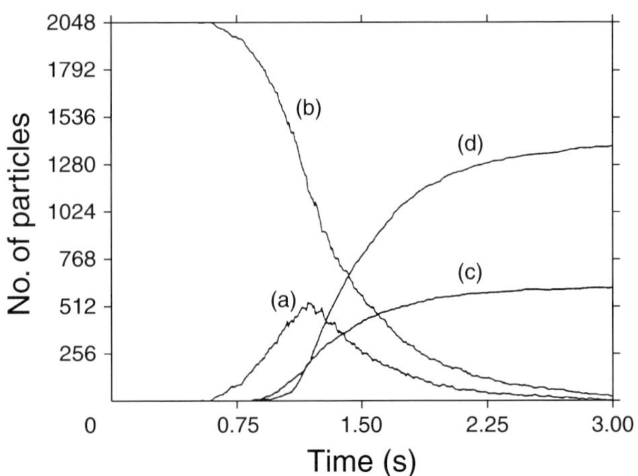

Figure 6 Monte Carlo simulation showing the time-dependent location of analyte atoms during a thermal heating cycle: (a) in the furnace gas phase; (b) on the furnace wall; (c) lost out of the dosing hole; and (d) lost out of the ends of the furnace. Data are for a small (3 mm diameter × 1 cm long) furnace. (Reproduced from Güell and Holcombe[20] by permission of the American Chemical Society.)

exponentially as the temperature ramp increases linearly. As the amount of copper on the furnace wall decreases, the generation rate falls off. The peak of the absorbance signal represents the point where generation and loss rates are equal, that is, $S(t) = R(t)$. It should be noted that a significant (ca. 20% in this case) amount of the analyte is lost from the dosing hole and that analyte loss occurs from the dosing hole before it is lost from the furnace ends. The decrease in the atoms found on the tube wall also does not conform to a simple exponential evolution as might be surmised from a linear temperature ramp and the assumption of "bulk amounts" of metal on the surface (i.e. activity of metal is unity). In the case of copper, this is a result of the release rate being dependent on the amount of copper (i.e. surface coverage) on the graphite, which changes with time (i.e. its activity decreases as material is vaporized). As the surface coverage drops, the rate of evolution decreases below that expected when surface coverage is not taken into account.

3.4 Preatomization Reactions

Many reactions occur on the surface prior to atom production. These can be as simple as prevolatilization of metal compounds and migration of material below the surface. Some processes only impact the appearance temperature and the shape of the analytical signal while others can affect the number of free atoms produced. The latter, of course, directly impacts analytical accuracy while the former may have little impact on accuracy when using the area under the absorbance signal. Most of these problems can be circumvented through the proper selection of temperature programming, a platform and chemical modifier.

Analyte prevolatilization involves the low temperature release of analyte-containing molecules at a temperature that is lower than that needed to thermally decompose these compounds. As a result, the absorbance signal is depressed. This is not uncommon for metal halides which are often relatively volatile but form stable metal halides in the gas phase. Some elements (e.g. selenium) also have volatile oxides that can also be lost from the furnace.

3.4.1 Decomposition

The form of the analyte salt residue that exists on the graphite surface after desolvation ("dry cycle") can often be predicted by the dominant matrix anions and relative solubilities of the possible salts that could form. For example, the metal nitrate salt is likely in a nitric acid matrix, metal halides in a HCl matrix, etc. Depending on the solution pH, metal hydroxides are not unexpected. During the dry or char cycle, many of these salts decompose to the metal or metal oxides as shown in Equations (4) and (5):

$$M(NO_3)_{2(s)} \longrightarrow MO_{(s)} + 2NO_2 + \tfrac{1}{2}O_2 \qquad (4)$$

$$M(OH)_{2(s)} \longrightarrow MO_{(s)} + H_2O \qquad (5)$$

Easily reducible metals (e.g. Au, Ag, Cu, etc.) are likely present on the surface in the elemental form before atomization. There is little doubt that the graphite also serves as a reducing agent to assist this process for a number of metals as the furnace is heated.

Graphite also serves as a potentially problematic reactant as well as a reductant for elemental formation. Many metals bind very strongly to carbon to form stable surface species or even stoichiometric metal carbide compounds. These species are often very refractory and can be difficult to vaporize completely unless very high atomization temperatures are employed. Examples of these metals include W, Mo, Ta and Si. These stable carbides often precluded the use of platform technology (see section 4.4) since the platform may be incapable of reaching the wall temperature even with sustained heating times. Several workers have successfully used metal liners (e.g. Ta, W, etc.) placed inside the graphite furnace to circumvent this problem. However, metal liners cannot be permanently used because they limit the final atomization temperature to values substantially lower than that for graphite and can react with some analyte metals to form relatively stable solid solutions (e.g. alloys).

While the pyrolytic coating used on most commercial furnaces greatly reduces permeation of the sample liquid and gaseous vapor through the furnace walls, it is not single crystalline in character and has a considerable number of surface imperfections, dislocations and "cracks" in the surface. These features permit samples to migrate below the surface, perhaps as early as during the drying cycle and often as far as several micrometers.[21,22] Majidi et al. have also shown that all metals do not behave the same. In many instances it is thought that the analyte forms intercalation compounds that are located in the interlamellar space between the graphitic sheets.

When metals that form intercalation compounds are present in large amounts such as quantities that might be present in a matrix or used in a chemical modifier, extensive damage to the furnace can occur. This is a result of these compounds causing an expansion in the interplanar spacing of the graphite which ultimately causes delamination and exfoliation of the material. The extreme result can include bending/warping of platforms and destruction of the furnace.

It should be noted in closing that it is not uncommon for materials to migrate on surfaces. Depending on the forces of attraction present, it is possible that adsorbed atoms, analyte-containing molecules and even small particles can move on the surface by surface diffusion. This diffusion

For references see page 9467

can bring about the formation of small microdroplets of metals even if the metals were initially quite dispersed on the surface. Similarly, metals present as small droplets or islands initially may disperse on the surface as the temperature increases and entropy provides a driving force.

In some instances the appearance of migration occurs as material vaporizes (or desorbs) and is readsorbed at some distance away from the original site. This process occurs quite readily and can precede the appearance of the absorbance signal by several hundred degrees. The net result is a picture of the release of atoms into the gas phase that is not as simple as "vaporize, enter the gas phase and diffuse from the ETA". With a mean free path of fractions of millimeters, there is ample opportunity for multiple collisions with the surface and readsorption at a measurable distance from the original sample deposition site.

3.4.2 Graphite–Oxygen Reaction

The importance of oxygen in forming metal oxides on the surface and in the gas phase has made studies of this "interferent" of special importance. Sources of oxygen include decomposition of oxyanion salts, the sample solvent (water), impurities in the sheath gas and diffusion of atmospheric oxygen into the furnace through the dosing hole. Reduction of oxygen by the graphite furnace is thermodynamically favorable at all temperatures and should produce CO and CO_2, with the $CO:CO_2$ ratio increasing at elevated temperatures if equilibrium is achieved. However, Sturgeon et al.(23) were the first to suggest that this equilibrium was not likely achieved in the ETA until temperatures exceeded ca. 1500 K. This was later confirmed and presented in more detailed models.(24) Sturgeon and Falk(25) were the first to present experimental evidence of the partial pressures of oxygen in the furnace which was not based on inferences from the metal AA signal. Interestingly, the values agreed well with values chosen earlier by Cedergren et al.(26) in their thermodynamic modeling of reactions within the furnace. In their studies they selected values that produced reasonable appearance temperatures for the metals under study.

As seen in the work of Gilmutdinov et al.(24) the reactions of oxygen with the graphite present a complex time and spatial picture within the furnace. However, when combined with the authors' spatial viewing of free atom densities within the ETV,(27) it explains many of the unusual distributions that were observed. For example, the disappearance of many AA signals near the top of the furnace could be attributed to the infusion of air into the furnace through the dosing hole. Similarly, with the platform in place, some metals showed a particularly strong absorbance signal under the platform, which is consistent with a rarified oxygen atmosphere that would favor metal oxide dissociation.

In some analytical situations, oxygen is intentionally added to the sheath gas during the dry and thermal pretreatment stage. This is referred to as "oxygen ashing" and is most commonly done when there is a significant amount of a combustible material in the sample, e.g. a biological or low volatility organic matrix. Pyrolysis (i.e. heating to high temperatures in the absence of oxygen) of a biological matrix, for example, can leave copious amounts of carbon in the furnace even after the atomization cycle. Using oxygen admixed with the Ar, this matrix undergoes combustion and consumes the carbon as CO and CO_2. It is obviously critical that the oxygen be removed from the furnace if it is heated above ca. 800 °C since oxidation (and destruction!) of the graphite atomizer will occur.

3.4.3 Gas-phase Reactions

The presence of gas-phase reactions with the analyte will effectively reduce the free atom density and produce a signal depression. The depressed signal is of analytical importance if the source of the reacting species originates in the sample and is not present in the standard solutions. In many instances, standard addition is still capable of producing accurate results under these circumstances. The gas-phase reactions that are significant are, of course, metal-dependent, but as a general guide one must be cautious of halides and oxides, both of which can form stable gas-phase diatomic species with many metals. In most instances, dissociation of these diatomic species is favored at elevated temperatures. As a consequence platform technology is often recommended. Another alternative is to "thermally stabilize" the analyte with the use of a matrix modifier. This accomplishes two objectives. By delaying the analyte vaporization until a higher temperature, dissociation of analyte molecules on the surface or in the gas phase is enhanced. Additionally, if the analyte is not released until a higher temperature, then thermal pretreatment can be conducted at a higher temperature without fear of analyte loss. In many cases, this may permit the removal of the problematic matrix during this thermal stage prior to the atomization heating cycle.

The other option to reduce gas-phase interferences is to minimize the concentration of interferents in the sample through the use of matrix modifiers. As an example, the simple addition of HNO_3 to a chloride-containing sample results in a significant loss of the chloride as HCl during the dry cycle. Ammonium salts (e.g. $NH_4H_2PO_4$) are also used since the resulting $NH_4Cl_{(s)}$ is relatively volatile and provides a pathway to remove the chloride at relatively low thermal pretreatment temperatures.

4 INSTRUMENTATION

Modern ETAAS instrumentation includes a light source, background correction, the atomizer, the spectrometer and signal processing. Accessories would include autosampling systems and other attachments that maximize the performance and flexibility of ETAAS in the analytical laboratory. While other articles also discuss some aspects of the instrumentation (**Atomic Spectroscopy: Introduction; Background Correction Methods in Atomic Absorption Spectroscopy; Flow Injection Analysis Techniques in Atomic Spectroscopy**), this section will highlight aspects specific to ETAAS and provide some indication of current instrumentation as well as approaches that are on the horizon.

4.1 Light Sources

The traditional light source is the HCL which is a glow discharge source whose cathode is fabricated from the analyte metal(s) of interest and is contained within a sealed envelope surrounded by a few torr of an inert gas. The choice of fill gas is element-dependent. The glow (or Grimm) discharge is a low current (i.e. few milliamps) source that relies on sputtering and electron impact excitation for sampling and excitation, respectively. The source is not under local thermodynamic equilibrium and exhibits a very low kinetic temperature but a large electron temperature. As a consequence, a large number of elemental lines are emitted but they have a very narrow half width, an ideal situation as a source for AA. Generally, a lamp provides the spectra for a single element. While multielement lamps are available, they usually provide reduced intensity for the elements within the lamp due to compromised excitation conditions and reduced sputtering efficiency because of fractional composition of the cathode material.

When low intensity from the HCL at the preferred wavelength presents a relatively high noise level, EDLs often become the source of choice. These devices are radiofrequency or microwave excited discharges which generally emit a more intense spectra. This is particularly useful for elements whose resonance lines are at shorter wavelengths such as As and Se.

More recently, tunable diode lasers have emerged as replacements for HCLs or EDLs. Because of their improved stability, narrow line width and reduced noise, diode laser sources have been used to record absorbance signals as low as 1×10^{-6} absorbance units. As a consequence, larger dynamic ranges and significantly reduced detection limits appear possible. A commercial source of these lamps is currently available. Widespread use may be contingent on the development of diode lasers with fundamental wavelengths deeper into the blue region of the spectra so that simple frequency doubling is only needed to reach the short wavelengths (ca. 180–200 nm) where the resonance lines of many elements of interest lie.

A continuum source with a high resolution spectrometer has also been explored. Uniquely, it offers the possibility of simultaneous multielement analysis for ETAAS. However, the weak intensity of the source at low wavelengths has traditionally made this source less sensitive than traditional HCLs or EDLs, especially at shorter wavelengths. Consequently, its use has been primarily confined to the research laboratory. Recent work with diode array detectors suggests that comparable sensitivities are available even at the shorter wavelengths. This approach may become another viable alternative to traditional line sources.

4.2 Background Correction

This topic has been discussed in greater detail in the article **Background Correction Methods in Atomic Absorption Spectroscopy** in this publication. In general, background correction is much more critical in ETAAS than in FAA because of the higher local density of sample vaporized and the presence of cooler areas near the dosing hole and furnace ends where molecular species may form and/or condensation may take place. Additionally, ETAAS is often employed with samples that have a very high concentration of matrix components, with the extreme case of using the method for direct solids analysis.

4.3 Atomizer

The atomizer has been discussed in general detail earlier in this chapter. In most cases the furnaces are fabricated from a high density "amorphous graphite", often referred to as "electrographite". This material is made from graphitized carbon at elevated temperatures but the crystals in the material are small and their orientation is random. While the density of the material is high (i.e. low porosity), it is not as great as pure graphite. This material is then coated with a 50 to 100-μm thick coating of "pyrolytic graphite" which is laid down by chemical vapor deposition techniques at elevated temperatures using a hydrocarbon gas as the source. (The temperatures and gas composition are usually proprietary information.) The coating makes the surface less permeable to diffusion by sample liquids and gaseous vapors. The coating is also more crystalline in nature with general orientation of the graphitic planes parallel to the surface. The reactivity of the coating is generally significantly less than the electrographite substrate.

Most furnaces are of a Massmann-type design with power supplied to the furnace via contacts at the tube ends. As noted earlier, Frech et al.[8] proposed an "isothermal cuvette" to minimize a temperature gradient that

For references see page 9467

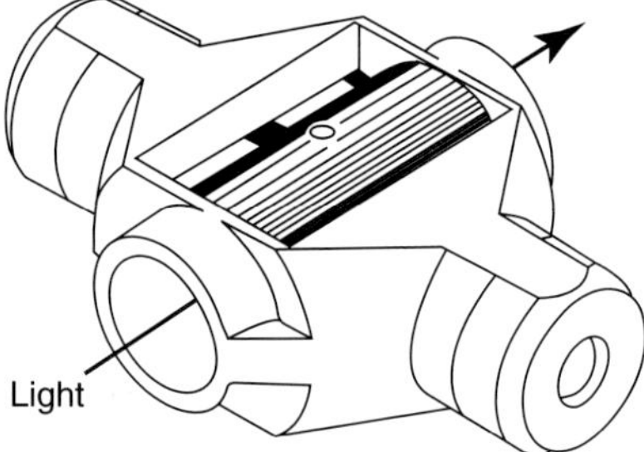

Figure 7 Transversely heated graphite atomizer where the current to heat the furnace is not passed axially through the furnace from the ends. The design is intended to eliminate thermal gradients along the length of the tube to minimize cool spots in the tube where condensation refractory metals and/or molecular formation might occur. This particular design also has a platform machined into the furnace. (Reproduced by permission of PerkinElmer Instruments.)

is known to persist from center-to-end in the more traditional furnace. A modification of the concept is commercially available as a "transversely heated graphite atomizer". Figure 7 shows an example of these two furnace types. As noted in earlier sections, platforms or probes are a routine part of modern analysis by ETAAS. Depending on the design and intimacy of the contact between the platform and the tube wall, refractory metals may or may not be vaporized completely from the platform. In those designs where the platform is incapable of reaching the needed temperature, wall vaporization is used.

More recently, the addition to the ends of the furnaces of circular caps with holes in the centers has been suggested.[28] These caps would reduce the rate of diffusive loss of analyte from the furnace and provide a more isothermal environment that may enhance sensitivity and reduce interferences. An even more dramatic change in furnace design has been proposed by Katskov et al.[29] who use a "filter furnace". In brief, the sample is deposited in a small chamber between the conventional filter wall and an insert within the tube that is fabricated from porous graphite. Upon heating, the sample diffuses through the hot porous graphite into the analytical volume. Results suggest fewer interferences while maintaining good sensitivity for the limited number of elements that they have considered.

4.4 Platforms and Probes

As was discussed at the start of this chapter and shown in Figure 4, the use of inserts (e.g. platforms or probes) is a common part of modern use of ETAAS. Platforms that have a more intimate thermal contact with the furnace wall will exhibit less of a thermal lag and may not prove as effective in dissociation of analyte-containing gas-phase molecules. However, this type of platform can be used for most analytes, including the more refractory elements.

4.5 Spectrometer and Optics

In general, the spectrometer is a medium or low resolution monochromator since it is only needed to isolate the resonance line of interest from other lines of the same element emitted from the line source, typically an HCL. Spectral bandpasses in the range of 0.1–5 nm are generally all that are required. The exception to this occurs when a continuum source is being considered in place of the HCL or other line sources. In this instance, a high throughput, high resolution spectrometer is preferred. To this end, an echelle spectrometer used in very high orders is the spectrometer of choice.

There are instruments which use several line sources which are alternately passed through the ETA and detection is made at several wavelengths. In this fashion the ETAAS system becomes capable of multielement analysis for a limited (four to eight) number of elements with a single firing of the atomizer.

Most systems are operated in a double beam mode where a chopper is used to alternately pass the source radiation through and around the ETA. In the case of the pulsed background correction, chopping is not needed since the same source provides on-line and off-line measurements (see **Background Correction Methods in Atomic Absorption Spectroscopy**). The path around the ETA is used primarily to correct for any drift in the source intensity. Additionally, the source may be modulated between the line source and the source used to register a background correction signal. For example, if an HCL is used with a D_2 continuum source, the signals measured are typically: (i) the line source through the ETA measuring line and background absorbance; (ii) the continuum source through the ETA measuring background equivalent absorbance (and scatter); (iii) the line source around the ETA; and (iv) the continuum around the ETA to use as a measure of source stability and to register a zero absorbance level. Additionally, there is a fraction of the measurement cycle when no source radiation passes. During this period any emission from the furnace (blackbody, scatter, line or molecular) is registered and corrected for.

A photomultiplier tube is employed as the detector in most cases because of its low noise and high sensitivity. Solid-state detectors and array-type detectors may find greater use in the future but are minimally used in current AA systems.

4.6 Signal Processing

Because of the transient nature of the ETAA signal, faster detection and processing is required than might be needed when the flame is used for atomization. Millisecond response times are often needed from the system to accurately reflect the transient signal. Collection rates of 20–120 Hz are typical for most modern instruments, yielding more than 100 measurement points to define a typical ETA signal of ca. 5 s duration. While faster modulation and collection could be used, the added noise from the more rapid collection generally does not compensate for the improved time resolution.

Most instruments provide the user with output of the time varying background-corrected atomic signal as well as the background signal. This can be useful in making an evaluation of whether the background level is of sufficient level that one should question the accuracy of the correction made by the instrument. Again, the article **Background Correction Methods in Atomic Absorption Spectroscopy** discusses the use and interpretation of the background signals.

Output is generally sent to a computer or "data station" where elaborate work up of the data takes place. Most instruments with a computer interface and high level software package permit the user to specify a number of analytical options, for example number and concentration of standards, number of replicates, options for standard additions and frequency of QA/QC checks. Additionally, most have capabilities of adding matrix modifiers from a separate reservoir to each standard and sample. Most software packages also provide complete reports on the analysis including information on the analytical precision as well as the accuracy and precision of the calibration curve. Additionally, these same software packages furnish full statistical analysis of the results and permit use of such analytical tools as standard additions.

4.7 Autosamplers and Other Accessories

ETAAS can be highly automated. The key ingredient to such automation is the use of an autosampler for sample introduction. The autosampler (under computer control) not only frees up the operator, but it also provides one of the most precise means of micropipetting the sample into the furnace (i.e. $<\pm3\%$). Many modern systems permit user input to the autosampler regarding the number of elements to be determined and number of replicates in each sample. The instrument then analyzes each sample and controls the spectrometer wavelength drive and source selector to cover all the elements of interest. In such an instrument, of course, the spectrometer must be set-up with a multilamp turret so that the lamps can be changed as one moves from sample to sample. Because of the improved precision and flexibility provided by the autosampler, most laboratories consider it as an essential part of a modern ETAAS system.

Other analytical schemes including flow injection analysis (see **Flow Injection Analysis Techniques in Atomic Spectroscopy**), hydride generation techniques (see **Flame and Vapor Generation Atomic Absorption Spectrometry**), slurries (see **Slurry Sampling Graphite Furnace Atomic Absorption Spectrometry in Environmental Analyses**) and solids directly are all available for use with ETAAS. In all the cases cited above, including direct solids analysis, high degrees of automation and sample handling are commercially available.

5 MATRIX MODIFIERS

A "matrix modifier" originally referred to a species that was added to the sample to improve the analytical accuracy and/or precision. It could be as simple as the addition of HNO_3 to assist in elimination of chloride as HCl during the dry cycle. The use of modifiers represents a cornerstone in practical utilization of ETAAS for chemical analysis. While there are literally hundreds of papers expounding on the virtues of various modifiers for particular analyte/matrix scenarios, most have been arrived at empirically and in most cases there is a minimal amount known about the actual physico-chemical interactions responsible for the observed improvements. Regardless of this unfortunate fact, many modifiers have sufficient documentation that one can feel comfortable that they will assist – if not ensure – the achievement of reliable analytical results. In all cases there are one or more fundamental functions that the modifier is intended to serve.

Matrix Removal. Some modifiers assist in the prevolatilization of the matrix prior to the appearance temperature of the analyte in the gas phase. The removal may minimize coincidence of analyte and a matrix component in the vapor that would otherwise form a stable analyte-containing molecule and lead to a depressed signal. Examples include HNO_3 or NH_4^+ for halide removal or use of oxygen ashing for combustion of biological compounds.

Analyte "Stabilization". By forming a more stable condensed phase species on the atomizer surface, higher thermal pretreatment temperatures may be used to remove unwanted matrix components. Examples include the addition of phosphate salts or the use of Pd (usually as the nitrate) to the sample and standards.

One of the more commonly used modifiers is Pd (usually as the nitrate salt). This appears to serve as a stabilizer for a number of metals. Obviously, it should not "stabilize" all species or it really serves no useful purpose except to delay vaporization of every sample component. The inclusion of $Mg(NO_3)_2$ in this solution has also been suggested to assist in the performance of the Pd modifier by acting as a "bulking agent".

For references see page 9467

Surface Modification. A number of studies have modified the graphite surface in an attempt to enhance the atomization conditions of various analyte metals. Surface modification often involves the addition of large amounts of a metal salt (as a solution) to the furnace followed by thermal pretreatment to form, for example, a metal carbide coating on the surface. Metals have also been sputtered onto the graphite surface to accomplish the same purpose. This modified surface has been reported to improve atomization efficiencies and provide immunity to matrix problems. Metals that have been used include Ta and Ir. In contrast to the use of more conventional modifiers, these surface modification techniques need not be done prior to each firing of the atomizer.(30)

There are certainly a large number of modifiers that have been used to satisfy certain analyte/matrix combinations. Some modifiers appear to be useful only for selected analyte/matrix combinations (e.g. Ni for semimetals) while others (e.g. Pd) have a broader application base. Tsalev and Slaveykova(31) used multivariate techniques in an attempt to place some order in those that are reported to have worked. There is little doubt that many can be understood based on their chemical properties, but the mechanism for others remains elusive.

6 PRACTICAL APPLICATION OF GRAPHITE FURNACE ATOMIC ABSORPTION SPECTROMETRY TO ANALYSIS

Most instruments come equipped with a methods manual or "cookbook", which provides information on analyte-specific determinations. In many instances, procedures are often included for specific matrices; and in the newer instruments, this manual is on-line, that is, part of the data station or computer system that operates the system. Additionally, there are a large number of method development papers in the literature discussing specific analyte determinations in specific matrices that can be employed. These procedures can be used as an excellent starting point to initiate the analysis. This encyclopedia also presents some key analytical areas where ETAAS (also known as GFAAS) are employed (see **Slurry Sampling Graphite Furnace Atomic Absorption Spectrometry in Environmental Analyses; Atomic Spectroscopy: Introduction; Background Correction Methods in Atomic Absorption Spectroscopy; Flame and Vapor Generation Atomic Absorption Spectrometry;** and **Flow Injection Analysis Techniques in Atomic Spectroscopy**.)

As applied and fundamental research progressed, several instrumentation and methodological characteristics were generally acknowledged as adding significantly to analytical accuracy and precision. Slavin et al.(32) attached the label of 'stabilized temperature platform furnace' conditions to these features. The embodiment of the general approach advocated the use of pyrocoated furnaces with a platform, rapid heating, fast responding digital electronics to accurately capture the transient signal, peak area (in place of peak height) for quantitation, Zeeman-effect background correction, and chemical modifiers. Some debate may persist regarding the universal applicability for all of these guidelines; but they are, in general, quite sound.

To follow is a generic approach to the set-up of a new method.

6.1 Setting Up Temperature Programs

6.1.1 Sample Deposition and Dry Cycle

While depositing the sample, one can look through the end of the furnace toward the light source and see a shadow of the autosampler tubing that can be manually positioned in its deposition position. The tube should not make contact with the dosing hole sides, graphite wall or platform. When the droplet is deposited it should be "pulled" away from the tubing so that the tubing is not resting in the liquid after the sample is dispensed. An angled cut on the tubing often assists in improved sample deposition. The drying temperature should be set to a ramp of 5–10 s duration to ca. 80 °C (10–20 °C higher with platform) to begin and held at this temperature for 20–30 s. The criterion for adequate drying is to see the droplet gradually disappear over a period of 5–10 s with no noticeable boiling or bumping. Variations in temperature and time can be adjusted to achieve this. Drying is one of the more important steps since improper drying will lead to accuracy and precision problems as well as tube degradation. Bearing in mind the temperature setting and display on most instruments is not measured during the cycle, the error could be several tens of degrees at the low temperature settings.

6.1.2 Thermal Pretreatment ("Char")

It is almost always advisable to have a thermal pretreatment step if for no other reason than to eliminate any residual water and acids as well as some of the volatile matrix components in the sample. Additionally, this step is useful in decomposing many of the inorganic salts, for example, nitrates.

It is most critical that the analyte is not lost during this step. To evaluate this, "char curves" can be constructed. Simply set the dry to the predetermined setting and the atomization to a sufficiently high value to feel confident in achieving atomization, for example from standard procedure. Next, make a series of firings using your sample where the thermal pretreatment temperature

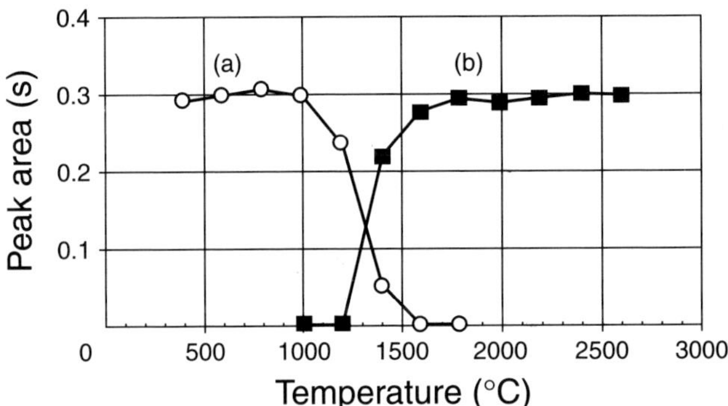

Figure 8 (a) Char and (b) atomization curves collected for the purpose of determining the correct thermal pretreatment (i.e. "char") and atomization temperature settings.

is changed by ca. 100 °C but the time program is kept constant. A typical program might be a 5 s ramp to the temperature and hold for 20 s. The absorbance peak area is then recorded for each char temperature used. A curve similar to that shown in Figure 8(a) will result. The highest temperature is then selected where there is no attenuation of the peak area for the thermal pretreatment, ∼900–1000 °C in this case.

6.1.3 Atomization Step

While a small gas flow through the atomizer is generally employed during the dry and thermal pretreatment steps, the atomization step is generally used in a "gas stop" mode (i.e. no flow through the furnace) to maximize residence time and to ensure maximum gas-phase temperatures.

The temperature setting for this step is critical. Too low an atomization temperature can produce a weak signal with memory effects while too high a temperature can cause excessive degradation of the furnace. Signs of too low a temperature include a badly tailing absorption peak. In most instances "step heating" (0 s ramp) is used for the atomization cycle. If a platform is used it is best to cool the atomizer to ca. 100 °C before initiating the atomization cycle. This provides the maximum time delay in the heating of the platform and should provide optimal results. After the ramp, the atomization temperature is usually held for ca. 5 s, sometimes longer if a platform is used and there is difficulty in removing the sample. If possible, it is always good to check the time-dependent temperature and absorbance signals to verify that a sufficient but not excessive hold time is being used.

Similar to the construction of a char curve, an atomization curve can be constructed. Using the previously determined dry and thermal pretreatment values and varying the atomization temperature, and noting the peak area, Figure 8(b) will result. An atomization temperature has to be selected where a minimum temperature yields the maximum area, ∼1900–2000 °C in this case. If the area continues to rise even with the maximum temperature used, it is possible that the analyte is sticking to the graphite. If a platform is being used, you may have to use wall atomization (i.e. remove the platform). If wall atomization is used and the problem seems to persist at the highest temperature available, you can verify complete vaporization by checking for memory effects by running a blank after a sample (or standard solution). The presence of a signal in the blank suggests that the analyte is not being completely removed, assuming that the blank is analyte-free. Some modifier or change in sample preparation may be needed.

6.1.4 Clean Cycle

For some samples a short (5–20 s) clean cycle near the maximum temperature of the atomizer is advisable. This is essential if an analyte signal or background signal persists when a blank is fired immediately after a sample solution. Many analysts use this cycle as a routine part of their thermal program.

It is not uncommon to conduct a quick check of the instrument and its settings at the start of the analysis by computing the experimentally determined characteristic mass, m_o, with that provided by the manufacturer. The characteristic mass is the mass of analyte needed to register a peak area of 0.0044 s. Some manufacturers quote other figures of merit (e.g. sensitivity values), which can be used in a similar fashion to that of characteristic mass. Most instrument manuals or on-line documentation provide these values for a large number of elements, and the value is element and instrument-dependent. As an example, m_o values range from a few tenths of a picogram to several tens of picograms, depending on the element and the instrument. The easiest way to determine m_o is to analyze a standard that provides a peak area of ca. 0.05–0.2. By multiplying the analyte mass in this sample by 0.0044 and dividing by the peak area of the signal, the resulting value will be m_o. While a value checking sensitivity often lies within ±20% of the manufacturer's tabulated value, a value that is no more than about two times the quoted value is reasonable assurance that there is no significant error in the methodology, instrument settings or lamp performance. Values that are much too high suggest that something is not quite right, for example lamp current is too high, temperature program is not correct, etc.

6.2 Sample Types

In most instances, ETAAS is used with 0–50 µL aliquots (20 µL being typical) of aqueous solutions, although

For references see page 9467

analysis of slurries (see **Slurry Sampling Graphite Furnace Atomic Absorption Spectrometry in Environmental Analyses**) and solids are routinely conducted using this technique. Both of these approaches minimize sample handling and minimize or eliminate contamination because they minimize reagents needed to prepare a homogeneous solution sample. As alluded to earlier, gases (e.g. hydrides, as alluded to and discussed in **Flame and Vapor Generation Atomic Absorption Spectrometry**) and even aerosols have been analyzed by direct deposition into the furnace.

6.3 Calibration Curves

When a large number of samples are anticipated, the use of calibration curves is generally preferred over the method of standard additions. The method of standard additions can correct for enhancements or depressions that are multiplicative but cannot correct for additive problems such as background correction errors. It is not uncommon to encounter nonlinear calibration curves and most exhibit a dynamic working range of two to three orders of magnitude. With Zeeman background correction, roll-over may limit this range even more. Like FAA, the nonlinearity in the calibration curves is a result of a combination of factors including the finite width of the absorbing and emitting line, secondary lines within the bandpass of the spectrometer and stray light reaching the detector. It has been suggested that contributions from nonuniform analyte distribution in the furnace combined with nonuniform spatial intensities from the line source may also contribute.[15] Because of the nonlinearity, one must exercise caution in using standard additions since this analytical approach assumes linear extrapolation of the standard additions data to the intercept on the concentration axis. Most commercial software packages permit curve fitting to nonlinear curves, which optimizes the useable dynamic range.

Critical to all analyses by ETAAS is the use of high purity reagents (e.g. acids, water, etc.) because of the sensitivity inherent to the technique. As a standard analytical procedure, it is also advisable to employ standard reference materials (SRMs) as a quality control check during methods development and as a routine part of the analyses.

6.4 Absolute Analysis

The general approach of absorption spectrometry has possibilities of conducting analysis without the use of standards, that is, absolute analysis. As an example, ultraviolet/visible solution spectrophotometry with known solvent and solution environment can be used for analysis working with Beer's law and knowledge of the proportionality between absorbance and concentration, that is, the molar absorptivity. It is acknowledged that there are certain caveats to do this successfully, but it can be done both in theory and practice. The success rests, in part, because absorption spectroscopy is a ratioing technique of the incident intensity with and without an absorber in the light path, that is, transmittance, and concentration is proportional to $\log(1/T)$.

Variables in nebulizer efficiency, flame gas flows (as associated dispersion of analyte in the flame), optical path of source light through the flame, etc., eliminated the application of absolute analysis to FAA. However, L'vov[33] recognized in his earliest work that this may be possible using ETAAS since a fixed volume was deposited in the furnace, atomization efficiency could be very nearly 100% and the full analyte could be retained in the analysis volume. (The latter fact was later modified to account for the continuous loss, and replaced by an assumption that the time spent by the analyte in the optical path could be constant and dependent only on furnace geometry if the analysis was conducted properly.) In essence, the approach suggested that one could calculate the peak area of the absorbance signal from fundamental spectroscopic constants combined with furnace geometry, thereby alleviating the need for preparing standard solutions and calibration curves. The potential of this concept was supported by experimental data from several laboratories.[34,35] Agreement between calculated absorbance values and experimental values were generally within 20% of each other. This should be very encouraging especially when such accuracy may really be all that is needed in many samples where parts per billion levels are being determined. However, the conventional use of calibration curves with ETAAS often yields accuracy and precision values of <5%, which many regulatory agencies and "customers" of the analyst demand. To date, this potentially powerful feature of absolute analysis by ETAAS has not been routinely employed in any major segment of the analytical community.

7 CONCLUSION

Graphite furnace AA exists in the modern trace metal arsenal as a reliable and rugged analytical tool. With modern use of the technique, high precision (3–5% relative standard deviation) and accurate results can be obtained on a routine basis. With absolute LODs in the low picogram to subpicogram range (low parts per billion for relative LODs), it also becomes one of the most sensitive elemental analytical techniques. The ability to handle micro amounts of sample as well as complex samples (e.g. slurries, high dissolved salt matrices and even solids) has made ETAAS a very attractive option when limited sample amounts or complex matrices are encountered.

Like all atomic spectroscopic techniques, the production of free atoms is essential to the analytical success of ETAAS. With modern furnace designs, matrix modifiers and a wide range of published procedures, the analyst has the key ingredients for obtaining reliable results.

Since its inception, a large number of fundamental studies have been conducted on the ETA and considerable knowledge gained on the processes preceding the formation of the free atoms. Many of the improvements in this technique have resulted from judicious use of this information. Similarly, this basic information is often of use in efficiently extracting solutions to analytical problems that are encountered with a new analyte or matrix.

ACKNOWLEDGMENTS

I would like to thank N.J. Miller-Ihli, Scott Baker and Debra Bradshaw for the comments and suggestions provided during the preparation of this manuscript.

ABBREVIATIONS AND ACRONYMS

AA	Atomic Absorption
EDL	Electrodeless Discharge Lamp
ETA	Electrothermal Atomizer
ETAAS	Electrothermal Atomic Absorption Spectrometry
FAA	Flame Atomic Absorption
GFAAS	Graphite Furnace Atomic Absorption Spectrometry
HCL	Hollow Cathode Lamp
LOD	Limit of Detection
QA/QC	Quality Assurance/Quality Control
SRM	Standard Reference Material

RELATED ARTICLES

Clinical Chemistry **(Volume 2)**
Atomic Spectrometry in Clinical Chemistry

Coatings **(Volume 2)**
Atomic Spectroscopy in Coatings Analysis

Environment: Water and Waste **(Volume 3)**
Flame and Graphite Furnace Atomic Absorption Spectrometry in Environmental Analysis • Heavy Metals Analysis in Seawater and Brines

Environment: Water and Waste cont'd **(Volume 4)**
Sample Preparation for Elemental Analysis of Biological Samples in the Environment • Sample Preparation for Environmental Analysis in Solids (Soils, Sediments, and Sludges)

Food **(Volume 5)**
Atomic Spectroscopy in Food Analysis

Industrial Hygiene **(Volume 6)**
Metals in Blood and Urine: Biological Monitoring for Worker Exposure

Steel and Related Materials **(Volume 10)**
Atomic Absorption and Emission Spectrometry, Solution-based in Iron and Steel Analysis

Atomic Spectroscopy **(Volume 11)**
Atomic Spectroscopy: Introduction • Background Correction Methods in Atomic Absorption Spectroscopy • Flame and Vapor Generation Atomic Absorption Spectrometry • Flow Injection Analysis Techniques in Atomic Spectroscopy

REFERENCES

1. A. Walsh, 'The Application of Atomic Absorption Spectra to Chemical Analysis', *Spectrochim. Acta*, **7**, 108–117 (1955).
2. A.S. King, 'The Production of Spectra by an Electrical Resistance Furnace in Hydrogen Atmosphere', *Astrophys. J.*, **27**, 353 (1908).
3. B.V. L'vov, 'The Analytical Use of Atomic Absorption Spectra', *Spectrochim. Acta*, **17**, 761–770 (1961).
4. R. Woodriff, G. Ramelow, 'Atomic Absorption Spectroscopy with a High-temperature Flame', *Spectrochim. Acta, Part B*, **23**, 665–671 (1968).
5. R.W. Woodriff, 'Atomization Chambers for Atomic Absorption Spectrochemical Analysis: A Review', *Appl. Spectrosc.*, **28**, 413–416 (1974).
6. H. Massmann, 'Vergleich von Atomabsorption und Atomfluoreszenz in der Graphitkuvette', *Spectrochim. Acta, Part B*, **23**, 215–226 (1968).
7. B. Welz, M. Sperling, G. Schlemmer, N. Wenzel, G. Marowsky, 'Spatially and Temporally Resolved Gas Phase Temperature Measurements in a Massmann-type Graphite Tube Furnace Using Coherent Anti-stokes Raman Scattering', *Spectrochim. Acta, Part B*, **43**, 1187–1207 (1988).
8. W. Frech, D.C. Baxter, B. Hutsch, 'Spatially Isothermal Graphite Furnace for Atomic Absorption Spectrometry Using Side-heated Cuvettes with Integrated Contacts', *Anal. Chem.*, **58**, 1973–1977 (1986).
9. B.V. L'vov, L.A. Pelieva, A.I. Sharnopolsky, *Zh. Prikl. Spektrosk.*, **27**, 395 (1977).
10. R.D. Ediger, 'Atomic Absorption Analysis with the Graphite Furnace Using Matrix Modification', *At. Absorpt. Newsl.*, **14**, 127–130 (1975).

11. R.F. Lonardo, A.I. Yuzefovsky, J.X. Zhou, J.T. McCaffrey, R.G. Michel, 'Extension of Working Range in Zeeman Graphite Furnace Atomic Absorption Spectrometry by Nonlinear Calibration with Prior Correction for Stray Light', *Spectrochim. Acta, Part B*, **51**, 1309–1323 (1996).
12. M.T.C. d. Loos-Vollebregt, L. de Galan, 'Stray Light in Zeeman and Pulsed Hollow Cathode Lamp Atomic Absorption Spectrometry', *Spectrochim. Acta, Part B*, **41**, 597–610 (1986).
13. P.L. Larkins, 'Atomic Line Profiles – Their Measurement and Importance in Analytical Atomic Spectroscopy', *J. Anal. At. Spectrom.*, **7**, 265 (1992).
14. R.J. Lovett, 'The Influence of Temperature on Absorbance in Graphite Furnace Atomic Absorption Spectrometry. I. General Considerations', *Appl. Spectrosc.*, **39**, 778–786 (1996).
15. A.K. Gilmutdinov, K.Y. Nagulin, Y.A. Zakharov, 'Analytical Measurement in Electrothermal Atomic Absorption Spectrometry. How Correct is It?', *J. Anal. At. Spectrom.*, **9**, 643–650 (1994).
16. C. Schnurer-Patschan, A. Zybin, H. Groll, K. Niemax, 'Improvement in Detection Limits in Graphite Furnace Diode Laser Atomic Absorption Spectrometry by Wavelength Modulation Technique', *J. Anal. At. Spectrom.*, **8**, 1103–1107 (1993).
17. H. Groll, K. Niemax, 'Multielement Diode Laser Atomic Absorption Spectrometry in Graphite Tube Furnaces and Analytical Flames', *Spectrochim. Acta, Part B*, **48**, 633–641 (1993).
18. R.E. Sturgeon, C.L. Chakrabarti, C.H. Langford, 'Studies on the Mechanism of Atom Formation in Graphite Furnace Atomic Absorption Spectrometry', *Anal. Chem.*, **48**, 1792–1807 (1976).
19. B. Smets, 'Atom Formation and Dissipation in Electrothermal Atomization', *Spectrochim. Acta, Part B*, **35**, 33–41 (1980).
20. O.A. Güell, J.A. Holcombe, 'Analytical Applications of Monte Carlo Techniques', *Anal. Chem.*, **62**, 529A–542A (1990).
21. V. Majidi, J.D. Robertson, 'Investigation of High Temperature Reactions on Solid Substrates with Rutherford Backscattering Spectrometry: Interaction of Palladium with Selenium on Heated Graphite Surfaces', *Spectrochim. Acta, Part B*, **46**, 1723–1733 (1991).
22. C.C. Eloi, J.D. Robertson, V. Majidi, 'Rutherford Backscattering Spectrometry Investigation of the Effects of Oxygen and Hydrogen Pretreatment of Pyrolytically Coated Graphite on Pb Atomization', *Anal. Chem.*, **67**, 335–340 (1995).
23. R.E. Sturgeon, K.W.M. Siu, G.J. Gardner, S.S. Berman, 'Carbon–Oxygen Reactions in Graphite Furnace Atomic Absorption Spectrometry', *Anal. Chem.*, **58**, 42–50 (1986).
24. A.K. Gilmutdinov, C.L. Chakrabarti, J.C. Hutton, 'Three-dimensional Distributions of Oxygen in Graphite and Metal Tube Atomizers for Analytical Atomic Spectrometry', *J. Anal. At. Spectrom.*, **7**, 1047–1062 (1992).
25. R.E. Sturgeon, H. Falk, 'Spectroscopic Measurement of Carbon Monoxide in a Graphite Furnace', *Spectrochim. Acta, Part B*, **43**, 421–428 (1988).
26. A. Cedergren, W. Frech, E. Lundberg, 'Estimation of Oxygen Pressure in Graphite Furnaces for Atomic Absorption Spectrometry', *Anal. Chem.*, **56**, 1382–1387 (1984).
27. A.K. Gilmutdinov, Y.A. Zakharov, V.P. Ivanov, A.V. Voloshin, K. Dittrich, 'Shadow Spectral Filming: A Method of Investigating Electrothermal Atomization. Part 2. Dynamics of Formation and Structure of the Absorption Layer of Aluminum, Indium and Gallium Molecules', *J. Anal. At. Spectrom.*, **7**, 675–683 (1992).
28. N. Hadgu, W. Frech, 'Performance of Side-heated Graphite Atomizers in Atomic Absorption Spectrometry Using Tubes with End Caps', *Spectrochim. Acta, Part B*, **49**, 445–457 (1994).
29. D.A. Katskov, R.I. McCrindle, R. Schwarzer, P.J.J.G. Marais, 'The Graphite Filter Furnace: A New Atomization Concept for Atomic Spectroscopy', *Spectrochim. Acta, Part B*, **50**, 1543–1555 (1995).
30. C.J. Rademeyer, B. Radziuk, N. Romanova, N.P. Skaugset, A. Skogstad, Y. Thomassen, *J. Anal. At. Spectrom.*, **10**, 739–745 (1995).
31. D.L. Tsalev, V.I. Slaveykova, 'Chemical Modification in Electrothermal Atomic Absorption Spectrometry. Organization and Classification of Data by Multivariate Methods', *J. Anal. At. Spectrom.*, **7**, 147–153 (1992).
32. W. Slavin, D.C. Manning, G.R. Carnrick, 'The Stabilized Temperature Platform Furnace', *At. Spectrosc.*, **2**, 137–143 (1981).
33. B.V. L'vov, 'Potentialities of the Graphite Crucible Method in Atomic Absorption Spectroscopy', *Spectrochim. Acta, Part B*, **24**, 53–70 (1969).
34. W. Frech, D.C. Baxter, 'Temperature Dependence of Atomization Efficiencies in Graphite Furnaces', *Spectrochim. Acta, Part B*, **45**, 867–886 (1990).
35. W. Slavin, G.R. Carnrick, 'The Possibility of Standardless Furnace Atomic Absorption Spectroscopy', *Spectrochim. Acta, Part B*, **39B**, 271–282 (1984).

Inductively Coupled Plasma/Optical Emission Spectrometry

Xiandeng Hou and Bradley T. Jones
Wake Forest University, Winston-Salem, USA

Inductively coupled plasma/optical emission spectrometry (ICP/OES) is a powerful tool for the determination of metals in a variety of different sample matrices. With this technique, liquid samples are injected into a radiofrequency (RF)-induced argon plasma using one of a variety of nebulizers or sample introduction techniques. The sample mist reaching the plasma is quickly dried, vaporized, and energized through collisional excitation at high temperature. The atomic emission emanating from the plasma is viewed in either a radial or axial configuration, collected with a lens or mirror, and imaged onto the entrance slit of a wavelength selection device. Single element measurements can be performed cost-effectively with a simple monochromator/photomultiplier tube (PMT) combination, and simultaneous multielement determinations are performed for up to 70 elements with the combination of a polychromator and an array detector. The analytical performance of such systems is competitive with most other inorganic analysis techniques, especially with regards to sample throughput and sensitivity.

1 INTRODUCTION

ICP/OES is one of the most powerful and popular analytical tools for the determination of trace elements in a myriad of sample types (Table 1). The technique is based upon the spontaneous emission of photons from atoms and ions that have been excited in a RF discharge. Liquid and gas samples may be injected directly into the instrument, while solid samples require extraction or acid digestion so that the analytes will be present in a solution. The sample solution is converted to an aerosol and directed into the central channel of the plasma. At its core the inductively coupled plasma (ICP) sustains a temperature of approximately 10 000 K, so the aerosol is quickly vaporized. Analyte elements are liberated as free atoms in the gaseous state. Further collisional excitation within the plasma imparts additional energy to the atoms, promoting them to excited states. Sufficient energy is often available to convert the atoms to ions and subsequently promote the ions to excited states. Both the atomic and ionic excited state species may then relax to the ground state via the emission of a photon. These photons have characteristic energies that are determined by the quantized energy level structure for the atoms or ions. Thus the wavelength of the photons can be used to identify the elements from which they originated. The total number of photons is directly proportional to the concentration of the originating element in the sample.

The instrumentation associated with an ICP/OES system is relatively simple. A portion of the photons emitted by the ICP is collected with a lens or a concave mirror. This focusing optic forms an image of the ICP on the entrance aperture of a wavelength selection device such as a monochromator. The particular wavelength exiting the monochromator is converted to an electrical signal by a photodetector. The signal is amplified and processed by the detector electronics, then displayed and stored by a personal computer.

The characteristics of the ICP as an analytical atomic emission source are so impressive that virtually all other emission sources [such as the flame, microwave-induced plasma (MIP), direct current plasma (DCP), laser-induced plasma (LIP), and electrical discharge] have been relegated to specific, narrowly defined application niches. Indeed, even much of the application field originally assigned to atomic absorption spectrometry (AAS), using both the flame and graphite furnace atomic absorption spectrometry (GFAAS), has been relinquished to the ICP. Compared to these other techniques,

For references see page 9484

Table 1 Survey of elemental application areas of ICP/OES[1–3]

Categories	Examples of samples
Agricultural and food	Animal tissues, beverages, feeds, fertilizers, garlic, nutrients, pesticides, plant materials, rice flour, soils, vegetables, wheat flour
Biological and clinical	Brain tissue, blood, bone, bovine liver, feces, fishes, milk powder, orchard leaves, pharmaceuticals, pollen, serum, urine
Geological	Coal, minerals, fossils, fossil fuel, ore, rocks, sediments, soils, water
Environmental and water	Brines, coal fly ash, drinking water, dust, mineral water, municipal wastewater, plating bath, sewage sludge, slags, seawater, soil
Metals	Alloys, aluminum, high-purity metals, iron, precious metals, solders, steel, tin
Organic	Adhesives, amino acids, antifreeze, combustion materials, cosmetics, cotton cellulose, dried wood, dyes, elastomers, epoxy, lubricant, organometallic, organophosphates, oils, organic solvent, polymers, sugars
Other materials	Acids, carbon, catalytic materials, electronics, fiber, film, packaging materials, paints and coatings, phosphates, semiconductors, superconducting materials

ICP/OES enjoys a higher atomization temperature, a more inert environment, and the natural ability to provide simultaneous determinations for up to 70 elements. This makes the ICP less susceptible to matrix interferences, and better able to correct for them when they occur. In cases where sample volume is not limited, ICP/OES provides detection limits as low as, or lower than its best competitor, GFAAS, for all but a few elements. Even for these elements, the simplicity with which the ICP/OES instrument is operated often outweighs the loss in sensitivity.

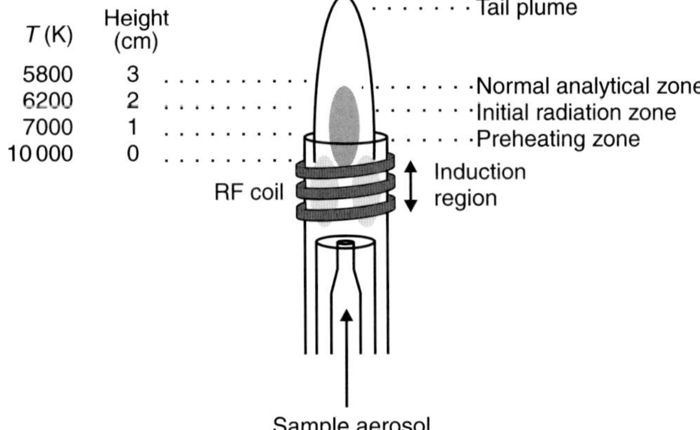

Figure 1 Schematic diagram of an ICP assembly showing the three concentric tubes composing the torch, the RF coil, the different plasma regions, and the temperature as a function of height above the load coil.

2 THEORY

The ICP was developed for optical emission spectrometry (OES) by Fassel et al. at Iowa State University in the US and by Greenfield et al. at Albright & Wilson, Ltd. in the UK in the mid-1960s.[1,4,5] The first commercially available ICP/OES instrument was introduced in 1974. The ICP is now not only the most popular source for OES but it is also an excellent ion source for inductively coupled plasma mass spectrometry (ICPMS).[6] ICP/OES is a proven commercial success, and the future is still bright for ICP-based spectroscopic techniques. Detectability has been continuously and dramatically improved over the past 35 years. Detection limits, for example, have improved by a factor of four to six orders of magnitude for many elements. Nevertheless, research and commercial opportunities for the further development of ICP/OES remain intriguing.[6]

2.1 Inductively Coupled Plasma Operation

As shown in Figure 1, the so-called ICP "torch" is usually an assembly of three concentric fused-silica tubes. These are frequently referred to as the outer, intermediate, and inner gas tubes. The diameter of the outer tube ranges from 9 to 27 mm. A water-cooled, two- or three-turn copper coil, called the load coil, surrounds the top section of the torch, and is connected to a RF generator. The outer argon flow (10–15 L min^{-1}) sustains the high-temperature plasma, and positions the plasma relative to the outer walls and the induction coil, preventing the walls from melting and facilitating the observation of emission signals. The plasma under these conditions has an annular shape. The sample aerosol carried by the inner argon flow (0.5–1.5 L min^{-1}) enters the central channel of the plasma and helps to sustain the shape. The intermediate argon flow (0–1.5 L min^{-1}) is optional and has the function of lifting the plasma slightly and diluting the inner gas flow in the presence of organic solvents.

The ICP is generated as follows. RF power, typically 700–1500 W, is applied to the load coil and an alternating current oscillates inside the coil at a rate corresponding to the frequency of the RF generator. For most ICP/OES instruments, the RF generator has a frequency of either

27 or 40 MHz. The oscillation of the current at this high frequency causes the same high-frequency oscillation of electric and magnetic fields to be set up inside the top of the torch. With argon gas flowing through the torch, a spark from a Tesla coil is used to produce "seed" electrons and ions in the argon gas inside the load coil region. These ions and electrons are then accelerated by the magnetic field, and collide with other argon atoms, causing further ionization in a chain reaction manner. This process continues until a very intense, brilliant white, teardrop-shaped, high-temperature plasma is formed. Adding energy to the plasma via RF-induced collision is known as inductive coupling, and thus the plasma is called an ICP. The ICP is sustained within the torch as long as sufficient RF energy is applied.[1] In a cruder sense, the coupling of RF power to the plasma can be visualized as positively charged Ar ions in the plasma gas attempting to follow the negatively charged electrons flowing in the load coil as the flow changes direction 27 million times per second.

Figure 1 shows the temperature gradient within the ICP with respect to height above the load coil. It also gives the nomenclature for the different zones of the plasma as suggested by Koirtyohann et al.[7] The induction region (IR) at the base of the plasma is "doughnut-shaped" as described above, and it is the region where the inductive energy transfer occurs. This is also the region of highest temperature and it is characterized by a bright continuum emission. From the IR upward towards to the tail plume, the temperature decreases.

An aerosol, or very fine mist of liquid droplets, is generated from a liquid sample by the use of a nebulizer. The aerosol is carried into the center of the plasma by the argon gas flow through the IR. Upon entering the plasma, the droplets undergo three processes. The first step is desolvation, or the removal of the solvent from the droplets, resulting in microscopic solid particulates, or a dry aerosol. The second step is vaporization, or the decomposition of the particles into gaseous-state molecules. The third step is atomization, or the breaking of the gaseous molecules into atoms. These steps occur predominantly in the preheating zone (PHZ). Finally, excitation and ionization of the atoms occur, followed by the emission of radiation from these excited species. These excitation and ionization processes occur predominantly in the initial radiation zone (IRZ), and the normal analytical zone (NAZ) from which analytical emission is usually collected.[1]

2.2 Inductively Coupled Plasma Characteristics

The main analytical advantages of the ICP over other excitation sources originate from its capability for efficient and reproducible vaporization, atomization, excitation, and ionization for a wide range of elements in various sample matrices. This is mainly due to the high temperature, 6000–7000 K, in the observation zones of the ICP. This temperature is much higher than the maximum temperature of flames or furnaces (3300 K). The high temperature of the ICP also makes it capable of exciting refractory elements, and renders it less prone to matrix interferences. Other electrical-discharge-based sources, such as alternating current and direct current arcs and sparks, and the MIP, also have high temperatures for excitation and ionization, but the ICP is typically less noisy and better able to handle liquid samples. In addition, the ICP is an electrodeless source, so there is no contamination from the impurities present in an electrode material. Furthermore, it is relatively easy to build an ICP assembly and it is inexpensive, compared to some other sources, such as a LIP. The following is a list of some of the most beneficial characteristics of the ICP source.

- high temperature (7000–8000 K)
- high electron density (10^{14}–10^{16} cm^{-3})
- appreciable degree of ionization for many elements
- simultaneous multielement capability (over 70 elements including P and S)
- low background emission, and relatively low chemical interference
- high stability leading to excellent accuracy and precision
- excellent detection limits for most elements (0.1–100 ng mL^{-1})
- wide linear dynamic range (LDR) (four to six orders of magnitude)
- applicable to the refractory elements
- cost-effective analyses.

3 SAMPLE INTRODUCTION

A sample introduction system is used to transport a sample into the central channel of the ICP as either a gas, vapor, aerosol of fine droplets, or solid particles. The general requirements for an ideal sample introduction system include amenity to samples in all phases (solid, liquid, or gas), tolerance to complex matrices, the ability to analyze very small amount of samples (<1 mL or <50 mg), excellent stability and reproducibility, high transport efficiency, simplicity, and low cost.[2] A wide variety of sample introduction methods have been developed, such as nebulization, hydride generation (HG), electrothermal vaporization (ETV), and laser ablation.[1,8,9]

3.1 Nebulizers

Nebulizers are the most commonly used devices for solution sample introduction in ICP/OES. With a

For references see page 9484

nebulizer, the sample liquid is converted into an aerosol and transported to the plasma. Both pneumatic and ultrasonic nebulizers (USNs) have been successfully used in ICP/OES. Pneumatic nebulizers make use of high-speed gas flows to create an aerosol, while the USN breaks liquid samples into a fine aerosol by the ultrasonic oscillations of a piezoelectric crystal. The formation of aerosol by the USN is therefore independent of the gas flow rate.

Only very fine droplets (about 8 µm in diameter) in the aerosol are suitable for injection into the plasma. A spray chamber is placed between the nebulizer and the ICP torch to remove large droplets from the aerosol and to dampen pulses that may occur during nebulization. Thermally stabilized spray chambers are sometimes adopted to decrease the amount of liquid introduced into the plasma, thus providing stability especially when organic solvents are involved. Pneumatic nebulization is very inefficient, however, because only a very small fraction (less than 5%) of the aspirated sample solution actually reaches the plasma. Most of the liquid is lost down the drain in the spray chamber. However, the pneumatic nebulizer retains its popularity owing to its convenience, reasonable stability, and ease of use. Efficiency may only be a concern when sample volumes are limited, or measurements must be performed at or near the detection limit.

Three types of pneumatic nebulizers are commonly employed in ICP/OES: the concentric nebulizer, the cross-flow nebulizer, and the Babington nebulizer (Figure 2). The concentric nebulizer is fashioned from fused silica. Sample solution is pumped into the back end of the nebulizer by a peristaltic pump. Liquid uptake rates may be as high as $4\,\mathrm{mL\,min^{-1}}$, but lower flows are more common. The sample solution flows through the inner capillary of the nebulizer. This capillary is tapered so that flexible tubing from the pump is attached at the entrance (4 mm outer diameter) and the exit has a narrow orifice approaching 100 µm or less in inner diameter. Ar gas ($0.5-1.5\,\mathrm{L\,min^{-1}}$) is supplied at a right angle into the outer tube. This tube is also tapered so that the exit internal diameter approaches the outer diameter for the sample capillary. As the Ar passes through this narrow orifice, its velocity is greatly increased, resulting in the shearing of the sample stream into tiny droplets. Concentric nebulizers have the advantages of excellent sensitivity and stability, but the small fragile fused-silica orifices are prone to clogging, especially when aspirating samples of high salt content. Concentric nebulizers also require a fairly large volume of sample, given the high uptake rate. The microconcentric nebulizer (MCN) is designed to solve this problem. The sample uptake rate for the MCN is less than $0.1\,\mathrm{mL\,min^{-1}}$. The compact MCN employs a smaller diameter capillary (polyimide or Teflon®) and poly(vinylidine difluoride) body to minimize the formation of large droplets and to facilitate the use of hydrofluoric acid.

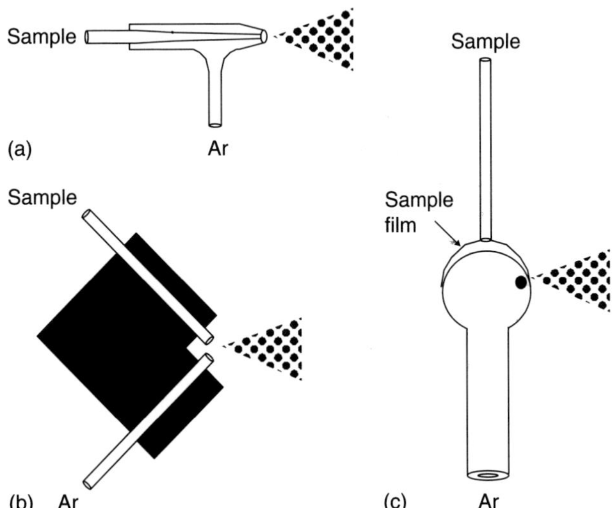

Figure 2 Schematic diagrams of three types of pneumatic nebulizer: (a) the concentric nebulizer; (b) the cross-flow nebulizer; and (c) the Babington nebulizer.

A second type of pneumatic nebulizer, the cross-flow nebulizer, is designed to reduce the clogging problem. In contrast to concentric nebulizers, cross-flow nebulizers use a high-speed stream of argon perpendicular to the tip of the sample capillary. Again the sample solution is broken into an aerosol, as shown in Figure 2. The drawbacks of the cross-flow nebulizer include lower sensitivity and potential capillary misalignment.

The third type of pneumatic nebulizer used for ICP/OES is the Babington nebulizer that allows a film of the sample solution to flow over a smooth surface having a small orifice (Figure 2).[10] High-speed argon gas emanating from the hole shears the sheet of liquid into small droplets. The essential feature of this type of nebulizer is that the sample solution flows freely over a small aperture, rather than passing through a fine capillary, resulting in a high tolerance to dissolved solids. In fact, even slurries can be nebulized with a Babington nebulizer. This type of nebulizer is the least susceptible to clogging and it can nebulize very viscous liquids.[1]

The Hildebrand grid nebulizer (HGN) (Figure 3) may be considered a specialized version of the Babington nebulizer with many orifices. Often the nebulizer has a screw-cap design. The outer member (cap) of the nebulizer holds two parallel platinum screens or grids. The grids are separated by approximately 2 mm. The inner body of the nebulizer (screw) has a single sample channel. Liquid is pumped through this channel at rates up to $1\,\mathrm{mL\,min^{-1}}$. The inner body also has a circular V-groove that allows the liquid to contact the entire perimeter of the inner platinum grid. In this fashion, the liquid completely wets both grids. A high-velocity stream

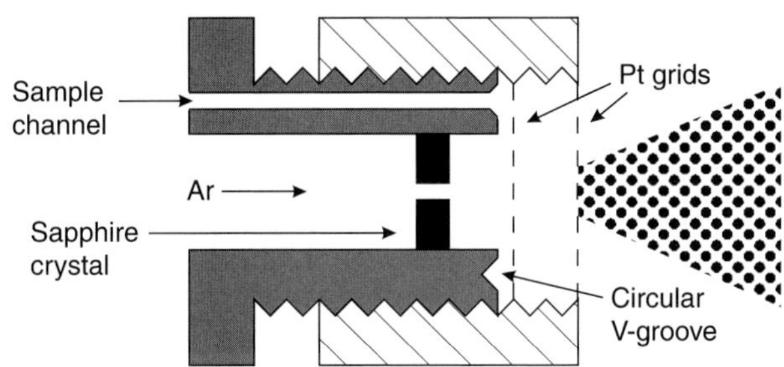

Figure 3 The HGN.

of argon ($1\,L\,min^{-1}$) blows through the center of the nebulizer. Often the velocity of the argon is increased by placing a sapphire crystal containing a small orifice (0.2 mm) in the center of the gas stream. The high-velocity argon forces the liquid through the tiny openings in the screens producing a fine aerosol. The HGN is characterized by clog-free operation, high efficiency, and excellent stability.

With the USN, sample solution is first introduced onto the surface of a piezoelectric transducer that is operated at a frequency between 0.2 and 10 MHz. The longitudinal wave, which is propagated in the direction perpendicular to the surface of the transducer towards the liquid–air interface, produces pressure that breaks the liquid into an aerosol.(11) The efficiency of an USN is typically between 10 and 20%. This nebulizing efficiency is greater than that of a pneumatic nebulizer, and it is independent of argon flow rate. Therefore, a slower gas flow rate can be used to transport the aerosol to the plasma, thus prolonging the residence time of analyte in the plasma. This results in improved sensitivity, and the limit of detection (LOD) is usually lowered by a factor of 8–200, depending upon the element. However, the USN is more complicated, more expensive, and more susceptible to matrix effects, memory effects, and high solid loading. The USN is not compatible with hydrofluoric acid. Various other means of nebulization have been tested with limited degrees of success.(11)

3.2 Hydride Generation

HG is a very effective sample introduction technique for some elements. These elements include arsenic, bismuth, germanium, lead, antimony, selenium, tin, and tellurium. In this method, the sample in diluted acid solution is mixed with a reducing agent, usually a solution of sodium borohydride in dilute sodium hydroxide. The reaction of sodium borohydride with the acid produces hydrogen. The hydrogen then reduces the analyte metal ions to hydrides, which are gaseous at ambient temperatures. The chemical reactions are shown in Equations (1) and (2).

$$NaBH_4 + 3H_2O + HCl \longrightarrow H_3BO_4 + NaCl + 8H \quad (1)$$

$$8H + E^{m+} \longrightarrow EH_n + H_2\ (\text{excess}) \quad (2)$$

where E is the hydride forming element of interest and m may or may not equal n.

The advantages of the HG technique include:

- physical separation of the analyte from possible matrix interferents;
- higher efficiency than conventional pneumatic nebulization;
- preconcentration of the analyte for better LOD;
- capability for inorganic and/or organic speciation;
- ease of automation when used with flow injection techniques.

On the other hand, several disadvantages of this technique may include:

- interference from those contaminants that reduce HG efficiency;
- slow reactions, necessitating hydride trapping prior to introduction;
- critical control of experimental conditions such as pH and reagent concentrations;
- extra influential factors, such as the oxidation state of the analyte of interest.

LODs for the hydride generating elements are listed in Table 2, together with those observed for conventional nebulization techniques. The LODs of HG/ICP/OES in Table 2 are the best LOD cited in Nakahara's review paper.(8) Compared with conventional pneumatic nebulization techniques, LODs achieved by the HG technique are enhanced by a factor between 10 and 1000, depending upon the element.

3.3 Electrothermal Vaporization

ETV has also been used to solve problems associated with pneumatic nebulization. Graphite furnaces or other

Table 2 LOD observed by hydride generating, HG/ICP/OES(1,8)

Elements	LOD by HG ($ng\,mL^{-1}$)	LOD by conventional nebulization ($ng\,mL^{-1}$)	Ratios (conventional/HG)
As	0.03	20	667
Bi	0.06	20	333
Ge	0.3	20	67
Pb	1.0	10	10
Sb	0.07	10	143
Se	0.04	50	1250
Sn	0.2	30	150
Te	0.04	10	250

For references see page 9484

electrothermal devices, such as carbon rods, carbon cups, graphite boats, graphite tubes, tungsten wire, and other metal filaments, have been used in research laboratories to electrothermally vaporize a liquid or solid sample for introduction into the ICP.[12] Other vaporization methods, such as arc/spark vaporization and laser ablation/vaporization, have also been used as means for sample introduction in ICP/OES. Even the ICP itself has been used to vaporize samples into a second ICP for analytical measurements.[13] In a typical experiment, a low current is applied to the ETV to remove the sample solvent. A small portion of the sample is then vaporized by the device through the application of a high current. An optional "ash" step may be used to remove some of the matrix prior to the analyte vaporization step. The resulting dense cloud of the analyte vapor is then efficiently swept into the center of the plasma by a flow of argon gas.

A commercial graphite furnace designed for AAS is most frequently used in ETV/ICP/OES. The major advantage of ETV as a means of sample introduction is that the transportation efficiency is dramatically improved over a pneumatic nebulizer, from less than 5% to over 60%. Consequently, the LODs are improved by at least an order of magnitude. Some difficult to analyze samples, those with high total dissolved solids (TDS) for example, can be introduced by the ETV. However, since these devices are generally not of a continuous-flow nature, the ICP instrument has to be capable of recording transient signals. Also, the simultaneous multielement capability of the system could be limited due to this transient nature of signals. Furthermore, when graphite material is used for the ETV, carbide formation could be a problem for some elements, resulting in lowered sensitivity and memory effects for refractory elements.

In an attempt to eliminate the problems associated with graphite, metal filaments have been employed for ETV/ICP/OES. For example, a tungsten coil from a commercial slide projector bulb can be used to vaporize liquid samples prior to their introduction into the ICP.[12] A small volume, typically 20 µL, of sample solution is delivered to the tungsten coil and dried at low current. Then, a higher current is applied to atomize the sample from the coil. The vapor is then rapidly introduced into the plasma as a dense plug by a flow of argon/hydrogen gas. The LOD is typically improved by 100–1500 times compared with pneumatic nebulization. These LODs are comparable to those obtained by GFAAS, but with the capability of simultaneous multielement measurement, and at a low cost. A tungsten loop has also been used as an in-torch vaporization (ITV) means for sample introduction to the ICP, and the operation can possibly be automated.[14] These approaches still share some of the other disadvantages associated with graphite furnace ETV, and commercial systems have not yet appeared.

3.4 Chromatographic Couplers

The combination of the separation power of chromatography and the detection power of atomic emission spectroscopy results in many advantages. One of the primary advantages of chromatography over conventional sample introduction is the ability to obtain speciation information.[15] When used as a detector for chromatographic methods, the ICP offers good sensitivity, wide LDR and multielement detection capability. The multielement capability of ICP, in turn, enhances the performance of chromatographic methods. Both gas chromatography (GC) and high-performance liquid chromatography (HPLC) can be coupled with ICP/OES.[16] Compared with HPLC, however, the GC/ICP coupling is less common because the analytical performance of ICP/OES for nonmetals is often not adequate. Undoubtedly, GC/ICP/OES is still useful in the analyses of volatile organometallics, as demonstrated in the determination of methylmercury species.[17] The successful combination of these two techniques is realized through the use of chromatographic couplers. Fortunately, most of the interface systems currently in use are relatively inexpensive and easy to construct, and they require few, if any, modifications to commercial ICP/OES instruments.

Direct connections between the end of the HPLC column and the nebulizer suffer from poor transport efficiency and low tolerance to many of the organic solvents commonly employed in mobile phases for HPLC, particularly when a pneumatic nebulizer is used. To improve the transport efficiency and to minimize the influence of organic solvents on the stability of the ICP, USNs, water-cooled thermospray chambers, and glass-frit nebulizers have been utilized for sample introduction in HPLC/ICP/OES. The solvent load on the plasma can also be decreased by aerosol thermostating, increasing the incident RF power, application of a condenser, or by use of a micro-HPLC column.

Other major sample introduction methods, such as thermospray, direct sample insertion, and laser ablation, have also been used for ICP/OES.[18–20] Each of these has its advantages and disadvantages. For example, laser ablation can be used to vaporize any solid samples into the ICP, but generally it has poor reproducibility and high cost.

4 TORCH CONFIGURATION

The atomic emission from the NAZ, as shown in Figure 1, is sampled for spectrometric measurements. Two configurations may be employed for observing emission from the ICP. One is referred to as a radial or side-on viewing of the plasma, and the other is known as an axial or end-on viewing of the plasma. A third viewing mode is the combination of these two basic modes, and is known as dual view. These are all commercially available, and each of them has advantages and disadvantages.

4.1 Radial View

The radial view is the classical operation mode for ICP/OES. With radial viewing, the plasma is operated in a vertical orientation, and the analytical zone is observed from the side of the plasma. Radial viewing constrains the observation volume in the NAZ, and thus limits the effect of potential spectral and background interferences.

4.2 Axial View

With the axial view, the plasma is rotated to a horizontal position and the NAZ of the ICP is observed from the end of the plasma. The axial view provides better LODs than radial view.[21] This may be attributed to the longer viewing path available down the axis of the plasma. Thus, a better sensitivity and a 5- to 10-fold improvement in the LOD can be achieved. The disadvantages of the axial view include the increased potential for spectral interference and matrix-induced interferences. Moreover, self-absorption effects can be quite severe because the observations are made through the much cooler tail plume of the plasma. These effects can be significantly reduced by use of a shear gas.[22] The shear gas displaces the tail plume from the optical path, and thus reduces the self-absorption. Spectral interferences may be either corrected or minimized by improving spectral resolution, using an alternate analytical line with less or no interference, or by applying an interelement correction (IEC) factor.[1]

4.3 Dual View

In cases of very complicated sample matrices having a wide range of elemental concentrations, the axial view may be inappropriate. Recent commercial instruments combine the axial view and radial view configurations into a single unit, known as dual view. This dual view system allows the user to optimize the appropriate configuration for the type of sample without the expense of two separate ICP/OES systems.[1]

5 DETECTION OF EMISSION

5.1 Gratings

ICP/OES is characterized by remarkably rich spectra. For example, the 70 elements most commonly determined by the technique give rise to at least 70 000 total emission lines in the 200–600 nm wavelength range. A consequence of this high density of spectral information is the need for high resolving power. The low-resolution dispersive systems typically employed with atomic absorption spectrometers will not suffice. Spectral interferences will occur in this case if only a small number of elements are present at moderate concentrations in the sample. Much higher resolution is desirable in ICP/OES, with spectral bandpass ($\Delta\lambda_s$) 0.01 nm or lower if possible. Traditionally, this degree of resolution has been accomplished using plane grating monochromators with large focal lengths ($f = 0.5$ m or more).

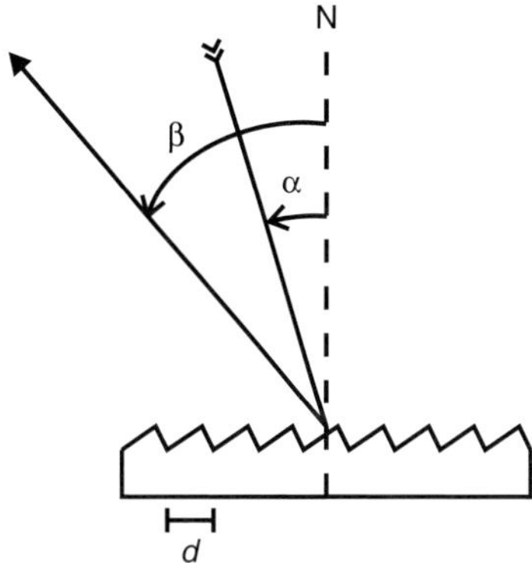

Figure 4 Diagram of the plane-ruled grating. N is the normal to the grating surface, α is the angle of incidence measured with respect to N, and β is the angle of diffraction measured with respect to N. d is the width of a single groove.

Figure 4 depicts a plane ruled grating. The normal to the grating surface (N) is shown as a dashed line. A light ray incident to the grating approaches at angle α measured with respect to N. The diffracted ray leaves the grating surface at angle β. Parallel rays striking the grating on different adjacent groove facets will travel a different distance before reaching a common position beyond the grating. If the difference in distance traveled is a multiple of the wavelength of the light incident upon the grating surface, then the rays will undergo constructive interference. Otherwise, destructive interference will occur. Relatively simple geometrical considerations result in the grating formula shown in Equation (3):

$$d(\sin\alpha + \sin\beta) = m\lambda \qquad (3)$$

This equation shows the relationship between α, β, the groove spacing (d), the wavelength of light (λ) and the order of diffraction (m). The order of diffraction may take any integer value including zero. At zero order, all wavelengths undergo constructive interference at the same diffraction angle. In the first order, one particular wavelength will undergo constructive interference at the angle β that corresponds to the specular reflection angle

For references see page 9484

for rays incident at angle α. This wavelength is called the blaze wavelength for the grating, and it is determined by the angle at which the grooves are cut with respect to the surface of the grating. A grating is most efficient at its blaze wavelength. Typically a grating may have an efficiency as high as 70% at its blaze wavelength, so the intensity measured at the blaze wavelength will be 70% of the intensity that would be measured at the specular reflectance angle for a polished mirror of the same coating and material as the grating.

Normally, when a plane grating is employed, the angle of incidence is nearly 0, so $\sin\alpha$ approaches 0. In this case, the grating formula may be further reduced, as shown in Equation (4):

$$\sin\beta = \frac{m\lambda}{d} \tag{4}$$

The angular dispersion of the grating ($\mathrm{d}\beta/\mathrm{d}\lambda$) may be found, as shown in Equations (5) and (6), by taking the derivative of both sides of the above equation with respect to λ:

$$(\cos\beta)\frac{\mathrm{d}\beta}{\mathrm{d}\lambda} = \frac{m}{d} \tag{5}$$

$$\frac{\mathrm{d}\beta}{\mathrm{d}\lambda} = \frac{m}{d\cos\beta} \tag{6}$$

The angular dispersion therefore increases with larger order, smaller groove spacing, or larger β. Angular dispersion may be converted to linear dispersion along the exit focal plane of the monochromator by simply multiplying by the focal length (f) of the monochromator. The spectral bandpass ($\Delta\lambda_s$) in wavelength units is then determined by dividing the slit width of the monochromator by the linear dispersion. So $\Delta\lambda_s$ gets smaller (higher resolution) for larger f, larger order, larger β, smaller d, and smaller slit width. A typical plane grating is operated in the first order. The groove density of the grating might be as high as 3600 grooves per millimeter, so the groove spacing (d) might be as small as 0.0003 mm. Therefore, assuming β is 45° and the slit width is 25 µm, a monochromator with a focal length of 500 mm will provide a spectral bandpass of 0.01 nm. Such a system will effectively isolate most ICP emission lines. Simultaneous multielement determinations are not performed effectively with such a system however. The focal plane for this type of monochromator is 5 cm long at best, so the entire spectral window will be approximately 20 nm in width. Unless the analytes of interest exhibit emission lines within 20 nm of one another, the wavelength must be scanned to detect multiple elements. So two other optical approaches have become more popular for simultaneous determinations: the concave grating placed on a Rowland circle, and the echelle grating coupled with a prism order-sorting device.

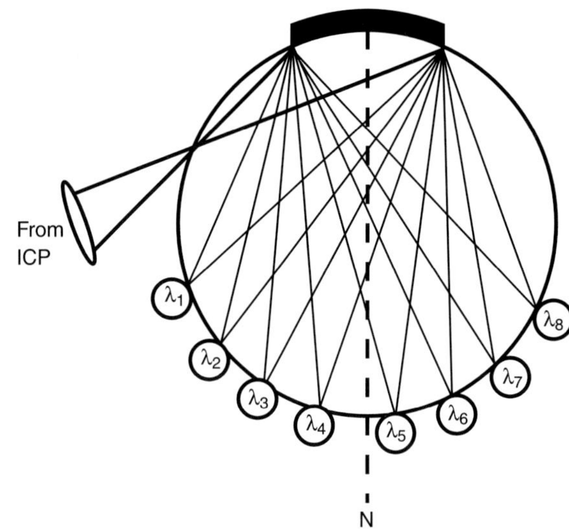

Figure 5 The Paschen–Runge mounting of a concave grating on a Rowland circle. The grating, entrance slit, and focused images of the diffracted wavelengths are all positioned on the perimeter of the circle. N is the grating normal.

Shortly after designing his grating ruling engine in 1881, Rowland first conceived the idea of ruling gratings on a spherical mirror of speculum metal.[23] The most important property of such a concave grating was also observed by Rowland. If the source of light and the grating are placed on the circumference of a circle, and the circle has a diameter equal to the radius of curvature of the grating, then the spectrum will always be brought to a focus on the circle. Hence the focal "plane" is curved, and of considerable length (Figure 5). In this case one entrance slit is placed on the circle for introduction of the source radiation, and multiple exit slits may be placed around the circle at the analytical wavelengths of interest. Hence Rowland's circle is ideally suited for multielement ICP emission spectrometry. A further advantage of the Rowland geometry is the elimination of the need for any collimating or focusing lenses or mirrors. A Rowland circle spectrometer with same groove density, slit width, and focal length as the plane grating system described above will provide similar spectral bandpass but with a much larger spectral window.

The echelle grating is a coarsely ruled grating, typically having a groove density of 70 grooves per millimeter, so $d = 0.014$ mm. The increase in spectral bandpass due to the increase in d is overcome by operating the grating in higher orders ($m = 25$ to 125), and by using steeper angles of diffraction ($\beta > 45°$). Figure 6 demonstrates how the steeper side of the groove facets are used with the echelle grating. If the steep sides of the grooves are blazed such that specular reflectance occurs when $\alpha = 60°$ and $\beta = 50°$, then each wavelength will exhibit a peak in grating efficiency at a particular order as determined by

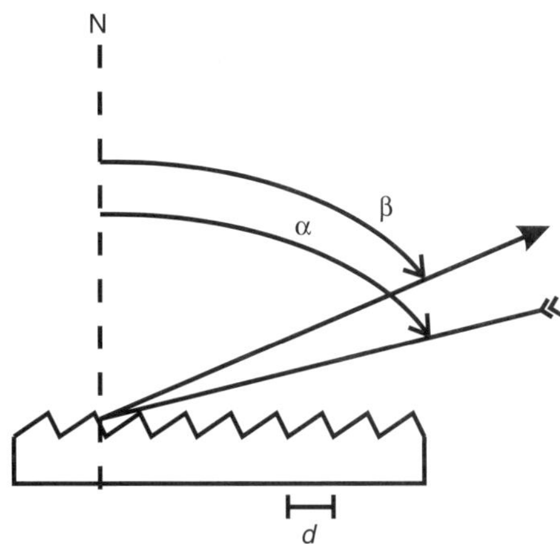

Figure 6 Diagram of the echelle grating. N is the normal to the grating surface, α is the angle of incidence measured with respect to N, and β is the angle of diffraction measured with respect to N. The light is incident to the steeper sides of the groove facets, and approaches the grating at nearly grazing angles (62°). d is the width of a single groove.

the grating formula. For example, for the 70 grooves per millimeter grating described above, the order of maximum efficiency (m_{max}) occurs as shown in Table 3. The efficiency of the grating for a given wavelength at its optimum order can be as high as 65%. This level of efficiency is typically attained across the free spectral range for a given order. The free spectral range ($\Delta\lambda_f$) is defined as the range of wavelengths over which no overlap from adjacent orders occurs, and is given by Equation (7):

$$\Delta\lambda_f = \frac{\lambda}{m+1} \quad (7)$$

As indicated in Table 3, $\Delta\lambda_f$ is very small for large values of m. Obviously, then, severe spectral overlap occurs with an echelle grating. The overlap does not simply involve adjacent orders, but all orders are dispersed in multiple layers along the same focal plane. This overlap is corrected most often with an order-sorting prism. This prism is placed between the echelle grating and the focal plane (Figure 7). The prism is positioned so that it disperses the light in a direction perpendicular to the

Table 3 The order of maximum efficiency and the free spectral range at that order for an echelle grating having 70 grooves per millimeter, angle of incidence (α) of 60°, and blaze angle (β) of 50°

λ (nm)	m_{max}	$\Delta\lambda_f$ (nm)
200	117	1.7
250	93	2.7
300	78	3.8
350	67	5.2
400	58	6.7

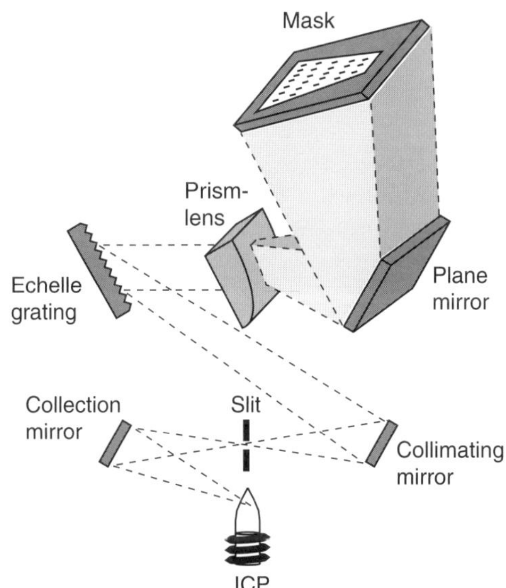

Figure 7 Schematic diagram of an echelle polychromator. (Adapted from the schematic diagram of the DRE/ICP provided by Leeman Labs, Inc., Hudson, NH.)

direction of dispersion of the grating. As a result, the focal plane has wavelength dispersed in the horizontal direction and order sorted in the vertical direction. The free spectral range (and the region of maximum efficiency) has a roughly triangular shape centered horizontally on the plane (Figure 8). An echelle monochromator often provides a spectral bandpass nearly 10 times smaller than that of a typical grating monochromator with a similar focal length. In addition, the echelle system provides high efficiency at many wavelengths rather than a single blaze wavelength. Finally, both the high efficiency and superior resolution are available over a very broad spectral window.

5.2 The Photomultiplier Tube

Figure 9 is a schematic representation of the PMT. Like its predecessor, the vacuum phototube, the heart of the PMT consists of two electrodes sealed in a fused-silica envelope. The cathode has a relatively large surface area, usually in the shape of a vertical, hollow "half cylinder". The cathode is made from a photoemissive material such as an alkali metal oxide. The anode is simply an electron collection wire or grid. Unlike the phototube, however, the PMT has up to 14 secondary emission dynodes placed between the cathode and the anode. Typically the anode is fixed to ground potential and the dynodes are at potentials that are successively more negative, by about 100 V per dynode. The potential of the cathode is typically −1000 V.

A photon generated in the ICP and passing through the wavelength selection device may pass through the fused silica envelope of the PMT, through a baffle-type grill,

For references see page 9484

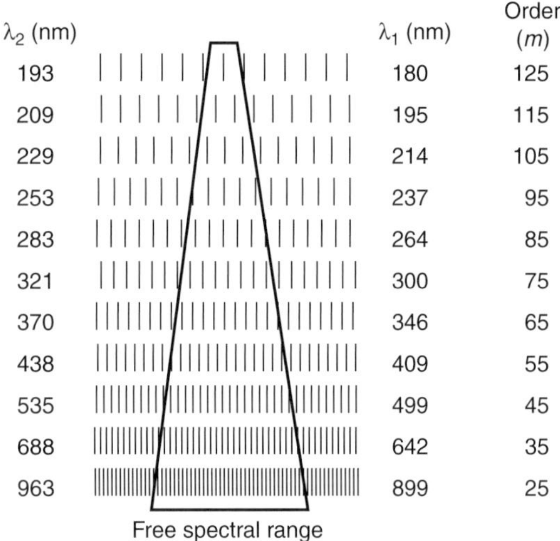

Figure 8 The two-dimensional focal plane provided by the echelle polychromator. The triangle-shaped free spectral range is the region of highest grating efficiency. The distance between adjacent vertical lines is 1 nm, and the beginning (λ_1) and ending (λ_2) wavelengths depicted by the lines are listed for each order. Only one out of every 10 orders is shown for clarity.

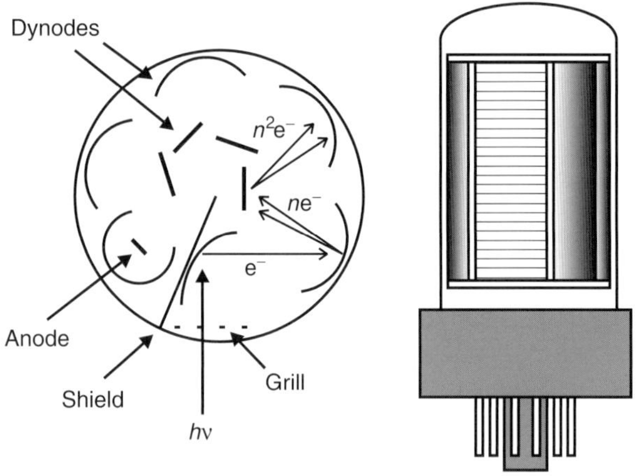

Figure 9 Cross-sectional and front views of a side-on PMT. In the cross-section, two types of dynodes are depicted: curved (outer) and flat (inner). The outside surfaces of the curved dynodes are seen in the front view.

and strike the photocathode. If the energy of the photon is higher than the work function of the photocathode material, then an electron may be ejected from the cathode. The fraction of photons with energy greater than the work function that actually produce a photoelectron is called the quantum efficiency of the photocathode. The quantum efficiency may be as high as 0.5, and it depends upon the photocathode material and upon wavelength. A plot of quantum efficiency versus wavelength is called the spectral response curve for the PMT. The spectral response curve is usually supplied by the manufacturer.

Once an electron has been ejected by the photocathode, it is accelerated towards the first dynode. Upon impact, the first dynode releases x secondary electrons (where x is typically between 2 and 5). This process continues at each dynode, with each electron impact imparting x new secondary electrons. So by the time that the pulse generated by a single photon reaches the anode it has been greatly multiplied, hence the name "multiplier" phototube or PMT. The gain, G, for a PMT can be defined as shown in Equation (8):

$$G = x^n \tag{8}$$

where n is the number of dynodes. The gain depends upon the voltage across the PMT and it may be as high as 10^8. One of the best features of the PMT is that the gain is acquired with almost no increase in noise. Thus the PMT is ideally suited for the detection of small analytical signals against a relatively dark background, as is the case near the detection limit in high resolution atomic emission spectrometry.

Usually the entrance aperture as defined by the grill on the PMT is very large compared to a single resolution element at the focal plane of a monochromator. This disparity is corrected by placing a mechanical slit on the focal plane in front of the PMT. The width of this slit then defines the range of wavelengths ($\Delta\lambda_s$) that is allowed to strike the PMT. Sequential detection of multiple elements can be accomplished by changing the grating angle in a conventional monochromator and thus scanning the wavelengths that are detected. A second approach is to move the PMT rapidly along the focal curve of a Rowland circle spectrometer. Such a spectrometer may have several pre-aligned exit slits along the focal curve, each corresponding to a particular element (Figure 5). A still more efficient method for sequential multielement determinations involves the echelle polychromator (Figure 7). A mask with many slits, each corresponding to a different element, is placed upon the two-dimensional focal plane. The PMT is held in a mechanical arm that quickly positions the detector at the appropriate x–y coordinates for a given element. This design allows very fast sequential determinations by "wavelength hopping" or direct reading rather than scanning linearly through all wavelengths to reach a select few. One commercial instrument using this design is called a Direct Reading Echelle (DRE) ICP instrument.

Simultaneous multielement determinations may be performed with multiple PMTs, but such designs quickly become limited by the size of the PMTs and the geometry of the polychromator. Multiple PMTs may be positioned either along the focal curve of a Rowland circle spectrometer or along the exit plane of an echelle polychromator. Solid-state detectors, with their relatively small size and their intrinsic multielement nature, are usually more effective for simultaneous determinations.

5.3 Array Detectors

Charge transfer devices (CTDs) include a broad range of solid-state silicon-based array detectors.[24] They include the charge injection device (CID) and the charge-coupled device (CCD). The CCD has found extensive use in nonspectroscopic devices such as video cameras, bar code scanners, and photocopiers. With the CTDs, photons falling on a silicon substrate produce electron–hole pairs. The positive electron holes migrate freely through the p-type silicon semiconductor material, while the electrons are collected and stored temporarily by an array of metal oxide semiconductor (MOS) capacitors (Figure 10). Each MOS capacitor is composed of a small metal electrode and a thin layer of insulating SiO_2 material on top of the p-type silicon substrate. A positive potential is applied to the metal electrode, so the electrons generated in a given region are trapped just below the insulating layer. Each MOS capacitor (or pixel) has a width in the 5- to 50-µm range, and a height that may be as large as 200 µm. A two-dimensional array of pixels is easily prepared by proper placement of the metal electrodes. Such arrays may vary in size up to 4096 pixels on an edge. The CCD differs from the CID mainly in the readout scheme. The CCD is read out in a sequential charge shifting manner towards the output amplifier. The CID on the other hand may be read out in a nondestructive manner by shifting charge between adjacent electrodes, and then shifting it back again. The CID thus benefits from quick random access, even during long integration periods.

Spectroscopic applications of CTDs has been hampered by the physical mismatch between the relatively small surface area of the detector and the large sometimes two-dimensional focal plane associated with polychromators. This mismatch may be overcome, however, and one commercial ICP spectrometer employs a CID detector having more than 250 000 pixels positioned upon an echelle focal plane.

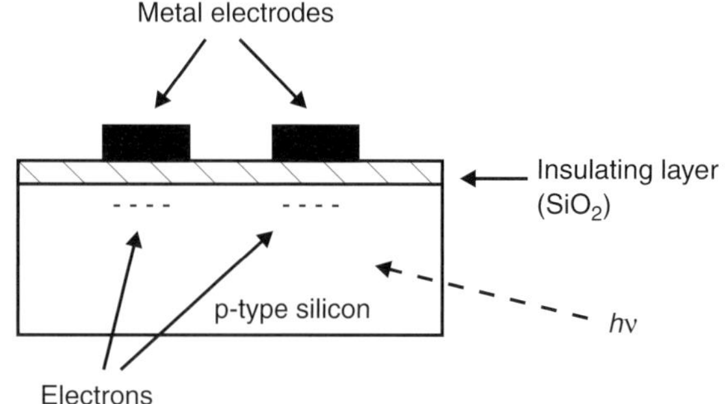

Figure 10 Cross-sectional diagram of two adjacent pixels in a CCD detector.

Alternative approaches have been successful with the CCD detector. In one case, a group of several CCD arrays are arranged around a Circular Optical System (CIROS) based upon a Rowland circle design. Rather than monitoring discrete wavelengths as is the case with the multiple PMT Rowland circle systems, the CIROS system provides total wavelength coverage from 120 to 800 nm, with resolution on the order of 0.009 nm.

A second multiple CCD array detector has become very popular commercially. This detector, called a segmented array charge-coupled device detector (SCD), employs over 200 small subarrays of 20–80 pixels each.[24] The subarrays are positioned along the two-dimensional focal plane of an echelle polychromator. The position of each subarray corresponds to one of the 236 most prominent ICP emission lines of the most commonly determined 70 ICP elements. This design allows for discrete wavelength determinations as seen with the multiple PMT designs, but it also provides additional spectral information around the vicinity of each emission line without exhaustively recording data at all wavelengths.

Another approach to correcting the mismatch between conventional imaging CCD arrays and the focal plane of an echelle spectrometer, is to specifically design a CCD array to exactly match the spectrometer image.[25] This process, called image-mapping the detector, has recently been accomplished and is available commercially in the form of the VistaChip. The VistaChip consists of a series of 70 diagonalized linear arrays (DLAs) of pixels that are designed to exactly match the individual diffraction orders present in the focal plane of the echelle spectrometer. The 70 DLAs correspond to orders 19–88, and the length of an individual DLA is set to match the free spectral range for the corresponding order. In this fashion, continuous wavelength coverage is provided across the range 167–363 nm, and selected coverage is provided in the range 363–784 nm where the diffraction orders are wider than the selected width of the detector (although no atomic emission lines of relevance miss the detector). The overall dimensions of the VistaChip are 15 by 19 mm, and a total of 70 908 pixels are packed inside the 70 DLAs.

5.4 Simultaneous Versus Sequential Detection

In the final analysis, the detection system most appropriate for an ICP emission system depends upon the application. In cases where only one or two elements will be determined routinely, the traditional scanning sequential detection system may be sufficient. The high sensitivity provided by a PMT coupled with the flexibility of interrogating any wavelength region may outweigh the

For references see page 9484

need for rapid determinations. On the other hand, if the application may vary between the determination of a few elements to the determination of many, the reasonable cost and high sensitivity of the PMT-based direct reading systems may be attractive. Finally, if a large suite of elements must be determined on a routine basis, one of the array-based detection systems might be most suitable.

6 ANALYTICAL PERFORMANCE

6.1 Analytical Wavelength

The ultraviolet and visible regions (160–800 nm) of the electromagnetic spectrum are most commonly used for analytical atomic spectrometry. In ICP/OES, the number of elements that can be determined is related to the wavelength window that can be covered by both the collimating and the dispersive optical system. Wavelengths above 500 nm should be used when alkali metals need to be determined, whereas wavelengths below 190 nm or even below 160 nm should be used when elements such as chlorine, bromine, nitrogen, arsenic must be determined. Spectral overlap must also be kept in mind in the selection of analytical lines. There are several criteria for selecting analytical lines. First, the wavelengths must be accessible by both the dispersive system and the detector. Second, the wavelengths must exhibit signal levels appropriate for the concentrations of the respective elements in the sample. Third, the wavelengths selected must be free from spectral interferences. When this is not possible, emission lines whose intensities can be corrected to account for spectral interferences should be chosen. Fourth, if an internal standard scheme is used, it may be preferable to match the analyte ion lines with an internal standard ion line, and analyte neutral atom lines with an internal standard neutral atom line.

6.2 Analytical Figures of Merit

For ICP/OES, the analytical figures of merit include the number of elements that can be determined, selectivity, reproducibility, long-term stability, susceptibility to matrix interferences, LOD, and accuracy.[26] The number of elements that can be measured by ICP/OES is often more than 70 out of a total of 92 naturally occurring elements, as listed in Table 4. Routine determination of 70 elements can be accomplished by ICP/OES at concentration levels below $1\,\mathrm{mg\,L^{-1}}$. As can be seen from Table 4, almost all naturally occurring elements, with the exception of hydrogen, oxygen, fluorine, and inert gases, can be determined by ICP/OES. The elements that are not usually determined by ICP/OES fall into three basic categories. The first category includes those elements that occur either as trace contaminants in the argon gas used in the ICP/OES (C from CO_2), constituents of the sample solvent (C, O, H), or as contaminants from the environment or atmosphere (N for example). The second category encompasses those elements that require high excitation energy, such as the halogens. These elements could be determined with poor LOD, however. The third category is the family of short-lived radioactive elements that are commonly determined by γ-ray spectrometry.[1]

Selectivity is important to minimize the spectral overlap interferences resulting from elements with rich line-emission spectra (tungsten, cobalt, niobium, molybdenum, tantalum, and rear earth elements) and to improve the signal-to-background ratio (SBR).[26] Selectivity is largely decided by the practical resolution of the wavelength dispersive system of the ICP/OES instrument. High selectivity is usually achieved with a sacrifice in sensitivity and the wavelength coverage range. For the best commercial ICP/OES instruments, a resolution of less than 5 pm is possible.[26]

The LODs of ICP/OES are generally in the nanogram per milliliter range. The LOD is usually defined as the analyte concentration that produces an analytical signal equivalent to three times the standard deviation observed for 16 measurements of a blank solution.[27] Another definition for the LOD of ICP/OES is related to the SBR of the analyte line at a given concentration, c, and the relative standard deviation (RSD) of the background, $\mathrm{RSD_B}$ as shown in Equation (9):[28]

$$\mathrm{LOD} = \frac{3 \times c \times \mathrm{RSD_B}}{\mathrm{SBR}} \tag{9}$$

The LOD is determined, therefore, by the sensitivity of the measurement and the noise level, or stability, of the ICP/OES instrument. The high degree of stability of an ICP was identified when Greenfield et al.[5] first

Table 4 A list of elements that can be determined by ICP/OES

Alkaline and alkaline earth	Rare earth	Transition metal	Others
Li, Na, K, Rb, Cs, Be, Mg, Ca, Sr, Ba	Ce, Pr, Nd, Sm, Eu, Gd, Tb, Dy, Ho, Er, Tm, Yb, Lu, Th, U	Sc, V, Ti, Cr, Mn, Fe, Co, Ni, Cu, Zn, Y, Nb, Zr, Mo, Ru, Th, Pd, Ag, Cd, La, Hf, Ta, W, Re, Os, Ir, Pt, Au, Hg	B, C, N, Al, Si, P, S, Cl, Ga, Ge, As, Se, Br, In, Sn, Sb, Te, I, Tl, Pb, Bi

used an ICP for analytical atomic spectrometry. For the best ICP/OES instruments, a long-term stability of less than 1% RSD has been achieved.[26] The atomic emission signals from the ICP are larger than those from other sources, such as a flame. This occurs because the high-temperature and inert-argon environment of the ICP leads to more efficient atomization, ionization, and excitation. In fact, the temperature of the ICP is so high that the largest signals are usually from the ionic lines. There are many other factors that may influence LOD, such as nebulizer type, view mode, and sample matrix.[29,30] Table 5 shows ICP/OES LODs obtained for various nebulizers and different viewing modes. LODs using an axially viewed plasma are typically better than those observed by radial viewing by a factor of 5–10,[1] and this is even true for instruments from different manufacturers, as shown in Table 5. Table 6 shows LODs for ETV/ICP/OES in comparison with those of GFAAS. LODs for many elements by ETV/ICP/OES are better than or equivalent to those achieved by GFAAS.[31,32] Notice that some elements that cannot be determined by GFAAS can be measured by ICP/OES.

Often the background equivalent concentration (BEC) is also used to check instrumental performance in ICP/OES. The BEC is defined as the concentration of a solution that results in an analyte emission signal equivalent in intensity to that of the background emission signal at the measurement wavelength. The BEC can be used as an indicator of relative sensitivity for an emission line. An unusually high BEC often indicates problems with the efficiency of the sample introduction system.

The LDR of calibration curves for ICP/OES is usually four to six orders of magnitude wide, starting from the LOD on the low concentration side. These LDRs are significantly larger than the two to three orders of magnitude observed for competing techniques such as AAS and arc/spark OES. The wide LDRs in ICP/OES translate into simple preparation of calibration curves. Very often a single standard together with a blank is enough to produce an accurate curve. Multiple sample dilutions are seldom needed prior to the analysis.

In general, the accuracy of the ICP/OES technique gets poorer as the analyte concentration approaches the LOD. For semiquantitative analysis (accuracy ±10%) the analyte concentration should be at least five times higher than the LOD. For accurate quantitation (±2%), the concentration should be 100 times greater than the LOD. At this concentration level, the precision is typically better than 1% RSD. This precision is considered sufficient for most trace element determinations. Better precision can be achieved, if necessary, by sacrificing analysis speed or with a more complex instrumental design. Recently, the concept of limit of quantitation (LOQ) has been defined as a concentration for which the precision, expressed as RSD, would be below a given threshold, for instance 10% or 5%.[33] A 5%-based LOQ would normally occur at concentrations approximately 10 times greater than the LOD (based upon three standard deviations).

Table 5 LOD (μg L^{-1}) observed for ICP/OES with different nebulizers and different viewing modes[1,3]

Element	HGN[3]	USN[3]	Ratio (HGN/USN)	Radial view[1]	Axial view[3]
Ag	6.1	0.71	9	1	0.6
Al	10.5	3.85	3	3	1.9
As	12.8	2.05	6	20	3.8
B	7.4	NA	NA	1	NA
Ba	0.28	0.11	3	0.1	0.12
Be	0.11	0.04	3	0.1	0.09
Ca	14.5	1.38	11	0.02	NA
Cd	1.3	0.59	2	1	0.2
Co	1.7	0.56	3	1	0.8
Cr	1.6	0.49	3	2	0.4
Cu	1.8	0.50	4	0.4	0.7
Fe	2.5	0.38	7	2	0.5
K	152	17.39	8	20	5
Li	4	0.40	10	0.3	NA
Mg	24.8	6.65	4	0.1	NA
Mn	0.61	0.09	7	0.4	0.07
Mo	2.2	0.58	4	3	0.8
Na	11.7	2.23	5	3	2.2
Ni	3.9	1.66	2	5	1.6
P	34.2	9.44	4	30	NA
Pb	10.7	1.60	7	10	1.6
Sb	15.6	2.65	6	10	2.6
Se	23	2.03	11	50	3.8
Si	14.3	3.48	4	4	NA
Sn	8.6	0.90	10	30	NA
Sr	0.3	0.05	6	0.06	NA
Ti	0.65	0.17	4	0.5	NA
Tl	14.2	2.05	7	30	4.8
V	2.8	0.41	7	0.5	0.4
Zn	1.5	0.40	4	1	0.4

6.3 Interferences

Among all commonly used analytical atomic spectrometry techniques, ICP/OES is probably the one with the fewest interferences. The argon plasma is inert when compared to the chemical reactivity of a flame. Also, the high temperature of the plasma helps to reduce chemical interferences. The temperature is high enough to break down most species into atoms or ions for excitation and subsequent emission. In contrast, in a low temperature flame, chemical interferences can be a severe problem. For example, a small amount of aluminum will interfere with the determination of calcium in flame AAS, but even at aluminum concentration 100 times higher than this, interference is not observed in ICP/OES. However, chemical interferences do exist in the ICP. Sometimes

For references see page 9484

Table 6 LODs (pg) for ETV/ICP/OES compared with those of GFAAS[2,31,32]

Elements	ETV/ICP/OES	GFAAS
Ag	1	0.5
Al	0.5	4
As	60	20
Au	10	10
Ba	0.3	10
Be	1	1
Bi	200	10
B	4	NA
Ca	150	5
Cd	1	0.3
Co	12	1
Cu	2	1
Er	34	300
Eu	12	10
Fe	10	2
Ga	10	10
Ge	10	20
Hg	4	100
In	20	5
K	1200	2
Li	2	5
Lu	54	4000
Mg	0.1	0.4
Mn	0.3	1
Mo	7	4
Na	400	5
Ni	27	10
P	100	3000
Pb	4	5
Pt	25	50
Rb	2800	5
Re	100	NA
Ru	95	40
Sb	100	20
Sc	17	40
Se	450	20
Si	100	40
Sn	20	20
Sr	5	2
Te	50	10
Ti	6	100
Tl	300	10
U	3	2.4 μg
V	6	20
W	160	NA
Y	25	NA
Yb	18	4
Zn	0.6	1

NA, not available.

higher RF power and/or lower inner argon flow rates are used to reduce these interferences. One particular type of chemical interference is the so-called easily ionized element (EIE) effect. The EIEs are those elements that have low ionization potentials, such as alkaline elements. High concentrations of EIEs can suppress or enhance emission signals, depending upon the analyte species. One way to reduce the EIE effect is to dilute the sample solution to the point that the EIE effect is not measurable. Sometimes, higher RF power or mathematical correction may be used to compensate for EIE interference. Instrumental conditions such as observation width, viewing height, and viewing volume can be chosen to minimize such interference and to optimize emission signal collection in either axial or radial configuration.[34]

The most common interference problem in ICP/OES is spectral interference (also referred to as background interference). Ironically, this type of interference arises because of the multielement nature of the plasma. Since the ICP is capable of exciting almost any element that is introduced into the plasma, spectra are likely to be rich especially for highly complex and concentrated samples. The solution to the spectral interference problem, as discussed previously, is the use of high-resolution spectrometers. Some spectral overlap may even exist with the best commercial system. In these cases advanced background correction techniques are employed or a different analytical wavelength for the element(s) of interest is chosen. Spectral interferences can be categorized into four categories: simple background shift, sloping background shift, direct spectral overlap, and complex background shift. The simple background shift is defined as a shift in background intensity that is essentially constant over a given wavelength range on either side of the analytical line. The background may shift up or down. There are two approaches to deal with this background problem. The first is to select a different analytical line at a wavelength with no background interference. The second is to correct for the background by measuring it somewhere near, but not falling on, the profile of the analytical line of the analyte element. Two background correction points, one on each side of the profile of the analytical line, are used to correct for the sloping background shift. This means that the average signal measured at the two points is subtracted from the total signal measured at the analytical line. In the worst case, direct spectral overlap occurs. This can be corrected if the magnitude of the interference is known as a function of the concentration of the interfering element. A correction factor can be calculated and used to correct the signal measured at the analyte wavelength. This can be best achieved by making simultaneous measurements of both the signal at the analyte wavelength and at a different wavelength for the interfering element. This method is referred to as the concentration ratio method or IEC. The basic requirement is that the concentration of the interfering element can be accurately measured at another wavelength. With advanced detector systems, all spectral lines are present, so IEC is possible. A complex background shift is a shift in a background intensity that varies significantly on both sides of the analytical line.

This is usually caused by the occurrence of a number of intense, closely spaced emission lines nearby, and perhaps directly overlapping the analyte wavelength. In this case, a different analytical wavelength should be chosen if possible.[1]

ACKNOWLEDGMENTS

The authors acknowledge that the preparation of the manuscript was partially done in Professor Robert G. Michel's research laboratory, and are grateful for access to all the facilities for the writing of this article. This work was funded in part by a grant from the National Science Foundation GOALI program (CHE-9710218).

ABBREVIATIONS AND ACRONYMS

AAS	Atomic Absorption Spectrometry
BEC	Background Equivalent Concentration
CCD	Charge-coupled Device
CID	Charge Injection Device
CIROS	Circular Optical System
CTD	Charge Transfer Device
DCP	Direct Current Plasma
DLA	Diagonalized Linear Array
DRE	Direct Reading Echelle
EIE	Easily Ionized Element
ETV	Electrothermal Vaporization
GC	Gas Chromatography
GFAAS	Graphite Furnace Atomic Absorption Spectrometry
HG	Hydride Generation
HGN	Hildebrand Grid Nebulizer
HPLC	High-performance Liquid Chromatography
ICP	Inductively Coupled Plasma
ICPMS	Inductively Coupled Plasma Mass Spectrometry
ICP/OES	Inductively Coupled Plasma/Optical Emission Spectrometry
IEC	Interelement Correction
IR	Induction Region
IRZ	Initial Radiation Zone
ITV	In-torch Vaporization
LDR	Linear Dynamic Range
LIP	Laser-induced Plasma
LOD	Limit of Detection
LOQ	Limit of Quantitation
MCN	Microconcentric Nebulizer
MIP	Microwave-induced Plasma
MOS	Metal Oxide Semiconductor
NAZ	Normal Analytical Zone
OES	Optical Emission Spectrometry
PHZ	Preheating Zone
PMT	Photomultiplier Tube
RF	Radiofrequency
RSD	Relative Standard Deviation
SBR	Signal-to-background Ratio
SCD	Segmented Array Charge-coupled Device Detector
TDS	Total Dissolved Solids
USN	Ultrasonic Nebulizer

RELATED ARTICLES

Clinical Chemistry **(Volume 2)**
Atomic Spectrometry in Clinical Chemistry

Coatings **(Volume 2)**
Atomic Spectroscopy in Coatings Analysis

Environment: Trace Gas Monitoring **(Volume 3)**
Laser-induced Breakdown Spectroscopy, Elemental Analysis

Environment: Water and Waste **(Volume 3)**
Environmental Analysis of Water and Waste: Introduction • Atomic Fluorescence in Environmental Analysis • Capillary Electrophoresis Coupled to Inductively Coupled Plasma-Mass Spectrometry for Elemental Speciation Analysis • Flame and Graphite Furnace Atomic Absorption Spectrometry in Environmental Analysis • Gas Chromatography with Atomic Emission Detection in Environmental Analysis • Heavy Metals Analysis in Seawater and Brines • Hydride Generation Sample Introduction for Spectroscopic Analysis in Environmental Samples • Inductively Coupled Plasma Mass Spectrometry in Environmental Analysis

Environment: Water and Waste cont'd **(Volume 4)**
Laser Ablation Inductively Coupled Plasma Spectrometry in Environmental Analysis • Microwave-assisted Techniques for Sample Preparation in Organic Environmental Analysis • Optical Emission Inductively Coupled Plasma in Environmental Analysis • Sample Preparation for Elemental Analysis of Biological Samples in the Environment • Sample Preparation for Environmental Analysis in Solids (Soils, Sediments, and Sludges) • Sample Preparation Techniques for Elemental Analysis in Aqueous Matrices • Sampling Considerations for Biomonitoring

Food **(Volume 5)**
Atomic Spectroscopy in Food Analysis

For references see page 9484

Forensic Science **(Volume 5)**
Atomic Spectroscopy for Forensic Applications

Industrial Hygiene **(Volume 6)**
Metals in Blood and Urine: Biological Monitoring for Worker Exposure • Surface and Dermal Monitoring

Petroleum and Liquid Fossil Fuels Analysis **(Volume 8)**
Metals, Nitrogen and Sulfur in Petroleum Residue, Analysis of

Steel and Related Materials **(Volume 10)**
Atomic Absorption and Emission Spectrometry, Solution-based in Iron and Steel Analysis • Metal Analysis, Sampling and Sample Preparation in

Atomic Spectroscopy **(Volume 11)**
Atomic Spectroscopy: Introduction

REFERENCES

1. C.B. Boss, K.J. Fredeen, *Concept, Instrumentation and Techniques in Inductively Coupled Plasma Optical Emission Spectrometry*, 2nd edition, Perkin-Elmer, Norwalk, CT, 1997.
2. J.M. Carey, J.A. Caruso, 'Electrothermal Vaporization for Sample Introduction in Plasma Source Spectrometry', *Crit. Rev. Anal. Chem.*, **23**(5), 397–439 (1992).
3. Leeman Labs, Inc., Hudson, NH, 1999.
4. R.H. Wendt, V.A. Fassel, 'Induction-coupled Plasma Spectrometric Excitation Source', *Anal. Chem.*, **37**(7), 920–922 (1965).
5. S. Greenfield, I.L. Jones, C.T. Berry, 'High-pressure Plasma as Spectroscopic Emission Sources', *Analyst*, **89**(11), 713–720 (1964).
6. V.A. Fassel, 'Analytical Inductively Coupled Plasma Spectroscopies – Past, Present, and Future', *Fresenius Z. Anal. Chem.*, **324**(6), 511–518 (1986).
7. S.R. Koirtyohann, J.S. Jones, C.P. Jester, D.A. Yates, 'Use of Spatial Emission Profiles and a Nomenclature System as Acids in Interpreting Matrix Effects in the Low-power Argon Inductively Coupled Plasma', *Spectrochim. Acta*, **36B**(1), 49–59 (1981).
8. T. Nakahara, 'Hydride Generation Techniques and Their Applications in Inductively Coupled Plasma–Atomic Emission Spectrometry', *Spectrochim. Acta Rev.*, **14**(1/2), 95–109 (1991).
9. R.E. Russo, 'Laser Ablation', *Appl. Spectrosc.*, **49**(9), 14A–28A (1995).
10. R.S. Babington, 'It's Superspray', *Popular Science*, May, 102–104 (1973).
11. J.G. Williams, 'Instrument Options', in *Inductively Coupled Plasma Mass Spectrometry*, eds. K.E. Jarvis, A.L. Gray, R.S. Houk, Blackie, New York, 58–80, 1992.
12. K. Levine, K.A. Wagner, B.T. Jones, 'Low-cost, Modular Electrothermal Vaporization System for Inductively Coupled Plasma Atomic Emission Spectrometry', *Appl. Spectrosc.*, **52**(9), 1165–1171 (1998).
13. G.M. Allen, D.M. Coleman, 'Segregated Sampling and Excitation with a Dual Inductively Coupled Plasma', *Anal. Chem.*, **56**(14), 2981–2983 (1984).
14. V. Karanassios, K.P. Bateman, 'Electrically Heated Wire-loop, In-torch Vaporization Sample Introduction System for Inductively Coupled Plasma Atomic Emission Spectrometry with Photodiode Array Detection', *Spectrochim. Acta*, **49B**(9), 847–865 (1994).
15. R. Lobinski, F.C. Adams, 'Speciation Analysis by Gas Chromatography with Plasma Source Spectrometric Detection', *Spectrochim. Acta*, **52B**(13), 1865–1903 (1997).
16. S.J. Hill, M.J. Bloxham, P.J. Worsfold, 'Chromatography Coupled with Inductively Coupled Plasma Atomic Emission Spectrometry and Inductively Coupled Plasma Mass Spectrometry', *J. Anal. At. Spectrom.*, **8**(6), 499–515 (193).
17. T. Kato, T. Uehiro, A. Yasuhara, M. Morita, 'Determination of Methylmercury Species by Capillary Column Gas Chromatography with Axially Viewed Inductively Coupled Plasma Atomic Emission Spectrometric Detection', *J. Anal. At. Spectrom.*, **7**(2), 15–18 (1992).
18. T.S. Conver, J. Yang, J.A. Koropchak, 'New Developments in Thermospray Sample Introduction for Atomic Spectrometry', *Spectrochim. Acta*, **52B**(8), 1087–1104 (1997).
19. V. Karanassios, G. Holick, 'Direct Sample Insertion Devices for Inductively Coupled Plasma Spectrometry', *Spectrochim. Acta Rev.*, **13**(2), 89–166 (1990).
20. L. Moenke-Blankenburg, 'Laser-ICP-spectrometry', *Spectrochim. Acta Rev.*, **15**(1), 1–38 (1993).
21. F.E. Lichte, S.R. Koirtyohann, 'Induction Coupled Plasma Emission from a Different Angle', Paper 26, Federation of Analytical Chemistry and Spectroscopy Society, Philadelphia, PA, 1976.
22. D.R. Demers, 'Evaluation of the Axially Viewed (End-on) Inductively Coupled Argon Plasma Source for Atomic Emission Spectrometry', *Appl. Spectrosc.*, **33**(6), 584–591 (1979).
23. E.C.C. Bailey, *Spectroscopy*, Longman, Green and Co., London, Chapter 1, 1905.
24. C.W. Earle, M.E. Baker, M. Bonner Denton, R.S. Pomeroy, 'Imaging Applications for Chemical Analysis Utilizing Charge Coupled Device Array Detectors', *Trends Anal. Chem.*, **12**(10), 395–403 (1993).
25. A.T. Zander, R.L. Chien, C.B. Cooper, P.V. Wilson, 'An Image Mapped Detector for Simultaneous ICP-AES', *Anal. Chem.*, **71**(16), 3332–3340 (1999).
26. J.M. Mermet, E. Poussel, 'ICP Emission Spectrometers: 1995 Analytical Figures of Merit', *Appl. Spectrosc.*, **49**(10), 12A–18A (1995).
27. G.L. Long, J.D. Winefordner, 'Limit of Detection: A Closer Look at the IUPAC Definition', *Anal. Chem.*, **55**(7), 712A–724A (1983).

28. P.W.J.M. Boumans, 'Detection Limits and Spectral Interferences in Atomic Emission Spectrometry', *Anal. Chem.*, **66**(8), 459A–467A (1994).
29. M. Thompson, R.M. Barnes, 'Analytical Performance of Inductively Coupled Plasma–Atomic Emission Spectrometry', in *Inductively Coupled Plasma in Analytical Atomic Spectrometry*, 2nd edition, eds. A. Montaser, D.W. Golightly, VCH Publishers, New York, 249–297, 1992.
30. T.C. Johnson, R.S. Perry, L.J. Fick, H.B. Fannin, 'An Examination of Relative Trends of Atomic Detection Limits in the Inductively Coupled Plasma', *Spectrochim. Acta*, **52B**(1), 125–129 (1997).
31. *The Guide to Techniques and Applications for Atomic Spectroscopy*, Perkin-Elmer, Norwalk, CT, 1990.
32. W. Slavin (ed.), *Graphite Furnace AAS: A Source Book*, Perkin-Elmer, Ridgefield, CT, 1984.
33. M. Carre, S. Excoffier, J.M. Mermet, 'A Study of the Relation between the Limit of Detection and the Limit of Quantitation in Inductively Coupled Plasma Spectrochemistry', *Spectrochim. Acta*, **52B**(14), 2043–2049 (1997).
34. P.J. Galley, G.M. Hieftje, 'Easily Ionizable Element (EIE) Interference in Inductively Coupled Plasma Atomic Emission Spectrometry-II. Minimization of EIE Effects by Choice of Observation Volume', *Spectrochim. Acta*, **49B**(7), 703–724 (1994).

Laser Ablation in Atomic Spectroscopy

Richard E. Russo, Xianglei Mao, Oleg V. Borisov, and Haichen Liu
Lawrence Berkeley National Laboratory, Berkeley, USA

Laser ablation (LA) is a unique technique to transform a solid sample into vapor-phase constituents, which then can be chemically analyzed by atomic spectroscopy. Ablation brings many exciting capabilities to the field of chemical analysis, primarily because of the laser-beam properties. The ability to analyze directly any solid sample without sample preparation and minimal sample-quantity requirements are just some of the unique capabilities. This article discusses current issues related to using LA in atomic spectroscopy. A general introduction to LA sampling is presented along with a comparison with other solid sampling techniques. The critical issues for all analytical techniques are calibration, accuracy, and sensitivity. Techniques that have been demonstrated to address these analytical characteristics and define these parameters for LA are investigated in detail. Finally, many unique applications are described, ranging from dating geological materials to providing crime-scene evidence. Most of these applications could not be performed without the use of a laser beam. The inductively coupled plasma (ICP) will be emphasized in this article because it is currently the most prevalent excitation and ionization source for chemical analysis using LA.

1 INTRODUCTION

Want to know the composition of an unknown solid sample? Don't want to deal with acids, wastes, and

For references see page 9502

laborious procedures to conduct the analysis? What is the best way to transform the unknown solid sample into vapor-phase constituents for chemical analysis? The answer: explode the sample with a high-power laser beam! This is precisely the role of LA in atomic spectroscopy. Complex solid samples, including environmental wastes, geochemical materials, coatings, extraterrestrial samples, and others dictate the development of a sampling approach for chemical analysis that does not rely on complicated, laborious dissolution procedures; a direct approach for analyzing unknown samples has been a quest of analytical chemistry for a long time. In some cases, the desire is to analyze a sample chemically without destruction, especially if that sample is the Shroud of Turin! These are some of the characteristics that make ablation sampling attractive for atomic spectroscopy.

This article will discuss current issues related to LA in atomic spectroscopy, including calibration and optimization, accuracy, sensitivity, and particle transport. In addition, many unique applications will be described, in some cases applications that cannot be performed without the use of a laser. The ICP is emphasized in this article because it is the most prevalent excitation and ionization source for chemical analysis using LA at this time.

1.1 Characteristics of Laser Ablation

One of the most unique characteristics of LA is that any solid sample can be directly ablated into vapor-phase constituents (vapor and particles), be it a rock, hair sample, extraterrestrial sample, or priceless piece of artwork. Only a very small portion of the sample is required for the analysis – sometimes of the order of picograms or less. Damage at these levels is generally not visible without the use of a microscope. In addition, there is no requirement for sample preparation; the laser beam can be used to ablate the surface contamination, as well as dig a crater and provide bulk analysis.

Because the laser beam directly converts the solid sample into the vapor phase, liquid reagents are not required to digest the sample. Therefore, there is no solution waste, no worries about loss of volatile species, and minimal sample handling. In addition, when liquid sample introduction is used, a larger portion of the sample is required because of dilution. Another significant and unique advantage of LA is the ability to perform spatial (micrometer) characterization. Since the laser beam can be guided to an exact location on a surface, the analyst can choose where the explosion and, therefore, the sampling occur. If the sample of interest is the microscopic inclusions in a bulk material, or particles on a filter paper, it is not necessary to analyze the entire sample; most of the mass in this case is not related and leads to large background signals in the analytical source. Finally, if one wants the analysis performed immediately, LA is the way to go – there are no time-consuming sample preparation or dissolution procedures; just place the sample in a simple chamber and hit it with the laser beam. With capabilities and advantages such as these, LA atomic spectroscopy is now utilized routinely in many industries, using commercially available and laboratory-built systems.

List of selected abbreviations appears in Volume 15

1.2 Solid Sampling Techniques: Comparison

The development of any "new" technique for elemental analysis requires parameters such as sampling, excitation, ionization, and detection to be critically investigated and optimized. An "ideal" sample introduction technique has to satisfy certain criteria, which include:

- remove a sufficient quantity of sample;
- reproducible sampling;
- minimal interference on the performance of the detection system;
- sampled material must be representative of the bulk composition (stoichiometry);
- absence of matrix effects (no variation of analyte signal in different matrices);
- sample transport to the excitation or ionization source without losses;
- absence of memory effects and sample carryover;
- adjustment of sampling parameters to satisfy detection requirements.

Among the most popular direct solid sampling methods are LA, glow discharge (GD), and spark/arc ablation. Direct analysis of solids with GD sources has been used with both atomic emission and mass spectrometric detection. GD converts a solid sample directly into the atomic phase by sputtering processes. Because sputtering is a primary sampling mechanism, heating is minimal and preferential vaporization is minimized. GD is best suited for conductive samples, although nonconductive samples can be analyzed with a radiofrequency discharge. A GD source coupled to a high-resolution mass spectrometer is a powerful technique for achieving excellent depth profiling and detection characteristics with a limit of detection (LOD) in the range 10–100 ng g^{-1}.

For arc/spark ablation, mass is eroded from the sample in the form of atoms, molecules, vapor, droplets, solid flakes, and large particles. Normally, only conductive samples can be used with a spark discharge; however, nonconductive materials can be analyzed by mixing the sample with a conductive matrix.

Because of unique benefits and capabilities for direct solid sampling, LA has outpaced development of these other techniques. Laser sampling has been employed with flame and graphite furnace atomic absorption spectroscopy, with typical LODs in the micrograms per gram range. A description of these methods can be found in a review by Darke and Tyson.[1] The direct current plasma (DCP) and microwave-induced plasma (MIP) have been used for the analysis of solid samples with LA sampling.[2,3] The DCP and MIP offer good detection limits in the micrograms per gram range, with linear calibration curves over several orders of magnitude of concentration.

The ICP has become the prevalent source for chemical analysis with LA sampling.[4] The ICP plasma is robust, with high temperature and electron number densities, and is in most cases unperturbed by small amounts of sample. A separate article (**Inductively Coupled Plasma/Optical Emission Spectrometry**) is dedicated to the ICP. Slurry nebulization of powders, electrothermal vaporization (ETV), direct sample insertion (DSI), spark/arc and LA techniques are among the most popular methods investigated for solid sampling into the ICP. Each of these techniques has unique features: sensitive analysis of powders (slurry nebulization, ETV); analysis of small volumes of solids and liquids with selective solvent or matrix removal (ETV); direct sampling and atomization in the ICP plasma (DSI); analysis of conductive samples (spark ablation); and analysis of any sample (LA). A comparison of slurry nebulization, ETV, and LA methods was reviewed by Darke and Tyson.[5] DSI for ICP was recently reviewed by Sing.[6]

2 FUNDAMENTAL CHARACTERISTICS

Because of the numerous capabilities and advantages for atomic spectroscopy, there is a tremendous demand to understand fundamental LA processes. LA is complex, consisting collectively of many nonlinear mechanisms, each operative on different timescales, from femtoseconds to seconds. When the laser beam irradiates the surface, electrons will be released on the femtosecond to picosecond timescales, followed by atomic and molecular mass on the nanosecond timescale, and then the eruption of large particles, microseconds after the laser pulse has ceased (Figure 1a–c). With different processes occurring over so many orders of magnitude in time, it is no wonder the collective phenomena have never been combined into a unifying theory. However, it is critical to point out before discussing fundamental issues that LA sampling is easier than liquid nebulization. In fact, there have been many years of science devoted to proper dissolution procedures in order to achieve accurate analysis – and still one must be very careful when dissolving a sample that is truly 'unknown'. The underlying fundamentals of LA should not preclude its development as the method of choice for direct solid sampling in atomic spectroscopy.

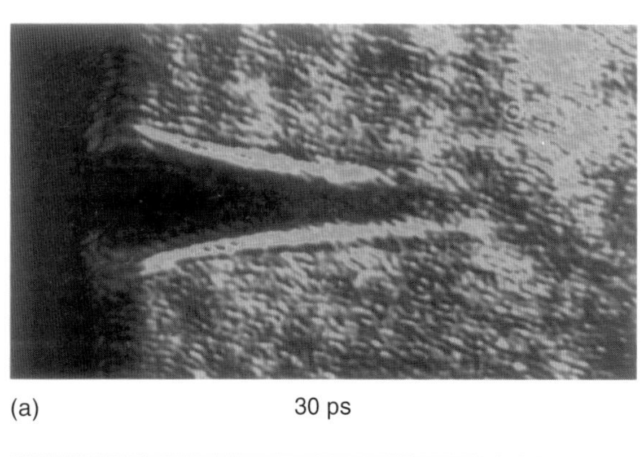

(a) 30 ps

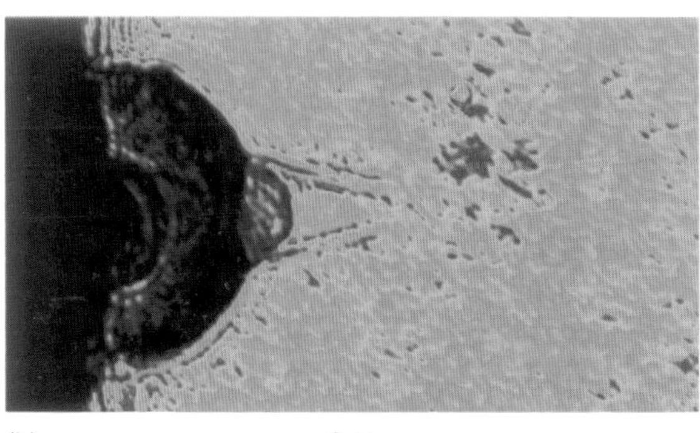

(b) 3 ns

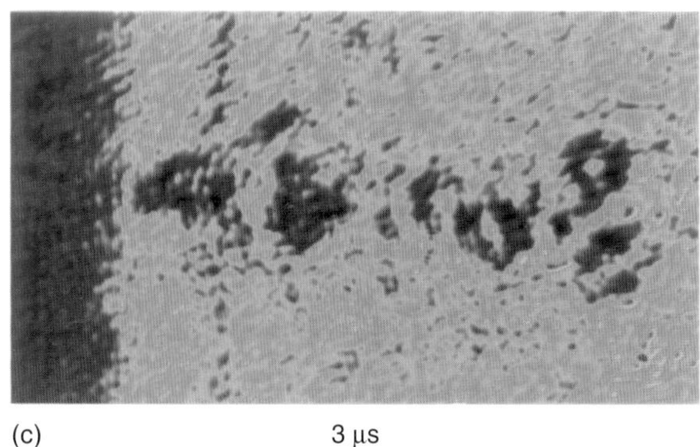

(c) 3 μs

Figure 1 LA involves complex and collective phenomenon, exhibiting significant nonlinearity in space and time. The images show the evolution of the laser explosion from a solid surface. (a) At 30 ps after the laser hits the sample, electrons leave the surface and collide with air to form the plasma; (b) at 3 ns, atomic vapor escapes from the surface; and (c) at 3 μs, large (>20 μm) particles are spalled from the bulk.

Studying fundamental LA mechanisms will provide further improvements and benefits for atomic spectroscopy. By understanding ablation, it will be possible to control process variables, such as the ability to couple the laser beam efficiently into the sample, ablate a reproducible quantity of mass, control the amount of mass ablated, minimize preferential ablation, produce stoichiometric ablation, and control the particle size distribution. By controlling these variables, LA in atomic spectroscopy will become a routine chemical analysis technology for

For references see page 9502

environmental, health, forensics, nonproliferation, and other applications where the primary sample is in the solid phase.

2.1 Ablation Processes

Many studies have been dedicated to understanding fundamental ablation mechanisms. Research studies have shown that LA of a solid sample consists of several stages, in which different kinds of 'vapor products' are ejected. The initial stage is electronic excitation inside the solid, accompanied by ejection of electrons at the sample surface, due to both photoelectric and thermionic emission. During this time, energetic electrons in the bulk of the solid also transfer energy to the lattice through a variety of scattering mechanisms; the sample target then undergoes melting and vaporization, followed by ionization and formation of a plasma plume consisting of the sample constituents. The expanding plume interacts with the surrounding gas to form a shock wave, causing the ambient gas to become further ionized. The expanding high-pressure plasma exerts a force back to the target, which flushes out the melted volume. This recoil pressure and flushing mechanism can produce large-sized particles (several micrometers). There is evidence of explosive boiling, occurring microseconds after the laser is finished, which also can produce micrometer-sized particles. Atomic- and micrometer-sized particles will both be transported to and digested in the ICP. Therefore, it is important to understand mechanisms of atomic- and micrometer-sized particle generation; particles in each of these size ranges can have different chemistries, affecting analytical accuracy. The particle size distribution will also affect transport efficiency and therefore analytical sensitivity.

To achieve accurate LA sampling, the composition of the ablated mass should be the same as the sample composition. To achieve high sensitivity, all of the ablated mass should be transported to and digested in the ICP. There are four primary processes that influence the accuracy and sensitivity in laser ablation/inductively coupled plasma (LA/ICP). The first is the LA process itself. For different laser conditions, the composition and quantity of the ablated mass can change significantly. The particle size distribution and particle composition can also change with laser conditions. The second process is transport, which includes the sample chamber design and tubing used to carry the ablated mass to the ICP. Transport efficiency will be different for different sized particles. The third process is sample digestion and excitation in the ICP. The conditions in the ICP (electron temperature and number density) which are controlled by ICP power, gas flow rate, and height above the load coil will influence vaporization, atomization, and ionization

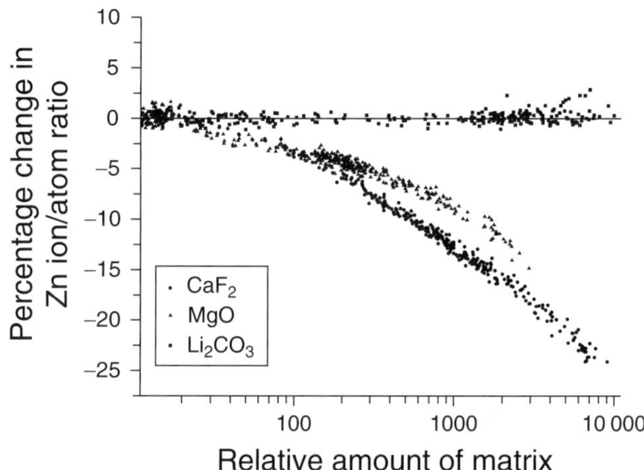

Figure 2 Percentage ratio change of Zn ionic to atomic emission lines in the ICP versus amount of matrix. The ratio of ionic to atomic line intensity correlates to the ICP conditions. LiF, CaF_2 and MgO were the matrices.

of the analyte. However, the amount and composition of the analyte and the matrix can change the ICP conditions (Figure 2). The fourth process includes detection, using the mass spectrometer for inductively coupled plasma mass spectrometry (ICPMS) or the spectrometer and photon detector for inductively coupled plasma atomic emission spectroscopy (ICPAES). These processes are discussed separately below as to their influence on accuracy and sensitivity.

2.2 Processes Affecting Accuracy

An expectation of any analytical procedure is accurate analysis; the detected mass composition must be the same as the sample composition. For LA sampling, fractionation (preferential mass removal during LA), which can cause inaccurate analysis, can occur under some conditions. Numerous studies have shown that fractionation can be minimized or eliminated, depending on the sample and laser properties.[7–13] The distribution of elements in the matrix, their status (inclusions), and zone migration will influence fractionation.[11,12] Fractionation at the sample is a function of wavelength; ultraviolet (UV) LA has been found to provide better stoichiometry than infrared (IR) LA.[14–16] The number of laser pulses at a fixed location on the sample, the laser irradiance, and laser beam spot size (related to focus position) also influence fractionation (Figure 3).[13,16,17] For fundamental purposes to describe fractionation at the sample, attempts have been made to correlate ablation behavior with melting, boiling, vapor pressure, atomic or ionic radius, charge, and speciation.[1,15,18]

The mechanism for atomic mass ejection can be thermal vaporization and/or plasma sputtering. The sputtering

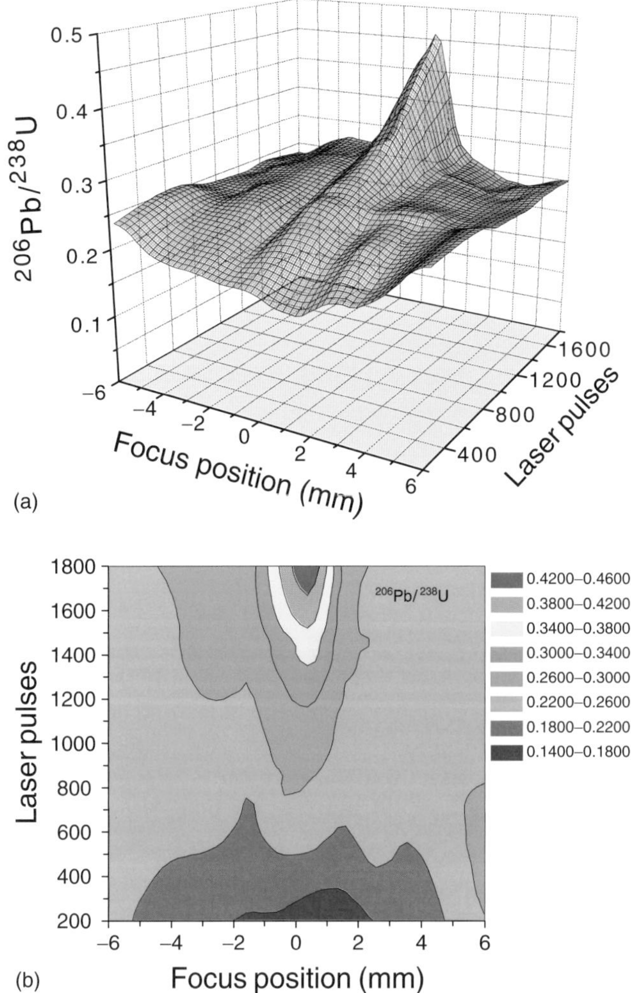

Figure 3 Pb–U fractionation versus number of laser pulses and lens-to-sample focal position (a) three-dimensional plot; (b) contour plot. Neodymium : yttrium aluminum garnet (Nd : YAG) laser, $\lambda = 213$ nm; pulse duration = 6 ns; energy = 1.15 mJ; sample: NIST SRM 610 glass in which the $^{206}Pb/^{238}U$ stoichiometric ratio is 0.2249. The positive and negative values represent the laser focus above and below the sample surface, respectively.

process involves the high-temperature, high-pressure plasma colliding with the sample surface. The ablation rate in these two processes will depend on the bond energy; the stronger the bond, the less mass is removed. Fractionation on the nanosecond timescale can be related to thermal properties of the sample constituents. Fractionation in alloys depends on the latent heat of vaporization; the higher the latent heat, the more difficult it is to vaporize a constituent. For example, in brass, the latent heat of vaporization of Cu is greater than that of Zn;[10] Zn can be preferentially ablated compared with Cu in these samples. By using a calibrated ICPAES instrument, fractionation based on thermal vaporization was demonstrated by measuring Zn/Cu mole ratios during ablation of brass. With a 30-ns, 248-nm excimer laser, the Zn/Cu mole ratio initially decreased with increasing laser irradiance, then stabilized at irradiance greater than approximately 0.3 GW cm^{-2}. The initial decreasing Zn/Cu ratio was due to thermal vaporization.[14] For higher laser irradiance, there are likely several competing mechanisms involved in the ablation processes. Melt flushing from the crater, plasma shielding, and/or radiative heating by the laser-induced plasma can contribute to mass ejection.[7,10,19,20] Picosecond LA is even more esoteric; nonthermal mechanisms must be operative. Using a 35-ps pulse (Nd : YAG, $Nd^{3+} : Y_3Al_5O_{12}$) laser, some type of nonthermal mechanism appears to govern the LA process because Cu is enriched in the vapor at lower irradiance. The Zn/Cu ratio approaches the stoichiometric value at higher irradiance. These experiments demonstrate that different mechanisms contribute to fractionation, although identification of the mechanisms is still a large research effort.

Fractionation during transport and/or in the ICP can occur.[21,22] Fractionation during transport involves selective vapor condensation on the tubing walls or the selective nucleation of species on different-sized particles.[12] These effects are governed by the vapor-phase morphology (droplet/particle shape and size) and chemical composition. Because transport efficiency and chemical composition are particle size dependent, fractionation can occur during transport. Figg et al.[22] demonstrated this effect by inserting a coiled Tygon tube into the transport path; both particle size distribution and fractionation changed significantly. It is believed that many of the large particles come from flushing out the molten liquid layer. The composition of the molten liquid could be significantly different from the bulk because of preferential vaporization and re-deposition from the plasma plume. For "large" particles entering the ICP and undergoing sublimation (vaporization), fractionation may exist in the ICP itself.[21,22] It is important to state here that fractionation is not a problem that precludes the use of LA for accurate chemical analysis, especially when matrix-matched standards exist. This section was designed to make the reader aware that such issues can exist. There are many applications of LA, as will be discussed in a later section, in which fractionation does not influence the analysis.

2.3 Processes Affecting Sensitivity

The signal intensity in atomic emission spectroscopy (AES) or mass spectrometry (MS) corresponds directly to the quantity of mass ablated and transported to the ICP. Depending on the concentration of the elements in the sample, it may be necessary to increase the quantity of ablated sample to achieve better signal-to-noise

For references see page 9502

ratios. Laser-beam properties (such as wavelength, pulse duration, energy, fluence, irradiance, and temporal and spatial profiles) and the ambient gas influence the quantity of mass ablated per laser pulse.[7,15,16,21,23–33]

2.3.1 *Laser Energy*

The ablated mass can be increased by increasing the laser-beam energy or fluence (energy per unit area). However, the relationship of ablated mass to laser-beam energy or irradiance is not linear;[7,18,19,34] mass was found to follow a power law with irradiance (I^m).[18,19,34] Using nanosecond laser pulses, with irradiance less than approximately $0.3\,\mathrm{GW\,cm^{-2}}$, m had values ranging from 2 to 5 for many samples. When the irradiance was $>0.3\,\mathrm{GW\,cm^{-2}}$, m became <1. However, as the irradiance was increased further, another change in the mass ablation rate occurred, in which m increased to >2. In the lower irradiance region, thermal vaporization was found to be a dominant process,[14] as evidenced also by fractionation (discussed earlier). In the middle laser irradiance range, plasma shielding may be a major factor governing ablation. The laser-induced plasma formed above the target surface can absorb and/or reflect incident laser energy, thereby decreasing the efficiency of laser energy available for mass ablation.[35] A plasma can transmit just enough energy to the surface to sustain itself. The density and temperature of the plasma can adjust in such a manner that the optical thickness remains constant; the final proportion of the laser energy absorbed and transmitted is constant. It is important to point out that this plasma is the basis of another chemical analysis technique, discussed in the article **Laser-induced Breakdown Spectroscopy**.

In the third irradiance region, mechanisms such as phase explosion and spallation may be dominant.

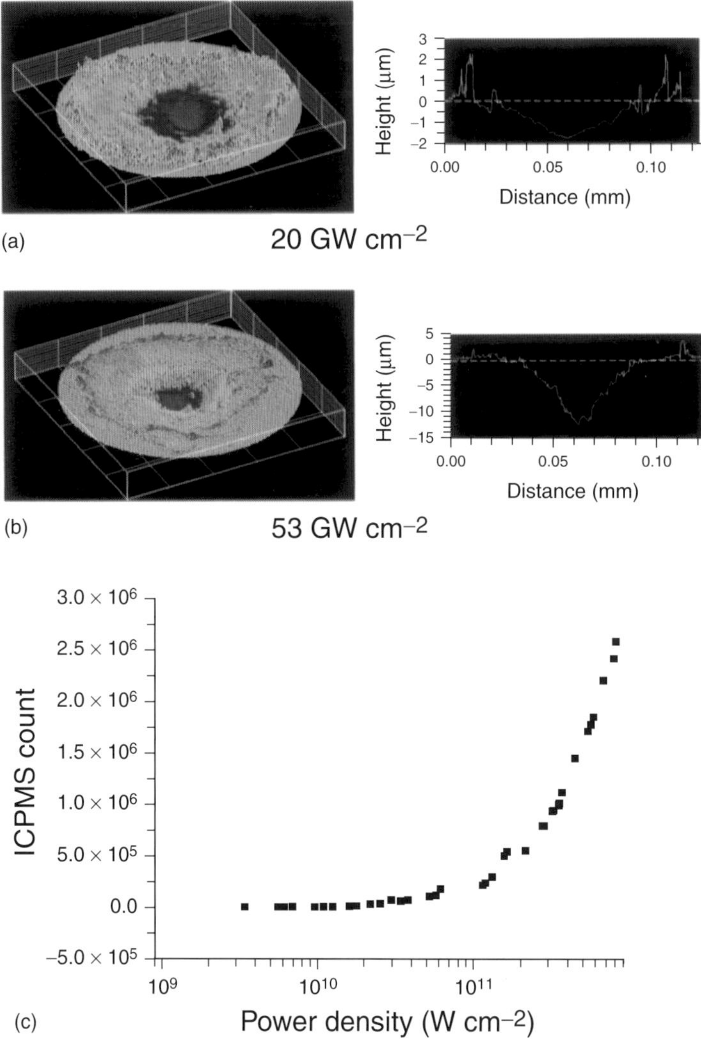

Figure 4 Crater profiles (a) before and (b) after an explosive boiling threshold. The crater depth changes dramatically from 1 μm to more than 10 μm. (c) ICPMS signal versus laser power density. The ICPMS intensity increases dramatically across the threshold.

Both crater depth and ICPMS intensity (Figure 4a–c) show dramatic increases at a threshold irradiance of 20–30 GW cm^{-2}. This dramatic change may be caused by explosive boiling.[36] The superheated molten liquid can experience an increased fluctuation in its density[37] when the temperature and pressure approach the critical point. Near the critical point, this fluctuation can generate vapor bubbles in the superheated liquid. For vapor bubbles larger than a critical radius, bubble growth will occur; bubbles smaller than the critical radius will collapse.[37] Once bubbles of a critical radius have been generated in the superheated liquid, the volume undergoes a rapid transition into a mixture of vapor and liquid droplets. During explosive boiling, rapid expansion of the high-pressure bubbles in the liquid leads to a violent ejection of the molten droplets from the sample. The shadowgraph images of liquid droplets ejected from the silicon surface (see Figure 1a–c) indicate that the onset of the explosive boiling is at an irradiance of approximately 2.2×10^{10} W cm^{-2}.

2.3.2 Laser Pulse Width

Laser wavelength and pulse duration also influence the quantity of ablated mass; the shorter the wavelength and pulse duration, the better the mass removal efficiency.[15,16,23–26,28–30,38,39] Using an Nd:YAG laser with a 3-ns pulse duration, the ICPAES intensity was found to be almost an order of magnitude greater for UV than IR ablation sampling, with the same fluence. The absolute enhancement was found to be a function of laser irradiance.[39] When plasma shielding does exist, a lower wavelength is better because there is less absorption by the plasma; more energy is used to remove mass instead of heating the plasma. The shorter the laser pulse duration, the more efficient is the ablation process; picosecond laser pulses provided an order of magnitude greater signal intensity in ICPAES compared with a nanosecond pulsed laser with the same fluence. An explanation for increased mass ablation using picosecond pulses is that plasma shielding may be weaker; more laser energy is coupled to the sample than is absorbed by the plasma. Another possible mechanism is that more laser energy is converted to ablated mass instead of being lost in the sample through thermal dissipation, which is a function of pulse duration.[40] If the pulse duration is femtoseconds, the ablation process should be even more efficient.[32,33] For femtosecond LA, the generation of vapor and a plasma will occur after the laser pulse is finished. Therefore, there should be no plasma shielding on the femtosecond timescale. Thermal diffusion to the solid also should be negligible. Currently, there are no reports of using femtosecond LA for chemical analysis. However, as femtosecond laser technology becomes more reliable, it is expected that new benefits to chemical analysis (and other applications) will be forthcoming.

2.3.3 Gas Environment

The gas atmosphere in the sample chamber can have a dramatic effect on ablation behavior.[7,30,31,34,41,42] The use of different gases has been found to enhance sensitivity either by reducing plasma shielding or by influencing the ablated particle size distribution. With He in the ablation chamber, the mass ablation rate was 2–5 times and 10 times greater than that with Ar in the chamber for nanosecond and picosecond laser pulses, respectively.[13,43] He and Ne provided an increased mass ablation efficiency compared with a decrease with Xe and Kr, relative to Ar for the nanosecond-pulsed LA. The enhancement or depression was found to be dependent on the laser irradiance. For picosecond LA sampling, only He provided an enhancement; there was very little influence on the mass ablation behavior by the other noble gases.

3 ANALYTICAL CHARACTERISTICS

A typical experimental system for LA sampling with the ICP is shown in Figure 5. In general, mass ablated from a target surface inside the sample chamber is entrained into an Ar gas flow and transferred into the ICP source, where particles are first vaporized and then atomized and ionized. In most cases, the sample chamber is placed on a motorized or manually controlled micrometer translation stage. This is especially convenient for the analysis of heterogeneous samples, where different areas on a sample surface are to be analyzed. The individual components are described separately below.

3.1 Ablation Chamber and Transport Tubing

Various types of chambers have been described in the literature; a summary can be found in a review by Moenke-Blankenburg.[44] Typically, the sample is placed inside a chamber. However, there are cases in which the sample itself was the bottom of the chamber.[45] The simplest chamber can be a glass tube with two ports for gas flow, and a quartz window (or other transparent material at the laser wavelength) for laser beam delivery. Argon gas flow dynamics inside the chamber play a significant role in particle entrainment. With laminar flow in the chamber, entrainment efficiency and thus detection characteristics are improved; turbulent flow contributes to losses by trapping particles in stagnant flow regions.[46] The volume of the chamber is an important parameter; a larger internal volume can lead to sample

For references see page 9502

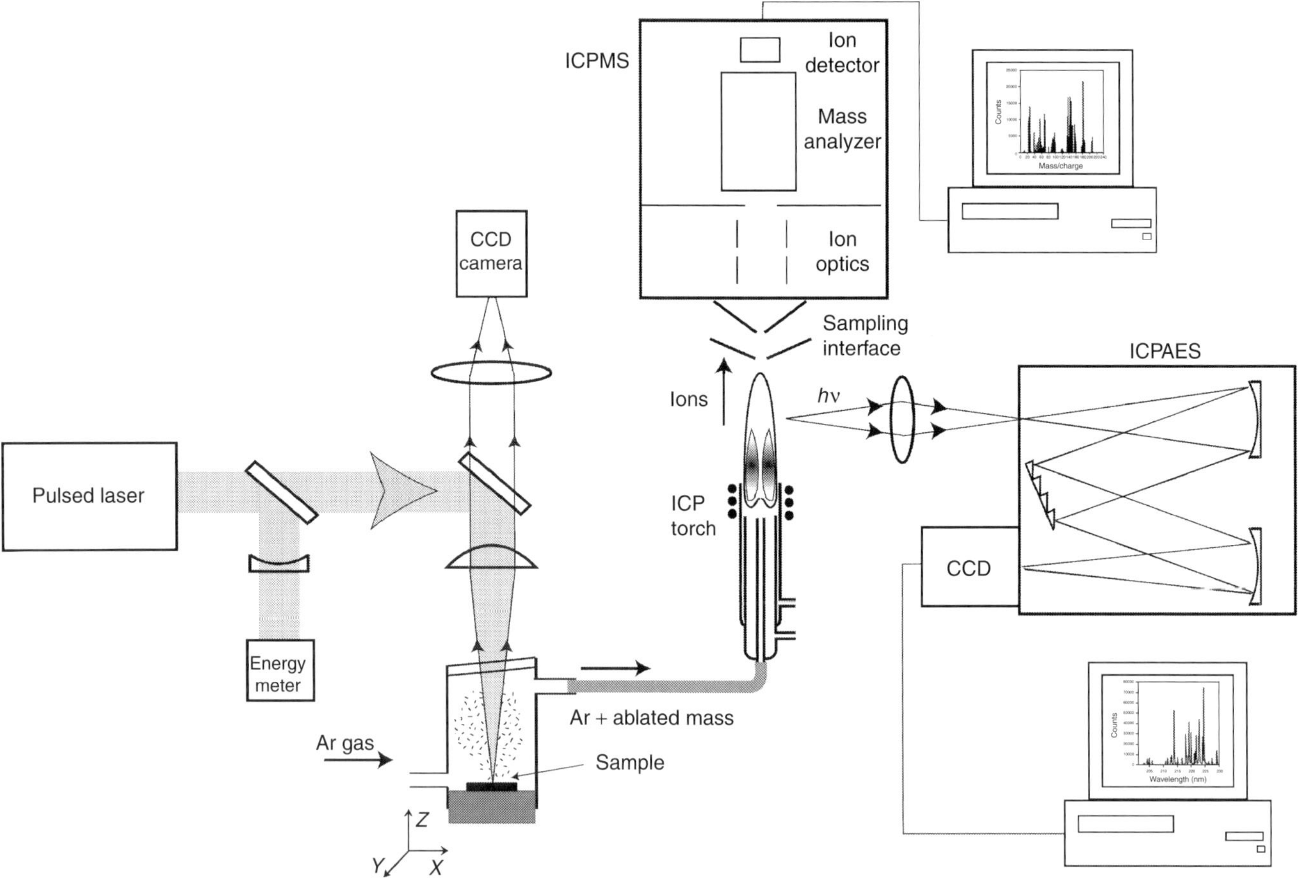

Figure 5 Schematic diagram of LA sampling into an ICP source. CCD, charge-coupled device.

dilution and memory effects. Because LA processes are transient in nature, larger chamber volumes may be advantageous for signal averaging during repetitive sampling experiments. Another important parameter is chamber length; if the laser beam window is located too close to the sample surface, vapor deposition can occur, reducing transmission and therefore the laser energy. Another issue concerning the chamber length is that different times are required for particles of different sizes to be entrained by the flow; larger particles will travel further before becoming entrained.(47) Depending on the flow velocity inside the chamber, large particles can be lost owing to collisions with the walls before entrainment.

Dispersion of different-sized particles in the ablation chamber depends on the Ar gas flow velocity in the chamber and can contribute to broadening of signal peaks (recorded during single laser pulse experiments). Further dispersion of ablated particles in the transport tube can occur and depends on the transfer tube length and internal diameter. To minimize dispersion, a short, narrow tube should be used. Although particle dispersion in the chamber and transport tube may be important for analysis during single laser pulse experiments, there are no significant advantages to using a short tube during LA sampling with repetitive pulses.

3.1.1 Mass Transport Efficiency

For a given ablation chamber/tube configuration, particle transport efficiency depends on the ablated particle size distribution. The number of particles and their size distribution depends on the laser and sample properties.(22,47,48) The volume distribution of laser-generated particles changes with respect to the laser wavelength; there are more large particles generated with an IR laser than with a UV laser.(48) Since melting and melt-fusing may be responsible for the generation of particles, a thicker molten layer may produce a greater fraction of large particles. The optical absorption depth in most solid samples is greater for IR than UV wavelengths. Therefore, a larger molten liquid volume may be responsible for large particle production using IR LA.

Entrainment of ablated mass into the gas stream and transport to the ICP are particle size dependent. Large particles may not be entrained and those that are may not completely vaporize in the ICP. Particle sizes should be less than about 2 µm for efficient transport to and excitation in the ICP.[45,49,50] Particle entrainment efficiency can be defined as the ratio of mass entering the ICP to the total mass ablated from the sample. Particle entrainment efficiency has been found to decrease with increasing laser irradiance. Entrainment efficiency was about 25% at low irradiance and decreased to about 5% at high irradiance.[47] One possible reason for the small entrainment efficiency at high irradiance may be the formation of excessively large particles (>5 µm); large particle ejection is observed in Figure 1(a–c). The removal of large fractured pieces (>~50 µm), possibly due to increased thermal stress and pressure on the sample surface, was also observed after ablation at high irradiances (Figure 6). These very large particles will not be entrained into the argon gas flow, but instead will settle in the ablation chamber due to gravity. Transport efficiency as a function of particle-size distribution needs to be critically studied for improving LA sampling in atomic spectroscopy.[45,51] A few studies have addressed particle transport, with preliminary data suggesting that particles in the 0.1–1.0 µm range reach and vaporize in the ICP.[22,45,48]

Literature values for LA transport efficiency are in the range of about 5–40%.[45,47] The large variation represents the effects of laser beam conditions, sample material properties, and ablation chamber/transfer tube geometries. In general, particle losses in the sample chamber and transport tube are mainly due to gravitational settling or inertial impact to the walls for large particles, and diffusion to the walls for smaller particles. These processes depend on the tube length and internal

Figure 6 Scanning electron microscope image of an ablated glass surface. A large piece (~50 µm) of the cracked sample was removed from the ablation spot.

diameter. Transport efficiency for larger particles, which represent most of the ablated mass and thus are responsible for most of the ICP signal, can be improved by utilization of a short, narrow tube. Generally, particles in the range 0.1–1.0 µm are most efficiently transported to the ICP. For a given tubing length and diameter, theoretical calculations predict losses to be flow dependent. In practice, however, no significant differences in the ICPMS signal count rates were observed when the Ar flow rate was varied from 0.1 to 0.9 L min^{-1} during ablation of Zr metal. Particle losses inside the sample chamber and transport tube can contribute to memory effects. To minimize these effects, it is recommended that the transport tube be cleaned regularly or replaced, and the system flushed with Ar gas at increased flow rates between measurements.

3.2 Laser Systems and Optics

Over the years, several pulsed lasers have been tested for ablation. Ruby (694.3 nm), CO_2 (10.6 µm), free-running Nd : YAG and excimer lasers have been shown to efficiently ablate solid samples.[1] However, Q-switched Nd : YAG lasers with nanosecond pulses have become the most prevalent systems used for LA today. These lasers are relatively inexpensive, easy to operate, and compact, have good pulse-to-pulse and long-term stability characteristics, and deliver sufficient energy for ablation. The fundamental harmonic of 1064 nm can be easily doubled (532 nm), tripled (355 nm), quadrupled (266 nm), and quintupled (213 nm). The Q-switched Nd : YAG laser operated at the fourth harmonic is currently the most popular for LA chemical analysis. Recently, it has been shown that samples with relatively low absorption at 266 nm, such as calcite and garnet, can be more efficiently ablated with the Nd : YAG at 213 nm[52] or the ArF excimer (193 nm).[53]

The laser beam is transferred to the sample chamber by means of mirrors, beam splitters, and/or prisms. In the simplest system, focusing is achieved with a plano-concave singlet lens, which has a low degree of spherical aberration. Laser beam spot size at a sample surface can be easily adjusted by translating the lens relative to the sample. The minimum spot size with radius, w, that can be achieved by focusing a Gaussian diffraction-limited laser beam can be approximated by Equation (1):

$$w \propto \frac{\lambda f}{\pi w_0} \quad (1)$$

where λ is the laser wavelength, f is the lens focal distance, and w_0 is the initial laser beam radius. By using a lens with a short focal distance and a wide initial beam diameter, the smallest LA spots (best spatial resolution) can be achieved. An optical microscope equipped with

For references see page 9502

a CCD camera is an excellent addition to experimental systems, allowing easy focusing of the laser beam on to a selected sample location, with beam spot sizes in the order of several micrometers. Such systems are widely used, especially for the analysis of geological samples where a high degree of spatial resolution is required.

3.3 Inductively Coupled Plasma Atomic Emission Spectroscopy and Inductively Coupled Plasma Mass Spectrometry Instrumentation

In contrast to liquid nebulization sample introduction, LA sampling is transient in nature. Thus, simultaneous detection of elements is needed. Single-channel instruments equipped with a photomultiplier tube (PMT) require scanning of a monochromator grating. Such systems are not very useful for transient signal detection when detection of more than one element is required. Direct-read polychromators can employ more than 60 exit slits with PMTs for detection of numerous wavelengths, providing simultaneous multielement capability. The photodiode array (PDA) and more recently the charge transfer device (CTD) used with a conventional Czerny–Turner-type spectrometer can cover spectral windows of several tens of nanometers simultaneously. Spectrometers with Paschen–Runge polychromators and with échelle gratings, equipped with solid-state detectors, have multielement capability and provide simultaneous UV and visible spectral coverage.

Most commercially available ICPMS instruments utilize radiofrequency quadrupole mass analyzers. Although quadrupole mass analyzers typically have a low resolution capability (ca. 0.5 u), their low cost, ease of coupling with an ICP source, and pseudosimultaneous mass detection make them attractive for elemental analysis with LA applications. The complete mass range from 1 to 250 u can be scanned in about 0.1 s.

Many LA applications require high-precision isotopic ratio measurements; high-resolution double-focusing ICPMS instruments equipped with multicollector array detectors are well suited for this purpose. ICPMS instruments with time-of-flight (TOF) analyzers were recently demonstrated with LA sampling.[(54)] The TOF approach allows the collection of several thousand complete mass spectra per second. Owing to the transient nature of LA, ICPMS with a TOF analyzer is potentially advantageous for multielemental determinations, although to date only limited research has been conducted.

The detection capabilities of LA with ICPAES and ICPMS techniques depend significantly on the experimental conditions and equipment. ICPMS is typically used for minor and trace elemental analysis, whereas ICPAES has lower sensitivity and is primarily used for the analysis of major and minor constituents. Most geological applications, which require a high degree of spatial resolution for accurate microanalysis of inclusions and grains in minerals, utilize ICPMS instrumentation.

3.4 Detection Limits

One of the advantages of dry ICP conditions from LA sample introduction is that plasma excitation/ionization temperatures and electron number densities are typically higher than with a wet plasma produced during liquid nebulization. Dry plasma conditions enhance ionization and excitation processes. This benefit, along with simplified mass spectra (due to reduction of polyatomic interferences), provides improved detection characteristics. Absolute detection limits (absolute detectable amount of analyte mass) for rare-earth elements in silicate samples were shown to be two orders of magnitude better for LA ICPMS than liquid nebulization sample introduction.[(55)] However when relative detection limits (detectable analyte concentration in a sample) are compared, liquid nebulization sample introduction is better. This situation only exists because the amount of mass nebulized into the ICP is much greater than that ablated by the laser. Assuming a 2% efficiency for conventional pneumatic nebulizers with a $1\,\mathrm{mL\,min^{-1}}$ sampling rate, roughly $20\,\mathrm{mg\,min^{-1}}$ of solution is introduced into the ICP. In contrast, the amount of ablated mass per laser pulse is typically only 1 ng–1 µg. Hence for a laser repetition rate of 10 Hz and a particle transfer efficiency of 40% (maximum value reported in the literature), the amount of mass introduced into the ICP is only from $0.2\,\mathrm{\mu g\,min^{-1}}$ to $0.2\,\mathrm{mg\,min^{-1}}$. Hence the mass per unit time introduced into the ICP as a result of LA is about 10^2–10^5 times less than that from liquid nebulization. Relative detection limits depend on the amount of mass and are therefore better for liquid nebulization sample introduction.

Typical detection limits, determined as three times the standard deviation of the blank, are in the low micrograms per gram and even nanograms per gram range for ICPMS detection with LA, in contrast to the picograms per gram level for liquid nebulization. For LA sampling with ICPAES, laser energies of 10–100 mJ are required, compared with only few milli- or even microjoules for sampling into the ICPMS. Typical LODs for LA sampling with ICPAES are in the micrograms per gram range and are comparable to those available from liquid nebulization sampling with ICPAES. Better detection limits can be achieved by increasing the amount of ablated mass per laser pulse, which can be realized by improvement in the laser energy coupling efficiency to the sample and/or improvement in entrainment/transport efficiency. When a sample was ablated inside the ICP torch, just below the discharge, particle losses associated with mass entrainment and transport were eliminated, and a significant improvement in detection characteristics was achieved.[56]

3.5 Calibration and Optimization

The signal intensity recorded by ICPAES and ICPMS during sample introduction (both liquid nebulization and solid sampling) is proportional to the concentration of an element of interest in the sample. Absolute measurements require calibration procedures to be established, which remains an essential issue for chemical analysis.[1,5,44,57] For liquid sample introduction, standards are relatively easy to obtain in the form of single- or multielemental solutions; linear calibration curves can be generated over three to six orders of magnitude for ICPAES and eight orders of magnitude for ICPMS.[57] Assuming similar viscosities for the diluted sample and the standard solutions, the amount of aerosol aspirated into the ICP is determined by the nebulizer/spray chamber parameters. In contrast, during LA the amount of mass ablated and transported into the ICP may be different for samples and standards, if the standards are not matrix matched. For LA, an internal or external standardization procedure is required to compensate for changes in the quantity of mass ablated, even when the analyte concentration remains constant. Matrix-matched solid standards are generally required for instrument calibration.

3.5.1 Internal Standardization and Matrix Matching

The ablated mass is determined by the properties of the matrix, and close matching of calibration standards to samples is preferred. However, the chemical form in which the analyte is present in the sample was shown to influence the elemental response.[58] Compensation for this effect often can be performed by normalization of analyte intensity to that of another element (internal standard). This procedure requires that the concentration of the internal standard, usually the matrix element, be either known [determined from independent methods, such as X-ray fluorescence (XRF) spectroscopy] or be constant in the sample and standards. In most cases, an internal standard provides excellent compensation for differences in ablation behavior for samples and standards. For trace elements in homogeneous samples, this approach has been used to improve the measurement precision to better than 1%.[18,49,59] Internal standardization is especially useful for bulk analysis with samples and standards pressed into pellets,[60] fused into glass beads,[49] or preconcentrated in an NiS button[61] where the internal standard is added during sample preparation. A single calibration graph was used for analysis of silicate rocks and limestones fused with Sc and Y oxides as internal standards.[62] For cases in which relatively "good" standards were available, LA proved to be a reliable and accurate chemical analysis procedure. In some cases, however, internal standardization can be limited by differences in the ablation behavior of the sample and the standards, especially when standards are not matrix matched. For example, elemental fractionation of W relative to the internal standard Ca was shown to be different for a scheelite sample and a silicate glass calibration standard.[63]

Calibration without an internal standard has been shown to be possible, especially in cases where only trace level impurities differ among the standards and sample; the trace impurities do not effect ablation behavior. Linear calibration curves were established and accurate trace elemental analysis was demonstrated for U,[64] Au and Ag,[65] and glass[66] without internal standardization. In contrast, if the matrix properties change because of changes in the analyte concentration, the LA behavior is affected. For example, for the analysis of Zn in brasses, where Zn is a major element, linear calibration cannot be achieved, and other signal normalization procedures are needed for accurate calibration.[67]

When internal standardization procedures are used for analyte characterization in an unknown sample, Equation (2) is applied:

$$C_{\mathrm{sa}}^{\mathrm{M}} = C_{\mathrm{sa}}^{\mathrm{IS}} \times \frac{I_{\mathrm{sa}}^{\mathrm{M}}}{I_{\mathrm{sa}}^{\mathrm{IS}}} \times \frac{C_{\mathrm{st}}^{\mathrm{M}}}{C_{\mathrm{st}}^{\mathrm{IS}}} \times \frac{I_{\mathrm{st}}^{\mathrm{IS}}}{I_{\mathrm{st}}^{\mathrm{M}}} \tag{2}$$

where C is the concentration and I is the ICP signal intensity. Subscripts sa and st denote sample and standard and superscripts M and IS correspond to the unknown analyte and the internal standard, respectively. This relationship assumes that the concentration of the internal standard is known and that fractionation between the analyte and internal standard is not significant, or is the same for analyte and standards.

3.5.2 Calibration with Liquid Standards

For many samples, matrix-matched standards will not be available. In these cases, external calibration can be performed using nebulization of liquid standards; a dual sample introduction method can be used to establish instrumental calibration with a series of standard solutions. To analyze the unknown sample by LA using liquid standards, wet ICP plasma conditions must be maintained.[68] However, the use of liquids for calibration defeats two important advantages of LA and dry ICP: elimination of isobaric interferences in ICPMS and the generation of solvent waste.[12] In addition, optimum ICP conditions will not be the same for LA and solution nebulization. For nebulization, the analyte dries from liquid droplets to form small particles in the plasma. For LA sampling, larger dry particles are introduced directly into the ICP.[45,48] Atomization and excitation processes in

For references see page 9502

the ICP are expected to be different for these two cases. Even if the water content is the same in the ICP, it is still possible that the excitation characteristics will be different because of the different vaporization mechanisms for nebulized solution and laser-ablated particles. On the other hand, calibration with dried solution aerosol[69] or by direct LA of liquids[70,71] does not significantly perturb dry ICP conditions.

LA of liquids has been shown to be effective for the analysis of microscopic fluid inclusions in minerals. In this method, direct ablation of a standard solution was used for external ICPMS calibration with Na as an internal standard.[71] In another study, instrument calibration was done by using artificial fluid inclusions prepared by drawing a standard solution into a microcapillary tube[72] or with microwells containing aqueous solutions.[73]

3.5.3 External Standardization

For some samples, internal standardization may not be a viable option. In such cases, external standardization must be used to compensate for differences in ablation of the sample versus the standards. Several external procedures have been demonstrated to compensate for changes in the quantity of ablated mass. They include light scattering,[48,74,75] acoustic emission in the sample[19,76,77] or in the ambient medium,[76,78] the use of a mass monitor to collect a portion of the laser-ablated aerosol,[79] and the measurement of spectral emission intensity in the ICP and laser-induced plasma simultaneously.[34,77] An absolute method to quantify the amount of mass ablated is to weigh the sample before and after ablation.[80] However, such direct mass measurements are tedious and may not be accurate because of the small quantity ($<1\,\mu g$) of mass ablated for each laser pulse, and because transport efficiency is not included.[22,45] Determination of mass ablation and transport efficiencies is possible by collecting particles on a nonporous filter, which is an indication of the total mass transported to the ICP.

4 APPLICATIONS

Because of the unique properties of the laser, many novel applications for chemical analysis using the ICP have been demonstrated. This section presents a brief overview of applications that would be difficult or impossible to perform without the use of LA. Most of the applications utilize ICPMS because of its enhanced sensitivity.

4.1 Environmental and Oceanography

4.1.1 Tree Rings

Spatial patterns in the chemical content of tree rings can be used as a monitor for changes in atmospheric conditions, changes in soil chemistry, and the pollution history of an area. Laser ablation inductively coupled plasma mass spectrometry (LA/ICPMS) is an excellent approach to determine chemical content in tree samples because of the high spatial resolution provided by the focused laser beam, and the excellent sensitivity of ICPMS to measure very low detection levels for many elements.[81–84] Watmough et al.[83] obtained the quantitative multielement analysis of 11 elements in red maple tree rings; significant changes in these elements were measured for trees grown in contaminated soils adjacent to a metal smelter versus trees grown in unpolluted soils. GarbeSchongberg et al.[84] analyzed pine and birch tree rings from Norway and Russia and discussed the elemental relationship with the pollution history of these areas.

4.1.2 Sea Shells

Trace-element fluctuations in sea shells represent environmental changes and major pollution events. In the work of Raith et al.,[85] the inner to the outer walls of a shell were analyzed; the elemental changes between growth bands of the shell showed a history of heavy-metal pollution over the years. Vander Putten et al.[86] measured spatial variations of Mg, Mn, Sr, Ba, and Pb in the calcite layer of *Mytilus edulis* sea shells.

4.1.3 Airborne Particulates

The analysis of trace elements in airborne particulates provides unique signature information for monitoring air quality and air pollution. For example, arsenic, considered the major "marker element" of air pollution, is one of the most hazardous anthropogenic air pollutants affecting humans globally. Traditionally, membrane filters with small pore size have been used for collecting particulate samples. The entire filter is then digested and the total volume analyzed. LA is a perfect alternative for such analyses; the particulates can be ablated directly from the filter media. Tanaka et al.[87] and Wang et al.[88–90] analyzed airborne particulates for more than 20 major, minor, and trace elements using this approach.

4.1.4 High-resolution Analysis of Coral

The concentration of trace elements in coral skeletons can provide information about changes in seawater properties; calcification in reef-building corals is strongly affected by environmental factors such as temperature, light, water motion, and pollution.[91] As coral grows, it secretes a calcareous skeleton into which trace elements are partitioned from the ambient seawater. Spatial analysis of the coral skeleton allows a detailed investigation of seasonal composition changes.[92,93] Compared with the

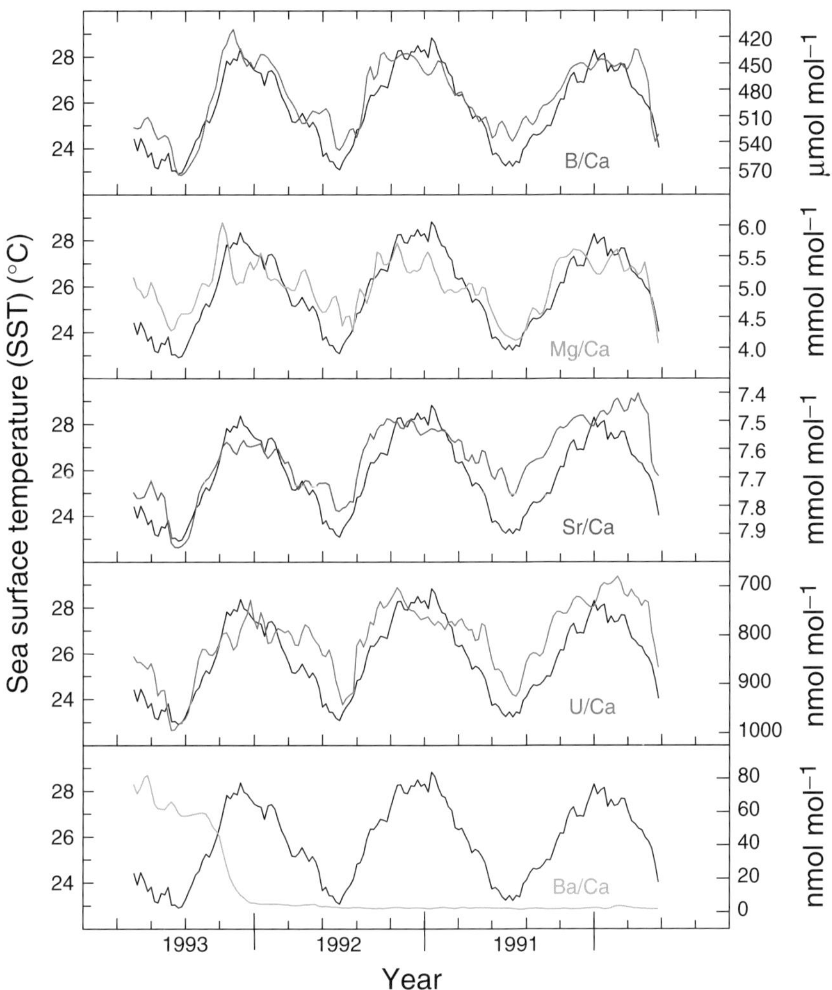

Figure 7 Trace element and sea surface temperature (SST) profiles for coral from Australia's Great Barrier Reef. Colored lines represent an average of three trace element profiles taken over the same coral track. Black lines are the instrumental SST data taken from the weather station. B, Mg, Sr, and U all display clear seasonal variation. Ba does not show a seasonality but displays a marked enrichment in the tissue zone. (Figure reproduced from original data with permission by Dr Daniel J. Sinclair[92].)

traditional method of sample milling, with processing and analysis by solution nebulization, LA/ICPMS provides in situ analysis of corals with spatial resolution less than 20 μm. Sinclair et al.[92] analyzed corals collected from Australia and showed that the elements B, Mg, Sr, and U exhibited seasonal variations, as shown in Figure 7. These fluctuations coincided with the changes of sea-surface temperature.

4.2 Geochemistry and Cosmochemistry

4.2.1 Uranium–Lead Geochronology

Zircon U–Pb geochronology is one of the principal dating tools used in the earth sciences; ages are calculated by measuring $^{206}Pb/^{238}U$, $^{207}Pb/^{235}U$, and $^{207}Pb/^{206}Pb$ ratios. The conventional method for U–Pb isotopic analysis has been by thermal ionization mass spectrometry (TIMS) with chemical separation. However, very low blank values are required and the data represent the "average" of the bulk grains. LA/ICPMS has the ability to perform spatially resolved in situ analysis of U–Pb isotopic compositions in zircons and similar minerals. Several groups[94–97] have studied zircon and monazite samples using LA/ICPMS. Most of these studies only show $^{207}Pb/^{206}Pb$ data and a few include the $^{206}Pb/^{238}U$ ratio, because of a fractionation problem. Without the $^{206}Pb/^{238}U$ ratio, the analysis cannot be extended to "young" (<600 Ma) zircon dating. In order to minimize Pb–U fractionation, methods such as active focusing, line scanning, and soft ablation (increasing the laser power when ablation is progressing) were attempted. Solving Pb–U fractionation is necessary to the success of LA/ICPMS for this application.

For references see page 9502

4.2.2 Inclusion Analysis

Microscopic inclusions in minerals contain direct evidence of the composition of fluids associated with large-scale material transport in the Earth's interior. Quantitative knowledge of the elements and isotopic composition of these fluid inclusions is a prerequisite for understanding and modeling fluid–rock interactions. Detailed chemical information is difficult to obtain because of the very small size of these inclusions (typically 10^{-11}–10^{-9} g).(70) Crush–leach analysis or bulk analysis of quartz containing fluid inclusions can provide concentration ratios averaged over many inclusions, but most samples contain multiple generations of fluid inclusions of different compositions. LA/ICPMS for the analysis of individual microscopic inclusion has, therefore, received considerable attention. Good quantitative results have been published mainly for synthetic fluid inclusions,(73,98) where a heavy trace element of a known concentration (e.g. Sr, U) was added as an internal standard. Günther et al.(70) reported a method for measuring complex polyphase inclusions using a stepwise opening procedure. A series of inclusions representing the fluid before, during, and after the deposition of cassiterite (SnO_2) in a tin deposit in Australia were analyzed by this method; physical and chemical mechanisms of ore precipitation were proposed based on these data.(99)

4.2.3 Precise In Situ Analysis of Hafnium, Tungsten, Strontium, Lead, and Osmium Isotopes

Many elements are of significant interest for isotope geochemistry and need to be measured at trace levels with excellent precision. Multiple collector inductively coupled plasma mass spectrometry (MC/ICPMS) with a magnetic sector is a new technology for the measurement of isotopic compositions with very high precision. It is particularly suited for elements with high first ionization potentials such as Hf, W, and Os, which cannot be measured with good precision using conventional TIMS. Combined with LA sampling, in situ isotopic measurements at the microscopic scale are possible. Although this technology is still in its infancy, diverse applications have already led to a number of important scientific developments.(100)

The initial Hf isotopic composition is more reliable than initial Nd as a geochemical tracer owing to the immobility of Hf. The Hf isotopic composition in zircon samples was analyzed by Thirlwall and Walder.(101) In low-temperature geochemistry, Hf isotopes may provide the most reliable and sensitive isotopic proxy for hydrothermal activity in the ocean. The Hf isotopic compositions in iron–manganese nodules and crusts were studied by Godfrey et al.(102) and the data reflected the concomitant growth from seawater and pore fluids. There is strong interest in measuring W and Hf isotopic compositions with high precision using LA with MC/ICPMS. The ^{182}Hf–^{182}W system is a method for constraining timescales of accretion and metal–silicate differentiation in planets; the age of the Earth's core, the Moon, and Mars have been measured using this new chronometer.(100)

Sr and Pb isotopic compositions are significant for geochemistry and oceanography research. Analysis of these elements traditionally requires complicated chemical separation procedures. LA with MC/ICPMS provided accurate and precise measurement of ^{87}Sr/^{86}Sr isotopic ratios in geological materials.(103) Pb isotopic compositions in a ferromanganese crust from the Pacific ocean were analyzed by Christensen et al.(104) and the authors concluded that the Pb isotopic data could be used to probe climate-driven changes in ocean circulation. Analysis of Os isotopic ratios for iridosmine samples was reported using LA with MC/ICPMS.(105) The mineral iridosmine has been used for the definition of the Os isotopic evolution of the mantle.

4.2.4 Bulk and Microbeam Analysis of Rocks and Minerals

Bulk and spatial analyses of rocks and minerals for elemental and isotopic compositions can provide fundamental information to help solve diverse geological and environmental problems. The application of LA/ICPMS for the analysis of whole-rock geological samples, such as pressed power pellets and lithium metaborate fusions, has been described by several authors.(106,107) In most of these studies, the measurement of rare-earth elements (Zr, Hf, U, Th, Sr Rb, Ba, Nb, Ta) was emphasized. LA/ICPMS was also used to determine the platinum group elements (Ru, Rh, Pd, Os, Ir and Pt) and gold, which have very low natural abundance but great economic and geological importance.(108)

The spatially resolved analysis of elements within minerals provides crystal-growth information and the variation in the physical and chemical nature of environments in which they grew. LA/ICPMS analysis, either on a single mineral grain or on a thin section, can provide such information. LA/ICPMS analyses of minerals such as calcite, zircon, olivine, plagioclase, feldspar, titanite, apatite, clinopyroxence, amphibole, and garnet have been reported.(109–111)

4.3 Forensics and Authentication

4.3.1 Authentication of Antique Objects

Chemical analysis is an excellent approach for verifying the authenticity of valuable artifacts. Obviously, the analytical method for authenticity verification of precious antiques should be either nondestructive or require extremely small sample quantities. Using LA/ICPMS, visible damage can be restricted to an acceptable minimum. Such studies were reported by Devos et al.,[112] in which a specially designed sample chamber was used, and almost invisible 100-μm craters were produced on antique silver objects. The elements Zn, Cd, Sn, Sb, Au, Pb, and Bi were measured, and their contents were used to distinguish forgery in silver antiques. Similar studies were reported by Wanner et al.[113] for trace element analyses of archaeological samples such as ancient coins, various antique silver items, and ancient iron. Owing to surface roughness of the samples, an autofocus system was used to achieve reproducible ablation conditions. In these experiments, a lateral resolution of 50 μm and absolute detection limits of 1–1.4 pg were achieved.

4.3.2 *Fingerprinting Crime Scene Evidence*

Many criminal activities result in the generation of debris or other materials, which become available to investigating authorities as physical evidence of the crime. However, the generation of traditional analytical and forensic chemical data is often costly and time-consuming. LA/ICPMS offers the potential for producing fast, definitive, and cost-effective forensic chemical analysis for use in identifying physical evidence that relates a suspect to the scene of a crime. Watling et al.[114] examined several kinds of glass and steel samples as physical evidence of a crime. They developed software to facilitate an intercomparison of three elements simultaneously (ternary plots) for large groups of samples. This approach established both the reproducibility of the "fingerprint" and the uniqueness of interelement associations. A similar idea was used to source the provenance of cannabis. Certain elements from a specific area or geological environment characterize uniquely the source of the plant. Watling[115] showed the trace element association "fingerprint" patterns of cannabis crops and the potential tracing of these crops to specific geological environments. The association of elements formed the basis for determining the provenance of cannabis crops and samples recovered during police drug raids.

4.4 Waste-sample Analysis

LA has many advantages when used to analyze radiological contaminated samples. For example, organic solvents or concentrated acids that are required for classical separation procedures will not be needed for LA sampling. Also, with LA, much less total sample (<1 μg) will be required, greatly reducing the risks associated with sample handling and contamination, and personnel exposure. Finally, elemental and isotopic analysis can be obtained entirely within a hot cell environment, further reducing the risk of contamination.[116] High-level waste analysis using LA/ICPMS was detailed by Smith et al.[117] A unique LA facility has been established in a hot cell environment at the Hanford Site for direct characterization of tank waste samples. Applications of LA and high-resolution ICPAES in the nuclear industry, especially high-resolution isotopic analysis of U and determination of lanthanides, were reported by Giglio et al.[116] and Zamzow et al.[118]

5 NONINDUCTIVELY COUPLED PLASMA APPLICATIONS

There are many LA applications that do not rely on particle digestion by an analytical source (ICP, flame, etc.), but instead directly measure the ablated atomic mass. Laser-induced breakdown spectroscopy (LIBS), described in more detail in the article **Laser-induced Breakdown Spectroscopy**, involves monitoring spectroscopic emission intensity directly in the laser plume. Other applications involve direct MS detection of the ablated mass, with and without secondary ionization, e.g. resonance ionization mass spectrometry (RIMS) and matrix-assisted laser desorption/ionization (MALDI). It would be impossible to discuss these numerous applications without doubling the length of this article. Instead, one particular technique employed by our research group is presented.

5.1 Laser Ablation Ion-storage Time-of-flight Mass Spectrometry

A new technique for single particle analysis was developed that uses ion-trap and TOF mass spectrometers. Commercial LA systems with imaging capabilities involve a chamber in which the ablated mass is entrained into a gas stream and transported to the analytical source by several meters of tubing, thereby diluting the mass vapor and influencing detection sensitivity. For single-particle analysis, excellent sensitivity must be available because of the limited absolute mass from a micrometer-sized particle. To overcome the dilution and sensitivity limitations, an instrument was developed using LA inside an ion-storage time-of-flight mass spectrometry (IS/TOFMS) system with an imaging camera to observe, select, vaporize, ionize, and analyze individual particles or spatial locations on a solid surface.

An ion trap was designed such that a pulsed Nd : YAG laser beam could ablate a sample affixed to a probe inserted radially through the ring electrode (Figure 8). Once the particle is selected using the imaging system, it is directly ablated/ionized by the laser beam. A primary advantage of this system is that the ions are generated

For references see page 9502

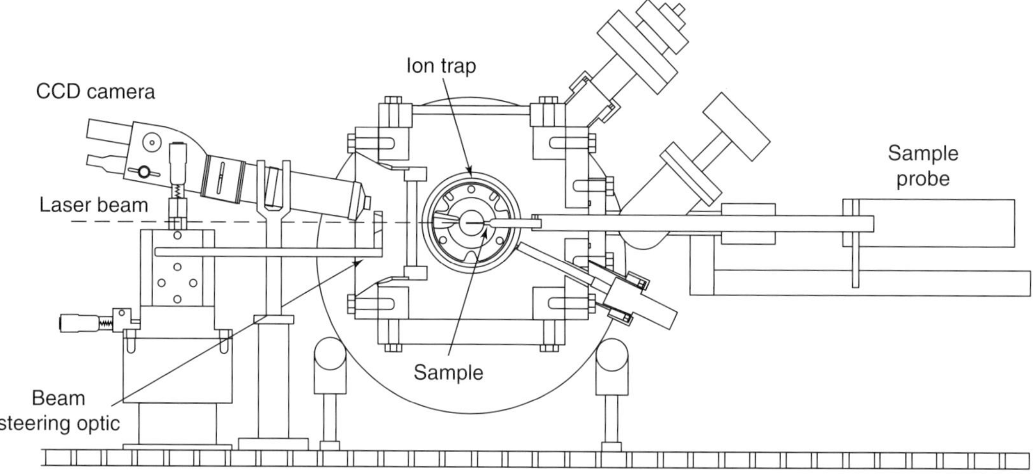

Figure 8 Schematic diagram of ion-trap TOF mass spectrometer with imaging capabilities. The system was designed such that a pulsed Nd : YAG laser beam could ablate a selected sample inside the ion trap mass spectrometer.

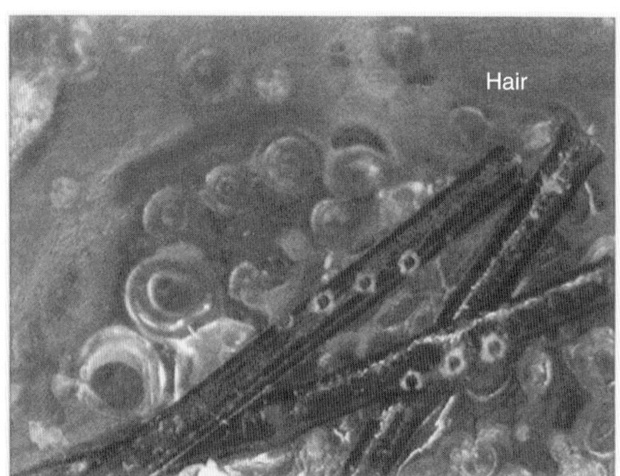

(a)

(b)

Figure 9 Images of the laser-induced craters in (a) hair and (b) alloy samples.

directly inside the ion trap. The ion trap is not used as the mass spectrometer, because space-charge effects would critically influence the mass resolution. Instead, the ion trap is used only as a storage device and the TOF performs the MS. In this way, space charging is significantly reduced except in cases when using extremely high laser fluence, which is not necessary for excellent sensitivity. In preliminary experiments, trace contaminants of Ag, Sn, and Sb were measured in a Pb target with a single laser shot. Reproducible spectra could be measured with only approximately 10 µJ of laser energy. The laser-beam focusing system provided a spatial resolution of approximately 12 µm, with imaging capability to guide the laser beam to a specific location on the sample. The photographs in Figure 9(a) and (b) shows images of the laser-induced craters in hair and alloy samples, respectively.

6 PERSPECTIVES AND FUTURE TRENDS

LA and atomic spectroscopy have become a mature partnership providing significant benefits to both techniques. On the one hand, LA provides the best method for directly converting a solid-phase sample into gas-phase constituents. On the other hand, atomic spectroscopy brings to the relationship a powerful method for understanding the fundamental mechanisms of ablation processes. For atomic spectroscopy, the relationship is continuing to flourish, as can be seen by the increasing number of applications, number of published papers and conference symposia, and new commercial systems. This article has primarily addressed LA as a sampling technique for atomic spectroscopy. There are numerous papers in the literature in which atomic spectroscopy is used to study the fundamental behavior of ablation processes; an example is the work by Bushaw and Alexander using high-resolution time-resolved atomic absorption spectroscopy for the investigation of LA plume dynamics.[119]

A literature search using the two keywords "laser" and "ablation" provided approximately 2500 published papers in the *Current Contents* database, increasing continuously from 1990 to the present. Interest in LA continues to flourish because of applications in atomic spectroscopy, and also medical, semiconductor, materials, and other areas. When matrix-matched standards are available, LA is an ideal quantitative method with excellent accuracy and precision. In cases when matrix-matched standards are not available, LA with atomic spectroscopy is suitable for semiquantitative analysis. Further development of LA and its maturation into an accurate and precise sampling technique for atomic spectroscopy will occur through understanding of fundamental processes and the means to control them. The benefits of LA sampling and atomic spectroscopy warrant success.

ACKNOWLEDGMENTS

The authors gratefully acknowledge the efforts of the Lawrence Berkeley National Laboratory (LBNL) researchers whose papers are referenced in this work, with special thanks to Sungho Jeong, Sam Mao, George Chan, and Jong Yoo for contributing data. R.E.R. also acknowledges G.L. Klunder, P.G. Grant, and B.D. Andresen for the work on IS/TOFMS at Lawrence Livermore National Laboratory (LLNL). This work was supported by the US Department of Energy, Office of Basic Energy Sciences, Division of Chemical Sciences, and by the Environmental Waste Management Science Program (EMSP) under a joint grant from the Office of Energy Research and Office of Environmental Management, through the LBNL under Contract No. DE-AC03-76SF00098.

ABBREVIATIONS AND ACRONYMS

AES	Atomic Emission Spectroscopy
CCD	Charge-coupled Device
CTD	Charge Transfer Device
DCP	Direct Current Plasma
DSI	Direct Sample Insertion
EMSP	Environmental Waste Management Science Program
ETV	Electrothermal Vaporization
GD	Glow Discharge
ICP	Inductively Coupled Plasma
ICPAES	Inductively Coupled Plasma Atomic Emission Spectroscopy
ICPMS	Inductively Coupled Plasma Mass Spectrometry
IR	Infrared
IS/TOFMS	Ion-storage Time-of-flight Mass Spectrometry
LA	Laser Ablation
LA/ICP	Laser Ablation/Inductively Coupled Plasma
LA/ICPMS	Laser Ablation Inductively Coupled Plasma Mass Spectrometry
LBNL	Lawrence Berkeley National Laboratory
LIBS	Laser-induced Breakdown Spectroscopy
LLNL	Lawrence Livermore National Laboratory
LOD	Limit of Detection
MALDI	Matrix-assisted Laser Desorption/Ionization
MC/ICPMS	Multiple Collector Inductively Coupled Plasma Mass Spectrometry
MIP	Microwave-induced Plasma
MS	Mass Spectrometry
Nd : YAG	Neodymium : Yttrium Aluminum Garnet
PDA	Photodiode Array
PMT	Photomultiplier Tube
RIMS	Resonance Ionization Mass Spectrometry
SST	Sea Surface Temperature
TIMS	Thermal Ionization Mass Spectrometry
TOF	Time-of-flight
UV	Ultraviolet
XRF	X-ray Fluorescence

RELATED ARTICLES

Environment: Trace Gas Monitoring **(Volume 3)**
Laser-induced Breakdown Spectroscopy, Elemental Analysis • Laser Mass Spectrometry in Trace Analysis • Laser- and Optical-based Techniques for the Detection of Explosives

Environment: Water and Waste **(Volume 3)**
Flame and Graphite Furnace Atomic Absorption Spectrometry in Environmental Analysis • Inductively Coupled Plasma Mass Spectrometry in Environmental Analysis

Environment: Water and Waste cont'd **(Volume 4)**
Laser Ablation Inductively Coupled Plasma Spectrometry in Environmental Analysis • Optical Emission Inductively Coupled Plasma in Environmental Analysis • Slurry Sampling Graphite Furnace Atomic Absorption Spectrometry in Environmental Analyses • Soil Sampling for the Characterization of Hazardous Waste Sites

Forensic Science **(Volume 5)**
Atomic Spectroscopy for Forensic Applications

For references see page 9502

Atomic Spectroscopy **(Volume 11)**
Atomic Spectroscopy: Introduction • Inductively Coupled Plasma/Optical Emission Spectrometry • Laser Spectrometric Techniques in Analytical Atomic Spectrometry • Laser-induced Breakdown Spectroscopy

REFERENCES

1. S.A. Darke, J.F. Tyson, 'Interaction of Laser Radiation with Solid Materials and Its Significance to Analytical Spectrometry', *J. Anal. At. Spectrom.*, **8**, 145–209 (1993).
2. A. Ciocan, J. Uebbing, K. Niemax, 'Analytical Applications of the Microwave Induced Plasma Used with Laser Ablation of Solid Samples', *Spectrochim. Acta, Part B*, **47**(5), 611–617 (1992).
3. P.G. Mitchell, J. Sneddon, L.J. Radziemski, 'Direct Determination of Copper in Solids by Direct Current Argon Plasma Emission Spectrometry with Sample Introduction Using Laser Ablation', *Appl. Spectrosc.*, **41**(1), 141–148 (1987).
4. A. Montaser, 'Assessment of the Potentials and Limitations of Plasma Sources Compared to ICP Discharges for Analytical Spectrometry', in *Inductively Coupled Plasmas in Atomic Spectrometry*, 2nd edition, eds. A. Montaser, D.W. Golightly, VCH, New York, 1–8, 1992.
5. S.A. Darke, J.F. Tyson, 'Review of Solid Sample Introduction for Plasma Spectrometry and a Comparison of Results for Laser Ablation, Electrothermal Vaporization, and Slurry Nebulization', *Microchem. J.*, **50**, 310–336 (1994).
6. R. Sing, 'Direct Sample Insertion for Inductively Coupled Plasma Spectrometry', *Spectrochim. Acta, Part B*, **54**, 411–441 (1999).
7. R.E. Russo, 'Laser Ablation', *Appl. Spectrosc.*, **49**, A14–A28 (1995).
8. L.A. Allen, H.M. Pang, A.R. Warren, R.S. Houk, 'Simultaneous Measurement of Isotope Ratios in Solids by Laser Ablation with a Twin Quadrupole Inductively Coupled Plasma Mass Spectrometer', *J. Anal. At. Spectrom.*, **10**(3), 267–271 (1995).
9. W.-T. Chan, X.L. Mao, R.E. Russo, 'Differential Vaporization During Laser Ablation/Deposition of Bi–Sr–Ca–Cu–O Superconducting Materials', *Appl. Spectrosc.*, **46**(6), 1025–1031 (1992).
10. X.L. Mao, W.T. Chan, M. Caetano, M.A. Shannon, R.E. Russo, 'Preferential Vaporization and Plasma Shielding during Nanosecond and Picosecond Laser Ablation', *Appl. Surf. Sci.*, **96–98**, 126–130 (1996).
11. E.F. Cromwell, P. Arrowsmith, 'Fractionation Effects in Laser Ablation Inductively Coupled Plasma Mass Spectrometry', *Appl. Spectrosc.*, **49**(11), 1652–1660 (1995).
12. P.M. Outridge, W. Doherty, D.C. Gregoire, 'The Formation of Trace Element-enriched Particulates During Laser Ablation of Refractory Materials', *Spectrochim. Acta, Part B*, **51**(12), 1451–1462 (1996).
13. S.M. Eggins, L.P.J. Kinsley, J.M.G. Shelley, 'Deposition and Element Fractionation Processes During Atmospheric Pressure Laser Sampling for Analysis by ICPMS', *Appl. Surf. Sci.*, **127–129**, 278–286 (1998).
14. X.L. Mao, A.C. Ciocan, R.E. Russo, 'Preferential Vaporization During Laser Ablation Inductively Coupled Plasma/Atomic Emission Spectroscopy', *Appl. Spectrosc.*, **52**, 913–918 (1998).
15. T.E. Jeffries, N.J.G. Pearce, W.T. Perkins, A. Raith, 'Chemical Fractionation During Infrared and Ultraviolet Laser Ablation Inductively Coupled Plasma Mass Spectrometry – Implications for Mineral Microanalysis', *Anal. Commun.*, **33**(1), 35–39 (1996).
16. D. Figg, M.S. Kahr, 'Elemental Fractionation of Glass Using Laser Ablation Inductively Coupled Plasma Mass Spectrometry', *Appl. Spectrosc.*, **51**(8), 1185–1192 (1997).
17. V. Kanicky, J. Musil, J.-M. Mermet, 'Determination of Zr and Ti in 3-μm-thick ZrTiN Ceramic Coating Using Laser Ablation Inductively Coupled Plasma Atomic Emission Spectroscopy', *Appl. Spectrosc.*, **51**(7), 1037–1041 (1997).
18. R.E. Russo, X.L. Mao, W.-T. Chan, M.F. Bryant, W.F. Kinard, 'Laser Ablation Sampling with Inductively Coupled Plasma Atomic Emission Spectrometry for the Analysis of Prototypical Glasses', *J. Anal. At. Spectrom.*, **10**(4), 295–301 (1995).
19. M.A. Shannon, X.L. Mao, A. Fernandez, W.-T. Chan, R.E. Russo, 'Laser Ablation Mass Removal Versus Incident Power Density During Solid Sampling for Inductively Coupled Plasma Atomic Emission Spectrometry', *Anal. Chem.*, **67**(24), 4522–4529 (1995).
20. S. Witanachchi, K. Ahmed, P. Sakthivel, P. Mukherjee, 'Dual-laser Ablation for Particulate-free Film Growth', *Appl. Phys. Lett.*, **66**(12), 1469–1471 (1995).
21. P. Goodall, S.G. Johnson, E. Wood, 'Laser Ablation Inductively Coupled Plasma Atomic Emission Spectrometry of a Uranium–Zirconium Alloy – Ablation Properties and Analytical Behavior', *Spectrochim. Acta, Part B*, **50**(14), 1823–1835 (1995).
22. D.J. Figg, J.B. Cross, C. Brink, 'More Investigations into Elemental Fractionation Resulting from Laser Ablation/Inductively Coupled Plasma/Mass Spectrometry on Glass Samples', *Appl. Surf. Sci.*, **127–129**, 287–291 (1998).
23. C. Geertsen, A. Briand, F. Chartier, J.-L. Lacour, P. Mauchien, S. Sjostrom, J.-M. Mermet, 'Comparison Between Infrared and Ultraviolet Laser Ablation at Atmospheric Pressure – Implications for Solid Sampling Inductively Coupled Plasma Spectrometry', *J. Anal. At. Spectrom.*, **9**(1), 17–22 (1994).
24. W. Sdorra, J. Brust, K. Niemax, 'Basic Investigations for Laser Microanalysis. 4. The Dependence on the Laser Wavelength in Laser Ablation', *Mikrochim. Acta*, **108**(1–2), 1–10 (1992).

25. Y.-I. Lee, K. Song, H.-K. Cha, J.-M. Lee, M.-C. Park, G.-H. Lee, J. Sneddon, 'Influence of Atmosphere and Irradiation Wavelength on Copper Plasma Emission Induced by Excimer and Q-switched Nd:YAG Laser Ablation', *Appl. Spectrosc.*, **51**(7), 959–964 (1997).
26. A. Dupont, P. Caminat, P. Bournot, J.P. Gauchon, 'Enhancement of Material Ablation Using 248, 308, 532, 1064 nm Laser Pulse with a Water Film on the Treated Surface', *J. Appl. Phys.*, **78**(3), 2022–2028 (1995).
27. S. Nakamura, K. Midorikawa, H. Kumagai, M. Obara, K. Toyoda, 'Effect of Pulse Duration on Ablation Characteristics of Tetrafluoroethylene Hexafluoropropylene Copolymer Film Using Ti-sapphire Laser', *Jpn. J. Appl. Phys., Part 1*, **35**(1A), 101–106 (1996).
28. W. Sdorra, K. Niemax, 'Basic Investigations for Laser Microanalysis. 3. Application of Different Buffer Gases for Laser-produced Sample Plumes', *Mikrochimica Acta*, **107**(3–6), 319–327 (1992).
29. Y.-I. Lee, T.L. Thiem, G.-H. Kim, Y.-Y. Teng, J. Sneddon, 'Interaction of an Excimer-laser Beam with Metals. 3. The Effect of a Controlled Atmosphere in Laser-Ablated Plasma Emission', *Appl. Spectrosc.*, **46**(11), 1597–1604 (1992).
30. Y. Iida, 'Effects of Atmosphere on Laser Vaporization and Excitation Processes of Solid Samples', *Spectrochim. Acta, Part B*, **45**(12), 1353–1367 (1990).
31. X.L. Mao, W.T. Chan, M.A. Shannon, R.E. Russo, 'Influence of Gas Medium on Plasma Shielding During Laser–Material Interactions in the Picosecond Time Regime', *J. Appl. Phys.*, **74**, 4915–4922 (1993).
32. A. Semerok, C. Chaleard, V. Detalle, J.L. Lacour, P. Mauchien, P. Meynadier, C. Nouvellon, B. Salle, P. Palianov, M. Perdrix, G. Petite, 'Experimental Investigations of Laser Ablation Efficiency of Pure Metals with Femto-, Pico- and Nanosecond Pulses', *Appl. Surf. Sci.*, **139**, 311–314 (1999).
33. C. Momma, B.N. Chichkov, S. Nolte, F. Vonalvensleben, A. Tunnermann, H. Welling, B. Wellegehausen, 'Short-pulse Laser Ablation of Solid Targets', *Opt. Commun.*, **129**(1–2), 134–142 (1996).
34. A. Fernandez, X.L. Mao, W.-T. Chan, M.A. Shannon, R.E. Russo, 'Correlation of Spectral Emission Intensity in the Inductively Coupled Plasma and Laser-induced Plasma During Laser Ablation of Solid Samples', *Anal. Chem.*, **67**(14), 2444–2450 (1995).
35. M. von Allmen, 'Evaporation and Plasma Formation', in *Laser-beam Interactions with Materials – Physical Principles and Applications*, 1st edition, ed. A. Mooradian, Springer, Berlin, 193–197, 1987.
36. J.H. Yoo, S.H. Jeong, R. Greif, R.E. Russo, 'Explosive Growth of Crater Volume during High Power Nanosecond Laser Ablation of Silicon', *J. Appl. Phys.*, to be published.
37. V.P. Carey, 'Phase Stability and Homogeneous Nucleation', in *Liquid–Vapor Phase Change Phenomena: An Introduction to the Thermophysics of Vaporization and Condensation Processes in Heat Transfer Equipment*, ed. V.P. Carey, Hemisphere, Washington, DC, 127, 1992.
38. M. Caetano, X.L. Mao, R.E. Russo, 'Laser Ablation Solid Sampling: Vertical Spatial Emission Intensity Profiles in the Inductively Coupled Plasma', *Spectrochim. Acta, Part B*, **51**, 1473–1485 (1996).
39. X.L. Mao, O.V. Borisov, R.E. Russo, 'Enhancements in Laser Ablation Inductively Coupled Plasma/Atomic Emission Spectrometry Based on Laser Properties and Ambient Environment', *Spectrochim. Acta, Part B*, **53**(5), 731–739 (1998).
40. A.M. Prokhorov, V.I. Konov, I. Ursu, I.N. Mihilescu, in *Laser Heating of Metals*, Adam Hilger, New York, 42, 1990.
41. N. Bloembergen, 'Fundamentals of Laser–Solid Interactions', in *Laser–Solid Interactions and Laser Processing – 1978*, eds. S.D. Ferris, H.J. Leamy, J.M. Poate, American Institute of Physics, New York, 1–9, 1979.
42. J.F. Ready, 'Gas Breakdown', in *Effects of High-power Laser Radiation*, Academic Press, New York, 213–272, 1971.
43. A.P.K. Leung, W.T. Chan, X.L. Mao, R.E. Russo, 'Influence of Gas Environment on Picosecond Laser Ablation Sampling Efficiency and ICP Conditions', *Anal. Chem.*, **70**(22), 4709–4716 (1998).
44. L. Moenke-Blankenburg, 'Laser–ICP Spectrometry', *Spectrochim. Acta Rev.*, **15**(1), 1–37 (1993).
45. P. Arrowsmith, S.K. Hughes, 'Entrainment and Transport of Laser Ablated Plumes for Subsequent Elemental Analysis', *Appl. Spectrosc.*, **42**(7), 1231–1239 (1988).
46. G. Su, S. Lin, 'Studies on the Complete Laser Vaporization of the Powdered Solid Samples into an Inductively Coupled Plasma for Atomic Emission Spectrometry', *J. Anal. At. Spectrom.*, **3**, 841 (1988).
47. S.H. Jeong, O.V. Borisov, J.H. Yoo, X.L. Mao, R.E. Russo, 'Effects of Particle Size Distribution on Inductively Coupled Plasma Mass Spectrometry Signal Intensity During Laser Ablation of Glass Samples', *Anal. Chem.*, **71**, 5123–5130 (1999).
48. M.L. Alexander, M.R. Smith, J.S. Hartman, A. Mendoza, D.W. Koppenaal, 'Laser Ablation Inductively Coupled Plasma Mass Spectrometry', *Appl. Surf. Sci.*, **127–129**, 255–261 (1998).
49. A.A. van Heuzen, 'Analysis of Solids by Laser Ablation/Inductively Coupled Plasma/Mass Spectrometry (LA/ICPMS) – I. Matching with Glass Matrix', *Spectrochim. Acta, Part B*, **46**(14), 1803–1817 (1991).
50. M. Thompson, S. Chenery, L. Brett, 'Nature of Particulate Matter Produced by Laser Ablation – Implications for Tandem Analytical Systems', *J. Anal. At. Spectrom.*, **5**(1), 49–55 (1990).
51. K.N. DeSilva, R. Guevremont, 'Direct Powder Introduction Inductively Coupled Plasma Atomic Emission Spectrometry with a Photodiode Array Spectrometer', *Spectrochim. Acta, Part B*, **46**(11), 1499–1515 (1991).

52. T.E. Jeffries, S.E. Jackson, H.P. Longerich, 'Application of a Frequency Quintupled Nd:YAG Source ($\lambda =$ 213 nm) for Laser Ablation Inductively Coupled Plasma Mass Spectrometric Analysis of Minerals', *J. Anal. At. Spectrom.*, **13**, 935–940 (1998).
53. D. Günther, R. Frischknecht, C.A. Heinrich, H.-J. Kahlert, 'Capabilities of an Argon Fluoride 193 nm Excimer Laser for Laser Ablation Inductively Coupled Plasma Mass Spectrometry Microanalysis of Geological Materials', *J. Anal. At. Spectrom.*, **12**, 939–944 (1997).
54. P.P. Mahoney, G. Li, G.M. Hieftje, 'Laser Ablation/Inductively Coupled Plasma Mass Spectrometry with a Time-of-flight Mass Analyzer', *J. Anal. At. Spectrom.*, **11**, 401–405 (1996).
55. T. Mochizuki, A. Sakashita, H. Iwata, 'Laser Ablation for Direct Elemental Analysis of Solid Samples by ICP–Atomic Emission Spectrometry and ICP–Mass Spectrometry', *Nippon Kagaku Kaishi*, **58**, 19–27 (1990).
56. X.R. Liu, G. Horlick, 'In Situ Laser Ablation Sampling for Inductively Coupled Plasma Atomic Emission Spectroscopy', *Spectrochim. Acta, Part B*, **50**(4–7), 537–548 (1994).
57. M. Thompson, R.M. Barnes, 'Analytical Performance of Inductively Coupled Plasma/Atomic Emission Spectrometry', in *Inductively Coupled Plasmas in Analytical Atomic Spectrometry*, 2nd edition, eds. A. Montaser, D.W. Golighty, VCH, New York, 276, 1992.
58. M. Motelica-Heino, O.F.X. Donard, J.-M. Mermet, 'Laser Ablation of Synthetic Geological Powders Using ICP/AES Detection: Effects of the Matrix, Chemical Form of the Analyte and Laser Wavelength', *J. Anal. At. Spectrom.*, **14**, 675–682 (1999).
59. M. Ducreux-Zappa, J.-M. Mermet, 'Analysis of Glass by UV Ablation Inductively Coupled Plasma Atomic Emission Spectrometry. Part 2. Analytical Figures of Merit', *Spectrochim. Acta, Part B*, **51**, 333–341 (1996).
60. A.A. van Heuzen, J.B. Morsink, 'Analysis of Solids by Laser Ablation/Inductively Coupled Plasma/Mass Spectrometry (LA/ICPMS) – II. Matching with a Pressed Pellet', *Spectrochim. Acta, Part B*, **46**(14), 1819–1828 (1991).
61. K.E. Jarvis, J.G. Williams, S.J. Parry, E. Bertalan, 'Quantitative Determination of the Platinum-group Elements and Gold Using NiS Fire Assay with Laser Ablation/Inductively Coupled Plasma/Mass Spectrometry (LA/ICPMS)', *Chem. Geol.*, **124**, 37–46 (1995).
62. V. Kinicky, J.-M. Mermet, 'Use of a Single Calibration Graph for the Determination of Major Elements in Geological Materials by Laser Ablation Inductively Coupled Plasma Atomic Emission Spectrometry with Added Internal Standard', *Fresenius' J. Anal. Chem.*, **363**, 294–299 (1999).
63. P.J. Sylvester, M. Ghaderi, 'Trace Element Analysis of Scheelite by Excimer Laser Ablation/Inductively Coupled Plasma/Mass Spectrometry (ELA/ICPMS) Using a Synthetic Silicate Glass Standard', *Chem. Geol.*, **141**, 49–65 (1997).
64. C. Leloup, P. Marty, D. Dall'ava, M. Perdereau, 'Quantitative Analysis for Impurities in Uranium by Laser Ablation Inductively Coupled Plasma Mass Spectrometry: Improvements in the Experimental Set-up', *J. Anal. At. Spectrom.*, **12**, 945–950 (1997).
65. V.V. Kogan, M.W. Hinds, G.I. Ramendik, 'The Direct Determination of Trace Metals in Gold and Silver Materials by Laser Ablation Inductively Coupled Plasma Mass Spectrometry Without Matrix Matched Standards', *Spectrochim. Acta, Part B*, **49**(4), 333–343 (1994).
66. A. Raith, J. Godfrey, R.C. Hutton, 'Quantitation Methods Using Laser Ablation ICPMS Part 2: Evaluation of New Glass Standards', *Fresenius' J. Anal. Chem.*, **354**, 163–168 (1996).
67. O.V. Borisov, X.L. Mao, A. Fernandez, M. Caetano, R.E. Russo, 'Inductively Coupled Plasma Mass Spectrometric Study of Non-linear Calibration Behavior During Laser Ablation of Binary Cu–Zn Alloys', *Spectrochim. Acta, Part B*, **54**(9), 1351–1365 (1999).
68. H.F. Falk, B. Hattendorf, K. Krengel-Rothensee, N. Wieberneit, S.L. Dannen, 'Calibration of Laser-ablation ICPMS. Can We Use Synthetic Standards with Pneumatic Nebulization?', *Fresenius' J. Anal. Chem.*, **362**, 468–472 (1998).
69. J.J. Leach, L.A. Allen, D.B. Aeschliman, R.S. Houk, 'Calibration of Laser Ablation Inductively Coupled Plasma Mass Spectrometry Using Standard Additions with Dried Solution Aerosols', *Anal. Chem.*, **71**, 440–445 (1999).
70. D. Günther, A. Audétat, R. Frischknecht, C.A. Heinrich, 'Quantitative Analysis of Major, Minor, and Trace Elements in Fluid Inclusions Using Laser Ablation/Inductively Coupled Plasma/Mass Spectrometry (LA/ICPMS)', *J. Anal. At. Spectrom.*, **13**, 263–270 (1998).
71. D. Günther, R. Frischknecht, H.-J. Müschenborn, C.A. Heinrich, 'Direct Liquid Ablation: A New Calibration Strategy for Laser Ablation/ICPMS Microanalysis of Solids and Liquids', *Fresenius' J. Anal. Chem.*, **359**, 390–393 (1997).
72. A.M. Ghazi, T.E. McCandless, D.A. Vanko, J. Ruiz, 'New Quantitative Approach in Trace Elemental Analysis of Single Fluid Inclusions: Applications of Laser Ablation Inductively Coupled Plasma Mass Spectrometry (LA/ICPMS)', *J. Anal. At. Spectrom.*, **11**, 667–674 (1996).
73. A. Moissette, T.J. Shepherd, S.R. Chenery, 'Calibration Strategies for the Elemental Analysis of Individual Aqueous Fluid Inclusions by Laser Ablation Inductively Coupled Plasma Mass Spectrometry', *J. Anal. At. Spectrom.*, **11**, 177–185 (1996).
74. T. Tanaka, K. Yamamoto, T. Nomizu, H. Kawaguchi, 'Laser Ablation/Inductively Coupled Plasma Mass

Spectrometry with Aerosol Density Normalization', *Anal. Sci.*, **11**(6), 967–971 (1995).
75. R.J. Watling, 'In-line Mass Transport Measurement Cell for Improving Quantification in Sulfide Mineral Analysis Using Laser Ablation Inductively Coupled Plasma Mass Spectrometry', *J. Anal. At. Spectrom.*, **13**, 927–934 (1998).
76. M.A. Shannon, B. Rubinsky, R.E. Russo, 'Mechanical Stress Power Measurements During High-power Laser Ablation', *J. Appl. Phys.*, **80**(8), 4665–4672 (1996).
77. C. Chaléard, P. Mauchien, N. Andre, J. Uebbing, J.L. Lacour, C. Geertsen, 'Correction of Matrix Effects in Quantitative Elemental Analysis with Laser Ablation Optical Emission Spectrometry', *J. Anal. At. Spectrom.*, **12**(2), 183–188 (1997).
78. V. Kanicky, V. Otruba, J.-M. Mermet, 'Characterization of Acoustic Signals Produced by Ultraviolet Laser Ablation Inductively Coupled Plasma Atomic Emission Spectroscopy', *Fresenius' J. Anal. Chem.*, **363**, 339–346 (1999).
79. D.P. Baldwin, D.S. Zamzow, A.P. D'Silva, 'Aerosol Mass Measurement and Solution Standard Additions for Quantitation in Laser Ablation Inductively Coupled Plasma Atomic Emission Spectroscopy', *Anal. Chem.*, **66**(11), 1911–1917 (1994).
80. J. Perez, B.R. Weiner, 'The Laser Ablation of Gold Films at the Electrode Surface of a Quartz Crystal Microbalance', *Appl. Surf. Sci.*, **62**(4), 281–285 (1992).
81. S.A. Watmough, T.C. Hutchinson, R.D. Evans, 'The Quantitative Analysis of Sugar Maple Tree Rings by Laser Ablation in Conjunction with ICPMS', *J. Environ. Qual.*, **27**(5), 1087–1094 (1998).
82. T. Prohaska, C. Stadlbauer, R. Wimmer, G. Stingeder, C. Latkoczy, E. Hoffmann, H. Stephanowitz, 'Investigation of Element Variability in Tree Rings of Young Norway Spruce by Laser-ablation–ICPMS', *Sci. Total Environ.*, **219**(1), 29–39 (1998).
83. S.A. Watmough, T.C. Hutchinson, R.D. Evans, 'Development of Solid Calibration Standards for Trace Elemental Analyses of Tree Rings by Laser Ablation Inductively Coupled Plasma/Mass Spectrometry', *Environ. Sci. Technol.*, **32**(14), 2185–2190 (1998).
84. C.D. GarbeSchonberg, C. Reimann, V.A. Pavlov, 'Laser Ablation ICPMS Analyses of Tree-ring Profiles in Pine and Birch From N Norway and NW Russia – a Reliable Record of the Pollution History of the Area?', *Environ. Geol.*, **32**(1), 9–16 (1997).
85. A. Raith, W.T. Perkins, N.J.G. Pearce, T.E. Jeffries, 'Environmental Monitoring on Shellfish Using UV Laser Ablation ICPMS', *Fresenius' J. Anal. Chem.*, **355**(7/8), 789–792 (1996).
86. E. Vander Putten, F. Dehairs, L. Andre, W. Baeyens, 'Quantitative In Situ Microanalysis of Minor and Trace Elements in Biogenic Calcite Using Infrared Laser Ablation/Inductively Coupled Plasma Mass Spectrometry: A Critical Evaluation', *Anal. Chim. Acta*, **378**(1–3), 261–272 (1999).
87. S. Tanaka, N. Yasushi, N. Sato, T. Fukasawa, S.J. Santosa, K. Yamanaka, T. Ootoshi, 'Rapid and Simultaneous Multi-element Analysis of Atmospheric Particulate Matter Using Inductively Coupled Plasma Mass Spectrometry with Laser Ablation Sample Introduction', *J. Anal. At. Spectrom.*, **13**(N2), 135–140 (1998).
88. C.J. Chin, C.F. Wang, S.L. Jeng, 'Multi-element Analysis of Airborne Particulate Matter Collected on PTFE-membrane Filters by Laser Ablation Inductively Coupled Plasma Mass Spectrometry', *J. Anal. At. Spectrom.*, **14**(4), 663–668 (1999).
89. C.F. Wang, S.L. Jeng, F.J.R. Shieh, 'Determination of Arsenic in Airborne Particulate Matter by Inductively Coupled Plasma Mass Spectrometry', *J. Anal. At. Spectrom.*, **12**(1), 61–67 (1997).
90. C.F. Wang, C.J. Chin, S.K. Luo, L.C. Men, 'Determination of Chromium in Airborne Particulate Matter by High Resolution and Laser Ablation Inductively Coupled Plasma Mass Spectrometry', *Anal. Chim. Acta*, **389**(N1–3), 257–266 (1999).
91. R. Vago, E. Gill, J.C. Collingwood, 'Laser Measurements of Coral Growth', *Nature (London)*, **386**(6620), 30–31 (1997).
92. D.J. Sinclair, L.P.J. Kinsley, M.T. McCulloch, 'High Resolution Analysis of Trace Elements in Corals by Laser Ablation ICPMS', *Geochim. Cosmochim. Acta*, **62**(11), 1889–1901 (1998).
93. S. Fallon, 'Corals at Their Limits: Laser Ablation Trace Element Systematics in Porites from Shirigai Bay, Japan', *Earth Planet. Sci. Lett.*, to be published.
94. R. Feng, N. Machado, J.N. Ludden, 'Lead Geochronology of Zircon by Laser Probe–Inductively Coupled Plasma Mass Spectrometry (LP/ICPMS)', *Geochim. Cosmochim. Acta*, **57**, 3479–3486 (1993).
95. B.J. Fryer, S.E. Jackson, P. Longerich, 'The Application of Laser Ablation Microprobe–Inductively Coupled Plasma Mass Spectrometry (LAM/ICPMS) to In Situ (U)–Pb Geochronology', *Chem. Geol.*, **109**, 1–8 (1993).
96. T. Hirata, R.W. Nesbitt, 'U–Pb Isotope Geochronology of Zircon: Evaluation of the Laser Probe–Inductively Coupled Plasma Mass Spectrometry Technique', *Geochim. Cosmochim. Acta*, **59**(12), 2491–2500 (1995).
97. H.C. Liu, B.Q. Zhu, Z.X. Zhang, 'Single Zircon Dating by LAM/ICPMS Technique', *Chin. Sci. Bull.*, **44**(2), 182–186 (1999).
98. T.J. Shepherd, S.R. Chenery, 'Laser Ablation ICPMS Elemental Analysis of Individual Fluid Inclusions – An Evaluation Study', *Geochim. Cosmochim. Acta*, **59**(19), 3997–4007 (1995).
99. A. Audetat, D. Günther, C.A. Heinrich, 'Formation of a Magmatic–Hydrothermal Ore Deposit: Insights with LA/ICPMS Analysis of Fluid inclusions', *Science*, **279**(27), 2091–2094 (1998).

100. A.N. Halliday, D.-C. Lee, J.N. Christensen, M. Rehkamper, W. Yi, X.Z. Luo, C.M. Hall, C.J. Ballentine, T. Pettke, C. Stirling, 'Application of Multiple Collector–ICPMS to Cosmochemistry, Geochemistry, and Paleoceanography', *Geochim. Cosmochim. Acta*, **62**(6), 919–940 (1998).
101. M.F. Thirlwall, A.J. Walder, 'In Situ Hafnium Isotope Ratio Analysis of Zircon by Inductively Coupled Plasma Multiple Collector Mass Spectrometry', *Chem Geol.*, **122**, 241–247 (1995).
102. L.V. Godfrey, D.-C. Lee, W.F. Sangrey, A.N. Halliday, V.J.M. Satlers, J.R. Hein, W.M. White, 'The Hf Isotopic Composition of Ferromanganese Nodules and Crusts and Hydrothermal Manganese Deposits: Implications for Seawater Hf', *Earth Planet. Sci. Lett.*, **151**, 91–105 (1997).
103. J N. Christensen, A.N. Halliday, D.-C. Lee, C.M. Hall, 'In Situ Sr Isotopic Analyses by Laser Ablation', *Earth Planet. Sci. Lett.*, **136**, 79–85 (1995).
104. J.N. Christensen, A.N. Halliday, L.V. Godfrey, J.R. Hein, D.K. Rea, 'Climate and Ocean Dynamics and the Lead Isotopic Records in Pacific Ferro-Manganese Crust', *Science*, **277**, 913–918 (1997).
105. T. Hirata, M. Hattori, T. Tanaka, 'In-situ Osmium Isotope Ratio Analyses of Iridosmines by Laser Ablation Multiple Collector Inductively Coupled Plasma Mass Spectrometry', *Chem. Geol.*, **144**(3–4), 269–280 (1998).
106. S.A. Baker, M. Bi, R.Q. Aucelio, B.W. Smith, J.D. Winefordner, 'Analysis of Soil and Sediment Samples by Laser Ablation Inductively Coupled Plasma Mass Spectrometry', *J. Anal. At. Spectrom.*, **14**(1), 19–26 (1999).
107. M. Odegard, S.H. Dundas, B. Flem, A. Grimstvedt, 'Application of a Double-focusing Magnetic Sector Inductively Coupled Plasma Mass Spectrometer with Laser Ablation for the Bulk Analysis of Rare Earth Elements in Rocks Fused with $Li_2B_4O_7$', *Fresenius' J. Anal. Chem.*, **362**(5), 477–482 (1998).
108. E.K. Shibuya, J.E.S. Sarkis, J. Enzweiler, A.P.S. Jorge, A.M.G. Figueiredo, 'Determination of Platinum Group Elements and Gold in Geological Materials Using an Ultraviolet Laser Ablation High-resolution Inductively Coupled Plasma Mass Spectrometric Technique', *J. Anal. At. Spectrom.*, **13**(9), 941–944 (1998).
109. F. Bea, P. Montero, G. Garuti, F. Zacharini, 'Pressure-dependence of Rare Earth Element Distribution in Amphibolite- and Granulite-grade Garnets. A LA/ICPMS Study', *Geostand. Newsl. J. Geostand. Geoanal.*, **21**(2), 253–270 (1997).
110. L.P. Bedard, 'Re-evaluation of the Homogeneity of REE, Th And U in Geochemical Reference Zircon 61.308b', *Geostand. Newsl.*, **20**(2), 289–293 (1996).
111. M.D. Norman, N.J. Pearson, A. Sharma, W.L. Griffin, 'Quantitative Analysis of Trace Elements in Geological Materials by Laser Ablation ICPMS – Instrumental Operating Conditions and Calibration Values of NIST Glasses', *Geostand. Newsl.*, **20**(2), 247–261 (1996).
112. W. Devos, C. Moor, P. Lienemann, 'Determination of Impurities in Antique Silver Objects for Authentication by Laser Ablation Inductively Coupled Plasma Mass Spectrometry (LA/ICPMS)', *J. Anal. At. Spectrom.*, **14**(4), 621–626 (1999).
113. B. Wanner, C. Moor, P. Richner, R. Bronnimann, B. Magyar, 'Laser Ablation Inductively Coupled Plasma Mass Spectrometry (LA/ICPMS) for Spatially Resolved Trace Element Determination of Solids Using an Autofocus System', *Spectrochim. Acta, Part B*, **54**(2), 289–298 (1999).
114. R.J. Watling, B.F. Lynch, D. Herring, 'Use of Laser Ablation Inductively Coupled Plasma Mass Spectrometry for Fingerprinting Scene of Crime Evidence', *J. Anal. At. Spectrom.*, **12**(2), 195–203 (1997).
115. R.J. Watling, 'Novel Application of Laser Ablation Inductively Coupled Plasma Mass Spectrometry in Forensic Science and Forensic Archaeology', *Spectroscopy*, **14**(6), 16–34 (1999).
116. J.J. Giglio, P.S. Goodall, S.G. Johnson, 'Application of High-resolution Inductively Coupled Plasma Atomic Emission Spectroscopy in the Nuclear Field', *Spectroscopy*, **12**(7), 26–37 (1997).
117. M.R. Smith, J.S. Hartman, M.L. Alexander, A. Mendoza, E.H. Hirt, M.R. Stewart, M.A. Hansen, W.R. Park, T.J. Peters, B.J. Burghard, J.W. Ball, C.T. Narquis, D.M. Thornton, R.L. Harris, 'Laser Ablation/Inductively Coupled Plasma Mass Spectrometry: Analysis of Hanford High-level Waste Materials', in *Science and Technology for Disposal of Radioactive Tank Wastes*, eds. W.W. Schullz, N.J. Lombardo, Plenum Press, New York, 135–158, 1998.
118. D.S. Zamzow, D.P. Baldwin, S.J. Weeks, S.J. Bajic, A.P. D'Silva, 'In Situ Determination of Uranium by Laser Ablation/Inductively Coupled Plasma Atomic Emission Mass Spectrometry', *Environ. Sci. Technol.*, **28**(2), 352–358 (1994).
119. B.A. Bushaw, M.L. Alexander, 'Investigation of Laser Ablation Plume Dynamics by High-resolution Time-resolved Atomic Absorption Spectroscopy', *Appl. Surf. Sci.*, **129**, 935–940 (1998).

Laser Spectrometric Techniques in Analytical Atomic Spectrometry

Ove Axner
Umeå University, Umeå, Sweden

Laser light has a number of spectacular properties that make it useful for analytical spectrometry. One is that it has a high directionality (i.e. it looks like a real "beam"). This implies, among other things, that it can be focused down to micrometer-sized spots. Another is that pulsed lasers can emit large amounts of light in very short pulses (often with a duration of 10^{-9}–10^{-8} s). These two properties imply that laser light often can reach high irradiance ($W\,m^{-2}$), which is of importance for a number of applications, not least when laser light is used for vaporization and/or atomization purposes of solid material. The most important attribute for spectroscopic applications, however, is that it often has a narrow frequency width (in the MHz–GHz range). This implies that laser light can induce one specific transition in one particular species at a time. The narrow frequency width is thus the basis for the high species selectivity that laser spectroscopic techniques possess. In addition, the combination of high irradiance and a narrow frequency width often gives laser light such staggering spectral irradiance ($W\,m^{-2}\,Hz^{-1}$) that a significant fraction of the atoms under illumination can produce at least one detectable event during the interaction time (one or several photons or an ion–electron pair). This explains why laser spectroscopic techniques can benefit from high species sensitivity.

The main purpose of this review article is to describe the theory and instrumentation for the field of laser spectroscopic techniques for analytical atomic spectrometry. This implies that techniques that use laser light for nonspectroscopic purposes, such as vaporization, and/or atomization purposes, e.g. laser ablation, laser-induced plasma spectrometry, laser mass spectrometry and laser-induced breakdown spectroscopy, will not be covered here. In addition, because atomic spectra in general consist

For references see page 9575

of a few strong and narrow-band transitions (whose widths often are comparable to those of the light from tunable laser systems and therefore seldom overlap with those from other atomic species) whereas those of molecules comprise a few broader, weaker and to a certain extent structured transition bands (which thus overlap more often with those of other molecular species), laser spectrometric techniques often show the highest sensitivity and selectivity when atomic species are detected. This is the main reason why this review focuses upon the use of laser spectrometric techniques for analytical atomic spectrometry. This implies, in turn, that laser-based spectrometric techniques that predominantly detect molecules, e.g. Raman, thermal lensing and photoacoustic spectrometry, and spectrofluorometry, will not be discussed.

This theory and instrumentation review focuses upon the most useful and versatile laser spectroscopic techniques for analytical atomic spectrometry. These techniques are based upon the concepts of fluorescence, ionization or absorption. Those based upon fluorescence are often referred to as either laser-induced fluorescence (LIF) or laser-excited atomic fluorescence spectrometry (LEAFS), whereas those based upon ionization are termed either laser-enhanced ionization (LEI) or resonance ionization spectrometry (RIS). The working principles, instrumentation and present status of laser-induced fluorescence/laser-excited atomic fluorescence spectrometry (LIF/LEAFS) and LEI are covered in some detail in two separate sections, whereas RIS is covered more briefly. The reason for this is the broader versatility and applicability of LEI: it can be performed in a variety of atomizers, including those working under atmospheric pressure, whereas RIS has to be carried out in a vacuum and thus with significantly more complex instrumentation. However, because the laser instrumentation and the theoretical basis for all these techniques are quite similar, a short introduction to laser instrumentation and a theory section that outlines the common basic features of excitation of atoms by laser light precede the detailed descriptions. Those laser spectroscopic techniques that are based upon absorption and used for analytical atomic spectrometry are, nowadays, all performed in conjunction with some sort of modulation methodology (often wavelength-modulation (WM)), frequently by the use of diode lasers owing to their rapid tunability. They are therefore often referred to as WM diode laser (atomic) absorption techniques and are covered in a separate section.

Typical analytical qualities, e.g. limits of detection (LOD) and selectivity, as well as the typical strengths and limitations of each of these techniques, are given or discussed. It is concluded among other things that the most impressive performance of the LIF/LEAFS technique has been obtained together with the graphite furnace (GF) as atomizer, resulting in detection limits in the femtogram range and a very high selectivity for a large variety of elements. LEI, which has found its best use with the flame as atomizer, can provide LOD in the pg mL^{-1} range for many elements. It does not, however, show the same high selectivity as the LIF/LEAFS technique because it suffers from background effects when samples with high concentrations of easily ionized elements (EIEs) are analyzed. RIS combines a high sensitivity with an extraordinary selectivity, but instead has to pay the price of a more complex instrumentation. It was demonstrated earlier, for example, that the RIS technique is able to provide single atom detection (SAD). Because RIS in general is performed in a vacuum it can also provide good isotopic selectivity. Although not yet applied to a broad range of elements for analytical applications, it has shown impressive LOD (in the attogram range) and excellent isotopic selectivity (>10^9) under a few specific conditions. The wavelength-modulation diode laser absorption spectrometry (WM-DLAS) technique shows good promise for becoming a user-friendly and widespread detection technique because it requires less complex instrumentation than most other laser spectroscopic techniques. Used with the GF as atomizer, LOD in the femtogram range have been achieved. The WM-DLAS technique also has good ability to correct for various types of unstructured background absorption signals that might appear when samples with complex matrices are analyzed. One drawback, however, is that its applicability is still restricted owing to the limited availability of diode lasers that emit light in the visible and ultraviolet (UV) region. The WM-DLAS technique, therefore, has been applied to only a limited number of elements so far. A common denominator for all these laser spectroscopic techniques is that they often have a large linear dynamic range (LDR). The LIF/LEAFS technique, for example, has demonstrated an LDR of 5–7 orders of magnitude. Finally, in addition to being used as sensitive and selective tools for analytical assessments, the laser spectroscopic techniques often show an excellent applicability to diagnostic studies (e.g. for the determination of processes such as atomization, diffusion, collision or ionization).

1 INTRODUCTION

Analytical spectrometry refers to a family of techniques by which a sample is characterized with regard to its content of atoms (or molecules) by spectroscopic means. It is customary to distinguish between optical (e.g. laser) and mass spectrometric techniques. The underlying foundation of optical spectrometric techniques is the quantization of energy levels, which gives each type of atomic or molecular species its own unique set of transition

List of selected abbreviations appears in Volume 15

wavelengths, whereas mass spectrometric techniques rely on the fact that each type of species has a unique mass. The present review is solely concerned with the use of optical spectrometric techniques. The reader is referred to other articles in this encyclopedia for a description of mass spectrometric techniques (**Mass Spectrometry: Overview and History; Time-of-flight Mass Spectrometry; Secondary Ion Mass Spectrometry as Related to Surface Analysis; Tandem Mass Spectrometry: Fundamentals and Instrumentation; Quadrupole Ion Trap Mass Spectrometer; Literature of Mass Spectrometry**).

1.1 Conventional Techniques for Analytical Atomic Spectrometry – Emission and Absorption

The two most commonly used conventional (optical) spectrometric techniques for analytical atomic spectrometry are emission and absorption spectroscopy.[1] In both of these techniques, the sample to be analyzed is introduced into a hot environment in which it is vaporized and subsequently atomized. Emission spectroscopy relies upon the fact that the heat of the atomizer thermally excites free atoms, which subsequently decay to lower-lying states by spontaneous emission. The wavelength distribution and the intensity of the spontaneously emitted light are measures of the analytical content of the sample. In absorption spectrometry, the wavelength distribution and the amount of the light absorbed when it is passing the atomized sample are measures of the analytical content of the sample.

Although both of these techniques have been developed to a considerable level of sophistication and applicability throughout the years (with a variety of atomizers), they suffer from a few inevitable inherent limitations. Emission spectrometry has a limited sensitivity (the thermal population of the excited states is usually rather low) and selectivity (given solely by the dispersive power of the detection instrumentation). In addition, the precision and accuracy are limited (the amount of light emitted for a given amount of analyte varies strongly with temperature, so any minor fluctuation of the atomizer conditions, e.g. from the matrix composition of the sample, can affect the sensitivity). The absorption technique is limited by influences from unspecific background absorption and stray light (although a variety of detection procedures, e.g. the D_2 lamp and Zeeman background correction, have been developed throughout the years in order to overcome these limitations). The present status of the emission and absorption spectrometry techniques can be found elsewhere in the literature[2–7] or in this encyclopedia (**Atomic Spectroscopy: Introduction; Inductively Coupled Plasma/Optical Emission Spectrometry; Flame and Vapor Generation Atomic Absorption Spectrometry; Graphite Furnace Atomic Absorption Spectrometry; Background Correction Methods in Atomic Absorption Spectroscopy; Flame and Graphite Furnace Atomic Absorption Spectrometry in Environmental Analysis**).

1.2 Lasers as Powerful Light Sources for Analytical Atomic Spectrometry

With the advent of the first laser (an acronym for light by amplified spontaneous emission (ASE) of Radiation) in 1960,[8] many of the limitations of the conventional analytical spectrometry techniques were expected to be overcome. Laser light has a number of spectacular properties that make it exceptionally useful for analytical spectrometry, the most important of which are: high directionality (i.e. it looks like a real "beam" of light), which implies that it has a high irradiance ($W\,m^{-2}$); and an extremely narrow frequency width (i.e. it can often be considered to consist of only "one" wavelength), implying that it has an even more impressive spectral irradiance ($W\,m^{-2}\,Hz^{-1}$). In addition, pulsed lasers can produce very short laser pulses, thus temporarily producing very high peak powers and thereby sometimes even staggering spectral irradiances (which is often of importance for analytical applications).

These properties give the laser-based spectrometric techniques both high sensitivity and excellent selectivity. The high sensitivity, which predominantly stems from the high irradiance of the laser light, implies that a significant fraction of the atoms under illumination can produce a detectable event (e.g. a high production of photons/ions/electrons per illuminated atom). Laser-based techniques, for example, can be used successfully for SAD.[9–11] The high selectivity, which mainly originates from the narrow bandwidth of the light, comes from the fact that it is often possible to induce one (and often only one) specific transition in one particular type of species at a time.

Several types of lasers with a variety of qualities have been developed since the first laser saw the light of day. The fact that some types of lasers can be scanned (or tuned) over the absorption profile (referred to as tunable lasers) makes them useful for analytical spectrometry (as well as indispensable for atomic physics[12–15]). With access to the tunable high-irradiance light sources that lasers constitute, a whole new family of detection techniques for analytical spectrometry has emerged during the last decades.[15–21]

1.3 Use of Lasers for Analytical Spectrometry

Laser light can be used in a variety of ways for analytical spectrometric purposes. It is customary to distinguish between spectrometric and nonspectrometric applications.

For references see page 9575

1.3.1 Lasers for Spectrometric Applications – Detection of Atoms versus Molecules

Laser-based spectrometric techniques refer to those techniques in which the narrow bandwidth of the laser light is used to provide a species-selective interaction with the sample. Although laser spectrometric techniques can be used successfully for the detection of both atoms and molecules, they show the highest sensitivity and selectivity when atomic species are detected because atomic spectra generally consist of a few strong and narrow-band transitions (whose widths often are comparable to those of the light from a tunable laser system) as opposed to those of molecules, which comprise a few broader, weaker and to a certain extent structured transition bands. This is the main reason why this review is devoted to the use of laser spectrometric techniques for analytical atomic spectrometry. This implies, in turn, that laser-based spectrometric techniques that predominantly detect molecules, e.g. Raman,[22] thermal lensing and photoacoustic spectrometry,[23–28] spectrofluorometry,[29–31] optoacoustic detection[32] and degenerate four-wave mixing,[33] will not be covered in this review. The reader is referred to the literature or other articles in this encyclopedia for further information about these types of technique (**Raman Spectroscopy: Introduction; Raman Scattering, Fundamentals; Dispersive Raman Spectroscopy, Current Instrumental Designs; Fourier Transform Raman Instrumentation; Raman Microscopy and Imaging**).

1.3.2 Lasers for Nonspectrometric Analytical Applications

Lasers can also be used as a tool for vaporization and/or atomization of the sample in conjunction with various types of conventional detection techniques. There are a number of closely related techniques in this field: e.g. laser ablation, laser-induced plasma spectrometry, laser mass spectrometry and laser-induced breakdown spectroscopy. However, because this use of lasers does not contribute to the elemental selectivity through their narrow bandwidth, none of these types of technique will be covered in this review. The reader is again referred to other articles in this encyclopedia (e.g. **Laser-induced Breakdown Spectroscopy**) or to the literature for this type of application of lasers in analytical spectrometry.[34]

1.4 Laser Spectrometric Techniques for Analytical Atomic Spectrometry – Fluorescence, Ionization and Absorption

Although a number of different laser-based spectroscopic techniques have been scrutinized for their analytical applicability throughout the years, the techniques based upon *fluorescence*, *ionization* and *absorption* have been shown to be the most useful and versatile for analytical atomic spectrometry.

1.4.1 Laser-induced Fluorescence Spectrometry

Tuned to a particular transition in one specific element, the narrow-band laser light will selectively increase the number of excited analyte species far above the thermal levels. This implies that the emission will be increased considerably. Such laser-enhanced emission is generally termed LIF. By detecting this fluorescent light by suitable means (most often a spectrometer and a light-sensitive detector), an assessment of the amount (or concentration) of the analyte can be made.[35] LIF is a very useful and versatile technique that has found applications in a number of areas (e.g. atomic physics, quantum optics, environmental monitoring, combustion analysis and analytical spectrometry[36]). When applied to analytical spectrometry for the detection of atoms or molecules, it appears frequently under the name LEAFS[37,38] or laser-excited molecular fluorescence spectrometry (LEMOFS).[38]

1.4.2 Laser Ionization Spectrometry

Laser-excited atoms cannot only decay to lower-lying states by emission of photons (i.e. fluorescence) or by collisions with surrounding species (i.e. by inelastic or so-called quenching collisions), but they can also be ionized as a direct (or indirect) consequence of the laser illumination. Techniques that are based on a monitoring of the production of charged species (i.e. ions or electrons) that result because of laser illumination are therefore termed *laser ionization spectrometry techniques*. A laser ionization technique can thus be seen as an optical-to-electrical transducer whose output (number of charges produced) depends on the amount (or concentration) of the analyte to be assessed. In analytical applications, the ionization techniques are most often referred to as either LEI spectrometry[39,40] or RIS.[11,19,41] The main difference between these two types of technique is that RIS is performed under a vacuum but LEI needs to be performed in an environment in which there is already an omnipresent thermal ionization. The reason for this is that the laser light is used for excitation as well as for ionization of the analytes in RIS, whereas the atoms only need to be promoted to an excited state by the laser light in the LEI technique (because the ionization is provided by an existing thermal ionization process in the atomic reservoir in this technique; see further discussion below). LEI has therefore the advantage over RIS that it can be performed in a variety of atomizers suitable for analytical purposes (flames, furnaces, etc.).

1.4.3 Laser Absorption Spectrometry

Tuned to a particular transition in an element, the laser light will not only increase the number of excited species but it will also experience an attenuation (one photon will be absorbed for each excitation event). This implies that the light will be attenuated as it passes a cloud of absorbing species. This is the basis for absorption techniques.

Laser-based absorption techniques have found best use for analytical spectrometry in combination with various modulation methodologies. Correctly used, the modulation reduces the noise in the system. Hence, although fluorescence and ionization techniques benefit from a high production of photons or ions/electrons, the laser-based absorption techniques benefit from a reduction of the noise in the system that the modulation brings. In addition, because diode lasers have the unique property that they can be tuned (i.e. modulated) at significantly higher rates than other tunable lasers, modulation absorption techniques often incorporate diode lasers and are therefore often referred to as wavelength-modulation (*WM*) or *frequency-modulation (FM) diode laser absorption techniques*.[42,43]

1.4.4 Coverage of this Review

Because these three laser spectroscopic techniques (fluorescence, ionization and absorption spectrometry) combine high sensitivity, high selectivity and broad versatility for the detection of a large variety of elements in trace concentrations/amounts in various types of samples, this review will focus upon their foundations, theory and instrumentation.

1.5 Organization of this Review

Section 2 gives some typical properties of modern laser systems suitable for analytical spectrometry. Section 3 covers the most important basic theoretical concepts related to the excitation of atoms by laser light and gives an introduction to the nomenclature and terminology of the field. Section 4 reviews the LIF/LEAFS technique in analytical applications. Section 5 deals with ionization techniques. After a short comparison of the RIS and the LEI techniques, most of the section is concerned with LEI owing to its broader versatility, whereas RIS is described more briefly at the end. Section 6 first makes some comments about FM and WM techniques in general, before it focuses the presentation on the WM diode laser absorption technique due to its suitability for analytical spectrometry. Finally, section 7 makes some concluding remarks about the field of laser spectrometric techniques for analytical atomic spectrometry and speculates briefly about its future.

2 LASER INSTRUMENTATION

Virtually all lasers used for analytical spectrometry are tunable (i.e. their wavelength can be tuned continuously over at least a certain part of the spectrum). With the exception of diode laser systems, a tunable laser system consists of a pump laser (producing nontunable high-power light) and a tunable laser (converting the pump laser light to preferably narrow-band tunable laser light).

Most tunable laser systems can be classified as either *continuous-wave* or *pulsed* systems, dictated by the qualities of the pump laser. Tunable continuous-wave laser systems most often consist of ion-laser-pumped dye laser or Ti : sapphire laser systems, or the diode lasers that have gained increasing interest during latter years. Tunable pulsed laser systems, which have dominated the area of laser analytical spectrometry for a number of years, consist most often now of excimer- or neodymium : yttrium aluminum garnet (Nd : YAG)-pumped dye laser or optical parametric oscillator (OPO) systems (although nitrogen- or flashlamp-pumped dye laser systems were common during the first years). Atomic-vapor-based systems, which have been used only occasionally for analytical spectrometry, take an intermediate position with pulse frequencies significantly exceeding and pulse energies significantly falling short of those of ordinary pulsed laser systems. A thorough and well-updated overview of the properties of laser systems in terms of the requirements of the field of analytical spectrometry has been given recently by Sneddon et al.[44]

2.1 Tunable Continuous-wave Laser Systems

2.1.1 Continuous-wave Dye Laser or Ti : Sapphire Laser Systems

Typical continuous-wave pump lasers produce light of a few (or tens of) watts. A tunable continuous-wave dye laser or Ti : sapphire laser system can therefore typically produce narrow-band laser light in the milliwatt to watt range in the visible part of the spectrum. Light in the UV region, which often is needed for excitation of atomic species, can be produced at microwatt powers by use of frequency-doubling instrumentation (frequency-doubling is done in nonlinear crystals in which two photons of the incoming light beam are being combined to one photon of twice the energy, i.e. half the incoming wavelength).

Although laser beams most often are produced and used with typical diameters of a few millimeters, they can be focused down to spots of micrometer-size dimensions. This implies that laser irradiances from tunable continuous-wave laser systems can, in theory, range from fractions of $\mathrm{W\,m^{-2}}$ to $10^{12}\,\mathrm{W\,m^{-2}}$, with

For references see page 9575

Table 1 Some "typical" (order-of-magnitude estimates) irradiances and spectral irradiances for three different light sources

	Irradiance[a] ($W m^{-2}$)	Spectral irradiance[a] ($W m^{-2} Hz^{-1}$)
The Sun[b]	10^3	10^{-11}
Continuous-wave laser system[c]	$10^2 - 10^3$ (10^{12})	$10^{-6} - 10^{-4}$ (10^5)
Pulsed laser system[d]	$10^8 - 10^{10}$ (10^{18})	$10^{-1} - 10^1$ (10^9)

[a] The values for the laser systems are given for a "typical" experimental situation with a beam diameter of a few millimeters. Values in parentheses refer to strongly focused situations (with a spot size of micrometer size dimensions).
[b] At the surface of the Earth.
[c] Values are given for a "typical" argon ion laser-pumped dye laser system.
[d] Values are given for a "typical" Nd:YAG- or excimer-pumped dye laser system with nanosecond pulse duration but are also in reasonable agreement with those for OPO systems.

"typical" numbers in the $10^2 - 10^3$ $W m^{-2}$ range (for the case of 1–10 mW of light and a beam area of 10 mm^2).

In addition, monochromatic continuous-wave laser light can be produced with a variety of bandwidths, often in the megahertz range (from a few to some hundreds of megahertz, depending on the type of system, type of stabilization, etc.). This implies that tunable continuous-wave laser light can be produced with spectral irradiances up to 10^5 $W m^{-2} Hz^{-1}$ under strongly focused conditions, with "typical" values in the $10^{-6} - 10^{-4}$ $W m^{-2} Hz^{-1}$ range for unfocused situations.

Comparing these numbers with those of an ordinary light source, e.g. the light from the Sun on the surface of the Earth (around 10^3 $W m^{-2}$), as is done in Table 1, one finds that the irradiances in fact are comparable. However, because sunlight is distributed over a large part of the spectrum (over approximately 10^{14} Hz), the typical spectral irradiance of tunable continuous-wave laser light ($10^{-6} - 10^{-4}$ $W m^{-2} Hz^{-1}$) is several orders of magnitude larger than that of the sunlight (10^{-11} $W m^{-2} Hz^{-1}$).

2.1.2 Diode Lasers

Modern diode lasers suitable for analytical spectrometry (i.e. single-mode lasers) can produce light at powers of tens of milliwatts directly (mostly in the red or infrared region). By using external frequency-doubling instrumentation, powers in the microwatt region (and occasionally in the milliwatt region) can be produced.

In addition to the aforementioned properties, diode lasers possess yet another unique property: namely, an extraordinary rapid and versatile tunability. Diode lasers are controlled by their temperature and injection current.[45] A modulation of the wavelength can be accomplished by a rapid modulation of the injection current. Although modern diode laser systems can be modulated at gigahertz frequencies for communication purposes, most diode laser systems for analytical purposes are modulated in the megahertz range (for FM techniques) and the kilohertz range (for WM techniques).[42,46]

2.2 Tunable Pulsed Laser Systems

2.2.1 Pump Lasers

The comparison between laser light and conventional light becomes even more spectacular when pulsed lasers are considered. Nd:YAG lasers can in general produce light with higher pulse energies (about 1 J per pulse) than can excimer lasers (typically a few hundred millijoules per pulse). However, because the fundamental wavelength of Nd:YAG-produced light is 1064 nm, it has to be converted to more suitable wavelengths in order to make pumping of dye lasers or OPOs possible. Light at 532- and 355-nm wavelengths can be produced by frequency doubling of the 1064-nm light and frequency-mixing of the 532- and 1064-nm light, respectively, in nonlinear crystals. Crystals dedicated for frequency conversion of Nd:YAG laser light have rather high conversion efficiencies, implying that frequency-doubled and frequency-tripled light can be produced at pulse energies comparable to that of excimer laser systems (which produce light useful for dye laser pumping directly, most often at 308 nm when XeCl is used as the laser firing medium).

Typical pulse lengths for these types of laser are in the nanosecond range (typically 5 ns for Nd:YAG-pumped systems and often around 10–30 ns for excimer-pumped systems). The shorter pulse lengths of Nd:YAG lasers therefore give them an edge over excimer lasers when it comes to irradiances but also a handicap when it comes to the illumination time (as will be discussed in the following section). The excimer systems, however, have an irrefutable advantage when it comes to repetition rates. Although many Nd:YAG lasers can produce light only at a repetition rate of 10 or 30 Hz (and some at 50 Hz), it is not uncommon for excimer lasers to run at several hundred hertz (some even up to 500 Hz).

2.2.2 Pulsed Dye Laser Systems

A pulsed dye laser system can produce tunable laser light in the low to medium millijoule range in the visible part of the spectrum. Because one particular dye can produce light only over a certain part of the spectrum (typically a few tens of nanometers), a number of different dyes are needed to cover the entire spectrum.[44] Typical conversion efficiencies of pulsed dye lasers depend on the choice of dye and the position within the laser firing range of the dye. Peak conversion efficiencies of various dyes range normally from a few per cent to a few tens per cent, with the highest efficiencies for coumarins (in the 450–600 nm range) for excimer-pumped dye lasers and for rhodamines (in the 560–650 nm range) for Nd : YAG-pumped systems. The fundamental wavelength ranges covered by these systems (i.e. using no nonlinear frequency-doubling or frequency-mixing instrumentation after the dye laser) are roughly 330–980 nm for XeCl-based excimer laser systems and 400–900 nm for Nd : YAG laser-based systems.

In order to create tunable UV light, the dye laser light has to be frequency-doubled in a nonlinear crystal. The conversion efficiencies of such crystals range from a fraction of a per cent to some tens per cent. Hence, UV light can most often be produced by pulsed dye laser systems with pulse energies in the micro- to low-millijoule range. The total wavelength ranges of pulsed dye laser systems depend on the particular nonlinear frequency-doubling (or frequency-mixing) schemes used. In general, however, virtually the entire wavelength region from below 200 nm up to around 1 μm can nowadays be covered with pulsed dye laser systems.

The irradiance from excimer- or Nd : YAG-pumped dye laser systems can be as high as $10^{18}\,\mathrm{W\,m^{-2}}$ (assuming 25 mJ of light distributed over 5 ns and focused down to an area of $5\,\mu\mathrm{m}^2$), with "typical" values in the 10^{8}–$10^{10}\,\mathrm{W\,m^{-2}}$ range (for 10 μJ–1 mJ of light distributed over 10 ns and a beam area of $10\,\mathrm{mm}^2$). Unless any special precautions are taken, tunable light produced by pulsed dye laser systems has typical spectral bandwidths of a few gigahertz. This implies that the corresponding spectral irradiance can become as high as $10^{9}\,\mathrm{W\,m^{-2}\,Hz^{-1}}$, but is under more normal conditions around 0.1–$10\,\mathrm{W\,m^{-2}\,Hz^{-1}}$, i.e. several orders of magnitude larger than from tunable continuous-wave laser systems (see Table 1).

2.2.3 *Pulsed Optical Parametric Oscillator Systems*

Tunable OPO laser systems have been developed during the latter years as a viable alternative to dye lasers. This all-solid-state laser has considerably improved the creation of pulsed tunable laser light. An OPO needs to be pumped by a pump laser in a similar way to dye lasers. An OPO does not, however, need any organic dye or wavelength-dispersive element because the laser light is being produced in a nonlinear crystal that splits the incoming photons into two (whose energies thus sum up to that of the incoming photon).[(21)] The two beams of light produced are referred to as signal and idler beams and have wavelengths that are above and below twice the wavelength of the incoming light, respectively. Pumped with the third harmonics of a Nd : YAG laser (355 nm), an OPO can therefore produce tunable laser light in two wavelength regions (approximately 410–690 nm and 730 nm–2 μm). Equipped with a frequency-doubling crystal, these lasers are tunable by computer control over the entire wavelength region of 220–2000 nm (with no need for any change of dye solution, etc.).

The output characteristics of an OPO depend greatly on the pump laser but in most situations are superior to those of dye lasers. Laser pulses with energies of several tens of millijoules can normally be produced in the entire visible part of the spectrum, whereas pulse energies of several millijoules can be produced in the UV region. Because there is no dispersive element in an OPO, the bandwidth is mainly determined by the bandwidth of the pump laser. In order to produce laser light with a sufficiently narrow bandwidth, the pump laser should therefore also have a narrow bandwidth. This requires the use of injection-seeded pump lasers (e.g. a Nd : YAG laser whose light is originally produced by a narrow-band diode laser system placed in the cavity).

2.3 Conclusions

In conclusion, modern tunable laser sources can produce narrow-band light with significantly higher spectral irradiance than any conventional light source. Because a narrow frequency width and a high spectral irradiance are the basis for the high selectivity and sensitivity, respectively, of laser-based spectroscopic techniques, laser light can, if correctly used, enhance the signal-to-noise (S/N) ratio and increase the detection power in many types of applications,[(47–52)] not least for analytical spectrometry.[(53–55)]

3 INTERACTION OF ATOMS WITH LASER LIGHT

Although all laser-based spectroscopic techniques have in common the fact that narrow-band laser light is used to interact selectively with the analyte atoms, the interaction has a different objective for the three techniques scrutinized in this review: fluorescence, ionization and absorption. Although the main aim of the fluorescence technique is to produce as many fluorescence photons as possible, the ionization techniques strive for ionizing (selectively) as large a fraction of the analyte atoms as possible. The common denominator for these two techniques is therefore to excite as many analyte species as possible. The absorption techniques, on the other hand, strive for minimizing, by the use of various modulation techniques, the amount of noise that affects the measurements, rather than maximizing the number of excited analyte species. This necessarily leads to different theoretical descriptions of, on the one hand, the fluorescence and

For references see page 9575

ionization techniques and, on the other hand, the absorption technique. In this section we will give a theoretical basis for the fluorescence and ionization techniques (in terms of the number of atoms excited); the theoretical description of the absorption techniques is referred to in section 6.

3.1 Fluorescence and Ionization

Before scrutinizing in detail how free atoms interact with laser light, let us first conclude which entities are of importance for the fluorescence and ionization techniques.

For the fluorescence technique, fluorescence is assumed to be measured from an excited level, here denoted level 2 for simplicity, down to a lower-lying level, denoted 1 (see Figure 1). The rate of photon emission from level 2 to level 1 in a volume V of excited atoms, $\dot{N}_{ph}(t)$, is directly proportional to the number density of atoms in the excited state, $n_2(t)$ (in units of m^{-3}), as shown in Equation (1):

$$\dot{N}_{ph}(t) = VA_{21}n_2(t) \quad (1)$$

where A_{21} represents the spontaneous emission rate from level 2 to level 1. The number of fluorescence photons emitted over a time t, $N_{ph}(t)$, can then be written as in Equation (2):

$$N_{ph}(t) = \int_0^t \dot{N}_{ph}(t)\,dt' = VA_{21}\int_0^t n_2(t')\,dt' \quad (2)$$

When ionization techniques are used, the entity of interest is the number of atoms that are ionized as a consequence of the laser illumination. As will be discussed in detail below, a prerequisite for the LEI technique is that there exists a thermal collisional ionization rate in the medium.[39] In such a system, species in all states, i (and not only those in one particular state, as is the case for the fluorescence technique), will contribute to the thermal ionization rate, $\dot{N}_{ion}(t)$, by their own (state-specific) collisional ionization rate, $k_{i,ion}$. This implies that

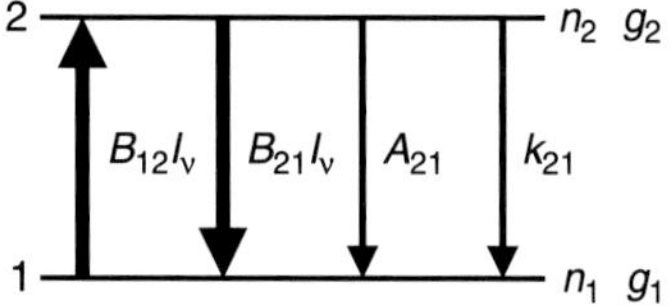

Figure 1 Schematic representation of excitation and de-excitation processes in a two-level atom: $B_{12}I_\nu$ and $B_{21}I_\nu$ represent the light-induced absorption and stimulated emission rates; A_{21} represents the spontaneous emission rate; k_{21} is the collisional de-excitation rate; and n_1 and n_2 are the populations (in units of m^{-3}) and g_1 and g_2 the degeneracies of the two levels respectively.

the total ionization rate can be written as in Equation (3):

$$\dot{N}_{ion}(t) = V\dot{n}_i(t) = V\sum_i k_{i,ion}n_i(t) \quad (3)$$

where $\dot{n}_i(t)$ is the ionization rate density (in units of $m^{-3}\,s^{-1}$) and where the sum goes over all levels in the analyte.[56,57] The total number of ions produced during a time t, $N_{ion}(t)$, can therefore be expressed by Equation (4):

$$N_{ion}(t) = \int_0^t \dot{N}_{ion}(t')\,dt' = V\sum_i k_{i,ion}\int_0^t n_i(t')\,dt' \quad (4)$$

For the RIS technique, for which photoionization most often is the dominating ionization mechanism,[11,41] Equations (3) and (4) are still valid, with the only difference that the state-specific collisional ionization rate, $k_{i,ion}$, needs to be interpreted as a state-specific photoionization rate, $\phi_{i,ion}$, given by Equation (5):

$$\phi_{i,ion} = I\frac{\sigma_{i,ion}}{h\nu} \quad (5)$$

where I is the irradiance of the laser light (in units of $W\,m^{-2}$), $\sigma_{i,ion}$ is the cross-section for photoionization from state i, and $h\nu$ is the photon energy. The sum in Equation (4) should in this case run over all states that have an energy deficit to ionization limit that is smaller that the photon energy.

The presentation below has, however, mostly been restricted to techniques utilizing atmospheric pressure atomizers, such as flames, furnaces or plasmas, owing to their versatility and user-friendliness. This presentation will therefore focus more upon the situation for the LEI technique than the RIS technique (although the two techniques have many aspects in common) and thereby consider the ionization to be mainly caused by collisions. The reader is referred to the literature for a more detailed presentation of the theory for the RIS technique.[19]

Hence, in order to estimate the number of photons emitted, $N_{ph}(t)$, or the number of ions created, $N_{ion}(t)$, from a given atomic system we need to determine the number densities (often referred to as the "populations") of the excited states ($n_2(t)$ for the LIF technique and all $n_i(t)$ for the LEI technique).

As will be shown by the treatment below, in order to get a high sensitivity for the LIF technique it is often sufficient to excite the analyte with light from one laser (commonly referred to as one-step excitation). One-step excitations in two- or three (multi)-level atoms will therefore be covered first. A one-step excitation is, however, normally not the optimum situation for the LEI technique. The highest performance (sensitivity as well as selectivity) is normally achieved when the atoms are excited in two separate steps using two individually tuned laser beams (using so-called two-step excitations). A two-step excitation process will therefore be modeled briefly thereafter. As will be discussed further below, two-step excitations can, under certain conditions, be beneficial also for the LIF technique. In addition, this presentation will be based upon the rate equation formalism and be restricted to the cases when the laser light is tuned exactly on resonance. This implies, among other things, that there will be virtually no discussion about the spectral shape of various excitation profiles.

3.2 Two-level Atoms

3.2.1 General Description of a Two-level System Exposed to a Resonant Laser Field

When laser light is tuned to a transition between the ground state and an excited state in an atom and the atomic structure is such that it is warranted to neglect the presence of other excited levels, the analyte atoms can often be described as two-level atoms. The two-level model can well be used, for example, for descriptions of excitations of the first resonant transition in elements that have no fine-structure splitting of any of the laser-connected levels and very few other excited states close to the laser-connected state (as, for example, is the case for Zn, Cd and Hg). The two-level model is sometimes appropriate also for other more complicated atomic systems.

3.2.1.1 Rate Equations for a Two-level System The excitation process in a two-level atom exposed to resonant laser light (as displayed in Figure 1) can be modeled in the so-called rate equation formalism as depicted in Equations (6) and (7):

$$\frac{dn_2}{dt} = B_{12}I_\nu n_1 - (B_{21}I_\nu + R_{21})n_2 \tag{6}$$

$$n_{\text{atom}} = n_1 + n_2 \tag{7}$$

where $B_{12}I_\nu$ and $B_{21}I_\nu$ represent the light-induced excitation and de-excitation rates between the two laser-connected states, respectively; R_{21} represents all the nonlight-induced de-excitation processes (in this case the sum of the spontaneous emission rate, denoted A_{ij}, and the collisional de-excitation rate, denoted k_{ij}); and n_{atom} is the total number density of (neutral) analyte atoms in the interaction region.

The light-induced excitation and de-excitation rates, $B_{12}I_\nu$ and $B_{21}I_\nu$, consist of a product of the Einstein coefficients for absorption and stimulated emission (B_{12} and B_{21}, respectively, in units of $\text{m}^2\,\text{Hz}\,\text{J}^{-1}$) and the spectral irradiance of the light (in units of $\text{W}\,\text{m}^{-2}\,\text{Hz}^{-1}$). The Einstein coefficients for absorption and emission are related to the rate of spontaneous emission, A_{21}, by the relations shown in Equation (8):

$$B_{12} = \frac{g_2}{g_1} B_{21} = \frac{g_2}{g_1}\frac{\lambda^3}{8\pi hc} A_{21} \tag{8}$$

where λ is the wavelength of the light, h is Planck's constant, c is the speed of light, and g_1 and g_2 are the degeneracies of the two levels.

3.2.1.2 Solution of the Rate Equations When solving rate equations, it is often convenient to work with normalized entities, i.e. to define the fraction of neutral atoms in a state i as $\bar{n}_i$ according to Equation (9):

$$\bar{n}_i = \frac{n_i}{n_{\text{atom}}} \tag{9}$$

Using this nomenclature, the fraction of neutral atoms in state 2 can be written according to Equation (10):

$$\bar{n}_2(t) = \bar{n}_2^{\text{sat}} \frac{B_{12}I_\nu}{B_{12}I_\nu + \bar{n}_2^{\text{sat}} R_{21}} \left[1 - \exp\left(\frac{-t}{\tau}\right)\right] \tag{10}$$

where $\bar{n}_2^{\text{sat}}$ and τ are the saturation population fraction of level 2 (Equation 11) and the time constant for establishment of a steady state (i.e. a time-independent situation; Equation 12) respectively.

$$\bar{n}_2^{\text{sat}} = \frac{g_2}{g_1 + g_2} \tag{11}$$

$$\tau = \frac{\bar{n}_2^{\text{sat}}}{B_{12}I_\nu + \bar{n}_2^{\text{sat}} R_{21}} \tag{12}$$

This implies that when the two-level system is exposed to a laser excitation, the fraction of excited atom will increase with time (in an exponential manner with a time constant τ) towards a steady-state value given by the product of the two first factors of Equation (10). The product of these two entities is therefore referred to as the *steady-state excitation population fraction*. In order to estimate a typical time constant for the establishment of a steady-state condition as well as the steady-state excitation population fraction of an atomic system exposed to laser light under conditions typically prevailing in the most commonly used atomizers, the various rates involved need to be estimated.

3.2.1.3 Typical Excitation and De-excitation Rates Typical values for the spontaneous emission rate, A_{21}, range generally from $10^7\,\text{s}^{-1}$ to a few times $10^8\,\text{s}^{-1}$. This implies that the Einstein coefficients for absorption and stimulated emission, B_{12} and B_{21}, are of the order of 3×10^{10}–$3 \times 10^{11}\,\text{m}^2\,\text{Hz}\,\text{J}^{-1}$ for a UV transition (at 250 nm), assuming equal degrees of degeneracy of the two levels. Because typical spectral irradiances of UV light from pulsed laser systems are in the 0.1–1 $\text{W}\,\text{m}^{-2}\,\text{Hz}^{-1}$

For references see page 9575

range (see Table 1), typical absorption and laser-induced stimulated emission rates for pulsed laser excitations are often in the gigahertz range. For example, excitation of atoms using a UV transition with a spontaneous emission rate of $10^8\,s^{-1}$ using laser pulses with a 10-μJ pulse energy, a 10-ns duration and a spectral width of 10 GHz, focused down to a diameter of 6 mm, gives rise to an excitation rate around 1 GHz. The corresponding excitation rate for a similar transition in the visible region can often become one to three orders of magnitude larger, owing to larger *B*-factors and higher laser pulse energies.

The collisional de-excitation rates depend on factors such as pressure, temperature and type of collision partner, and differ therefore among various atomizers. Typical collision rates between radiatively coupled levels in atoms in inert gas atmospheric pressure atomizers, e.g. flames with Ar as buffer gas,[58] often referred to as a high quantum efficiency flame, or electrothermal atomizers (ETAs), are often slightly below the gigahertz range. In air-supported atomizers, the rates are often in the low gigahertz range, with particularly high values for rates between closely spaced states.[59] This implies that collisional de-excitation rates often can dominate over spontaneous emission (especially in high collisional media such as flames). On the other hand, because light-induced excitation and de-excitation rates often can be well into the gigahertz range, they can, in turn, often exceed all nonlaser-connected de-excitation rates in the system.

3.2.1.4 Justification of the Steady-state Approximation Because excitation rates often are in the gigahertz range when pulsed laser systems are used, Equation (12) shows that the time constant for establishment of a steady-state condition in a two-level system becomes very short, often significantly less than 1 ns. A couple of examples of the time dependence of the population of the upper level of a two-level system are shown in Figure 2(a) for two different laser excitation rates (0.5 and 5 GHz, respectively) for a "typical" situation with a nonlaser-induced de-excitation rate of 1 GHz. Because most lasers used for LIF and LEI spectrometry are pulsed with pulse lengths in the 5–30 ns range, and the saturated excited fraction of atoms reaches its steady-state value within about 1 ns, one can conclude that any two laser-connected levels will reach their steady-state values within a fraction of the pulse length. This warrants the often-used approximation (valid for pulses whose duration is significantly longer than the time constant τ) that the number densities of atoms in the two laser-connected levels can be considered to be *locked* to each other (to their steady-state population fractions) during the laser illumination period.

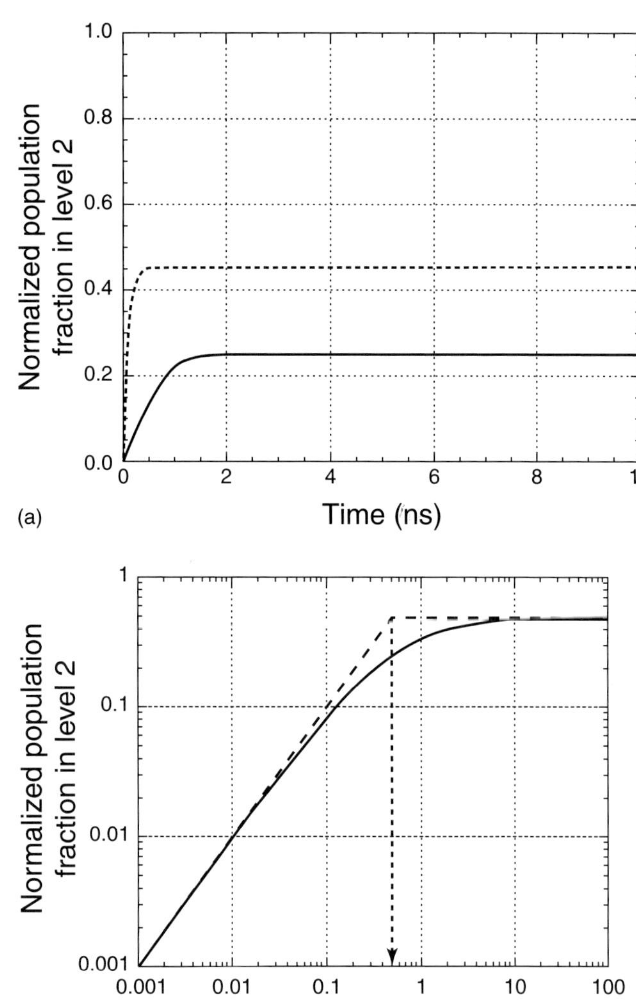

Figure 2 (a) The time dependence of the excited fraction of atoms of a two-level system exposed to laser-induced excitation rates of 0.5 (——) and 5 GHz (- - - -). (b) A saturation curve for the excited fraction of atoms (under steady-state conditions) as a function of laser-induced excitation rate. We have, for simplicity, assumed equal degeneracy of the two levels and a total nonlight-induced de-excitation rate of 1 GHz. The saturation rate for the system (0.5 GHz) is specifically marked.

3.2.1.5 Optical Saturation Moreover, Equation (10) above shows that the steady-state number density of excited atoms in the interaction volume will be proportional to the laser spectral irradiance (i.e. to the intensity) for low light levels and that it will level off towards a plateau for higher light levels, i.e. the saturated excited fraction of atoms (given by $\bar{n}_2^{sat}$) shown in Figure 2(b). This particular type of population density versus irradiation curve is called a *saturation curve*. The reason for the leveling off of the population density for high light levels is that the rate of stimulated emission dominates the combined rate of spontaneous emission and collisional de-excitation. When this takes place, the system is said to be *optically saturated*. The second term in the denominator of Equation (12), i.e. $\bar{n}_2^{sat} R_{21}$, is called the *saturation rate* (or *saturation parameter*). This rate has an important role when laser-based techniques are used for diagnostic purposes.[60] The low- and high-intensity asymptotes are drawn with dashed lines in Figure 2(b). The saturation rate (which is the rate at which these two asymptotes meet, or where the saturated excited fraction of atoms reaches half of its maximum value) is also indicated for clarity.

The intensity (or spectral irradiance) for which the excitation rate is equal to the saturation parameter is termed *saturation intensity* (or *spectral saturation irradiance*).[61–68] The spectral saturation irradiance for a two-level system can be written according to Equation (13):

$$I_\nu^{sat} = \bar{n}_2^{sat} \frac{R_{21}}{B_{12}} = \frac{g_1}{g_1 + g_2} \frac{8\pi hc}{\lambda^3} \frac{1}{Y_{21}} \tag{13}$$

where Y_{21} represents the *fluorescence yield* (also referred to as *quantum yield* in the literature). The fluorescence yield can be defined as the ratio of the spontaneous emission rate to the total nonlaser-induced de-excitation rate (Equation 14), i.e.

$$Y_{21} = \frac{A_{21}}{R_{21}} = \frac{A_{21}}{A_{21} + k_{21}} \tag{14}$$

and is an entity of importance for the amount of fluorescence emitted from a system (see discussion below).

Because the collisional de-excitation rate differs among various atomizers (and depends on the running conditions to a certain extent), the fluorescence yield and thereby the saturation spectral irradiance will differ. Depending on details of the transitions and the collision partners, the fluorescence yield can range from a few percent to values close to unity in atmospheric pressure atomizers.[59] As an example, Hannaford reported on fluorescence yields in the 0.24–0.87 range for the strongest resonance lines of a number of elements (Li, Na, K, Rb, Cs, Ga, In, Tl, Au, Sn, Pb, etc.) in an argon-diluted stoichiometric oxygen–hydrogen flame.[69] Fluorescence yields in this range give "typical" spectral saturation irradiances for a two-level system in the 10^{-4}–10^{-3} W m^{-2} Hz^{-1} range for UV transitions and in the 10^{-3}–10^{-2} W m^{-2} Hz^{-1} range for transitions in the visible spectrum. By comparison with the "typical" spectral irradiance from pulsed as well as continuous-wave laser systems (see Table 1), one can conclude that optical saturation of a two-level system in atmospheric pressure atomizers can easily be obtained with light from pulsed laser systems. For a continuous-wave system, on the other hand, optical saturation is normally only achieved with some degree of focusing of the laser beam.

3.2.2 Fluorescence from a Two-level System

Let us define the number of photons emitted per atom under a time t, $\overline{N}_{ph}(t)$, according to Equation (15):

$$\overline{N}_{ph}(t) = \frac{N_{ph}(t)}{N_{atom}} \tag{15}$$

where N_{atom} is the number of illuminated atoms (given by Vn_{atom}). The rate of photon emission per atom, $\dot{\overline{N}}_{ph}(t)$, can then be written (using Equations 1, 9 and 15) according to Equation (16):

$$\dot{\overline{N}}_{ph}(t) = A_{21}\bar{n}_2(t) \tag{16}$$

3.2.2.1 Fluorescence from an Unsaturated Two-level System The rate of photon emission per atom, for a two-level system exposed to low light levels (i.e. for laser light intensities below the saturation intensity, $I_\nu \ll I_\nu^{sat}$), under steady-state conditions (i.e. for $t \gg \tau$), can then (using Equations 10, 14 and 16) be expressed as shown in Equation (17):

$$\dot{\overline{N}}_{ph} = Y_{21} B_{12} I_\nu \tag{17}$$

This shows that the rate of photon emission per atom from a two-level system (illuminated by a laser pulse whose duration is longer than τ) is independent of time and that it depends, in the unsaturated case, on the product of the excitation rate, $B_{12}I_\nu$, and the fluorescence yield, Y_{21}. The number of photons emitted per atom under a time t is therefore given by Equation (18):

$$\overline{N}_{ph}(t) = Y_{21} B_{12} I_\nu t \tag{18}$$

Although these expressions look handy, they show that the fluorescence technique indeed has a weakness. The fact that the signal depends on the fluorescence yield implies that the sensitivity of the LIF technique (the number of photons produced from a given amount of atoms) varies, under unsaturated conditions, not only with the amount of laser light but also with the quenching collisional rate. This implies, in turn, that the sensitivity will vary in time and in space in a manner that is not fully controlled if the local environment in the atomizers differs between various parts of the interaction region or if it is dependent on time (e.g. if it is affected by the matrix in the sample).

3.2.2.2 Fluorescence from a Saturated Two-level System For a high laser light irradiance (i.e. for $I_\nu \gg I_\nu^{sat}$), the fraction of atoms excited (and thereby also the number of emitted photons) becomes independent of the laser light intensity as well as of the quenching rate. This implies that the rate of emission of fluorescence photons per

For references see page 9575

atom under saturated conditions can be written as shown in Equation (19):

$$\dot{\overline{N}}_{\text{ph}} = \bar{n}_2^{\text{sat}} A_{21} \tag{19}$$

The number of photons emitted per atom from a saturated two-level system under a time t is therefore given by Equation (20):

$$\overline{N}_{\text{ph}}(t) = \bar{n}_2^{\text{sat}} A_{21} t \tag{20}$$

This implies that the number of photons emitted per (two-level) atom can be made independent of the fluorescence yield as well as the laser irradiation under saturated conditions.

3.2.2.3 Number of Photons Emitted per Two-level Atom Using "typical" numbers from the discussion above, one finds that the number of photons emitted per two-level atom, excited by light from an excimer system under saturated conditions, can be close to unity (e.g. for a situation with a spontaneous emission rate of $10^8\,\text{s}^{-1}$, equal degeneracies of the two levels, and a pulse duration of 20 ns).

The number of photons emitted per atom under saturated conditions will exceed unity for sufficiently long laser illumination times (i.e. for $t > (\bar{n}_2^{\text{sat}} A_{21})^{-1}$).

The interaction time for continuous-wave illumination is often determined by the time the atoms spend in the interaction volume, which, in turn, depends on properties of the particular instrumentation used: primarily the atom drift velocity and dimensions of the laser beam. The number of photons emitted per two-level atom can therefore significantly exceed unity for continuous-wave excitations (even if optical saturation is not always reached). However, because illumination with continuous-wave light implies that detection has to be performed over long periods of time, more noise will be detected than under pulsed conditions. This does not necessarily imply, therefore, that the detectability of atoms with continuous-wave light is better than with pulsed laser sources.

3.2.3 Practical Aspects of Optical Saturation

3.2.3.1 Analytical Aspects of Optical Saturation The advantages of optical saturation are that it maximizes the fluorescence signal, minimizes the signal fluctuations (which implies that it improves on the precision), and makes the signal independent of the fluorescence yield (improves on the accuracy). Although optical saturation seems to solve many problems related to quantitative spectrometry, it has some drawbacks. The main disadvantage is that optical saturation can deteriorate the S/N ratio by introducing high levels of scattered light or background signals.[68] The phenomenon of optical saturation also leads to a broadening of the profiles when scanning the laser wavelength. This is referred to as saturation broadening, as has been reviewed in detail in the literature.[65] It has therefore been suggested that a suitable working methodology for the fluorescence technique (when applied to analytical applications) is to work at laser intensities close to (but not really at) optical saturation.[66]

3.2.3.2 Diagnostic Aspects of Optical Saturation It was prophesied earlier that the LIF technique, owing to the existence of optical saturation as well as the tunability, narrow bandwidth and high directionality of the laser light, also has a unique diagnostic capability.[35,67] Numerous publications have therefore analyzed the properties of the optical saturation phenomenon in detail.[35,61–63,70] It should be possible, for example, to use the concept of saturated fluorescence to determine a variety of physical or chemical entities in the system under study. Such an experiment would consist of operating the laser at a sufficiently high spectral irradiance (so that optical saturation is guaranteed) and measuring the amount of fluorescence. By then attenuating the laser light in known fractions (e.g. using neutral density filters) and measuring the fluorescence, it should be possible to determine the saturation spectral irradiance. Entities such as the fluorescence yield, the de-excitation rate, or the quenching cross-section could then be evaluated from the saturation spectral irradiance through Equation (13).

History has shown, however, that the analysis of the experimental data is not always as unambiguous as suggested above. The reason is that the saturation curves very seldom conform to the predicted ideal saturation curve predicted by Equation (10).[61–63,71] Instead, experimentally measured saturation curves often continue to increase with increasing light levels instead of reaching the irradiance-independent plateau. There are numerous reasons for this.[62] Two important ones are spatial and temporal nonhomogeneities of the laser beam. A laser beam with a higher intensity in its center part than in its periphery will saturate the atoms in its center parts to a larger extent that in its periphery.[62,71] A pulse with a temporal inhomogeneity will saturate the atoms more strongly during its high-intensity part than in its weak-intensity parts. Other possible reasons for nonideal saturation curves are that the atomic system is not sufficiently well represented by simple two-level atoms or that the steady-state condition is not valid. For further details, the reader is referred to an excellent review by Alkemade about anomalous saturation curves in LIF.[62]

3.2.4 Limitations of the Two-level Representation

The situation described by this two-level model, in which the fluorescence is measured at the same transition wavelength as is used for the excitation, is referred to as *resonance fluorescence*. However, this means of detecting fluorescence has little practical use for the most demanding analytical applications because the detection system, which necessarily has to be tuned to detect fluorescence at the laser wavelength, has no possibility to reject scattered laser light (which originates from laser light being reflected at various surfaces or by particles or molecules in the atomizer). It is therefore of importance to find viable alternatives to resonance fluorescence. One such alternative is to excite the atoms to levels from which they can fluoresce at more than one wavelength. This implies that it is of importance to investigate to which extent other levels (i.e. nonlaser-connected levels) interact with the laser-connected levels. The following section is therefore concerned with an analysis of a multilevel atomic system, represented by various types of three-level system for simplicity, exposed to one laser field. This issue is thus of special importance to all modes of detection that are not well represented by a two-level system. A three-level model will also enable us to describe the most basic phenomena for the LEI technique. As will be apparent, the existence of various nonlaser-connected levels will significantly affect the atomic system (and thereby also the signal strength and the detection capability).[72–74]

3.3 Multilevel Atoms Exposed to One Laser Field

3.3.1 General Description of a Three-level System Exposed to One Laser Field

3.3.1.1 Rate Equation System for a Three-level System Exposed to One Laser Field The rate equations for a three-level atomic system exposed to one laser field connecting levels 1 and 2, and with the nonlaser-connected level denoted 2′, can be written as Equations (21–23):

$$\frac{dn_2}{dt} = (B_{12}I_\nu + R_{12})n_1 - (B_{21}I_\nu + R_{21} + R_{22'})n_2 + R_{2'2}n_{2'} \tag{21}$$

$$\frac{dn_{2'}}{dt} = R_{12'}n_1 + R_{22'}n_2 - (R_{2'1} + R_{2'2})n_{2'} \tag{22}$$

$$n_{\text{atom}} = n_1 + n_2 + n_{2'} \tag{23}$$

where the various R_{ij}, as before, represent the sum of the collisional de-excitation rate, k_{ij}, and the spontaneous emission rate, A_{ij}, for the case of dipole-allowed exothermic transitions, but only the collisional excitation or de-excitation rates (i.e. k_{ij}) for the other transitions.

3.3.1.2 Various Types of Three-level Systems Although it is possible to solve the set of rate equations once and for all,[59,63,73,75] it is more illustrative and often more convenient to solve the system under a few typical conditions so that the specific influence of each type of nonlaser-connected level can be appraised clearly. There are three conceptually different three-level systems of specific interest for the LIF and LEI techniques, namely when the nonlaser-connected state is:

1. representative of the huge number of highly excited states (Rydberg levels[76]) that exist in all atoms (see Figure 3a);
2. an excited level in close proximity to the upper laser-connected level, which also is radiatively coupled to the ground state, e.g. the ^{2}P-states in alkali atoms (see Figure 3b);
3. a metastable state, i.e. one that is radiatively coupled to the upper laser-connected level but not to the ground state, e.g. the upper fine-structure component of the ground configuration in elements such as Al, In, Ga and Tl (see Figure 3c).

These three systems will be referred to below as cases 1, 2 and 3 respectively.

3.3.1.3 Three-level System with a High-lying Nonlaser-connected State The influence of a highly excited state (case 1) or a state in close proximity to the upper laser-connected level (case 2) can be investigated by solving the set of rate equations (Equations 21–23) under steady-state conditions. The fraction of atoms in the two excited states (under steady-state conditions) can be written according to Equations (24) and (25):

$$\bar{n}_2 = \bar{n}_{2(3)}^{\text{sat}} \frac{B_{12}I_\nu}{B_{12}I_\nu + \bar{n}_{2(3)}^{\text{sat}}(R_{21} + R_{2'1}\chi_{2'})} \tag{24}$$

$$\bar{n}_{2'} = \bar{n}_{2'(3)}^{\text{sat}} \frac{B_{12}I_\nu}{B_{12}I_\nu + \bar{n}_{2(3)}^{\text{sat}}(R_{21} + R_{2'1}\chi_{2'})} \tag{25}$$

where $\bar{n}_{2(3)}^{\text{sat}}$ and $\bar{n}_{2'(3)}^{\text{sat}}$ are the saturation population fractions of levels 2 and 2′ in a three-level system, respectively, given by Equations (26) and (27):

$$\bar{n}_{2(3)}^{\text{sat}} = \frac{g_2}{g_1 + g_2(1 + \chi_{2'})} \tag{26}$$

$$\bar{n}_{2'(3)}^{\text{sat}} = \chi_{2'}\bar{n}_{2(3)}^{\text{sat}} = \frac{g_2\chi_{2'}}{g_1 + g_2(1 + \chi_{2'})} \tag{27}$$

and where $\chi_{2'}$, in turn, represents the ratio of the filling and emptying rates of the nonlaser-connected level (Equation 28):

$$\chi_{2'} = \frac{R_{22'}}{R_{2'1} + R_{2'2}} \tag{28}$$

These expressions show clearly that not only the density of atoms in the upper laser-connected level is being affected by the laser excitation but also the densities

For references see page 9575

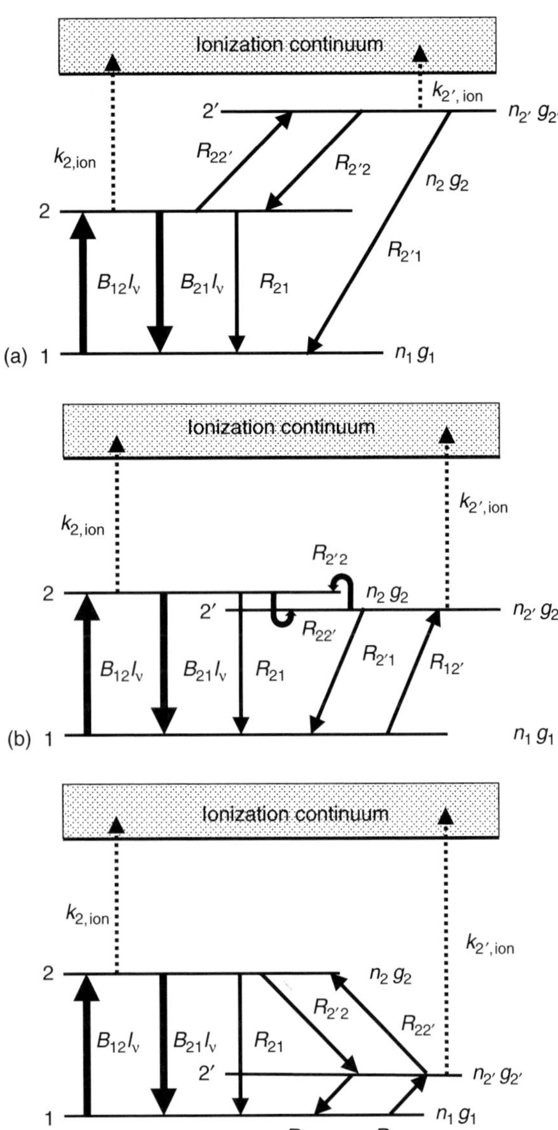

Figure 3 Three different three-level systems exposed to one laser field: (a) the nonlaser-connected level represents a highly excited state (a Rydberg state); (b) the nonlaser-connected state represents an excited state in close proximity to the upper laser-connected level; (c) the nonlaser-connected level represents a low-lying metastable state. The nonlaser-connected rates, R_{ij}, represent collisional excitation rates, collisional de-excitation rates or the sum of collisional de-excitation rates and the spontaneous emission rate, depending on the type of atom. The dashed lines represent collisional ionization rates that are assumed to interact weakly with the three-level systems.

of atoms in the nonlaser-connected levels (represented by level 2′) are being influenced.[77–80] This implies, among other things, that although the laser excitation only induces a transition to one specific state, LIF will in fact be emitted from atoms in virtually all excited states.[78,80] This opens up viable alternatives to resonance fluorescence.

The fraction of atoms in a specific nonlaser-connected excited state can be calculated by inserting the appropriate values for all the quantities involved. The collisional excitation rates can be related to the corresponding de-excitation rates by the concept of detailed balance, as shown in Equation (29):

$$k_{ij} = \frac{g_j}{g_i} \exp\left(\frac{-\Delta E_{ij}}{kT}\right) k_{ji} \tag{29}$$

where ΔE_{ij} is the energy difference between the two states for which $E_j > E_i$ (in J), k is Boltzmann's constant and T is the temperature (in K). However, because not all such rates used to be known, some approximations or generalizations are often used.

One way to estimate the fraction of atoms in a specific nonlaser-connected excited state for cases 1 and 2 is to assume that the collisional rates from level 2′ dominate over any possible spontaneous emission rate from that particular level and that the decay of the nonlaser-connected level to the ground state is smaller than that to the excited state (i.e. $R_{2'1} < R_{2'2}$). One then finds that $\chi_{2'}$ takes a particularly simple form, as shown in Equation (30):

$$\chi_{2'} = \frac{g_{2'}}{g_2} \exp\left(\frac{-\Delta E_{22'}}{kT}\right) \tag{30}$$

Equation (27) then shows that the fraction of atoms in highly excited states (i.e. those that have a large energy deficit to the upper laser-connected level; case 1 atoms) will be significantly lower than that of the upper laser-connected level (because $\chi_{2'} \ll 1$).

For atoms with an excited level in close proximity to the upper laser-connected level (for which the energy deficit normally is small), on the other hand, i.e. for case 2 atoms, the fraction of atoms in this nonlaser-connected level can be comparable to that of the upper laser-connected level (because $\chi_{2'} \approx 1$).

3.3.1.4 Three-level System with a Metastable State The third case of special importance (case 3) is when the nonlaser-connected state represents a metastable state (i.e. a state that is radiatively coupled to the upper laser-connected level but not to the ground state). Elements that have a metastable state are, for example, those with fine-structure splitting of the ground configuration, e.g. Al, In, Ga and Tl, or other low-lying states of the same parity as the ground state, e.g. Cu, Ag and Au. It has been found that the actual lifetimes of metastable states are significantly longer than the typical pulse lengths from the most commonly used laser systems (µs versus ns), even in such highly collisional media as atmospheric pressure atomizers.[72,81–85] These long lifetimes preclude any steady-state condition to be established with the nonlaser-connected level within the laser pulse duration.[59,63] The system of rate equations (Equations 21–23) therefore needs to be solved with its full temporal behavior for atoms with metastable states.(59,63,72,73,85)

A simplified solution for the population of the uppermost laser-connected level in the presence of a lower-lying metastable level can be obtained by assuming that a steady-state condition first is established between the two laser-connected levels within a fraction of the pulse length (following the discussion in section 3.2.1.4). Under this assumption, atoms in the upper laser-connected level will decay to the metastable state at such a rate that the relation between the populations of the two laser-connected levels is not altered.(72,73) The fraction of atoms in the upper laser-connected level then has a time dependence that can be written as shown in Equation (31):

$$\bar{n}_2(t) = \bar{n}_2^{\rm ss}\left[1 + \left(\frac{\bar{n}_2^{\rm pss}}{\bar{n}_2^{\rm ss}} - 1\right)\exp\left(\frac{-t}{\tau_{\rm m}}\right)\right] \tag{31}$$

where $\bar{n}_2^{\rm pss}$ is the fraction of atoms in the second level directly after the onset of steady state between the two laser-connected levels but before any substantial population has been transferred to the metastable state (i.e. $\bar{n}_2(t) \approx \bar{n}_2^{\rm pss}$ when $t \approx 0^+$), $\bar{n}_2^{\rm ss}$ is the steady-state value of the fraction of atoms in level 2 after infinite time (i.e. $\bar{n}_2(t) \approx \bar{n}_2^{\rm ss}$ when $t \to \infty$) and $\tau_{\rm m}$ is the decay time of the laser-connected pair of transitions into the metastable state. These three entities are given by Equations (32–34):

$$\bar{n}_2^{\rm pss} = \bar{n}_2^{\rm sat}\frac{B_{12}I_\nu}{B_{12}I_\nu + \bar{n}_2^{\rm sat}(R_{21} + R_{22'})} \tag{32}$$

$$\bar{n}_2^{\rm ss} = (\chi_{2'})^{-1}\frac{B_{12}I_\nu}{B_{12}I_\nu + (\chi_{2'})^{-1}(R_{21} + R_{22'})} \tag{33}$$

$$\tau_{\rm m} = \frac{1}{\bar{n}_2^{\rm sat}R_{22'}} \tag{34}$$

where $\bar{n}_2^{\rm sat}$ and $\chi_{2'}$ are the saturation population fraction in a two-level system (Equation 11) and the ratio of the filling and emptying rates of the nonlaser-connected level (Equation 28), respectively. It important to realize that $\chi_{2'} \gg 1$ (and thereby $\chi_{2'}^{-1} \ll 1$) for a metastable state, which implies the condition in Equation (35):

$$\bar{n}_2^{\rm pss} \gg \bar{n}_2^{\rm ss} \tag{35}$$

which, in turn, and in practice, implies that the fraction of atoms in the upper laser-connected level rapidly (in this approximation, instantaneously after the onset of the laser excitation) reaches a value similar to what would be the case if there were not any metastable state $\bar{n}_2^{\rm pss}$ (for saturated condition, the fraction of atoms in this state is momentarily $\bar{n}_2^{\rm sat}$). The decay to the metastable state will, however, rapidly drain this population of the upper laser-connected state. This is done with a rate roughly given by the decay rate from the upper laser-connected level to the metastable state ($R_{22'}$). Because this rate can be substantial, this implies that the fraction of atoms in the upper laser-connected level can decay to significantly lower values even within the duration of a laser pulse from a pulsed laser system. For excitation with a continuous laser light, a steady-state condition given by $\bar{n}_2^{\rm ss}$ will prevail. The fraction of atoms in the upper laser-connected level will, under these conditions, only be a fraction ($\chi_{2'}^{-1}$) of the value reached momentarily directly after onset of the excitation.

3.3.1.5 Conclusions This analysis of a three-level system exposed to a single laser field shows clearly that each nonlaser-connected level (irrespective of whether it is a Rydberg state, a closely situated state, or a lower-lying metastable state) influences the populations of the laser-connected levels as well as the requirement of optical saturation (and thereby also the fluorescence and ionization rates). It is not possible, however, to investigate the consequences of this in any detail here. The reader is referred to the literature for a more thorough treatment.(62,63,73,85,86) It can be concluded, however, that the multilevel structure of some atoms explains some (but far from all) of the unusual findings about optical saturation in the literature. Bolshov et al., for example, found that the $6p^2\,{}^3P_0$–$6p\,7s\,{}^3P_1$ transition (at 283.31 nm) in Pb atoms in a GF became saturated at a laser irradiance of 20 kW cm^{-2}, whereas the $3d^6\,4s^2\,{}^5D_4$–$3d^7\,4p\,{}^5F_5$ transition (at 296.69 nm) in Fe showed no sign of saturation up to 300 kW cm^{-2}.(85) As can be concluded from a comparison with the treatment of the two-level system above, both of these saturation irradiances are larger than expected from a simple two-level atom model, possibly indicating influences of nonlaser-connected levels.

The fact that atoms will be promoted also to nonlaser-connected states by the laser excitation has significant but different implications for the fluorescence and ionization techniques. The effects on the fluorescence technique therefore will be discussed separately from those on the ionization techniques. More detailed analysis of these situations can again be found in the literature.(63,72–74)

3.3.2 Implications of Nonlaser-connected Levels for Laser-induced Fluorescence

3.3.2.1 Laser-induced Fluorescence from a Three-level System with a Nonlaser-connected Rydberg State (Case 1) Equations (24–27) and (30) show that the number density of atoms in highly excited nonlaser-connected states is significantly smaller than that of the upper laser-connected level. The density of atoms decreases by one order of magnitude for each 4000 cm^{-1} that separates the

For references see page 9575

two excited states in an atomizer with a temperature of 2500 K. Hence, although LIF in principle will be emitted from virtually all excited states, very few photons will be emitted from excited levels significantly above the upper laser-connected level. Hence, highly excited nonlaser-connected levels generally play a small or insignificant role in the LIF technique.

3.3.2.2 Laser-induced Fluorescence from a Three-level System with a Split Upper Level (Case 2) The situation is different for those atoms that have a nonlaser-connected level in close proximity to the upper laser-connected level. Two levels in close proximity not only have a small energy difference, i.e. $\Delta E_{ij} \ll kT$, but often they also have a fast mixing rate, i.e. $R_{2'2}$ and $R_{22'} \gg R_{2'1}$. This justifies the assumption above, leading to Equation (30), which, in turn, implies that $\chi_{2'}$ often takes a value equal to the ratio of the degeneracy factors of the two levels, i.e. $g_{2'}/g_2$. This implies, furthermore, that the population density on the nonlaser-connected level, in fact, can be comparable to (or even exceed) that of the upper laser-connected level. If this nonlaser-connected level also is radiatively coupled to the ground state, this implies that LIF often can be detected from levels in close proximity to the upper laser-connected level without any significant loss of signal strength. This is of special importance for those elements that have a ground state with no fine-structure splitting, or no other suitable low-lying levels to which fluorescence can be measured from the upper laser-connected level (for which resonant fluorescence otherwise would have been the only possibility). In conclusion, detection of fluorescence from nonlaser-connected excited states in close proximity to the upper laser-connected level might often be a viable alternative to resonance detection.

3.3.2.3 Laser-induced Fluorescence from a Three-level System with a Metastable State (Case 3) The rate of emission of photons per atom for a system with a metastable state is given directly by the time evolution of the fraction of atoms in the upper laser-connected level (Equation 31) and can therefore, in general, be written according to Equation (36):

$$\dot{\overline{N}}_{\mathrm{ph}}(t) = \bar{n}_2^{\mathrm{ss}} A_{22'} \left[1 + \left(\frac{\bar{n}_2^{\mathrm{pss}}}{\bar{n}_2^{\mathrm{ss}}} - 1 \right) \exp\left(\frac{-t}{\tau_{\mathrm{m}}} \right) \right] \quad (36)$$

This simplifies considerably under saturated conditions to the form given in Equation (37):

$$\dot{\overline{N}}_{\mathrm{ph}}(t) = \frac{A_{22'}}{\chi_{2'}} \left[1 + \bar{n}_2^{\mathrm{sat}} \chi_{2'} \exp\left(\frac{-t}{\tau_{\mathrm{m}}} \right) \right] \quad (37)$$

where $\chi_{2'} \gg 1$ (and hence $\bar{n}_2^{\mathrm{pss}} \gg \bar{n}_2^{\mathrm{ss}}$) has been assumed for a metastable state. This implies that the rate of emission of fluorescence from a metastable system rapidly

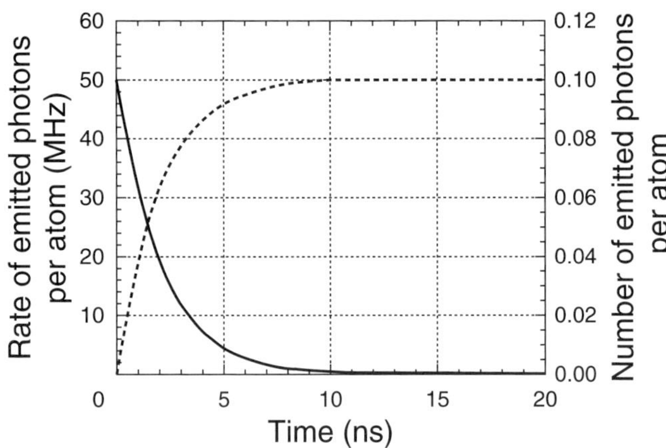

Figure 4 The rate of emitted photons per atom, $\dot{\overline{N}}_{\mathrm{ph}}(t)$ (———), and the number of emitted photons per atom, $\overline{N}_{\mathrm{ph}}(t)$ (- - - -), for the three-level system with a metastable state displayed in Figure 3(c) as a function of exposure time (according to Equations 37 and 41 respectively). It is assumed that the populations on the laser-connected levels become locked to each other instantaneously (to their relative steady-state population fractions). The following parameters have been used: $g_1 = g_2 = g_{2'}$, $A_{22'} = 10^8\,\mathrm{s}^{-1}$, $R_{22'} = 1\,\mathrm{GHz}$ (implying that $\tau_{\mathrm{m}} = 2\,\mathrm{ns}$ and $Y_{22'} = 0.1$), $R_{2'1} = 1\,\mathrm{MHz}$ and $R_{2'2} \ll R_{2'1}$ (implying that $\chi_{2'} = 10^3$).

decreases following the onset of laser illumination, as shown in Figure 4.

Although the initial rate of emission of fluorescence photons from a saturated metastable system (i.e. for $0 < t \ll \tau_{\mathrm{m}}$) is given by Equation (38):

$$\dot{\overline{N}}_{\mathrm{ph}}(t \approx 0^+) = \bar{n}_2^{\mathrm{sat}} A_{22'} \qquad (\approx A_{22'}) \quad (38)$$

which is similar to that of a saturated two-level system, the rate of emission becomes significantly lower when the system has reached steady-state conditions (i.e. for $t \gg \tau_{\mathrm{m}}$), according to Equation (39):

$$\dot{\overline{N}}_{\mathrm{ph}}(t \gg \tau_{\mathrm{m}}) = \frac{A_{22'}}{\chi_{2'}} \qquad (\ll A_{22'}) \quad (39)$$

This implies that the photon emission rate will be significantly smaller for continuous-wave excitations than for pulsed excitations. Because no steady-state condition will be established within the time period of a laser excitation from a pulsed laser, it is more appropriate to discuss the concept of the number of emitted fluorescence photons per atom within a time period t of laser excitation, $\overline{N}_{\mathrm{ph}}(t)$, according to Equation (2). One then finds that $\overline{N}_{\mathrm{ph}}(t)$ can be written as shown in Equation (40):

$$\overline{N}_{\mathrm{ph}}(t) = \bar{n}_2^{\mathrm{pss}} \left\{ \left(1 - \frac{\bar{n}_2^{\mathrm{ss}}}{\bar{n}_2^{\mathrm{pss}}} \right) \tau_{\mathrm{m}} \left[1 - \exp\left(\frac{-t}{\tau_{\mathrm{m}}} \right) \right] + \frac{\bar{n}_2^{\mathrm{ss}}}{\bar{n}_2^{\mathrm{pss}}} t \right\} A_{22'} \quad (40)$$

for a system with a metastable state. This expression again simplifies under pulsed, saturated conditions, as shown in Equation (41):

$$\overline{N}_{ph}(t) = Y_{22'}\left[1 - \exp\left(\frac{-t}{\tau_m}\right)\right] \quad (41)$$

where τ_m, as above, is the decay time of the laser-connected pair of transitions given by Equation (34) and we again have used $\chi_{2'} \gg 1$ for a metastable state, assumed that the fluorescence is measured primarily over the most intensive part of the fluorescence pulse (i.e. over a time period t that does not substantially exceed the lifetime of the metastable state) and introduced $Y_{22'}$ as the fluorescence transition yield for the transition between the upper laser-connected level and the metastable state, defined according to Equation (42):

$$Y_{22'} = \frac{A_{22'}}{R_{22'}} \quad (42)$$

Equation (41) (see Figure 4) shows that the number of photons emitted per atom from this atomic system under pulsed conditions is limited to $Y_{22'}$. This value is reached (within 63%) after a time equal to the decay time of the laser-connected pair of transitions, i.e. τ_m, which, in turn, is roughly given by the inverse of the decay rate from the upper laser-connected state to the metastable state. Because many decay rates are slightly below or in the low gigahertz range, it implies that the actual length of typical pulses from pulsed laser systems (5–30 ns) corresponds fairly well to an efficient detection of fluorescence from atoms with metastable states in atmospheric pressure atomizers. This implies that no atom with a metastable state can emit more than one photon per laser pulse under excitation with pulsed laser light. It also implies that the fluorescence yield for the entire detection process of atoms with metastable states is given by the fluorescence transition yield for the transition between the upper laser-connected level and the metastable state, $Y_{22'}$. For continuous-wave excitations, on the other hand, atoms with metastable states give significantly poorer detectability because the number of photons per atom does not increase as much with increased illumination time as does the amount of noise detected.

3.3.3 *One-step Excitation Laser-enhanced Ionization*

The influence of the various nonlaser-connected levels will be somewhat different for the LEI technique.

3.3.3.1 Ionization Through Nonlaser-connected Rydberg States Although the high-lying levels play a minor role in the fluorescence technique, they are of major importance for the LEI technique. As stated above (near Equations 3 and 4), the total ionization signal depends on the ionization rate density, $\dot{n}_{ion}$, which in turn is given by a sum over all state-specific ionization rates. Adapting this reasoning to the current model with two excited states (one laser-connected and one not), we find that the ionization rate density can be written as shown in Equation (43):

$$\dot{n}_{ion}(t) = k_{2,ion}n_2(t) + k_{2',ion}n_{2'}(t) \quad (43)$$

It is commonplace to assume that the collisional ionization rate scales in a similar way to that of exciting collisions between two bound states, as was discussed in proximity to Equation (29), i.e. as $\exp(-\Delta E_{i,ion}/kT)$, where $\Delta E_{i,ion}$ is the energy deficit to the ionization limit.[57,87,88] This implies that even without any detailed knowledge about each of the two ionization rate constants in the expression above, they can be related to each other (assuming that $E_{2'} > E_2$) according to Equation (44):

$$k_{2,ion} \approx k_{2',ion}\exp\left(\frac{-\Delta E_{22'}}{kT}\right) \quad (44)$$

This implies that although the collisional ionization rate constant from level 2 is smaller than that of level 2′ by the factor $\exp(-\Delta E_{22'}/kT)$, Equations (24–27) and (30) show that the population of level 2′ is smaller than that of level 2 by almost the same factor (given by $\chi_{2'}$), which is expressed as Equation (45):

$$\bar{n}_{2'} \approx \bar{n}_2\frac{g_{2'}}{g_{2'}}\exp\left(\frac{-\Delta E_{22'}}{kT}\right) \quad (45)$$

This implies, in turn, that all excited levels whose populations have an approximate relation to the upper laser-connected level by the activation energy factor (i.e. as given by Equation 30) also contribute approximately equally to the total ionization rate. This can be interpreted as if the ionization simultaneously takes place through a large number of "channels". Following the treatment above, it would be easy to conclude that all levels whose energy is higher than that of the upper laser-connected level would qualify for being such a channel.

However, this argument gives rise to a contradiction. Because there are, in principle, an infinite number of excited states above the upper laser-connected level, this model predicts that the ionization rate would be infinite. In reality this is, of course, not the case. It has been argued in the literature that atoms that are not too highly excited can (or will) exist in atmospheric pressure atomizers (one reason is that their size makes them very vulnerable in the highly collisional environment). It has therefore been found suitable to introduce a certain cutoff principal number, representing the highest excited state that can contribute to the ionization.[74] This is equivalent to introducing an *effective* collisional ionization rate that would act from only one excited state, most conveniently

For references see page 9575

the upper laser-connected level.[57] Hence, although the excited atoms will redistribute rapidly among a large number of excited states due to collisions,[78,80] all of which will have their own state-specific ionization rate, we will model LEI by an effective collisional ionization rate, $k_{i,\text{ion}}^{\text{eff}}$, acting only from the upper laser-connected level, as shown in Equation (46):

$$\dot{n}_{\text{ion}}(t) = k_{2,\text{ion}}^{\text{eff}} n_2(t) \tag{46}$$

3.3.3.2 Ionization Yield for One-step Laser-enhanced Ionization The electrical field across the interaction region in LEI will rapidly separate ions from electrons. The recombination rate will therefore be insignificant during the laser pulse. If we allow for a substantial ionization during the laser pulse (so that the number density of neutral atoms, n_{atom}, will decrease during the laser pulse), the number density of ions, n_{ion}, needs to be expressed in terms of the number density of neutral atoms, n_{atom}, as shown in Equation (47):

$$n_{\text{tot}} = n_{\text{atom}} + n_{\text{ion}} \tag{47}$$

where n_{tot} is the total number density of analyte species. This implies that the equation for the number density of ions can be written according to Equation (48):

$$\dot{n}_{\text{ion}}(t) = k_{2,\text{ion}}^{\text{eff}} \bar{n}_2 [n_{\text{tot}} - n_{\text{ion}}(t)] \tag{48}$$

where $\bar{n}_2$ is given by Equation (10) for atoms lacking metastable states.

Following the convention above of working with normalized entities, it is convenient to define an ionization yield, $\overline{N}_{\text{ion}}(t)$ (often also denoted Y^{ion} in the literature) as the fraction of analyte species that ionize as a consequence of the laser illumination, as shown in Equation (49):

$$\overline{N}_{\text{ion}}(t) = \frac{n_{\text{ion}}(t)}{n_{\text{tot}}} \tag{49}$$

Solving Equation (48) for the number density of ions, one finds that the ionization yield in this case is given by Equation (50):

$$\overline{N}_{\text{ion}}(t) = 1 - \exp(-k_{2,\text{ion}}^{\text{eff}} \bar{n}_2 t) \tag{50}$$

(under the assumption that $\bar{n}_2$ takes its steady-state value within a time that is short with respect to the ionization depletion time).

We can see here that the ionization yield depends strongly on the effective collision rate, which in turn depends on the energy deficit to the ionization limit of the upper laser-connected level, and that a substantial degree of ionization can be obtained when the interaction time is equal to, or larger than, the inverse of this effective collision rate.

Although the ionization limit of various elements differs, we can conclude that most elements have an ionization limit that is around 50 000–70 000 cm^{-1} above the ground state (with an important exception for the alkali elements). Typical UV light (with 250–330 nm wavelength) excites ground-state atoms to states with energies around 30 000–40 000 cm^{-1}. This implies that the energy deficit of atoms excited by UV light often is in the 20 000–30 000 cm^{-1} region (i.e. a few electronvolts). It is reasonable to assume that the effective collisional ionization rate is in the 10^4–$10^5\,\text{s}^{-1}$ range for states a couple of (or a few) electronvolts below the ionization limit.[57,88] Because $\bar{n}_2$ is close to unity for a saturated transition (on an orders of magnitude basis) and a typical laser illumination time is 10 ns, we get an ionization yield of about 10^{-4}–10^{-3} for a typical one-step LEI excitation. This shows that it is not possible in general to reach any high degrees of ionization from one-step LEI (unless the atoms have unusually low ionization limits).

3.3.3.3 Means to Improve on the Low Ionization Yield for One-step Laser-enhanced Ionization An obvious remedy to this is to excite the atoms to levels closer to the ionization limit.[88–90] Although this can be done by a one-step excitation using far-UV light, it has been found to be less suitable. One reason is that it is getting increasingly difficult to create far-UV photons the further down in wavelength one goes. Another reason is that far-UV photons have a greater tendency to give rise to background signals from flame molecules and concomitant elements than visible photons have (which thus would counteract the expected increase in S/N ratio). It has been found more convenient (and considerably more beneficial in terms of selectivity) to excite the atoms by light from two simultaneously pumped tunable lasers, alternatively referred to as stepwise excitation or two-step excitation (although also a rich variety of more complex denotations exists in the literature).[91]

3.4 Three-level Atoms Exposed to Two Laser Fields

There are several treatments of two-step (or even multistep) excitation of atoms in the literature, some of them with slightly different approaches.[73,74,92–94] Although the presence of metastable states might require a more sophisticated description and solution than those given here, we will describe two-step excitations of atoms in the simplest possible manner that can still correctly account for ionization depletion. This implies that we will only consider direct two-step excitations (i.e. those that share a common intermediate level), neglect the influence of metastable trapping levels, and solve the system of rate equations solely under steady-state conditions. The reader is referred to the literature for analyses of other configurations.[94]

3.4.1 General Description of a Three-level System Exposed to Two Laser Fields

Consider a situation in which the atoms are described as three-level atoms (with levels denoted 1, 2 and 3 as shown in Figure 5) between which two resonantly tuned laser fields are acting. In line with the discussion above, it is often a good approximation to assume that the laser excitations rapidly (i.e. within the first nanosecond of the laser pulse) create a population equilibrium among the three laser-connected states. The ionization is then assumed to act on this strongly coupled three-level system.

3.4.1.1 Rate Equation System for a Three-level System Exposed to Two Laser Fields The rate equations for a three-level system exposed to two resonant laser fields, as depicted in Figure 5, can in general be written as in Equations (51–53):

$$\frac{dn_3}{dt} = R_{13}n_1 + (R_{23} + B_{23}I_v^{23})n_2 - (R_{31} + R_{32} + B_{32}I_v^{23})n_3 \quad (51)$$

$$\frac{dn_2}{dt} = (R_{12} + B_{12}I_v^{12})n_1 - (B_{21}I_v^{12} + R_{21} + B_{23}I_v^{23} + R_{23})n_2 + (R_{32} + B_{32}I_v^{23})n_3 \quad (52)$$

$$n_{atom} = n_1 + n_2 + n_3 \quad (53)$$

3.4.1.2 Solution of the Rate Equation System for a Three-level System Exposed to Two Laser Fields Because ionization rates most often are considerably smaller than typical de-excitation and laser-induced rates, this three-level system can be solved under steady-state conditions. Under the assumption that smaller exciting collisions can be neglected in comparison with large de-exciting collision rates or laser-induced rates, the fraction of neutral atoms in the uppermost laser-connected state (level 3) can be written according to Equation (54):

$$\bar{n}_3 = \bar{n}_3^{sat} \frac{B_{12}I_v^{12}B_{23}I_v^{23}}{(B_{12}I_v^{12} + a)(B_{23}I_v^{23} + b) + c} \quad (54)$$

where the various coefficients are given in Table 2.

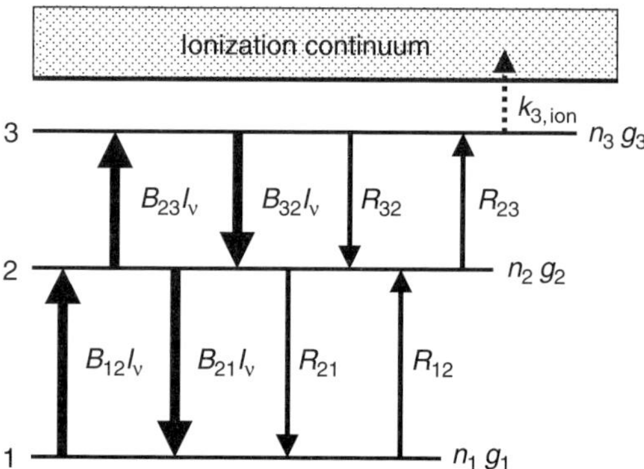

Figure 5 A three-level system exposed to two resonant laser fields.

Table 2 Coefficients for Equation (54)

Coefficient	Expression
$\bar{n}_3^{sat}$	$\frac{g_3}{g_1 + g_2 + g_3}$
a	$\bar{n}_3^{sat}R_{31} + \bar{n}_2^{sat}R_{21}$
b	$\frac{\bar{n}_3^{sat}}{\bar{n}_{2,off}^{sat}}(R_{31} + R_{32})$
c	$(\bar{n}_3^{sat})^2(R_{31} + R_{32})\left(R_{21} - \frac{R_{31}}{\bar{n}_{2,off}^{sat}}\right)$
$\bar{n}_{2,off}^{sat}$	$\frac{g_2}{g_1 + g_2}$

3.4.2 Two-step Excitation Laser-enhanced Ionization

As long as there is any appreciable laser illumination on the upper transition, the main contribution to the ionization rate will be from the uppermost laser-connected level. Hence, for a two-step excitation, we can write the ionization rate as shown in Equation (55):

$$\dot{n}_{ion}(t) = k_{3,ion}^{eff}n_3(t) \quad (55)$$

This implies that the ionization yield for a two-step excitation, $\overline{N}_{ion}^{2s}(t)$, can be written in an analogous way to that for a one-step excitation, namely according to Equation (56):

$$\overline{N}_{ion}^{2s}(t) = 1 - \exp(-k_{3,ion}^{eff}\bar{n}_3 t) \quad (56)$$

Because the second excitation step often can excite the atoms to states a couple of electronvolts closer to the ionization continuum than can be achieved with one-step excitation (a second-step photon with a wavelength of 500 nm corresponds to a decreased energy deficit of 2.5 eV), the (effective) ionization rate can be increased by two to three orders of magnitude by using two-step excitation. In addition, because the fraction of neutral atoms in the uppermost laser-connected state can still take values close to unity under saturated conditions, this implies that the ionization yield can approach unity within the time of a laser pulse excitation.[72,95] Two-step excitation is therefore the most efficient means to achieve strong LEI signals. As will be discussed below, two-step excitation not only increases the sensitivity but also improves the selectivity.

For references see page 9575

3.4.3 Two-step Excitation Laser-induced Fluorescence

Although one-step excitation often gives rise to a sufficiently high sensitivity (number of photons produced per atom) for the fluorescence technique, there are situations when a one-step excitation is not appropriate. One such example is when the atomic structure is such that resonance fluorescence is the only possibility when the atom is excited by light from one laser (as is the case for Cd, for example). Another example is when the fluorescence appears in the visible part of the region, a part of the spectrum in which substantial amounts of background emission often exist (e.g. from the blackbody radiation in ETAs). In this case the technique will not reach its optimum performance. A remedy for these situations is to use two-step excitation also for the LIF technique.(41,63) Because the fluorescence from a two-step excitation often appears in the UV or far-UV regions (thus, far from the exciting wavelengths, as well as blackbody radiation from the atomizer), two-step excitation LIF used to be characterized by freedom from scattered laser light and background radiation (see below).(72,94)

The expressions above (Equation 54 and Table 2) show that a strong fluorescence signal is expected from the upper laser-connected state (or from any collisionally coupled excited state) directly following the onset of laser excitation. However, as is shown by Equations (55) and (56), the atomic system can also suffer from an appreciable ionization depletion during the laser pulse if the uppermost laser-connected level has too small an energy deficit compared to the ionization limit. Taking this drain of the number of neutral atoms during the laser pulse into account, the rate of emission of fluorescence photons per atom (in this case, from the uppermost laser-connected level to any lower-lying level, here denoted i) can be expressed as shown in Equation (57):

$$\dot{\overline{N}}_{\text{ph}}(t) = A_{3i}\bar{n}_3(t)\exp(-k_{3,\text{ion}}^{\text{eff}}\bar{n}_3 t) \qquad (57)$$

This shows that the emission rate of fluorescence photons is highest directly following the onset of laser illumination, after which it decays with a decay time given by the inverse of the product of the effective ionization rate and the fraction of atoms in the uppermost excited level. This behavior resembles that for the LIF signal from a three-level system with a metastable state exposed to one laser field (Equation 37). This is not unexpected because the drain to ionization acts in an equivalent manner to that to a metastable state on these timescales.

Using Equation (2), the total number of photons emitted per atom from such a system (in a time t) can then be expressed as shown in Equation (58):

$$\overline{N}_{\text{ph}}(t) = \frac{A_{3i}}{k_{3,\text{ion}}^{\text{eff}}}[1 - \exp(-k_{3,\text{ion}}^{\text{eff}}\bar{n}_3 t)] \qquad (58)$$

This shows that for laser excitation to states whose (effective) collisional ionization rate is smaller than the inverse of the laser pulse illumination time (i.e. $k_{3,\text{ion}}^{\text{eff}} < t^{-1}$), the number of photons emitted per atom is approximately given by Equation (59):

$$\overline{N}_{\text{ph}}(t) = A_{3i}\bar{n}_3 t \qquad (59)$$

which is equivalent to neglecting ionization depletion during the laser excitation.

If the excitation is performed to a state whose (effective) collisional ionization rate is significantly larger than the inverse of the laser pulse illumination (i.e. $k_{3,\text{ion}}^{\text{eff}} \gg t^{-1}$), the number of photons emitted per atom cannot exceed that shown in Equation (60):

$$\overline{N}_{\text{ph}}\left(t \gg (k_{3,\text{ion}}^{\text{eff}})^{-1}\right) = \frac{A_{3i}}{k_{3,\text{ion}}^{\text{eff}}} \qquad (60)$$

due to ionization losses, which again is a similar behavior to that of the metastable three-level system exposed to one laser field (Equation 42). This shows that if the upper laser-connected level experiences too large an effective ionization rate, ionization losses can put a limitation on the number of photons emitted per atom from two-step excitation.

3.4.4 Fluorescence Dip

If the system of rate equations for a three-level system exposed to two laser fields (sharing a common intermediate level; Equations 51–53) is solved for the population of the intermediate state, one can see clearly that this population (and hence the fluorescence from the intermediate state) will experience a decrease with the onset of the second-step laser.(94,96) This phenomenon, which is similar to the ionization depletion phenomenon discussed above, has been termed *fluorescence dip* and has been prophesied to have extraordinary diagnostic capabilities, e.g. for the determination of transition probabilities between excited states or photoionization cross-sections.(97) The first investigations of the fluorescence dip phenomenon were made by Omenetto et al.(97,98)

The relative decrease of the fluorescence from the intermediate state is termed relative fluorescence dip, ζ_2. It was concluded that the maximum relative fluorescence dip for an optically saturated two-level system (such as that depicted in Figure 5) under steady-state conditions (i.e. neglecting the influence of ionization depletion) can be written as shown in Equation (61):

$$\zeta_2 = \frac{\bar{n}_2[I_\nu^{12} > (I_\nu^{12})^{sat}, I_\nu^{23} = 0) - \bar{n}_2(I_\nu^{12} > (I_\nu^{12})^{sat}, I_\nu^{23} > (I_\nu^{23})^{sat}]}{\bar{n}_2[I_\nu^{12} > (I_\nu^{12})^{sat}, I_\nu^{23} = 0]} = \frac{g_3}{g_1 + g_2 + g_3} \quad (61)$$

For the case of Sr (with a $5s^2$–5s 5p–5s 6d excitation, an upper state about $6000\,cm^{-1}$ below the ionization limit, and degeneracy factors of 1, 3 and 5 for the three states, respectively), the optically saturated relative fluorescence dip should be 55% under the assumption that a steady-state situation prevails. It was found experimentally by Axner et al., however, that significantly larger optically saturated relative fluorescence dips could be obtained in an acetylene–air flame (up to 85%).(99) Simultaneous measurement of the fluorescence dip and the increase in the ionization signal, coupled to an analysis based upon time-dependent rate equations for the fluorescence dip, revealed that the system was exposed to significant ionization depletion during the laser illumination time. This finding does not only show that substantial depletion of the intermediate laser-connected level follows the onset of the second excitation step, but it is also an example of the importance of taking ionization depletion into account when excitations of atomic systems to high-lying states in high-collision media are considered. This is in line with the theoretical predictions of ionization and fluorescence depletion made by Omenetto et al.(100)

3.5 Conclusions – Justification of the Use of the Rate Equation Formalism

Section 3 has given the basic mechanisms for the production of fluorescence photons and charges (ions and electrons) primarily for the LIF and the LEI techniques, respectively, for both one- and two-step excitations in high-collision media. The description was given in terms of the number of photons emitted per atom in time t, $\overline{N}_{ph}(t)$, and the fraction of analyte species that ionizes as a consequence of the laser illumination, $\overline{N}_{ion}(t)$, commonly also termed the ionization yield. It has been shown that both of these entities can reach values close to unity under a rather broad range of optimized pulsed conditions.

The description above was given in the rate equation formalism with no particular justification of its applicability. A correct representation of the excitation and de-excitation processes of atoms by narrow-band laser light can normally only be accomplished by using the so-called density matrix formalism.(101–103) In this formalism, the atomic system is described by its quantum mechanical wavefunction. Solved with its full temporal dependence, the density matrix formalism is capable of describing both the time dependence of the dipole moments as well as the atomic level populations. The density matrix formalism is therefore able to describe effects that appear as a result of coherence between the induced dipole moments and the populations, e.g. rapid population oscillations between various states, commonly termed Rabi oscillations. Solved under steady-state conditions for the dipole moments, the density matrix equations (often termed Bloch equations in this context) revert to pure population equations of similar type to those of the rate equation formalism. These population equations are able to describe how an atomic system responds to narrow-band laser light of an arbitrary wavelength (i.e. including wing excitations) and can therefore predict, among other things, profile shapes of atomic transitions. They can also describe coherent contributions in the excitation process, e.g. two-photon excitations and dynamic Stark effects, such as broadening, splitting, and shifts of transitions induced by the laser light.(73,102–109)

The density-matrix-based population equations, in turn, revert to the rate equations for sufficiently large laser bandwidths and collisional rates in the system (rather rapidly for one-step excitations but more reluctantly for two-step excitations).(73,103,107–109) The rate equation formalism assumes, among other things, that there is a constant irradiance of the light across the width of the transition. This implies, in turn, that the formalism assumes tacitly that the light is fully in resonance with the transition. It is therefore unable to predict excitation profiles correctly. The high collisional rates in atmospheric pressure atomizers, together with the fact that light with bandwidths larger than that of the analytes often is used, often justify (although not always) the use of the simpler rate equations for calculations of various-level populations.(73,103,107) It has been commonplace, however, to use the rate equation formalism for description of excitation processes in more or less all fields of analytical spectrometry, even under situations when not all of the requirements for its use are fully fulfilled. The most common justification of the use of the rate equation formalism seems too often to be its simplicity rather than its appropriateness. As a result, the main part of the theoretical description in the literature of the interactions between laser light and atoms for analytical applications (as well as the description given above) has been cast in the framework of the rate equation formalism.(35,59–63,71–74,94,99,103,110–112) One exception to this is the work by Boudreau et al.(113) These authors have developed a user-friendly program, based upon a fully time-dependent density matrix model that describes stepwise excitations of atoms with degenerate states under collision-dominated conditions,(108) that can predict phenomena such as dynamic Stark effects (splitting shifting and broadening) and two-photon transitions.(113) The

For references see page 9575

reader is referred to the literature for a proper description of how density matrix formalism is adapted to the conditions applicable to analytical spectrometry.[103–109,114,115]

4 LASER-INDUCED FLUORESCENCE SPECTROMETRY

4.1 Introduction

LIF spectrometry, commonly also termed LEAFS when applied to analytical spectrometry, is based upon resonant absorption of laser radiation by the analyte and detection of the subsequently emitted fluorescence radiation.

Because laser light has a high spectral irradiance, the population of excited analytes can be made significantly larger when laser light is being used for excitation than when other light sources are used. This implies a high sensitivity and low detection limits for laser-based spectroscopic techniques in general and for LIF/LEAFS in particular. LOD in the low picogram per milliliter range, and in the femtogram range, have been obtained for a number of elements when used with a GF as atomizer. The technique has also been applied to other atomizers (e.g. flames and plasmas) but not with the same impressive LOD. In addition, because laser light consists of narrow bands, laser-based spectroscopic techniques generally have good selectivity. Because the LIF/LEAFS technique often uses a dispersive element for detection of the fluorescence (most often a spectrometer), the elemental selectivity is excellent. The LIF/LEAFS technique also has a large dynamic range (often five to seven orders of magnitude).

The first time fluorescence was used for analytical purposes was in 1964 by Winefordner et al.[116,117] The first laser-based instrumentation for fluorescence was constructed by Fraser and Winefordner in 1971 and 1972.[118,119] The first time frequency-doubling was used to extend the applicability of laser-based spectroscopic techniques for analytical applications was in 1973 when Kuhl and Spitschan detected Mg, Ni, and Pb in a flame by the LIF/LEAFS technique.[120] Since then, a large number of papers have been published in the field, with significant contributions by the groups of Winefordner, Omenetto and Michel as well as others.

The range of elements that can be detected conveniently by the LIF/LEAFS technique has been restricted mostly to metals, although a few nonmetal elements also have been studied. In addition, the fluorescence technique has been used for the detection of molecules; it is then referred to as LEMOFS[38] or occasionally LEMFS[121]). Moreover, in hot plasmas, where the degree of ionization is fairly high, ionic fluorescence has been used as an alternative to atomic fluorescence in order to improve the sensitivity or extend the number of accessible elements. The isotope selectivity of the LIF/LEAFS technique, on the other hand, is in general not very impressive under normal conditions because most optical isotope shifts are smaller than typical laser bandwidths or Doppler and collisional broadening mechanisms in most atomizers (although a few exceptions exist, e.g. among the actinides and lanthanides[122]). Pulsed laser systems (primarily those with pulse duration in the nanosecond range, i.e. excimer-pumped dye laser systems and, to some extent, Nd:YAG-pumped dye laser or OPO systems) have shown the highest applicability for the LIF/LEAFS technique, although some impressive works also have been done with continuous-wave diode lasers. In addition, a variety of excitation–detection schemes have been investigated. Although the most common excitation–detection scheme is one-step excitation followed by Stokes direct-line or stepwise detection (mainly due to its simplicity and high fluorescence yield), some elements require two-step excitation in order to obtain the most favorable conditions.

This implies that in order to characterize fully the LIF/LEAFS technique with respect to its properties and applicability to analytical spectrometry, one can thus envisage a huge number of possible combinations of laser, atomization and detection instrumentation, excitation and detection schemes, as well as detection procedures. This naturally leads to a large spread in performance capabilities. Numerous detection limits of the LIF/LEAFS technique for a variety of elements have therefore been published in the literature throughout the years. It is not possible to cite all individual contributions to the continuously ongoing development of the LIF/LEAFS technique in a presentation like this so we have chosen to present the specific analytical attributes and characteristics of the LIF/LEAFS technique (sensitivity, types of noise, LOD, background reduction techniques, LDR, etc.) by referring to only a selection of papers in the field (but including those that report on the lowest LOD for each element).

Prior to that, however, a survey of previous reviews of the LIF/LEAFS technique is given, followed by a description and discussion of "typical" LIF/LEAFS instrumentation. Because the LIF/LEAFS technique can be used also as a diagnostic probe for specific physical or chemical phenomena (e.g. state-specific ionization rates, fluorescence yields, or recombination rates), some comments on the diagnostic capabilities of the LIF/LEAFS technique are given in a special section.

4.2 Previous Reviews

The field of conventional-source-excited atomic fluorescence spectrometry (i.e. not using laser light for excitation) has been reviewed in detail by Butcher et al.[123] Smith et al. have made an extensive compilation of a variety of atomic fluorescence spectrometry results (up to 1989) arranged by elements.[124] The field of laser-induced/excited fluorescence, i.e. LIF/LEAFS, has been reviewed by several authors. An early review (from 1978) that serves as a good introduction to the practical aspects and early applicability of the LIF/LEAFS technique is that of Winefordner.[36] Contemporary reviews that are focused on the basic properties of the fluorescence technique are those of Omenetto and Winefordner.[35,125] Another early landmark review paper is that of Butcher et al.[37] Omenetto has written many interesting papers about various aspects of the LIF/LEAFS technique[35,48,66,125–129] but of special importance is the work by Omenetto and Human[66] in which many important issues for the analytical use of LEAFS in flames, furnaces and inductively coupled plasmas (ICPs) are discussed in detail (e.g. the influence of peak power, pulse duration, spectral bandwidth and repetition rate on the detectability, together with the choice of excitation line and the optical arrangement of the detection system). This paper also contains many "handy recipes" for the LIF/LEAFS technique. Sjöström[130] has written a comprehensible review of LIF/LEAFS in ETAs and, together with Mauchien, has given a critical view of the entire field of laser-based spectrometric techniques for analytical spectrometry.[131] Interesting reviews not only of LIF/LEAFS but also of other laser-based spectroscopic techniques have been given by Falk,[132,133] as well as by Omenetto.[134] The most recent review of the LIF/LEAFS technique is that by Hou et al.[38] The group of Winefordner has presented a series of papers critically discussing the ultimate possibilities of laser-based techniques in general and LIF/LEAFS in particular for analytical spectrometry, as well as their possibilities to achieve SAD.[53–55,129,135]

4.3 Instrumentation

4.3.1 Typical Set-up

A typical experimental set-up for LIF/LEAFS is shown in Figure 6. Laser light is directed into an atomizer in which the sample is atomized. The emitted fluorescence is collected by an optical system and directed into a spectrometer. The transmitted light is detected by a sensitive detector – most often a photomultiplier tube (PMT), whose signal is subsequently amplified, and infrequently an intensified charge-coupled device (ICCD). When pulsed lasers are employed, gated detection, most often synchronously triggered with light from the laser, needs to be used (accomplished by a boxcar integrator for the case of a PMT). The output signal from the boxcar integrator is finally fed into a computer for data storage and handling.

4.3.2 Atomizers

4.3.2.1 Flames The first LIF/LEAFS measurements for analytical purposes during the 1970s used flames as atomizers.[35,36,70,71,86,118–120,136–147] The prime reasons were the suitability of flames for analysis of liquid samples, their simplicity and reliability, and the fact that they have a continuous mode of operation (and hence provide a continuously refreshing distribution of

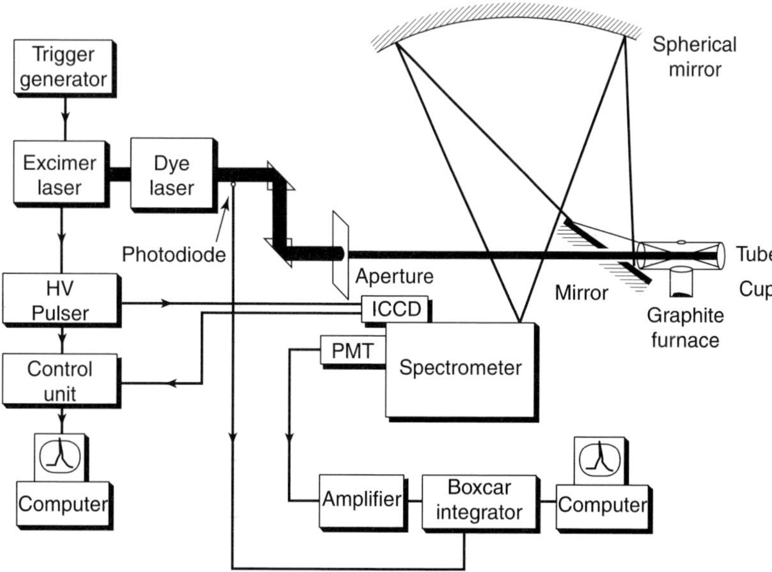

Figure 6 A typical experimental set-up for LIF/LEAFS in a GF. (Reproduced from A. Marunkov, N. Chekalin, J. Enger, O. Axner, *Spectrochimica Acta*, **49B**, 1385–1410, Copyright 1994, with permission from Elsevier Science.)

For references see page 9575

atoms). However, although LIF/LEAFS in flames was prophesied early on to be a technique with extraordinary detectability and selectivity,[35] the actual detection limits turned out not to be as impressive as first anticipated.[37,38,60,64–67,78,81,115,148–158] The main reason is that the flame itself is far from the ideal atomizer for the LIF/LEAFS technique. Flames give rise to a rather large amount of background radiation and act as a source of scattering of laser light, which reduce the detection limits (see discussion below). However, LIF/LEAFS has been found to be a useful tool for the diagnostics of flames.

4.3.2.2 Electrothermal Atomizers The ETA, mostly represented by the GF, has several advantages in comparison with flames. The one that affects sensitivity the most is that the vaporized atomic cloud is held more concentrated, which implies that the atomic residence time (and thereby the interaction time) is longer. Other advantages are that only microliter volumes of samples are needed and that staged heating in an inert atmosphere makes analysis of samples with high concentrations of concomitant elements possible. A specific advantage with the ETA for the LIF/LEAFS technique is that it provides a rather quiet environment with a minimum amount of background radiation. It was therefore found early on that the LIF/LEAFS technique had its largest potential for analytical spectrometry in combination with the ETA.[37,66,77,79,80,85,121,130,132,156–227]

4.3.2.3 Other Atomizers The development of the ICP during the last decade has considerably facilitated the determination of trace elements in many types of matrices using conventional detection techniques. The most significant advantages with the ICP are that it provides an efficient atomization, excitation and ionization environment. It has therefore taken over the role of the flame as the workhorse for a large variety of analytical investigations in many laboratories. This has implied that the ICP has been of interest also for the community of laser spectroscopists. LIF/LEAFS has therefore been applied to various types of plasmas.[37,60,66,94,122,157,228–230] In addition, other atomizers, e.g. discharges[231–235] and cells,[178] have been of interest for analytical laser-based spectrometric techniques. The analytical performance of LIF/LEAFS systems containing these types of atomizers, however, has not been found to rival those of ETA systems. Therefore, only a limited number of studies in this area have been pursued.[37,60,66,94,122,157,178,228–235]

4.3.3 Laser Sources

A variety of lasers have been used for the LIF/LEAFS technique over the years. Pulsed laser systems (primarily those with pulse duration in the nanosecond range, i.e. excimer-pumped dye laser systems,[37,66,67,80,120,137,153,155–189,191–199] N_2-pumped dye laser systems,[118,119,138–141,148,152,214–224] and Nd:YAG-pumped dye laser[37,77,79,85,190,200–209] or OPO systems[154,227,236]) have shown the highest applicability for the LIF/LEAFS technique, although some impressive studies also have been done with Cu-vapor lasers,[209–213] ion-pumped dye laser systems,[142–145,151] flashlamp-pumped dye lasers[146,147,149,150] and continuous-wave diode lasers.[225]

The prime reasons for the success of pulsed lasers for LIF/LEAFS (as well as the ionization techniques, see below) are: versatility in producing UV light (because tunable UV laser light is normally produced by frequency-doubling visible light in a nonlinear crystal using processes that are strongly nonlinear, then, with a conversion efficiency that increases with irradiance, pulsed lasers, which have a significantly higher irradiance than continuous-wave systems, can more easily produce UV wavelengths); the possibility of producing a large signal over a short period of time (thereby maximizing the S/N ratio); and the fact that they can saturate the transitions more easily than continuous-wave systems.

A disadvantage with pulsed laser systems is that the duty cycle is rather poor, so that only a small fraction of all atoms entering the atomic reservoir will be exposed to laser light. Consider, for example, the case of a 10-Hz Nd:YAG laser illuminating an air–acetylene flame (whose rise time is around $10\,\mathrm{m\,s^{-1}}$ and width is around 1 cm). The analyte atoms are being distributed over an area of $100\,\mathrm{cm^2}$ in-between each laser firing. If the laser beam diameter is between 3 and 4 mm, it covers an area of about $10\,\mathrm{mm^2}$. Hence, not more than 10^{-3} of all the atoms in the flame can interact with the laser light. The situation improves if excimer-pumped laser systems are used. Modern excimer lasers can produce laser pulses at repetition rates of up to a few hundred hertz (although often at a reduced pulse energy). An excimer-pumped dye laser system running at 100 Hz can then potentially still interact with only a few percent of all the atoms in the flame using an enlarged beam. This is, however, not such a devastating drawback for LIF/LEAFS as might be expected because the pulsed detection also reduces the noise in the system (noise is measured only over $2\,\mathrm{\mu s\,s^{-1}}$ using a 100-Hz system and a 20-ns fluorescence detection gate).[237]

Copper vapor lasers, on the other hand, which produce pulses at kilohertz repetition rates, can potentially interact with all atoms in a sample. However, a severe drawback of metal-vapor-pumped laser systems is that not the entire spectral region can be accessed.[238]

The most commonly used type of laser for LIF/LEAFS is the XeCl excimer-pumped dye laser, mainly due to a combination of a high repetition rate (up to 500 Hz), a high peak power (typically up to 1 MW in the visible spectrum and tens of kilowatts in the UV range) and pulse durations that are a few times longer than those of Nd:YAG lasers.[191] Another issue of importance in this respect is that there are commercial excimer-pumped dye laser systems in which the dye laser cuvette and dye pump form a combined unit that can be removed/exchanged rather easily. With access to a few sets of dye laser cuvette and pump units, a change of dye can be accomplished rather conveniently, i.e. within a few minutes (a procedure that often takes many hours with other dye laser systems).

The recent development of new pulsed tunable laser sources has had some impact on the field of analytical spectrometry. The development of OPO systems, with a rapidly scanned wide spectral tuning range (220–2200 nm) and pulse energies of tens of millijoules per pulse in the visible region and several millijoules in the UV region, takes the laser-based spectroscopic techniques one step further towards multi-element determinations.[154,227,236] It is not clear, a priori, however, that pulsed dye laser systems with nanosecond pulse lengths provide the optimum excitation source for the LIF/LEAFS technique for *all* elements in *all* instances. As will be discussed further below, they can, under certain conditions, be rivaled by other types of laser systems (depending on the dominating type of noise).

The development of diode lasers has, in particular, opened up the possibility of constructing new, powerful and compact LIF/LEAFS instrumentation. Although most applications of diode lasers for analytical purposes so far have been pursued within the field of FM or WM absorption techniques (which is described in section 6), Zybin et al. established 10- and 20-fg LOD for Li and Rb in a graphite tube, respectively, detected by diode-laser-excited LIF/LEAFS.[225]

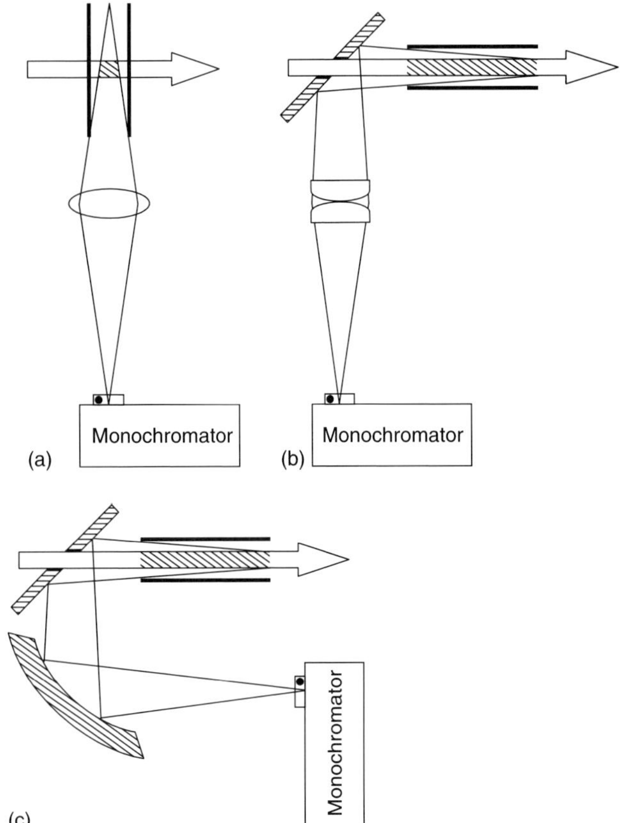

Figure 7 Three different types of optical systems that have been used with the LIF/LEAFS technique: (a) right-angle detection using refractive optics; (b) front-surface detection using refractive optics; (c) front-surface detection using reflective-only optics (i.e. a curved mirror).

4.3.4 Detection Equipment

4.3.4.1 Optical Systems As was concluded by Yuzefovsky et al.[239] and clearly summarized by Hou et al.,[38] there are five main requirements for an efficient optical system, namely: a high light collection efficiency in both the UV and visible regions; a good ability to discriminate between background and analytical signals; minimal losses of the analytical signal at the detection wavelength; easy-to-align optics; and inexpensive and commercially available optical components.

The optical systems that have been used the most with the LIF/LEAFS technique over the years are the three that are depicted schematically in Figure 7 (exemplified using a graphite tube as the atomizer). Most early experimental set-ups directed the fluorescence into the spectrometer in a right-angle mode of detection, as displayed in Figure 7(a) (see references 35, 37, 65, 66, 77, 85, 115–121, 132, 137–153, 157–174, 203–208, 213–222). Later set-ups have instead had front-surface detection using a pierced mirror (see references 37, 79, 80, 154–156, 158, 171, 173–202, 209–213, 223–227), either flat (together with a suitable set of lenses) (see references 37, 79, 156, 158, 173–185, 190–198, 200–202, 209–212, 224–226) as in Figure 7(b), or curved[80,155,171,186–193,199,209,223] (ellipsoidal[171,191–193,199,209] or spherical[80,155,186–189]) as in Figure 7(c). The front-surface mode of detection has been found to increase the volume of illuminated atoms seen by the spectrometer and hence the sensitivity of the system.[37,156,209] A front-surface mode of detection also has the advantage that it reduces so-called post-filtering effects (see below).

Most early experimental set-ups used a single biconvex lens for directing the fluorescence light into the spectrometer, whereas later ones have used either multi-lens systems (Figure 7b) or reflective optics (Figure 7c). It was calculated by Farnsworth et al. (using computer

For references see page 9575

modeling of the collection efficiency of LIF) that the use of a single biconvex lens for collection of fluorescence (as shown in Figure 7a) would not provide the best collection of the fluorescence owing to aberrations (mainly spherical aberration).(240) They found that single-point collection efficiencies in general differ significantly from those that would be expected based on the solid angle subtended by the lens. An optical system consisting of a pair of planoconvex lenses, as indicated in Figure 7(b), was found to be far more effective in imaging the interaction volume onto the entrance slit of the spectrometer than a single biconvex lens. The best performance was obtained by a pair of achromatic doublets. The single-point collection efficiency was found to range from below 15% of the ideal value for a biconvex lens to nearly 100% for a pair of the achromats. Some of these predictions have been confirmed experimentally by Liang et al.(183)

In addition, because all optical material has dispersion (i.e. a wavelength-dependent index of refraction), instrumentation using refractive components (e.g. lenses) for the collection of fluorescence light will experience chromatic aberrations (implying that the light collection efficiency will have a wavelength dependence). The use of achromatic lenses therefore decreases significantly the chromatic aberrations in the system. Because there is no dispersion in fully reflective optical systems, the use of curved mirrors instead of lenses for focusing the light onto the spectrometer slit has the advantage that there are no chromatic aberrations.(187)

4.3.4.2 Detectors The light filtered by the spectrometer/monochromator is being detected by a sensitive detector. Because the vast majority of all the LIF/LEAFS measurements have been done using pulsed excimer- or Nd:YAG-pumped laser systems, and typical de-excitation rates in atmospheric pressure atomizers are in the range of hundreds of megahertz to a few gigahertz, the duration of the emitted fluorescence will be of the same order of magnitude as the duration of the laser pulses (i.e. most often a few to some tens of nanoseconds). Furthermore, because these types of laser have repetition rates of some tens or hundreds of hertz, the duty cycle is low (typically 10^{-6}–10^{-5}). Gated detection therefore has to be used in order to obtain the optimum S/N conditions.

By far the most common detector used is the PMT,(37,65,66,77,79,85,116–121,132,137–150,152–182,190–224,227) because it has a number of useful properties: it is sensitive, inexpensive and easy to use. The only disadvantage with a PMT is that it is not capable of providing any fluorescence spectrum – it can only give rise to one data point per laser firing (corresponding to the fluorescence integrated over the bandpass of the spectrometer at the center wavelength chosen). When PMTs are used the gating is done by a boxcar integrator, whereas diode arrays and charge-coupled device (CCD) detectors, which sometimes have been used with the LIF/LEAFS technique, need to be equipped with an intensifier in order to provide the necessary gating.

ICCD detectors(80,185–190) or diode arrays(151,226) have the capability of detecting an entire fluorescence spectrum for each laser shot, which can be an advantage when samples with complex matrices are to be detected. One example of the versatility of an ICCD detector is shown in Figure 8. The figure shows a pair of fluorescence spectra from Ti atoms excited by pulsed 264.108-nm light in a GF. Figure 8(a) displays the situation in a normal Ar atmosphere whereas Figure 8(b) shows the corresponding spectrum when Ar was substituted by N_2.(80) As can be

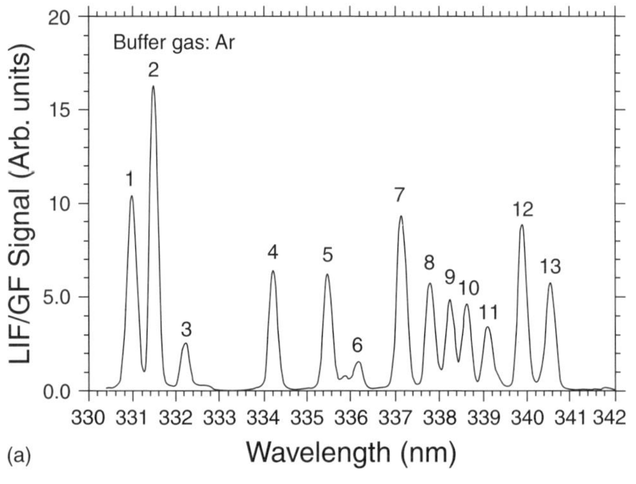

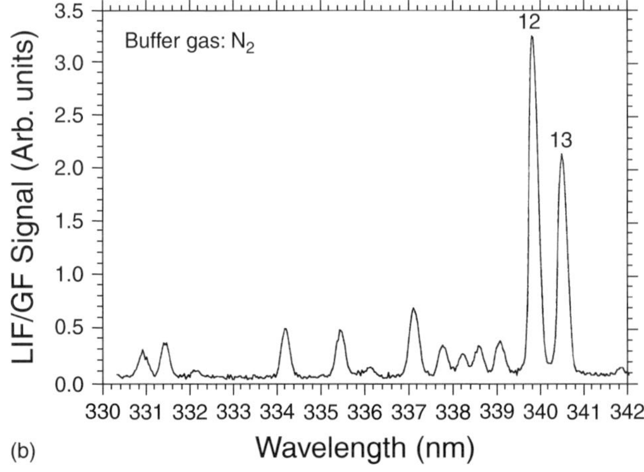

Figure 8 Two LIF/LEAFS/ICCD spectra from Ti atomized in a GF excited by 264.108-nm light to the $3d^2\,4s\,4p\,^3D_1$ state at 37 852.021 cm^{-1} when Ar (a) and N_2 (b) have been used as buffer gas. A full identification of the fluorescence lines is given in Ljung et al.(80) N_2 is a stronger quencher than Ar, so peaks 12 and 13 can be referred to as "direct" transitions. Note that several of the indirect peaks are equally as strong as the direct ones in the Ar atmosphere. (Reproduced from P. Ljung, E. Nyström, J. Enger, P. Ljungberg, O. Axner, *Spectrochimica Acta*, **52B**, 675–701, Copyright 1997, with permission from Elsevier Science.)

concluded from the figure, the fluorescence spectrum is fairly rich. To aid in the identification of the fluorescence peaks, however, Ar was occasionally substituted for N_2, which is a better quencher of excited atoms. The peaks that decrease the most (labeled 1–11 in Figure 8) originate from "indirect" transitions (i.e. originating from upper levels that are different from that accessed by the laser), whereas the others refer to "direct" transitions (i.e. those that originate from the upper laser-connected level). This study led to the conclusion that most of the fluorescence spectra from Ti are dominated by "indirect" transitions. In addition, many direct transitions were found to be inferior to other indirect transitions, making it virtually impossible even to make a qualified guess of the optimum excitation–detection combination prior to the experiment. The results indicate also that the collisional redistribution processes amongst the excited levels are, in general, faster than typical fluorescence rates for Ti in GFs (which supports the discussion about rapid collisional redistribution processes among excited states in section 3.3.1.1 above).

4.3.5 Types of Fluorescence Transitions

A variety of different excitation–detection modes have been used throughout the years, mainly dictated by the atomic structure of the analyte to be investigated. Although attempts have been made to define a clear nomenclature for these various excitation–detection modes,[119,241] a slight disorder has been prevalent. Lately, however, a new International Union of Pure and Applied Chemistry (IUPAC) notation for the description of various excitation–detection processes, applicable to fluorescence as well as ionization techniques, has been released.[242] A few of the most commonly used excitation–detection schemes for the LIF/LEAFS technique are given in Figure 9. The reader is referred to the literature, e.g. any of the works by Omenetto and Winefordner,[35,241] for a more extensive description of possible excitation–detection schemes for the LIF/LEAFS technique.

4.4 Analytical Performance

The detectability of a certain species in a given atomizer depends on a number of factors. On the one hand, we have the sensitivity, i.e. the relation between the measured signal and the concentration (or amount) of the analyte in the sample. On the other hand, there is always noise in the system. The relation between the two determines the LOD for the system. Because both the sensitivity and limiting noise will differ between various types of instrumentation, the LOD for a given element will depend on the particular atomizer and laser system used, as well as their mode of operation.

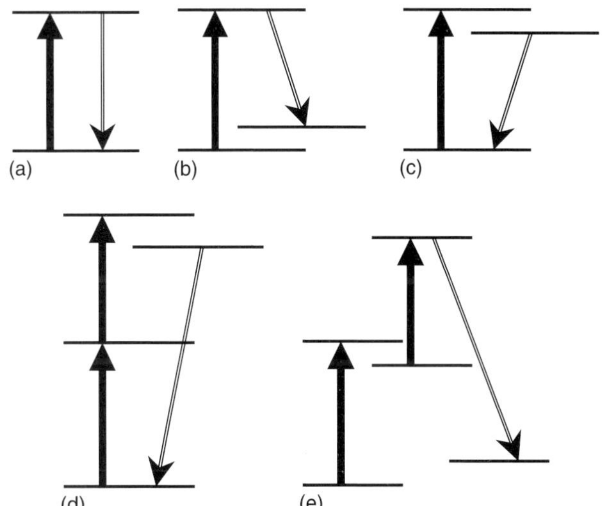

Figure 9 The most commonly used excitation–detection schemes for the LIF/LEAFS technique. Thick and thin arrows represent laser excitations and fluorescence, respectively. (a) Resonance fluorescence ($R_{12}F_{21}$); (b) Stokes direct line fluorescence ($R_{13}F_{32}$); (c) Stokes step-wise line fluorescence ($R_{13}F_{21}$); (d) connected double-resonance stepwise line fluorescence ($R_{12}R_{24}F_{31}$); (e) disconnected double-resonance direct line fluorescence ($R_{14}R_{35}F_{52}$). Both conventional notation[241] and the new IUPAC terminology (within parentheses)[242] are used.

4.4.1 Sensitivity

The sensitivity for a given instrumentation is normally defined as the rate of change of the signal with amount/concentration of analyte. Because PMTs are often capable of detecting single events and LIF/LEAFS measurements very seldom are limited by detector noise (see below), it is often more convenient to define an overall efficiency of a system: as the number of events detected per concentration unit of the analyte in the sample for continuous-mode atomizers; and as the number of events detected per mass unit of the analyte or per atom in the atomizer for discontinuous atomizers (furnaces, discharges etc.). There are, however, several closely related entities defined and discussed in the literature (e.g. the efficiency of detection and efficiency of measurement).[37,65,129,174,214,243,244] The detection efficiency, for example, introduced by Alkemade,[243,244] refers to the probability that a given atom appearing in the probed volume produces an event during the probing time. However, we will not scrutinize these issues in detail here. It is sufficient to conclude that the overall sensitivity can be written as a product of several individual efficiency terms, e.g. the sample introduction efficiency, the atomization efficiency, the illumination efficiency (the fraction of atomized atoms that is illuminated by the laser light), the number of photons emitted per atom (as defined in section 3), the solid angle over which fluorescence is

For references see page 9575

collected, the light collection efficiency of the optical system, the transmission of the spectrometer, and the quantum efficiency of the detector.[37,129,174,243,244]

One illustrative example of the concept of sensitivity is the determination by Wei et al.[174] of the overall efficiency for detection of Pb atoms in an ETA by the LIF/LEAFS technique for three different experimental set-ups: right-angle detection combined with dispersive detection; front-surface detection combined with dispersive detection; and front-surface detection combined with nondispersive detection.[174] These authors found overall efficiencies for their three set-ups of 4.3×10^{-7}, 2.0×10^{-6} and 1.2×10^{-5}, respectively. The major reason for the relatively low numbers for the overall efficiency is the small solid angle over which fluorescence is collected by the detection system. Their results show that although front-surface detection has a higher overall efficiency than right-angle detection, nondispersive detection systems can, in turn, have a higher overall efficiency than dispersive detection.

4.4.2 Noise and Background Signals

Noise in a measurement can be classified as either extrinsic or intrinsic. Extrinsic noise is the noise that arises from a nonspecific background signal that is present even in the absence of analyte. Extrinsic sources of noise comprise, for example, thermal noise, background emission, stray light and concomitant scatter. The intrinsic noise, on the other hand, is due to the inherent statistical fluctuations in the number of atoms present in the volume probed by the laser.[35,132,244] Under the most favorable conditions, only the intrinsic noise will limit the detectability of a certain species. Although discussed in the literature in the assessment of the ultimate detection power of laser-based techniques (in particular, in connection with the concept of SAD)[54,129,243,244] and because of the existence of many other types of noise (see below), the intrinsic noise will only occasionally limit an actual measurement under real conditions.

4.4.2.1 Thermal Noise Thermal noise consists of electronic noise from the instrumentation, e.g. detector noise and preamplifier noise, and is an omnipresent source of noise that only occasionally (when a small number of events is to be detected) is the limiting factor for the LIF/LEAFS technique. Because PMTs can detect a single event above its noise level, a low number of signal photons more often creates a situation limited by the intrinsic noise than by the thermal noise in experiments in which very few background photons are created. Hence, it is the consensus that PMT-based systems are seldom limited by the thermal noise.

A gated CCD detector (i.e. an ICCD), on the other hand, which consists of many thousands of individual detector elements each with its own noise, can, when used in a binned or integrating mode (as is most often the case when hooked up to a spectrometer), give rise to a higher amount of detector noise. Hence, although not yet fully confirmed, especially regarding the most modern ICCD systems, it seems most plausible that the detection of weak signals using ICCDs in binned modes can be limited by the thermal noise from the detector rather than the intrinsic noise.[187]

4.4.2.2 High-frequency (Radio-) Interferences High-power pulsed laser systems can produce significant amounts of high-frequency (radio-) interferences that can interfere with the detection equipment. Although this type of noise is more of a nuisance for the LEI technique (in which minuscule signal currents are measured in unshielded environments), it will only occasionally limit the detectability for the LIF/LEAFS technique (especially when laser or detector systems are used that are not fully shielded electrically).

4.4.2.3 Nonlaser-induced Background Signals Nonlaser-induced background signals from the atomizer comprise emission from constituents in the atomizer or blackbody radiation. Although flame emission can be significant at the wavelength regions where the most frequent flame species emit (e.g. OH and C_2), the general trend is that the atomizer background signals decrease with wavelength.[38] For ETAs, the nonlaser-induced background signals consist mostly of the blackbody radiation, which can limit the detectability primarily at wavelengths above 300 nm.[130] A remedy for this is thus to strive to obtain detection wavelengths as far as possible into the UV region.[176]

4.4.2.4 Stray Light Stray light is defined as laser light reflected into the detection system from various parts of the instrumentation (but excluding scattering from the sample). Most of the stray light originates from scattering from various optical surfaces. This source of noise is therefore most severe when closed atomizers (e.g. a GF) are used (in which the light has to pass through a window before it enters the interaction volume). Various attempts to design the instrumentation (i.e. the atomizer[132,137,164] or the optical system[239,240]) so that light originating from the entrance and exit windows does not reach the detector have therefore been made. One possibility is to use an optical system with a short depth-of-focus and the atomizer with its windows at the Brewster angle as far from the interaction region as possible. This implies that only a small part of the stray light will be imaged onto the spectrometer slit.

Stray light is always the dominating source of noise for resonant LIF/LEAFS but can also affect the measurements when other types of excitation–detection modes are used (especially when the detection wavelength is close to the excitation wavelength).[144,149,159,176,207,214,245] Remedies for this are to use excitation–detection wavelengths that are as widely separated as possible and to use a double monochromator (which has a better stray-light rejection ratio than ordinary monochromators/spectrometers).[200] However, these measures are insufficient if the stray light originates from ASE from the laser (which is a broad emission background from dye lasers that appears in addition to the stimulated light that is produced in the system) or if it originates from so-called "environmental" (or "white") fluorescence. Environmental fluorescence arises when UV light impinges upon optical components that generate fluorescence photons in the 290–420 nm region.[209] An efficient solution to this problem is to use double-resonance fluorescence (as was shown in Figures 9d and 9e), because the fluorescence light then can have a significantly shorter wavelength than the excitation light (described in more detail below).[94,176] The ordinary spectrometer instrumentation can then be complemented with either a suitable filter (which thus only transmits the fluorescence wavelength and not the laser excitation wavelengths) or with a solar-blind photomultiplier (which is sensitive only to UV light).[72,176]

4.4.2.5 Concomitant Scatter Scattering of laser radiation off particles introduced by the sample is referred to as concomitant scatter.[245] This phenomenon is closely related to the concept of molecular fluorescence, which may be generated either by the species commonly present in the atomizer (e.g. OH, NO, or C_2 molecules in flames) or by those being produced by the sample (metal chlorides, metal oxides or metal hydroxides).[38,139–141,145,246,247] These types of background can therefore be significant (and severely limit the detectability) when samples containing large amounts of nonvolatile and undissociated matrices are detected. Concomitant scatter is generally more severe in flames than in ETAs because the flame, with its chemical environment, has a larger tendency to produce nonvolatile species than has an inert gas atmosphere.[38] Atomizers with large temperature gradients (e.g. graphite filaments or rods) also give rise to potentially larger concomitant scatter and molecular fluorescence than those with good temperature homogeneity (e.g. transversely heated graphite atomizers), because the cooler region may allow recombination of matrix elements or the formation of scattering species from the analyte.[38,208]

One of the very few documented examples of the existence of concomitant scatter in a GF is shown in Figure 10 in which Sb is detected in riverine and estuarine water by LIF/LEAFS in a GF using an ICCD detector.[189] Figure 10(a) shows fluorescence spectra from (tenfold-diluted) estuarine water (containing $540\,\mathrm{pg\,mL^{-1}}$ Sb and 2×10^7 times more Na, i.e. $11\,600\,\mathrm{\mu g\,mL^{-1}}$) for three different laser wavelengths (in resonance with the Sb transition as curve A, and detuned −30 and +30 pm as curves B and C, respectively). The peak in curve A at around 289 nm, marked with an arrow in the figure, originates from Sb, whereas the other regularly spaced peaks in Figure 10(a) belong to some unidentified molecular species in the sample. Figure 10(b), in which riverine water (containing $275\,\mathrm{pg\,mL^{-1}}$ Sb and about 10^4 times less Na, i.e. $1.86\,\mathrm{\mu g\,mL^{-1}}$) is investigated, shows no such background, indicating that it originates from the matrix in the estuarine water. Despite the clearly visible background signal from the matrix in the estuarine water at this particular excitation wavelength, the most prevalent situation for the LIF/LEAFS technique is the situation shown in Figure 10(b), i.e. in which there is no matrix background signal.

4.4.2.6 Nonanalyte Atomic Fluorescence Nonanalyte atomic background fluorescence is a rather uncommon phenomenon for the LIF/LEAF technique in ETAs (although it exists to a certain degree in flames owing to their richer chemical composition).[140] The main reason for this is the inherent double selectivity of the LIF/LEAFS technique. A potentially interfering species needs to have a wavelength overlap in both the excitation and detection steps in order to interfere with the analyte signal. Because atoms generally have very narrow excitation profiles, background signals from nonanalyte atoms are extremely rare, although molecular background signals occasionally have been found (as was exemplified in Figure 10).

4.4.2.7 Conclusions It can be concluded that the laser stray light (e.g. "environmental" fluorescence), atomizer emission, or concomitant scatter limits the detectability of LIF/LEAFS in most practical situations.

4.4.3 Means to Reduce Background Signals

In order to reduce the influence of background signals, a variety of background-reducing or background-correction techniques have been developed and investigated.

4.4.3.1 Double-resonance Fluorescence As was concluded above, the limiting noise in many situations is laser stray light, concomitant scattering or atomizer background emission noise (e.g. blackbody radiation in furnaces). A possible approach to decrease the influence of atomizer background emission noise is to bring the fluorescence far into the UV region. This often requires laser photons of very short wavelength (sometimes even below 220 nm), which can be difficult or cumbersome

For references see page 9575

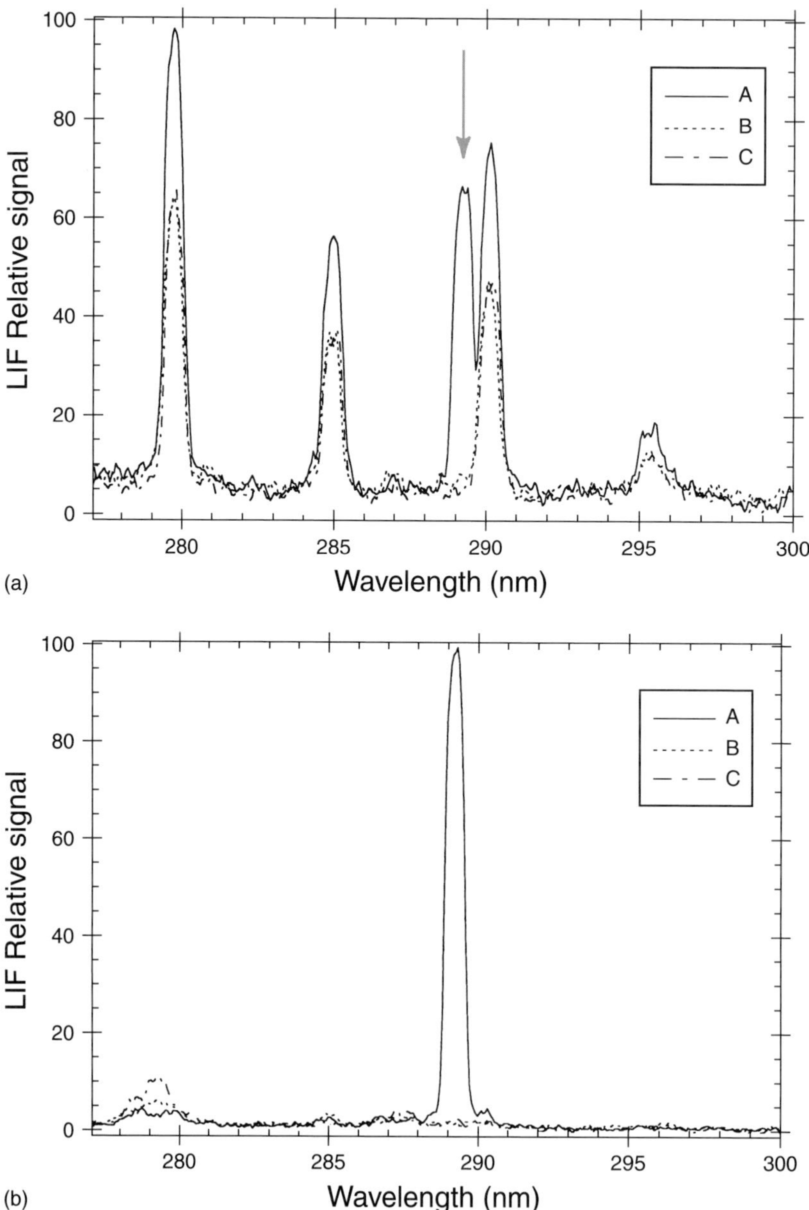

Figure 10 Two LIF/LEAFS/GF fluorescence spectra from tenfold-diluted estuarine water (a) and riverine water (b), respectively. The laser wavelength is 231.147 nm, which is fully resonant with the $5p^3\,{}^4S_{3/2}$–$5p^2\,6s\,{}^4P_{1/2}$ transition in Sb for the two curves A. Curves B and C correspond to detuned laser excitation (−30 and +30 pm, respectively). The Sb contents in the two samples were assessed to be 0.275 ± 0.01 and $0.54 \pm 0.07\,\text{ng}\,\text{mL}^{-1}$, whereas the Na contents were 1.86 and $11\,600\,\mu\text{g}\,\text{mL}^{-1}$, respectively. The peak in curve A at around 289 nm originates from Sb, whereas the other regularly spaced peaks in (a) belong to some unidentified molecular species in the estuarine sample. (Reproduced by permission of the Royal Society of Chemistry from J. Enger, A. Marunkov, N. Chekalin, O. Axner, *Journal of Analytical Atomic Spectrometry*, **10**, 539–549 (1995).)

to produce. An alternative approach, which can reduce the influence of all these sources of noise significantly, is to use a double-resonance (two-step) excitation (i.e. to excite the atoms simultaneously with two laser pulses), as was shown in Figure 9(d) and (e). The greatest advantage of double-resonance excitation is that the fluorescence light can have a significantly shorter wavelength than the excitation light.(72,94,176)

Vera et al.(79) defined the ideal excitation–detection conditions, as shown in Equation (62):

$$\begin{cases} \lambda_1, \lambda_2 > 320\,\text{nm} \\ \lambda_F < 320\,\text{nm} \end{cases} \tag{62}$$

where λ_1 and λ_2 are the two excitation wavelengths and λ_F is the wavelength at which the fluorescence is detected. This allows for the use of a solar-blind

PMT (i.e. one that does not respond to light above 320 nm) and implies that the detector will respond neither to any laser stray light nor to any concomitant scatter. Double-resonance excitation has thus a greater potential to bring the LIF/LEAFS technique closer to the predicted intrinsic LOD than ordinary one-step excitation–detection schemes.

The first double-resonance excitation LIF/LEAFS measurement was made on Pb by Miziolek and Willis in 1981 under the name of saturated optical nonresonant-emission spectrometry (SONRES).(77) Lead atoms were excited by 283.3- and 600.2-nm light. They concluded that very short wavelength photons (e.g. down to 202 nm) could be generated by collisional energy transfer among excited states. They studied a number of transitions and found the 261.4-nm transition to be the strongest. They quoted an order-of-magnitude LOD of 1 fg for Pb. Because this work was performed under an unconventional name, it took the community of scientists several years to realize the importance of this mode of detection for LIF/LEAFS. A number of two-step excitation LIF/LEAFS measurements were made in 1988 by Omenetto et al. (establishing LOD for Tl, Pb and Cd in a GF of 2, 5 and 18 fg, respectively),(176) by Leong et al. (who studied both direct and disconnected double-resonance excitations in Pb and made a comparison with one-step excitation),(200) and in 1989 by Vera et al. (who assessed LOD for In, Ga and Yb of 2, 1, and 220 fg, respectively).(79) Additional two-step excitation experiments have been performed by several authors thereafter, e.g. Sjöström et al. when detecting V,(194) Axner and Rubinsztein-Dunlop when investigating Cr,(173) and Petrucci et al. when detecting Au.(158)

4.4.3.2 Multichannel Background Monitoring Various approaches for the monitoring of background signals have been developed and scrutinized.

A multichannel background correction technique for pulsed LEAFS with a GF was developed by Sjöström (a poor man's multichannel analyzer but with the benefit of using PMTs).(130) The background was measured simultaneously with the analyte signal using three optical fibers positioned at the exit slit of the monochromator and coupled to two different PMTs (one for the center fiber and one for the two flanking fibers). Blackbody radiation, scattered laser light and molecular fluorescence could be corrected for by this simultaneous background-correction technique. Gallium was detected with an LOD of 50 fg.

An alternative approach was taken by Remy et al.(201) They split the fluorescence light, using a beamsplitter, into two different monochromators – one tuned to the analytical transition and one slightly detuned so as to measure the background from scattered laser light, furnace emission, concomitant scatter, etc. The authors conclude that this leads to improvements in LOD and precision. The drawback is that the beamsplitter reduces the analytical signal by a factor of 2.

A more powerful approach to this concept is to use an ICCD to detect the fluorescence.(80,186–189) An ICCD detector has several advantages over a PMT. It can detect several wavelengths simultaneously, so it can be used for fast and convenient investigations of the fluorescence spectra to find the most sensitive excitation–detection wavelength combination from atoms with complex atomic structure. It can be used to increase the absolute sensitivity of the LIF technique in a GF as compared to PMT detection by allowing simultaneous detection of the fluorescence at several wavelengths. It can monitor and correct for background signals from matrix interferences, blackbody radiation and scattered laser light both at and around the wavelength of detection, which is of special importance when samples with high concentrations of matrices are being analyzed. This leads to an improved spectral selectivity. In addition, owing to the delayed read-out of the ICCD detector (delayed in time with respect to the laser pulse), it has a high immunity from radiofrequency pick-ups emitted from the pump laser. Moreover, by storing a number of consecutive fluorescence spectra from within one furnace heating, the time development of spectra from the analyte as well as from any matrix constituent in the sample giving rise to background signals can be studied. Finally, a two-dimensional ICCD detector can be used for spatial studies of atomization and diffusion processes in the GF if combined with an imaging spectrometer. There are a few disadvantages of ICCD detectors, the most significant of which are: they can have a lower sensitivity than a PMT, resulting from a lower quantum efficiency of the photocathode; and they have a limited read-out rate (around one or a fraction of a hertz for a full two-dimensional read-out, or up to a few hundred hertz for binned read-outs). They produce a huge amount of data, which necessitates the use of computers with large data-storage capabilities, and they are relatively expensive.

The first use of an ICCD together with LIF/LEAFS in an ETA was the investigation of Ni.(187) The most sensitive and versatile excitation and detection wavelengths were identified. The LOD of Ni by LIF/LEAFS in an ETA could thereby be improved by two orders of magnitude. The ICCD detector was also compared with an ordinary PMT. The LOD were found to be 15 and 10 fg for ICCD and PMT detection, respectively. The simultaneous monitoring of entire fluorescence spectra by the use of an ICCD made possible the detection of Ni in various aqueous standard reference samples with sodium concentrations ranging from micrograms to tens of milligrams per milliliter (riverine water and estuarine

For references see page 9575

water) with good accuracy and precision.[187] ICCD detectors have also been used for studies of Ti (as was shown in Figure 8 above)[80,188] and Sb in environmental samples and human blood.[189]

4.4.3.3 Zeeman Background Correction Another technique for background correction is the Zeeman technique, extensively used for atomic absorption spectrometry (AAS) in an ETA. Use of the Zeeman background-correction technique together with LIF/LEAFS in an ETA (termed Zeeman electrothermal atomization/laser-excited atomic fluorescence spectrometry (ZETA/LEAFS)[171]) has been studied by the group of Michel. ZETA/LEAFS was first investigated with Co as a pilot element by Dougherty et al.[161] ZETA/LEAFS was found to correct for furnace blackbody radiation, scatter and stray light. However, the Zeeman effect has a drawback in that it degrades sensitivity. Dougherty et al. found an LOD of 0.3 pg with and 0.7 pg without Zeeman correction. It was later found that the Zeeman correction degraded the LOD for elements such as Ag, Co, In, Mn, Pb, and Tl by a factor between 1 and 10, whereas no successful Zeeman correction could be done for Cu.[171] The Zeeman correction technique did not affect the LOD but enabled correction for scatter in resonance fluorescence determination of Mn in a zinc chloride matrix and brain tissue. A diagnostic study of the Zeeman effect was made by Preli et al.[163] They investigated the influence of Zeeman splitting, applied field strength, laser excitation line width, and atomic spectral profile on the LOD and LDR for six elements (Ag, Co, Cu, In, Pb, Tl). All these investigations used a longitudinal Zeeman correction system. Transverse ZETA/LEAFS has been investigated by Irwin et al. on Pb and Co.[193]

4.4.3.4 Wavelength-modulating Laser-induced Fluorescence/Laser-excited Atomic Fluorescence Spectrometry Techniques Various WM methodologies have been investigated for the LIF/LEAFS techniques by Goff et al.[143] and by Su et al.[248] The light from a dye laser was modulated periodically either by rapid tuning of an etalon or by tilting a mirror in the cavity by electrical means. Both of these groups reported on the reduction of the laser scatter background in various types of measurements. A potential drawback of this type of wavelength-jumping method is that it cannot correctly compensate for a structured background.

4.4.4 Limits of Detection

LOD are normally defined as the concentration (or absolute amount) of analyte in a sample that produces a signal equivalent to three times the standard deviation of the noise associated with the blank. For steady-state atomizers (e.g. flames and plasmas) LOD are expressed in concentration units (e.g. $ng\,mL^{-1}$), whereas for nonsteady-state atomizers (e.g. ETAs and vapor cells) absolute amounts are used (e.g. pg). A convenient way to determine LOD for a given instrumentation is to divide the amount of noise (three times the standard deviation of the noise associated with the blank) by the sensitivity. This implies that LOD depend both on the sensitivity and the noise of the technique. Because both of these two entities in turn depend on most other parts of the instrumentation (type of laser, atomizer, detector, as well as sample constituents), LOD can vary widely between various elements and instrumentation. This also implies that one given instrumentation does not necessarily provide the optimum conditions for all elements. This can be exemplified by the works by Wei et al.[174] and Vera et al.[202,209]

Wei et al. determined the LOD for two elements (Tl and Pb) atomized in a GF for the three different optical systems examined for their overall efficiency above (i.e. right-angle detection combined with dispersive detection; front-surface detection combined with dispersive detection; and front-surface detection combined with nondispersive detection).[174] They found that although the sensitivity increased with front-surface detection (in comparison with right-angle detection) for Tl (for dispersive detection), the noise increased by almost the same amount. Hence, the front-surface and the right-angle approaches gave the same LOD for Tl (3 fg). For Pb, on the other hand, the sensitivity increased significantly more than did the noise, and LOD were improved by almost one order of magnitude for front-surface detection (from 7 to 1 fg). They attributed this difference to the fact that the dominating source of noise for Pb was blackbody emission, whereas scattered laser light noise dominated for Tl. They therefore concluded that the front-surface mode of detection will offer a significant improvement of the LOD over right-angle mode detection for situations where blackbody emission dominates but not necessarily when noise from scattered laser light is dominating.

The same authors also found that nondispersive detection (e.g. narrow-bandpass filters) had an even better light-gathering power, and hence a better sensitivity, than dispersive detection (using a spectrometer).[174] They therefore claimed that nondispersive detection should reduce noise and lead to improved LOD. It remains to be proven, however, whether nondispersive systems can be as versatile as dispersive systems because they generally have a lower selectivity.

Vera et al.[209] evaluated three different laser systems (nitrogen-, copper-vapor- and Nd:YAG-pumped dye laser systems, respectively) for their Pb detection power by LIF/LEAFS in a graphite tube. They found that the best LOD of Pb could be achieved by the high-repetition-rate (6 kHz) Cu-vapor laser (0.5 fg, laser scatter limited), whereas a 20-Hz repetition-rate nitrogen laser and a Nd : YAG laser yielded virtually identical LOD (3 fg).

However, in an accompanying work by the same authors, concerned with the detection of Fe and Ga, the Cu-vapor laser did not yield the same advantage over the other systems.(202) The reason was attributed to the low pulse energy of the Cu-vapor laser (a few hundred nanojoules per pulse), which did not allow for optical saturation of the transitions in these elements. LOD for Ga, for example, were found to be almost two orders of magnitude lower when detected by the Nd : YAG laser than with the Cu-vapor laser (25 versus 2000 fg). Despite the appealingly high repetition rate, Cu-vapor lasers have not been used extensively for LIF/LEAFS in an ETA, with the powerful exception of the direct determination of Pb in Great Lakes waters by Cheam et al.(210)

The above examples show that it is *not* clear, a priori, that pulsed dye laser systems of nanosecond pulse length *always* provide the most optimum excitation source for the LIF/LEAFS. They can, under certain conditions, be rivaled by other types of laser systems. Theoretical considerations for (and discussions of) S/N ratios and LOD for the LIF/LEAFS technique have been given by a number of authors.(53–55,129,243,244) The actual LOD for the LIF/LEAFS technique that have been achieved in various experiments are presented below (with respect to the type of atomizer used).

4.4.4.1 Laser-induced Fluorescence/Laser-excited Atomic Fluorescence Spectrometry in Flames Table 3 gives a compilation of the best LOD for the flame LIF/LEAFS technique.(36,136,144,145,149–152,154,157) As was mentioned above, although the LIF/LEAFS technique was first applied to flames as atomizers and a large number of studies have been done in the field,(35–38,60,64–67,70,71,78,81,86,115,118–120,136–158) the actual detection limits have turned out not to be as impressive as first anticipated.(35,36,125) The assessment of subnanogram per milliliter detection limits of Pb and Tl in an acetylene–air flame (0.02 and 0.8 ng mL^{-1}, respectively) by Human et al. in 1984 is still among the most impressive results by flame LIF/LEAFS.(157) Most other elements show detection limits in the ng mL^{-1} range (ranging from some parts of ng mL^{-1} to several μg mL^{-1}).

One important reason for the rather high LOD is that flames give rise to significant amounts of background radiation, which acts as a source of scattering for laser light. Many types of flames, supported on a variety of burners, have therefore been scrutinized for their applicability to LIF/LEAFS. Even if slot burners (which were specially designed for AAS by providing a long pathlength) were handy to use, they were found not to be suitable for the LIF/LEAFS technique.(35) A variety of alternative burners have therefore been developed for the LIF/LEAFS technique – some with small holes or capillary tubes, and some surrounded by a sheet of inert gas (most often N_2 or Ar) in order to reduce the effects of quenching that can result from the entraining of air and reduce the fluorescence yield.(150)

Table 3 LOD for LIF/LEAFS in flames (ng mL^{-1})

Element	LOD	Refs.	Element	LOD	Refs.
Ag	4	145	Mo	12	145
Al	0.6	145	Na	0.1	144, 145
Au	4	152	Nb	1500	36, 136
Ba	0.7	151	Nd	2000	36, 144
Bi	3	145	Ni	0.5	150
Ca	0.08	145	Os	150 000	36, 136
Cd	8	145	Pb	0.02	157
Ce	500	36	Pd	1	152
Co	2	154	Pr	1000	36
Cr	1	145	Pt	0.7	152
Cu	1	145	Rh	100	36
Dy	300	36	Ru	2	152
Er	500	36	Sb	50	36
Eu	20	36	Sc	10	36, 136
Fe	0.2	149	Sm	100	36
Ga	0.9	145	Sn	3	150
Gd	800	36	Sr	0.3	145
Hf	100 000	136	Tb	500	36
Ho	100	36	Ti	2	145
In	0.2	145	Tl	0.8	157
Ir	9	152	Tm	100	36
Li	0.5	145	U	500 000	144
Lu	3000	36	V	30	145
Mg	0.2	145	Yb	10	36
Mn	0.4	145			

Another way to improve the sensitivity and LOD of conventional flame LIF/LEAFS is to use a multipass configuration (using a flat mirror with a pierced hole in the perpendicular position together with a toroidal mirror).(149) Using a flashlamp-pumped dye laser for excitation, Epstein et al. demonstrated that the LOD of Fe could be improved from 0.6 to 0.2 ng mL^{-1}.

The field of flame LIF/LEAFS has been quite stagnant during latter years but an impressive exception is the multi-element analysis of river sediment for five elements using a modern tunable OPO system. Comparable or improved flame LEAFS detection limits over previous literature values were obtained for Co, Cu, Pb, Mn and Tl (2, 2, 0.4, 0.2 and 0.9 ng mL^{-1}, respectively).(154)

4.4.4.2 Laser-induced Fluorescence/Laser-excited Atomic Fluorescence Spectrometry in Graphite Furnaces If the actual detection limits turned out not to be as impressive as first anticipated when flames were used with the LIF/LEAFS technique, the results were soon found

For references see page 9575

to be significantly more impressive when the ETA was used as atomizer. Detection limits well into the picogram per milliliter range and in the femtogram range were rapidly established. There are a number of developmental achievements related to furnaces, the optical system, detectors and detection strategies that have brought the LIF/LEAFS technique in an ETA to where it stands today.

Following the first LIF/LEAFS measurements in ETAs, which were made on graphite rods (see references 85, 157, 216, 218, 220–222), filaments[222] and boats,[217,219] open cups rapidly became popular (see references 132, 160, 165–170, 178, 198, 203–208, 213–215, 221) owing to their better performance. For example, Goforth and Winefordner compared a graphite rod, a plain graphite cup and a slotted graphite cup for a number of elements and found that the rod worked well for volatile elements but gave insufficient atomization for less-volatile elements, and that the plain graphite cup gave the best overall results of these atomizers. They also compared the LOD for pyrolytic coating, a tantalum foil liner and a tantalum carbide coating of a graphite cuvette, and found the pyrolytic coating to give the best results.[221]

It was meanwhile found by AAS, however, that open atomizers (e.g. rods, boats and cups) are impractical due to diffusion losses[249] and vapor-phase interferences[250] that occur in the cool zone above the atomizer. Similar effects were soon found for the LIF/LEAFS technique, for example by Bolshov et al.[203] It was therefore concluded, first by Human et al.[157] and Dittrich and Stärk,[218] and later by others (e.g. Falk and Tilch[214]), that graphite tubes provided the best atomization environment for analysis (see references 77, 79, 80, 121, 158–164, 171–177, 179–197, 199–202, 209–212, 214, 218, 220, 224–226). Dittrich and Stärk,[218] and shortly after, Falk and Tilch,[214] for example, demonstrated that the tube atomizer had an increased sensitivity over the rod atomizer. They attributed this to an increased residence time and a higher degree of atomization. Other advantages of the tube are a more homogeneous temperature distribution and a higher heating rate.[174,218]

The most important development of the optical system for LIF/LEAFS in an ETA is the front-surface detection (as was shown in Figure 7), which not only provides freedom from post-filter effects (see below) and the imaging of a larger interaction volume but it also has yet another distinct advantage over right-angle detection when it comes to ETAs – it can be applied directly to the atomizer without any modification of the furnace system. In addition, the use of reflective instead of refractive optics has also improved the performance of LIF/LEAFS in an ETA (because it provides chromatic-free light collection).

Regarding detectors, most LOD have been determined using PMTs. The ICCDs have not proven themselves to yield any better LOD than the PMTs under normal conditions. Their main advantage instead is that they can monitor the fluorescence background significantly more effectively than the PMTs and hence yield more reliable results when samples with complex matrices are detected.[189] The ICCDs have also proven to be indispensable when efficient excitation–detection transitions are to be found in atoms with complicated atomic structure.[80,187]

The state-of-the art performance of the LIF/LEAFS technique in an ETA is shown in Table 4, which displays the best LOD published so far for each element (see references 79, 158–160, 169, 170, 173, 174, 178, 179, 183, 187–190, 194–196, 199, 201, 203–206, 224, 225). Table 4 is based upon a compilation of around 200 LOD from the literature for the LIF/LEAFS technique in an ETA.

As can be seen from Table 4, quite a number of elements have LOD in the low femtogram range. For many elements, improvements of two to four orders of magnitude (with respect to the conventional AAS technique in an ETA) have been obtained. Elements that have poor LOD in the AAS technique because of furnace atomization problems (in general those that require high atomization temperatures, normally above 2500 °C) also have relatively poor LOD in the LIF/LEAFS technique. The LOD for these elements are limited either by memory effects in the graphite cuvette[80,173,188,194] or by blackbody radiation from the atomizer. Although the majority of the LOD have been measured using one-step excitation, there are a few elements whose LOD have been obtained by using double-resonance fluorescence.

Molecular detection has found a certain (although a limited) use for the assessments of some elements whose excitation wavelengths are difficult to access with laser light, e.g. halogens.[121,177,192,195,223,251] In the case of fluorine, for example, a matrix of Mg was added to the sample (analyzed in a GF), leading to the formation of MgF, which could be detected by LEMOFS.[177,192]

4.4.4.3 Laser-induced Fluorescence/Laser-excited Atomic Fluorescence Spectrometry in Other Atomizers LIF/LEAFS has also been used in conjunction with other types of atomizers. Owing to the success of the ICP with conventional detection techniques, a number of investigations have been devoted to investigations of the use of this atomizer together with LIF/LEAFS.[37,60,66,94,122,157,228–230] An early, mostly diagnostic, investigation of the use of the LIF/LEAFS technique with an ICP was performed by Omenetto et al.[228] A more extensive demonstration of the analytical capability of the LIF/LEAFS technique applied to an ICP was made by Human et al. in 1984.[157] The

Table 4 LOD for the LIF/LEAFS technique in ETAs (fg)

Element	First-step excitation	Second-step excitation	Fluorescence wavelength	ETA[a]	Mode of detection[b]	Excitation source[c]	LOD	Refs.
Ag	328		338	T/P	90°	E	20[d]	159, 160
Al	308		394/6	T/W	180°	E	100	174
As	193.7		245.7	T/P	180°	Y	54	190
Au	267.6	406.5	200	T/W	180°	E	3	158
Ba	597.2		611.1			F	40 000 000	205
Bi	223.1		299.3	C	90°	E	2.5	170
Cd	228.8		228.8	C	90°	E	0.5	169
Co	308.3		345.4	T/W	180°	Y	4	201
Cr	427.5	529.8	360	T/W	180°	E	1400	173
Cu	324.7		510.5	C	90°	Y	150	203
Eu	287.9		536.1	C	90°	Y	300 000	203, 206
Fe	296.7		373.5	T/W	180°	E	70	174
Ga	403.3	641.4	250.0	T	180°	Y	1	79
Hg	253.7	435.8	546.1	C	180°	E	90	178
Ir	237.3		254.4	T/W	180°	E	18	179
Li	670.8		670.8	T/W	180°	D	10	225
Mn	279.5		279.5	T/P	90°	E	80	159
Mo	313.3		317.0	T/W	180°	N	100 000	224
Ni	224.5		231.4	T/W	180°	E	10	187
P	213.6		253.4/6	T/W	180°	E	8000	195, 199
Pb	283.3		405.8	T/W	180°	E	0.2	174
Pd	244.8		343.3	T/W	180°	E	4	179
Pt	264.7		270.2/6	T/W	180°	E	70	179
Rb	780.0		780.0	T/W	180°	D	20	225
Rh	236.2		381.5	T/W	180°	E	190	179
Ru	287.5		366.3	C/LP	90°	Y	3000[e]	204
Sb	287.8		372.2	T/W	180°	E	5	189
Se	196		204/6	T/P	180°	Y	15	190
Sn	286		318	T/W	180°	E	30	174
Te	214.3		238.3/6	T/W	180°	E	20	183
Ti	264.6		295	T/W	180°	E	1000	188
Tl	277		353	T/W	180°	E	0.1	196
V	458.0	578.6	256	T/W	180°	E	10 000[f]	194
Yb	398.8	666.8	246.4	T/W	180°	Y	220	79

[a] T = tube; P = platform; W = wall; C = cup.
[b] Direction of detection of the fluorescence.
[c] E = excimer-pumped dye laser system; N = nitrogen-pumped dye laser system; Y = Nd : YAG-pumped dye laser system; D = diode laser.
[d] Contamination-limited LOD reported: estimated LOD without contamination: 8 and 10 fg.
[e] A volume of 30 μL has been assumed.
[f] Contamination-limited LOD reported: estimated LOD without contamination: 2500 fg.

authors assessed LOD for a range of elements (Al, B, Ba, Ga, Mo, Pb, Si, Sn, Ti, Tl, V, Y, Zr and U). The LOD were found to be between 0.4 and 20 ng mL^{-1}. Atomic fluorescence from Ag, Au, Hf, Ir, Mo, Nb, Pd, Pt, Ru, Ta and Zr in an ICP was measured by Huang et al.[(229)] LOD found were in the 1.3–58 ng mL^{-1} range, with an LDR of over four orders of magnitude for most elements. Double-resonance *ionic* fluorescence has been used as an alternative to atomic fluorescence from Ca, Sr, Ba and Mg atoms in an ICP in order to improve on the sensitivity or to extend the number of accessible elements.[(94)] The LOD for these four elements were 0.007, 1, 1, and 0.05 ng mL^{-1}, respectively. Simeonsson et al.[(230)] used both single- and double-resonance excitation for detection of atomic or ionic fluorescence from Ag(I), Au(I), Co(I), Cu(I), Ni(I), Pb(I), Pd(I), Pt(I) and Sc(II). The LOD for single excitations for these elements were found to be 0.8, 4.2, 15, 3, 5, 0.7, 3.9, 3.3 and 0.2 ng mL^{-1}, respectively. The corresponding LOD for double excitation were 1.7, 6.0, 530, 5, 170, 8, 45, 29 and 13 ng mL^{-1}, respectively. Their results show that double-resonance excitation in fact yields higher LOD than single-resonance excitation. The authors stressed, however, that the spectral selectivity for the double-resonance excitation technique should widely supersede that of the single-resonance excitation technique.

These studies show that the LOD for the LIF/LEAFS technique with an ICP are generally in the low nanogram

For references see page 9575

per milliliter range, with a few exceptions such as ionic detection of Ca and Mg, which are well into the picogram per milliliter range.[94]

Although the LIF/LEAFS technique is generally not very impressive for isotope detection, because most optical isotope shifts are smaller than the typical laser bandwidths or Doppler and collisional broadening mechanisms in most atomizers (with the exception for some actinides and lanthanides), it has been used for isotopic detection of uranium isotopes in complex matrices in an ICP.[122]

LIF/LEAFS has also been applied to a hollow-cathode glow discharge, which served the purpose of being the atomizer (aqueous solutions were dried in graphite electrodes used as disposable hollow cathodes).[231] LOD for Pb and Ir were 500 fg and 20 pg, respectively. A commercial hollow-cathode lamp (which has a very low background emission and excellent atomization characteristics) has also been investigated for its use for SAD (referring to the number of atoms in the interaction volume).[232] An LOD for Pb of 1.8 ag (i.e. 0.0018 fg) in the laser beam was established. Work with a hot, hollow-cathode atomizer has also been performed.[233] LOD for dried solutions of Co and Ni were 15 pg mL^{-1} and 5 ng mL^{-1}, respectively. LIF/LEAFS measurements for trace detection in conducting solids (Fe in brass) have been performed by Travis et al.[234] The LOD were below the microgram per gram level. LIF/LEAFS has also been applied to direct detection of U(VI) in solution, with a LOD of 40 pg mL^{-1}.[235]

4.4.5 Linear Dynamic Range

Although the LIF/LEAFS signal is linear with analyte concentration for small concentrations, pre-filter, self-absorption, and post-filter effects can degrade the linearity at higher concentrations.[35,37,70,137,164,241] Pre-filter effects refer to the reduction in fluorescence signal that occurs because of absorption of *laser light* by analyte atoms on its way to (i.e. prior to) the interaction volume (the interaction volume is defined as the overlapping volumes of laser irradiation, atomic density and detection), whereas self-absorption refers to reabsorption of *fluorescence light* by analyte atoms *within* the laser-irradiated volume. Finally, post-filter effects refer to reabsorption of *fluorescence light* by atoms *outside* the irradiated volume (i.e. absorption of fluorescence by atoms within the detection volume but outside the illuminated volume). The risk for self-absorption and post-filter effects to occur is naturally higher for transitions that terminate in highly populated states (e.g. the ground state or in low-lying excited states) than for those transitions that terminate in short-lived excited states.[38,164] Self-absorption can occur also for transitions terminating in relatively long-lived excited states mainly when continuous-wave excitation is used.

Because both pre-filter and self-absorption originate from absorption of light by atoms within the irradiated volume of atoms, these effects can be reduced by optically saturating the transition because optical saturation occurs when the rates of absorption and stimulated emission balance each other, a medium with optically saturated atoms is transparent at that particular wavelength.[145] Post-filter effects, on the other hand, are not affected by the laser irradiation. Hence, they can cause significant reduction in the LDR even when the atomic transition is saturated. On the other hand, post-filter effects can be eliminated by using front-surface detection (because all atoms viewed by the detector are illuminated by laser light).[35,37,38,156,164,174,176,195,221,224,248] This implies that the front-surface mode of detection yields smaller post-filter effects and hence a larger LDR than do systems with right-angle detection.[156,164,174,176,195,221,224,248] Dynamic ranges of five to seven orders of magnitude have been demonstrated for several elements using front-surface detection.[164,175,212,221]

4.4.6 Selectivity

The selectivity of the LIF technique is excellent due to a double selectivity for the conventional single-resonance fluorescence spectrometry technique and a triple selectivity for the double-resonance technique. As was discussed in connection with concomitant scatter and nonanalyte atomic fluorescence above, a concomitant species needs to have a spectral overlap both in the excitation as well as in the de-excitation processes in order to yield a background signal. There are very few such examples documented in the literature. One exception was the study of Sb in estuarine water, as shown above in Figure 10.

4.4.7 Applications

Although the limited space in this theory and instrumentation review precludes any detailed analysis of the applicability of LIF/LEAFS, a few examples can be given.

4.4.7.1 Laser-induced Fluorescence/Laser-excited Atomic Fluorescence Spectrometry in Flames The flame LIF/LEAFS technique has been used for a limited number of applications. Compilations can be found in the literature.[36,68] One example is Fe detected in simulated fresh water, unalloyed copper and fly ash. Other examples are the detection of Pb in blood[155] and the determination of Ni and Sn in various reference materials (riverine water, unalloyed copper and fly ash).[150] A multi-element LIF/LEAFS analysis of river sediment in flames for Co, Cu, Pb, Mn and Tl using a modern tunable OPO system, encompassing 640 measurements in 6 h with triplicate measurements of all solutions and aqueous calibration curves, with a relative standard deviation (RSD) precision better than 5%, was done by Zhou et al.[154] Flame LIF/LEAFS has also been coupled to high-performance liquid chromatography (HPLC) for speciation. Walton et al., for example, detected various organomanganese and organotin species by HPLC and flame LIF/LEAFS. The detection limits for organomanganese species ranged from 8 to 22 pg of Mn.[156]

4.4.7.2 Laser-induced Fluorescence/Laser-excited Atomic Fluorescence Spectrometry in Eletrothermal Atomizers There are a large number of applications of the LIF/LEAFS technique in ETAs. A recent compilation is given in the review by Hou et al.[38] An early example is the detection of Cd and Pb in Antarctic and Greenland snow and ice by Bolshov et al.[169] (with an LOD of 0.5 fg). Thallium, manganese and lead were detected by slurry sampling in food and agriculture standard reference materials by Butcher et al.[175] The technique has also been used for the assessment of trace levels of gold in size-segregated, atmospheric particulate samples.[158] The simultaneous monitoring of entire fluorescence spectra by the use of an ICCD made possible the detection of Ni in various aqueous standard reference samples with sodium concentrations ranging from micrograms to tens of milligrams per milliliter (riverine water and estuarine water) with good accuracy and precision.[187] The Zeeman correction technique has been used for assessment of Mn in a zinc chloride matrix and brain tissue using resonance fluorescence.[171]

4.5 Diagnostic Capabilities

The LIF/LEAFS technique has been used by numerous authors for a variety of diagnostic studies.[60,252] Example of entities that have been studied are temperatures,[60,253–256] number densities,[35,87,257–259] quantum efficiencies,[260] lifetimes of atomic states in flames,[69,261] spatially resolved atomic distributions in flames,[67] laser-induced flame chemistry processes,[262,263] collisional effects in flames,[264] collisional mixing rates of excited atomic states in flames,[78,86] the role of chemical reactions on saturation curves,[265] excitation–detection wavelength combinations of atoms with complex atomic structure by the use of an ICCD,[80] memory effects in GFs,[188] and the Zeeman effect.[163,193] Of special importance is the fluorescence dip,[97–99,266] which has been used for studies of quantum efficiencies,[267] radiative transfer,[268] ionization depletion[99] and autoionizing cross-sections.[269] A thorough study of the capabilities of performing time-resolved investigations for diagnostic purposes was made by Omenetto and Matveev.[270] Further discussions about the diagnostic capability of the LIF/LEAFS technique are given by Omenetto.[127,266]

4.6 Conclusions and Outlook

LIF/LEAFS is a technique that has clearly demonstrated its usefulness for analytical applications. The most impressive performance has been achieved with the GF as an atomizer. The most versatile detection systems for LIF/LEAFS today are based upon a combination of front-surface detection and chromatic-free reflective optics (either by a pierced curved mirror or a pierced flat mirror followed by a curved mirror). Double-resonance fluorescence can bring high detection powers to some elements. The use of an ICCD when samples with complex matrices are analyzed is advantageous. LOD in the low femtogram range have been obtained for a number of elements. Owing to its double (or triple) selectivity, the LIF/LEAFS technique shows an extraordinary selectivity. Moreover, the LDR has been found to be large (five to seven orders of magnitude has been demonstrated). This implies that the LIF/LEAFS technique in an ETA can rival most other analytical techniques for trace element analysis.

Work has also been done with regard to the SAD capability of the LIF/LEAFS technique. It is clear that the LIF/LEAFS technique is capable of detecting the presence of one single atom in quiet surroundings (e.g. in a vacuum). However, the SAD capability of the technique seems more difficult to address when the technique is coupled to practically useful atomizers. It is clear that the graphite tube (and in particular the transversely heated tube) so far provides the best (although not the ideal) atomizer for practical analytical work. Work with other, quieter atomizers, e.g. a hollow-cathode lamp, might bring down the LOD of the technique, although at the price of less versatile and less user-friendly instrumentation.

Looking back at the development of the LIF/LEAFS technique in an ETA – with the improved knowledge about the underlying principles for the technique that has been gained throughout the years, which in turn has resulted in more efficient experimental set-ups and detection methodologies – and speculating into the future, one can conclude that it would be surprising if the technique were not to be quoted with even better LOD, selectivity and LDR values a few years from now.

5 LASER IONIZATION SPECTROMETRY

5.1 Laser-enhanced Ionization versus Resonance Ionization Spectrometry

As was alluded to in the Introduction (section 1.4.2), there are basically two different laser-based spectroscopic ionization techniques with good potential for analytical applications: LEI and RIS. The main difference between the two is that LEI is performed in conventional

For references see page 9575

atmospheric pressure atomizers (flames, furnaces, etc.) and RIS is performed under vacuum conditions.

Because atoms in atmospheric pressure atomizers are exposed to a high collision rate, they also experience a certain thermal ionization. As was discussed earlier, this thermal ionization ionizes excited atoms at a rate faster than ground-state atoms. Hence, this thermal ionization can be increased significantly by exciting atoms in high-collision media by laser light. By applying an electrical field across the interaction volume, the increased ionization rate is then detected as an increase in the current through the atomizer, which constitutes the signal in LEI spectrometry.(40) In the environment to which RIS is applied, there is no such thermal ionization. This implies that although the laser light has only to excite selectively the analytes in the LEI techniques (the ionization is provided by the existing thermal ionization process in the atomic reservoir), the laser light has to be used for both selective excitation and ionization of the analytes in RIS. Although the RIS technique has shown some extraordinary properties (i.e. SAD and extremely high elemental selectivity, especially when coupled to a mass spectrometer for detection of the ions produced), the fact that atomization of the sample in general is performed under vacuum conditions puts some restrictions on the practical applicability of the technique. It will therefore be covered in less detail in section 5.8. The reader is referred to the literature for a detailed description of the technique.(10,11,19,41,271–285)

5.2 Laser-enhanced Ionization Spectrometry – An Introduction

LEI is a technique that originates from a finding by a group of scientists at the National Institute of Standards and Technology (NIST, formerly National Bureau of Standards NBS), Gaithersburg, Maryland, USA, in 1976.(286) Green et al. found that atoms in a flame irradiated with laser light experience an enhanced ionization rate if the laser light is resonant with a transition in the atoms. This increased ionization could be detected most easily by applying an electrical field across the flame and measuring the corresponding increase in current. This effect was originally termed "optogalvanic detection" for a few years by the American community,(286–291) a name that rapidly became used exclusively for the phenomenon in discharges.(288) The name LEI was used for the "optogalvanic" phenomenon in flames for the first time in 1979.(91,292,293) The technique has also appeared under a variety of other names throughout the years, e.g. (stepwise) photoionization in flames (primarily by the Russian community),(294) dual laser ionization,(295–297) laser-assisted ionization,(298) laser-induced ionization,(299,300) laser step-photoionization,(301) or just stepwise photoionization.(294)

As was alluded to previously, it was found early on that laser light in fact does not need to ionize the atoms under study in LEI spectrometry; it is sufficient to *excite* them (e.g. promote ground-state atoms to an excited state) in order to create an increased rate of ionization. This led to the conclusion that the laser light enhances an already existing thermal ionization process, sustained by the heat of the flame, which ionizes excited atoms at a higher rate than it does for unexcited (ground-state) atoms. A more concise terminology of the technique would therefore have been laser-enhanced collisional ionization,(302) an expression that never came into use.

The common denominator for all atomic reservoirs (atomizers) to which LEI can be applied is, therefore, that they must possess a certain omnipresent thermal ionization that converts photons absorbed by the atoms under study into freely movable charges (electrons and ions). This implies, in turn, that the atomizer performs two functions simultaneously in LEI spectrometry. It renders an atomization of the sample and provides a collisional excitation of the laser-excited atoms.

The LEI technique has most often been used with flames as atomic reservoirs, although GFs, plasmas, discharges, cells and thermionic diodes also have been used but to a significantly lesser extent than flames. We will therefore direct most of this presentation of LEI to its use in flames and only briefly mention its use together with other atomizers (e.g. GFs and plasmas). Moreover, in a perspective of analytical chemistry, LEI spectrometry with optogalvanic detection in discharges or thermionic diodes is of less interest because these types of atomizers do not provide for any simple means of introducing and atomizing an unknown sample into the atomic reservoir. Optogalvanic detection in discharges has become of more interest for spectroscopic investigations and as a technique for wavelength calibration, and therefore will not be discussed further here. The only exception is the development by Petrucci and Winefordner of a sensitive photon detector based upon the optogalvanic effect.(303) The reader is therefore directed to the literature for more information about optogalvanic spectrometry.(304,305)

5.3 Previous Reviews

The field of LEI spectrometry has been reviewed previously by a number of authors.(39,40,237,289,304,306–320) An early but very illustrative review is that by Travis et al.(39) The work by Green and Seltzer is a thorough and easy-reading review of the status of the field up to 1989.(314) A contemporary review, although somewhat more focused upon the principles of the technique, is that of Axner and Rubinsztein.(309) Butcher has recently provided an updated complement to the review of Green and Seltzer.(315) The most comprehensive review, however, is given by a recent book dedicated to the field.(40) This book covers the fundamental mechanisms of LEI in terms of the production of ions(57) and the signal detection;(316) the analytical performance of LEI in flames;(237) applications of LEI;(317) nonflame reservoirs for LEI spectrometry;(318) and the interplay of LIF and LEI in different atomic reservoirs.(319) Although mostly used for the detection of atoms, the LEI technique has also been used for the detection of molecules, which has been reviewed specifically by Webster and Rettner.(304)

5.4 Instrumentation

5.4.1 Typical Set-up

Because LEI mostly has been used with flames as atomizers, the flame LEI set-up shown in Figure 11 will serve the purpose of illustrating a "typical" LEI instrumentation. In short, light from a laser system, most often a pulsed dye laser system, is directed into a flame at a height where there is a sufficient atomic population of the analyte species to be detected. An electrical field is established in the interaction volume by applying a potential (most often ranging from -1 to -2 kV) between an electrode immersed in the flame (or alternatively between two external electrodes placed on each side of the flame) and the burner head (which is the head at ground potential). In order to be able to detect the charges created, the burner head has to be isolated electrically with respect to ground, with only a controlled electric path (accomplished by a resistor) to ground potential. When pulsed excitation sources are used, as is most often the case, a pulsed LEI current will result in this path. This current pulse is extracted from the direct current (DC) (originating from the thermal ionization) by a capacitor before it is converted to a voltage pulse by a current-to-voltage amplifier of suitable trans-impedance gain. The output signal is most often monitored by using a boxcar integrator whose detection window is set equal to or slightly longer than the resulting voltage pulse (often around 1 µs). The output from the boxcar is most conveniently fed into a laboratory computer.

5.4.2 Atomizers

5.4.2.1 Flames Most flames used for LEI have been supported on commercially available slot-burner units (including nebulizer units) developed for the

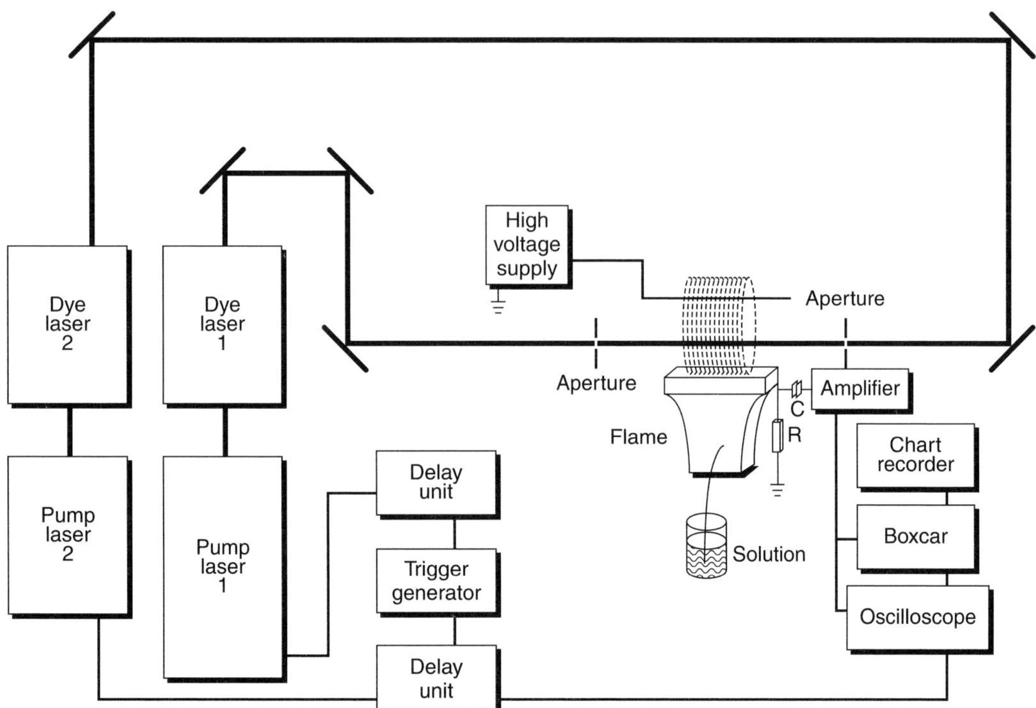

Figure 11 A set-up for LEI spectrometry in flames. The solution to be analyzed is aspirated into a flame. The atomic vapor is illuminated by light from either one or two tunable lasers. In conventional one-step LEI only one laser system is needed. For two-step excitation, the pump laser from the pulse laser is normally split and directed into two dye lasers simultaneously (not shown in the figure). By using two independent laser systems, as shown here, pump-and-probe experiments (e.g. for measurements of the lifetime of excited states) can be performed. (Reproduced by permission of Springer Verlag from O. Axner, P. Ljungberg, Y. Malmsten, *Applied Physics B*, **54**, 144–155 (1992).)

For references see page 9575

AAS technique. The elongated form of the slot-burner-supported flame (normally 5 or 10 cm) is suitable for LEI because it gives a long interaction region (which in turn makes interaction with a large number of analytical species per laser pulse possible). In addition, the use of commercially available slot-burner units has also meant that the LEI technique has been able to benefit from the development of burner and nebulizer units. A typical pneumatic nebulizer, which converts the sample solution into a liquid aerosol, has a liquid sample uptake rate of around $5\,mL\,min^{-1}$ and an efficiency of about 10% (the rest being drained to waste).

All standard types of flame used for AAS have been used for LEI (e.g. air–acetylene, nitrous oxide–acetylene, air–hydrogen), with the air–acetylene flame being the most popular. The reason for this is that this flame constitutes a good compromise between a high degree of atomization, relatively low occurrence of chemical interferences and a high degree of thermal ionization.

The higher-temperature flames (e.g. the nitrous oxide–acetylene flame) have a better degree of atomization of refractory elements and a higher degree of thermal ionization (and thereby collisional ionization), which would justify their use in LEI spectrometry. A disadvantage is that the higher degree of thermal ionization also produces a larger background signal (from which the LEI signal has to be extracted). This implies that the detection power of atoms that atomize readily in air–acetylene or air–hydrogen flames decreases when being detected in a nitrous oxide–acetylene flame.[237] For nonvolatile elements, however, the hotter nitrous oxide–acetylene flame offers an advantage.[321]

The cooler flames (e.g. hydrogen flames), on the other hand, benefit from a lower background ionization, which in principle should give opportunities to lower detection limits for those elements that atomize readily in such flames. The lower background ionization rate of the hydrogen flames, however, is not such an advantage in reality as might be expected. The reason is that when samples containing "normal" concentrations of EIEs are detected, the background ionization rate is dominated by thermal ionization from the EIEs rather than the flame background ionization. The cooler flames are instead limited by a lower degree of atomization and a lower collisional ionization rate, resulting in a poorer LEI detection power for most elements. In addition, the cooler hydrogen flames are more susceptible to chemical interferences, which limits their use for practical purposes. The cooler flames, however, are useful for LEI detection of alkali metals, because these ionize readily also in the cooler flames (and in fact exist with a larger fraction of neutral atoms).[91,237,322]

5.4.2.2 Plasmas Only a limited number of studies concerned with the use of plasmas as atomizers for LEI have been carried out.[300,323–327] The reason is that the plasma does not offer the same suitable environment for the LEI technique as does the flame. The main reasons for this are the following: firstly, the ion fractions in ICPs are much larger than in flames, which implies that the neutral atom population (i.e. those that are available for excitation) is smaller than in a flame for the atoms that already atomize readily in a flame; secondly, the electrical environment around a plasma is harsher than around a flame, implying that the radiofrequency pick-ups become more severe; thirdly, the electron density in a plasma is several orders of magnitude larger than in the flame, which makes it more difficult to detect the laser-induced creation of charges. LEI detection in an ICP has therefore been performed either in the tail end of the plasma[323] or in a power-modulated plasma.[324] Attempts have also been pursued to detect the laser-produced ions with a mass spectrometer rather than electrically.[325,327]

5.4.2.3 Electrothermal Atomizers The increased LOD that followed when the flame was substituted with the GF for the AAS and the LIF/LEAFS techniques also encouraged the field of LEI spectroscopists to move to the GF. Early attempts using the GF as an atomizer for LEI were therefore made by Gonchakov et al.,[301] Bykov et al.,[328] Magnusson et al.[329–331] and Sjöström.[332]

One drawback with LEI in a GF is that the GF does not allow for any apparently efficient geometry for the charge collection as does the flame in flame LEI (see section 5.4.4.2). Therefore, although the discussion about electrode design for flame LEI has been conveniently placed in section 5.4.4.3, together with a discussion about the principles for the charge collection process, this is not possible in the case of LEI in a GF because the performance of this technique is strongly dependent on the interplay between the GF and the charge-collection electrode(s).

Gonchakov et al., who used a graphite cup as an atomizer,[301] used a circular wire probe placed above the cup. Magnusson et al., on the other hand, who used a large graphite tube (HGA-72), placed the charge-collecting probe inside the tube parallel with the laser light.[329,330]

Both Magnusson et al.[329,330] and Gonchakov et al.[301] investigated the influence of various voltages and potentials applied to the probe. Although the first work by Magnusson et al. concluded that the best reproducibilities and S/N ratios were obtained with a negative potential applied to the electrode,[329] they found contrary results in their second study.[330] With a negative voltage, the signal duration was determined by the time constant of the resistance–capacitance (RC) high-frequency cut-off filter (1 µs or 100 ns, depending on the resistor value). The total signal (integrated over time), however, was found to be independent of the RC cut-off frequency. With a positive voltage, the signal lasted around 20 µs, and was found to be significantly larger than that obtained with a negative probe potential. It is as yet unclear as to whether this can be explained by the conventional model for charge collection and signal apportionment (based upon the difference in mobility between electrons and ions) developed by Travis and Turk (section 5.4.4.2).[40] Magnusson et al. suggested that signal-enhancing space–charge effects, similar to what exists in thermionic diodes, might play a role. Also Gonchakov et al. found a signal-enhancing effect. They state that: "The signal amplification coefficient arising from the avalanche ionization of the probe was about 100. This was verified by replacing the tungsten probe with a stainless steel probe, for which the signal proved to be 100 times smaller. A polarity change at the probe reduced the signal by four to five orders."[301]

In addition, it was concluded early on that the long-term reproducibility was not satisfactory for LEI in a GF. Magnusson concluded, for example, that the reproducibility could be improved by replacing the wire electrode with one of a solid graphite rod.[331] In this work, the laser beam was enlarged so that it filled the entire tube, and the electrode was placed in the center and held at zero potential. They reported on a large variation in the duration of the signal pulses when a positive potential was applied. These and similar results indicate that the signal collection process is not yet fully understood in LEI spectrometry in a GF.

It was found early on that the large currents from the heating system induced currents to the probe. In order to minimize these effects, Magnusson et al. synchronized the triggering of the laser with the 50-Hz power line. In this way, the fluctuations of the induced disturbances on the signal could be virtually eliminated.[329]

Both thermionic and photoelectric emission of electrons from surfaces (e.g. the electrode) had been noticed previously from flame LEI measurements when an uncooled metal surface was inserted into the flame or when intense UV light struck a surface, respectively. The same effects were found also for the LEI technique in a GF. Because a large electronic emission rate can lead to suppression of the analytical signal or electrical breakdown, it is not impossible that this might be one reason for the poor reproducibility of the LEI in a GF. Irrespective of whether this is the case, it will limit the applicability of the technique considerably.

Magnusson et al. showed that the effects of photoelectric emission could be reduced by carefully aligning the laser beam sufficiently far from the electrode.[330] It was noticed, however, that the heating of the furnace made the atmosphere act as a thermal lens, thus focusing the light during heating and defocusing it during cool-off, which can be a source of signal fluctuation.[330] Further investigations reported on the existence of dual peaks for LEI in a GF.[330] The same work also investigated, among other things, the influence of different protective gases on the thermionic emission of electrons and found significantly lower thermionic emission in N_2 than in Ar.[330]

An alternative approach to the careful aligning of the laser beam away from the electrode, first suggested by Magnusson[331] and then investigated practically by Sjöström,[332] was to separate spatially the atomization volume from the charge detection volume so as to overcome some of the problems occurring in the previous construction. A T-shaped furnace was constructed in which the atomization took place in the "base" of the T while the detection was carried out in the "bar" part of the furnace. This measurement section consisted of two stainless-steel plates, separated by Teflon insulation, with the outer plate held at −500 V. By separating the atomization and detection regions, less thermionic emission of electrons resulted.

Chekalin and Vlasov achieved a separation of the atomization and detection regions by placing the charge-collecting electrode above the dosing hole of a graphite tube while the laser beams were directed outside the tube (between the electrode and tube).[333]

The common denominator of all these LEI measurements in a GF is that none of them made use of the most modern and up-to-date furnace technology. Butcher et al. used a modern HGA-500 (Perkin-Elmer) furnace and constructed a system with a moveable probe that was introduced automatically into the furnace shortly after the onset of the atomization so as to minimize the amount of thermionic emission.[334] They investigated the LEI and GF instrumentation with respect to a number of parameters in order to find the optimum conditions. Optimization of the atomization temperature for iron showed that no signal was obtained at temperatures exceeding 2000 °C. The loss of signal was attributed to the presence of a large thermionic emission of electrons. As a consequence, only temperatures up to 1900 °C could be used. This implied that relatively involatile elements such as Li, Mg, Fe and Mn required a long atomization period (20 s was used but this still did not ensure complete atomization).

5.4.2.4 Rod–Flame and Graphite Furnace–Flame Systems Attempts to combine the properties of the GF with the flame for LEI measurements have been pursued by Chekalin et al.[333,335–338] as well as by Winefordner, Smith et al.,[339,340] but with different approaches.

Chekalin et al. development a rod–flame system (see Figure 12).[335,336] The sample is placed onto a graphite rod placed in close proximity to a flame. By electrically heating the rod, the sample can be dried and charred

For references see page 9575

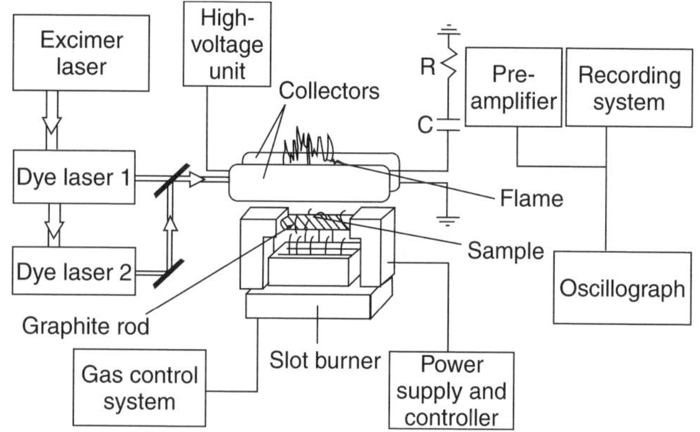

Figure 12 An early version of a set-up for flame–rod LEI spectrometry. In a more developed system, the rod can be swung in and out of the flame to gain better control of the vaporization process. (Reproduced by permission of the Royal Society of Chemistry from N.V. Chekalin, I.I. Vlasov, *Journal of Analytical Atomic Spectrometry*, **7**, 225–228 (1992).)

before the rod is swung into the flame where the dried sample is evaporated. Because evaporation normally takes place at a lower temperature than atomization, atomization is delayed until the sample is well into the flame. Excitation and ionization then take place at a suitable height above the rod (determined by the height of the laser beams). By this procedure, the regions of evaporation, atomization and excitation–ionization–detection are separated from each other. This reduces, among other things, problems associated with thermionic emission from the graphite rod that otherwise limit the detection power and matrix effects.

Smith et al. pursued an alternative route in order to separate the vaporization from the ionization detection.[339] The output of a GF was coupled into the argon flow of a miniature flame system.

5.4.2.5 Cells Detection of mercury differs from most other elements in that it already has a high vapor pressure at room temperature. No heated atomizer is therefore needed for mercury – a room-temperature atomic reservoir suffices. Mercury can therefore be measured directly in a cell. The groups of Winefordner and Omenetto have made extensive studies of the applicability of detection of mercury by the LEI technique (section 5.5.3.6).[178,270,341–346]

5.4.2.6 Thermionic Diodes Yet another alternative to the detection of laser-excited atoms by electrical means is the thermionic diode. Thermionic diodes are known to be extremely sensitive detectors for ions,[347] having a large dynamic range (over four orders of magnitude) and a very high gain (10^6 or larger).[318] The thermionic diode consists basically of a cylindrical anode in which an axially mounted cathode filament, heated by a DC, is mounted. No bias is needed to run the thermionic diode in a space-charge-limited mode. The ions produced by laser irradiation will be trapped within the negative space-charge cloud. Their presence reduces the space charge, which can be detected as a change in the diode current.[348–352]

Owing to its construction, the thermionic diode is most suitable for continuous-wave laser excitations. Although the first experiments were made with continuous-wave dye lasers, diode lasers have been used more frequently lately.[353–359] An improved type of thermionic diode that is suitable for analytical trace element analysis in samples with low vapor pressure has been constructed by Franzke et al.[354] By inserting a grid between the atomizer and the thermionic diode detector, and with appropriate biasing of grid and atomizer, the perturbation of the space charge due to the emission of thermal ions and electrons from the surface of the atomizer could be eliminated. The thermionic diode is also suitable for Doppler-free spectroscopy, which allows for precise isotope determinations.[355–357] It has also been found suitable for diagnostic purposes, i.e. the determination of collisional cross-sections and energy transfer rates.[353–363]

5.4.3 Laser Sources

As the selectivity of the LEI technique lies solely in the excitation step(s), narrow-band tunable laser systems must be used. Although the first demonstration of the LEI technique utilized a continuous-wave dye laser system (based upon an argon ion laser),[277] most subsequent work has been done using pulsed dye laser systems pumped by excimer lasers (see references 72, 74, 82–84, 89, 90, 95, 99, 109, 173, 181, 327, 329, 336, 364–381), Nd:YAG lasers (see references 294, 322, 328, 365, 382–386), nitrogen lasers (see references 91, 322, 328, 387, 388), flashlamps (see references 74, 293, 306, 389, 390) or copper vapor lasers[238] (with the two first types being by far the most used). A smaller number of investigations have used continuous-wave lasers (tunable diode lasers,[391] argon- or krypton-ion-laser-pumped dye lasers,[392] or atomic line lasers[368]). As was alluded to earlier, the main reasons for the popularity of pulsed lasers is that they produce UV light (most atoms require UV light for efficient excitation) and saturate the transitions more easily than do continuous-wave systems, and they can produce a large signal over a short period of time (thereby maximizing the S/N ratio). A disadvantage with pulsed laser systems is the poor duty cycle.

5.4.4 Detection Equipment

5.4.4.1 Principles The most distinctive aspect of LEI is the charge-collection process. In LEI, the charge created as a consequence of laser illumination (i.e. the enhanced ionization rate) is detected electrically by applying an electrical potential across the volume of interaction. Hence, the atomic reservoir (e.g. the flame) is part of an electrical circuit. The applied potential gives rise to an electrical field that separates the charges created. The movement of the charges in the electrical field, in turn, gives rise to an induced current in the rest of the electrical circuit. This current is converted to a voltage by a current-to-voltage transducer (often termed amplifier), which thus constitutes the output signal of the LEI instrumentation. This implies that LEI differs from most other laser-based spectrometric techniques (e.g. LIF/LEAFS) because it does not require any light-detecting equipment.

5.4.4.2 Charge-collecting Electrical Field Although the electrical field strength between a cathode and an anode is constant under vacuum conditions, the thermal production of charges in a flame will give rise to a nonuniform electrical field strength. The cathode will attract the thermally created positively charged ions and the anode will attract the electrons. However, because the mobility of the electrons is about two to three orders of magnitude larger than that of the ions, they will be extracted more rapidly from the flame than the ions, leaving a certain density of positively charged ions behind (distributed with their highest density close to the cathode, decreasing in the direction of the anode). Hence, the flame will exhibit a certain positive charge. The presence of a net charge will give rise to a gradient in the electrical field. Because the highest density of ions is found close to the cathode, the largest field gradients will also be found there. This implies that the electrical field strength will decrease monotonically in a direction towards the anode. Depending on the potential applied, the distance between the burner head and the immersed cathode, and the amount of thermally produced charges, there will either be a finite but monotonically decreasing electrical field strength in all positions between the cathode and the burner head, or an electrical field strength that will be reduced to zero at a position before the anode. In the latter case, which can occur when there is a high thermal production of charges in the flame, the flame can be considered to be composed of two distinct regions: a positive ion space-charge region, often referred to as the cathode sheath or simply the sheath, which extends from the cathode to some intermediate position in the flame at which the electrical field strength becomes zero; and a part extending from the anode (the burner head) to the position in the flame in which there is no electrical field (hence constituting an unperturbed region in the flame). In order to be able to detect the increased ionization rate, the charges have to be created in a region where there is an electrical field that can separate the charges. This implies that only charges created in the sheath will contribute to the detectable signal.[316]

In addition, and as discussed in the literature in more detail, the LEI signal consists of two parts: one related to the movement of the electrons and one to that of the ions.[307,316] Because the mobility of the electrons is two or three orders of magnitude larger that of the ions, the electronic signal will last only a fraction of the ionic signal. Moreover, because the area of the signal versus time curve represents the relative potential passed by each type of particle, they are of roughly equal size, implying that the electronic signal is not only shorter but also about two or three orders of magnitude higher than the ionic signal. Detection of only the electronic part of the LEI signal (whose duration is about 1 μs) has been shown to yield better S/N than detection of the entire signal (with a microsecond duration) owing to the significantly smaller amount of noise detected with the smaller gate.

5.4.4.3 Electrode Design A variety of electrode designs for flame LEI have been used and evaluated throughout the years. The first set-ups used electrodes (plates or rods) placed in close proximity to, but outside, the flame. The electrodes had either different polarity (giving rise to a mainly horizontally directed electrical field across the flame) or the same polarity (acting as a common cathode with respect to the burner head so as to produce mainly a vertically directed electrical field in the flame). It was soon found, however, that the best practical performance (for best immunity to interferences from so-called EIEs; see section 5.5.4.1 for details) could be obtained using a cathode immersed in the flame and directing the laser beam close to the electrode. It has been demonstrated repeatedly that the immersed cathode clearly outperforms the external cathodes as soon as samples with any substantial amounts of EIEs are analyzed.[316]

5.5 Analytical Performance

5.5.1 Sensitivity

5.5.1.1 Optimum Laser-enhanced Ionization Transitions A great deal of work in the field of LEI has been performed with the aim of finding the best/most suitable excitation wavelengths for various elements. It is obvious that whenever there are a number of transitions that all can be optically saturated by the laser light, the highest sensitivity will be obtained for the one that excites the atoms to the highest possible energy level (i.e. as close as possible to the ionization limit). However, because the transition probability (and hence the laser-induced excitation rate) rapidly decreases as a function of

For references see page 9575

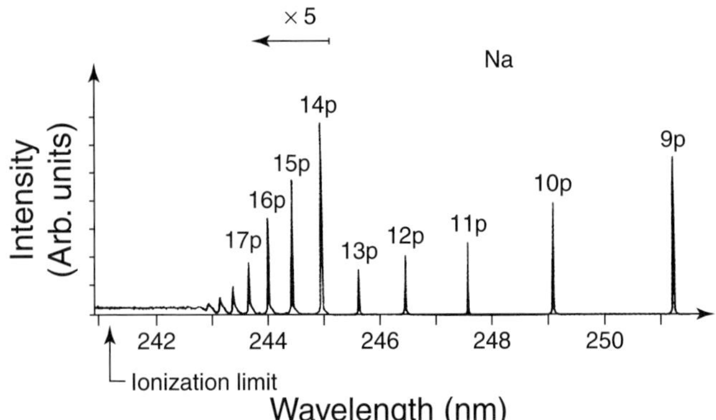

Figure 13 An LEI scan over a number of transitions (3s to higher-lying *n*p states, $n > 9$) in Na in a flame. (Reproduced by permission of the Society for Applied Spectroscopy from O. Axner, T. Berglind, *Applied Spectroscopy*, **43**(6), 940–951 (1989).)

principal quantum number of the upper state, transitions to high-lying states will in general not be saturated by light from normal pulsed dye laser systems (which is illustrated in Figure 13). The question of which transition is the most sensitive in a given element then arises.

A systematic study of this concept was made by Axner.[89] Based upon a simple model (assuming that the transition probability from the ground state to excited states decreases with the principal quantum number cubed, as it does for transitions to Rydberg states, assuming no optical saturation, and a particularly simple form of the ionization efficiency, namely a pure Boltzmann factor), he predicted that the transitions to states positioned approximately $1.7\,kT$ below the ionization limit should show the highest LEI sensitivity for one-step excitations. In an acetylene–air flame with a temperature of 2500 K, this corresponds roughly to $3000\,\text{cm}^{-1}$, which for the case of Na and Li (the pilot elements in many investigations) corresponds to the 7p and 6p states, respectively. Measured LEI sensitivities confirm these simple estimates fairly well. The highest LEI sensitivity (for unsaturated transitions) for Na has been found for the 3s–6p transition (i.e. to a state approximately $4150\,\text{cm}^{-1}$ below the ionization limit), whereas the 2s–5p and 2s–6p transitions for Li have been shown to have approximately the same sensitivity. The small discrepancy between prediction and findings was explained by the use of an oversimplified expression for the effective ionization rate for excited atoms.[89] The same work also predicted the most sensitive unsaturated one-step transitions for a number of other elements (K, Rb, Cs, Mg, Ca, Sr, Ba, Al, Ga, In and Tl) using real transition probabilities rather than asymptotically extrapolated values. Because many elements have approximately the same level structure among the excited states, one can conclude that the most sensitive one-step transitions for LEI (for unsaturated transitions) are those that excite the atoms to states approximately 2–3 kT below the ionization limit.

5.5.1.2 Ionization Yield The ionization yield, defined as the fraction of illuminated atoms that are ionized as a consequence of laser irradiation, Y^{ion}, is a convenient quantitative measure of the sensitivity of the LEI excitation–ionization process. It has been found that the ionization yield can approach unity for pulsed as well as continuous-wave excitations for elements with low ionization potential.

For example, Smith et al. determined the ionization yield of Li atoms excited to the 2p and 4d states by pulsed lasers in an acetylene–air flame.[96] The authors concluded that approximately 60% of all the atoms were ionized following a stepwise excitation to the 4d state. Axner et al. measured ionization yields of up to 80% for Na atoms excited to the 7d state in a two-step process (3s–3p–7d) using a pulsed excimer laser system.[95] Unity (or close to unity) ionization depletion from states further away from the ionization limit (the 3p state in Na in this particular case) has been demonstrated using continuous-wave excitation by Schenck et al.[393] The increased illumination time for continuous-wave excitations compensates for the lower ionization rate of the 3p state.

5.5.1.3 Two-step versus One-step Laser-enhanced Ionization As was alluded to earlier, a one-step excitation is not sufficient to obtain a high ionization yield for elements with higher ionization potential. An alternative then is to excite the atoms under study in a two-step excitation process, as was discussed in section 3.4. The two-step versus one-step signal enhancement in LEI, defined as S_{12}/S_1, where S_{12} and S_1 are the two-step and one-step LEI signals, respectively, has been found to range from unity (i.e. no enhancement) to a few thousand, depending on element, choice of transitions, irradiances, etc. The use of two-step LEI will thus not only provide a larger sensitivity to the elements with a moderate or a high ionization limit but will also improve on the selectivity because it enhances the signal from the analyte significantly more than that of the background (see section 5.5.5.2).

5.5.1.4 Ionization Efficiency Axner and Berglind determined a closely related entity, referred to as the ionization efficiency (defined as the probability that an excited atom will ionize before returning to the ground state), for excited *n*p states in Na and Li.[90] They found that the ionization efficiencies for these elements were considerably higher than those given by a simple Boltzmann factor (close to unity for states within $1\,kT$

from the ionization limit and around 50% for states 2.5 kT below the ionization limit), thus supporting the discussion above (section 3) about an effective ionization rate that is larger than a single state-specific ionization rate.

5.5.1.5 Overall Efficiency There have been several attempts to model the overall efficiency of LEI based upon the basic assumptions given above. The total number of charges created (per laser pulse), for example, can, in general, be written as a product of the density of analyte species, the interaction volume, and the ionization yield. It has been found, however, that such descriptions do not always agree with the experimental findings with expected accuracy.[(74)] Although the dependence of most parameters (e.g. laser pulse energy and analyte concentration) is well understood, not all dependencies have yet found a satisfactory explanation. The area dependence is such an unexplained feature. In fact, the discrepancy between predicted and experimental behavior has been so extensive that not one single work has yet been able to address successfully this seemingly simple matter for the LEI technique.

It is rather straightforward to argue that there should be an optimum laser beam area for a given laser pulse energy.[(57)] As long as the transition is saturated, and the cross-section of the laser beam is smaller than the distribution of the analyte atoms in the part of the atomizer in which there is a sufficient charge collection efficiency, the total number of charges created can be increased by enlarging the laser beam area. As the beam area becomes larger, the degree of optical saturation will decrease, eventually implying that the linear excitation regime will be reached. In this case, any further defocusing will no longer increase the number of charges created (for a one-step LEI case) because the increasing number of atoms illuminated will be balanced by a decreasing fraction of atoms excited in the interaction volume. Eventually, when the beam area becomes larger than the distribution of the analyte atoms in the atomizer (or the part of the atomizer in which there is a sufficient electrical field for collection of the charges created), the total number of charged created or detected should decrease. However, this behavior has never been verified experimentally.

5.5.1.6 Influence of "Scattered Laser Light" A possible explanation as to why this discrepancy exists was given by the findings of Sjöström and Axner when they found that the signal strength of the LEI technique could be influenced significantly by what they referred to as "scattered laser light".[(394)]

Figure 14 shows the results from an experiment on Sr atoms in an acetylene–air flame. A laser with a small beam diameter (in this case 2 mm) was scanned across the $5s^2\,{}^1S_0$–$5s\,5p\,{}^1P_1$ transition around 460.733 nm for four different laser pulse energies in Figure 14(a). A distinct peak from Sr atoms is clearly visible in all the four scans. Worth noticing is that even though the irradiance of the light was increased by a factor of 45 from 12 to 550 kW cm^{-2}, the peak signal only increased by a factor of 2. This behavior is normally interpreted as a manifestation of optical saturation. However, a simple modeling of the behavior of a two-level system shows a clear discrepancy with respect to the experimental findings. The reason is that an optically saturated system shows saturation broadening when the laser wavelength is being scanned over the transition. Saturation broadening originates from the fact that a laser beam that strongly saturates the transition on resonance also saturates the transition when being slightly detuned. This implies that the profile broadens. The result of a modeling of a two-level system (with some photoionization) that replicates the peak values of the experimental curves in Figure 14(a) are shown in Figure 14(b). It can be seen clearly that there is a significant discrepancy between prediction and experimental findings when it comes to the width of the curves.

In order to understand this discrepancy, additional experiments were made. Figure 14(c) shows the results from a corresponding set of experiments for a larger beam area. In this case, the beam area was enlarged 16 times (the beam diameter was expanded from 2 to 8 mm), as was the beam pulse energy (so as to preserve the irradiance of the light and degree of optical saturation). It was then found unexpectedly that the signals were only marginally larger than with the 16 times smaller beam area (approximately 40–50% larger) and that a significant broadening of the lower parts of the peaks could be discerned. This gave some insight into the true physical origin of the LEI signal in this particular experiment. Analysis of the data showed that the LEI signal, in fact, can have several origins. It was found that each curve could be decomposed into a sum of two curves with significantly different widths, as is shown by Figure 14(d). The top experimental curve can be decomposed into a curve with broad shoulders and one narrow-band feature (next two curves). The curve with broad shoulders was found to follow rather well the predicted behavior of the LEI signal from an optically saturated system, but the narrow-band feature did not. This led to the conclusion that the signal with broad shoulders originates from atoms within the geometrically confined interaction region (referred to as the "true" interaction region), whereas the narrow-band peak has to be attributed to the signal from nonsaturated atoms in another part of the atomizer (i.e. outside the "true" interaction region) affected by what was referred to as "scattered laser light". It was thus concluded that a significant (sometimes even a dominating) part of the signal from Sr can originate from "scattered laser

For references see page 9575

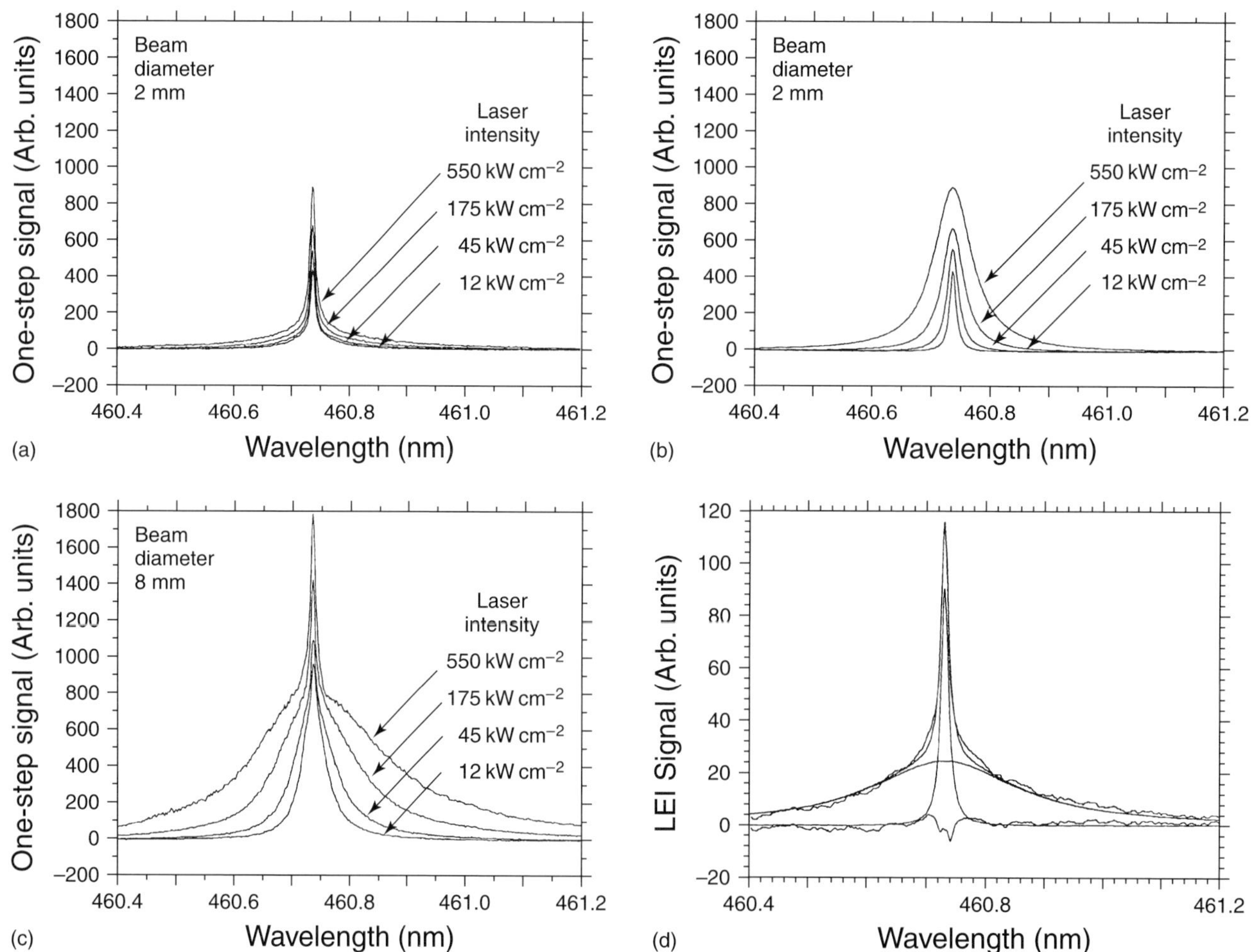

Figure 14 A set of one-step LEI scans over the $5s^2\,^1S_0-5s\,5p\,^1P_1$ transition in Sr in an acetylene–air flame. (a) The situation when light with four different laser irradiances (intensities) was scanned over the transition (12, 45, 175, and 550 kW cm^{-2} as the four curves, from bottom to top, respectively) with a beam diameter of 2 mm. (b) The results from the simplest possible model of the situation that can reproduce the peak values (a two-level model of a Lorentzian absorption profile with some photoionization). (c) A situation corresponding to (a), with a 16-fold larger area (a beam diameter of 8 mm) but with the same irradiances, from which one can see clearly that the LEI one-step signal from Sr is composed of at least two parts. (d) A one-step spectrum (top curve) decomposed into the sum of two Lorentzian-shaped peaks (next two curves), with the bottom curve being the residual error of the decomposition. It can be concluded from a comparison between the widths of the curves in (a–d) that it is the broad peak that behaves as expected from theory. Hence, the narrow peak, which dominates the curves in (a), originates from "scattered" laser light. (Parts (a) and (c) are reproduced by permission of the Society of Applied Spectroscopy from O. Axner, S. Sjöström, *Applied Spectroscopy*, **44**(5), 864–870 (1990).)

light" under a variety of conditions, as is exemplified by Figure 14(a).

A number of possible mechanisms for the existence of "scattered laser light" outside the "true" interaction region have been proposed: the light might be due to diffraction of laser light from an aperture positioned outside the flame; it might consist of fluorescence light from excited analyte atoms in the interaction region; or it might originate from light scattered from particles of flame molecules in the "true" interaction region of the flame. Despite attempts to understand the physical origin of this "scattered laser light", no clear conclusion about its origin has yet been drawn. Furthermore, it has not been investigated to what extent this affects other elements detected by LEI. It is clear, however, that the influence of "scattered laser light" is smaller in two-step than in one-step experiments.[394] This also explains why there has been a better agreement between measured and predicted signal strengths for two-step excitations than for one-step excitations in LEI spectrometry.[74]

5.5.2 Noise

The most thorough description of the various sources of noise in LEI spectrometry has been given by Turk.[237]

He divides the various types of noise into two categories: multiplicative and additive noise. Multiplicative noise is defined as the noise that is proportional to the concentration of the analyte, and additive noise comes from the other components. This implies that as the concentration of the analyte decreases, the influence of the multiplicative noise decreases accordingly, so that only the additive noise determines the limit of detection. Multiplicative noise, on the other hand, can affect the precision.

5.5.2.1 Multiplicative Noise Multiplicative noise can be due to fluctuations in the atomic population (affected by properties of the nebulizer/flame), the ionization yield (affected by properties of the laser), and the detection efficiency (affected by properties of the applied potential and the flame). Although pulse-to-pulse variation in the laser power can be substantial, it has been found that there is little correlation between the fluctuations of the LEI signal and those of the laser power.[395] The main reason is attributed to optical saturation. Instead, in a simultaneous measurement of LEI and LIF in a flame, it was found that the correlation between the noise in the LEI and LIF signals was substantial. This provides evidence that the major source of multiplicative noise in this particular case resulted from fluctuations of the atomic population within the laser-irradiated volume.[395]

5.5.2.2 Additive Noise Additive noise can consist of electronic noise, thermal background ionization noise and laser-induced background ionization noise. Electronic noise can in turn consist of either Johnson noise in resistors or radiofrequency interferences emitted from pulsed lasers synchronous with the laser pulses. As was concluded by Turk, the electrodes, burner head and preamplifier seem to form an excellent antenna for radiofrequency interferences.[237] Care in shielding and grounding the LEI instrumentation is therefore essential in LEI spectrometry. A comparison between the nonlaser-induced sources of noise for two flames (acetylene–air and hydrogen–air) and two different samples (synthetic drinking water with a few to some tens of ppm of EIEs and citrus leaves, diluted and digested to $10\,\mathrm{mg\,mL^{-1}}$ with one to two orders of magnitude higher concentrations of EIE) were made by Turk.[237] It was found that although the root mean square of the electronic noise (no flame on) was 1.3 nA (unfiltered), the flames themselves contributed very little to this noise level. When drinking water was aspirated, however, the noise increased to approximately twice the value of the electronic noise (to 2.7 and 1.7 nA for the two flames, respectively). When the high-EIE samples were aspirated (citrus leaves), on the other hand, the noise increased considerably (to 23 and 6.2 nA, respectively). This shows that the dominating source of noise originates from the increased thermal ionization from EIE when samples with high concentrations of such elements are detected. In the absence of EIE most of the noise can be attributed to either electronic noise or flame current noise. Turk concluded furthermore that when the noise is dominated by flame noise, the noise magnitude is consistent with shot noise from the flame DC. When also considering the effect of signal averaging, the noise levels given above were found to be decreased significantly. Assuming a time constant of 1 s and exponential averaging over 10 Hz (as typical for Nd:YAG systems) the noise levels in the acetylene–air flame decreased to 0.45 and 3 nA (for the two samples, respectively), whereas the corresponding levels for a 300-Hz excimer system become 0.07 and 0.52 nA.

The relative importance of laser-induced background ionization noise is more difficult to assess because this type of noise depends on the amount of background signal generated by the laser (which normally is considered to be a type of interference). Irrespective of its origin (it can, for example, be due to wing excitation or multiphoton ionization of flame molecules), one can conclude, however, that this type of noise is particularly destructive for the LEI technique because in general it is affected by the fluctuations of the laser power (because wing excitation and multiphoton ionization are seldom optically saturated) as well as the atomic population in the flame, two entities that are known to carry a significant amount of noise.

5.5.3 Limits of Detection

5.5.3.1 Laser-enhanced Ionization in Flames As was alluded to above, the most commonly used atomizer for LEI has been the flame. A rather large number of detection limits for LEI in flames have therefore been published throughout the years. The best LOD for each element have been collected in Table 5 (see references 72, 293, 306, 321, 322, 328, 336, 364, 366, 367, 370, 371, 383–385, 387, 388, 392, 396–399). The LOD have been stated as given in the original references. This implies that although some elements have LOD that are limited by the actual noise of the background (i.e. including background signals from contamination, etc.), some have been estimated from the instrumentation noise (i.e. obtained when detuning the laser(s) from resonance so as to avoid the influence of contamination). In addition, although most LOD have been given with an S/N of 3, measured over 1 s, some are taken under slightly different conditions (or under unspecified conditions).

As can be seen from Table 5, the majority of the LOD are in the picogram per milliliter region with a few even below, i.e. in the femtogram per milliliter range (In, Li,

For references see page 9575

Table 5 LOD for the LEI technique in flames (ng mL^{-1})

Element	First-step excitation (nm)	Second-step excitation (nm)	Laser[a]	Flame[b]	LOD	Refs.	Additional Refs.[c]
Ag	328.1	421.1	E	AA	0.05	72	
Al	308.2/309.2		F	AN	0.2	321	371
As	278.0		E	AA	3000	364	
Au	242.8	479.3	Y	AA	1	383	364
Ba	307.2		F	AA	0.2	293	364
Bi	227.7		E	AA	0.2	371	
Ca	227.6		E	AA	0.006	371	387, 396
Cd	228.8	466.2	Y	AA	0.1	383	271
Co	240.8		E	AA	0.06	371	383
Cr	240.9		E	AA	0.2	371	
Cs	455.5		N	AA	0.002	328	322, 388
Cu	324.8	453.1	Y	AA	0.07	383	336
Eu	459.4	564.0	N	AA	4000	397	
Fe	271.9		E	AA	0.1	364	364, 371
Ga	241.9		E	AA	0.03	371	306, 364, 366
In	303.9	532	Y	AA	0.0004	336, 398	328, 364
Ir	266.5	562.0 + 642.0	E	AA	0.3	367	
K	404.4		N	PBA	0.1	322	392
Li	670.8	460.3	E	AA	0.0003	72	293
Mg	285.2		E	AA	0.005	366	
Mn	279.5	521.5	Y	AA	0.02	384	364
Mo	319.4		F	AN	10	321	
Na	589.0	568.8	N	AA	0.0006	328	399
Ni	229.0/232.0		E	AA	0.02	371	
Pb	282.3	600.2 + 1064	Y	AA	0.0007	398	
Rb	420.2		N	AA	0.0006	387	
Sb	287.8		E	AA	50	364	
Si	288.2		F	AN	40	321	
Sn	284.0	597.0	Y	HA	0.3	383	321
Sr	230.7		E	AA	0.006	371	
Ti	318.6/319.2/320.0		F	AN	1	321	
Tl	276.8		E	AA	0.006	364	366, 370
V	318.4/318.5		F	AN	0.9	321	
W	283.1		E	AA	300	364	
Yb	555.6	581.2	Y	AA	0.1	328	
Zn	213.9	396.5	Y	AA	1	385	

[a] E = excimer-pumped dye laser system; N = nitrogen-pumped dye laser system; Y = Nd : YAG-pumped dye laser system; F = flashlamp-pumped dye laser system.
[b] AA = acetylene–air; AN = acetylene–nitrous oxide; PBA = propane–butane–air; HA = hydrogen–air.
[c] Additional references with LOD that are within a factor of three from the best. These works in general do not use the same type of laser, wavelength or flame as that specified.

Na, Pb and Rb), and a limited number above, i.e. in the nanogram per milliliter range (As, Eu, Mo, Sb, Si and W).

The best LOD correspond fairly well to the realistically estimated best LOD for the LEI technique. Turk calculated the best possible LOD for two different situations.[(337)] Under the somewhat unrealistic condition that the detection power should be limited by the shot noise of the thermal ionization, he found that the LOD should be 3 fg mL^{-1}. This is far below any experimentally found LOD. When basing the calculation upon a more realistic situation (assuming that the detection power is limited by the actually measured noise values), the LOD was found to be of the order of 0.1 pg mL^{-1}. As can be seen from Table 5, there are a few elements that have their LOD within one order of magnitude from this estimated optimum LOD (primarily In, Li, Na, Pb and Rb). This indicates that there is still room for improvements for most of the detected elements for LEI in flames.

5.5.3.2 Laser-enhanced Ionization in Plasmas The detection power of laser-based ionization techniques has not been improved as much as for conventional techniques when going from flames to ICPs, for reasons given in the instrumentation section (section 5.4).

The first attempts to combine an ICP with a laser-based ionization technique were made by Turk and Watters. In order to circumvent these problems, the authors used an extended plasma torch and placed two charge-collecting electrodes 19 cm above the load coil (in the tail flame, so as to benefit from the higher neutral atom fraction that exists in the cooler part of the plasma and to minimize the influence of radiofrequency interferences).[323] The detection powers were clearly inferior to those previously attained by the LEI technique in flames (e.g. LOD for Cu of 7 μg mL^{-1}).

In order to increase the detection power of LEI in an ICP and to perform measurements in the analytically most preferable position in the plasma, Turk et al. power-modulated an ICP by temporarily turning the plasma off (reducing the influence of radiofrequency interferences and decreasing the plasma background ion/electron concentrations).[324] Ionization measurements were made 1.4 ms after complete interruption of plasma. LOD in the nanogram per milliliter range were achieved: for Fe and Ga, 80 and 20 ng mL^{-1}, respectively. It was concluded that the primary mode of ionization of the laser-excited atoms was photoionization rather than collisional ionization (i.e. more in-line with RIS).

Other atomizers have been investigated for possible application in LEI spectrometry, but with limited success. The microwave-induced plasma, for example, has been investigated by Seltzer and Green.[326] Another example is the use of an atmospheric pressure micro-arc plasma with LEI, which was investigated by Churchwell et al.[300]

5.5.3.3 Laser-enhanced Ionization in Graphite Furnaces
As was alluded to above (in section 5.4), a variety of experimental configurations of furnaces and probes have been scrutinized for LEI in a GF. The common experiences are that the sensitivity is not as large as expected and that signal fluctuations are considerable.

For example, already the first measurements indicate that there is a somewhat lower collisional ionization rate (or yield) in the furnace than in the flame. One such indication was the finding by Gonchakov et al. that the LEI signal from two-step excitation of Na could be increased significantly (by a factor of 20–40) by actively photoionizing the atoms with infrared light from a Nd : YAG laser.[301] Magnusson et al. compared signal strengths (on a sensitivity basis) from flame LEI and LEI in a GF and concluded that the Na signal from the latter was smaller than expected and that from Mg was larger.[329] The smaller Na signal strength was attributed to a lower collisional ionization rate in the furnace than in the flame, whereas the larger Mg signal was explained partly as photoionization.

Moreover, many authors have noticed that there was always a background signal, fully resonant with the species detected, as soon as any liquid was injected into the furnace (corresponding to that from a few picograms of analyte).[301,329,330] Magnusson et al. showed explicitly that there were no background signals when the furnace was heated empty (with the lasers on) and concluded that the background signal originates from contamination.

Many LOD for LEI in a GF are in the picogram range. Magnusson et al. quoted contamination-limited LOD that are in the picogram range (Co, 5 pg; Cr, 5 pg; Mg, 1 pg; Mn, 1 pg; Na, 5 pg; Ni, 7 pg; Pb, 0.5 pg).[329,330] The contamination-free LOD for Mg and Na were estimated to be 0.03 and 0.5 pg, respectively (obtained from measurements when no sample was injected into the furnace).[329] Gonchakov et al. had previously estimated the LOD for Na in the absence of contamination to be in the low femtogram range.[301] Moreover, the LOD for Mn and Sr measured with the T-furnace by Magnusson et al. were assessed to be 1 and 2 pg, respectively.[331] Detection of Yb and In, using two-step excitation with the probe above the dosing hole, yielded LOD of 1 pg and 0.8 fg, respectively.[333]

Butcher et al., finally, found that the limiting noise for their LEI system in a modern HGA-500 furnace with a moveable probe was furnace noise and thermionic emission. LOD were established for a number of elements (Tl, 2 pg; In, 0.7 pg; Pb, 60 pg; Li, 1 pg; Mg, 10 pg; Mn, 30 pg; Fe, 50 pg; the LOD were determined on-line for all elements except Li and Mg). In agreement with previous authors, they found that the precision was poor (between 12 and 16%). They attributed this to poor precision of the manual pipetting, but gave no conclusive evidence as to whether this originated from any parts of the LEI or GF instrumentation (e.g. fluctuations in the charge-collection efficiency) as had been suggested earlier.

It can be concluded that the furnace does not bring the same advantages for the LEI technique as it does for the LIF/LEAFS technique, for the reasons described in section 5.4.2.3.

5.5.3.4 Laser-enhanced Ionization in Rod–Flame Systems
As was alluded to above, an alternative means of atomization is the rod–flame system developed by Chekalin et al.[333,335–338] The most significant advantages with this system come from its ability to handle samples with high concentrations of concomitant elements and for analyses of solid materials. It was, for example, concluded that the background signal from the matrix decreased by over two orders of magnitude when samples in solid form were analyzed in comparison with samples in liquid form. This allowed for detection limits in solid samples down to the picogram per gram level. LOD for the rod–flame system are generally low (Au, 200 fg;[336] Co, 1 pg;[338] Cr, 2 pg;[338] Cu, 2 pg;[335,336] In, 4 fg;[335,336] Mn, 3 pg;[338] Na, 2 fg;[336] Ni, 800 fg[338]).

The rod–flame instrumentation has been used for detection of: In in CdHgTe solutions,[335,336] Au in

For references see page 9575

$AgNO_3$,[333] Cu in Ge,[335,336] Na and Cu in orthophosphoric acid,[333] Li, K, Na, Sb, Cu and Ag in phosphoric acid, acetone and isopropanol,[337] Ni, Co, Cr and Mn in fiber-optic fluoride materials,[337] and Cr, Co, Mn and Ni in NH_4F and NaF.[338]

5.5.3.5 Laser-enhanced Ionization in Graphite Furnace–Flame Systems Two-step excitation from a pulsed Nd:YAG-pumped dye laser system was used to detect Mg, Tl and In in the previously described combined GF–flame system developed by Smith et al.[339] After optimization of various parameters (e.g. carrier gas flow), the LOD for these three elements could be assessed as 1, 120 and 260 fg, respectively.

Riter et al. determined experimentally the vaporization, transport, atomization, probing, and detection efficiencies of Mg atoms in an LEI system in which a GF was coupled to a miniature flame.[340] The authors estimated the overall efficiency of the system to be 0.0025%. The experimental LOD was 20 pg. The LOD for Mg was estimated to be 6 fg in the absence of the blank and after a reduction of radiofrequency noise.[340] The same group has subsequently used an LEI–GF–flame system for the determination of Pb in whole blood. An LOD of 90 pg mL^{-1} (0.9 pg absolute) was found for lead in whole blood.[400]

5.5.3.6 Laser-enhanced Ionization in Cells Mercury has been studied by ionization techniques directly in a cell.[178,270,341–346] Clevenger et al. studied the temporal behavior of the LEI signal of Hg in a quartz cell under low buffer gas pressure. Using fast electronics and a short laser pulse, it was possible to distinguish, in one single time-resolved ionization waveform, the nonselective photoionization component of the signal from that due to collisional ionization from selected levels. Experimental results were shown to agree with those obtained by computer simulations, and optimal conditions for deconvolution of the two components were assessed.[343] A spectroscopic study of Hg Rydberg states ($n = 10–42$) was also made. Broadening and splitting caused by the influence of Stark effects and of increasing buffer gas pressure were observed. The optimal operating conditions, in terms of pressure and applied high voltage, for obtaining the best sensitivity and LOD by analytical LEI spectroscopy were identified.[344]

A new method for the detection of Hg atoms in an inert gas atmosphere by LEI, based on the avalanche amplification of the signal resulting from the ionization from a selected Rydberg level reached by a three-step laser excitation of mercury vapor in a simple quartz cell, was developed by Clevenger et al.[345] An avalanche amplification effect was found when Ar and P-10 gases were used at atmospheric pressure. The authors estimate that an LOD of 15 Hg atoms per laser pulse in the interaction region is achievable under amplifier noise-limited conditions.[345] A comparison of various means to detect Hg (including laser-based techniques) is given by Clevenger et al.[346]

5.5.4 Interferences

Interferences are phenomena that adversely affect the performance of a technique. There are two types of interferences in LEI spectrometry: space-charge and laser-induced interferences. Although space-charge interferences are exclusively associated with EIEs and affect the charge-collection process, spectral interferences can originate from either matrix elements (to a large extent EIEs) or flame native molecules and therefore affect primarily the excitation–ionization process.

5.5.4.1 Space-charge Interferences Space-charge interferences originate from the fact that (for a given potential applied across the flame) the electrical field distribution depends strongly on the thermal ionization rate.[307,316,401]

As was described in the instrumentation section (section 5.4), applying a potential across a weakly ionized plasma (as LEI atomizers can be characterized as) gives rise to two distinctly different regions: the sheath, in which there is a finite electrical field; and a fieldless region (close to the anode). Only charges created in the sheath will contribute to the signal. Because the extent of the sheath depends on the amount of thermally produced charges, the active interaction volume (i.e. the volume in the flame in which the laser beam overlaps with a nonzero external field) will decrease (i.e. shrink towards the cathode) with increasing amount of EIEs (because EIEs contribute significantly to the thermal ionization in the flame). Hence, when samples with high concentrations of EIEs are investigated, only the charges created close to the cathode will contribute to the signal. Moreover, because normally only the electronic part of the signal is being measured, the creation of a pair of charges close to the cathode (from which the electronic part of the total LEI signal is larger than the ionic part) will give rise to a larger LEI signal than if the charges were created further away from the cathode. This implies that the laser beam preferably should be directed close to (i.e. just below) the immersed electrode in order to have the highest immunity to EIEs and yield the highest signal. This also explains why the externally positioned cathodes are not capable of handling samples with high concentrations of EIEs (they have their cathode sheath across a cold region in which no free atoms exist). Although the influence of space-charge interference can be reduced by a proper choice of electrode and probe volume, this type of interference is a considerable limitation to the applicability of the LEI technique.

A flow injection system has been interfaced to an LEI apparatus to handle samples with high concentrations of EIEs by Wang et al.[402] With the use of the flow injection/laser-enhanced ionization (FI/LEI) apparatus, the electrical interference induced by the matrix could be effectively diminished. It was found that the FI/LEI technique is capable of detecting indium in a solution containing an Na matrix of more than 40 ppm ($\mu g\,mL^{-1}$), which was about 20 times higher than what the conventional LEI apparatus could tolerate. The authors state that it also has a much larger LDR than conventional LEI. Additional advantages are good reproducibility and a rapid sampling rate.[402,403]

5.5.4.2 Spectral Interferences Spectral interferences can be separated into the following categories: ionization of native flame molecules; matrix spectral interferences (i.e. wing excitations or direct overlap between the spectral line of the analyte species with those of the matrix elements); and ionization of matrix elements by laser ASE.

Although there are several molecular species present in a flame, there are only a few that give rise to ionization signals. The NO molecule is the flame species that gives rise to the most severe flame background signals. With its strongest bandhead around 227 nm, and with others at around 237 and 248 (etc.) nm, it is a source of background signals primarily in the UV region. It has been found, however, that the ionization signal from NO is due to a two-photon excitation, in which case the signals can be held at moderate levels by avoiding focusing too strongly in the flame. Most other flame molecules that contribute to background signals in fluorescence or emission techniques, e.g. OH, have a larger tendency to dissociate than to ionize when being excited under the conditions typically prevailing in LEI spectroscopy, so they do not contribute to the LEI background signal.

Matrix spectral interferences, on the other hand, are in general a more severe problem in LEI spectrometry. Direct overlaps between a transition in an analyte and that of a matrix are rather scarce. Wing excitations of transitions in matrix elements, on the other hand, are much more frequent. It has been found that wing excitations of ubiquitous alkali and alkaline earth elements give rise to substantial background signals in LEI spectrometry. One such example is seen in Figure 15, which shows the LEI signal from a 20 µg mL^{-1} Na solution in the vicinity of the 3s–3p transition. As can be seen, wing excitations exist even several nanometers from the transition wavelength.

In addition, alkali elements often have quite a low ionization limit. When such an element is exposed to laser light with wavelengths below their threshold wavelength for direct photoionization from the ground state, a

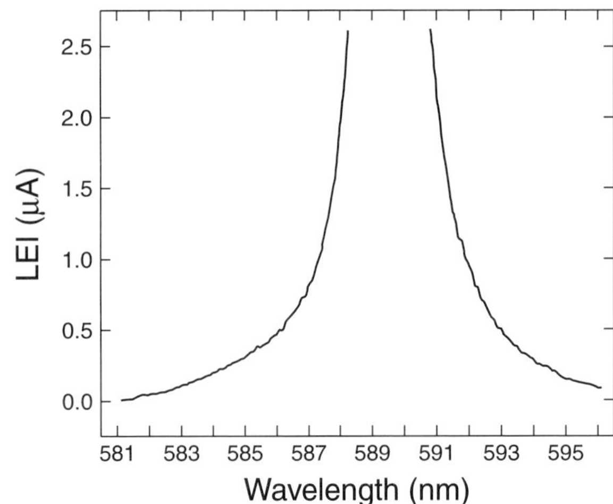

Figure 15 One-step LEI signal from 20 $\mu g\,mL^{-1}$ of Na in the vicinity of the 3s–3p transition. (Reproduced by permission of John Wiley & Sons, from G.C. Turk, in *Laser-enhanced Ionization Spectrometry*, eds. J. Travis, G. Turk, 161–211, 1996.)

structureless LEI background signal appears.[372,377] The threshold wavelengths for direct photoionization for Li, Na, K, Rb and Cs are 230, 241, 286, 297 and 318 nm, respectively. Hence, whenever UV light is used for investigating samples with large amounts of alkali matrix elements, a significant and unstructured LEI background signal will appear. As with all background signals, this will add to the existing noise in the system, thus reducing the detection power of the LEI technique.

A thorough study of background signals (direct overlap, wing excitations and photoionization) from Na in the entire 200–450 nm region was conducted by Axner et al.[377] They found that even when the laser was detuned as much as possible from the discrete transitions, there was an Na signal from wing excitation that will create a background signal whenever samples with large amounts of Na are detected, as is shown in Figure 16. It was found that the background signal originates mainly from direct photoionization from the ground state below 240 nm; excitation to a number of closely spaced Rydberg levels gives rise to a strongly structured background signal in the 240–260 nm region; direct overlap or wing excitations of the strong 3s–4p, 3s–5p, 3s–6p and 3s–7p transitions dominate between 260 and 360 nm, whereas photoionization of thermally excited 3p atoms predominantly takes place slightly below 420 nm.

Moreover, a part of the light emitted from a dye laser consists of ASE. This broad-band light, which occurs within the tuning curve of the dye used, is usually of much lower power than the laser light but nevertheless can be responsible for excitation of concomitant elements. Because the amount of ASE emitted from a laser is not strongly dependent on the laser light wavelength (except

For references see page 9575

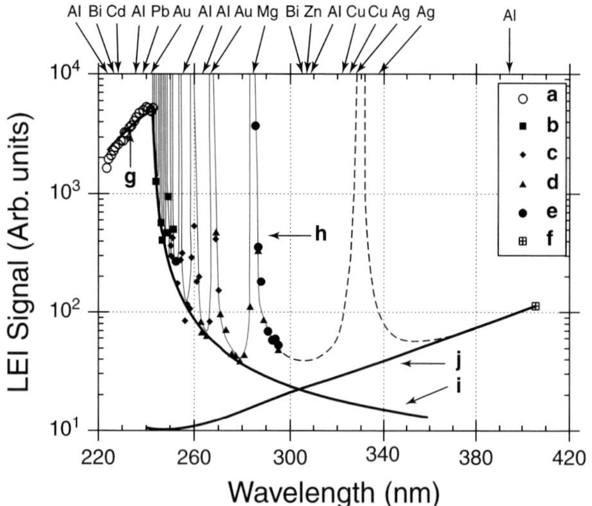

Figure 16 Experimentally measured (intensity- and concentration-normalized) LEI signal strengths for Na versus wavelength. The open and solid markers represent individual measurements done by various dyes in the dye laser: **(a)** Coumarin 47; **(b)** Coumarin 102; **(c)** Coumarin 307; **(d)** Coumarin 153; **(e)** Rhodamine 6G; **(f)** DPS. The predicted LEI signal from direct photoionization from the ground state (below 240 nm) is indicated by solid curve **(g)**. Various discrete resonance transitions are indicated with the solid and dashed line **(h)** (the peak at around 185 nm corresponds to the 3s–5p transition). The lowest contribution from wing excitations of the 3s–*n*p transitions in each wavelength region is indicated by solid line **(i)**. The predicted contribution from photoionization of the thermally populated 3p state is indicated by the solid curve **(j)**. (Reproduced by permission of the Society of Applied Spectroscopy from O. Axner, M. Norberg, H. Rubinsztein-Dunlop, *Applied Spectroscopy*, **44**(7), 1124–1133 (1990).)

at the edges of the tuning curves where it increases), it gives rise to a rather wavelength-independent background signal.

5.5.5 Selectivity

It has been corroborated frequently that LEI is a technique that has a high sensitivity (implying that the number of charges created per photon absorbed is high, in some cases close to unity). This is especially the case for the alkali elements because they have a rather low ionization potential and are easily atomized in a flame. This is, however, also the Achilles heel of the LEI technique. As was shown for Na above, all alkali elements will give rise to various types of spectral interferences (wing excitations or photoionization from lower-lying states). Any analyte to be detected in a matrix of alkali element therefore has to be detected on top of a background signal. The high sensitivity of the alkali elements thus reduces the detection power of other analytes and reduces the selectivity of the LEI technique. The selectivity of LEI therefore has been found to be lower than that of the LIF/LEAFS technique. This is an unrelenting consequence of the fact that LEI is a technique in which the selectivity provided by the laser source is limited by the amount of resolution it can provide when coupled with an undiscriminating charge-collecting detector.

5.5.5.1 Means to Improve Selectivity – Proper Choice of Laser Irradiance One possible remedy to the aforementioned problem of interferences and limited selectivity is to excite the analyte atoms with as low an irradiance as possible in order to avoid excess optical saturation of the transitions of the analyte (and minimize the effect of background signals from matrix elements).

An example is shown in Figure 17. So far, the most severe spectral interference encountered for LEI spectrometry occurs when Mg is to be detected in the presence of Na. The reason is that Mg has its only practically accessible transition ($3s^2$–3s 3p) at 285.21 nm whereas Na has one of its strongest transitions (3s–5p) at 285.28 nm, which is only 0.07 nm away. It is therefore clear that significant wing excitations of Na atoms will occur when the laser is tuned to the Mg resonance. Figure 17(a) shows the situation for one-step excitation of 10 ng mL^{-1} Mg in a matrix of 10 times more Na, i.e. 100 ng mL^{-1}, using a laser irradiance of 25 kW cm^{-2}. As can be seen from Figure 17(a), there is a spectral interference at the wavelength of the Mg transition already for this rather moderate amount of Na matrix under these conditions.

One way to increase the selectivity is to adjust the laser irradiance so as to avoid excess optical saturation of the Mg transition. Figure 17(b) shows the situation when the laser irradiance has been attenuated to 0.5 kW cm^{-2}. The interference from Na at the Mg transition has decreased considerably.

The increase in selectivity is not sufficient, however, when samples with higher Na concentrations are to be analyzed. Figure 17(c) shows the situation when the Na concentration has been increased by one order of magnitude (i.e. to 1 μg mL^{-1}).

5.5.5.2 Means to Improve Selectivity – Use of Two-step Excitations By illuminating the atoms from the example above also with 470.30-nm light, in order to excite the Mg atoms closer to the ionization limit by a two-step process ($3s^2$–3s 3p–3s 5d), the Mg signal will be increased and thereby also the selectivity. In this particular case, the signal enhancement was about 300, as is shown in Figure 17(d), and thus virtually all of the Na interference at the Mg transition has been eliminated.

However, although an optimization of the working conditions (regarding laser irradiance, laser beam area, etc.) and the use of two-step excitation can increase the

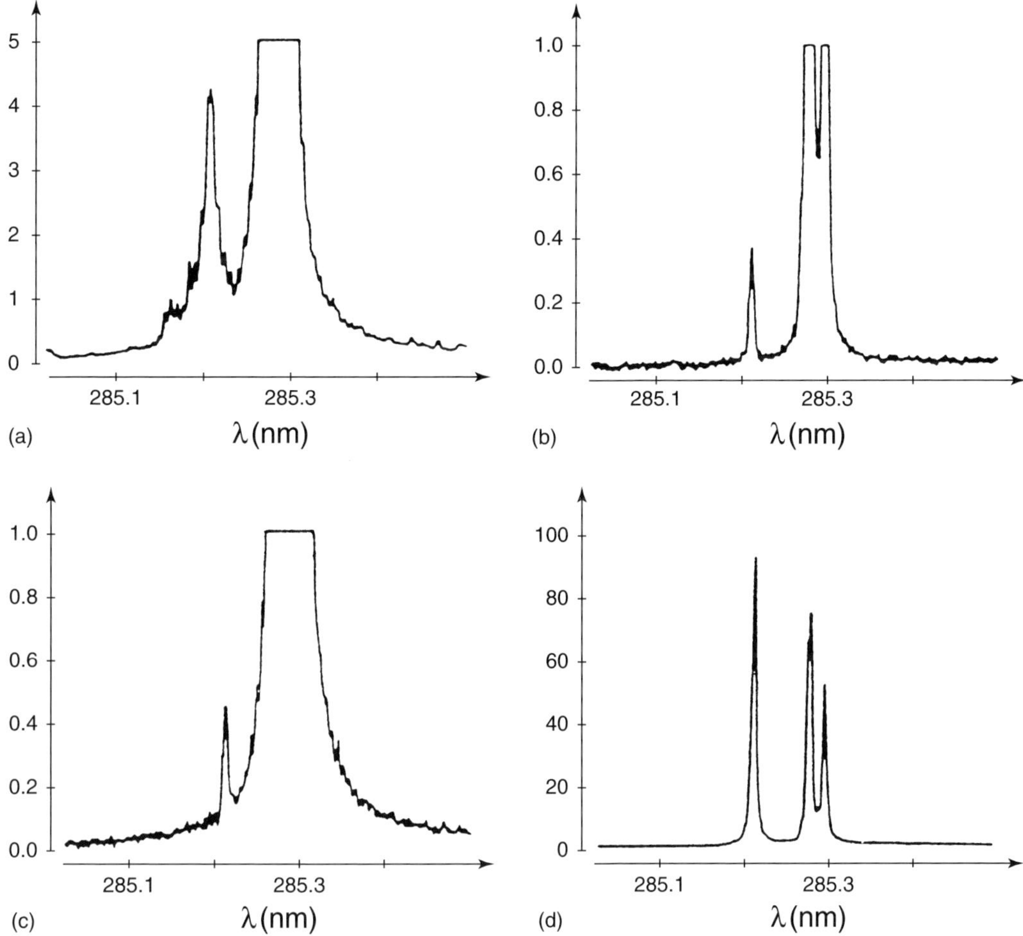

Figure 17 LEI signals from solutions of $10\,\mathrm{ng\,mL^{-1}}$ Mg in various matrices of Na for a variety of experimental conditions: (a) one-step excitation using $25\,\mathrm{kW\,cm^{-2}}$ irradiance laser light in a matrix of $100\,\mathrm{ng\,mL^{-1}}$ Na; (b) one-step excitation using $0.5\,\mathrm{kW\,cm^{-2}}$ irradiance laser light in the same Na matrix as in (a); (c) one-step excitation using the same laser light irradiance as in (b) in a matrix of $1\,\mathrm{\mu g\,mL^{-1}}$ Na; (d) two-step excitation using the same laser light irradiance in the first step as in (b) and (c) and an irradiance of approximately $4\,\mathrm{MW\,cm^{-2}}$ fixed on the second-step transition in Mg (at 470.30 nm) as the second step in the same Na matrix as in (c). (Reproduced by permission of The Royal Swedish Academy of Sciences from I. Magnusson, O. Axner, H. Rubinsztein-Dunlop, *Physica Scripta*, **3**, 429–433 (1986).)

selectivity considerably, they cannot completely eliminate the spectral interferences. This is illustrated in Figure 18, where the influence of wing excitation on the detection of Ni in an Na matrix is displayed. Nickel atoms are excited by a two-step excitation (at 300.3 and 561.5 nm) for best sensitivity. The second-step laser is scanned across the second-step excitation in Ni for two water solutions containing $100\,\mathrm{ng\,Ni\,mL^{-1}}$ with and without the presence of $100\,\mathrm{\mu g\,Na\,mL^{-1}}$ of matrix. A background signal is clearly visible for the sample containing the Na matrix. In this case, the background signal originates from a wing excitation for the 3p–4d transition with a central wavelength of 568.8 nm. Although the presence of $100\,\mathrm{\mu g\,Na\,mL^{-1}}$ does not influence the charge collection efficiency (the Ni peaks are virtually the same), it considerably decreases the S/N ratio.

5.5.6 Applications

The LEI technique has been used for a variety of applications, both analytical as well as diagnostic. A thorough review of the analytical applications of LEI has been given by Green.[317] A common analytical application is the assessment of traces in standard reference samples. Another is to determine the metal content in rocks.[404] The LEI technique has also been used with liquid chromatography in order to create a very sensitive method for speciation: organolead species were studied by Epler et al.,[405] and Cr speciation was studied by Paquet et al.[381]

Another example of an application of LEI is the development of a resonance photon ionization detector,[303,406–410] which is a narrow-band detector for photons or collisional energy transfer. In short, the

For references see page 9575

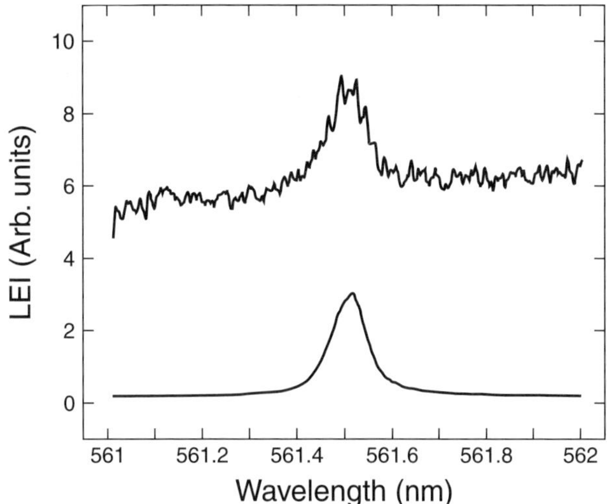

Figure 18 Two-step LEI spectra for $10\,ng\,mL^{-1}$ Ni in pure deionized water and in a matrix of $100\,\mu g\,mL^{-1}$ Na as the lower and upper curves, respectively. The first-step laser is fixed on the 300.249 nm transition in Ni and the second-step laser is scanned over the second-step excitation at around 561.5 nm. (Reproduced by permission of John Wiley & Sons, from G.C. Turk, in *Laser-enhanced Ionization Spectrometry*, eds. J. Travis, G. Turk, 161–211 (1996).)

detector consists of a two-step LEI instrumentation in which the second-step excitation is provided with light from a laser system. This system then acts as a sensitive and narrow-band detector of light whose wavelength corresponds to that of the first excitation step. One area of application for this technique is to serve as a detector for Raman light;[303,406,407,411,412] another area is to measure temperatures in a flame with high spatial resolution.[408,409]

A field-deployable flame-based LEI spectrometry system (termed laser optogalvanic spectrometry) has been developed to measure the concentration of metal species present in the near-atmospheric pressure off-gases of mixed-waste thermal treatment systems.[413]

The technique has also been used for assessment of unidentified transitions from Fe(I) from solar spectra, resulting in the possibility for several high-lying states in Fe(I) to be identified and labeled correctly.[380]

5.6 Diagnostic Capabilities

The high sensitivity of the LEI technique makes it useful as a diagnostic tool. A number of various entities and phenomena have been investigated by the LEI technique throughout the years and a few examples in this field follow.

5.6.1 Laser-enhanced Ionization in Flames

Studies concerned with the excitation and ionization processes (including energy pooling effects, charge exchange rates, etc.) in a variety of atomic reservoirs have been performed by a number of authors.[270,357,360–362,414] For example, Turk developed a methodology that can determine where within a flame LEI actually takes place.[414a,414b] He found that spatial control of the flame volume sampled by the ionization detection electrodes could be obtained by varying the applied voltage across the flame. This procedure was used to observe anomalous contributions to LEI caused by scattering or fluorescence from the laser beam to other regions in the flame. Axner and Berglind demonstrated that the LEI technique can be used to measure the electrical field distribution in a flame by monitoring the Stark-splitting Rydberg states that are exposed by the electrical field.[373] The ionization efficiency of Na atoms in an acetylene–air flame (defined as the probability than an excited atom ionizes rather than returning to the ground state) was measured by the same authors in another work.[90] The ionization yield (defined as the fraction of atoms in the interaction region that ionize) has been measured for a few elements in acetylene–air flames.[415] The LEI technique has also been used for measuring the atomization efficiency of atoms in an acetylene–air flame,[416,417] ionization rates,[418] flame temperatures[418,419] and dynamic Stark effects (i.e. effects on the energy levels induced by the laser light).[109]

A demonstration of the applicability of a delayed two-step LEI technique for measurement of the lifetimes of metastable states (in Pb and Tl) in an acetylene–air flame was first demonstrated by Omenetto et al.[81] Axner et al. made a more through study of nine different metastable states under a variety of conditions in an acetylene–air flame.[82] A few such examples are given in Figure 19. They concluded, among other things, that metastable states that are forbidden to decay to lower states because of a parity violation generally have longer lifetimes (85 ns–3.1 µs) than those that are forbidden to decay by violation of spin conservation (5–33 ns).[82] They also showed that the lifetime of a metastable state depends strongly on the local stoichiometry in the interaction volume (the lifetime of a specific metastable state in Au was found to range between 600 ns and 8.8 µs, depending on the local stoichiometry), which led to the finding that the LEI technique could be used to map the local stoichiometry in an acetylene–air flame by monitoring the lifetimes of atomic metastable states.[82–84]

The two-step LEI technique has also been used to measure flame flow velocities or gas velocities in a shock tube.[420–423] One laser system is used to deplete the density of neutral atoms locally in the medium while another is used to probe the time development of this depleted region. The fluorescence dip technique, finally, has been used to study laser-enhanced emission processes.[268]

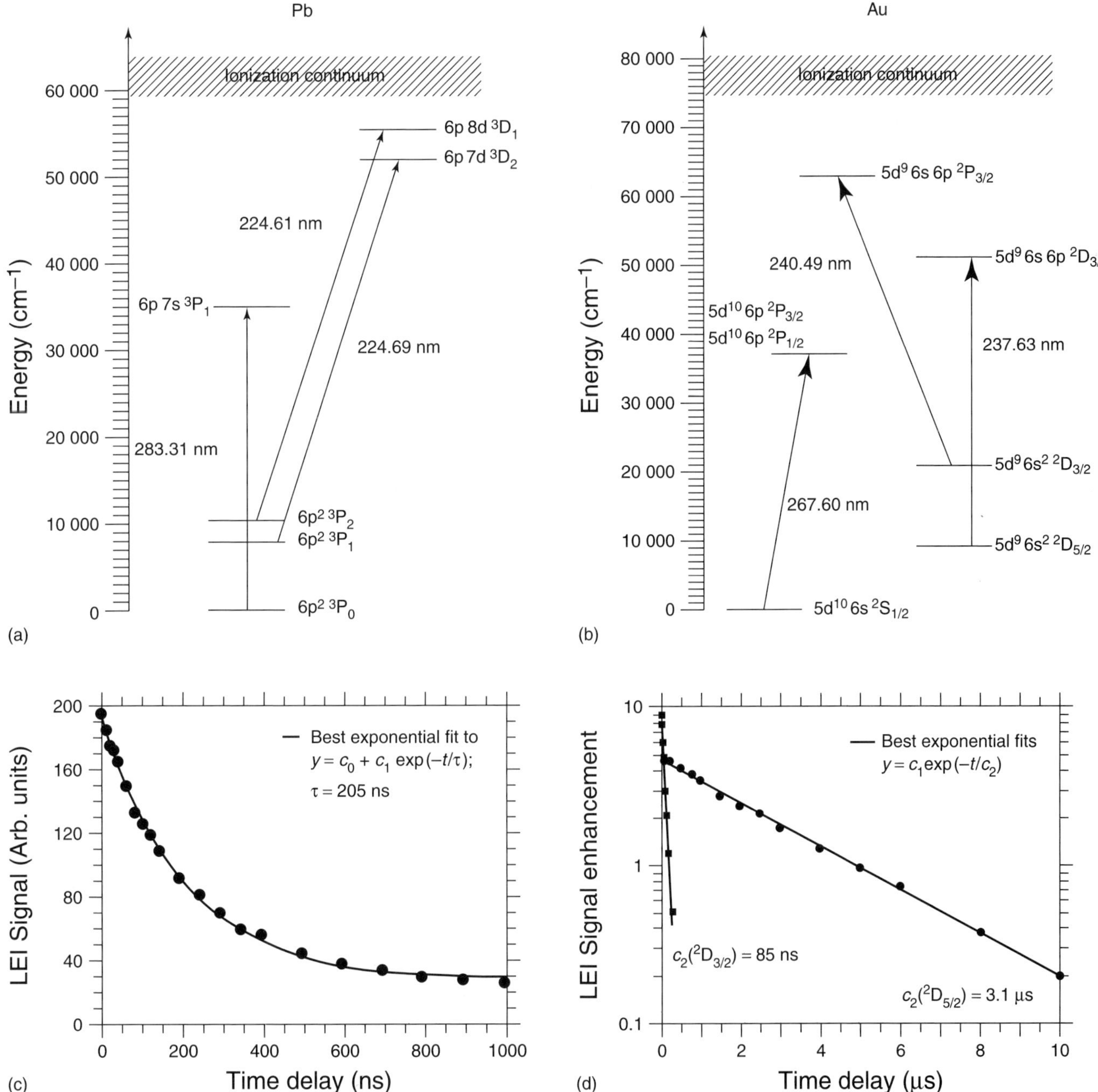

Figure 19 (a, b) Simplified energy level diagrams of Pb and Au in which some pump-and-probe excitations are marked. (c, d) Display decay curves for the $6p^2\ ^3P_1$ metastable state of Pb and the $5d^9\ 6s^2\ ^2D_{3/2,5/2}$ metastable states of Au, respectively, in an acetylene–air flame. (Reproduced by permission of Springer Verlag from O. Axner, P. Ljungberg, Y. Malmsten, *Applied Physics B*, **54**, 144–155 (1992).)

5.6.2 Laser-enhanced Ionization in Plasma

Because ICP provides a rather different ionization environment than do flames, it is not clear, a priori, how excited atoms ionize in this type of atomizer. Laser power studies were therefore made on several elements (Cu, Na, Fe and Mn) to investigate the ionization mechanisms of the excited atoms.(323) However, the results were not fully conclusive. Both Fe and Mn, which were excited to states whose energy is slightly above half of the ionization energy, showed a linear laser power dependence. Assuming that the optical transition was saturated at the intensities used, this indicates that the major ionization route is photoionization for these elements. Copper was found to be excited in two-photon

For references see page 9575

transitions to states close to the ionization limit from which ionization is assumed to proceed, with no further involvement of any light-induced process. For Na (excited to the 3p state), collisional ionization was the most probable ionization mechanism.

Laser-induced ionization in an ICP has also been performed by Turk et al. in order to measure the ionization yield and the fate of the laser-produced Sr ions in the ICP. Neutral Sr atoms were excited and photoionized by one laser system, while the number of Sr ions produced was monitored by another laser system (using the LIF technique).(327) By inducing a delay between the two ionizing and probing laser systems, it was found, among other things, that the recombination time constant of Sr ions in the ICP was 15.5 µs (but this decreased with increasing numbers of EIEs).

In another study concerned with the fate of ions in an ICP, Yu et al. found that the laser-induced enhancement of Sr^+ ions in a modified inductively coupled plasma mass spectrometry (ICPMS) instrument lasted for approximately 0.25 ms and peaked 0.2 ms after the laser pulse at a value of 11% relative to the Sr^+ ion signal due to the plasma ionization alone. The authors found that the laser-induced signal was limited by the high degree of direct plasma ionization and ion–electron recombination of laser-produced ions.(325)

With the use of a simplified model of the magnesium atom, several possible approaches to the measurement of charge exchange rates between Mg and Ar in the ICP have been developed.(414) Two of the approaches, which used pulsed dye lasers to populate the Mg(II) levels that are close to resonance for charge exchange, were tested experimentally. The experiments yielded an effective rate constant for the transfer of charge from Mg^{2+} to Ar of $1.1 \times 10^8\,s^{-1}$. This value is a factor of 50 lower than previously published estimates.(414)

A commercial ICPMS instrument, modified for use with a flame rather than an ICP, was used to sample and detect the LEI ions. Using double-resonance LEI with pulsed dye lasers, an Na^+ signal that was 350 times larger than that induced by thermal ionization could be achieved. Using a 5-mm laser beam diameter, the LEI signal ion pulse was found to last for 0.53 ms. Spatial studies, in which the position of the laser beam relative to the mass spectrometer sampler cone was varied, demonstrated that the ions produced by LEI travel with the flame velocity into the mass spectrometer, with no significant losses due to recombination from as far as 13 mm from the interface.(424)

5.7 Conclusions and Outlook for Laser-enhanced Ionization

It is clear that LEI is a highly sensitive laser-based spectroscopic technique that benefits from using a simple flame as atomizer (which allows for an uncomplicated sample introduction). LOD in the picogram per milliliter range have been obtained for a large number of elements (and with a few LOD even below this range). The most significant drawback of the technique is its sensitivity to EIEs. When samples with considerable numbers of EIEs are analyzed, significant background effects that limit the detection power of the analyte appear. This has limited the use of LEI primarily to situations when samples with uncomplicated matrices are to be analyzed. The technique has a great potential for diagnostics (e.g. of atomizer-specific physical and chemical processes) as well as for basic studies.

5.8 Resonance Ionization Spectrometry

5.8.1 Historical Development of Resonance Ionization Spectrometry

RIS was historically developed as a means of probing excited (metastable)-state populations in He as part of a study of the intricate interplay between various types of energy transfer processes in noble gas systems in the middle of the 1970s.(19) At that time it was conceived that photoionization could provide a more unambiguous probing of the excited-state populations than emission because the emission processes from metastable states in noble gas atoms are indirect. It was soon found, however, that the real importance of the photoionization approach was that it could provide a quantum-state selectivity if a tunable pulsed laser were used to first excite the metastable atom to another excited state, after which photoionization could proceed by absorption of a second photon from the same laser. The use of an intermediate state ensured that the ionization process could take place only if the laser wavelength was resonant with an existing transition in the atoms, which thus gave the process its name: RIS.(19)

5.8.2 Early Acquaintances for Resonance Ionization Spectrometry with the Field of Analytical Applications – Single Atom Detection

It was soon recognized that the RIS technique had an extraordinary potential for analytical applications by tuning the laser light to transitions originating from the highly populated ground state. It was even prophesied early on that SAD capabilities were within reach. The SAD capability of RIS was demonstrated by Hurst et al. in 1977 in a proportional counter detecting individual Cs atoms.(9,10) Since then the field developed rapidly, with the aim of meeting the various requirements of an analytical technique.(10,11,19,41,271–285) The RIS technique was first applied to species like alkali atoms, whose ionization limit was so low that the RIS process was feasible with the laser systems available at that time. As tunable laser systems became more advanced and more prevalent, however, RIS was also applied to other types of species.[9–11,19,41,271–285]

5.8.3 *Estimated Analytical Properties of Resonance Ionization Spectrometry*

The greatest advantage of RIS is undoubtedly the unmatched combination of high sensitivity (a high ionization yield for the laser-illuminated atoms) and high selectivity. It has been found, for example, that almost 100% of the illuminated atoms can be ionized. The selectivity (calculated as a product of the estimated selectivity of each excitation step) was estimated to be of the order of $10^9 - 10^{12}$ under quite general conditions (e.g. from RIS using two resonant excitation steps followed by photoionization from the uppermost laser-connected level), and even higher under most favorable conditions. In addition, if the laser-produced ions were detected by a mass spectrometer (most often a time-of-flight mass spectrometer for pulsed excitations and a quadrupole mass spectrometer for continuous-wave excitations), the selectivity could be even larger. The technique is then often referred to as resonance ionization mass spectrometry (RIMS). This high selectivity gives the resonance ionization spectrometry/resonance ionization mass spectrometry (RIS/RIMS) technique good applicability to isotopic selective assessments, an application area in which most other optical techniques are not as amenable owing to the increased transition widths in finite-pressure atomizers (collisionally broadened). This extraordinary selectivity has truly given the RIS/RIMS technique extraordinary promise for analytical applications.

5.8.4 *Practical Limitations of Resonance Ionization Spectrometry/Resonance Ionization Mass Spectrometry*

Reality showed that not all of the expected properties of RIS/RIMS could be realized readily as first anticipated. The main limitation of the technique is a poor temporal and/or spatial overlap between a thermally atomized sample and the interaction volume. This is particularly evident when pulsed laser sources are used for excitation (which was the prevalent situation for the first years, mainly dictated by the availability of UV laser light) together with thermal atomization. For example, the spatial and temporal overlap of a thermal atomic beam and the light from a pulsed laser system is of the order of 10^{-5} using a 25-Hz repetition rate laser (and typical values for other parameters). Means to improve on this consist mainly of using a high-duty-cycle laser system (i.e. atomic-vapor-pumped dye laser systems[425] or continuous-wave lasers, either ion-laser-pumped dye laser systems[426] or, lately, diode laser systems[427]) or pulsed atomization sources (e.g. laser ablation[428,429] and sputter-initiated atomization[430]).[431] The use of a 6.5-kHz atomic-vapor-pumped dye laser system instead of a conventional pulse laser with a 25-Hz repetition rate, for example, increases the spatial and temporal overlap by two or three orders of magnitude to a few tenths of a percent.[432]

5.8.5 *Practical Applicability of Resonance Ionization Spectrometry/Resonance Ionization Mass Spectrometry*

An assessment of the analytical capabilities of the RIS/RIMS technique can most easily be performed by studying the practical applicability of the technique. The RIS/RIMS technique has been used for a variety of applications in both the basic and applied sciences throughout the years.

5.8.5.1 Basic Studies with Resonance Ionization Spectrometry/Resonance Ionization Mass Spectrometry The basic studies have often been concerned with measurements of atomic physics entities, e.g. hyperfine structures,[433] ionization potentials,[434] photoionization and continuum structure[435,436] (including studies of negative ions[437,438]), and for high-precision measurements,[439] etc. but also for studies of basic quantum phenomena,[440] e.g. dynamic Stark shift[441,442] and population dynamics.[443]

5.8.5.2 Analytical Applications of Resonance Ionization Spectrometry/Resonance Ionization Mass Spectrometry The analytical applications of RIS/RIMS were originally devoted to studies of atomic species but molecular species have been assessed also during latter years.[285] The most recent activity within the field of analytical atomic spectrometry has been on isotopic issues. For example, double-resonance RIMS using single-frequency continuous-wave dye lasers has demonstrated detection limits in the attogram range for ^{90}Sr[444] and an optical isotope selectivity above 10^9 in measurements of Pb.[445]

One particular problem of using RIS/RIMS for isotope determinations, however, is that various isotopes can have different sets of (i.e. different numbers of) hyperfine components of a given atomic level and thereby different degeneracy factors of the laser-connected states. Because the ionization rate depends on details of the level structure of the atoms under study, the ionization fraction will vary between various isotopes. This is often referred to as the odd–even effect because isotopes with an even mass number generally lack nuclear spin ($I = 0$) and thereby hyperfine structure, whereas isotopes with

For references see page 9575

odd mass number have a finite nuclear spin (i.e. $I \neq 0$) and show hyperfine structure.(446) This effect is now understood (theoretical predictions have been verified by experiments(447)) but the effect has to be taken into account properly when analyzing experimental data. There are several ways to reduce the odd–even effect in practical RIS/RIMS measurements, e.g. reducing the resonance laser intensity well below saturation, increasing the ionization intensity to full saturation so that all isotopes are ionized, or working with depolarized laser light or short pulses.(448)

One of the most important uses of RIMS applies its ability to detect low concentrations of long-lived radioisotopes, prompted by the fact that radiometry faces various problems in the detection of some of the most important and hazardous radioisotopes. For example, for very long-lived α- and β-emitters (e.g. ^{239}Pu with $t_{1/2} = 2.41 \times 10^4$ years and ^{99g}Tc with $t_{1/2} = 2.1 \times 10^5$ years), the measuring time is very long using radiometric methods. In addition, detection limits are impaired by the background. Some isotopes cannot be distinguished by α-spectroscopy owing to the very similar energies of the α-lines (e.g. the pairs ^{239}Pu/^{240}Pu and ^{238}Pu/^{241}Am). Furthermore, because the continuous energy spectra of the β-emitters ^{89}Sr and ^{90}Sr overlap, they cannot be measured separately by β-spectroscopy. ^{90}Sr ($t_{1/2} = 28.5$ years) therefore often has to be detected via the daughter isotope ^{90}Y, which includes a waiting time of about 10–14 days.(432) This situation has justified the RIS/RIMS technique to seek applications in the detection of long-lived radioisotopes. Erdmann et al., for example, demonstrated detection of Pu with LOD of only 10^6 atoms using RIMS.(448) Isotopic Pu compositions could be determined with sufficient accuracy to assess the isotopic composition of Pu in soil from the Chernobyl area.(448) Passler used RIMS to detect long-lived radioisotopes like Pu, Tc, and ^{89,90}Sr.(432) The LOD of ^{89}Sr and ^{90}Sr were 5×10^7 and 3×10^6 atoms, respectively.(432)

RIS/RIMS has also been used for precise determination of Ca isotope ratios, primarily ^{41}Ca,(427) for purposes of geological and anthropological dating. Other uses of techniques capable of a precise assessment of widely different Ca isotope ratios include the evaluation of exposure histories of extraterrestrial materials, measurements of ratios of the minor stable isotopes in meteorite inclusions for tests of nucleosynthetic models, and investigations of isotope traces in medical studies.(427) The analytical requirements for these applications vary widely, but for dating measurements with ^{41}Ca they include, among other things, an isotopic selectivity of 10^{15}.(427) In order to accomplish this, and therefore with the objective of finding the most optimal excitation schemes for precise determination of Ca isotope ratios, a series of studies concerned with isotope shifts and hyperfine structure,(449) as well as line shapes and optical selectivity issues in Ca,(450) have recently been performed.

Although many laser sources produce light with an extremely narrow bandwidth, the combination of small optical isotope shifts and finite broadening mechanisms (Doppler shifts and natural broadening) restricts the isotopic selectivity of the RIS/RIMS technique in certain applications. As a means to improve on this, collinear RIS/RIMS on accelerated atoms has been performed. The use of accelerated atoms eliminates Doppler broadening and enlarges the isotope shifts, whereas the collinear approach increases the interaction volume. After a long series of methodology improvements, LOD of 3×10^6 atoms of ^{90}Sr (corresponding to an activity of 2 mBq) in the presence of up to 10^{17} atoms of stable ^{88}Sr could be achieved.(426)

The combination of pulsed atomization and RIS/RIMS not only provides a higher temporal and spatial overlap between atomized sample and the interaction volume, but also provides the possibility to perform microanalysis on solid samples. RIS/RIMS has been used successfully for selective post-ionization of sputtered neutral species, proving powerful ultratrace analysis capabilities below atomic fractions of 10^{-9} by removing only a few monolayers of the substrate.(451) One such application is the detection of Al in brain tissue homogenates with a spatial resolution of 100 μm.(452) When RIMS is combined with techniques for pulsed atomization, however, it becomes closely related to many other analytical techniques and thus under these situations has appeared under a large variety of names, e.g. sputter-initiated RIS or laser ablation RIS.

In addition to the few examples of the applicability of the RIS/RIMS technique to analytical atomic spectrometry given above, there are also a number of other applications of the RIS/RIMS technique that prove its applicability to ultrasensitive trace element or isotope analysis. The latest conference proceedings of the series of "International Symposia on Resonance Ionization Spectroscopy and its Applications" provide an unusually good coverage of the status of the field, and the reader is referred to these for a more in-depth description of the RIS/RIMS technique.(281–285)

5.9 Conclusions and Outlook for Resonance Ionization Spectrometry/Resonance Ionization Mass Spectrometry

Because RIS/RIMS normally is performed under vacuum conditions (thus with a minimum of broadening mechanisms), it can make full use of the narrow frequency width of laser light. By the use of double (or sometime even triple) resonant excitations followed by detection of the produced ions using mass spectrometry, it can provide impressive selectivity. This huge selectivity also makes isotopic determinations possible, even in cases when one isotope has an abundance that exceeds another by nine or ten orders of magnitude. One drawback is that the RIS/RIMS technique has difficulties in providing a high temporal and spatial overlap between a thermally atomized sample and the interaction volume, which can put practical restrictions on the sensitivity of the technique. Combined with pulsed atomization (e.g. laser ablation or ion sputtering) it can, however, be used for a variety of applications (e.g. surface analysis) with high sensitivity. Another drawback is that the RIS/RIMS technique requires rather complex instrumentation. The current development of continuous-wave diode lasers might be a useful solution to this problem. On the other hand, the technique has an excellent potential for studies of basic phenomena, e.g. within the fields of quantum optics or atomic physics.

6 FREQUENCY- AND WAVELENGTH-MODULATION DIODE LASER ABSORPTION SPECTROMETRY

6.1 Introduction to Diode Lasers and Their Use for Analytical Applications

6.1.1 Historical Development of Diode Lasers

Although the diode laser was invented before the dye laser,[453–456] for a long period of time it was not able to compete successfully with the dye laser for spectroscopic applications. The reason was the poor performance of the early diode lasers, which made them unattractive in comparison with dye lasers. They needed to be operated at liquid nitrogen temperatures, the spatial profile of their emission was not stable, their spectral emission was spread over many cavity modes, the output intensity was not always linear with injection current, and their lifetimes were not very impressive.[457] The development of diode lasers during the 1970s and 1980s has implied, however, that most of the problems associated with early diode lasers have been overcome. Today, there are handy single-mode room-temperature diode lasers with a number of unique and attractive properties, of which rapid tunability and low noise are two of the most important.

6.1.2 Spectroscopic Applications of Diode Lasers

Diode lasers have found use mainly for fundamental studies – atomic and molecular spectroscopy;[352,458–466] environmental monitoring, particularly remote sensing;[467–475] and analytical applications[43,45,225,476–480] – but also for a variety of other purposes (e.g. studies of effects of collisions[462,463] and evaluation of absolute number densities of atoms in atomic reservoirs[481]). Although the general properties of diode lasers have been reviewed repeatedly,[457,482–489] the spectroscopic properties of commercial laser diodes firing in the visible and near-infrared regions have been specially addressed by a handful of authors,[45,391,457,459,474,486,490–493] of which the reviews by Franzke et al.[45] and Fox et al.[391] are of particular interest for spectrochemical analysis. Pioneering work towards bringing diode lasers into the field of spectrochemistry has been done primarily and extensively (but not exclusively) by the groups of Niemax et al.[43,225,352,359,460,477–479,494–503] and Winefordner et al.[504–507]

Despite a number of advantages of diode lasers, a crucial drawback is the lack of lasers that can operate directly at visible or UV wavelengths. The number of elements today that can be studied efficiently by diode laser spectroscopy is therefore somewhat limited. Only a few elements can be detected using diode laser light directly and around 50 can be detected utilizing frequency-doubled diode laser light.[43] The atomic species that have been detected so far by diode laser spectroscopy in various atomizers (primarily flames or furnaces) for analytical purposes are Ba,[477,479,494] Br,[498] Ca,[479] Cl,[497,498] Cr,[508] Cs,[494,508] Eu,[479] K,[494] Li,[479,494,509] Pb,[479] Rb,[477,494,510–513] Sr[479,494] and Ti,[508] amongst which Li and Rb are the most frequently studied, primarily due to strong transitions from the ground state at wavelengths that can be produced rather easily by diode lasers. Issues that have been addressed are the multi-element capability of diode laser spectrometry,[225,494] the isotope selectivity of the technique,[359,461,479,492,509] and the use of diode laser techniques for assessment of the elemental content in samples with complex matrices.[513] Also, the concept of frequency-doubling of diode laser light has received a certain amount of attention lately.[514–516]

A number of molecular species have been detected by diode laser light for remote-sensing applications.[467,468,473,517,518] A somewhat more limited number of molecules have been detected for analytical applications. They have been detected either directly (e.g. water vapor[519,520]) or indirectly (e.g. $C_2F_4Cl_2$, CCl_4, CHF_3 and O_2 by detection of Cl, F and O in a DC discharge[496]). A good review of the field of analytical molecular spectroscopy with diode lasers has been given by Imasaka.[480]

The detection capabilities for various species differ considerably, depending on the type of atomizer, the transition (i.e. population of the lower laser-connected level and transition probability) and the detection technique used. Diode lasers have been used for spectrochemical purposes with a variety of techniques, e.g. fluorescence techniques,[225,504,510] ionization techniques,[278,352,460–463,476,479,509,521] noise detection techniques[511] and, last but not least, absorption

For references see page 9575

techniques.[477,478,481,494,496–498,505,508,512,522] Although very low detection limits can be obtained by applying diode lasers to the RIS/RIMS technique (which has to be performed under low-pressure conditions in rather complex instrumental set-ups),[11,19,274,278] it has been found that FM or WM techniques significantly improve the detectability of various species using absorption spectrometry under far more versatile conditions.[42,46] The following presentation of the use of diode lasers for analytical spectrometric applications will therefore be devoted to modulated (mostly WM) diode laser absorption techniques.

6.2 Frequency-modulation versus Wavelength-modulation

The general approach of modulation techniques comprises modulating the frequency (or the wavelength) of the laser light (most conveniently done by modulating the injection current to the laser diode), transmitting the light through the species cloud to be analyzed, and analyzing the detector signal at the modulation frequency (or at some frequency related to this). This process significantly reduces noise from the laser (primarily $1/f$ noise) by shifting the detection to higher frequencies where the noise is less significant. It also removes baseline offsets, slopes or even some curvature of the background (depending on the mode of operation).

Although being basically the same, the techniques utilizing a modulation of the frequency (or wavelength) are referred to as WM or FM techniques, depending on the relative sizes of the various modulation parameters and the means of detection.[42,46]

WM techniques are characterized by a modulation frequency, f_m, that is much *smaller* than the half-width at half-maximum of the absorption peak, $\Delta\nu_p$ (i.e. $f_m \ll \Delta\nu_p$).[42,523] The signal is detected at a suitable harmonic of the modulation frequency (most often at $2f_m$) using a lock-in amplifier. The optimum conditions for WM techniques include, among other things, a modulation amplitude, ν_a, that is a few times larger than the half-width at half-maximum of the absorption profile ($\nu_a = 2.2\Delta\nu_p$ for $2f$ detection, $\nu_a = 3.9\Delta\nu_p$ for $4f$ detection, etc.).[42] This implies, in turn, that the modulation frequency is smaller than the modulation amplitude (i.e. $f_m \ll \nu_a$).

When atmospheric pressure-broadened absorption profiles are detected (whose widths often are in the gigahertz range), the modulation amplitude therefore needs to be in the gigahertz range but the modulation frequency (often limited by the performance of the lock-in amplifier) is in the 10–100 kHz range. The use of modulation frequencies in the mid-kilohertz range implies that the influence of laser excess noise has been reduced significantly (although not necessarily fully eliminated).

FM techniques, on the other hand, pioneered by Bjorklund in 1980,[524] and investigated by numerous authors since then,[46,524–532] utilize, under optimum conditions, a modulation frequency that is *larger* than the width of the absorption peak (i.e. $f_m > \Delta\nu_p$). Such a high modulation frequency will impose sidebands (in frequency space) whose spacing is equal to the modulation frequency. By detuning the center frequency of the laser light by an amount equal to the sideband spacing, a sideband will be placed atop the absorption profile. One of the two sidebands will then be absorbed by the analyte. Detection is then achieved by feeding the signal from the detector to a mixer in which this signal is mixed with a reference signal (at frequency f_m), thus producing the sums and differences of these two signals (often referred to as heterodyning). The resulting signal, which is proportional to the difference in electrical field amplitude of the two sidebands, and therefore also the density of analyte species, can then often be detected directly by ordinary (fast) electronics. In order to optimize the signal strength, preferably only one sideband on each side of the carrier should be produced. This can be achieved if the modulation frequency also is *larger* than the FM amplitude (i.e. $f_m \gg \nu_a$).

This technique has found its best use for the detection of various molecular species under low-pressure conditions (often in multipass cells). Typical modulation frequencies for the FM techniques are in the high megahertz range (or occasionally around a few gigahertz), which is larger than the typical (often Doppler-broadened) absorption profiles of molecules.[533] Because excess laser noise is negligible at such detection frequencies, it should, in principle, be possible to obtain shot-noise-limited detection limits with FM techniques.[528,529] FM techniques in most cases should therefore outperform WM techniques when it comes to detection power.[46] However, this has not always been the case in practice. The high modulation frequencies involved put large demands on the instrumentation and increase the complexity of the system considerably. The instrumentation used therefore does not always fulfill all simultaneous requirements for optimum detectability. This is particularly the case when atomic species are to be detected under atmospheric pressure conditions (in which absorption profiles can have widths in the 5–10 GHz range). The rather complex instrumentation is also the prime drawback to the widespread use of FM techniques. Another reason why FM techniques do not outperform WM techniques (despite a lower inherent noise) is the existence of background signals (which, in many practical cases, is the limiting factor for analytical applications). Hence, despite the inherently higher detectability of FM techniques, WM techniques are still often used for the detection of species for spectrochemical purposes[495–498,508,510,513,522,523,534–539] whereas

FM techniques have been used almost exclusively for gas species analysis.(517,520,526–533,540) We will therefore only discuss the WM technique below. The reader is referred to the literature for a discussion of FM techniques.(42,43,520,524–533,540) Thorough investigations and comparisons of (theoretical as well as experimental) various FM and WM techniques have been done by Silver(42) as well as Bomse et al.(46)

6.3 Wavelength-modulation Spectrometry

6.3.1 Instrumentation

A typical experimental set-up for WM spectrometry is shown in Figure 20. Wavelength-modulated light from a diode laser is directed through an atomizer onto a detector. The detector output is fed to a lock-in amplifier (through a preamplifier when applicable). The output of the lock-in amplifier constitutes the final signal to be registered and stored by a computer.

There are a few different modes of detection for the WM techniques. The most common is $2f$ detection (i.e. detecting the absorption at twice the modulation frequency, $2f_m$),(43,495–499,508,510,513,522,531,534–538,541–544) although occasionally higher harmonics ($4f_m$ and $6f_m$) have been used.(467) Reasons for this are that: the $2f$ output is intuitively the harmonic that carries the main part of the analytical signal in WM spectrometry (see discussion below); it produces the largest WM signals; and it eliminates (or reduces significantly) both constant and linearly sloping baselines. Another attractive feature is that the signal peaks at the absorption line center.

6.3.2 Theory

6.3.2.1 Absorption Process and Nomenclature For absorption measurements, the detector signal, $S_D(\nu, \nu_0)$, can be written as shown in Equation (63):

$$S_D(\nu, \nu_0) = \kappa I_0 e^{-\rho L \sigma(\nu-\nu_0)} \tag{63}$$

where ν is the frequency of the laser light, ν_0 is the center frequency of the absorption profile, κ is an instrumentation constant (including a photodiode response and an amplification function) that relates the light intensity impinging upon the photodiode to the measured signal, I_0 is the laser light intensity prior to the interaction region, ρ is the density of species to be detected (in m^{-3}), L is the interaction length, and σ is the absorption cross-section of the species to be detected at the particular wavelength (frequency) used. Because the absorption cross-section depends on the wavelength of the light (i.e. the frequency), it can be written conveniently as in Equation (64):

$$\sigma(\nu - \nu_0) = \sigma_0 \bar{\chi}(\nu - \nu_0) \tag{64}$$

where σ_0 represents the absorption cross-section at the peak of the profile, i.e. $\sigma(\nu = \nu_0)$, and $\bar{\chi}(\nu - \nu_0)$ is the (peak-normalized) line-shape function for the species to be detected.

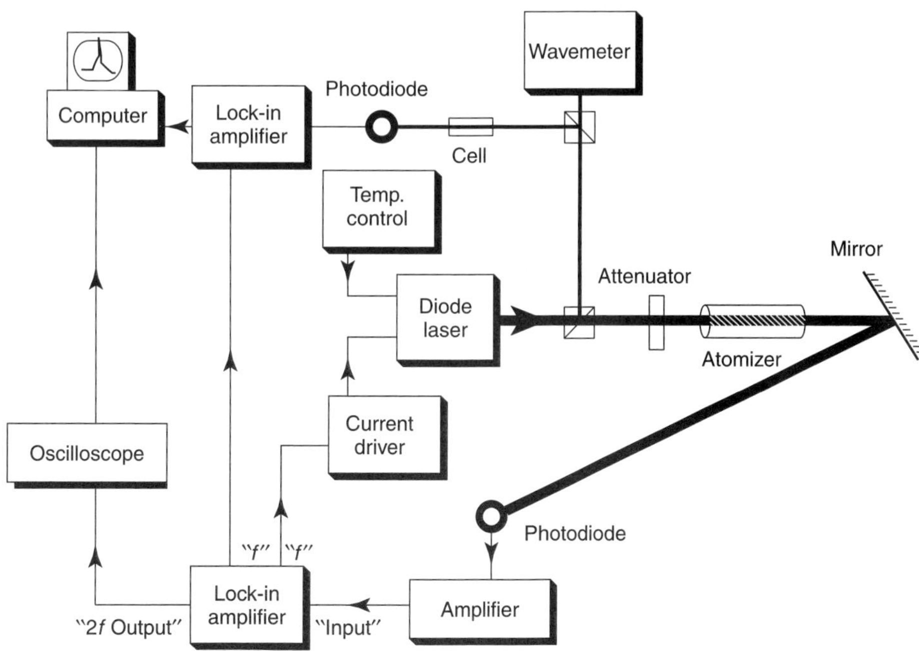

Figure 20 A typical set-up for WMDLAS spectrometry in a GF. (Reproduced from P. Ljung, O. Axner, *Spectrochimica Acta*, **52B**, 305–319, Copyright 1997, with permission from Elsevier Science.)

For references see page 9575

6.3.2.2 Wavelength Modulation WM techniques build upon modulation of the laser light wavelength, λ, by a sinusoidal modulation frequency, f_m, of the injection current, as shown in Equation (65):

$$\lambda(t) = \lambda_c + \lambda_a \sin(2\pi f_m t) \tag{65}$$

where λ_c and λ_a are the center wavelength and amplitude of the WM, respectively. Because it is more convenient to work with broadening and distribution functions in frequency space, the modulation can be written in terms of frequency (i.e. in Hz), as in Equation (66):

$$\nu(t) = \nu_c + \nu_a \sin(2\pi f_m t) \tag{66}$$

where ν_c and ν_a are the corresponding entities in frequency space, i.e. the center frequency and the modulation amplitude, respectively.

6.3.2.3 The nf *Wavelength-modulation Signal* This implies that the detector signal has a rather complex time dependence. An intuitive understanding of the origin of the 2*f* WM signal in resonance can be obtained from Figure 21. Figure 21(a) displays: the detector signal as a function of frequency for an analyte species with its absorption profile centered around ν_0; the time evolution of the laser light frequency (modulated at a frequency f_m); and the time-dependent detector signal constructed from both of the former. The absorption profile has been turned through 90° to correlate the frequency axis with that of the laser light modulation. Because the center frequency of the laser light coincides with that of the analyte absorption profile (at λ_0), the detector signal will experience an absorption maximum twice per modulation period, as shown in Figure 21(a). This explains clearly why the detector output signal will contain a significant amount of $2f_m$ frequency components. Figure 21(b) shows the same entities for a linear background absorption. As can be seen, a linear (or constant) background absorption produces no $2f_m$ frequency components.

There is, however, no need to study the time dependence of the detector signal directly. A lock-in amplifier is normally used to extract the *n*th harmonic of this detector signal, $\overline{S}_{nf}(\nu_d, \nu_a)$, according to Equation (67):

$$\overline{S}_{nf}(\nu_d, \nu_a) = \frac{1}{\tau}\int_0^{\tau} S_D[\nu_c + \nu_a \sin(2\pi f_m t), \nu_0] \times \sin(2\pi n f_m t + \phi)\,dt \tag{67}$$

where ϕ is a suitable chosen phase and where we, for convenience, have introduced ν_d as the detuning of the modulation center frequency with respect to that of the absorption profile, $\nu_d = \nu_c - \nu_0$. This implies in practice that $\overline{S}_{nf}(\nu_d, \nu_a)$ is the *n*th Fourier component of the time-dependent detector signal, $S_D(\nu_c + \nu_a \sin(2\pi f_m t), \nu_0)$.[523]

Arndt[545] has shown that the signal strength of the *n*th harmonic can be expressed in terms of an integral as shown in Equation (68):

$$\overline{S}_{nf}(\nu_d, \nu_a) = \frac{(-1)^n(2-\delta_{n0})}{\pi}\int_{-\nu_a}^{\nu_a} S(\nu_c - \nu', \nu_0) \times \frac{T_n(\nu'/\nu_a)}{\sqrt{(\nu_a)^2 - (\nu')^2}}\,d\nu' \tag{68}$$

where δ_{n0} is the Kronecker delta and T_n is the Chebyshev polynomial of degree *n*.

An *nf* WM spectrum of the species to be detected, $\overline{S}_{nf}(\nu_d)$, can be created by slowly scanning the center frequency across the absorption profile (i.e. the detuning) for a given modulation amplitude.

For the case when the medium is optically thin, the exponential factor in Equation (63) can be approximated with the first two terms in its series expansion; thus, the *nf* WM signal, $\overline{S}_{nf}(\nu_d, \nu_a)$, can be written in terms of the *n*th harmonic of the line-shape function, $\bar{\chi}_n(\nu_d, \nu_a)$, according to Equation (69):

$$\overline{S}_{nf}(\nu_d, \nu_a) = -\kappa I_0(\rho L \sigma_0)\bar{\chi}_n(\nu_d, \nu_a) \tag{69}$$

The *nf* lock-in output signal depends not only on the detuning, the amplitude of the modulation, ν_a, and the width of the absorption profile, $\Delta\nu_p$, but also on the particular form of the absorption profile. The general form of a single absorption transition has a Voigt form, originating from a convolution of a Lorentzian-broadened homogeneous profile (resulting from collision broadening) and a Gaussian-broadened inhomogeneous profile (resulting from Doppler broadening).

There are analytical expressions for the *nf* WM signal for only two limiting cases: when small modulation amplitudes are being used ($\nu_a \ll \Delta\nu_p$); and when homogeneous broadening dominates the inhomogeneous profite under optically thin conditions (so that the *nf* WM signal is directly proportional to the *n*th harmonics of a Lorentzian function). Equation (67) or (68) therefore has to be evaluated by numerical means for all other conditions (i.e. either by numerical integration[546] or by simulations[535–539]).

For sufficiently small modulation amplitudes (i.e. significantly smaller than the width of the absorption profile), the *nf* WM signal is directly proportional to the *n*th derivative of the line-shape function, as shown in Equation (70):

$$\lim \overline{S}_{nf}(\nu_c)|_{\nu_a \to 0} = -\kappa I_0(\rho L \sigma_0)\frac{(\nu_a)^n}{n!2^{n-1}}\left.\frac{\partial^n \bar{\chi}(\nu)}{\partial \nu^n}\right|_{\nu=\nu_c} \tag{70}$$

This property has given the WM technique its alternative name of *derivative spectroscopy*. This description describes the shape of the WM spectra for small modulation amplitudes and also shows that the *nf* signal strength

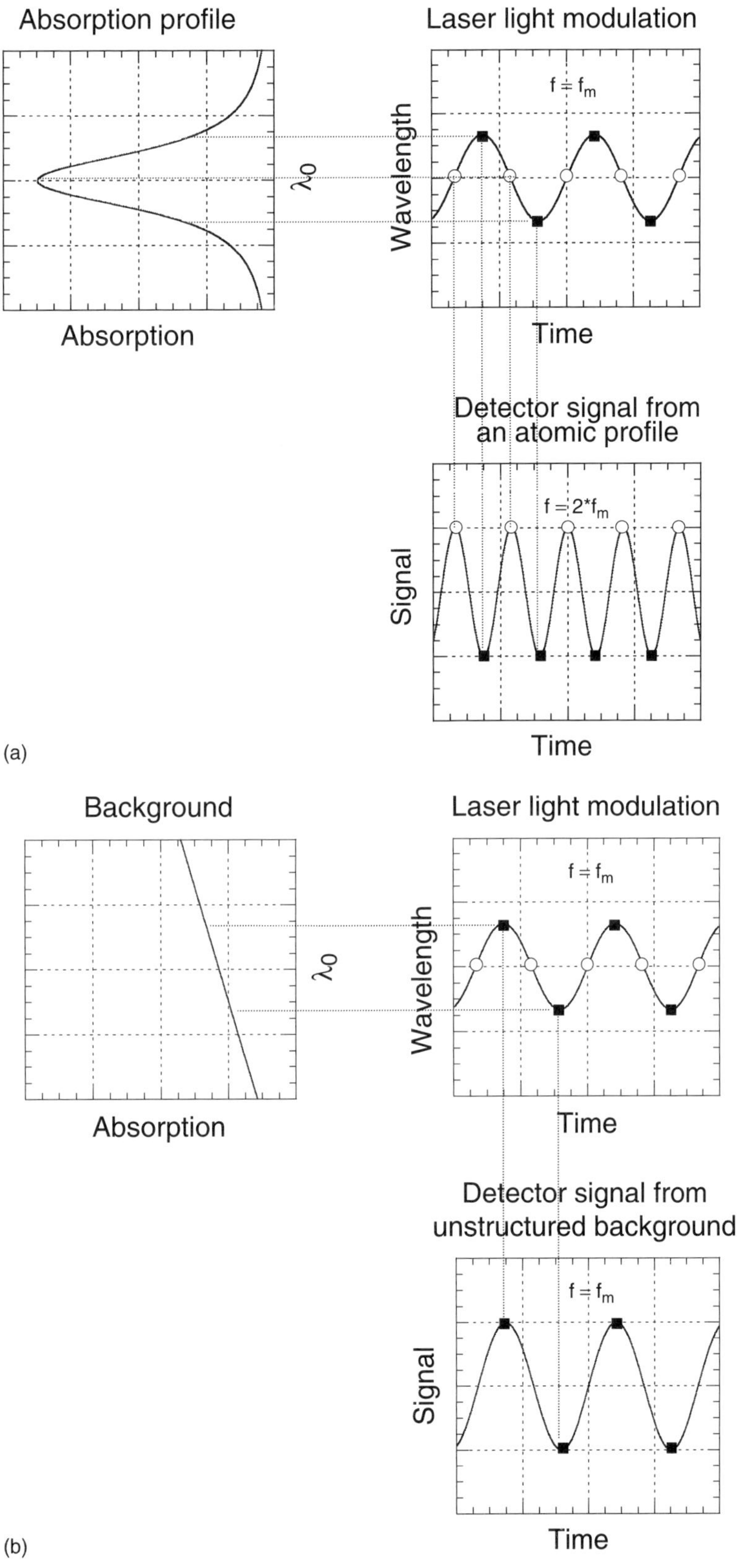

Figure 21 Absorptions, modulated laser light frequencies and resulting detector signals for the absorption profile (a) and for linear background absorption (b).

rapidly decreases with decreasing modulation amplitude (i.e. as $(\nu_a)^n$). This implies that it is not particularly applicable to the conditions under which the WM technique is used for analytical applications (which is to use a modulation amplitude that is a few times larger than the width of the absorption profile).

It is also possible to give an analytical description of the WM signal when the absorption profile is purely Lorentzian. In the description of the WM technique it has been found convenient to work with all frequency entities normalized with respect to the half-width at half-maximum of the atomic profile (i.e. $\bar{\nu}_d = \nu_d/\Delta\nu_p$ and $\bar{\nu}_a = \nu_a/\Delta\nu_p$, the latter often referred to as the modulation index, m). An analytical expression for the various nth harmonics of a peak-normalized Lorentzian-shaped line-shape function (expressed in normalized units) is given in Equation (71):

$$\bar{\chi}(t) = \frac{1}{1 + [\bar{\nu}_d + \bar{\nu}_a \sin(2\pi f_m t)]^2} \quad (71)$$

which is valid for arbitrary modulation amplitudes and has been derived previously by Arndt,[545] as shown in Equation (72):

$$\bar{\chi}_n(\bar{\nu}_d, \bar{\nu}_a) = \frac{2 - \delta_{n0}}{2} \frac{\left\{ \begin{array}{c} \sqrt{[(1 - i\bar{\nu}_d)^2 + \bar{\nu}_a^2]} \\ -(1 - i\bar{\nu}_d) \end{array} \right\}^n}{\bar{\nu}_a^n \sqrt{[(1 - i\bar{\nu}_d)^2 + \bar{\nu}_a^2]}} i^n + c.c. \quad (72)$$

where i is the imaginary unit and *c.c.* represents the complex conjugate.

Although the above expression is exact, it is not always suitable for direct use because it is cast in terms of complex numbers. Kluczynski and Axner showed recently that this expression can be written more conveniently as shown in Equation (73):

$$\bar{\chi}_n(\bar{\nu}_d, \bar{\nu}_a) = \frac{A_n}{(\bar{\nu}_a)^n} \left[B_n + \frac{C_n S_+ + D_n S_-}{\sqrt{2}R} \right] \quad (73)$$

where $S_+ = \sqrt{R + M}$, $S_- = \sqrt{R - M}$, $R = \sqrt{M^2 + 4(\bar{\nu}_d)^2}$, $M = 1 + (\bar{\nu}_a)^2 - (\bar{\nu}_d)^2$, and the four parameters A_n–D_n take different values (in terms of $\bar{\nu}_a$ and $\bar{\nu}_d$) depending on the order of the harmonics.[523] The components for the first five harmonics are given in Table 6.

Typical $1f, 2f, 4f$, and $6f$ WM spectra from a single-peak (Lorentzian-broadened) transition are exemplified in Figure 22. Each spectrum is evaluated for the modulation amplitude that maximizes that particular nf WM component (2.0, 2.2, 3.9 and 7.4, respectively; see section 6.3.3 below). As can be seen from the graph, the even harmonics have their maximum value at the position of the peak of the absorption profile, whereas the odd harmonics have it elsewhere, and the $2f$ signal gives rise to the largest signal of all even components.

6.3.3 Maximum Signal Conditions

The conditions for maximization of the WM signal have been scrutinized by a number of authors. Cassidy and Reid calculated relative peak heights for the 2nd, 4th, 6th and 8th harmonics of a pure Lorentzian profile.[467] Wilson used numerical integration to obtain the three first harmonics of both Gaussian and Lorentzian profiles.[546] Silver determined the optimum values of the modulation index (m) for both Gaussian and Lorentzian profiles for the 1st, 2nd, 4th and 6th harmonics.[42] He found the optimum modulation indices to be: 1.6 and 2.0; 2.1 and 2.2; 3.6 and 3.9; and 5.2 and 7.4, respectively. Rojas et al.[535] and Gustafsson et al.[536–539] investigated three different issues of the $2f$ WM technique: when atoms are detected under not necessarily optically thin conditions;[535,538,539] for atoms with complicated (multi-line) atomic structure (a special study was performed of the 780-nm transition in Rb, which constitutes two isotopes each with six individual close-lying transitions);[535–539] and the extent to which a low-pressure cell can be used as a wavelength reference for $2f$ WM experiments made under conditions of atmospheric pressure and high temperature (i.e. a GF).[536–538] The same authors also investigated the temperature dependence of the $2f$ WM technique.[539]

6.3.4 Signal-to-noise Ratios

The most extensive analysis of the S/N conditions for diode-laser-based modulation techniques has been given by Silver[42] and Bomse et al.[46] They discussed the concept of S/N for both WM and FM techniques and

Table 6 The A_n–D_n coefficients for the five lowest harmonics (i.e. $n = 0$–4) of the Fourier components of a Lorentzian absorption profile (Equation 73)

	$n=0$	$n=1$	$n=2$	$n=3$	$n=4$
A_n	1	−2	−2	2	2
B_n	0	0	−2	$-8\bar{\nu}_d$	$-4(1 + M - 5\bar{\nu}_d^2)$
C_n	1	$\bar{\nu}_d$	$1 + M - \bar{\nu}_d^2$	$[9 + 3M - \bar{\nu}_d^2]\bar{\nu}_d$	$(1 + 6M + M^2) - 2(17 + 3M)\bar{\nu}_d^2 + \bar{\nu}_d^4$
D_n	0	$\frac{-\bar{\nu}_d}{\lvert\bar{\nu}_d\rvert}$	$4\lvert\bar{\nu}_d\rvert$	$\frac{-(1 + 3M - 9\bar{\nu}_d^2)\bar{\nu}_d}{\lvert\bar{\nu}_d\rvert}$	$16(1 + M - \bar{\nu}_d^2)\lvert\bar{\nu}_d\rvert$

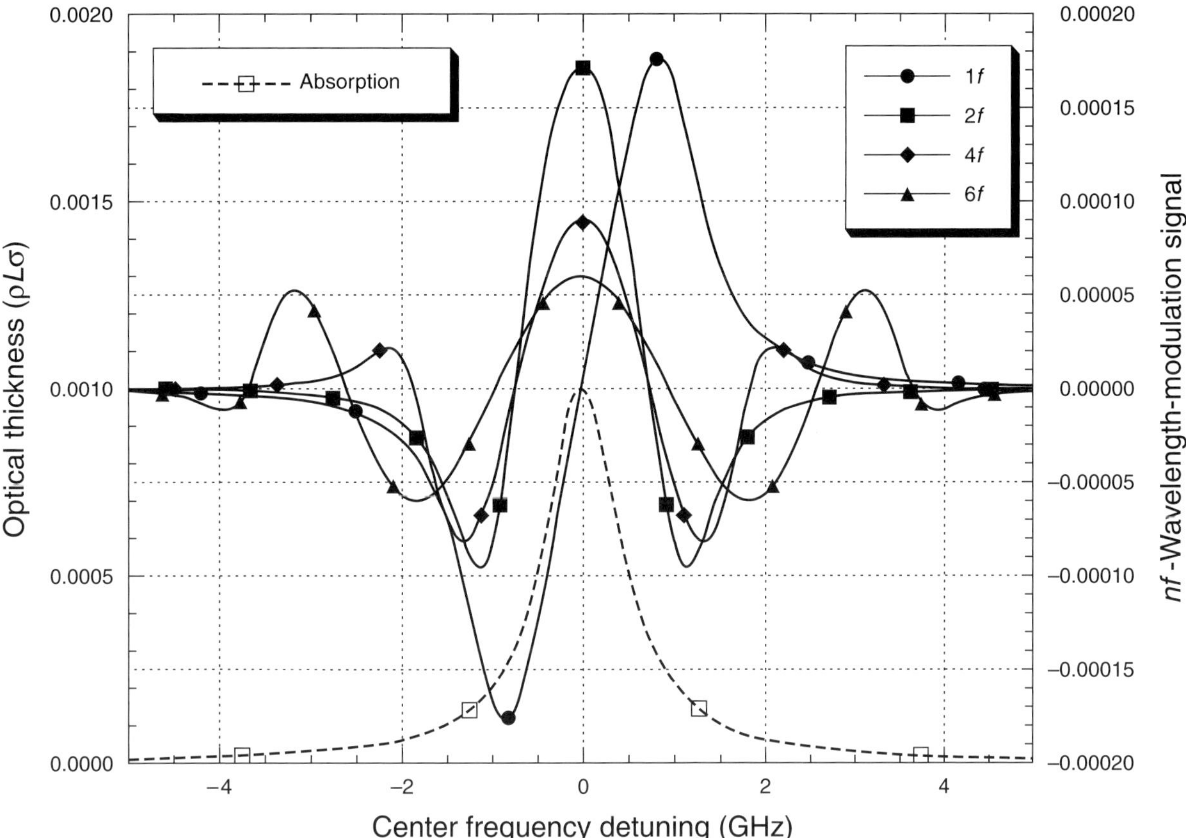

Figure 22 Schematic representation of a Lorentzian-shaped absorption profile (dashed line, left axis) and the corresponding 1*f*, 2*f*, 4*f* and 6*f* WM spectra (solid lines, right axis). The half-width at half-maximum of the absorption profile ($\Delta\nu_p$) has been taken as 0.5 GHz, for simplicity. The four WM spectra are evaluated for the FM amplitudes that maximize each *nf*-WM curve, which are 2.0, 2.2, 3.9, and 7.4 times $\Delta\nu_p$ for the four WM curves, respectively. The density of absorbers has been chosen to create an optical thickness of 10^{-3} at resonance.

their conclusions are based upon experiments made with a liquid-nitrogen-cooled infrared lead–salt diode laser, working in the 8-µm wavelength range, used for detection of N_2O in a multipass cell. Although the conditions are not typical of those when diode-laser-based modulation techniques are used for analytical atomic spectrometry, some of their conclusions are of importance.

They defined an intrinsic S/N (power) ratio as the ratio of the square of the time-averaged analytical signal and the sum of the squares of the noise contribution from internal sources of noise, i.e. laser-induced detector shot-noise, detector–preamplifier thermal noise, residual amplitude modulation (RAM)-induced noise, and laser excess noise. They analyzed their results, however, in terms of extrinsic sources of noise, in this case that originating from etalon background signals (see section 6.3.5).

In the absence of noise from RAM and background signals, the authors concluded that under their experimental conditions (a laser with 11 µW of output power) the WM signals ought to be dominated by laser excess noise at low modulation frequencies and by detector noise at higher frequencies. Shot-noise-dominated situations should, under optimum conditions (i.e. in the absence of RAM and background signals), be obtained for a combination of higher laser powers (in their case above 40 or 560 µW, depending on the detector used) and high modulation frequencies (in the megahertz range). At low or medium modulation frequencies the detectability of WM techniques should therefore (again under optimum conditions) be limited by laser excess noise (which is the dominating noise in the kilohertz range).[42]

6.3.5 *Background Signals*

In reality, there are background signals that can compete with the analytical signal when small enough quantities/concentrations of analyte are to be detected.

RAM (originating from an associated modulation of the intensity with the laser diode injection current) is a type of background signal frequently referred to. Although predicted by the theory presented by Silver,[42]

For references see page 9575

and used by a number of authors since then for describing their background signals, it was first shown by Zhu and Cassidy[547] and then discussed further by Kluczynski and Axner[523] that there is no *nf* RAM signal for $n \geq 2$ (including 2*f* RAM) as long as the modulation of the intensity is linear with that of the injection current. The incorrect prediction of a 2*f* RAM signal from a linear intensity modulation originates from an uncritical application of a theoretical formalism developed for FM spectrometry (in which the intensity modulation is described as a modulation of the electrical field strength of the laser light instead of the intensity) to the WM technique.

The dominating cause of background signals is instead multiple reflections between various optical surfaces (which thus act as etalons). Because the transmission through an etalon is strongly wavelength dependent, the detector will experience a wavelength-dependent modulation of the intensity that will contain the same higher harmonics as the analytical signal.[523] The drift and fluctuations of these background signals are often the limiting factors to the WM technique. Therefore, unless any special concerns are taken, the detection power of modulation techniques is often limited by the noise and drift in the background signal originating from multiple reflections in the system rather than any of the intrinsic sources of noise, giving rise to detection limits (fractional absorbencies) in the 10^{-5} range.[43,46,467,495,498,513,517,530,544,548]

6.3.6 Means to Reach Optimum Conditions

Remedies for reducing etalon effects comprise: eliminating the use of optical surfaces that are perpendicular to the direction of the laser beam; antireflection coating; replacing parallel plates with wedges; and modulating either the position of the surfaces[499,520,549–551] or the atomization at a frequency different from that modulating the laser wavelength.[498,550]

6.3.6.1 Dithering of Optical Surfaces Bomse et al.[46] concluded that with piezocrystal dithering of appropriate optical components the optimum conditions for 1*f* and 2*f* WM spectrometry comprise modulation of the frequency at a rate between 100 kHz (which were sufficient for bringing the 1/*f* noise down to levels below that of the background signals in their experiment) and 20 MHz (at which instrumentation difficulties arose). They also concluded that neither of the FM techniques investigated could outperform the high-frequency WM techniques (although their theoretical treatment predicted that the one-tone FM technique should provide an almost fivefold improvement over high-frequency WM spectrometry). They attributed this to instrumental deficiencies (low detector sensitivity and inefficient radiofrequency coupling of the modulation current to the laser). The authors concluded finally that, owing to the high sensitivity of the etalon background signals to small changes in optical alignment, the reproducibility of the final detection power of the WM techniques ($<10^{-6}$) depended strongly on the details and stability of the optical alignment and variations in the effectiveness of the etalon reduction scheme.

It is important to point out, however, that dithering the position of various surfaces has the drawback that it can only eliminate background signals from etalons created *between* two different optical components. Any possible etalon created *inside* a component (i.e. between the front and back surface of the same optical component) cannot be eliminated by that technique.[523] Such etalons can therefore set the ultimate level for the background signal in conventional WM spectrometry and thereby also the S/N level.[523]

6.3.6.2 Modulation of the Atomization An efficient way around these residual problems is to modulate also the atomization process and detect the signal at a combination of the ordinary *nf* frequency and that of the atomization. Although modulation of an optical surface is rather straightforward to achieve, modulation of the atomization process is more difficult in thermal atomizers such as flames and furnaces, although it can be used with advantage using various types of plasmas. The most impressive demonstration of a double-modulation technique that has been performed so far is the detection of Cl atoms in a modulated low-pressure microwave-induced plasma by Liger et al.[499] The authors showed that spurious etalon effects, background absorption, residual diode-laser-amplitude modulation and the noise that accompanies these entities could be suppressed by the use of a double-modulation technique in which both the diode laser wavelength and the plasma were modulated (although at different frequencies). This double-modulation technique enabled a detection limit of 1×10^{-6} AU to be obtained for Cl atoms in a modulated low-pressure microwave-induced plasma (with a time constant of 1 s).

6.3.6.3 Double-modulation, Dual-beam, and Logarithmic Detection In order to eliminate laser excess noise and signal variations due to changes/drifts of the optical transmittance, a double-beam arrangement with logarithmic subtraction of sample and reference detector currents was developed by the same set of authors.[499] As is exemplified in Figure 23, the logarithmic amplification not only makes the system independent of variations in the radiation intensity but it also enables suppression of the noise by decoupling the modulations of the intensity and wavelength. A detection limit of 2×10^{-7} AU limited by shot-noise could finally be established. The authors

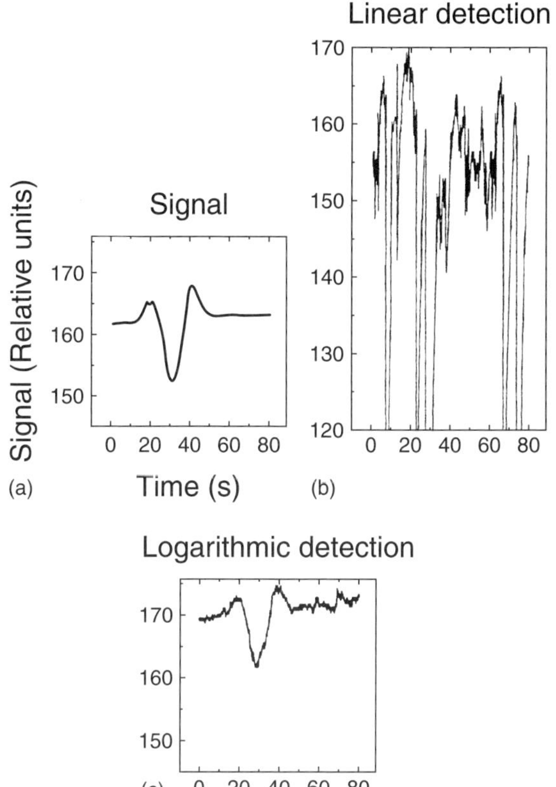

Figure 23 The $2f$ WM spectra from the 833-nm line in Cl measured in a directly coupled plasma: (a) single-beam absorption and linear amplification under optimum conditions; (b) the same instrumentation as in (a) but with turbulent smoke in the absorption path; (c) double-beam absorption with logarithmic conversion measured simultaneously with the curve displayed in (b). (Reproduced from V. Liger, A. Zybin, Y. Kuritsyn, K. Niemax, *Spectrochimica Acta*, **52B**, 1125–1138, Copyright 1997, with permission from Elsevier Science.)

conclude finally that preliminary experiments showed that shot-noise was the dominating source of noise up to time constants of about 1000 s in their experiments and that as a consequence it should be possible to reach detection limits as low as 6×10^{-9} AU. If such detection powers can be realized, WM techniques would surely find a much broader applicability in spectrochemical analysis than they have today.

6.4 Conclusions and Outlook

Frequency- and wavelength-modulation diode laser absorption spectrometry (FM/WMDLAS) has the great potential to outperform conventional AAS in the future. There is no doubt that the technique has better sensitivity, LDR and selectivity than the AAS technique. The most significant drawback of FM/WMDLAS today is the lack of diode lasers producing light in the visible and UV regions.

7 CONCLUSIONS

It is clear that laser light, with its narrow bandwidth and high spectral irradiance, is ideal for selective interactions with atomic and molecular species. Extraordinary results have been achieved in a number of application areas,[47–52] of which the cooling and trapping of single atoms or ions and the creation of Bose–Einstein condensates[552] in the fields of atomic and molecular spectroscopy and quantum optics are just a few impressive examples. Despite many impressive demonstrations in the field of analytical spectrometry,[21,48] laser-based spectrometric techniques so far have been techniques only for the physical or chemical laboratories around the world. The reasons for this are that: the production of laser light has not been as carefree as first anticipated; the cost of tunable laser systems has been too high; and laser-based techniques are basically single-element detection techniques.

With the advent of single-mode, tunable diode lasers (and maybe also new OPO-based systems), however, and the new application areas that these handy types of lasers now start to pervade, the prospects for a more regular use of laser-based spectroscopic techniques for analytical spectrometry seems to be better than ever before. If such techniques can be developed and commercialized in a proper manner in the future, then the regular analytical laboratory will be able to benefit from the extraordinary properties of laser-based spectroscopic techniques regarding sensitivity and selectivity (including various types of powerful means of background correction).

Even if laser-based spectroscopic techniques do not find their way onto the commercial market as regular techniques for trace species determinations, they will always have their role in certain types of applications. The nonintrusiveness of the techniques, for example, makes them attractive in certain situations, e.g. when samples with long-lived but toxic radioactive species are to be analyzed. In addition, as has been demonstrated repeatedly, their capabilities as diagnostic tools are indeed unmatched. Their applicability for remote-sensing applications is also unparalleled.

Looking back to what the joint community of laser spectroscopists has achieved during the last 20 years, it is difficult not to be impressed. Even in such a perspective, it is more difficult than ever to try to prophesy what the future will bring in this area: where will this field of science stand 20 years from now? Can we then, maybe, really count the atoms in an unknown sample?

ACKNOWLEDGMENTS

The author is indebted to all his colleagues in the Laser Physics Group at the Department of Experimental

For references see page 9575

Physics at Umeå University for their continuous support during the writing of this review article. The author would also like to thank the Swedish Natural Science Research Council, the Swedish Research Council for Engineering Sciences (under project nos 251-97-733 and 288-98-40) and the Faculty of Science and Technology at Umeå University for various types of economic support, without which this review would never have become a reality.

ABBREVIATIONS AND ACRONYMS

AAS	Atomic Absorption Spectrometry
ASE	Amplified Spontaneous Emission
CCD	Charge-coupled Device
DC	Direct Current
EIE	Easily Ionized Element
ETA	Electrothermal Atomizer
FI/LEI	Flow Injection/Laser-enhanced Ionization
FM	Frequency-modulation
FM/WMDLAS	Frequency- and Wavelength-modulation Diode Laser Absorption Spectrometry
GF	Graphite Furnace
HPLC	High-performance Liquid Chromatography
ICCD	Intensified Charge-coupled Device
ICP	Inductively Coupled Plasma
ICPMS	Inductively Coupled Plasma Mass Spectrometry
IUPAC	International Union of Pure and Applied Chemistry
LDR	Linear Dynamic Range
LEAFS	Laser-excited Atomic Fluorescence Spectrometry
LEI	Laser-enhanced Ionization
LEMOFS	Laser-excited Molecular Fluorescence Spectrometry
LIF	Laser-induced Fluorescence
LIF/LEAFS	Laser-induced Fluorescence/ Laser-excited Atomic Fluorescence Spectrometry
LOD	Limits of Detection
NBS	National Bureau of Standards
Nd:YAG	Neodymium:Yttrium Aluminum Garnet
NIST	National Institute of Standards and Technology
OPO	Optical Parametric Oscillator
PMT	Photomultiplier Tube
RAM	Residual Amplitude Modulation
RC	Resistance–Capacitance
RIMS	Resonance Ionization Mass Spectrometry
RIS	Resonance Ionization Spectrometry
RIS/RIMS	Resonance Ionization Spectrometry/Resonance Ionization Mass Spectrometry
RSD	Relative Standard Deviation
SAD	Single Atom Detection
S/N	Signal-to-noise
SONRES	Saturated Optical Nonresonant-emission Spectrometry
UV	Ultraviolet
WM	Wavelength-modulation
WM-DLAS	Wavelength-modulation Diode Laser Absorption Spectrometry
ZETA/LEAFS	Zeeman Electrothermal Atomization/Laser-excited Atomic Fluorescence Spectrometry

List of selected abbreviations appears in Volume 15

RELATED ARTICLES

Coatings **(Volume 2)**
Atomic Spectroscopy in Coatings Analysis

Environment: Trace Gas Monitoring **(Volume 3)**
Environmental Trace Species Monitoring: Introduction • Automotive Emissions Analysis with Spectroscopic Techniques • Differential Optical Absorption Spectroscopy, Air Monitoring by • Diode Laser Spectroscopic Monitoring of Trace Gases • Fourier Transform Infrared Spectrometry in Atmospheric and Trace Gas Analysis • Infrared LIDAR Applications in Atmospheric Monitoring • Laser Absorption Spectroscopy, Air Monitoring by Tunable Mid-infrared Diode • Laser-induced Breakdown Spectroscopy, Elemental Analysis • Laser Mass Spectrometry in Trace Analysis • Laser-based Combustion Diagnostics • Laser- and Optical-based Techniques for the Detection of Explosives • Optical Gas Sensors in Analytical Chemistry: Applications, Trends and General Comments • Photoacoustic Spectroscopy in Trace Gas Monitoring • Ultraviolet/Visible Light Detection and Ranging Applications in Air Monitoring

Environment: Water and Waste **(Volume 3)**
Atomic Fluorescence in Environmental Analysis • Flame and Graphite Furnace Atomic Absorption Spectrometry in Environmental Analysis

Environment: Water and Waste cont'd **(Volume 4)**
Laser Ablation Inductively Coupled Plasma Spectrometry in Environmental Analysis • X-ray Fluorescence Spectroscopic Analysis of Liquid Environmental Samples

Food **(Volume 5)**
Atomic Spectroscopy in Food Analysis • Fluorescence Spectroscopy in Food Analysis

Forensic Science **(Volume 5)**
Fluorescence in Forensic Science • X-ray Fluorescence in Forensic Science

Peptides and Proteins **(Volume 7)**
Fluorescence Spectroscopy in Peptide and Protein Analysis

Remote Sensing **(Volume 10)**
Remote Sensing: Introduction

Atomic Spectroscopy **(Volume 11)**
Background Correction Methods in Atomic Absorption Spectroscopy • Flame and Vapor Generation Atomic Absorption Spectrometry • Glow Discharge Optical Spectroscopy and Mass Spectrometry • Graphite Furnace Atomic Absorption Spectrometry • Inductively Coupled Plasma/Optical Emission Spectrometry • Laser Ablation in Atomic Spectroscopy • Laser-induced Breakdown Spectroscopy

Electronic Absorption and Luminescence **(Volume 12)**
Fluorescence Imaging Microscopy • Fluorescence Lifetime Measurements, Applications of

Infrared Spectroscopy **(Volume 12)**
Cavity Ringdown Laser Absorption Spectroscopy

Raman Spectroscopy **(Volume 15)**
Raman Spectroscopy: Introduction

X-ray Photoelectron Spectroscopy and Auger Electron Spectroscopy **(Volume 15)**
X-ray Photoelectron Spectroscopy and Auger Electron Spectroscopy: Introduction • X-ray Photoelectron and Auger Electron Spectroscopy

X-ray Spectrometry **(Volume 15)**
X-ray Techniques: Overview • Absorption Techniques in X-ray Spectrometry • Energy Dispersive, X-ray Fluorescence Analysis • Sample Preparation for X-ray Fluorescence Analysis • Total Reflection X-ray Fluorescence • Ultrafast Diffraction Techniques • Wavelength-dispersive X-ray Fluorescence Analysis

General Articles **(Volume 15)**
Ultrafast Laser Technology and Spectroscopy

REFERENCES

1. K.W. Jackson, S. Lu, 'Atomic Absorption, Atomic Emission, and Flame Emission Spectrometry', *Anal. Chem.*, **70**(12), 363R–383R (1998).
2. J. Marshall, E.H. Evans, A. Fisher, S. Chenery, 'Atomic Spectrometry Update – Atomic Emission Spectrometry', *J. Anal. At. Spectrom.*, **12**(6), 263R–290R (1997).
3. J. Marshall, S. Chenery, E.H. Evans, A. Fisher, 'Atomic Spectrometry Update – Atomic Emission Spectrometry', *J. Anal. At. Spectrom.*, **13**(6), 107R–128R (1998).
4. E.H. Evans, S. Chenery, A. Fisher, J. Marshall, K. Sutton, 'Atomic Spectrometry Update – Atomic Emission Spectrometry', *J. Anal. At. Spectrom.*, **14**(6), 977–1004 (1999).
5. S.J. Hill, J.B. Dawson, W.J. Price, I.L. Shuttler, C.M.M. Smith, J.F. Tyson, 'Atomic Spectrometry Update – Advances in Atomic Absorption and Fluorescence Spectrometry and Related Techniques', *J. Anal. At. Spectrom.*, **12**(8), R327–R379 (1997).
6. S.J. Hill, J.B. Dawson, W.J. Price, I.L. Shuttler, C.M.M. Smith, J.F. Tyson, 'Atomic Spectrometry Update – Advances in Atomic Absorption and Fluorescence Spectrometry and Related Techniques', *J. Anal. At. Spectrom.*, **13**(8), 131R–170R (1998).
7. S.J. Hill, J.B. Dawson, W.J. Price, I.L. Shuttler, C.M.M. Smith, J.F. Tyson, 'Atomic Spectrometry Update – Advances in Atomic Absorption and Fluorescence Spectrometry and Related Techniques', *J. Anal. At. Spectrom.*, **14**(8), 1245–1285 (1999).
8. T.H. Maiman, 'Stimulated Optical Radiation in Ruby', *Nature*, **187**, 493–494 (1960).
9. G.S. Hurst, M.H. Nayfeh, J.P. Young, 'A Demonstration of One-atom Detection', *Appl. Phys. Lett.*, **30**(5), 229–231 (1977).
10. G.S. Hurst, M.H. Nayfeh, J.P. Young, 'One-atom Detection Using Resonance Ionization Spectroscopy', *Phys. Rev. A*, **15**(6), 2283–2292 (1977).
11. G.S. Hurst, M.G. Payne, S.D. Kramer, J.P. Young, 'Resonance Ionization Spectrometry and One-atom Detection', *Rev. Mod. Phys.*, **51**, 767–819 (1979).
12. C.T.J. Alkemade, 'Atomic Physics and Atomic Spectroscopy: Mother and Daughter', *Spectrochim. Acta*, **38B**, 1395–1409 (1983).
13. A.L. Schawlow, 'Laser Spectroscopy of Atoms and Molecules', *Science*, **202**, 141–147 (1978).
14. W. Demtröder, *Laser Spectroscopy – Basic Concepts and Instrumentation*, 2nd edition, Springer-Verlag, Berlin, 1996.
15. L.J. Radziemski, R.W. Solarz, J.A. Paisner, *Laser Spectroscopy and its Applications*, Marcel Dekker, New York, 1987.

List of selected abbreviations appears in Volume 15

16. N. Omenetto (ed.), *Analytical Laser Spectroscopy*, Wiley, New York, 1979.
17. V.S. Letokhov (ed.), *Laser Analytical Spectrochemistry, Optics and Optoelectronics*, eds. E.R. Pike, W.T. Welford, Adam Hilger, Bristol, 1985.
18. V.S. Letokhov, *Laser Photoionization Spectroscopy*, Academic Press, Orlando, 1987.
19. G.S. Hurst, M.G. Payne, *Principles and Applications of Resonance Ionization Spectroscopy*, Adam Hilger, Bristol, 1988.
20. D.L. Andrews (ed.), *Applied Laser Spectroscopy*, VCH Publishers, New York, 1992.
21. J. Sneddon, T.L. Thiem, Y.-I. Lee (eds.), *Lasers in Analytical Atomic Spectroscopy*, VCH Publishers, New York, 1997.
22. L.A. Lyon, C.D. Keating, A.P. Fox, B.E. Baker, L. He, S.R. Nicewarner, S.P. Mulvaney, M.J. Natan, 'Raman Spectroscopy', *Anal. Chem.*, **70**(12), 341R–361R (1998).
23. A.G. Abroskin, T.V. Belyaeva, V.A. Filichkina, E.K. Ivanova, M.A. Proscurnin, V.M. Savostina, Y.A. Barbalat, 'Thermal Lensing Spectrometry in Trace Metal Analysis', *Analyst*, **117**, 1957–1962 (1992).
24. C. Moulin, N. Delorme, T. Berthoud, P. Mauchien, 'Double Beam Thermal Lens Spectroscopy for Actinides Detection and Speciation', *Radiochim. Acta*, **44/45**, 103–106 (1988).
25. C.D. Tran, 'Helium–Neon Laser Intracavity Photothermal Beam Deflection Spectrometry', *Anal. Chem.*, **58**, 1714–1716 (1986).
26. N. Omenetto, P. Cavalli, G. Rossi, G. Bidoglio, G.C. Turk, 'Thermal Lensing Spectrophotometry of Uranium(VI) with Pulsed Laser Excitation', *J. Anal. At. Spectrom.*, **2**, 579–583 (1987).
27. T. Berthoud, N. Delorme, 'Differential Dual-beam Thermal Lensing Spectrometry: Determination of Lanthanides', *Appl. Spectrosc.*, **41**(1), 15–19 (1987).
28. C.M. Phillips, S.R. Crouch, G.E. Leroi, 'Matrix Effects in Thermal Lensing Spectrometry: Determination of Phosphate in Saline Solutions', *Anal. Chem.*, **58**, 1710–1714 (1986).
29. T. Berthoud, P. Decambox, B. Kirsch, P. Mauchien, C. Moulin, 'Direct Uranium Trace Analysis in Plutonium Solutions by Time-resolved Laser-induced Spectrofluorometry', *Anal. Chem.*, **60**, 1296–1299 (1988).
30. T. Berthoud, P. Decambox, B. Kirsch, P. Mauchien, C. Moulin, 'Direct Determination of Traces of Lanthanide Ions in Aqueous Solutions by Laser-induced Time-resolved Spectrofluorimetry', *Anal. Chim. Acta*, **220**, 235–241 (1989).
31. P. Decambox, P. Mauchien, C. Moulin, 'Direct Trace Determination of Curium by Laser-induced Time-resolved Spectrofluorometry', *Radiochim. Acta*, **48**, 23–28 (1989).
32. G.P. Smith, M.J. Dyer, D.R. Crosley, 'Pulsed Laser Optoacoustic Detection of Flame Species', *Appl. Opt.*, **22**(24), 3995–4003 (1983).
33. W.G. Tong, J.M. Andrews, Z. Wu, 'Laser Spectrometry Based on Phase Conjugation by Resonant Degenerate Four-wave Mixing in an Analytical Flame', *Anal. Chem.*, **59**, 896–899 (1987).
34. K. Song, Y.-I. Lee, J. Sneddon, 'Applications of Laser-induced Breakdown Spectrometry', *Appl. Spectrosc. Rev.*, **32**, 183–235 (1997).
35. N. Omenetto, J.D. Winefordner, 'Atomic Fluorescence Spectrometry Basic Principles and Applications', *Prog. Anal. Atom. Spectrosc.*, **2**, 1–183 (1979).
36. J.D. Winefordner, 'Principles, Methodologies, and Applications of Atomic Fluorescence Spectrometry', *J. Chem. Educ.*, **55**(2), 72–78 (1978).
37. D.J. Butcher, J.P. Dougherty, F.R. Preli, A.P. Walton, G.-T. Wei, R.L. Irwin, R.G. Michel, 'Laser-excited Atomic Fluorescence Spectrometry in Flames, Plasmas and Electrothermal Atomisers, a Review', *J. Anal. At. Spectrom.*, **3**, 1059–1078 (1988).
38. X. Hou, S.-J.J. Tsai, J.X. Zhou, K.X. Yang, R.F. Lonardo, R.G. Michel, 'Laser-excited Atomic Fluorescence Spectrometry: Principles, Instrumentation and Applications', in *Lasers in Analytical Spectroscopy*, eds. J. Sneddon, T.L. Thiem, Y.-I. Lee, VCH Publishers, New York, 83–123, 1997.
39. J.C. Travis, G.C. Turk, R.B. Green, 'Laser-enhanced Ionization Spectrometry', *Anal. Chem.*, **54**, 1006A–1018A (1982).
40. J.C. Travis, G.C. Turk (eds.), *Laser-enhanced Ionization Spectrometry. Analytical Chemistry and its Application*, ed. J.D. Winefordner, John Wiley & Sons, New York, Vol. 136, 1996.
41. G.S. Hurst, 'Resonance Ionization Spectroscopy', *Anal. Chem.*, **53**(13), 1448A–1456A (1981).
42. J.A. Silver, 'Frequency-modulation Spectroscopy for Trace Species Detection: Theory and Comparison Among Experimental Methods', *Appl. Opt.*, **31**(6), 707–717 (1992).
43. K. Niemax, H. Groll, C. Schnürer-Patschan, 'Element Analysis by Diode Laser Spectroscopy', *Spectrochim. Acta Rev.*, **15**(5), 349–377 (1993).
44. J. Sneddon, Y.-I. Lee, X. Hou, J.X. Zhou, R.G. Michel, 'Lasers', in *Lasers in Analytical Spectroscopy*, eds. J. Sneddon, T.L. Thiem, Y.-I. Lee, VCH Publishers, New York, 41–81, 1997.
45. J. Franzke, A. Schnell, K. Niemax, 'Spectroscopic Properties of Commercial Laser Diodes', *Spectrochim. Acta Rev.*, **15**, 379–395 (1993).
46. D.S. Bomse, A.C. Stanton, J.A. Silver, 'Frequency Modulation and Wavelength Modulation Spectroscopies: Comparison of Experimental Methods Using

a Lead–Salt Diode Laser', *Appl. Opt.*, **31**(6), 718–731 (1992).

47. V.S. Letokhov, 'Laser-induced Chemistry – Basic Nonlinear Processes and Applications', *Appl. Phys. B*, **46**, 237–251 (1988).
48. N. Omenetto, 'The Impact of Several Atomic and Molecular Laser Spectroscopic Techniques for Chemical Analysis', *Appl. Phys. B*, **46**, 209–220 (1988).
49. J. Wolfrum, 'Laser Spectroscopy for Studying Chemical Processes', *Appl. Phys. B*, **46**, 221–236 (1988).
50. J.A. Paisner, 'Atomic Vapor Laser Isotope Separation', *Appl. Phys. B*, **46**, 253–260 (1988).
51. D. Bäuerle, 'Chemical Processing with Lasers: Recent Developments', *Appl. Phys. B*, **46**, 261–270 (1988).
52. S. Svanberg, *Atomic and Molecular Spectroscopy – Basic Aspects and Practical Applications*, 2nd edition, eds. G. Ecker, P. Lambropoulus, I.I. Sobelman, H. Walther, Springer-Verlag, Berlin, Vol. 6, 1992.
53. J.D. Winefordner, B.W. Smith, N. Omenetto, 'Theoretical Considerations of Laser Induced Fluorescence and Ionization Spectrometry: How Close to Single Atom Detection', *Spectrochim. Acta*, **44B**(12), 1397–1403 (1989).
54. J.D. Winefordner, C.L. Stevenson, 'Linking Principles with Absolute Detection Power in Atomic Spectrometry: How Far Can We Go?', *Spectrochim. Acta*, **48B**, 757–767 (1993).
55. J.D. Winefordner, G.A. Petrucci, C.L. Stevenson, B.W. Smith, 'Theoretical and Practical Limits in Atomic Spectroscopy Plenary Lecture', *J. Anal. At. Spectrom.*, **9**(3), 131–143 (1994).
56. G.N. Fowler, T.W. Preist, 'Ionization Cross-sections in Flames', *J. Chem. Phys.*, **56**, 1601–1605 (1972).
57. O. Axner, H. Rubinsztein-Dunlop, 'Fundamental Mechanisms of Laser-enhanced Ionization: The Production of Ions', in *Laser-enhanced Ionization Spectrometry*, eds. J. Travis, G. Turk, John Wiley & Sons, New York, 1–98, 1996.
58. R. Kelly, P.J. Padley, 'Measurement of Collisional Ionization Cross-sections for Metal Atoms in Flames', *Proc. R. Soc. London, A*, **327**, 345–366 (1972).
59. G. Zizak, J.D. Bradshaw, J.D. Winefordner, 'Rate Equation Solution for the Temporal Behavior of a Three-level System', *Appl. Opt.*, **19**(21), 3631–3639 (1980).
60. N. Omenetto, 'Analytical and Diagnostic Applications of Laser-induced Fluorescence in Flames and Plasmas', in *Analytical Laser Spectroscopy*, eds. S. Martellucci, A.N. Chester, Plenum Press, New York, 131–146, 1985.
61. N. Omenetto, C.A.v. Dijk, J.D. Winefordner, 'Some Considerations on the Saturation Parameter for 2- and 3-Level Systems in Laser-excited Fluorescence', *Spectrochim. Acta*, **37B**, 703–711 (1982).
62. C.T.J. Alkemade, 'Anomalous Saturation Curves in Laser-induced Fluorescence', *Spectrochim. Acta*, **40B**, 1331–1368 (1985).
63. D.R. de Olivares, G.M. Hieftje, 'Saturation of Energy Levels in Analytical Atomic Fluorescence Spectrometry – I. Theory', *Spectrochim. Acta*, **33B**, 79–99 (1978).
64. D.R. de Olivares, G.M. Hiefte, 'Saturation of Energy Levels in Analytical Atomic Fluorescence Spectrometry – II. Experimental', *Spectrochim. Acta*, **36B**(11), 1059–1079 (1981).
65. N. Omenetto, J. Bower, J. Bradshaw, C.A.v. Dijk, J.D. Winefordner, 'A Theoretical and Experimental Approach to Laser Saturation Broadening in Flames', *J. Quant. Spectrosc. Radiat. Transfer*, **24**, 147–158 (1980).
66. N. Omenetto, H.G.C. Human, 'Laser Excited Analytical Atomic and Ionic Fluorescence in Flame, Furnaces and Inductively Coupled Plasmas: I. – General Considerations', *Spectrochim. Acta*, **39B**, 1333–1343 (1984).
67. P.E. Walters, J. Lanauze, J.D. Winefordner, 'Spatially Resolved Concentration Studies of Ground State Atoms in a Flame: Saturated Absorption Spectroscopic Method', *Spectrochim. Acta*, **39B**(1), 125–129 (1984).
68. J.D. Winefordner, N. Omenetto, 'Laser-excited Atomic and Ionic Fluorescence in Flames and Plasmas', in *Analytical Applications of Lasers*, ed. E.H. Piepmeier, John Wiley & Sons, New York, 31–73, 1986.
69. P. Hannaford, 'Time-resolved Atomic Fluorescence in Flames', in *21st Colloquium Spectroscopicum Internationale and 8th International Conference on Atomic Physics*, Heyden, Cambridge, 250–261, 1979.
70. N. Omenetto, L.P. Hart, P. Benetti, J.D. Winefordner, 'On the Shape of Atomic Fluorescence Analytical Curves with a Laser Excitation Source', *Spectrochim. Acta*, **28B**, 301–307 (1973).
71. J.W. Daily, 'Saturation of Fluorescence in Flames with a Gaussian Laser Beam', *Appl. Opt.*, **17**(2), 225–229 (1978).
72. N. Omenetto, B.W. Smith, L.P. Hart, 'Laser Induced Fluorescence and Ionization Spectroscopy: Theoretical and Analytical Considerations for Pulsed Sources', *Fresenius' Z. Anal. Chem.*, **324**, 683–697 (1986).
73. O. Axner, P. Ljungberg, 'A Tutorial Review of the Rate-equation and Density-matrix Formalisms for Two- and Three-level Atomic and Molecular Systems in High Collisional Media Exposed to Pulsed Laser Light with Arbitrary Bandwidths', *Spectrochim. Acta Rev.*, **15**(4), 181–287 (1993).
74. O. Axner, T. Berglind, J.L. Heully, I. Lindgren, H. Rubinsztein-Dunlop, 'Improved Theory of Laser-enhanced Ionization in Flames: Comparison with Experiment', *J. Appl. Phys.*, **55**(9), 3215–3225 (1984).
75. M.A. Bolshov, A.V. Zybin, V.G. Koloshnikov, K.N. Koshelev, 'Some Characteristics of Laser-excited Atomic

Fluorescence in a Three-level Scheme', *Spectrochim. Acta*, **32B**, 279–286 (1977).

76. D. Kleppner, M.G. Littman, M.L. Zimmerman, 'Highly Excited Atoms', in *Sci. Am.*, **245**, 130–149 (1981).
77. A.W. Miziolek, R.J. Willis, 'Saturated Double-resonance Emission Spectroscopy of Lead for Sensitive Atomic Analysis', *Opt. Lett.*, **6**, 528–530 (1981).
78. C.A. van der Wijngaart, P. Kuik, H.A. Dijkerman, T. Hollander, C.T.J. Alkemade, 'Collisional Repopulation of Excited Na-states upon Laser-pumping in an H_2-O_2-Ar Flame', *J. Quant. Spectrosc. Radiat. Transfer*, **37**(3), 267–281 (1987).
79. J.A. Vera, C.L. Stevenson, B.W. Smith, N. Omenetto, J.D. Winefordner, 'Laser-excited Atomic Fluorescence Spectrometry Using Graphite Tube Electrothermal Atomization and Double-resonance Excitation', *J. Anal. At. Spectrom.*, **4**, 619–623 (1989).
80. P. Ljung, E. Nyström, J. Enger, P. Ljungberg, O. Axner, 'Detection of Titanium in Electrothermal Atomizers by Laser-induced Fluorescence: Part 1 – Determination of Optimum Excitation and Detection Wavelengths', *Spectrochim. Acta*, **52B**(6), 675–701 (1997).
81. N. Omenetto, T. Berthoud, P. Cavalli, G. Rossi, 'Lifetime Measurements of Metastable Levels of Thallium and Lead in the Air–Acetylene Flame by Laser-enhanced Ionization Spectrometry', *Appl. Spectrosc.*, **39**, 500–503 (1985).
82. O. Axner, P. Ljungberg, Y. Malmsten, 'Lifetime Measurements of Metastable States of Mg, Sr, Cd, Au, Pb and Bi in an Acetylene–Air Flame by Laser-enhanced Ionization', *Appl. Phys. B*, **54**, 144–155 (1992).
83. O. Axner, P. Ljungberg, Y. Malmsten, 'Monitoring of the Lifetimes of the Metastable States in Au in an Air–Acetylene Flame as a Probe for Local Stoichiometric Conditions by the Laser-enhanced Ionization Technique', *Appl. Phys. B*, **56**, 355–362 (1993).
84. O. Axner, P. Ljungberg, Y. Malmsten, 'Laser-based Spectroscopic Technique for Fast and Reliable Measurements of Lifetimes of Atomic Metastable States in Combustive Situations', *Appl. Opt.*, **32**(6), 899–906 (1993).
85. M.A. Bolshov, A.V. Zybin, L.A. Zybina, V.G. Koloshnikov, I.A. Majorov, 'The Use of a Dye-laser for the Detection of Sub-picogram Amounts of Lead and Iron by Atomic Fluorescence Spectrometry', *Spectrochim. Acta*, **31B**, 493–500 (1976).
86. N. Omenetto, M.S. Epstein, J.D. Bradshaw, S. Bayer, J.J. Horvath, J.D. Winefordner, 'Fluorescence Ratio of the Two D Sodium Lines in Flames for D_1 and D_2 Excitation', *J. Quant. Spectrosc. Radiat. Transfer*, **22**, 287–291 (1979).
87. C.T.J. Alkemade, T. Hollander, W. Snelleman, P.J.T. Zeegers, *Metal Vapors in Flames*, Pergamon Press, Oxford, 1982.
88. K.C. Smyth, P.K. Schenck, W.G. Mallard, 'What Really Does Happen to Electronically Excited Atoms in Flames?', in *Laser Probes of Chemistry*, Chapter 12, American Chemical Society, Washington, DC, 1980.
89. O. Axner, 'Determination of Optimal Conditions for Laser-enhanced Ionization Spectrometry in Flames. Part 1. One-step Signal Strength versus Excitation Transition', *Spectrochim. Acta*, **45B**, 561–579 (1990).
90. O. Axner, T. Berglind, 'Determination of Ionization Efficiencies of Excited Atoms in a Flame by Laser-enhanced Ionization', *Appl. Spectrosc.*, **43**, 940–952 (1989).
91. G.C. Turk, W.G. Mallard, P.K. Schenck, K.C. Smyth, 'Improved Sensitivity for Laser-enhanced Ionization Spectrometry in Flames by Stepwise Excitation', *Anal. Chem.*, **51**, 2408–2410 (1979).
92. J.R. Ackerhalt, J.H. Eberly, 'Coherence versus Incoherence in Stepwise Laser Excitation of Atoms and Molecules', *Phys. Rev. A*, **14**(5), 1705–1710 (1976).
93. R.M. Measures, P.G. Cardinal, 'Laser Ionization Based on Resonance Saturation – a Simple Model Description', *Phys. Rev. A*, **23**(2), 804–815 (1981).
94. N. Omenetto, B.W. Smith, L.P. Hart, P. Cavalli, G. Rossi, 'Laser-induced Double-resonance Ionic Fluorescence in an Inductively Coupled Plasma', *Spectrochim. Acta*, **40B**, 1411–1422 (1985).
95. O. Axner, M. Norberg, M. Persson, H.R. Dunlop, 'Reduction of Spectral Interferences from Na in Laser-enhanced Ionization Spectrometry by Laser-preionization', *Appl. Spectrosc.*, **44**, 1117–1123 (1990).
96. B.W. Smith, L.P. Hart, N. Omenetto, 'Measurement of the Laser-induced Ionization Yield for Lithium in an Air Acetylene Flame', *Anal. Chem.*, **58**, 2147–2151 (1986).
97. N. Omenetto, G.C. Turk, M. Rutledge, J.D. Winefordner, 'Atomic and Ionic Fluorescence Dip Spectroscopy as a Tool for Flame and Plasma Diagnostics', *Spectrochim. Acta*, **42B**(6), 807–817 (1987).
98. N. Omenetto, M. Leong, C. Stevensson, J. Vera, B.W. Smith, J.D. Winefordner, 'Fluorescence Dip Spectroscopy of Sodium Atoms in an Inductively Coupled Plasma', *Spectrochim. Acta*, **43B**, 1093–1100 (1988).
99. O. Axner, M. Norberg, H. Rubinsztein-Dunlop, 'Simultaneous Measurements of Laser-induced Fluorescence (LIF Dip) and Laser-enhanced Ionization (LEI Enhancement) in Flame – Investigation of Physical Properties', *Spectrochim. Acta*, **44B**, 693–712 (1989).
100. N. Omenetto, B.W. Smith, B.T. Jones, J.D. Winefordner, 'Considerations on the Simultaneous Behavior of the Resonance Fluorescence and Ionization Signals versus Laser Intensity in Flames', *Appl. Spectrosc.*, **43**, 595–598 (1989).
101. M. Sargent, III, P. Horwitz, 'Three-level Rabi Flopping', *Phys. Rev. A*, **13**(5), 1962–1964 (1976).
102. J.W. Daily, 'Coherent Optical Transient Spectroscopy in Flames', *Appl. Opt.*, **18**(3), 360–367 (1979).
103. J.W. Daily, 'Use of Rate Equations to Describe Laser Excitation in Flames', *Appl. Opt.*, **16**(8), 2322–2327 (1977).

104. R. Salomaa, S. Stenholm, 'Two-photon Spectroscopy: Effects of a Resonant Intermediate State', *J. Phys. B*, **8**(11), 1795–1805 (1975).
105. R. Salomaa, S. Stenholm, 'Two-photon Spectroscopy II. Effects of Residual Doppler Broadening', *J. Phys. B*, **9**(8), 1221–1235 (1976).
106. R. Salomaa, 'Effects of Finite Laser Bandwidths in Two-photon Spectroscopy', *J. Phys. B*, **11**, 3745–3755 (1978).
107. O. Axner, P. Ljungberg, 'A Density Matrix Treatment of Step-wise Laser Excitations of Atoms with Degenerate States in High Collisional Media', *J. Quant. Spectrosc. Radiat. Transfer*, **50**(3), 277–292 (1993).
108. P. Ljungberg, D. Boudreau, O. Axner, 'Full Temporal Density Matrix Treatment of Dual-wavelength, Arbitrary Bandwidth, Pulsed-laser Excitation of Atoms with Degenerate States in High Collisional Media', *Spectrochim. Acta*, **49B**(12–14), 1491–1505 (1994).
109. O. Axner, S. Sjöström, 'Experimental and Theoretical Investigation of Pulsed Step-wise Laser Excitations of Atoms in Flames – The Role of Two-step versus Two-photon Excitations and Dynamical Stark Effects', *Spectrochim. Acta*, **47B**(2), 245–273 (1992).
110. R.P. Lucht, D.W. Sweeney, N.M. Laurendeau, 'Balanced Cross-rate Model for Saturated Molecular Fluorescence in Flames Using a Nanosecond Pulse Length Laser', *Appl. Opt.*, **19**(19), 3295–3300 (1980).
111. R.P. Lucht, N.M. Laurendeau, 'Two-level Model for Near Saturated Fluorescence in Diatomic Molecules', *Appl. Opt.*, **18**(9), 856–861 (1979).
112. W.G. Mallard, J.H. Miller, K.C. Smyth, 'Resonantly Enhanced Two-photon Photoionization of NO in an Atmospheric Flame', *J. Chem. Phys.*, **76**, 3483–3492 (1982).
113. D. Boudreau, P. Ljungberg, O. Axner, 'DENSMAT: Fully Time Resolved Simulation of Two-step Pulsed Laser Excitation of Atoms in Highly Collisional Media', *Spectrochim. Acta*, **51B**, 413–428 (1996).
114. R. Salomaa, 'Two-photon Spectroscopy III. General Strong-field Aspects', *J. Phys. B*, **10**(15), 3005–3021 (1977).
115. J.W. Daily, 'Laser-induced Fluorescence Spectroscopy in Flames', in *Laser Probes for Combustion Chemistry, ACS Symp. Ser.*, **134**, 61–83, 1980.
116. J.D. Winefordner, T.J. Vickers, 'Atomic Fluorescence as a Means of Chemical Analysis', *Anal. Chem.*, **36**(1), 161–165 (1964).
117. J.D. Winefordner, R.A. Staab, 'Determination of Zinc, Cadmium, and Mercury by Atomic Fluorescence Flame Spectrometry', *Anal. Chem.*, **36**(1), 165–168 (1964).
118. L.M. Fraser, J.D. Winefordner, 'Laser-excited Atomic Fluorescence Flame Spectrometry', *Anal. Chem.*, **43**(12), 1693–1696 (1971).
119. L.M. Fraser, J.D. Winefordner, 'Laser-excited Atomic Fluorescence Flame Spectroscopy as an Analytical Method', *Anal. Chem.*, **44**(8), 1444–1451 (1972).
120. J. Kuhl, H. Spitschan, 'Flame-fluorescence Detection of Mg, Ni, and Pb with a Frequency-doubled Dye Laser as Excitation Source', *Opt. Comm.*, **7**(3), 256–259 (1973).
121. K. Dittrich, H.-J. Stark, 'Laser-excited Molecular Fluorescence Spectrometry for the Determination of Traces on Nonmetals. Part 1. Determination of Traces of Fluoride, Chloride and Bromide Based on Diatomic Molecules in a Graphite Furnace', *Anal. Chim. Acta*, **200**, 581–591 (1987).
122. G.M. Murray, S.J. Weeks, M.C. Edelson, 'Determination of Uranium Isotopes in a Complex Matrix by Optical Spectroscopy', *J. Alloys Compd.*, **181**, 57–62 (1992).
123. D.J. Butcher, J.P. Dougherty, J.T. McCaffrey, F.R. Preli, A.P. Walton, R.G. Michel, 'Conventional Source Excited Atomic Fluorescence Spectrometry', *Prog. Anal. Atom. Spectrosc.*, **10**, 359–506 (1987).
124. B.W. Smith, K.N. Spears, J.D. Winefordner, M.R. Glick, 'A Comprehensive Table of Atomic Fluorescence Detection Limits and Experimental Conditions', *Appl. Spectrosc.*, **43**(3), 376–414 (1989).
125. N. Omenetto, J.D. Winefordner, 'Atomic Fluorescence Spectroscopy with Laser Excitation', in *Analytical Laser Spectroscopy*, ed. N. Omenetto, Wiley, New York, Chapter 4, 1979.
126. N. Omenetto, J.D. Winefordner, 'Lasers in Analytical Spectroscopy', *Crit. Rev. Anal. Chem.*, **13**(1), 59–115 (1981).
127. N. Omenetto, 'Laser-induced Atomic Fluorescence Spectroscopy: A Personal Viewpoint on its Status, Needs, and Perspectives', *Spectrochim. Acta*, **44B**(2), 131–146 (1989).
128. N. Omenetto, 'Laser-induced Fluorescence in a Furnace: A Viable Approach to Absolute Analysis?', *Mikrochim. Acta*, **II**, 277–285 (1991).
129. N. Omenetto, B.W. Smith, J.D. Winefordner, 'Laser-induced Fluorescence and Electrothermal Atomization: How Far from the Intrinsic Level of Detection?', *Spectrochim. Acta*, **43B**, 1111–1118 (1988).
130. S. Sjöström, 'Laser-excited Atomic Fluorescence Spectrometry in Graphite Furnace Electrothermal Atomizer', *Spectrochim. Acta Rev.*, **13**, 407–465 (1990).
131. S. Sjöström, P. Mauchien, 'Laser Atomic Spectroscopic Techniques – The Analytical Performance for Trace Element Analysis of Solid and Liquid Samples', *Spectrochim. Acta Rev.*, **15**(3), 153–180 (1993).
132. H. Falk, 'Analytical Capabilities at Atomic Spectrometric Methods Using Tunable Lasers: A Theoretical Approach', *Prog. Anal. Atom. Spectrosc.*, **3**, 181–208 (1980).
133. H. Falk, 'Methods for the Detection of Single Atoms Using Optical and Mass-spectrometry', *J. Anal. At. Spectrom.*, **7**, 255–260 (1992).
134. N. Omenetto, 'Role of Lasers in Analytical Atomic Spectroscopy: Where, When and Why – Plenary Lecture', *J. Anal. At. Spectrom.*, **13**(5), 385–399 (1998).

List of selected abbreviations appears in Volume 15

135. C.L. Stevenson, J.D. Winefordner, 'Estimating Detection Limits in Ultra Trace Analysis. 3. Monitoring Atoms and Molecules with Laser-induced Fluorescence', *Appl. Spectrosc.*, **46**(5), 715–724 (1992).
136. N. Omenetto, N.N. Hatch, L.M. Fraser, J.D. Winefordner, 'Laser-excited Atomic Fluorescence of Some Transition Elements in the Nitrous Oxide–Acetylene Flame', *Anal. Chem.*, **45**(1), 65–78 (1973).
137. N. Omenetto, P. Benetti, L.P. Hart, J.D. Winefordner, C.T.J. Alkemade, 'Nonlinear Optical Behavior in Atomic Fluorescence Flame Spectrometry', *Spectrochim. Acta*, **28B**, 289–300 (1973).
138. N. Omenetto, N.N. Hatch, L.M. Fraser, J.D. Winefordner, 'Laser-excited Atomic and Ionic Fluorescence of the Rare Earths in the Nitrous Oxide–Acetylene Flame', *Anal. Chem.*, **45**(1), 195–197 (1973).
139. S.J. Weeks, H. Haraguchi, J.D. Winefordner, 'Laser-excited Molecular Fluorescence of CaOH in an Air–Acetylene Flame', *J. Quant. Spectrosc. Radiat. Transfer*, **19**, 633–640 (1978).
140. K. Fujiwara, N. Omenetto, J.B. Bradshaw, J.N. Bower, S. Nikdel, J.D. Winefordner, 'Laser-induced Molecular Background Fluorescence in Flames', *Spectrochim. Acta*, **34B**, 317–329 (1979).
141. M.E. Blackburn, J.-M. Mermet, J.D. Winefordner, 'CW-Laser-excited Molecular Fluorescence of Species in Flames', *Spectrochim. Acta*, **34A**, 847–852 (1978).
142. R.B. Green, J.C. Travis, R.A. Keller, 'Resonance Flame Atomic Fluorescence Spectrometry with Continuous Wave Dye Laser', *Anal. Chem.*, **48**(13), 1954–1959 (1976).
143. D.A. Goff, E.S. Yeung, 'Atomic Fluorescence Spectrometry with a Wavelength-modulated Continuous Wave Dye Laser', *Anal. Chem.*, **50**(4), 625–627 (1978).
144. B.W. Smith, M.B. Blackburn, J.D. Winefordner, 'Atomic Fluorescence Flame Spectroscopy with a Continuous Wave Dye Laser', *Can. J. Spectrosc.*, **22**, 57–61 (1977).
145. S.J. Weeks, H. Haraguchi, J.D. Winefordner, 'Improvement of Detection Limits in Laser-excited Atomic Fluorescence Flame Spectrometry', *Anal. Chem.*, **50**, 360–368 (1978).
146. C.A.v. Dijk, 'Excitation Profiles of the Sodium D-doublet in an H_2–O_2–Ar Flame', *Opt. Commun.*, **22**(3), 343–345 (1977).
147. R.A. van Calcar, M.J. van de Ven, B.K. van Uitert, K.J. Biewenga, T. Hollander, C.H. Alkemade, 'Saturation of Sodium Fluorescence in a Flame Irradiated with a Pulsed Tunable Dye Laser', *J. Quant. Spectrosc. Radiat. Transfer*, **21**, 11–18 (1979).
148. J.J. Horvath, J.D. Bradshaw, J.N. Bower, M.S. Epstein, J.D. Winefordner, 'Comparison of Nebulizer–Burner Systems for Laser-excited Atomic Fluorescence Flame Spectrometry', *Anal. Chem.*, **53**, 6–9 (1981).
149. M.S. Epstein, S. Bayer, J. Bradshaw, E. Voigtman, J.D. Winefordner, 'Application of Laser-excited Fluorescence Spectrometry to the Determination of Iron', *Spectrochim. Acta*, **35B**, 233–237 (1980).
150. M.S. Epstein, S. Bayer, J. Bradshaw, E. Voigtman, J.D. Winefordner, 'Application of Laser-excited Fluorescence Spectrometry to the Determination of Nickel and Tin', *Appl. Spectrosc.*, **34**, 372–376 (1980).
151. F.E. Hovis, J.A. Gelbwachs, 'Determination of Barium at Trace Levels by Laser-induced Ionic Fluorescence Spectrometry', *Anal. Chem.*, **56**, 1392–1394 (1984).
152. S.V. Kachin, B.W. Smith, J.D. Winefordner, 'Laser-excited Atomic Fluorescence of some Precious Metals in the Air/Acetylene Flame', *Appl. Spectrosc.*, **39**(4), 587–590 (1985).
153. M.D. Seltzer, M.S. Hendrick, R.G. Michel, 'Photomultiplier Gating for Improved Detection in Laser-excited Atomic Fluorescence Spectrometry', *Anal. Chem.*, **57**, 1096–1100 (1985).
154. J.X. Zhou, X.D. Hou, S.J.J. Tsai, K.X. Yang, R.G. Michel, 'Characterization of a Tunable Optical Parametric Oscillator Laser System for Multielement Flame Laser-excited Atomic Fluorescence Spectrometry of Cobalt, Copper, Lead, Manganese, and Thallium in Buffalo River Sediment', *Anal. Chem.*, **69**(3), 490–499 (1997).
155. N. Omenetto, H.G.C. Human, P. Cavalli, G. Rossi, 'Direct Determination of Lead in Blood by Laser-excited Flame Atomic Fluorescence Spectrometry', *Analyst*, **109**, 1067–1070 (1984).
156. A.P. Walton, G.-T. Wei, Z. Liang, R.G. Michel, 'Laser-excited Atomic Fluorescence in a Flame as a High Sensitivity Detector for Organomanganese and Organotin Compounds Following Separation by High-performance Liquid Chromatography', *Anal. Chem.*, **63**, 232–240 (1991).
157. H.G.C. Human, N. Omenetto, P. Cavalli, G. Rossi, 'Laser-excited Analytical Atomic and Ionic Fluorescence in Flames, Furnaces and Inductively Coupled Plasmas – II. Fluorescence Characteristics and Detection Limits for Fourteen Elements', *Spectrochim. Acta*, **39B**, 1345–1363 (1984).
158. G.A. Petrucci, H. Beisler, O. Matveev, P. Cavalli, N. Omenetto, 'Analytical and Spectroscopic Characterization of Double-resonance Laser-induced Fluorescence of Gold Atoms in a Graphite Furnace and in a Flame', *J. Anal. At. Spectrom.*, **10**(10), 885–890 (1995).
159. J.P. Dougherty, F.R. Preli, R.G. Michel, 'Laser-excited Atomic Fluorescence Spectrometry in an Atomic Absorption Graphite Tube Furnace', *J. Anal. At. Spectrom.*, **2**, 429–434 (1987).
160. F.R. Preli, J.P. Dougherty, R.G. Michel, 'Laser-excited Atomic Fluorescence Spectrometry with a Laboratory Constructed Tube Electrothermal Atomizer', *Anal. Chem.*, **59**, 1784–1789 (1987).
161. J.P. Dougherty, F.R. Preli, Jr, J.T. McCaffrey, M.D. Seltzer, R.G. Michel, 'Instrumentation for Zeeman

Electrothermal Atomizer Laser-excited Fluorescence Spectrometry', *Anal. Chem.*, **59**, 1112–1119 (1987).

162. J.P. Dougherty, J.A. Costello, R.G. Michel, 'Determination of Tl in Bovine Liver and Mouse Brains by ETA LEAFS', *Anal. Chem.*, **60**, 336–340 (1988).
163. F.R. Preli, J.P. Dougherty, R.G. Michel, 'Diagnostic Studies of the Zeeman Effect for Laser-excited Atomic Fluorescence Spectrometry in an Electrothermal Atomizer', *Spectrochim. Acta*, **43B**, 501–517 (1988).
164. J.P. Dougherty, F.R. Preli, G.-T. Wei, R.G. Michel, 'Nonlinearity of Calibration Graphs for Laser-excited Atomic Fluorescence in Graphite-tube Atomizers', *Appl. Spectrosc.*, **44**(6), 934–944 (1990).
165. V.M. Apatin, B.V. Arkhangel'skii, M.A. Bol'shov, V.V. Ermolov, V.G. Koloshnikov, O.N. Kompanetz, N.I. Kuznetsov, E.L. Mikhailov, V.S. Shishkovskii, 'Automated Laser-excited Atomic Fluorescence Spectrometer for Determination of Trace Concentration of Elements', *Spectrochim. Acta*, **44B**(3), 253–262 (1989).
166. M.A. Bolshov, V.G. Kolosnikov, S.N. Rudnev, C.F. Boutron, U. Görlach, C.C. Patterson, 'Detection of Trace Amounts of Toxic Metals in Environmental Samples by Laser-excited Atomic Fluorescence Spectrometry', *J. Anal. At. Spectrom.*, **7**, 99–104 (1992).
167. C.L. Boutron, M.A. Bolshov, V.G. Koloshnikov, C.C. Patterson, N.I. Barkov, 'Direct Determination of Lead in Vostok Antarctic Ancient Ice by Laser-excited Atomic Fluorescence Spectrometry', *Atmos. Environ.*, **24A**(7), 1797–1800 (1990).
168. M.A. Bolshov, C.F. Boutron, A.V. Zybin, 'Determination of Lead in Antarctic Ice at the Picogram-per-gram Level by Laser Atomic Fluorescence Spectrometry', *Anal. Chem.*, **61**, 1758–1762 (1989).
169. M.A. Bolshov, C.F. Boutron, F.M. Ducroz, U. Görlach, O.N. Kompanetz, S.N. Rudniev, B. Hutch, 'Direct Ultratrace Determination of Cadmium in Antarctic and Greenland Snow and Ice by Laser Atomic Fluorescence Spectrometry', *Anal. Chim. Acta*, **251**, 169–175 (1991).
170. M.A. Bolshov, S.N. Rudnev, J. Brust, 'Analytical Characterization of Laser-excited Atomic Fluorescence of Bismuth', *Spectrochim. Acta*, **49B**(12–14), 1437–1444 (1994).
171. J.P. Dougherty, F.R. Preli, R.G. Michel, 'Laser-excited Atomic-fluorescence Spectrometry in an Electrothermal Atomizer with Zeeman Background Correction', *Talanta*, **36**(1/2), 151–159 (1989).
172. Z. Liang, G.-T. Wei, R.L. Irwin, A.P. Walton, R.G. Michel, J. Sneddon, 'Determination of Subnanogram per Cubic Meter Concentrations of Metals in the Air of a Trace Metal Clean Room by Impaction Graphite Furnace Atomic Absorption and Laser-excited Atomic Fluorescence Spectrometry', *Anal. Chem.*, **62**, 1452–1457 (1990).
173. O. Axner, H. Rubinsztein-Dunlop, 'Detection of Trace Amounts of Cr by Two Laser-based Spectroscopic Techniques: Laser-enhanced Ionization in Flames and Laser-induced Fluorescence in Graphite Furnace', *Appl. Opt.*, **32**(6), 867–884 (1993).
174. G.-T. Wei, J.P. Dougherty, F.R. Preli, R.G. Michel, 'Signal and Noise Considerations of Nondispersive Laser-excited Atomic Fluorescence in a Graphite Tube Atomizer with Front-surface Illumination', *J. Anal. At. Spectrom.*, **5**, 249–259 (1990).
175. D.J. Butcher, R.L. Irwin, J. Takahashi, G. Su, G.-T. Wei, R.G. Michel, 'Determination of Thallium, Manganese, Lead in Food, Agricultural Standard Reference Materials by Electrothermal Atomizer Laser-excited Atomic Fluorescence, Atomic Absorption Spectrometry with Slurry Sampling', *Appl. Spectrosc.*, **44**(9), 1521–1533 (1990).
176. N. Omenetto, P. Cavalli, M. Broglia, P. Qi, G. Rossi, 'Laser-induced Single Resonance and Double Resonance Atomic Fluorescence Spectrometry in a Graphite Tube Atomizer', *J. Anal. At. Spectrom.*, **3**, 231–235 (1988).
177. D.J. Butcher, R.L. Irwin, J. Takahashi, R.G. Michel, 'Determination of Fluorine in Urine and Tap Water by Laser-excited Molecular Fluorescence Spectrometry in a Graphite Tube Furnace with Front-surface Illumination', *J. Anal. At. Spectrom.*, **6**, 9–18 (1991).
178. W. Resto, R.G. Badini, B.W. Smith, C.L. Stevenson, J.D. Winefordner, 'Two-step Laser-excited Atomic Fluorescence Spectrometry Determination of Mercury', *Spectrochim. Acta*, **48B**(5), 627–632 (1993).
179. E. Masera, P. Mauchien, Y. Lerat, 'Electrothermal Atomization – Laser-induced Fluorescence Determination of Iridium, Rhodium, Palladium, Platinum and Gold at the ng/l Level in Pure Water', *Spectrochim. Acta*, **51B**, 543–548 (1996).
180. N. Chekalin, A. Marunkov, O. Axner, 'Laser-induced Fluorescence in Graphite Furnaces under Low Pressure Conditions as a Powerful Technique for Studies of Atomization Mechanisms: Investigation of Ag', *Spectrochim. Acta*, **49B**(12–14), 1411–1435 (1994).
181. O. Axner, N. Chekalin, P. Ljungberg, Y. Malmsten, 'Direct Determination of Thallium in Natural Waters by Laser-induced Fluorescence in a Graphite Furnace', *Int. J. Environ. Anal. Chem.*, **53**, 185–193 (1993).
182. R.L. Irwin, D.J. Butcher, J. Takahashi, G.-T. Wei, R.G. Michel, 'Determination of Thallium and Lead in Nickel-based Alloys by Direct Solid Sampling with Graphite Furnace Laser-excited Atomic Fluorescence', *J. Anal. At. Spectrom.*, **5**, 603–610 (1990).
183. Z. Liang, R.F. Lonardo, R.G. Michel, 'Determination of Tellurium and Antimony in Nickel Alloys by Laser-excited Atomic Fluorescence Spectrometry in a Graphite Furnace', *Spectrochim. Acta*, **48B**(1), 7–23 (1993).
184. E. Masera, P. Mauchien, B. Remy, Y. Lerat, 'Characterization and Reduction of Silver Matrix Induced Effects in the Determination of Gold, Iridium, Palladium, Platinum

and Rhodium by Graphite Furnace Laser-induced Fluorescence Spectrometry', *J. Anal. At. Spectrom.*, **11**(3), 213–223 (1996).

185. E. Masera, P. Mauchien, Y. Lerat, 'Silver Matrix Effects on Gold Atomization in a Graphite-furnace Investigated by 2-Dimensional Laser Imaging with a Gated Charge-coupled-device Camera', *J. Anal. At. Spectrom.*, **10**(2), 137–144 (1995).
186. J. Enger, Y. Malmsten, P. Ljungberg, O. Axner, 'Laser-induced Fluorescence in a Graphite Furnace as a Sensitive Technique for Assessment of Traces in North Arctic Atmospheric Aerosol Samples', *Analyst*, **120**, 635–641 (1995).
187. A. Marunkov, N. Chekalin, J. Enger, O. Axner, 'Detection of Trace Amounts of Ni by Laser-induced Fluorescence in Graphite Furnace with ICCD Detection', *Spectrochim. Acta*, **49B**(12–14), 1385–1410 (1994).
188. P. Ljung, E. Nyström, O. Axner, W. Frech, 'Detection of Titanium in Electrothermal Atomizers by Laser-induced Fluorescence: Part 2 – Investigation of Various Types of Atomizers', *Spectrochim. Acta*, **52B**(6), 703–716 (1997).
189. J. Enger, A. Marunkov, N. Chekalin, O. Axner, 'Direct Detection of Sb in Environmental and Biological Samples at pg/ml Concentrations by Laser-induced Fluorescence in Graphite Furnace with Intensified Charge-coupled Device', *J. Anal. At. Spectrom.*, **10**, 539–549 (1995).
190. U. Heitmann, T. Sy, A. Hese, G. Schoknecht, 'High-sensitivity Detection of Selenium and Arsenic by Laser-excited Atomic Fluorescence Spectrometry Using Electrothermal Atomization', *J. Anal. At. Spectrom.*, **9**, 437–442 (1994).
191. A.I. Yuzefovsky, R.F. Lonardo, M. Wang, R.G. Michel, 'Determination of Ultra-trace Amounts of Cobalt in Ocean Water by Laser-excited Atomic Fluorescence Spectrometry in a Graphite Electrothermal Atomizer with Semi On-line Flow Injection Preconcentration', *J. Anal. At. Spectrom.*, **9**, 1195–1202 (1994).
192. A.I. Yuzefovsky, R.G. Michel, 'Role of Barium Chemical Modifier in the Determination of Fluoride by Laser-excited Molecular Fluorescence of Magnesium Fluoride in a Graphite Tube Furnace', *J. Anal. At. Spectrom.*, **9**, 1203–1207 (1994).
193. R.L. Irwin, G.T. Wei, D.J. Butcher, Z. Liang, E.G. Su, J. Takahashi, A.P. Walton, R.G. Michel, 'Transverse Zeeman Background Correction for Graphite-furnace Laser-excited Atomic Fluorescence Spectrometry – Determination of Lead and Cobalt in Standard Reference Materials', *Spectrochim. Acta*, **47B**(13), 1497–1515 (1992).
194. S. Sjöström, O. Axner, M. Norberg, 'Detection of Vanadium by Laser-excited Atomic Fluorescence Spectrometry in a Side-heated Graphite Furnace', *J. Anal. At. Spectrom.*, **8**(2), 375–378 (1993).
195. Z.W. Liang, R.F. Lonardo, J. Takahashi, R.G. Michel, F.R. Preli, 'Laser-excited Fluorescence Spectrometry of Phosphorus Monoxide and Phosphorus in an Electrothermal Atomizer – Determination of Phosphorus in Plant and Biological Reference Materials and Nickel-alloys', *J. Anal. At. Spectrom.*, **7**(6), 1019–1028 (1992).
196. B.W. Smith, P.B. Farnsworth, P. Cavalli, N. Omenetto, 'Optimization of Laser-excited Atomic Fluorescence in a Graphite Furnace for the Determination of Thallium', *Spectrochim. Acta*, **45B**(12), 1369–1373 (1990).
197. S. Sjöström, 'Multi-channel Background Correction Technique for Pulsed Laser-excited Atomic Fluorescence with a Graphite Furnace', *J. Anal. At. Spectrom.*, **5**, 261–267 (1990).
198. M.A. Bolshov, S.N. Rudnev, J.-P. Candelone, C.F. Boutron, S. Hong, 'Ultratrace Determination of Bi in Greenland Snow by Laser-excited Atomic Fluorescence Spectrometry', *Spectrochim. Acta*, **49B**(12–14), 1445–1452 (1994).
199. R.F. Lonardo, A.I. Yuzefovsky, K.X. Yang, R.G. Michel, E.S. Frame, J. Barren, 'Electrothermal Atomizer Laser-excited Atomic Fluorescence Spectrometry for the Determination of Phosphorus in Polymers by Direct Solid Analysis and Dissolution', *J. Anal. At. Spectrom.*, **11**(4), 279–285 (1996).
200. M. Leong, J. Vera, B.W. Smith, N. Omenetto, J.D. Winefordner, 'Laser-induced Double Resonance Fluorescence of Lead with Graphite Tube Atomization', *Anal. Chem.*, **60**, 1605–1610 (1988).
201. B. Remy, I. Verhaeghe, P. Mauchien, 'Real Sample Analysis by ETA/LEAFS with Background Correction: Application to Gold Determination in River Water', *Appl. Spectrosc.*, **44**, 1633–1638 (1990).
202. J.A. Vera, M.B. Leong, C.L. Stevenson, G. Petrucci, J.D. Winefordner, 'Laser-excited Atomic-fluorescence Spectrometry with Electrothermal Tube Atomization', *Talanta*, **36**(12), 1291–1293 (1989).
203. M.A. Bolshov, A.Z. Zybin, I.I. Smirenkina, 'Atomic Fluorescence Spectrometry with Laser Excitation', *Spectrochim. Acta*, **36B**, 1143–1152 (1981).
204. M.A. Bolshov, A.V. Zybin, V.G. Koloshnikov, I.I. Smirenka, 'Analytical Applications of a LEAFS ETA Method', *Spectrochim. Acta*, **43B**, 519–528 (1988).
205. L.K. Denisov, A.F. Loshin, N.A. Kozlov, V.G. Nikiforov, 'Analysis of Na and Ba in an Atomic Fluorescence Method with Excitation by a Pulsed Dye Laser with Lamp Pumping', *Zh. Prikl. Spektrosk.*, **43**(4), 1092–1096 (1985).
206. M.A. Bolshov, A.V. Zybin, V.G. Koloshnikov, A.V. Pisarskii, A.N. Smirnov, 'Atom-fluorescence Analysis of Pt, Ir, Eu under Excitation by Pulsed Dye Lasers', *Zh. Prikl. Spektrosk.*, **28**, 45–49 (1978).
207. M.A. Bolshov, A.V. Zybin, V.G. Koloshnikov, M.V. Vasnetsov, 'Detection of Extremely Low Lead Concentrations by Laser Atomic Fluorescence Spectrometry', *Spectrochim. Acta*, **36B**(4), 345–350 (1981).

208. M.A. Bolshov, A.V. Zybin, V.G. Koloshnikov, I.A. Mayorov, I.I. Smirenkina, 'Laser-excited Fluorescence Analysis with Electrothermal Sample Atomization in Vacuum', *Spectrochim. Acta*, **41B**, 487–492 (1986).
209. J.A. Vera, M.B. Leong, N. Omenetto, B.W. Smith, B. Womack, J.D. Winefordner, 'Evaluation of Three Different Laser Systems for the Determination of Lead by Laser-excited Atomic Fluorescence Spectrometry with Graphite Tube Atomizers', *Spectrochim. Acta*, **44B**(10), 939–948 (1989).
210. V. Cheam, J. Lechner, I. Sekerka, J. Nriagu, G. Lawson, 'Development of a Laser-excited Atomic Fluorescence Spectrometer and a Method for the Direct Determination of Lead in Great Lakes Waters', *Anal. Chim. Acta*, **269**, 129–136 (1992).
211. V. Cheam, J. Lechner, I. Sekerka, R. Desrosiers, 'Direct Determination of Lead in Seawaters by Laser-excited Atomic Fluorescence Spectrometry', *J. Anal. At. Spectrom.*, **9**, 315–320 (1994).
212. V. Cheam, J. Lechner, R. Desrosiers, I. Sekerka, 'Direct Determination of Dissolved and Total Thallium in Lake Waters by Laser-excited Atomic Fluorescence Spectrometry', *Int. J. Environ. Anal. Chem.*, **63**(2), 153–165 (1996).
213. J.B. Womack, C.A. Ricard, B.W. Smith, J.D. Winefordner, 'Evaluation of a Continuous-flow Furnace Atomizer for Laser-excited Atomic Fluorescence Spectrometry', *Spectrosc. Lett.*, **22**(10), 1333–1345 (1989).
214. H. Falk, J. Tilch, 'Atomization Efficiency and Over-all Performance of Electrothermal Atomizers in Atomic Absorption, Furnace Atomization Non Thermal Excitation and Laser-excited Atomic Fluorescence Spectroscopy', *J. Anal. At. Spectrom.*, **2**, 527–531 (1987).
215. H. Falk, H.-J. Paetzold, K.P. Schmidt, J. Tilch, 'Analytical Application of Laser-excited Atomic Fluorescence Using a Graphite Cup Atomizer', *Spectrochim. Acta*, **43B**, 1101–1109 (1988).
216. S. Neumann, M. Kriese, 'Sub-picogram Detection of Lead by Nonflame Atomic Fluorescence Spectrometry with Dye Laser Excitation', *Spectrochim. Acta*, **29**, 127–137 (1974).
217. J.P. Hohimer, P.J. Hargis, Jr, 'Automatic Fluorescence Spectrometry of Thallium with a Frequency-doubled Dye-laser and Vitreous Carbon Atomizer', *Anal. Chim. Acta*, **97**, 43–49 (1978).
218. K. Dittrich, H.-J. Stärk, 'Laser-excited Atomic Fluorescence Spectrometry as a Practical Analytical Method: Part 1 Design of a Graphite Tube Atomizer for the Determination of Trace Amounts of Lead', *J. Anal. At. Spectrom.*, **1**, 237–241 (1986).
219. J.P. Hohimer, P.I. Hargis, Jr, 'Picogram Detection of Cesium in Aqueous Solution by Nonflame Atomic Fluorescence Spectroscopy with Dye Laser Excitation', *Appl. Phys. Lett.*, **30**(7), 344–346 (1977).
220. K. Dittrich, H.-J. Stärk, 'Laser-excited Atomic Fluorescence Spectrometry as a Practical Analytical Method: Part 2 Evaluation of a Graphite Tube Atomizer for the Determination of Trace Amounts of Indium, Gallium, Aluminum, Vanadium and Iridium by LEAFS', *J. Anal. At. Spectrom.*, **2**, 63–66 (1987).
221. D. Goforth, J.D. Winefordner, 'Laser-excited Atomic Fluorescence of Atoms Produced in a Graphite Furnace', *Anal. Chem.*, **58**, 2598–2602 (1986).
222. P. Wittman, J.D. Winefordner, 'Laser-excited Atomic Fluorescence Spectrometry with Graphite Filament Atomization', *Can. J. Spectrosc.*, **29**(3), 75–78 (1983).
223. J. Anwar, J.M. Anzano, G. Petrucci, J.D. Winefordner, 'Determination of Chloride at Picogram Levels by Molecular Fluorescence in a Graphite Furnace', *Analyst*, **116**(10), 1025–1028 (1991).
224. D. Goforth, J.D. Winefordner, 'A Graphite Tube Furnace for Use in Laser-excited Atomic-fluorescence Spectrometry', *Talanta*, **34**(2), 290–292 (1987).
225. A. Zybin, C. Schnürer-Patschan, K. Niemax, 'Simultaneous Multi-element Analysis in a Commercial Graphite Furnace by Diode Laser-induced Fluorescence', *Spectrochim. Acta*, **47B**(14), 1519–1524 (1992).
226. A. Mellone, J.D. Winefordner, 'Graphite-furnace Vaporization of Polycyclic Aromatic-compounds with Laser-induced Fluorescence of Vapors to Fingerprint Complex Environmental Materials', *Microchem. J.*, **42**(1), 126–137 (1990).
227. J.X. Zhou, X.D. Hou, K.X. Yang, R.G. Michel, 'Laser-excited Atomic Fluorescence Spectrometry in a Graphite Furnace with an Optical Parametric Oscillator Laser for Sequential Multi-element Determination of Cadmium, Cobalt, Lead, Manganese and Thallium in Buffalo River Sediment', *J. Anal. At. Spectrom.*, **13**(1), 41–47 (1998).
228. N. Omenetto, S. Nikdel, J.D. Bradshaw, M.S. Epstein, R.D. Reeves, J.D. Winefordner, 'Diagnostic and Analytical Studies of the Inductively Coupled Plasma by Atomic Fluorescence Spectrometry', *Anal. Chem.*, **51**(9), 1521–1525 (1979).
229. X. Huang, J. Lanauze, J.D. Winefordner, 'Laser-excited Atomic Fluorescence of Some Precious Metals and Refractory Elements in the Inductively Coupled Plasma', *Appl. Spectrosc.*, **39**(6), 1042–1047 (1985).
230. J.B. Simeonsson, K.C. Ng, J.D. Winefordner, 'Single-resonance and Double-resonance Atomic Fluorescence Spectrometry with Inductively Coupled Plasma Atomization and Laser Excitation', *Appl Spectrosc.*, **45**(9), 1456–1462 (1991).
231. M. Glick, B.W. Smith, J.D. Winefordner, 'Laser-excited Atomic Fluorescence in a Pulsed Hollow-cathode Glow-discharge', *Anal. Chem.*, **62**(2), 157–161 (1990).
232. B.W. Smith, J.B. Womack, J.D. Winefordner, N. Omenetto, 'Approaching Single Atom Detection with Atomic Fluorescence in a Glow-discharge Atom Reservoir', *Appl. Spectrosc.*, **43**(5), 873–876 (1989).
233. O.S. Lunyov, S.V. Oshemkov, 'Determination of Trace Metals by Laser-excited Fluorescence with Hot Hollow

Cathode Atomization', *Spectrochim. Acta*, **47B**(1), 71–81 (1992).
234. J.C. Travis, G.C. Turk, R.L. Watters, Jr, L.-J. Yu, J.L. Blue, 'Trace Detection in Conducting Solids Using Laser-induced Fluorescence in a Cathodic Sputtering Cell', *J. Anal. At. Spectrom.*, **6**, 261–271 (1991).
235. G.I. Romanovska, V.I. Pogonin, A.K. Chibisov, 'Fluorescence Determination of Trace Amounts of Uranium(VI) in Various Materials by a Repetitive Laser Technique', *Talanta*, **34**(1), 207–210 (1987).
236. J.X. Zhou, X.D. Hou, K.X. Yang, S.-J.J. Tsai, R.G. Michel, 'Lasers Based on Optical Parametric Devices: Wavelength Tunability Empowers Laser-based Techniques in the UV, VIS, and near-IR', *Appl. Spectrosc.*, **52**, 176A–189A (1998).
237. G.C. Turk, 'Analytical Performance of Laser-enhanced Ionization in Flames', in *Laser-enhanced Ionization Spectrometry*, eds. J. Travis, G. Turk, John Wiley & Sons, New York, 161–211, 1996.
238. M.J. Rutledge, M.E. Tremblay, J.D. Winefordner, 'Measurement Methods for Atomic Fluorescence and Laser-enhanced Ionization Spectrometries with a Copper-vapor Pumped Dye Laser', *Appl. Spectrosc.*, **41**(1), 5–9 (1987).
239. A.I. Yuzefovsky, R.F. Lonardo, R.G. Michel, 'Spatial Discrimination Against Background with Different Optical-systems for Collection of Fluorescence in Laser-excited Atomic Fluorescence Spectrometry with a Graphite Tube Electrothermal Atomizer', *Anal. Chem.*, **67**(13), 2246–2255 (1995).
240. P.B. Farnsworth, B.W. Smith, N. Omenetto, 'Computer Modeling of Collection Efficiency of Laser-excited Fluorescence from a Graphite Furnace', *Spectrochim. Acta*, **45B**, 1151–1166 (1990).
241. N. Omenetto, J.D. Winefordner, 'Types of Fluorescence Transitions in Atomic Fluorescence Spectrometry', *Appl. Spectrosc.*, **26**, 555–557 (1972).
242. N. Omenetto, J.M. Mermet, G.C. Turk, D.S. Moore, 'Nomenclature, Symbols, Units and their Usage in Spectrochemical Analysis. XIV. Laser-based Atomic Spectroscopy: A New Notation for Spectrochemical Processes (IUPAC recommendations 1997)', *Pure Appl. Chem.*, **70**(2), 517–526 (1998).
243. C.T.J. Alkemade, 'Detection of Small Numbers of Atoms and Molecules', in *Analytical Applications of Lasers*, ed. E.H. Piepmeier, John Wiley & Sons, New York, 107–162, 1986.
244. C.T.J. Alkemade, 'Single-atom Detection', *Appl. Spectrosc.*, **35**, 1–14 (1981).
245. M.S. Epstein, J.D. Winefordner, 'Summary of the Usefulness of Signal-to-noise Treatment in Analytical Spectrometry', *Prog. Anal. Atom. Spectrosc.*, **7**, 67–137 (1984).
246. P.A. Bonczyk, J.A. Shirley, 'Measurement of CH and CN Concentration in Flames by Laser-induced Saturated Fluorescence', *Combust. Flame*, **34**, 253–264 (1979).
247. H. Haraguchi, S.J. Weeks, J.D. Winefordner, 'Selective Excitation of Molecular Species in Flames by Laser-excited Molecular Fluorescence', *Spectrochim. Acta*, **35A**, 391–399 (1979).
248. E.G. Su, R.L. Irwin, Z.W. Liang, R.G. Michel, 'Background Correction by Wavelength Modulation for Pulsed-laser-excited Atomic Fluorescence Spectrometry', *Anal. Chem.*, **64**(15), 1710–1720 (1992).
249. R.D. Reeves, B.M. Patel, C.T. Molnar, 'Decay of Atom Population Following Graphite Rod Atomization in Atomic Absorption Spectrometry', *Anal. Chem.*, **45**, 246–249 (1973).
250. J. Aggett, T.S. West, 'Atomic Absorption and Fluorescence Spectroscopy with a Carbon Filament Atom Reservoir', *Anal. Chim. Acta*, **55**, 349–357 (1971).
251. K. Dittrich, B. Hanisch, H.-J. Stärk, 'Molecule Formation in Electrothermal Atomizers: Interferences and Analytical Possibilities by Absorption, Emission and Fluorescence', *Fresenius' Z Anal. Chem.*, **324**, 497–506 (1986).
252. G. Zizak, 'Laser Diagnostics in Flames by Fluorescence Techniques', in *Analytical Laser Spectroscopy*, eds. S. Martellucci, A.N. Chester, Plenum Press, New York, 147–158, 1985.
253. N. Omenetto, P. Benetti, G. Rossi, 'Flame Temperature Measurements by Means of Atomic Fluorescence Spectrometry', *Spectrochim. Acta*, **27B**, 453–461 (1972).
254. J.D. Bradshaw, N. Omenetto, G. Zizak, J.N. Bower, J.D. Winefordner, 'Five Laser-excited Fluorescence Methods for Measuring Spatial Flame Temperatures. 1: Theoretical Basis', *Appl. Opt.*, **19**(16), 2709–2716 (1980).
255. G. Zizak, J.D. Winefordner, 'Application of the Thermally Assisted Atomic Fluorescence Technique to the Temperature Measurement in a Gasoline–Air Flame', *Combust. Flame*, **44**, 35–41 (1982).
256. R.G. Joklik, J.W. Daily, 'Two-line Atomic Fluorescence Temperature Measurement in Flames: An Experimental Study', *Appl. Opt.*, **21**(22), 4158–4162 (1982).
257. D.R. Crosley, 'Laser Probes for Combustion Chemistry', *ACS Symp. Ser.*, **134**, 3–18 (1980).
258. J.W. Daily, 'Pulsed Resonance Spectroscopy Applied to Turbulent Combustion Flows', *Appl. Opt.*, **15**, 955–960 (1976).
259. J.W. Daily, 'Saturation Effects in Laser-induced Fluorescence Spectroscopy', *Appl. Opt.*, **16**(3), 568–572 (1977).
260. H. Uchida, M.A. Kosinski, J.D. Winefordner, 'Laser-excited Atomic and Ionic Fluorescence in an Inductively Coupled Plasma', *Spectrochim. Acta*, **38B**, 5–13 (1983).
261. N.S. Ham, P. Hannaford, 'Direct Observation of Lifetimes of Excited Atoms in a Flame at Atmospheric Pressure', *J. Phys. B*, **12**, L199–L204 (1979).
262. C.H. Muller, III, K. Schofield, M. Steinberg, 'Laser-induced Flame Chemistry of Li (2 $^2P_{1/2,3/2}$) and Na (3 $^2P_{1/2,3/2}$). Implications for Other Saturated Mode Measurements', *J. Chem. Phys.*, **72**(12), 6620–6631 (1980).

263. C.H. Muller, III, K. Schofield, M. Steinberg, 'Near Saturation Laser-induced Chemical Reactions of Na (3 $^2P_{1/2,3/2}$) in $H_2/O_2/N_2$ Flames', *Chem. Phys. Lett.*, **72**(12), 6620–6631 (1980).
264. D.R. Crosley, 'Collisional Effects on Laser-induced Fluorescence Flame Measurements', *Opt. Eng.*, **20**, 511–521 (1981).
265. M. Iino, H. Yano, Y. Takubo, M. Shimazu, 'Saturation Characteristics of Laser-induced Na Fluorescence in a Propane–Air Flame: The Role of Chemical Reaction', *J. Appl. Phys.*, **52**, 6025–6031 (1981).
266. J.B. Simeonsson, B.W. Smith, J.D. Winefordner, N. Omenetto, 'On the Possibility of Observing Negative Fluorescence Dips in Laser-induced Fluorescence Experiments', *Appl. Spectrosc.*, **45**(4), 521–523 (1991).
267. J.B. Simeonsson, K.C. Ng, J.D. Winefordner, 'Fluorescence-dip Spectroscopy as a Method for Measuring the Quantum Efficiency of Fluorescence', *J. Quant. Spectrosc. Radiat. Transfer*, **48**(2), 131–139 (1992).
268. L. Xu, Y.Y. Zhao, G.Y. Wang, M.Q. He, Z.Y. Wang, 'Observation of Laser-enhanced Radiative-transfer of Barium 6s 6p $^1P_1$ to 6s 5d $^1D_2$ by Means of Fluorescence-dip Detection', *J. Opt. Soc. Am. B*, **9**(7), 1017–1019 (1992).
269. G.A. Petrucci, C.L. Stevenson, B.W. Smith, J.D. Winefordner, N. Omenetto, 'Experimental Evaluation of the Autoionization Cross-section of the Magnesium Transition at 300.9 nm by Laser-induced Fluorescence', *Spectrochim. Acta*, **46B**(6/7), 975–981 (1991).
270. N. Omenetto, O.I. Matveev, 'Time-resolved Fluorescence as a Direct Experimental Approach to the Study of Excitation and Ionization Processes in Different Atom Reservoirs', *Spectrochim. Acta*, **49B**(12–14), 1519–1535 (1994).
271. M.H. Nayfeh, 'Laser Detection of Single Atoms', *Am. Sci.*, **67**(2), 204–213 (1979).
272. G.S. Hurst, M.G. Payne, S.D. Kramer, C.H. Chen, 'Counting the Atoms', *Phys. Today*, **24 Sept.**, 24–29 (1980).
273. G.I. Bekov, V.S. Lethokhov, 'Laser Atomic Photoionization Spectral Analysis of Element Traces', *Appl. Phys. B*, **30**, 161–176 (1983).
274. J.D. Fassett, L.J. Moore, J.C. Travis, J.R. DeVoe, 'Laser Resonance Ionization Mass Spectrometry', *Science*, **230**, 263–267 (1985).
275. T. Whitaker, 'Isotopically Selective Laser Measurements', *Laser Appl.*, **5**(8), 67–73 (1986).
276. G.S. Hurst, M.G. Payne, 'Elemental Analysis Using Resonance Ionization Spectroscopy', *Spectrochim. Acta*, **42B**, 715–726 (1987).
277. J.D. Fassett, R.J. Walker, J.C. Travis, F.C. Ruegg, 'Measurement of Low Abundance Isotopes by Laser Resonance Ionization Mass Spectrometry (RIMS)', *Anal. Instrum.*, **17**(1/2), 69–86 (1988).
278. J.D. Fassett, J.C. Travis, 'Analytical Applications of Resonance Ionization Mass Spectrometry (RIMS)', *Spectrochim. Acta*, **43B**, 1409–1422 (1988).
279. J.P. Young, R.W. Shaw, D.H. Smith, 'Resonance Ionization Mass Spectrometry', *Anal. Chem.*, **61**(22), 1271A–1279A (1989).
280. G.I. Bekov, V.N. Radayev, V.S. Lethokhov, 'Laser Photoionization Spectroscopy of Atomic Traces at Part per Trillion Levels', *Spectrochim. Acta*, **43B**, 491–499 (1988).
281. T.B. Lucatorto, J.E. Parks (eds.), Proceedings of the 4th International Symposium on Resonance Ionization Spectroscopy and its Applications: *Resonance Ionization Spectroscopy, 1988*, Institute of Physics Conference Series Number 94, National Bureau of Standards, Gaithersburg, MD, April 10–15, 1988.
282. J.E. Parks, N. Omenetto (eds.), Proceedings of the 5th International Symposium on Resonance Ionization Spectroscopy and its Applications: *Resonance Ionization Spectroscopy, 1990*, Institute of Physics Conference Series Number 114, Congress Center Villa Ponti, Varese, Italy, Sept. 16–21, 1990.
283. C.M. Miller, J.E. Parks (eds.), Proceedings of the 6th International Symposium on Resonance Ionization Spectroscopy and its Applications: *Resonance Ionization Spectroscopy, 1992*, Institute of Physics Conference Series Number 128, Los Alamos National Laboratory, Santa Fe, NM, May 24–29, 1992.
284. H.-J. Kluge, J.E. Parks, K. Wendt (eds.), Proceedings of the 7th International Symposium on Resonance Ionization Spectroscopy and its Applications: *Resonance Ionization Spectroscopy, 1994*, American Institute of Physics Conference Proceeding 329, Bernkasel-Kues, Germany, July 3–8, 1994.
285. N. Winograd, J.E. Parks (eds.), Proceedings of the 8th International Symposium on Resonance Ionization Spectroscopy and its Applications: *Resonance Ionization Spectroscopy, 1996*, American Institute of Physics Conference Proceeding 338, State College, PA, June 30–July 5, 1996.
286. R.B. Green, R.A. Keller, P.K. Schenck, J.C. Travis, G.G. Luther, 'Opto-galvanic Detection of Species in Flames', *J. Am. Chem. Soc.*, **98**, 8517–8518 (1976).
287. R.A. Keller, R. Engleman, Jr, E.F. Zalewski, 'Opto-galvanic Spectroscopy in a Uranium Hollow Cathode Discharge', *J. Opt. Soc. Am. B*, **69**(5), 738–742 (1979).
288. A. Rosenfeld, S. Mory, R. König, 'Observation of the Opto-galvanic Effect in Neon and Argon Using a Nanosecond Dye Laser', *Opt. Commun.*, **30**(3), 394–396 (1979).
289. D.S. King, P.K. Schenck, 'Optogalvanic Spectroscopy', *Laser Focus*, **14**, 50–57 (1978).
290. P.K. Schenck, W.C. Mallard, J.C. Travis, K.C. Smyth, 'Absorption Spectra of Metal Oxides Using Optogalvanic Spectroscopy', *J. Chem. Phys.*, **69**(11), 5147–5150 (1978).

291. D.S. King, P.K. Schenck, K.C. Smyth, J.C. Travis, 'Direct Calibration of Laser Wavelength and Bandwidth Using the Optogalvanic Effect', *Appl. Opt.*, **16**(10), 2617–2619 (1977).
292. J.C. Travis, P.K. Schenck, G.C. Turk, W.G. Mallard, 'Effect of Selective Laser Excitation on the Ionization of Atomic Species in Flames', *Anal. Chem.*, **51**(9), 1516–1520 (1979).
293. G.C. Turk, J.C. Travis, J.R. DeVoe, T.C. O'Haver, 'Laser-enhanced Ionization Spectrometry in Analytical Flames', *Anal. Chem.*, **51**, 1890–1896 (1979).
294. A.S. Gonchakov, N.B. Zorov, Y.Y. Kuzyakov, O.I. Matveev, 'Determination of Picogram Concentrations of Sodium in Flame by Stepwise Photoionization of Atoms', *Anal. Lett.*, **12**, 1037–1048 (1979).
295. F.M. Curran, K.C. Lin, G.E. Leroi, P.M. Hunt, S.R. Crouch, 'Energy Considerations in Dual Laser Ionization Processes in Flames', *Anal. Chem.*, **55**, 2382–2387 (1983).
296. F.M. Curran, C.A.v. Dijk, S.R. Crouch, 'Dual Laser Ionization in Flames: a Search for Electrical Interferences', *Appl. Spectrosc.*, **37**, 385–389 (1983).
297. Y.-Y.J. Wu, P.M. Hunt, G.E. Leroi, S.R. Crouch, 'Dual Laser Ionization (DLI) Measurements of Ion Mobilities in a H_2-O_2-Ar Flame', *Chem. Phys. Lett.*, **155**(1), 69–76 (1989).
298. C.A.v. Dijk, F.M. Curran, K.C. Lin, S.R. Crouch, 'Two-step Laser-assisted Ionization of Sodium in a Hydrogen–Oxygen–Argon flame', *Anal. Chem.*, **53**, 1275–1279 (1981).
299. M.F. Hineman, S.R. Crouch, 'Modeling Signals, Background and Noise in Flame Laser-induced Ionization Spectroscopy', *Spectrochim. Acta*, **43B**(9–11), 1119–1131 (1988).
300. M.E. Churchwell, T. Beeler, J.D. Messman, R.B. Green, 'Laser-induced Ionization in an Atmospheric-pressure Microarc-induced Plasma', *Spectrosc. Lett.*, **18**(9), 679–693 (1985).
301. A.S. Gonchakov, N.B. Zorov, Y.Y. Kuzyakov, O.I. Matveev, 'Detection of Subpicograms of Sodium by Laser Step-photoionization of Atoms and Laser Atomic Fluorescence with Electrothermal Atomization', *J. Anal. Chem. (USSR)*, **34**, 1792–1795 (1980).
302. J.C. Travis, G.C. Turk, 'Preface', in *Laser-enhanced Ionization Spectrometry*, eds. J.C. Travis, G.C. Turk, John Wiley & Sons, New York, xv–xvi, 1996.
303. G.A. Petrucci, J.D. Winefordner, 'The Double-resonance Optogalvanic Effect of Neon as a Sensitive Photon Detector', *Spectrochim. Acta*, **47B**(3), 437–447 (1992).
304. C.R. Webster, C.T. Rettner, 'Laser Optogalvanic Spectroscopy of Molecules', *Laser Focus*, **19**, 41–53 (1983).
305. P.K. Schenck, J.W. Hastie, 'Optogalvanic Spectroscopy – Application to Combustion Systems', *Opt. Eng.*, **20**, 522–528 (1981).
306. J.C. Travis, G.C. Turk, R.B. Green, 'Laser-enhanced Ionization for Trace Metal Analysis in Flames', *ACS Symp. Ser.*, **85**, 91–101 (1978).
307. J.C. Travis, G.C. Turk, J.R. DeVoe, P.K. Schenck, C.A.v. Dijk, 'Principles of Laser-enhanced Ionization Spectrometry in Flames', *Prog. Anal. At. Spectrosc.*, **7**, 199–241 (1984).
308. O. Axner, 'Laser-enhanced Ionization Spectrometry and its Principles – a Powerful Tool for Ultra-sensitive Trace-element Analysis', PhD Thesis, Department of Physics, Chalmers University of Technology, Göteborg, Sweden, 1987.
309. O. Axner, H. Rubinsztein-Dunlop, 'Laser-enhanced Ionization Spectrometry in Flames – a Powerful and Versatile Technique for Ultra-sensitive Trace Element Analysis', *Spectrochim. Acta*, **44B**, 835–866 (1989).
310. O. Axner, S. Sjöström, H. Rubinsztein-Dunlop, 'Analytical Applications of LEI Spectrometry in Flames and Furnaces', *Mikrochim. Acta*, **III**, 197–214 (1989).
311. K. Niemax, 'LIF, LEI, RIMS etc. – New Promising Techniques in Elemental Micro- and Trace Analysis', *Fresenius' J. Anal. Chem.*, **337**, 551–556 (1990).
312. G.C. Turk, 'Laser-enhanced Ionization Spectroscopy in Flames and Plasmas – Plenary Lecture', *J. Anal. At. Spectrom.*, **2**, 573–577 (1987).
313. Y.Y. Kuzyakov, N.B. Zorov, 'Atomic Ionization Spectrometry: Prospects and Results', *Crit. Rev. Anal. Chem.*, **20**(4), 221–290 (1988).
314. R.B. Green, M.D. Seltzer, 'Laser-induced Ionization Spectrometry', in *Advances in Atomic Spectrometry*, ed. J. Sneddon, JAI Press, Greenwich, CT, 37–79, 1992.
315. D.J. Butcher, 'Laser-enhanced Ionization Spectrometry', in *Lasers in Analytical Spectroscopy*, eds. J. Sneddon, T.L. Thiem, Y.-I. Lee, VCH Publishers, New York, 237–272, 1997.
316. J.C. Travis, G.C. Turk, 'Fundamental Mechanisms of Laser-enhanced Ionization: Signal Detection', in *Laser-enhanced Ionization Spectrometry*, eds. J.C. Travis, G.C. Turk, John Wiley & Sons, New York, 99–160, 1996.
317. R.B. Green, 'Applications of Laser-enhanced Ionization', in *Laser-enhanced Ionization Spectrometry*, eds. J. Travis, G. Turk, John Wiley & Sons, New York, 213–232, 1996.
318. N.B. Zorov, 'Nonflame Reservoirs for Laser-enhanced Ionization Spectrometry', in *Laser-enhanced Ionization Spectrometry*, eds. J. Travis, G. Turk, John Wiley & Sons, New York, 233–264, 1996.
319. N. Omenetto, P.B. Farnsworth, 'Ions and Photons: Interplay of Laser-induced Ionization and Fluorescence Techniques in Different Atomic and Molecular Reservoirs', in *Laser-enhanced Ionization Spectrometry*, eds. J. Travis, G. Turk, John Wiley & Sons, New York, 265–326, 1996.

320. R.B. Green, 'Laser-enhanced Ionization in Flames', in *Analytical Applications of Lasers*, ed. E.H. Piepmeier, John Wiley & Sons, New York, 75–105, 1986.
321. J.D. Messman, N.E. Schmidt, J.D. Parli, R.B. Green, 'Laser-enhanced Ionization of Refractory Elements in a Nitrous Oxide–Acetylene Flame', *Appl. Spectrosc.*, **39**(3), 504–507 (1985).
322. V.I. Chaplygin, Y.Y. Kuzyakov, O.A. Novodvorsky, N.B. Zorov, 'Determination of Alkali Metals by Laser-induced Atomic-ionization in Flames', *Talanta*, **34**(1), 191–196 (1987).
323. G.C. Turk, R.L. Watters, Jr, 'Resonant Laser-induced Ionization of Atoms in an Inductively Coupled Plasma', *Anal. Chem.*, **57**, 1979–1983 (1985).
324. G.C. Turk, L.J. Yu, R.L. Watters, J.C. Travis, 'Laser-induced Ionization of Atoms in a Power-modulated Inductively Coupled Plasma', *Appl. Spectrosc.*, **46**(8), 1217–1222 (1992).
325. L.J. Yu, S.R. Koirtyohann, G.C. Turk, M.L. Salit, 'Selective Laser-induced Ionization in Inductively-coupled Plasma-mass Spectrometry', *J. Anal. Atom. Spectrom.*, **9**(9), 997–1000 (1994).
326. M.D. Seltzer, R.B. Green, 'An Active Nitrogen Plasma Atom Reservoir for Laser-induced Ionization Spectrometry', *Spectrosc. Lett.*, **20**, 601–617 (1987).
327. G.C. Turk, O. Axner, N. Omenetto, 'Optical Detection of Laser-induced Ionization in the Inductively Coupled Plasma for the Study of Ion–Electron Recombination and Ionization Equilibrium', *Spectrochim. Acta*, **42B**, 873–881 (1987).
328. I.V. Bykov, A.B. Skvortsov, T.G. Tatsii, N.V. Chekalin, 'Metal Trace Analysis by Flame/Graphite Furnace OG Spectroscopy', *J. Phys. Colloq.*, **44**(C7), 345–352 (1983).
329. I. Magnusson, O. Axner, I. Lindgren, H. Rubinsztein-Dunlop, 'Laser-enhanced Ionization Detection of Trace Elements in a Graphite Furnace', *Appl. Spectrosc.*, **40**(7), 968–971 (1986).
330. I. Magnusson, S. Sjöström, M. Lejon, H. Rubinsztein-Dunlop, 'Trace Element Analysis by Two-color Laser-enhanced Ionization Spectrometry in a Graphite Furnace', *Spectrochim. Acta*, **42B**(5), 713–718 (1986).
331. I. Magnusson, 'The Applicability to Trace Element Analysis of Laser-enhanced Ionization Spectroscopy in a Graphite Furnace', *Spectrochim. Acta*, **43B**, 727–735 (1988).
332. S. Sjöström, I. Magnusson, M. Lejon, H. Rubinsztein-Dunlop, 'Laser-enhanced Ionization Spectrometry in a T-furnace', *Anal. Chem.*, **60**(15), 1629–1631 (1988).
333. N.V. Chekalin, I.I. Vlasov, 'Direct Analysis of Liquid and Solid Samples without Sample Preparation Using Laser-enhanced Ionization', *J. Anal. At. Spectrom.*, **7**(2), 225–228 (1992).
334. D.J. Butcher, R.L. Irwin, S. Sjöström, A.P. Walton, R.G. Michel, 'Probe Atomization for Laser-enhanced Ionization in a Graphite Tube Furnace', *Spectrochim. Acta*, **46B**(1), 9–33 (1991).
335. N.V. Chekalin, V.I. Pavlutskaya, I.I. Vlasov, 'A "Rod Flame" System in Direct Atomic-ionization Analysis of High-purity Substances', *Inst. Phys. Conf. Ser.*, **114**, 283–288 (1990).
336. N.V. Chekalin, V.I. Pavlutskaya, I.I. Vlasov, 'A "Rod Flame" System in Direct Laser-enhanced Ionization Analysis of High-purity Substances', *Spectrochim. Acta*, **46B**(13), 1701–1709 (1991).
337. N.V. Chekalin, A.G. Marunkov, I.I. Vlasov, A.T. Khalmanov, 'Multipurpose Atomic-ionization Spectrometer – Analysis of High-purity Substances', *High Energy Chem.*, **28**(5), 412–416 (1994).
338. N.V. Chekalin, A. Khalmanov, A.G. Marunkov, I.I. Vlasov, Y. Malmsten, O. Axner, V.S. Dorofeev, E. Glukhan, 'Determination of Co, Cr, Mn and Ni Traces in Fluorine-containing Materials for Optical Fibers Using Laser-enhanced Ionization Techniques with Flame and Rod–Flame Atomizers', *Spectrochim. Acta*, **50B**(8), 753–761 (1995).
339. B.W. Smith, G.A. Petrucci, R.G. Badini, J.D. Winefordner, 'Graphite-furnace Vaporization with Laser-enhanced Ionization Detection', *Anal. Chem.*, **65**(2), 118–122 (1993).
340. K.L. Riter, W.L. Clevenger, L.S. Mordoh, B.W. Smith, O.I. Matveev, J.D. Winefordner, 'Laser-enhanced Ionization Detection of Magnesium Atoms by a Combination of Electrothermal Vaporization and Flame Atomization', *J. Anal. At. Spectrosc.*, **11**(6), 393–399 (1996).
341. N. Omenetto, O.I. Matveev, W. Resto, R. Badini, B.W. Smith, J.D. Winefordner, 'Nonlinear Behavior of Atomic Fluorescence in Mercury Vapors Following Double-resonance Laser Excitation', *Appl. Phys. B*, **58**(4), 303–307 (1994).
342. O.I. Matveev, B.W. Smith, N. Omenetto, J.D. Winefordner, 'Single Photo-electron and Photon Detection in a Mercury Resonance Ionization Photon Detector (RID)', *Spectrochim. Acta*, **51B**(6), 563–567 (1996).
343. W.L. Clevenger, L.S. Mordoh, O.I. Matveev, N. Omenetto, B.W. Smith, J.D. Winefordner, 'Analytical Time-resolved Laser-enhanced Ionization Spectroscopy. 1. Collisional Ionization and Photoionization of the Hg Rydberg States in a Low Pressure Gas', *Spectrochim. Acta*, **52B**(3), 295–304 (1997).
344. W.L. Clevenger, O.I. Matveev, N. Omenetto, B.W. Smith, J.D. Winefordner, 'Laser-enhanced Ionization Spectroscopy of Mercury Rydberg States', *Spectrochim. Acta*, **52B**(8), 1139–1149 (1997).
345. W.L. Clevenger, O.I. Matveev, S. Cabredo, N. Omenetto, B.W. Smith, J.D. Winefordner, 'Laser-enhanced Ionization of Mercury Atoms in an Inert Atmosphere with Avalanche Amplification of the Signal', *Anal. Chem.*, **69**(13), 2232–2237 (1997).

346. W.L. Clevenger, B.W. Smith, J.D. Winefordner, 'Trace Determination of Mercury: a Review', *Crit. Rev. Anal. Chem.*, **27**(1), 1–26 (1997).
347. K. Niemax, 'Spectroscopy Using Thermionic Diode Detectors', *Appl. Phys. B*, **38**, 147–157 (1985).
348. K. Niemax, C.J. Lorenzen, 'The Thermionic Double-diode, an Efficient Detector in Excited-state Laser Spectroscopy', *Opt. Commun.*, **44**(3), 165–169 (1983).
349. K. Niemax, J. Lawrenz, A. Obrebski, K.-H. Weber, 'Isotope-selective Trace-element Detection with the Thermionic Diode', *Anal. Chem.*, **58**, 1566–1571 (1986).
350. K. Niemax, 'Investigations on the Thermionic Diode: The Ionization Probability of Rb ^2P_J Atoms by Noble Gas Collisions', *Appl. Phys. B*, **32**, 59–62 (1983).
351. A. Obrebski, R. Hergenröder, K. Niemax, 'Narrowing of Spectral Lines by Energy Pooling Collisions in Laser-enhanced Ionization Spectroscopy', *Z. Phys. D*, **14**, 289–292 (1989).
352. J. Lawrenz, A. Obreski, K. Niemax, 'Measurement of Isotope Ratio by Doppler-free Laser Spectroscopy Applying Semiconductor Diode Laser and Thermionic Diode Detection', *Anal. Chem.*, **59**, 1232–1236 (1987).
353. J. Brust, D. Veza, M. Movre, K. Niemax, 'Collisional Excitation Transfer between Lithium Isotopes', *Z. Phys. D*, **32**(4), 305–309 (1995).
354. J. Franzke, D. Veza, K. Niemax, 'An Improved Thermionic Diode Detector for Analytical Laser Spectroscopy', *Spectrochim. Acta*, **47B**(5), 593–599 (1992).
355. V. Horvatic, D. Veza, M. Movre, K. Niemax, C. Vadla, 'Collision Cross-sections for Excitation-energy Transfer in $Na^*(^3P_{1/2}) + K(^4S_{1/2})$ to $Na^*(^3P_{3/2})$', *Z. Phys. D*, **34**(3), 163–170 (1995).
356. V. Horvatic, C. Vadla, M. Movre, K. Niemax, 'The Collision Cross-sections for the Fine-structure Mixing of Caesium 6P Levels Induced by Collisions with Potassium Atoms', *Z. Phys. D*, **36**(2), 101–104 (1996).
357. C. Vadla, D. Veza, M. Movre, K. Niemax, 'Fine-structure Excitation Transfer between the Lithium d-lines by Collisions with Cesium Atoms', *Z. Phys. D*, **22**(3), 591–595 (1992).
358. D. Veza, C. Vadla, K. Niemax, 'Excitation-energy Transfer in the Li–Cs Collision $Li^*(2p) + Cs^-(6s) \longrightarrow Li(2s) + Cs^{-*}(5d)$', *Z. Phys. D*, **22**(3), 597–601 (1992).
359. H.D. Wizemann, K. Niemax, 'Isotope Selective Element Analysis by Diode Laser Atomic Absorption Spectrometry', *Mikrochim. Acta*, **129**(3/4), 209–216 (1998).
360. C. Vadla, K. Niemax, V. Horvatic, R. Beuc, 'Population and Deactivation of Lowest Lying Barium Levels by Collisions with He, Ar, Xe and Pa Ground-state Atoms', *Z. Phys. D*, **34**(3), 171–184 (1995).
361. C. Vadla, K. Niemax, J. Brust, 'Energy Pooling in Cesium Vapor', *Z. Phys. D*, **37**(3), 241–247 (1996).
362. C. Vadla, K. Niemax, V. Horvatic, 'Energy Pooling to the Ba 6s 6p $^1P_1^0$ Level Arising from Collisions between Pairs of Metastable Ba 6s 5d ^3D_J Atoms', *Z. Phys. D*, **1**(2), 139–147 (1998).
363. J. Brust, M. Movre, K. Niemax, 'Measurement and Calculation of the Fine-structure Changing Collision Cross-sections in the Mg^+ and Ca^+ Resonance States by Helium', *Z. Phys. D*, **27**(3), 243–248 (1993).
364. O. Axner, I. Magnusson, J. Petersson, S. Sjöström, 'Investigation of the Multi-element Capability of Laser-enhanced Ionization Spectrometry in Flames for Analysis of Trace Elements in Water Solutions', *Appl. Spectrosc.*, **41**, 19–26 (1987).
365. J.E. Hall, R.B. Green, 'Laser-enhanced Ionization Spectrometry with a Total Consumption Burner', *Anal. Chem.*, **55**, 1811–1814 (1983).
366. O. Axner, I. Lindgren, I. Magnusson, H. Rubinsztein-Dunlop, 'Trace Element Determination in Flames by Laser-enhanced Ionization Spectrometry', *Anal. Chem.*, **57**, 773–776 (1985).
367. O.I. Matveev, P. Cavalli, N. Omenetto, 'Three-step Laser-induced Ionization of Ir and Hg Atoms in an Air–Acetylene Flame and a Gas Cell', in *Resonance Ionization Spectroscopy 1994*, American Institute of Physics, New York, 269–272, 1994.
368. D.J. Ehrlich, R.M. Osgood, Jr, G.C. Turk, J.C. Travis, 'Atomic Resonance-line Lasers for Atomic Spectrometry', *Anal. Chem.*, **52**, 1354–1356 (1980).
369. L.P. Hart, B.W. Smith, N. Omenetto, 'Laser-induced Stepwise and Two-photon Ionization Studies of Strontium in the Air Acetylene Flame', *Spectrochim. Acta*, **40B**, 1637–1649 (1985).
370. N. Omenetto, T. Berthoud, P. Cavalli, G. Rossi, 'Analytical Laser-enhanced Ionization Studies of Thallium in the Air–Acetylene Flame', *Anal. Chem.*, **57**, 1256–1261 (1985).
371. O. Axner, I. Magnusson, 'Determination of Trace Elements in Water Solution by Laser-enhanced Ionization Using Coumarin 47', *Phys. Scr.*, **31**, 587–591 (1985).
372. O. Axner, T. Berglind, S. Sjöström, 'Laser-enhanced Ionization Spectroscopy Around the Ionization Limit', *Phys. Scr.*, **34**, 18–23 (1986).
373. O. Axner, T. Berglind, 'Stark Structure Observation in Rydberg States of Li in Flames by Laser-enhanced Ionization – a New Method for Probing Local Electrical Fields in Flames', *Appl. Spectrosc.*, **40**, 1224–1231 (1986).
374. O. Axner, M. Lejon, I. Magnusson, H. Rubinsztein-Dunlop, S. Sjöström, 'Detection of Traces in Semiconductor Materials by Two-color Laser-enhanced Ionization Spectroscopy in Flames', *Appl. Opt.*, **26**, 3521–3525 (1987).
375. O. Axner, S. Sjöström, 'Anomalous Contributions to the Signal in Laser-enhanced Ionization Spectrometry from "Scattered" Laser Light', *Appl. Spectrosc.*, **44**(5), 864–870 (1990).
376. O. Axner, S. Sjöström, 'Anomalous Lineshapes and Signals in Laser-enhanced Ionization Spectrometry in Flames – The Role of "Scattered" Light, Dynamic Stark Effects and Two-step versus Two-photon Excitations', *Inst. Phys. Conf. Ser.*, **114**(6), 31–36 (1990).

377. O. Axner, M. Norberg, H.R. Dunlop, 'Investigation of Background Signals from Na as a Source of Interference in Laser-enhanced Ionization Spectrometry in Flames', *Appl. Spectrosc.*, **44**, 1124–1133 (1990).
378. P. Ljungberg, Y. Malmsten, O. Axner, 'Monitoring of Atomic Metastable State Lifetimes by the Laser-enhanced Ionization Technique – Development of a Method for Probing Local Stoichiometric Combustive Conditions', in *Resonance Ionization Spectroscopy 1994*, American Institute of Physics, New York, 273–276, 1994.
379. I. Magnusson, O. Axner, H. Rubinsztein-Dunlop, 'Elimination of Spectral Interferences Using Two-step Excitation Laser-enhanced Ionization', *Phys. Scr.*, **33**, 429–433 (1986).
380. G. Nave, S. Johansson, O. Axner, P. Ljungberg, Y. Malmsten, B. Baschek, 'Analysis of the $3d^6$ 4s (^{6}D) 6d Subconfiguration of Fe I by Laser-enhanced Ionization and Grating Spectroscopy', *Phys. Scr.*, **49**, 581–587 (1994).
381. P.M. Paquet, J.-F. Gravel, P. Norbert, D. Boudreau, 'Speciation of Chromium by Ion Chromatography and Laser-enhanced Ionization: Optimization of the Excitation–Ionization Scheme', *Spectrochim. Acta*, **53**(12), 1907–1917 (1998).
382. G.J. Havrilla, C.C. Carter, 'Laser-enhanced Ionization Detection of Trace Copper in High Salt Matrices', *Appl. Opt.*, **17**, 3511–3515 (1987).
383. G.C. Turk, J.R. DeVoe, J.C. Travis, 'Stepwise Excitation Laser-enhanced Ionization Spectrometry', *Anal. Chem.*, **54**, 643–645 (1982).
384. G.C. Turk, J.C. Travis, J.R. DeVoe, 'Laser-enhanced Ionization Spectrometry for Trace Metal Analysis', *J. Phys. Colloq.*, **44**(11 : C7), 301–309 (1983).
385. G.J. Havrilla, K.-J. Choi, 'Detection and Spectroscopic Study of Zinc by Laser-enhanced Ionization Spectrometry', *Anal. Chem.*, **58**, 3095–3100 (1986).
386. M.D. Seltzer, R.B. Green, 'Direct Laser Ionization in Analytically Useful Flames', *Appl. Spectrosc.*, **43**, 257–263 (1989).
387. Y.Y. Kuzyakov, N.B. Zorov, V.I. Chaplygin, A.A. Gorbatenko, 'The Role of Resonance Photoionization of Atoms in Seeded Flames', in *Resonance Ionization Spectroscopy 1988*, IOP Publishing, Bristol, 179–182, 1988.
388. V.I. Chaplygin, N.B. Zorov, Y.Y. Kuzyakov, 'Laser Atomic-ionization Determination of Caesium in Flames', *Talanta*, **30**, 505–508 (1983).
389. G.C. Turk, J.C. Travis, J.R. DeVoe, T.C. O'Haver, 'Analytical Flame Spectrometry with Laser-enhanced Ionization', *Anal. Chem.*, **50**, 817–820 (1978).
390. O. Axner, T. Berglind, J.L. Heully, I. Lindgren, H. Rubinsztein-Dunlop, 'Theory of Laser-enhanced Ionization in Flames – Comparison with Experiments', *J. Phys. Colloq.*, **44**(C7), 311–317 (1983).
391. R.W. Fox, C.S. Weimer, L. Hollberg, G.C. Turk, 'The Diode-laser as a Spectroscopic Tool', *Spectrochim. Acta Rev.*, **15**(5), 291–299 (1993).
392. G.J. Havrilla, S.J. Weeks, J.C. Travis, 'Continuous Wave Excitation in Laser-enhanced Ionization Spectrometry', *Anal. Chem.*, **54**, 2566–2570 (1982).
393. P.K. Schenck, J.C. Travis, G.C. Turk, 'Studies of Physical Mechanisms in Laser-enhanced Ionization in Flames', *J. Phys. (Paris) Colloq.*, **C7**, 75–84 (1983).
394. O. Axner, S. Sjöström, 'Anomalous Contributions to the Signal in Laser-enhanced Ionization Spectrometry from "Scattered" Laser Light', *Appl. Spectrosc.*, **44**, 864–870 (1990).
395. G.C. Turk, J.C. Travis, 'Simultaneous Detection of Laser-enhanced Ionization and Laser-induced Fluorescence in Flames: Noise Correlation Studies', *Spectrochim. Acta*, **45B**, 409–419 (1990).
396. A.A. Gorbatenko, N.B. Zorov, S.Y. Karopova, Y.Y. Kuzyakov, V.I. Chapplygin, 'Determination of Trace Amounts of Calcium by Laser-enhanced Ionization Spectrometry in Flames', *J. Anal. At. Spectrom.*, **3**, 527–530 (1988).
397. N.B. Zorov, Y.Y. Kuzyakov, O.I. Matveev, 'Atomic Ionization Analysis by Tunable Lasers', *Zh. Anal. Khim.*, **37**(3), 520–533 (1982).
398. A.G. Marunkov, N.V. Chekalin, 'The Limiting Performance of a Flame Atomic-ionization Spectrometer', *J. Anal. Chem. USSR (Engl. Transl.)*, **42**, 506–508 (1987).
399. L.C. Chandola, P.P. Khanna, M.A.N. Razvi, 'Sub-picogram Detection of Sodium by One-step Laser-enhanced Ionization (LEI) Spectrometry Using Two-photon Transition', *Anal. Lett.*, **24**(9), 1685–1693 (1991).
400. K.L. Riter, O.I. Matveev, B.W. Smith, J.D. Winefordner, 'The Determination of Lead in Whole Blood by Laser-enhanced Ionization Using a Combination of Electrothermal Vaporizer and Flame', *Anal. Chim. Acta*, **333**(1/2), 187–192 (1996).
401. I. Magnusson, 'On the Signal Collection in Laser-enhanced Ionization Spectrometry', *Spectrochim. Acta*, **42B**, 1113–1124 (1987).
402. S.C. Wang, K.C. Lin, 'Automatic-determination of Optimum Dilution Levels for Laser-enhanced Ionization Detection of Matrix-interfered Sample by Flow-injection', *Analyst*, **120**(10), 2593–2599 (1995).
403. S.C. Wang, K.C. Lin, 'Novel Technique to Reduce Electrical Interference Inherent in Laser-enhanced Ionization Detection by Using Flow-injection Analysis', *Anal. Chem.*, **66**(13), 2180–2186 (1994).
404. N.V. Chekalin, A.G. Marunkov, V.I. Pavlutskaya, S.V. Bachin, 'Determination of Traces of Cs, Li and Rb in Rocks by Laser-enhanced Ionization Spectrometry without Preconcentration', *Spectrochim. Acta*, **46B**, 551–558 (1991).
405. K.S. Epler, T.C. O'Haver, G.C. Turk, 'Liquid-chromatography Laser-enhanced Ionization Spectrometry for the Speciation of Organolead Compounds', *J. Anal. At. Spectrom.*, **9**(2), 79–82 (1994).
406. G.A. Petrucci, R.G. Badini, J.D. Winefordner, 'Photon Detection based on Pulsed Laser-enhanced Ionization

and Photoionization of Magnesium Vapor – Experimental Characterization', *J. Anal. At. Spectrom.*, **7**(3), 481–491 (1992).

407. G.A. Petrucci, R.G. Badini, J.D. Winefordner, 'Resonance Detection of Photons by Atomic Ionization', *Inst. Phys. Conf. Ser.*, **128**, 333–336 (1992).

408. G.A. Petrucci, D. Imbroisi, B.W. Smith, J.D. Winefordner, 'Detection of OH in an Atmospheric-pressure Flame via Laser-enhanced Ionization of Indium', *Spectrochim. Acta*, **49B**(12–14), 1569–1578 (1994).

409. G.A. Petrucci, D. Imbroisi, R.D. Guenard, B.W. Smith, J.D. Winefordner, 'High-spatial-resolution OH Rotational Temperature-measurements in an Atmospheric-pressure Flame Using an Indium-based Resonance Ionization Detector', *Appl. Spectrosc.*, **49**(5), 655–659 (1995).

410. G.A. Petrucci, J.D. Winefordner, N. Omenetto, 'The Relaxation Oscillator as a Resonance Photon Detector', *Appl. Phys. B*, **62**(5), 457–464 (1996).

411. B.W. Smith, N. Omenetto, J.D. Winefordner, 'A New, Sensitive, High Resolution Raman Detector Based on Ionization', in *Future Trends in Spectroscopy*, Pergamon Press, Oxford, and Pontificia Academica Scientiarum, The Vatican, 101–111, 1989.

412. N. Omenetto, B.W. Smith, J.D. Winefordner, 'Theoretical Considerations on the Optogalvanic Detection of Laser-induced Fluorescence in Atmospheric Pressure Atomizers', in *Future Trends in Spectroscopy*, Pergamon Press Oxford and Pontificia Academica Scientiarum, The Vatican, 91–99, 1989.

413. D.L. Monts, Abhilasha, S.C. Qian, D. Kumar, X. Yao, S.P. McGlynn, 'Comparison of Atomization Sources for a Field-deployable Laser Optogalvanic Spectrometry System', *J. Thermophys. Heat Transfer*, **12**(1), 66–72 (1998).

414a. P.B. Farnsworth, N. Omenetto, 'The Kinetics of Charge-exchange between Argon and Magnesium in the Inductively-coupled Plasma', *Spectrochim. Acta*, **48B**(6/7), 809–816 (1993).

414b. G.C. Turk, 'Imaging the Active Flame Volume for Pulsed Laser-enhanced Ionization Spectroscopy', *Anal. Chem.*, **64**(17), 1836–1839 (1992).

414c. G.C. Turk, 'Spatially Selective Detection of Anomalous Contributions to Laser-enhanced Ionization in Flames', *Inst. Phys. Conf. Ser.*, **128**(4), 157–160 (1992).

415. I.I. Vlasov, N.V. Chekalin, 'A New Approach to the Determination of the Ionization Yield of Atoms by Laser-enhanced Ionization', *Spectrochim. Acta*, **48B**(4), 597–603 (1993).

416. K.C. Lin, 'Applications of Laser-enhanced Ionization in Analytical Chemistry', *J. Chin. Chem. Soc.*, **41**(3), 293–308 (1994).

417. K.D. Su, K.C. Lin, W.T. Luh, 'Application of Laser-enhanced Ionization–Atomization Efficiency Determination', *Appl. Spectrosc.*, **46**(9), 1370–1375 (1992).

418. O.I. Matveev, N. Omenetto, 'On the Possibility of Direct Evaluation of Ionization Rates Induced by Laser-radiation in Flames: Part 1 – Time-resolved Ionization Measurements', *Spectrochim. Acta*, **49B**(7), 691–702 (1994).

419. C.B. Ke, K.C. Lin, 'Spatially Resolved Temperature Determination of an Air/Acetylene Flame Using the Two-step Laser-enhanced Ionization Technique', *Appl. Spectrosc.*, **52**(2), 187–194 (1998).

420. P. Barker, A. Thomas, H. Rubinsztein-Dunlop, P. Ljungberg, 'Velocity Measurements by Flow Tagging Employing Laser-enhanced Ionization and Laser-induced Fluorescence', *Spectrochim. Acta*, **50B**(11), 1301–1310 (1995).

421. P. Barker, H. Rubinsztein-Dunlop, 'Measurements of Neutral Atom Diffusion and Electron–Ion Recombination by Laser-enhanced Ionization and Planar Laser-induced Fluorescence in an Air–Acetylene Flame', *Spectrochim. Acta*, **52B**(4), 459–469 (1997).

422. P. Barker, A. Bishop, H. Rubinsztein-Dunlop, 'Supersonic Velocimetry in a Shock Tube Using Laser-enhanced Ionization and Planar Laser-induced Fluorescence', *Appl. Phys. B*, **64**(3), 369–376 (1997).

423. P.F. Barker, A.M. Thomas, T.J. McIntyre, H. Rubinsztein-Dunlop, 'Velocimetry and Thermometry of Supersonic Flow around a Cylindrical Body', *AIAA J.*, **36**(6), 1055–1060 (1998).

424. G.C. Turk, L.J. Yu, S.R. Koirtyohann, 'Laser-enhanced Ionization Spectroscopy of Sodium Atoms in an Air Hydrogen Flame with Mass-spectrometric Detection', *Spectrochim. Acta*, **49B**(12–14), 1537–1543 (1994).

425. N. Trautmann, 'Ultratrace Analysis of Long-lived Radioisotopes in the Environment', in *7th International Symposium on Resonance Ionization Spectroscopy and its Applications*, American Institute of Physics Conference Proceeding 329, Bernkastel-Kues, Germany, 209–214, 1994.

426. K. Wendt, B.A. Bushaw, G. Bhowmick, V.A. Bystrow, N. Kotovski, J.V. Kratz, J. Lantzsch, P. Muller, W. Nörtershäuser, E.W. Otten, A. Seibert, N. Trautmann, A. Waldek, Y. Yushkevich, '^{89,90}Sr-determination in Various Environmental Samples by Collinear Resonance Ionization Spectroscopy', in *8th International Symposium on Resonance Ionization Spectroscopy and its Applications*, American Institute of Physics Conference Proceeding 388, State College, PA, 361–364, 1996.

427. B.A. Bushaw, F. Juston, W. Nörtershäuser, N. Trautmann, P. Voss-de-Haan, K. Wendt, 'Multiple Resonance RIMS Measurements of Calcium Isotopes Using Diode Lasers', in *8th International Symposium on Resonance Ionization Spectroscopy and its Applications*, American Institute of Physics Conference Proceeding 388, State College, PA, 115–118, 1996.

428. M.I.K. Santala, H.M. Lauranto, R.R.E. Salomaa, 'RIS and Laser-induced Fluorescence of Nb from Laser-ablated Metal Surfaces', in *8th International Symposium*

on Resonance Ionization Spectroscopy and its Applications, American Institute of Physics Conference Proceeding 388, State College, PA, 191–194, 1996.

429. J.E. Anderson, T.M. Allen, A.W. Garret, C.G. Gill, P.H. Hemberg, P.B. Kelly, N.S. Nogar, 'Resonant Laser Ablation: Mechanisms and Applications', in *8th International Symposium on Resonance Ionization Spectroscopy and its Applications*, American Institute of Physics Conference Proceeding 388, State College, PA, 195–198, 1996.
430. H.F. Arlinghaus, X.Q. Guo, T.J. Whitaker, M.N. Kwoka, 'Fish & Chips: Analytical Applications of Resonance Ionization Mass Spectrometry', in *8th International Symposium on Resonance Ionization Spectroscopy and its Applications*, American Institute of Physics Conference Proceeding 388, State College, PA, 139–144, 1996.
431. G.I. Bekov, 'Resonance Ionization Spectroscopy', in *7th International Symposium on Resonance Ionization Spectroscopy and its Applications*, American Institute of Physics Conference Proceeding 329, Bernkastel-Kues, Germany, 9–14, 1994.
432. G. Passler, 'Application of Resonance Ionization Mass Spectrometry for Trace Analysis and in Fundamental Research', in *8th International Symposium on Resonance Ionization Spectroscopy and its Applications*, American Institute of Physics Conference Proceeding 388, State College, PA, 209–214, 1996.
433. B.A. Bushaw, H.-J. Kluge, J. Lantsch, R. Schwalbach, M. Schwarz, J. Stenner, H. Stevens, K. Wendt, K. Zimmer, 'Hyperfine Structure of ^{87,89}Sr 5s 4d ^{3}D–5s *n*f Transitions in Collinear Fast Beam RIMS', in *7th International Symposium on Resonance Ionization Spectroscopy and its Applications*, American Institute of Physics Conference Proceeding 329, Bernkastel-Kues, Germany, 9–14, 1994.
434. S. Köhler, F. Albus, R. Deissenberger, N. Erdmann, H. Funk, H.-U. Hasse, G. Herrmann, G. Huber, H.-J.N. Kluge, G. Passler, P.M. Rao, J. Riegel, N. Trautmann, F.-J. Urban, 'Determination of the First Ionization Potential of Actinides by Resonance Ionization Mass Spectrometry', in *7th International Symposium on Resonance Ionization Spectroscopy and its Applications*, American Institute of Physics Conference Proceeding 329, Bernkastel-Kues, Germany, 377–380, 1994.
435. S. Cavalieri, R. Eramo, L. Fini, M. Matera, 'Experimental Studies of Laser-induced Continuum Structure in the Atomic Ionization Continuum', in *7th International Symposium on Resonance Ionization Spectroscopy and its Applications*, American Institute of Physics Conference Proceeding 329, Bernkastel-Kues, Germany, 165–170, 1994.
436. S.V. Bobashev, N.A. Cherepkov, A.Y. Elizarov, V.K. Prilipko, V.V. Korshunov, 'Linear and Circular Dichroism in Two-step Photoionization of Barium Atoms', in *7th International Symposium on Resonance Ionization Spectroscopy and its Applications*, American Institute of Physics Conference Proceeding 329, Bernkastel-Kues, Germany, 399–402, 1994.
437. V.V. Petrunin, H.H. Andersen, P. Balling, P. Kristensen, T. Andersen, 'Resonance Ionization Spectroscopy of Negative Ions', in *8th International Symposium on Resonance Ionization Spectroscopy and its Applications*, American Institute of Physics Conference Proceeding 388, State College, PA, 103–108, 1996.
438. D. Hanstorp, U. Ljungblad, U. Berzinsh, D.J. Pegg, 'Observation of Doubly Excited States in Negative Ions Using Resonance Ionization Spectroscopy', in *8th International Symposium on Resonance Ionization Spectroscopy and its Applications*, American Institute of Physics Conference Proceeding 388, State College, PA, 97–102, 1996.
439. S.D. Bergeson, A. Balakrishnan, K.G.H. Baldwin, T.B. Lucatorto, J.P. Marangos, T.J. McIlrath, T.R. O'Brian, S.L. Rolston, N. Vansteenkiste, 'Doppler-free Resonance Ionization Spectroscopy of the He 1s^2 ^{1}S–1s 2s ^{1}S Transition at 120.3 nm', in *8th International Symposium on Resonance Ionization Spectroscopy and its Applications*, American Institute of Physics Conference Proceeding 388, State College, PA, 109–112, 1996.
440. H. Walther, 'Laser-experiments with Single Atoms and the Test of Basic Quantum Phenomena', in *8th International Symposium on Resonance Ionization Spectroscopy and its Applications*, American Institute of Physics Conference Proceeding 388, State College, PA, 3–12, 1996.
441. D.-Y. Jeong, K.S. Lee, A.S. Choe, J. Lee, B.K. Rhee, 'Dynamic Stark Shift in a 3-Level Atomic System', in *8th International Symposium on Resonance Ionization Spectroscopy and its Applications*, American Institute of Physics Conference Proceeding 388, State College, PA, 89–92, 1996.
442. M.G. Payne, L. Deng, J.Y. Zhang, W.R. Garrett, 'Effect of Pressure Dependent Quantum Interference on the AC Stark Shifting of a Four-photon Resonance', in *7th International Symposium on Resonance Ionization Spectroscopy and its Applications*, American Institute of Physics Conference Proceeding 329, Bernkastel-Kues, Germany, 373–376, 1994.
443. M. Shapiro, P. Brumer, 'Quantum Control of Dynamics', in *8th International Symposium on Resonance Ionization Spectroscopy and its Applications*, American Institute of Physics Conference Proceeding 388, State College, PA, 69–88, 1996.
444. B.A. Bushaw, 'Attogram Measurements of Rare Isotopes by CW Resonance Ionization Mass Spectrometry', in *6th International Symposium on Resonance Ionization Spectroscopy and its Applications*, Institute of Physics Conference Series Number 128, Los Alamos National Laboratory, Santa Fe, NM, 31–36, 1992.
445. B.A. Bushaw, J.T. Munely, 'Isotopically Selective RIMS of Rare Radionuclides by Double-resonance Excitation with Single-frequency CW Lasers', in *5th International Symposium on Resonance Ionization Spectroscopy and*

its Applications, Institute of Physics Conference Series Number 114, Congress Center Villa Ponti, Varese, Italy, 387–392, 1990.

446. W.M. Fairbank, Jr, 'A Seat-of-the pants Approach to RIS Theory', in *8th International Symposium on Resonance Ionization Spectroscopy and its Applications*, American Institute of Physics Conference Proceeding 388, State College, PA, 457–464, 1996.
447. X. Xiong, J.M.R. Hutchinson, J. Fassett, W.M. Fairbank, Jr, 'Measurements of the Odd–Even Effect in the Resonance Ionization of Tin as a Function of Laser Intensity', in *6th International Symposium on Resonance Ionization Spectroscopy and its Applications*, Institute of Physics Conference Series Number 128, Los Alamos National Laboratory, Santa Fe, NM, 123–126, 1992.
448. N. Erdmann, G. Herrmann, G. Huber, S. Köhler, J.V. Kratz, A. Mansel, M. Nunnemann, G. Passler, N. Trautmann, A. Waldek, 'Trace Analysis of Plutonium in Environmental Samples by Resonance Ionization Mass Spectrometry (RIMS)', in *8th International Symposium on Resonance Ionization Spectroscopy and its Applications*, American Institute of Physics Conference Proceeding 388, State College, PA, 205–208, 1996.
449. W. Nörtershäuser, N. Trautmann, K. Wendt, B.A. Bushaw, 'Isotope Shifts and Hyperfine Structure in the $4s^2\,{}^1S_0$–$4s\,4p\,{}^1P_1$–$4s\,4d^1\,D_2$ Transitions of Stable Calcium Isotopes and Calcium-41', *Spectrochim. Acta*, **53B**(5), 709–721 (1998).
450. B.A. Bushaw, W. Nörtershäuser, K. Wendt, 'Lineshapes and Optical Selectivity in High-resolution Double-resonance Ionization Mass Spectrometry', *Spectrochim. Acta*, **54B**(2), 321–332 (1999).
451. C.S. Hansen, W.F. Calaway, M.J. Pellin, R.C. Wiens, D.S. Burnett, 'Three-color Resonance Ionization Spectroscopy of Zr in Si', in *8th International Symposium on Resonance Ionization Spectroscopy and its Applications*, American Institute of Physics Conference Proceeding 388, State College, PA, 215–218, 1996.
452. O.R. Jones, C.J. Abraham, H.H. Telle, A.E. Oakley, 'The Feasibility of RIMS for the Analysis of Potentially Toxic Elements Accumulation in Neural Tissue', in *8th International Symposium on Resonance Ionization Spectroscopy and its Applications*, American Institute of Physics Conference Proceeding 388, State College, PA, 289–292, 1996.
453. R.N. Hall, G.E. Fenner, J.D. Kingsley, T.J. Soltys, R.O. Carlson, 'Coherent Light Emission from GaAs Junctions', *Phys. Rev. Lett.*, **9**, 366–368 (1962).
454. M.I. Nathan, W.P. Dumke, G. Burns, F.H. Dill, Jr, G. Lasher, 'Stimulated Emission of Radiation from GaAs p–n Junctions', *Appl. Phys. Lett.*, **1**(3), 62–64 (1962).
455. N. Holonyak, S.F. Bevacqua, 'Coherent (visible) Light Emission from $Ga(As_{1-x}P_x)$ Junctions', *Appl. Phys. Lett.*, **1**(4), 82–83 (1962).
456. T.M. Quist, R.H. Rediker, R.J. Keyes, W.E. Krag, B. Lax, A.L. McWhorter, H.J. Zeigler, 'Semiconductor Maser of GaAs', *Appl. Phys. Lett.*, **1**(4), 91–92 (1962).
457. A.W. Mantz, 'A Review of Spectroscopic Applications of Tunable Semiconductor Lasers', *Spectrochim. Acta*, **51A**(13), 2211–2236 (1995).
458. J.C. Camparo, C.M. Klimack, 'Laser Spectroscopy on a "Shoestring"', *Am. J. Phys.*, **51**(12), 1077–1081 (1983).
459. J.C. Camparo, 'The Diode Laser in Atomic Physics', *Contemp. Phys.*, **26**(5), 443–477 (1985).
460. K.-H. Weber, J. Lawrenz, A. Obreski, K. Niemax, 'High-resolution Laser Spectroscopy of Aluminum, Gallium and Thallium', *Phys. Scr.*, **35**, 309–312 (1987).
461. C. Vadla, A. Obreski, K. Niemax, 'Isotope Shift of the $3s\,{}^2S_{1/2}$ and $3p\,{}^2P_J$ Levels in ${}^{6,7}Li$', *Opt. Commun.*, **63**(5), 288–292 (1987).
462. C. Vadla, J. Lawrenz, K. Niemax, 'Measurement of Velocity Dependence of Collisional Excitation Transfer Applying Resonant Doppler-free 2-photon Laser Spectroscopy', *Opt. Commun.*, **63**(5), 293–297 (1987).
463. D. Veza, J. Lawrenz, K. Niemax, 'Velocity Dependence of Impact Line-broadening Studied by Resonant Doppler-free Two-photon Laser Spectroscopy', *Z. Phys. D*, **9**, 135–141 (1988).
464. A.S. Zibrov, R.W. Fox, R. Ellingsen, C.S. Weimer, V.L. Velichansky, G.M. Tino, L. Hollberg, 'High-resolution Diode-laser Spectroscopy of Calcium', *Appl. Phys. B*, **59**, 327–331 (1994).
465. F.S. Pavone, 'Diode Lasers and their Applications in Spectroscopy', *Riv. Nuovo Cimento*, **19**(9), 1–42 (1996).
466. C.E. Wieman, L. Hollberg, 'Using Diode Laser for Atomic Physics', *Rev. Sci. Instrum.*, **62**(1), 1–20 (1991).
467. D.T. Cassidy, J. Reid, 'Atmospheric Pressure Monitoring of Trace Gases Using Tunable Diode Lasers', *Appl. Opt.*, **21**(7), 1185–1190 (1982).
468. D.T. Cassidy, J. Reid, 'High-sensitivity Detection of Trace Gases Using Sweep Integration and Tunable Diode Lasers', *Appl. Opt.*, **21**(14), 2527–2530 (1982).
469. P. Werle, R. Mucke, F. Slemr, 'The Limits of Signal Averaging in Atmospheric Trace-gas Monitoring by Tunable Diode-laser Absorption-spectroscopy (TDLAS)', *Appl. Phys. B*, **57**(2), 131–139 (1993).
470. P. Werle, B. Scheumann, J. Schandl, 'Real-time Signal-processing Concepts for Trace-gas Analysis by Diode-laser Spectroscopy', *Opt. Eng.*, **33**(9), 3093–3105 (1994).
471. P. Werle, 'Tunable Diode Laser Absorption Spectroscopy: Recent Findings and Novel Approaches', *Infrared Phys. Technol.*, **37**(1), 59–66 (1996).
472. P. Werle, 'Analytical Applications of Infrared Semiconductor-lasers in Atmospheric Trace Gas Monitoring', *J. Phys. IV*, **4**(C4), 9–12 (1994).
473. P. Werle, 'Spectroscopic Trace Gas Analysis Using Semiconductor Diode Lasers', *Spectrochim. Acta*, **52A**(8), 805–822 (1996).

474. P. Werle, 'A Review of Recent Advances in Semiconductor Laser Based Gas Monitors', *Spectrochim. Acta*, **54A**(2), 197–236 (1998).
475. R. Schieder, 'High Resolution Diode Laser and Heterodyne Spectroscopy with Applications toward Remote Sensing', *Infrared Phys. Technol.*, **35**(2/3), 477–486 (1994).
476. K. Niemax, 'Hochauflösende und hochempfindliche Laserspektroskopie in der Elementanalytik', *Naturwissenschaften*, **74**, 474–481 (1987).
477. R. Hegenröder, K. Niemax, 'Laser Atomic Absorption Spectroscopy Applying Semiconductor Diode Laser', *Spectrochim. Acta*, **43B**(12), 1443–1449 (1988).
478. R. Hegenröder, K. Niemax, 'Atomic Absorption Spectroscopy with Tunable Semiconductor Diode Lasers', *Trends Anal. Chem.*, **8**(9), 333–335 (1989).
479. A. Obreski, J. Lawrenz, K. Niemax, 'On the Potential and Limitations of Spectroscopic Isotope Ratio Measurements by Resonant Doppler-free Two-photon Laser-enhanced Ionization Spectroscopy', *Spectrochim. Acta*, **45B**(1/2), 15–36 (1990).
480. T. Imasaka, 'Analytical Molecular Spectroscopy with Diode Lasers', *Spectrochim. Acta Rev.*, **15**(5), 329–348 (1993).
481. T.E. Barber, P.E. Walters, J.D. Winefordner, N. Omenetto, 'Evaluation of Absolute Number Densities by Diode Laser Atomic Absorption Spectroscopy', *Appl. Spectrosc.*, **45**(4), 524–526 (1991).
482. P.W. Milonni, J.H. Eberly, *Lasers*, John Wiley & Sons, New York, 1988.
483. J. Hecht, 'Long-wavelength Diode Lasers are Tailored for Fiber optics', *Laser Focus World*, **Aug.**, 79–89 (1992).
484. J. Hecht, 'Diode-laser Performance Rises as Structures Shrink', *Laser Focus World*, **May**, 127–143 (1992).
485. W.R. Trutna, P. Zorabedian, 'Research on External-cavity Lasers', *Hewlett-Packard J.*, 35–38 (1993).
486. J. Hecht, 'Semiconductor Diode Lasers Span the Rainbow', *Laser Focus World*, **April**, 199–211 (1993).
487. J. Hecht, 'Gallium Arsenide Lasers offer an Array of Options', *Laser Focus World*, **July**, 83–92 (1993).
488. D.W. Nam, R.G. Waarts, 'Advanced Laser Diodes bring Compact Blue-green Sources to Light', *Laser Focus World*, **Aug.**, 49–55 (1994).
489. T.V. Higgins, 'The Smaller, Cheaper, Faster World of the Laser Diode', *Laser Focus World*, **April**, 65–76 (1995).
490. J. Lawrenz, K. Niemax, 'A Semiconductor Diode Laser Spectrometer for Laser Spectrochemistry', *Spectrochim. Acta*, **44B**(2), 155–164 (1989).
491. H.R. Telle, 'Stabilization and Modulation Schemes of Laser Diodes for Applied Spectroscopy', *Spectrochim. Acta Rev.*, **15**(5), 301–327 (1993).
492. A.W. Mantz, 'A Review of the Applicability of Tunable Diode-laser Spectroscopy at High-sensitivity', *Microchem. J.*, **50**(3), 351–364 (1994).
493. R.U. Martinelli, 'Mid-infrared Wavelengths Enhance Trace-gas Sensing', *Laser Focus World*, **March**, 77–81 (1996).
494. H. Groll, K. Niemax, 'Multielement Diode Laser Atomic Absorption Spectrometry in Graphite Tube Furnace and Analytical Flames', *Spectrochim. Acta*, **48B**(5), 633–641 (1993).
495. C. Schnürer-Patschan, A. Zybin, H. Groll, K. Niemax, 'Improvement in Detection Limit in Graphite Furnace Diode Laser Atomic Absorption Spectrometry by Wavelength Modulation Technique', *J. Anal. At. Spectrom.*, **8**, 1103–1107 (1993).
496. A. Zybin, C. Schnürer-Patschan, K. Niemax, 'Measurements of $C_2F_4Cl_2$, CCl_4, CHF_3 and O_2 by Wavelength Modulated Laser Absorption Spectroscopy of Excited Cl, F and O in a DC Discharge Applying Semiconductor Diode Lasers', *Spectrochim. Acta*, **48B**(14), 1713–1718 (1993).
497. C. Schnürer-Patschan, K. Niemax, 'Elemental Selective Detection of Chlorine in Capillary Gas Chromatography by Wavelength Modulation Diode Laser Atomic Absorption Spectrometry in a Microwave Induced Plasma', *Spectrochim. Acta*, **50**, 963–969 (1995).
498. A. Zybin, C. Schnürer-Patschan, K. Niemax, 'Wavelength Modulation Diode Laser Atomic Spectrometry in Modulated Low-pressure Helium Plasmas for Element-selective Detection in Gas Chromatography', *J. Anal. At. Spectrom.*, **10**, 563–567 (1995).
499. V. Liger, A. Zybin, Y. Kuritsyn, K. Niemax, 'Diode-laser Atomic-absorption Spectrometry by the Double-beam–Double-modulation Technique', *Spectrochim. Acta*, **52B**(8), 1125–1138 (1997).
500. H.D. Wizemann, K. Niemax, 'Cancellation of Matrix Effects and Calibration by Isotope Dilution in Isotope-selective Diode Laser Atomic Absorption Spectrometry', *Anal. Chem.*, **69**(20), 4291–4293 (1997).
501. A. Zybin, K. Niemax, 'GC Analysis of Chlorinated Hydrocarbons in Oil and Chlorophenols in Pliant Extracts Applying Element-selective Diode Laser Plasma Detection', *Anal. Chem.*, **69**(4), 755–757 (1997).
502. A. Zybin, K. Niemax, 'Improvement of the Wavelength Tunability of Etalon-type Laser Diodes and Mode Recognition and Stabilization in Diode Laser Spectrometers', *Spectrochim. Acta*, **52B**(8), 1215–1221 (1997).
503. A. Zybin, C. Schnürer-Patschan, M.A. Bolshov, K. Niemax, 'Elemental Analysis by Diode Laser Spectroscopy', *Trends Anal. Chem.*, **17**(8–9), 513–520 (1998).
504. P.A. Johnson, J.A. Vera, B.W. Smith, J.D. Winefordner, 'Determination of Rubidium Using Laser Diode Excited Atomic Fluorescence', *Spectrosc. Lett.*, **21**(7), 607–612 (1988).
505. K.C. Ng, A.H. Ali, T.E. Barber, J.D. Winefordner, 'Multiple Mode Semiconductor Diode Laser as a Spectral Line Source for Graphite Furnace Atomic Absorption Spectroscopy', *Anal. Chem.*, **62**, 1893–1895 (1990).

506. K.C. Ng, A.H. Ali, T.E. Barber, J.D. Winefordner, 'Flame Atomic Absorption Spectroscopy Using a Single-mode Laser Diode as the Line Source', *Appl. Spectrosc.*, **44**(6), 1094–1096 (1990).
507. K.C. Ng, A.H. Ali, T.E. Barber, J.D. Winefordner, 'The Applicability of a Multiple-mode Diode Laser in Flame Atomic Absorption Spectroscopy', *Appl. Spectrosc.*, **44**(5), 849–852 (1990).
508. H. Groll, C. Schnürer-Patschan, Y. Kuritsyn, K. Niemax, 'Wavelength Modulation Diode Laser Atomic Absorption Spectrometry in Analytical Flames', *Spectrochim. Acta*, **49B**(12–14), 1463–1472 (1994).
509. R. Hegenröder, D. Veza, K. Niemax, 'Detection Limit and Selectivity for Lithium Isotopes in Continuous Wave Field Ionization Laser Spectroscopy', *Spectrochim. Acta*, **48B**(4), 589–596 (1993).
510. P.E. Walters, T.E. Barber, M.W. Wensing, J.D. Winefordner, 'A Diode Laser Wavelength Reference System Applied to the Determination of Rb in Atomic Fluorescence Spectroscopy', *Spectrochim. Acta*, **46B**(6/7), 1015–1020 (1991).
511. D.H. McIntyre, C.E. Fairchild, J. Cooper, R. Walser, 'Diode-laser Noise Spectroscopy of Rubidium', *Opt. Lett.*, **18**(21), 1816–1818 (1993).
512. T.E. Barber, P.E. Walters, M.W. Wensing, J.D. Winefordner, 'Diode Laser Atomic Absorption Using a New Reference Method', *Spectrochim. Acta*, **46B**(6/7), 1009–1014 (1991).
513. P. Ljung, O. Axner, 'Measurements of Rubidium in Standard Reference Samples by Wavelength-modulation Diode Laser Absorption Spectrometry in a Graphite Furnace', *Spectrochim. Acta*, **52B**(3), 305–319 (1997).
514. K. Hayasaka, M. Watanabe, H. Imajo, R. Ohmukai, S. Urabe, 'Tunable 397-nm Light Source for Spectroscopy Obtained by Frequency Doubling of a Diode Laser', *Appl. Opt.*, **33**(12), 2290–2293 (1994).
515. W.P. Risk, W.J. Kozlovsky, S.D. Lau, 'Generation of 425-nm Light by Waveguide Frequency Doubling of a GaAlAs Laser Diode in an Extended-cavity Configuration', *Appl. Phys. Lett.*, **63**(23), 3134–3136 (1993).
516. J. Franzke, J. Brust, C. Vadla, H.D. Wizemann, K. Niemax, 'Second Harmonic Generation by Applying Injection-locked Radiation from a High-power, Broad-stripe Laser Diode', *Spectrochim. Acta*, **53B**(5), 763–768 (1998).
517. F. Slemr, G.W. Harris, D.R. Hastie, G.I. Mackay, H.I. Schiff, 'Measurement of Gas Phase Hydrogen Peroxide in Air by Tunable Diode Laser Absorption Spectrometry', *J. Geophys. Res.*, **91**, 5371–5378 (1986).
518. A. Stanton, C. Hovde, 'Near-infrared Diode Lasers Measure Greenhouse Gases', *Laser Focus World*, **Aug.**, 117–120 (1992).
519. L.-G. Wang, H. Riris, C.B. Carlisle, T.F. Gallagher, 'Comparison of Approaches to Modulation Spectroscopy with GaAlAs Semiconductor Lasers: Application to Water Vapor', *Appl. Opt.*, **27**(10), 2071–2077 (1988).
520. L.-G. Wang, D.A. Tate, H. Riris, T.F. Gallagher, 'High-sensitivity Frequency-modulation Spectroscopy with a GaAlAs Diode Laser', *J. Opt. Soc. Am. B*, **6**(5), 871–876 (1989).
521. G.S. Hurst, M.G. Payne, 'Elemental Analysis Using Resonance Ionization Spectroscopy', *Spectrochim. Acta*, **43B**, 715–726 (1988).
522. H. Groll, G. Schaldach, H. Berndt, K. Niemax, 'Measurement of Cr(III)/Cr(VI) Species by Wavelength Modulation Diode Laser Flame Atomic Absorption Spectrometry', *Spectrochim. Acta*, **50B**, 1293–1298 (1995).
523. P. Kluczynski, O. Axner, 'Theoretical Description of Wavelength Modulation Spectrometry in Terms of Analytical and Background Signals based upon Fourier Analysis', *Appl. Opt.*, **38**, 5803–5815 (1999).
524. G.C. Bjorklund, 'Frequency-modulation Spectroscopy: A New Method for Measuring Weak Absorption and Dispersion Lineshapes', *Opt. Lett.*, **5**(1), 15–17 (1980).
525. J.L. Hall, L. Hollberg, T. Baer, H.G. Robinson, 'Optical Heterodyne Saturation Spectroscopy', *Appl. Phys. Lett.*, **39**, 680–682 (1981).
526. W. Lenth, 'High Frequency Heterodyne Spectroscopy with Current-modulated Diode Lasers', *IEEE J. Quantum Electron.*, **QE-20**(9), 1045–1050 (1984).
527. M. Gehrtz, W. Lenth, A.T. Young, H.S. Johnston, 'High-frequency-modulation Spectroscopy with a Lead-salt Diode Laser', *Opt. Lett.*, **11**(3), 132–134 (1986).
528. M. Gehrtz, G.C. Bjorklund, E.A. Whittaker, 'Quantum-limited Laser Frequency-modulation Spectroscopy', *J. Opt. Soc. Am. B*, **2**(9), 1510–1526 (1985).
529. C.B. Carlisle, D.E. Cooper, H. Preier, 'Quantum Noise-limited FM Spectroscopy with a Lead-salt Diode Laser', *Appl. Opt.*, **28**(13), 2567–2576 (1989).
530. P. Werle, F. Slemr, M. Gehrtz, C. Bräuchle, 'Quantum-limited FM-spectroscopy with a Lead-salt Diode Laser', *Appl. Phys. B*, **49**, 99–108 (1989).
531. J.M. Supplee, E.A. Whittaker, W. Lenth, 'Theoretical Description of Frequency Modulation and Wavelength Modulation Spectroscopy', *Appl. Opt.*, **33**(27), 6294–6302 (1994).
532. G.R. Janik, C.B. Carlisle, T.F. Gallagher, 'Two-tone Frequency-modulation Spectroscopy', *J. Opt. Soc. Am. B*, **3**(8), 1070–1074 (1986).
533. G.C. Bjorklund, M.D. Levenson, W. Lenth, C. Ortiz, 'Frequency Modulation (FM) Spectroscopy – Theory of Lineshapes and Signal-to-noise Analysis', *Appl. Phys. B*, **32**, 145–152 (1983).
534. N. Hadgu, J. Gustafsson, W. Frech, O. Axner, 'Rubidium Atom Distribution and Nonspectral Interference Effects in Transversely Heated Graphite Atomizers Evaluated by Wavelength Modulated Diode Laser Absorption Spectrometry', *Spectrochim. Acta*, **53B**, 923–943 (1998).
535. D. Rojas, P. Ljung, O. Axner, 'An Investigation of the $2f$-wavelength Modulation Technique for Detection of

Atoms under Optically Thin as well as Thick Conditions', *Spectrochim. Acta*, **52B**(11), 1663–1686 (1997).

536. J. Gustafsson, D. Rojas, O. Axner, 'The Influence of Hyperfine Structure and Isotope Shift on the Detection of Rb Atoms in Atmospheric Pressure Atomizers by the 2*f*-wavelength Modulation Technique', *Spectrochim. Acta*, **52B**, 1937–1953 (1997).
537. J. Gustafsson, O. Axner, 'The Influence of Hyperfine Structure and Isotope Shift on the Detection of Rb by 2*f*-wavelength Modulation Diode Laser Absorption Spectrometry – Experimental Verification of Simulations', *Spectrochim. Acta*, **53B**, 1895–1905 (1998).
538. J. Gustafsson, N. Chekalin, D. Rojas, O. Axner, 'Extension of the Dynamic Range of the 2*f*-wavelength Modulated Diode Laser Absorption Spectrometry Technique – Detection of Atoms under Optically Thick Conditions', *Spectrochim. Acta*, **55B**, 237–262 (2000).
539. J. Gustafsson, O. Axner, 'Theoretical Investigation of the Temperature Dependence of the 2*f*-wavelength Modulation Diode-laser Absorption Signal', *Spectrochim. Acta*, **53B**, 1827–1846 (1998).
540. D.E. Cooper, R.E. Warren, 'Frequency Modulation Spectroscopy with Lead-salt Diode Lasers: A Comparison of Single-tone and Two-tone Techniques', *Appl. Opt.*, **26**, 3726–3732 (1987).
541. J. Reid, D. Labrie, 'Second-harmonic Detection with Tunable Diode Lasers – Comparison of Experiment and Theory', *Appl. Phys. B*, **26**, 203–210 (1981).
542. M.L. Olsen, D.L. Grieble, P.R. Griffiths, 'Second Derivative Tunable Diode Laser Spectrometry for Line Profile Determination. I. Theory', *Appl. Spectrosc.*, **34**(1), 50–56 (1980).
543. L.C. Philippe, R.K. Hanson, 'Laser Diode Wavelength-modulation Spectroscopy for Simultaneous Measurement of Temperature, Pressure, and Velocity in Shock-heated Oxygen Flows', *Appl. Opt.*, **32**(30), 6090–6103 (1993).
544. E.I. Moses, C.L. Tang, 'High-sensitivity Laser Wavelength-modulation Spectroscopy', *Opt. Lett.*, **1**, 115–117 (1977).
545. R. Arndt, 'Analytical Line Shapes for Lorentzian Signals Broadened by Modulation', *J. Appl. Phys.*, **36**(8), 2522–2524 (1965).
546. G.V.H. Wilson, 'Modulation Broadening of NMR and ESR Line Shapes', *J. Appl. Phys.*, **34**(11), 3276–3285 (1963).
547. X. Zhu, D.T. Cassidy, 'Modulation Spectroscopy with a Semiconductor Diode Laser by Injection-current Modulation', *J. Opt. Soc. Am. B*, **14**(8), 1945–1950 (1997).
548. D.M. Bruce, D.T. Cassidy, 'Detection of Oxygen Using Short-extended-cavity GaAs Semiconductor Diode Laser', *Appl. Opt.*, **29**, 1327–1332 (1990).
549. J.A. Silver, A.C. Stanton, 'Optical Interference Fringe Reduction in Laser Absorption Experiments', *Appl. Opt.*, **27**(10), 1914–1916 (1988).
550. C.R. Webster, 'Brewster-plate Spoiler: A Novel Method for Reducing the Amplitude of Interference Fringes that Limit Tunable-laser Absorption Sensitivities', *J. Opt. Soc. Am. B*, **2**(9), 1464–1470 (1985).
551. H.C. Sun, E.A. Whittaker, 'Novel Etalon Fringe Rejection Technique for Laser Absorption Spectroscopy', *Appl. Opt.*, **31**(24), 4998–5002 (1992).
552. M.H. Anderson, J.R. Ensher, M.R. Matthews, C.E. Wieman, E.A. Cornell, 'Observation of Bose–Einstein Condensation in a Dilute Atomic Vapor', *Science*, **269**, 198–199 (1995).

Laser-induced Breakdown Spectroscopy

David A. Cremers and Andrew K. Knight
Los Alamos National Laboratory, Los Alamos, USA

For references see page 9612

Laser-induced breakdown spectroscopy (LIBS) is a novel method of elemental analysis based on a laser-generated plasma. Pulses from a laser are focused on a sample to atomize a small amount of material resulting in the formation of a microplasma. Because of the high plasma temperature, the resulting atoms are electronically excited to emit light. The plasma light is spectrally resolved and detected to determine the elemental composition of the sample based on the unique emission spectrum of each element. Because of the simplicity of the method, it is suited for analyses that cannot be carried out using conventional methods of atomic emission spectroscopy (AES). This is particularly true for measurements that must be conducted outside of an analytical laboratory. A particular advantage of LIBS is the ability to analyze most types of samples without any preparation. This means that samples can be interrogated in situ, providing rapid measurement capability and permitting the method to be used in the analysis of gases, liquids, and solids in a variety of different sampling configurations. Although LIBS provides sensitive detection for many elements, it is not an ultrasensitive detection technique. In addition, under field conditions, the method typically does not provide the high accuracy and precision offered by laboratory-based methods of AES.

1 INTRODUCTION

LIBS is a method of elemental analysis based on AES.[1–3] Three basic steps are involved in AES measurements. These are (1) vaporization of the sample to produce free atoms, (2) electronic excitation of the atoms to induce optical emissions at discrete wavelengths indicative of the atoms, and (3) collection, recording and analysis of the optical emissions. Atoms in the sample are identified by their unique spectral emissions or 'fingerprint' spectra. In conventional AES methods, a plasma source is used for steps (1) and (2). Conventional plasma sources include arcs and sparks and the inductively coupled plasma (ICP). These sources all require the use of some physical device to deliver electrical energy to form and sustain the plasma such as metal electrodes (for sparks and arcs) and a metal coil (ICP). In LIBS, however, the plasma or laser spark is formed by intense optical radiation focused on the sample to produce dielectric breakdown leading to plasma formation.

List of selected abbreviations appears in Volume 15

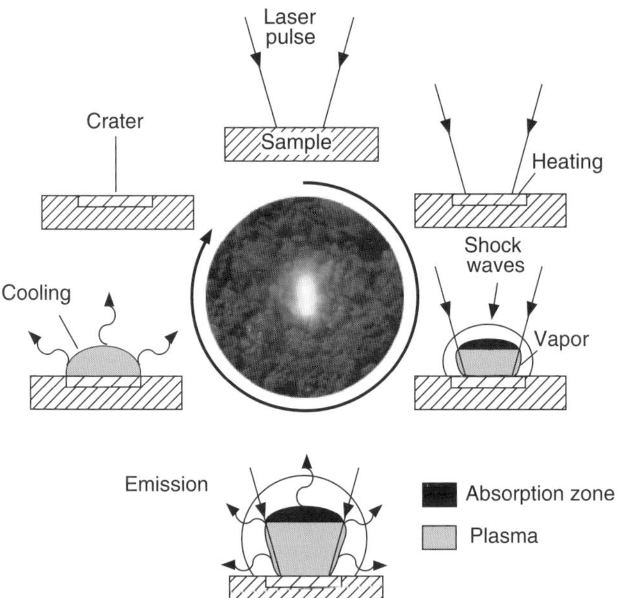

Figure 1 Diagram showing the main events in the interrogation of a solid surface by a laser pulse during a LIBS measurement.

The basic steps involved in a LIBS measurement are shown in Figure 1. Focusing a laser pulse on the sample produces local heating of material in the focal volume resulting in the ejection of a small mass of material in the form of solid particles, liquid drops, and atomic vapor. Via interactions between the incident laser pulse, electrons, and free atoms, a hot plasma is formed above the target surface. Through collisions between the different species in the plasma the atoms are electronically excited to emit light at discrete frequencies unique to each species. These unique spectral 'fingerprints' permit identification of the species through collection and analysis of the plasma light. Because the laser plasma is formed by a laser pulse of short duration (typically 5–20 ns), the plasma decays and cools rapidly, usually within tens of microseconds after formation. The ejected material results in a small crater formed on the sample surface.

Because the LIBS plasma is formed by light, it represents a unique method of elemental analysis having many advantages over conventional methods of AES. These advantages include (1) in situ analysis (only optical access to the sample is required), (2) little or no sample preparation, and (3) the ability to sample gases, liquids, and solids directly. On the other hand, like other plasma-based methods of AES, it provides for (1) simultaneous multi-element detection and (2) exhibits good sensitivity for many elements. The main limitations of LIBS relate to the sampling characteristics of the technique. Specifically, that the amount of sample vaporized and the plasma characteristics (e.g. temperature, electron density) are strongly dependent on the parameters of the laser pulse, the physical characteristics of the target sample, and the sampling geometry. These limitations are manifested in terms of limited accuracy and precision compared to more conventional analytical methods.

2 THEORY

2.1 Fundamentals of Laser-induced Breakdown

The LIBS method is based on the formation of a plasma, or highly ionized gas, within the focus of the laser pulse directed either into a material (gas or liquid) or onto the surface of a solid target. Plasma formation is accompanied by a bright flash of light and a loud 'snap' emanating from the focal region. Detailed analysis of the physics of laser-induced breakdown can be found elsewhere.[4,5] Briefly, the plasma is formed by the transfer of optical energy from the laser pulse to atoms, ions, and electrons resulting from deposition of energy into the target. This energy transfer occurs through the process of inverse bremmstrahlung absorption involving interactions between free electrons and atoms and ions. In order for the inverse bremmstrahlung process to be of sufficient strength to produce breakdown, a high density of electrons (and ions) must be present. Two mechanisms accounting for electron (e^-) production are described by two processes involving atoms (M) (Equations 1 and 2):

$$e^- + M \longrightarrow 2e^- + M^+ \tag{1}$$

$$M + nh\nu \longrightarrow M^+ + e^- \tag{2}$$

Mechanism (1) involves absorption of laser radiation by an electron during a collision with a neutral atom. If sufficient energy is absorbed to ionize the atom then two free electrons result. If this process continues generating two electrons for each one electron involved in the collision then what is termed cascade electron growth results. Mechanism (2) is termed multiphoton ionization. In this process a number of photons (n) each of energy $h\nu$ are absorbed simultaneously by an atom resulting in a free electron and an ionized species. In order for this process to occur, it must be that $nh\nu > Ip$, where Ip is the ionization potential of the atom.

2.2 Post-breakdown Phenomena

Following formation of the initial plasma by the laser pulse, the plasma passes through several different phases. Detailed descriptions of these can be found in the literature.[4,6] Here a brief overview is presented. Once free electrons have been generated in the focal volume, the plasma grows through the process of inverse bremsstrahlung (IB) of which two types can be identified, electron–neutral and electron–ion, each being important during different stages of plasma growth. When few electrons are present, electron–neutral IB (IB_{e-n}) predominates but as plasma formation continues to the point at which about 1% ionization occurs, electron–ion IB (IB_{e-i}) becomes the chief absorption process. The absorption coefficients (α) for each type are given in Equations (3) and (4):

$$\alpha(IB_{e-n}) = \left[1 - \exp\left(\frac{-hc}{\lambda kT}\right)\right] \sum Q_j n_e n_j \tag{3}$$

$$\alpha(IB_{e-i}) = \left[1 - \exp\left(\frac{-hc}{\lambda kT}\right)\right] \left(\frac{4e^6\lambda^3}{3hc^4 m_e}\right) \times \left(\frac{2\pi}{3m_e kT}\right)^{1/2} n_e \sum z_i^2 n_i g_i \tag{4}$$

where:

Q_j = average cross-section for absorption of a photon (wavelength λ) as a result of collisions with the jth species
n_e = electron number density
n_i and n_j = number density of ith and jth species
z_i = charge of the ith species
g_i = Gaunt factor.

The temperature regime over which the change from the two types of IB absorption occurs is between 6000 and 10 000 K, which is also the temperature regime of a LIBS plasma. Both absorption coefficients decrease with increased temperature with the coefficient $\alpha(IB_{e-i})$ having the stronger dependence. Because this coefficient depends on the square of the electron density compared to the linear dependence for $\alpha(IB_{e-n})$, electron–ion IB predominates as the temperature and hence the electron density increase.

Following the formation of an absorbing plasma within the focal zone, the plasma will continue to grow away from the surface towards the direction of the incoming laser pulse. This growth occurs because the laser pulse feeds energy into the plasma with most of the absorption occurring in the upper boundary region of the plasma as shown in Figure 1. This process accounts for the inverted cone shape of the plasma observed in the focal volume. The exact character of the evolving plasma will depend on parameters of the laser pulse (irradiance, spot size, wavelength) and the surrounding atmosphere. Three distinct evolutionary paths can be described in terms of the characteristics of absorption and shock or pressure waves emanating from the focal zone. The three types are classed as either laser-supported combustion (LSC), laser-supported detonation (LSD) or laser-supported radiation (LSR) waves. The interested reader is directed to the

For references see page 9612

literature for a discussion of the characteristics of each type.[4,6]

2.3 Dynamics of Laser Ablation

The interaction between a laser pulse of high power density and a material is a complicated process. A thorough consideration of the interaction depends on characteristics of the laser pulse and the material. Several authors have considered the interaction in detail.[7,8] Here we present a brief description of the main processes involved.

The laser pulse incident on the surface interacts mainly with the electrons in the bulk material. These interactions involve absorption of energy from the pulse by electrons that in turn collide with other electrons and with other constituents of the matrix. These collisions result in the transfer of energy from the laser pulse to the material in the focal volume resulting in heating. This heating occurs rapidly and results in melting of a thin surface layer with some of the heat being conducted away into the bulk matrix. Because of the high power densities on the surface however, the melted material continues to be heated and when the deposited energy exceeds the latent heat of vaporization, atomized material is ejected from the surface. Some solid and liquid particles may also be ejected via the strong shock/pressure waves generated at the surface.

An estimate of the minimum laser pulse-power density required to produce vaporization ($P_{\min}$) can be obtained from Equation (5):

$$P_{\min} = \rho L_{\mathrm{v}} \kappa^{1/2} \Delta t^{-1/2} \quad (5)$$

where:

ρ = mass density of target
L_{v} = latent heat of vaporization
κ = thermal diffusivity of target
Δt = pulse length.

For pure aluminum metal, $P_{\min} \approx 1.75 \times 10^8\,\mathrm{W\,cm^{-2}}$. Note that this calculation assumes all incident pulse energy goes toward heating therefore representing a lower bound on the minimum power density required for vaporization.

Although ejection of material can proceed over a wide range of power densities, two distinct operating regimes can be identified, one most favorable for quantitative analysis. For low power densities ($<10^8\,\mathrm{W\,cm^{-2}}$), the ejection of positive ions of volatile elements such as Na and K is readily observed from the surface whereas more refractory elements (Mo, W, etc.) remain in the bulk material. In this case the ejected material will not be representative of the bulk composition and therefore operation at these low power densities is not appropriate for quantitative analysis.

At higher power densities ($>10^9\,\mathrm{W\,cm^{-2}}$) significantly more laser energy is transferred into the material resulting in a greater degree of vaporization and atomization of a more representative distribution of elements from the molten pool. In addition, because of the greater mass of ablated material, the incident laser pulse also interacts with electrons and ions in the plume above the surface resulting in increased excitation of atomized species. Because laser ablation is more pronounced at higher power densities, a visible depression or crater is left on the target surface in the interaction zone. Typically, the mass of material ejected from the surface is very small. Assuming again that all energy goes toward the ablation process, a simple calculation shows that the upper limit to the amount of material that can be vaporized by a 100 mJ pulse is about 6.6 µg. Because of reflection losses at the surface and the conversion of some of the incident energy into light and sound, the amount of material actually vaporized will be significantly less. Measurements indicate, for example, that about 10 ng are ablated from aluminum metal using a laser pulse of 175 mJ.

3 LASER-INDUCED BREAKDOWN SPECTROSCOPY INSTRUMENTATION

3.1 General

A diagram of a generalized LIBS instrument is shown in Figure 2. The main components include a laser, a method of spectrally resolving or of spectrally selecting a certain narrow region of the spectrum to monitor, and a method of detecting the spectrally selected light. The specifications of each component as well as the method of sampling used will depend on the application. Factors to consider include (1) the elements to be monitored (number and type), (2) the characteristics of the sample (complexity, homogeneity, etc.), (3) the type of analysis (e.g. a qualitative versus quantitative measurement), and (4) the state of the sample (e.g. gas, liquid, or solid).

3.2 Laser Systems

The parameters important in the specification of the laser to be used for LIBS include (1) pulse energy, (2) pulse repetition rate, (3) beam mode quality, (4) size/weight, and (5) cooling and electrical power requirements. The wavelength of the laser beam is not an important factor in most cases.

Solid state lasers, in particular pulsed and Q-switched Nd : YAG lasers having pulsewidths less than 15 ns, are typically used for LIBS measurements because these

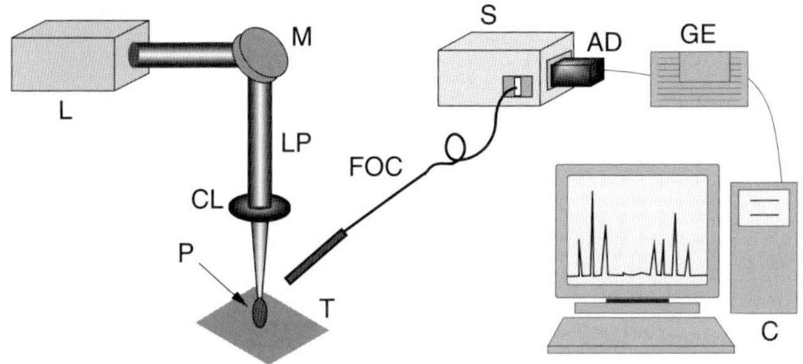

Figure 2 Diagram of a typical LIBS apparatus. L = laser; M = mirror; LP = laser pulse; CL = converging lens; P = plasma; T = target; FOC = fiber optic cable; S = spectrograph; AD = array detector; GE = gating electronics; C = computer.

lasers are a compact and convenient source of the powerful pulses needed to generate the laser plasmas.[9] The fundamental wavelength of the Nd:YAG laser (1064 nm) can easily be converted to shorter wavelengths (532, 355, and 266 nm) via passive harmonic generation techniques which may have certain advantages in terms of increased energy coupling into a particular sample, but typically the 1064 nm wavelength is used because this provides the highest power density. Other types of lasers, most notably the pulsed CO_2 laser (10 600 nm wavelength) and the excimer laser (typical wavelengths of 193, 248, 308 nm) have been used for LIBS. In comparison to solid state lasers, however, these lasers require more maintenance (e.g. change in gases) and special optical materials because their wavelengths lie in the far-infrared and ultraviolet spectral regions, respectively. For this reason these lasers are not widely used.

Another advantage of the Nd : YAG laser is the variety of different sizes available commercially. These range from laboratory-based models which can output a Joule or more of pulse energy at repetition rates between 10 and 50 Hz to small hand-held versions with a repetition rate of 1 Hz and a pulse energy of about 17 mJ. The laboratory models require 208 VAC electrical services of at least 20 A and may require external water cooling or at least a heat exchanger. The hand-held versions are air cooled, and can be operated from batteries or low voltage direct current sources.

3.3 Methods of Spectral Resolution

The basis of a LIBS measurement is the collection and analysis of an emission spectrum. The emission lines of the elements are tabulated in various sources.[10,11] Examples of emission spectra from three different samples are shown in Figure 3.

All samples contained the element Si having a strong emission at 288.1 nm but each sample differed in the number of other elements present as major and minor

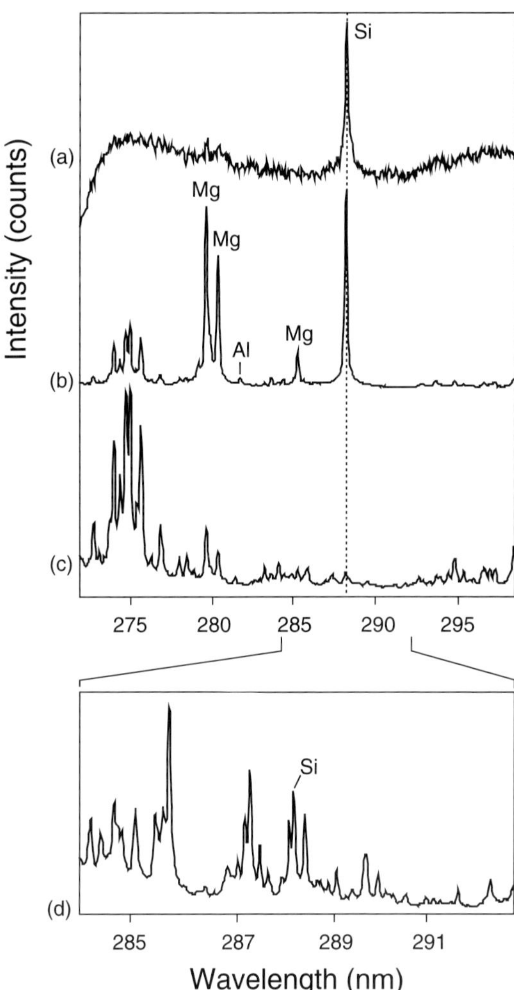

Figure 3 Emission line of Si (288.1 nm) observed from different matrices using LIBS. Spectra (a), (b), and (c) were obtained using a spectrograph of moderate resolution. (a) Si in water. (b) Si in aluminum alloy. (c) Si in steel. (d) Si in steel obtained using 3× greater spectra resolution than shown in (c).

species. In general, the greater the number of elements the more complicated the spectrum. The simplest spectrum is that of Si in water [Figure 3a]. Here the Si line appears alone without interferences from either H or O which have few emission lines making the emission spectrum particularly simple. Silicon in aluminum metal is readily observed in Figure 3(b) although lines due to Mg, Al, and Fe are also present. Steel represents a complex matrix because of the large number of Fe lines and steels typically contain a large number of other elements. The Si line from a steel sample is apparent in Figure 3(c) but because it is adjacent to an Fe line it is not completely resolved in this spectrum. Using a spectral resolution method having greater resolving power, however, the Si can be separated to some extent from the adjacent Fe line as shown in Figure 3(d). The spectra shown in Figure 3 demonstrate

For references see page 9612

that the complexity of the sample will determine the method of spectral resolution needed to monitor the element(s) of interest. Methods of providing spectral discrimination of the plasma light are listed below.[12]

- Narrow bandpass (<1 nm) fixed wavelength line filter.
- An acousto-optic tunable filter (AOTF) consisting of a crystalline material (e.g. TeO_2) to which a radio-frequency wave is applied.[13] By adjusting the frequency of the wave, the bandpass of the AOTF can be varied continuously over a certain range.
- A monochromator that is tuned to monitor a single wavelength and this wavelength is presented at the exit slit of the device for detection.
- A spectrograph is similar to a monochromator except it has an exit plane at which a continuous range of wavelengths is presented for detection using some type of AD or a series of single wavelength detectors positioned behind individual slits.

Diagrams of methods (1) and (2) are shown in Figure 4(a). In the case of the narrow bandpass filter or AOTF, only a single narrow wavelength band is passed through the wavelength selective element. The transmitted light is then detected using some sort of light detector (section 3.4). The advantage of the fixed wavelength filter is very small size and low weight and cost. The AOTF, on the other hand, is somewhat larger and requires a power supply, but it can be tuned to monitor different wavelength regions. The use of a monochromator or spectrograph is diagrammed in Figure 4(b). The monochromator can be considered a spectrograph with a slit at the exit focal plane so that only a single narrow wavelength region is monitored at one time. The spectrograph, on the other hand, provides simultaneous detection over a wide spectral range because the single exit slit is removed and an AD is used to record the spectrum or a series of exit slits each with its own detector is placed along the focal plane.

The most suitable method of spectral resolution depends mainly on the analysis requirements. Factors that must be considered include (1) the complexity of the sample (i.e. how many elements are in the sample and do these elements have many emission lines or only a few strong lines?), (2) the number of elements to be monitored, (3) whether the elements are to be monitored simultaneously or is sequential detection adequate, and (4) the location of the emission lines in the spectrum.

Although not all possible combinations of the spectral resolution methods can be considered here because of the wide range of factors that will influence the final choice, Table 1 summarizes the capabilities of each method. It should be noted that in the case of the fixed wavelength

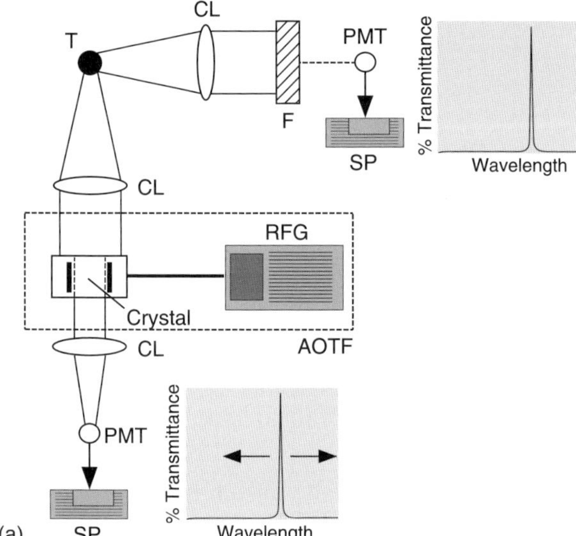

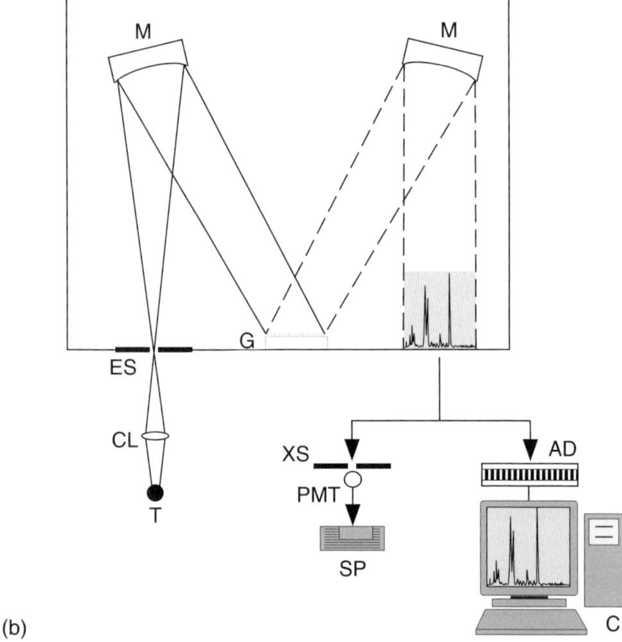

Figure 4 Methods of spectral resolution of the plasma light using (a) filter-based methods and (b) spectrometer-based methods. T = target; CL = converging lens; F = filter; photomultiplier tube (PMT); SP = signal processing; RFG = radiofrequency generator; AOTF = acousto-optic tunable filter; M = mirror; G = grating; ES = entrance slit; XS = exit slit; AD = array detector; C = computer.

methods a second, third, etc. fixed wavelength system may be added to provide simultaneous multiple wavelength detection capabilities.

If only a single element or a few elements are to be monitored then narrow bandpass fixed wavelength filters may be the method of choice in terms of simplicity and small size. These filters can be manufactured having a narrow

Table 1 Comparison of methods of spectral resolution

Method	Wavelength region		Simultaneous detection of multiple wavelengths	Detector
	Fixed	Tunable		
Filter	X		X (using multiple systems)	PD
AOTF	X	X	X (using multiple systems)	PD
Monochromator	X	X		PD
Spectrograph	X	X	X	AD or multiple PDs

PD, photodetector.

bandpass (down to 0.1 nm) to selectively monitor an element line in complex spectra as shown in Figures 3(c) and 3(d). If, on the other hand, multiple wavelengths are to be monitored using a spectrograph, then the spectral resolution and spectral coverage provided by the instrument must be considered. As spectral resolution increases, spectral coverage decreases. For example, in Figure 3(c) a fairly wide spectral range is shown. However, if Si is the element of interest, there appears to be an interference with an adjacent line. Using a spectrograph having resolution three times greater, results in the spectrum shown in Figure 3(d) with the degree of overlap between the Si line and the adjacent line reduced. There is, however, a loss in the spectral coverage so that element lines at both extremes of the spectrum shown in Figure 3(c) will not be recorded with the instrument of higher resolution. In the case of the monochromator and spectrograph, the spectral resolution or ability to resolve adjacent emission lines is determined by the focal length of the device and the number of lines per millimeter on the grating.

3.4 Detectors

The method of spectral selection determines the type of detector used for LIBS measurements.[14,15] The simplest detector consists of a photosensitive material that generates a signal proportional to the amount of light incident on the device. These PDs include PMTs and photodiodes. These devices are used with spectral selection methods such as fixed filters, AOTFs, and monochromators. By placing small photosensitive elements (pixels) in either a linear or 2-dimensional array, an AD is produced that provides spatial information concerning the light pattern incident on the array. Common examples of these ADs include photodiode arrays (PDA), charge coupled devices (CCD) and charge injection devices (CID). ADs are used with spectrographs to record the continuous spectrum presented at the focal plane of the instrument. The spectra shown in Figure 3 were obtained using a CCD detector. Each firing of the laser produced a spectrum.

The type of detector determines the method of signal processing. The signal from the PMT, for example, consists of a current that is converted to a voltage by the recording device. The response of a PMT is typically very fast, less than a few nanoseconds, so it can be used to record the temporal variation of the plasma light at the selected wavelength. If multiple wavelengths must be monitored simultaneously, several slit/PMT assemblies can be placed in the focal plane of a spectrograph with the slit positions aligned to the emission peaks of the elements of interest. Time-resolved detection of an element signal is obtained by electronic processing of the PMT signal using a sample-and-hold circuit.

The spectral coverage provided by an AD is determined by the physical size of the array and the characteristics of the spectrograph. The PDA is a one-dimensional arrangement of diodes that provides spatial intensity information in one dimension. Typical spacing between individual photodiodes is 25 microns so an array of 1024 pixels has a physical length of about 25 mm. The CCD and CID, on the other hand, are two-dimensional arrays of photodiodes that can provide intensity information along two axes. Typical pixel sizes range from 9×9 microns up to 24×24 microns and array formats range from 576×384 pixels up to 3072×2048 pixels with a large number of other formats in between these extremes.

A generalized method of processing the plasma light using a detector is shown in Figure 5. The short duration laser pulse generates the plasma that typically lives for many microseconds after formation. Using either a PD or AD, the plasma light can be integrated over the entire emission period of the plasma. The resultant signal is shown in Figure 5.

In the case of a PMT, the detector current is stored on a capacitor that charges up to a certain voltage as light strikes the detector. In the case of an AD, the charge is stored on the pixel and then is read out at a later time. The magnitude of the voltage or stored charge is proportional to the number of photons incident on each detector. At the end of the period of light collection, the voltage or charge is read out and the signal levels are reset to zero for the next pulse. Instead of integrating the plasma light over the entire plasma lifetime, it may be useful to only monitor a certain period of light emission as the plasma decays. An example would be to discriminate against

For references see page 9612

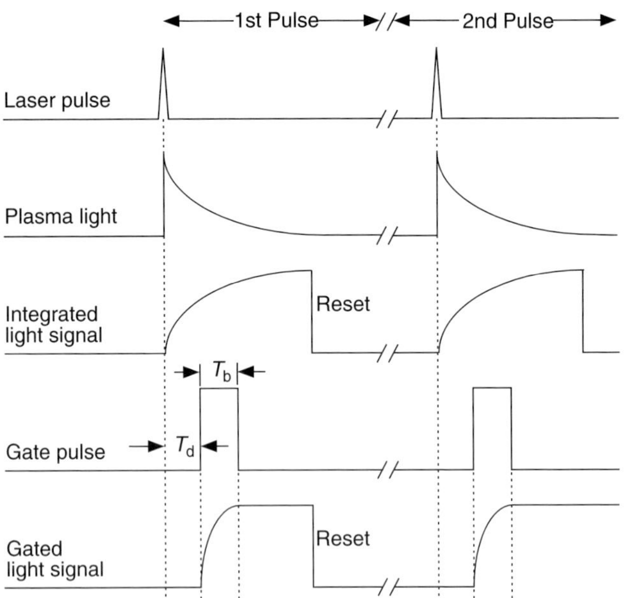

Figure 5 Methods of recording the plasma light during a LIBS measurement. The plasma light decays slowly after the laser pulse. The integrated light signal is the total light recorded by the detector. The gated signal only monitors the light during the gate pulse. After recording the light signals the signal levels are reset to zero to await the signal from the next pulse.

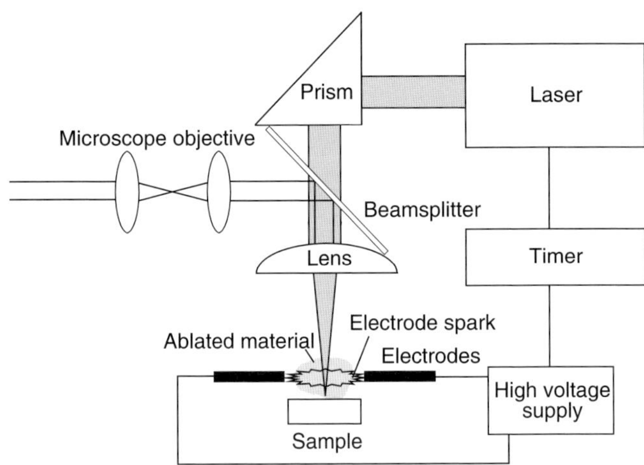

Figure 6 Laser microprobe instrument in which the laser pulse is used to ablate material from a surface that is subsequently analyzed using the spark produced by a pair of electrodes.

the intense white light that emanates from the plasma during the first 1–2 μs. Generally, a gate pulse is applied to the detection electronics that determines the time after plasma formation at which recording of the plasma light is to begin (T_d) and the length of time over which the light is collected (T_b). In the case of a PMT, the signal is processed by a simple circuit such as a gated integrator that only stores the current from the PMT according to the timing designated by an applied gate pulse. When the gate pulse is high, the current is stored. In the case of an AD, the gate pulse is directed to a microchannel plate image intensifier that only transmits light through to the photosensitive area of the PDA, CCD, or CID when the gate pulse is present.

3.5 Examples of Instruments

3.5.1 *Laser Microprobe*

The first instrument developed for the commercial market based on the laser spark was the laser microprobe.(16) This device combined laser ablation of a solid sample with the spark produced by conventional metal electrodes (Figure 6). In this laboratory-based instrument, the ablating laser pulse produced an aerosol of material above the sample surface that was then interrogated by the spark produced by the electrodes. These types of instruments were made by several manufacturers and were on the market for several years. Detailed reviews of applications of these instruments can be found in the literature.(16) Some additional information on the capabilities of the technique can be found in section 6.2.

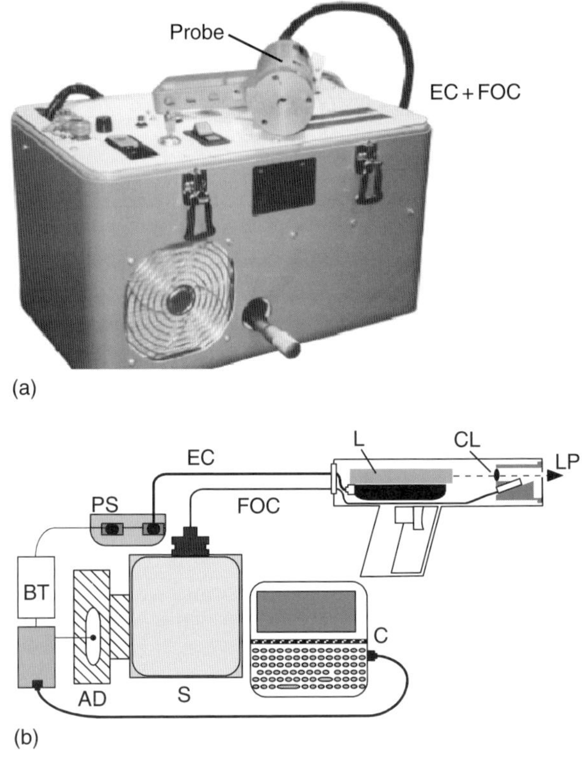

Figure 7 (a) Photograph of field-portable LIBS analyzer and (b) diagram showing main components of device. PS = power supply; EC = electrical cable; L = laser; CL = converging lens; LP = laser pulse; BT = battery; AD = array detector; S = spectrograph; FOC = fiber optic cable; C = computer. (Photograph supplied courtesy of CMEI, Inc.)

3.5.2 Field-portable Instrument

Because of recent developments in miniaturization of components used for LIBS, it has been possible to fabricate truly portable LIBS-based instruments.[17] An example is shown in Figure 7. This device consists of a main analysis unit connected to a hand-held probe. The probe contains the laser, focusing optics, and a fiber optic cable to collect the plasma light. The light is transported back to the analysis unit that houses the small spectrograph, detector, and laser power supply. The instrument is operated through the use of a micro or laptop computer. An analysis is carried out by placing the probe on the sample and then firing the laser. The laser can be repetitively fired to average the spectra from many shots and increase measurement accuracy and precision. The resulting spectrum is then analyzed via software to determine the element signals. Quantitative analysis is possible through calibration of the instrument using calibration standards of known composition.

4 CHARACTERISTICS OF THE LASER PLASMA

The characteristics of the LIBS plasma (e.g. temperature, electron density, size) depend on the state of the sample (gas, liquid, solid) and other parameters such as the pressure and composition of the surrounding atmosphere. Here we present a brief overview of the properties of the laser plasma for different samples.

4.1 On Solids

Solid samples were among the first to be quantitatively analyzed with LIBS. Because most solid targets are opaque at the laser wavelength, the laser plasma is formed on the surface of the solid even though the distance between the target and the focusing lens may be different from the focal length of the lens. As long as the power density is above a threshold value, a laser plasma will be formed. This situation is different from that which occurs in the analysis of a transparent gas or liquid. In these cases, the laser plasma is always formed in the focal volume, the position of which is determined by the focal length of the lens. Because there are a range of lens-to-sample distances (LTSD) over which the plasma can be formed on a solid, the plasma parameters such as temperature and electron density and the amount of material ablated by the pulse have a range of values determined by the LTSD.[18]

For a Nd:YAG laser pulse (186 mJ), typical plasma temperatures range from 6000 to 9500 K with electron densities in the range of 10^{17} to $10^{19}\,cm^{-3}$ immediately after plasma formation. The mass of material ablated from aluminum metal ranges from 5 to 80 ng $pulse^{-1}$. These parameters vary within the specified limits and depend on factors such as ambient pressure and LTSD. As the plasma cools, the temperature and electron density both decrease. Using the white light from the plasma as an indicator, the temporal duration of the plasma ranges out to about 10–20 µs if formed on a metal in air at atmospheric pressure.

4.2 In Gases

The laser spark phenomenon was first discovered by the breakdown of gases shortly after the discovery of the laser. However, it was not until time-resolution was applied that quantitative measurements could be carried out. The LIBS technique allows for the analysis of gases and combustion products as well as the analysis of aerosols and suspended particles in a gas. The typical precision for the analysis of gases is around 5–8% RSD (relative standard deviation). This value is considerably elevated in the analysis of aerosols and airborne particles to around 30% and is probably due to shot-to-shot differences in sampling particles of different sizes and having different locations in the plasma volume.

The spark formed in a gas has a much higher temperature than that formed on a solid. Although the temperature is dependent upon the ambient gas, values up to 20 000 K have been measured as the initial temperature of the plasma. The plasma cools with time to values as low as 12 000 K at 10 µs after plasma formation. The electron density of this high temperature plasma is about the same as measured for solid samples.

4.3 In Liquids

The spark produced in a bulk liquid is significantly different from that produced in a gas or on a solid.[19] The plasma is visibly much smaller and the temporal decay of light emission occurs more rapidly such that useful spectral emissions are absent for times >2 µs. In addition, there is significant temporal overlap between emissions from neutral and ionized species and also simple molecules formed in the recombining plasma so that time resolved detection is less useful for the spark in a liquid.

The electron density of the plasma formed in water is on the order of 10^{18} to $10^{19}\,cm^{-3}$ with temperatures in the range of 7000–12 000 K.

5 ANALYTICAL CAPABILITIES

5.1 Advantages

The analytical capabilities of a method are usually specified by a set of measurable parameters called

For references see page 9612

figures-of-merit. The parameters we use here are (1) the limit of detection (LOD) for an element in a specified matrix and measurement, (2) precision, and (3) accuracy. Methods of computing the LOD can be found in the literature.[20,21] Here we use LOD = 3(SD)/m where SD is the standard deviation of replicate measurements of the element signal using the same sample and m is the slope of the linear calibration curve of element signal or referenced element signal (see section 5.2.2) vs. element concentration. Precision refers to how reproducibly an element signal or concentration can be measured and it is specified in terms of % RSD determined by repeating a measurement several times (preferably > 6), determining the average of the element signals (S_{ave}), and then computing %RSD = (SD/S_{ave}) × 100%. Accuracy refers to how close the predicted concentration of an element in a sample (C_{pred}) is to the actual concentration (C_{act}). The figure-of-merit here is %Accuracy where

$$\%\text{Accuracy} = \frac{100\%|C_{act} - C_{pred}|}{C_{act}}.$$

5.1.1 *Range of Analysis Scenarios*

Because the laser spark is formed by focused optical radiation, in addition to the conventional LIBS apparatus shown in Figure 2, several different analysis scenarios can be implemented. Some of these are shown in Figure 8 and have been incorporated into instruments.

5.1.1.1 Direct Analysis This configuration, shown in Figure 2, is used for most LIBS measurements. Here a short focal length lens is used to focus the laser pulses onto the sample (solid, liquid) or into a liquid or gas to form the plasma. The plasma light can be collected using a lens which focuses the light onto a spectrograph slit or collimates the light to pass it directly through a bandpass filter or other frequency selective device (e.g. AOTF). As an alternative, a fiber optic cable can collect and transport the light to a remotely located frequency selective device, typically a spectrograph. This analysis method represents the simplest embodiment of a LIBS apparatus.

5.1.1.2 Fiber Optic Delivery With the development of improved fiber optic materials it is now possible to focus power densities on the order of MW cm^{-2} onto the end of a fiber, without damage, and inject and then transport tens of millijoules of energy through the fiber [Figure 8a].[22] By placing a lens system at the distal end of the fiber, the laser pulse can be focused to produce a spark. Transport of laser pulses over distances up to 100 meters has been demonstrated. Depending on the analysis requirements (e.g. the elements to be monitored, concentrations, etc.) the plasma light can be collected and transported back to the detection system using either the same fiber or a second fiber optic cable.

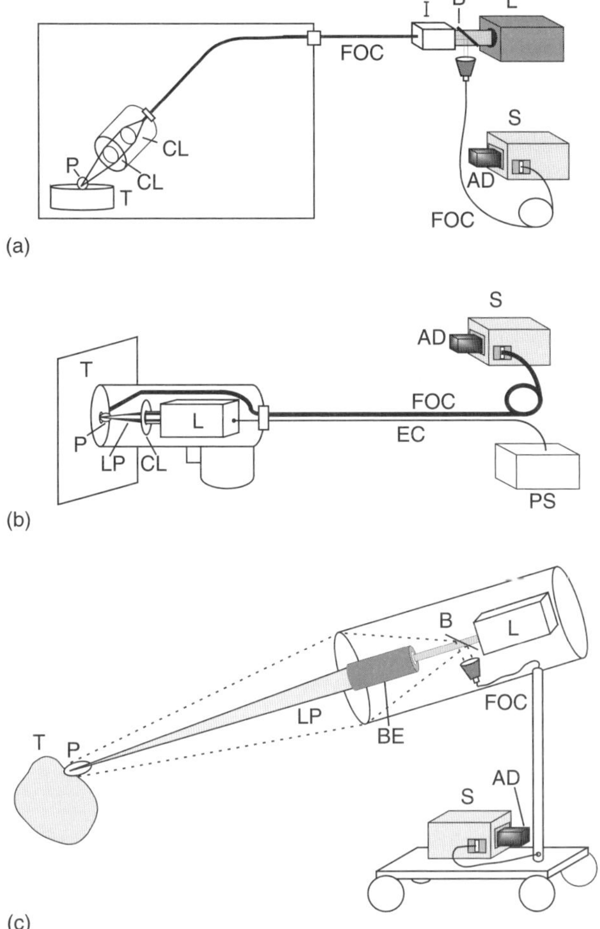

Figure 8 LIBS analysis of materials using different sampling methods. (a) Transport of the laser pulse to the remotely located sample using a fiber optic cable. (b) Incorporation of a small laser into a probe connected to the main analysis unit of a LIBS instrument through cabling. (c) Stand-off analysis in which the laser pulses are focused at a distance onto the remotely located sample. L = laser; B = beamsplitter; I = injector; FOC = fiber optic cable; CL = converging lens; P = plasma; T = target; S = spectrograph; AD = array detector; EC = electrical cable; LP = laser pulse; PS = power supply; BE = beam expander.

5.1.1.3 Compact Probe By combining a very compact laser with fiber optics it is possible to construct a small probe to use for remote LIBS measurements [Figure 8b].[17] The laser power supply and detection system can be located remotely from the probe, connected by an umbilical cable containing electrical cables for the laser and a fiber optic cable to transport the plasma light. This configuration is used in at least one commercial LIBS unit. Compared to fiber optic delivery, this method has the advantage that spot sizes of small diameter and hence greater power density can be delivered to the target enhancing the element signal.

5.1.1.4 Stand-off Analysis Because the laser plasma is formed by focused light, it is possible to generate the laser spark at a distance on a remotely located sample [Figure 8c].[23] The distances that can be achieved are a function of many parameters including the laser pulse energy and power, the beam divergence, and the optical system used to focus the pulses at a distance. With good quality components, for example, a laser plasma can be formed on soil at a distance of 19 meters using only 35 mJ pulse^{-1}. In such a system, efficient collection of the plasma light is critical to obtain useful signals. Unlike the other analysis scenarios described above in which a bare fiber can be aimed at the plasma to collect sufficient light, in the case of stand-off analysis, a lens is required to increase the solid angle over which the plasma light is collected and then directed into the fiber. The light collection system can be either adjacent to or collinear with the optical axis of the system used to focus the pulses on the sample. The latter configuration is shown in Figure 8(c).

5.1.2 Minimal Sample Preparation

Because of the high power densities used to form the laser spark, all types of materials can be atomized and a plasma formed in the focal region. Only optical access to the sample is required. Within the laser plasma additional atomization of ablated material continues and the resulting atoms are excited to emit light. For this reason, in general, no sample preparation is required for a LIBS measurement which is a distinct advantage compared to most other forms of analysis. For example, analysis using the ICP generally requires that the sample be chemically ashed to produce a solution that is then nebulized into the ICP plasma. Chemical preparation is a time-consuming and sometimes labor intensive process that precludes rapid sample analysis. Using LIBS however, a measurement can be carried out immediately merely by focusing the laser pulses directly on the sample.

5.1.3 Speed of Analysis

LIBS measurements can be carried out in what is considered real-time because of the lack of sample preparation and the simplicity of the method which make it amenable to deployment in the field. Due to the short lifetime of the plasma (few tens of microseconds), the time required to record a spectrum is less than 100 milliseconds and using automated instrumentation driven by computer software, the analysis of the resulting spectrum is immediate. In this case, the main factor determining the speed of a measurement is the number of laser shots required to obtain a representative spectrum. Many shots may be needed to:

(1) obtain an average composition reading due to compositional inhomogeneity of the sample;
(2) ablate away an overlying surface layer having a composition that may not be representative of the underlying bulk material;
(3) average out shot-to-shot variations in the plasma characteristics.

The use of multiple pulses to overcome sampling problems associated with (1) and (2) above is obvious and is discussed in section 5.2.1. Advantages to be gained from using many laser shots in the analysis of even a uniform sample when plasma characteristics are changing shot-to-shot can be demonstrated by performing replicate measurements, each consisting of 50, 200, and 1600 laser pulses, and then computing the precision of the measurements. For uranium in solution, the results obtained in one set of measurements were 13.3, 7.2, and 1.8% RSD, respectively.[24] These measurements were carried out with repetitive laser plasmas formed on the surface of a liquid in which the lens-to sample distance and hence plasma characteristics changed on each shot due to strong pressure waves generated at the liquid surface. The strong dependence of precision on the number of averaged laser pulses demonstrates the advantage of repetitive measurements.

LIBS measurements made using uniform geological samples (i.e. certified reference materials such as those available from the National Institute of Standards and Technology), metals, and liquid samples indicates that 100 laser shots can produce measurement precision on the order of 10%.

5.2 Considerations in the Use of Laser-induced Breakdown Spectroscopy

5.2.1 Sample Homogeneity

One of the more appealing aspects of LIBS is the ability to analyze samples with little or no preparation. For samples such as mixed gases and liquids containing well mixed and dissolved materials, sample homogeneity may be assumed so every plasma interrogates a small volume having a composition representative of the bulk sample. In this case, the number of laser shots to be averaged for a measurement is determined by factors such as the method of sampling and perturbations of the laser plasma. Other samples, solids in particular, cannot always be assumed to be homogeneous and, in fact, except for metals and plastics, etc. inhomogeneity should be assumed. Two types of inhomogeneity likely to be encountered are listed below.

For references see page 9612

5.2.1.1 Bulk Non-uniformity The small area interrogated by the laser pulse represents point detection of an area of 0.1 to 1 mm diameter involving a very small mass of material (i.e. tens of nanograms, section 2.3). Surface features on rocks, for example, may display visual irregularities in the distribution of materials that are on the order of the area sampled on each shot. These non-uniformities may be averaged out using a number of laser plasmas to repetitively interrogate a sample with the results then averaged.

5.2.1.2 Non-representative Surface Composition Some samples, such as metal alloys, and rocks exposed to the elements may have a surface layer composition that is not representative of the underlying bulk composition. For example, weathered rocks usually have a desert varnish layer ranging from 30 to 100 microns thick that has a composition different from the bulk rock matrix. Depending on the laser parameters, each interrogation by the laser pulse will produce a sampling depth ranging from a few microns up to perhaps 10 to 20 microns. Therefore, to obtain a more representative analysis, repetitive sampling by the laser spark at the same location on the sample can be used to ablate away the outer layers revealing the true bulk composition underneath.

For samples that are significantly inhomogeneous either as a result of bulk or surface non-uniformity, it may be advisable to grind the samples and then press the resulting particles to produce a flat surface for analysis. Although this procedure eliminates the real-time and in-situ analysis advantages of LIBS, it still preserves analysis capability without the need for chemical ashing of the sample.

5.2.2 Matrix Effects

LIBS, as other analytical methods, displays so-called matrix effects. That is, the composition of the sample affects the element signal such that changes in concentration of one or more of the elements forming the matrix alter an element signal even though the element concentration remains constant. For example, the signals from Si in water, in steel, and in soil appear much different even though the concentration of the element is the same in all three matrices. Even for samples that are more closely allied in matrix composition such as soils and stream sediments, significant differences in signal levels are observed for elements at identical concentrations in these materials. Because the laser spark both ablates and excites the sample, these effects can be more pronounced than in other methods that require sample preparation. Matrix effects can be divided into two kinds, physical and chemical.

Physical matrix effects depend on the physical properties of the sample and generally relate to the ablation step of LIBS. That is, differences between the specific heat, latent heat of vaporization, thermal conductivity, absorption, etc. of different matrices can change the amount of an element ablated from one matrix compared to another matrix even though the properties of the ablation laser pulse remain constant. Changes in the amount of material ablated can often be corrected for by computing the ratio of the element emission signal to some reference element known to be in the matrix at a fixed or known concentration. In this case, it is assumed that the relative ablated masses of the element and reference elements remain constant although the total mass of ablated material may change on a shot-to-shot basis. In this case, calibration is provided by using the ratio of the element signal to the signal produced by the reference element.

Chemical matrix effects occur when the presence of one element affects the emission characteristics of another element. This can complicate calibration of the technique and hence the ability to obtain quantitative results. These effects can be calibrated out of the analysis if the concentration and effect of the interfering element(s) are known but changes in the concentration of the interfering species from sample-to-sample can be a difficult correction procedure. An example of a chemical matrix effect is the reduction in emission intensity of an ionized species [e.g. Ba(II)] upon a significant increase in the concentration of an easily ionizable species in the matrix. The easily ionizable species perturbs the electron density thereby decreasing the concentration of Ba(II).[25] In addition, there are indications that the compound form of an element [e.g. $PbNO_3$, $PbCl_2$, etc.] may result in different emission signal strengths for the same element concentrations.

Extensive work remains to be done to characterize chemical and physical matrix effects for all types of samples and to develop methods to correct for their effects and increase the quantitative ability of LIBS.

5.2.3 Sampling Geometry

In the analysis of a solid, a plasma will be formed on the surface if the power density is sufficiently high even though the distance between the sample and the lens may be different from the focal length of the lens. These changes in the LTSD can result in changes in the mass ablated as well as changes in the temperature and electron density of the plasma which in turn affect the element emission signals.[18] Keeping the sampling geometry constant is important to achieve the best analytical results. In the interrogation of some samples, such as soil or rocks on a conveyor belt, for example, maintaining the LTSD constant may not be possible to a high degree. This can be dealt with in several ways including the use of a lens of long focal length to focus the pulses on the sample so that relative changes in the LTSD are less important or developing an active feedback system to automatically change the lens position to keep the LTSD constant. In addition, it is often possible to compute the ratio of the element emission to the emission of some reference element (e.g. Fe in steel or in soil) known to be in the sample at a fixed concentration. In this case, relative changes in both signals are the same and the ratio remains constant as described in section 5.2.2.

5.2.4 Safety

There are three areas of safety that must be considered in the use of LIBS. These are (1) the ocular hazard posed by the laser pulse, (2) the potentially lethal high voltage circuits used by the laser, and (3) the explosive potential of the laser spark for certain materials. The first two hazards are discussed at length in the industry safety standard ANSI Z136.2 to which one is referred for further information.[26]

The lasers used for LIBS are almost exclusively typed as Class IV indicating that both eye and skin exposure should be avoided. Established safety procedures require that precautions be taken to ensure that personnel are not exposed to the laser radiation through proper administrative and engineering controls. Engineering controls include interlocks connected to an enclosure that if opened cause immediate deactivation of the laser. The safest procedure is to completely enclose the laser radiation so that operations can be carried out as if the LIBS system contained a Class I laser in which case no extraordinary safety measures are needed outside the enclosure. In most cases, it should be possible to completely enclose the laser to protect the operator against the laser light. An exception may be the stand-off sampling geometry shown in Figure 8(c). For Nd : YAG lasers of the powers normally used for LIBS measurements, the main hazard is eye exposure. Exposure of the skin should be avoided but rarely results in significant damage at the pulse energies used for LIBS. On the other hand, the use of lasers producing radiation in the ultraviolet spectral region, such as the excimer laser, can pose a significant hazard to the skin. If interlocks must be defeated to work on an operating system or for alignment purposes, eye protection is easily provided for by the use of goggles appropriate for the laser wavelength in use.

The high voltages used by all lasers represent a lethal hazard that is most easily avoided by operating the laser with all interlocks to electrical circuits maintained in place. Only qualified personnel should be permitted to work on the high voltage power supplies in either an energized or non-energized configuration. Industry procedures such as lockout/tagout are called for in many cases involving work on electrical systems of this magnitude.

The laser plasma, as any other type of plasma source, has the potential to ignite flammable gases or to cause detonation of explosives. For this reason, the environment in which LIBS measurements will be carried out must be evaluated regarding this risk.

5.3 Representative Figures-of-merit

Because LIBS analyzes a sample directly with little or no preparation, the analytical figures-of-merit of the method are strongly dependent on sample characteristics. Listed in Table 2 are some representative LOD values for elements in different matrices. For gases, liquids, and solids the detection limits are in units of ppm (parts-per-million wt wt^{-1}) whereas for materials on filters, the LOD values refer to a surface detection limit with units of mass per surface area.

6 HYBRID TECHNIQUES

6.1 Double-pulse Techniques

A typical LIBS measurement is made by repetitively sampling a surface with a specified number of laser pulses and then averaging the element signals together. The time separation between the repetitive pulses, determined by the maximum repetition rate of the laser, is large compared to the lifetime of the plasma. This type of measurement is carried out using the RSS (repetitive single sparks) method, defined as such when the time separation between laser pulses is so large (e.g. of 1 ms) that interpulse effects are negligible. The RSS method is used most widely for both field and laboratory LIBS analyses. However, in some cases, the element signal can be strongly enhanced by the use of a technique termed RSP (repetitive spark pair).[19] The RSP method is defined as interrogation of a target by dual laser pulses having a time separation on the order of tens of microseconds or less so that some interpulse effects are present. An interpulse effect is defined as some effect induced by the first pulse of the pulse pair that still resides in the focal volume interrogated by the second pulse. An example is ablated material from the first pulse still residing in the focal volume during formation of a plasma by the second pulse of the pulse pair. In general, the dual pulses of the RSP are focused by the same optical elements so the focal volumes of the two pulses are coincident.

The timing between the laser pulse(s) and the gate pulse to the detector for the RSS and the RSP techniques is shown in Figure 9. Here ΔT is the time between

For references see page 9612

Table 2 Representative detection limits for elements in selected matrices

Element	Gas (ppm)	Liquid (ppm)	Filter (ng cm^{-2})	Solid/Matrix (ppm)
Ag		20	17	
Al				130 (Fe ore)
As	0.5 (aerosol)		44	
B		80		
Ba		6.8; 130	1	3 (soil)
Be	0.0006 (aerosol)	10	1	0.5 (soil)
Ca		0.13		1300 (organic material) 300 (Fe ore)
Cd	0.018 (aerosol)	0.8; 500	300	60; 90 (soil) 20 (silica gel)
Cl	8 (freon)			
Cr			160	10; 150 (steel alloy) 30 (soil) 40 (silica gel)
Cs		1		
Cu			11	10; 38 (soil) 7; 100 (steel alloy)
F	38 (freon)			
Fe		0.16		
Hg	0.5 (aerosol)		15	300 (soil)
K	1.5 (aerosol)	1.2		
Li		0.006; 0.013		2.5 (soil)
Mg		25; 100		230 (Fe ore)
Mn			115	160; 210 (steel alloy)
Mo				120; 200 (steel alloy)
Na	0.006 (aerosol)	0.0075; 0.014		
Ni			185	12; 20 (soil) 64; 150 (steel alloy) 310 (silica gel)
P	1.2 (aerosol)			
Pb	0.2 (aerosol)	12.5	450	17; 40 (soil) 96 (leaded paint)
Rb		0.2		
S	200 (aerosol)			
Sb				73 (soil)
Si				1190 (Al alloy) 380 (steel alloy) 1500 (Fe ore)
Sn				26 (soil)
Sr			5	1.2 (soil)
Ti				44 (Al alloy) 230 (Fe ore)
Tl			40	
U		100		1000 (soil)
V				200 (steel alloy)
Zn	0.23 (aerosol)		135	250 (soil)

pulses of the RSP. In the analysis of metals, the greatest enhancements were observed for ΔT of about 5 μs.[27] In the RSS technique, the time delay T_d is referenced to the single pulse used to excite the sample at the laser repetition rate. In an RSP measurement, the parameters T_d and T_b are referenced to the second pulse of the pulse pair as shown. For soil and metal samples at atmospheric pressure, when identical timing parameters T_d and T_b were used to compare the RSS to the RSP, enhancements in element signals up to factor of 15 were obtained. The mechanism responsible for the signal enhancement seen with the RSP in these cases is that the first laser pulse ablates particles into the focal volume of the second pulse that can then more efficiently excite the ablated material.

Another use of the RSP is in the analysis of a bulk metal surface under water.[28] Using the RSS, no element signal is obtained as shown in Figure 9(b). However, interrogation of the metal using the RSP results in the appearance of a strong spectrum shown in Figure 9(d). The reason for the enhancement is that a large bubble is

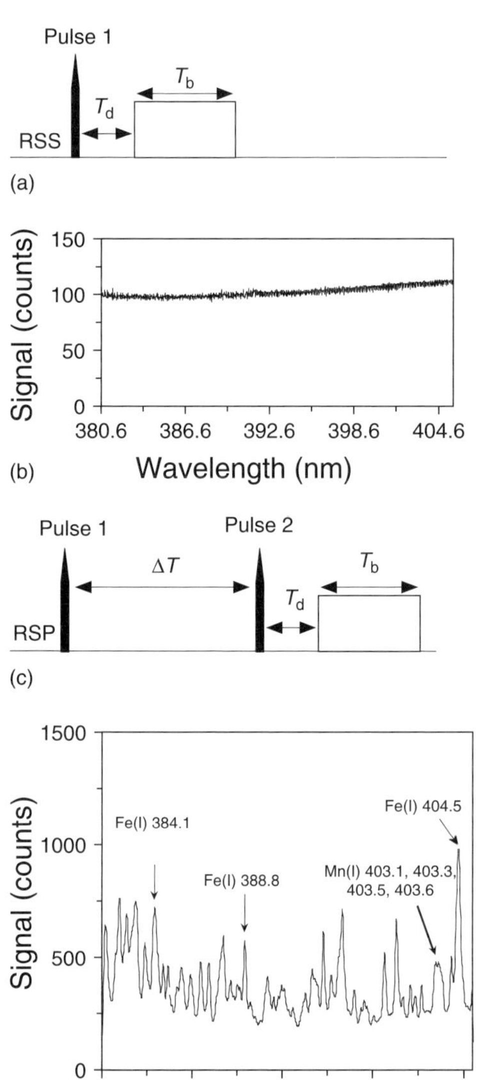

Figure 9 Use of double-pulse excitation to enhance detection capabilities of LIBS. Timing used in (a) RSS and (c) RSP measurements. Spectra obtained using (b) RSS and (d) RSP excitation of metal located under water. Here ΔT is the time between dual pulses, T_d is the delay time between a pulse and the start of the gate pulse and T_b is the duration of the gate pulse.

formed on the metal surface by the first pulse allowing the second pulse to form inside the bubble creating conditions similar to an analysis conducted in air. The cavity created from the first pulse depends upon laser energy but is typically about 8 mm in diameter.

6.2 Combined with Other Methods

The use of the laser spark alone as an analytical technique has been discussed above. The ability of the laser pulse to ablate a wide range of materials resulting in atomized material has led to its combining with other analytical techniques. In most combined techniques, the laser spark serves to introduce the sample to the other technique in a vaporized form. This introduction of the vaporized sample eliminates the time-consuming sample preparation step allowing for rapid analysis. Another useful characteristic is that the vaporized sample material can be transported in a gas stream to a desired location. In this case, analysis of a large specimen can be completed without having to introduce the large sample into the analytical equipment since the laser spark ablates only a few nanograms of material. Furthermore, all types of solid materials can be introduced.

The first of the techniques discussed here is that of the laser microprobe with cross excitation referred to earlier (section 3.5.1).[16] This technique was developed in 1962 and is the oldest laser plasma analytical device. A typical microprobe apparatus is shown in Figure 6. A laser pulse is focused onto the sample surface using a microscope objective. A collinear visual imaging system is incorporated into the instrument design to allow for accurate sample placement under the laser pulse. Directly above the sample surface is a pair of electrodes used to excite the ablated material. The light from the electrical discharge is transmitted to the spectrograph.

While the light from the laser plasma alone can be used to make analytical determinations in the microprobe devices, the combination of the electrode spark offers some spectrochemical advantages. The spark produced between the electrodes is more intense than the laser spark by a factor of 100 to 1000 times. This results in greater element emission line intensities resulting in greater sensitivity over the excitation provided by the laser spark alone. Also, because the electron density of the electrode spark is lower than that of the laser spark, the line widths are narrower and are less likely to be self-reversed. A third advantage of using the coupled technique is that the electrode spark does not contain as strong a broadband background continuum light.

The laser microprobe with cross excitation has been applied to a greater variety of materials than any other laser plasma-based method. The list of samples analyzed includes minerals, ceramics, glasses, biological materials, oils, crystals, alloys, powders, and geologic materials. The absolute detection capabilities depend on the element and the sample matrix. Typical mass detection limits for many elements lie in the range of 0.1 pg to 500 pg and concentration LOD values range from 10 ppm to 500 ppm for most elements.[16]

The laser microprobe with cross excitation is a well-researched technique and several disadvantages have become evident when compared to other analytical techniques. Since the method uses an electrode spark, the disadvantages associated with electrodes are present such

For references see page 9612

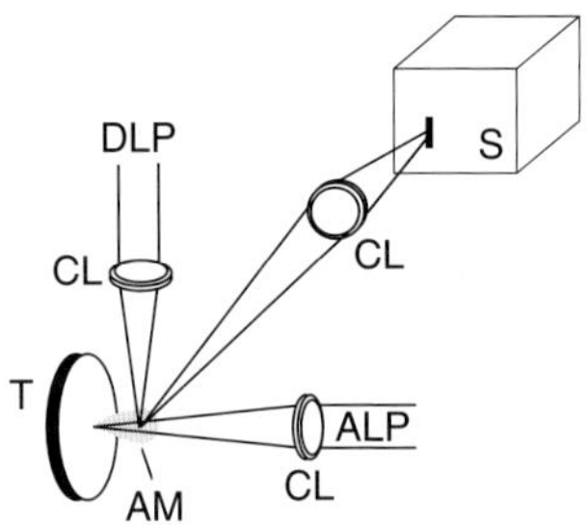

Figure 10 Diagram of the geometry used for LIF detection of laser-ablated material. DLP = dye laser pulse; CL = converging lens; T = target; AM = ablated material; ALP = ablating laser pulse; S = spectrograph.

as possible interference from the electrode material and wear of electrodes with use. However, the most important drawback of this technique is poor measurement precision. Typically around 25–40% RSD. Although creative attempts have been made to increase precision, the best achievable precision lies in the range of 3.6 to 16% RSD.

Laser-induced fluorescence (LIF) has been used to probe the atomic constituents of the laser plume.(29,30) In LIF, the free atoms of interest are electronically excited using a dye laser pulse tuned to a specific atomic transition. The relaxation of the excited species produces fluorescence. The geometry of a typical laser ablation/LIF set-up is shown in Figure 10. Some advantages of this method compared to monitoring the plasma light directly include (1) greater concentration and absolute mass detection sensitivity, (2) fewer matrix effects, (3) wider dynamic range, and (4) isotope selectivity.

In one study, Cr was analyzed from steel, flour, and skim milk powder.(29) The calibration curves were linear over 10 ppm to 1000 ppm and no chemical matrix effects were observed. The detection limit for Cr was determined to be 1 ppm corresponding to 0.1 pg of atomized Cr. The uncertainty in the measurements was 20–30% but was attributed to correctable experimental difficulties. Foremost was Mie scattering of the laser radiation from ablated particles in the laser plasma. This can be corrected by using filters to block the laser wavelength if the fluorescence is shifted from the laser wavelength or in the case of resonance fluorescence, by time gating the fluorescence detection. The effects of atmosphere, pressure, and time delay all have an effect on the analytical capabilities of this technique. At lower pressures, the longer lifetimes of the free atoms allow for greater resolution in time gating when resonance fluorescence is being detected.

6.3 Continuous Optical Discharge

The plasma formed by a laser pulse has a relatively low duty cycle (10^{-6}) given that the plasma lifetime is only about 20 μs and a typical laser repetition rate is 20 Hz. It is possible to generate a continuously operating laser plasma, called the continuous optical discharge (COD), by focusing a cw-CO_2 laser beam into air or another gas.(31) Whereas typically megawatts of pulsed energy are required to form the pulsed laser plasma, only hundreds or a few thousand watts are required to form the COD. The main reason for this is that absorption of the incoming laser energy by the plasma is proportional to the square of the laser wavelength making the power requirements of the long wavelength CO_2 laser (10 600 nm) a factor of 100 less then would be required using the 1064 nm wavelength of the Nd : YAG laser. Even though the CO_2 laser provides sufficient power to sustain the COD, a pulsed laser or other method of introducing a spark into the focal volume of the CO_2 laser is needed to initiate the COD. A diagram of a cell used to generate the COD is shown in Figure 11.

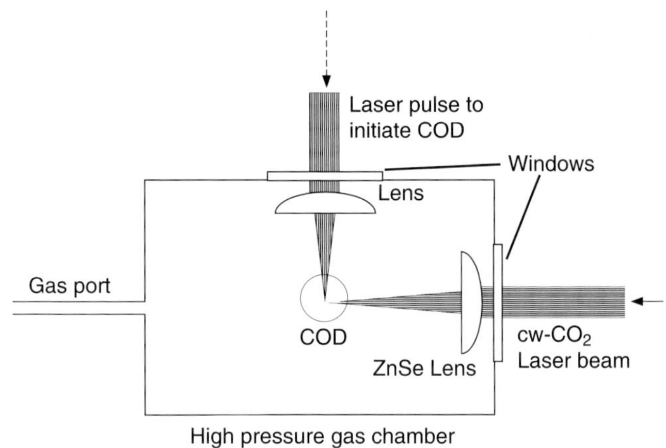

Figure 11 Diagram of the method used to generate the COD in a high-pressure cell.

The laser power required to sustain the COD depends on several factors including type of gas, pressure, and quality of the laser beam. For Xe gas, which has a comparatively low ionization potential, 20 W are needed to sustain the COD at a pressure of 10 atm. While in air, which contains molecules and hard to ionize species, 2 kW are needed to maintain the plasma. For a given laser power, the COD can only be sustained over a certain pressure range for a specific gas. For instance, using a 45 W CO_2 beam in Xe, the operating pressures were between 1.8 atm and 4.2 atm.

The plasma temperature and electron density depend upon the operating parameters and the gas used. For a CO_2 laser power of 45 W, the temperature of the COD sustained in Ar at 16 atm was 13 000 K. The temperature of the COD increased to 18 000 K when a pressure of 2 atm in Ar was used. The type of gas greatly affects the plasma temperature as temperatures as high as 30 000 K have been measured in 5 atm of He. While the temperature is highly depend upon the gas and the pressure, the electron density is somewhat less variant and is typically in the range of 10^{17} to 10^{18} cm^{-3}.

The analytical capabilities of the COD are most easy to evaluate for analytical gas samples because these are easily introduced into a high-pressure gas cell. A linear relationship of the calibration curves for many gases has been demonstrated over several orders of magnitude. However, there seems to be an upper limit on the concentration of the analytical gas that the COD can tolerate without extinguishing. Also, some analytical gases absorb at the wavelength of the CO_2 beam resulting in a significant change in the excitation parameters of the COD. Another problem is that some gases have decomposition products that coat the optical elements used to form the COD. To compensate for these problems, the COD should operate as a plasmatron in which the gases are flowed through the laser beam focus and carried away without effect on the COD. Although the COD may be best suited to analyze gases, solids can be introduced into the COD via laser ablation. In conjunction with the gas stream formerly mentioned, solid samples can be ablated and carried into the COD for analysis. This process has yielded semi-quantitative analytical results using thin coatings of solid samples on a metal substrate ablated with a Nd:YAG laser pulse. Although spectrochemical analysis with the COD is possible, further development is needed to devise a method to reproducibly introduce samples into the COD without perturbing the plasma characteristics.

7 PERSPECTIVE AND FUTURE DEVELOPMENTS

The LIBS technique has been under investigation for almost four decades. Practical application of the method has been concurrent with much fundamental work aimed at characterizing and understanding the laser plasma and laser ablation. A variety of systems have been investigated and in some cases the processes are well understood. In other areas, much work remains. The realization of LIBS instrumentation has a less successful history. Although the laser microprobe as a laboratory instrument was offered commercially, the technology has fallen into disuse being displaced by more capable analytical methods meeting the increasing needs for better analytical results including highly sensitive detection. In the setting of an analytical laboratory, where sample preparation and the size of an instrument are not of significant concern, LIBS does not appear to have a viable role. In recent years there has been a resurgence of interest in the technology, however, for a variety of reasons the most notable of which has been environmental regulation. Often such measurements must be performed outside the laboratory in less than ideal conditions not suitable for more conventional instrumentation. In addition, often such measurements do not require the high accuracy and precision characteristic of conventional methods. In these cases, LIBS may have something to offer. Because of the simplicity of the method, the lack of sample preparation, and the non-invasive nature of the sampling step, LIBS may be the ideal technique. In addition, recent technological developments in computers, lasers, spectrographs, and detectors, resulting in rugged and compact field-portable LIBS devices, have given rise to renewed interest in the method. The activity in the development of commercial LIBS devices over the past five years is probably greater than the combined activity over the first thirty years of the method. Whether this interest will continue and is realized on a large scale remains to be seen. What is certain, however, is that the method can uniquely meet some analysis requirements and does have a place, perhaps limited, in the arsenal of modern analytical techniques. The key to increasing these applications is a more thorough investigation and hopefully more thorough understanding of the method.

ACKNOWLEDGMENTS

This work was done under the auspices of the US Department of Energy. Los Alamos National Laboratory is operated by the University of California for the US DOE.

ABBREVIATIONS AND ACRONYMS

AOTF	Acousto-optic Tunable Filter
AD	Array Detector
AES	Atomic Emission Spectroscopy
CCD	Charge Coupled Devices
CID	Charge Injection Devices
COD	Continuous Optical Discharge
ICP	Inductively Coupled Plasma
IB	Inverse Bremsstrahlung
LIBS	Laser-induced Breakdown Spectroscopy
LIF	Laser-induced Fluorescence
LSC	Laser-supported Combustion
LSD	Laser-supported Detonation
LSR	Laser-supported Radiation
LTSD	Lens-to-sample Distances
LOD	Limit of Detection
PD	Photodetector
PDA	Photodiode Array

For references see page 9612

PMT	Photomultiplier Tube
RSD	Relative Standard Deviation
RSS	Repetitive Single Sparks
RSP	Repetitive Spark Pair
SD	Standard Deviation

RELATED ARTICLES

Environment: Trace Gas Monitoring **(Volume 3)**
Laser-induced Breakdown Spectroscopy, Elemental Analysis

Environment: Water and Waste **(Volume 4)**
Laser Ablation Inductively Coupled Plasma Spectrometry in Environmental Analysis • Optical Emission Inductively Coupled Plasma in Environmental Analysis

Atomic Spectroscopy **(Volume 11)**
Atomic Spectroscopy: Introduction • Laser Ablation in Atomic Spectroscopy • Laser Spectrometric Techniques in Analytical Atomic Spectrometry

REFERENCES

1. *Laser-induced Plasmas and Applications*, eds. L.J. Radziemski, D.A. Cremers, Marcel Dekker, New York, 1989.
2. L. Moenke-Blankenburg, *Laser Microanalysis*, John Wiley, New York, 1989.
3. D.A. Rusak, B.C. Castle, B.W. Smith, J.D. Winefordner, 'Fundamentals and Applications of Laser-induced Breakdown Spectroscopy', *Crit. Rev. Anal. Chem.*, **27**(4), 257–290 (1997).
4. T.P. Hughes, *Plasmas and Laser Light*, John Wiley and Sons, New York, 1975.
5. G.M. Weyl, 'Physics of Laser-induced Breakdown: An Update', in *Laser-induced Plasmas and Applications*, Marcel Dekker, New York, 1989.
6. R.G. Root, 'Modeling of Post-breakdown Phenomena', in *Laser-induced Plasmas and Applications*, Marcel Dekker, New York, 1989.
7. J.F. Ready, *Effects of High Power Laser Radiation*, Academic Press, New York, 1971.
8. C.R. Phipps, R.W. Dreyfus, 'The High Laser Irradiance Regime', in *Laser Ionization Mass Analysis*, John Wiley and Sons, New York, 1993.
9. O. Svelto, *Principles of Lasers*, Plenum Press, New York, 1976.
10. *Wavelengths and Transition Probabilities for Atoms and Atomic Ions Part II. Transition Probabilities*, NSRDS-NSB 68, U.S. Government Printing Office, Washington, DC, 1980.
11. A.R. Striganov, N.S. Sventitskii, *Tables of Spectral Lines of Neutral and Ionized Atoms*, IFI/Plenum, New York, 1968.
12. K.I. Tarasov, *The Spectroscope*, John Wiley & Sons, Inc., New York, 1974.
13. I.C. Chang, 'Acousto-optic Tunable Filters', *Opt. Eng.*, **20**, 824–829 (1981).
14. *The Photonics Design and Applications Handbook, Book 3*, Laurin Publishing Company, Inc., Pittsfield, MA, 1997.
15. Y. Talmi, (ed.), *Multichannel Image Detectors*, ACS Symp. Series No. 102, ACS, Washington, DC, 1979.
16. H. Moenke, L. Moenke-Blankenburg, *Laser Micro-spectrochemical Analysis*, Crane, Russak and Company, Inc., New York, 1973.
17. K.Y. Yamamoto, D.A. Cremers, L.E. Foster, M.J. Ferris, 'Detection of Metals in the Environment Using a Portable Laser-induced Breakdown Spectroscopy (LIBS) Instrument', *Appl. Spectrosc.*, **50**, 222–233 (1996).
18. R.A. Multari, L.E. Foster, D.A. Cremers, M.J. Ferris, 'The Effects of Sampling Geometry on Elemental Emissions in Laser-induced Breakdown Spectroscopy', *Appl. Spectrosc.*, **50**, 1483–1499 (1996).
19. D.A. Cremers, L.J. Radziemski, T.R. Loree, 'Spectrochemical Analysis of Liquids Using the Laser Spark', *Appl. Spectrosc.*, **38**, 721–729 (1984).
20. C.Th.J. Alkemade, W. Snelleman, G.D. Boutilier, B.D. Pollard, J.D. Winefordner, T.L. Chester, N. Omenetto, 'Review and Tutorial Discussion of Noise and Signal-to-noise Ratios in Analytical Spectrometry. I. Fundamental Principles of Signal-to-noise Ratios', *Spectrochim. Acta*, **33B**, 383–399 (1978).
21. G.D. Boutilier, B.D. Pollard, J.D. Winefordner, T.L. Chester, N. Omenetto, 'Review and Tutorial Discussion of Noise and Signal-to-noise Ratios in Analytical Spectrometry. II. Fundamental Principles of Signal-to-noise Ratios', *Spectrochim. Acta*, **33B**, 401–416 (1978).
22. C.M. Davies, H.H. Telle, D.J. Montgomery, R.E. Corbett, 'Quantitative Analysis Using Remote Laser-induced Breakdown Spectroscopy (LIBS)', *Spectrochim. Acta*, **50B**, 1059–1075 (1995).
23. J.D. Blacic, D.R. Pettit, D.A. Cremers, 'Laser-induced Breakdown Spectroscopy for Remote Elemental Analysis of Planetary Surfaces', in *Proc. Int. Symp.* Spectral Sensing Research, Maui, HI, November, 15–20, 1992.
24. J.R. Wachter, D.A. Cremers, 'Determination of Uranium in Solution Using Laser-induced Breakdown Spectroscopy', *Appl. Spectrosc.*, **41**, 1042–1048 (1987).
25. A.S. Eppler, D.A. Cremers, D.D. Hickmott, M.J. Ferris, A.C. Koskelo, 'Matrix Effects in the Determination of Pb and Ba in Soils Using Laser-induced Breakdown Spectroscopy', *Appl. Spectrosc.*, **50**(9), 1175–1181 (1996).
26. *American National Standard for the Safe Use of Lasers ANSI standard ANSI Z136.2*, American National Standards Institute (most recent issue).

List of selected abbreviations appears in Volume 15

27. R. Sattman, V. Sturm, R. Noll, J., 'Laser-induced Breakdown Spectroscopy of Steel Samples Using Multiple Q-switched Nd:YAG Laser Pulses', *Appl. Phys. D*, **28**, 2181–2187 (1995).
28. A.E. Pichahchy, D.A. Cremers, M.J. Ferris, 'Elemental Analysis of Metals Under Water Using Laser-induced Breakdown Spectroscopy', *Spectrochim. Acta*, **52B**, 25–39 (1997).
29. H.S. Kwong, R.M. Measures, 'Trace Element Laser Microanalyzer with Freedom from Chemical Matrix Effect', *Anal. Chem.*, **51**, 428–432 (1979).
30. I. Gobernado-Mitre, A.C. Prieto, V. Zafiropulos, Y. Spetsidou, C. Fotakis, 'On-line Monitoring of Laser Cleaning of Limestone by Laser-induced Breakdown Spectroscopy and Laser-induced Fluorescence', *Appl. Spectrosc.*, **51**, 1125–1129 (1997).
31. D.R. Keefer, 'Laser-sustained Plasmas', in *Laser-induced Plasmas and Applications*, Marcel Dekker, New York, 1989.

Microwave-induced Plasma Systems in Atomic Spectroscopy

José A.C. Broekaert
University of Leipzig, Leipzig, Germany

Ulrich Engel
University of Dortmund, Dortmund, Germany

The use of microwave plasmas as radiation sources for optical atomic emission (AES), absorption (AAS) and fluorescence (AFS) spectroscopy and for laser ionization spectroscopy is treated and reference is also made to the use of microwave-induced plasmas (MIPs) as ion sources for mass spectrometry (MS). Devices for producing both single-electrode and electrodeless microwave plasmas are treated, in addition to methods for their diagnostics, and results for the analytically relevant plasma parameters are presented. Methods for sample introduction are discussed. They include dry aerosol generation techniques (hydride generation (HG), electrothermal evaporation, spark and laser ablation (LA)) and possibilities for the uptake of wet aerosols. Special reference is made to coupling with gas chromatography (GC) and also to the potential for coupling with high-performance liquid chromatography (HPLC) Further, the use of microwave plasmas for

For references see page 9645

cross-excitation in the case of glow discharges (GDs) is treated. The analytical figures of merit in the case of AES with low-power microwave plasmas, single-electrode microwave plasmas, plasma torch sources and stabilized capacitively coupled plasmas are given also in the case of atomic absorption, fluorescence and laser ionization with these sources. The developments in MS in the case of both low-power and high-power microwave plasmas and in the case of various types of sample introduction are discussed.

Applications of MIP atomic spectroscopy are in the fields of biological samples with special reference to microanalysis, and of environmental samples with special emphasis on metal speciation, on-line monitoring and direct solids analysis. A critical comparison of the methodology with other methods for the determination of the elements and their species is given.

1 INTRODUCTION

For elemental determinations at widely differing concentration levels, since the 1950s plasma sources have played an important role as radiation sources for atomic spectrometry. The latter methodology, especially in the case of optical emission spectroscopy (OES) and MS, enables many elements to be determined simultaneously in very different types of sample materials. The use of plasma sources started with the classical arc and spark sources, as known from direct solids analysis. Experiments were particularly designed aiming at a constant supply of analyte in these sources, be it for the case of liquids, finely dispersed solids or gases. This led to procedures dealing with dc plasma jets, which are still in use. Plasma atomic spectrometry in this sense, however, was hampered by the need to use electrodes for energy supply, causing blanks as a result of electrode burn-off and poor long-term stability. Therefore, there was intensive research on the availability of electrodeless discharges. Work at atmospheric pressure was found to have an important advantage, as then a number of techniques such as solution nebulization could be used easily and successfully for sample introduction into the sources. This explains the considerable attention which single-electrode microwave plasmas attracted towards the end of the 1950s, despite the fact that they were not electrodeless, as shown by papers by Mavrodineanu and Hughes,(1) Jecht and Kessler(2) and Tappe and Van Calker.(3) Here plasmas were mostly produced in noble gases by the interaction of microwave energy at gigahertz frequencies with gas flows. The hot plasmas permitted efficient drying of wet aerosols, evaporation of solid particles, atomization, excitation and ionization. They were operated at a metal electrode (W, Au-coated Cu, brass) and they could be operated with Ar, He and even with air at an electrical power ranging from 200 to 600 W. The discharge has a bush form and the hottest zones are in the center of the plasma. In the case of high concentrations of alkali, however, the plasma geometry drastically changed, which led to large concomitant effects as compared with the inductively coupled plasma (ICP).(4) Accordingly, these sources could be used successfully as radiation sources in atomic emission and, provided that ways are found to enter the ions in a mass spectrometer, also as ion sources for MS.

The trends in the development of plasma spectroscopy in general were reflected in regular surveys in the literature. They started with the conference proceedings of the first Winter Conference on Plasma Spectrochemistry,(5) and reviews on commercial instrumentation (see e.g. Broekaert(6)) and on the trends in developments in plasma spectrometry (see e.g. Broekaert(7)) regularly occur. Also, the *Journal of Analytical Atomic Spectrometry* updates are very informative to the field (see e.g. Sharp et al.(8)).

The success of plasma atomic spectroscopy first started with the availability of the ICP, first described by Greenfield et al.(9) and by Wendt and Fassel.(10) Here a toroidal plasma with a diameter of up to 20 mm and a height of up to 50 mm could be realized. Further, both wet and dry aerosols containing even large particulates could be entered, guaranteeing residence times of several milliseconds at temperatures above 4000 K. These sources, however, also necessitate working with gas flows higher than 10 L min^{-1} and at an electrical power in the kilowatt range, both of which make the operating costs high. Apart from the single-electrode microwave plasmas, considerable effort was put into low-pressure microwave plasmas, as they became available towards the end of the 1950s also (see e.g. Uden(11)). Here excitation in sealed tubes could be used for drinking water analysis, but the analysis times were long.(12)

The breakthrough in microwave plasmas first took place when electrodeless plasma discharges, which can be operated at atmospheric pressure, became available. This was realized in 1976 by the work of Beenakker,(13) who described a cavity in which both He and Ar microwave discharges could be produced with a power of up to ca. 100 W with a gas consumption of below 1 L min^{-1}. These sources became very popular for the excitation of dry analyte vapors, as obtained in the electrothermal vaporization (ETV) of dry solution residues and also in GC. For the latter case, low-power electrodeless microwave discharges are now widely used.

To bridge the gap towards high-power plasmas with and without electrodes, in which wet aerosols can be taken up and where air operation is also possible, intensive research was carried out. This resulted in a wide variety of devices such as the surfatron,(14) which will

List of selected abbreviations appears in Volume 15

be discussed further, and the recent microwave plasma torch (MPT) developed by Jin et al.,[15] and microwave plasma atomic spectroscopy became a powerful tool for elemental analysis.

The special features of the microwave plasma itself have been the subject of several review and tutorial papers by experts in the field. Jin et al.[16] reviewed work up to 1997 with 255 references. Winefordner et al.[17] surveyed microwave plasmas and GDs with respect to their typical lines of development. Reviews on microwave plasmas have been regularly published, e.g. for the work before 1980 by Zander and Hieftje,[18] in 1988 by Abdillahi,[19] in 1995 by Culp and Ng,[20] in 1984 by Matousek et al.[21] and in 1990 and 1996 by Broekaert.[22,23]

2 DEVICES FOR THE GENERATION OF MICROWAVE PLASMAS

Microwave generators operated at the allowed frequency of 2.45 GHz make use of standard technology in fields such as microwave heating and microwave sputtering devices, whereas the development of microwave cavities and resonators and properly designed torches was and still is a field of continuous innovation, as it can be seen from reviews such as that by Goode and Baughman.[24]

2.1 Microwave Generators

For the production of microwave energy, use is made of clystrons and magnetrons, the latter being the more important when a power level above ca. 100 W is required. In a magnetron (Figure 1), a filament current is amplified and flows to an annular anode, which contains a certain number of cavity resonators.[26] Perpendicular to this plane, a very strong stationary magnetic field is applied. Through the action of the secondary cavities, an ultrahigh frequency (UHF) current with a characteristic frequency is produced, which can be coupled out with a loop and transported through a coaxial connector to a waveguide end or to a resonator. The power output of the magnetron can be regulated and must be stabilized and smoothed so as to allow, at any level, stable operation of the plasmas produced. Means for converting pulsed into continuous-wave operation in generators for microwave plasma production were described by Brandl et al.[27] Microwave generators operating at power levels from a few watts to 1 kW with sufficient stability (ripple and drift <1%) are now commercially available. Miniaturization of microwave generators, especially with the aim of realizing small size and low power as required when generating smaller plasmas is possible through the use of transistor technology.

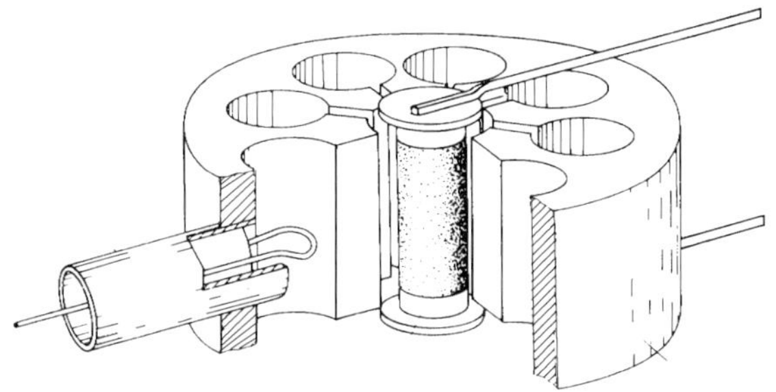

Figure 1 Schematic diagram of a magnetron.[25] (Reproduced by permission from O. Zinke, H.L. Hartnagel, *Lehrbuch der Hochfrequenztechnik*, Springer, Berlin, Band 2, 1987. Copyright Springer Verlag, 1987.)

When transporting microwave energy into suitable structures, only part of the energy will be taken up in the device, part being reflected to the generator. The forward and reflected power can be measured with the aid of a bidirectional coupling loop. When transporting microwaves to a coaxial or cavity waveguide, the dimensions of the latter must be tuned to the wavelength of the microwave. Cavity conductors are rigid metal tubes with a rectangular or circular cross-section in which the microwaves propagate and their dimensions are standardized, e.g. for the case of 2.54 GHz, 91 mm internal width and 42 mm internal height. Individual waveguides must be tightly connected through suitable flanges to prevent microwave leakage. Each slit acts as a secondary source. Hence holes and slits for observation must be provided with a chimney so as to prevent the sorting of microwave radiation by damping. The resonance conditions required to optimize the ratio of forward to reflected power can be realized by fine tuning with stubs or displaceable walls and with screws perturbing the microwave field.

2.2 Cavities and Resonant Structures

A plasma can be operated at the top of the internal conductor of a waveguide. To couple microwave energy inductively into a gas, a microwave cavity or resonator is used. The waveguide then should end at a length where the microwave amplitude maximum is located, whereas a cavity consists of a closed metal tube of rectangular or circular cross-section and its internal dimensions allow the formation of standing waves. Tapered rectangular cavities, Evenson quarter-wave cavities and Broida quarter- or three-quarter-wave cavities are commonly used. Anyhow, as the plasma formed disturbs the field, fine tuning with screws and stubs is required.

At reduced pressure many cavities could be used to obtain a microwave discharge, even in the case of gas mixtures (see e.g. McKenna et al.[28]). They include Evanson cavities and other devices. They played an

For references see page 9645

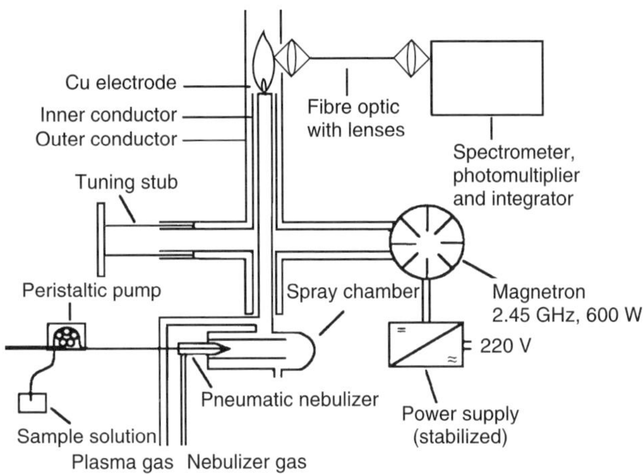

Figure 2 Schematic diagram of a set-up for capacitively coupled microwave plasma optical emission spectroscopy (CMPOES).[32]

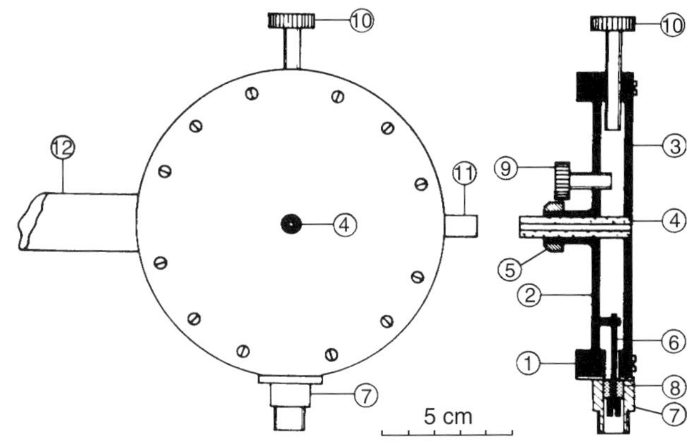

Figure 3 Schematic diagram of the Beenakker cavity.[38] (1) Cylindrical wall with (2) a fixed bottom and (3) a removable lid; (4) fused-silica discharge tube; (5) tube holder; (6) coupling loop; (7) connector; (8) sealing kit; (9,10) tuning screws; (11) air cooling; (12) cavity holder. (Reprinted from C.I.M. Beenakker, *Spectrochim. Acta, Part B*, **39**(7), 931–937 (1984), Copyright 1984, with permission from Elsevier Science.)

important role in the excitation of analytes enclosed in quartz ampules, which was developed as a technique for dry solution residue analyses for volatile elements by OES[29–31] and especially also to element-specific detection in GC (see section 4.4).

Microwave plasmas operated at the tip of a metal electrode connected with the internal conductor were the oldest microwave structures used for plasma spectrochemical analysis. Here the analyte flows around the tip of the electrode and enters the plasma axially (Figure 2). Although this method is not optimal as the plasma has a bush form and maximum temperatures in the center lead to poorer signal-to-noise ratios (S/Ns), it found widespread use. Indeed, it is stable, can be operated with both noble and molecular gases and it even accepts wet aerosols. Attempts were made to improve the sampling capacity further by using a gas flow around the tip of the electrode to stabilize and to concentrate the analyte flow.[33] It could be further decreased in size and in this form it still attracts interest (for the analytical figures of merit, see e.g. Wünsch et al.[34]).

Microwave discharges first became popular when it became possible to produce them at atmospheric pressure and low power, as was first possible with the TM_{010} resonator described by Beenakker and was improved especially with respect to coupling.[35–37] As shown in Figure 3, in this device a standing wave is produced in a circular cavity into which the microwave energy is entered with the aid of a loop or with the aid of the end of the inner conductor acting as an antenna. Coarse tuning can be accomplished through the positioning of the antenna and for fine tuning a ceramic stub is provided, with which distortions of the field in the cavity can be produced. The latter was found to be no longer required in the case of delivering the microwave power with the aid of a loop.[39]

In this cavity the power may vary from a few tens of watts to more than 300 W, depending on a cooling of the cavity, the burner and the respective gas flows. These parameters also determine the ways of sample introduction possible and sources operated at moderate power especially were designed with the aim of taking up wet aerosols. (see e.g. Haas et al.[40]).

At the end of the 1970s the surfatron was proposed as a source for analytical atomic spectroscopy. This microwave plasma, as shown in Figure 4, is produced in a microwave structure with varying depth and side-on coupling of the microwave power through the internal conductor.[41] Here, a microwave expands through the slit between the front plate and the plasma, through which a slightly hollow plasma with improved sampling capacity is realized. Selby and Hieftje[42] optimized the construction of the surfatron as a device for analytical work by making it of the same dimensions as the TM_{010} resonator according to Beenakker. The working power and pressure then become very similar to those of the resonator according to Beenakker. However, the plasma in a surfatron seemed to be more robust, e.g. when entering gases such as H_2 accompanying the volatile hydrides in HG work.[43]

Rectangular cavities, as described by Matusiewicz,[44] offer good possibilities for spectrochemical analysis. They can be operated from 100 W to several hundred watts. Even solvent-loaded aerosols can be entered in these sources after careful optimization. Further, the slab-line cavity described by Estes et al.[45] is also of use, but for the excitation of gaseous substances only. The different devices described in the literature were often compared when using one sample introduction technique (in the

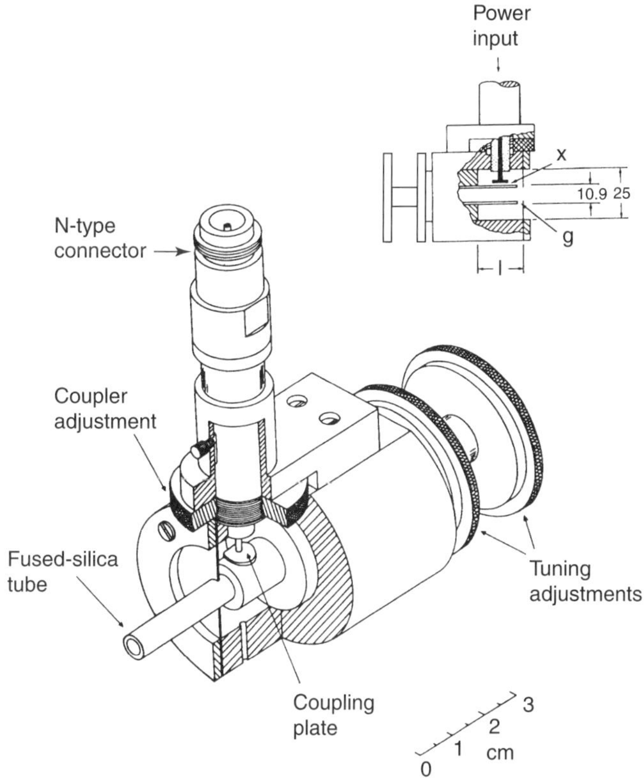

Figure 4 Schematic diagram of a surfatron set-up.[42] (Reprinted from M. Selby, G.M. Hieftje, *Spectrochim. Acta, Part B*, **42**(1–2), 285–298 (1987), Copyright 1987, with permission from Elsevier Science.)

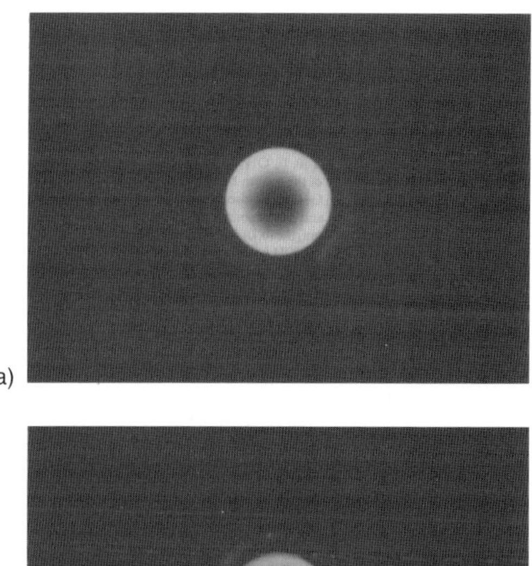

Figure 5 Pictures of (a) toroidal Ar MIP and (b) diffuse He MIP. Quartz torch tube: internal diameter 4 mm, external diameter 6 mm. (a) Pneumatic nebulization (PN), 1.2 L min^{-1} argon, −110 W; (b) without aerosol introduction, 0.2 L min^{-1} He, −130 W.[52]

case of HG, see e.g. Mulligan et al.[46]) or with other plasmas[47] also for the case of solution analysis.

2.3 Plasma Torches

The analytical features of different types of microwave discharges mainly depend on the type of torch in which they are produced. The latter especially determines the admissible power and whether a filament-type, a delocalized or a toroidal plasma can be obtained.

In Ar or He microwave discharges, up to about 80 W can easily be produced in a quartz capillary with an internal diameter of up to 1 mm, as described in the original version according to Beenakker. However, they also can be operated with cooling by an outer gas flow. Through observation of the Si lines, it could be shown that the erosion of the quartz tube can be considerable and it could be substantially decreased by using a torch where the discharge capillary is liquid cooled.[48] This tube erosion could even be used as a technique of sample introduction, namely by supplying a liquid aliquot in the capillary of the MIP prior to initiating the discharge and ablating the dry solution residue with the aid of the discharge.[49]

Extensive efforts have been made to produce toroidal MIP plasma sources for OES. Axial observation then enables it to obtain high signal-to-background ratios and accordingly low detection limits. Success in this respect was achieved by Kollotzek et al.[50,51] by using wetted Ar and carefully centering a quartz capillary of 4 mm internal diameter in the resonator according to Beenakker. Later on it was found that such a plasma (Figure 5a) at a power of about 100 W could be obtained when the working gas only was led through a water-filled washing flask.[52] With dry He a diffuse delocalized discharge, which homogeneously fills a quartz tube of 4 mm internal diameter, could be obtained at 100–300 W (Figure 5b). However, here the cooling of the discharge tube is much more critical.

Another approach especially makes use of special torch constructions. Here Bollo-Kamara and Codding[53] were successful in using a dual-tube arrangement with a threaded inset, by means of which a tangential gas flow could be realized. More work on tangential flow torches included extensive diagnostic studies of hydrodynamic flows and spatially resolved temperature measurements[54] and studies on self-centering plasmas[55] and their use in GC detection.[56] For the study of different arrangements, demountable torches in

For references see page 9645

a polytetrafluoroethylene (PTFE) socket are very useful, as then the induividual tubes can easily be changed in position or exchanged after attack by the discharge.[57] A detailed study and an improvement of the torch proposed by Bollo-Kamara and Codding was realized with a such demountable torch; here a PTFE gas-swirl modifier was used and also a PTFE socket for the tubes, making glass-blowing superfluous.[58]

Further work was also done on laminar-flow plasmas with the aim of obtaining a plasma with a high discharge stability and robustness with respect to incoming analyte clouds. This is very important so as to keep baseline changes in GC detection to the minimum.[59,60] It was found that because of improved stability the detection limits obtainable with laminar flow torches were considerably lower than those with tangential flow torches. It was reported that stabilities even better than 2% over 2 days could be realized.[61]

A significant innovation in MIP work with respect to burner design was made by Jin et al.[15] They used a concentric arrangement of two metal tubes mounted in an enclosure and the power was coupled in by means of the intermediate tube (Figure 6). Power coupling initially was realized with the aid of a gliding ring moving over the intermediate tube. Effective power coupling could be achieved by varying the depth of the outer enclosure and by changing the coupler position. It was found that with both Ar and He stable discharges could be realized at a gas consumption below 1 L min^{-1} and with a power of up to 200 W depending on the efficiency of the cooling. As a wide variety of sample introduction techniques could be used, this source is now one of the prominent microwave structures.

The design of the MPT was considerably improved by Pack and Hieftje.[62] First, it was possible to close the adjustment slit completely, so that air no longer enters between the intermediate and the outer tubes and microwave leakages also are minimized. Further, it was possible to use a quartz instead of a metal tube as the internal tube. This considerably lowers the risks for contamination when analyzing real samples. It was also found that sheathing of the plasma with N_2 greatly improved its stability. Further improvements of the torch design resulted from modeling of the electromagnetic fields without and with plasma. As a result, it was possible to design a torch which no longer requires tuning and with which ignition and stable operation of the MPT was also possible (Figure 7).[63] This significantly improves the reproducibility of the device with respect to ignition and operation and makes the MPT a robust spectrochemical source. Here the internal conductor of the coaxial cable is directly connected to the coupler, which is rigidly fixed to the intermediate tube. Further, it was no longer required to use a sheating gas in the metal part of the torch. A quartz enclosure around the plasma itself, however, was found to stabilize it considerably against air turbulence in the laboratory. Further developments

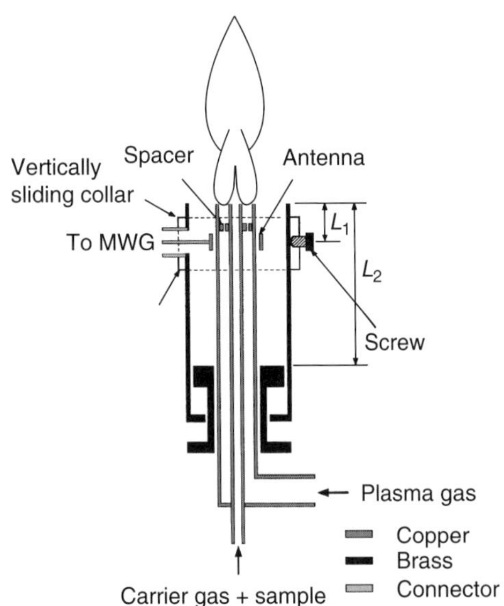

Figure 6 Schematic diagram of the MPT.[15] (Reprinted from Q. Jin, C. Zhu, W. Borer, G.M. Hieftje, *Spectrochim. Acta, Part B*, **46**(3), 417–430 (1991), Copyright 1991, with permission from Elsevier Science.)

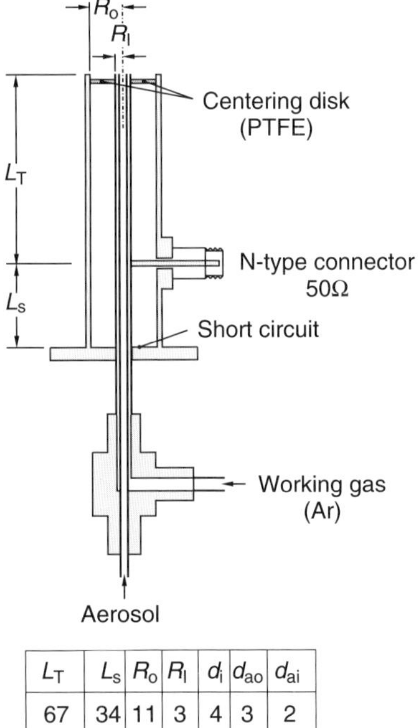

L_T	L_s	R_o	R_i	d_i	d_{ao}	d_{ai}
67	34	11	3	4	3	2

Figure 7 Modified MPT set-up for atomic spectroscopy.[63] Dimensions of the MPT are in millimeters; inner diameter of center conductor d_i and outer diameter d_{ao} together with inner diameter of analyte introduction tube.

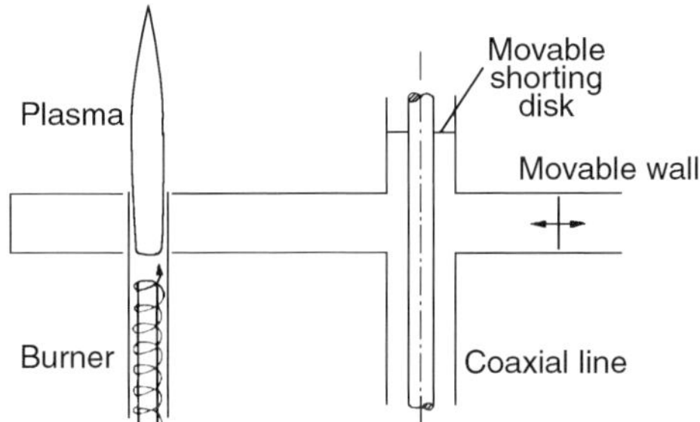

Figure 8 Resonant rectangular TE_{102} resonator and plasma torch for the 800 W MIP.[67]

include pulsed operation of a microwave discharge at higher power (500 W) and its coupling with spark ablation (SA) sampling devices.[64]

High-power inductively coupled microwave plasmas could be operated in various MIP resonant structures. This first is possible in the TM_{010} resonator according to Beenakker where, at powers of up to 500 W, Ar, He and air plasmas can be obtained, again by using modular cooled torches.[65] Such devices were often compared in the case of well-defined sampling techniques such as GC detection.[66]

Another possibility is the use of ICP-like torches in a rectangular cavity (Figure 8). Here a power up to the kilowatt range can be used and at several liters per minute gas consumption a very stable plasma can be obtained with Ar and also with air. This MIP is a powerful source for AES and for MS.[68]

3 DIAGNOSTICS OF MICROWAVE PLASMAS

The diagnostics of microwave plasmas deal with both the measurements of the plasma physical properties, including the flow dynamics, temperatures and particle number densities, and their spectroscopic properties with respect to optical emission, absorption and fluorescence, their properties as atom reservoirs and their properties as ion sources. The results obtained here differ for the different types of microwave discharges used for spectrochemical purposes; however, some trends are common.

As long as MIPs at atmospheric pressure are considered, the plasma is near to thermal equilibrium (TE); however, there might already be much greater differences between the different temperatures than in the case of the ICP. The main processes taking place in the plasma are as follows:

- electron impact for the excitation of working gas species (Equation 1):

$$\mathrm{A + e \longrightarrow A^*} \tag{1}$$

- electron impact for the ionization of working gas species (Equation 2):

$$\mathrm{A + e \longrightarrow A^+} \tag{2}$$

- electron impact for the excitation of analyte species (Equation 3):

$$\mathrm{S + e \longrightarrow S^*} \tag{3}$$

- electron impact for the direct ionization of the analyte species (Equation 4):

$$\mathrm{S + e \longrightarrow S^+} \tag{4}$$

- radiative recombination (Equation 5):

$$\mathrm{A^* \text{ or } S^* + e \longrightarrow A \text{ or } S} + h\nu \tag{5}$$

- excitation of analyte through collisions with excited working gas atoms (Equation 6):

$$\mathrm{S + A^* \longrightarrow S^* + A} \tag{6}$$

- ionization of analyte through collisions with excited working gas atoms (Equation 7):

$$\mathrm{S + A^* \longrightarrow S^+ + A} \tag{7}$$

A special case is excitation through collisions with metastable noble gas species (Ar^m). These are species for which a decay through the emission of radiation is forbidden as the multiplicity of the lower states differs from that of the excited states. As the energy of these species is high (e.g. Ar^m at 11.7 eV, He_2^m at 14.6–17.4 eV and $He_2{}^{+m}$ at 18.8–21.6 eV[69]), this process may be relevant for the one-step excitation of ion lines and it is highly selective.

- Charge-transfer processes (Equation 8):

$$\mathrm{A^{+(*)} + S \longrightarrow A + S^{+(*)}} \tag{8}$$

- radiation trapping (Equation 9):

$$\mathrm{A} + h\nu \longrightarrow \mathrm{A^*} \tag{9}$$

may be important for energy exchange and excitation as well.

With respect to diagnostics, many papers have described the determination of excitation temperatures from the relative intensities of atomic emission lines originating from the same stage of ionization. The intensity of an atomic emission line I_{qp}, corresponding with a transition from the higher level q to the lower level p, is

For references see page 9645

proportional to the population of the atoms a in level q (n_{aq}) and is given by Equation (10):

$$I_{qp} = A_{qp} n_{aq} h\nu_{qp} \tag{10}$$

After substitution of the population of the excited level according to Boltzmann's equation (Equation 11):

$$\frac{n_{aq}}{n_a} = \left(\frac{g_q}{Z_a}\right) \exp\left(\frac{-E_q}{kT}\right) \tag{11}$$

where n_a is the total density of atoms a over all levels, g_q the statistical weight of the excited level q, Z_a the partition function for the atoms (being a function of the temperature), E_q the excitation energy for the excited level q and T the excitation temperature (often denoted T_{ex}), one obtains Equation (12):

$$I_{qp} = A_{qp} h\nu_{qp} n_a \left(\frac{g_q}{Z_a}\right) \exp\left(\frac{-E_q}{kT}\right) \tag{12}$$

T then can be determined from the intensity ratio for two lines (a and b) with wavelength λ_a and λ_b of the same ionization state of an element according to Equation (13):

$$T = \frac{5040(V_a - V_b)}{\log[(gA)_a/(gA)_b] - \log(\lambda_a/\lambda_b) - \log(I_a/I_b)} \tag{13}$$

Often the line pair Zn 307.206/Zn 307.59 nm is used. This line pair is very suitable because ionization of zinc is low as a result of its relatively high ionization energy. Also, the wavelengths are close to each other, which minimizes errors introduced by changes in the spectral response of the detector, and the ratio of the gA values is well known.

The excitation temperatures can also be determined from the slope of the plot of $\ln[I_{qp}/(g_q A_{qp} \nu_{qp})]$ or $\ln(I_{qp}\lambda/gA_{qp})$ versus E_q, which is $-1/kT$. The λ/gA values for a large number of elements and lines are available. Spectroscopic measurements of temperatures from line intensity ratios may be hindered by deviations from thermodynamic behavior in real radiation sources and by inaccuracies in the transition probabilities. The determination of excitation temperatures in spatially inhomogeneous plasmas, which have a cylindrical symmetry, can be performed with the aid of Abel inversion, as treated extensively in classical textbooks on atomic spectroscopy (see e.g. the book by Boumans[(70)]).

For the measurement of the gas kinetic temperatures, both measurements of the rotational temperatures from the band spectra of species such as the OH radical or the N_2^+ ion and measurements of Doppler widths have been proposed. In the first case one makes use of the rotational fine structure of the vibration bands.

Molecules or radicals have various electronic energy levels ($^1\Sigma$, $^2\Sigma$, $^2\Pi$, etc.) which have a vibrational fine structure ($v = 0, 1, 2, 3, \ldots$), and these levels, in turn, have a rotational hyperfine structure ($J = 0, 1, 2, 3, \ldots$). The total energy of a state may be written as Equation (14):

$$E_i = E_{el} + E_{vib} + E_{rot} \tag{14}$$

E_{el} is of the order of 1–10 eV, the energy difference between two vibrational levels of the same electronic state is of the order of 0.25 eV and the separation of rotational levels is of the order of 0.005 eV. When the rotational levels considered belong to the same electronic level, the emitted radiation is in the infrared (IR) region. When they belong to different electronic levels, they occur in the ultraviolet (UV) or visible (VIS) region. Transitions are characterized by the three quantum numbers of the states involved: n', v', j' and n'', v'', j''. All lines which originate from transitions between rotational levels belonging to different vibrational levels of two electronic states form the band: n', $v' \rightarrow n''$, v''. For these band spectra, the selection rule is $\Delta j = j' - j'' = \pm 1, 0$. Transitions for which $J'' = j' + 1$ give rise to the P-branch, $j'' = j' - 1$ to the R-branch and $j' = j''$ to the Q-branch of the band. The line corresponding to $j' = j'' = 0$ is the zero line of the band. When $v' = v'' = 0$ it is also the zero line of the system. The difference between the wavenumber of a rotational line and the wavenumber of the zero line in the case of the P- and the R-branch is a function of the rotational quantum number j and the rotation constant B_v, for which (Equation 15):

$$\frac{E_j}{hc} = B_v j(j+1) \tag{15}$$

The functional relation is quadratic and known as the Fortrat parabola.

As in the case of atomic spectral lines, the intensity of a rotational line can be written as Equation (16):

$$I_{nm} = \frac{N_m A_{nm} h\nu_{nm}}{2\pi} \tag{16}$$

where N_m is the population of the excited level and ν_{nm} the frequency of the emitted radiation. The transition probability for dipole radiation is given by Equation (17):

$$A_{nm} = \frac{64\pi^4 \nu_{nm}^3}{3k} \left[\frac{1}{g_m}\right] \sum |Rn_i m_k|^2 \tag{17}$$

where i and k are the degenerancies of the upper (m) and the lower state (n). $Rn_i m_k$ is a matrix element of the electrical dipole moment and g_m is the statistical weight of the upper state; N_m is given by the Boltzmann equation (see Equation 11) where the energy to be included is E_j (see Equation 15) with $j = J'$, being the rotational term of the upper level. For a $^2\Sigma_g$–$^2\Sigma_u$ transition, between a so-called "gerade" (g) and "ungerade" (u) level, the term $\Sigma|Rn_im_k|^2 = J' + J'' + 1$, where J' and J'' are the rotational quantum numbers of the upper and lower state. Accordingly (Equation 18):

$$I_{\mathrm{nm}} = \left[\frac{16\pi^3 cN\nu_{\mathrm{nm}}^4}{3Z(T)}\right](J' + J'' + 1) \times \exp\left[\frac{-hcB_{v'}J'(J'+1)}{kT}\right] \quad (18)$$

or

$$\ln\left[\frac{I_{\mathrm{nm}}}{J' + J'' + 1}\right] = \ln\frac{16\pi^3 cN\nu_{\mathrm{nm}}^4}{3Z(T)} - \frac{hcB_{v'}J'(J'+1)}{kT} \quad (19)$$

By plotting $\ln[I_{\mathrm{nm}}/(J' + J'' + 1)]$ versus $J'(J'+1)$ for a series of rotational lines, a so-called rotational temperature can be determined from the slope. It reflects the kinetic energy of neutrals and ions in the plasma. For the determination of the gas kinetic temperatures, both measurements of the rotational temperatures from the band spectra of species such as the OH radical or the N_2^+ ion as well as measurements of Doppler widths have been used. For the measurements of electron temperatures Thomson scattering has also been used.(71) The basics of these methods are described in classical textbooks(70,72) and papers.(73) Also, efforts have been made to describe the flow dynamics of microwave discharges, which especially were found to be very useful for improving the design of microwave structures used for spectrochemical analysis.(63)

3.1 Low-power Microwave-induced Plasmas

In early MIP work, Beenakker MIP(69) mentioned that the electron temperatures in an MIP operated in a TM_{010} cavity should be high whereas rotational temperatures, being a good approximation to the gas kinetic temperatures, as described by Heltai et al.,(52) are only of the order of 1500 –2400 K (Figure 9a and b) both for He and Ar MIPs. Similar measurements were made in the surfatron plasma,(74) in the modulated surfatron(75) and also in MIPs operated in CO_2 and in He.(76) For the case of a surfatron the electromagnetic properties were studied(77) in addition and also their fundamental parameters.(78) In the case of the MIP the importance of charge transfer was shown by Brandl and Carnahan.(79) The diagnostic data are important for the dissociation of molecular species, being especially studied with respect to combinations of chromatography and MIP excitation, as described for hydrocarbons.(80)

Diagnostics have been performed not only for some widely used types of MIPs but also for a low-power Ar MIP used for the analysis of solutions(81) and for an MIP operated in a liquid-cooled discharge tube.(82) Studies on the energetic balance in low-pressure MIPs

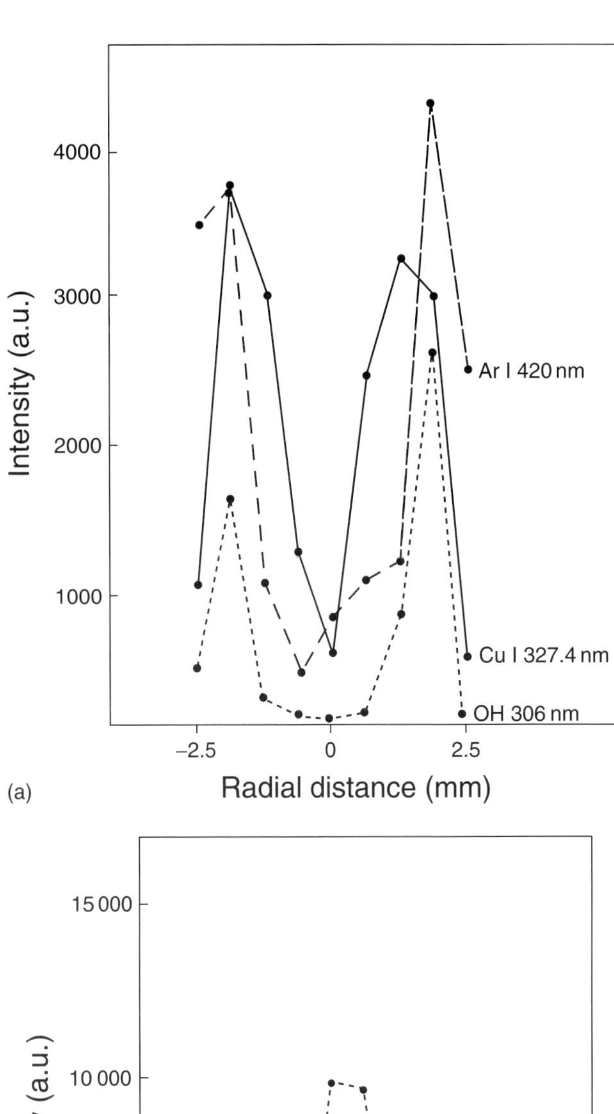

Figure 9 (a) Radial intensity profiles of a toroidal Ar MIP. Discharge tube: external diameter 7 mm, internal diameter 5 mm; PN of H_2O or of a 100 µg mL^{-1} solution of copper, −100 W. (b) Radial intensity profiles measured for a cylindrical He MIP. Discharge tube: external diameter 7 mm, internal diameter 5 mm; 0.2 L min^{-1} helium; without aerosol introduction, −150 W. (○) He 388.7 nm; (•) (0, 0) N_2^+ band head at 390 nm; (▲) (0, 0) OH band head at 306 nm.(52)

For references see page 9645

have also been made.[83] With special plasmas, such as a low-density He electron cyclotron resonance microwave plasma, techniques for temperature measurements, e.g. with line broadening, have been developed.[84]

In addition to plasma parameter measurements, noise analysis of the reflected power of MIPs,[85] near-infrared (NIR) characteristics of MIPs with respect to the determination of nonmetals[86] and studies on MS of MIPs[87] were also performed.

3.2 Single-electrode Plasmas

For single-electrode plasmas, electron number densities and excitation temperatures were determined by Kirsch et al.,[88] who had already extensively used this plasma for analytical purposes (see e.g. Hanamura et al.[89]), as did Boumans et al. (see e.g. Boumans et al.[4]). Here excitation temperatures of 5000–7000 K were reported for a 500 W capacitively coupled microwave plasma (CMP) operated in Ar or He with N_2 as wall-stabilizing gas. The flow pattern changes as the result of the addition of the alkali metals, as reflected in the tomographic studies reported by Bings and Broekaert.[90] Here it could be shown that the plasma blows up as a result of ambipolar diffusion and becomes broader when Cs or Na is added (Figure 10). Ethanol addition could also be shown to thermalize the plasma, as excitation and rotational temperatures measured then came nearer to each other. The plasma temperatures for CMPs operated with air only could be measured and tomographic studies performed. As shown in Table 1, these temperatures did not differ much from these obtained by Kirsch et al.[88] with noble gases and neither did the electron number densities or the rotational temperatures. Anyhow, it could be shown that CMPs, despite the fact that they are often operated at a power of 500 W which is near to the power of low-power analytical ICPs, have a larger deviation from local thermal equilibrium (LTE) than the ICP. This can be understood from the higher operational frequencies of microwave discharges, which even more hamper energy exchange as compared with the ICP.

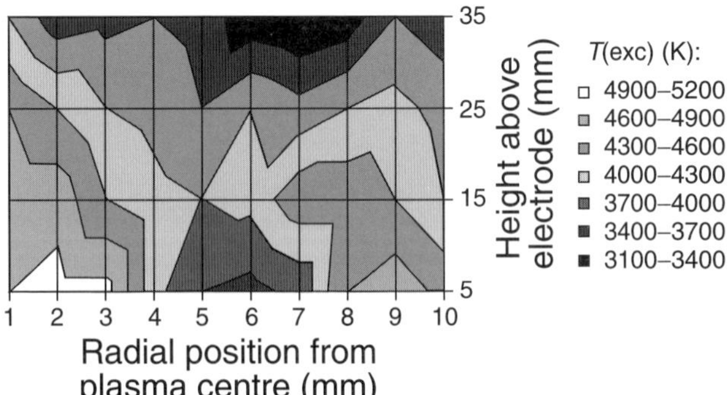

Figure 10 Mapping for excitation temperature (T_{exc}) in an air CMP.[32]

Table 1 Ranges of T_{exc}, T_{rot} and n_e obtained for 600 W N_2, air and Ar CMPs[32]

Working gas	T_{exc} (K)	T_{rot} (K)	n_e (cm^{-3})
N_2	5500–3400	4300–2500	10^{11}–10^{14}
Air	4600–3400	3400–2800	10^{10}–10^{14}
Ar	4900–3100	4300–2800	10^{9}–10^{14}
Standard deviation	300	300	10^{7}–10^{12}

3.3 Plasma Torch Discharges

Since the first publications by Jin et al.[15,91] on the MPT, it became clear that a microwave plasma with a toroidal structure can be operated in an ICP-like torch and that the moderate power and the possibility of using both He and Ar as working gases made it a very powerful atomic spectrometric source. This MIP could be well operated in combination with flow-cell HG. There was no need to remove the excess of H_2 produced before leading the reaction gases, including H_2 and the hydrides, into the plasma, which was found to be impossible with an MIP according to Beenakker. From noise power spectra obtained with the MPT in the case of HG, as shown in Figure 11(a) and (b), it was found that even with an excess of H_2 the frequency component stemming from the plasma swirl still is the important one, as it is in the case of the ICP. Further, however, flicker noise was found to occur at frequencies of 0.2–0.5 Hz and white noise was found to dominate below 100 Hz.[93] Temperatures were also measured in the microwave plasma operated in the MPT. For electron and gas kinetic temperatures, values at different heights in the plasma were obtained by Thomson scattering.[94] A complete temperature and electron number density mapping in the case of a modified MPT was performed recently.[95] It showed that in the case of wet plasmas the electron temperature increased similarly as in the case of the ICP. Further, mappings of the rotational temperatures, as possible with a charge-coupled device (CCD) spectrometer, were performed by Engel et al.[96] They showed that the rotational temperature in the MPT increases with the water uptake, which might even improve the robustness of the source. This may explain why it easily seemed to be possible to take up water- and even acetonitrile-loaded aerosols into the MPT, as described by Prokisch and Broekaert.[97]

Also for other types of MIPs, spatial mappings of analyte intensities and plasma parameters were made as an aid in optimization studies.[98] Further, studies were not confined to noble gas plasmas, but also plasma torches

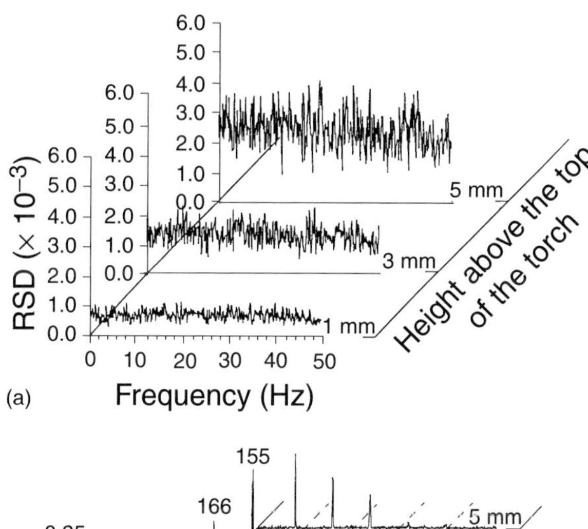

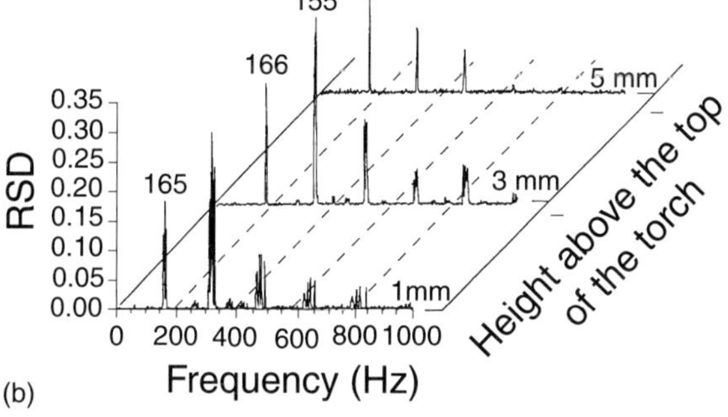

Figure 11 Effect of H_2 addition ($13.5\,mL\,min^{-1}$) on the noise spectra of the He I 501.56 nm line at different heights above the top of the torch.(92)

obtained in the case of N_2 at a higher power were operated in a TM_{010} resonator.(99) These devices turned out to be of use for solution analysis and could use an ICP-like torch. This discharge, as well as that referred to by Ohata and Furuta(100) and later proposed as a source for MIP/MS, is similar to that described by Leis and Broekaert.(67) It is operated in a rectangular cavity and is very stable; however, it is much more prone to easily ionized element interferences than the ICP.

Thorough diagnostic studies also have been performed with the so-called "torche à injection axiale" (TIA) by the group of Van der Mullen as described by Jonkers et al.(71,101) recently. This plasma has a toroidal shape, it can take up water-loaded aerosols and it can be operated with both molecular and noble gases.

For this plasma it has been reported that the Saha equation applies. This equation describes the ionization in a thermal plasma as (Equation 20):

$$S_{pj}(T) = \frac{p_{ij}p_e}{p_{aj}} = \frac{(2\pi m)^{\frac{3}{2}}(kT)^{\frac{5}{2}}}{h^3(2Z_{ij}/Z_{aj})[\exp(-E_{ij}/kT)]} \quad (20)$$

where S_{pj} is the Saha function expressed in terms of the partial pressures for the component j, p_i, p_e and p_a are the partial pressures for ions, electrons and atoms, respectively, the factor 2 is the statistical weight of the free electron (two spin orientations), m is the mass of the electron (9.11×10^{-28} g), k is the Boltzmann constant ($k = 1.38 \times 10^{-23}\,J\,K^{-1}$), T is the absolute temperature (K), h is Planck's constant ($h = 6.62 \times 10^{-34}$ J s), Z_{ij} and Z_{aj} are the partition functions of the ions and the atoms of the element j, respectively, and E_{ij} is the ionization energy of the element j. As $1\,eV = 1.6 \times 10^{-19}$ J, this leads to Equation (21):

$$\log S_{pj} = \frac{5}{2}\log\left[\frac{T-5040}{T}V_{ij}\right] + \log\left(\frac{Z_{ij}}{Z_{aj}}\right) - 6.18 \quad (21)$$

where V_{ij} is the ionization energy (eV). It should be mentioned, however, that the Saha equation is only valid for plasmas in LTE, which strictly does not apply in the case of many MIPs. The latter is understandable from the fact that the light electrons can easily take up energy at microwave frequencies, but heavy ions much less so. This results in different kinetic energies and different temperatures for the light electrons and the heavy ions and, accordingly, in departures from LTE. By Thomson scattering, the electron temperature can be determined irrespective of whether the plasma is in LTE or not and for the TIA, for instance, electron temperatures in the case of Ar were reported to be of the order of >8700 K.(71) This was the case even for a He plasma operated in air.(102) Atom state densities for this plasma have also been calculated.(101)

4 SAMPLE INTRODUCTION

As in any spectrochemical source, the analyte has to be brought as an atomic cloud in the zones where excitation and/or ionization takes place. Because of the relatively low gas kinetic temperatures in the microwave discharges, their atomization capacity is not as high as in the ICP, which is generally operated at higher power than microwave discharges. Accordingly, the analyte must be atomized more efficiently or at least it must be present in a very fine form before it enters the microwave plasma.

4.1 Techniques for Sample Introduction

Sample introduction is the "Achilles' heel" of plasma spectrochemical analysis(103,104) and many techniques which can be used in the case of microwave plasmas have been described in the plasma spectrometry literature, as shown in a monograph by Sneddon,(105) a special issue of *Spectrochimica Acta, Part B,*(106) and various reviews treating aspects such as flow injection,(107) coupling techniques,(108) on-line separation(109) and sample introduction in atmospheric pressure MIPs(110) or MIPs

For references see page 9645

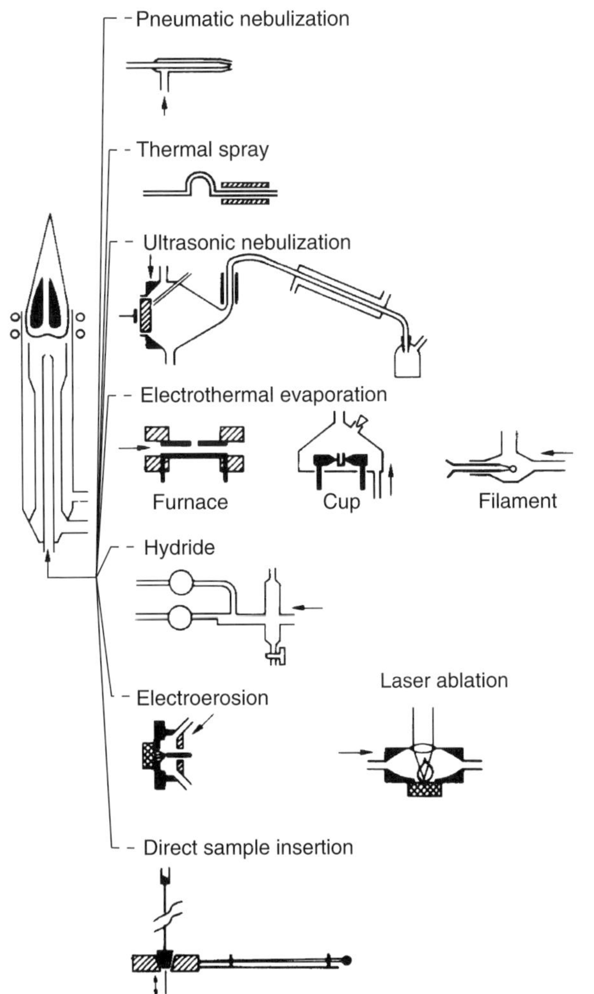

Figure 12 Different techniques for sample introduction on inductively coupled plasma optical emission spectroscopy (ICPOES).[113]

in general.[111,112] These techniques include PN, ultrasonic nebulization (USN), ETV of microsamples, HG for the elements having volatile hydrides and direct solids sampling methods such as slurry nebulization, SA and LA and direct sample insertion, as are well known from the ICP literature (Figure 12). In the case of microwave plasmas at low power, especially techniques where a dry atom cloud is formed will be favorable. This explains, why tandem systems using an MIP for the excitation of the exhaust aerosols of an ICP were of use,[114] in which the MIP even could be operated at reduced pressure.[115]

4.2 Sample Introduction in Low-power Microwave Plasmas

In the case of very weak discharges such as the MIP according to Beenakker, especially ETV and HG[116] have been used for sample introduction. In the case of ETV, as reviewed by Matusiewicz,[117,118] different types of electrothermal devices can be used. They include a commercial graphite furnace (GF) as known from AAS[119–121] as well as a graphite rod atomizer,[122] even for the case of a low-pressure MIP,[123] wire-loop devices entered into a microtube[124,125] and Ta boat[126] and Ta strip devices.[127] In the case of the resonator according to Beenakker, it could be shown that the detection limits with GF sample evaporation can be considerably improved by using a toroidal instead of a filament discharge.[120] Special attention was also given to the acquisition of the transient signals, for which a rapid scanning monochromator can be very helpful.[128] The technique could be used for the determination of I in HCl,[129] for the determination of nonmetals in the case of an He discharge,[130,131] for the determination of S,[132,133] of halogenated compounds[134] and of Zn[135] and for the determination of Ni and Pb in bones.[136] A useful way to increase further the detection power of ETV coupled to OES with the MIP consists of preconcentration of the trace elements to be determined by electrolysis on a graphite tube filled with vitreous C, as shown for the case of Pb.[137] Chromatography on a vitrous C-filled column can also be used for this purpose.[138] Many phenomena, however, still have to be studied, such as the aerosol transport mechanisms[139] and also vaporization at low pressure. The latter, for example, has been studied for the case of chlorides, sulfates and nitrates.[140] Special ETV techniques may also make use of moving bands.[141]

The generation of hydrides and of other volatile compounds has found widespread use in MIP work. In a number of cases, however, the vapors generated have to be freed from an excess of molecular gases such as H_2 or even water vapor, as these species may change the plasma impedance so much that the circuit leaves the resonance frequency and the plasma is extinguished. In the case of a surfatron MIP, the excess of H_2 envolved in continuous HG was found not to hamper the discharge stability,[142] whereas in the case of the MIP in a Beenakker resonator the excess of H_2 is mostly removed by fixing the hydrides in a cold trap and releasing them at once into the MIP.[143] In this way, detection limits at the sub nanograms per milliliter level can be obtained (Tables 2 and 3), and also by hot-trapping in a preheated GF.[143,152,153] Also a Nafion® membrane can be used for separating moisture or H_2.[144] Further, one can work with microtechniques, where microaliquots are deposited on an $NaBH_4$ pellet.[154,155] At well selected operating conditions, continuous HG could even be directly coupled with microwave plasmas, also for the case of real samples, such as for the determination of Se in soils.[156] With low-pressure microwave plasmas, the introduction of an excess of H_2 was found to hamper the discharge stability less.[157] Also, the determination of

Table 2 Comparison of the concentration detection limits in microwave-induced plasma optical emission spectroscopy (MIPOES) for As, Sb and Se obtained with different sample introduction techniques[a] [143]

Procedure[b]	As ($ng\,mL^{-1}$)	Sb ($ng\,mL^{-1}$)	Se ($ng\,mL^{-1}$)
PN/T/MIP	500 (4.8)	250 (4.0)	540 (4.2)
GF/3F/MIP	50 (1.8)	20 (2.0)	46 (1.5)
HG/CT/3F/MIP	0.8 (6.7)	0.4 (6.8)	0.5 (7.2)
HG/GFT/3F/MIP	0.4 (4.5)	0.35 (4.8)	0.25 (4.6)

[a] At 10 × detection limit; data in parentheses are RSD, relative standard deviation (%).
[b] T/MIP, toroidal microwave-induced plasma; 3F/MIP, three-filament/microwave-induced plasma; GFT, graphite furnace trapping. CT, cold trapping.

Table 3 Comparison of literature values for the detection limits ($ng\,mL^{-1}$) for As, Sb and Se in AAS, AES and MS[143]

Technique[a]	Reference	As	Sb	Se
PN/ICP	145	50	32	75
PN/MIP	51	300	–	–
HG/ICP	146	1.0	2.4	1.3
HG/ICP	147	1.0	–	–
HG/ICP	148	0.06	0.18	–
HG/MIP	144	0.32	6.1	–
HG/GF/MIP	149	0.12	–	–
PN/ICP/MS	150	0.14	0.019	1.5
HG/ICP/MS	150	0.017	0.031	0.17
HG/AAS	151	0.16	0.08	0.18

[a] PN/ICP, pneumatic nebulization/inductively coupled plasma; PN/MIP, pneumatic nebulization/microwave-induced plasma; HG/ICP, hydride generation/inductively coupled plasma; HG/MIP, hydride generation/microwave-induced plasma; HG/GF/MIP, hydride generation/graphite furnace/microwave-induced plasma; PN/ICPMS, pneumatic nebulization/inductively coupled plasma mass spectrometry; HG/ICPMS, hydride generation/inductively coupled plasma mass spectrometry; HG/AAS, hydride generation atomic absorption spectroscopy.

As and Se subsequent to microwave sample dissolution and hot hydride trapping has been described.[158] Further, miniature hydride systems for the determination of As, Sb, Pb and Sn[155] and the determination of Pb by HG[159] have been described. For their optimization, spatially resolved studies of Ga, In, Se, Te, As and Sb lines[160] were very useful. Not only the hydrides but also further volatile compound-forming elements can be easily determined with the aid of an MIP at a very high power of detection. This first applies to the determination of Hg with the aid of the cold vapor (CV) technique, known from AAS work. This can be done at low pressure[161] or at atmospheric pressure[162] and also down to ultratrace levels, as already shown in early work.[163] By this method I also can be determined, simply by a "reverse titration",[164] as already shown for the case of seawater.[165] Much work has also been done on the determination of the halogens after volatilization into an He MIP.[166]

Cl could be determined as HCl,[167,168] all halogens could be determined with the aid of detected ion with a surfatron MIP[169] and also Br[170–173] and I[174,175] could be determined using an MIP. In the case of I, methods for differentiation between iodate and iodide in brines have been described.[176] The influences of N_2 impurities on the F and Cl signals were also studied.[177] Cl and Br could be volatilized chemically through the use of strong oxidants such as a solution of $K_2Cr_2O_7$ in highly concentrated sulfuric acid.[178] Also further nonmetals can be determined by volatilization methods,[179] including S after H_2S/SO_2 generation.[180] Generation of volatile species can be applied for the determination of Ni by using the formation of volatile carbonyls.[181,182] The introduction of solid halocarbons,[183] the determination of C, H, N and O for elemental analysis purposes with the aid of a diode-array spectrometer and atomic emission lines in the IR region[184] and organic microanalysis have further been reported.[185] Special attention was also given to the determination of the N species in the MIP[186] and to the emission of H for radiation standardization purposes.[187]

Wet aerosols are mostly generated by PN, USN or by newer techniques such as hydraulic high-pressure nebulization (HHPN). In PN with cross-flow, concentric, grid or Babington nebulizers (Figure 13a–d), the aerosol is generated by splitting off small droplets from a liquid surface by viscosity drag forces and a good description of the relation between the gas flow nebulizing a liquid flow and the pressure decay at the nebulizer nozzle is given by the Nukuyama–Tanasawa equation. The Sauter diameter, being the diameter of the droplets for which the volume-to-surface ratio equals that of the complete aerosol, is given by Equation (22):

$$d_0 = \left(\frac{C}{v_G}\right)\left(\frac{\sigma}{\rho}\right)^{\frac{1}{2}} + C'\left\{\frac{\eta}{(\sigma\rho)^{\frac{1}{2}}}\right\}^{C''}\left[1000\left(\frac{Q_L}{Q_G}\right)\right]^{C'''} \quad (22)$$

where v_G is the gas velocity, Q_G the gas flow, Q_L the liquid flow, η the viscosity, ρ the density, σ the surface tension of the liquid, and C, C', C'' and C''' are constants. When the nebulizer gas flow increases d_0 becomes smaller, the sample introduction efficiency increases and so do the signals. However, as more gas is blown through the plasma, it is cooled and the residence time of the droplets decreases so that atomization, excitation and ionization also decrease. These facts counteract the increase in the signals as a result of the improved sampling. Maximum signal intensity and power of detection are thus achieved at a compromise gas flow. The size of the aerosol droplets obtained is in the low micrometer range and their diameter depends on the physical properties of the liquid as described above. Apart from PN, HHPN, as

For references see page 9645

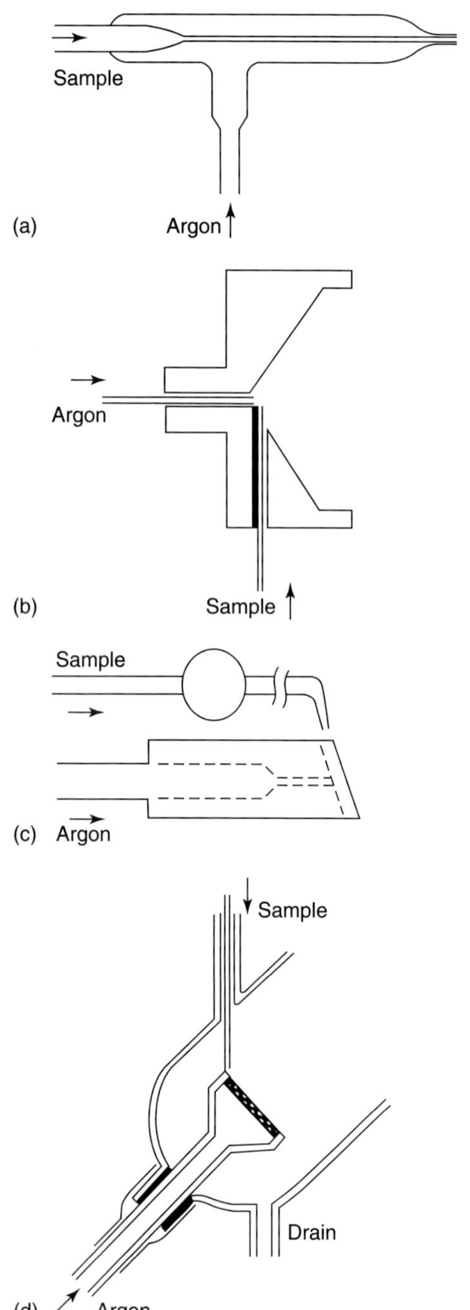

Figure 13 Pneumatic nebulizers for plasma atomic spectrometry: (a) concentric glass nebulizer; (b) cross-flow nebulizer; (c) Babington-type nebulizer; (d) fritted-disc nebulizer.[188]

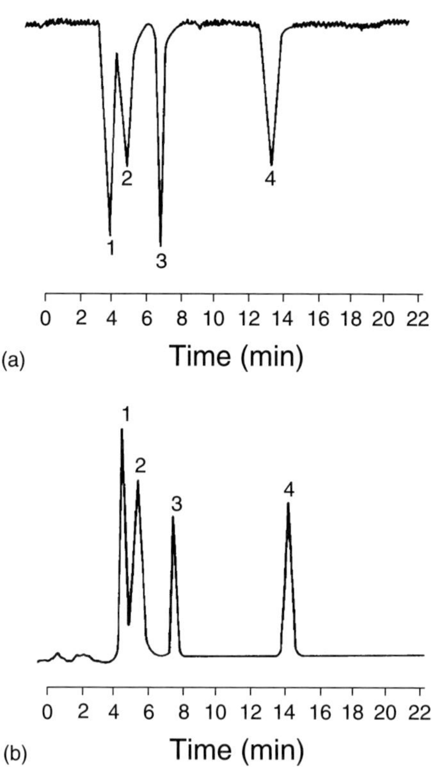

Figure 14 Ultraviolet/visible (UV/VIS) and MIP detection of different Hg species after their isocratic separation as 2-mercaptoethanol complexes by reversed-phase HPLC. Peaks: 1 = Hg^{2+}; 2 = CH_3Hg^+; 3 = $C_2H_5Hg^+$; 4 = phenyl-Hg^+; 300 ng Hg in each case. Column: RP-6, 5 µm film thickness, 250 µm × 4.6 mm i.d. Mobile phase: methanol–water (35 : 65, v/v). (a) MIP detection: Hg I 253.6 nm. (b) UV/VIS detection: absorption at 250 nm, reference wavelength 430 nm.[195] (Reproduced by permission from D. Kollotzek, D. Oechsle, G. Kaiser, P. Tschöpel, G. Tölg, *Fresenius' Z. Anal. Chem.*, **318**(7), 485–489 (1984). Copyright Springer Verlag, 1984.)

developed by Berndt[189] and based on the expansion of a liquid jet, can also be used subsequent to desolvation.[190] This, of course, also applies for USN, of which the state-of-the-art was described by Fassel and Bear.[191] The introduction of wet aerosols into the MIP requires the use of higher power and special geometries.

Some papers have described the use of PN subsequent to aerosol desolvation, e.g. in the case of a glass frit nebulizer and an He MIP.[192] At moderate microwave power, desolvation even might be superfluous, as shown in the case of a Hildebrand grid nebulizer.[193] With a Babington nebulizer it was even possible to analyze slurries directly using an MIP.[194] Toroidal plasmas were shown to well accept wet aerosols[51] and they are suitable for direct coupling with liquid chromatography (LC) (Figure 14a and b). In this field, refinement in HPLC and capillary zone electrophoresis (CZE) is still needed. Also, here a plasma with moderate power is advantageous.[196,197] With the MPT the introduction of wet aerosols generated from aqueous and even acetonitrile-containing solutions without desolvation is possible, but with lower detection limits compared with dry aerosols (Table 4). The concomitant effects were found to be low[199] and the system could be used for the determination of noble metals.[200] The use of USN in MIP work was described e.g. by Michlewicz and Carnahan.[201]

Table 4 Limits of detection (c_L) and upper limit of linear dynamic range (*ldr*) obtained with OES using an MPT for different sample solutions[97]

Element/line	Wavelength (nm)	Sample[a]	Without desolvation (this work)[97]		Without desolvation[198]	With desolvation[198]
			c_L (ng mL^{-1})	*ldr* (ng mL^{-1})	c_L (ng mL^{-1})	c_L (ng mL^{-1})
Li I	670.78	Water	20	200 000	–	0.99
		100 µg mL^{-1} Cs solution	20	200 000	–	–
		30% AcN	4	10 000	–	–
Cd I	228.80	Water	400	50 000	100	18
		100 µg mL^{-1} Cs solution	300	50 000	–	–
		30% AcN	2000	100 000	–	–
Cr I	359.35	Water	200	1 000 000	100	9.2
		100 µg mL^{-1} Cs solution	300	600 000	–	–
		30% AcN	300	100 000	–	–
Pb I	368.35	Water	500	500 000	800	27
		100 µg mL^{-1} Cs solution	600	500 000	–	–
		30% AcN	2000	500 000	–	–

[a] AcN = acetonitrile.

4.3 Sample Introduction in Medium- and High-power Microwave Plasmas

In microwave discharges operated at a power level of at least 150 W, many more techniques for sample introduction can be easily applied. This applies both to electrodeless MIPs and to single-electrode plasmas.

In the case of the MPT, it is possible to apply flow-cell HG, where the excess of the H_2 produced is also led into the discharge (Figure 15). The detection limits for the volatile hydride-forming elements As, Se, etc. are in the nanograms per milliliter range and in the case of hot trapping the absolute detection limits of Hg, As, Se and Sb with both He and Ar discharges are 0.1–0.2 ng.[202] Medium-power plasmas were also found to be useful for trace determinations in gases.[203] Also for element-specific detection in supercritical fluid chromatography (SFC), it was found that the high-power MIP can well cope with the excess of CO_2 produced.[204]

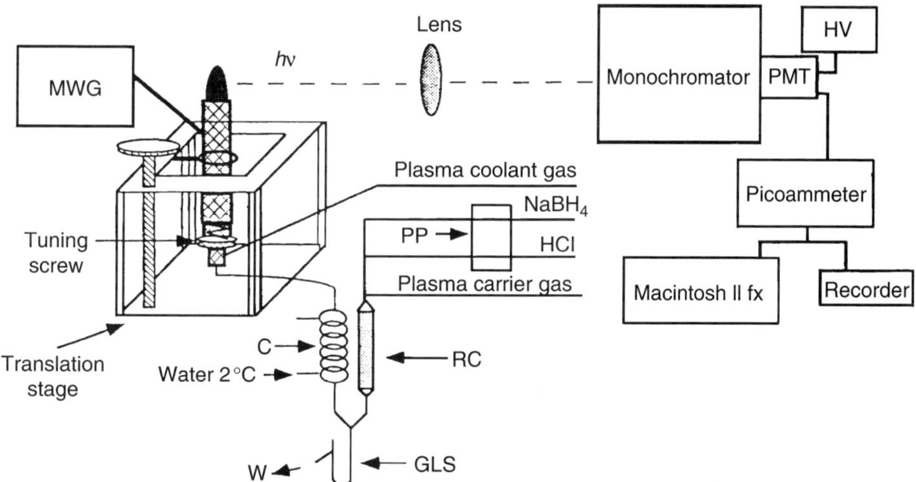

Figure 15 Schematic diagram of instrumentation for OES with an MPT and HG. PP = peristaltic pump; C = cooler; RC = reaction column (25 cm × 1.5 cm i.d.); GLS = gas–liquid separator (25 cm × 0.5 cm i.d.); W = waste; MWG = microwave generator; PMT = photomultiplier; HV = high-voltage power supply.[92]

In the case of ETV the switching of gas lines for plasmas at medium and high power becomes less critical than with low-power MIPs. This has been shown for the case of the MPT, where detection limits of 3–100 pg for elements such as Ag, Au, Ge, Pb, S and Te were reported.(205) In the case of the CMP, the detection limits for direct volatilization of microsamples from a W cup are in the same range.(206,207)

With PN of solutions desolvation is often still necessary, but for the MPT(208) it becomes superfluous at powers higher than ca. 100 W.(209–211) This allows the use of direct injection nebulization (DIN)(212) and of flow injection analysis (FIA)(213) also for on-line preconcentration.(214) USN, because of the small droplet size, is very suitable, even for the the determination of nonmetals such as S(215) and C, S and P at powers above 1 kW.(216) Inexpensive air humidifiers can also be used for this purpose.(217) Even for the determination of F in solutions an He MIP was found suitable.(218) The high-power plasma torch (Figure 8) in a rectangular cavity operated with air was found to allow the direct introduction of wet aerosols, but with a lower power of detection and higher concomitant effects than the ICP.(67)

4.4 Element-specific Detection in Chromatography

Microwave plasmas, owing to their compactness and the low operating costs of the plasma source and also owing to the possibility of constructing generators with lower costs than ICP generators, are valuable sources for element-specific detection in chromatography. This applies both when they are used as radiation sources for AES and as ion sources for elemental MS. A further advance lies in the dimensions of these sources, which are very small (some a few cubic millimeters) compared with ICPs (in the cubic centimeter range). The latter is advantageous both for preserving the obtained chromatographic resolution and for realizing a high absolute power of detection.

The features of MIPs both for LC and GC were outlined in reviews such as that by Jia et al.(219) and for the He plasmas especially that by Long et al.(220) In GC, He is often used as the working gas, which then is very suitable for entry into microwave plasmas as they also can be operated in He and then allow the detection of nonmetals including the halogens with a high absolute power of detection. The latter is due to the possible formation of $He_2{}^{+m}$, having energies of 18.8–21.6 eV.(221) In the case of LC, nebulization of the effluent from the chromatographic system has to be applied before excitation with microwave plasmas is possible. This is complicated by the fact that wet aerosols, at the present state-of-the-art, cannot be entered in small microwave plasma sources without significantly influencing their stability and excitation conditions. Here research efforts have to be made, either by the design of suitable nebulization systems, working with minimal effluent flows, as is now possible with CZE, or by using alternative microwave sources still to be developed. For SFC, which is becoming of increasing interest for the detection of metal chelates,(222,223) tolerance against the introduction of CO_2 must be ensured.

The possibilities of using MIPs for element-specific detection in GC have been described in several reviews(224–227) and their features as compared with AAS and MS have been outlined.(228) Many papers have described the use of microwave plasmas operated at reduced pressure for element-specific GC detection [see e.g. Hobbs(229)]. The commercially available microwave plasma detector (MPD) at reduced pressure in combination with a conventional polychromator was very useful. More recently, the MIP according to Beenakker was also used at reduced pressure for C, H and S detection.(230) However, the breakthrough first occurred with the availability of low-power microwave plasmas at atmospheric pressure coupled with GC(231) and of commercial equipment combining a gas chromatograph through a heated transfer line with an MIP and a suitable diode-array spectrophotometer, such as available from Hewlett-Packard (Figure 16).(233,234) Herewith, a system using an MIP for element-specific detection also of organometallic compounds at the trace level after GC is available.(235–237) However, this system makes use of a plasma as described by Beenakker, which is not the only possible way, as shown by work using a surfatron MIP(238) with the same spectrometer or by systems constructed from different components.(239) Here both glass capillaries(240) and fused-silica capillaries(45) can be used.

Comparisons were made between MIPs operated with He only and those operated with He–Ar mixtures.(241) With respect to the types of MIPs used, the surfatron,(242) the MIP according to Beenakker and other types were compared.(243) Such studies also include noise studies, as was done in the case of the surfatron(244) and the MPT.(245) With respect to torch design, both a tangential flow torch and a capillary water-cooled MIP torch can be used, where the latter was found to result in the best sensitivity.(246) Further, different torch tube materials can be used, such as BN and quartz. The former material is brittle but is more resistant to etching by chromatographic effluents such as acetone and acetonitrile. Laminar-flow torches were found to give better results in the determination of empirical formulas than the other types, as shown in the case of dioxins and pyrethroids.(247) Apart from the already mentioned types of MIPs, microwave plasmas generated at about 15 W(248) between the tip of a hollow metal electrode (e.g. the GC capillary end) and an annular electrode surrounding the tip have also been used.(249) Power modulation of the plasma at 10 Hz has also been applied and found useful to

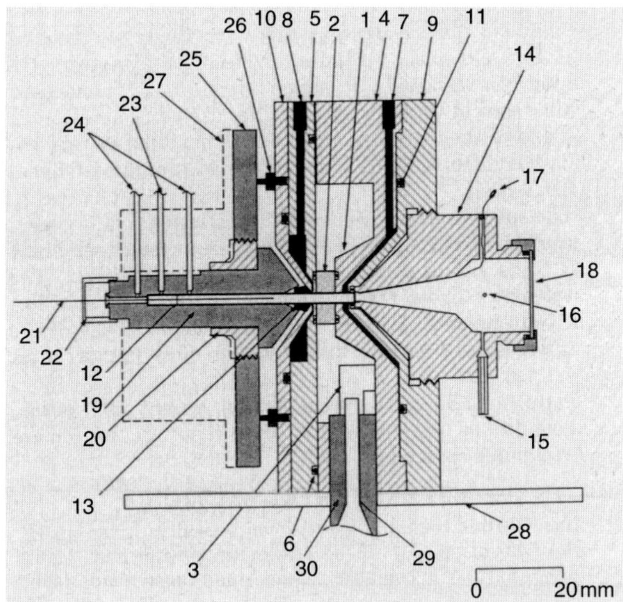

Figure 16 Re-entrant cavity for gas chromatography/microwave-induced plasma atomic emission detector (GC/MIPAED) system. (1) Pedestal; (2) quartz jacket; (3) coupling loop; (4) main cavity body; (5) cavity cover plate; (6) gasket; (7,8) cooling water inlet and outlet; (9,10) water plates; (11) O-ring; (12) silica discharge tube; (13) polyimide ferrule; (14) exit chamber; (15,16) window purge inlet and outlet; (17) sparker wire; (18) window; (19) gas union; (20) threaded collar; (21) column; (22) capillary column fitting; (23) makeup and reagent gas inlet; (24) purge flow outlets; (25) stainless steel plate; (26) standoff; (27) heater block; (28) mounting flange; (29) brass center conductor; (30) PTFE coaxial insulator.[232] (Reprinted with permission from B.D. Quimby, J.J. Sullivan, *Anal. Chem.*, **62**(10), 1027–1034 (1990). Copyright 1990, American Chemical Society.)

maximize the excitation efficiency of the MIP.[250] For the detection of dioxins the use of a tangential torch, of low flow rates and of low power, was found to give the best sensitivity.[251] It was even found possible to use the reflected power of the MIP for registration of the chromatogram.[252] Also the VIS wavelength range often is used in the case of nonmetals.[253,254] Here Fourier transform spectrometry is very useful and allows multiwavelength measurements.[255] Apart from rather expensive dispersive spectrometers using a diode array, interference filtering[256] can also be used. Attention must also be given to the acquisition of the spectral background, which may be influenced by the effluent. This can be done by using rapid scanning techniques, as has long been known.[257] Multicapillary GC is very useful for obtaining high throughput at high resolution, as shown for hydrocarbons and organotin compound separations.[258] On-column detection is also useful in elemental ratio detection,[259] as then the chromatographic resolution is maintained. This has been described for the characterization of an organomercury reference material held on microcolumns.[260] Also the use of zero-dead-volume cross and PTFE transfer lines in headspace work for organomercury detection has been shown to be useful in minimizing memory effects.[261] Further, provisions for solvent venting have to be realized in the coupling between GC and MIP, as already known from low-pressure MPD.[262] Here heated Al tubes are used and near to the plasma one employs all-glass open-split connections[263] and zero-transfer line systems.[264] The He MIP can even be operated as a very soft electron source and used as an alternative to a β-emitter in electron-capture detection.[265]

Many papers have dealt with the optimization and use of the gas chromatography/atomic emission detector (GC/AED) system for element-specific detection. It has been increasingly used for speciation in environmental samples for both metals and nonmetals [for a state-of-the-art report, see e.g. Lobinski and Adams.[266]]. During optimization special attention has to be paid to memory effects arising from deposits and also from the detector response, which could become of relevance in high-speed GC.[267] For the optimization factorial analysis was applied and response surfaces were studied.[268] Influences of the compound structures on the elemental response obtained were also investigated.[269] Further, noise studies were performed and showed some flicker noise contributions of the plasma depending on the concentration range of the analytes.[270] Detection with a microwave-induced plasma/atomic emission detector (MIP/AED) was found to facilitate considerably the structure of gas chromatograms as compared with those obtained with conventional detection, as shown for polychlorinated biphenyl (PCB) mixtures[271] and also for amines[272] and carboxylic acids[273] both after derivatization. As one works with an He plasma nonmetals can be detected with low detection limits.[274]

In view of the possibility of elemental analysis of organic compounds, efforts were made to determine O selectively. This was found to be difficult,[275] because of the need to exclude contributions from the atmosphere. This necessitates purification of the scavenger and carrier gas[276] but also optimization of the viewing position in the plasma,[277] of the gas flow rates[278] and of the addition of Cu as catalyst and of I_2 to reduce the spectral background for O.[279] Also the detection of H was optimized.[280] As the signals of deuterated and nondeuterated compounds can often be separated by GC/AED this is of interest for the control of deuteration experiments. This can be performed both in solution and in the gas phase with suitable provisions,[281] but unfortunately the spectral lines of ^{1}H and ^{2}H partially overlap.[282] The method was shown to be suitable in the separation of the deuteroisotopomers of caffeine.[283] The detection of

For references see page 9645

C has also been carefully studied. After optimization ^{13}C can be detected down to the sub-picograms per second level.[284] Owing to the experience with the detection of C and H, alkanes could be unambiguously identified in gas chromatograms, based on the C/H ratios[285] and empirical formula determinations can also be performed.[286] This has been shown, for example, for polymer pyrolyzates.[287,288] As N can also be well determined,[289] the GC/AED system could be applied to a wide diversity of organics.

The halogens certainly can be detected with an He MIP, as shown with a system using an oscillating filter-based detection system, developed by Mueller and Cammann.[290] Accordingly, empirical formulas can also be determined for chlorinated hydrocarbons[291] even including the fluorinated compounds[292,293] and quantitative determinations of dioxins and dibenzofurans were shown to be possible in the micrograms per milliliter range.[294] The results agreed with those of MS detection and also in the case of determinations of chlorophenols.[295] However, it was shown that the structure of the halogenated compounds can influence the elemental signals obtained.[296] Further, brominated compounds can also be detected, as shown for hydrocarbons[297] or after derivatization with phenylboronic acid.[298] For S the power of detection of the MIP was shown to be better than that with chemiluminescence detection.[299] With the S 525.45-nm line a high power of detection and linearity were reported.[300] The element-specific detection of B was shown to permit the determination of organoboron compounds in motor and lubricating oils by GC/AED.[301] Detection of Si down to a few picograms as required for the determination of organosilicon compounds was shown to necessitate the use of plasma discharge tubes made of BN or alumina.[302]

Organoarsenicals can also be determined by GC/AED. This could be shown for the determination of the methylated arsines together with organomercurials in oil shale[198] and for the analysis of complex organoarsenical mixtures,[303] also after derivatization.[304] Futher interesting applications of GC/AED analysis are in the analysis of petroleum-related products for dioxane,[305] the analysis of gasoline for oxygenated compounds,[306] the determination of the halogenated products of humic substances in water,[307] pesticide residue analysis[308] and dioxin detection.[309] The use of a GC/MIPAED for the determination of molecular impurities such as CH_4 in Ar[310] or for the determination of organomercury compounds in air, after trapping them on Chromosorb®, has been described.[311]

Apart from nonmetals, metal chelates can also be determined with GC/MIPAED and its application to metal speciation has been widely studied. The latter field especially developed through coupling of chromatography and atomic spectroscopy (for a review see Lobinski[223]). Such work was initiated in 1980[312] and has now been performed for Fe, Ge, Hg, Ni, Pb, Se, Sn and V species, as discussed by Lobinski and Adams.[313] Organotin determinations in water samples and sediments were performed, including a thorough optimization of the sample preparation and the measurement conditions.[314] Also a derivatization by a Grignard reaction often is applied [see e.g. Tutschku et al.[315]]. The determination of alkyllead compounds was described very early.[316,317] A review on the optimization of microwave-induced plasma/gas chromatography (MIP/GC) determinations of organolead and organomercury compounds with the aid of simplex methods was given by Greenway and Barnett.[318] For the speciation of Hg, however, cold vapor atomic absorption spectroscopy (CVAAS) and AFS are very powerful. The determination of methylmercury in biological samples has been performed with headspace techniques.[261,319–321] After derivatization, methylmercury could be determined in fish subsequent to microwave-assisted extraction.[322] In addition to speciation, pre-enrichment for inorganic analysis can also be performed,[323] as shown by the preconcentration of Be from natural waters with the aid of acetylacetonates.[324] The use of acetylacetonates could also be applied in the determination of Ni and Cu in ores.[325] Further fluorinated and non-fluorinated tetradentate β-ketoaminate ligands could be used in the determination of Pd.[326] Se(V) could be determined by volatilization after ethylation with tetraethylborate[327] and volatile Ni, V and Fe porphyrins could be determined directly in crude oils.[328]

Also for element-specific detection in LC, carefully optimized types of MIPs already have some prospects. This is certainly true when using desolvated aerosols, as shown by Billiet et al.[329] The latter especially makes sense with USN, after which halides and oxyhalide salts subsequent to anion exchange could be determined.[330] Further, dry aerosol generation with a moving-wheel system is also possible.[331]

PN without desolvation also could be shown to be of use. Kollotzek et al.[195] used a toroidal mixed-gas MIP in a Beenakker cavity to obtain element-specific signals for Hg. Also the modified MPT can take up wet aerosols without applying desolvation. This is even possible for acetonitrile-containing solutions, but with deterioration in the power of detection, as shown e.g. also for Cr.[97] In the case of Hg the problems can be solved by an on-line CV technique between the exit of the HPLC system and the MIP. However, care must be taken to minimize dead volumes so as not to deteriorate the chromatographic resolution.[332] Wet aerosols produced with a frit nebulizer in the case of microcolumn HPLC with some restrictions can be led into MIPs.[333] In CZE work the effluent rates are low. When using an ion-exchange membrane as an electrical junction in front of the capillary, dead volumes can be minimized and with high-efficiency nebulizers sensitive detection becomes possible.(334) Also for element-specific detection in ion chromatography (IC)(335) the use of MIPOES has been described.

For element-specific detection in SFC,(336) the gases used are very important. Mostly CO_2 is used and it was shown that in the case of an He MIP a sensitive detection of S, Cl and P is possible.(337) The use of Xe both for atmospheric- and low-pressure plasma sources cannot be justified.(338) Also the use of N_2O as mobile phase was reported to allow the sensitive detection of S.(339) Both CO_2 and H_2 addition, however, were shown to decrease the emission intensities of Cl, Br, I, S and P considerably.(340) Binary mobile phases such as CO_2 with methanol addition also have prospects especially for the extraction of metal chelates and these mixed phases were shown to be tolerated by the MIP.(341) Further, the use of medium-power MIPs was shown to be advantageous so as to reduce the influence of the mobile phase of the stability of the MIP.(342) Jin et al. showed that the MPT was very useful in the case of SFC and in OES they reported a detection limit for Cl of 300 pg.(343) A surfatron can also be used for element-specific detection in SFC for the determination of S-containing aromatics.(344) The NIR lines can also be used, as shown for the detection of S and Cl.(345) With CO_2 as mobile phase and an He MIP, Zhang et al. reported a detection limit for Cl of 40 pg s^{-1}.(346) Webster and Carnahan also proposed the use of internal standards to cope with the influence of the effluent gases on the plasma in the case of nonmetal detection.(347) With the example of ferrocene, it could be shown that the approach is also viable for the isolation and determination of organometallics.(348) In the case of packed columns, the detection limit of supercritical fluid chromatography/microwave-induced plasma (SFC/MIP) detection of Fe was reported to be 30 pg.(349)

4.5 Glow Discharges

Since the work of Paschen on hollow cathodes,(350) the GD has become well known as a source for atomic spectroscopic analyses, especially in the case of solid samples. GDs are operated at low pressure. In a classical GD there are different zones, where the formation of excited species takes place. Near the cathode positive ions are accelerated and impact on the cathode practically without any energy exchange with gas ions (dark zone), whereas in the negative glow the free electron concentrations are high and considerable ionization of the gas takes place. Near the anode the energy conversion is again very low. Accordingly, volatilization, excitation and ionization processes differ widely from those in plasma sources operated at atmospheric pressure. As outlined in state-of-the-art reviews,(351,352) the samples to be analyzed mostly constitute or are placed in one of the electrodes, preferably the cathode. From here they can be volatilzed by cathodic sputtering resulting from the impact of heavy gas ions formed in the negative glow of the discharge. Often heating of the cathode also occurs and this still enhances sample volatilization and in some cases it is the process mainly responsible for material volatilization. Excitation and ionization mainly result from electron impact; however, collisions with metastables, which have a long lifetime in a discharge at reduced pressure, and charge transfer(353) may also play a role. GDs may be produced by dc and also with rf energy, as proposed for the spectrochemical analysis especially of non-conducting samples by Winchester et al.(354) and other groups. Indeed, as the electrons can take up energy in the high-frequency field much better than the heavy positive ions do, a bias potential is built up in front of the sample, through which the field is coupled into the working gas. Accordingly, sputtering through the presence of this field can occur as in a dc field, but also for electrically nonconducting samples. Although GDs are now routinely used for solids analysis, especially when it comes to depth profiling as is now required for new surface-improved materials (for a thorough review see Bengtson(355)), its potential is not fully used. Indeed, much of the sputtered material is not excited and certainly not ionized in the optically thin low-pressure discharge. Therefore, already in early work endeavors to achieve cross-excitation were made. Leis et al.(356) successfully designed a GD incorporating cross-excitation of the sputtered material in an MIP produced at low pressure in a resonator according to Beenakker, which is constructed just in front of the GD source (Figure 17). The system was optimized and it could be shown that owing to the increased excitation efficiencies the line intensities in OES were considerably improved, but also the spectral background intensities. It could be shown by Fourier transform high-resolution studies of line profiles(357,358) that the self-reversal in the case of microwave boosting diminishes. This proves the increase in the excitation and ionization efficiency in the plasma especially in front of the sample. For steels a gain in power of detection by up to an order of magnitude could be realized (Table 5) and the system also could be used successfully for the analysis of Al, Cu and Pb samples.(359) Similar results were also reported by other groups.(360) The analyte signal enhancements due to microwave boosting seemed to be at least as strong for a microsecond-pulsed GD.(361)

Hollow cathodes have also been widely applied as atomic emission spectrochemical sources. Here the analyte residence time in the plasma is very high, leading to high excitation and ionization efficiencies. Also here

For references see page 9645

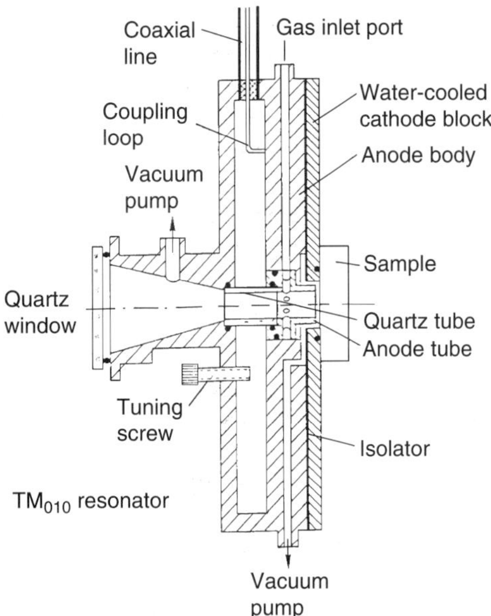

Figure 17 Schematic representation of GD lamp with integrated microwave cavity.[356]

Table 5 Detection limits for some elements in steel obtained with a microwave-supported GD lamp, calculated with an RSD of the background signal equalized to 1%; for comparison the values obtained with a conventional GD lamp are also presented[356]

Element/ line	Wavelength (nm)	Excitation energy (eV)	Detection limits ($\mu g\,g^{-1}$)	
			Without microwaves	With microwaves
Al I	396.2	3.1	0.4	0.1
B I	209.1	5.9	0.8	0.3
B I	208.9	5.9	0.9	0.4
Cr I	425.4	2.9	0.2	0.05
Cu I	327.4	3.8	1.5	0.3
Cu I	324.8	3.8	2	0.9
Mg I	285.2	4.3	2	1.5
Mg II	279.6	4.4	1.3	0.9
Mn I	403.1	3.1	1	0.2
Mo I	386.4	3.2	1.5	0.8
Nb I	405.9	3.2	4	0.6
Ni I	232.0	5.3	0.5	0.1
Si I	288.2	5.1	3	0.4
Ti I	364.3	3.4	3	0.6
V I	318.4	3.9	3	1
Zr I	360.1	3.6	8	1.5

microwave boosting was found to increase the excitation and ionization efficiencies and accordingly the analytical signals.[362] These sources have been found especially useful for dry solution residue analysis; however, they also might have potential for elemental detection in GC, because of the low gas consumption and simple power supply systems.

5 OPTICAL ATOMIC SPECTROSCOPY WITH MICROWAVE PLASMAS

Many method developments in optical atomic spectrochemical procedures making use of MIPs have been described. For this purpose conventional grating spectrometers using gratings with constants down to 1/3600 mm and focal lengths of 0.5–1 m with a Czerny–Turner or Ebert set-up are used and sequential systems or simultaneous spectrometers including polychromators mainly use a Paschen–Runge mounting (Figure 18a–d). Radiation measurements are made with photomultipliers but also more and more with diode-array detection, as then line and background intensities for the atomic spectral lines can be recorded simultaneously. The latter systems also are very useful for measurements of band intensities, as shown in the case of the OH bands[96] (Figure 19a and b), which are required for the determination of rotational temperatures. Here illumination of the entrance slit with a 90° rotated image of the plasma and measurements at different heights along each rotational line allow tomography of the plasma for the rotational temperature, as shown as an example in Figure 20.

More advanced spectrometers can also be used, such as those including an interferometer and a photodiode-array detector[364] or a Fourier transform spectrometer (for a tutorial discussion, see Faires[365]).

Particularly interesting also is the use of échelle spectrometers with crossed dispersion (see Figure 21). Here an échelle grating with a small number of lines (often down to 50 lines mm^{-1}), a high order (30 and more) and a prism is employed, which has its dispersion direction perpendicular to that of the grating so as to separate the different spectral orders. Accordingly, a two-dimensional spectrum is obtained. As échelle systems require the use of a small slit height, high radiation density sources are required so as not to come into shot-noise limitations with the detector. As microwave plasmas also at low power fulfill this condition, they easily can be used together with échelle spectrometers realizing both high resolution with short focal length spectrometers, abscence from shot-noise limitations and high resolving power. This combination is particularly promising as two-dimensional array detectors then also allow simultaneous multielement detection with true background correction.

Systems with N_2-purged optics have also been used, with which the lines at vacuum-ultraviolet (VUV) wavelengths for I become accessible.[366] Method development depends, of course, on the type of microwave plasma used in particular and it will be outlined for the low- and medium-power MIPs, the single-electrode plasmas and the related stabilized capacitively coupled high-frequency

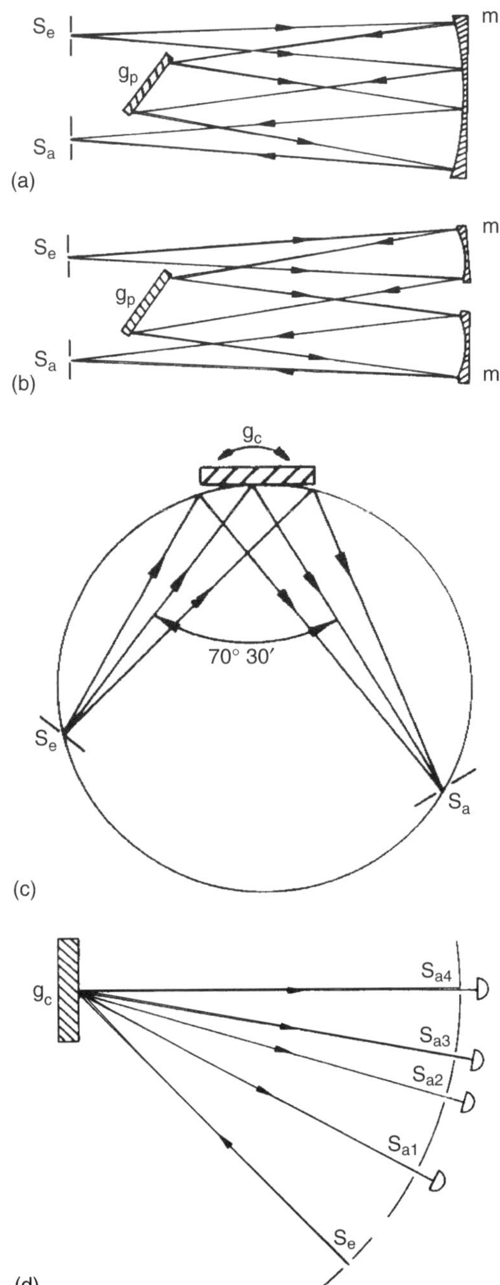

Figure 18 Optical mountings for optical spectrometers with a plane grating.[363] (a) Ebert; (b) Czerny–Turner; (c) Seya–Namioka; (d) Paschen–Runge. S_e = Entrance slit; S_a = exit slit; g_p = plane grating; g_c = concave grating; m = mirror.

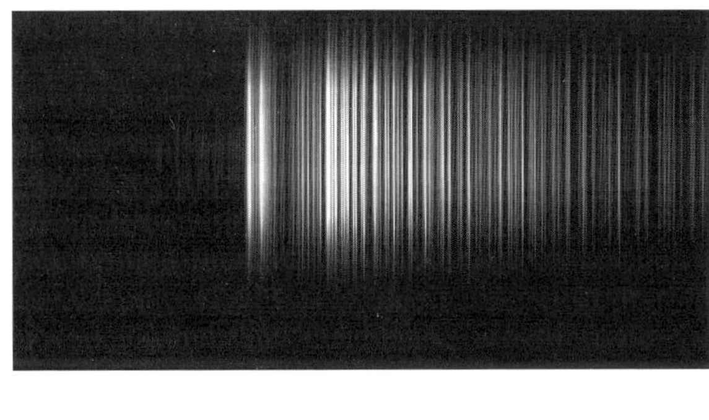

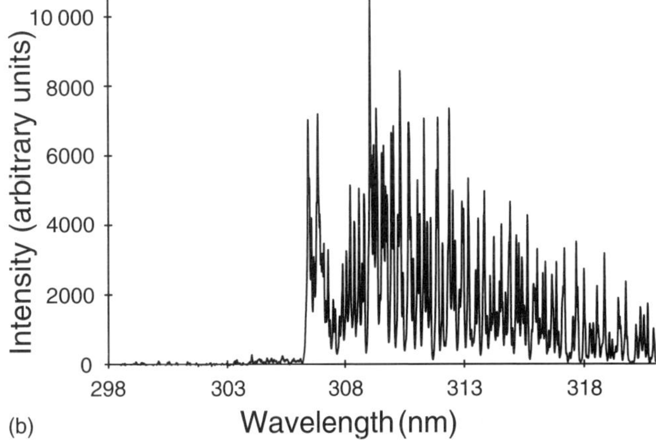

Figure 19 OH band at 306.4 nm. (a) CCD image; (b) signal of one row of the CCD image.[96]

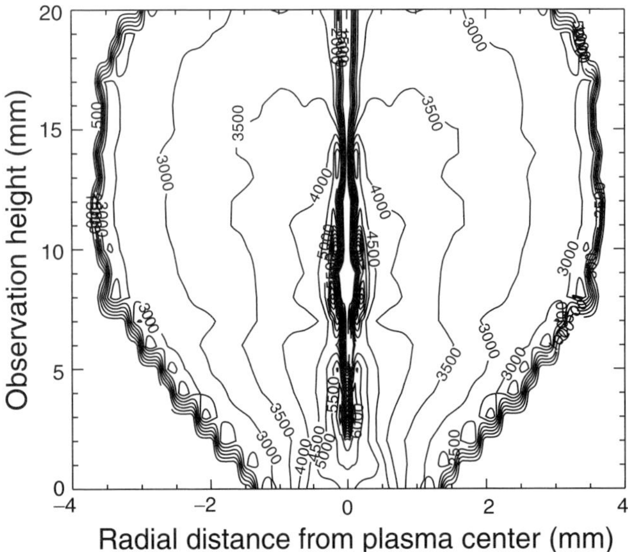

Figure 20 Radially resolved rotational temperatures in an Ar MPT (forward power 100 W, 150 mL min^{-1} working gas, 450 mL min^{-1} carrier gas.)[96]

plasma (SCP), as well as for atomic emission, absorption, fluorescence and laser-induced ionization.

5.1 Atomic Emission Spectroscopy with Low-power Microwave-induced Plasmas

The special features of the low-power MIPs are their low instrument and operating costs, but they are restricted in their use to those techniques where gaseous and dry analyte clouds are entered into the plasma. This has been discussed for ETV and HG and also includes the use of volatile compound formation, as developed, for instance,

For references see page 9645

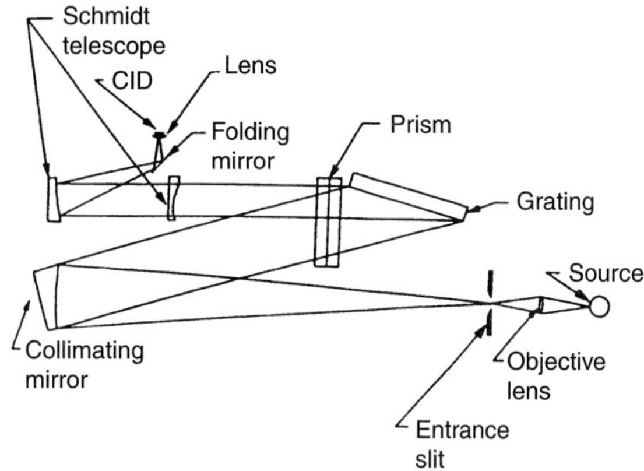

Figure 21 Échelle spectrometer with crossed dispersion and charge injection device (CID) detector for plasma spectrometery (courtesy of Perkin-Elmer Corporation).

for the determination of I.[367] Here detection limits below 100 ng mL^{-1} can easily be obtained. The determination of Hg can also be performed with extremely low detection limits, as here the small volume of the source, the absence of an excess of H_2 when reducing with $SnCl_2$ and the possibility of using He are very favorable conditions. Here the use of a surfatron source was shown to allow the highest sensitivity.[368] For the determination of N in natural gas, a low-pressure microwave plasma was found to permit detection limits down to 0.01 ppm by using the band heads at 336 and 337 nm[369] and also O could be determined from the OH band emission.[370] For the low-power microwave plasma sources dopants were found to considerably influence the atomic line intensities, as shown for the case of Sn analyte lines in the GC of alkylated Sn compounds.[371] Low-power microwave plasmas in the case of Ar as working gas further seemed particularly useful for the excitation of elements with low excitation potentials such as alkali metals.[372]

5.2 Single-electrode Microwave Plasmas

Single-electrode CMPs are very stable in the case of many techniques for sample introduction. They can be operated between about 200 and 600 W with different gases. It might be problematic, however, to enter the analytes in an efficient way into the hot plasma center. For this purpose the aerosols are often led into the CMP through holes near the plasma tip and wall stabilization with the aid of a mantle gas flow is used, as described by Patel et al.[373] In the case of a high-power plasma a graphite cup which contains the sample can even be an integrated part of the torch.[374] At a sufficiently high power the CMP can easily be operated with air only and, as shown by Bings et al.,[32] the excitation and rotational temperatures in this source are still in the 5000 and 4000 K range, respectively. As shown already for many elements by Zhang et al.,[375] the detection limits in aqueous solutions with this source are still in the sub-micrograms per liter range. A CMP can be operated also in He/H_2 at the tip of a graphite electrode[376] with a power of up to 800 W. The addition of H_2 was found to remove typical volatilization interferences such as the $Ca^{2+}–PO_4{}^{3-}$ interference. With an He CMP Cl and Br can also be determined down to detection limits of 1 and 0.4 µg mL^{-1}, respectively, as required in the case of organohalogenated compounds.[377] With ETV Si can be determined with a detection limit as low as 0.03 µg mL^{-1} and in the case of PN 0.3 µg mL^{-1}.[378] With an He CMP organotin compounds can determined in GC effluents with a detection limit of 1 ng.[379]

When combining OES using the CMP with HG, As can be determined with absolute detection limits of down to 60 pg in the case of CT.[380] In GC, element-specific detection of the halogens can be performed.[381,382] H and O can be determined in metals by heating the samples and bringing the released gases into the CMP.[383] The plasma even is so stable that insertion of samples in microcups placed on the tip of the electrode is possible. Then extremely low absolute detection limits are obtained, e.g. 8 pg in the case of Cd.[384,385] From an external GF several heavy metals in National Institute of Science and Technology (NIST) fly ash and tomato leaf samples could be evaporated and determinend with the CMP.[386] For Hg, thermal vaporization with steam led to a detection limit of 1 µg L^{-1} with the CMP.[387] In the case of wet aerosols desolvation can be applied and just as in ICPOES it may improve the detection limits.[388] However, wet aerosols can easily be entered directly into the plasma, especially when using a tubular electrode of Ta or Al.[33,389] The plasma can also take up solvent-loaded aerosols obtained by the PN of organic solutions, which made it possible to use the CMP to perform analysis in oils eventually after dilution with the appropriate organic solvent[390] (Table 6).

As in the plasma the ground-state atom population is still considerable, the CMP can also be used as an atom reservoir for atomic absorption work.[34]

5.3 Plasma Torch Sources

In a review, Blades[391] showed the state of development of He MPTs and other microwave plasmas and compared them with the ICP. The so-called microwave-induced nitrogen discharge at atmospheric pressure (MINDAP) source described by Deutsch et al.[392,393] operates at high power and it is shown that also in a Beenakker cavity an N_2 discharge can accept wet aerosols with low detection limits, especially for the alkali metals. The MPT, however, became a well-studied source. A comparison

Table 6 Analysis of a waste motor oil, a standard oil and a fresh motor oil by ICPOES subsequent to digestion and by CMPOES (± standard deviations from five replicate analyses)[390]

Method	Sample	Co ($\mu g\,g^{-1}$)	Cr ($\mu g\,g^{-1}$)	Fe ($\mu g\,g^{-1}$)	Ni ($\mu g\,g^{-1}$)
ICPOES	Standard oil	<0.1	1.2 ± 0.1	2.1 ± 0.3	0.5 ± 0.1
	Fresh motor oil	0.6 ± 0.1	0.8 ± 0.1	2.6 ± 0.1	1.1 ± 0.1
	Waste motor oil	1.0 ± 0.1	7.1 ± 0.1	52.0 ± 2.2	4.0 ± 0.3
CMPOES	Waste motor oil	–	6.0 ± 0.9	51.4 ± 1.0	3.9 ± 0.5

with the surfatron showed that for Hg determinations the two sources have similar performances.[394] However, in the case of HG the MPT allows it to enter the excess of H_2 produced in chemical HG together with the hydrides into the plasma,[92] which also was found to be possible with a TE_{010} cavity MIP in the case of electrochemical HG.[395] In the case of the MPT, volatilization also was used for the determination of the halogens in aqueous solutions.[396] For the nonmetals also the sensitive lines in the IR region are accessible with Fourier transform spectroscopy, as shown for a high-power MIP.[397] Flow injection can be coupled on-line with OES using an MPT in the case of USN including desolvation[398] or of PN and desolvation and Cd, Cu, Mn and Zn can be determined at concentrations down to some nanograms per milliliter.[399] The necessity for desolvation, however, is not always given.[400] This could be confirmed by measurements with a Légère nebulizer, where desolvation proved to be not necessary at all.[97] Also, ion exchange can be very efficiently coupled on-line with an MPT, as shown by the determination of B using cation exchange[401] and also for metals forming chloro complexes in the case of anion exchange.[402] Ion exchange could be applied for a removal of the matrix element Fe in the determination of Si in steel.[403] The analytical performance of the MPT was also demonstrated by the determination of the rare earths in solutions using USN.[404] It was shown that with He as discharge gas, the halogens, P and S could be determined at the $100\,ng\,mL^{-1}$ level.[405] At high power (>500 W), it was shown by Urh and Carnahan[406] that MPT discharges can take up wet aerosols without any problem. Then also Cl[407] could be determined with an ICP like torch positioned in a rectangular resonator. Leis and Broekaert[67] showed that with wet aerosols sub-$\mu g\,mL^{-1}$ detection limits can be obtained but that the interferences by easily ionized elements are less favorable than in ICPOES.

5.4 Stabilized Capacitively Coupled Plasma

The SCP described by Gross et al.,[408] through its low operating power, is very similar in its features to low-power microwave plasmas. It is operated between two ring-shaped electrodes at 200 W and has a diameter of 1 mm, defined by the water-cooled plasma capillary (Figure 22). Originally designed for element-specific detection in GC, it also can be well used for the detection of nonmetals after volatilization[409] and in combination with all dry vapor generation devices, including HG without removal of the H_2, ETV and PN using desolvation.[410]

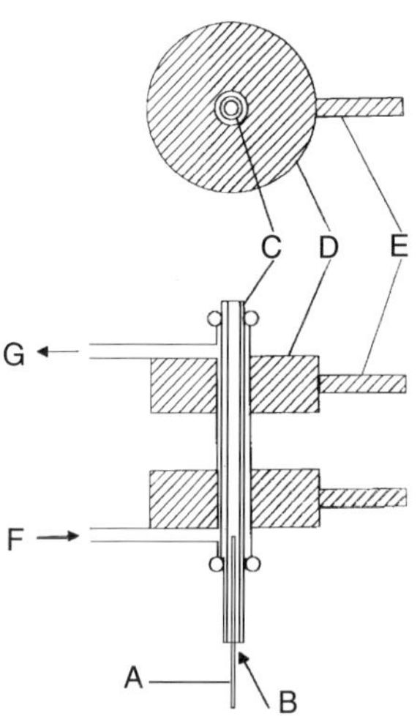

Figure 22 Cross-section of torch unit for SCP. (A) chromatographic capillary column; (B) plasma gas (He); (C) fused-silica discharge tube; (D) annular electrode (gold plated brass); (E) rf connector; (F) cooling water in; (G) cooling water out.[408] (Reprinted from R. Gross, B. Platzer, E. Leitner, A. Schalk, H. Sinabell, H. Zach, G. Knapp, *Spectrochim. Acta, Part B*, **47**(1), 95–106 (1992), Copyright 1992, with permission from Elsevier Science.)

5.5 Microwave-induced Plasmas as Atom Reservoirs for Atomic Absorption Spectroscopy, Atomic Fluorescence Spectroscopy and Laser-enhanced Ionization

As the ground-state atom population densities in the case of microwave plasma discharges are still high and the plasma volumes small, microwave discharges in a number of cases could be very well used as atom reservoirs. This was shown for the case of the MPT in which samples were introduced by USN. Here with AAS lower detection limits could be obtained than in the case of the MIP.[411] For sample introduction GF evaporation could

For references see page 9645

also be combined with the MPT[412] and in the case of a Beenakker MIP PN with a glass frit nebulizer also was applied for AAS measurements.[413] The performance in the case of desolvation is improved,[414] as also for an MPT as described by Duan et al.[415] Because of the longer pathlength and decreases in matrix effects in the latter case, axial measurements are worth considering.[416]

Apart from hollow-cathode lamps, diode lasers can also be used as primary radiation sources for AAS. For the halogens a low-pressure He MIP can be used as atom reservoir, thus allowing the detection of Cl compounds through the metastable Cl levels in GC effluents.[417–419] Applications of microwave plasma torch atomic absorption spectrometry (MPTAAS) include the determination of Cd.[420]

For atomic fluorescence both line (hollow-cathode lamps) and continuous sources (Xe arc) were applied in the case of a high-power Ar MIP according to Beenakker,[421] but the detection limits obtained were poorer than those in AAS and OES. In the case of an He MIP nonmetals could also be determined[422] and detection limits of 0.25 ng mL^{-1} Cd and 120 ng mL^{-1} As can be obtained with AFS[423] and 3 ng mL^{-1} for Hg.[424] For the determination of down to 10 µg L^{-1} Na, ETV from a W filament and AFS in an MIP could be successfully applied.[425] For the determination of Fe, Ni and Co in pure water, introduction of their volatile carbonyls could be applied.[426] The full power of detection of AFS for the analysis of pure waters can only be obtained with laser excitation; only then can saturation of the excited levels be obtained, as shown in work using ETV from a W filament and an MIP.[427]

LEI (laser-enhanced ionization) can also be performed with an MIP as atom reservoir. Here the ion currents resulting from increases in ionization when populating the excited states of analytes with the aid of laser radiation of a suitable wavelength are measured and the technique is well known from flame work.[428] As the method needs no spectrometric equipment at all (Figure 23), its instrumentation costs are low but nevertheless the method is very sensitive and even isotope-selective determinations are possible. The features of a set-up using a nitrogen MIP and an Ar ion laser were shown in the case of Na, for which a detection limit of a few nanograms per milliliter was obtained.[429] Especially, the electrical characteristics of the plasma and their effect on ion detection have to be studied.[430] With a microarc atomizer and LEI in a nitrogen plasma using a flashlamp-pumped dye laser, In could be determined down to the picogram level.[431]

Comparative studies of AAS, AFS and OES with an MIP for the case of Hg revealed differences in detection limits of less than a factor of two.[432] The plasma afterglow can also be used for atomization.[433]

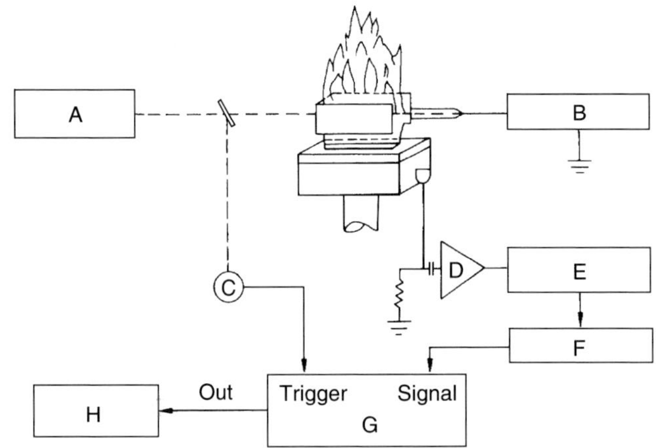

Figure 23 Flame LEI spectroscopy.[428] (A) Flashlamp/dye laser; (B) high voltage; (C) trigger photodiode; (D) pre-amplifier; (E) pulse amplifier; (F) active filter; (G) boxcar averager; (H) chart recorder. (Reprinted with permission from G.C. Turk, J.C. Travis, J.R. DeVoe, T.C. O'Haver, *Anal. Chem.*, **51**(12), 1890–1896 (1979). Copyright 1979, American Chemical Society.)

6 MASS SPECTROMETRY WITH MICROWAVE PLASMAS

Also for plasma MS, MIPs have been used as ion source.[434,435] They even were used earlier than the ICP. The latter became more succesful, as with aqueous solution analysis desolvation was not required. However, MIPs in their whole variety at low and high power and at atmospheric and reduced pressure, just as the ICP, are of use as ion sources for MS. This could be shown for the detection of metals and nonmetals in aqueous solutions.[436] Microwave-induced plasma mass spectrometry (MIPMS) especially found use for element-specific detection in chromatography.[437–440] Compared with OES, MS not only has the advantage of increased power of detection but it also permits the determination of isotope ratios, with a precision of better than 3% in the case of comparable abundances.[441]

6.1 Low-power Microwave Plasmas

Especially at low pressure and in the case of He as discharge gas, MIPs have been shown to be very useful ionization sources for MS and both dry vapors and as solvent-loaded aerosols[442] and organics[443] can be introduced. These working conditions especially promote microwave discharges to soft ionization sources, yielding molecular and radical species, as does electron impact ionization. Their use facilitates the identification of chromatographic peaks[444] and in this respect MIPs have similar potential to low-pressure rf plasmas.[445]

At atmospheric pressure the use of a nitrogen MIP and USN including desolvation has been described. In this case, however, contributions of N species to the background were found to be considerable.(446) With a high-power MIP operated in a surfatron, desolvation even in the case of USN was found not to be necessary.(447) Particularly with low-power MIPs in a TM_{010} cavity, fragmentation spectra could be obtained, when working under reduced pressure and at the lowest possible power.(448)

Especially in the case of He the MPT also became of use as ion source for MS. When operated at 150 W and using USN and subsequent desolvation, the nonmetals could be determined with detection limits below 1 μg mL^{-1}.(449)

Especially for GC detection, MIP/MS became recognized as very powerful.(450) This is due to the small volume of the MIP, which keeps the deterioration of the high resolution in GC work to the minimum. The He MPT became of use for the detection of many elements by GC, including the halogens,(451) O and H.(452) For this application microwave plasma torch time-of-flight mass spectrometry (MPTTOFMS) is very useful. Here the chromatographic speed and resolution can remain high while acquiring all element and radical signals for a peak simultaneously. Accordingly, complete mass spectra, except for the range around the He ion and the N species stemming from the surrounding atmosphere and which cannot be completely removed by deflection (Figure 24), can be recorded. As the mass spectrometer ion conductance remains high compared with quadrupoles, one can achieve very low absolute detection limits, as found with halogenated hydrocarbons (see Table 7). Elemental ratio determinations and isotopic dilution in principle can also be performed with high precision compared with sequentially operating quadrupoles and sector feld instruments.

Many studies have been devoted to the detection of the halogens, because of the environmental interest in the detection of halogenated hydrocarbons. The detection of Br, Cl and I down to the picogram level is possible with a low-pressure MIP in a resonator according to Beenakker(455) when working in the positive ion mode.(456) This value is still lower in the case of MPTTOFMS as discussed.(453) The detection of P and S, which is especially interesting from the point of view of pesticide residue analysis, could be perfomed with a water-cooled low-pressure MIP.(457) The atmospheric pressure plasma in this case was found to give more background species.(458) For such applications a tangential flow torch was often found advantageous.(454) The mass spectra for organics in the case of plasmas in O_2 or N_2 were also studied(459) and clusters identified.(460) The system can accordingly be very useful for elemental analysis.(434) However, organometallics can also be well determined, as impressively shown for the case of the alkylated Sn compounds.(461)

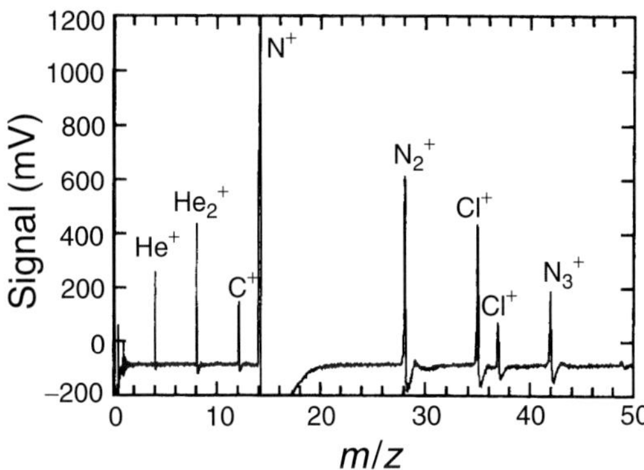

Figure 24 TOF (time-of-flight mass spectrum of CCl_4 obtained with the MPT and a flow cell sample introduction technique with the use of a digital oscilloscope for data acquisition. Peaks arising from air entrainment have been eliminated by a flowing sheath of N_2. Peak undershoot is caused by the high-speed amplifier used in this study.(453)

Table 7 Detection limits for halogenated compounds(453) using MPTTOFMS GC

Compound	MPTTOFMS					Detection limit as total compound (Mohamad et al.(454)) (fg)
	Amount injected (pg)	S/N	Detection limit[a]			
			As halogen (fg)	As total compound (fg)	As total compound (fg)	
Chlorobenzene	10.1	118.6	75.4	2.21	260	9200
1-Chloropentane	8.8	91.6	95.2	2.72	290	–[b]
p-Chlorotoluene	12.7	114.9	91.2	2.61	330	11000
Bromobenzene	14.9	287.0	74.0	0.93	150	–[b]
Bromoform	28.0	320.7	254	1.01	270	–[b]
Iodobenzene	18.2	199.7	168	1.32	270	1500
1-Iodobutane	15.0	272.0	110	0.87	160	–[b]

[a] Detection limits were calculated at 3σ and were based on peak heights.
[b] Detection limits for these compounds were not reported.

For references see page 9645

GC often has the disadvantage of requiring derivatization of the analytes so as to obtain volatile compounds. This is not necessary when applying SFC. In the case of halogenated compounds this could be well applied in combination with element-specific detection by MIPMS when using CO_2 and a low-power plasma.(462) HPLC could be applied for the case of halogenated compounds and also organoarsenicals, but requires the use of a 300 W MIP.(463)

Apart from its use in combination with chromatography, MIPMS in the case of vapor generation can also be applied for the determiantion of nonmetals. Here Cl, Br, I and S compounds(464,465) can be determined down to the nanograms per milliliter level. Se in water can also be determined after its volatilization as a complex into MIPMS(466) and halogens and Br-containing species after volatilizing their derivatives.(467)

6.2 High-power Microwave Plasmas

The introduction of high-power microwave plasmas as ion sources for MS is aimed at the use of conventional nebulization of solutions without need for desolvation. This is possible at 150 W when using a micronebulizer: for Pd, Ru and Rh detection limits down to 0.04 $\mu g\,mL^{-1}$ could be achieved in this way with AES(468) and with an He MIP at 350 W detection limits for Se and Cd of 19 and 0.9 $ng\,mL^{-1}$, respectively, could be obtained with MIPMS, when applying direct solution analysis using DIN.(469) At a moderate power of 500 W, an N_2-operated MIP can easily accept nondesolvated aerosols. When using a water-cooled torch(470) or an Al_2O_3 tube,(471) the detection limits for Ca and K were shown to be lower than those with the ICP. The use of an annular nitrogen microwave plasma at 1.3 kW removes all restrictions for solution work and in MS detection limits down to the picograms per milliliter level have been reported.(472) The source was used for the analysis of biological samples using calibration by isotopic dilution(473) and even work with high salt concentrations was found to be possible.(474) The source also can easily cope with organics when using O_2 as the working gas, as shown in the analysis of photoresists.(475)

7 APPLICATIONS OF MICROWAVE PLASMA ATOMIC SPECTROSCOPY

The applications of microwave plasma atomic spectroscopy narrowly hang together with the types of available instrumentation. Therefore, element-specific detection in GC for residue analysis and now also for metal speciation is a major field of study. A review dealing with topics from the coupling of AAS to GC to the use of the MIP illustrates this development(476) and the importance of coupled techniques for speciation in general is now obvious.(477) As shown e.g. by the determination of organomercurials in biological materials,(478) the coupling of the MIP to GC is now well developed. There are, however, precautions to be taken in calibrating for organometallic compounds as a result of their stability [for organomercury compounds, see e.g. Lansens et al.(479)] and there is certainly room for advanced spectroscopic concepts(480) making the system cheaper.

7.1 Analysis of Biological Samples

As shown in a review,(481) most MIP work for the analysis of biological samples makes use of volatile compound formation using hydrides or GC subsequent to derivatization [see Krull et al.(482,483)] or ETV. For biological studies ^{15}N detection is very important, as discussed by Deruaz et al.,(484) and makes thorough optimization of column separation and MIP detection necessary.

With respect to plants, volatile elements have been determined in tobacco using sampling of fumes on Tenax® tubes and thermal desorption with subsequent gas chromatography/mass spectrometry (GC/MS) or GC/AED.(485) Pesticide residue analyses,(486) e.g. in the case of fruit,(487) also are of crucial importance and may make use of Cl-specific MIP detection. Further, the volatile hydride-forming elements As, Sb and Se have been determined in Chinese tea, and electrochemical HG was found to give detection limits as low as the chemical conventional method using $NaBH_4$.(395) In the case of the speciation of Se, GC has to be included, as shown for the determination of selenomethionine in wheat.(488) The power of detection of these procedures is well illustrated by the analysis of human garlic breath for organoselenium compounds.(489) Also in beer, S compounds and coffee can be determined with gas chromatography/microwave-induced plasma optical emission spectrometry (GC/MIPOES) subsequent to suitable trapping.(490) With respect to organometallic species, especially the determination of organolead in wine has been studied(491) and for very volatile species such as the organomercurials headspace techniques were optimized.(492) ETV from a GF, as already shown by Aziz(119) (Figure 25), can be very well combined even with a low-power MIP and used for determinations in plant materials. This also applies to the direct determination of very volatile elements in solids such as As(493) in leaves and for Pd, Cd. With selective volatilization speciation also can be realized, as shown for Hg and As compounds.(494)

In blood(495) and in water and urine,(496) F-containing compounds could be determined with GC coupled to MIPOES. As metals, and especially total Hg concentrations(497) and organomercurials,(498) and also Pb have

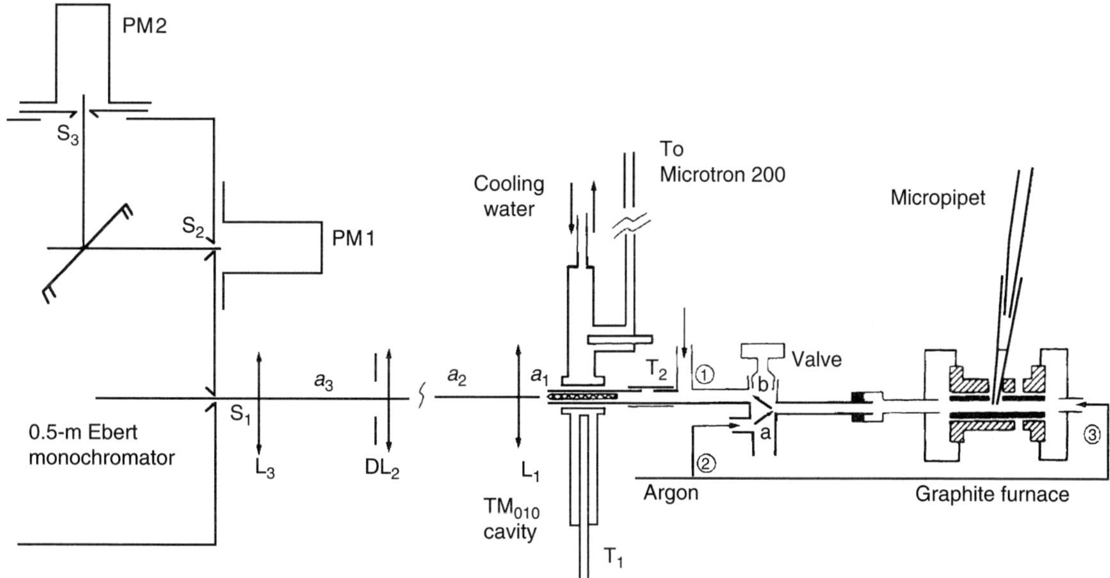

Figure 25 GF MIP instrumentation. Ar gas flows: (1) plasma gas flow; (2) for exhaust of solvent vapor; (3) carrier gas flow. (T_1) Coupling rod; (T_2) fine tuning rod. Spherical quartz lenses (f = focal length, d = diameter): (L_1) $f = 35$ mm, $d = 19$ mm, $a_1 = 41$ mm; (L_2) $f = 66$ mm, $d = 19$ mm, $a_2 = 249$ mm; (L_3) $f = 82$ mm, $d = 19$ mm, $a_3 = 90$ mm. (S_1) Entrance slit; (S_2) fixed exit slit; (S_3) movable exit slit (wavelength coverage: $\lambda_a \pm 2$ nm); (PM_1, PM_2) photomultipliers; (D) diaphragm (6 × 6 mm).[119]

been determined. For the determination of Pb in blood, the CMP could be used very successfully when applying dry residue analysis in a W cup placed on top of the electrode.[499–501] In fish tissue especially organomercury compounds have to be determined[502,503] and also the element Se [see e.g. Tsunoda et al.[504]]. As has been determined in human air, pig kidney and mussel tissue by electrochemical HG combined with hot trapping and CT.[505] For the determination of organotin compounds in tissue, microwave-assisted leaching procedures have been shown to be very helpful,[506] but clean-up of the extracts and derivatization of the organotin compounds are required.[507]

7.2 Analysis of Environmental Samples

MIPs have been used both for the determination of the elements and for the determination of the species in water, soil and air. For this purpose hyphenated techniques, coupling chromatography to atomic spectroscopic methods, are often used so as to realize on-line pre-enrichment or separation and the determination of elemental species. For this purpose, GC, LC and SFC have been used (for a review see Dai and Jia[508]). With this approach organics, metals and organometallic species have been determined in various types of water samples.

For organics, GC coupled to MS and to MIP AES have been shown to be complementary for screening tests. This has been shown e.g. for the case of atrazine, where a peak identification was possible from the elemental signal ratios for C, H, Cl, Br, N and S on the one hand and from library mass spectra on the other.[509] Further, the power of detection of the GC/MIPAED technique for organics using empirical formula determinations, as required for screening in river water, could be considerably increased by preconcentration on LiChrolut EN® microcartridges.[510] Especially organohalogens have long been determined in waters by MIPOES.[511] Cl-containing organics, such as trihalomethane[512] and trichloroacetic acid,[134] have long been determined by precolumn enrichment and microwave plasma detection. "Purge-and-trap" prior to pyrolysis with subsequent GC using MIP detection was used for the determination of 20 halogenated hydrocarbons in drinking water.[513] For pesticide residue analysis, element-specific detection for C, H, N, O, S, P, F, Cl and Br with a commercially available GC/MIPAED system (Hewlett-Packard) was shown to be very useful.[514] Pyrolysis, known from GC/MS work, could be shown to be useful also in the case of MIP/AED for improved elemental detection both for chlorinated[515] and for S-containing species.[516] In the latter case proteins and S-containing amino acids could be detected very sensitively after converting S into H_2S and trapping the latter by freezing.

In synthetic ocean water, metal determinations (Cr, Mn, In, V, Pb and Sr) could be performed by the direct introduction of aerosols produced by a MAK pneumatic nebulizer without the need for desolvation.[517] For elements such as Pb a pre-enrichment as dithiocarbamate and direct GC/MIPAED were found to be

For references see page 9645

very useful in improving the power of detection as required for the analysis of rain and tap water.[518] For volatile and volatile compound-forming elements very sensitive detection also is possible. This has been shown for the case of I, which chemically could be volatilized from brines or seawater,[519] for Hg, which could be determined with the CV technique[520] and after Au amalgamation down to the nanograms per liter level in lake water,[521] and for As, which could be determined in sewage sludge when using a CMP and HG.[522]

Especially more recently much work has been done on the determination of organometallic species in water samples. Especially for organotin, organolead and organomercury compounds, derivatization was found very useful.[523] Both ethylation and phenylation with sodium tetraethylborate or tetraphenylborate, respecively, and Grignard reactions are useful for converting the analytes into thermally better stable and more volatile compounds. Organolead species in this way could be determined in River Danube water down to the nanograms per milliliter level[524] and organotin compounds in water from the Thermaikos Gulf.[525] In the case of Sn, derivatization with tetraethyl borate could be performed on-line for the case of river water.[526] For drinking water an extraction of 30 mL with hexane was found to be sufficient to allow the detection of organo-lead and -tin compounds, the latter down to the $40\,\mathrm{ng\,L^{-1}}$ level.[527] Much work has been done on the detection of the organomercurials in water samples, as already mentioned for the case of seawater in early MIP work.[528] Owing to the labile character of many organomercurials and organolead compounds, filtration through a dithiocarbamate-containing cartridge has been proposed and tested for pre-enrichment in the case of natural waters[529] and seawater;[530] however, it was found to be problematic for long storage. For integrated sample preparation and speciation, purge-and-trap injection GC is very useful. Here sodium tetraethylborate is added to the water sample and the mixture is purged with He while trapping the analytes at $-100\,°\mathrm{C}$ and applying subsequent flash desorption and GC analysis. Accordingly, organo-tin, -lead and -mercury compounds could be determined in estuarine waters down to the nanograms per liter range.[531] For organomercury determinations in water samples, adsorption of the derivatives on a nonpolar microcolumn also was applied.[532] For the case of organoselenium compounds in water samples, purge-and-trap procedures were also found to be very useful.[533]

The determination of organolead species in ice samples was also found to be interesting as it delivers archive data on the nature of Pb pollution. When using derivatization and solvent venting procedures, the analysis of Greenland ice was shown to be possible down to the femtograms per gram level.[534]

The determination of traces of organometallic compounds in soils and sediments puts high requirements on the sample preparation. The determination of Pd in spent catalysts in waste from a production plant was found to be possible by acid extraction and analysis of the extracts by MIPOES using USN. In the extraction of soils with organic solvents, recoveries of organotin species were determined[535] and the organo-lead, -tin and -mercury analytes were derivatized with a Grignard reaction. As a soft extraction method for methylmercury determinations in sediments, supercritical fluid extraction with CO_2 was found to be promising.[536] ^{15}N is an important tracer for studies on the N cycle in agriculture and it was shown that both ^{14}N and ^{15}N can be determined separately in soils by MIPOES on converting NH_3 nitrogen into N_2 with NaOBr and measuring the intensities of the N_2 bandheads.[537] Volatile arsenic compounds could be determined in soils after their thermal volatilization and MIPOES.[538]

With respect to gas analysis, organobromine compounds and hydrogen bromide could be determined in car exhaust gases after trapping them on Tenax® and applying thermal desorption with subsequent GC/MIPAED.[539] Also low-molecular-weight S compounds could be determined in air with MIPOES.[540] A direct determination of gaseous and particulate Pb in air could be performed with an air CMP[541] and a detection limit of $5\,\mathrm{\mu g\,m^{-3}}$ was reported.

For the analysis of petroleum- and natural gas-related products, MIPOES is also of use.[542] Organomercury species, for example, could be determined in natural gas using on-line amalgamation for trapping or solid-phase microextraction prior to GC/MIPAED,[543] and detection limits at the $0.2\,\mathrm{\mu g\,L^{-1}}$ level could be obtained for the various species.

7.3 On-line Monitoring

Because of the low capital and operating costs of microwave plasma discharges compared with ICPs, they are very useful for on-line monitoring purposes where metal concentrations in gas flows or in flowing liquids have to be monitored continuously. This growing need is introduced by the requirements for process control so as to optimize industrial processes with respect to product quality, costs and environmentally relevant emissions. Especially the cost effectiveness but also the fact that microwave plasmas in a number of cases can be operated with air makes them very attractive for on-line monitoring.

Microwave discharges have long been proposed for air quality monitoring.[544] Their possibilities for the determination of Hg in molecular gases were investigated in the case of an MIP operated in a TM_{010} cavity according to Beenakker.(545) When using a tangential flow torch, N_2 loaded with Hg can be entered as outer gas and detection limits in the case of OES are in the $10\,\mu g\,m^{-3}$ range. It was found that in view of possible applications of such a system for the on-line analysis of flue gases, considerable deterioration of the analytical signals occurred as a consequence of the presence of water vapor or molecular gases such as CO_2 and SO_2. It also could be shown that on-line speciation for Hg is possible. Indeed, by the use of a KCl-coated denuder (for $HgCl_2$) and an Au- or Ag-coated denuder (for metallic and oxidized Hg) placed in-line, the most relevant species can be separated. MIPOES, however, must still be optimized further so as to reach the detection limits of filter-based methods using determinations in decomposed filter residues by CVAAS, where a detection limit of $0.1\,\mu g\,m^{-3}$ was obtained for Hg.(546) Also for the determination of Hg in wastewater, MIPOES subsequent to the volatilization of metallic Hg can be used and here a detection limit of $0.31\,\mu g\,L^{-1}$ has been reported.(547)

The CMP, owing its stability, is much more appropriate for on-line measurements, as shown by lead determinations in air.(541) This plasma can be operated with air only at a power of some ca. 100 W and the gas flows can be varied from 0.2 to several liters per minute, allowing for flexibility in sampling, which should be performed isokinetically with the flowing medium to be monitored. The air CMP also tolerated the uptake of large amounts of organics and with both wet and dry aerosols the excitation temperatures were above 5000 K. Rotational temperatures were about 4000 K,(32) while the electron number densities were at the $10^{14}\,cm^{-3}$ level. With an air CMP operated at 600 W, detection limits for Cr, Cd, Pb, Co, Fe and Ni are at the $0.03–10\,\mu g\,m^{-3}$ level but they increase by up to 10-fold when the water loading increases above $100\,g\,m^{-3}$ or the CO_2 concentration increases to 20% (Table 8). As the air sampling certainly can still be improved, this especially shows that the CMP is a strong tool for on-line monitoring not only in complex gas mixtures such as flue gases but also for liquids, as the uptake of wet aerosols without the need for desolvation is possible.

Table 8 Detection limits ($\mu g\,m^{-3}$) using OES with an air CMP and different CO_2 concentrations; total gas flow = $1.2\,L\,min^{-1}$(548)

Element	CO_2 (%)				
	0	3	5	11	22
Cd	0.03	0.1	0.1	0.2	0.2
Cr	1.1	0.8	1.0	1.8	3.2
Pb	9.6	15	15	22	84

7.4 Direct Solids Analysis

Different direct solid sampling techniques known from ICP work can also be directly used in the case of microwave plasmas. As the relevant techniques, such as SA, LA and ETV, deliver dry analyte vapors or aerosols, they do not even require any precautions in the case of low-power microwave discharges.

Direct sample insertion requires a more voluminous plasma to accommodate nonmicrosamples. For this technique the CMP could be used very successfully. In biological samples such as soils, tomato leaves and bovine liver, Cd could be determined by placing the solid samples in a W cup loaded in a graphite holder in the center tube of a quartz torch with two concentric tubes. With suitable temperature programming, absolute detection limits of 40 pg for Cd could be obtained.(549) Steel samples could also be analyzed with the same set-up and Cr, Mn, Sn and Pb could be determined, Sn and Pb with detection limits of 0.08 and $5\,\mu g\,g^{-1}$, respectively. The water content of solid samples could also be determined with the aid of OES using a CMP.(550) Here the solid samples are heated in a quartz crucible and the water evolved is carried with a He flow into the CMP, while measuring the line intensities of O and H. The technique has been applied to water determinations in biological samples (tuna fish, freeze-concentrated and dried orange juice), coal and $CuSO_4 \cdot 5H_2O$.

Because of its small volume and low power, the MIP is only capable of evaporating very small particles (micrometer range) or of exciting vapors. This capacity can even be used for sizing particles in aerosol flows.(551) Here particles are first sampled on a filter and then aspirated into an He MIP with the aid of a scanning nozzle. The emission signals are proportional to the mass of the particles and the wavelengths present in the spectrum provide information on their composition. As the particles enter the MIP one by one, the number of signals counted provides information on their number. The apparatus is commercially available and a number of industrial and environmental applications have been reported.

ETV can easily be used for the analysis of powdered biological samples with MIPOES. Here a conventional GF can be used for sample volatilization and the lead vapor without overpressures suddenly occurring into the MIP.(552) For sample dispensing, a powder pipet can be used with which samples of down to 2 mg can be dispensed reproducibly and a calibration can be performed by standard addition with solutions. After performing a simultaneous and time-resolved measurement of line and background intensities, Mn could be determined reliably in the $10–150\,\mu g\,g^{-1}$ range and detection limits of $0.07\,\mu g\,g^{-1}$ Cu, $0.5\,\mu g\,g^{-1}$ Fe and $0.2\,\mu g\,g^{-1}$ Mn could

For references see page 9645

be obtained. A real-time background correction was found necessary as small sample amounts entering the plasma could be observed visually to change its form and excitation conditions.

SA is known to produce aerosols with particle sizes in the micrometer range, as described by Raeymaekers et al.[553] The concentrations of analytes in the aerosol as a function of the sparking conditions are well representative of the composition of the solids, as investigated for steels, brass and aluminum samples. However, as shown by single-particle analysis with an electron microprobe, a redistribution of the elements over the different particle classes may occur. This may lead to matrix effects as shown in early spark ablation/inductively coupled plasma optical emission spectroscopy (SA/ICPOES) work on aluminum samples.[554] Referring to Helmer and Walters,[555] the production of particles in SA is very complex, as both large particles, which are deposited immediately besides the spark gap, and small transportable particles are formed in a spark. The formation results both from direct ablation and melting in the spark plasma and from vapor condensation. With an Ar carrier gas flow the aerosols obtained by SA can be transported reproducibly and with low memory effects into an MIP discharge. With the aid of vidicon detection, spectra from the material produced by microwave plasma discharge excitation could be observed, which were less complex than the spark spectra.[556] In the case of a moderate-power MIP, the technique could be shown to be useful for the determination of Cr, Ni, Mn, Cu and Si in steel and of Fe, Si, Mg, Mn, Cu and Zn in aluminum alloys.[557] The spark-ablated aerosols also could be led into an MPT, which is viewed axially or laterally with an optical spectrometer. After optimizing the gas flows in the ablation chamber and the MIP, detection limits were obtained which were higher than in the case of SA/ICPOES using a sequential spectrometer. However, probably the difference by one order of magnitude is due to the differences in the spectrometers used. Further, it could be shown that the precision and the linearity of the calibration curves could be considerably improved by using a matrix line as reference.[558] A pulsed MIP could also be operated very successfully for the analysis of steels.[559]

As a direct solids sampling technique, LA has the advantage that also for electrically nonconducting samples aerosols can be produced and that at the same time laterally resolved information can be obtained. Because of the availability of very robust Nd:YAG lasers with a reproducible power output from one laser pulse to another and the growing need for the direct characterization of electrically nonconducting samples such as plastics and ceramics, the technique has been investigated in detail and combined with various radiation and ion sources including microwave plasma discharges, as described as early as 1980.[560] In the case of brass, it must be mentioned, however, that without carefully optimizing the working pressure and the laser parameters, selective volatilization can easily occur, as reflected by electron probe microanalysis in the crater, at the crater wall and of the ablated particles. As shown in Table 9, this contrasts with steel samples.[561]

LA has developed into a mature procedure. When combining it with an MIP operated in a TM_{010} resonator and when using a simultaneous échelle emission spectrometer with crossed dispersion and accordingly high resolving power, Be, Mg, Cr and Al could be determined in Cu, Ni and borax glass. High precision could be obtained when using internal standardization. Even from a single laser shot Na and Li could be determined in quartz with detection limits of 35 and 70 ng g^{-1}, respectively.[562] Fluctuations in LA resulting from power changes or surface effects could be corrected for by applying scattering light measurements and on comparing an Ar ICP with an He MIP the latter was found to give the lowest background intensities.[563] As for a few microseconds the plasma is briefly perturbed as a result of the laser pulse whereas the emission lasts for milliseconds in the case of an MIP

Table 9 Selective ablation by LA and SA for brass and steel[561]

Sample	LA Type	LA Concentration (%)			SA Type	SA Concentration (%)		
		Cu	Zn			Cu	Zn	
Brass (S 40/2)	Bulk	58 ± 1	42 ± 1		Bulk	30 ± 3	40 ± 3	
	Crater wall	66 ± 1	34 ± 1		Burning spot	71 ± 3	29 ± 2	
	Droplet	65 ± 1	35 ± 1		Particles	51 ± 6	49 ± 7	
		Concentration (%)				Concentration (%)		
		Fe	Ni	Cr		Fe	Ni	Cr
Steel (E-106)	Bulk	51 ± 2	20 ± 1	28 ± 1	Bulk	69 ± 2	18 ± 1	13.2 ± 0.6
	Crater wall	50 ± 2	20 ± 1	30 ± 1	Burning spot	70 ± 2	17 ± 1	13.4 ± 0.7

operated at reduced pressure, it was proposed to perform time-resolved measurements.[564] When applying time-gated detection of the emission spectra combined with internal standardization of the line intensities for both the main and trace elements in steel, Cu, Al and borax glass, a matrix-independent analysis was found to be possible.[565] With LA coupled to MIPOES using a low-pressure plasma, traces of Mg, Al, Si and Fe also could be determined in high-temperature superconductors ($YBa_2Cu_3O_7$) and Na could be determined in GeO_2-doped quartz tubes down to $40\,ng\,g^{-1}$. Care must be taken to ensure an air-leak-free construction of the MIP source.[566]

When using MIPs at higher power, direct powder injection could also be applied succesfully, as shown by the example of coal analysis.[567] C/H elemental ratios could be determined, but not Cl and S concentrations, when flushing the vials containing ground coal with He.

8 COMPARISON WITH OTHER METHODS AND OUTLOOK

Compared with other plasma sources for chemical analysis, such as plasma arcs and ICPs, MIPs have some differences which make their analytical features understandable. First they are operated at frequencies in the gigahertz range. This makes it understandable that deviations from TE are more pronounced than in high-frequency or even in arc sources. Indeed, the electrons can easily take up the microwave energy rather than giving it to the much slower gas atoms and ions. These deviations from LTE are known to lead to higher line to background intensity ratios as described in pertinent discussions in the literature. This is responsible for their high absolute power of detection. Further, microwave plasmas can be small and can often be contained in capillaries. This leads to high analyte number densities. This, however, is not always most favorable in AES as the background intensities in the analytical zone are then relatively high. Therefore, the development of toroidal plasmas, such as that realized in a resonator according to Beenakker with quartz tubes having a 4 mm internal diameter, and the use of the MPT are very important, as in their sources the analytical zone has a lower temperature than the plasma-generating zones which are outside. Microwave discharges have been shown to be very useful atom reservoirs for AAS and AFS or LEI, because of the high analyte number densities. Indeed, these "zero-background" types of spectroscopy are not hampered by spectral background emission in the analytical zones. In the case of MS analyses, the very small plasmas may not suffer from this drawback that limits OES, but also here larger plasmas with a better stabilized analytical zone may be advantageous so as to keep matrix influences resulting from changes in the plasma geometry low. Microwave plasmas also have the advantage over ICPs that they can easily be operated with He, which makes economic sense because of their low gas consumption. This makes microwave plasmas very useful for the determination and detection of nonmetals, which explains their success in speciation and in empirical formula determinations for organics. The fact that MIPs and especially the CMP can easily be operated on an air base only makes it very useful for on-line measurements. Indeed, here the operation costs really become low and as both wet and dry aerosols and gases can be sampled, both in process flows and exhaust gas flows, and also in wastewater, metals and nonmetals can be determined on-line. This makes microwave discharges very promising for environmental monitoring, e.g. in flue gases from waste incineration.

Microwave plasmas still have considerable potential for further development. Especially the variety of cavities which can be used for the production of different types of discharges with varying properties has hardly been exploited up to now. Here, both the demand for robust discharges with good sample uptake capacities also in the case of solvent-loaded aerosols, operation with both noble and molecular gases, the costs of operation and the ease of handling are important criteria. With respect to the last point, it was interesting how the need for retuning could be made superfluous in the case of the MPT by redesigning the device with respect to the coupler and dimensions as revealed by modeling.[63]

A further trend in development lies in the miniaturization of MIP devices together with their use in combination small spectrometers. Ideally, both the source and the spectrometer than could become plug-in accessories to be housed in computer racks. This point is of key importance for the wider use of microwave plasma detection in chromatography. Then the MIP surely could replace the conventional flame ionization detector (FID) and electron capture detector (ECD) much more easily and their capabilities for element-specific detection making the interpretation of chromatograms much easier could be more widely used.

The microwave discharges certainly should be improved with respect to their uptake of wet aerosols. Here, however, more powerful aerosol generation devices, which produce aerosols with a smaller droplet size, a narrower droplet size distribution and a better independence from the physical properties of the liquids are desirable, in addition to optimal adaption to the microwave plasmas. In this respect nebulization systems such as the DIN, HHPN, as introduced by Berndt,[189] and also the straightforward design of nebulization chambers to be used are important. Indeed, these are prerogatives for making use

For references see page 9645

of the small volume and thus the less disturbing properties of MIPs in LC detection as compared with the ICP.

Finally, as ionization sources for MS, microwave plasmas also have great potential, as here He can easily be used. This reduces the problems of interferences, especially in the lower mass range as compared with inductively coupled plasma mass spectrometry (ICPMS). With molecular gases toroidal plasmas can successfully be operated, as shown by Leis and Broekaert.[67] In atomic emission they suffer from a number of limitations as a result of background intensities, but this is not the case to the same extent when using them as ion sources for MS.

ABBREVIATIONS AND ACRONYMS

AAS	Atomic Absorption Spectroscopy
AES	Atomic Emission Spectroscopy
AFS	Atomic Fluorescence Spectroscopy
CCD	Charge-coupled Device
CID	Charge Injection Device
CMP	Capacitively Coupled Microwave Plasma
CMPOES	Capacitively Coupled Microwave Plasma Optical Emission Spectroscopy
CT	Cold Trapping
CV	Cold Vapor
CVAAS	Cold Vapor Atomic Absorption Spectroscopy
CZE	Capillary Zone Electrophoresis
DIN	Direct Injection Nebulization
ECD	Electron Capture Detector
ETV	Electrothermal Vaporization
FIA	Flow Injection Analysis
FID	Flame Ionization Detector
GC	Gas Chromatography
GC/AED	Gas Chromatography/Atomic Emission Detector
GC/MIPAED	Gas Chromatography/Microwave-induced Plasma Atomic Emission Detector
GC/MIPOES	Gas Chromatography/Microwave-induced Plasma Optical Emission Spectrometry
GC/MS	Gas Chromatography/Mass Spectrometry
GD	Glow Discharge
GF	Graphite Furnace
GFT	Graphite Furnace Trapping
HG	Hydride Generation
HG/AAS	Hydride Generation Atomic Absorption Spectroscopy
HG/GF/MIP	Hydride Generation/Graphite Furnace/Microwave-induced Plasma
HG/ICP	Hydride Generation/Inductively Coupled Plasma
HG/ICPMS	Hydride Generation/Inductively Coupled Plasma Mass Spectrometry
HG/MIP	Hydride Generation/Microwave-induced Plasma
HHPN	Hydraulic High-pressure Nebulization
HPLC	High-performance Liquid Chromatography
IC	Ion Chromatography
ICP	Inductively Coupled Plasma
ICPMS	Inductively Coupled Plasma Mass Spectrometry
ICPOES	Inductively Coupled Plasma Optical Emission Spectroscopy
IR	Infrared
LA	Laser Ablation
LC	Liquid Chromatography
LEI	Laser-enhanced Ionization
LTE	Local Thermal Equilibrium
MINDAP	Microwave-induced Nitrogen Discharge at Atmospheric Pressure
MIP	Microwave-induced Plasma
MIP/AED	Microwave-induced Plasma/Atomic Emission Detector
MIP/GC	Microwave-induced Plasma/Gas Chromatography
MIPMS	Microwave-induced Plasma Mass Spectrometry
MIPOES	Microwave-induced Plasma Optical Emission Spectroscopy
MPD	Microwave Plasma Detector
MPT	Microwave Plasma Torch
MPTAAS	Microwave Plasma Torch Atomic Absorption Spectrometry
MPTTOFMS	Microwave Plasma Torch Time-of-flight Mass Spectrometry
MS	Mass Spectrometry
NIR	Near-infrared
NIST	National Institute of Science and Technology
OES	Optical Emission Spectroscopy
PCB	Polychlorinated Biphenyl
PN	Pneumatic Nebulization
PN/ICP	Pneumatic Nebulization/Inductively Coupled Plasma
PN/ICPMS	Pneumatic Nebulization/Inductively Coupled Plasma Mass Spectrometry
PN/MIP	Pneumatic Nebulization/Microwave-induced Plasma
PTFE	Polytetrafluoroethylene
RSD	Relative Standard Deviation
SA	Spark Ablation

List of selected abbreviations appears in Volume 15

SA/ICPOES	Spark Ablation/Inductively Coupled Plasma Optical Emission Spectroscopy
SCP	Stabilized Capacitively Coupled High-frequency Plasma
SFC	Supercritical Fluid Chromatography
SFC/MIP	Supercritical Fluid Chromatography/ Microwave-induced Plasma
S/N	Signal-to-noise Ratio
TE	Thermal Equilibrium
TIA	Torche à Injection Axiale
T/MIP	Toroidal Microwave-induced Plasma
TOF	Time-of-flight
UHF	Ultrahigh Frequency
USN	Ultrasonic Nebulization
UV	Ultraviolet
UV/VIS	Ultraviolet/Visible
VIS	Visible
VUV	Vacuum-ultraviolet
3F/MIP	Three-filament/Microwave-induced Plasma

RELATED ARTICLES

Environment: Water and Waste **(Volume 3)**
Atomic Fluorescence in Environmental Analysis • Capillary Electrophoresis Coupled to Inductively Coupled Plasma-Mass Spectrometry for Elemental Speciation Analysis • Hydride Generation Sample Introduction for Spectroscopic Analysis in Environmental Samples • Inductively Coupled Plasma Mass Spectrometry in Environmental Analysis

Forensic Science **(Volume 5)**
Atomic Spectroscopy for Forensic Applications

Atomic Spectroscopy **(Volume 11)**
Atomic Spectroscopy: Introduction • Background Correction Methods in Atomic Absorption Spectroscopy • Flame and Vapor Generation Atomic Absorption Spectrometry • Flow Injection Analysis Techniques in Atomic Spectroscopy • Glow Discharge Optical Spectroscopy and Mass Spectrometry • Graphite Furnace Atomic Absorption Spectrometry • Inductively Coupled Plasma/Optical Emission Spectrometry • Laser Spectrometric Techniques in Analytical Atomic Spectrometry

Gas Chromatography **(Volume 12)**
Gas Chromatography: Introduction • Instrumentation of Gas Chromatography

Liquid Chromatography **(Volume 13)**
Liquid Chromatography: Introduction • Capillary Electrophoresis • Supercritical Fluid Chromatography

Mass Spectrometry **(Volume 13)**
Quadrupole Ion Trap Mass Spectrometer • Time-of-flight Mass Spectrometry

REFERENCES

1. R. Mavrodineanu, R.C. Hughes, 'Excitation in Radio Frequency Discharges', *Spectrochim. Acta, Part B*, **19**(8), 1309–1317 (1963).
2. U. Jecht, W. Kessler, 'Uber den Anregungsmechanismus einer HF-Fackelentladung bei 2400 MHz', *Z. Phys.*, **178**, 133–145 (1964).
3. W. Tappe, J. van Calker, 'Quantitative Spektrochemische Untersuchungen mit hochfrequenten Plasmaflammen', *Z. Anal. Chem.*, **198**, 13–20 (1963).
4. P.W.J.M. Boumans, F.J. de Boer, F.J. Dahmen, H. Hoelzel, A. Meier, 'A Comparative Investigation of Some Analytical Performance Characteristics of an ICP and a CMP for Solution Analysis by Emission Spectrometry', *Spectrochim. Acta, Part B*, **30**(10–11), 449–469 (1975).
5. R.M. Barnes, *Developments in Atomic Plasma Spectrochemical Analysis*, Heyden, London, 1982.
6. J.A.C. Broekaert, 'Plasma Atomic Spectrometry', *Nachr. Chem. Tech. Lab.*, **37**, 57–80 (1989).
7. J.A.C. Broekaert, 'Trends in Optical Spectrochemical Trace Analysis with Plasma Sources', *Anal. Chim. Acta*, **196**, 1–21 (1987).
8. B.L. Sharp, S. Chenery, R. Jowitt, S.T. Sparkes, A. Fisher, 'Atomic Spectrometry Update – Atomic-emission Spectrometry', *J. Anal. At. Spectrom.*, **9**(6), 171R–188R (1994).
9. S. Greenfield, I.L. Jones, C.T. Berry, 'High-pressure Plasmas as Spectroscopic Emission Sources', *Analyst*, **89**(1064), 713–720 (1964).
10. R.H. Wendt, V.A. Fassel, 'Induction-coupled Plasma Spectrometric Excitation Source', *Anal. Chem.*, **37**(7), 920–922 (1965).
11. P.C. Uden, *Element-specific Chromatographic Detection by Atomic Emission Spectroscopy*, American Chemical Society, Washington, DC, 1992.
12. D.W. Holman, T.J. Vickers, 'Optical Emission Spectroscopy with a Microwave-induced Plasma in a Sealed Micro-tube', *Talanta*, **29**(5), 419–421 (1982).
13. C.I.M. Beenakker, 'A Cavity for Microwave-induced Plasmas Operated in Helium and Argon at Atmospheric Pressure', *Spectrochim. Acta, Part B*, **31**(8–9), 483–486 (1976).
14. J. Hubert, M. Moisan, A. Ricard, 'A New Microwave Plasma at Atmospheric Pressure', *Spectrochim. Acta, Part B*, **34**(1), 1–10 (1979).
15. Q. Jin, C. Zhu, W. Borer, G.M. Hieftje, 'Microwave-plasma Torch Assembly for Atomic-emission Spectrometry', *Spectrochim. Acta, Part B*, **46**(3), 417–430 (1991).

List of selected abbreviations appears in Volume 15

16. Q.H. Jin, Y.X. Duan, J.A. Olivares, 'Development and Investigation of Microwave Plasma Techniques in Analytical Atomic Spectrometry', *Spectrochim. Acta, Part B*, **52**(2), 131–161 (1997).
17. J.D. Winefordner, E.P. Wagner, B.W. Smith, 'Status of and Perspectives on Microwave and Glow Discharges for Spectrochemical Analysis. Plenary Lecture', *J. Anal. At. Spectrom.*, **11**(9), 689–702 (1996).
18. A.T. Zander, G.M. Hieftje, 'Microwave-supported Discharges', *Appl. Spectrosc.*, **35**(4), 357–371 (1981).
19. M.M. Abdillahi, 'Microwave-induced Plasmas for Trace Elemental Analysis', *Int. Lab.*, **18**(5), 16–24 (1988).
20. R.C. Culp, K.C. Ng, 'Recent Developments in Analytical Microwave-induced Plasmas', in *Advances in Atomic Spectrometry*, ed. J. Sneddon, JAI Press, Greenwich, CT, 215–283, Vol. 2, 1995.
21. J.P. Matousek, B.J. Orr, M. Selby, 'Microwave-induced Plasmas: Implementation and Application', *Prog. Anal. At. Spectrosc.*, **7**(3), 275–314 (1984).
22. J.A.C. Broekaert, 'Current Developments in Optical-emission Spectrometry with High-frequency and Microwave Induced Plasmas', *J. Chin. Chem. Soc.*, **37**(1), 1–9 (1990).
23. J.A.C. Broekaert, 'Microwave Plasmas for Atomic Spectrometry – State of the Art and Analytical Applications', *GIT Fachz. Lab.*, **40**(4), 323–327 (1996).
24. S.R. Goode, K.W. Baughman, 'Review of Instrumentation Used to Generate Microwave-induced Plasmas', *Appl. Spectrosc.*, **38**(6), 755–763 (1984).
25. O. Zinke, H.L. Hartnagel, *Lehrbuch der Hochfrequenztechnik*, Springer, Berlin, Band 2, 1987.
26. A.J. Baden Fuller, *Mikrowellen*, Vieweg & Sohn, Braunschweig, 1974.
27. P.G. Brandl, C.L. Arnow, C.L. Caldwell, J.W. Carnahan, 'Conversion of a Pulsed 120 Hz, 3 kW, 2.45 GHz Generator for Continuous-wave Output for Microwave-induced Plasma Maintenance', *Appl. Spectrosc.*, **51**(2), 280–283 (1997).
28. M. McKenna, I.L. Marr, M.S. Cresser, E. Lam, 'Analysis of Some Diving Gas Mixtures by Microwave-induced Plasma Optical Emission Spectroscopy', *Spectrochim. Acta, Part B*, **41**(7), 669–676 (1986).
29. P. Brassem, F.J.M.J. Maessen, 'The Influence of Flow Rate and Pressure on the Excitation Conditions in Low-pressure Microwave-induced Plasmas', *Spectrochim. Acta, Part B*, **30**(12), 547–556 (1975).
30. P. Brassem, F.J.M.J. Maessen, L.de Galan, 'Excitation Conditions in a Mixed-gas Low-pressure Microwave-induced Plasma', *Spectrochim. Acta, Part B*, **31**(10–12), 537–545 (1976).
31. P. Brassem, F.J.M.J. Maessen, 'Electron Temperatures and Electron Concentrations at Low Pressure in Microwave-induced Plasmas', *Spectrochim. Acta, Part B*, **29**(7–8), 203–210 (1974).
32. N.H. Bings, M. Olschewski, J.A.C. Broekaert, 'Two-dimensional Spatially Resolved Excitation and Rotational Temperatures as well as Electron Number Density Measurements in Capacitively Coupled Microwave Plasmas Using Argon, Nitrogen and Air as Working Gases by Spectroscopic Methods', *Spectrochim. Acta, Part B*, **52**(13), 1965–1981 (1997).
33. A. Disam, P. Tschöpel, G. Tölg, 'Emission-spectrometric Determination of Trace Elements in Aqueous Solutions by Means of a Sheath-gas-stabilized Capacitatively Coupled Microwave Plasma (CMP)', *Fresenius' Z. Anal. Chem.*, **310**(1–2), 131–143 (1982).
34. G. Wünsch, G. Hegenberg, N. Czech, 'Capacitively Coupled Microwave Plasma (CMP) as Source of Atomization in Atomic-absorption Spectroscopy', *Spectrochim. Acta, Part B*, **38**(8), 1135–1141 (1983).
35. C.I.M. Beenakker, P.W.J.M. Boumans, P.J. Rommers, 'Microwave-induced Plasma as Excitation Source for Atomic-emission Spectrometry', *Philips Tech. Rev.*, **39**(3–4), 65–77 (1980).
36. C.I.M. Beenakker, P.W.J.M. Boumans, P.J. Rommers, 'Microwave-induced Plasma as Excitation Source for Atomic-emission Spectrometry. II', *GIT Fachz. Lab.*, **25**(3), 179–189 (1981).
37. C.I.M. Beenakker, P.W.J.M. Boumans, P.J. Rommers, 'Microwave-induced Plasma as Excitation Source for Atomic-emission Spectrometry', *GIT Fachz. Lab.*, **25**(2), 82–87 (1981).
38. C.I.M. Beenakker, 'A Cavity for Microwave-induced Plasmas Operated in Helium and Argon at Atmospheric Pressure', *Spectrochim. Acta, Part B*, **31**(7), 483–486 (1976).
39. N. Rait, D.W. Golightly, C.J. Massoni, 'Improved Beenakker-type Cavity for Microwave-induced Plasma Spectrometry', *Spectrochim. Acta, Part B*, **39**(7), 931–937 (1984).
40. D.L. Haas, J.W. Carnahan, J.A. Caruso, 'Internally Tuned TM_{010} Microwave Resonant Cavity for Moderate-power Microwave-induced Plasmas', *Appl. Spectrosc.*, **37**(1), 82–85 (1983).
41. J. Hubert, S. Bordeleau, K.C. Tran, S. Michaud, B. Milette, R. Sing, J. Jalbert, D. Boudreau, M. Moisan, J. Margot, 'Atomic Spectroscopy with Surface-wave Plasmas', *Fresenius' J. Anal. Chem.*, **355**(5–6), 494–500 (1996).
42. M. Selby, G.M. Hieftje, 'Taming the Surfatron', *Spectrochim. Acta, Part B*, **42**(1–2), 285–298 (1987).
43. S. Luge, J.A.C. Broekaert, 'Use of Optical Emission Spectrometry with Microwave-induced Plasma (MIP) Discharges in a Surfatron Combined to Different Types of Hydride Generation for the Determination of Arsenic', *Mikrochim. Acta*, **113**(3–6), 277–286 (1994).
44. H. Matusiewicz, 'Novel Microwave Plasma Cavity Assembly for Atomic-emission Spectrometry', *Spectrochim. Acta, Part B*, **47**(10), 1221–1227 (1992).
45. S.A. Estes, P.C. Uden, R.M. Barnes, 'Microwave-Excited Atmospheric-pressure Helium-plasma Emission Detection Characteristics in Fused-silica Capillary

Gas Chromatography', *Anal. Chem.*, **53**(12), 1829–1837 (1981).

46. K.J. Mulligan, M.H. Hahn, J.A. Caruso, F.L. Fricke, 'Comparison of Several Microwave Cavities for Simultaneous Determination of Arsenic, Germanium, Antimony and Tin by Plasma Emission Spectrometry', *Anal. Chem.*, **51**(12), 1935–1938 (1979).
47. A.T. Zander, 'Atomic-emission Sources for Solution Spectrochemistry', *Anal. Chem.*, **58**(11), 1139A–1149A (1986).
48. H. Matusiewicz, R.E. Sturgeon, 'Liquid Cooling of a Torch for Microwave-induced Plasma Spectrometry', *Spectrochim. Acta, Part B*, **48**(4), 515–519 (1993).
49. R.F. Wandro, H.B. Friedrich, 'Investigation of the Interactions of a Microwave-induced Plasma with a Quartz Capillary Wall', *Anal. Chem.*, **56**(14), 2727–2731 (1984).
50. D. Kollotzek, P. Tschöpel, G. Tölg, 'Lösungsemissionsspektrometrie mit mikrowelleninduzierten Mehrfaden- und Hohlzylinderplasmen – I: Die radiale Verteilung von Signal- und Untergrundintensitäten', *Spectrochim. Acta, Part B*, **37**, 91–96 (1982).
51. D. Kollotzek, P. Tschöpel, G. Tölg, 'Three-filament and Toroidal Microwave-induced Plasmas as Radiation Sources for Emission-spectrometric Analysis of Solutions and Gaseous Samples. II. Analytical Performance', *Spectrochim. Acta, Part B*, **39**(5), 625–636 (1984).
52. G. Heltai, J.A.C. Broekaert, F. Leis, G. Tölg, 'Study of a Toroidal Argon and a Cylindrical Helium Microwave-induced Plasma for Analytical Atomic-emission Spectrometry. I. Configurations and Spectroscopic Properties', *Spectrochim. Acta, Part B*, **45**(3), 301–311 (1990).
53. A. Bollo-Kamara, E.G. Codding, 'Considerations in the Design of a Microwave-induced Plasma Utilizing the TM_{010} Cavity for Optical Emission Spectroscopy', *Spectrochim. Acta, Part B*, **36**(10), 973–982 (1981).
54. G.K. Webster, J.W. Carnahan, 'Improved Supercritical-fluid Chromatography Mobile-phase Tolerance with a Moderate-power Helium Microwave-induced Plasma', *Appl. Spectrosc.*, **45**(8), 1285–1290 (1991).
55. S.R. Goode, B. Chambers, N.P. Buddin, 'Use of a Tangential-flow Torch with a Microwave-induced Plasma Emission Detector for Gas Chromatography', *Spectrochim. Acta, Part B*, **40**(1–2), 329–333 (1985).
56. S.R. Goode, B. Chambers, N.P. Buddin, 'Critical Evaluation of the Tangential-flow Torch Microwave-induced Plasma Detector for Gas Chromatography', *Appl. Spectrosc.*, **37**(5), 439–442 (1983).
57. K.C. Ng, M.J. Brechmann, 'Demountable Torch for Combination-gas Microwave-induced Plasmas', *Spectroscopy*, **2**(1), 23–25 (1987).
58. G.S. Sobering, T.D. Bailey, T.C. Farrar, 'Simple Torch Design for a Microwave-induced Plasma', *Appl. Spectrosc.*, **42**(6), 1023–1025 (1998).
59. P.R. Fielden, M. Jiang, R.D. Snook, 'Laminar-flow Microwave-plasma Detector for Gas Chromatography', *Appl. Spectrosc.*, **43**(8), 1444–1449 (1989).
60. M.R. Jiang, P.R. Fielden, R.D. Snook, 'Performance of a Laminar-flow Torch – Microwave Plasma Detector for Gas Chromatography', *Appl. Spectrosc.*, **45**(2), 227–230 (1991).
61. M.L. Bruce, J.M. Workman, J.A. Caruso, D.J. Lahti, 'Low-flow Laminar-flow Torch for Microwave-induced Plasma Emission Spectrometry', *Appl. Spectrosc.*, **39**(6), 935–942 (1985).
62. B.W. Pack, G.M. Hieftje, 'An Improved Microwave Plasma Torch for Atomic Spectrometry', *Spectrochim. Acta, Part B*, **52**(14), 2163–2168 (1997).
63. A.M. Bilgic, C. Prokisch, J.A.C. Broekaert, E. Voges, 'Design and Modelling of a Modified 2.45 GHz Coaxial Plasma Torch for Atomic Spectrometry', *Spectrochim. Acta, Part B*, **53**(5), 773–777 (1998).
64. M.M. Mohamed, T. Uchida, S. Minami, 'Direct Sample Introduction of Solid Material into a Pulse-operated MIP [Microwave-induced Plasma]', *Appl. Spectrosc.*, **43**(5), 794–800 (1989).
65. K.G. Michlewicz, J.J. Urh, J.W. Carnahan, 'Microwave-induced Plasma System for Maintenance of Moderate-power Plasmas of Helium, Argon, Nitrogen and Air', *Spectrochim. Acta, Part B*, **40**(3), 493–499 (1985).
66. H.A. Dingjan, H.J. De Jong, 'Comparative Study of Two Cavities for Generating a Microwave-induced Plasma in Helium or Argon as a Detector in Gas Chromatography', *Spectrochim. Acta, Part B*, **36**(4), 325–331 (1981).
67. F. Leis, J.A.C. Broekaert, 'High-power Microwave-induced Plasma for Analysis of Solutions', *Spectrochim. Acta, Part B*, **39**(9–11), 1459–1463 (1984).
68. M. Ohata, N. Furuta, 'Evaluation of the Detection Capability of a High Power Nitrogen Microwave-induced Plasma for Both Atomic Emission and Mass Spectrometry', *J. Anal. At. Spectrom.*, **5**(13), 447–453 (1998).
69. C.I.M. Beenakker, 'Evaluation of a Microwave-induced Plasma in Helium at Atmospheric Pressure as an Element-selective Detector for Gas Chromatography', *Spectrochim. Acta, Part B*, **32**(3–4), 173–187 (1977).
70. P.W.J.M. Boumans, *Theory of Spectrochemical Excitation*, Hilger and Watts, London, 1966.
71. J. Jonkers, J.M. De Regt, J.A.M. Van der Mullen, H.P.C. Vos, F.P.J. De Groote, E.A.H. Timmermans, 'On the Electron Temperatures and Densities in Plasmas Produced by the "torche à Injection Axiale"', *Spectrochim. Acta, Part B*, **51**(11), 1385–1392 (1996).
72. H.R. Griem, *Plasma Spectroscopy*, McGraw-Hill, New York, 1964.
73. D. Robinson, P.D. Lenn, 'Plasma Diagnostics by Spectroscopic Methods', *Appl. Opt.*, **6**, 983–1000 (1967).
74. A. Garnier, E. Bloyet, P. Leprince, J. Marec, 'Microwave Plasma in Argon Produced by a Surface Wave: Study of the Effect of Pressure on the Optical Emission and the Potentials for Analysis of Gaseous Samples', *Spectrochim. Acta, Part B*, **43**(8), 963–970 (1988).

List of selected abbreviations appears in Volume 15

75. B. Riviere, J.M. Mermet, D. Deruaz, 'Behaviour and Analytical Applications of a Modulated-power Microwave-induced Plasma (Surfatron)', *J. Anal. At. Spectrom.*, **4**(6), 519–523 (1989).
76. L.K. Olson, W.C. Story, J.T. Creed, W.L. Shen, J.A. Caruso, 'Fragmentation of Organic Compounds Using Low-pressure Microwave-induced Plasma Mass Spectrometry', *J. Anal. At. Spectrom.*, **5**(6), 471–475 (1990).
77. M. Moisan, R. Grenier, Z. Zahrzewski, 'The Electromagnetic Performance of a Surfatron-based Coaxial Microwave Plasma Torch', *Spectrochim. Acta, Part B*, **50**(8), 781–789 (1995).
78. J. Cotrino, M. Saez, M.C. Quinero, A. Menendez, E. Sanchez-Uria, A. Sanz-Medel, 'Spectroscopic Determination of Fundamental Parameters in an Argon Microwave-induced Plasma (Surfatron) at Atmospheric Pressure', *Spectrochim. Acta, Part B*, **47**(3), 425–435 (1992).
79. P.G. Brandl, J.W. Carnahan, 'Charge Transfer in Analytical Helium Plasmas', *Spectrochim. Acta, Part B*, **49**(1), 105–115 (1994).
80. M.P. Dziewatkoski, C.B. Boss, 'Study of Hydrocarbon Dissociation in a Microwave-induced Plasma Sustained in Ar Gas', *Spectrochim. Acta, Part B*, **49**(2), 117–135 (1994).
81. W. Zyrnicki, 'Spectroscopic Study of Argon MIP (Microwave-induced Plasma) Discharge Used for Analysis of Solutions', *Microchem. J.*, **46**(3), 356–351 (1992).
82. J. Mierzwa, R. Brandt, J.A.C. Broekaert, P. Tschöpel, G. Tölg, 'Performance of a Microwave-induced Plasma (MIP) Operated in a Liquid-cooled Discharge Tube for Atomic-emission Spectrometry', *Spectrochim. Acta, Part B*, **51**(1), 117–126 (1996).
83. J. Roepcke, A. Ohl, M. Schmidt, 'Comparison of Optical-emission Spectrometric Measurements of the Concentration and Energy of Species in Low-pressure Microwave and Radio-frequency Plasma Sources', *J. Anal. At. Spectrom.*, **8**(6), 803–808 (1983).
84. H. Jimenez-Dominguez, S. Cruz-Jimenez, A. Cabral-Prieto, 'Spectroscopic Applications of the Plasma Dispersion Function', *Spectrochim. Acta, Part B*, **51**(1), 165–174 (1996).
85. R.M. Alvarez-Bolainez, C.B. Boss, 'Noise Analysis of the Reflected Power Signal from a Microwave-induced Plasma Gas-chromatographic Detector', *Anal. Chim. Acta*, **269**(1), 89–97 (1992).
86. J.E. Freeman, G.M. Hieftje, 'Analytical Characteristics of Near-infrared Non-metal Atomic Emission from a Helium Microwave-induced Plasma', *Spectrochim. Acta, Part B*, **40**(3), 475–492 (1985).
87. D.M. Chambers, J.W. Carnahan, Q. Jin, G.M. Hieftje, 'Fundamental Studies of the Sampling Process in an Inductively Coupled Plasma Mass Spectrometer. IV. Replacement of the Inductively Coupled Plasma with a Helium Microwave-induced Plasma', *Spectrochim. Acta, Part B*, **46**(13), 1745–1765 (1991).
88. B. Kirsch, B.S. Hanamura, J.D. Winefordner, 'Diagnostical Measurements in a Single-electrode, Atmospheric-pressure, Microwave Plasma', *Spectrochim. Acta, Part B*, **39**(8), 955–963 (1984).
89. S. Hanamura, B.W. Smith, J.D. Winefordner, 'Single-electrode Atmospheric-pressure Microwave Discharge System for Elemental Analysis', *Can. J. Spectrosc.*, **29**(1), 13–18 (1984).
90. N.H. Bings, J.A.C. Broekaert, 'The Use of Different Plasma Gases (Argon, Nitrogen and Air) for Capacitively Coupled Microwave-plasma Optical Emission Spectrophotometry (CMP OES): Figures of Merit and Temperature Measurements', *Fresenius' J. Anal. Chem.*, **355**(3–4), 242–243 (1996).
91. Q. Jin, G. Yang, A. Yu, J. Liu, H. Zhang, Y. Ben, 'A Novel Plasma Emission Source', in *Abstracts Book of Pittcon '85*, Abstract No. 1171, 1985.
92. R. Pereiro, M. Wu, J.A.C. Broekaert, G.M. Hieftje, 'Direct Coupling of Continuous Hydride Generation with Microwave Plasma Torch Atomic-emission Spectrometry for the Determination of Arsenic, Antimony and Tin', *Spectrochim. Acta, Part B*, **49**(1), 59–73 (1994).
93. Y. Madrid, M.W. Borer, C. Zhu, Q.H. Jin, G.M. Hieftje, 'Noise Characterization of the Microwave Plasma Torch (MPT) Source', *Appl. Spectrosc.*, **48**(8), 994–1002 (1994).
94. M. Huang, D.S. Hanselman, Q. Jin, G.M. Hieftje, 'Non-thermal Features of Atmospheric-pressure Argon and Helium Microwave-induced Plasmas Observed by Laser-light Thomson Scattering and Rayleigh Scattering', *Spectrochim. Acta, Part B*, **45**(12), 1339–1352 (1990).
95. C. Prokisch, A.M. Bilgic, E. Voges, J.A.C. Broekaert, J. Jonkers, M. van Sande, J.A.M. van der Mullen, 'Photographic Plasma Images and Electron Number Density as well as Electron Temperature Mappings of a Plasma in a Modified Argon Microwave Plasma Torch (MPT) Measured by Spatially Resolved Thomson Scattering', *Spectrochim. Acta, Part B*, **54**(9), 1253–1266 (1999).
96. U. Engel, C. Prokisch, E. Voges, G.M. Hieftje, J.A.C. Broekaert, 'Spatially Resolved Measurements and Plasma Tomography with Respect to the Rotational Temperatures for a Microwave Plasma Torch', *J. Anal. At. Spectrom.*, **13**(9), 955–961 (1998).
97. C. Prokisch, J.A.C. Broekaert, 'Element Determination in Aqueous and Acetonitrile Containing Solutions by Atomic Emission Spectrometry Using a Microwave Plasma Torch', *Spectrochim. Acta, Part B*, **52**(6–8), 1909–1119 (1998).
98. S.R. Goode, J.N. Emily, 'Measuring the Spatial Distribution of Properties and Species in Microwave-induced Helium Plasmas', *Spectrochim. Acta, Part B*, **49**(1), 31–45 (1998).
99. Y. Okamoto, 'Annular-shaped Microwave-induced Nitrogen Plasma at Atmospheric Pressure for Emission Spectrometry of Solutions', *Anal. Sci.*, **7**(2), 283–288 (1991).

100. M. Ohata, N. Furuta, 'Spatial Characterization of Emission Intensities and Temperatures of a High-power Nitrogen Microwave-induced Plasma', *J. Anal. At. Spectrom.*, **12**(3), 341–347 (1997).
101. J. Jonkers, H.P.C. Vos, J.A.M. Van der Mullen, E.A.H. Timmermans, 'On the Atomic State Densities of Plasmas Produced by the "Torche à Injection Axiale"', *Spectrochim. Acta, Part B*, **51**(5), 457–465 (1996).
102. A. Rodero, M.C. Quintero, A. Sola, A. Gamero, 'Preliminary Spectroscopic Experiments with Helium Microwave-induced Plasma Produced in Air by Use of a New Structure: The Axial Injection Torch', *Spectrochim. Acta, Part B*, **51**(5), 467–479 (1996).
103. R.F. Browner, A.W. Boorn, 'Sample Introduction: The Achilles' Heel of Atomic Spectroscopy?', *Anal. Chem.*, **56**(7), 786A–798A (1984).
104. R.F. Browner, A.W. Boorn, 'Sample-introduction Techniques for Atomic Spectroscopy.', *Anal. Chem.*, **56**(7), 875A–888A (1984).
105. J. Sneddon, *Sample Introduction in Atomic Spectroscopy*, Elsevier, Amsterdam, 1990.
106. Special Issue: 'Sample Introduction in Atomic Spectrometry', *Spectrochim. Acta, Part B*, **50**(4–7), 271–654 (1995).
107. Special Issue: 'Flow Injection Analysis – State of the Art Applied to Atomic Spectroscopy', *Spectrochim. Acta, Part B*, **51**(14), 1733–1941 (1996).
108. L. Ebdon, 'The Joys of Coupling', *Lab. Pract.*, **38**(3), 15–20 (1989).
109. A. Sanz-Medel, A. Menendez, M.L. Fernandez, E. Sanchez-Uria, 'Tandem Online Separations: An Alternative Sample Presentation in Atomic Spectrometry for Ultratrace Analysis', *Acta. Chim. Hung.*, **128**(4–5), 551–558 (1991).
110. G. Heltai, J.A.C. Broekaert, 'Sample Introduction Problems of Atmospheric MIPs (Microwave-induced Plasmas)', *Acta. Chim. Hung.*, **128**(4–5), 599–611 (1991).
111. J.A. Caruso, 'Microwave-induced Plasmas as Sources for Atomic Spectrometry', *J. Res. Natl. Bur. Stand. (US)*, **93**(3), 447–449 (1988).
112. E.I. Brooks, K.J. Timmins, 'Sample-introduction Device for Use with a Microwave-induced Plasma', *Analyst*, **110**(5), 557–558 (1985).
113. J.A.C. Broekaert, G. Tölg, 'Recent Developments in Atomic Spectrometry Methods for Elemental Trace Determinations', *Fresenius' Z. Anal. Chem.*, **326**(6), 495–509 (1987).
114. W.M. Borer, G.M. Hieftje, 'Inductively Coupled Plasma–Microwave-induced Plasma Tandem Source for Atomic-emission Spectrometry', *J. Anal. At. Spectrom.*, **8**(2), 339–348 (1993).
115. W.M. Borer, G.M. Hieftje, 'Design Considerations for a Pressure-differential Tandem Source for Use in Atomic Spectrometry', *J. Anal. At. Spectrom.*, **8**(2), 333–338 (1993).
116. N.W. Barnett, 'Some Experience with Various Sample Introduction Techniques for Microwave-induced Plasma Atomic-emission Spectrometry [MIP/AES]', *Vestn. Slov. Kem. Drus.*, **36**(1), 1–23 (1989).
117. H. Matusiewicz, 'A Microwave Plasma Cavity Assembly for Atomic-emission Spectrometry', *Fresenius' J. Anal. Chem.*, **355**(5–6), 623–625 (1996).
118. H. Matusiewicz, 'Thermal Vaporization for Sample Introduction in Microwave-induced Plasma Analytical Emission Spectrometry', *Spectrochim. Acta Rev.*, **13**(1), 47–68 (1990).
119. A. Aziz, J.A.C. Broekaert, F. Leis, 'Contribution to the Analysis of Microamounts of Biological Samples Using a Combination of Graphite-furnace and Microwave-induced Plasma Atomic-emission Spectroscopy', *Spectrochim. Acta, Part B*, **37**(5), 381–389 (1982).
120. G. Heltai, J.A.C. Broekaert, P. Burba, F. Leis, P. Tschöpel, G. Tölg, 'Study of a Toroidal Argon MIP (Microwave-induced Plasma) and a Cylindrical Helium MIP for Atomic-emission Spectrometry. II. Combination with Graphite-furnace Vaporization and Use for Analysis of Biological Samples', *Spectrochim. Acta, Part B*, **45**(8), 857–866 (1990).
121. J.F. Yang, J.Y. Zhang, C. Schickling, J.A.C. Broekaert, 'Study of a Microwave-induced Argon Plasma Sustained in a TE_{101} Cavity as Spectrochemical Emission Source Coupled with Graphite-furnace Evaporation', *Spectrochim. Acta, Part B*, **51**(6), 551–562 (1996).
122. M.M. Abdillahi, 'Analytical Feasibility of Graphite-rod Vaporization–Helium Microwave-induced Plasma (GRV–He-MIP) for Some Non-metal Determinations', *Appl. Spectrosc.*, **47**(3), 366–374 (1993).
123. J.F. Alder, M.T.C. Da Cunha, 'Evaluation of a Low-pressure Microwave-induced Argon Plasma with Carbon-rod Sample Introduction for Trace-metal Analysis', *Can. J. Spectrosc.*, **25**(2), 32–38 (1980).
124. C.E. Pfluger, T. Nessel, 'Wire-loop Microfurnace–Power Supply Controller System for the Miniature Helium Direct-current Discharge Tube Emission Method', *Analyst*, **109**(5), 593–596 (1984).
125. R.G. Stahl, L. Brett, K.J. Timmins, 'Automation of a Wire-filament Atomizer Helium Microwave-induced Plasma System', *J. Anal. At. Spectrom.*, **4**(4), 337–340 (1989).
126. K. Chiba, M. Kurosawa, K. Tanabe, H. Haraguchi, 'Micro-sampling Technique Utilizing an Electrothermal Tungsten-boat Vaporization Device for Atmospheric-pressure Helium Microwave-induced Plasma Emission Spectrometry', *Chem. Lett.*, 75–78 (1984).
127. W.H. Evans, J.A. Caruso, R.D. Satzger, 'Evaluation of a Tantalum-tip Electrothermal-vaporization Sample-introduction Device for Microwave-induced Plasma Mass Spectrometry and Atomic-emission Spectrometry', *Appl. Spectrosc.*, **45**(9), 1478–1484 (1991).
128. M. Zerezghi, K.J. Mulligan, J.A. Caruso, 'Simultaneous Multi-element Determination in Microlitre Samples by

Rapid-scanning Spectrometry Coupled to a Microwave-induced Plasma', *Anal. Chim. Acta*, **154**, 219–226 (1983).
129. N.W. Barnett, G.F. Kirkbright, 'Electrothermal-vaporization Sample Introduction into an Atmospheric-pressure Helium Microwave-induced Plasma for the Determination of Iodine in Hydrochloric Acid', *J. Anal. At. Spectrom.*, **1**(5), 337–342 (1986).
130. J.P. Matousek, B.J. Orr, M. Selby, 'Spectrometric Analysis of Nonmetals Introduced from a Graphite Furnace into a Microwave-induced Plasma', *Talanta*, **33**(11), 875–882 (1986).
131. M. Wu, J.W. Carnahan, 'Trace Determination of Cadmium, Copper, Bromine and Chlorine with Electrothermal Vaporization into a Helium Microwave-induced Plasma', *Appl. Spectrosc.*, **44**(4), 673–678 (1990).
132. K. Dittrich, H. Fuchs, H. Berndt, J.A.C. Broekaert, G. Schaldach, 'Determination of Traces of Sulfur by Electrothermal Evaporation and Non-thermal Excitation of Sulfur-containing Species in a Hollow-cathode Discharge (FANES/MONES) [Furnace Non-thermal Excitation Spectrometry/molecular Non-thermal Excitation Spectrometry] and in a Microwave-induced Plasma (MIP)', *Fresenius' J. Anal. Chem.*, **336**(4), 303–310 (1990).
133. H.P.J. Van Dalen, B.G. Kwee, L. De Galan, 'Selective Determination of Halogens and Sulfur in Solution by Atmospheric-pressure Helium Microwave-induced Plasma Emission Spectrometry Coupled to an Electrothermal Introduction System', *Anal. Chim. Acta*, **142**, 159–171 (1982).
134. J.W. Carnahan, J.A. Caruso, 'Determination of Halogenated Organic Compounds by Electrothermal Vaporization into a Helium Microwave-induced Plasma at Atmospheric Pressure', *Anal. Chim. Acta*, **136**, 261–267 (1982).
135. T. Kumamaru, J.F. Riordan, B.L. Vallee, 'Low-pressure Microwave-induced Helium Plasma Emission Spectrometry: Determination of Sub-nanogram Quantities of Zinc by Use of a Tungsten-filament Vaporization System', *Anal. Biochem.*, **126**(1), 208–213 (1982).
136. N.W. Barnett, 'Determination of Lead and Nickel in Animal Bone by Microwave-induced Plasma Atomic-emission Spectrometry with Sample Introduction by Electrothermal Vaporization', *Anal. Chim. Acta*, **198**, 309–314 (1987).
137. E. Beinrohr, E. Bulska, P. Tschöpel, G. Tölg, 'Determination of Lead by Electrothermal-vaporization Microwave-induced-plasma Atomic-emission Spectrometry After Flow-through Electrolytic Deposition in a Graphite Tube Packed with Reticulated Vitreous Carbon', *J. Anal. At. Spectrom.*, **8**(7), 965–968 (1993).
138. K. Kitagawa, A. Mizutani, M. Yanagisawa, 'Separative Column Atomizer Attached to Atmospheric-pressure Helium Microwave-induced Plasma for Direct Trace-element Analysis by Atomic-emission Spectrometry', *Anal. Sci.*, **5**(5), 539–544 (1989).
139. H. Matusiewicz, I.A. Brovko, R.E. Sturgeon, V.T. Luong, 'Aerosol Transport Interface for Electrothermal Vaporization–Microwave-induced Plasma Emission Spectrometry', *Appl. Spectrosc.*, **44**(4), 736–739 (1990).
140. M. Yanagisawa, H. Kawaguchi, B.L. Vallee, 'Low-pressure Microwave-induced Helium-plasma Emission Spectrometry: Vaporization Characteristerics of Metal Chlorides, Nitrates and Sulfates', *Anal. Biochem.*, **95**(1), 8–13 (1979).
141. P.B. Mason, L. Zhang, J.W. Carnahan, R.E. Winans, 'Helium Microwave-induced Plasma Atomic-emission Detection for Liquid Chromatography Utilizing a Moving Band Interface', *Anal. Chem.*, **65**(19), 2596–2600 (1993).
142. S. Luge, J.A.C. Broekaert, 'Use of Optical Emission Spectrometry with Microwave-induced Plasma (MIP) Discharges in a Surfatron Combined to Different Types of Hydride Generation for the Determination of Arsenic', *Mikrochim. Acta*, **113**(3–6), 277–286 (1994).
143. E. Bulska, P. Tschöpel, J.A.C. Broekaert, G. Tölg, 'Different Sample Introduction Systems for the Simultaneous Determination of Arsenic, Antimony and Selenium by Microwave-induced Plasma Atomic-emission Spectrometry', *Anal. Chim. Acta*, **271**(1), 171–181 (1993).
144. H. Tao, A. Miyazaki, 'Determination of Germanium, Arsenic, Antimony, Tin and Mercury at Trace Levels by Continuous Hydride-generation Helium Microwave-induced Plasma Atomic-emission Spectrometry', *Anal. Sci.*, **7**(1), 55–59 (1991).
145. A. Montaser, D.W. Golightly, *Inductively Coupled Plasmas in Analytical Atomic Spectrometry*, Verlag Chemie, Weinheim, 1987.
146. G.S. Pyen, R.F. Browner, 'Comparison of Flow Injection and Continuous Sample Introduction for Automatic Simultaneous Determination of Arsenic, Antimony and Selenium by Hydride Generation and Inductively Coupled Plasma Optical Emission Spectrometry', *Appl. Spectrosc.*, **42**(3), 508–512 (1988).
147. J.D. Hwang, H.P. Huxley, J.P. Diomiguardi, W.J. Vaughn, 'Determination of Arsenic in Environmental Samples by Inductively Coupled Plasma Atomic-emission Spectrometry with an In Situ Nebulizer/Hydride Generator', *Appl. Spectrosc.*, **44**(3), 491–496 (1990).
148. H. Tao, A. Miyazaki, K. Bansho, 'Hydride-generation System with a Hydrogen Separation Membrane for Low-power Inductively Coupled Plasma Emission Spectrometry', *Anal. Sci.*, **6**(2), 195–199 (1990).
149. H. Matusiewicz, R.E. Sturgeon, S.S. Bermann, 'In Situ Hydride Generation Preconcentration of Arsenic in a Graphite Furnace with Sample Vaporization into a Microwave-induced Plasma for Emission Spectrometry', *Spectrochim. Acta, Part B*, **45**(1–2), 209–214 (1990).
150. D.T. Heitkemper, J.A. Caruso, 'Continuous Hydride Generation for Simultaneous Multi-element Detection

with Inductively Coupled Plasma Mass Spectrometry', *Appl. Spectrosc.*, **44**(2), 228–234 (1990).

151. K. Dittrich, R. Mandry, 'Investigations into Improvement of Analytical Application of the Hydride Technique in Atomic-absorption Spectrometry by Matrix Modification and Graphite-furnace Atomization. I. Analytical Results', *Analyst*, **111**(3), 269–275 (1986).

152. E. Bulska, E. Beinrohr, P. Tschöpel, J.A.C. Broekaert, G. Tölg, 'Determination of Arsenic, Antimony and Selenium After Preconcentration onto the Reticulated Vitreous Carbon with MIP AES Detection', *Chem. Anal. (Warsaw)*, **41**(4), 615–623 (1996).

153. H. Matusiewicz, R.E. Sturgeon, S.S. Berman, 'In Situ Hydride-generation Pre-concentration of Arsenic in a Graphite Furnace with Sample Vaporization into a Microwave-induced Plasma for Emission Spectrometry', *Spectrochim. Acta, Part B*, **45**(1–2), 209–214 (1990).

154. N.W. Barnett, L.S. Chen, G.F. Kirkbright, 'Rapid Determination of Arsenic by Optical Emission Spectroscopy Using a Microwave-induced Plasma Source and a Miniature Hydride-generation Device', *Spectrochim. Acta, Part B*, **39**(9–11), 1141–1147 (1984).

155. N.W. Barnett, 'Further Experience with a Miniature Hydride Generation Device Used in Conjunction with Microwave-induced Plasma Atomic-emission Spectrometry (MIP/AES) for the Determination of Antimony, Arsenic, Lead and Tin', *Spectrochim. Acta, Part B*, **42**(6), 859–864 (1987).

156. K.C. Ng, X. Xu, M.J. Brechmann, 'Direct, Continuous Hydride Generation Coupled with Microwave-induced Plasma Atomic-emission Spectrometry for the Determination of Selenium', *Spectrosc. Lett.*, **22**(9), 1251–1262 (1989).

157. F. Lunzer, R. Pereiro-Garcia, N. Bordel-Garcia, A. Sanz-Medel, 'Continuous Hydride Generation Low-pressure Microwave-induced Plasma Atomic Emission Spectrometry for the Determination of Arsenic, Antimony and Selenium', *J. Anal. At. Spectrom.*, **10**(4), 311–315 (1995).

158. H. Matusiewicz, Z. Kurzawa, 'Electrothermal-vaporization Microwave-induced Plasma Atomic-emission Spectrometric Determination of Arsenic and Selenium in Sulfur Following Teflon (PTFE) Bomb Microwave Acid Dissolution and In Situ Hydride Generation', *Acta Chim. Hung.*, **128**(3), 401–410 (1991).

159. K.C. Ng, W. Shen, 'Direct, Continuous Lead Hydride Generation into a Low-power Helium–Hydrogen Plasma for Atomic-emission Spectrometry', *Spectroscopy*, **2**(1), 50–53 (1987).

160. K.J. Timmins, 'Excitation of Gallium, Indium, Selenium, Tellurium, Arsenic and Antimony Microwave-induced Plasma', *J. Anal. At. Spectrom.*, **2**(2), 251–252 (1987).

161. J.M. Costa-Fernandez, R. Pereiro-Garcia, A. Sanz-Medel, N. Bordel-Garcia, 'Effect of Plasma Pressure on the Determination of Mercury by Microwave-induced Plasma Atomic-emission Spectrometry', *J. Anal. At. Spectrom.*, **10**(9), 649–653 (1995).

162. T. Nakahara, K. Nakanishi, T. Walsa, 'Atmospheric-pressure Helium Microwave-induced Plasma Atomic-emission Spectrometry for Determination of Mercury', *Chem. Express*, **1**(3), 149–152 (1986).

163. K. Tanabe, K. Chiba, H. Haraguchi, K. Fuwa, 'Determination of Mercury at the Ultra-trace Level by Atmospheric-pressure Helium Microwave-induced Plasma Emission Spectrometry', *Anal. Chem.*, **53**(9), 1450–1453 (1981).

164. T. Nakahara, T. Wasa, 'Feasibility Study of Indirect Determination of Iodine by Helium Microwave-induced Plasma Atomic-emission Spectrometry of Mercury', *Chem. Express*, **4**(8), 495–498 (1989).

165. T. Nakahara, T. Walsa, 'Indirect Determination of Iodine in Seawater and Brine by Atmospheric-pressure Helium Microwave-induced Plasma Atomic-emission Spectrometry Using Continuous-flow Cold-vapor Generation of Mercury', *Microchem. J.*, **41**(2), 148–155 (1990).

166. F. Camuna, J.E. Sanchez-Uria, A. Sanz-Medel, 'Continuous-flow and Flow Injection Halogen Generation for Chloride, Bromide and Iodide Determinations by Microwave-induced Plasma Atomic-emission Spectroscopy', *Spectrochim. Acta, Part B*, **48**(9), 1115–1125 (1993).

167. J.F. Alder, Q. Jin, R.D. Snook, 'Determination of Chloride by Microwave Helium-plasma Emission Spectrometry Using a Hydrogen Chloride-generation Technique', *Anal. Chim. Acta*, **123**, 329–333 (1981).

168. J.F. Alder, Q. Jin, R.D. Snook, 'Determination of Traces of Chloride in Solution by Microwave-induced Plasma Emission Spectrometry Using Chlorine Generation', *Anal. Chim. Acta*, **120**, 147–154 (1980).

169. M.D. Calzada, M.C. Quintero, A. Gamero, M. Gallego, 'Chemical Generation of Chlorine, Bromine and Iodine for Sample Introduction into a Surfatron-generated Argon Microwave-induced Plasma' *Anal. Chem.*, **64**(13), 1374–1378 (1992).

170. T. Nakahara, S. Morimoto, T. Wasa, 'Analyte Volatilization Procedure for Continuous-flow Determination of Bromine by Atmospheric-pressure Helium Microwave-induced Plasma Atomic-emission Spectrometry', *J. Anal. At. Spectrom.*, **7**(2), 211–218 (1992).

171. M.D. Calzada, M.C. Quintero, A. Gamero, J. Cotrino, J.E. Sanchez-Uria, A. Sanz-Medel, 'Determination of Bromide by Low Power Surfatron Microwave-induced Plasma After Bromine Continuous Generation', *Talanta*, **39**(4), 341–347 (1992).

172. M.M. Abdillahi, R.D. Snook, 'Determination of Bromide Using a Helium Microwave-induced Plasma with Bromine Generation and Electrothermal Vaporization for Sample Introduction', *Analyst*, **111**(3), 265–267 (1986).

173. M.M. Abdillahi, W. Tschanen, R.D. Snook, 'Microwave-induced Plasma Emission-spectrometric Determination of Bromide', *Anal. Chim. Acta*, **172**, 139–145 (1985).
174. M.C. Quintero-Ortega, J. Cotrino-Bautista, M. Saez, A. Menendez-Garcia, J.E. Sanchez-Uria, A. Sanz-Medel, 'Determination of Iodide by Low Power Surfatron Microwave-induced Plasma After Iodine Continuous Generation', *Spectrochim. Acta, Part B*, **47**(1), 79–87 (1992).
175. T. Nakahara, T. Wasa, 'Effect of Acidity and Reductant Concentration on the Indirect Determination of Iodine by Helium Microwave-induced Plasma Atomic-emission Spectrometry', *Chem. Express*, **5**(3), 121–124 (1990).
176. T. Nakahara, S. Yamada, T. Wasa, 'Selective Determination of Iodide and Iodate in Brine Waters by Atmospheric Pressure Helium Microwave-induced Plasma Atomic-emission Spectrometry with Continuous-flow Generation of Volatile Iodine', *Chem. Express*, **6**(1), 5–8 (1991).
177. S.R. Koirtyohann, 'Effect of Nitrogen Impurity on Fluorine and Chlorine Emission from an Atmospheric-pressure Helium Microwave Plasma', *Anal. Chem.*, **55**(2), 374–376 (1983).
178. N.W. Barnett, 'Improvements in the Chemical Generation of Chlorine and Bromine, and their Respective Hydrides, as a Means of Sample Introduction into an Atmospheric-pressure Helium Microwave-induced Plasma', *J. Anal. At. Spectrom.*, **3**(7), 969–972 (1988).
179. J. Alvarado, J.W. Carnahan, 'Direct Detection of Vacuum-ultra-violet Radiation for Non-metal Determinations with a Helium Microwave-induced Plasma', *Appl. Spectrosc.*, **47**(12), 2036–2043 (1993).
180. T. Nakahara, T. Mori, S. Morimoto, H. Ishikawa, 'Continuous-flow Determination of Aqueous Sulfur by Atmospheric-pressure Helium Microwave-induced Plasma Atomic-emission Spectrometry with Gas-phase Sample Introduction', *Spectrochim. Acta, Part B*, **50**(4–7), 393–403 (1995).
181. W. Drews, G. Weber, G. Tölg, 'Trace Determination of Nickel by Microwave-induced Plasma Atomic-emission Spectrometry After Pre-concentration of Nickel Tetracarbonyl on Chromosorb®', *Anal. Chim. Acta*, **231**(2), 265–271 (1990).
182. W. Drews, G. Weber, G. Tölg, 'Flow Injection System for the Determination of Nickel by Means of MIPOES [Microwave-induced Plasma Optical Emission Spectrometry] After Conversion to Nickel Tetracarbonyl', *Fresenius' J. Anal. Chem.*, **332**(8), 862–865 (1989).
183. G.K. Webster, J.W. Carnahan, 'Non-metal Analyte Introduction Device for Atomic Spectrometry', *Anal. Chem.*, **61**(1), 790–793 (1989).
184. J.M. Keane, D.C. Brown, R.C. Fry, 'Red and Near-infrared Photodiode Array Atomic-emission Spectrograph for the Simultaneous Determination of Carbon, Hydrogen, Nitrogen and Oxygen', *Anal. Chem.*, **57**(13), 2526–2533 (1985).
185. D.B. Hooker, J. DeZwaan, 'Organic Micro-analysis of Sub-microgram Samples', *J. Res. Natl. Bur. Stand.*, **93**(3), 245–249 (1988).
186. K. Tanabe, K. Matsumoto, H. Haraguchi, K. Fuwa, 'Determination of Ultra-trace Ammonium-, Nitrite- and Nitrate-nitrogen by Atmospheric-pressure Helium Microwave-induced Plasma Emission Spectrometry with Gas-generation Technique', *Anal. Chem.*, **52**(14), 2361–2365 (1980).
187. F.P. Schwarz, 'Characterization of Emission from Atomic Hydrogen in a Microwave-induced Plasma', *Anal. Chem.*, **51**(9), 1508–1512 (1979).
188. J.A.C. Broekaert, P.W.J.M. Boumans, in *Inductively Coupled Plasma Emission Spectroscopy*, ed. P.W.J.M. Boumans, Wiley, New York, 296–357, Chapter 6, Vol. 1, 1987.
189. H. Berndt, 'High-pressure Nebulization: A New Way of Sample Introduction for Atomic Spectroscopy', *Fresenius' Z. Anal. Chem.*, **331**(3–4), 321–323 (1988).
190. G. Heltai, T. Jozsa, K. Percsich, 'A Possibility of Element-specific Detection in HPLC by Means of MIP/AES Coupled with Hydraulic High-pressure Nebulization', *Fresenius' J. Anal. Chem.*, **355**(5–6), 638–641 (1996).
191. V.A. Fassel, B.R. Bear, 'Ultrasonic Nebulization of Liquid Samples for Analytical Inductively Coupled Plasma–Atomic Spectroscopy: An Update', *Spectrochim. Acta, Part B*, **41**(10), 1089–1113 (1986).
192. R.G. Stahl, K.J. Timmins, 'Use of a Glass-frit Nebulizer with a Helium Microwave-induced Plasma', *J. Anal. At. Spectrom.*, **2**(6), 557–559 (1987).
193. H. Matusiewicz, 'Use of the Hildebrand Grid Nebulizer as a Sample-introduction System for Microwave-induced Plasma Spectrometry', *J. Anal. At. Spectrom.*, **8**(7), 961–964 (1993).
194. H. Matusiewicz, R.W. Sturgeon, 'Slurry Sample Introduction with Microwave-induced Plasma Atomic-emission Spectrometry', *Spectrochim. Acta, Part B*, **48**(5), 723–727 (1993).
195. D. Kollotzek, D. Oechsle, G. Kaiser, P. Tschöpel, G. Tölg, 'Application of a Mixed-gas Microwave-induced Plasma as an On-line Element-specific Detector in High-performance Liquid Chromatography', *Fresenius' Z. Anal. Chem.*, **318**(7), 485–489 (1984).
196. G.L. Long, L.D. Perkins, 'Direct Introduction of Aqueous Samples into a Low-powered Microwave-induced Plasma for Atomic-emission Spectrometry', *Appl. Spectrosc.*, **41**(6), 980–985 (1987).
197. P.G. Brown, D.L. Haas, J.M. Workman, J.A. Caruso, F.L. Fricke, 'Moderate-power Argon Microwave-induced Plasma for the Detection of Metal Ions in Aqueous Samples of Complex Matrix', *Anal. Chem.*, **59**(10), 1433–1436 (1987).
198. K.B. Olsen, D.S. Sklarew, J.C. Evans, 'Detection of Organomercury, Selenium and Arsenic Compounds by a Capillary-column Gas Chromatography–Microwave

Plasma Detector System', *Spectrochim. Acta, Part B*, **40**(1–2), 357–365 (1985).

199. V.P. Baluda, L.M. Filimonov, 'Effect of Accompanying Components on Spectrographic Analyses with Use of the PVS-1 Microwave Plasmatron', *Zh. Anal. Khim.*, **35**(6), 1061–1073 (1980).
200. F. Liang, D.X. Zhang, Y.H. Lei, H.Q. Zhang, Q.H. Jin, 'Determination of Selected Noble Metals by MPT/AES Using a Pneumatic Nebulizer', *Microchem. J.*, **52**(2), 181–187 (1995).
201. K.G. Michlewicz, J.W. Carnahan, 'Determination of Aqueous Bromide, Iodide and Chloride with Pneumatic and Ultrasonic Nebulization into a Helium Microwave-induced Plasma', *Anal. Chem.*, **58**(14), 3122–3125 (1986).
202. E. Bulska, J.A.C. Broekaert, P. Tschöpel, G. Tölg, 'Comparative Study of Argon and Helium Plasmas in a TM_{010} Cavity and a Surfatron and Their Use for Hydride-generation Microwave-induced-plasma Atomic-emission Spectrometry', *Anal. Chim. Acta*, **276**(2), 377–384 (1993).
203. S. Kirschner, A. Golloch, U. Teigheder, 'First Investigations for the Development of a Microwave-induced Plasma Atomic-emission Spectrometry System to Determine Trace Metals in Gases', *J. Anal. At. Spectrom.*, **9**(9), 971–974 (1994).
204. G.K. Webster, J.W. Carnahan, 'Improved Supercritical-fluid Chromatography Mobile-phase Tolerance with a Moderate-power Helium Microwave-induced Plasma', *Appl. Spectrosc.*, **45**(8), 1285–1290 (1991).
205. Q.H. Jin, H.Q. Zhang, W.J. Yang, Q. Jin, Y.H. Shi, 'Determination of Trace Silver, Gold, Germanium, Lead, Tin and Tellurium by Microwave Plasma Torch Atomic Emission Spectrometry Coupled with an Electrothermal Vaporization Sample Introduction System', *Talanta*, **44**(9), 1605–1614 (1997).
206. A.H. Ali, J.D. Winefordner, 'Micro-sample Introduction by Tungsten-filament Electrode into Capacitively Coupled Microwave Plasma for Atomic-emission Spectroscopy: Analytical Figures of Merit', *Anal. Chim. Acta*, **264**(2), 327–332 (1992).
207. A.H. Ali, J.D. Winefordner, 'Micro-sample Introduction by Tungsten-filament Electrode into Capacitively Coupled Microwave Plasma for Atomic-emission Spectroscopy: Diagnostics', *Anal. Chim. Acta*, **262**(2), 319–325 (1992).
208. Q.H. Jin, H.Q. Zhang, Y. Wang, X.L. Yuan, W.J. Yang, 'Study of Analytical Performance of a Low-powered Microwave Plasma Torch in Atomic-emission Spectrometry', *J. Anal. At. Spectrom.*, **9**(8), 851–856 (1994).
209. Y.N. Pak, S.R. Koirtyohann, 'Fundamental Characterization of the Moderate-power Helium Microwave-induced Plasma', *Spectrochim. Acta, Part B*, **49**(6), 593–606 (1994).
210. Y. Okamoto, M. Yasuda, S. Murayama, 'High-power Microwave-induced Plasma Source for Trace-element Analysis', *Jpn. J. Appl. Phys., Part 2*, **29**(4), L670–L672 (1990).
211. D.L. Haas, J.A. Caruso, 'Characterization of a Moderate-power Microwave-induced Plasma for Direct Solution Nebulization of Metal Ions', *Anal. Chem.*, **56**(12), 2014–2019 (1984).
212. K.C. Ng, R.C. Culp, 'Direct Injection Nebulizer-coupled Microwave-induced Plasma for Atomic-emission Spectroscopy', *Appl. Spectrosc.*, **51**(10), 1447–1452 (1997).
213. M. Wu, Y. Madrid, J.A. Auxier, G.M. Hieftje, 'New Spray Chamber for Use in Flow Injection Plasma Emission Spectrometry', *Anal. Chim. Acta*, **286**(2), 155–167 (1994).
214. H.Q. Zhang, X.L. Yuan, X.J. Zhao, Q.H. Jin, 'On-line Preconcentration with Activated Carbon for Microwave Plasma Torch Atomic-emission Spectrometry', *Talanta*, **44**(9), 1615–1623 (1997).
215. J. Alvarado, M.G. Wu, J.W. Carnahan, 'Electrothermal Vaporization and Ultrasonic Nebulization for the Determination of Aqueous Sulfur Using a Kilowatt-plus Helium Microwave-induced Plasma', *J. Anal. At. Spectrom.*, **7**(8), 1253–1256 (1992).
216. M. Wu, J.W. Carnahan, 'Direct Determination of Aqueous Carbon, Phosphorus and Sulfur Using a Kilowatt-plus Helium Microwave-induced Plasma System with Ultrasonic Nebulization', *J. Anal. At. Spectrom.*, **7**(8), 1249–1252 (1992).
217. Q. Jin, C. Zhu, K. Brushwyler, G.M. Hieftje, 'Efficient and Inexpensive Ultrasonic Nebulizer for Atomic Spectrometry', *Appl. Spectrosc.*, **44**(2), 183–186 (1990).
218. J.M. Gehlhausen, J.W. Carnahan, 'Determination of Aqueous Fluoride with a Helium Microwave-induced Plasma and Flow Injection Analysis', *Anal. Chem.*, **61**(7), 674–677 (1989).
219. C.R. Jia, X.H. Wang, Q.Y. Ou, 'Microwave-induced Plasma Atomic-emission Spectrometer Used as the Detector for Online Coupled High-performance Liquid Chromatography–Gas Chromatography', *Anal. Commun.*, **34**(2), 53–56 (1997).
220. G.L. Long, G.R. Ducatte, E.D. Lancaster, 'Helium Microwave-induced Plasmas for Element Specific Detection in Chromatography', *Spectrochim. Acta, Part B*, **49**(1), 75–87 (1994).
221. J.P.J. van Dalen, P.A. de Lezenne-Coulander, L. de Galan, 'Optimization of the Microwave-induced Plasma as an Element-selective Detector for Non-metals', *Anal. Chim. Acta*, **94**(1), 1–19 (1977).
222. N.P. Vela, L.K. Olson, J.A. Caruso, 'Elemental Speciation with Plasma Mass Spectrometry', *Anal. Chem.*, **65**(13), 585A–597A (1993).
223. R. Lobinski, 'Elemental Speciation and Coupled Techniques', *Appl. Spectrosc.*, **51**(7), 260A–278A (1997).
224. D. Deruaz, J.M. Mermet, 'Gas Chromatography–Microwave-induced Plasma Atomic-emission Spectrometry', *Analusis*, **14**(3), 107–118 (1986).

225. P.C. Uden, 'Element-selective Chromatographic Detection by Plasma Atomic-emission Spectroscopy', *Trends Anal. Chem.*, **6**(9), 238–246 (1987).
226. P.C. Uden, Y. Yoo, T. Wang, Z. Cheng, 'Element-selective Gas-chromatographic Detection by Atomic Plasma Emission Spectroscopy. Review and Developments', *J. Chromatogr.*, **468**, 319–328 (1989).
227. W. Yu, 'Development and Application of a Microwave-induced Plasma Emission-spectrometric Detector for Gas Chromatography in China', *J. Anal. At. Spectrom.*, **3**(6), 893–900 (1988).
228. P.C. Uden, 'Element-specific Chromatographic Detection by Atomic Absorption, Plasma Atomic Emission and Plasma Mass Spectrometry', *J. Chromatogr. A*, **703**(1–2), 393–416 (1995).
229. J. Hobbs, 'Microwave Plasma Detection [in Gas Chromatography] with the MPD 80', *Eur. Spectrosc. News*, **26**, 40 (1979).
230. J.C. Evans, K.B. Olsen, D.S. Sklarew, 'Low-pressure Beenakker-type Microwave-induced Helium Plasma Source as a Simultaneous Multi-element Gas-chromatographic Detector', *Anal. Chim. Acta*, **194**, 247–260 (1987).
231. K.J. Slatkavitz, P.C. Uden, L.D. Hoey, R.M. Barnes, 'Atmospheric-pressure Microwave-induced Helium Plasma Spectroscopy for Simultaneous Multi-element Gas-chromatographic Detection', *J. Chromatogr.*, **302**, 277–287 (1984).
232. B.D. Quimby, J.J. Sullivan, 'Evaluation of a Microwave Cavity, Discharge Tube and Gas Flow System for Combined Gas Chromatography–Atomic Emission Detection', *Anal. Chem.*, **62**(10), 1027–1034 (1990).
233. T. Bandemer, 'Element-selective GC Detection by AES', *Int. Lab.*, **20**(8), 28–35 (1990).
234. R.L. Firor, 'Multi-element Detection Using GC–Atomic-emission Spectroscopy', *Int. Lab.*, **19**(7), 44–52 (1989).
235. K. Tanabe, H. Haraguchi, K. Fuwa, 'Application of an Atmospheric-pressure Helium Microwave-induced Plasma as an Element-selective Detector for Gas Chromatography', *Spectrochim. Acta, Part B*, **36**(7), 633–639 (1981).
236. C.F. Bauer, D.F.S. Natusch, 'Speciation at Trace Levels by Helium Microwave-induced Plasma Emission Spectrometry', *Anal. Chem.*, **53**(13), 2020–2027 (1981).
237. E. Bulska, 'Microwave-induced Plasma as an Element-specific Detector for Speciation Studies at the Trace Level', *J. Anal. At. Spectrom.*, **7**(2), 201–210 (1992).
238. Y. Takigawa, T. Hanai, J. Hubert, 'Microwave-induced Plasma Emission Spectrophotometer Combined with a Photodiode-array Monitor for Capillary-column Gas Chromatography', *J. High Resolut. Chromatogr. Chromatogr. Commun.*, **9**(11), 698–702 (1986).
239. C.S. Cerbus, S.J. Gluck, 'Design, Optimization and Utilization of a Microwave-induced Helium-plasma Gas-chromatographic Detector in an Industrial Laboratory', *Spectrochim. Acta, Part B*, **38**(1–2), 387–397 (1983).
240. Q.Y. Ou, G.C. Wang, K.W. Zeng, W.L. Yu, 'Investigation of a Microwave-plasma Emission Spectrometer as a Quantitative Detector for Glass-capillary Gas Chromatography', *Spectrochim. Acta, Part B*, **38**(1–2), 419–425 (1983).
241. A.L. Pires-Valente, P.C. Uden, 'Comparison of a Combined Helium–Argon Plasma with Pure Helium Plasmas for Gas Chromatography with Atomic-emission Detection', *Analyst*, **120**(2), 419–421 (1995).
242. G. Chevrier, T. Hanai, K.C. Tran, J. Hubert, 'Development of a Microwave Plasma Detector for Gas Chromatography', *Can. J. Chem.*, **60**(7), 898–903 (1982).
243. K.A. Forbes, E.E. Reszke, P.C. Uden, R.M. Barnes, 'Comparison of Microwave-induced Plasma Sources', *J. Anal. At. Spectrom.*, **6**(1), 57–71 (1991).
244. R.L.A. Sing, J. Hubert, 'Noise Characterization of a 'Surfatron' Microwave-induced Plasma and the Implications for Fourier-transform-based Detection in GC–Microwave-induced Plasma AES', *Appl. Spectrosc.*, **44**(10), 1605–1612 (1990).
245. J.F. Camuna-Aguilar, R. Pereiro-Garcia, J.E. Sanchez-Uria, A. Sanz-Medel, 'Comparative Study of Three Microwave-induced Plasma Sources for Atomic-emission Spectrometry. II. Evaluation of Their Atomization/Excitation Capabilities for Chlorinated Hydrocarbons', *Spectrochim. Acta, Part B*, **49**(6), 545–554 (1994).
246. R.M. Alvarez-Bolainez, M.P. Dziewatkoski, C.B. Boss, 'Sensitivity Comparison in a Microwave-induced Plasma Gas-chromatographic Detector: Effect of Plasma Torch Design', *Anal. Chem.*, **64**(5), 541–544 (1992).
247. M.L. Bruce, J.A. Caruso, 'Laminar-flow Torch for Gas-chromatographic Helium Microwave Plasma Detection of Pyrethroids and Dioxins', *Appl. Spectrosc.*, **39**(6), 942–949 (1985).
248. G.W. Jansen, F.A. Huf, H.J. De Jong, 'Low-power Microwave Plasma Source for Chromatography Detection', *Spectrochim. Acta, Part B*, **40**(1–2), 307–316 (1985).
249. F.A. Huf, G.W. Jansen, 'Microwave-induced Plasma Between Electrodes as a Detector for Gas and Liquid Chromatography', *Spectrochim. Acta, Part B*, **38**(7), 1061–1064 (1983).
250. K. Cammann, L. Lendero, H. Feuerbacher, K. Ballschmiter, 'Power-modulated Microwave-induced Plasma with Enhanced Sensitivity and Practicability as an Element-specific GC Detector', *Fresenius' Z. Anal. Chem.*, **316**(2), 194–200 (1983).
251. D.L. Haas, J.A. Caruso, 'Moderate-power Helium Plasma as an Element-selective Detector for Gas Chromatography of Dioxins and Other Halogenated Compounds', *Anal. Chem.*, **57**(3), 846–851 (1985).

252. R.M. Alvarez-Bolainez, C.B. Boss, 'Microwave-induced-plasma Reflected-power Detector for Gas Chromatography', *Anal. Chem.*, **63**(2), 159–163 (1991).
253. J.E. Freeman, G.M. Hieftje, 'Near-infrared Nonmetal Atomic Emission from a Helium Microwave-induced Plasma: Element Ratio Determinations', *Spectrochim. Acta, Part B*, **40**(4), 653–664 (1985).
254. D.E. Pivonka, W.G. Fateley, R.C. Fry, 'Simultaneous Determination of Carbon, Hydrogen, Nitrogen, Oxygen, Fluorine, Chlorine, Bromine and Sulphur in Gas-chromatographic Effluent by Fourier-transform Red–Near-infrared Atomic-emission Spectroscopy', *Appl. Spectrosc.*, **40**(3), 291–297 (1986).
255. C. Lauzon, K.C. Tran, J. Hubert, 'Multi-wavelength Detection in Gas Chromatography with Microwave-induced Plasma Atomic-emission Fourier Transform Spectrometry', *J. Anal. At. Spectrom.*, **3**(6), 901–905 (1988).
256. D. Rieping, J. Bettmer, W. Buscher, K. Cammann, 'The Development of a Simple Gas Chromatography Detector Based on Atomic-emission Spectrophotometry', *GIT Fachz. Lab.*, **39**(2), 95–98 (1995).
257. S.R. Koirtyohann, 'Effect of Nitrogen Impurity on Fluorine and Chlorine Emission from an Atmospheric-pressure Helium Microwave Plasma', *Anal. Chem.*, **55**(2), 374–376 (1983).
258. V.O. Schmitt, I. Rodriguez-Pereiro, R. Lobinski, 'Flash Species-selective Analysis of Multicapillary Gas Chromatography with Microwave-induced-plasma Atomic-spectrometric Detection', *Anal. Commun.*, **34**(5), 141–143 (1997).
259. T.N. Asp, S. Pedersen Bjergaard, T. Greibrokk, 'Calculation of Elemental Ratios by On-column Radio-frequency Plasma Atomic-emission Detection Coupled with Capillary Gas Chromatography', *J. Chromatogr. A*, **736**(1–2), 157–164 (1996).
260. M.L. Mena, C.W. McLeod, 'Mercury Species Immobilized on Sulphydryl Cotton: A New Candidate Reference Material for Mercury Speciation', *Mikrochim. Acta*, **123**(1–4), 103–108 (1996).
261. P. Lansens, W. Baeyens, M. Termonia, 'Modification of the Perkin-Elmer HS-6 Headspace Sampler for the Determination of Organomercury Compounds', *J. High Resolut. Chromatogr. Chromatogr. Commun.*, **12**(3), 132–133 (1989).
262. M.E. Birch, 'Solvent-venting Technique for Gas Chromatography with Microwave-induced Plasma Atomic-emission Spectrometry', *Anal. Chim. Acta*, **282**(2), 451–458 (1993).
263. S.W. Jordan, B.L. Karger, I.S. Krull, S.B. Smith Jr, 'All-glass-lined, Open-split, Solvent-venting Interface for Gas Chromatography–Microwave-induced Plasma Emission Spectroscopy (GC/MIP)', *Chem. Biomed. Environ. Instrum.*, **12**(4), 263–274 (1983).
264. L. Zhang, J.W. Carnahan, R.E. Winans, P.H. Neill, 'Solvent-venting Interface for Capillary Gas Chromatography and a Microwave-induced Plasma', *Anal. Chim. Acta*, **233**(1), 149–154 (1990).
265. W.E. Wentworth, T. Limero, C.F. Batten, E.C.M. Chen, 'Non-radioactive Electron-capture Detector for Gas Chromatography', *J. Chromatgr.*, **468**(5), 215–224 (1989).
266. R. Lobinski, F.C. Adams, 'Application of Capillary Gas Chromatography with Atomic Emission Detection (CGC–AED) in Ultratrace Speciation Analysis of Organometals and Metalloids', *ICP Inf. Newsl.*, **19**(3), 164–166 (1993).
267. A.L.P. Valente, P.C. Uden, 'Memory Effects in Gas-chromatographic Detection with a Microwave-induced Plasma', *J. High Resolut. Chromatogr.*, **16**(5), 275–278 (1993).
268. M. Caetano, R.E. Golding, E.A. Key, 'Factorial Analysis and Response Surface of a Gas Chromatography Microwave-induced Plasma System for the Determination of Halogenated Compounds', *J. Anal. At. Spectrom.*, **7**(6), 1007–1011 (1992).
269. H. Yieru, O. Qingyu, Y. Weile, 'Effect of Compound Structure on the Elemental Responses in Gas Chromatography–Microwave-induced Plasma Atomic-emission Spectrometry', *J. Chromatogr. Sci.*, **28**(11), 584–588 (1990).
270. S.R. Goode, L.K. Kimbrough, 'Experimental Study of the Signal-to-noise Ratio in the Microwave-induced Plasma Gas-chromatographic Detector', *Spectrochim. Acta, Part B*, **42**(1–2), 309–322 (1987).
271. K. Cammann, H. Mueller, 'Reduced Calibration Efforts in Gas Chromatography by the Use of an Element-selective Plasma-emission Detector', *Fresenius' Z. Anal. Chem.*, **331**(3–4), 336–341 (1988).
272. D.F. Hagen, J.S. Marhevka, L.C. Haddad, 'Multi-element Taggant and Dual Homolog Derivatization Concepts in G.C.–MED [Microwave-sustained Helium Emission Detector]', *Spectrochim. Acta, Part B*, **40**(1–2), 335–347 (1985).
273. J.H. Brill, B.A. Narayanan, J.P. McCormick, 'Selective Determination of Pentafluorobenzyl Ester Derivatives of Carboxylic Acids by GC Using Microwave Plasma and Mass Selective Detection', *Appl. Spectrosc.*, **45**(10), 1617–1620 (1991).
274. D. Deruaz, J.L. Brazier, 'Applications of Gas Chromatography–Helium Microwave Plasma Detection', *Spectra 2000*, **161**, 8–13 (1991).
275. C. Webster, M. Cooke, 'Use of Microwave-induced Plasma Atomic-emission Detection for the Quantification of Oxygen Containing Compounds', *Anal. Proc.*, **31**(8), 237–240 (1994).
276. K. Zeng, Q. Ou, G. Wang, W. Yu, 'Analysis of Oxygenated Compounds by Gas Chromatography–Microwave

Plasma Emission Spectrometry', *Spectrochim. Acta, Part B*, **40**(1–2), 349–356 (1985).
277. S.R. Goode, C.L. Thomas, 'Characterizing the Factors that Influence Oxygen Selectivity in Gas Chromatography–Microwave-induced Plasma Atomic-emission Spectrometry', *J. Anal. At. Spectrom.*, **9**(9), 965–970 (1994).
278. S.R. Goode, L.K. Kimbrough, 'Study of the Factors Influencing the Detection of Oxygen [in Organic Compounds] with Gas Chromatography–Microwave-induced Plasma Atomic-emission Spectrometry', *J. Anal. At. Spectrom.*, **3**(6), 915–918 (1988).
279. K.J. Slatkavitz, P.C. Uden, R.M. Barnes, 'Consideration of an Atmospheric-pressure Microwave-induced Helium Plasma as an Oxygen-selective Gas-chromatographic Detector', *J. Chromatogr.*, **355**(1), 117–126 (1986).
280. P. Stilkenboehmer, K. Cammann, 'Plasma Emission GC Detector for Selective Determination of Hydrogen-containing Organic Compounds', *Fresenius' Z. Anal. Chem.*, **335**(7), 764–768 (1989).
281. D.F. Hagen, L.C. Haddad, J.S. Marhevka, 'Solution and Gas-phase Deuterium Derivatization Reactions for Gas Chromatography–Microwave Emission Detection', *Spectrochim. Acta, Part B*, **42**(1–2), 253–267 (1987).
282. S.R. Goode, J.J. Gemmill, B.E. Watt, 'Determination of Deuterium by Gas Chromatography with a Microwave-induced Plasma Emission Detector', *J. Anal. At. Spectrom.*, **5**(6), 483–486 (1990).
283. A. Bannier, D. Deruz, C. Weber, J.L. Brazier, 'Analysis of Caffeine Deutero-isotopomers by Gas Chromatography and Atomic-emission Detection', *Anal. Lett.*, **25**(6), 1073–1085 (1992).
284. F. Leclerc, D. Deruaz, A. Bannier, J.L. Brazier, 'Limit of Detection of Carbon-13 Using an Atomic-emission Detector (MIP) Coupled to Gas Chromatography', *Anal. Lett.*, **27**(7), 1325–1338 (1994).
285. J.T. Jelink, A. Venema, 'Investigations into the Use of Capillary GC with Atomic-emission Detection', *J. High Resolut. Chromatogr.*, **13**(6), 447–450 (1990).
286. H.A. Dingjan, H.J. De Jong, 'Determination of Ratio Formulae for Organic Compounds Using a Microwave-induced Plasma', *Spectrochim. Acta, Part B*, **38**(5–6), 777–781 (1983).
287. H.J. Perpall, P.C. Uden, R.L. Deming, 'Empirical and Molecular Formula Determination of Polymer Pyrolysates by Multi-referencing GC–Microwave-induced Plasma Spectroscopy', *Spectrochim. Acta, Part B*, **42**(1–2), 243–251 (1987).
288. L.G. Sarto Jr, S.A. Estes, P.C. Uden, S. Siggia, R.M. Barnes, 'Fused-silica Capillary GC with Boron-specific Microwave-excited Plasma Detection in Derivatization and Polymer Pyrolysis', *Anal. Lett.*, **14**(3), 205–218 (1981).
289. J.J. Sullivan, B.D. Quimby, 'Detection of Carbon, Hydrogen, Nitrogen and Oxygen by Atomic-emission in Capillary Gas Chromatography', *Chem. Labor. Betr.*, **41**(4), 200–207 (1990).
290. H. Mueller, K. Cammann, 'Development of a Plasma Emission Gas-chromatographic Detector for the Simultaneous Selective Detection of Chlorine, Bromine and Carbon', *J. Anal. At. Spectrom.*, **3**(6), 907–913 (1988).
291. P.C. Uden, K.J. Slatkavitz, R.M. Barnes, R.L. Deming, 'Empirical and Molecular Formula Determination by Gas Chromatography–Microwave-induced Plasma Atomic-emission Spectrometry', *Anal. Chim. Acta*, **180**, 401–416 (1986).
292. N. Kovacic, T.L. Ramus, 'Application of a Microwave-induced Plasma Atomic-emission Detector for Quantification of Halogenated Compounds by Gas Chromatography', *J. Anal. At. Spectrom.*, **7**(6), 999–1005 (1992).
293. J.H. Brill, B.A. Narayanan, J.P. Doom, J.P. McCormick, 'Selective Determination of Organofluorine Compounds by Capillary Column Gas Chromatography with an Atmospheric Pressure Helium Microwave-injected Plasma Detector', *J. High Resolut. Chromatogr. Chromatogr. Commun.*, **11**(5), 368–374 (1988).
294. H. Schimmel, B. Schmidt, R. Bacher, K. Ballschmiter, 'Molar Response of Polychlorinated Dibenzo-*p*-dioxins and Dibenzofurans by the Mass-spectrometric Detector', *Anal. Chem.*, **65**(5), 640–644 (1993).
295. C. Webster, M.J. Cooke, 'Use of an Atomic-emission Detector to Study the Variation in Elemental Response for Chlorine, Carbon and Oxygen in Phenols', *J. High Resolut. Chromatogr.*, **18**(5), 319–322 (1995).
296. Y. Huang, Q. Ou, W. Yu, 'Study of gas Chromatography–Microwave-induced Plasma Atomic-emission Spectrometry. I. Effect of the Structure of a Compound on the Determination of its Empirical Formula', *J. Anal. At. Spectrom.*, **5**(2), 115–120 (1990).
297. M.M. Abdillahi, 'Gas Chromatography–Microwave-induced Plasma for the Determination of Halogenated Hydrocarbons', *J. Chromatogr. Sci.*, **28**(12), 613–616 (1990).
298. A. Sarafraz-Yazdi, M.Y. Khuhawar, P.C. Uden, 'Gas Chromatography of Chloride and Bromide as Phenylboronic Acid–Mercuric Nitrate Derivatives with Microwave-induced Plasma Atomic-emission Detection', *J. Chromatogr.*, **594**(1–2), 395–399 (1992).
299. S.E. Eckert-Tilotta, S.B. Hawthorne, D.J. Miller, 'Comparison of Commercially Available Atomic-emission and Chemiluminescence Detectors for Sulphur-selective Gas Chromatographic Detection', *J. Chromatogr.*, **591**(1–2), 313–323 (1992).
300. B. Schmidt, J.T. Anderson, K. Ballschmiter, 'Sulfur-selective Detection in Gas Chromatography with a Microwave-induced Plasma Detector', *Fresenius' Z. Anal. Chem.*, **333**(7), 765–766 (1989).
301. J.A. Seeley, P.C. Uden, 'Element-selective Gas-chromatographic Detection and Determination of Organoboron Compounds (in Motor and Lubricating Oils)', *Analyst*, **116**(2), 1321–1326 (1991).

302. K.J. Slatkavitz, L.D. Hoey, P.C. Uden, R.M. Barnes, 'Element-specific Detection of Organosilicon Compounds by Gas Chromatography–Atmospheric-pressure Microwave-induced Helium Plasma Spectrometry', *Anal. Chem.*, **57**(9), 1846–1853 (1985).
303. G.B. Limentani, P.C. Uden, 'High-resolution Gas and Liquid Chromatography of Organoarsenic Compounds', *J. Chromatogr.*, **325**(1), 53–60 (1985).
304. H. Haraguchi, A. Takatsu, 'Derivative Gas Chromatography of Methylarsenic Compounds with Atmospheric-pressure Helium Microwave-induced Plasma Atomic-emission Spectrometric Detection', *Spectrochim. Acta, Part B*, **42**(1–2), 235–241 (1987).
305. C. Bradley, J.W. Carnahan, 'Oxygen-selective Microwave-induced Plasma Gas Chromatography Detector for Petroleum-related Samples', *Anal. Chem.*, **60**(9), 858–863 (1988).
306. S.R. Goods, C.L. Thomas, 'Determination of Oxygen-containing Additives in Gasoline by Gas Chromatography–Microwave-induced Plasma Atomic-emission Spectrometry', *J. Anal. At. Spectrom.*, **9**(2), 73–78 (1994).
307. M.P. Italia, P.C. Uden, 'Multiple-element Emission-spectral-detection Gas-chromatographic Profiles of Halogenated Products from Chlorination of Humic Acid and Drinking Water', *J. Chromatogr.*, **438**(1), 35–43 (1988).
308. B. Riviere, J.M. Mermet, D. Deruaz, 'Study of a Microwave-induced Plasma (Surfatron) as a Detector in Capillary-column Gas Chromatography with Reference to Pesticides', *J. Anal. At. Spectrom.*, **2**(7), 705–709 (1987).
309. A.H. Mohamad, M. Zerezghi, J.A. Caruso, 'Determination of Polychlorinated Dibenzo-*p*-dioxins Using Capillary Gas Chromatography with Microwave-induced Plasma Detection', *Anal. Chem.*, **58**(2), 469–471 (1986).
310. M.A. George, J.P. Hessler, J.W. Carnahan, 'Determination of Impurities in Argon by Gas Chromatography with a Microwave-induced Plasma Detector', *J. Anal. At. Spectrom.*, **4**(1), 51–54 (1989).
311. D.S. Ballantine Jr, W.H. Zoller, 'Collection and Determination of Volatile Organic Mercury Compounds in the Atmosphere by Gas Chromatography with Microwave Plasma Detection', *Anal. Chem.*, **56**(8), 1288–1293 (1984).
312. S.A. Estes, P.C. Uden, M.D. Rausch, R.M. Barnes, 'Fused-silica Capillary GC Separation and Element-selective Microwave Plasma-emission Detection of Volatile Organometallics', *J. High Resolut. Chromatogr. Chromatogr. Commun.*, **3**(9), 471–472 (1980).
313. R. Lobinski, F.C. Adams, 'Recent Advances in Speciation Analysis by Capillary Gas Chromatography –Microwave-induced-plasma Atomic-emission Spectrometry', *Trends Anal. Chem.*, **12**(2), 41–49 (1993).
314. R. Lobinski, W.M.R. Dirkx, M. Ceulemans, F.C. Adams, 'Optimization of Comprehensive Speciation of Organotin Compounds in Environmental Samples by Capillary Gas Chromatography–Helium Microwave-induced Plasma Emission Spectrometry', *Anal. Chem.*, **64**(2), 159–165 (1992).
315. S. Tutschku, S. Mothes, K. Dittrich, 'Determination and Speciation of Organotin Compounds by Gas Chromatography–Microwave-induced Plasma Atomic-emission Spectrometry', *J. Chromatogr. A.*, **683**(1), 269–276 (1994).
316. S.A. Estes, P.C. Uden, R.M. Barnes, 'Determination of *n*-butylated Trialkyl-lead Compounds by Gas Chromatography with Microwave Plasma-emission Detection', *Anal. Chem.*, **54**(14), 2402–2405 (1982).
317. S.A. Estes, P.C. Uden, R.M. Barnes, 'High-resolution Gas Chromatography of Trialkyl-lead Chlorides with an Inert-solvent-venting Interface for Microwave-excited Helium-plasma Detection', *Anal. Chem.*, **53**(9), 1336–1340 (1981).
318. G.M. Greenway, N.W. Barnett, 'Optimization of an Atmospheric-pressure Helium Microwave-induced Plasma Coupled with Capillary Gas Chromatography for the Determination of Alkyl-lead and Alkylmercury Compounds', *J. Anal. At. Spectrom.*, **4**(8), 783–787 (1989).
319. G. Decadt, W. Baeyens, D. Bradley, L. Goeyens, 'Determination of Methylmercury in Biological Samples by Semi-automated Headspace Analysis', *Anal. Chem.*, **57**(14), 2788–2791 (1985).
320. P. Lansens, C. Meuleman, M. Leermakers, W. Baeyens, 'Determination of Methylmercury in Natural Waters by Headspace Gas Chromatography with Microwave-induced Plasma Detection After Pre-concentration on a Resin Containing Dithiocarbamate Groups', *Anal. Chim. Acta*, **234**(2), 417–424 (1990).
321. P. Lansens, W. Baeyens, 'Improvement of the Semi-automated Headspace Analysis Method for the Determination of Methylmercury in Biological Samples', *Anal. Chim. Acta*, **228**(1), 93–99 (1990).
322. E. Bulska, D.C. Baxter, W. Frech, 'Capillary Column Gas Chromatography for Mercury Speciation (in Environmental and Biological Materials)', *Anal. Chim. Acta*, **249**, 545–554 (1991).
323. P.C. Uden, 'Element Specific Chromatographic Detection for Trace Inorganic Analysis', *Anal. Proc.*, **30**(10), 405–408 (1993).
324. H. Tao, A. Miyazaki, K. Bansho, 'Determination of Trace Beryllium in Natural Waters by Gas Chromatography–Helium Microwave-induced Plasma Emission Spectrometry', *Anal. Sci.*, **4**(3), 299–302 (1988).
325. M.Y. Khuhawar, A. Sarafraz-Yazdi, P.C. Uden, 'Capillary Gas-chromatographic Determination of Copper and Nickel Using Microwave-induced Plasma Atomic-emission Detection.', *J. Chromatogr.*, **636**(2), 271–276 (1993).
326. M.Y. Khuhawar, A. Sarafraz-Yazdi, Y. Zeng, P.C. Uden, 'Capillary Gas-chromatographic Determination of Palladium Chelates Using Microwave-induced Plasma

Atomic-emission Detection', *Chem. Anal. (Warsaw)*, **40**(3), 271–279 (1995).

327. M.B. De la Calle Guntinas, R. Lobinski, F.C. Adams, 'Interference-free Determination of Selenium(IV) by Capillary Gas Chromatography–Microwave-induced Plasma Atomic-emission Spectrometry After Volatilization with Sodium Tetraethylborate', *J. Anal. At. Spectrom.*, **10**(2), 111–115 (1995).
328. B.D. Quimby, P.C. Dryden, J.J. Sullivan, 'Selective Detection of Volatile Nickel, Vanadium and Iron Porphyrins in Crude Oils by Gas Chromatography–Atomic-emission Spectroscopy', *J. High Resolut. Chromatogr.*, **14**(2), 110–116 (1991).
329. H.A.H. Billiet, J.P.J. Van Dalen, P.J. Schoenmakers, L. De Galan, 'Measurement of (Column Volume from) Deuterium Oxide Elution Data in Reversed-phase Liquid Chromatography with Microwave-induced-plasma Detection', *Anal. Chem.*, **55**(6), 847–851 (1983).
330. K.G. Michlewicz, J.W. Carnahan, 'Directly Coupled Microwave-induced Plasma Atomic-emission Liquid Chromatography Detector for Non-metals: Preliminary Characterization with Halides and Oxyhalogen Salts', *Anal. Lett.*, **20**(8), 1193–1205 (1987).
331. L. Zhang, J.W. Carnahan, R.E. Winans, P.H. Neill, 'Moving-wheel Liquid Chromatography–Helium Microwave-induced Plasma Interface', *Anal. Chem.*, **61**(8), 895–897 (1989).
332. J. Costa-Fernandez, F. Lunzer, R. Pereiro-Garcia, A. Sanz-Medel, N. Bordel-Garcia, 'Direct Coupling of High-performance Liquid Chromatography to Microwave-induced Plasma Atomic-emission Spectrometry Via Volatile-species Generation and its Application to Mercury and Arsenic Speciation', *J. Anal. At. Spectrom.*, **10**(11), 1019–1025 (1995).
333. M. Ibrahim, W. Nisamaneepong, J. Caruso, 'Micro-column High-pressure Liquid Chromatography with a Glass-frit Nebulizer Interface for Plasma Emission Detection', *J. Chromatogr. Sci.*, **23**(4), 144–150 (1985).
334. Y. Liu, V. Lopez-Avila, 'Online Microwave-induced Helium Plasma Atomic-emission Detection for Capillary Zone Electrophoresis', *J. High Resolut. Chromatogr.*, **16**(12), 717–720 (1993).
335. L.J. Galante, D.A. Wilson, G.M. Hieftje, 'Detection of Ions by Replacement-ion Chromatography Coupled to a Microwave-induced Nitrogen Discharge at Atmospheric Pressure', *Anal. Chim. Acta*, **215**(1–2), 99–109 (1988).
336. J.M. Carey, J.A. Caruso, 'Plasma-spectrometric Detection for Supercritical-fluid Chromatography', *Trends Anal. Chem.*, **11**(8), 287–293 (1992).
337. C.B. Motley, G.L. Long, 'Examination of a Helium Highly Efficient Microwave-induced Plasma as an Element-selective Detector for Supercritical-fluid Chromatography', *J. Anal. At. Spectrom.*, **5**(6), 477–482 (1990).
338. M. Montes-Bayon, F. Camuna-Aguilar, R. Pereiro, J.E. Sanchez-Uria, A. Sanz-Medel, 'The Effect of Two Gases Forming Supercritical Fluids (Xenon and Carbon Dioxide) on the Spectral Characteristics and Analytical Capabilities of Microwave-induced Plasmas', *Spectrochim. Acta, Part B*, **51**(7), 685–695 (1996).
339. L.J. Galante, M. Selby, D.R. Luffer, G.M. Hieftje, M. Novotny, 'Characterization of Microwave-induced Plasma as a Detector for Supercritical-fluid Chromatography', *Anal. Chem.*, **60**(14), 1370–1376 (1988).
340. G.R. Ducatte, G.L. Long, 'Effect of Carbon Dioxide and Hydrogen on Non-metal Emission Intensities in a Helium Microwave-induced Plasma', *Appl. Spectrosc.*, **48**(4), 493–501 (1994).
341. Y. Wang, J.W. Carnahan, 'Binary Mobile Phases for Supercritical-fluid Chromatography with Helium Microwave-induced Plasma Detection', *Anal. Chem.*, **65**(22), 3290–3294 (1993).
342. G.K. Webster, J.W. Carnahan, 'Atomic-emission Detection for Supercritical-fluid Chromatography Using a Moderate-power Helium Microwave-induced Plasma', *Anal. Chem.*, **64**(1), 50–55 (1992).
343. Q. Jin, F. Wang, C. Zhu, D.M. Chambers, G.M. Hieftje, 'Atomic-emission Detector for Gas Chromatography and Supercritical-fluid Chromatography', *J. Anal. At. Spectrom.*, **5**(6), 487–494 (1990).
344. D.R. Luffer, L.J. Galante, P.A. David, M. Novotny, G.M. Hieftje, 'Evaluation of a Supercritical-fluid Chromatograph Coupled to a Surface-wave-sustained Microwave-induced Plasma Detector', *Anal. Chem.*, **60**(14), 1365–1369 (1988).
345. D.R. Luffer, M. Novotny, 'Element-selective Detection After Supercritical-fluid Chromatography by Means of a Surfatron Plasma in the Near-infrared Spectral Region', *J. Chromatogr.*, **517**, 477–489 (1990).
346. L. Zhang, J.W. Carnahan, R.E. Winans, P.H. Neill, 'Supercritical-fluid Chromatography with a Helium Microwave-induced Plasma for Chlorine-selective Detection', *Anal. Chem.*, **63**(3), 212–216 (1991).
347. G.K. Webster, J.W. Carnahan, 'Simulated Supercritical-fluid Chromatography Mobile Phase Introduction to a Helium Microwave-induced Plasma: Effects on Non-metal Emission and Spectral Behaviour', *Appl. Spectrosc.*, **44**(6), 1020–1027 (1990).
348. C.B. Motley, M. Ashraf-Khorassani, G.L. Long, 'Microwave-induced Plasma as an Elemental Detector for Packed-column Supercritical-fluid Chromatography', *Appl. Spectrosc.*, **43**(5), 737–741 (1989).
349. C.B. Motley, G.L. Long, 'Evaluation of Sample Introduction Techniques of Packed-column SFC into an MIP [Microwave-induced Plasma]', *Appl. Spectrosc.*, **44**(4), 667–672 (1990).
350. F. Paschen, 'Bohrs Heliumlinien', *Ann. Phys.*, **50**(14), 901–940 (1916).
351. R.K. Marcus, *Glow Discharge Spectroscopies*, Plenum Press, New York, 1993.

352. J.A.C. Broekaert, 'State of the Art of Glow Discharges Lamp Spectrometry–Plenary Lecture', *J. Anal. At. Spectrom.*, **2**(6), 537–542 (1987).
353. E.B.M. Steers, F. Leis, 'Observations on the Use of the Microwave-boosted Glow-discharge Lamp and the Relevant Excitation Processes', *J. Anal. At. Spectrom.*, **4**(2), 199–204 (1989).
354. M.R. Winchester, C. Lazik, R.K. Marcus, 'Characterization of a Radio Frequency Glow Discharge Emission Source', *Spectrochim. Acta, Part B*, **46**(4), 483–499 (1991).
355. A. Bengtson, 'Quantitative Depth-profile Analysis by Glow Discharge', *Spectrochim. Acta, Part B*, **49**(4), 411–429 (1994).
356. F. Leis, J.A.C. Broekaert, K. Laqua, 'Design and Properties of a Microwave Boosted Glow-discharge Lamp', *Spectrochim. Acta, Part B*, **42**(11–12), 1169–1176 (1987).
357. E.B.M. Steers, A.P. Thorne, 'Application of High-resolution Fourier-transform Spectrometry to the Study of Glow-discharge Sources. I. Excitation of Iron and Chromium Spectra in a Microwave-boosted Glow-discharge Source', *J. Anal. At. Spectrom.*, **8**(2), 309–315 (1993).
358. F. Leis, E.B.M. Steers, 'Boosted Glow Discharges for Atomic Spectroscopy–Analytical and Fundamental Properties', *Spectrochim. Acta, Part B*, **49**(3), 289–325 (1994).
359. F. Leis, J.A.C. Broekaert, 'Optical-emission Spectroscopic Measurements on Aluminium, Copper and Lead Samples with a Microwave-boosted Glow-discharge Lamp', *Spectrochim. Acta, Part B*, **46**(2), 243–251 (1991).
360. Y.X. Duan, Y.M. Li, Z.H. Du, Q.H. Jin, J.A. Olivares, 'Instrumentation and Fundamental Studies on Glow-discharge–Microwave-induced Plasma (GD/MIP) Tandem Source for Optical Emission Spectrometry', *Appl. Spectrosc.*, **50**(8), 977–984 (1996).
361. Y.X. Su, P.Y. Yang, D.Y. Chen, Z.G. Zhang, Z. Zhou, X.R. Wang, B.L. Huang, 'Microwave-induced-plasma-boosted Microsecond-pulse Glow-discharge Optical Emission Spectrometry', *J. Anal. At. Spectrom.*, **12**(8), 817–822 (1997).
362. S. Caroli, O. Senofonte, N. Violante, L. Di Simone, 'Novel Version of the Microwave-coupled Hollow-cathode Lamp for Atomic-emission Spectrometry', *Appl. Spectrosc.*, **41**(4), 579–583 (1987).
363. J.A.C. Broekaert, 'Atomic Spectroscopy', in *Ullmann's Encyclopedia of Industrial Chemistry*, eds. B. Elvers, S. Hawkins, W. Russey (Volume ed. H. Günzler), Verlag Chemie, Weinheim, 559–652, Vol. B5, 1994.
364. J.J. Sullivan, B.D. Quimby, 'Detection of Carbon, Hydrogen, Nitrogen and Oxygen by Atomic-emission in Capillary Gas Chromatography', *Chem. Labor. Betr.*, **41**(4), 200–207 (1990).
365. L.M. Faires, 'Fourier Transforms for Analytical Atomic Spectroscopy', *Anal. Chem.*, **58**(9), 1023A–1032A (1986).
366. T. Nakahara, S. Yamada, T. Wasa, 'Simple Nitrogen-purged Optical System for Atmospheric-pressure Helium Microwave-induced Plasma Atomic-emission Spectrometry in the Vacuum Ultra-violet Spectral Region', *Bunko Kenkyu*, **38**(3), 205–208 (1989).
367. T. Nakahara, S. Yamada, T. Wasa, 'Continuous-flow Determination of Trace Iodine by Atmospheric-pressure Helium Microwave-induced Plasma Atomic-emission Spectrometry Using Generation of Volatile Iodine from Iodide', *Appl. Spectrosc.*, **44**(10), 1673–1678 (1990).
368. J.F. Camuna-Aguilar, R. Pereiro-Garcia, J.E. Sanchez-Uria, A. Sanz-Medel, 'Comparative Study of Three Microwave-induced Plasma Sources for Atomic-emission Spectrometry. I. Excitation of Mercury and its Determination After On-line Continuous Cold Vapor Generation', *Spectrochim. Acta, Part B*, **49**(5), 475–484 (1994).
369. E. Denkhaus, A. Golloch, H.M. Kuss, 'Determination of Gaseous Nitrogen in Gas Mixtures Using Low Pressure Microwave-induced Plasma Emission Spectrometry', *Fresenius' J. Anal. Chem.*, **353**(2), 156–161 (1995).
370. K. Dittrich, H. Fuchs, J.M. Mermet, B. Riviere, 'Determination of Nitrogen and Oxygen and Species Containing Nitrogen by Molecular Non-thermal Excitation Spectrometry (MONES) Using Microwave-induced Plasma (MIP) and Furnace Atomization Non-thermal Excitation Spectrometry (FANES) Sources', *J. Anal. At. Spectrom.*, **6**(4), 313–316 (1991).
371. A. Besner, J. Hubert, 'Effects of Dopants on Tin Emission in a Helium Microwave-induced Plasma', *J. Anal. At. Spectrom.*, **3**(2), 381–385 (1988).
372. D.F. Hagen, L.C. Haddad, J.S. Marhevka, 'Solution and Gas-phase Deuterium Derivatization Reactions for Gas Chromatography–Microwave Emission Detection', *Spectrochim. Acta, Part B*, **42**(1–2), 253–267 (1987).
373. B.M. Patel, E. Heithmar, J.D. Winefordner, 'Tubular Electrode Torch for Capacitatively Coupled Helium Microwave Plasma as a Spectrochemical Excitation Source', *Anal. Chem.*, **59**(19), 2374–2377 (1987).
374. J.D. Hwang, W. Masamba, B.W. Smith, J.D. Winefordner, 'Development of High-power, Capacitatively Coupled Microwave Helium Plasma for Atomic-emission Spectroscopy (CMP/AES)', *Can. J. Spectrosc.*, **33**(6), 156–160 (1988).
375. Y.K. Zhang, S. Hanamura, J.D. Winefordner, 'Evaluation of a Microwave-induced Air Plasma as an Excitation Source', *Appl. Spectrosc.*, **39**(2), 226–230 (1985).
376. W.R.L. Masamba, J.D. Winefordner, 'Analytical Characteristics of a Helium–Hydrogen Capacitively Coupled Microwave Plasma', *Spectrochim. Acta, Part B*, **48**(4), 521–529 (1993).
377. B.M. Spencer, A.R. Raghani, J.D. Winefordner, 'Investigation of Halogen Determination in a Helium Capacitively-coupled Microwave Plasma Atomic-emission Spectrometer', *Appl. Spectrosc.*, **48**(5), 643–646 (1994).

378. B.M. Spencer, J.D. Winefordner, 'Determination of Silicon in Organic Samples by Atomic-emission Spectrometry by Using a Capacitively-coupled Microwave Plasma', *Can. J. Spectrosc.*, **39**(2), 43–53 (1994).
379. H. Uchida, P.A. Johnson, J.D. Winefordner, 'Evaluation of the Capacitatively Coupled Helium Microwave Plasma as an Excitation Source for the Determination of Inorganic and Organic Tin', *J. Anal. At. Spectrom.*, **5**(1), 81–85 (1990).
380. D.M. Hueber, W.R.L. Masamba, B.M. Spencer, J.D. Winefordner, 'Application of Hydride Generation to the Determination of Trace Concentrations of Arsenic by Capacitively Coupled Microwave Plasma', *Anal. Chim. Acta*, **278**(2), 279–285 (1993).
381. H. Uchida, A. Berthod, J.D. Winefordner, 'Determination of Non-metallic Elements by Capacitively Coupled Helium Microwave Plasma Atomic-emission Spectrometry with Capillary Gas Chromatography', *Analyst*, **115**(7), 933–937 (1990).
382. D. Huang, D.C. Liang, M.W. Blades, 'Capacitatively Coupled Plasma Detector for Gas Chromatography', *J. Anal. At. Spectrom.*, **4**(8), 789–791 (1989).
383. S. Hanamura, W.J. Wang, J.D. Winefordner, 'Determination of Hydrogen and Oxygen in Metals with the Aid of a Helium Single-electrode Microwave-plasma Emission Technique', *Can. J. Spectrosc.*, **30**(2), 46–49 (1985).
384. A.M. Pless, B.W. Smith, M.A. Bolshov, J.D. Winefordner, 'A Capacitively Coupled Microwave-plasma Atomic-emission Spectrometer for the Determination of Trace Metals in Micro Samples', *Spectrochim. Acta, Part B*, **51**(1), 55–64 (1996).
385. A.H. Ali, K.C. Ng, J.D. Winefordner, 'Micro-sampling in Graphite-cup Capacitively Coupled Microwave-plasma Atomic-emission Spectroscopy', *Spectrochim. Acta, Part B*, **46**(8), 1207–1214 (1991).
386. A.H. Ali, K.C. Ng, J.D. Winefordner, 'Direct Solid Sampling in Capacitively Coupled Microwave-plasma Atomic-emission Spectrometry', *J. Anal. At. Spectrom.*, **6**(3), 211–213 (1991).
387. K. Kitagawa, N. Nishimoto, 'Thermal Vaporizer–Capacitively Coupled Microwave Plasma System for Trace Mercury Determination', *Bunko Kenkyu*, **38**(4), 282–287 (1989).
388. H. Uchida, W.R. Masamba, T. Uchida, B.W. Smith, J.D. Winefordner, 'New Desolvation System for Use with Capacitatively Coupled Microwave Plasma and Inductively Coupled Plasma Atomic-emission Spectrometry', *Appl. Spectrosc.*, **43**(3), 425–430 (1989).
389. B.M. Patel, J.P. Deavor, J.D. Winefordner, 'Solution Nebulization of Aqueous Samples into the Tubular-electrode Torch Capacitatively Coupled Microwave Plasma', *Talanta*, **35**(8), 641–645 (1988).
390. M. Seelig, N.H. Bings, J.A.C. Broekaert, 'Use of a Capacitively Coupled Microwave Plasma (CMP) with Ar, N_2 and Air as Working Gases for Atomic Spectrometric Elemental Determinations in Aqueous Solutions and Oils', *Fresenius' J. Anal. Chem.*, **360**(2), 161–166 (1998).
391. M.W. Blades, 'Atmospheric-pressure, Radio-frequency, Capacitively Coupled Helium Plasmas', *Spectrochim. Acta, Part B*, **49**(1), 47–57 (1994).
392. R.D. Deutsch, J.P. Keilson, G.M. Hieftje, 'Analytical Characteristics of the Microwave-induced Nitrogen Discharge at Atmospheric Pressure (MINDAP)', *Appl. Spectrosc.*, **39**(3), 531–534 (1985).
393. R.D. Deutsch, G.M. Hieftje, 'Development of a Microwave-induced Nitrogen Discharge at Atmospheric Pressure (MINDAP)', *Appl. Spectrosc.*, **39**(2), 214–222 (1985).
394. Y.X. Duan, X.G. Du, Q.H. Jin, 'Comparative Studies of Surfatron and Microwave Plasma Torch Sources for Determination of Mercury by Atomic-emission Spectrometry', *J. Anal. At. Spectrom.*, **9**(5), 629–633 (1994).
395. J.F. Yang, C. Schickling, J.A.C. Broekaert, P. Tschöpel, G. Tölg, 'Evaluation of Continuous Hydride Generation Combined with Helium and Argon Microwave-induced Plasmas Using a Surfatron for Atomic-emission Spectrometric Determination of Arsenic, Antimony and Selenium', *Spectrochim. Acta, Part B*, **50**(11), 1351–1363 (1995).
396. K.G. Michlewicz, J.W. Carnahan, 'Quantitation of Trace Aqueous Halides by Volatilization into a Microwave-induced Helium Plasma', *Anal. Chim. Acta*, **183**, 275–280 (1986).
397. J.E. Freeman, G.M. Hieftje, 'Interferometric Detection of Near-infrared Non-metal Atomic Emission from a Microwave-induced Plasma', *Appl. Spectrosc.*, **39**(2), 211–214 (1985).
398. Y. Madrid, M. Wu, Q.H. Jin, G.M. Hieftje, 'Evaluation of Flow Injection Techniques for Microwave Plasma Torch Atomic-emission Spectrometry', *Anal. Chim. Acta*, **277**(1), 1–8 (1993).
399. D.M. Ye, H.Q. Zhang, Q.H. Jin, 'Flow Injection Online Column Preconcentration for Low Powered Microwave Plasma Torch Atomic-emission Spectrometry', *Talanta*, **43**(4), 535–544 (1996).
400. H.Q. Zhang, D.M. Ye, J.H. Zhao, J.L. Yu, R.Z. Men, Q.H. Jin, D.F. Dong, 'Flow Injection Sample Introduction into Microwave Plasma', *Microchem. J.*, **53**(1), 69–78 (1996).
401. Q. Jin, H.Q. Zhang, F. Liang, W.J. Yang, Q.H. Jin, 'Determination of Trace Amounts of Boron by Microwave Plasma Torch Atomic-emission Spectrometry Using an Online Separation and Preconcentration Technique', *J. Anal. At. Spectrom.*, **11**(5), 331–337 (1996).
402. Q.H. Jin, H.Q. Zhang, F. Liang, Q. Jin, 'Some Observations on the On-line Anion-exchange Preconcentration for Microwave Plasma-torch Atomic-emission Spectrometry', *J. Anal. At. Spectrom.*, **10**(10), 875–879 (1995).

403. F. Liang, H.Q. Zhang, Q. Jin, D.X. Zhang, Y.H. Lei, Q.H. Jin, 'Use of Microwave Plasma Torch Atomic Emission Spectrometry for the Determination of Silicon', *Fresenius' J. Anal. Chem.*, **357**(4), 384–388 (1997).
404. Y. Duan, Y. Li, X. Tian, H. Zhang, Q. Jin, 'Analytical Performance of the Microwave Plasma Torch in the Determination of Rare-earth Elements with Optical-emission Spectrometry', *Anal. Chim. Acta*, **295**(3), 315–324 (1994).
405. Q.H. Jin, H.Q. Zhang, D.M. Ye, J.S. Zhang, 'Study of a Low-powered Helium Microwave-induced Plasma for Determination of Phosphorus, Chlorine, Bromine and Iodine by Atomic-emission Spectrometry', *Microchem. J.*, **47**(3), 278–286 (1993).
406. J.J. Urh, J.W. Carnahan, 'Determination of Metals in Aqueous Solution by Direct Nebulization into an Air Microwave-induced Plasma', *Anal. Chem.*, **57**(7), 1253–1255 (1985).
407. K.G. Michlewicz, J.W. Carnahan, 'Determination of Aqueous Chloride by Direct Nebulization into a Helium Microwave-induced Plasma', *Anal. Chem.*, **57**(6), 1092–1095 (1985).
408. R. Gross, B. Platzer, E. Leitner, A. Schalk, H. Sinabell, H. Zach, G. Knapp, 'Atomic-emission Gas-chromatographic Detection – Chemical and Spectral Interferences in the Stabilized Capacitive Plasma (SCP)', *Spectrochim. Acta, Part B*, **47**(1), 95–106 (1992).
409. J.F. Camuna, M. Montes, R. Pereiro, A. Sanz-Medel, C. Katschthaler, R. Gross, G. Knapp, 'Determination of Halides by Microwave-induced Plasma and Stabilized Capacitive Plasma Atomic-emission Spectrometry After Online Continuous Halogen Generation', *Talanta*, **44**(4), 535–544 (1997).
410. S. Luge, J.A.C. Broekaert, A. Schalk, H. Zach, 'The Use of Different Sample-introduction Techniques in Combination with the Low-power Stabilized Capacitive Plasma (SCP) as a Radiation Source for Atomic-emission Spectrometry', *Spectrochim. Acta, Part B*, **50**(4–7), 441–452 (1995).
411. Y.X. Duan, M.Y. Hou, Z.H. Du, Q.H. Jin, 'Evaluation of the Performance of Microwave-induced Plasma Atomic-absorption Spectrometry (MIP/AAS)', *Appl. Spectrosc.*, **47**(11), 1871–1879 (1993).
412. Y.X. Duan, X.Y. Li, Q.H. Jin, 'Electrothermal Vaporization for Sample Introduction in Microwave-induced-plasma Atomic-absorption Spectrometry', *J. Anal. At. Spectrom.*, **8**(8), 1091–1096 (1992).
413. K.C. Ng, R.S. Jensen, M.J. Brechmann, W.C. Santos, 'Microwave-induced Plasma Atomic-absorption Spectrometry with Solution Nebulization', *Anal. Chem.*, **60**(24), 2818–2821 (1988).
414. K.C. Ng, T.J. Garner, 'Microwave-induced Plasma Atomic-absorption Spectrometry with Solution Nebulization and Desolvation–Condensation', *Appl. Spectrosc.*, **47**(2), 241–242 (1993).
415. Y.X. Duan, H.Q. Zhang, M.Y. Huo, Q.H. Jin, 'Desolvation Effect on the Analytical Performance of Microwave-induced-plasma Atomic-absorption Spectrometry (MIP/AAS)', *Spectrochim. Acta, Part B*, **49**(6), 583–592 (1994).
416. Y.X. Duan, M.Y. Huo, J. Liu, Q.H. Jin, 'Study on Microwave-induced Plasma Atomic-absorption Spectrometry (MIP/AAS) for the Determination of Calcium in Axial Viewing Mode', *Fresenius' J. Anal. Chem.*, **349**(4), 277–282 (1994).
417. A. Zybin, K. Niemax, 'GC Analysis of Chlorinated Hydrocarbons in Oil and Chlorophenols in Plant Extracts Applying Element-selective Diode Laser Plasma Detection', *Anal. Chem.*, **69**(4), 755–757 (1997).
418. C. Schnuerer-Patschan, K. Niemax, 'Element-selective Detection of Chlorine in Capillary Gas Chromatography by Wavelength-modulation Diode-laser Atomic-absorption Spectrometry in a Microwave-induced Plasma', *Spectrochim. Acta, Part B*, **50**(9), 963–969 (1995).
419. A. Zybin, C. Schnuerer-Patschan, K. Niemax, 'Wavelength Modulation Diode Laser Atomic-absorption Spectrometry in Modulated Low-pressure Helium Plasmas for Element-selective Detection in Gas Chromatography', *J. Anal. At. Spectrom.*, **10**(9), 563–567 (1995).
420. H. Lu, Y.L. Ren, H.Q. Zhang, Q. Jin, 'Quantitation of Cadmium by Microwave-induced-plasma Atomic-absorption Spectrometry', *Microchem. J.*, **44**(1), 86–92 (1991).
421. L.D. Perkins, G.L. Long, 'Evaluation of Line and Continuum Sources for Atomic-fluorescence Spectrometry Using a Low-powered Argon Microwave-induced Plasma', *Appl. Spectrosc.*, **42**(7), 1285–1289 (1988).
422. L.D. Perkins, G.L. Long, 'Characterization of a High-efficiency Helium Microwave-induced Plasma as an Atomization Source for Atomic-spectrometric Analysis', *Appl. Spectrosc.*, **43**(3), 499–504 (1989).
423. Y.X. Duan, X.G. Du, Y.M. Li, Q.H. Jin, 'Characterization of a Modified, Low-power Argon Microwave Plasma Torch (MPT) as an Atomization Cell for Atomic-fluorescence Spectrometry', *Appl. Spectrosc.*, **49**(8), 1079–1085 (1995).
424. Y. Duan, X. Kong, H. Zhang, J. Liu, Q. Jin, 'Evaluation of a Low-powered Argon Microwave Plasma Discharge as an Atomizer for the Determination of Mercury by Atomic-fluorescence Spectrometry', *J. Anal. At. Spectrom.*, **7**(1), 7–10 (1992).
425. Y. Oki, H. Uda, C. Honda, M. Maeda, J. Izumi, T. Morimoto, M. Tanoura, 'Laser-induced-fluorescence Detection of Sodium Atomized by a Microwave-induced Plasma with Tungsten-filament Vaporization', *Anal. Chem.*, **62**(7), 680–683 (1990).
426. V. Rigin, 'Simultaneous Atomic-fluorescence-spectrometric Determination of Traces of Iron, Cobalt and Nickel After Conversion to Their Carbonyls and Gas-phase Atomization by Microwave-induced Plasma', *Anal. Chim. Acta*, **283**(2), 895–901 (1993).

427. Y. Oki, E. Tashiro, M. Maeda, C. Honda, Y. Hasegawa, H. Futami, J. Izumi, K. Matsuda, 'Sensitive Detection of Trace Elements in Pure Water by Laser-induced Atomic-fluorescence Spectroscopy in Microwave Discharge Atomization', *Anal. Chem.*, **65**(15), 2096–2101 (1993).
428. G.C. Turk, J.C. Travis, J.R. DeVoe, T.C. O'Haver, 'Laser-enhanced Ionization Spectrometry in Analytical Flames', *Anal. Chem.*, **51**(12), 1890–1896 (1979).
429. R.S. Lysakowski Jr, R.E. Dessey, G.L. Long, 'Laser-enhanced Ionization in Microwave-induced Plasmas', *Appl. Spectrosc.*, **43**(7), 1139–1145 (1989).
430. M.D. Seltzer, R.B. Green, 'Electrical Characteristics of Microwave-induced Plasmas for Laser-induced-ionization Spectrometry', *Spectrosc. Lett.*, **22**(4), 461–470 (1989).
431. M.D. Seltzer, R.B. Green, 'Active Nitrogen Plasma Atom Reservoir for Laser-induced Ionization Spectrometry', *Spectrosc. Lett.*, **20**(8), 601–617 (1987).
432. M. Okumura, K. Fukushi, S.N. Willie, R.E. Sturgeon, 'Evaluation of Atomic Fluorescence, Absorption and Emission Techniques for the Determination of Mercury', *Fresenius' J. Anal. Chem.*, **345**(8–9), 570–574 (1993).
433. H. Brouwers, S. De Jaegere, 'Plasma Afterglow Atomization in Electrothermal Atomic-absorption Spectrometry', *Spectrochim. Acta, Part B*, **43**(8), 901–915 (1988).
434. D.J. Douglas, J.B. French, 'Elemental Analysis with a Microwave-induced Plasma/Quadrupole Mass Spectrometer System', *Anal. Chem.*, **53**(1), 37–41 (1981).
435. D.J. Douglas, E.S.K. Quan, R.G. Smith, 'Elemental Analysis with an Atmospheric-pressure Plasma (Microwave-induced Plasma, ICP)/Quadrupole Mass Spectrometer System', *Spectrochim. Acta, Part B*, **38**(1–2), 39–48 (1983).
436. J.T. Creed, T.M. Davidson, W.L. Shen, P.G. Brown, J.A. Caruso, 'Helium Microwave-induced Plasma Mass Spectrometry for Detection of Metals and Non-metals in Aqueous Solutions', *Spectrochim. Acta, Part B*, **44**(9), 909–924 (1989).
437. E.H. Evans, J.J. Giglio, T.M. Castillano, J.A. Caruso, N.W. Barnett, *Inductively Coupled and Microwave-induced Plasma Sources for Mass Spectrometry*, Royal Society of Chemistry, Cambridge, 1995.
438. F.A. Byrdy, J.A. Caruso, 'Elemental Analysis of Environmental Samples. Using Chromatography Coupled with Plasma Mass Spectrometry', *Environ. Sci. Technol.*, **28**(12), 528A–534A (1994).
439. B.S. Sheppard, J.A. Caruso, 'Plasma Mass Spectrometry: Consider the Source', *J. Anal. At. Spectrom.*, **9**(3), 145–149 (1994).
440. N.P. Vela, L.K. Olson, J.A. Caruso, 'Elemental Speciation with Plasma Mass Spectrometry', *Anal. Chem.*, **65**(13), 585A–597A (1993).
441. C.J. Park, Y.N. Pak, K.W. Lee, 'Isotope-ratio Measurements of Iron, Bromine and Selenium in Aqueous Solutions by Helium Microwave-induced-plasma Mass Spectrometry', *Anal. Sci.*, **8**(3), 443–448 (1992).
442. L.K. Olson, J.A. Caruso, 'Helium Microwave-induced Plasma: An Alternative Ion Source for Plasma Mass Spectrometry', *Spectrochim. Acta, Part B*, **49**(1), 7–30 (1994).
443. W.L. Shen, R.D. Satzger, 'Development of a Low-power Microwave Atmospheric-pressure Molecular Ionization Source for Mass Spectrometry with Direct Introduction of Gaseous and Liquid Organic Samples', *Anal. Chem.*, **63**(18), 1960–1964 (1991).
444. E. Poussel, J.M. Mermet, D. Deruaz, C. Beaugrand, 'Evaluation of a Microwave-induced Plasma as a Soft Ionization Source in Mass Spectrometry', *Anal. Chem.*, **60**(9), 923–927 (1988).
445. A.G. Kouzmin, L.N. Gall, M.Z. Muradimov, 'About Some Peculiarities of Ionization in Radio-frequency Glow-discharge (r.f. GD)', *Fresenius' J. Anal. Chem.*, **355**(7–8), 866–867 (1996).
446. D.A. Wilson, G.H. Vickers, G.M. Hieftje, 'Use of Microwave-induced Nitrogen Discharge at Atmospheric Pressure as an Ion Source for Elemental Mass Spectrometry', *Anal. Chem.*, **59**(13), 1664–1670 (1987).
447. D. Boudreau, J. Hubert, 'Atmospheric-pressure Argon Surface-wave Plasma (SWP) as an Ion Source in Elemental Mass Spectrometry', *Appl. Spectrosc.*, **47**(5), 609–614 (1993).
448. N.P. Vela, J.A. Caruso, R.D. Satzger, 'Potential for an Atmospheric-pressure Low-power Microwave-induced Plasma Ionization Source for Mass Spectrometry', *Appl. Spectrosc.*, **51**(10), 1500–1503 (1997).
449. M. Wu, Y.X. Duan, Q.H. Jin, G.M. Hieftje, 'Elemental Mass Spectrometry Using a Helium Microwave Plasma Torch as an Ion Source', *Spectrochim. Acta, Part B*, **49**(2), 137–148 (1994).
450. J.T. Creed, A.H. Mohamad, T.M. Davidson, G. Ataman, J.A. Caruso, 'Helium Source Microwave-induced Plasma Mass-spectrometric Detection in the Analysis of Gas-chromatographic Eluates', *J. Anal. At. Spectrom.*, **3**(6), 923–926 (1988).
451. P.G. Brown, T.M. Davidson, J.A. Caruso, 'Application of Helium Microwave-induced Plasma Mass Spectrometry to the Detection of High Ionization Potential Gas-phase Species', *J. Anal. At. Spectrom.*, **3**(6), 763–769 (1988).
452. Q. Jin, G. Yang, Z. Guo, A. Yu, J. Liu, 'Some Observations on the Development of a Gas-chromatographic Microwave-plasma Ionization Detector', *Microchem. J.*, **35**(3), 281–287 (1987).
453. B. Pack, J.A.C. Broekaert, J. Guzowski, J. Poehlman, G.M. Hieftje, 'Determination of Halogenated Hydrocarbons by Helium Microwave Plasma Torch Time-of-flight Mass Spectrometry Coupled to Gas Chromatography', *Anal. Chem.*, **70**(18), 3957–3963 (1998).
454. A.H. Mohamad, J.T. Creed, T.M. Davidson, J.A. Caruso, 'Detection of Halogenated Compounds by Capillary Gas Chromatography with Helium-plasma Mass

Spectrometry Detection', *Appl. Spectrosc.*, **43**(7), 1127–1131 (1989).
455. J.T. Jelink, A. Venema, 'Investigations into the Use of Capillary GC with Atomic-emission Detection', *J. High Resolut. Chromatogr.*, **13**(6), 447–450 (1990).
456. R.D. Satzger, F.L. Fricke, P.G. Brown, J.A. Caruso, 'Detection of Halogens as Positive Ions Using a He Microwave-induced Plasma as an Ion Source for Mass Spectrometry', *Spectrochim. Acta, Part B*, **42**(5), 705–712 (1987).
457. W.C. Story, J.A. Caruso, 'Gas-chromatographic Determination of Phosphorus, Sulfur and Halogens (in Pesticides) Using a Water-cooled Torch with Reduced-pressure Helium Microwave-induced Plasma Mass Spectrometry', *J. Anal. At. Spectrom.*, **8**(4), 571–575 (1993).
458. W.C. Story, L.K. Olson, W.L. Shen, J.T. Creed, J.A. Caruso, 'Reduced-pressure Microwave-induced Plasma Mass Spectrometric Detection of Phosphorus and Sulphur in Gas Chromatographic Eluates', *J. Anal. At. Spectrom.*, **5**(6), 467–470 (1990).
459. T. Shirasaki, K. Yasuda, 'Mass Spectra of Oxygen–Nitrogen Plasma by Introduction of Organic Solvent in a Microwave-induced-plasma Mass Spectrometer', *Anal. Sci.*, **8**(3), 375–376 (1992).
460. R.A. Heppner, 'Elemental Detection with a Microwave-induced Plasma–Gas Chromatograph–Mass Spectrometer System', *Anal. Chem.*, **55**(13), 2170–2174 (1983).
461. H. Suyani, J. Creed, J. Caruso, R.D. Satzger, 'Helium Microwave-induced Plasma Mass Spectrometry for Capillary Gas Chromatographic Detection: Speciation of Organotin Compounds', *J. Anal. At. Spectrom.*, **4**(8), 777–782 (1989).
462. L.K. Olson, J.A. Caruso, 'Determination of Halogenated Compounds with Supercritical-fluid Chromatography–Microwave-induced Plasma Mass Spectrometry', *J. Anal. At. Spectrom.*, **7**(6), 993–998 (1992).
463. D. Heitkemper, J. Creed, J.A. Caruso, 'Helium Microwave-induced Plasma Mass-spectrometric Detection for Reversed-phase High-performance Liquid Chromatography', *J. Chromatogr. Sci.*, **28**(4), 175–181 (1990).
464. Y.X. Duan, M. Wu, Q.H. Jin, G.M. Hieftje, 'Vapor Generation of Non-metals Coupled to Microwave Plasma-torch Mass Spectrometry', *Spectrochim. Acta, Part B*, **50**(9), 1095–1108 (1995).
465. Y.X. Duan, M. Wu, Q.H. Jin, G.M. Hieftje, 'Vapor-generation-assisted Nebulization for Non-metal Determination', *Spectrochim. Acta, Part B*, **50**(9), 971–974 (1995).
466. M. Moini, G. Li, F. Perez, F.E. Ibarra, D. Sandoval, 'Selective Detection of Selenium in Water Utilizing Chemical Reaction Interface Mass Spectrometry', *J. Mass Spectrom.*, **32**(4), 420–424 (1997).
467. J.T. Morre, M. Moini, 'Selective Detection and Characterization of Chlorine- and Bromine-containing Compounds in Complex Mixtures Using Microwave-induced Plasma–Chemical-reaction-interface Mass Spectrometry', *Biol. Mass. Spectrom.*, **21**(12), 693–699 (1992).
468. K. Jankowski, 'Application of the Microwave-induced Plasma Atomic-emission Spectrometry to Platinum-group Metals Analysis', *Chem. Anal. (Warsaw)*, **39**(3), 381–387 (1994).
469. J.J. Giglio, J.S. Wang, J.A. Caruso, 'Evaluation of a Direct-injection Nebulizer (DIN) for Liquid Sample Introduction in Helium Microwave-induced Plasma Mass Spectrometry (MIP/MS)', *Appl. Spectrosc.*, **49**(3), 314–319 (1995).
470. W.S. Shen, T.M. Davidson, J.T. Creed, J.A. Caruso, 'Moderate-power Nitrogen Microwave-induced Plasma as an Alternative Ion Source for Mass Spectrometry', *Appl. Spectrosc.*, **44**(6), 1003–1010 (1990).
471. R.D. Satzger, F.L. Fricke, J.A. Caruso, 'Elemental Mass Spectrometry Using a Moderate-power Microwave-induced Plasma as an Ion Source', *J. Anal. At. Spectrom.*, **3**(2), 319–323 (1988).
472. Y. Okamoto, 'High-sensitivity Microwave-induced Plasma Mass Spectrometry for Trace Element Analysis', *J. Anal. At. Spectrom.*, **9**(7), 745–749 (1994).
473. J. Yoshinaga, T. Shirasaki, K. Oishi, M. Morita, 'Isotope Dilution Analysis of Selenium in Biological Materials by Nitrogen Microwave-induced Plasma Mass Spectrometry', *Anal. Chem.*, **67**(9), 1568–1574 (1995).
474. P.A. Fecher, A. Nagengast, 'Trace Analysis in High Matrix Aqueous Solutions Using Helium Microwave-induced Plasma Mass Spectrometry', *J. Anal. At. Spectrom.*, **9**(9), 1021–1027 (1994).
475. T. Shirasaki, K. Hiraki, 'Determination of Trace Elements in a Photoresist Solution by Microwave-induced Plasma Mass Spectrometry', *Bunseki Kagaku*, **43**(1), 25–29 (1994).
476. M. Ceulemans, F.C. Adams, 'GC-based Hyphenated Techniques Applied to the Speciation Analysis of Organometals', *LC-GC Int.*, **7**(12), 694–697 (1994).
477. S.J. Hill, 'Advances in Coupled Techniques for Speciation Studies', *Anal. Proc.*, **29**(9), 399–401 (1992).
478. A.M. Carro-Diaz, R.A. Lorenzo-Ferreira, R. Cela-Torrijos, 'Speciation of Organomercurials in Biological and Environmental Samples by Gas Chromatography with Microwave-induced Plasma Atomic-emission Detection', *J. Chromatogr. A*, **683**(1), 245–252 (1994).
479. P. Lansens, C. Meuleman, W. Baeyens, 'Long-term Stability of Methylmercury Standard Solutions in Distilled, Deionized Water', *Anal. Chim. Acta*, **229**(2), 281–285 (1990).
480. B. Rosenkranz, C.B. Breer, W. Buscher, J. Bettmer, K. Cammann, 'The Plasma Emission Detector – a Suitable Detector for Speciation and Sum Parameter Analysis', *J. Anal. At. Spectrom.*, **12**(9), 993–996 (1997).
481. A. Sanz-Medel, 'Atomic Spectrometry in the Analysis and Research of Biological Systems', *Egypt. J. Anal. Chem.*, **3**(1), 60–67 (1994).

482. I.S. Krull, S.W. Jordan, S. Kahl, S.B. Smith, Jr, 'Trace Analysis for Steroidal Carboranes Via Gas Chromatography–Microwave-induced Plasma Emission-spectroscopic Detection', *J. Chromatogr. Sci.*, **20**(11), 489–498 (1982).

483. S.W. Jordan, I.S. Krull, S.B. Smith Jr, 'Trace Analysis for Catechol Derivatives Via Boronate Ester Formation and Gas Chromatographic–Microwave-induced Plasma-emission Spectroscopic Detection (GC/MIP)', *Anal. Lett.*, **15**(A14), 1131–1148 (1982).

484. D. Deruaz, A. Bannier, C. Pionchon, M.S. Boukraa, W. Elbast, J.L. Brazier, 'Limit of Detection of Nitrogen-15 by Gas-chromatography Atomic-emission Detection. Optimization Using an Experimental Design', *Anal. Lett.*, **28**(11), 2095–2113 (1995).

485. P. Clarkson, M. Cooke, 'The Identification of an Unusual Volatile Component in Processed Tobacco by Gas Chromatography with Mass Spectrometry and Atomic-emission Detection', *Anal. Chim. Acta*, **335**(3), 253–259 (1996).

486. J. Hajslova, P. Cuhra, M. Kempny, J. Poustka, K. Holadova, V. Kocourek, 'Determination of Polychlorinated Biphenyls in Biotic Matrices Using Gas Chromatography–Microwave-induced Plasma Atomic-emission Spectrometry', *J. Chromatogr. A*, **699**(1–2), 231–239 (1995).

487. K.C. Ting, P. Kho, 'GC–Microwave-induced Plasma Atomic-emission Detection Method for Pesticide Residue Determination in Fruits and Vegetables', *J. Assoc. Off. Anal. Chem.*, **74**(6), 991–998 (1991).

488. M.B. De La Calle Guntinas, C. Brunori, R. Scerbo, S. Chiavarini, P. Quevauviller, F.C. Adams, R. Morabito, 'Determination of Selenomethionine in Wheat Samples: Comparison of Gas Chromatography–Microwave-induced Plasma Atomic-emission Spectrometry, Gas Chromatography–Flame-photometric Detection and Gas Chromatography–Mass Spectrometry', *J. Anal. At. Spectrom.*, **12**(9), 1041–1046 (1997).

489. E. Block, X.J. Cai, P.C. Uden, X. Zhang, B.D. Quimby, J.J. Sullivan, 'Allium Chemistry: Natural Abundance of Organoselenium Compounds from Garlic, Onion and Related Plants and in Human Garlic Breath', *Pure Appl. Chem.*, **68**(4), 937–944 (1996).

490. C. Gerbersmann, R. Lobinski, F.C. Adams, 'Determination of Volatile Sulfur Compounds in Water Samples, Beer and Coffee with Purge and Trap Gas Chromatography–Microwave-induced Plasma Atomic-emission Spectrometry', *Anal. Chim. Acta*, **316**(1), 93–104 (1995).

491. R. Lobinski, J. Szpunar Lobinska, F.C. Adams, P.L. Teissedre, J.C. Cabanis, 'Speciation Analysis of Organolead Compounds in Wine by Capillary Gas Chromatography Microwave-induced Plasma Atomic-emission Spectrometry', *J. AOAC Int.*, **76**(6), 1262–1267 (1993).

492. P. Lansens, C. Casais Laino, C. Meuleman, W. Baeyens, 'Evaluation of Gas-chromatographic Columns for the Determination of Methylmercury in Aqueous Headspace Extracts from Biological Samples', *J. Chromatogr.*, **586**(2), 329–340 (1991).

493. B. Sarx, K. Baechmann, 'Speciation of Arsenic Compounds by Volatilization from Solid Samples', *Fresenius' Z. Anal. Chem.*, **316**(6), 621–626 (1983).

494. S. Hanamura, B.W. Smith, J.D. Winefordner, 'Speciation of Inorganic and Organometallic Compounds in Solid Biological Samples by Thermal Vaporization and Plasma Emission Spectrometry', *Anal. Chem.*, **55**(13), 2026–2032 (1983).

495. D.F. Hagen, J. Belisle, J.S. Marhevka, 'Capillary GC Helium Microwave Emission Detector Characterization of Fluorine-containing Metabolites in Blood Plasma', *Spectrochim. Acta, Part B*, **38**(1–2), 377–385 (1983).

496. K. Chiba, K. Yoshida, K. Tanabe, M. Ozaki, H. Haraguchi, J.D. Winefordner, 'Determination of Ultra-trace Levels of Fluorine in Water and Urine Samples by a Gas-chromatographic–Atmospheric-pressure Helium Microwave-induced Plasma Emission-spectrometric System', *Anal. Chem.*, **54**(4), 761–764 (1982).

497. B. Lind, R. Body, L. Friberg, 'Mercury Speciation in Blood and Brain Tissue from Monkeys. Inter-laboratory Comparison of Magos' Method with Other Spectroscopic Methods, Using Alkylation and Gas Chromatography Separation as well as RNAA (Radiochemical NAA) in Combination with Westoo's Extraction Methods', *Fresenius' J. Anal. Chem.*, **345**(2–4), 314–317 (1993).

498. E. Bulska, H. Emteborg, D.C. Baxter, W. Frech, D. Ellingsen, Y. Thomassen, 'Speciation of Mercury in Human Whole Blood by Capillary Gas Chromatography with a Microwave-induced Plasma Emission Detector System Following Complexometric Extraction and Butylation', *Analyst*, **117**(3), 657–663 (1992).

499. A.D. Besteman, N. Lau, D.Y. Liu, B.W. Smith, J.D. Winefordner, 'Determination of Lead in Whole Blood by Capacitively-coupled Microwave Plasma Atomic-emission Spectrometry', *J. Anal. At. Spectrom.*, **11**(7), 479–481 (1996).

500. M.W. Wensing, D.Y. Liu, B.W. Smith, J.D. Winefordner, 'Determination of Lead in Whole Blood Using a Capacitively Coupled Microwave Plasma Atomic-emission Spectrometer', *Anal. Chim. Acta*, **299**(1), 1–7 (1994).

501. M.W. Wensing, B.W. Smith, J.D. Winefordner, 'Capacitively Coupled Microwave Plasma Atomic-emission Spectrometer for the Determination of Lead in Whole Blood', *Anal. Chem.*, **66**(4), 531–535 (1994).

502. S.C. Hight, M.T. Corcoran, 'Rapid Determination of Methylmercury in Fish and Shellfish: Method Development', *J. Assoc. Off. Anal. Chem.*, **70**(1), 24–30 (1987).

503. K. Fukushi, S.N. Willie, R.E. Sturgeon, 'Subnanogram Determination of Inorganic and Organic Mercury by Helium-microwave-induced Plasma Atomic-emission Spectrometry', *Anal. Lett.*, **26**(2), 325–340 (1993).

504. A. Tsunoda, K. Matsumoto, H. Haraguchi, K. Fuwa, 'Application of Microwave-induced Helium Plasma Atomic-emission Spectrometry in Gas-chromatographic Detection of Dimethyl Selenide in Biological Samples', *Anal. Sci.*, **2**(1), 99–100 (1986).
505. C. Schickling, J. Yang, J.A.C. Broekaert, 'Optimization of Electrochemical Hydride Generation Coupled to Microwave-induced Plasma Atomic-emission Spectrometry for the Determination of Arsenic and its Use for the Analysis of Biological Tissue Samples', *J. Anal. At. Spectrom.*, **11**(9), 739–745 (1996).
506. J. Szpunar, V.O. Schmitt, R. Lobinski, J.L. Monod, 'Rapid Speciation of Butyltin Compounds in Sediments and Biomaterials by Capillary Gas Chromatography–Microwave-induced Plasma Atomic-emission Spectrometry After Microwave-assisted Leaching-digestion', *J. Anal. At. Spectrom.*, **11**(3), 193–199 (1996).
507. T. Suzuki, R. Matsuda, Y. Saito, H. Yamada, 'Application of Helium Microwave-induced Plasma Emission Detection System to Analysis of Organotin Compounds in Biological Samples', *J. Agric. Food Chem.*, **42**(1), 216–220 (1994).
508. S.G. Dai, C.R. Jia, 'Applications of Hyphenated Techniques in Environmental Analysis', *Anal. Sci.*, **12**(2), 355–361 (1996).
509. H. Frischenschlager, C. Mittermayr, M. Peck, E. Rosenberg, M. Grasserbauer, 'The Potential of Gas Chromatography with Microwave-induced Plasma Atomic-emission Detection (GC/MIP/AED) as a Complementary Analytical Technique in Environmental Screening Analysis of Aqueous Samples', *Fresenius' J. Anal. Chem.*, **359**(3), 213–221 (1997).
510. H. Frischenschlager, M. Peck, C. Mittermayr, E. Rosenberg, M. Grasserbauer, 'Improved Screening Analysis of Organic Pollutants in River Water Samples by Gas Chromatography with Atomic Emission Detection (GC/MIPAED)', *Fresenius' J. Anal. Chem.*, **357**(8), 1133–1141 (1997).
511. R. Roehl, H.J. Hoffmann, 'Determination of Organic Halogen Compounds by MIP/AES [Microwave-induced Plasma–a.e.s.]', *Vom Wasser*, **63**, 225–237 (1984).
512. B.D. Quimby, M.F. Delaney, P.C. Uden, R.M. Barnes, 'Determination of Trihalomethanes in Drinking Water by Gas Chromatography with a Microwave-plasma Emission Detector', *Anal. Chem.*, **51**(7), 875–880 (1979).
513. K. Chiba, H. Haraguchi, 'Determination of Halogenated Organic Compounds in Water by Gas Chromatography–Atmospheric-pressure Helium Microwave-induced Plasma Emission Spectrometry with a Heated Discharge Tube for Pyrolysis', *Anal. Chem.*, **55**(9), 1504–1508 (1983).
514. P.L. Wylie, R. Oguchi, 'Pesticide Analysis by Gas Chromatography with a Novel Atomic-emission Detector', *J. Chromatogr.*, **517**, 131–142 (1990).
515. C. Flodin, M. Ekelund, H. Boren, A. Grimvall, 'Pyrolysis–GC/AED and Pyrolysis–GC/MS Analysis of Chlorinated Structures in Aquatic Fulvic Acids and Chlorolignins', *Chemosphere*, **34**(11), 2319–2328 (1997).
516. J.S. Alvarado, J.W. Carnahan, 'Reductive Pyrolysis for the Determination of Aqueous Sulfur Compounds with a Helium Microwave-induced Plasma', *Anal. Chem.*, **65**(22), 3295–3298 (1993).
517. K.C. Ng, W. Shen, 'Solution Nebulization into a Low-power Argon Microwave-induced Plasma for Atomic-emission Spectrometry Study of Synthetic Ocean Water', *Anal. Chem.*, **58**(9), 2084–2087 (1986).
518. R. Lobinski, F.C. Adams, 'Sensitive Speciation Analysis of Lead in Environmental Waters by Capillary Gas Chromatography–Microwave-induced Plasma Atomic-emission Spectrometry', *Anal. Chim. Acta*, **262**(2), 285–297 (1992).
519. T. Nakahara, S. Yamada, T. Wasa, 'Determination of Total Iodine in Brines and Sea-waters by Atmospheric-pressure Helium Microwave-induced Plasma Atomic-emission Spectrometry with Continuous-flow Gas-phase Sample Introduction', *Appl. Spectrosc.*, **45**(9), 1561–1563 (1991).
520. S. Natarajan, 'Determination of Parts-per-trillion Levels of Mercury with Low-power Microwave-induced Argon-plasma Emission Spectrometry', *At. Spectrosc.*, **9**(2), 59–62 (1988).
521. Y. Nojiri, A. Otsuki, K. Fuwa, 'Determination of Sub-nanogram-per-litre Levels of Mercury in Lake Water with Atmospheric-pressure Helium Microwave-induced Plasma Emission Spectrometry', *Anal. Chem.*, **58**(3), 544–547 (1986).
522. I. Atsuya, K. Akatsuka, 'Determination of Trace Amounts of Arsenic Using a Capacitively Coupled Microwave Plasma Combined with Hydride Generation: Application to Sewage Sludge, Iron and Steels', *Spectrochim. Acta, Part B*, **36**(7), 747–755 (1981).
523. V. Minganti, R. Capelli, R. De Pellegrini, 'Evaluation of Different Derivatization Methods for the Multi-element Detection of Mercury, Lead and Tin Compounds by Gas Chromatography–Microwave-induced Plasma–Atomic Emission Spectrometry in Environmental Samples', *Fresenius' J. Anal. Chem.*, **351**(4–5), 471–477 (1995).
524. M. Paneli, E. Rosenberg, M. Grasserbauer, M. Ceulemans, F. Adams, 'Assessment of Organolead Species in the Austrian Danube Basin Using GC/MIPAED', *Fresenius' J. Anal. Chem.*, **357**(6), 756–762 (1997).
525. S. Girousi, E. Rosenberg, A. Voulgaropoulos, M. Grasserbauer, 'Speciation Analysis of Organotin Compounds in Thermaikos Gulf by GC/MIPAED', *Fresenius' J. Anal. Chem.*, **358**(7–8), 828–832 (1997).
526. J. Szpunar Lobinska, M. Ceulemans, R. Lobinski, F.C. Adams, 'Flow Injection Sample Preparation for Organotin Speciation Analysis of Water by Capillary Gas Chromatography–Microwave-induced Plasma Atomic-emission Spectrometry', *Anal. Chim. Acta*, **278**(1), 99–113 (1993).

527. A. Sadiki, D.T. Williams, 'Speciation of Organotin and Organolead Compounds in Drinking Water by Gas Chromatography–Atomic-emission Spectrometry', *Chemosphere*, **32**(10), 1983–1992 (1996).
528. K. Chiba, K. Yoshida, K. Tanabe, H. Haraguchi, K. Fuwa, 'Determination of Alkylmercury in Sea-water at the Nanogram per Litre Level by Gas Chromatography–Atmospheric-pressure Helium Microwave-induced Plasma Emission Spectrometry', *Anal. Chem.*, **55**(3), 450–453 (1983).
529. H. Emteborg, D.C. Baxter, W. Frech, 'Speciation of Mercury in Natural Waters by Capillary Gas Chromatography with a Microwave-induced Plasma Emission Detector Following Pre-concentration Using a Dithiocarbamate Resin Micro-column Installed in a Closed Flow Injection System', *Analyst*, **118**(8), 1007–1013 (1993).
530. M. Johansson, H. Emteborg, B. Glad, F. Reinholdsson, D.C. Baxter, 'Preliminary Appraisal of a Novel Sampling and Storage Technique for the Speciation Analysis of Lead and Mercury in Seawater', *Fresenius' J. Anal. Chem.*, **351**(4–5), 461–466 (1995).
531. M. Ceulemans, F.C. Adams, 'Integrated Sample Preparation and Speciation Analysis for the Simultaneous Determination of Methylated Species of Tin, Lead and Mercury in Water by Purge-and-trap Injection Capillary Gas Chromatography–Atomic-emission Spectrometry', *J. Anal. At. Spectrom.*, **11**(3), 201–206 (1996).
532. M.L. Mena, C.W. McLeod, P. Jones, A. Withers, V. Minganti, R. Capelli, P. Quevauviller, 'Microcolumn Preconcentration and Gas Chromatography–Microwave-induced Plasma Atomic-emission Spectrometry (GC/MIP/AES) for Mercury Speciation in Waters', *Fresenius' J. Anal. Chem.*, **351**(4–5), 456–460 (1995).
533. M.B. De La Calle Guntinas, M. Ceulemans, C. Witte, R. Lobinski, F.C. Adams, 'Evaluation of a Purge-and-trap Injection System for Capillary Gas Chromatography–Microwave-induced Plasma Atomic-emission Spectrometry for the Determination of Volatile Selenium Compounds in Water', *Mikrochim. Acta*, **120**(1–4), 73–82 (1995).
534. R. Lobinski, J. Szpunar Lobinska, F.C. Adams, 'Ultratrace Analysis for Organolead Compounds in Greenland Snow', *Analusis*, **22**(7), M54–M57 (1994).
535. Y. Liu, V. Lopez-Avila, M. Alcaraz, W.F. Beckert, 'Simultaneous Determination of Organotin, Organolead and Organomercury Compounds in Environmental Samples Using Capillary Gas Chromatography with Atomic-emission Detection', *J. High Resolut. Chromatogr.*, **17**(7), 527–536 (1994).
536. H. Emteborg, E. Bjorklund, F. Odman, L. Karlsson, L. Mathiasson, W. Frech, 'Determination of Methylmercury in Sediments Using Supercritical Fluid Extraction and Gas Chromatography Coupled with Microwave-induced Plasma Atomic Emission Spectrometry', *Analyst*, **121**(1), 19–29 (1996).
537. G. Heltai, T. Jozsa, 'Nitrogen-15-tracer Technique with the MIPOES Detection of Stable Nitrogen-isotopes for Soil Ecological Studies', *Microchem. J.*, **51**(1–2), 245–255 (1995).
538. B. Pohl, K. Baechmann, 'Comparative Investigations of the Concentrations of Dimethylarsinic Acid (Hydroxydimethylarsine Oxide), Arsenic(III) and Arsenic(V) in Soil and Air-borne Dust Samples', *Fresenius' Z. Anal. Chem.*, **323**(8), 859–864 (1986).
539. H. Baumann, K.G. Heumann, 'Analysis of Organobromine Compounds and Hydrogen Bromide in Motor Car Exhaust Gases with a GC–Microwave Plasma System', *Fresenius' Z. Anal. Chem.*, **327**(2), 186–192 (1987).
540. J.L. Genna, W.D. McAninch, R.A. Reich, 'Atmospheric Microwave-induced Plasma Detector for the Gas-chromatographic Analysis of Low-molecular-weight Sulfur Gases [in Air]', *J. Chromatogr.*, **238**(1), 103–112 (1982).
541. H. Vermaak, O. Kujirai, S. Hanamura, J.D. Winefordner, 'Potential Method for Determination of Gaseous and Particulate Lead in Exhaust Gas by Microwave induced Air-plasma Emission Spectrometry and Zeeman Furnace Atomic-absorption Spectrometry', *Can. J. Spectrosc.*, **31**(4), 95–99 (1986).
542. E.R. Adlard, *Chromatography in the Petroleum Industry*, Elsevier, Amsterdam, 1995.
543. J.P. Snell, W. Frech, Y. Thomassen, 'Performance Improvements in the Determination of Mercury Species in Natural Gas Condensate Using Online Amalgamation Trap or Solid Phase Microextraction with Capillary Gas Chromatography–Microwave-induced Plasma Atomic-emission Spectrometry', *Analyst*, **121**(8), 1055–1060 (1996).
544. *Proceedings [of the] 76th APCA [Air Pollution Control Association] Annual Meeting, June 19–24, 1983, Atlanta, Georgia*, Air Pollution Control Association, Pittsburgh, PA, 1983.
545. V. Siemens, T. Harju, T. Laitinen, K. Larjava, J.A.C. Broekaert, 'Applicability of Microwave-induced Plasma Optical Emission Spectrometry (MIPOES) for Continuous Monitoring of Mercury in Flue Gases', *Fresenius' J. Anal. Chem.*, **351**(1), 11–18 (1995).
546. K. Larjava, T. Laitinen, T. Vahlman, S. Artmann, V. Siemens, J.A.C. Broekaert, D. Klockow, 'Measurement and Control of Mercury Species in Flue Gases from Liquid Waste Incineration', *Int. J. Environ. Anal. Chem.*, **49**(1–2), 73–85 (1992).
547. T. Nakahara, K. Kawakami, T. Wasa, 'Continuous Determination of Low Concentrations of Mercury by Atomic-emission Spectrometry with Helium Microwave-induced Plasma', *Chem. Express.*, **3**(11), 651–654 (1988).
548. M. Seelig, J.A.C. Broekaert, Unpublished Work.
549. A.M. Pless, A. Croslyn, M.J. Gordon, B.W. Smith, J.D. Winefordner, 'Direct Determination of Cadmium in Solids Using a Capacitively Coupled Microwave Plasma

Atomic-emission Spectrometer', *Talanta*, **44**(1), 39–46 (1997).

550. S. Hanamura, B. Kirsch, J.D. Winefordner, 'Determination of Trace Levels of Water in Solid Samples by Evolved Gas Analysis–Helium Microwave Plasma Emission Spectrometry', *Anal. Chem.*, **57**(1), 9–13 (1985).
551. E.M. Skelly Frame, Y. Takamatsu, T. Suzuki, 'Characterization of Solid Particles by Helium Microwave-induced Plasma Atomic-emission Spectrometry', *Spectroscopy*, **11**(1), 17–22 (1996).
552. J.A.C. Broekaert, F. Leis, 'Application of Electrothermal Evaporation Using Direct Solids Sampling Coupled with Microwave-induced Plasma Optical Emission Spectroscopy to Elemental Determinations in Biological Matrices', *Mikrochim. Acta*, **II**(3–4), 261–272 (1985).
553. B. Raeymaekers, P. Van Espen, F. Adams, J.A.C. Broekaert, 'A Characterization of Spark-produced Aerosols by Automated Electron-probe Micro-analysis', *Appl. Spectrosc.*, **42**(1), 142–150 (1988).
554. A. Aziz, J.A.C. Broekaert, K. Laqua, F. Leis, 'A Study of Direct Analysis of Solid Samples Using Spark Ablation Combined with Excitation in an Inductively Coupled Plasma', *Spectrochim. Acta, Part B*, **39**(9–11), 1091–1103 (1984).
555. D.J.C. Helmer, J.P. Walters, 'Microscopic Investigations of the Material Eroded by a High-voltage Spark Discharge', *Appl. Spectrosc.*, **38**(3), 399–405 (1984).
556. D.J.C. Helmer, J.P. Walters, 'Analysis of the Effluent from a High-voltage Spark Discharge with a Microwave-induced Plasma', *Appl. Spectrosc.*, **38**(3), 392–398 (1984).
557. Yong Nam Pak, S.R. Koirtyohann, 'Direct Solid Sample Analysis in a Moderate-power Argon Microwave-induced Plasma with Spark Generation', *J. Anal. At. Spectrom.*, **9**(11), 1305–1310 (1994).
558. U. Engel, A. Kehden, E. Voges, J.A.C. Broekaert, 'Direct Solid Atomic Emission Spectrometric Analysis of Metal Samples by an Argon Microwave Plasma Torch Coupled to Spark Ablation', *Spectrochim. Acta, Part B*, **54**(9), 1279–1289 (1999).
559. M.M. Mohamed, T. Uchida, S. Minami, 'Direct Sample Introduction of Solid Material into a Pulse-operated MIP [Microwave-induced Plasma]', *Appl. Spectrosc.*, **43**(5), 794–800 (1989).
560. T. Ishizuka, Y. Uwamino, 'Atomic-emission Spectrometry of Solid Samples with Laser Vaporization–Microwave-induced Plasma System', *Anal. Chem.*, **52**(1), 125–129 (1980).
561. A. Kehden, J.A.C. Broekaert, Unpublished Work.
562. L. Hiddemann, J. Uebbing, A. Ciocan, O. Dessenne, K. Niemax, 'Simultaneous Multi-element Analysis of Solid Samples by Laser Ablation–Microwave-induced-plasma Optical Emission Spectrometry', *Anal. Chim. Acta*, **283**(1), 152–159 (1993).
563. P. Richner, M.W. Borer, K.R. Brushwyler, G.M. Hieftje, 'Comparison of Different Excitation Sources and Normalization Techniques in Laser-ablation AES Using a Photodiode-based Spectrometer', *Appl. Spectrosc.*, **44**(8), 1290–1296 (1990).
564. J. Uebbing, A. Ciocan, K. Niemax, 'Optical Emission Spectrometry of a Microwave-induced Plasma Used with Laser Ablation of Solid Samples', *Spectrochim. Acta, Part B*, **47**(5), 601–610 (1992).
565. A. Ciocan, J. Uebbing, K. Niemax, 'Analytical Application of the Microwave-induced Plasma Used with Laser Ablation of Solid Samples', *Spectrochim. Acta, Part B*, **47**(5), 611–617 (1992).
566. A. Ciocan, L. Hiddemann, J. Uebbing, K. Niemax, 'Measurement of Trace Elements in Ceramic and Quartz by Laser Ablation Microwave-induced Plasma Atomic-emission Spectrometry. Plenary Lecture', *J. Anal. At. Spectrom.*, **8**(2), 273–278 (1993).
567. J.M. Gehlhausen, J.W. Carnahan, 'Simultaneous Determination of Elemental Ratios in Coal by Direct Powder Injection into a Helium Microwave-induced Plasma', *Anal. Chem.*, **63**(21), 2430–2434 (1991).

Chemometrics

SECTION CONTENTS

S.D. Brown
University of Delaware,
Newark,
USA

Chemometrics

Steven D. Brown
University of Delaware, Newark, USA

1 INTRODUCTION

The mathematical and statistical analysis of chemical data is generally discussed collectively under the term "chemometrics." Chemometrics is an interfacial field, lying between the more established fields of chemistry, chemical engineering, statistics, electrical engineering, and computer science. Like many interfaces, there is uncertainty in the exact location of the borders. Many of the borders between chemometrics and the other disciplines are unclear or even disputed, and there are even differences of opinion on the correct pronunciation of the word "chemometrics," but most agree that the goals of chemometrics include the development of mathematical and statistical methods for extracting and representing chemically relevant information from chemical data. The most important applications of chemometrics come in the discovery of the quantitative relationships between chemical measurements and properties and the qualitative modeling and classification of the behavior of chemical systems. A sizeable fraction of the applications of chemometrics have been reported in analytical chemistry and in the closely related field of process analytical chemistry, where the main goal is often the quantitative estimation of chemical concentrations of mixtures from multivariate physical or chemical measurements obtained on those mixtures. Applications of chemometrics to problems in the fields of chemical and electrical engineering are far less numerous, and most of these have dealt with applications of chemometric methods in connection with process monitoring and control or for signal processing.

The field of chemometrics has a similarly uncertain beginning. Several prominent workers in chemometrics, when interviewed by Geladi and Esbensen,[1,2] reported a wide range of dates and people as being the first to do what is now called chemometrics. The dates cited by these workers ranged from about 1976 to as early as 1908. The early date cited here is important, because it focuses our attention on the fact that researchers have done chemometrics from the point when substantial chemical data began to become available. Gossett published his work for Guinness in 1908. Youden and Wernimont did data analysis and wrote articles and books on quality control in the 1940s and 1950s. There was some interest in their work in chemistry for a good while, but much of that died away by the early 1960s. Box, now a famous statistician, also started as a research chemist who got involved with analysis of plant data. Substantial support by chemists for analysis of chemical data as a sub-field of chemistry failed to occur during these times, and interest died out in chemistry. The individual workers doing such analyses mostly went in the direction of statistics because there they found numbers, interest and acceptance. Statisticians then were a relatively small group too, but they were vastly more numerous than the chemical data analysis community, and they had enough desire to broaden their field that these more applied people were welcomed.[3]

Only in the late 1960s and early 1970s, when automated data collection and computation had both gotten much easier, did the need in chemistry again arise for the field. Factors that led to the formation of the field were the press of too much data, the increasing availability of improved mainframe computers with new statistical software packages, the changing attitudes in the statistics and analytical chemistry communities, and the need for answers.[4] A few researchers realized very early that mathematical methods for finding the relations between measured variables and relations between measurements and latent variables could help in finding the information embedded in large amounts of data. These ideas were revolutionary, in that they offered completely new ways of examining the complex, multivariate data sets that were beginning to appear in chemical studies. Because these methods provided useful answers to questions posed by chemists studying large data sets, they attracted a great deal of interest. Very quickly, a sizeable number of

For references see page 9677

researchers were exploring the limits of this new approach to data analysis, and the field now known as chemometrics came to be. The field's very name, the publication of the early Chemometrics Newsletter and the formation of the Society for Chemometrics were early indicators of the eagerness with which the field was accepted at first. Howery and Hirsch[3] have briefly summarized the early days of chemometrics.

Changes to the fields of analytical chemistry and statistics also helped shape the new sub-field's goals and make-up. The huge increase in the number and sophistication of chemical instruments made available the chemical data that drives chemometrics. It also changed analytical chemistry into an instrument-oriented field, with more emphasis on the physical chemistry of measurements and less on the classical elements of equilibrium theory. Analytical chemists went from being practitioners of the meticulous art of classical analysis, where the quirks of solution-phase chemical reactions were either used or avoided, to specialist workers focusing on the construction and use of a single, increasingly complex instrument that, in ideal cases, made measurement of some analyte species concentration simpler and more sensitive through the application of ideas from physical chemistry, as well as physics and engineering.[5] In that process, these chemists subscribed – perhaps unknowingly – to the idea of latent relationships in data. Finding these latent relationships in data is often the goal of a chemometric study.

The increased sophistication of the computer also changed statistics, though not as much as might have been expected. The field of statistics has always had an uneasy time with its close relationship with mathematics. The 1930s and 1940s saw rapid progress in the field of statistics. Mathematical statisticians made great progress by using properties of distributions to solve a wide range of problems concerned with classical modeling. The more applied statisticians tended to take a Bayesian approach to problems when developing their applications of these newer methods in the many areas where data analysis was applied. The analysis of mostly chemical data from the medical sciences was an area of particular focus because a large amount of univariate data was routinely available. Economics and psychology, two areas whose data sets also involve latent relationships, but with far fewer variables than were routinely being generated by the new chemical instruments, were, like chemometrics, left to those with an interest in nonclassical modeling methods.

2 SOFT MODELING IN LATENT VARIABLES

Many of the methods employed in chemometrics are based on the concept of soft modeling, a linear modeling method that originated in the field of multivariate statistical analysis but which has become synonymous with the term "chemometrics." The focus of the soft modeling method on the properties of the signal rather than on the noise help to distinguish chemometrics from statistics, where the emphasis is usually on the structure and properties of the error term. Chemists often confuse the two fields. It is worthy of note that, while chemometrics brought many statistical concepts to chemical applications early in its history, in recent years the transfer of information and technology has been from chemometrics to other fields, including statistics. Soft modeling is an area where advances from chemometrics are now aiding a wide range of diverse fields, from remote sensing to psychology.

Because of the heavy emphasis on soft modeling in chemometrics, the field developed around an algorithmic rather than theoretical framework, an attribute that is only now beginning to change. Discussions of soft modeling in current literature are more likely to focus on the linear algebraic theory of the modeling than on the specific steps needed to form the model.

To understand the field's present status and direction, it is necessary to have an appreciation for some of the key approaches and assumptions of soft modeling, as these underlie the logic of many of the chemometric methods.[6] More details are available in the Encyclopedia article **Soft Modeling of Analytical Data** to follow, so it is appropriate to provide only an overview here.

Traditionally, modeling in chemistry and chemical engineering has been done using "first-principles" (hard) models. A hard model is one that describes the system in terms of mathematical relationships developed using the measurement variables as independent variables and the desired outputs as dependent variables. Because chemical systems studied are complex, the hard modeling used in chemistry has either been applied to simplified systems or has involved either limiting "laws" or other approximations and restrictions to the region of application of the hard model. Soft modeling sees the problem from an entirely different logical perspective: it presumes that the chemical system under study is complex, and that it is not possible or economically feasible to adequately describe the behavior of the system using a hard model. The soft model is based on variation and correlation in the data, as captured in a covariance matrix – which can be thought of as a measure of the overall fluctuation in each independent variable present in the data set, as well as the variable–variable interactions.

A set of orthogonal components made from linear combinations of the independent variables is created to describe independent sources of the observed variation in the covariance matrix created from the data set analyzed, according to Equation (1)

$$\mathbf{X} = \mathbf{U}\mathbf{V}^{\mathrm{T}} \quad (1)$$

where matrix **V**, the loadings of **X**, contains the linear combinations of the original measurement variables that define the new variation-based coordinate system spanning the data in **X**, and matrix **U**, called the scores of **X**, contains the coordinates of the data **X** in that variation-based coordinate system.(7)

These linear combinations of the measured variables are called latent variables because they are derived rather than measured. The latent variables extracted to describe a data set are ordered in terms of the size of the independent sources of variation that they explain: the first latent variable explains the largest independent source of variance in the data, the second latent variable the second largest, and so on, until all variation in the data set is accounted for by one of the linear combinations of the measurements. Generally, re-expression of the data set **X** in terms of latent variables is not useful unless a decision is also made on the *number* of latent variables that are needed to adequately explain the systematic variation in the data **X**.

Any correlation between the dependent variable(s) and these latent variables is captured by means of a regression model. The regression step can be done after the creation of latent variables for the independent variables and the removal of sources of variation believed to be unrelated to the systematic effects under study. Building a regression model by first soft-modeling and truncating the independent (measured) variables is known as principal components regression (PCR) in statistics and chemometrics (or it can be done in concert with the extraction of the latent variables using a modeling method known as partial least squares (PLS) regression).

The strength of soft modeling with latent variables is that systematic sources of variation in the data do not go unmodeled, as they might under hard models developed with inadequate theory. Relationships in data can be modeled even when the existing theory is incomplete, wrong, or even missing. The ability to model linear relationships in the absence of suitable theory has made soft modeling an attractive approach for a wide range of problems in applied and analytical chemistry, including modeling of quantitative relationships in spectroscopy, quantitative structure–property relationships, and many other quantitative modeling scenarios. Soft modeling with latent variables also has uses in semi-quantitative modeling, where latent variables can be used to group or to classify samples by some latent property. Examination of the latent variables can often provide insights into the chemistry behind the observations. And, because latent variables are linear combinations of what may be many measurements, they offer a way to screen measurements for appropriateness and impact on the modeling. For this reason, soft modeling in latent variables is often used to discover relationships in data.

Soft modeling comes with several inherent defects, however. The most serious is that the latent variables created in the soft modeling are *local* to the data set analyzed. The soft model developed explains the variation seen in the data set used to create it. Unfortunately, quantitative modeling often requires a predictive model, and there is strong incentive to take a soft model developed on one set of data for use on an entirely different set. Any use of this model on new data is an extrapolation and, unless great care is taken in collecting and analyzing the data, the soft model may have little predictive use. Because the soft model defines a truncated mathematical basis for a series of measurements, it is possible to create a soft model on one set of data and to use it on a closely related set. To make this possible, though, the truncated basis defining the latent variables developed in the soft model must span the bases defining any new set of experimental data subjected to the model. The subtle and not-so-subtle consequences of this requirement force the user of soft models to consider the experimental design of data sets and the sensitivity of the models to individual samples to a degree that goes far beyond that done with more traditional modeling methods.

A second difficulty with soft modeling arises from the fact that the latent variables describe orthogonal mathematical effects, and not physical effects. Most physical effects show high correlation between measurements, especially between adjacent measurement channels (e.g. individual variable "channels" in spectra or chromatograms). For example, suppose that in a chemical calibration, the varying amounts of two different compounds with different spectra are responsible for the two different sources of variation in the analytical response. These two chemically distinct sources may not be *mathematically* independent, however, because of the redundant nature of most multichannel responses measured with chemical instrumentation. Described in terms of latent variables, though, the sources must be mathematically independent. The result is the mixing of separate, chemically significant effects in each of the latent variables and the spread of these effects over several latent variables. The latent variables describe *variation* in measurements rather than some easily understood measurement phenomenon, such as the absorbance at a particular wavelength. Direct interpretation of the chemical significance of these latent variables is difficult or impossible. Usually, conversion of mathematically significant effects as evidenced in the latent variables surviving after truncation of the soft model back to physically significant effects is needed to extract chemically relevant

For references see page 9677

information. In essence, the conversion needed is an oblique rotation of orthogonal basis defined by the latent variables into the nonorthogonal basis defined by the nature of the physically significant information desired. Without external constraints or some prior knowledge of the solution, constraining the large number of possible rotations is impossible in many cases. The conversion of a mathematically useful latent variable set to a chemically relevant basis is therefore a stumbling block in getting usable chemical information from the soft modeling. This fact more than anything limits the use of chemometrics to those secure enough in the mathematics to apply it without many of the usual "safety lines" associated with visual interpretation or simple statistical inference (confidence intervals, etc.).

Another defect in soft modeling is the lack of an easy route to distinguish the sources of variation attributable to systematic, chemical effects in the data from other sources of variation. Accidental correlations arising between systematic variation in the data and various kinds of noise present either in the measurements themselves or in the taking of the data can corrupt the latent variables. One consequence of this accidental correlation is the aliasing of measurement variables weighted heavily in the latent variable; it is possible to see increased or decreased weights given measurements in a latent variable as a consequence of noise-induced artifactual correlation with another measurement variable with large variation. These artifacts in the correlation and variation present in the data set frustrate the modeling effort in several ways. First, when present, accidental correlation induced by noise effects can make difficult the direct interpretation of the latent variables, an important part in extracting information from the soft model. Second, the presence of the incorrect weighting of the measurement variables in the latent variables can corrupt the rotation of latent variables back into chemically significant information presented in terms of the original measurement variables. Third, the noise-related artifacts complicate statistical decisions on the number of latent variables needed to describe a systematic effect in data, a key part of the development of the soft model.

3 CURRENT DIRECTIONS IN CHEMOMETRICS

In view of the importance of soft modeling to the study of complex chemical systems, most standard chemometric methods rely heavily on soft modeling in one form or another and make full use of its advantages. Advances in these chemometric methods often come in the development of new algorithms or new ways of using older methods that can either ameliorate or avoid the above-mentioned difficulties associated with soft modeling.

4 MULTIVARIATE (FIRST-ORDER) CALIBRATION

The best-known examples of the use of new algorithms to make use of soft modeling practical come in the area of multivariate calibration. This is the subject of the Encyclopedia article **Multivariate Calibration of Analytical Data**. An overview is provided here.

Here, the PCR algorithm and later the PLS algorithm were modified and put to use in improving the extraction of quantitative information such as concentrations from analytical responses. One modification came in that the soft modeling was done on an inverse relationship between responses and concentration, namely Equation (2)

$$\mathbf{C} = \mathbf{RB} \tag{2}$$

where $\mathbf{C}$ is an $m \times n$ matrix of concentrations, $\mathbf{R}$ is an $m \times p$ matrix of responses, and $\mathbf{B}$ is a $p \times n$ matrix describing the calibration relationship. Equation (2) is an inverse Beer's law relation, as it implies that concentration $\mathbf{C}$ carries the error and is dependent on the response variables in $\mathbf{R}$. By soft modeling the response $\mathbf{R}$ (and $\mathbf{C}$ with PLS), then truncating the soft model to decrease the variance unrelated to the calibration relationship, a useful calibration model results. This model has the benefit of being suited to predict one component in a calibrated mixture.[(8)]

The most widely known success of first-order calibration methods has come in near-infrared (NIR) spectrometry, a fairly general analytical measurement that is often useless for direct quantitative measurements because of the lack of specificity of NIR bands.[(9,10)] Efforts to enhance the specificity or selectivity of analytical instrumentation, especially spectroscopic and chromatographic instrumentation, continue to make up a sizable fraction of chemometric research. Multivariate calibration using PLS for the modeling of the inverse response–concentration relationship has become commonplace in the last 10 years, and these techniques have begun to be embraced by industrial chemists and chemical engineers.[(11)] Published applications of multivariate calibration abound, as is clear from the number of citations in the Fundamental Reviews reported over the last 10 years,[(12)] but it is the author's experience that many more applications of multivariate calibration are put into practice than are reported. A good, short introduction to PLS theory and coding is also available.[(13,14)]

The current trend in first-order calibration is to explain the calibration in terms of the net analyte signal.[15,16]

5 SECOND- AND HIGHER-ORDER CALIBRATION

Second-order calibration, where the analytical response gives rise to a matrix rather than a vector of data, is a very active area of research at present. This area is the subject of the Encyclopedia article **Second-order Calibration and Higher**, so again, only an overview is provided here. There are two reasons for the interest. One is practical: a large number of analytical measurements give rise to a matrix of data (e.g. a liquid chromatographic separation monitored with a diode-array detector). There is also a theoretical "second-order advantage" in using these data in a calibration, in that the mathematics permits analysis of a calibrated component in the presence of an uncalibrated interference.[17]

The analysis of higher-order chemical data – that is, data with dimension higher than one – involves a choice of the way that the data are unfolded to make matrices of data (where the rows are defined by the individual samples and the columns by the components of the one-dimensional measurements) for more conventional matrix-oriented analysis. Smilde has reviewed different unfolding methods, including the various Tucker unfolding schemes and parallel factor analysis (PARAFAC) modeling, and offered an explanation of the history and applications of higher-order analysis.[18] As in first-order data analysis, early chemometrics work on higher-order data has benefited from porting of the substantial work on three-way methods published in statistics, psychometrics, and related fields. But, as with analysis of first-order data, the maturation of chemometrics research has led to significant contributions by chemometricians to this approach to data analysis.

Like multivariate calibration of first-order data, higher-order analysis is also based on soft modeling. It is no surprise that it carries some of the defects inherent in soft modeling, including sensitivity to noise, difficulty in getting the correct model size (the rank here) and problems in assuring the match of bases. There are some important differences, however: the rank annihilation method deals with information in multiple measurement dimensions, and it is far less susceptible to the effect of an interference that shows in a prediction sample but not in any calibration sample. Higher-order calibration is very dependent on a good match of the calibration model with the response data to be determined.

6 NONLINEAR METHODS

The history of measurement chemistry is filled with studies of linear systems and of efforts to make those nonlinear systems studied linear enough for examination by linear methods. It is therefore not surprising that relatively little effort has been placed on the systematic study of nonlinear methods as applied to chemical calibrations. The large majority of studies of nonlinear systems have been done using nonlinear causal models, because methods of nonlinear soft modeling developed to date require the underlying presence of specific nonlinear functional forms (e.g. quadratic relationships for loadings) and have not proved useful in practice.

Recent interest in soft modeling nonlinear relationships has centered on the use of neural networks. Artificial neural networks (ANNs), a modeling method that does not presume any underlying relationship in the data, have received some attention for the calibration of nonlinear chemical systems and have been used for modeling a number of chemical processes with success. An ANN models the relationship between the input independent variables and the output dependent variables by using weighted sums of transforms usually based on a sigmoid function or the tanh function, a close relative. These transforms are capable of describing very complex relationships between the independent and the dependent variables. According to theory, a three-layer net with sigmoid transfer function in the hidden layer and linear transfer function in the output layer can model any continuous function.[19] The modeling is done by first establishing the topology of the network model, i.e. number of layers of nodes to be used, the number of nodes in each layer, and the transform operation used in each layer. Finding an appropriate topology is a fairly straightforward but time-consuming process. Generally, a number of possible candidate topologies are tried – these can be found by simple enumeration or by more sophisticated methods such as genetic algorithms – and the topologies giving the best results are kept for further analysis and optimization.

7 SEMI-QUANTITATIVE METHODS USING SOFT MODELING

In many studies, the goal is not quantitative, indirect measurement of one or more known, calibrated chemical species in a series of mixtures. The goal may be discovery of what chemical species are involved in a dynamic system, or possibly discovery of the number of species changing in that system. It may just be an objective measure that the dynamic system has not changed in its overall make-up. Chemometric methods based on soft-modeling of the

For references see page 9677

multivariate sensor measurements taken on these systems are well suited to obtaining information on the identity and number of chemical species.

When less is known about a data set, self-modeling can be attempted. The loadings used to describe a set of data are rotated to find another set of axes that has physical significance. In this case, less may be known about the composition of the system under study, and the number of possible rotations is infinite. External constraints such as all positive spectra and all positive amounts of species are needed to help limit the search to the space between the convex hulls defined by the different constraints. It is now known that the solutions are found near the apexes of the convex hulls, but there still is no systematic means to map out convex hulls and locate their apexes in the presence of noise and confounding effects. SIMPLISMA, one popular method for self-modeling, uses a visual examination to help find suitable rotations of the soft model.[20] The solutions that result from this self-modeling are subject to uncertainty unless each species has a signature that is independent of the others at one or more of the sensors used to collect the multivariate set. The more similar the species' true responses over the sensor set used, the larger the uncertainty in the self-modeled responses for these species will tend to be.

Often, self-modeling is needed because of a lack of solid information on the composition of the system, but the data are collected in such a way that ensures their serial correlation in both the sensor axis and another axis, often time. In these cases, the additional correlation offers an additional constraint that helps to remove ambiguity in the rotated solution. Now, it is possible to develop a series of soft models that evolve as the data evolves in time, in both the forward (increasing time) and backward (decreasing time) directions. Tracking of the number of significant principal components in the soft models over the time axis permits an estimate of the number of species that vary over the data set, as well as where it is in the data that the different species enter into the chemical system's response. Rotation of the soft model to simultaneously describe both the response and the time-dependent data axes, subject to constraints in spectra and amounts, gives semiquantitative estimates of pure responses and relative amounts. Malinowski reviewed early work showing that this modeling approach can be used to discover relationships in data.[21]

8 CLASSIFICATION AND CLUSTERING OF CHEMICAL DATA

It may be that there is grouping of the samples in the data according to some latent property. Often, a set of data is heterogeneous, in that the samples clump together rather than spread evenly over the multivariate space. Building a single PLS regression model on the full set of multivariate data is inappropriate because the variation captured in the modeling will be heavily influenced by variation between groups of data and will be less influenced by the variation between the independent and dependent variables. In these cases, a grouping step is needed before regression modeling to ensure that a homogeneous structure is present in some subset of the data prior to modeling this portion of the data. If a multivariate grouping structure present in the data can be guessed or determined, separate soft models can be developed for each group in the data.

Once grouping is decided, separate soft models can be built for each of the groups. Soft modeling can be used to develop class regression models for different values of the latent property. New data can be classified into the appropriate group on the basis of the distance of the new data to each of the class models. An F-test is used to decide on membership, again using residual variance in the fit and in the model. The SIMCA method focuses on modeling the classes rather than on finding an optimal classifier.[22,23] By examining the mathematics behind SIMCA, it has become apparent that there may be better classifiers, depending on the structure of the data.[24,25]

Current research in soft modeling for semiquantitative modeling has as an emphasis the removal of many of the problems associated with soft modeling that carry over to self-modeling. Making the evolving factor analysis methods work more reliably is a major focus of research, as are improved methods for getting self-modeling to work on many systems. Background and current research in classification methods are discussed in the Encyclopedia article **Clustering and Classification of Analytical Data**.

9 SIGNAL PROCESSING METHODS

Since the initial days of chemometrics, signal processing has long been a focus. Initially, the aim was noise reduction through smoothing off-line, but in recent times, the aim of signal processing has focused more on real-time data enhancement through filtering or on real-time control of a process through use of real-time data. Details are available in the Encyclopedia article **Signal Processing in Analytical Chemistry**. As this area has evolved, the interest of analytical chemists has waned while interest from chemical engineers and process chemists has increased substantially.[26,27]

10 CLASSICAL AND NONCLASSICAL OPTIMIZATION METHODS

Chemometrics is mostly focused on data analysis but one aim has been the improvement of a process or system producing the data. Chemometrics has long been involved in process optimization, from the evolutionary optimization schemes of Box to the Simplex optimization methods perfected by Deming. Newer work on genetic algorithms and simulated annealing continue the advancement of optimization methods. The Encyclopedia article **Classical and Nonclassical Optimization Methods** discusses these optimizers.

11 KEY REVIEWS OF THE FIELD OF CHEMOMETRICS

Because of its early, close association with the field of analytical chemistry, the field of chemometrics has received frequent and thorough review in that field's literature. Beginning in 1976, a detailed review of the methodology and practice of data analysis in chemistry has appeared in the biennial Fundamental Reviews issue of the journal *Analytical Chemistry* (the most recent as of this writing is that of Lavine[(12)]). These reviews provide a comprehensive analysis of the state of chemometrics and, when examined together, also provide a glimpse at the evolution of concepts in chemometrics over the past 10 years. Along with the reviews of basic research in chemometrics, three major reviews on chemometrics applied to problems arising in process analytical chemistry also have appeared in the Applications Reviews issue of *Analytical Chemistry*.

ABBREVIATIONS AND ACRONYMS

ANN	Artificial Neural Network
NIR	Near-infrared
PCR	Principal Components Regression
PLS	Partial Least Squares

REFERENCES

1. P. Geladi, K. Esbensen, 'The Start and Early History of Chemometrics: Selected Interviews', *J. Chemometrics*, **4**, 337–354 (1990).
2. P. Geladi, K. Esbensen, 'The Start and Early History of Chemometrics: Selected Interviews. Part 2', *J. Chemometrics*, **4**, 389–412 (1990).
3. D.G. Howery, R.L. Hirsch, 'Teaching Chemometrics', *J. Chem. Educ.*, **60**, 656–659 (1983).
4. S.A. Borman, 'Math is Cheaper than Physics', *Anal. Chem.*, **54**, 1379A–1380A (1982).
5. B.G.M. Vandeginste, 'Teaching Chemometrics', *Anal. Chim. Acta*, **150**, 199–206 (1983).
6. S. Wold, 'Chemometrics: What do we mean with it, and what do we want from it?', *Chemom. Intell. Lab. Syst.*, **30**, 109–115 (1995).
7. J.E. Jackson, *A User's Guide to Principal Components*, Wiley-Interscience, New York, 1991.
8. P.J. Brown, 'Multivariate Calibration (with Discussion)', *J. Roy Stat. Soc., Ser. B*, **44**, 287–321 (1982).
9. H. Martens, T. Karstang, T. Næs, 'Improved Selectivity in Spectroscopy by Multivariate Calibration', *J. Chemometrics*, **1**, 201–219 (1987).
10. T. Næs, H. Martens, 'Principal Component Regression in NIR Analysis: Viewpoints, Background Details and Selection of Components', *J. Chemometrics*, **2**, 155–168 (1988).
11. H. Martens, T. Næs, *Multivariate Calibration*, John Wiley & Sons, Chichester, 1989.
12. B.K. Lavine, 'Chemometrics: Fundamental Review', *Anal. Chem.*, **70**, 209R–218R (1998).
13. P. Geladi, B.R. Kowalski, 'Partial Least Squares Regression: A Tutorial', *Anal. Chim. Acta*, **185**, 1–17 (1986).
14. P. Geladi, B.R. Kowalski, 'An example of 2-Block Predictive Partial Least Squares Regression with Simulated Data', *Anal. Chim. Acta*, **185**, 19–32 (1986).
15. K.S. Booksh, B.R. Kowalski, 'Theory of Analytical Chemistry', *Anal. Chem.*, **66**, 782A–791A (1994).
16. K. Faber, A. Lorber, B.R. Kowalski, 'Net Analyte Signal Calculation in Multivariate Calibration', *Anal. Chem.*, **69**, 1620–1626 (1997).
17. B.R. Kowalski, M.B. Seasholz, 'Recent Developments in Multivariate Calibration', *J. Chemometrics*, **5**, 129–146 (1992).
18. A.K. Smilde, 'Three-way Analyses: Problems and Prospects', *Chemom. Intell. Lab. Syst.*, **15**, 143–157 (1992).
19. G. Cybenko, 'Approximation by Superpositions of a Sigmoidal Function', *Math. Control Signals Systems*, **2**, 303–314 (1989).
20. W.Windig, 'Mixture Analysis of Spectral Data by Multivariate Methods', *Chemom. Intell. Lab. Syst.*, **4**, 201–213 (1988).
21. E.R. Malinowski, *Factor Analysis in Chemistry*, 2nd edition, Wiley-Interscience, New York, 1991.
22. S. Wold, 'Pattern Recognition by Means of Disjoint Principal Component Models', *Patt. Recog.*, **8**, 127–139 (1976).
23. S. Wold, C. Albano, W.J. Dunn III, U. Edlund, K. Esbensen, P. Geladi, S. Hellberg, E. Johansson, W. Lindberg, M. Sjöström, 'Multivariate Data Analysis in Chemistry', in *Chemometrics: Mathematics and Statistics in Chemistry*, ed. B.R. Kowalski, NATO ASI Series, Reidel, Dordrecht, 17–95, Vol. 138, 1984.

List of selected abbreviations appears in Volume 15

24. I.E. Frank, J.H. Friedman, 'Classification: Oldtimers and Newcomers', *J. Chemometrics*, **3**, 463–475 (1989).
25. G.J. McLachlan, *Discriminant Analysis and Statistical Pattern Recognition*, Wiley-Interscience, New York, 1992.
26. J.F. MacGregor, 'Statistical Process Control of Multivariable Processes', *Control Eng. Practice*, **3**, 403–414 (1995).
27. C.E. Miller, 'The Use of Chemometric Techniques in Process Analytical Method Development and Operation', *Chemom. Intell. Lab. Syst.*, **30**, 11–22 (1995).

Classical and Nonclassical Optimization Methods

Ron Wehrens and Lutgarde M.C. Buydens
University of Nijmegen, The Netherlands

Optimization problems are abundant in analytical chemistry, examples being the determination of optimal conditions for experiments or optimal settings for instruments. In general, all the required information should be obtained from as few experiments as possible. Classical techniques such as response surface models or simplex optimization are often used. These techniques, which can be very efficient in cases where the underlying assumptions are fulfilled, are called "strong" methods.

With the advent of the computer in the laboratory, a new class of optimization problems arose which could not be tackled with the standard methodologies. For these search-type problems, new strategies such as simulated annealing (SA) and genetic algorithms (GA) are applied. Although these are not guaranteed to give the optimal result, in almost all cases they are able to find very good solutions where other techniques fail completely. These methods find themselves in an intermediate position between the strong methods, mentioned above, and weak methods, where hardly any assumptions are made.

This article provides an overview of classical and nonclassical optimization techniques, and stresses the differences in their areas of application. The key ideas are highlighted and references to important publications are given.

List of selected abbreviations appears in Volume 15

1 INTRODUCTION

As in all empirical sciences, optimization problems are abundant in chemistry, and have been since the very first alchemist experiments. In the synthesis of gold from lesser substances, every conceivable variation of ingredients and experimental conditions was tried and eventually rejected. Sometimes these experiments were conducted in a very systematic way, sometimes random, sometimes somewhere in between. Although the main goal was never achieved, a lot was learned from these early experiments.

In the modern laboratory, the advent of complicated instruments controlled by computers has changed the nature of the problems faced by the chemist. Whereas the actual doing of the experiment used to be a major limiting factor in obtaining relevant information, and therefore warranted a large investment in time and energy to be designed and executed, in the modern laboratory experiments can be performed in a fast, reproducible way. For the sake of ever higher performance requirements, dictated by economic reasons, systematic optimizations are performed in all chemical laboratories, often consisting of hundreds or thousands of experiments.

Optimization is the search for a maximum or minimum in the value of a certain response function. For example, the yield of a chemical reaction is a function of several variables, such as concentrations of components in the mixture and physical characteristics such as temperature and pressure. The relationship between these variables

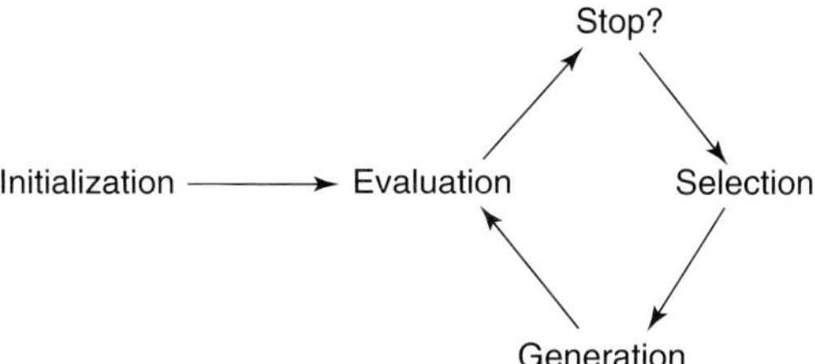

Figure 1 The basic iterative optimization cycle.

and the response function (the yield) is unknown, but it can be sampled by performing experiments. The conditions in these experiments can be carefully selected by using knowledge (section 2.2) and, depending on the outcome, new experiments may be performed.

Such optimizations are typically performed in an iterative fashion (Figure 1). The search is started either from one or more random positions or from a set of points, picked according to some criterion. In the evaluation stage, the quality of the current point(s) is assessed by experiments. Several criteria may be used to stop the optimization, such as the number of experiments, or the quality of the solutions found so far. If no stopping criterion applies, the next stage is to accept the solution(s) as a starting point for new solutions, or to reject and to proceed from another solution. This process is called selection. Finally, the optimization method generates one or more new candidate solutions. These in turn are evaluated and the cycle enters its next iteration.

Many different optimization methods exist and they can be classified in a number of ways. Among the classical methods[1] are methods relying on a model of the response surface. The sampling of the space of possible solutions is meant either to describe this surface or to define the direction in which the largest improvement will be found. Typical examples are steepest-descent and response-surface methods. Many assumptions about the relationship between problem parameters and response are made, but if they are correct then the performance of these methods is very good. These "strong" methods are treated in section 2. The other category of classical optimization methods consists of the weak methods. These methods (section 3) make no assumptions about the response surface, and include the well-known but rarely used random search technique.

The newest optimization techniques, i.e. the nonclassical techniques, place themselves between these two extremes. Whereas weak methods rely on chance to find the optimum and strong methods rely on problem-specific assumptions, these "intermediate" methods seek a balance between chance and information. As they typically require a large number of experiments, they are often applied in computational optimization problems. The most well-known method is SA.[2]

The differences between the methods discussed here and the problems for which they are applied can be found at all levels of the basic cycle depicted in Figure 1. Whereas weak and intermediate methods typically rely on random choices for the starting points, strong methods utilize knowledge about the problem domain to identify promising regions. Some methods stop as soon as no improvement is found in the last step, whereas others just proceed with the best move available. Almost all methods use a quality-weighted selection – bad solutions are more often rejected than good solutions, but traditionally each optimization method uses its own particular selection scheme. Finally, in the generation of new trial solutions, strong methods use moves defined a priori by the user; weak methods randomly generate the next trial set; and intermediate methods usually apply a method-specific and/or problem-specific mechanism.

In cases where the response is an error value, as in curve-fitting applications, the optimal value is zero and all other response values are larger than zero. In many other applications the response variable is maximized, such as the yield of a chemical reaction. This article does not distinguish between these two cases. The examples given always relate to minimization problems without loss of generality.

1.1 Local and Global Optimality

To clarify the concepts of local and global optimality, the map of a mountain landscape, in which lines indicate regions of similar response, is often used as a metaphor. The peak of a mountain always is a local optimum, because in its direct neighborhood there are no higher places. However, only the peak of the highest mountain is the global optimum. In many optimization problems, finding the global optimum is the challenge, and the presence of many local optima complicates the problem significantly. In the same vein, local optimizers are methods that always find the (usually local) optimum near the starting position, and global optimizers are methods that end up at the highest peak, no matter where the search started from.

Many methods utilize a sense of direction in which to proceed. Steepest ascent/descent methods (discussed below) choose the direction in which most short-term gains can be made. These invariably end up at the nearest local optimum, and are therefore very bad global optimizers. However, techniques that have a better search behavior can also be led astray. Problems in which the global optimum is located in another area, as suggested by the slope of the landscape, are described as deceptive, and much research has been devoted to these. It should be noted that no method guarantees finding the global solution in finite time.

For references see page 9688

1.2 Problem Types

Several different problem types can be distinguished. By far the most common in analytical chemistry is the numerical problem, where for a number of parameters optimal values must be found. These parameters might be settings for a chromatograph or another instrument, or coefficients in a nonlinear equation describing a curve. The parameters may interact, or may be constrained to certain regions, and the resulting problem landscape may therefore be quite complex.

The second type of problem is formed by combinatorial problems. The complexity of this kind of problem often is huge; most are nonpolynomial (NP)-complete, an indication meaning that when the size of the problem grows, the time needed to evaluate all possible solutions increases faster than the polynomial. In practice this means that even for small problems evaluating all possibilities is completely impossible. This also implies that only in special cases it is possible to know whether an optimum is truly a global optimum. Two basic forms can be distinguished: nonordered and ordered problems. A well-known example of nonordered problems is subset selection, which is a common issue in calibration and classification procedures. The problem relates to which subset of all available variables should be included in the statistical model. Not only does a small number of variables in model building lead to more robust and parsimonious models, it also may decrease experimental effort. Examples include wavelength selection from IR (infrared) spectra and variable selection in quantitative structure–activity relationship (QSAR) modeling, where a small set of relevant descriptors is to be selected from a potentially very large list.

Ordered combinatorial problems are also called sequential or scheduling problems. They are not so common in (analytical) chemistry but of tremendous importance in operations research, for example. The question is to put a set of objects in the optimal order, or in more than one dimension, at the optimal location. The traveling salesperson problem is a well-known example; microchip design and scheduling of analyses in a routine laboratory belong to this category. Just as problems from the previous class, sequential problems are more often than not NP-complete.

1.3 Example Problem: Fitting Laser-induced Fluorescence Spectra

An example used throughout this article concerns the description of laser-induced fluorescence (LIF) spectra of organic molecules.[3] This kind of spectroscopy can provide much information on the structure and dynamics of molecules and molecular complexes in ground states as well as excited states. The experimental spectrum of benzimidazole is shown in Figure 2.

Hundreds of well-resolved peaks can be seen, whose positions and intensities depend on a small number of parameters. Some of these parameters are related to experimental conditions such as temperature; others are related to quantum-chemical properties such as rotational constants. For small molecules, the spectra can be interpreted and in this way a set of valid parameters is found. However, for larger molecules the number of possibilities quickly becomes too large and one is forced to resort to search methods. For a given set of parameters the theoretical spectrum for a molecule, can be calculated, and so the agreement between theory and experiment can be used to assess the validity of the parameter set. The goal of optimization is to find parameters that yield a good approximation to the experimental spectrum. The difference is expressed as the root mean square (RMS) error.

In Figure 3 the RMS response surface is plotted if two parameters, the rotational constants around the two main axes, A and B, are systematically varied. All other parameters are kept constant at their true levels. This response surface is used throughout the article. Note that the set-up where only A and B are optimized is only a toy problem, although the contour lines are not very smooth.

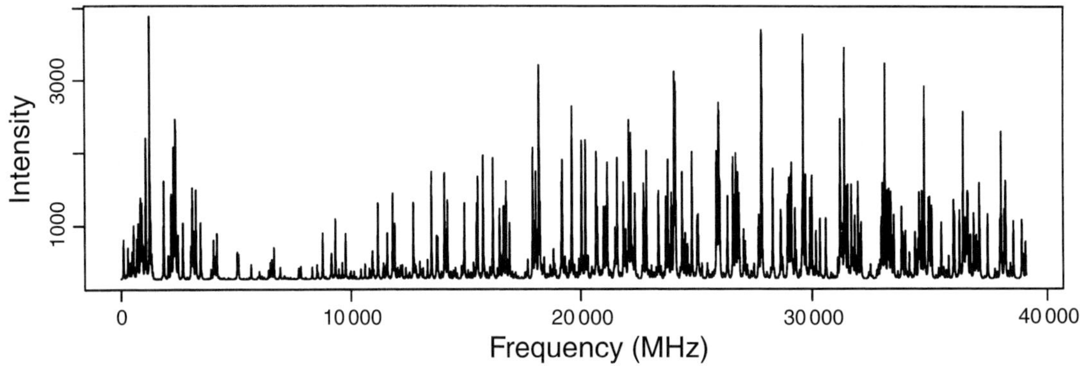

Figure 2 Part of the LIF spectrum of benzimidazole (22 797 data points).

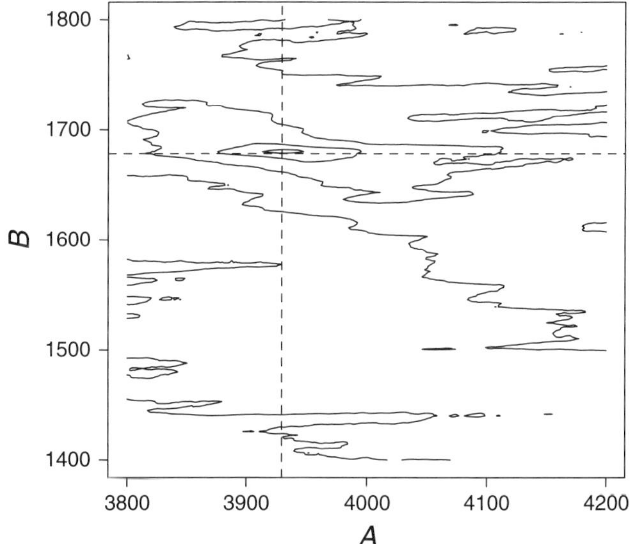

Figure 3 True response surface showing the RMS error between experimental and simulated LIF spectra of benzimidazole, depending on two parameters A and B. The true values for A and B are indicated by the dotted lines.

The real, high-dimensional optimization problem is much more difficult.

1.4 Criteria for Optimization

In an optimization problem a minimum or a maximum value is sought for a response variable as a function of problem parameters that can be controlled by the user. The extent to which these parameters can be controlled determines the precision of the outcome. It is of course senseless to try and optimize beyond experimental error; nevertheless, it is easy to forget the obvious.

Whereas the previous considerations concerned noise in the response value, there is also the case where the parameters can be influenced by noise. Moreover, if a small change in optimal parameter settings (e.g. noise) leads to a drastic decrease in product quality, it may be useful to look not for the optimal solution but for solutions which are acceptable and not so sensitive to small deviations from optimality. In an industrial context robust and good solutions are often of greater interest than the very best solution, which may be unstable.

In other cases solutions proposed by an optimization algorithm may be physically impossible or economically undesirable. These constraints are not always easy to define. If that is the case, one alternative is to generate several solutions and pick from those the ones that least violate the boundary conditions.

Good optimization procedures should yield the desired information in a fast, reproducible, and reliable way. This means that repeated application should yield more or less the same answers or answers of the same quality. In some cases it may be important that all acceptable solutions are found, in other cases it is more important that the global optimum is found or that at least one good solution is found within time or cost constraints. These aspects are to some extent determined by the problem at hand.

1.5 Multicriteria Optimization

If additional constraints can be defined, often in terms of costs, it may be a good idea to include them in the response from the experiments. There are several ways of doing this, the simplest of which is a weighted sum of several response variables. However, the weights need to be set to realistic values and in some cases this requires quite a bit of fine tuning. Another approach is to consider several response variables at a time. One solution then dominates another if it is better at one or more of the response variables while not being worse for all other criteria. This is called Pareto optimality, after the Italian economist Vilfredo Pareto. The result is a set of solutions that dominate other solutions at least for a number of variables. The user than has to make the trade-off between optimality in one variable versus optimality in the other. Especially with a small number of responses it may be that a useful Pareto-optimal set of solutions can be found.

Optimization problems where the robustness of the solution is important form a good example of the possibilities of multicriteria optimization. Then one criterion is the quality of the optimum and the other the sensitivity of the response to small changes in the parameters (Figure 4). Fifty random samples were taken from the contour surface of Figure 3, and the RMS error (x-axis) is plotted against the difference in RMS

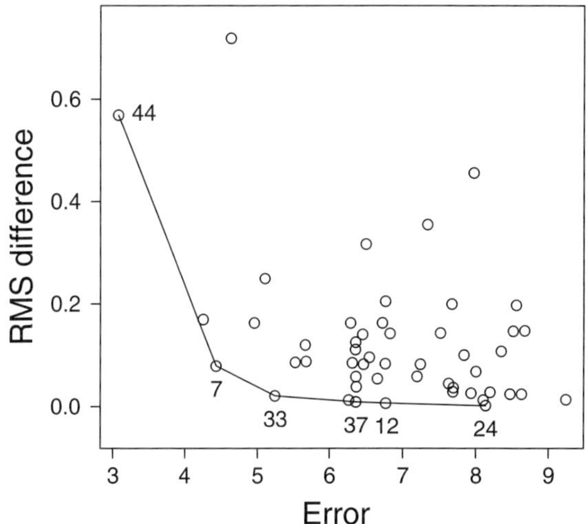

Figure 4 Pareto-optimality. Solutions 44, 7, 33, 37, 12, and 24 are on the Pareto-optimal front, where attention shifts from a small error to a smaller difference with neighboring points.

For references see page 9688

value with the surrounding points (y-axis). The smaller this RMS difference, the more robust the solution. A front of Pareto-optimal solutions delimits a cloud of possible solutions. Note that the point with the lowest error (number 44) is not very robust, so in cases where robustness is important solution 7 might be a better alternative.

2 STRONG METHODS

Strong methods assume a certain structure in the solution space, such as a quadratic response surface or only one optimum. If the assumptions are correct, these methods are fast and reliable. The main problem is that if the assumptions are not correct, these methods will not find the global optimum even if the number of experiments is increased drastically. In difficult optimization problems they are often used for the last part of the optimization, where the approximate global optimum has been identified and the exact location still must be found. Especially in low-dimensional optimization problems these methods are often very suitable. Examples include optimization in organic synthesis[4,5] and method development in high-performance liquid chromatography (HPLC).[6,7]

2.1 Gradient-based Optimization

The most intuitive methods of this class rely on an accurate estimate of the local gradient, and proceeding in the steepest direction. Depending on whether we are dealing with a minimization or maximization problem, the method is called "steepest descent" or "steepest ascent", respectively.

The simplest and rather naive form is to assess the quality of neighboring points and, if there are better points, to proceed towards the best of these. Then, new neighboring points are evaluated and this process continues until there is no neighbor better than the current point. This process is depicted in Figure 5, using the error surface from Figure 3. Clearly, only from one of the 16 starting points the global optimum is located. The other starting points lead to (sometimes even quite bad) local optima. Each optimum has its own basin of attraction, indicating that any steepest descent search starting from a point within that basin will end up at that particular optimum. This, of course, is the definition of a local search method.

In some cases, especially those where the landscape being searched looks like a long narrow valley (in the case of minimization), steepest descent does very poorly. Unless the valley is approached at a perfect right-angle, it takes many small steps to reach the optimum. The more sophisticated conjugate gradient (CG) methods[8] determine the new direction based not solely on the current gradient but on old gradients as well. This leads to a much better search behavior.

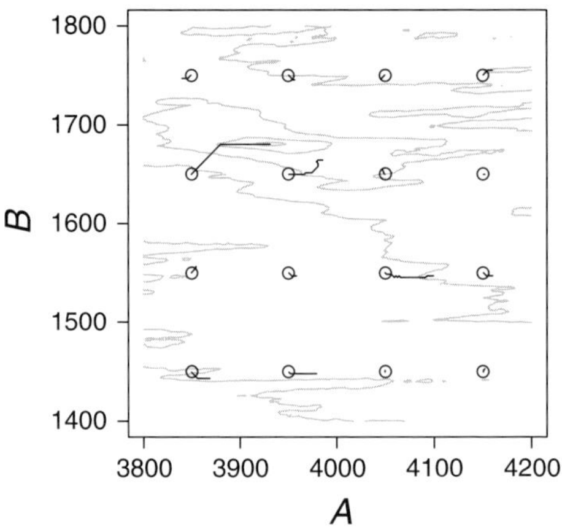

Figure 5 Gradient optimization (naive form) starting from 16 points. The global optimum was reached in 52 steps from the starting point at (3850, 1650). The mean number of steps before the gradient optimizations stopped was just above 12.

Inclusion of the second derivative (the Hessian matrix) is often used in curve-fit problems, or nonlinear least-squares routines. The quantity that is minimized must follow a χ^2 distribution, which is often the case with errors from a fit. Calculation of the Hessian is possible precisely because of the assumption of the χ^2 distribution. The standard method, which combines steepest descent and Hessian approaches, is the Marquardt–Levenberg method,[8] using the former far from the optimum and the latter in the neighborhood of the optimum.

2.2 Response Surface Methods

Response surface methods stem from the area of experimental design,[4,9] where in as few experiments as possible the maximum amount of information is extracted from a system. It is assumed that the response surface can be parameterized by a simple function with one optimum. The aim is to estimate the response function surface by choosing the experimental conditions intelligently. An often-used set-up is central composite design (CCD) depicted in Figure 6(a). Four points, arranged in a square, are combined with five points in the form of a star. In cases where the response is subject to error, one or more points (often the central point) may be replicated to obtain an estimate of the standard deviation. The fitted surface, using all terms to second degree, is given by Equation (1):

$$z = 433.7 - 0.061x - 0.371y - 1.1210^{-5}x^2 - 3.8810^{-6}y^2 + 9.4010^{-5}xy \quad (1)$$

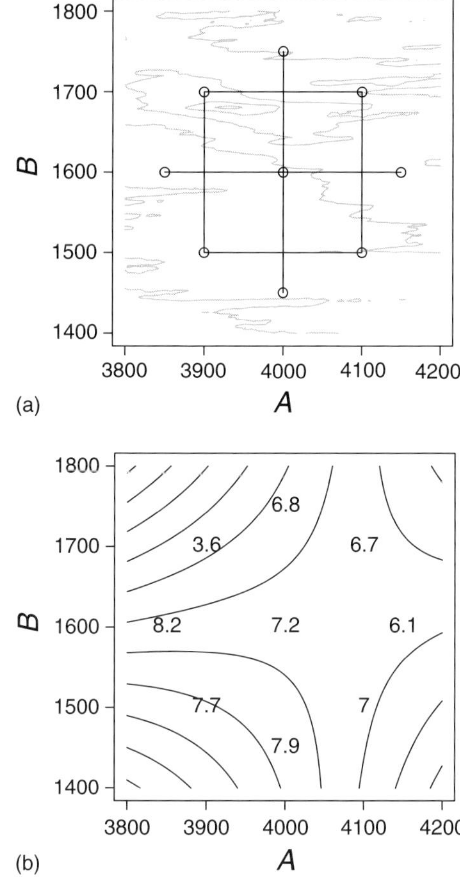

Figure 6 (a) Application of a CCD to the response surface of Figure 3. (b) Contour lines of the fitted surface; error values at the experimental points are indicated. Optimal values according to the model are the upper left and lower right corners.

where z is the error, x is A and y is B. The resulting surface is depicted in Figure 6(b).

This figure also clearly demonstrates the danger of extrapolating. According to the model, the optimal values are to be found in the upper left or lower right corners. One should only use these designs in the neighborhood of an optimum where one can be reasonably sure that the surface behaves as expected. Even very close to the global optimum this may not be the case.

2.3 Simplex Methods

Repeated application of small experimental designs can lead to the accurate and precise estimate of the optimum. Although sampling points can be reused, significantly decreasing the experimental effort, most response surface methods become impractical in spaces with more than three or four dimensions.

A simple extension that is more practical in spaces of moderate dimension is the simplex method.[10] Instead of picking points in the search space according to an experimental design, one starts with the smallest number of experiments defining a surface in space; for example, for a two-dimensional space three points are chosen or, for a three-dimensional space, four points are chosen. This set of experiments, or points, is called the initial simplex. Next, one iterates by repeatedly selecting the worst point of the simplex and replacing it with a new point, obtained by mirroring the worst point in the side defined by the remaining points.

In two dimensions, this yields a series of adjoining triangles, as depicted in Figure 7(a), where the simplex method is applied to the error surface of Figure 3. The simplex is able to locate the global optimum approximately. The size of the simplex remains constant. This prohibits the method from finding an optimum with a better accuracy, and leads to an inefficient search behavior in smooth and monotonous error surfaces. An extension is the so-called modified simplex method,[11] where the size of the simplex can be increased for more rapid convergence, or decreased for a more precise search. The results of this method are depicted in Figure 7(b). However, the optimum is found much more precisely,

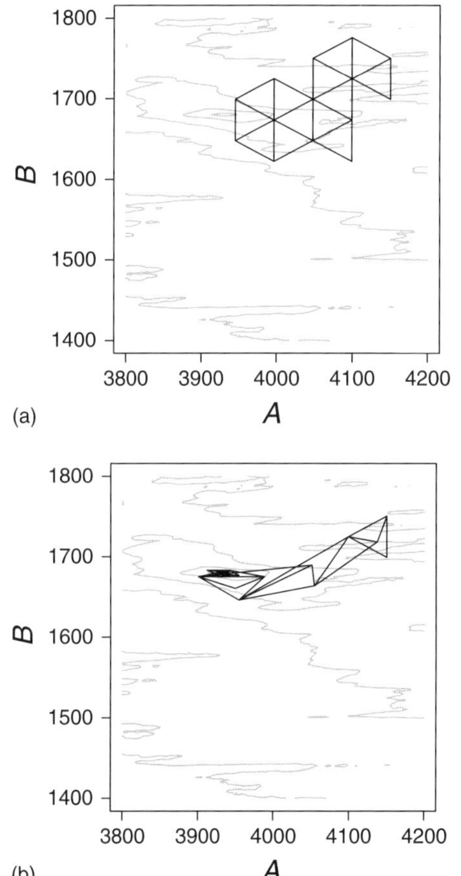

Figure 7 (a) Simplex method and (b) modified simplex method.

For references see page 9688

and in several cases the method is able to take big steps in the right direction, thus speeding up the optimization considerably.

In conclusion, the simplex method (and especially the modified simplex method) is an attractive optimization method in problems where experimentation is difficult or expensive, and where a number of other constraints are satisfied. The most important of these is the requirement that there is only one optimum, because otherwise the simplex method may easily get stuck in a local optimum. This also implies that there should not be too much noise on the response variable, because this may lead to a projection of the wrong simplex vertex. In some cases this merely decreases the efficiency of the algorithm, in other cases it leads to completely wrong answers.[12]

Although a number of experiments are performed, the results cannot readily be used to calculated response surfaces, because the experiments may be concentrated in one area of the search space. However, response-surface methods explicitly cover the relevant space. Simplex methods are very useful if improvement rather than an optimal solution is sought.

3 WEAK METHODS

Weak methods make almost no assumptions, but at a certain cost. In many cases they are not very effective, and they should only be used in very specific circumstances. The ultimate weak technique, evaluating all possible solutions, is practically infeasible for all but the most simple problems. Therefore, weak techniques sample the solution space in order to locate areas of good solutions. A strong random component is used in either selecting the sampling locations or defining the orientation of a group of samples. Repeated application of these methods may lead to quite different results, so the reproducibility of these searches is low.

3.1 Random Search

Random search is not a strategy many people would use, unless there is no other alternative. Yet in some molecular mechanics software packages it is implemented to serve as a reference point for other optimization procedures. The strategy is simple: just keep on trying new candidate solutions and keep the best one(s) until the time is up. The only situation in which this strategy is expected to be not worse than other methods is for "needle-in-a-haystack" problems – the landscape is completely flat and does not give any indication that one is near the optimum until the optimum is actually found. In such cases an exhaustive search is the only method guaranteed to find a solution, but is often not feasible. A typical example is molecular recognition – seemingly small changes to an active molecule can lead to a drastic decrease in activity. It is therefore not surprising that, especially in combinatorial chemistry, the random search strategy has led to the discovery of several interesting and completely new drug leads, although also in this field more sophisticated techniques are being investigated.

3.2 Sampling Methods

A more systematic way of sampling a landscape is to use a grid of a predefined size. Additional grids with smaller spacings may be placed in promising areas. An example of this is depicted in Figure 8(a), where the grid of 16 points used in the gradient search is followed by two other sampling steps. A disadvantage of this method is that phenomena smaller than the grid spacing are

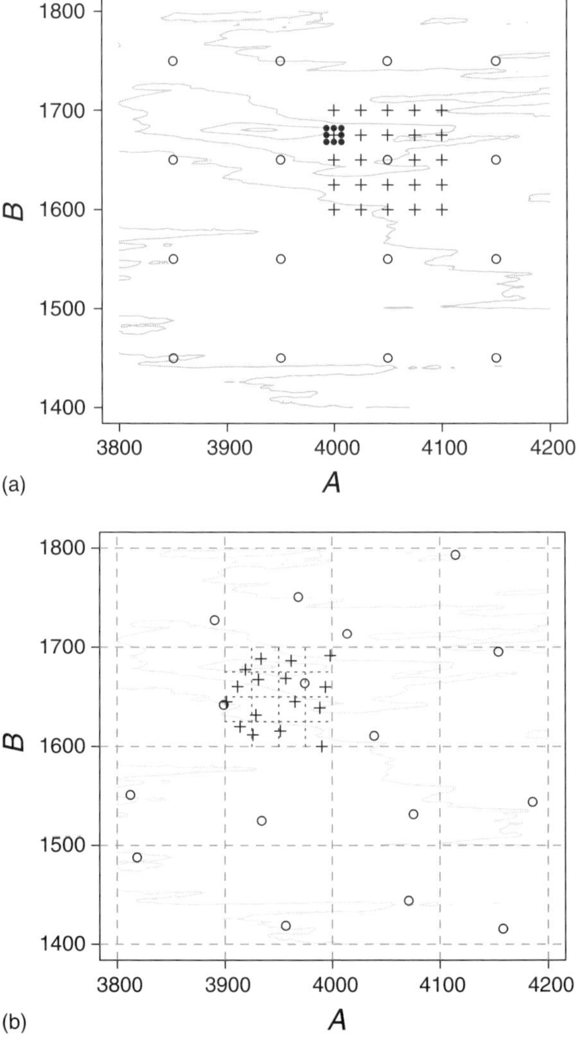

Figure 8 (a) Grid sampling and (b) stratified sampling. The spacing in the grids is 100, 25, and 7 units, respectively. The strata sizes are 100 × 100 and 25 × 25 units.

easily overlooked. Moreover, in search spaces with more dimensions the number of samples required increases drastically. In some well-understood problems of low dimension, however, this method can be used very well. The main advantage is its versatility – no attempt is made to parameterize the response surface and, if more than one optimum is found, several grids can be evaluated in parallel.

The stratified random search combines grid search with random search, by dividing the search space into strata and taking a random sample from each stratum. The problem that phenomena with a size smaller than the grid spacing may be overlooked is somewhat less acute in this approach. However, especially with large strata, some samples may be very close to each other leaving large parts of the search space empty. The number of samples per stratum may be increased if this is a problem. Again, a sequential approach may be used in which larger strata are followed by smaller ones. An example is indicated in Figure 8(b). There, the square in the original grid with the best sample value is further subdivided into smaller strata. Alternatively, a new set of strata can be centered at the location of the best sample.

Although sampling methods are grouped with the class of weak optimization methods, for a repeated application with decreasing intersample distance one must assume a more or less continuous form of response surface. This is not a very demanding assumption, however. The main problem with sampling approaches is the chance of missing important optima with a basin of attraction smaller than the intersample distance. In more than two or three dimensions this will frequently occur, because the number of sampling points must be kept manageable.

4 INTERMEDIATE METHODS

Stochastic optimization methods have been developed to counter the weaknesses of classical optimization methods in high-dimensional search problems. Instead of following a fixed path that is only determined by the choice of the starting point, they feature a strong random component, mainly to avoid getting trapped in local optima. Although finding the optimal solution is not guaranteed for all methods, in almost all cases these methods will find very good solutions. The most important methods are Monte Carlo simulated annealing (MCSA, or SA for short), GA and evolutionary strategies (ES), and tabu search (TS). The principles behind these methods are analogous to the physical cooling of a liquid (SA), Darwin's theories of evolution (GA, ES) and learning processes (TS), respectively.

Although the roots of these methods lie in the 1960s, it was not until the mid-1980s (SA and GA) or later (TS) that these methods were applied in chemistry. The results clearly indicated that the methods offered solutions not achievable with the classical optimization methods.

4.1 Monte Carlo Simulated Annealing

MCSA, also known as the Metropolis algorithm, is probably the most used optimization method in chemistry today. Although the principle of the method was used in 1953 by Metropolis et al.,[13] it only became popular after 1983 when Kirkpatrick et al.[2] described its use as a general optimization scheme. The method is implemented in a large number of software packages and is the de facto standard in structure optimization problems.[14]

The principle of the method is very simple indeed. A random walk is performed in the search space, accepting all moves that lead to a better solution, and accepting moves that lead to a worse solution with a probability $e^{-\Delta E/T}$. This is analogous to the well-known Boltzmann distribution. If T, a control parameter often called the temperature is large, the chance of accepting a bad move is large as well and the search truly resembles a random walk. The search characteristics of SA are obtained by gradually lowering the temperature so that bad moves are more and more often rejected. It is clear that the cooling scheme is of prime importance for the success of the method. Indeed, proofs have been published guaranteeing convergence to the global optimum provided the cooling is done slowly enough. In practice, this proof is not very useful because adequate cooling may be infinitely slow. Many variants exist, including schemes that take into account the distance to the optimal evaluation function value (e.g. error = 0.0; this particular form of SA is called generalized simulated annealing (GSA)) or algorithms that use a Cauchy rather than a Boltzmann distribution.

The only parts that must be provided by the user are an evaluation function, a procedure to generate a new trial state and a cooling scheme. In particular the latter two have a profound influence on the performance of the algorithm and in some cases may be difficult to define. The algorithm is stopped when either the optimal solution has been found, a predefined number of evaluations is reached, or no improvement has been obtained for a specified number of evaluations.

A simple example using the error surface of Figure 3 is depicted in Figure 9(a). The SA is started from the center of the permissible region. A step consists of moving to one of the 24 closest points in the neighborhood. The temperature factor was set to 575 energy units at the beginning and gradually lowered to 525 units at the last of the 300 iterations. Nearly 40% of all proposed moves was accepted. As can be seen in the right plot in Figure 9(b), the algorithms wander around aimlessly for

For references see page 9688

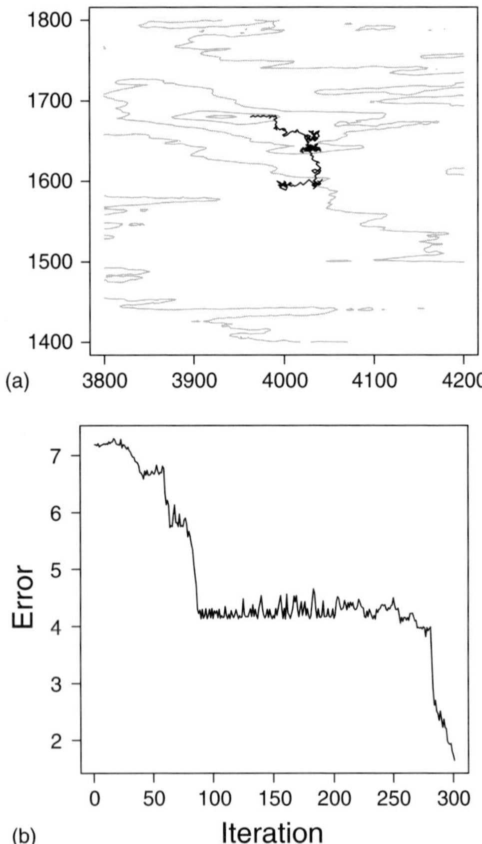

Figure 9 (a) SA to locate the optimum. Starting from the center of the admissible region, 300 iterations are performed. (b) The development of the error is depicted. Note the long period between iterations 80 and 260 where no improvement is made.

several periods of time, whereas in other periods rapid improvements are made.

Although in this case the SA finds the optimum without much trouble, in real problems a fair amount of fine tuning is usually necessary, especially in the move-generating and cooling procedures. Many practitioners use cooling schedules that keep the ratio of rejected versus accepted moves at around 1. Because of the stochastic nature of SA, it is advisable to perform several runs from different starting points. There is no guarantee of finding the global optimum, especially in more difficult problems.

4.2 Tabu Search

Whereas SA performs an optimization by randomly generating a new state and applying the Boltzmann rule to decide whether to accept or reject, TS uses sophisticated history mechanisms to avoid evaluating solutions that are already known.[15] In its simplest form, TS generates a new state by selecting the best available move, excluding those states that are members of the tabu list T. To prevent memory problems, however, a tabu list usually contains moves, or components of moves, instead of states. This list of tabu moves is continually updated and consists of those moves (or move components) that would undo the most recent iterations in the search process. The net effect of the tabu list is that cycles with a length smaller than the length of the tabu list are avoided.

The TS paradigm is made more sophisticated by inclusions of aspiration criteria. These play the opposite role of the tabu moves – a move is regarded as admissible if its aspiration criteria are satisfied. Examples include situations where a move would lead to a better solution than all previously visited solutions, or where a move (component) has not been executed for a long time. Furthermore, intermediate- and long-term memory functions are used to identify attribute values that are clearly important for good solutions, and to diversify the search, respectively. Thus, the search can be focused to particular regions in space or the algorithm can escape from local optima.

The separate components of TS methods are much more problem dependent than is usually the case with other optimization methods. However, they are intuitively appealing and with expert knowledge about the problem not too difficult to define. The number of applications in chemistry is still limited, mainly because the method has been used primarily for sequential problems. Recently, several applications have been reported on structure optimization problems,[15] assessing chemical similarity,[16] and molecular docking.[17]

4.3 Evolutionary Optimization

A large number of different techniques fall under the header of evolutionary optimization, most notably evolutionary algorithms (EA)[18] and GAs.[19,20] Their main characteristic is that they use a *population* of trial solutions, instead of proceeding from one trial solution to the next. This may have several advantages. For example, in cases where one is interested not only in the global optimum but also in other good solutions, population-based methods may prevent the same optimum being found continuously. Thus, in principle, a more diverse set is found. Examples where this is very important are abundant – in chemistry, molecular structure optimization is used to find not only the global energy minimum but also all other structures that are feasible at room temperature.

The second important aspect of the use of a population is that there is no need to define how new solutions are obtained. In methods such as TS or SA one has to define how to proceed from the current solution to the next, but in EA this is typically performed by standard operators such as cross-over and mutation. Where the latter in most applications is a more or less random disruption of a member of the population, the former is specific for evolutionary optimization. In analogy with nature, two (or more) members of the population mate and exchange genetic material to produce children. The genetic material consists of parts of the solutions, and is coded either in binary form or as real numbers. Cross-over is also governed by chance, so no attempt is made to identify "good" parts of a solution for reproduction. The best genetic material is more or less automatically preserved because solutions of a high quality will be selected for parenthood more often than bad solutions. The "genes" of the latter will gradually be removed from the population as the evolution goes on for a number of generations.

The success of EAs depends on the combination of representation and evolutionary operators. In general, evolutionary operators that yield invalid solutions (such as parameter values outside a permissible range, or impossible chemical structures in a structure-manipulation application) are to be avoided. Many applications have been described in molecular structure optimization, spectrum interpretation, subset selection for QSAR, and multivariate calibration and other fields.[21,22]

4.4 Hybrid Methods

Hybrid methods combine the characteristics of different optimization methods. Typically, the basic optimization cycle of Figure 1 is composed of elements from the individual methods. Of course, many methods differ only in one aspect. The main difference between TS and steepest descent methods, for example, is in the generation stage – some solutions are forbidden by the tabu list or explicitly allowed, but each method selects the best of the admissible moves. One further difference is that TS may accept a worse solution if that is the best one available, whereas steepest descent then stops.

Of course, hybrid methods are most useful when they combine the best features of their components. One approach which has been used in practice is a hybrid form of a GA and SA. The GA is used to maintain a pool of trial solutions and to generate offspring; the SA governs the selection part by accepting all improving solutions and accepting worse solutions according to the Boltzmann distribution. The advantages of the GA are that inadmissible regions of the search space are handled gracefully and do not constitute a barrier that a vintage SA may find difficult to cross. The advantage of the SA is the added control that the user may exert through the cooling scheme. For the selection and generation stages, typical strategies are summarized in Table 1. Other possible hybrids are GA/TS (a pool of solutions is maintained and the random reproduction mechanism is supervised by TS methods) and SA/TS (probabilistic TS, where the chance of accepting a new solution is governed by the SA Boltzmann distribution).

Table 1 Selection and generation characteristics for SA, TS, and GA

Method	Characteristic
Selection	
SA	Accept new solution if better, else accept with probability $e^{-\Delta E/T}$
TS	Accept best of all admissible new solutions, even if it is worse than the previous solution
GA	Accept solutions according to their relative quality in the population
Generation	
SA	Randomly pick one solution out of the neighborhood
TS	Generate all solutions in neighborhood
GA	Use random-based procedures such as cross-over and mutation to generate new solutions

The main strengths of all these methods – their versatility and their ability to adapt to all kinds of problems – are also their weakness. No generally accepted strategies for selecting, adapting and fine tuning these global search methods exist. If one is lucky, the methods work straight away. However, one may not be lucky, and although there are a million ways to alter and tweak the algorithms, no guidelines exist as to which one is best. Fortunately, in many cases suitable starting points and standard settings can be found in the literature for analogous problems.

4.5 Comparison of Methods

The question as to which of the intermediate methods mentioned so far is most suited for a particular problem is not easily answered, for several reasons. Although they are often called "global" optimization methods, in many practical problems there is no guarantee that the true global optimum will be found. Replicate runs may find radically different solutions, either because of the stochastic nature of the search performed (all but TS) or because of the influence of the starting position (all methods). This means that for a thorough comparison many replicate runs should be performed on a representative set of example problems. As these problems usually take a fair amount of computing time, extensive comparisons between algorithms are rarely performed.

A second problem is that each algorithm comes in many variants, each with a set of optimization parameters that can be fine tuned. This tuning may be critical to the performance of the algorithm and constitutes a meta-optimization problem. It is also difficult to give each of the

For references see page 9688

algorithms tested the same level of attention – often, one of the techniques is already known to the implementors and this may give it a decisive advantage.

Finally, there is no clear-cut criterion with which to judge the performance of the optimization methods. Simply looking at the quality of the best solution clearly does not provide all the information. Therefore, other criteria are needed.[17,23] If these are not explicitly based on the value of the evaluation function, the form of this function can be tuned as well. For a successful application of these intermediate search methods, the ensemble of representation, the search operations, and the evaluation function must be adjusted to each other.

5 CONCLUSIONS AND OUTLOOK

The field of optimization is broad and has applications in all areas of chemistry. Naturally, many different methods have been used and in some cases even developed by chemists. In this article, an overview is given where the methods have been classified as "classical" or "nonclassical". The former category comprises methods that are very good in dealing with low-dimensional problems. An example is the optimization of a chromatographic experiment, where optimal separation conditions should be found in as few experiments as possible. These methods are said to be strong because they incorporate much knowledge from the user about the problem at hand. The classical category also embraces weak methods where knowledge is not a prerequisite. As the efficiency of these methods is in many cases very low, they are not often used in chemistry.

The nonclassical optimization techniques described in this article have been specifically developed for those cases where the classical techniques were not suitable – high-dimensional search problems with many local optima. Because the number of evaluations may be quite high they usually are applied in connection with computer experiments rather than with laboratory experiments. A notable exception is found in the field of combinatorial chemistry, where GAs have been used to guide the synthesis of new promising lead molecules.[24]

The classical and nonclassical methods have complementary roles in chemistry. With the increasing role of computational chemistry, the nonclassical methods will continue to flourish. However, due to their complicated nature and great versatility, they will remain the playground of experts in the near future. With classical methods the situation is different. Many standard computer software packages are available for experimental design, and are sometimes even included in laboratory instruments. Although the principles of the methods should still be understood by the user, no programming is required, thereby significantly lowering the threshold for their use.

ABBREVIATIONS AND ACRONYMS

CCD	Central Composite Design
CG	Conjugate Gradient
EA	Evolutionary Algorithms
ES	Evolutionary Strategies
GA	Genetic Algorithms
GSA	Generalized Simulated Annealing
HPLC	High-performance Liquid Chromatography
IR	Infrared
LIF	Laser-induced Fluorescence
MCSA	Monte Carlo Simulated Annealing
NP	Nonpolynomial
QSAR	Quantitative Structure–Activity Relationship
RMS	Root Mean Square
SA	Simulated Annealing
TS	Tabu Search

RELATED ARTICLES

Pharmaceuticals and Drugs **(Volume 8)**
Quantitative Structure–Activity Relationships and Computational Methods in Drug Discovery

Process Instrumental Methods **(Volume 9)**
Chemometric Methods in Process Analysis

Chemometrics **(Volume 11)**
Chemometrics • Clustering and Classification of Analytical Data • Multivariate Calibration of Analytical Data • Second-order Calibration and Higher • Signal Processing in Analytical Chemistry

Infrared Spectroscopy **(Volume 12)**
Spectral Data, Modern Classification Methods for

REFERENCES

1. T. Schlick, 'Optimization Methods in Computational Chemistry', in eds. K.B. Lipkowitz, D.B. Boyd, *Reviews in Computational Chemistry*, VCH, New York, Chapter 1, Vol. 3, 1992.
2. S. Kirkpatrick, C.D. Gelatt, M.P. Vecchi, 'Optimization by Simulated Annealing', *Science*, **220**, 671–680 (1983).

3. G. Berden, W. Leo Meerts, E. Jalviste, 'Rotationally Resolved UV Spectroscopy of Indole, Indazole and Benzimidazole, Inertial Axis Reorientation in the $S_1(^1L_b) \leftarrow S_0$ Transitions', *J. Chem. Phys.*, **103**, 9596–9606 (1995).
4. G.E.P. Box, W. Hunter, J. Hunter, *Statistics for Experimenters. An Introduction to Design, Data Analysis and Model Building*, John Wiley & Sons, New York, 1978.
5. R. Carlson, *Design and Optimization in Organic Synthesis*, Elsevier, Amsterdam, 1992.
6. J.C. Berridge, *Techniques for the Automated Optimization of HPLC Separations*, Wiley, New York, 1985.
7. J.L. Glajch, L.R. Snyder, *Computer-assisted Method Development for HPLC*, Elsevier, Amsterdam, 1990.
8. W.H. Press, B.P. Flannery, S.A. Teukolsky, W.T. Vetterling, *Numerical Recipes in C*, Cambridge University Press, Cambridge, 1988.
9. D.L. Massart, B.G.M. Vandeginste, L.M.C. Buydens, S. de Jong, P.J. Lewi, J. Smeyers-Verbeke (eds.), *Handbook of Chemometrics and Qualimetrics: Part A, Data Handling in Science and Technology*, Elsevier Science, Vol. 20A, 1998.
10. W. Spendley, G.R. Hext, F.R. Himsworth, 'Sequential Application of Simplex Designs in Optimization and Evolutionary Operations', *Technometrics*, **4**, 441–461 (1962).
11. J.A. Nelder, R. Mead, 'A Simplex Method for Function Optimization', *Computer J.*, **7**, 308–313 (1965).
12. F.H. Walter, L.R. Parker, S.L. Morgan, S.N. Deming, *Sequential Simplex Optimization*, CRC Press, Boca Raton, 1991.
13. N. Metropolis, A.W. Rosenbluth, M.N. Rosenbluth, A.H. Teller, E. Teller, 'Equation for State Calculation by Fast Computing Machines', *J. Chem. Phys.*, **21**, 1087–1091 (1953).
14. J.H. Kalivas (ed.), *Adaption of Simulated Annealing to Chemical Optimization Problems: Data Handling in Science and Technology*, Elsevier Science, Vol. 15, 1995.
15. F. Glover, M. Laguna, *Tabu Search*, Kluwer, Boston, 1997.
16. V. Kvasnicka, J. Pospichal, 'Fast Evaluation of Chemical Distance by Tabu Search Algorithm', *J. Chem. Inf. Comput. Sci.*, **34**, 1109–1112 (1994).
17. D.R. Westhead, D.E. Clark, C.W. Murray, 'A Comparison of Heuristic Search Algorithms for Molecular Docking', *J. Comput.-Aided Mol. Des.*, **11**, 209–228 (1997).
18. H.P. Schwefel, *Numerical Optimization of Computer Models*, Wiley, Chichester, 1981.
19. D.E. Goldberg, *Genetic Algorithms in Search, Optimization and Machine Learning*, Addison-Wesley, 1989.
20. L. Davis (ed.), *Handbook of Genetic Algorithms*, Van Nostrand Reinhold, New York, 1991.
21. J. DeVillers (ed.), *Genetic Algorithms in Molecular Modeling, Principles of QSAR and Drug Design*, Academic Press, 1996.
22. R.S. Judson, 'Genetic Algorithms and Their use in Chemistry', in eds. K.B. Lipkowitz, D.B. Boyd, *Reviews in Computational Chemistry*, VCH, 1–73, 1997.
23. R. Wehrens, E. Pretsch, L.M.C. Buydens, 'Quality Criteria of Genetic Algorithms for Structure Optimization', *J. Chem. Inf. Comput. Sci.*, **38**, 151–157 (1998).
24. J.L. Fauchere, J.A. Boutin, J.M. Henlin, N. Kucharczyk, J.C. Ortuno, 'Combinatorial Chemistry for the Generation of Molecular Diversity and the Discovery of Bioactive Leads', *Chemometr. Intell. Lab. Systems*, **43**, 43–68 (1998).

Clustering and Classification of Analytical Data

Barry K. Lavine
Clarkson University, Potsdam, USA

Clustering and classification are the major subdivisions of pattern recognition techniques. Using these techniques, samples can be classified according to a specific property by measurements indirectly related to the property of interest (such as the type of fuel responsible for an underground

For references see page 9708

spill). An empirical relationship or classification rule can be developed from a set of samples for which the property of interest and the measurements are known. The classification rule can then be used to predict the property in samples that are not part of the original training set. The set of samples for which the property of interest and measurements is known is called the training set. The set of measurements that describe each sample in the data set is called a pattern. The determination of the property of interest by assigning a sample to its respective category is called recognition, hence the term pattern recognition.

For pattern recognition analysis, each sample is represented as a data vector $\boldsymbol{x} = (x_1, x_2, x_3, x_j, \ldots, x_n)$, *where component* x_j *is a measurement, e.g. the area* a *of the* j*th peak in a chromatogram. Thus, each sample is considered as a point in an* n*-dimensional measurement space. The dimensionality of the space corresponds to the number of measurements that are available for each sample. A basic assumption is that the distance between pairs of points in this measurement space is inversely related to the degree of similarity between the corresponding samples. Points representing samples from one class will cluster in a limited region of the measurement space distant from the points corresponding to the other class. Pattern recognition (i.e. clustering and classification) is a set of methods for investigating data represented in this manner, in order to assess its overall structure, which is defined as the overall relationship of each sample to every other in the data set.*

1 INTRODUCTION

Since the early 1980s, a major effort has been made to substantially improve the analytical methodology applied to the study of environmental samples. Instrumental techniques such as gas chromatography, high-performance liquid chromatography (HPLC) and X-ray fluorescence spectroscopy have dramatically increased the number of organic and inorganic compounds that can be identified and quantified, even at trace levels, in the environment. This capability, in turn, has allowed scientists to attack ever more complex problems, such as oil and fuel spill identification, but has also led to an information-handling problem.[1]

The reason for this problem is that in any monitoring effort it is necessary to analyze a large number of samples in order to assess the wide variation in composition that an environmental system may possess. The large number of samples that must be analyzed and the number of constituents that must be measured per sample give rise to data sets of enormous size and complexity. Often, important relationships in these data sets cannot be uncovered, when the data are examined one variable at a time, because of correlations between measurement variables, which tend to dominate the data and prevent information from being extracted.

Furthermore, the relationships sought in the data often cannot be expressed in quantitative terms, such as the source of a pollutant in the environment. These relationships are better expressed in terms of similarity or dissimilarity among groups of multivariate data. The task that confronts the scientist when investigating these sorts of relationships in multivariate data, is twofold:

- Can a useful structure based on distinct sample groups be discerned?
- Can a sample be classified into one of these groups for the prediction of some property?

The first question is addressed using principal component analysis (PCA)[2] or cluster analysis,[3] whereas the second question is addressed using pattern recognition methods.[4]

PCA is the most widely used multivariate analysis technique in science and engineering.[5] It is a method for transforming the original measurement variables into new variables called principal components. Each principal component is a linear combination of the original measurement variables. Often, only two or three principal components are necessary to explain all of the information present in the data. By plotting the data in a coordinate system defined by the two or three largest principal components, it is possible to identify key relationships in the data, that is, find similarities and differences among objects (such as chromatograms or spectra) in a data set.

Cluster analysis[6] is the name given to a set of techniques that seek to determine the structural characteristics of a data set by dividing the data into groups, clusters, or hierarchies. Samples within the same group are more similar to each other than samples in different groups. Cluster analysis is an exploratory data analysis procedure. Hence, it is usually applied to data sets for which there is no a priori knowledge concerning the class membership of the samples.

Pattern recognition[7] is a name given to a set of techniques developed to solve the class-membership problem. In a typical pattern recognition study, samples are classified according to a specific property using measurements that are indirectly related to that property. An empirical relationship or classification rule is developed from a set of samples for which the property of interest and the measurements are known. The classification rule is then used to predict this property in samples that are not part of the original training set. The property in question may be the type of fuel responsible for a spill, and the measurements are the areas of selected gas chromatographic

List of selected abbreviations appears in Volume 15

(GC) peaks. Classification is synonymous with pattern recognition, and scientists have turned to it and PCA and cluster analysis to analyze the large data sets typically generated in monitoring studies that employ computerized instrumentation.

This article explores the techniques of PCA, cluster analysis, and classification. The procedures that must be implemented to apply these techniques to real problems are also enumerated. Special emphasis is placed on the application of these techniques to problems in environmental analysis.

2 PRINCIPAL COMPONENT ANALYSIS

PCA is probably the oldest and best known of the techniques used for multivariate analysis. The overall goal of PCA is to reduce the dimensionality of a data set, while simultaneously retaining the information present in the data. Dimensionality reduction or data compression is possible with PCA because chemical data sets are often redundant. That is, chemical data sets are not information rich. Consider a gas chromatogram of a JP-4 fuel (Figure 1), which is a mixture of alkanes, alkenes, and aromatics. The gas chromatogram of a JP-4 fuel is characterized by a large number of early-eluting peaks, which are large in size. There are a few late-eluting peaks, but their size is small. Clearly, there is a strong negative correlation between the early- and late-eluting peaks of the JP-4 fuel. Furthermore, many of the alkane and alkene peaks are correlated, which should not come as a surprise as alkenes are not constituents of crude oil but instead are formed from alkanes during the refining process. In addition, the property of a fuel most likely to be reflected in a high resolution gas chromatogram is its distillation curve, which does not require all 85 peaks for characterization.

Redundancy in data is due to collinearity (i.e. correlations) among the measurement variables. Collinearity diminishes the information content of the data. Consider a set of samples characterized by two measurements, X_1 and X_2. Figure 2 shows a plot of these data in a two-dimensional measurement space, where the coordinate axes (or basis vectors) of this measurement space are the variables X_1 and X_2. There appears to be a relationship between these two measurement variables, which suggests that X_1 and X_2 are correlated, because fixing the value of X_1 limits the range of values possible for X_2. If the two measurement variables were uncorrelated, the enclosed rectangle in Figure 2 would be fully populated by the data points. Because information is defined as the scatter of points in a measurement space, it is evident that correlations between the measurement variables decrease the information content of this space. The data points, which are restricted to a small region of the measurement space due to correlations among the variables, could even reside in a subspace if the measurement variables are highly correlated. This is shown in Figure 3. Here X_3 is

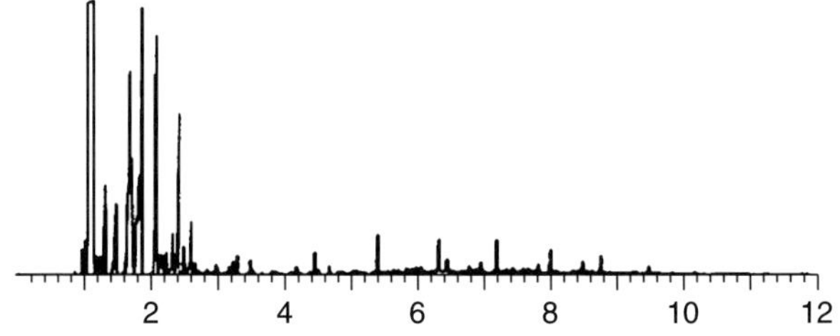

Figure 1 A high-resolution capillary column gas chromatogram of a JP-4 fuel.

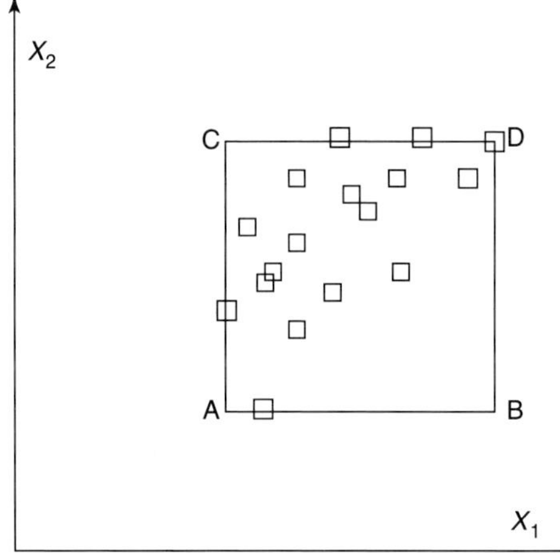

Figure 2 Seventeen hypothetical samples projected onto a two-dimensional measurement space defined by the measurement variables X_1 and X_2. The vertices, A, B, C, and D, of the rectangle represent the smallest and largest values of X_1 and X_2. (Adapted from Mandel.[8])

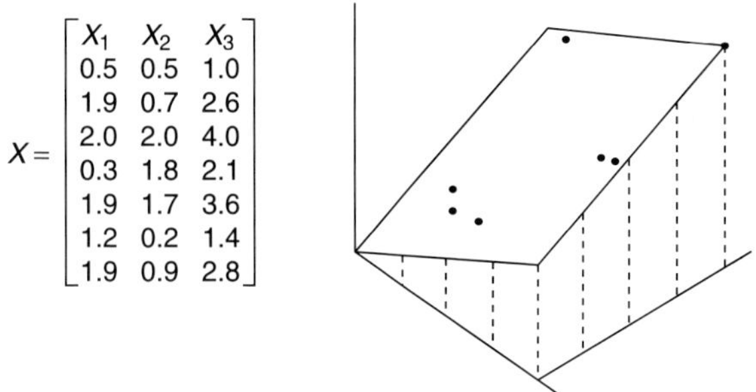

Figure 3 In the case of strongly correlated variables, the data points may reside in a subspace of the original measurement space. (Adapted from *Multivariate Pattern Recognition in Chemometrics*.[4] Copyright 1992, with permission from Elsevier Science.)

For references see page 9708

perfectly correlated with X_1 and X_2 because X_1 plus X_2 equals X_3. Hence, the seven sample points lie in a plane even though each data point has three measurements associated with it.

2.1 Variance-based Coordinate System

Variables that have a great deal of redundancy or are highly correlated are said to be collinear. High collinearity between variables is a strong indication that a new set of basis vectors can be found that will be better at conveying the information content present in data than axes defined by the original measurement variables. The new basis set that is linked to variation in the data can be used to develop a new coordinate system for displaying the data. The principal components of the data define the variance-based axes of this new coordinate system. The largest or first principal component is formed by determining the direction of largest variation in the original measurement space and modeling it with a line fitted by linear least squares (Figure 4) that passes through the center of the data. The second largest principal component lies in the direction of next largest variation – it passes through the center of the data and is orthogonal to the first principal component. The third largest principal component lies in the direction of next largest variation – it also passes through the center of the data, it is orthogonal to the first and second principal components, and so forth. Each principal component describes a different source of information because each defines a different direction of scatter or variance in the data. (The scatter of the data points in the measurement space is a direct measure of the data's variance.) Hence, the orthogonality constraint imposed by the mathematics of PCA ensures that each variance-based axis will be independent.

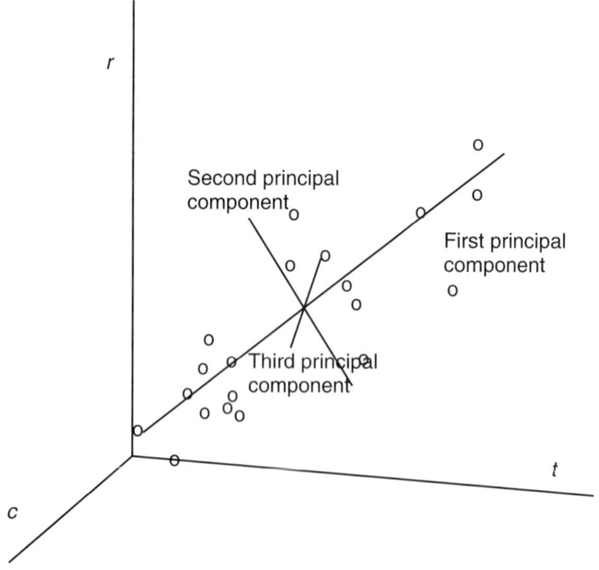

Figure 4 Principal component axes developed from the measurement variables *r*, *c*, and *t*. (Reproduced by permission from Brown[(5)] courtesy of Society of Applied Spectroscopy.)

List of selected abbreviations appears in Volume 15

2.2 Information Content of Principal Components

One measure of the amount of information conveyed by each principal component is the variance of the data explained by the principal component. The variance explained by each principal component is expressed in terms of its eigenvalue. For this reason, principal components are usually arranged in order of decreasing eigenvalues or waning information content. The most informative principal component is the first and the least informative is the last. The maximum number of principal components that can be extracted from the data is the smaller of either the number of samples or number of measurements in the data set, as this number defines the largest number of independent variables in the data.

If the data are collected with due care, one would expect that only the first few principal components would convey information about the signal, as most of the information in the data should be about the effect or property of interest being studied. However, the situation is not always this straightforward. Each principal component describes some amount of signal and some amount of noise in the data because of accidental correlation between signal and noise. The larger principal components primarily describe signal variation, whereas the smaller principal components essentially describe noise. When smaller principal components are deleted, noise is being discarded from the data, but so is a small amount of signal. However, the reduction in noise more than compensates for the biased representation of the signal that results from discarding principal components that contain a small amount of signal but a large amount of noise. Plotting the data in a coordinate system defined by the two or three largest principal components often provides more than enough information about the overall structure of the data. This approach to describing a data set in terms of important and unimportant variation is known as soft modeling in latent variables.

PCA takes advantage of the fact that a large amount of data is usually generated in monitoring studies when sophisticated chemical instrumentation, which is commonly under computer control, is used. The data have a great deal of redundancy and therefore a great deal of collinearity. Because the measurement variables are correlated, 85 peak gas chromatograms do not necessarily require 85 independent axes to define the position of the sample points. Utilizing PCA, the original measurement variables that constitute a correlated axis system can be converted into a system that removes correlation by forcing the new axes to be independent and orthogonal.

This requirement greatly simplifies the data because the correlations present in the data often allow us to use fewer axes to describe the sample points. Hence, the gas chromatograms of a set of JP-4 and Jet-A fuel samples may reside in a subspace of the 85-dimensional measurement space. A plot of the two or three largest principal components of the data can help us to visualize the relative position of the Jet-A and JP-4 fuel samples in this subspace.

2.3 Case Studies

With PCA, we are able to plot the data in a new coordinate system based on variance. The origin of the new coordinate system is the center of the data, and the coordinate axes of the new system are the principal components of the data. Employing this new coordinate system, we can uncover relationships present in the data, that is, find distinct samples subgroups within the data. This section shows, by way of two published studies, how principal components can be used to discern similarities and differences among sample within a data set.

2.3.1 *Troodos Data Set*

In the first study, 143 rock samples collected in the Troodos region of Cyprus were analyzed by X-ray fluorescence spectroscopy for 10 metal oxides, which contained information about the formation of these rocks. If the formation of the entire Troodos region occurred at the same time, one would expect all of the rocks to be similar in composition. However, if there are distinct subgroups in the data, other conclusions may have to be drawn about the formation of the Troodos region. This study[9] was initiated to settle a controversy about the geological history of Cyprus.

Figure 5 shows a plot of the two largest principal components of the 143 rock samples. (The original Troodos data set was modified for the purpose of this principal component mapping exercise.) Samples 65 and 66 appear to be outliers in the plot as they are distant from the other samples. As a general rule, outliers should deleted because of the least-squares property of principal components. In other words, a sample that is distant from the other points in the measurement space can pull the principal components towards it and away from the direction of maximum variance. Figure 6 shows the results of a principal component mapping experiment with samples 65 and 66 removed from the data. It is evident from the plot that samples 129 and 130 are also outliers. Figure 7 summarizes the results of a principal component mapping experiment with samples 65, 66, 129, and 130 removed. Although samples 19 and 20 are probably outliers and are also candidates for removal, it is evident from the principal component plot that the rock samples can be divided into two groups, which would suggest that other conclusions should be drawn about the geological history of the Troodos region. The clustering of the rocks samples into two distinct groups was not apparent until the four outliers were removed from the data.

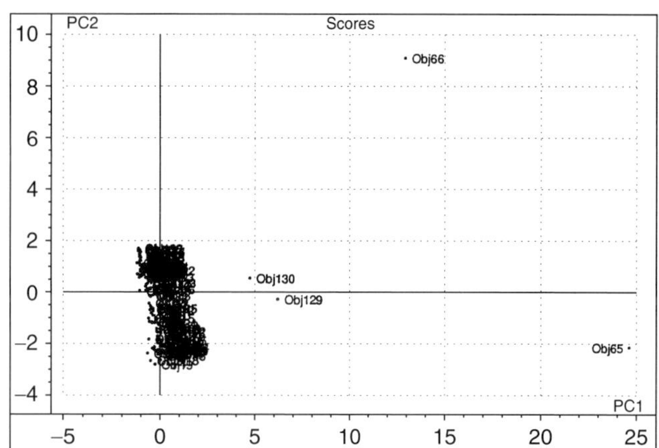

Figure 5 A principal component map of the 143 rock samples. Samples 65 and 66 are outliers. (The original data set was modified for this principal component mapping exercise.) The principal component map was generated using the program UNSCRAMBLER.

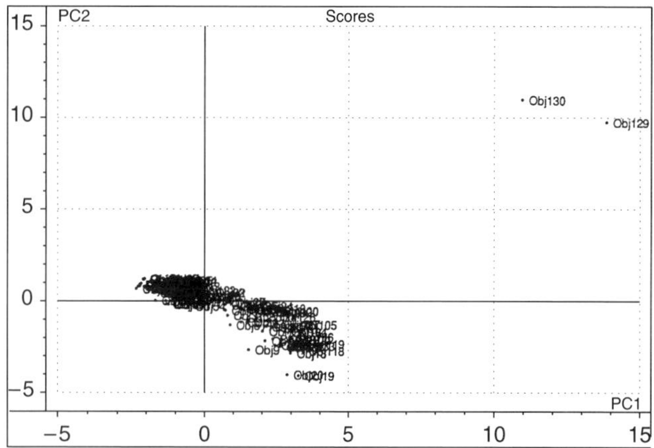

Figure 6 A principal component map of the Troodos rock samples with samples 65 and 66 removed. Samples 129 and 130 appear as outliers in the plot. (The original data set was modified for this principal component mapping exercise.) The principal component map was generated using the program UNSCRAMBLER.

2.3.2 *Obsidian Data Set*

The second study[10] also involves X-ray fluorescence data. Sixty-three Indian artifacts (such as jewelry, weapons, and tools) made from volcanic glass, were collected from four quarries in the San Francisco Bay

For references see page 9708

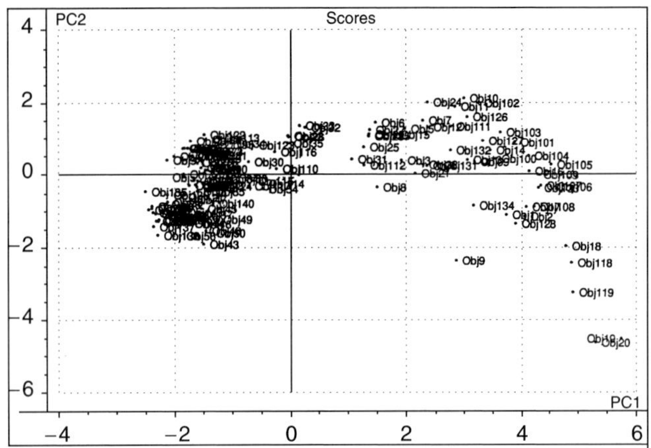

Figure 7 A principal component map of the Troodos rock samples with samples 65, 66, 129, and 130 removed. The principal component map was generated using the program UNSCRAMBLER.

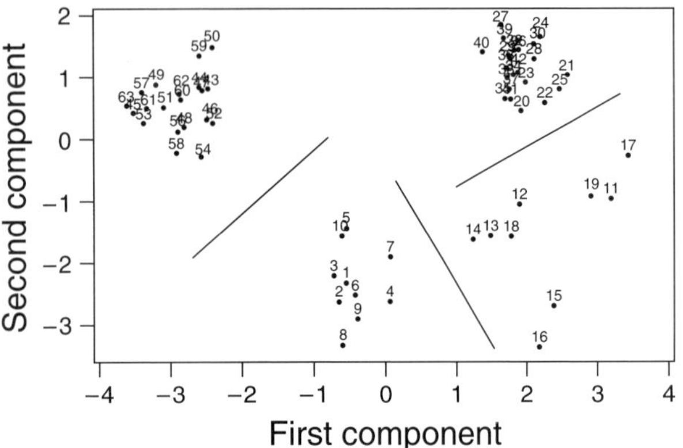

Figure 8 A plot of the two largest principal components for the 63 Indian artifacts developed from the concentration data of 10 metals. The principal component map was generated using the program SCAN.

area. (Samples 1–10 are from quarry 1, samples 11–19 are from quarry 2, samples 20–42 are from quarry 3, and samples 43–63 are from quarry 4.) Because the composition of volcanic glass is characteristic of the site and tends to be homogeneous, it is reasonable to assume that it should be possible to trace these artifacts to their original source material. In this study, the investigators attempted to do this by analyzing the 63 glass samples for 10 elements: Fe, Ti, Ba, Ca, K, Mn, Rb, Sr, Y, and Zn. Next, a PCA was performed on the data (63 artifacts with 10 features per artifact). The goal was to identify the overall trends present in the data. Figure 8 shows a plot of the two largest principal components of the data. From the principal component map, it is evident that the 63 Indian artifacts can be divided into four groups, which correspond to the quarry sites from which the artifacts were collected. Evidently, the artifacts in each quarry were made from the same source material. This result is significant because it provides the archaeologists with important information about the migration patterns and trading routes of the Indians in this region. Further details about the obsidian data can be found elsewhere.(10)

3 CLUSTER ANALYSIS

Cluster analysis is a popular technique whose basic objective is to discover sample groupings within data. The technique is encountered in many fields, such as biology, geology, and geochemistry, under such names as unsupervised pattern recognition and numerical taxonomy. Clustering methods are divided into three categories, hierarchical, object-functional, and graph theoretical. The focus here is on hierarchical methods, as they are the most popular.

For cluster analysis, each sample is treated as a point in an n-dimensional measurement space. The coordinate axes of this space are defined by the measurements used to characterize the samples. Cluster analysis assesses the similarity between samples by measuring the distances between the points in the measurement space. Samples that are similar will lie close to one another, whereas dissimilar samples are distant from each other. The choice of the distance metric to express similarity between samples in a data set depends on the type of measurement variables used.

Typically, three types of variables – categorical, ordinal, and continuous – are used to characterize chemical samples. Categorical variables denote the assignment of a sample to a specific category. Each category is represented by a number, such as 1, 2, 3, etc. Ordinal variables are categorical variables, in which the categories follow a logical progression or order, such as 1, 2, and 3 denoting low, middle, and high, respectively. However, continuous variables are quantitative. The difference between two values for a continuous variable has a precise meaning. If a continuous variable assumes the values 1, 2, and 3, the difference between the values 3 and 2 will have the same meaning as the difference between the values 2 and 1, because they are equal.

Measurement variables are usually continuous. For continuous variables, the Euclidean distance is the best choice for the distance metric, because interpoint distances between the samples can be computed directly (Figure 9). However, there is a problem with using the Euclidean distance, which is the so-called scaling effect. It arises from inadvertent weighting of the variables in the analysis that can occur due to differences in magnitude among the measurement variables. For example, consider a data set where each sample is described by two variables:

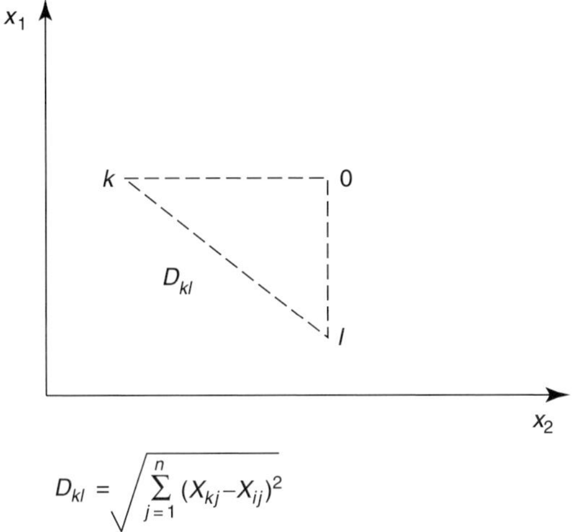

Figure 9 Euclidean distance between two data points in a two-dimensional measurement space defined by the measurement variables x_1 and x_2. (Reprinted from ref. 3.)

the concentration of Na and the concentration of K as measured by atomic flame emission spectroscopy. The concentration of Na varies from 50 to 500 ppm, whereas the concentration of K in the same samples varies from 5 to 50 ppm. A 10% change in the Na concentration will have a greater effect on Euclidean distance than a 10% change in K concentration. The influence of variable scaling on the Euclidean distance can be mitigated by autoscaling the data, which involves standardizing the measurement variables, so that each variable has a mean of zero and a standard deviation of 1 (Equation 1):

$$x_{i,\text{standardized}} = \frac{x_{i,\text{orig}} - m_{i,\text{orig}}}{s_{i,\text{orig}}} \quad (1)$$

where $x_{i,\text{orig}}$ is the original measurement variable i, $m_{i,\text{orig}}$ is the mean of the original measurement variable i, and $s_{i,\text{orig}}$ is the standard deviation of the original measurement variable i. Thus, a 10% change in K concentration has the same effect on the Euclidean distance as a 10% change in Na concentration when the data is autoscaled. Clearly, autoscaling ensures that each measurement variable has an equal weight in the analysis. For cluster analysis, it is best to autoscale the data, because similarity is directly determined by a majority vote of the measurement variables.

3.1 Hierarchical Clustering

Clustering methods attempt to find clusters of patterns (i.e. data points) in the measurement space, hence the term cluster analysis. Although several clustering algorithms exist, e.g. K-means, K-median, Patrick-Jarvis, FCV (fuzzy clustering varieties), hierarchical clustering is by far the most widely used clustering method. The starting point for a hierarchical clustering experiment is the similarity matrix which is formed by first computing the distances between all pairs of points in the data set. Each distance is then converted into a similarity value (Equation 2):

$$s_{ik} = 1 - \frac{d_{ik}}{d_{\max}} \quad (2)$$

where s_{ik} (which varies from 0 to 1) is the similarity between samples i and k, d_{ik} is the Euclidean distance between samples i and k, and $d_{\max}$ is the distance between the two most dissimilar samples (i.e. the largest distance) in the data set. The similarity values are organized in the form of a table or matrix. The similarity matrix is then scanned for the largest value, which corresponds to the most similar point pair. The two samples constituting the point pair are combined to form a new point, which is located midway between the two original points. The rows and columns corresponding to the old data points are then removed from the matrix. The similarity matrix for the data set is then recomputed. In other words, the matrix is updated to include information about the similarity between the new point and every other point in the data set. The new nearest point pair is identified, and combined to form a single point. This process is repeated until all points have been linked.

There are a variety of ways to compute the distances between data points and clusters in hierarchical clustering (Figure 10). The single-linkage method assesses similarity between a point and a cluster of points by measuring the distance to the closest point in the cluster. The

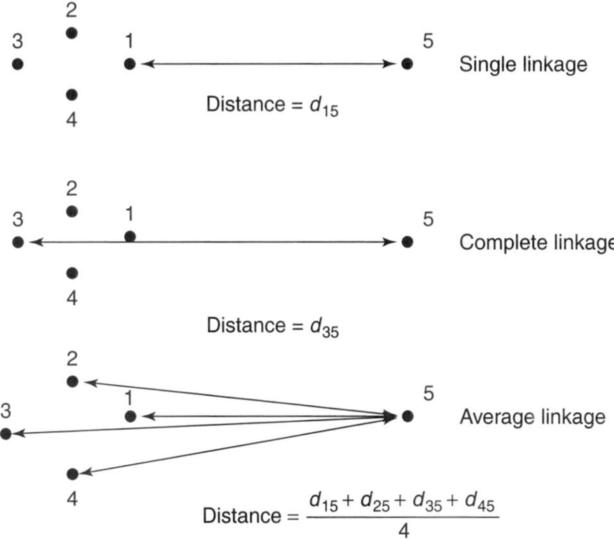

Figure 10 The distance between a data cluster and a point using single linkage, complete linkage, and average linkage. (Reproduced by permission from Lavine[20] courtesy of Marcel Dekker, Inc.)

Table 1 HPLC data set

Column	k_1	k_2	k_3
A	0.31	17.8	3
B	0.10	9.30	3
C	0.11	21.5	1
D	0.58	22.0	2
E	0.50	16.0	1

(Reprinted from *Multivariate Pattern Recognition in Chemometrics*.[4] Copyright 1992, by permission of Elsevier Science.)

Table 2 Similarity matrix

Columns	A	B	C	D	E
A	1.00	0.79	0.58	0.69	0.61
B	0.79	1.00	0.36	0.17	0.34
C	0.58	0.36	1.00	0.51	0.72
D	0.69	0.17	0.51	1.00	0.75
E	0.61	0.34	0.72	0.75	1.00

Table 3 Updated similarity matrix

Columns	A, B	C	D	E
A, B	1.00	0.58	0.69	0.61
C	0.58	1.00	0.51	0.72
D	0.69	0.51	1.00	0.75
E	0.61	0.72	0.75	1.00

Table 4 Updated similarity matrix

Columns	A, B	C	D, E
A, B	1.00	0.58	0.61
C	0.58	1.00	0.72
D, E	0.61	0.72	1.00

Table 5 Updated similarity matrix

Columns	A, B	C, D, E
A, B	1.00	0.69
C, D, E	0.69	1.00

complete linkage method assesses similarity by measuring the distance to the farthest point in the cluster. Average linkage assesses the similarity by computing the distances between all point pairs where a member of each pair belongs to the cluster. The average of these distances is used to compute the similarity between the data point and the cluster.

To illustrate hierarchical clustering, consider the data shown in Table 1. (The example shown here is an adaptation of the exercise described in Chapter 6 of reference 4.) Five HPLC columns were characterized by the capacity factor values obtained from three substances, which served as retention probes. To perform single-linkage hierarchical clustering on this chromatographic data, it is necessary to first compute the similarity matrix for the data, given as Table 2.

The similarity matrix (Table 2) is then scanned for the largest value, which corresponds to the two HPLC columns that are most similar. An examination of the similarity matrix suggests that chromatographic columns A and B with a score of 0.79 are the most similar. Hence, chromatographic columns A and B should be combined to form a new point. The rows and columns corresponding to the two original points (A and B) are removed from the similarity matrix. The similarity matrix for the data set is then updated to include information about the similarity between the new point and every other point (C, D, and E) in the data set. In this study, the investigators chose to use the single-linkage criterion for assessing the similarity between a data point and a point cluster, see Table 3. (Using the single-linkage method, the similarity between point D and the cluster consisting of columns A and B is the larger of the values of 0.69 and 0.17, see Table 2. For complete linkage, the similarity between this cluster and point D is the smaller of the two values.)

The updated similarity matrix is then scanned for the largest value; the new nearest pair is combined to form a single point, which is points D and E. The rows and columns corresponding to points D and E are deleted from the similarity matrix. The similarity matrix for the data set is then updated (Table 4) to include information about the similarity between the new point (D + E) and every other point in the data set. This process is repeated (Table 5) until all points are merged into a single cluster.

The results of a hierarchical clustering study are usually displayed as a dendogram, which is a tree-shaped map of the intersample distances in the data set. The dendogram shows the merging of samples into clusters at various stages of the analysis and the similarities at which the clusters merge, with the clustering displayed hierarchically. The dendogram for the single-linkage analysis of the HPLC data is shown in Figure 11. Interpretation of the results is intuitive, which is the major reason for the popularity of these methods.

3.2 Practical Considerations

A major problem in hierarchical clustering (or cluster analysis for that matter) is defining a cluster. Contrary to many published reports, there is no cluster validity measure that can serve as an indicator of the quality of a proposed partitioning of the data. Hence, clusters are defined intuitively, depending on the context of the problem, not mathematically, which limits the utility of this technique. Clearly, prior knowledge about the problem is essential when using these methods. The

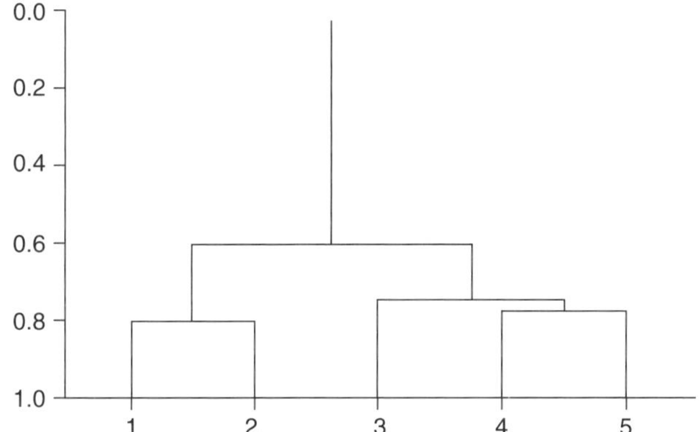

Figure 11 Single-linkage dendogram of the HPLC data set. There are two groups of HPLC columns: 1, 2 and 3, 4, 5. (Reprinted from *Multivariate Pattern Recognition in Chemometrics*.[4] Copyright 1992, with permission from Elsevier Science.)

criterion for determining the threshold value for similarity is often subjective and depends to a large degree on the nature of the problem investigated – for example, the goals of the study, the number of clusters sought, previous experience, and common sense.

All clustering procedures yield the same results for data sets with well-separated clusters. However, the results will differ when the clusters overlap. That is why it is a good idea to use at least two different clustering algorithms, such as single and complete linkage, when studying a data set. If the dendograms are in agreement, then a strong case can be made for partitioning the data into distinct groups as suggested by the dendograms. If the cluster memberships differ, the data should be further investigated using average linkage or PCA. The results from average linkage or PCA can be used to gauge whether the single or farthest linkage solution is the better one.

All hierarchical clustering techniques suffer from so-called space distorting effects. For example, single-linkage favors the formation of large linear clusters instead of the usual elliptical or spherical clusters. As a result, poorly separated clusters are often chained together. However, complete linkage favors the formation of small spherical clusters. Because of these space-distorting effects, hierarchical clustering methods should be used in tandem with PCA to detect clusters in multivariate data sets.

All hierarchical methods will always partition data, even randomly generated data, into distinct groups or clusters. Hence, it is important to ascertain the significance level of the similarity value selected by the user. For this task, a simple three-step procedure is proposed. First, a random data set is generated with the same correlation structure, the same number of samples, and the same number of measurements as the real data set that is currently being investigated. Second, the same clustering technique(s) is applied to the random data. Third, the similarity value, which generates the same number of clusters as identified in the real data set, is determined from the dendogram of the random data. If the similarity value is substantially larger for the real data set, the likelihood of having inadvertently exploited random variation in the data to achieve clustering is probably insignificant.

3.3 Case Studies

Hierarchical clustering methods attempt to uncover the intrinsic structure of a multivariate data set without making a priori assumptions about the data. This section, shows, by way of two published studies, how clustering methods can be used to find clusters of points in data.

3.3.1 Obsidian Data Set

The first study is the obsidian data set, discussed in section 2.3.2. A principal component map of the data (Figure 8) revealed four distinct clusters, which correspond to the sites from which these artifacts were obtained. To confirm the four-cluster hypothesis, the investigators also analyzed their data using single-linkage analysis. The resulting dendogram (Figure 12) indicated that it is reasonable to divide the glass samples into four categories based on the quarry sites from which these artifacts were obtained. (At similarity 0.40, samples 1–10 form a cluster, samples 11–19 form another cluster, samples 20–42 form a third cluster, and samples 43–63 form the fourth cluster.) Because the dendogram and principal component map of the data are in strong agreement, the investigators decided to partition their data into four groups based on the quarry labels of the samples.

3.3.2 Artificial Nose Data Set

The second study involves data from an artificial nose.[11] A salesman claims that an electronic nose can successfully sniff odor. The salesman obtained data from the literature to support his claim. Of the 41 compounds in the data set, compounds 1–21 are etheral, compounds 22–32 are pungent, and compounds 33–41 are minty. Each compound was characterized by six electronic measurements. Using a back-propagation neural network algorithm, the salesman was able to correctly classify all of the compounds in the data set. Should we then accept the salesman's claim that an electronic nose can sniff odor?

In order to validate the salesman's claim, the odor data was analyzed using single and complete linkage hierarchical clustering. Figures 13 and 14 show dendograms of the odor data. It is evident from the dendograms that dividing the 41 compounds into three categories based on odor type cannot be justified. (Although the data can

For references see page 9708

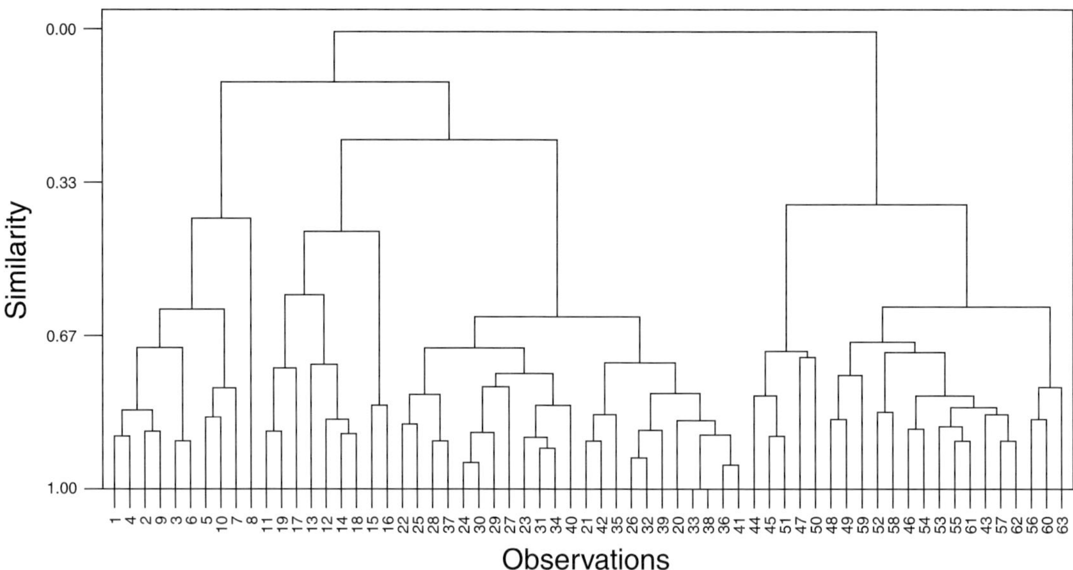

Figure 12 A single-linkage dendogram of the obsidian data set. The dendogram was generated using the program SCAN.

be divided into three clusters by complete linkage at a 0.56 similarity, the cluster memberships cannot be correlated to compound odor – why? Cluster 1 consists of samples 41, 21, 19, 16, 13, 39, 31, 36, 29, 8, 34, and 32. Cluster 2 consists of samples 38, 5, 26, 14, 24, 15, 37, 6, 33, 35, 20, 30, and 7. Cluster 3 consists of samples 40, 17, 4, 22, 2, 25, 10, 28, 18, 27, 23, 12, 11, 1, 9, and 3.) Evidently, the six electronic measurements from the nose do not contain sufficient discriminatory information to force the compounds to cluster on the basis of odor type. Therefore, the salesman's claim about the efficacy of the proposed artificial nose should not be accepted at face value.

4 PATTERN RECOGNITION

So far, only exploratory data analysis techniques, i.e. cluster analysis and PCA, have been discussed. These techniques attempt to analyze data without directly using information about the class assignment of the samples. Although cluster analysis and PCA are powerful methods for uncovering relationships in large multivariate data sets, they are not sufficient for developing a classification rule that can accurately predict the class-membership of an unknown sample. In this section, pattern recognition techniques will be discussed. These techniques were originally developed to categorize a sample on the basis of regularities in observed data. The first applications of pattern recognition to chemistry were studies involving low-resolution mass spectrometry.[12] Since then, pattern recognition techniques have been applied to a wide variety of chemical problems, such as chromatographic fingerprinting,[13–15] spectroscopic imaging,[16–18] and data interpretation.[19–21]

Pattern recognition techniques fall into one of two categories: nonparametric discriminants, and similarity-based classifiers. Nonparametric discriminants,[22–24] such as neural networks, attempt to divide a data space into different regions. In the simplest case, that of a binary classifier, the data space is divided into two regions. Samples that share a common property (such as fuel type) will be found on one side of the decision surface, whereas those samples comprising the other category will be found on the other side. Nonparametric discriminants have provided insight into relationships contained within sets of chemical measurements. However, classification based on random or chance separation[24] can be a serious problem if the data set is not sample rich. Because chemical data sets usually contain more variables than samples, similarity-based classifiers are generally preferred.

Similarity-based classifiers, e.g. k-nearest neighbor (KNN)[25] and soft independent modeling by class analogy (SIMCA),[26–30] treat each chromatogram or spectrum as a data vector $\mathbf{x} = (x_1, x_2, x_3, \ldots, x_j, \ldots, x_p)$ where component x_j is the area of the jth peak or the absorbance value of the jth wavelength. Such a vector can also be viewed as a point in a high-dimensional measurement space. A basic assumption is that distances between points in the measurement space will be inversely related to their degree of similarity. Using a similarity-based classifier we can determine the class-membership of a sample by examining the class label of the data point closest to it or from the principal component model of the class, which lies closest to

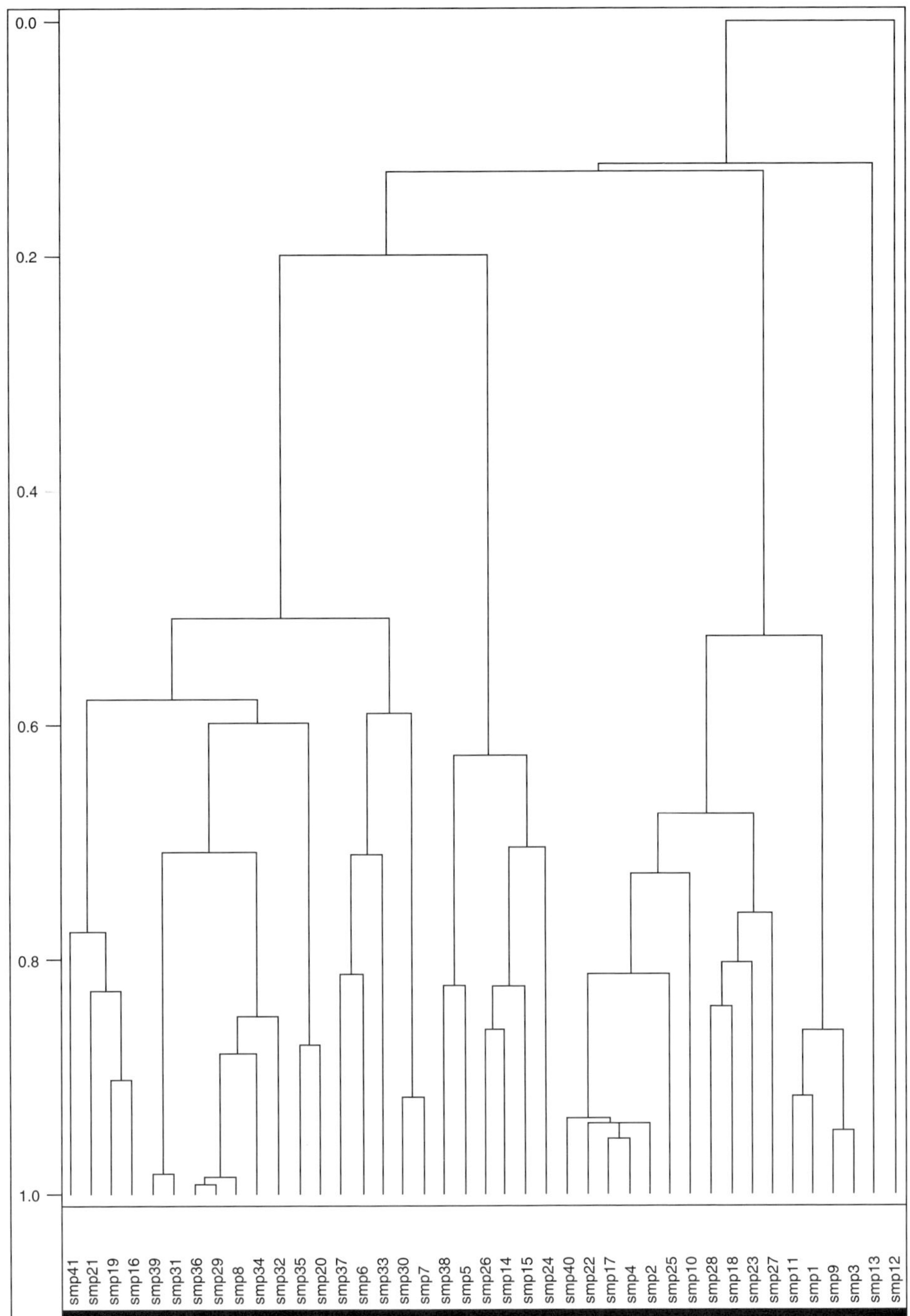

Figure 13 A single-linkage dendogram of the nose data set. The dendogram was generated using the program Pirouette.

the sample in the measurement space. In chemistry, similarity-based classification rules are implemented using either KNN or SIMCA.

4.1 *k*-Nearest Neighbor

For its simplicity, KNN is a powerful classification technique. A sample is classified according to the majority vote of its KNNs, where k is an odd integer (one, three, or five). For a given sample, Euclidean distances are first computed from the sample to every other point in the data set. These distances arranged from smallest to the largest are used to define the sample's KNNs. A poll is then taken by examining the class identities among the point's KNNs. Based on the class identity of the majority of its KNNs, the sample is assigned to a class

For references see page 9708

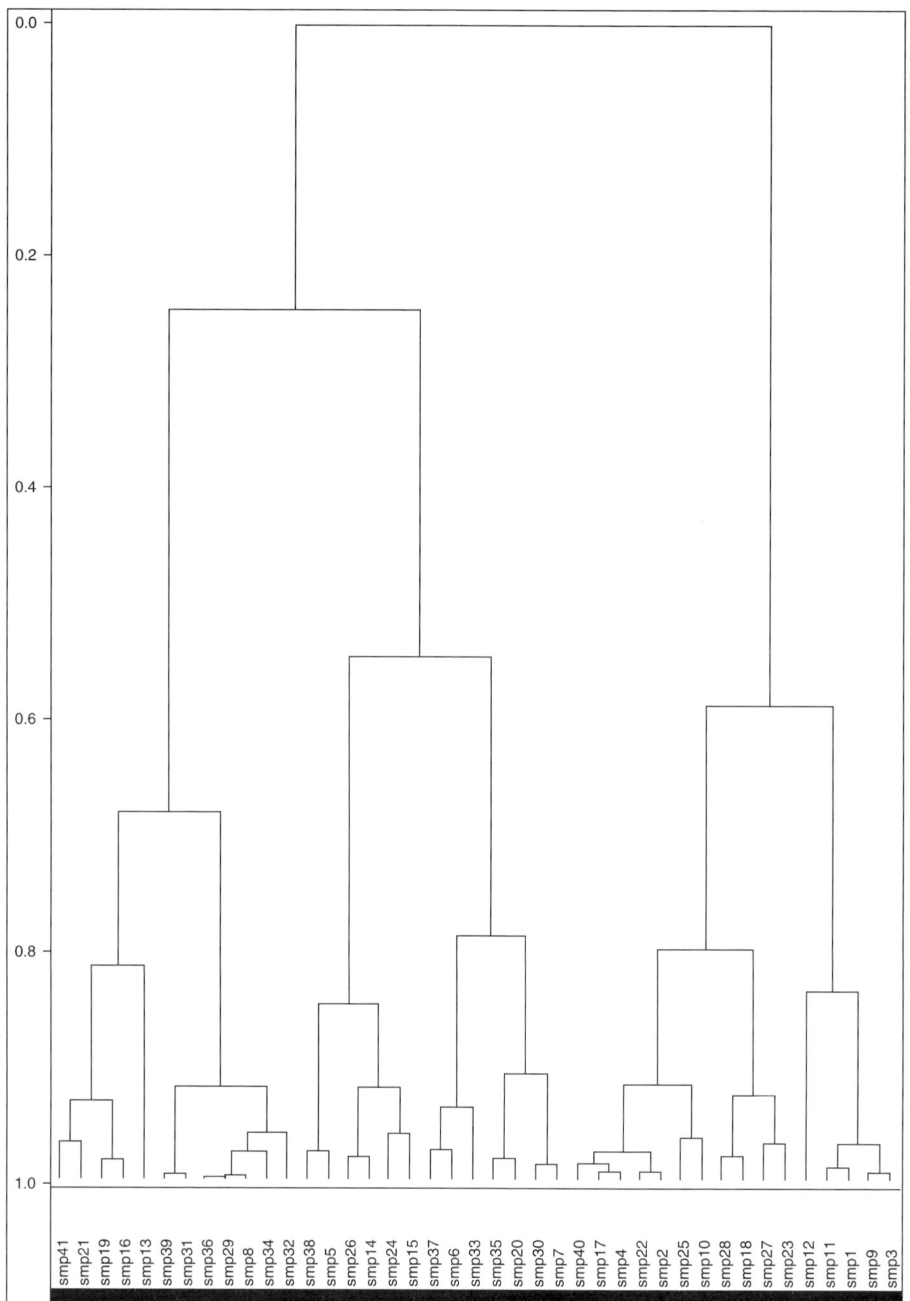

Figure 14 A complete-linkage dendogram of the nose data set. The dendogram was generated using the program Pirouette.

in the data set. If the assigned class and the actual class of the sample match, the test is considered a success. The overall classification success rate, calculated over the entire set of points, is a measure of the degree of clustering in the set of data. Clearly, a majority vote of the KNNs can only occur if the majority of the measurement variables concur, because the data is usually autoscaled.

KNN cannot furnish a statement about the reliability of a classification. However, its classification risk is bounded. In other words, the Bayes classifier will generate the optimal classification rule for the data, and 1-nearest neighbor has an error rate which is twice as large as the Bayes classifier. (To implement the Bayes classifier, one must have knowledge about all the statistics of the data set including the underlying probability distribution function for each class. Usually, this knowledge is not available.) Hence, any other classification method, no matter how sophisticated, can at best only improve on the performance of KNN by a factor of two.

4.2 Soft Independent Modeling by Class Analogy

In recent years, modeling approaches have become popular in analytical chemistry for developing classification rules because of the problems with nonparametric discriminants. Although there are a number of approaches to modeling classes, the SIMCA method, based on PCA, has been developed by Wold for isolating groups of multivariate data or classes in a data set. In SIMCA, a PCA is performed on each class in the data set, and a sufficient number of principal components are retained to account for most of the variation within each class. Hence, a principal component model is used to represent each class in the data set. The number of principal components retained for each class is usually different. Deciding on the number of principal components that should be retained for each class is important, as retention of too few components can distort the signal or information content contained in the model about the class, whereas retention of too many principal components diminishes the signal-to-noise. A procedure called cross-validation(31) ensures that the model size can be determined directly from the data. To perform cross-validation, segments of the data are omitted during the PCA. Using one, two, three, etc., principal components, omitted data are predicted and compared to the actual values. This procedure is repeated until every data element has been kept out once. The principal component model that yields the minimum prediction error for the omitted data is retained. Hence, cross-validation can be used to find the number of principal components necessary to describe the signal in the data while ensuring high signal-to-noise by not including the so-called secondary or noise-laden principal components in the class model.

The variance that is explained by the class model is called the modeled variance, which describes the signal, whereas the noise in the data is described by the residual variance or the variance not accounted for by the model. (The residual variance is explained by the secondary principal components, which have been truncated or omitted from the principal component model.) By comparing the residual variance of an unknown to the average residual variance of those samples that make up the class, it is possible to obtain a direct measure of the similarity of the unknown to the class. This comparison is also a measure of the goodness of fit of the sample to a particular principal component model. Often, the F-statistic is used to compare the residual variance of a sample with the mean residual variance of the class.(32) Employing the F-statistic, an upper limit for the residual variance can be calculated for those samples belonging to the class. The final result is a set of probabilities of class-membership for each sample.

An attractive feature of SIMCA is that a principal component mapping of the data has occurred. Hence, samples that may be described by spectra or chromatograms are mapped onto a much lower dimensional subspace for classification. If a sample is similar to the other samples in the class, it will lie near them in the principal component map defined by the samples representing that class. Another advantage of SIMCA is that an unknown is only assigned to the class for which it has a high probability. If the residual variance of a sample exceeds the upper limit for every modeled class in the data set, the sample would not be assigned to any of the classes because it is either an outlier or comes from a class that is not represented in the data set. Finally, SIMCA is sensitive to the quality of the data used to generate the principal component models. As a result, there are diagnostics to assess the quality of the data, such as the modeling power(33) and the discriminatory power.(34) The modeling power describes how well a variable helps the principal components to model variation, and discriminatory power describes how well the variable helps the principal components to classify the samples in the data set. Variables with low modeling power and low discriminatory power are usually deleted from the data because they contribute only noise to the principal component models.

SIMCA can work with as few as 10 samples per class, and there is no restriction on the number of measurement variables, which is an important consideration, because the number of measurement variables often exceeds the number of samples in chemical studies. Most standard discrimination techniques would break down in these situations because of problems arising from collinearity and chance classification.(24)

4.3 Feature Selection

Feature selection is a crucial step in KNN or SIMCA, because it is important to delete features or measurements that contain information about experimental artifacts or other systematic variations in the data not related to legitimate chemical differences between classes in a data set. For profiling experiments of the type that are being considered (see section 4.4) it is inevitable that relationships may exist among sets of conditions used to generate the data and the patterns that result. One must realize this in advance when approaching the task of analyzing such data. Therefore, the problem is utilizing information contained in the data characteristic of the class without being swamped by the large amount of qualitative and quantitative information contained in the chromatograms or spectra about the experimental conditions used to generate the data. If the basis of classification for samples in the training set is other than desired group differences, unfavorable classification results for the prediction set will be obtained despite a linearly separable training set. The existence of these

For references see page 9708

confounding relationships is an inherent part of profiling data. Hence, the goal of feature selection is to increase the signal-to-noise ratio of the data by discarding measurements on chemical components that are not characteristic of the source profile of the classes in the data set. Feature selection in the context of pattern recognition is described in greater detail in the next section by way of the two worked examples.

4.4 Case Studies

Pattern recognition is about reasoning, using the available information about the problem to uncover information contained within the data. Autoscaling, feature selection, and classification are an integral part of this reasoning process. Each plays a role in uncovering information contained within the data.

Pattern recognition analyses are usually implemented in four distinct steps: data preprocessing, feature selection, classification, and mapping and display. However, the process is iterative, with the results of a classification or display often determining a further preprocessing step and reanalysis of the data. Although the procedures selected for a given problem are highly dependent upon the nature of the problem, it is still possible to develop a general set of guidelines for applying pattern recognition techniques to real data sets. In this section, a framework for solving the class-membership problem is presented by way of two recently published studies on chromatographic fingerprinting of complex biological and environmental samples.

4.4.1 Fuel Spill Identification

The first study[(35)] involves the application of gas chromatographic and pattern recognition (GC/PR) methods to the problem of typing jet fuels, so that a spill sample in the environment can be traced to its source. The test data consisted of 228 gas chromatograms of neat jet fuel samples representing the major aviation fuels (JP-4, Jet-A, JP-7, JPTS, and JP-5) found in the USA. The neat jet fuel samples used in this study were obtained from Wright Patterson Air Force Base or Mulkiteo Energy Management Laboratory (Table 6). They were splits from regular quality control standards, which were purchased by the United States Air Force (USAF) to verify the authenticity of the manufacturer's claims.

The prediction set consisted of 25 gas chromatograms of weathered jet fuels (Table 7). Eleven of the 25 weathered fuels were collected from sampling wells as a neat oily phase found floating on top of the well water. Eleven of the 25 weathered fuel samples were extracted from the

Table 6 Training set

Number of samples	Fuel type
54	JP-4 (fuel used by USAF fighters)
70	Jet-A (fuel used by civilian airliners)
32	JP-7 (fuel used by SR-71 reconnaissance plane)
29	JPTS (fuel used by TR-1 and U-2 aircraft)
43	JP-5 (fuel used by Navy jets)

Table 7 Prediction set

Sample	Identity	Source	Sample	Identity	Source
PF007	JP-4	A[a]	MIX1	JP-4	C[c]
PF008	JP-4	A[a]	MIX2	JP-4	C[c]
PF009	JP-4	A[a]	MIX3	JP-4	C[c]
PF010	JP-4	A[a]	MIX4	JP-4	C[c]
PF011	JP-4	A[a]	STALE-1	JP-4	D[d]
PF012	JP-4	A[a]	STALE-2	JP-4	D[d]
PF013	JP-4	A[a]	STALE-3	JP-4	D[d]
KSE1M2	JP-4	B[b]	PIT1UNK	JP-5	E[e]
KSE2M2	JP-4	B[b]	PIT1UNK	JP-5	E[e]
KSE3M2	JP-4	B[b]	PIT2UNK	JP-5	E[e]
KSE4M2	JP-4	B[b]	PIT2UNK	JP-5	E[e]
KSE5M2	JP-4	B[b]			
KSE6M2	JP-4	B[b]			
KSE7M2	JP-4	B[b]			

[a] Sampling well at Tyndall AFB: the sampling well was near a previously functioning storage depot. Each well sample was collected on a different day.
[b] Soil extract near a sampling well: dug with a hand auger at various depths. Distance between sampling well Tyndall and soil extract was approximately 80 yards.
[c] Weathered fuel added to sand.
[d] Old JP-4 fuel samples that had undergone weathering in a laboratory refrigerator.
[e] Sampling pit at Keywest Air Station: two pits were dug near a seawall to investigate a suspected JP-5 fuel leak.

soil near various fuel spills. The other three fuel samples had been subjected to weathering in a laboratory.

The neat jet fuel samples were stored in sealed containers at $-20\,°C$. Prior to chromatographic analysis, each fuel sample was diluted with methylene chloride and injected onto a fused silica capillary column (10 m × 0.10 mm) using a split injection technique. The fused silica capillary column was temperature programmed from 60 °C to 270 °C at $18\,°C\,min^{-1}$. High-speed gas chromatograms representative of the five fuel types (JP-4, Jet-A, JP-7, JPTS, and JP-5) are shown in Figure 15.

The gas chromatograms were peak matched using a computer program[36] that correctly assigned peaks by first computing the Kovats retention index (KI)[37] for compounds eluting off the GC column. Because the n-alkane peaks are the most prominent features present in the gas chromatograms of these fuels, it is a simple matter to compute KI values. The peak-matching program then developed a template of peaks by examining integration reports and adding peaks to the template that did not match the retention indices of previously observed peaks. By matching the retention indices of the peaks in each chromatogram with the retention indices of the features in the template, it was possible to produce a data vector for each gas chromatogram. Each feature in a peak-matched chromatogram was assigned a value corresponding to the normalized area of the peak in the chromatogram. (If the peak was present, its normalized area from the integration report was assigned to the corresponding element of the vector. If the peak was not present, the corresponding feature was assigned a value of zero.) The number of times a particular peak was found to have a nonzero value was also computed, and features below a user-specified number of nonzero occurrences (which was set equal to 5% of the total number of fuel samples in the training set) were deleted from the data set. This peak-matching procedure yielded a final cumulative reference file containing 85 features, though not all peaks were present in all chromatograms.

Because outliers have the potential to adversely influence the performance of pattern recognition methods, outlier analysis was performed on each fuel class in the training set prior to pattern recognition analysis. The generalized distance test[38] at the 0.01 significance level was implemented via SCOUT[39] to identify discordant observations in the data. Three Jet-A and four JP-7 fuel samples were found to be outliers and were subsequently removed from the data set. The training set, comprising 221 gas chromatograms of 85 peaks each, was analyzed using pattern recognition methods. Prior to pattern recognition analysis, the data were autoscaled so that each feature had a mean of zero and standard deviation of one within the set of 221 gas chromatograms. Hence, each gas chromatogram was initially represented as an 85-dimensional

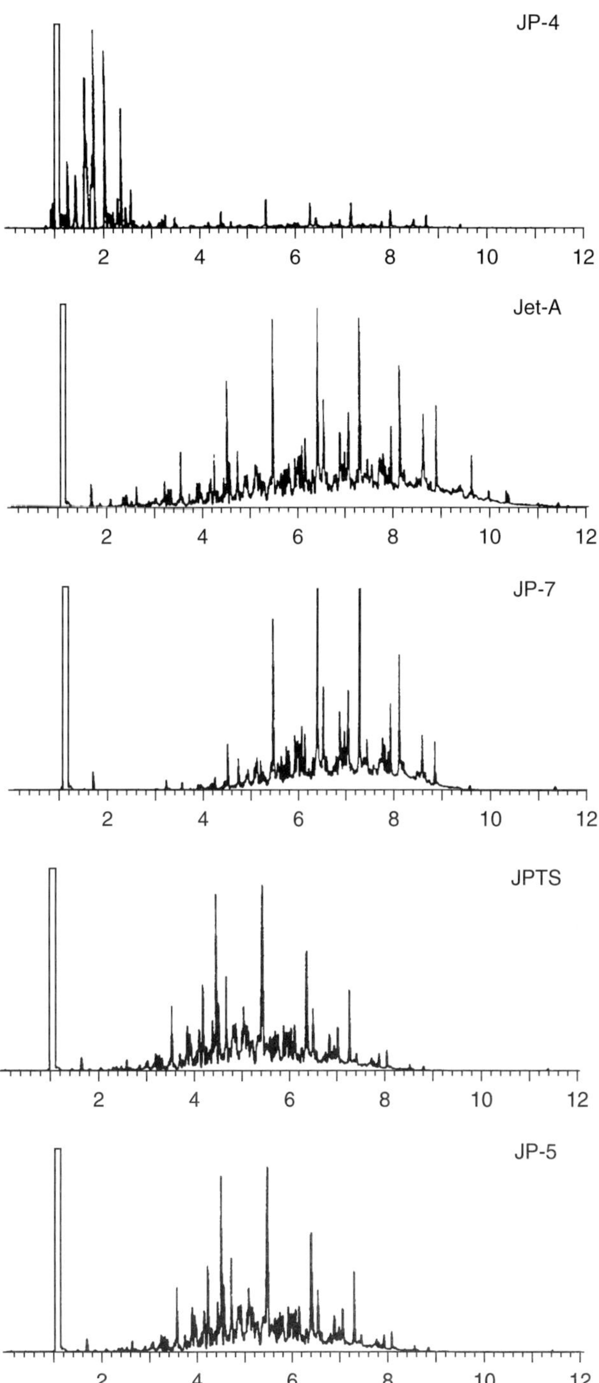

Figure 15 High speed gas chromatograms of JP-4, Jet-A, JP-7, JPTS, and JP-5. (Reproduced with permission from Lavine et al.[35] Copyright 1995 American Chemical Society.)

data vector, $\mathbf{x} = (x_1, x_2, x_3, \ldots, x_j, \ldots, x_{85})$, where x_j is the normalized area of the jth peak.

The first step in the study was to apply PCA to the training set data. Figure 16 shows a plot of the two largest principal components of the 85 GC peaks obtained from

For references see page 9708

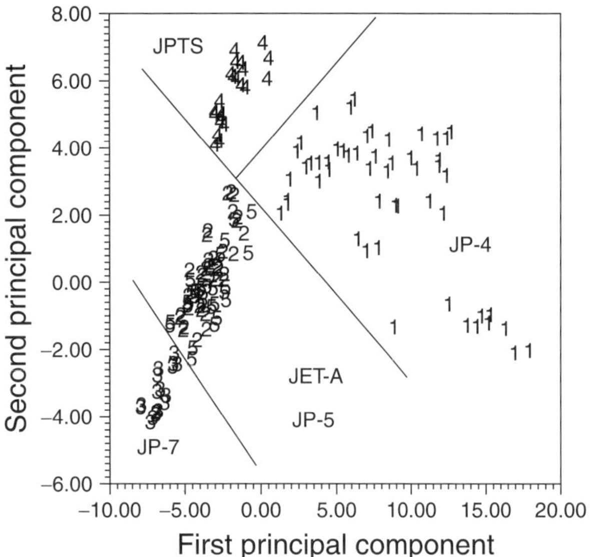

Figure 16 A plot of the two largest principal components of the 85 GC peaks for the 221 neat jet fuel samples. The map explains 72.3% of the total cumulative variance: 1 = JP-4, 2 = Jet-A, 3 = JP-7, 4 = JPTS, 5 = JP-5. (Reproduced with permission from Lavine et al.[35] Copyright 1995 American Chemical Society.)

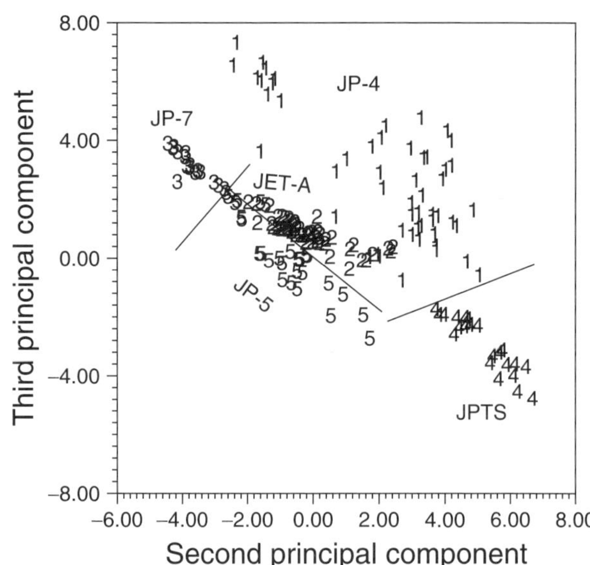

Figure 17 A plot of the second and third largest principal components of the 85 GC peaks for the 221 neat jet fuel samples. The map explains 23.1% of the total cumulative variance: 1 = JP-4, 2 = Jet-A, 3 = JP-7, 4 = JPTS, 5 = JP-5. (Reproduced with permission from Lavine et al.[35] Copyright 1995 American Chemical Society.)

the 221 neat jet fuel samples. Each fuel sample or gas chromatogram is represented as a point in the principal component map of the data. The JP-7 and JPTS fuel samples are well separated from one another and from the gas chromatograms of the JP-4, Jet-A, and JP-5 fuel samples, suggesting that information about fuel type is present in the high speed gas chromatograms of the neat jet fuels. However, the overlap of the JP-5 and Jet-A fuels in the principal component map suggests that gas chromatograms of these two fuels share a common set of attributes, which is not surprising in view of their similar physical and chemical properties.[40] Mayfield and Henley[41] have also reported that gas chromatograms of Jet-A and JP-5 fuels were more difficult to classify than gas chromatograms of other types of jet fuels. Nevertheless, they concluded that fingerprint patterns exist within GC profiles of Jet-A and JP-5 fuels characteristic of fuel type, which is consistent with a plot of the second and third largest principal components of the training set data. The plot in Figure 17 indicates that differences do indeed exist between the GC profiles of Jet-A and JP-5 fuels. However, the second and third largest principal components do not represent the direction of maximum variance in the data. (In fact, they only represent 23.1% of the total cumulative variance or information content of the data.) Hence, it must be concluded that the bulk of the information contained within the 85 GC peaks is not about differences between the GC profiles of Jet-A and JP-5 fuels.

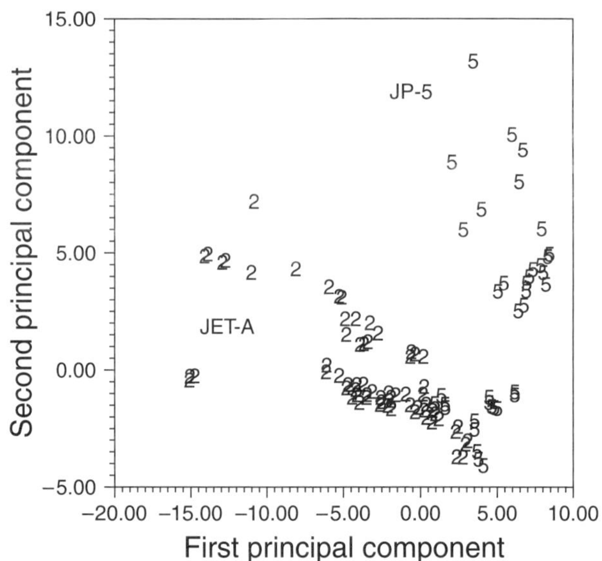

Figure 18 A principal component map of the 110 neat Jet-A and JP-5 fuel samples developed from the 85 GC peaks. The JP-5 fuel samples can be divided into two distinct groups: fuel samples that lie close to the Jet-A fuels and fuel samples distant from the Jet-A fuels. (Reproduced with permission from Lavine et al.[35] Copyright 1995 American Chemical Society.)

To better understand the problems associated with classifying the gas chromatograms of Jet-A and JP-5 fuels, it was necessary to reexamine this classification problem in greater detail. Figure 18 shows a plot of the two largest principal components of the 85 GC peaks of the 110 Jet-A and JP-5 neat fuel samples. It is evident from an examination of the principal component map that Jet-A and JP-5 fuel samples lie in different regions, suggesting that Jet-A and JP-5 fuels can be differentiated from each other on the basis of their GC profiles. However, the points representing the JP-5 fuels form two distinct subgroups in the map, which could pose a problem as an important requirement for a successful pattern recognition study is that each class is represented by a collection of samples that are in some way similar. The subclustering suggests a lack of homogeneity among the samples representing the JP-5 fuels. Therefore, it is important to identify and delete the GC peaks responsible for the subclustering of the JP-5 fuels.

The following procedure was used to identify the GC peaks strongly correlated with the subclustering. First, the JP-5 fuel samples were divided into two categories on the basis of the observed subclustering. Next, the ability of each GC peak alone to discriminate between the gas chromatograms from the two JP-5 subclusters was assessed. The dichotomization power of each of the 85 GC peaks was also computed for the following category pairs: JP-5 versus JP-4, JP-5 versus Jet-A, JP-5 versus JP-7, and JP-5 versus JPTS. A GC peak was retained for further analysis only if its dichotomization power for the subclustering dichotomy was lower than for any of the other category pairs. Twenty-seven GC peaks that produced the best classification results when the chromatograms were classified as Jet-A, JPTS, JP-7, or JP-5 were retained for further study.

Figure 19 shows a plot of the two largest principal components of the 27 GC peaks obtained from the 221 neat jet fuel samples. It is evident from the principal component map of the 27 features that the five fuel classes are well separated. Furthermore, the principal component map of the data does not reveal subclustering within any class. This indicates that each fuel class is represented by a collection of samples that are in some way similar when the 27 GC peaks are used as features.

A five-way classification study involving JP-4, Jet-A, JP-7, JPTS, and JP-5 fuels was also undertaken using SIMCA. A principal component model for each fuel class in the training set was developed from the 27 GC peaks. The complexity of the principal component model was determined directly from the data using the technique of cross-validation. For each class, a single principal component was used to model the data. The gas chromatograms in the training set were then fitted to these models, and the residual – that is the sum of the squares difference between the original gas chromatogram and the chromatogram reproduced by the model – was computed for each gas chromatogram.

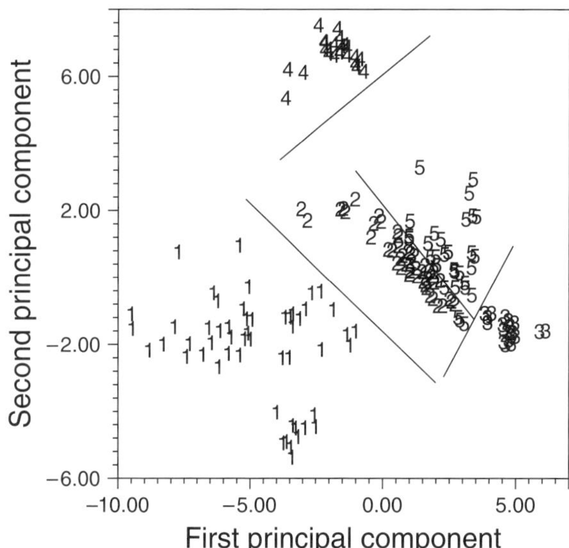

Figure 19 A principal component map of the 27 GC peaks: 1 = JP-4, 2 = Jet-A, 3 = JP-7, 4 = JPTS, 5 = JP-5. (Reproduced with permission from Lavine et al.[35] Copyright 1995 American Chemical Society.)

Each gas chromatogram in the training set was then classified on the basis of its goodness of fit. The probability of a gas chromatogram belonging to a particular fuel class was determined from its residual for the corresponding principal component model by way of an F-test. A gas chromatogram was assigned to the fuel class for which it had the lowest variance ratio. However, if the variance ratio exceeded the critical F-value for that class, then the sample would not be assigned to it. Results from the five-way classification study involving the training set samples are summarized in Table 8. The recognition rate for JP-4, Jet-A, JP-7, and JPTS fuels is very high. However, Jet-A is more difficult to recognize because of its similarity to JP-5, which is undoubtedly the reason for SIMCA classifying 16 Jet-A fuel samples as JP-5.

The ability of the principal component models to predict the class of an unknown fuel was first tested using a method called internal validation. The training set of 221 gas chromatograms was subdivided into 13 training set – prediction set pairs. Each training set had 204 gas chromatograms and each prediction set had 17 gas chromatograms. The members of the sets were chosen randomly. Furthermore, a particular chromatogram was present in only 1 of the 13 prediction sets generated. Principal component models were developed for each of the training sets and tested on the corresponding prediction set. The mean classification success rate for these so-called prediction sets was 90.5%.

To further test the predictive ability of the 27 GC peaks and the classification models associated with them, an external prediction set of 25 gas chromatograms was

For references see page 9708

Table 8 SIMCA training set results

Class	F-criterion[a]			
	Principal components	Number in class	Correct	Percentage
JP-4	1	54	54	100
Jet-A[b]	1	67	51	76.1
JP-7[b]	1	28	27	96.4
JPTS	1	29	29	100
JP-5	1	43	43	100
Total		221	204	92.3

[a] Classifications were made on the basis of the variance ratio $F = [s_p/S_o]^2[N_q - NC - 1]$, where s_p^2 is the residual of sample p for class i, S_o^2 is the variance of class i, N_q is the number of samples in the class, and NC is the number of principal components used to model the class. A sample is assigned to the class for which it has the lowest variance ratio. However, if the sample's variance ratio exceeds the critical F-value for the class, then the sample cannot be assigned to the class. The critical F-value for each training set sample is $F_\alpha = 0.975$ $[(M - NC),(M - NC)(N_q - NC - 1)]$ where M is the number of measurement variables or GC peaks used to develop the principal component model.

[b] Misclassified Jet-A and JP-5 fuel samples were categorized as JP-5.

Table 9 Prediction set results

Samples	F-values[a]				
	JP-4	JET-A	JP-7	JPTS	JP-5
PF007	3.82	10.04	58.9	12.4	7.43
PF008	3.69	9.62	57.6	12.5	7.14
PF009	3.71	9.84	57.6	12.6	7.32
PF010	3.30	16.7	73.7	11.8	10.8
PF011	3.57	9.64	58.9	12.8	7.39
PF012	4.11	7.74	78.2	13.5	12.04
PF013	4.33	8.19	79.8	12.6	12.3
KSE1M2	2.83	24.4	63.9	30.4	11.21
KSE2M2	2.25	16.2	70.8	21.6	11.09
KSE3M2	2.51	9.41	71.0	17.3	10.2
KSE4M2	2.40	10.11	71.3	17.83	10.4
KSE5M2	2.33	7.76	56.4	17.9	7.61
KSE6M2	1.87	13.4	69.3	20.8	10.4
KSE7M2	2.21	9.85	67.3	18.3	9.78
MIX1	1.33	34.9	71.3	38.2	13.3
MIX2	1.33	11.93	53.3	20.9	7.37
MIX3	1.44	12.3	55.2	20.6	7.71
MIX4	1.59	9.51	48.6	19.9	6.27
STALE-1	1.72	73.7	151.9	54.7	31.5
STALE-2	0.58	28.7	123.8	30.9	22.6
STALE-3	0.541	28.7	127.3	29.9	22.6
PIT1UNK	6.62	1.19	6.106	33.02	0.504
PIT1UNK	6.57	1.15	6.03	32.9	0.496
PIT2UNK	6.51	1.14	6.14	32.8	0.479
PIT2UNK	6.51	1.14	6.27	32.7	0.471

[a] An object is assigned to the class for which it has the lowest variance ratio. However, if the variance ratio exceeds the critical F-value for that class, then the object cannot be assigned to it. Critical F-values of prediction set samples are obtained using one degree of freedom for the numerator and $N_q - NC - 1$ degrees of freedom for the denominator.(32) For JP-4, the critical F-value at $\alpha = 0.975$ is $F(1, 52) = 5.35$, and it is $F(1, 41) = 5.47$ for JP-5.

employed. The gas chromatograms in the prediction set were run a few months before the neat jet fuel gas chromatograms were run. The results of this study are shown in Table 9. All the weathered fuel samples were correctly classified. This is an important result, as the changes in composition that occur after a jet fuel is released into the environment constitute a major problem in fuel spill identification. These changes may arise from evaporation of lower-molecular-weight alkanes, microbial degradation, and the loss of water-soluble compounds due to dissolution.(42) Because the weathered fuel samples used in this study were recovered from a subsurface environment, loss of lower alkanes due to evaporation will be severely retarded. Furthermore, dissolution of water-soluble components should not pose a serious problem as only a small fraction of the fuel's components is soluble in water.(43) Hence, the predominant weathering factor in subsurface fuel spills is probably biodegradation, which does not appear to have a pronounced effect on the overall GC profile of the fuels. Clearly, weathering of aviation turbine fuels in a subsurface environment will be greatly retarded compared to surface spills, thereby preserving the fuel's identity for a longer period of time.

4.4.2 Africanized Honeybees

GC/PR has also been used to develop a potential method for differentiating Africanized honeybees from European honeybees.(44–47) The test data consisted of 109 gas chromatograms (49 Africanized and 60 European) of cuticular hydrocarbons obtained from bodies of Africanized and European honeybees. Cuticular hydrocarbons were obtained by rinsing the dry or pinned bee specimens in hexane for approximately 15 min. The cuticular hydrocarbon fraction analyzed by gas chromatography was isolated from the concentrated washings by means of a silicic acid column. Hexane was used as the eluent. The extracted hydrocarbons (equivalent to 4% of a bee) were coinjected with authentic n-alkane standards. KIs were assigned to compounds eluting off the column. These indices were used for peak identification.

Each gas chromatogram contained 40 peaks corresponding to a set of standardized retention time windows. A typical GC trace of the cuticular hydrocarbons from an Africanized honeybee sample is shown in Figure 20. The GC column had about 5000 plates. The hydrocarbon extract was analyzed on a glass column (1.8 m × 2 mm) packed with 3% OV-17 on Chromosorb® WAW DMCS packing (120–140 mesh).

The gas chromatograms were translated into data vectors by measuring the area of the 40 GC peaks. However, only 10 of the GC peaks were considered

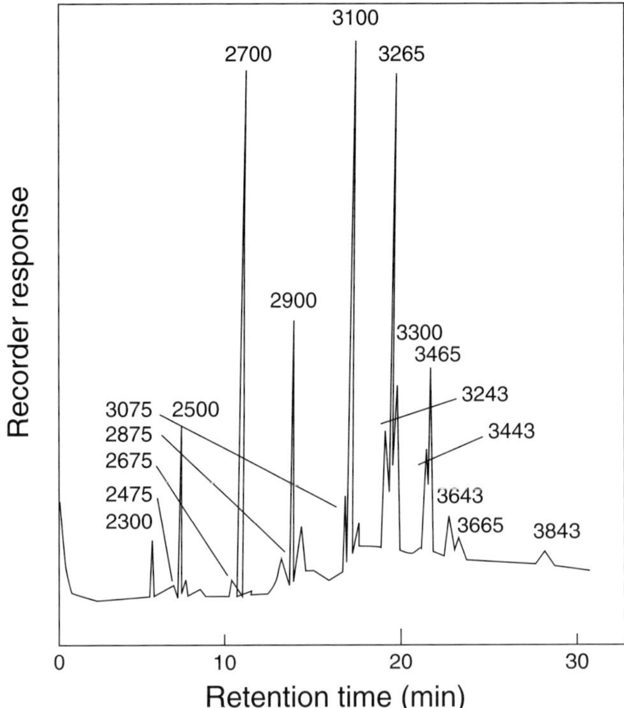

Figure 20 A gas chromatogram of the cuticular hydrocarbons from an Africanized honey bee. The KI values for the peaks used in the pattern recognition study were 2300, 2675, 2875, 3075, 3100, 3243, 3265, 3300, 3443, and 3465. (Reproduced with permission from Lavine and Carlson.[15] Copyright 1987 American Chemical Society.)

for pattern recognition analysis. Compounds comprising these peaks were found in the wax produced by nest bees, and the concentration pattern of the wax constituents is believed to convey genetic information about the honeybee colony. Because the feature selection process was carried out on the basis of a priori considerations, the probability of inadvertently exploiting random variation in the data was minimized.

Each gas chromatogram was normalized to constant sum using the total integrated area of the 40 GC peaks. Also, the training set data were autoscaled to ensure that each feature had equal weight in the analysis. The normalized and autoscaled data were then analyzed using KNN, which classifies the data vectors in the training set according to a majority vote of its KNNs. Hence, a sample will be classified as an Africanized or European bee only if the majority of its KNNs in the measurement space are Africanized bees. When the 1-nearest neighbor classification rule was applied to the 10 GC peaks, it could correctly classify every chromatogram in the training set. This result indicates that Africanized and European bee specimens are well separated from each other in the feature space defined by the 10 GC peaks.

To test the predictive ability of these descriptors and the classifier associated with them, a prediction set of 55 gas chromatograms (15 Africanized and 40 European) was employed. The distances between the prediction set samples and the samples in the training set were calculated, with class assignments computed in the same manner as in the training phase. Using the 1-nearest neighbor classification rule, a classification success rate of 100% was achieved for the gas chromatograms in the prediction set. This result is important because it demonstrates that information derived solely from cuticular hydrocarbons can categorize bees as to subspecies. This suggests a direct relationship between the concentration pattern of these compounds and the identity of the subspecies (Africanized or European). Clearly, these results imply that GC/PR can be used to identify the presence of the African genotype in honeybees.

5 SOFTWARE

There are a number of Windows 95/98 software packages sold by commercial vendors that can be used for clustering and classification. UNSCRAMBLER (Camo A/S, Olav Tryggvasonsgt. 24, N-7011 Trondheim, Norway) offers data preprocessing, PCA, SIMCA classification, and graphics in a flexible package. Pirouette (Infometrix Inc., P.O. Box 1528, 17270 Woodinville-Redmond Road NE, Suite 777, Woodinville, WA 98072-1528) has a nice user interface, with good quality graphics. The package has data preprocessing, hierarchical clustering, PCA, KNN, and SIMCA classification. Pirouette, which has been validated according to the United States Food and Drug Administration Standard Operating Procedure, is a good introductory package because of its broad functionality.

SCAN (Minitab Inc., 3081 Enterprise Drive, State College, PA 16801-3008) has PCA, hierarchical clustering, KNN, SIMCA, and discriminant analysis (quadratic, linear, regularized, and DASCO, which is an advanced classification method in chemometrics). The user interface, which is similar to the popular Minitab statistics package, has many advanced editing features, such as brushing. The package is a good mix of statistical and pattern recognition methods. SIRIUS (Pattern Recognition Associates, P.O. Box 9280, The Woodlands, TX 77387-9280) is a graphics oriented package intended for modeling and exploratory data analysis, such as SIMCA and PCA. The PLS TOOLBOX (Eigenvector Technologies, P.O. Box 483, 196 Hyacinth, Manson, WA 89931) is for Matlab and contains routines for PCA, discriminant analysis, and cluster analysis.

For references see page 9708

6 CONCLUSION

In this article, a basic methodology for analyzing large multivariate chemical data sets is described. A chromatogram or spectrum is represented as a point in a high-dimensional measurement space. Exploratory data analysis techniques (PCA and hierarchical clustering) are then used to investigate the properties of this measurement space. These methods can provide information about trends present in the data. Classification methods can then be used to further quantify these relationships. The techniques, which have been found to be most useful, are nonparametric in nature. As such, they do not attempt to fit the data to an exact functional form; rather, they use the data to suggest an appropriate mathematical model, which can identify structure within the data. Hence, the approach described in this article relies heavily on graphics for the presentation of results, because clustering and classification methods should be used to extend the ability of human pattern recognition to uncover structure in multivariate data. Although the computer can assimilate more data at any given time than can the chemist, it is the chemist, in the end, who must make the necessary decisions and judgements about their data.

ABBREVIATIONS AND ACRONYMS

FCV	Fuzzy Clustering Varieties
GC	Gas Chromatographic
GC/PR	Gas Chromatographic and Pattern Recognition
HPLC	High-performance Liquid Chromatography
KI	Kovats Retention Index
KNN	*k*-Nearest Neighbor
PCA	Principal Component Analysis
SIMCA	Soft Independent Modeling by Class Analogy
USAF	United States Air Force

RELATED ARTICLES

Environment: Water and Waste **(Volume 4)**
Solid-phase Microextraction in Environmental Analysis • Underground Fuel Spills, Source Identification

Field-portable Instrumentation **(Volume 5)**
Solid-phase Microextraction in Analysis of Pollutants in the Field

Chemometrics **(Volume 11)**
Chemometrics

List of selected abbreviations appears in Volume 15

Gas Chromatography **(Volume 12)**
Data Reduction in Gas Chromatography

Infrared Spectroscopy **(Volume 12)**
Spectral Data, Modern Classification Methods for

REFERENCES

1. B.R. Kowalski, 'Analytical Chemistry as an Information Science', *TRACS*, **1**, 71–74 (1988).
2. I.T. Jolliffee, *Principal Component Analysis*, Springer-Verlag, New York, 1986.
3. D.L. Massart, L. Kaufman, *The Interpretation of Analytical Chemical Data by the Use of Cluster Analysis*, John Wiley & Sons, New York, 1983.
4. R.G. Brereton (ed.), *Multivariate Pattern Recognition in Chemometrics*, Elsevier, Amsterdam, 1992.
5. S.D. Brown, 'Chemical Systems Under Indirect Observation: Latent Properties and Chemometrics', *Appl. Spectrosc.*, **49**(12), 14A–31A (1995).
6. L. Kaufman, P.J. Rousseeuw, *Finding Groups in Data*, John Wiley & Sons, New York, 1990.
7. G.L. McLachlan, *Discriminant Analysis and Statistical Pattern Recognition*, John Wiley & Sons, New York, 1992.
8. J. Mandel, 'The Regression Analysis of Collinear Data', *J. Res. Natl. Bur. Stand.*, **90**(6), 465–476 (1985).
9. P. Thy, K. Esbensen, 'Seafloor Spreading and the Ophiolitic Sequence of the Troodos Complex: A Principal Component Analysis of Lava and Dike Compositions', *J. Geophysical Research*, **98**(B7), 11799–11805 (1993).
10. B.R. Kowalski, T.F. Schatzki, F.H. Stross, 'Classification of Archaeological Artifacts by Applying Pattern Recognition Techniques to Trace Element Data', *Anal. Chem.*, **44**, 2176–2180 (1972).
11. H. Abe, S. Kanaya, Y. Takahashi, S. Sasaki, 'Extended Studies of the Automated Odor-sensing Systems Based on Plural-semiconductor Gas Sensors with Computerized Pattern Recognition Techniques', *Anal. Chim. Acta*, **215**, 155–168 (1988).
12. P.C. Jurs, B.R. Kowalski, T.L. Isenhour, C.N. Reilley, 'Computerized Learning Machines Applied to Chemical Problems: Molecular Structural Parameters from Low Resolution Mass Spectrometry', *Anal. Chem.*, **42**(12), 1387–1394 (1970).
13. J.A. Pino, J.E. McMurry, P.C. Jurs, B.K. Lavine, A.M. Harper, 'Application of Pyrolysis/Gas Chromatography/Pattern Recognition to the Detection of Cystic Fibrosis Heterozygotes', *Anal. Chem.*, **57**(1), 295–302 (1985).
14. A.B. Smith, A.M. Belcher, G. Epple, P.C. Jurs, B.K. Lavine, 'Computerized Pattern Recognition: A New Technique for the Analysis of Chemical Communication', *Science*, **228**(4696), 175–177 (1985).

15. B.K. Lavine, D. Carlson, 'European Bee or Africanized Bee? Species Identification Through Chemical Analysis', *Anal. Chem.*, **59**(6), 468A–470A (1987).
16. P. Geladi, H. Grahn, *Multivariate Image Analysis*, John Wiley and Sons, New York, 1996.
17. W.H.A. van den Broek, D. Wienke, W.J. Melssen, R. Feldhoff, T. Huth-Fehre, T. Kantimm, L.M.C. Buydens, 'Application of a Spectroscopic Infrared Focal Plane Array Sensor for On-line Identification of Plastic Waste', *Appl. Spectrosc.*, **51**(6), 856–865 (1997).
18. P. Robert, D. Bertrand, M.F. Devaux, A. Sire, 'Identification of Chemical Constituents by Multivariate Near-infrared Spectral Imaging', *Anal. Chem.*, **64**, 664–667 (1992).
19. D.D. Coomans, D.I. Broeckaert, *Potential Pattern Recognition in Chemical and Medical Decision-making*, Research Studies Press Ltd, Letchworth, England, 1986.
20. B.K. Lavine, 'Signal Processing and Data Analysis', in *Practical Guide to Chemometrics*, ed. S.J. Haswell, Marcel Dekker, New York, 1992.
21. F.W. Pjipers, 'Failures and Successes with Pattern Recognition for Solving Problems in Analytical Chemistry', *Analyst*, **109**(3), 299–303 (1984).
22. P.D. Wasserman, *Neural Computing*, Van Nostrand Reinhold, New York, 1989.
23. J. Zupan, J. Gasteiger, *Neural Networks for Chemists*, VCH Publishers, New York, 1993.
24. B.K. Lavine, P.C. Jurs, D.R. Henry, 'Chance Classifications by Nonparametric Linear Discriminant Functions', *J. Chemom.*, **2**(1), 1–10 (1988).
25. B.R. Kowalski, C.F. Bender, 'Pattern Recognition. A Powerful Approach to Interpreting Chemical Data', *J. Am. Chem. Soc.*, **94**, 5632–5639 (1972).
26. B.R. Kowalski, S. Wold, 'Pattern Recognition in Chemistry', in *Classification, Pattern Recognition and Reduction of Dimensionality*, eds. P.R. Krishnaiah, L.N. Kanal, North Holland, Amsterdam, 1982.
27. M. Sjostrom, B.R. Kowalski, 'A Comparison of Five Pattern Recognition Methods based on the Classification Results from Six Real Data Bases', *Anal. Chim. Acta*, **112**, 11–30 (1979).
28. B. Soderstrom, S. Wold, G. Blomqvist, 'Pyrolysis Gas Chromatography Combined with SIMCA Pattern Recognition for Classification of Fruit-bodies of Some Ectomycorrhizal Suillus Species', *J. Gen. Microbiol.*, **128**, 1783–1794 (1982).
29. S. Wold, 'Pattern Recognition: Finding and Using Regularities in Multivariate Data', in *Food Research and Data Analysis*, eds. H. Martens, H. Russwurm, Applied Science, Essex, England, 1983.
30. S. Wold, 'Pattern Recognition by Means of Disjoint Principal Component Models', *Patt. Recog.*, **8**, 127–139 (1976).
31. S. Wold, 'Cross-validatory Estimation of the Number of Components in Factor and Principal Components Models', *Technometrics*, **20**, 397–406 (1978).
32. S. Wold, C. Albano, U. Edlund, K. Esbensen, S. Hellberg, E. Johansson, W. Lindberg, M. Sjostrom, 'Pattern Recognition by Means of Disjoint Principal Component Models (SIMCA). Philosophy and Methods', in *Proc. Symp. Applied Statistics, NEUCC*, eds. A. Hoskuldsson, K. Esbensen, RECAU and RECKU, Copenhagen, 183–218, 1981.
33. S. Wold, C. Albano, W.J. Dunn III, U. Edlund, K. Esbensen, S. Hellberg, E. Johansson, W. Lindberg, M. Sjostrom, 'Multivariate Data Analysis in Chemistry', in *Chemometrics, Mathematics and Statistics in Chemistry*, eds. B.R. Kowalski, D. Reidel Publishing Company, Dordrecht, 1984.
34. S. Wold, M. Sjostrom, 'SIMCA, a Method for Analyzing Chemical Data in Terms of Similarity and Analogy', in *Chemometrics, Theory and Application*, ed. B.R. Kowalski, American Chemical Society, Washington, DC, 1977.
35. B.K. Lavine, H. Mayfield, P.R. Kroman, A. Faruque, 'Source Identification of Underground Fuel Spills by Pattern Recognition Analysis of High-speed Gas Chromatograms', *Anal. Chem.*, **67**, 3846–3852 (1995).
36. H.T. Mayfield, W. Bertsch, 'An Algorithm for Rapidly Organizing Gas Chromatographic Data into Data Sets for Chemometric Analysis', *Comput. Appl. Lab.*, **1**, 13–137 (1983).
37. E. Kovats, in *Advances in Chromatography*, eds. J.C. Giddings, R.A. Keller, Marcel Dekker, New York, Vol. 1, 230, 1965.
38. R.A. Johnson, D.W. Wichern, 'Applied Multivariate Statistical Analysis', Prentice Hall, Englewood Cliffs, New Jersey, 1982.
39. M.A. Stapanian, F.C. Garner, K.E. Fitzgerald, G.T. Flatman, J.M. Nocerino, 'Finding Suspected Causes of Measurement Error in Multivariate Environmental Data', *J. Chemom.*, **7**, 165–176 (1993).
40. *Handbook of Aviation Fuel Properties*, Coordinating Research Council, Inc., Atlanta, GA, 1983.
41. H.T. Mayfield, M. Henley, in *Monitoring Water in the 1990's: Meeting New Challenges*, eds. J.R. Hall, G.D. Glayson, American Chemical Society for Testing of Materials, Philadelphia, PA, 1991.
42. J.C. Spain, C.C. Sommerville, L.C. Butler, T.J. Lee, A.W. Bourquin, 'Degradation of Jet Fuel Hydrocarbons in Aquatic Communities', USAF Report ESL-TR-83-26, AFESC, Tyndall AFB, 1983.
43. W.E. Coleman, J.W. Munch, R.P. Streicher, H.P. Ringhand, W.F. Kopfler, 'Optimization of Purging Efficiency and Quantification of Organic Contaminants from Water using a 1-L Closed Looped Stripping Apparatus and Computerized Capillary Columns', *Arch. Environ. Contam. Toxicol.*, **13**, 171–180 (1984).
44. B.K. Lavine, D.A. Carlson, D.R. Henry, P.C. Jurs, 'Taxonomy Based on Chemical Constitution: Differentiation of Africanized Honeybees from European Honeybees', *J. Chemomet.*, **2**(1), 29–38 (1988).

45. B.K. Lavine, A.J.I. Ward, R.K. Smith, O.R. Taylor, 'Application of Gas Chromatography/Pattern Recognition Techniques to the Problem of Identifying Africanized Honeybees', *Microchem. J.*, **39**, 308–316 (1989).
46. B.K. Lavine, D.A. Carlson, 'Chemical Fingerprinting of Africanized Honeybees by Gas Chromatography/Pattern Recognition Techniques', *Microchem. J.*, **42**, 121–125 (1990).
47. B.K. Lavine, A.J.I. Ward, J.H. Han, R.K. Smith, O.R. Taylor, 'Taxonomy Based on Chemical Constitution: Differentiation of Heavily Africanized Honeybees from Moderately Africanized Honeybees', *Chemolab.*, **8**, 239–243 (1990).

Multivariate Calibration of Analytical Data

Svante Wold
Research Group for Chemometrics, Umeå University, Sweden

Mats Josefson
AstraZeneca R&D, Mölndal, Sweden

List of selected abbreviations appears in Volume 15

Multivariate calibration (MVC) is a methodology for using multiple signals, for instance a digitized spectrum, to determine the levels of concentrations of chemical compounds in analytical samples. MVC can also be used to determine other properties of interest, for instance viscosity, particle size distribution, energy content, or taste. MVC is made in two phases. In the first, the "training" or "calibration" phase, samples with known concentration (property) values and their signal profiles are used to develop a model of their relationship, a multivariate standard curve. In the second phase, this model is used with new samples to determine their concentration (property) values from their signal profiles.

1 INTRODUCTION

The calibration of an analytical instrument means the construction of a quantitative relationship between, on the one hand the instrument "signals" as they are measured on analytical samples, and on the other hand one or several properties of the samples, usually concentrations of analytes in the samples.

In univariate calibration (UVC) the amplitude of one signal, e.g. the absorption of light at a certain wavelength, is related to the concentration of one analyte. In MVC, several signals, for instance from a whole spectrum digitized at regularly spaced wavelengths, are used to derive a multivariate model (a generalized standard curve).[1] This model simultaneously relates the amplitudes of *all* the signals, or a substantial part of them, to the concentrations of one or several analytes in the samples. The multivariate model is then used to estimate the analyte concentrations in new samples from the multiple signals measured on these. The use of MVC and spectral profiles to "measure" concentrations or other properties is often referred to as "indirect measurement" in contrast to the direct measurement of, say, concentration by means of titration, or precipitation and weighing.

In both UVC and MVC, the deviations between the model and the data are used to derive statistical measures of uncertainty of the model and of the estimated concentrations of the new samples. In MVC, the multivariate information can be used for additional statistical diagnostics such as the similarity between a new sample and the calibration set (CS), the presence of clusters in the data, and the relative information content of the "signals" (predictor variables).

The relation between spectroscopic signals and concentrations is fairly linear as long as the variation of concentration is moderate. This is often referred to as Lambert–Beer's law. Most calibrations are limited to concentration intervals where a linear standard curve is adequate, but the extension to mildly nonlinear curves are straightforward using, for instance, quadratic polynomials as the functional basis.

MVC has six steps, the first five of which comprise the training phase, and the sixth is the prediction phase. These steps are:

1. Specification of the analytes with concentration ranges. Selection of the instrumental method, including the range of wavelengths, reflectance or transmission mode, etc.
2. Selection of a representative set of calibration samples – the training set (TS) or CS. This CS should span the range of analyte concentrations and also the concentration ranges of interferents.
3. The multivariate signals (typically the digitized spectra) are recorded for the calibration samples and stored in an appropriate data base. The analyte concentrations are measured by a reference method.
4. The data are investigated for the presence of outliers and other anomalies. Thereafter, the data are preprocessed and transformed to a form suitable for the subsequent data analysis.
5. The calibration model is developed and optimized. This includes checking for linearity, and the determination of selectivity, detection limits, precision, accuracy, and other measures of performance. Statistical measures of uncertainty are calculated and used to construct confidence intervals for predicted values. Also, the model is interpreted chemically, important variables (wavelength regions) looked at, interferences are identified, etc.
6. The model is used to estimate the analyte concentrations in new samples (prediction set), including confidence intervals. Diagnostics for dissimilarity (outliers) are checked.

For references see page 9734

MVC has a much improved precision and selectivity compared to UVC. Also MVC can handle complicated samples with unknown interfering compounds, and it works even when there is no selective wavelength region for the analyte. This has made MVC particularly useful with nonselective spectral methods, such as near-infrared spectroscopy (NIR). In a slightly generalized sense, MVC can be applied to the relation between any instrumental "profile", including chromatograms, kinetic curves, thermo-gravimetric curves, sound spectra, etc. and any properties of the analyzed sample. The properties can be other than concentrations, for instance viscosity and molecular weight of polymer samples, the energy content of oil, gasoline, coal, or peat, or the taste of cheese, wine, or beer. Hence the MVC methodology provides exciting possibilities for the indirect measurement of complicated properties of complicated samples.

1.1 Univariate Calibration

The idea of calibration is simple and powerful. The instrument is given a training or CS of samples with known concentration (c_i) of the analyte, and the amplitude of the instrument signal (z_i) is recorded for each sample, i. The resulting data ($z_i, c_i, i = 1, 2, \ldots, n$) are then used to construct a standard curve (usually a straight line), which is used for new samples to predict the concentration for new samples from the amplitude of the signal measured on these samples (Figure 1).

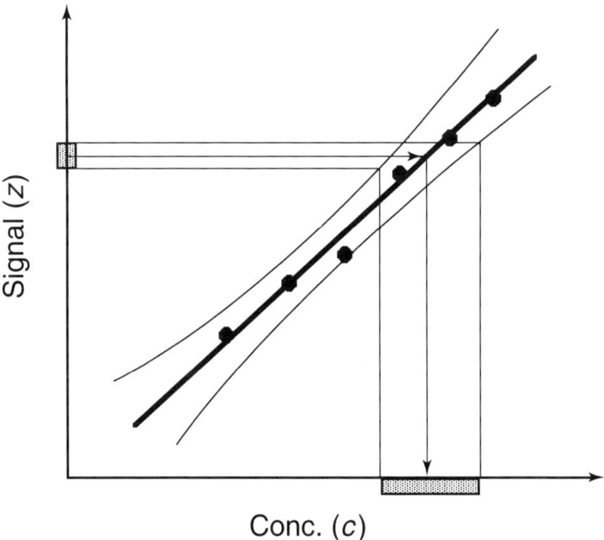

Figure 1 The CS (filled circles) of samples with known concentrations are used to derive a standard curve (here straight line). For new samples, the concentrations can be predicted from the amplitude of the signal measured on these samples by moving horizontally from the signal axis to the standard curve and then vertically down to the concentration axis. Owing to the noise in the data, the standard curve has some uncertainty in slope and intercept, which together with the imprecision of the measured signal, causes an uncertainty in the predicted concentration value.

A prerequisite for calibration to work is that the signal amplitude really changes in a reproducible way with the analyte concentration. The greater this change is for a unit change of concentration, the greater is the sensitivity of the instrument with respect to the analyte. Moreover, the signal must not be much affected by other components in the sample, it must be fairly selective.

1.2 Standard Curve, Predictions, Measures of Uncertainty

Normally in UVC, the signal amplitude (z) is used as the y-variable and the concentration (c) used as $\boldsymbol{x}$ in the calculation of the standard curve with intercept a and slope b (Equation 1).

$$z_i = a + bc_i + e_i \tag{1}$$

Least squares (linear regression) is normally used for this calculation.(2–4) The residuals e_i are calculated, with their variance (s^2) and standard deviation (SD), s, (Equations 2 and 3).

$$e_i = z_{i_{\text{observed}}} - z_{i_{\text{model}}} \tag{2}$$

$$s = \sqrt{\sum_i \frac{e_i^2}{n-1}} \tag{3}$$

This residual standard deviation (RSD) should be the same as the precision of measurement within statistical uncertainty. Regression(2–4) is then used to calculate standard errors and confidence intervals for the parameters, as well as for the concentrations of new samples (calculated from their signal amplitudes and the standard curve). If the RSD is substantially different at different concentrations, a weighted regression may be needed for an efficient data analysis. It should be noted that the estimation of the parameters (a and b in Equation 1) by regression, is not based on the assumption of constant and normally distributed noise. The sizes of their confidence intervals depend, however, on the distribution of the noise.

1.3 Multivariate Calibration; Many Signals

In MVC the *whole* spectral profile is used for the calibration instead of just the signal at one single wavelength.(1,5) The signal data now consists of a matrix, $\mathbf{Z}$ (dimension $n \times K$). In MVC, one can calibrate on a single analyte, i.e. $\mathbf{y}$ is an $(n \times 1)$ vector, or one can simultaneously calibrate on several (M) analytes. In the latter case, the concentrations of the CS form an $(n \times M)$ matrix, $\mathbf{Y}$. The latter is discussed in sections 3.5 and 8.3.

1.4 Interferences

In most calibrations, the samples contain compounds other than the actual analyte, so-called interferents. These other compounds will most likely also affect the signal, and hence the measured signal will have a systematic bias, an interference. UVC thus works well only when the signal is close to selective, and is affected only by the change of the analyte concentration. This often necessitates the use of chromatography or some other separation method to clean up the sample from interfering compounds. MVC is affected much less by the presence of other compounds due to its ability to find nearly selective combinations of the multiple signals that model and predict the analyte concentration(s). A preprocessing of the data to enhance this selectivity may still be needed (see section 7).

A second problem is that any calibration model is valid only for samples that are similar to the calibration samples. Thus for example, a model developed for gold in seawater samples, does not give accurate results for fresh water samples, and even less well with samples of orange juice. Hence one should check that each new sample is indeed similar to the calibration samples. Unfortunately, there is nothing in the single signal employed in UVC that is helpful for this matter, and the similarity must be checked in some other way, or accepted by faith. In MVC, however, the profile of the multiple signals contains information also about the similarity between the samples. Hence, MVC provides an "autodiagnosis" for the similarity between the new samples and the calibration samples, which greatly improves the reliability of the results.(1)

1.5 Direct and Indirect Calibration

If the spectra of all the pure constituents are available (denoted by r_{jk} for the spectral amplitude of constituent j at wavelength k), the concentrations c_{ij} in sample i can be estimated from the spectrum z_i' of a mixture of the analytes by means of multiple regression using the linear model (Equation 4):

$$z_i = \sum_j r_j c_{ij} + e_j \tag{4}$$

This is often called direct calibration because the causal physical model (concentration → spectrum) and the statistical model ($z = \mathrm{f}(c)$) coincide.(1) This direct approach works when (a) *all* constituents in a given sample are known together with their spectra, and (b) the spectrum of any constituent is the same when it is pure and when it is present in the mixture (sample). This makes direct calibration work for simple and rather dilute gas samples and some simple dilute liquid samples, but less well or not at all with complicated samples where many constituents are unknown (interferents) and where the spectra change with concentration. This is exemplified by samples of grain, blood, tissue, lake sediment, polymer formulations, pharmaceutical formulations, wood, pulp, etc. where indirect MVC is the only feasible approach.

The MVC model is formulated inversely to the direct model above, with the explicit objective to predict the concentrations (**Y**) from the digitized and preprocessed spectra (**X**), (Equation 5):

$$\mathbf{Y} = \mathbf{XB} + \mathbf{F} \tag{5}$$

Here the information flow (from **X** to **Y**) is the opposite to the causal model (**Y** → **X**), which made this be called the "indirect calibration" model. This inverse formulation of the calibration model was first proposed by Krutchoff,(6) causing much initial consternation and controversy. With the work of Hoadley,(7) Naes,(8) and others, however, we have learnt to distinguish between the direction of the information flow and the direction of causality, and (indirect) MVC is today established and accepted as a good working approach.

The mechanism of inverse MVC is, quite simply, that as long as the correlation structure between **X** (signals) and **Y** (concentrations, properties) remains, then the predictions can be made in either direction **X** → **Y**, or **Y** → **X**. The principle of MVC is to utilize this fact in the direction **X** → **Y** (signals predict concentrations or properties). The correlation structure of (**X**, **Y**) is constant as long as the samples are similar to each other, but it changes when very different samples are brought in. Hence the CS must be representative for the future samples for which the concentrations or other properties are to be determined, see section 4.

1.6 The Concentrations of Two Drug Compounds Determined by Ultraviolet Spectroscopy

To illustrate the calculations and the interpretation of resulting parameters, we shall use a small example with ultraviolet (UV) spectra of samples containing two drug substances. There is a set of $n = 10$ spectra of mixed standards, where all combinations of three levels of two substances (felodipine and metoprolol) are included. One of the samples is included in duplicate (#6 and 7). Three additional samples are included as a prediction set. In addition, the spectra of the two pure substances are included, showing overlap with no selective region for metoprolol, but a selective region of felodipine with a maximum at around variable 58, corresponding to around 363 nm (Figure 2).

UV spectra in the wavelength range 250 to 450 nm have been recorded, and digitized at every 2 nm giving $K = 101$ spectral variables per sample. The concentration intervals

For references see page 9734

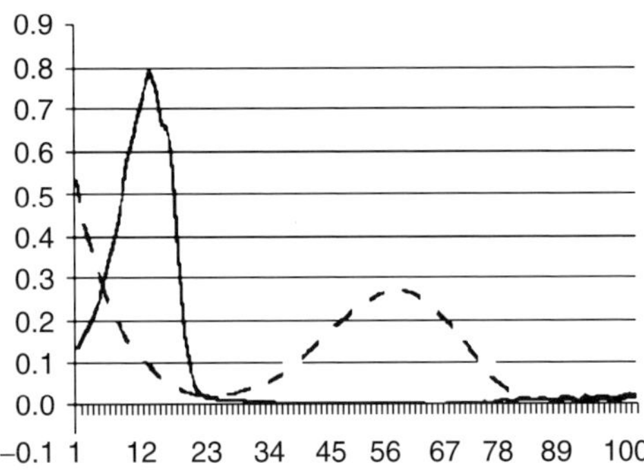

Figure 2 UV Spectra of felodipine (- - - - -) and metoprolol (——) (pure samples 11 and 12).

are 20–29 µg L^{-1} for the first substance (felodipine) and 191–239 µg L^{-1} for the second (metoprolol).

1.7 Mathematical and Statistical Formulation

Assuming that the signal amplitudes change proportionally to the concentrations (Lambert–Beer's law), the calibration model will be linear, with **B** a matrix of parameters to be determined in the calibration stage.

$$\mathbf{Y} = \mathbf{ZB} + \mathbf{F} \tag{6}$$

The model (Equation 6) works as a multivariate standard curve, where the parameters **B** define directions in the *Z*-space (the space of the spectral variables) which are linearly related to the **Y**-variables, i.e. the analyte concentrations or other properties of interest.

The **Y**-residual vector or matrix **F** contains the deviations between model and the concentration/property values (**Y**), and should have small elements everywhere. The size of the elements in **F** is determined by the precision of the instrument, the experimental control, the adequacy of the linear model and the method of data analysis. These residuals can be inspected for large values (outliers) and systematic patterns, which indicate model inadequacies. The SD of these **Y**-residuals should not greatly exceed the known precision of the used method and instrument.

In the case where Equation (6) is estimated by a projection model (section 8) this allows also the estimation of residuals of the (preprocessed) signals, **X**. These residuals, denoted **E**, provide indications of spectral areas with large and small information content.

2 NOTATION AND ABBREVIATIONS USED IN MULTIVARIATE CALIBRATION

2.1 Mathematical Notation

Vectors are denoted by bold lower-case letters, e.g. **t**, and matrices by bold capital letters, e.g. **X**. Vectors are assumed to be column vectors unless indicated by a prime (transpose), e.g. $\mathbf{p}'$. The length (norm) of a vector or a matrix is denoted by $\|\mathbf{v}\|$ or $\|\mathbf{X}\|$ respectively. This length divided by the square root of the number of elements, often corrected for the degrees of freedom, is the SD of the vector or the matrix.

For notation see the list of symbols.

3 PROBLEM FORMULATION AND CHOICE OF ANALYTICAL METHOD

MVC provides an improved set of tools for analytical chemistry compared to UVC. There is no longer a need to find the "ultimate variable" with as little interference as possible. One or a few analytical techniques yielding several variables that carry relevant information for the analysis are preferred. Interferences can be compensated for, a measure of the appropriateness of using an MVC for a given sample is available, and the correlation patterns among the variables are a starting point for new knowledge.

By treating analytical data from a given sample as a profile of variables, where the profile is used as a total signature of the sample, there is less need for special optimization regarding the protection of the single variable from interfering compounds and processes. For efficient use, the focus should be on analytical methods that give several variables with a short measurement time. The measurements should be made using best analytical chemistry practice.

A number of questions are closely related to the MVC problem, but we will here only discuss the first, namely (1) How much of a known compound is present in the given sample? Another, more demanding question is: (2) Which compounds are present in the given sample? An even more elusive type of questions is: (3) Will this intermediate product yield a good-quality end product?; e.g. Will this wine be better with storage? Finally, there is a related question of classification that in some cases is easier to answer if the amount of reference samples is limited: (4) Is the analytical profile in the present sample similar to that of the previous samples?

3.1 How Much of a Known Compound is Present in the Given Sample?

The type 1 question has traditionally been answered with UVC. The use of UVC in this case implies that the signal is selective, i.e. the interfering compounds in the sample matrix must have a low influence on the magnitude of the signal. This is today often accomplished by a chromatographic separation step. By using MVC it is possible to omit the chromatographic separation before a spectrometric detection. This will save run time for a long series of determinations and even allow for direct measurements on solid materials, thus avoiding the time spent for sample preparation. The recognition of aberrant samples usually attained by visual inspection of chromatograms, can be automated in MVC. The spectral residual profile provides this ability.

3.2 Prepared Reference Samples, or Use of a Reference Method

We can distinguish two calibration situations, the first where reference samples can be prepared with desired levels of analytes and interferents, and the second where samples are collected from nature or some process of interest, and analyzed by a reference method. In statistics this is referred to as the concentrations being fixed or random, respectively.

The first situation is illustrated by the example (section 1.6), where two known substances are weighed and dissolved to make reference samples of known composition. The concentration levels in these samples should then be laid out according to a statistical design, typically a full or fractional grid design with five or more levels. The second situation is typical for complicated samples, such as samples of grain where one wishes to determine the concentrations of protein, fat, and water. Then many samples are needed in the TS to ensure the proper spanning of all important directions in the concentration space, and to compensate for the imprecision of the reference method.

The reference method should have good accuracy and precision, providing accurate reference values for the TS and the validation set. Effectively the second type of MVC involves an inter-calibration of two different analytical methods that puts both the MVC and the reference method in a critical light. This is not always in favor of the reference method and it may need improvement in order to yield high-quality data.

3.3 Multivariate Calibration for Explorative Work

Often the objective is not to build a precise calibration but rather to explore the correlations between variables investigated in development work. In this case the MVC is used in an inductive way to make steps forward in the building of knowledge during explorative work. Here the resulting information is the structure of the calibration model rather than the specific amount of a compound. This use of MVC is less obvious but very important in industrial development work.

A much more rough calibration can then be made based on a minimal number of samples, just to give an idea about the feasibility of the methodology. Here the minimum number of samples is five, and these samples must also be well distributed over the range of interest. With so few samples and with many spectral variables, no extensive preprocessing can be afforded, and the partial least squares (PLS) projection to latent structures approach is preferred.

3.4 Multivariate Classification of Samples in Narrow Concentration Intervals

In some cases it less practical to make a calibration, for instance, when the properties of an industrial product are going to be monitored. The product is produced according to a specification aiming at a constant set of properties. Then a large set of samples is available with concentrations close to the specification in a very narrow distribution. Samples outside this narrow distribution may not be available from the production. Then calibration samples outside the specification have to be manufactured at lab scale in order to span a variation of the concentration that is needed for the TS. This may be feasible for liquid solutions with few components, but it is more involved for solid samples such as powders and pharmaceutical tablets.

An alternative approach in this case is to use multivariate classification.[9] Then the classification is used as a positive verification of the presence of the correct concentration in a certain concentration interval. In the method validation step, concentrations outside this interval have to be tested and should fail a positive identity. Analogously to the residual variance in MVC, the classification residuals indicate aberrant spectra resulting from deviations from the specified composition profile.[10,11]

Using multivariate classification in this way may save a lot of development time when it is costly to create concentration variations for the calibration samples, while it is still possible to get the benefit of using spectral residuals to find the sources for variations when they occur.

3.5 One or Several Analytes

When PLS is used in the MVC, it is possible to have more than one response from the same calibration model. This is useful in cases where the responses are correlated. When the reference method for one or a few correlated responses is performing less well, the use of a joint model can support the prediction of these responses. This is useful in explorative work when the most information has to be extracted from limited amounts of data.

When uncorrelated responses are used in the same model, this will need more PLS components than the corresponding single-response calibration models. Hence it is better in this case to develop separate PLS models for each response. Typical situations with correlated

For references see page 9734

responses are examplified by isocratic chromatography where the peak widths are correlated with the peak retention time, by environmental analysis where pollutants often vary together, and by kinetic measurements where the amount of reacted compound at time j is also correlated with the amount of reacted compound at time $j-1, j-2$, etc.

3.6 The Chemical Analytical Method

Analyte concentrations can be determined in various types of phases, i.e. transparent and turbid liquids, solids, gases, and plasma. Traditionally, most analyses have been performed in the liquid state and in some cases in the gas phase. If the sample came as a solid, it would be dissolved in an appropriate solvent and then analyzed. In chromatographic and electrophoretic separations the sample is also transported in solution or in the gas phase.

The technology for measuring absorbance by light passing through a liquid or gas sample cell is well established and yields close to linear responses in the appropriate concentration range. These optical transmission techniques are applicable for ultraviolet/visible (UV/VIS), NIR, infrared (IR), and their combinations with separation methods.

Transmission spectroscopy is limited by the amount of light that is coming through the sample. If the sample in the sample cell is absorbing too much light in some wavelength regions, the measurement at those wavelengths will be of little analytical value since stray light in the instrument will be the dominating feature. This may be adjusted by the cell path length. Technically it is more difficult to get a close to linear response at short wavelengths in UV below 210 nm due to absorption of glass and Rayleigh scattering due to dust. However, this is a range that is commonly used for amino acids and proteins.

When water is a solvent, IR spectrometry has to be done with very short path lengths in attenuated total reflection (ATR). For NIR transmission in water solutions, the proportion between the ≈1440 nm and the ≈1930 nm water peaks may be used as an indicator of nonlinearity. In many cases the longer wavelengths including the 1900 nm peak have to be discarded to get a close to linear response.

For trace amounts of substances it is advantageous to employ fluorescence measurements to get a high sensitivity. Fluorescence data with both excitation and emission wavelengths collected is also a case where two-dimensional (2-D) MVC is working well.

With MVC it is also possible to use UV/VIS spectrometry in turbid solutions.[12] For NIR the turbidity is less of a problem due to the smaller spectral response for turbidity at longer wavelengths. If larger particles are present in the sample, the optics of the instrument has to have a wide enough beam to pass beside the particles. For optical fibers this means that when larger particles are present it is better with fiber bundles than single fibers. When the sample becomes too turbid it is advantageous to change from transmission to diffuse reflectance.

Solid samples also have polymorphic properties. These are destroyed if the sample is dissolved. Hence it is often of value to measure directly on solid samples. This can be done by IR, NIR or UV/VIS diffuse reflectance, NIR transmission, and Raman. X-ray diffraction is a possible reference method for MVC with polymorphic samples. For most complex solid samples measured with NIR, the MVC is the only way to get a calibration. NIR is working well with most organic solids but an exception is black plastics, which are better characterized with diffuse reflectance IR. Raman is also working for some inorganic solids.

Color measurement is a special form of UV/VIS diffuse reflectance with the colors as a standardized set of responses in a fixed instrument calibration.

Measurements in the gas phase are theoretically simpler, since there are less interactions between molecules than in the solid and liquid state. However, the equipment may become bulkier because of the longer cell paths needed to get sufficient sensitivity.

Measurements in plasma are done for example in inductively coupled plasma (ICP) and inductively coupled plasma/mass spectrometry (ICP/MS) where metal salts in solution are burnt in a plasma and the emission spectrum is measured, or the plasma is subsequently put into a mass spectrometer. The plasma is used to eliminate molecules and only keep the elemental responses from the excited atoms. Spectra are obtained in parallel with many elements. A potential for MVC exists with ICP and ICP/MS.

Spectroscopy is ideal for creating variables for MVC. One reason is the stability of the wavelength scale. MVC is dependent upon having the same information sorted into the same variables (wavelengths) over all measured samples. This generally holds when a single instrument is used for both calibration and prediction. For dispersive spectrometers such as UV/VIS, NIR, IR, and fluorescence and Raman, there are several levels of methods adapted for wavelength-scale verification and correction, all the way from using, for instance, a rare-earth oxide or polystyrene transmission or reflection standard manually during maintenance, to an automatic wavelength correction preceding each obtained spectrum. A frequent and automatic wavelength correction is preferable.

The photo diode array spectrometer (PDA) is a special case of the dispersive spectrometers where the grating is mounted in a fixed position. Due to the nonmoving parts, the PDA is suitable for fast data acquisition but is in general limited to narrower wavelength ranges and is less sensitive than the corresponding nonarray detector. For use with NIR the InGaAs array is presently emerging as a sensitive detector with high speed data acquisition in the 800–1700 nm range, and also in extended versions up to 2300 nm.

For nuclear magnetic resonance (NMR) it is possible to use compounds with known resonating frequencies as internal standards for the frequency scale. In mass spectrometry (MS), mass standards also have to be employed in order to verify and adjust the mass scale.

The wavelength scale may also be controlled by fixed or variable filters. Arrays of fixed filters have been designed for specific measurements, such as protein in wheat by NIR diffuse reflectance. In general, fixed filter instruments are considered to be less repeatable between instruments and filters tend to age faster than gratings. Another line of filter instruments uses acousto-optical tuneable filters (AOTF). This type of instrument is suitable when fast spectral acquisition is needed in process applications.

Even if MVC can cope with mild instrumental nonlinearities, it is preferable to obtain spectra of good quality from analytical instruments with good specifications. If a certain variation is not present in the spectra, it will not be caught by the MVC either. However, the presence of a variation need not be visible for the human eye in, e.g. a set of NIR spectra. MVC performs better than the human eye in catching small systematic variations that occur in combinations over a large number of wavelengths.

3.7 Spectral or Time Domain

The IR, NIR, NMR, and Raman spectrometers have developed to a state where spectra are often obtained as interferograms. Then Fourier transform (FT) is applied to translate the measured signals into spectra. For IR, NIR, and Raman the raw data for the wavelength or frequency scale is generated by a laser beam that passes through the same path as the analytical light in the spectrometer. By the use of interference fringes between the laser beams passing through the two different pathways in the spectrometer, it is possible to accurately define the relative position of the instrument mirrors and thus the frequency scale in the calculated spectrum. The precision of the wavelength scale from this type of instrument is generally high.

From an MVC point of view there may be alternative ways to treat the interferograms. They could be used directly as variables in an MVC or be preprocessed by, for instance, a wavelet transform.[13,14] The main reason for the use of FT methods is that the spectra should be convenient to inspect one by one for the human eye. To accomplish this, the interferogram is usually premultiplied by a filter that has a slight smoothing effect before the FT is applied. Since MVC works also for patterns that are less interpretable by the human eye, this smoothing may not be necessary and may even remove some of the important spectral information.

When using MVC based on FT generated spectra, it is important to keep track of the filter used before the FT calculation. This filter should be the same for all spectra in a calibration. With a less smoothing filter, wiggles may occur on the spectral baseline. These may not show up in the spectra but will still be visible in PLS loading vectors.

3.8 Chromatography

Chromatographic and electrophoretic methods also yield a vector (i.e. a chromatogram) for each sample with peaks representing different compounds separated in time. Such vectors are useful for characterizing biological samples and their relation to properties such as flavor, taste, and degree of disease.[15–17] Examples are gas chromatography (GC), liquid chromatography (LC) and thin-layer chromatography (TLC).

Just treating the chromatogram as a spectrum will not work well since the retention time axis of chromatograms varies with e.g. the column temperature, the mobile phase, the column age, and the column to column variation. Thus it is not straightforward to align chromatographic peaks. When there are few and known peaks, the peak areas or peak heights can be used provided that they are assigned correctly. Internal standards may be used as retention time reference points and also as a basis for the magnitude scale. When the whole chromatographic profile is important, such as for characterization of natural materials, e.g. oil, peat, and wine, it is more appropriate to use the entire chromatogram as a set of variables. Successful approaches for aligning the retention times have been reported.[18] Alternatively, Fourier or wavelet transformation of the chromatographic profile may give a representation less sensitive to shifts in the chromatogram.

When quantitative measurements are based on image scanned TLC plates or electrophoretic gels it is important to recognize the limited resolution in magnitude which may cause a nonlinear response.

Often a separation is optimized to give an even separation of the peaks. This is not the optimal way to run a separation method when molecular properties are wanted. Instead the conditions should be selected to show interactions of different kinds such as hydrogen bonding, η–η interaction, acid–base properties, etc. The interactions of a compound with a set of mobile and stationary phases may be used in an MVC for molecular properties, such as lipophilicity, polarity and other related chemical properties for a class of chemical compounds.

For references see page 9734

3.9 Combination Methods

The combination methods often use one method to separate the sample into purer fractions, followed by a spectrometric method to give as much information as possible about the separated fractions. The fractions may not always be pure compounds, they can also be combinations of related compounds. GC may be combined with MS or IR; LC may be combined with MS or UV; tandem MS may be used.

In these cases spectral data may be directly used in MVC. Automated multivariate classification is a good primary step in order to align retention times. This will now be easier since each chromatographic peak also has a spectral profile.

Hyphenated methods also lend themselves to the next level of MVC, working with variables as 2-D arrays instead of vectors. Special MVC methods have been developed for this situation.[19,20] However, when a chromatographic or electrophoretic separation method is involved, the time alignment problem in section 3.8.4 still persists.

3.10 Sensor Arrays (Electronic Noses, Crystal Sensors)

Sensor arrays are often constructed with silicon chip technology as resonators with a layer of substrate that is catching target molecules. The idea behind sensor arrays is that the individual sensor does not need to be totally selective, but through a combination of slightly different substrates, MVC is given the possibility to distinguish between different compounds. There are no alignment problems, but memory effects may be a problem, as the substrate may age and become chemically modified during harsh chemical conditions. When used for MVC these type of calibrations should have some means for shutting down individual sensors without too much degradation of the calibration performance.

3.11 Kinetic Curves

Kinetic measurements give a profile of variables that changes with time. Kinetic raw variables can be used in MVC as they are.

Another approach is to fit a physical kinetic model to the time traces and then use the coefficients of this model as variables in an MVC. In this case it is important that these coefficients really reflect a measured property and not just constitute an extrapolation. For example, an LD50 value will be an extrapolation in a toxicological study if the substance had such a low toxicity that the LD50 conditions were never approached in the study. In this case the LD10 or LD20 values are bracketed by measured values, and hence are more suitable for MVC.

3.12 Images and Other Measured Data

Two- or three-dimensional images from microscopes, digital cameras, tomography, etc. can all be transformed to data vectors and hence be used as data for MVC. Special preprocessing may be needed to make the data comparable between samples, such as 2-D FT or 2-D wavelet transforms. Actually, ordinary spectra UV, IR, etc.) can be considered as one-dimensional images, and quantitative image analysis can be considered as just a special case of MVC.[21]

Any quantitative multivariate data measured on analytical samples can be used as the basis for MVC provided that they contain relevant information. Recently, profiles of sound and other vibrations have been shown to contain interesting information about concentrations, the degree of mixing and other properties of complex material flows.[22]

4 SELECTION OF THE CALIBRATION SET

The calibration results in a model that relates the signal amplitudes to concentration, the so-called standard curve. All models have a domain of validity which depends on the mathematical stability of the model and the range of the data on which the model is based. The uncertainty of the predicted concentrations for a given signal amplitude increase rapidly outside the concentration range defined by the calibration samples (see Figure 1). Hence the calibration model (the standard curve) should not be applied far outside this range. Consequently, if one knows the future operating range of the analytical method, the calibration samples must span this range. The calibration sample concentrations should also be distributed fairly uniformly over the range of interest.

The precision of the standard curve improves with an increasing number of calibration samples, n.[2–4] With large n, the uncertainty of the predicted concentrations is dominated by the uncertainty of the signal amplitude of the new samples. In the situation where reference samples are prepared from pure substances, there is therefore little gain in increasing n above around $M \times 10$ (M is the number of analytes), except for validation of linearity and for the estimation of the precision of measurement in different parts of the concentration interval.

In the situation where the concentration levels in the reference samples cannot be directly controlled, however, one needs a fairly large number of such samples for the calibration of the model to ensure representativity, typically thirty to several hundreds.[1,5] In the subsequent data analysis one must also confirm that indeed the concentration space is appropriately represented.

The reference samples provide a TS of data where both the "signal matrix" **X** – usually consisting of a set of digitized spectra from NMR, NIR, FTIR, ... – and the property matrix **Y** for the same samples (the concentrations or other property values) are defined. These data provide the basic structure of the model, i.e. how the signals (digitized spectra) co-vary with the concentrations (or other properties) in the **Y** matrix.

4.1 Validation Set (Test Set)

To ensure that the calibration model works well in the desired concentration range(s), it is essential to validate the model by using it on a representative set of **new** samples – the validation set, often also called the test set. This is further discussed in section 9.

4.2 Prediction Set

The samples on which the developed model finally is used are often collectively called the prediction set. This set is often open ended and in principle infinite, and is usually not known or available at the time of the model development, the calibration.

5 MAKING THE MEASUREMENTS

This part is of course a most important part of the calibration. MVC is by no way a substitute for sloppy measurements. On the contrary, MVC shows the weakness of the data, and is negatively affected by imprecision, bad sampling, etc. just as UVC. Hence, the multivariate measurements must be made in a reproducible and precise way. This means, in turn, that the measuring conditions should be controlled to be as constant as possible, or at least carefully recorded, solvents should be checked, etc. i.e. normal analytical precautions should be taken also with MVC. See section 5.1.

5.1 Uncontrolled Variables

Uncontrolled variables such as temperature, humidity, time order, powder packing density, cell path length, etc. may affect the spectral response and in some cases the chromatographic retention. MVC can often compensate for variations in uncontrolled variables. This is achieved if the uncontrolled variable is allowed to vary in the calibration TS. The best form of such variation is according to a statistical experimental design. The variations could also be introduced as random variations. It is bad practice to try to keep all influencing variables constant in the calibration training and test sets, and thereafter relax the conditions in routine use. On the contrary, the calibration should contain slightly larger variations than expected in routine use in order to enter compensations for the interfering variations in the calibration model. If, for example, means to control the climate and precise powder packing devices are available during calibration time for a diffuse reflectance NIR method for powders, these should be used to induce known designed or random variations rather than keeping the conditions unrealistically constant.

5.2 Ambiguous Spectral Regions (Variables)

Together with high-performing variables, in general analytical instruments produce variables with bad performance in spectral regions that are technically available but of little analytical value. One example is when optical fibers are used together with a spectrometer. The original design of the spectrometer usually allows for measurements at wavelengths that are not coming through the optical fiber. Rather than resorting to automatic variable selection schemes, the analyst should take decisions in these cases and avoid using variables that are out of the analytical range.

5.3 Randomization of Measuring Order

It is good practice to make the measurements in random order when the training and test sets are made. It is not advisable to order the samples in the same or reversed order as the magnitude of the *y*-variable to be calibrated for. If the samples are ordered, changes that occur with time, such as instrument warming, operator learning, etc. will introduce a bias in the calibration.

When trace amounts of compounds are to be measured, this randomized measuring order is not always practical. In these cases the samples may have to be grouped to reduce carry over. However, extra diagnostics must be applied to verify that the measurements are not correlated with time. This can be done with an explorative MVC with the actual time or the run order for the measurement as the response to the measured variables.

5.4 Data Collection and Administration

Data suitable for import into for MVC software are usually organized as tables or matrices where each row holds the measured values for one sample. These row vectors are also commonly called objects or observations. Each column in the table holds the measurements of a specific variable. Each object and variable should be identified by a name (label) in ASCII text format. When the objects are spectra, it is useful to attach the spectral scale to the data table. This may be done in the name fields of the variables or as a separate object.

For references see page 9734

6 INITIAL QUALITY CONTROL OF THE DATA

In any data set there is a risk of bad measurements, fouled samples, etc. giving outlying observations. Unless these problems are identified and properly dealt with, they will badly affect the calibration model.

6.1 Deviations of Individual Spectra from the Average

During work with small changes in spectra, it is beneficial to inspect the spectral shapes after subtraction of the mean spectrum. An example is shown in Figures (3a–c) where the diffuse reflectance NIR spectra from three homogeneous metoprolol tablets are shown, first (a) as original spectra, then (b) with the mean from the three tablets subtracted, and finally (c) a spectrum from the pure metoprolol tartrate. From Figure 3(a) it is not possible to see the metoprolol variation. However, in Figure 3(b) the same main patterns are visible as in the pure substance spectrum in Figure 3(c). Also note that the scale in Figure 3(b) covers 0.05 $\log(1/R)$ units while the scale in Figure 3(a) covers 0.6 units. This inspection serves well to give a first impression of the main spectral variation. The spectral interpretation aids in the confidence of the method. When the TS is generated according to a design there may be more than one distinct spectral variation. In that case this inspection may be complemented by inspection of the principal components analysis (PCA) spectral loadings as described below. This type of inspection is a good alternative to the use of derivatives as inspection tool. The original shape of the spectral variation is preserved and no information is removed.

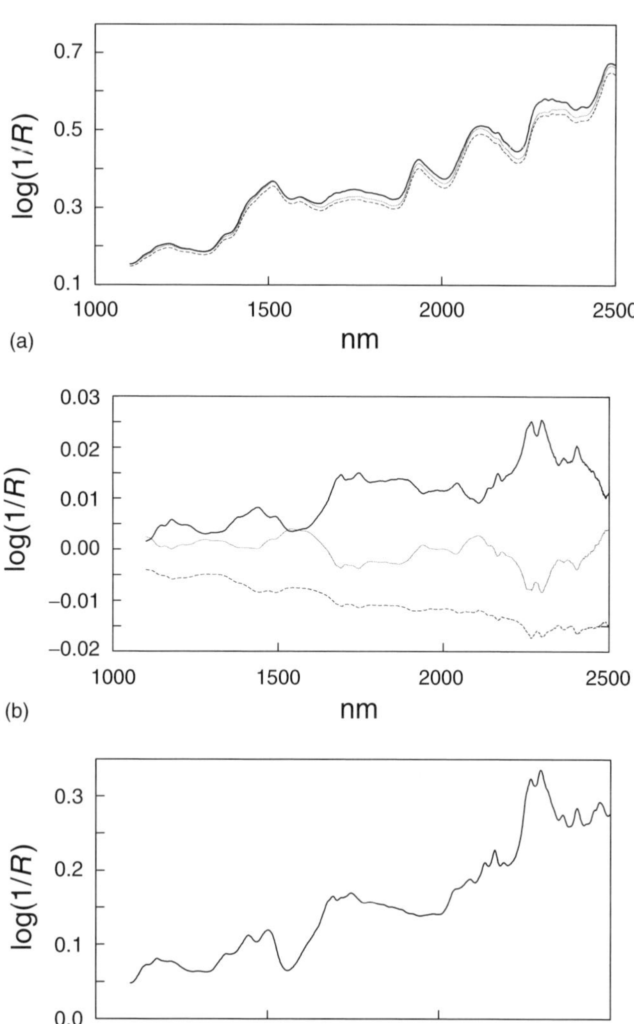

Figure 3 NIR spectra from three homogeneous metoprolol tablets: (a) original spectra, (b) with the average spectrum subtracted, and (c) a spectrum from the pure metoprolol tartrate.

6.2 Multivariate Quality Control of the Data by Principal Components Analysis

The correlation structure in multivariate data (here digitized spectra) allows the straightforward identification of outliers by means of PCA. Also "univariate outliers", deviating strongly in only a single variable, are seen in the PCA, particularly in the RSDs, s_i. The same analysis provides indications also about the balance and representativity of the signal matrix, **X**, as well as about the rank of **X**, i.e. the inherent dimensionality (complexity) of **X**. The latter is closely related to the number of detectable constituents in the investigated set of samples.

In the example, a PCA of the training data (centered but unscaled, see section 7.3.4) gives three significant components indicating the presence of three constituents. However, only two components are expected with two analytes. The third, very small component ($R^2 = 1\%$) indicates some kind of deviation from Lambert–Beer behavior. The loading vector of the third component (p_3) looks similar to a linear combination of the first two loadings (Figure 4), but the third score vector (t3) does not have any clear relation neither to felodipine, nor to metoprolol (Figure 5b).

6.3 Outliers in X and Unbalanced X-data

The t1, t2 score plot of the TS (Figure 5a) does not show any strongly deviating observations, but displays a pattern corresponding to the concentration design matrix. This indicates that the dominating systematic part of the spectroscopic data is related to the concentrations, **Y**. The RSDs, s_i (section 8.4.2) also are within the normal limits.

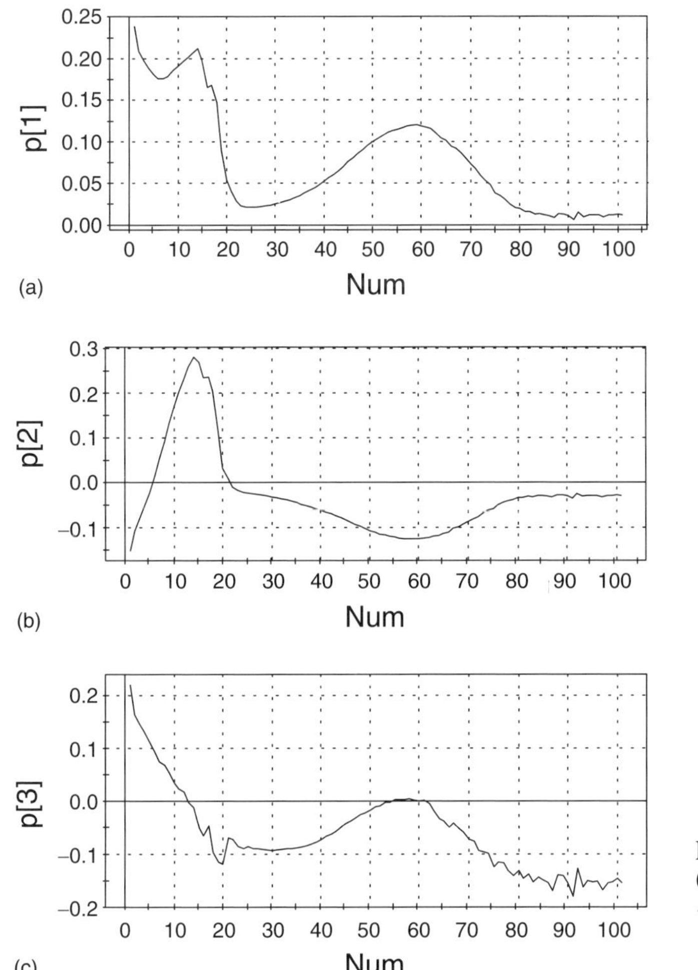

Figure 4 The three PC loading vectors (p1 to p3) of ex.1 plotted vs variable number.

The t1, t2 score plot also shows a well balanced TS without any empty regions or strong clusters. This further corroborates the good distribution of the data, and hence these training data sets have passed the first "quality control".

6.4 Outliers in Y

With many **Y**-variables, a PCA of **Y** shows that, analogously to above, the balance of the **Y**-data, the presence/absence of outliers and clusters, etc. This analysis is relevant when the data has more than four or five y's. For a partial least-squares multivariate calibration (PLS/MVC) the **Y**-outliers will also be visible in t vs. u plots for the significant components. If several outliers are present the most deviating will show up first. When these are removed additional milder outliers may show up. A common cause for **Y**-outliers is that the value from the reference method is wrong.

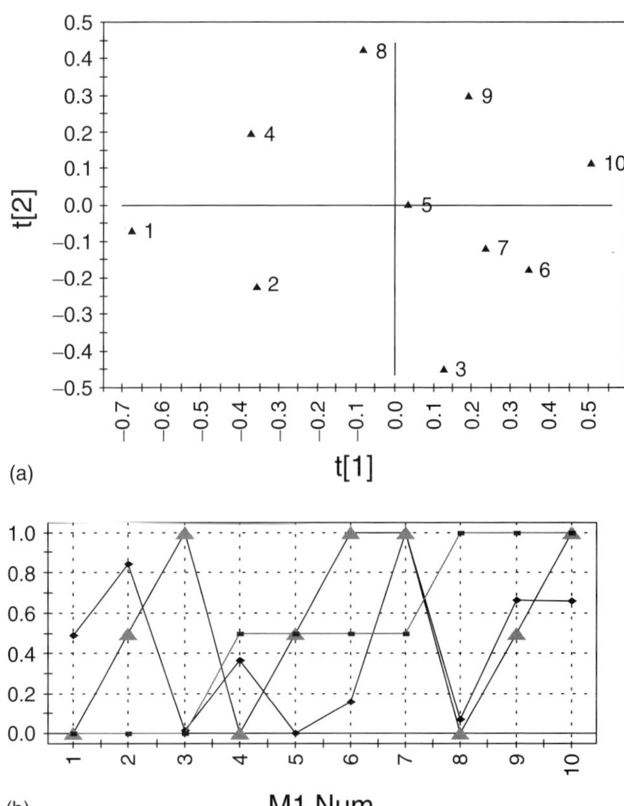

Figure 5 (a) The t1, t2 score plot of a PCA of the TS of ex.1. (b) The third scores (t3) and Y1 and Y2 vs obs number ♦ M1.t(3); ▲ DS1. Felodipi (DS); ■ DS1. Metoprol (DS).

7 DATA PREPROCESSING

The raw data coming out of the measuring instrument (typically the spectrometer) are not necessarily expressed in the form that is most suitable for the mathematical calibration. Hence there is often a motivation for modifying the raw signals, both by simple transformations such as logarithmic, or by complex modifications such as Fourier transformation, filtering, variable selection, baseline correction, and differentiation.

The preprocessing of the TS and of new samples (the prediction set) should be the same. When the preprocessing contains adjustable parameters, e.g. averages, SDs, orthogonal signal correction (OSC) components, etc. these parameters are based on the TS, and the same parameter values are then used for the new samples.

7.1 Choice of Representation (Time or Frequency Domain)

As discussed in section 3.7, multivariate signal profiles come out of the instruments usually either in the form of

For references see page 9734

a spectrum (light absorption at different wavelengths or frequencies) or, with a FT spectrometer, as an interferogram (spectral amplitude as a function of mirror distance or time). As humans we usually prefer the first representation, but nothing indicates that either is more useful for MVC. Hence, with FT spectroscopy it may sometimes be preferable to use the digitized interferograms directly as objects, and then back-transform the resulting model coefficient profiles to a spectral representation for the interpretation of the results. Our experience is still not sufficient to provide any guidelines.

7.2 Logarithms and Other Stabilizing Transformations

If the signals are exponentially distributed, taking their logarithms is recommended before the data analysis. This may also stabilize the influence of the noise on the model if the size of the noise is approximately proportional to the concentration. This is common when the latter varies over several magnitude of ten. If the value zero occurs frequently in the signal matrix, the fourth root is an alternative transformation which has a similar "compressing" effect as the logarithm on large values, but it does not inflate very small values, and the fourth root of zero remains zero.

If a linear relation is expected between the original signals (**Z**) and the concentrations (**Y**), then both **Z** and **Y** should be transformed in the same way. Hence, if log(**Z**) is used, this implies that also log(**Y**) should be used in the calibration.

7.3 Spectral Correction Methods

Particularly with diffuse reflectance spectroscopy, there is often a need of a spectral preprocessing to remove baseline variation and other systematic variations in the data that are not related to the analyte concentrations or other properties of interest. Then so-called signal correction methods are applied, or, alternatively, first or second derivatives of the spectra along the wavelength axis are calculated (spectral differentiation). We emphasize that there is presently not sufficient experience with data preprocessing and MVC to select the appropriate approach a priori, and a comparison between the results of different approaches including the use of unmodified data should always be made.

7.3.1 Additive and Multiplicative Signal Corrections

Large interfering additive and multiplicative variations are often present in spectra along with the desired information. The additive effects are visible in spectra as offsets. The multiplicative effects are visible as a scaling factor where proportions between spectral peaks are preserved while the spectral magnitude varies.

It is possible to include these variations in the MVC, but at the cost of an increased complexity of the calibration model. With principal components regression (PCR) and partial least-squares regression (PLSR) several extra components may be needed. This often affects the predictive precision negatively, and also makes the model interpretation more difficult.

This type of variation is common with diffuse reflectance methods, since the effective cell path is not exactly defined, as in transmission spectrometry. In NIR diffuse reflectance the cell path variations will cause multiplicative effects and e.g. a smaller particle size will give a lower baseline of a $\log(1/R)$ spectrum, i.e. the spectrum will appear "lighter" for the spectrophotometer.

In diffuse reflectance NIR spectrometry these effects are often the most dominating, which means that they are modeled in the first component if PCR or PLSR is used. When the objective is to model chemical properties with a simple MVC model, the corrections can be used. As a consequence of the simple model, the scores and loadings will in many cases be easier to interpret. The correction will, however, not cure bad data.

In cases when the MVC is aimed at physical properties such as particle size, no correction should usually be made, however. This because these physical properties may be reflected in light scatter, etc. i.e. components little related to concentration. As an alternative, the sizes of the additive and multiplicative corrections may be added as two appropriately scaled extra variables in the calibration.[1]

Multiplicative signal correction (MSC), or mean centering of the objects followed by a scaling of each row to unit SD, are two variants of the same correction except for a scaling factor.[23,24] The object centering and auto-scaling has also been named the standard normal variate (SNV) transform.[25] Both transforms remove the additive and multiplicative effects. Spectra may also be detrended by fitting and subtraction of a low order polynomial. Dhanoa et al. recommends that this should be done before the SNV correction.[24]

The MSC algorithm uses univariate linear regression with a plot of the TS mean spectrum against each individual spectrum. The resulting intercept is used for the additive correction and the slope is used as the multiplicative correction of each spectrum in the TS. Further measured samples, in the test and prediction sets, have to be corrected with the saved TS mean spectrum.

Denoting the row vector of the average spectrum of the TS by $\mathbf{m}'$, each training and prediction set observation row vector, $\mathbf{z}_i'$ is corrected as follows. First the slope, a_i, and the intercept b_i, are calculated by least squares (**e** is a residual vector), Equation (7):

$$\mathbf{z}_i = a_i + b_i\mathbf{m} + \mathbf{e} \tag{7}$$

Then, the "corrected" vector, $\mathbf{x_i}$, is calculated as Equation (8):

$$\mathbf{x_i} = \frac{\mathbf{z_i} - a_i}{b_i} \tag{8}$$

The relation to a mean spectrum is a disadvantage with MSC since the mean spectrum has to be recalculated for each selected TS. That is avoided with SNV.

The MSC has been applied both to the entire spectral range present in the MVC and to parts thereof. One way to correct measured variables that contains an internal standard, would be to correct the whole spectrum with an MSC that get its slope and intercept only at the variable interval where an internal standard is present.[1]

7.3.2 Optimal Scaling

Optimal scaling is preprocessing that corrects for the nonlinearity created when spectra are normalized before the calibration.[26] Optimal scaling can also correct for multiplicative effects. This is done by the addition of a multiplier m_i for each sample in $\mathbf{Y}$. Then Equation (9) will be complemented by a diagonal multiplier matrix $\mathbf{M}$:

$$\mathbf{MY} = \mathbf{XB} + \mathbf{F} \tag{9}$$

The MVC is then constituted as a solution to Equation (9) including the optimized $m_i:s$. The regression coefficients B may be solved as

$$\mathbf{B} = \mathbf{X}^{+}\mathbf{MY} \tag{10}$$

where $\mathbf{X}^{+}$ is the generalized inverse of $\mathbf{X}^{T}\mathbf{X}$ for the cases when $\mathbf{X}$ is singular.

Optimal scaling improves PLS calibration on normalized X-ray diffraction spectra.[26] This scaling is, however, not appropriate for the correction of additive effects. Raw data with additive anomalies should be corrected with another method before the application of optimal scaling.

7.3.3 Orthogonal Signal Correction

OSC[27,28] is a recently published correction method. The main idea is to remove systematic variation in the $\mathbf{X}$-variables that is not correlated to the variation in the y-variables. This is another way to remove variations introduced by variations in the apparent cell pathlength. In contrast to the MSC and SNV, OSC needs the y-variable to guide the correction. It is also possible to use OSC to reduce the size of interferences such as the switch of instruments.[28,29]

The OSC operates as a reversed PLS calibration. A multivariate model of the TS is built, but the objective of the model is to describe variations that are orthogonal to the Y-variable(s). The model is composed of OSC-components, just as a PLS-model is composed of PLS-components, (Equation 11):

$$\mathbf{X} = \mathbf{T}_{\mathbf{OSC}}\mathbf{P}'_{\mathbf{OSC}} + \mathbf{E} \tag{11}$$

The residual matrix $\mathbf{E}$ contains the OSC-filtered data, which are thereafter used as the TS signal matrix in the MVC. Just like ordinary PLS, OSC also gives a weight matrix, $\mathbf{W}^{*}_{\mathbf{OSC}}$ such that (Equation 12):

$$\mathbf{T}_{\mathbf{OSC}} = \mathbf{X}\mathbf{W}^{*}_{\mathbf{OSC}} \tag{12}$$

Spectra of new samples (centered and scaled) are then filtered using the same OSC model before they are entered the calibration model to predict the values of $\mathbf{Y}$ (the concentrations). First their "OSC-score" values are calculated, Equation (13):

$$\mathbf{t}'_{\mathbf{OSC}} = \mathbf{x}'_{\mathbf{new}}\mathbf{W}^{*}_{\mathbf{OSC}} \tag{13}$$

and thereafter the "filtered" values are calculated as Equation (14):

$$\mathbf{x}'_{\mathbf{new,\ filtered}} = \mathbf{x}'_{\mathbf{new}} - \mathbf{t}'_{\mathbf{OSC}}\mathbf{P}_{\mathbf{OSC}} \tag{14}$$

The limited experience is that one or two OSC components usually are adequate. The score plots of both the OSC model and of the final PLS/MVC should be used as diagnostic tools. When the score-pattern that is being corrected for disappears from the final MVC, no further OSC components should be used. As always with a multivariate modelling, the OSC may give overfitting if too many components are used.

7.3.4 Centering and Scaling

After transformations (log, etc.), the variables are usually centered by subtracting the average values (a_k, see below) from each column in $\mathbf{X}$ and $\mathbf{Y}$. Thereafter, one can optionally scale the variables by multiplying each by a scaling weight, v_k or v_m. The most common is to scale both the $\mathbf{X}$ and $\mathbf{Y}$ variables to unit variance. This is done by calculating the column averages (a_k) and column SDs (s_k), and then make the linear transformation, Equation (15):

$$x_{ik} = \frac{z_{ik} - a_k}{s_k} \tag{15}$$

This corresponds to using the inverse SDs as scaling weights (i.e. $v_k = 1/s_k$). The result is that each variable gets the same initial importance in the calculation of the calibration model.

When the variables comprise digitized spectra and all the variables are in the same unit, another common "scaling" is to keep the spectrum as it is, except for centering. With a projection method such as PCR and

For references see page 9734

PLSR this means that the variables get the importance for the calibration proportional to their variation. Variables should still be centered, and it is the variation after centering that gives the weight (see Figure 3b). This is sometimes a good approach for spectra if large sections of spectra are only reflecting baseline noise. If scaling to unit variance is applied to such data, it will magnify the amount of noise that is entered into the calibration model. However, sometimes the variations in the baseline may be significant for the analysis. This may occur with e.g. fluorescence in Raman spectrometry and in gradient LC. A compromise is to weight the variables proportionally to the inverse of the square root of their SD, often called Pareto scaling. Thus the large variations are still dominating, but the small variations are not scaled down as much.

When blocks of variables give less than optimal response for the instrument type, but these blocks still contain information that is important for the calibration, an option is to down-weight these blocks. This will enhance the detection of aberrant measurements and the calibration performance may increase.

When the **X**-variables comprise combinations of spectra from several analytical instruments, the blocks of variables from the different instruments may be combined with weights that let the instruments have "equal votes" in the calibration (block scaling).

It may be also possible to automatically optimize the individual variable weights in a calibration. This is in general not a good practice, since the risk of over-fitting will increase. This method is related to variable selection methods where a large set of selections are tried and the "best" is chosen. This greatly increases the risk of producing spurious correlations.

7.3.5 *First and Second Derivatives*

When the x-variables for an MVC have the same unit and are arranged as a continuous function (e.g. in spectra), the first or second derivatives may replace the original variables as **X**-variables for the MVC. Using first derivatives removes the linear slope of the baseline and changes all peak maxima to zero-crossings. Using second derivatives reduces the background further and leaves narrower negative peaks at the original peak positions. However, the differentiation also removes all variation in the variables that is not carried by differences between adjacent variables. In the rare case where the majority of the variation in the **X**-variables remains after the differentiation, this may be a good pretreatment.

Differentiation removes a sizable amount of variation that may be important for the calibration. Noise will then be a larger proportion of the signal, and will be more visible in spectra. This has made it possible for most derivatives in instrument software to be calculated with smoothing, for instance using Savitsky–Golay approach,[30] or by subsequent differences between moving averages.[31]

Care should be taken not to smooth away information. When digital spectra are collected, there is often a digital over-sampling in relation to the bandwidth of the spectrometer. A good rule of thumb is to keep the smoothing parameters in a range that is not degrading the efficient bandwidth of the instrument significantly.

A common misuse of derivatives to use it as a replacement for mean-centering of the variables. MVC models with and without differentiated **X**-variables should always be calculated with mean centered **X**-variables before comparison.

8 THE ANALYSIS OF THE TRAINING SET DATA; THE CALIBRATION MODEL DEVELOPMENT

With many spectral variables (more than one), a computerized analysis is necessary to use the data efficiently. The basis for this analysis is to develop a model, a mathematical description, of the relationship between the raw or preprocessed spectral variables (**Z** or **X**) and the concentrations (**Y**). There are two basic approaches to this data analysis, multiple linear regression (MLR), or a multivariate projection. The most commonly used projection methods are PCR and PLSR.

8.1 The Mathematical Formulation of Multivariate Calibration

The general form of the calibration model can be derived from Lambert–Beer's law, which states that at any wavelength the light absorption is linearly related to the concentrations of constituents that absorb at that wavelength.

Hence, according to Lambert–Beer's law, spectral data z_{ik} – the spectral absorbance of sample i at the kth wavelength – is a sum of contributions from chemical constituents with the concentrations c_{im} and the spectral unit absorbance d_{km}, plus noise, e_{ik} (Equation 16):

$$z_{ik} = \sum_{m} c_{im} d_{km} + e_{ik} \qquad (16)$$

In matrix form, this is Equation (17):

$$\mathbf{Z} = \mathbf{C}\mathbf{D}' + \mathbf{E} \qquad (17)$$

After monotonic transformations, centering, scaling, and preprocessing (section 7), giving the signal data matrix **X**, this matrix is still decomposable in the same way (the resulting residual matrix, $\mathbf{E}$, is not the same as above, however), Equation (18):

$$\mathbf{X} = \mathbf{TP}' + \mathbf{E} \tag{18}$$

An important conclusion is that the signal matrix $\mathbf{X}$ can be modeled by this PC-like model, also called a bilinear model.

The data analytical problem is now basically to find the parameter matrices $\mathbf{T}$ and $\mathbf{P}'$ which are related to the concentrations of the analytes of interest. There are several complications, however. First, the number of constituents in a complicated sample is often large, meaning that the rank of $\mathbf{T}$ and $\mathbf{P}$ is potentially large. Second, there is a rotation problem, which makes $\mathbf{T}$ and $\mathbf{P}$ nonunique. Inserting any nonsingular matrix $\mathbf{R}$ and its inverse $\mathbf{R}^{-1}$ between $\mathbf{T}$ and $\mathbf{P}'$ in Equation (18) gives Equation (19) (remembering that $\mathbf{R}^{-1}\mathbf{R} = \mathbf{1}$):

$$\mathbf{X} = \mathbf{TP}' + \mathbf{E} = \mathbf{TR}^{-1}\mathbf{RP}' + \mathbf{E} = \mathbf{VQ}' + \mathbf{E} \tag{19}$$

Finally, the presence of the noise, $\mathbf{E}$, presents further problems. The rank of $\mathbf{X}$ cannot be determined with full certainty, and any parameter estimation based on $\mathbf{X}$ will also result in noise in the parameters.

Hence it is not sufficient to just decompose $\mathbf{X}$ by PCA because the resulting score matrix $\mathbf{T}$ cannot be identified with the desired analyte concentrations $\mathbf{Y}$. Instead one has to find a way to find the loadings $\mathbf{P}$ so that at least the resulting scores ($\mathbf{T} = \mathbf{XP}$) are linearly related to the concentrations $\mathbf{Y}$ (PCR and PLSR, below) by coefficients $\mathbf{Q}$, Equation (20):

$$\mathbf{Y} = \mathbf{TQ}' + \mathbf{F} \tag{20}$$

Equation (17) can be multiplied from the right with $\mathbf{D}'^{-1}$, giving Equation (21), after rearrangement:

$$\mathbf{C}(\text{or } \mathbf{Y}) = \mathbf{ZD}'^{-1} + \mathbf{F} = \mathbf{ZB} + \mathbf{F} \tag{21}$$

Hence, we can, in principle, find coefficients $\mathbf{B}$ that directly relate the signal matrix to the concentrations by a MLR model (Equation 5). This might seem more attractive than going via the bilinear decomposition of $\mathbf{X}$, but actually leads to a loss of information since no modelling is made of the signal matrix $\mathbf{X}$ and its correlation structure. This information loss makes both the interpretation of the model and the recognition of anomalous samples model difficult. Also, usually the variables in $\mathbf{X}$ are strongly correlated, which makes the MLR solution impossible to compute without an extensive variable reduction. This can be made in simple cases, but is difficult with complicated samples without distinct spectral regions. The different approaches – PCR, PLSR, and MLR – are discussed below (section 8.4).

8.2 Linear or Nonlinear Calibration Model

Relationships between measured data are, in principle, nonlinear. However, over limited intervals these relationships are usually very close to linear, and can hence be adequately approximated by a linear model. However, if the relationship appears to be curved, the linear models can easily be extended to quadratic or cubic models, handling polynomial (mild) nonlinearities (section 8.4). These nonlinear models work as long as there is a monotonic relationship between signals ($\mathbf{X}$) and concentrations ($\mathbf{Y}$), i.e. the spectral variables always increase (or decrease) when the concentrations or other calibration properties increase in value. Stronger nonlinearities are impossible to deal with even in principle, since then the calibration model becomes noninvertable and a given signal profile will correspond to several different concentrations.

8.3 One or Several Analytes

With several analytes and a multivariate $\mathbf{Y}$-matrix of concentrations, there are two principally different ways to perform the MVC. Either one can develop a separate model for each analyte, or one can develop a single model for all analytes. Each approach has its merits. The separate models are often easier to understand and use, and often give slightly better predictions, especially if the y-variables are close to orthogonal. This situation typically arises when calibration samples are prepared from pure substances and their mixtures varied according to a statistical design as discussed in section 3.2 above.

The multi-analyte model is preferable when the number of analytes is large and the concentrations (or other properties used as $\mathbf{Y}$) are correlated. This is typical for noncontrolled samples of natural origin, i.e. the second case discussed in section 3.2. Then the simplification of a single model is strongly desirable, and possible small gains in predictive precision in one or a few of the analytes by using separate calibration models are offset by the ensuing increase in complexity of developing and maintaining a large number of different calibration models.

With noncontrolled samples, it often happens that the analytes are correlated groupwise. This grouping is usually seen in the loading plots of a PCA of the concentration matrix, $\mathbf{Y}$. Making a separate model for each group of correlated analytes is then a natural approach.

8.4 The Calibration Model

The choice of the calibration model depends mainly on the complexity of the calibration problem and the corresponding data structure. One of the great advantages

For references see page 9734

with projection methods is that they can handle data sets with very many and correlated **X**-variables, even many more than the number of calibration samples, *N*. For such data, the ordinary MLR cannot even be computed. With many **X**-variables, K, and most of them containing information about **Y**, projection methods are hence the choice, while for very simple problems where a few wavelengths can be selected to give an adequate calibration model, MLR is often used. As indicated in the end of section 8.1, however, projection methods may still be preferable also in simpler problems, because they give additional information beyond the results of MLR.

8.4.1 Projection Methods

As discussed in section 8.1, we expect that the signal matrix **X** can be decomposed as a product of a matrix **T** (the score matrix) and a matrix **P′** (the loading matrix), Equation (18). Moreover, the scores (the columns of **T**) can be used as predictor variables of the concentrations, **Y** (Equation 20). This is the basis for the two most commonly used MVC models, namely PCR[32] and PLSR.[33,34]

The former PCR is based on first making a PCA of the signal matrix **X** according to Equation (18), and then developing a second model of the relation between the score matrix (**T**) and the concentration matrix **Y** (Equation 20). Each PC consists of one score vector ($\mathbf{t_a}$) and one loading vector ($\mathbf{p_a}$), and the component index (*a*) runs between one and A (the number of components of the model). The components, i.e. the columns of **T** and **P**, are mutually orthogonal, and the latter (the **P** columns) are normalized to length one. Moreover, the score vectors ($\mathbf{t_a}$, i.e. the columns of **T**) are sorted in order of the importance for **X**, i.e. $\mathbf{t_1 p_1'}$ describes more of **X** (its sum of squares (SS)) than does $\mathbf{t_2 p_2'}$, etc. The PC loadings ($\mathbf{p_a}$) and score vectors ($\mathbf{t_a}$) are eigenvectors to the variance covariance matrix (**X′X**) and the association matrix (**XX′**), respectively.

Due to the orthonormality of the PCA loading matrix **P**, and the orthogonality between **P′** and the residual matrix **E**, Equation (18) multiplied from the right by **P′** gives Equation (22):

$$\mathbf{T} = \mathbf{X}\mathbf{P}' \tag{22}$$

For an additional sample (in the prediction set) with the preprocessed data vector $\mathbf{x}'_{\mathbf{new}}$, the predicted PC score values are calculated from this relation, which when inserted in the calibration model (Equation 20) gives the predicted values of **Y**, Equations (23) and (24):

$$\mathbf{t_{new}} = \mathbf{x}'_{\mathbf{new}}\mathbf{P} \tag{23}$$

$$\mathbf{y}'_{\mathbf{new}} = \mathbf{x}'_{\mathbf{new}}\mathbf{P}\mathbf{Q}' \tag{24}$$

Standard errors ($\mathrm{serr_y}$) and confidence intervals (approximately 2 or 3 times the standard errors, according to the t-distribution) can be calculated in many ways for the predicted *y*-values, the simplest being based on the regression formalism and the fairly accurate assumption of a fixed (exact) X-score matrix, **T**. The square of the standard error (error variance) including the variability of the new measure *y*-values, and assuming a variance of the measurement errors of σ^2, is Equation (25):[35]

$$\mathrm{serr_y} = \left(1 + \frac{1}{\mathrm{N}} + \mathbf{x}'_{\mathbf{new}}(\mathbf{T}'\mathbf{T})^{-1}\mathbf{x}_{\mathbf{new}}\right)\sigma^2 \tag{25}$$

PCR presents some complications, however. In particular in complicated samples, there is often a strong systematic variation in **X** that is not related to **Y**. Also, especially in reflectance spectroscopy, there is often a baseline variation and a multiplicative effect due to light scatter and varying light path length. This makes the first few PCs little related to **Y**, with the ensuing loss of information. This may be remedied by appropriate data preprocessing (section 7). However, the fact that the components of **X** are derived without using the information in **Y**, makes PCR often need many components to capture the parts of **X** that are relevant to **Y**, with a resulting loss of predictive precision.

PLS projection to latent structure PLS[33–36] is a method where the projection of **X** is made to enhance the information content about **Y** in the score matrix, **T**. This decreases the risk for losing information about **Y** in the projection of **X**, and with complicated samples PLSR usually needs fewer components than PCR to achieve the same explanation of **Y**. With few calibration samples and many **X**-variables, this is often an advantage, and PLS usually gives a smaller prediction error than PCR. In many applications, however, where the information in **X** about **Y** sits in the first few components like in the present example, the differences are miniscule and chemically and statistically insignificant.

The PLS model is developed with two objectives, namely to model **X**, and to predict (model) **Y** from **X**. The former is accomplished by finding new "variables", the scores ($\mathbf{t_a}$), that well summarize **X**, and the latter objective is achieved by focusing the projection on the parts of **X** that are related to **Y**.

The PLSR model has identical form as the PCR model (Equations 18 to 20), but the actual values of the parameters (**T** and **P**) are different, as well as the dimensionality (A). The computation of the PLS parameters is made in such a way that the PLS score matrix, **T**, is an optimal compromise between modelling **X** and predicting **Y**. This compromise is sometimes called the "H-principle" which points to the analogy with Heisenberg's principle of uncertainty.[36] This is achieved by having the first PLS score vector ($\mathbf{t_1}$) be the largest eigenvector of the extended association matrix, $\mathbf{XX'YY'}$ in contrast to the first PC score vector which is the largest eigenvector to just $\mathbf{XX'}$. The subsequent PLS score vectors ($\mathbf{t_a}$) are eigenvectors of $\mathbf{E_{a-1}E_{a-1}YY'}$, where $\mathbf{E_{a-1}}$ is the matrix of X-residuals ($\mathbf{X} - \mathbf{T_{a-1}P'_{a-1}}$) after a – 1 components. The PLS model of $\mathbf{Y}$ is formulated as a linear model between the $\mathbf{X}$-scores and the $\mathbf{Y}$-variables, with a coefficient matrix $\mathbf{C}$, Equation (26):

$$\mathbf{Y} = \mathbf{TC'} + \mathbf{F} = \mathbf{XW^*C'} + \mathbf{F} = \mathbf{XB} + \mathbf{F} \quad (26)$$

In PLS there is a second matrix of "loading vectors", denoted by $\mathbf{W}$, with one column per PLS component. A variant of $\mathbf{W}$ denoted by $\mathbf{W^*}$ contains the weights that directly combine the $\mathbf{X}$-variables to form the scores, $\mathbf{t_a}$. Thus, analogously to Equation (22) we obtain Equations (27) and (28):

$$\mathbf{t_a} = \mathbf{Xw_a^*} \quad (27)$$

$$\mathbf{T} = \mathbf{XW^*} \quad (28)$$

This is used for calculating the predicted values of $\mathbf{Y}$ for new samples with X-data $\mathbf{x'_{new}}$ according to Equation (29):

$$\mathbf{y'_{new}} = \mathbf{x'_{new}W^*C'} \quad (29)$$

The standard errors and confidence intervals for these predicted values are calculated in the same way as for PCR, i.e. by Equation (25).

8.4.2 Diagnostics Common to Both Regression Methods

The diagnostics and the interpretation of PLSR is the same as for PCR, as well as the way to calculate predicted y-values (e.g. concentrations) for new samples. The only difference is that in PCR, the X-scores for new samples are calculated by means of the loadings, $\mathbf{P}$, while in PLS, these scores are calculated by means of the weight matrix $\mathbf{W^*}$ (Equations 22 and 28 respectively). We note that the X-residuals are calculated in the same way for both PCR and PLS (from Equation 18) by Equation (30):

$$\mathbf{E} = \mathbf{X} - \mathbf{TP'} \quad (30)$$

The X-residuals, $\mathbf{E}$, express the deviations between the data ($\mathbf{X}$) and the model ($\mathbf{TP'}$). For new observations (samples), the X-residuals ($\mathbf{e'_{new}}$) are calculated in the same way as the preprocessed sample x-vector minus the calculated scores times the loading matrix of the model, Equation (31):

$$\mathbf{e'_{new}} = \mathbf{x'_{new}} - \mathbf{t_{new}P'} \quad (31)$$

A great advantage of projection methods is that data of new samples in the prediction set can be "quality controlled" by inspection of their RSD values (s_{new}) (Equation 32). The RSD of observation i shows how far this observation is from the model, however dissimilar it is to the model.

$$s_{new} = \left\| \frac{\mathbf{e'_{new}}}{\mathrm{K - A}} \right\| \quad (32)$$

If the value of this RSD for a new sample is larger than a certain limit (d below), the sample is an outlier not being similar to the TS. This, in turn, indicates that the predictions of the concentrations ($\mathbf{y}$) are unreliable for the sample. The limit (d) for the RSD is calculated from the RSD of the whole training matrix (s_0) and a critical value of the F-distribution (Equations 33 and 34).

$$s_0 = \mathrm{sqrt}\left[\sum_i \sum_k \frac{e_{ik}^2}{(\mathrm{N - A - 1})(\mathrm{K - A})}\right] \quad (33)$$

$$d = s_0 \times \mathrm{sqrt}(\mathrm{F_{crit}}) \quad (34)$$

8.4.3 The Number of Model Components (A)

The number of components used in the PCA of $\mathbf{X}$ (A) is determined either by cross-validation (CV), or by using components with eigenvalues significantly larger than some limit, usually between 1.0 and 2.0. The eigenvalue of a component is the fraction of the SS of the signal matrix $\mathbf{X}$ that is explained by the component, multiplied by the number of variables (K). A component with an eigenvalue larger than one explains the worth of at least one variable. One further alternative (not recommended) is to sufficiently use many components to make the SS of the residuals $\mathbf{E}$ smaller than a fraction of the SS of the signal matrix $\mathbf{X}$.

With CV one estimates the number of components (A) by dividing the elements in the data matrix $\mathbf{X}$ in a number of groups, usually seven or nine. One then makes a number of separate analyses, each time keeping out one of the groups of the data (setting them as "missing"). The model is then used to predict the deleted data, and the ensuing prediction errors (predicted value minus actual values) are calculated. The resulting SS of these prediction errors is called predictive residuals sum of squares (PRESS), and one selects the model dimensionality (A), which gives the smallest PRESS. This approach is called full CV. Alternatively, one calculates one PC after another with a renewed CV for each component, until PRESS stops to decrease in size, so-called partial CV. Full and partial CV usually indicate the same number of components or very close.

Like with PCR, the number of significant PLS components is determined by CV. Since there is a $\mathbf{Y}$-matrix in PLS, the CV of PLSR is different from that of PCA, and the PRESS relates to the predictive residuals of the $\mathbf{Y}$-matrix. In CV of PLS, the deleted groups of data contain

For references see page 9734

whole objects (**X** and **Y**-data). A PLS model is developed without this group, and the predicted Y-values are then calculated for the deleted objects. These Y-values are compared with the actual values, and PRESS equals the SS of the difference between observed and predicted **Y**, the predictive residuals. Hence the CV in PLS relates directly to the objective of predicting **Y**, while the CV of PCA tells us only about the significant structure of **X**. Also for PLS, CV can be made in a full or partial way, giving similar results in most practical cases.

8.4.4 Computational Algorithms for Both Regression Methods

PCA is mathematically equivalent to singular value decomposition (SVD) as it is usually called in numerical analysis and computation science, and to principal factor analysis, Karhunen-Loève decomposition, and eigenvector analysis as this is often called in physics and engineering. Hence, PCR starts with a PCA of **X**, followed by a simple MLR using the first few PC score vectors, $\mathbf{t_a}$, and predictor variables.

Many different algorithms exist for the calculation of the principal components parameters, where, historically, the first were based on the diagonalization of the variance covariance matrix $\mathbf{X'X}$. The resulting eigenvectors equal the loadings, $\mathbf{p_a}$. Today, however, the most commonly used algorithms are based on SVD,[37] or on the NIPALS method.[38] The latter is very fast if just a few PCs are needed, and also works for incomplete data matrices.

Like for PCR, many algorithms exist for calculating the PLS parameters. The most common is based on the NIPALS algorithm[39] which has the advantage of also operating with incomplete **X** and **Y** matrices, and to extract only as many components as actually are needed in the model. Other algorithms are based on the calculation of eigenvectors of the matrices $\mathbf{XX'YY'}$ or $\mathbf{X'YY'X}$.[40–42] These two matrices have the dimensions $(n \times n)$ and $(K \times K)$. The first is small for few samples, and hence advantageous for fast computations.

8.4.5 The Example

In the example, a PCA of the first 10 samples (the TS) gives three components as shown above in section 6. The two PCR calibration models for y1 = felodipine and y2 = metoprolol based on the corresponding score vectors explain the two Y-variables well ($R^2 = 0.9996$ and 0.9991, and $Q^2 = 0.9817$ and 0.9153 respectively). The fitted and predicted values for the second model are shown in Figure 6, as well as in Table 1.

8.4.6 Multiple Linear Regression

The first method used in MVC was MLR used with a small selected subset of variables. These variables are usually selected by inspecting spectra of fairly pure constituents. The number of variables should be at least two per constituent, i.e. four in the example. In the absence of impurities, the 4 wavelengths 11, 2, 39, and 57 seem like a good selection in this example. A number of automatic variable selection procedures exist for MLR, but the experience shows that a selection based on chemical and spectroscopic knowledge is preferable. In the case where such knowledge is weak or absent, full spectrum methods such as PCR and PLSR should be used at least as starting points to identify important regions in the spectrum.

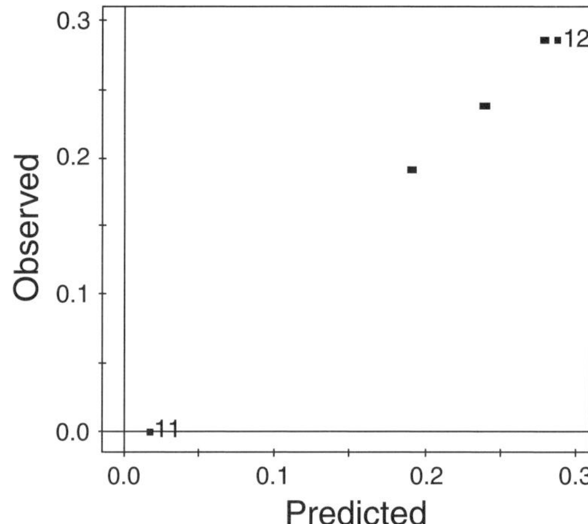

Figure 6 The observed versus fitted or predicted values for y2 = metoprolol in the example. The calibration model was one of PCR (A = 3 components) based on samples 1 to 10 (not labeled). Samples 11 and 12 are prediction samples.

Table 1 Resulting R^2 and Q^2 values of the PCA of the spectral data of the example. Symbols explained in 10.1

A	R^2X	M6 R^2X(cum)	$N = 15$ Q^2	Q^2(cum)
1	0.659	0.659	0.625	0.625
2	0.331	0.989	0.957	0.984
3	0.010	1.000	0.956	0.999

MLR starts directly with the model in the form of Equation (5). The coefficients **B** that minimize the SS of the residuals in **F** are given by Equation (35):

$$\mathbf{B} = (\mathbf{X'X})^{-1}\mathbf{X'Y} \tag{35}$$

Predicted values of **Y** for new samples are obtained as Equation (36):

$$\mathbf{y'_{new}} = \mathbf{x'_{new}B} \tag{36}$$

The standard errors and ensuing confidence intervals are given in Equation (37) which is analogous to Equation (25), but with **X** instead of **T**, because in MLR all

X-variables are assumed to be independent:

$$\mathrm{serr_y} = \left(1 + \frac{1}{\mathrm{N}} + \mathbf{x}'_{\mathbf{new}}(\mathbf{X}'\mathbf{X})^{-1}\mathbf{x}_{\mathbf{new}}\right)\sigma^2 \quad (37)$$

9 METHOD PERFORMANCE AND VALIDATION

The use of analytical method validation is crucial for the operation of many regulated businesses such as the pharmaceutical industry, and for quality-rated analytical laboratories. Method validation is an activity that is meant to prove that an analytical method is indeed doing what it is supposed to do. The requirements on the analytical method should correspond to the needs of the operation. For process analysis, speed of analysis may be more important than the lowest possible error in each analysis. When product bulk properties are measured, other requirements are more important than when environmental trace analysis is performed. This means that the first step in method validation is to determine what we want and what requirements that will lead to.

The validation is usually broken down into a set of performance measures such as selectivity, sensitivity, limit of detection, linearity or model fit, repeatability, reproducibility, accuracy and robustness. The possible lowest limits for these performance measures vary with the analytical method. The validation task for a MVC is often to prove that the MVC performs on the same level as the reference method. It is important to notice that the model is not the method. The analytical method has to be a description of the development of calibration models.

9.1 Measures of Model Performance

Today the pharmaceutical industry has a well regulated set of performance measures that has to be reported for an analytical method. Hence these can be used as a framework for MVC. A set of performance measures in pharmaceutical analytical method validation mainly taken from the suggestion list of the International Harmonization Committee (IHC) (International Conference on Harmonization of Technical Requirements for Registration of Pharmaceuticals for Human Use IHC secretariat c/o IFPMA, 30 rue de St-Jean, PO Box 9, CH-1211 Geneva 18, Switzerland.)[(43)] will be covered. These measures are mostly developed with UVCs in mind. Their application for MVC will be discussed.

The calibration model can be evaluated according to very many criteria. According to our present knowledge of analytical chemistry, the following set of criteria cover most practical needs.

9.2 Selectivity

The selectivity of an analytical method is defined as the extent to which it can determine particular analytes in a complex mixture without interference from other components in the mixture.

MVC is able to make a selective analysis by the use of mathematics. The requirements for this to work is that the other compounds in a mixture have different magnitude relations for the same spectral variables and that the amounts of the interfering compounds are varied in the TS independently of the analyte amount. This can be tested by the variation of physical properties, matrix components, and interfering compounds to samples in the test set. Then the selectivity as a deviation in the analytical result can be estimated as Q_{test} for this test set, by Equation (38):

$$Q_{\mathrm{test}} = \sqrt{1 - \frac{S^2_{F,\mathrm{test}}}{S^2_{Y,\mathrm{test}}}} \quad (38)$$

Here $S_{F,\mathrm{test}}$ is the SD of the y-residual for the selected test set and $S_{Y,\mathrm{test}}$ is the SD for the total variation in the y-variables.

When the amounts of interfering compounds are outside the scope of the model, the residual error for the x-variables (**E**) should be large enough to rise beyond a decided limit of acceptance. The verification of the residual diagnostics should be included in the method validation.

Selectivity is related to the concept of "net analyte signal" (NAS) of Lorber et al.[(44)] It is claimed that NAS can be used as a basis for a broad range of MVA diagnostics. In its present form, however, the NAS estimation needs the pure spectra of the constituents of interest, making NAS work only in simple concentration calibrations. Also, the experience with the NAS concept in MVC is too short to allow an evaluation.

9.3 Accuracy

The accuracy of a method can be defined as the closeness between the analytical result and an accepted reference value, sometimes termed the true value.

The estimation of the accuracy should be applied across the range of the analytical method. Reference materials, or a separate well-characterized procedure with known accuracy, may be used for this purpose. The reference method for the MVC is a valid alternative for the comparison if it is well characterized. A report of the accuracy may include a root mean square error of prediction (RMSEP) for the test set in Equation (39):

$$\mathrm{RMSEP} = \sqrt{\frac{\sum_{\mathrm{i}=1}^{n}(y_{\mathrm{i,ref}} - y_{\mathrm{i,pred}})^2}{n}} \quad (39)$$

For references see page 9734

where $y_{i,ref}$ denotes the values obtained from the reference method and $y_{i,pred}$ the values predicted by the MVC. This may be complemented by an estimate of the average bias between the reference method and the MVC results for the test set, in Equation (40):

$$\text{bias} = \frac{1}{n}\sum_{i=1}^{n}(y_{i,ref} - y_{i,pred}) \qquad (40)$$

For pharmaceutical-method validation with one y-variable, a minimum of nine determinations is recommended by IHC over a minimum of three concentration levels. For the MVC with more than one y-variable this may be extended to minimum three levels in each y-variable.

9.4 Precision

The precision of an analytical method can be defined as the closeness between repeated measurement of subsamples taken from the same homogeneous sample. The sample should be authentic. Three levels of precision measurements are defined by IHC: repeatability, intermediate precision, and reproducibility. The precision is often reported as the variance, SD, or coefficient of variation of a series.

9.4.1 Repeatability

The repeatability expresses the variation during a short interval of time with the same operating conditions. For pharmaceutical use, a minimum is six determinations at a 100% level of the analyte, or nine determinations covering the analytical range.

9.4.2 Intermediate Precision

The intermediate precision should include possible variations within a single laboratory, such as different days, different analysts, and different equipment. It is not necessary to study these effects one by one. For pharmaceutical use, the experimental design approach is encouraged. Then the sizes of e.g. the day, analyst and equipment effects are quantified as coefficients in a regression for evaluation of the experimental design. See section 9.10 below.

9.4.3 Reproducibility

The reproducibility can be defined as a case when both within and between laboratory variations are included in the evaluation of the precision. This involves an inter-laboratory trial. For certain types of MVC, e.g. NIR calibrations, this level of precision involves the topic of calibration transfer or the use of well-accepted standards. As with intermediate precision, the results can be evaluated using an experimental design.

9.5 Sensitivity

Sensitivity, or how low are the concentrations that an instrument can detect, can be defined in many ways, where the most common are detection and quantitation limits.

9.5.1 Detection Limit

The detection limit can be defined as the lowest amount of analyte that is still detectable by the method but not necessarily quantifiable with good precision.

The detection limit may be estimated in ways such as visual evaluation, signal-to-noise ratio (S/N) (often 3 : 1 or 2 : 1), SD of the blank, a calibration curve on the same level as the detection limit, or an extrapolation of the calibration curve. When extrapolation is used it should be justified with test samples.

The multivariate spectral S/N can be defined (Equation 41) as the net signal ($S_X^2 - S_E^2$) divided by the noise:

$$\text{S/N} = \frac{\sqrt{S_X^2 - S_E^2}}{\sqrt{S_E^2}} \qquad (41)$$

where S_X is the SD of the gross signal i.e. the SD over all measured samples and variables, and S_E is the SD of the noise i.e. the spectral x-residuals.

The MVC detection limit is, however, more directly dependent of the S_F. which is the SD of the y-residuals. Hence a multivariate detection limit can be estimated as $3S_F$.

For pharmaceutical use this estimation is recommended for nonquantitative limit tests, i.e. classification methods, while the quantitation limit is recommended in use of quantitative calibrations at low analyte levels.

9.5.2 Quantitation Limit

The quantitation limit can be defined as the lowest level of the analyte where the calibration is still working with a suitable accuracy and precision. The same methods as for the detection limit are applicable but a recommended S/N is now increased to 10 : 1. Analogously the multivariate quantitation limit can be estimated as $10 \cdot S_F$.

For pharmaceutical use, this estimation is recommended for low analyte levels, but it is not necessary if the calibration works in a range well above the low levels for the given analytical method.

9.6 Model Fit or Linearity

The linearity of an analytical procedure may be defined as the ability to give results that are proportional to the analyte concentration in the sample within a given range. This works well with univariate linear regression. For MVC there are additional measures of performance.

The model fit may be based on diagnostics such as percent explained variation, CV, t vs. u score plots, calibration y-residuals vs. y-size, multivariate correlation coefficient, etc.

9.6.1 Degree of Fit

The size of the "leftovers" after modelling, i.e. the residuals, is useful as a measure of how well the model fits the data. This is often re-expressed as the amount of variation explained, and denoted as R^2. This can, in projection models, be calculated both for the Y and the X parts of the data, Equations (42) and (43):

$$R_{\mathrm{Y}}^2 = \sqrt{1 - \frac{S_F^2}{S_Y^2}} \tag{42}$$

$$R_{\mathrm{X}}^2 = \sqrt{1 - \frac{S_E^2}{S_X^2}} \tag{43}$$

Here the residual variances, S_F^2 and S_E^2, as well as the variances of the data, S_Y^2 and S_X^2, are uncorrected for degrees of freedom, i.e. the sums of squares divided by $\mathrm{N} \times \mathrm{M}$ and $\mathrm{N} \times \mathrm{K}$, respectively. If one uses the variance corrected for degrees of freedom, which for the residuals **F** and **E** are $(\mathrm{N} - \mathrm{A} - 1) \times (\mathrm{M} - \mathrm{A})$ and $(\mathrm{N} - \mathrm{A} - 1) \times (\mathrm{K} - \mathrm{A})$, respectively, the R^2s are called "adjusted".

The square roots of these variances, then usually corrected for the degrees of freedom, are called the RSDs, of Y and X, as given above in Equation (33).

9.6.2 Cross-validation

If the *predictive* residuals from CV (8.4.3) are used instead of the "fitted" residuals in Equations (42) and (43) we get the cross-validated R^2s. These are often denoted by Q_{Y}^2 and Q_{X}^2. Here Q_{Y}^2 is usually the only one of interest, and it is often denoted just by Q^2. The corresponding RSD based on predictive residuals is often called SECV (standard error of cross-validation). One should note that the denominator in this is always N, since there is no loss of degrees of freedom in the CV process.

9.6.3 Prediction of Validation or Prediction Set

Analogously, if the residuals from a real prediction with a new prediction or test set are available, they can be used to calculate a predictive Q^2, denoted by Q_{pred}^2, with the corresponding predictive RSD, denoted by SEP (standard error of prediction).

9.7 Range

The range of an analytical method can be defined as the interval between the upper and the lower levels of analyte present in the samples, where the precision, accuracy and model fit are suitable for the analysis.

Since MVC involves the use of a local model, extrapolations will not be accurate. This means that the range will be limited approximately to the range of levels entered in the CS used during method validation. Since samples outside the validated range may occur, it is advisable to include a path for extension of the range in the analytical MVC method together with its validation e.g. using the diagnostic tools for model fit above.

9.8 Robustness

The robustness can be defined as the measure of the ability of the analytical method to remain unaffected by small, but purposely made disturbances in method parameters. These are: measurement errors during sample pretreatment, temperature, humidity, variations in grinding, the stability of analytical solutions, extraction times, influence of chromatographic mobile-phase composition, flow rate, different lots of columns, etc. The evaluation of the robustness indicates the level of reliability of the method. Individual factors that may harm the robustness can be identified and compensated for, if the robustness test is made according to an experimental design.

9.9 System Suitability Testing

The system suitability testing can be made as an integral part of an analytical method. The system suitability test is aimed to make an overall test of the entire function of the analytical equipment used.

During the use of a MVC model for prediction, the RSD from the calibration x-variables is useful as a system suitability test.

9.10 Validating the Calibration Model (Including its Robustness)

The investigation of all points above is very time-consuming unless an experimental design is made taking into factors that may influence the results of the MVC. The ranges of the design factors should be small and correspond to around three times the range of experimental control of these factors. The center of the design should be the conditions specified for the analytical method. Since the variations are small, a reduced design aimed at a linear model for evaluation is expected.

If these design factors include "day of analysis", "analyst", "instrument" (equipment), and "laboratory", the results will be informative both for intermediate

For references see page 9734

precision and robustness. If the center point of the design is repeated six times the results also indicate the repeatability of the method.

We must remember that this validation (and robustness study) indicates the performance of the whole analytical process, from sample preparation to the stability of the instrument, and that the MVC is just a part of this process. A common and good result from robustness studies will be that none or few factors are affecting the analytical result. This is reflected in the evaluation of the design as no model present or only a few significant factors.

10 CALIBRATION TRANSFER (BETWEEN INSTRUMENTS, LABS, AND OVER TIME)

Calibration transfer is a process where a calibration model developed for one or more instruments is going to be applied to data from additional instruments not included in the primary calibration model. If the variations are large and the instrument is linear, external standards may be applied. This often holds for UV/VIS spectrometry in liquids and LC. In e.g. NIR diffuse reflectance measurements the magnitude range that is actually used is much smaller (see Figure 3b). For these types of cases, a very local calibration in one instrument has to be moved to another instrument. To preserve the accuracy of the analysis, the local magnitude range of the secondary instrument has to be accurately mapped. This is best done with the same type of samples as the ones present in the MVC.[45]

When the calibration samples are stable with time, the most straightforward solution is to build a new calibration for the secondary instrument but with the same samples and reference analysis results.

Another approach that works for e.g. the same brand and type of NIR instruments even with non-stable samples, is to complement a previous PLS-calibration model by adding samples measured with the new instrument. When the calibration for the primary instrument is well developed also for the interfering variations, it may be sufficient with 10–20% of the original number of samples to span the calibration model also for the secondary instrument. This updating scheme will also work for the same instrument over time.

A variation of this is to make deliberate variations of the instrument parameters and develop the MVC for a set of instruments of the same type from the beginning.[46] This is common in the agricultural sector with calibrations for e.g. protein in wheat. It is also common to adjust this type of calibration for bias and slope on the y-side for the individual instruments.

Another approach is to make the instruments as similar as possible. This route has been taken e.g. by Perkin Elmer for an oil analysis NIR spectrometer.[47] Here the optical bandwidth is artificially degraded to a preset level in order to be constant for all instruments. In addition to this, corrections are made to the wavelength scale and the absorbance scale.

When working in diffuse reflectance mode, the blank spectrum is usually defined by a material that has a flat NIR spectrum. This material should be stable with time. Often a PTFE standard is used.[48] Ceramic is another common material. The batch to batch variations in these materials may lead to different zero levels for the obtained spectra. In some cases is sufficient to bias correct all variables in the spectra[28,29] in such a way that they appear to have been measured with the primary instrument, and then to use the primary calibration model for the adjusted spectra. The relative differences of the zero levels may be established by a set of standards that is measured with both instruments.

Shenk has developed another transfer method.[49,50] Here the wavelength scale is scaled by a wavelength standard, then the intensity scale is scaled in an univariate way for each wavelength using approximately 30 standards.

Direct standardization (DS) involves the use of a normal MVC with spectra from standards measured on the secondary instrument as x-variables and spectra from the same standards measured with the primary instrument as y-variables. This technique has been shown to work by Hanssen.[51] One problem here is that the random variations in the translated spectra are lost. Thus the variations that result in large residual errors for the primary MVC will be masked in the translation step. To compensate for this, the residual error should be monitored with the MVC for translation.

Piecewise direct standardization (PDS) is another spectrum modifying technique.[52] Adjacent x-variables have to be continuous and correlated, as they are e.g. in spectra or chromatograms. Here the main idea is to correct both for magnitude and wavelength differences at the same time. This is done by measuring a set of transfer standards that are similar to the measured samples. Then a sliding window for calculations is used along with the standards to build a series of PCR or PLS calibration models with the window range in the secondary instrument as x-variables to the window center of the primary instrument as y-variable. The transfer mechanism is then to apply the array of models to translate spectra from e.g. the secondary spectrometer to spectra as measured by the primary spectrometer.

With PDS the transfer models for some variables are likely to perform less well. This is visible as discontinuities in the translated spectra. In many cases these discontinuities are sufficiently small to be ignored but in other cases, certain regions of the spectra may have to be removed. This may occur for grating spectrometers at the wavelengths where the Wood's anomaly occurs,[53] and for sections of the spectra that have less dynamic range.

It is also possible to use the OSC[27] for the removal of spectral instrument differences.[28,29] In this case an OSC correction model is developed by measuring a set of samples of the same type as in the primary calibration. They are measured on both instruments in the same manner as for PDS. Spectra from both instruments are then put together in the OSC calculation with the corresponding reference method y-values. Since the instrumental variation is uncorrelated with the y-value variation, the OSC will be able to correct for the instrument effects.

11 SUMMARY AND CONCLUSIONS

MVC is a recently developed approach to use the multidimensional data emerging from spectrometers and other analytical instruments to indirectly estimate properties of interest in chemical, biological, environmental, and other complicated samples. These properties often are concentrations, but, particularly in process analytical chemistry, MVC is often used to calibrate for other qualities such as strength, film thickness, and viscosity. Due to its novelty, the approach is still not fully developed, and interesting research remains regarding both the applicability of MVC in various areas, and the further development of the methodology toolbox. However, MVC is starting to be used extensively in the chemical industry, which is seen in the appearance of introductory books for the industrial analytical chemist.[5]

The idea of using "all" signals instead of one selected frequency or wavelength is simple but powerful. Simple mathematics is used to combine the signals so that the resulting combinations are much more selective, and usually also much more precise, than the original individual signals. Provided that the instrumental data, the signals X, are *relevant* for the actual measurement problem, and provided that the TS is representative, and the TS Y-values are sufficiently accurate, MVC works very well.

MVC needs a computer for the computations, which is a major reason why this approach was not used before about 1970. Today, when most analytical instruments come integrated with a computer, MVC is the natural choice for most calibrations.

LIST OF SYMBOLS

a	component index ($a = 1, \ldots, A$)
$\mathbf{b}$	vector of coefficients ($K \times 1$) that combine the $\mathbf{X}$ columns to map $\mathbf{y}$
$\mathbf{B}$	matrix of coefficients ($K \times M$) that combine the $\mathbf{X}$ columns to map $\mathbf{Y}$
$\mathbf{c}, \mathbf{C}$	concentration vector and matrix
$\mathbf{E}$	matrix of $\mathbf{X}$-residuals; $\mathbf{E_a}$ denotes the $\mathbf{X}$-residuals after a components
$\mathbf{F}$	matrix of $\mathbf{Y}$-residuals
$\mathbf{H}$	a general mathematical function, i.e. $\mathbf{Y} = \mathbf{H}(\mathbf{X}, \mathbf{B}) + \mathbf{F}$
i	sample index ($i = 1, 2, \ldots, n$)
k	index of $\mathbf{X}$-variables, columns ($k = 1, 2, \ldots, K$)
K	number of $\mathbf{X}$-variables
m	index of $\mathbf{Y}$-variables, columns ($m = 1, 2, \ldots, M$)
M	number of $\mathbf{Y}$-variables
n, M	number of calibration (CS/TS) samples
$\mathbf{p_a}$	$\mathbf{X}$-loading vector ($K \times 1$)
$\mathbf{P}$	$\mathbf{X}$-loading matrix ($K \times A$)
$\mathbf{s_i}$	residual SD of one row in $\mathbf{E} = \lVert \mathbf{e_i}/(\mathrm{K} - \mathrm{A}) \rVert$
$\mathbf{t_a}$	$\mathbf{X}$-score vector ($n \times 1$) of component a
$\mathbf{T}$	matrix of $\mathbf{X}$-score vectors ($n \times A$)
$\lVert \mathbf{v} \rVert$	the length (norm) of the vector $\mathbf{v}$
$\mathbf{w_a}$	implicit $\mathbf{X}$-weight vector in PLS models ($K \times 1$): $\mathbf{t_a} = \mathbf{E_{a-1} w_a}$
$\mathbf{w_a^*}$	explicit $\mathbf{X}$-weight vector in PLS models ($K \times 1$); $\mathbf{t_a} = \mathbf{X w_a^*}$
$\mathbf{W}$	implicit $\mathbf{X}$-weight matrix in PLS models ($K \times A$)
$\mathbf{W^*}$	explicit $\mathbf{X}$-weight matrix in PLS models ($K \times A$)
$\mathbf{x_i'}$	row vector ($1 \times K$) of one observation in $\mathbf{X}$ (e.g. one digitized spectrum)
$\mathbf{X}$	TS matrix ($n \times K$) of preprocessed spectra (or other signal profiles)
$\mathbf{y}$	TS column vector ($n \times 1$) of concentrations or other property
$\mathbf{Y}$	TS matrix of concentrations ($n \times M$) or other property
$\mathbf{z}, \mathbf{Z}$	raw data of digitized spectra (before transformation and preprocessing)
$'$	denotes the transpose of a vector or a matrix, e.g. $\mathbf{x}'$

ABBREVIATIONS AND ACRONYMS

AOTF	Acousto-optical Tuneable Filters
ATR	Attenuated Total Reflection
CS	Calibration Set
CV	Cross-validation
DS	Direct Standardization
FT	Fourier Transform
GC	Gas Chromatography
ICP	Inductively Coupled Plasma
ICP/MS	Inductively Coupled Plasma/Mass Spectrometry

For references see page 9734

IHC	International Harmonization Committee
IR	Infrared
LC	Liquid Chromatography
MLR	Multiple Linear Regression
MS	Mass Spectrometry
MSC	Multiplicative Signal Correction
MVC	Multivariate Calibration
NAS	Net Analyte Signal
NIR	Near-infrared Spectroscopy
NMR	Nuclear Magnetic Resonance
OSC	Orthogonal Signal Correction
PCA	Principal Components Analysis
PCR	Principal Components Regression
PDA	Photo Diode Array Spectrometer
PDS	Piecewise Direct Standardization
PLS	Partial Least Squares
PLS/MVC	Partial Least-squares Multivariate Calibration
PLSR	Partial Least-squares Regression
PRESS	Predictive Residuals Sum of Squares
RMSEP	Root Mean Square Error of Prediction
RSD	Residual Standard Deviation
SD	Standard Deviation
SECV	Standard Error of Cross-validation
SEP	Standard Error of Prediction
S/N	Signal-to-noise Ratio
SNV	Standard Normal Variate
SS	Sum of Squares
SVD	Singular Value Decomposition
TLC	Thin-layer Chromatography
TS	Training Set
UV	Ultraviolet
UVC	Univariate Calibration
UV/VIS	Ultraviolet/Visible
2-D	Two-dimensional

RELATED ARTICLES

Biomedical Spectroscopy **(Volume 1)**
Magnetic Resonance, General Medical • Multinuclear Magnetic Resonance Spectroscopic Imaging

Food **(Volume 5)**
Flavor Analysis in Food • Infrared Spectroscopy, Gas Chromatography/Infrared in Food Analysis • Near-infrared Spectroscopy in Food Analysis • Water Determination in Food

Forensic Science **(Volume 5)**
Nuclear Magnetic Resonance Spectroscopy for the Detection and Quantification of Abused Drugs

Petroleum and Liquid Fossil Fuels Analysis **(Volume 8)**
Near-infrared Spectroscopy in Analysis of Crudes and Transportation Fuels • Petroleum Residues, Characterization of

Pharmaceuticals and Drugs **(Volume 8)**
Nuclear Magnetic Resonance Spectroscopy in Pharmaceutical Analysis • Quantitative Structure–Activity Relationships and Computational Methods in Drug Discovery

Polymers and Rubbers **(Volume 9)**
Infrared Spectroscopy in Analysis of Polymer Structure–Property Relationships

Process Instrumental Methods **(Volume 9)**
Process Analysis: Introduction • Chemometric Methods in Process Analysis • Infrared Spectroscopy in Process Analysis • Near-infrared Spectroscopy in Process Analysis • Raman Spectroscopy in Process Analysis

Chemometrics **(Volume 11)**
Chemometrics • Classical and Nonclassical Optimization Methods • Clustering and Classification of Analytical Data • Second-order Calibration and Higher • Signal Processing in Analytical Chemistry • Soft Modeling of Analytical Data

Infrared Spectroscopy **(Volume 12)**
Spectral Data, Modern Classification Methods for

General Articles **(Volume 15)**
Quantitative Spectroscopic Calibration

List of selected abbreviations appears in Volume 15

REFERENCES

1. H. Martens, T. Naes, *Multivariate Calibration*, Wiley, New York, 1989.
2. N.R. Draper, H. Smith, *Applied Regression Analysis*, 2nd edition, Wiley, New York, 1981.
3. O.L. Davies, P.L. Goldsmith (eds.), *Statistical Methods in Research and Production*, 4th edition, Longman, London, 1980.
4. J.C. Miller, J.N. Miller, *Statistics for Analytical Chemistry*, 3rd edition, Ellis Horwood, New York, 1993.
5. K.R. Beebe, R.J. Pell, M.B. Seasholtz, *Chemometrics, a Practical Guide*, Wiley, New York, 1998.
6. R.G. Krutchoff, 'Classical and Inverse Regression Models of Calibration', *Technometrics*, **9**, 425–439 (1967).
7. B. Hoadley, 'A Bayesean Look at Inverse Linear Regression', *J. Am. Stat. Assoc.*, **65**, 356–369 (1970).
8. T. Naes, 'Comparison of Approaches of Multivariate Linear Calibration', *Biometrical J.*, **27**, 267–275 (1985).

9. W. Plugge, C. van der Vlies, 'Near-infrared Spectroscopy as an Alternative to Assess Compliance of Ampicillin Trihydrate with Compendial Specifications', *J. Pharm. Biomed. Anal.*, **11**(6), 435–442 (1993).
10. S. Wold, 'Pattern Recognition by Means of Disjoint Principal Components Models', *Pattern Recognition*, **8**, 127–139 (1976).
11. J.V. Kresta, J.F. MacGregor, T.E. Marlin, 'Multivariate Statistical Monitoring of Process Operating Performance', *Can. J. Chem. Eng.*, **69**, 35–47 (1991).
12. M. Josefson, E. Johanson, A. Torstensson, 'Optical Fiber Spectroscopy in Turbid Solutions by Multivariate Calibration Applied to Tablet Dissolution Testing', *Anal. Chem.*, **60**, 2666–2671 (1988).
13. B. Walczak, B. Bogaert, D.L. Massart, 'Application of Wavelet Packet Transform in Pattern Recognition of Near-IR Data', *Anal. Chem.*, **68**, 1742–1747 (1996).
14. J. Trygg, S. Wold, 'PLS Regression on Wavelet Compressed NIR Spectra', *Chemom. Intell. Lab. Syst.*, **42**, 209–220 (1998).
15. E. Jellum, I. Björnson, R. Nesbakken, E. Johansson, S. Wold, 'Classification of Human Cancer Cells by Means of Capillary Gas Chromatography and Pattern Recognition Analysis', *J. Chromatogr.*, **217**, 231–237 (1981).
16. J.S. Swan, D. Howie, 'Correlation of Sensory and Analytical Data in Flavor Studies into Scotch Malt Whisky', *Found. Biotech. Industr. Fermentation*, **3**, 291–309 (1984).
17. M. Forina, G. Drava, 'Chemometrics for Wine Applications', *Analusis*, **25**(3), 38–42 (1997).
18. G. Malmquist, R. Danielsson, 'Alignment of Chromatographic Profiles for Principal Component Analysis: a Prerequisite for Fingerprinting Methods', *J. Chromatogr. A*, **687**, 71–88 (1994).
19. L. Ståhle, 'Three-way PLS Regression', *Chemom. Intell. Lab. Syst.*, **7**, 95–100 (1989).
20. R. Bro, 'Multiway Calibration, Multi-linear PLS', *J. Chemometrics*, **10**, 47–62 (1996).
21. P. Geladi, H. Grahn, *Multivariate Image Analysis*, John Wiley & Sons, Chichester, UK, 1996.
22. K.H. Esbensen, M. Halstensen, et al. 'Acoustic Chemometrics – from Noise to Information', *Chemom. Intell. Lab. Syst.*, **44**, 61–76 (1998).
23. I.S. Helland, T. Naes, T. Isaksson, 'Related Versions of the Multiplicative Scatter Correction Method for Preprocessing Spectroscopic Data', *Chemom. Intell. Lab. Syst.*, **29**, 233–241 (1995).
24. M.S. Dhanoa, S.J. Lister, R. Sanderson, R.J. Barnes, 'The Link between Multiplicative Scatter Correction (MSC) and Standard Normal Variate (SNV) Transformation of NIR Spectra', *J. Near Infrared Spectrosc.*, **2**, 43–47 (1994).
25. R.J. Barnes, M.S. Dhanoa, S.J. Lister, 'Standard Normal Variate Transformation and De-trending of Near Infrared Diffuse Reflectance Spectra', *Appl. Spectrosc.*, **43**, 772–777 (1989).
26. T.V. Karstang, R. Manne, 'Optimized Scaling. A Novel Approach to Linear Calibration with Closed Data Sets', *Chemom. Intell. Lab. Syst.*, **14**, 165–173 (1992).
27. S. Wold, H. Antti, F. Lindgren, J. Öhman, 'Orthogonal Signal Correction of Near-infrared Spectra', *Chemom. Intell. Lab. Syst.*, **44**, 175–185 (1998).
28. J. Sjöblom, O. Svensson, M. Josefson, H. Kullberg, S. Wold, 'An Evaluation of Orthogonal Signal Correction Applied to Calibration Transfer of Near-infrared Spectra', *Chemom. Intell. Lab. Syst.*, **44**, 229–244 (1998).
29. P. Geladi, H. Barring, E. Dabakk, J. Trygg, H. Antti, S. Wold, A. Karlberg, 'Calibration Transfer for pH Prediction in Lake Sediment Samples', *J. NIR Spectrosc.*, (1999) (in press).
30. A. Savitzky, M.J.E. Golay, 'Smoothing and Differentiation of Data by Simplified Least Squares Procedures', *Anal. Chem.*, **36**, 1627–1639 (1964).
31. K.H. Norris, P.C. Williams, *Cereal Chem.*, **61**, 158 (1984).
32. Ref. 2, 327.
33. S. Wold, H. Martens, H. Wold, 'The Multivariate Calibration Method in Chemistry Solved by the PLS Method', in *Proc. Conf. Matrix Pencils*, eds. A. Ruhe, B. Kågström, Lecture Notes in Mathematics, Springer Verlag, Heidelberg, 286–293, 1983.
34. W. Lindberg, J.-Å. Persson, S. Wold, 'Partial Least-squares Method for Spectro-fluorimetric Analysis of Mixtures of Humic Acid and Ligninsulfonate', *Anal. Chem.*, **55**, 643–648 (1983).
35. A. Höskuldsson, 'PLS Regression Methods', *J. Chemometrics*, **2**, 211–228 (1988).
36. A. Höskuldsson, *Prediction Methods in Science and Technology*, Thor Publishing, Copenhagen, Vol. 1, 1996.
37. G.H. Golub, W. Kahan, 'Calculating the Singular Values and Pseudoinverse of a Matrix', *SIAM J. Numer. Anal.*, **B2**, 205–224 (1965).
38. H. Wold, 'Nonlinear Estimation by Iterative Least Squares Procedures', in *Research Papers in Statistics*, ed. F. David, John Wiley & Sons, New York, 411–444, 1966.
39. S. Wold, A. Ruhe, H. Wold, W.J. Dunn III, 'The Collinearity Problem in Linear Regression. The Partial Least Squares Approach to Generalized Inverses', *SIAM J. Sci. Stat. Comput.*, **5**, 735–743 (1984).
40. F. Lindgren, P. Geladi, S. Wold, 'The Kernel Algorithm for PLS', *J. Chemometrics*, **7**, 45–59 (1993).
41. S. Rännar, F. Lindgren, P. Geladi, S. Wold, 'A PLS Kernel Algorithm for Data Sets with Many Variables and Less Objects. Part I. Theory and Algorithms', *J. Chemometrics*, **8**, 111–125 (1994).
42. B.S. Dayal, J.F. MacGregor, 'Improved PLS Algorithms', *J. Chemometrics*, **11**, 73–85 (1997).
43. The International Harmonization Committee (IHC) Q2: Validation of Analytical Procedures A: Definitions and Terminology http://www.ifpma.org/pdfifpma/q2a.pdf Q2: Validation of Analytical Procedures B: Methodology http://www.ifpma.org/pdfifpma/q2b.pdf

List of selected abbreviations appears in Volume 15

44. A. Lorber, K. Faaber, B.R. Kowalski, 'Net Analyte Signal Calculation in Multivariate Calibration', *Anal. Chem.*, **69**, 1620–1626 (1997).
45. O. de Nord, 'Multivariate Calibration Standardization', *Chemom. Intell. Lab. Syst.*, **25**, 85–97 (1994).
46. N.B. Büchmann, S. Runfors, 'The Standardisation of Infratech 1221 Near Infrared Transmission Instruments in the Danish Network used for the Determination of Protein and Moisture in Grains', *J. Near Infrared Spectrosc.*, **3**, 35–42 (1995).
47. J. Workman, J. Coates, 'Multivariate Calibration Transfer. The Importance of Standardizing Instrumentation', *J. Spectrosc.*, **8**(9), 36–42 (1993).
48. S.V. Hammond, 'Networking Qualitative Databases of Pharmaceutical Raw Materials', Lecture at International Conference on Pharmaceutical Applications of NIR Spectroscopy, Stockholm, Sweden, June 13–15, 1996.
49. J.S. Shenk, M.O. Westerhaus, US Patent No. 4866644, Sept 12, 1989.
50. Y. Wang, D.J. Veltkamp, B.R. Kowalski, 'Multivariate Instrument Standardization', *Anal. Chem.*, **63**, 2750–2756 (1991).
51. W.G. Hansen, 'Production Control by Near Infrared Spectroscopy', in *Leaping Ahead Near Infrared Spectrosc.*, [Proc. Int. Conf. Near Infrared Spectrosc.], 6th Int. Conf., NIR, 376–81, 1994.
52. Y. Wang, D.J. Veltkamp, B.R. Kowalski, 'Multivariate Instrument Standardization', *Anal. Chem.*, **63**, 2750–2756 (1991).
53. R.W. Wood, 'On a Remarkable Case of Uneven Distribution of Light in a Diffraction Grating Spectrum', *Philos. Mag.*, **4**, 396–402 (1902).

Quantitative Spectroscopy Calibration

see Quantitative Spectroscopic Calibration (*General Articles* – **Volume 15**)

Second-order Calibration and Higher

Cliona M. Fleming and Bruce R. Kowalski
University of Washington, Seattle, USA

List of selected abbreviations appears in Volume 15

Most analytical chemists recognize calibration as the quantitation of analytes by relating the analyte signal and analyte concentration, although it is also possible to obtain information about other characteristics of the sample through calibration. In the calibration process, a number of samples that have known values of the characteristic to be determined, e.g. concentration, are measured. These are reference measurements, and are known as calibration standards. Traditionally, the measurements have consisted of a single number (a univariate or scalar measurement); they are plotted against the known concentration, to generate a "calibration curve". (An example of a scalar measurement is a pH value or the area under the curve of a chromatographic peak.) When unknown (test) samples are measured under the same experimental conditions as the standards, their concentrations may be found by regressing their measured signals against the standard calibration curve. However, a significant disadvantage of univariate calibration is that a sensor must be fully selective for the analyte of interest, since any interfering species that may be present cannot be detected when measurements are scalar. This has led to the development of "first-order" calibration methods. These techniques use the relationship between an array of measurements (such as a chromatogram or spectrum) and the analyte concentration to develop a calibration model from which analyte concentrations in test samples can be determined. The array of measurements is a vector, and is known as a first-order tensor, while the calibration model is equivalent to the calibration curve. Since first-order calibration uses many measurements per sample to generate a model, rather than just a single measurement as in the univariate case, it has a number of advantages. For example, multianalyte analysis is made possible as long as there is a standard available for each analyte present in a sample, and outlier detection is also feasible. These types of methods have become popular in recent years, and include partial least squares (PLS) and principal component regression (PCR). It makes sense that when more data are available per sample, more information may be extracted, and this is the source of the success of first-order methods. It therefore follows that, if an analyte signal consists of a matrix of data (also known as a second-order tensor), even more advantages exist. [Examples of instruments that produce matrices of data for each sample analyzed include liquid chromatography/ultraviolet detection (LC/UV) and gas chromatography/mass spectrometry (GC/MS).] This situation allows powerful calibration techniques, known as second-order methods, to be used. The primary advantage of these methods is known as the "second-order advantage"; it allows analytes to be quantitated even in the presence of unmodeled interferents. In other words, the calibration standards do not have to contain any information about the interfering species in the sample. This is of particular value in situations where complex samples are being analyzed, as they do not have to be fully characterized for quantitation of a single analyte. The disadvantage of these methods, however, is their complexity: there are often decisions to be made, such as the values to assign to various parameters, and each method has analysis situations which are more favorable than others. The most common second-order analysis methods are alternating least-squares (ALS) methods, such as parallel factor analysis (PARAFAC) and multivariate curve resolution (MCR), and eigenvalue–eigenvector based methods, such as the generalized rank annihilation method (GRAM) and direct trilinear decomposition (DTD).

1 INTRODUCTION

Calibration is the mathematical and statistical process of extracting information, usually analyte concentration, from an instrument signal. The analytical instruments can be classified according to the tensor order of the data they are capable of acquiring.[1–3] A single sensor (e.g. ion-selective electrode) generates a single number, a scalar or zero-order tensor, per sample. Calibration of zero-order data requires the use of univariate methods, where a calibration curve is generated by plotting standard concentrations versus the instrument response. However, the sensor must be fully selective, since if an interferent is present it cannot be detected. An array of sensors, or a spectrometer, generates an array of measurements, a vector or first-order tensor, per sample. In this case, the existence of many measurements per sample makes multivariate analysis and outlier detection possible. Interferents are not a problem as long as they are included in the calibration design. The calibration methods that are suited to this type of data include PLS and PCR, which have become common analysis tools in today's laboratories.

Recent years have seen the increased use of analytical instruments that produce so-called multiway data, classed as having a tensor order of two (a matrix) or higher.[3] "Multiway" data consist of at least a matrix of numbers per sample analyzed. Powerful calibration methods exist specifically for the analysis of such data, and these methods produce the so-called "second-order advantage".[4,5] This advantage allows analytes to be quantified even in the presence of unknown interferents, along with the resolution of the analyte signal in each order. Another advantage of higher-order tensor analysis is the possibility of automatic correction of instrument malfunction and/or drift or baseline changes.

When a matrix of data is produced per sample, the data are known as "second-order" data, because it is produced by a second-order instrument. Some examples of techniques which produce second-order data are the GC/MS, liquid chromatography/mass spectrometry (LC/MS), tandem mass spectrometry (MS/MS) and LC/UV. Fluorescence excitation/emission is also an example of such an instrument. Figure 1(a) and (b) show the three-dimensional image of a sample that was measured by comprehensive two-dimensional gas chromatography (GC × GC).

When multiple samples are analyzed simultaneously, the matrices are combined to form a block of data, often referred to as a three-way data set. For LC/MS data, use of a second-order calibration method would result in the resolution of the chromatographic profile for each analyte, along with the appropriate mass spectrum. With knowledge of the analyte concentrations in the calibration sample, their quantitation in the unknown samples is also possible.

Instruments also exist that produce even higher order data, although they are less common. Liquid chromatography/tandem mass spectrometry (LC/MS/MS) is an example of the type of data one might obtain from such an instrument. The resulting data from this instrument are in the form of a block of data per sample, creating a third-order tensor, and therefore a collection of samples analyzed together will produce a fourth-order tensor. Some of the same calibration methods can be used to analyze both second- and higher-order data. The methods used, however, depend on the underlying structure of the data.

A second-order data matrix may be classed as bilinear or nonbilinear. Data can be described as being bilinear if the pseudo-rank of the matrix (i.e. rank in the absence of noise) equals the number of chemical components in the sample. This type of data results when one order modulates the other, such as in techniques that

For references see page 9762

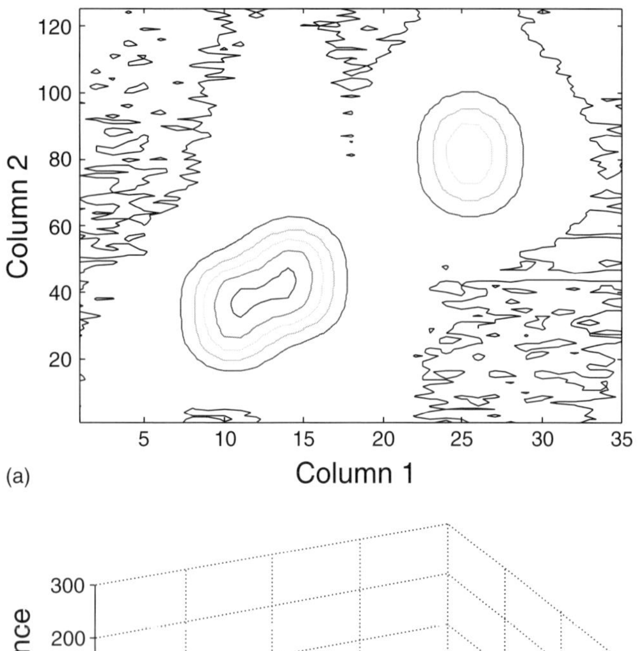

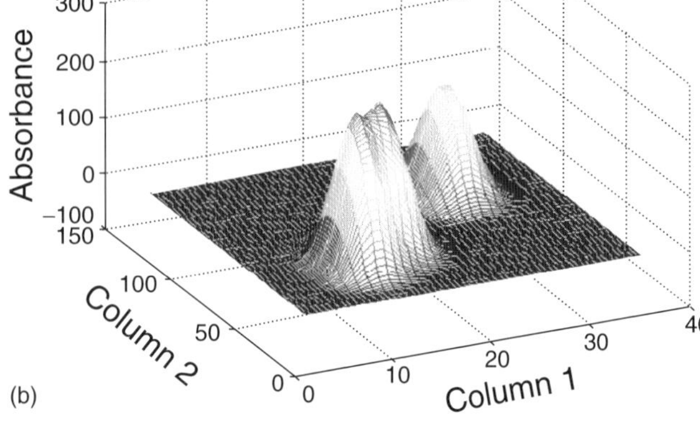

Figure 1 GC × GC instrument response for a sample containing three analytes. (a) Contour plot, where the x- and y-axes represent the separations performed by the first and second columns, respectively; (b) three-dimensional view, showing the intensity of the responses on the z-axis.

combine separation with spectrometry. In GC/MS data, for example, the mass spectrum of an analyte will not change as it elutes except in magnitude. Equation (1) represents the nth analyte in a bilinear matrix:

$$\mathbf{M}_{b,n} = \mathbf{x}_n c_n \mathbf{y}_n^T + \mathbf{E} \tag{1}$$

where $\mathbf{M}_{b,n}$ is the bilinear data matrix for the nth analyte, $\mathbf{x}$ is the analyte profile in the first-order (e.g. pure elution profile), $\mathbf{y}$ is the analyte profile in the second-order (e.g. pure spectrum), c is the concentration of the analyte and $\mathbf{E}$ represents the unmodeled error. The superscript T represents the transpose operator. A mixture of chemical components is therefore represented by Equation (2):

$$\mathbf{M}_b = \sum_{n=1}^{N} \mathbf{M}_{b,n} + \mathbf{E} = \sum_{n=1}^{N} \mathbf{x}_n c_n \mathbf{y}_n^T \tag{2}$$

List of selected abbreviations appears in Volume 15

where N is the number of chemical components present in the mixture. The decomposition of a bilinear data matrix is depicted in Figure 2.

Nonbilinear data, on the other hand, may be represented by $\mathbf{M}_{nb}$ (Equation 3):

$$\mathbf{M}_{nb} = \sum_{p=1}^{P} \mathbf{q}_p r_p \mathbf{s}_p^T + \mathbf{E} \tag{3}$$

which describes a matrix that contains a single pure component,[6] where P is the rank of $\mathbf{M}_{nb}$, $(\mathbf{q}, \mathbf{s})$ is a set of vectors which span the column and row spaces of $\mathbf{M}_{nb}$ and each r is a scaling factor that normalizes the corresponding $(\mathbf{q}, \mathbf{s})$ vectors. In other words, a pure component response has a pseudo-rank >1 when the data are nonbilinear. The reason for this difference lies in the way the data are constructed; for bilinear data, one order modulates the other, whereas nonbilinear data do not have this characteristic. In the case of MS/MS data, the mass spectra in the second-order change independently over the mass to charge (m/z) range of the first-order. Consequently, there is no correlation between the mass spectrum at m/z i (in the first-order) and the mass spectrum at m/z $i+1$, so that a pure component cannot be represented by the outer product of two mass spectra.

The different characteristics of bilinear and nonbilinear data call for different calibration methods. There

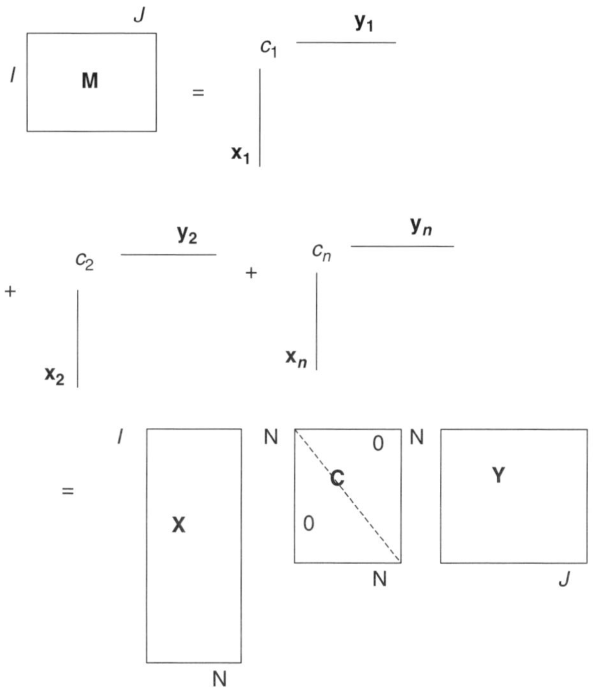

Figure 2 Decomposition of a bilinear data matrix: one dyad [$(\mathbf{x}, \mathbf{y})$ pair] exists for each component/chemical species.

have been some relatively successful attempts to develop calibration methods for nonbilinear data, such as the nonbilinear rank annihilation method (NBRA) which is based on an eigenvalue decomposition.[6] However, analysis methods are far more common for bilinear data, for a number of reasons. First, bilinear data are more commonly produced by instruments in analytical chemistry, particularly through the combination of separation and spectrometric techniques such as LC/UV. Second, the bilinear model is more amenable to mathematical solutions using linear models.

Bilinear data analysis methods can calibrate for analytes with as few as two response matrices, where one is acquired from a standard sample with known analyte concentrations. Of course, owing to the second-order advantage that has been previously mentioned, either the standard or the sample may contain unknown interferents without adversely affecting the calibration. The particular method that operates on two matrices is named the GRAM, and is based on an eigenvalue–eigenvector decomposition.[4]

In cases where multiple standards and/or multiple test samples of second-order bilinear data are available, blocks of data, or data cubes, can be created. This is commonly referred to as three-way data, and the calibration methods that are utilized may be termed either second-order methods or three-way methods. These techniques decompose the blocks into three "ways", where one way corresponds to the sample concentration information. This allows calibration when a standard is available as part of the data set. One such method is an iterative procedure known as PARAFAC, which was developed in the discipline of psychometrics in the early 1970s for decomposition of three-way data.[7] This method uses alternating least squares, and is suitable for data that have low signal-to-noise ratios (S/N). Since its development, other methods have been reported that also utilize ALS algorithms, and they have a similar advantage with low S/Ns.[8,9] A second group of methods, which includes DTD, uses the generalized eigenvalue problem to decompose three-way data.[5,10] These methods are non-iterative, and are especially effective when the data have a high S/N.

In some cases, the accuracy of the calibration may be threatened when unknown components in the sample interfere with the analyte response so that the analyte signal actually depends on the interferent concentration. In such situations, many of the methods mentioned above will be unable to predict the analyte concentrations reliably; this requires the use of a method for standard additions. Such a method exists that is analogous to the standard addition method commonly used with zero-order data: it is termed the second-order standard addition method (SOSAM),[11] and has been shown to increase the reliability of the calibration when matrix effects exist in the sample.

Another tool that can improve calibration accuracy is data pretreatment. In some cases, where the raw data do not exactly fit the model upon which a particular method is based, a preprocessing technique may be of some assistance in improving the fit. For example, in cases where instrumental variations cause minute changes in analyte response from sample to sample, methods that rely on identical responses may fail. This is especially true in cases involving time series data, as the response in the time domain may not be exactly the same from run to run. One approach to solving this problem is to standardize the responses so that the instrumental variations are reduced. Second-order standardization methods are available that perform such an operation.[12,13]

Other work that has been carried out in the second-order area and is related to calibration involves the figures of merit.[14] These include net analyte signal, selectivity, sensitivity, limit of detection and S/N, and they can be used to aid in the development of better second-order instruments which will ultimately result in improved calibration due to improved data.

2 THREE-WAY MODELS

Calibration methods are based upon different models that describe the structure of the data. In calibration, the most commonly used model for second-order (three-way) data is the trilinear model.[15] However, the trilinear model is actually one specific case of a broader model, the Tucker3 model. Tucker proposed his model for use in multiway psychology, in cases where the data contained three modes.[16] His example was of individuals × traits × raters, and the aim of the analysis was to determine the underlying trends in each mode. The Tucker3 decomposition can be written as Equation (4):

$$r_{ijk} = \sum_{a=1}^{A}\sum_{b=1}^{B}\sum_{c=1}^{C} x_{ia}\, y_{jb}\, z_{kc}\, g_{abc} + e_{ijk} \tag{4}$$

where r_{ijk} is a single data point in the three way data block under analysis, x_{ia} is an element of the loading matrix from the first mode, y_{jb} is an element of the loading matrix from the second mode and z_{kc} is an element of the loading matrix from the third mode; g_{abc} is an element of Tucker's core matrix, and represents the magnitude of any interactions between the modes, while e_{ijk} is an element of a matrix of residuals. This equation is written in matrix form as Equation (5):

$$\underline{\mathbf{R}} = \mathbf{X}\underline{\mathbf{G}}\mathbf{YZ} \tag{5}$$

For references see page 9762

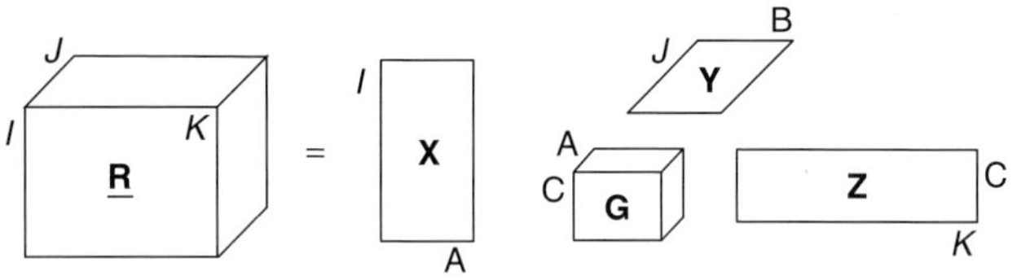

Figure 3 The Tucker3 decomposition: a core matrix, $\underline{\mathbf{G}}$, models the interactions between different factors.

Another form (Equation 6) was suggested by Geladi to eliminate confusion in the placement of the matrices:[17]

$$\mathbf{Y}$$

$$\underline{\mathbf{R}} = \mathbf{X}\underline{\mathbf{G}}\mathbf{Y} + \mathbf{E} \qquad (6)$$

The general decomposition described by these equations is visually represented in Figure 3. It can be seen that this decomposition allows a different number of factors to be present in each mode (a, b and c), a fact that can complicate analysis. The core models interactions between factors by permitting nonzero elements on the off-diagonal. A unique solution is not obtained by this decomposition, however, as rotations can provide different solutions.

When the core is constrained to have nonzero elements only on the superdiagonal, a specific case of the Tucker3 model results. This model, the trilinear model, is much more restricted and can provide unique solutions. The calibration methods described later in this paper (e.g. DTD and PARAFAC) will generally use this trilinear model as in Equation (7):

$$r_{ijk} = \sum_{n=1}^{N} x_{in}\, y_{jn}\, z_{kn} + e_{ijk} \qquad (7)$$

where N is the number of underlying factors in the data, x_{in} is an element of the loading matrix from the first mode, y_{jn} is an element of the loading matrix from the second mode and z_{kn} is an element of the loading matrix from the third mode. This is also shown in matrix form in Equation (8a):

$$\underline{\mathbf{R}} = \mathbf{XYZ} + \underline{\mathbf{E}} \qquad (8a)$$

or Equation (8b):

$$\underline{\mathbf{R}} = \mathbf{X}\underline{\mathbf{I}}\mathbf{YZ} + \underline{\mathbf{E}} \qquad (8b)$$

The visual representation of this model is shown in Figure 4. Because the core matrix now just contains "ones" on the superdiagonal, and is, in effect an identity matrix, this is a trilinear model that does not allow interactions between modes. Another fact that simplifies analysis is that the number of factors in each mode

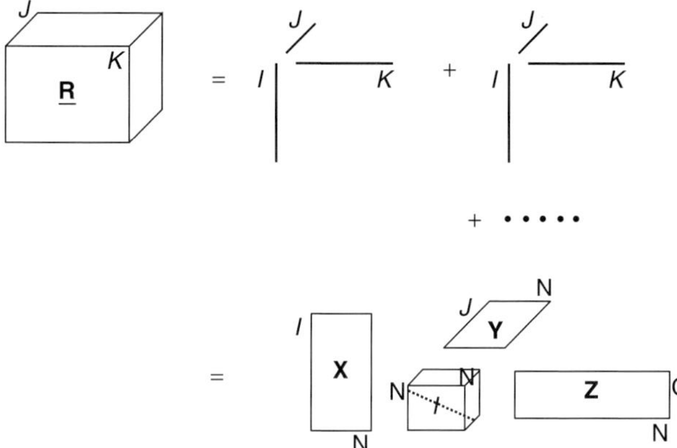

Figure 4 The trilinear model: **X**, **Y** and **Z** are the loading matrices for each mode.

must be the same: this requires a single rank estimation rather than three, which was the case with the Tucker3 model. In addition, the decomposition does not require the loading vectors **x**, **y** and **z** to be orthogonal, unlike the Tucker3 model, allowing the solutions to approximate more closely the true solutions found in analytical data which may not necessarily be orthogonal. The trilinear model is considered to be the most restricted model of all three-way models, and is the only one that allows a unique solution, provided the following condition is met: if $\underline{\mathbf{R}}$ is decomposed into factor matrices **X**, **Y** and **Z**, and every I column of **X** is linearly independent, every J column of **Y** is linearly independent and every K column of **Z** is also independent, then $I + J + K \leq 2M + 2$, where M is the number of columns in each of A, B and C.

There are other special cases of the Tucker model, one of which is the basis of unfold principal component analysis (Unfold-PCA) and unfold partial least squares (Unfold-PLS) techniques. In this special case, the core matrix is not calculated, but is multiplied with two of the loading matrices. Only one of the loading matrices is explicitly calculated.[18] This is performed by unfolding the data block along one of its ways, and placing each of the resulting "slices" side-by-side to form a new matrix. There are three possible ways to unfold the cube, but usually if two of the modes are linked or associated in some way, they are kept together. The possible models that underlie this procedure are specified in Equation (9a–c):

$$r_{ijk} = \sum_{n=1}^{N} x_{in}\, h_{jkn} + e_{ijk} \qquad (9a)$$

$$r_{ijk} = \sum_{n=1}^{N} y_{jn}\, h_{ikn} + e_{ijk} \qquad (9b)$$

$$r_{ijk} = \sum_{n=1}^{N} z_{kn}\, h_{ijn} + e_{ijk} \qquad (9c)$$

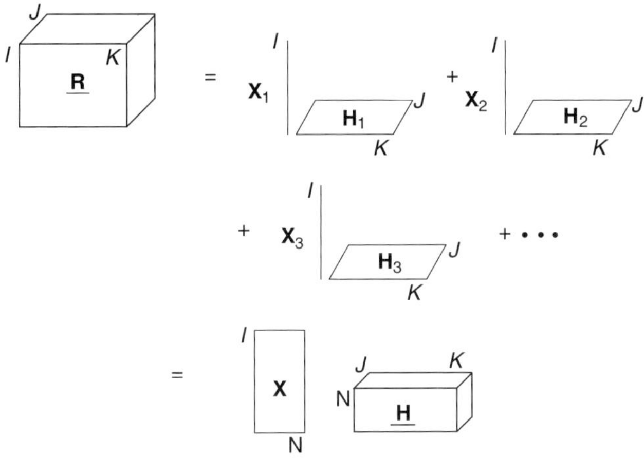

Figure 5 Unfold-PCA model: only one of the loading matrices (**X**) is explicitly calculated.

where x_{in} is an element of the (I × N) score matrix **X** and h_{jkn} is an element of the loading matrix $\mathbf{H}_n$.

The loading matrices $\mathbf{H}_1$ to $\mathbf{H}_n$ can be stacked and the three-way loadings matrix $\underline{\mathbf{H}}$ is obtained. Figure 5 shows one form of this unfolded model, corresponding to Equation (9a). This model does not produce a unique solution, as rotations will produce different solutions. In addition, there will more than likely be different residual matrices depending on which way the data block was unfolded.

3 CALIBRATION METHODS

There are many methods that have been published on solving three-way decompositions, which then allow quantitative analysis. Each model requires different techniques, and choosing the correct calibration method therefore requires that the user know the type of data structure present. Booksh et al. discussed ways of choosing the calibration method, based on the theoretical instrument response function.[19] If this is known, it may be compared with the models described below, and then the appropriate method may be used.

3.1 Alternating Least-squares Techniques

ALS methods have been used for second-order analysis for some years. PARAFAC is one such method that solves the trilinear model that is shown in Equation (4). This algorithm was first introduced in the psychometric literature in 1970.[7] At the same time, an equivalent algorithm was introduced by a separate research group, who named their approach canonical decomposition (CANDECOMP).[20] The approaches are equivalent, and use an ALS algorithm in their solution. The first application of this procedure in the chemical literature was by Appeloff and Davidson in 1980.[21]

Table 1 The PARAFAC algorithm

0	Initialize X and Y
1	$\widehat{\mathbf{Z}} = \mathbf{R}^T\mathbf{H}(\mathbf{H}^T\mathbf{H})^{-1}$
2	$\widehat{\mathbf{Y}} = \mathbf{R}^T\mathbf{H}(\mathbf{H}^T\mathbf{H})^{-1}$
3	$\widehat{\mathbf{X}} = \mathbf{R}^T\mathbf{H}(\mathbf{H}^T\mathbf{H})^{-1}$
4	Repeat steps 1–3 until convergence criterion is satisfied

The first step in PARAFAC is to obtain initial estimates for two of its ways, for example, **X** and **Y** (Step 0 below). These guesses are then used to estimate the third way, $\widehat{\mathbf{Z}}$. In the next step, $\widehat{\mathbf{Z}}$ and **X** are used to estimate $\widehat{\mathbf{Y}}$, and then $\widehat{\mathbf{Z}}$ and $\widehat{\mathbf{Y}}$ are used to estimate $\widehat{\mathbf{X}}$. The process continues in an iterative fashion until a specified ending criterion is met. The algorithm is given in Table 1.

3.1.1 Step 0: Choosing the Starting Estimates

Choosing a starting estimate for two of the ways is the initialization step in the algorithm. A number of approaches are possible, but it must be noted that this choice will greatly affect how the algorithm proceeds, since the PARAFAC algorithm can become trapped in local minima so that convergence is slow and erroneous solutions can be produced.[22] If there is a priori knowledge of the analyte profiles, these may be used. A random number generator is also a frequent choice.[23] In that case in particular, a wise decision is to consider many starting values, although this will considerably delay the process. If a number of different starting values give the same solution, then confidence in the accuracy of the result is greater. The closer the starting guesses are to the true solution, the faster the algorithm will proceed, and the probability of encountering local minima is reduced. Speed is a big consideration when using PARAFAC analyses, as convergence is usually very slow. Results obtained from DTD provide another option for starting estimates: when these results are not complex, this choice can speed up the convergence process.[24]

3.1.2 Step 1: Estimating the First Set of Profiles

In each of steps 1–3, the **R** and **H** matrices are working matrices, redefined at each step. **R** is the data block, $\underline{\mathbf{R}}$, rearranged in matrix form, and **H** is created from the parameters not being estimated in that step. When calculating the **Z** profiles, therefore, the current **X** and **Y** information is employed to define **H**, as in Equation (10):

$$\mathbf{H}^T\mathbf{H} = (\mathbf{X}^T\mathbf{X}) \otimes (\mathbf{Y}^T\mathbf{Y}) \tag{10}$$

For references see page 9762

and Equation (11):

$$\mathbf{R}^T\mathbf{H} = \begin{bmatrix} diag(\mathbf{X}^T\mathbf{R}_1\mathbf{Y})^T \\ diag(\mathbf{X}^T\mathbf{R}_2\mathbf{Y})^T \\ \vdots \\ diag(\mathbf{X}^T\mathbf{R}_k\mathbf{Y})^T \end{bmatrix} \quad (11)$$

where $\otimes$ represents the Kronecker operator (element-wise multiplication), $\mathbf{R}$ is the data block unfolded in the K direction, so that $\mathbf{R}_1$ is the $I \times J$ submatrix of $\underline{\mathbf{R}}$, and $diag(\mathbf{X}^T\mathbf{R}_1\mathbf{Y})^T$ is a column vector containing the diagonal of $\mathbf{X}^T\mathbf{R}_1\mathbf{Y}$. At this step, Equation (12) is being minimized as

$$\min_z \left\|\mathbf{R}^T - \mathbf{H}\mathbf{Z}^T\right\|_2^2 \quad (12)$$

3.1.3 *Steps 2 and 3: Estimating the Second and Third Sets of Profiles*

The procedures for estimating $\mathbf{X}$ and $\mathbf{Y}$ are similar to that for determining $\mathbf{Z}$, but in each case the $\mathbf{H}$ and $\mathbf{R}$ matrices are newly defined, using the parameters not being estimated.

3.1.4 *Step 4: Convergence*

There are two main approaches that are used to determine when the solution has been reached and the iterations should be stopped. One of these is to compare changes in the fit after each iteration, stopping when the change in the residuals is less than a specified tolerance, generally close to 10^{-6}. The residual change is calculated using the root of squared residuals. An alternative is to compare the loadings obtained after each iteration, stopping when they do not change more than a specified amount. This comparison is performed by calculating the angle between the predicted profile from successive iterations, which is found in each case by unfolding the $\mathbf{X}$, $\mathbf{Y}$ and $\mathbf{Z}$ matrices into column vectors and then calculating (Equation 13):

$$\cos\theta_x \cos\theta_y \cos\theta_z = \left(\frac{\mathbf{x}_i\mathbf{x}_{i+1}}{\sqrt{\mathbf{x}_i^2\mathbf{x}_{i+1}^2}}\right)\left(\frac{\mathbf{y}_i\mathbf{y}_{i+1}}{\sqrt{\mathbf{y}_i^2\mathbf{y}_{i+1}^2}}\right) \times \left(\frac{\mathbf{z}_i\mathbf{z}_{i+1}}{\sqrt{\mathbf{z}_i^2\mathbf{z}_{i+1}^2}}\right) \quad (13)$$

where $\mathbf{x}_i$ is from the ith iteration, and $\mathbf{x}_{i+1}$ is from the following iteration. When this term is less than a specified criterion, for example 10^{-6} once again, convergence has been reached.

A difficulty that exists with PARAFAC is the occasional occurrence of two-factor degeneracies.[25] This is signified when the estimated profiles for two components are highly correlated in all three modes. The uncorrelated correction coefficient (UCC) has been introduced to detect such a situation. The UCC for two components, A and B, is (Equation 14):

$$\mathrm{UCC} = \left(\frac{\mathbf{x}_A\mathbf{x}_B}{\sqrt{\mathbf{x}_A^2\mathbf{x}_B^2}}\right)\left(\frac{\mathbf{y}_A\mathbf{y}_B}{\sqrt{\mathbf{y}_A^2\mathbf{y}_B^2}}\right)\left(\frac{\mathbf{z}_A\mathbf{z}_B}{\sqrt{\mathbf{z}_A^2\mathbf{z}_B^2}}\right) \quad (14)$$

where $\mathbf{x}_A, \mathbf{x}_B$ are the $\mathbf{x}$ profiles, $\mathbf{y}_A, \mathbf{y}_B$ are the $\mathbf{y}$ profiles and $\mathbf{z}_A$, $\mathbf{z}_B$ are the $\mathbf{z}$ profiles of the two components. When the UCC is close to -1, a two-factor degeneracy has occurred. In that case, the profiles will be either positively correlated in two or more of the modes or negatively correlated in one or all three of the modes. Effectively, the results suggest that there is one factor fewer present than is actually the case. Sometimes it happens that a solution is close to degeneracy; this is indicated when the UCC is close to -0.85. This is not considered to be a true degeneracy, and further iterations may result in an improvement in the UCC as the solution moves closer to the real one. The temporary degeneracy can then be considered to be due to a local minimum, most likely due to a poor choice of initial estimates, and is known as a "swamp". A true degeneracy may arise if the algorithm has been asked to calculate more factors than exist in the data. It may also occur if the wrong model has been chosen, i.e. if the trilinear model does not fit the data.

The existence of swamps is a source of difficulty, as extremely slow convergences may result.[22] These are areas where the convergence slows dramatically, making it appear as though the solution has been reached. However, if the iterations are allowed to proceed, the slow convergence will once again become faster and the true solution will emerge. While the algorithm is in a swamp, the residuals will not change dramatically, whereas the loadings will. If a swamp is mistakenly thought to indicate convergence, therefore, the solution provided will not be accurate. Mitchell and Burdick[24] have proposed that swamps tend to occur simultaneously with two-factor degeneracies, and caution must be exercised when these play a role in convergence. A solution that uses a stabilization method has been proposed, so that the number of iterations spent in a swamp is greatly reduced.[25] It is suggested that a high positive UCC may be indicative of serious problems due to instabilities in the estimation steps.

Constraints are generally imposed on the parameters to be estimated in order to increase interpretability. As was mentioned previously, PARAFAC was used in the psychometric field for some years before its applicability to chemical data was recognized. Psychometricians have employed a number of constraints, including orthogonal constraints, to overcome problems with unstable solutions to help convergence to the correct solution. (An orthogonal constraint helps with scaling: although scaling does not affect the uniqueness of the solution, it can be an issue in determining how long it takes to converge.) In chemistry, constraints are chosen based on a priori knowledge of the data being analyzed. For example, knowing that mass spectra, chromatographic profiles and concentrations are always positive, non-negativity constraints can be applied to all modes of a set of LC/MS samples. In addition, it is known that chromatographic profiles should be unimodal, allowing the imposition of a unimodality constraint. While non-negativity constrained least-squares techniques are common, however, the unimodality constraint is more difficult to impose.

The results that are obtained when PARAFAC is applied to excitation–emission data demonstrate the improvement that can occur when constraints are imposed. Samples that contained tyrosine, tryptophan and phenylalanine were measured by fluorescence, and unconstrained PARAFAC was used in an attempt to resolve the excitation and emission spectra.(15) Figure 6(a) shows the emission spectra that were obtained. It is obvious that the profiles are not correct, as the spectra should be non-negative; however, when the concentrations and spectra are constrained to be positive, PARAFAC produces emission profiles (Figure 6b) that are very close to the true spectra (Figure 6c).

The only deviation from the true spectra is a slight hump below 300 nm, and this is reported to be due to non-multilinear Rayleigh scatter. Using the knowledge that the spectra should be positive enabled the analysts to use PARAFAC in such a way (i.e. with a non-negativity constraint) that the difficulties encountered could be overcome. These problems were most likely due to the small number of samples used, which in this case was just two. A related area of research in PARAFAC analysis is in increasing the algorithm's efficiency by reducing the computations required;(26–29) constrained algorithm's tend to reach convergence more quickly than the traditional versions, thereby resulting in research that combines the two.(30)

PARAFAC has been applied to various analytical problems, and has resulted in the successful calibration of naphthalene and styrene in ocean water.(31) In that case, an excitation–emission matrix imaging spectrofluorimeter was used, and the data contained a complex background signal in addition to overlapping spectral interferents that were not included in the calibration set. It was found that a two-factor PARAFAC model could be used to decompose the data so as to produce estimates of the excitation and emission profiles of naphthalene and background intensities in the samples. The factor that corresponded to naphthalene had a 2.6 ppb standard deviation of the blank, which equaled a limit of determination of 8 ppb, so that the measured linear dynamic range for naphthalene extended from 8 to 2000 ppb. When

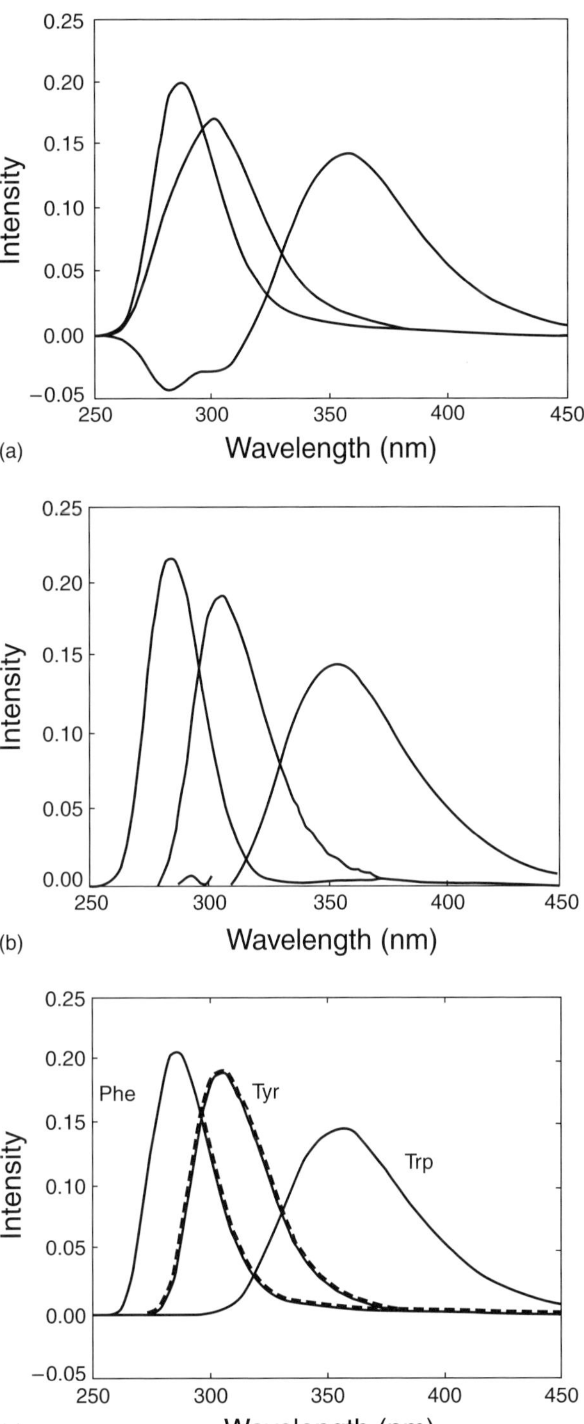

Figure 6 Emission spectra from (a) unconstrained PARAFAC, (b) PARAFAC with non-negativity constraints and (c) true emission profiles. (Reprinted with permission from R. Bro, *Chemom. Intell. Lab. Syst.*, **38**, 149–171 (1997). Copyright 1997 Elsevier Science.)

gasoline/naphthalene/ocean water samples were included in the decomposition, one extra factor was needed to

Table 2 Prediction errors obtained from PARAFAC quantitation of naphthalene: (i) two-factor model (calibration set contains naphthalene/ocean water standards only) and (ii) three-factor model (calibration set includes naphthalene/gasoline/ocean water). (Reprinted with permission from K.S. Booksh, A.R. Muroski, M.L. Myrick *Anal. Chem.*, **68**, 3539–3544 (1996). Copyright 1996 American Chemical Society.)

Naphthalene added (ppb)	Prediction error	
	(i) 2-factor	(ii) 3-factor
28	−9.5(4.0)[a]	−10.8(2.6)[a]
28	−13.3(−0.1)	−14.3(−1.0)
56	−6.7(7.4)	−4.1(10.0)
84	−11.5(3.1)	−12.8(1.8)
110	−18.3(−3.2)	−17.5(−2.5)
286	−14.0(4.7)	−16.0(2.7)
571	−30.2(−5.9)	−31.9(−7.6)
666	78.3	77.7
800	113.0	112.2
1000	40.8	42.7
1333	−90.3(−51.5)	−91.0(−52.0)
2000	−18.9(35.1)	−18.8(36.2)
RMSE[b]	47.2(19.2)	47.3(19.7)

[a] Errors derived after discarding the 666, 800 and 1000 ppb naphthalene/ocean water standards.
[b] Root mean square error, including blanks.

account for the fluorescence spectrum of the gasoline; the question to be answered was whether or not the quantitation would be affected by the inclusion of these samples. Calibration curves were generated for both the two- and three-factor decompositions, where standard additions of naphthalene had been made to generate the standards. It was found that the quality of the calibration curve formed by the naphthalene/ocean water standards was not degraded by the presence of the gasoline as long as an extra factor was included to account for its variation. Table 2 displays the prediction errors obtained in each calibration, and it is apparent that they are very similar, each having a root mean square error of prediction (RMSEP) of 47 ppb. However, three of the naphthalene standards deviate from linearity on the calibration curve: when these standards are excluded from the calibration, the RMSEP drops to <20 ppb in each case. The important conclusion that can be drawn from this analysis is that when second-order calibration techniques are employed, accurate analyte quantitation is possible even in the presence of large background signals and interfering chemical species.

An accurate calibration of a model system using flow-injection analysis/diode-array data has also been reported.[32] Yet another investigation used PARAFAC to predict enzymatic activity and substrate consumption during the enzymatic browning of vegetables,[33] while kinetic studies have also been investigated using this technique.[34]

A recently published algorithm, the alternating trilinear decomposition (ATLD) algorithm, also solves the trilinear model with the aid of ALS, and is in fact an improvement of the traditional PARAFAC algorithm without constraints.[8] While it does not involve the use of constraints, it still converges faster than the original PARAFAC algorithm (ATLD is reported as converging in less than ten iterations), and its iterative procedure uses the Moore–Penrose generalized inverse with singular value decomposition (SVD) in the trilinear sense. As mentioned previously, some of the main drawbacks to the use of PARAFAC are (a) the occurrence of two-factor degeneracies and (b) the presence of swamps. ATLD was developed with the aim of regularizing the procedure, thus avoiding these traps. The loss function to be minimized in this case is (Equation 15):

$$\sigma = \sum_{i=1}^{I}\sum_{j-1}^{J}\sum_{k-1}^{K}\left(r_{ijk} - \sum_{n=1}^{N} x_{in}\, y_{jn}\, z_{kn}\right)^2 \tag{15}$$

where the symbols have the same meanings as before. In order to solve the loss function, the algorithm alternates over **Y** for fixed **X** and **Z**, over **X** for fixed **Y** and **Z**, and over **Z** for fixed **X** and **Y**. The equations that are solved at each step are (Equations 16a–c)

$$\mathbf{x}_i^T = diag\left(\mathbf{Y}^+\mathbf{R}_{i..}(\mathbf{Z}^T)^+\right), \quad i = 1, \ldots, I \tag{16a}$$

$$\mathbf{y}_j^T = diag\left(\mathbf{Z}^+\mathbf{R}_{.j.}(\mathbf{X}^T)^+\right), \quad j = 1, \ldots, J \tag{16b}$$

$$\mathbf{z}_k^T = diag\left(\mathbf{X}^+\mathbf{R}_{..k}(\mathbf{Y}^T)^+\right), \quad k = 1, \ldots, K \tag{16c}$$

where $\mathbf{R}_{i..}$ is the data cube $\underline{\mathbf{R}}$ unfolded along the $J \times K$ axis. These equations involve the use of the Moore–Penrose generalized inverse, denoted by the + superscript, whereas the original PARAFAC algorithm did not (see Equation 11). Because this computation is based on the SVD, singular values less than a certain tolerance are treated as zero. This makes it possible to perform the calculations even when N is greater than the number of chemical species present in the sample, reducing the rank deficiency problem that exists with PARAFAC and that can cause two-factor degeneracies. The importance of this result stems from the fact that rank determination is not always straightforward, especially in the presence of noise, and an overestimation of the number of chemical species in the sample may sometimes occur.

A comparison of the ATLD and traditional PARAFAC algorithm has been carried out on liquid chromatography/diode-array detection (LC/DAD) data for overlapped chlorinated aromatic hydrocarbons. The analytes were *p*-chlorotoluene and *o*-chlorotoluene, with *o*-dichlorobenzene being an interfering species (i.e. not included in the calibration set). Four samples were included in the calibration set, while there were three

Table 3 Comparison of PARAFAC and ATLD quantitation of overlapped chlorinated aromatic hydrocarbons. (Reprinted with permission from H.-L. Wu, M. Shibukawa, K. Oguma, *J. Chemom.*, **12**, 1–26 (1998). Copyright 1998 John Wiley & Sons Limited.)

Unknown sample no.	*p*-Chlorotoluene (μg ml^{-1})					*o*-Chlorotoluene (μg ml^{-1})				
	Known	PARAFAC		ATLD		Known	PARAFAC		ATLD	
		Predicted	Recovery (%)	Predicted	Recovery (%)		Predicted	Recovery (%)	Predicted	Recovery (%)
7	25.2	24.2 ± 3.9	95.8 ± 15.5	25.2 ± 0.2	100.0 ± 0.7	91.2	88.1 ± 18.2	108.7 ± 19.9	90.8 ± 0.9	99.6 ± 1.0
8	50.4	54.9 ± 2.8	109.3 ± 5.5	53.7 ± 0.5	106.6 ± 1.0	30.4	30.7 ± 12.0	101.1 ± 39.5	29.5 ± 0.8	97.2 ± 2.7
9	75.6	75.3 ± 6.1	99.6 ± 8.1	78.2 ± 0.1	103.4 ± 0.1	60.8	67.7 ± 9.9	111.3 ± 16.3	64.7 ± 0.9	106.4 ± 1.4

test samples. Table 3 shows the true concentrations of the analytes in each sample, along with the predicted concentrations that were given by both the traditional PARAFAC algorithm and the ATLD algorithm.

The concentration estimates are, in general, closer to the true values than the PARAFAC predictions; moreover, convergence was reached much more quickly with the ATLD algorithm. It must also be noted that the standard deviations of the ATLD predictions are much smaller than those of the PARAFAC predictions. However, it is known that PARAFAC performance can be improved through the use of constraints, and such an algorithm may have performed better.

Another method that uses an ALS algorithm is MCR. As the name suggests, this technique was originally developed for resolution of the profiles in second-order data.(9,35,36) However, further development of the method has allowed the calibration of three-way data in addition to profile resolution.(37)

There are three main steps that are carried out during this procedure. First, the chemical rank of the data, N, must be determined. Generally, SVD may be used. By examining the singular values in regions where no chemical components exist, variance due to noise can be estimated. Once the noise estimate is known, singular values due to the chemical components can be determined in different regions of the data, thereby allowing pseudo-rank estimation.

As with other techniques that employ ALS, initial estimates of the profiles are required. In this case, however, only a concentration profile estimate is needed. Evolving factor analysis (EFA) is used,(38) and for a specific reason: it is known that, in order to use curve-resolution methods, assumptions must be made about the signal. Common assumptions include bilinearity, non-negativity, unimodality and closure. However, even with such constraints, a unique solution (for the analysis of a single sample) is not guaranteed, as rotational and intensity ambiguities may still exist.(9) The use of EFA allows curve resolution to overcome, in part, the rotational ambiguity, as it provides information regarding the windows of existence of components in the data. Of course, when multiple samples are analyzed simultaneously, the rotational ambiguity no longer exists, but EFA remains a good choice for initial estimates. The third step to be taken when performing MCR is to calculate iteratively new profiles using constrained ALS. When a single sample is being analyzed (Equation 17):

$$\mathbf{D} = \mathbf{CS} \tag{17}$$

where $\mathbf{D}$ is a bilinear data matrix, dimensioned $m \times n$, $\mathbf{C}$ is a matrix of concentration profiles, dimensioned $m \times N$, and $\mathbf{S}$ is a matrix of spectral profiles, dimensioned $N \times n$. It can be seen that this is not the trilinear model used by PARAFAC and the eigenvalue–eigenvector-based methods, as the concentration information is contained within the concentration profiles rather than being separately calculated.

The initial estimates of the concentration profiles in $\mathbf{C}$, obtained by EFA, are then used to calculate the corresponding spectra (Equation 18):

$$\widehat{\mathbf{S}} = \mathbf{C}^{+}\overline{\mathbf{D}} \tag{18}$$

where $\overline{\mathbf{D}}$ has been truncated to rank using SVD. The following step involves a new estimation of $\mathbf{C}$, using the new spectral profiles, $\widehat{\mathbf{S}}$ (Equation 19):

$$\widehat{\mathbf{C}} = \overline{\mathbf{D}}\widehat{\mathbf{S}}^{+} \tag{19}$$

The algorithm iterates back and forth between Equations (18) and (19) until the stopping criterion is met. This criterion is similar to PARAFAC ending criteria.

There are a number of constraints that may be used, based on the previously mentioned assumptions. Constraining the spectra and concentration profiles to have non-negative values is common, while the shape of the signal may also be a constraint. For example, a unimodality constraint is frequently imposed upon chromatographic profiles. As mentioned already, however, the use of such constraints will not always remove the ambiguity problems. Analyzing two or more samples simultaneously, on the other hand, often allows a unique solution. In addition, quantitation in the presence

For references see page 9762

of unknown analytes becomes possible: the second-order advantage.[39,40]

In order to use MCR for calibration, the data matrices to be analyzed must be arranged in one matrix. They may be concatenated in three ways: row-wise, column-wise or along the tubes. Samples are concatenated so that the columns, or spectra, are in common (Equation 20a):

$$\mathbf{D}_s = \mathbf{C}_s\mathbf{S} \tag{20a}$$

or (Equation 20b)

$$\begin{bmatrix} \mathbf{D}_1 \\ \mathbf{D}_2 \\ \vdots \\ \mathbf{D}_s \end{bmatrix} = \begin{bmatrix} \mathbf{C}_1 \\ \mathbf{C}_2 \\ \vdots \\ \mathbf{C}_s \end{bmatrix} \mathrm{S} \tag{20b}$$

$\mathbf{D}_s$ is dimensioned $m^*s \times n$, $\mathbf{C}_s$ is $m^*s \times N$ and $\mathbf{S}$ is $N \times s$, where s is the number of samples, N is the number of principal components required to describe the data and $m \times n$ are the dimensions of each individual $\mathbf{D}$ matrix. It can be seen that $\mathbf{S}$, which contains the spectral profiles, is common to each individual sample, while the concentration profiles vary.

The quantitative results are obtained through the resolved concentration profiles in $\mathbf{C}$. Since the concentration information is embedded in the profiles, the area under the curves can be related to concentration. When calibration standards are included in the MCR analysis, the unknown analyte concentrations can be found via Equation (21):

$$c_{\mathrm{un}} = \frac{A_{\mathrm{un}}}{A_{\mathrm{std}}} c_{\mathrm{std}} \tag{21}$$

where c_{un} and c_{std} are the unknown and standard concentrations for a particular analyte, and A_{un}, A_{std} are the areas under the resolved profiles for that analyte in the unknown and standard samples.

With MCR analysis, it is possible to analyze data matrices that do not strictly follow a trilinear data structure when concatenated. Although each individual matrix must be bilinear, the profiles of one mode may vary. For example, if the data are chromatographic/spectrometric, the chromatographic profiles may vary in retention time and/or shape from one sample to the next. Such data do not have a trilinear data structure, and methods that rely on a trilinear model will fail. By constraining the spectra to be equal, but not the concentration profiles, MCR can resolve the profiles and also obtain adequate quantitative results.[37]

For example, quantitative analysis using second-order MCR has been applied to mixtures of amino acids that were measured with flow-injection analysis and diode-array multiwavelength detection.[39] The system monitored the reaction of lysine and proline with

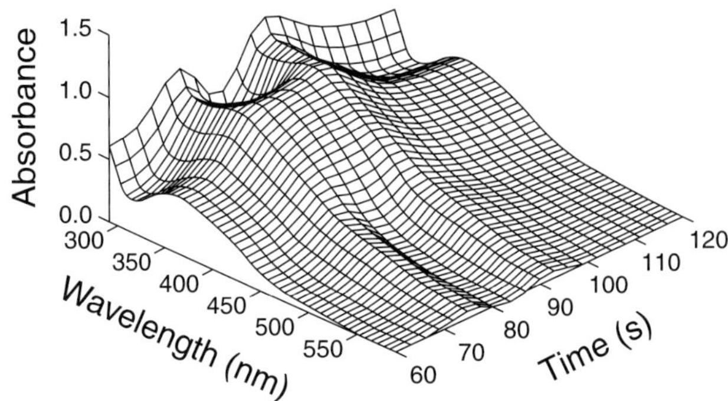

Figure 7 Flow-injection analysis/diode-array detector response for a mixture of lysine and proline. (Reprinted with permission from J. Saurina, S. Hernandez-Cassou, R. Tauler, *Anal. Chim. Acta*, **335**, 41–49 (1996). Copyright 1996 Elsevier Science.)

1,2-naphthoquinone-4-sulfonate (NQS). Four samples were analyzed, water, lysine, proline and a mixture of lysine and proline, all of which were injected into a system containing NQS. This data is not trilinear; whereas the concentration profiles of the amino acid derivatives retain their shapes over the course of the reaction, the shapes of unreacted NQS species vary depending on the amino acid concentration in the mixture. The three-dimensional data from the mixture of lysine and proline are shown in Figure 7.

The sample data matrices were concatenated and analyzed as described above, and Table 4 shows the predicted concentrations that were obtained under a number of different constraints (options a–d). The fitting error of the resolved profiles is also reported.

Because of the nontrilinear nature of the data, option c (which forces both concentration and spectral profiles to be identical and therefore assumes trilinearity) does not give the best quantitative results, producing a quantitation error of 3.34%. Conversely, option d constrains only the concentration profiles of the amino acid derivatives to be the same, and this more flexible option produces much better quantitative results with a quantitation error of 0.36%.

In contrast, when data with a trilinear structure are analyzed, the concentration profiles can be constrained to be identical in addition to the spectral profiles. In that case, the additional constraint improves both quantitative and qualitative results. It has been found that when data follow the trilinear data structure, the predictive error is within the level of measurement noise.[41] In cases where only one order is identical from one sample to another, the error is higher, but still within acceptable limits, provided that there is some selectivity present for the analyte in question. If complete overlap exists, then the results will be ambiguous.[40] Comparison studies have been carried out between MCR and DTD, an eigenproblem-based method that operates specifically on trilinear data, and

Table 4 Predicted concentrations for lysine and proline obtained from MCR under four different constraints. (Reprinted with permission from J. Saurina, S. Hernandez-Cassou, R. Tauler, *Anal. Chim. Acta*, **335**, 41–49 (1996). Copyright 1996 Elsevier Science.)

Option[a]	Recovered Lys concentration ($M \times 10^4$)	Recovered Pro concentration ($M \times 10^4$)	Fitting error(%)[b]	Quantitation error(%)[c]
a	2.04	1.89	3.75	4.21
b	1.97	1.93	4.77	2.76
c	1.93	1.94	6.08	3.34
d	2.01	2.00	5.36	0.36

[a] Applied constraints: (a) no closure or equal shape constraint; (b) closure constraint with respect to NQS; (c) closure constraint with respect to NQS and equal shape constraint to all concentration profiles; (d) closure constraint with respect to NQS and equal shape constraint to only the derivative concentration profiles.

[b] Fitting error (%) $= \left[\sum(A_{ij,\text{experimental}} - A_{ij,\text{reproduced}})^2\right]^{1/2} \Big/ \left[(\sum A_{ij,\text{experimental}})^2\right]^{1/2} \times 100$, where $A_{ij,\text{experimental}}$ is the experimental absorbance at wavelength i and time j and $A_{ij,\text{reproduced}}$ is the reproduced absorbance obtained by MCR.

[c] Quantitation error (%) $= \left[\sum(C_{\text{known}} - C_{\text{calc}})^2\right]^{1/2} \Big/ \left[(\sum C_{\text{known}}{}^2)^{1/2}\right] \times 100$, where C_{known} is the known concentration of each analyte, Lys and Pro, in the unknown sample and C_{calc} is the calculated concentration of each analyte recovered from MCR.

the results indicate that each method's performance is comparable for trilinear data.[(41)] It has also been suggested that a simple method for determining the presence or absence of trilinearity might be to compare the lack of fit obtained with DTD and MCR. This is because data that fit the trilinear model will be fitted well with DTD, whereas deviations from trilinearity will result in a better fit with MCR.[(42)]

3.2 Rank Annihilation Methods

Calibration methods that use rank annihilation have been used in the chemical field for almost 20 years, and they have undergone many changes and improvements in that time. The first form of rank annihilation factor analysis (RAFA) was introduced by Ho et al. in 1980.[(43–45)] That method suffered from being very slow, as it employed iterative methods for its solution. When an improved version of RAFA was proposed by Lorber 4 years later, the problem was transformed into an eigenvalue problem that could be directly solved in a very short time.[(46)] In this method, two samples were required for calibration. One was the unknown or test matrix that was to be analyzed, **M**, while the second was the calibration sample, **N**, that could contain just one of the analytes present in **M**. Both **M** and **N** must follow the bilinear data format, where the mixture matrix, **M**, can be represented by Equation (2), and the pure standard matrix, **N**, can be represented by Equation (1). The kth component in **M** is common to both samples.

These equations can be rearranged to form a generalized eigenvalue problem (Equation 22):

$$\mathbf{N}_k \mathbf{z}_k = \lambda_k \mathbf{M} \mathbf{z}_k \tag{22}$$

where λ_k is an eigenvalue that is equivalent to the concentration of the kth analyte in **N** divided by its concentration in **M**. $\mathbf{z}_k$ is an eigenvector that was later shown by Sanchez and Kowalski to be related to the response profile of the analyte.[(4)] However, because **M** and **N** are not necessarily square, the problem as written may not be solved. **M** is therefore truncated using SVD (Equation 23):

$$\mathbf{M} = \mathbf{USV}^T \tag{23}$$

Equation (22) is rewritten as Equation (24):

$$\mathbf{x}_k\, \mathbf{y}_k^T\, \mathbf{z}_k = \lambda_k\, \overline{\mathbf{USV}}^T \mathbf{z}_k \tag{24}$$

and an eigenvalue problem is formed (Equation 25):

$$\overline{\mathbf{U}}^T \mathbf{x}_k\, \mathbf{y}_k^T\, \overline{\mathbf{VS}}^{-1} \mathbf{z}' = \lambda_k\, \mathbf{z}' \tag{25}$$

where $\overline{\mathbf{U}}$, $\overline{\mathbf{S}}$ and $\overline{\mathbf{V}}$ are the SVD factors truncated to the number of principal components in **M**, $\mathbf{z}' = \overline{\mathbf{SV}}^T \mathbf{z}_k$, $\overline{\mathbf{U}}^T \mathbf{x}_k\, \mathbf{y}_k^T\, \overline{\mathbf{VS}}^{-1}$ is square, and $\mathbf{z}'$ is the eigenvector to be found. The column and row profiles, $\mathbf{y}_k$ and $\mathbf{x}_k$, may be found (Equation 26):

$$\mathbf{y}_k = (\overline{\mathbf{VS}}^{-1} \mathbf{z}')^{T+}; \qquad \mathbf{x}_k = \mathbf{N} \mathbf{z}_k\, c_{n,k} \tag{26}$$

This formulation of RAFA (minus the use of Equation 26), when introduced, decreased computation time, and it also overcame a problem that had existed with the original version. In using that particular method, it was necessary to decide which eigenvalues should be examined to find the minimum which indicates rank annihilation; the use of an eigenvalue problem eliminates that need.

For references see page 9762

The form of RAFA that is generally applied today is called generalized RAFA, also known as GRAM.[4,47] This technique allows the simultaneous quantitation of multiple analytes, whereas the earlier RAFA formulations permit the quantitation of just one analyte at a time. Again, one standard matrix, **N**, is required for the calibration of one sample matrix, **M**. However, in this case, the standard and the test samples may contain unknown interferents that will not affect the calibration; this is the second-order advantage previously mentioned.

The first step that must be taken in a GRAM analysis is to find joint basis sets for the analyte signals in **M** and **N**. The fact that interferents may be present in either **M** or **N** means that projection of one matrix on to the principal components of the other (as in Lorber's method, Equation 25) will change the information present and result in an inaccurate calibration. GRAM overcomes this by finding the principal components from both matrices. There have been a few suggestions on the best way to calculate the principal components: Sanchez and Kowalski proposed calculating them from the sum of **M** and **N**,[4] thereby producing (Equation 27):

$$\mathbf{W} = \mathbf{M} + \mathbf{N} = \mathbf{U}_W\mathbf{S}_W\mathbf{V}_W^T \tag{27}$$

where $\mathbf{U}_W$ and $\mathbf{V}_W$ are the principal components of the joint row and column spaces. Wilson et al. found the principal components as[47] (Equation 28):

$$\mathbf{M}|\mathbf{N} = \mathbf{QSV}^T \tag{28a}$$

$$\frac{\mathbf{M}}{\mathbf{N}} = \mathbf{PSV}^T \tag{28b}$$

M|N is obtained by concatenating the matrices row-wise, whereas **M/N** is obtained through the column-wise concatenation of **M** and **N**. **Q** and **P** are the joint row and column spaces found by this technique.

Following the principal component calculation, the eigenvalue problem must be formulated. There are a number of ways to calculate the eigenvalues and eigenvectors from which the concentration ratios and row and column profiles are determined. Either a standard eigenproblem or a generalized eigenproblem may be solved. The standard eigenproblem is similar to Lorber's formulation for a single component, as it involves projecting the test sample matrix onto the principal components of the calibration matrix, and is shown using the singular vectors from Equation (27) (Equation 29):

$$(\overline{\mathbf{U}}_W^T\mathbf{M}\overline{\mathbf{V}}_W\overline{\mathbf{S}}_W^{-1})\mathbf{Z}'_W = \mathbf{Z}'_W\mathbf{\Lambda} \tag{29}$$

where **U**, **S** and **V** calculated in Equation (27) have been truncated to the number of principal components present in **W** and $\mathbf{Z}'_W = \overline{\mathbf{S}}_W\overline{\mathbf{V}}_W^T\mathbf{Z}$ and **Z** is a matrix of eigenvectors. In this case, $\mathbf{\Lambda}$ is a diagonal matrix that consists of the concentration ratios for all of the components common to **M** and **N**. It should be noted that the ratios that result from this formulation are (Equation 30):

$$\lambda_k = \frac{c_{m,k}}{c_{n,k} + c_{m,k}} \tag{30}$$

where $c_{n,k}$ and $c_{m,k}$ are the concentrations of analyte k in **N** and **M**, respectively. This means that if a component is not present in the calibration sample (**N**), $\lambda_k = 0$, while $\lambda_k = 1$ if a component is not present in the unknown sample (**M**). From this fact, the analyst can determine which components are present in either or both samples. The row and column profiles of each analyte common to both matrices may also be found (Equation 31):

$$\mathbf{Y}^T = (\overline{\mathbf{V}}_W\overline{\mathbf{S}}_W^{-1}\mathbf{Z}'_W)^+; \qquad \mathbf{X} = \overline{\mathbf{U}}_W\mathbf{Z}'_W\mathbf{C}_m^{-1} \tag{31}$$

An alternative formulation of GRAM uses the generalized eigenvalue problem.[47] It has a similar form to Equation (22), but in this case there are multiple analytes in both calibration matrices (Equation 32):

$$\mathbf{MZ} = \mathbf{NZ\Lambda} \tag{32}$$

where **Z** is a matrix of eigenvectors and $\mathbf{\Lambda}$ is a diagonal matrix of eigenvalues. If **M** and **N** are square, this equation can be solved directly using the QZ algorithm. However, **M** and **N** are generally rectangular matrices, and therefore a projection operation is needed to make them square. The principal components for the joint row and column spaces from either Equation (27) or (28) may be used, and this then leads to Equations (33a) and (33b):

$$\overline{\mathbf{M}}_{PQ} = \overline{\mathbf{P}}^T\mathbf{M}\overline{\mathbf{Q}} \tag{33a}$$

$$\overline{\mathbf{N}}_{PQ} = \overline{\mathbf{P}}^T\mathbf{N}\overline{\mathbf{Q}} \tag{33b}$$

where $\overline{\mathbf{M}}_{PQ}$ and $\overline{\mathbf{N}}_{PQ}$ are square and $\overline{\mathbf{P}}$ and $\overline{\mathbf{Q}}$ are truncated principal components from Equation (28). The eigenproblem can then be rewritten as Equation (34):

$$\overline{\mathbf{M}}_{PQ}\mathbf{Z}_Q = \overline{\mathbf{N}}_{PQ}\mathbf{Z}_Q\mathbf{\Lambda} \tag{34}$$

where $\mathbf{Z}_Q = (\mathbf{Y}_Q^T)^+$. $\mathbf{\Lambda}$ is a diagonal matrix that contains eigenvalues of the form $c_{k,m}/c_{k,n}$. Use of the QZ algorithm has the advantage that it yields two matrices α and β where $\Lambda = \alpha/\beta$, rather than calculating $\mathbf{\Lambda}$ directly. This means that there is no danger of running into difficulties when the calibration matrix, **N**, is missing an analyte that is present in **M**. With the original method (Equation 29), this scenario would have led to a calculation of 1 divided by 0. The analyte concentrations in the test matrix, **M**, may be calculated from this relation if the concentrations in **N** are known, and vice versa. As with the previous description of GRAM above, the row and column profiles can then be found (Equation 35):

$$\mathbf{Y} = \overline{\mathbf{Q}}(\mathbf{Z}^{+})^{T}; \qquad \mathbf{X} = \overline{\mathbf{P}}(\mathbf{N}_{PQ} + \mathbf{M}_{PQ})\mathbf{Z} \qquad (35)$$

The use of the generalized eigenvalue problem and QZ algorithm is preferable in situations where construction of the factor space has not been performed with maximum precision.[48]

There is another modification to the GRAM algorithm that in some instances can prevent the occurrence of complex solutions. It applies to the GRAM formulation that uses the QZ algorithm, and simply involves projecting the sum of **M** and **N** (i.e. **W**) rather than **M** alone when formulating the eigenproblem.[49] This makes the QZ formulation more similar to the original Sanchez–Kowalski method. In the case of the generalized eigenvalue problem, Equation (33a) becomes Equation (36):

$$\overline{\mathbf{W}}_{PQ} = \overline{\mathbf{P}}^{T}\mathbf{W}\overline{\mathbf{Q}} \qquad (36)$$

and Equation (34) is rewritten as Equation (37):

$$\overline{\mathbf{W}}_{PQ}\mathbf{Z}_{Q} = \overline{\mathbf{N}}_{PQ}\mathbf{Z}_{Q}\mathbf{\Lambda} \qquad (37)$$

such that $\mathbf{\Lambda}$ now contains eigenvalues as per Equation (30) rather than the $c_{k,m}/c_{k,n}$ ratios that are obtained from the eigenproblem in Equation (34). The advantage of this variant is that it is numerically more stable, because the eigenvectors that are calculated are common to both of the matrices under analysis, **W** and **N**. In cases where the analyte profiles of the common components in **M** and **N** vary, such as in situations where retention time shifts are present, the use of this version avoids the complex solutions that can result from dissimilarities in the matrices analyzed.

The utility of GRAM is demonstrated in an application that quantitates ethylbenzene and *m*-xylene, in a white gas mixture that was analyzed by GC × GC.[50] This technique produces bilinear data matrices, and the samples analyzed together give a trilinear structure, allowing GRAM to be used. Data analysis was carried out only on the section of data that contained the analytes of interest; the ethylbenzene and *m*-xylene responses were severely overlapped, with a resolution of 0.46 on the first axis of the gas chromatogram and 0.2 on the second axis: these data can be seen in Figure 8.

Five samples with varying concentrations of the analytes were used as test samples in the analysis, and a sixth sample was available as a standard. Despite the low resolution of the analytes, use of GRAM produced the pure compound elution profiles shown in Figure 9(a) and (b), in addition to the quantitative results shown in Table 5.

It can be seen that, for most of the samples, the GRAM results compare favorably with those of the reference method (where the gasoline samples were run on a conventional one-dimensional gas chromatograph so that

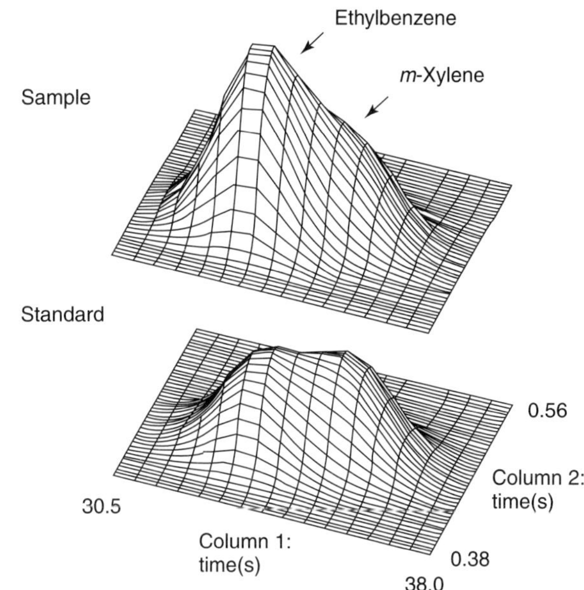

Figure 8 GC × GC responses for two mixtures of ethylbenzene and *m*-xylene. The sample is set no. 4 and the standard is set no. 6 as defined in Table 5. (Reprinted with permission from C.A. Bruckner, B.J. Prazen, R.E. Synovec, *Anal. Chem.*, **70**, 2796–2804 (1998). Copyright 1998 American Chemical Society.)

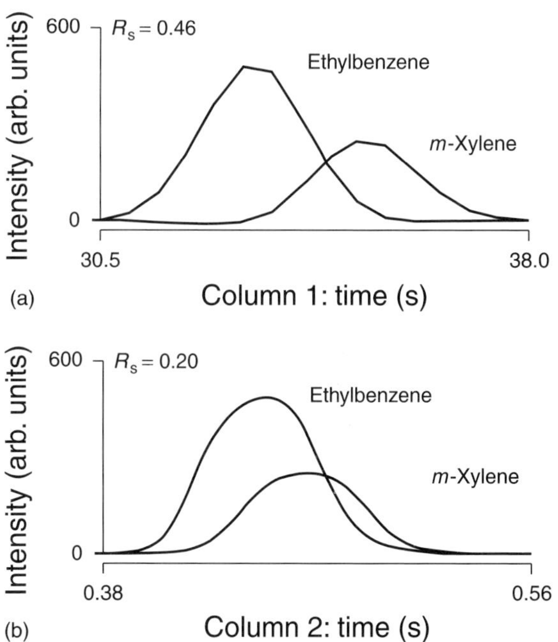

Figure 9 Resolved profiles of ethylbenzene and *m*-xylene on (a) the first GC column and (b) the second GC column after application of GRAM. (Reprinted with permission from C.A. Bruckner, B.J. Prazen, R.E. Synovec, *Anal. Chem.*, **70**, 2796–2804 (1998). Copyright 1998 American Chemical Society.)

Table 5 GRAM concentration estimates for ethylbenzene and *m*-xylene, along with the relative standard deviation in each case. (Reprinted with permission from C.A. Bruckner, B.J. Prazen, R.E. Synovec, *Anal. Chem.*, **70**, 2796–2804 (1998). Copyright 1998 American Chemical Society.)

Set no.	Ethylbenzene (%, w/w)		*m*-Xylene (%, w/w)	
	Ref.[a]	GRAM[b] (RSD, %)[c]	Ref.[a]	GRAM[b] (RSD, %)[c]
1	0.031	0.024(7.8)	0.090	0.094(2.9)
2	1.10	1.09 (4.1)	0.090	0.097(5.0)
3	0.031	0.016(42)	1.11	1.08 (2.7)
4	2.20	2.19 (5.9)	1.11	1.07 (3.0)
5	1.12	1.06 (5.9)	2.13	2.20 (4.7)
6	1.11[d]		1.12[d]	

[a] Quantifying resolved peaks by peak area using a reference single-column gas chromatography (GC) method.
[b] Quantifying incompletely resolved components in GC × GC by GRAM.
[c] RSD is the relative standard deviation of predictions, expressed as a percentage.
[d] Set 6 used as standard. Compound concentrations were determined gravimetrically during sample preparation.

all components were fully resolved). For the samples with low concentrations, however, the prediction error is fairly high. For ethylbenzene in sample 1, for example, the prediction is 22% lower than that of the reference method. The reason for this may be due to the large difference in ethylbenzene concentration between the sample and standard, as the level of ethylbenzene in the standard was 35 times greater than that in the sample, implying that analyte concentrations in the sample and standard should be similar in order to obtain accurate and precise quantitation with GRAM.

As was mentioned previously, complex results from GRAM are occasionally obtained. In fact, the eigenvalues and eigenvectors from the generalized eigenvalue problem are guaranteed to be real only when **M** and **N** are positive definite and nonsingular. As they are generally real, unsymmetrical rectangular matrices, complex solutions will result at times. In particular, when the shape and retention time of any analyte vary between **M** and **N**, perhaps owing to the presence of a high-noise signal, or at other times to variations in instrumental conditions, complex results are produced. In such cases, the imaginary part of the estimated row and column profiles is too large to be attributed to round-off error. However, when such a problem arises, there is a solution in the form of a similarity transform.[51]

The success of the similarity transform is due to the fact that it can rotate the imaginary part of the complex solution to the real plane, and this can be achieved when complex eigenvalues and their corresponding eigenvectors occur in conjugate pairs. However, because eigenvalues and eigenvectors from the QZ algorithm do not always occur in this fashion, a preliminary step is required before the transform is applied. This first step following the solution of the eigenproblem is to order the eigenvalues and eigenvectors in descending order: the complex conjugate pairs of eigenvalues are grouped together by this operation. A preliminary transform, $\mathbf{T}_1$, is then carried out to group real eigenvalues with their corresponding eigenvectors. The direction of a complex number in the complex plane can be changed by multiplication with $e^{i\theta}$ without changing its magnitude. When eigenvalues occur in complex conjugate pairs but eigenvectors do not, it is therefore possible to find two arguments α and β that transform the eigenvectors simultaneously to complex conjugate pairs. The elements of $\mathbf{T}_1$ that perform this operation are $e^{i\alpha}$ and $e^{i\beta}$. The operation on the eigenvalues and eigenvectors is (Equation 38a and 38b):

$$\mathbf{Z}^* = \mathbf{Z}(\mathbf{T}_1)^{-1} \tag{38a}$$

$$\mathbf{\Lambda}^* = \mathbf{T}_1\mathbf{\Lambda}(\mathbf{T}_1)^{-1} \tag{38b}$$

Finally, a second transform (the similarity transform, $\mathbf{T}_2$) is found which transforms the complex eigenvalues and eigenvectors to real eigenvalues and eigenvectors. The elements of $\mathbf{T}_2$ for a complex conjugate pair are (Equation 39):

$$\begin{bmatrix} t_{2_{j,j}} & t_{2_{j,j+1}} \\ t_{2_{j+1,j}} & t_{2_{j+1,j+1}} \end{bmatrix} = \begin{bmatrix} 1 & 1 \\ 1 & -1 \end{bmatrix} \tag{39}$$

The $\mathbf{T}_2$ matrix is applied to the eigenvalues and eigenvectors that are produced by the first operation displayed in Equation (38), to give Equations (40a) and (40b):

$$\mathbf{Z}^{**} = \mathbf{Z}^*(\mathbf{T}_2)^{-1} \tag{40a}$$

$$\mathbf{\Lambda}^{**} = \mathbf{T}_2\mathbf{\Lambda}^*(\mathbf{T}_2)^{-1} \tag{40b}$$

The effectiveness of this transform has been shown in the determination of unbound hydrocortisone in urine. LC/DAD was used to measure a real urine sample in addition to a standard hydrocortisone sample. When these data were analyzed by GRAM, complex values appeared in both the eigenvectors, **V**, and the eigenvalues, **Λ** (see Table 6) thereby requiring the use of the transform.

Following this procedure, the imaginary parts of the eigenvalues and the eigenvectors had disappeared, as can be seen in Table 7.

The validity of the results is evident upon examination of the true hydrocortisone concentration ratio (33.3%) versus that produced by the modified GRAM algorithm (34.6%). The source of the imaginary numbers in this particular example may be due in part to the large rank of the system; owing to the complicated nature of the urine sample, it was found that six principal components were required to describe the data. As mentioned previously,

Table 6 Eigenvectors, **V**, and eigenvalues, **Λ**, obtained from GRAM before application of the similarity transform. (Reprinted with permission from S. Li, J.C. Hamilton, P.J. Gemperline, *Anal. Chem.*, **64**, 599–608 (1992). Copyright 1992 American Chemical Society.)

Parameter	Value					
V	0.32051	0.03541 + 0.01033*i*	0.01033 + 0.03541*i*	−0.08109	−0.00536	−0.04901
	0.53393	−0.01334 + 0.01781*i*	0.01781 − 0.01334*i*	0.04569	0.02561	0.34098
	−0.62034	0.03839 + 0.04811*i*	−0.04811 + 0.03839*i*	−0.26059	0.03452	0.76677
	−0.03676	0.43400 − 0.37987*i*	−0.37987 + 0.43400*i*	0.63946	0.02958	0.47519
	0.44596	0.22458 + 0.74070*i*	0.74070 + 0.22458*i*	−0.28228	−0.23033	−0.22309
	−0.16473	−0.07732 − 0.23799*i*	−0.23799 − 0.07732*i*	0.65942	−0.97170	−0.13352
Λ	490.86	0	0	0	0	0
	0	16.9 − 18.54*i*	0	0	0	0
	0	0	16.90 + 18.54*i*	0	0	0
	0	0	0	−4.30	0	0
	0	0	0	0	0.81	0
	0	0	0	0	0	0.35

Table 7 After using the similarity transform, the eigenvectors, $\mathbf{V}^{**}$, and the eigenvalues, $\mathbf{\Lambda}^{**}$, no longer have imaginary parts. (Reprinted with permission from S. Li, J.C. Hamilton, P.J. Gemperline, *Anal. Chem.*, **64**, 599–608 (1992). Copyright 1992 American Chemical Society.)

Parameter	Value					
$\mathbf{V}^{**}$	0.32051	−0.01864	−0.03183	−0.08109	−0.00536	−0.04901
	0.53393	−0.01402	−0.01728	0.04569	0.025661	0.34098
	−0.62034	0.03731	−0.04895	−0.26059	0.03452	0.76677
	−0.03676	0.26265	−0.51349	0.63946	0.02958	0.47519
	0.44596	−0.77310	−0.03731	−0.28228	−0.23033	−0.22309
	−0.16473	0.24966	0.01699	0.65942	−0.97170	−0.13352
$\mathbf{\Lambda}^{**}$	490.86	0	0	0	0	0
	0	16.90	18.54	0	0	0
	0	−18.54	16.90	0	0	0
	0	0	0	−4.3	0	0
	0	0	0	0	0.81	0
	0	0	0	0	0	0.35

slight retention time shifts and the presence of noise can cause complex results to be produced by GRAM, and in a case such as this where the samples contain many components, the adverse effects of shifts can be exacerbated. This is especially true if the shifts are nonlinear.

GRAM theory has been discussed at length in various publications since its introduction.[48,52,53] Degenerate solutions, for example, can arise when two components have the same eigenvalues and, in such a case, the corresponding eigenvectors do not have a fixed direction.[52] In calibration, this case can occur when the ratio of analyte A (between samples **M** and **N**) is the same as the concentration ratio of analyte B. In such a situation, the profiles determined by GRAM are not unique. If each of the signals from A and B are unimodal chromatographic peaks, both profiles will appear to be bimodal. Closer inspection will reveal that both profiles display maxima/minima at the retention times of A and B, as they are in fact linear combinations of these analytes. However, quantitation in this situation is not affected, as the ratios of the analytes are still valid. This has been demonstrated through the use of both GRAM and trilinear decomposition (TLD) on simulated emission–excitation data.[5]

Effects of model errors in GRAM have also been investigated.[53–55] Matrix effects can be minimized through the use of standard addition,[11] while interaction effects cannot (because a cross-term between concentrations is present in such a case). GRAM is very sensitive to retention time shifts, as stated above, thereby producing a bias in the estimated concentration ratios. A theoretical study of bias and variance in the estimated eigenvalues in a case where heteroscedastic noise exists found that a negative bias is generally obtained,[54] and these results have also been shown with realistic simulations.[56] A similar study in the case of homoscedastic noise, however, found a positive error to occur.[53] The agreement of the study with heteroscedastic noise to the simulation lends credence to the theory that the presence of homoscedastic noise is an unrealistic assumption, and does not adequately describe real systems.

For references see page 9762

GRAM can be extended to allow a TLD, also known as DTD.[5,10] GRAM is actually the simplest case of a third-order tensor, with only two slices in the third-order. When more than two slices are present, DTD can be applied to project all slices to just two, followed by SVD which finds a joint row, column and sample space.

The input to the algorithm is an $I \times J \times K$ block of data. The data are unfolded along each order, and in each case the resulting matrices are decomposed by SVD, as mentioned above. The left singular vectors for the first n principal components are retained from the decomposition of the joint row and column spaces, but in the case of the joint sample space, only the first two singular vectors are retained, to give Equations (41a–c):

$$[\mathbf{R}_1|\mathbf{R}_2|\mathbf{R}_3\cdots|\mathbf{R}_K] = \mathbf{U}_r\mathbf{S}_r\mathbf{V}_r^T \qquad \mathbf{U} = \mathbf{U}_r(I \times N) \quad (41a)$$

$$[\mathbf{R}_1|\mathbf{R}_2|\mathbf{R}_3\cdots|\mathbf{R}_I] = \mathbf{U}_c\mathbf{S}_c\mathbf{V}_c^T \qquad \mathbf{V} = \mathbf{U}_c(J \times N) \quad (41b)$$

$$[\mathbf{R}_1|\mathbf{R}_2|\mathbf{R}_3\cdots|\mathbf{R}_J] = \mathbf{U}_w\mathbf{S}_w\mathbf{V}_w^T \qquad \mathbf{W} = \mathbf{U}_w(K \times 2) \quad (41c)$$

Two representative matrices are then generated using Equations (42a) and (42b):

$$\mathbf{G}_1 = \sum_{k=1}^{K} w_{k,2}\mathbf{U}^T\mathbf{R}_k\mathbf{V} \quad (42a)$$

$$\mathbf{G}_2 = \sum_{k=1}^{K} w_{k,1}\mathbf{U}^T\mathbf{R}_k\mathbf{V} \quad (42b)$$

The first two columns of $\mathbf{W}$ are suitable for calculating representative matrices because they are the "scores" of the first two principal components, and are therefore likely to contain contributions from all components. GRAM is then carried out on $\mathbf{G}_1$ and $\mathbf{G}_2$ to find $\mathbf{X}$, $\mathbf{Y}$ and $\mathbf{Z}$, where $\mathbf{Z}$ is generally the concentration order. The generalized eigenproblem that is solved is[10] (Equation 43):

$$\mathbf{G}_1\boldsymbol{\Psi} = \mathbf{G}_2\boldsymbol{\Psi}\boldsymbol{\Lambda} \quad (43)$$

$\mathbf{X}$ and $\mathbf{Y}$ are then determined from the calculated eigenvectors (Equation 44):

$$\widehat{\mathbf{Y}} = \mathbf{V}(\boldsymbol{\Psi}^{\mathbf{T}})^{-1} \qquad \widehat{\mathbf{X}} = \mathbf{U}\mathbf{G}_2\boldsymbol{\Psi} \quad (44)$$

The concentrations in $\mathbf{Z}$ are determined by least squares. For each slice in the instrument response, $\mathbf{R}_k$, $\mathbf{Z}$ is calculated to minimize the error term (Equation 45):

$$\mathbf{R}_k = \sum_{n=1}^{N} \mathbf{z}_n\mathbf{x}_n\mathbf{y}_n^T + \mathbf{E}_k \quad (45)$$

The full $\widehat{\mathbf{Z}}$ $(K \times N)$ matrix can be calculated simultaneously by Equation (46):

$$\widehat{\mathbf{Z}} = \mathbf{P}\mathbf{Q}^{-1} \quad (46)$$

where (Equation 47):

$$p_{kn} = \sum_{i=1}^{I}\sum_{j=1}^{J} r_{ijk}\,\hat{x}_{in}\,\hat{y}_{jn} \quad (47)$$

and (Equation 48):

$$\mathbf{Q} = (\widehat{\mathbf{X}}\widehat{\mathbf{X}}^T)\cdot(\widehat{\mathbf{X}}\widehat{\mathbf{X}}^T) \quad (48)$$

where the $\cdot$ symbol signifies element-by-element multiplication.

The original DTD algorithm was slightly different, in that there were two eigenproblems to be solved,[10] where $\widehat{\mathbf{Y}}$ was calculated from Equation (49):

$$\mathbf{G}_1\boldsymbol{\Psi}_a = \mathbf{G}_2\boldsymbol{\Psi}_a\boldsymbol{\Lambda}_a \quad (49)$$

and $\widehat{\mathbf{X}}$ from Equation (50):

$$\mathbf{G}_1^T\boldsymbol{\Psi}_b = \mathbf{G}_2^T\boldsymbol{\Psi}_b\boldsymbol{\Lambda}_b \quad (50)$$

However, the resulting $\mathbf{X}$ and $\mathbf{Y}$ are often mismatched when this version is implemented, and therefore it is advised to calculate these parameters using a single eigenproblem. In other words, the nth eigenvector in Ψ_a does not always correspond to the nth eigenvector in Ψ_b. This has been demonstrated with two samples of simulated LC/DAD data, where the test sample contained the analyte of interest plus four interfering species, and the calibration standard contained only the analyte of interest.[10] The true relative analyte concentration in this case was 1.0000, and Table 8 shows the results that were obtained with different formulations of TLD in addition to the result given by GRAM. The prediction given by the original TLD algorithm (Equations 49 and 50) is low by 2%, owing to mismatched eigenvalues and

Table 8 Concentration estimates from GRAM and various formulations of TLD. (Reprinted with permission from K.S. Booksh, Z. Lin, Z. Wang, B.R. Kowalski, *Anal. Chem.*, **66**, 2561–2569 (1994). Copyright 1994 American Chemical Society.)

True	GRAM[a]	TLD[b]	TLD [Equation 49]	TLD [Equation 50]
1.0000	1.0009	0.9802	1.009	1.009

[a] From Li et al.[51]
[b] Using the algorithm from Sanchez and Kowalski.[5]

eigenvectors; however, when they are rearranged so that they correspond, the error, although improved, is still higher than that given by GRAM plus the similarity transform. This is because, even though the pair of eigenvectors (ψ_{a1} and ψ_{a2}) calculated from Equation (49) span the same space as the pair of eigenvectors (ψ_{b1} and ψ_{b2}) calculated from Equation (50), different linear combinations of the eigenvectors yield different intrinsic profile estimates. In other words (Equation 51):

$$\hat{\mathbf{x}}_a \neq \hat{\mathbf{x}}_b \quad \text{and} \quad \hat{\mathbf{y}}_a \neq \hat{\mathbf{y}}_b \tag{51}$$

Therefore, Equation (52):

$$\mathbf{R}_1 + \mathbf{R}_2 \neq \hat{\mathbf{x}}_{b1}\hat{\mathbf{y}}_{a1} + \hat{\mathbf{x}}_{b2}\hat{\mathbf{y}}_{a2}^T \tag{52}$$

On the other hand, when Equations (43) and (44) are used, it is found that the quantitation results are identical with GRAM, and the prediction error is only 0.09%.

Similarly to GRAM, DTD occasionally has problems with complex solutions. This has been addressed with the use of similarity transforms, as with GRAM.[57] In this case, the transform is carried out on $\boldsymbol{\Psi}$ and $\boldsymbol{\Lambda}$ from the eigenvalue problem. However, the similarity transform, when applied to DTD decomposition factors, contains a rotational ambiguity. This means that when eigenvectors have occurred in complex pairs, the resulting chromatograms and spectra are not uniquely identified. The use of similarity transforms to eliminate the problem of complex solutions is therefore not as widely applicable to DTD data as to GRAM.

The rank annihilation methods have been applied in many different situations. GRAM has been successfully applied to LC/UV data, and also to bimodal chromatography data.[58,59] A unique application involved nuclear magnetic resonance (NMR) data, where a single multicomponent pulsed-gradient spin-echo nuclear magnetic resonance (PGSENMR) data set was analyzed by GRAM.[60] This was possible because PGSENMR data consist of exponential decaying profiles, and the authors were able to "create" two data sets from a single one by using spectra 1 to $n-1$ for the first set and spectra 2 to n for the second set. Despite the highly overlapped nature of the data, GRAM could resolve the spectra of the two components present. Another extension or application of DTD is direct trilinear decomposition with matrix reconstruction (DTDMR).[61] This method is suitable for cases where one of the responses is actually a matrix. GRAM and DTD have both been used to predict heavy metal concentrations, from data obtained from a second-order fiber-optic heavy metal sensor.[62] The signals in each order were time and wavelength, respectively, and unknown interferents were present. The GRAM and DTD predictions both gave improvements over a zero-order analysis, which gave a positive bias. It was found, however, that when the sensor response deviated from linearity, the prediction errors increased, as is to be expected from these methods when the trilinear model is not held. In a separate study, DTD was used to resolve mathematically porphyrins from emission–excitation fluorescent spectra of canine and feline dental calculus deposits.[63] The small quantity of samples in that study had made physical separation and analysis unfeasible.

Although the trilinear model is the most commonly employed model for second-order calibration, there are cases where data do not fit this model. For example, when an instrument response is such that a pure analyte has a rank of >1, the data are nonbilinear. In such a case, application of any of the methods that specifically solve that type of data will not be of benefit, and may in fact result in an unreliable calibration. NBRA is a method that enables calibration of nonbilinear data when the response matrix of the pure analyte is available as a calibration standard.[6,64] The method is based on the model in Equation (53):

$$\mathbf{N} = \sum_{p=1}^{P} \mathbf{x}_p c_p \mathbf{y}_p^T \tag{53}$$

where $\mathbf{N}$ is a data matrix of a single pure component from a second-order instrument that produces nonbilinear data, P is the total number of detected components in the mixture, $\mathbf{x}$ and $\mathbf{y}$ are factors relating to the response profiles and c is related to concentration. The factors can be approximated using any three-way decomposition method, such as an eigenvalue–eigenvector decomposition, but it must be noted that the bilinear components from the decomposition do not contain the chemical and physical meaning that they do for bilinear data. (Approximating nonbilinear data using an eigenvalue–eigenvector decomposition is analogous to using PLS or PCR to approximate nonlinear first-order data with extra factors.)

It can be seen, therefore, that the greatest difference between nonbilinear and bilinear data lies in P, the number of factors in the data. For a pure component in a bilinear response matrix, there is one detectable factor present, as previously explained and shown in Equation (1). In other words, the rank of the matrix in the absence of noise is 1. On the other hand, the rank in a nonbilinear response matrix for a pure component is >1, so that when a second-order decomposition is carried out more than one concentration estimate is obtained.

A figure of merit that may be calculated to determine the reliability of the calibration on a particular data set is the rank linear additivity (RLA). It can be said that RLA holds if the rank of a mixture, $\mathbf{M}$, is equal to the sum of the ranks of individual components in the mixture. In that case, different components do not overlap with respect to

For references see page 9762

their rank. For example, if **M** consists two components, a and b, whose instrument responses can be represented as in Equation (54a):

$$\mathbf{M}_a = c_a(\mathbf{x}_1\mathbf{y}_1^T + \mathbf{x}_2\mathbf{y}_2^T) \tag{54a}$$

and Equation (54b):

$$\mathbf{M}_b = c_b(\mathbf{x}_3\mathbf{y}_3^T + \mathbf{x}_4\mathbf{y}_4^T) \tag{54b}$$

where c_a and c_b are the concentrations of the components, and (Equation 55):

$$\mathbf{M} = \mathbf{M}_a + \mathbf{M}_b \tag{55}$$

then the calibration is similar to the case of bilinear components with identical concentrations. In such a case, the response profiles are not resolved because they are linearly combined, but the concentration prediction is accurate.[10] In this case, $(\mathbf{x}_1, \mathbf{x}_2)$ and $(\mathbf{y}_1, \mathbf{y}_2)$ would be linear combinations of one another, as would $(\mathbf{x}_3, \mathbf{x}_3)$ and $(\mathbf{y}_4, \mathbf{y}_4)$, but the four concentration estimates will simply be c_a repeated twice, and similarly for c_b.

If RLA does not hold, then the calibration becomes more complicated, because the components are now overlapping in rank. For example, if (Equation 56):

$$\mathbf{M} = c_a\mathbf{x}_1\mathbf{y}_1^T + c_b\mathbf{x}_3\mathbf{y}_3^T + \mathbf{x}_2(c_a\mathbf{y}_2^T + c_b\mathbf{y}_4^T) \tag{56}$$

then the rank of **M** equals 3 rather than 4.

Here it can be seen that $\mathbf{x}_2 = \mathbf{x}_4$, so that these factors, from different components, are not independent. What will happen when a decomposition is carried out? Will it be possible to calculate the concentrations of these analytes? In a case such as this, it is important to calculate the net analyte rank (NAR).

NAR may be defined as (Equation 57):

$$\text{NAR}(\mathbf{N}) = rank(\mathbf{M}) - rank(\mathbf{M}|\overline{\mathbf{N}}) \tag{57}$$

where $rank(\mathbf{M}|\overline{\mathbf{N}})$ is the rank of mixture **M** in the absence of the pure analyte **N**. If the NAR is equal to at least one, then calibration is still possible: it has been determined that NBRA retains the second-order advantage as long as the "NAR" of the data is ≥ 1.[65] In this case, shown in Equation (54), if the NAR is calculated for component a, then $\text{NAR} = 3 - 2 = 1$. When the decomposition is carried out, therefore, the concentration estimate corresponding to the $\mathbf{x}_1\mathbf{y}_1$ factor will be c_a. The same situation holds for analyte b in this case. There is another way to determine the correct concentration estimate when there is just one analyte in the calibration sample, **N**. This procedure requires NBRA to be repeated using different multiples of N in place of N $(\varepsilon \times N)$, and the resulting concentrations which are proportional to $1/\varepsilon$ are chosen as the correct concentration estimates. The corresponding $\mathbf{x}_i$, $\mathbf{y}_i$ pairs then comprise the net analyte signal.

Whether or not the data can be calibrated depends considerably on the instrument on which the data is collected. If the instrument produces data whose nonlinearities are polynomial in nature, then the data in general can be described by a low rank, and the NAR will generally be >1. If the nonlinearities are more complex, the rank will be higher and the bilinear model is less likely to be able to model the data. In such cases, the rank of a single component will be very high, and it is likely that there will be no NAR, thereby eliminating the ability to perform calibration. It has been found that MS/MS data generally fall into the first category, so that calibration of MS/MS data is possible with NBRA.

The NBRA method has been applied to two-dimensional NMR and to MS/MS data.[66] It was found that whereas NBRA tends to have a negative bias, the quantitative results were better than those provided by a first-order technique, multiple linear regression (MLR). The NMR data were for a six-component sugar mixture, while the MS/MS data were for warfarin, 3′-, 4′-, 5-, 6- and 7-hydroxywarfarin plus phenylbutazone. The average quantitation error over the test sets was 2.5%.

3.3 Multiway Partial Least Squares

PLS is a calibration technique that is widely used on first-order data. It is a method for building regression models between independent and dependent variables. Because of its success with first-order data, there has been interest in extending this technique to the calibration of second-order data.

The most obvious procedure might be to take the three-way data block, unfold it into a matrix and continue with regular PLS. However, this process has no advantage over two-way methods as far as prediction errors are concerned.[67] A second approach that may be taken is to divide the data block into two sets, one of which is the predictor set, the other being the test set. The model being estimated by Unfold-PLS is that represented by Equation (9a–c). However, because two sets of data are used, both must be modeled as Equation (58a) and (58b):

$$r_{ijk} = \sum_{n=1}^{N} x_{in}\, h_{njk} + e_{r,ijk} \tag{58a}$$

$$s_{ijk} = \sum_{n=1}^{N} x_{in}\, c_{njk} + e_{s,ijk} \tag{58b}$$

where s_{ijk} is an element of the dependent variable data set, $\underline{\mathbf{S}}$, and r_{ijk} is an element of the independent variable set, $\underline{\mathbf{R}}$. The model estimates the x_{in}, h_{njk} and c_{njk} parameters such that a compromise is made between minimizing $\sum\sum\sum e_{r,ijk}^2$ and $\sum\sum\sum e_{s,ijk}^2$, by maximizing the covariance of $\mathbf{x}_n$ and a linear combination of the variables in $\underline{\mathbf{S}}$.[(68)]

A third PLS method is named n-PLS, so-called because it is applicable to second-order and higher data.[(69,70)] In the case of three-way data, it is termed tri-PLS. PLS actually consists of two steps. In the first, the data array is decomposed, and in the second, a relationship between the independent ($\underline{\mathbf{R}}$) and dependent ($\mathbf{y}$) variables is established. While the array decomposition is trilinear, the part of $\underline{\mathbf{R}}$ relevant for describing $\mathbf{y}$ does not have to be of rank one. This enables calibration in the presence of matrix and other nonlinear effects. The tri-PLS algorithm is based on a decomposition of the data block such that the successively computed score factors have the property of maximum paired covariance with the unexplained part of the dependent variable; this constraint is not met in a trilinear sense with the other Unfold-PLS techniques. In this case, it is assumed that the dependent variables are in the form of a vector, $\mathbf{s}$, rather than a block of data. This tri-PLS model boils down to the problem of finding the weight vectors $\mathbf{w}_1$ and $\mathbf{w}_2$ that satisfy Equation (59):

$$\max_{\mathbf{w}_1\mathbf{w}_2}\left\{\operatorname{cov}(\mathbf{x},\mathbf{s})\middle|\min\left[\sum_{i=1}^{I}\sum_{j=1}^{J}\sum_{k=1}^{K}(r_{ijk}-x_i w_{1,j} w_{2,k})^2\right]\right\} \quad (59)$$

where $\mathbf{x}$, $\mathbf{w}_1$ and $\mathbf{w}_2$ correspond to the factors for each mode in the decomposition. The least-squares solution to the tri-PLS model is given by Equation (60):

$$\max_{\mathbf{w}_1,\mathbf{w}_2}\left\lfloor(\mathbf{w}_1)^T\mathbf{Z}\mathbf{w}_2\right\rfloor \quad (60)$$

where $\mathbf{Z}$ is a $J\times K$ matrix formed by $\underline{\mathbf{R}}$ and $\mathbf{s}$. The $\mathbf{w}_1$ and $\mathbf{w}_2$ vectors can be found by performing an SVD on $\mathbf{Z}$, and then the scores matrix, $\mathbf{X}$, can also be determined. The regression coefficients are calculated according to Equation (61):

$$\mathbf{b}=(\mathbf{X}^T\mathbf{X})^{-1}\mathbf{X}^T\mathbf{S} \quad (61)$$

This method can be extended to higher order arrays with an additional weight vector introduced for each extra way.

The applicability of Unfold-PLS and n-PLS can be seen in an example that determines the ash content of sugar.[(69)] In the analysis, the fluorescence of 67 samples of sugar was measured at four excitation and 63 emission wavelengths – this produced the $\underline{\mathbf{R}}$ block of data. The $\mathbf{y}$ block contained data regarding the ash content of each sample, as determined by a standard European Union method (no. 1265/09). The $\underline{\mathbf{R}}$ and $\mathbf{y}$ blocks were divided into a calibration and test set, and mean centering was carried out. The unfolded $\underline{\mathbf{R}}$ block and the $\mathbf{y}$ block are shown in Figure 10(a) and (b).

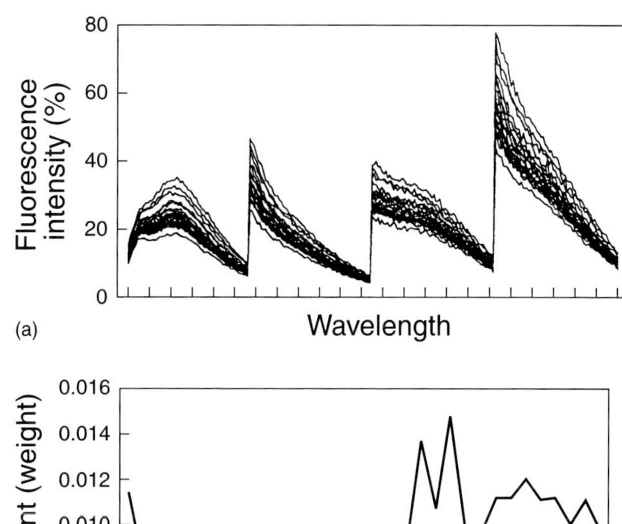

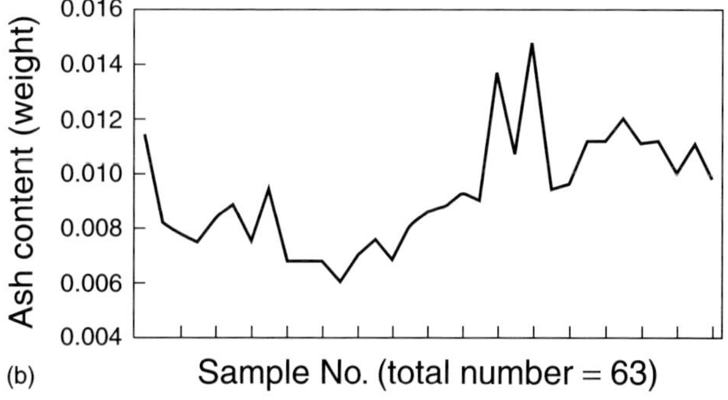

Figure 10 (a) Unfolded excitation–emission spectra from sugar samples and (b) the ash content in each sample. (Reprinted with permission from R. Bro, *J. Chemom.*, **10**, 47–61 (1996). Copyright 1996 John Wiley & Sons Limited.)

Table 9 Results from (i) Unfold-PLS and (ii) tri-PLS. (Reprinted with permission from R. Bro, *J. Chemom.*, **10**, 47–61 (1996). Copyright 1996 John Wiley & Sons Limited.)

Method	R^{2a}	RSD[b]
Tri-PLS	0.9396	6.95%
Unfold-PLS	0.9401	7.01%

[a] Correlation between concentrations and predictions.
[b] Relative standard deviation.

During the analysis, both Unfold-PLS and n-PLS used two components to describe the variations in $\mathbf{y}$, and the predictions of ash in the test sets were practically identical for both methods (see Table 9) This is due to the fact that the number of samples was large compared with the number of components. However, the difference between the methods is seen in the weight vectors, which show that the solution is much more stabilized in the case of n-PLS. (see Figures 11 and 12a and b). The Unfold-PLS weight vectors, especially that of the second component, have incorporated a lot of noise, unlike the tri-PLS vectors, which had fewer parameters to calculate.

For references see page 9762

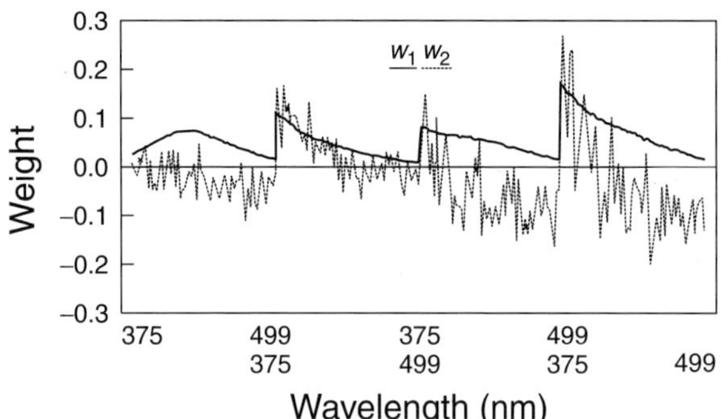

Figure 11 Weight vectors from Unfold-PLS. (Reprinted with permission from R. Bro, *J. Chemom.*, **10**, 47–61 (1996). Copyright 1996 John Wiley & Sons Limited.)

4 CALIBRATION TRANSFER/ STANDARDIZATION

Standardization and calibration transfer are important topics that must often be addressed in order to obtain accurate calibration results.

Many of the multivariate calibration methods that have been discussed have strict requirements when it comes to the characteristics of the data employed and, in particular, these methods perform better when the signal from all analytes in different samples remain constant. Often, however, this is not the case. Consider, for example, an analyte that has been analyzed by LC/MS. If that analyte is analyzed again at a later time, its profile in the new data will not be identical with that in the earlier sample. Its retention time on the chromatographic axis may have changed, its peak shape may have changed or there may be a combination of both. In order for the second-order calibration to be effective, the differences between the analyte signals in the samples must first be minimized.

The first solution to this problem was proposed by Wang and Kowalski, and was named "second-order standardization".[12] This method standardizes two-dimensional responses measured on multiple instruments, or on a single instrument under different operational conditions. This technique does not, however, address the case where analyte response is constantly changing from sample to sample.

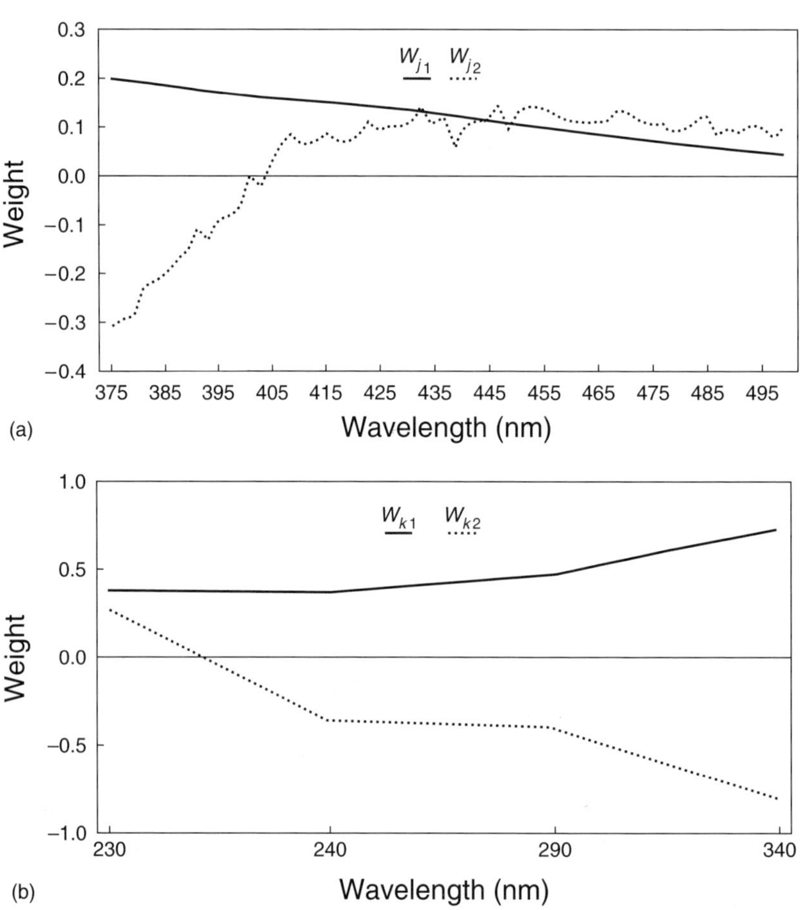

Figure 12 Weight vectors from trilinear-PLS in order of (a) excitation and (b) emission. (Reprinted with permission from R. Bro, *J. Chemom.*, **10**, 47–61 (1996). Copyright 1996 John Wiley & Sons Limited.)

Second-order standardization is related to a method that exists for first-order (two-way) data, namely piecewise direct standardization (PDS).[71] PDS relates responses of samples measured on different instruments through the use of a banded diagonal matrix, **F** (Equation 62):

$$\overline{\mathbf{R}}_1 = \overline{\mathbf{R}}_2\mathbf{F} \tag{62}$$

where $\overline{\mathbf{R}}_1$ and $\overline{\mathbf{R}}_2$ are a small transfer set of samples measured on instruments 1 and 2 ($\overline{\mathbf{R}}$ refers to a truncated data set). This method assumes that the response at each channel is related to the response in a small window surrounding that channel on the second instrument. However, in second-order data, an analyte response may vary in both orders from instrument 1 to instrument 2, requiring a second banded matrix to account for changes in the second order. The placement of the banded matrices is (Equation 63):

$$\mathbf{R}_{1,\alpha} = \mathbf{A}\mathbf{R}_{2,\alpha}\mathbf{B} \tag{63}$$

where $\mathbf{R}_{1,\alpha}$ is the response from sample α on instrument 1 and $\mathbf{R}_{2,\alpha}$ is the response on instrument 2. **A** and **B** are banded left and right transformation matrices that correct for differences between the responses: these are estimated from a set of simultaneous nonlinear equations via the Gauss–Newton method. It must be noted that the left and right transformation matrices may not be found separately, as their solutions depend upon one another.

Following the estimation of **A** and **B**, they may then be used to transfer other samples that have also been run on the second instrument, so that it appears as if they have been run on the first, as (Equation 64):

$$\mathbf{R}_{1,\beta} = \mathbf{A}\mathbf{R}_{2,\beta}\mathbf{B} \tag{64}$$

where $\mathbf{R}_{1,\beta}$ and $\mathbf{R}_{2,\beta}$ are the responses of sample β run on the first and second instruments respectively.

In this way, samples run on instrument 2 can be calibrated using samples run on instrument 1, allowing a laboratory to calibrate their samples using a calibration set from a second laboratory, for example. However, one sample must be run on both instruments, and this transfer sample must have information in all channels, or the transfer will not be optimal. This is because the banded transformation matrices must contain information at all channels that are in new samples to be transferred. An additive correction matrix may be required in addition to Equations (62–64) to correct for additive differences between $\mathbf{R}_1$ and $\mathbf{R}_2$. However, a method to find the additive matrix does not currently exist in the literature.

This technique has been tested on LC/UV data, and variation from the ultraviolet (UV) data was found to be reduced from 0.2 AU before the transfer to 0.03 AU after the standardization.

A second standardization method that has been reported for second-order data is "second-order chromatographic standardization".[13] This technique, introduced by Prazen et al., is specifically directed toward data in which one of the orders is a chromatographic separation, while the other order is not chromatographic, e.g. GC/MS. The method corrects for retention shifts between a standard sample and a calibration sample, which allows calibration then to be performed.

In the first step of this method, the pseudo-rank of the sample matrix is estimated, and the two data sets are then augmented. SVD is carried out on the augmented matrix. The remaining eigenvalues beyond the pseudo-rank of the sample are used to calculate the percentage residual variance ($\%RV$). The retention time of the samples is shifted and the process is repeated until the point is identified where the residual variance is at a minimum. The residual variance calculation is (Equation 65):

$$\%RV = 100\left(\frac{\sum_{j=n+1}^{c}\lambda_j^0}{\sum_{j=1}^{c}\lambda_j}\right)\left[\frac{c \times r}{(c-n)(r-n)}\right] \tag{65}$$

where n is the number of chemical components, λ is an eigenvalue and c and r are the dimensions of the augmented matrix.

In order to use this standardization method effectively, the pseudo-rank of the sample matrix must be determined, which is not always a straightforward matter. In addition, the test sample must contain the same analytes as the standard sample, which reduces the generality of the procedure. However, the correct use of this technique allows calibration to be carried out with greater accuracy, as was demonstrated on LC/DAD data: methyl vinyl ketone–dinitrophenylhydrazine (MVK–DNPH) in a mixture of five DNPH derivatives was measured, along with two calibration standards, each of which contained only the MVK–DNPH analyte at different concentrations. When GRAM was carried out on the mixture with each of the standards, quantitation was inaccurate (see Table 10). Upon calculation of the $\%RV$ at different time shifts, it was found that one of the analyses required a shift of 0.2 s in the retention time, while the other required a 0.3 s shift, as can be seen in Figure 13.

After the correction, GRAM was performed once again, and the new concentration estimates were drastically improved, as is evident in Table 10. This shows the importance of data quality for accurate GRAM quantitation and the value of standardization in this type of case.

For references see page 9762

Table 10 Quantitative results obtained from GRAM for two different MVK–DNPH standards, before and after standardization. (Reprinted with permission from B.J. Prazen, R.E. Synovec, B.R. Kowalski, *Anal. Chem.*, **70**, 218–226 (1998). Copyright 1998 American Chemical Society.)

Experiment	Volumetric ratio of concentrations	GRAM ratio of concentrations without standardization	GRAM ratio of concentrations with standardization	Error without standardization (%)	Error with standardization (%)
MVK–DNPH standard 1	1.44	2.14	1.67	48	15
MVK–DNPH standard 2	0.722	0.871	0.0713	21	1.3

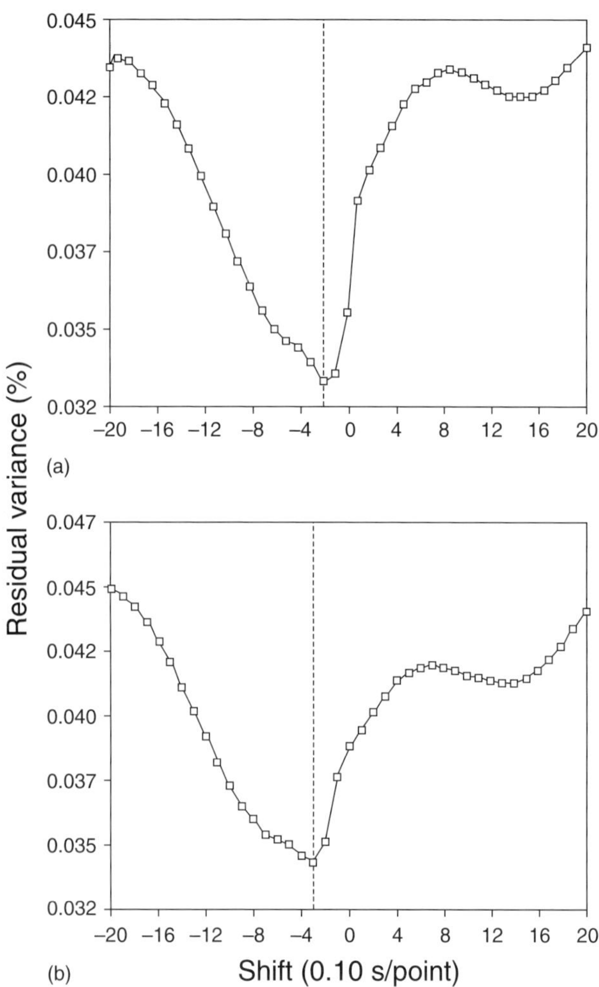

Figure 13 Standardization plots for two different MVK–DNPH analyses. The minimum shows the shift needed to align the matrices for GRAM analysis. (Reprinted with permission from B.J. Prazen, R.E. Synovec, B.R. Kowalski, *Anal. Chem.*, **70**, 218–226 (1998). Copyright 1998 American Chemical Society.)

5 STANDARD ADDITION

Standard addition is a method that is commonly employed in zero- and first-order analyses to reduce the effects of matrix effects on calibration. In order for this method to be effective for zero-order data, there are two assumptions that must be fulfilled. First, the instrument response must vary linearly with analyte concentration. Second, for an analyte response of zero, the instrument response must be zero. The extension to first-order data, generalized standard addition, involves sequential spiking of the sample with the analyte plus each interferent that is affecting the analyte signal. In other words, each source of instrumental signal must be modeled and calibrated. When the data are from an instrument that produces second-order data, however, second-order methods can be used that eliminate this constraint. While the second-order advantage allows calibration in the presence of unmodeled interferents, this is only true when the data are completely bilinear. When there are interferents present in the sample that change the way in which the instrument responds to the analyte, then the data are no longer bilinear. For example, the analyte signal may change in scale or shape depending on the other components that are present in the sample, making traditional calibration unreliable. (Traditional in this sense refers to the second-order calibration methods that rely on the trilinear model.)

A SOSAM that mathematically separates the instrument response of the analyte from the instrument response of any interfering species was proposed by Booksh et al. in order to solve this problem.[11] While trilinear data follow the trilinear model (see Equation 7) which depends on the number of chemical species in the sample (N), "fragmented" data follow the model (Equation 66):

$$R_{ijk} = \sum_{m=1}^{M} X_{im}\, Y_{jm}\, Z_{km} + E_{ijk} \qquad (66)$$

where M is the total number of detectable species formed by the N chemical components in the sample. This model is applicable in cases where two or more components in a sample mixture combine to form an additional component that can be detected. Matrix effects, on the other hand, occur when two or more sample components interact such that the profiles of the analyte depend on the concentration of the interferent in the sample. Both scenarios result in the data requiring standard addition for reliable calibration.

There are three steps to the SOSAM. The data block, $\underline{\mathbf{R}}$, must first be decomposed. This can be achieved using any of the second-order methods already mentioned, such as DTD or PARAFAC. Following this step, the N columns in the sample space of $\underline{\mathbf{R}}$ (i.e. $\mathbf{Z}$) must be regressed against the standard additions to obtain N concentration estimates. For zero-order standard addition, a least-squares model is used (Equation 67):

$$r_k = \delta_k m + b \tag{67}$$

where r_k is the instrument response for the kth sample, δ_k is the change in analyte concentration from the first to the kth sample (from the standard additions) and m and b are the slope and intercept of the regression line, respectively. Using this model, the estimated analyte concentration in the original (first) sample is given by Equation (68):

$$\hat{c} = \frac{b}{m} \tag{68}$$

For the extension to second-order data, $\mathbf{Z}$ is the equivalent of r_k in Equation (65), because whether the data are bilinear or fragmented, there will be one column of $\mathbf{Z}$ that is linearly related to the analyte of interest. Equation (69) shows how the slope and intercept may be calculated in the second-order case, so that Equation (68) may then be used to calculate the analyte concentration:

$$\hat{\mathbf{z}}_n = \delta m + b \tag{69}$$

Following the regression, a decision may have to be made as to which concentration estimate is correct. This will be the case when the data are fragmented, as there will be a number of concentration estimates, each of which will relate to a different factor. For example, there will be unique factors for both the interferent and analyte, but there may also be factors that relate to the interactions between them, in addition to factors that describe nonlinearities. The analyte concentration estimate may be determined as the smallest estimate that does not change when additional standard additions are included in $\underline{\mathbf{R}}$.

When applied to bilinear data, this algorithm is robust to shifting errors, and the bias in concentration estimates is low. Calibration in the case of fragmented data results in a low bias when the data are nondegenerate. When the algorithm is applied to degenerate data, however, the effect of random errors is increased. This may occur, for example, when two components have a common profile in one of the orders and therefore cannot be resolved.

This technique has been applied to a data set that resulted from the collection of UV diode-array spectra over a 30 min time period, over the course of a Fujiwara reaction. Trichloroethylene (TCE) in the presence of chloroform ($CHCl_3$) was quantified, and six standard additions of TCE were made. It had previously been discovered that TLD fails to predict the concentration of TCE in the presence of large quantities of $CHCl_3$ owing to an interaction between the two reagents.[72] In that case, the prediction error was found to be 29.5%. However, the data were determined to be fragmented and degenerate, therefore the model in Equation (64) applies. The prediction error was 10.8% when all six standard additions were included in the calculation, which is an improvement over the TLD error of 29.5%. Furthermore, when only the four most linear standard additions were used in the analysis, the error decreased to 2.6%, while TLD produced an error of 30.0% for the same standards.

6 FIGURES OF MERIT

Figures of merit provide a means of determining how well analytical determinations, including calibrations, perform. These performance characteristics include sensitivity, selectivity and S/N, in addition to net analyte signal. For good calibration, a high sensitivity and selectivity are desirable, as is a high S/N. A high sensitivity indicates that the analyte response changes significantly with concentration, which is important since small concentration changes may then be detected. Selectivity, on the other hand, is a measure of how well the method can distinguish the analyte in question from other analytes. This characteristic is related to the net analyte signal, which calculates the proportion of the signal unique to the analyte. A high net analyte signal translates into a high selectivity. Taken together, these characteristics can be used as criteria to decide whether or not a given instrument is suitable for attacking a particular analytical problem, or they can be used to determine the optimal set of analytical conditions for the problem.

The figures of merit mentioned above are well defined for zero-order data: the net analyte signal can be obtained simply be performing a background subtraction, while the sensitivity is determined from the slope of the calibration curve. Work by Lorber allowed them to be calculated for first-order data,[73,74] and there has also been research in the area of second-order figures of merit.[14,75] The derivation of the second-order figures of merit is much more complicated than the zero- and first-order versions, owing to the added complexity of the data. To date, there have been derivations only for bilinear data matrices.

Most figures of merit are related to the net analyte response, $\mathbf{R}_k^*$, which for the kth analyte in a second-order data matrix, $\mathbf{R}$, can be defined as (Equation 70):

$$\mathbf{R}_k^* = \mathbf{P}_{k,x}\mathbf{R}\mathbf{P}_{k,y} \tag{70}$$

For references see page 9762

where (Equations 71a and 71b):

$$\mathbf{P}_{k,x} = \mathbf{I} - \mathbf{X}_{-k}\mathbf{X}^*_{-k} \tag{71a}$$

$$\mathbf{P}_{k,y} = \mathbf{I} - \mathbf{Y}_{-k}\mathbf{Y}^*_{-k} \tag{71b}$$

While the net analyte response is a matrix that contains the profile of that part of the analyte signal that is orthogonal to every other component in the matrix, the net analyte signal, r_k^*, is a scalar that is derived from the net analyte response (Equation 72):

$$r_k^* = \left\| vec\mathbf{R}_k^* \right\| \tag{72}$$

where $\| \|$ refers to the Euclidean norm, and the *vec* operator corresponds to "vectorizing" the matrix. Selectivity, l_k, and sensitivity, s_k, are both found using the net analyte signal as (Equations 73 and 74):

$$l_k = \frac{r_k^*}{\left\| vec\underline{\mathbf{R}}_k \right\|} \tag{73}$$

$$s_k = \frac{r_k^*}{c_k} \tag{74}$$

A selectivity of zero implies that an analyte signal cannot be distinguished from other analytes – in other words, the net analyte signal is completely overlapped by other analyte signals. A selectivity of 1, on the other hand, means that the net analyte signal is not at all overlapped by other analyte signals.

These equations can be extended to *n*th-order data by using the *vec* operator to string out the data. Carrying this out on second-order data (e.g. Equation 2) produces Equation (75):

$$vec\mathbf{R} = \sum \mathbf{y}_k \otimes \mathbf{x}_k \tag{75}$$

where $\otimes$ is the N-fold Kronecker product.[14] Similarly, the *vec* operator can be applied to three-way arrays such that (Equation 76):

$$vec\underline{R} = \sum_{k=1}^{K} \bigotimes_{n=1}^{N} \mathbf{a}_{k,n} \tag{76}$$

where $\mathbf{A}_n$ $(n = 1, \ldots, N)$ is a factor matrix of $\underline{\mathbf{R}}$ instead of $\mathbf{X}$ and $\mathbf{Y}$. The net analyte response may then be denoted by Equation (77):

$$vec\underline{\mathbf{R}}_k^* = \left(\bigotimes_{n=1}^{N} \mathbf{P}_{k,n} \right) vec\underline{\mathbf{R}} = \mathbf{P}_k vec\underline{\mathbf{R}} \tag{77}$$

As is the case with the second-order data, all of the figures of merit may be calculated from the net analyte response.

The second-order selectivity is one of the most important figures of merit for calibration, as a low value means that quantitation will be very difficult in the presence of interfering species. Selectivities of analytes in LC/DAD spectrochromatograms have been calculated, and serve to demonstrate typical values based on overlapped chromatograms and spectra in a data matrix. The analytes are toluene, naphthalene and *m*-xylene, all of which are spectrally similar, and biphenyl, which is spectrally dissimilar. The pure spectra and chromatograms are shown in Figure 14(a) and (b).

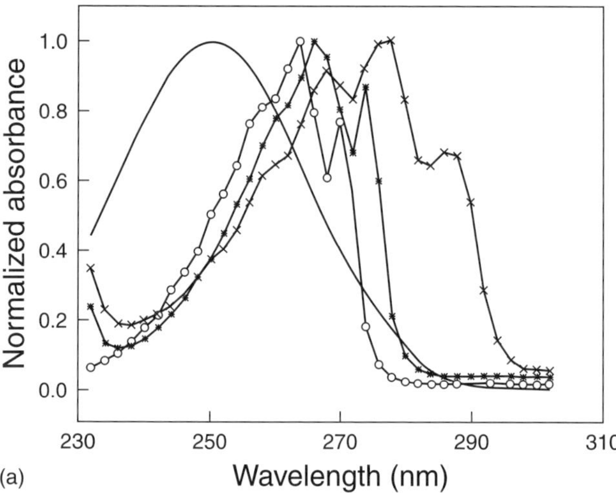

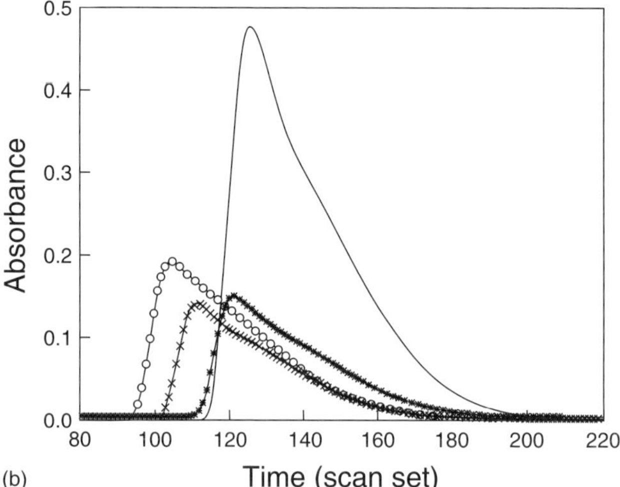

Figure 14 (a) Pure spectra and (b) chromatograms of toluene (○), naphthalene (×), *m*-xylene (∗) and biphenyl (no symbol). (Reprinted with permission from N.J. Messick, J.H. Kalivas, P.M. Lang, *Anal. Chem.*, **68**, 1572–1579 (1996). Copyright 1996 American Chemical Society.)

Upon calculation with Equation (73) for each analyte, the values in Table 11 were obtained. The first-order selectivities for the spectra of these analytes are shown in Table 12 for comparison.

The second-order selectivities can be explained based on the signals in the two orders (chromatographic and spectroscopic). For example, the largest value is for the comparison of naphthalene and biphenyl. When the

Table 11 Second-order selectivities for LC/DAD spectra for each pair of analytes. (Reprinted with permission from N.J. Messick, J.H. Kalivas, P.M. Lang, *Anal. Chem.*, **68**, 1572–1579 (1996). Copyright 1996 American Chemical Society.)

Component	Compared component	Second-order selectivity
Toluene	Naphthalene	0.7758
Toluene	*m*-Xylene	0.7893
Toluene	Biphenyl	0.8873
Naphthalene	*m*-Xylene	0.7272
Naphthalene	Biphenyl	0.8920
m-Xylene	Biphenyl	0.6916

Table 12 Spectral first-order selectivities for each pair of analytes. (Reprinted with permission from N.J. Messick, J.H. Kalivas, P.M. Lang, *Anal. Chem.*, **68**, 1572–1579 (1996). Copyright 1996 American Chemical Society.)

Component	Compared component	First-order selectivity
Toluene	Naphthalene	0.6877
Toluene	*m*-Xylene	0.3727
Toluene	Biphenyl	0.5530
Naphthalene	*m*-Xylene	0.5287
Naphthalene	Biphenyl	0.7684
m-Xylene	Biphenyl	0.6514

first-order selectivity for this combination is examined, it can be seen that it is high, meaning that these analytes are the most dissimilar spectrally; moreover, the chromatographic profiles show that they are also the most dissimilar in that axis, resulting in the high second-order selectivity. On the other hand, *m*-xylene and biphenyl have the lowest selectivity, implying that the signals from these analytes are more similar than any of the others. While their spectral similarity is not the highest (as seen by the relatively low first-order selectivity), their chromatographic profiles are very much overlapped, and this is what leads to their low second-order selectivity.

7 FUTURE WORK

While there have been tremendous advances in the area of second-order calibration, much work remains. Some of that work will pertain to improving the robustness and speed of the algorithms, while another part of the work should focus on increasing the awareness of the analytical community as to the value of such tools. Generally, while two-dimensional instruments are commonly used in today's laboratories, analysts often reduce the huge amount of data produced to one-dimension, and consequently they also reduce the quantity of information present. For example, when LC/MS is used, a two-dimensional data array is produced, but these data are often only viewed as a total ion chromatogram plot, thereby losing the mass spectral information. Often, the analysts are not aware of the existence of second-order methods. However, when this is not the case, the techniques are still not used owing to the perceived difficulties in their application. The great difficulty, therefore, is with the transfer of technology from the research arena to industry. Once this occurs, the use of second-order tools, including calibration, will become common place.

ACKNOWLEDGMENTS

C.M.F. acknowledges the financial support of the Endowed Analytical Professorship.

ABBREVIATIONS AND ACRONYMS

ALS	Alternating Least-squares
ATLD	Alternating Trilinear Decomposition
CANDECOMP	Canonical Decomposition
DTD	Direct Trilinear Decomposition
DTDMR	Direct Trilinear Decomposition with Matrix Reconstruction
EFA	Evolving Factor Analysis
GC	Gas Chromatography
GC × GC	Comprehensive Two-dimensional Gas Chromatography
GC/MS	Gas Chromatography/Mass Spectrometry
GRAM	Generalized Rank Annihilation Method
LC/DAD	Liquid Chromatography/Diode-array Detection
LC/MS	Liquid Chromatography/Mass Spectrometry
LC/MS/MS	Liquid Chromatography/Tandem Mass Spectrometry
LC/UV	Liquid Chromatography/ Ultraviolet Detection
MCR	Multivariate Curve Resolution
MLR	Multiple Linear Regression
MS/MS	Tandem Mass Spectrometry
NAR	Net Analyte Rank
NBRA	Nonbilinear Rank Annihilation
NMR	Nuclear Magnetic Resonance

For references see page 9762

NQS	1,2-Naphthoquinone-4-sulfonate
PARAFAC	Parallel Factor Analysis
PCR	Principal Component Regression
PDS	Piecewise Direct Standardization
PGSENMR	Pulsed-gradient Spin-echo Nuclear Magnetic Resonance
PLS	Partial Least Squares
RAFA	Rank Annihilation Factor Analysis
RLA	Rank Linear Additivity
RMSEP	Root Mean Square Error of Prediction
S/N	Signal-to-noise Ratio
SOSAM	Second-order Standard Addition Method
SVD	Singular Value Decomposition
TCE	Trichloroethylene
TLD	Trilinear Decomposition
UCC	Uncorrelated Correction Coefficient
Unfold-PCA	Unfold Principal Component Analysis
Unfold-PLS	Unfold Partial Least Squares
UV	Ultraviolet

RELATED ARTICLES

Chemometrics **(Volume 11)**
Chemometrics • Classical and Nonclassical Optimization Methods • Clustering and Classification of Analytical Data • Multivariate Calibration of Analytical Data • Signal Processing in Analytical Chemistry • Soft Modeling of Analytical Data

Gas Chromatography **(Volume 12)**
Multidimensional Gas Chromatography

Infrared Spectroscopy **(Volume 12)**
Gas Chromatography/Infrared Spectroscopy • Liquid Chromatography/Infrared Spectroscopy

Mass Spectrometry **(Volume 13)**
Gas Chromatography/Mass Spectrometry • Liquid Chromatography/Mass Spectrometry

REFERENCES

1. K.S. Booksh, B.R. Kowalski, 'Theory of Analytical Chemistry', *Anal. Chem.*, **66**(15), 782A–791A (1994).
2. E. Sanchez, B.R. Kowalski, 'Tensorial Calibration: I. First-order Calibration', *J. Chemom.*, **2**(4), 247–264 (1988).
3. E. Sanchez, B.R. Kowalski, 'Tensorial Calibration: II. Second-order Calibration', *J. Chemom.*, **2**(4), 265–280 (1988).
4. E. Sanchez, B.R. Kowalski, 'Generalized Rank Annihilation Factor Analysis', *Anal. Chem.*, **58**(2), 496–499 (1986).
5. E. Sanchez, B.R. Kowalski, 'Tensorial Resolution: a Direct Trilinear Decomposition', *J. Chemom.*, **4**(1), 29–45 (1990).
6. B.E. Wilson, W. Lindberg, B.R. Kowalski, 'Multicomponent Quantitative Analysis Using Second-order Nonbilinear Data: Theory and Simulations', *J. Am. Chem. Soc.*, **111**(11), 3797–3804 (1989).
7. R.A. Harshman, 'Foundations of the PARAFAC Procedure: Models and Methods for an "Explanatory" Multi-Mode Factor Analysis', *UCLA Working Papers Phonetics*, **16**, 1–84, 1970.
8. H.-L. Wu, M. Shibukawa, K. Oguma, 'An Alternating Trilinear Decomposition Algorithm with Application to Calibration of HPLC–DAD for Simultaneous Determination of Overlapped Chlorinated Aromatic Hydrocarbons', *J. Chemom.*, **12**(1), 1–26 (1998).
9. R. Tauler, B.R. Kowalski, S. Fleming, 'Multivariate Curve Resolution Applied to Spectral Data from Multiple Runs of an Industrial Process', *Anal. Chem.*, **65**(15), 2040–2047 (1993).
10. K.S. Booksh, Z. Lin, Z. Wang, B.R. Kowalski, 'Extension of Trilinear Decomposition Method with an Application to the Flow Probe Sensor', *Anal. Chem.*, **66**(15), 2561–2569 (1994).
11. K. Booksh, J.M. Henshaw, L.W. Burgess, B.R. Kowalski, 'A Second-order Standard Addition Method with Application to Calibration of a Kinetics–Spectroscopic Sensor for Quantitation of Trichloroethylene', *J. Chemom.*, **9**(4), 263–282 (1995).
12. Y. Wang, B.R. Kowalski, 'Standardization of Second Order Instruments', *Anal. Chem.*, **65**(9), 1174–1180 (1993).
13. B.J. Prazen, R.E. Synovec, B.R. Kowalski, 'Standardization of Second-order Chromatographic/Spectroscopic Data for Optimum Chemical Analysis', *Anal. Chem.*, **70**(2), 218–226 (1998).
14. K. Faber, A. Lorber, B.R. Kowalski, 'Analytical Figures of Merit for Tensorial Calibration', *J. Chemom.*, **11**(5), 419–461 (1997).
15. R. Bro, 'PARAFAC. Tutorials and Applications', *Chemom. Intell. Lab. Syst.*, **38**(2), 149–171 (1997).
16. L.R. Tucker, 'Relations between Multidimensional Scaling and Three-mode Factor Analysis', *Psychometrika*, **37**(1), 3–27 (1972).
17. P. Geladi, 'Analysis of Multi-way (Multi-mode) Data', *Chemom. Intell. Lab. Syst.*, **7**(1–2), 11–30 (1989).
18. A.K. Smilde, 'Three-way Analyses. Problems and Aspects', *Chemom. Intell. Lab. Syst.*, **15**(2–3), 143–157 (1992).
19. K.S. Booksh, B.R. Kowalski, 'Calibration Method Choice by Comparison of Model Basis Functions to the

List of selected abbreviations appears in Volume 15

Theoretical Instrument Response Function', *Anal. Chim. Acta*, **348**(1–3), 1–9 (1997).
20. J.D. Carroll, J.J. Chang, 'Analysis of Individual Differences in Multidimensional Scaling via an *N*-way Generalization of "Eckart–Young" Decomposition', *Psychometrika*, **35**(3), 283–319 (1970).
21. C.J. Appellof, E.R. Davidson, 'Strategies for Analyzing Data from Video Fluorometric Monitoring of Liquid Chromatographic Effluents', *Anal. Chem.*, **53**(13), 2053–2056 (1981).
22. B.C. Mitchell, D.S. Burdick, 'Slowly Converging PARAFAC Sequences: Swamps and Two-factor Degeneracies', *J. Chemom.*, **8**(2), 155–168 (1994).
23. R.A. Harshman, M.E. Lundy, 'The PARAFAC Model for Three-way Factor Analysis and Multidimensional Scaling', in *Research Methods for Multimode Data Analysis*, eds. H.G. Law, C.W. Snyder Jr, J.A. Hattie, R.P. McDonald, Praeger, New York, 122–215, 1984.
24. B.C. Mitchell, D.S. Burdick, 'An Empirical Comparison of Resolution Methods for Three-way Arrays', *Chemom. Intell. Lab. Syst.*, **20**(2), 149–161 (1993).
25. W.S. Rayens, B.C. Mitchell, 'Two-factor Degeneracies and a Stabilization of PARAFAC', *Chemom. Intell. Lab. Syst.*, **38**(2), 173–181 (1997).
26. C.A. Andersson, R. Bro, 'Improving the Speed of Multiway Algorithms. Part I: Tucker3', *Chemom. Intell. Lab. Syst.*, **42**(2), 93–103 (1998).
27. R. Bro, C.A. Andersson, 'Improving the Speed of Multiway Algorithms. Part II: Compression', *Chemom. Intell. Lab. Syst.*, **42**(2), 105–113 (1998).
28. H.A.L. Kiers, W.P. Krijnen, 'An Efficient Algorithm for PARAFAC of Three-way Data with Large Numbers of Observation Units', *Psychometrika*, **56**(1), 147–152 (1991).
29. R. Bro, S. de Jong, 'A Fast Non-negativity Constrained Least Squares Algorithm', *J. Chemom.*, **11**(5), 393–401 (1997).
30. P. Paatero, 'A Weighted Non-negative Least Squares Algorithm for Three-way 'PARAFAC' Factor Analysis', *Chemom. Intell. Lab. Syst.*, **38**(2), 223–242 (1997).
31. K.S. Booksh, A.R. Muroski, M.L. Myrick, 'Single-measurement Excitation/Emission Matrix Spectrofluorometer for Determination of Hydrocarbons in Ocean Water. 2. Calibration of Naphthalene and Styrene', *Anal. Chem.*, **68**(20), 3539–3544 (1996).
32. I.E. Bechmann, 'Second-order Data by Flow Injection Analysis with Spectrophotometric Diode-array Detection and Incorporated Gel-filtration Chromatographic Column', *Talanta*, **44**(4), 585–591 (1997).
33. R. Bro, H. Heimdal, 'Enzymatic Browning of Vegetables. Calibration and Analysis of Variance by Multiway Methods', *Chemom. Intell. Lab. Syst.*, **34**(1), 85–102 (1996).
34. P. Hindmarch, K. Kavianpour, R.G. Brereton, 'Evaluation of Parallel Factor Analysis for the Resolution of Kinetic Data by Diode-array High Performance Liquid Chromatography', *Analyst*, **122**(9), 871–877 (1997).
35. R. Tauler, A. Izquierdo-Ridorsa, E. Casassas, 'Simultaneous Analysis of Several Spectroscopic Titrations with Self-modeling Curve Resolution', *Chemom. Intell. Lab. Syst.*, **18**(3), 293–300 (1993).
36. R. Tauler, S. Lacorte, D. Barcelo, 'Application of Multivariate Self-modeling Curve Resolution to the Quantitation of Trace Levels of Organophosphorus Pesticides in the Natural Waters from Interlaboratory Studies', *J. Chromatogr. A*, **730**(1–2), 117–183 (1996).
37. R. Tauler, 'Multivariate Curve Resolution Applied to Second-order Data', *Chemom. Intell. Lab. Syst.*, **30**(1), 133–146 (1995).
38. M. Maeder, 'Evolving Factor Analysis for the Resolution of Overlapping Chromatographic Peaks', *Anal. Chem.*, **59**(3), 527–530 (1997).
39. J. Saurina, S. Hernandez-Cassou, R. Tauler, 'Second-order Curve Resolution Applied to the Flow Injection Analysis of Mixtures of Amino Acids', *Anal. Chim. Acta*, **335**(1–2), 41–49 (1996).
40. R. Tauler, A. Smilde, B. Kowalski, 'Selectivity, Local Rank, Three-way Data Analysis and Ambiguity in Multivariate Curve Resolution', *J. Chemom.*, **9**(1), 31–58 (1995).
41. J. Saurina, S. Hernandez-Cassou, R. Tauler, 'Multivariate Curve Resolution and Trilinear Decomposition Methods in the Analysis of Stopped-flow Kinetic Data for Binary Amino Acid Mixtures', *Anal. Chem.*, **69**(13), 2329–2336 (1997).
42. A. de Juan, S.C. Rutan, R. Tauler, D.L. Massart, 'Comparison Between the Direct Trilinear Decomposition and the Multivariate Curve Resolution–Alternating Least Squares Methods for the Resolution of Three-way Data Sets', *Chemom. Intell. Lab. Syst.*, **40**(1), 19–32 (1998).
43. C.-N. Ho, G.D. Christian, E.R. Davidson, 'Application of the Method of Rank Annihilation to Quantitative Analysis of Multicomponent Fluorescence Data from the Video Fluorometer', *Anal. Chem.*, **50**(8), 1108–1113 (1978).
44. C.-N. Ho, G.D. Christian, E.R. Davidson, 'Application of the Method of Rank Annihilation to Fluorescent Multicomponent Mixtures of Polynuclear Aromatic Hydrocarbons', *Anal. Chem.*, **52**(7), 1071–1079 (1980).
45. C.-N. Ho, G.D. Christian, E.R. Davidson, 'Simultaneous Multicomponent Rank Annihilation and Applications to Multicomponent Fluorescent Data Acquired by the Video Fluorometer', *Anal. Chem.*, **53**(1), 92–98 (1981).
46. A. Lorber, 'Quantifying Chemical Composition from Two-dimensional Data Arrays', *Anal. Chim. Acta*, **164**, 293–297 (1984).
47. B.E. Wilson, E. Sanchez, B.R. Kowalski, 'An Improved Algorithm for the Generalized Rank Annihilation Method', *J. Chemom.*, **3**(3), 493–498 (1988).

48. N.M. Faber, L.M.C. Buydens, G. Kateman, 'Generalized Rank Annihilation Method. III: Practical Implementation', *J. Chemom.*, **8**(4), 273–285 (1994).
49. C.M. Fleming, B.R. Kowalski, unpublished work, 1998.
50. C.A. Bruckner, P.J. Prazen, R.E. Synovec, 'Comprehensive Two-dimensional High-speed Gas Chromatography with Chemometric Analysis', *Anal. Chem.*, **70**(14), 2796–2804 (1998).
51. S. Li, J.C. Hamilton, P.J. Gemperline, 'Generalized Rank Annihilation Method using Similarity Transformations', *Anal. Chem.*, **64**(6), 599–607 (1992).
52. N.M. Faber, L.M.C. Buydens, G. Kateman, 'Generalized Rank Annihilation Method. I: Derivation of Eigenvalue Problems', *J. Chemom.*, **8**(2), 147–154 (1994).
53. N.M. Faber, L.M.C. Buydens, G. Kateman, 'Generalized Rank Annihilation Method. II: Bias and Variance in the Estimated Eigenvalues', *J. Chemom.*, **8**(3), 181–203 (1994).
54. K. Faber, A. Lorber, B.R. Kowalski, 'Generalized Rank Annihilation Methods: Standard Errors in the Estimated Eigenvalues if the Instrumental Errors are Heteroscedastic and Correlated', *J. Chemom.*, **11**(2), 95–109 (1997).
55. K. Booksh, B.R. Kowalski, 'Error Analysis of the Generalized Rank Annihilation Method', *J. Chemom.*, **8**(1), 45–63 (1994).
56. R.B. Poe, S.C. Rutan, 'Effects of Resolution, Peak Ratios and Sampling Frequency in Diode-array Fluorescence Detection in Liquid Chromatography', *Anal. Chim. Acta*, **283**(2), 845–853 (1993).
57. S.S. Li, P.J. Gemperline, 'Eliminating Complex Eigenvectors and Eigenvalues in Multiway Analyses Using the Direct Trilinear Decomposition Method', *J. Chemom.*, **7**(2), 77–88 (1993).
58. E. Sanchez, L.S. Ramos, B.R. Kowalski, 'Generalized Rank Annihilation Method. I. Application to Liquid Chromatography–Diode Array Ultraviolet Detection Data', *J. Chromatogr.*, **385**, 151–164 (1987).
59. L.S. Ramos, E. Sanchez, B.R. Kowalski, 'Generalized Rank Annihilation Method. II. Analysis of Bimodal Chromatographic Data', *J. Chromatogr.*, **385**, 165–180 (1987).
60. W. Windig, B. Antalek, 'Direct Exponential Curve Resolution Algorithm (DECRA): A novel Application of the Generalized Rank Annihilation Method for a Single Spectral Mixture Data Set with Exponentially Decaying Contribution Profiles', *Chemom. Intell. Lab. Syst.*, **37**(2), 241–254 (1997).
61. Y.S. Zeng, P.K. Hopke, 'A New Receptor Model: A Direct Trilinear Decomposition Followed by a Matrix Reconstruction', *J. Chemom.*, **6**(2), 65–83 (1992).
62. Z.H. Lin, K.S. Booksh, L.W. Burgess, B.R. Kowalski, 'Second-order Fiber Optic Heavy Metal Sensor Employing Second-order Tensorial Calibration', *Anal. Chem.*, **66**(15), 2552–2560 (1994).
63. M.M.C. Ferreira, M.L. Brandes, I.M.C. Ferreira, K.S. Booksh, W.C. Dolowy, M. Gouterman, B.R. Kowalski, 'Chemometric Study of the Fluorescence of Dental Calculus by Trilinear Decomposition', *Appl. Spectrosc.*, **49**(9), 1317–1325 (1995).
64. A. Lorber, 'Features of Quantifying Chemical Composition from Two-dimensional Data Array by the Rank Annihilation Factor Analysis Method', *Anal. Chem.*, **57**(12), 2395–2397 (1985).
65. Y.D. Wang, O.S. Borgen, B.R. Kowalski, M. Gu, F. Turecek, 'Advances in Second-order Calibration', *J. Chemom.*, **7**(2), 117–130 (1993).
66. B.E. Wilson, B.R. Kowalski, 'Quantitative Analysis in the Presence of Spectral Interferents using Second-order Non-bilinear Data', *Anal. Chem.*, **61**(20), 2277–2284 (1989).
67. C. Demir, R.G. Brereton, 'Calibration of Gas Chromatography–Mass Spectrometry of Two-component Mixtures using Univariate Regression and Two- and Three-way Partial Least Squares', *Analyst*, **122**(7), 631–638 (1997).
68. A.K. Smilde, D.A. Doornbos, 'Simple Validatory Tools for Judging the Predictive Performance of PARAFAC and Three-way PLS', *J. Chemom.*, **6**(1), 11–28 (1992).
69. R. Bro, 'Multiway Calibration. Multilinear PLS', *J. Chemom.*, **10**(1), 47–61 (1996).
70. A.K. Smilde, 'Comments on Multilinear PLS', *J. Chemom.*, **11**(5), 367–377 (1997).
71. Y.D. Wang, B.R. Kowalski, 'Calibration Transfer and Measurement Stability of Near-infrared Spectrometers', *Appl. Spectrosc.*, **46**(5), 764–771 (1992).
72. J.M. Henshaw, 'Reaction-based Chemical Sensing for the Multicomponent Determination of Chlorinated Hydrocarbons', Doctoral Dissertation, University of Washington, 1993.
73. A. Lorber, 'Error Propagation and Figures of Merit for Quantification by Solving Matrix Equations', *Anal. Chem.*, **58**(6), 1167–1172 (1986).
74. A. Lorber, A. Harel, Z. Goldbart, I.B. Brenner, 'Curve Resolution and Figures of Merit Estimation for Determination of Trace Elements in Geological Materials by Inductively Coupled Plasma Atomic Emission Spectrometry', *Anal. Chem.*, **59**(9), 1260–1266 (1987).
75. N.J. Messick, J.H. Kalivas, P.M. Lang, 'Selectivity and Related Measures for *n*th Order Data', *Anal. Chem.*, **68**(9), 1572–1579 (1996).

Signal Processing in Analytical Chemistry

Peter D. Wentzell and Christopher D. Brown
Dalhousie University, Halifax, Canada

Signal processing refers to a variety of operations that can be carried out on a continuous (analog) or discrete (digital) sequence of measurements in order to enhance the quality of information it is intended to convey. In the analog domain, electronic signal processing can encompass such operations as amplification, filtering, integration, differentiation, modulation/demodulation, peak detection, and analog-to-digital (A/D) conversion. Digital signal processing can include a variety of filtering methods (e.g. polynomial least-squares smoothing, differentiation, median smoothing, matched filtering, boxcar averaging, interpolation, decimation, and Kalman filtering) and domain transformations (e.g. Fourier transform (FT), Hadamard transform (HT), and wavelet transform (WT)). Generally the objective is to separate the useful part of the signal from the part that contains no useful information (the noise) using either explicit or implicit models that distinguish these two components. Signal processing at various stages has become an integral part of most modern analytical measurement systems and plays a critical role in ensuring the quality of those measurements.

1 INTRODUCTION

The reliability of analytical results is vitally dependent on the quality of the measurements leading to their determination. Signal processing refers to a variety of operations that can be carried out on a continuous or discrete sequence of measurements in order to enhance the quality of information they are intended to convey. The term 'signal' is ordinarily applied to a sequence of measurements that are related by some ordinal variable, such as time or wavelength, and usually obtained via electrical transduction. Operations can be carried out on a continuous electrical signal (analog signal processing) or on discretely sampled numerical values (digital signal processing). Generally, the objective is to separate the desired part of the signal (the pure signal, which is correlated to some physical or chemical property of interest) from the unwanted part of the signal, or noise. Often, the term 'signal processing' implies that operations are carried out in real time, or as the data are acquired, but this is not a requirement. The distinction between signal processing and other forms of data analysis is often open to interpretation, but usually signal processing emphasizes alternative representations of the sequence of measurements as opposed to the direct extraction of secondary information such as analyte concentrations or chemical structures.

The goal of this article is to provide an overview of signal processing methods used in analytical chemistry with an emphasis on their capabilities, weaknesses, and practical implementation. Although both analog and digital signal processing are discussed, a much greater emphasis is placed on the latter because of its greater relevance to the practicing analytical chemist. Owing to the scope of the subject area, some topics will no doubt be neglected or underemphasized, but an attempt has been made to balance coverage with the importance to the field.

2 OVERVIEW

2.1 History

Historically, signal processing in analytical chemistry can be regarded as originating with the first quantitative analytical measurements, such as the end-point of a titration, the potential of an electrochemical cell, or

For references see page 9798

the absorbance of a solution at a single wavelength. Such scalar quantities are now sometimes referred to as zero-order measurements, with reference to the fact that a scalar is a zero-order tensor. Rudimentary signal processing consisted of averaging replicate measurements or damping the response of electrical signals, but these were rather simplistic approaches when compared to today's more sophisticated methods.

The evolution of modern signal processing in analytical chemistry can be traced to three parallel developments. The first was the emergence of analytical instruments capable of producing first-order data, or a vector (first-order tensor) of measurements. The development of instruments such as chromatographs and scanning spectrometers meant that the signal from an instrument could no longer be considered static, but rather changed in a regular fashion with some variable such as time or wavelength, thereby requiring more flexible signal processing methods that could remove the noise without distorting the pure signal. A second important influence was the appearance of the analog-to-digital converter (ADC) and the computers that drove them. This opened the door to more versatile digital signal processing methods. Finally, there was the parallel development of efficient digital signal processing algorithms, such as the Savitzky–Golay (SG) implementation of polynomial least-squares filters,[(1)] and the Cooley–Tukey algorithm for the fast Fourier transform (FFT),[(2)] which are still among the most highly cited papers in the literature. Such algorithms made the practical implementation of signal processing methods a reality.

Although first-order instruments are still the mainstay of analytical chemistry, second-order instruments which provide a matrix of data, such as chromatographs with multichannel detection and tandem mass spectrometers, are now routinely employed, and higher-order instruments are commonplace. These, combined with ever more powerful computational platforms, have advanced modern signal processing to yet another level.

2.2 Definitions and Notation

Throughout this article, the term 'signal' will be used to refer to either a continuous or discrete measurement sequence which consists of a pure or undistorted signal corrupted by noise. The signal is implicitly measured as a function of some other variable which will be referred to as the ordinal variable because it correlates directly with the sequence order. Traditionally, this variable is time, but other variables such as wavelength or applied voltage (which may or may not be correlated with time), can be employed without loss of generality. References to representation of the signal in the 'time domain' will therefore refer to this the original measurement sequence even if the ordinal variable is not time. Likewise, references to 'frequency domain' representations (section 3) will refer to a FT (section 6) into the inverse domain of the ordinal variable.

The term 'noise' is used to refer to unwanted fluctuations from the pure signal that obscure its measurement. This definition is quite general and means that what is considered noise can vary with the situation. If a signal consists of contributions from two sources, A and B, then B is considered noise if one is looking for A, and vice versa. Noise, especially random noise, is often characterized by its root-mean-square (rms) amplitude (continuous signals) or by its standard deviation (discrete signals).

Most of this article concerns digital signal processing methods where the signal is a vector of discrete measurements, usually made on equal intervals of the ordinal variable referred to as the sampling interval, t_s. The reciprocal of this interval will be referred to as the sampling frequency, f_s. (Again, analogous definitions hold when the ordinal variable is not time.) In vector notation, the signal $\mathbf{x}$ is the result of combining the pure signal vector $\mathbf{x^0}$ with the noise vector $\mathbf{e}$ (Equation 1):

$$\mathbf{x} = \mathbf{x^0} + \mathbf{e} \quad (1)$$

Throughout this article, boldface lower case letters are used to represent vectors (assumed to be column vectors unless otherwise indicated). Scalar quantities will be represented by lower case italic letters and matrices by boldface upper case letters. Transposes of vectors or matrices will be indicated with a superscript 'T' and inverses of matrices with a superscript '−1'. The identity matrix will be designated '**I**'.

3 SIGNALS AND NOISE

3.1 Signals and Signal Domains

The features that distinguish a signal from a random series of measurements are: the measurements have a definite order, and the measurements are generally correlated in the time domain. For a discrete signal undistorted by noise, $\mathbf{x^0} = [x_1^0 x_2^0 x_3^0 \ldots]$, the correlation of two measurements is described by their covariance (Equation 2),

$$\mathrm{cov}(x_1^0, x_2^0) = \mathrm{E}(x_1^0 x_2^0) \quad (2)$$

where 'E' denotes the expectation value. To say that two measurements are correlated means that knowing one allows us to say something about the other. Although the covariance among measurements in a sequence is not generally known, some knowledge of its characteristics may be known, and it is these characteristics that are often used to distinguish the pure signal from the noise. For example, if the pure signal changes relatively slowly, it exhibits long-range correlations which may not be present in the noise.

Although signals are usually presented in the time domain, the same information can be conveyed by transforming them in alternate domains. The most useful of these is the frequency domain. For a continuous signal, this transformation can be accomplished by using a spectrum analyzer, a device that consists of a continuous series of electronic bandpass filters. The signal, which must be repetitively applied to the input of the device, is filtered by the spectrum analyzer and the rms signal at each frequency is determined. In essence, the power of the signal is plotted as a function of frequency to give a power spectrum. In the age of discrete signals, this approach is rarely used anymore, instead being replaced by the discrete FT, which is the digital counterpart of the analog spectrum analyzer. FTs are discussed in more detail in section 6 but, because of their importance to signal processing, they are introduced here. All of the information conveyed in the original signal is carried in the FT, but it is represented in the frequency domain rather than the time domain. This allows the composition of the signal in terms of sinusoidal frequencies to be analyzed, providing more direct information about the correlations in the time domain. Slowly varying signals in the time domain will have significant low-frequency components, and the shape of the power spectrum conveys important information about optimal signal processing methods. Figure 1 shows some simple signals and their FTs. The first example shows that a pure sinusoid gives rise to a single peak in the frequency domain. In contrast, the sharp edges of the square wave in the second example lead to high-frequency components in the FT. The last two examples demonstrate that more slowly varying signals have fewer high frequency components.

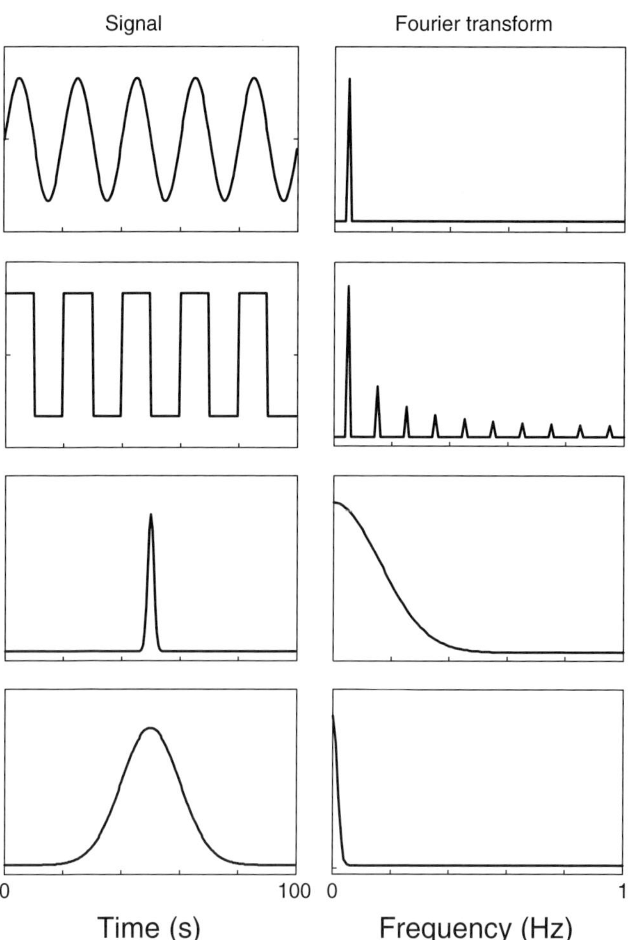

Figure 1 FTs of some simple signals.

3.2 Noise

The noise in an analytical signal can be classified in a number of ways, including (a) its distribution, (b) its source, (c) its characteristics in the time domain, and (d) its characteristics in the frequency domain. Because different classifications are used in different situations and they are not all mutually exclusive, it is necessary to understand the different cases and how they relate to signal processing. In doing so, it is helpful to imagine a signal from which we could subtract the pure signal component leaving only the noise or, alternatively, a situation in which the pure signal is zero so that we are measuring only noise, as shown in Figure 2(a).

If we were to plot a histogram of the magnitude of the noise for a large number of measurements, we might obtain a distribution such as that shown in Figure 2(b). By far the most common noise distribution (assumed or measured) for analytical measurements is the normal, or Gaussian, distribution, shown by the solid line in the figure. The reason for this is the central limit theorem which, simply put, states that if a measurement is the sum of a series of values drawn from arbitrary distributions, the distribution of the measurement will approach a normal distribution as the length of the series approaches infinity. As, in an analytical instrument, the observed noise is a consequence of many smaller random events, the central limit theorem can be rationalized to hold. Other noise distributions (e.g. uniform, log-normal) are also observed but are much less common. One other type which is common, however, is the Poisson distribution, which is observed in cases where the signal arises from a collection of discrete events, such as photons striking a photomultiplier tube. However, in reality a histogram constructed from Poisson noise would look essentially the same as the normal distribution in most cases. The distinction is in how the magnitude of the noise (i.e. its standard deviation) changes with signal intensity. The

For references see page 9798

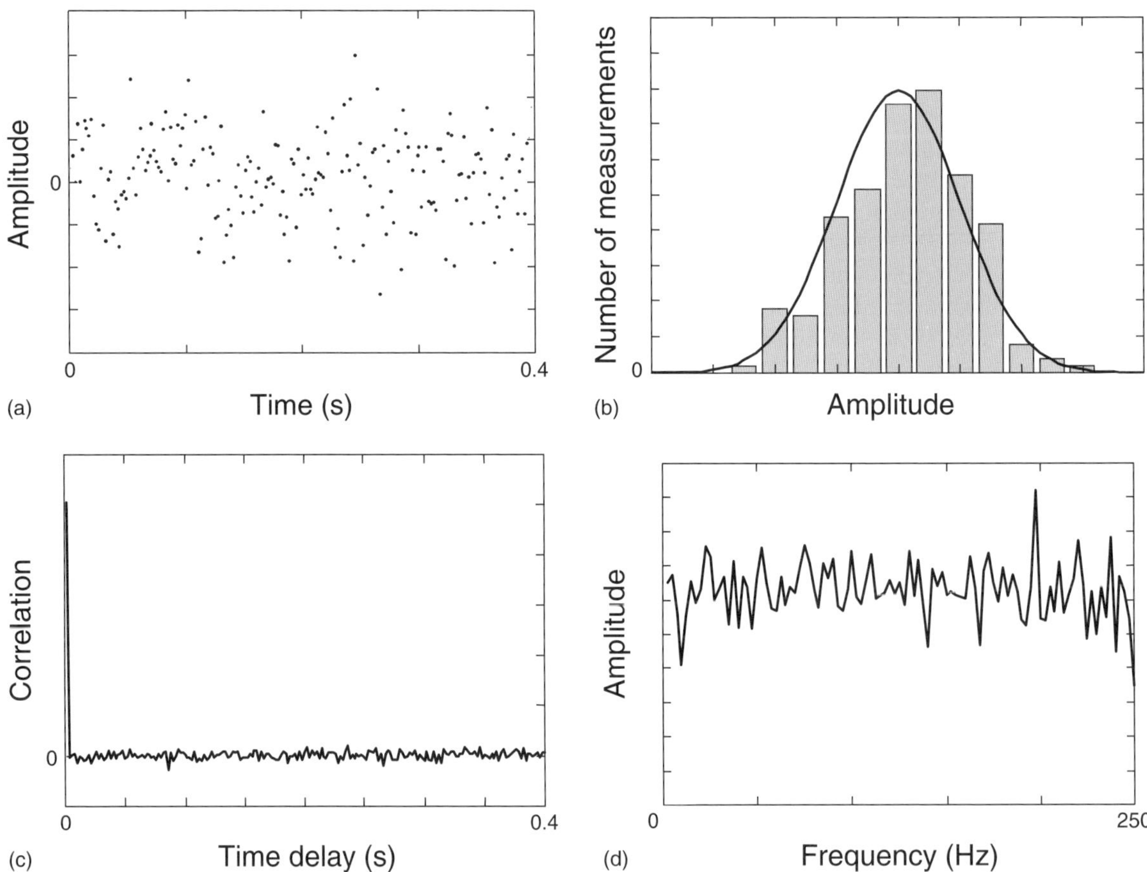

Figure 2 Representations of white noise: (a) noise sequence in the time domain, (b) distribution, (c) autocorrelation function, (d) FT.

standard deviation of Poisson noise will increase with the square root of the signal intensity. To say that noise has a normal distribution, however, does not imply anything about how its magnitude changes with signal intensity. In this sense, the Poisson distribution can be regarded as a special case of the normal distribution.

In some cases noise is classified according to its dominant source. Such classifications often imply information about the distribution or temporal characteristics of the noise. For example shot noise, also called Schottky noise or quantum noise, arises in detectors based on discrete events, such as photons striking a photomultiplier, and exhibits a Poisson distribution. Shot noise is a type of fundamental noise because it originates from the random statistical nature of the events themselves and not from any deficiencies of the instrument. This type of noise dominates in cases where the number of events is relatively small, such as in fluorescence measurements. Johnson noise is another type of fundamental noise that arises from the random thermal motion of electrons in resistors. Flicker noise is considered to be a type of nonfundamental or excess noise in which the magnitude of the noise is directly proportional to the signal amplitude, hence it is often referred to as proportional or multiplicative noise. Flicker noise is often associated with variations in source intensity in absorption spectroscopy and can have distinctive frequency characteristics (see below). Other types of noise include interference noise (electrical, optical or other interferences that arise at specific frequencies, such as 60 Hz line interference), detector noise (a general term referring to instruments in which the limiting noise, such as shot noise or thermal noise, occurs at the detector), amplifier-readout noise (the noise observed for the readout circuitry when the input signal is zero) and quantization noise (observed when measurement precision is limited by the A/D conversion step).

In the time domain, noise can be classified in two ways: correlated and uncorrelated. Uncorrelated, or independent, noise implies that the noise observed at one point in the series of measurements is not related in any way to that at other points. If we consider the sequence of noise to be represented by the vector $[e_1\ e_2\ e_3\ \ldots]$, then the covariance between the first two noise elements can be represented by Equation (3),

$$\sigma_{12} = \mathrm{E}(e_1 e_2) \qquad (3)$$

and will be zero for uncorrelated noise. In many applications, independent and identically distributed noise with a normal distribution, or *iid normal* noise, is assumed, implying uncorrelated noise with a Gaussian distribution and equal variances at all channels. However, in reality correlated noise is commonplace, arising from varied sources such as temporal variations in spectroscopic source intensities, spatial correlations (cross-talk) in array detectors, thermal variations and electronic filtering. Signal processing itself can turn uncorrelated noise into correlated noise, a fact that can be important for subsequent data analysis methods. The correlation of a noise sequence can be examined through its autocorrelation function. This is obtained through an element-by-element multiplication of the noise sequence by a time shifted version of itself and averaging the results of the n multiplications. This is repeated for each time shift. Uncorrelated noise should give a single spike at zero time delay, with the products averaging to zero everywhere else. This is demonstrated in Figure 2(c).

The complete characterization of the correlation among elements of a noise vector of length n is given by the error covariance matrix Σ (Equation 4):

$$\Sigma = \begin{bmatrix} \sigma_1^2 & \sigma_{12} & \cdots & \sigma_{1n} \\ \sigma_{12} & \sigma_2^2 & \cdots & \sigma_{2n} \\ \vdots & \vdots & \ddots & \vdots \\ \sigma_{1n} & \sigma_{2n} & \cdots & \sigma_n^2 \end{bmatrix} \quad (4)$$

where the diagonal elements represent the noise variances and the off-diagonal elements (zero for uncorrelated noise) represent the covariance. If a noise sequence is a stationary process, then its statistical properties remain constant throughout the sequence, which means $\sigma_{12} = \sigma_{23} = \sigma_{34}$ and so on. The terms 'homoscedastic' and 'heteroscedastic' are also used to indicate whether the variance of the noise remains constant or changes with the position in the series, respectively.

As with the signal, the characteristics of the noise can be examined in the frequency domain by using a spectrum analyzer or the FT. The result is called the noise power spectrum (NPS) and it conveys important information about the time domain correlations of the noise. Uncorrelated noise is referred to as white noise and, analogous to white light, it contains equal contributions at all frequencies. This is the type of noise that is often assumed or hoped for when designing or implementing signal processing methods, but exceptions are very common. The NPS for the white noise in Figure 2(a) is shown in Figure 2(d). Note that because of the stochastic nature of noise, the NPS is not perfectly flat.

Often purely white noise is corrupted by pink noise or $1/f$ noise. This noise, also known as drift in the time domain and arising from flicker noise, has a NPS which varies as the reciprocal of the intensity. The dominance at low frequencies is indicative of correlated noise in the time domain. The third type of noise readily identified in the NPS is interference noise, which appears as spikes at the corresponding frequencies and often higher harmonics. Figure 3 illustrates a typical correlated noise sequence and its representative autocorrelation function and NPS. Note the slow decay of the autocorrelation function indicating that the noise is not white. Both $1/f$ noise and interference noise are apparent in the NPS. It should be noted that, in order to obtain a clear representation of the autocorrelation function and the NPS in this case, it was necessary to average results from 20 noise sequences, because a single noise sampling does not usually give a clear indication of its characteristics.

3.3 Signal Averaging

One of the most effective ways to separate the pure signal from the noise is through the use of signal averaging. True signal averaging, as opposed to boxcar averaging or smoothing (section 5), requires a repetitive signal sequence. A common example is the fluorescence or phosphorescence decay curve generated by pulsing a laser or flash lamp repetitively. Each experiment (pulse) results in a signal that can be added to the next, and the total sum can be averaged over the number of experiments. Unlike most of the methods described here, which take advantage of signal correlation within a single experiment, signal averaging exploits signal correlation at a given time channel for repeated experiments. For this reason, it is discussed here, separately from the other techniques.

In order for signal averaging to be effective and useful, a number of conditions need to be met. First, the noise should be uncorrelated between corresponding time channels for successive experiments or else it will not be effectively removed. (Note that correlation of noise within an experiment is not important.) A second requirement is that the shape of the signal needs to be truly repetitive in the time domain. Translation of the signal or changes in its profile (other than scaling) due to poor synchronization or changes in the process will lead to an unrepresentative average, although noise should be reduced. Finally, the duration of the experiment needs to be short enough to make signal averaging practical.

When these conditions are met, signal averaging can be particularly advantageous because: (a) the signal-to-noise ratio (SNR) improves by a factor of $\sqrt{n}$, where n is the number of repetitions, and (b) there is no distortion in the profile of the signal. In contrast, most of the methods discussed in the sections that follow have the potential to distort the shape of the signal.

For references see page 9798

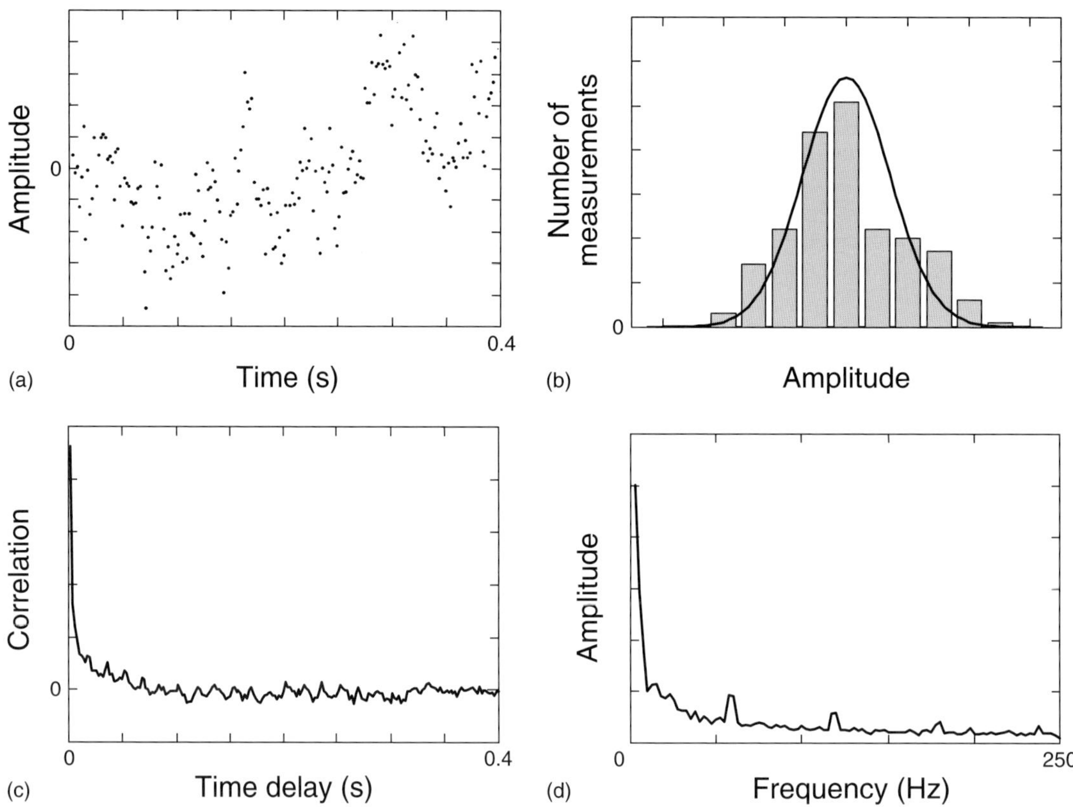

Figure 3 Representations of correlated noise: (a) noise sequence in the time domain, (b) distribution, (c) autocorrelation function, (d) FT.

4 ANALOG SIGNAL PROCESSING

4.1 Overview

As virtually all instrumental methods involve some form of electrical transduction of a particular phenomenon into a continuous signal, analog signal processing is as universal as it is diverse. Such processing begins the moment the quantity being measured is converted into some electrical property such as current or voltage (assuming that it did not originate in that form) and continues until the final measurements are recorded in digital or analog form. This obviously opens up a tremendous range of topics, the detailed coverage of which is beyond the scope of this article. However, there has been a gradual shift over the years which has placed an increasing emphasis on digital signal processing over analog signal processing. One reason for this is that access to the analog signal in modern instruments has become more restricted and more often the chemist is presented with data that have already been digitized, as evidenced by the demise of the chart recorder from most analytical labs. A second reason is that improvements in the speed and storage capacity of digital components have removed many of the limitations of early devices. Finally, digital processing of results has the advantage that it can be carried out any time after the data are required, whereas this is not true for analog signal processing.

On this basis, it could be argued that a detailed comprehension of analog signal processing is less important now than it once was, although it is still essential to understand the basic capabilities and limitations of this stage of processing, because it will always precede the generation of digital information and can be the 'weak link in the chain' if care is not taken. For this reason, a brief coverage of electronic signal processing methods is presented here, with a special emphasis on two aspects which are closely related to digital processing, analog filters and A/D conversion.

The electronic manipulation of signals can be divided into methods which are based on passive or active circuits. Passive circuits (those consisting only of simple components such as resistors, capacitors or diodes) do not require an external power source (other than the signal itself) whereas active circuits do. Although passive circuits are normally much simpler than active circuits, their capabilities are severely limited by comparison. Most complex manipulations of electronic signals are based on operational amplifiers, active circuit elements employed for a wide variety of linear and nonlinear

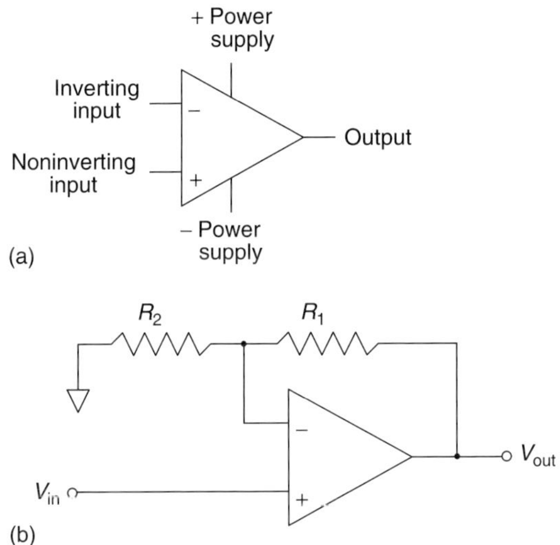

Figure 4 (a) Symbolic representation of an operational amplifier. (b) A noninverting amplifier with gain.

operations. A basic operational amplifier, represented symbolically in Figure 4(a), is essentially a high gain (typically 10^5) differential amplifier. By taking advantage of the high gain in negative feedback, the two inputs (referred to as the inverting (−) and noninverting (+) inputs) and the output can be configured into a wide range of useful circuits using passive components. As an example, the configuration for a simple fixed-gain voltage amplifier (gain $= V_{out}/V_{in} = (R_1 + R_2)/R_2$) is shown in Figure 4(b).

The number of electronic operations that can be carried out on electrical signals is too extensive to cover here, but a brief summary of some important operations is given in Table 1. More details on analog signal processing can be found in appropriate references on the subject.[3–5] Analog filters and A/D conversion are also covered in more detail in sections 4.2 and 4.3.

4.2 Analog Filters

One of the simplest operations that can be carried out to improve signal quality is the application of an electronic filter. As the object of signal processing is to distinguish the pure signal from the noise, a means is generally sought to distinguish the two. As noted in section 3, this distinction can often be made in the frequency domain. Typically, white noise has a flat NPS, whereas a slowly varying signal will have most of the information at low frequencies. Therefore, a filter that removes high-frequency components from a noisy signal, called a low-pass filter, will retain most of the information about the pure signal while eliminating much of the noise. However, drift and offset noise (low frequency) may be dominant features in the noise, and a high-pass

Table 1 Summary of some common analog signal processing applications

Operation or circuit	Description
Domain conversion	Conversion of an analog signal between domains, such as current-to-voltage, resistance-to-voltage, time-to-amplitude, A/D.
Amplification	Multiplication of an analog signal by a constant factor called the gain. Sometimes coupled with domain conversion.
Inversion	Changes the sign of an analog signal.
Addition/subtraction	Two or more signals are added/subtracted.
Multiplication/division	Two signals are multiplied/divided by one another. Often used with modulation (see below).
Other mathematical operations	Includes logarithm, antilogarithm, absolute value, reciprocal, etc.
Integration/differentiation	Integrals or derivatives of analog signals.
Modulation	Process by which the property (e.g. amplitude or frequency) of a carrier wave, typically a high frequency sinusoid or square wave, is modified to convey information about an analog signal of interest. Demodulation is used to recover the original analog signal.
Comparator	Circuit that compares two voltages and produces one of two outputs, depending on which signal is larger.
Pulse height discriminator	Circuit that detects the presence of pulses with a peak amplitude within a certain threshold region.
Peak detector	Circuit that follows an analog signal until it reaches a peak value and then holds that value for a certain time period.
Sample-and-hold amplifier	Circuit that samples an analog voltage at a particular point in time and holds the value until further processing, normally A/D conversion, is completed.
Boxcar integrator	Circuit used to measure a rapidly changing but repetitive signal by sampling it after a particular delay time and holding the sampled measurement. Often the delay time is scanned to obtain the time profile of the signal.
Lock-in amplifier	One type of circuit used to demodulate analog signals (see above).

For references see page 9798

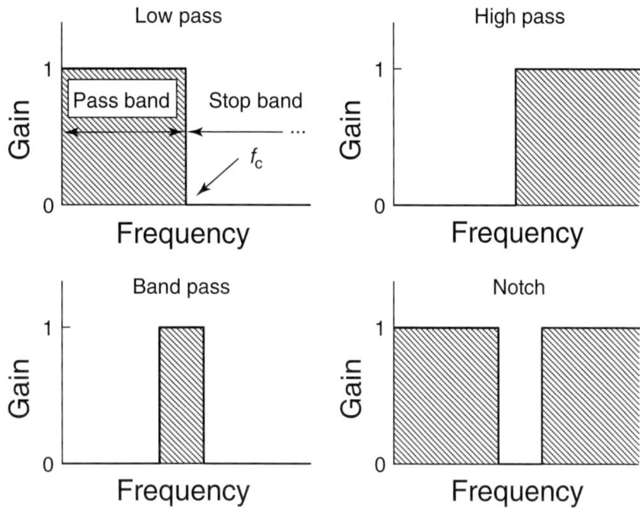

Figure 5 Transfer functions for ideal filters of various types.

filter may be more effective, provided there are sufficient high-frequency components in the signal. In another scenario, the signal may be modulated at a particular frequency and it may be necessary to use a band pass filter to isolate the components of interest. Finally, if interference noise, such as 60 Hz noise, is a problem, it may be removed using a notch filter. The ideal transfer functions for each of these types of filter is shown in Figure 5. The transfer function gives the amplification or gain of the filter as a function of the frequency of the input and ideally should be unity within the pass band and zero in the stop band.

As shown in section 5, analog filters bear many similarities to digital filters, but it is important to understand the former for several reasons. First, analog filters evolved before digital filters and there is substantial overlap of terminology. Second, although digital filters are becoming more widely used, analog filters are often a more effective way to eliminate noise near the source and are essential to limit the noise bandwidth in any digital data acquisition system. Finally, although they share similarities in their characteristics, transfer functions for digital and analog filters are significantly different.

Analog filters can be classified as either active or passive. Passive filters use only resistive (i.e. resistors) and reactive (i.e. capacitors and inductors) components, whereas active filters employ operational amplifiers as well. Figure 6 shows examples of simple low-pass and high-pass resistor–capacitor (RC) filters. In principle, filters could also be constructed using resistors and inductors, but inductors tend to be more bulky, expensive and less ideal, so in practice capacitors are more commonly used. The transfer function for a simple low-pass filter is shown in Figure 7(a). The gain is given by the ratio of the capacitive reactance to the total impedance

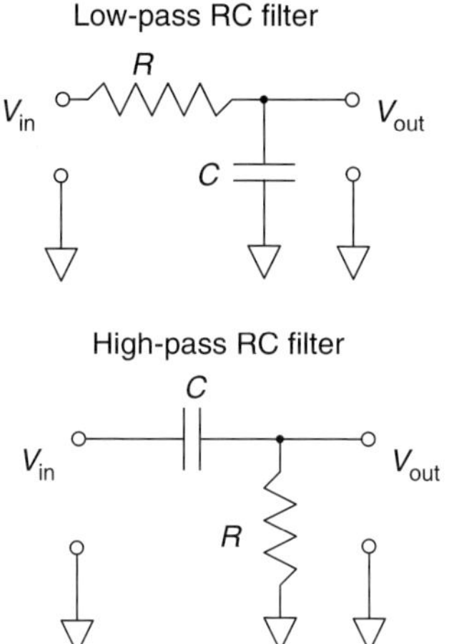

Figure 6 Simple passive RC low-pass and high-pass filters.

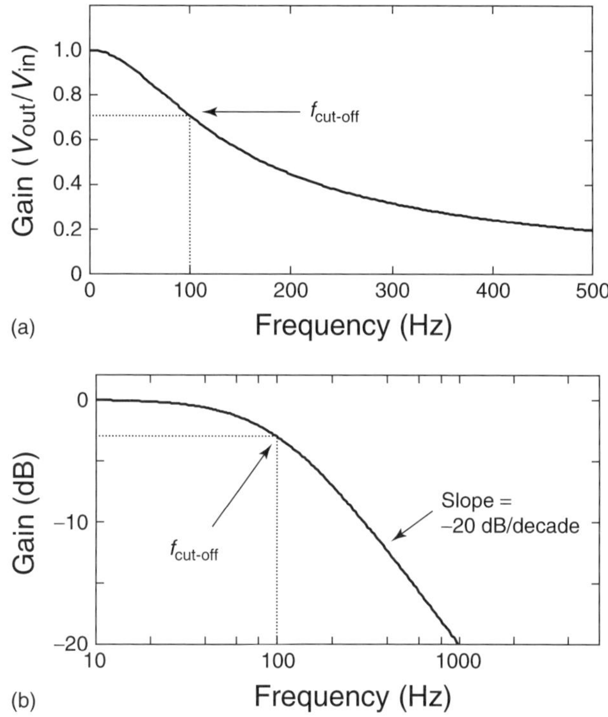

Figure 7 (a) Transfer function and (b) Bode plot for a simple low-pass RC filter with a cut-off frequency of 100 Hz.

(Equation 5):

$$\text{gain} = \frac{V_{out}}{V_{in}} = \frac{X_C}{Z} = \frac{1/(2\pi fC)}{\sqrt{R^2 + 1/(2\pi fC)^2}} = \frac{1}{\sqrt{(2\pi fRC)^2 + 1}} \tag{5}$$

Here X_C is the capacitive reactance, Z is the total impedance of the circuit, f is the frequency in hertz, R is the resistance in ohms, and C is the capacitance in farads. Note that the voltages referred to in this equation are not instantaneous voltages, but rather peak or rms values for inputs of a fixed frequency, as analog filters usually impose a phase shift on the original signal. More commonly, the frequency response of a filter is shown with a Bode plot, in which logarithmic scales are used on both axes. The voltage gain on the vertical axis is normally expressed in decibels (dB), given by Equation (6):

$$\text{gain(dB)} = 20 \log \left(\frac{V_{\text{out}}}{V_{\text{in}}} \right) \tag{6}$$

A Bode plot for a simple low-pass filter is shown in Figure 7(b). The frequency cut-off is usually taken to be the point at which the capacitive reactance equals the resistance, i.e. $V_{\text{out}}/V_{\text{in}} = 1/\sqrt{2}$. This corresponds to a gain of −3.01 dB, and so the operational cut-off frequency is usually referred to as the '3 dB point'. The cut-off frequency depends on the product RC and, for the simple low pass filter shown in Figure 6, this is given by Equation (7):

$$f_{\text{cut-off}} = \frac{1}{2\pi RC} \tag{7}$$

After the cut-off frequency, the Bode plot shows a linear region which has a slope of −20 dB per decade, more gradual than the ideal case shown in Figure 5. In order to provide a steeper slope, the order of a filter has to be increased. The filter order, also referred to as the number of poles, is the number of reactive components required for each cut-off frequency. Therefore, a first-order low- or high-pass filter such as the ones above requires only one capacitor, whereas a band-pass filter would require two. The higher the order of the filter, the more one can approach the ideal filter with a sharp transition between the pass band and the stop band. Practical limitations related to loading and other factors restrict the number of poles that can be used with passive filters, however, and active filters are normally used when more sophisticated filters are required.

The use of operational amplifiers in active filters allows greater flexibility in filter design. The design of active filters is well beyond the scope of this article, but fortunately, in those cases where a chemist may wish to incorporate an active filter, standard designs have been developed.[3–7] An active filter is classified not only according to its order, but also according to the way it optimizes a variety of other parameters, such as the steepness of the transition region, the flatness of the pass band, and its phase characteristics. Two of the most common types are the Butterworth, or maximally flat, filter and the Chebyshev, or equal ripple, filter. As shown in Figure 8, the frequency response of the Butterworth filter is very flat in the pass band, but its transition region is not as steep as the Chebyshev filter, which exhibits ripples in the pass band. The design for this active filter is shown in Figure 9. Note that the difference between the two filter types lies simply in the selection of the components and that the ripple can always be decreased at the expense of sharpness. The implementation of these filters is made even simpler through the use of integrated circuits incorporating switched capacitor filters that allow the cut-off frequency to be set simply with a reference clock signal.

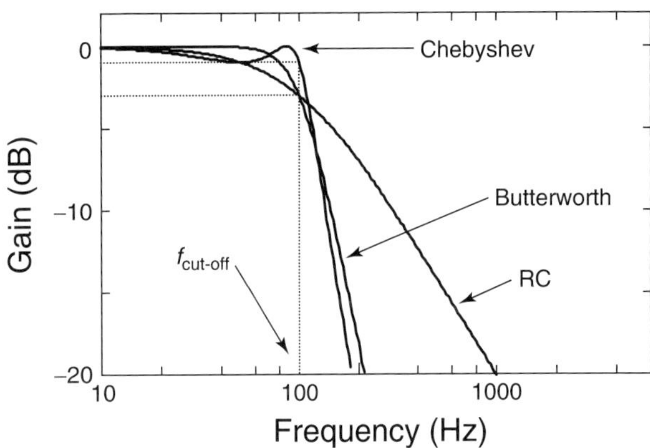

Figure 8 Frequency response (Bode plot) for passive (RC) and active (Butterworth and Chebyshev) filters.

From a digital signal acquisition perspective, perhaps the most important kind of filter is the anti-aliasing filter. When a signal is converted from the analog domain into the digital domain, it must be sampled at discrete and normally equally spaced time intervals, Δt. The sampling frequency is given by $f_s = 1/\Delta t$, and restricts the upper limit of signal frequencies that can be accurately represented in the transition from A/D information. More specifically, any components of the signal with frequencies above the Nyquist frequency will have their components aliased to lower frequencies. The Nyquist frequency, f_N, is given by Equation (8):

$$f_N = \frac{f_s}{2} = \frac{1}{2\Delta t} \tag{8}$$

The phenomenon of aliasing is shown in Figure 10, where four different sinusoids are sampled at a frequency of

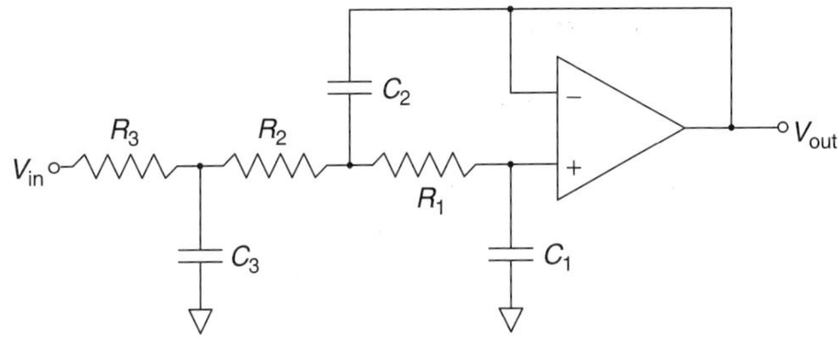

Figure 9 A third-order active filter.

For references see page 9798

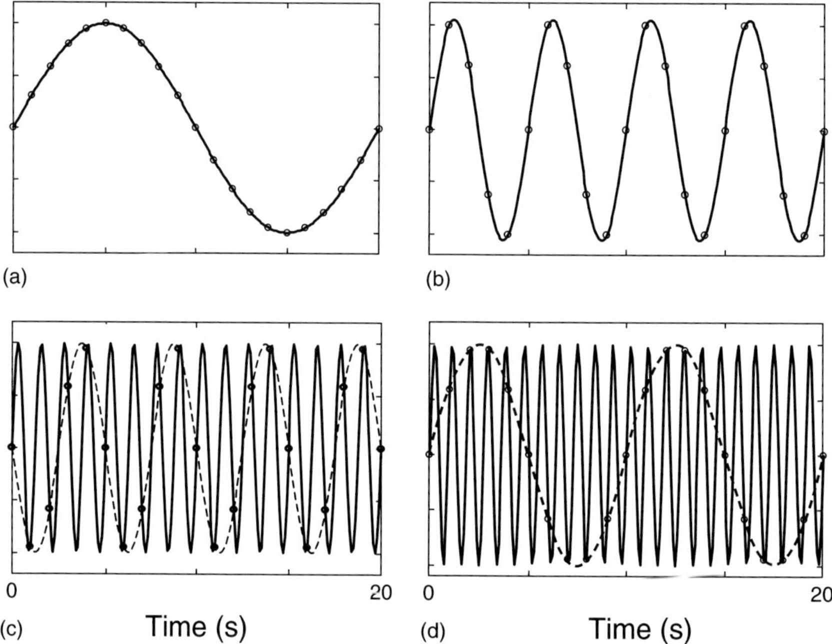

Figure 10 An illustration of aliasing in which open circles indicate sampled points. Parts (a) and (b) represent signals below the Nyquist frequency which are not aliased. The solid lines in (c) and (d) represent signals above the Nyquist frequency which are aliased to lower frequencies (dashed lines).

1 Hz (i.e. $f_N = 0.5$ Hz). The first two signals shown are below the Nyquist frequency and the sampling reflects their variations accurately. However, the last two signals are above the Nyquist frequency and aliased back to lower frequencies (Figure 10).

The importance of an anti-aliasing filter, which is simply a low-pass filter with a cut-off near the Nyquist frequency, has to do with the noise bandwidth of the analog signal. If noise components are present above the Nyquist frequency and no filtering is applied, this noise will be aliased back to lower frequencies and appear in the acquired signal. As no information at frequencies higher than the Nyquist frequency can be accurately extracted anyway, it is wise to always use an anti-aliasing filter to reduce noise in a digitally acquired signal. Also note that this noise can only be removed in the analog domain and digital filtering cannot help once the noise is aliased to lower frequencies.

4.3 Analog-to-digital Conversion

A/D conversion refers to the process by which a continuously variable analog signal, usually a voltage, is converted into a discrete numerical value with a fixed precision. In most modern analytical instruments, A/D conversion is a key step in the signal processing sequence. Although this process is generally transparent to the user, an understanding of the principles involved can be useful in practice.

As digital logic circuits are based on binary states, the digital representation of a measurement is made in the binary, or base 2, number system, consisting of a series of binary digits, or bits. Thus the number 27 in base 10 would be represented by the following 8 bits, or byte, in base 2:

$$\begin{aligned} 27_{10} &= 0001\,1011_2 \\ &= 1(2^4) + 1(2^3) + 1(2^1) + 1(2^0) \end{aligned}$$

Alternatively, binary coded decimal (BCD) can be used in which each series of 4 bits (sometimes called a nibble) represents a decimal digit:

$$27_{10} = 0010\,0111 \text{ (BCD)}$$

For n bits, a binary representation gives the numbers from 0 to $2^n - 1$, whereas a BCD representation gives a smaller range, from 0 to $10^{n/4} - 1$. To include the sign with a binary number several strategies can be employed. In the offset binary notation, a value of 2^{n-1} is subtracted from the binary number to give the signed result. In the sign magnitude notation, the most significant bit (MSB) (the left-most bit) is used to represent the sign (0 = positive, 1 = negative). Finally, the most practical representation from a mathematical point of view is the 2's complement representation. In this case, a number is negated by inverting the original bits and adding 1. The 8-bit representations of the number −27 using each of these notations is given below.

$$\begin{aligned} -27_{10} &= 0110\,0101 \text{ (offset binary)} \\ &= 1001\,1011 \text{ (sign magnitude binary)} \\ &= 1110\,0101 \text{ (2's complement binary)} \end{aligned}$$

A/D conversion is facilitated using a largely self-contained circuit, the ADC. In addition to binary coding issues, important ADC parameters from a signal-processing perspective include precision, accuracy, linearity, monotonicity, and speed. The precision of an ADC is directly related to the number of bits in the digital output, but increased precision also means decreased conversion speed for a given type of ADC. Typically, 8-bit converters are used in applications where speed is more critical than precision, whereas precision applications can use as many as 16 or 20 bits. For many scientific applications, a 12-bit ADC is a good compromise, with a precision of 1 part in 4095, or 0.02%. It is important that the precision of the ADC is better than the standard deviation of the noise in the measurement, or else the dominant source of noise will be the quantization (or digitization) noise arising from rounding of the result. As the quantization noise is always fixed at a value corresponding to the least significant bit (LSB), its relative contribution increases for small values. In some cases, this problem can be addressed through autoranging, in which small signals are amplified in the analog domain prior to conversion. This is most important when the amplitude of the noise increases with the signal.

The accuracy of an ADC refers to the closeness of the converted value to the expected value based on the range of the ADC and the reference voltage. Linearity is an indication of the constancy of the proportionality of the digital output to the analog voltage over the full range of conversion, and in some cases may be more important than the actual accuracy. In a plot of the digital output versus the analog input, linearity is often specified as the maximum deviation from the straight line drawn between zero and the full scale output, or alternatively the best-fit straight line. Monotonicity is a specification that requires an increasing analog input to give an increasing digital output with no missing codes over the full range, and likewise for decreasing inputs. These specifications (accuracy, linearity, and monotonicity) are primarily a function of the ADC design and the quality of the components used.

In addition to being discrete in the measurement domain, analog signals are also discrete in the time domain, and the required conversion speed, or sampling frequency, will depend on the application. It is essential that all of the important signal characteristics in the frequency domain fall below the Nyquist frequency, f_N (section 4.2). The speed of an ADC is primarily a function of the type of converter used, although it is also dependent on factors such as the number of bits and clock frequency. Although the design of new ADCs is an ongoing process driven by consumer electronics, most are variations on five basic types: (1) parallel, (2) tracking, (3) successive approximation, (4) integration, and (5) voltage-to-frequency (V/F) conversion.

The fastest and conceptually simplest ADC is the parallel ADC, often referred to as a flash converter. In this circuit, the input voltage is simultaneously compared with 2^n reference values and logic circuits use the closest match to produce the digital output. This brute-force approach can provide conversion speeds under 10 ns, but is not component efficient, requiring 2^n voltage comparators. As a consequence, this type of ADC is usually expensive and limited to 8 bits, although hybrid circuits referred to as half-flash converters, can increase the precision.

Other relatively fast circuits are based on the use of a digital-to-analog converter (DAC) to provide an analog voltage for comparison with the input voltage. The digital value passed to the DAC is systematically changed until its analog output matches as closely as possible the input voltage. At that point, the digital input to the DAC is taken to be the digital output of the ADC. Differences exist in the way that these devices change the digital values for comparison. In a tracking ADC, a simple counter is used to increment or decrement the digital value until a match is found. A block diagram for a simple tracking ADC is shown in Figure 11. Depending on the change in the voltage, the conversion time for this type of ADC could range from 0 to 2^n clock cycles for an n-bit converter. The successive approximation ADC operates on the same basic principle as the tracking

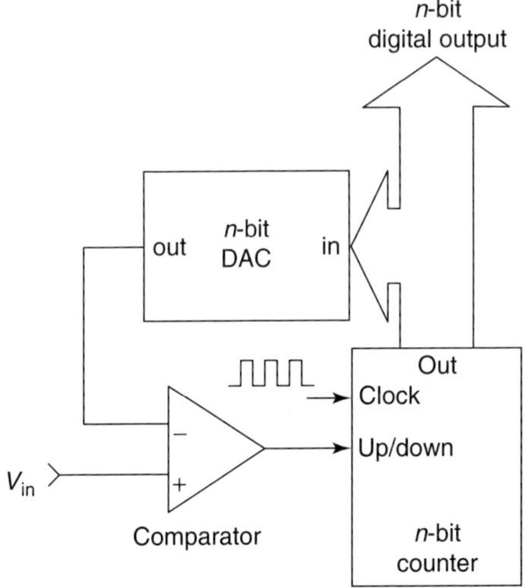

Figure 11 Block diagram of a simple tracking ADC.

For references see page 9798

ADC, but has a fixed conversion time of only n clock cycles. This is accomplished by replacing the counter with circuitry that uses an efficient binary search algorithm to split the digital range in two for each comparison. Successive approximation ADCs have typical conversion times ranging from 1 to 100 μs and are among the most common converters in use.

The remaining two types of ADCs are relatively slow by comparison. Integrating ADCs are based on the use of a fixed current to charge a capacitor through a resistor. The digital result is obtained by timing the period necessary for the capacitor to reach the input voltage. This basic strategy, termed a single slope integrating ADC, suffers from a dependence on accurate and stable circuit components and has been replaced by the dual slope (or even quad slope) integrating ADC. The dual slope ADC is also based on charge integration, but uses a charge–discharge cycle to cancel the effects of component variations. This type of ADC is known for its accuracy, stability, monotonicity, low cost and noise rejection characteristics and has been used in a large number of precision applications where speed is not critical. Typical conversion rates are around 10 conversions per second. In many applications, the dual slope ADC has been replaced by the V/F converter. This device simply produces a series of pulses whose frequency is proportional to the input voltage applied, with typical maximum frequencies in the range from 10 kHz to 1 MHz. To complete the conversion to a binary number, a counter is attached to the pulsed output and counted for a fixed period of time. Precision can be improved by counting for a longer period. In addition to low cost, simplicity, and good linearity, the V/F converter has the advantage that its output can be transmitted in serial over a single line, simplifying remote data acquisition. Like the integrating ADCs, however, this type of ADC is relatively slow.

Although the analyst typically has very little control over the type of A/D conversion that is used in a particular instrument, this is the first stage of all subsequent digital signal processing, so it is important to recognize the strengths and weaknesses of these devices. Further details on the design and application can be found in a number of references.[3–5,7–8]

5 DIGITAL FILTERING

5.1 Introduction

Because digital filters are among the most widely used methods for signal processing in analytical chemistry, much of this article is dedicated to describing their implementation and operation. The term 'filter' is a reference to the similarities they share with their electronic counterparts. In both cases, the data are presented to the filter in a sequential fashion and the distinction between pure signal and noise is often made on the basis of differences in power spectra. Strictly speaking, however, a filter uses only information from the past up to and including the current point in obtaining an estimate of the current point. Although this is true for electronic filters, it does not hold for most digital filters, which use points before and after the measurement of interest to form an estimate. Thus, they may be more properly classified as smoothers or smoothing filters, but all three terms are used in the literature. The term 'filter' is used in a general way throughout this section, incorporating both smoothers and other types of filters.

Digital filtering can be performed either in real-time or in a post-acquisition mode. The advantage of the former is that it can be transparent to the user and optimized at the time of instrument design. Fast digital signal processors (DSPs) are available for performing common signal processing operations and can be built into the instrument itself. A disadvantage of this approach is that it removes some flexibility and may obscure some of the features of the original data. For these reasons, real-time digital filtering is often kept to a minimum and much signal processing is carried out in post-acquisition mode. Therefore, it is important to appreciate the advantages and limitations of different types of digital filters. A complete coverage of the subject of digital filters is well beyond the scope of this article. The objective here is to present some of the terminology and describe some of the digital filters commonly used in analytical chemistry. For readers seeking more information on the subject, there is abundant literature available.[9–19]

5.2 Filter Types

In general terms, a digital filter could be defined simply as an operation that is carried out on a contiguous subset of the original signal sequence to produce an estimate of a value in the filtered signal sequence. This is illustrated in Figure 12. For conventional digital filters, the filtering operation consists of the convolution of a series of filter coefficients with the signal, but this approach is by no

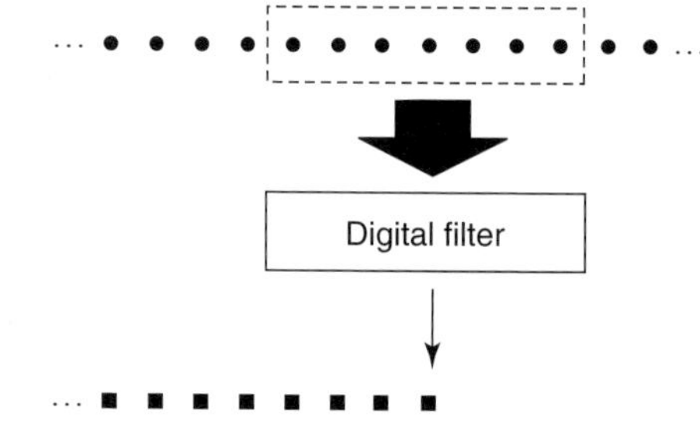

Figure 12 General operation of a digital filter.

List of selected abbreviations appears in Volume 15

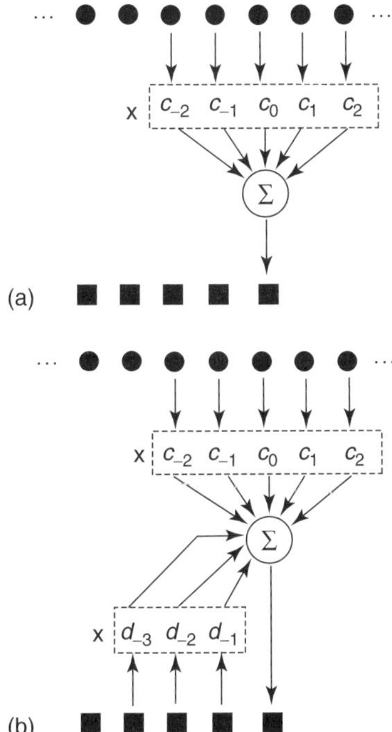

Figure 13 Operation of (a) nonrecursive and (b) recursive digital filters.

means universal. Digital filters are therefore classified in a number of ways according to the manner of their operation.

The most common type of digital filter employed in analytical chemistry is the nonrecursive filter, also referred to as the finite impulse response (FIR) filter because its response to an impulse (delta) function will always fall to zero at some point in time. As illustrated in Figure 13(a), nonrecursive filters use the conventional approach of convoluting a set of filter coefficients with the sequence of measurements to produce the filtered signal. If z_i represents the filtered value for measurement i, it is determined mathematically by Equation (9):

$$z_i = \sum_{j=p}^{q} c_j y_{i+j} \qquad (9)$$

where the c terms are the filter coefficients and the y terms are the original measurements. Most often, the filter coefficients are arranged symmetrically around the point to be estimated so that $p = -q$, but this is not a requirement. Nonrecursive filters have good stability and are relatively easy to design. A simple example of a nonrecursive filter is a five-point moving average filter, which averages the five points around a central value (two on either side plus the point itself) to obtain its estimate. The coefficients in this case would be $c_{-2} = c_{-1} = c_0 = c_1 = c_2 = 0.2$.

Recursive filters differ from nonrecursive filters in that they make use of previously filtered values to estimate the current measurement, as shown in Figure 13(b). Mathematically, this can be represented by Equation (10),

$$z_i = \sum_{j=p}^{q} c_j y_{i+j} + \sum_{k=r}^{s} d_k z_{i+k} \qquad (10)$$

where z represents the filtered measurement, y is the unfiltered measurement, and c and d are the filter coefficients. Note that the indices r and s must be less than zero, as the filter coefficients d_k can only be applied to previously filtered values. In practice, many nonrecursive filters are designed to function in real time and so also limit the maximum value of q to be zero to make them physically realizable (i.e. they do not make use of future values). A simple example of a recursive digital filter is integration using the trapezoid rule, which gives (Equation 11):

$$z_i = \tfrac{1}{2}(y_{i-1} + y_i) + z_{i-1} \qquad (11)$$

This corresponds to coefficients $c_{-1} = 0.5$, $c_0 = 0.5$ and $d_{-1} = 1$. Recursive digital filters are also known as infinite impulse response (IIR) filters because it is possible for a single impulse input to influence filter output values indefinitely, as illustrated in this example. These filters can be used for normal smoothing operations in addition to integration, and have the advantage that they provide more efficient filters with fewer coefficients than nonrecursive filters. However, recursive filters are more difficult to design and have more complex properties than their nonrecursive counterparts, so they have not been widely used in analytical chemistry.

It is not required that digital filter coefficients remain constant throughout a sequence of measurements. In some cases, the coefficients may change in response to variations in another external input or the signal itself. These kinds of filters are called adaptive filters. Such filters are used, for example, in removing noise from audio signals using external measurements of noise characteristics. Chemical applications of this type of filter are rare, although digital processing of modulated signals, such as modulated sources in atomic absorption spectroscopy, could be considered to be a simple form of adaptive filtering.

Filters can also be classified according to the types of results they are intended to produce. Smoothing filters are intended to reproduce the pure signal and suppress the noise, whereas derivative filters are intended to estimate the derivative (first, second, or higher) of the pure signal. In other cases, one may wish to increase the apparent sampling frequency of a set of data, filling in points for esthetic or practical reasons (e.g. locating a

For references see page 9798

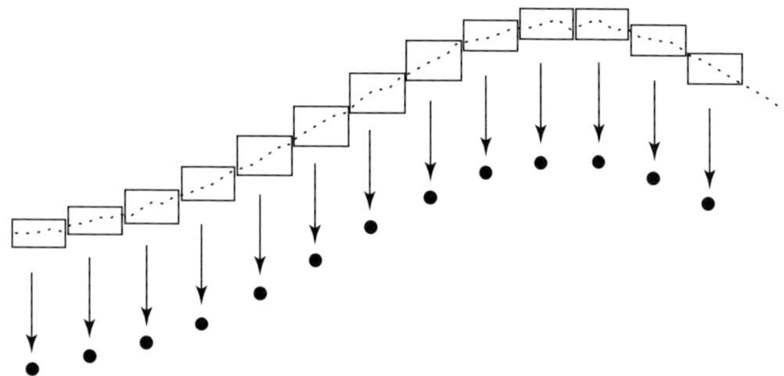

Figure 14 Illustration of boxcar averaging.

peak maximum). This can be done using an interpolation filter, although care should be taken when applying such approaches. In contrast, in cases where the signal is highly oversampled, one may wish to reduce the volume of data while improving its quality by using a decimation filter.

Some types of digital filters do not follow the usual pattern of smooth convolution of a set of filter coefficients with the data, but are considered digital filters nevertheless. In boxcar averaging, for example, the signal is divided into subsets of n measurements which are averaged to produce a single result, as shown in Figure 14. As this reduces the total number of points, it is one type of decimation filter. Another rarely used but very useful filter is the median filter. Although this filter should be used with care because of its nonlinear transformation of the data, it is particularly effective at removing spikes or outliers from a measurement sequence. These outliers may arise, for example, from cosmic rays striking a photodetector or bubbles passing through a detector flow-cell. The median filter works by sorting the data within a window of length n and choosing the median value as the filtered estimate, thus automatically eliminating outliers unless they occur in clusters. An example of the application of the median filter is shown in Figure 15.

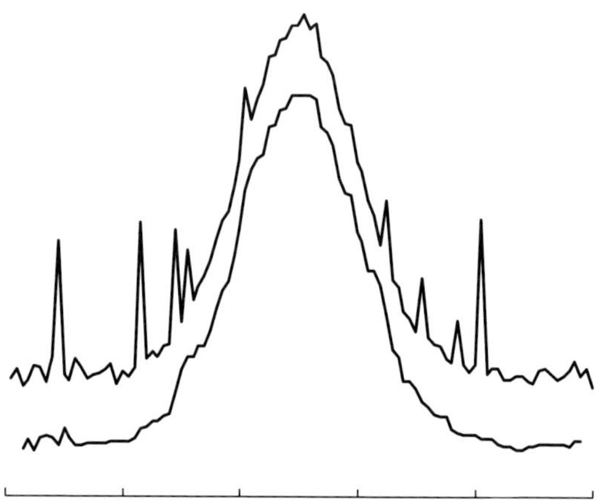

Figure 15 A noisy peak (upper curve) before and (lower curve) after filtering with a median filter. Note the removal of spikes.

List of selected abbreviations appears in Volume 15

Because of their dominance in chemical applications the main emphasis of this article is on nonrecursive digital filters, particularly polynomial least-squares filters, although some discussion of Kalman filters is also presented.

5.3 Polynomial Least-squares Smoothing Filters

By far the most widely applied digital filter in analytical chemistry is the polynomial least-squares smoothing filter. These filters are more commonly known to analytical chemists as SG filters, a reference to their introduction into the analytical chemistry literature by Savitzky and Golay in 1964.(1) Although these filters were known in the field of signal processing prior to this, the SG paper made their utility known to chemists at a time when the digital acquisition of signals was becoming more commonplace. In addition, the paper presented tables of precalculated coefficients for different types of filters. At the time, computational efficiency was poor, so the authors presented the coefficients as integers with a normalization factor rather than as a series of floating point numbers.

Among the advantages of the SG filters are their simplicity and versatility. Application of these filters assumes that a local region of the data set (i.e. the filter window) can be fit to a low-order polynomial, and the central point within that window is estimated by performing such a fit. This is illustrated in Figure 16 for first- and second-order polynomials. Fitting the data in this way should model the correlations in the pure signal while reducing the influence of random noise fluctuations. Furthermore, when the measurements are evenly spaced in the time domain, the fitted estimate of the central point can be obtained simply by multiplying the points in the

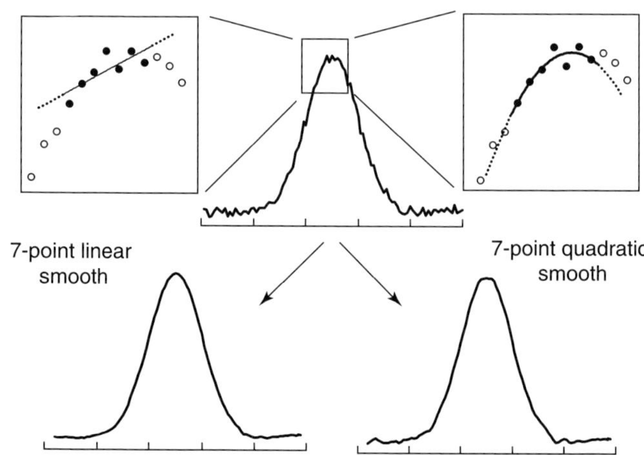

Figure 16 Illustration of polynomial smoothing as a least-squares fitting procedure.

window by a set of precalculated coefficients, thus making the fit equivalent to a nonrecursive digital filter.

The original tables published by SG had a number of errors that were later corrected in the literature,[20] but it is now just as simple to calculate the coefficients for a given application, so the tables are seldom used. To illustrate how this is done, consider the design of a second-order polynomial smoothing filter using a five-point window. The model to be fit is Equation (12):

$$y = b_0 + b_1 x + b_2 x^2 \tag{12}$$

Equation (13) is the equivalent matrix form:

$$\begin{bmatrix} y_1 \\ y_2 \\ y_3 \\ y_4 \\ y_5 \end{bmatrix} = \begin{bmatrix} 1 & x_1 & x_1^2 \\ 1 & x_2 & x_2^2 \\ 1 & x_3 & x_3^2 \\ 1 & x_4 & x_4^2 \\ 1 & x_5 & x_5^2 \end{bmatrix} \begin{bmatrix} b_0 \\ b_1 \\ b_2 \end{bmatrix} \tag{13}$$

which can be expressed as (Equation 14):

$$\mathbf{y} = \mathbf{X}\mathbf{b} \tag{14}$$

In these expressions, x represents the time or other ordinal variable, while $\mathbf{X}$ is the matrix containing the basis functions for the polynomial fit. It is important to note that the fitted values obtained for y are independent of the scale of x and, if the time interval between each measurement is equal (as is usually the case), we can arbitrarily set $\mathbf{x} = [-2 \quad -1 \quad 0 \quad 1 \quad 2]$, giving Equation (15):

$$\mathbf{X} = \begin{bmatrix} 1 & -2 & 4 \\ 1 & -1 & 1 \\ 1 & 0 & 0 \\ 1 & 1 & 1 \\ 1 & 2 & 4 \end{bmatrix} \tag{15}$$

The least-squares solution for the vector of regression coefficients, $\mathbf{b}$, is well known from linear algebra to be (Equation 16):

$$\mathbf{b} = (\mathbf{X}^T\mathbf{X})^{-1}\mathbf{X}^T\mathbf{y} = \mathbf{A}\mathbf{y} \tag{16}$$

The matrix $\mathbf{A}$ is a 3×5 matrix which can be regarded as being composed of three row vectors, $\mathbf{a}_1$, $\mathbf{a}_2$, and $\mathbf{a}_3$; as in Equation (17):

$$\mathbf{A} = \begin{bmatrix} a_{11} & a_{12} & a_{13} & a_{14} & a_{15} \\ a_{21} & a_{22} & a_{23} & a_{24} & a_{25} \\ a_{31} & a_{32} & a_{33} & a_{34} & a_{35} \end{bmatrix} = \begin{bmatrix} \leftarrow \mathbf{a}_1 \rightarrow \\ \leftarrow \mathbf{a}_2 \rightarrow \\ \leftarrow \mathbf{a}_3 \rightarrow \end{bmatrix} \tag{17}$$

Note that the intercept coefficient for the fit, b_0, is obtained from Equation (18),

$$b_0 = \mathbf{a}_1\mathbf{y} = a_{11}y_1 + a_{12}y_2 + \cdots + a_{15}y_5 \tag{18}$$

Also note that, as $x = 0$ for the central point in the five point sequence, Equation (19) holds:

$$\hat{y}_3 = b_0 + b_1(0) + b_2(0)^2 = b_0 = \mathbf{a}_1\mathbf{y} \tag{19}$$

Therefore, because of the way the problem has been set-up, the estimate of the central point in the sequence is obtained simply by multiplying each measurement by the corresponding element in $\mathbf{a}_1$. In other words, the digital filter coefficients are simply the first row of the matrix $(\mathbf{X}^T\mathbf{X})^{-1}\mathbf{X}^T$, i.e. $\mathbf{c} = \mathbf{a}_1$.

The above reasoning holds for polynomial smoothing filters of any length and any order. All that is required to determine the filter coefficients is to set up the matrix of basis functions, $\mathbf{X}$, perform the calculation in Equation (16) using a spreadsheet or other software, and extract the first row of the resulting matrix. Polynomial smoothing filters are convenient for improving the appearance and SNR of many signals and have the added advantage over electronic filters that different types of filters can be applied after the signal has been recorded.

One of the drawbacks of polynomial smoothing filters is sometimes referred to as the edge effect. As the filters are designed to obtain an estimate of the central point in a window, there will be points at the beginning and end of a measurement vector that cannot be estimated with the symmetric filter. For example, with the five-point filter described above, two points could not be filtered at each end of the data sequence. Several options are available to deal with this problem. The simplest is either to drop these points from the data set, or leave them in the data set unfiltered. Another possibility, if only baseline data occurs at the limits of the measurement vector, is to use the points at one end of the data set to filter those at the other. For example, to estimate y_1 with the five-point filter, we could use the sequence (y_{n-2}, y_{n-1}, y_1, y_2, y_3). Finally, we could employ what are sometimes referred to as initial point filters or extended sliding window filters,[21,22] designed to estimate values other than the central point of a sequence. The coefficients for these filters are easily obtained by simply shifting the $\mathbf{X}$ matrix accordingly. For example, to obtain coefficients to estimate the first point of a five point sequence with a second-order smooth, we would use Equation (20):

$$\mathbf{X} = \begin{bmatrix} 1 & 0 & 0 \\ 1 & 1 & 1 \\ 1 & 2 & 4 \\ 1 & 3 & 9 \\ 1 & 4 & 16 \end{bmatrix} \tag{20}$$

All other aspects of the problem are the same. Although this is an elegant way to solve the problem, it requires several sets of filter coefficients to handle the points at the edges of the data and the noise rejection and signal

For references see page 9798

distortion characteristics of these filters are not identical to the symmetric filters.[23]

The selection of smoothing filter parameters (order, number of points) for a given application is often a matter of trial and error and intuition. Obviously, one would like to obtain the maximum noise reduction with the minimum amount of distortion. Although the best noise reduction occurs with wider filters (more points), wider filters also limit the ability of the chosen function to obtain a good local model for a changing signal. In general, noise rejection improves and signal distortion increases as the width of the signal increases and the order of the filter decreases. It should be noted, however, that as a consequence of the mathematics, smoothing filters for orders 0 and 1 are identical, as are those for orders 2 and 3, and so on. To determine the amount of noise reduction a filter will provide, there is a very simple relationship (Equation 21):

$$\frac{\sigma^2_{\text{filtered}}}{\sigma^2_{\text{unfiltered}}} = \sum_{j=p}^{q} c_j^2 \qquad (21)$$

Thus the ratio of the variance of the noise in the filtered signal to that in the unfiltered signal is simply the sum of the squared filter coefficients. This equation applies to any type of nonrecursive digital filter, but is only valid in cases where the signal exhibits white noise.

Although Equation (21) is useful for describing the amount of noise reduction, it provides no information about the extent of signal distortion. To this end, it is often more useful to examine the response of a digital filter in the frequency domain in much the same way as for an electronic filter. For a symmetric smoothing filter, the frequency response is given by Equation (22),

$$H(f) = \left| \sum_{k=-m}^{m} c_k \cos\left(\frac{k\pi f}{f_{\text{N}}}\right) \right| \qquad (22)$$

where $H(f)$ is the amplitude gain of the filter at frequency f, c_k represents the filter coefficients, and f_{N} is the Nyquist frequency. Note that this does not apply to unsymmetric smoothing filters, such as the initial point filters, because these filters also involve a phase shift, but alternative expressions are available.[23] Figure 17 shows the frequency response for an 11-point quadratic smoothing filter, plotted as a function of f/f_{N} to make it universal. As expected, the amplitude gain is unity at low frequencies and drops off at higher frequencies. Unlike simple electronic filters, however, the gain of these filters does not smoothly approach zero, but instead oscillates around a number of nodes that are related to the size and order of the filter. Furthermore, if the frequency response were plotted beyond the Nyquist frequency, the function would simply reflect itself as signals were aliased

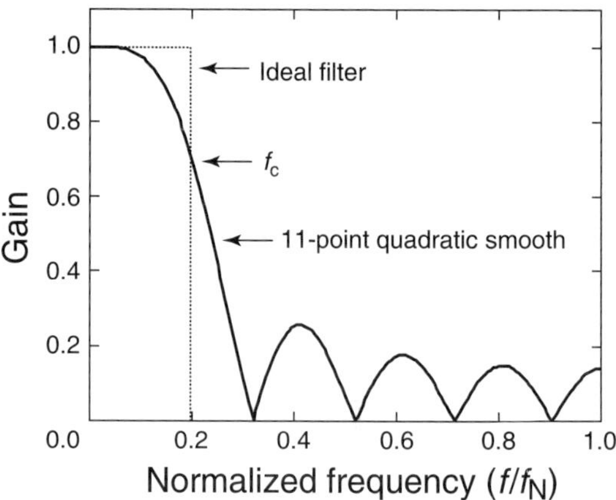

Figure 17 Frequency response (transfer function) for an 11-point quadratic smoothing filter.

to lower frequencies, reaching a gain of unity once again at the sampling frequency. If one has some idea of the amplitude spectrum of the signal to be filtered, plots such as Figure 17 can be very useful in assessing the degree of signal distortion that will result from filter application. The amount of noise rejected can be also ascertained from the ratio of the area under the NPS before and after multiplication by the filter frequency response. Clearly, this type of filter will be effective when white noise is present, but will be less effective for $1/f$ (drift) noise, as the noise exists predominantly at low frequencies. It should also be noted that even if white noise was present before filtering, measurement noise will become correlated after filtering.

Of course, whenever a digital filter is applied to experimental data, there will be changes in the shape of the signal. In certain cases where parameters of the signal such as peak height, area or width are important in themselves, consideration must be given to the consequences of applying a digital filter. Generally, for signals which exhibit the same shape, the effects in the time domain (e.g. width at half-maximum) will be the same, whereas effects on the amplitude (e.g. peak height/area) will be linear with the magnitude of the signal. The reader is referred to a useful, although somewhat empirical, study by Enke and Nieman for more details on these effects.[16]

5.4 Derivative Filters

The numerical differentiation of signals with respect to time (or other ordinal variable) is a common practice in analytical chemistry. This procedure can be used, for example, to locate the position of a peak maximum, to determine the end-point of a titration, or to highlight poorly defined features in a signal sequence (e.g. the shoulder on a peak). These can be regarded as qualitative applications in the sense that one is looking for the location of specific points in the signal derivative (e.g. the maximum or zero values) but not using the derivative sequence for further calculations. Quantitative applications, in which the differentiated signal is used for purposes such as calibration, have become more common in recent years. Because the derivative of a function is unaffected by the addition of a constant, differentiation proves valuable for methods which exhibit a baseline shift (offset) between samples. Likewise, if sample measurements are plagued with a baseline that changes linearly with the ordinal variable (drift), calculation of the second derivative can solve the problem.

One serious problem with signal differentiation is the selective amplification of high-frequency noise. Because the derivative of a signal is, by definition, its rate of change, the more rapidly varying components of a signal, including noise, are amplified to a greater extent than the more slowly changing features typically associated with the pure signal. Because of this, derivative filters (first, second or higher) are most useful for signals which exhibit relatively small amounts of high-frequency noise compared to the low-frequency contributions, i.e. cases where $1/f$ or drift noise dominates. A classic example of this is near infrared (NIR) spectroscopy, where second derivatives are routinely calculated prior to quantitative analysis. Although NIR measurements are widely characterized in the literature as having very high SNRs, these measurements suffer from serious noise problems in the form of baseline offset and drift, but traditional SNR calculations normally do not incorporate these components. Fortunately, the characteristics of NIR spectra make them almost ideal benefactors of derivative filtering.

The noise amplification characteristics of derivative filters can be better understood by examining the transfer function in the frequency domain. As shown in Figure 17, the gain of a smoothing filter is typically unity at low frequencies and falls off at high frequencies. In contrast, Figure 18 shows that the gain of a true derivative filter increases linearly with frequency. This can be easily confirmed by recognizing that (Equation 23),

$$\frac{\mathrm{d}(\sin wt)}{\mathrm{d}t} = w \cos wt \qquad (23)$$

where $w = 2\pi f$ is the angular frequency. As a signal in the time domain can be represented as the sum of a series of sines and cosines (section 6.2) it is clear the calculation of the derivative amplifies a signal component by a factor of w and also gives rise to a 90° phase shift. The consequence of this is that the excessive amplification of high-frequency noise components generally makes the calculation of the true derivative for practical measurements useless (and is the reason why the term ‘true’ is used instead of ‘ideal’).

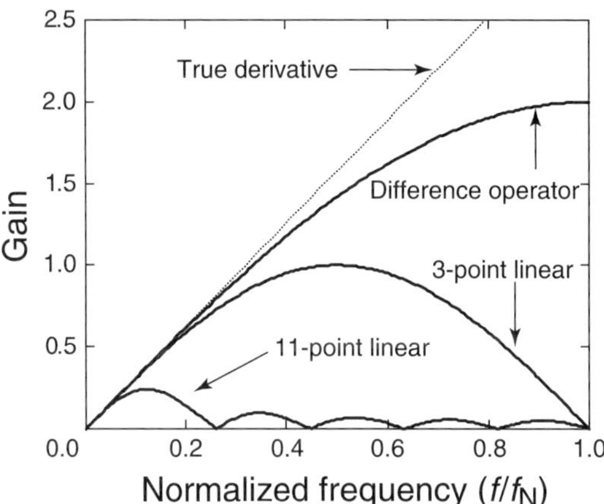

Figure 18 Frequency response (transfer function) for various types of derivative filters.

In practice, most derivative filters implicitly combine the derivative calculation with a low-pass filter to reduce the contribution of high frequency components.

The simplest method for calculating a signal derivative, and one which can be described as a simple digital filter, is the difference operator, defined by Equation (24),

$$\hat{y}_i' = \frac{y_{i+1} - y_i}{\Delta t} \qquad (24)$$

where $\hat{y}_i'$ represents the estimate of the derivative of the function at point i, and Δt is the sampling interval. The transfer function for this type of filter is shown in Figure 18. It is clear that the frequency response for a simple difference operator matches that of the true derivative closely except at very high frequencies. This type of filter provides no low-pass filtering, however.

An alternative way of calculating derivatives is to use a slight modification of the polynomial least-squares filters described in section 5.2. If we carry out the least-squares fit in the same manner as the previous example for a five-point quadratic model, the estimate of the derivative for the central point is given by Equation (25),

$$\begin{aligned}\hat{y}_0' &= \frac{\mathrm{d}}{\mathrm{d}x}(b_0 + b_1x_0 + b_2x_0^2) = b_1 + 2b_2x_0 \\ &= b_1 + 2b_2(0) = b_1\end{aligned} \qquad (25)$$

Therefore, the derivative estimate of the central point is simply the first-order coefficient of the fit. In a manner analogous to Equation (18), this is obtained by simply multiplying the second row of the $\mathbf{A}$ matrix by the windowed measurement vector $\mathbf{y}$. This means that the coefficients of the derivative filter are given by the second row of $\mathbf{A}$ ($\mathbf{c} = \mathbf{a}_2$) as opposed to the first row of $\mathbf{A}$ for the smoothing filter. Likewise, the filter coefficients for the second derivative are given by the third row of $\mathbf{A}$, and so

For references see page 9798

on. It is clear then, that there is a simple, common path to the calculation of polynomial filters of various types.

A number of characteristics of derivative filters calculated in this way should be noted. First, unlike smoothing filters, the calculation of numerically correct derivatives requires consideration of the sampling interval. The adjusted coefficients necessary to obtain the correct scale are given by Equation (26),

$$\mathbf{c}_{\text{adjusted}} = \frac{\mathbf{c}_{\text{original}}\, p!}{(\Delta t)^p} \quad (26)$$

where p is the order of the derivative. In many cases, this scaling is ignored, because it is only the relative changes in the derivative that are important. A second characteristic of these derivative filters that should be apparent from the mathematics is that the determination of coefficients for a pth order derivative requires at least a pth order polynomial. Also, as with smoothing filters, there is a duplication of coefficients for adjacent polynomial orders, although the pairing shifts with each higher derivative. For example, for the first derivative, the filter coefficients for the linear and quadratic polynomials are the same, whereas the quadratic and cubic coefficients are the same for the second derivative.

The polynomial filters described here are symmetric in the sense that there are an equal number of coefficients on either side of the central point (in fact, a first-derivative filter is better described as antisymmetric, as $c_{-i} = -c_i$). As with smoothing filters, it is possible to develop derivative filters for the edges of a window, but the usual precautions regarding the quality of estimates apply. For symmetric first-derivative filters, the frequency response is given by,

$$H(f) = \left| \sum_{k=-m}^{m} c_k \sin\left(\frac{k\pi f}{f_{\text{N}}}\right) \right| \quad (27)$$

Equation (27) is a slight modification of Equation (22), where the substitution of sine for cosine results from the 90° phase shift brought about by the derivative filter. The frequency responses for 3-point and 11-point first-order derivative filters are shown in Figure 18 for comparison with the true derivative and difference operator. Note that although the transfer functions for the polynomial filters match the true derivative at low frequencies, there is significant attenuation at high frequencies due to the low-pass filtering. The effects of this low-pass filtering are clearly seen in Figure 19, which shows the application of three types of derivative filters to a noisy signal. Because of the effect of high-frequency noise on derivative filter response, it is a common practice by some to first apply a smoothing filter to the data, but Figures 18 and 19 demonstrate that if the derivative filter is properly designed, this practice is redundant.

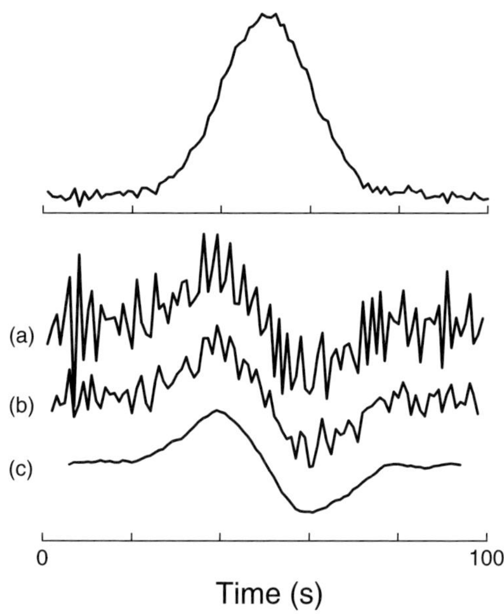

Figure 19 Result of the application of various types of derivative filters to a noisy peak, shown above: (a) difference operator, (b) 3-point linear derivative filter, (c) 11-point linear derivative filter.

5.5 Kalman Filters

The Kalman filter is a recursive linear least-squares estimator with the capability of estimating the parameters associated with a system model in real time.[17–19] It is not so much a filter in the conventional sense as it is a means for carrying out linear least-squares in a recursive fashion. The estimates it provides are not the smoothed measurements, but rather the parameters associated with the linear model, or the state parameters. These parameters are sometimes considered to represent a state vector in an n-dimensional state space (n = number of parameters). Once the state parameters have been estimated, it is possible to generate a smooth curve for the measurements from the model, but this is usually a secondary objective.

A simple example of recursive estimation in a manner similar to the Kalman filter is the calculation of a mean from a series of measurements as the measurements are being acquired. It is usual to start with an estimate of the mean equal to the first measurement, i.e. $\hat{m}_1 = x_1$. Once a second measurement was acquired, the estimate could be improved by using $\hat{m}_2 = 1/2\hat{m}_1 + 1/2x_2$. In general, after the ith measurement, Equation (28) would hold:

$$\hat{m}_i = \frac{i-1}{i}\hat{m}_{i-1} + \frac{1}{i}x_i \quad (28)$$

As each new measurement is assimilated into the estimation, the quality of the parameter estimate improves. It is apparent that Equation (28) has the form of a recursive filter whose coefficients are changing with each measurement. An advantage of recursive estimation is that continuous updates of the parameter(s) of interest are obtained with each new measurement. Although this could also be done in batch mode, the recursive formulation is computationally more efficient.

The general model which is covered by the Kalman filter can be described by Equations (29) and (30):

$$\mathbf{x}_{k+1} = \mathbf{F}_k\mathbf{x}_k + \mathbf{w}_k \quad (29)$$

$$\mathbf{z}_k = \mathbf{H}_k\mathbf{x}_k + \mathbf{v}_k \quad (30)$$

In these equations, $\mathbf{x}_k$ represents the $n \times 1$ vector of parameters to be estimated (the state vector) at measurement interval k and $\mathbf{z}_k$ represents the $m \times 1$ vector of measurements at interval k. The first equation describes how the state vector is expected to change from one measurement interval to the next and contains both systematic and stochastic terms. The $n \times n$ state transition matrix, $\mathbf{F}$, describes the systematic linear transformation, whereas the vector of random variables, $\mathbf{w}$, represents the stochastic change. Each element of $\mathbf{w}$ is assumed to be derived from a zero-mean white-noise sequence and $\mathbf{w}$ is characterized by an $n \times n$ covariance matrix $\mathbf{Q}$.

Equation (30) describes how the state parameters are translated into a measurement or observation vector. The linear relationship is described by the $m \times n$ observation matrix, $\mathbf{H}$. There is also a random noise component assumed for the observations, represented by the vector $\mathbf{v}$. The elements of this vector are also assumed to comprise a white-noise sequence and the covariance of noise in the measurement vector is described by the $m \times m$ covariance matrix, $\mathbf{R}$.

The basic algorithm for the Kalman filter is shown in Figure 20, although several variants exist. The application of this algorithm is best described through a simple example. Suppose, for the purposes of illustration, we are using an absorption spectrometer to monitor a reaction in which two absorbing species, A and B, are reacting independently to form nonabsorbing products by first-order kinetics. The defining Equation (31) is:

$$\mathbf{z}(t) = A(t) = C_A\varepsilon_A e^{-k_A t} + C_B\varepsilon_B e^{-k_B t} + v(t) \quad (31)$$

where $A(t)$ is the absorbance at time t, C_A and C_B represent the initial concentrations of the two species, ε_A and ε_B are their molar absorptivities, and k_A and k_B are their first-order decay constants. Assuming that all quantities are known except the initial concentrations which are to be estimated, this is a linear problem (Equations 32):

$$\mathbf{x}(t) = \begin{bmatrix} C_A \\ C_B \end{bmatrix}; \quad \mathbf{H}(t) = [\varepsilon_A e^{-k_A t} \quad \varepsilon_B e^{-k_B t}] \quad (32)$$

A simulated data set was generated using $C_A = 0.2\,\text{mM}$, $C_B = 0.5\,\text{mM}$, $\varepsilon_A = 1000\,\text{M}^{-1}\,\text{cm}^{-1}$, $\varepsilon_B = 1500\,\text{M}^{-1}\,\text{cm}^{-1}$, $k_A = 0.1\,\text{s}^{-1}$, $k_B = 0.3\,\text{s}^{-1}$, and a noise level of 0.02 absorbance units (AU). These data, showing the measured points and the curve with no error, are presented in Figure 21(a). The objective of the Kalman filter is to

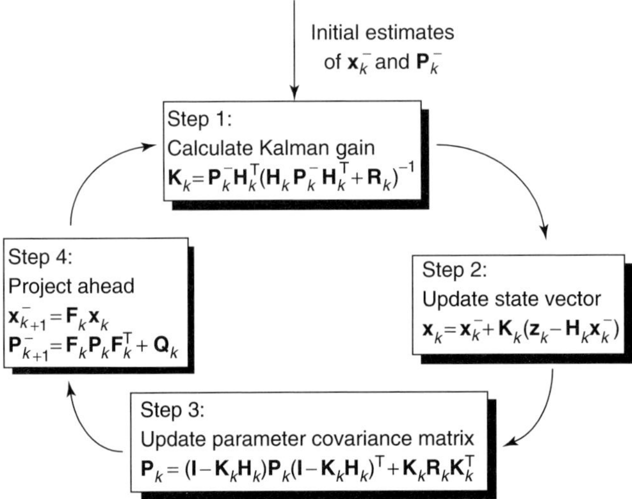

Figure 20 Basic algorithm for the Kalman filter.

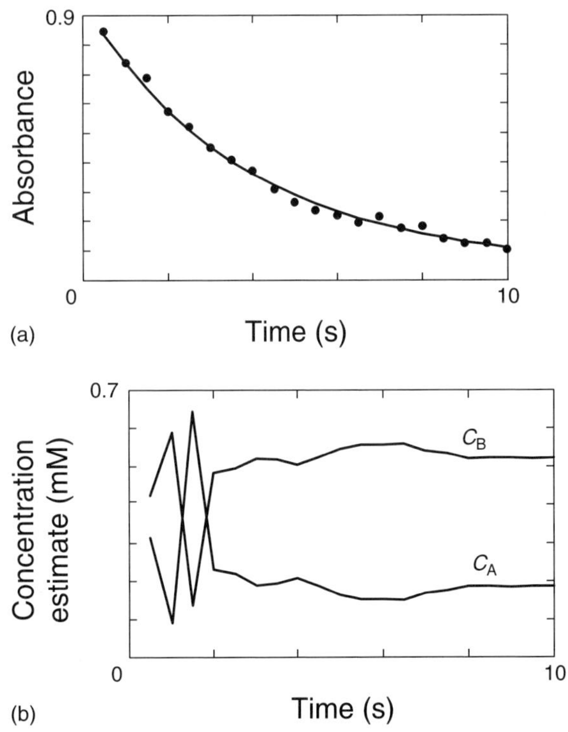

Figure 21 Results from the application of the Kalman filter to the first-order kinetic data described in the text. (a) The solid line shows the true absorbance decay curve and the points indicate the noisy measurements made. (b) The evolution of estimates for the initial concentrations of components A and B by the Kalman filter.

For references see page 9798

provide estimates of C_A and C_B as each new measurement is made.

To initiate the Kalman filter prior to step 1, we need an estimate of the state parameters and the error covariance matrix, **P**, describing the uncertainty in those parameters. Assuming we have no prior knowledge, we use Equations (33):

$$\mathbf{x}_1^- = \begin{bmatrix} 0 \\ 0 \end{bmatrix}; \quad \mathbf{P}_1^- = \begin{bmatrix} 10^{10} & 0 \\ 0 & 10^{10} \end{bmatrix} \tag{33}$$

The superscript '−' indicates that these are the estimates before we have assimilated the first measurement. As we have no prior knowledge of the parameters, the diagonal elements (variances) of the covariance matrix are set to very large values. In order to estimate the Kalman gain, **K**, in step 1, the observation matrix, **H**, can be calculated from Equation (32). As the measurement, **z**, is a scalar in this case, the measurement error covariance matrix, **R**, is simply equal to the variance of the measurements. Thus, at $t = 0.5$ s, Equations (34) hold:

$$\mathbf{H}_1 = [951.2 \quad 1291.1]; \quad \mathbf{R}_1 = (0.02)^2 = 0.0004 \tag{34}$$

With these values, the $n \times 1$ Kalman gain vector for this iteration is Equation (35):

$$\mathbf{K}_1 = \begin{bmatrix} 3.699 \times 10^{-4} \\ 5.020 \times 10^{-4} \end{bmatrix} \tag{35}$$

In step 2 of the algorithm, the difference between the actual observation, $\mathbf{z}_k$, and the observation predicted by the current state parameters, $\mathbf{H}_k\mathbf{x}_k$, is calculated. This difference is sometimes called the innovation and is like an ordinary residual except that it is calculated using the current rather than the final parameter estimates. The Kalman gain vector, **K**, determines how much the innovation is weighted in updating the state parameter estimates. It is also used in the third step of the algorithm to update the error parameter error covariance matrix, **P**, following integration of the new measurement. Using the first observation of 0.846 AU gives Equations (36):

$$\mathbf{x}_1 = \begin{bmatrix} 0.313 \\ 0.425 \end{bmatrix} \times 10^{-3}; \quad \mathbf{P}_1 = \begin{bmatrix} 6.5 & -4.8 \\ -4.8 & 3.5 \end{bmatrix} \times 10^9 \tag{36}$$

Note that because only one measurement has been processed and there are two parameters to be estimated, neither **x** nor **P** can be regarded as reliable at this point.

In step 4, the state vector and its covariance matrix are projected ahead to the next measurement interval. This requires a knowledge of **F** and **Q**, which are trivial in this example. As the state parameters here are static (i.e. $\mathbf{x} \neq \mathbf{f}(t)$), the state transition matrix, **F**, is simply the identity matrix. This would not be the case, for example, if the state parameters were the concentrations at time t, rather than the initial concentrations, but the modifications to **H** and **F** would be straightforward in that case. Likewise, we are assuming no random variation in the initial concentrations, so the state vector covariance matrix, **Q**, is equal to zeros. The role of **Q** in a more complex application is to allow for random variation in the state parameters over time. As an example, suppose that instead of absorbance, we were measuring total pressure in a gas phase reaction which was subject to random temperature fluctuations between measurements. This would effectively change the initial pressures we were trying to estimate.

The iterations of the Kalman filter continue in this way until all of the measurements have been processed. For the example presented here, Figure 21(b) shows the concentration estimates as a function of time with the final estimates:

$$C_A = 0.186 \pm 0.019\,\text{mM}$$

$$C_B = 0.522 \pm 0.023\,\text{mM}$$

The uncertainties are the standard deviations of the parameters from **P**. Note that these converge to values close to the true concentrations. It should also be noted that these are essentially the same estimates that would have been obtained by linear least-squares implemented in batch mode.

Although Kalman first introduced this filter in 1960,[(24)] applications in chemistry were not abundant until the late 1970s and early 1980s. However, many of these applications employed the Kalman filter mainly as a recursive implementation of simple least-squares, such as the example above, and did not exploit its full capabilities. For instance, given spectra of mixture components, the Kalman filter can be used to estimate component concentrations as a spectrum was being scanned. Although this offered certain advantages such as speed and the ability to terminate an experiment when the desired precision was achieved, developments in instrumental and computational efficiency have made these benefits less significant. The potential exists for more effective utilization of the algorithm, however.

At least two modifications of the basic Kalman filter have also appeared in the analytical chemistry literature. The extended Kalman filter[(25)] has been used to model nonlinear systems (e.g. the estimation of rate constants in the example above) through a linearization of the equations but, like most nonlinear methods, convergence can be slow and subject to initial estimates. Often, several passes are needed, defeating the advantages of recursion. A more successful application has been the adaptive Kalman filter,[(26)] which examines the innovations sequence to detect model errors and effectively turns the filter off in those regions by using an inflated measurement variance estimate. This allows the filter to be applied in situations where strict adherence to the model is not a certainty.

5.6 Other Filters

This section has only scratched the surface of digital filter design, focusing on those filters which are most commonly implemented in analytical chemistry. The reader should be aware that nonrecursive filters with more desirable transfer characteristics, such as a flatter stop band, can be designed with relatively little additional effort, and is referred to appropriate texts on the subject.(9–11) Even more flexibility can be achieved with recursive filters, with characteristics analogous to the Butterworth and Chebyshev designs described earlier for analog signal processing. The popularity of polynomial least-squares filters appears to be a consequence of their intuitive simplicity and the fact that, although not necessarily optimal, they are sufficient for many applications.

The subject of optimal filtering is revisited in the next section with the Weiner filter in the Fourier domain. In terms of optimal filtering in the time domain, however, one additional filter, the matched filter, deserves mention because it often appears in the analytical literature. With a matched filter, the filter coefficients are obtained simply by normalizing the shape of the pure signal. This is illustrated in Figure 22 with a noisy Gaussian. For white noise, the matched filter is optimal in the sense that it produces the largest SNR, interpreted as the maximum value divided by the baseline noise. Unfortunately, it requires an advance knowledge of the signal shape and has the undesirable consequence of broadening the peak. The optimality of the matched filter derives from its connection to regression. This connection, as well as the relationship between Kalman filtering and regression, has been described by Erickson et al.(27)

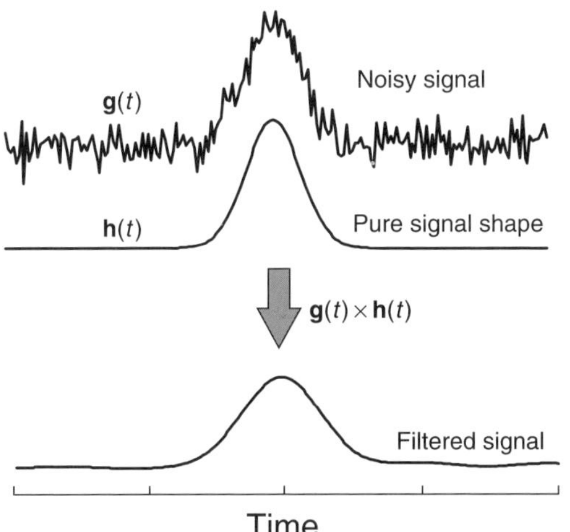

Figure 22 Illustration of the application of a matched filter (**h**(t)) to a noisy signal (**g**(t)).

6 DOMAIN TRANSFORMATIONS

6.1 Introduction

In the context of signal processing, a domain transformation can be defined as a mathematical or physical process that converts a sequence of measurements into an alternative representation which retains all of the information in the original sequence. A domain transformation is distinguished from a simple domain conversion, such as scaling or current-to-voltage conversion, in that it involves a redefinition of the ordinal variable. As such, the procedures are comparatively complex.

There are two principal reasons why domain transformations are used in chemistry. The first is so that information can be represented in a form commonly used for interpretation. A familiar example is Fourier transform infrared (FTIR) spectroscopy in which the signal, collected by means of an interferometer in the time domain, must undergo a transformation in order to represent it as the familiar plot of transmittance versus wavenumber. The second use of domain transformations is to allow certain operations to be carried out on signals with greater ease. As the objective of signal processing is to separate the pure signal from the noise, transformations which provide a better distinction between these two elements of the signal are useful.

Although there are a large number of possible domain transformations that can be employed, this section will focus on three which have been particularly useful in analytical chemistry: the FT, the WT and the HT.

6.2 Fourier Transforms

Without a doubt, the most widely encountered domain transformation in chemistry is the FT. In addition to being a useful stand-alone signal processing tool, the FT has become an integral part of many instrumental methods (FTIR, FT/NMR (nuclear magnetic resonance), FTRS (Fourier transform Raman Spectroscopy) and FTMS (Fourier transform mass spectrometry)).(28,29) Although the FT can be applied to both continuous and discretely sampled functions, it is the latter which dominates instrumental applications and will be the focus of this section. The section begins with a basic description and simple illustration of the principles of the FT and concludes with some examples of its application to signal processing. Abundant supplementary information can be found in the literature.(13,14,28–34)

The fundamental principle behind the discrete FT is that any signal sampled at equal intervals in the time

For references see page 9798

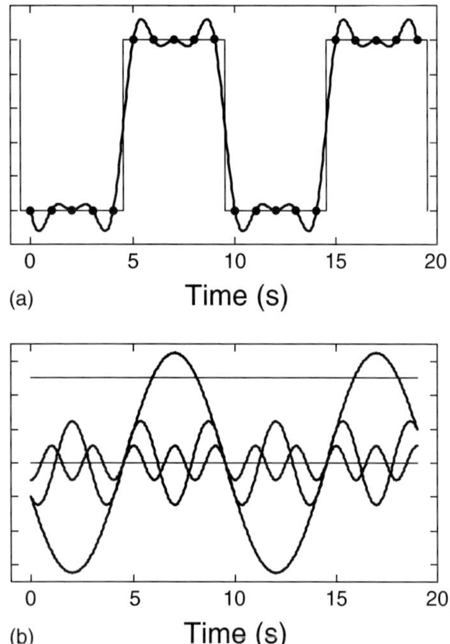

Figure 23 (a) Illustration of the reconstruction of a sampled square wave with a sum of sinusoids. (b) The individual sinusoids used for the reconstruction.

domain can have the sampled points reproduced by the addition of a finite number of sinusoids at defined frequency intervals with variable amplitude and phase. This is illustrated in Figure 23 with the simple example of a sampled square wave. Figure 23(a) shows the square wave with the sampled points and the reconstruction using the combination of sinusoids. Although the reconstruction does not match the square wave exactly, it does reproduce the sampled points exactly, which is its only requirement. If the square wave were sampled more frequently, a larger number of sinusoids would be required for reconstruction. Figure 23(b) shows the individual sinusoids added to give the reconstruction, including the DC (direct current) offset (sine wave with frequency of zero).

Although two cycles of the square wave are shown in Figure 23 for clarity, the FT is based on a single cycle of 10 points. The reconstruction of the sampled points from the sinusoid terms can be represented as

$$h(t) = \sum_{n=0}^{N/2} C_n \cos\left(\frac{2\pi n f_s t}{N} + \phi_n\right) \tag{37}$$

In Equation (37), $h(t)$ represents the reconstructed signal at time t, f_s is the sampling frequency, N is the number of points sampled, and C_n and ϕ_n represent the amplitude and phase of the nth sinusoid, which has a frequency of $f_n = (n/N)f_s$. Note that an equivalent representation using sines rather than cosines could have been written simply by adding 90° to ϕ_n, but a cosine expansion is

Table 2 Amplitudes and phase angles for simple FT example

n	0	1	2	3	4	5
f_n (Hz)	0	0.1	0.2	0.3	0.4	0.5
C_n	0.5	−0.647	0	−0.247	0	−0.1
ϕ_n (°)	0	−72	0	−36	0	0
C_n	0.5	0.647	0	0.247	0	0.1
ϕ_n (°)	0	108	0	144	0	180
A_n	0.5	−0.2	0	−0.2	0	−0.1
B_n	0	0.616	0	0.145	0	0

more consistent with FT calculations. For the example in Figure 23, the coefficients C_n and phase angles ϕ_n are given in Table 2 (a sampling interval of 1 s was assumed). Several points should be noted here. First, two sets of amplitudes and angles are given in the table to illustrate an ambiguity in this type of representation – the same result can be obtained by changing the sign of any of the coefficients and shifting the corresponding phase angle by 180°. A second important point is that, whichever set of values is used, there are 12 parameters provided to describe the sinusoids (6 amplitudes and 6 phase angles). Given that we are representing 10 points in the time domain, it would seem that an excessive number of parameters is needed to describe the signal in the Fourier (frequency) domain. However, this is misleading, because the mathematical restrictions can always fix the phase angles for 0 Hz and the Nyquist frequency (0.5 Hz in this case) to be 0°. Therefore, an equal number of values can be used to represent the signal in both domains. Finally, it should be noted that the periodic nature of the sinusoidal basis functions will give rise to a periodic reconstruction even if the original signal is not periodic. This does not mean that nonperiodic signals cannot be transformed, but it should be kept in mind that the FT will treat them as if they are periodic. Discontinuities in amplitude between the beginning and the end of a signal sequence will be reflected in the high frequency components of the FT.

The ambiguity which arose in the amplitude/phase representation of the FT can be resolved by exploiting the fact that a phase shifted sinusoid can be represented as a linear combination of sine and cosine terms. Therefore, an equivalent form of Equation (37) is Equation (38):

$$h(t) = \sum_{n=0}^{N/2} A_n \cos\left(\frac{2\pi n f_s t}{N}\right) + B_n \sin\left(\frac{2\pi n f_s t}{N}\right) \tag{38}$$

For the square wave example, the coefficients A_n and B_n are also given in Table 2. Although the phase angle has been removed, the same number of parameters as before is required to describe the signal, but there is no ambiguity. The basic objective of the FT is to obtain the coefficients A_n and B_n. Mathematically, this is done by separating the sine and cosine terms through complex arithmetic, recalling Euler's relationship (Equation 39):

$$e^{-i\theta} = \cos\theta - i\sin\theta \quad (39)$$

where $i = \sqrt{-1}$.

The discrete FT is mathematically defined as follows. Given a series of N measurements in the time domain, h_k, where $k = 0 \ldots N-1$, the N complex coefficients of the FT, H_n, where $n = 0 \ldots N-1$, are given by Equation (40):

$$H_n = \sum_{k=0}^{N-1} h_k e^{2\pi ikn/N} \quad (40)$$

This calculation results in N real coefficients and N imaginary coefficients which are related to the coefficients A_n and B_n in Equation (38). However, the total number of coefficients here is $2N$, whereas the total number in Equation (38) was $N+2$, which suggests that there is some redundancy. This arises because the FT produces coefficients at both positive and negative frequencies. This is illustrated in Figure 24 which shows the actual FT of the sampled square wave in Figure 23. Both the real and imaginary parts of the FT are shown and the labels on the x axis indicate the correspondence between the coefficient number, n, and the frequency. Note that 0 Hz (DC) and the Nyquist frequency (f_N) are represented only once in the mapping, whereas both positive and negative values are shown for other frequencies. Furthermore, there is a symmetry between the positive and negative frequencies such that Equation (41) holds:

$$H(-f) = H(f)^* \quad (41)$$

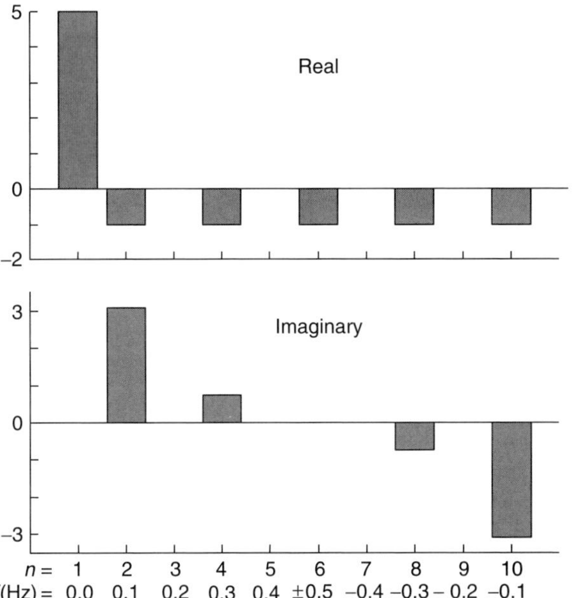

Figure 24 FT of the sampled square wave in Figure 23(a) (one cycle of 10 points only) showing the real and imaginary components and the mapping of points to the frequency domain.

where the asterisk indicates the complex conjugate. This symmetry arises from the fact that h_k is a real function and has no imaginary components.

The representation of the real and imaginary parts of the FT as shown in Figure 24 is an unambiguous presentation of the transform and is the one used in calculations, but it is not normally the one shown in practice. Typically, figures show an amplitude spectrum or a power spectrum and (less frequently) a phase spectrum. Unfortunately, there is a great variability in the scaling and presentation of these spectra, so caution needs to be employed in their interpretation. One way to calculate the amplitude spectrum is to calculate the modulus of H directly (Equation 42):

$$\begin{aligned}\mathrm{Amp}(f) &= \sqrt{\mathrm{real}(H(f))^2 + \mathrm{imag}(H(f))^2}\\ &= \sqrt{H(f)H(f)^*} = |H(f)|\end{aligned} \quad (42)$$

The results of this calculation for the square wave example are shown in Figure 25(a), where the frequency values have also been properly ordered. Because of the symmetry of the figure for real data, the amplitude spectrum is often represented as simply the right-hand side, excluding negative frequencies. If this were done here, it is clear that the amplitude spectrum would not be consistent with the values given in Table 2. There are three reasons for this. First, the amplitude spectrum calculated in this way will always give positive values because of the ambiguity in the sign of the square root. Second, there is a scaling factor of $1/N$ needed to go between Figure 25(a) and Table 2. This scaling factor normally appears in the *inverse Fourier transform* (IFT), defined by Equation (43):

$$h_k = \frac{1}{N}\sum_{n=0}^{N-1} H_n e^{-2\pi ikn/N} \quad (43)$$

(Note that the IFT is essentially the same as the forward FT except for the sign change and the scaling factor.) Finally, in order to arrive at an amplitude consistent with Table 2, it is necessary to combine positive and negative frequencies (except for DC and f_N). Thus an alternative definition of the amplitude spectrum (nonnegative frequencies only) is Equations (44):

$$\begin{aligned}\mathrm{Amp}(0) &= \frac{1}{N}|H(0)|\\ \mathrm{Amp}(f) &= \frac{1}{N}(|H(f)| + |H(-f)|) = \frac{2}{N}|H(f)|\\ \mathrm{Amp}(f_N) &= \frac{1}{N}|H(f_N)|\end{aligned} \quad (44)$$

This representation of the amplitude spectrum is shown in Figure 25(b) and, except for the signs, is consistent with the data in Table 2.

For references see page 9798

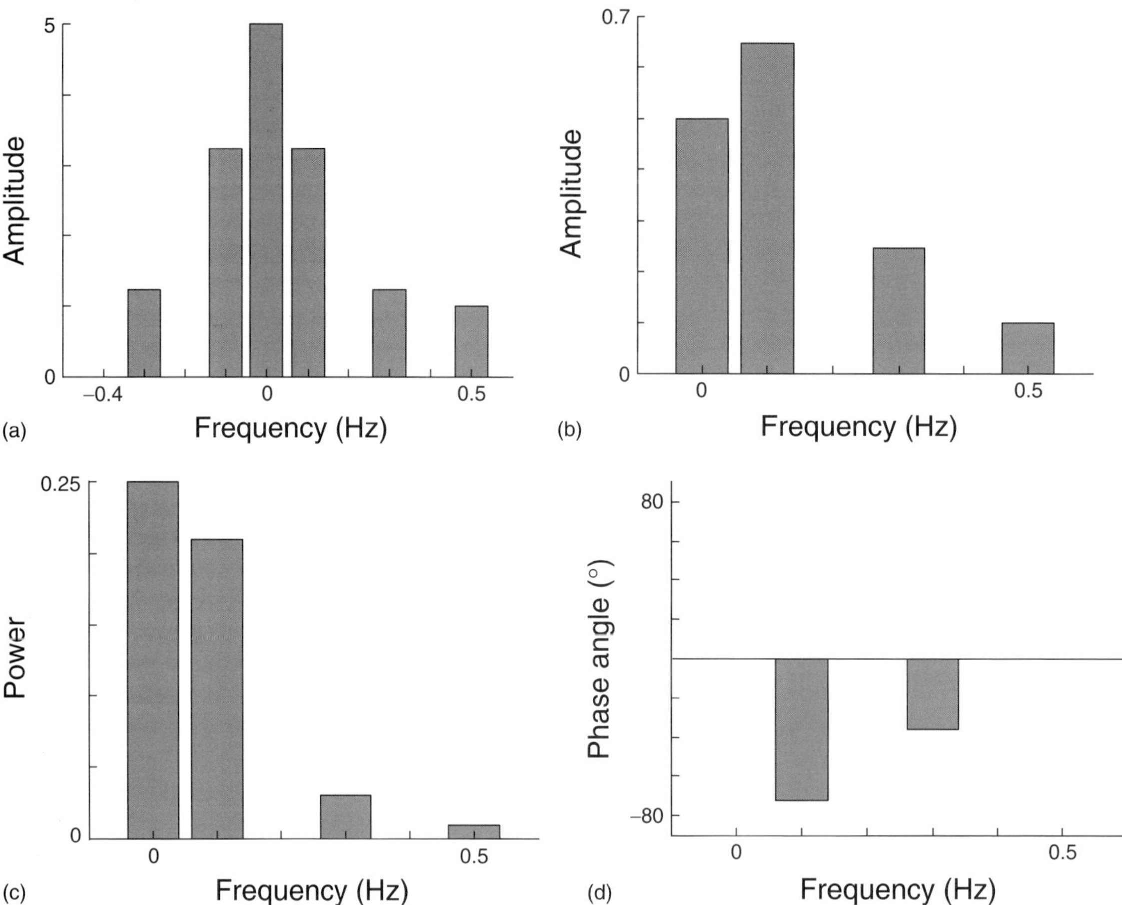

Figure 25 Alternative representations of the FT of the sampled square wave in Figure 23(a). Parts (a) and (b) are different representations of the amplitude spectrum sometimes used; (c) is the power spectrum and (d) is the phase spectrum.

The power spectrum (or power spectral density function) is often used in place of the amplitude spectrum. As its calculation involves squaring the amplitudes, the same scaling inconsistencies exist here as for the amplitudes, so care should be taken. The power of the signal in the two domains is related through Parseval's theorem (Equation 45):

$$\sum_{k=0}^{N-1} |h_k|^2 = \frac{1}{N}\sum_{n=0}^{N-1} |H_n|^2 \qquad (45)$$

One method to arrive at a valid power spectrum is to use a set of equations similar to those given in Equation (44) for the amplitude, replacing amplitude with power by squaring each of the modulus terms and each N. This gives the power spectrum in Figure 25(c). It can be verified that the sum of the elements is equal to 0.5, which is equal to the mean squared value of the signal in the time domain. There are other aspects to the calculation of power spectra, such as windowing methods to prevent leakage among frequencies and improve the quality of the spectral estimation, but these will not be described here.

Although there are some variations in the manner of calculation of amplitude and power spectra, these are not especially serious when the FT is used for descriptive purposes. The phase spectrum, which plots the phase angle as a function of frequency, is less useful than the amplitude spectrum in most instances. The phase spectrum for the current example is shown in Figure 25(d). As the negative frequencies contain redundant information by symmetry, only the positive half of the phase spectrum is shown. The phase angles are calculated from Equation (46):

$$\phi(f) = \tan^{-1}\left(\frac{\mathrm{imag}(H(f))}{\mathrm{real}(H(f))}\right) = \arg(H(f)) \qquad (46)$$

As in the amplitude calculation, this equation will have an ambiguity due to the fact that the angle calculated will always be between −90° and +90°. Table 2 indicates that, in this case, positive amplitudes should produce phase angles outside this range, but this is clearly not reflected in the phase spectrum. Therefore, although the amplitude/power and phase spectra are the most common descriptive forms of the FT, the indeterminacy of the resultant parameters means that they cannot actually be used to regenerate the original signal in most cases. It should also be noted that Equation (46) involves the ratio of two numbers and this can cause problems in the phase angle calculation when both terms are very close to zero, since round-off error leads to an arbitrary phase angle. Although this is of no importance in the final result (the amplitude of the frequency component is zero), it can complicate the interpretation of the phase spectrum.

As already noted, the calculation of the discrete FT can be carried out using Equation (40), but for most real applications involving a substantial number of data points, the application of this equation is impractical due to the large number of operations required (on the order of N^2). For this reason, few applications employed the FT until the mid-1960s, when the FFT algorithm was popularized by Cooley and Tukey.[2] The FFT greatly reduced the number of operations required (of the order of $N \log N$) and made transformations practical for a wide variety of problems. Although beautifully elegant in its partitioning of the problem, a somewhat annoying requirement of the original algorithm was that it required the number of points to be equal to a power of two. Improvements on the original algorithm have largely removed this restriction (although they are not quite as efficient), and FTs are now calculated with ease for most signals of moderate length using a variety of software packages.

One of the principal applications of the FT is the conversion of data recorded by instruments such as FTIR spectrometers, where it is needed to transform the measurements from the time domain to the frequency domain (or vice versa) before it can be interpreted in the conventional way. In addition to this, the FT is used for a great many signal manipulation purposes, such as smoothing, deconvolution, and interpolation. Some of these applications are now illustrated.

A great deal of the utility of the FT in signal processing derives from the convolution theorem, which states that the convolution of two signals in the time domain is equivalent to the element-by-element multiplication of the functions in the frequency domain. Mathematically, if g and h are functions in the time domain and G and H are the corresponding functions in the frequency domain, Equation (47) holds:

$$g(t) * h(t) = \mathrm{IFT}[G(f)H(f)] \tag{47}$$

where the asterisk indicates the convolution of the two functions and IFT indicates the inverse transform. As digital filtering is the convolution of filter coefficients with a noisy signal, this immediately leads to an application in Fourier smoothing. The difference here is that we can specify the transfer function of the filter exactly through its FT. This is illustrated in Figure 26 with the smoothing of a noisy Gaussian. The FT of the noisy signal is first calculated and then both the real and imaginary parts are multiplied by the ideal transfer function which sets all of the high frequencies where there is no significant signal contribution to zero. Note that both positive and negative frequencies must be included in this multiplication, which is why the transfer function looks somewhat different than that shown earlier. After the multiplication is carried out, an IFT is applied to the result to give the smoothed signal in the time domain.

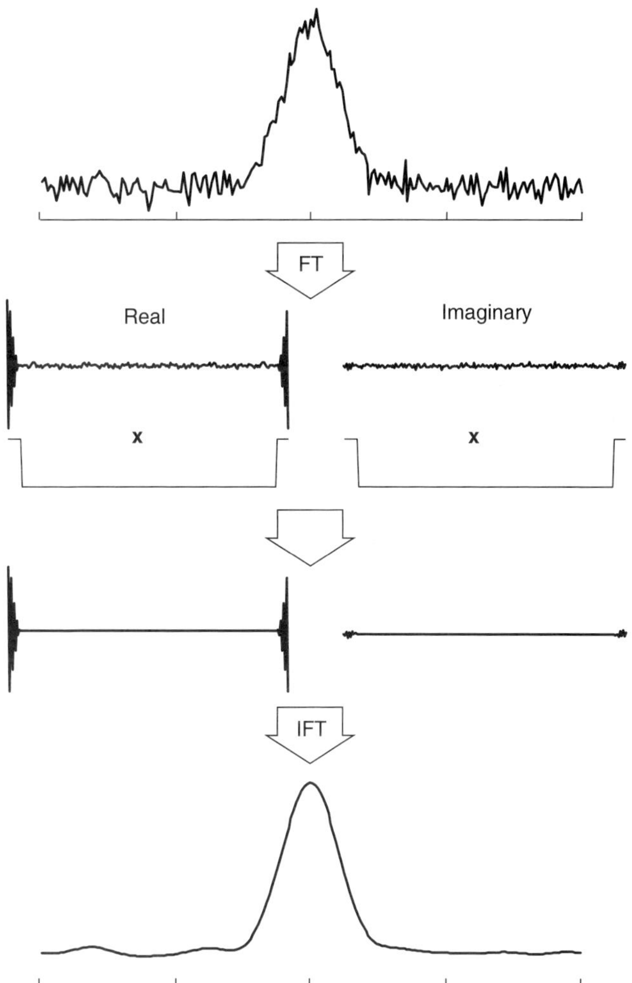

Figure 26 Fourier filtering of a noisy signal.

Although this procedure works very well, it has some drawbacks. First, it is slower than a digital filter and cannot be done in real time because the entire signal is required. Second, artifacts such as the oscillations near the tails of the peak are often observed due to the sharp transition of the transfer function. More severe distortion can result if the cut-off frequency is moved closer to the signal components, but there will be less noise reduction if it is moved to higher frequencies. To avoid this characteristic

For references see page 9798

of the ideal filter, a more gradual decrease in the transfer function is often employed. Such a function is sometimes referred to as an apodization function. If the FT of the pure signal is designated as S and that of the noise as N, it can be shown that the transfer function of the optimal, or Weiner, filter Φ is given by Equation (48):

$$\Phi(f) = \frac{|S(f)|^2}{|S(f)|^2 + |N(f)|^2} \tag{48}$$

This is the transfer function that will give the optimal reproduction of the true signal in the least-squares sense. The difficulty with applying this filter is in the estimation of $S(f)$ and $N(f)$ for the pure signal and noise, but that is beyond the scope of this article.

In addition to convolution, the FT can aid in the deconvolution of two signals. If h represents the convolution of two signals, f and g in the time domain ($h = f^*g$), where g is known, the deconvoluted signal f can be obtained through an element-by-element division in the frequency domain (Equation 49):

$$f(t) = \mathrm{IFT}\left[\frac{H(f)}{G(f)}\right] \tag{49}$$

This is illustrated in Figure 27 where a simulated spectral doublet has been convoluted with the slit function of the spectrometer which smears the two peaks. Through Fourier deconvolution, it is seen that the original line shape can be recovered. Practically speaking, there are a number of difficulties associated with this procedure. First, one of the convolution functions needs to be known in advance. Second, the division presents problems when the denominator in Equation (49) is close to zero and adjustments need to be made in this case. Depending on how this is done, artifacts such as those apparent in the baseline of the deconvoluted spectrum of Figure 27 can result. Finally, the presence of noise can lead to additional complications in the deconvolution. In a best-case scenario, however, this type of deconvolution can be used to improve the resolution of the instrument after the measurements have been obtained.

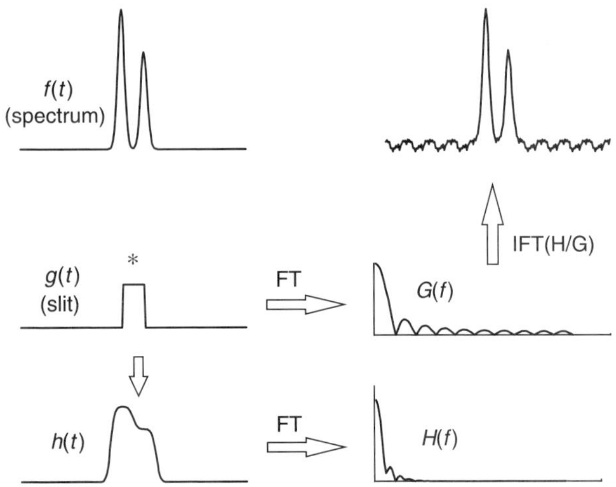

Figure 27 Convolution of a spectral doublet with a slit function followed by Fourier deconvolution to recover the original line shapes.

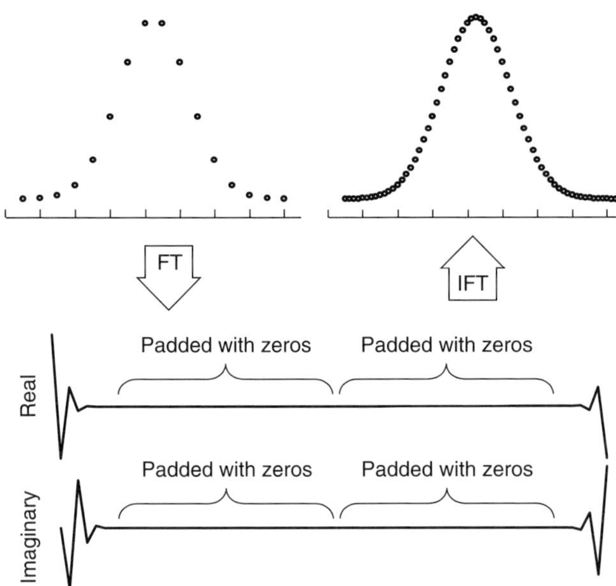

Figure 28 Fourier interpolation of an undersampled signal by a factor of four. Note that the FT has been padded with 24 zeros in both the positive and negative frequency regions to give a total of 64 points.

As a final example of the use of the FT in signal processing, Figure 28 shows an example of function interpolation. This procedure, which has been referred to as the zoom FT, is based on the fact that the Nyquist frequency is directly related to the sampling frequency. If the FT of a signal is padded with zeros at the frequency limit where there is little signal contribution, the Nyquist frequency can be increased, and consequently the sampling interval decreased. This essentially interpolates the function between existing measurements. Although this approach has been used in certain applications, such as locating the peak maxima in undersampled mass spectrometry peaks, it should be used with caution because the zero-padding makes implicit assumptions about the form of the function between the original points which may be erroneous.

6.3 Wavelet Transforms

Although a more recent development than the FT, the WT approach is gaining increased acceptance as a signal processing tool for the analytical chemist. The first applications in the chemical literature of the WT as a denoising, smoothing and data compression procedure appeared in the early 1990s, and their frequency of mention has increased steadily. The utility of the WT for noise reduction purposes rests largely on its decomposition of the signal into successive levels of high- and low-frequency components. In data compression applications the decimation filter property of the WT is useful, effectively reducing the number of elements needed to represent the signal with minimal loss of information. Numerous algorithms for the WT have been devised, with the most popular being the recursive form of the discrete wavelet transform (DWT) attributable to Mallat,(35) the generalization of which is known as the wavelet packet transform (WPT).(30,36) An increasing number of software packages are now available for performing DWTs and WPTs as well as related functions in a relatively straightforward manner.

Like the FT, the WT converts the data into a more useful domain for signal processing by projecting the observed signal onto a set of orthogonal basis functions. In the FT, the signal is projected into the frequency domain using sinusoids as the basis functions. In the frequency domain, the basis functions are localized, but when transformed to the time domain the functions extend globally along the time axis. In contrast, the WT uses basis functions that are both localized in the frequency and time domains to project the data into the wavelet domain. The WT, therefore, has perhaps a more intuitive appeal for some who routinely deal with signals that are time localized, such as chromatographic or spectroscopic measurements. A very readable introduction to denoising and compression of chemical signals using wavelets has been written in tutorial fashion by Walczak and Massart.(37)

In the popular DWT pyramid algorithm of Mallat,(35) a recursive decomposition of the signal is performed using both high- (**H**) and low-pass (**L**) filter matrices which are rectangular (each $n/2 \times n$, if the observed signal vector $\mathbf{x}$ has n elements). The coefficients of these decimation filters depend on the family of wavelets that are used; the Daubechies family appears to be the most popular in chemical applications. The portion of the signal that passes through the low-pass filter is typically called the mth-level approximation to $\mathbf{x}$, $\mathbf{a}_m$, and the portion of the signal that is rejected by **L** (passes through the high-pass filter) is referred to as the mth-level detail in $\mathbf{x}$, $\mathbf{d}_m$ ($\mathbf{d}_m$ can be considered to be the information in $\mathbf{a}_{m-1}$ not included in the approximation $\mathbf{a}_m$). As **H** and **L** are decimation matrices with a down-sampling rate of two, the number of elements in both **a** and **d** drops by a half at each level of approximation (an effective doubling of the sampling interval). The algorithm requires the number of measured channels in the signal vector to be a power of two, although this requirement can be side-stepped with zero-padding. Edge effects can often result from the filter convolutions; however, signals that are not of a length 2^J (J is an integer) can be zero-padded on both ends of the signal vector to sterilize the distortion.

The DWT algorithm is logistically summarized in Figure 29(a). Clearly, the detail at a given level is not used in subsequent approximation steps of the pyramid algorithm, because $\mathbf{a}_m$ is only approximated from $\mathbf{a}_{m-1}$, and the possibility exists that important information was rejected by the low-pass filter and resides in the detail vector. When the detail vectors are also incorporated into the decomposition, the WPT algorithm results (illustrated in Figure 29b). The structure of the **L** and **H** filters used in the decomposition is shown in Figure 30 for the Daubechies-4 wavelet (four coefficients). Note that the coefficients for **H** are the same as for **L** except that the order is reversed and alternate signs are changed. Normally, the **L** and **H** matrices are concatenated and

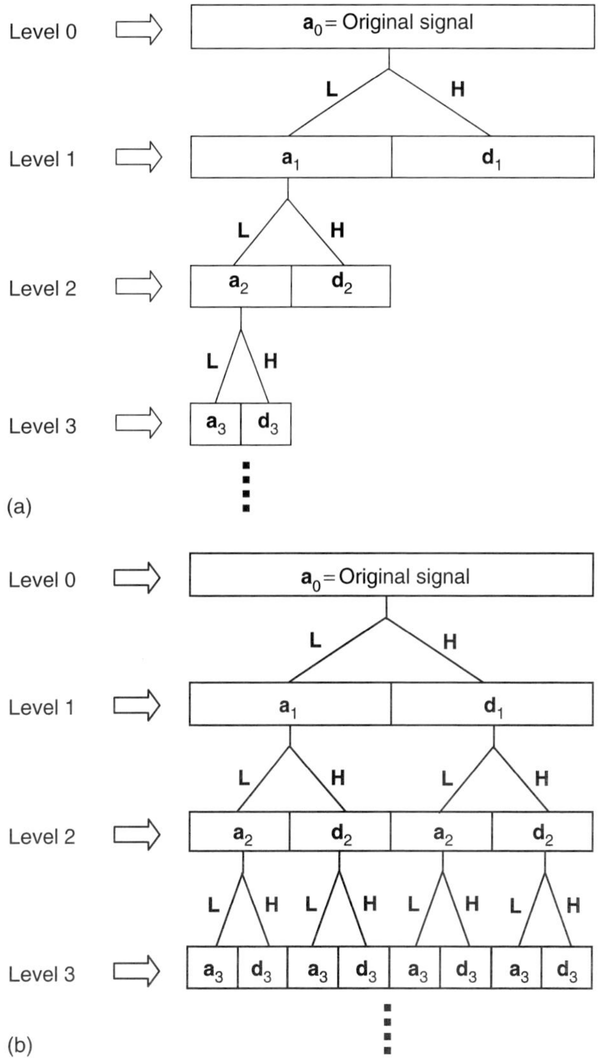

Figure 29 Illustration of (a) the DWT and (b) the WPT.

For references see page 9798

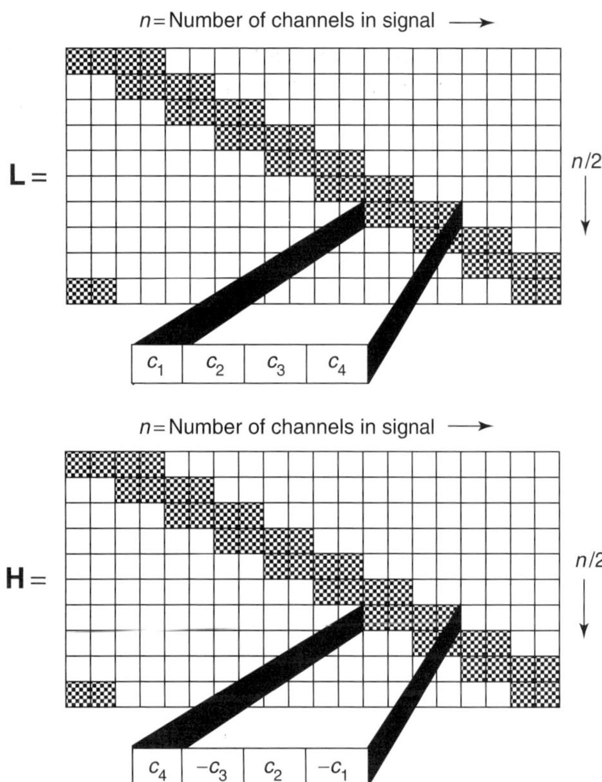

Figure 30 Illustration of the **L** and **H** matrices (four coefficients) used in the WPT to generate the approximation (**a**) and detail (**d**) vectors, respectively, at successive levels of resolution. For the Daubechies-4 wavelet, the coefficients are $c_1 = 0.4830$, $c_2 = 0.8365$, $c_3 = 0.2241$, and $c_4 = -0.1294$. Note that the coefficients will not change for different levels of resolution, but the size of the filter matrix will depend on n, the number of channels in the signal at the previous level of resolution.

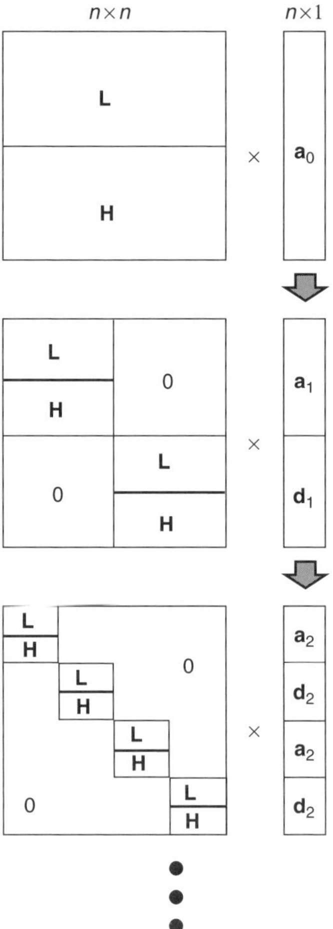

Figure 31 Arrangement of the **L** and **H** filter matrices through the first three levels of the WPT.

multiplied by the concatenated basis vectors at each level. With this approach, the arrangement of **L** and **H** matrices will change at each level, as shown in Figure 31.

With the decomposition of the original signal accomplished, it is possible to examine the coefficients of the approximation and detail levels. This is analogous to examining the coefficients of a Fourier decomposition, except that the wavelet decomposition is arranged in two dimensions – time and frequency. As in Fourier denoising applications, it is desirable to eliminate or reduce coefficients believed to be associated with noise, and retain coefficients reflecting information in the original signal. In the WPT, however, not only must coefficients be selected in the frequency realm, but also in the level of approximation. In order to determine the appropriate basis vectors for reconstruction of the signal, it is useful to examine the approximation and detail vectors resulting from the transform. A simple example of denoising using the WPT is presented in Figure 32. In this example, the WPT has only been carried out to two levels for purposes of illustration. At level 2, it is apparent that three of the four vectors contain little useful information, so these are set to zero before the inverse wavelet transform (IWT) is carried out, resulting in a reduction in the noise. It should be noted that no attempt was made to optimize the denoising in this example, and decompositions to additional levels may have allowed further improvement. Denoising with the WT involves more options than that for the FT, such as the selection of a set of basis functions, the level of decomposition used, and the choice of basis vectors to set to zero.

Because of the wide range of possibilities, algorithmic methods of basis selection have been proposed, with the most intuitive and successful using the minimum entropy (or maximum information) condition of Coifman and Wickerhauser.[38] Best-basis selection according to the minimum entropy condition proceeds on the principle that the most useful basis vectors will be those that contain the most information, and informative vectors will tend to be those which have some large coefficients and some small ones. The coefficients of uninformative basis

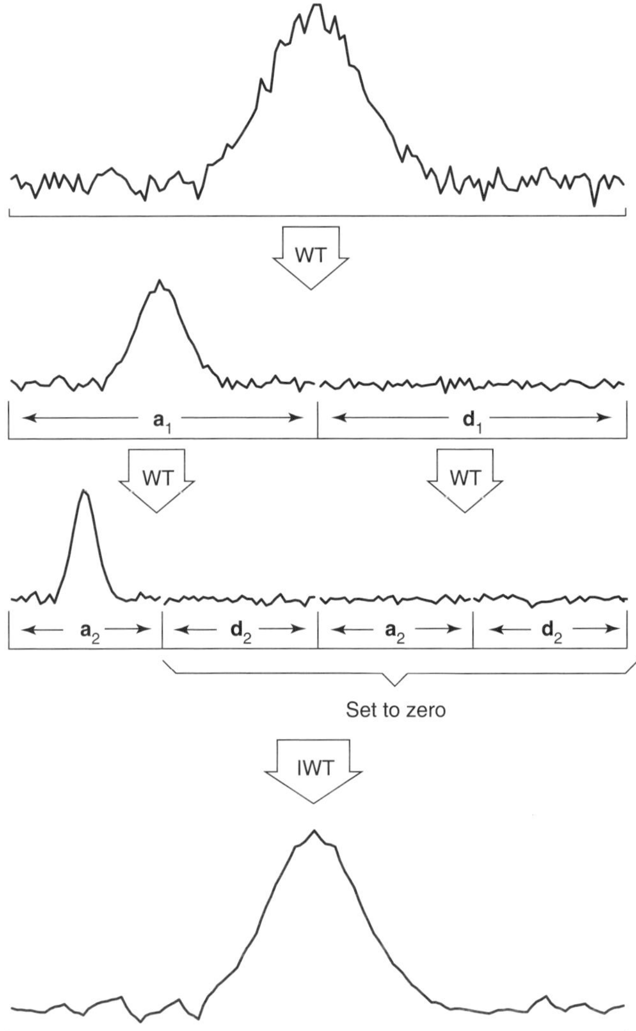

Figure 32 Simple illustration of signal denoising using the WPT.

vectors will largely be the same. The minimum entropy condition in best-basis selection is typically applied by seeking the basis vectors which have the greatest number of coefficients above a preset threshold value.[39] Another method of selecting the best basis vectors is the minimum description length (MDL) method.[40,41] The MDL is primarily used when data compression is desired, and proceeds on a version of the principle of parsimony, seeking the basis vectors which contain the most information in the fewest coefficients.

With the best set of basis vectors selected from the full decomposition, additional signal adjustment can be made by using hard or soft thresholding measures on the remaining coefficients. In hard thresholding, coefficients above a preset threshold are retained, whereas coefficients below this level are discarded. Soft thresholding can also entail zeroing of coefficients below the threshold, but coefficients above the threshold are also typically shrunk towards zero by an amount inversely proportional to their magnitude. Evidently the selection of the threshold value is crucial in these procedures, and several methods exist for estimating the optimal threshold value, including estimating the threshold based on the level of noise,[42] and setting the threshold as a percentage of the largest coefficient.[43] Typical hard, and soft thresholding functions are shown in Equations (50) and (51):

hard thresholding:

$$c_{\text{new}} = \begin{cases} 0, & \text{if } |c_{\text{old}}| < T \\ c_{\text{old}}, & \text{if } |c_{\text{old}}| \geq T \end{cases} \tag{50}$$

soft thresholding:

$$c_{\text{new}} = \begin{cases} 0, & \text{if } |c_{\text{old}}| \leq T \\ \text{sign}(c_{\text{old}})(|c_{\text{old}}| - T) & \text{if } |c_{\text{old}}| > T \end{cases} \tag{51}$$

As an alternative to using threshold values to select relevant coefficients, wavelet smoothing can be achieved by simply discarding detail vectors and performing the inverse WPT from the desired approximation vectors. Although this technique has the potential to achieve greater compression ratios, it is a perilous operation when one lacks knowledge of the location of relevant information in the wavelet decomposition – some useful information may well be contained in some detail vectors. With this possibility looming, it is generally recommended that wavelet smoothing by discarding detail vectors be reserved for situations in which extensive knowledge of the signal allows for educated detail removal.

With basis selection, and coefficient adjustment complete, it is possible to approximate the original signal in the original resolution. This domain is revisited by passing the selected basis vectors back through the high- and low-pass filters. To make the filters interpolation rather than decimation filters, the conjugates of **H**, and **L** are employed. Based on the orthonormality of the two matrices, the conjugates of **H** and **L** are equivalent to the transposes. Therefore the IWT proceeds straightforwardly through the up-sampling filters $\mathbf{H}^{\mathrm{T}}$ and $\mathbf{L}^{\mathrm{T}}$ (transposes of the matrices in Figure 31) until the desired resolution is achieved, with the denoised signal resulting.

Figure 33 illustrates the utility of the WT in signal denoising using a simulated spectrum. White noise was added at a level corresponding to $\sigma_{\text{noise}} = 0.1$, giving the noisy spectrum in Figure 33(a). The first wavelet chosen for the transform was the Daubechies-4 ($c_1 = 0.4829$, $c_2 = 0.8365$, $c_3 = 0.2241$, $c_4 = -0.1294$). Best-basis selection was performed using the minimum entropy procedure of Coifman, and hard thresholding was used on the chosen basis vectors. The results of the overall wavelet denoising with the Daubechies-4 wavelet are shown in Figure 33(b). To illustrate the effect that different wavelet families can

For references see page 9798

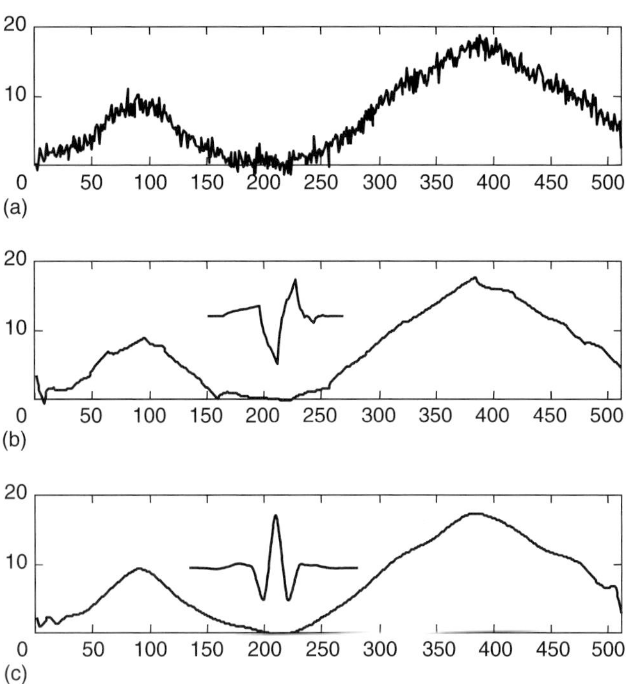

Figure 33 Illustration of signal denoising using Coifman's best-basis selection algorithm and coefficient adjustments using hard thresholding: (a) original simulated noisy spectrum, (b) denoised spectrum using the Daubechies-4 wavelet (inset), and (c) denoised spectrum using the Coiflet-3 wavelet (inset).

have, the Coiflet-3 wavelet was also used, and the results shown in Figure 33(c). In practice, a comparison would more probably be made between different members of a family to find the best result.

6.4 Hadamard Transforms

Like other transform methods, the HT can be thought of as a transformation from one space to another, with a Hadamard matrix acting as the transformation matrix.(44–47) HTs are one method of gaining the Felgett advantage, or multiplex advantage, as it is often called. The multiplex advantage is a statistical gain in SNR as a result of simultaneously measuring multiple spectral resolution elements. In contrast to dispersive methods, in which a single spectral element is measured at a time, multiplex methods measure several coincident spectral elements simultaneously. In order for the multiplex design to prove beneficial from an SNR perspective, the noise in the signal must be considered to be independent of the strength of the incident radiation (i.e. detector noise is the overwhelming noise source). If this condition is met, simple propagation of error reveals that the SNR of a multiplex instrument, relative to a dispersive instrument is $\sqrt{N}$, where N is the number of spectral elements that impinge on the detector at any one time.

Although FT spectrometers are perhaps the best known instruments to utilize the multiplex advantage, the HT spectrometer is also a valuable option. The principles of the HT are based on the concept of Hadamard matrices. As noted above, the benefit of the HT stems from propagation of measurement error into the estimated spectral values. If one observation is made with an inherent detector error of e, then the error in the estimated value is e. However when we wish to estimate several unknowns we can reduce the error associated with a particular estimate by measuring groups of unknowns together in a well-designed fashion. We can subsequently use systems of linear equations to solve for the estimates, and decrease the error in those estimates in the process. The classic analogy is to a weighing scheme for several unknown objects. In the example sketched here, four objects are weighed in experiment A one at a time, with a detector noise level of 0.1. (Experiment A is analogous to a dispersive spectrometer.) In experiment B, two or three objects are weighed together at any one time, although we still only have four total measurements to estimate each individual object's mass. (Experiment B is analogous to a single detector HT spectrometer.)

Experiment A

$$\begin{bmatrix} 1 & 0 & 0 & 0 \\ 0 & 1 & 0 & 0 \\ 0 & 0 & 1 & 0 \\ 0 & 0 & 0 & 1 \end{bmatrix} \begin{bmatrix} m_1 \\ m_2 \\ m_3 \\ m_4 \end{bmatrix} = \begin{bmatrix} x_1 \\ x_2 \\ x_3 \\ x_4 \end{bmatrix} + \begin{bmatrix} e_1 \\ e_2 \\ e_3 \\ e_4 \end{bmatrix}$$

$$\mathbf{W}_{\mathrm{A}}\mathbf{m} = \mathbf{x} + \mathbf{e}$$

$$\mathbf{m} = \mathbf{W}_{\mathrm{A}}^{-1}\mathbf{x} + \mathbf{W}_{\mathrm{A}}^{-1}\mathbf{e}$$

Experiment B

$$\begin{bmatrix} 0 & 1 & 1 & 1 \\ 1 & 1 & 0 & 0 \\ 1 & 0 & 1 & 0 \\ 1 & 0 & 0 & 1 \end{bmatrix} \begin{bmatrix} m_1 \\ m_2 \\ m_3 \\ m_4 \end{bmatrix} = \begin{bmatrix} x_1 \\ x_2 \\ x_3 \\ x_4 \end{bmatrix} + \begin{bmatrix} e_1 \\ e_2 \\ e_3 \\ e_4 \end{bmatrix}$$

$$\mathbf{W}_{\mathrm{B}}\mathbf{m} = \mathbf{x} + \mathbf{e}$$

$$\mathbf{m} = \mathbf{W}_{\mathrm{B}}^{-1}\mathbf{x} + \mathbf{W}_{\mathrm{B}}^{-1}\mathbf{e}$$

Here the x_i terms represent the observed reading on the scales, and the m_i terms represent the estimated mass of the ith object. Through propagation of error, it is relatively easy to show that if the measurement uncertainties for the x_i terms are independent and given by σ_x, then the uncertainties in the masses will be given by Equation (52),

$$\begin{bmatrix} \sigma_{m_1}^2 \\ \sigma_{m_2}^2 \\ \vdots \end{bmatrix} = \sigma_x^2 \mathrm{diag}\,[(\mathbf{W}\mathbf{W}^{\mathrm{T}})^{-1}] \tag{52}$$

where 'diag' indicates extraction of the diagonal elements. Solution of this equation using $\sigma_x = 0.1$ gives $\sigma_m = 0.1$

for all masses in experiment A, whereas the values for experiment B are 0.067, 0.088, 0.088, 0.088 for m_1–m_4, respectively. Clearly, noise reduction in the estimates has occurred via the multiplex advantage.

In HT spectrometers, the weighing design matrix as shown above is embodied by a mask (Hadamard mask) that physically impedes the incidence of some spectral elements while letting others pass through to the detector. Whereas early HT instruments used a moving mask, the current inclination is toward stationary masks whose codes are changed using electrooptical devices. In true HT spectrometers, light is not only blocked from the coagulating detector, but it is also reflected back to a subtracting detector, such that the measured total intensity is the difference of the adding and subtracting detectors. The weighing matrix in these scenarios, **H**, is a series of 1 and (−1) values representing which elements are subtracted and which are added. These matrices are designed based on Hadamard mathematics. When this arrangement is used, the SNR enhancements observed in FT instruments can be achieved. In practice the HT instruments are difficult to construct to the required specifications and thus single-detector instruments are principally used. The weighing matrix used in these systems is the **S** matrix, and the elements are similar to the weighing matrices shown above (zeros and ones). **S** matrices can be easily constructed from Hadamard matrices by removing the first row and column of **H** and changing all −1 elements in **H** to zeros in the **S** matrix. Although closely related to the Hadamard matrices, **S**-matrix methods do not afford the same enhancement in the SNR as **H**-matrix methods because, with N spectral elements, only $(N+1)/2$ may be measured at any one time.

Like the interferogram resulting from the Michelson interferometer, the encodegram is the resulting signal output from a Hadamard mask experiment. The encodegram relates the radiative flux reaching the detector with the position of the Hadamard mask. To convert this signal in the Hadamard domain to the desired frequency domain the inverse HT is used. Given the properties of **S** (orthonormal rows/columns and square) this is easily accomplished by convolution of the encodegram with the inverse of **S**, i.e. $\mathbf{S}^{-1} = \mathbf{S}^{\mathrm{T}}$.

With the use of electrooptic Hadamard masks come new problems with the standard HT. Although these stationary masks remove the problem associated with the continuously moving parts of the FT instruments, nonidealities in the opacity or transmissiveness of the mask require adjustments to the weighing matrices.[47]

When noise is independent of the signal intensity, as is the case when detector noise dominates, Hadamard multiplexing can prove a useful method of improving the SNR of the spectral estimates. In true Hadamard multiplexing noise reduction follows the general formula of Equation (53):

$$\sigma_{\mathrm{HT}} = \frac{\sigma}{\sqrt{N}} \tag{53}$$

where σ_{HT} is the standard deviation of the estimated elements using HT methods, σ is the standard deviation of the detector output (equivalent to the noise level in the same experiment using a monochromator), and N is the number of spectral elements to be estimated. However, most HT instruments employ **S** matrix methods which, at best, allow reduction of the uncertainty of the estimate according to Equation (54):

$$\sigma_{\mathrm{HT}} = \frac{2\sqrt{N}}{N+1}\sigma \approx \frac{2\sigma}{\sqrt{N}} \tag{54}$$

Although the HT has found some utility in analytical applications and is likely to continue to do so in situations for which multiple channel detection or FT methods are unfeasible, its implementation has not been extensive. With the increasing prominence and quality of multichannel detection systems, the multiplex approach of the HT to signal processing is likely to have limited future utility.

7 HIGHER-ORDER SIGNAL PROCESSING

The bulk of this article centers on signal processing methods for first-order data sets – those cases where the signal can be represented as a vector of measurements. In recent years, however, there has been an increased emphasis on the use of higher order data in analytical chemistry. This is particularly true for second-order data sets (matrices of measurements), but increased use of third-order data is also apparent in the literature. This phenomenon can be attributed to three main factors: (a) the demand for new kinds of analytical information and more efficient analytical methods; (b) the increased availability of multichannel detectors, such as photodiode arrays and charge-coupled devices, as well as rapid scanning instruments; and (c) the development of chemometric methods capable of dealing with multidimensional data. Techniques such as multivariate calibration and pattern recognition are now used routinely, and their application has led to the increased need for signal processing for higher-order data.

In discussing signal processing for higher-order data, it is necessary to make the distinction between the order of the data and the order of the signal. A defining characteristic of a signal is that it exhibits correlation in the ordinal variable for some domain, so a higher-order signal should exhibit correlation in the ordinal variable for each dimension. For example, a collection of spectra from

For references see page 9798

different samples for a multivariate calibration or pattern recognition study would not normally exhibit correlations among the samples, and so this second-order data set can be regarded as a collection of first-order signals. In contrast, spectra obtained during a chromatography or kinetics experiment would result in second-order signals in a second-order data set, because there would be a relationship among the spectra in the time domain. Other combinations are also possible. For instance, if fluorescence emission–excitation spectra were collected for an arbitrary series of samples, we would have second-order signals composing a third-order data set.

For data sets that are composed of first-order signals, signal processing is generally restricted to first-order methods such as those already described. Nevertheless, such signal processing can still have effects across multiple orders and for that reason may be regarded as even more important for higher-order data than for the first-order case. For example, the presence of a variable baseline offset or drift between sample spectra can be detrimental to multivariate calibration methods, but this effect can be minimized by derivative filtering in the spectral domain. The application of such techniques prior to data analysis falls under the subject area of data preprocessing, and includes such methods as mean-centering, baseline subtraction, scaling, smoothing, differentiation, and domain transformation. Choice of an appropriate preprocessing method can be critical and often determines the success or failure of a multivariate analysis application. A complete discussion of these methods is beyond the scope of this article; however, many appropriate texts on chemometrics give more information.[13,48–50]

For signals that are truly higher order, first-order signal processing methods can still be used, but other options are also available. In large part, these are extensions of the first-order methods which have already been discussed. In the case of second-order signals, for example, there are two-dimensional (2D) smoothing methods, 2D FTs and 2D WTs. The application of these techniques can offer greater power and flexibility since the characteristics of the signals in both dimensions can be exploited. For example, in the case of spectra collected during a chromatography experiment (a spectrochromatogram), filtering using a nine-point moving average filter in either the time or spectral dimension requires convoluting

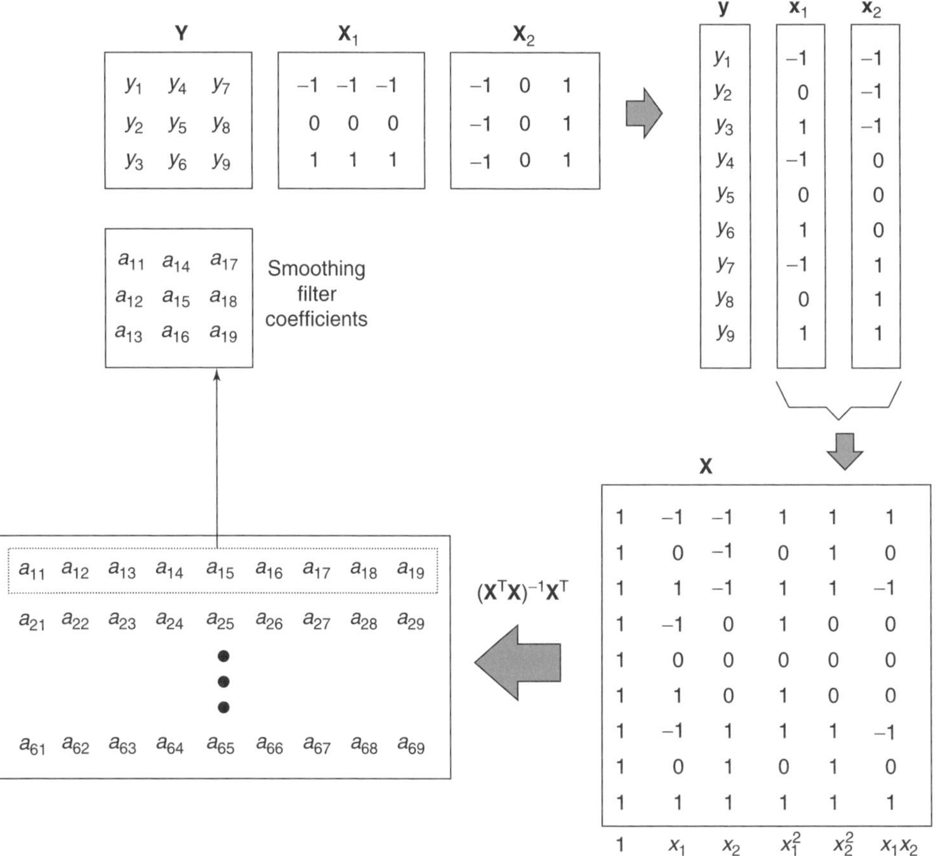

Figure 34 Illustration of the calculation of filter coefficients for a 3 × 3 quadratic filter with an interaction term.

each signal vector with a 1×9 smoothing vector. However, 2D smoothing could use a convolution of the full matrix with a 3×3 smoothing matrix and the same level of noise reduction would be achieved (in the case of a moving average filter) with less distortion. Understandably, the use of 2D techniques introduces greater complexities in terms of computation, implementation, interpretation and optimization than their one-dimensional (1D) counterparts, but these can be overcome.

As most higher-order signal processing methods are extensions of their 1D counterparts, a detailed discussion is not presented here. However, one example of 2D smoothing is presented as an illustration. In this example, a 3×3 polynomial smoothing filter is used. To demonstrate the design, a quadratic filter with an interaction term was chosen, with the corresponding Equation (55):

$$\hat{y} = b_0 + b_1x_1 + b_2x_2 + b_3x_1^2 + b_4x_2^2 + b_5x_1x_2 \quad (55)$$

where x_1 and x_2 represent the two ordinal variables. The generation of filter coefficients requires the unfolding of matrices representing the ordinal variables and the process is illustrated in Figure 34. The resulting smoothing coefficients are (Equation 56)

$$\mathbf{C} = \begin{bmatrix} -0.111 & 0.222 & -0.111 \\ 0.222 & 0.556 & 0.222 \\ -0.111 & 0.222 & -0.111 \end{bmatrix} \quad (56)$$

The result of the application of this filter to the noisy fluorochromatogram of a mixture of pyrenes in Figure 35(a) is shown in Figure 35(b). Although some noise reduction results, it is not as great as that obtained with a simple 3×3 moving average filter, as shown in Figure 35(c). It is clear that the 2D filter involves the same trade-off between noise reduction and distortion as the 1D filters, but the optimization in the 2D case involves a greater number of options, such as the size and order in each dimension and the inclusion of interaction terms. So far, unlike first-order methods, there have not been extensive studies on the relationship between second-order signal processing methods and the signals they are applied to in chemistry, but this is likely to change as higher-order data become more prevalent.

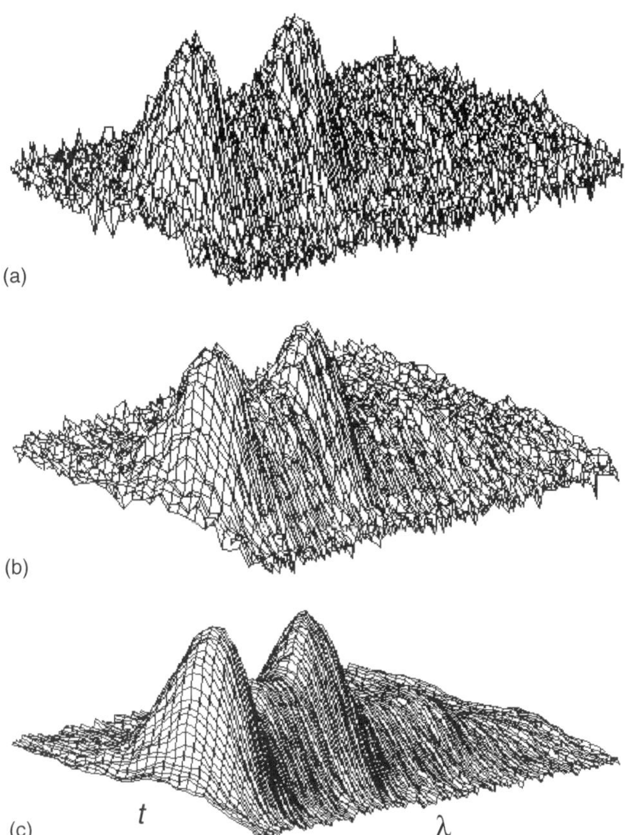

Figure 35 Application of a 2D smoothing filter to noisy data from a chromatogram with multiwavelength fluorescence detection: (a) original data, (b) data filtered with a 3×3 quadratic filter, (c) data filtered with a 3×3 moving average filter.

8 CONCLUSIONS

This article provides a general description of some of the signal processing tools commonly employed in analytical chemistry. As a general principle, it is apparent that all signal processing methods make assumptions about the models for signals and for noise in order to distinguish the two. The power of a particular method in a given application depends on the nature of the assumptions made (very general or very restrictive) and the extent to which they are valid. It is also true that the use of signal processing methods is a double-edged sword. Although the quality of information may be enhanced, it is also possible to distort the signal to the point where results become unreliable. Clearly, a knowledge of the nature of signals, noise and the capabilities of signal processing methods is essential. For this reason, a significant portion of this article is dedicated to the practical aspects of implementing different methods and their effects on signals.

Developments in signal processing applications to analytical measurements will no doubt continue, particularly for digital signals. Although some methods, such as polynomial smoothing and FT-related techniques, will continue to permeate all areas of analytical chemistry, other methods, such as Kalman filtering and HTs, have found more specialized niches. The impact of WTs is evidence of the ongoing research in signal processing applications. Undoubtedly, future developments will

For references see page 9798

exploit greater computational abilities and present new challenges in application and interpretation. As noted in the previous section, applications to higher-order methods will be a focus of research. In any case, it is apparent that, whether it is the enhancement of fuzzy images from atomic microscopes or the removal of background signals in remote sensing from space, signal processing methods will continue to play a key role in all aspects of analytical chemistry.

ACKNOWLEDGMENTS

The authors gratefully acknowledge the research support of the Natural Sciences and Engineering Research Council of Canada (NSERC) and the Dow Chemical Company. Prof. L. Ramaley is thanked for his helpful comments.

ABBREVIATIONS AND ACRONYMS

A/D	Analog-to-digital
ADC	Analog-to-digital Converter
AU	Absorbance Units
BCD	Binary Coded Decimal
DAC	Digital-to-analog Converter
DC	Direct Current
DSP	Digital Signal Processor
DWT	Discrete Wavelet Transform
FFT	Fast Fourier Transform
FIR	Finite Impulse Response
FT	Fourier Transform
FTIR	Fourier Transform Infrared
FTMS	Fourier Transform Mass Spectrometry
FTRS	Fourier Transform Raman Spectroscopy
HT	Hadamard Transform
IFT	Inverse Fourier Transform
IIR	Infinite Impulse Response
IWT	Inverse Wavelet Transform
LSB	Least Significant Bit
MDL	Minimum Description Length
MSB	Most Significant Bit
NIR	Near Infrared
NMR	Nuclear Magnetic Resonance
NPS	Noise Power Spectrum
RC	Resistor–Capacitor
rms	root-mean-square
SG	Savitzky–Golay
SNR	Signal-to-noise Ratio
V/F	Voltage-to-frequency
WPT	Wavelet Packet Transform
WT	Wavelet Transform
1D	One-dimensional
2D	Two-dimensional

List of selected abbreviations appears in Volume 15

RELATED ARTICLES

Process Instrumental Methods **(Volume 9)**
Chemometric Methods in Process Analysis

Chemometrics **(Volume 11)**
Chemometrics • Multivariate Calibration of Analytical Data • Second-order Calibration and Higher

Electronic Absorption and Luminescence **(Volume 12)**
Ultraviolet and Visible Molecular Absorption and Fluorescence Data Analysis

Gas Chromatography **(Volume 12)**
Data Reduction in Gas Chromatography

Infrared Spectroscopy **(Volume 12)**
Spectral Data, Modern Classification Methods for

Kinetic Determinations **(Volume 12)**
Data Treatment and Error Analysis in Kinetics

General Articles **(Volume 15)**
Multivariate Image Analysis

REFERENCES

1. A. Savitzky, M.J.E. Golay, 'Smoothing and Differentiation of Data by Simplified Least Square Procedures', *Anal. Chem.*, **36**, 1627–1639 (1964).
2. J.W. Cooley, J.W. Tukey, 'An Algorithm for the Machine Calculation of Complex Fourier Series', *Math. Comput.*, **19**, 297–301 (1965).
3. A.J. Diefenderfer, B.E. Holton, *Principles of Electronic Instrumentation*, 3rd edition, Saunder's College Publishing, Philadelphia, 1994.
4. P. Horowitz, W. Hill, *The Art of Electronics*, 2nd edition, Cambridge University Press, New York, 1989.
5. H.V. Malmstadt, C.G. Enke, S.R. Crouch, *Electronics and Instrumentation for Scientists*, Benjamin/Cummings, Menlo Park, CA, 1981.
6. A.B. Williams, *Active Filter Design Handbook: LC, Active and Digital Filters*, 2nd edition, McGraw-Hill, New York, 1988.
7. Z.H. Meiksin, P.C. Thackray, *Electronic Design with Off-the-shelf Integrated Circuits*, 2nd edition, Prentice-Hall, Englewood Cliffs, NJ, 1984.

8. K.M. Daugherty, *Analog-to-digital Conversion: A Practical Approach*, McGraw-Hill, New York, 1995.
9. R.W. Hamming, *Digital Filters*, 2nd edition, Prentice-Hall, Englewood Cliffs, NJ, 1983.
10. T.J. Terrell, *Introduction to Digital Filters*, 2nd edition, Wiley, New York, 1988.
11. L.R. Rabiner, B. Gold, *Theory and Application of Digital Signal Processing*, Prentice-Hall, Englewood Cliffs, NJ, 1975.
12. A.V. Oppenheim, R.W. Schafer, *Digital Signal Processing*, Prentice-Hall, Englewood Cliffs, NJ, 1975.
13. D.L. Massart, B.G.M. Vandeginste, S.N. Deming, Y. Michotte, L. Kaufman, *Chemometrics: A Textbook*, Elsevier, Amsterdam, 1988.
14. S.D. Brown, 'Signal Processing and Data Enhancement', in *Practical Guide to Chemometrics*, ed. S.J. Haswell, Marcel Dekker, New York, 1992.
15. M.U.A. Bromba, H. Ziegler, 'Application Hints for Savitzky–Golay Digital Smoothing Filters', *Anal. Chem.*, **53**, 1583–1586 (1981).
16. C.G. Enke, T.A. Nieman, 'Signal-to-noise Ratio Enhancement by Least-squares Polynomial Soothing', *Anal. Chem.*, **48**, 705A–712A (1976).
17. R.G. Brown, *Introduction to Random Signal Analysis and Kalman Filtering*, Wiley, New York, 1983.
18. S.D. Brown, 'The Kalman Filter in Analytical Chemistry', *Anal. Chim. Acta*, **181**, 1–26 (1986).
19. H.N.J. Poulisse, 'Multicomponent Analysis Computations Based on Kalman Filtering', *Anal. Chim. Acta*, 361–374 (1979).
20. J. Steinier, Y. Termonia, J. Deltour, 'Comments on Smoothing and Differentiation of Data by Simplified Least Square Procedures', *Anal. Chem.*, **44**, 1906–1909 (1972).
21. A. Proctor, P.M.A. Sherwood, 'Smoothing of Digital X-ray Photoelectron Spectra by an Extended Sliding Least-squares Approach', *Anal. Chem.*, **52**, 2315–2321 (1980).
22. R.A. Leach, C.A. Carter, J.M. Harris, 'Least-squares Polynomial Filters for Initial Point and Slope Estimation', *Anal. Chem.*, **56**, 2304–2307 (1984).
23. P.D. Wentzell, T.P. Doherty, S.R. Crouch, 'Frequency Response of Initial Point Least Squares Polynomial Filters', *Anal. Chem.*, **59**, 367–371 (1987).
24. R.E. Kalman, 'A New Approach to Linear Filtering and Prediction Problems', *Trans. ASME, Ser. D*, **82**(Mar.), 35–45 (1960).
25. S.C. Rutan, S.D. Brown, 'Estimation of First-order Kinetic Parameters by Using the Extended Kalman Filter', *Anal. Chim. Acta*, **167**, 23–37 (1985).
26. S.C. Rutan, S.D. Brown, 'Adaptive Kalman Filtering Used to Compensate for Model Errors in Multicomponent Methods', *Anal. Chim. Acta*, **160**, 99–119 (1984).
27. C.L. Erickson, M.J. Lysaght, J.B. Callis, 'Relationship Between Digital Filtering and Multivariate Regression in Quantitative Analysis', *Anal. Chem.*, **64**, 1155A–1163A (1992).
28. A.G. Marshall, F.R. Verdun, *Fourier Transforms in NMR, Optical and Mass Spectrometry: A User's Handbook*, Elsevier, Amsterdam, 1990.
29. M. Cartwright, *Fourier Methods for Mathematicians, Scientists and Engineers*, Ellis Horwood Ltd, Chichester, 1990.
30. W.H. Press, B.P. Flannery, S.A. Teukolsky, W.T. Vetterling, *Numerical Recipes in C*, Cambridge University Press, New York, 1992.
31. R.P. Wayne, 'Fourier Transformed', *Chem. Britain*, **23**, 440–446 (1987).
32. R.B. Lam, R.C. Wieboldt, T.L. Isenhour, 'Practical Computations with Fourier Transforms for Data Analysis', *Anal. Chem.*, **53**, 889A–901A (1981).
33. G. Horlick, 'Digital Data Handling of Spectra Utilizing Fourier Transforms', *Anal. Chem.*, **44**, 943–947 (1972).
34. G. Horlick, 'Fourier Transform Approaches to Spectroscopy', *Anal. Chem.*, 61A–66A (1971).
35. S. Mallat, 'A Theory for Multiresolution Signal Decomposition: The Wavelet Representation', *IEEE Trans. Pattern Anal. Machine Intell.*, **11**, 674–693 (1989).
36. C.K. Chui, *Introduction to Wavelets*, Academic Press, Boston, 1991.
37. B. Walczak, D.L. Massart, 'Noise Suppression and Signal Compression Using the Wavelet Packet Transform', *Chemom. Intell. Lab. Syst.*, **36**, 81–94 (1997).
38. R.R. Coifman, M.V. Wickerhauser, 'Entropy-based Algorithms for Best Basis Selection', *IEEE Trans. Inform. Theory*, **38**, 713–719 (1992).
39. M.V. Wickenhauser, *Adapted Wavelet Analysis from Theory to Software*, A.K. Peters, Ltd, Wellesley, MA, 1994.
40. J. Rissanen, 'A Universal Prior for Integers and Estimation by Minimum Description Length', *Ann. Stat.*, **11**, 416–431 (1983).
41. J. Rissanen, *Stochastic Complexity in Statistical Inquiry*, World Scientific Publishing, Singapore, 1989.
42. D.L. Donoho, 'De-noising by Soft Thresholding', *IEEE Trans. Inform. Theory*, **41**, 613–627 (1995).
43. V.J. Barclay, R.F. Bonner, I.P. Hamilton, 'Application of Wavelet Transforms to Experimental Spectra: Smoothing, Denoising, and Data Set Compression', *Anal. Chem.*, **69**, 78–90 (1997).
44. A.G. Marshall, *Fourier, Hadamard, and Hilbert Transforms in Chemistry*, Plenum Press, New York, 1982.
45. D.K. Graff, 'Fourier and Hadamard: Transforms in Spectroscopy', *J. Chem. Ed.*, **73**, 304–309 (1995).
46. P.J. Treado, M.D. Morris, 'A Thousand Points of Light: The Hadamard Transform in Chemical Analysis and Instrumentation', *Anal. Chem.*, **61**, 723A–734A (1989).
47. M. Harwit, N.J.A. Sloane, *Hadamard Transform Optics*, Academic Press, New York, 1979.

List of selected abbreviations appears in Volume 15

48. K.R. Beebe, R.J. Pell, M.B. Seasholtz, *Chemometrics: A Practical Guide*, Wiley, New York, 1998.
49. H. Martens, T. Næs, *Multivariate Calibration*, Wiley, New York, 1989.
50. M.A. Sharaf, D.L. Illman, B.R. Kowalski, *Chemometrics*, Wiley, New York, 1986.

Soft Modeling of Analytical Data

A. de Juan, E. Casassas, and R. Tauler
University of Barcelona, Barcelona, Spain

In this contribution different methods for soft modeling data analysis are reviewed. Soft modeling approaches attempt the description of a system without the need of an a priori model postulation, physical and/or chemical. The goal of these methods is the explanation of data variance using the minimal or softer assumptions about data. Most of these soft modeling approaches are based on factor analysis (FA) decompositions of experimental data matrices. These decompositions are done by pure mathematical means and allow the identification of the number of data variance sources, their qualitative and, eventually, quantitative estimation. Results of soft modeling data analysis are useful to validate hard modeling results and also for investigation of complex chemical systems. In this contribution, the soft modeling data analysis methods described can be applied to one data matrix or to several data matrices (three-way data sets). The purpose of these methods are mainly exploratory analysis and resolution of mixture data sets. Within the latter group, special attention is devoted to multivariate curve resolution (MCR) techniques and their extension to three-way data analysis. Three examples of application are given covering the chromatographic coelution of mixtures of pesticides using liquid chromatography/diode-array detection (LC/DAD), the infrared (IR) spectral data analysis from multiple runs of an industrial process and the interpretation of thermodynamic and conformational transitions of polynucleotides using spectrometric titrations.

1 INTRODUCTION

Data modeling and data fitting in chemical sciences has been traditionally done by hard modeling techniques, i.e. data are tested against a model based on physical and chemical laws and the parameters of this model are obtained by least-squares curve fitting optimization techniques. This approach is valid for well-known phenomena and laboratory data, where the variables of the model are under control during the experiments and only the phenomena under study affect the data. However, this ideal situation is not obeyed in many circumstances in chemistry, specially in analytical chemistry, when natural samples or unknown processes are investigated. Complex phenomena like those involving macromolecular compounds or industrial processes, where physical parameters cannot be appropriately fixed, are typical examples not solved by the traditional model-based data treatments. Alternative approaches to hard modeling have been proposed. In particular, soft modeling approaches attempt the description of a system without the need of an a priori model postulation, physical and/or chemical. The goal of these methods is the explanation of data variance using the minimal or softer assumptions about data. Most of these soft modeling approaches are based on FA decompositions of experimental data. These decompositions are done by pure mathematical means and allow the identification of the number of data variance sources and often their qualitative and, eventually, quantitative estimation. Results of soft modeling data analysis are useful to validate hard modeling results and also for investigation of complex chemical systems. In this contribution, soft modeling data analysis based on FA techniques and, especially, MCR techniques are reviewed and some examples of application are given at the end of the chapter.

2 SOFT-MODELING DATA ANALYSIS BASED ON FACTOR ANALYSIS

Dealing with huge data sets is a common situation in analytical chemistry. Indeed, the characterization of a sample involves usually the measurement of several independent parameters (e.g. concentrations of constituents, pH, temperature, etc.) or the recording of a multivariate instrumental response (e.g. a spectrum). Modern instrumentation able to acquire and store data easily and powerful computers that can handle efficiently large amounts of information have promoted the transition from the univariate to the multivariate domain. However, enormous amounts of raw experimental measurements reveal little about a chemical problem unless they are treated with the appropriate data analysis method.

In this and subsequent sections, the explanations are focused on data sets that can be organized in a table or, in mathematical words, in a data matrix. These data matrices can be the result of organizing in rows the multivariate response of a series of samples or the responses recorded at different stages of a chemical reaction (see Figure 1). The general terminology adopted calls *objects* the parameters that define the rows and *variables* the parameters that define the columns in a data matrix.

FA started as a method to find out the underlying structure in these data matrices, i.e. to know how many sources of variation (*factors*) are needed to describe the data sets and, if possible, to identify the chemical nature of these factors.(1) The first step in this process is called abstract FA or principal component analysis (PCA) and establishes a bilinear model of variation in the data set by using abstract variables. The second and more important step links these abstract factors with chemical sources of variation and can be done using several procedures that will be described below.

The first problem associated with large data sets is "how to look at them". The observation of hundreds or even thousands of numerical values does not provide any straightforward information, but displaying the raw data in plots does not seem possible due to the size of the data space (e.g. displaying samples whose spectra have 100 absorbance readings would imply drawing a plot with 100 axes, when the human eye can cope with, at most, three-dimensional figures). Trying to find out methods to visualize multidimensional data spaces, the next question

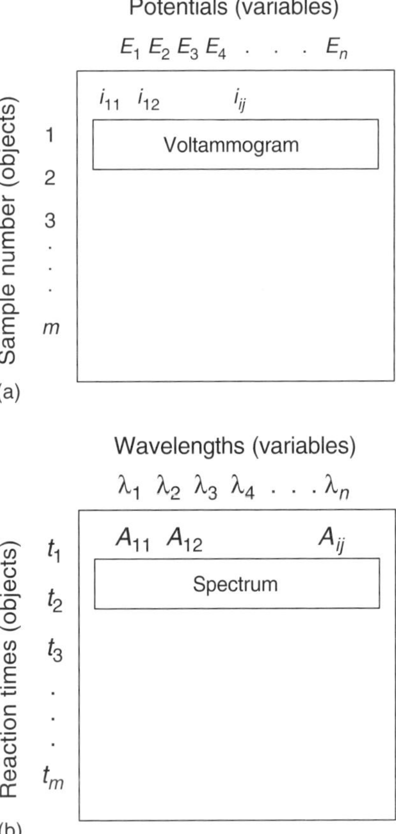

Figure 1 Examples of multivariate data sets. (a) Matrix of samples analyzed voltammetrically. i_{ij} is an intensity measurement collected for the ith sample at the jth potential. (b) Spectrometric monitoring of a kinetic process. A_{ij} is an absorbance reading recorded at the ith reaction time for the jth wavelength.

is if all the variables are necessary to describe properly the variation in our data set. There are often many original variables that are clearly correlated and, as a consequence, most of the information that they provide is redundant. In view of this fact, a plot of the data set could be drawn either using the most representative and uncorrelated original variables[2] or replacing the pool of original variables by a reduced set of efficient abstract variables, calculated as a linear combination of the original ones. These new variables, which are completely uncorrelated and describe the main directions of variation of the data set, are the so-called *principal components*.

Figure 2 shows graphically how principal components are designed with a simple example. The data set used consists of a series of spectra for solutions of a pure substance with different concentrations recorded at three different wavelengths. All the points plotted lie in a straight line because spectra of the same pure substance differ only in intensity and not in shape. It can be seen that the information contained in the plot of the

For references see page 9834

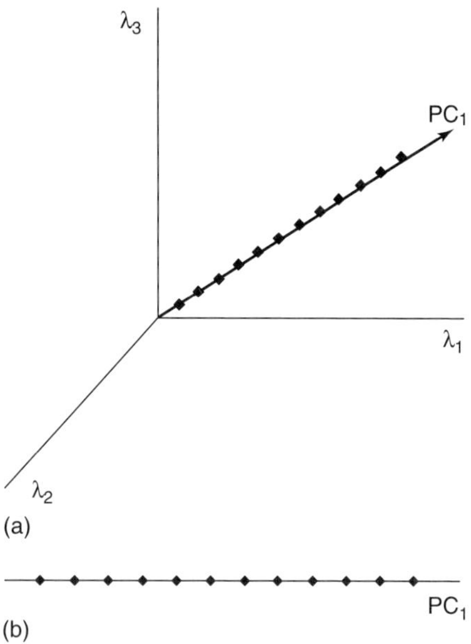

Figure 2 Graphical representation of a data set formed by solution spectra of a pure substance recorded at three wavelengths. (a) Plot in the original data space. (b) Plot in the space of principal components.

spectra in the original space could be kept using only one axis drawn in the direction of maximum variation of the data set. This axis is the first principal component (PC_1) and allows the reduction of the dimensions of the data space from three to one (note that if the spectra for the pure substance had been recorded at m wavelengths, the reduction in dimensions would have gone from m to one). In general, for complex systems, more than one principal component is needed, but the reduction of dimensions from the original space of variables to the space of principal components is always drastic.

The principal components are axes (vectors) which are orthogonal to each other, i.e. completely uncorrelated (eigenvectors). Each new principal component is calculated so as to describe the largest direction of variation in the data set that has not yet been described by other principal components. This means that they decrease gradually in importance because every time the remaining variation in the data set is less. Given a data set with m rows and n columns ($m \times n$), the number of principal components that can be calculated is m or n, whichever is the smallest; however, only a few of them are needed to obtain a good picture of the variation in the original matrix. Thus, for any data matrix, the principal components can be divided into two groups: those which account for the chemical variations in the data set, which are calculated first and are usually few; and those which are related to the noise description. Only the first group will be used to model the data matrix.

The concepts that relate the objects and the variables of the original data matrix to the space of principal components are the *scores* and the *loadings*, respectively. The scores are the coordinates of each object in the principal component axes. The loadings express how large the contribution of each original variable is to the principal components. As mentioned above, principal components are obtained as weighted sums of the original variables. These weights are the loadings and their magnitude (in absolute value) is proportional to the importance they have in the description of the principal component. A variable with large loadings for a certain principal component is significantly correlated to this abstract variable, i.e. their directions in the data space are similar.

Figure 3 shows the PCA results of a real example related to an environmental problem.[3] The data set in this case is formed by 22 sediment samples collected in the northwestern Mediterranean Sea (objects) for which the concentration of 96 compounds of anthropogenic and biogenic origin (variables) has been determined. The first two principal components explain around 70% of the total variation in the data set. Samples close to each other in the *score plot* of PC_2 versus PC_1 are similar. The PCA allowed the identification of several groups of samples related, in this case, to diverse sampling areas with clearly different compound apportionment. The *loading plot* linked principally the concentrations of compounds coming from land-based pollution sources to the first principal component, whereas those due to atmospheric deposition were mainly related to the second principal component. Correlated variables were close to each other or placed along the same straight line through the origin in the loading plot. The information in the scores plot and the loadings plot can be put together in a *biplot*, which relates the objects and variables of a data set. A biplot helps to characterize the different clusters of objects found in the data set. Thus, in this example, the cluster of samples collected in the Ebro prodelta (region 3 in Figure 3), which extends primarily along the PC_1, is mostly related to the variables associated with this first principal component, namely those compounds coming from land-based pollution sources. The other two groups of samples, collected in the open sea, have an expected larger contribution of compounds coming from the deposition of airborne particles.

Mathematically, the relationship between the original data space and the principal component space is expressed by Equation (1):

$$\mathbf{D} = \mathbf{T}\mathbf{P}^{\mathrm{T}} \tag{1}$$

where $\mathbf{D}$ ($m \times n$) is the original data matrix, $\mathbf{T}$ ($m \times npc$) is the matrix of scores and $\mathbf{P}^{\mathrm{T}}$($npc \times n$) is the loading matrix.

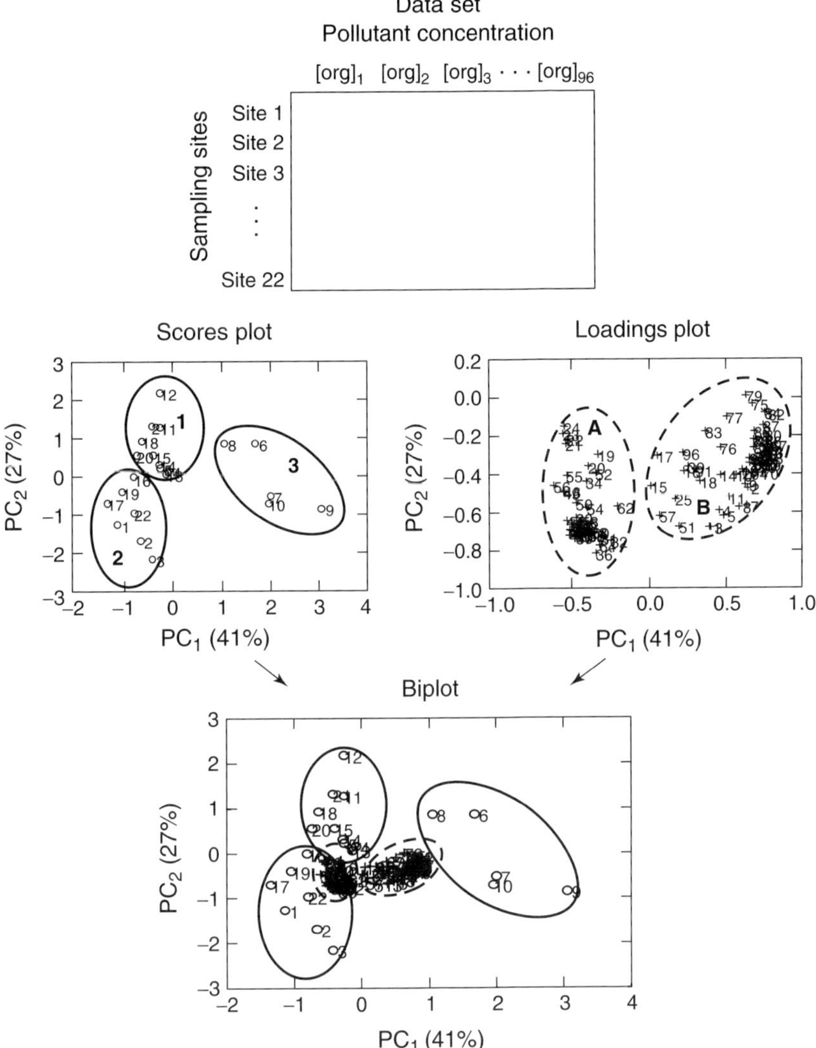

Figure 3 Scores plot, loadings plot and biplot of an environmental data set. Each row in the original data matrix refers to a sampling site (○) and each column to the concentration of an organic compound (+). Figures in the scores plot indicate different geographical zones (1 and 2 are open sea zones, whereas 3 is a river delta). Letters in the loadings plot indicate the origin of the contaminants (A refers mainly to deposited atmospheric pollutants and B mainly to anthropogenic or biogenic pollutants).

npc is the number of principal components related to chemical sources of variation in the data set. All columns in the **T** matrix are orthogonal to each other and so are all rows in $\mathbf{P}^T$. This is the data decomposition obtained using the nonlinear iterative partial least squares (NIPALS) algorithm,[(4)] by far the most used when FA was introduced in chemistry (it is important to note that the rows of the $\mathbf{P}^T$ matrix are the eigenvectors obtained from the diagonalization of the covariance matrix $\mathbf{Z} = \mathbf{D}\mathbf{D}^T$). Nowadays, the most popular algorithm is singular value decomposition (SVD),[(5)] which provides a matrix decomposition closely related to Equation (1), though formally different. Equation (2) is provided by SVD:

$$\mathbf{D} = \mathbf{U}\mathbf{S}\mathbf{V}^T \tag{2}$$

where the columns in **U** and the rows in $\mathbf{V}^T$ are orthonormal vectors, i.e. unit length orthogonal vectors. The information contained in **U** and in $\mathbf{V}^T$ is related to the scores matrix **T** and the loading matrix $\mathbf{P}^T$ respectively. **S** is a diagonal matrix, whose diagonal elements are the singular values of **D**, i.e. the square root of the eigenvalues. The mathematical expressions that are generally used to link the NIPALS and SVD results are given in Equations (3) and (4):

$$\mathbf{T} = \mathbf{U}\mathbf{S} \tag{3}$$

$$\mathbf{V} = \mathbf{P} \tag{4}$$

though some other combinations have also been reported.[(6)]

For references see page 9834

The singular values are ranked in descending order of magnitude along the diagonal of the **S** matrix and their numerical values are related to the importance of their associated principal component. When the singular values (or the logarithm of the eigenvalues) of a matrix are plotted vs the principal component number, there is often a clear cut-off point between the eigenvalues connected with chemically meaningful factors and those describing noise contributions. This kind of plot is frequently used to find out how many principal components (*factors*) are needed to describe the sources of variation in a data set.

Deduction of this number of principal components is sometimes not an easy task because of the presence of nonrandom experimental noise. Besides this visual method, several methods have been proposed:[1] methods based on the previous knowledge of experimental error; and approximate methods which do not require previous knowledge of experimental noise. Obviously, the first type of method is preferred when the experimental error is known. However, this knowledge is not available in many circumstances and the methods of the second group should be used. One of the methods which has been found to be useful and does not require previous knowledge of the experimental error is the indicator function (IND) proposed by Malinowski.[7] This function reaches a minimum when the correct number of components has been chosen. It is more sensitive than other functions proposed for the same purpose. There are also other statistically sounder methods, like cross-validation[8] methods. These methods consist of the elimination of a reduced data subset of the whole data set, for instance one row (or several rows) of the data matrix, and the determination of the eigenvalues and eigenvectors of the reduced data matrix which does not have the eliminated rows. All these eliminated rows are then estimated using different numbers of components and the differences between the reproduced values and those which were eliminated are calculated and expressed in a table. The process is repeated for all the rows of the matrix and the prediction errors are summed, giving the function PRESS (prediction error sum of squares). Once this function has been calculated, different approaches can be applied; the simplest one is just plotting it for different numbers of components and looking for a minimum or for a nonstatistical difference between two consecutive numbers of components. Malinowski[9] has also proposed other methods based on the concept of reduced eigenvalue, which theoretically should follow a normal distribution. The significant reduced eigenvalues will be statistically larger than the average variance associated with the error eigenvalues, and therefore, distinguishable from them. Malinowski[1] gives more details about these methods.

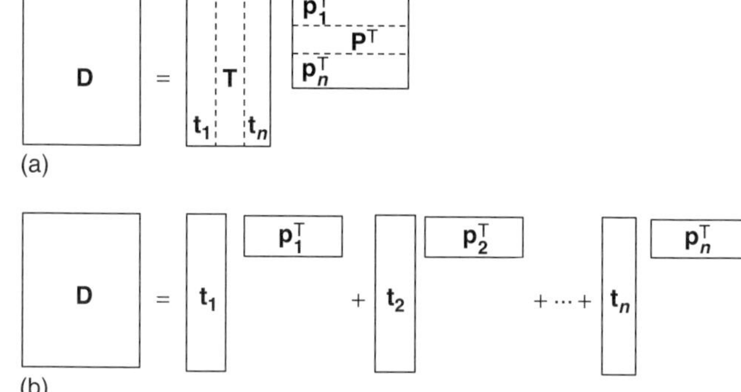

Figure 4 PCA decomposition of a data matrix written as: (a) the product of the matrix of scores (**T**) and the loadings matrix ($\mathbf{P}^T$); and (b) a bilinear model formed by terms described as the outer product of each score vector by its related loading.

The value of PCA for exploratory data analysis has been shown in this section and in many examples published along the history of chemometrics; however, the abstract model provided by PCA is useful for other purposes. Figure 4 shows the bilinear PCA model in detail.

According to Figure 4, the PCA decomposition of a matrix can be written as in Equation (1) or as the sum of bilinear terms as shown in Equation (5):

$$\mathbf{D} = \mathbf{t}_1\mathbf{p}_1^T + \mathbf{t}_2\mathbf{p}_2^T + \cdots + \mathbf{t}_n\mathbf{p}_n^T \quad (5)$$

In using PCA as the starting point to find the underlying model of a chemical data set, the most important requirement is that this set should be intrinsically bilinear, i.e. should have the model form of Equation (5). Happily, many chemical models follow this pattern. A known example is a spectrometric data set, which behaves according to the Beer–Lambert law as shown in Equation (6):

$$\mathbf{D} = \boldsymbol{\varepsilon}_1\mathbf{c}_1 + \boldsymbol{\varepsilon}_2\mathbf{c}_2 + \cdots + \boldsymbol{\varepsilon}_i\mathbf{c}_i \quad (6)$$

where **D** is a matrix of composite spectra, $\mathbf{c}_i$ is the vector that contains the concentrations of the *i*th component in the different samples and $\boldsymbol{\varepsilon}_i$ is the vector of absorptivities for the *i*th component. As can be seen, Equations (5) and (6) are formally identical. Such a similarity suggests that PCA can be useful to decide which size the simplest bilinear model to describe a data set should have, but once this information is obtained, the chemist may be more interested in a model equally sized where the scores and loadings be replaced by terms with a chemical meaning.

At the beginning of this section, it was pointed out that principal components are orthogonal axes that describe the directions of largest variation in a data set. These directions, however, are not necessarily the most informative. In these cases, the best option is performing a *rotation* of the original axes so that the new axes span the same data space, this being more easily interpretable. For example, for a system with two principal components, these components can be replaced by any other couple of axes which lie on the same plane defined by the two principal components. Some rotation procedures are based on mathematical criteria,[10] such as the VARIMAX rotation, which replaces the principal components by the set of orthogonal axes of maximal simplicity.[11] When two axes are compared, e.g. [0.3 0.3 0.3] and [0 0 1], the simplest one, [0 0 1], has its nonnull loadings grouped over a smaller number of variables. In chemistry, these simple factors are often closer to the real ones and, when not, they help to simplify the interpretation of the data set.

Despite the benefits provided by VARIMAX and other abstract rotations, the rotated axes can seldom be identified as chemical factors. Whenever this is possible, the use of rotation procedures which take into account chemical information is preferred. This is the basis of target factor analysis (TFA), which replaces the principal component axes by real factors with chemical meaning.[1] This is done after testing individually a pool of potential real factors (*targets*) which may be useful to build a chemical model for the data set.

TFA works by following the steps below:

1. determination of the number of factors of the data set (PCA);
2. selection of potential real factors (targets);
3. target testing;
4. data reconstruction with the targets accepted.

A crucial step in this method is the selection of the targets. The appropriate choice of these vectors and, consequently, the quality of the final model obtained, would largely depend on the chemical knowledge of the data set. The criterion to choose a target can be theoretical or empirical, or simply rely on previous experience but, in all cases, the target selected should potentially justify part of the variation in the data set. There is no limitation on the number of targets to be tested.

Figure 5 shows two real examples of data sets with their associated targets. The example proposed in (a) is a matrix whose columns are mixture spectra. The targets, which are proposed to replace the score vectors, would be pure spectra of compounds that may be present in the mixtures. TFA would link the models in Equations (5) and (6).[12] The example in (b) belongs to the field of linear solvation energy relationships (LSER).[13] LSER are linear models that explain solvent-dependent variations in solute properties as a function of changes in certain solvent properties (see Equation 7).

$$\mathbf{XYZ} = (\mathbf{XYZ})_o + \mathbf{s_1a_1} + \mathbf{s_2a_2} + \cdots + \mathbf{s_ia_i} \quad (7)$$

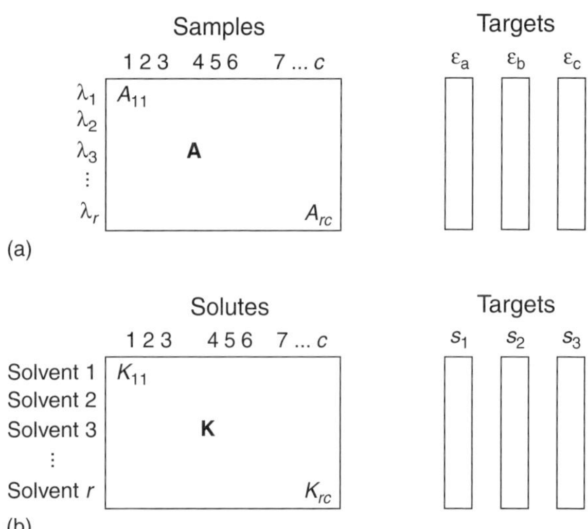

Figure 5 Examples of data sets with their related targets. (a) Matrix of mixed spectra. Targets: pure spectra of potential compounds in the mixture. (b) Matrix of solute properties measured in different solvents. Targets: vectors of solvent properties.

$\mathbf{XYZ}$ is the solvent-dependent solute property, $(\mathbf{XYZ})_o$ is the value of the $\mathbf{XYZ}$ property in a hypothetical solvent with null interaction with the solute, $\mathbf{s_i}$ is a solvent property responsible for the variation of the solute property and $\mathbf{a_i}$ is the weight coefficient related to the contribution of $\mathbf{s_i}$ in the variation of the solute property $\mathbf{XYZ}$. Each column in the data matrix of Figure 5(b) contains the equilibrium constants of a solute determined in different solvents. The targets are vectors of solvent properties that may cause the variation in the equilibrium constants of solutes. TFA links the model in Equation (5) with the model in Equation (7). Apart from the targets with chemical meaning, the *unity target*, which is a vector whose elements are all equal to one, is always tested to account for the possible presence of an offset in the bilinear model.

Once the selection of targets is finished, the next step is assessing if they can be included in the model that describes the variation of the data set. Till this point in the modeling process, PCA has indicated how many real factors should be present in the model and has perfectly defined the data space by means of the score vectors. Any real factor used to replace the scores must belong to the space defined by these abstract vectors. Therefore, testing the goodness of a target is simply finding out how far the vector related to this potential factor is from the space defined by the score vectors.

From a geometrical point of view, target testing consists of projecting the target selected, called hereafter *input target* ($\mathbf{x}_i$), onto the scores space. The projected target, the so-called *output target* ($\mathbf{x}_o$), lies on the scores space.

For references see page 9834

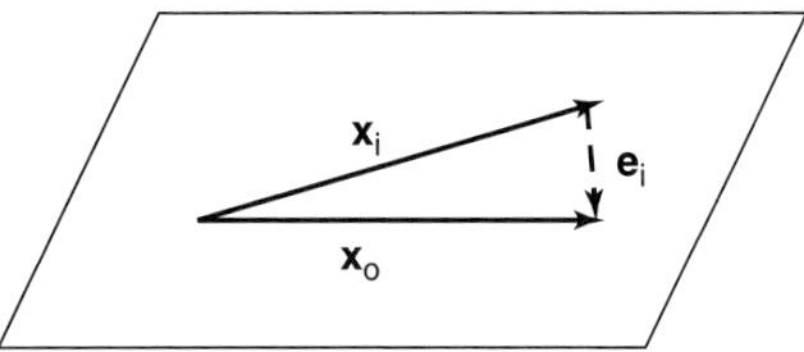

Figure 6 Geometrical representation of the target testing process for a data set described by two factors.

Figure 6 shows the process of target testing for a data set described with two factors. The two principal components define, in this case, a plane.

The perfect target would give $\mathbf{x}_i = \mathbf{x}_o$. However, this equality is never fulfilled in practice due to the experimental error in both the data matrix and the target vector. As a consequence, a target is accepted when the input target is close enough to the output target, i.e. when the length of the vector $\mathbf{e}_i$ (*apparent error in the target*) is small enough.[1] The difference between analogous elements of $\mathbf{x}_i$ and $\mathbf{x}_o$ is related to the quality of the target; thus, the smaller the difference the better the target. Nevertheless, this observation alone is not reliable enough to decide whether a target must be accepted. Empirical methods, such as the SPOIL function,[14] and statistical tests, such as an F-test that compares the error in the target with the error in the data matrix,[9] have been proposed to decide more soundly the acceptance or rejection of a target.

After the target testing step, several targets have been confirmed as suitable candidates to build the real model of variation of the data set. From all the targets that passed this test, a number equal to the number of factors determined by PCA should be chosen. When the number of targets accepted exceeds the size of the model, the best among them are selected.

In the definite model, the targets accepted replace the scores. To explain how to replace the loadings ($\mathbf{P}^T$) by a matrix of weight coefficients associated with the real targets ($\mathbf{C}^T$ for the model in Equation 6 and $\mathbf{A}^T$ for the model in Equation 7), the close relationship between a PCA model and a real model should be recalled. Taking the LSER example in Equation (7) yields Equation (8)

$$\mathbf{TP}^T = \mathbf{SA}^T \tag{8}$$

where $\mathbf{S}$ is the matrix whose columns are the solvent properties ($\mathbf{s}_i$) accepted in the target testing step. This equation can also be written as Equation (9):

$$\mathbf{TY}(\mathbf{Y}^{-1}\mathbf{P}^T) = \mathbf{SA}^T \tag{9}$$

where

$$\mathbf{TY} = \mathbf{S} \tag{10}$$

$$\mathbf{Y}^{-1}\mathbf{P}^T = \mathbf{A}^T \tag{11}$$

From Equation (10), where $\mathbf{T}$ and $\mathbf{S}$ are known, $\mathbf{Y}$ can be calculated. Once $\mathbf{Y}$ is known, $\mathbf{A}^T$ can be calculated using Equation (11) and the real model is completed. The reproduction of the original data set, $\mathbf{D}$, would be as shown in Equation (12):

$$\mathbf{D}_{\mathbf{TFA}} = \mathbf{SA}^T \tag{12}$$

If $\mathbf{D}_{\mathbf{TFA}}$ and $\mathbf{D}$ are sufficiently similar, i.e. the differences can be considered as due to experimental error, the real model proposed is suitable to describe the data set. When this does not happen, some targets in $\mathbf{S}$ can be correlated and, then, new models with other accepted targets should be tried.

All the data sets that can be organized in matrices do not have the same features. This is clearly seen when the two data sets in Figure 1 are compared. The data set in Figure 1(a) is formed by voltammograms related to different samples. There is no loss of information if the order of the rows (samples) is exchanged. Figure 1(b) contains the data set obtained in the spectrometric monitoring of a kinetic process. If the rows of the data matrix, which are related to each reaction time, are organized differently, the information about the kinetic evolution is lost. Data sets like the latter describe dynamic systems and are often found in chemistry, e.g. high-performance liquid chromatography/diode-array detector (HPLC/DAD) data, or spectrometric monitoring of thermodynamic, thermal or any other parameter-dependent reaction. For these cases, there are some FA-based techniques, namely evolving factor analysis (EFA) and fixed-size moving window/evolving factor analysis (FSMW/EFA), which are designed to model the evolution of the factors in the data set. The evolution of these abstract factors is connected with the development of the chemical process.

EFA was first proposed by Maeder and co-workers[15–17] to treat spectrometrically monitored equilibria and was afterwards applied to any other kind of data set coming from dynamic chemical processes,[18] such as HPLC/DAD, kinetic transformations, or structural transitions.

The EFA method calculates the eigenvalues obtained from the PCA of gradually growing submatrices of the original data set, $\mathbf{D}$ ($m \times n$). PCA is performed repeatedly in submatrices generated by increasing the size of the previous submatrix by one row. This size increase is performed from the top to the bottom of the original data matrix $\mathbf{D}$ (*forward EFA*) and also in the opposite sense (*backward EFA*). Thus, forward EFA first carries out PCA on rows 1 and 2 of the data matrix, then on rows 1, 2, and 3 and so on until the final PCA on the m rows of the whole data matrix is performed. Backward EFA starts by performing PCA on the rows m and $(m-1)$, then adds the row $(m-2)$ to the analysis and so forth until the final PCA is carried out with the whole data matrix.

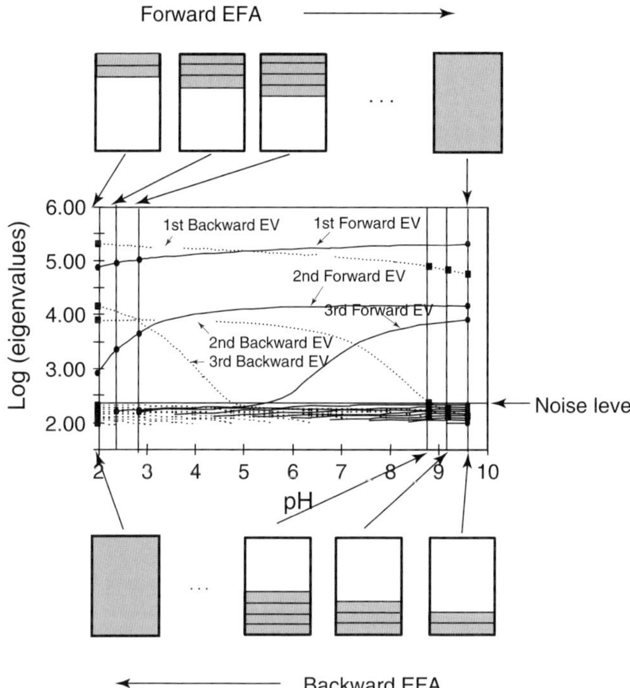

Figure 7 Construction of an EFA plot related to the spectrometric titration of a diprotic acid (simulated data). The continuous lines join the eigenvalues (circles) obtained in each PCA of forward EFA and the dashed lines connect the eigenvalues (squares) obtained in each PCA of backward EFA. EV stands for eigenvalue.

The EFA plot displays the evolution of the eigenvalues as the submatrix analyzed grows, i.e. as the evolutionary process in the data matrix goes from the beginning to the end (forward EFA) or vice versa (backward EFA). The log(eigenvalues) are plotted versus the row number of the last row included in the PCA or, better, versus the value of the variable responsible for the evolutionary process related to this row (e.g. a pH value in an acid–base equilibrium or a retention time in an HPLC/DAD data set). Figure 7 shows the EFA plot related to a matrix formed by data coming from the spectrometric monitoring of the acid–base equilibria of a diprotic acid. Each row in the matrix is a spectrum collected at a certain pH value.

As PCA does, an EFA plot indicates the total number of components in the data set. The eigenvalues related to chemical components are clearly higher than the eigenvalues associated with noise. At the bottom of the plot, the many overlapping noise eigenvalues define a graphical threshold that can be considered the noise level of the system. All the eigenvalues above this limit are chemically significant. The total number of chemical components can be determined from the number of significant eigenvalues obtained in the last PCA performed in both forward and backward data analyses (i.e. with the results in the right and left extremes of the forward and backward EFA plot respectively). In the example of Figure 7, three chemical components, related to each of the species of a diprotic acid, are detected.

In contrast to PCA, an EFA plot also marks the zones of appearance and disappearance of chemical components. The appearance of a new species is detected in the forward EFA through the emergence of a new significant eigenvalue. Thus, when the row added to a certain submatrix includes a new chemical component, the different information introduced by this component will cause a significant increase in the 'until then' first noise eigenvalue. The chemical variable (e.g. pH, time, etc.) corresponding to the new row included in the analysis indicates the chemical conditions in which the new component appears. When performing backward EFA, the process is followed from the end to the beginning. Therefore, the presence of a new significant eigenvalue indicates the row, and consequently the chemical conditions, in which a component disappears. According to the forward EFA results in Figure 7, the two most protonated forms of the diprotic acid are present at pH 2, whereas the third one appears at a pH close to 5.5. Looking at the backward EFA plot, the most deprotonated form is still present at pH 10, the next one disappears at a pH close to 9 and the most protonated at a pH around 5.

The evolution of the eigenvalues in the forward and backward EFA is connected with the formation and decay curves of the chemical components present in the data matrix. When the appearance and disappearance of the chemical components are sequential, as in the example of Figure 7 and other reaction processes, or as in most of the chromatographic elutions, the suitable connections between the lines of the forward EFA plot and the backward EFA plot can provide abstract profiles related to the evolution of each of the chemical components in the data matrix. Thus, for a system with n components, the profile of the first component appearing is built connecting the line of the first eigenvalue of the EFA forward plot with the line of the nth significant eigenvalue of the EFA backward plot (related to the first compound disappearing); the profile of the second component connects the second eigenvalue in the EFA forward plot with the $(n-1)$th eigenvalue of the EFA backward plot and so on. The zone of the matrix (rows) where the concentration profile of one compound shows values different from zero is the *concentration window* related to this compound (see Figure 8). The abstract profiles obtained with EFA are often used as a starting point in the iterative resolution methods described in sections 3 and 4 of this chapter.

For references see page 9834

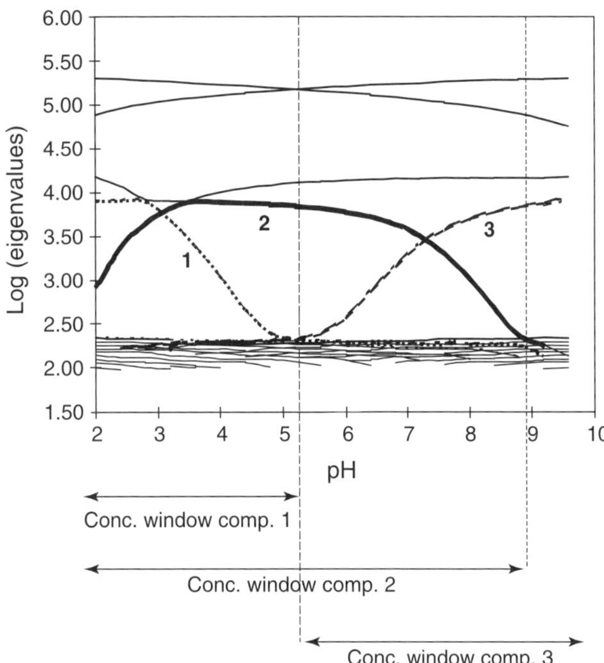

Figure 8 Abstract concentration profiles of a diprotic acid obtained using EFA, with their respective concentration windows.

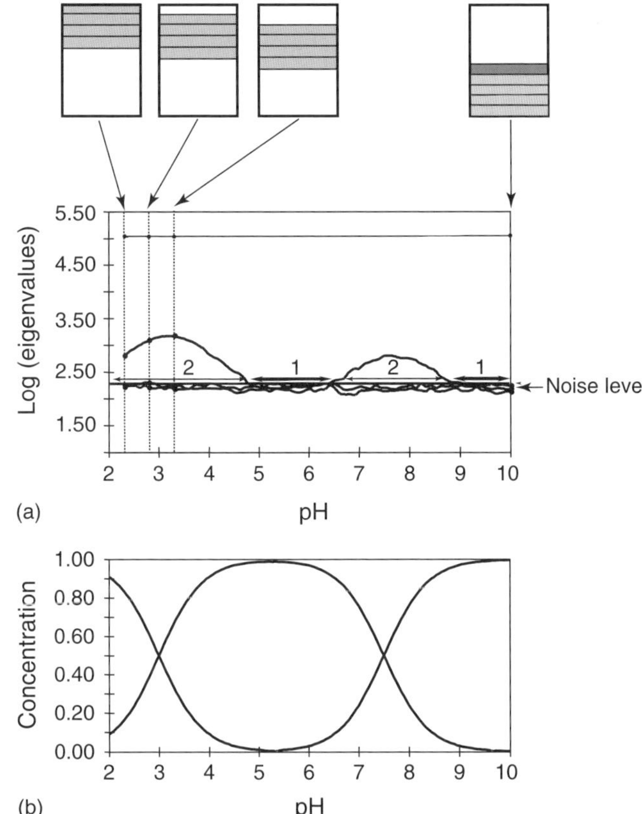

Figure 9 (a) FSMW/EFA plot related to the spectrometric titration of a diprotic acid. The window size is four rows. The figures within the plot indicate the number of species in the zones defined by the lines. The thick lines mark the selective zones. Circles are the EV obtained in the PCA of each submatrix. (b) True distribution plot of the diprotic acid.

The EFA method can be applied to both directions in the data matrix (i.e. the successive PCA can be performed in row- or column-wise growing submatrices of the original data matrix). In any case, the most useful results will always be obtained in the direction where the overlap of the different chemical components is less.

FSMW/EFA was proposed by Keller and Massart[18] as a derivation of EFA. In contrast to EFA, FSMW/EFA scans the original data matrix from top to bottom by performing PCA on "equally sized" submatrices. A new submatrix is built by removing the first row of the previous submatrix and adding the following row in the original data matrix, i.e. moving a window of a fixed number of rows one row downwards. The size of the window usually exceeds the number of chemical components by one, though more information can be obtained if FSMW/EFA is applied several times using different window sizes.[19]

A local rank map, i.e. a graph with information about the number of significant principal components (chemical components) in each zone of the data matrix, is obtained by plotting the eigenvalues obtained in each PCA vs the variable responsible for the evolution of the process, as shown in Figure 9. In this plot, the noise level is also defined graphically by the zone in which the noise eigenvalues appear together. The emergence of one significant eigenvalue from the noise level zone indicates the presence of only one species in that zone; the presence of two significant eigenvalues defines the zone where two species coexist, three significant eigenvalues would detect the overlap of three species, and so on.

The local rank map obtained through the application of FSMW/EFA allows the *detection of selective zones*, i.e. zones where only one chemical component is present. The zones where only one significant eigenvalue is present are the selective zones of the species (the detection and location of these regions is essential in the resolution of a data matrix). In Figure 9, two selective zones are detected, one for the most deprotonated form of the diprotic acid and another for the intermediate form of this compound. FSMW/EFA also contributes to the *detection of minor constituents*. Actually, FSMW/EFA was initially proposed to deal with peak purity problems in chromatography. The local analysis of the data matrix and the clear graphical information help in the detection of minor species, although they are embedded under major constituents.

3 MULTIVARIATE CURVE RESOLUTION: GENERAL BACKGROUND

All the resolution methods were born as a tool to analyze multivariate experimental data coming from multicomponent dynamic systems.[20–27] The common goal for all these methods is to mathematically decompose the global instrumental response into the pure contributions due to each of the components in the system. The use of such methods has become valuable when obtaining individual signals experimentally is not possible or when this process is too complex or too time-consuming.

The multivariate output of an experiment monitoring a dynamic process is organized in a double-structured matrix **D** containing mixed information about the evolution of all the components present in the successive stages of the chemical process. The ultimate goal of the curve resolution (CR) methods is the decomposition of the initial mixture data matrix **D** into the product of two data matrices **C** and **S**, each of them including the pure response profiles of the *n* mixture components associated with one of the directions of the initial data matrix (see Figure 10).

In matrix notation, Equation (13) gives the general expression valid for all CR procedures:

$$\mathbf{D} = \mathbf{C}\mathbf{S}^{\mathrm{T}} + \mathbf{E} \qquad (13)$$

where $\mathbf{D}$ $(r \times c)$ is the original data matrix, $\mathbf{C}$ $(r \times n)$ and $\mathbf{S}^{\mathrm{T}}(n \times c)$ are the matrices containing the pure response profiles related to the data variation in the row direction and in the column direction respectively, and $\mathbf{E}$ $(r \times c)$ is the error matrix, i.e. the residual variation of the data set that is not related to any chemical contribution. n is the number of chemical components in matrix **D**.

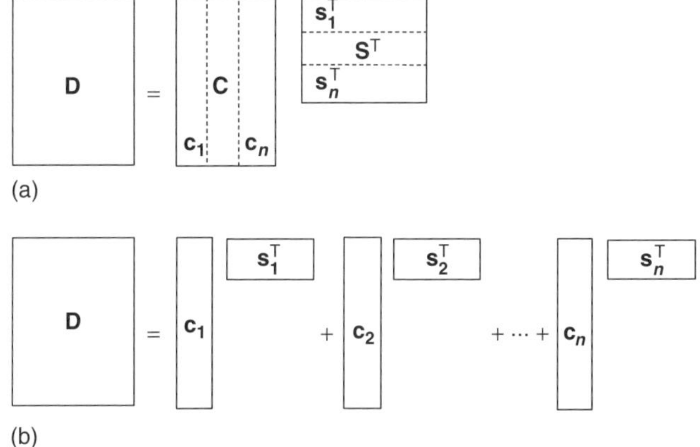

Figure 10 Two graphical views of the decomposition of a multicomponent data matrix: (a) as the product of the matrices including the pure response profiles; (b) as the sum of the products related to each pure contribution of the mixture components.

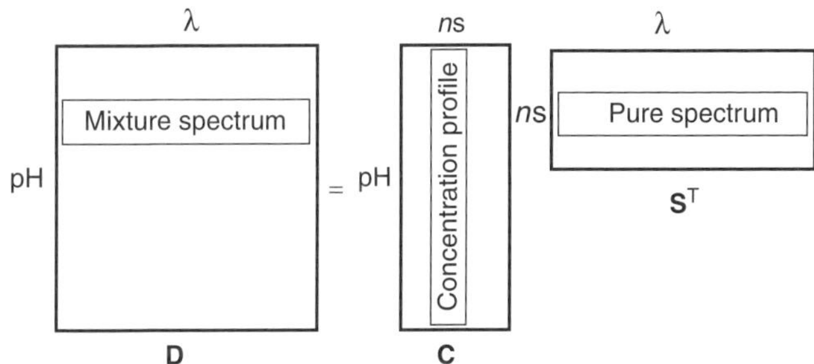

Figure 11 Results obtained after the application of a CR method on a data matrix coming from a pH-dependent process monitored spectrometrically. *ns* means number of absorbing species.

Taking as an example a pH-dependent process monitored spectrometrically, the **C** matrix would contain the pure concentration profiles of all the absorbing species and the $\mathbf{S}^{\mathrm{T}}$ matrix would be formed by their related pure spectra (see Figure 11). For the sake of simplicity, **C** and $\mathbf{S}^{\mathrm{T}}$ are referred to as concentration profile matrix and spectra matrix, though this does not mean that the applicability of CR methods is restricted to this kind of chemical data.

The mathematical decomposition of a single data matrix is inherently affected by two sources of ambiguity: rotational ambiguity and intensity ambiguity. Whereas the former accounts for the possibility of reproducing correctly the original data matrix by using **C** and **S** matrices containing linear combinations of the true profiles, the latter warns about the possibility of having profiles equal in shape to the true ones, though different in magnitude. In other words, the correct reproduction of the original data matrix can be achieved by using response profiles differing in shape (rotational ambiguity) or in magnitude (intensity ambiguity) from the true ones.

The explanation of these two ambiguities is simple. The basic equation associated with CR methods, $\mathbf{D} = \mathbf{C}\mathbf{S}^{\mathrm{T}}$, can be easily transformed as shown in Equations (14–16):

$$\mathbf{D} = \mathbf{C}(\mathbf{T}\mathbf{T}^{-1})\mathbf{S}^{\mathrm{T}} \qquad (14)$$

$$\mathbf{D} = (\mathbf{C}\mathbf{T})(\mathbf{T}^{-1}\mathbf{S}^{\mathrm{T}}) \qquad (15)$$

$$\mathbf{D} = \mathbf{C}'\mathbf{S}'^{\mathrm{T}} \qquad (16)$$

where $\mathbf{C}' = \mathbf{C}\mathbf{T}$ and $\mathbf{S}'^{\mathrm{T}} = (\mathbf{T}^{-1}\mathbf{S}^{\mathrm{T}})$ describe the **D** matrix as correctly as the true **C** and $\mathbf{S}^{\mathrm{T}}$ matrices do, though $\mathbf{C}'$ and $\mathbf{S}'^{\mathrm{T}}$ lack chemical sense. On the basis of the transformation shown in these equations, the mathematical formulation of the rotational ambiguity problem indicates that the possible solutions of a resolution method are as numerous as the **T** matrices can be, i.e infinite. However, the inclusion of information related to the internal structure of the data (e.g. the presence of selective zones) and to their chemical properties in the resolution process

For references see page 9834

often allows the suppression of this ambiguity or, at least, a large decrease in the number of feasible solutions.

When a system lacking rotational ambiguity is considered, the basic CR equation can still be rewritten as shown in Equations (17) and (18):

$$\mathbf{D} = \left(\frac{1}{k}\right)\mathbf{C}k\mathbf{S}^{\mathrm{T}} \tag{17}$$

$$\mathbf{D} = \mathbf{C}'\mathbf{S}'^{\mathrm{T}} \tag{18}$$

where k is a scalar. The concentration profiles of the new $\mathbf{C}' = (1/k)\mathbf{C}$ matrix have the same shape as the real ones, but are k times smaller, whereas the spectra of the new $\mathbf{S}' = k\mathbf{S}$ matrix are shaped like the $\mathbf{S}$ spectra, though k times more intense. This ambiguity cannot be solved unless external information is introduced in the resolution process. Both rotational and intensity ambiguities are drastically diminished when several matrices (*three-way data sets*) are analyzed together.

The correct performance of the CR methods depends strongly on the internal features of the data set being analyzed, specially on selectivity and local rank.[28] Regardless of the quality of each CR method, the two following conditions must be fulfilled if the true concentration profile and spectrum of each compound in the data matrix are to be recovered:[29]

- The true concentration profile of a compound can be recovered when all the compounds inside its concentration window are also present outside.
- The true spectrum of a compound can also be recovered if its concentration window is not completely embedded inside the concentration window of a different compound.

The same formulation written above holds when, instead of looking at the concentration windows (windows of rows), the 'spectral' windows (windows of columns) are considered. The content of these theorems supports the fact that the goodness of the resolution result depends more strongly on the features of the data set (particularly those related to selectivity and local rank) than on the mathematical background of the CR method. Therefore, a good knowledge of the properties of the data sets before carrying out a resolution process provides a clear idea about the quality of the results that can be expected.

Seeing how close the PCA and the CR decomposition of a matrix are formally, it is not surprising that some CR methods, such as window factor analysis (WFA)[30] or heuristic evolving latent projections (HELP),[31] work with the abstract variables obtained by PCA in the resolution process. Some other methods transform initial estimates into real solutions projecting them iteratively onto the scores space, such as iterative transformation target factor analysis (ITTFA).[24,32] Methods like multivariate curve resolution/alternating least squares (MCR/ALS)[33] are linked to PCA because they often use initial estimates obtained by PCA-based techniques, such as EFA. Initial estimates can be also obtained by some other methods that use real variables, such as the simple-to-use interactive self-modeling analysis (SIMPLISMA)[26] or the orthogonal projection approach (OPA).[34] The CR methods can also be classified according to the data sets they can deal with. Thus, some methods are applied to the resolution of single matrices, such as HELP, WFA and ITTFA, whereas others can deal with two (generalized rank annihilation method, GRAM)[35] or more matrices together (direct trilinear decomposition (DTD),[36] parallel factor analysis (PARAFAC),[37–39] and restricted Tucker models).[40] Methods like MCR/ALS are adapted to one or more matrices.[41–43]

4 NONITERATIVE AND ITERATIVE CURVE RESOLUTION TECHNIQUES

As mentioned in section 3, the goal of all CR is the decomposition of a mixture data matrix $\mathbf{D}$ into the product of the small matrices, $\mathbf{C}$ and $\mathbf{S}^{\mathrm{T}}$, which contain profiles related to the evolution along the rows and along the columns of each pure component in $\mathbf{D}$. The noniterative resolution methods obtain $\mathbf{C}$ and $\mathbf{S}^{\mathrm{T}}$ in one calculation step, whereas iterative resolution methods refine the profiles in $\mathbf{C}$, in $\mathbf{S}^{\mathrm{T}}$ or in both matrices at each iterative cycle till an optimal solution is attained.

Noniterative resolution methods are the fastest, but they often require that the data set has certain features for them to be applied (e.g. WFA is meant to work with evolutionary data where each chemical compound has a unique maximum in its concentration profile). The user intervention is less, but usually more critical; wrong decisions cannot be changed in a one-step calculation method. Nevertheless, when the data sets fulfill the conditions needed by the CR method, these resolution procedures are a good alternative. The information provided by the global and local application of PCA to the data set is essential in most of these noniterative methods; thus, the quality of the final solutions depends basically on the correct determination of the total number of chemical components in the data set and on the construction of a reliable local rank map.

Within the group of noniterative resolution methods, HELP[31] and WFA[30] are probably the most well known. The HELP method is based on the local rank analysis of the data set and focuses on finding selective concentration and/or spectral windows. When these selective zones exist, the resolution of the system is clear. Thus, for a data set related to the spectrometric monitoring of a kinetic process (see Figure 1b), the row related to a selective reaction time directly provides the shape of the spectrum of the only component present at that stage of the reaction and the column related to a selective wavelength directly provides the kinetic concentration profile of the only absorbing compound at that wavelength. The main contributions of HELP have been offering a sophisticated graphical tool (the so-called datascope) to visually detect potential selective zones in the score plot of the data matrix and a statistical method to confirm the presence of selectivity in the concentration and/or spectral windows graphically chosen. The statistical method is based on the use of an F-test to compare the magnitude of eigenvalues related to selective zones of the data set with eigenvalues related to noise zones of the data matrix, i.e. those where no chemical components are supposed to be present. The confirmation of a selective zone in the data set, which is actually a rank-one window in the data matrix, will then be obtained when no significant differences are found among the first eigenvalue of a noise-related zone of the data matrix and the second eigenvalue of the potential selective zone. The statistical part of the method can be negatively affected by the presence of nonchemical significant variations in the data, such as instrumental drift or nonrandom noise. It should be also noted that the application of this method requires that the data set has zero-component zones, which is easy in an HPLC/DAD data set (time windows with no compounds eluting are commonly found) but not so evident when a reaction monitoring is carried out (there is always some chemical compound in the reaction vessel), for instance.

WFA seems to have a wider field of applications than HELP and improved versions of the algorithm have recently been published.[44] This method inspired in the rank annihilation evolving factor analysis (RAEFA)[25] was proposed by Malinowski.[30,45,46] WFA is applicable to data sets where the formation and decay of the compound profiles in one of the directions of the matrix (usually the concentration direction) is sequential, i.e. the compound appearing in the nth position disappears in the nth position. As a consequence, out of the concentration window of a certain compound all the other compounds in the data set are present.

WFA recovers each true concentration profile as follows (see Figure 12):

1. PCA of the original data matrix, **D**.
2. Determination of the concentration window of each component.
3. PCA of a matrix, $\mathbf{D^o}$, where the rows related to the concentration window of the nth compound have been removed.

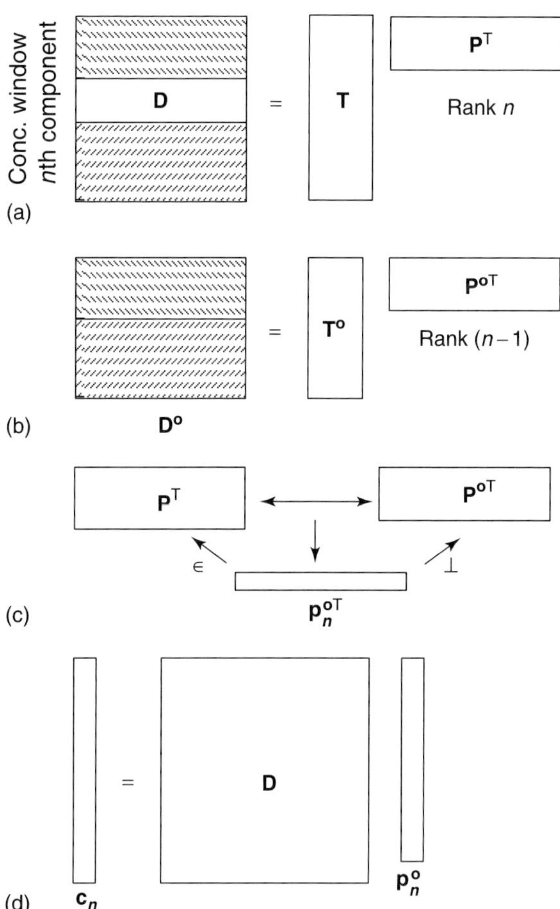

Figure 12 Recovery of the concentration profile of the nth compound by WFA. (a) PCA of the raw data matrix, **D**; (b) PCA of the matrix formed by suppression of the concentration window of the nth component, $\mathbf{D^o}$; (c) recovery of the component of the spectrum of the nth component orthogonal to all the spectra in $\mathbf{D^o}$, $\mathbf{p}_n^{o\mathrm{T}}$; and (d) recovery of the concentration profile of the nth component.

4. Calculation of the vector, $\mathbf{p}_n^{o\mathrm{T}}$, which is the part of the spectrum of the nth component orthogonal to the spectra of the other components.
5. Recovery of the true concentration profile of the nth component using $\mathbf{p}_n^{o\mathrm{T}}$ and **D**.
6. Calculation of the spectra matrix $\mathbf{S}^\mathrm{T}$ by least squares using **D** and **C**.

The mathematical formulation associated with this method is not trivial and only a rough intuitive explanation will be given in this section. A detailed description can be found in the literature.[30]

WFA starts with the PCA decomposition of the **D** matrix into the product $\mathbf{TP}^\mathrm{T}$ (remember Equation 1). In a general case, the **D** matrix can have n components, i.e. rank n. The definition of the concentration windows for each component is carried out later using EFA (see

For references see page 9834

Figure 8) or other methods. Steps 3 to 5 are the core of the WFA method and should be performed as many times as compounds are present in matrix **D** to recover every time one concentration profile of the **C** matrix.

For each component, a $\mathbf{D^o}$ matrix is built removing the rows related to its concentration window. Then, a PCA is performed and the product $\mathbf{T^oP^{oT}}$ is obtained. Note that the $\mathbf{D^o}$ matrix has rank $n-1$ because the variation due to one of the components in the data set disappears when the related rows of the data matrix **D** are deleted. The loading matrices, $\mathbf{P^T}$ and $\mathbf{P^{oT}}$, describe the space of the n pure spectra in **D** and the $(n-1)$ pure spectra in $\mathbf{D^o}$ respectively. The rows in these loading matrices are actually "abstract spectra" and the real spectra can be expressed as a linear combination of them. The vectors either in $\mathbf{P^T}$ or in $\mathbf{P^{oT}}$ could be interchangeably used to build these linear combinations if the information related to the removed component was present in $\mathbf{P^{oT}}$. Using these two loading matrices, it is possible to calculate a vector $\mathbf{p}_n^{oT}$, which is orthogonal to the $(n-1)$ $\mathbf{p}_i^{oT}$ vectors and which belongs to the space defined by $\mathbf{P^T}$. This vector completes the set of vectors in $\mathbf{P^{oT}}$ and contains the part of the spectra of the removed component which is uncorrelated to the spectra of the other $(n-1)$ components in the data matrix. Using this vector with information exclusively related to the removed component, the true concentration profile of this compound can be calculated as shown in Equation (19):

$$\mathbf{D}\mathbf{p}_n^o = \mathbf{c_n} \quad (19)$$

The complete **C** matrix is then obtained by appending row-wise the concentration profiles found for each compound in the **D** matrix. Using the basic equation of CR methods, $\mathbf{D} = \mathbf{CS^T}$, the matrix of spectra, $\mathbf{S^T}$, is obtained by least squares using the **D** and **C** matrices as shown in Equation (20):

$$\mathbf{S^T} = (\mathbf{C^TC})^{-1}\mathbf{C^TD} \quad (20)$$

The main drawbacks of WFA are the impossibility of solving data sets with nonsequential profiles (e.g. data sets with embedded profiles, for instance) and the dangerous effects of a bad definition of the concentration windows. Tackling this last point has been the main goal of recent modifications of this algorithm.[44]

Despite the limitations associated with the noniterative methods, the information provided by local PCA performed in selected windows of the data matrix still inspires new algorithms for resolution, such as the very recent subwindow factor analysis (SFA), proposed by Manne et al.[47] based on the comparison of matrix windows sharing one compound in common.

Iterative resolution methods are in general more versatile than noniterative methods. They apply to more diverse data (e.g. with sequential and nonsequential profiles, with different degrees of selectivity, etc.) and the previous knowledge about the data set (chemical or related to mathematical features) can be used in the optimization process. The main complaint about iterative resolution methods has often been the larger calculation times required to obtain the optimal results; however, improved fast algorithms[48,49] and more powerful personal computers have overcome this historical limitation.

All iterative resolution methods refine the profiles in **C**, in $\mathbf{S^T}$ or in both matrices at each step of the optimization process. The profiles in **C** and/or $\mathbf{S^T}$ are "tailored" according to the chemical properties and the mathematical features of each particular data set. The iterative process stops when a convergence criterion is fulfilled (e.g. a preset number of iterative cycles is exceeded or the lack of fit goes below a certain value).

Starting the iterative optimization of the profiles in **C** or $\mathbf{S^T}$ requires a matrix sized as **C** or as $\mathbf{S^T}$ with more or less rough approximations of the concentration profiles or spectra that will be obtained as the final results. This matrix contains the *initial estimates* of the resolution process. Though some authors are in favor of using random initial estimates,[50] the most generalized opinion tends to recommend the use of nonrandom estimates to shorten the iterative optimization and to avoid the convergence to local optima different from the solution searched. If the initial estimates are either a **C**-type or an $\mathbf{S^T}$-type matrix can depend on which kind of profiles are less overlapped, on which direction of the matrix (rows or columns) has more available information or simply on the will of the chemist. There are many chemometric methods to build initial estimates: some are especially suitable when the data consist of evolutionary profiles, such as EFA (see Figure 8), whereas some others mathematically select the purest rows or the purest columns of the data matrix as initial profiles. Within the latter, key set factor analysis (KSFA)[51] works in the FA abstract domain and some other procedures, such as SIMPLISMA[26] and OPA,[34] work with the real variables in the data set and select the rows or the columns most representative and most dissimilar to each other. Besides the use of chemometric methods, a matrix of initial estimates can always be formed including the rows or columns of the data set that the researcher considers most representative for chemical reasons.

The general process in an iterative resolution method goes on with the optimization of the initial profiles with the help of some selected constraints. A constraint can be defined as any systematic feature in the data set (mathematical or chemical) translated into mathematical language. Therefore, the effectiveness of a constraint can change depending on the way it has been implemented and active research is in progress to optimize this point.[52–55] In any profile, a constraint works by updating the elements that do not allow the fulfillment of a certain condition by some others that do. The use of constraints is optional and should be adapted to each particular data set. Small departures from the conditions imposed by some constraints are often allowed.

The most essential constraint in the resolution process is *selectivity*.[28] The selective zones in a data matrix are those regions (row windows or column windows) where only one species is present, i.e. those with rank one; therefore, this is a constraint related to a mathematical feature. As mentioned in the brief comment about the HELP method (see this section, above), the presence of selective zones for all the species in a data matrix eliminates the rotational ambiguity and ensures the recovery of the real response profiles of the chemical system. Local rank analysis methods, such as EFA and FSMW/EFA, are the most suitable techniques to detect and locate the selective zones of a system, as shown in section 2. The selectivity for one compound can be forced by setting to zero the elements of all the other response profiles in the selective regions of this compound.

The most frequent constraints related to chemical features of the profiles are described below:

1. Non-negativity. This constraint forces the values in a profile to be equal to or greater than zero and is applied to all the concentration profiles and to some experimental responses, such as ultraviolet (UV) absorbances.
2. Unimodality. This constraint allows the presence of only one maximum per response profile. It is applied to chromatographic peaks, to the concentration profiles of some chemical reactions and to some peak-shaped instrumental responses, such as voltammograms.
3. Closure. This constraint is applied to closed reaction systems for which the sum of the concentrations of all the species involved in the reaction or the sum of some of them is forced to be constant at each stage of the reaction. Actually, closure is a mass balance constraint.

A possible modification of some simulated profiles after the application of the constraints above is graphically shown in Figure 13.

Two of the most representative iterative resolution methods are ITTFA[24,32,56] and MCR/ALS.[27,28,41–43] Whereas ITTFA optimizes either the **C** profiles or the $\mathbf{S}^T$ profiles in each iterative cycle, MCR/ALS modifies the profiles in both matrices in each step of the optimization. An additional difference is that ITTFA works by optimizing one profile at a time, whereas MCR/ALS refines together all the profiles in the constrained data matrices.

As the name suggests, ITTFA is based on TFA (see section 2). In TFA, some vectors (targets) are tested to see if they can be used to describe real sources of variation in the data set. These vectors have a chemical meaning and are perfectly characterized. In facing a resolution problem, the appropriate targets should be either potential concentration profiles or instrumental responses (spectra). However, in practice, the user can, at most, know exactly some of these real profiles and, in some situations, none of them. Therefore, the straightforward application of TFA to solve completely a resolution problem is not possible. ITTFA borrows two main ideas from TFA: the fact that the space of either the concentration profiles or the spectra can be perfectly known and the initial use of a target to finally obtain a true profile of the **C** or the $\mathbf{S}^T$ matrix. Essentially, what ITTFA does is to modify a target until it lies on the real space of concentrations or spectra and fulfills the appropriate constraints. This process is repeated with as many targets as the data set has components.

It has been pointed out that ITTFA works by optimizing one at a time the profiles in either the **C** matrix or the $\mathbf{S}^T$ matrix. The direction of optimization will depend on which information is considered most important by the chemist: e.g. in an HPLC/DAD run, getting the chromatographic peaks of each compound may be the priority; whereas in a process, monitoring the knowledge of the identity of the compounds involved (spectra) can be the main goal. The explanation given below holds for a data set where ITTFA is applied to obtain the profiles in the **C** matrix. Transposing the original data matrix, this process would be useful to get the profiles in $\mathbf{S}^T$. ITTFA gets each concentration profile following the steps below:

1. Calculation of the score matrix by PCA.
2. Use of an estimated concentration profile as initial target.
3. Projection of the target onto the score space.
4. Constraint of the target projected.
5. Projection of the constrained target.
6. Return to step 4 until convergence is achieved.

ITTFA starts performing PCA in the original data matrix, **D**. There is a formal analogy between the PCA decomposition, i.e. $\mathbf{D} = \mathbf{TP}^T$, and the CR decomposition, i.e. $\mathbf{D} = \mathbf{CS}^T$, of a data matrix. The scores matrix, **T**, and the loadings matrix, $\mathbf{P}^T$, span the same data space as the **C** and the $\mathbf{S}^T$ matrix do and their profiles can be described as abstract concentration profiles and abstract spectra respectively. This means that any real concentration profile of **C** belongs to the score space and can be described as a linear combination of the abstract concentration profiles in the **T** matrix.

For references see page 9834

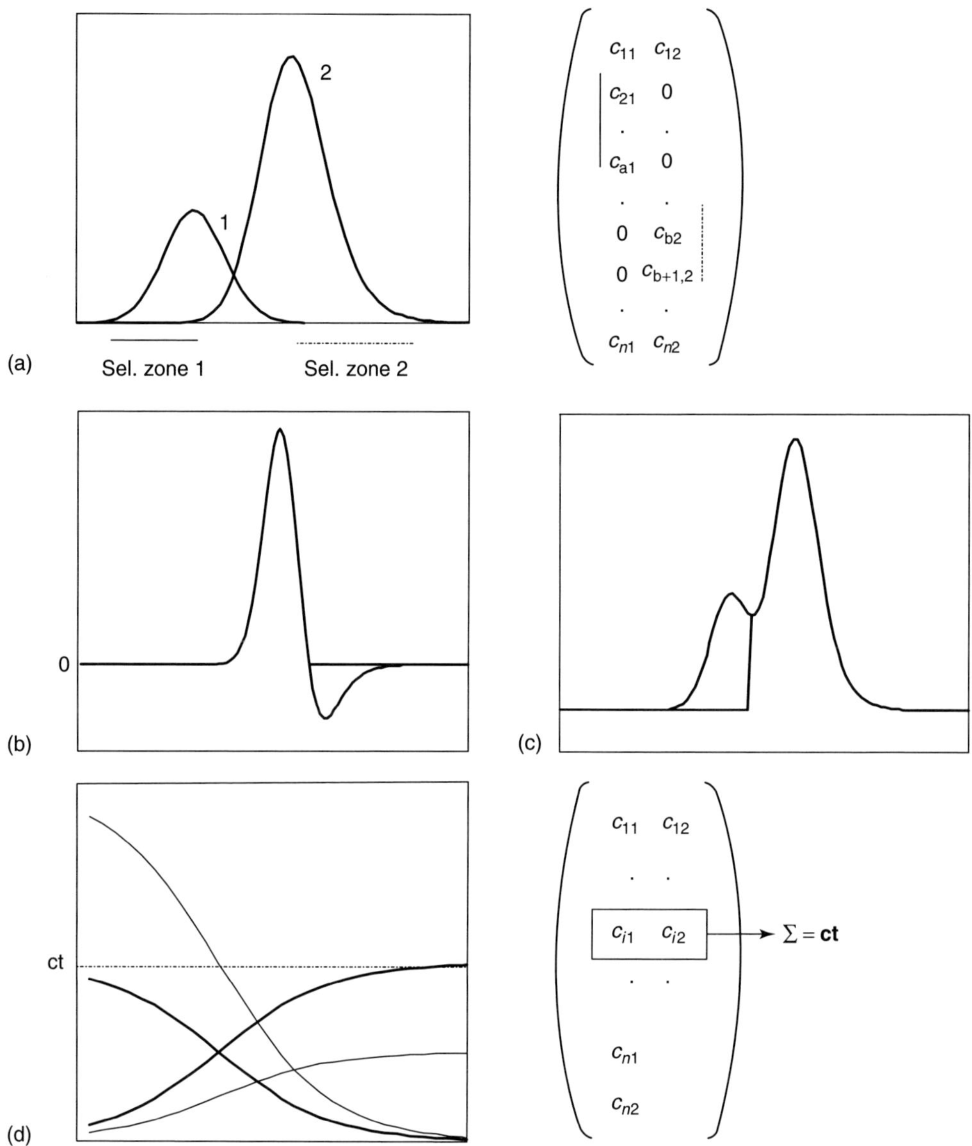

Figure 13 Performance of the constraints in the MCR/ALS method. (a) Selectivity. The solid line and the dashed line indicate the selective zones for compounds 1 and 2, respectively. On the right, the constrained matrix is shown. (b) Non-negativity, (c) unimodality and (d) closure: the normal line and the thick line show a response profile before and after the application of the constraint, respectively. (d) also presents the constrained concentration matrix.

The next step would be choosing approximate concentration profiles as targets. There are many ways to select these initial vectors; actually, any method used to provide initial CR estimates can be useful for this purpose. Historically, the vectors obtained after performing VARIMAX rotation onto the scores were used[24] and also the needle targets (i.e. vectors with only one nonnull element equal to 1), which are the simplest representation of a peak-shaped profile.[32,57]

The next step is the projection of the initial target ($\mathbf{x}_{1\ \mathrm{in}}$) onto the score space. The projected target ($\mathbf{x}_{1\ \mathrm{out}}$) belongs to the space of the real concentration profiles and, from a pure mathematical point of view, could be accepted to describe the data set. However, when this profile is plotted, the chemist may not like some of the features that present (e.g. negative parts, secondary maxima, etc.). When this happens, the projected target $\mathbf{x}_{1\ \mathrm{out}}$ is modified by using the appropriate constraints. The application of constraints satisfies the chemical features demanded by the profile but, as a consequence of the modification of its elements, lifts the target from the score plane. The constrained target, $\mathbf{x}_2$, is projected again onto the score space and the new projected target, $\mathbf{x}_{2\ \mathrm{out}}$, is constrained if necessary. The process goes on until the projected target

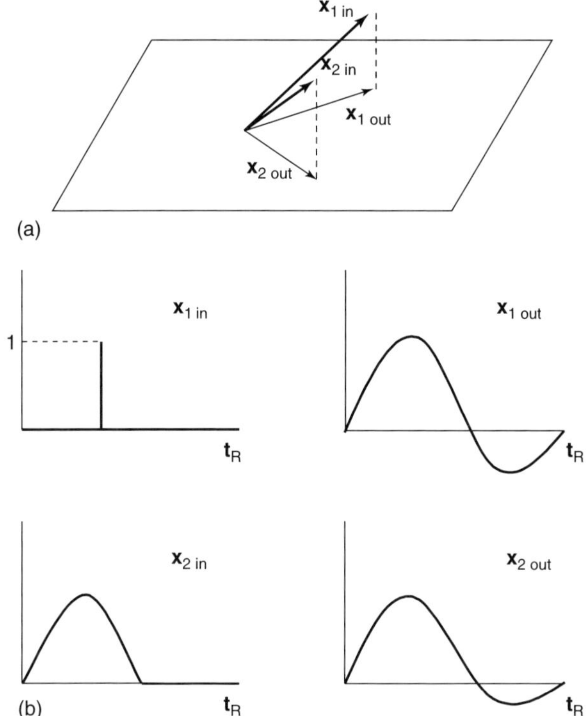

Figure 14 (a) Geometrical representation of the optimization of a chromatographic profile by ITTFA from an initial needle target. The example represents a two-compound data set. Thick lines represent targets out of the score plane; thinner lines are targets on the score plane. (b) Evolution in the shape of the chromatographic profile through the ITTFA process.

makes sense from both mathematical and chemical points of view, i.e. until the constrained profile belongs to the score space or until it is very close to it. Figure 14 shows the ITTFA optimization of a chromatographic profile for a system with two compounds.

Once all the concentration profiles obtained are appended to form the **C** matrix, the $\mathbf{S}^{\mathrm{T}}$ matrix can be calculated by least squares from **D** and **C** (see Equation 13).

MCR/ALS uses an alternative approach to find iteratively the matrices of concentration profiles and instrumental responses. In this method, neither the **C** nor the $\mathbf{S}^{\mathrm{T}}$ matrix have priority over each other and both are optimized at each iterative cycle. The general operating procedure of MCR/ALS includes:

1. Determination of the number of compounds in **D**.
2. Calculation of initial estimates (e.g. **C**-type matrix).
3. Calculation and constraint of the $\mathbf{S}^{\mathrm{T}}$ matrix.
4. Calculation and constraint of the **C** matrix.
5. Return to step 3 until convergence is achieved.

The number of compounds in **D** can be determined using PCA or can be known beforehand. In any case, the number obtained must not be considered a fixed parameter and resolution of the system considering different numbers of components is a usual and recommendable practice. In contrast to ITTFA, MCR/ALS uses complete **C**- or $\mathbf{S}^{\mathrm{T}}$-type matrices during the alternating least squares (ALS) optimization. The two least-squares problems shown in Equations (21) and (22) are solved under restrictions such as non-negativity, unimodality or closure.

$$\min_{\widehat{\mathbf{C}}} \|\widehat{\mathbf{D}}_{\mathbf{PCA}} - \widehat{\mathbf{C}}\widehat{\mathbf{S}}^{\mathrm{T}}\| \tag{21}$$

and

$$\min_{\widehat{\mathbf{S}}^{\mathrm{T}}} \|\widehat{\mathbf{D}}_{\mathbf{PCA}} - \widehat{\mathbf{C}}\widehat{\mathbf{S}}^{\mathrm{T}}\| \tag{22}$$

In these two equations the norm of the residuals between the PCA reproduced data, $\widehat{\mathbf{D}}_{\mathbf{PCA}}$, using the selected number of components and the ALS reproduced data using the least-squares estimates of **C** and $\mathbf{S}^{\mathrm{T}}$ matrices, $\widehat{\mathbf{C}}$ and $\widehat{\mathbf{S}}^{\mathrm{T}}$, is alternatively minimized keeping constant $\widehat{\mathbf{C}}$ (Equation 21) or $\widehat{\mathbf{S}}^{\mathrm{T}}$ (Equation 22). This is equivalent to the minimization of the least squares function $\sum\sum(d_{ij} - \hat{c}_{ik}\hat{s}_{kj})^2$, where d_{ij} are the experimental data and $\hat{c}_{ik}$ and $\hat{s}_{kj}$ are the current ALS estimations of the concentrations and spectra values (spectra i, wavelength j and species k). The unconstrained least-squares solution of Equation (21) is given by Equation (23)

$$\widehat{\mathbf{S}}^{\mathrm{T}} = \widehat{\mathbf{C}}^{+}\mathbf{D}_{\mathbf{PCA}} \tag{23}$$

where $\widehat{\mathbf{C}}^{+}$ is the pseudoinverse of the concentration matrix which for a full rank matrix gives Equation (24)

$$\widehat{\mathbf{S}}^{\mathrm{T}} = (\widehat{\mathbf{C}}^{\mathrm{T}}\widehat{\mathbf{C}})^{-1}\widehat{\mathbf{C}}^{\mathrm{T}}\mathbf{D}_{\mathbf{PCA}} \tag{24}$$

and the unconstrained least-squares solution of Equation (22) is given by Equation (25):

$$\widehat{\mathbf{C}} = \mathbf{D}_{\mathbf{PCA}}(\widehat{\mathbf{S}}^{\mathrm{T}})^{+} \tag{25}$$

where $(\widehat{\mathbf{S}}^{\mathrm{T}})^{+}$ is the pseudoinverse of the spectra matrix which for a full rank matrix gives Equation (26):

$$\widehat{\mathbf{C}} = \mathbf{D}_{\mathbf{PCA}}\widehat{\mathbf{S}}(\widehat{\mathbf{S}}^{\mathrm{T}}\widehat{\mathbf{S}})^{-1} \tag{26}$$

Equations (23) and (24) are solved sequentially, i.e. from a given estimation of the concentration matrix **C** obtained in the previous cycle of the ALS optimization a new estimation of the spectra matrix $\mathbf{S}^{\mathrm{T}}$ (Equation 24) is calculated, and from this a new estimation of the concentration matrix **C** is then calculated (Equation 26). These solutions may be constrained to fulfill particular requirements like non-negativity, unimodality and closure immediately after they are calculated using Equations (23) and (25) (i.e. before the next equation is applied). In this way, at each iteration of the ALS procedure, the solutions are improved not only from a least-squares sense but also to

For references see page 9834

fulfill a particular set of constraints. Direct non-negative least squares solutions of Equations (21) and (22) can also be found using either the Lawson and the Hanson algorithm[58] or recent faster modifications.[54] Likewise, special algorithms have been proposed for the implementation of unimodality constraints.[55] Initial estimates to start the ALS optimization can be obtained for **C** or for $\mathbf{S}^T$ matrices, using either pure variable detection methods, EFA-derived methods[59] or previously known profiles. In the current implementations of the MCR/ALS method,[60] different constraints may be selected for the **C** and the $\mathbf{S}^T$ matrix and, within each of those matrices, all or some of the profiles can be constrained.

The convergence criterion in the ALS optimization is based on the comparison of the fit obtained in two consecutive iterations. When the relative difference in fit is below a threshold value, the optimization is finished. Sometimes a maximum number of iterative cycles is used as the stop criterion. This method is very flexible and may be adapted to very diverse real examples, as shown in section 6.

5 THREE-WAY RESOLUTION METHODS

The methods presented in previous sections are suitable to work with a data matrix and give results related to the two different directions of the data matrix, i.e. profiles related to the variation along the rows and along the columns of the data matrix. This is the reason why a data matrix is also called a two-way data set.

A data matrix is not the most complex data set that can be found in chemistry. Let us consider a kinetic process monitored fluorimetrically; at each reaction time, a series of emission spectra recorded at different excitation wavelengths are obtained. This means that we collect a data matrix at each stage of the reaction and if the goal is getting a picture of the global kinetic process, the matrices should be considered altogether. The information about the whole kinetic process should now be organized in a cube of data (tensor) with three informative directions, i.e. in a three-way data set. Another usual example is coupling data matrices from different samples that share all or some of their compounds, e.g. several HPLC/DAD runs. In this case, the third direction of the data set accounts for the quantitative differences among samples. Figure 15 shows both examples mentioned.

Though there is a clear gain in quality and quantity of information when going from two- to three-way data sets, the mathematical complexity associated with the treatment of three-way data sets can seem, at first sight, a drawback. To overcome this problem, most of the three-way data analysis methods transform the original cube of data into a stack of matrices, where simpler

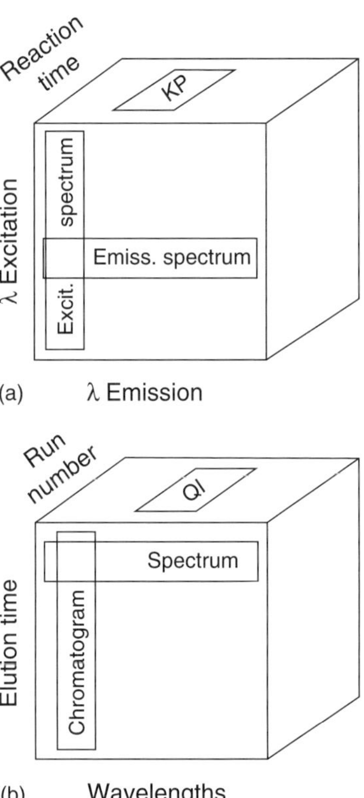

Figure 15 Examples of three-way data sets. (a) Fluorimetric monitoring of a kinetic process. KP, kinetic profile. (b) Coupling of several HPLC/DAD runs. QI, quantitative information.

mathematical methods can be applied. This process is often known as *unfolding*. A cube of data sized $(m \times n \times p)$ can be unfolded in three different directions: along the row space, along the column space and along the third direction of the cube, also called the tube space. The three unfolding procedures give a row-wise augmented matrix $\mathbf{D_r}(m \times np)$, a column-wise augmented matrix $\mathbf{D_c}(n \times mp)$ and a tube-wise augmented matrix $\mathbf{D_t}(p \times mn)$, respectively (see Figure 16). When the rank analysis of the three augmented matrices is carried out, the number of components obtained for the three different directions (*modes*) of the data set may be the same or not. When $\mathbf{D_r}$, $\mathbf{D_c}$ and $\mathbf{D_t}$ have the same rank, the three-way data set is said to be *trilinear* and when their ranks are different from each other, the data set is *nontrilinear*. (Please note that this definition holds for by far most of the chemical data sets, except those for which phenomena of rank deficiency or rank overlap are present).[61,62] The resolution of a three-way data set into the matrices **X**, **Y** and **Z**, which contain the pure profiles related to each of the directions of the three-way data set, changes for trilinear and nontrilinear systems, as can be seen in Figure 17. For trilinear systems, **X**, **Y** and **Z** have the same number of profiles (nc) and the three-way core, **C**, is an identity cube $(nc \times nc \times nc)$ whose

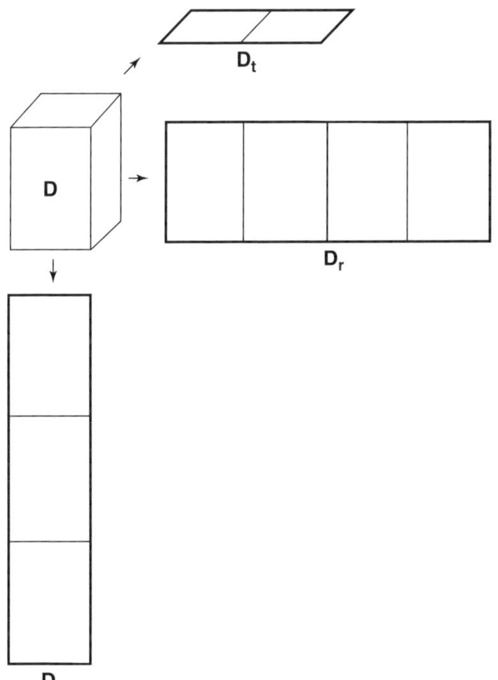

Figure 16 Unfolding of a three-way data set into a row-wise data matrix, $\mathbf{D_r}$, a column-wise data matrix, $\mathbf{D_c}$, and a tube-wise data matrix, $\mathbf{D_t}$.

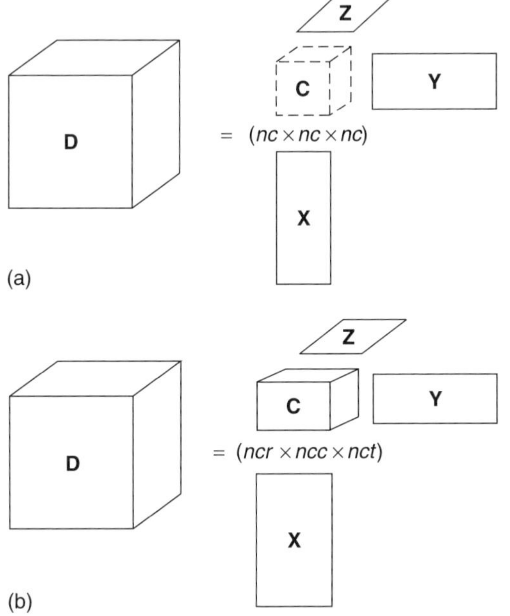

Figure 17 Decomposition of three-way data sets. (a) With trilinear structure. The dashed core can be omitted and is a regular identity cube. (b) With nontrilinear structure. The core is necessary and has a different number of components in each direction.

unity elements are placed in the superdiagonal. In this case, the three-way core is often omitted because it does not modify numerically the reproduction of the original tensor. Each element in the original three-way data set can be reproduced as follows:

$$d_{ijk} = \sum_{f=1}^{nc} x_{if}\, y_{if}\, z_{kf} \tag{27}$$

Equation (27) is the fundamental expression of the PARAFAC model,[37–39] which is used to describe the decomposition of trilinear data sets.

For nontrilinear systems, the core **C** is no longer a regular cube ($ncr \times ncc \times nct$) and the nonnull elements are spread out in different ways depending on each particular data set. *ncr*, *ncc* and *nct* hold for the rank in the row-wise, column-wise and tube-wise augmented data matrices respectively. Each element in the original data set can now be obtained as shown in Equation (28):

$$d_{ijk} = \sum_{f=1}^{ncr} \sum_{g=1}^{ncc} \sum_{h=1}^{nct} x_{if}\, y_{jg}\, z_{kh}\, c_{fgh} \tag{28}$$

Equation (28) defines the decomposition of nontrilinear data sets and is the underlying expression of the Tucker3 model.[40]

Decompositions of three-way arrays into these two different models require different data analysis methods; therefore, finding out if the internal structure of a three-way data set is trilinear or nontrilinear is essential to ensure the selection of the suitable chemometric treatment.

In this section, the concept of trilinearity has been tackled as an exclusively mathematical matter. However, the chemical information is often enough to determine if a three-way data set presents this feature. How to link the chemical knowledge with the mathematical structure of a three-way data set can be easily seen with real examples. Let us consider the three-way data sets in Figure 15. If a data set is trilinear, **X**, **Y** and **Z** will have as many profiles as chemical compounds in the original data set and this number will be equal to the rank of the data set. For each chemical compound, there will be only one profile in **X**, in **Y** and in **Z** common to all the appended matrices in the original data set. In the HPLC/DAD example, the decomposition of the three-way array gives an **X** matrix with chromatographic profiles, a **Y** matrix with pure spectra and a **Z** matrix with the quantitative information about each compound in the different chromatographic runs. In this case, a trilinear structure would imply that the pure spectrum and the pure chromatogram of a compound remain invariant in the different chromatographic runs. If the experimental conditions in the runs analyzed are similar enough, the UV spectrum of a pure compound should not change; however, run-to-run differences in peak shape

For references see page 9834

and position are commonly found in practice. Assuming that the elution process of the same compound in the different runs yields always the same chromatographic profile does not make sense from a chemical point of view and, therefore, the data set should be considered nontrilinear. In the example related to the fluorimetric monitoring of a kinetic process, the decomposition of the original data set gives a matrix **X** with pure excitation spectra, a matrix **Y** with pure emission spectra and a matrix **Z** with the kinetic profiles of the process. A trilinear structure would indicate that the shape of the excitation spectrum and the emission spectrum of a compound does not change at the different reaction times of the kinetic process. This invariability of the spectra is an acceptable statement if the experimental conditions during the process are not modified. Therefore, this data set may be considered trilinear.

In practice, most of the systems are nontrilinear due either to the underlying chemical process (e.g. UV reaction monitoring coupling experiments with different reagent ratios) or to the instrumental lack of reproducibility in the response profiles (e.g. chromatographic profiles in different HPLC/DAD runs). Therefore, section 6 is mainly focused on the study of real examples lacking the trilinear structure.

Despite the higher abundance of nontrilinear data sets, many of the algorithms proposed to study three-way arrays rely on the assumption of trilinear structure. This is the case of GRAM,[35] designed to work with two matrices, or its natural extension, DTD,[36,63] which can handle larger data sets with more appended matrices. Both GRAM and DTD are noniterative methods and use latent variables to resolve the profiles in **X**, **Y** and **Z**. When these methods are applied to nontrilinear data sets, the profiles obtained often belong to the imaginary domain. Iterative methods are also used in three-way data and the scheme followed in their application is the same as for a single data matrix, i.e. determination of number of components, use of initial estimates, application of constraints and iterative optimization until convergence. Most of the iterative algorithms are based on least-squares calculations. As for two-way data sets, three-way iterative methods are more flexible and can deal with more diverse data sets.

It has been commented that three-way resolution methods generally work with the unfolded matrices. Depending on the algorithm used, the three unfolded matrices are used or only some of them. As examples, the PARAFAC decomposition of a trilinear data set using the three unfolded data matrices and the resolution of a nontrilinear data set by applying the MCR/ALS method to only one of the unfolded matrices will be described.

The PARAFAC decomposition of a trilinear data set can be obtained by using a constrained least-squares algorithm.[37–39] The steps of the method are listed below:

1. Determination of the number of chemical compounds (nc) in the original three-way array.
2. Calculation of initial estimates for **X** and **Y**.
3. Estimation of **Z**, given $\mathbf{D_t}$, **X** and **Y**.
4. Estimation of **X**, given $\mathbf{D_r}$, **Y** and **Z**.
5. Estimation of **Y**, given $\mathbf{D_c}$, **X** and **Z**.
6. Return to step 3 until convergence is achieved.

The number of chemical compounds (rank) in the three-way array is estimated for the unfolded matrices in the three directions of the data set ($\mathbf{D_t}$, $\mathbf{D_r}$ and $\mathbf{D_c}$). If the data set is really trilinear, the rank values should coincide unless situations of rank deficiency or rank overlap occur.

Since the original data set is decomposed into three matrices, **X**, **Y** and **Z**, the initial estimates should be a pair of matrices sized as two of the matrices resulting from the decomposition. Any combination, e.g. **X**- and **Y**-type matrices, and **X**- and **Z**-type matrices, is accepted. The initial estimates can be calculated using any of the methods explained for resolution of two-way data sets.

From the original data set and the two matrices proposed as initial estimates, e.g. **X** ($m \times nc$) and **Y** ($n \times nc$), the third matrix of the PARAFAC model, **Z** ($p \times nc$), may be calculated. To do so, the unfolded matrix in the direction of the matrix to be calculated, $\mathbf{D_t}(p \times mn)$, is used (mathematical details related to this step are in the literature[39]). In each iterative cycle, this operation is performed analogously to obtain **X** and **Y** and every time the suitable unfolded matrix, $\mathbf{D_r}(m \times np)$ or $\mathbf{D_c}(n \times mp)$, participates in the calculation. The constraints explained in section 4 may be optionally applied to the profiles in **X**, **Y** and **Z**. The original three-way array is reproduced according to Equation (27) and the end of the iterative process arrives when changes in the fit among two consecutive iterations are small enough.

Three-way data sets have been presented as cubes of data formed by appending several matrices together. This means implicitly that all the data matrices in a tensor should be equally sized; otherwise, the cube cannot be constructed. Besides, the information in the rows, in the columns and in the third direction of the array must be synchronized for each of the layers of the cube, e.g. in the HPLC/DAD example, if the columns are wavelengths, all the runs should span the same wavelength range and if the rows are retention times, the elution time range should also coincide. In experimental measurements, it may not be easy or convenient to fulfill these two requirements. Actually, synchronization can be difficult when the parameter that changes in one of the directions of the array cannot be controlled in a simple manner and getting matrices equally sized may also be inconvenient if this condition forces the inclusion of irrelevant information in some of the two-way appended arrays. An example of difficult synchronization is the coupling of experiments related to UV monitoring of pH-dependent processes since pH variations may not be easily reproducible among experiments. The inconvenience of appending equally sized two-way arrays is evident when HPLC/DAD runs related to a mixture and to matrices of single standards are treated together; if the runs of standards should cover the same elution time range as the run of the mixture, most of the information in these standard matrices will be formed by baseline spectra which are not relevant for the resolution of the mixture.

When building a typical three-way data set is not possible, there is no need to give up the simultaneous analysis of a group of matrices that have something in common. Some methods, such as MCR/ALS, are designed to work with only one of the three possible unfolded matrices. This operating procedure greatly relaxes the demands to the two-way arrays that are treated together. Indeed, MCR/ALS requires only one common direction in all the matrices analyzed, e.g. in the previous examples, the wavelength range of the spectra collected (see Figure 18).

The MCR/ALS decomposition of the three-way data set shows the ability of this method to deal with nontrilinear systems. Whereas the profile of each compound related to the common order of the column-wise augmented matrix (spectra) is considered to be invariant for all the matrices, the unfolded **C** matrix allows the profile of each compound in the concentration direction to be different for each appended data matrix. This freedom in the shape of the **C** profiles agrees with the nontrilinear structure of the data array. The least-squares problems solved by MCR/ALS when applied to a three-way data set are the same as those in Equations (23) and (25); the only difference is that **D** and **C** are now augmented matrices. Though the operating procedure of the method has been shown in section 5, some particularities connected with the treatment of three-way data sets deserve a comment.

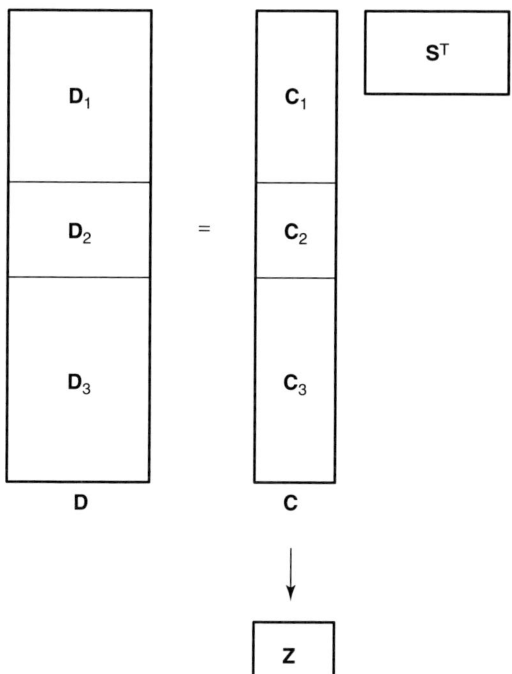

Figure 18 MCR/ALS decomposition of a three-way data set formed by three appended matrices.

In the resolution of a column-wise augmented data matrix, the initial estimates can be either a single $\mathbf{S}^T$ matrix or a column-wise augmented **C** matrix. The column-wise concentration matrix is built by placing the initial **C**-type estimates obtained for each data matrix in the three-way data set one on top of the other. The appended initial estimates must be sorted as the initial data matrices are in **D** and must keep a correct correspondence of species, i.e. each column in the augmented **C** matrix must be formed by appended concentration profiles related to the same chemical compound. When no prior information about the identity of the compounds in the different data matrices is available, the correct correspondence of species can be known from the resolution results of each single matrix.

The same constraints used in the resolution of a data matrix can be applied to three-way data sets. Selectivity and non-negativity affect the spectrum and the augmented concentration profile of each species, whereas unimodality is applied separately to each of the profiles appended to form the augmented concentration profile. The closure constraint operates by applying the corresponding closure constant to each of the single matrices in the column-wise concentration matrix. Another constraint specific of three-way data sets is the so-called *correspondence among species*. In each single matrix of a three-way data set, the concentration profiles of absent compounds are set equal to zero after each iterative cycle.

Though MCR/ALS is specially relevant to cope with nontrilinear data sets formed by matrices of varying sizes, it can also work with trilinear data sets. Because of the inherent freedom in the modeling of the profiles of the augmented **C** matrix, the trilinear feature is introduced in the MCR/ALS method as an optional constraint.(41) The application of this constraint is performed separately on the concentration profile of each species. Thus, the profiles appended to form the augmented concentration profile of a certain species are placed one beside the other to form a new data matrix and PCA is performed on it. If the system is trilinear, the score vector related to the first principal component must show the real shape of the concentration profile and the rest of principal components must be related to noise contributions. The loadings related to the first principal component are

For references see page 9834

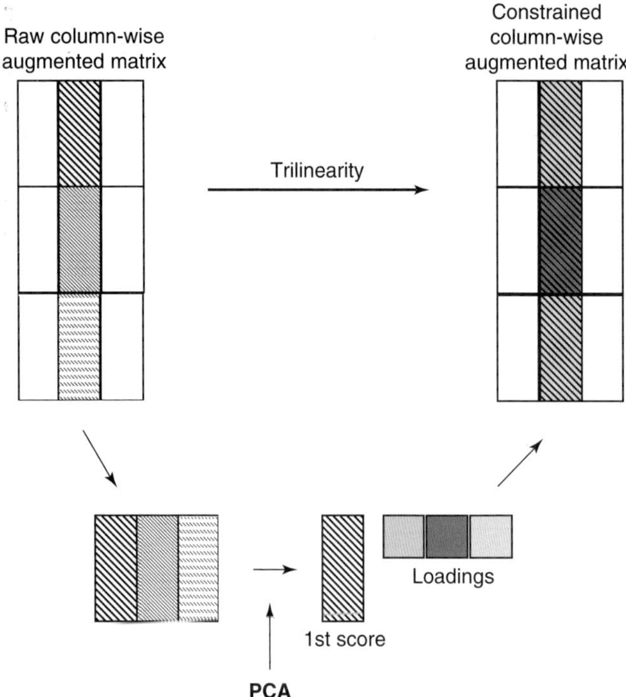

Figure 19 Application of the trilinearity constraint in the MCR/ALS method.

scaling factors accounting for the species concentration level in the different appended matrices. Therefore, the new single profiles will be calculated as the product of the score vector and their corresponding scaling factor. The constrained single profiles are finally appended to form the new augmented concentration profile. All of this process is shown graphically in Figure 19. In contrast to some other three-way resolution methods specially designed to work with trilinear systems, the implementation of this constraint in MCR/ALS should not necessarily be complete, i.e. all or some of the compounds can be forced to have common profiles in the **C** matrix. This flexibility allows a more representative modeling of some real situations, like those of systems with trilinear profiles related to the evolution of chemical compounds and a free modeled profile related to an important background contribution.

The information in the third direction of the array, i.e. the **Z** matrix, is directly extracted from the augmented matrix **C** in MCR/ALS.[41] This dimension of the data set is usually the smallest in size and is connected with scaling differences among the matrices appended. Since the $\mathbf{S}^T$ profile of one compound is common to all the appended data matrices, the area of the concentration profiles of this compound is scaled according to the concentration level of the species in each single data matrix. Thus, the profile of a compound in the **Z** matrix accounts for the relative concentration of a particular compound in each of the appended matrices and can be obtained from the ratio between the area of its concentration profile in a given matrix and the area related to the concentration profile of the same compound in a matrix taken as reference.

Since MCR/ALS has been proven to be a very versatile method to deal with any kind of three-way data set, this method has been used to work with all the diverse examples shown in section 6.

6 EXAMPLES OF APPLICATION

Different examples have been selected showing the possibilities of soft modeling methods in the analysis of chemical data. The first example refers to a chromatographic co-elution using LC/DAD. This is a typical example of analytical chemistry where the motivation of the data analysis covers most of the aspects analyzed in previous sections: estimation of the number of components coeluting in a chromatographic peak using FA-derived methods; estimation of the elution windows of each of these co-eluting components using EFA-derived methods; resolution of these components by means of multivariate resolution methods; and finally the eventual quantitation of these components. The second example will be the analysis of IR spectral data from multiple runs of an industrial process. In this case the goal of the data analysis is the development of a soft model explaining how the industrial process evolves, resolving the different components of the system and estimating the relative concentrations of the different constituents at any stage of the process from its time spectrum. The third example refers to the study and interpretation of thermodynamic and conformational transitions of polynucleotides using UV absorption and circular dichroism (CD) spectrometric methods.

6.1 Example 1: Chromatographic Co-elution Using Liquid Chromatography/Diode-array Detection[41]

Soft modeling FA-based methods have been proposed for the analysis of unresolved peaks in LC/DAD and particular attention has been focused in the use of FA and MCR methods.[23] CR methods, like ITTFA,[24,32] EFA,[15] WFA,[30] SIMPLISMA[26] or HELP,[31] have been used as qualitative tools for resolution of co-eluted peaks and peak purity, and they are not intended to provide fully quantitative information about the overlapped components. However if several chromatographic runs are simultaneously analyzed using MCR, relative quantitative information may be obtained similarly to three-way data analysis and second-order calibration methods.[36] The key aspects for this quantitation to be possible are that the spectrum of the same chemical component in

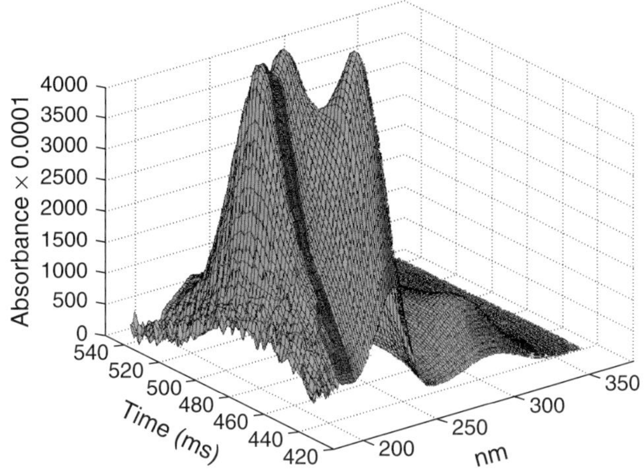

Figure 20 Three-dimensional plot of the LC/DAD peak of the unresolved mixture naphthol–pirimicarb.

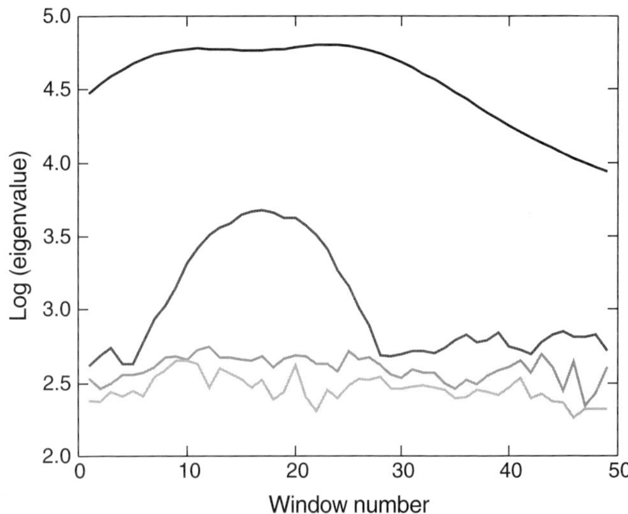

Figure 21 FSMW/EFA of the data given in Figure 20. During the experiment two distinctive components are clearly distinguished with an overlapped range between 450×10^{-3} and 510×10^{-3} min.

the different chromatographic runs does not change and, eventually, that the elution profile of a component in the different chromatographic runs also has the same shape and is synchronized. In case this second condition be not totally obeyed, resolution and quantitation of the co-eluted compounds is still possible, but the accuracy of the obtained solutions will depend on local rank and resolution conditions.

As an example of chromatographic co-elution, Figure 20 shows a three-dimensional plot of the data matrix obtained in the co-elution of two pesticides, pirimicarb and 1-naphthol, using LC/DAD. Apparatus, reagents, chromatographic conditions and other experimental details are given in Tauler et al.[(64)] Figure 21 shows the results of EFA using a fixed-size moving window of the data given in Figure 20. Along the experiment, two components were detected and distinguished from the noise level. The FSMW/EFA results suggested that the system might be resolved with few ambiguities since selectivity and local rank conditions were present. This is a common situation in chromatography, except for embedded peaks (one peak inside another[(29)]). Embedded unresolved peaks can still be resolved when several chromatographic runs are simultaneously analyzed and at least in one of the runs, the local rank and resolution conditions are also achieved for that embedded component.[(65,66)]

Application of MCR/ALS to the data of Figure 20, together with the data from the individual analyses of the two pesticides, gave the resolved pure spectra and elution profiles of Figures 22 and 23. Component 1 refers to pirimicarb, and component 2 refers to naphthol which elutes second. This resolution was obtained using non-negativity constraints for elution and spectral profiles and using unimodality constraint for the elution profiles.

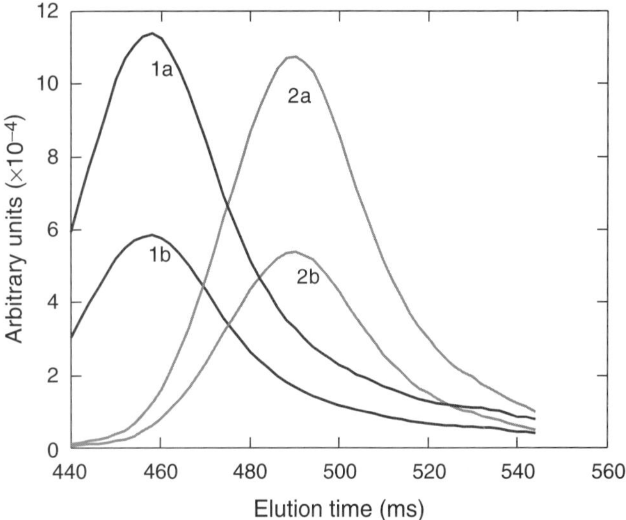

Figure 22 Resolved elution–concentration profiles of the components in the three samples simultaneously analyzed: (1a) pirimicarb in the pure analyte sample; (1b) pirimicarb in the mixture; (2a) naphthol in the pure analyte sample; (2b) naphthol in the mixture.

Each component is characterized by a pure spectrum equal in all the data matrices where this component is present. The pure spectra recovered (Figure 23) in the simultaneous numerical treatment were found to be equal to the pure spectra obtained in the individual chromatographic analysis of the pure analyte samples. In Figure 22, the two elution (concentration) profiles for each of the two components are given, one for the component in the pure analyte sample and the

For references see page 9834

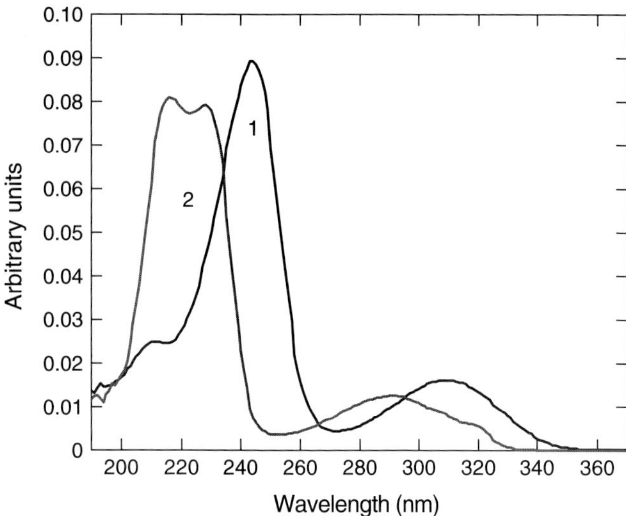

Figure 23 Resolved pure spectra of the two overlapped components. Spectrum 1 corresponds to pirimicarb and spectrum 2 corresponds to naphthol.

Table 1 Results of the quantitation of the unresolved LC/DAD peak mixture of pirimicarb and naphthol[a]

Method[b]	Pirimicarb	Naphthol	N[c]
1	0.267 (4%)[d]	0.039 (17%)	1
2a	0.270 (6%)	0.045 (3%)	3
2b	–	0.043 (8%)	2
2c	0.189 (26%)	–	2
3a	0.263 (3%)	0.047 (1%)	3
3b	–	0.049 (5%)	2
3c	0.252 (1%)	–	2
4a	–	0.050 (2%)	2
4b	0.256 (0%)	–	2

[a] Quantitation of the unresolved pirimicarb–naphthol mixture shown in Figure 20. Correct concentrations are, respectively (in ppm), pirimicarb 0.256 and naphthol 0.047.
[b] Method used in the quantitation of the unresolved mixture: (1) least-squares estimation using known pure spectra of the two unresolved components; (2) simultaneous analysis of multiple chromatographic runs using MCR and nonnegative constraint; (2a) mixture sample and two pure analyte samples; (2b) mixture sample and naphthol pure analyte samples; (2c) mixture sample and pure pirimicarb sample; (3) As in (2) but with the additional constraint of equal shape elution profiles of common components between runs (trilinearity constraint, Figure 19); (4) GRAM;[35] (4a) mixture sample and pure naphthol sample; (4b) mixture sample and pure pirimicarb sample.
[c] Number of data matrices used in the analysis.
[d] Error percentage.

other for the same component in the unknown mixture. The resolution of the two co-eluted components in the unknown mixture was equal to 0.489.

Once the concentration profiles of the different components in the mixture are recovered using the proposed method, the concentration of the analytes in the samples can be estimated. If the conditions of linearity hold, the area under the concentration profile of a certain component is proportional to the concentration of this analyte. The ratio between these areas for a particular component gives the ratio between the concentrations of that particular analyte in the different samples, recovering directly the relative quantitative information. If, in addition, in one or some of the experiments the concentration of the analyte is known, the concentration of the analytes in the unknown samples can be calculated also in absolute values. In this example, the method of quantitation is improved when the constraint of equal shape over the concentration profiles is applied (trilinear data).

In Table 1 a summary of the results obtained when different methods of quantitation are applied to the unresolved mixture of pirimicarb and naphthol (Figure 20) are given. For comparison, the results given first (method 1) correspond to those obtained by classical least-squares regression when the pure unit spectra of all the eluted components are given in the input. The analysis is performed only over the unresolved mixture data matrix. Whereas the results for pirimicarb are rather good, those for naphthol are poorer. The results of the simultaneous analysis of the data matrices obtained in the chromatographic analysis of an unknown mixture together with one or two pure analyte samples are also given. When the unknown mixture is analyzed together with these two samples, the results (method 2a) are better than before, although still with some error in the estimation of the pirimicarb concentration. When one of the two co-eluted components is considered an interferent and the unknown mixture is analyzed together with a pure analyte sample containing the other co-eluted component, the results (methods 2b and 2c) are worse, specially for pirimicarb with an error of 26% (method 2c). Conversely, when the constraint of equal shape in the elution profiles is added (method 3), the results of the quantitation are good in all the cases. When naphthol is considered an interferent and only the pure pirimicarb sample is included in the simultaneous analysis, the quantitation of pirimicarb is still very good (1% error). Although method 2 has the advantage that it did not require synchronization in the time order, it has the disadvantage that it requires more knowledge about the interferents to resolve the overlapped components in quantitative terms. Method 3 is the best choice for quantitation when interferents are present and it is closely related to second-order calibration and trilinear resolution methods.[35,36] However, it requires the data to be trilinear, which is rather unusual in real chromatographic co-elution conditions.

The results of Table 1 show that MCR can be adapted to handle LC/DAD data of different complexity. For those cases where the elution of the components is not completely reproducible because of experimental or instrumental limitations, the method can still be used for quantitation, although now the error caused by the presence of unknown interferents not present in the known analyte samples will be higher. Other recent examples of application of the MCR method to chromatographic co-elution problems have proved the utility of the proposed method and have been extended to more difficult resolution problems. Examples are the resolution of mixtures of traces of herbicides and pesticides,[66] in interlaboratory studies[67] in the resolution of liquid chromatography/mass spectrometry (LC/MS) co-eluted peaks with common mass spectrometric ions,[68] as well as in the validation of the MCR/ALS method for peak purity and quantitation in chromatographic analysis of mixtures of unresolved hormones.[65]

6.2 Example 2: Infrared Spectral Data Analysis from Multiple Runs of an Industrial Process[69]

Recent advances in process instrumentation and in data collection techniques have resulted in a rapid increase in the amount of data that can be acquired from chemical processes. Extracting the significant information from the data produced by modern instrumentation is in many circumstances a nontrivial task. The description and modeling of the evolution of a chemical process is important for both practical and economic reasons. In many of these cases multivariate calibration and regression methods[70] cannot be applied because there is no previous information available to perform the calibration of the system. Examples of this situation in process analysis are abundant and include monitoring the evolution of chemical processes where one or more parameters are changed, like time, temperature, pH, the concentration of a reagent or any other parameter, and there is no previous quantitative information about the evolution of the process. The multivariate data acquired with spectroscopic probes produce continuous data which can be arranged in an ordered data matrix according to the variation of the parameter changed during the process. One way to address this problem is by CR methods.

In the case of process analytical chemistry[71] the final goal is the estimation of the concentration of the constituents in the mixture change along the process and simultaneously the estimation of the unit responses (pure spectra) of those constituents. To have a meaningful solution from the CR decompositions it is necessary to make some assumptions about the signals obtained apart from their bilinearity, such as non-negativity, unimodality or closure. Taking advantage of the ordered structure of the process analysis data, EFA provides very valuable information concerning the windows of existence and relative importance of every component in the unknown mixtures existing at any time during the evolution of the process.

In addition, if several runs of the same industrial process are available and as usually at least one of the two orders is common between them (e.g. the scanned spectral range), the intensity ambiguities associated with the analysis of a single process run can be resolved in relative terms. Assuming that the same component in the mixture has the same pure spectrum in the different process runs, the simultaneous analysis of different process runs can give the correct relative amounts of the common components in the different mixtures analyzed in the different process runs at any particular stage of the process.

Data for the present example are spectra obtained from successive runs of an industrial chemical process. The spectra are evenly spaced in time, but may represent different elapsed times from beginning to end. The spectral intensities are stored in a matrix **D** with one spectrum per row. The dimensions of **D** are the number of spectra by the number of spectral channels (e.g. wavelength). The object is to determine the number of unique chemical components included during the full process run along with their pure spectra and concentration/time profiles. Eight different runs of the same industrial chemical process at different days of production were analyzed. Every run contained between 75 and 125 spectra, 795 in total, measured along a broad IR spectral range of 66 channels. An example of the data collected in one of the runs of the process is given in Figure 24. As seen from Figure 24, the spectra change with time, starting from a very weak and flat background absorption, increasing the absorption giving two main absorption bands at channel numbers 10–20 and 50–60 and a broader absorption band around channel number 30, and finally decreasing the absorption very fast on all those bands when the process is terminated. All the runs show a similar pattern but with slight differences between them in the timing and in the position of the

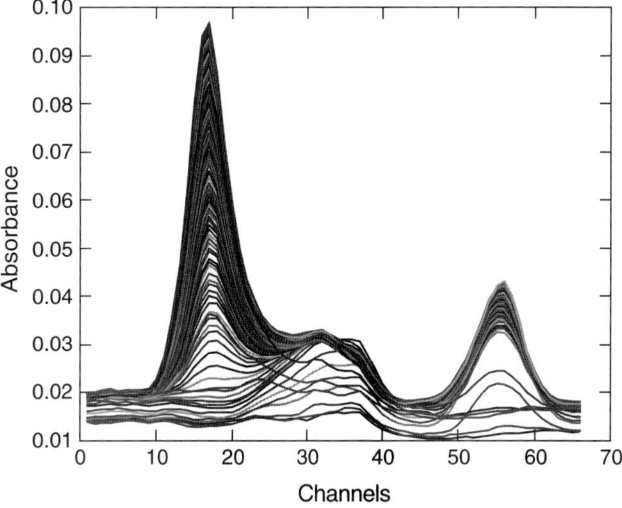

Figure 24 Example of the spectra acquired in one process run.

For references see page 9834

maxima (shifting), which show that there should be also some differences in the chemistry of the different runs of the process.

As also seen in Figure 24, the background absorption and baseline can also change during the process and are different among different runs. Therefore some data pretreatment is needed. First, to account for the differences in the initial baseline absorption among different process runs, subtraction of the first spectrum of each run from the following spectra in the same run removes these differences. This is true because in the first spectrum only the baseline or initial background absorption is present and the spectral bands of interest for the process itself have not appeared yet. With this treatment the first spectrum in each run always will be zero. Second, to account for the changes in the baseline or background during a particular run, the first and second derivatives of the raw spectra are calculated. That pretreatment allows the minimization of the contributions which are constant along a particular spectrum (first derivative) as well as those contributions which produce constant slopes within every spectrum (second derivative). In Figure 25, the second derivative spectra of the data set of Figure 24 are given. Most of the baseline changes are now removed. Pretreatment, second derivative and subtraction of the first spectrum were all analyzed and also compared with the analysis of the raw experimental spectra without any pretreatment.

The estimation of the singular values related with noise is performed using the first channels of the second derivative spectra where no band is present (Figure 25). As mentioned before, in second derivative spectra, the background and baseline contributions to the data variance are considerably diminished. The singular values obtained in this narrow spectral range are estimated for all the process runs together, to include the variations between process runs. At the same time, the singular values of the complete data set comprising the 64 channels of measurement, the process runs individually, and all together, were also calculated. In the comparison (Table 2) it was taken into account that the dimensions of the data matrix in one case and another were not the same and therefore the reduced singular values[9] were used. When the complete set of runs and spectra were analyzed, the maximum number of contributions associated with different chemical contributions was estimated as two or three, since the value of the third singular value is similar to the first singular value associated with the noise in the nonabsorbing parts of the spectra. When the analysis was performed over the individual process runs, it was found also that the number of chemical contributions was between two and three. The number of components obtained in this way was only a first estimation to start the resolution procedure. The number of three components was finally confirmed from the results of EFA and of the ALS optimization procedures (see below).

Table 2 Comparison of the reduced singular values[9] obtained in the analysis of the complete second derivative data set to the reduced singular values obtained in the analysis of the spectral regions where there is no contribution of the components of interest

795 spectra 64 channels	795 spectra 5 blank channels
1.5381559×10^{-6}	1.4242273×10^{-7}
4.4253437×10^{-7}	1.1659144×10^{-7}
1.3784506×10^{-7}	1.3244814×10^{-7}
9.2429634×10^{-8}	1.6341295×10^{-7}
4.8231603×10^{-8}	2.0548819×10^{-7}

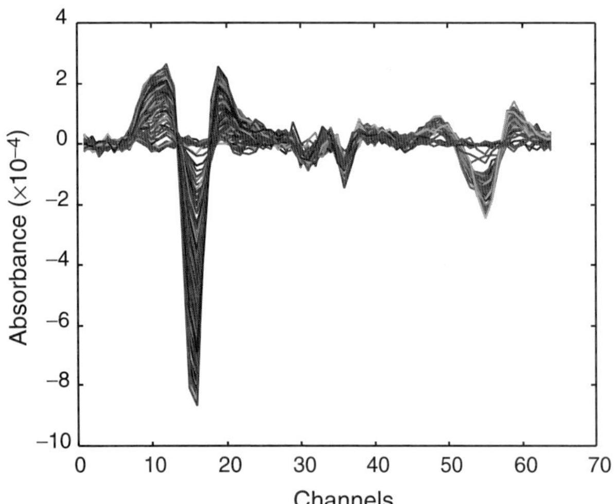

Figure 25 Second derivative spectra of one process run.

Figure 26 shows the spectra shown in Figure 25 recalculated using PCA. The principal components used in that reproduction are the three more significant found by the PCA of the whole augmented data matrix. As seen by comparison of Figures 25 and 26, most of the dominant spectral features are described by the three principal components (albeit obtained in the analysis of the 795 spectra of the eight different runs of the process, each one with 66 channels). Noise filtering is also achieved in this way and very little information is lost by considering only these three components. If only two components are considered a poorer reproduction of the original data is observed and if four components are considered very little improvement is achieved. The standard deviations of the residuals between the eight-run experimental matrix and the reproduced matrix considering two, three and four principal components are given in Table 3.

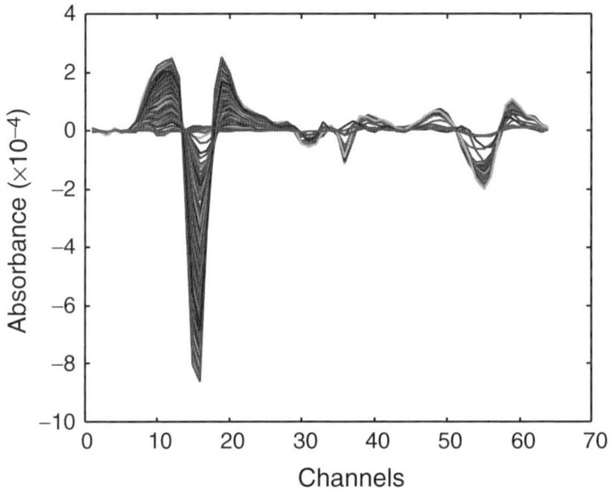

Figure 26 Reproduced second derivative spectra of the process run given in Figure 24 using three principal components.

Table 3 Standard deviation of the residuals obtained in the data analysis[a]

	Two components	Three components	Four components
exp–PCA[b]	2.82×10^{-3}	1.18×10^{-3}	1.16×10^{-3}
PCA–calc[c]	5.50×10^{-3}	3.30×10^{-3}	6.76×10^{-3}
exp–calc[d]	6.19×10^{-3}	3.80×10^{-3}	6.86×10^{-3}

[a] Results are given for two, three and four components in the simultaneous analysis of the eight runs of the process.
[b] Standard deviation of the residuals between the experimental data matrix and the PCA reproduced data matrix.
[c] Standard deviation of the residuals between the PCA reproduced data matrix and the calculated data matrix using the MCR/ALS optimized set of pure spectra and concentration profiles given in Figures 28 and 29.
[d] Standard deviation of the residuals between the experimental data matrix and the calculated data matrix using the MCR/ALS optimized set of pure spectra and concentration profiles given in Figures 28 and 29.

In Figure 27 an example of the plot of the concentration profiles obtained by EFA when applied to one run of the process is given. From EFA, three components were detected and differentiated from the other contributions. The fourth and fifth components emerged significantly from the error contributions only at the very end of the process when the reaction was terminated and therefore not of interest for the present study. Similarly, the EFA of each of the eight other process runs provides an initial estimation of the concentration profiles of the components in each process run. These concentration profiles were used as initial values in the ALS optimization. They can be used in the individual analysis of each set of data, or better, to build up the initial estimation of the augmented concentration matrix to be used in the simultaneous analysis of the eight runs of the process.

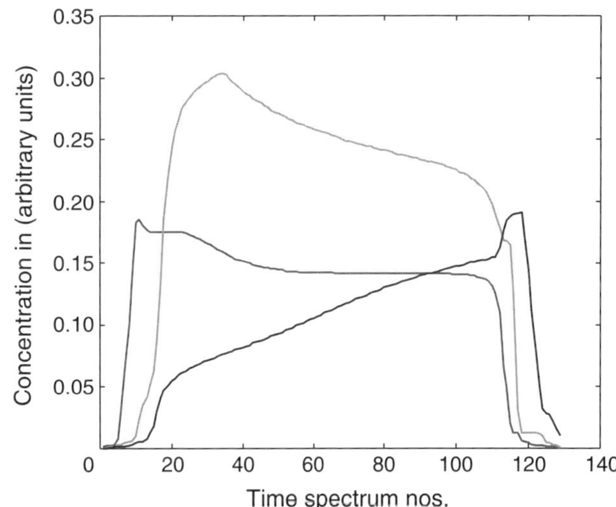

Figure 27 Initial concentration changes estimated using EFA of a single process run.

Although each process run was extensively analyzed individually, the results presented here refer only to those obtained in the simultaneous analysis. These solutions were improved by the added constraint of equality of spectra between process runs. The alternating and constrained least-squares optimization method was applied to the augmented data matrix containing the eight runs of the process arranged in the following three forms: without any pretreatment; with the first spectrum of each run subtracted; and with the second derivative augmented data matrix. The concentration matrix used initially in the optimization was the augmented concentration matrix containing the concentrations obtained in the individual EFA of each run. Of the three arrangements of the data matrix, the one which gave better results is the second case where the spectra of each run were corrected by subtracting the first spectrum of the same run. The reason for the better results in this case is that the subtraction of the first spectrum of each run removes the arbitrary offset between data from different process runs and that the non-negativity constraint can also be applied over the unit pure spectra. Conversely, when the optimization is performed over the second derivative spectra the non-negativity constraint is lost and it cannot be applied over the unit spectra. While the results are still in agreement with those obtained with the subtracted data matrix, the shapes of the recovered concentration profiles and pure spectra are less reliable. Of the three cases, the poorest results were obtained when no pretreatment was performed. The reason for this degradation of the resolution is because in that case the effects of the baseline (background absorption) between process runs and within process runs is rather high and the optimization is more difficult. In order to summarize the

For references see page 9834

large amount of calculations performed during the work, only the results obtained in the simultaneous analysis of the eight runs of the process will be given. In the ALS optimization of the complete data set, the number of three components was again reconfirmed. If another number of components is considered, not only is the fit worse (see Table 3) but also the shapes of the recovered unit spectra and concentration profiles do not make chemical sense.

Figure 28 shows the concentration profiles obtained in the analysis of the eight runs of the process after applying the constrained ALS optimization. Component one rises very fast at the beginning of the process, decreases a little and then keeps stable until the process is terminated. Component two is present except in process run 8, and always shows the same pattern of growth. The reason why this component does not appear in process run 8 is because the data analyzed pertain only to the first part of the process before it became present in appreciable concentrations. Component three does not appear at appreciable concentrations in all the runs. It is nearly nonexistent in process runs 1, 2 and 3, but it is the dominant contribution in process runs 4, 7 and 8.

The two constraints applied on the shape of the concentration profiles, non-negativity and unimodality, have an important role during the optimization. It was a little more difficult to handle the unimodality constraint because it has to account also for the small random oscillations and changes in the concentration of the components during the process; this means that although the global shape has to be unimodal and smooth, locally some small oscillations of the unimodality condition have to be allowed.

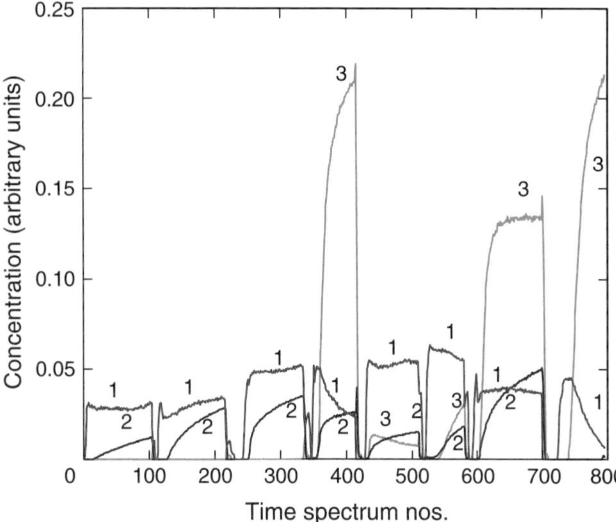

Figure 28 Concentration changes calculated by constrained ALS of eight process runs. The *x*-axis is a time axis and gives an indication of the spectrum number within the total number of spectra analyzed (795 in total).

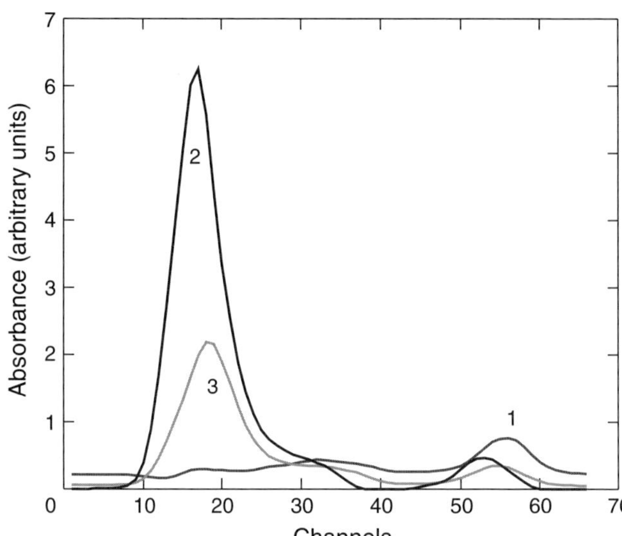

Figure 29 Pure spectra calculated by constrained ALS of eight process runs.

Figure 29 gives the pure spectra of the three common components deduced from the analysis of the eight runs of the process with the ALS optimization procedure applied to the augmented data matrix describing the eight runs of the process. The shapes obtained for these three unit spectra explain very well the changes observed in the shapes of the raw process experimental spectra of every run. For instance the first component only has the band around channels 50–60; this is in agreement with the first experimental spectra of every run which only show that band. Conversely, the second and third components have two bands approximately in the same locations but shifted between them. The strong absorption around channels 10–20 is more important for the third component than for the second, but the latter becomes the dominant contribution in some experiments. This is in agreement with what is observed from the detailed comparison of the spectra of the different runs. Moreover, the recovered pure spectra show a different baseline as a consequence of the observed baseline changes in the experimental spectra.

When the analysis is performed using the second derivative spectra the results are very similar to the ones given in Figures 28 and 29 but with a poorer description of the concentration profiles in some parts of the process, especially for component 2 in the first three runs. The reason for this degradation of the resolution is that the non-negativity constraint cannot be applied over the second derivative unit spectra. The non-negativity constraints applied over both the concentrations and unit spectra imply that the raw spectra contain only positive values, which is not always the case; therefore in the presence of negative absorbance the model will not be correct.

6.3 Example 3: Interpretation of Thermodynamic and Conformational Transitions of Polynucleotides[72]

Nowadays, the experimental monitoring of biochemical processes is relatively straightforward due to the instrumental techniques and data acquisition systems available to the researcher. The typical output of an instrument used in monitoring the evolution of a process according to the variation of a certain chemical variable consists of arrays of data (e.g. spectra) recorded at certain stages during the reaction (e.g. pH values, temperature, solvent polarity, etc.). These data can be organized in a data matrix, where the rows contain the instrumental responses and the columns reflect the relationship between the variation of the chemical variable and the evolution of the concentration of the species in the process. Despite the availability of such experimental data, univariate monitoring is still widely used in many chemical fields, such as biochemistry. Traditional biochemical studies tend to focus either on obtaining structural information under fixed conditions (e.g. physiological conditions) or on studying dynamic processes by recording univariate measurements (e.g. melting studies using single-wavelength absorbance readings).

Experiments conducted with small molecules have been traditionally interpreted by applying classical iterative least-squares methods based on the refinement of a postulated chemical model to obtain the optimal fit to the experimental data.[73] This approach is often applied to the data coming from the spectrometric monitoring of chemical equilibria and is also known under the name of global analysis.[74] The clear understanding of these simple equilibria allows the following assumptions in the process of model building: the fulfillment of the mass action law (i.e. the validity of a fixed equilibrium constant throughout the reaction); and the one-to-one correspondence between instrumental response and chemical species.

Nevertheless, most of these procedures are unable to interpret many biochemical processes due to the macromolecular nature of many biomolecules (e.g. polynucleotides, proteins, etc.) that causes a more complex evolution of the processes. The inapplicability of the classical methods can be explained by:

1. The existence of polyelectrolyte effects. The mass action law is no longer valid when either important changes in the electric field on the surface of the macromolecule or other effects related to conformational transitions modify the tendency of analogous sites to react. If this happens, Equation (29) gives

$$\log K = f(\alpha) \tag{29}$$

   where α denotes the extent of the reaction. Mathematical expressions which might model this effect cannot be devised since no information concerning its existence and its pattern (i.e. linear or nonlinear) for an unknown macromolecular process is available beforehand.
2. The existence of conformational transitions. Macromolecular biomolecules can show conformational transitions associated with a chemical reaction or with a spatial rearrangement of the molecule due to physical reasons. If the latter phenomenon occurs, the number of structural species will exceed the number of chemical species and the assumption of a one-to-one correspondence between chemical species and spatial configurations will be erroneous.

Of these techniques available for analyzing multivariate data sets, the self-modeling CR methods are the most appropriate for the analysis of evolutionary chemical processes when neither models nor prior information about the number and identity of the species involved are available, as is the case with biomacromolecular equilibria.

To illustrate the usefulness of the soft-modeling CR methods in the interpretation of biomacromolecular processes, the spectrometric (UV and CD) study of the acid–base behavior of the homopolynucleotides poly(uridylic acid) (polyU),[75] poly(cytidylic acid) (polyC)[76] and poly(adenylic acid) (polyA) in dioxane–water (30% v/v) is shown by way of example. Information about the identification and evolution of the chemical species, the existence and pattern of the polyelectrolyte effect and the conformational transitions associated with the studied reactions is presented. The solvent effect on these equilibria can be determined by comparing the results obtained in the hydroorganic mixture with those previously reported in aqueous solution using the same resolution method.

All the experiments were carried out in a dioxane–water (30% v/v) mixture. This hydroorganic mixture is often employed in biocoordination studies to emulate low-polar biological microenvironments.[75,77,78] See the references for experimental details of these studies.

6.3.1 Concentration Profiles and Pure Unit Spectra of the Individual Species Involved in Acid–Base Equilibria

The number of species involved in each acid–base equilibrium and the initial estimates were determined using EFA. The acid–base equilibria of polyC and polyU were explained with two species, related to the protonated and the deprotonated species, whereas the polyA protonation process needed three species to be described. Figures 30 and 31 show the concentration profiles and spectra obtained in the MCR/ALS optimization for the

For references see page 9834

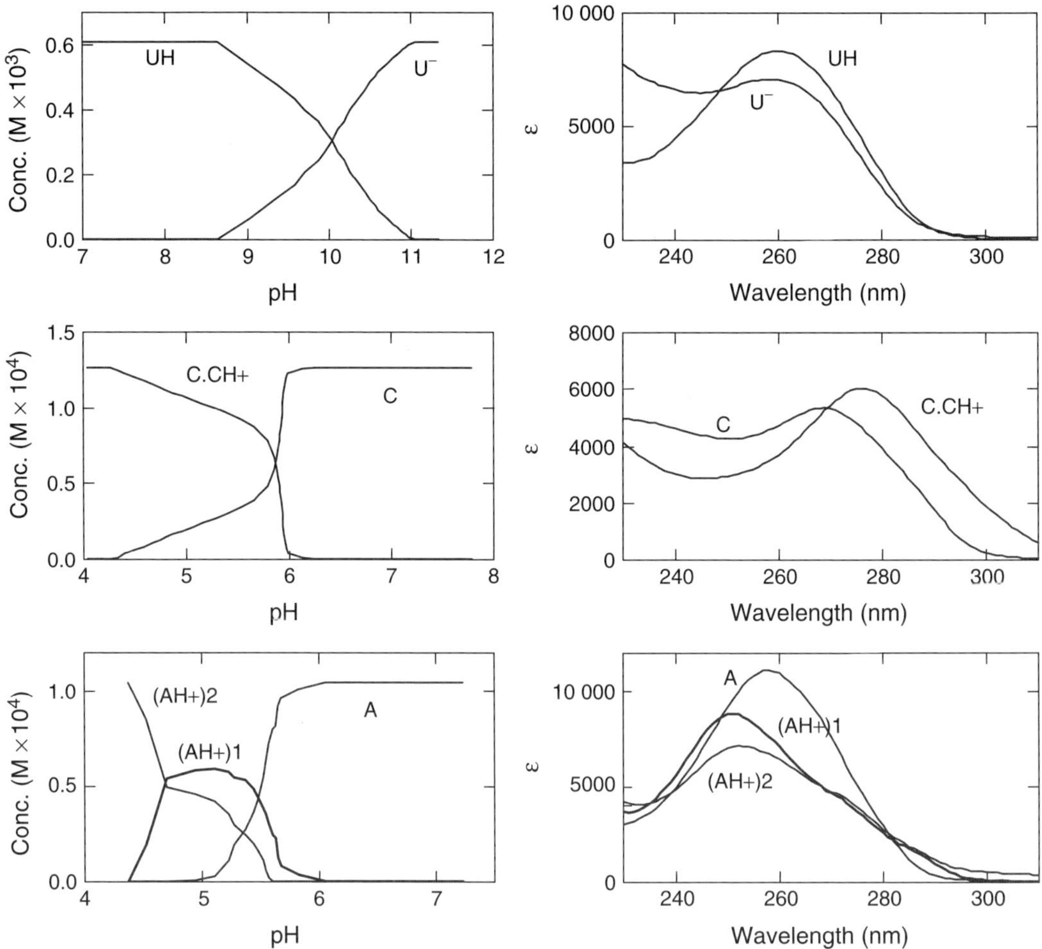

Figure 30 MCR/ALS results from UV titrations in dioxane–water (30% v/v): concentration profiles and UV spectra related to the polyU–H, polyC–H and polyA–H systems. ε values are molar absorptivities.

three polynucleotides from the UV and CD titrations, respectively. The agreement between the distribution plots obtained with both spectrometric techniques for the same polynucleotide protonation process reflects the reproducibility of the experimental work and the good performance of the MCR/ALS method on this kind of biochemical data.

6.3.2 Detection and Description of Conformational Transitions

The concentration profiles diagram obtained with ALS provides complete information about the evolution of all the species present in the acid–base equilibria. This includes transitions between chemical species which do not involve changes in the spatial structure of the molecule, conformational changes associated with the proton uptake process and changes in the spatial configuration of a single chemical species which do not alter the protonation state of the molecule. Identifying which kind of transition takes place and which conformations are involved depends on the chemical knowledge about the process being analyzed. polyU, polyC and polyA are good examples of the three transitions mentioned above.

Chemical literature on polyU generally accepts the existence of random coil structures related to both protonated and deprotonated species.[79,80] This is confirmed by the low ellipticities in polyU experiments compared with the ellipticities obtained for polyC and polyA experiments (Figure 31). The similar shape of the UV and CD spectra related to poly(UH) and poly(U^-), where the shift between their absorption maxima is the main difference, is in agreement with the hypothesis of a protonation process without conformational changes.

Literature studies of the polyC structure in aqueous solution suggested three different chemical species and conformations depending on the protonation form of the molecule, namely a single-helical deprotonated polyC, a double-stranded helical half-protonated [poly(C) · poly(CH^+)] and a fully protonated random coiled poly(CH^+).[79] Only two species were detected

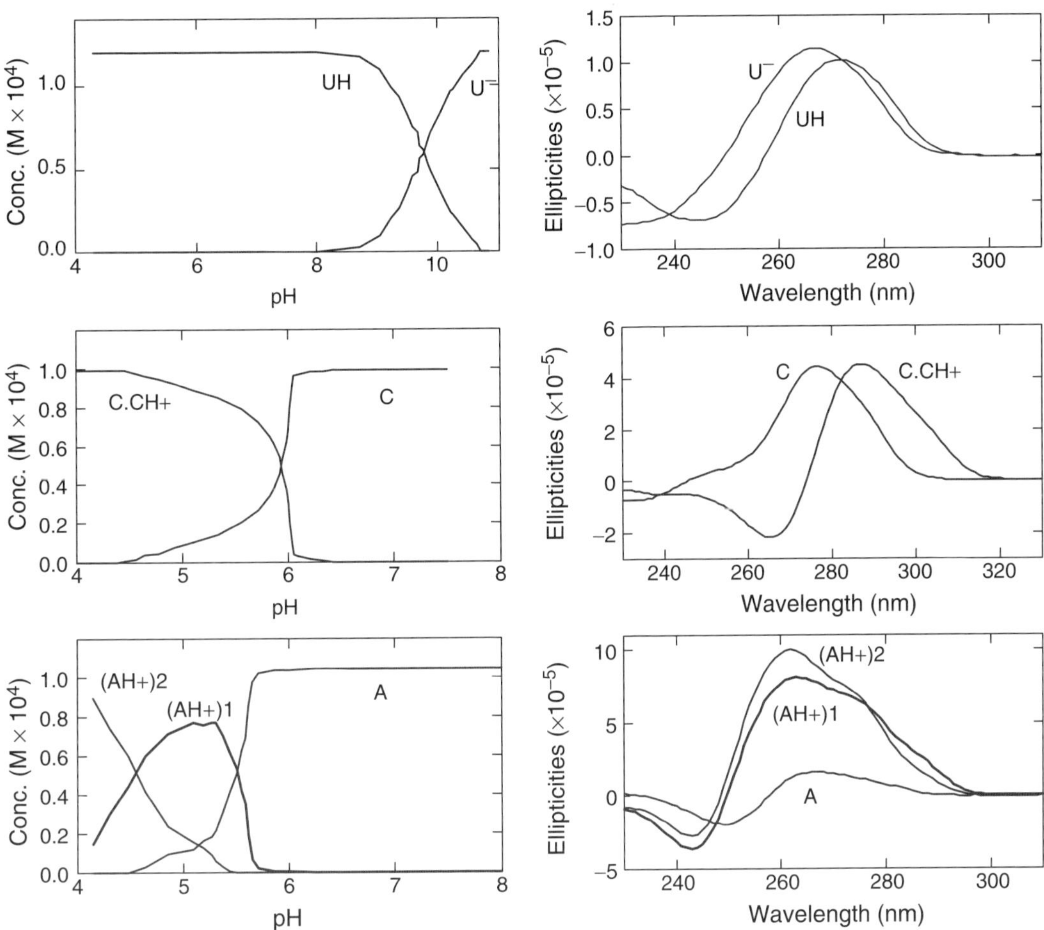

Figure 31 MCR/ALS results from CD titrations in dioxane–water (30% v/v): concentration profiles and CD spectra related to the polyU–H and polyA–H systems.

when this equilibrium took place in the hydroorganic mixture because of the narrower pH working range, limited by the precipitation of the charged polynucleotide at more acidic pH values. The most plausible identification of the two existing species includes the presence of the deprotonated and half-protonated polyC species. Apart from the similar shape of the spectra and the similar pH region of existence of the species presented in Figures 30 and 31 and those found in aqueous solution, the absence of the characteristic decrease in the CD ellipticities associated with a helix to random coil transition, and the more favorable situation of a charged species partially stabilized with the formation of an interstrand hydrogen bond $N-H^+ \cdots N$ in comparison with a species with a net charge NH^+ in a low-polar solution, support the identification proposed. Thus, the polyC–H system is a good example of acid–base equilibrium with a conformational change associated with the proton uptake process.

PolyA behaves in a way which is somewhat more complex than that of the two previous polynucleotides. Though the single-helical conformation of deprotonated polyA and its first transition to double-stranded protonated polyA is widely accepted,[79,81–83] different conformational transitions between double-helical configurations of the protonated polynucleotide have also been proposed.[81–83] Figures 30 and 31 show three different species associated with the polyA–H system in the dioxane–water mixture. The similar shape of the two species occurring at more acidic pH values seem to identify them as different double-helical protonated configurations, whereas the species present at higher pH values can probably be attributed to the single-helical deprotonated polyA. The transition between the deprotonated and the first protonated species, poly(AH^+)1, is associated with the proton uptake process, and therefore very reproducible. The transformation between the two protonated forms, poly(AH^+)1 $\rightarrow$ poly(AH^+)2, presents more irregular concentration profiles owing to the lack of chemical reaction and to the probable time-dependency of the process.[83] Nevertheless, CD and UV spectra show coherent hyperchromism and hypochromism, respectively, when going from poly(AH^+)1 to poly(AH^+)2. Such

For references see page 9834

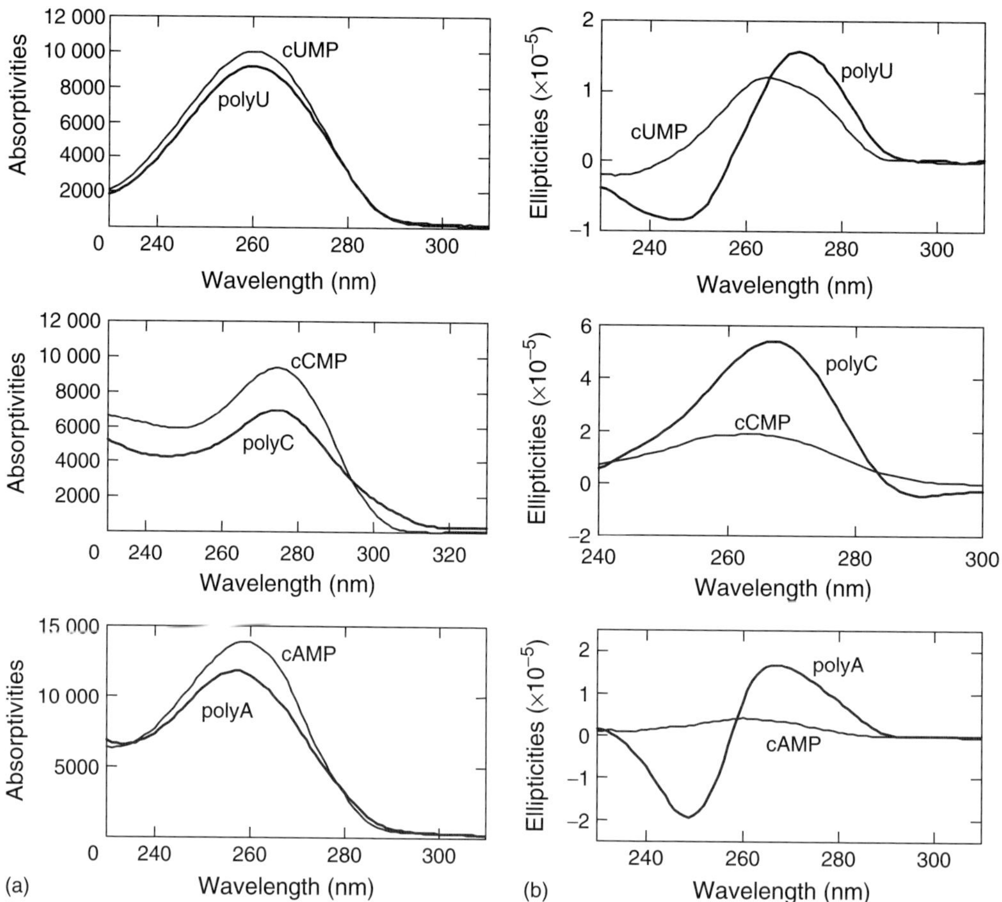

Figure 32 (a) UV and (b) CD spectra of the polymer and the respective cyclic nucleotide (polyU/cUMP, polyC/cCMP and polyA/cAMP) in dioxane–water (30% v/v).

a phenomenon could be explained by the gradual minimization of the electrostatic repulsion of neighboring phosphate groups due to the increase in the stabilizing electrostatic interactions between the negative charges of these groups and the protonated sites of the adenine bases as the protonation proceeds. Such a stabilizing effect would probably allow the formation of a more compact and ordered structure with the phosphate groups occurring closer together and with a consequent stronger base stacking.

The presence of highly ordered polymeric structures could be confirmed by comparing the polynucleotide spectra with the spectra of their respective cyclic nucleotides, as shown in Figure 32. The marked hyperchromism in the CD spectra and hypochromism in the UV spectra of polyA and polyC with respect to cyclic adenosine monophosphate (cAMP) and cyclic cytidine monophosphate (cCMP) indicate a significant stacking in both polynucleotides and, therefore, the existence of ordered structures. The slight differences in terms of intensity between the polyU and cyclic uridine monophosphate (cUMP), and their respective low ellipticities compared with the other two previous cases, support the hypothesis of random structure for this polymer.

6.3.3 *Existence and Pattern of Polyelectrolyte Effect*

Once the species in the polynucleotide protonation process are identified, the equilibrium constant of this reaction can be properly evaluated. The distribution plot obtained by applying the MCR/ALS method provides an estimation of the concentration values of all the protonated and the deprotonated species for each pH measured. A $\log K$ value may be then calculated for each titration point, bearing in mind that a fixed $\log K$ value for each functional site of the polynucleotide during the whole protonation process cannot be obtained unless no polyelectrolyte effect exists. Plotting the $\log K$ values versus their corresponding protonation degrees (α) is a graphical way of studying both the existence and the pattern of a polyelectrolyte effect. If this effect is revealed, the so-called apparent constant (K_{app}) of analogous sites of the polynucleotide changes as the protonation process advances and the $\log K$ value usually given is not a thermodynamic constant, but an intrinsic constant (K_{int}) defined as the extrapolated K_{app} value for a protonation degree equal to zero, i.e. for the theoretical point where no effects of neighboring protonated sites are present.

Bearing in mind that the equilibrium process followed is the protonation of the site in the monomer unit of the macromolecule, the concentration of the polynucleotide is expressed as moles of monomer per volume unit and, therefore, K is defined by Equation (30):

$$K = \frac{[\text{protonated monomer}]}{[\text{deprotonated monomer}]\,[\text{H}^+]} \tag{30}$$

and consequently, α is given by Equation (31):

$$\alpha = \frac{[\text{protonated monomer}]}{[\text{total monomer}]} \tag{31}$$

In the expressions below referring to the protonation constants of each polynucleotide, the names between brackets indicate the form of the macromolecule where the protonated or deprotonated monomers are placed.

Figure 33 shows in thick lines the $\log K$ versus α plots for polyU, polyC and polyA in dioxane–water (30% v/v). The plots include the theoretical lines obtained after the polynomial fitting of the experimental (α, $\log K$) values. The related equations $\log K = f(\alpha)$ can be found in Table 4.

According to the identification of species in the MCR/ALS resolved distribution plots, the polyU protonation constant may be determined as shown in Equation (32):

$$K = \frac{[\text{poly(U–H)}]}{[\text{poly(U}^-)]\,[\text{H}^+]} \tag{32}$$

No polyelectrolyte effect has been detected and, therefore, a thermodynamic protonation constant can be given for all the analogous protonation sites in the macromolecule. The absence of polyelectrolyte effect is probably due to previously detected random coiled structures associated with the protonated and the deprotonated species. These disordered structures are more flexible and allow spatial rearrangements of the macromolecule to minimize the between-sites effect during the protonation process.

The polyC protonation constant has been calculated using Equation (33):

$$K = \frac{[\text{poly(C)}\cdot\text{poly(CH}^+)]/2}{\{[\text{poly(C)}] + [\text{poly(C)}\cdot\text{poly(CH}^+)]/2\}\,[\text{H}^+]} \tag{33}$$

Please note that the [poly(C) · (poly(CH$^+$)] concentration is always divided by two when included as protonated and deprotonated forms due to the existence of one protonated base and one deprotonated base per base

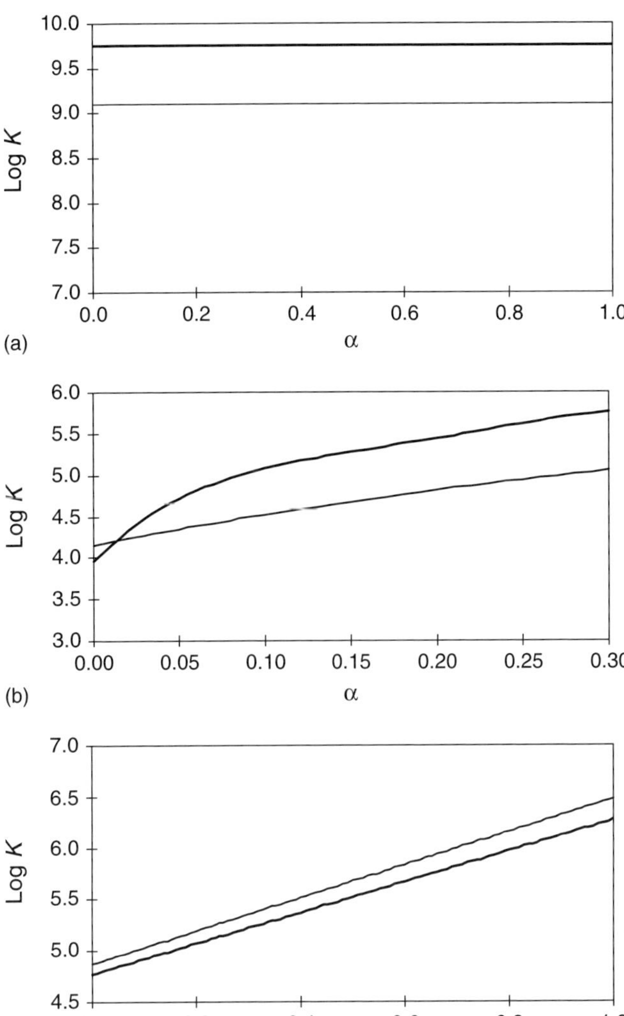

Figure 33 Theoretical models, $\log K = f(\alpha)$, of the polyelectrolyte effect related to the (a) polyU, (b) polyC and (c) polyA protonation processes in dioxane–water (30% v/v) and in aqueous solution (thin line).

pair. There is a nonlinear pattern in the polyelectrolyte effect owing to the cooperative action between the protonation process and the formation of the double-stranded helical structure. Thus, the formation of this helix stabilizes the protonated base because of the interstrand hydrogen bond $\text{N–H}^+\cdots\text{N}$ and, at the same time, this base pair arrangement is responsible for the growth and stabilization of the helical structure. As the protonation process advances, the intensity of the positive polyelectrolyte effect decreases, becoming negative for α_p values higher than 0.3[76] because the increase of charge density in the macromolecular structure means that the repulsive effect between the protonated sites is more important than the stabilization caused by the formation of the double-helical structure.

For references see page 9834

Table 4 Results of the experiments performed

System	Solvent	Technique	Data treatment	$\log K$	σ[a]	χ^{2}[a]	Lack of fit (%)[b]	Fit $\log K = f(\alpha)$	
								Model	r^2
cUMP–H	Dioxane–water (30%)	Potentiometry	CV[c]	9.250(4)[d]	2.14	60.09	–	–	–
cCMP–H	Dioxane–water (30%)	Potentiometry	CV	3.766(3)	2.41	34.03	–	–	–
cAMP–H	Dioxane–water (30%)	Potentiometry	CV	3.195(8)	1.29	30.55	–	–	
polyU–H	Water[e]	Potentiometry	CV	9.364	–	–	–	$\log K = 9.364$	–
	Water	CD	ALS	9.1(2)	–	–	5.9	$\log K = 9.1$	–
	Dioxane–water (30%)	Potentiometry	CV	9.756(4)	1.40	80	–	$\log K = 9.756$	–
	Dioxane–water (30%)	UV	ALS	9.9(2)	–	–	6.2	$\log K = 9.9$	–
	Dioxane–water (30%)	CD	ALS	9.75(6)	–	–	5.8	$\log K = 9.75$	–
polyC–H[f]	Water[e]	UV	ALS	4.21(5)	–	–	1.3	$\log K = 4.21 + 4.6\alpha - 4.5\alpha^2$	0.980
	Water	CD	ALS	4.16(4)	–	–	3.6	$\log K = 4.16 + 3.9\alpha - 3.1\alpha^2$	0.985
	Dioxane–water (30%)	UV	ALS	4.04(2)	–	–	1.6	$\log K = 4.04 + 17\alpha - 110\alpha^2 + 350\alpha^3 - 400\alpha^4$	0.996
	Dioxane–water (30%)	CD	ALS	3.96(2)	–	–	3.5	$\log K = 3.96 + 21\alpha - 140\alpha^2 + 480\alpha^3 - 600\alpha^4$	0.996
polyA–H[f]	Water[e]	CD, UV	ALS	4.87(4)	–	–	2.0	$\log K = 1.61\alpha + 4.87$	0.94
	Dioxane–water (30%)	UV	ALS	4.74(6)	–	–	3.3	$\log K = 1.57\alpha + 4.74$	0.97
	Dioxane–water (30%)	CD	ALS	4.78(4)	–	–	4.9	$\log K = 1.46\alpha + 4.78$	0.99

[a] Figures of merit related to the curve fitting program SUPERQUAD:[85] σ, ratio between the root mean square of the weighted residuals and the estimated error in the working conditions, and the statistical parameter χ^2 based on the distribution of the weighted residuals.

[b] Figure of merit lack of fit related to ALS: lack of fit $= \sqrt{\sum_{ij}(d_{ij} - \widehat{d}_{ij})^2 \big/ \sum_{ij} d_{ij}^2}$ where d_{ij} and $\widehat{d}_{ij}$ are, respectively, the experimental and ALS calculated data values.

[c] CV, classical least-squares curve fitting procedure (SUPERQUAD program[85]).

[d] Numbers in parentheses are the errors associated with the last numeral.

[e] Results in aqueous solution are taken from Casassas et al.[80] for polyU, Casassas et al.[84] for polyC, and Casassas et al.[81] for polyA.

[f] $\log K$ values for polyC and polyA are always intrinsic constants.

The equation related to the polyA protonation process is given by Equation (34):

$$K = \frac{([\mathrm{poly(AH^+)1}] + [\mathrm{poly(AH^+)2}])}{[\mathrm{poly(A)}][\mathrm{H^+}]} \tag{34}$$

A linear positive polyelectrolyte effect caused by the stabilizing electrostatic interactions between the negatively charged phosphate groups and the protonated adenine sites is shown. The difference between the polyelectrolyte effect patterns of polyC and polyA depends on the role of the protonation sites in the formation of the double-stranded helix of the polynucleotide. PolyC protonated sites are responsible for the formation of one of the interstrand hydrogen bonds, whereas polyA protonated sites, though having a positive effect on the stabilization of the double-helical structure because of the minimization of the electrostatic repulsion of the phosphates, do not participate directly in the interstrand hydrogen bonding.[82] Thus, a nonlinear pattern of the polyelectrolyte effect can in all likelihood be attributed to the presence of a cooperative mechanism associated with the protonation process, as occurs in polyC, whereas the more common linear pattern would appear to be caused mainly by purely electrostatic interactions, as shown in polyA and in other simpler polyelectrolytes, such as polyacrylic acid.

A good agreement between the numerical values of polyU and of cUMP protonation constants and between the intrinsic protonation constant of polyC and the protonation constant of cCMP is shown, the slight differences between the values of K_{polyU} and K_{cUMP} being due to hindrance effects (see Table 4). No such agreement is noticed when the cAMP protonation constant and the polyA intrinsic protonation constant are compared. The significantly lower value of the cAMP protonation constant can probably be explained by the absence of the positive influence of the surrounding phosphate groups, present on the protonation of all the adenine sites of polyA, on the protonation process of the cyclic nucleotide, cAMP.

6.3.4 *Solvent Effect on the Acid–Base Equilibria of Polynucleotides*

The acid–base equilibria of polyA, polyC and polyU were also studied in aqueous solution.[80,81,84] The comparison of the results obtained in water with those obtained in dioxane–water (30% v/v) allows the inference of solvent effects on the protonation processes of the polynucleotides. The system dioxane–water was used because, in comparison with some other hydroorganic mixtures, it provides a larger decrease of polarity with the inclusion of smaller amounts of organic cosolvent. No experiments were performed at dioxane proportions higher than 30% (v/v) because some of the polynucleotides studied presented precipitation phenomena.

When comparing the experiments mentioned above, a general observation is the greater instability of charged macromolecular structures in the hydroorganic medium. Indeed, some species detected in aqueous solutions are apparently not formed in the lower-polar medium because the macromolecule is precipitated at suitable pH values. This is the case for the fully protonated poly(CH^+), whose formation involves the appearance of two net charges in solution, one from the new protonated site and the other from the breaking of the interstrand hydrogen-bonding $N{-}H^+ \cdots N$. The precipitation of the macromolecule from dioxane–water solution also takes place for the protonated polyA at pH values lower than 4 and this prevents the formation of other postulated double-helical configurations at more acidic pH values. Disordered charged structures, such as deprotonated polyU, are more stable than ordered forms in dioxane–water solution due to the greater ability of the macromolecule to reach spatial arrangements suitable for accommodating the net charges.

All the species detected in dioxane–water (30% v/v) are present in water with the same spatial structure. The only difference is the formation of more relaxed configurations in the dioxane–water mixtures because of the weakening of the base stacking interactions caused by the low-polar solvents.

The polyelectrolyte effect, if absent or linear, maintains the same behavior in both solvents studied, as shown in Table 4 and in Figure 33. Thus, the polyU acid–base behavior does not present a polyelectrolyte effect either in water or in the working dioxane–water mixture and the only difference between the polyU equilibrium in both media is the higher value of the protonation constant in the hydroorganic mixture owing to the lower stabilizing effect of this low-polar solvent on the negatively charged deprotonated polynucleotide. Linear models with fairly similar slopes describe the polyelectrolyte effect associated with the polyA protonation process in hydroorganic and in aqueous solutions. Nevertheless, there is not enough information to determine whether the small difference between slopes is due to the intrinsic similarity of the polyelectrolyte effects in both media or whether the stabilizing effect of the inert salt counterions around the negatively charged phosphates conceals the real solvent effect on the polyA protonation process. To clarify this point, it would be necessary to work at lower ionic strengths.

There is a clear solvent influence on the pattern of the polyelectrolyte effect when a cooperative mechanism is involved in the polynucleotide protonation process, as can be seen in the polyC protonation process. Though not visible in Figure 33, where only the mathematically fitted positive polyelectrolyte effect is shown, there is a change in the sign of the polyelectrolyte effect associated with the polyC protonation process in both water and water–dioxane solutions.[76] Solvent effect is first noticed in the ranges of existence of the positive and the negative polyelectrolyte effects. Whereas the change of sign appears in α values around 0.3 in the hydroorganic solution, this negative behavior is not seen until α values higher than 0.5 in water. The greater stability of charged structures in water explains why the negative polyelectrolyte effect does not appear in this solvent until the breakdown of the stable double-stranded helical structure takes place, whereas an increase in the charge density of the double-stranded helix is enough to change the tendency of the polyelectrolyte effect in less polar media. The second difference concerns the pattern of the positive effect in both media: in the hydroorganic mixture, a fourth-order polynomial is needed to explain the variation of $\log K$ with α, whereas a second-order polynomial is enough to explain the same data in water, as shown in Table 4. A much steeper effect for low α values is detected in water–dioxane because of the easier formation of the interstrand hydrogen bond $N{-}H^+ \cdots N$ due to the weaker competition of the solvent molecules in the development of these interactions in a hydroorganic mixture less polar than water. As the protonation degree increases, this tendency is inverted and the polyelectrolyte effect in water–dioxane becomes less and less pronounced, being close to a plateau for α values next to 0.3. The smoothing in the evolution of the polyelectrolyte effect comes from

For references see page 9834

the gradual balance between the favorable tendency to form the hydrogen bond $N-H^{+}\cdots N$ and the destabilizing influence associated with the increase of charge density in the macromolecular structure.

ABBREVIATIONS AND ACRONYMS

ALS	Alternating Least Squares
cAMP	Cyclic Adenosine Monophosphate
cCMP	Cyclic Cytidine Monophosphate
CD	Circular Dichroism
CR	Curve Resolution
cUMP	Cyclic Uridine Monophosphate
CV	Classical Least-squares Curve Fitting Procedure
DTD	Direct Trilinear Decomposition
EFA	Evolving Factor Analysis
FA	Factor Analysis
FSMW/EFA	Fixed-size Moving Window/ Evolving Factor Analysis
GRAM	Generalized Rank Annihilation Method
HELP	Heuristic Evolving Latent Projections
HPLC/DAD	High-performance Liquid Chromatography/Diode-array Detector
IND	Indicator Function
IR	Infrared
ITTFA	Iterative Transformation Target Factor Analysis
KSFA	Key Set Factor Analysis
LC/DAD	Liquid Chromatography/Diode-array Detection
LC/MS	Liquid Chromatography/Mass Spectrometry
LSER	Linear Solvation Energy Relationships
MCR	Multivariate Curve Resolution
MCR/ALS	Multivariate Curve Resolution/ Alternating Least Squares
NIPALS	Nonlinear Iterative Partial Least Squares
OPA	Orthogonal Projection Approach
PARAFAC	Parallel Factor Analysis
PCA	Principal Component Analysis
polyA	Poly(adenylic Acid)
polyC	Poly(cytidylic Acid)
polyU	Poly(uridylic Acid)
PRESS	Prediction Error Sum of Squares
RAEFA	Rank Annihilation Evolving Factor Analysis
SFA	Subwindow Factor Analysis
SIMPLISMA	Simple-to-use Interactive Self-modeling Analysis
SVD	Singular Value Decomposition
TFA	Target Factor Analysis
UV	Ultraviolet
WFA	Window Factor Analysis

List of selected abbreviations appears in Volume 15

RELATED ARTICLES

Chemometrics **(Volume 11)**
Chemometrics • Classical and Nonclassical Optimization Methods • Clustering and Classification of Analytical Data • Multivariate Calibration of Analytical Data • Second-order Calibration and Higher • Signal Processing in Analytical Chemistry

Infrared Spectroscopy **(Volume 12)**
Spectral Data, Modern Classification Methods for

Kinetic Determinations **(Volume 12)**
Data Treatment and Error Analysis in Kinetics

General Articles **(Volume 15)**
Multivariate Image Analysis

REFERENCES

1. E.R. Malinowski, *Factor Analysis in Chemistry*, 2nd edition, Wiley-Interscience, New York, 1992.
2. W. Wu, Y. Mallet, B. Walczak, W. Penninckx, D.L. Massart, S. Heuerding, F. Erni, 'Comparison of Regularized Discriminant Analysis, Linear Discriminant Analysis and Quadratic Discriminant Analysis, Applied to NIR Data', *Anal. Chim. Acta*, **329**, 257–265 (1996).
3. J.S. i Salau, R. Tauler, J.M. Bayona, I. Tolosa, 'Input Characterization of Sedimentary Organic Contaminants and Molecular Markers in the Northwestern Mediterranean Sea by Exploratory Data Analysis', *Environ. Sci. Technol.*, **37**, 3482–3490 (1997).
4. S. Wold, 'Soft-modeling by Latent Variables: the Nonlinear Iterative Partial Least Squares (NIPALS) Algorithm', in *Perspectives in Probability and Statistics*, ed. J. Garni, Academic Press, London, 117–142, 1975.
5. G.H. Golub, C. Reinsch, 'Singular Value Decomposition and Least Squares Solutions', *Numer. Math.*, **14**, 403–420 (1970).
6. D.L. Massart, B.G.M. Vandeginste, L.M.C. Buydens, S. de Jong, P.J. Lewi, J. Smeyers-Verbeke, 'Handbook of Chemometrics and Qualimetrics', in *Data Handling in Science and Technology*, Elsevier, Amsterdam, Vol. 20, 1997.
7. E.R. Malinowski, 'Theory of the Distribution of the Error Eigenvalues Resulting from Principal Component

Analysis with Application to Spectroscopic Data', *J. Chemom.*, **1**, 33–40 (1987).
8. S. Wold, 'Cross-validatory Estimation of the Number of Components in Factor and Principal Component Models', *Technometrics*, **20**, 397–405 (1978).
9. E.R. Malinowski, 'Statistical F-tests for Abstract Factor Analysis and Target Testing', *J. Chemom.*, **3**, 49–60 (1988).
10. R.J. Rummel, *Applied Factor Analysis*, Northwestern University Press, Evanston, IL, 1970.
11. H.F. Kaiser, 'The Varimax Criterion for Analytic Rotation in Factor Analysis', *Psychometrika*, **23**, 187–200 (1958).
12. S.Q. Xia, P.L. Su, J.H. Sun, Z.X. Pan, M.S. Zhang, L.M. Shi, 'Determination of Six Aromatic Compounds by Using Target Factor Analysis UV Spectrometry', *Fresenius' J. Anal. Chem.*, **351**, 325–327 (1995).
13. E. Casassas, G. Fonrodona, A. de Juan, R. Tauler, 'Assessment of Solvent Parameters and their Correlation with Protonation Constants in Dioxane–Water Mixtures Using Factor Analysis', *Chemom. Intell. Lab. Syst.*, **12**, 29–38 (1991).
14. E.R. Malinowski, 'Theory of the Error for Target Factor Analysis with Applications to Mass Spectrometry and Nuclear Magnetic Resonance Spectrometry', *Anal. Chim. Acta*, **103**, 339–354 (1978).
15. H. Gampp, M. Maeder, C.J. Meyer, A.D. Zuberbühler, 'Calculation of Equilibrium Constants from Multiwavelength Spectroscopic Data', III. Model-free Analysis of Spectrophotometric and ESR Titrations', *Talanta*, **32**, 1133–1139 (1985).
16. M. Maeder, 'Evolving Factor Analysis for the Resolution of Overlapping Chromatographic Peaks', *Anal. Chem.*, **59**, 527–530 (1987).
17. R. Tauler, E. Casassas, 'Principal Component Analysis Applied to the Study of Successive Complex Formation Data in the Cu(II) Ethanolamine Systems', *J. Chemom.*, **3**, 151–161 (1988).
18. H.R. Keller, D.L. Massart, 'Peak Purity Control in Liquid Chromatography with Photodiode Array Detection by Fixed Size Moving Window Evolving Factor Analysis', *Anal. Chim. Acta*, **246**, 379–390 (1991).
19. J. Toft, O.M. Kvalheim, 'Eigenstructure Tracking Analysis for Revealing Noise Patterns and Local Rank in Instrumental Profiles: Application to Transmittance and Absorbance IR Spectroscopy', *Chemom. Intell. Lab. Syst.*, **19**, 65–73 (1993).
20. W.H. Lawton, E.A. Sylvestre, 'Self Modeling Curve Resolution', *Technometrics*, **13**, 617–633 (1971).
21. E. Spjotvoll, H. Martens, R. Volden, 'Restricted Least Squares Estimation of the Spectra and Calibration of Two Unknown Constituents Available in Mixtures', *Technometrics*, **24**, 173–180 (1982).
22. O.S. Borgen, B.R. Kowalski, 'An Extension of the Multivariate Component-resolution Method to Three Components', *Anal. Chim. Acta*, **174**, 1–26 (1985).
23. J.C. Hamilton, P.J. Gemperline, 'Mixture Analysis Using Factor Analysis. II. Self-modeling Curve Resolution', *J. Chemom.*, **4**, 1–13 (1990).
24. B.G.M. Vandeginste, W. Derks, G. Kateman, 'Multicomponent Self-modeling Curve Resolution in High Performance Liquid Chromatography by Iterative Target Transformation Factor Analysis', *Anal. Chim. Acta.*, **173**, 253–264 (1985).
25. H. Gampp, M. Maeder, Ch.J. Meyer, A.D. Zuberbühler, 'Quantification of a Known Component in an Unknown Mixture', *Anal. Chim Acta*, **193**, 287–292 (1987).
26. W. Windig, J. Guilment, 'Interactive Self-modeling Mixture Analysis', *Anal. Chem.*, **63**, 1425–1432 (1991).
27. R. Tauler, E. Casassas, 'Application of Principal Component Analysis to the Study of Multiple Equilibria Systems. Study of Copper(II) Salicylate Monoethanolamine, Diethanolamine and Triethanolamine Systems', *Anal. Chim. Acta*, **223**, 257–268 (1989).
28. R. Tauler, A.K. Smilde, B.R. Kowalski, 'Selectivity, Local Rank, Three-way Data Analysis and Ambiguity in Multivariate Curve Resolution', *J. Chemom.*, **9**, 31–58 (1995).
29. R. Manne, 'On the Resolution Problem in Hyphenated Chromatography', *Chemom. Intell. Lab. Syst.*, **27**, 89–94 (1995).
30. E.R. Malinowski, 'Window Factor Analysis: Theoretical Derivation and Application to Flow Injection Analysis Data', *J. Chemom.*, **6**, 29–40 (1992).
31. O.M. Kvalheim, Y.Z. Liang, 'Heuristic Evolving Latent Projections–Resolving 2-way Multicomponent Data. 1. Selectivity, Latent Projective Graph, Datascope, Local Rank and Unique Resolution', *Anal. Chem.*, **64**, 936–946 (1992).
32. P.J. Gemperline, 'A Priori Estimates of the Elution Profiles of the Pure Components in Overlapped Liquid Chromatography Peaks Using Target Factor Analysis', *J. Chem. Inf. Comput. Sci.*, **24**, 206–212 (1984).
33. R. Tauler, 'Multivariate Curve Resolution Applied to Second Order Data', *Chemom. Intell. Lab. Syst.*, **30**, 133–146 (1995).
34. F.C. Sánchez, J. Toft, B. van den Bogaert, D.L. Massart, 'Orthogonal Projection Approach Applied to Peak Purity Assessment', *Anal. Chem.*, **68**, 79–85 (1996).
35. E. Sánchez, B.R. Kowalski, 'Generalized Rank Annihilation Method', *Anal. Chem.*, **58**, 496–499 (1986).
36. E. Sánchez, B.R. Kowalski, 'Tensorial Resolution: a Direct Trilinear Decomposition', *J. Chemom.*, **4**, 29–45 (1990).
37. R.A. Harshman, 'Foundations of the PARAFAC Procedure: Models and Conditions for an "Explanatory" Multi-modal Factor Analysis', *UCLA Working Papers in Phonetics*, 1–84 (1970).
38. J.D. Carroll, J. Chang, 'Analysis of Individual Differences in Multidimensional Scaling via an N-way Generalization of "Eckart-Young" Decomposition', *Psychometrika*, 283–319 (1970).

39. R. Bro, 'PARAFAC: Tutorial and Applications', *Chemom. Intell. Lab. Syst.*, 149–171 (1997).
40. A.K. Smilde, R. Tauler, J.M. Henshaw, L.W. Burgess, B.R. Kowalski, 'Multicomponent Determination of Chlorinated Hydrocarbons Using a Reaction-based Chemical Sensor. 3. Medium-rank 2nd-order Calibration with Restricted Tucker Models', *Anal. Chem.*, **66**, 3345–3351 (1994).
41. R. Tauler, D. Barceló, 'Multivariate Curve Resolution Applied to Liquid Chromatography–Diode Array Detection', *Trends Anal. Chem.*, **12**, 319–327 (1993).
42. R. Gargallo, R. Tauler, A. Izquierdo-Ridorsa, 'Application of a Soft Modeling Curve Resolution Method to the Analysis of Second-order Melting Data of Synthetic and Natural Polynucleotides', *Anal. Chem.*, **69**, 1785–1792 (1997).
43. R. Tauler, A. Izquierdo-Ridorsa, E. Casassas, 'Simultaneous Analysis of Several Spectroscopic Titrations with Self-modeling Curve Resolution', *Chemom. Intell. Lab. Syst.*, **18**, 293–300 (1993).
44. E.R. Malinowski, 'Automatic Window Factor Analysis. A More Efficient Method for Determining Concentration Profiles from Evolutionary Spectra', *J. Chemom.*, **10**, 273–279 (1996).
45. W. Den, E.R. Malinowski, 'Investigation of Copper(II)–Ethylenediaminetetraacetate Complexation by Window Factor Analysis of Ultraviolet Spectra', *J. Chemom.*, **7**, 89–98 (1993).
46. M.M. Darj, E.R. Malinowski, 'Complexation Between Copper(II) and Glycine in Aqueous Acid Solutions by Window Factor Analysis of Visible Spectra', *Anal. Chem.*, **68**, 1593–1598 (1996).
47. R. Manne, H. Shen, Y. Liang, 'Subwindow Factor Analysis', *Chemom. Intell. Lab. Syst.*, **45**, 171–176 (1999).
48. C.A. Andersson, R. Bro, 'Improving the Speed of Multiway Algorithms: Part I. Tucker3', *Chemom. Intell. Lab. Syst.*, **42**, 93–103 (1998).
49. R. Bro, C.A. Andersson, 'Improving the Speed of Multiway Algorithms: Part II. Compression', *Chemom. Intell. Lab. Syst.*, **42**, 105–113 (1998).
50. E.J. Karjalainen, 'The Spectrum Reconstruction Problem. Use of Alternating Regression for Unexpected Spectral Components in Two-dimensional Spectroscopies', *Chemom. Intell. Lab. Syst.*, **7**, 31–38 (1989).
51. E.R. Malinowski, 'Obtaining the Key Set of Typical Vectors by Factor Analysis and Subsequent Isolation of Component Spectra', *Anal. Chim. Acta*, **134**, 129–137 (1982).
52. A. de Juan, Y. Vander Heyden, R. Tauler, D.L. Massart, 'Assessment of New Constraints Applied to the Alternating Least Squares Method', *Anal. Chim. Acta*, **346**, 307–318 (1997).
53. J. Mendieta, M.S. Díaz-Cruz, R. Tauler, M. Esteban, 'Application of Multivariate Curve Resolution to Voltammetric Data. Part 2: Study of Metal-binding Properties of the Peptides', *Anal. Biochem.*, **240**, 134–141 (1996).
54. R. Bro, S. de Jong, 'A Fast Nonnegativity-constrained Least Squares Algorithm', *J. Chemom.*, **11**, 393–401 (1997).
55. R. Bro, N.D. Sidiropoulos, 'Least Squares Algorithms Under Unimodality and Nonnegativity Constraints', *J. Chemom.*, **12**, 223–247 (1998).
56. P.K. Hopke, 'Target Transformation Factor Analysis', *Chemom. Intell. Lab. Syst.*, **6**, 7–19 (1989).
57. A. de Juan, B. van den Bogaert, F. Cuesta Sánchez, D.L. Massart, 'Application of the Needle Algorithm for Exploratory Analysis and Resolution of HPLC–DAD Data', *Chemom. Intell. Lab. Syst.*, **33**, 133–145 (1996).
58. R.J. Hanson, C.L. Lawson, *Solving Least-squares Problems*, Prentice-Hall, Englewood Cliffs, NJ, 1974.
59. H. Gampp, M. Maeder, Ch.J. Meyer, A. Zuberbühler, 'Calculation of Equilibrium Constants from Multiwavelength Spectroscopic Data IV. Model Free Least Squares Refinement by Use of Evolving Factor Analysis', *Talanta*, **33**, 943–951 (1986).
60. http://www.ub.es/gesq/eq1_eng.htm
61. A. Izquierdo-Ridorsa, J. Saurina, S. Hernández-Cassou, R. Tauler, 'Second-order Multivariate Curve Resolution Applied to Rank Deficient Data Obtained from Acid–Base Spectrophotometric Titrations of Mixtures of Nucleic Bases', *Chemom. Intell. Lab. Syst.*, **38**, 183–196 (1997).
62. M. Amrhein, B. Srinivasan, D. Bonvin, M.M. Schumacher, 'On the Rank Deficiency and Rank Augmentation of the Spectral Measurement Matrix', *Chemom. Intell. Lab. Syst.*, **33**, 17–33 (1996).
63. M. Gui, S.C. Rutan, A. Agbodjan, 'Kinetic Detection of Overlapped Aminoacids in Thin-layer Chromatography with a Direct Trilinear Decomposition Method', *Anal. Chem.*, **67**, 3293–3299 (1995).
64. R. Tauler, G. Durand, D. Barceló, 'Deconvolution and Quantitation of Unresolved Mixtures in Liquid Chromatography–Diode Array Detection Using Evolving Factor Analysis', *Chromatographia*, **33**, 244–254 (1992).
65. R. Gargallo, F. Cuesta-Sànchez, D.L. Massart, R. Tauler, 'Validation of Alternating Least Squares Multivariate Curve Resolution for the Resolution and Quantitation of Overlapped Peaks Obtained in Liquid Chromatography with Diode Array Detection', *Trends Anal. Chem.*, **15**, 279–286 (1996).
66. S. Lacorte, D. Barceló, R. Tauler, 'Quantitation and Validation of Trace Herbicides Using On-line Solid-phase Extraction Followed by Liquid Chromatography Diode-array and Multivariate Curve Resolution', *J. Chromatogr. A*, **697**, 345–355 (1995).
67. R. Tauler, S. Lacorte, D. Barceló, 'Application of Multivariate Curve Self-modeling Curve Resolution for the Quantitation of Trace Levels of Organophosphorous Pesticides in Natural Waters from Interlaboratory Studies', *J. Chromatogr. A*, **730**, 177–183 (1996).
68. J.M. Salou, M. Honing, R. Tauler, D. Barceló, 'Resolution and Quantitative Determination of Coeluted Pesticide

Mixtures in Mass Spectrometry Liquid Chromatography by Multivariate Curve Resolution', *J. Chromatogr. A*, **795**, 3–12 (1998).
69. R. Tauler, B.R. Kowalski, S. Fleming, 'Multivariate Curve Resolution Applied to Spectral Data from Multiple Runs of an Industrial Process', *Anal. Chem.*, **65**, 2040–2047 (1993).
70. H. Martens, T. Naes, *Multivariate Calibration*, Wiley & Sons, New York, 1989.
71. F.C. McLennan, B.R. Kowalski (eds.), *Process Analytical Chemistry*, Chapman & Hall, London, 1995.
72. A. de Juan, A. Izquierdo-Ridorsa, R. Tauler, G. Fonrodona, E. Casassas, 'A Soft Modeling Approach to Interpret Thermodynamic and Conformational Transitions of Polynucleotides', *Biophys. J.*, **73**, 2937–2948 (1997).
73. D.J. Legget (ed.), *Computational Methods for the Determination of Formation Constants*, Modern Inorganic Chemistry Series, Plenum Press, New York, 1985.
74. R.M. Dyson, S. Kardeli, G.A. Lawrence, M. Maeder, A. Zuberbühler, 'Second Order Global Analysis: the Evaluation of Series of Spectrophotometric Titrations for Improved Determination of Equilibrium Constants', *Anal. Chim. Acta*, **353**, 381–393 (1997).
75. A. de Juan, G. Fonrodona, R. Gargallo, A. Izquierdo-Ridorsa, R. Tauler, E. Casassas, 'Application of a Self-modeling Curve Resolution Approach to the Study of Solvent Effects on the Acid–Base and Copper(II)-complexing Behaviour of Polyuridylic Acid', *J. Inorg. Biochem.*, **63**, 155–173 (1996).
76. A. de Juan, A. Izquierdo-Ridorsa, R. Gargallo, R. Tauler, G. Fonrodona, E. Casassas, 'Three-way Curve Resolution Applied to the Study of Solvent Effect on the Thermodynamic and Conformational Transitions Related to the Protonation of Polycytidylic Acid', *Anal. Biochem.*, **249**, 174–183 (1997).
77. H. Sigel, 'Interactions of Metal Ions with Nucleotides and Nucleic Acids and their Constituents', *Chem. Soc. Rev.*, **22**, 255–267 (1993).
78. H. Sigel, R.B. Martin, R. Tribolet, U.K. Häring, R. Malini-Balakrishnan, 'An Estimation of the Equivalent Solution Dielectric Constant in the Active-site Cavity of Metalloenzymes. Dependence of Carboxylate–Metal Ion Complex Stabilities on the Polarity of Mixed Aqueous/Organic Solvents', *Eur. J. Biochem.*, **152**, 187–193 (1985).
79. W. Saenger, *Principles of Nucleic Acid Structure*, Springer-Verlag, New York, 1988.
80. E. Casassas, R. Gargallo, I. Giménez, A. Izquierdo-Ridorsa, R. Tauler, 'Application of an Evolving Factor Analysis-based Procedure to Speciation Analysis in the Copper(II)–Polyuridylic Acid System', *Anal. Chim. Acta*, **283**, 538–547 (1993).
81. E. Casassas, R. Tauler, I. Marqués, 'Interactions of H^+ and Cu(II) ions with Poly(adenylic) Acid: Study by Factor Analysis', *Macromolecules*, **27**, 1729–1737 (1994).
82. V.P. Antao, D.M. Gray, 'CD Spectral Comparisons of the Acid-induced Structures of poly[d(A)], poly[r(A)], poly[d(C)], and poly[r(C)]', *J. Biomol. Struct. Dyn.*, **10**, 819–838 (1993).
83. R. Maggini, F. Secco, M. Venturini, H. Diebler, 'Kinetic Study of Double-helix Formation and Double-helix Dissociation of Polyadenylic Acid', *J. Chem. Soc., Faraday Trans.*, **90**, 2359–2363 (1994).
84. E. Casassas, R. Gargallo, A. Izquierdo-Ridorsa, R. Tauler, 'Application of a Multivariate Curve Resolution Procedure for the Study of the Acid–Base and Copper(II) Complexation Equilibria of Polycytidylic Acid', *React. Polym.*, **27**, 1–14 (1995).
85. P. Gans, A. Sabatini, A. Vacca, 'SUPERQUAD: an Improved General Program for Computation of Formation Constants from Potentiometric Data', *J. Chem. Soc., Dalton Trans.*, 1195–1200 (1985).

ELECTROANALYTICAL METHODS

SECTION CONTENTS

H.S. White
University of Utah,
Salt Lake City,
USA

R.M. Crooks
Texas A&M University,
College Station,
USA

Electroanalytical Methods: Introduction

Henry S. White
University of Utah, Salt Lake City, USA
Richard M. Crooks
Texas A&M University, College Station, USA

This section focuses on recent advances in the theory, instrumentation, and methods of electroanalytical chemistry. Electroanalytical methods are well-suited for analysis of a wide range of chemical systems and problems of both fundamental and technological interest. These methods tend to have excellent sensitivity, are versatile and relatively inexpensive, and in a growing number of recent examples, exhibit the high spatial resolution required for modern analysis in the arenas of biological chemistry and nanotechnology. For these reasons, electroanalysis remains at the forefront of modern analytical methods. Advances in sensitivity and range of electroanalytical methods during the past decade, including single-molecule and single-reaction event detection, high-speed electrochemical measurements for investigations of fast kinetics, and the growing role of electrochemistry in bioanalysis have had an especially high impact on the field.

The authors contributing to this section have produced a well-rounded perspective of electroanalytical chemistry at the turn of the century. For example, the importance of electroanalytical chemistry to the field of biology and biological chemistry is represented in the article by Ewing, **Neurotransmitters, Electrochemical Detection of**. Here, the focus of the methods is on very high spatial resolution and time-resolved detection of biological signaling events.

The theme of detection and chemical sensing is continued in the chapters by Bard, Peters, and Young. Bard pioneered the field of electrogenerated chemiluminescence, and over the years has played a major role in elucidating the theory and developing new analytical applications of this technique. The article **Chemiluminescence, Electrogenerated** focuses on analytical applications, and readers interested in the use of this high-sensitivity, zero-background method for bioassays are especially fortunate to have this review at their disposal. Peters has been working in the field of electroanalytical chemistry for many years, and in **Selective Electrode Coatings for Electroanalysis** he discusses recent advances and applications of electrochemistry to detection. As the title suggests, the focus of this chapter is on the use of electrodes that have been engineered to respond selectively to target analytes. This theme is continued in **Ion-selective Electrodes: Fundamentals** by Young. The remarkable attributes of electrodes modified with well-defined organic monolayers, and especially their application to electroanalytical chemistry, are described by Finklea in **Self-assembled Monolayers on Electrodes**. Such materials are useful for fundamental studies of electrode processes, but also for designing chemical detection systems.

Similarly to other analytical methods, the usefulness of electroanalytical methods are greatly enhanced when they are coupled to instrumental techniques. The articles **Ion-selective Electrodes: Fundamentals** and **Ultraviolet/Visible Spectroelectrochemistry** by Anderson and Scherson, respectively, provide excellent examples of how optical methods can be coupled with electroanalytical chemistry to learn more about the system under study than either technique could provide if applied independently. Similarly, in **Microbalance, Electrochemical Quartz Crystal** Ward clearly demonstrates that gravimetric methods are extremely powerful for studying both electroactive films sorbed to electrode surfaces and physical properties of electrolyte solutions. In addition to spectroscopic and gravimetric methods, scanning probe microscopies can provide detailed information about the structure and electronic properties of electrode surfaces. The article by Itaya, **Self-assembled Monolayers on Electrodes**, deals with this subject and shows that the structure of solid electrodes, as well as the molecular structure of adsorbed organic monolayers, can be elucidated as a function of electrode potential. Finally, the article by Herrero, **X-ray Methods for the Study of Electrode Interaction**, describes how powerful diffraction methods can be applied in situ for the study of electrode processes.

In-situ methods are, of course, most desirable for studying electrode processes, but often it is either impractical or impossible to use such technique. So it is with ultrahigh vacuum surface spectroscopy. However, in the article **Surface Analysis for Electrochemistry: Ultrahigh Vacuum Techniques** Soriaga shows how well-defined electrodes, combined with carefully developed emersion procedures, provide detailed chemical and structural information about electrode surfaces.

The study of organic reaction mechanisms at electrode surfaces has been a central focus of electroanalytical chemistry for many years. In the article **Organic Electrochemical Mechanisms**, Andrieux provides an excellent update of the field. Of course not all electrochemical reactions occur at solid electrodes, and in recent years there has been much interest in electron transfer at liquid–liquid interfaces, for among other reasons the key role played by such interfaces in catalytic and biological reactions. This subject is covered in the article **Liquid/Liquid Interfaces, Electrochemistry at** by Girault.

Carter provides an excellent review on the many techniques that comprise the more traditional workhorse family of methods that fall under the category

List of selected abbreviations appears in Volume 15

of Pulse Voltammetry. Much faster methods, based on the application of ultramicroelectrodes, are described by Forster in the article **Ultrafast Electrochemical Techniques**. These techniques offer great promise in improving the temporal aspects of electroanalytical methods in the coming years.

Taken together, this section of the *Encyclopedia of Analytical Chemistry* provides the reader with an insightful overview of modern electroanalytical chemistry and its application to real-world problems. The remarkable improvements in the sensitivity, spatial resolution, and speed of these methods is sure to expand the scope of their use in the years ahead.

Chemiluminescence, Electrogenerated

Allen J. Bard
University of Texas at Austin, Austin, USA
Jeff D. Debad, Jonathan K. Leland, George B. Sigal, James L. Wilbur, and Jacob N. Wohlstadter
IGEN International, Inc., Gaithersburg, USA

Electrogenerated chemiluminescence (ECL) is the process in which electrogenerated species undergo electron transfer reactions to form excited states that emit light. Many molecules have the potential to produce ECL, however $Ru(bpy)_3^{2+}$ (bpy = 2,2′-bipyridine) is the most common emitter used for analytical applications. Application of a voltage to an electrode in the presence of an emitter induces light production and allows for the detection of the emitter at very low concentrations. Advantages over other analytical methods include low backgrounds, precise spatial and temporal control over the emission, and the possibility of signal amplification. Commercial systems exist that use ECL to detect numerous clinically relevant analytes with high sensitivity using a variety of assay formats.

List of selected abbreviations appears in Volume 15

1 INTRODUCTION

ECL (also called electrochemiluminescence) is a term commonly applied to systems in which energetic electron transfer reactions of electrogenerated species occur to produce excited states, usually with regeneration of luminescent species. An important distinction between ECL and chemiluminescence (CL) is control of the light generation process. CL processes are initiated and controlled by mixing and fluid flow, while ECL processes are initiated and controlled by switching an electrode voltage.

ECL reactions were first investigated in the early 1960s and were carried out with systems composed of aromatic compounds in highly purified and carefully deaerated aprotic solvents such as acetonitrile (MeCN). In 1972 ECL of luminescent inorganic species, mainly $Ru(bpy)_3^{2+}$, was introduced. Much of this early work has been reviewed.[1,2] The use of aqueous ECL in 1981 was made possible by the introduction of the concept of coreactants in 1977, and this led to the use of ECL labels in clinical diagnostics.[3–5]

ECL has been used for a wide range of analytical applications.[6] The most important commercial application to date for ECL is its use in diagnostic assays. ECL provides for highly sensitive and precise diagnostic assays and offers several advantages when compared with other methods. Unlike fluorescent methods, ECL does not require expensive and complicated light sources (i.e. lasers) to generate luminescence. Moreover, because ECL excitation is very selective, it does not suffer from the high background signals that are typical of fluorescent methods. ECL has the advantage over analytical CL applications in having high spatial and temporal resolution, since the ECL reaction is initiated at the surface of an electrode and only when the system is electrochemically activated. Moreover, unlike most CL systems, the luminescent species can be regenerated in ECL, allowing the possibility of amplification.

2 PRINCIPLES AND THEORY

2.1 Annihilation Electrogenerated Chemiluminescence

The ECL annihilation mechanism describes a process wherein a compound is electrochemically reduced and oxidized at an electrode (Equations 1 and 2, where E^0_{R,R^-} and $E^0_{R^+,R}$ are reduction and oxidation potentials, respectively). The resulting products can react (Equation 3, a "radical annihilation reaction") to produce excited species ($^1R^*$ in Equation 3) capable of emitting light.[7]

$$R + e \longrightarrow R\bullet^- \qquad E^0_{R,R^-} \tag{1}$$

$$R - e \longrightarrow R\bullet^+ \qquad E^0_{R^+,R} \tag{2}$$

$$R\bullet^+ + R\bullet^- \longrightarrow {}^1R^* + R \tag{3}$$

If the enthalpy, ΔH^0, of the electron transfer reaction shown in Equation (3) is larger than the energy required to produce the excited singlet state from the ground state, E_S (Equation 4), then one of the products of the reaction is produced with excess energy ($^1R^*$). This excess energy can be emitted as light (Equation 5). This process is known as the "S-route", indicating that the luminescence is emitted by a species in an excited singlet state.

$$\Delta H^0 = \Delta G^0 + T\Delta S^0 = E^0_{R^+,R} - E^0_{R,R^-} - 0.1\,\text{eV} > E_S \tag{4}$$

$$^1R^* \longrightarrow R + h\nu \tag{5}$$

It is also possible to use two different precursors, A and D, in an ECL annihilation process (Equations 6–8):

$$A + e \longrightarrow A\bullet^- \tag{6}$$

$$D - e \longrightarrow D\bullet^+ \tag{7}$$

$$D\bullet^+ + A\bullet^- \longrightarrow {}^1A^* + D \quad \text{or} \quad {}^1D^* + A \tag{8}$$

Light may be emitted from either of the possible high-energy products $^1A^*$ or $^1D^*$.

Even if the energetic condition of Equation (4) is not satisfied, it is possible for light emission to occur via the formation of lower energy triplet states, followed by triplet–triplet annihilation (the T-route, Equations 9 and 10):

$$D\bullet^+ + A\bullet^- \longrightarrow {}^3A^* + D \tag{9}$$

$$^3A^* + {}^3A^* \longrightarrow {}^1A^* + A \tag{10}$$

Direct emission from the triplet state is usually not observed for organic species, because they have long radiative lifetimes and are quenched before emitting. Quenching of the triplets by the radical ions or other species in the system leads to lower emission yields for the T-route than the S-route systems.

The ECL mechanisms shown above generate oxidized and reduced species at a single electrode. It is also possible to obtain emission through the steady generation of the two reactants at two different electrodes that are sufficiently close to allow the reactants to interdiffuse and react.[8,9]

2.2 Coreactant Electrogenerated Chemiluminescence

Although analytical ECL is possible with annihilation systems,[10] most systems require impractical procedures like the use of aprotic solvents and rigorous deaeration of solvents. Annihilation ECL is generally not convenient in aqueous solutions, because the stability range for water oxidation and reduction is too small to generate both species needed for the radical ion annihilation reaction. Moreover, one or both of the precursors are usually unstable in water. It is possible, however, to obtain ECL in aqueous solution through the use of a coreactant.[3,11] ECL coreactants are species that, upon electrochemical oxidation or reduction, produce species that react with other compounds to produce excited species capable of emitting light. Oxalate ion, $C_2O_4^{2-}$, the first ECL coreactant discovered, represents a typical oxidative coreactant.[3] Consider an ECL system composed of species D and oxalate in an aqueous solution. An oxidizing potential at an electrode oxidizes D to produce the radical cation $D\bullet^+$. It also oxidizes the oxalate, which decomposes to produce the reducing agent $CO_2\bullet^-$ as shown in Equations (11) and (12):

$$C_2O_4^{2-} - e \longrightarrow C_2O_4^- \longrightarrow CO_2\bullet^- + CO_2 \tag{11}$$

$$D\bullet^+ + CO_2\bullet^- \longrightarrow D^* + CO_2 \tag{12}$$

In ECL systems that use coreactants, the electrode typically only oxidizes or reduces the reagents; chemical transformations of the electrochemical oxidation or reduction products provide the other required species. For example, in the oxalate system the electrode only oxidizes oxalate and the ECL reactant D; the reductant, $CO_2\bullet^-$, is generated via Equation (11). This strategy is used in most analytical applications, with the reactant D being $Ru(bpy)_3^{2+}$ (oxidized to $Ru(bpy)_3^{3+}$).

A variety of different coreactants, most notably aliphatic amines, can be used with the $Ru(bpy)_3^{2+}$ system[6,12,13] and form the basis of many analytical schemes described in the following section. Commercial applications typically use $Ru(bpy)_3^{2+}$ as the emitter and tripropylamine (TPA) as a coreactant. In this ECL system, developed by IGEN International Inc.[14] in the 1980s, $Ru(bpy)_3^{2+}$ is excited repeatedly and emits a photon many times (Figure 1).[3,13] This process – often referred to as an "amplification" process – allows a single label to generate many photons and contributes to the remarkable sensitivity of ECL systems.

For references see page 9849

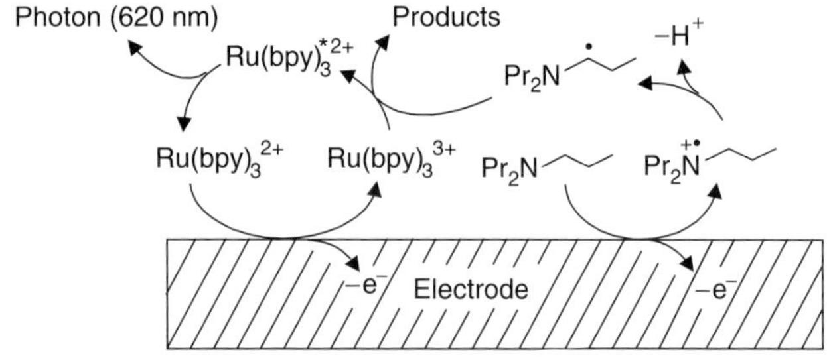

Figure 1 Proposed reaction mechanism for $Ru(bpy)_3^{2+}$/TPA ECL system. The coreactant forms a strong reducing agent upon oxidation that reacts with the oxidized form of the emitter ($Ru(bpy)_3^{3+}$) to generate light.

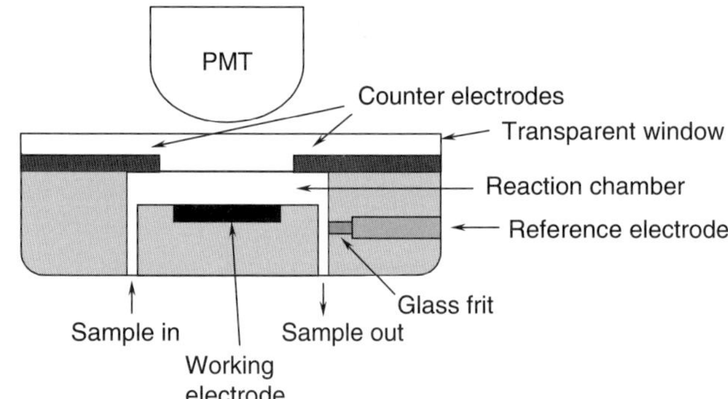

Figure 2 Schematic of a simple electrochemical flow cell used for ECL.

Other systems use coreactants that are reduced to generate reactive species. For example, ECL emission occurs on reduction of a mixture of $Ru(bpy)_3^{2+}$ and peroxydisulfate ($S_2O_8^{2-}$) via the sequence shown in Equations (13–15):[15]

$$Ru(bpy)_3^{2+} + e \longrightarrow Ru(bpy)_3^{+} \quad (13)$$

$$S_2O_8^{2-} + e \longrightarrow SO_4^{\bullet -} + SO_4^{2-} \quad (14)$$

$$Ru(bpy)_3^{+} + SO_4^{\bullet -} \longrightarrow Ru(bpy)_3^{2+*} + SO_4^{2-} \quad (15)$$

Here, reduction of $S_2O_8^{2-}$ generates the oxidant, $SO_4^{\bullet -}$, which reacts with the ECL label to form the excited state of the label.

3 INSTRUMENTATION

The design of an instrument based on ECL depends on the intended application and the mechanism used for light generation. Three major components are found in most devices: an electrochemical cell, a potentiostat, and a light detection system. Here we discuss these basic elements and other important aspects of ECL; examples of integrated instruments are presented in later sections.

The electrochemical cell is the center of the ECL process. A basic description of electrochemical cells can be found in any introductory electrochemistry text; Figure 2 shows a diagram of a typical electrochemical flow cell used for ECL. Generation of ECL takes place in the reaction chamber, which holds a solution containing the reagents required for ECL. This chamber also houses three electrodes that, during an ECL measurement, are in contact with the solution. The working and counter electrodes are essential for ECL, because they, upon passage of the current, generate the energetic reactants needed for ECL. A reference electrode controls the electrical potential of the working electrode and thus the reactions that occur there. The reference electrode is often separated from the reaction chamber by a porous frit that prevents fluid flow between the chamber and the reference electrode compartment. A potentiostat controls and applies potentials in this three-electrode system. Computer controlled potentiostats are readily available that have the flexibility to deliver any waveform required. Solutions within the electrochemical cell must contain an electrolyte to support the passage of current. This requirement may be met in aqueous systems simply by the presence of a buffer, while quaternary ammonium salts are common electrolytes for nonaqueous solvents.

Electrodes used in electrochemical ECL cells are typically made of gold or platinum. Even these robust electrodes may require periodic cleaning, and many cells are designed for easy dismantling to aid in electrode replacement or mechanical polishing. Highly reproducible methods of cleaning the electrodes without removing them from the cell have also been introduced, for example by electrochemical oxidation of water at the electrode surface in basic media.

Light generated by ECL is typically detected with a photodetector such as a photomultiplier tube (PMT), a photodiode, or a charge-coupled device (CCD). The choice of detectors depends on several factors including size, cost, sensitivity, power requirements and the ability to image. The spectral sensitivity of the detector may also be important because different ECL emitters can have very different emission wavelengths. For example, $Ru(bpy)_3^{2+}$ emits red light at 610 nm, $Os(bpy)_3^{2+}$ emits infrared light (720 nm), while diphenylanthracene's emission is blue (420 nm).

Efficient light detection helps to maximize the sensitivity of ECL measurements. Part of the electrochemical cell is transparent so that the light generated at the electrode(s) can escape, and the detector is positioned to maximize collection of light from the electrochemical cell. The cell and photodetector are housed in a light-tight box to reduce background light.

Virtually all integrated systems that use ECL for diagnostic measurements use the ORIGEN™ technology developed and patented by IGEN International, Inc. The essential component of the ORIGEN™ system is the measurement module, which consists of a flow cell containing an electrode and a light detection means such as a PMT or a photodiode. The current design of the ORIGEN™ measurement module includes a resusable platinum or gold electrode to generate ECL. ORIGEN™ assays typically employ magnetic microparticles as a binding phase; these microparticles are concentrated at the surface of the electrode to improve the efficiency and selectivity of the ECL process. More detailed descriptions of certain aspects of the ORIGEN™ technology are described in subsequent sections.

4 ANALYTICAL APPLICATIONS

For convenience, we divide analytical measurements that use ECL into two broad categories: assays for ECL coreactants; and assays for species capable of ECL (emitters). In the first category, biologically or pharmacologically important compounds act as ECL coreactants and react with ECL labels to generate luminescent species. The intensity of ECL measured in the presence of these coreactants is a function of the concentration of the coreactant as well as that of the emitter. ECL measurements in the presence of high, predetermined concentrations of ECL emitters can be used, therefore, as a means to assay for compounds that act as coreactants. The second category of analytical measurements uses ECL emitters as labels in affinity binding assays that attach the ECL emitter to the analyte of interest. The coreactant, typically TPA, is present in high concentrations; the amount of luminescence is thus dependent on the amount of the ECL emitter present in the assay. Since the emitters are bound to the analyte of interest, the amount of luminescence can be correlated with the concentration of the analyte.

4.1 Assays for Electrogenerated Chemiluminescence Coreactants

ECL from $Ru(bpy)_3^{2+}$ has been used to measure the concentration of coreactants such as peroxydisulfate, oxalate, and a variety of amines.[6] ECL assays for amines find many applications because amine groups are prevalent in biologically and pharmacologically important compounds and because amines absorb weakly in the ultraviolet/visible spectrum. ECL assays have been reported for numerous amine-containing compounds including alkylamines, antibiotics, antihistamines, opiates, nicotinamide adenine dinucleotide (reduced form) (NADH) and β-blockers.[6,16] As a general rule, the ECL signal from alkylamine coreactants follows the order: $3^\circ > 2^\circ > 1^\circ$.[13]

In a typical solution-based assay, a sample is combined with a solution containing $Ru(bpy)_3^{2+}$; the resulting solution is mixed and introduced into the electrochemical cell (see Figure 2). An oxidizing potential is applied to the working electrode and the ECL generated at the working electrode is measured using a light detector coupled to the flow cell. Similar methods have been used as detectors for the chromatographic separation of amines and in high-performance liquid chromatography (HPLC) systems that separate and detect amino acids for amino acid analysis.[17] Alternatively, $Ru(bpy)_3^{2+}$ may be immobilized in a thin polymer film deposited directly on the working electrode, which eliminates the need for a constant stream of $Ru(bpy)_3^{2+}$.[18]

ECL has been used to monitor enzymatic reactions by coupling the enzymatic reaction to the generation or consumption of an ECL coreactant. The coenzyme NADH contains an amine moiety and acts as a coreactant for $Ru(bpy)_3^{2+}$, however its oxidized form (NAD^+) is not a coreactant.[18] Scheme 1 illustrates an ECL measurement used to assay for glucose.[19] Numerous NADH-dependent enzymes are known, allowing for the adaptation of this technique to assay for a variety of different analytes.

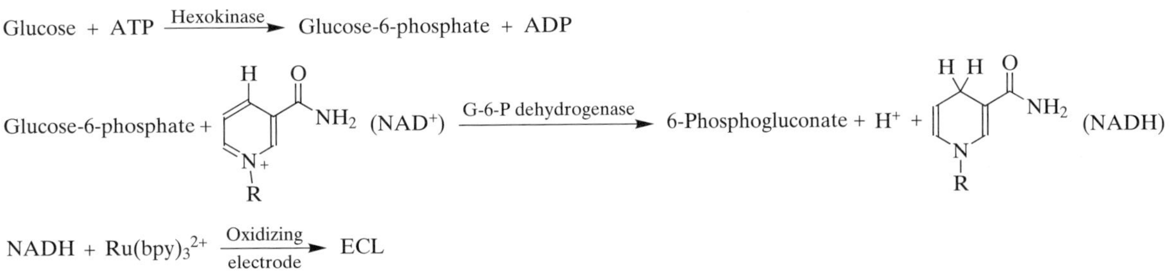

$$\text{Glucose} + \text{ATP} \xrightarrow{\text{Hexokinase}} \text{Glucose-6-phosphate} + \text{ADP}$$

$$\text{Glucose-6-phosphate} + (\text{NAD}^+) \xrightarrow{\text{G-6-P dehydrogenase}} \text{6-Phosphogluconate} + \text{H}^+ + (\text{NADH})$$

$$\text{NADH} + Ru(bpy)_3^{2+} \xrightarrow[\text{electrode}]{\text{Oxidizing}} \text{ECL}$$

Scheme 1 Schematic of reactions used to measure glucose by ECL. NADH, which is produced enzymatically, is used as a coreactant for $Ru(bpy)_3^{2+}$ ECL.

For references see page 9849

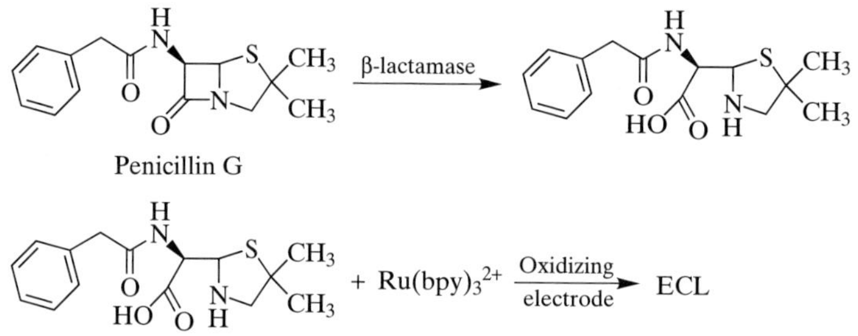

Penicillin G

β-lactamase

+ $Ru(bpy)_3^{2+}$ Oxidizing electrode ECL

Scheme 2 ECL assay that measures the presence of β-lactamase and/or its activity. Enzymatic hydrolysis of penicillin by β-lactamase creates a coreactant that can be used to produce ECL from $Ru(bpy)_3^{2+}$.

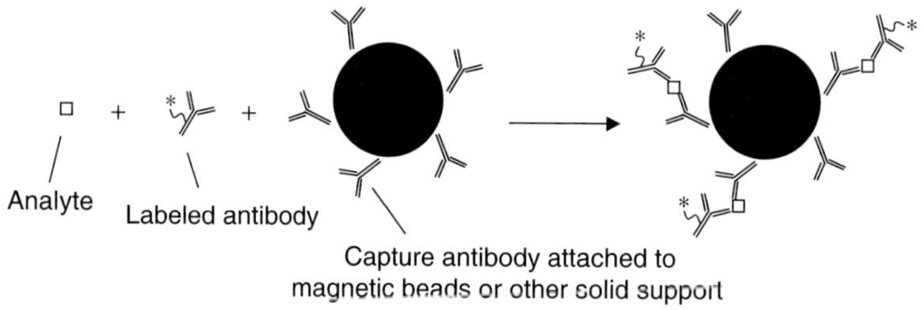

Analyte

Labeled antibody

Capture antibody attached to magnetic beads or other solid support

Figure 3 An affinity assay using an antibody–antigen–antibody "sandwich" assay. Once the analyte and labeled antibody are bound, the support is washed, and ECL is initiated after addition of a coreactant solution (the asterisk is an ECL label such as $Ru(bpy)_3^{2+}$, and the open square is an analyte of interest).

ECL has also been used to detect β-lactamase activity.[20] Penicillin does not act as a coreactant with $Ru(bpy)_3^{2+}$ to produce ECL but β-lactamase catalyzed hydrolysis of penicillin forms a molecule with a secondary amine that can act as a coreactant (Scheme 2). The efficiency of the ECL process can be increased by direct covalent attachment of the β-lactamase substrate to a $Ru(bpy)_3^{2+}$ derivative.[21]

The ECL detection of enzyme activity has also been used to develop highly sensitive assays in which an enzyme that generates or consumes an ECL coreactant is used as the detectable label on an antibody. The presence of the antibody is determined by measuring the activity of the enzyme label via ECL.

4.2 Assays for Electrogenerated Chemiluminescence Emitters

The most common analytical uses of ECL employ the emitter as a label in affinity binding assays.[5] Affinity binding assays detect an analyte of interest by monitoring the binding of the analyte to one or more binding partners specific for the analyte. The label is physically linked to one of the binding partners in the assay and provides the means for detecting the coupling of the binding partner to the analyte. Several classes of binding partners are used in these assays: antibody/antigen, enzyme/inhibitor, carbohydrate/lectin, and nucleic acid/complementary nucleic acid are the most common.[22]

The "sandwich assay" format (Figure 3) is often used for affinity binding assays. In a sandwich immunoassay, a sample is combined with two antibodies specific for different regions of the analyte: one of these antibodies is immobilized on a solid phase support and the other is linked to a detectable label. The presence of analyte leads to the formation of "sandwich" complexes that link the detectable label to the solid phase; the concentration of the label on the solid phase is directly proportional to the concentration of analyte. In the ECL version of the sandwich assay, the number of ECL labels on the solid phase are determined through an ECL measurement.

ECL-based affinity assays typically employ coreactants for compatibility with the aqueous conditions used in biological assays. The concentrations of the coreactants are kept constant and high to maximize the sensitivity of the detection and to prevent fluctuations in the concentration of the coreactant from changing the ECL.

Commercial instruments that use ECL and affinity binding assays utilize $Ru(bpy)_3^{2+}$ as the emitter and TPA as coreactant. The coreactant TPA dissolves in buffers, does not interfere with binding reactions, and is very stable. $Ru(bpy)_3^{2+}$ is also very stable, and is a versatile label that can be readily linked to biological molecules using various linking groups, for example, activated *N*-hydroxysuccinamide esters.

Magnetic microparticles are used in most commercial ECL assays as supports for affinity binding assays. In a sandwich assay format, the microparticles are coated

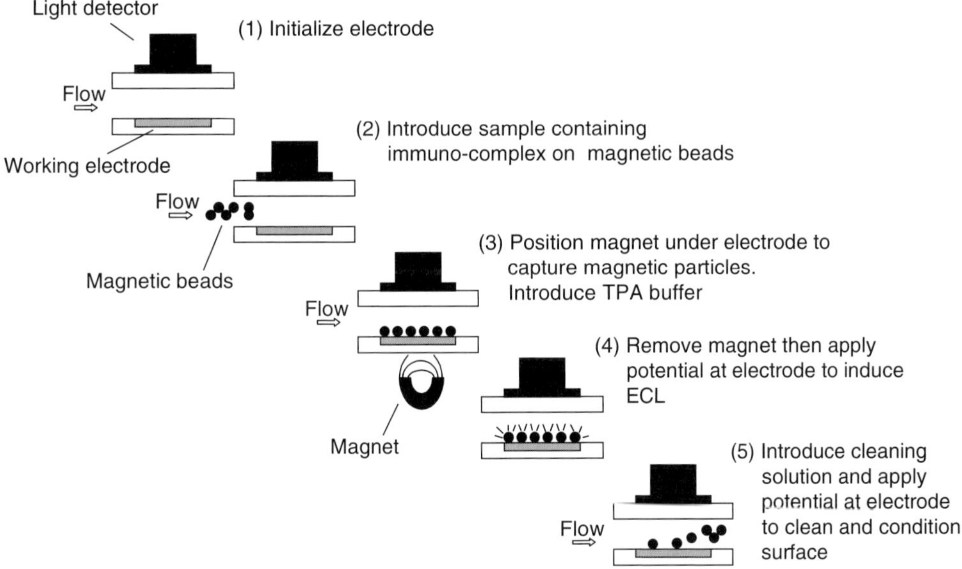

Figure 4 Schematic of process used for ECL assays employing magnetic microparticles. The microparticles are captured by positioning a magnet underneath the working electrode, and the ECL is initiated upon introduction of a coreactant solution. Once the electrode is electrochemically cleaned, another sample can be introduced.

with an antibody specific for the analyte to be measured and mixed with a solution that contains both the sample and a second antibody that is linked to one or more ECL labels. Analytes present in the sample bind to the antibodies on the microparticles, and the labeled antibody binds to these analytes to complete the sandwich complex. After the binding reactions are complete, the magnetic microparticles, (and the associated sandwich complexes) are collected on the working electrode by positioning a magnet beneath the working electrode (Figure 4). The electrode is then used to initiate ECL and the intensity of the emitted light is measured. This intensity can be correlated to the amount of emitter, and hence analyte, that is bound to the magnetic beads, which in turn can be correlated with the amount of analyte that was present in the sample. The use of magnetic microparticles enables very sensitive assay measurements by concentrating analyte and labels bound to the particles on the electrode.

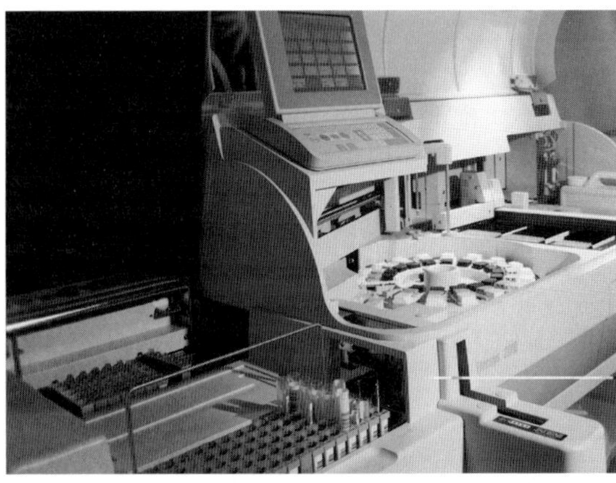

Figure 5 The ORIGEN™ analyzer (IGEN International, Inc.) was the first commercial ECL instrument (a). The ELECSYS™ 2010 (Roche Diagnostics) was designed for use in centralized testing and reference laboratories (b).

The first commercial instrument that used ECL was the ORIGEN™ analyzer (IGEN International, Inc., Figure 5). This ECL analyzer provides highly sensitive and precise assays in an automated format. Typically, the assays use magnetic microparticles as a solid support and $Ru(bpy)_3^{2+}$/TPA as the label and coreactant, respectively. A personal computer controls the instrument and aids in the processing and storage of data. The sample and assay reagents are combined in plastic tubes; these tubes are placed in a carousel (located on the ORIGEN™ analyzer) that can accommodate 50 tubes at one time. The carousel agitates the tubes to mix the sample and reagents and to enhance the speed of the assay binding reactions. After the reactions are complete, the contents of the tubes are combined with other assay reagents and transported into the electrochemical cell. ECL is induced, the light is measured with a PMT, the contents of the cell are removed and the cell is reconditioned for the next measurement. A typical read cycle requires approximately 1 min. The ORIGEN™ analyzer is capable of conducting a broad menu of assays; different assays require only different assay reagents and, occasionally, modest modification of the instrumental procedures.

For references see page 9849

Roche Diagnostics, a licensee of IGEN International, Inc.'s technology, developed the ELECSYS™ instrument for conducting immunoassays in centralized hospital and reference laboratories. This highly automated instrument offers some of the most sensitive and precise assays available to the medical diagnostics market. (For example, the lower detection limit for the total concentration of the tumor marker prostate specific antigen (PSA) is less than 0.01 ng ml^{-1}.) Its fundamental technology and operation is similar, in principle, to the ORIGEN™ analyzer. The ELECSYS™ offers a menu of more than 40 assays, including assays for tumor markers, cardiac markers, and analytes relevant to infectious diseases, fertility therapies, thyroid disease, and many other clinical conditions. The instrument can operate in random access mode and has the capability to process STAT samples.

ECL has also been used to detect cryptosporidium in water, *Escherichia coli* in foods, and in the very precise quantitation of polymerase chain reaction (PCR) products for DNA-based diagnostic testing.[23] All these detection schemes rely on affinity binding mechanisms, and typically use $Ru(bpy)_3^{2+}$/TPA as the ECL reagents.

Recently, IGEN International, Inc. developed a new generation of ECL-based instruments. These instruments are based on IGEN's ECL module, a self-contained, fully functional diagnostic operating system that approximates 1/20th the size of the first generation of ORIGEN™ operating systems and which has accuracy and sensitivity equal to the first generation. The ECL module forms the foundation for a wide variety of diagnostic instruments and allows for flexible design and development of ECL-based instrumentation. ECL module-based systems (Figure 6) should have a variety of applications in patient point-of-care, high-throughput drug screening, food, water and animal health testing.

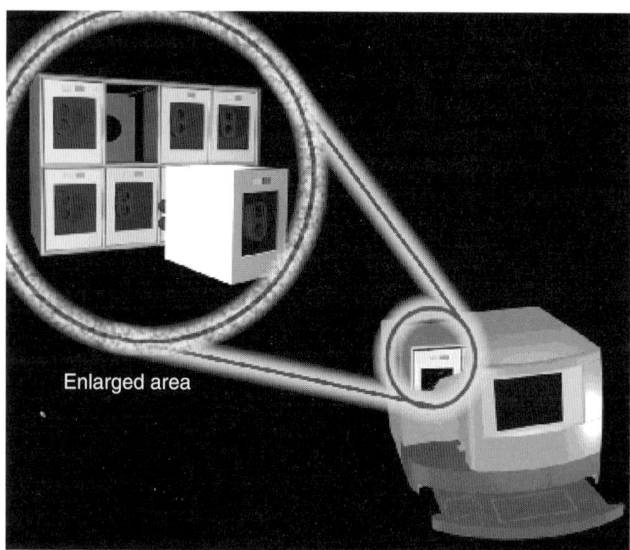

Figure 6 Schematic of an instrument based on the ECL module technology (IGEN International, Inc.). This instrument can be used for high-throughput diagnostic testing and drug discovery. Multiple self-contained, fully functional diagnostic modules (inset) are incorporated into an integrated device. The modules are interchangeable and easily removed and installed. Instruments intended for other applications (i.e. instruments that may have different requirements for operation, throughput and supporting infrastructure) also use ECL modules.

5 FUTURE CONSIDERATIONS

ECL will continue as an important area of exploration for basic research. It provides a rich platform for understanding fundamental questions in chemistry, biology and physics. Applications for ECL will also continue to expand in scope and importance. Analytical and diagnostic technology based on ECL currently provides a route to simple, low-cost, highly sensitive and very precise measurements. New ECL-based systems will provide sensitive measurements in local doctor's offices, small clinics, emergency rooms, and even in the home. These new developments will expand the impact of ECL in medicine, drug discovery, analytical science, environmental testing and industrial diagnostics.

ABBREVIATIONS AND ACRONYMS

CCD	Charge-coupled Device
CL	Chemiluminescence
ECL	Electrogenerated Chemiluminescence
HPLC	High-performance Liquid Chromatography
NADH	Nicotinamide Adenine Dinucleotide (Reduced Form)
PCR	Polymerase Chain Reaction
PMT	Photomultiplier Tube
PSA	Prostate Specific Antigen
TPA	Tripropylamine

List of selected abbreviations appears in Volume 15

RELATED ARTICLES

Biomolecules Analysis **(Volume 1)**
Fluorescence-based Biosensors

Clinical Chemistry **(Volume 2)**
Clinical Chemistry: Introduction • Immunochemistry • Phosphorescence, Fluorescence, and Chemiluminescence in Clinical Chemistry

Electroanalytical Methods **(Volume 11)**
Electroanalytical Methods: Introduction

REFERENCES

1. A.J. Bard, L.R. Faulkner, 'Techniques of Electrogenerated Chemiluminescence', in *Electroanalytical Chemistry*, ed. A.J. Bard, Marcel Dekker, New York, 1–95, Vol. 10, 1977.
2. R.S. Glass, L.R. Faulkner, 'Electrochemiluminescence', in *Chemical and Biological Generation of Excited States*, eds. W. Adam, G. Cilento, Academic Press, New York, Chapter 6, 1982.
3. I. Rubinstein, A.J. Bard, 'ECL 37. Aqueous ECL Systems Based on $Ru(bpy)_3^{2+}$ and Oxalate or Organic Acids', *J. Am. Chem. Soc.*, **103**, 512–516 (1981).
4. A.J. Bard, G.M. Whitesides, 'Luminescent Metal Chelate Labels and Means for Detection,' US Patent No. 5 221 605 (June 22, 1993).
5. G.F. Blackburn, H.P. Shah, J.H. Kenten, J.K. Leland, R.A. Kamin, J. Link, J. Peterman, M.J. Powell, A. Shah, D.B. Talley, S.K. Tyagi, E. Wilkins, T.-G. Wu, R.J. Massey, 'Electrochemiluminescence Detection for Development of Immunoassays and DNA Probe Assays for Clinical Diagnostics', *Clin. Chem.*, **37**, 1534–1539 (1991).
6. A.W. Knight, G.M. Greenway, 'Occurrence, Mechanisms and Analytical Applications of Electrogenerated Chemiluminescence', *Analyst*, **119**, 879–890 (1994).
7. K.M. Maness, J.E. Bartelt, R.M. Wightman, 'Effects of Solvent and Ionic Strength on the Electrochemiluminescence of 9,10-Diphenylanthracene', *J. Phys. Chem.*, **98**, 3993–3998 (1994).
8. J.T. Maloy, K.B. Prater, A.J. Bard, 'Electrogenerated Chemiluminescence. V. The Rotating Ring-disk Electrode-digital Simulation and Experimental Evaluation', *J. Am. Chem. Soc.*, **93**, 5959 (1971).
9. G.H. Brilmyer, A.J. Bard, 'Electrogenerated Chemiluminescence. XXXVI. The Production of Steady Direct Current ECL in Thin Layer and Flow Cells', *J. Electrochem. Soc.*, **127**, 104 (1980).
10. S.A. Cruser, A.J. Bard, 'Concentration–Intensity Relationships in Electrogenerated Chemiluminescence', *Anal. Lett.*, **1**, 11–17 (1967).
11. M.-M. Chang, T. Saji, A.J. Bard, 'ECL 30. Electrochemical Oxidation of Oxalate in the Presence of Luminescers in Acetonitrile Solution', *J. Am. Chem. Soc.*, **99**, 5399–5403 (1977).
12. J.B. Noffsinger, N.D. Danielson, 'Generation of CL upon Reaction of Aliphatic Amines with $Ru(bpy)_3^{2+}$', *Anal. Chem.*, **59**, 865–868 (1987).
13. J.K. Leland, M.J. Powell, 'ECL: An Oxidative–Reduction Type ECL Reaction Sequence Using Tripropyl Amine', *J. Electrochem. Soc.*, **137**, 3127–3131 (1990).
14. See www.igen.com.
15. H.S. White, A.J. Bard, 'ECL 41. ECL and CL of the $Ru(bpy)_3^{2+}-S_2O_8^{2-}$ System in Acetonitrile–Water Solutions', *J. Am. Chem. Soc.*, **104**, 6891–6895 (1982).
16. A.W. Knight, G.M. Greenway, 'Relationship between Structural Attributes and Observed Electrogenerated Chemiluminescence (ECL) Activity of Tertiary Amines as Potential Analytes for the Tris(2,2′-Bipyridine)Ruthenium(II) ECL Reaction', *Analyst*, **121**, 101R–106R (1996).
17. W.A. Jackson, D.R. Bobbitt, 'Chemiluminescent Detection of Amino Acids Using In Situ Generated $Ru(bpy)_3^{3+}$', *Anal. Chem. Acta*, **285**, 309–320 (1994).
18. T.M. Downey, T.A. Nieman, 'Chemiluminescence Detection Using Regenerable Tris(2,2′-Bipyridyl)Ruthenium(II) Immobilized in Nafion', *Anal. Chem.*, **64**, 261–268 (1992).
19. F. Jameison, R.I. Sanchez, L. Dory, J.K. Leland, D. Yost, M.T. Martin, 'Electrochemiluminescence-based Quantitation of Classical Clinical Chemistry Analytes', *Anal. Chem.*, **68**, 1298–1302 (1996).
20. P. Liang, R.I. Sanches, M.T. Martin, 'Electrochemiluminescence-based Detection of β-Lactam Antibiotics and β-Lactamases', *Anal. Chem.*, **68**, 2426–2431 (1996).
21. P. Liang, L. Dong, M.T. Martin, 'Light Emission from Ruthenium-labeled Penicillins Signaling their Hydrolysis by β-Lactamase', *J. Am. Chem. Soc.*, **118**, 9198–9199 (1996).
22. D. Wild (ed.), *The Immunoassay Handbook*, Macmillan Press Ltd, 1994.
23. J. DiCesare, B. Grossman, E. Katz, E. Picozza, R. Ragusa, T. Woudenberg, 'A High-Sensitivity ECL Based Detection System for Automated PCR Product Quantitation', *Biotechniques*, **15**, 152–157 (1993).

Infrared Spectroelectrochemistry

Mark R. Anderson and C. Douglas Taylor
Virginia Polytechnic Institute and State University, Blacksburg, USA

For references see page 9875

The combination of infrared (IR) spectroscopy and electrochemistry in a single measurement has been an area of active research interest since the late 1960s. IR spectroelectrochemical experiments provide insight about interfacial electrochemical phenomena and/or oxidation–reduction reaction mechanisms that cannot be obtained using either technique individually. The power of the measurement is in the complementary nature of the spectroscopic and electrochemical information available when obtained simultaneously. The combination of measurements, however, requires compromise between the IR spectroscopy and the electrochemistry because of their different sampling requirements. The need for compromise illustrates the unique strengths and weaknesses of the combined spectroelectrochemical measurement. Three spectral sampling methods are commonly used: transmission, internal reflection, and specular reflection sampling of the electrochemical experiment. Additional experimental perturbations, such as polarization and electrode potential modulation, are frequently applied to increase the spectral sensitivity toward changes confined to the electrode interfacial region. This article presents the instrumental requirements for, and a theoretical background of, different types of IR spectroelectrochemical measurements.

1 INTRODUCTION

The combination of two or more unique measurements conducted simultaneously provides an opportunity to more completely characterize a chemical system without dramatically increasing the time of analysis. This concept has driven many instrumental developments in analytical chemistry over the last 30 years, and is a principal force behind the explosive growth of the field of spectroelectrochemistry. The combination of molecular structural/electronic information that is provided by spectroscopy with the analytical sensitivity and ability to control interfacial properties afforded by electrochemistry suggests the utility of spectroelectrochemical measurements. When combining these measurements, however, care must be taken to ensure that the efforts to conduct the simultaneous determinations do not dramatically compromise the quality of either the spectroscopic or the electrochemical data. This issue is particularly important to the field of IR spectroelectrochemistry because of the strict experimental requirements of both the IR and electrochemical measurements.

The article comprises four sections. Initially, background information about the field of spectroelectrochemistry is presented. This discussion provides a perspective on the issues of spectral sampling of the electrochemical experiment, and on the instrumental challenges faced by combined IR spectroscopy and electrochemical experiments. The focus then is on the three principal sampling methods commonly used in spectroelectrochemistry: transmission, internal reflection, and specular reflection. Here, discussion centers on the experimental requirements for acquiring spectral data during the course of an electrochemical experiment.

1.1 Historical Background

The field of spectroelectrochemistry has its origins in the mid-1960s.[1–4] The first publication that demonstrated the feasibility of combined spectroscopy and electrochemistry involved the spectral monitoring of the oxidation of *o*-toluidine using an optically transparent tin oxide conducting glass electrode (CGE).[1] Use of the semiconducting SnO_2 compromised the electrochemistry slightly, but the transparency of the CGE to UV/VIS (ultraviolet/visible) radiation was crucial so that the incident visible radiation could pass through the working electrode and probe the electrolyte solution during the heterogeneous redox reaction. Although the spectral monitoring in this application was with visible radiation, this demonstration of spectroelectrochemical measurements illustrated the potential utility of combining electrochemistry with spectroscopic monitoring. This measurement also illustrated that instrumental compromises, such as the introduction of resistance by the use of the semiconducting electrode, are frequently required when combining these measurements.

Several other publications in which the method of spectral sampling of the electrogenerated/consumed material was altered followed this initial publication. Hansen et al. demonstrated the ability of the SnO_2-coated glass to serve as the electrode as well as an internal reflection element (IRE)/waveguide in a spectroelectrochemical measurement.[2] Here, due to the limited penetration depth of the evanescent field, they

List of selected abbreviations appears in Volume 15

were able to restrict spectral monitoring to the region of solution adjacent to the electrode, effectively eliminating spectral response from the bulk of the solution. Murray et al. developed the optically transparent thin-layer electrode (OTTLE) by sandwiching a conducting grid, prepared from fine-mesh Au wires, between two optically flat glass or quartz microscope slides.[3] Unlike the transmission experiment by Kuwana et al.,[1] the thin-layer cell restricted the amount of solution monitored by the spectroscopy.

Although these initial efforts in the field of spectroelectrochemistry utilized the visible region of the spectrum, they address many of the fundamental issues central to all spectroelectrochemical measurements. The initial CGE configuration used by Kuwana indicated the need to compromise the requirements of the electrochemical measurement so that spectroscopy could be conducted simultaneously. The internal reflection measurements by Hansen addressed the issue of spectral selectivity for the electrode interfacial region. The OTTLE cell design by Murray considered minimizing the interference from a complex sample by eliminating most absorbances from species in the bulk solution that are not influenced by the applied potential. These issues, which are pivotal to the development of spectroelectrochemistry in general, are particularly important as the technique migrated to the IR region of the spectrum.

1.1.1 *Infrared Spectroelectrochemistry*

In 1966, Mark and Pons published the first application of spectroelectrochemistry to the IR region of the electromagnetic spectrum.[5] The experimental set-up for this measurement made use of a semiconducting Ge plate as both the working electrode and as an IRE for the IR spectroscopy. As with the UV/VIS measurements by Hansen, this IR spectroscopic measurement isolated the interfacial region by using the evanescent wave emanating from the semiconducting working electrode as the incident radiation. Minimizing the spectral sampling of the bulk solvent was a driving force to the internal reflection method. Unlike in the visible region of the spectrum, typical solvents used in electrochemistry experiments are strong absorbers of IR radiation and can easily dominate the IR spectrum. In the internal reflection experiment, the evanescent wave extends approximately one-tenth of a wavelength into the solution, effectively minimizing the amount of solvent that is probed by the IR radiation. As was the case in the application by Hansen et al., use of a semiconducting working electrode compromised the electrochemical response in favor of IR transparency. Despite the poor signal-to-noise, this result clearly demonstrated the possibility of extending spectroelectrochemical measurements to the IR region of the spectrum.

In 1968 Heineman, Burnett, and Murray designed an OTTLE cell for use in the IR region of the spectrum.[6] This cell used an Au-mesh minigrid electrode sandwiched between two IR transparent NaCl plates. This design parallels earlier OTTLE cells developed for use in the UV/VIS regions of the spectrum. As with the research by Mark and Pons, the OTTLE cell designed by Heineman et al. compromised the electrochemistry by increased resistance of the thin layer in order to accommodate the IR spectroscopic measurement. Of particular importance, both experimental configurations by Mark and by Heineman had low signal-to-noise, limiting the utility of these early IR spectroelectrochemistry measurements.

These initial demonstrations of IR spectroelectrochemistry illustrate the principal issues faced when conducting this combined measurement: (a) low spectral signal-to-noise for solutes, and (b) absorption by the background electrolyte. Several other papers appeared in the late 1960s and early 1970s that tried to address the experimental limitations with qualified success.[7,8] The potential of IR spectroelectrochemistry to obtain structural information about changes that accompany electrode reactions was intriguing and provided sufficient reason for the area to be revisited as instrumentation improved over the years.

1.1.2 *Instrumental Developments of Importance to Infrared Spectroelectrochemistry*

Two developments, independent of electrochemical measurements, in the mid-1960s and early 1970s made significant contributions to the field of IR spectroelectrochemistry. In 1966, Greenler published the first of two papers that describe the IR spectroscopy of molecules adsorbed to metallic substrates using a specular reflection geometry.[9] Greenler showed that the absorbance of a thin film was nearly 5000 times greater at a glancing angle for IR light polarized parallel to the plane of reflection compared to the absorbance obtained with light reflecting from the surface with normal incidence. Further, he calculated that the absorbance from the glancing angle reflection experiment was 25 times larger than the absorbance for a transmission experiment with a free-standing film of similar dimensions. This result provided the theoretical basis for the infrared reflection–absorption spectroscopy (IRRAS) experiment that has become a routine measurement today. As was the case with the early efforts in the field of IR spectroelectrochemistry, the external reflection measurements of supported thin films showed promise but were ultimately limited by the poor spectral signal-to-noise obtainable at the time.[10]

Throughout the 1960s and 1970s, instruments for IR spectroscopy were traditional spectrometers which based wavelength selection on prism or grating dispersion and

For references see page 9875

slits. These instruments suffered from low source throughput, slit width limited resolution, grating/prism dispersion, detector sensitivity, and slow detector-response time. Each of these instrumental properties contributed to the low signal-to-noise of the initial IR spectroelectrochemistry measurements and the initial specular reflection measurements. It was not until Fourier transform infrared (FTIR) spectroscopy and the associated accessories (e.g. detectors and electronics) became widely available that the field of IR spectroelectrochemistry became instrumentally feasible.

During the 1970s and early 1980s FTIR equipment began replacing instruments based on the dispersion of light for collecting IR spectra.[(11)] Instrumental advances that accompanied the development of FTIR addressed many of the signal-to-noise limitations found in the early IR spectroelectrochemistry measurements. The poor signal-to-noise of the initial spectroelectrochemical measurements may be attributed to the combination of low concentration of solutes that are being measured, and to the low intensity of the IR source that probes the sample at any given wavelength using the dispersive instrument. The issue of low signal-to-noise can be addressed by taking advantage of the high throughput of a Fourier transform instrument. Because of the design and use of an interferometer with a Fourier transform instrument, all frequencies are measured simultaneously. Compared to a dispersive instrument, therefore, the Fourier transform instrument may obtain the same resolution, but it does so in a fraction of the time. Consequently, multiple spectra may be obtained and co-added to increase the signal-to-noise in the same amount of time that it would take to monitor the wavelength region of interest with a dispersive instrument. This multiplex advantage leads to a substantial improvement in the signal-to-noise of the spectrum (the signal-to-noise increases by a factor of $\sqrt{N}$ where N is the number of co-added spectra). Similarly, use of the interferometer to encode the wavelength data, as opposed to separating the light by dispersion and slits, leads to greater source throughput. The dispersion of the source radiation and selection of individual wavelengths by slits is inefficient and lowers the total source intensity directed through the sample. The Fourier transform instrument directs most of the source intensity through the sample and separates the individual frequencies by mathematical transformation of the interferometer encoded source intensity. In this manner, less of the typically weak IR source is needlessly wasted. Each of these characteristics of the FTIR improves the signal-to-noise of the IR spectroscopy, providing the opportunity to acquire spectra of more challenging, less concentrated samples typically encountered in spectroelectrochemistry experiments.

An additional development in FTIR offers new opportunities for IR spectroelectrochemistry.[(12–15)] Step-scan FTIR instruments provide the same advantages as the continuous-scan instruments described previously; in addition, step-scan instruments introduce time as an accessible experimental parameter. In a step-scan instrument, the interferogram is built up point-by-point by moving the mirror of the interferometer in discrete steps. The timing of the experiment can be gated to the interferometer mirror steps, allowing acquisition of spectra as a function of time. With continuous-scan FTIR instruments, limited control of the time-frame is available through the interferometer scan speed and the overall instrumental resolution (e.g. the scan distance of the moving mirror). While individual scans of the interferometer are being co-added, the system being interrogated should be in some equilibrium condition. With the step-scan instrument, the time frame of the experiment is established by the dwell time at each step position of the interferometer. The dwell time may be set, providing an additional degree of flexibility for monitoring phenomena that occur at both short and long times. In this manner, one can easily monitor chemical systems at defined intervals after the interferometer mirror step, providing the opportunity to make measurements on dynamic systems whose processes occur over tens of microseconds. The utility of step-scan instruments for monitoring dynamic electrochemical processes is discussed later.

Although advances in FTIR instrumentation addressed some of the signal-to-noise limitations of IR spectroelectrochemistry measurements, these advances could not resolve all the fundamental issues faced when combining IR spectroscopy and electrochemical measurements:

- The general incompatibility of the solvent types required for these two measurements.
- The small quantity of material that must be detected during the experiment.

Most typical solvents for electrochemistry are poor solvents for IR spectroscopy (and vice versa). The polar solvents required to dissolve the supporting electrolyte generally have strong absorptions in the IR region of the spectrum such that the solvent would dominate the measured spectrum.

Two methods designed to reduce the solvent contribution to the IR spectrum have been developed:

- The use of thin solution layers to minimize the amount of solvent that the incident radiation samples.
- Modulating the interfacial properties and using phase-sensitive detection to selectively detect the interfacial region at the frequency of the modulation.

Advances in instrumentation over the years provide sufficient spectral sensitivity that IR spectroelectrochemical measurements have become common measurements despite the restrictions confronting them. In the following, the instrumental requirements for conducting IR spectroelectrochemical experiments are reviewed and discussed. The focus is on the strict conditions placed on sampling methods, and on how the experiment is used to obtain both electrochemical and spectroscopic data with good signal-to-noise. Where required, fundamental principles are addressed to illustrate why methods have certain restrictions, and why some experimental compromises must be made to accommodate the simultaneous spectroscopy and electrochemistry. The following sections are organized according to the main sampling geometries in common use today: transmission using the OTTLE cell, internal reflection, and external reflection. The discussion in each section focuses on the instrumentation requirements, the type of information available from the experiment, the limitations of the experiment, and the theory behind the measurements.

2 TRANSMISSION INFRARED SPECTROELECTROCHEMISTRY

Perhaps the most common applications of IR spectroelectrochemistry use the transmission mode for spectral sampling. In this case an OTTLE cell is used. Here the source radiation is directed through the transparent working electrode, sampling solution throughout the entire dimension of the thin layer. The relatively large surface area of the transparent minigrid electrode, combined with the small solution volume within the thin layer, allow complete electrochemical conversion of the solute. This method is therefore a convenient way to spectroscopically probe the structure and identity of products and long-lived intermediates (e.g. those with lifetimes of the order of minutes) of electrode reactions. In addition, the short path length of the thin-layer cell helps to minimize the absorbance by the background electrolyte. Despite the amount of use that the OTTLE cell finds today, the basic design has not significantly changed from that published by Heineman in 1968 (Figure 1).[6]

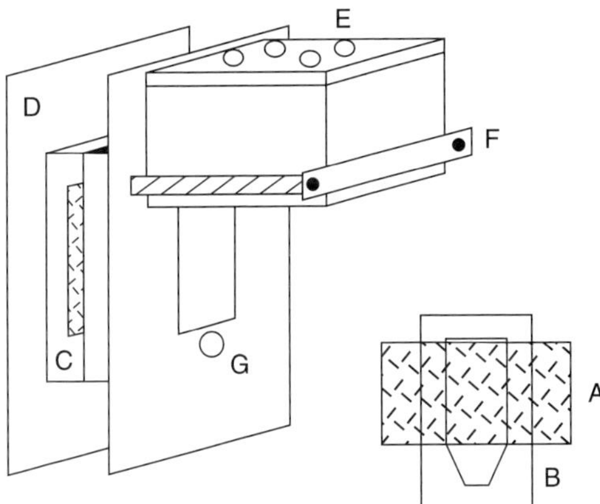

Figure 1 Prototype IR OTTLE spectroelectrochemical cell where: A is the gold minigrid electrode; B is a Teflon® spacer; C are NaCl plates; D is the steel frame; E are holes for reference and auxillary electrodes, and nitrogen inlet and outlet; F is a Teflon® cup and support bar; G is an outlet to draw solution from the cell. (Reprinted from W.R. Heineman, J.N. Burnett, R.W. Murray, 'Optically Transparent Thin-layer Electrodes: Ninhydrin Reduction in an Infrared Transparent Cell', *Anal. Chem.*, **40**(13), 1974–1978, copyright (1968) American Chemical Society.)

2.1 Optically Transparent Thin-layer Electrode Cell

Using the concepts first illustrated with UV/VIS thin-layer cells, the OTTLE cell for use in the IR region of the electromagnetic spectrum sandwiches an Au minigrid working electrode between two NaCl plates. A Teflon® gasket of known thickness spaces the two IR-transparent NaCl plates from each other and defines the thickness of the thin layer. A hole near the top of the front NaCl plate provides a path for solution to enter a larger-volume reservoir. The secondary and reference electrodes are placed in this larger solution reservoir to complete the electrochemical circuit.

Heineman used what is now commonly called an optically thick thin-layer design by placing multiple Au minigrids parallel to and spaced from each other with Teflon® gaskets. Although not required, the optically thick thin-layer design helps in the spectral detection of weak IR absorptions by allowing a larger quantity of solute to experience the applied electrode potential. The closely spaced minigrid electrodes allow efficient conversion of the electroactive material within the thin layer because the distance required for material to diffuse to the nearest electrode is diminished in this arrangement.

The principal advantage of the thin-layer design is the ability to quantitatively convert material within the thin layer between oxidation states. The depletion of electroactive species in the thin-layer cell avoids most bulk mass-transport effects. This occurs because the size of the thin layer is small compared to that of the diffusion layer thickness over the time-frame of the measurement (minutes). The acquired IR spectrum is therefore not complicated by the presence of unreacted analyte, nor is the response complicated by species diffusing into or out of the spectral observation window. The electrochemical behavior of thin-layer cells has been described previously.[16,17]

The open nature of the minigrid electrode allows IR source radiation to shine through the working electrode while maintaining electrochemical behavior that approximates a continuous conductor. Electrochemical behavior

For references see page 9875

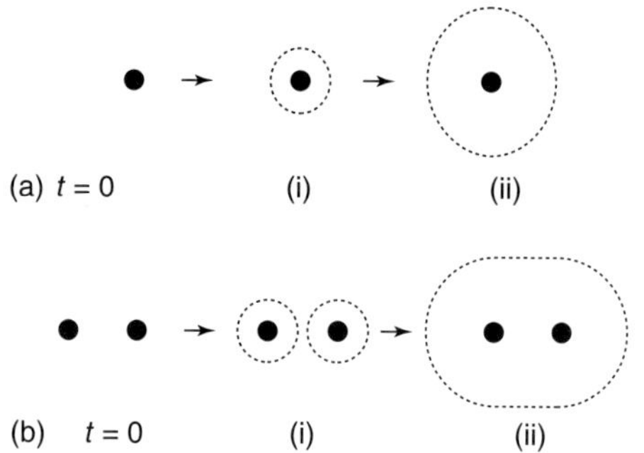

Figure 2 (a) Illustration of the diffusion layer generated at an isolated wire, cross-sectional view. (b) Model of diffusion at two isolated wires illustrating that, after some period of time, the individual diffusion layers overlap.

of the minigrid electrode can be understood using simple diffusion mass-transport arguments.[18] Consider the diffusion behavior of electroactive species at an isolated wire (Figure 2a). When a potential sufficient to oxidize or reduce electroactive material is applied, the species adjacent to the electrode are converted, establishing a concentration gradient in the adjacent solution. Under the influence of this gradient, electroactive reactant diffuses toward the electrode where it will then react, whereas the product of the electrode reaction will diffuse away from the electrode surface into the bulk solution. Consequently, the thickness of this concentration gradient expands in a radial direction away from the isolated wire with time after application of the potential.

Now consider two wires of the same dimension placed a small distance from each other. After application of the same potential to both wires the concentration gradient associated with the electrode reaction at each wire will similarly expand in time (Figure 2b). Ultimately, the diffusion layers from these adjacent wires will overlap. At that point, the diffusion approximates the behavior of a single electrode, not a pair of independent electrodes.

The minigrid electrodes commonly used in OTTLE cells have many closely spaced wires arranged parallel and perpendicular to each other. By analogy to the simple two-wire model given above, shortly after application of the potential to the minigrid, the diffusion layers associated with any given portion of the minigrid will overlap with that of its nearest neighbors. Consequently, the depletion of electroactive material due to oxidation or reduction at the working electrode will rapidly approximate linear diffusion to the macroscopic dimension of the minigrid. The electrochemical response, therefore, has the appearance of a reaction occurring at a planar electrode despite the open structure of the minigrid. Under these conditions, the electrochemical behavior of the thin-layer cell can be simplified to a one-dimensional model of linear diffusion over a finite distance to an infinitely large planar electrode.[18] Because the walls to the vessel are a finite distance from each other, depletion of the solute will extend to the dimensions of the thin layer, provided that the time-frame of the experiment is long compared to the time required for analytes to traverse the diffusion distance. Once this happens, the faradaic current due to the electrode reaction will return to the baseline level because all of the electroactive material within the thin layer has been converted into product. This thin-layer behavior has been described in detail previously.[16,17]

2.2 Cell Characteristics

An advantage of thin-layer electrochemical behavior is that experimental limitations due to mass transport are eliminated. This thin-layer configuration allows quantitative conversion between reactants, long-lived intermediates, and products for spectral monitoring. From an electrochemical point of view, this allows separation of the effects due to the heterogeneous reaction from the effects due to mass transport phenomena. Spectroscopic determination, therefore, provides the opportunity to acquire information regarding the mechanism of electrode reactions that is not available from the electrochemical data alone.

Although the response and planar electrode behavior of the OTTLE cell provide for a conceptually simple experiment, the thin-layer minigrid working electrode introduces some experimental complications. In electrochemical measurements, a three-electrode electrochemical cell is generally used to maintain accurate control of the working electrode potential. Accurate potential control, however, can be compromised by the OTTLE cell design due to the nonideal electrode arrangement.

Without accurate potential control, the experiment runs the risk of unevenly polarizing the working electrode (e.g. having an unequal potential distribution along the surface of the working electrode). Uneven working electrode polarization is generally avoided by maintaining the surface area of the secondary electrode large compared to that of the working electrode, and by the arrangement of the working, secondary, and reference electrodes. Typically, the surface area of the secondary electrode is 10–100 times larger than that of the working electrode. Similarly, the ideal arrangement of the working and secondary electrodes is opposite and parallel to each other with the reference electrode placed between these electrodes and close to the surface of the working electrode.[19] These conditions prevent the generation of current paths that differ significantly from each other between the secondary electrode and points along the surface of the working electrode. Under these conditions an equipotential distribution exists along the surface of the working electrode.

If these conditions are not met, the working electrode may have a distribution of potentials along the surface, introducing the risk that a distribution of phenomena will occur at the working electrode. This would complicate the electrochemistry and make spectral interpretation difficult. Clearly the gold minigrid electrode has a large surface area, which requires a secondary electrode of large area so as to avoid uneven electrode polarization. Unfortunately, the thin-layer cell prevents placement of the secondary electrode in the ideal geometry. It is therefore likely that the working electrode experiences some uneven polarization.

Placement of the reference electrode in the reservoir away from the working electrode may also have some deleterious effects. Under normal bulk-solution experimental conditions, the solution resistance that exists along the current path between the working and secondary electrodes may be partially eliminated by placing the reference electrode between these two electrodes and close to the working electrode. In an OTTLE thin-layer cell, it is difficult to place the reference electrode in this arrangement. Consequently, substantial solution resistance may exist within the thin layer. In addition, the total solution resistance experienced will vary along the surface of the working electrode, depending on the distance between points along the working electrode and the secondary electrode. In the thin-layer geometry, it is impossible to avoid different distances between points along the working electrode and the secondary electrode, leading to variable levels of resistance within the thin layer. This variable solution resistance leads to uneven electrode polarization and potential gradients along the working electrode within the thin layer. Murray et al. illustrated this when they were able to visually monitor a color change migrating along the length of the working electrode in an OTTLE cell following the application of potential.[3]

The presence of solution resistance and potential gradients principally influences the time response of the electrochemical measurement. Electrochemical experiments may be simply modeled as an equivalent circuit of resistors and capacitors. Because of the nature of the electrochemical experiments, the time response (e.g. how quickly the working electrode potential responds to the application of a new value) is an important parameter determined by values of the solution resistance and interfacial capacitance. Experiments in which the potential is rapidly altered in time need to have a fast response to accurately represent the electrochemical phenomena. With high resistance, OTTLE cells generally have a slow time response, resulting in distortion of the electrochemical data. For example, Heineman et al. obtained cyclic voltammetry data for the reduction of ninhydrin in their thick-layer OTTLE cell.[6] In this experiment there is a clear distortion of the current–potential response when measured in the thin layer compared to the current potential behavior in the bulk solution, suggestive of substantial resistance effects within the thin layer.

However, for most applications of transmission IR spectroelectrochemistry the slow time response of the OTTLE cell is not debilitating. Commonly, the working electrode potential is set to some value suitable for quantitatively oxidizing or reducing the analyte. When the current decays to the baseline level, indicating that quantitative transformation has occurred and equilibrium established, the sample is spectroscopically probed. In this manner, IR spectra of the reactants, products, or intermediates of the electrode reaction are obtained at their equilibrium concentrations determined by the electrode potential.

2.3 New Cell Designs

Although the basic design of the OTTLE cell for IR spectroelectrochemistry has not appreciably changed since the initial publication of Heineman, some alterations aimed at improving the time response, and at improving the compatibility of the cell with organic solvents, have appeared.

Hartl et al. prepared an OTTLE cell in which all three electrodes are placed within the solution thin layer. In this design a gold minigrid serves as the working electrode, a platinum mesh serves as the secondary electrode, and a silver wire serves as a pseudo-reference electrode (Figure 3).[20] These three electrodes are melt sealed into a 200-µm-thick polyethylene spacer to establish the small dimension of the thin-layer cell. In this arrangement, all three electrodes sample the same small portion of the solution. Although this design is intended to minimize solution resistance and increase the slow current response, cyclic voltammetry of 0.018 M ferrocene (in CH_2Cl_2 containing 0.10 M tetra-*n*-butylammonium hexafluorophosphate (TBAF)) had substantial current–potential distortion.[20]

Placement of the working and secondary electrodes within the same volume of solution may introduce some problems. The electrode reaction that occurs at the working electrode has a corresponding but opposite reaction at the secondary electrode to maintain electroneutrality (e.g. if a reduction occurs at the working electrode, then an oxidation must occur at the secondary electrode). Typically, when the solution is completely electrolyzed, the working and the secondary electrodes are placed in separate compartments to prevent mixing of the products of the separate electrode reactions. Mixing of the products generated at these two electrodes can result in undesired homogeneous reactions occurring. Brisdon et al. suggest that mixing of the counter-electrode

For references see page 9875

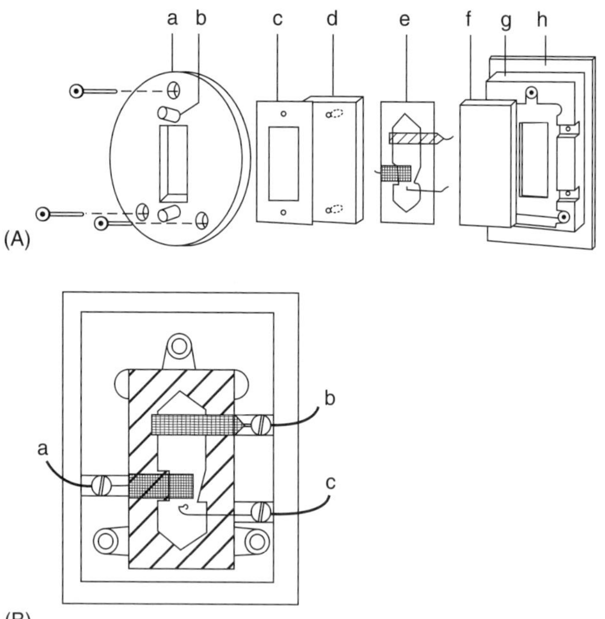

Figure 3 IR OTTLE cell in which the working, secondary and reference electrodes are all melt sealed into a polyethylene gasket. Part (A) shows: (a) steel pressure plate; (b) inlet; (c) Teflon® gasket; (d, f) front and back KBr windows; (e) polyethylene spacer with melt sealed electrodes; (g) Teflon® holder; (h) back plate. Part (B) shows: (a) Au minigrid working electrode; (b) Pt mesh auxillary electrodes; (c) Ag wire pseudo-reference electrode. (Reprinted from M. Krejcik, M. Danek, F. Hartl, 'Simple Construction of an Infrared Optically Transparent Thin-layer Electrochemical Cell: Applications to the Redox Reactions of Ferrocene, $Mn_2(CO)_{10}$ and $Mn(CO)_3$(3,5-di-*t*-butyl-catecholate)', *J. Electroanal. Chem.*, **317**, 179–187, copyright (1991) with permission from Elsevier Science.)

products in a thin-layer cell is negligible on the time-frame of the spectroelectrochemical measurement.[21] This is probably due to the relatively large distances that exist between the anode and cathode within the thin layer. However, repeated and prolonged use of this type of OTTLE cell can lead to the eventual build-up of residue at the edges of the working and counter electrodes. Permanently sealed cells therefore need to be discarded after several uses because of this residue build-up. Designs that rely on gaskets and pressure to seal the thin layer have also appeared.[21–23] These designs allow disassembly of the thin layer for cleaning and subsequent re-use.

The design described by Hartl allows maintenance of anaerobic conditions over the relatively long time-frame of the spectroelectrochemical experiment. The exclusion of oxygen from the electrolyte solution is particularly important when studying reductive electrochemistry. Oxygen is easily reduced and can interfere with the desired electrode reactions under these conditions. In most electrochemical measurements, the electrolyte solution is purged with an inert gas immediately prior to conducting the electrochemical reduction. However, with the OTTLE cell the solution within the thin layer cannot be easily purged. In this case, the analyte solution is purged prior to placement within the thin layer. Maintaining anaerobic conditions within the thin layer over the long time-frame of the transmission spectroelectrochemical measurement is required to avoid oxygen interference with the desired electrode reaction. As they demonstrate, this feature is particularly important for spectroscopically identifying oxygen-sensitive reaction intermediates frequently encountered in inorganic chemistry.[20]

The ease and overall utility of transmission spectroelectrochemical measurement makes it one of the more widely applied spectroelectrochemical measurements. Despite the many applications of this technique that have appeared in the literature, the basic sampling is largely the same as described by Heineman et al. in 1968.[6] The method is most useful for identifying bulk phase products and reactants of oxidation–reduction reactions. The thin layer of the OTTLE cell compromises the time response of electrochemical reactions. Consequently, transmission IR spectroelectrochemical measurements are best suited for identifying solution components after the solution has reached its equilibrium composition at a given applied potential. Recognizing the limitations allows one to study redox reaction mechanisms using transmission IR spectroelectrochemistry and the OTTLE cell.

3 INTERNAL REFLECTION INFRARED SPECTROELECTROCHEMISTRY

Although IR spectroelectrochemical applications using an OTTLE cell are numerous, the restrictions placed on electrochemical measurement in this sampling configuration are significant. The resistance within the thin layer prevents rapid changes of the electrode potential; consequently, the observation of short-lived intermediates is virtually impossible. Mass transport into or out of the thin layer is also compromised by the geometry of the thin layer. This behavior is described as diffusionally decoupling the electrolyte which is trapped within the thin layer from the rest of the solution.[24] This may result in a number of deleterious effects including electrolyte concentration gradients, pH gradients, and slow diffusional replenishment of materials within the thin layer. A configuration of the spectroelectrochemical cell for improved electrochemical response would probably have to eliminate the thin layer.

The thin solution layer was required for these spectroelectrochemical measurements to minimize the absorbance contribution by the bulk electrolyte. Mark and Pons, in the first publication in the field of IR spectroelectrochemistry, showed in 1966 that absorption by the bulk electrolyte could also be reduced by sampling the interface with the evanescent wave that develops at the interface where IR light is internally reflected.(5) In this application, the evanescent wave emanates from a Ge plate configured as both the working electrode and as a waveguide for IR radiation (Figure 4). Although the spectroscopic signal-to-noise was poor, this measurement demonstrated the potential for using the evanescent wave for IR spectroelectrochemical measurements.

Hansen(25) and Harrick(26) demonstrated that light incident on an interface between a phase of high refractive index and a phase of lower refractive index reflects at the interface with a small portion of the incident light transmitted through the interface when the incident angle is greater than a critical value. The transmitted light, called the evanescent wave, decays exponentially into the lower-index medium with the depth of penetration being dependent on the wavelength of the incident light and on the angle of incidence. Typically, this decay length is approximated as one-tenth of the wavelength of the incident light. For IR spectroscopy, therefore, the spectral sampling by the evanescent wave will be of the order of hundreds of nanometers. The penetration of the evanescent wave is smaller than the typical diffusion layer thickness in electrochemical experiments, indicating that internal reflection spectroscopy will sample only the interfacial region during the IR spectroelectrochemical experiment. Because of the very small interfacial region probed by the internal reflection method, IR absorption by the bulk solvent is not significant and the need for a thin-layer configuration is eliminated. Eliminating the thin layer removes many of the electrochemical restrictions imposed on the experiment by the OTTLE cell (e.g. slow time response and resistance effects), and provides the opportunity to observe transient electrochemical phenomena on faster timescales.

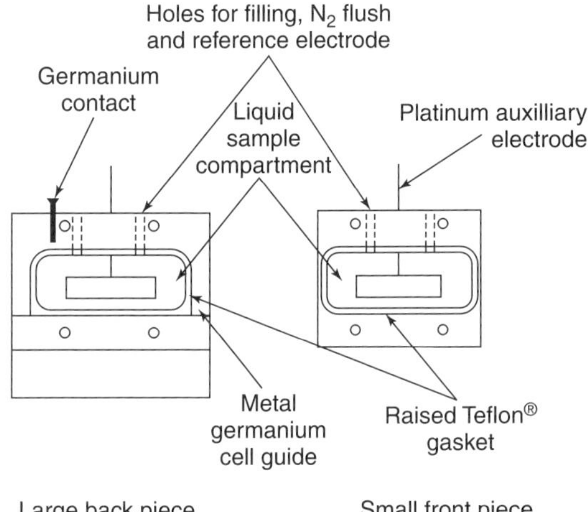

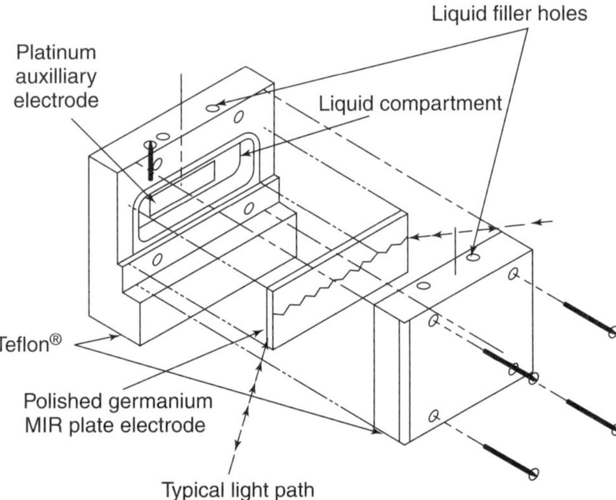

Figure 4 IR spectroelectrochemical cell with a Ge working electrode–waveguide combination for internal reflection spectroelectrochemistry. MIR = multiple internal reflections. (Reprinted by permission from H.B. Mark, B.S. Pons, 'An In Situ Spectrophotometric Method for Observing the Infrared Spectra of Species at the Electrode Surface during Electrolysis', *Anal. Chem.*, **38**(1), 119–121, copyright (1966) American Chemical Society.)

Unfortunately, the combination of the IRE and the working electrode into a single package introduces some new compromises to the combined spectroelectrochemical experiment. Mark and Pons,(5) as well as subsequent work by Tallant and Evans(7) and by Reed and Yeager,(8) used semiconducting *n*-doped Ge as both the working electrode and the IR-transparent IRE in their spectroelectrochemical experiments. Because of the semiconducting nature of Ge, there was significant internal resistance across the Ge plates (100–2000 Ω as measured across a 5-cm dimension by Tallant and Evans(7)). Any resistance in the electrochemical measurement, whether it comes from the electrode or the solution, places restrictions on the time response of the measurement. Additional doping can decrease the resistance of the semiconducting Ge, but improvements in the electrochemical performance in this case are offset by the diminished IR transparency of the more highly doped Ge. Tallant and Evans suggested the use of metallic films deposited on the surface of the IR-transparent Ge as a way of increasing the conductivity of the working electrode.(7) In this case, the evanescent wave must penetrate through the metal film to sample the electrolyte solution. When using a metallic overlayer as the working electrode, the metal layer must be sufficiently thin to allow the evanescent wave to penetrate, because IR radiation is strongly absorbed by metallic films. To function as an electrode, however, the metallic film must be

For references see page 9875

sufficiently thick to ensure a good, continuous conductor. As with the OTTLE cell, the different requirements of the spectroscopy and electrochemistry elements of the combined measurement require a compromise that ultimately may degrade the performance of both portions of the measurement.

It is evident that the strengths of the IRE spectroelectrochemical experiment are complementary to those of transmission IR spectroelectrochemistry with the OTTLE cell. Although the OTTLE cell provides a convenient method for acquiring spectra of reactants, intermediates, and products in the bulk solution, measurements that use IRE sampling isolate the interface and might be useful for identifying species at the electrode interface or within the electrochemical double layer during electrode reactions.

3.1 Evanescent Waves

A detailed discussion of the theory of internal reflection and evanescent wave properties is found elsewhere and is not reproduced here. Hansen[25,27] and Harrick[26] provide details for solving the boundary value problem for transmission and reflection of electromagnetic radiation at interfaces between media of different refractive indices using Fresnel's equations. Hansen in particular explored systems of relevance to spectroelectrochemical applications (e.g. two-layer and three-layer systems in which the middle layer is a thin metallic film).[25,27] He shows that the evanescent wave has a field strength that decays exponentially into the electrolyte medium. For a three-layer system (Figure 5), Equations (1–3) describe the field strength of the evanescent wave:

$$\langle E_{\parallel z}^2 \rangle = \left| \frac{n_1 \sin\theta_1}{\hat{n}_3} t_{\parallel} \right|^2 \exp\left[\left(\frac{4\pi}{\lambda} \right) \mathrm{Im}\,\xi_3 (z-h) \right] \quad (1)$$

$$\langle E_{\parallel x}^2 \rangle = \left| \frac{\xi_3}{\hat{n}_3} t_{\parallel} \right|^2 \exp\left[\left(\frac{4\pi}{\lambda} \right) \mathrm{Im}\,\xi_3 (z-h) \right] \quad (2)$$

$$\langle E_{\perp}^2 \rangle = |t_{\perp}|^2 \exp\left[\left(\frac{4\pi}{\lambda} \right) \mathrm{Im}\,\xi_3 (z-h) \right] \quad (3)$$

In these equations, $\langle E_{\parallel z}^2 \rangle$ and $\langle E_{\parallel x}^2 \rangle$ are the mean electric field strengths in the z and x directions resulting from light polarized parallel to the plane of reflection, $\langle E_{\perp}^2 \rangle$ is the mean electric field strength resulting from light polarized perpendicular to the plane of reflection, n_1 and $\hat{n}_3$ are the refractive index and complex refractive index of the incident and final phases, h is the thickness of the metallic layer, θ_1 is the incident angle, ξ_3 is the complex refractive index of the third phase multiplied by the cosine of the incident angle, $t_{\perp}$ and $t_{\parallel}$ are symbols representing Fresnel's equations for the transmitted light polarized perpendicular and parallel to the plane of reflection, and λ is the wavelength of the incident light. The Im function

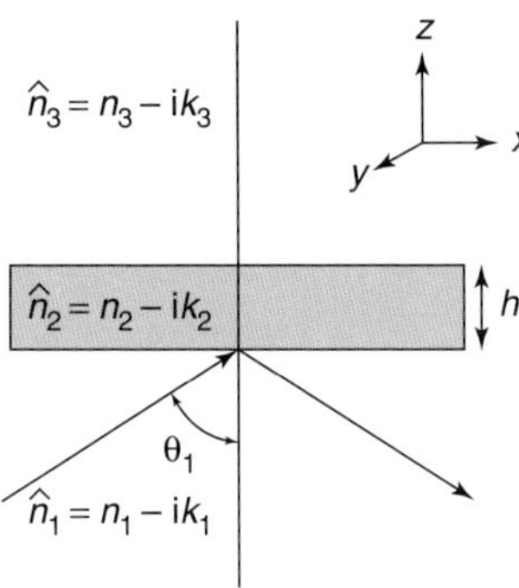

Figure 5 Diagram of the three-layer model of the interface between an IRE, a thin metallic layer and the electrolyte solution used to define the Cartesian axes of the boundary value problem for calculating the mean electric field strength of the evanescent wave.

indicates that the imaginary part of the complex refractive index is taken, and z is defined in Figure 5.

Because the imaginary portion of ξ_3 is negative, the field strength decays exponentially into the lower index medium. It is estimated that the penetration depth, d_p, of the evanescent wave is equal to the distance away from the interface where the amplitude of the evanescent electric field strength decays to 1/e of its initial value. This is given by[28]

$$d_p = \frac{\lambda}{2\pi \left(\sin^2\theta - (n_2/n_1)^2 \right)} \quad (4)$$

Equation (4), strictly applied, is for a two-layer system; however, it is frequently used to approximate the penetration depth of three-layer systems where the assumption is made that the metallic film is sufficiently thin so as not to dramatically attenuate the evanescent wave. This is generally thought to be the case if the thickness of the layer is much smaller than the wavelength of the incident light. Equation (4) illustrates that the sampling depth of the evanescent wave in an internal reflection experiment depends explicitly on several factors:

- the refractive indices of the incident (e.g. the waveguide with high refractive index) and final layer (e.g. the electrolyte solution);
- the angle of incidence;
- the wavelength of the incident radiation.

Equation (4) illustrates that, over a broad range of angles, the sampling depth of the internal reflection experiment can probe different total dimensions of the interface. As the angle approaches the critical angle, the penetration depth increases, as expected. This behavior suggests that internal reflection methods may be used to spectroscopically probe interfacial electrolyte concentration gradients at the electrode interface by varying the angle of the incident radiation. Equation (4) also shows that over the mid-IR region of the electromagnetic spectrum, the sampling depth changes dramatically. For example, d_p is calculated to change from 600 nm at 1000 cm^{-1} to only 200 nm at 3000 cm^{-1}. This result suggests that quantitative spectroscopic analysis may be distorted if using spectral features having different wavelengths for analysis, a problem when using FTIR spectroscopy.

3.2 Sampling in Internal Reflection Measurements

Although not as widely applied as other IR spectroelectrochemical methods, several different sampling methods have been reported since the pioneering work of Mark and Pons.[5] The following examines these different methods.

3.2.1 Multiple Internal Reflections

The initial spectroelectrochemical experiments of Mark and Pons,[5] Tallant and Evans,[7] and Reed and Yeager[8] used a multiple reflection IRE for acquiring the IR spectrum. In this arrangement, the Ge waveguide is much longer than it is thick. As a consequence, the incident light reflects multiple times within the Ge working electrode as it propagates down the length of the waveguide. At each reflection of the incident light with the interface between the Ge plate and the electrolyte solution, attenuation of the evanescent wave due to absorbance by species on the electrolyte side of the interface is possible. Consequently, the multiple reflections can lead to greater spectroscopic signal-to-noise for species at the electrode interface.

Ashley et al. demonstrate the utility of multiple internal reflection Fourier transform infrared spectroscopy (MIRFTIRS) for studying material adsorbed onto a metallic interface.[29] Here they investigate the adsorption of SCN^- on silver and gold films by MIRFTIRS, and compare the results to other spectroelectrochemical experiments. Their spectroelectrochemical cell is similar in construction to Mark and Pons'; however, Ashley et al. used a 200-nm thin film of Au or Ag deposited on Si substrates as the working electrode. The sampling depth of the IR radiation in this cell was estimated by Ashley et al. to be approximately of 350 nm for 2000 cm^{-1} incident radiation. This estimate of the sampling depth apparently does not consider the attenuation of the incident radiation that may occur due to the presence of a relatively thick layer of metal between the Si IRE and the electrolyte solution. The thickness of the metal layer is approximately 5% of the wavelength of the incident light in this experiment; consequently, estimating the evanescent wave penetration depth with the expression for a two-layer system is not adequate for this system. The 200 nm metal thin film used by Ashley et al. is approximately 10 times larger than metal thin films used in other internal reflection experiments.[24,30–33] The presence of the thick metal film probably attenuates the penetration depth of the evanescent wave (considering the dependence on h, i.e. the metal film thickness, in the mean evanescent field strength given by Equations 1–3). This probably makes the actual sampling depth in this application somewhat less than this estimate of h. If the sampling depth is assumed to be much smaller than the initial estimate, the spectral results with the thick metal layer correlate with the observation that these measurements detect only SCN^- species adsorbed directly to the metal (e.g. those species within approximately 10 Å of the interface). In this case, the multiple internal reflections provide sufficient sampling of the interface to allow adequate signal-to-noise for observation of the adsorbed species. The evanescent wave attenuation by the thick metal film apparently prevents the observation of species in solution.

Sherson et al. also compared MIRFTIRS with other sampling techniques for IR spectroelectrochemistry.[34] Unlike Ashley et al., they prepared their metal film by patterning a thin metal section (thickness estimated to be approximately 4 nm) down the middle of the IRE surrounded by a thicker metal layer on the IRE periphery (Figure 6). A problem that has been observed with very thin metal layers on IREs is the formation of nonisotropic, nonideal interfacial structures (e.g. island formation) during the deposition step.[30,35] Island formation may result in electronic isolation of portions of the thin metal layer, making electrochemical phenomena inhomogeneous across the surface of the

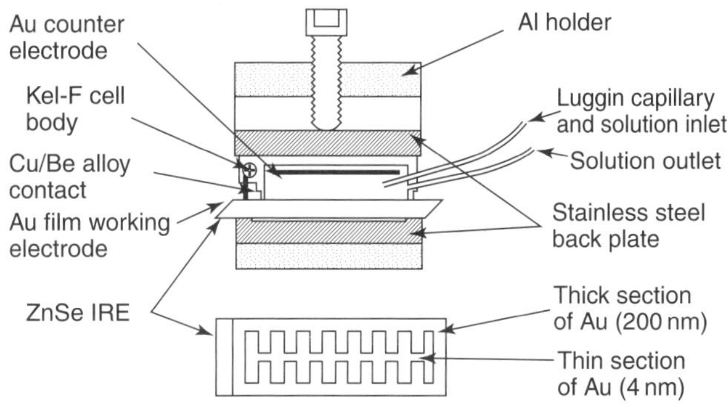

Figure 6 Patterned electrode and spectroelectrochemical cell used by Sherson in MIRFTIRS to study the electrochemical adsorption of SCN^-. (Reprinted by permission from I.T. Bae, M. Sandifer, Y.W. Lee, D.A. Tryk, C.N. Sukenik, D.A. Sherson, 'In Situ Fourier Transform Infrared Spectroscopy of Molecular Adsorbates at Electrode–Electrolyte Interfaces: A Comparison between Internal and External Reflection Modes', *Anal. Chem.*, **67**(24), 4508–4513, copyright (1995) American Chemical Society.)

For references see page 9875

metal. The intent of the patterned electrode was to maintain a thin metallic region for efficient evanescent wave penetration into the solution while also ensuring that the bulk metal film was sufficiently thick to maintain good conductivity for the electrochemistry. This cell demonstrates almost no electrochemical distortion due to resistance effects, and no obvious spectral irregularities from a nonisotropic surface. IR spectra with excellent signal-to-noise that are qualitatively equivalent to those obtained by IRRAS methods are obtained using this spectroelectrochemical cell.

Multiple internal reflection methods demonstrate excellent signal-to-noise characteristics. This is probably due to the many times that the interface is sampled as the incident light propagates down the waveguide. Ashley et al. showed that MIRFTIRS measurements can use thicker metallic films than others predict, probably due to the many times that the interface is sampled. Sherson demonstrated that patterned interfaces can provide continuous electrical contact across extremely thin metallic films. These results demonstrate the applicability of MIRFTIRS for studying interfacial electrochemical phenomena.

3.2.2 Single Reflection Interfaces

Single reflection interfaces have also been used as IREs for IR spectroelectrochemical measurements. A typical spectroelectrochemical cell for a single internal reflection experiment is shown in Figure 7.(36) In these applications, the IRE is a prism rather than a thin waveguide. Many different materials have been used for the prism (e.g. Ge, GaAs, Si, KRS-5, ZnSe, CdTe, BaF_2, CaF_2).(30) Dependent on the optical properties of the prism material and the adjacent solution, it is found that allowable dimensions of the metallic layer for adequate penetration of the evanescent wave through the metal film vary from

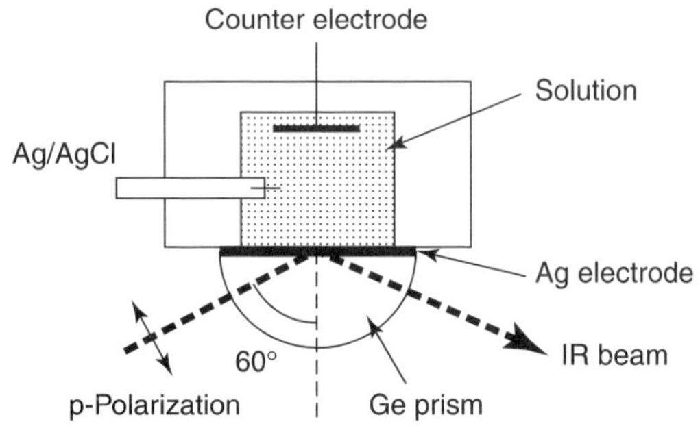

Figure 7 A Typical Kretschmann spectroelectrochemical cell. (Reprinted from M. Osawa, K. Ataka, K. Yoshii, T. Yotsuyanagi, 'Surface-enhanced Infrared ATR Spectroscopy for In Situ Studies of Electrode–Electrolyte Interfaces', *J. Electron Spectrosc. Relat. Phenom.*, **64/65**, 371–379, copyright (1993) with permission from Elsevier Science.)

10 nm up to 40 nm.(30) This dependence of the allowable metal thickness on the optical properties of the prism is evident from Hansen's treatment of the three-layer system (Equations 1–3), yet it introduces complications when trying to compare results that are obtained with different IRE–metal overlayer combinations.

Johnson et al. attempted to rationalize behavioral differences observed based upon the prism–metal film combinations, particularly with regard to the inability of Hansen's treatment to make accurate quantitative predictions for some of these IRE–metal overlayer combinations.(30) They found that the optical properties of the metal films are different from those of the bulk metals, making predictions based on bulk metal properties invalid. They proposed that the origin of these property differences comes from two effects, namely the inhomogeneous nature of the metallic thin film and the interdiffusion of semiconducting metalloids and the metal overlayer. The three-layer model used for theoretical description of the internal reflection behavior at an IRE, as originally developed by Hansen,(25,27) assumes that all layers are isotropic. However, when depositing a thin metal layer for use as an electrode it is known that the metal film is not isotropic.(30,35) Instead, it forms initially as isolated islands that merge into a rough, continuous film as more material is deposited. Because of the inhomogeneities in the dimensions of the metal film, the isotropic model that is generally used does not accurately describe the system.

The role of interdiffusion is clearly demonstrated when comparing spectral results obtained with thin metal films deposited on an ionic reflection element (such as BaF_2) and a semiconducting metal (such as Ge).(30,37) With Ge substrates, diffusion of the Ge into the metal overlayer may occur and change the optical properties of the metal overlayer. With the ionic crystal, interdiffusion of the substrate into the metal overlayer will not occur. When measurements were made with BaF_2 substrates coated with metallic thin films, close agreement between experimental results and Hansen's theory was obtained.(30,35) Agreement between experimental spectral data with the metal layer on a Ge substrate and theoretical predictions was not as close.

The single-reflection arrangement using a prism IRE with a thin metallic layer for IR spectral sampling is generally known as the Kretschmann arrangement, and has been widely applied for studying the structure of materials adsorbed on metallic films. Kretschmann IRE cells possess several characteristics that are potentially useful for spectroelectrochemical applications. The metallic thin film can be used as the working electrode in a spectroelectrochemical experiment, provided that it is thick enough to conduct. The film attenuates the evanescent wave penetration into the electrolyte, serving to further isolate the interface for spectral observation. As demonstrated in MIRFTIRS applications, this provides improved selectivity for adsorbed materials relative to bulk species. Finally, the extremely thin metallic films (generally 10–20 nm thick) typically used in this application are found to enhance the IR spectrum of species adjacent to the metallic layer.[31,32,38]

The enhancement of the IR spectrum of adsorbed material when using the Kretschmann arrangement is generally known as surface-enhanced infrared absorption spectroscopy (SEIRAS), and was first demonstrated by Hartstein et al. in 1980.[38] Like the related technique of surface-enhanced Raman spectroscopy (SERS), SEIRAS has been associated with the roughness of the thin metallic layer. However, unlike SERS, the phenomenon is observed using a variety of metallic substrates including Ag, Au, Cu, In, Pt, and Pd.[30,36]

Osawa et al. have studied applications and the theory of the SEIRAS technique.[35,36,39] They find that the enhancement is influenced by the optical properties of the IR-transparent substrate, the conditions used in depositing the metal film on the substrate, and the overall morphology of the film. Although each of these factors plays a role in the SEIRAS effect, the effects are not independent of each other. For example, when comparing BaF_2 and Ge IRE substrates ($n = 1.42$ and 4.0, respectively), Osawa found the SEIRAS intensity to be larger for the metal films deposited on BaF_2. This result can be partially understood from the optical properties of the materials and the field strength of the evanescent wave in a three-phase system (Equations 1–3). The closer the optical properties of the waveguide and the phase on the opposite side of the metal film, the larger will be the strength of the evanescent electric field. This result corresponds with those of Johnson.[37]

Different substrate–metal combinations generate different metal film morphologies when the film is vapor deposited. Wetting of the substrate surface during deposition refers to the ability of the deposited material to spread over the substrate surface and form an isotropic layer. The ability of a metal vapor to wet the substrate is related to the relative surface energies of the substrate and the deposition metal. In general, metal deposition initially occurs as isolated metal islands that eventually merge together as additional metal is deposited. The amount of metal required before the isolated islands merge into a continuous film depends on the efficiency with which the metal spreads across the substrate. Inefficient wetting of the substrate suggests that the metal will not deposit as a continuous film; rather, it will deposit and grow as discrete ellipsoids. These ellipsoids generate morphology/roughness of the metal film at the interface, a property that has been identified as an important contributor to the SEIRAS effect.[40,41] Different IRE materials have different surface energies; consequently, it is expected that there will be morphological differences between substrates resulting from the vapor deposition of thin metal films.

Osawa et al. report enhancement of the IR spectrum by a factor of 1000 when using a thin metal film made up of discontinuous islands of prolate ellipsoids.[39] In general, the deposition of metals on substrates begins by forming individual islands at very small coverage (5 nm films and smaller), and these islands merge into a continuous film as the thickness of the layer increases. The term "thickness" for a vapor-deposited metallic film is perhaps a misnomer because it is assumed that all the evaporated metal deposits on the substrate to form a uniformly thick film. The film thickness is calculated based upon the rate of thermal evaporation of the source metal and not on the actual thickness of the deposited film. When the individual metallic islands begin to merge, the amount of spectral enhancement decreases sharply (Figure 8).[40] This result is unfortunate for spectroelectrochemical applications because the electrochemistry needs to have a continuous metallic film to ensure uniform conductivity across the surface of the IRE. For spectroelectrochemical applications, therefore, the amount of spectral enhancement does not approach the factor of 1000 reported by Osawa et al. Tenfold sensitivity increases are observed in spectroelectrochemical applications where a continuous film

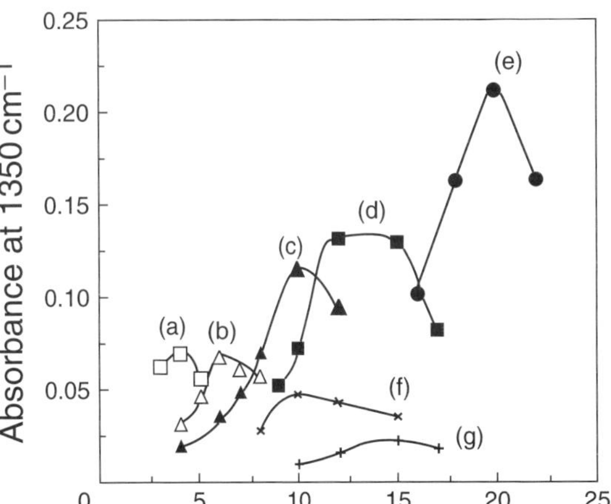

Figure 8 Plot of peak intensity of the 1350 cm^{-1} band as a function of Ag film thickness for different substrates. (a) Glow-discharge treated carbon-BaF_2; (b) untreated carbon-BaF_2; (c–g) bare BaF_2 prepared with different metal evaporation rates: (a–c, f) 0.10 nm s^{-1}, (e) 0.001 nm s^{-1}; and with different substrate temperatures: (a–e) 27 °C, (f, g) 100 °C. (Reprinted from Y. Nishikawa, T. Nagasawa, K. Fujiwara, M. Osawa, 'Silver Island Films for Surface-enhanced Infrared Absorption Spectroscopy: Effect of Island Morphology on the Absorption Enhancement', *Vib. Spectrosc.*, **6**, 43–53, copyright (1993) with permission from Elsevier Science.)

For references see page 9875

is required (the sensitivity enhancement is calculated by comparison to spectra acquired by other IR methods).

Several groups have considered the mechanism of SEIRAS theoretically, with the most detailed descriptions given by Osawa et al.(35,36,39,40) and by Suetaka et al.(31,41,42) Osawa et al. have shown that the enhancement of the IR absorption is general for molecules near small metallic particles regardless of the spectral probe used (e.g. transmission, internal reflection, or specular reflection).(39) Regardless of the method of interrogating the sample, the general premise is that the oscillating field from the incident light induces an oscillating dipole in the metallic particle (illustrated in Figure 9).(39) The induced dipole of the particle generates an electric field with a principal direction perpendicular to the particle surface. Because of the small size of the particle (estimated to be approximately 10–20 nm along the major axis of the ellipsoid), the molecular vibrations of the adsorbed material modulate the polarizability of the metallic particle. Consequently, maximum field enhancement will occur when the system (incident radiation, particle dipole moment change, and vibrational motion of adsorbed material) is in resonance. Calculations by Osawa et al. using this model are in excellent agreement with experimental data.

Suetaka et al. investigated different contributions to the enhancement of the IR absorption.(31,35,41) They found that the thickness of the metal played a significant role in determining the type of enhancement mechanism operating. At low metal thickness (approximately 5 nm), both p- and s-polarized light (that is, light whose electric field is linearly polarized and aligned with the plane of reflection and perpendicular to the plane of reflection, respectively) produced an enhanced IR spectrum. This was attributed to an electromagnetic effect. At higher coverage (200 nm), only the p-polarized light was found to enhance the IR absorption. This observation is important to spectroelectrochemical applications because these generally use thick metal film electrodes. The p-polarized light introduces a directional aspect to the IR experiment that may be used to obtain information about the structure of the molecule adsorbed to the interface. Originally, the thick film behavior was attributed to an excitation of delocalized surface plasmons.(41) Osawa et al. showed later that the surface plasmons could not explain the broad range of incident angles that lead to the observed enhancement.(39) They assert that the polarization dependence of the enhancement with the thick films is likely a function of the continuous nature of the metal layer. That is, the p-polarized light can more efficiently induce polarization changes of the metal film in a direction normal to the macroscopic dimension of the film than can the s-polarized light.

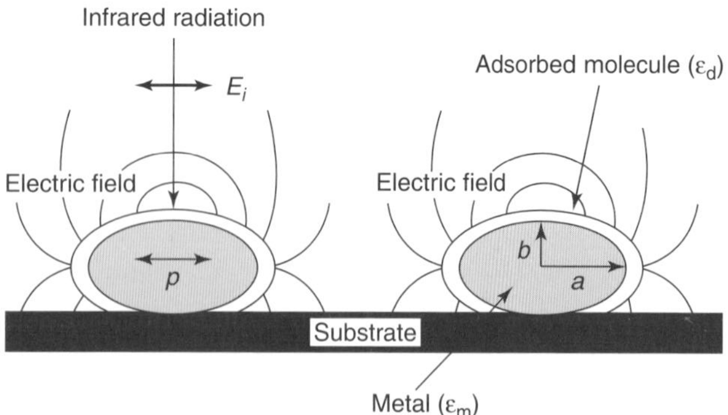

Figure 9 Diagram of the electromagnetic enhancement mechanism for SEIRAS experiments. (Reprinted from M. Osawa, K. Ataka, K. Yoshii, Y. Nishikawa, 'Surface-enhanced Infrared Spectroscopy: The Origin of the Absorption Enhancement and Band Selection Rule in the Infrared Spectra of Molecules Adsorbed on Fine Metal Particles', *Appl. Spectrosc.*, **47**(9), 1497–1502, copyright (1993) with permission from the Society for Applied Spectroscopy.)

For spectroelectrochemical applications, the field enhancement at the metal–particle interface has several benefits. The large magnitude of the electric field at the interface increases the absorbance by material adjacent to the metallic thin film. This leads to a spectral response that is approximately 10 times larger than that obtained by other IR spectroelectrochemical techniques. Hatta et al. showed that the SEIRAS enhancement was dependent upon the incident polarization as well as the thickness of the metallic layer.(41) For thicker metal layers (approximately 20 nm) where the islands have merged into a continuous film, only incident light that is p-polarized generated the enhanced field at the interface. Because the metallic film is continuous in electrochemical applications, and because the incident radiation probes the interface by internal reflection, the field enhancement on the solution side of the interface has a principal direction largely normal to the macroscopic dimension of the IRE interface. The directional aspect of the induced electric field results in selective differentiation of the vibrational modes for molecules adsorbed on the surface. Vibrational modes whose transition dipole moment change is aligned with the enhanced field are observed, whereas those modes whose transition dipole moment change is perpendicular to the field enhancement are not observed. This behavior is generally known as the surface selection rule and provides information about the orientation that adsorbed molecules make with respect to the interface. Inspection of the equations for the evanescent wave field strength given by Hansen (Equations 1–3) shows that the evanescent wave has polarization components both perpendicular and parallel to the plane of reflection. For the internal reflection experiment, therefore, the surface selection rules must arise from the field enhancement due to the SEIRAS effect.

Taking advantage of this polarization dependence of the field enhancement, Hatta et al. demonstrated a polarization modulation method used in conjunction with the Kretschmann configuration.[43] In this method, alternating p- and s-polarized light is reflected from the prism interface. The reflected light is detected and demodulated with a phase-sensitive lock-in amplifier at the frequency of the polarization modulation. Because only the p-polarized light efficiently couples into the continuous metallic film to enhance the IR spectrum of adsorbed material, modulating the incident radiation between p- and s-polarization has the effect of introducing time dependence to the spectral signal. Consequently, lock-in detection improves the signal-to-noise ratio for the IR spectrum obtained in this fashion. Because this polarization effect is present only with continuous metallic films, the polarization modulation technique can easily be applied to electrochemical systems.

The polarization modulation used with the Kretschmann arrangement illustrates a method where modulating the analytical signal improves the signal-to-noise of the experiment. In this application, the polarization modulation relies on the properties of the interface to differentiate the effects measured.

3.3 Dynamic Measurements

Osawa et al. recently demonstrated how SEIRAS measurements using the Kretschmann arrangement could be used to characterize dynamic (e.g. short time-frame) electrochemical phenomena.[44] Typical IR spectroelectrochemical measurements yield spectra under equilibrium conditions; that is, conditions at some time after the potential alteration has occurred and the system current has returned to baseline levels. In making spectral measurements in this fashion, the dynamics of the electrode process are lost. This restriction is largely due to the incompatible time-frames of the electrochemical phenomena (milliseconds) and the IR spectral acquisition time when using FTIR (seconds to minutes).

The advent of step-scan FTIR instrumentation makes dynamic FTIR measurements possible. With this instrument, the moving mirror of the interferometer is not continuously scanned; rather, it is sequentially stepped and held at discrete positions along the scan dimension. In scanning interferometry, each scan of the moving mirror represents the acquisition of one spectrum, and the signal-to-noise of the final spectrum is improved by co-adding many individual spectra. Depending on the scan distance and the mirror velocity, acquiring one spectrum requires some defined time. In a step-scan instrument, the signal is integrated not as a function of the number of scans, but as a function of time at each step position. With this instrument, therefore, the interferogram is built up point-by-point with the signal-to-noise of the final spectrum being related to the dwell time at each mirror position. If used effectively, there is the opportunity to gate an experiment to the individual steps of the interferometer so that time-resolved spectral data might be acquired. In this manner, one may obtain IR spectra at a defined duration after application of some experimental perturbation. For example, if the interferometer mirror position step were synchronized with a change in the applied potential at the working electrode, then it would be possible to acquire IR spectra at some defined time after the potential change. Alternatively, one could use the interferometer mirror steps to initiate a linear potential scan of the working electrode. In this case, one could acquire spectra of the interface at discrete potentials (because time and potential are linearly related in a linear potential scan experiment) during the potential scan.

To conduct these synchronized experiments, the system being studied must exhibit reversible behavior with respect to the perturbation applied. This requirement is easily understood because the experiment must be repeated many times at each mirror position so that the corresponding interferogram can be built up point-by-point. Consider that each mirror position is providing only a portion of the total interferogram. Only after each portion of the interferogram (e.g. at each mirror position) is collected and assembled can the final spectrum be calculated. If the system being studied is not reversible with respect to the experimental perturbation, then the chemical system is different at each mirror position. To obtain a sensible spectrum, then the condition of the sample must be the same at every interval following the mirror position change, requiring that the associated chemistry be reversible and reproducible.

Osawa et al. have demonstrated the utility of step-scan coupled with the Kretschmann arrangement for monitoring electrochemical processes at short times.[44] In this application, they monitor the deposition of heptaviologen radical cation as a function of time after the electrode is stepped from a value where no electrochemistry occurs (-0.2 V) to a value where the mass-transport-limited one-electron reduction of heptaviologen occurs (-0.55 V). After the potential step, spectral data are acquired at discrete times (every 100 µs) at each mirror position. From this spectral data, individual IR spectra of the interface corresponding to every 100 µs after the potential step are constructed (Figure 10). This provides a dynamic spectral picture of the interfacial electrochemical phenomena with time resolution acceptable for following dynamic electrode processes. The combination of step-scan FTIR and the rapid response, low resistance spectroelectrochemical

For references see page 9875

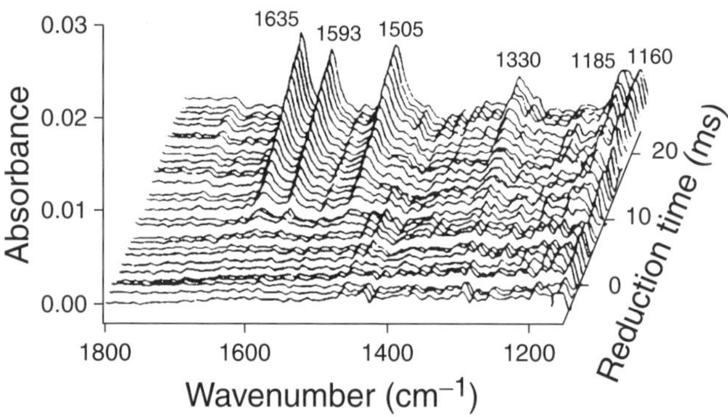

Figure 10 IR spectra acquired using a step-scan FTIR at 100-µs intervals after the application of a potential to the working electrode in a Kretschmann spectroelectrochemical cell. (Reprinted with permission from M. Osawa, K. Yoshii, K. Ataka, T. Yotsuyanagi, 'Real-time Monitoring of Electrochemical Dynamics by Submillisecond Time-resolved Surface-enhanced Infrared Attenuated-total-reflection Spectroscopy', *Langmuir*, **10**(3), 640–642, copyright (1994) American Chemical Society.)

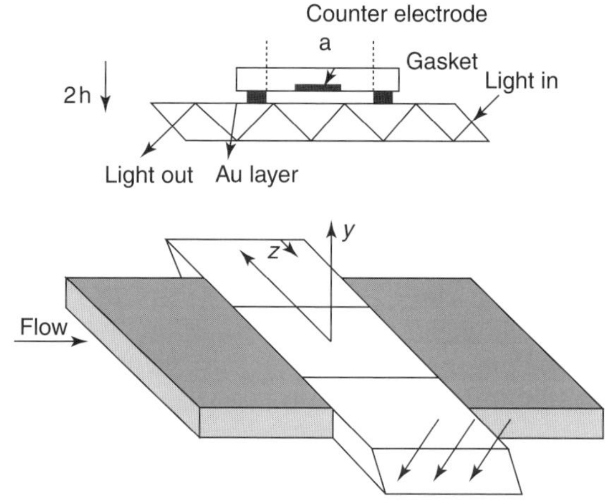

Figure 11 Schematic diagram of a channel-type spectroelectrochemical cell for ATR/FTIR measurements of solution phase electrogenerated species. (2h) is the full height of the cell and (a) is the width of the flow channel. (Reprinted with permission from R. Barbour, Z. Wang, I.T. Bac, Y.V. Tolmachev, D.A. Sherson, 'Channel Flow Cell for Attenuated Total Reflection Fourier Transform Infrared Spectroelectrochemistry', *Anal. Chem.*, **67**(21), 4024–4027, copyright (1995) American Chemical Society.)

cell used in the internal reflection sampling mode provide the instrumental characteristics necessary for this experiment.

3.4 Other Sampling Configurations

Internal reflection IR spectroelectrochemical measurements in which the working electrode is a thin metal film deposited on the IRE are shown to have excellent sensitivity for material adsorbed to the metal thin film. However, the technique is poor at monitoring species present only in the solution phase. Sherson et al. address this shortcoming by using a channel flow electrochemical cell (Figure 11).[34,45] With this cell, the electrochemical process and the spectral monitoring are separated in space. Using a flowing solution through a thin channel, electroactive material undergoes an electrode reaction at the working electrode and then it is swept into the spectroscopic observation region by the flowing solution. The electrochemical response to the flowing stream follows standard steady-state electrochemical behavior, with the flowing stream maintaining the concentration of the electroactive material constant at the diffusion layer boundary.[16] Because the evanescent wave penetrates only a small distance into the solution phase, only species within the diffusion layer are observed spectroscopically. The concentration determined by the spectroscopic measurement is a function of only the down stream distance from the near edge of the working electrode. This method holds great promise for extending the internal reflection sampling geometry for studying solution phase reactants and products of electrode reactions.

4 SPECULAR REFLECTION

The strengths and weaknesses of transmission spectroelectrochemistry with an OTTLE cell and internal reflection spectroelectrochemical measurements are somewhat complementary to each other. The OTTLE cell allows spectral observation of solution phase products and intermediates of electrode reactions. However, this method does not provide any information about material at the solution–electrode interface. Because of the shallow depth of penetration, internal reflection methods are most sensitive to the interface and provide almost no information about material in the bulk solution. Resistance effects limit transmission measurements to static, equilibrium conditions whereas, as demonstrated by Osawa et al. with step-scan FTIR,[44] sampling by internal reflection allows acquisition of spectra during dynamic electrochemical phenomena.

Ideally, a single IR spectroscopic sampling method would be able to provide information about both solution phase and adsorbed materials, as well as information about dynamic electrode processes. One sampling method that may be able to provide this information uses a specular, or external, reflection sampling geometry. In this geometry for spectral sampling, the IR source passes through a small thickness of the bulk solution before and after reflection, sampling the solution phase species. According to Greenler, the phase behavior of reflected light provides an enhanced sensitivity for species at the metallic surface.[9,10] The specular reflection sampling method therefore appears to be able to provide information about both solution phase and surface adsorbed material.

4.1 Spectroelectrochemical Cells for Specular Reflection Experiments

Although specular reflection has the potential to provide more spectral information than either transmission or internal reflection methods, it also has limitations. Primary among these limitations is the required use of a solution thin layer. The thin layer, as in transmission experiments, is required to improve the total throughput of the source radiation through the highly absorbing solvent. As with the OTTLE cell, use of the thin layer creates resistance effects and possible uneven polarization of the working electrode interface, limiting the opportunity for dynamic measurements. Designs of the specular reflection IR spectroelectrochemical cells provide additional flexibility, compared to OTTLE cells, which help to reduce resistance effects and improve the time response of the measurements.

The cell design used by Bewick et al. in the first demonstration of specular reflection IR spectroelectrochemistry is shown in Figure 12.[46] Here the working electrode was a Pt disk mounted on a syringe plunger. The cell body was fashioned from the syringe barrel with the front sealed with an IR-transparent window. When assembled, the working electrode is inserted through the syringe barrel until it is pressed against the IR-transparent window, forming the solution thin layer with a thickness estimated to be approximately 10–20 µm. To complete the electrochemical circuit, the secondary electrode is a Pt coil placed close to the IR-transparent window and surrounding the working electrode. The reference electrode tip is

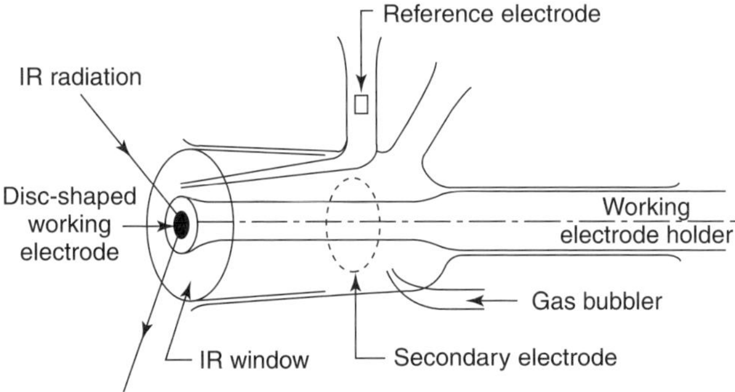

Figure 12 Diagram of the spectroelectrochemical cell used for specular reflection experiments. (Reprinted from A. Bewick, K. Kunimatsu, B.S. Pons, J.W. Russell, 'Electrochemically Modulated Infrared Spectroscopy (EMIRS): Experimental Details', *J. Electroanal. Chem.*, **160**, 47–61, copyright (1984) with permission from Elsevier Science.)

also placed close to the edge of the working electrode using a Luggin capillary. Although the electrochemistry is conducted on a thin layer of solution, in this specular reflection cell design the placement of the secondary and reference electrodes is more ideal than in the OTTLE cell, helping to reduce some of the resistance effects.

The cell design by Bewick et al. is similar to the full-edge current flow thin-layer spectroelectrochemical cell described by Lin and Kadish.[47] Although Lin and Kadish's cell is designed for use in the UV/VIS region of the spectrum, the placement of the secondary electrode surrounding the working electrode and just outside of the thin layer is like the design of Bewick et al's. Similarly, placement of the reference electrode tip just outside the thin layer is similar to the cell of Bewick et al. Lin and Kadish's thin-layer cell has excellent current–voltage response, suggestive of reduced resistance effects (compared to that of a traditional OTTLE cell). Indeed, equivalent circuit analysis suggests that the more ideal arrangement of the electrodes, particularly the reference electrode, make significant contributions toward reducing the resistance effects.[47]

In addition to possible thin-layer solution resistance effects, a problem introduced by this design that should be addressed is the need to carefully align the IR-transparent window with the working electrode. Unlike the OTTLE cell, where the boundaries of the thin layer are held apart and parallel to each other by a Teflon® gasket of known thickness, the specular reflection cell does not define the thin-layer dimension. Consequently, the thin-layer size is difficult to reproduce experiment to experiment. Defining the thin-layer dimensions with partial gaskets at the edge of the working electrode is possible; however, this may increase the resistance of the thin layer by impeding current paths. The IR-transparent window and the surface of the electrode must also be parallel to each other. If they are not parallel, then the thickness of the thin layer will vary along the surface of the working electrode. The irreproducibility of the solution thin layer may result in variation in the spectral signal-to-noise from one experiment to another.

Although the basic design principles used by Bewick et al. in constructing their specular reflection cell have not changed, many alternative designs have appeared. These designs address the sampling issues identified above.

Mosier-Boss et al. describe a cell, similar to that of Bewick et al., that is machined from a glass ceramic material.[48] The arrangement of the working, secondary, and reference electrodes is similar to the glass cell of Bewick et al. By machining the cell body from a ceramic, higher tolerances for the electrode and window alignments are possible. These higher construction tolerances improve the reproducibility for generating the solution thin layer, for ensuring that the IR-transparent window

For references see page 9875

and the working electrode are parallel to each other, and for decreasing the dimension of the thin layer. Using this cell with a 10-μm thin-layer thickness, Mosier-Boss et al. were able to obtain current versus potential responses in cyclic voltammetry experiments that are largely undistorted by resistance effects.(48) This particular current versus potential response is characteristic of semi-infinite diffusion, not thin-layer electrochemistry. At these potential scan rates, the 10-μm dimension of the thin layer is much larger than the diffusion layer thickness. This cell can respond quickly to changing potentials (e.g. potential scans of $100\,mV\,s^{-1}$) and is not subject to high thin-layer resistance effects; consequently, it is not limited to static, equilibrium electrochemical measurements and could be used in dynamic electrochemical measurements. This result illustrates the effectiveness of using more ideal electrode placements on reducing resistance effects within the thin layer.

Reproducing the thin-layer thickness is a difficult, yet important, parameter to the specular reflection spectroelectrochemistry experiments. Frequently, one needs to collect spectral data, pull the electrode away from the window to allow replenishment of the thin layer, and then re-establish the thin layer for additional spectroscopic measurements. If the thin-layer thickness is not reproduced, spectra acquired before and after the thin-layer disruption cannot be reliably compared with each other. Several spectroelectrochemical cells have been designed in which the working electrode is mounted on a micrometer.(49–51) The micrometer allows exact and reproducible placement of the working electrode surface based on the micrometer setting.

Zhang and Lin sought to reduce even further the resistance within the thin layer by using an electrode microarray as their working electrode.(52) Here, rather than having a continuous disk, they use a 3×3 array of 1-mm diameter disks (Figure 13) as their working electrode. In this design, nine closely spaced, full-edge current-flow thin-layer cells(47) with dimensions much smaller than that of a single disk electrode with an equivalent surface area act as a single working electrode. Using equivalent circuit analysis, they show that the resistance of this microarray is one-ninth the resistance of a single electrode with an equivalent surface area. Likewise, the time-constant of the thin-layer cell using the microarray working electrode will improve by a factor of nine compared to the single disk. These results suggest that a fast electrochemical response would be expected when using this microarray in a thin-layer cell. Indeed, when cyclic voltammetry experiments are conducted with the microarray in the thin-layer arrangement, negligible distortion of the current versus potential curve is observed. Importantly, peak current heights obtained with the microarray electrode in the thin-layer arrangement are the same as that obtained under bulk solution conditions. This result indicates that, under the thin-layer condition, the entire surface area of the microarray electrode is available for electrochemical reaction and is participating in the reaction. If the electrode were unevenly polarized, one would expect that only a portion of the electrode surface area would be involved in the electrochemical reaction, and the observed peak currents would be reduced or the current–voltage behavior distorted compared to the bulk electrochemical behavior. This electrochemical result is important because the reflection IR measurement samples the entire surface area of the electrode. To ensure that the spectroscopy results properly represent the electrochemical phenomena, the electrochemical process must be homogeneous across the entire spectral sampling area (e.g. generally the entire electrode surface area in a specular reflection experiment).

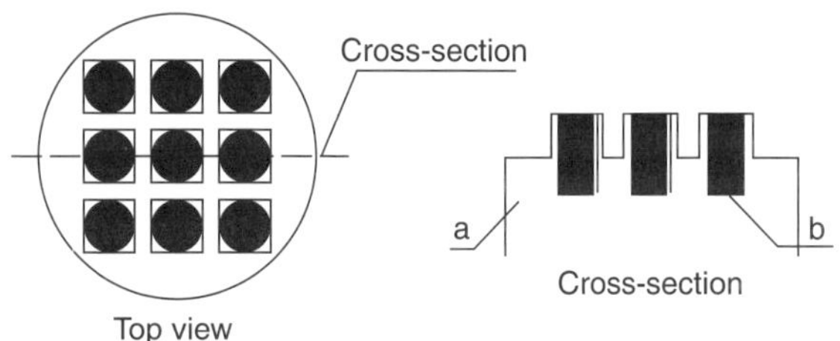

Figure 13 Diagram of the 3×3 microarray working electrode; (a) is a Teflon® rod and (b) are the embedded 1-mm Pt wires. (Reprinted from H.Q. Zhang, X.Q. Lin, 'In Situ FTIR Spectroelectrochemistry of a Microarray Electrode', *J. Electroanal. Chem.*, **434**, 55–59, copyright (1997) with permission from Elsevier Science.)

Li and Lin describe a reflection IR spectroelectrochemical cell that uses a 10-μm Pt disk as the working electrode.(53) The advantage of using a microelectrode is the small influence that resistance effects have on microelectrode applications. The small surface area of the working electrode generates very small faradaic currents (nanoamperes to picoamperes). With small currents, even a large thin-layer resistance will have only small influence on potential polarization of the working electrode; consequently, microelectrodes are frequently used in high-resistance applications. This microelectrode cell, therefore, should demonstrate little uneven polarization of the working electrode and have a fast time response. As with the OTTLE thin-layer cells designed by Hartl et al.,(20,54) this microspectroelectrochemical cell places all three electrodes within the solution thin layer (Figure 14). In this arrangement, electrochemistry and spectroscopy measurements could be made on less than $20\,\mu L$ of solution, making this a useful sampling configuration when the sample size is limited. Because of the small dimension of the working electrode, this cell required the tight focus

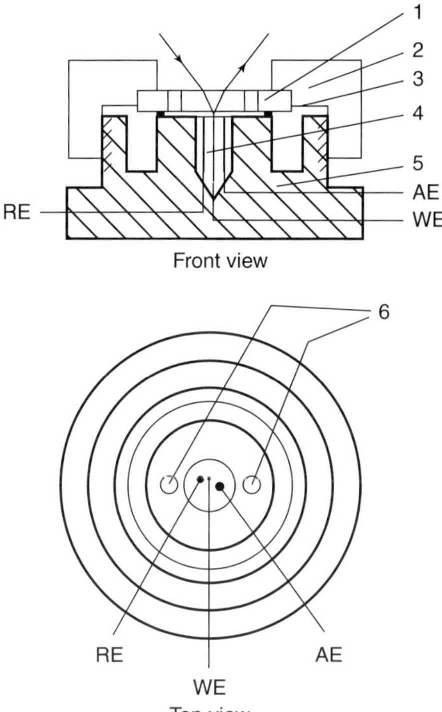

Figure 14 Diagram of the thin-layer microelectrode reflection cell where (1) is the IR optical window, (2) is the Teflon® cell top, (3) is a ring spacer, (4) is an epoxy adhesive, (5) is the Teflon® cell body, (6) are the inlet and outlet, WE is the working electrode, RE is the reference electrode and AE is the auxiliary electrode. (Reprinted from Z. Li, X. Lin, 'Electrochemistry and In Situ IR Microspectroelectrochemistry in a Versatile Thin Layer Reflection Fourier Transform IR Microspectroelectrochemical Cell', *J. Electroanal. Chem.*, **386**, 83–87, copyright (1995) with permission from Elsevier Science.)

of the IR beam offered by an IR microscope to conduct the spectroelectrochemical experiment.

Although each of these specular reflection cell designs requires the use of a thin layer of solution, all have demonstrated resistance characteristics that are much improved compared to that of a typical OTTLE thin-layer cell. This result suggests that the time response for specular reflection experiments will be sufficient for measurements of dynamic electrochemical phenomena. A rapid electrochemical response is important to specular reflection experiments because electrode potential modulation is one technique that is frequently utilized to improve the signal-to-noise ratio of the spectroelectrochemical measurement.

4.2 Potential Modulation

In specular reflection IR spectroelectrochemistry, one is trying to detect chemical species that are trapped within a solution thin layer, or are present at the electrode–electrolyte interface. Generally this is a very small amount of material, particularly in comparison with the amount of solvent and supporting electrolyte also present in the thin layer. Bewick demonstrated that the signal-to-noise for electrochemically active solutes could be substantially improved by introducing a time variation to the IR spectroelectrochemical experiment.(55) In this first application of specular reflection IR spectroelectrochemistry, a 10-Hz potential modulation was applied to the working electrode. The potential modulation created a continuous oscillation of the identity of the chemical species present within the solution thin layer that was monitored spectroscopically. In many respects, this procedure is analogous to using an optical chopper to introduce a time dependence on the electromagnetic radiation as it arrives at the detector. With the potential modulation, the radiation is being turned on and off at a defined frequency because the interaction between the sample (e.g. the solution thin layer) and the light changes in phase with the potential modulation. The output of the IR detector is directed to a lock-in amplifier that demodulates the detector signal at the frequency of the potential modulation. Consequently, the final output from the lock-in detection is representative of the changes in the IR spectrum of the interface that are in phase with the electrochemical perturbation.

With the potential modulation, the phase-sensitive detection will detect only spectral changes that occur at the frequency of the potential changes. This aspect of the measurement has several important implications to spectroelectrochemical measurements made in this manner. Because detection is at the frequency of the potential modulation, the acquired spectra result from the interfacial reflectance differences at the two potentials of the modulation. The magnitude of the reflectance spectrum is normally given by Equation (5):

$$\frac{\Delta R}{R} = \frac{R_2(\nu) - R_1(\nu)}{R_1(\nu)} \tag{5}$$

where $R_1(\nu)$ is the interfacial reflectance at applied potential E_1 and $R_2(\nu)$ is the reflectance at potential E_2. The reflectance spectrum is related to the absorbance by Equation (6):

$$A = -\log\left(\frac{\Delta R}{R} - 1\right) \tag{6}$$

As a difference spectrum, the appearance is not the same as an absolute spectrum of the thin layer at any given potential. Only those vibrational features that undergo some change caused by the changing value of the applied potential have the same time variance as the potential modulation; therefore, they are the only features that are detected in the IR spectrum. As a result, the baseline of the measurement is at $\Delta R/R = 0$. This reflectance

For references see page 9875

baseline suggests that, for most of the spectral region, there are no changes in the sample brought about by the potential alteration. Considering the spectral contributions from the solvent over most of the available potential window for a given solvent–electrolyte combination, the vibrational features of the solvent will be invariant with the value of the applied potential. Consequently, the solvent contributions to the IR spectrum are effectively cancelled out by measuring the difference spectrum.

A second implication of the potential modulation spectrum is that the final reflectance spectrum will have features with both positive and negative values. The positive and negative features of the difference spectrum arise from changes in the thin-layer reflectances brought about by the two potentials used in the modulation. To illustrate this concept, consider a potential modulation about the standard potential for the oxidation/reduction of some solute. At one potential (E_1), the oxidized form is favored within the thin layer. At the other potential (E_2), the reduced form is favored. From the $\Delta R/R$ expression (Equation 5), it can be seen that the difference spectrum will have negative-pointing features corresponding to vibrations characteristic of the oxidized form and positive features characteristic of the reduced form. Correlation of the observed spectral features with the potential allows diagnosis of mechanistic aspects of the electrochemical response of the system under investigation. Unlike transmission spectroelectrochemical measurements with the OTTLE cell, the potential changes do not have to convert all of the material within the thin layer. The potential modulation need only create enough of a time-dependent difference in the vibrational modes of species within the thin layer to be observed by the IR spectroscopy. How much of a change is enough depends on the absorptivities of the individual species.

Only those portions of the molecule whose vibrational features are altered by the potential changes are observed. For complex molecules, therefore, the potential modulation method offers the opportunity to identify the site of oxidation/reduction within the molecule (e.g. by observing which vibrational features are most influenced by the potential changes). Although redox conversion of electroactive solutes represents an obvious application of the potential modulation, other more subtle changes brought about by the potential changes may also be observed. Other example applications of processes associated with a changing applied potential include: studies of potential-induced adsorption or desorption of molecules;(56) orientational changes of molecules adsorbed to the electrode interface;(57,58) changes in the type or strength of adsorption interactions at the interface;(59) changes brought about by the changing electric field strength at the interface;(60) and ion migration into or out of the thin layer.(58,61)

4.2.1 *Electrochemically Modulated Infrared Spectroscopy*

The initial demonstration of specular reflection IR spectroelectrochemistry by Bewick et al. used a continuous 10 Hz potential modulation and a high-throughput dispersive-IR spectrometer.(55) In this case, the potential modulation was much faster than the rate of wavelength scanning, providing improved signal-to-noise for the measurement. They demonstrated the utility of this method by modulating the oxidation state of thianthrene, and by spectroscopically monitoring the potential-induced adsorption/desorption of indole.(55) The method, which they called electrochemically modulated infrared spectroscopy (EMIRS), provided improved signal-to-noise by having relatively long dwell times at each resolution element of the dispersive spectrometer, and by introducing the time dependence of the spectral signal.

Shortly after its initial demonstration, the EMIRS technique was used to study the electrosorption of methanol.(59) Modulating the electrode potential at 8.5 Hz, two IR features were observed. The first, at $1850\,cm^{-1}$, was attributed to an adsorbed CO at a bridging site, and a complex bipolar band centered at approximately $2070\,cm^{-1}$ was attributed to adsorbed CO bonded to a linear site. The bipolar nature of the band at $2070\,cm^{-1}$ is representative of the changing interfacial properties as the potential is altered. In this case, changes in the IR spectrum of the adsorbed CO with potential alterations provide a probe of the interfacial properties. The shift in the absorption position of the linearly bonded CO illustrates the subtle changes that arise from the potential modulation that can be monitored by EMIRS.

The power of EMIRS measurements comes from the ability to measure subtle changes within the thin layer that are brought about by the potential change. Although other techniques use the EMIRS method with a dispersive spectrometer, FTIR instruments offer signal-to-noise advantages beyond those obtained by using a potential modulation. When using a scanning FTIR instrument, however, the method of spectral data collection must be modified from that used with EMIRS.

4.2.2 *Potential Difference Infrared Spectroscopy*

With scanning FTIR instruments, a continuous 10-Hz potential modulation is not useful because the sample must be at equilibrium during each interferometer scan. However, potential difference infrared spectroscopy (PDIRS) may still be acquired by synchronizing the potential alteration with the interferometer scans. Two general methods have been used.

In the first potential difference FTIR technique, a small number of interferometer scans are collected at one applied potential, then the potential is changed to a second value and a second set of interferometer data collected. The electrode potential is returned to the first value and this sequence is repeated until a sufficient total number of interferometer scans are collected and co-added at each potential value. The interferograms for each potential value are then Fourier transformed, and the spectrum corresponding to the first potential is ratioed against the spectrum obtained at the second potential and then normalized. In this sequence, the potential is modulated between two values at a frequency that depends on the number of interferometer scans and the individual interferometer scan times. The difference spectra acquired by this technique are equivalent to those obtained by the EMIRS method. As with EMIRS, because the potential is altered between two values repeatedly, the electrochemical response must be reversible with respect to those potential values.

If the electrode response is not reversible with respect to the potential, difference spectra can still be measured, but spectra at individual potentials must be acquired in sequence. Here, the potential is systematically altered and individual spectra are acquired at each applied potential before measuring spectra at the second, third, fourth, etc. potentials. The catalytic decomposition of small organic molecules on Pt has been extensively studied in this manner.[62–64] After Fourier transformation, any two of these spectra may be ratioed against each other to obtain spectral information regarding changes that occur in the system between these different applied potentials.

An alternative strategy, used by Weaver et al., is to collective spectra as a very slow potential ramp is applied to the working electrode.[63–65] In this manner, IR spectra are acquired during different time blocks along the potential ramp. Assuming that the spectral acquisition is fast enough (depending on resolution, and interferometer frequency) and/or the potential scan slow enough, during the individual blocks of time the electrode potential changes only slightly. Using this technique, one can obtain a measure of the interfacial dynamics during a continuous, albeit slow, potential scan.

With the exception of the slow potential ramp method used by Weaver et al., continuous-scan FTIR instrumentation requires that spectra be acquired with the chemical system at equilibrium. Consequently, spectral information about fast, dynamic electrochemical phenomena cannot be studied using PDIRS. As described for internal reflection, however, step-scan FTIR instruments provide the opportunity to study dynamic phenomena on relatively short time-frames.

4.2.3 *Electrochemically Modulated Infrared Reflection Spectroscopy by Step-scan Fourier Transform Infrared*

Griffiths et al. have demonstrated the utility of step-scan methods for obtaining spectral information about dynamic electrochemical processes using specular reflection sampling.[13–15] One application studied the potential dependence of the IR spectrum of adsorbed CO. CO adsorbed on Pt has been extensively studied, providing a wealth of literature for comparison. They found that the signal-to-noise in their step-scan measurements was comparable to data collected by other EMIRS and PDIRS techniques. In addition, the opportunity to conduct a double modulation experiment with the step-scan instrument allowed Griffiths et al. to identify and compensate for the slow time response of the thin-layer cell – effectively using the spectral data and computational methods to correct for the resistance within the thin layer. Although not a critical parameter in these measurements, the opportunity to use the step-scan instrument to determine the level of resistance effects on the electrochemical response is a unique application of the spectral data. As step-scan instrumentation becomes more widely available this may become an important method if quantitative dynamic information is to be extracted from the measurements (e.g. reaction kinetics).

Griffiths et al. also demonstrated that EMIRS experiments using step-scan FTIR can be used to differentiate between species in the bulk solution and those confined to the interface.[14,15] When demodulating the spectral response due to the continuous potential modulation, the signal may be separated into in-phase and quadrature components. From these components, a magnitude and a phase spectrum are calculated. When compared, the magnitude spectrum indicates when absorption of the source radiation is occurring and the phase spectrum indicates whether the absorption arises from species in an isotropic (e.g. solution) or a nonisotropic (e.g. interface) environment. Knowing the phase behavior of solution species allows the separation of all of the solution vibrational bands into one of the detection channels (e.g. the quadrature channel), and all the other IR absorptions arising from interfacial species in the other detection channel.[14,15] This method provides an opportunity to determine both solution and interfacial species during the same spectroelectrochemical experiment using phase-selective detection. In this case, the contributions from the solution and the interface are separated from each other in the final spectrum.

The combination of the more ideal electrode geometry to reduce the thin-layer resistance, along with the ability to measure and compensate for electrode time delay, suggests that step-scan instrumentation with specular

For references see page 9875

reflection sampling will be a powerful method for studying the dynamics of electrode processes. Although the Griffiths' method clearly distinguishes solution and interfacial species, alternative methods for separating the interfacial contributions from the solution contributions are also available.

4.3 Polarization Effects

Greenler in 1966 demonstrated that specular reflection IR spectroscopy could be used to obtain spectra of thin organic films on metallic substrates.(9) Although not initially applied to spectroelectrochemistry, this work established the fundamental principles that govern the experimental aspects of specular reflection IR spectroelectrochemistry today. As with internal reflection methods, the behavior of IR light on reflection from a metallic substrate is determined by solving the boundary value problem for reflection with Fresnel's equations. This calculation has been conducted for many different systems of relevance to spectroelectrochemical applications, and are not reproduced here.(66,67) Only the principal results are discussed, in terms of their influence on sampling and instrumentation in the spectroelectrochemical experiment.

Greenler showed that the mean feld strength for the incident radiation maximizes when the IR radiation is incident on the substrate at glancing angles.(9) Quantitatively, the mean field strength is calculated by Greenler to be nearly 5000 times larger when the incident light reflects from the surface at 88°, compared to the value obtained for a normal-incidence reflection. In addition, Greenler showed that the sensitivity of the reflection experiment is approximately 25 times larger than that for a transmission experiment using a free-standing film of the same dimension. These calculations were for a thin isotropic organic layer on top of a metallic substrate and exposed to an air atmosphere. Although these results illustrate the utility of the specular reflection technique, they do not realistically model electrochemical systems.

Faguy and Fawcett reproduced Greenler's calculations assuming a multilayer model that more closely resembles an electrochemical system.(66) Their calculations not only accounted for many of the electrochemical parameters, they also considered real instrumental limitations of the spectroelectrochemical experiment; that is, parameters such as the angular divergence of the focused IR beam and the beam focus size were included in the model. In this model, a thin organic layer exists on top of the metallic substrate. The ambient atmosphere is a thin solution layer of some defined thickness (6 μm of acetonitrile was used in their calculations). On top of the solution layer is an initial layer that represents the IR-transparent window (e.g. CaF_2). This four-layer model allows calculation of the mean electric field strength at the metal interface using conditions that closely approximate the electrochemical experiment.

A result of particular interest from this work was the optimum angle of incidence for maximum field strength at the interface calculated by Faguy and Fawcett for this electrochemical system.(66) They found that the optimum angle of incidence for maximum reflectance for these electrochemical systems was not as glancing as reported by Greenler (Figure 15).(66) When the angular divergence due to a focusing beam of incident IR radiation is considered, the optimum incident angle is reduced even further. Faguy and Fowcett's more realistic electrochemical model also found that the mean field strength is reduced at the interface, compared to values calculated by Greenler. This result indicates that the reflectance for these multilayer systems will be smaller than predicted by Greenler's model. These results are important because a glancing angle of incidence is a nearly impossible sampling geometry for spectroelectrochemical cells designed for external reflection experiments to attain. With a finite thickness of the solution thin-layer and refraction effects at the boundary between the window and the solution, experimental incident angles of 60–65° are more common. Although Faguy and Fawcett's calculations predict lower reflectances for real systems, they also show that the angle of incidence typically used in spectroelectrochemical measurements is not as debilitating as expected from Greenler's work.

In his calculations, Greenler also showed that on reflection from the metallic substrate the phase behavior

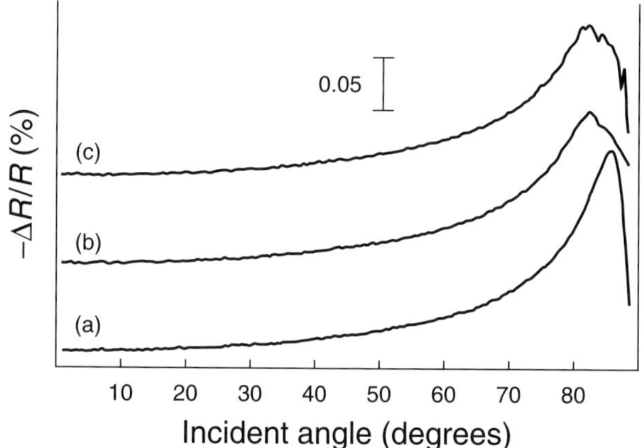

Figure 15 Calculated $-\Delta R/R$ versus incident angle at 2350 cm^{-1} for an adsorbed layer thickness of 0.5 mm (a) with noise injected, (b) as with (a) but with Gaussian averaging for angular divergence and (c) as with (b) but with beam attenuation. (Reprinted from P.W. Faguy, W.R. Fawcett, 'Infrared Reflection–Absorption Spectroscopy of the Electrode–Electrolyte Solution Interface: Optical Considerations', *Appl. Spectrosc.*, **44**(8), 1309–1316, copyright (1990) with permission from the Society for Applied Spectroscopy.)

of the incident light depended on the polarization of the incident light.[9] Light that is polarized perpendicular to the plane of reflection, commonly called s-polarized light, has a near 180° phase shift at the reflection plane for all angles of incidence. Consequently, the incident field strength from the reflected light with perpendicular polarization goes to zero at the interface. Light which is polarized parallel to the plane of reflection, commonly called p-polarized light, has a phase shift that depends on the angle of incidence. For glancing angles, this phase shift is nearly 90°, resulting in an electric field that oscillates normal to the substrate. The result of this phase behavior is that, on reflection, the incident radiation generates a field at the interface that is oriented normal to the metallic substrate. This polarization dependence at the interface generates what has commonly been referred to as the surface selection rules. That is, only those vibrational motions for confined species with a component of their transition dipole moment change that lies perpendicular to the substrate are observed. This is due to the electric field from the incident radiation being present only in a direction normal to the substrate. There are a number of important experimental consequences that come from this result.

Discrimination between solution-phase material and adsorbed species can be obtained by comparing spectra collected using polarized incident radiation. When the incident light is fixed so that its electric field is oriented parallel to the plane of incidence (p-polarized), the resultant spectrum contains information about species present in both the bulk solution and those confined to the interface. Repeating the experiment with the incident light polarized perpendicular to the plane of reflection results in spectra that contain information only about species in the bulk solution. These results are due to the directional aspect of the interfacial electric field. Species present in the solution are randomly oriented and can interact equally with the p- and s-polarized light. When confined to the interface, species can only interact with incident radiation that has p-polarization. Comparison of spectra, obtained under identical conditions with the exception of the polarization of incident light, allows identification of those vibrational modes that arise from confined versus solution-phase material. Foley et al. demonstrated this effect when investigating the potential induced adsorption of SCN^- by specular reflection PDIRS.[56]

For systems in which only species confined to the interface are studied, the spectral signal-to-noise ratio may be increased by a factor of two by using incident radiation that is only p-polarized. This is understood by considering the behavior of p- and s-polarized light at the interface. p-Polarized light is able to interact with the confined material; consequently, it will possess spectral information as well as random noise. Incident light that is s-polarized has no interaction with species confined to the electrochemical interface. Consequently, it will contain no spectral information about the confined species, but will carry random noise. By filtering the s-polarized light, 50% of the noise contribution has been eliminated without affecting the spectral signal.

Evaluation of the reflection IR spectrum of a material confined to the interface relative to its isotropic bulk spectrum can also provide information about the orientation/structure of the molecule at the interface. Although not an electrochemical application, Allara and Nuzzo illustrate the principle of this analysis when characterizing the structure of long-chain surfactants adsorbed on metallic substrates.[68] From the results of a Kramers–Kronig evaluation of an IR spectrum of a randomly distributed isotropic sample, a theoretical reflection spectrum of the sample can be obtained. The calculated reflection spectrum assumes that the adsorbed layer is homogeneous, isotropic and has the same structure as found in the bulk. If the interfacial structure is anisotropic, the experimental reflection spectrum will differ from the theoretical spectrum by the intensity of the vibrational modes. These intensity differences arise from the anisotropy of the sample and the directional aspect of the incident field strength at the interface. Comparison of this calculated, theoretical spectrum with the actual experimental reflection spectrum allows evaluation of the average orientation/structure of the molecules when they are confined to the interface.

4.3.1 Reflection Losses

Although polarization behavior on reflection from the electrode surface provides valuable information regarding surface-confined species, care must be taken to ensure that the experiment does not introduce polarization artifacts. Just as the two polarizations have different reflectivities from the electrode, they also reflect with different efficiencies from other surfaces. It is possible, therefore, that the optical path of the instrument may introduce some differential polarization of the source radiation. For example, the different polarizations of light may reflect with different efficiencies from the front surface of the IR-transparent window of the spectroelectrochemical cell. For a flat disk-shaped window, the amount of light that is transmitted through the window depends on the incident polarization as well as the angle of incidence. When using a glancing angle, light polarized perpendicular to the reflection plane reflects from the window and is not as efficiently transmitted through to probe the electrochemical interface as is light polarized parallel to the reflection plane. Using a trapezoidal or hemicylindrical prism as the window may circumvent

For references see page 9875

this problem. With prism windows, the IR light has normal incidence at the ambient–window interface and each polarization is transmitted with equal efficiency. Differential polarization induced by the instrument or sample cell is of particular concern when using a continuous polarization modulation for spectral sampling.

4.3.2 Polarization Modulation

Like potential modulation, one can use a continuous polarization modulation to effectively cancel out those portions of the spectrum that are invariant to the polarization of light. Unlike potential modulation, however, the incident light polarization may be modulated at a frequency high enough that this modulation method may be coupled to scanning FTIR spectral data acquisition. Conceptually, polarization modulation takes advantage of the surface selection rules to differentiate between the interphase and the bulk. Species in the bulk solution are equally likely to interact with p- and s-polarized light because their vibrational motions are randomly oriented in space and not confined to a defined direction over the time of the experiment. Due to the phase behavior at the reflection, interfacial material only interacts with the p-polarized light. If the reflected light is detected at the frequency of the polarization modulation, the resultant reflection spectrum is given by Equation (7),

$$-\frac{\Delta R}{R} = \frac{R_p - R_s}{R_p + R_s} \tag{7}$$

where R_p is the sample reflectance obtained with the p-polarized incident radiation and R_s is the sample reflectance obtained with the s-polarized radiation. It is obvious from this equation that species that have equal interaction with both p- and s-polarized light (e.g. bulk-phase material) will cancel in the final reflectance spectrum, whereas those with differential reflectivity will be observed. Consequently, the final reflectance spectrum will contain spectral information about only the interfacial region.

Typically the phasing behavior is thought to predominate in the region of space extending a distance approximately one-quarter of a wavelength away from the substrate into the solution. In the mid-IR region of the spectrum, therefore, the polarization modulation experiment samples approximately the first 1.5–2.5 µm into the solution away from the substrate. For a well-aligned spectroelectrochemical cell, this sampling dimension can be up to 50% of the total thin-layer thickness. Although sampling by polarization modulation does not discriminate entirely against solvent in the thin layer, it does isolate a smaller region of the interphase. This results in an overall enhancement of the spectral signal-to-noise. This provides the possibility, under the correct experimental conditions, of monitoring by IR spectroscopy the diffusion behavior at the interface. Combining polarization modulation with a potential modulation scheme, as Kunimatsu et al. demonstrate,[49] provides excellent signal-to-noise for monitoring the potential-dependent behavior of adsorbed molecules.

There have been several experimental demonstrations of the utility of polarization modulation to electrochemical systems. Kunimatsu et al. used the technique to study the adsorption of CO to Pt electrodes.[49] Anderson and Gatin investigated the structural behavior of self-assembled monolayers when potential is applied to the substrate.[57] Saez and Corn investigated the structure of polymer films deposited on electrodes as a function of electrode potential.[69]

4.3.3 Photoelastic Modulator

To conduct the polarization modulation experiment, the polarization of the incident radiation must be rapidly alternated between the linear p- and s-polarization extremes. The frequency of this polarization modulation must be fast relative to the frequency of the interferometer frequency in order to separate the signal modulation due to the polarization from that due to the scanning interferometer. This separation of the two time-dependent signals is possible if the light polarization is modulated at a high frequency relative to the frequency of the interferometer. An instrument called a photoelastic modulator provides this high-frequency dynamic change in the incident light polarization state.

Construction of the photoelastic modulator was first described in 1966.[70] Hipps and Crosby have described the principle of the dynamic polarization of light by photoelastic modulation.[71] The optically active portion of the photoelastic modulator is an isotropic crystal of some material transparent in the spectral region of interest. For IR measurements, instruments that use CaF_2 and ZnSe crystals are commercially available. A periodic stress is applied to the isotropic crystal with a physical transducer, generally a piezoelectric material bonded to the stress axis of the IR-transparent crystal (Figure 16).[72] This mechanical stress induces a strain in the crystal along this axis. The strain in the crystal induces a variation of the optic ellipsoid of the crystal material at a resonance frequency.

As seen in Figure 16, the incident radiation is linearly polarized prior to passing through the photoelastic modulator. The electric field of the incident linear polarized light is aligned 45° away from the crystal strain axis. The strain is driven sinusoidally at the resonance frequency of the crystal, so the refractive index of the crystal along this axis varies sinusoidally as well.

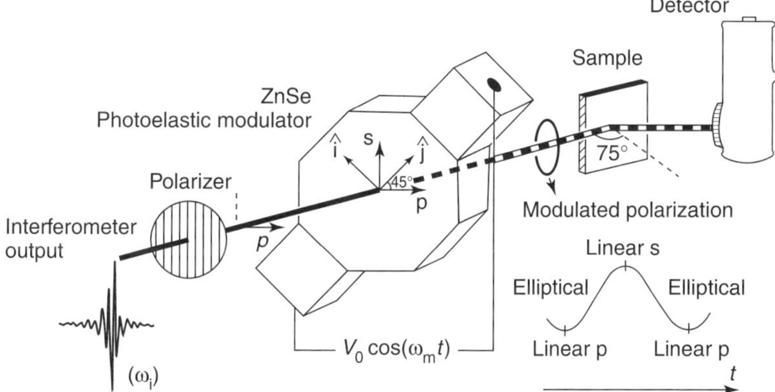

Figure 16 Diagram of the optical path used in the polarization modulation FTIR experiment. Illustration of the linear to elliptical polarization change as the linearly polarized light passes through the photoelastic modulator. (Reprinted from T. Buffeteau, B. Desbat, J.M. Turlet, 'Polarization Modulation FTIR Spectroscopy of Surfaces and Ultra-thin Films: Experimental Procedure and Quantitative Analysis', *Appl. Spectros.*, **45**(3), 380–389, copyright (1991) with permission from the Society for Applied Spectroscopy.)

The refractive index change creates a time dependent phase shift of the incident light, which is given by Equation (8),

$$\phi = \phi_0 \cos(wt) \tag{8}$$

where ϕ_0 is the maximum phase shift (180°), w is the resonant frequency, and t is time. When linearly polarized light passes through the photoelastic modulator crystal, the light emerges elliptically polarized due to the strain-induced phase shift. Dependent on when during the stress–strain cycle the light emerges, the polarization of the emerging light changes between linear polarization and different degrees of elliptical polarization. If the maximum phase shift (ϕ_0) is 180°, then during a complete strain cycle the polarization of the light emerging from the modulator goes through two complete cycles of linear p- to linear s-polarized light (which are 90° out of phase from each other).

The intensity of the light that reaches the detector after passing through the photoelastic modulator and reflecting from the metallic substrate is given by Equation (9):[72]

$$I_d = C\frac{I_o(w)}{2}\{(R_p + R_s) - 2(R_p - R_s)J_2(\phi_0)\cos(2wt)\} \tag{9}$$

where C is a constant, $I_o(w)$ is the light intensity as it enters the photoelastic modulator, $J_2(\phi_0)$ is a second-order Bessel function, and all other terms have their normal meaning. The second-order Bessel function originates from the wavelength dependence of the peak efficiency from the polarization modulation. Strictly, the photoelastic modulator will only modulate the polarization of the light between the linear p- and s-polarized extremes for a single wavelength. This wavelength is determined by the potential used to drive the strain of the modulator crystal. That is, the maximum phase change, ϕ_0, is given by Equation (10):[71]

$$\phi_0 = \frac{GV}{\lambda} \tag{10}$$

where G is a constant value, V is the strain driving voltage, and λ is the wavelength of peak polarization retardation. For a given drive voltage, therefore, only one wavelength will have the maximum phase shift. For all other wavelengths, the polarization states during the mechanical strain cycle are elliptical.

Seki et al. investigated this wavelength dependence of the polarization state and its influence on distorting the measured IR spectrum.[49] Although the polarization modulation is only accurate for a single wavelength, they found that the elliptical light does not dramatically distort the measured spectrum in the wavelength region surrounding modulation wavelength. For example, if the retardation of the photoelastic modulator is set at $2100\,\mathrm{cm^{-1}}$, the multiplication factor from the Bessel function varies from 1 (at $2100\,\mathrm{cm^{-1}}$) to 0.8 (at $1450\,\mathrm{cm^{-1}}$ and $2600\,\mathrm{cm^{-1}}$) within the range $1450–2600\,\mathrm{cm^{-1}}$. Over this wavenumber region with this limited distortion, the Bessel function can be approximated by a parabola, fit to the baseline, and then subtracted from the polarization modulation spectrum.

4.3.4 *Detection of Polarization Modulation/Fourier Transform/Infrared Reflection–Absorption Spectroscopy*

As stated before, the different polarizations of light will reflect from surfaces with different efficiencies. In the polarization modulation experiment, after the source radiation passes through the photoelastic modulator it

For references see page 9875

should reflect from only one surface (e.g. the surface being characterized). This ensures that the spectral information encoded by the differential reflectivity at the interface of interest is not mixed with polarization effects at other reflections. Consequently, lenses are used to collect and focus the light after reflecting from the interface of interest in the polarization modulation experiment. In cases where mirrors are used after the sample reflection, a plate of some IR-transparent material (e.g. KBr) is placed into the optical path and at an angle relative to the direction of light propagation.[73] This plate offsets any polarization discrimination due to reflection differences from the optics of the instrument. Although this optical compensation works to balance polarization differentiation that occurs within the optics of the instrument, it is inefficient because it throws away some of the source radiation.

Polarization modulation reflection–absorption spectroscopy using FTIR instrumentation is a double modulation experiment. As such, the frequency components of these two modulations need to be sufficiently disparate so that they may be effectively separated by lock-in detection. A typical IR photoelastic modulator operates at a frequency of 74 kHz. The spectral information from an FTIR instrument operating under normal conditions contains frequencies ranging from a few hertz to tens of kilohertz. The upper range of the spectral frequencies is too close to the frequency of the polarization modulation by the photoelastic modulator for efficient separation by lock-in detection. Lowering the scan speed of the interferometer increases the difference between the spectral frequency and the polarization modulation frequency to allow lock-in detection. The ultimate decrease in the interferometer frequency is accomplished using a step-scan interferometer where the discrete steps of the interferometer moving mirror reduce the instrument frequency to essentially zero.[15] Highly sensitive detectors, such as HgCdTe and InSb, operate most efficiently when there are fast changes of the IR light intensity incident on them. Decreasing the interferometer scan speed lowers the efficiency of these detectors. Lowering the scan speed of the interferometer also lowers the stability of the interferometer and increases the level of noise in the measured spectrum. Although lowering the interferometer scan rate is often required for detection with a lock-in amplifier, it does come at the cost of lower spectral signal-to-noise. To efficiently collect the double modulation spectra with a continuous scan FTIR instrument, the rise time of the lock-in amplifier must be short, requiring a short time constant of the lock-in amplifier instrument (below 1 ms). This requirement places additional constraints on the instrumentation required for these measurements. Nevertheless, polarization modulation FTIR methods using lock-in detection have been demonstrated, and have been used in several spectroelectrochemical applications.[49,57,69]

Corn et al. developed electronics that allow real-time sampling for the acquisition of FTIR spectra during the double modulation experiment.[74,75] Unlike demodulating the frequency components of the experiment with a lock-in amplifier, these electronics do not require that the two modulation frequencies have a large difference. The instrument provides an accurate separation of the high-frequency component of the polarization modulation from the slower spectral frequency of the interferometer. Consequently, the interferometer can be operated using normal scan rates during the spectroelectrochemical experiment. This improves the overall signal-to-noise of the polarization modulation experiment because the interferometer is stable at the higher scan rates, and because the IR detectors operate more efficiently at the higher scan rates. Several spectroelectrochemical applications of the real-time sampling electronics have appeared. Richmond et al. acquired spectrochemical data for the adsorption of thiocyanate, imidazole, and glucose using real-time sampling electronics with the polarization modulation method as well as a static polarization technique. They found that the spectra were virtually identical, but that the real-time electronics provided better rejection of the solvent and better overall signal-to-noise.[76] Real-time sampling of polarization modulation FTIR experiments has also been used to monitor water and hydroxide vibrations associated with the interface.[77] This is an extremely challenging spectroscopic measurement because of the presence of water as the bulk solvent in the thin layer. The properties of the polarization modulation experiment provide the sensitivity and ability to discriminate against the bulk required to make these types of measurements.

5 SUMMARY

As illustrated throughout this article, the combination of IR spectroscopy and electrochemistry is an area of active research interest. The two experiments provide complementary information. Conducting the two measurements simultaneously provides insight about the chemical system that could not be obtained otherwise. The principal issues faced when conducting these simultaneous determinations are spectral sampling of the electrochemical system, and the overall signal-to-noise of the measurement. The three dominant sampling methods have unique strengths and weaknesses. Interestingly, these three methods tend to be complementary to each other, each having strengths in different areas. The type of information desired from the spectroelectrochemical experiment therefore dictates the sampling method used.

Sensitivity advances in the field of IR spectroelectrochemistry have followed improvements in the instrumentation available for the measurements, and this is likely to continue to drive the area. Modulation techniques introduce a time-variant signal that allows use of phase-sensitive detection methods to improve the sensitivity of the measurement. Polarization modulation methods help to isolate the interface, enhancing the sensitivity for double-layer phenomena. The availability of step-scan FTIR instrumentation allows use of a wider range of modulation frequencies that were not previously available for spectroelectrochemical experiments with traditional scanning FTIR instrumentation. These developments have made accessible for spectroelectrochemical measurement those phenomena that occur at the electrode interface and at short times after initiating the electrochemical experiment. New challenges continue to appear that can benefit from IR spectroelectrochemical measurements.

ACKNOWLEDGMENTS

M.R.A. would like to acknowledge Dr Marilyn Gatin, Dr John Roush, Dr Jimin Huang, Susanne Dana, Minhui Zhang, C. Douglas Taylor, Alison Grieshaber, Joshua Joseph, Mark Scalf, Cynthia Kraft, Richard Anderson, and Michaiah Parker for the many contributions that they have made to our research group over the years.

ABBREVIATIONS AND ACRONYMS

CGE	Conducting Glass Electrode
EMIRS	Electrochemically Modulated Infrared Spectroscopy
FTIR	Fourier Transform Infrared
IR	Infrared
IRE	Internal Reflection Element
IRRAS	Infrared Reflection–Absorption Spectroscopy
MIRFTIRS	Multiple Internal Reflection Fourier Transform Infrared Spectroscopy
OTTLE	Optically Transparent Thin-layer Electrode
PDIRS	Potential Difference Infrared Spectroscopy
SEIRAS	Surface-enhanced Infrared Absorption Spectroscopy
SERS	Surface-enhanced Raman Spectroscopy
TBAF	Tetra-*n*-butylammonium Hexafluorophosphate
UV/VIS	Ultraviolet/Visible

RELATED ARTICLES

Coatings **(Volume 2)**
Infrared and Raman Spectroscopy and Imaging in Coatings Analysis

Surfaces **(Volume 10)**
Infrared and Raman Spectroscopy in Analysis of Surfaces

Electroanalytical Methods **(Volume 11)**
Electroanalytical Methods: Introduction • Microbalance, Electrochemical Quartz Crystal • Selective Electrode Coatings for Electroanalysis • Self-assembled Monolayers on Electrodes • Surface Analysis for Electrochemistry: Ultrahigh Vacuum Techniques • Ultraviolet/Visible Spectroelectrochemistry

Infrared Spectroscopy **(Volume 12)**
Infrared Spectroscopy: Introduction • Infrared Reflection–Absorption Spectroscopy

REFERENCES

1. R. Kuwana, R.K. Darlinton, D.W. Leedy, 'Electrochemical Studies Using Conducting Glass Indicator Electrodes', *Anal. Chem.*, **36**(10), 2023–2025 (1964).
2. W.N. Hansen, T. Kuwana, R.A. Osteryoung, 'Observation of Electrode–Solution Interface by Means of Internal Reflection Spectroscopy', *Anal. Chem.*, **38**(13), 1810–1821 (1966).
3. R.W. Murray, W.R. Heineman, G.W. O'Dom, 'An Optically Transparent Thin-layer Electrochemical Cell', *Anal. Chem.*, **39**(13), 1666–1668 (1967).
4. A. Yildiz, P.T. Kissinger, C.N. Reilley, 'Evaluation of an Improved Thin-layer Electrode', *Anal. Chem.*, **40**(7), 1018–1024 (1968).
5. H.B. Mark, B.S. Pons, 'An In Situ Spectrophotometric Method for Observing the Infrared Spectra of Species at the Electrode Surface During Electrolysis', *Anal. Chem.*, **38**(1), 119–121 (1966).
6. W.R. Heineman, J.N. Burnett, R.W. Murray, 'Optically Transparent Thin-layer Electrodes: Ninhydrin Reduction in an Infrared Transparent Cell', *Anal. Chem.*, **40**(13), 1974–1978 (1968).
7. D.R. Tallant, D.H. Evans, 'Application of Infrared Internal Reflection Spectrometry to Studies of the Electrochemical Reduction of Carbonyl Compounds', *Anal. Chem.*, **41**(6), 835–838 (1969).
8. A.H. Reed, E. Yeager, 'Infrared Internal Reflexion Studies of the Germanium/Electrolyte Interface', *Electrochim. Acta*, **15**, 1345–1354 (1970).
9. R.G. Greenler, 'Infrared Study of Adsorbed Molecules on Metal Surfaces by Reflection Techniques', *J. Chem. Phys.*, **44**(1), 310–315 (1966).

List of selected abbreviations appears in Volume 15

10. R.G. Greenler, 'Reflection Method for Obtaining the Infrared Spectrum of a Thin Layer on a Metal Surface', *J. Chem. Phys.*, **50**(5), 1963–1968 (1969).
11. S.A. Borman, 'Fourier Transform IR: Are the Older Grating Instruments Going the Way of the Dinosaurs?', *Anal. Chem.*, **55**(11), 1054A–1056A (1983).
12. R.A. Palmer, J.L. Chao, R.M. Dittmar, V.G. Gregoriou, S.E. Plunkett, 'Investigation of Time-dependent Phenomena by Use of Step-scan FTIR', *Appl. Spectrosc.*, **47**(9), 1297–1310 (1993).
13. B. Budevska, P.R. Griffiths, 'Step-scan FTIR External Reflection Spectrometry of the Electrode–Electrolyte Interface by Potential Modulation', *Anal. Chem.*, **65**(21), 2963–2971 (1993).
14. C.M. Pharr, P.R. Griffiths, 'Infrared Spectroelectrochemical Analysis of Adsorbed Hexacyanoferrate Species Formed During Potential Cycling in the Ferrocyanide/Ferricyanide Redox Couple', *Anal. Chem.*, **69**(22), 4673–4679 (1997).
15. C.M. Pharr, P.R. Griffiths, 'Step-scan FTIR Spectroelectrochemical Analysis of Surface and Solution Species in the Ferricyanide/Ferrocyanide Redox Couple', *Anal. Chem.*, **69**(22), 4665–4672 (1997).
16. A.J. Bard, L.R. Faulkner, *Electrochemical Methods: Fundamentals and Applications*, John Wiley and Sons, New York, 1980.
17. A.T. Hubbard, F.C. Anson, 'The Theory and Practice of Electrochemistry with Thin Layer Cells', in *Electroanalytical Chemistry*, ed. A.J. Bard, Marcel Dekker, New York, 129–214, Vol. 4, 1970.
18. W.R. Heineman, F.M. Hawkridge, H.N. Blount, 'Spectroelectrochemistry at Optically Transparent Electrodes. II. Electrodes Under Thin-layer and Semi-infinite Diffusion Conditions and Indirect Coulometric Titrations', in *Electroanalytical Chemistry*, ed. A.J. Bard, Marcel Dekker, New York, 1–113, Vol. 13, 1984.
19. F.M. Hawkridge, 'Electrochemical Cells', in *Laboratory Techniques in Electroanalytical Chemistry*, eds. P.T. Kissinger, W.R. Heineman, Marcel Dekker, New York, 267–291, 1996.
20. M. Krejcik, M. Danek, F. Hartl, 'Simple Construction of an Infrared Optically Transparent Thin-layer Electrochemical Cell: Applications to the Redox Reactions of Ferrocene, $Mn_2(CO)_{10}$ and $Mn(CO)_3$(3,5-di-*t*-butylcatecholate)', *J. Electroanal. Chem.*, **317**, 179–187 (1991).
21. B.J. Brisdon, S.K. Enger, M.J. Weaver, R.A. Walton, 'Infrared Spectroelectrochemistry of η^3-Allyl Dicarbonyl Complexes of Molybdenum(II) and Tungsten(II)', *Inorg. Chem.*, **26**, 3340–3344 (1987).
22. J.P. Bullock, D.C. Boyd, K.R. Mann, 'In Situ Infrared Spectroelectrochemistry Studies of Ferrocene, $[Rh_2(dimen)_2(dppm)_2](PF_6)_2$, and $(mes)Cr(CO)_3$(dimen = 1,8-Diisocyanomenthane; dppm = Bis(diphenylphosphinomethane); mes = Mesitylene). A Useful Technique for the Characterization of Electrochemically Generated Organometallic Species', *Inorg. Chem.*, **26**, 3084–3086 (1987).
23. W.A. Nevin, A.B.P. Lever, 'Reusable Thin-layer Spectroelectrochemical Cell for Nonaqueous Solvent Systems', *Anal. Chem.*, **60**, 727–730 (1988).
24. I.T. Bae, M. Sandifer, Y.W. Lee, D.A. Tryk, C.N. Sukenik, D.A. Sherson, 'In Situ Fourier Transform Infrared Spectroscopy of Molecular Adsorbates at Electrode–Electrolyte Interfaces: A Comparison between Internal and External Reflection Modes', *Anal. Chem.*, **67**(24), 4508–4513 (1995).
25. W.N. Hansen, 'Internal Reflection Spectroscopy in Electrochemistry', in *Advances in Electrochemistry and Electrochemical Engineering*, eds. P. Delahay, C.W. Tobias, John Wiley & Sons, New York, 1–60, Vol. 9, 1973.
26. N. Harrick, *Internal Reflection Spectroscopy*, Interscience Publishers, New York, 1967.
27. W.N. Hansen, 'Electric Fields Produced by the Propagation of Plane Coherent Electromagnetic Radiation in a Stratified Medium', *J. Opt. Soc. Am.*, **58**(3), 380–390 (1968).
28. B.D. MacCraith, 'Enhanced Evanescent Wave Sensors Based on Sol-gel-derived Porous Glass Coating', *Sens. Actuators, B*, **11**, 29–34 (1993).
29. D.B. Parry, J.M. Harris, K. Ashley, 'Multiple Internal Reflection Fourier Transform Infrared Spectroscopic Studies of Thiocyanate Adsorption on Silver and Gold', *Langmuir*, **6**(1), 209–217 (1990).
30. B.W. Johnson, J. Bauhofer, K. Doblhofer, B. Pettinger, 'Practical Considerations of the IR Attenuated Total Reflection (IR-ATR) Technique for Electrochemical Investigations', *Electrochim. Acta*, **37**(12), 2321–2329 (1992).
31. A. Hatta, T. Ohshima, W. Suëtaka, 'Observation of the Enhanced Infrared Absorption of *p*-Nitrobenzoate on Ag Island Films with an ATR Technique', *Appl. Phys. A*, **29**, 71–75 (1982).
32. A. Hatta, Y. Chiba, W. Suëtaka, 'Infrared Absorption Study of Adsorbed Species at Metal/Water Interfaces by use of the Kretschmann Configuration', *Surf. Sci.*, **158**, 616–623 (1985).
33. T. Wadayama, O. Suzuki, Y. Suzuki, A. Hatta, 'Infrared Absorption Enhancement of *p*-Cyanobenzoic Acid on Silver Island Films Deposited on Oxidized and Hydrogen-terminated Si(100)', *Appl. Phys. A*, **64**(501), 506 (1997).
34. R. Barbour, Z. Wang, I.T. Bae, Y.V. Tolmachev, D.A. Sherson, 'Channel Flow Cell for Attenuated Total Reflection Fourier Transform Infrared Spectroelectrochemistry', *Anal. Chem.*, **67**(21), 4024–4027 (1995).
35. M. Osawa, M. Kuramitsu, A. Hatta, W. Suëtaka, 'Electromagnetic Effect in Enhanced Infrared Absorption of Adsorbed Molecules on Thin Metal Films', *Surf. Sci.*, **175**, L787–L793 (1986).
36. M. Osawa, K. Ataka, K. Yoshii, T. Yotsuyanagi, 'Surface-enhanced Infrared ATR Spectroscopy for In Situ Studies

of Electrode–Electrolyte Interfaces', *J. Electron Spectrosc. Relat. Phenom.*, **64/65**, 371–379 (1993).
37. B.W. Johnson, K. Doblhofer, 'Barium Fluoride as an Internal Reflection Element of IR-ATR Spectroelectrochemistry', *Electrochim. Acta*, **38**(5), 695–701 (1993).
38. A. Hartstein, J.R. Kirtley, J.C. Tsang, 'Enhancement of the Infrared Absorption from Molecular Monolayers with Thin Metal Overlayers', *Phys. Rev. Lett.*, **45**(3), 201–204 (1980).
39. M. Osawa, K. Ataka, K. Yoshii, Y. Nishikawa, 'Surface-enhanced Infrared Spectroscopy: The Origin of the Absorption Enhancement and Band Selection Rule in the Infrared Spectra of Molecules Adsorbed on Fine Metal Particles', *Appl. Spectrosc.*, **47**(9), 1497–1502 (1993).
40. Y. Nishikawa, T. Nagasawa, K. Fujiwara, M. Osawa, 'Silver Island Films for Surface-enhanced Infrared Absorption Spectroscopy: Effect of Island Morphology on the Absorption Enhancement', *Vib. Spectrosc.*, **6**, 43–53 (1993).
41. A. Hatta, Y. Suzuki, W. Suëtaka, 'Infrared Absorption Enhancement of Monolayer Species on Thin Evaporated Ag Films by Use of a Kretschmann Configuration: Evidence for Two Types of Enhanced Surface Electric Fields', *Appl. Phys. A*, **35**, 135–140 (1984).
42. T. Wadayama, T. Sakurai, S. Ichikawa, W. Suëtaka, 'Charge-transfer Enhancement in Infrared Absorption of Thiocyanate Ions Adsorbed on a Gold Electrode in the Kretschmann ATR Configuration', *Surf. Sci.*, **198**, L359–L364 (1988).
43. A. Hatta, Y. Sasaki, W. Suëtaka, 'Polarization Modulation Infrared Spectroscopic Measurements of Thiocyanate and Cyanide at the Silver Electrode/Aqueous Electrolyte Interface by Means of Kretschmann's ATR Prism Configuration', *J. Electroanal. Chem.*, **215**, 93–102 (1986).
44. M. Osawa, K. Yoshii, K. Ataka, T. Yotsuyanagi, 'Real-time Monitoring of Electrochemical Dynamics by Submillisecond Time-resolved Surface-enhanced Infrared Attenuated Total Reflection Spectroscopy', *Langmuir*, **10**(3), 640–642 (1994).
45. Y.V. Tolmachev, Z. Wang, D.A. Sherson, 'Theoretical Aspects of Laminar Flow in a Channel-type Electrochemical Cell as Applied to In Situ Attenuated Total Reflection/Infrared Spectroscopy', *J. Electrochem. Soc.*, **143**(10), 3160–3166 (1996).
46. A. Bewick, K. Kunimatsu, B.S. Pons, J.W. Russell, 'Electrochemically Modulated Infrared Spectroscopy (EMIRS): Experimental Details', *J. Electroanal. Chem.*, **160**, 47–61 (1984).
47. X.Q. Lin, K.M. Kadish, 'Characteristics of a Full Edge Current Flow Thin-layer Electrochemical Cell that Uses Both Internal (Real) and External (Auxillary) Reference Points', *Anal. Chem.*, **58**(7), 1493–1497 (1986).
48. P.a. Mosier-Boss, R. Newbery, S. Szpak, S.H. Lieberman, 'Versatile, Low-volume, Thin-layer Cell for In Situ Spectroelectrochemistry', *Anal. Chem.*, **68**(18), 3277–3282 (1996).
49. H. Seki, K. Kunimatsu, W.G. Golden, 'A Thin-layer Electrochemical Cell for Infrared Spectroscopic Measurements of the Electrode–Electrolyte Interface', *Appl. Spectrosc.*, **39**(3), 437–443 (1985).
50. D.S. Bethune, A.C. Luntz, J.K. Sass, D.K. Roe, 'Optical Analysis of Thin-layer Electrochemical Cells for Infrared Spectroscopy of Adsorbates', *Surf. Sci.*, **197**, 44–66 (1988).
51. D.K. Roe, J.K. Sass, D.S. Bethune, A.C. Luntz, 'Prospects for Transient IR Reflection–Absorption Spectroscopy of Adsorbed Species on Electrode Surfaces: Cell Design for a Laser Source', *J. Electroanal. Chem.*, **216**, 293–301 (1987).
52. H.Q. Zhang, X.Q. Lin, 'In Situ FTIR Spectroelectrochemistry of a Microarray Electrode', *J. Electroanal. Chem.*, **434**, 55–59 (1997).
53. Z. Li, X. Lin, 'Electrochemistry and In-situ IR Microspectroelectrochemistry in a Versatile Thin Layer Reflection Fourier Transform IR Microspectroelectrochemical Cell', *J. Electroanal. Chem.*, **386**, 83–87 (1995).
54. F. Hartl, H. Luyten, H.A. Nieuwenhuis, G.C. Shoemaker, 'Versatile Cryostated Optically Transparent Thin-layer Electrochemical (OTTLE) Cell for Variable-temperature UV/VIS IR Spectroelectrochemical Studies', *Appl. Spectrosc.*, **48**(12), 1522–1528 (1994).
55. A. Bewick, K. Kunimatsu, B.S. Pons, 'Infrared Spectroscopy of the Electrode–Electrolyte Interphase', *Electrochim. Acta*, **25**, 465–468 (1980).
56. J.K. Foley, B.S. Pons, J.J. Smith, 'Fourier Transform Infrared Spectroelectrochemical Studies of Anodic Processes in Thiocyanate Solutions', *Langmuir*, **1**(6), 697–701 (1985).
57. M.R. Anderson, M. Gatin, 'Effects of Applied Potential upon the In Situ Structure of Self-assembled Monolayers on Gold Electrodes', *Langmuir*, **10**(6), 1638–1641 (1994).
58. F.C. Nart, T. Iwasita, 'On the Adsorption of $H_2PO_4^-$ and H_3PO_4 on Platinum: An In Situ Study', *Electrochim. Acta*, **37**(3), 385–391 (1992).
59. B. Beden, C. Lamy, A. Bewick, K. Kunimatsu, 'Electrosorption of Methanol on a Platinum Electrode. IR Spectroscopic Evidence for Adsorbed CO Species', *J. Electroanal. Chem.*, **121**, 343–347 (1981).
60. M.R. Anderson, J. Huang, 'The Influence of Cation Size upon the Infrared Spectrum of Carbon Monoxide Adsorbed on Platinum Electrodes', *J. Electroanal. Chem.*, **318**, 335–347 (1991).
61. V.B. Paulissen, C. Korzeniewski, 'Vibrational Analysis of Interfacial Phosphate Equilibria', *J. Electroanal. Chem.*, **290**, 181–189 (1990).
62. T. Iwasita, F.C. Nart, B. Lopez, W. Vielstich, 'On the Study of Adsorbed Species at Platinum from Methanol, Formic Acid, and Reduced Carbon Dioxide via In Situ FTIR Spectroscopy', *Electrochim. Acta*, **37**(12), 2361–2367 (1992).

63. D.S. Corrigan, M.J. Weaver, 'Adsorption and Oxidation of Benzoic Acid, Benzoate, and Cyanate at Gold and Platinum Electrodes as Probed by Potential-difference Infrared Spectroscopy', *Langmuir*, **4**(3), 599–606 (1988).
64. D.S. Corrigan, M.J. Weaver, 'Mechanism of Formic Acid, Methanol, and Corbon Monoxide Electrooxidation at Platinum as Examined by Single Potential Alteration Infrared Spectroscopy', *J. Electroanal. Chem.*, **241**, 143–162 (1988).
65. L.H. Leung, M.J. Weaver, 'Reactant Structural Effects on the Formation and Electro-oxidation of Adsorbed Carbon Monoxide from Small Organic Molecules at Platinum as Studied by Time-resolved FTIR Spectroscopy', *J. Electroanal. Chem.*, **240**, 341–348 (1988).
66. P.W. Faguy, W.R. Fawcett, 'Infrared Reflection–Absorption Spectroscopy of the Electrode–Electrolyte Solution Interface: Optical Considerations', *Appl. Spectrosc.*, **44**(8), 1309–1316 (1990).
67. D.D. Popenoe, S.M. Stole, M.D. Porter, 'Optical Considerations for Infrared Reflection Spectroscopic Analysis in the C–H Stretching Region of Monolayer Films at an Aqueous/Metal Interface', *Appl. Spectrosc.*, **46**(1), 79–87 (1992).
68. D.L. Allara, R.G. Nuzzo, 'Spontaneously Organized Molecular Assemblies. 2. Quantitative Infrared Spectroscopic Determination of Equilibrium Structures of Solution-adsorbed *n*-Alkanoic Acids on an Oxidized Aluminum Surface', *Langmuir*, **1**, 52–66 (1985).
69. E.I. Saez, R.M. Corn, 'In Situ Polarization Modulation/Fourier Transform Infrared Spectroelectrochemistry of Phenazine and Phenothiazine Dye Films at Polycrystalline Gold Electrodes', *Electrochim. Acta*, **38**(12), 1619–1625 (1993).
70. S.N. Jasperson, S.E. Schnatterly, 'An Improved Method for High Reflectivity Ellipsometry Based on a New Polarization Modulation Technique', *Rev. Sci. Instrum.*, **40**(6), 761–767 (1969).
71. K.W. Hipps, G.A. Crosby, 'Applications of the Photoelastic Modulator to Polarization Spectroscopy', *J. Phys. Chem.*, **83**(5), 555–562 (1979).
72. T. Buffeteau, B. Desbat, J.M. Turlet, 'Polarization Modulation FTIR Spectroscopy of Surfaces and Ultra-thin Films: Experimental Procedure and Quantitative Analysis', *Appl. Spectrosc.*, **45**(3), 380–389 (1991).
73. W.G. Golden, D.S. Dunn, J. Overend, 'A Method for Measuring Infrared Reflection–Absorption Spectra of Molecules Adsorbed on Low-area Surfaces at Monolayer and Submonolayer Concentrations', *J. Catal.*, **71**, 395–404 (1981).
74. M.J. Green, B.J. Barner, R.M. Corn, 'Real-time Sampling Electronics for Double Modulation Experiments with Fourier Transform Infrared Spectrometers', *Rev. Sci. Instrum.*, **62**(6), 1426–1430 (1991).
75. B.J. Barner, M.J. Green, E. Sáez, R.M. Corn, 'Polarization Modulation Fourier Transform Infrared Reflectance Measurements of Thin Films and Monolayers at Metal Surfaces Utilizing Real-time Sampling Electronics', *Anal. Chem.*, **63**(1), 55–60 (1991).
76. W.N. Richmond, P.W. Faguy, R.S. Jackson, S.C. Weibel, 'Comparison Between Real-time Polarization Modulation and Static Linear Polarization for In Situ Infrared Spectroscopy at Electrode Surfaces', *Anal. Chem.*, **68**(4), 621–628 (1996).
77. P.W. Faguy, W.N. Richmond, 'Real-time Polarization Modulation Infrared Spectroscopy Applied to the Study of Water and Hydroxide Ions at Electrode Surfaces', *J. Electroanal. Chem.*, **410**, 109–113 (1996).

Ion-selective Electrodes: Fundamentals

Vaneica Young
University of Florida, Gainesville, USA

This article begins with an introduction to ion-selective electrodes (ISEs). The classification protocol based on the most recent International Union of Pure and Applied Chemistry (IUPAC) recommendations is given. The historical development of ISEs is summarized, starting with the pH electrode. For the inorganic cations and anions, a periodic table "time-line" is given. Recent history has seen the rise of polymer membrane ISEs, which are important in clinical analysis. The historical section concludes with a brief look at these ISEs. The general theory of the potential generation process is explained, and then aspects of the potential generation process that are specific for the different classes of ISEs. A short review of the instrumentation needed to make the measurements is given, followed by an overview of the latest commercially available instrumentation. One aspect of ISE research is the development of selective probes for determining the analyte content of individual cells. This requires miniaturization, and this has been achieved by the development of microelectrode ISEs. An overview of this area is given. Finally, some recent developments that represent the directions of research in the ISE field are given. Included are the development of sensors for patient monitoring, the development of ISE sensor arrays, and the development of more sophisticated electronic and mathematical methods of data analysis.

1 INTRODUCTION

Potentiometry is a static, interfacial electroanalytical technique that has found acceptance across a wide range of disciplines. This is due in large part to the development of ISEs. In the most recent recommendations of IUPAC, an ISE is defined as follows:[1]

> This is an electrochemical sensor, based on thin films or selective membranes as recognition elements, and is an electrochemical half-cell equivalent to other half-cells of zeroth (inert metal in a redox electrolyte), 1st, 2nd and 3rd kinds. These devices are distinct from systems that involve redox reactions (electrodes of zeroth, 1st, 2nd and 3rd kinds), although they often contain a 2nd kind electrode as the 'inner' or 'internal' reference electrode. The potential difference response has, as its principal component, the Gibbs energy change associated with permselective mass transfer (by ion exchange, solvent extraction or some other mechanism) across a phase boundary.

As indicated by this definition, one surface of the thin film or membrane is in contact with a bulk phase whose composition may be varied. The "heart" of an ISE is its thin film or membrane. Thus, much of the research in this area has involved a search for or the fabrication of thin films or membranes that are highly selective for a single ion. The ideal thin film or membrane is one that is specific for a single ion. The actual thin film or membrane, designed to target a single ion, called the analyte, is selective for a small set of ions. The ions left in the set after the analyte ion is removed are called electrode interferences. For example, the lithia-based glass membrane pH electrode targets H^+, but is selective for $\{H^+, Li^+, Na^+\}$. Its electrode interferences are $\{Li^+, Na^+\}$. For the glass membrane pH electrodes, the electrode interferences are serious problems only when they are at fairly high concentrations. This is the exception, rather than the rule.

Electrode interferences are by no means the only interferences that one encounters in working with these devices. Chemical interferences may be of equal or even greater importance than electrode interferences. Chemical interferences are species that either tie up the analyte ion or that change the membrane so that it can no longer effectively interact with the analyte. These types of interferences are very dependent on the type of active material constituting the membrane. Thus a second area of research in the field of ISEs has involved identifying various interferences and quantifying their effects on an ISE's signal, since interferences affect how and when various ISEs may be used.

ISEs are now being exploited as detectors in flow injection analysis systems. In such systems, the dynamic behavior of the ISE assumes great importance, since that will determine the signal shape and the sampling rate. Characterizing the dynamic behavior of ISEs constitutes a third area of research in the field.

For the most part, research in these three areas has unfolded in the order given. For example, when a new membrane was developed, the obvious electrode interferences were examined. However, many interferences were identified only after researchers tried to develop applications in various disciplines. Because of the way in which early analyses were done, the dynamic response of an ISE was not of extreme importance. The response rate had to be rapid enough that potentiometric titration curves would not be distorted. For direct analysis, one could simply wait for the signal to stabilize.

This field has just passed its golden anniversary. So many membranes have been developed that a classification scheme had to be recommended. The membranes will be introduced by means of the latest classification scheme. After a short delineation of the historical development, the theory of the electrode response will be presented. Both the equilibrium and the dynamic response behavior in the absence and presence of interferences will be discussed. Instrumentation has played a very large role in the ascendency of this field. The instrumentation requirements will be discussed, and then some of the most recent instrumentation will be presented. Physiological and biological applications of ISEs have been a goal since the inception of the field. Now, miniaturization in instrumentation seems to be a current paradigm. ISEs are part of this effort, as researchers have and continue to develop ion-selective microelectrodes for intracellular measurements of ion activities. The progress in this area will be summarized. Finally, some recent, novel developments in this field will be reported, particularly those which may find commercial applications within the next 10 years. In this

For references see page 9904

article, only a very small part of a vast literature can be included. However, the journal *Analytical Chemistry* has published a series of review articles that include ISEs.[2–24] They are an excellent place to start a search for more detail on a specific topic. For the most part, physiological and biological applications of ISEs involve the use of carrier-based microelectrode ISEs. These have been comprehensively reviewed by Bakker et al.[25,26]

In the new classification scheme, the old scheme is retained as a single category, *primary ISEs*. Two new categories, *compound or multiple membrane ISEs* and *metal contact or all-solid-state ISEs*, have been added. None of these categories include the ISFET (ion-selective field effect transistor) or CHEMFET (chemical-sensing field effect transistor), because they are considered to be purely capacitatively coupled devices.[27] However, some of these devices are commercially available, and they are called "electrodes" by their vendors.[28] It should be noted that some researchers have argued that the glass membrane ISEs are capacitative devices.[29,30] The experimental results that are the basis of their conclusions have not been independently verified. If true, the glass membrane ISE would be a half-cell for an electrochemical capacitor. The development of electrochemical capacitors is a fairly new field; the United States Department of Energy has had such a program since 1992.[31] The electrochemical capacitor is just a third type of electrochemical cell. Should it be shown that at least one ISE is a half-cell for an electrochemical capacitor, then we might expect a future classification scheme to include ISFETs or CHEMFETs. Reproduced below is the classification scheme according to the IUPAC 1994 recommendations.[1]

A. Primary ISEs

1. *Crystalline electrodes*
   - (a) Homogeneous membrane electrodes
   - (b) Heterogeneous membrane electrodes
2. *Non-crystalline electrodes*
   - (a) Rigid, self-supporting, matrix electrodes
   - (b) Electrodes with mobile charged sites
     - Positively charged, hydrophobic cations
     - Negatively charged, hydrophobic anions
     - Uncharged carrier electrodes
     - Hydrophobic ion-pair electrodes

B. Compound or Multiple Membrane ISEs

1. *Gas-sensing electrodes*
2. *Enzyme substrate electrodes*

C. Metal Contact or All-solid-state ISEs

Examples of the constructions used for the three main categories are shown in Figures 1–3. Gas-sensing electrodes consist of a reference electrode and indicator electrode that are in electrolytic contact. Some researchers say that these devices should not be called electrodes, because they are really electrochemical cells.[32] An ordinary ISE is used by placing it in a solution containing the analyte ion and measuring the potential, using a high-impedance potentiometer circuit, with respect to a reference electrode, placed in the same solution. The reference electrode, analyte solution and ISE constitute an electrochemical cell.

Many ISEs may now be purchased as combination electrodes. A combination electrode consists of an ISE and a reference electrode in a single package. An example of a combination electrode, the Orion glass pH electrode, is shown in Figure 4. Some researchers would probably argue that a combination electrode is not really an electrode. If one immerses it in an analyte solution,

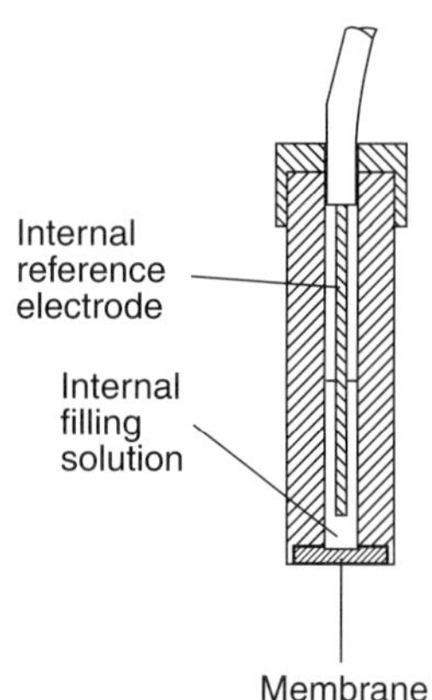

Figure 1 Schematic of a primary ISE.

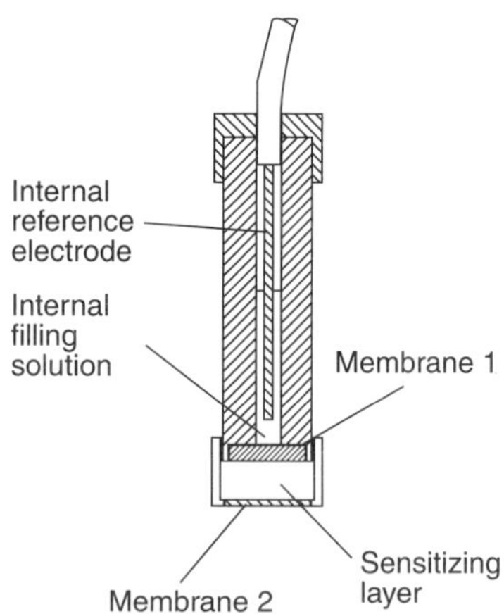

Figure 2 Schematic of a compound or multiple membrane ISE.

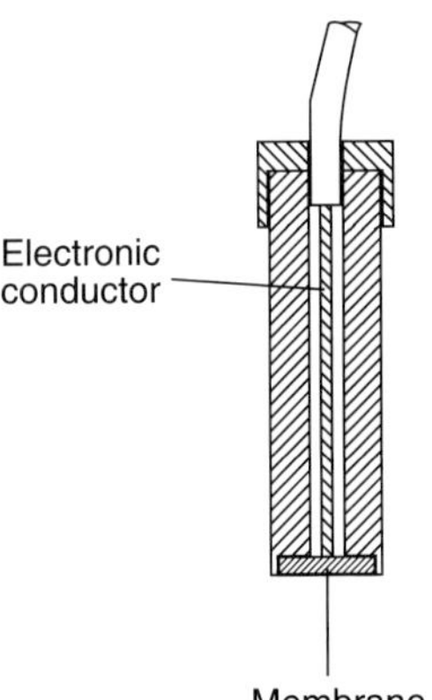

Figure 3 Schematic of a metal contact or all-solid-state ISE.

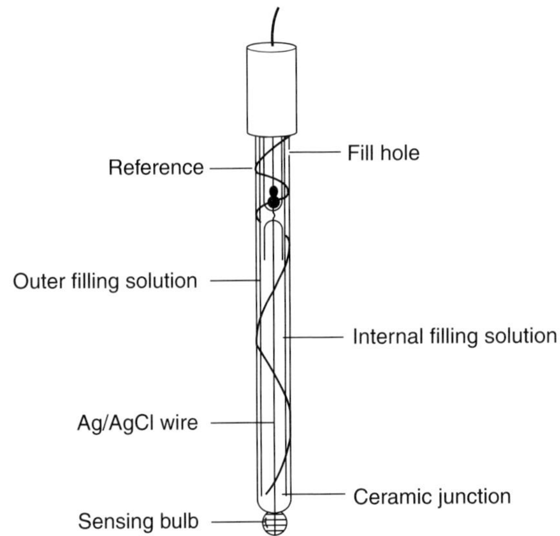

Figure 4 Orion combination glass pH electrode. (Reproduced from *A Concise Guide to Ion Analysis*,(28) by permission of Orion Research Inc.)

then one has an electrochemical cell. An analogy is the conventional car battery without the electrolyte. We do not regard it as an electrode just because the electrolyte is missing. For the foreseeable future, the use of the term combination electrode will undoubtedly continue. Technically correct and noncumbersome nomenclature for these devices does not exist at present.

2 HISTORICAL DEVELOPMENT

It is generally agreed that the progenitor of ISEs is the glass membrane acidity electrode. Cremer is credited with making the discovery that a large potential difference exists across a thin glass membrane in contact with solutions with large differences in acidity on opposite sides of the membrane.(33) It is fascinating to note that Cremer's work was actually the culmination of many studies on the electrical properties of glasses starting as far back as 1761!(34) At that time, glasses were considered to be materials that consisted of silicon dioxide and various other oxides. The properties of the glasses were known to depend on which other oxides were added and on the stoichiometry of the resultant material. Thus, some glasses were shown to be conductors of electricity, while others were shown to be electrical insulators. Cremer, who was interested in the potential differences generated across certain biological tissues, recognized that the Warburg model of glass meant that it should be an ideal semipermeable membrane. He performed several experiments using a flask containing an electrolyte solution immersed in a beaker, also containing an electrolyte solution. A platinum wire was in contact with each solution. He found that when the two electrolytes differed in their acidities, a large potential difference could be measured across the two platinum electrodes.

Systematic, quantitative studies on this phenomenon were made by Haber and Klemensiewicz on glass bulb electrodes.(35) In that paper, the theoretical model for the electrode response was developed by Haber, and the experimental confirmation of the model was performed by Klemensiewicz. For the theoretical model, Haber extended the work of H.W. Nernst.(36) Of particular relevance to the development of the mathematical form for the electrode potential of an ISE is Nernst's treatment of the potential difference across the interface between two electrolyte solutions, the well-known liquid junction potential, generated by nonFaradaic processes. He showed that the steady-state potential difference across an interface between a dilute solution of HCl (concentration c_1) and a more concentrated solution of HCl (concentration c_2) has the form of Equation (1):(36)

$$E = \frac{RT}{\Im}(t_+ - t_-)\ln\left(\frac{c_1}{c_2}\right) \tag{1}$$

where E is the potential of the interface on the side of the more concentrated solution relative to the more dilute solution, t_+ is the transport number for the cation, t_- is the transport number for the anion and $\Im$ is Faraday's constant. (Note that in this and other equations, the expressions for E have been written in modern form. In the original equations, R, the gas constant, is given in electrical dimensions. In modern usage, R is given in energy dimensions. R, in electrical units, is equal to R in energy dimensions divided by the Faraday constant $\Im$.) To verify his theory, Nernst studied concentration cells of the form in Equation (2):

$$\mathrm{M|MX}(c_1)\mathrm{|MX}(c_2)\mathrm{|M} \tag{2}$$

For references see page 9904

where MX is a strong electrolyte and $c_1 \neq c_2$. Here, Nernst proposed that the mathematical relationship for the potential difference across the metal electrolyte interface under conditions of zero current flow is given by Equation (3):[36]

$$E = \frac{RT}{\mathfrak{F}} \ln\left(\frac{C}{c}\right) \tag{3}$$

where C is the concentration of ion cores in the metal and c is the concentration of the electrolyte. According to Nernst, dissolution of the metal may occur, such that ions cross to the solution side and electrons remain in the metal, or metal ions from the electrolyte may specifically absorb on the metal surface. These processes are also nonFaradaic. Thus Nernst clearly proposed that his mathematical treatment was a general one valid for all electrified interfaces under conditions of equilibrium or steady-state charge transport. Haber extended Nernst's treatment to the case of a concentration cell involving a solid electrolyte MX and an aqueous electrolyte MX, with the cell notation as in Equation (4):

$$M|MX_{(s)}|MX_{(aq)}|M \tag{4}$$

He showed that the potential at the interface between the aqueous and the solid electrolytes is given by Equation (5):

$$E = -\frac{RT}{\mathfrak{F}} \ln(C_M) + K \tag{5}$$

where C_M is the concentration of the metal ion in the aqueous electrolyte solution and K is a constant. Haber then proposed that a hydrated glass membrane is a type of solid water phase. By considering the cell in Equation (6):

$$H_2|H_2O_{(s)}|H_2O_{(l)}|H_2 \tag{6}$$

he arrived at Equation (7) for the potential difference across the glass membrane – water interface:

$$E = -\frac{RT}{\mathfrak{F}} \ln(C_{H^+}) + K \tag{7}$$

where C_{H^+} is the concentration of hydronium ions and K is a constant. Similarly, by considering the cell in Equation (8):

$$O_2|H_2O_{(s)}|H_2O_{(l)}|O_2 \tag{8}$$

he arrived at a second expression for the potential difference across the glass membrane – water interface (Equation 9):

$$E = +\frac{RT}{\mathfrak{F}} \ln(C_{OH^-}) + K' \tag{9}$$

where C_{OH^-} is the concentration of hydroxide ions and K' is a constant. We see that Haber recognized that the glass membrane can sense either hydronium ions or hydroxide ions. He further indicated that Equation (9) could be written in terms of C_{H^+} by making use of the relationship $C_{OH^-} = K_w/C_{H^+}$, giving Equation (10):

$$E = -\frac{RT}{\mathfrak{F}} \ln(C_{H^+}) + \frac{RT}{\mathfrak{F}} \ln(K_w) + K' \tag{10}$$

Haber then stated that Equations (7) and (10) are identical, which meant that K and K' are related by Equation (11):

$$K = \frac{RT}{\mathfrak{F}} \ln(K_w) + K' \tag{11}$$

Haber's statement, given without proof, was widely accepted by subsequent analytical chemists. Later, in describing the microscopic behavior giving rise to the electrode response, many analytical chemists only considered Equation (7), which has led to some misconceptions that are only now being addressed in the literature. Haber used his results to derive an expression for the potential difference across a glass membrane electrode which had an acidic solution on one side and an alkaline solution on the other. Thus, for the scheme in Equation (12):

$$\text{1. alkaline water } |H_2O_{(s)}| \quad \text{2. acidic water} \tag{12}$$

he wrote Equation (13):

$$E = \frac{RT}{\mathfrak{F}} \ln\left(\frac{C_{H^+,2}}{C_{H^+,1}}\right) \tag{13}$$

where $C_{H^+,1}$ is the hydronium ion concentration in the alkaline water, $C_{H^+,2}$ is the hydronium ion concentration in the acidic water and E is the potential difference between the alkaline water – $H_2O_{(s)}$ interface and the acidic water – $H_2O_{(s)}$ interface.

Klemensiewicz performed an eloquent set of experiments that constituted the proof of Haber's theory. The electrochemical cells used resemble the type of set-up we use today, in that a reference electrode, a 1 N calomel electrode, and an indicator electrode, the glass membrane electrode, were used. The set-up is shown in Figure 5. The glass membrane electrodes were hand-blown from Thüringer glass rods to give a bulb approximately 2.5 cm in diameter and a thickness of 0.06–0.1 mm. Inside the glass membrane electrode, a dilute solution of HCl or KCl was used. Various strong acid – strong base titrations were performed, as follows: 0.1 N HCl with 1 N KOH, 0.1 N HCl with 1 N NaOH, 0.1 N H_2SO_4 with 0.1 N KOH, 0.1 N H_2SO_4 with 0.1 N NaOH, and 0.1 N KOH with 1 N HCl. The potential of the glass membrane electrode was measured with respect to the normal calomel electrode using a quadrant electrometer with reading telescope. Plots of the potential difference versus volume of titrant gave the typical sigmoidal curves

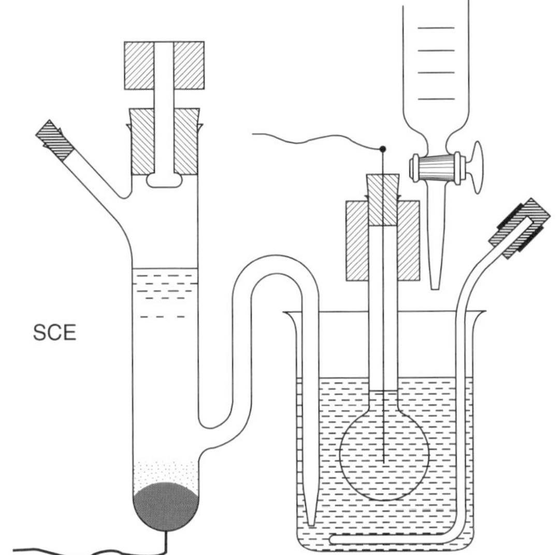

Figure 5 Experimental set-up for the pioneering studies of Klemensiewicz. SCE = saturated calomel electrode.

that are well known to us today. The magnitude of the potential difference between the initial point of the titration and the final point of the titration, which corresponded to an extent of titration = 2, is expected to be given by Equation (13), with the final point corrected for dilution effects and the ratio of the concentrations of hydronium ion being that for the acidic solution to that for the basic solution. For titrations involving KOH, the experimental error for titrations performed at 20 °C averaged −7%, which is fairly good. When KOH was replaced with NaOH, much larger relative errors were obtained, −15 to −22%. This came to be known as the sodium error. Titration of a weak acid with a strong base was also performed; it showed the characteristic sigmoidal curve with the buffer region so familiar to us today.

The effect of temperature was also studied, as was the effect of the glass on the potential generation. Two other easily fusible glasses, one a blue-colored, silicic acid-poor and sodium-rich glass and the other a colorless, silicic acid-rich and alkali-poor glass, were studied. The freshly blown blue-colored glass gave a relative error of −14%, but after steam treatment the error increased to −51%. The error for the colorless glass was even worse, being −57% for freshly blown and −71% for steam treated. When difficult to fuse Jena glass was used, no potential difference developed. Although the glass studies were qualitative at best, since the compositions were not measured, these results opened the way to what soon became an explosion in the development of ISEs.

Around this same time, Sørenson proposed the pH scale for acidity measurements(37) and the glass membrane acidity electrode ultimately came to be known as the pH electrode. From 1920 to 1935, quantitative studies on the effect of glass composition on the potential generation for the acidity electrode proliferated.(38) The compositions of Thuringian glass and the Jena glasses were determined.(39) This period was capped by the development of the first commercial pH glass manufactured by Corning as Corning 015 glass.(40) Because this glass composition was not ideal for the entire range of the pH scale, studies on glass composition for pH electrodes continued, with the results for over 500 different compositions appearing in the literature.(41) The sodium error results seem to have led some researchers to search for glass compositions that would give glass membranes sensitive to other ions. Thus, in 1934, Lengyel and Blum(42) reported on a glass membrane electrode that gave a Nernstian response to sodium ions. Subsequent glass composition studies by Eisenman,(43) reported in 1957, led to commercial glass membrane electrodes for sodium and potassium ions.

It is interesting that materials other than glass were reported as active materials for potentiometric ion sensors in the period from 1930 to 1960. These included inorganic sparingly soluble salts, surface-modified collodion, and cation-exchange resins for ions such as Ca^{2+}, Na^+, K^+, NH_4^+, Mg^{2+}, Ag^+, Cl^-, Br^-, I^-, Ba^{2+}, ClO_4^-, NO_3^-, and SO_4^{2-}.(44) At the time, this research seems to have been ignored, largely because of irreproducible results and lack of a Nernstian response. Glass membrane studies continued, and additional glass membrane electrodes were developed for Li^+, Ag^+, NH_4^+, Rb^+ and Cs^+. In the 1960s, studies by Pungor et al.(45) led to the first nonglass ISEs to give thermodynamically reversible responses – sparingly soluble inorganic salts in silicone rubber. These were commercialized and marketed by Radelkis in Hungary in 1965. Up to this point, cation ISEs with acceptable responses existed only for univalent cations, so the next breakthrough was a very big one – the development in 1966 of the Ca^{2+} liquid membrane ISE by Ross.(45) As a result of this development and the development of the solid-state F^- ISE by Ross and Frant,(45) a small company in the USA, Orion Research, experienced tremendous growth.

Another development in 1966 was the introduction of the first neutral carrier-based membrane ISE by Stefanac and Simon.(46,47) The electrode was a K^+ ISE based on the macrotetrolide homologs of the antibiotic nonactin. The membrane supports studied were sintered-glass disks, Millipore filter disks, nylon mesh, polyethylene films and gel formers. In 1967, Shatkay described a neutral carrier-based polymeric ISE for calcium based on poly(vinyl chloride) (PVC), tributyl phosphate and theonyltrifluoroacetone.(48) This period of time is regarded by many as the incipient point for

For references see page 9904

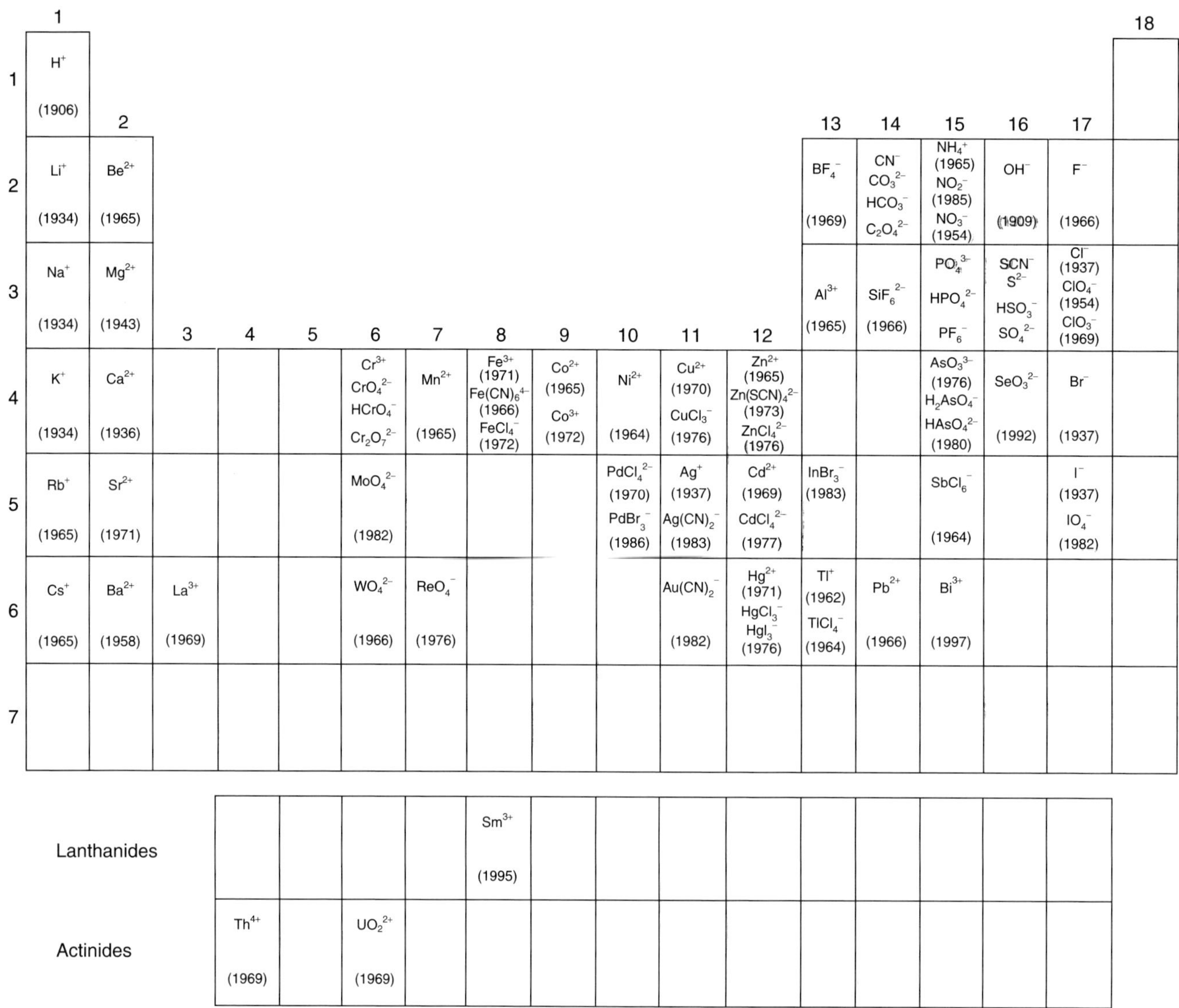

Figure 6 Periodic table profile for the historical development of ISE membranes for inorganic ions.

what Ruzicka[49] calls the "golden age" for ISEs, with obvious reference to ISEs which target inorganic cations and anions. In Figure 6, a timeline periodic table for ISE membranes for inorganic ions is given. The year of the first literature report is given, although many years often elapsed before a useful ISE was fabricated for an ion. Of the 79 ions in the table, membranes for 27 of them were reported from 1960 to 1969 and membranes for 24 of them were reported from 1970 to 1979.

Most research on membranes for inorganic ions now involves developing more selective membranes for ions already shown in Figure 6. This is reflected by the fact that only three "brand new" ISEs for inorganic ions were developed in the period from 1990 to 1999. ISEs for direct potentiometry are available commercially for only 24 of these ions, as shown in Figure 7. These are electrodes for which the analyte determinations have a high volume demand, and for which ISE potentiometry is competitive with other methods of analysis.

By contrast, the development of ISEs for organic and bio-organic analysis is now experiencing the kind of growth that was observed for inorganic ions from 1960 to 1979. This era seems to have commenced with the development in 1969 of the urea ISE by Guilbault and Montalvo,[50] the development in 1972 of coated-wire ISEs for amino acid anions and salicylate by James, Carmack and Freiser[51] and the development in 1972 of PVC matrix membrane ISEs for large onium ions by Scholer and Simon.[52] Electrodes have now been reported for glucose, L- and D-amino

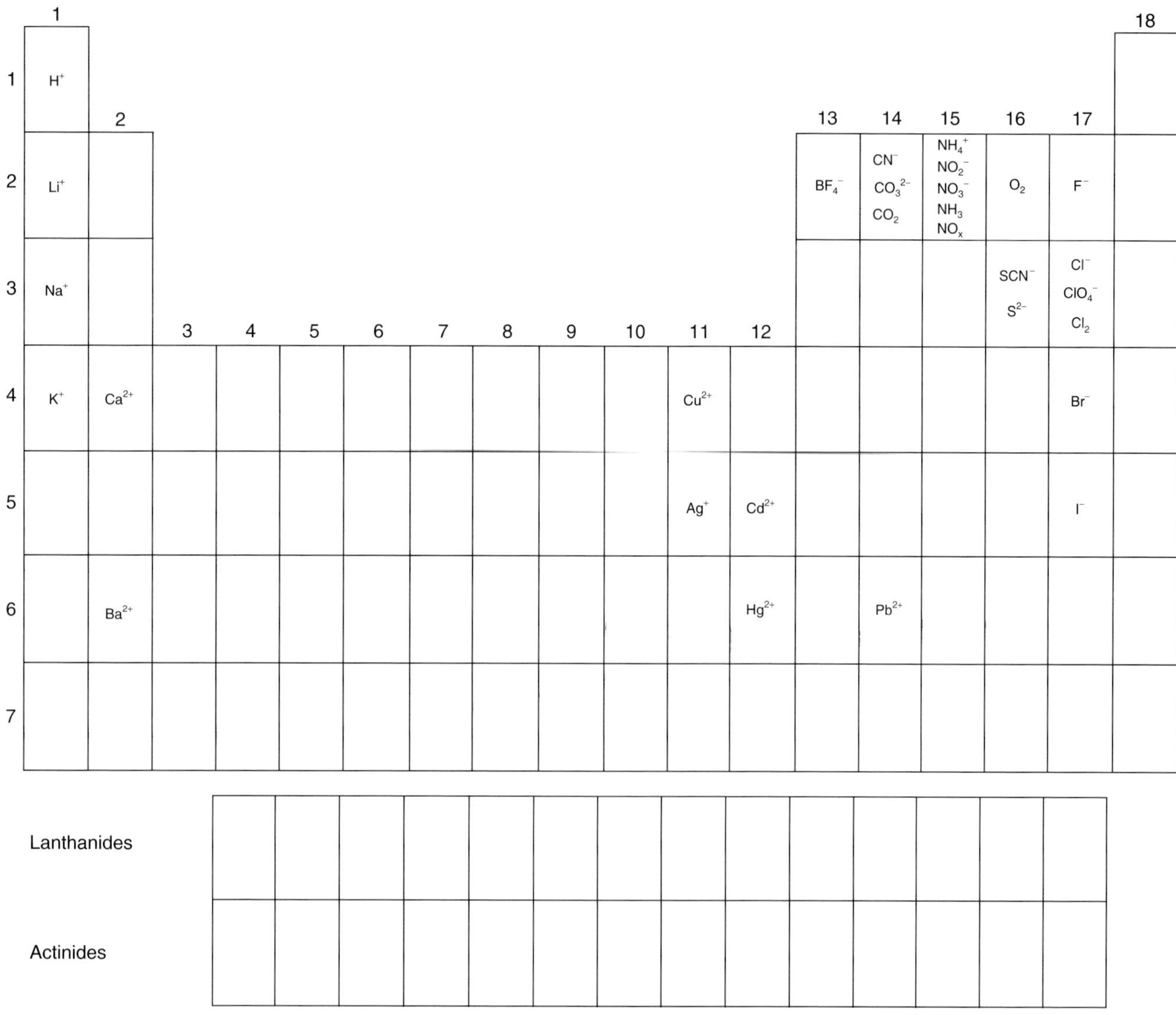

Figure 7 Periodic table diagrams for commercially available inorganic ions.

acids and their methyl esters, 1-phenylethylamine, 1-(1-naphthyl)ethylamine, ephedrine, norephedrine, pseudoephedrine, amphetamine, propanolol, α-amino-ε-caprolactam, amino acid amides, benzylamine, alkylamines, dopamine, procaine, prilocaine, lidocaine, bupivacaine, lignocaine, diquat, paraquat, acetylcholine, albumin, amygdalin, pencillin, protamine, catechol, cholesterol, creatinine, uric acid, 5′-AMP, glutamine, glutamate, pyruvate, vitamin B_6, guanine, guanidine, amprolium, heparin, methotrexate, mexiletine, metformin, phenformin, clotrimazole, bifonazole, human immunoglobulin G, organohosphate nerve agents, cationic surfactants, anionic surfactants, and organosulfonates.[25,50–63] Only the glucose, urea and creatinine electrodes are available commercially.

The first gas-sensing ISE was the carbon dioxide gas electrode, conceptualized in 1957 by Stow, Baer and Randall.[64] After a 10-year gap, the ammonia gas-sensing electrode was developed, and by 1976 electrodes had been developed additionally for NO_x, SO_2, H_2S and CH_3COOH.[64] As shown in Figure 7, 10 gas sensing ISEs are now available commercially. Further development in this area is not likely, because acoustic wave devices with higher sensitivities, faster responses, and simpler construction are likely to become the sensors of choice for gas analysis.[65]

For references see page 9904

3 THEORY OF ELECTRODE RESPONSE

3.1 General

All of the ISEs share in common that the component of the measured signal which is related to the activity of the analyte to be determined is the potential difference across interface 1 in Figure 8(a) or (b). Figure 8(a) is the configuration for the primary ISEs. Figure 8(b) is the configuration for the metal contact or all-solid-state ISEs. The compound or multiple membrane ISE has an intermediate compartment which controls the activity of analyte in the external solution of an electrode of type shown in Figure 8.

The formation of a potential difference across the interface between two different phases is a general process. The potential difference can be written as the sum of two or more potential differences resulting from essentially independent processes occurring at the interface. Let us assume that the membrane is a perfectly ordered solid which fills all space (infinite three-dimensional lattice). If the solid is cut along the *yz* plane, leaving a semi-infinite three-dimensional solid with a face in contact with a vacuum, then surface states whose wave functions penetrate a short distance into the vacuum are created. The surface states are not empty. Since electron density is proportional to the product of a wave function and its complex conjugate, this means that some electrons have spilled across the solid surface and into the vacuum. However, these electrons belong to the solid, and not the vacuum (i.e. ionization has not occurred). Thus, all solids in a vacuum have a surface dipole barrier at the interface between the surface and the vacuum.[66] The magnitude of the surface dipole barrier would be largest for a free electron metal and smallest for an insulator. In the case of an ionic solid, surface defects are the source of the equilibrium space charge potential.[67] If a semi-infinite electrolyte solution can be placed in contact with a vacuum, even in this case a surface dipole barrier would be formed. It would be due to the reorientation of water molecules to give a net dipole moment at the surface of the solution.[68] One would expect that more water molecules are orientated so that a lone pair on the oxygen points toward the vacuum than are oriented with the hydrogens pointed toward the vacuum. Thus the picture for the surface dipole barrier is analogous to that for the solid. These surface dipole potentials are often called the "chi potentials".[67]

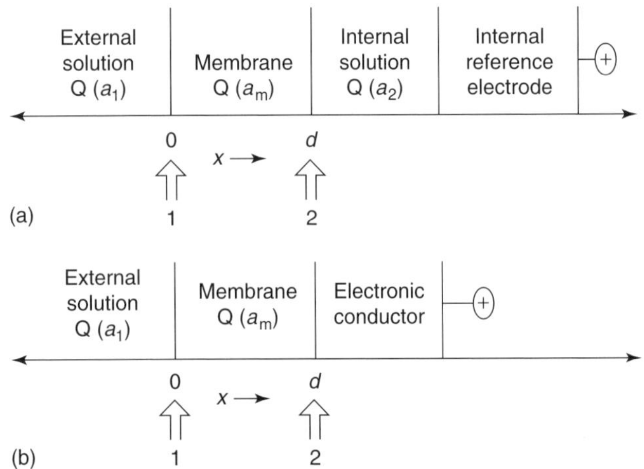

Figure 8 Simplest interface diagram for the physical interface locations of (a) primary ISEs and (b) metal contact or all-solid-state ISEs.

Now, if the semi-infinite solid and the semi-infinite electrolyte solution are brought together and no transfer of charge occurs between them, the two phases remain neutral, but the interface between them is electrified owing to the difference between their surface potentials. The potential difference will be smaller than the numerical difference between the two vacuum referenced values, because there will be a readjustment of the electron density on the solid side, the distribution of defects, the orientation of water dipoles, etc. Nevertheless, there will be a finite potential difference across the interface. However, because this potential difference across the interface is not due to any potential-determining ion, it cannot provide a useful analytical signal. An analytically useful signal can be obtained only if a charge-transfer process can occur which results in a nonzero charge on each phase across the interface. Charge transfer may be caused by or influenced by concentration gradients, potential gradients, temperature gradients and/or pressure gradients.

It is usual to classify charge-transfer processes as either Faradaic or nonFaradaic. A Faradaic process involves an oxidation or a reduction reaction. This type of process will alter the nature of certain ions or molecules on both sides of the interface. It has been shown that this type of process is frequently the basis of chemical interference for some type of membranes, but it does not usually lead to a potential difference across an interface that can be related to the activity of a potential-determining ion. One exception is the homogeneous Cu(II) membrane ISE, which will be discussed later. Thus, the relevant charge-transfer processes are mostly nonFaradaic. Examples of nonFaradaic chemical or physical processes that can lead to charge transfer are ion adsorption/desorption, complex formation, ion diffusion, ion extraction, and ion exchange.

A complete description of a membrane system involves the determination of various concentrations and potentials in both space and time. This requires that a set of partial differential equations for pertinent fluxes and a set of conservation equations be solved for some set of initial and boundary conditions. For the most general solution, only those initial and boundary conditions necessary to prevent the violation of physical laws may be specified. No such solution exists. Instead, solutions have been obtained for a number of special cases. In all of these special cases, temperature is kept constant.

We cannot examine all of these special cases here. Instead, it is instructive to examine the thermodynamic treatment for a number of simple cases in which the material properties of the nonaqueous phase are different. In every case, we consider that the aqueous phase and the nonaqueous phase have at least a univalent cation in common. Using in part the notation of Koryta,[69] we consider first the interface for zero current flow in Equation (14):

$$\begin{array}{ccc} \alpha & & \beta \\ P^+X^- & | & P^+e^- \end{array} \qquad (14)$$

where α is an aqueous solution of the strong electrolyte P^+X^- and β is a free electron metal P with ion cores P^+ and free electrons e^-. An example would be silver metal in contact with $AgNO_{3(aq)}$. Since X^- is not potential determining, on the α side of the interface, we have (Equation 15)[70]

$$\bar{\mu}_{P^+}(\alpha) = \mu_{P^+}(\alpha) + \mathcal{F}\phi_\alpha \qquad (15)$$

where $\bar{\mu}$ is the electrochemical potential, μ is the chemical potential and ϕ is the Galvani potential. On the metal side of the interface, we have (Equation 16)

$$\bar{\mu}_P = \bar{\mu}_{P^+}(\beta) + \bar{\mu}_{e^-}(\beta) = \mu_P \qquad (16)$$

where

$$\bar{\mu}_{e^-}(\beta) = \mu_{e^-}(\beta) - \mathcal{F}\phi_\beta \qquad (17)$$

Substituting and rearranging gives Equation (18):

$$\mathcal{F}(\phi_\beta - \phi_\alpha) = \mu_{P^+}(\alpha) + \mu_{e^-}(\beta) - \mu_P \qquad (18)$$

The chemical potentials may be written in terms of the standard chemical potential and the activity. For example (Equation 19):

$$\mu_{P^+}(\alpha) = \mu^0_{P^+}(\alpha) + RT\ln[a_{P^+}(\alpha)] \qquad (19)$$

Making this substitution, and given that $a_{e^-}(\beta) = a_P = 1$, we obtain Equation (20):

$$\phi_\beta - \phi_\alpha = K + RT\ln[a_{P^+}(\alpha)] \qquad (20)$$

where K is given by

$$K = \frac{[\mu^0_{P^+}(\alpha) + \mu^0_{e^-}(\beta) - \mu^0_P]}{\mathcal{F}} \qquad (21)$$

Next, consider the case in Equation (22):

$$\begin{array}{ccc} \alpha & & \beta \\ P^+X^- & | & P^+X^- \end{array} \qquad (22)$$

where α is an aqueous phase containing the electrolyte P^+X^- and β is an immiscible, nonaqueous phase containing the same electrolyte. At equilibrium, we have (Equation 23):

$$\mathcal{F}(\phi_\beta - \phi_\alpha) = [\mu_{P^+}(\alpha) - \mu_{P^+}(\beta)] - [\mu_{X^-}(\alpha) - \mu_{X^-}(\beta)] \qquad (23)$$

If there is differential transport of P^+ and X^-, then a Galvani potential difference results, but both ions are potential determining. How do we obtain an interface in which a single ion is potential determining? The result for the aqueous solution/metal phase gives the clue. We see that X^- cannot cross into the metal side of the interface. Likewise, electrons cannot move from the metal into the solution phase. Therefore, if the negative ions in α are excluded from β, and the negative ions in β are excluded from α, only the cations will determine the Galvani potential difference. This would correspond to the case in Equation (24):

$$\begin{array}{ccc} \alpha & & \beta \\ P^+X^- & | & P^+Y^- \end{array} \qquad (24)$$

where X^- is hydrophilic and Y^- is lipophilic. This case represents the simplest model for the polymeric membrane-type ISEs. With the anion exclusions, Equation (23) becomes Equation (25):

$$\mathcal{F}(\phi_\beta - \phi_\alpha) = [\mu_{P^+}(\alpha) - \mu_{P^+}(\beta)] \qquad (25)$$

$$\mathcal{F}(\phi_\beta - \phi_\alpha) = \mu^0_{P^+}(\alpha) - \mu^0_{P^+}(\beta) + RT\ln\left[\frac{a_{P^+}(\alpha)}{a_{P^+}(\beta)}\right] \qquad (26)$$

If $a_{P^+}(\beta)$ is constant, an equation similar in form to Equation (20) is obtained, with K given by Equation (27):

$$K = \frac{\mu^0_{P^+}(\alpha) - \mu^0_{P^+}(\beta) + RT\ln[a_{P^+}(\beta)]}{\mathcal{F}} \qquad (27)$$

Notice that the interface in Equation (24) could just as easily represent an aqueous phase containing the electrolyte P^+X^- in contact with a sparingly soluble salt P^+Y^-. Clearly, at equilibrium there will be some Y^- in α, but owing to the common ion effect, its activity will be very small over some range of activities for P^+. The interface can be approximated as one in which Y^- is essentially confined to β, and the equation for the Galvani potential difference across the interface has the same form as Equation (25). Again, Equation (20) gives the final form of the mathematical expression for the Galvani potential difference. Equation (20) is a form of the Nernst equation for an interface at equilibrium. The Galvani potential difference across an interface cannot be measured. What can be measured is the potential difference across a cell. By coupling the indicator electrode to a reference electrode, we obtain an electrochemical cell whose potential difference is the

For references see page 9904

sum of the potential differences across many interfaces. If all interfaces except for the interface between the external membrane surface and the sample solution are constant, then $E_{\mathrm{cell}} = (\phi_\beta - \phi_\alpha) + K'$. Furthermore, if we write $E_{\mathrm{cell}} = E_{\mathrm{ind}} - E_{\mathrm{ref}}$, the potential difference between the indicator electrode (metal or ISE in the cases above) and the reference electrode, then the electrode potential for the indicator may be written in its usual form (Equation 28):

$$E_{\mathrm{IND}} = K_{\mathrm{electrode}} + \frac{RT}{\mathcal{F}} \ln[a_{\mathrm{P}^+}(\alpha)] \qquad (28)$$

where any liquid junction potentials in the cell are assumed to be negligible. Converting to common logarithms and generalizing to a potential-determining ion with charge z, sign and magnitude, Equation (28) becomes Equation (29):

$$E_{\mathrm{IND}} = K_{\mathrm{electrode}} + \frac{2.303RT}{z\mathcal{F}} \log[a_{\mathrm{Q}}] \qquad (29)$$

where Q is the target ion shown in Figure 8. At 25 °C, the magnitude of the multiplicand of the logarithm terms is 59.2, 29.6, and 19.7 mV for $z = \pm1$, ±2, and ±3, respectively. In practice, these ideal values are frequently not observed for ISEs, so it is common practice to add a so-called efficiency factor, β, to that term and to treat the multiplicand as a constant which must be measured. Thus, most textbooks give Equation (30) for the electrode response of an ISE:

$$\begin{aligned} E_{\mathrm{ISE}} &= K_{\mathrm{electrode}} + \frac{\beta \times 2.303RT}{z\mathcal{F}} \log[a_{\mathrm{Q}}] \\ &= K_{\mathrm{electrode}} + S\log[a_{\mathrm{Q}}] \end{aligned} \qquad (30)$$

If Q can be transported through the membrane, i.e. the membrane is an ionic conductor for Q, then there may be a membrane diffusion potential difference across the bulk membrane.[71] Frequently, it is the case that the ion transported through the membrane is not the potential-determining ion responsible for the potentials generated at the boundaries. Even in that case, a membrane diffusion potential difference across the bulk membrane may be generated. The transport equations have been solved for several applicable cases.[71–73] The mathematical formulation is at best quasi-thermodynamic. Fortunately, it has been shown that the diffusion potential difference across the bulk membrane is negligible for most practical cases.[25,74,75]

The time-variable properties of ISEs determine the upper limit to the rate of sampling. They are particularly important in flow analysis applications or in monitoring fast processes. These properties are determined by subjecting an ISE to activity steps. The time-variable properties of ISEs are referred to as their dynamic characteristics. In Figure 9(a) and (b) are shown the responses of a Corning I^-/CN^- ISE to forward and reverse iodide steps.[76]

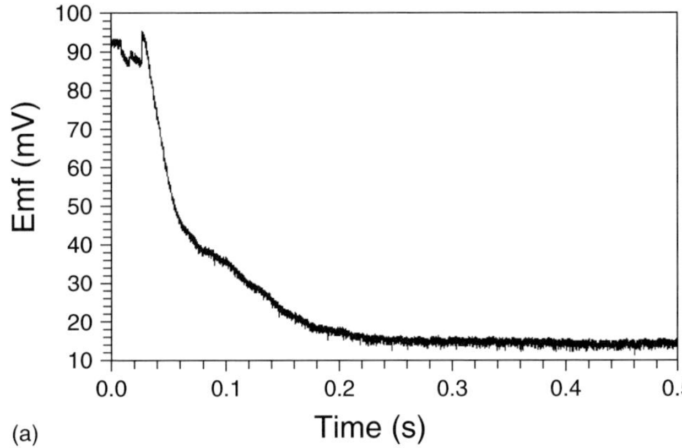

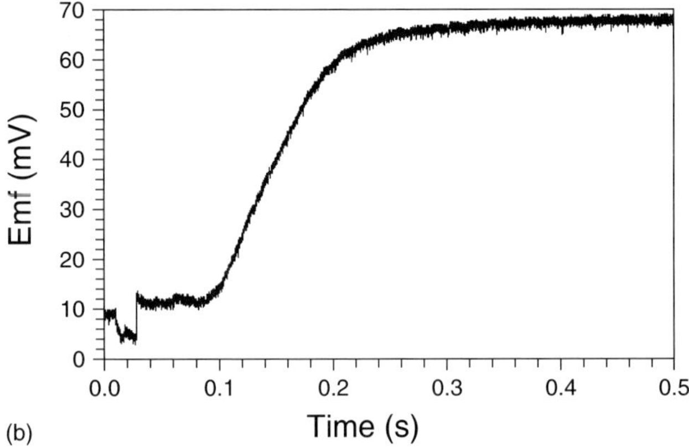

Figure 9 Dynamic responses of a Corning iodide/cyanide ISE to (a) forward and (b) reverse iodide steps. (a) 0.001 to 0.01 M I^-, ionic strength = 0.15 M; (b) 0.01 to 0.001 M I^{-1}, ionic strength = 0.15 M.

The characterization of the dynamic response behavior of ISEs seems to have commenced in the mid-1960s,[77] but progress has been slow owing to the complexity of the membrane response mechanisms. The figure of merit for the dynamic characteristics of an ISE is the response time. This has variously been defined in the literature as $t_{1/2}$, t_{90}, t_{95}, or $t_{99.5}$ (times to reach 50%, 90%, 95%, or 99.5% of the total potential change) for a 10-fold activity change. The IUPAC recommendation for the definition of this parameter is the following: "The time which elapses between the instant when an ISE and a reference electrode (ISE cell) are brought into contact with a sample solution (or at which the activity of the ion of interest in a solution is changed) and the first instance at which the electromotive force (emf)/time slope ($\Delta E/\Delta t$) becomes equal to a limiting value selected on the basis of the experimental conditions and/or requirements concerning the accuracy (e.g. 0.6 mV min^{-1}). ... In clinical applications (the physiological activity range corresponds to a small emf span), a smaller slope, e.g. 0.1 mV min^{-1}, may be chosen, provided the standard deviation of the response is less than the required slope."[1] The response time defined as the time from which the activity step is initiated until the slope becomes zero is identical to t_{100}.

Both of these definitions are impractical, because many electrodes never reach a constant potential after the step. Unfortunately, it is not usually possible to convert from one definition to another, because the mathematical equation for the potential as a function of time must be known. There does not seem to be a single equation that adequately represents the time response for all ISEs. Berube et al.[78] examined eight different models from the literature for a study on the homogeneous, crystalline iodide ISE. If the measured time response is not corrected for drift, the determined response time will not be accurate. Drift is defined as "the slow non-random change with time in the emf of an ion-selective electrode cell assembly maintained in a solution of constant composition and temperature".[1] The drift may be measured on the front end before the activity step, and we have no choice but to assume that it does not change as a result of the step. If precise response times need to be known, then independent studies on drift as a function of the activity of the potential-determining ion should be performed. It is necessary to define the initial and final solution conditions for the step, since the response curves depend on these.

Finally, this subsection concludes with a discussion of the general treatment of electrode interferences. The 1994 IUPAC recommendations specify that the general expression for electrode response is the modified Nikolsky–Eisenman equation, given by Equation (31):[1]

$$E_{\mathrm{ISE}} = K_{\mathrm{electrode}} + \frac{2.303RT}{z\mathcal{F}} \log \left[a_{\mathrm{A}} + \sum K^{\mathrm{pot}}_{\mathrm{A,B}} (a_{\mathrm{B}})^{z_{\mathrm{B}}} \right] \tag{31}$$

where $K^{\mathrm{pot}}_{\mathrm{A,B}}$ are selectivity coefficients.

Over the years, several different methods have been proposed for the evaluation of selectivity coefficients. The IUPAC recommendation is that the fixed interference method is preferred and the separate solution method is acceptable, but less desirable. When the analyte ion and the electrode interferences have the same charge (sign and magnitude), and when the response is Nernstian for solutions of each type of ion, then the electrode interference is adequately described by the Nikolsky–Eisenman formalism. This formalism fails when either of these conditions is not met. Bakker et al.[79] examined critically the Nikolsky–Eisenman formalism for polymer membrane-based ISEs, demonstrated its general inadequacy and proposed a new formalism, which they showed corresponded to the matched potential method proposed earlier by Gadzekpo and Christian.[80] Umezawa et al.[81] critically evaluated the validity of Equation (28) for two crystalline membrane ISEs and three liquid membrane ISEs. They also recommended that selectivity coefficients be obtained using the matched potential method. More recently, Nägele et al.[82] developed a general formalism for electrode interferences for polymer membrane-based ISEs. Thus, the formalism for the response of polymeric membrane ISEs in the presence of electrode interferences has been completely and rigorously established. Since it is based on ion extraction/exchange equilibria, the formalism is not necessarily a general one applicable to all ISEs. These results will be discussed in more detail under the specific heading of noncrystalline electrodes.

By using statistical mechanics and equations obtained from nonequilibrium thermodynamics, Hall[83] developed a general limiting expression valid for electrode interferences whose charges differ in both sign and magnitude. In this treatment, the total potential difference across the membrane is not partitioned into phase boundary potentials and a diffusion potential. With reference to Figure 8(a), Hall writes the potential difference across the membrane as $\phi^{(1)} - \phi^{(2)}$, which can be written as Equation (32):

$$\phi^{(1)} - \phi^{(2)} = \frac{\left[\bar{\mu}^{(1)}_{\mathrm{Q}} - \bar{\mu}^{(2)}_{\mathrm{Q}}\right]}{z\mathcal{F}} - \frac{\left[\mu^{(1)}_{\mathrm{Q}} - \mu^{(2)}_{\mathrm{Q}}\right]}{z\mathcal{F}} \tag{32}$$

where z is the charge on Q. The first term in Equation (32) is the deviation of the membrane response from Nernstian behavior; it represents the contribution to the cell emf due to interference. In the equilibrium thermodynamic treatment, this term is zero. Considering both ionic and nonionic interferences, Hall developed a general expression for the interference term. Unfortunately, from a practical perspective, the mathematical expression is not useful. If it were possible to correlate these results with those obtained for the polymeric membranes, then it might be that modifications of those equations could be used for ISEs with other types of membranes.

3.2 Specific

The electrical properties of these membranes determine to a great extent the nature of their response to potential-determining ions. Both the electronic properties and the electrolytic properties must be considered. It is common knowledge that electronic materials may be classified as conductors, semiconductors, or insulators, based on the magnitude of their conductivities. It is not as widely appreciated that a similar classification scheme exists for electrolytic materials. They may be fast-ion conductors, electrolytic semiconductors, or electrolytic insulators. The boundaries between classes are determined by the magnitude of their ionic conductivities. In Table 1, we show these two classification schemes.[84]

For references see page 9904

Table 1 Electrical properties of materials

Type	Electronic	Electrolytic
Conductor	$10^2\ \Omega^{-1}\,cm^{-1} < \sigma_e$	$10^1\ \Omega^{-1}\,cm^{-1} < \sigma_I$
Semiconductor	$10^{-9}\ \Omega^{-1}\,cm^{-1} < \sigma_e < 10^2\ \Omega^{-1}\,cm^{-1}$	$10^2\ \Omega^{-1}\,cm^{-1} < \sigma_I < 10^1\ \Omega^{-1}\,cm^{-1}$
Insulator	$\sigma_e < 10^{-9}\ \Omega^{-1}\,cm^{-1}$	$\sigma_I < 10^{-9}\ \Omega^{-1}\,cm^{-1}$

3.2.1 Primary Ion-selective Electrodes

3.2.1.1 Crystalline Electrodes The active materials in these electrodes are sparingly soluble, inorganic salts. They vary widely in their properties. Commercial electrodes with membranes in this category are the fluoride, chloride, bromide, iodide, sulfide, cyanide, thiocyanate, cadmium ion, cupric ion, lead ion, silver ion and mercuric ion ISEs. The fluoride ISE is the most important member of this group. Direct potentiometry for the in vitro determination of fluoride is the method of choice. Official methods for the determination of fluoride by use of the fluoride ISE appear in the *Official Methods of Analysis* Manual of the Association of Official Analytical Chemists, and also manuals of analysis published by the American Society for Testing and Materials and the Environmental Protection Agency. The active material of the fluoride ISE is lanthanum fluoride, a wide band gap electronic insulator, but an electrolytic semiconductor. Fluoride ion is the mobile charge carrier in this defect ionic conductor. Impedance spectroscopy[85] and X-ray photoelectron spectroscopy (XPS)[86] show that it forms a surface-hydrolyzed layer on contact with solution, hence a better model for this ISE is the segmented membrane model shown in Figure 10(a) and (b).

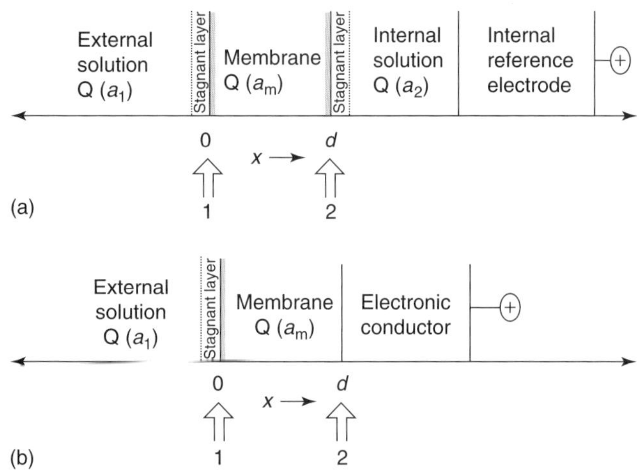

Figure 10 The segmented membrane model for (a) primary ISE and (b) a metal contact or all-solid-state ISE. In the latter case, the space-charge region in the membrane at the membrane/electronic conductor interface is not shown.

The solution in contact with the membrane may also be regarded as segmented into a bulk solution and a stagnant layer in contact with the electrode. In a stirred solution, there is convection in the bulk, but no convection in the stagnant layer. Although a gel layer has been considered critical for the response of glass membrane electrodes, or the fluoride ISE, the gel layer may not be essential for the response, and is sometimes actually detrimental. The gel layer, described as $LaF_3 \cdot 3H_2O$, is produced by diffusion of water into the membrane; it reaches a thickness of 20 nm in less than 1 week. The dynamic response of the electrode is not significantly affected by the gel layer, but with increasing time, leaching of fluoride is observed, concomitant with the incorporation of hydroxide. The aged electrode shows a sluggish response to a decade fluoride activity step and the change in emf becomes significantly less than Nernstian. After 18 weeks of aging, the first 5 nm of the 20-nm-thick gel layer has the stoichiometry $LaF_{3-n}(OH)_n \cdot mH_2O$. Fortunately, the aging effects may be reverse by polishing. At room temperature, hydroxide is the only electrode interferent; the mechanism of interference has been shown to be ion exchange in the gel layer.[87] A very significant result of that study is that carbonate and phosphate, both of which form sparingly soluble salts with lanthanum, adsorbed on the membrane but did not affect the potential response at 20 °C, but at 60 °C the two ions did interfere with the potential response. At that temperature, they undergo ion exchange in the gel layer. There is some debate as to whether adsorption/desorption or ion exchange is the mechanism for potential generation of crystalline membrane ISEs. Experiments such as these may allow the debate to be resolved. The lanthanum fluoride membrane has not been used for the direct determination of lanthanum, so the membrane apparently does not give a useful potential response to this ion. The useful concentration range is from saturated to 10^{-6} M fluoride.

The other commercial crystalline membrane ISEs are based on sparingly soluble silver salts or a homogeneous mixture of silver sulfide and a sparingly soluble metal sulfide. The silver salts are mixed conductors, so that chemical interference from redox active species, in addition to electrode interferences, may adversely affect an analysis. For example, a chloride ISE may have silver chloride as its active material. Silver chloride is a mixed conductor; it is an electronic semiconductor and an electrolytic semiconductor (silver ion is the mobile ion). Redox interference is determined by the electronic band gap of the material. The band gap of the silver salts is large enough that redox interference is not usually a problem in practice.

The silver chloride membrane ISE differs from the fluoride ISE in other ways: the electrode does not have a gel layer and it gives a Nernstian response to both silver ions and chloride ions. An example of the response behavior is shown in Figure 11.

Note that the electrode gives a constant response whenever the activity of chloride or the activity of silver ion is less than the square root of the K_{sp} of silver chloride. By calling chloride ion a silver ion buffer and drawing the collinear line segments as a single straight line, the linear dynamic range for silver activity has been said to extend from 0.10 to 1.0×10^{-9} M. Mechanistically, these curves indicate that when silver ion is the dominant ion in solution, its interaction with the membrane determines the potential, and vice versa for chloride ion. It is well known that when solid silver chloride is present in a solution with excess silver ion (cation common ion effect), silver ion adsorbs on the surface of the solid. When the solid is in a solution with excess chloride ion (anion common ion effect), chloride ion adsorbs on the solid.

Adsorption/desorption equilibrium at a membrane–electrolyte interface will alter the distribution of defects in the space-charge layer of the membrane. For example, it has been shown that the space-charge potential of a silver chloride membrane in contact with an aqueous solution containing silver ion depends on pAg.[88,89] A diffusion-controlled transport model with $D \approx 5 \times 10^{-8}\,\mathrm{cm^2\,s^{-1}}$ gave a good fit to experimental conductance

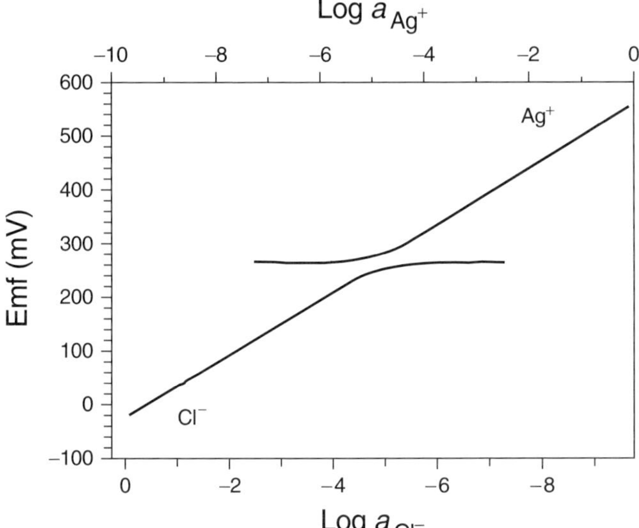

Figure 11 Response of an AgCl ISE to solutions of silver ions and solutions of chloride ions.

and capacitance data obtained on a frequency range of $30-10^5$ Hz. Using data from Hoyen et al.,[89] the thickness of the space-charge layer is calculated to be 25 nm. An estimate of the time constant needed to achieve a new equilibrium in the space-charge region after changing the activity of silver ion or chloride ion at the membrane surface is given by Equation (33):

$$\tau = \frac{4\delta}{\pi^2 D} \tag{33}$$

for first-order diffusion transport, where δ is the space-charge layer thickness. A value of 64 µs is obtained, which corresponds to a t_{95} response time of 192 µs ($t_{95} = 3\tau$). Rangarajan and Rechnitz[90] found $t_{95} \approx$ 200 ms for an Orion-type 94-17-00 electrode subjected to a 10-fold concentration step-up in an electrode flow chamber. Pungor[91] reported a response time of 20 ms, which is probably τ, using a switched wall-jet measuring cell arrangement. The experimental response times include diffusion across the Nernst layer plus whatever process gives rise to the interfacial potential. Adsorption/desorption leading to modulation of the space-charge defect distribution in the space-charge layer is fast enough to account for these experimental results.

Buck[84] found that the impedance plots for silver chloride were independent of pCl. This would seem to suggest that chloride adsorption did not modulate the space-charge distribution, so that adsorption alone must be responsible for the interfacial potential at the membrane. In essence, the potential is changed only at the top monolayer of the membrane, which is equivalent to causing a jump discontinuity in the space-charge potential distribution. This is consistent with the coupled diffusion/adsorption model used by Berube et al.[78] to interprete dynamic response data for the iodide ISE perturbed by iodide steps. Here, diffusion refers to the transport that occurs through the Nernst layer. The bromide ISE based on silver bromide is in all respects similar to the chloride and iodide ISEs.

Silver sulfide has a much smaller band gap than the silver halides. At room temperature, the most stable crystallographic form is β-Ag_2S, which has a silver ion conductivity of $3.60 \times 10^{-4}\,\Omega^{-1}\,\mathrm{cm^{-1}}$ and an electronic conductivity of only $4.00 \times 10^{-6}\,\Omega^{-1}\,\mathrm{cm^{-1}}$. Hence, its electrolytic properties dominate, and it is this form that is useful as an ion sensor. Not only is it a sensor for silver ions and sulfide ions, but it is also a component of the homogeneous mixture membranes of the cadmium ISE (CdS/Ag_2S), the lead ISE (PbS/Ag_2S), the copper ISE (Cu_2S/Ag_2S and CuS/Ag_2S), the mercury(II) ISE (HgS/Ag_2S), and the iodide/cyanide ISE (AgI/Ag_2S). Other metal sulfide/silver sulfide homogeneous mixture membranes have been studied, but they have not been commercialized.

For references see page 9904

Like lanthanum fluoride, the surface of silver sulfide becomes hydrated on soaking in water.[92] However, the concentration profile for water shows an approximate exponential decay, $e^{-\lambda d}$, with depth d into the membrane. The reciprocal of λ is ~2 nm, and at $d \approx 8$ nm, the boundary with dry silver sulfide is reached. The interfacial potential is believed to be generated by a space-charge mechanism.[92] The divalent sulfides are electronic semiconductors but electrolytic insulators. In the homogeneous mixtures, they provide absorption centers for the divalent cations, while the silver sulfide provides the electrolytic conductivity. However, in an elegant experiment, Uosaki et al.[93] showed that a single crystal of CdS functions as an ISE. The configuration was one of all solid state, and the external surface was either an (0001) Cd face or an (0001) S face. The (0001) Cd face showed a Nernstian response to HS^- concentration (Na_2S solutions buffered at pH 9) and the (0001) S face showed a Nernstian response to Cd^{2+} concentration. The results have been plotted in Figure 12 using the same format as Figure 11.

The equilibrium expression allowing the concentrations of Cd^{2+} and HS^- to be related is given by Equation (34):

$$CdS_{(s)} + H_2O \rightleftharpoons Cd^{2+} + HS^- + OH^- \quad (34)$$

with $K = 6.4 \times 10^{-28}$. Since pH = 9.0, $C_{cd} \times C_{HS} = 6.4 \times 10^{-19}$, where activity effects have been ignored. The plot is analogous to those obtained for silver iodide and silver sulfide.[94] Most significantly, the slopes of the emf versus log C plots were identical with the slopes of the flat band potential versus log C. This is, perhaps, the strongest evidence yet for an adsorption mechanism for interfacial potential generation. As discussed in that paper, a general equation for the variation of the flat band potential with activity of the adsorbate, assuming a Frumkin adsorption isotherm, reduces to the Nernst equation when there are a large number of adsorption sites and there is no interaction between adsorbates. Because repulsion will keep adsorbates from approaching close to each other on the surface, this approximation will almost always be valid.

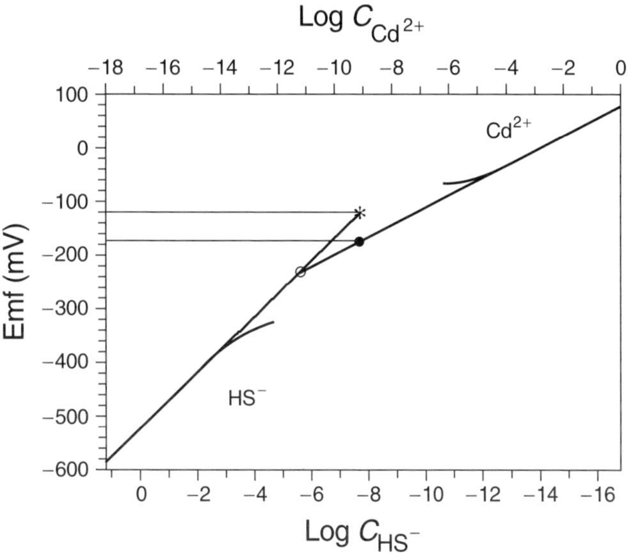

Figure 12 Response of a single-crystal cadmium sulfide ISE to solutions of cadmium ions and solutions of hydrogen sulfide ions. (○) Point of intersection of the extended linear regions; (∗) point on extended hydrogen sulfide line where the hydrogen sulfide concentration equals 8×10^{-10} M; (●) point on extended cadmium line where the cadmium concentration equals 8×10^{-10} M.

The electrode interferences for these ISEs are easy to predict. Any cation or anion which forms a sparingly soluble salt with the counterion of the membrane is an electrode interference. A metathetical reaction, which alters the membrane surface composition and modifies the electrode response, will occur. The selectivity order may be predicted by calculating the equilibrium constant of the metathetical reaction using pertinent activity product constants. Selectivity will be of the order of the K values: the larger is K, the more serious is the interference by the ion. For quantitative work the selectivity coefficients should be measured.

These electrodes are also subject to numerous chemical interferences. The lead, cadmium and copper ISEs are subject to oxidation or reduction by sufficiently strong redox-active species. If the band gap is small, then photodegradation may also occur. Photodegradation may activate oxidation or reduction reactions which are energetically unfavorable in the absence of light. It can also catalyze the degradation of chemical interferences. All of the cations in these membranes form complex ions with some ligands. These are usually charged species, and therefore have a high solubility in aqueous media. The net result is that etching or leaching of the membrane will occur. Mild etching will cause so-called sub-Nernstian responses and longer response times. Severe etching may cause loss of response. In many cases, the responses of the electrodes may be restored by mechanical polishing. However, since membrane material is lost with each polish, the lifetime of the electrode will be significantly diminished. It is best to avoid exposing the electrodes to solutions containing known interferences. The interference problem has somewhat limited the use of these electrodes as detectors for real-time analysis of complex samples by methods such as flow injection analysis. Taken as a set, the quantitation detection limits of these electrodes range from 10^{-8} to 5×10^{-5} M, with 6×10^{-6} M as the median value. The upper quantitation detection limits range from 10^{-2} M to saturation, with 1.0 M as the median value.

3.2.1.2 Noncrystalline Electrodes The glass membrane electrodes and the electrodes with mobile charged sites, the various ionophore-containing polymer membranes, belong to this class. There are many types of inorganic glasses, but the silicate glasses are the only ones used to make commercial ion-sensing membranes. The most important ISEs of this type are the pH electrode and the sodium ion electrode. Silicate glass is formed from SiO_2, a network former, and one or more metal oxides, network modifiers. For example, in a sodium silicate glass, some of the oxygen atoms are bonded to two silicon atoms (bridging oxygen), other oxygen atoms are bonded to only one silicon atom [nonbridging oxygen (NBO) atoms, which carry a charge of −1], and sodium ions are either electrostatically bonded to NBOs or isolated in percolation channels. Percolation channels are tunnels through which alkali metal ions may move. Some network modifiers, e.g. aluminum oxide, give rise to substitutional impurities in the network itself, e.g. an aluminum substitutes for a silicon. When the valence of the substitutional impurity is not the same as that of silicon, a second network modifier, an alkali or alkaline earth metal oxide, must be used to provide for charge compensation.

Although silica glass membrane electrodes have been studied since the beginning of this century, it is easy to understand why their behavior is still being characterized. The properties of these materials are not monolithic; they depend strongly on the chemical composition of the dry glass, on the history of its preparation, and on its state of hydration during utilization. Depending on its composition, a dry silicate glass can be an electronic insulator and an electrolytic semiconductor or an electronic insulator and an electrolytic insulator. For example, at 27 °C NAS 25-25 glass has a conductivity of $1.0 \times 10^{-8}\,\Omega^{-1}\,cm^{-1}$ and is an electrolytic semiconductor, while Corning 015 glass has a conductivity of $8.6 \times 10^{-12}\,\Omega^{-1}\,cm^{-1}$ and is an electrolytic insulator. The former types might be expected to exhibit some of the behavior that we have observed for crystalline membrane ISEs. In the dry state, the latter types can only behave as capacitative devices. However, sufficient hydration of a dry glass can cause it to undergo a transition from an electrolytic insulator to an electrolytic semiconductor. Thus, conflicting reports on the behavior of glasses with the same nominal chemical composition can probably be traced to a failure to consider their preparation and utilization histories.

All glass membrane ISEs respond to monovalent cations and NH_4^+, and to a lesser extent to divalent cations. The dry glass chemical composition determines which ion gives the greatest response, i.e. it defines the target, and also defines the selectivity order for electrode interferences.

In Table 2, some important glass compositions which are used as membranes to target H^+, Na^+ and K^+ are shown. Note that a gel layer forms on Corning 015 glass when it comes in contact with water or aqueous solutions. A glass scientist would say that Corning 015 glass is subject to "glass corrosion". A glass subject to corrosion will begin to corrode immediately after water contact. As can be seen from Table 2, not all glasses are subject to significant glass corrosion.

Using infrared (IR) spectroscopy, Raman spectroscopy and magic-angle spinning (MAS) nuclear magnetic resonance (NMR) spectroscopy, Pandya et al.[95] examined water speciation in alkali silicate glasses with the compositions shown in rows 3, 5 and 7 of Table 2. These had been hydrated over a period of 6 years under ambient conditions, which apparently meant exposure to moisture-laden air at room temperature. The lithia glass had <0.01 wt% water, present as silanol, and no gel layer. The sodium glass had about 5.2 wt% water, present as monomeric, dimeric, and polymeric water. No silanol was observed. The potassium glass was essentially all gel, with 32 wt% water present as monomeric, dimeric, and polymeric water. Again, no silanol was observed. The absence of silanols in the latter cases was probably due to the conditions of hydration. Diffusion of water into glass initiates stage 1 corrosion, which involves the leaching of alkali metal ions from the glass, and which requires that the glass be in contact with bulk water. In stage 1 corrosion, alkali metal ions are leached as water reacts with ≡SiONa groups to form silanol, ≡SiOH, and release Na^+ and OH^- to water or to the solution. The process has been characterized for Corning 015 glass, both for pure

Table 2 Examples of glasses used to form membranes for H^+, Na^+ and K^+ (the numerical values give the composition in mol%)

Target	Li_2O	Na_2O	K_2O	CaO	BaO	Al_2O_3	SiO_2	Gel layer?	Comments
H^+	–	21.3	–	6.4	–	–	72.3	Yes	Corning 015
H^+	14.3	–	–	–	7.0	–	68.7		Lithia glass
–	30.0	–	–	–	–	–	70.0	No	
Na^+	–	11.0	–	–	–	18.0	70.0		NAS 11-18
–	–	45.0	–	–	–	–	55.0	Yes	
K^+	–	27.0	–	–	–	6.0	67.0		NAS 27-6
–	–	–	35.0	–	–	–	65.0	Yes	

For references see page 9904

water and for electrolyte solutions and for temperatures of 27, 46, and 57 °C.[96]

If pure water is used, the pH increases to ≥9, and then stage 2 corrosion begins. This involves hydration of the Si−O−Si groups, i.e. breakdown of the network. Stage 2 corrosion was not characterized for Corning 015 glass, but Zhang et al.[97] measured both sodium and silicon loss for densified and undensified $Na_2O \cdot 3SiO_2$ glasses in distilled water at 5 °C for both stage 1 and stage 2 corrosion. These researchers showed a scanning electron micrograph of the gel layer formed on the undensified glass. Similar behavior is expected for the Corning 015 glass, and other sodium- and potassium-containing glasses.

According to Buck,[98] gel layers with thicknesses between 60 and 1000 nm may form on pH-, sodium- and potassium-sensitive glasses. Since these gel layers may have diffusivities that are several orders of magnitude greater than dry glass,[99] the gel layers may exhibit electrolytic semiconductivity. These glass electrodes will exhibit a space-charge layer, and the mechanism of response should be similar to that of the crystalline membrane electrodes. The results on the lithia glass may indicate that the pH electrodes based on lithia glasses have no gel layer. In that case, the potential must be due to adsorption of hydronium ions or hydroxide ions limited to the surface layer. The adsorption model developed for CdS should apply. This is probably why pH electrodes based on lithia glasses have a quantitation concentration range from pH 0 to 14. The electrode probably has an upper quantitation detection limit of 1.0 M and a lower quantitation detection limit of 10^{-7} M for H^+ and also for OH^-. Hence the detection limits are similar to those observed for the crystalline membrane ISEs. For the sodium ISE, the upper quantitation detection limit is a saturated solution, and the lower quantitation detection limit is 10^{-6} M. More studies need to be carried out on these glasses, using techniques such as NMR, Raman spectroscopy, and scanning electron microscopy to characterize their behavior more fully.

The polymer-based liquid membrane ISEs come in three varieties: the neutral carrier-based ISE, the charged carrier-based ISE and the ion exchanger-based ISE.[25] The ion exchanger-based ISE membrane has an ion exchanger dissolved in an organic solvent, and the resultant solution is dispersed in an inert polymer support. Traditionally, long-chain alkylammonium salts or nonlabile metal complexes have been considered to be anion exchangers.[100] The electroactive carrier is dissolved in an organic solvent, and the organic solution formed is usually dispersed in an inert polymer support. For example, Nielsen et al.[101] constructed a nitrate ISE from tetraoctylammonium nitrate, dialkyl phthalate or dialkyl adipate, and PVC. When such a membrane is placed in contact with an aqueous phase, the tetraoctylammonium ion is confined to the membrane, but the nitrate ion is free to move into or out of the membrane. An example of a nonlabile metal complex is $Fe(II)(o\text{-phen})_3(ClO_4)_2$.[100] The term nonlabile means that the Fe(II) ion is not present as the free ion in the membrane. The anion, which is the counter ion, is free to move. Cation exchangers are typically long-chain metal sulfonate salts or labile metal ion complexes dissolved in an organic solvent, with this organic solution dispersed in an inert polymeric support. For example, the first Ross calcium ISE membrane was formed from calcium bis(2-ethylhexyl phosphate) dissolved in dioctyl phenylphosphonate and supported in a porous, inert polymer.[45] The calcium bis(2-ethylhexyl phosphate) can ionize, and the free metal ion can move between the membrane and a contacting aqueous phase. However, the 2-ethylhexyl phosphate ligands, which carry a charge of −1, cannot leave the membrane.

Ion-exchanger ISE membranes are electronic insulators and electrolytic semiconductors. Potential generation for these membrane systems is similar to that for other electrolytic semiconductors. For example, Li and Harrison[102] characterized transport in a nitrite ion-selective membrane using spatial imaging photometry. The system consisted of bromo(pyridine)(5,10,15,20-tetraphenylporphyrinato)cobaltate in dioctyl adipate supported in PVC. They found that on exposure to water, hydration occurs, giving a gel layer that is about 50 µm thick. When the membrane was conditioned in a nitrite solution, the nitrite ion replaced the bromide ion. Nitrite ion diffused across the membrane, to give ultimately a nearly uniform distribution, but the diffusion coefficient was only about $5 \times 10^{-9}\,cm^2\,s^{-1}$. The potential is established in a thin (∼100 Å thick) region at the membrane/solution interface, i.e. a space-charge layer exists.

NMR studies have revealed that the uptake of water by polymer-based liquid membranes occurs in two stages.[103] The first stage has been attributed to water miscibility in the membrane phase. This is equivalent to the existence of monomeric water in the silicate glasses. In the second stage, droplets of water are formed. With the charged-carrier ISE membrane, the carrier is charged in its uncomplexed state, and forms a neutral complex with the target ion. Such membranes also contain ion-exchanger sites, which are needed to insure permselectivity for the target ion.[25]

An example of a charged-carrier anion ISE is the nitrite electrode based on the aquocyanocobalt(III) heptakis(phenylethyl)cobyrinate ion.[104] This membranes has the following composition: 1.0 wt% aquocyanocobalt(III) heptakis(phenylethyl)cobyrinate perchlorate, 37 mol% potassium tetrakis[3,5-bis(trifluoromethyl)phenyl]borate (KTFPB), 65 wt% 2-nitrophenyl octyl ether (NPOE) and 33 wt% PVC.[26] The ligand, confined to the membrane, carries a positive charge when not complexed to the target ion. Hence it excludes cations from the membrane. KTFPB is used to improve the selectivity of the membrane. It dissociates to give a large, lipophilic anion in the membrane which functions to exclude anions. Only an anion which forms a neutral complex with the carrier can enter the membrane.

A charged-carrier cation ISE will have the complementary membrane composition, i.e. a ligand which is negative when not complexed, a salt which contains a lipophilic cation, the solvent, and the polymer support. In their review, Bühlmann et al.[26] included examples of attempts to form such electrodes, but it seems that none of them have given adequate responses or selectivity. For example, Midgley[105] tried to make an ion-exchanger Fe(III) ISE based on the charged-carrier myobactin S. It failed to respond to Fe(III), but did respond to salicylate. The predominant mechanism responsible for potential generation for the charged-carrier ISEs is extraction by means of complex formation.

The third type of polymer-based liquid membrane is the neutral-carrier type. A neutral carrier is an uncharged, lipophilic complexing ligand. Some are large macrocycles with hydrophobic exteriors and internal cavities lined with polar groups. A metal ion which has a diameter such that a "lock and key" fit may occur between the metal and the ligand will form a strong complex with the ligand. Smaller or larger ions form weak or no complexes with the ligand. An example of a neutral-carrier macrocycle is shown in Figure 13(a). Many of the neutral carriers being used today are not macrocycles. Examples are the various Nile Blue oxazone derivatives (ETH *wxyz*, where *w*, *x*, *y*, and *z* are integers)[106] and *para*-substituted trifluoroacetyl benzenes. The former are chromoionophores originally developed for optodes, but have now been used as neutral carriers for several different cations.[26] The latter are neutral carriers used in commerical carbonate ISEs. An example of a neutral-carrier ISE membrane for a cation is the valinomycin potassium ion ISE. Valinomycin and potassium ion give a "lock and key" fit, as shown in Figure 13(b).

An example of a recipe for such a membrane is as follows: 1.0 wt% valinomycin, 0.01 wt% potassium tetraphenylborate ($KBPh_4$), 66.0 wt% dioctyl sebacate (DOS) and 33.0 wt% PVC. $KBPh_4$ is completely ionized in the organic solution, with the potassium ions all complexed by valinomycin and the tetraphenylborate ions constituting mobile negative ion sites confined to the membrane. A schematic of such a membrane is shown in Figure 14. The tetraphenylborate ions function to exclude sample anions from the membrane.

It has been shown that functional ISEs may be prepared without adding the ion-exchanger salt. Ye et al.[107] showed by XPS and time-of-flight (TOF) secondary ion mass spectrometry (SIMS) that there are carboxylate and

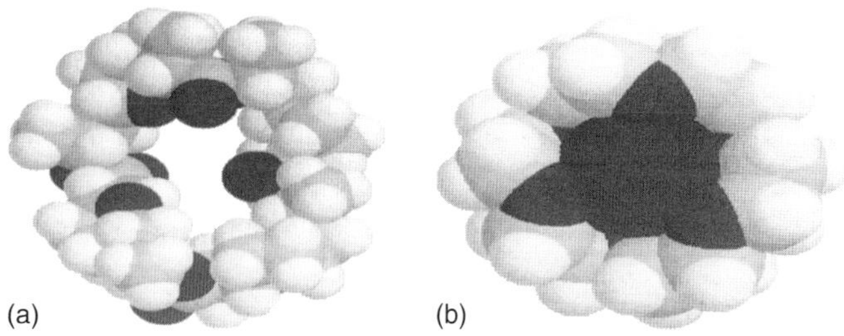

Figure 13 (a) A macrocyclic neutral ligand, nonactin. (b) The K^+–valinomycin complex ion. (Reproduced by permission of David Woodcock.)

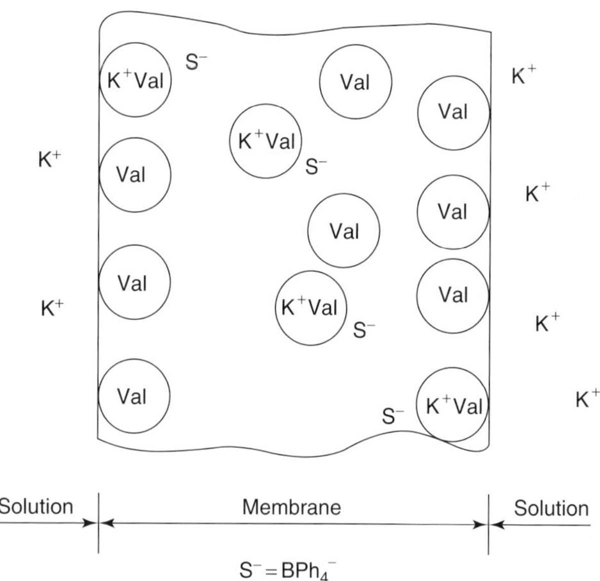

Figure 14 Schematic of a polymeric neutral-carrier liquid membrane.

sulfate groups covalently attached to the polymer support. Where these are in contact with water in the membrane, they will form fixed charge sites in the membrane. The number of such sites, and their distribution, are unknown. The addition of ion-exchanger salts allows the number of sites to be controlled, and this leads to improvements in selectivity. The potassium ions from the $KBPh_4$ are complexed with the neutral carrier, giving lipophilic cations. These will be excluded from the membrane cations which do not complex with the neutral carrier. DOS is the solvent.

An example of a neutral-carrier anion ISE is one whose membrane contains *N*,*N*-dioctyl-4-trifluoroacetylbenzamide, a neutral carrier for carbonate. The membrane composition is as follows:[26] 3.4 wt% *N*,*N*-dioctyl-4-trifluoroacetylbenzamide, 41 mol% tridodecylmethylammonium chloride (TDDMACl), 54 wt%

For references see page 9904

DOS and 41 wt% PVC. Here, TDDMACl is the ion-exchanger salt, which gives large, lipophilic cations in the membrane. With the neutral carrier-type ISEs, the mechanism for potential generation is ion extraction. An excellent discussion of the potential generation mechanism for the valinomycin potassium electrode has been published by Sandifer.[108]

While the polymeric membrane ISEs are being used to analyze for lithium, potassium, magnesium and carbonate in commercial clinical analyzers, the greatest potential for these electrodes lies in their ability to be tailored for the analysis of biologically important organic cations and anions. The key is to find carriers that selectively target an analyte present at a low concentration in a complex sample with high concentrations of other ions. The ability to achieve selectivity by design has been greatly advanced by the recent development of a precise selectivity model for these electrodes. It began with a demonstration of the inability of the Nicolsky–Eisenman formalism to give self-consistent results for a target ion with charge +1 and an ion interference with charge +2. If we consider the electrode response for an ISE immersed in series of mixed solutions with a fixed activity of the ion interference of charge +2 in which the activity of the target ion of charge +1 varies, we expect the response to be described by a single curve which increases linearly as a function of the logarithm of the target activity, a_Q, when a_Q is much larger than the activity of the ion interference and is constant when a_Q is much smaller than the activity of the ion interference. Thus, the latter part of the curve is described by a straight line of zero slope. In the intermediate region, a single curve must be obtained, because the electrode will give a single measured value for each solution. However, the Nicolsky–Eisenman formalism gives different curves, depending on whether the electrode is viewed as a sensor for Q or as a sensor for the ion interference.[79] By considering the response mechanisms for the three types of polymeric membrane electrodes and using appropriate electroneutrality conditions, complex formation constants and mass balances, the authors derived explicit electrode response functions for the case of $z_Q = +1$ and z ion interference $= +2$. This involved the introduction of a selectivity factor, k^{sel} = the ratio of the activity of the target to the activity of the ion interference which gives the same potential when measurements are made in solutions of the single ions. For the ion-exchanger polymeric membrane, k^{sel} can be expressed in terms of the so-called "single ion distribution coefficients". For the charged-carrier polymeric membrane, k^{sel} can be expressed in terms of respective single ion distribution coefficients and complex formation constants. For the neutral-carrier polymeric membrane, k^{sel} depends on the total neutral carrier, as well as the single ion distribution coefficients and complex formation constants. The explicit functions are obtained by expressing the conventional Nicolsky selectivity coefficient in terms of k^{sel}.

In a subsequent paper,[25] explicit response functions were given for the following coupled charge magnitudes: (1,2), (2,1), (1,3), (2,3), (3,2) and (3,1), where the first element of the pair corresponds to the target ion and the second to the ion interference. Most recently, a general implicit function valid for any number of sample ions with any type of charge has been developed.[82] Using E_B as the phase boundary potential difference (this is directly proportional to $\phi_\beta - \phi_\alpha$ in our treatment in the general section), the implicit function is given as Equation (35):[82]

$$1 = \sum_i a_i(\mathrm{org}) K_i \exp\left(\frac{-z_i E_B}{s}\right) \tag{35}$$

where i represents the ions, a is activity, z_i is the ion charge, $s = RT/\Im$ and K_i is a weighing factor, which combines a number of constants that depend on the type of polymeric membrane under consideration. For the ion-exchanger membrane, it is given by Equation (36):[82]

$$K_i = \frac{z_i k_i}{\gamma_i R_T} \tag{36}$$

where k_i is the single ion distribution coefficient, γ_i is the activity coefficient and R_T is the concentration of nonexchangeable ions in the membrane. For the ionophore-based membrane (neutral carrier or charged carrier), K_i is given by Equation (37):[82]

$$K_i = \frac{z_i k_i \beta_{i,n} c_L^0(\mathrm{org})^n}{R_T \gamma_{i,n}} \tag{37}$$

where $\beta_{i,n}$ is the complex formation constant for ion i forming a $1:n$ complex with the ligand, $\gamma_{i,n}$ is the corresponding activity coefficient, c_L^0 (org) is the initial concentration of the uncomplexed carrier, and the other terms have been defined above. Potentiometric selectivity coefficients are expressed in terms of these weighing factors as Equation (38):[82]

$$K_{I,i}^{pot} = \frac{K_I^{\;z_I/z_i}}{K_i} \tag{38}$$

In turn, k_{Ii}^{sel} can be related to the potentiometric selectivity coefficients, and this leads to a simple expression for the boundary potential (Equation 39):[82]

$$E = E_I^0 + \frac{RT}{z_I \mathcal{F}} \ln\left[a_I(\mathrm{aq}) + \sum_{i \neq I} k_{Ii}^{sel} a_i(\mathrm{aq})\right] \tag{39}$$

The usefulness of the k^{sel} is obvious. First, by measuring single ion distribution coefficients and complex formation constants, and by determining the appropriate concentration of lipophilic ion sites needed, the selectivity of the membrane can be designed. Optimization techniques can be used to design membranes to give the best response for a set of specified ion interferences. This will allow the designer to generate quickly a set of membrane compositions likely to be suitable detectors for a given problem. The experimental work would then be limited to those compositions. Intelligent design allows time to be used more efficiently and minimizes waste of reagents.

Just as the interference order may be qualitatively predicted for crystalline membrane ISEs by using K_{ap} values, the selectivity order for charged-carrier ionophore anion ISEs often obeys the Hofmeister lyophilic series (Equation 40):[109]

$$R^- > ClO_4^- > I^- > NO_3^- > Br^- > Cl^- > F^- \quad (40)$$

where R^- is an organic anion. For charged-carrier ionophore cation ISEs, the Hofmeister series is [Equation 41][109]

$$R^+ > Cs^+ > Rb^+ > K^+ > Na^+ > Li^+ \quad (41)$$

where R^+ is an organic cation. There does not appear to be a recurring selectivity order for divalent cations, etc. For the neutral carrier-type liquid membranes, the size and shape of the ion cavity determine the selectivity order to a large extent. It is possible to use molecular modeling techniques to investigate selectivity orders for any given neutral carrier. Since the organic solvent also influences the selectivity order, these methods are a kind of zeroth-order approximation. Commercially available macroelectrode ISEs with plastic membranes have an average quantitation lower detection limit of about 4×10^{-6} M and an average quantitation upper detection limit of about 1.0 M. With crystalline membrane ISEs, the selectivity is determined by the properties of the solid. By buffering the sample solution, the lowest quantitation detection limits may be realized, but there can be no further improvement. With the polymeric membrane ISEs, the composition of the membrane can be changed to improve the selectivity, which, when coupled with buffering, can push the lower quantitation detection limit down. One way to improve membrane selectivity is to find more selective ionophores. This is why so much research in this area involves synthesizing and evaluating new ionophores, not only for new target ions, but also for ions that already have polymeric membrane ISE sensors. For example, neutral-carrier polymeric membrane ISEs for calcium and lead with lower quantitation detection limits of 10^{-9} M have been described.[110,111] In comparison with the commercial calcium polymeric membrane ISE, this respresents an improvement by a factor of 500, and in comparison with the commercial crystalline membrane lead ISE, an improvement by a factor of 1000.

3.2.2 Compound or Multiple Membrane Ion-selective Electrodes

These include gas-sensing electrodes (called Severinghaus-type potentiometric detectors) and enzyme electrodes. A gas-sensing electrode probes the concentration (activity) of a dissolved gas in an external solution as the result of admission of the dissolved gas into a compartment containing a solution which can interact with the gas to form the target ion of the internal ISE. The compartment and the external solution are separated by a microporous, semipermeable membrane, so the gas must diffuse across this membrane. If the gas of interest is in the vapor phase, then its solution concentration (activity) will be determined by its solubility equilibrium. The process that occurs in the compartment is kinetically fast, so the response time of the electrode will be determined by the kinetics of the diffusion process and the kinetics of the ISE response. The former is always much slower. The measurement should be taken under steady-state conditions. The concentration (activity) of the gas to be measured can be related to the activity of the target ion by means of a mathematical expression based on the multiple equilibria which occur. No general expression is possible, but an example might be instructive. Suppose that we choose to monitor the partial pressure of ammonia in a stack gas using an ammonia gas ISE. The ammonia gas electrode contains a pH glass electrode, a compartment which contains either 10^{-3} M NH_4Cl or 0.1 M KNO_3, and a 0.1-mm micropore Teflon® membrane.[112] The electrochemical cell consists of the ammonia gas electrode, a reference electrode and an external solution in contact with the stack gas. The concentration of dissolved ammonia is related to the partial pressure of ammonia gas in the stack gas by means of Henry's law, $P_{gas} = k[NH_3]_{aq}$. Since a concentration gradient for dissolved ammonia exists across the membrane, ammonia diffuses into the compartment. Diffusion must occur until the concentration gradient is zero, so the concentration of dissolved ammonia in the compartment will be the same as that in the external solution at steady state. If the comparment contains 0.1 M KNO_3, then base hydrolysis of the ammonia fixes the hydroxide ion activity, which fixes the hydrogen ion activity. The ISE's electrode response is related to P_{gas} by Equation (42):

$$E_{ISE} = K_{electrode} + S\log\left(\sqrt{kK_wK_a}\right) - \frac{S}{2}\log(P_{gas}) \quad (42)$$

where S is the slope factor for the pH electrode and the relationship $K_b = K_w/K_a$ has been used. The electrode gives a linear response as long as the approximation $[NH_4^+] = [OH^-]$ is valid, which is so as long as the hydrolysis of water can be neglected. If 10^{-3} M NH_4Cl is in the compartment, then a buffer is formed in

For references see page 9904

the compartment when ammonia diffuses across the membrane. In that case, the ISE's response is related to P_{gas} by Equation (43):

$$E_{ISE} = K_{electrode} + S\log(10^{-3}kK_a) - S\log(P_{gas}) \quad (43)$$

The electrode response is linear so long as the equilibrium concentrations of ammonia and ammonium ion are given by their analytical concentrations. In both cases, ionic strength effects must be adequately addressed. With these electrodes, the recovery times may be long (many minutes), so care must be taken to avoid memory effects in successive measurements. Interferences arise from other gases which can diffuse across the membrane and either control the activity of the target ion or form an ion which is an electrode interference for the internal ISE. For example, for the ammonia gas electrode, volatile amines interfere, since they can diffuse across the Teflon® membrane and they are weak bases, like ammonia. The quantitation lower detection limits for the commercial macroelectrode ISEs for NO_x and ammonia are 4×10^{-6} and 5×10^{-7} M, respectively. For the commercial carbon dioxide macroelectrode ISE, the quantitation lower detection limit is only 10^{-4} M. The detection ranges for the NO_x and for the carbon dioxide macroelectrode ISEs are fairly small; the quantitation upper detection limits are 5×10^{-3} and 10^{-2} M, respectively. By contrast, the ammonia macroelectrode ISE has a quantitation upper detection limit of 1.0 M. Recently, polymeric membrane-type ISEs have been used to assemble carbon dioxide and ammonia sensors that are analogs to the Severinghaus-type sensors. The carbon dioxide sensor uses a neutral-carrier hydrogen ion polymeric membrane,[26,113] whereas the ammonia electrode uses a neutral-carrier ammonium ion polymeric membrane.[26,114–116] The neutral-carrier ammonium ion polymeric membrane shows an improved quantitation lower detection limit of 25, relative to the macroelectrode ISE. A neutral-carrier polymeric membrane-based ISE has been developed for sulfur dioxide.[26] At present, there is no commercial macroelectrode ISE for this gas. Oxygen sensors have also been described,[26] but it is not clear that the response is due entirely to ion transport and not electron transfer. These electrodes may not be ISEs in the traditional sense.

An enzyme electrode has a gel layer containing an enzyme interposed between the indicator ISE membrane and an external solution. The enzyme catalyzes the conversion of a substrate to a set of products, one of which is a target of the ISE. Targets are frequently small molecules or ions, e.g. ammonia, carbon dioxide, hydronium ion, cyanide ion.[50] The selectivity of these electrodes is governed by the specificity of the enzyme. Because enzymes are stereoselective, these ISEs come the closest to being electrodes that give an activity-dependent response for a single type of molecule or ion, the substrate of the enzyme. The generation of the signal requires three steps: (1) the substrate must diffuse from the external solution into the enzyme gel layer, (2) the enzyme must catalysis the conversion to the set of products and (3) the product which is the ISE target must diffuse to the surface of the ISE membrane. In principle, the kinetics associated with step 2 can be complex. Also, while the concentration gradient is such that the substrate diffuses into the gel, the products will diffuse both toward the membrane surface and toward the external solution. Practitioners have chosen enzyme systems which obey Michaelis–Menten kinetics, which is the simplest type. Even so, the transport equations are inherently nonlinear. In order to obtain an equation which relates the potential of the enzyme electrode to the substrate activity (concentration), the equations are solved for steady-state conditions. At the steady state, the diffusion processes may be ignored, and only step 2 is important. In such a case, a set of simultaneous first-order differential equations need to be solved. An example of such a treatment can be found in the book by Morf.[117] Brady and Carr[118] actually solved numerically the coupled set of nonlinear, second-order differential equations for a set of reasonable boundary conditions. Since an expression for the electrode potential as a function of substrate activity (concentration) cannot be obtained in analytical form, this method is not useful to the practitioner. However, it has allowed certain approximations, which constitute general guidelines for the use of such electrodes, to be validated. In general, a linear response region is observed, the lower bound being determined by the ionic background in the sample, and the upper bound being determined by the Michaelis constant for the enzyme system.

3.2.3 Metal Contact or All-solid-state Electrodes

The behavior of these electrodes is similar to that of the primary electrodes. One type of failure which we have seen for microelectrodes of this type is the decoration of the membrane surface with silver, when silver is used to make the internal contact. Silver atoms have fairly large diffusion constants, in comparison with other metal atoms, and thus even microcracks will result in its ultimate transport to the surface. When this happens, the electrode response gives sub-Nernstian slope factors.

4 INSTRUMENTATION

An ISE measurement involves the determination of the potential difference between an ISE and an external reference electrode. The instrumentation used to make this voltage measurement must not load the circuit. There are two types of instrumentation that may be used to make voltage measurements – a potentiometer or a voltmeter. A potentiometer is a null device, whose limitations depend on the current sensitivity of the null detector. The classical voltmeter is a direct measuring device, e.g. the D'Arsonval meter, in which a small current causes a meter deflection. For a given electrochemical cell, the absolute error in the voltage measurement will depend on both the magnitude of the actual voltage to be measured and on the meter resistance (impedance). The electrochemical cell resistance is determined essentially by the resistance of the ISE, which in turn depends on the geometry of the sensing membrane. In Table 3 are shown some typical values of ISE resistances for various types of electrodes.(119,120) The meter resistance needed to give an absolute error of no more than 1 mV for the measurement of pH values from 1 to 14 may be calculated using the extremes of the output potential range, +414 to −414 mV.(119) A value of $8.0 \times 10^{11}\,\Omega$ is calculated. Thus, a meter resistance (impedance) of 10^3–10^4 times that of the cell resistance is needed.

Classical voltmeters, so-called VOM instruments, cannot be used to measure the voltages of electrochemical cell using ISEs, because the meter resistances (impedances) are too small (typically $1 \times 10^5\,\Omega$). Thus, early pH meters made use of potentiometers, wherein the null detectors evolved from galvanometers to vacuum tube configurations. However, the classical instrumentation era culminated with the development of the vacuum tube voltmeter by A. O. Beckman. This interesting history has been discussed by Jaselskis et al.(121)

The semiconductor revolution introduced the next innovation, as vacuum tubes were replaced by semiconductors, giving rise to the electronic voltmeter. The early electronic ion meters were meter readout instruments, but today's electronic ion meters are digital readout instruments, which utilize either a light-emitting diode (LED) display or a liquid crystal display (LCD). These instruments can have meter resistances (impedances) in excess of $10^{13}\,\Omega$. The heart of a modern electronic ion meter is the operational amplifier. The operational amplifier is a small-scale integration device which has a high input impedance ($\geq 10^{12}\,\Omega$), a low output impedance (≤ 1–$10\,\Omega$), and a large open-loop gain (10^4–10^6).(122) Many circuit design are possible; in Figure 15 is shown a simple circuit design which uses a voltage follower configuration with manual adjustment for the slope factor. Slope factor adjustment is needed to adjust for measurements made at different temperatures.

Table 3 Typical ISE resistances

Type of ISE	R (Ω)
Conventional glass membrane	5.0×10^7–2.0×10^8
Conventional crystalline membrane	2.0×10^5–3.0×10^7
Conventional liquid membrane	1.0×10^7–2.5×10^8
Conventional gas sensing	$\sim 5.0 \times 10^9$
Conventional enzyme electrode	1.0×10^7–2.0×10^9
Microelectrode	Up to 10^{12}

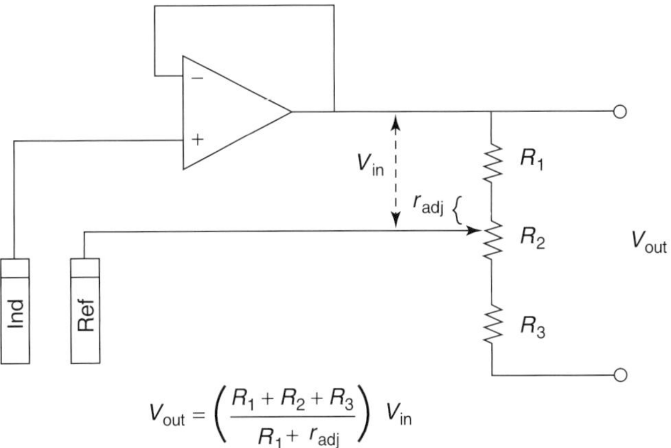

Figure 15 E_{cell} measurement with manual slope adjustment using a voltage follower operational amplifier circuit.

Modern digital instruments have many more features. Most are microprocessor control instruments with multiple measuring modes (e.g. millivolts, pX, concentration mode in various units), with temperature compensation probes, and connectors for interfacing the instrument with auxiliary instruments. In Figure 16 is shown a block diagram of the Fisher Accumet Model 750 selective ion analyzer. The important circuits are shown in blocks. For example, the temperature compensation may be done manually (by entering information into the system via the keyboard) or automatically by means of a probe. The instrument has an analog-to-digital converter, and the digital data are available in binary coded decimal at an external connector and in decimal form at an LED display. This meter is a desktop model, and the technology dates from about 1981.

The new benchtop Accumet research meters are microprocessor-based instruments with backlit LCD touchscreens, prompts and context-specific help screens. In both the pH mode and ion mode, they offer up to five points of calibration. There are also portable Accumet meters that offer the same calibration options, but these do not have the help screens.

In Figures 17–19 are shown examples of the most recent Orion technology. In Figure 17 is shown the Orion Model 710A ion meter, a basic benchtop ISE/pH meter. It also uses an LCD and a touchpad input module with

For references see page 9904

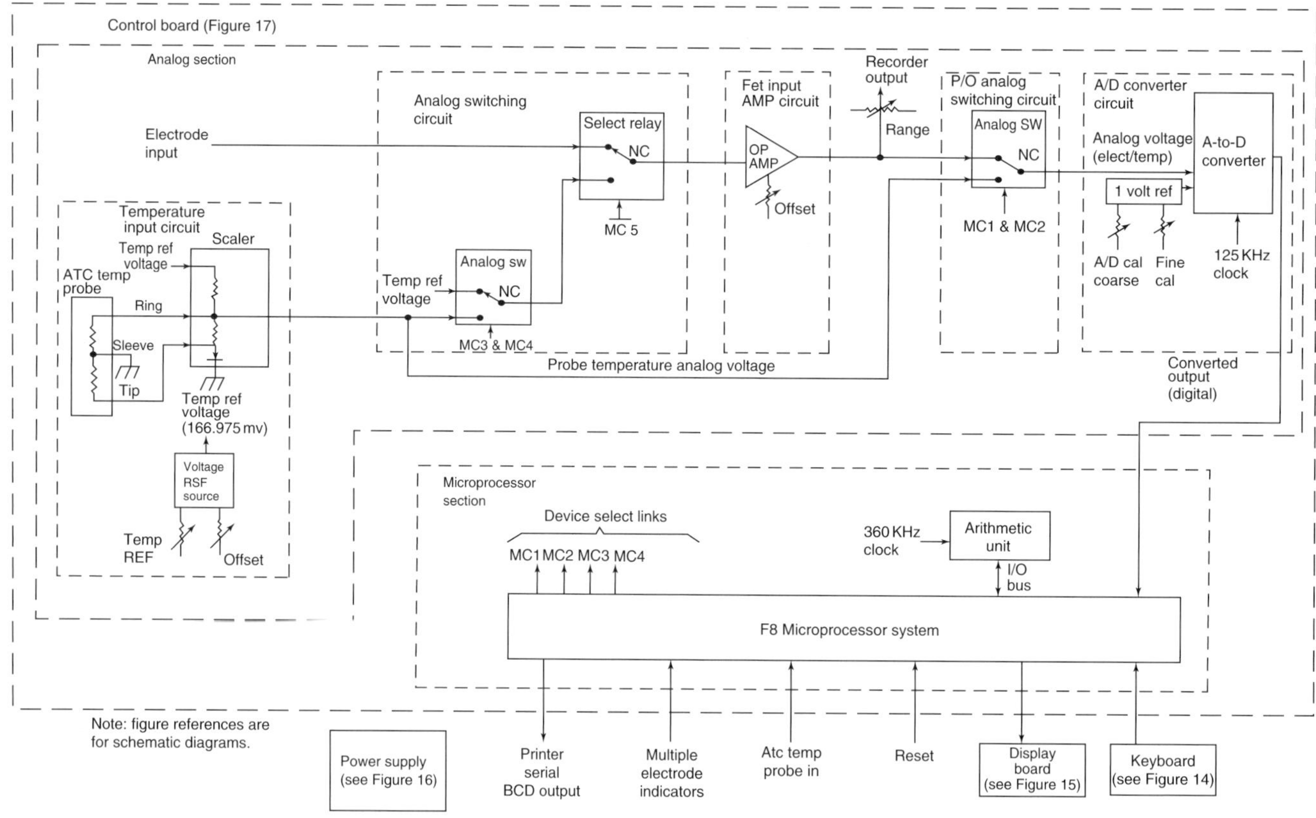

Figure 16 Block diagram for the Fisher Accumet Model 750 selective ion analyzer. (Reproduced by permission of Fisher Scientific.)

Figure 17 The Orion Model 710A pH/ISE meter. (Reproduced by permission of Orion Research, Inc.)

a large number of functions. It covers a pX range of −2.000 to 19.999, but the resolution depends on the pX value. For example, with pX = 7.00, the resolution is two

Figure 18 A portable pH/ISE meter, the Orion Model 290A. (Reproduced by permission of Orion Research, Inc.)

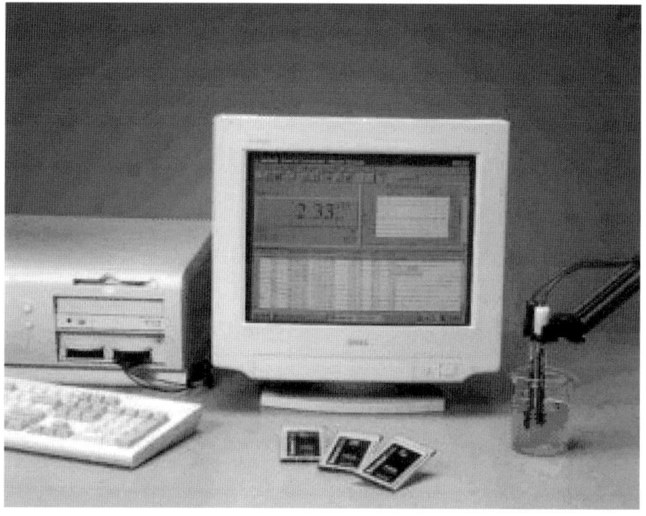

Figure 19 The Orion SensorLink® system interfaced to a PC. (Reproduced by permission of Orion Research, Inc.)

decimal places. The meter has auto-buffer recognition for five-point pH standardization. There are built-in buffer/temperature tables. By contrast, the older technology, e.g. the Accumet Model 750 meter mentioned above, uses two-point calibration, with the data entered into the system manually. In Figure 18, the Orion Model 290A meter is shown. This is an advanced, portable pH/ISE meter which is not much larger than a pocket calculator. Portable instruments are ideal for field measurements; the two-way RS232 communication and datalogging capabilities allow results to be downloaded later into a laboratory computer. In Figure 19 is shown one of the Orion SensorLink® systems. This is a PCMCIA card-based system, shown here interfaced with a PC; systems are also available for laptop computers. A standard Windows interface, using either Windows 95 or Windows 3.1, is used. The laptop systems are ideal for traveling scientific consultants or for Environmental Protection Agency compliance officers. Data can be obtained and analyzed and reports generated all within 1 day. The portable pH/ISE meters and the card-based systems are truly the wave of the future, being realized today.

5 ION-SELECTIVE MICROELECTRODES

As we have seen, the development of the pH electrode was initiated by researchers interested in making physiological measurements of acidity. Although the Beckman pH meter revolutionized the routine measurement of pH in wine, pickling baths, etc., the first pH meter for measuring the pH of blood was not introduced until 1954.[123] ISE technology has replaced many of the clinical analysis techniques based on colorimetric analysis, flame photometry, and atomic absorption. Flame photometry was once the method of choice for the determination of lithium, sodium and potassium in samples. However, the inorganic material had to be separated from its biological matrix, a time-consuming process. Using ISE analyzers, whole blood samples can be analyzed for these ions.[123] Atomic absorption used to be the method of choice for the clinical determination of calcium and magnesium. ISE analyzers are now used for these determinations. The trend has been to replace multiple instruments with a single analyzer capable of providing analysis for multiple targets. For example, the Nova 16 Stat chemical analyzer, an all-electrode (but not all-ISE) instrument, provides for the analysis of blood urea nitrogen, glucose, creatinine, sodium, potassium, chloride, and total carbon dioxide in less than 1 mL of whole blood.

The development of ISEs for both in vivo and in vitro biological measurements and medical applications has always been a goal of researchers in the health-related fields and in the biological sciences. Since in vivo measurements must be made in very tiny compartments, very tiny probes – microelectrodes – have to be developed. Ammann[124] presented an excellent chart of the size ranges of various biological compartments, against which we may assess how far we have come and how much further we need to go in developing such electrodes. In Table 4, that information is shown using the median size of the biological compartments. In Figure 20 the three most useful geometries, in our opinion, for making in vivo measurement are shown. Coated-wire electrodes can be inserted into egg cells, but they are too large to be inserted into any of the other cells. The catheter electrodes are about the same size as the coated-wire electrodes. Just as catheters are useful for the continuous injection of drugs intravenously, catheter electrodes could be used for the continuous monitoring of molecules and ions that circulate in the bloodstream. However, only the microelectrode has been made small enough that the contents

Table 4 Median sizes (equivalent spherical diameter) of selective biological samples

Biological sample	Median size
Egg cell	820 μm
Animal cell	53 μm
Plant cell	53 μm
Vacuole	18 μm
Nucleus	6.9 μm
Mitochondrion	1.8 μm
Bacterium	770 nm
Lysosome	350 nm
Ribosome	46 nm
Virus	46 nm

For references see page 9904

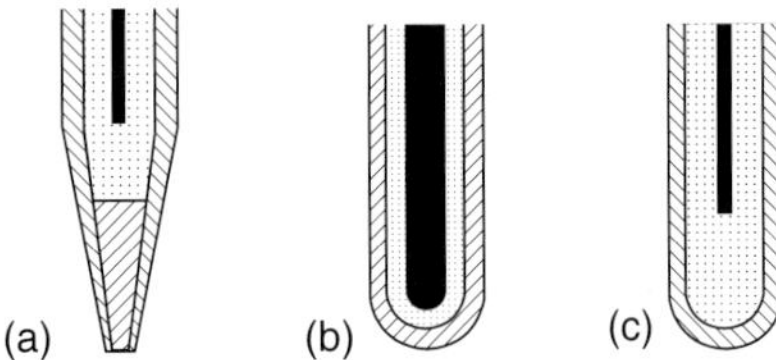

Figure 20 Some useful microelectrode geometries: (a) microelectrode; (b) coated-wire electrode; (c) catheter electrode.

of cells and some of their organelles may, in principle, be measured. The smallest tip diameter of an enzyme microelectrode is 2 µm, which means that these electrodes may be used to probe the cytoplasmic regions of plant and animal cells, vacuoles and the nucleus. With the neutral carrier microelectrodes, with the smallest tip diameter of 300 nm, the above compartments, and also mitochondria, bacteria, and lysosomes, may be probed. For the ion-exchanger and charged-carrier-type polymeric membranes, microelectrodes with minimum tip diameters of 100 nm have been constructed. These can interrogate the same compartments as the neutral carrier-type microelectrodes, but are not small enough to insert into ribosomes or viruses.

The prospect of using enzyme microelectrodes for the in vivo, selective monitoring of biological molecules is extremely attractive. What is needed is another factor of 10 decrease in the smallest tip diameter of these microelectrodes. However, obtaining useful signals from these smaller microelectrodes would require concomitant improvements in instrumentation capabilities. At present, microelectrodes may have resistances of the order of $1 \times 10^{12}\,\Omega$. A decrease in diameter by a factor of 10 with no decrease in thickness would cause the resistance to increase by a factor of 100. This means that for comparable measurement errors, the meter resistances also need to increase by the same factor. There are other problems. The resistance of the cable cladding needs to be much larger than the resistance of the ISE, so that voltage drops between the cable core and the cladding are negligible compared with the magnitude of the signal. When the cladding resistance and the ISE resistance are similar in magnitude, electrode cable shielding must be performed in such a way as to reduce the voltage drop between the cable core and the cable cladding to virtually zero.[120] The cladding problem is already being encountered for ISE resistances of $1 \times 10^{12}\,\Omega$. Commercially available neutral carrier microelectrodes exist for the monitoring of lithium, sodium, potassium, magnesium, calcium, and hydrogen ions. Since the uptake or release of various of these ions often accompanies metabolic processes, these are useful for in vivo studies. Numerous examples of intracellular measurements of ions using neutral carrier microelectrodes exist.[125–129] Recently, a neutral-carrier polymeric ISE microelectrode was developed to study the uptake of cadmium in plant cells.[130]

6 NOVEL DEVELOPMENTS IN ION-SELECTIVE ELECTRODE RESEARCH

In this section, some recent developments which show particular promise in advancing the field of ISEs are discussed. This section is not intended to be comprehensive, and may unintentionally reflect the biases of the author! The development of polymer membrane-type ISEs for clinical uses and for monitoring the condition of patients during surgical intervention or during their recuperation in critical care units is a very active area. Several specific reviews have appeared.[131–133] In particular, the development of polymeric ISEs for polyion analysis must be mentioned. One such polyion is heparin. Heparin is a glycosaminoglycan that occurs naturally in mast cells and cells that line the walls of arteries. It functions as an anticoagulant, and is used in hemodialysis, during heart surgery, and postsurgically. A heparin ISE has been developed by Meyerhoff et al. and is being improved continuously.[134–136]

Of course, the monitor itself must not induce thrombosis, so it is important that the polymeric membrane be biocompatible. Therefore, studies on modified PVC and other polymer matrices have been carried out.[137] A novel development is a heparin ISE which slowly releases NO, a platelet antiaggregation agent.[138] A disadvantage of these ISEs is that heparin must be removed from the membrane between measurements. An improved cleaning method has recently been developed.[139] The development of polyion-sensitive membrane ISEs for nonseparation immunoassays also seems promising.[63] The ability to analyze economically and quickly undiluted blood samples for illicit drugs would be useful for sports compliance monitoring and also for police monitoring for drug use.

Another development is ion imaging by means of scanning ion-selective potentiometric microscopy.[140] It was shown that this technique could be used to image ions over conducting and insulating targets. This comes very close to being a measuring instrument that does not perturb the system that is being measured. This instrument will make it possible to use microelectrodes in order to advance our understanding of the detailed behavior of macroelectrodes, which can be used as targets.

More developments in the area of simultaneous analyte monitoring are expected. This area began with the development of sensors for the simultaneous monitoring of two analytes, e.g. pH and carbon dioxide.[141,142]

Table 5 Ion-selective sensors in the WCL

Sensor	Type	Analyte
pH	Polymer membrane	pH
pH	Iridium dioxide	pH
Silver/sulfide	Crystalline membrane	Silver or sulfide
Cadmium	Crystalline membrane	Cadmium
Chloride	Crystalline membrane	Chloride
Bromide	Crystalline membrane	Bromide
Iodide	Crystalline membrane	Iodide
Lithium	Polymer membrane	Used as reference
Sodium	Polymer membrane	Sodium
Potassium	Polymer membrane	Potassium
Magnesium	Polymer membrane	Magnesium
Calcium	Polymer membrane	Magnesium
Ammonium	Polymer membrane	Ammonium
Nitrate	Polymer membrane	Nitrate
Perchlorate	Polymer membrane	Perchloride
Carbon dioxide	Membrane gas sensor	Carbon dioxide/ hydrogencarbonate

The further development of ISE arrays for simultaneous multitarget determinations will continue.(143,144) This also includes methods of microfabrication.(145,146) The development of data acquisition cards, which allow as many as 14 different channels to be interfaced to a computer, and sampling under computer control make these types of arrays attractive. The main problem with ISE arrays is that many of the ion mixtures that one might want to determine give rise to a complex electrode interference problem. The development of data acquisition designs and chemometric approaches to the analysis of such data are steps toward solving that problem.(147–149)

Finally, ISEs are heading to Mars! A project called MECA (Mars Environmental Compatibility Assessment) is scheduled to fly aboard the Mars Surveyor 2001 Lander.(150) MECA contains a wet chemistry laboratory (WCL) that consists of four beakers, which contain sensor arrays that are mostly ISEs. Both solid-state and polymeric membrane electrodes are being used. In Table 5, the ISE sensors being used in the WCL are shown. Conspicuously absent are glass ISEs. Cracking due to thermal cycling and degradation of the insulation resistance led to catastrophic electrical leakage.

LIST OF SYMBOLS

a	activity
β	efficiency factor in electrode potential equation
$\beta_{i,n}$	complex formation constant
c, C	concentration, usually in $mol\,L^{-1}$
γ	activity coefficient
δ	space-charge layer thickness. SI units m, but usually expressed in μm, nm, etc.
D	diffusion coefficient. SI units $m^2\,s^{-1}$, but frequently given in $cm^2\,s^{-1}$ in the literature
E	cell emf, electrode emf, in V or mV; phase boundary potential difference
$\mathfrak{F}$	Faraday constant, 96 500 J C^{-1}
k	Henry's constant
k_i	single ion distribution coefficient
k^{sel}	selectivity coefficient
K	constant in the Nernst equation in V or mV
K_a	acid dissociation constant
K_b	base hydrolysis constant
K_i	weighing factor
K_{sp}	solubility product constant
K_w	autoprotolysis constant of water
$K_{A,B}^{pot}$	selectivity coefficient
μ	chemical potential, J mol^{-1}
p	osmotic pressure
P	partial pressure of a gas, atm
Φ, ϕ	Galvani potential, V or mV
R_T	concentration of nonexchangeable ions
S	empirical slope factor, V or mV
τ	time constant, s^{-1}
t_+, t_-	transport number
X, χ	chi potential, V or mV

ABBREVIATIONS AND ACRONYMS

CHEMFET	Chemical-sensing Field Effect Transistor
DOS	Dioctyl Sebacate
emf	Electromotive Force
IR	Infrared
ISE	Ion-selective Electrode
ISFET	Ion-selective Field Effect Transistor
IUPAC	International Union of Pure and Applied Chemistry
KTFPB	Potassium Tetrakis[3,5-bis(trifluoromethyl)phenyl]borate
LCD	Liquid Crystal Display
LED	Light-emitting Diode
MAS	Magic-angle Spinning
MECA	Mars Environmental Compatibility Assessment
NBO	Nonbridging Oxygen
NMR	Nuclear Magnetic Resonance
NPOE	2-Nitrophenyl Octyl Ether
PVC	Poly(vinyl chloride)
SIMS	Secondary Ion Mass Spectrometry
TDDMACl	Tridodecylmethylammonium Chloride
TOF	Time-of-flight
WCL	Wet Chemistry Laboratory
XPS	X-ray Photoelectron Spectroscopy

For references see page 9904

RELATED ARTICLES

Biomolecules Analysis **(Volume 1)**
Biomolecules Analysis: Introduction

Clinical Chemistry **(Volume 2)**
Clinical Chemistry: Introduction • Automation in the Clinical Laboratory • Biosensor Design and Fabrication • Electroanalysis and Biosensors in Clinical Chemistry • Electroanalytical Chemistry in Clinical Analysis • Electrolytes, Blood Gases, and Blood pH • Laboratory Instruments in Clinical Chemistry, Principles of

Environment: Water and Waste **(Volume 4)**
Ion-selective Electrodes in Environmental Analysis

Field-portable Instrumentation **(Volume 4)**
Electrochemical Sensors for Field Measurements of Gases and Vapors

Field-portable Instrumentation cont'd **(Volume 5)**
Solid-state Sensors for Field Measurements of Gases and Vapors

Remote Sensing **(Volume 10)**
Remote Sensing: Introduction

Surfaces **(Volume 10)**
X-ray Photoelectron Spectroscopy in Analysis of Surfaces

General Articles **(Volume 15)**
Analytical Problem Solving: Selection of Analytical Methods • Titrimetry

REFERENCES

1. R.P. Buck, E. Lindner, 'Recomendations for Nomenclature of Ion-selective Electrodes', *Pure Appl. Chem.*, **66**(12), 2527–2536 (1994).
2. N.H. Furman, 'Potentiometric Titrations', *Anal. Chem.*, **22**, 33–41 (1950).
3. N.H. Furman, 'Potentiometric Titrations', *Anal. Chem.*, **23**, 21–24 (1951).
4. N.H. Furman, 'Potentiometric Titrations', *Anal. Chem.*, **26**, 84–90 (1954).
5. C.N. Reilley, 'Potentiometric Titrations', *Anal. Chem.*, **28**, 671–678 (1956).
6. C.N. Reilley, 'Potentiometric Titrations', *Anal. Chem.*, **30**, 185R–193R (1960).
7. R.W. Murray, C.N. Reilley, 'Potentiometric Titrations', *Anal. Chem.*, **34**, 313R–380R (1962).
8. R.W. Murray, C.N. Reilley, 'Potentiometric Titrations', *Anal. Chem.*, **36**, 370R–380R (1964).

List of selected abbreviations appears in Volume 15

9. D.K. Roe, 'Potentiometric Titrations', *Anal. Chem.*, **38**, 461R–469R (1966).
10. E.C. Toren, Jr, 'Potentiometric Titrations', *Anal. Chem.*, **40**, 402R–412R (1968).
11. E.C. Torens, Jr, P.M. Gross, R.P. Buck, 'Potentiometric Titrations', *Anal. Chem.*, **42**, 284R–304R (1970).
12. R.P. Buck, 'Ion-selective Electrodes, Potentiometry, and Potentiometric Titrations', *Anal. Chem.*, **44**, 270R–295R (1972).
13. R.P. Buck, 'Ion-selective Electrodes, Potentiometry, and Potentiometric Titrations', *Anal. Chem.*, **46**, 28R–51R (1974).
14. R.P. Buck, 'Ion-selective Electrodes', *Anal. Chem.*, **48**, 23R–39R (1976).
15. R.P. Buck, 'Ion-selective Electrodes', *Anal. Chem.*, **50**, 17R–29R (1978).
16. G.H. Fricke, 'Ion-selective Electrodes', *Anal. Chem.*, **52**, 259R–275R (1980).
17. M.E. Meyerhoff, Y.M. Fraticelli, 'Ion-selective Electrodes', *Anal. Chem.*, **54**, 27R–44R (1982).
18. M.A. Arnold, M.E. Meyerhoff, 'Ion-selective Electrodes', *Anal. Chem.*, **56**, 20R–48R (1984).
19. M.A. Arnold, R.L. Solsky, 'Ion-selective Electrodes', *Anal. Chem.*, **58**, 84R–101R (1986).
20. R.L. Solsky, 'Ion-selective Electrodes', *Anal. Chem.*, **60**, 106R–113R (1988).
21. R.L. Solsky, 'Ion-selective Electrodes', *Anal. Chem.*, **62**, 21R–33R (1990).
22. J. Janata, 'Chemical Sensors', *Anal. Chem.*, **64**, 196R–219R (1992).
23. J. Janata, M. Josowicz, D.M. DeVaney, 'Chemical Sensors', *Anal. Chem.*, **66**, 207R–228R (1994).
24. J. Janata, M. Josowicz, P. Vanýsek, D.M. DeVaney, 'Chemical Sensors', *Anal. Chem.*, **70**, 179R–208R (1998).
25. E. Bakker, P. Bühlmann, E. Pretsch, 'Carrier-based Ion-selective Electrodes and Bulk Optodes. 1. General Characteristics', *Chem. Rev.*, **97**, 3083–3132 (1997).
26. P. Bühlmann, E. Pretsch, E. Bakker, 'Carrier-based Ion-selective Electrodes and Bulk Optodes. 2. Ionophores for Potentiometric and Optical Sensors', *Chem. Rev.*, **98**, 15933–1687 (1998).
27. R.P. Buck, 'Theory and Principles of Membrane Electrodes', in *Ion-selective Electrodes in Analytical Chemistry*, ed. H. Freiser, Plenum Press, New York, 1–141, Vol. 1, 1978.
28. *A Concise Guide to Ion Analysis*, Laboratory Products Catalog & Electrochemistry Handbook, Orion Research, Cambridge, MA, 85–92, 1998.
29. K.L. Cheng, 'pH Glass Electrode and Its Mechanism', in *Electrochemistry, Past and Present*, eds. J.S. Stock, M.V. Orna, American Chemical Society, Washington, DC, 286–302, 1989.
30. K.L. Cheng, 'Capacitor Theory for Nonfaradaic Potentiometry', *Microchem. J.*, **42**, 5–24 (1990).
31. A.F. Burke, T.C. Murphy, 'Materials Characteristics and the Performance of Electrochemical Capacitors for

Electric/Hybrid Vehicle Applications', *Materials for Electrochemical Energy Storage and Conversion – Batteries, Capacitors and Fuel Cells*, Materials Research Society Symposium Proceedings, Materials Research Society, Pittsburg, PA, 375–395, Vol. 393, 1995.
32. D.A. Skoog, J.J. Leary, 'Potentiometric Methods', *Principles of Instrumental Analysis*, 4th edition, Saunders College Publishing, Fort Worth, TX, 503–506, 1992.
33. M. Cremer, 'Über die Ursache der Elektromotorischen Eigenschaften der Gewebe Zugleich ein Betrag zur Lehre von dem Polyphasichen Elekolytketten', *Z. Biol.*, **47**, 562–607 (1906).
34. C.E. Moore, B. Jaselskis, A. Smolinski, 'Development of the Glass Electrode', in *Electrochemistry, Past and Present*, eds. J.S. Stock, M.V. Orna, American Chemical Society, Washington, DC, 272–285, 1989.
35. F. Haber, Z. Klemensiewicz, 'Über Elektrische Phasengrenzkräfte', *Z. Phys. Chem.*, **67**, 385–431 (1909).
36. M.D. Archer, 'Genesis of the Nernst Equation', in *Electrochemistry, Past and Present*, eds. J.S. Stock, M.V. Orna, American Chemical Society, Washington, DC, 115–126, 1989.
37. S.P.L. Sørenson, 'Enzyme Studies. II. The Measurement and Importance of the Hydrogen Ion Concentration in Enzyme Reactions', *Biochem. Z.*, **21**, 131–200 (1909).
38. C.E. Moore, B. Jaselskis, A. Smolinski, 'Development of the Glass Electrode', in *Electrochemistry, Past and Present*, eds. J.S. Stock, M.V. Orna, American Chemical Society, Washington, DC, 272–285, 1989.
39. J.O. Isard, 'The Dependence of Glass-electrode Properties on Composition', in *Glass Electrodes for Hydrogen and Other Cations*, ed. G. Eisenman, Marcel Dekker, New York, 51–100, 1967.
40. G.A. Perley, 'Composition of pH-responsive Glasses', *Anal. Chem.*, **21**, 391–394 (1949).
41. G.A. Perley, 'Glasses for Measurement of pH', *Anal. Chem.*, **21**, 394–401 (1949).
42. B. Lengyel, E. Blum, 'The Behaviour of the Glass Electrode in Connection with Its Chemical Composition', *Trans. Faraday Soc.*, **30**, 461–471 (1934).
43. M.S. Frant, 'History of the Early Commercialization of Ion-selective Electrodes', *Analyst*, **119**, 2293–2301 (1994).
44. A.K. Covington, 'Introduction: Basic Electrode Types, Classification, and Selectivity Considerations', in *Ion-selective Electrode Methodology*, ed. A.K. Covington, CRC Press, Boca Raton, FL, 1–20, Vol. I, 1979.
45. M.S. Frant, 'Where Did Ion-selective Electrodes Come From?', *J. Chem. Educ.*, **74**, 159–166 (1997).
46. Z. Stefanac, W. Simon, 'In-vitro-verhalten von Makrotetroliden in Membranen als Grundlage für Hochselektive Kationenspezifische Elektrodensysteme', *Chimia*, **20**, 436 (1966).
47. Z. Stefanac, W. Simon, 'Ion Specific Electrochemical Behavior of Macrotetrolides in Membranes', *Microchem. J.*, **12**, 125–132 (1967).
48. A. Shatkay, 'Ion Specific Membranes as Electrodes in Determination of Activity of Calcium', *Anal. Chem.*, **39**, 1056–1065 (1967).
49. J. Ruzicka, 'The Seventies – Golden Age for Ion-selective Electrodes', *J. Chem. Educ.*, **74**, 167–170 (1997).
50. P. Vadgama, 'Enzyme Electrodes', in *Ion-selective Electrode Methodology*, ed. A.K. Covington, CRC Press, Boca Raton, FL, 23–40, Vol. II, 1979.
51. R.W. Cattrall, 'Heterogeneous-membrane, Carbon-supported, and Coated-wire Ion-selective Electrodes', in *Ion-selective Electrode Methodology*, ed. A.K. Covington, CRC Press, Boca Raton, FL, 131–1730, Vol. I, 1979.
52. G.J. Moody, J.D.R. Thomas, 'Polyvinyl Chloride Matrix Membrane Ion-selective Electrodes', in *Ion-selective Electrode Methodology*, ed. A.K. Covington, CRC Press, Boca Raton, FL, 111–130, Vol. I, 1979.
53. M.A. Arnold, G.A. Rechnitz, 'Optimization of a Tissue-based Membrane Electrode for Guanine', *Anal. Chem.*, **54**, 777–782 (1982).
54. P. Rolfe, M.J. Martin, 'Medical Sensors and Biosensors', *Chem. Br.*, **24**, 1026–1028 (1988).
55. S.M.S. Hassan, G.A. Rechnitz, 'Enzyme Amplification for Trace Level Determination of Pyridoxal 5′-Phosphate with a pCO_2 Electrode', *Anal. Chem.*, **53**, 512–515 (1981).
56. A. Mulchandani, P. Mulchandani, I. Kaneva, W. Chen, 'Biosensor for Direct Determination of Organophosphate Nerve Agents Using Recombinant *Escherichia coli* with Surface-expressed Organophosphorus Hydrolase. 1. Potentiometric Microbial Electrode', *Anal. Chem.*, **70**, 4140–4145 (1998).
57. P. Seegopaul, G.A. Rechnitz, 'Enzyme-amplified Determination of Methotrexate with a pCO_2 Membrane Electrode', *Anal. Chem.*, **56**, 852–854 (1994).
58. S.B. Amemiya, P. Buhlmann, Y. Umezawa, R.C. Jagessar, D.H. Burns, 'An Ion-selective Electrode for Acetate Based on a Urea-functionalized Porphyrin as a Hydrogen-bonding Ionophore', *Anal. Chem.*, **71**, 1049–1054 (1999).
59. T. Katsu, Y. Mori, K. Furuno, Y. Gomita, 'Mexiletine-sensitive Membrane Electrode for Medical Application', *J. Pharm. Biomed. Anal.*, **19**, 585–593 (1999).
60. O.V. Mushik, V.I. Tkach, N.I. Karandeeva, O.I. Glukhova, L.P. Tsyganok, 'Electrochemical and Analytical Properties of Solid-contact Ion-selective Electrodes Reversible to the Imidazole Derivatives Clotrimazole and Bifonazole', *J. Anal. Chem.*, **53**, 1110–1112 (1998).
61. R. Koncki, A. Owczarek, W. Dzwolak, S. Glab, 'Immunoenzymatic Sensitisation of Membrane Ion-selective Electrodes', *Sens. Actuators B*, **47**, 246–250 (1998).
62. Y.M. Issa, M.S. Rizk, A.F. Shoukry, E.M. Atia, 'Plastic Membrane Electrodes for Amprolium', *Mikrochim. Acta*, **129**, 195–200 (1998).

63. S. Dai, J.M. Esson, O. Lutze, N. Ramamurthy, V.C. Yang, M.E. Meyerhoff, 'Bioanalytical Applications of Polyion-sensitive Electrodes', *J. Pharm. Biomed. Anal.*, **19**, 1–14 (1999).
64. P.L. Bailey, *Analysis with Ion-selective Electrodes*, 2nd edition, Heyden, London, 158–181, 1980.
65. S.Z. Yao, X.L. Su, Y.J. Xu, 'New BAW Sensors for Dissolved Gases', *Instrum. Sci. Technol.*, **26**, 473–490 (1998).
66. A. Zangwill, *Physics at Surfaces*, Cambridge University Press, Cambridge, 54–109, 1988.
67. J.O'M. Bockris, A.K.N. Reddy, *Modern Electrochemistry*, 1st Edition, Plenum Press, New York, 667–669, Vol. 2, 1970.
68. R.P. Poeppel, J.M. Blakely, 'Origin of Equilibrium Space Charge Potentials in Ionic Crystals', *Surf. Sci.*, **15**, 507–523 (1969).
69. J. Koryta, 'Theoretical Treatment and Experimental Investigation of Ion Transfer Across Liquid–Liquid Interfaces', in *Ion-transfer Kinetics: Principles and Applications*, ed. J.R. Sandifer, VCH Publishers, New York, 1–17, 1995.
70. J.O'M. Bockris, A.K.N. Reddy, *Modern Electrochemistry*, 1st Edition, Plenum Press, New York, 897–901, Vol. 2, 1970.
71. W.E. Morf, 'Part A – Theory of Membrane Potentials and Membrane Transport', *The Principles of Ion-selective Electrodes and of Membrane Transport*, Elsevier, Amsterdam, 27–161, 1981.
72. R.P. Buck, 'Theory and Principles of Membrane Electrodes', in *Ion-selective Electrodes in Analytical Chemistry*, ed. H. Freiser, Plenum Press, New York, 1–141, Vol. 1, 1978.
73. G. Eisenman, 'The Origin of the Glass-electrode Potential', in *Glass Electrodes for Hydrogen and Other Cations*, ed. G. Eisenman, Marcel Dekker, New York, 133–173, 1967.
74. E. Pungor, 'Working Mechanism of Ion-selective Electrodes', *Pure Appl. Chem.*, **64**, 503–507 (1992).
75. E. Bakker, M. Nägele, U. Schaller, E. Pretsch, 'Applicability of the Phase Boundary Potential Model to the Mechanistic Understanding of Solvent Polymeric Membrane-based Ion-selective Electrodes', *Electroanalysis*, **7**, 817–822 (1995).
76. M.L. Clay, R.E. Pierce, V.Y. Young, S. Hoke, 'A Converging Wall-jet Instrument for Ion Selective Electrode Dynamic Response Studies', *Instrum. Sci. Technol.*, **26**, 461–472 (1998).
77. E. Lindner, K. Toth, E. Pungor, *Dynamic Characteristics of Ion-selective Electrodes*, CRC Press, Boca Raton, FL, 1988.
78. T.R. Berube, R.P. Buck, E. Lindner, K. Toth, E. Pungor, 'Coupled Diffusion/Adsorption Model for Response of Precipitate-based Iodide-selective Electrodes to Primary-ion Activity Steps', *Anal. Chem.*, **63**, 946–953 (1991).
79. E. Bakker, R.K. Meruva, E. Pretsch, M.E. Meyerhoff, 'Selectivity of Polymer Membrane-based Ion-selective Electrodes: Self-consistent Model Describing the Potentiometric Response in Mixed Ion Solutions of Different Charge', *Anal. Chem.*, **66**, 3021–3030 (1994).
80. V.P.Y. Gadzekpo, G.D. Christian, 'Determination of Selectivity Coefficients of Ion-selective Electrodes by a Matched-potential Method', *Anal. Chim. Acta*, **164**, 279–282 (1984).
81. Y. Umezawa, K. Umezawa, H. Sato, 'Selectivity Coefficients for Ion-selective Electrodes: Recommended Methods for Reporting $K_{A,B}^{pot}$ Values', *Pure Appl. Chem.*, **67**(3), 507–518 (1995).
82. M. Nägele, E. Bakker, E. Pretsch, 'General Description of the Simultaneous Response of Potentiometric Ionophore-based Sensors to Ions of Different Charge', *Anal. Chem.*, **71**, 1041–1048 (1999).
83. D.G. Hall, 'Ion-selective Membrane Electrodes: A General Limiting Treatment of Interference Effects', *J. Phys. Chem.*, **100**, 7230–7236 (1996).
84. R.P. Buck, 'The Impedance Method Applied to the Investigation of Ion-selective Electrodes', *Ion-Sel. Electrode Rev.*, **4**, 3–74 (1982).
85. M.J.D. Brand, G.A. Rechnitz, 'Mechanistic Studies on Crystal-membrane Ion-selective Electrodes', *Anal. Chem.*, **42**, 478–483 (1970).
86. R. De Marco, P.C. Hauser, R.W. Catrall, J. Liesegang, G.L. Nyberg, I.C. Hamilton, 'XPS Studies of the Fluoride Ion-selective Electrode Membrane LaF_3: Evidence for a Gel Layer on the Surface', *Surf. Interface Anal.*, **14**, 463–468 (1989).
87. R. De Marco, R.W. Catrall, J. Liesegang, G.L. Nyberg, I.C. Hamilton, 'XPS Studies of the Fluoride Ion-selective Electrode Membrane LaF_3: Ion Interferences', *Surf. Interface Anal.*, **14**, 457–462 (1989).
88. H.A. Hoyen, Jr, J.A. Strozier, Jr, C.-Y. Li, 'Evidence for Space Charge Limited Ionic Transport at the Silver Chloride–Aqueous Solution Interface', *Appl. Phys. Lett.*, **14**, 104–106 (1969).
89. H.A. Hoyen, Jr, J.A. Strozier, Jr, C.-Y. Li, 'Space Charge Limited AC Conduction with Non-blocking Electrodes', *Surf. Sci.*, **20**, 258–268 (1970).
90. R. Rangarajan, G.A. Rechnitz, 'Dynamic Response of Ion-selective Membrane Electrodes', *Anal. Chem.*, **47**, 324–326 (1975).
91. E. Pungor, 'The Theory of Ion-selective Electrodes', *Anal. Sci.*, **14**, 249–256 (1998).
92. R. De Marco, R.W. Cattrall, J. Liesegang, G.L. Nyberg, I.C. Hamilton, 'Surface Studies of the Silver Sulfide Ion-selective Electrode Membrane', *Anal. Chem.*, **62**, 2339–2346 (1990).
93. K. Uosaki, Y. Shigematsu, H. Kita, Y. Umezawa, 'Crystal-face Specific Response of a Single-crystal Cadmium Sulfide Based Ion-selective Electrode', *Anal. Chem.*, **61**, 1980–1983 (1989).

94. W.E. Morf, 'Part B – Ion-selective Electrodes', *The Principles of Ion-selective Electrodes and of Membrane Transport*, Elsevier, Amsterdam, 178–198, 1981.
95. N. Pandya, D.W. Muenow, S.K. Sharma, B.L. Sherriff, 'The Speciation of Water in Hydrated Alkali Silicate Glasses', *J. Non-Cryst. Solids*, **176**, 140–146 (1994).
96. H. Dunken, R.H. Doremus, 'Short Time Reactions of a $Na_2O-CaO-SiO_2$ Glass with Water and Salt Solutions', *J. Non-Cryst. Solids*, **92**, 61–72 (1987).
97. Z. Zhang, N. Soga, K. Hirao, 'Water Corrosion Behavior of Densified Glass. I. Silicate Glass', *J. Non-Cryst. Solids*, **135**, 55–61 (1991).
98. R.P. Buck, 'Transient Electrical Behavior of Glass Membranes. Part I. Theory of D.C. Pulse Processes', *J. Electroanal. Chem.*, **18**, 363–380 (1968).
99. W.M. Mularie, W.F. Furth, A.R.C. Westwood, 'Influence of Surface Potential on the Kinetics of Glass Reactions with Aqueous Solutions', *J. Mater. Sci.*, **14**, 2659–2664 (1979).
100. A.K. Covington, P. Davison, 'Liquid Exchanger Types', in *Ion-selective Electrode Methodology*, ed. A.K. Covington, CRC Press, Boca Raton, FL, 85–110, Vol. I, 1979.
101. I. Nielsen, H. Jorgen, E.H. Hansen, 'Nitrate Ion-selective Electrode', US Pat., 4 059 499, 1977.
102. X. Li, D.J. Harrison, 'Measurement of Concentration Profiles Inside a Nitrite Ion Selective Electrode Membrane', *Anal. Chem.*, **63**, 2168–2174 (1991).
103. A.D.C. Chan, D.J. Harrison, 'NMR Studies of the State of Water in Ion-selective Electrode Membranes', *Anal. Chem.*, **65**, 32–36 (1993).
104. R. Stepánek, B. Kräutler, P. Schulthess, B. Lindemann, D. Ammann, W. Simon, 'Aquocyanocobalt(III) Hepta(2-phenylethyl)cobyrinate as a Cationic Carrier for Nitrite-selective Liquid-membrane Electrodes', *Anal. Chim. Acta*, **182**, 83–90 (1986).
105. D. Midgley, 'Ion-selective Electrodes Based on Siderophores', *J. Chem. Soc., Faraday Trans. I*, **82**, 1187–1193 (1986).
106. E. Bakker, M. Lerchi, T. Rosatzin, B. Rusterholz, W. Simon, 'Synthesis and Characterization of Neutral Hydrogen Ion-selective Chromoionophores for Use in Bulk Optodes', *Anal. Chim. Acta*, **278**, 211–225 (1993).
107. Q. Ye, G. Horvai, A. Tóth, I. Bertóti, M. Botreau, T.M. Duc, 'Studies of Ion-selective Solvent Polymeric Membranes by X-ray Photoelectron Spectroscopy and Time-of-flight Static Secondary Ion Mass Spectroscopy', *Anal. Chem.*, **70**, 4241–4246 (1998).
108. J.R. Sandifer, 'Ion Transfer Into and Ion Transport within Plastic Ion-selective Membranes', in *Ion-transfer Kinetics: Principles and Applications*, ed. J.R. Sandifer, VCH Publishers, New York, 115–138, 1995.
109. W.E. Morf, 'Part B – Ion-selective Electrodes', *The Principles of Ion-selective Electrodes and of Membrane Transport*, Elsevier, Amsterdam, 211–218, 1981.
110. U. Schefer, D. Ammann, E. Pretsch, U. Oesch, W. Simon, 'Neutral Carrier Based Ca^{2+}-selective Electrode with Detection Limit in the Sub-nanomolar Range', *Anal. Chem.*, **58**, 2282–2285 (1986).
111. E. Bakker, M. Willer, E. Pretsch, 'Detection Limit of Ion-selective Bulk Optodes and Corresponding Electrodes', *Anal. Chim. Acta*, **282**, 265–271 (1993).
112. K. Cammann, 'Gas Sensors for CO_2, NH_3, SO_2, NO_2, HF, H_2S, HCN, etc.', *Working with Ion-selective Electrodes*, Springer, New York, 93–98, 1979.
113. U. Oesch, E. Malinowska, W. Simon, 'Bicarbonate-sensitive Electrode Based on Planar Thin Membrane Technology', *Anal. Chem.*, **59**, 2131–2135 (1987).
114. M.E. Meyerhoff, 'Polymer Membrane Electrode Based Potentiometric Ammonia Gas Sensor', *Anal. Chem.*, **52**, 1532–1534 (1980).
115. Y.M. Fraticelli, M.E. Meyerhoff, 'Selectivity Characteristics of Ammonia-gas Sensors Based on a Polymer Membrane Electrode', *Anal. Chem.*, **53**, 1857–1861 (1981).
116. D.M. Pranitis, M.E. Meyerhoff, 'Continuous Monitoring of Ambient Ammonia with a Membrane-electrode-based Detector', *Anal. Chem.*, **59**, 2345–2350 (1987).
117. W.E. Morf, 'Part B – Ion-selective Electrodes', *The Principles of Ion-selective Electrodes and of Membrane Transport*, Elsevier, Amsterdam, 406–413, 1981.
118. J.E. Brady, P.W. Carr, 'Theoretical Evaluation of the Steady-state Response of Potentiometric Enzyme Electrodes', *Anal. Chem.*, **52**, 977–980 (1980).
119. P.R. Burton, 'Instrumentation for Ion-selective Electrodes', in *Ion-Selective Electrode Methodology*, ed. A.K. Covington, CRC Press, Boca Raton, FL, 21–41, Vol. I, 1979.
120. K. Cammann, 'Measurements of Intracellular Ion Activities', *Working with Ion-selective Electrodes*, Springer, New York, 174–178, 1979.
121. B. Jaselskis, C.E. Moore, A. Smolinski, 'Development of the pH Meter', in *Electrochemistry, Past and Present*, eds. J.S. Stock, M.V. Orna, American Chemical Society, Washington, DC, 254–271, 1989.
122. D.A. Skoog, J.J. Leary, 'Operational Amplifiers in Chemical Instrumentation', *Principles of Instrumental Analysis*, 4th edition, Saunders College Publishing, Fort Worth, TX, 10–27, 1992.
123. C.C. Young, 'Evolution of Blood Chemistry Analyzers Based on Ion-selective Electrodes', *J. Chem. Educ.*, **74**, 177–182 (1997).
124. D. Ammann, 'Classification of Ion-selective Electrodes', *Ion-selective Microelectrodes*, Springer, Berlin, 3–8, 1986.
125. D. Ammann, 'General Aspects of Intracellular Measurements of Ions', *Ion-selective Microelectrodes*, Springer, Berlin, 154–193, 1986.
126. K. Venkova, J. Krier, 'Postjunctional α_1- and β-Adrenoceptor Effects of Noradrenaline on Electrical Slow Waves and Phasic Contractions of Cat Colon Circular Muscle', *Br. J. Pharmacol.*, **116**, 3265–3273 (1995).

List of selected abbreviations appears in Volume 15

127. Z. Hu, T. Buehrer, M. Mueller, B. Rusterholz, M. Rouilly, W. Simon, 'Intracellular Magnesium Ion-selective Microelectrode Based on a Neutral Carrier', *Anal. Chem.*, **61**, 574–576 (1989).
128. V. Lyall, T.L. Croxton, W.M. Armstrong, 'Measurement of Intracellular Chloride Activity in Mouse Liver Slices with Microelectrodes', *Biochim. Biophys. Acta*, **903**, 56–67 (1987).
129. C.O. Lee, 'Ionic Activities in Cardiac Muscle Cells and Application of Ion-selective Microelectrodes', *Am. J. Physiol.*, **241**, H459–H478 (1981).
130. M.A. Pineros, J.E. Shaff, L.V. Kochian, 'Development, Characterization, and Application of a Cadmium-selective Microelectrode for the Measurement of Cadmium Fluxes in Roots of Thlaspi Species and Wheat', *Plant Physiol.*, **116**, 1393–1401 (1998).
131. U. Oesch, D. Ammann, W. Simon, 'Ion-selective Membrane Electrodes for Clinical Use', *Clin. Chem.*, **32**, 1448–1459 (1986).
132. D.M. Pranitis, M. Telting-Diaz, M.E. Meyerhoff, 'Potentiometric Ion-, Gas-, and Bio-selective Membrane Electrodes', *Crit. Rev. Anal. Chem.*, **23**, 163–186 (1992).
133. M.E. Collison, M.E. Meyerhoff, 'Chemical Sensors for Bedside Monitoring of Critically Ill Patients', *Anal. Chem.*, **62**, 425A–437A (1990).
134. V.C. Yang, S.C. Ma, M.E. Meyerhoff, 'Thinning Blood Safely', *CHEMTECH*, **23**, 25–32 (1993).
135. M.E. Meyerhoff, V.C. Yang, J.A. Wahr, L.M. Lee, J.H. Yun, B. Fu, E. Bakker, 'Potentiometric Polyion Sensors: New Measurement Technology for Monitoring Blood Heparin Concentrations During Open Heart Surgery', *Clin. Chem.*, **41**, 1355–1356 (1995).
136. T.M. Ambrose, M.E. Meyerhoff, 'Photo-cross-linked Decyl Methacrylate Films for Electrochemical and Optical Polyion Probes', *Anal. Chem.*, **69**, 4092–4098 (1997).
137. C. Espadas-Torre, M.E. Meyerhoff, 'Thrombogenic Properties of Untreated and Poly(ethylene oxide)-modified Polymeric Matrixes Useful for Preparing Intraarterial Ion-selective Electrodes', *Anal. Chem.*, **67**, 3108–3114 (1995).
138. C. Espadas-Torre, V. Oklejas, K. Mowery, M.E. Meyerhoff, 'Thromboresistant Chemical Sensors Using Combined Nitric Oxide Release/Ion Sensing Polymeric Films', *J. Am. Chem. Soc.*, **119**, 2321–2322 (1997).
139. S. Mathison, E. Bakker, 'Renewable pH Cross-sensitive Potentiometric Heparin Sensors with Incorporated Electrically Charged H^+ Ionophores', *Anal. Chem.*, **71**, 4614–4621 (1999).
140. C. Wei, A.J. Bard, G. Nagy, K. Toth, 'Scanning Electrochemical Microscopy. 28. Ion-selective Neutral Carrier-based Microelectrode Potentiometry', *Anal. Chem.*, **67**, 1346–1356 (1995).
141. M.E. Collison, G.V. Aebli, J. Petty, M.E. Meyerhoff, 'Potentiometric Combination Ion–Carbon Dioxide Sensors for In Vitro and In Vivo Blood Measurements', *Anal. Chem.*, **61**, 2365–2372 (1989).

142. M. Telting-Diaz, M.E. Collison, M.E. Meyerhoff, 'Simplified Dual-lumen Catheter Design for Simultaneous Potentiometric Monitoring of Carbon Dioxide and pH', *Anal. Chem.*, **66**, 576–583 (1994).
143. K. Beebe, D. Uerz, J. Sandifer, B. Kowalski, 'Sparingly Selective Ion-selective Electrode Arrays for Multicomponent Analysis', *Anal. Chem.*, **60**, 66–71 (1988).
144. A. Lynch, D. Diamond, P. Lemoine, J. McLaughlin, M. Leader, 'Solid-state Ion-selective Arrays', *Electroanalysis*, **10**, 1096–1100 (1998).
145. R.S. Glass, R.G. Musket, K.C. Hong, 'Preparation of Solid Membrane Chloride Ion Selective Electrodes by Ion Implantation', *Anal. Chem.*, **63**, 2203–2206 (1991).
146. E. Lindner, V.V. Cosofret, T.M. Nahir, R.P. Buck, 'Characterization of Stability of Modified Poly(vinyl chloride) Membranes for Microfabricated Ion-selective Electrode Arrays in Biomedical Applications', *ACS Symp. Ser.*, **556**, 149–157 (1994).
147. K.R. Beebe, B.R. Kowalski, 'Nonlinear Calibration Using Projection Pursuit Regression: Application to an Array of Ion-selective Electrodes', *Anal. Chem.*, **60**, 2273–2278 (1988).
148. R.J. Forster, D. Diamond, 'Nonlinear Calibration of Ion-selective Electrode Arrays for Flow Injection Analysis', *Anal. Chem.*, **64**, 1721–1728 (1992).
149. A. Akhmetshin, V. Baranovsky, A. Akhmetshina, 'Use of Four-factorial Design in Ion-selective Potentiometry for Analysis of Multi-ionic Solutions', *Fresenius' J. Anal. Chem.*, **361**, 282–284 (1998).
150. S.J. West, M.S. Frant, X. Wen, R. Geis, J. Herdan, T. Gillette, M.H. Hecht, W. Schubert, S. Grannan, S.P. Kounaves, 'Electrochemistry on Mars', *Am. Lab.*, **31**(20), 48–54 (1999).

Liquid/Liquid Interfaces, Electrochemistry at

Frédéric Reymond and Hubert H. Girault
Ecole Polytechnique Fédérale de Lausanne, Lausanne, Switzerland

This article outlines the electrochemical methodology at the interface between two electrolyte solutions (ITIES). The fundamental concepts of the thermodynamics in biphasic systems are presented in order to show how ions are distributed between the two adjacent phases, and hence how a Galvani potential difference is established at an ITIES. Polarizable and nonpolarizable ITIES are then characterized, and it is further evidenced that the classical electroanalytical methodology at a solid electrode can be directly transposed to the ITIES, thereby allowing reversible charge-transfer reactions to be easily monitored and interpreted.

This theoretical approach is completed by a review of the analytical methods used at ITIES, namely cyclic voltammetry, dropping electrolyte electrodes, optical techniques and microinterfaces (which are the biphasic analogs of microelectrodes). The last part deals with the practical applications that electrochemistry at ITIES has attracted during the last decade in the development of amperometric ion sensors and detectors, in the extraction of metal ions by interfacial formation of a complex and in the assessment of the lipophilicity of ionizable drugs.

1 INTRODUCTION

Electrochemistry at the ITIES is concerned with three main types of charge-transfer reactions, namely:

- ion-transfer reactions from one phase to another
- ion-transfer reactions assisted by the presence of a complexing agent
- electron-transfer reactions between a hydrophilic and a hydrophobic redox couple.

So far, the electroanalytical applications of electrochemistry at liquid/liquid interfaces have been limited to simple or assisted ion transfers,(1) and this article will be restricted to this subject.

The amperometric study of ion-transfer reactions is a relatively recent area which was developed in the 1970s in France by Gavach et al.(2–5) and in Prague by Koryta et al.(6–10) who demonstrated that the ITIES were polarizable in the same way as the interface between a metallic electrode and an electrolyte solution. It ensued that techniques commonly used for the measurement of electron transfer at solid/liquid interfaces could be applied to study transfer processes through the ITIES, that is, cyclic voltammetry,(11) chronoamperometry,(12) polarography,(13) differential-pulse stripping voltammetry(14) and ac voltammetry.(15)

The route to modern studies of both the interfacial structure of the ITIES and of charge-transfer reactions started only with the introduction of the concept of the four-electrode potentiostat with ohmic drop compensation,(6) which permitted potentiostatic control of the interfacial potential difference. Most publications have reported mainly experimental results,(16–26) and few theoretical models of charge-transfer processes(27–35) and of the interfacial structure(36–43) have been proposed recently. Although the potential distribution at the ITIES is now well established,(44) the structure of the interface and the kinetics of transfer are still controversial subjects.(24,45)

Numerous efforts have been made to acquire a better understanding of the molecular mechanisms involved in ionic motion in liquids,(1,46) and recent results suggest that the rate-limiting step in ion transport is the necessary interchange of the solvation shell from one liquid to the other. The roughness of the interface is likely to manifest itself as capillaries or fingers of one liquid protruding into another.(47,48) This "fingering" resulting from the long-range ion–dipole interactions plays a major role in the change of the solvation shell,(49) implying that ion transfer may be an activated process(50) This has been corroborated by molecular dynamics calculations(49,51) which tend to confirm that the interface is not a sharp but an extended region in which the two solvents mix. One of the greatest challenges in the theory of ITIES is to understand the spatial distribution of the driving forces. From an electroanalytical viewpoint, the majority of the charge-transfer reactions studied are reversible (i.e. kinetically fast) and the development of

For references see page 9921

their applications is not hindered by these theoretical limitations.

This article is intended to describe the general theory of charge-transfer reactions at ITIES and to outline the various methodologies that electrochemistry at liquid/liquid interfaces offers for analytical purposes. It does not deal with interfacial structure and kinetics, but focuses on the various applications of ITIES and on recent advances in this field. Also, the very important field of the ion-selective electrode (ISE),[52–55] which is a significant application of electrochemistry at liquid/liquid interfaces is omitted, as it is a subject of its own.

2 THEORETICAL BACKGROUND

2.1 Thermodynamics for the Partition of Ions

2.1.1 *Gibbs Energy of Transfer and Nernst Equation at the Interface Between Two Immiscible Electrolyte Solutions*

The standard transfer Gibbs energy of a species from one phase (say water) to another phase (say the organic solvent), $\Delta G_{\mathrm{t}}^{o,\mathrm{w}\rightarrow\mathrm{o}}$, is by definition equal to the difference between the standard Gibbs energy of solvation, $\mu^{o,\mathrm{o}}$, and the standard Gibbs energy of hydration, $\mu^{o,\mathrm{w}}$ (Equation 1):

$$\Delta G_{\mathrm{t}}^{o,\mathrm{w}\rightarrow\mathrm{o}} = \mu^{o,\mathrm{o}} - \mu^{o,\mathrm{w}} \tag{1}$$

In the case of an ionic species i, we have to consider the electrochemical potentials which are equal at equilibrium. In developing this equality, we can write Equation (2):

$$\mu_{\mathrm{i}}^{o,\mathrm{w}} + RT\ln a_{\mathrm{i}}^{\mathrm{w}} + z_{\mathrm{i}}F\phi^{\mathrm{w}} = \mu_{\mathrm{i}}^{o,\mathrm{o}} + RT\ln a_{\mathrm{i}}^{\mathrm{o}} + z_{\mathrm{i}}F\phi^{\mathrm{o}} \tag{2}$$

from which we can express the Galvani potential difference between the two phases $\Delta_{\mathrm{o}}^{\mathrm{w}}\phi$ according to Equation (3):

$$\Delta_{\mathrm{o}}^{\mathrm{w}}\phi = \Delta_{\mathrm{o}}^{\mathrm{w}}\phi_{\mathrm{i}}^{o} + \frac{RT}{z_{\mathrm{i}}F}\ln\left(\frac{a_{\mathrm{i}}^{\mathrm{o}}}{a_{\mathrm{i}}^{\mathrm{w}}}\right) = \Delta_{\mathrm{o}}^{\mathrm{w}}\phi_{\mathrm{i}}^{o'} + \frac{RT}{z_{\mathrm{i}}F}\ln\left(\frac{c_{\mathrm{i}}^{\mathrm{o}}}{c_{\mathrm{i}}^{\mathrm{w}}}\right) \tag{3}$$

where a_{i} and c_{i} are the activity and the concentration of the ion, respectively, in both phases and $\Delta_{\mathrm{o}}^{\mathrm{w}}\phi_{\mathrm{i}}^{o}$ and $\Delta_{\mathrm{o}}^{\mathrm{w}}\phi_{\mathrm{i}}^{o'}$ are called the standard and the formal transfer potentials, respectively.

This equation is often called the Nernst equation for ion transfer at liquid/liquid interfaces, and the term $\Delta_{\mathrm{o}}^{\mathrm{w}}\phi_{\mathrm{i}}^{o}$ corresponds to the standard Gibbs energy of transfer expressed on a voltage scale (Equation 4):

$$\Delta_{\mathrm{o}}^{\mathrm{w}}\phi_{\mathrm{i}}^{o} = \frac{\Delta G_{\mathrm{t,i}}^{o,\mathrm{w}\rightarrow\mathrm{o}}}{z_{\mathrm{i}}F} \tag{4}$$

It is important to realize that although the Nernst equation for ion transfer resembles the classical Nernst equation for redox reactions on an electrode, there is no redox reaction involved in the definition of Equation (3).

If a ligand or an ionophore able to complex the transferring ion is present in the organic phase, then the complexation equilibrium can be taken into account. In the case of 1 : 1 stoichiometry, the association constant is simply given by Equation (5):

$$K_{\mathrm{a}} = \frac{a_{\mathrm{ML}}^{\mathrm{o}}}{a_{\mathrm{M}}^{\mathrm{o}}a_{\mathrm{L}}^{\mathrm{o}}} \tag{5}$$

where M refers to the transferring ion and L to the ligand. The Galvani potential difference now reads as Equation (6):

$$\begin{aligned}\Delta_{\mathrm{o}}^{\mathrm{w}}\phi &= \Delta_{\mathrm{o}}^{\mathrm{w}}\phi_{\mathrm{M}}^{o} + \frac{RT}{z_{\mathrm{M}}F}\ln\left(\frac{a_{\mathrm{M}}^{\mathrm{o}}}{a_{\mathrm{M}}^{\mathrm{W}}}\right)\\ &= \Delta_{\mathrm{o}}^{\mathrm{w}}\phi_{\mathrm{ML}}^{o} + \frac{RT}{z_{\mathrm{M}}F}\ln\left(\frac{a_{\mathrm{ML}}^{\mathrm{o}}}{a_{\mathrm{M}}^{\mathrm{W}}}\right)\end{aligned} \tag{6}$$

with the apparent standard transfer potential given by Equation (7):

$$\Delta_{\mathrm{o}}^{\mathrm{w}}\phi_{\mathrm{ML}}^{o} = \Delta_{\mathrm{o}}^{\mathrm{w}}\phi_{\mathrm{M}}^{o} - \frac{RT}{z_{\mathrm{M}}F}\ln\left(K_{\mathrm{a}}a_{\mathrm{L}}^{\mathrm{o}}\right) \tag{7}$$

This equation shows that the presence of an ionophore in the organic phase can shift the apparent solvation energy and therefore facilitate the transfer of hydrophilic species from water to oil. We shall call this type of ion-transfer reaction facilitated ion transfer.

2.1.2 *Polarizable and Non-polarizable Interface Between Two Immiscible Electrolyte Solutions*

If a salt such as tetrabutylammonium bromide is dissolved in two immiscible solvents in contact, the distribution of the salt induces a polarization of the interface. The resulting Galvani potential difference is then called a distribution potential and is defined by applying Equation (3) to both the cation and the anion (Equation 8):

$$\Delta_{\mathrm{o}}^{\mathrm{w}}\phi = \Delta_{\mathrm{o}}^{\mathrm{w}}\phi_{+}^{o} + \frac{RT}{F}\ln\left(\frac{a_{+}^{\mathrm{o}}}{a_{+}^{\mathrm{w}}}\right) = \Delta_{\mathrm{o}}^{\mathrm{w}}\phi_{-}^{o} - \frac{RT}{F}\ln\left(\frac{a_{-}^{\mathrm{o}}}{a_{-}^{\mathrm{w}}}\right) \tag{8}$$

In the case of dilute solutions, this equation simplifies to Equation (9):

$$\Delta_{\mathrm{o}}^{\mathrm{w}}\phi = \tfrac{1}{2}(\Delta_{\mathrm{o}}^{\mathrm{w}}\phi_{+}^{o} + \Delta_{\mathrm{o}}^{\mathrm{w}}\phi_{-}^{o}) \tag{9}$$

This simple example illustrates that as soon as we partition salts between two adjacent phases, the interface becomes polarized at a fixed potential defined by the standard transfer potentials of the different ionic species. Because this polarization potential is fixed, we shall say that the interface is nonpolarizable in the sense that it is not possible to polarize the interface without modifying the chemical composition of the two phases.

In the case where a hydrophilic salt is dissolved in water and a hydrophobic salt is dissolved in the organic phase such that the concentration of the hydrophilic salt in the organic phase is negligible compared with that of the hydrophobic salt and, conversely, the concentration of the hydrophobic salt in water is negligible compared with that of the hydrophilic salt, then the interface will be called polarizable. This definition means that it is now possible to polarize the interface from an external potential source without modifying the chemical composition of the adjacent phases. In this way, there is what is called a potential window such that it is possible to polarize the interface up to a point where the applied Galvani potential difference is enough for an ion to transfer.

To illustrate the principle of the potential window, let us consider an interface between an aqueous solution of Li_2SO_4 and a solution of tetrabutylammonium (TBA^+) tetraphenylborate (TPB^-) in an organic solvent, e.g. 1,2-dichloroethane (1,2-DCE).

At the potential of zero charge, the two adjacent phases are by definition uncharged. If a positive Galvani potential difference (water vs oil) is applied from an external source, two back-to-back Gouy–Chapman diffuse layers will be established with an excess of cations in the aqueous phase and an excess of anions in the organic phase.(20,46,56) As shown in Figure 1, we can polarize the interface until the Galvani potential difference reaches the standard transfer potential of either Li^+ or TPB^-. As it happens, the standard transfer potential of TPB^- is less than that of Li^+ ($\Delta_o^w\phi_{TPB^-}^o = 340\,mV$ and $\Delta_o^w\phi_{Li^+}^o = 580\,mV$), and TPB^- then starts to transfer as soon as the Galvani potential difference approaches 200 mV. The chemical composition of the adjacent phases is then altered by

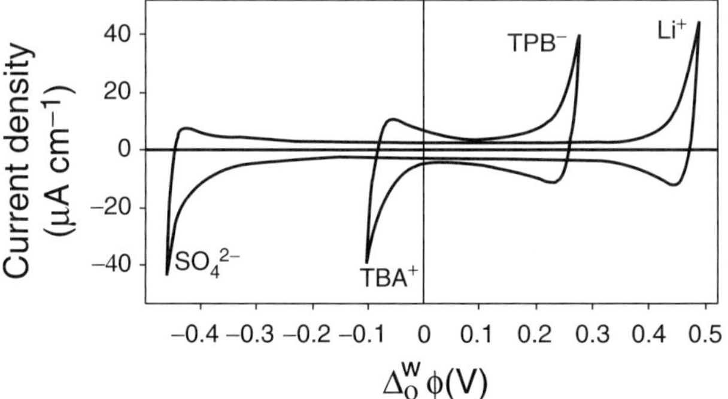

Figure 1 Potential window for 1,2-DCE/water systems containing 5 mM Li_2SO_4 as aqueous supporting electrolyte and 5 mM tetrabutylammonium tetraphenylborate (TBATPB) or 5 mM bis(triphenylphosphoranylidene) ammonium tetrakis(4-chlorophenyl borate) (BTTPATPBCl) as organic supporting electrolyte.

the Faradaic current across the interface and the redox reactions at the two electrodes connected to the external source. If instead of using TPB^- as the organic anion we choose a more hydrophobic anion such as tetrakis(4-chlorophenyl)borate ($TPBCl^-$) for which the standard transfer potential is very large, then a positive polarization of the interface will result in the transfer of Li^+ from water to oil (see Figure 1). It can be concluded that $TPBCl^-$ is more hydrophobic than Li^+ is hydrophilic.

When a negative polarization is applied, then the interface is polarized until the Galvani potential difference reaches the standard transfer potential of either TBA^+ or SO_4^{2-} ($\Delta_o^w\phi_{TBA^+}^o = -230\,mV$ and $\Delta_o^w\phi_{SO_4^{2-}}^o < -600\,mV$, respectively). Since the standard transfer potential of SO_4^{2-} is more negative than that of TBA^+, the potential window is limited by the transfer of TBA^+. However, if we use in the organic phase a more hydrophobic cation such as bis(triphenylphosphoranylidene) ammonium ($BTPPA^+$), which has a very negative standard transfer potential, then the potential window is limited by the transfer of SO_4^{2-}. Again, it can be concluded that $BTPPA^+$ is more hydrophobic that sulfate is hydrophilic.

Standard Gibbs energies of ion transfer have been tabulated in different reviews.(22,57–59)

2.2 Structure of the Interface and Potential Distribution

A liquid/liquid interface is by definition a molecular interface between two condensed media. The solvent dynamics results in an interface fluctuating to a certain limit following the capillary wave theory. If we take a time average view of the interface, we can say that it is composed of a thin (i.e. about 1 nm) mixed solvent layer. Snapshots from molecular dynamics computer simulation show that the local structure of the interface is greatly influenced by the presence of ionic charges.(38,47,49,51,60,61)

As discussed above, the interface between two immiscible electrolytes can be polarized using an external source. From an experimental viewpoint, it is usual to operate with a four-electrode potentiostat comprising two reference electrodes and two counter electrodes to provide the current,(6,11) as illustrated in Figure 2.

At the interface, the polarization is distributed over the two back-to-back Gouy–Chapman diffuse layers. One of the key unresolved issues is the dependence of the interfacial electric field on the applied polarization. At the point of zero charge, it is usually accepted that there is no specific orientation of dipoles resulting from the molecular interactions between the solvent molecules. However, it is difficult to estimate if the applied polarization influences the orientation of the solvent molecules and to quantify the strength of the field. Some authors choose, as in bioelectrochemistry, a constant-field approach to model the polarized interface, as shown in Figure 3.

For references see page 9921

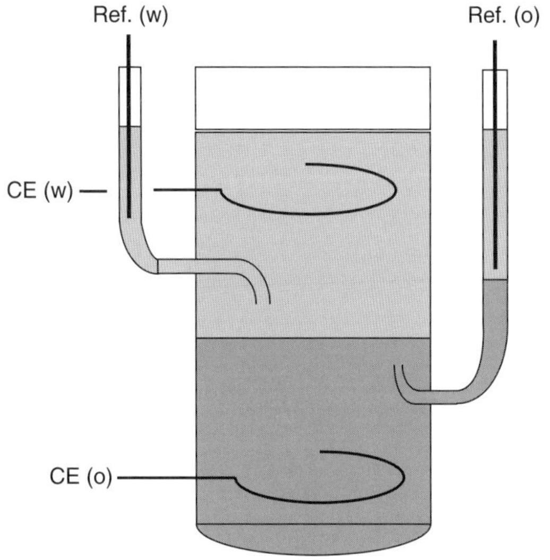

Figure 2 Schematic diagram of an electrochemical cell for the study of charge-transfer reactions at an ITIES. For the system in Figure 1, the aqueous reference electrode can be $Ag|Ag_2SO_4$. The organic reference electrode includes an unpolarized liquid junction using the organic cation or anion as the common ion. If the organic supporting electrolyte is TBATPB, the liquid junction can comprise either tetrabutyl ammonium chloride (TBACL) or sodium tetrabutylborate (NaTPB) so that the reference electrode can be $TBATPB_o|TBACL_w|AgCl|Ag$ or $TBATPB_o|NaTPB|AgTPB|Ag$.

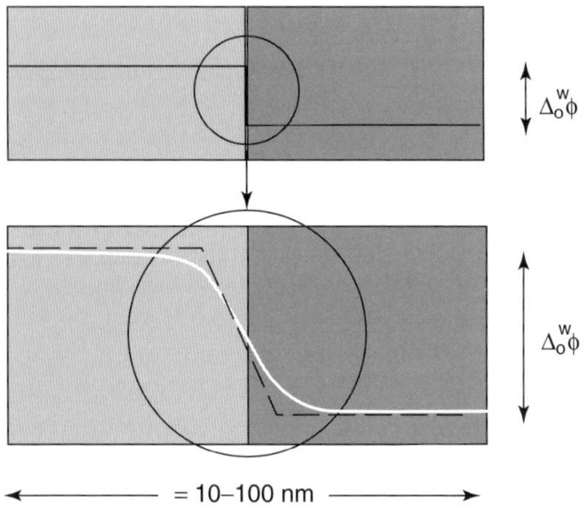

Figure 3 Potential distribution for a positive polarization (water vs oil). The top curve is a macroscopic representation, whereas the bottom curve illustrates the potential drop across the two back-to-back diffuse layers. The dashed line represents the constant-field model.

2.3 Charge-transfer Reactions

2.3.1 Simple Ion Transfer

From a practical viewpoint, the kinetics of ion transfer can be considered as very fast, such that it can be assumed that the surface concentrations always follow the Nernst Equation (4). In electrochemical nomenclature, it is then said that ion-transfer reactions are reversible. Similarly to a reversible redox reaction on an electrode that is limited by the mass transfer of the reactants to the electrode and by that of the products away from the electrode, an ion-transfer reaction is limited by the mass transfer of ions to the interface and away from it. Hence the mass transport differential equations and boundary conditions are similar in both cases, and all the electroanalytical methodology can therefore be transposed to the study of ion transfer reactions.

As in classical amperometry, the response of the system stems from the resolution of the diffusion equations of the ion in the two adjacent phases (Equation 10):

$$\frac{\partial c_i^w}{\partial t} = D_i^w \frac{\partial^2 c_i^w}{\partial x^2} \quad \text{and} \quad \frac{\partial c_i^o}{\partial t} = D_i^o \frac{\partial^2 c_i^o}{\partial x^2} \tag{10}$$

By taking the interface as the origin, the current is then simply given by the flux of i across the interface of area A (Equation 11):

$$I = z_i FA \left(\frac{\partial c_i^w}{\partial x}\right)_{x=0} \tag{11}$$

The boundary conditions are the Nernst Equation (4) and the equality of the fluxes (Equation 12):

$$D_i^w \left(\frac{\partial c_i^w}{\partial x}\right)_{x=0} + D_i^o \left(\frac{\partial c_i^o}{\partial x}\right)_{x=0} = 0 \tag{12}$$

2.3.2 Facilitated Ion Transfer

Assisted ion-transfer reactions are also very fast and can be considered reversible in most cases. However, mass transport is more complicated and, as shown schematically in Figure 4, we can distinguish four types of reactions:[62]

- aqueous complexation followed by transfer (ACT)
- transfer by interfacial complexation (TIC)
- transfer by interfacial decomplexation (TID)
- transfer followed by organic-phase complexation (TOC).

In each case, we have to consider the mass transport of the ions and of the ligands to and away from the interface. This leads to many specific cases, which have been treated in the literature (see section 4.2). More interesting from an electroanalytical aspect are the two limiting cases where either the ions or the ligands are in excess in their respective phases. In these two cases, the mass transport is limited by that of the ligand or that of the ions, respectively.

Assisted, or facilitated, ion transfer was first reported in 1979 by Koryta,[7] who observed that the transfer of potassium and sodium ions in the aqueous phase

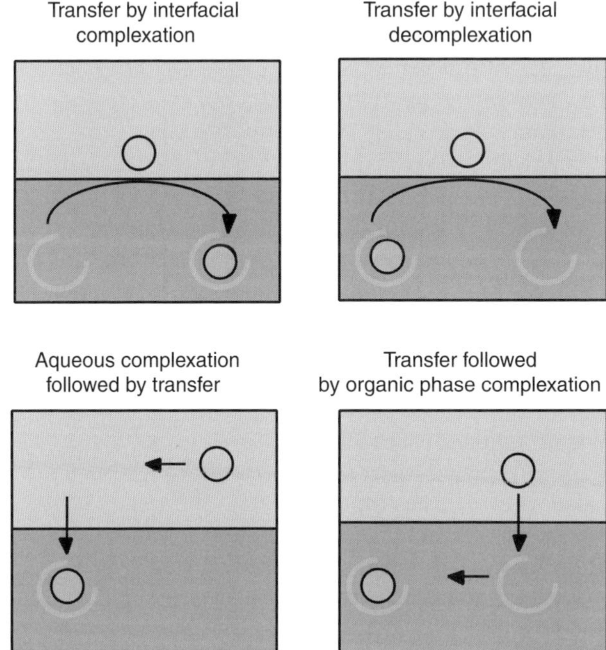

Figure 4 Schematic representation of assisted ion-transfer reactions.

was facilitated by the formation of a complex in the organic phase with the synthetic polyether dibenzo-18-crown-6 (DB18C6) and with the natural antibiotic valinomycin, respectively. This work was a landmark in ITIES research, because it meant that both potential window-limiting species and neutral ionophore molecules were amenable to study. As a result, this field spread quickly to solvent extraction and purification, detection of trace ions, assisted transfer of proton and development of amperometric sensors (see also section 4).

3 ANALYTICAL METHODS

3.1 Cyclic Voltammetry

Cyclic voltammograms produced by reversible ion-transfer reactions are similar to those obtained for reversible electron-transfer reactions at a metal/electrolyte solution interface,[63] as shown in Figure 5. Thus, for reversible transfer reactions of an ion across a large (or planar) interface, the maximum forward peak current I_p^{FWD} may then be expressed by the Randles–Sevcik Equation (13):[64]

$$I_p^{FWD} = 0.4463\, z_i F A c_i^w \sqrt{\frac{z_i F D_i^w v}{RT}} \tag{13}$$

where A is the interfacial area, v the rate of a potential sweep and c_i^w the aqueous bulk concentration of i.

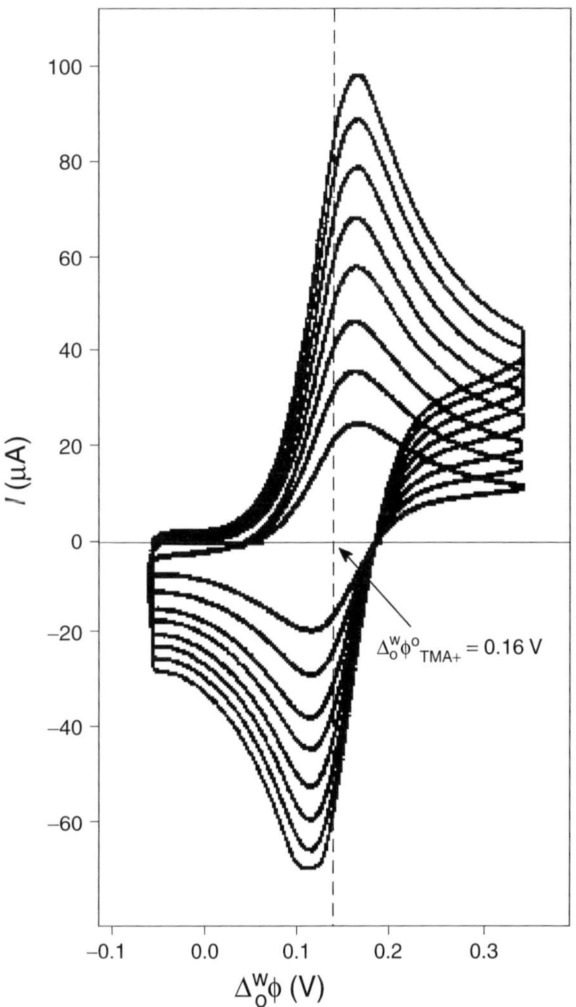

Figure 5 Cyclic voltammogram for the transfer of tetramethylammonium across the water/1,2-DCE interface at different sweep rates (the interfacial area is 1.13 cm^2).

Together with the evaluation of the diffusion coefficient of the transferring ion, the determination of the formal transfer potential of an ion (and thus of its Gibbs energy of transfer) is the most important application of cyclic voltammetry. For a reversible ion-transfer reaction at a large planar interface, $\Delta_o^w\phi_i^{o'}$ may be expressed in terms of the half-wave potential, $\Delta_o^w\phi_{i,1/2}$, by Equation (14):

$$\Delta_o^w\phi_{i,1/2} = \Delta_o^w\phi_i^{o'} + \frac{RT}{2z_iF}\ln\left(\frac{D_i^w}{D_i^o}\right) \tag{14}$$

Experimentally, $\Delta_o^w\phi_{i,1/2}$ is considered equal to the mid-peak potential, and is directly deduced from the voltammograms. However, as the diffusion coefficients in the organic phase are rarely known because of experimental difficulties, the ratio D_i^o/D_i^w in Equation (14) is

For references see page 9921

usually approximated to the inverse ratio of the solvent viscosities, η, by application of Walden's rule[64] (Equation 15):

$$\frac{D_i^o}{D_i^w} = \frac{\eta^w}{\eta^o} \tag{15}$$

Finally, for a reversible charge transfer, the classical following conditions also apply:

- the peak potentials must be independent of the scan rate
- the peak-to-peak separation is $59/z_i$ mV at 25 °C.

3.2 Dropping-electrolyte Electrode

In the same way that the mercury-drop electrode permits very reproducible measurements owing to the continuous renewal of the interface for every drop, the dropping-electrolyte electrode allows the study of both ion-transfer and facilitated ion-transfer reactions.[13,65–77] This approach has been widely used to study facilitated ion transfer. The principle consists in either dropping down or dropping up an electrolyte solution and recording the current as in polarography. The methodology of polarography can then be transposed directly.

3.3 Micro Interface Between Two Immiscible Electrolyte Solutions

3.3.1 *Micropipets and Microholes*

Microelectrodes benefit from diffusion fields controlled by the geometry of the interface and from reduced ohmic loss. Because studies at ITIES always involve the use of an organic solvent, much effort has been dedicated to supporting micro liquid/liquid interfaces. Using the micropipet technology developed for electrophysiology, liquid/liquid interfaces were first supported at the tip of micropipets, the water phase being usually located inside the pipet.[62,78–87] Micropipets provide an asymmetry of diffusion fields. Ingress motion of ion occurs by pseudo-spherical diffusion to a microdisk interface whereas egress motion out of the pipet occurs by linear diffusion. This yields asymmetric voltammograms as shown in Figure 6. The ingress current reaches a steady-state value proportional to the radius, the diffusion coefficient and to a geometric factor.

Another approach is to support micro liquid/liquid interfaces in a microhole in a thin polymer film or in a silicon structure.[88–95] A technique often used to micromachine polymers is UV laser photo-ablation,[96,97] whereas that used to machine silicon is anisotropic

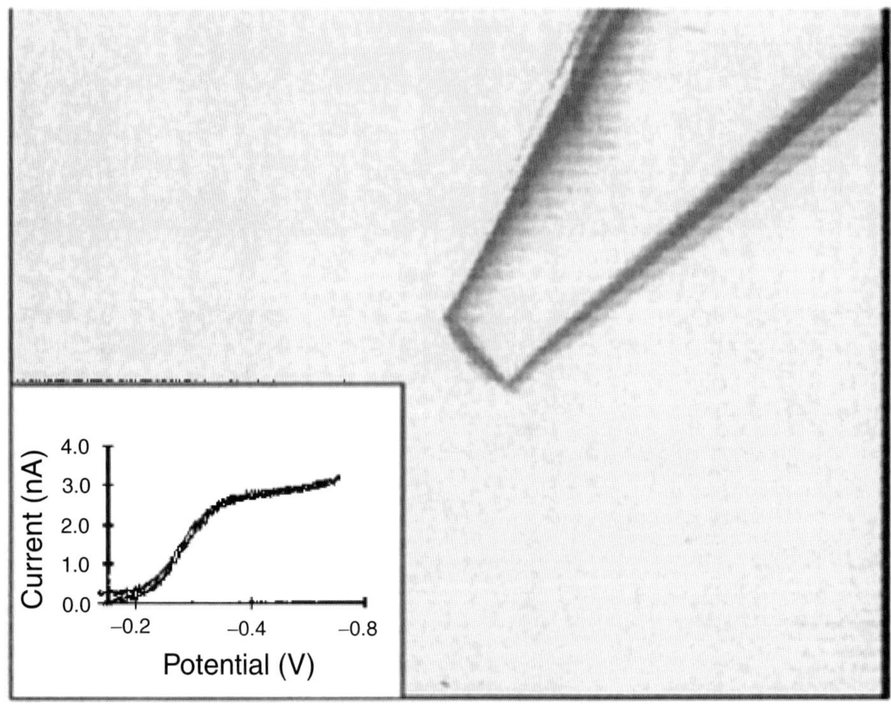

Figure 6 Video micrograph of a 15.5 µm radius micropipet filled with an aqueous KCl solution and immersed in a 1,2-DCE solution of DB18C6. No external pressure was applied, and the micro ITIES is flat. The insets show the corresponding steady-state voltammograms of facilitated transfer of potassium. (Reprinted with permission from Y. Shao, M.V. Mirkin, 'Voltammetry at Micropipet Electrodes', *Anal. Chem.*, **70**, 3155–3161 (1998). Copyright 1998 American Chemical Society.)

etching. For microinterfaces supported in microholes, the thinner is the supporting film the more symmetric are the diffusion fields. For very thin films, the mass transport equations are similar to those for a microdisk electrode of radius r.[88,93] In this case, the half-wave potential is given by Equation (16):

$$\Delta_o^w\phi_{i,1/2} = \Delta_o^w\phi_i^{o'} + \frac{RT}{z_iF}\ln\left(\frac{D_i^w}{D_i^o}\right) \tag{16}$$

and the steady-state current by Equation (17):

$$I_{SS} = 4z_iFD_i^w c_i^w r \tag{17}$$

With thicker films, however, the microhole supporting the ITIES modifies the mass transport, and both the ingress and egress geometries in and out of the microhole interface must be introduced.[98,99] Following the shape given in Figure 7, the expression of the half-wave potential then becomes Equation (18):[100]

$$\Delta_o^w\phi_{i,1/2} = \Delta_o^w\phi_i^{o'} + \frac{RT}{z_iF}\ln\left(\frac{D_i^w}{D_i^o}\cdot\frac{h^w + \pi r/4}{h^o + \pi r/4}\right) \tag{18}$$

The main advantage of working with microliquid/liquid interfaces stems from the low IR drop, which provides clean voltammetric responses.

3.3.2 Ionodes

The major barriers to the electroanalytical exploitation of amperometry at liquid/liquid interfaces have been the mechanical stability of the interface and the resistivity of the organic phase. To circumvent these difficulties, the first approach has been to "gelify" one of the phases (usually the organic phase) using the methodology developed to produce polymer membranes for the design of ISEs [see Figures 8 and 9a and b].

Recently, new composite membranes comprising an organic electrolyte gel supported on a thin polymer micromachined so as to include a regular array of

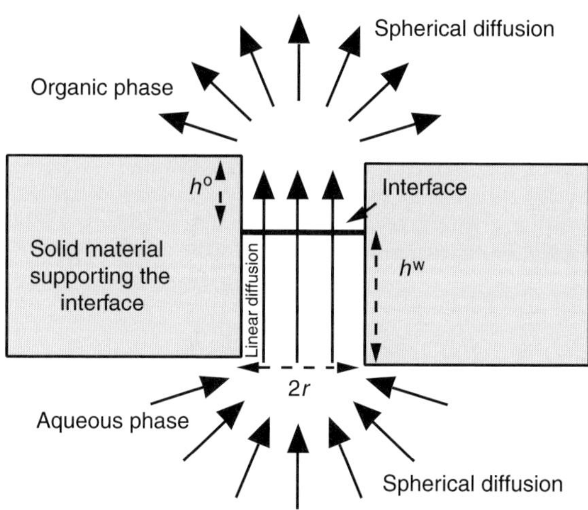

Figure 7 Schematic diagram showing the diffusion processes of an ion passing from water to the organic phase through a microhole supported liquid/liquid interface. r is the radius of the microhole, h^w the penetration depth of the aqueous phase and h^o that of the organic phase into the microhole.

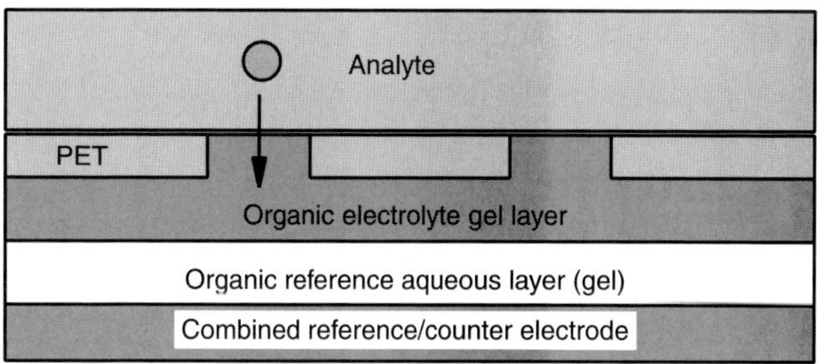

Figure 8 Simplified diagram showing the common composition of an ionode. The interface is supported in a microhole array pierced through poly(ethylene terephthalate) (PET) by laser photoablation.

(a)

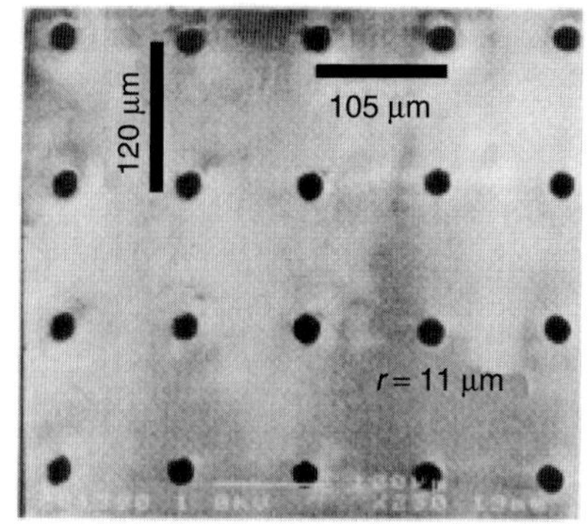

(b)

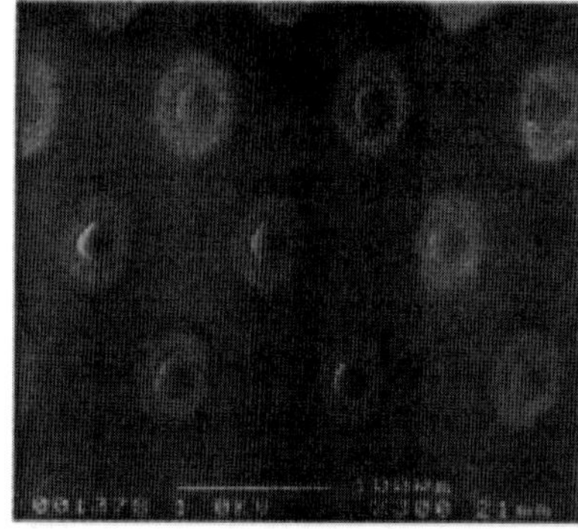

Figure 9 Scanning electron micrographs of the entrance-side holes in a PET film. (a) Empty holes; (b) array filled with poly(vinyl chloride) (PVC) at 70 °C. (Adapted with permission from H.J. Lee, P.D. Beattie, B.J. Seddon, M.D. Osborne, H.H. Girault, 'Amperometric Ion Sensors Based on Laser-patterned Composite Polymer Membranes', *J. Electroanal. Chem.*, **440**, 73–82 (1997). Copyright 1998 American Chemical Society.)

For references see page 9921

microholes have been developed and named ionodes for simplicity.[101,102]

3.4 Optical Techniques

Optical techniques are interesting alternatives to classical electrochemical methods, because they have the advantage of providing in situ spectroscopic measurements at liquid/liquid interfaces. Two main types of experiments (namely voltfluorimetry or voltabsorptometry on the one hand and sum-frequency generation (SFG) or second harmonic generation (SHG) on the other) are commonly conducted at ITIES depending on whether linear or, non-linear optics, respectively, are used. For simplicity, however, only the linear optical techniques are presented here, and readers interested in SHG or SFG techniques should consult reviews.[103–109]

Ultraviolet/visible (UV/VIS) absorption measurements can be carried out when the incident light beam impinges on the interface from the phase of larger optical index in total internal reflection (TIR) geometry, i.e. with an incidence angle greater than the critical angle, so that the transmitted light wave in the adjacent phase cannot propagate. Voltabsorptometry measures then the changes of light intensity of the reflected beam due to absorption of charge-transfer products. Assuming that the absorbance A_{TIR} is proportional to the integral of the bulk concentration of the absorbing species, its time derivative is again proportional to the Faradaic current.[110] As an example, Figure 10(a) and (b) show the evolution of the absorbance in 1,2-DCE upon transfer of methyl orange. The evolution of the UV/VIS spectra follows the current change in the corresponding voltammogram. The band at 420 nm increases monotonically during the forward sweep owing to deprotonated methyl orange transfer from water to 1,2-DCE, and decreases only after the corresponding isobestic point after which the transfer is inverted.

This technique has also been used to study Cu(II) transfer assisted by 6,7-dimethyl-2,3-di(2-pyridyl)quinoxaline, the absorption spectrum of which changes when it forms a complex. The transfer of rose bengal and eosin B has been studied by voltfluorometry,[111–114] where, similarly to UV/VIS experiments, the change in fluorescence intensity from an exciting beam is monitored as a function of applied interfacial potential.

The advantage of fluorescence and luminescence measurements is that the technique is extremely sensitive, and that small changes in interfacial concentration can be readily monitored. In contrast to cyclic voltammetry, these techniques are totally specific to the transferring species and are not influenced by the transfer of undesired species, such as supporting electrolyte ions.[113,115] Furthermore, they are insensitive to double-layer charging and ohmic drop, so that chronoabsorptometry and

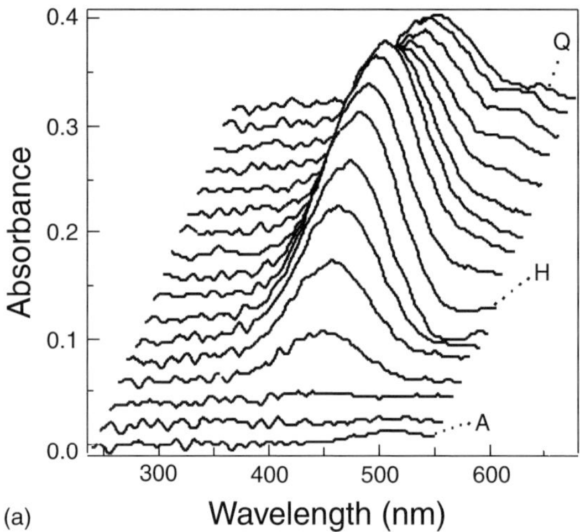

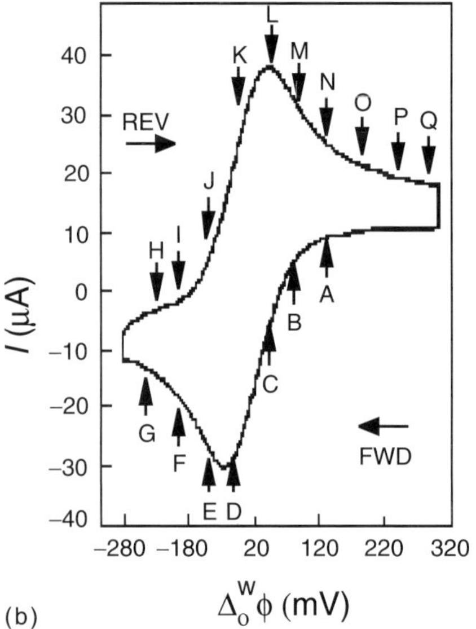

Figure 10 In situ UV/VIS spectra (a) and cyclic voltammogram (b) for the transfer of deprotonated methyl orange from water to 1,2-DCE at a scan rate of 36 mV s^{-1}. (Reprinted from Z. Ding, F. Reymond, P. Baumgartner, D. Fermin, P.-F. Brevet, P.-A. Carrupt, H.H. Girault, 'Mechanism and Dynamics of Methyl and Ethyl Orange Transfer across the Water/1,2-Dichloroethane Interface', *Electrochim. Acta*, **44**, 3–13 (1998). Copyright 1998, with permission from Elsevier Science.)

chronofluorimetry[112,114] studies proved to be very attractive in measuring the kinetics of ion-transfer,[115–119] for which reliable data were previously difficult to obtain. Other spectroscopic techniques are also of interest in this field, such as time-resolved laser-induced fluorescence for monitoring the lifetimes of excited species[120] and potential-modulated reflectance spectroscopy,[117] where the transfer kinetics of a species are estimated by measuring its frequency-dependent absorption following an ac potential perturbation.

4 APPLICATIONS

4.1 Amperometric Ion Sensors and Detectors

Nonredox ionic species can be detected amperometrically by measuring the current associated with ion-transfer reactions across a polarized ITIES.(19,24) However, the major difficulties in designing transducers with liquid/liquid systems stem from the mechanical instability of the ITIES and from the resistive nature of the organic phase. As can be deduced from section 3.3, these two problems can be circumvented by using microinterfaces such as micropipets and microholes in thin polymer films or by gelifying one or two of the phases (although the diffusion coefficient of ions in a gel is much reduced).(121–123)

Plasticized polymers have been used to solidify the organic phase(121,124–128) and efficient sensors for sodium,(129–131) ammonium,(132) choline(133) and urea(134–136) have been obtained by immobilizing DB18C6 within the organic phase to facilitate the transfer of Na^+ and $NH_4{}^+$, which were detected amperometrically. Moreover, micro liquid/liquid interface arrays(93) have been developed for the assay of urea(89) and creatinine,(91) where the transducer relied again on the amperometric detection of ammonium.

Amperometric detection of choline using either direct ion-transfer(137) or stripping ion-transfer reactions(138) has been achieved with a water/*o*-nitrophenyl octyl ether (NPOE)–PVC gel interface supported on an array of microholes. Otherwise, the electrochemical behavior of NPOE gelified with 1,3 : 2,4-dibenzylidene sorbitol (DBS) has been characterized(139) and a NPOE–PVC gel has been used to manufacture an amperometric sensor for alkali metal ions by incorporation in the organic gel of ionophores such as DB18C6 and valinomycin,(140) which has been shown to facilitate discriminantly the transfer of lithium, sodium, ammonium and potassium.(101,102) This principle has also been used to detect lithium in samples such as blood serum that contains a large excess of sodium (dibenzyl-14-crown-4 and *o*-nitrophenyl phenyl ether were used as ionophore and organic phase, respectively).(141) These last two results are very promising, because they constitute a powerful alternative to conductimetric detection of nonredox species in ion-exchange chromatography,(142) which suffers from a lack of selectivity compared with optical or direct amperometric detectors (see Figure 11a and b).

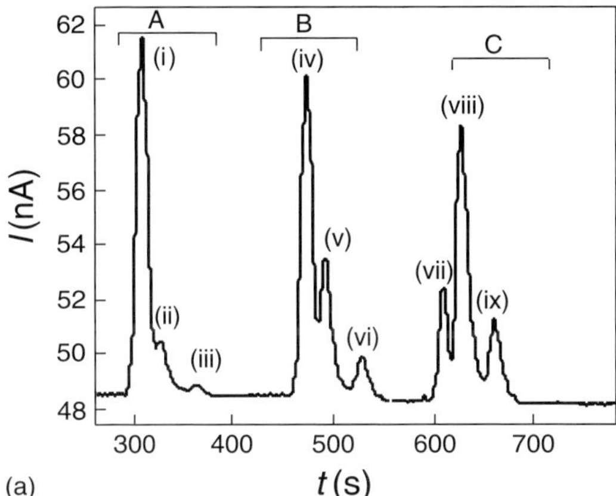

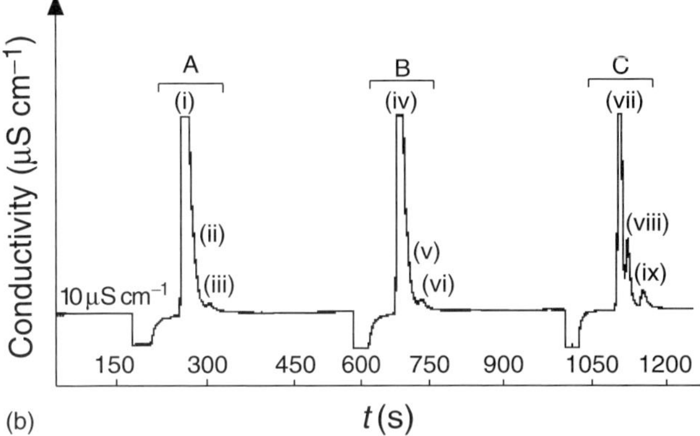

Figure 11 (a) Pulse amperograms recorded for monovalent cations facilitated by 1 mM valinomycin using a PVC–NPOE microgel membrane and tetrabutylammonium cation as organic supporting electrolyte. (b) Chromatogram obtained for the monovalent cations based on a conductimetric detector. The concentrations of the standard solutions are (A) (i) 100 ppm Na^+, (ii) 0.5 ppm $NH_4{}^+$ and (iii) 0.5 ppm K^+; (B) (iv) 100 ppm Na^+, (v) 1 ppm $NH_4{}^+$ and (vi) 1 ppm K^+; and (C) (vii) 20 ppm Na^+, (viii) 4 ppm $NH_4{}^+$ and (ix) 4 ppm K^+. Flow rate, 0.85 mL min^{-1}; 5 mM tartaric acid is used as the eluent. (Reprinted with permission from H.J. Lee, H.H. Girault, 'Amperometric Ion Detector for Ion Chromatography', *Anal. Chem.*, **70**, 4280–4285 (1998). Copyright 1998 American Chemical Society.)

4.2 Analytical Aspects of Metal Extraction

The chemical analysis of liquid solutions has for a long time been a very important domain of analytical chemistry.(143) Numerous applications in biology, medicine and environmental studies require the sensitive and selective detection of chemical constituents in different media. Heavy metals are an example where the need for a precise analytical tool is of major importance owing to their high toxicity towards life in general and people in particular.

Electrochemistry at the ITIES has attracted much attention in this field during the last decade, because it provides a simple way of measuring the stoichiometry and the association constants of ion–ionophore complexes in organic solvents.(140,144,145) With the rapid development of coordination chemistry, numerous ligands with specific binding properties have become available, extending complexation reactions where the transfer of a monocharged metal ion is facilitated by the formation

For references see page 9921

of a complex of 1 : 1 ion-to-ligand stoichiometry to more complicated systems.

Theoretical studies on facilitated ion-transfer reactions now provide a reliable framework to analyze experimental data and to determine the physicochemical parameters governing complexation reactions at ITIES. Matsuda et al.(146) published a general theoretical equation for the polarographic response of reversible facilitated ion-transfer reactions, leading to a prediction of the half-wave potential dependence on the initial concentrations of both the metal and the ionophore (denoted c_{Minit} and c_{Linit}). In ligand excess, the transfer is limited by the diffusion of the free metal ions towards the interface, whereas it is limited by the diffusion of the complex away from the interface in metal excess. Between these two limiting cases, a mixed diffusion regime is established and, for 1 : 1 stoichiometry, the authors determined criteria separating these three regions in which the dependence of $\Delta_o^w\phi_{1/2}$ on the initial concentrations changes (see Figure 12a and b). They further described a simple method to calculate the association constants, and this work has been widely used to interpret both polarographic and cyclic voltammetric experiments.

Few studies have been carried out to model cyclic voltammetric experiments(147–150) and the approach followed by Matsuda et al. has recently been generalized to $1:m$ ion-to-ligand stoichiometries, showing that variations of c_{Minit} and c_{Linit} do not lead to a similar evolution of $\Delta_o^w\phi_{1/2}$.(151,152) This is illustrated in Figure 12(a) and (b) and demonstrated by the relationships obtained for the TIC, TID and TOC mechanisms (Equations 19–21):

when $c_{Linit} \gg c_{Minit}$:

$$\Delta_o^w\phi_{1/2} = \Delta_o^w\phi_{M^{z+},1/2} - \frac{RT}{zF}\ln\left[\sum_{j=0}^{m}\beta_j^o(c_{Linit})^j\right] \quad (19)$$

when $c_{Minit} \gg c_{Linit}$:

$$\Delta_o^w\phi_{1/2} = \Delta_o^w\phi_{M^{z+}}^{o'} - \frac{RT}{zF}\ln\left[\sum_{j=0}^{m}j\beta_j^o c_{Minit}\left(\frac{c_{Linit}}{2}\right)^{(j-1)}\right] \quad (20)$$

where β_j^o are the reduced association constants, defined as

$$\beta_j^o = \frac{c^o_{ML_j^{z+}}}{c^o_{ML_j^{z+}}(c_L^o)^j} = \prod_{k=1}^{j}K_{ak}^o \quad (21)$$

The transfer of alkali metal cations such as Li^+, Na^+, K^+, Rb^+, Cs^+ and Tl^+ is facilitated by DB18C6(71,131,153,154) and some of them by monensin, valinomycin, nigericin,(155) nonactin(156) and various crown ethers;(149,157) the transfer of alkaline earth metal ions such as Mg^{2+}, Ca^{2+}, Sr^{2+} and Ba^{2+} is facilitated by various water-soluble and -insoluble crown ethers and by poly(oxyethylene)octylphenyl ethers(158) and calix[4]-arene.(159) The transfer of the alkali and alkaline earth metal ions is also facilitated by other synthetic neutral ionophores.(160,161) Various studies were carried out with 1,10-phenanthroline, which forms complexes with heavy metal ions such as Cd^{2+}, Zn^{2+}, Co^{2+}, Ni^{2+},(67)

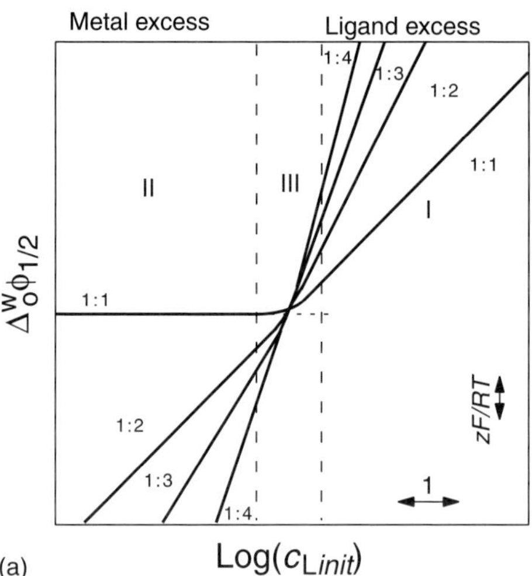

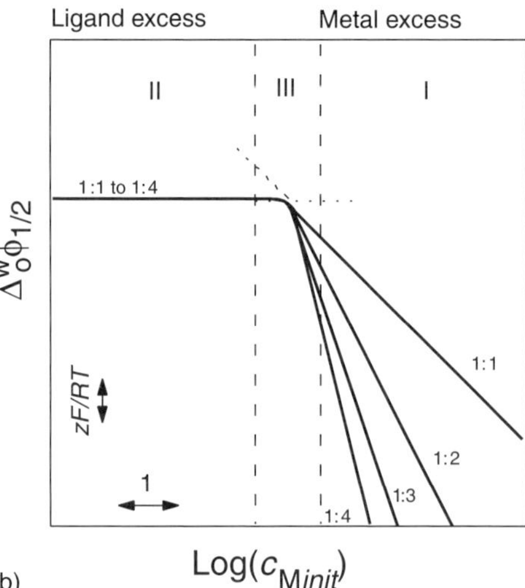

Figure 12 Schematic diagram showing the half-wave potential dependence on (a) the initial ligand concentration and (b) the initial metal concentration for TIC/TID facilitated ion-transfer reactions of 1 : 1 to 1 : 4 ion-to-ligand stoichiometry. An intersection point has been chosen arbitrarily to facilitate comparison between the various stoichiometries. These schemes show that these curves do not have equal slopes in metal or ligand excess and that a transition domain separates these two extreme cases [the roman numbers indicate the domains defined by Matsuda et al.,(146) who derived criteria defining these three regions].

Mn^{2+}[72] and Cu^{2+},[76] and also with protons[68,162] and alkali and alkaline earth metal cations in various stoichiometries.[163] 8-Quinolinol (Ni^{2+}), hydroxy oximes (Cu^{2+}) and dithizones (Ni^{2+}, Zn^{2+} and Cd^{2+}) have been used for solvent extraction of these metals.[164,165] Thiocrown ligands such as 1,4,7,10-tetrathiacyclododecane facilitate the transfer of Cu^{2+},[166–168] Cu^{+},[168,169] Hg^{2+},[168] Ag^{+},[168,170] Ni^{2+}[171] and Pb^{2+} and Cd^{2+}[172] and extraction of lanthanide ions has been achieved with 1,4,10,13-tetrathia-7,6-diazacyclooctadecane.[173]

Likewise, ethylenediamines of different hydrophobicities have been shown to form complexes of various stoichiometries with Ni(II).[174] Complexes of 1 : 1 to 1 : 3 stoichiometries have been clearly identified for the transfer of Ni(II), Zn(II), Fe(II) and Fe(III) assisted by the bidentate ionophores *o,o′*-bipyridine and *o*-phenanthroline[175] and for that of Co^{2+}[176] and Cd^{2+}[177] assisted by 2,2′-bipyridine. Similar investigations were made with the terdentate ligands 2,2′,2″-terpyridine[178] and 2,2′,6′,2″-terpyridine[179] that form stable 1 : 1 and 1 : 2 complexes with transition metals[178] and with the tetradentate phosphorus–nitrogen ligand *N,N′*-bis[2-(diphenylphosphino)phenyl]propane-1,3-diamine for the complexation of Co^{2+}.[180] Dihydroxynaphthalene has also been used as a complexing agent to detect Mo[181] and cupferron to detect Al,[182] Cr,[183] U,[184] V[185] and Eu.[186]

The facilitation effect of drug molecules such as valinomycin,[7,140,187] monensin,[188] nigericin,[155] piroxicam,[189] creatinine,[91] terramycin,[190] rifamycin,[191] lidocaine and dicaine[192] has also been studied, as well as the transfer of U, Np and Pu ions facilitated by phosphine oxide.[193] Finally, phosphine ligands have been used successfully to separate simultaneously Pb^{2+}, Zn^{2+}, Co^{2+}, Fe^{2+}, Cd^{2+}, Mn^{2+} and Ca^{2+}[92]. Polarographic and molecular dynamics studies have also been introduced as methods for the calculation of complex association constants in polar solvents[194–198] and for the determination of the structural conformation of various crown ether complexes, respectively.[199–201] Recently, complex formation at ITIES has also been studied by SHG suggesting that a reorientation of the ionophore occurs before the complexation and the transfer of the metal ion take place.[202,203]

This listing shows the variety of the systems studied, and a complete review and discussion of the advances in assisted ion-transfer reactions at the ITIES up to 1993 have been published.[23,24] However, most of the above reactions follow a TIC or TID[62] mechanism, and only a few studies show evidence of an ACT mechanism,[158,204] because most ligands used in electrochemistry are poorly soluble in water, so that complexes may only be formed in the organic phase. Furthermore, the TOC mechanism has not been differentiated from the TIC mechanism, because this necessitates kinetic data that still remain unmeasurable, the diffusion being the factor limiting the transfer and not the energy required to allow the ions to cross the interface.

4.3 Lipophilicity of Ions

Because of their intrinsic nature, interfaces between two immiscible liquids may serve as simple artificial models of biological membranes. Therefore, studies on drug transfer characteristics and mechanisms in such systems are of great importance to understand better the behavior of drugs in their pharmacokinetic phase,[205,206] their distribution in vivo[207] and hence the delivery problems that limit their efficiency.[208–211] The transport of exogeneous chemicals (and hence of the majority of common drugs) is a passive process,[212] for which it is commonly assumed that ionizable compounds can only cross biological membranes in their neutral form.[213] However, recent studies suggest a significant passive transfer of ions.[214–217] As many drugs are organic compounds that are thus partly or largely ionized at physiological pH, membrane transport can be deeply affected by the lipophilicity of charged species.[218] The lipophilicity of a species is generally evaluated by its partition coefficient, defined as the logarithm of the ratio of the activity of a species in the organic phase to that in the aqueous phase and denoted $\log P$. This parameter is widely used in medicinal chemistry to relate the structure and the physicochemical properties of a drug to its biological activity,[219] which is the objective of all quantitative structure–activity relationship (QSAR) studies.[220]

Dealing only with ions for which $\log P$ is directly given by the formal transfer potential,[221] electrochemistry at the ITIES appears to be a method of choice for assessing the lipophilicity of ionizable drugs. The literature is still very scarce on this topic, and ion-transfer voltammetry has only been applied to determine the transfer potentials of a few compounds of biological interest [1,10-phenanthroline,[68,222] acetylcholine,[125,223–225] various amines,[77,226,227] phosphorylation uncouplers,[228,229] pyrazolone derivatives,[230] picrate,[63,231–233] cinchonidine[75] and quinine[74,221,234,235]] and to investigate the transfer of several antibiotics[236–239] and of a series of hypnotic, anesthetic, cholinergic and adrenergic agents.[192,240–243] However, the recent introduction of ionic partition diagrams[234] (which are a transposition at liquid/liquid interfaces of Pourbaix's pH–potential diagrams[244] for metals in solution) has improved the understanding of the partition processes of ionizable compounds, since they allow reliable predictions and interpretations of their transfer mechanisms across ITIES, as exemplified by Figure 13 for the case of trimetazidine. Further insight into the influence of electronic

For references see page 9921

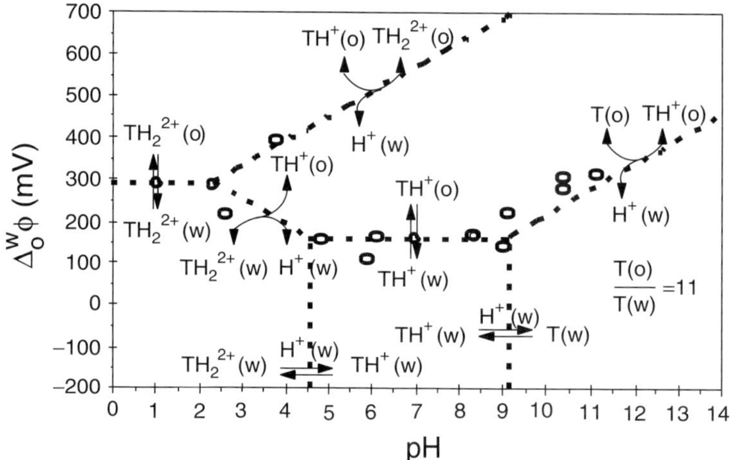

Figure 13 Ionic partition diagram of trimetazidine at 25 °C in water–1,2-DCE. The figure shows the formal transfer potentials obtained by cyclic voltammetry as a function of aqueous pH (bold circles), the equiconcentration lines between two adjacent species (bold dashed lines) and the corresponding transfer mechanisms (chemical equilibria). T, TH^+ and TH_2^{2+} stand for the neutral, the singly protonated and the doubly protonated forms of trimetazidine, respectively. (Adapted from Reymond et al.[247] by permission of Plenum Publishing Corporation.)

structure on lipophilicity has revealed the importance of intramolecular charge delocalization to stabilize ions in the organic phase.[245–248] Although this effect has not yet been quantified, electrochemistry at ITIES is an easy methodology to assess the pH–lipophilicity profiles of ionizable drugs, and it should soon become increasingly popular.

It must finally be noted that octanol is not polarizable,[249] so that most electrochemical studies use 1,2-DCE or nitrobenzene as the organic phase. In order to interpret the lipophilicity of ionizable drugs in pharmacological terms, the thermodynamic parameters obtained with these solvents must be correlated to the octanol–water system commonly used in pharmacology. This correspondence has been established,[250] offering more relevance to the results presented above and opening ideal perspectives for extending electrochemical measurements to medicinal chemistry.

5 CONCLUSION

The increasing number of experiments conducted at liquid/liquid interfaces shows that electrochemical methodologies at such boundaries can be an efficient and versatile tool to probe ions. The great demand for innovative analytical techniques with accurate sensitivity and high selectivity offers a brilliant future for ITIES. Solvent extraction, separation, transfer-phase catalysis and biomembrane studies are other areas where ITIES are expected to develop rapidly.

Further, the study of charge-transfer processes gives an insight into the fundamental problems of physical chemistry in biphasic systems. Thanks to the improved theoretical understanding of interfacial structure, the transfer process of solvated ions from one phase to the other becomes clearer, but it is still very demanding because future applications of ITIES will depend on our theoretical knowledge of the motion of ions in biphasic systems.

Nevertheless, the above applications still suffer from a lack of experimental information compared with other techniques, and further investigations are required.

List of selected abbreviations appears in Volume 15

LIST OF SYMBOLS

a	Activity
A	Absorbance
A	Interfacial Area
c	Concentration
D	Diffusion Coefficient
F	Faraday Constant
h	Penetration Depth of a Phase into a Microhole
I	Current
I_p^{FWD}	Maximum Forward Peak Current
K_a	Association Constant
r	Radius of the Interface
R	Gas Constant
t	Time
T	Temperature
x	Position
z	Charge

Greek Letters

β_j	Reduced Association Constant of the ML_j Complex
$\Delta_o^w\phi$	Galvani Potential Difference Between the w and o Phases
$\Delta_o^w\phi_i^o$	Standard Transfer Potential of i
$\Delta_o^w\phi_i^{o'}$	Formal Transfer Potential of i
$\Delta_o^w\phi_{i,1/2}$	Half-wave Potential
$\Delta G_{t,i}^{o,w\to o}$	Standard Transfer Gibbs Energy of a Species i from Phase w to Phase o
η	Solvent Viscosity
μ^o	Standard Chemical Potential or Gibbs Energy of Solvation
v	Rate of a Potential Sweep

Superscripts

o	Organic Phase
w	Aqueous Phase
o	Standard Value
o′	Formal Value
FWD	Forward

Subscripts

i	Ion i
L	Ligand or Ionophore
init	Initial
M	Metal Ion
ML	Metal–Ionophore Complex
+	Positive Ion
−	Negative Ion
SS	Steady State
t	Transfer

ABBREVIATIONS AND ACRONYMS

ACT	Aqueous Complexation followed by Transfer
BTTPATPBCl	Bis(triphenylphosphoranylidene) Ammonium Tetrakis(4-chlorophenyl Borate)
DB18C6	Dibenzo-18-crown-6
DBS	1,3:2,4-Dibenzylidene Sorbitol
ISE	Ion-selective Electrode
ITIES	Interface Between Two Electrolyte Solutions
NaTPB	Sodium Tetrabutylborate
NPOE	*o*-Nitrophenyl Octyl Ether
PET	Poly(ethylene terephthalate)
PVC	Poly(vinyl chloride)
QSAR	Quantitative Structure–Activity Relationship
SFG	Sum-frequency Generation
SHG	Second Harmonic Generation
TBACL	Tetrabutyl Ammonium Chloride
TBATPB	Tetrabutylammonium Tetraphenylborate
TIC	Transfer by Interfacial Complexation
TID	Transfer by Interfacial Decomplexation
TIR	Total Internal Reflection
TOC	Transfer followed by Organic-phase Complexation
UV/VIS	Ultraviolet/Visible
1,2-DCE	1,2-Dichloroethane

RELATED ARTICLES

Environment: Water and Waste **(Volume 3)**
Detection and Quantification of Environmental Pollutants

Pharmaceuticals and Drugs **(Volume 8)**
Pharmaceuticals and Drugs: Introduction • Quantitative Structure–Activity Relationships and Computational Methods in Drug Discovery

Electroanalytical Methods **(Volume 11)**
Electroanalytical Methods: Introduction • Chemiluminescence, Electrogenerated • Ion-selective Electrodes: Fundamentals • Pulse Voltammetry

Electronic Absorption and Luminescence **(Volume 12)**
Electronic Absorption and Luminescence: Introduction

REFERENCES

1. A.G. Volkov, D.W. Deamer (eds.), *Liquid–Liquid Interfaces: Theory and Methods*, CRC Press, Boca Raton, FL, 1996.
2. C. Gavach, 'Kinetics of Electroadsorption and Polarization at the Interface Between Certain Nonmiscible Ionic Solutions', *Experientia, Suppl.*, **18**, 321–331 (1971).
3. C. Gavach, A. Savajols, 'Biionic Potentials of Strongly Dissociated Liquid Membranes', *Electrochim. Acta*, **19**, 575–581 (1974).
4. C. Gavach, P. Seta, F. Henry, 'A Study of the Ionic Transfer Across an Aqueous Solution/Liquid Membrane Interface by Chronoamperometric and Impedance Measurements', *Bioelectrochem. Bioenerg.*, **1**, 329–342 (1974).
5. C. Gavach, F. Henry, 'Chronopotentiometric Investigation of the Diffusion Overvoltage at the Interface

Between two Nonmiscible Solutions. Aqueous Solution-Tetrabutylammonium Ion Specific Liquid Membrane', *J. Electroanal. Chem.*, **54**, 361–370 (1974).

6. Z. Samec, V. Marecek, J. Koryta, M.W. Khalil, 'Investigation of Ion Transfer Across the Interface Between Two Immiscible Electrolyte Solutions by Cyclic Voltammetry', *J. Electroanal. Chem.*, **83**, 393–397 (1977).
7. J. Koryta, 'Electrochemical Polarization Phenomena at the Interface of Two Immiscible Electrolyte Solutions', *Electrochim. Acta*, **24**, 293–300 (1979).
8. D. Homolka, L.Q. Hung, A. Hofmanova, M.W. Khalil, J. Koryta, V. Marecek, Z. Samec, S.K. Sen, P. Vanysek, 'Faradaic Ion Transfer Across the Interface of Two Immiscible Electrolyte Solutions: Chronopotentiometry and Cyclic Voltammetry', *Anal. Chem.*, **52**, 1606–1610 (1980).
9. J. Koryta, M. Brezina, A. Hofmanova, D. Homolka, H. Le Quoc, W. Khalil, V. Marecek, Z. Samec, S.K. Sen, P. Vanysek, J. Weber, 'A New Model of Membrane Transport: Electrolysis at the Interface of Two Immiscible Electrolyte Solutions', *Bioelectrochem. Bioenerg.*, **7**, 61–68 (1980).
10. J. Koryta, 'Ion Transfer Across Water/Organic Phase Boundaries and Analytical Applications', *Ion-Sel. Electrode Rev.*, **5**, 131–164 (1983).
11. Z. Samec, J. Weber, V. Marecek, 'Charge Transfer Between Two Immiscible Electrolyte Solutions. Part II. The Investigation of Cs+ Ion Transfer Across the Nitrobenzene/Water Interface by Cyclic Voltammetry with IR Drop Compensation', *J. Electroanal. Chem.*, **100**, 841–852 (1979).
12. T. Kakutani, T. Osakai, M. Senda, 'A Potential-step Chronoamperometric Study of Ion Transfer at the Water/Nitrobenzene Interface', *Bull. Chem. Soc. Jpn.*, **56**, 991–996 (1983).
13. Z. Samec, V. Marecek, J. Weber, D. Homolka, 'Charge Transfer Between Two Immiscible Electrolyte Solutions. Advances in Method of Electrolysis with the Electrolyte Dropping Electrode (EDE)', *J. Electroanal. Chem.*, **99**, 385–389 (1979).
14. V. Marecek, Z. Samec, 'Determination of Calcium, Barium and Strontium Ions by Differential Pulse Stripping Voltammetry at a Hanging Electrolyte Drop Electrode', *Anal. Chim. Acta*, **151**, 265–269 (1983).
15. B. Hundhammer, T. Solomon, H. Alemu, 'Investigation of the Ion Transfer Across the Water/Nitrobenzene Interface by AC Cyclic Voltammetry', *J. Electroanal. Chem.*, **149**, 179–183 (1983).
16. P. Vanysek, *Electrochemistry on Liquid–Liquid Interfaces, Lecture Notes in Chemistry*, Springer-Verlag, Berlin, 1985.
17. H.H. Girault, 'Electrochemistry at the Interface Between Two Immiscible Electrolyte Solutions', *Electrochim. Acta*, **32**, 383–385 (1987).
18. J. Koryta, 'Electrochemical Polarization Phenomena at the Interface of Two Immiscible Electrolyte Solutions-III. Progress since 1983', *Electrochim. Acta*, **33**, 189–197 (1988).
19. V. Marecek, Z. Samec, J. Koryta, 'Charge Transfer Across the Interface of Two Immiscible Electrolyte Solutions', *Adv. Colloid Interface Sci.*, **29**, 1–78 (1988).
20. Z. Samec, 'Electrical Double Layer at the Interface Between Two Immiscible Electrolyte Solutions', *Chem. Rev.*, **88**, 617–632 (1988).
21. V.S. Markin, A.G. Volkov, 'The Gibbs Free Energy of Ion Transfer Between Two Immiscible Liquids', *Electrochim. Acta*, **34**, 93–107 (1989).
22. H.H. Girault, D.J. Schiffrin, 'Electrochemistry of Liquid–Liquid Interfaces', in *Electroanalytical Chemistry*, ed. A.J. Bard, Marcel Dekker, New York, 1–141, Vol. 15, 1989.
23. M. Senda, T. Kakiuchi, T. Osakai, 'Electrochemistry at the Interface Between Two Immiscible Electrolyte Solutions', *Electrochim. Acta*, **36**, 253–262 (1991).
24. H.H. Girault, 'Charge Transfer Across Liquid–Liquid Interfaces', in *Modern Aspects of Electrochemistry*, eds. J.O.M. Bockris, B. Conway, R. White, Plenum Press, New York, 1–62, Vol. 25, 1993.
25. P. Vanysek, 'Charge Transfer Processes on Liquid–Liquid Interfaces: The First Century', *Electrochim. Acta*, **40**, 2841–2847 (1995).
26. M. Senda, 'Voltammetry at Liquid–Liquid Interface', *Anal. Sci. Technol.*, **8**, 95A–100A (1995).
27. H.H. Girault, D.J. Schiffrin, 'Theory of the Kinetics of Ion Transfer Across Liquid–Liquid Interfaces', *J. Electroanal. Chem*, **195**, 213–227 (1985).
28. Y.Y. Gurevich, Y.I. Kharkats, 'Ion Transfer through a Phase Boundary: A Stochastic Approach', *J. Electroanal. Chem.*, **200**, 3–16 (1986).
29. Z. Samec, Y.I. Kharkats, Y.Y. Gurevich, 'Stochastic Approach to the Ion Transfer Kinetics across the Interface Between Two Immiscible Electrolyte Solutions. Comparison with Experimental Data', *J. Electroanal. Chem*, **204**, 257–266 (1986).
30. G.M. Torrie, J.P. Valleau, 'Double Layer Structure at the Interface Between Two Immiscible Electrolyte Solutions', *J. Electroanal. Chem.*, **206**, 69–79 (1986).
31. F. Milner, M.J. Weaver, 'A Simulation Analysis of Errors in the Measurement of Standard Electrochemical Rate Constants from Phase-selective Impedance Data', *J. Electroanal. Chem.*, **222**, 21–33 (1987).
32. Y.I. Kharkats, J. Ulstrup, 'The Electrostatic Gibbs Energy of Finite-size Ions Near a Planar Boundary Between Two Dielectric Media', *J. Electroanal. Chem.*, **308**, 17–26 (1991).
33. T. Kakiuchi, 'Current–Potential Characteristic of Ion Transfer Across the Interface Between Two Immiscible Electrolyte Solutions Based on the Nernst–Planck Equation', *J. Electroanal. Chem.*, **322**, 55–61 (1992).
34. A.V. Indenbom, 'Interpretation of Ion Transfer Reactions Across the Interface Between Two Immiscible

Electrolyte Solutions by the Model of "Elastic" Inner Layer', *Electrochim. Acta*, **40**, 2985–2991 (1995).

35. W. Schmickler, 'A Unified Model for Electrochemical Electron and Ion Transfer Reactions', *Chem. Phys. Lett.*, **237**, 152–160 (1995).
36. H.H. Girault, D.J. Schiffrin, 'Thermodynamics of a Polarized Interface Between Two Immiscible Electrolyte Solutions', *J. Electroanal. Chem.*, **170**, 127–141 (1984).
37. I. Benjamin, 'Dynamics of Ion Transfer Across a Liquid–Liquid Interface: A Comparison Between Molecular Dynamics and Diffusion Model', *J. Chem. Phys.*, **96**, 577–585 (1992).
38. I. Benjamin, 'Theoretical Study of Water/1,2-Dichloroethane Interface: Structure, Dynamics, and Conformational Equilibria at the Liquid–Liquid Interface', *J. Chem. Phys.*, **97**, 1432–1445 (1992).
39. A. Vincze, G. Horvai, F.A.M. Leermakers, J.M.H.M. Scheutjens, 'Mathematical-modeling of the Interface of Two Immiscible Electrolyte-Solutions', *Sens. Actuators B*, **18**, 42–46 (1994).
40. T. Wandlowski, K. Holub, V. Marecek, Z. Samec, 'The Double-layer at the Interface Between 2 Immiscible Electrolyte-solutions. IV. Solvent Effect', *Electrochim. Acta*, **40**, 2887–2895 (1995).
41. S. Toxvaerd, J. Stecki, 'Density Profiles at a Planar Liquid–Liquid Interface', *J. Chem. Phys.*, **102**, 7163–7168 (1995).
42. I. Benjamin, 'Theory and Computer Simulations of Solvation and Chemical Reactions at Liquid Interfaces', *Acc. Chem. Res.*, **28**, 233–239 (1995).
43. D.J. Henderson, W. Schmickler, 'Simple Model for Liquid/Liquid Interfaces', *J. Chem. Soc., Faraday Trans.*, **92**, 3839–3842 (1996).
44. Y. Shao, J.A. Campbell, H.H. Girault, 'Kinetics of Transfer of Acetylcholine Across the Water/Nitrobenzene-Tetrachloromethane Interface. The Gibbs Energy of Transfer Dependence of the Standard Rate Constant', *J. Electroanal. Chem.*, **300**, 415–429 (1991).
45. T. Kakiuchi, J. Noguchi, M. Senda, 'Double-layer Effect on the Transfer of Some Monovalent Ions Across the Polarized Oil–Water Interface', *J. Electroanal. Chem.*, **336**, 137–152 (1992).
46. A.G. Volkov, D.W. Deamer, D.L. Tanelian, V.S. Markin, 'Electrical Double Layers at the Oil/Water Interface', *Prog. Surf. Sci.*, **53**, 1–134 (1996).
47. I. Benjamin, 'Mechanism and Dynamics of Ion Transfer Across a Liquid–Liquid Interface', *Science*, **261**, 1558–1560 (1993).
48. K.J. Schweighofer, I. Benjamin, 'Electric Field Effects on the Structure and Dynamics at a Liquid–Liquid Interface', *J. Electroanal. Chem.*, **391**, 1–10 (1995).
49. K.J. Schweighofer, I. Benjamin, 'Transfer of Small Ions Across the Water/1,2-Dichloroethane Interface', *J. Phys. Chem.*, **99**, 9974–9985 (1995).
50. H.H. Girault, 'The Potential Dependence of the Rate of Ion Transfer Reactions Across a Liquid/Liquid Interface', *J. Electroanal. Chem.*, **257**, 47–55 (1988).
51. W. Schmickler, 'A Model for Ion Transfer through Liquid/Liquid Interfaces', *J. Electroanal. Chem.*, **426**, 5–9 (1997).
52. J. Koryta, 'Theory and Applications of Ion-selective Electrodes', *Anal. Chim. Acta*, **183**, 1–46 (1986).
53. E. Wang, Z. Sun, 'Development of Electroanalytical Chemistry at the Liquid–Liquid Interface', *Trends Anal. Chem.*, **7**, 99–106 (1988).
54. M. Senda, H. Katano, M. Yamada, 'Amperometric Ion-selective Electrode. Voltammetric Theory and Analytical Applications at High Concentration and Trace Levels', *J. Electronal. Chem.*, **468**, 34–41 (1999).
55. M. Senda, Y. Yamamoto, 'Amperometric Ion-selective Electrode Sensors', in *Liquid–Liquid Interfaces. Theory and Methods*, eds. A.G. Volkov, D.W. Deamer, CRC Press, Boca Raton, FL, 277–293, 1996.
56. T. Kakiuchi, 'Partition Equilibrium of Ionic Components in Two Immiscible Electrolyte Solutions', in *Liquid–Liquid Interfaces, Theory and Methods*, eds. A.G. Volkov, D.W. Deamer, CRC Press, Boca Raton, FL, 1–18, 1996.
57. Y. Marcus, 'Single Ion Gibbs Free Energies of Transfer from Water to Organic and Mixed Solvents', *Anal. Chem.*, **5**, 53–137 (1980).
58. Y. Marcus, 'Thermodynamic Functions of Transfer of Single Ions from Water to Nonaqueous and Mixed Solvents: Part 1 – Gibbs Free Energies of Transfer to Nonaqueous Solvents', *Pure Appl. Chem.*, **55**, 977–1021 (1983).
59. Y. Marcus, 'Ion Solvation', in *Liquid–Liquid Interfaces. Theory and Methods*, eds. A.G. Volkov, D.W. Deamer, CRC Press, Boca Raton, FL, 39–61, 1996.
60. I. Benjamin, 'Chemical Reactions and Solvation at Liquid Interfaces: A Microscopic Perspective', *Chem. Rev.*, **96**, 1449–1475 (1996).
61. I. Benjamin, 'Molecular Structure and Dynamics at Liquid–Liquid Interfaces', *Annu. Rev. Phys. Chem.*, **48**, 407–451 (1997).
62. Y. Shao, M.D. Osborne, H.H. Girault, 'Assisted Ion Transfer at Micro-ITIES Supported at the Tip of Micropipettes', *J. Electroanal. Chem.*, **318**, 101–109 (1991).
63. D. Homolka, V. Marecek, 'Charge Transfer Between Two Immiscible Electrolyte Solutions. Part VI. Polarographic and Voltammetric Study of Picrate Ion Transfer Across the Water/Nitrobenzene Interface', *J. Electroanal. Chem.*, **112**, 91–96 (1980).
64. A.J. Bard, L.R. Faulkner, *Electrochemical Methods: Fundamentals and Applications*, John Wiley & Sons, New York, 1980.
65. J. Koryta, P. Vanysek, M. Brezina, 'Electrolysis with an Electrolyte Dropping Electrode', *J. Electroanal. Chem.*, **67**, 263–266 (1976).

66. J. Koryta, P. Vanysek, M. Brezina, 'Electrolysis with Electrolyte Dropping Electrode. II. Basic Properties of the System', *J. Electroanal. Chem.*, **75**, 211–228 (1977).
67. Z. Yoshida, H. Freiser, 'Ascending Water Electrode Studies of Metal Extractants. Role of Kinetic Factors in the Faradaic Ion Transfer of Metal–Phenanthroline Complex Ions Across an Aqueous–Organic Solvent Interface', *Inorg. Chem.*, **23**, 3931–3935 (1984).
68. Z. Yoshida, H. Freiser, 'Ascending Water Electrode Studies of Metal Extractants. Faradaic Ion Transfer of Protonated 1,10-Phenanthroline and Its Derivatives Across an Aqueous 1,2-Dichloroethane Interface', *J. Electroanal. Chem.*, **162**, 307–319 (1984).
69. S. Kihara, M. Suzuki, K. Maeda, K. Ogura, S. Umetani, M. Matsui, Z. Yoshida, 'Fundamental Factors in the Polarographic Measurements of Ion Transfer at the Aqueous/Organic Solution Interface', *Anal. Chem.*, **58**, 2954–2961 (1986).
70. L. Sinru, Z. Zaofan, H. Freiser, 'Potassium Ion Transport Processes Across the Interface of an Immiscible Liquid Pair in the Presence of Crown Ethers', *J. Electroanal. Chem.*, **210**, 137–146 (1986).
71. Z. Samec, P. Papoff, 'Electrolyte Dropping Electrode Polarographic Studies. Solvent Effect on Stability of Crown Ether Complexes of Alkali-metal Cations', *Anal. Chem.*, **62**, 1010–1015 (1990).
72. H. Doe, H. Freiser, 'Ion-transfer Current-scan Polarographic Studies of Metal Extractants with Ascending Water Electrode: Manganese(II)-1,10-Phenanthroline and Related Ligand Systems', *Anal. Sci.*, **7**, 303–311 (1991).
73. H. Doe, H. Freiser, 'Ion-transfer Current-scan Polarographic Studies of Metal Extractants with Ascending Water Electrode: Transfer of Divalent Metal Ion Complexes of Phenanthrolines Across a Liquid–Liquid Interface', *Anal. Sci.*, **7**, 313 (1991).
74. J. Xiao, L. Nie, S.-Z. Yao, 'Transfer Mechanism of Quinine Drug Across the Oil/Water Interface', *Sci. China (Ser. B)*, **34**, 42–53 (1991).
75. S. Lin, D. Shen, L. Nie, S. Yao, 'The Transfer Mechanism and Analytical Properties of the Variable-valency Drug Cinchonidine Across the Liquid/Liquid Interface', *Electrochim. Acta*, **38**, 207–213 (1993).
76. H. Doe, M. Hoshiyama, L. Jian, 'Cooperative Ion-transfer of Copper(II)-1,10-Phenanthroline-trifluoroacetylacetone Complex Across a Water/1,2-Dichloroethane Interface', *Electrochim. Acta*, **40**, 2947–2951 (1995).
77. Y. Kubota, H. Katano, K. Maeda, M. Senda, 'Voltammetric Study of the Partition of Amines Between Water and an Organic Solvent', *Electrochim. Acta*, **44**, 109–116 (1998).
78. G. Taylor, H.H. Girault, 'Ion Transfer Reaction across a Liquid–Liquid Interface Supported on a Micropipette Tip', *J. Electroanal. Chem.*, **208**, 179–183 (1986).
79. J.A. Campbell, H.H. Girault, 'Steady-state Current for Ion Transfer Reactions at a Micro Liquid/Liquid Interface', *J. Electroanal. Chem.*, **266**, 465–469 (1989).
80. J.A. Campbell, A.A. Stewart, H.H. Girault, 'Determination of the Kinetics of Facilitated Ion Transfer Reactions Across the Micro Interface Between Two Immiscible Electrolyte Solutions', *J. Chem. Soc., Faraday Trans. 1*, **85**, 843–853 (1989).
81. A.A. Stewart, G. Taylor, H.H. Girault, J. McAleer, 'Voltammetry at MicroITIES Supported at the Tip of a Micropipette. Part I. Linear Sweep Voltammetry', *J. Electroanal. Chem.*, **296**, 491–515 (1990).
82. A.A. Stewart, Y. Shao, C.M. Pereira, H.H. Girault, 'Micropipette as a Tool for the Determination of the Ionic Species Limiting the Potential Window at Liquid/Liquid Interfaces', *J. Electroanal. Chem.*, **305**, 135–139 (1991).
83. T. Ohkouchi, T. Kakutani, T. Osakai, M. Senda, 'Voltammetry with Ion Selective Microelectrode Based on Polarizable Oil/Water Interface', *Anal. Sci.*, **7**, 371–376 (1991).
84. P.D. Beattie, A. Delay, H.H. Girault, 'Investigation of the Kinetics of Ion and Assisted Ion Transfer by the Technique of ac Impedance of the Micro-ITIES', *Electrochim. Acta*, **40**, 2961–2969 (1995).
85. V. Cunnane, D.J. Schiffrin, D.E. Williams, 'Micro-cavity Electrode: A New Type of Liquid–Liquid Microelectrode', *Electrochim. Acta*, **40**, 2943–2946 (1995).
86. A.-K. Kontturi, K. Kontturi, L. Murtomäki, B. Quinn, V.J. Cunnane, 'Study of Ion Transfer across Phospholipid Monolayers Adsorbed at Micropipette ITIES', *J. Electroanal. Chem.*, **424**, 69–74 (1997).
87. Y. Shao, M.V. Mirkin, 'Voltammetry at Micropipet Electrodes', *Anal. Chem.*, **70**, 3155–3161 (1998).
88. M.C. Osborne, Y. Shao, C.M. Pereira, H.H. Girault, 'Micro-hole Interface for the Amperometric Determination of Ionic Species in Aqueous Solutions', *J. Electroanal. Chem.*, **364**, 155–161 (1994).
89. M.D. Osborne, H.H. Girault, 'The Liquid–Liquid Micro-Interface for the Amperometric Detection of Urea', *Electroanalysis*, **7**, 714–721 (1995).
90. C.M. Pereira, F. Silva, 'Square Wave Voltammetry with Arrays of Liquid/Liquid Microinterfaces', *Electroanalysis*, **6**, 1034–1039 (1994).
91. M.D. Osborne, H.H. Girault, 'The Micro Water/1,2-Dichloroethane Interface as a Transducer for Creatinine Assay', *Mikrochim. Acta*, **117**, 175–185 (1995).
92. S. Wilke, H. Wang, M. Muraczewska, H. Müller, 'Amperometric Detection of Heavy Metal Ions in Ion Pair Chromatography at an Array of Water/Nitrobenzene Micro Interfaces', *Fresenius' J. Anal. Chem.*, **356**, 233–236 (1996).
93. S. Wilke, M.D. Osborne, H.H. Girault, 'Electrochemical Characterization of Liquid/Liquid Microinterface Arrays', *J. Electroanal. Chem.*, **436**, 53–64 (1997).

94. C. Beriet, H.H. Girault, 'Electrochemical Studies of Ion Transfer at Micro-machined Supported Liquid Membranes', *J. Electroanal. Chem.*, **444**, 219–229 (1998).
95. W. Mickler, A. Mönner, E. Uhlemann, S. Wilke, H. Müller, 'Transfer of β-Diketone and 4-Acylpyrazolone Anions Across the Electrified Water/Nitrobenzene Interface', *J. Electroanal. Chem.*, **469**, 91–96 (1999).
96. B.J. Seddon, Y. Shao, J. Fost, H.H. Girault,'The Application of Excimer Laser Micromachining for the Fabrication of Disc Microelectrode', *Elecrochim. Acta*, **39**, 783–791 (1994).
97. M.A. Roberts, J.S. Rossier, P. Bercier, H. Girault, 'UV Laser Machined Polymer Substrates for the Development of Microdiagnostic Systems', *Anal. Chem.*, **69**, 2035–2042 (1997).
98. S. Wilke, T. Zerihun, 'Diffusion Effects at Microhole Supported Liquid/Liquid Interfaces', *Electrochim. Acta*, **44**, 15–22 (1998).
99. C. Beriet, H.H. Girault, 'Simulation of the Diffusion Process to a Regular Array of Micro-disc Electrodes', submitted.
100. J. Josserand, J. Morandini, H.J. Lee, R. Ferrigno, H.H. Girault, 'Finite Element Simulation of Ion Transfer Reactions at a Single Micro-Liquid/Liquid Interface Supported on a Thin Polymer Film, *J. Electroanal. Chem.*, **468**, 42–52 (1999).
101. H.J. Lee, C. Beriet, H.H. Girault, 'Amperometric Detection of Alkali Metal Ions on Micro-fabricated Composite Polymer Membranes', *J. Electroanal. Chem.*, **453**, 211–219 (1998).
102. H.J. Lee, H.H. Girault, 'Amperometric Ion Detector for Ion Chromatography', *Anal. Chem.*, **70**, 4280–4285 (1998).
103. P. Guyot-Sionnest, J.H. Hunt, Y.R. Shen, 'Sum-frequency Vibrational Spectroscopy of a Langmuir Film: Study of Molecular Orientation of a Two-dimensional System', *Phys. Rev. Lett.*, **59**, 1597–1600 (1987).
104. Y.R. Shen, 'Optical Second Harmonic Generation at Interfaces', *Annu. Rev. Phys. Chem.*, **40**, 327–350 (1989).
105. K. Wolfrum, H. Graener, A. Laubereau, 'Sum-frequency Vibrational Spectroscopy at the Liquid–Air Interface of Methanol. Water Solutions', *Chem. Phys. Lett.*, **213**, 41–46 (1993).
106. C.D. Stanners, Q. Du, R.P. Chin, P. Cremer, G.A. Somorjai, Y.R. Shen, 'Polar Ordering at the Liquid/Vapor Interface of *n*-Alcohol (C1–C8)', *Chem. Phys. Lett.*, **232**, 407–413 (1995).
107. K.B. Eisenthal, 'Liquid Interfaces Probed by Second-harmonic Generation and Sum-frequency Spectroscopy', *Chem. Rev.*, **96**, 1343–1360 (1996).
108. P.-F. Brevet, *Surface Second Harmonic Generation*, Presses Polytechniques et Universitaires Romandes, Lausanne, 1996.
109. P.F. Brevet, H.H. Girault, 'Second Harmonic Generation at Liquid/Liquid Interfaces', in *Liquid–Liquid Interfaces: Theory and Methods*, eds. A.G. Volkov, D.W. Deamer, CRC Press, Boca Raton, FL, 103–137, 1996.
110. Z. Ding, R.G. Wellington, P.F. Brevet, H.H. Girault, 'Spectroelectrochemical Studies of $Ru(bpy)_3^{2+}$' at the Water/1,2-Dichloethane Interface', *J. Phys. Chem.*, **100**, 10658–10663 (1996).
111. T. Kakiuchi, Y. Takasu, M. Senda, 'Voltage-scan Fluorometry of Rose Bengal Ion at the 1,2-Dichloroethane–Water Interface', *Anal. Chem.*, **64**, 3096–3100 (1992).
112. T. Kakiuchi, Y. Takasu, 'Differential Cyclic Voltfluorometry and Chronofluorometry of the Transfer of Fluorescent Ions Across the 1,2-Dichloroethane–Water Interface', *Anal. Chem.*, **66**, 1853–1859 (1994).
113. T. Kakiuchi, Y. Takasu, 'Ion Selectivity of Voltage-scan Fluorometry at the 1,2-Dichloroethane/Water Interface', *J. Electroanal. Chem.*, **365**, 293–297 (1994).
114. T. Kakiuchi, Y. Takasu, 'Potential-step Chronofluorometric Response of Fluorescent-ion Transfer Across a Liquid/Liquid Interface', *J. Electroanal. Chem.*, **381**, 5–9 (1995).
115. Z. Ding, F. Reymond, P. Baumgartner, D. Fermin, P.-F. Brevet, P.-A. Carrupt, H.H. Girault, 'Mechanism and Dynamics of Methyl and Ethyl Orange Transfer Across the Water/1,2-Dichloroethane Interface', *Electrochim. Acta*, **44**, 3–13 (1998).
116. H.D. Duong, P.-F. Brevet, H.H. Girault, 'Heterogeneous Electron Transfer Reactions at Liquid–Liquid Interfaces Studied by Time Resolved Absorption Spectroscopy', *J. Photochem. Photobiol. A*, **117**, 27–33 (1998).
117. D.J. Fermin, Z. Ding, P.-F. Brevet, H.H. Girault, 'Potential Modulated Reflectance Spectroscopy of the Methyl Orange Transfer Across the Water/1,2-Dichloroethane Interface', *J. Electroanal. Chem.*, **447**, 125–133 (1998).
118. Z. Ding, D.J. Fermin, P.-F. Brevet, H.H. Girault, 'Spectroelectrochemical Approaches to Heterogeneous Electron Transfer Reactions at the Polarized Water/1,2-Dichloroethane Interfaces', *J. Electroanal. Chem.*, **458**, 139–148 (1998).
119. L. Tomaszewski, Z. Ding, D.J. Fermin, H.M. Caçote, C.M. Pereira, F. Silva, H.H. Girault, 'Spectroelectrochemical Study of Copper(II) Transfer Assisted by 6,7-Dimethyl-2,3-di(2-pyridyl)quinoxaline at the Water/1,2-Dichloroethane Interface', *J. Electroanal. Chem.*, **453**, 171–177 (1998).
120. R.A.W. Dryfe, Z. Ding, R.G. Wellington, P.-F. Brevet, A.M. Kuznetzov, H.H. Girault, 'Time-resolved Laser-induced Fluorescence Study of Photoinduced Electron Transfer at the Water/1,2-Dichloroethane Interface', *J. Phys. Chem. A*, **101**, 2519–2524 (1997).
121. T. Osakai, T. Kakutani, M. Senda, 'Ion-transfer Voltammetry with the Interfaces Between Polymer–Electrolyte Gel and Electrolyte Solutions', *Bunseki Kagaku*, **33**, E371–E377 (1984).

122. O. Dvorak, V. Marecek, Z. Samec, 'Ion Transfer Across Polymer Gel/Liquid Boundaries. Electrochemical Kinetics by Faradaic Impedance', *J. Electroanal. Chem.*, **284**, 205–215 (1990).
123. V. Marecek, M.P. Colombini, 'Charge Transfer Across a Polymer Gel/Liquid Interface: The Polyvinyl Chloride+ Nitrobenzene Gel/Water Interface', *J. Electroanal. Chem.*, **241**, 133–141 (1988).
124. T. Kakutani, T. Ohkouchi, T. Osakai, T. Kakiuchi, M. Senda, 'Ion-transfer Voltammetry and Potentiometry of Acetylcholine with the Interface Between Polymer–Nitrobenzene Gel and Water', *Anal. Sci.*, **1**, 219–225 (1985).
125. V. Marecek, H. Janchenova, M.P. Colombini, P. Papoff, 'Charge Transfer Across a Polymer Gel/Liquid Interface. A Voltammetric Detector for a Flow System', *J. Electroanal. Chem.*, **217**, 213–219 (1987).
126. E. Wang, H. Ji, W. Hou, 'The Use of Chemically Modified Electrodes for Liquid Chromatography and Flow-injection Analysis', *Electroanalysis*, **3**, 1–11 (1991).
127. V. Marecek, H. Janchenova, M. Brezina, M. Betti, 'Charge Transfer Across a Polymer Gel/Liquid Interface: Determination of Ionophores', *Anal. Chim. Acta*, **244**, 15–19 (1991).
128. S. Wilke, H. Franzke, H. Müller, 'Simultaneous Determination of Nitrate and Chloride by Means of Flow-injection Amperometry at the Membrane-stabilized Water/Nitrobenzene Interface', *Anal. Chim. Acta*, **268**, 285–292 (1992).
129. T. Kakutani, Y. Nishiwaki, T. Osakai, M. Senda, 'On the Mechanism of Transfer of Sodium Ion Across the Nitrobenzene/Water Interface Facilitated by Dibenzo-18-crown-6', *Bull. Chem. Soc. Jpn.*, **59**, 781–788 (1986).
130. Y. Yamamoto, T. Osakai, M. Senda, 'Potassium and Sodium Ion Sensor Based on Amperometric Ion-selective Electrode', *Bunseki Kagaku*, **39**, 655 (1990).
131. B. Hundhammer, T. Solomon, T. Zerihun, M. Abegaz, A. Bekele, K. Graichen, 'Investigation of Ion Transfer Across the Membrane-stabilized Interface of Two Immiscible Electrolyte Solutions. Part III. Facilitated Ion Transfer', *J. Electroanal. Chem.*, **371**, 1–11 (1994).
132. T. Osakai, T. Kakutani, M. Senda, 'A Novel Amperometric Ammonia Sensor', *Anal. Sci.*, **3**, 521–526 (1987).
133. E. Wang, H. Ji, 'Ion Transfer Across Water/Solidified Nitrobenzene Interface as Amperometric Flow Detector', *Electroanalysis*, **1**, 75–80 (1989).
134. Y. Yamamoto, M. Senda, 'Amperometric Ammonium Ion Sensor and its Application to Biosensors', *Sens. Actuators B*, **13–14**, 57–60 (1993).
135. T. Osakai, T. Kakutani, M. Senda, 'A Novel Amperometric Urea Sensor', *Anal. Sci.*, **4**, 529–530 (1988).
136. M. Senda, Y. Yamamoto, 'Urea Biosensor Based on Amperometric Ammonium Ion Electrode', *Electroanalysis*, **5**, 775–779 (1993).
137. H.J. Lee, P.D. Beattie, B.J. Seddon, M.D. Osborne, H.H. Girault, 'Amperometric Ion Sensors Based on Laser-patterned Composite Polymer Membranes', *J. Electroanal. Chem.*, **440**, 73–82 (1997).
138. H.J. Lee, C. Beriet, H.H. Girault, 'Stripping Voltammetry Determination of Choline Based on Micro-fabricated Composite Membrane', *Anal. Sci.*, **14**, 71–77 (1998).
139. F. Silva, M.J. Sousa, C.M. Pereira, 'Electrochemical Study of Aqueous–Organic Gel Micro-interfaces', *Electrochim. Acta*, **42**, 3095–3103 (1997).
140. P. Vanysek, W. Ruth, J. Koryta, 'Valinomycin Mediated Transfer of Potassium Across the Water/Nitrobenzene Interface. A Study by Voltammetry at the Interface Between Two Immiscible Electrolyte Solutions', *J. Electroanal. Chem.*, **148**, 117–121 (1983).
141. S. Sawada, H. Torii, T. Osakai, T. Kimoto, 'Pulse Amperometric Detection of Lithium in Artificial Serum Using a Flow Injection System with a Liquid/Liquid-Type Ion-selective Electrode', *Anal. Chem.*, **70**, 4286–4290 (1998).
142. J. Weiss, *Ion Chromatography*, VCH, Weinheim, 1995.
143. B.A. Moyer, Y. Sun, 'Principles of Solvent Extraction of Alkali Metal Ions: Understanding Factors Leading to Cesium Selectivity in Extraction by Solvation, in *Ion Exchange and Solvent Extraction*, eds. J.A. Marinsky, Y. Marcus, Marcel Dekker, New York, 295–391, Vol. 13, 1997.
144. D. Guo, A. Hofmanova, J. Koryta, 'Voltammetric Study of Ion Transfer Across the Oil/Water Interface Facilitated by an Ionophore', *Yingyong Huaxue*, **2**, 29–33 (1985).
145. F. Heitz, C. Gavach, G. Spach, Y. Trudelle, 'Analysis of the Ion Transfer Through the Channel of 9,11,13,15-Phenylalanylgramicidin A', *Biophys. Chem.*, **24**, 143–148 (1986).
146. H. Matsuda, Y. Yamada, K. Kanamori, Y. Kudo, Y. Takeda, 'On the Facilitation Effect of Neutral Macrocyclic Ligands on the Ion Transfer Across the Interface Between Aqueous and Organic Solutions. I. Theoretical Equation of Ion-transfer–Polagraphic Current–Potential Curves and Its Experimental Verification', *Bull. Chem. Soc. Jpn.*, **64**, 1497–1508 (1991).
147. D. Homolka, K. Holub, V. Marecek, 'Facilitated Ion Transfer Across the Water/Nitrobenzene Interface. Theory for Single-scan Voltammetry Applied to a Reversible System', *J. Electroanal. Chem.*, **138**, 29–36 (1982).
148. T. Kakiuchi, M. Senda, 'Current–Potential Curves for Facilitated Ion Transfer Across Oil/Water Interfaces in The Presence of Successive Complex Formation', *J. Electroanal. Chem.*, **300**, 431–445 (1991).
149. Y. Kudo, Y. Takeda, H. Matsuda, 'On the Facilitating Effect of Neutral Macrocyclic Ligands on Ion Transfer Across the Interface Between Aqueous and Organic Solutions. 2. Alkali-metal Ion Complexes with Hydrophilic Crown-ethers', *J. Electroanal. Chem.*, **396**, 333–338 (1995).

150. P.D. Beattie, R.G. Wellington, H.H. Girault, 'Cyclic Voltammetry for Assisted Ion Transfer at ITIES', *J. Electroanal. Chem.*, **396**, 317–323 (1995).
151. F. Reymond, P.-A. Carrupt, H.H. Girault, 'Facilitated Ion Transfer Reactions Across Oil/Water Interfaces. Part I. Algebraic Development and Calculation of Cyclic Voltammetry Experiments for Successive Complex Formation', *J. Electroanal. Chem.*, **449**, 49–65 (1998).
152. F. Reymond, G. Lagger, P.-A. Carrupt, H.H. Girault, 'Facilitated Ion Transfer Reactions Across Oil/Water Interfaces. Part II. Use of the Convoluted Current for the Calculation of the Association Constants and for an Amperometric Determination of the Stoichiometry of ML_j^{z+} Complexes', *J. Electroanal. Chem.*, **451**, 59–76 (1998).
153. G. Rounaghi, Z. Eshaghi, E. Ghiamati, 'Thermodynamic Study of Complex Formation between 18-Crown-6 and Potassium Ion in Some Binary Non-aqueous Solvents Using a Conductometric Method', *Talanta*, **44**, 275–282 (1997).
154. S.A. Dassie, L.M. Yudi, A.M. Baruzzi, 'Voltammetric Analysis of Cs^+-DB18C6 Complex Stoichiometry at the Water/1,2-Dichloroethane Interface', *J. Electroanal. Chem*, **464**, 54–60 (1999).
155. A. Sabela, J. Koryta, O. Valent, 'Ion Carrier Properties of Nigericin Studied by Voltammetry at the Interface of Two Immiscible Electrolyte Solutions', *J. Electroanal. Chem.*, **204**, 267–272 (1986).
156. M.D. Osborne, H.H. Girault, 'Amperometric Detection of the Ammonium Ion by Facilitated Ion Transfer Across the Interface Between Two Immiscible Electrolyte Solutions', *Electroanalysis*, **7**, 425–434 (1995).
157. T. Okada, T. Usui, 'Role of Anions in the Complex Formation of Crown Ethers with Ammonium Ions Chemically Bonded on Silica Gel', *J. Chem. Soc., Faraday Trans.*, **92**, 4977–4981 (1996).
158. Z. Yoshida, S. Kihara, 'The Role of Non-ionic Polyoxyethylene Ether Surfactants on Ion Transfer Across Aqueous/Organic Solutions Interfaces Studied by Polarography with the Electrolyte Dropping Electrode', *J. Electroanal. Chem.*, **227**, 171–181 (1987).
159. V. Marecek, A. Lhotsky, K. Holub, I. Stibor, 'Surface Complex Formation at the Water/1,2-Dichloroethane Interface', *Electrochim. Acta*, **44**, 155–159 (1998).
160. Z. Samec, D. Homolka, V. Marecek, 'Charge Transfer Between Two Immiscible Electrolyte Solutions. Part VIII. Transfer of Alkali and Alkaline Earth-metal Cations Across the Water/Nitrobenzene Interface Facilitated by Synthetic Neutral Carriers', *J. Electroanal. Chem.*, **135**, 265–283 (1982).
161. E. Wang, Z. Yu, D. Qi, C. Xu, 'Alkali and Alkaline Earth metal Ion Transfer Across the Liquid/Liquid Interface Facilitated by Ionophore', *Electroanalysis*, **5**, 149–154 (1993).
162. N. Ogawa, H. Freiser, 'Study of Ion Transfer at a Liquid–Liquid Interface by Current Linear Sweep Voltammetry. 1. The 1,10-Phenanthroline–Phenanthrolinium System', *Anal. Chem.*, **65**, 517–522 (1993).
163. S.A. Dassie, L.M. Yudi, A.M. Baruzzi, 'Comparative Analysis of the Transfer of Alkaline and Alkaline-earth Cations Across the Water/1,2-Dichloroethane Interface', *Electrochim. Acta*, **40**, 2953–2959 (1995).
164. H. Freiser, 'Metal Complexation at the Liquid–Liquid Interface', *Chem. Rev.*, **88**, 611–616 (1988).
165. W. Yu, H. Freiser, 'Electrochemical Study of the Mechanism of Cadmium Extraction with Dithizone', *Anal. Chem.*, **61**, 1621–1623 (1989).
166. T.E. Jones, L.L. Zimmer, L.L. Diaddario, D.B. Rorabacher, L.A. Ochrymowycz, 'Macrocyclic Ligand Ring Size Effects on Complex Stabilities and Kinetics. Copper(II) Complexes of Cyclic Polythiaethers', *J. Am. Chem. Soc.*, **97**, 7163–7165 (1975).
167. L.L. Diaddario, L.L. Zimmer, T.E. Jones, L.S.W.L. Sokol, R.B. Cruz, E.L. Yee, L.A. Ochrymowycz, D.B. Rorabacher, 'Macrocyclic Ligand Complexation Kinetics. Solvent, Ring Size, and Macrocyclic Effect on the Formation and Dissociation Reactions of Copper(II)-Cyclic Polythiaether Complexes', *J. Am. Chem. Soc.*, **101**, 3511–3520 (1979).
168. K. Saito, S. Murakami, A. Muromatsu, E. Sekido, 'Liquid–Liquid Extraction of Copper(II) with Cyclic Tetrathioethers', *Anal. Chim. Acta*, **294**, 329–335 (1994).
169. K. Saito, S. Murakami, A. Muromatsu, E. Sekido, 'Liquid–Liquid Extraction of Copper(I) by Cyclic tetrathio Ethers', *Anal. Chim. Acta*, **237**, 245–249 (1990).
170. S. Tanaka, H. Yoshida, 'Stripping Voltammetry of Silver(I) with a Carbon-paste Electrode Modified with Thiacrown Compounds', *Talanta*, **36**, 1044–1046 (1989).
171. W. Rosen, D.H. Busch, 'Octahedral Nickel(II) Complexes of some Cyclic Polyfunctional Thioethers', *Inorg. Chem.*, **9**, 262–265 (1970).
172. G. Lagger, L. Tomaszewski, M.D. Osborne, B.J. Seddon, H.H. Girault, 'Electrochemical Extraction of Heavy Metal Ions Assisted by Cyclic Thioether Ligands', *J. Electroanal. Chem.*, **451**, 29–37 (1998).
173. Y. Masuda, Y.W. Zhang, C.H. Yan, B.G. Li, 'Studies on the Extraction and Separation of Lanthanide Ions with a Synergistic Extraction System Combined with 1,4,10,13-Tetrathia-7,16-Diazacyclooctadecane and Lauric Acid', *Talanta*, **46**, 203–213 (1998).
174. A.M. Baruzzi, H. Wendt, 'Interfacial Phase Transfer of Amino-complexed Ni(II) Cations. A Model for Solvent Extraction', *J. Electroanal. Chem.*, **279**, 19–30 (1990).
175. D. Homolka, H. Wendt, 'Transfer of Multicomplexed Ions Across the Interface Between Two Immiscible Electrolyte Solutions; Fe(II,III), Ni(II) and Zn(II) Ions Complexed by Bidentate Nitrogen Bases', *Ber. Bunsenges. Phys. Chem.*, **89**, 1075–1082 (1985).
176. R. Fan, X. Wang, 'Semi-differential Cyclic Voltammetry of Cadmium Ion Transfer Across the Water/Nitrobenzene Interface Facilitated by 2,2′-Bipyridine', *J. Electroanal. Chem.*, **261**, 77–88 (1989).

177. E. Wang, Y. Liu, 'Electrochemistry of Cadmium Ion at the Water/Nitrobenzene Interface', *J. Electroanal. Chem.*, **214**, 465–472 (1986).
178. H. Alemu, B. Hundhammer, T. Solomon, 'Transfer of Transition Metal–terpyridine Complexes Across the Water/Nitrobenzene Interface', *J. Electroanal. Chem.*, **294**, 165–177 (1990).
179. I. Bustero, Y. Cheng, J.C. Mugica, T. Fernandez-Otero, F. Silva, D.J. Schiffrin, 'Electro-assisted Solvent Extraction of Cu^{2+}, Ni^{2+} and Cd^{2+}', *Electrochim. Acta*, **44**, 29–38 (1998).
180. Y. Cheng, D.J. Schiffrin, 'Copper(II) Ion Complexation by the Tetradentate Phosphorus–Nitrogen Ligand P_2N_2 at the Water/1,2-DCE Interface', *Inorg. Chem.*, **33**, 765–769 (1994).
181. H. Li, R.B. Smart, 'Square Wave Catalytic Stripping Voltammetry of Molybdenum Complexed with Dihydroxynaphthalene', *J. Electroanal. Chem.*, **429**, 169–174 (1997).
182. J. Wang, J. Lu, R. Setiadji, 'Adsorptive Stripping Measurements of Trace Aluminum in the Presence of Cupferron', *Talanta*, **40**, 351–354 (1993).
183. J. Wang, J. Lu, K. Olsen, 'Measurement of Ultratrace Levels of Chromium by Adsorptive-catalytic Stripping Voltammetry in the Presence of Cupferron', *Analyst*, **117**, 1913–1917 (1992).
184. J. Wang, R. Setiadji, 'Selective Determination of Trace Uranium by Stripping Voltammetry Following Adsorptive Accumulation of the Uranium–Cupferron Complex', *Anal. Chim. Acta*, **264**, 205–211 (1992).
185. J. Wang, B. Tian, J. Lu, 'Adsorptive-catalytic Stripping Measurements of Ultratrace Vanadium in the Presence of Cupferron and Bromate', *Talanta*, **39**, 1273–1276 (1992).
186. O. Abollino, M. Aceto, E. Mentasti, C. Sarzanini, C.M.G. van den Berg, 'Determination of Trace Europium by Adsorptive Cathodic Stripping Voltammetry after Complexation with Cupferron', *Electroanalysis*, **9**, 444–448 (1997).
187. J. Koryta, Y.N. Kozlov, M. Skalicky, 'Potassium Ion Transfer Across the Nitrobenzene/Water Interface Facilitated by Valinomycin', *J. Electroanal. Chem.*, **234**, 355–360 (1987).
188. G. Du, J. Koryta, W. Ruth, P. Vanysek, 'Diversity of Ion Carrier Functions of Monensin: A Study Using Voltammetry at the Interface of Two Immiscible Electrolyte Solutions', *J. Electroanal. Chem.*, **159**, 413–420 (1983).
189. F. Reymond, G. Steyaert, A. Pagliara, P.-A. Carrupt, B. Testa, H. Girault, 'Transfer Mechanism of Ionic Drugs: Piroxicam as an Agent Faciliting Proton Transfer', *Helv. Chim. Acta*, **79**, 1651–1669 (1996).
190. E. Wang, Y. Liu, 'Cyclic Voltammetry and Chronopotentiometry with Cyclic Linear Current Scanning of Terramycin at the Water/Nitrobenzene Interface', *J. Electroanal. Chem.*, **214**, 459–464 (1986).
191. S.M. Hassan, W.H. Mahmoud, M. Othman, 'A Novel Potassium Ion Membrane Sensor Based on Rifamycin Neutral Ionophore', *Talanta*, **44**, 1087–1094 (1997).
192. E. Wang, Z. Yu, N. Li, 'Anaesthetic Lidocaine and Dicaine Transfer across Liquid–Liquid Interfaces', *Electroanalysis*, **4**, 905–909 (1992).
193. Z. Yoshida, H. Aoyagi, Y. Meguro, Y. Kitatsuji, S. Kihara, 'Voltammetric Study on the Transfer of U, Np, and Pu Ions at the Aqueous–Organic Interface Facilitated by Phosphine Oxides', *J. Alloys Compd.*, **213/214**, 324–327 (1994).
194. D.P. Zollinger, E. Bulten, A. Christenhusz, M. Bos, W.E. Van Der Linden, 'Computerized Conductometric Determination of Stability Constants of Complexes of Crown Ethers with Alkali Metal Salts and with Neutral Molecules in Polar Solvents', *Anal. Chim. Acta*, **198**, 207–222 (1987).
195. R.D. Hancock, I. Cukrowski, J. Baloyi, J. Mashishi, 'The Affinity of Bismuth(III) for Nitrogen-donor Ligands', *J. Chem. Soc., Dalton Trans.*, 2895–2899 (1993).
196. A. Ohki, J.-P. Lu, R.A. Bartsch, 'Effect of Side-arm Variation in Dibenzo-16-crown-5 Compounds on the Potentiometric Selectivity for Sodium Ion', *Anal. Chem.*, **66**, 651–654 (1994).
197. I. Cukrowski, E. Cukrowska, R.D. Hancock, G. Anderegg, 'The Effect of Chelate Ring Size on Metal Ion Size-based Selectivity in Polyamine Ligands Containing Pyridyl and Saturated Nitrogen Donor Groups', *Anal. Chim. Acta*, **312**, 307–321 (1995).
198. I.Cukrowski, 'A Polarographic Method of Speciation of Labile Metal–Ligand Systems Based on Mass-balance Equations. A Differential Pulse Polarographic Study at Fixed Ligand to Metal Ratio and Varied pH', *Anal. Chim. Acta*, **336**, 23–36 (1996).
199. M.H. Mazor, J.A. McCammon, T.P. Lybrand, 'Molecular Recognition in Nonaqueous Solvents. 2. Structural and Thermodynamic Analysis of Cationic Selectivity of 18-Crown-6 in Methanol', *J. Am. Chem. Soc.*, **112**, 4411–4419 (1990).
200. L. Troxler, G. Wipff, 'Conformation and Dynamics of 18-Crown-6, Cryptand 222, and their Cation Complexes in Acetonitrile Studied by Molecular Dynamics Simulations', *J. Am. Chem. Soc.*, **116**, 1468–1480 (1994).
201. L.X. Dang, P.A. Kollman, 'Free Energy of Association of the K+:18-Crown-6 Complex in Water: A New Molecular Dynamics Study', *J. Phys. Chem.*, **99**, 55–58 (1995).
202. K. Tohda, Y. Umezawa, S. Yoshiyagawa, S. Hashimoto, M. Kawasaki, 'Cation Permselectivity at the Phase Boundary of Ionophore-incorporated Solvent Polymeric Membranes as Studied by Optical Second Harmonic Generation', *Anal. Chem.*, **67**, 570–577 (1995).
203. M.J. Crawford, J.G. Frey, T.J. VanderNoot, Y. Zhao, 'Investigation of Transport Across an Immiscible Liquid–Liquid Interface. Electrochemical and Second

Harmonic Generation Studies', *J. Chem. Soc., Faraday Trans.*, **92**, 1369–1373 (1996).
204. S.N. Tan, H.H. Girault, 'Potassium Transfer Facilitated by Monoaza-18-crown-6 Across the Water–1,2-Dichloroethane Interface', *J. Electroanal. Chem.*, **332**, 101–112 (1992).
205. L.B. Kier, B. Testa, *Complexity and Emergence in Drug Research*, Academic Press, London, 1995.
206. L.Z. Benet, B.Y.T. Perotti, 'Drug Absorption, Distribution and Elimination', in: *Burger's Medicinal Chemistry and Drug Discovery*, ed. M.E. Wolff, John Wiley & Sons, New York, 113–128, Vol. 1, 1995.
207. R.A. Conradi, P.S. Burton, R.T. Borchardt, 'Physicochemical and Biological Factors that Influence a Drug's Cellular Permeability by Passive Diffusion', in: *Lipophilicity in Drug Action and Toxicology*, eds. V. Pliska, B. Testa, H. van de Waterbeemd, VCH, Weinheim, 233–252, Vol. 4, 1996.
208. J.B. Taylor, P.D. Kennewell, *Modern Medicinal Chemistry*, Ellis Horwood, New York, 1993.
209. V.H.L. Lee, *Peptide and Protein Delivery*, Marcel Dekker, New York, 1991.
210. K.L. Audus, T.J. Raub, *Biological Barriers to Protein Delivery*, Plenum Press, New York, 1993.
211. C.J.H. Porter, W.N. Charman, 'Uptake of Drugs into the Intestinal Lymphatics after Oral Administration', *Adv. Drug Del. Rev.*, **25**, 71–89 (1997).
212. P.D.Kennewell, 'The Architecture of the Cell', in *Comprehensive Medicinal Chemistry: The Rational Design, Mechanistic Study and Therapeutic Application of Chemical Compounds*, eds. C. Hansch, P.G. Sammes, J.B. Taylor, Pergamon Press, Oxford, 169–194, Vol. 1, 1990.
213. H. van de Waterbeemd, B. Testa, *The Parametrization of Lipophilicity and Other Structural Properties in Drug Research*, Academic Press, London, 1987.
214. M.G. Davis, C.N. Manners, D.W. Payling, D.A. Smith, C.A. Wilson, 'Gastrointestinal Absorption of the Strongly Acidic Drug Proxicromil', *J. Pharm. Sci.*, **73**, 949–953 (1984).
215. W.A. Banks, A.J. Kastin, 'Peptides and the Blood–Brain Barrier: Lipophilicity as a Predictor of Permeability', *Brain Res. Bull.*, **15**, 287–292 (1985).
216. D.A. Smith, K. Brown, M.G. Neale, 'Chromone-2-carboxylic Acids: Roles of Acidity and Lipophilicity in Drug Disposition', *Drug Metab. Rev.*, **16**, 365–388 (1986).
217. C.J. Alcorn, R.J. Simpson, D.E. Leahy, T.J. Peters, 'Partition and Distribution Coefficients of Solutes and Drugs in Brush Border Membrane Vesicles', *Biochem. Pharmacol.*, **45**, 1775–1782 (1993).
218. A. Pagliara, P.-A. Carrupt, G. Caron, P. Gaillard, B. Testa, 'Lipophilicity Profiles of Ampholytes', *Chem. Rev.*, **97**, 3385–3400 (1997).
219. V. Pliska, B. Testa, H. van de Waterbeemd, (eds.), *Lipophilicity in Drug Action and Toxicology. Methods and Principles in Medicinal Chemistry*, VCH, Weinheim, Vol. 4, 1996.
220. C. Hansch, A. Leo, *Exploring QSAR: Fundamentals and Applications in Chemistry and Biology*, American Chemical Society, Washington, DC, 1995.
221. F. Reymond, G. Steyaert, P.-A. Carrupt, B. Testa, H.H. Girault, 'Mechanism of Transfer of a Basic Drug Across the Water/1,2-Dichloroethane Interface: The Case of Quinidine', *Helv. Chim. Acta*, **79**, 101–117 (1996).
222. H. Doe, K. Yoshioka, T. Kitagawa, 'Voltammetric Study of Protonated 1,10-Phenanthroline Cation Transfer Across the Water/Nitrobenzene Interface', *J. Electroanal. Chem.*, **324**, 69–78 (1992).
223. P. Vanysek, M. Behrendt, 'Investigation of Acetylcholine, Choline and Acetylcholinesterase at the Interface of the Two Immiscible Electrolyte Solutions', *J. Electroanal. Chem.*, **130**, 287–292 (1981).
224. V. Marecek, Z. Samec, 'Electrolysis at the Interface Between Two Immiscible Electrolyte Solutions: Determination of Acetylcholine by Differential Pulse Stripping Voltammetry', *Anal. Lett.*, **14**, 1241–1253 (1981).
225. Z. Samec, V. Marecek, D. Homolka, 'Charge Transfer Between Two Immiscible Electrolyte Solutions. Part IX. Kinetics of the Transfer of Choline and Acetylcholine Cations Across the Water/Nitrobenzene Interface', *J. Electroanal. Chem.*, **158**, 25–36 (1983).
226. D. Homolka, V. Marecek, Z. Samec, K. Base, W. Wendt, 'The Partition of Amines Between Water and an Organic Solvent Phase', *J. Electroanal. Chem.*, **163**, 159–170 (1984).
227. O. Dvorak, V. Marecek, Z. Samec, 'Selective Complexation of Biogenic Amines by Macrocyclic Polyethers at a Liquid/Liquid Interface', *J. Electroanal. Chem.*, **300**, 407–413 (1991).
228. T. Ohkouchi, T. Kakutani, M. Senda, 'Electrochemical Study of the Transfer of Uncouplers Across the Organic/Aqueous Interface', *Bioelectrochem. Bioenerg.*, **25**, 71–80 (1991).
229. T. Ohkouchi, T. Kakutani, M. Senda, 'Electrochemical Theory of the Transfer of Protons Across a Biological Membrane Facilitated by Weak Acid Uncouplers Added to the Medium', *Bioelectrochem. Bioenerg.*, **25**, 81–89 (1991).
230. E. Wang, H. Wang, Z. Yu, 'Electrochemical Study of Pyrazolone Derivatives at the Liquid/Liquid Interface', *Electroanalysis*, **6**, 1020–1023 (1994).
231. T. Osakai, T. Kakutani, M. Senda, 'Kinetics of the Transfer of Picrate Ions at the Water/Nitrobenzene Interface', *Bull. Chem. Soc. Jpn.*, **58**, 2626–2633 (1985).
232. T. Wandlowski, V. Marecek, Z. Samec, 'Kinetic Analysis of the Picrate Ion Transfer Across the Interface Between Two Immiscible Electrolyte Solutions from Impedance Measurements at the Equilibrium Potential', *J. Electroanal. Chem.*, **242**, 291–302 (1988).
233. O. Shirai, Y. Yoshida, M. Matsui, K. Maeda, S. Kihara, 'Voltammetric Study on the Transport of Ions of

Various Hydrophobicity Types through Bilayer Lipid Membranes Composed of Various Lipids', *Bull. Chem. Soc. Jpn.*, **69**, 3151–3162 (1996).
234. F. Reymond, G. Steyaert, P.-A. Carrupt, B. Testa, H.H. Girault, 'Ionic Partition Diagrams: A Potential-pH Representation', *J. Am. Chem. Soc.*, **118**, 11951–11957 (1996).
235. F. Reymond, P.F. Brevet, P.-A. Carrupt, H.H. Girault, 'Cyclic Voltammetry for the Transfer of Multiple Charged Ions at Large ITIES: General Computational Methodology and Application to Simple and Facilitated Ion Transfer Reactions', *J. Electroanal. Chem.*, **424**, 121–139 (1997).
236. Y.N. Kozlov, J. Koryta, 'Determination of Tetracycline Antibiotics by Voltammetry at the Interface of Two Immiscible Electrolyte Solutions', *Anal. Lett.* **B16**, 255–263 (1983).
237. L.M. Yudi, A.M. Baruzzi, V.M. Solis, 'Quantitative Determination of Erythromycin and its Hydrolysis Products by Cyclic Voltammetry at the Interface Between Water and 1,2-Dichloroethane', *J. Electroanal. Chem.*, **360**, 211–219 (1993).
238. L.M. Yudi, E. Santos, A.M. Baruzzi, V.M. Solis, 'Erythromycin Transfer across the Water/1,2-Dichloroethane Interface Modified by a Phospholipid Monolayer', *J. Electroanal. Chem.*, **379**, 151–158 (1994).
239. J. Di, X. Xu, J. Luo, 'Determination of Minocycline by Semi-differential Cyclic Voltammetry at a Liquid/Liquid Interface', *Anal. Lett.*, **29**, 2691–2700 (1996).
240. K. Arai, M. Ohsawa, F. Kusu, K. Takamura, 'Drug Ion Transfer Across an Oil/Water Interface and Pharmacological Activity', *Bioelectrochem. Bioenerg.*, **31**, 65–76 (1993).
241. K. Arai, F. Kusu, N. Tsuchiya, S. Fukuyama, 'Ion Transfer of Weak Electrolytes Across the Oil/Water Interface. Ion Transfer of Scopolamine and Lidocaine', *Denki Kagaku*, **62**, 840–845 (1994).
242. K. Arai, F. Kusu, K. Takamura, 'Electrochemical Behavior of Drugs at Oil/Water Interfaces', in *Liquid–Liquid Interfaces. Theory and Methods*, eds. A.G. Volkov, D.W. Deamer, CRC Press, Boca Raton, FL, 375–400, 1996.
243. Z. Samec, J. Langmaier, A. Trojanek, E. Samcova, J. Malek, 'Transfer of Protonated Anesthetics Across the Water/*o*-Nitrophenyl Octyl Ether Interface: Effect of the Ion Structure on the Transfer Kinetics and Pharmacological Activity', *Anal. Sci.*, **14**, 35–41 (1998).
244. M. Pourbaix, *Atlas d'Equilibres Electrochimiques*, Gauthier-Villars, Paris, 1963.
245. F. Reymond, P.-A. Carrupt, B. Testa, H.H. Girault, 'Charge and Delocalization Effects on the Lipophilicity of Protonable Drugs', *Chem. Eur. J.*, **5**, 39–47 (1999).
246. F. Reymond, V. Chopineaux-Courtois, G. Steyaert, G. Bouchard, P.-A. Carrupt, B. Testa, H.H. Girault, 'Ionic Partition Diagrams of Ionizable Solutes: pH Lipophilicity Profiles, Transfer Mechanisms and Charge Effects on Solvation', *J. Electroanal. Chem.*, **462**, 235–250 (1999).
247. F. Reymond, G. Steyaert, P.-A. Carrupt, D. Morin, J.-P. Tillement, H.H. Girault, B. Testa, 'The pH-Partition Profile of the Anti-ischemic Drug Trimetazidine May Explain Its Reduction of Intracellular Acidosis', *Pharm. Res.*, **16**, 616–624 (1999).
248. V. Chopineaux-Courtois, F. Reymond, G. Bouchard, P.-A. Carrupt, B. Testa, H.H. Girault, 'Effects of Charge and Intramolecular Structure on the Lipophilicity of Nitrophenols', *J. Am. Chem. Soc.*, **121**, 1743–1747 (1999).
249. K. Kontturi, L. Murtomaki, 'Electrochemical Determination of Partition Coefficients of Drugs', *J. Pharm. Sci.*, **81**, 970–975 (1992).
250. G. Steyaert, G. Lisa, P.-A. Carrupt, B. Testa, F. Reymond, H.H. Girault, 'Solvatochromic Analysis of Partition Coefficients in the Water/1,2-Dichloroethane System', *J. Chem. Soc., Faraday Trans.*, **93**, 401–406 (1997).

Microbalance, Electrochemical Quartz Crystal

Michael D. Ward
University of Minnesota, Minneapolis, USA

The need for direct measurement of interfacial events at solid surfaces under ambient conditions has provoked surface scientists and electrochemists to develop direct in situ methods. Such processes include electroless or electrochemical metal deposition, insertion of ions into polymer ion-exchange films, growth of oxide films on metals, or the loss of material from corrosion processes. An increasing interest in molecular films has prompted development of methods for measuring adsorption of molecules from vapor or liquid phases. Many of these processes share a common feature, namely they are accompanied by changes in mass at the solid surface.

During the 1990s a new analytical method for the in situ examination of interfacial electrode processes, the electrochemical quartz crystal microbalance (EQCM), has emerged that has substantially influenced electrochemical science. This method relies on a single crystal of quartz that has been cut into a thin wafer and coated with gold electrodes on both sides. These electrodes are used to provide an alternating electric field so that the crystal is vibrated at a specific resonant frequency, while one of the electrodes is simultaneously used as a working electrode in an electrochemical cell. The resonant frequency shifts upon changes in mass that occur on the working electrode during an electrochemical process, with sensitivity as high as 100 pg cm^{-2} of electrode area. Recent developments also have illustrated that the quartz crystal microbalance (QCM) is capable of measuring the viscosity and density of liquids near the QCM surface. The purpose of this article is to provide the reader with a fundamental understanding of the EQCM and several illustrative examples of applications that demonstrate its unique capabilities, as well as the technical details required to build this apparatus.

1 INTRODUCTION

In the 1980s and 1990s surface scientists and electrochemists have made considerable strides in sophisticated instrumental techniques for probing the structure and composition of solid interfaces and electrodes, motivated by the need to measure and understand interfacial processes. However, many of these techniques involve examination of solid surfaces outside the medium in which they are typically used. For example, ultrahigh vacuum studies of electrode and catalyst surfaces provide invaluable data concerning structure and composition. The relevance of data obtained from such ex situ methods to properties and applications under ambient conditions, particularly applications involving immersion of the surface in solution, can be questionable.

The need for direct measurement of interfacial events at solid surfaces under ambient conditions has provoked surface scientists and electrochemists to develop direct in situ methods. In addition to providing data that can be used for understanding fundamental mechanistic details, in situ approaches can be useful in sensor applications where measurement under specific, and sometimes harsh, conditions is required. Such processes include electroless or electrochemical metal deposition, insertion of ions into polymer ion-exchange films, growth of oxide films on metals, or the loss of material from corrosion processes. An increasing interest in molecular films has prompted development of methods for measuring adsorption of molecules from vapor or liquid phases. Many of these processes share a common feature, namely they are accompanied by changes in mass at the solid surface.

In the 1990s, a new analytical method for the in situ measurement of interfacial electrode processes, the EQCM, has emerged that has substantially influenced electrochemical science. This method relies on the piezoelectric properties of a single crystal of quartz that has been cut into a thin wafer. The word piezoelectric derives from the word *piezein*, meaning to press. Hence, the piezoelectric effect hinges on "pressure electricity," a phenomenon first observed by Jacques and Pierre Curie when they discovered that mechanical stress applied to the surfaces of certain crystals, including quartz, resulted in an electrical potential across the crystal. Shortly afterwards, the converse piezoelectric effect – a mechanical strain produced by application of a electric potential across the crystal – was discovered. This effect is sometimes referred to as the converse piezoelectric effect. The motor generator properties have long been associated with underwater sound transducers (sonar), and electromechanical devices such as speakers, microphones, and phonograph pickups.

The QCM earns its name from its ability to measure the mass of thin films that have adhered to its surface. The QCM generally comprises a thin quartz wafer with a diameter of 0.25–1.0 in (1 in = 2.54 × 10^{-2} m), sandwiched between two metal electrodes which are used to establish an electric field across the crystal. If an alternating electric field and appropriate electronics

For references see page 9954

are used, the crystal can be made to oscillate at its resonant frequency. Most crystals of current interest resonate between 5 and 30 MHz. The measured frequency is dependent upon the combined thickness of the quartz wafer, metal electrodes, and material deposited on the QCM surface. Because the resonance is very sharp, high-precision frequency measurements allow the detection of minute amounts of deposited material, as small as $100\,\mathrm{pg\,cm^{-2}}$.

Early applications of the QCM involved the well-documented measurement of metal deposition in high-vacuum metal evaporators, which is still widely practiced. This allows for real-time rapid measurement of film thicknesses with ångström resolution ($1\,\text{Å} = 10^{-10}\,\mathrm{m}$). Advances in QCM methodology in the 1990s allow for dynamic measurements of minute mass changes at surfaces, thin films, and electrode interfaces prepared on the quartz crystal, while the surface is immersed in liquid. The capability for direct real-time highly sensitive mass measurement in the liquid phase offers opportunities not available by other means. Developments also have illustrated that the QCM is capable of measuring the viscosity and density of liquids near the QCM surface. Combining this technique with electrochemical instrumentation allows simultaneous measurement of mass and electrochemical variables such as electrochemical potential, current, and charge.

The EQCM has emerged as a powerful technique capable of detecting very small mass changes at the electrode surface that accompany electrochemical processes. This relatively simple technique requires, in addition to conventional electrochemical equipment, an inexpensive radiofrequency (RF) oscillator, a frequency counter, and commercially available AT-cut quartz crystals. EQCM has evolved into a routine experimental method used in numerous electrochemical laboratories. The purpose of this article is to provide the reader with a fundamental understanding of the EQCM and several illustrative examples of applications that demonstrate its unique capabilities. It is hoped that this will enable readers to add the EQCM to their battery of electrochemical methods. Other review articles can be consulted for different emphasis or greater detail.[1–3]

2 THEORY AND PRINCIPLES

In order to understand the operation of the EQCM a fundamental understanding of the piezoelectric effect is required. In 1880, Jacques and Pierre Curie discovered that a mechanical stress applied to the surfaces of various crystals, including quartz, rochelle salt ($NaKC_4H_4O_6 \cdot 4H_2O$), and tourmaline, resulted in an electrical potential across the crystal whose magnitude was proportional to the applied stress.[4] This behavior is referred to as the piezoelectric effect which is derived from the Greek word *piezein* meaning "to press." This property only exists in materials that are acentric, that is, those that crystallize into noncentrosymmetric space groups. A single crystal of an acentric material will possess a polar axis due to dipoles associated with the arrangement of atoms in the crystalline lattice. The charge generated in a quartz crystal under mechanical stress is a manifestation of a change in the net dipole moment because of the physical displacement of the atoms and a corresponding change in the net dipole moment. This results in a net change in electrical charge on the crystal faces, the magnitude and direction of which depends upon the relative orientation of the dipoles and the crystal faces. Following their initial discovery, the Curies discovered the converse piezoelectric effect, in which the application of a potential across these crystals resulted in a corresponding mechanical strain. It is this effect that is the operational basis of the EQCM.

The EQCM is actually the electrochemical version of the QCM, which has long been used for frequency control and mass sensing in vacuum and air. The QCM consists of a thin, AT-cut quartz crystal with very thin metal electrode "pads" on opposite sides of the crystal. The terminology "AT" simply refers to the orientation of the crystal with respect to its large faces; this particular crystal is fabricated by slicing through a quartz rod at an angle of approximately 35° with respect to the crystallographic x axis. The electrode pads overlap in the center of the crystal with tabs extending from each to the edge of the crystal where electrical contact is made. When an electrical potential is applied across the crystal using these electrodes, the AT-cut quartz crystal experiences a mechanical strain in the shear direction. Crystal symmetry dictates that the strain induced in a piezoelectric material by an applied potential of one polarity will be equal and opposite in direction to that resulting from the opposite polarity (Figure 1). Therefore, an alternating potential across the crystal causes vibrational motion of the quartz crystal with the vibrational amplitude parallel to the crystal surface and in the x-direction. This oscillatory behavior and the electromechanical "motor generator" properties are the basis of numerous applications, including the QCM, sonar transducers, speakers, microphones, phonograph pickups, and quartz digital watches. It is important to note that the direction of the crystal vibration is critical for liquid-phase applications. An AT-cut crystal vibrates in the shear mode, parallel to the crystal–liquid interface. Consequently, damping of the crystal vibration by a contacting fluid are minimized (see below).

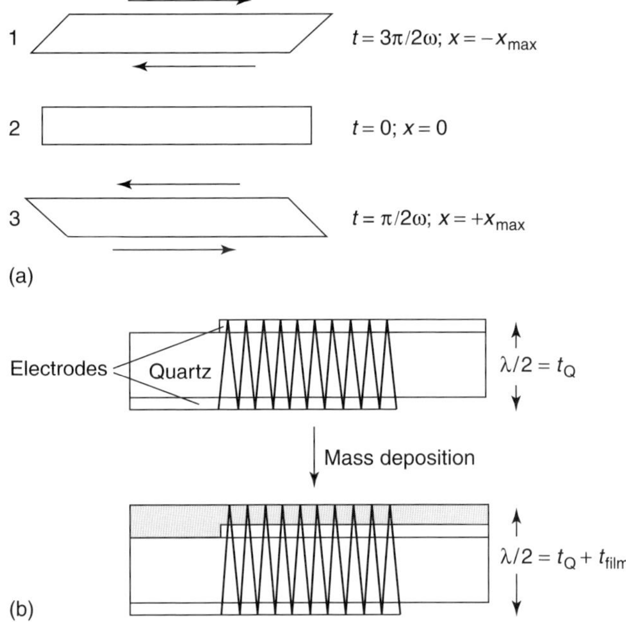

Figure 1 (a) Schematic representation of the shear vibration of an AT-cut quartz resonator. The time at which the crystal achieves maximum strain during oscillation is indicated. The crystal has maximum kinetic energy at $x = 0$, but maximum potential energy at $x = \pm x_{max}$, similar to classical oscillators. (b) Schematic representation of the transverse shear wave in a quartz crystal with excitation electrodes and a composite resonator comprising the quartz crystal, electrodes, and a thin layer of a foreign material. The acoustic wavelength is longer in the composite resonator because of the greater thickness, resulting in a lower frequency compared to the quartz crystal.

The shear vibrational motion of the quartz crystal results in a transverse acoustic wave that propagates back and forth across the thickness of the crystal between the crystal faces. Accordingly, a standing wave condition is established in the quartz resonator when the acoustic wavelength is equal to twice the combined thickness of the crystal and electrodes. The frequency, f_0, of the acoustic wave fundamental mode is given by Equation (1), where v_{tr} is the transverse velocity of sound in AT-cut quartz ($3.34 \times 10^4\,\mathrm{m\,s^{-1}}$) and t_Q is the resonator thickness. Some useful parameters and relationships for quartz resonators are provided in Table 1. An assumption is commonly made that the velocity of sound in quartz and the electrodes is identical. While this is not rigorously true, for small electrode thicknesses the error introduced by this approximation is negligible. The acoustic velocity is dependent upon the modulus and density of the crystal. The quartz crystal surface is at an antinode of the acoustic wave, and therefore the acoustic wave propagates across the interface between the crystal and a foreign layer on its surface. If it is assumed that the acoustic velocity in the foreign layer, and its density,

Table 1 Some useful parameters and relationships for quartz resonators

Velocity of sound in quartz (v_Q)	$3340\,\mathrm{m\,s^{-1}}$
Shear modulus of quartz (μ_Q)	$2.947 \times 10^{11}\,\mathrm{g\,cm^{-1}\,s^{-2}}$
Density of quartz (ρ_Q)	$2.648\,\mathrm{g\,cm^{-3}}$
Thickness–frequency relationship	$t_Q = v_Q/2f_0$
Amplitude of oscillation (thickness shear mode)	10–100 nm
Shear velocity	$v(z,t) = v_0 \exp(k_1 z) \times \cos(k_1 z - 2pf_0 t)$
Propagation constant	$k_1 = (\pi f_0 \rho_1/\eta_1)^{1/2}$
Decay length of shear wave	$1/k_1$
Decay length of 5-MHz shear wave in water	250 nm
Typical equivalent circuit values	
C_1 (energy stored during oscillation)	$23 \times 10^{-15}\,\mathrm{F}$
R_1 (energy lost during oscillation, air)	ca. $100\,\Omega$
R_1 (energy lost during oscillation, water)	ca. $1000\,\Omega$
L_1 (inertial component related to mass displaced during oscillation)	$45 \times 10^{-3}\,\mathrm{H}$
C_0 (dielectric capacitance of quartz)	$10^{-12}\,\mathrm{F}$
Series resonance, f_s	$f_s = [2\pi(L_1 C_1)^{1/2}]^{-1}$
Quality factor	$Q = (2pf_s C_1 R_1)^{-1} = 2\pi f_s L_1/R_1$
Quality factor	Q = Peak width at half height of conductance/f_s

are identical to those for quartz (cf. the assumption for the metal electrodes), a change in thickness of the foreign layer is tantamount to a change in the thickness of the quartz crystal. Under these conditions, a fractional change in thickness results in a fractional change in the resonant frequency; appropriate substitutions yield the well-known Sauerbrey equation, Equation (2), where Δf is the measured frequency change, f_0 the frequency of the quartz resonator prior to a mass change, Δm the mass change, A the piezoelectrically active area, ρ_Q the density of quartz ($2.648\,\mathrm{g\,cm^{-3}}$) and μ_Q the shear modulus of AT-cut quartz ($2.947 \times 10^{11}\,\mathrm{dyn\,cm^{-2}}$) ($1\,\mathrm{dyn} = 10^{-5}\,\mathrm{N}$).

$$f_0 = \frac{v_{tr}}{2t_Q} \tag{1}$$

$$\Delta f = \frac{-2f_0^2 \Delta m}{A\sqrt{\mu_Q \rho_Q}} \tag{2}$$

This equation is the primary basis of most QCM and EQCM measurements wherein mass changes occurring at the electrode interface are evaluated directly from the frequency changes of the quartz resonator. It is generally considered to be accurate as long as the thickness of the film added to the QCM is less than

For references see page 9954

2% of the quartz crystal thickness. With this constraint, the errors resulting from the discrepancy between the acoustic propagation characteristics in quartz and the film are minimal. Deviations from Equation (2) due to higher mass loadings may be compensated, however, by use of the "*Z*-match" method.[5] While this method has been used for vacuum applications, it has yet to be employed in EQCM applications. Typical operating frequencies of the EQCM of the EQCM lie within the range 5–10 MHz, although recently the operation of 30 MHz quartz crystals in EQCM applications has been achieved.[6] These operating frequencies provide for mass detection limits approaching 1 ng cm^{-2}.

The reader may better understand the mass sensing properties of the QCM by comparing the motion of the quartz crystal to other oscillating systems, such as a vibrating string, a pendulum, or a mass on a spring. In all cases the amplitude is defined by the initial energy input and the resonant frequencies are defined by characteristics of mass and length. The quartz crystal motion can be described as moving about the $x = 0$ rest point between limits of $-x_{max}$ and $+x_{max}$. The magnitude of x_{max} will depend upon the applied alternating voltage across the crystal. As with the more familiar oscillating systems, the potential energy of the crystal is at a maximum at $x = \pm x_{max}$, whereas the kinetic energy is at a maximum at $x = 0$. The effect of mass (or thickness) changes on the quartz resonant frequency can be understood by analogy to a classical system such as a vibrating string. Standing waves can exist in a vibrating string if their wavelengths are integral divisors of $2l$, where l is the length of the string. The fundamental frequency, f_0, is given by Equation (3), where S is the tension on the string and m_l is the mass per unit length.

$$f_0 = \frac{(S/m_l)^{1/2}}{2l} \qquad (3)$$

An increase in the mass or length of the string therefore results in a decrease in f_0. This is identical to increasing the thickness of a quartz crystal, with a dimensional increase resulting in a longer standing wave propagation distance and a corresponding reduction in the frequency. The stress applied to the vibrating string is analogous to the modulus of the quartz crystal; an increase in either of these quantities increases the velocity of the standing wave.

The key distinguishing feature of quartz resonators is their negligible energy dissipation during oscillation. While a pendulum may lose considerable energy during oscillation because of friction, a quartz crystal loses only a minute amount due to phonon interactions that produce heat, vibrational damping by the mounting components, and acoustical losses to the environment. This property is generally characterized by the quality factor, Q, which is the ratio of the energy stored to energy lost during a single oscillation. For quartz crystals, this quantity can exceed 100 000. Low energy losses in oscillating systems are manifested as high accuracy. As a system loses more energy during oscillation the period of the oscillation becomes less well defined. This is the basis for the widespread use of quartz crystals in timepieces and frequency control elements. In fact, the frequency of a typical quartz crystal can be determined to an accuracy of 1 part in 10^8. It is this precision that makes the QCM and EQCM so useful. In liquid applications, including the EQCM, Q will generally have values of 1000–3000, indicative of energy damping by the fluid. Nevertheless, the quartz crystals will still perform acceptably at these levels.

The vibration of the quartz crystal parallel to the QCM–liquid interface results in the radiation of a shear wave into the liquid (Figure 2). The instantaneous shear wave velocity decays as an exponentially damped cosine function according to Equation (4), where k is the propagation constant, z the distance from the resonator surface, A the maximum amplitude of the shear wave, and w the angular frequency. The inverse of the propagation constant k is the decay length, δ, which is given by Equation (5), where ρ_L and η_L are the liquid density and viscosity, respectively. This leads to Equation (6), which gives the dependence of the resonant frequency on $(\rho_L\eta_L)^{1/2}$.[7]

$$V_x(z,t) = A\mathrm{e}^{-kz}\cos(kz - wt) \qquad (4)$$

$$\delta = \sqrt[z]{\frac{\eta_L}{\pi f_0 \rho_L}} \qquad (5)$$

$$\Delta f = -f_0^{3/2}\left(\frac{\rho_L\eta_L}{\pi\rho_Q\mu_Q}\right)^{1/2} \qquad (6)$$

This effect of a contacting fluid medium is evidenced by the approximately 750-Hz decrease in resonant frequency that occurs when a QCM is immersed in water. This decrease can be attributed to the effective mass of liquid contained in this decay length. It should be noted that Equations (4–6) rely on the assumption of a "no-slip" condition at the QCM–liquid interface. That is, the molecular layer of liquid directly in contact with the vibrating region of the QCM surface moves with the same velocity and amplitude as the vibrating region. The shear wave then decays in the liquid because of the velocity gradient along the direction normal to the surface. This is a consequence of the fluid viscosity. While there are some reports claiming deviations from the no-slip condition, these are sparse and not yet conclusive.[8,9,111]

Inspection of Equation (6) reveals that an increase in either the density or viscosity results in a decrease in the resonant frequency. This can be better understood by

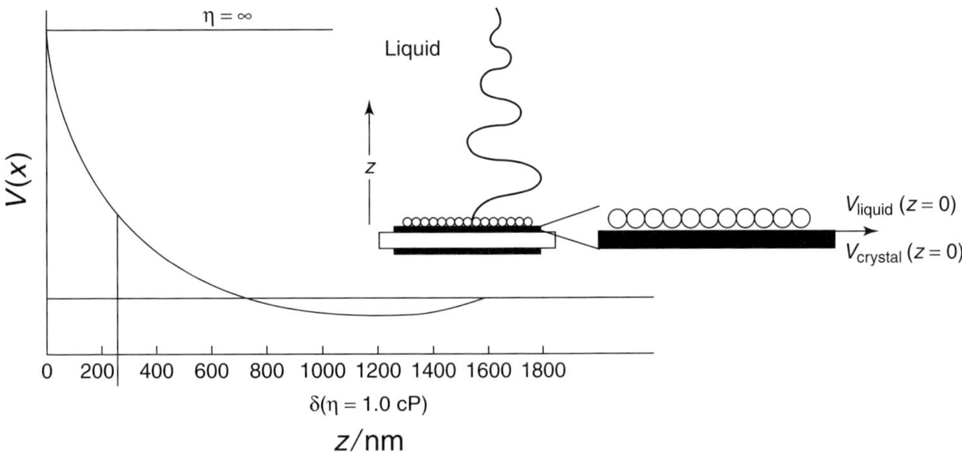

Figure 2 Description of the shear wave propagation in a Newtonian fluid in terms of the shear velocity in the x-direction (parallel to the resonator–liquid interface) as a function of distance from the resonator–liquid interface. If the viscosity of the film in contact with the resonator is infinitely large, the acoustic wave will propagate without loss. In water, the amplitude of the acoustic wave decays, with the decay length $\delta = 2500$ Å. The decay length represents the distance at which the amplitude of the shear wave decays to 1/e of its maximum value at the interface. The circles represent the layer of liquid molecules at the interface, at which a "no-slip" condition generally is assumed. That is, these molecules move with the same velocity and amplitude as the resonator ($V_{liquid} = V_{crystal}$).

considering two extremes: in air, the quantity $(\rho_L\eta_L)^{1/2}$ is negligible, but if the liquid was actually a rigid solid, $(\rho_L\eta_L)^{1/2}$ would be substantial. In the case of a rigid solid with $\eta = \infty$, the acoustic wave will propagate indefinitely and the frequency change would reflect the true thickness of the solid. If the thickness of this rigid film was large, crystal oscillation would be difficult because of the significant damping by the large mass. Therefore, it is the decay of the shear wave in liquids that enables operation of the QCM and EQCM in liquid media.

As the viscosity of the fluid becomes larger, the accuracy and performance of the submerged quartz resonator diminishes. These effects can be especially important when the EQCM is modified with polymer films, which may undergo changes in $(\rho_L\eta_L)^{1/2}$ during measurements. In addition, polymer films may be viscoelastic, which can further complicate interpretation of EQCM data (see below). Indeed, there are several examples of significant QCM responses to these changes under conditions where mass changes are not operative.(10)

Another effect that needs to be considered when quartz resonators are immersed in liquids is the microscopic roughness of the resonator surface. The cavities on a rough surface can trap liquid which will be manifested as an additional mass on the surface. The amount of trapped liquid will depend upon the cavity geometry and size. This effect has been inferred in gold and copper surfaces on the EQCM, in which the electrode surfaces were roughened during electrochemical cycling through the oxide regions of the metal electrodes.(11–13) An extensive study of these gold surfaces involved comparison of the frequency shift with that expected for liquid trapped in surface cavities whose dimensions were measured by scanning electron microscopy (SEM). These studies indicated that the frequency changes were smaller than expected based on the SEM measurements, suggesting that the trapped liquid did not behave as a rigid mass. Therefore, surface roughness effects can be very difficult to quantify even when the exact roughness is known. These effects may be significant in many published EQCM studies because of the wide use of unpolished quartz crystals, which are commonly used because of their ready availability and low cost. These are widely sold for vacuum thickness monitor applications where good adhesion of metal films is required, but liquid trapping is not a factor. Our laboratories have always used polished quartz crystals in order to minimize roughness effects. In any case, any published work employing the EQCM should contain a description of the quartz crystals and their roughness.

3 ELECTROCHEMICAL METHODS AND EXAMPLES OF APPLICATIONS

The EQCM has been used to examine a wide variety of electrochemical processes. Unfortunately, space does not permit here a comprehensive review of the area and the reader is referred to other reviews cited earlier. Rather, this section will highlight some illustrative examples of EQCM applications which are meant to provide the reader with a general understanding of the scope of this method with respect to its capability, phenomena

For references see page 9954

that affect measurements, and the types of system that are amenable to examination.

3.1 Experimental Apparatus and Operation

Several versions of EQCM instrumentation have been described,[14–17] differing mostly in minor details. The system described below is one that has been developed in our laboratory (Figure 3). Commercial systems are now available (EG&G PAR, Elchema), but it has been our experience that the practitioner gains a better awareness of the strengths and limitations of the method by building the homemade system. EQCM quartz crystals commonly have diameters of 0.5 in and 1.0 in, with appropriately sized excitation electrodes. The Bechmann numbers[18–20] dictate that, for a crystal diameter that is 50 times the crystal thickness, the electrode diameter must be 18 times the crystal thickness in order to avoid interference from other acoustic modes. Accordingly, a reduction in crystal and electrode diameter must be accompanied by a corresponding reduction in the crystal thickness to maintain frequency stability.

The crystals can be mounted at the bottom of a glass cylinder that assumes the role of the working electrode compartment of an otherwise conventional electrochemical cell. The crystal is mounted so that the excitation electrode that is to be employed as the working electrode is facing the solution; the opposite electrode is therefore facing air in order to avoid electrical shorting between the electrodes. If the quartz crystal is immersed under a column of the electrolyte solution, hydrostatic pressure results in a stress on the quartz crystal that affects the resonant frequency. Equation (7) describes a parabolic dependence of f_0 on the hydrostatic pressure.[21,22]

$$f_0 - f_0^{\max} = A(p - p_{\max})^2 \qquad (7)$$

However, the significance of this effect is probably minimal as the hydrostatic pressure is generally constant (barring evaporation) for most experiments. This problem can be avoided completely by configurations in which the crystal is simply mounted vertically. Crystals can be mounted between O-rings or with epoxy, the former being more convenient as the crystals can be easily demounted for reuse or further surface studies. Our laboratory has found that Teflon™-coated O-rings give the best performance.

The two excitation electrodes are electrically connected to an oscillator circuit that contains a broadband RF amplifier. Several oscillator designs are available, although standard commercial oscillators, such as those sold for metal evaporation control, typically need modification in order to supply the crystal with sufficient gain to sustain oscillation in liquids. Therefore, it is usually simpler to build the oscillator. The schematic for the oscillator circuit used in our laboratory is illustrated in Figure 4. For the convenience of the reader, a parts manifest is provided in Table 2. A commercially fabricated printed circuit board is also recommended in order to avoid noise pickup that accompanies circuit built on crude breadboards. The circuit is designed so that the crystal is in a feedback loop, therefore driving the crystal at a frequency at which the maximum current can be sustained in a zero-phase angle condition. The electrode

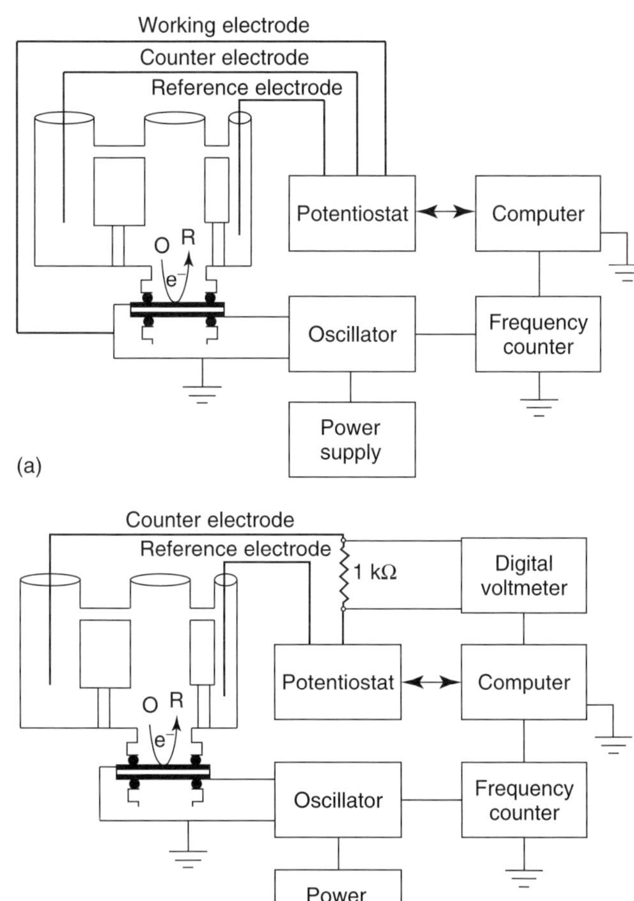

Figure 3 (a) Schematic representation of typical EQCM apparatus in which a Wenking potentiostat is employed when the working electrode is at hard ground. (b) An alternative EQCM in which a conventional commercially available potentiostat is used. In this arrangement, the working electrode lead of the potentiostat is not connected to the EQCM working electrode. Rather, the current is measured by the voltage drop across a 1 kW resistor in series with the counter electrode. This arrangement is required when using commercially available potentiostats because these potentiostats operate with the working electrode at virtual ground; connection of this virtual-ground to the hard-grounded EQCM working electrode can result in oscillator instabilities.

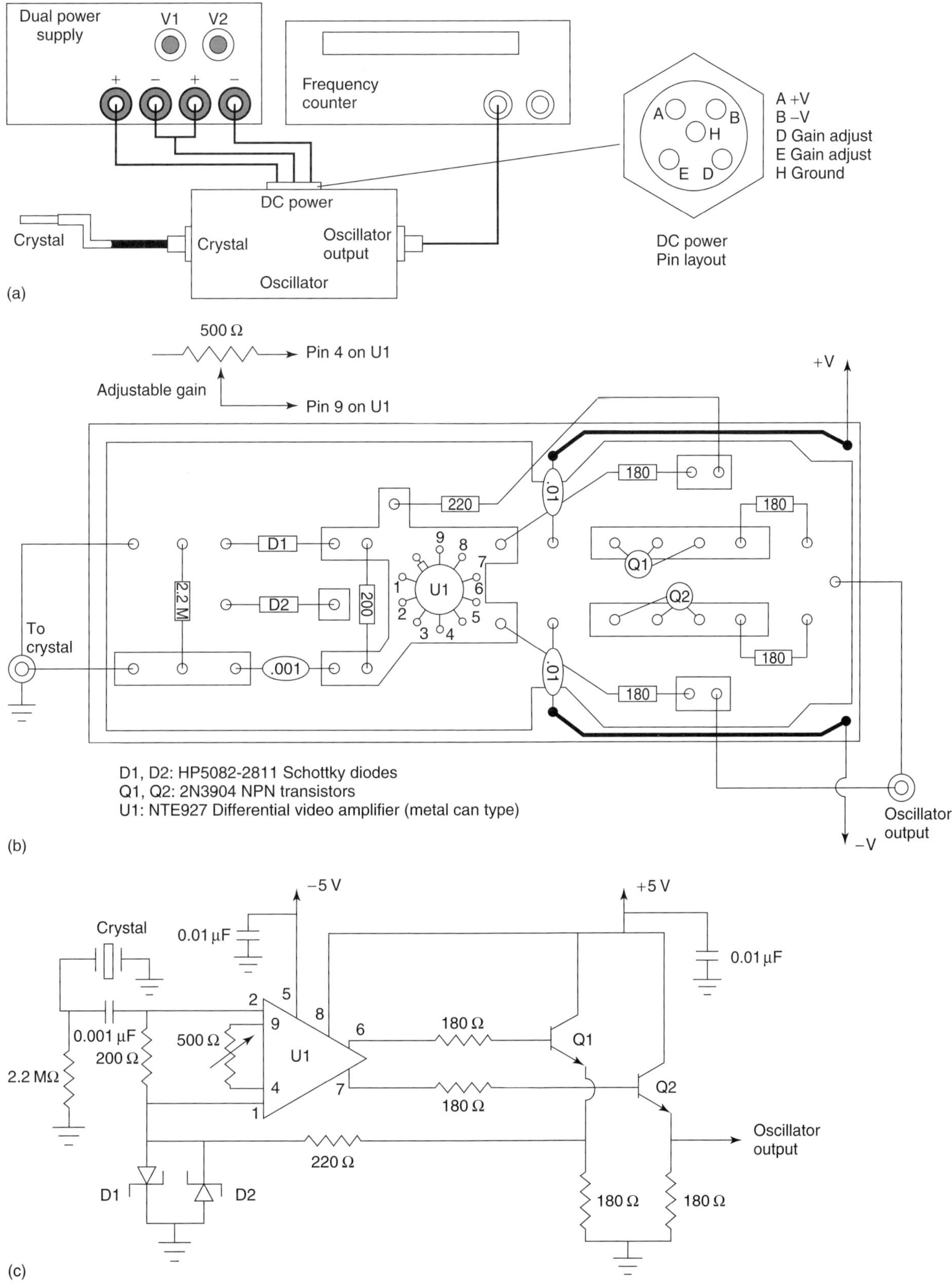

Figure 4 (a) Schematic representation of the hardware configuration for the EQCM, neglecting the electrochemical apparatus depicted in Figure 3. (b) Schematic of an oscillator circuit and related components. (c) Detailed schematic of the component layout for the oscillator circuit. The components are listed in Table 1.

For references see page 9954

Table 2 Hardware and circuit components for a user-constructed EQCM

Hardware	Quantity	Vendor[a]	Part no.
Dual output power supply	1	1	6234A
Frequency counter	1	1(2)	5384A ()
IEEE interface board (data acquisition)	1	3	Any compatible version
Cable (oscillator to crystal)	1	3	RG 58 A/U (with two BNC connectors)
Cable (oscillator to power supply)	1	3	RG 174/U coaxial (with BNC connector)
Shielded IEEE cable	1	3	
Type 3302 component box	1	4	35F3515
H-90 trimmer tool (optional)	1	4	12F8866
5-pin jack	1	5	126-218
Printed circuit board	1	Custom	
Circuit components			
Cemet trimmer 500 Ω	1	4	12F9636
Panel adapter type 6-2-0	1	4	81F009
Differential video amplifier	1	6	NTE927
ULP-IC socket	1	4	65F1881
RN-55D type resistor, 182 Ω, 1/8 W	4	4	58F001
RN-55D type resistor, 200 Ω, 1/8 W	1	4	58F001
RN-55D type resistor, 221 Ω, 1/8 W	1	4	58F001
YX series polyester film capacitor, 0.001 μF	1	4	89F3228
YX series polyester film capacitors, 0.01 μF	2	4	89F3232
RC07 type CB resistor, 2.2 MΩ, 1/4 W		4	10F305
Schottky diodes	2	6	HP5082-2811
NPN transistors	2	3	2N3904

[a] Suggested vendors: (1) Hewlett-Packard, (2) Phillips Electronics, (3) any vendor, (4) Newark Electronics, (5) Wire-Pro, Inc., (6) Allied Electronics.

facing solution is at hard ground. The output of the oscillator is connected to a conventional frequency meter for measurement. A critical feature of the EQCM is the type of potentiostat: the Wenking potentiostat functions with the working electrode at hard ground whereas most commercially available potentiostats generally function with the working electrode at virtual ground. Commercial potentiostats can only be used if the working electrode is not connected to the potentiostat and the potential difference between the reference and hard ground is used to control the working-electrode potential. Because the current is generally measured at the working-electrode side, this format requires that the current be measured by the voltage drop across a resistor in series with the counter-electrode connection. Because this equipment generally is available in most electrochemical laboratories, the EQCM practitioner need not build, or buy, a custom-made Wenking potentiostat. Finally, a computer is used to collect frequency and electrochemical data simultaneously, as well as control the waveform applied to the working electrode. This arrangement allows simultaneous measurement of the electrochemical charge, current, voltage, and EQCM frequency. The timescale of the analysis is in a particularly useful domain for electrochemists. The time constant of a quartz resonator is fixed by $Q/\pi f_0$. The quality factor for a 5-MHz resonator is $Q \approx 10^3$, and therefore the minimum sampling time is in the millisecond range. Frequency counters are capable of sampling the frequency output of the oscillator at 100 ms intervals. This capability enables analysis of the kinetics of a wide range of electrochemical processes, including electrodeposition and dissolution, nucleation and growth, and ion/solvent insertion in redox polymer films. It is important to stress that this equipment is not prohibitively expensive and is well within the reach, economically and technically, of most electrochemical investigators.

3.2 Electrochemical Measurements of Crystal Resonant Properties in Fluids

There are other aspects of the EQCM that distinguish their operation in liquids from that of the QCM in vacuum or the gas phase, namely the effect of the liquid on the actual vibrating area of the quartz crystal. The Sauerbrey equation, Equation (2), assumes that the frequency shift resulting from a localized deposit is equivalent to the contribution of that deposit when it is a portion of a thin film of identical thickness distributed over the entire active EQCM area. However, a general expression that accounts for localized or nonuniform mass deposits covering the EQCM electrode to $r = r_e$ (where r_e is the ideal radius of the excitation electrodes, between which the electric field induces crystal motion) is given by

Equation (8),

$$\Delta f = \left(\frac{1}{\pi r_e^2}\right) \int_0^{2\pi} \int_0^{r_e} S(r,\theta)\, m(r,\theta)\, r\, dr\, d\theta \quad (8)$$

where $S(r,\theta)$ is the differential mass sensitivity (df/dm) and $m(r,\theta)$ is the mass distribution with respect to r, the distance of the deposit from the center of the crystal and θ, the angle in the crystal plane with respect to the x-axis. It has long been accepted, on the basis of theory and experiment, that $S(r,\theta)$ has a near-Gaussian form in which the crystal vibration is maximum in the center and negligible at the edges.[23–27] Indeed, Equation (8) and the Sauerbrey equation are based on the assumption that the vibrating area, and therefore the mass sensitive area, is limited to within the region where the excitation electrodes overlap. Any changes in $S(r,\theta)$ resulting from the liquid, or any crystal motion beyond the electrode edges, will cause deviations from the ideal response.

It has been shown in liquids that $S(r,\theta)$ and the actual vibrational area depend significantly upon the liquid and its properties, including its viscosity.[28–31] EQCM measurements performed in conjunction with SEM demonstrated that $S(r,\theta)$ was gaussian-like (Figure 5). However, the shape and the maximum value in the center were dramatically affected by the viscosity of the liquid in contact with the EQCM. Increasing the viscosity resulted in a suppression of the maximum and an increase in the mass-sensitive area due to a phenomenon referred to as field fringing, in which the crystal vibrations extend beyond the electrode area. This results in an overall reduction in the sensitivity of the EQCM. The details of these effects will depend upon crystal contour, electrode geometry, and the liquid properties. Trapping of the crystal vibrations within the excitation electrode boundaries is much more efficient for plano–convex crystals, in which one side of the crystal is contoured by polishing such that its shape resembles a convex lens. The difference in mass between the center and outer regions of the crystal results in a focusing of the acoustic energy toward the center of the resonator, as is evident from an increase in the value of $S(r,\theta)$ at the center and negligible mass sensitivity at $r > r_e$. Therefore, there appear to be distinct advantages to using plano–convex crystals. Understanding of these effects is crucial if precise quantitative interpretation of EQCM data is required. At the very least, these effects should be appreciated in a qualitative way.

3.3 Data Interpretation for Ideally Behaved Systems

Interpretation of EQCM data is accomplished in a rather straightforward manner. Since the electrochemical charge represents the total number of electrons transferred in a given electrochemical process, it corresponds to mass

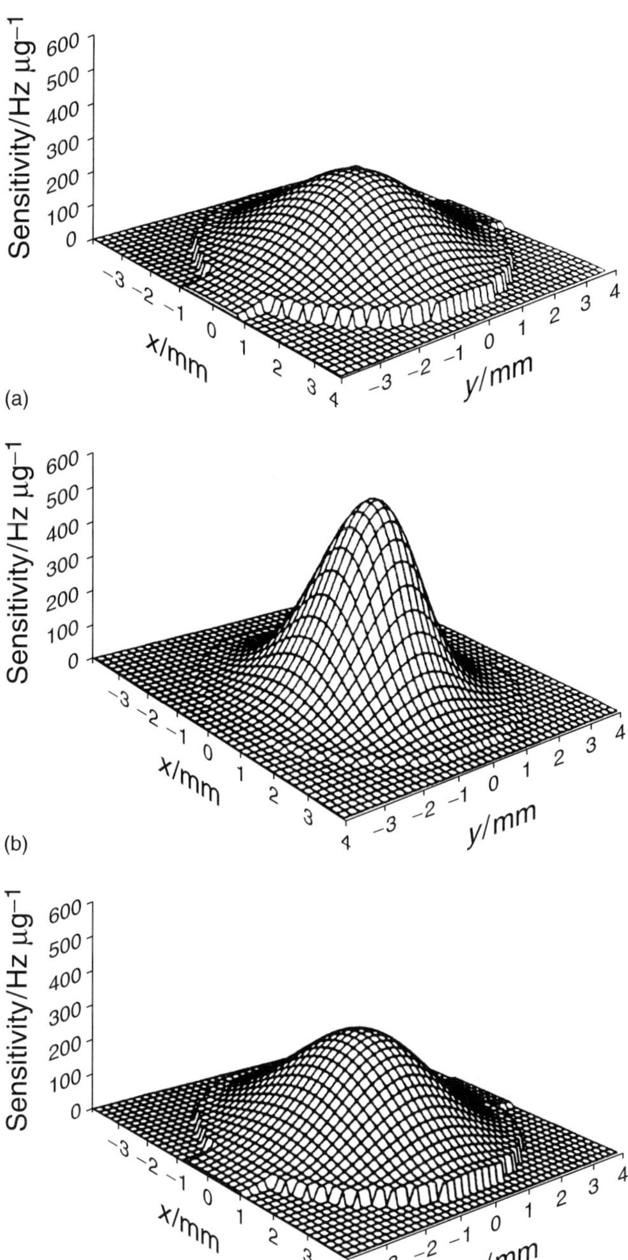

Figure 5 Sensitivity distributions $S(r,\theta)$ for a 5-MHz AT-cut quartz crystal determined by a SEM method in which 100-µm diameter circular copper features are electrodeposited at different values of r,θ while simultaneously measuring electrochemical charge and frequency. The amplitude of these plots at a given value of r,θ represents the sensitivity of the EQCM to mass changes at that location, and the area under the plot represents the total, or integral, sensitivity C_f. The $S(r,\theta)$ plots shown were measured on (a) plano–plano and (b) plano–convex quartz crystals in aqueous 20-mM $CuSO_4$ ($\eta_L = 0.8904 \times 10^{-2}\,g\,cm^{-1}\,s^{-1}$), and (c) plano–convex quartz crystals in aqueous 20-mM $CuSO_4$ containing 15% sucrose ($\eta_L = 1.469 \times 10^{-2}\,g\,cm^{-1}\,s^{-1}$). (Reproduced with permission from Hillier and Ward,[28] Copyright (1992) American Chemical Society.)

For references see page 9954

changes occurring at the electrode surface. Accordingly, under ideal conditions, the frequency change measured with the EQCM will be proportional to the electrochemical charge, and will be related to the apparent molar mass by Equation (9), where MW is the apparent molar mass ($g\,mol^{-1}$), Q is the electrochemical charge, n is the number of electrons involved in the electrochemical process, F is the Faraday constant, and C_f ($Hz\,g^{-1}$) the sensitivity constant derived from the Sauerbrey relationship.

$$\Delta f = \frac{\mathrm{MW}C_f Q}{nF} \tag{9}$$

Inspection of Equation (9) reveals that plots of Δf versus Q are particularly useful in the determination of MW/n, which represents the molar mass per electron transferred. This calculation, of course, depends upon knowledge of C_f. Because this value can differ for different crystal contours, electrode geometries, and solution conditions (see above), it is important to calibrate the EQCM with a well-behaved electrochemical reaction under conditions similar to those present during the experiments of interest. This is generally accomplished with copper or silver electrodeposition, for which all terms on the right side of Equation (9) are known except for C_f. The term apparent molar mass is stressed because in many cases the measured value of MW may not be that expected based on a simple stoichiometric relationship, but may involve solvent or co-adsorbed species that can reveal considerable insight into the electrochemical behavior. Nonlinearity in plots of Δf versus Q can be particularly useful for diagnosing nonideal behavior such as roughness and viscoelastic effects that may become evident over the range of frequency changes examined. An alternative approach to data analysis involves the relationship between the electrochemical current and the first derivative of the frequency change with respect to time, as given in Equation (10), where ν is the scan rate in units $V\,s^{-1}$. This format is particularly useful for cyclic voltammetry experiments, as $d\Delta f/dt$ should appear similar in form to the voltammograms if the electrochemical events are accompanied by corresponding mass changes.

$$i = \frac{d(\Delta f)}{dE}\,\frac{n\nu F}{\mathrm{MW}C_f} \tag{10}$$

The utility of the EQCM method stems from its capability of measuring electrochemical charge and current while simultaneously measuring mass changes with extraordinary sensitivity. A typical operating resonant frequency of 5 MHz provides a theoretical sensitivity of $0.5666\,Hz\,cm^{-2}\,ng^{-1}$. Since the frequency can generally be measured to within an accuracy of 1 Hz, the EQCM can detect approximately $10\,ng\,cm^{-2}$. This translates roughly into 10% of a monolayer of Pb atoms. Much higher sensitivity can be realized with quartz crystals that operate at a higher fundamental frequency, as the sensitivity increases with f_0^2. Alternatively, the quartz crystal can be driven at one of its odd harmonic modes with appropriate circuitry, which provides an n-fold increase in sensitivity, where n is the harmonic number. While the third harmonic of a 5-MHz crystal (i.e. an operating frequency of 15 MHz) has been employed successfully in the examination of underpotential deposition (UPD) on metal electrodes,[32] general use of this approach can be limited by the lower stability of the harmonic modes relative to the fundamental mode.

It is much more beneficial to employ crystals that can be operated at higher fundamental frequencies. However, higher fundamental frequencies can only be achieved by fabricating thinner crystals. The operating frequency range of the EQCM has been extended as far as 30 MHz. While these crystals are rather thin ($t_Q = 50\,\mu m$), they could be used conveniently by constructing the electrochemical cell with heat-shrinkable Teflon™ tubing constricted about the periphery of the crystal. In liquid media the frequency of a 30-MHz resonator responds to changes in the liquid viscosity and density according to Equation (6). That is, these resonators exhibit a linear relationship between Δf and $(\eta_L \rho_L)^{1/2}$, identical to behavior expected and observed for lower frequency resonators. Electrodeposition of copper on the EQCM surface (Figure 6) affords sensitivity constants in exact agreement with Equation (9), indicating that energy trapping of the fundamental mode is very efficient for the 30-MHz resonators (lower frequency resonators commonly exhibit sensitivity constants that are lower than expected owing to acoustic field fringing). The higher frequency resonators have many potential advantages, including greater sensitivity to mass, better signal-to-noise characteristics than other acoustic wave devices,[33] and less energy dissipation during oscillation. The origin of the improved stability is evident from admittance analysis of 30-MHz resonators (see below). Higher frequency EQCMs are feasible if chemically milled crystals are employed. Such a crystal is obtained by chemically etching the center of a quartz crystal to obtain a thin quartz "membrane" in the center of a thicker outer quartz ring, the outer ring providing improved mechanical stability.

Changes in the depletion layer that accompany redox processes also can influence EQCM measurements. Our laboratory demonstrated that cyclic voltammetry in solutions containing the $Fe(CN)_6^{3-/4-}$ or $Ru(NH_3)_6^{2+/3+}$ redox couples was accompanied by potential-dependent changes in the resonator frequency even though no mass changes occurred at the electrode surface (Figure 7).[34] This behavior can be attributed to changes in the density of the depletion layer that accompany the change in redox

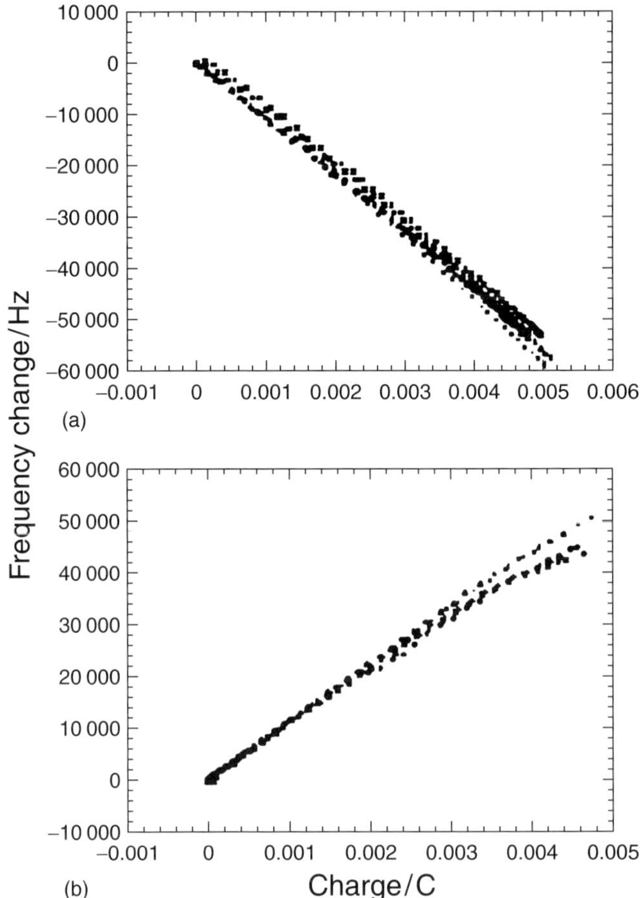

Figure 6 Frequency dependence of the charge consumed at the electrode of a 30-MHz EQCM during electrodeposition and removal of copper in 0.1-M $CuSO_4$ solution. (a) Electrodeposition of copper, slope $= 1.135 \pm 0.036 \times 10^7$ Hz C^{-1}. (b) Electrochemical removal of copper, slope $= 1.064 \pm 0.043 \times 10^7$ Hz C^{-1}. (Reproduced with permission from Lin et al.,[6] Copyright (1993) American Chemical Society.)

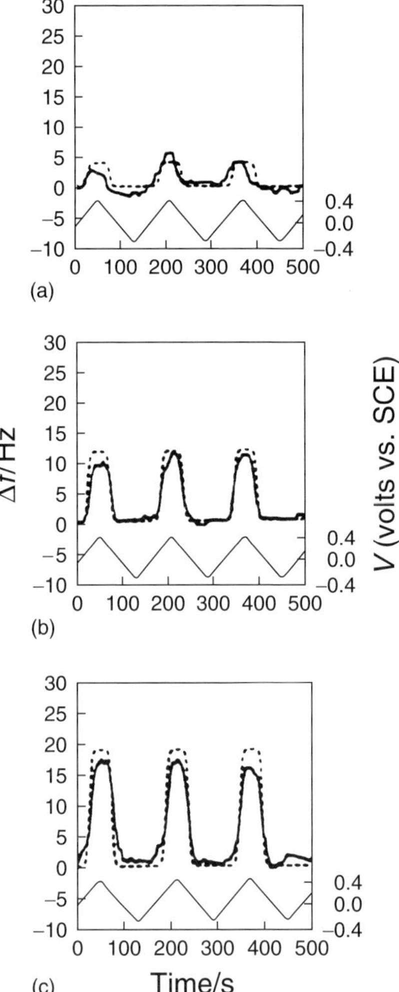

Figure 7 Comparison of measured (——) and simulated (- - - - -) frequency changes at a 5-MHz EQCM during potential cycling between −0.3 V and 0.5 V at 10 mV s^{-1} for solutions of $K_4Fe(CN)_6$ at concentrations of (a) 20, (b) 60, and (c) 100 mM in 1.0 M Na_2SO_4. (Reproduced with permission from Lee et al.,[34] Copyright (1993) American Chemical Society.)

state. Modeling of the depletion-layer characteristics using known densities and viscosities of each redox species have confirmed that the frequency changes during cyclic voltammetry behaved according to Equation (6).

3.4 Electrochemical Quartz Crystal Microbalance Investigations of Metal Electrodes and Films

As stated above, the most common technique for calibrating the EQCM involves simultaneous measurement of the electrochemical charge and resonant frequency during the electrochemical deposition or dissolution of a metal film. This capability was demonstrated by Bruckenstein and Shay in which the deposition of 10 layers of Ag metal on a 10-MHz EQCM gave frequency shifts that were within 3% of the amount expected based on the Sauerbrey equation.[35] Therefore, the EQCM can be an effective method for examining the Faradaic efficiencies of electroplating and dissolution processes such as corrosion.[36]

The high sensitivity of the EQCM has enabled examination of UPD processes in situ. For example, examination of the UPD of Pb on the Au working electrode of the EQCM revealed mass and electrochemical charge changes in the UPD region corresponding to a hexagonally closest packed monolayer. The ability to measure mass and charge simultaneously allowed determination of the electrovalency number for this process, which was found to be $\gamma = 2.08 \pm 0.10$. Similar experiments revealed that other UPD processes occurred with γ values lower than expected for complete electron transfer, with $\gamma_{Bi} = 2.7$, $\gamma_{Cu} = 1.4$.[37]

For references see page 9954

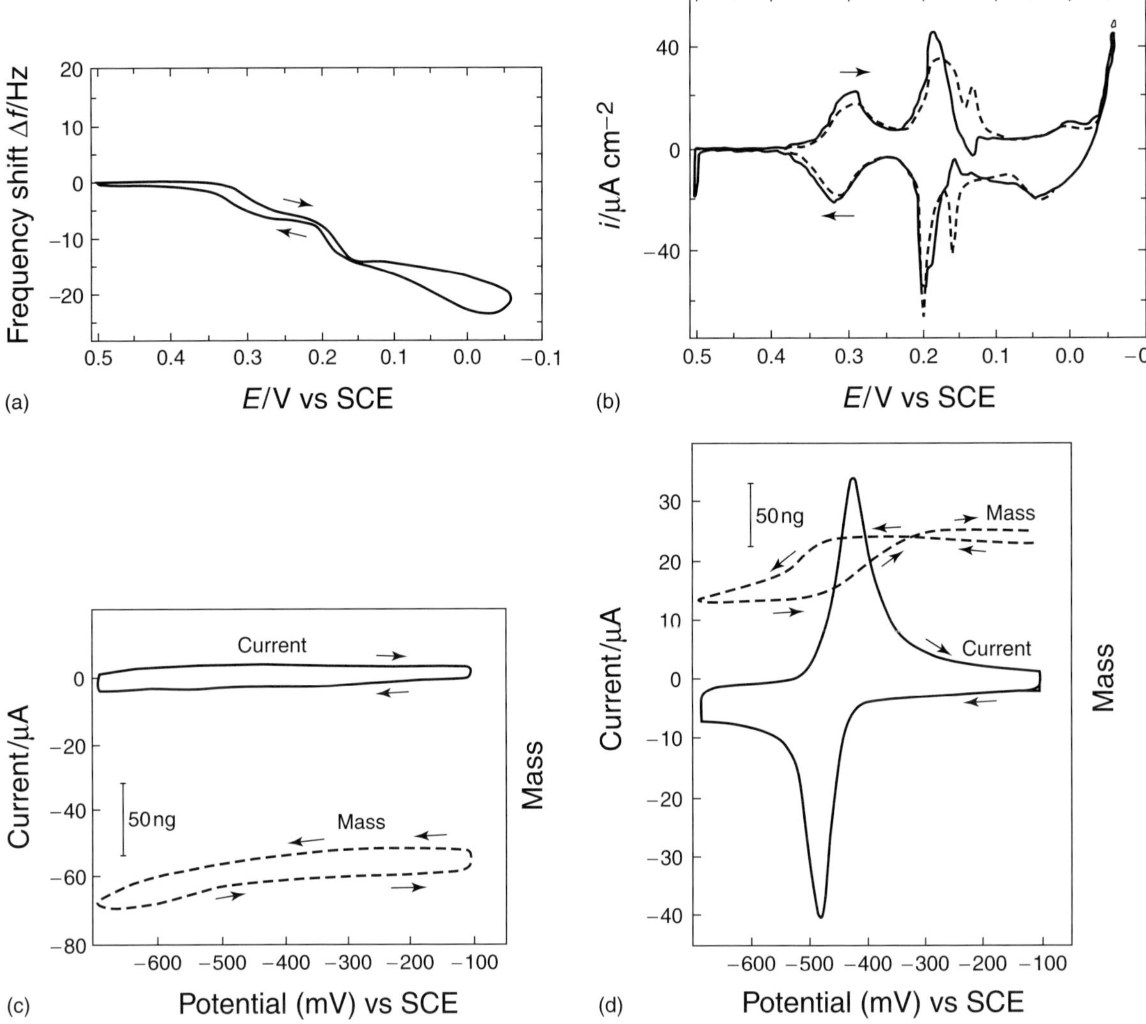

Figure 8 (a) Frequency shift of the EQCM during Bi UPD on a Au electrode (1.0-mM Bi in 0.1-M $HClO_4$). (b) Current response during Bi UPD on the Au EQCM electrode (- - - - -) and current response expected from the observed frequency response (———). (Reproduced from Deakin and Melroy, *J. Electroanal. Chem.*[37] Copyright (1998) with permission from Elsevier Science.) (c) Cyclic voltammetric scan showing the current and EQCM response obtained at an EQCM silver electrode in 0.1-M borate buffer at pH = 9.15; scan rate = 50 mV s^{-1}. The electrode was conditioned at −100 mV (vs SCE (saturated calomel electrode)) for 30 s before initiating the potential scan. (d) Cyclic voltammetric scan showing the current and EQCM response obtained at an EQCM silver electrode in 0.1-M borate buffer at pH = 9.15 containing 2.7×10^{-5} M Pb(II); scan rate = 50 mV s^{-1}; starting potential = −100 mV. (Reproduced from Hepel et al., *J. Electroanal. Chem.*[39] Copyright (1989) with permission from Elsevier Science.)

The EQCM has revealed several interesting features in UPD processes that otherwise would be difficult to detect. For example, in the aforementioned investigations with bismuth, it was discovered that a precipitous increase in electrochemical charge accompanied the third UPD peak, with a slight decrease in mass (Figure 8). This behavior may be associated with the loss of electrolyte ions that are adsorbed during the first two UPD events. More detailed studies of Pb UPD on Au and Ag revealed behavior that demonstrated the importance of the underlying substrate in UPD processes.[38–40] Whereas Pd UPD on Au electrodes in borate buffer was accompanied by a mass increase, UPD on Ag electrodes exhibited a mass decrease due to the desorption of BO_2^- ligands from a previously adsorbed anionic Pb(II) species. The adsorption of the latter was detected by the frequency decrease upon addition of Pb(II) to the electrolyte solution prior to electrochemical experiments. The data reflected the desorption of three BO_2^- ions per adsorbed Pb(II) species during the UPD process. These studies clearly reveal the value of the EQCM in probing electrode surface processes, providing information critical to complete understanding of rather complicated mechanistic schemes.

The EQCM has also enabled determination of the electrovalency numbers for anion adsorption. Simultaneous measurement of electrochemical charge and frequency during anodic adsorption of Br^- and I^- under conditions where monolayer coverage previously had been demonstrated revealed $\gamma_{Br^-} = -0.39 \pm 0.03$ and $\gamma_{I^-} = -1.01 \pm 0.03$.[41] These studies indicate that, while adsorption of iodide occurs with complete charge transfer, bromide ions partially retain their negative charge.

The capabilities of the EQCM provide a unique approach for examining electrochemical dissolution, which is important in processes such as corrosion and electrochemical machining. While EQCM studies of dissolution have been rather limited, these reports indicate that important mechanistic information can be obtained readily with this method. For example, EQCM investigations of the anodic dissolution of nickel films revealed two maxima in the Δf versus potential plots, indicating a potential-dependent dissolution of the α and β phases of NiH_x.[42] Analysis of the frequency changes and electrochemical charge in EQCM studies of the anodic dissolution of a nickel–phosphorus film revealed that two different Ni–P compositions were present in the film prior to dissolution. This was consistent with the known Ni–P phase diagram.[43] EQCM studies of the electrochemical dissolution of copper films in oxygenated sulfuric acid revealed that the dissolution rate was linearly dependent upon $[O_2]$ and $[H^+]$, enabling the authors to conclude that a heterogeneous surface reaction was operative.[44]

Several groups have demonstrated that the EQCM is also an ideal method for examining the electrochemically induced adsorption of hydrogen and deuterium in metal films such as palladium.[45–47] Such studies can have significant impact on understanding the commercially important isotopic separation of H and D. In addition to measurement of H/D adsorption, these EQCM studies have revealed the important role of stress in EQCM measurements. Cheek and O'Grady reported a rather novel approach to this issue by using a "double resonator" technique that previously had been reported for measuring stresses in Si films resulting from ion implantation (in vacuum).[48,49] This method involves the comparison of frequency responses from palladium-coated AT-cut and BT-cut quartz crystals during electrochemically induced ingress and egress of H or D (Figure 9). Both crystals have identical sensitivities to mass changes, but a compressive stress in a film on an AT-cut crystal results in a frequency decrease while an identical stress in a film on a BT-cut crystal results in an increase in frequency of similar magnitude. The amount of stress can be determined by Equation (11),

$$S = (K^{AT} - K^{BT})\left(\frac{t_Q^{AT}\Delta f^{AT}}{f_0^{AT}} - \frac{t_Q^{BT}\Delta f^{BT}}{f_0^{BT}}\right) \quad (11)$$

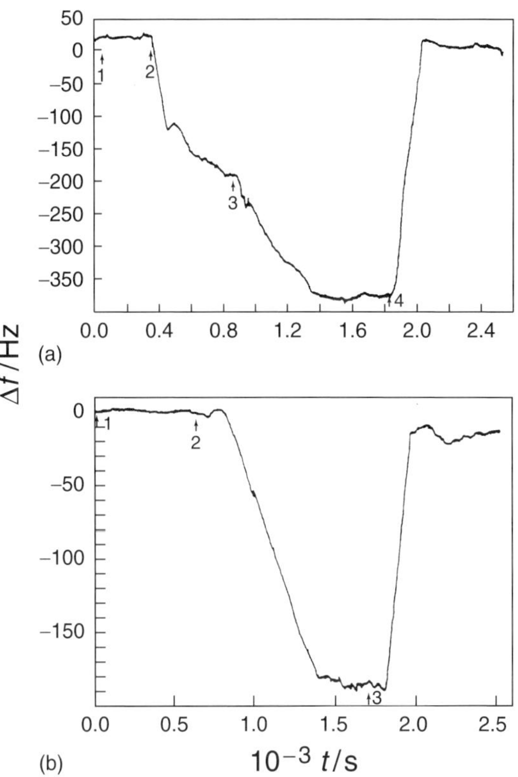

Figure 9 (a) Frequency shift versus time for a 366.6-nm thick Pd film on an AT-cut quartz resonator in 0.1-M $LiOH/H_2O$. The numbered arrows indicate the potential applied to the film: (1) 0.00 V; (2) E scanned to and held at −1.14 V; (3) E scanned to and held at −1.27 V; (4) E scanned to and held at 0.0 V. The frequency decrease is larger than that expected based on the mass of the cathodic hydrogen absorption process, and is attributed to compressive stress in the Pd film upon hydrogen absorption which decreases the frequency. (b) Frequency shift versus time for a 475-nm thick Pd film on an AT-cut quartz resonator in 0.1-M $LiOH/D_2O$. The numbered arrows indicate the potential applied to the film: (1) 0.00 V; (2) E scanned to and held at −1.20 V; (3) E scanned to and held at 0.0 V. The frequency decrease is smaller than that expected based on the mass of the cathodic deuterium absorption process, and is attributed to compressive stress in the Pd film upon deuterium absorption that increases the frequency. (Reproduced from Cheek and O'Grady, *J. Electroanal. Chem.*[45] Copyright (1990) with permission from Elsevier Science.)

where K^{AT} and K^{BT} are the stress coefficients for the different resonators ($K^{AT} = 2.75 \times 10^{-12}\,cm^2\,dyn^{-1}$ and $K^{BT} = -2.65 \times 10^{-12}\,cm^2\,dyn^{-1}$). Stress in thin films is manifested in a stress in the quartz crystal, which is tantamount to a change in the elastic modulus, μ_q. Because the acoustic velocity depends upon this quantity, these stresses result in changes in the resonant frequency (this effect is similar to tightening a violin string to increase the frequency of vibration). Thus, hydrogen adsorption in a Pd film on an AT-cut crystal gave a frequency decrease that was much larger than expected, indicating

For references see page 9954

compressive stresses in the Pd film upon H adsorption, −351 Hz for a 367-nm thick Pd film, compared to −185 Hz expected for the mass change due to H adsorption. However, the same experiment on a BT-cut crystal resulted in a very small increase in frequency, +20 Hz for a 475 nm thick Pd film.

3.5 Electrochemical Quartz Crystal Microbalance Investigations of Thin-film Growth

The EQCM provides an extremely useful approach to the in situ study of the nucleation and growth of a wide variety of thin films on electrodes, ranging from oxide films to molecular solids and monolayers. In principle, the ability to measure mass and electrochemical current simultaneously allows determination of chemical stoichiometries, Faradaic efficiencies, and reaction kinetics. The EQCM therefore complements current transient methods typically used to study nucleation and growth, providing information which otherwise cannot be obtained readily. The time constant for EQCM measurements (see above) is within the timescale of many electrochemical processes.

The growth of metal-oxide films has been investigated at copper and gold electrodes, the EQCM enabling determination of the stoichiometry of the oxides formed.[11–13] In addition, these studies revealed frequency responses that were consistent with morphological changes of the electrode surface accompanying oxide formation. This resulted in much larger changes in the frequency during oxide growth that was attributed to surface roughening and subsequent trapping of water in the cavities of the roughened surface. The surface roughness was retained initially after reduction of the oxide back to the metal, but the frequency gradually returned to its original value, indicating dynamic changes in the morphology of the roughened metal electrodes (Figure 10). In addition to providing valuable mechanistic insight into oxide growth, these studies illustrate the sensitivity of the EQCM to surface roughness effects.

EQCM studies of the electrochemical deposition of molecular films have also been studied. The adsorption of surfactants containing redox-active ferrocene (Fc) groups (referred to as C_{12} and C_{14}, which designates the length of the alkyl chain) could be induced electrochemically, because adsorption occurred when these species were in the reduced (Fc^0) state (Figure 11).[50,51] Upon oxidation to the oxidized form (Fc^+), the surfactants rapidly desorbed. Interestingly, the desorption rate was dependent upon the chain length of the surfactant. Whereas short-chain-length surfactants resulted in rapid desorption following electrochemical oxidation, the EQCM frequency changes indicated that long-chain surfactants desorbed much more slowly following oxidation. Mechanistic and thermodynamic information for these films was

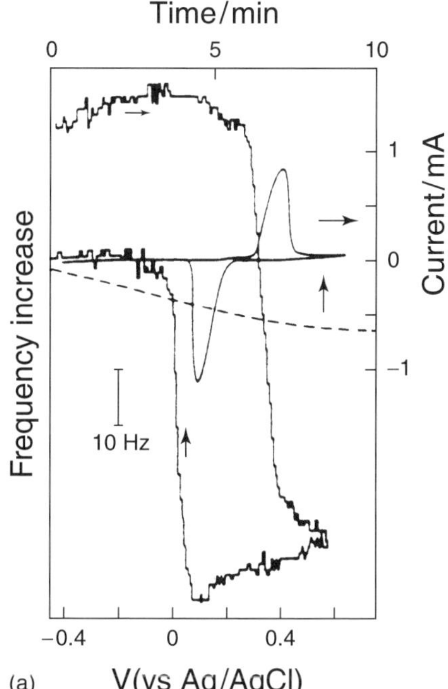

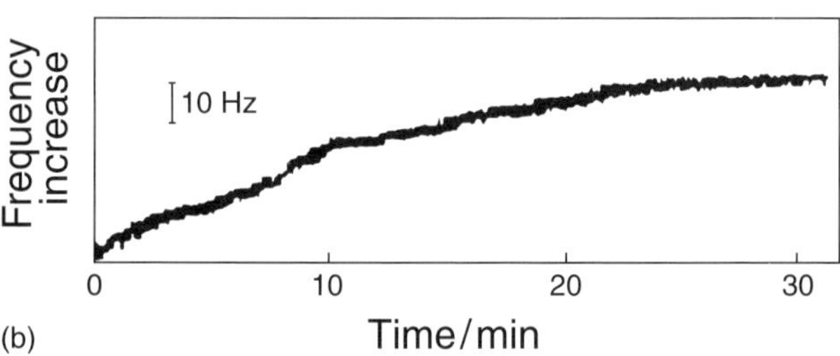

Figure 10 (a) Current versus potential (- - - - -) and frequency versus potential (jagged line) for a silver EQCM electrode during cyclic voltammetry from an initial potential of −0.4 V (vs Ag/AgCl). Scan rate = 20 mV s^{-1}, surface area = 0.33 cm^2. (———) Frequency versus time with the electrode held at −0.4 V at the end of the cycle. Note that the frequency at the end of the cycle is approximately 30 Hz less than at the beginning, and this difference increases to 45 Hz after 10 min at −0.4 V. This indicates a significant change in the morphology of the electrode. (b) Frequency versus time response for the silver electrode held at −0.4 V after the 10 min period depicted in (a). The frequency increase is attributed to a morphological relaxation of the silver electrode. (Reproduced from Schumacher et al.[12] Copyright (1987) with permission from Elsevier Science.)

thereby attainable.[52] The nucleation, growth, and dissolution of thicker films of related species, namely diheptyl viologen, was also examined with the EQCM. Examination of these processes at different potentials provided insight into the nucleation behavior of this system.[53]

The EQCM has also been useful in the investigation of the nucleation and growth of "electronic" materials. For example, the deposition of thin semiconductor films

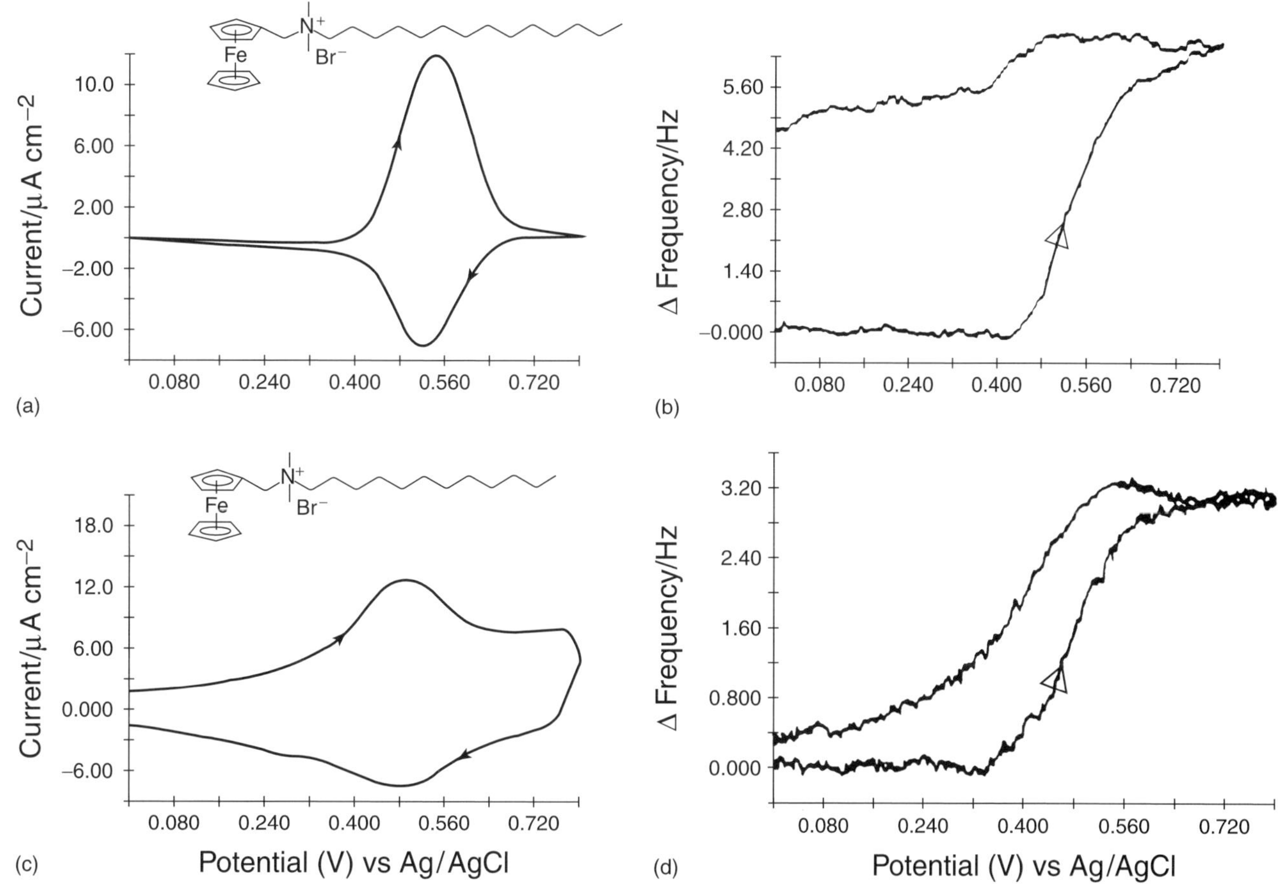

Figure 11 (a) Cyclic voltammetric scan for a 7-μM solution of C_{14} compound from 0.0 to 0.80 V (vs Ag/AgCl) in 1-M H_3PO_4; 50 mV s^{-1}. The current was digitally smoothed and corrected for background. (b) EQCM frequency response for (b). (c) Cyclic voltammetric scan for a 22-μM solution of C_{12} from 0.0 to 0.80 V in 1-M H_3PO_4; 50 mV s^{-1}. (d) EQCM frequency response for (c). Higher concentrations of the C_{12} compound were required to obtain the same coverage as for the C_{14}. Note that the frequency response in (d) is less than that in (b), consistent with the lower molecular weight of C_{12}. (Reproduced with permission from Donahue and Buttry,(50) Copyright (1989) American Chemical Society.)

was examined, and the rather complicated mechanism involved in the deposition of thin films of Te was determined by comparison of the EQCM frequency changes and electrochemical charge.(54) Similarly, the electrodeposition of B-doped β-PbO_2 thin films was studied with the EQCM and details of the film composition and catalytic activity toward oxygen atom transfer reactions were realized.(55)

The growth and redox chemistry of Prussian blue and related films has been examined by EQCM, with particular attention paid to the degree of solvation of these films and the ion transport during electrochemical cycling. These studies would be expected to have an impact on the use of these films in sensors and electrochromic displays. The initial EQCM study of Prussian blue films found that the frequency decrease during electrochemical deposition was consistent with a high degree of hydration.(56) Subsequent EQCM studies of these films revealed that cation transport accompanied changes in the redox state during potential cycling, although at low pH conditions proton transport was also involved. Ion and solvent transport in related nickel ferrocyanide films during potential cycling in 0.1-M CsCl solutions also has been examined. The results were consistent with transport of Cs^+ ions upon change in the redox state of the film.(57) The mass change calculated from the Sauerbrey equation, Equation (2), which assumes rigid-layer behavior, suggested that H_2O was expelled from the film as Cs^+ was incorporated during reduction. This was verified by a clever experiment in which the process was studied in H_2O and D_2O; the mass change associated with the solvent was found to increase by 10% in D_2O. This verified the participation of solvent and provided corroborative evidence of rigid-layer behavior.

The Prussian blue and nickel ferrocyanide films consist of a low density pseudocrystalline lattice in which the metal atoms are organized by cyanide ligand bridges

For references see page 9954

between octahedral metal centers. As a result, metal ion intercalation into the interstices in the lattice is affected by the size and hydration of the cation. The different amount of work required for intercalation of the different ions is manifested as a dependence of the formal potential of the films on the identity and concentration of the metal ion in the supporting electrolyte. Analysis of EQCM data acquired during redox cycling of Prussian blue films in propylene carbonate containing different relative amounts of $NaClO_4$ and $LiClO_4$ revealed that intercalation of Na^+ was favored over Li^+ by a factor of 15.[58] This capability was demonstrated to be useful for the analytical identification of metal ions in a flow injection mode.[59] Simultaneous electrochemical and EQCM measurements were performed in which the Prussian blue film was held at a potential known to be selective for a given cation, and the charge and mass changes associated with ion transport were measured. Comparison of the charge and mass changes enabled determination of the molar mass and, therefore, identification of the cation.

The electrodeposition of polymer films has also been investigated with the EQCM, in particular the electrodeposition of poly(vinylferrocene) (PVF) films.[60] These studies indicated that in oxidative electrodeposition of PVF in CH_2Cl_2, initially more polymer was deposited than expected based on the deposition of one monomer unit per electron. It was suggested that during the initial stages partially oxidized polymer was deposited (i.e. large Faradaic efficiencies), possibly with solvent or electrolyte trapped in the polymer. However, the results indicated that the total frequency change was smaller than that expected for the amount of charge passed during the deposition. This apparent loss of mass sensitivity suggests, as with some of the examples above, that the deposited film is not ideally rigid.

3.6 Electrochemically Active Polymer Films

Without a doubt, the EQCM has found its most extensive use in the investigation of electrochemically active polymer films, including redox and conducting polymers. Indeed, to a large extent it was the first report of an EQCM study of polypyrrole films by Kaufman et al. that heightened interest in the EQCM method.[61] The benefits of the EQCM were immediately obvious, as it became clear that the identity of the counterions exchanging between the films and electrolyte could be determined. In addition, the amount of solvent accompanying ion exchange could be determined by calculating the mass in excess of that expected from the electrochemical charge. Since these initial studies, numerous polymer films have been investigated with the EQCM and many mechanistic and thermodynamic insights into their behavior have been realized. The practitioner should be cautious when using the EQCM with polymer films, as changes in the mechanical properties of the films are not uncommon when ion population and solvent swelling are involved. The corresponding changes in viscoelasticity can result in dramatic frequency changes which are unrelated to mass changes. In many cases, this warrants experimental verification that rigid-layer behavior is present over the thickness and composition ranges examined. This can be accomplished by performing experiments with different polymer thicknesses over a reasonable range and by using impedance analysis to elucidate the contribution from changes in mechanical properties. If these factors are taken into account, the EQCM can be an especially valuable tool in examining electrochemically active polymer films.

It is reasonable to claim that PVF is ubiquitous in the subdiscipline of polymer-modified electrodes. Similarly, it has been among the polymer films most extensively examined with the EQCM. One of the initial investigations involved measurement of the frequency changes associated with ion transport required for electroneutrality and the amount of solvent accompanying ion transport in PVF films.[62] Based on a frequency decrease that accompanied oxidation of the PVF film, it was determined that ClO_4^- and PF_6^- were inserted without accompanying solvent, with one equivalent of anions inserted for each equivalent of electrons (Figure 12). The process was reversible, as evidenced by the frequency increase upon electrochemical reduction. Electrochemical oxidation of PVF films in the presence of other counterions, however, occurred with varying amounts of solvent incorporation, with the amount of incorporated water per ion decreasing in the order $Cl^- > IO_3^- \gg BrO_3^-(5) > ClO_3^-(1) \approx NO_3^-(1) > CH_3{-}C_6H_4{-}SO_3^-(0.5)$.[63] Indeed, cross-linking of the PVF film was necessary to prevent its dissolution upon oxidation. Unfortunately, determination of the amount of incorporated solvent is not feasible in the case of Cl^- or IO_3^- because of the dramatic changes in viscoelasticity that accompany the large degree of swelling when these ions are present. On the other hand, solvent incorporation for the other ions could be measured reliably (the amount of solvent incorporated per counterion are indicated above within the parentheses).

The EQCM has been employed to investigate transport in redox polymer films such as PVF and poly(thionine) of mobile neutral species such as water, particularly with regard to thermodynamic changes in solvent activity in the polymer film and transport kinetics.[64–66] These studies indicate that the number of solvent molecules per counterion need not be integral, and that it is feasible that neutral ion pairs are also involved in the transport. Comparison of EQCM frequency changes at different scan rates also revealed that mass changes

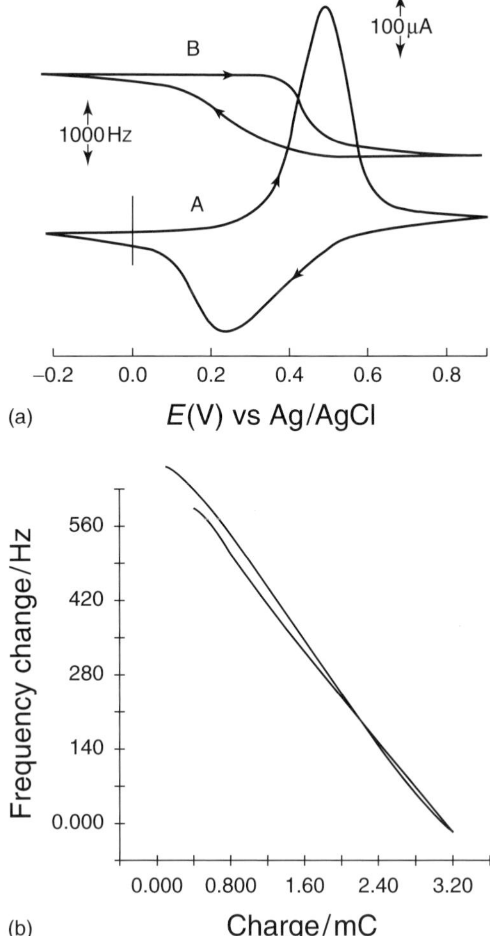

Figure 12 (a) (A) cyclic voltammogram of PVF on a gold EQCM electrode in 0.1-M KPF_6. Scan rate = $10\,mV\,s^{-1}$. (B) EQCM frequency response obtained simultaneously with (A). (b) Plot of frequency versus charge for a scan from 0.0 to 0.60 V and back for a PVF film in 0.1-M $NaClO_4$ + 0.1-M $HClO_4$. Scan rate = $25\,mV\,s^{-1}$. The linearity suggests ideal behavior and the slope gives the molar mass of the transport species. (Reproduced with permission from Varineau and Buttry,[62] Copyright (1987) American Chemical Society.)

and electrochemical charge changes do not always occur simultaneously.[67,68] That is, counterion motion to maintain electroneutrality during redox changes must always be established, but global equilibrium may lag behind. In particular, these studies revealed that electroneutrality would be achieved initially during redox by using counterions already present in the film, followed by transport of ions from the solution. Similar behavior was observed for redox-active poly(nitrostyrene) films.[69] The behavior has been attributed to potential gradients within the film that affect ion transport, and to the low dielectric constant of the films.[70–72] In the case of poly(thionine) experiments, comparison of current transients following a series of potential steps, suggested that transport of different species occurs on different timescales (Figure 13).[73] It was surmised that proton transport necessary to achieve electroneutrality was rapid, but that global equilibrium occurred more slowly by transport of the neutral ion pair $H_3O^+ClO_4^-$ and H_2O. These conclusions were supported by experiments performed in D_2O, which exhibited trends consistent with the difference in mass of H and D. These results have important consequences in the design and synthesis of polymer films in applications where charging and discharging rates are critical, for example in sensors or energy storage.

EQCM studies revealed that ion transport into PVF films upon oxidation was slower than their transport

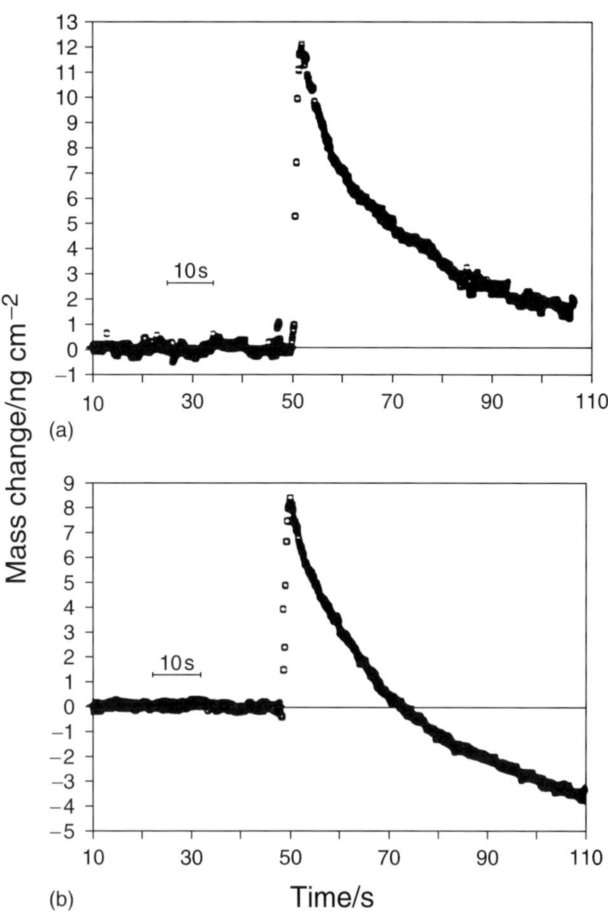

Figure 13 Transient mass changes following the application of a potential step from −0.1 V to +0.5 V (vs SCE) to a poly(thionine) film in $HClO_4$ solution. (a) Solvent in H_2O, pH = 1.6; (b) Solvent is D_2O, pD = 2.1. The potential step converts a fully reduced film to a fully oxidized film at long times. The pH and pD values were adjusted to achieve near zero net mass change at long times after switching. Mass increases upwards. The same polymer film was used in H_2O and D_2O. Note that the initial mass change is larger than that after global equilibrium is achieved at longer times. (From Bruckenstein et al.[73])

For references see page 9954

out of the film during reduction.[74] This is presumably because of the change in density accompanying the swelling of the oxidized film. It was estimated that in the first oxidation cycle of electrochemically deposited PVF films, water uptake began only after the film was approximately 40% oxidized and that 50% of the water was retained after subsequent reduction.[75] Water transport in following cycles was reversible, apparently with some water always retained in the polymer film. Transport in PVF films can also be severely affected by the nature of the counterion. For example, EQCM studies indicated that ion exchange of ferro- or ferricyanide into PVF during electrochemical cycling resulted in irreversible incorporation of $Fe(CN)_6^{3-/4-}$ in the film.[76] This process led to a slow decrease in the electroactivity of the film, presumably due to an electrostatic "crosslinking" by the multiply charged ion that inhibited further transport of $Fe(CN)_6^{3-/4-}$. It should be noted that variability in these results may be expected in different laboratories because of differences in film preparation.

The EQCM can also be used to examine the kinetics of chemical reactions of solution species with redox polymer films. The chemical oxidation of a PVF film by KI_3 was monitored by measurement of the frequency decrease associated with insertion of the I_3^- counterion following oxidation (Figure 14). These experiments were possible because the PVF film could be held in its reduced form prior to the chemical reaction in the presence of the oxidant by holding the potential of the electrode negative of E^0 (PVF/PVF^+). This established the initial conditions for this measurement. The results yielded the stoichiometry of the reaction (1 equivalent I_3^- per PVF^+ formed) as well as the pseudo-first-order rate constant for the reaction.

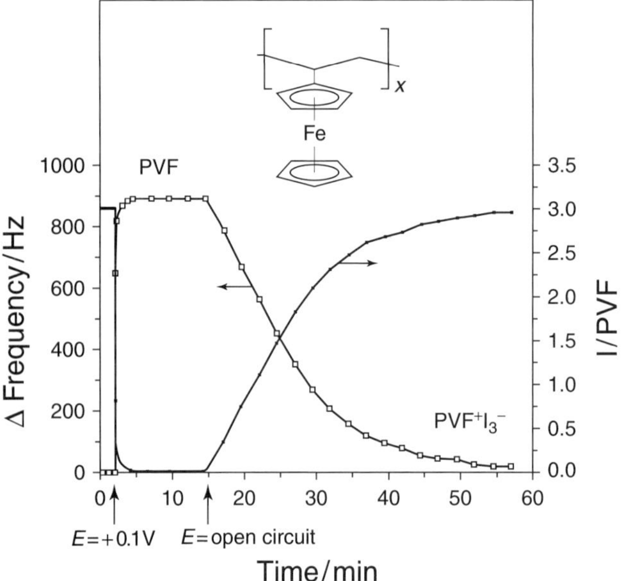

Figure 14 Frequency response of a PVF film in the presence of 5-mM KI_3 and 1.0-M KNO_3. The right hand ordinate refers to the number of equivalents of I incorporated into the film normalized to the amount of PVF coverage on the piezoelectrically active area (0.28 cm^2). $[\Gamma_{PVF}]_0 = 4.1 \times 10^{-8}$ equiv. cm^{-2}. (Reproduced with permission from Ward,[17] Copyright (1988) American Chemical Society.)

3.7 Electrochemical Quartz Crystal Microbalance Investigations of Conducting Polymers

Interest in the EQCM method heightened considerably after it was reported in the investigation of polypyrrole films.[61] Since that time, numerous reports have appeared describing attempts to deconvolute the typically complicated behavior involved in the preparation of conducting polymer films and their subsequent doping behavior. The value of the EQCM was immediately evident in those initial polypyrrole studies in which it was determined that reduction of electrochemically (by oxidation of pyrrole in $LiClO_4$) prepared polypyrrole films resulted in Li^+ insertion to maintain electroneutrality rather than expulsion of ClO_4^-. This behavior was attributed to strong ionic pairing between cationic sites on the polymer and ClO_4^-. It was also discovered that subsequent cycling of these films in the presence of n-$Bu_4N^+$$p$-$CH_3C_6H_4{-}SO_3^-$ resulted in frequency changes consistent with anion insertion during the oxidative doping step, with its expulsion during reduction (Figure 15). This behavior clearly revealed that the nature of ion transport in conducting polymers was strongly dependent upon the nature of the counterions. The impact of this behavior on the design of energy storage systems with regard to weight, power density, and energy density prompted several other EQCM investigations of conducting polymers, including poly(pyrrole), poly(aniline), and poly(thiophene). Decreasing the mass of the transport species would lead to higher energy densities, whereas higher power densities would be realized by faster transport rates.

Poly(pyrrole) film growth was also examined with the EQCM. It was concluded that film deposition involved the initial formation of soluble oligomers, which upon further polymerization precipitated on the electrode.[77,78] EQCM frequency changes that corresponded to the mass of deposited polypyrrole revealed a second-order dependence of the electropolymerization rate on pyrrole concentration, suggesting the bimolecular coupling of oxidatively formed radical cations was involved in the rate determining step. Subsequent EQCM studies of poly(pyrrole) revealed more details of the ion transport and its dependence upon film preparation. Electropolymerization of pyrrole in the presence of large polymeric anions such as poly(4-styrenesulfonate)[79–82] and poly(vinylsulfonate) resulted in films whose transport

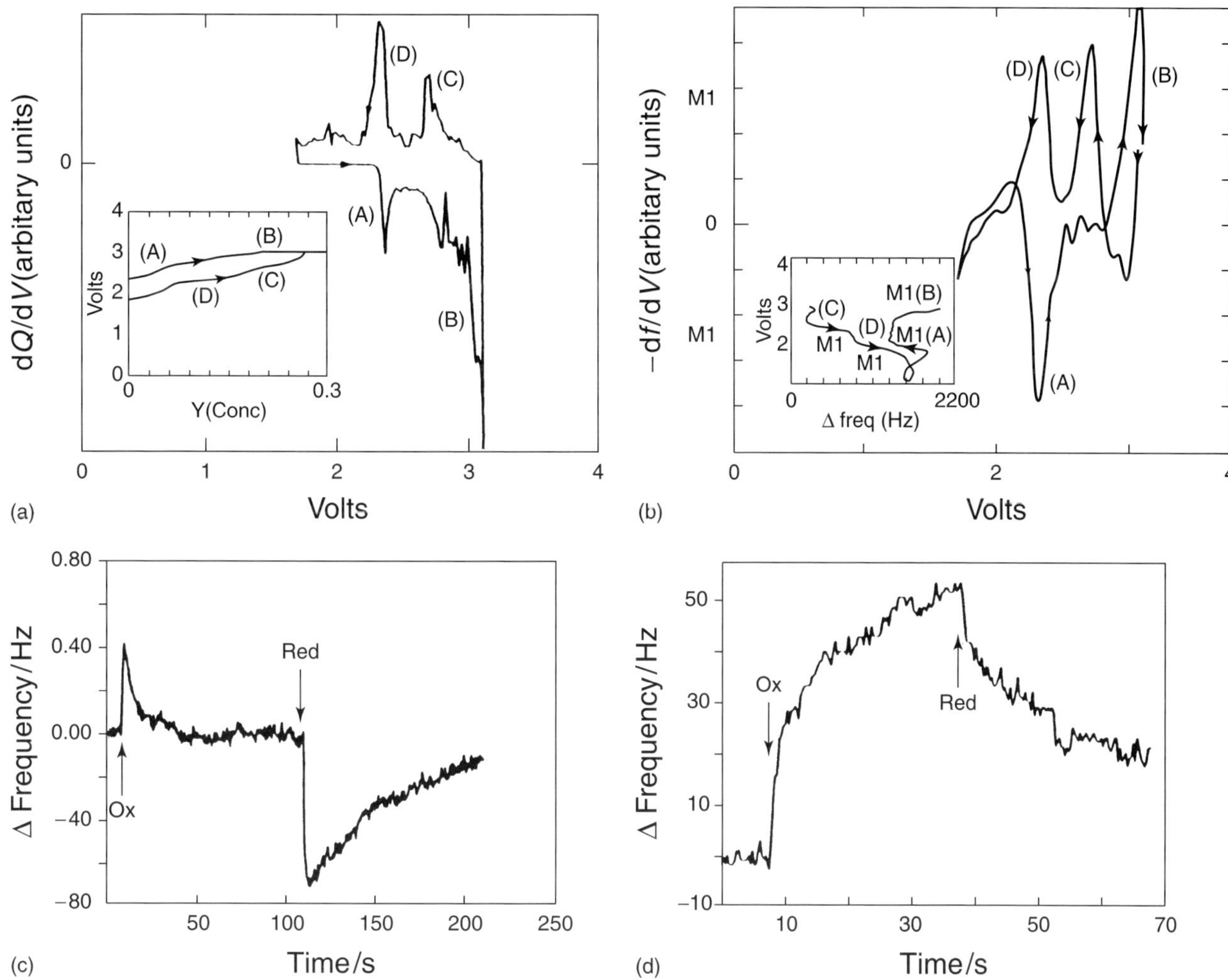

Figure 15 (a) Cyclic voltammetry data for a polypyrrole film immobilized on an EQCM in 1.0-M $LiClO_4$/THF, presented as dQ/dV versus V. Oxidation occurs at peaks A and B while reduction occurs at C and D. (inset shows V vs Q) (b) EQCM data presented as df/dV versus V showing differential mass changes associated with the oxidation and reduction features in (a). The up arrow (↑) indicates increasing mass and the down arrow indicates decreasing mass. (Reproduced with permission from Varineau and Buttry,[62] Copyright (1987) American Chemical Society.) (c) EQCM response upon oxidation of a polypyrrole film followed by reduction in tetraethylammonium tosylate electrolyte. The polypyrrole film was prepared by electrochemical oxidation of pyrrole. (d) EQCM response upon oxidation of a polypyrrole-co-[3-(pyrrol-1-yl)propanesulfonate] film followed by reduction in tetraethylammonium tosylate electrolyte. In both cases, the increase in frequency upon oxidation and decrease in frequency upon reduction are associated with transport of cations. (Reproduced with permission from Kaufman et al.[61] *Phys. Rev. Lett.*, Copyright (1984) by the American Physical Society.)

properties differed significantly from films prepared in the presence of more conventional anions. In these studies conventionally prepared films exhibited frequency changes during doping–undoping cycles that indicated anion transport for ClO_4^-, BF_4^-, or PF_6^-. However, in electrolytes containing poly(4-styrenesulfonate) mixed transport of both poly(4-styrenesulfonate) anions and cations was observed in conventionally prepared films. If the films prepared in poly(4-styrenesulfonate) were used, then the transport during doping–undoping was completely dominated by the cation of the supporting electrolyte. Cation transport also dominated in the doping–undoping cycles of self-doped conducting copolymers of pyrrole and 3-(pyrrol-1-yl)propanesulfonate.

Poly(thiophene)-based polymers, conducting polymers closely related to poly(pyrrole) have also been examined with the EQCM.[83–85] These studies revealed the doping levels of poly(3-methylthiophene), as well as a nonlinear dependence of the charge on film thickness, the latter being inferred from the frequency change. Subsequent studies provided a thermodynamic model based on the EQCM response for transport of the ions, neutral ion pairs, and solvent based on the thermodynamic activity of the polymer film, similar to the model mentioned

For references see page 9954

above. The transport of ions during doping–undoping cycles of poly(aniline), a conducting polymer film with doping behavior that is quite different from that of the aforementioned examples, has also been investigated with the EQCM.[86] These studies provided the Faradaic efficiencies during oxidative film growth by comparison of the coulometric charge and the frequency decrease. Interestingly, it was discovered that the Faradaic efficiencies for films prepared by cycling the potential were higher (ca. 40%) than those prepared at constant potential (ca. 10%). More importantly, mass changes during doping–undoping cycles revealed the extent of protonation of the polymer in both the doped and undoped states. Based on the EQCM results, it was concluded that each aniline ring contributed one electron to the conduction band of the polymer, in agreement with a proposed model.[87] Subsequent EQCM studies of poly(aniline) in nonaqueous solvents revealed that protons on the imine nitrogens of the polymer chains were retained during doping–undoping cycles, and that electroneutrality was maintained by transport of anions from the supporting electrolyte.[88]

3.8 Electrochemical Quartz Crystal Microbalance Investigations of Biologically Relevant Films

The EQCM has been used increasingly to examine the formation and properties of biologically relevant redox active films. One of the first such studies examined the electrochemically induced adsorption of the proteins human IgG and anti-IgG on Ag electrodes.[89] Upon oxidation of the Ag electrodes, the frequency of the EQCM decreased owing to formation of an insoluble layer of the proteins. These effects suggested that initial protein adsorption was related to interactions between oxidized Ag^+ ions on the electrode surface and functional groups on the protein.

Recent efforts have further demonstrated the utility of the EQCM in the characterization of electrode processes involving biologically relevant species. The adsorption of organosulfur reagents, followed by alternate electrostatic adsorption of ferrocene-modified poly(allylamine) and anionic glucose oxidase was monitored with the QCM in order to quantify the amount of redox-active material deposited on the gold electrode after each step.[90] Impedance analysis was employed to characterize viscoelastic contributions from the poly(allylamine) polymer solution in contact with the QCM. Viscoelastic contributions were also observed during the deposition of the glucose oxidase layer, although at long times these effects vanished so that the Sauerbrey approximation could be used reliably. The ability to measure the amount of material deposited in each step was important for elucidation of the electrochemical properties of these films. In a similar manner, the EQCM was employed to quantify the amount of electroactive and nonelectroactive protein in multilayer films prepared by alternate layer-by-layer deposition of cationic myoglobin or cationic cytochrome $P450_{cam}$ with either poly(styrenesulfonate) or poly(dimethyl diallyl) ammonium chloride.[91] The EQCM data reflected a linear increase of film mass with increasing number of layers.

The adsorption of flavin adenine dinucleotide (FAD) on the gold electrode of the EQCM revealed deposition of FAD upon reduction and its reversible desorption upon oxidation.[92] The voltammetric adsorption and desorption of cysteine at the gold electrode of an EQCM has been examined. Cysteine was adsorbed following cathodic reduction of the surface gold oxide, but the EQCM data indicated that the apparent surface mass upon reduction was less when cysteine was present than in solutions without cysteine. This was attributed to a decrease in surface hydration accompanying the loss of the oxide and adsorption of cysteine.[93]

The adsorption of green fibrous polymeric products formed by reaction of electrochemically generated dications of the carotenoid (7*E*,7′*Z*)-diphenyl-7,7′-diapocarotene was monitored with the EQCM. Coulometric measurements performed during cyclic voltammetry, combined with the EQCM measurement of the amount of polymer adsorbed, enabled determination of the average molar mass per electron ($5400\,g\,mol^{-1}$ electrons).[94]

4 IMPEDANCE ANALYSIS METHODS

Impedance analysis is being used increasingly to evaluate the contributions from viscous and viscoelastic effects associated with films on the quartz resonator. There is not sufficient space in this article to give a detailed explanation of this approach, but it can be described briefly as a technique in which a voltage within a specified range of frequencies is broadcast across the crystal and the current measured. The impedance or admittance is measured, and the Butterworth–van Dyke (BVD) equivalent circuit parameters determined by numerical fitting of the data. The mechanical properties corresponding to the electrical parameters can then be assessed. Commercially available impedance analyzers allow a complete set of parameters to be measured within 1 min, enabling dynamic measurements, albeit on a rather slow timescale.

Quartz crystals are electromechanical devices, and therefore their mechanical vibrations can be described in terms of electrical equivalents.[95] This also serves to enhance understanding of the EQCM, particularly the conditions under which the Sauerbrey equation, Equation (2), is valid. The quartz resonator can be described according to a mechanical model with elements of mass, compliance (the ability of an object to yield elastically under an applied force), and friction. The electrical equivalent of this system is an electrical circuit that has an inductor, a capacitor, and a resistor connected in series (Figure 16). In this equivalent circuit, the inductor, L_1, represents the mass displaced during oscillation, C_1 the energy stored during oscillation (the compliance is the inverse of the elastic, or Hooke's constant), and R_1 the energy dissipation due to losses that are tantamount to internal friction. In order to describe the quartz crystal behavior accurately, a parallel capacitance must also be included that represents the static capacitance of the quartz plate with its electrodes and any stray parasitic capacitances. The complete circuit is commonly referred to as the BVD circuit.[96,97] The series branch of the circuit is referred to as the motional branch since it reflects the vibrational behavior of the crystal.

The relationship between this circuit and the quartz crystal is especially useful because the *LCR* branch is identical to a "tank circuit", in which oscillations can be sustained by cycling of current between the capacitor and the inductor. When the capacitor in this circuit discharges through the inductor, a magnetic field is established around the inductor as it opposes the current. When the capacitor discharge is complete and the current falls to zero, the electromotive force in the inductor creates a current in the direction opposite to the original current and the capacitor recharges. Repetition of this cycle results in electrical oscillation, with the oscillations dampened by an amount proportional to R. In the case of quartz crystals, the R values are rather small and sustained oscillations are favored.

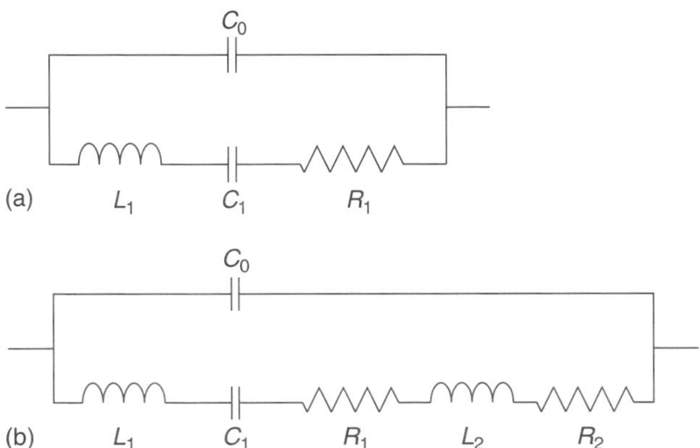

Figure 16 (a) BVD equivalent electrical circuit used to describe the mechanical properties of a quartz resonator. The components L_1, C_1 and R_1 in the motional branch of the circuit represent the inertial mass, compliance, and energy dissipation in the crystal, and C_0 represents the static capacitance of the quartz crystal. (b) Equivalent electrical circuit used to describe the mechanical properties of a quartz resonator immersed in a liquid. The inductance L_2 and resistance R_2 represent the mass and viscosity components of the liquid.

As a result of the electromechanical relationship between quartz crystals and electrical circuits, the equations of harmonic motion of the quartz crystals are closely related in form to the expressions describing the properties of the *LCR* tank circuit. This has been reviewed elsewhere[3] and a detailed description is not given here. The equivalent electrical parameters in terms of crystal properties are given in Equations (12–15), along with typical experimental values for these parameters. In these relationships, D_Q is the dielectric constant of quartz, ε_0 the permittivity of free space, r a dissipation coefficient corresponding to the energy losses during oscillation, ε the piezoelectric stress constant and c the elastic constant. Note that while L_1, C_1, and R_1 depend upon ε, C_0 does not participate directly in piezoelectricity. It should also be noted that L_1 depends upon the density; in fact, the quantity $t_Q^3\rho/A$ is equivalent to the mass per unit area in the Sauerbrey equation. These equivalent representations provide a quantitative approach to examining the properties of the EQCM, the role of the liquid environment and thin films on the resonant frequency response, and the design of quartz resonators.

$$C_0 = \frac{D_Q \varepsilon_0 A}{t_Q} \approx 10^{-12}\,\mathrm{F} \tag{12}$$

$$C_1 = \frac{8A\varepsilon^2}{\pi^2 t_Q c} \approx 10^{-14}\,\mathrm{F} \tag{13}$$

$$R_1 = \frac{t_Q^3 r}{8A\varepsilon^2} \approx 100\,\Omega \tag{14}$$

$$L_1 = \frac{t_Q^3 \rho}{8A\varepsilon^2} \approx 0.075\,\mathrm{H} \tag{15}$$

The use of impedance analysis, or its inverse admittance analysis, for characterization of crystal properties and performance can be demonstrated by the rather simple comparison of the admittance plots for 5-MHz and 30-MHz quartz crystals mentioned above (Figure 17). The smaller thickness of the 30-MHz quartz crystal leads to an increase in C_0 and the expected shift of the admittance locus upwards on the admittance plot (by an amount equal to wC_0). This increases the likelihood that the admittance locus will not cross the real axis. Under this condition, operation of the EQCM at its resonant frequency using a feedback-mode oscillator, which relies on maintaining a frequency at zero phase angle, can be difficult. The smaller area of the electrode overlap for the 30-MHz resonators somewhat diminishes this effect. More important, the 30-MHz resonators have smaller values of R_1 and L_1,

For references see page 9954

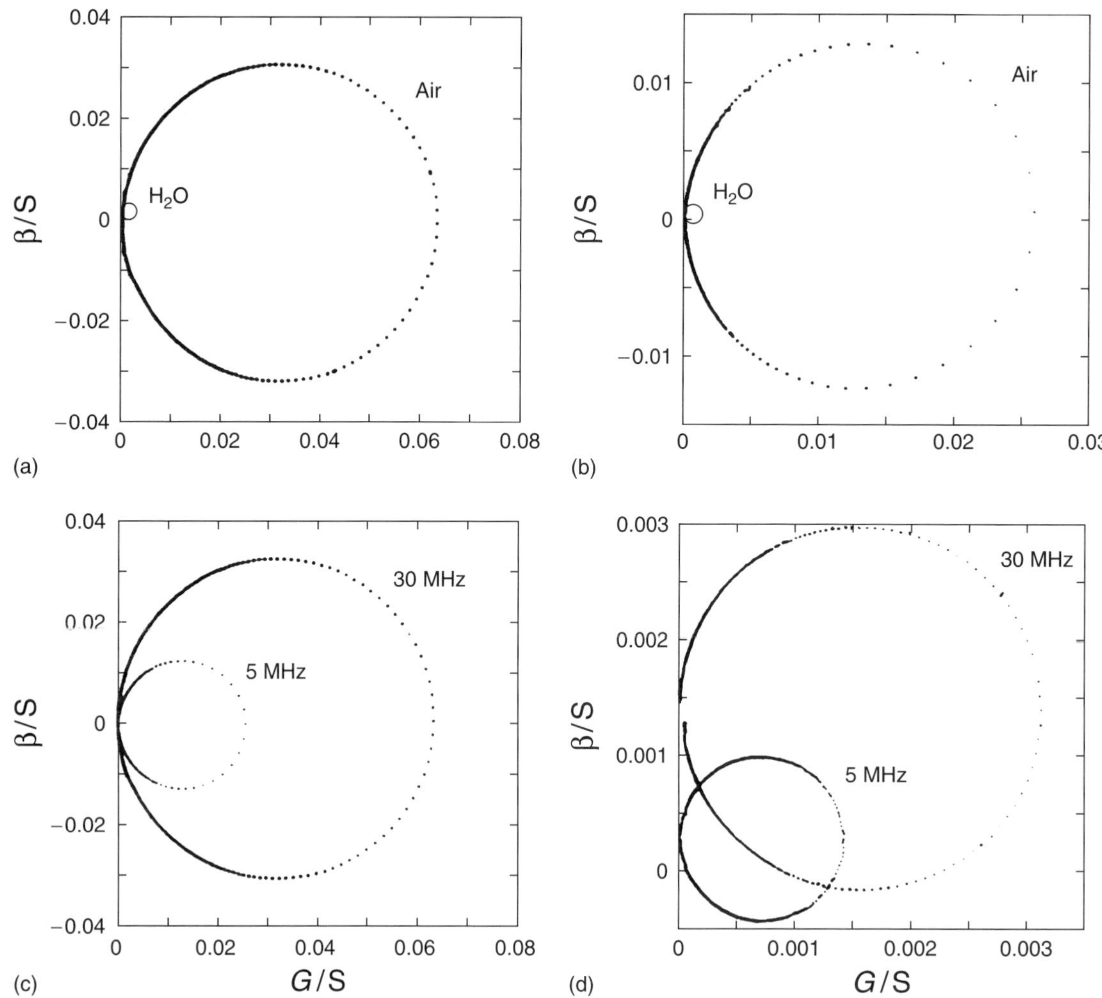

Figure 17 Admittance analysis data for 5-MHz and 30-MHz quartz resonators in aqueous media. The increased capacitance of the thinner 30-MHz resonator lifts the admittance circle upward along the imaginary susceptance (jB) axis, but the decreased resistance associated with the decreased crystal thickness increases the diameter of the admittance circle. The latter effect is critical as it compensates for the capacitance effect such that the admittance circle crosses the real axis, which is required for establishing resonance. The admittance data was collected with a Hewlett Packard Model 4194A Impedance Analyzer. (Reproduced with permission from Lin et al.,[6] Copyright (1993) American Chemical Society.)

owing to the smaller thickness of these crystals compared to their lower frequency counterparts. The decrease in R_1 is evident from the 30-MHz crystal admittance locus, which is significantly larger than the 5-MHz locus. The admittance locus crosses the real axis for both resonators, but the conductance value of the series resonant frequency f_s, where the locus crosses the real axis, is larger for the 30-MHz resonator. Because EQCM experiments using a feedback-mode oscillator operate at f_s, the quartz crystals can be operated with lower loss in liquid media or in air than the more commonly used 5-MHz crystals.

While the BVD circuit accurately describes the operation of the QCM in vacuum or the gas phase, it is not a sufficient description when the QCM or EQCM is used in liquids where the liquid density and viscosity alter the resonator characteristics. The density of the liquid effectively adds to the mass of the resonator, while its viscosity provides additional energy damping. The effect of a Newtonian liquid can be described by adding an additional inductance and resistance L_2 and R_2 in series with the motional branch of the BVD circuit, in which L_2 and R_2 are related to the extra mass of the liquid and its viscosity, respectively (Figure 17).[98]

Our laboratory devised a method based on an electromechanical model that relates the electrical parameters to the mechanical characteristics of a composite resonator consisting of a AT-cut thickness shear mode quartz crystal and a viscoelastic polymer film. The theoretical component of this method is based on a previously reported relationship Equation (16),[99] where μ_F is the film shear modulus (or equivalently, the storage modulus G'), $\bar{\bar{\mu}}_F$ is the complex film shear modulus, η_F is the film viscosity (or equivalently, the loss modulus-frequency quotient G''/w), ρ_F is the film density, l_F is the film thickness, ρ_Q is the quartz density, w is the angular frequency ($w = 2\pi f$), l_Q is the quartz thickness, A_{ACT} is the piezoelectric active area, ε_{22} is the quartz dielectric constant, e_{26} is the quartz piezoelectric constant, c_{66} is the quartz shear modulus, $\bar{\bar{c}}_{66}$ is the complex quartz shear modulus, η_Q is the quartz viscosity, and Y is the complex resonant admittance.

$$Y = \frac{\dfrac{iwA_{ACT}\varepsilon_{22}}{l_Q}\left[k_Q\bar{\bar{c}}_{66}\sin(k_Q l_Q) + k_F\bar{\bar{\mu}}_F\tan(k_F l_F)\cos(k_Q l_Q)\right]}{k_Q\sin(k_Q l_Q) + k_F\bar{\bar{\mu}}_F\tan(k_F l_F)\cos(k_Q l_Q) - \dfrac{2e_{26}^2}{l_Q\varepsilon_{22}} \times \left[1 - \cos(k_Q l_Q) + \dfrac{k_F\bar{\bar{\mu}}_F}{2k_Q\bar{\bar{c}}_{66}}\tan(k_F l_F)\sin(k_Q l_Q)\right]} \tag{16}$$

where

$$k_Q = w\sqrt{\frac{\rho_Q}{\bar{\bar{c}}_{66}}} = w\sqrt{\frac{\rho_Q}{c_{66} + \dfrac{e_{26}^2}{\varepsilon_{22}} + iw\eta_Q}}$$

and

$$k_F = w\sqrt{\frac{\rho_F}{\bar{\bar{\mu}}_F}} = w\sqrt{\frac{\rho_F}{\mu_F + iw\eta_F}}$$

This electromechanical equation is derived from linear equations of motion relating the electrical admittance and the mechanical properties of a composite resonator. An algorithm based on this relationship was limited to measurement of the mechanical properties of polymer films containing various amounts of solvent that affected the viscoelastic properties. These measurements require somewhat cumbersome and time-consuming numerical analysis during measurement that inhibits the use of complete impedance analysis techniques in EQCM applications.[100–102] Even direct impedance analysis requires long analysis times because the measurement involves sweeping a range of frequencies to evaluate the resonator conductance and numerically fit the data to obtain the BVD parameters. Consequently, experimental strategies based on impedance analysis have not been used extensively in EQCM applications. However, changes in R can be rapidly inferred from changes in the amplitude of the voltage output of the oscillator during operation with a conventional feedback oscillator.

5 EMERGING METHODS

Several modifications of the quartz microbalance have appeared in which the oscillation is induced by electrodes that are near, but not in contact with the quartz surface.[103–107] These resonators, which can be referred to as electrode-separated quartz crystal resonators, allow examination of interfacial processes occurring directly on the quartz surface or on other films deposited on the quartz. The capabilities of these methods have been extended to a scanning configuration in which the position of a small conducting probe, serving as one of the excitation electrodes, is controlled by computer-activated positioning of the sample stage.[108] In principle, this procedure enables qualitative mapping of the mass distribution and viscoelastic properties of films on the quartz resonator. Two configurations have actually been demonstrated. In an "overscanning" mode the conducting probe is scanned over the upper surface, onto which a film of interest has been deposited, and the bottom surface has the conventional metal electrode covering a large region of the crystal. An alternative "underscanning" mode relies on the conducting probe scanning the lower surface of the quartz crystal while the upper surface is coated conventionally with the large area excitation electrode. Both configurations allow introduction of an electrochemical working electrode. In the overscanning mode the working electrode is independent of the excitation electrodes, whereas in the underscanning mode the excitation and working electrode are the same.

The format of the EQCM is amenable to combination with other in situ techniques. This was illustrated for the SEM/EQCM experiments described earlier in this article. This capability is further demonstrated by ellipsometric studies of the nucleation and growth of poly(aniline) films performed in conjunction with the EQCM,[109,110] which allowed complete characterization of the polymer films (Figure 18). Whereas the EQCM provided the total mass of the deposit, ellipsometry provided the thickness. Accordingly, the combination of these techniques allows determination of the density, which is not realized by either technique alone. The major conclusions of these studies were that the kinetics for growth of the poly(aniline) films depended upon the electrochemical conditions, and that self-assembled monolayers of aniline derivatives on the electrode surface promoted nucleation and films with higher densities.

6 SUMMARY

The increasing use of the EQCM and the examples described in this article demonstrate the power of this

For references see page 9954

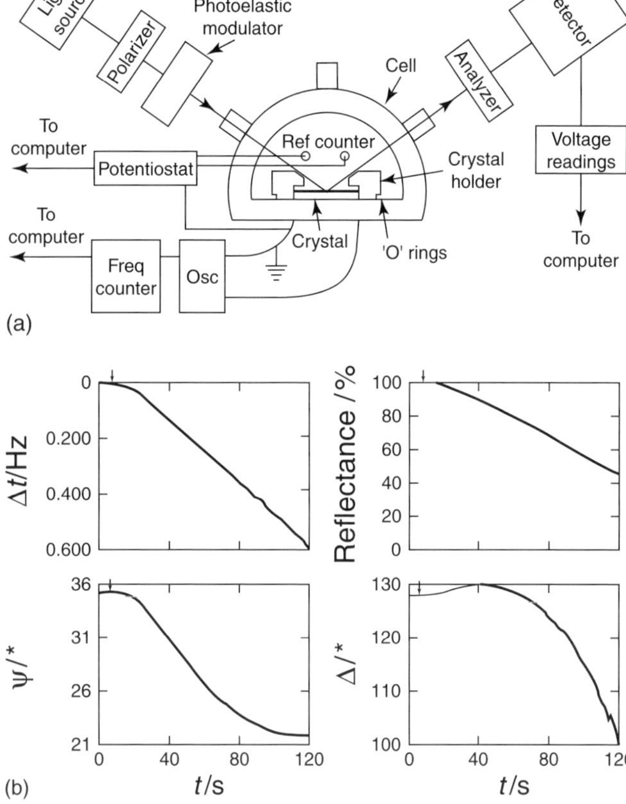

Figure 18 (a) Schematic representation of the experimental apparatus for simultaneous ellipsometry–EQCM measurements. (b) Optical and EQCM data for poly(aniline) growth on a Pt electrode in 1-M aniline containing 2-M HCl at 0.7 V (vs SCE). Wavelength = 550 nm, angle of incidence = 65°. Δ and ψ are the ellipsometric parameters derived from the 50 kHz, 100 kHz, and DC (direct current) components of the signal generated at the detector by the reflected beam. (Reproduced from Rishpon et al.[109] Copyright (1990) with permission from Elsevier Science.)

method in elucidating the fundamental interfacial processes occurring at electrode surfaces. The rather simple concept and low cost of the equipment necessary to perform EQCM experiments should encourage physical, analytical, and electrochemists to use this technique routinely in the laboratory. Although numerous laboratories use the EQCM regularly, the contributions of interfacial slip, stress, surface roughness, viscosity, viscoelasticity, and influence of the liquid phase to propagation of the acoustic energy are difficult to quantify. The new practitioner (and even experienced ones) should expend some effort to grasp the basic principles of EQCM operation in order to avoid pitfalls and misinterpretations that can occur if these effects are ignored. Generally, verification of ideal behavior is possible through fairly simple experiments, as discussed in this article, or through more sophisticated impedance analysis techniques. It is anticipated that as the EQCM becomes more widely appreciated and used, many of these effects will be better understood. Nevertheless, the EQCM is well within the reach of most electrochemical laboratories and properly used can provide details of electrode processes that were previously unattainable.

ABBREVIATIONS AND ACRONYMS

BVD	Butterworth–van Dyke
DC	Direct Current
EQCM	Electrochemical Quartz Crystal Microbalance
FAD	Flavin Adenine Dinucleotide
PVF	Poly(vinylferrocene)
QCM	Quartz Crystal Microbalance
RF	Radiofrequency
SCE	Saturated Calomel Electrode
SEM	Scanning Electron Microscopy
UPD	Underpotential Deposition

RELATED ARTICLES

Coatings **(Volume 2)**
Rheology in Coatings, Principles and Methods

Electroanalytical Methods **(Volume 11)**
Electroanalytical Methods: Introduction • Surface Analysis for Electrochemistry: Ultrahigh Vacuum Techniques

REFERENCES

1. M.D. Ward, D.A. Buttry, 'In Situ Interfacial Mass Detection with Piezoelectric Transducers', *Science*, **249**, 1000 (1990).
2. D.A. Buttry, M.D. Ward, 'Measurement of Interfacial Processes at Electrode Surfaces with the Electrochemical Quartz Crystal Microbalance, *Chem. Rev.*, **92**, 1355 (1992).
3. D.A. Buttry, 'Applications of the Quartz Crystal Microbalance to Electrochemistry', *Electroanalytical Chemistry*, ed. A.J. Bard, Marcel Dekker, New York, 2, Vol. 17, 1991.
4. P. Curie, J. Curie, 'Cristallophysique-Development de L'electricite Polaire dans les Cristaux Hemiedes a Faces Inclines', *C.R. Acad. Sci.*, **91**, 294 (1880).
5. C.S. Lu, O. Lewis, 'Investigation of Film-thickness Determination by Oscillating Quartz Resonators with Large Mass Load', *J. Appl. Phys.*, **43**, 4385 (1972).

6. Z. Lin, C. Yip, I.S. Joseph, M.D. Ward, 'Operation of an Ultrasensitive 30 MHz Quartz Crystal Microbalance in Liquids', *Anal. Chem.*, **65**, 1646 (1993).
7. K.K. Kanazawa, J.G. Gordon III, 'Frequency of a Quartz Microbalance in Contact with Liquid Sensors to the Liquid Phase by Interfacial Viscosity', *Anal. Chem.*, **57**, 1770 (1985).
8. L.V. Rajakovic, B.A. Cavic-Vlasak, V. Ghaemmaghami, K.M.R. Kallury, A.L. Kipling, M. Thompson, 'Mediation of Acoustic Energy Transmission from Acoustic Wave', *Anal. Chem.*, **63**, 615 (1991).
9. A.L. Kipling, M. Thompson, 'Network Analysis Method Applied to Liquid Phase Acoustic Sensors', *Anal. Chem.*, **62**, 1514 (1990).
10. J. Wang, L.M. Frostman, M.D. Ward, 'Self-assembled Thiol Monolayers with Carboxylic Acid Functionality, Measuring pH-dependent Phase Transitions with the Quartz Crystal Microbalance', *J. Phys. Chem.*, 96 (1992).
11. R. Schumacher, G. Borges, K.K. Kanazawa, *Surf. Sci.*, **163**, L621 (1985).
12. R. Schumacher, J. Gordon, O. Melroy, 'Observation of Morphological Relaxation of Copper and Silver Electrodes in Solution Using a Quartz Crystal Microbalance', *J. Electroanal. Chem.*, **216**, 127 (1987).
13. A. Muller, M. Wicker, R. Schumacher, R.N. Schindler, *Ber. Bunsenges. Phys. Chem.*, **92**, 1395 (1988).
14. T. Nomura, M. Iijima, 'Electrolytic Determination of Nanomolar Concentrations of Silver in Solution with a Piezoelectric Quartz Crystal', *Anal. Chim. Acta*, **131**, 97 (1981).
15. S. Bruckenstein, M. Shay, 'Experimental Aspects of Use of the Quartz Crystal Microbalance in Solution', *Electrochim. Acta*, **30**, 1295 (1985).
16. R. Schumacher, 'The Quartz Microbalance, a Novel Approach to the In-situ Investigation of Interfacial Phenomena at the Solid/Liquid Junction', *Angew. Chem. Int. Ed. Engl.*, **29**, 329 (1990).
17. M.D. Ward, 'Investigation of Open Circuit Reactions of Polymer Films Using the Quartz Crystal Microbalance. Reactions of Poly(vinylferrocene) Films', *J. Phys. Chem.*, **92**, 2049 (1988).
18. R. Bechmann, 'Single Response Thickness-shear Mode Resonators Using Circular Bevelled Plates', *J. Sci. Instrum.*, **29**, 73 (1952).
19. R. Bechmann, 'Quartz AT-type Filter Crystals for Frequency Range 0.7 to 60 MC', *Proc. Inst. Radio Engineers*, **49**, 523 (1961).
20. H.F. Tiersten, 'Analysis of Trapped Energy Resonators Operating in Overtones of Thickness-shear', *Proc. Annu. Freq. Control Symp.*, **28**, 44 (1974).
21. K.E. Heusler, A. Grzegorzewski, L. Jackel, J. Pietrucha, 'Measurement of a Mass and Surface Stress at One Electrode of a Quartz Oscillator', *Ber. Bunsenges. Phys. Chem.*, **92**, 1395 (1988).
22. E.P. EerNisse, *Methods and Phenomena*, eds. C. Lu, A. Czanderna, Elsevier Sequoia, Lausanne, Vol. 4, 1984.
23. G. Sauerbrey, 'The Use of a Quartz Crystal Oscillator for Weighing Thin Layers and for Microweighing Applications', *Z. Physik*, **155**, 206 (1959).
24. D.M. Ullevig, J.F. Evans, M.G. Albrecht, 'Effects of Stressed Materials on the Radial Sensitivity Function of a Quartz Crystal Microbalance', *Anal. Chem.*, **54**, 2341 (1982).
25. L. Koga, Y. Tsuzuki, S.N. Witt, A.L. Bennet, 'Measurements of the Vibrations of Quartz Plates', *Proc. Annu. Freq. Control Symp.*, **14**, 53 (1960).
26. H. Fukuyo, A. Yokoyama, N. Ooura, S. Nonaka, 'Vibration of Biconvex Circular AT-Cut Plate', *Bull. Tokyo Inst. Technol.*, **72**, 1 (1965).
27. K.S. van Dyke, 'Strain Patterns in Thickness-shear Quartz Resonators', *Proc. Annu. Freq. Control Symp.*, **111**, 1 (1957).
28. A.C. Hillier, M.D. Ward, 'Scanning Electrochemical Mass Sensitivity Mapping of the Quartz Crystal Microbalance in Liquid Media', *Anal. Chem.*, **64**, 2539 (1992).
29. M.D. Ward, E.J. Delawski, 'Radial Sensitivity of the Quartz Crystal Microbalance in Liquid Media', *Anal. Chem.*, **63**, 886 (1991).
30. C. Gabrielli, M. Keddam, R. Torresi, 'Calibration of the Electrochemical Quartz Crystal Microbalance', *J. Electrochem. Soc.*, **138**, 2657 (1991).
31. B.A. Martin, H.E. Hager, 'Velocity Profile On Quartz Crystals Oscillating in Liquids', *J. Appl. Phys.*, **65**, 2630 (1989).
32. O. Melroy, K.K. Kanazawa, J.G. Gordon, D.A. Buttry, 'Direct Determination of the Mass of an Underpotentially Deposited Monolayer of Lead On Gold', *Langmuir*, **2**, 697 (1987).
33. G.K. Guttwein, A.D. Ballato, T.J. Lukaszek, US Patent, 3, 694, 677; Sept. 26, 1972.
34. W.W. Lee, H.S. White, M.D. Ward, 'Depletion Layer Effects on the Response of the Electrochemical Quartz Crystal Microbalance', *Anal. Chem.*, **65**, 3232 (1993).
35. S. Bruckenstein, M. Shay, 'Experimental Aspects of Use of the Quartz Crystal Microbalance in Solution', *Electrochim. Acta*, **30**, 1295 (1985).
36. H.E. Hager, R.D. Ruedisueli, M.E. Buehler, 'The Use of Piezoelectric Crystals as Electrode Substrates in Iron Corrosion Studies, the Real-time, In Situ Determination of Dissolution and Film Formation Reaction Rates', *Corrosion*, **42**, 345 (1986).
37. M.R. Deakin, O. Melroy, 'Underpotential Metal Deposition on Gold, Monitored In Situ with a Quartz Crystal Microbalance', *J. Electroanal. Chem.*, **239**, 321 (1988).
38. M. Hepel, S. Bruckenstein, 'Tracking Anion Expulsion During Underpotential Deposition of Lead at Silver Using the Quartz Microbalance', *Electrochim. Acta*, **34**, 1499 (1989).
39. M. Hepel, K. Kanige, S. Bruckenstein, 'In Situ Underpotential Deposition Study of Lead on Silver Using the Electrochemical Quartz Crystal Microbalance. Direct

Evidence for Lead(II) Adsorption Before Spontaneous Charge Transfer', *J. Electroanal. Chem.*, **266**, 409 (1989).
40. M. Hepel, K. Kanige, S. Bruckenstein, 'Expulsion of Borate Ions from the Silver/Solution Interfacial Region During Underpotential Deposition Discharge of Pb(II) in Borate Buffers', *Langmuir*, **6**, 1063 (1990).
41. M. Deakin, T. Li, O. Melroy, 'A Study of the Electrosorption of Bromide and Iodide Ions on Gold Using the Quartz Crystal Microbalance', *J. Electroanal. Chem.*, **243**, 343 (1988).
42. M. Benje, M. Eiermann, U. Pitterman, K.G. Weil, 'An Improved Quartz Crystal Microbalance. Applications to the Electrocrystallization and -dissolution of Nickel', *Ber. Bunsenges. Phys. Chem.*, **90**, 435 (1986).
43. M. Benje, U. Hofmann, U. Pitterman, K.G. Weil, 'Anodic Dissolution of Thin Nickel–Phosphorus Films', *Ber. Bunsenges. Phys. Chem.*, **92**, 1257 (1988).
44. R. Schumacher, A. Muller, W. Stockel, 'An In Situ Study of the Mechanism of the Electrochemical Dissolution of Copper in Oxygenated Sulfuric Acid. An Application of the Quartz Microbalance', *J. Electroanal. Chem.*, **219**, 311 (1987).
45. G.T. Cheek, W.E. O'Grady, 'Measurement of Hydrogen Uptake by Palladium Using a Quartz Crystal Microbalance', *J. Electroanal. Chem.*, **277**, 341 (1990).
46. L. Grasjo, M. Seo, 'Measurement of Absorption of Hydrogen and Deuterium into Palladium During Electrolysis by a Quartz Crystal Microbalance', *J. Electroanal. Chem.*, **296**, 233 (1990).
47. N. Yamamoto, T. Ohsaka, T. Terashima, N. Oyama, 'In Situ Electrochemical Quartz Crystal Microbalance Studies of Water Electrolysis at a Palladium Cathode in Acidic Aqueous Media', *J. Electroanal. Chem.*, **274**, 313 (1989).
48. E.P. EerNisse, 'Simultaneous Thin Film Stress and Mass-change Measurements Using Quartz Resonators', *J. Appl. Phys.*, **43**, 1330 (1972).
49. E.P. EerNisse, 'Extension of the Double Resonator Technique', *J. Appl. Phys.*, **44**, 4482 (1973).
50. J.J. Donahue, D.A. Buttry, 'Adsorption and Micellization Influence the Electrochemistry of Redox Surfactants Derived from Ferrocene', *Langmuir*, **5**, 671 (1989).
51. L. Nordyke, D.A. Buttry, 'Redox Surfactants are Chemical Probes of Electrode Surface Functionalization Derived from Disulfide Immobilization on Gold', *Langmuir*, **7**, 380 (1991).
52. H.C. De Long, J.J. Donahue, D.A. Buttry, 'Ionic Interactions in Electroactive Self-assembled Monolayers of Ferrocene Species', *Langmuir*, **7**, 2196 (1991).
53. G.S. Ostrom, D.A. Buttry, 'Quartz Crystal Microbalance Studies of Deposition and Dissolution Mechanisms of Electrochromic Films of Diheptylviologen Bromide', *J. Electroanal. Chem.*, **256**, 411 (1988).
54. E. Mori, C.K. Baker, J.R. Reynolds, K. Rajeshwar, 'Aqueous Electrochemistry of Tellurium at Glassy Carbon and Gold. A Combined Voltammetry–Oscillating Quartz Crystal Microgravimetry Study', *J. Electroanal. Chem.*, **252**, 441 (1988).
55. L.A. Larew, J.S. Gordon, Y.-L. Hsiao, D.C. Johnson, D.A. Buttry, 'Application of an Electrochemical Quartz Crystal Microbalance to a Study of Pure and Bismuth-doped B-Lead Dioxide Film Electrodes', *J. Electrochem. Soc.*, **137**, 3071 (1990).
56. B.J. Feldman, O.R. Melroy, 'Ion Flux During Electrochemical Charging of Prussian Blue Films', *J. Electroanal. Chem.*, **234**, 213 (1987).
57. S.J. Lasky, D.A. Buttry, 'Mass Measurements Using Isotopically Labeled Solvents Reveal the Extent of Solvent Transport During Redox in Thin Films on Electrodes', *J. Am. Chem. Soc.*, **110**, 6258 (1988).
58. K. Aoki, T. Miyamoto, Y. Ohsawa, 'The Determination of the Selectivity Coefficient of Na^+ Versus Li^+ on Prussian Blue Thin Film in Propylene Carbonate by Means of a Quartz Crystal Microbalance', *Bull. Chem. Soc. Jpn.*, **62**, 1658 (1989).
59. M.R. Deakin, H. Byrd, 'Prussian Blue Coated Quartz Crystal Microbalance as a Detector for Electroinactive Cations in Aqueous Solution', *Anal. Chem.*, **61**, 290 (1989).
60. A.R. Hillman, D.C. Loveday, S. Bruckenstein, 'Electrochemical Quartz Crystal Microbalance Monitoring of Poly(vinylferrocene) Electrodeposition', *Langmuir*, **7**, 191 (1991).
61. J.N. Kaufman, K.K. Kanazawa, G.B. Street, 'Gravimetric Electrochemical Voltage Spectroscopy, In Situ Mass Measurements During Electrochemical Doping of the Conducting Polymer Polypyrrole', *Phys. Rev. Lett.*, **53**, 2461 (1984).
62. P.T. Varineau, D.A. Buttry, 'Applications of the Quartz Crystal Microbalance to Electrochemistry. Measurement of Ion and Solvent Populations in Thin Films of Poly(vinylferrocene) as Functions of Redox State', *J. Phys. Chem.*, **91**, 1292 (1987).
63. P.T. Varineau, 'Determination of the Ion and Solvent Content of Thin Films of Poly(vinylferrocene) on Gold [Electrode] Using the Quartz Crystal Microbalance', PhD Thesis, University of Wyoming, 1989.
64. S. Bruckenstein, A.R. Hillman, 'Consequences of Thermodynamic Restraints on Solvent and Ion Transfer During Redox Switching of Electroactive Polymers', *J. Phys. Chem.*, **92**, 4837 (1988).
65. A.R. Hillman, D.C. Loveday, M.J. Swann, S. Bruckenstein, C.P. Wilde, 'Transport of Neutral Species in Electroactive Polymer Films', *J. Chem. Soc., Faraday Trans.*, **87**, 2047 (1991).
66. A.R. Hillman, D.C. Loveday, S. Bruckenstein, 'A General Approach to the Interpretation of Electrochemical Quartz Crystal Microbalance Data, Part II. Chronoamperometry, Temporal Resolution of Mobile Species Transport in Poly(vinylferrocene) Films', *J. Electroanal. Chem.*, **300**, 67 (1991).

67. S. Bruckenstein, C.P. Wilde, M. Shay, A.R. Hillman, D.C. Loveday, 'Observation of Kinetic Effects During Interfacial Transfer at Redox Polymer Films Using the Quartz Microbalance', *J. Electroanal. Chem.*, **258**, 457 (1989).
68. A.R. Hillman, D.C. Loveday, S. Bruckenstein, C.P. Wilde, 'Criteria Governing Ion and Solvent Transport Rates in Electroactive Polymers, the Existence of Kinetic Permselectivity', *J. Chem. Soc., Faraday Trans.*, **86**, 437 (1990).
69. R. Borjas, D.A. Buttry, 'Solvent Swelling Influences the Electrochemical Behavior and Stability of Thin Films of Nitrated Poly(styrene)', *J. Electroanal. Chem.*, **280**, 73 (1990).
70. S. Bruckenstein, C.P. Wilde, M. Shay, A.R. Hillman, 'Experimental Observations on Transport Phenomena Accompanying Redox Switching in Polythionine Films Immersed in Strong Acid Solutions', *J. Phys. Chem.*, **94**, 787 (1990).
71. A.R. Hillman, M.J. Swann, S. Bruckenstein, 'General Approach to the Interpretation of Electrochemical Quartz Crystal Microbalance Data. 1. Cyclic Voltammetry, Kinetic Subtleties in the Electrochemical Doping of Polybithiophene Films', *J. Phys. Chem.*, **95**, 3271 (1991).
72. S. Bruckenstein, C.P. Wilde, A.R. Hillman, 'Transport Phenomena Accompanying Redox Switching in Polythionine Films Immersed in Aqueous Acetic Acid Solutions', *J. Phys. Chem.*, **94**, 6458 (1990).
73. S. Bruckenstein, A.R. Hillman, M.J. Swann, 'Transient Mass Excursions During Switching of Polythionine Films Between Equi-mass Equilibrium Redox States', *J. Electrochem. Soc.*, **137**, 1323 (1990).
74. A.R. Hillman, D.C. Loveday, M.J. Swann, R.M. Eales, A. Hamnett, S.J. Higgins, S. Bruckenstein, C.P. Wilde, 'Charge Transport in Electroactive Polymer Films', *Faraday Discuss. Chem. Soc.*, **88**, 151 (1989).
75. A.R. Hillman, N.A. Hughes, S. Bruckenstein, 'Solvation Phenomena in Poly(vinylferrocene) Films, Effect of History and Redox State', *J. Electrochem. Soc.*, **139**, 74 (1992).
76. M.D. Ward, 'Ion Exchange of Ferro(Ferricyanide) into Poly(vinylferrocene) Films', *J. Electrochem. Soc.*, **135**, 2747 (1988).
77. C.K. Baker, J.R. Reynolds, 'Use of the Quartz Microbalance in the Study of Polyheterocycle Electrosynthesis', *Synth. Met.*, **28**, C21 (1989).
78. C.K. Baker, J.R. Reynolds, 'A Quartz Microbalance Study of the Electrosynthesis of Polypyrrole', *J. Electroanal. Chem.*, **251**, 307 (1988).
79. C.K. Baker, Y.-J. Qiu, J.R. Reynolds, 'Electrochemically Induced Charge and Mass Transport in Polypyrrole/Poly(styrenesulfonate) Molecular Composites', *J. Phys. Chem.*, **95**, 4446 (1991).
80. K. Naoi, M.M. Lien, W.H. Smyrl, 'Quartz Crystal Microbalance Study, Ionic Motion Across Conducting Polymers', *J. Electrochem. Soc.*, **138**, 440 (1991).
81. K. Naoi, M.M. Lien, W.H. Smyrl, 'Quartz Crystal Microbalance Analysis. Part 1. Evidence of Anion or Cation Inversion into Electropolymerized Conducting Polymers', *J. Electroanal. Chem.*, **272**, 273 (1989).
82. J.R. Reynolds, N.S. Sundaresan, M. Pomerantz, S. Basak, C.K. Baker, 'Self-doped Conducting Copolymers, a Charge and Mass Transport Study of Poly{pyrrole-co[3-(pyrrol-1-yl)propanesulfonate]}', *J. Electroanal. Chem.*, **250**, 355 (1988).
83. R. Borjas, D.A. Buttry, 'EQCM Studies of Film Growth, Redox Cycling and Charge Trapping of a N-Doped and P-Doped Poly(thiophene)', *Chem. Mater.*, **3**, 872 (1991).
84. S. Servagent, E. Vieil, 'In Situ Quartz Microbalance Study of the Electrosynthesis of Poly(3-Methylthiophene)', *J. Electroanal. Chem.*, **280**, 227 (1990).
85. A.R. Hillman, M.J. Swann, S. Bruckenstein, 'Ion and Solvent Transfer Accompanying Polybithiophene Doping and Undoping', *J. Electroanal. Chem.*, **291**, 147 (1990).
86. D.O. Orata, D.A. Buttry, 'Determination of Ion Populations and Solvent Content as Functions of Redox State and pH in Polyaniline', *J. Am. Chem. Soc.*, **109**, 3574 (1987).
87. S.H. Glarum, J.H. Marshall, 'In Situ Potential Dependence of Poly(aniline) Paramagnetism', *J. Phys. Chem.*, **90**, 6076 (1986).
88. H. Daifuku, T. Kawagoe, N. Yamamoto, T. Ohsaka, N. Oyama, 'A Study of the Redox Reaction Mechanisms of Polyaniline Using a Quartz Crystal Microbalance', *J. Electroanal. Chem.*, **274**, 313 (1989).
89. E.S. Grabbe, R.P. Buck, O.R. Melroy, 'Cyclic Voltammetry and Quartz Microbalance Electrogravimetry of Igg and Anti-igg Reactions on Silver', *J. Electroanal. Chem.*, **223**, 67 (1987).
90. J. Hodak, R. Etchenique, E.J. Calvo, K. Singhal, P.N. Bartlett, 'Layer-by-layer Self-assembly of Glucose Oxidase with a Poly(allylamine)ferrocene Redox Mediator', *Langmuir*, **13**, 2708 (1997).
91. Y.M. Lvov, Z. Lu, J.B. Schenkman, X. Zu, J.F. Rusling, 'Direct Electrochemistry of Myoglobin and Cytochrome $P450_{cam}$ in Alternate Layer-by-layer Films with DNA and Other Polyions', *J. Am. Chem. Soc.*, **120**, 4073 (1998).
92. Y. Wang, G. Zhu, E. Wang, 'Electrochemical Behavior of FAD at a Gold Electrode Studied by Electrochemical Quartz Crystal Microbalance', *Anal. Chim. Acta*, **338**, 97 (1997).
93. A.J. Tudos, P.J. Vandeberg, D.C. Johnson, 'Evaluation of EQCM Data from a Study of Cysteine Adsorption on Gold Electrodes in Acidic Media', *Anal. Chem.*, **67**, 552 (1995).
94. G. Gao, D.B. Wurm, Y.-T. Kim, L.D. Kispert, 'Electrochemical Quartz Crystal Microbalance, Voltammetry, Spectroelectrochemical, and Microscopic Studies

of Adsorption Behavior for (7*E*,7′*Z*)-Diphenyl-7,7′-diapocarotene Electrochemical Oxidation Product', *J. Phys. Chem. B*, **101**, 2038 (1997).

95. V.G. Bottom, *Introduction to Quartz Crystal Unit Design*, Van Nostrand Reinhold, New York, 1982.
96. W.P. Mason, *Electromechanical Transducers and Wave Filters*, Van Nostrand, New York, 1948.
97. W.G. Cady, *Piezoelectricity*, Dover, New York, 1964.
98. S.J. Martin, V.E. Granstaff, G.C. Frye, 'Characterization of a Quartz Microbalance with Simultaneous Mass and Liquid Loading', *Anal. Chem.*, **63**, 2272 (1991).
99. C.E. Reed, K.K. Kanazawa, J.H. Kaufman, 'Physical Description of a Viscoelastically Loaded AT-cut Quartz Resonator', *J. Appl. Phys.*, 68, 1993 (1990).
100. V.E. Granstaff, S.J. Martin, 'Characterization of a Thickness-shear Mode Quartz Resonator with Multiple Nonpiezoelectric Layers', *J. Appl. Phys.*, **75**, 1319 (1994).
101. K. Tozaki, M. Kimura, N. Komatsu, S. Itou, 'An Improved Method to Determine Complex Acoustic-impedance Using a Piezoelectric Resonator', *Jpn. J. Appl. Phys.*, **31**, 2592 (1992).
102. K.K. Kanazawa, ANTEC'90, 1049 (1990).
103. W.G. Cady, 'Piezoelectric Resonator and Effect of Electrode Spacing upon Frequency', *Physics*, **7**, 237 (1936).
104. T. Nomura, F. Tanaka, 'Frequency Characteristics of Piezoelectric Quartz Crystals Having a Separated Electrode in Liquids', *Bunseki Kagaku*, **39**, 773 (1990).
105. T. Nomura, T. Yanagihara, T. Mitsui, 'Electrode-separated Piezoelectric Quartz Crystal and its Application as a Detector for Liquid Chromatography', *Anal. Chim. Acta*, **248**, 329 (1991).
106. Z. Mo, L. Nie, S.J. Yao, 'A New Type of Piezoelectric Detector in Liquid. 1. Theoretical Considerations and Measurements of Resonance Behavior Dependent on Liquid Properties', *Electroanal. Chem.*, **316**, 79 (1991).
107. K. Takada, T. Tatsuma, N. Oyama, T. Nomura, 'Impedance and Frequency Characteristics of Electrode-separated Piezoelectric Quartz Crystals in Liquids', *Electroanal. Chem.*, **370**, 103 (1994).
108. N. Oyama, T. Tatsuma, S. Yamaguchi, M. Tsukahara, 'Scanning Electrode Quartz Crystal Analysis', *Anal. Chem.*, **69**, 1023 (1997).
109. J. Rishpon, A. Redondo, C. Derouin, S. Gottesfeld, 'Simultaneous Ellipsometric and Microgravimetric Measurements During the Electrochemical Growth of Polyaniline', *J. Electroanal. Chem.*, **294**, 73 (1990).
110. I. Rubinstein, J. Rishpon, E. Sabatini, A. Redondo, S. Gottesfeld, 'Morphology Control in Electrochemically Grown Conducting Polymer Films. 1. Precoating the Metal Substrate with an Organic Monolayer', *J. Am. Chem. Soc.*, **112**, 6135 (1990).
111. W. Stockel, R. Schumacher, 'In Situ Microweighing at the Metal/Electrolyte Junction', *Ber. Bunsenges. Phys. Chem.*, **345**, 91 (1987).

Neurotransmitters, Electrochemical Detection of

Thomas L. Colliver and Andrew G. Ewing
Pennsylvania State University, PA, USA

In this article, various electrochemical methods used in neuronal cell studies are examined. Electrochemical methods useful for investigating neuronal communication directly in the brain and in ex vivo brain slices are discussed. Additionally, we describe how voltammetric methods can be used to make intracellular and extracellular measurements at single cells. Technical aspects of each method are emphasized and selected applications are highlighted.

Although techniques such as electron microscopy and patch clamp can be used to visualize and monitor cellular events, electrochemical methods can provide chemical information about neurotransmitters being released and transported in brain tissue or even at single cells. The main competing technique for in vivo measurements of neurotransmitters is microdialysis. Although microdialysis coupled to separation techniques can be more selective and more sensitive than electrochemical methods, the latter methods are considerably faster and more accurate in determining the in vivo concentration. At single cells, electrochemistry provides extremely small probes compared to microdialysis thus facilitating measurements of single-cell events. Here, again, the response time associated with electrochemical methods is important for measurements of millisecond exocytosis events.

Electrochemical methods are clearly a powerful means to measure rapid changes in neurotransmitters that are electroactive and found in tissue or cellular microenvironments.

1 INTRODUCTION

The study of the brain and its function has been an area of research for many years. Analytical techniques have provided unique information regarding the neurophysiology and neuropharmacology of the brain. One technique, electroanalysis, has found extensive applications since many neurochemicals are easily oxidized or reduced. Electroanalytical techniques possess unique characteristics that make them useful for studying chemical processes occurring in complex biological matrices. Attributes that are particularly advantageous include the ability to provide qualitative and quantitative information for electroactive substances in the brain.

The fundamental building block of the brain is the single nerve cell or neuron.(1,2) It has been estimated that the human brain contains 10^{10} to 10^{12} neurons.(1) Nerve cells are unique in that they are specialized to communicate with each other. A typical neuron is comprised of a cell body, dendrites and an axon (Figure 1).(3) According to the classical model of neurotransmission, neuronal signals are received at the dendrites, integrated at the cell body, and then sent down the axon to the terminal where they are relayed to the next cell or cells. The site of information transfer between two cells is called the synapse. This structure is essentially a very small gap between cells which ranges in size from 1 to 100 nanometers.

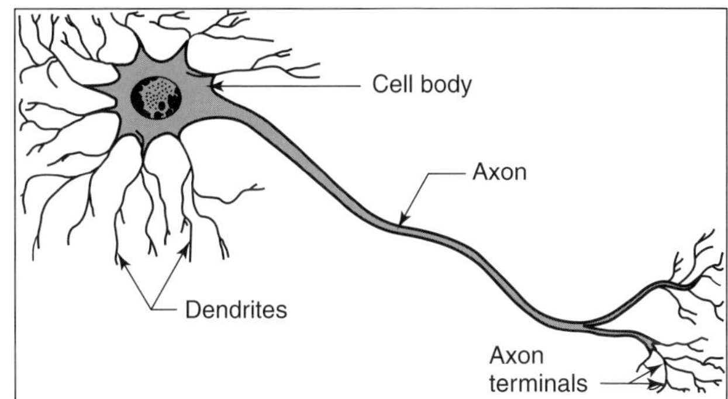

Figure 1 "Typical" neuron in mammalian brain. (Reproduced with permission from ref. 3. © 1976 American Chemical Society.)

Electron microscopy has revealed that storage granules or vesicles aggregate at the synapse near the cell membrane inside the presynaptic neuron (i.e. the neuron sending the signal). These storage compartments are where neurotransmitters are packaged, stored, and sometimes synthesized. When an electrical signal, called an action potential, travels down the axon and reaches the axon terminal, a series of events are initiated which cause the vesicles to transiently fuse with the cell membrane and expel their contents into the synapse. This process is termed exocytosis (Figure 2).(4,5)

After exocytosis, neurotransmitters diffuse across the synapse and bind to receptors on the postsynaptic cell. In some instances, the binding of neurotransmitters to these receptors causes a change in the permeability of the postsynaptic membrane to certain ions leading to

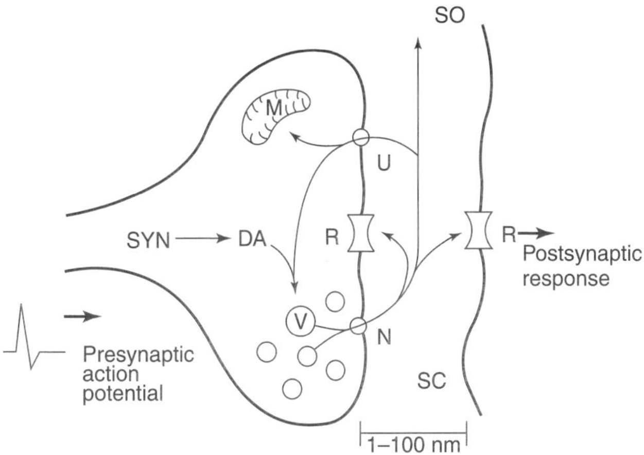

Figure 2 The terminal (or synaptic) region coupling two cells. (SYN = synthesis; DA = dopamine; V = dopamine storage vesicles; N = neuronal release; SC = synaptic cleft; R = pre- and postsynaptic dopamine receptors; U = dopamine uptake; SO = synaptic overflow; M = metabolic enzymes.) (Reproduced with permission from ref. 5. © 1988 American Chemical Society.)

For references see page 9978

the generation of a small postsynaptic potential. The magnitude of this potential is directly dependent on the amount of neurotransmitter that has been released, and can be either excitatory or inhibitory. At the cell body, postsynaptic potentials are integrated and the net sum of these potential changes determines whether the receiving neuron initiates a new action potential.

Once released, the excess neurotransmitter needs to be deactivated to prevent continual stimulation of the postsynaptic cell. Deactivation usually occurs via three mechanisms: diffusion of the neurotransmitter out of the synapse, uptake of the neurotransmitter into the presynaptic cell (or surrounding glial cells), and/or metabolic breakdown of the neurotransmitter.

Neurons utilize a variety of different transmitters and in some cases an individual neuron can release more than one type of transmitter. The general classes of neurotransmitters include acetylcholine, catecholamines, indoleamines, amino acids, and peptides.[1] Since biogenic amines are electroactive, electroanalytical techniques have mainly been used in the study of these transmitters. In this category are the catecholamines; dopamine, norepinephrine, epinephrine, and the indoleamine serotonin.

In the 1970s, Adams et al. began to develop voltammetric methods to monitor dopamine in a dopamine-rich portion of the mammalian brain.[6] This work sparked interest in the use of electroanalysis in neuroscience-related research and has led to the continuing development of methods to monitor neurotransmitter action more effectively in neuronal microenvironments. Throughout the course of this article, various electrochemical methods used in neuronal cell studies will be examined. Specifically, the development of electrochemical techniques useful for investigating neuronal communication directly in the brain and in ex vivo brain slices are discussed. Additionally, the final two sections will examine voltammetric methods used for intracellular and extracellular measurements at single cells, both in vivo and in vitro. At the end of relevant sections, selected applications of these methods will be highlighted.

2 OVERVIEW OF ELECTROCHEMICAL METHODS

2.1 Potentiometric Compared with Voltammetric Measurements

Potentiometry and voltammetry are two of the basic electroanalysis methods available for chemical analysis. These are broad categories that include many different techniques developed for use in different experimental situations. Potentiometric ion-selective electrodes (ISEs)[7] are commonly used to measure concentrations of ions both outside and inside cells. The most common ISEs are used to measure pH, K^+, Na^+, or Ca^{2+}[8] levels in resting and stimulated cells.[9,10] These types of electrodes have been used to study diffusion characteristics as well as to determine the volume of the extracellular space in the brain.[11,12] Although potentiometric methods can be used to study ion fluxes and concentrations, these techniques are not capable of detecting neurotransmitters at neuronal cells. For these types of measurements voltammetry must be employed.

Several neuroactive species are easily oxidized (Figure 3) and are therefore amenable to detection by voltammetry. The current created by the oxidation of these species in response to an applied voltage is the common operating principle behind the various voltammetric techniques. Voltammetric techniques can be put into one of two broad categories: potential step (or pulse) and continuously applied potentials. The pulse techniques include chronoamperometry, normal pulse voltammetry, and differential pulse voltammetry, while linear sweep voltammetry, cyclic voltammetry, and amperometry can be categorized as continuous potential methods. A comparison of the various electrochemical methods reveals that each method has specific advantages and disadvantages in particular protocols. Generally, the voltammetric method used depends on the needs of the user.

2.2 Electrode Types

Platinum wire and carbon fibers are just two of the many materials that have been used in the construction of

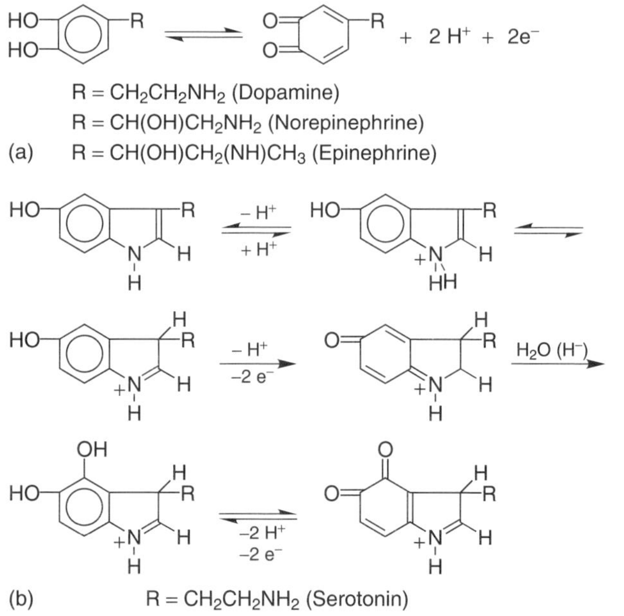

(a) $R = CH_2CH_2NH_2$ (Dopamine)
$R = CH(OH)CH_2NH_2$ (Norepinephrine)
$R = CH(OH)CH_2(NH)CH_3$ (Epinephrine)

(b) $R = CH_2CH_2NH_2$ (Serotonin)

Figure 3 Oxidation reactions for (a) catecholamines and (b) indoleamines. (Reproduced from ref. 104, 225. © 1998 with permission from Marcel Dekker Inc.)

microelectrodes. Both platinum and carbon can be fabricated into durable electrodes with micron or smaller size tips and used in minute environments.[13,14] Platinum and carbon have also been combined to construct platinized carbon microelectrodes.[15–17] Selectivity, sensitivity, stability, and reproducibility of the signals are the major differences between these two types of electrodes, especially when working in biological environments such as the brain or with cells in culture. The electrode material of choice for a given experiment depends upon the nature of the substance(s) being investigated.

2.2.1 Micro Platinum Electrodes

Platinum possesses partially unsaturated surface d-orbitals which facilitate the adsorption stabilization of free radical products in slow electrooxidation reactions and are beneficial in the detection of aliphatic alcohols and amines. Platinum electrodes are also very sensitive to the presence of hydrogen and oxygen.[15] This is advantageous if one wishes to detect H_2 or O_2, but is detrimental if these gases are background signals that obscure detection of the analyte. The strong adsorption characteristics of Pt makes it useful in a "clean" environment, but much less so in biological matrices where proteins and other biological molecules can adsorb to the surface. Ultimately, electrode fouling decreases the sensitivity, stability, and reproducibility of the measurements. These issues have been addressed by depositing thin layers of platinum on carbon surfaces as well as by coating the platinum with various permselective membranes. Although these modifications have improved the performance of platinum electrodes, there is still room for further improvements.

2.2.2 Carbon-fiber Electrodes

Owing to their resistance to drift[18] and inert nature in biological environments, carbon fibers have been used to develop carbon-fiber microelectrodes with dimensions in the micrometer range. The small size of these electrodes provides many advantages in terms of electrochemistry and the practical aspects of their use in biological environments. The improved faradaic to nonfaradaic current ratio at microelectrodes compared to macroelectrodes, for example, enhances the signal-to-noise ratio.[19] Additionally, currents generated at microelectrodes typically lie in the picoampere to nanoampere (pA to nA) range. Such small currents facilitate the use of simpler two-electrode configurations in voltammetric studies.[20] In terms of the biological environment, the smaller currents also mean fewer products are generated at the electrode. This is beneficial since many of these products are detrimental to cells at high concentrations.[21] Small currents created at microelectrodes are not as disruptive to normal cellular function and are less likely to interfere with the normal activity of the surrounding tissue or cells.[22]

Like platinum, some adsorption of large biomolecules does occur on carbon-fiber electrodes[23] causing the electrode's response to deteriorate over time. Usually there is a 30–50% decrease in electrode sensitivity following implantation into brain tissue. This decrease levels off after approximately two hours and reproducible, stable measurements can be collected for six to twelve hours.[21,24,25,26] In culture, sensitivity loss is much less.[18] In both instances, the decrease in electrode sensitivity requires that electrodes be calibrated after experiments.

3 IN VIVO VOLTAMMETRY

3.1 Introduction

The brain is a chemically heterogeneous organ with a fluid environment containing various chemical substances at different concentrations that can change on the second to millisecond timescale. An ideal technology to investigate the neurochemistry of the brain should be able to sample from spatially discrete locations without perturbing the local neuronal environment and have the sensitivity and selectivity to detect and discriminate between chemicals that are present at a wide range of concentrations. This technology should also be able to make measurements on a timescale similar to that on which neuronal signaling occurs. Over the last three decades, two methods have been developed which are complementary to each other and collectively satisfy these requirements. These methods are microdialysis and in vivo voltammetry.

3.2 Microdialysis Compared with In Vivo Voltammetry

In vivo voltammetry and microdialysis sampling have played a large role in detecting neurotransmitter release and turnover. For both these techniques, stereotaxic coordinates are used to implant the sensors into a specific region of the brain. Probes for microdialysis consist of two concentric tubes. The tips of these probes are covered with a low-molecular-weight cut-off dialysis membrane. Small molecules are removed from the extracellular fluid (ECF) by diffusing across this membrane. Fluid containing the analytes is pumped through the probe and out of the brain for subsequent analysis. High-performance liquid chromatography (HPLC) can be used to simultaneously detect many analytes in the dialysis samples. The combination of a sample volume requirement of microliters for HPLC and slow diffusion of molecules across the dialysis membrane results in sampling rates that are generally five to ten minutes,

For references see page 9978

although this has been pushed into the range of a few seconds in recent years.[27,28]

Voltammetry permits measurements to be made more rapidly by applying a small potential to an electrode and measuring the current resulting from the oxidation of electroactive species. In contrast to microdialysis, voltammetric techniques analyze the sample at the surface of the probe. The rate at which molecules can be detected is limited only by diffusion and the response time of the electrode. Another distinction between in vivo voltammetry and microdialysis is the small size of the recording electrodes used in voltammetry. Most electrodes used in vivo have a 10–20 µm total tip diameter. This is approximately ten times smaller than the outer diameter of a microdialysis probe. The size of both types of probes is large compared to cells and synapses. As a result, measurements obtained from both of these techniques reflect the dynamics of chemical events occurring in the extracellular space of the brain. However, the smaller electrochemical probes make it possible to get closer to the synaptic region while minimizing damage to the surrounding brain tissue.

The fast acquisition rates and small size of voltammetric probes allow more rapid and spatially discrete measurements of chemical dynamics. Although microdialysis, in contrast, is better suited to monitor concentration changes for a wider dynamic range of compounds at low levels over prolonged periods of time, the temporal and spatial resolution of this technique are not as good as for voltammetry at microelectrodes.

Since its inception, in vivo voltammetry has been used to measure the dynamic changes of many different substances in the brain that could be considered "informational". One goal which has driven the development of this technology has been to measure directly, in real time, changes in the concentration of neurotransmitter(s) that result from neural activity. Throughout this section of the article, key developments that have helped to push this technology toward this goal will be highlighted and findings from select studies using in vivo voltammetry will be discussed.

3.3 Development of In Vivo Voltammetry to Detect Stimulated Neurotransmitter Release

3.3.1 *Selectivity of In Vivo Voltammetry*

In vivo voltammetry is a general term used to describe a group of loosely related electrochemical techniques developed to measure electroactive compounds in the brain. These techniques can vary with the type of scanning method (i.e. pulse or continuous), the rate at which they scan, and the type of electrode used. As a result of these different experimental approaches, each technique varies in selectivity, rate of data acquisition, and sensitivity. Although it is not the intent here to compare the merits of every technique, (this has been done elsewhere)[21,29] in general, the voltammetric method used depends on the needs of the user. Despite the varied approaches, one of the main disadvantages of using voltammetry is its limited selectivity. The lack of chemical selectivity in voltammetric techniques results mainly from the fact that electrodes respond to any substance that can be electrolyzed at the applied potential.

Initial studies reporting the voltammetric detection of neurotransmitters in the rat brain attributed the increase in oxidation current following intraperitoneal (i.p.) amphetamine injection to an increase in the catecholamine dopamine.[30,31] Gonon et al. later exploited the ability of electrochemically modified electrodes to clearly resolve ascorbic acid and dihydroxyphenylacetic acid (DOPAC) to provide evidence that ascorbic acid, not dopamine, was responsible for the increased electrochemical signal observed following administration of amphetamine.[32] Using an electrode that could discriminate between ascorbic acid, dopamine, and DOPAC, investigators from Wightman's laboratory also concluded that ascorbic acid increased in the ECF of the brain following amphetamine injection.[33,34] The conclusions from these in vivo voltammetry studies were later independently confirmed using push–pull perfusion and microdialysis methods. Using the superior selectivity of HPLC, Salamone et al. showed that ascorbic acid increases in concentration in response to amphetamine.[35] It should be noted that further technical developments did make it possible to detect the amphetamine-induced release of dopamine using voltammetry (see below). Additionally, voltammetry continues to be a very useful method for following changes in ECF levels of ascorbic acid in vivo.[36,37]

Ascorbic acid and DOPAC are both electroactive and are always present in ECF at concentrations 10^4 to 10^3 greater than biogenic amines. Detecting these neurotransmitters in the presence of ascorbic acid and metabolites such as DOPAC therefore remains an analytical challenge. Another way to enhance the selectivity of voltammetric electrodes is to coat the electrode with a permeation selective film. Nafion™ is a sulphonated polymer that is selectively permeable to cations, but repels anions. When coated onto the tip of an electrode, Nafion™ permits cations such as dopamine and serotonin to reach the electrode surface, but rejects anions such as ascorbic acid and DOPAC. Thus, the development of electrodes coated with Nafion™[38] has been a critical component of selective measurements with in vivo voltammetry.

3.3.2 *Fast Compared with Slow Voltammetric Methods*

By utilizing the above techniques to improve selectivity, several investigators have been able to successfully demonstrate that neurotransmitters can be directly detected in the brain using voltammetry. However, exocytosis occurs on a millisecond timescale. In these early examples, data was acquired on the second to minute timescale.[32,33,39] Therefore, these relatively slower voltammetric methods were not capable of detecting faster signals that could have been present. The merits of fast and slow in vivo voltammetric methods have been reviewed elsewhere.[40] In general, slower voltammetric methods, and microdialysis, are best suited to monitor slow changes in concentration of metabolites and related species that result from pharmacological or behavioral stimuli.[41] However, these types of changes are slow and long lasting (minutes to hours) compared to release processes and do not require that high-resolution sampling methods be used. Furthermore, changes in metabolite levels do not reliably reflect release of the parent neurotransmitter.[42,43]

To accurately measure neurotransmitter dynamics following direct rapid stimulation of nerve terminals, voltammetric methods need to acquire data on a sub-second timescale. Owing to their high rate of data acquisition, chronoamperometry and fast cyclic voltammetry (FCV) are both well suited to measure the acute kinetic characteristics of release processes in "real time". Before the advent of very small electrodes, which facilitated faster scanning methods, the most popular voltammetric methods for rapid electrochemical measurements were potential step techniques such as chronoamperometry.[5] For typical chronoamperometry, data acquisition usually requires 50 to 100 ms. Compared to pulse techniques, however, scanning methods provide a higher degree of selectivity and are more effective at discriminating possible interferents (see below). Data for an entire FCV can be acquired in about 10 to 15 ms using a scan rate of $300\,V\,s^{-1}$. Since small carbon-fiber microelectrodes can be used in combination with FCV, this method provides excellent temporal and spatial resolution. Consequently, FCV has been the most widely utilized electrochemical technique for in vivo voltammetry. In the following sections, issues related to the selectivity and sensitivity of in vivo voltammetry will be discussed with an emphasis on FCV.

3.3.3 *Selectivity of Fast Cyclic Voltammetry*

There are two ways FCV provides selectivity. FCV at microelectrodes is dominated by double-layer charging current which masks the faradaic current arising from the oxidation–reduction of electroactive neurotransmitters. This background current can be digitally subtracted from the faradaic current to give a "background subtracted" voltammogram. During periods between stimulation, the majority of the FCV signal is presumably due to the detection of ascorbic acid. Since individual fast cyclic voltammograms are obtained on a timescale similar to release processes, following neuronal stimulation, neurotransmitters are the most likely compounds to change during the time of each scan. Metabolite and ascorbic acid concentrations should change on a much slower timescale.

Generally, the shape of background corrected voltammograms vary with different chemical species. At the fast scan rates used in FCV, voltammogram shapes are kinetically controlled. As a result, most electroactive analytes in the ECF have a unique "fingerprint" which can be used to tentatively identify what has changed in concentration during the time course of the scan. Unfortunately, qualitative chemical analysis cannot be solely based on the shape of the voltammogram. The neurotransmitters serotonin, epinephrine, and dopamine for example, all oxidize at potentials between +0.2 and 0.5 V against Ag–AgCl with untreated electrodes, and the shape of the signal from these compounds is not always readily distinguishable. Additionally, interpreting the shape of the voltammogram can become complicated if more than one neurotransmitter changes in concentration. When using FCV to detect electroactive neurotransmitters in the brain, it is essential that independent neuropharmacological interventions be used to verify the identity of the detected neurotransmitter.[41,44]

3.3.4 *Sensitivity of In Vivo Voltammetry*

Realizing the limited sensitivity of voltammetric methods in general, Ewing et al.[33] used elevated levels of K^+ to stimulate massive amounts of dopamine release (Figure 4). This method of stimulation increased ECF levels of dopamine from basal levels of approximately 5–10 nanomolar to the low micromolar range. Although a relatively slower method of voltammetry (normal pulse voltammetry) was used, these results demonstrated that, with an adequate stimulus, voltammetry could be used to monitor the stimulated release of dopamine.

Owing to the limited spatial and temporal control provided by elevated levels of K^+, this method of stimulation was later replaced by the use of a bipolar stimulating electrode.[45] Stimulating electrodes can be used to artificially trigger action potentials within a specific group of neurons. Stimulating and recording electrodes can be placed in different locations in the brain of an animal during an experiment. For example, with the working electrode in the terminal field where release occurs, the stimulating electrode could be placed in the cell body region, along ascending fiber tracts, or directly in the terminal fields to stimulate release. Despite these options, most workers stimulate ascending pathways. One approach that has been used extensively

For references see page 9978

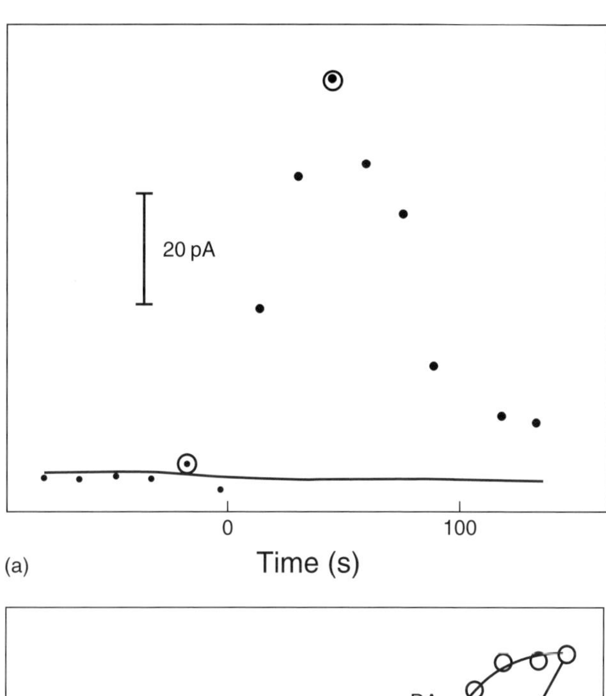

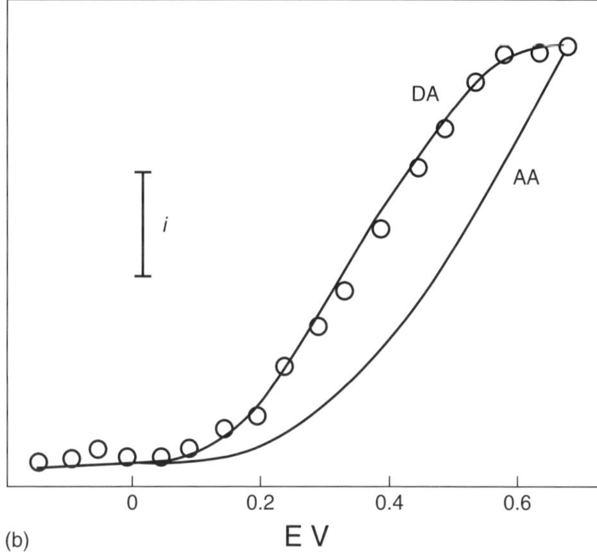

Figure 4 Electrochemical response to injection at 0 time of 0.6 μmol of potassium chloride in 1.0 μL of physiological buffer into the caudate nucleus. Distance from electrode to syringe tip is 0.8 mm. (a) Current vs time response at 0.5 V. Solid line, injection of physiological buffer only; points, injection of potassium chloride. (b) Difference voltammograms. Solid lines, dopamine (25 μM) and ascorbic acid (200 μM) in pH 7.4 buffer after in vivo use; circles, in vivo result from subtracting voltammograms obtained at the circles shown in (a). Current scales: $i = 25$ pA, dopamine; $i = 28$ pA, ascorbic acid; and $i = 22$ pA, in vivo. (Reproduced from Brain Research, A.G. Ewing, R.M. Wightman, M.A. Dayton 'In vivo Voltammetry with Electrodes that Discriminate Between Dopamine and Ascorbate', **249**, 361–370. © 1982 with permission from Elsevier Science.)

is to electrically stimulate the medial forebrain bundle (MFB) while detecting dopamine concentration changes in the rat striatum with a carbon-fiber electrode.

3.4 Fast Cyclic Voltammetry and Electrical Stimulation: Applications In Vivo

Combined, the use of local electrical stimulation and FCV make it possible to directly measure, in real time, changes in the concentration of neurotransmitter(s) that result from neural activity. Stamford et al.[46] were the first to realize the advantages of combining these two techniques and, since then, FCV and electrical stimulation have been the most widely used methods for examining the dynamics of electroactive neurotransmitters in vivo and in vitro (see below). Detailed and exhaustive discussion of the neurochemistry examined in all of these studies is beyond the scope of this article. However, selected findings are discussed here to illustrate the utility of this methodology.

Kuhr and Wightman[47] have used fast-scan voltammetry at Nafion™-coated, beveled, carbon-disk electrodes with electrical stimulation of the rat MFB to monitor the time course of stimulated dopamine release in the caudate nucleus. The maximum concentration of dopamine detected at a single electrode location during stimulation has been found to be directly proportional to the frequency of pulses in the stimulation train. Dopamine is observed in the ECF for approximately 1.5 s when short stimulation times are employed demonstrating the brief time that detectable dopamine levels are present following stimulation. The disappearance of dopamine appears to occur exponentially, which is characteristic of first-order uptake kinetics. Using the uptake inhibitor, nomifensine, a large increase in current is observed when cells are stimulated with a small electrical stimulus (15 pulses in 250 ms), which normally gives only a barely detectable signal (Figure 5). These results confirm the large effect that uptake has on the detected dopamine signal in vivo.

In another example, in vivo voltammetry has been used to suggest that the brain is not "hard-wired" meaning that some neurons communicate by pathways outside of the synapse. Garris et al.[48] have used FCV with Nafion™-coated, beveled, disk electrodes and electrical stimulation to examine dopamine efflux from the synaptic junction. It has been determined that binding of released dopamine to receptors and uptake sites does not appreciably alter the diffusion of dopamine from the synaptic cleft in the rat nucleus accumbens. This implies that the dopamine synapse is designed for efficient transmitter efflux from the synaptic cleft so that dopamine can escape the synapse and affect a greater number of cells. This proposed scheme contrasts with the established model of neurotransmission derived from experiments at the neuromuscular junction where communication occurs only inside the synapse (Figure 6).

The simultaneous detection of two neurochemically relevant substances in vivo has also been accomplished using FCV. Zimmerman and Wightman[49] have

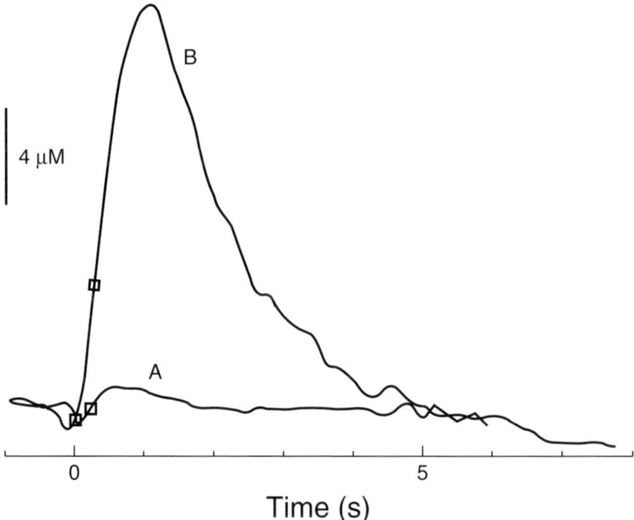

Figure 5 Time course of dopamine release before (a), and 8 min after (b), administration of $20\,mg\,kg^{-1}$ nomifensine. The release was induced by stimulation consisting of fifteen 300-μA bi-phasic pulses delivered at a frequency of 60 Hz. Although only one time course is shown, identical temporal profiles were obtained in 3 additional rats. (Reproduced from Brain Research, W.G. Kuhr, R.M. Wightman, 'Real-time Measurement of Dopamine Release in Rat Brain', **381**, 168–171. © 1986 with permission from Elsevier Science.)

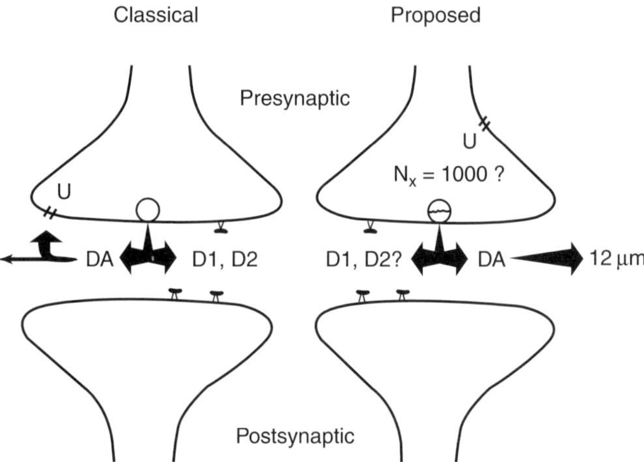

Figure 6 Schematic drawings of dopamine neurotransmission in the core of the nucleus accumbens. Schematics contrast the classical and proposed model of dopamine efflux from dopamine synapses in the nucleus accumbens of the rat brain. For the proposed model, it is suggested that the dopamine transporter (DAT) (U) mainly functions to regulate extrasynaptic dopamine levels as opposed to preventing the efflux of transmitter from the synaptic cleft. Note the difference in the size of the arrows representing the efflux of dopamine from the synaptic cleft. (Reproduced from *J. Neuroscience*, **14**(10). © 1994 with permission.)

demonstrated that both dopamine and O_2 can be independently and simultaneously monitored at a single Nafion™-coated disk electrode by scanning from 0.0 V to +0.8 V followed by a cathodic sweep to −1.4 V before returning to a resting potential of 0.0 V. Several laboratories have also used voltammetry at carbon-fiber electrodes, in combination with other techniques, to simultaneously acquire electrochemical and electrophysiological information in vivo and in vitro.(22,50,51) In one particular study, Ewing et al. used carbon-fiber microelectrodes combined with tungsten microelectrodes to simultaneously measure the effects of *d*-amphetamine and ascorbic acid on the electrochemical and unit activity of neostriatal neurons.(22) Neurotransmitter efflux and unit activity may be considered respectively as indices of presynaptic and postsynaptic neuronal function. The ability to record both of these phenomena makes it possible to correlate the release of neurotransmitter from a discrete population of release sites with changes in neuronal activity at a specific region in the brain.(51)

4 VOLTAMMETRY IN BRAIN SLICES

4.1 Introduction: Use of Brain Slices Compared with In Vivo Techniques

The first report of voltammetric techniques being used in combination with brain slices described experiments that were designed to shed light on data obtained in vivo.(52) Owing to the many technical advantages of using this approach, slice-voltammetry methods have become very popular and have nicely complemented work done in vivo. For these types of experiments, voltammetry is performed in slices of brain tissue that are created by taking "sections" from a whole brain. This approach has several advantages that make it attractive. It avoids any potential confounding effects that may arise from the use of anesthetics and inherently reduces/simplifies the complexities of the neuronal circuits present during an experiment. Additionally, when using brain slices one does not need to worry about the ability of a given agent to cross the blood brain barrier or the possible toxic side effects that an agent may have on the whole animal. Combined, these two factors widen the range of agents which can be investigated. Since the investigator can control the composition of the fluid perfusing the tissue, it is also possible to minimize potential interferents such as ascorbic acid.

4.2 Technical Issues

4.2.1 Slice Preparation

In general, when slices of brain tissue are prepared for an experiment, the brain is quickly removed from an animal and immediately placed in chilled physiological saline. A

For references see page 9978

block of tissue containing the desired region of the brain is then isolated from the whole brain and slices of tissue are created by sectioning this block using a vibratome. A slice is then placed into a tissue bath where the tissue has complete access to the physiological buffer. In the bath, the tissue is continuously superfused with warm physiological buffer saturated with oxygen and carbon dioxide (32–37 °C; some investigators have reported that the temperature of the slice affects the long-term stability of the electrochemical signal[51]). Slices are usually cut to a thickness of 150–400 µm and the physiological buffer is flowed through the bath at a rate of approximately 0.25–1 mL min^{-1}. The dimensions of the slice allow oxygen to easily diffuse into the tissue and the constant exposure to fresh buffer helps to establish a metabolically stable system.[53] Once situated within the tissue bath, the slice is allowed to incubate for 30 to 60 minutes prior to an experiment. This allows the preparation to stabilize and equilibrate with its surrounding environment.

4.2.2 *Experimental Set-up*

With the aid of a dissecting microscope, the working electrode is lowered into the slice, using a micromanipulator, to a position of 75 to 100 µm below the surface of the tissue. This ensures that the whole sensing area of the electrode is completely enclosed within the slice. Auxiliary and reference electrodes are placed in convenient positions within the tissue bath. While solutions of elevated K^+ have been used to stimulate neurotransmitter release in slice-voltammetry experiments,[54] as with experiments in vivo, the most popular method used in slices has been electrical stimulation. This type of stimulation is accomplished by placing a bipolar stimulating electrode in or on the slice approximately 100 to 400 µm away from the working electrode. The stimulating electrodes can be accurately placed, with the aid of the microscope, at a specific region within the slice. Interestingly, the relative efficacy of this method of stimulation varies with the area of the brain being studied. Single pulse stimulations (0.1 ms width, 20 V) for example, consistently elicit detectable levels of dopamine in slices of rat striatum[55] while trains of stimuli (i.e. 25 pulses, 50 Hz, 0.1 ms, 20 V) are required to give similar voltammetric signals for serotonin in slices of rat dorsal raphe or suprachiasmatic nuclei.[56] For a more detailed discussion of the technical issues concerning the use of voltammetry in brain slices, the reader is referred to an excellent review by Stamford et al.[57]

4.3 Applications

Although FCV is not the only voltammetric technique that has been used with slices,[58,59] it is the method that has been most widely applied to these preparations. Since voltammetry was first used in slices, there have been a variety of experiments performed using different voltammetric techniques. The focus of the majority of these studies can be loosely categorized as: examining neurotransmitter release and uptake in different regions of the brain,[55,56,60] comparing neurotransmitter release and uptake between different regions of the brain,[61,62] and investigating the presynaptic regulation of neurotransmitter release by pharmacologically manipulating autoreceptors.[63,64] Recent studies have even used voltammetry in slices to better understand the physiological mechanisms regulating ischaemia-induced dopamine release.[65,66] Since it is beyond the scope of this article to cover in detail all the findings from these studies, we have chosen to illustrate the utility of using voltammetry in slices by focusing on work recently done in brain slices from mice lacking the DAT. The DAT is associated with presynaptic dopamine neurons and mediates uptake processes to clear dopamine from the synaptic cleft (see Figure 2).

The central importance of cellular uptake mechanisms for regulating dopamine levels in the ECF has been investigated by disrupting the mouse DAT gene. Voltammetry experiments in brain slices from these "knockout" mice have shown that disruption of the DAT significantly increases the time required for dopamine to be cleared from the ECF. Indeed, when both alleles of the DAT are disrupted, dopamine remains in the ECF up to 100 times longer than normal (Figure 7). These results directly demonstrate that the DAT is critically involved in regulating dopamine ECF levels, and hence dopamine signaling, and provide a biochemical basis for the spontaneous hyperactivity observed in DAT knockout mice.[67]

Subsequent experiments in slices from DAT knockout mice have taken advantage of the unique genetic background of these animals to help unravel the complex actions of the psychostimulant amphetamine.[68] Amphetamine can cause the release of dopamine from neurons in a manner which is impulse independent and has little calcium dependence. The neurochemical mechanism responsible for the releasing action of amphetamine is attributed to both the drug's ability to redistribute dopamine from secretory vesicles and its ability to cause the release of cytoplasmic dopamine into the ECF via the DAT (a process that has been called reverse transport[69]). The relative importance of these two mechanisms in the releasing action of amphetamine, however, has been controversial.

Figure 8 shows how amphetamine affects the electrically stimulated and baseline levels of dopamine when applied to striatal slices from wild-type (DAT +/+, control) and DAT knockout (DAT −/−) mice. As shown

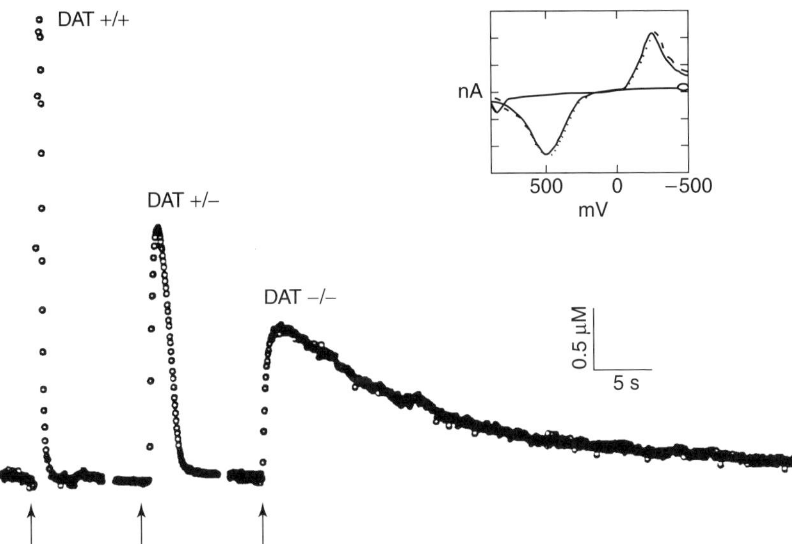

Figure 7 Typical recordings of electrically stimulated dopamine efflux in striatal slices from DAT +/+, DAT +/− and DAT −/− mice. DAT +/+ represents mice with both alleles of the DAT; DAT +/− represents mice missing one allele; DAT −/− represents mice missing both alleles. A single biphasic electrical pulse (4 ms, 350 μA) was applied at the time indicated by the arrows with a locally placed bipolar electrode. Data points (open circles) were collected from a recording electrode every 100 ms. Inset shows cyclic voltammograms recorded at the peak of dopamine efflux in each slice. Voltammograms recorded from DAT +/+ (solid line), DAT +/− (broken line) and DAT −/− (dotted line) mice were scaled to approximately the same peak amplitude for comparison. These cyclic voltammograms are not significantly different from those obtained with authentic dopamine. Measurements were made in at least 22 slices from at least 5 animals for each group. The average time to clear released dopamine was 1, 3, and 100 s in DAT +/+, DAT +/− and DAT −/−, respectively. (Reproduced from ref. 67. © 1996 with permission from *Nature*.)

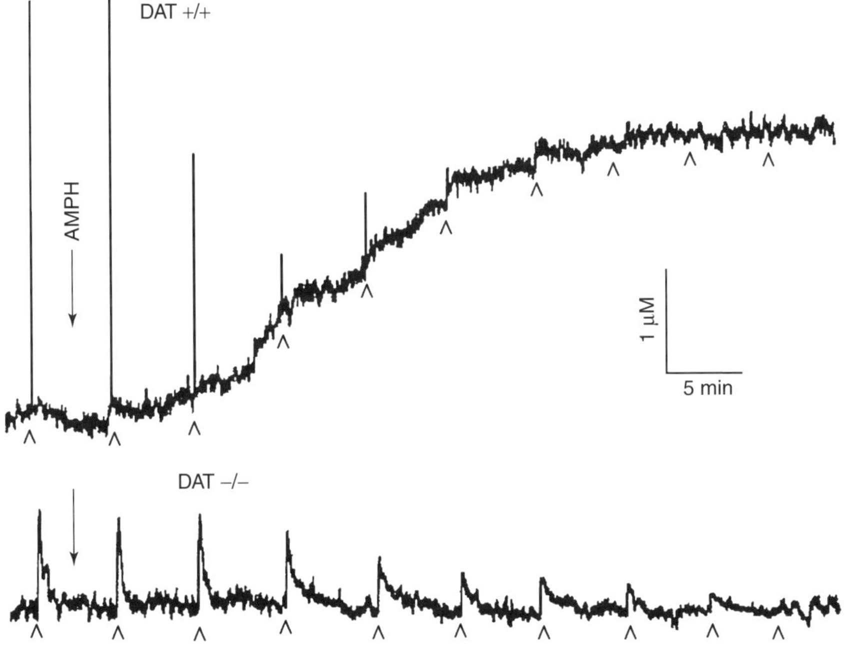

Figure 8 Effect of amphetamine (10 μM) on dopamine efflux in striatal slices from wild-type (DAT +/+) and homozygote DAT knockout (DAT −/−) mice. Current was measured by cyclic voltammetry at microelectrodes implanted in the slices. Stimulated dopamine release was elicited by single electrical pulses applied to the slice at the times indicated by small arrowheads. On the timescale shown, electrically stimulated dopamine release events appear as sharp spikes. Baseline release was monitored as any increase in the current measured between stimulations that was identified as dopamine. In the absence of pharmacological or electrical intervention, baseline and electrically stimulated dopamine recordings were stable for >3 h. (Reproduced from *J. Neuroscience*, **18**(16). © 1998 with permission.)

For references see page 9978

in Figure 8, approximately 10 min after amphetamine is applied to slices from wild-type mice there is a distinct rise in baseline dopamine levels. At the same time, as can be seen by the decrease in height of the sharp 'spikes' of dopamine efflux, amphetamine also decreases the amount of electrically stimulated dopamine release. Approximately 25 min after the drug has been administered, the steady increase in the baseline reaches a plateau. In contrast to these results, when amphetamine is applied to slices from DAT –/– mice, although there is also a decrease in the stimulated efflux of dopamine, no change in baseline dopamine levels are observed. These results directly demonstrate that the DAT is required for the releasing action of amphetamine.

To examine vesicle depletion and its effect on reverse transport independently, slices from control and DAT deficient mice were exposed to the vesicle depleting agent Ro4-1284. As can be seen in Figure 9, this drug, like amphetamine, caused a gradual reduction in the stimulated release of dopamine. Unlike amphetamine however, Ro4-1284 did not cause an increase in baseline dopamine levels. When amphetamine was applied to slices from wild-type mice, after Ro4-1284 had completely blocked stimulated dopamine release, there was a rapid increase in baseline dopamine levels that reached values similar to those observed when amphetamine was used alone. The rate of this increase, however, was much faster with baseline concentrations of dopamine reaching their maximum values 5 to 10 min after the application of amphetamine.

From these two sets of experiments it has been estimated that it takes approximately 25 min for amphetamine to exert its depleting action on synaptic vesicles and only 5 min for amphetamine to release cytoplasmic dopamine via reverse transport. These results demonstrate that both vesicular depletion and reverse transport play central roles in the releasing action of amphetamine. However, vesicle depletion appears to be the rate-limiting step.

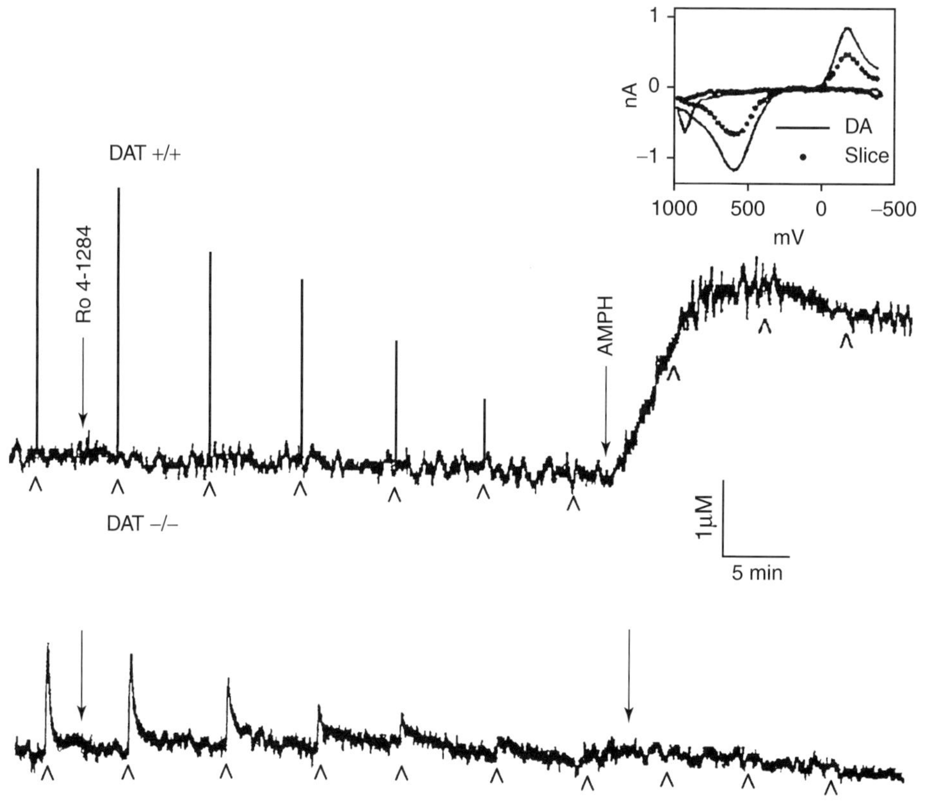

Figure 9 Effect of a fast-acting reserpine-like drug, Ro4-1284 (10 μM), and amphetamine (104 μM) on stimulated and baseline release of dopamine. Individual recordings of dopamine efflux were measured by cyclic voltammetry in striatal slices from wild-type (DAT +/+) and homozygote DAT knockout mice (DAT –/–). Drugs were applied to slices at the times indicated by the large arrows. Top, Single-pulse stimulations were applied to a slice from a wild-type mouse at the times indicated by the small arrowheads. Inset, Cyclic voltammograms, current (nA) versus potential (mV) plots, recorded during postcalibration with authentic dopamine (solid line) and during the plateau phase of the release induced by amphetamine (filled circles), are plotted for comparison. Bottom, Single-pulse stimulations were applied to a slice from a homozygote DAT knockout mouse at the times indicated by the small arrowheads. (Reproduced from *J. Neuroscience*, **18**(16). © 1998 with permission.)

5 INTRACELLULAR VOLTAMMETRY AT SINGLE CELLS

5.1 Overview

The ability to detect species inside single nerve cells is very important in order to gain further insight into the metabolism, function and regulation of individual neurons. Electrode tips 10 µm or less in diameter are small enough to impale fairly large cells (100–200 µm diam.) without causing excessive damage or interfering with cellular function. The development of carbon and platinum microelectrodes with total overall dimensions in the micron range has made it possible to directly detect several different electroactive substances in the cytoplasm of single nerve cells.[14,70,71]

Initial intracellular voltammetric studies, conducted by Meulmans et al.,[14] to measure the endogenous ascorbic acid levels in single *Aplysia* neurons, demonstrated the potential of electrochemical methods to intracellularly monitor electroactive neurochemicals. Using 0.5- to 2-µm tip diameter carbon-fiber or platinum electrodes in combination with differential pulse voltammetry, it has been determined that the endogenous ascorbic acid concentration in these single neurons is approximately 100 µM. Similar approaches have also been used to monitor the kinetics of uptake, clearance of two electroactive drugs,[72] and levels of serotonin in single *Aplysia* neurons.[14]

Development of glucose microsensors has provided the ability to qualitatively monitor glucose transients in the cytoplasm of a single cell. Glucose metabolism is of particular interest because glucose is a major energy source for the brain.[16] Abe et al.[73,74] have used 2-µm diameter platinized-carbon ring electrodes with amperometric detection (0.6 V versus SSCE (sodium saturated calomel electrode)) to monitor intentional manipulations of cytoplasmic glucose in the giant dopamine cell of the pond snail *Planorbis corneus*. Intracellular oxygen levels in the giant dopamine cell of *Planorbis* have also been measured using Nafion™-coated platinized-carbon ring electrodes with differential pulse voltammetry.[16] In the following section, experiments investigating intracellular dopamine in the giant dopamine cell of *Planorbis* will be described.

5.2 Monitoring Intracellular Dopamine

Carbon and platinum microelectrodes and various electrochemical techniques have been used to monitor dopamine inside the giant dopamine cell from the pond snail *Planorbis corneus*. Active transport of dopamine into the cell has been monitored using linear scan voltammetry and poly(ester sulfonic acid)-coated carbon ring electrodes. In these experiments, the electrode is implanted in the neuron and the cell is bathed in 0.5 mM dopamine.[70] When the oxidation current at the electrode is monitored at +0.78 V and plotted against time, distinct peaks can be seen following extracellular application of the dopamine solution (Figure 10). The rising portion of each peak represents the active transport of dopamine across the cell membrane, and the decreasing portion of each peak represents the clearance of cytoplasmic dopamine through metabolism and/or vesicularization.

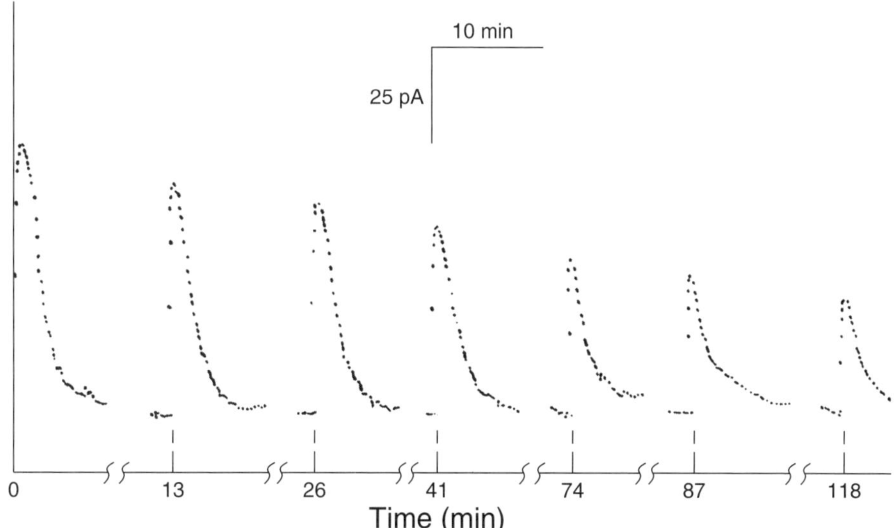

Figure 10 Observed time course plot of repeated dopamine bathings (0.5 mM, 30 µL each arrow) with the electrode placed in the cell body of the identified dopamine neuron of *Planorbis corneus*. The dopamine solution used for extracellular bathings was a modified snail saline solution. The current was monitored at +0.78 V vs SSCE with 7 s between measurements. (Reproduced from ref. 70. © 1991 with permission from Wiley-VCH.)

For references see page 9978

In these experiments, the peak current for repeated bathings with dopamine gradually decreases over time. The average loss of response in the cell is 36% after 20 min of continuous voltammetry. This loss is most likely due to a decrease in electrode sensitivity brought on by specific adsorption of large biomolecules. Loss of sensitivity from fouling of the electrode surface has been largely overcome by use of integrated pulse linear scan voltammetry, at platinized-carbon ring electrodes.(17) This method, also referred to as pulsed amperometric detection (PAD), minimizes electrode fouling while maintaining a reactive electrode surface. The applied waveform, based on the pulse techniques developed by Johnson's group,(75,76) consists of a rapid cyclic potential sweep followed by a large positive potential pulse to oxidatively clean the electrode surface and then a large negative potential pulse to restore the reactivity of the electrode.(17) This technique allows continuous amperometric detection that is both sensitive and reproducible. The average peak concentration of dopamine uptake in the *Planorbis* studies has been found to be $44 \pm 2\,\mu M$, and the average rate of dopamine clearance, estimated from the linear portion of the declining limiting current, is $0.29\,\mu M\,s^{-1}$.(17)

To gain insight into the distribution of dopamine inside the cell, intracellular studies measuring basal dopamine levels have also been carried out. Carbon ring electrodes with electrode tips of approximately 6 µm have been placed inside the giant dopamine cell of *Planorbis* to monitor dopamine using staircase voltammetry.(71) Basal levels of free dopamine are below the detection limit of this method, so 0.5 mM dopamine is added extracellularly. The uptake of dopamine by the cell causes a sharp increase in the oxidation current detected in the cytoplasm, results in agreement with those presented by Lau et al.(17,70) The extracellular application of nominfensine (700 µL, 0.6 mM), an uptake inhibitor, diminishes the oxidation current, verifying that the rise in current observed is due to the uptake of extracellular dopamine and not free intracellular dopamine.

Bathing the cell in 300 µL of 50% ethanol disrupts the cellular membranes and results in a large increase in intracellular dopamine. Capillary electrophoresis has been used to verify the identification of the electroactive compound released upon ethanol exposure. In the resting state, cytoplasmic dopamine levels are below the detection limits of this method. However, upon exposure to ethanol, 14 fmol of dopamine is detected. Assuming an injection volume of 100 to 300 pL, the cytoplasmic dopamine concentration following ethanol treatment is approximately 1.4×10^{-4} to 4.7×10^{-5} M.(71) This technique has also been coupled to off-column amperometric detection(77) with a detection limit of 10^{-8} M in 50 pL samples.(78) These lower detection limits are necessary for detection of basal levels of cytoplasmic free dopamine. Olefirowicz and Ewing(78) have determined basal dopamine concentration in *Planorbis* to be $2.2 \pm 0.5\,\mu M$.

6 EXTRACELLULAR VOLTAMMETRY AT SINGLE CELLS

6.1 Introduction

As described above, carbon-fiber microelectrodes have been used in a number of studies to monitor extracellular changes in neurotransmitter levels both in the brain and in brain slices. One limitation to these types of studies is that the observed signal arises from multiple cells. Furthermore, since the electrodes are much larger than the synapse, one can only infer what is going on at the site of neurotransmitter release. The next frontier for voltammetry has been to monitor the exocytotic release of neurotransmitters at the level of a single cell.

Several different examples are given below illustrating how carbon-fiber microelectrodes can be used to monitor exocytosis at a single cell. In all examples the experimental protocols are similar. In most cases, with the exception of two examples, tissue containing the cell(s) of interest is dissociated into a single-cell suspension and cultured. The distance between single cells in culture facilitates electrode placement onto a cell and minimizes possible interferences from neighboring cells. Carbon-fiber microelectrodes, which are usually beveled for these experiments, are positioned via a micromanipulator next to or directly against a single cell (see Figure 11). To stimulate exocytosis, nanoliter volumes of a chemical stimulant (i.e. elevated K^+ and/or nicotine) are administered to a

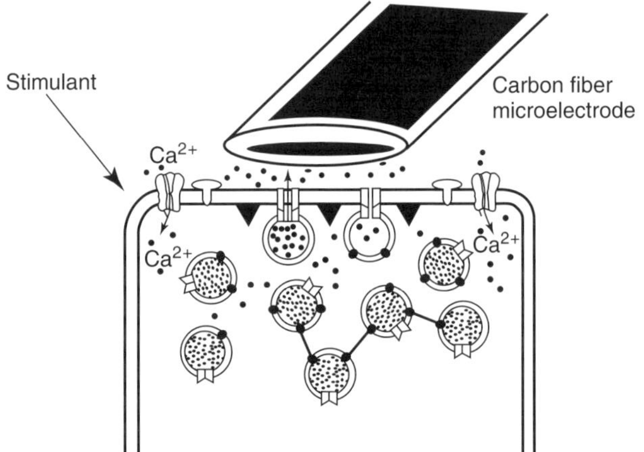

Figure 11 Schematic showing the experimental arrangement used for measuring neurotransmitter secretion from a single cell. (Reproduced from ref. 104, 225. © 1998 with permission from Marcel Dekker Inc.)

cell via pressure injection through a micropipette placed approximately 20–30 μm from the cell. The electrode is held at a constant potential and the current is sampled at high rates. This electrochemical technique is referred to as amperometry. When electroactive neurotransmitters are released from a cell and approach the electrode surface they are oxidized. The resulting current gives rise to a positive deviation in the baseline.

6.2 Overview of Systems Used for Single-cell Analysis

Single-cell amperometry has been used to investigate many types of cells that release easily oxidized substances by exocytosis. Some of the nonsynaptic cell systems that have been investigated include bovine adrenal chromaffin cells, rat pheochromocytoma (PC12) cells, beige mouse mast cells, and human pancreatic β-cells. Synaptic systems including invertebrate and mammalian neurons have also been investigated. A brief overview of some of the key studies using these different cell types will be presented in the following sections.

6.2.1 Bovine Adrenal Chromaffin Cells

The direct measurement of catecholamine secretion from a single cell was first reported by workers in Wightman's laboratory.[(79)] Bovine adrenal chromaffin cells were used in this study. Adrenal chromaffin cells are useful models of neurotransmitter biosynthesis, metabolism, and secretion owing to their neuroectodermal origin and biochemical and functional similarities with postganglionic sympathetic neurons.[(80)] Changes in the chemistry surrounding a single chromaffin cell are monitored using FCV at an electrode placed next to a cell. The application of 100 μM nicotine results in a series of sharp irregular concentration spikes superimposed on a secretion envelope (Figure 12d). Individual cyclic voltammograms confirm that the substances secreted and detected are catecholamines (Figure 12b). The secretion of catecholamines at a single cell has been shown to be induced by stimulants such as nicotine, the nicotinic agonist carbamylcholine, and K^+.[(81)] Furthermore, K^+- and nicotine-stimulated secretion could be inhibited by Cd^{2+}, a calcium channel blocker, while nicotine-stimulated secretion could be selectively inhibited by hexamethonium, a sympathetic ganglionic blocker.[(81)]

Many of these observations of catecholamine secretion are consistent with similar measurements made at multiple-cell preparations such as the intact adrenal gland, adrenal slices, and whole cultures of adrenal cells. Additionally, since vesicular fusion events should result in discrete packets of catecholamines being released from the cell surface, the observed chemical concentration spikes are the expected consequence of exocytosis.

Combined, the above evidence suggested that the single-cell electrochemical signals were the result of secretion of catecholamines via exocytosis. To improve the temporal resolution of the measurements made with FCV and to more accurately measure exocytosis, which can occur on the millisecond to microsecond timescale, single-cell measurements have been made in the amperometric mode. The faster sampling rates possible with this technique make it possible to more clearly resolve the chemical spikes. For each spike, the integrated current, or charge, is proportional to the mass of the analyte detected and can be used in Faraday's law, Equation (1), to calculate the number of molecules detected per each spike event. Faraday's law can be expressed as

$$\frac{Q}{nF} = N \tag{1}$$

where Q is the charge in coulombs, n is the number of moles of electrons transferred per mole of analyte oxidized, F is the Faraday constant (96 485 C equiv^{-1}) and N is the total number of moles of substance oxidized.

Several results from the above studies have demonstrated that the properties of the electrochemically detected spikes at single bovine adrenal cells correlated well with expectations based on the exocytotic secretion of catecholamines from individual vesicles. For example, cyclic voltammograms at individual concentration spikes confirm that the spikes are due to the detection of concentration packets of catecholamines rather than an electrical artifact.[(79,81)] Additionally, in agreement with the requirements of Ca^{2+} for exocytosis, spikes do not occur in the absence of this ion.[(80)] Within each vesicle there is a discrete or quantal amount of neurotransmitter. If each of the observed chemical spikes was due to the exocytotic release of neurotransmitter(s) from a single vesicle, the amount detected per each spike event should be independent of the method of stimulation. Different methods of stimulation have been used to test this hypothesis.[(80)] Stimulation with 100 μM nicotine, results in the highest frequency response (1.2 ± 0.2 Hz), whereas at a lower concentration (10 μM) the response is much slower (0.5 ± 0.2 Hz). Both the nicotinic agonist carbamylcholine (1 mM), and KCl (60 mM) give a similar response (0.7 ± 0.2 and 0.6 ± 0.2 Hz, respectively). The mean ± SEM (standard error of the mean) spike areas for the different stimulants is 1.19 ± 0.9 pC for K^+, 0.98 ± 0.13 pC for carbamylcholine, 1.04 ± 0.16 pC for 10 μM nicotine, and 1.19 ± 0.10 pC for 100 μM nicotine. Clearly, the frequency of the spikes increases with the intensity of the stimulus. However, the average quantity of catecholamine in each spike does not significantly differ with the intensity or type of stimulant used. While previous experiments in other laboratories have provided

For references see page 9978

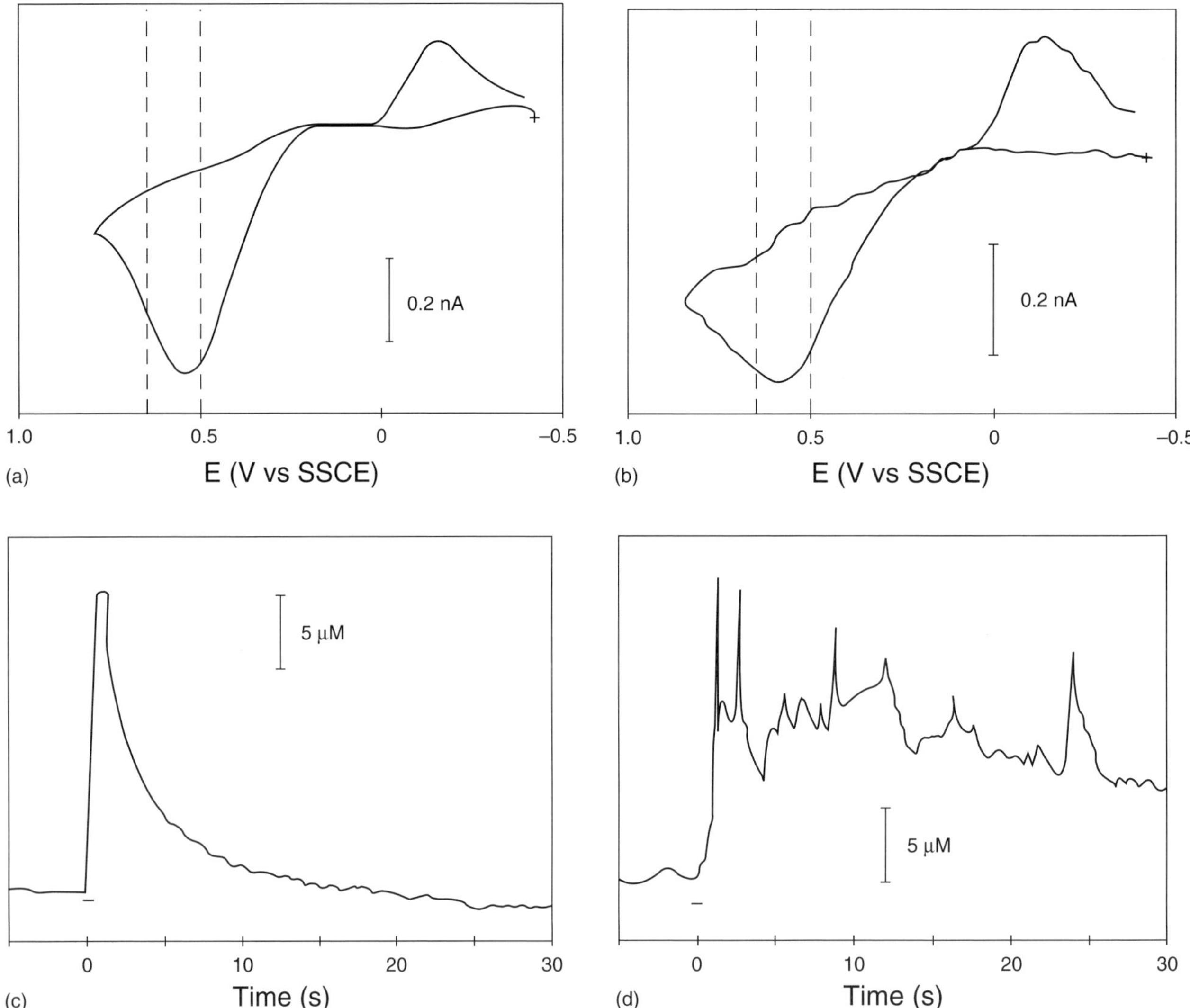

Figure 12 Cyclic voltammetric response ($200\,V\,s^{-1}$, repeated at 100-ms intervals) of a carbon-fiber electrode to norepinephrine ejection from an adjacent micropipette and to catecholamine secretion from a single chromaffin cell. Panels (a) and (b) are averaged background-subtracted voltammograms of the substances whose concentration changed during the measurement interval of panels (c) and (d). Each time point in panels c and d is the integrated current recorded from 0.5–0.6 V from individual voltammograms (hatched lines in panels (a) and (b)); bars to the right in panels (c) and (d) are the conversion of current to catecholamine concentration based on calibration curves constructed with standards. Panels (a) and (c) are the electrode response to a 1-s ejection (3 nL) of 20 μM norepinephrine applied at $t = 0$ with the ejection pipette 20 μm from the electrode. Panels (b) and (d) are the electrochemical response obtained with the electrode tip adjacent to a single cell; at $t = 0$, a 1-s ejection of nicotine (100 μM) was made 20 μm away from the cell. (Reproduced with permission from ref. 79. © 1990 The American Society for Biochemistry.)

evidence for the quantized nature of neurotransmitter release, results from these studies were the first to provide direct chemical evidence for exocytosis. Further support for these conclusions was provided by Chow et al.[82] In this study, it was found that the time-averaged signal of the carbon-fiber microelectrode closely resembled the derivative of the membrane capacitance trace (Figure 13). Membrane capacitance is proportional to the cell surface area and increases when individual vesicles fuse with the plasma membrane during exocytosis.[83]

Since the initial studies, bovine adrenal cells have been one of the most widely used cell types in single-cell amperometry experiments. From these experiments, an incredibly large amount of information has been generated which has helped further our basic understanding of exocytosis, the fundamental means by which chemical communication takes place between cells in the nervous system, at the single-cell level. Several studies have investigated how conditions of the extracellular environment, such as temperature,[84] osmolarity,[85] and pH,[86] affect

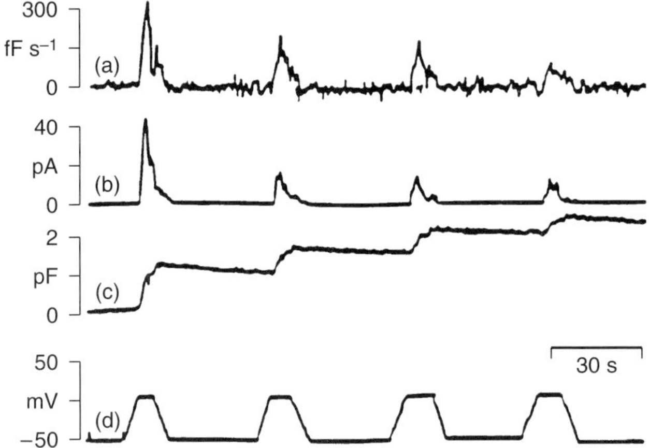

Figure 13 Correlation between the amperometric signal and the rate of capacitance change at a single adrenal cell. A cell was held at −60 mV and repetitively depolarized to 0 mV for about 8 s at a time (d). Capacitance (c) increased during depolarizations, and the amperometric signal (b) displayed peaks. The time derivative of the capacitance signal, which, like the amperometric signal, should be proportional to secretion rate, is displayed in the top trace (in femtofarads per s) for comparison with the latter. (Reproduced from ref. 82. © 1992 with permission of Nature.)

the release of catecholamines from single chromaffin vesicles. Amperometry has also been used to dissect different temporal stages of vesicular catecholamine release at bovine adrenal cells.[87–90] Chow et al.[82] for example, reported a delay in vesicle fusion compared to neuronal systems and a small peak or "foot" at the onset of vesicular release events. Schroeder et al.[91] also proposed that there are three distinct stages of exocytosis at bovine adrenal cells that include formation of a fusion pore, pore expansion to release catecholamine, and a final dissociation of the intravesicular matrix. Bovine adrenal cells have been used with amperometry and in conjunction with other single-cell techniques to learn more about the role of Ca^{2+} and Ca^{2+} channels in the exocytotic process.[92–97] By using small (1 µm) flame-etched carbon-fiber electrodes, it has been possible to map the surface of adrenal cells and identify the existence of spatially localized release sites.[98] Amperometry at single bovine adrenal cells has also been used to study the regulatory role of autoreceptors in exocytosis.[99] More detailed discussions of findings from these studies and those discussed in the following sections may be found in the literature.[100–104]

6.2.2 *Rat Pheochromocytoma (PC12) Cells*

PC12 cells are an immortalized cell line derived from rat adrenal chromaffin cells.[105] These cells have been widely used as a model for adrenal chromaffin cells and sympathetic neurons. PC12 cells are an attractive model neuronal system owing to their ability to differentiate into neuronal-type cells upon treatment with nerve growth factor (NGF). Additionally, like sympathetic neurons, PC12 cells can synthesize acetylcholine, as well as the catecholamines dopamine and norepinephrine.[106,107]

Using amperometric detection at carbon-fiber electrodes, catecholamine release has been observed at the zmol level from single PC12 cells in culture.[108] The amperometric response at a single PC12 cell following three successive stimulations with both 1 mM nicotine and 105 mM K^+ is compared to that of a control in Figure 14. The shape of each transient is similar to those observed from bovine adrenal cells[80] except that the average charge for each transient is significantly smaller. Individual current spikes have an average half-width of 9.3 ± 0.1 ms and the average catecholamine content calculated using the area under each current transient and Faraday's law is 190 ± 3.5 zmol.[108]

Following treatment with NGF, PC12 cells extend processes.[109] Along these processes, are bulbous regions (1–2 µm diam.), referred to as varicosities. Zerby and Ewing[110] have used PC12 cells treated with NGF to investigate how the site of exocytosis changes during the differentiation process. To investigate these changes, beveled carbon-fiber microelectrodes have been placed at several different locations on differentiated PC12 cells. Experiments carried out between days 10 and 14 of culture show no release from the cell body ($n = 3$), only occasional responses from the smooth regions of the neurites ($n = 5$), and frequent release when the electrode is placed at a varicosity ($n = 16$). Additionally, the average vesicular catecholamine content observed at varicosities is not significantly different from that observed at the cell body of undifferentiated cells. These results lend some insight into functional changes that occur during differentiation and indicate that during differentiation the relocation of release sites to PC12 varicosities does not significantly alter the mean catecholamine content of individual vesicles.

The time resolution of carbon-fiber microelectrodes not only allows the detection of single exocytotic events, but also provides an excellent means by which to monitor the time course of the stimulus–secretion process. PC12 cells possess both nicotinic and muscarinic receptors that trigger exocytosis through two different mechanisms. Nicotine causes a change in the conformation of the nicotine receptor opening channels permeable to Na^+. The influx of Na^+ causes depolarization of the cell membrane which opens voltage-sensitive Ca^{2+} channels that allow Ca^{2+} into the cell to cause exocytosis.[111] Muscarine activates exocytosis through muscarinic receptors that act through intracellular second messengers to release Ca^{2+} from intracellular stores.[112] In contrast to nicotine and muscarine, K^+ causes exocytosis through direct membrane

For references see page 9978

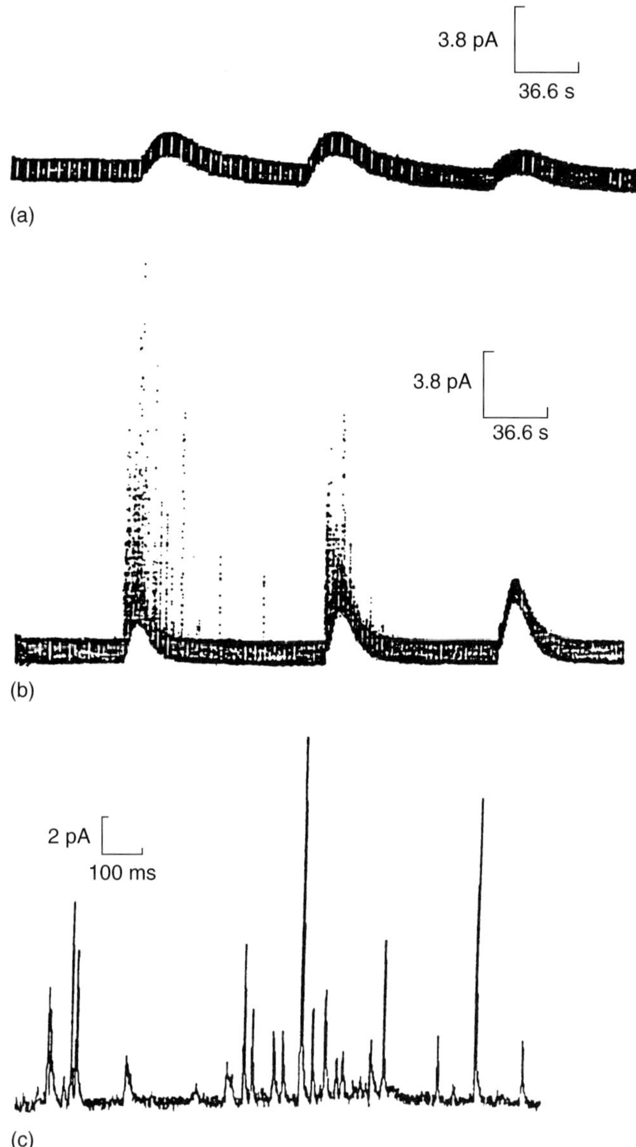

Figure 14 Amperograms of individual vesicular exocytosis events from PC12 cells. (a) Amperogram of a control experiment with the tip of the working electrode 200 μm from the cell. (b) Amperogram of vesicular exocytosis induced by bathing a cell with 1 mM nicotine in 105 mM K^+ balanced salt solution while the electrode was placed on top of the cell. (c) Enlargement of a 1-s period of the amperogram from the first stimulation in (b). Data displayed correspond to the time period near the middle of the first baseline disturbance of (b). Traces (a) and (b) were obtained from computer screen dumps; each printed line represents 600 data points. (Reproduced with permission from ref. 108. © 1994 American Chemical Society.)

depolarization. Zerby and Ewing have used amperometry at single PC12 cells to investigate how these different mechanisms affect the time course or latency of the release process.[113] Their results have shown that while the average vesicle catecholamine content is unaltered by the different mechanisms of release, their latencies (time between application of the stimulant and onset of exocytotic events) vary significantly. The mean latencies for each type of stimulation are reported as: 6 ± 1 s for K^+ (105 mM), 37 ± 5 s for nicotine (1 mM), and 103 ± 11 s for muscarine (1 mM). Figure 15 shows representative amperograms obtained using the three different stimulants. Interestingly, these results are in agreement with similar measurements made at single rat adrenal cells.[114]

Several studies have used PC12 cells to investigate how different pharmacological agents effect the average catecholamine content of PC12 vesicles. For example, PC12 vesicular dopamine levels have been shown to decrease when cells are exposed to amphetamine and reserpine.[69,115] Recently, Pothos et al. have used

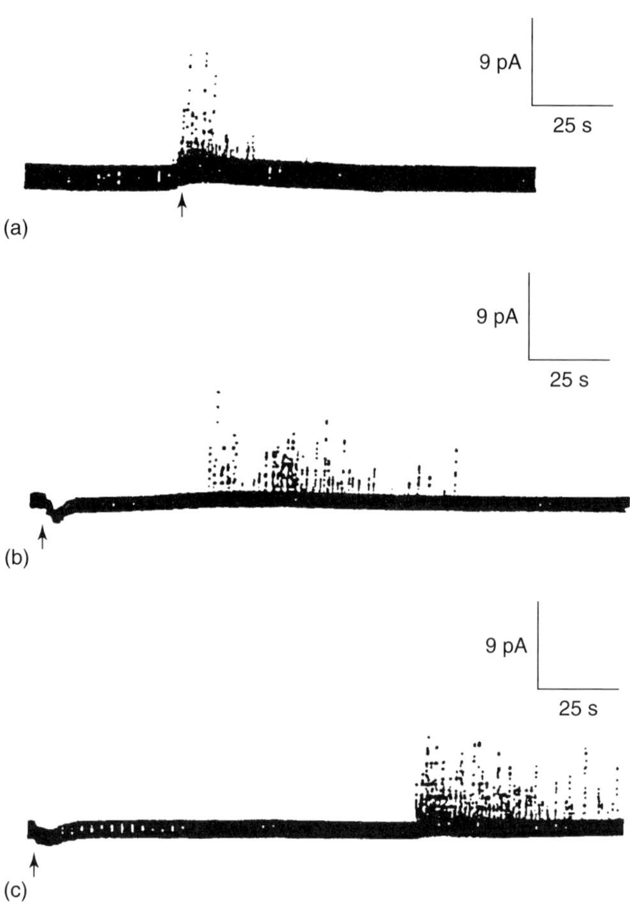

Figure 15 Current–time traces for exocytosis at single PC12 cells. A 6-s ejection of stimulant [105 mM K^+ (a), 1 mM nicotine (b), or 1 mM muscarine (c)] from a microinjector was administered at each arrow. The resulting current transients correspond to the oxidation of dopamine at the electrode tip as it is released from the cell. Detection was performed in the amperometric mode at 650 mV vs SSCE. (Reproduced from *J. Neurochem.*[113] © 1996 with permission from Lippincott, Williams and Wikins.)

amperometry at single PC12 cells to investigate the effects of D_2 autoreceptors on the exocytotic release of neurotransmitters. By treating PC12 cells with the D_2 agonist quinpirole, it was possible to decrease the quantal size of release events by approximately 50%.[116] In contrast to results obtained with reserpine and amphetamine, these results provide evidence for receptor-mediated mechanisms that can decrease quantal release. The size of release events detected at PC12 cells has also been modulated by exposing cells to the catecholamine precursor L-3,4-dihydroxyphenylalanine (L-Dopa). Following exposure to 50 μM L-Dopa for 40 to 90 min, the average size of quantal events is increased to 251% of control values.[117]

6.2.3 Mast Cells, Pancreatic β-cells, and Rat Melanotrophs

Beige mouse mast cells have been utilized for amperometric investigations of the secretory agents histamine and serotonin. Histamine and serotonin are easily oxidized molecules and are important in the nervous and immune systems of mammals. One advantage of using this cell system is the large size of the vesicles present. The larger vesicles allow release to be more easily correlated with simultaneous changes in membrane capacitance.[118] Using amperometric methods, histamine and serotonin have been shown to be co-released from single mast cell vesicles.[119]

Amperometric techniques have also been extended to pancreatic β-cells and rat melanotrophs, secretory cells not normally considered as model systems for studying neurotransmission. Pancreatic β-cells secrete insulin in response to glucose levels in the body. The common link between insulin secretion from β-cells and the other secretory cells described is that insulin is stored in vesicles and secreted via exocytosis. Membrane-potential experiments with intracellular microelectrodes[120] and amperometric monitoring[121] of insulin secretion, using chemically modified electrodes, have been carried out at individual pancreatic β-cells in an effort to better understand how these cells regulate blood sugar levels in the body. Kennedy et al. have further investigated secretion from β-cells using chromatographic methods and have confirmed that the secreted products are indeed insulin.[122]

Peptide hormones are necessary for many biological functions and understanding how they are stored and secreted at the single-cell level is extremely important. Kennedy et al. have demonstrated that it is possible to detect the exocytotic release of peptides at single rat melanotrophs. The secretion of peptide hormones can be detected electrochemically by the oxidation of tryptophan and tyrosine residues found in the proopiocortin cleavage products. The calcium-mediated release of ∂-melanocyte stimulating hormone has also been measured at a carbon-fiber electrode placed on a single rat melanotroph.[123]

6.2.4 Invertebrate and Mammalian Neurons

Exocytotic events at single neurons from the pond snail *Planorbis corneus*[124,125] and the leech *Hirudo medicinalis*[126] have been studied using extracellular voltammetry. Bruns and Jahn[126] have described electrochemical detection of serotonin release from isolated Retzius cells of the leech *Hirudo medicinalis*. Single leech Retzius neurons have been cultured and a carbon-fiber electrode placed at the axonal stump and the cell body. Two types of exocytotic responses have been observed, one interpreted as resulting from serotonin release from small clear synaptic vesicles and the other from large dense-core vesicles. Release from the small clear vesicles appears to occur more rapidly (faster time constant) than from the large dense-core vesicles, which are more randomly distributed throughout the cell. It has been suggested that the faster rate observed for the small clear vesicles is due to the discharge of their contents on a submillisecond timescale through an undilated fusion pore.

Using single-cell amperometry, Chen et al.[124] have found that stimulation of the giant dopamine neuron of the pond snail *Planorbis corneus*, with elevated KCl, results in massive exocytotic release from the cell body (Figure 16). Since norepinephrine and epinephrine have not been detected in the giant dopamine cell,[127] it appears that the observed current transients are due to the release of dopamine from the cell body. Capillary electrophoresis with electrochemical detection has been used to verify that dopamine, indeed, is being released. The current transients observed from the cell body have rise times between 2 and 5 ms, an average base width of 14 ± 0.8 ms ($n = 13$ cells, 12 324 transients), and are not observed in the absence of Ca^{2+}.

In addition to the large number of vesicular release events at the cell body, the amount of dopamine released from the vesicles has an interesting distribution.[124,125] When the area of each spike is converted into vesicular content and plotted as a histogram, a bimodal distribution is observed in which each phase drops off exponentially (Figure 17a). To correlate the content of release events with vesicle radii, it has been assumed that the catecholamine concentration within each vesicle is constant.[128] In contrast to results obtained from other cell types,[128] when release events from the giant dopamine cell are plotted against the cube root of vesicle content two distinct Gaussians can clearly be seen in the histogram. These results suggest that there are at least two distinct classes of vesicles or release events occurring at this cell (Figure 17b). Results from the *Planorbis*

For references see page 9978

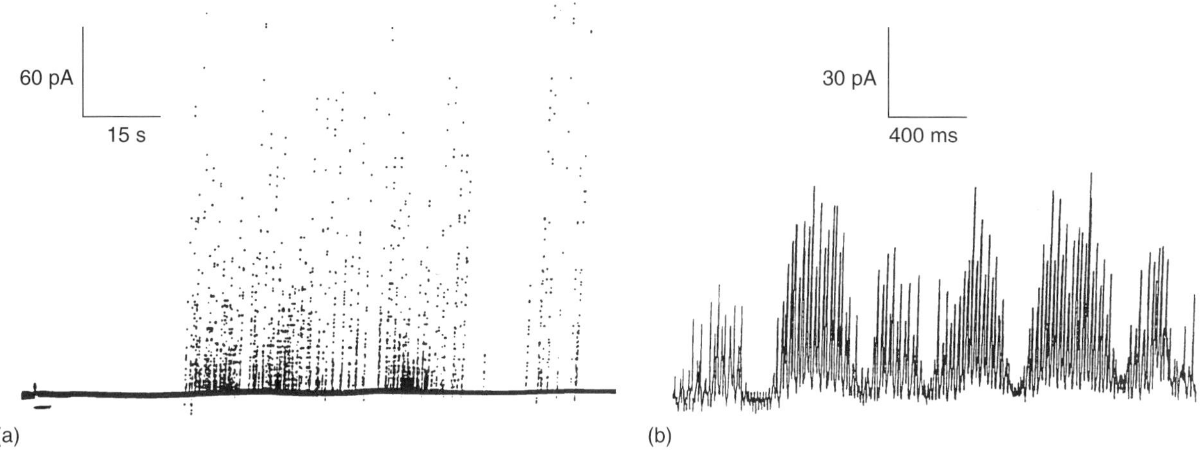

Figure 16 Dopamine current transients detected by a carbon-fiber disk electrode placed on the cell body of the dopamine neuron in *Planorbis corneus*. (a) An example of current transients recorded by amperometry. A large *Planorbis* dopamine cell (diameter about 100 µm) was stimulated with a 4-s potassium chloride (1 M) pulse (87 nL) delivered from a glass pipette that was placed about 15 µm from the cell body. The stimulation is shown by the horizontal bar below the trace. (b) Bursting release events were observed in 24 out of 29 cells that showed release transients. The overall success rate for observation of current transients from cells sampled was approximately 50%. (Reproduced from *J. Neuroscience*, **15**(11). © 1995 with permission.)

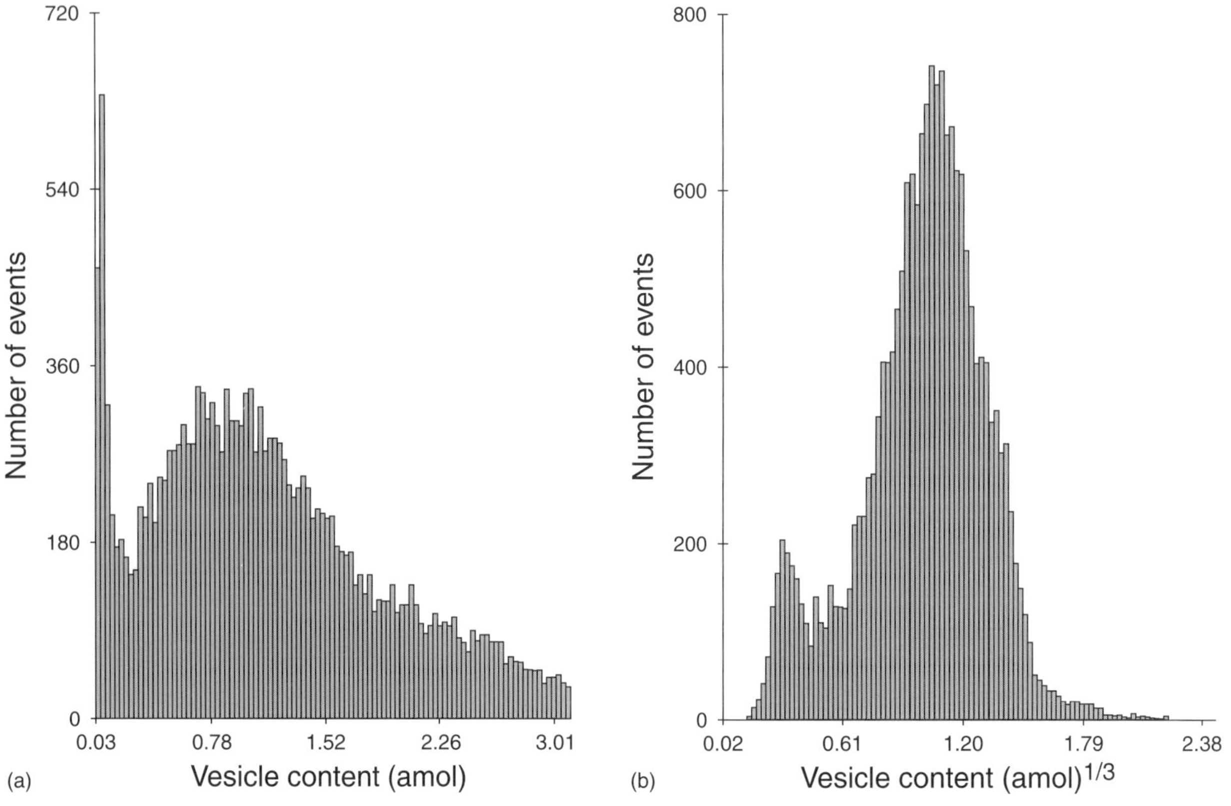

Figure 17 Histograms of the frequency of release events vs the amount of neurotransmitter released from the dopamine cell of *Planorbis*. (a) Frequency vs attomoles of dopamine released (16 cells, 18 456 events). Only release events with base widths less than 40 ms were considered. Wider transients, which made up less than 2% of the total number and mostly appeared to contain contributions from more than one vesicle (appearing as doublets and shoulders), were excluded in the calculations. Based on the results obtained from 16 cells, the average vesicle content was 1.36 ± 0.53 amol (means ± s.d.), equivalent to 818 000 ± 319 000 molecules of dopamine. (b) Frequency vs the cubed root of attomoles of dopamine released (16 cells and 18 456 events). (Reproduced from *J. Neuroscience*, **15**(11). © 1995 with permission.)

system have also shown that amphetamine differentially alters the catecholamine content of the two classes of vesicles[125] and that release events at the cell body occur in a bursting pattern with regular time intervals between individual events (Figure 16).[129]

While much has been learned by studying exocytosis at model neuronal systems and invertebrate neurons, efforts have also been focused on directly analyzing exocytosis at single mammalian neurons from the central nervous system (CNS) and peripheral nervous system (PNS). Developing superior cervical ganglion neurons from neonatal rats, for example, have been investigated with amperometric methods and vesicular release has been detected at varicosities along axons from these neurons following the application of potassium or black widow spider venom.[130] Zhou and Misler[130] report a median spike charge of 11.3 femtocoulombs, which corresponds to 35 000 catecholamine molecules per release event (58 zmol). These values are similar to those obtained for release events associated with small clear vesicles from leech Retzius neurons.[126]

Recently, two studies have used amperometry to investigate the quantal release of neurotransmitters from CNS neurons. Pothos et al. adapted amperometric methods to directly observe quantal dopamine release from axonal varicosites of midbrain dopamine neurons.[131] Events were elicited by high K^+ or α-latrotoxin, required extracellular Ca^{2+}, and were abolished by reserpine. The characteristics of these events indicated that, on average, 3000 molecules were detected over 200 µs, parameters much smaller and faster than those associated with release events detected at other vertebrate preparations. The number of dopamine molecules detected per event increased to 380% of control values after cells were exposed to glial-derived neurotrophic factor and to 350% of control values after exposure to L-Dopa. These results indicate that quantal size is not invariant in CNS neurons.

Substantia nigra neurons release dopamine from their somatodendritic regions. However, previous studies had not been able to determine whether dopamine release occurred via exocytosis or some nonvesicular mechanism(s). To shed light on this process, Jaffe et al.[132] have used amperometry at carbon-fiber microelectrodes to assay catecholamine secretion from the cell bodies of substantia nigra neurons in rat midbrain slices. Detected secretion events had charge integrals ranging from a few to several hundred femtocoulombs. While local application of glutamate enhanced the frequency of events, the mean area of the spikes was not changed. The addition of Cd^{2+}, a blocker of voltage-dependent Ca^{2+} channels, to the cells blocked the stimulatory effects of glutamate. Together, these results suggest that dopamine is released from the cell body region of substantia nigra neurons via exocytosis.

7 CONCLUDING REMARKS AND FUTURE DIRECTIONS

Understanding communication between mammalian neurons is the ultimate goal of most neuronal studies. Since the initial experiments in Adams' laboratory some 30 years ago, several developments have helped to push voltammetry to the forefront of available technologies. Today, voltammetry is being used in many different ways by chemists and neuroscientists to study fundamental processes related to neuronal communication. Fast scan rate cyclic voltammetry at microelectrodes combined with electrical stimulation has proven to be a very powerful means for detecting neurotransmitter dynamics in vivo and in vitro. These methods have been used to identify released neurochemicals, monitor how drugs affect neuronal communication, investigate the kinetics of neurotransmitter release and uptake, and further our understanding of the basic process of neurotransmission in the brain.

With the development of microelectrodes, many of the advantages of voltammetry have been exploited and used to study neurochemical processes both inside and at the surface of single cells and neurons. The small size of microelectrodes makes it easy to investigate spatially discrete regions of release on a single cell with the sensitivity and response time necessary to detect the rapid exocytotic release of minute amounts of neurotransmitters. The large amount of information that has been generated using single-cell electrochemical methods attests to the utility of this technology. In its current state, voltammetric methods are best suited to study the neurochemistry of biological systems that utilize easily oxidized neurotransmitters. Combined with enzyme-based electrodes and other chemically modified electrodes, a new group of electrochemical sensors promises to move this technology into the area of monitoring molecules that are normally not easily oxidized in biological microenvironments (key developments in this area have been published[104,133]). It seems likely that some of the important molecules to be examined in the future will be glutamate, acetylcholine, peptides, and NO.[134,135]

Future developments will probably also include the development of nanometer-size electrochemical probes to access the small synaptic gap. Looking back on the developments that have already taken place, it seems logical that intra-synaptic measurements will be the next frontier in which microelectrodes will be used for the electrochemical detection of neurotransmitters.

For references see page 9978

ACKNOWLEDGMENTS

The contributions by our coworkers that are referenced herein are gratefully acknowledged. This work was supported, in part, by grants from the National Science Foundation, and the National Institutes of Health. T.L.C. is a National Institute of Mental Health Pre-doctoral Fellow.

ABBREVIATIONS AND ACRONYMS

CNS	Central Nervous System
DAT	Dopamine Transporter
DOPAC	Dihydroxyphenylacetic Acid
ECF	Extracellular Fluid
FCV	Fast Cyclic Voltammetry
HPLC	High-performance Liquid Chromatography
i.p.	Intraperitoneal
ISEs	Ion-selective Electrodes
L-Dopa	L-3,4-dihydroxyphenylalanine
MFB	Medial Forebrain Bundle
NGF	Nerve Growth Factor
PAD	Pulsed Amperometric Detection
PNS	Peripheral Nervous System
SEM	Standard Error of the Mean
SSCE	Sodium Saturated Calomel Electrode

RELATED ARTICLES

Clinical Chemistry **(Volume 2)**
Biosensor Design and Fabrication • Electroanalysis and Biosensors in Clinical Chemistry • Electroanalytical Chemistry in Clinical Analysis

Electroanalytical Methods **(Volume 11)**
Electroanalytical Methods: Introduction • Pulse Voltammetry • Selective Electrode Coatings for Electroanalysis • Ultrafast Electrochemical Techniques

REFERENCES

1. G.M. Shepherd, *Neurobiology*, 2nd edition, Oxford University Press, Oxford, 1988.
2. J.R. Cooper, F.E. Bloom, R.H. Roth, *The Biochemical Basis of Neuropharmacology*, Oxford University Press, Oxford, 1986.
3. R.N. Adams, 'Probing Brain Chemistry with Electroanalytical Techniques', *Anal. Chem.*, **48**, 1128–1138A (1976).
4. F. Valtorta, R. Fesce, F. Grohovaz, C. Haimann, W.P. Hurlbut, N. Iezzi, F. Torri Tarelli, A. Villa, B. Ceccarelli, 'Neurotransmitter Release and Synaptic Vesicle Recycling', *Neuroscience*, **35**, 477–489 (1990).
5. R.M. Wightman, L.J. May, A.C. Michael, 'Detection of Dopamine Dynamics in the Brain', *Anal. Chem.*, **60**, 769–779A (1988).
6. P.T. Kissinger, J.B. Hart, R.N. Adams, 'Voltammetry in Brain Tissue: A New Neurophysiological Measurement', *Brain Res.*, **55**, 209–213 (1973).
7. C. Nicholson, 'Ion-selective Microelectrodes and Diffusion Measurements as Tools to Explore the Brain Cell Microenvironment', *J. Neurosci. Methods*, **48**, 199–213 (1993).
8. S. Baudet, L. Hove-Madsen, D.M. Bers, 'How to Make and Use Calcium-specific Mini- and Microelectrodes', *Methods Cell Biol.*, **40**, 93–113 (1994).
9. J.O. Schenk, E. Miller, R.N. Adams, 'Electrochemical Techniques for the Study of Brain Chemistry', *J. Chem. Ed.*, **60**, 311–314 (1983).
10. R.M. Wightman, 'In Vivo Electrochemistry', in *Electrochemistry in Research and Development*, eds. R. Kalvoda, R. Parsons, Plenum Press, New York, 189–202, 1985.
11. C. Nicholson, J.M. Phillips, 'Ion Diffusion Modified by Tortuosity and Volume Fraction in the Extracellular Microenvironment of the Rat Cerebellum', *J. Physiol.*, **321**, 225–257 (1981).
12. M.E. Rice, C. Nicholson, 'Diffusion Characteristics and Extracellular Volume Fraction During Normoxia and Hypoxia in Slices of Rat Neostriatum', *J. Neurophysiol.*, **65**, 264–272 (1991).
13. R.M. Penner, M.J. Heben, T.L. Longin, N.S. Lewis, 'Fabrication and Use of Nanometer-sized Electrodes in Electrochemistry', *Science*, **250**, 1118–1121 (1990).
14. A. Meulemans, B. Poulain, G. Baux, L. Tauc, 'Changes in Serotonin Concentration in a Living Neurone: A Study by On-line Intracellular Voltammetry', *Brain Res.*, **414**, 158–162 (1987).
15. N. Georgolios, D. Jannakoudakis, P. Karabinas, 'Pt Electrodeposition on PAN-based Carbon Fibers', *J. Electroanal. Chem.*, **264**, 235–245 (1989).
16. Y.Y. Lau, T. Abe, A.G. Ewing, 'Voltammetric Measurement of Oxygen in Single Neurons Using Platinized Carbon Ring Electrodes', *Anal. Chem.*, **64**, 1702–1705 (1992).
17. Y.Y. Lau, D.K.Y. Wong, A.G. Ewing, 'Intracellular Voltammetry at Ultrasmall Platinum Electrodes', *Microchem. J.*, **47**, 308–316 (1993).
18. K.T. Kawagoe, J.B. Zimmerman, R.M. Wightman, 'Principles of Voltammetry and Microelectrode Surface States', *J. Neurosci. Methods*, **48**, 225–240 (1993).
19. R.M. Wightman, D.O. Wipf, 'Voltammetry at Ultramicroelectrodes', in *Electroanalytical Chemistry*, ed. A.J. Bard, Marcel Dekker, New York, 267–353, Vol. 15, 1989.
20. A. Fitch, D.H. Evans, 'Use of Microelectrodes for the Study of a Fast Chemical Step in an Electrode Reaction', *J. Electroanal. Chem.*, **202**, 83–92 (1986).

21. J.A. Stamford, 'In Vivo Voltammetry: Some Methodological Considerations', *J. Neurosci. Methods*, **17**, 1–29 (1986).
22. A.G. Ewing, K.D. Alloway, S.D. Curtis, M.A. Dayton, R.M. Wightman, G.V. Rebec, 'Simultaneous Electrochemical and Unit Recording Measurements: Characterization of the Effects of D-Amphetamine and Ascorbic Acid on Neostriatal Neurons', *Brain Res.*, **261**, 101–108 (1983).
23. P.M. Plotsky, 'Differential Voltammetric Measurement of Catecholamines and Ascorbic Acid at Surface-modified Carbon Filament Microelectrodes', *Brain Res.*, **235**, 179–184 (1982).
24. A.G. Ewing, M.A. Dayton, R.M. Wightman, 'Pulse Voltammetry with Microvoltammetric Electrodes', *Anal. Chem.*, **53**, 1842–1847 (1981).
25. F.G. Gonon, F. Navarre, M.J. Buda, 'In Vivo Monitoring of Dopamine Release in the Rat Brain with Differential Normal Pulse Voltammetry', *Anal. Chem.*, **56**, 573–575 (1984).
26. P. Capella, B. Ghasemzadeh, K. Mitchell, R.N. Adams, 'Nafion-coated Carbon Fiber Electrodes for Neurochemical Studies in Brain Tissue', *Electroanalysis*, **2**, 175–182 (1990).
27. M.W. Lada, R.T. Kennedy, 'Quantitative In Vivo Monitoring of Primary Amines in Rat Caudate Nucleus Using Microdialysis Coupled by a Flow-gated Interface to Capillary Electrophoresis with Laser-induced Fluorescence Detection', *Anal. Chem.*, **68**, 2790–2797 (1996).
28. B.L. Hogan, S.M. Lunte, J.F. Stobaugh, C.E. Lunte, 'On-line Coupling of In Vivo Microdialysis Sampling with Capillary Electrophoresis', *Anal. Chem.*, **66**, 596–602 (1994).
29. C.A. Marsden, M.P. Brazell, N.T. Maidment, 'An Introduction to In Vivo Electrochemistry', in *Measurement of Neurotransmitter Release In Vivo*, ed. C.A. Marsden, J. Wiley and Sons, New York, 127–151, 1984.
30. J.C. Conti, E. Strope, R.N. Adams, C.A. Marsden, 'Voltammetry in Brain Tissue: Chronic Recording of Stimulated Dopamine and 5-Hydroxytryptamine Release', *Life Sciences*, **23**, 2705–2715 (1978).
31. R.F. Lane, A.T. Hubbard, K. Fukunaga, R.J. Blanchard, 'Brain Catecholamines: Detection In Vivo by Means of Differential Pulse Voltammetry at Surface-modified Platinum Electrodes', *Brain Res.*, **114**, 346–352 (1976).
32. F. Gonon, M. Buda, R. Cespuglio, M. Jouvet, J.F. Pujol, 'In Vivo Electrochemical Detection of Catechols in the Neostriatum of Anaesthetized Rats: Dopamine or DOPAC?', *Nature*, **286**, 902–904 (1980).
33. A.G. Ewing, R.M. Wightman, M.A. Dayton, 'In Vivo Voltammetry with Electrodes that Discriminate between Dopamine and Ascorbate', *Brain Res.*, **249**, 361–370 (1982).
34. M.A. Dayton, A.G. Ewing, R.M. Wightman, 'Evaluation of Amphetamine-induced In Vivo Electrochemical Response', *Eur. J. Pharmacol.*, **75**, 141–144 (1981).
35. J.D. Salamone, L.S. Hamby, D.B. Neill, J.B. Justice, 'Extracellular Ascorbic Acid Increases in Striatum Following Systemic Amphetamine', *Pharmacol. Biochem. Behav.*, **20**, 609–612 (1984).
36. R.L. Wilson, R.M. Wightman, 'Systemic and Nigral Application of Amphetamine Both Cause an Increase in Extracellular Concentration of Ascorbate in the Caudate Nucleus of the Rat', *Brain Res.*, **339**, 219–226 (1985).
37. R.D. O'Neill, 'The Measurement of Brain Ascorbate In Vivo and its Link with Excitatory Amino Acid Neurotransmission', in *Neuromethods, Voltammetric Methods in Brain Systems*, eds. A.B. Boulton, G.B. Baker, R.N. Adams, Humana Press, Totowa, NJ, 221–268, Vol. 27, 1995.
38. G.A. Gerhardt, A.F. Oke, G. Nagy, B. Moghaddam, R.N. Adams, 'Nafion-coated Electrodes with High Selectivity for CNS Electrochemistry', *Brain Res.*, **290**, 390–395 (1984).
39. G. Rose, G. Gerhardt, I. Stromberg, L. Olson, B. Hoffer, 'Monoamine Release from Dopamine-depleted Rat Caudate Nucleus Reinnervated by Substantia Nigra Transplants: An In Vivo Electrochemical Study', *Brain Res.*, **341**, 92–100 (1985).
40. J.A. Stamford, 'In Vivo Voltammetry: Prospects for the Next Decade', *Trends Neurosci.*, **12**, 407–412 (1989).
41. R.N. Adams, 'In Vivo Electrochemical Measurements in the CNS', *Prog. Neurobiol.*, **35**, 297–311 (1990).
42. F. Crespi, 'In Vivo Voltammetry with Micro-biosensors for Analysis of Neurotransmitter Release and Metabolism', *J. Neurosci. Methods*, **34**, 53–65 (1990).
43. J.W. Commissiong, 'Monoamine Metabolites: Their Relationship and Lack of Relationship to Monoaminergic Neuronal Activity', *Biochem. Pharmacol.*, **34**, 1127–1131 (1985).
44. R.M. Wightman, D.S. Brown, W.G. Kuhr, R.L. Wilson, 'Molecular Specificity of In Vivo Electrochemical Measurements', in *Voltammetry in the Neurosciences*, ed. J.B. Justice, Humana Press, Totowa, NJ, 103–138, 1987.
45. A.G. Ewing, J.C. Bigelow, R.M. Wightman, 'Direct In Vivo Monitoring of Dopamine Released from Two Striatal Compartments in the Rat', *Science*, **221**, 169–171 (1983).
46. J.A. Stamford, Z.L. Kruk, J. Millar, R.M. Wightman, 'Striatal Dopamine Uptake in the Rat: In Vivo Analysis by Fast Cyclic Voltammetry', *Neurosci. Lett.*, **51**, 133–138 (1984).
47. W.G. Kuhr, R.M. Wightman, 'Real-time Measurement of Dopamine Release in Rat Brain', *Brain Res.*, **381**, 168–171 (1986).
48. P.A. Garris, E.D. Ciolkowski, P. Pastore, R.M. Wightman, 'Efflux of Dopamine from the Synaptic Cleft in the Nucleus Accumbens of the Rat Brain', *J. Neurosci.*, **14**, 6084–6093 (1994).
49. J.B. Zimmerman, R.M. Wightman, 'Simultaneous Electrochemical Measurements of Oxygen and Dopamine In Vivo', *Anal. Chem.*, **63**, 24–28 (1991).

50. J. Millar, M. Armstrong-James, Z.L. Kruk, 'Polarographic Assay of Iontophoretically Applied Dopamine and Low-noise Unit Recording Using a Multibarrel Carbon Fibre Microelectrode', *Brain Res.*, **205**, 419–424 (1981).
51. J.A. Stamford, P. Palij, C. Davidson, C.M. Jorm, J. Millar, 'Simultaneous "Real-time" Electrochemical and Electrophysiological Recording in Brain Slices with a Single Carbon-fibre Microelectrode', *J. Neurosci. Methods.*, **50**, 279–290 (1993).
52. J.O. Schenk, E. Miller, M. Rice, R.N. Adams, 'Chronoamperometry in Brain Slices: Quantitative Evaluations of In Vivo Electrochemistry', *Brain Res.*, **277**, 1–8 (1983).
53. D.N. Middlemiss, P.H. Hutson, 'Measurement of the In Vitro Release of Endogenous Monoamine Neurotransmitters as a Means of Identification of Prejunctional Receptors', *J. Neurosci. Methods*, **34**, 23–28 (1990).
54. M. Rice, A.F. Oke, C.W. Bradberry, R.N. Adams, 'Simultaneous Voltammetric and Chemical Monitoring of Dopamine Release In Situ', *Brain Res.*, **340**, 151–155 (1985).
55. D.R. Bull, P. Palij, M.J. Sheehan, J. Millar, J.A. Stamford, Z.L. Kruk, P.P.A. Humphrey, 'Application of Fast Cyclic Voltammetry to Measurement of Electrically Evoked Dopamine Overflow from Brain Slices In Vitro', *J. Neurosci. Methods*, **32**, 37–44 (1990).
56. J.J. O'Connor, Z.L. Kruk, 'Fast Cyclic Voltammetry can be used to Measure Stimulated Endogenous 5-Hydroxytryptamine Release in Untreated Rat Brain Slices', *J. Neurosci. Methods*, **38**, 25–33 (1991).
57. J.A. Stamford, P. Palij, C. Davidson, C.M. Jorm, P.E.M. Phillips, 'Fast Cyclic Voltammetry in Brain Slices', in *Neuromethods, Voltammetric Methods in Brain Systems*, eds. A.B. Boulton, G.B. Baker, R.N. Adams, Humana Press, Totowa, NJ, 81–116, Vol. 27, 1995.
58. F. Crespi, C. Pietra, 'Middle Cerebral Artery Occlusion Alters Neurotransmitter Activities in Ipsilateral and Contralateral Rat Brain Regions: An Ex Vivo Voltammetric Study', *Neurosci. Lett.*, **230**, 77–80 (1997).
59. R.S. Kelly, R.M. Wightman, 'Detection of Dopamine Overflow and Diffusion with Voltammetry in Slices of Rat Brain', *Brain Res.*, **423**, 79–87 (1987).
60. P. Palij, J.A. Stamford, 'Real-time Monitoring of Endogenous Noradrenaline Release in Rat Brain Slices Using Fast Cyclic Voltammetry: 3. Selective Detection of Noradrenaline Efflux in the Locus Coeruleus', *Brain Res.*, **634**, 275–282 (1994).
61. M.A. Bunin, C. Prioleau, R.B. Mailman, R.M. Wightman, 'Release and Uptake Rates of 5-Hydroxytryptamine in the Dorsal Raphe and Substantia Nigra Reticulata of the Rat Brain', *J. Neurochem.*, **70**, 1077–1087 (1998).
62. S. Cragg, M.E. Rice, 'Heterogeneity of Electrically Evoked Dopamine Release and Reuptake in Substantia Nigra, Ventral Tegmental Area, and Striatum', *J. Neurophysiol.*, **77**, 863–873 (1997).
63. J.J. O' Connor, Z.L. Kruk, 'Frequency Dependence of 5-HT Autoreceptor Function in Rat Dorsal Raphe and Suprachiasmatic Nuclei Studied Using Fast Cyclic Voltammetry', *Brain Res.*, **568**, 123–130 (1991).
64. P. Palij, J.A. Stamford, 'Real-time Monitoring of Endogenous Noradrenaline Release in Rat Brain Slices Using Fast Cyclic Voltammetry. 2. Operational Characteristics of the α_2 Autoreceptor in the Bed Nucleus of Stria Terminalis, Pars Ventralis', *Brain Res.*, **607**, 134–140 (1993).
65. C.C. Toner, J.A. Stamford, 'Sodium Channel Blockade Unmasks Two Temporally Distinct Mechanisms of Striatal Dopamine Release during Hypoxia/Hypoglycaemia In Vitro', *Neuroscience*, **81**, 999–1007 (1997).
66. C.C. Toner, J.A. Stamford, 'Involvement of N- and P/Q-but not L- or T-type Voltage-gated Calcium Channels in Ischaemia-induced Striatal Dopamine Release In Vitro', *Brain Res.*, **748**, 85–92 (1997).
67. B. Giros, M. Jaber, S.R. Jones, R.M. Wightman, M.G. Caron, 'Hyperlocomotion and Indifference to Cocaine and Amphetamine in Mice Lacking the Dopamine Transporter', *Nature*, **379**, 606–612 (1996).
68. S.R. Jones, R.R. Gainetdinov, R.M. Wightman, M.G. Caron, 'Mechanisms of Amphetamine Action Revealed in Mice Lacking the Dopamine Transporter', *J. Neurosci.*, **18**, 1979–1986 (1998).
69. D. Sulzer, T.K. Chen, Y.Y. Lau, H. Kristensen, S. Rayport, A.G. Ewing, 'Amphetamine Redistributes Dopamine from Synaptic Vesicles to the Cytosol and Promotes Reverse Transport', *J. Neurosci.*, **15**, 4102–4108 (1995).
70. Y.Y. Lau, J.B. Chien, D.K.Y. Wong, A.G. Ewing, 'Characterization of the Voltammetric Response at Intracellular Carbon Ring Electrodes', *Electroanalysis*, **3**, 87–95 (1991).
71. J.B. Chien, R.A. Wallingford, A.G. Ewing, 'Estimation of Free Dopamine in the Cytoplasm of the Giant Dopamine Cell of *Planorbis corneus* by Voltammetry and Capillary Electrophoresis', *J. Neurochem.*, **54**, 633–638 (1990).
72. A. Meulemans, B. Poulain, G. Baux, L. Tauc, D. Henzel, 'Micro Carbon Electrode for Intracellular Voltammetry', *Anal. Chem.*, **58**, 2088–2091 (1986).
73. T. Abe, Y.Y. Lau, A.G. Ewing, 'Characterization of Glucose Microsensors for Intracellular Measurements', *Anal. Chem.*, **64**, 2160–2163 (1992).
74. T. Abe, Y.Y. Lau, A.G. Ewing, 'Intracellular Analysis with an Immobilized-enzyme Glucose Electrode Having a 2-μm Diameter and Subsecond Response Times', *J. Am. Chem. Soc.*, **113**, 7421–7423 (1991).
75. W.R. LaCourse, D.A. Mead, D.C. Johnson, 'Anion-exchange Separation of Carbohydrates with Pulsed Amperometric Detection Using a pH-selective Reference Electrode', *Anal. Chem.*, **62**, 220–224 (1990).
76. G.G. Neuburger, D.C. Johnson, 'Pulsed Coulometric Detection with Automatic Rejection of Background

Signal in Surface-oxide-catalyzed Anodic Detections at Gold Electrodes in Flow-through Cells', *Anal. Chem.*, **60**, 2288–2293 (1988).
77. R.A. Wallingford, A.G. Ewing, 'Capillary Zone Electrophoresis with Electrochemical Detection', *Anal. Chem.*, **59**, 1762–1766 (1987).
78. T.M. Olefirowicz, A.G. Ewing, 'Dopamine Concentration in the Cytoplasmic Compartment of Single Neurons Determined by Capillary Electrophoresis', *J. Neurosci. Methods*, **34**, 11–15 (1990).
79. D.J. Leszczyszyn, J.A. Jankowski, O.H. Viveros, E.J. Diliberto, Jr, J.A. Near, R.M. Wightman, 'Nicotinic Receptor-mediated Catecholamine Secretion from Individual Chromaffin Cells: Chemical Evidence for Exocytosis', *J. Biol. Chem.*, **265**(25), 14736–14737 (1990).
80. R.M. Wightman, J.A. Jankowski, R.T. Kennedy, K.T. Kawagoe, T.J. Schroeder, D.J. Leszczyszyn, J.A. Near, E.J. Diliberto, Jr, O.H. Viveros, 'Temporally Resolved Catecholamine Spikes Correspond to Single Vesicle Release from Individual Chromaffin Cells', *Proc. Natl. Acad. Sci. USA*, **88**, 10754–10758 (1991).
81. D.J. Leszczyszyn, J.A. Jankowski, O.H. Viveros, E.J. Diliberto, Jr, J.A. Near, R.M. Wightman, 'Secretion of Catecholamines from Individual Adrenal Medullary Chromaffin Cells', *J. Neurochem.*, **56**, 1855–1863 (1991).
82. R.H. Chow, L. von Rüden, E. Neher, 'Delay in Vesicle Fusion Revealed by Electrochemical Monitoring of Single Secretory Events in Adrenal Chromaffin Cells', *Nature*, **356**, 60–63 (1992).
83. E. Neher, A. Marty, 'Discrete Changes of Cell Membrane Capacitance Observed Under Conditions of Enhanced Secretion in Bovine Adrenal Chromaffin Cells', *Proc. Natl Acad. Sci. USA*, **79**, 6712–6716 (1982).
84. K. Pihel, E.R. Travis, R. Borges, R.M. Wightman, 'Exocytotic Release from Individual Granules Exhibits Similar Properties at Mast and Chromaffin Cells', *Biophys. J.*, **71**, 1633–1640 (1996).
85. R. Borges, E.R. Travis, S.E. Hochstetler, R.M. Wightman, 'Effects of External Osmotic Pressure on Vesicular Secretion from Bovine Adrenal Medullary Cells', *J. Biol. Chem.*, **272**, 8325–8331 (1997).
86. J.A. Jankowski, J.M. Finnegan, R.M. Wightman, 'Extracellular Ionic Composition Alters Kinetics of Vesicular Release of Catecholamines and Quantal Size During Exocytosis at Adrenal Medullary Cells', *J. Neurochem.*, **63**, 1739–1747 (1994).
87. Z. Zhou, S. Misler, R.H. Chow, 'Rapid Fluctuations in Transmitter Release from Single Vesicles in Bovine Adrenal Chromaffin Cells', *Biophys. J.*, **70**, 1543–1552 (1996).
88. R.M. Wightman, T.J. Schroeder, J.M. Finnegan, E.L. Ciolkowski, K. Pihel, 'Time Course of Release of Catecholamines from Individual Vesicles during Exocytosis at Adrenal Medullary Cells', *Biophys. J.*, **68**, 383–390 (1995).
89. A.F. Oberhauser, I.M. Robinson, J.M. Fernandez, 'Simultaneous Capacitance and Amperometric Measurements of Exocytosis: A Comparison', *Biophys. J.*, **71**, 1131–1139 (1996).
90. A. Albillos, G. Dernick, H. Horstmann, W. Almers, G. Alvarez de Toledo, M. Lindau, 'The Exocytotic Event in Chromaffin Cells Revealed by Patch Amperometry', *Nature*, **389**, 509–512 (1997).
91. T.J. Schroeder, R. Borges, J.M. Finnegan, K. Pihel, C. Amatore, R.M. Wightman, 'Temporally Resolved, Independent Stages of Individual Exocytotic Secretion Events', *Biophys. J.*, **70**, 1061–1068 (1996).
92. J.A. Jankowski, T.J. Schroeder, R.W. Holz, R.M. Wightman, 'Quantal Secretion of Catecholamines Measured from Individual Bovine Adrenal Medullary Cells Permeabilized with Digitonin', *J. Biol. Chem.*, **267**, 18329–18335 (1992).
93. J.M. Finnegan, R.M. Wightman, 'Correlation of Real-time Catecholamine Release and Cytosolic Ca^{2+} at Single Bovine Chromaffin Cells', *J. Biol. Chem.*, **270**, 5353–5359 (1995).
94. I.M. Robinson, J.M. Finnegan, J.R. Monck, R.M. Wightman, J.M. Fernandez, 'Colocalization of Calcium Entry and Exocytotic Release Sites in Adrenal Chromaffin Cells', *Proc. Natl Acad. Sci. USA*, **92**, 2474–2478 (1995).
95. R.H. Chow, J. Klingauf, E. Neher, 'Time Course of Ca^{2+} Concentration Triggering Exocytosis in Neuroendocrine Cells', *Proc. Natl Acad. Sci. USA*, **91**, 12765–12769 (1994).
96. A. Elhamdani, Z. Zhou, C.R. Artalejo, 'Timing of Dense-core Vesicle Exocytosis Depends on the Facilitation L-type Ca Channel in Adrenal Chromaffin Cells', *J. Neurosci.*, **18**, 6230–6240 (1998).
97. K.L. Engisch, N.I. Chernevskaya, M.C. Nowycky, 'Short-term Changes in the Ca^{2+}–Exocytosis Relationship during Repetitive Pulse Protocols in Bovine Adrenal Chromaffin Cells', *J. Neurosci.*, **17**, 9010–9025 (1997).
98. T.J. Schroeder, J.A. Jankowski, J. Senyshyn, R.W. Holz, R.M. Wightman, 'Zones of Exocytotic Release on Bovine Adrenal Medullary Cells in Culture', *J. Biol. Chem.*, **269**(25), 17215–17220 (1994).
99. R. Zhou, G. Luo, A.G. Ewing, 'Direct Observation of the Effect of Autoreceptors on Stimulated Release of Catecholamines from Adrenal Cells', *J. Neurosci.*, **14**, 2402–2407 (1994).
100. A.G. Ewing, T.K. Chen, G. Chen, 'Voltammetric and Amperometric Probes for Single-cell Analysis', in *Neuromethods, Voltammetric Methods in Brain Systems*, eds. A.B. Boulton, G.B. Baker, R.N. Adams, Humana Press, Totowa, NJ, 269–304, Vol. 27, 1995.
101. R.M. Wightman, J.M. Finnegan, K. Pihel, 'Monitoring Catecholamines at Single Cells', *Trends Anal. Chem.*, **14**, 154–158 (1995).
102. G. Chen, A.G. Ewing, 'Chemical Analysis of Single Cells and Exocytosis', *Crit. Rev. Neurobiol.*, **11**, 59–90 (1997).

103. R.A. Clark, A.G. Ewing, 'Quantitative Measurements of Released Amines from Individual Exocytosis Events', *Mol. Neurobiol.*, **15**, 1–16 (1997).
104. R.A. Clark, S.E. Zerby, A.G. Ewing, 'Electrochemistry in Neuronal Microenvironments', in *Electroanalytical Chemistry a Series of Advances*, eds. A.J. Bard, I. Rubinstein, Marcel Dekker, New York, 227–294, 1998.
105. L.A. Greene, A.S. Tischler, 'Establishment of a Noradrenergic Clonal Line of Rat Adrenal Pheochromocytoma Cells which Respond to Nerve Growth Factor', *Proc. Natl Acad. Sci. USA*, **73**, 2424–2428 (1976).
106. D. Schubert, F.G. Klier, 'Storage and Release of Acetylcholine by a Clonal Cell Line', *Proc. Natl Acad. Sci. USA*, **74**, 5184–5188 (1977).
107. J.A. Wagner, 'Structure of Catecholamine Secretory Vesicles from PC12 Cells', *J. Neurochem.*, **45**, 1244–1253 (1985).
108. T.K. Chen, G. Luo, A.G. Ewing, 'Amperometric Monitoring of Stimulated Catecholamine Release from Rat Pheochromocytoma (PC12) Cells at the Zeptomole Level', *Anal. Chem.*, **66**(19), 3031–3035 (1994).
109. R. Levi-Montalcini, P.U. Angeletti, 'Nerve Growth Factor', *Physiol. Rev.*, **48**, 534–569 (1968).
110. S.E. Zerby, A.G. Ewing, 'Electrochemical Monitoring of Individual Exocytotic Events from the Varicosities of Differentiated PC12 Cells', *Brain Res.*, **712**, 1–10 (1996).
111. W.B. Stallcup, 'Sodium and Calcium Fluxes in a Clonal Nerve Cell Line', *J. Physiol.*, **286**, 525–540 (1979).
112. M.J. Berridge, R.F. Irvine, 'Inositol Trisphosphate, a Novel Second Messenger in Cellular Signal Transduction', *Nature*, **312**, 315–321 (1984).
113. S.E. Zerby, A.G. Ewing, 'The Latency of Exocytosis Varies with the Mechanism of Stimulated Release in PC12 Cells', *J. Neurochem.*, **66**, 651–657 (1996).
114. P.S. Chowdhury, X. Guo, T.D. Wakade, D.A. Przywara, A.R. Wakade, 'Exocytosis from a Single Rat Chromaffin Cell by Cholinergic and Peptidergic Neurotransmitters', *Neuroscience*, **59**, 1–5 (1994).
115. K.D. Kozminski, D.A. Gutman, V. Davila, D. Sulzer, A.G. Ewing, 'Voltammetric and Pharmacological Characterization of Dopamine Release from Single Exocytotic Events at Rat Pheochromocytoma (PC12) Cells', *Anal. Chem.*, **70**, 3123–3130 (1998).
116. E.N. Pothos, S. Przedborski, V. Davila, Y. Schmitz, D. Sulzer, 'D_2-like Dopamine Autoreceptor Activation Reduces Quantal Size in PC12 Cells', *J. Neurosci.*, **18**, 5575–5585 (1998).
117. E. Pothos, M. Desmond, D. Sulzer, 'L-3, 4-Dihydroxyphenylalanine Increases the Quantal Size of Exocytotic Dopamine Release In Vitro', *J. Neurochem.*, **66**, 629–636 (1996).
118. G.A. deToledo, R. Fernández-Chacón, J.M. Fernández, 'Release of Secretory Products During Transient Vesicle Fusion', *Nature*, **363**, 554–558 (1993).
119. K. Pihel, S. Hsieh, J.W. Jorgenson, R.M. Wightman, 'Electrochemical Detection of Histamine and 5-Hydroxytryptamine at Isolated Mast Cells', *Anal. Chem.*, **67**, 4514–4521 (1995).
120. H.P. Meissner, 'Membrane Potential Measurements in Pancreatic Beta Cells with Intracellular Microelectrodes', *Methods Enzymol.*, **192**, 235–246 (1990).
121. R.T. Kennedy, L. Huang, M.A. Atkinson, P. Dush, 'Amperometric Monitoring of Chemical Secretions from Individual Pancreatic β-Cells', *Anal. Chem.*, **65**, 1882–1887 (1993).
122. L. Huang, H. Shen, M.A. Atkinson, R.T. Kennedy, 'Detection of Exocytosis at Individual Pancreatic β-Cells by Amperometry at a Chemically Modified Microelectrode', *Proc. Natl Acad. Sci. USA*, **92**, 9608–9612 (1995).
123. C.D. Paras, R.T. Kennedy, 'Electrochemical Detection of Exocytosis at Single Rat Melanotrophs', *Anal. Chem.*, **67**, 3633–3637 (1995).
124. G. Chen, P.F. Gavin, G. Luo, A.G. Ewing, 'Observation and Quantitation of Exocytosis from the Cell Body of a Fully Developed Neuron in *Planorbis corneus*', *J. Neurosci.*, **15**, 7747–7755 (1995).
125. G. Chen, A.G. Ewing, 'Multiple Classes of Catecholamine Vesicles Observed During Exocytosis from the Planorbis Cell Body', *Brain Res.*, **701**, 167–174 (1995).
126. D. Bruns, R. Jahn, 'Real-time Measurement of Transmitter Release from Single Synaptic Vesicles', *Nature*, **377**, 62–65 (1995).
127. B. Powell, G.A. Cottrell, 'Dopamine in an Identified Neuron of *Planorbis corneus*', *J. Neurochem.*, **22**, 605–606 (1974).
128. J.M. Finnegan, K. Pihel, P.S. Cahill, L. Huang, S.E. Zerby, A.G. Ewing, R.T. Kennedy, R.M. Wightman, 'Vesicular Quantal Size Measured by Amperometry at Chromaffin, Mast, Pheochromocytoma, and Pancreatic β-Cells', *J. Neurochem*, **66**, 1914–1923 (1996).
129. G. Chen, D.A. Gutman, S.E. Zerby, A.G. Ewing, 'Electrochemical Monitoring of Bursting Exocytotic Events from the Giant Dopamine Neuron of *Planorbis corneus*', *Brain Res.*, **733**, 119–124 (1996).
130. Z. Zhou, S. Misler, 'Amperometric Detection of Stimulus-induced Quantal Release of Catecholamines from Cultured Superior Cervical Ganglion Neurons', *Proc. Natl Acad. Sci. USA*, **92**, 6938–6942 (1995).
131. E.N. Pothos, V. Davila, D. Sulzer, 'Presynaptic Recording of Quanta from Midbrain Dopamine Neurons and Modulation of the Quantal Size', *J. Neurosci.*, **18**, 4106–4118 (1998).
132. E.H. Jaffe, A. Marty, A. Schulte, R.H. Chow, 'Extrasynaptic Vesicular Transmitter Release from the Somata of Substantia Nigra Neurons in Rat Midbrain Slices', *J. Neurosci.*, **18**, 3548–3553 (1998).
133. L. Huang, R.T. Kennedy, 'Exploring Single-cell Dynamics Using Chemically-modified Microelectrodes', *Trends Anal. Chem.*, **14**, 158–164 (1995).

134. T. Malinski, Z. Taha, S. Grunfeld, A. Burewicz, P. Tomboulian, F. Kiechle, 'Measurements of Nitric Oxide in Biological Materials Using a Porphyrinic Microsensor', *Anal. Chim. Acta*, **279**, 135–140 (1993).

135. F. Lantoine, S. Trevin, F. Bedioui, J. Devynck, 'Selective and Sensitive Electrochemical Measurement of Nitric Oxide in Aqueous Solution: Discussion and New Results', *J. Electroanal. Chem.*, **392**, 85–89 (1995).

Organic Electrochemical Mechanisms

Claude P. Andrieux
Université Paris 7 – Denis Diderot, Paris, France

The determination of electrochemical mechanisms by electrochemical techniques is very efficient in organic electrochemistry. In contrast to the electrochemistry of small molecules and ions, the electron transfers on organic compounds are mainly outer-sphere electron transfers following a Marcus–Hush law. Hence these fast electron transfers open up the possibility of obtaining kinetic limitation by the associated homogeneous chemical reactions.

The classical methods for these determinations are stationary techniques such as rotating disk electrode voltammetry (RDEV) and polarography but also more powerful methods, involving large-amplitude transient methods: linear sweep voltammetry (LSV), cyclic voltammetry (CV) and double-step chronoamperometry (DSC) allow the time window to be extended to thousands of microseconds; 1 µs can be reached with the use of ultramicroelectrodes. When the chemical reaction is the rate-determining step, the response, i.e. current as a function of time (or potential), provides information about the nature of this reaction and in many case allows the measurement of its rate constant. Characteristic behaviors are detailed for the principal chemical reactions preceding (C + E) or following (E + C) the electron transfer.

Organic electrochemical mechanisms often involve several electron transfers. The influence of homogeneous electron transfers is underlined in particular for two-electron reductions or oxidations and also in redox catalysis. The possibility of an electron transfer concerted with a chemical reaction is characterized. Some examples of mechanisms and kinetic determinations are given in the field of widely used organic electrosynthetic processes.

1 INTRODUCTION

An organic electrochemical process involves electron transfers on an organic molecule. Such processes are possible with a great number of organic compounds and are usually conducted in an electrolytic cell where electron transfers are located at the interface between an electrode

For references see page 10008

and a conducting solution containing the compound to be reduced (electron addition) or to be oxidized (electron removal).

In the field of organic chemistry, a simple electron transfer never occurs alone, with the exception of the electrochemical preparation of stable free radicals. An organic electrochemical mechanism is an association of electron transfers and chemical reactions. Hence an electrochemical reaction appears as a sequence of three kinds of simple phenomena:

- electron transfer on organic species at an electrode surface; these species are usually in the solution but could be located in somes cases in a particular layer (adsorption layer, for example) near the electrode;
- chemical reactions from organic species produced after one or several electron transfers; these reactions can involve other reactants present in the electrolytic solution;
- transport of reactants, intermediates and products from and to the electrode surface.

This article deals with the description of the sequence of electron transfers and chemical reactions for organic species when particular conditions of potential or current are fixed on an electrode. The nature of this electrode, the solvent used and the other reactants in the medium are important parameters in allowing a particular global electrochemical reaction to occur from the reactant to one or several products. Some books deal with the more interesting processes in this field.[1–3]

The purpose here is mainly to describe how it is possible to determine this sequence (or some part of this sequence) from electrochemical measurements by analyzing the electrical response when using electrochemical techniques at the level of small electrodes (microelectrodes) set in the initial solution. As described in several textbooks,[4,5] the kinetics of the overall process are the result of the kinetics of the above three kinds of phenomena, but in many cases the global kinetics are a function of only a few key steps. The analysis of the electrical response on the microelectrode, associated with the description of the final products, is an efficient tool in the determination of this mechanism.

2 ELECTROCHEMICAL TECHNIQUES USED FOR THE DETERMINATION OF ORGANIC ELECTROCHEMICAL MECHANISMS

2.1 Cells and Electrodes

As organic compounds are usually insoluble in water, most electrochemical studies of organic compounds are carried out in solvents other than water. These solvents must also permit the dissolution of organic species and of some strong electrolytes (supporting electrolytes) at about 0.1 or 1 M concentrations, leading to an electrolytic solution with sufficiently good conductivity. Thus, solvents with a dielectric constant ε_r in the range 20–50 are the most commonly used, e.g. acetonitrile $\varepsilon_r = 37.4$, dimethylformamide (DMF) $\varepsilon_r = 36.7$ and dimethyl sulfoxide (DMSO) $\varepsilon_r = 46.7$ as aprotic solvents and methanol $\varepsilon_r = 32.6$ and ethanol $\varepsilon_r = 24.3$ as protic organic solvents, the latter able to be used pure or mixed with water.[6] The possibilities opened up by the addition of a surfactant to an aqueous solution that allows the solubilization of organic compounds have often been described and pose a particular challenge at the level of industrial preparation.[7]

Because these electrolytes must remain inert within the potential range of interest, tetraalkylammonium or lithium cations are used when a large potential range is necessary in reduction; perchlorate, tetrafluoroborate and hexafluorophosphate are used when a large potential range is necessary in oxidation.

For the electrode material, the same metallic material must be used for the analytical probe and for the preparative electrolysis, so a wide number of conducting materials are used. For reduction processes, mercury or lead are often used because they allow a high hydrogen overvoltage. In oxidation processes the inertness to most chemical environments calls for the use of platinum or glassy carbon. The latter allow both oxidation and reduction.[1–3,8]

From an analytical point of view, a three-electrode device cell is generally used. The electrical potential of the working electrode is controlled regardless of what current is flowing through it, independent of the reactions occurring at the other electrode (counter electrode or auxiliary electrode) by a potentiostat.[5,9] This electronic device, built from operational amplifiers, assumes a potential difference automatically adjusted between the working and the auxiliary electrodes in such a way that the potential difference between the working electrode and the reference electrode is fixed at the desired value. The auxiliary electrode is commonly an inert, larger sized electrode compared with the working electrode.

The reference electrode should satisfy the following requirements: its potential should remain constant with time, and if small currents are passed through it, they must have little effect on its potential, with an immediate return to the initial value upon elimination of polarization. Therefore, its impedance must be high and fast electron transfer between two stable species is necessary.

The most common reference electrode is a saturated calomel electrode (SCE) in water. In fact, it is very often used in all media provided that a salt-conducting separator is introduced between the two solutions. In an aprotic medium, other reference electrodes are used, e.g. silver–silver ions in acetonitrile or silver–silver chloride.[10] The normal hydrogen electrode is never used in practical experiments.

According to readings from the potentiostat and from the electrochemical method used, it is possible to indicate suitable dimensions of the working electrode. For classical methods, millimeter-sized electrodes are used, but for very fast transient methods the dimensions can reach 1 μm. In order to interpret the electrochemical response, an analysis of the electrical circuit of an electrochemical cell with its connection to the potentiostat must be given (Figure 1).[4,5,11] The effect of the double-layer capacitance will be discussed later. Recent potentiostats permit the minimization (without being able to suppress completely the ohmic drop for stability reasons) of the value of the uncompensated resistance R_u by a positive feedback compensation.[12] A current measurer is also associated with the electronic device. The readings of these electronic devices are a function of the technique used. If rapidly varying working potentials are involved, the potentiostat response has to be short and requires the use of large bandpass amplifiers.

It must be noted that although in the preparative scale-up some separation between the working compartment and the auxiliary compartment often occurs, this separation is not necessarily at the level of the mechanism determination since the concentration perturbation stays in the vicinity of the working electrode and therefore the reactions at the auxiliary electrode do not affect the process. Consequently, the analysis that is developed here is not directly applicable to coupled reduction and oxidation reactions where products produced on one electrode interfere with the electrochemical reactions on the other electrode.

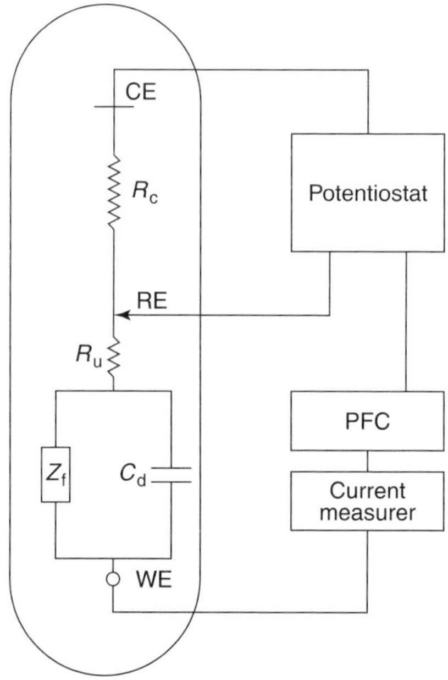

Figure 1 Scheme of the equivalent electrical circuit of an electrochemical cell and its control of potential by a three-electrode potentiostat with a positive feedback compensation (PFC). WE, CE and RE are the working, counter and reference electrodes, R_u and R_c the cell resistances, Z_f the faradaic impedance and C_d the double-layer capacitance.

In order to describe the different techniques that are used, we shall first describe, briefly, the electrical response for a single, fast, one-electron transfer neglecting secondary phenomena and more specifically ohmic drop and double-layer capacitance. Fast electron transfer[5,13] means that the kinetics of the electron transfer do not interfere with other limiting kinetics and that we can apply the Nernst law to this simple electron transfer (Equations 1 and 2):

$$A + e \rightleftharpoons B \tag{1}$$

$$E = E^0 + \frac{RT}{F}\left[\ln\left(\frac{C_A^0}{C_B^0}\right)\right] \tag{2}$$

where C_A^0 and C_B^0 are the concentrations of A and B at the electrode, E is the electrode potential and E^0 is the standard potential of the couple A/B. The initial concentration of only electroactive species A present in solution is C^0, with an order of magnitude for this concentration of 10^{-3} M. The example here is of a reductive process but transposition to oxidation is easy.

Because we are assuming that no chemical complications are involved, only transport phenomena are kinetically determinant. In fact, the transport phenomenon is mainly diffusion and under certain conditions convection for the reactants and the products. This is because in the solution as a large excess of supporting electrolyte is present (the concentration C^0 remains very low compared with the concentration of electrolyte), so the migration of the charged species involved in the electrode reaction is restricted to a very narrow double-layer region (a few ångstroms) and, as seen below, very narrow compared with the diffusion layer.[5,14] Hence we can neglect this migration effect for all the described techniques and the diffusion phenomenon can be explained by a simple relationship according to Fick's law.

2.2 Stationary Techniques

RDEV[5,14,15] is a method where the working electrode is a small conducting disk with a surface ranging from 1 to 10 mm^2. Thus a large amount of material can be used. Laminar flow is involved for moderate rotation speeds

For references see page 10008

and the concentration profiles for the reactant and the product are calculated, taking into account the forced convection produced by the rotation of the electrode. The current at the electrode is dependent on potential but reaches a stationary value for any imposed potential (Figure 2a and b).

The Nernst approximation applies and we can replace the actual profile by a linear gradient in a diffusion layer (Equation 3):

$$\delta = 4.98\, D^{1/3} \nu^{1/6} w^{-1/2} \tag{3}$$

where D is the diffusion coefficient (cm^2), ν the kinematic viscosity (cm^2 s^{-1}), w the angular rotation rate (rpm) and δ is expressed in centimeters. In this diffusion layer, a stationary profile is obtained and a linear variation of the two concentrations is observed. The possible rotation rate is between 50 and 20 000 rpm and the corresponding thickness of the diffusion layer is between 7×10^{-3} and 4×10^{-4} cm.

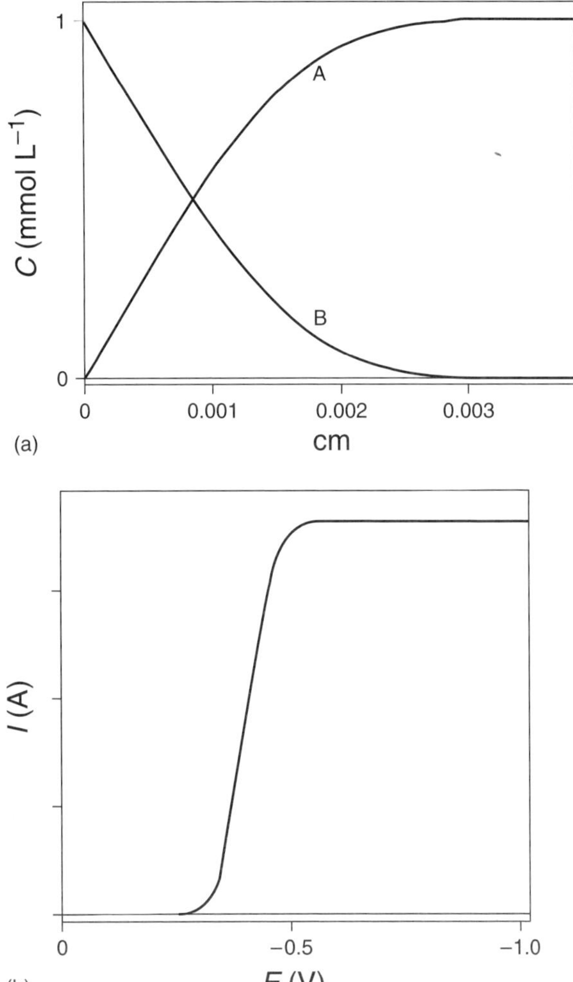

Figure 2 (a) Concentration profile for reactant A and product B in RDEV on the plateau current. (b) Current potential curve in RDEV for the reaction $A + e^- \rightleftharpoons B$ with $E^0 = -0.4$ V.

Polarography[5] is the oldest of all electrochemical techniques. A dropping mercury electrode serves as the working electrode, imposing severe limitations on reactions on this metal. The current law seems rather complex because this method involves a moving sphere, the surface area increasing until the drop falls. The current varies with time yet at the level of the maximum (or the average) current, a satisfactory approximation is a stationary regime in which the Nernst approximation (linear profiles) is applied, with a diffusion layer thickness in relation to the drop time θ (Equation 4):

$$\delta = \left(\frac{3\pi D\theta}{7}\right)^{1/2} \tag{4}$$

As the drop time is variable only within a narrow window (1–10 s), the thickness of the diffusion layer is only between 4×10^{-3} and 10^{-2} cm, leading to a very small variation in the experimental parameters.

With both stationary techniques, it is possible to scan the potential very slowly and thus obtain a current–potential curve (voltammetric curve). This gives a clear picture of the range of potential that must be applied in order to conduct the electrochemical reaction. The characteristic values of this plateau-shaped curve are the half-wave potential $E_{1/2} = E^0$ and plateau current $i_l = FSDC^0/\delta$.

As a complement to the RDEV and with the purpose of detecting the product obtained on the disk in the electrochemical reaction and to see if this intermediate is decomposed by a chemical reaction, rotating ring-disk electrode voltammetry (RRDEV) is used.[16] The working electrode is now a disk as in RDEV, surrounded by a concentric ring separated from the disk by an insulator. The gap between disk and ring must be small to ensure that species produced on the disk can easily reach the ring. The potentials of the disk and the ring are controlled independently. In the case of a simple electron exchange, if the potential of the ring is set at a value more positive than the reduction potential of A and if the potential of the disk is slowly changed from positive to negative as in an RDEV experiment, a wave of similar shape (with a plateau for a negative potential of the disk) is also obtained on the disk during this experiment. The current obtained on the ring is the current of reoxidation of B and the height of this wave depends on the geometric dimensions of the electrode but will always be smaller than the reduction current.

2.3 Transient Techniques

Two kinds of transient techniques are described. At the level of mechanism investigations, large-amplitude techniques are usually more adapted to the problem, since they have the same advantage as stationary techniques of easily visualizing the electrochemical process. These techniques also allow a large range for the time window.[11,17] LSV and CV are powerful tools in this area.

The working electrode is a stationary disk or sphere with an area in the range of $1\,mm^2$ and the working electrode potential is imposed as a linear function of time (LSV). Very often, the potential returns linearly to the initial value (CV). The sweep rate v is the most important parameter because it controls the rate of diffusion of the species at the electrode and thus plays a role analogous to the rotation speed in RDEV. This sweep rate can be varied between 0.1 and a few thousand $V\,s^{-1}$.

In the case of a simple electron transfer, when the potential is swept within the potential range where the electrochemical reaction is, at the beginning (initial potential), nonoperative and completely occurring (inversion potential), the curve shown in Figure 3 is obtained.

A convenient experimental characterization of the current–potential curve is measured according to the height, the location and the width of the peak for the first and second parts of the sweep.[11,17,18]

For the first scan, these cathodic peak characteristics for a single fast transfer are expressed by Equations (5–7):

$$\text{Peak current:}\quad i_p = 0.446FSC^0D^{1/2}\left(\frac{Fv}{RT}\right)^{1/2} \tag{5}$$

$$\text{Peak potential:}\quad E_p = E^0 - 1.11\left(\frac{RT}{F}\right) \tag{6}$$

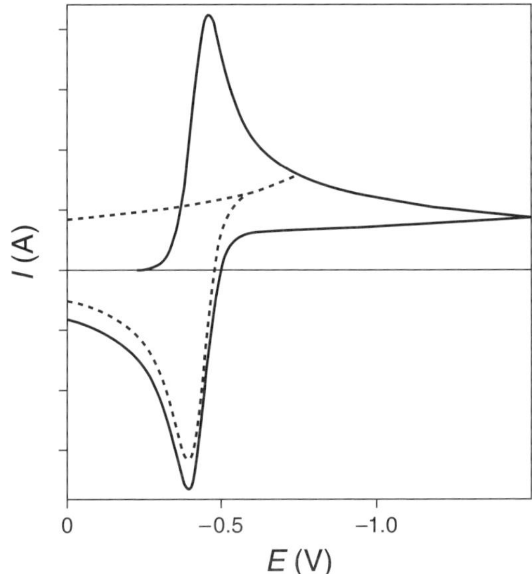

Figure 3 LSV and CV. Diffusion-controlled voltammogram (solid curve) and the measurement of the anodic part from the prolongation of the cathodic part (dashed curve).

$$\text{Peak width:}\quad E_{p/2} - E_p = 2.20\left(\frac{RT}{F}\right) \tag{7}$$

where $E_{p/2}$ is the potential corresponding to a current intensity equal to half that peak current.

For the reverse part of the potential sweep, if the curve is calculated from the prolongation of the diffusion part of the first part, the same peak current (anodic current) and peak width are found with an anodic peak potential (Equation 8) of

$$(E_p)_2 = E^0 + 1.11\left(\frac{RT}{F}\right) \tag{8}$$

leading to a cathodic and anodic peak separation (Equation 9) of

$$\Delta E_p = 2.22\left(\frac{RT}{F}\right) \tag{9}$$

Other useful large-amplitude techniques in the field of organic electrochemical mechanism determination are potential step chronoamperometry (PSC) and DSC.[5,11,17] In PSC, the stationary electrode is stepped from a rest potential (no reduction of A) to a more negative value, markedly more negative than the standard potential E^0 so that the concentration of A at the electrode is zero. The cathodic current is only controlled by diffusion and can be expressed by Equation (10):

$$i = \frac{FSC^0D^{1/2}}{(\pi t)^{1/2}} \tag{10}$$

Moreover, in DSC, after a time θ, the potential is stepped back to the initial potential and, as for a CV experiment, the reoxidation curve is exactly the same as the cathodic curve if the current is measured starting from the prolongation of the cathodic curve (Figure 4).

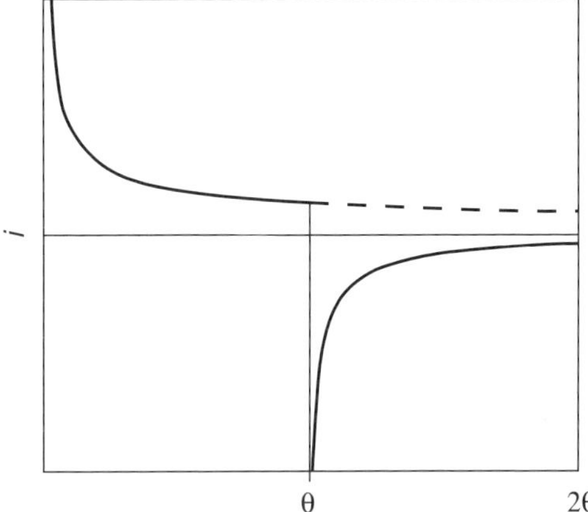

Figure 4 DSC. Current as a function of time with inversion time θ. Reconstruction of the anodic current from the prolongation of the cathodic part (dashed line).

For references see page 10008

An easy way to analyze the reoxidation curve is to measure the anodic current from the current axis at a time 2θ (double the inversion time). The ratio between the anodic and the cathodic currents for θ is 0.293.

Small-amplitude transient techniques are often used for the analytical aspects of these techniques.[19] For the purpose of determining the characteristics of the electron transfers, some aspects of square- or sinusoidal-superimposed potential [alternating current (ac) technique, also called the impedance technique][20] have been developed with a range of frequency of 10–10 000 Hz. It has been claimed that the second-harmonic ac technique is very efficient.[21] These techniques consist of applying on the electrode a small-amplitude sinusoidal potential (frequency ν) and a slow variation of a dc potential and detecting the amplitude and the phase of the first- (frequency ν) or second-harmonic (frequency 2ν) current amplitude as a function of the dc potential. The two components in-phase and out-of-phase with the input potential are then separated. It must be noted that the second-harmonic ac technique contains the same limitation as the simple ac technique.[22] At the level of mechanism determinations for complex sequences, these techniques are very rarely used.

2.4 Comparison of the Time Windows

For stationary techniques, the time window is determined from the mean transport (diffusion) time from the solution to the electrode and thus is D/δ^2. The limitations are dependent on the hydrodynamic limitation with respect to the rotation rate in RDEV, as mentioned earlier, and dependent on the drop time in polarography.

For transient techniques, the limitations that occur are essentially due to secondary phenomena.[11] In the preceding sections, we have assumed that the current at the electrode surface results only from diffusion in solution when an electron transfer occurs between an electrode and the organic substrate (faradaic current). However, several side effects can disturb this assumption.

To start with, a double layer is always present at the electrode interface, which leads to an additional capacitor SC_d in parallel with the faradaic impedance, as indicated in Figure 1 (C_d is the capacitance per unit of area). At the level of the stationary techniques, this capacitor does not give any additional current, but for all the transient techniques its presence imposes severe limitations.

For example, in LSV, a constant capacitive current $i_c = SC_d v$ is obtained and it should be noted that this current is a function of the sweep rate v as compared with the faradaic current (that is a function the square root of v). Hence, for large values for v, the capacitive component increases as compared with the faradaic current. For a concentration of substrate of about 10^{-3} M, these two components have the same order of magnitude for a sweep rate of a few thousand volts per second.

As previously remarked, an uncompensated resistance between the working and reference electrodes remains in a three-electrode potentiostat; even when this potentiostat is equipped with feedback compensation, part of this resistance remains.[12] The total capacitive current is expressed by Equation (11):

$$i_c = SC_d v\left[1 - \exp\left(\frac{-t}{R_u SC_d}\right)\right] \tag{11}$$

The resulting voltammetry is represented in Figure 5. The duration of the increasing part of the capacitive current is approximated by $R_u SC_d$ and a faradaic signal can be accurately measured if the time of the experiment is long compared with $R_u SC_d$. The uncompensated resistance also leads to important modifications on the faradaic curve itself owing to an effective potential applied where the ohmic drop $R_u i$ must be taken into account. Let us estimate the order of magnitude of these different effects. If a potentiostat with no feedback compensation is used in an aprotic medium, R_u is about 500 Ω, where $S = 0.01\ \text{cm}^2$, $C_d = 10\ \text{mF cm}^{-2}$ and $D = 10^{-5}\ \text{cm}^2\ \text{s}^{-1}$, using a sweep rate of $1000\ \text{V s}^{-1}$, then $i_F = 283$ mA, $i_c = 100$ mA, $R_u i = 161$ mV and $R_u SC_d = 50\ \mu\text{s}$. With a reduced resistance at 20 Ω by feedback compensation, the rising time becomes 2 μs and the residual ohmic drop falls at 8 mV. This value is not sufficient for an accurate determination of peak potential, which would require 1–2 mV. However, by taking into account the possibility of approximating the ohmic drop via a simple

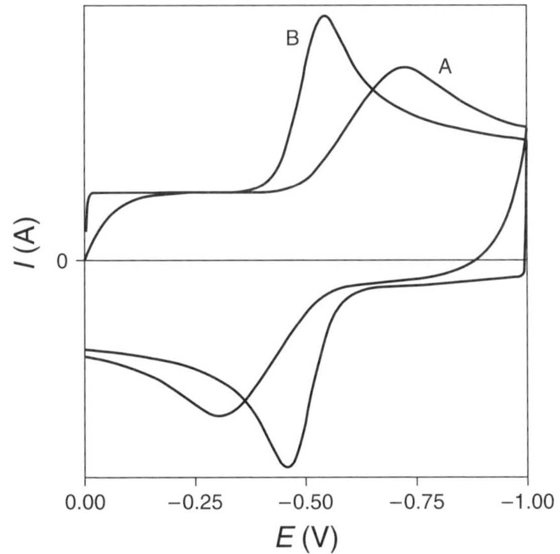

Figure 5 CV at $1000\ \text{V s}^{-1}$ (A) without positive feedback compensation, $R_u = 500\ \Omega$, and (B) with positive feedback compensation, $R_u = 20\ \Omega$ ($S = 0.01\ \text{cm}^2$, $C_d = 10\ \text{mF cm}^{-2}$, $D = 10^{-5}\ \text{cm}^2\ \text{s}^{-1}$).

calculation, since the residual resistance can be measured, we can determine this peak potential with the desired accuracy. It must be noted that a complete resistance compensation can never be obtained because when the resistance is reduced to a very small value, oscillations at the beginning of the capacitive current curve appear and it is therefore possible theoretically to show that for zero residual resistance a complete unstable system is involved.

For other transient techniques the same kind of limitation occurs, which prohibits a very short experiment time. For example, in PSC, even if the effect of ohmic drop can theoretically be neglected because the step value in potential is chosen far enough away from the standard potential value, the effect is still present on the shape of the curve, mainly for short observation times. Moreover, the capacitive and the faradaic currents in this short-time experiment are high enough that the limitation for short times cannot be estimated merely in terms of R_uSC_d but by about 10 or 20 times these values before which the diffusion current is not affected by the capacitive current. These high currents also prevent the use of very efficient ohmic drop compensation. Thus, the short-time limitation for this technique is 1 ms without ohmic drop compensation and 0.1 ms otherwise.

Another limitation is the effect of nonlinear diffusion on the small disk or on the sphere acting as the working electrode. This effect appears for long-time values or for low sweep rates and it can be neglected on millimeter-sized electrodes down to $0.1\,V\,s^{-1}$ (or up to 1 s in PSC). Thus the sweep rate window in LSV ranges from 0.1 to $1000\,V\,s^{-1}$. The time window corresponding to this sweep rate is RT/Fv, a value that is of the same order of magnitude as the time occurring between the standard potential and peak potential. Hence the time window for an LSV experiment on a millimeter-sized electrode ranges from 25 µs to 0.25 s.

Recently, a means of obtaining higher sweep rates was found using ultramicroelectrodes[23–25] provided that large-bandpass amplifiers were used. Disk electrodes between 1 and 10 µm are now commercially available for this purpose. As shown previously, three parameters must be minimized in order to shorten response times (or increase sweep rates): R_uSC_d, R_ui and i_c as compared with i_F. A possibility is to decrease the size of the electrode. In this way not only R_uSC_d is decreased but also the resistance near the working electrode (the important part of the resistance between the working and reference electrodes) is more or less a function of the inverse of the radius of this electrode. Thus R_u is decreased by a factor of 100 between a millimeter-sized and a 10-µm electrode. Figure 6 illustrates the possibility of the use of these ultramicroelectrodes, shown with a sweep rate up to $10^6\,V\,s^{-1}$.

The use of an ultramicroelectrode is not restricted to LSV; we can also use it in DSC. The gain obtained compared with a millimeter-sized electrode is about two orders of magnitude, which leads to short time values of about 1 µs.

With an ultramicroelectrode, a limitation due to spherical diffusion occurs at a higher sweep rate (about $100\,V\,s^{-1}$) but does not present a real problem because at lower sweep rates a stationary response is obtained even if a stationary electrode is used. From this change in the diffusion regime, an elegant method for determining the exchange number of electrons in a complex mechanism is developed.[25]

Table 1 summarizes the various possibilities of the more commonly used methods in the analysis of organic electrochemical mechanisms.

In addition to the classical electrochemical methods, an effective method using photoinjection of electrons by means of a laser flash has been described,[26] as shown in Figure 7.

The advantage of this method is that it produces solvated electrons that react with a reducible neutral compound very close to the surface. Therefore, analysis of the chemical reaction in this region near the electrode is possible. The potential of the electrode can be fixed in a potential region beneath the reduction potential of the initial compound. The measurement is the residual photoinduced charge transferred from the electrode to the solution. If the intermediate product (anion radical) is stable, the response is similar to the response of the recombination of the free electrons produced in the solution without any reducible compound (the solvated electron and the radical anion are reoxidized at the electrode). However, if the intermediate compound cleaves very quickly, yielding an unoxidizable compound, the residual photoinduced charge increases. According to the analysis time for the residual photoinduced charge at the end of the laser flash, a first-order kinetic constant can be obtained between 2×10^{-3} and 10^{-8}–$10^{-9}\,s^{-1}$, leading to the shortest time window of only a few nanoseconds.

However, with the preceding performance, this method is restricted to an aqueous solution and even in DMF this application is difficult. On the other hand, this method allows the study of the second electron transfer when a very fast cleavage occurs at the level of the first step.[27] The reduction of radicals giving anions can be characterized via its standard potential and its kinetics of electron transfer even in an aprotic medium such as DMF. Some results are given in the following section. At the level of the determination of electrochemical properties of transient radicals, another effective method is used, photomodulation voltammetry (PMV),[28] which, yields either cations in oxidation or anions in reduction. If these radicals are destroyed by a dimerization process, the

For references see page 10008

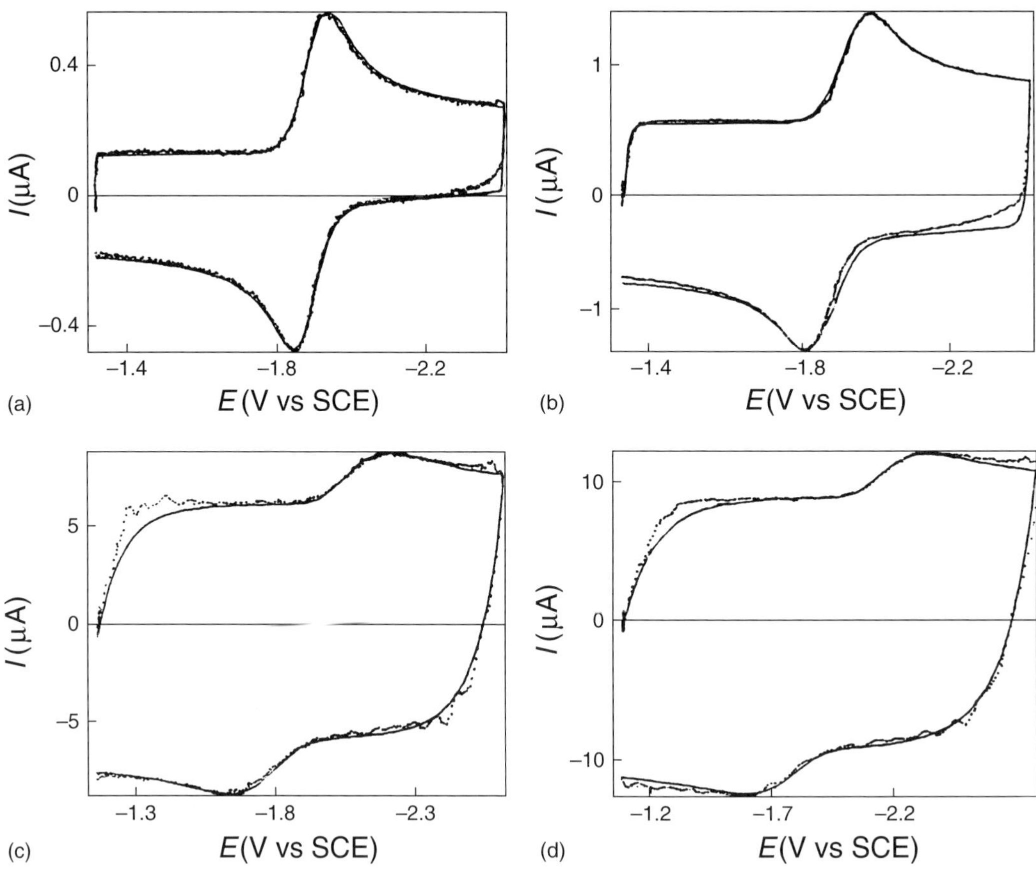

Figure 6 CV of anthracene (10 mM) in acetonitrile at a 5-μm diameter gold electrode. Scan rate: (a) 22 100, (b) 113 400, (c) 1 191 000 and (d) 1 724 000 V s^{-1}. [Reproduced by permission of the American Chemical Society from C.P. Andrieux, P. Hapiot, J.-M. Savéant, *Chem. Rev.*, **90**, 723–736 (1990).]

Table 1 Time window for stationary and transient techniques

Technique	Experimental parameter	Range of accessible values	Time window (s)
Polarography	Drop time	1–10 s	1–10
RDEV	Rotation speed w	50–20 000 rpm	2×10^{-2}–5
LSV (millimeter electrode)	Sweep rate	0.1–1000 V s^{-1}	3×10^{-5}–0.3
DSC (millimeter electrode)	Step time	1 s–1 ms	10^{-3}–1
Sinusoidal technique	Frequency	1–10 000 Hz	10^{-4}–1
LSV (ultramicroelectrode)	Sweep rate	10^2–10^6 V s^{-1}	3×10^{-8}–3×10^{-4}
DSC (ultramicroelectrode)	Step time	10 ms–1 μs	10^{-6}–10^{-2}

measurement is always possible even when this reaction reaches the diffusion limit.

In addition to the electrochemical methods described here, the determination of electrochemical reaction mechanisms is often assisted by spectrometric methods ex situ or in situ. In the latter case, spectroelectrochemical techniques with ultraviolet/visible (UV/VIS), infrared (IR) or electron spin resonance (ESR) spectroscopy are used. These techniques are described in other articles.

List of selected abbreviations appears in Volume 15

3 CHARACTERISTICS OF ELECTRON TRANSFERS FOR ORGANIC COMPOUNDS

Two characteristics of the electron transfer are of particular importance because they open up possibilities for the use of electrochemical techniques in mechanism determinations. The first is the weak adsorption of organic reactants, products and intermediates when aprotic solvents and classical inert electrodes are used. Hence

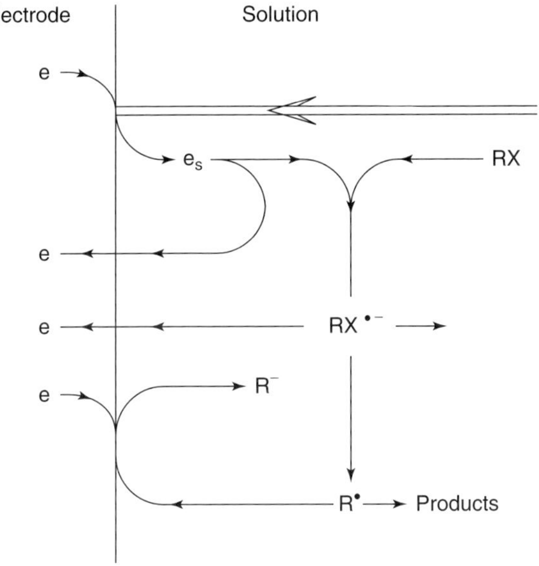

Figure 7 Principle of the photoinjection of electrons method in the case of the reduction of a substrate RX.

most organic mechanisms deal only with homogeneous chemical reactions and electron transfer reactions at the electrode. The second is related to the kinetics of electron transfer in organic electrochemistry.

3.1 Theory

Electron transfers constitute one or several elementary steps of the mechanism of electrochemical reactions. As organic species are molecules of significant size, the Marcus–Hush theory[29,30] is especially adapted to the description of the kinetic laws of this transfer. This theory was developed in the case of an electron transfer occurring without any breaking or any formation of chemical bonds, still taking into account changes in the solvation (changes in orientation of solvent molecules near the molecule) and changes in bond lengths or angles. The theory of electron transfer at the electrode is completely associated with the theory of homogeneous electron transfer when the electron exchange occurs between two molecules differing only by one oxidation degree.[30–32]

From a theory similar to that concerning activated complexes, the electron transfer rate can be written for an adiabatic process, and according to the Franck–Condon principle, which states that the slow step involves the heavier particle, leading to kinetics determined by the changes at the level of the reactants and the product. The kinetics can be expressed using Equations (12) and (13):

$$k_S = Z_{el} \exp\left(\frac{-\Delta G^*}{kT}\right) \tag{12}$$

with

$$\Delta G^* = \left(\frac{\lambda}{4}\right)\left(1 + \frac{\Delta G^0}{\lambda}\right)^2 \tag{13}$$

where Z_{el} is the collision number for the heterogeneous reaction, ΔG^* the free energy of activation, ΔG^0 the standard free energy of the reaction ($\Delta G^0 = 0$ at the standard potential) and λ a reorganization term composed of solvational (λ_0) and internal (λ_i) components. The standard activation free energy (intrinsic barrier) for $\Delta G^0 = 0$ is then given by Equation (14):

$$(\Delta G^*)^0 = \frac{\lambda_0 + \lambda_i}{4} \tag{14}$$

In an outer-sphere electron transfer with small internal reorganization, so that the solvent reorganization term λ_0 is the largest, this term can be estimated from the ionic radii of reactant and product and a distance d that expresses the location of the electron exchange (Equation 15):

$$\lambda_0 = (\Delta e)^2\left(\frac{1}{a} - \frac{1}{d}\right)\left(\frac{1}{\varepsilon_{op}} - \frac{1}{\varepsilon_S}\right) \tag{15}$$

where Δe is the change in charge between reactants and product, ε_{op} and ε_S are the optical and static dielectric constant of the solvent and the estimation of d is $2a$ according to Marcus[30] and $1/d = 0$ according to Hush.[29]

From this theory and for large-sized molecules such as aromatic compounds in aprotic solvents where only solvent reorganization is involved, the rate constant of the electron transfer can be estimated. For radii in the range 2–5 Å, values of k_S between 1 and 10 cm s^{-1} were estimated. It must be noted that the rate of the electron transfer is quadratic with the potential having a value of $\alpha(\mathrm{d}\Delta G^*/\mathrm{d}\Delta G^0)$ equal to 0.5 at the standard potential of the reaction.

The internal or vibrational term provides an important contribution when some bond lengths or bond angles are changed during the electron transfer. This is important, for example, in the reduction of an aliphatic radical in an anion.

It is also possible to estimate the kinetics of the electron transfer when this transfer is concerted with a bond cleavage (Equation 16):

$$\text{A–X} + \text{e}^- \rightleftharpoons \text{A}^\bullet + \text{X}^- \tag{16}$$

As with the Marcus–Hush theory, the Born–Oppenheimer approximation is assumed to be applicable and the reaction to be adiabatic. The potential energy surfaces for the reactants and products depend now on three kinds of reaction coordinates: (1) the solvent fluctuational configuration as in the simple outer-sphere electron transfer, (2) the vibration coordinates for bonds that are

For references see page 10008

not cleaved during the reaction and (3) the stretching of one bond cleaved during the reaction. These three parts are additive and for the last one the potential energy of the reactant is assumed to depend on the bond length according to a Morse curve whereas the product potential energy curve is assumed to be the same as the repulsive part of the reactant Morse curve.(33,34) A comparison of the potential energies for the outer-sphere mechanism of electron transfer and dissociative electron transfer is illustrated in Figure 8(a) and (b).

As a result, the standard activation free energy is now given by Equation (17):

$$(\Delta G^*)^0 = \frac{D_e + \lambda_0 + \lambda_i}{4} \tag{17}$$

where D_e represents the bond dissociation energy. It can easily be shown that $(\Delta G^*)^0$ can have large values compared with the outer-sphere case and that the standard potential of the redox couple involved

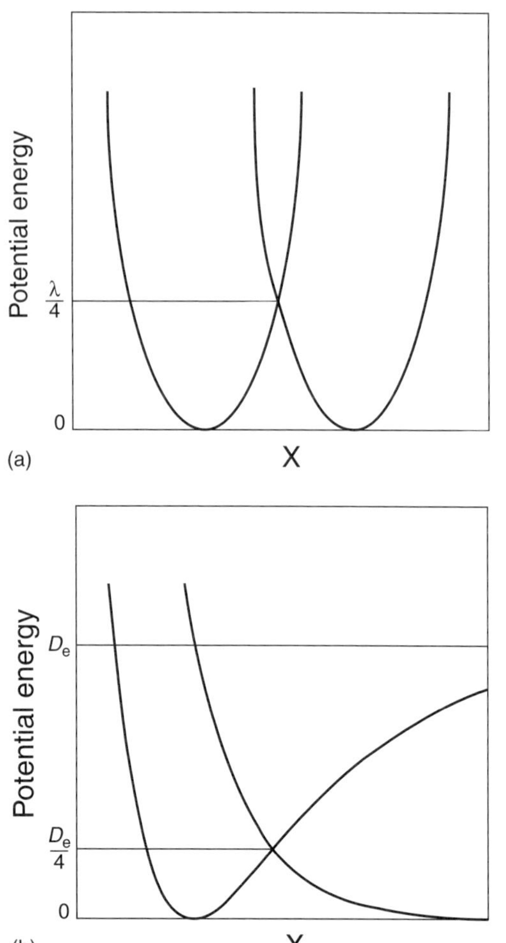

Figure 8 Potential energy (a) for an outer-sphere mechanism of electron transfer (Marcus–Hush theory) and (b) for a dissociative electron transfer.

(Equation 16) can be very different to the standard potential from the possible simple electron transfer (Equation 18):

$$\text{A–X} + e^- \rightleftharpoons \text{AX}^{\bullet -} \tag{18}$$

3.2 Characterization Procedures

3.2.1 Linear Sweep Voltammetry and Convolution

The analysis of the kinetics of the electron transfer must be taken into account in comparison with the always present rate of diffusion. This depends on the method used and on the time window of this method. It should be emphasized that there is a possible difficulty in ensuring that the kinetics are dependent on the electron transfer and not on other phenomena such as chemical evolution of an intermediate product.

The effect of the kinetics of the electron transfer on the curves obtained by the different electrochemical methods is illustrated via LSV and CV. As a first approach, the electronic rate law is often described with the empirical Butler–Volmer law (Equation 19):(5,14)

$$i = FSk_S \exp\left[\frac{-\alpha F(E - E^0)}{RT}\right] \times \left\{(C_A)_0 - (C_B)_0 \exp\left[\frac{F(E - E^0)}{RT}\right]\right\} \tag{19}$$

(for a reduction) where k_S is the standard rate constant and α the transfer coefficient. It must be noted that this approach can be related to the Marcus law and, for example, when the potential is approximately equal to the standard potential, $k_S = Z_{el} \exp(-(\Delta G^*)^0/kT)$ and $\alpha = 0.5$.

The change in the voltammetric curve occurs according to the change in k_S and to the time window of the method. The operational parameter is, in fact, $k_S/(DFv/RT)^{1/2}$. As the values of k_S become smaller and smaller, the more the cathodic peak potential becomes negative and the more the cathodic and anodic peaks are separated. A cathodic curve depending only on the shape of α is obtained for lower values of k_S.(5,11)

However, as mentioned earlier, the quadratic law is predicted by the Marcus–Hush theory and its extension. Under these conditions, some different evolutions are obtained that are characterized by the standard activation free energy. The value of the apparent transfer coefficient is then a function of the gap between peak potential and standard potential. Figure 9(a) and (b) shows a comparison between these two possible approaches. The apparent transfer coefficient decreases from 0.5 in the vicinity of the standard potential when the potential is set to a more negative value.

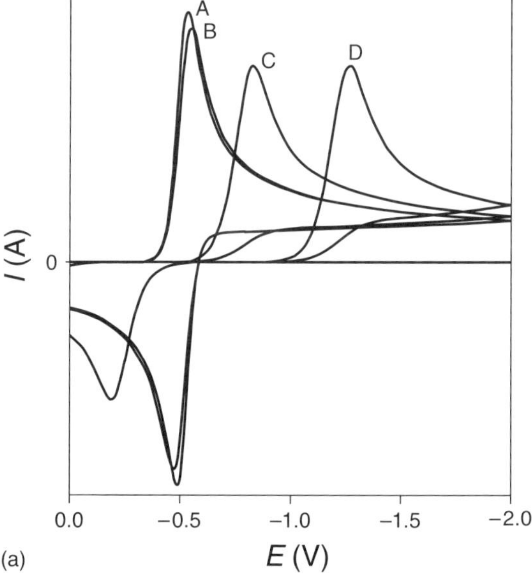

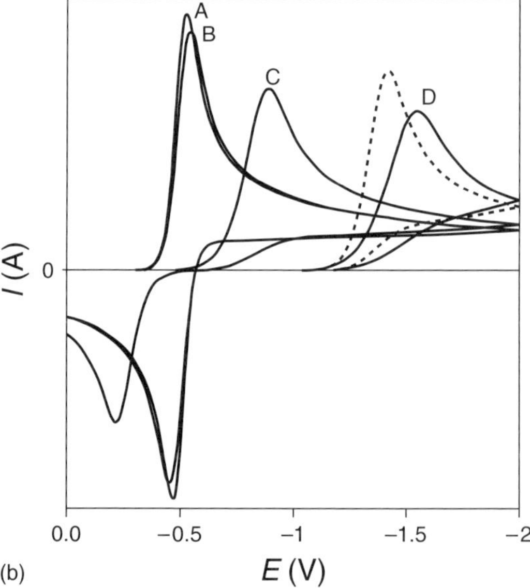

Figure 9 CV with a slow-rate electron transfer, (a) according a Butler–Volmer law with $k_S(\mathrm{cm\,s^{-1}}) = $ (A) ∞, (B) 0.03 (C) 5×10^{-5} and (D) 1×10^{-8} and (b) according a Marcus–Hush law. For curve D, comparison of the shape of the curve corresponding to the Butler–Volmer law (dashed line).

The shape of the curve is determined by the value of the peak width (gap between peak potential and half-peak potential) (Equation 20):

$$E_{p/2} - E_p = 1.85\left(\frac{RT}{\alpha F}\right) \tag{20}$$

and for all the values of $\alpha \leq 0.5$ this peak width is larger than for a fast electron transfer where only the diffusion is rate-limiting step. This measurement of the peak width is a good way to determine the experimental value of α_{ap} but not very accurate, particularly when this value is near 0.5.

A way in which to increase this accuracy is to take the convolution of the LSV curve (Equation 21):[35]

$$I = \frac{1}{\pi^{1/2}}\int_0^{\tau} \frac{i(\nu)}{(\tau - \nu)^{1/2}}\,\mathrm{d}\nu \tag{21}$$

This quantity is easily calculated and is directly proportional to the diffusing reactant concentration. Hence a plateau curve identical with what is obtained using a stationary technique is obtained regardless of the kinetic law of the electron transfer that is followed (Figure 10a).

From the shape of the curve and from the different convolution curves obtained using LSV curves at several sweep rates, the kinetics of the electron transfer are measured by determining the value of $\ln k(E)$

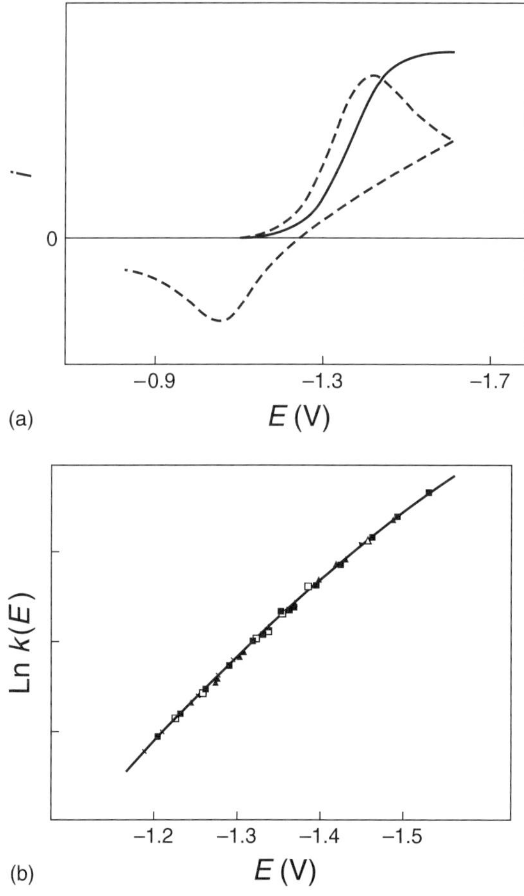

Figure 10 (a) CV (dashed line) and the corresponding convolution curve (solid line) for *tert*-nitrobutane in DMF. (b) Charge transfer rate $k(E)$ for *tert*-nitrobutane in DMF, using sweep rates between 0.59 and 189 $\mathrm{V\,s^{-1}}$. [Reproduced by permission of Elsevier Sequoia from J.-M. Savéant, D. Tessier, *J. Electroanal. Chem.*, **65**, 57–66 (1975).]

For references see page 10008

(Equation 22):

$$i = FSk(E)\left\{(C_A)_0 - (C_B)_0 \exp\left[\frac{F(E-E^0)}{RT}\right]\right\} \quad (22)$$

An example of the determination of this quantity is given in Figure 10(b), where a curvature of this relationship with potential, resulting from a Marcus-like law of the electron transfer, can be noted. The apparent transfer coefficient α_{ap} is obtained from Equation (23):

$$\alpha_{ap} = -\left(\frac{RT}{F}\right)\frac{d[\ln k(E)]}{dE} \quad (23)$$

The standard potential is the value of the potential when α_{ap} is exactly 0.5.

3.2.2 *Impedance Technique*

This kind of determination may also be conducted using an impedance technique,[20] mainly in the vicinity of the standard potential for fairly fast electron transfers (quasi-reversible system). By superimposing an alternating current on a slowly variable dc potential, the faradaic resistance R_f and the faradaic capacitance C_f are analyzed taking into account the double-layer capacitance and the cell resistance that are obtained from a test experiment. The apparent rate constant $k(E)$ is obtained from Equation (24):

$$\log\left[\frac{k(E)}{D^{1/2}}\right] = -\log\left[\frac{(R_fC_fw-1)[1+\exp\{(F/RT)(E-E_{1/2})\}]}{(2w)^{1/2}}\right] \quad (24)$$

The results obtained by this technique for a quasi-reversible system have comparable accuracy to those is obtained via convolution.[36]

3.3 Some Results

Organic species producing stable radicals (mainly anion radicals in reduction and cation radicals in oxidation) are for the most part aromatic and bulky molecules. Thus, the electron transfer for these molecules can be explained using the Marcus–Hush theory where the predominant effect is solvation reorganization. The transfers that occur are fairly fast and their measurement requires very short time window techniques.

Some examples of the determination of the kinetics of simple reduction processes as measured using the standard rate constant and the standard free energies of activation where DMF is the solvent are given in Table 2.

It is possible to show that a quadratic Marcus-type law is followed for the reduction of nitrodurene and *tert*-nitrobutane where the electron transfer is simple but slower than for the first ones. There the change in the solvation is localized at the level of the nitro group, which leads to a significant change in the solvation reorganization term.[38] For the reduction of aliphatic radicals a significant change in the internal reorganization term explains the large standard activation energy that results.[39]

Table 2 Kinetics of electron transfer for a simple reduction process

Compound	k_S ($cm\,s^{-1}$)	$(\Delta G^*)^0$ (eV)	Technique	Refs.
Anthracene	5	0.13	Impedance	32
Phthalonitrile	1.8	0.17	Impedance	32, 37
Nitrobenzene	2.2	0.16	Impedance	32
Nitrodurene	0.12	0.26	Impedance	37
tert-Nitrobutane	4×10^{-4}	0.36	Convolution	38
tert-Butyl radical	9×10^{-7}	0.58	R[a]	39
Benzyl radical	0.7	0.23	P[b]	27

[a] By analysis of the second step of reduction of *tert*-butyl iodide by LSV.
[b] By photoinjection of electrons.

When chemical reactions occur (the product of electron transfer is not stable), the kinetics of the electron transfer are difficult to determine, yet, as explained below, the measurement of the kinetics for very slow electron transfer is possible whatever the following chemical or electrochemical processes. For this reason, it is easy to characterize dissociative electron transfer where the $(\Delta G^*)^0$ values are high owing to the influence of the dissociation energy. Some examples of the determination of $(\Delta G^*)^0$, of values of α_{ap}, of transfer coefficients and of the order of magnitude of the gap between the standard potential and the effective reduction potential in a voltammetric experiment are presented for the case of reductive cleavage of organic halides[40,41] in Table 3.

All these electron transfers involve the following electrochemical elementary reaction (Equation 25):

$$R{-}X + e^- \rightleftharpoons R^\bullet + X^- \quad (25)$$

Table 3 Values of standard free energy of dissociative electron transfer $(\Delta G^*)^0$, of values of α_{ap} and of the gap between the standard potential and the effective reduction potential

Compound	$(\Delta G^*)^0$ (eV)	α_{ap}	$E^0(RX/RX^{\bullet-}) - E_m$ (V)
9-Chloromethylanthracene	0.77	0.38	1.14
4-Cyanobenzyl bromide	0.72	0.33	0.97
Benzyl bromide	0.74	0.34	1.03
n-Butyl bromide	0.93	0.25	1.60
n-Butyl iodide	0.80	0.30	1.10

E_m is the mean value of the reduction peak in an LSV experiment at $0.1\,V\,s^{-1}$.

with a standard potential $E^0(\mathrm{RX/R^\bullet + X^-})$. It is important to note that this standard potential is absolutely different from the standard potential $E^0(\mathrm{RX/RX^{\bullet-}})$ in the couple (Equation 26)

$$\mathrm{R{-}X + e^- \rightleftharpoons RX^{\bullet-}} \qquad (26)$$

which corresponds only to a simple electron exchange on the same molecule. For example, in the case of the reduction in DMF of 9-chloromethylanthracene,[41] $E^0(\mathrm{RX/R^\bullet + X^-}) = -0.15\,\mathrm{V}$ versus SCE and $E^0(\mathrm{RX/RX^{\bullet-}})$ is estimated by comparison with other anthracenyl compounds to be about −1.8 to −1.9 V versus SCE.[42] The fact that the reduction potential is −1.36 V in LSV at $1\,\mathrm{V\,s^{-1}}$ and that this value is less negative than for the possible simple reduction reaction is proof of the efficiency of the reduction via dissociative electron transfer.

4 SIMPLE ORGANIC ELECTROCHEMICAL MECHANISMS

4.1 Nature and Classification of Coupled Chemical Reactions

In this section, the different kinds of chemical reactions that occur in organic electrochemistry and their influence on the kinetics of the overall process will be presented. For this purpose, we have chosen some simple examples where the nature of these reactions is clearly demonstrated and not merely postulated. This is an important point because in complex mechanisms not all the steps are fully elucidated as they are out of reach of characterization techniques.[5,11]

The way in which these chemical reactions may be qualitatively characterized will be explained for each particular case.

4.1.1 Preceding Chemical Reaction

The reaction scheme in Scheme 1 is followed:

$$\mathrm{C \rightleftharpoons A}$$
$$\mathrm{A \pm e^- \rightleftharpoons B}$$

Scheme 1

Compound C present in solution is not electroactive in the potential range of interest and compound A must be formed through prior conversion of C. It must be noted that compound C is sometimes reducible or oxidizable directly in another range of potential. The reduction of carbonyl compound in acidic medium illustrates this behavior (Scheme 2):[1,2]

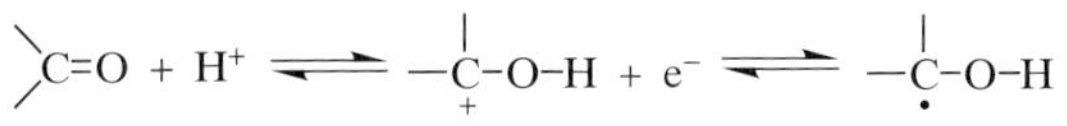

$$\mathrm{{>}C{=}O + H^+ \rightleftharpoons {-}\overset{|}{\underset{+}{C}}{-}O{-}H + e^- \rightleftharpoons {-}\overset{|}{\underset{\bullet}{C}}{-}O{-}H}$$

Scheme 2

Reactions that thermodynamically favor C are the only ones that are easily observable and the main manifestation of this process is the height of the wave in RDEV, the peak current in LSV or the current values in a single potential step technique. If the chemical reaction is slow compared with the diffusion, the result will be a measurement of the equilibrium concentration of A, whereas if the reaction is fast, the result will be proportional to the total concentration of A and C. So, for example, in RDEV the height of a wave does not follow $w^{1/2}$ when the time window is of the same order of magnitude as that of the rate of reaction.

4.1.2 Consecutive Reaction

A chemical reaction occurs after the electron transfer (Scheme 3):

$$\mathrm{A \pm e^- \rightleftharpoons B}$$
$$\mathrm{B \rightleftharpoons C}$$

Scheme 3

In the case where C is thermodynamically stable compared with B, this intermediate B is not stable and cannot accumulate near the electrode. As written above, this reaction is first order (or pseudo-first order if B reacts with solvent or a species in high concentration in the solution), but a second-order reaction (dimerization) is also possible. The characterization of such a process is not possible at the level of the wave or peak current but other techniques that allow the detection of the reverse current (reoxidation of B for a reduction of A) can help to visualize this type of process. Thus, with CV or RRDEV, if the lifetime of the intermediate is small compared with the time window of the method, no reoxidation current is observed. However, if the time window of the technique can be shortened compared with the lifetime of B, then nothing changes in the pattern obtained from a simple electron transfer.

Another way to detect a fast chemical reaction involves the displacement of the potential in response to the chemical reaction. The electrochemical reaction is facilitated in the presence of a chemical reaction (more positive reduction).

At the level where it is possible to determine the order of the reaction, an accurate determination of the peak potential in LSV or of the half-wave in RDEV provides an answer to the question. The displacement as a function of the initial concentration also provides information.

For references see page 10008

Some examples of consecutive reactions are given in Schemes 4–6.

First-order, bond cleavage reaction:[43]

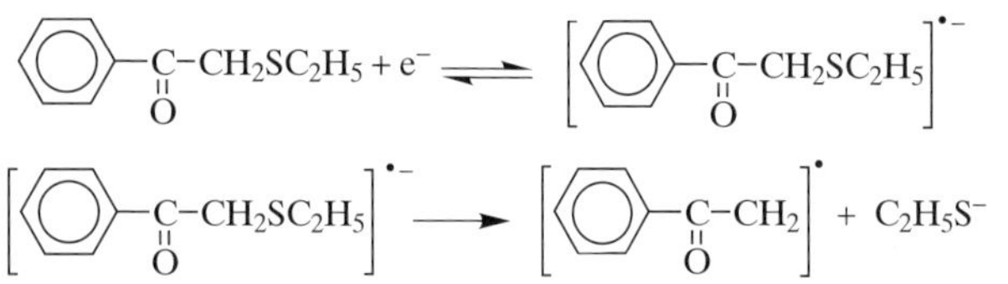

$$PhC(O){-}CH_2SC_2H_5 + e^- \rightleftharpoons [PhC(O){-}CH_2SC_2H_5]^{\bullet -}$$

$$[PhC(O){-}CH_2SC_2H_5]^{\bullet -} \longrightarrow [PhC(O){-}CH_2]^{\bullet} + C_2H_5S^-$$

Scheme 4

Pseudo-first-order reaction, protonation of a ketone by phenol:[44]

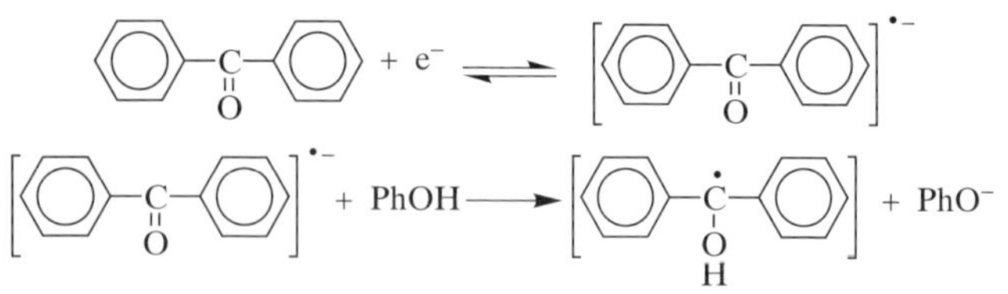

$$PhC(O)Ph + e^- \rightleftharpoons [PhC(O)Ph]^{\bullet -}$$

$$[PhC(O)Ph]^{\bullet -} + PhOH \longrightarrow [Ph\dot{C}(OH)Ph] + PhO^-$$

Scheme 5

Dimerization:[45]

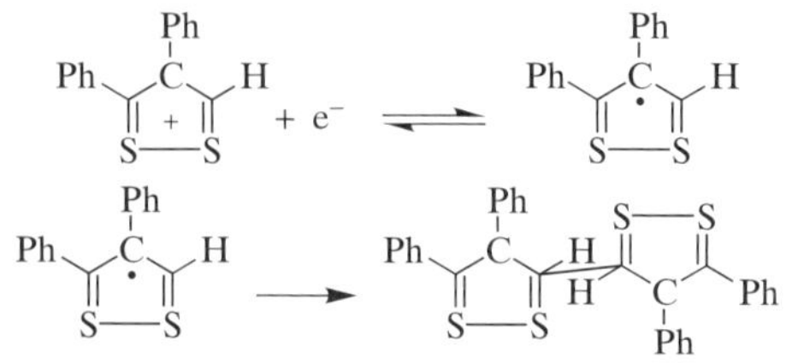

$$Ph(Ph)C_3H(S{-}S)^{+} + e^- \rightleftharpoons Ph(Ph)C_3H(S{-}S)^{\bullet}$$

$$Ph(Ph)C_3H(S{-}S)^{\bullet} \longrightarrow \text{dimer } [Ph(Ph)C_3H(S{-}S)]_2$$

Scheme 6

4.2 Kinetic Analysis for a Single Chemical Reaction Acting as the Determining Step

As explained earlier, in organic electrochemistry the electron transfers are fairly fast and homogeneous chemical reactions occur more frequently than surface chemical reactions. When electron transfer is fast, it is possible to express, using efficient methods, how the homogeneous chemical reactions (acting only with the diffusion), influence the electrochemical response. For each method used the same kind of kinetic limits are obtained but, as LSV and CV are often used, the results are presented for these methods.[46] The sequence of electrochemical and chemical reactions are expressed by the letters E for electrochemical and C for chemical; if necessary, the order of the reaction is indicated after the letter C.[13]

4.2.1 *First-order Preceding Chemical Reaction (C + E) (Scheme 7)*

As explained above, only cases where the equilibrium is displaced towards C are of interest. For each value of this equilibrium constant, three main kinetic behaviors are singled out according to the value of the forward rate constant k:

$$C \underset{}{\overset{k}{\rightleftharpoons}} A \quad \text{Equilibrium constant: } K$$
$$A \pm e^- \rightleftharpoons B$$

Scheme 7

- $k \approx 0$, curve corresponding to a simple diffusion of the A species in the thermodynamically low concentration: "pure diffusion";
- $k \approx \infty$, curve corresponding to the reduction of the whole concentration of species A + C: "pure extraordinary diffusion";
- intermediate value of k where a plateau-shaped curve is obtained at negative potential; this "pure kinetics" case depends on the parameter $K\sqrt{k}$ and under these conditions the platcau current obtained is independent of the sweep rate (Equation 27):

$$i_{pl} = FSC^0D^{1/2}K\sqrt{k} \quad (27)$$

These limiting cases are easily visualized in a kinetic diagram (Figure 11) where kinetic control intermediate zones depending on the accuracy of the determination

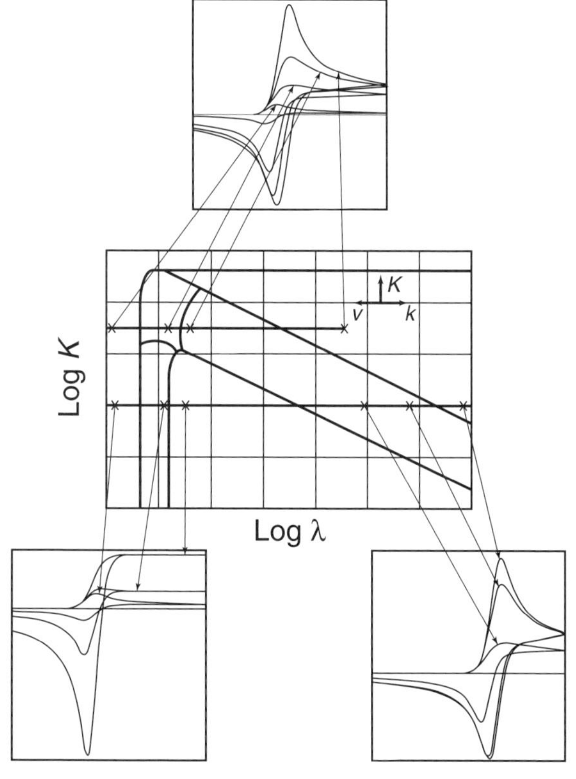

Figure 11 First-order C + E mechanism. Kinetic zone diagram for LSV with $\lambda = (RT/F)(k/v)$ and the characteristic cyclic voltammograms.

are present between the three pure limiting zones. The two parameters are the thermodynamic constant K and the kinetic constant k.

4.2.2 Consecutive Chemical Reaction

4.2.2.1 First-order Reaction $(E + C1)$ (Scheme 8)

$$A \pm e^- \rightleftharpoons B$$
$$B \overset{k}{\rightleftharpoons} C \quad \text{Equilibrium constant: } K$$

Scheme 8

The kinetic diagram with the two parameters k and K is shown in Figure 12. Here also a "pure kinetic" zone is obtained for intermediate values of k in the region of an equilibrium displaced towards C. In this pure kinetic zone, chemical irreversibility occurs (no reverse current in CV) and the characteristics of the peak current are expressed by Equations (28–30):

Peak current:

$$i_p = 0.496FSC^0D^{1/2}\left(\frac{Fv}{RT}\right)^{1/2} \quad (28)$$

Peak potential:

$$E_p = E^0 - 0.78\left(\frac{RT}{F}\right) + \left(\frac{RT}{2F}\right)\ln\left[\left(\frac{RT}{F}\right)\frac{k}{v}\right] \quad (29)$$

Peak width:

$$E_{p/2} - E_p = 1.85\left(\frac{RT}{F}\right) \quad (30)$$

Thus, the peak current in a "pure kinetic" condition is proportional to the square root of the sweep rate and the peak potential is displaced to $-(RT/2F)\ln 10$ (-30 mV at 302 K) when the sweep rate is increased by one order of magnitude. This value is for a reduction reaction as in the other following schemes. For an oxidation a change in the direction of the variations is obvious.

The displacement to -30 mV at 302 K when the sweep rate is increased by one order of magnitude without any variation with the concentration of the initial species is a characteristic of the order of the chemical reaction.

4.2.2.2 Second-order Reactions $(E + C2)^{(47)}$ If the chemical reaction is a second-order reaction, the same kind of kinetic zone diagram is obtained and a pure kinetic zone is also found. In fact, several coupling processes are possible and each of them can be characterized, in the "pure kinetic" zone, by the shape of the voltammetric curve in CV, by the peak potential variations with sweep rate and concentration in LSV.

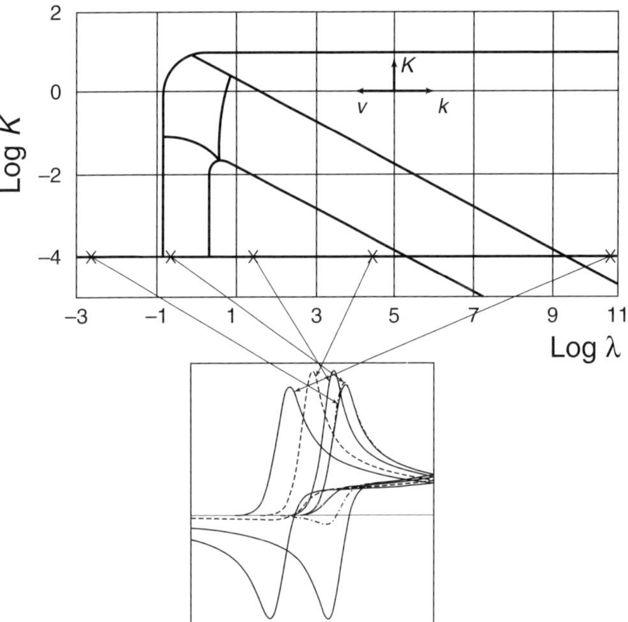

Figure 12 First-order E + C mechanism. Kinetic zone diagram for LSV with $\lambda = (RT/F)(k/v)$ and the characteristic cyclic voltammograms.

$$A \pm e^- \rightleftharpoons B$$
$$2B \overset{k}{\rightleftharpoons} D$$

Scheme 9

Dimerization Between Two Reduced (or Oxidized) Species $[E + C2_{(\mathrm{Arr})}]$ (Scheme 9). The characteristics of the peak current are expressed by Equations (31–33):

$$\text{Peak current:} \quad i_p = 0.527FSC^0D^{1/2}\left(\frac{Fv}{RT}\right)^{1/2} \quad (31)$$

$$\text{Peak potential:} \quad E_p = E^0 - 0.902\left(\frac{RT}{F}\right) + \left(\frac{RT}{3F}\right) \times \ln\left[\left(\frac{2RT}{3F}\right)\left(\frac{kC^0}{v}\right)\right] \quad (32)$$

$$\text{Peak width:} \quad E_{p/2} - E_p = 1.51\left(\frac{RT}{F}\right) \quad (33)$$

The displacements to -20 mV at 302 K when the sweep rate is increased by one order of magnitude and to 20 mV when the concentration of the initial species is increased by one order of magnitude characterize this dimerization process.

Dimerization Between One Reduced (or Oxidized) Species and the Substrate $[E + C2_{(\mathrm{Ars})}]$ (Scheme 10). The characteristics of the peak current are expressed by

For references see page 10008

$$\mathrm{A \pm e^- \rightleftharpoons B}$$
$$\mathrm{A + B \overset{k}{\rightleftharpoons} D}$$

Scheme 10

Equations (34–36):

Peak current: $$i_p = 0.430FSC^0D^{1/2}\left(\frac{Fv}{RT}\right)^{1/2} \quad (34)$$

Peak potential: $$E_p = E^0 - 0.457\left(\frac{RT}{F}\right) + \left(\frac{RT}{2F}\right) \times \ln\left[\left(\frac{RT}{F}\right)\left(\frac{kC^0}{v}\right)\right] \quad (35)$$

Peak width: $$E_{p/2} - E_p = 2.27\left(\frac{RT}{F}\right) \quad (36)$$

The displacements to −30 mV at 302 K when the sweep rate is increased by one order of magnitude and to 30 mV when the concentration of the initial species is increased by one order of magnitude characterize this dimerization process.

When only one compound is electroactive, from these four simple kinetic situations (C + E, E + C1 and the two E + C2), it is possible to analyze the complex mechanisms obtained in organic electrochemistry.

One question remains concerning the possibility of considering the kinetics of the electron transfer as a non-limiting kinetic phenomenon. For example, in chemical reactions following the electron transfer in reduction, the CV curve is displaced towards increasingly positive potentials as the chemical rate constant increases. Taking into account the law of electron transfer, it appears that the apparent electron rate constant is lower at more positive potential values than at standard potentials. Hence the faster the chemical reactions are, the more the limitation by electronic transfer becomes significant.(11) However, in many experiments, in a range up to 10^4–10^6 s for the first-order chemical reaction constant, the hypothesis of fast electron transfer still remains valid.

5 COMPLEX MECHANISMS

The mechanism for electrochemical reactions is the pathway leading from the initial reactants to the final products where at least one electron transfer occurs. In many cases the electrochemical reactions on a substrate can yield several products. On the one hand, the process is potential dependent and, according to the applied potential, the nature of the products is different. On the other hand, for a certain potential, competitive sequences of reactions are possible. In this section we limit ourselves to mechanisms where only one sequence is involved (no branching reactions) and also to cases where only one substrate is electroactive.

An electrochemical mechanism is written as a succession of electrochemical steps and chemical steps and we shall focus on two main problems: (1) how to understand the reaction sequence for complex reactions and (2) to elucidate what happens when the mechanism involves several electron transfers.

5.1 Sequence of Chemical Reactions After One Electron Transfer

Consider the formation of pinacol from the reduction of a ketone in a slightly acidic medium (the ketone cannot be protonated prior to its reduction). The first electrochemical reaction is (Equation 37):

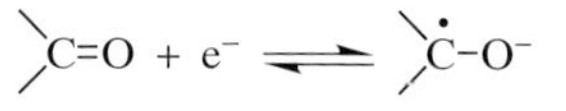

$$\mathrm{>C{=}O + e^- \rightleftharpoons >\dot{C}{-}O^-} \quad (37)$$

and successive reactions lead to the pinacol (Equation 38):

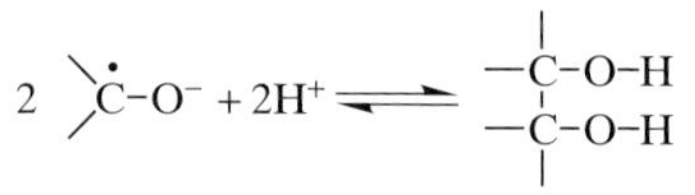

$$\mathrm{2\ >\dot{C}{-}O^- + 2H^+ \rightleftharpoons \begin{matrix}-C{-}O{-}H\\ -C{-}O{-}H\end{matrix}} \quad (38)$$

Because the dimerization of the two anions is slow, the possible sequence is as shown in Scheme 11:

$$\mathrm{A + e^- \rightleftharpoons B^-}$$
$$\mathrm{(i)\quad B^- + H^+ \underset{k_2}{\overset{k_1}{\rightleftharpoons}} BH\ (Pseudo\text{-}first\text{-}order\ reaction)}$$
$$\mathrm{(ii)\quad 2\,BH \overset{k_d}{\rightleftharpoons} D}$$

Scheme 11

and several kinetic limiting behaviors are possible according to what the slow chemical step (rate-determining step) is, as in chemical kinetics.(44)

If reaction (i) is slow compared with reaction (ii), the kinetics and the characteristics of the electrochemical curves are those of the mechanism E + C1. For example, in LSV, the characteristics are expressed by Equations (39–41):

Peak current: $$i_p = 0.496FSC^0D^{1/2}\left(\frac{Fv}{RT}\right)^{1/2} \quad (39)$$

Peak potential: $$E_p = E^0 - 0.78\left(\frac{RT}{F}\right) + \left(\frac{RT}{2F}\right) \times \ln\left[\left(\frac{RT}{F}\right)\frac{k}{v}\right] \quad (40)$$

Peak width: $$E_{p/2} - Ep = 1.85\left(\frac{RT}{F}\right) \quad (41)$$

Hence the peak current is, as in pure diffusion conditions, proportional to the square root of the sweep rate and the peak potential is displaced to $-(RT/2F)\ln 10$ (-30 mV at 302 K) when the sweep rate is increased by one order of magnitude.

If reaction (i) is fast compared with reaction (ii) ($k_2 > k_d$), the reaction acting as a pre-equilibrium for reaction (ii), the kinetics and the characteristics of the electrochemical curves are those of the mechanism $E + C2_{(Arr)}$ resulting in the pre-equilibrium that replaces k_d by $(k_d k_1/k_2)[H^+]$, which introduces a variation with pH and more generally with concentration of other species occuring before the slow step. In LSV, the characteristics are expressed by Equations (42–44):

$$\text{Peak current: } i_p = 0.430FSC^0D^{1/2}\left(\frac{Fv}{RT}\right)^{1/2} \tag{42}$$

$$\text{Peak potential: } E_p = E^0 - 0.457\left(\frac{RT}{F}\right) + \left(\frac{RT}{2F}\right) \times \ln\left[\left(\frac{RT}{F}\right)\left(\frac{k_d k_1}{k_2}\right)[H^+]\left(\frac{C^0}{v}\right)\right] \tag{43}$$

$$\text{Peak width: } E_{p/2} - E_p = 2.27\left(\frac{RT}{F}\right) \tag{44}$$

The displacement is to -30 mV at 302 K when the sweep rate is increased by one order of magnitude and to 30 mV when the concentration of the initial species or the concentration in protons is increased by one order of magnitude. The preceding results are applicable when the proton concentration is sufficiently high to keep its concentration constant in the process. This is also the case when a buffer medium is used.

More generally, all the reactions occurring after the rate-determining step do not play any role in the kinetics and all the reactions before the rate-determining step act as a pre-equilibrium for the rate-determining step. In an unbuffered medium, a more complicated law results. This is not considered here.

5.2 Two-electron Process: Electron Transfer at the Electrode or in Solution

In organic electrochemistry, the ability to transfer more than one electron during a single step is impossible without a chemical reaction between the two electron transfers. This result is explained as follows. When two electron transfers occur in the same part of the molecule, adding (or removing) an electron leads to a group that is more difficult to reduce (or to oxidize) (repulsion of charge). Hence the second step will occur at a potential that is distinct from the first one and the two steps can be separated by a technique such as LSV.

This result is illustrated by the reduction potentials for a series of dinitro compounds.[48] When the two groups are on the same phenyl ring, a separation of 300–500 mV is observed according to the position of these groups on the ring. When a saturated carbon chain is introduced this separation decreases and it must be noted that for a long chain (three or four CH_2 groups are sufficient) the curve obtained is exactly the same in shape and potential as for a one-electron transfer with a double current. In fact, the two standard potentials are not equal but separated by $(RT/F)\ln 4$ as a function of the statistical factors for the two steps.

However, in many cases a first single two-electron process is observed. This is due to the influence of a chemical reaction that occurs between the two-electron steps that provides for a second step easier than the first one.

This is often the case because injection of an electron into an organic molecule increases its Lewis basicity and thus increases its ability either to capture a proton (or an electrophile) or to lose a base. On the oxidation side, removal of an electron from the molecule increases its Lewis basicity and thus increases its ability either to capture a nucleophile or to lose an acid. This explains the choice of a weakly acidic solvent for reduction (and the use of weakly basic medium in oxidation) in order to stabilize the initially formed ion radical and to yield useful chemical reactions, provided that some nucleophiles are specially added.

An example of a two-electron process is illustrated by the reduction process of anthracene in the presence of an acid (phenol) leading to dihydroanthracene in only one step[49] (Equation 45):

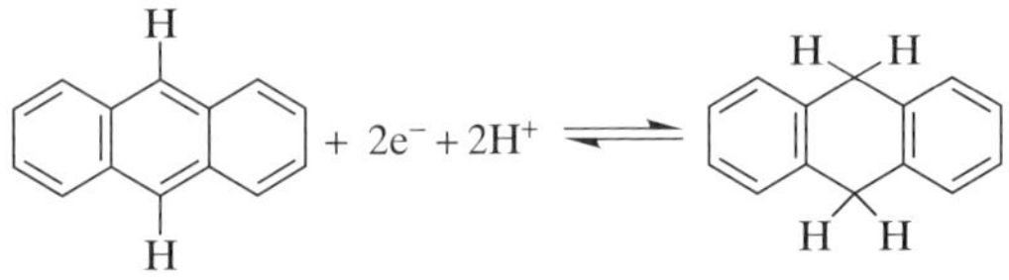

$$\text{Anthracene} + 2e^- + 2H^+ \rightleftharpoons \text{9,10-dihydroanthracene} \tag{45}$$

What are the possible mechanisms for this reaction, involving two-electron transfers? Often, a mechanism involving the sequence shown in Scheme 12 is postulated:

$$An + e^- \rightleftharpoons An^{\bullet -} \quad (E^0{}_1)$$
$$An^{\bullet -} + H^+ \rightleftharpoons AnH^{\bullet}$$
$$AnH^{\bullet} + e^- \rightleftharpoons AnH^- \quad (E^0{}_2)$$
$$AnH^- + H^+ \rightleftharpoons AnH_2$$

Scheme 12

This mechanism is a succession of electron transfers at the electrode and of chemical reactions in solution. It is an E + C + E + C (often called an ECE) mechanism.

For references see page 10008

By the fact that a two-electron transfer is obtained, the second electron transfer is easier than the first. This means that E_2^0 is more positive than E_1^0. The reaction, an electron transfer in solution (Equation 46):

$$\mathrm{AnH^{\bullet} + An^{\bullet -} \rightleftharpoons AnH^{-} + An} \qquad (46)$$

is a disproportionation of two once-reduced species. This is displaced to the right, thus making the sequence in Scheme 13 also possible:

$$\mathrm{An + e^- \rightleftharpoons An^{\bullet -}} \quad (E^0{}_1)$$
$$\mathrm{An^{\bullet -} + H^+ \rightleftharpoons AnH^{\bullet}}$$
$$\mathrm{AnH^{\bullet} + An^{\bullet -} \rightleftharpoons AnH^- + An}$$
$$\mathrm{AnH^- + H^+ \rightleftharpoons AnH_2}$$

Scheme 13

It appears that three limiting kinetic behaviors are possible, leading to characteristic variations within the experimental parameters (Table 4).[50,51]

As is easily explained by the fact that the same reaction is the rate-determining step, the first two mechanisms are not easily separable by electrochemical techniques and under pure kinetic conditions (as in Table 4) it is impossible to deduce this determination. Yet at the level of the transition between the two-electron process and the one-electron process when the effect of the chemical reaction is increasingly weakened, some discrepancies appear between these two mechanisms. For example, in DSC[52] the law of variation with respect to inversion time is given in Figure 13.

It is also possible to predict what the mechanism followed is according to the values of the kinetic constants of the two reactions and according to the method used. In LSV (Scheme 14) the two parameters governing the competition between these three mechanisms are k_2/k_1, the equilibrium constant $K = [\mathrm{B}]/[\mathrm{C}]$ and $p = k_\mathrm{d}C^0(RT/Fv)^{-1/2}(k)^{-3/2}$, where $k = k_1 + k_2$.

Figure 14 shows the predicted kinetic control diagram.[51] From this diagram and taking into account that the value of k_d is very high since the equilibrium has been displaced toward A + D, and therefore can reach the diffusion limit in the solvent, it is possible to show that a second-order disproportionation mechanism is only possible in the case where an equilibrium B/C is displaced towards B and where k_2 has very high values. An ECE mechanism is often written in the case of a two-electron mechanism without a real proof of the mechanism. It appears that this case is mainly possible for an equilibrium displaced towards C and where the

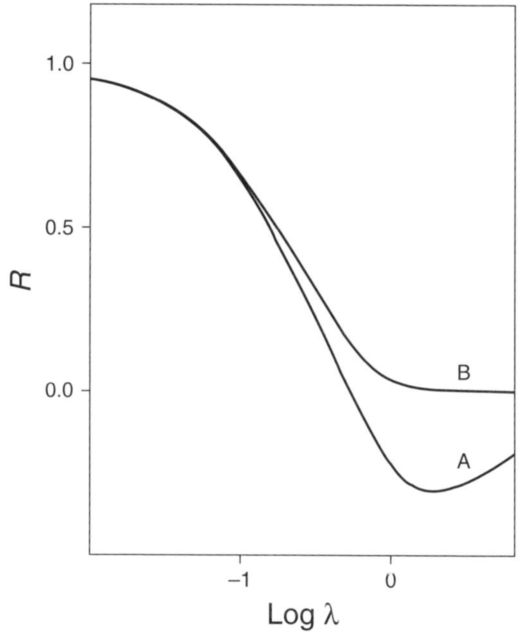

Figure 13 Two-electron process. Comparison of (A) ECE and (B) first-order disproportionation mechanisms in DSC ($\lambda = k_1\theta$, $R = [i(2\theta)/i(\theta)]/0.293$).

$$\mathrm{B} \underset{k_2}{\overset{k_1}{\rightleftharpoons}} \mathrm{C}$$
$$\mathrm{B + C} \xrightarrow{k_\mathrm{d}} \mathrm{D + A}$$

Scheme 14

Table 4 Two-electron process: mechanisms and diagnostic criteria in LSV

ECE	DISP 1	DISP 2
$\mathrm{A + e^- \rightleftharpoons B}$ $\mathrm{B \longrightarrow C}$ $\mathrm{C + e^- \rightleftharpoons D}$ $\mathrm{D \rightleftharpoons E}$ $\partial E_\mathrm{p}/\partial \log v$ in LSV = −30 mV at 302 K $\partial E_\mathrm{p}/\partial C^0$ in LSV = 0	$\mathrm{A + e^- \rightleftharpoons B}$ $\mathrm{B \longrightarrow C}$ $\mathrm{B + C \rightleftharpoons D + A}$ $\mathrm{D \rightleftharpoons E}$ $\partial E_\mathrm{p}/\partial \log v$ in LSV = −30 mV at 302 K $\partial E_\mathrm{p}/\partial C^0$ in LSV = 0	$\mathrm{A + e^- \rightleftharpoons B}$ $\mathrm{B \rightleftharpoons C}$ $\mathrm{B + C \longrightarrow D + A}$ $\mathrm{D \rightleftharpoons E}$ $\partial E_\mathrm{p}/\partial \log v$ in LSV = −20 mV at 302 K $\partial E_\mathrm{p}/\partial C^0$ in LSV = 20 mV at 302 K

DISP 1 and 2 represent first- and second-order disproportionation, respectively.

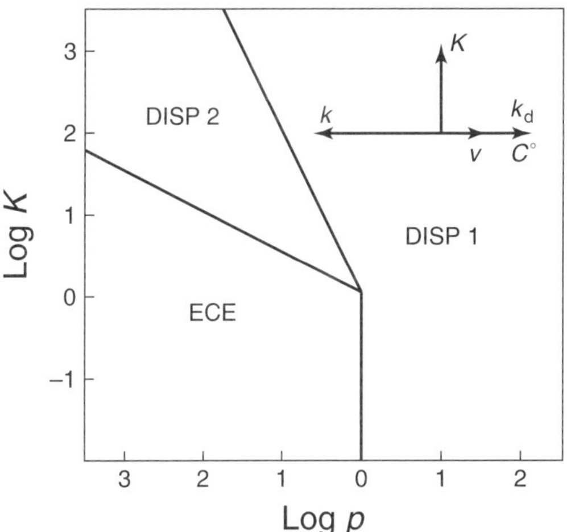

Figure 14 Two-electron process. Kinetic zone diagram showing the possibilities for the three mechanisms; $p = k_d C^0(RT/Fv)^{-1/2}(k)^{-3/2}$.

values of the kinetic constant k_1 are high. At a moderate sweep rate of $1\,\mathrm{V\,s^{-1}}$ at a millimolar concentration, if the disproportionation is at the diffusion limit ($\sim 10^{10}\,\mathrm{s^{-1}}$), the value of k_1 must be higher than $2 \times 10^5\,\mathrm{s^{-1}}$. This is explained qualitatively by the fact that if the reaction $B \rightarrow C$ is fast the production of C occurs near the electrode and the second transfer can take place another time at the electrode. However if the reaction is slow, the production of C takes place far from the electrode and the solution reaction of electron exchange becomes predominant.

5.3 Redox Catalysis

From a practical point of view, e.g. in order to avoid passivation of the electrodes, or for kinetic purposes, it is sometimes possible to realize an electrochemical reaction via the intermediary of a homogeneous electron transfer from a mediator acting as an electron carrier from the electrode to the solution.

The principle of this process is described in the example of a reductive first-order E + C reaction scheme[53] (Scheme 15):

Direct electrochemical process:

$$A \pm e^- \rightleftharpoons B$$

$$B \xrightarrow{k} C$$

Mediated process:

$$P + e^- \rightleftharpoons Q \quad (E^0_{PQ})$$

$$Q + A \underset{k_2}{\overset{k_1}{\rightleftharpoons}} B + P$$

$$B \xrightarrow{k} C$$

Scheme 15

The mediator couple (or catalyst couple) is chosen not to undergo any chemical reaction alone. Species P and A are only initially present at concentrations C^0_P and C^0_A. It is possible to obtain a mediated process at a potential where the direct process is not operative. This would be true when the direct process is kinetically slow, that is, either when the intrinsic rate constant is low and/or if the kinetic constant k is large.

Two limiting behaviors are reached according to the values of the kinetic constants and the parameters of concentration and diffusion rate. Thus, in LSV the system depends upon three parameters (Equations 47–49):

$$\lambda_1 = k_1 C^0_P \left(\frac{RT}{Fv}\right) \tag{47}$$

$$\sigma = \frac{k}{k_2 C^0_P} \tag{48}$$

$$\gamma = \frac{C^0_A}{C^0_P} \tag{49}$$

Two limiting behaviors are distinguished accordingly the value of σ. When $\sigma \gg 1$, the reaction of homogeneous electron exchange in the direct direction is the rate-determining step and the observed curve is a function of λ_1 and γ. An example of a curve and peak current evolution is presented in Figure 15(a–c). It must be noted that for large values of γ, a plateau-shaped curve is obtained for a wide range of values of λ_1. When $\sigma \ll 1$, the important parameters are $\lambda_1\sigma$ [$= (kk_1/k_2)(RT/Fv)$] and γ. It is interesting that as the parameter is experimentally varied by changing the mediator concentration, it is therefore possible, under favorable circumstances, to shift the kinetic control from the homogeneous forward reaction to an intervention of the deactivation of B by decreasing the concentration of the mediator. This fact is used to determine of the rate constant k in the range 10^6–$10^9\,\mathrm{s^{-1}}$, out of reach of all direct electrochemical techniques, even the most modern ones.

Another possible catalytic process, called chemical catalysis, is obtained when an adduct is formed between the catalyst (or a species obtained by electron transfer from the catalyst) and the substrate.[54] The reaction induced by this adduct leads to the products and to a regeneration of the catalytic species after several steps and perhaps after several electron transfers. The shape of the voltammetric curves obtained in these cases can be explained by a sequence of chemical and electrochemical reactions.

6 EXAMPLES OF MECHANISM ANALYSIS AND RATE CONSTANT DETERMINATION

The ability to reduce or to oxidize electrochemically organic molecules has been the topic of numerous books

For references see page 10008

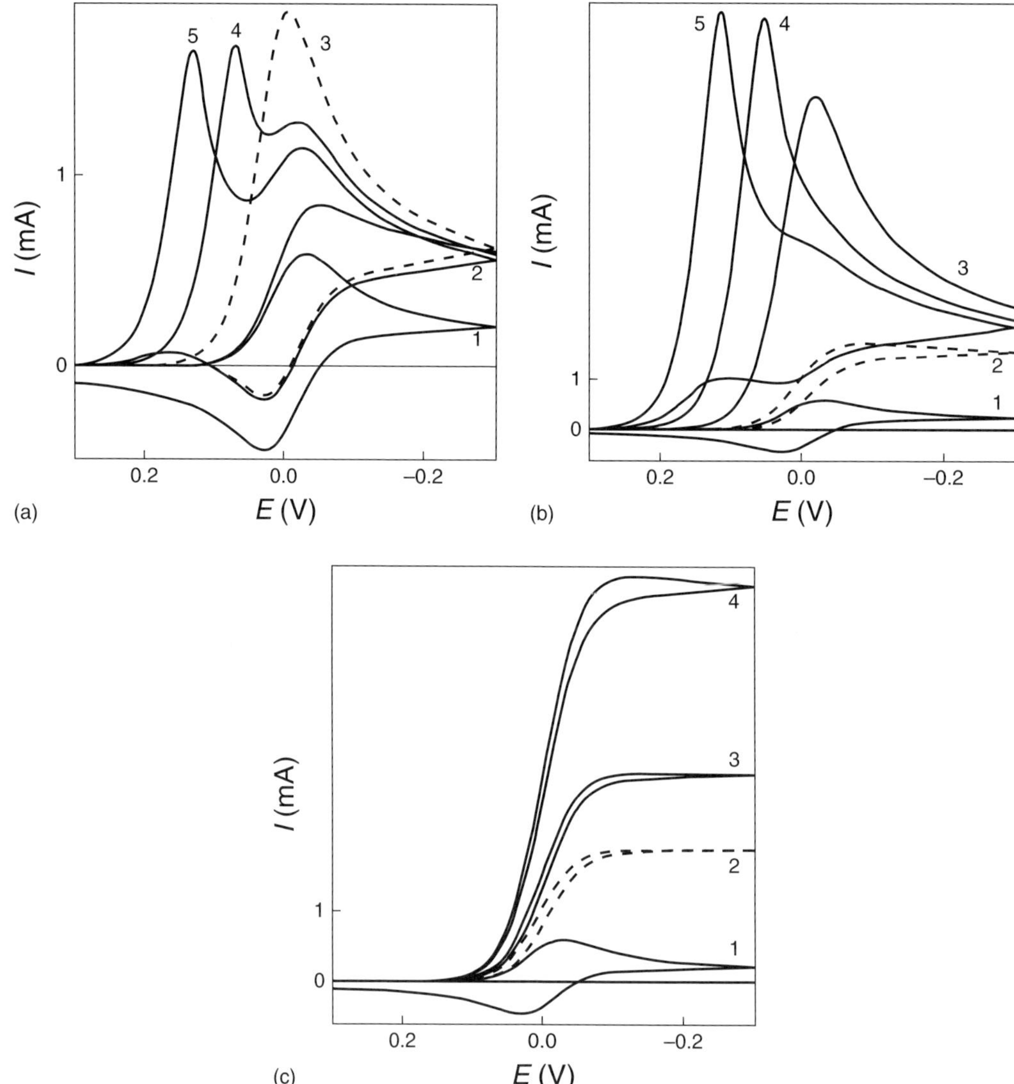

Figure 15 Redox catalysis. CV when the reaction of homogeneous electron exchange in the direct direction is the rate-determining step with $\lambda_1 = k_1 C_P^0(RT/Fv)$, $\gamma = C_A^0/C_P^0$. (a) $\gamma = 2$, $\gamma_1 =$ (1) 0, (2) 0.2, (3) 20, (4) 2000, (5) 200 000. (b) $\gamma = 10$, $\gamma_1 =$ (1) 0, (2) 0.2, (3) 20, (4) 2000, (5) 200 000. (c) $\gamma = 1000$, $\gamma_1 =$ (1) 0, (2) 0.02, (3) 0.05, (4) 0.2.

and even series of books. It is not the purpose of this article to study in detail the possible reactions for each class of compounds. In many cases, the mechanism is only described by the overall reactions leading from the substrates to the products. These products are analyzed using traditional methods of organic chemistry.

We present here only a selection of some important electrochemical reactions where the mechanism is analyzed by means of electrochemical methods, leading in some cases to the determination of kinetic constants.

6.1 Kolbe Reaction

This first example deals with the oldest electrosynthetic process in organic chemistry. In an aqueous alcoholic medium without any particular precautions, salts of organic acids by oxidation undergo a decarboxylation and then the formation of a carbon–carbon bond between the two organic groups[1,2,55] (Equation 50):

$$2\text{R–COO}^- \longrightarrow \text{R–R} + 2\text{CO}_2 + 2\text{e}^- \qquad (50)$$

This reaction is often used with long-chain carboxylates. At the level of the mechanism determination this reaction is very difficult to analyze because these salts are difficult to oxidize and furthermore the solvents are oxidized in the same range of potential. Two kinds of mechanisms have been postulated for the oxidation of these compounds:

- radical–radical coupling leading to a dimer compound after a one-electron transfer (Scheme 16):
- the formation of a carbocation after a two-electron transfer leading to several kinds of compounds such as esters, alcohols and ethers.

These results are supported by the analysis of the products obtained after electrolysis taking into account the structure of the carboxylate and the nature of the solvent.

$$R{-}COO^- \longrightarrow R^\bullet + CO_2 + e^-$$

$$2R^\bullet \longrightarrow RR$$

Scheme 16

However, information about the first step of the reaction or about the rate of dimerization of the two radicals is very difficult to obtain in most cases. In terms of the possibility of detecting the intermediate $RCOO^\bullet$, nothing was possible through electrochemistry for aliphatic or for unsubstituted aromatic carboxylates and the observation of a slow electron transfer prevents further kinetic determinations. For example, it is not possible to say either that the departure of CO_2 is concerted with the electron transfer or that it occurs in a second step. Even in a nonaprotic medium for easily oxidizable compounds such as aryl acetates an answer is not possible owing to the grafting process on the electrode, except in the case of the oxidation of *N*-dimethylaminophenyl acetate where a sweep rate of $20\,V\,s^{-1}$ led to a fully chemically reversible curve in CV.[56] In this last case the first two steps of the reaction are as shown in Scheme 17.

6.2 Reduction of Halo Compounds

These studies are conducted in an aprotic medium such as acetonitrile or DMF, and from the comparison of the heterogeneous and homogeneous electron transfer an accurate answer to the question could be given concerning the first steps of the reduction process according to the structure of the compounds.

For polyaromatic or easily reducible benzenic compounds bearing an electron-withdrawing group, a two-step mechanism is demonstrated by electrochemical methods or by a redox catalysis using as mediator a stable anion radical of aromatic compounds that permit the determination of the cleavage rate constant.[42] In Table 5 some determinations of standard potential and cleavage rate constants are shown.

For simple phenyl halides or halopyridines, the two-step mechanism is not directly determined without a redox catalysis study showing kinetic control by the forward homogeneous exchange for all the catalysts used. From the determination of these rate constants and by observing the law showing the predicted activation and diffusion lines, homogeneous electron exchange can be characterized as an equilibrium and the standard potential of the aromatic halide/aromatic halide anion couple can be determined (Equation 51):

$$\frac{1}{k_1} = \frac{1}{k_1^{\mathrm{act}}} + \frac{1}{k^{\mathrm{dif}}}\left\{1 + \exp\left[\left(\frac{F}{RT}\right)\left(E^0_{\mathrm{PQ}} - E^0_{\mathrm{AB}}\right)\right]\right\} \tag{51}$$

where k_1^{act} is the kinetic constant under activation control, which in the vicinity of the standard potential E^0_{PQ} is approximately a straight line with a slope of half the diffusion line when the backward reaction k_2 reaches k^{dif}, the diffusion limit.[57]

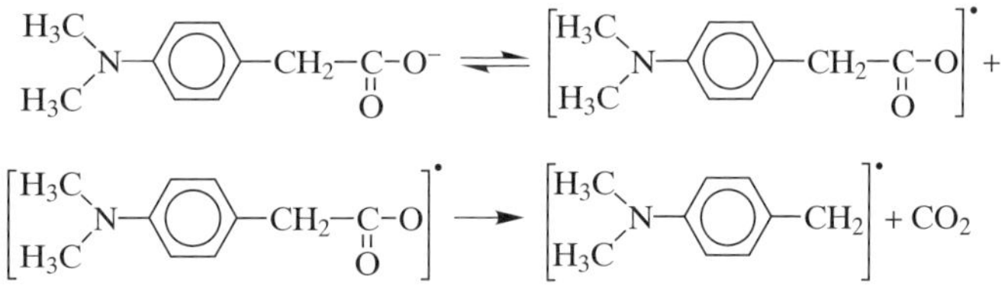

$$(H_3C)_2N{-}C_6H_4{-}CH_2{-}C(=O){-}O^- \rightleftharpoons [(H_3C)_2N{-}C_6H_4{-}CH_2{-}C(=O){-}O]^\bullet + e^-$$

$$[(H_3C)_2N{-}C_6H_4{-}CH_2{-}C(=O){-}O]^\bullet \longrightarrow [(H_3C)_2N{-}C_6H_4{-}CH_2]^\bullet + CO_2$$

Scheme 17

Table 5 Standard potentials and cleavage rate constants for halo compounds in DMF

Chloro compound	E^0 (V vs SCE)	Log k (s^{-1})	Bromo compound	E^0 (V vs SCE)	Log k (s^{-1})
4-Chloronitrobenzene	−1.05	−2	4-Bromonitrobenzene	−0.98	<−2
4-Chlorobenzophenone	−1.64	1.6	4-Bromobenzophenone	−1.63	5
9-Chloroanthracene	−1.73	2.2	9-Bromoanthracene	−1.70	6.4
1-Chloronaphthalene	−2.26	7.2	1-Bromonaphthalene	−2.19	7.8

For references see page 10008

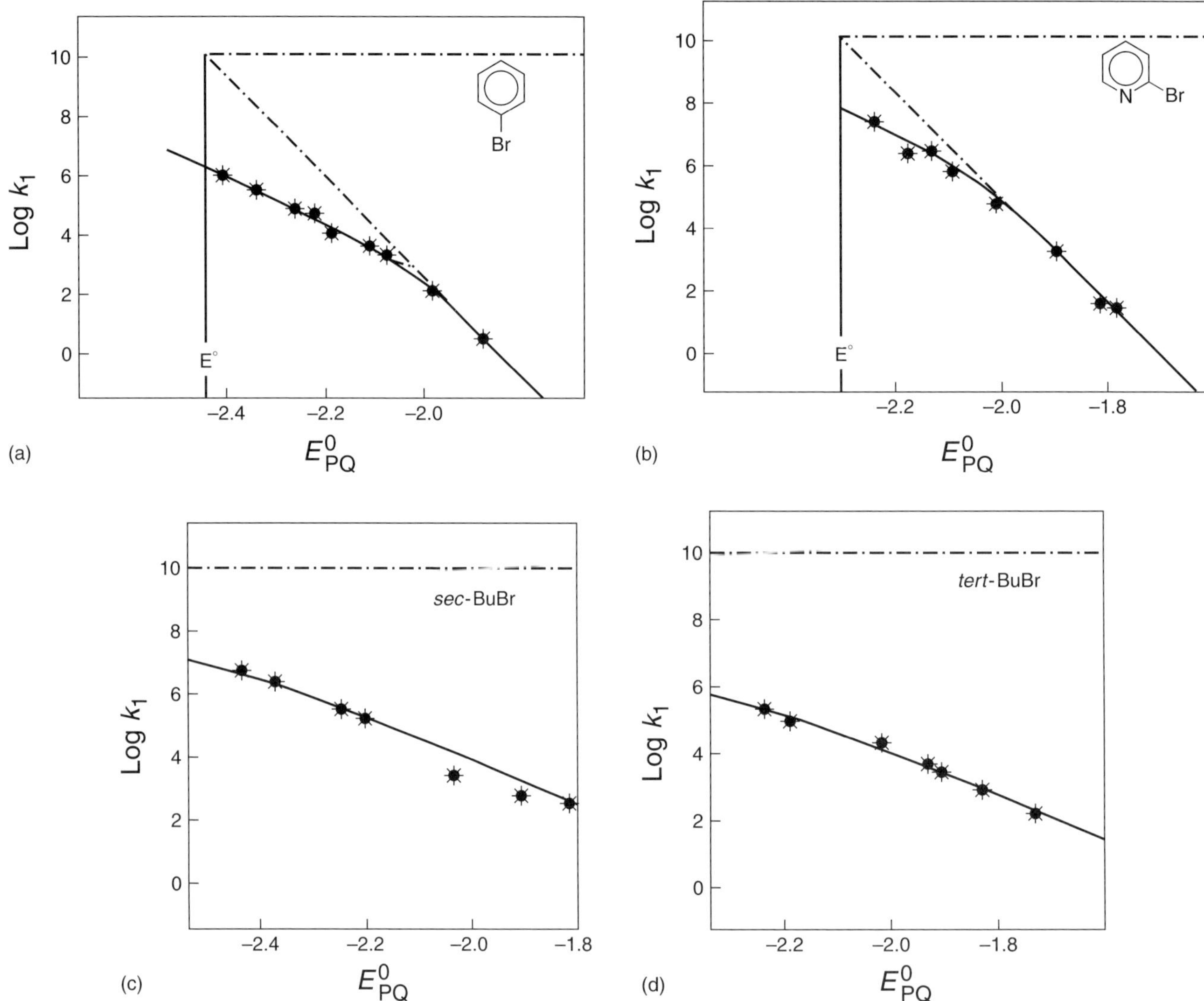

Figure 16 Forward homogeneous electron exchange in the redox catalysis of halo compounds. (a) Bromobenzene; (b) 2-bromopyridine; (c) *sec*-butyl bromide; (d) *tert*-butyl bromide.

Although these results clearly indicate that the rate constant of cleavage is not measurable, the two-step mechanism is still followed (Figure 16a and b).

In contrast to aliphatic halides such as butyl halides, only the activation part of the reaction is observed in the redox catalysis of the compounds.[40] The shape differs significantly from a straight line and the slope as compared with a reverse diffusion is less than 0.5 (Figure 16c and d). The shape of the curves obtained by direct electrochemical methods (for bromides and iodides; the chloro compounds are too difficult to reduce at the electrode) demonstrates a slow electron transfer, which is also attested by a transfer coefficient significantly less than 0.5. In the context of the Marcus–Hush theory, this leads to a large gap between the reduction potential and the standard potential. All these results are characteristic of a concerted reduction and cleavage for the first step of the mechanism.

For bromides, the electron stoichiometry is 2, since the radical produced at the electrode is more easily reducible than the initial substrate. For example, with *n*-butyl bromide (Scheme 18):

$$n\text{-}C_4H_9\text{–}Br + e^- \longrightarrow n\text{-}C_4H_9^{\bullet} + Br^-$$
$$n\text{-}C_4H_9^{\bullet} + e^- \rightleftharpoons n\text{-}C_4H_9^-$$
$$n\text{-}C_4H_9^- + AH \longrightarrow n\text{-}C_4H_{10} + A^-$$

Scheme 18

where AH could be an acidic compound or the solvent itself.

For the tertiary and secondary iodides the first step yields a dimer compound and only in a second step (with a more negative potential) is the radical reduced[39] (Scheme 19):

First step:

$$t\text{-}C_4H_9\text{–}I + e^- \longrightarrow t\text{-}C_4H_9^\bullet + I^-$$

$$2\ t\text{-}C_4H_9^\bullet \longrightarrow \begin{array}{c} t\text{-}C_4H_9 \\ | \\ t\text{-}C_4H_9 \end{array}$$

Second step:

$$t\text{-}C_4H_9^\bullet + e^- \rightleftharpoons t\text{-}C_4H_9^-$$

$$t\text{-}C_4H_9^- + AH \longrightarrow t\text{-}C_4H_{10} + A^-$$

Scheme 19

From these results, it is possible to estimate the standard potential of the anion radical–anion couple for *sec*-butyl (-1.4_4 V vs SCE) and *t*-butyl(-1.5_4 V vs SCE).

6.3 Organic Electrodimerization

This reaction is a powerful method for the industrial preparation of adiponitrile resulting from the reduction of acrylonitrile[1–3,55] (Equation 52):

$$2\ CH_2{=}CHCN \xrightarrow{+2e^- + 2H^+} \begin{array}{c} CH_2CH_2CN \\ | \\ CH_2CH_2CN \end{array} \qquad (52)$$

In fact, this is a hydrodimerization and with this sample compound, as with many other compounds, the mechanism is very difficult to analyze owing to a reduction that is very close to the solvent reduction.

For more conjugated compounds such as *p*-methylbenzylidenemalononitrile in an aprotic solvent, it is possible to elucidate the sequence of reactions. One important question that arises from this reaction is how to determine whether the dimerization occurs between two reduced species [two anion radicals, $E + C2_{(Arr)}$] or between one reduced species reacting on the substrate [$E + C2_{(Ars)}$]. It is also necessary to indicate at what level the protonations by the residual water are involved.

In the present case, via LSV and via convolution,[58] it is possible to demonstrate that the first reaction is the coupling of two radical anions followed by two protonations (Scheme 20):

$$Ph\text{–}CH{=}CHCN + e^- \rightleftharpoons Ph\text{–}\dot{C}H\text{–}\bar{C}HCN$$

$$2\ Ph\text{–}\dot{C}H\text{–}\bar{C}HCN \xrightarrow{k_d} \begin{array}{c} Ph\text{–}CH\text{–}\bar{C}HCN \\ | \\ Ph\text{–}CH\text{–}\bar{C}HCN \end{array}$$

$$k_d = 10^7\ M^{-1}\ s^{-1}$$

Scheme 20

The same kind of mechanism occurs during oxidation for the coupling of two cation radicals and during the oxidation of aromatic amines,[59] the loss of proton occurring in a second chemical step.

It is possible to obtain mixed coupling when two compounds are electroactive in the same range of potential and also to obtain intramolecular coupling, yielding cyclic compounds.

The formation of electroactive polymers with compounds such as pyrrole and thiophene is an important electrosynthetic process.[60] As the formation of long-chain oligomers is described, it has often been postulated that the propagation mechanism is the result of the addition of a cation radical or of a deprotonated radical to the monomer or to the polymer moieties. However, for example, polypyrrole is obtained as a long-chain insoluble product and even the use of the fastest method in the electrochemistry of pyrrole does not permit any way of visualizing the intermediates.

In contrast, when substituted pyrroles[61] or bipyrroles are used, it is possible to demonstrate by CV or DSC that the initial step of this polymerization is a dimerization of two cation radicals; the deprotonation steps occur consecutively[62] (Scheme 21):

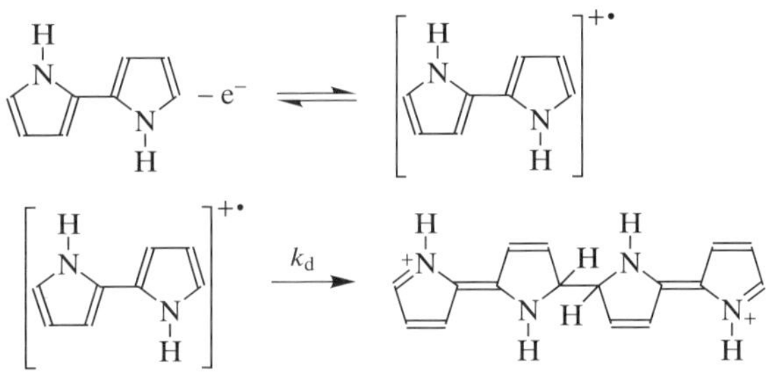

Scheme 21

The same mechanism was demonstrated in the polythiophene series for long-chain polythiophene (three or more units) owing to the strong reactivity of thiophenes as compared with that of pyrroles.

6.4 Chemical Reactions Induced by Electrochemistry

Electrochemistry is also used for inducing chemical reactions when these reactions are impossible or slow when unaccompanied by an activating process. Under these conditions, the electrochemical methods described earlier yield interesting results about the mechanism.

These processes can be induced electrochemically without any addition of other compounds (only by some change in the oxidation degree of one reactant) or with the addition of some chemical catalysts. In the first case the process can be described by Scheme 22.

The "zero-electron" reaction is a nucleophilic substitution induced by electrochemistry when the reduction of AY leads to the replacement of the Y substituent by L, which leads to the reoxidation of $AL^{\bullet-}$ in AL.[63,64] The reaction is called an $S_{RN}1$ reaction (first-order radical

For references see page 10008

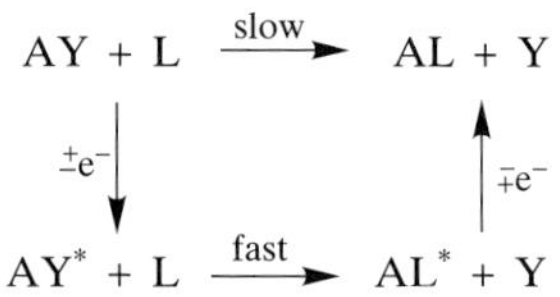

$$\begin{array}{ccc} AY + L & \xrightarrow{\text{slow}} & AL + Y \\ \downarrow \pm e^- & & \uparrow \mp e^- \\ AY^* + L & \xrightarrow{\text{fast}} & AL^* + Y \end{array}$$

Scheme 22

nucleophilic substitution) when a one-electron reduction is involved, e.g. in the substitution by phenoxy or phenylthio groups in halo compounds (Equation 53):

$$RX + Nu^- \longrightarrow RNu + X^- \qquad (53)$$

This reaction is easily studied via CV in the absence or presence of nucleophiles as illustrated in Figure 17(a) and (b). The mechanism of this reaction and good working conditions are determined by the following observations:

- the potential of initiation is the potential of reduction of RX;
- the potential of reduction of RNu must be more negative than the potential of reduction of RX;
- the "zero-electron" is more efficient for RX compounds that give a slow cleavage reaction because the faster the cleavage, the closer to the electrode the reaction occurs.

Taking into account the preceding results about the rate of cleavage of halo compounds, polyaromatic compounds must be suitable for this process. The reduction of the radical into an anion is a competing process that must be minimized, either by using chloride rather than bromide or iodide (the cleavage is faster for bromide or iodide than for chloride) or by redox catalysis of the process that also leads to radical formation far from the electrode. On the other hand, we must avoid the possibility of atom transfer from the solvent on the radical; for this purpose, liquid ammonia would be the best solvent but we can also use DMSO or acetonitrile[65] (Scheme 23):

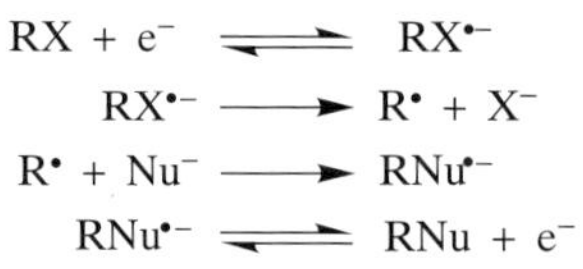

$$\begin{array}{rcl} RX + e^- & \rightleftharpoons & RX^{\bullet -} \\ RX^{\bullet -} & \longrightarrow & R^{\bullet} + X^- \\ R^{\bullet} + Nu^- & \longrightarrow & RNu^{\bullet -} \\ RNu^{\bullet -} & \rightleftharpoons & RNu + e^- \end{array}$$

Scheme 23

A good example is the following one shown in Equation (54):

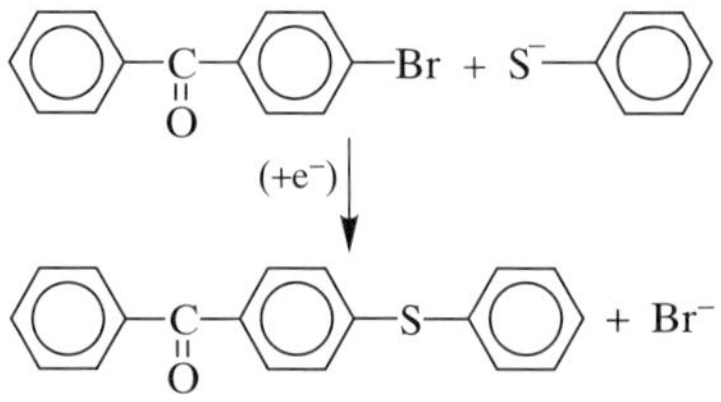

$$C_6H_5{-}C(=O){-}C_6H_4{-}Br + {}^-S{-}C_6H_5 \xrightarrow{(+e^-)} C_6H_5{-}C(=O){-}C_6H_4{-}S{-}C_6H_5 + Br^- \qquad (54)$$

In many cases the simple activation by reduction (or oxidation) of one reactant does not lead to a significant reaction. A catalyst acting by a chemical reactant on the substrate can be used. For example, in the same kind of reactions, by using palladium complexes,[66,67] some new reactions are possible. The electrochemistry can play on two registers: it is often necessary to prepare the active species at the right degree of oxidation and on the other hand the mechanism of the reaction is followed by electrochemical methods (Scheme 24).

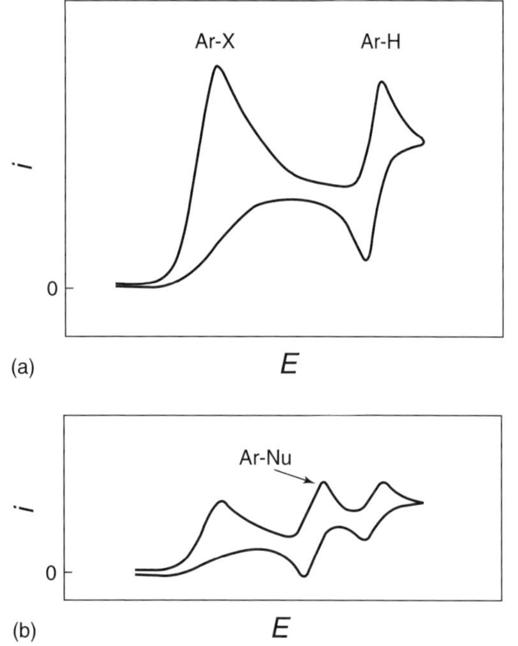

Figure 17 Schematic CV for an $S_{RN}1$ reaction on an aromatic halide. (a) CV without nucleophile and (b) CV in the presence of nucleophile.

7 CONCLUSION

In contrast to small-molecule electrochemistry, the mechanisms of organic electrochemical reactions are slightly influenced by the adsorption of reactants or of intermediates on the electrodes. Hence it is possible to determine a great number of mechanisms that involve a sequence of electrochemical reactions and homogeneous chemical reactions and to characterize the key steps of these processes.

Our aim was not to list the mechanisms of the electrochemistry of each class of organic compounds but to show what can be determined using different electrochemical methods.

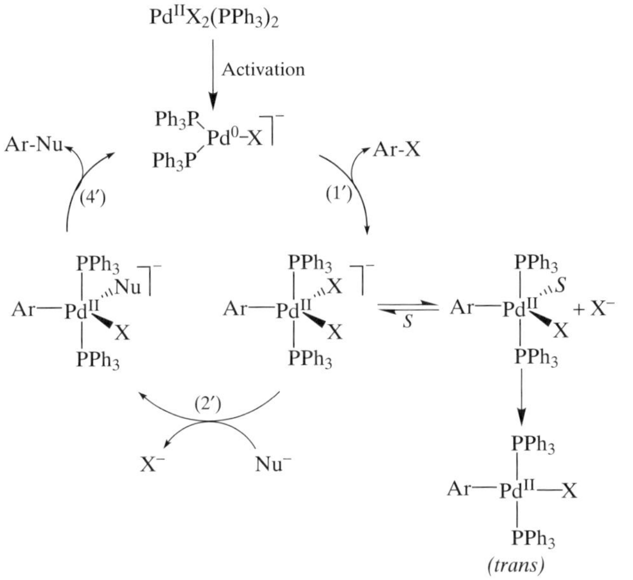

Scheme 24 Catalytic cycle for an $S_{RN}1$ reaction on an aromatic halide using a palladium catalyst. [Reproduced by permission of La Société Française de Chimie from C. Amatore, A. Jutand, L. Thouin, J.-N. Verpeaux, *Actual. Chim.*, (8–9), 43–62 (1998).]

Because simple electron transfer steps are rather fast for organic molecules, the chemical reactions acting before or after the electron transfers are easily characterized. Using double-step techniques, mainly CV and also RRDEV or DSC, when the follow-up chemical reactions are not too fast, it is easy first to visualize the stability of the intermediates with half-lives of about 10^{-4} s. Under these conditions the rate constants of these chemical reactions are accurately measured using either double-step or single-step techniques (LSV, RDEV). Recent advances in electronic devices and in the miniaturization of electrodes allow the detection of intermediates with a lower limit of the detectable half-life of a few microseconds.

Preceding reaction, catalytic reaction or multielectron reactions with the same range of half-life of intermediates are easier to study because the loss of stability of the intermediates can also be followed by the significant changes in the currents giving a clearer determination of the rate constants.

Even when intermediates cannot be detected, the nature of the chemical reactions is accurately demonstrated by the use of single-step techniques as measured by the order of the reaction owing to the possibility of changing the operational window time that these techniques afford.

Even if the limitations due to the electron transfer occur less for organic than for inorganic species, very fast chemical reactions nevertheless lead to kinetic control by electron transfer and prevent kinetic information about the chemical step from being obtained. An important question still remains regarding the fact that electron transfer and chemical reaction are either two separate reactions or only one concerted process. The kinetic behavior of each electron transfer is characteristic and leads to the conclusion that contrary to common knowledge, concerted electron transfer and chemical reaction exist in many cases.

Beyond the description of the reaction sequence, the electrochemical methods described here open up the possibility of obtaining accurate information about the elementary steps in organic chemistry and can contribute to the development of structure–reactivity relationships concerning the kinetics of both electron transfers and coupled chemical reactions.

ACKNOWLEDGMENTS

I am indebted to all the members of the Laboratoire d'Electrochimie Moléculaire who participated in the work on organic electrochemical mechanisms and whose names may be found in the reference list. The invaluable contribution of Dr Jean-Michel Savéant to this field should be particularly emphasized.

For the simulation of cyclic voltammograms, Digisim® from BAS was used.

ABBREVIATIONS AND ACRONYMS

CV	Cyclic Voltammetry
DMF	Dimethylformamide
DMSO	Dimethyl Sulfoxide
DSC	Double-step Chronoamperometry
ESR	Electron Spin Resonance
IR	Infrared
LSV	Linear Sweep Voltammetry
PMV	Photomodulation Voltammetry
PSC	Potential Step Chronoamperometry
RDEV	Rotating Disk Electrode Voltammetry
RRDEV	Rotating Ring-disk Electrode Voltammetry
SCE	Saturated Calomel Electrode
UV/VIS	Ultraviolet/Visible

RELATED ARTICLES

Electroanalytical Methods **(Volume 11)**
Electroanalytical Methods: Introduction • Infrared Spectroelectrochemistry • Pulse Voltammetry • Scanning Tunneling Microscopy, In Situ, Electrochemical • Ultrafast Electrochemical Techniques • Ultraviolet/Visible Spectroelectrochemistry

For references see page 10008

Kinetic Determinations **(Volume 12)**
Kinetic Determinations: Introduction • Catalytic Kinetic Determinations: Nonenzymatic

Kinetic Determinations cont'd **(Volume 13)**
Instrumentation for Kinetics

REFERENCES

1. H. Lund, M.M. Baizer, *Organic Electrochemistry*, 3rd edition, Marcel Dekker, New York, 1991.
2. A.J. Bard, H. Lund, *Encyclopedia of the Electrochemistry of the Elements*, Marcel Dekker, New York, Vols. 11–15, 1978–1984.
3. T. Shono, *Electroorganic Synthesis*, Academic Press, London, 1991.
4. P. Delahay, *New Instrumental Methods in Electrochemistry*, Interscience, New York, 1952.
5. A.J. Bard, L.R. Faulkner, *Electrochemical Methods*, John Wiley, New York, 1980.
6. C.K. Mann, 'Nonaqueous Solvents for Electrochemical Use', in *Electroanalytical Chemistry*, ed. A.J. Bard, Marcel Dekker, New York, 57–134, Vol. 3, 1969.
7. H. Feess, H. Wendt, 'Performance of Two-phase-electrolyte Electrolysis', in *Techniques of Electroorganic Synthesis, Part III*, eds. N.L. Weinberg, B.V. Tilak, John Wiley & Sons, New York, 81–177, 1982.
8. R.N. Adams, *Electrochemistry at Solid Electrodes*, Marcel Dekker, New York, 1969.
9. D.T. Sawyer, J.L. Roberts, *Experimental Electrochemistry for Chemists*, John Wiley, New York, 1974.
10. D.J.G. Ives, G.J. Janz, *Reference Electrodes*, Academic Press, New York, 1961.
11. C.P. Andrieux, J.-M. Savéant, 'Electrochemical Reactions', in *Investigations of Rates and Mechanisms of Reactions*, ed. C. Bernasconi, John Wiley, New York, 305–390, Vol. 6, 4/E, part 2, 1986.
12. D. Garreau, J.-M. Savéant, 'Linear Sweep Voltammetry. Compensation of Cell Resistance and Stability. Determination of the Residual Uncompensated Resistance', *J. Electroanal. Chem.*, **35**, 309–331 (1972).
13. C.P. Andrieux, 'Terminology and Notations for Multistep Electrochemical Reaction Mechanisms', *Pure Appl. Chem.*, **66**, 2245–2250 (1994).
14. W.J. Albery, *Electrode Kinetics*, Clarendon Press, Oxford, 1975.
15. F. Opekar, P. Beran, 'Rotating Disk Electrodes', *J. Electroanal. Chem.*, **69**, 1–115 (1976).
16. W.J. Albery, M.L. Hitchman, *Ring-disk Electrodes*, Clarendon Press, Oxford, 1971.
17. D.D. MacDonald, *Transient Techniques in Electrochemistry*, Plenum Press, New York, 1977.
18. R.S. Nicholson, I. Shain, 'Theory of Stationary Electrode Polarography. Single Scan and Cyclis Methods Applied to Reversible, Irrevesible and Kinetic Systems', *Anal. Chem.*, **36**, 706–723 (1964).
19. J. Osteryoung, J.J. O'Dea, 'Square-wave Voltammetry', in *Electroanalytical Chemistry*, ed. A.J. Bard, Marcel Dekker, New York, 209–308, Vol. 14, 1986.
20. D.E. Smith, 'AC Polarography and Related Techniques: Theory and Practice', in *Electroanalytical Chemistry*, ed. A.J. Bard, Marcel Dekker, New York, 1–148, Vol. 1, 1966.
21. E.M. Arnett, R.A. Flowers, 'Bond Cleavage Energies for Molecules and Their Associated Radical Ions', *Chem. Soc. Rev.*, **22**, 9–15 (1993).
22. C.P. Andrieux, P. Hapiot, J. Pinson, J.-M. Savéant, 'Determination of Formal Potentials of Chemically Unstable Redox Couples by Second Harmonic Alternating Current Voltammetry and Cyclic Voltammetry. Application to the oxidation of thiophenoxide ions', *J. Am. Chem. Soc.*, **115**, 7783–7788 (1993).
23. C.P. Andrieux, D. Garreau, P. Hapiot, J. Pinson, J.-M. Savéant, 'Fast Sweep Cyclic Voltammetry at Ultramicroelectrodes. Evaluation of the Method for Fast Electron-transfer Kinetic Measurements', *J. Electroanal. Chem.*, **243**, 321–335 (1988).
24. C.P. Andrieux, D. Garreau, P. Hapiot, J.-M. Savéant, 'Ultramicroelectrodes. Cyclic Voltammetry Above One Million Volts per Second', *J. Electroanal. Chem.*, **248**, 447–450 (1988).
25. I. Montenegro, M.A. Queiros, J.L. Daschbach (eds.), *Microelectrodes: Theory and Applications*, NATO ASI Series, Series E, Kluwer Academic, Dordrecht, Vol. 197, 1993.
26. V.V. Konavalov, A.M. Raitsimring, Y.D. Tsvetkov, I.I. Bilkis, 'Photoelectrochemical Study of the Radical Anion Cleavage Rates in Aromatic Molecules: Halobenzoic Acids', *Chem. Phys. Lett.*, **157**, 257–264 (1989).
27. D.D.M. Wayner, D.J. McPhee, D. Griller, 'Oxidation and Reduction Potentials of Transient Free Radicals', *J. Am. Chem. Soc.*, **110**, 132–137 (1988).
28. P. Hapiot, V. Konovalov, J.-M. Savéant, 'Application of Laser Pulse Photoinjection of Electrons from Metal Electrodes to the Determination of Reduction Potentials of Organic Radicals in Aprotic Solvents', *J. Am. Chem. Soc.*, **117**, 1428–1434 (1995).
29. N.S. Hush, 'Adiabatic Theory of Outer-sphere Electron-transfer Reactions in solution', *Trans. Faraday Soc.*, **57**, 557–580 (1961).
30. R.A. Marcus, 'On the Theory of Electron-transfer Reactions VI.Unified Treatment for Homogeneous and Electrode Reactions', *J. Chem. Phys.*, **43**, 679–701 (1965).
31. L. Eberson, *Electron Transfer Reactions in Organic Chemistry*, Springer, Berlin, 1987.
32. H. Kojima, A.J. Bard, 'Determination of Rate Constants for the Electroreduction of Aromatic Compounds and Their Correlation with Homogeneous Electron Transfer Rates', *J. Am. Chem. Soc.*, **97**, 6317–6324 (1975).

33. J.-M. Savéant, 'A Simple Model for the Kinetics of Dissociative Electron Transfer in Polar Solvents. Application to the Homogeneous and Heterogeneous Reduction of Alkyl Halides', *J. Am. Chem. Soc.*, **109**, 6788–6795 (1987).
34. J.-M. Savéant, 'Dissociative Electron Transfer', in *Adv. Electron Transfer Chem.*, **4**, 53–116 (1994).
35. J.C. Imbeaux, J.-M. Savéant, 'Convolution Potential Sweep Voltammetry. I. Introduction', *J. Electroanal. Chem.*, **44**, 169–187 (1973).
36. D. Garreau, J.-M. Savéant, D. Tessier, 'A Computer Coupled High Frequency AC Technique for Investigating Fast Non-volmerian Electron Transfers', *J. Electroanal. Chem.*, **103**, 321 (1979).
37. D. Garreau, J.-M. Savéant, D. Tessier, 'Potential Dependence of the Electrochemical Transfer Coefficient. An Impedance Study of the Reduction of Aromatic Compounds', *J. Phys. Chem.*, **83**, 3003–3007 (1979).
38. J.-M. Savéant, D. Tessier, 'Potential Dependence of the Electrochemical Transfer Coefficient. Reduction of Some Nitro-compounds in Aprotic Media', *J. Phys. Chem.*, **81**, 2192–2197 (1977).
39. C.P. Andrieux, I. Gallardo, J.-M. Savéant, 'Outer-sphere Electron-transfer Reduction of Alkyl Halides: A Source of Alkyl Radicals or of Carbanions? Reduction of Alkyl Radicals', *J. Am. Chem. Soc.*, **111**, 1620–1626 (1989).
40. C.P. Andrieux, I. Gallardo, J.-M. Savéant, K.B. Su, 'Dissociative Electron Transfer. Homogeneous and Heterogeneous Reductive Cleavage of the Carbon–Halogen Bond in Simple Aliphatic Halides', *J. Am. Chem. Soc.*, **108**, 638–647 (1986).
41. C.P. Andrieux, A. Le Gorande, J.-M. Savéant, 'Electron Transfer and Bond Breaking. Examples of Passage from a Sequential to a Concerted Mechanism in the Electrochemical Reductive Cleavage of Arylmethyl Halides', *J. Am. Chem. Soc.*, **114**, 6892–6904 (1992).
42. C.P. Andrieux, J.-M. Savéant, D. Zann, 'Relationship Between Reduction Potentials and Cleavage Rates in Aromatic Molecules', *Nouv. J. Chim.*, **8**, 107–115 (1984).
43. C.P. Andrieux, J.-M. Savéant, A. Tallec, R. Tardivel, C. Tardy, 'Concerted and Stepwise Dissociative Electron Transfers. Oxidizability of the Leaving Group and Strength of the Breaking Bond as Mechanism and Reactivity Governing Factors Illustrated by the Electrochemical Reduction of α-Substituted Acetophenones', *J. Am. Chem. Soc.*, **119**, 2420–2429 (1997).
44. L. Nadjo, J.-M. Savéant, 'Dimerization, Disproportionation and ECE Mechanism in the Reduction of Aromatic Carbonyl Compounds in Alkaline Media', *J. Electroanal. Chem.*, **33**, 419–451 (1971).
45. C.T. Pedersen, V.D. Parker, 'The Electrochemistry of Organic Sulfur Compounds. Part II. Cathodic Reduction of 1,2-Dithiolylium Ions. Redox Analogy to the Tropylium Bitrotenyl System', *Tetrahedron Lett.*, **9**, 767–770 (1972).
46. J.-M. Savéant, E. Vianello, 'Potential Sweep Voltammetry. General Theory of Chemical Polarization', *Electrochim. Acta*, **12**, 629–646 (1967).
47. C.P. Andrieux, L. Nadjo, J.-M. Savéant, 'Electrodimerization. I. One Electron Irreversible Dimerization. Diagnostic Criteria and Rate Determination Procedures for Voltammetric Studies', *J. Electroanal. Chem.*, **26**, 147–186 (1970).
48. F. Ammar, J.-M. Savéant, 'Thermodynamics of Successive Electron Transfers. Entropy Effects in the Series of Polynitro Compounds', *J. Electroanal. Chem.*, **47**, 115–125 (1973).
49. M.E. Peover, 'Electrochemistry of Aromatic Hydrocarbons and Related Substances', in *Electroanalytical Chemistry*, ed. A.J. Bard, Marcel Dekker, New York, 1–29, Vol. 2, 1967.
50. M. Mastragostino, L. Nadjo, J.-M. Savéant, 'Disproportionation and ECE Mechanisms. I. Theoretical Analysis for Linear Sweep Voltammetry', *Electrochim. Acta*, **13**, 721–749 (1968).
51. C. Amatore, J.-M. Savéant, 'ECE and Disproportionation. V. Stationary State General Solution. Application to Linear Sweep Voltammetry', *J. Electroanal. Chem.*, **85**, 27–46 (1977).
52. C. Amatore, J.-M. Savéant, 'Electrochemical Hydrogenation of Aromatic Hydrocarbons. Discrimination Between ECE and Disproportionation Mechanisms by Double Potential Step Chronoamperometry', *J. Electroanal. Chem.*, **107**, 353–364 (1980).
53. C.P. Andrieux, C. Blocman, J.-M. Dumas-Bouchiat, F. M'Halla, J.-M. Savéant, 'Homogeneous Redox Catalysis of Electrochemical Reactions. V. Cyclic Voltammetry', *J. Electroanal. Chem.*, **113**, 19–40 (1980).
54. D. Lexa, J.-M. Savéant, K.B. Su, D.L. Wang, 'Chemical Versus Redox Catalysis of Electrochemical Reactions. Reduction of *trans*-1,2-Dibromocyclohexane by Electrogenerated Aromatic Anion Radicals and Low Oxidation State Metallo-porphyrins', *J. Am. Chem. Soc.*, **109**, 6464–6470 (1987).
55. R.D. Little, M.K. Schwaebe, 'Reductive Cyclizations at the Electrode', *Top. Curr. Chem.*, **185**, 1–48 (1997).
56. C.P. Andrieux, F. Gonzales, J.-M. Savéant, 'Derivatization of Carbon Surfaces by Anodic Oxidation of Arylacetates. Electrochemical Manipulation of the Grafted Films', *J. Am. Chem. Soc.*, **119**, 4292–4300 (1997).
57. C.P. Andrieux, C. Blocman, J.-M. Dumas-Bouchiat, J.-M. Savéant, 'Heterogeneous and Homogeneous Electron Transfers to Aromatic Halides. A Redox Catalysis Study in the Halobenzene and Halopyridine Series', *J. Am. Chem. Soc.*, **101**, 3431–3441 (1979).
58. L. Nadjo, J.-M. Savéant, D. Tessier, 'Electrodimerization. XI. Coupling Mechanism of an Activated Olefin: *p*-Methylbenzylidenemalononitrile as Studied by Convolution Potential Sweep Voltammetry', *J. Electroanal. Chem.*, **64**, 143–154 (1975) .

59. D. Larumbe, I. Gallardo, C.P. Andrieux, 'Anodic Oxidation of Some Tertiary Amines', *J. Electroanal. Chem.*, **304**, 241–247 (1991).
60. H.S. Nalwa, *Handbook of Organic Conductive Molecules and Polymers*, John Wiley, New York, 1997.
61. C.P. Andrieux, P. Audebert, P. Hapiot, J.-M. Savéant, 'Identification of the First Steps of the Electrochemical Polymerization of Pyrroles by Means of Fast Potential-step Techniques', *J. Phys. Chem.*, **95**, 10 158–10 164 (1991).
62. L. Guyard, P. Hapiot, P. Neta, 'Redox Chemistry of Bipyrroles: Further Insights into the Oxidative Polymerization Mechanism of Pyrrole and Oligopyrroles', *J. Phys. Chem. B*, **101**, 5698–5706 (1997).
63. J. Pinson, J.-M. Savéant, 'Electrochemically Induced Aromatic Nucleophilic Substitution', *J. Am. Chem. Soc.*, **100**, 1506–1510 (1978).
64. R.A. Rossi, R.H. Rossi, *Aromatic Nucleophilic Substitution by the SRN1 Mechanism*, ACS Monograph 178, American Chemical Society, Washington, DC, 1983.
65. C. Amatore, J. Chaussard, J. Pinson, J.-M. Savéant, A. Thiébault, 'Electrochemically Induced Aromatic Nucleophilic Substitution in Liquid Ammonia. Competition with Electron Transfer', *J. Am. Chem. Soc.*, **101**, 6012–6020 (1979).
66. C. Amatore, M. Azzabi, A. Jutand, 'Rates and Mechanism of the Reversible Oxidative Addition of (*Z*)- and (*E*)-1,2 Dichloroethylene to Low Liganded Zerovalent Palladium', *J. Am. Chem. Soc.*, **113**, 1640–1677 (1991).
67. C. Amatore, A. Jutand, A. Suarez, 'Intimate Mechanism of Oxidative Addition of Zerovalent Palladium Complexes in Presence of Halide Ions and Its Relevance to the Mechanism of Palladium-catalyzed Nucleophilic Substitutions', *J. Am. Chem. Soc.*, **115**, 9531–9541 (1993).

Pulse Voltammetry

Michael T. Carter
Eltron Research Inc., Boulder, USA

Robert A. Osteryoung
North Carolina State University, Raleigh, USA

This article covers fundamentals and applications of pulse voltammetry. These techniques use a train of timed potential pulses to perturb the electrode–solution interface. Electron transfer reactions are driven at the electrode surface during each pulse. Current flowing as a result of electron transfer is measured at specified times during the pulse. Popular pulse voltammetric methods include normal pulse voltammetry (NPV), reverse pulse voltammetry (RPV), square wave voltammetry (SWV) and differential pulse voltammetry (DPV). Pulse voltammetry generally provides improved signal-to-noise ratio (S/N) and a decreased lower limit of detection (LLD) compared with continuous electrolysis methods such as cyclic voltammetry (CV), by controlling the extent of electrolysis prior to current measurement and discrimination against background processes. A concise discussion of the basic concepts behind the various pulse voltammetric methods is presented, followed by method-specific discussions of the most popular methods. Instrumentation and practical experimental considerations are covered. Finally, a discussion of selected illustrative applications of pulse voltammetry to analysis and other chemical problems is presented, demonstrating the use of these techniques as a laboratory tool.

1 INTRODUCTION

Pulse voltammetry comprises a family of electrochemical methods which use a train of potential pulses and corresponding current responses to generate a current–potential curve or voltammogram. Pulse voltammetry takes advantage of the difference between Faradaic (electron transfer) and nonFaradaic (interfacial charging) processes to discriminate against background, resulting in improved S/N compared with continuous electrolysis [direct current (DC)] methods, such as CV. A variety of specialized pulse sequences, conditioning potentials and other steps can be included in the complete experimental waveform, making these methods very versatile for electrochemical studies. The applied potential required to oxidize or reduce a particular species can be used to identify it in many cases. This is particularly useful in differential methods, including SWV and DPV, which provide a peaked voltammogram. The analytical utility of pulse voltammetry, as with other voltammetric methods, comes from proportionality between concentration of redox-active species in solution and measured current.

The predominant modern techniques of pulse voltammetry are SWV, NPV, RPV and DPV. SWV is a method which combines features of the pulse method with continuous electrolysis. Strictly, "voltammetry" refers to techniques applied at solid electrodes, e.g. platinum, carbon or gold and to mercury electrodes exclusive of the dropping mercury electrode (DME), such as the static mercury drop electrode (SMDE), the hanging mercury drop electrode (HMDE) and the thin-film mercury electrode (TFME). The term "polarography" strictly applies to the DME only. Normal and reverse pulse methods at the DME, for example, will be referred to as normal pulse polarography (NPP) and reverse pulse polarography (RPP), respectively, while the corresponding experiments performed at, say, a glassy carbon (GC) disk electrode are NPV and RPV. The reader will no doubt encounter many violations of this rule and a confusing array of specialized nonstandard jargon in the electrochemical literature, but one should keep in mind that all these techniques have a common basis.

2 THEORY AND OPERATING PRINCIPLES

Pulse voltammetric methods are potentially very powerful analysis tools, sometimes capable of detection limits rivaling more popular analytical methods such as atomic absorption spectroscopy. Voltammetric measurements may also provide information on the thermodynamics and kinetics of an interfacial electrochemical reaction. Information is contained in the redox potential at which an electron transfer reaction proceeds, which is often characteristic of the particular species reacting and in the current passed, which is a direct measure of the overall rate of the reaction and is commonly proportional to analyte concentration.

We shall concentrate on basic operating principles, which will therefore be limited to simple reversible cases to illustrate the main concepts, although common complications of which the reader should be aware are mentioned where appropriate. The reader is directed to the literature citations for additional detail.

2.1 Chronoamperometric Response to a Potential Step

Chronoamperometry, measurement of current flowing in response to a potential step as a function of time, is the fundamental experiment behind all pulse voltammetry. The most basic results for a simplified redox system are described here. For convenience, electrochemists are fond of discussing generic electron transfer reactions in terms of the O/R redox couple (Equation 1):

$$\mathrm{O} + ne^- \rightleftharpoons \mathrm{R} \qquad (1)$$

where O is the oxidized form of the couple, R is the reduced form and n is the number of electrons transferred in the reaction. In the simplest case, O and R differ by one electron, but are otherwise essentially identical, e.g. Fe^{3+}/Fe^{2+}. We also usually make assumptions that both O and R are soluble, i.e. neither adsorbs on the electrode, and that both are chemically stable on the *timescale of the measurement*. The experimental timescale is an important aspect of pulse voltammetry. Furthermore, the electron transfer reaction is assumed to be reversible, i.e. the concentrations of O and R at the electrode–solution interface obey the Nernst equation. Clearly, this is a lot to expect from a real system. We shall discuss caveats and exceptions to the simple case as appropriate.

2.1.1 Reversible Systems Under Diffusion Control

We shall consider a case in which the working electrode is immersed in a solution of O.[1–4] We generally ignore the rest of the electrochemical cell, concentrating on what happens at the working electrode, which is the detector in an analytical application. It is common to speak of reduction of O to R, but the reasoning applies equally to oxidations. During the electrochemical reaction, O is converted to R at the electrode surface. This conversion depends on time-dependent phenomena such as diffusion and heterogeneous reaction rate, and so pulsed electrochemical methods are characterized by currents which vary with time. As conversion proceeds, the total analytical concentration of O dissolved in solution, denoted C_O^*, is partitioned between O and R. These concentrations, denoted C_O and C_R, respectively, change with time and distance from the electrode surface, with the sum of C_O and C_R equaling C_O^*. Designation of concentration of O or R as a function of distance (x) and time (t) is often given in shorthand as $C_\mathrm{O}(x,t)$ and $C_\mathrm{R}(x,t)$, or some variation thereof.[1]

When the electrode–solution interface is perturbed by application of a potential, denoted E (volts), and E

For references see page 10032

is sufficiently extreme compared with $E^{o'}_{O/R}$, the standard formal potential of the O/R system, to drive redox Equation (1) at its maximum rate from left to right, it is common (but not necessarily always the case) that the current, usually denoted i (amperes), is controlled by diffusion of O to the electrode surface. A typical potential step traversing the $E^{o'}_{O/R}$ region and resulting current response are shown in Figure 1(a) and (b). C_O at the surface of the electrode rapidly decreases to zero after pulse application and C_R increases correspondingly, as shown in Figure 2(a). The exact concentrations of O and R at the electrode surface are fixed by the Nernst equation for a reversible reaction (Equation 2):

$$E = E^{o'} - \frac{0.059}{n}\log\left[\frac{C_R(x=0)}{C_O(x=0)}\right] \qquad (2)$$

where E is the applied potential, $E^{o'}$ is the standard formal potential of the O/R couple and $C_R(x=0)$ and $C_O(x=0)$ are the concentrations of O and R at the electrode surface. The term $0.059/n$ is RT/nF at 25 °C, where R is the gas constant, T is absolute temperature (kelvin) and F is the Faraday constant. Once C_O has reached zero at the electrode surface, a concentration gradient of O exists and material must diffuse to the electrode to sustain the reduction, hence "diffusion-limited current". When the reaction is fast, current is controlled by the diffusion gradient of O established at the electrode surface, which is characterized by the diffusion coefficient, D, of the diffusing species. The current, $i(t)$, at any time during the potential step is given by Equation (3):

$$i(t) = \frac{nFAD^{1/2}C^*_O}{\pi^{1/2}t^{1/2}} \qquad (3)$$

where A is electrode area and the other terms were defined previously. Equation (3) is commonly known as the Cottrell equation.[1–4] The main points to take from this are that diffusion-controlled currents decay with a $t^{-1/2}$ dependence and they are proportional to analyte concentration.

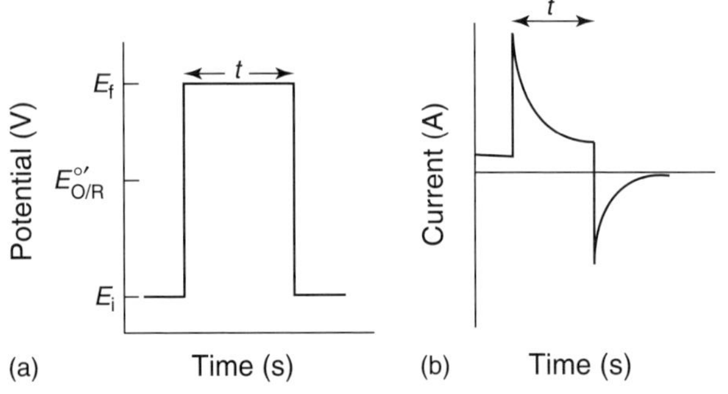

Figure 1 (a) A potential step experiment and (b) the resulting chronoamperometric response. $E^{o'}_{O/R}$ is the formal potential of the O/R redox couple; E_i and E_f denote initial and final values of the potential, respectively.

List of selected abbreviations appears in Volume 15

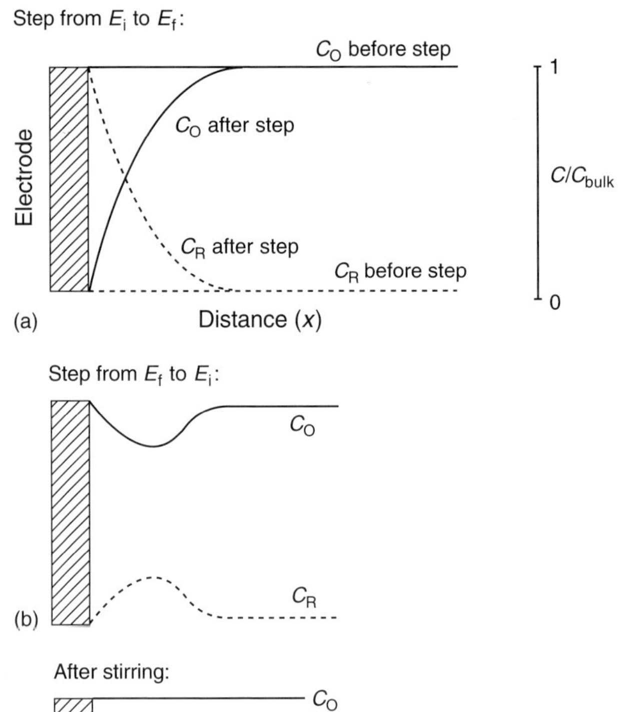

Figure 2 Concentration–distance profiles at various stages of a potential step: (a) changes in concentrations of oxidized (C_O) and reduced (C_R) forms following a potential step to a diffusion-limited value; (b) effect of returning the potential to the initial value; (c) re-establishment of initial conditions by convection.

Most pulse voltammetric experiments rely on the re-establishment of "initial boundary conditions" at the electrode surface between potential steps. Following the step from E_i to E_f, above, returning the potential to E_i (Figure 2b) drives R at the electrode surface back to O. A slight concentration gradient still exists, however, since some R has diffused away from the electrode surface during the step. Stirring or other modes of forced convection, as shown in Figure 2(c), or waiting for a sufficiently long time at E_i are commonly used to remove residual gradients and return the interface to its initial state.

2.1.2 Charging Current

An electrochemical interface has the properties of a capacitor.[5] When an electrode in contact with an aqueous medium containing dissolved ionic species is subjected to a potential step, the ions redistribute in response to change in interfacial potential. This redistribution results in a structure called the electrical double layer. Charging of the interface can be thought of conveniently as charge separation in a capacitor (Equation 4):

$$Q = CE \tag{4}$$

where charge built up on the plates of the capacitor (Q, coulombs) is proportional to the potential applied (E) and the capacitance (C, farads), which depends on the dielectric properties of the medium. Experimentally, a potential step applied to the interface results in current–time behavior described by a simple RC circuit (Equation 5):

$$i(t) = \frac{\Delta E}{R_s} e^{-t/R_s C_{dl}} \tag{5}$$

where $i(t)$ is the time-dependent current, t is time, ΔE is applied potential, R_s is solution resistance and C_{dl} is the double-layer capacitance. Typical capacitances are of the order of several $\mu F\,cm^{-2}$ for metal electrodes in contact with aqueous electrolytes. Charging current decays as e^{-t} whereas current from a Faradaic process decays with a slower $t^{-1/2}$ dependence, as shown in Figure 3. This is an important result for pulse voltammetry because it suggests that proper timing will allow significant Faradaic current to be measured at times when charging contributions are diminished. The ability to discriminate against background processes by such an approach is one feature that gives pulse methods significantly improved S/N over DC waveforms.

2.1.3 Kinetic Complications

It is convenient to think of most electrochemical processes as reversible, since this is the simplest case conceptually and mathematically. Reversibility depends on the rate of the redox reaction vs timescale of the experiment. In the real world there are not many perfectly reversible systems. Many nonreversible systems, however, can be treated as reversible without introducing significant error. It is also possible for a redox reaction to appear reversible by slowing the measurement speed, but this often comes at the cost of reduced signal.

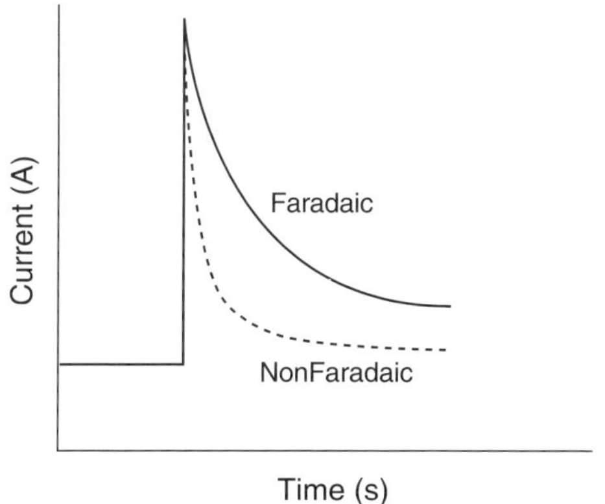

Figure 3 Chronoamperometry of Faradaic (electron transfer) and nonFaradaic (charging) processes.

The treatment of electrode kinetics is too involved for a detailed discussion here, but it is useful to understand the meaning of kinetic limitations in terms of analytical applications. A main concept of all voltammetric methods is that the current is a direct reflection of the rate of electrolysis of O to R[6] (Equation 6):

$$i = nFAv \tag{6}$$

where v is rate of conversion of O to R in $mol\,s^{-1}\,cm^{-2}$. The area term comes from the heterogeneous nature of the interfacial process, conversion per unit electrode area per unit time. Kinetic complications enter into v. If one thinks of the O/R conversion of Equation (1) as a general kinetic process (Equation 7):

$$O + ne^- \underset{k_b}{\overset{k_f}{\rightleftharpoons}} R \tag{7}$$

where k_f and k_b are the forward and backward rate constants, respectively, then the current, i_{net}, which reflects the net rate of conversion of O to R is given by Equation (8):

$$i_{net} = i_c - i_a = nFAk_0 \times \left[C_O(0,t) e^{\frac{-\alpha nF(E-E^{o'})}{RT}} - C_R(0,t) e^{\frac{(1-\alpha)nF(E-E^{o'})}{RT}} \right] \tag{8}$$

where i_c and i_a are currents for the forward and reverse parts of Equation (7) at the given potential, E, and k_0 is the standard heterogeneous rate constant ($cm\,s^{-1}$), the reaction rate constant when $E = E^{o'}$. $C_O(0,t)$ and $C_R(0,t)$ are the concentrations of O and R at the electrode surface at time t, α is the transfer coefficient (which usually has a value of ca. 0.5) and $E - E^{o'}$ is the overpotential, denoted η. Overpotential corresponds to reaction driving force (applied potential) in excess of the standard potential that must be supplied to drive the O/R conversion. The current depends on the net rate of conversion of O to R, which in turn depends on applied potential. The determination of kinetic parameters from voltammetric data has been greatly simplified by computer-assisted fitting of theory to experiment. Many modern electrochemical instruments incorporate such procedures as software options.

Many physical and chemical factors manifest themselves as slow electron transfer. Specific adsorption of ions

For references see page 10032

from solution on to an electrode surface, distance requirements for electron transfer at the electrode–solution interface, chemical reactions coupled to electron transfer and adsorption of macromolecular contaminants on the electrode can all result in sluggish kinetics.

Real samples often contain species which can strongly adsorb on an electrode surface, e.g. proteins in biological matrices or humic substances in fresh water. Adsorption is often strongly dependent on the nature of the electrode material. The effect of these adsorbates is to slow the rate of electron transfer at any applied potential by increasing the distance over which this event must take place. This means that more extreme potentials must be applied to drive the reaction at a rate comparable with that observed in the absence of the blockage. Such a kinetic effect typically broadens the volammetric wave, decreases its magnitude and shifts the wave to more extreme potentials. This has important implications for detection since it can degrade both S/N and resolution.

2.1.4 Surface-confined Redox Systems

While adsorption of macromolecular interferents is usually a hindrance to analytical detection, confinement of an analyte at the electrode surface by adsorption[7,8] or other types of intentional confinement[9] can actually be beneficial. Adsorption offers the opportunity to concentrate reactant at the interface and eliminate diffusion from the analytical signal. Anodic stripping voltammetry (ASV)[10] and cathodic stripping voltammetry (CSV)[11] both take advantage of surface effects and preconcentration of analyte to allow extremely low detection limits and high analytical sensitivity.

2.2 Pulse Voltammetry at Electrodes of Classical Dimensions

We make the distinction between "classically" sized electrodes, i.e. electrodes with dimensions of the order of 1–0.1 mm, and microelectrodes, having at least one microscopic dimension (i.e. <0.1 mm), because the nature of analyte mass transfer and hence current response change as the dimensions of the electrode decrease. Most common electrodes that one will encounter are the "macro" variety, but the use of microelectrodes, discussed in section 2.4, is increasing owing to their advantageous properties.

2.2.1 Normal and Reverse Pulse Voltammetry

The NPV[1,2,12–15] waveform is illustrated in Figure 4. The electrode is subjected to series of potential steps, each originating from the same initial potential (E_i), where O is stable. The potential step duration is t_p. At a point late in each step, the current is measured which

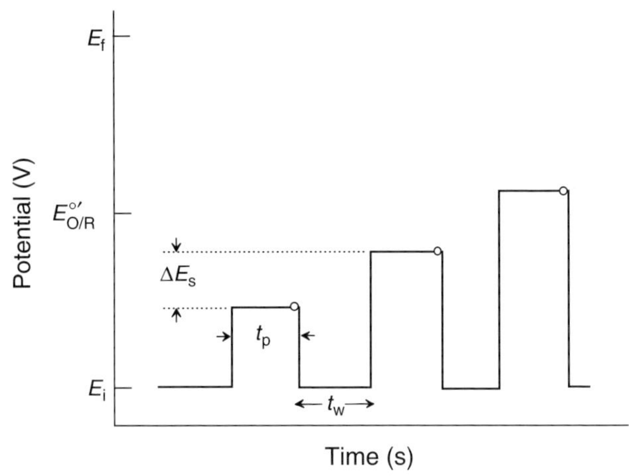

Figure 4 The NPV waveform with experimental parameters. Current measurement at the end of each pulse is denoted by an open circle.

is denoted by an open circle superimposed on the E–t waveform in Figure 4. The potential is then returned to E_i for a waiting period, t_w. Returning the potential to E_i drives any R at the electrode surface back to O, as shown in Figure 2(b). During t_w, any residual R further removed from the interface diffuses away from the electrode. It is often useful during this period to stir the solution or otherwise renew the initial boundary conditions, so that the concentration of O is everywhere equal to the original bulk concentration. If $t_w \gg t_p$, stirring may not be necessary. Following the first step, the potential is stepped incrementally to more extreme values, $E_i + 2\Delta E_s$, $E_i + 3\Delta E_s$ and so forth, until E_f is reached. This sequence of evenly spaced, progressively more extreme potential steps and corresponding current measurements span the region of $E^{o'}_{O/R}$. The early steps produce no electrolysis of O to R, as shown in Figure 5(a), and so generate only background current by double-layer charging. Successive potential steps eventually perturb the concentration profile of O at the electrode surface as the step approaches $E^{o'}$ and currents increase accordingly. At sufficiently extreme potentials, the diffusion layer is established quickly on the timescale of the potential step and the flux of O at the electrode surface is not changed appreciably at potentials well beyond $E^{o'}_{O/R}$. The current reaches a plateau in this regime.

The NPV curve (Figure 5d) has a sigmoid shape centered around the half-wave potential, $E_{1/2}$, which corresponds to $E^{o'}$ when the diffusion coefficients of the oxidized (D_O) and reduced (D_R) forms are equal (Equation 9):

$$E_{1/2} = E^{o'} + \frac{RT}{nF} \ln \left(\frac{D_O}{D_R}\right)^{1/2} \tag{9}$$

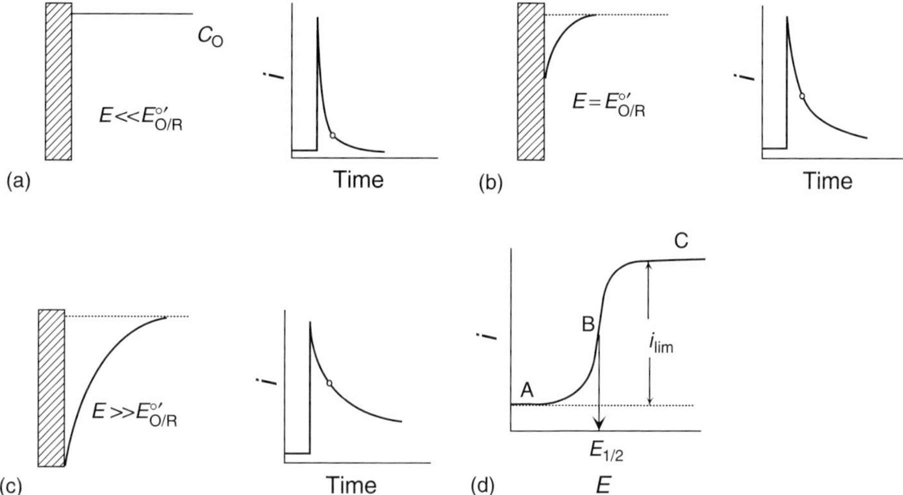

Figure 5 Origin of the NPV in a series of potential steps to various values vs $E^{o'}_{O/R}$. (a), (b) and (c) show potential steps to values positive of, equal to and negative of $E^{o'}_{O/R}$ and the corresponding chronoamperometric trace. Plotting currents sampled at t_p vs the step potential yields the sigmoid NPV curve (d). (Reproduced in part by permission of Marcel Dekker from W.R. Heineman, P.T. Kissinger, 'Large Amplitude Controlled-potential Techniques', in *Laboratory Techniques in Electroanalytical Chemistry*, eds. P.T. Kissinger, W.R. Heineman, Marcel Dekker, New York, Chapter 3, 51–127, 1984.)

For most practical circumstances, the difference between $E_{1/2}$ and $E^{o'}$ is negligible. The overall shape of the NPV trace can be cast in terms of $E_{1/2}$ and the limiting current, i_{lim}, at the plateau of the curve (Equation 10):

$$E = E_{1/2} + \frac{RT}{nF} \ln\left(\frac{i_{lim} - i}{i}\right) \quad (10)$$

A semilogarithmic plot of the rising portion of the NPV curve in the vicinity of $E_{1/2}$ therefore should have a slope of 59.1/n mV at 25 °C for a reversible system. This serves as a useful diagnostic criterion for reversibility, as does an absence of dependence of $E_{1/2}$ on t_p. Reversibility is not necessary for analytical detection methods based on pulse voltammetry, but resolution is maximized under these conditions.

The limiting current of an NPV curve is given by Equation (3), where t is replaced by the pulse time, t_p (Equation 11):

$$i_{lim} = \frac{nFAD^{1/2}C}{\pi^{1/2}t_p^{1/2}} \quad (11)$$

i_{lim} is linearly proportional to concentration, hence its analytical utility. For a fixed pulse time the simple Equation (12):

$$i_{lim} = (\text{slope})C \quad (12)$$

may be used to relate signal to concentration, where the slope is $nFAD^{1/2}/\pi^{1/2}t_p^{1/2}$. Often in analysis the nature of the slope will not be important, but it will vary with electrode area and diffusion coefficient. Since limiting current is proportional to $D^{1/2}$, the dependence of i_{lim} on t_p can also be used to determine the diffusion coefficient of an electroactive species.

Figure 6 shows consecutive one-electron transfers[16] (Equations 13 and 14):

$$\text{TTF} \rightleftharpoons \text{TTF}^+ + e^- \quad (13)$$

$$\text{TTF}^+ \rightleftharpoons \text{TTF}^{2+} + e^- \quad (14)$$

where TTF is tetrathiafulvalene (CAS 31366-25-3). This figure illustrates the NPV behavior of consecutive, simple, reversible one-electron transfer reactions. Circles denote currents measured at the end of t_p and crosses denote currents prior to the pulse. These latter values can be used to assess boundary condition renewal. Deviations from zero indicate inconsistent renewal of boundary conditions. The value of i_{lim} for each reaction varied linearly with $t_p^{-1/2}$. $E_{1/2}$ was essentially independent of t_p for each wave, suggesting reversibility of the electron transfers.

One drawback of NPV at solid electrodes, as in Figure 6, is the time needed for proper experimental control. Following each potential pulse, the working electrode was returned to 0 V and the solution stirred to renew the boundary conditions. Allowing 12 s for this process at each pulse took 6.6 min to acquire 33 data points. The use of renewable electrodes, such as the SMDE, can significantly speed data acquisition. Although NPV is useful for certain measurements, it is extremely slow when performed at solid, stationary electrodes. Of

For references see page 10032

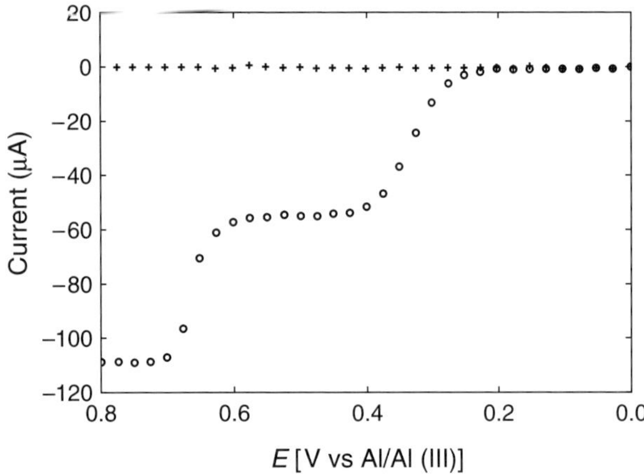

Figure 6 Two-step NPV oxidation of 4.8 mM tetrathiafulvalene at a GC disk electrode in an ambient temperature ionic liquid (25 °C). Circles are currents at the end of each pulse and crosses are currents at E_i prior to the pulse. Waiting time, t_w, at E_i was 12 s and pulse time, t_p, was 50 ms. (Reproduced by permission of The Electrochemical Society from M.T. Carter, R.A. Osteryoung, 'Heterogeneous and Homogeneous Electron Transfer Reactions of Tetrathiafulvalene in Ambient Temperature Chloroaluminate Molten Salts', *J. Electrochem. Soc.*, **141**, 1713–1720 (1994).)

course, if one decreases t_p then t_w can be decreased commensurately.

RPV[12,14,17,18] is complementary to NPV. Typical RPV waveforms are shown in Figure 7(a) and (b). The two common variants are RPV and reverse pulse voltammetry with waiting (RPW). RPV is essentially an inverted NPV experiment, whereas RPW entails additional potential steps in the nonFaradaic region positive of $E^{o'}_{O/R}$. Like NPV, the potential waveform is chosen to span the region of $E^{o'}_{O/R}$. However, unlike NPV, the experiment is initiated at a potential where R is generated, typically at a diffusion-controlled rate. After holding the potential for t_w at a value E_f where O is stable, the potential is stepped to E_i, where R is generated for time t_g. This generates a diffusion layer of R at the electrode surface and current flows according to Equation (3). This current is usually not of analytical interest. After generating R, the potential is pulsed for time t_p to successively more positive values with current measurement at the end of each pulse. The potential is finally returned to E_f and stirring initiated to renew the initial boundary conditions between pulses.

The RPV curve is shown in Figure 8. A corresponding NPV trace is shown for comparison. If both forms of the redox couple are stable, the RPV curve will have a limiting current, $i_{lim,RPV}$, equal to that of the corresponding NPV experiment. The full $i-E$ curve is displaced on the current axis from the NPV experiment, however, because of the current that flows at E_i owing to generation of R. This current, denoted i_{DC}, is given by Equation (11) for a reversible redox couple where the pulse time is t_g. Under ideal conditions, i_{DC} should be constant at all points in the experiment. The sum of i_{DC} and the reverse pulse current i_{RP} for electrolysis of R to O in the diffusion layer is equal to the corresponding NPV i_{lim} if all species are stable (Equation 15):

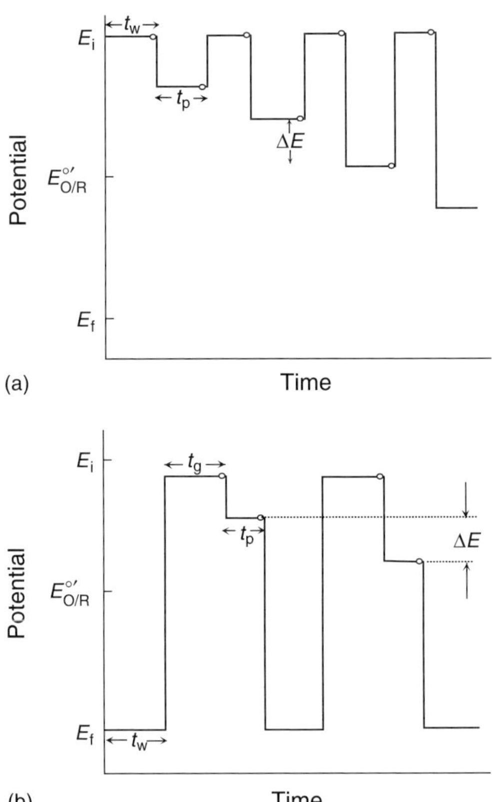

Figure 7 (a) Reverse pulse and (b) reverse pulse with waiting voltammetric waveforms and relevant parameters. Points of current measurement are denoted by open circles.

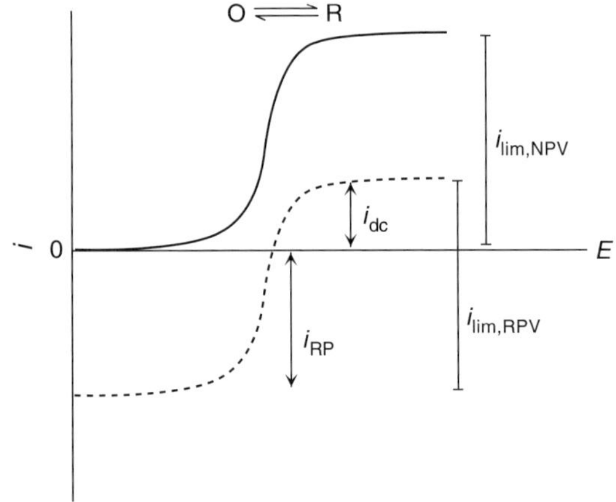

Figure 8 NPV and RPV curves for a reversible system.

$$i_{RP} + i_{DC} = i_{lim,RPV} = i_{lim,NPV} \tag{15}$$

The value of $E_{1/2}$ from RPV should equal the NPV value in the reversible case and the shape of the curve is described completely analogously to NPV.

RPV can be used to electrogenerate a species under controlled conditions and examine its stability. For example, if the value of t_p is tens of milliseconds, but the product R decays on a timescale of seconds, then NPV will behave reversibly, since R is essentially stable on the timescale of that experiment. However, in RPV, if R is generated for several seconds, i.e. long enough to allow measurable decay, a fraction of R will not be available for reoxidation during t_p, resulting in a decrease in i_{RP} compared with the case where R is completely stable. Instability of R on the timescale of the RPV measurement could be probed by systematic variation of t_g from short times, where R is completely stable to long times, where significant loss of R occurs and $i_{RP} + i_{DC} < i_{NPV}$. RPV applications have included detection of halogenated organics,(19) determination of OH^-,(20) evaluation of stability of organometallics(21) and characterization of chemical reactions coupled to electron transfer.(22)

The use of RPV to evaluate the stability of an organometallic complex is shown in Figure 9(a) and (b).(21) Here, decomposition of $[Ru^{III}(EDTA)]_2$ dimer, the product of electroreduction of a related starting material, was followed as a function of t_p (Equation 16):

$$[Ru^{III}(EDTA)]_2 \underset{k_{-1}}{\overset{k_1}{\rightleftharpoons}} 2Ru^{III}(EDTA)(OH)_2 \quad (16)$$

Removal of electroactive $[Ru^{III}(EDTA)]_2$ product by Equation (16) on a timescale comparable to t_p was reflected in decrease in the ratio i_{RPV}/i_{NPV} as t_p increased. A rate constant of $0.75\,s^{-1}$ at pH 4.8 was calculated based on this trend.

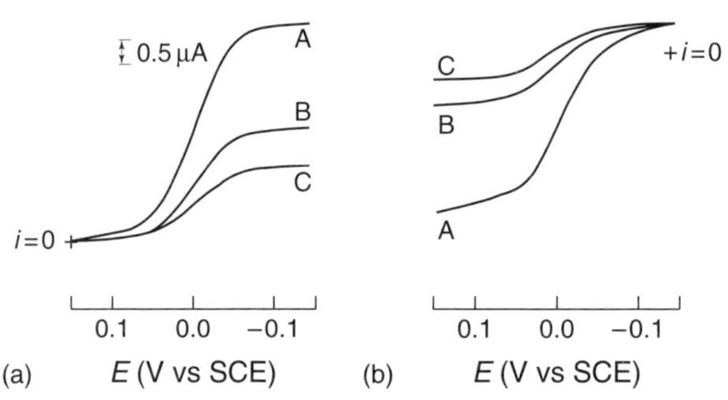

Figure 9 (a) NPV and (b) RPV of 0.25 mM $[Ru^{III\frac{1}{2}}(EDTA)_2]_2$ dimer for $t_p =$ (a) 10.4, (b) 39.7 and (c) 90 ms. (Reprinted from C.-L. Ni, F.C. Anson, J.G. Osteryoung, 'A Pulse Polarographic Measurement of the Kinetics of Dissociation of the $[Ru^{III}(EDTA)]_2$ Dimer', *J. Electroanal. Chem.*, **202**, 101–109 (1986), Copyright 1986, with permission from Elsevier Science.)

2.2.2 Differential Pulse Voltammetry

DPV is one of the most popular pulse voltammetric methods.(1,23–28) The DPV waveform is shown in Figure 10. Each pulse of time t_p and magnitude ΔE_p is preceded by a uniform waiting time, t_w, as in NPV, but instead of making one current measurement at the end of t_p, two measurements are made. The first (i_1) is made at τ' just prior to initiation of the pulse and the second (i_2) at τ toward the end of the pulse. After the pulse is completed, the potential does not return to E_i, as in NPV, but is incremented by a step ΔE_s, typically 1–10 mV. Pulse widths and waiting times are similar to those in NPV. In differential pulse polarography (DPP) a new drop is generated at the end of t_p, as discussed in section 2.3.

The output of the experiment is the difference $i_2 - i_1$ vs the base potential E_s, hence the name of the method. DPV yields a peaked output as shown in Figure 11(a) and (b). At the beginning of the experiment, no Faradaic current flows, so $i_2 - i_1$ is very close to zero. At the other extreme, i.e. well beyond the peak, the O/R reaction is driven at its diffusion-limited rate. Further pulses do not appreciably raise the rate beyond that driven by the base potential, so again $i_2 - i_1$ is near zero. It is only in the region near $E^{o'}$ that small changes in potential between the base value and the pulse value result in large changes in current and nonzero $i_2 - i_1$.

The peaked DPV output has distinct advantages for quantitation and speciation in analytical applications. The differential nature of DPV tends to flatten sloping baselines, in addition to increasing S/N. The height of the peak varies with pulse amplitude, ΔE_p(23) (Equation 17):

$$i_p = \frac{nFAD_O^{1/2}C_O}{\pi^{1/2}(\tau - \tau')^{1/2}}\frac{1-\sigma}{1+\sigma} \quad (17)$$

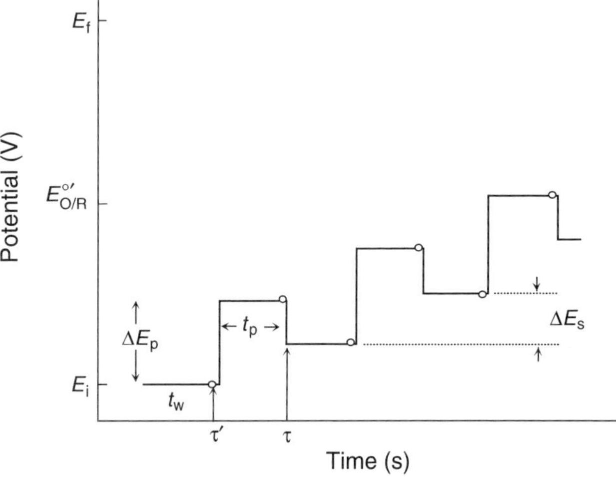

Figure 10 The DPV waveform and associated experimental parameters.

For references see page 10032

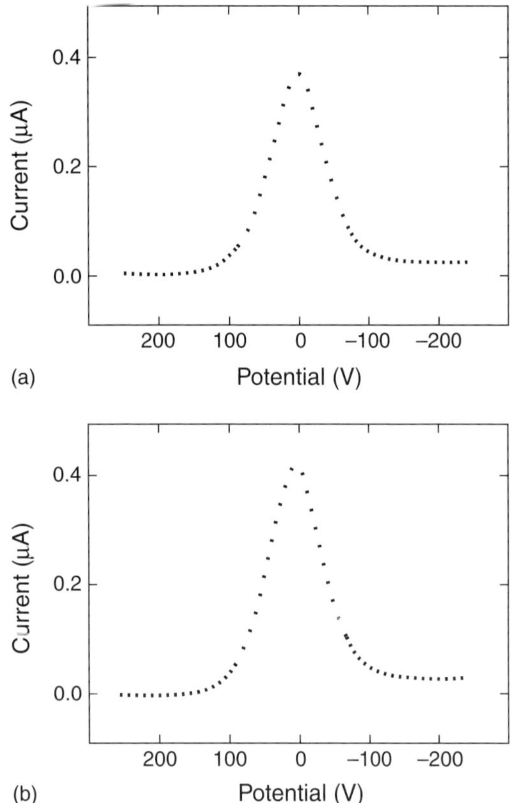

Figure 11 Simulated differential pulse voltammograms at (a) a planar electrode and (b) a DME with sphericity. Drop time, 0.5 s; pulse time, 40 ms; ΔE_s, 10 mV; ΔE_p, 10 mV. (Reproduced by permission of The American Chemical Society from J.W. Dillard, J.A. Turner, R.A. Osteryoung, 'Digital Simulation of Differential Pulse Polarography with Incremental Time Change', *Anal. Chem.*, **49**, 1246–1250 (1977).)

where $(1-\sigma)/(1+\sigma)$ is a function of ΔE_p and n.[23] At large pulse amplitudes $(1-\sigma)/(1+\sigma)=1$ and Equation (17) reduces to Equation (11). The DPV curve has a maximum current equal to that in the corresponding NPV experiment under these conditions. Usually, the DPV peak current is smaller than the corresponding NPV limiting current, at typical values of ΔE_p. The enhanced sensitivity and detection limit of DPV over NPV, as much as an order of magnitude, is not due to enhancement of the Faradaic signal but to a reduction in charging current.[1] Maximization of sensitivity is essentially a trade-off between larger currents obtained at large pulse amplitudes and the increased background component that accompanies it. Larger pulse widths also broaden the wave, which decreases the ability to resolve multiple analytes. DPV has been used most extensively in conjunction with the DME and SMDE (section 2.3).

The peak potential, E_p, corresponds approximately to the NPV $E_{1/2}$. Resolution of multicomponent mixtures is defined by peak width, $W_{1/2}$, which reaches a limiting value of $90.4/n$ mV at 25 °C, as ΔE_p approaches zero.

2.2.3 Square Wave Voltammetry

SWV[29–39] employs the waveform shown in Figure 12. Following an initial period of quiet time, t_w, a symmetrical train of pulses of amplitude $2\Delta E_{SW}$ is applied to the electrode, superimposed on a small DC potential increment, ΔE_s, changed at a frequency (f) of $1/\tau$, where τ is the period of the square wave. Current is sampled at the end of each forward (i_f) and reverse (i_r) half-cycle. The difference $i_f - i_r$, called the net current, is plotted against the DC potential to yield the net voltammogram, where $i_{net} = i_f - i_r$.

SWV can be performed very quickly compared with NPV and RPV since boundary conditions do not need to be renewed before each pulse. For example, a typical square wave frequency of 50 Hz and step height of 10 mV complete a 1-V scan in 2 s, whereas the corresponding NPV experiment would take 1000 s if a 10-s waiting period was used between pulses. Measurement speed makes SWV a very promising electrochemical technique for applications including chromatography detectors and on-line monitors where real-time data acquisition may be necessary.[40,41]

The differential nature of SWV imparts enhanced S/N and sensitivity. SWV is more sensitive than DPV by as much as 24% under optimum conditions since it uses signals from both the forward and reverse reactions in the output, whereas DPV only uses the forward component. Noise is reduced because charging components in the forward and reverse half-cycles are nearly identical, and are zeroed in the calculation of i_{net}.

Figure 13 shows a calculated reversible SWV curve with forward, reverse and net currents. Here the current is given as ψ vs $n(E - E_{1/2})$, which are simply dimensionless variables which correspond to current and potential,

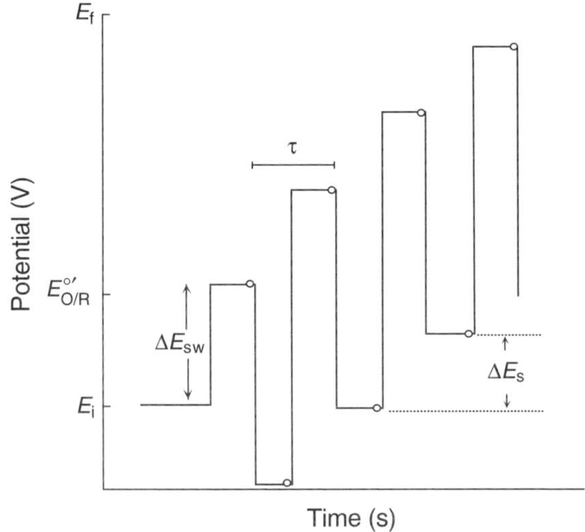

Figure 12 The SWV waveform and associated experimental parameters.

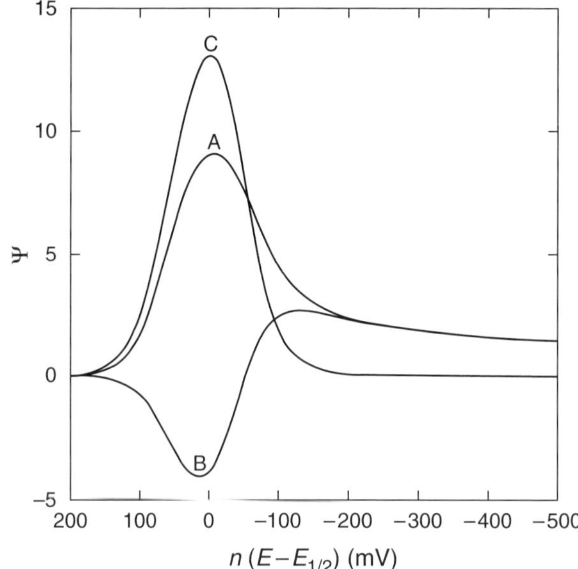

Figure 13 Calculated square wave voltammograms for (A) forward, (B) reverse and (C) net currents in dimensionless units. (Reproduced by permission of The American Chemical Society from J.J. O'Dea, J. Osteryoung, R.A. Osteryoung, 'Theory of Square-wave Voltammetry for Kinetic Systems', *Anal. Chem.*, **53**, 695–701 (1981).)

respectively. The peak current of a reversible SWV curve is given by Equation (18):[(29)]

$$i_p = \frac{nFAD_O^{1/2}C_O}{\pi^{1/2}(t_p)^{1/2}}\psi_p \quad (18)$$

where ψ_p is the dimensionless current function for SWV, which depends on nE_{SW} and $n\Delta E_s$. Values of ψ_p have been tabulated.[(29)] The peak height of the net voltammogram depends linearly on $f^{1/2}$ for reversible systems. The peak height of a DPV curve is approximately the same as that of the forward SWV component.

The width at half-height for a reversible system is $90.5/n$ mV, at 25 °C, for small values of E_{SW}. The height and width of the wave both depend on E_{SW}. For analytical applications, the peak width-to-height ratio is maximized using $nE_{SW} = 50$ mV.[(27)] Table 1 summarizes typical ranges of experimental parameters for commonly used pulse voltammetric methods.

2.3 Pulse Polarography and Voltammetry at Mercury Electrodes

Mercury electrodes have played an important role in the development of electroanalytical methods and are still used today.[(42,43)] The ease of oxidation of Hg in aqueous media limits its use at positive potentials. Nonetheless, Hg is an excellent material for the electrochemical detection of species with fairly negative redox potentials and for the detection of metal ions in particular. Hg also has an advantage over solid electrodes that the electrode surface can be renewed during an experiment by replacing an old drop with a new one. This is particularly useful in situations where the electrode surface may be degraded or fouled during the measurement. Recent advances in renewable mercury electrodes have been reviewed.[(44)]

2.3.1 Dropping Mercury Electrode

The DME was the original working electrode of polarography. A DME is a gravity-fed device in which Hg flows from a reservoir through a narrow-bore glass capillary having an inner diameter of the order of 0.1 mm. Hg drops form at the open end of the capillary, grow and finally fall from the capillary tip. Hg drops extruded from a capillary tip are spherical and their size generally falls

Table 1 Pulse voltammetry methods

Method	Acronym	Parameter	Designation	Typical ranges[a]
Normal pulse voltammetry	NPV	Pulse time	t_p	1–100 ms
		Waiting time	t_w	1–10 s
		Step height	ΔE_s	1–25 mV
Reverse pulse voltammetry	RPV[b]	Waiting time	t_w	As for NPV
		Pulse time	t_p	As for NPV
		Step height	ΔE_s	As for NPV
		Generation time	t_g	1–10 s
Differential pulse voltammetry	DPV	Pulse height	ΔE_p	1–10 mV
		Pulse width	t_p	As in NPV
		Step increment	ΔE_s	1–10 mV
Square wave voltammetry	SWV	Step height	ΔE_s	1–10 mV
		Square wave amplitude	ΔE_{sw}	1–100 mV
		Frequency	f	1–1000 Hz

[a] Ranges indicate order of magnitude for various parameters. Drop times for Hg electrodes have not been included in the table but are typically of the order of 1–5 s.
[b] RPV consists of two different methods, reverse pulse (RP) and reverse pulse with waiting (RPW).

For references see page 10032

in the macroelectrode range. The rate of drop formation and growth is dependent upon the height of the Hg reservoir above the tip and the diameter of the capillary. The ultimate size of the Hg drop formed depends on the surface tension between the drop and the surrounding aqueous medium into which it grows. A "drop knocker" is often used to dislodge mechanically the drop at a fixed point during growth, i.e. at constant surface area. Prior to the advent of pulse methods, continuous measurement of current was performed during the entire cycle of drop birth, growth and removal. Since the surface area of the electrode increased during the drop's lifetime, periodic current oscillations and low S/N were facts of life for polarographic methods. Developments in pulse voltammetry addressed this problem by timing potential application and current measurement to the cycle of the DME.(1,23) The basic strategy for pulse timing, e.g. in DPP, was to apply pulses and measure currents at the very end of a drop's life, so that surface area changes between measurements were minimized. This was possible using short pulse widths, 10–100 ms, which allowed the experiment to be performed under nearly constant electrode area conditions. Significant reduction in charging current was also realized.

An example of SWV at a DME(45) is given in Figure 14(a) and (b). Here, the redox reaction is (Equation 19):

$$Cd^{2+} + 2e^- \rightleftharpoons Cd(Hg) \quad (19)$$

in which aqueous Cd^{2+} (CAS 10022-68-1) is reduced by $2e^-$ to Cd metal, which dissolves into the Hg electrode. The entire SWV scan was performed on a single Hg drop after it had grown for 2 s. Drop surface area growth was compensated by normalizing currents to $t_d^{2/3}$, where t_d is the drop time, since the drop surface area changes with $t^{2/3}$. Today, the DME is mostly of historical interest.

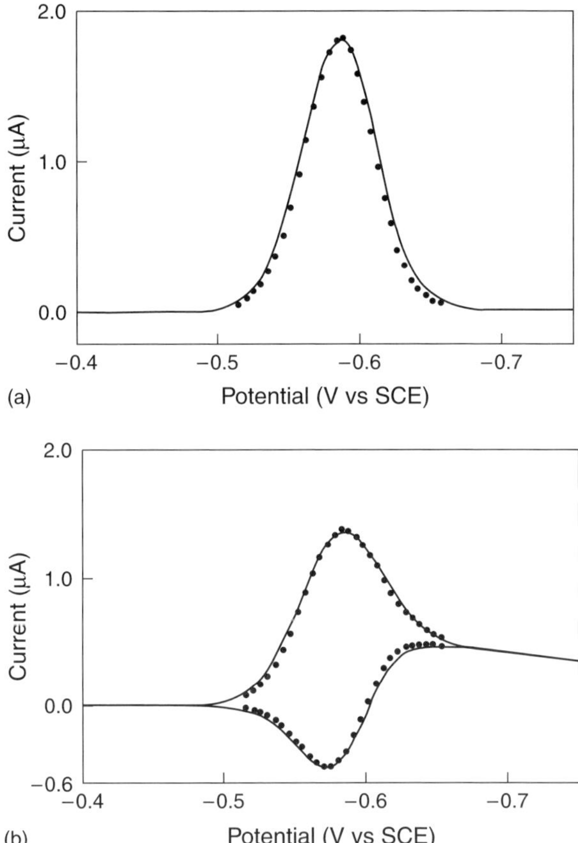

Figure 14 Experimental SWV of $Cd^{2+}/Cd(Hg)$ in 0.1 M HCl for 8.43×10^{-5} M Cd^{2+} at a single drop (t_d, 2 s) of a DME. (a) Net current. (b) Forward and reverse currents. $\Delta E_s = 5$ mV; $E_{SW} = 30$ mV; $f = 30$ Hz. (Reproduced by permission of The American Chemical Society from J.A. Turner, J.H. Christie, M. Vukovic, R.A. Osteryoung, 'Square-wave Voltammetry at the Dropping Mercury Electrode: Experimental', *Anal. Chem.*, **49**, 1904–1908 (1977).)

2.3.2 Static Mercury Drop Electrode

The SMDE(46) was developed by EG&G Princeton Applied Research (PAR) to address the problem of time-dependent electrode surface area and is still in widespread use today. This device can be used in either a repetitive mode analogous to the DME or as an HMDE where a single drop is used for an entire experiment. The HMDE is a spherical, stationary mercury electrode. The SMDE shares some characteristics of the DME in that it retains the Hg reservoir and uses a capillary tip to form the Hg drop.

When the SMDE is used in repetitive mode, Hg drops are extruded from the end of the large bore capillary by activation of a solenoid and plunger which prevents Hg flow when closed. When opened Hg flows, rapidly forming a drop. Deactivation of the solenoid closes the plunger and Hg flow ceases. The drop is grown to its final size and surface area very quickly compared with the DME. After a predetermined time, the drop is dislodged and the cycle repeated. Advantages realized by a renewable electrode are countered somewhat by accumulation of large amounts of waste mercury generated by the SMDE.

NPV and RPV of $Fe^{2+/3+}$ (CAS 10025-77-1) and $Pb^{2+}/Pb(Hg)$ (CAS 10099-74-8)/(CAS 7439-92-1) at an SMDE are shown in Figure 15.(18) For $Fe^{2+/3+}$, RPV was performed starting at −0.5 V vs a saturated calomel electrode (SCE) to generate Fe^{2+}. Pb was generated at −0.8 V vs SCE. The SMDE drop grew for 1.5 s at the initial potential, 0 V ($Fe^{2+/3+}$) or −0.3 V (Pb^{2+}), prior to the RPV run. Figure 15 shows that spherical diffusion complications were apparent for RPV of $Pb^{2+}/Pb(Hg)$, i.e. $i_{RP} + i_{DC} > i_{NP}$, when Pb dissolved in the Hg drop, but were absent for $Fe^{2+/3+}$ where neither oxidation state was soluble in Hg.

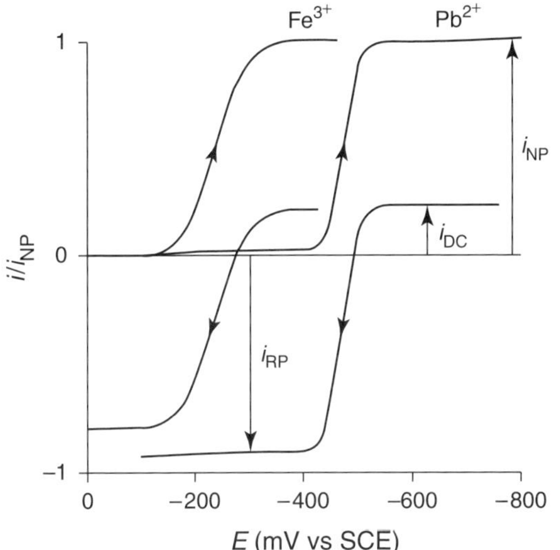

Figure 15 Normalized NPV and RPV of 0.639 mM $Fe^{3+/2+}$ in 0.1 M sodium oxalate–0.1 M $NaClO_4$ and 0.488 mM $Pb^{2+/0}$ in 0.5 M acetate buffer (pH 4.8) at an SMDE. (Reproduced by permission of The American Chemical Society from T.R. Brumleve, J. Osteryoung, 'Spherical Diffusion and Shielding Effects in Reverse Pulse Voltammetry', *J. Phys. Chem.*, **86**, 1794–1801 (1982).)

2.3.3 *Thin-film Mercury Electrode*

TFMEs are fabricated by plating Hg on solid substrate electrodes such as carbon, platinum or gold. TFMEs allow the use of Hg in a convenient, planar format when the use of Hg drop electrodes is not practical. Such applications could include chromatographic flow cell detectors and monitors for flowing streams. The TFME is not renewable, but imparts the cathodic advantages of Hg to a detection scheme.

Pulse voltammetry has been carried out at TFMEs in DPV[47,48] and SWV modes.[49–52] TFMEs have been shown to improve the analytical detection limit, sensitivity and resolution compared with Hg drops, particularly in ASV discussed in section 4.2.1, by virtue of exhaustive removal of electrodeposited material, which often is not achieved with a mercury drop.

2.4 Pulse Voltammetry at Ultramicroelectrodes

Ultramicroelectrodes (UMEs) have proven useful in electroanalytical pulse voltammetry. Once at the cutting edge of electrochemical research, UMEs are now fairly standard devices. A variety of geometries have been characterized, including disks, rings, bands, cylinders and arrays of these.[53] UMEs have been prepared from solid electrode materials such as Pt, Au and C. Hemispherical Hg drop UMEs have also been fabricated. We shall discuss the characteristics of UMEs that make them useful in electroanalytical work and illustrate these points with examples of pulse voltammetric applications.

UMEs are small, with at least one microscopic dimension. An ultramicrodisk electrode, for example, may have a diameter of 20 µm or less. An ultramicroband electrode, on the other hand, could have a width of 20 µm and a length of 1–2 mm. Small physical size makes the UME obviously useful in sampling small volumes and restricted spaces. Since UMEs have a small surface area, the current flowing at them is also small. This characteristic tends to reduce the effects of uncompensated resistance (iR_u) in the electrochemical cell[5,54] and allows undistorted voltammetry in media or physical vicinities of high resistance, e.g. a nonaqueous solvent or a very thin flow channel. UMEs feature reduced double-layer capacitance compared with larger electrodes and have smaller *RC* time constants which allow faster electrochemical measurements (microseconds or less) to be performed compared with their macroscale brethren.

The nature of diffusion of electroactive species to a very small electrode is different from that of its conventional counterparts. Generally UMEs show enhanced mass transfer in the form of significant radial diffusion contributions in addition to planar diffusion. This feature actually simplifies and speeds NPV and RPV experiments by eliminating the necessity for convection to achieve boundary condition renewal, and enables true steady-state limiting currents to be obtained.

SWV,[55,56] NPV and RPV[57–68] have been studied extensively at microelectrodes in both their theoretical and experimental aspects. Hg-based microelectrodes have also been characterized.[69–71] Reduced iR_u at microelectrodes has allowed pulse voltammetry to be applied to resistive media[67,68] and media with high concentrations of electroactive species.[70] The small *RC* time constant of microelectrodes has enabled fast pulse voltammetry on the microsecond timescale.[62,64] Radial diffusion contributions allow rapid renewal of initial boundary conditions in the NPV/RPV experiment.[57,58,63]

Radial diffusion contributions to mass transfer at a microdisk electrode have important implications for the timescale on which NPV/RPV experiments can be performed. A stationary 5 µm radius microdisk electrode, for example, has been shown to have a mass transfer coefficient comparable to that which would be obtained by rotating a conventional size electrode at 10 000 rpm.[57] Simply stated, this means that it is only necessary to wait sufficiently long during t_w for re-establishment of initial conditions. If $t_p/t_w \leq 0.1$, initial boundary conditions can be restored without stirring for electrode diameters up to 25 µm.[57]

Figure 16(a–f) shows RPV curves taken in the RPW mode for oxidation of 5 mM $Fe(CN)_6{}^{4-}$ (CAS 14459-95-1) in 0.4 M $Sr(NO_3)_2$.[57] The curves show variation of t_g between 0.01 and 5.00 s for a constant t_w of 5 s. The magnitudes of i_{dc} and i_{RP} were within 15% of theory

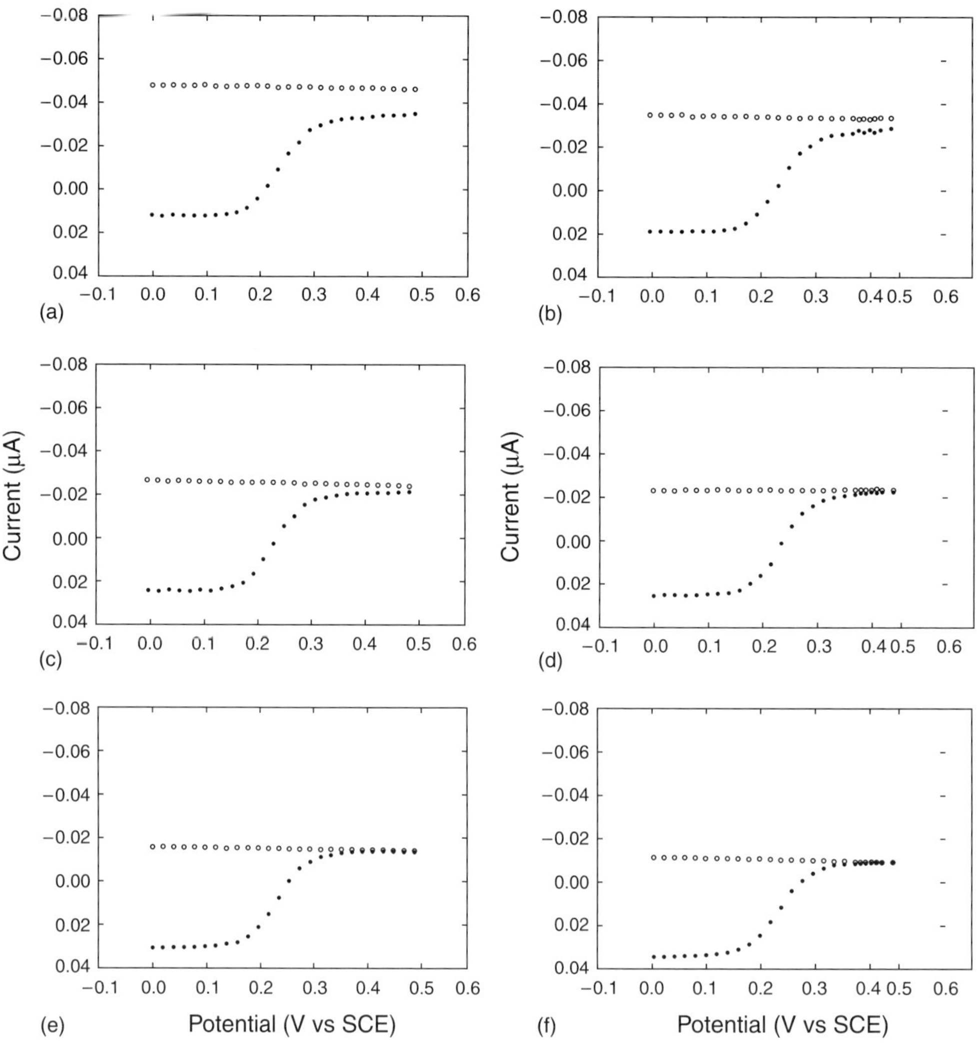

Figure 16 Reverse pulse voltammograms with waiting for 5 mM $Fe(CN)_6^{4-}$ in 0.4 M $Sr(NO_3)_2$ at a 12.98-μm radius Pt electrode with $t_w = 5$ s and $t_p = 0.1$ s; $t_g =$ (a) 0.01, (b) 0.02, (c) 0.03, (d) 0.10, (e) 1.00 and (f) 5 s. (Reproduced by permission of The American Chemical Society from L. Sinru, J. Osteryoung, J.J. O'Dea, R.A. Osteryoung, 'Normal and Reverse Pulse Voltammetry at Microdisk Electrodes', *Anal. Chem.*, **60**, 1135–1141 (1988).)

even though no stirring was used to renew the boundary conditions.

3 INSTRUMENTATION AND OPERATION

This section covers some basic considerations in carrying out pulse voltammetry experiments in the laboratory. The advent of widely available commercial benchtop instruments controlled by PCs has greatly simplified the implementation of pulse voltammetry in the laboratory. It has also virtually eliminated the need to construct one's own potentiostatic instrumentation. User friendliness has reached the stage where practically anyone can setup and use electrochemistry as a research tool. The reader will find much helpful information about the particulars of doing specific electrochemical experiments in the volumes by Heineman and Kissinger(2) and Sawyer and Roberts,(3) both of which have been updated in new editions within the past few years. With the appearance of microprocessor-controlled potentiostats in the early 1980s, the potential of analytical voltammetry as a working tool has increased significantly. Modern Windows®-based graphic user interfaces have made many sophisticated electrochemical techniques available to a wider spectrum of users from diverse disciplines.

3.1 Commercial Instrumentation

We mainly discuss commercially available instrumentation from US manufacturers and among these only several representative examples. A recent review of electrochemical literature by Anderson et al.[72] lists many useful Web sites for electrochemical instrument vendors. The reader is directed to these sources for additional information.

PAR (*http://www.egginc.com*) has long been a leading manufacturer of electrochemical instruments. PAR offers a line of no less than nine different potentiostat models, the most prevalent of which is probably Model 273A, which is scheduled to be discontinued. Its replacement is Model 283. Most PAR potentiostats are computer controlled from a Windows® environment. PAR offers two highly sophisticated software packages, M250 and M270, which allow the user to employ a wide array of electrochemical methods and exercise advanced data acquisition and analysis including curve fitting for mechanism and kinetics. All of the popular pulse voltammetric methods discussed in this article are represented. Two units targeted for analytical applications are the Model 394 Electrochemical Trace Analysis System and the Model 264B Polarographic Analyzer, both of which handle NPV and NPP. Model 394 also performs SWV, DPP and DPV.

Bioanalytical Systems (BAS) (*http://bioanalytical.com*) markets two multi-technique, research-grade potentiostats, the 100 B/W and the CV-50 W. BAS was the first company to attempt to make electrochemical methods more transparent and user friendly with introduction of the BAS 100 in the early 1980s. These units enable a wide range of methods to be performed, including NPV, NPP, DPV, DPP and SWV. Many automatic functions are included in these instruments, such as automatic curve analysis for peak potential and current calculation and quantitation from calibration curves. BAS also sells software for fitting of electrochemical data to mechanisms.

Cypress Systems (*http://www.cypresshome.com*) sells a variety of electrochemical instruments ranging from models suitable for an undergraduate laboratory to full research-grade instruments. The CS-1190 and CS-1090, for example, support SWV, DPV and DPP, but not NPV and NPP.

CH Instruments (*http://www.chinstruments.com*) is a relatively new company that offers a full line of electrochemical instruments in three different series, Models 600, 700 and 800. Within each series a huge array of choices may be made in terms of available techniques and other system capabilities, from units that are fairly limited for specific applications, to fully capable research-grade workstations. NPV, NPP, DPV, DPP and SWV are all supported. Many models incorporate extensive data analysis and curve fitting capabilities.

ECO CHEMIE (*http://www.ecochemie.nl*) is a very innovative and successful vendor of electrochemical instrumentation in Europe. Their Autolab instrument is particularly useful for implementation of modern pulse voltammetric methods.

3.2 General Considerations

Basic electrochemical experiments are actually fairly easy to setup once the electrochemical instrument is available. A real experiment could involve no more than a beaker or other suitable container, a few electrodes including a reference electrode for the three-electrode potentiostat and some solutions or samples. Pulse voltammetry will also require efficient stirring which can be actuated, e.g. by a transistor-transistor logic (TTL) pulse. SMDE units are also run via the potentiostat. The books by Heineman and Kissinger[2] and Sawyer and Roberts[3] cited above contain ample detail on specifics of electrochemical experiments for the interested reader.

Renewal of the initial boundary conditions is probably the most important factor in the successful implementation of analytical pulse voltammetry. If one uses an SMDE, this requirement is automatically satisfied and, as we saw above, microelectrodes also satisfy the condition easily, provided that appropriate time parameters are chosen. If solid electrodes of conventional size are employed, one must ensure that adequate stirring and time to achieve quiescence are incorporated into the pulse voltammetry protocol.

Selection of the appropriate pulse voltammetric method and experimental parameters depends on the task at hand. For a laboratory researcher wanting to evaluate $E^{\circ\prime}$ and diffusion coefficients for newly synthesized organometallic complexes, NPV will probably be the method of choice. The sigmoidal shape of the NPV waveform is naturally adapted to facile comparison of wave heights and half-wave potentials. For an environmental researcher wanting to determine what electroactive trace metals are present in river water samples, then a differential method such as DPV or SWV may be more appropriate because the peaked response of these techniques is more amenable to identification and quantification of components in multianalyte mixtures. Differential methods also have a certain tolerance for irreversible background electrochemical processes such as O_2 reduction. O_2 is present at ca. 25 ppm in air-equilibrated aqueous samples and is reducible at potentials where many common analytes are electroactive. In the laboratory, one may simply deoxygenate a solution with N_2 or Ar, but in field applications this practice should be avoided since it adds extra steps and cost to the analysis process. Differential methods are therefore the choice if one wishes to develop a fieldable unit.

Selection of the appropriate electrode for a particular application is often extremely important since electrode materials have widely disparate properties. Which electrode material will work acceptably will depend on the

For references see page 10032

particular problem to be solved. Among the solid metals, Pt and Au have been popular. Pt is harder than Au and is therefore easier to polish to a mirror finish. Both have relatively low overpotentials for H^+ reduction to H_2, which means that their use in acidic media can be problematic. Some relief from proton reduction background can be gained by using the differential methods for the same reasons as for O_2 reduction. Pt and Au both form surface oxide films at positive potentials which may interfere with some analytical applications.

Carbon electrodes, including GC and highly oriented pyrolytic graphite (HOPG), display high overpotential for H^+ reduction and are therefore useful in acidic media. Carbon is more difficult to fabricate as a microelectrode owing to its fragility compared with common metals, but the extra effort is often worthwhile. Carbon electrodes are used extensively in aqueous and nonaqueous electrochemistry. One problem that can be encountered with carbon is the chemical nature of the electrode surface. Carbon surfaces generally are covered with functional groups such as quinones, carboxylic acids and so forth, the presence of which can affect electron transfer reversibility for certain analytes. Electrochemical pretreatment of the electrode is often necessary to achieve repeatable results. These pretreatments change the chemical nature and distribution of functional groups on the carbon surface. The effect of pretreatment protocols is usually determined empirically for a particular problem.

Hg electrodes have a very large overpotential for H^+ reduction, which means that they are ideal for cathodic reduction processes in acidic media. However, Hg is easily oxidized, especially in the presence of complexing anions such as Cl^-. Therefore, one cannot very often study oxidative chemistry at an Hg electrode. In certain nonaqueous media, the anodic range of an Hg electrode can be extended to more positive values. A major drawback of Hg, aside from the negative environmental connotations involved in using this metal, is the large amount of waste Hg generated by devices such as the SMDE. One must store and either clean up or dispose of this Hg, which can be a nuisance if a large amount of activity is anticipated. Electrodes in conventional formats, such as the SMDE, are not useful for field work, but the TFME can be used in the field.

Selection of pulse voltammetric parameters, within reasonable limits, is not as critical to the success of a method as selection of appropriate technique and electrode provided that common sense is exercised. For example, if NPV of a dissolved species is being performed to determine its diffusion coefficient and the species is stable, one would not want to select $t_w = 500\,s$ to renew the initial conditions between pulses when $t_w = 5\,s$ will do. In SWV, $\Delta E_{sw} = 1\,V$ probably will not be useful since one will quickly exceed the available potential window of the aqueous medium. Values in range 25–100 mV are more useful. Common ranges of timing parameters for various pulse voltammetric methods are given in Table 1. Within these ranges, empirical adjustment of pulse amplitudes and timing can be performed to optimize the method for a particular problem, although the effects of the parameters on the shape of the resulting voltammogram are known quantitatively and exactly for the various pulse voltammetric methods, as noted in previous sections.

In general, one cannot expect more than a ca. $\pm 1\,V$ window for performing aqueous electrochemistry. The available potential window depends ultimately on the potential needed to oxidize or reduce the solvent, but other limiting processes, such as redox chemistry of dissolved electrolytes, usually occur prior to solvent discharge. Redox reactions occurring at potentials more extreme than the solvent limits are not generally accessible by electrochemical methods. Electrochemical processes which occur on the edge of the solvent limit, e.g. reduction of Zn(II) in aqueous media, do not typically provide reliable quantitative information unless careful background subtraction can be performed. Nonaqueous media have significantly extended potential ranges compared with water: as much as $\pm 2\,V$ or more.

4 APPLICATIONS AND EXAMPLES

In this section we present a variety of illustrative examples of the use of various pulse voltammetric methods for examination of problems in chemistry, analysis and monitoring. These problems range from the most basic to the most applied and hopefully will give the reader a flavor of the possibilities for using pulse voltammetry in their own work. In the course of reviewing the literature one of us (M.T.C.) found 1703 papers containing some use of pulse voltammetric methods published in the chemical literature between January 1991 and June 1998 using only a single bibliographic search source (Chemistry Citation Index, Institute for Scientific Information, Philadelphia, PA). This is an average of 227 papers per year! Examples presented here are therefore selective, but hopefully show a spectrum of uses for pulse voltammetry.

4.1 Electrochemical Studies

4.1.1 Mass Transport

A common theme in fundamental experimental studies in electrochemistry is comparison of the shape of a voltammogram with that predicted by theory. Knowledge of the shape of a voltammogram can provide significant information about the thermodynamics and rates of the electron transfer reactions, magnitudes of diffusion coefficients and coupled chemical reaction rates and equilibrium constants. The following examples show how pulse voltammetric methods have been applied to problems involving mass transport.

Steady-state pulse voltammetry at microelectrodes has been used to probe the properties of strong and weak acids, both in simple molecular form, such as phosphoric acid (H_3PO_4, CAS 7664-38-2), and in polyelectrolytes, e.g. poly(styrene sulfonate).[67,73–75] Steady-state current for proton reduction at an 11.25-μm radius platinum microelectrode was simply described by Equation (20):

$$i_{ss} = 4nFDCr \tag{20}$$

where i_{ss} is the steady-state limiting current, D is the diffusion coefficient of proton, C is the acid concentration and r is the microelectrode radius. In the case of a simple strong acid, e.g. $HClO_4$ (CAS 7601-90-3), the limiting current for proton reduction was controlled by the total acid concentration and displayed variation with ionic strength consistent with the influence of ionic strength on the diffusion coefficient of proton.[72] In the case of strong acids, which are completely dissociated in water, the only reaction of interest was proton reduction (Equation 21):

$$H^+ + e^- \longrightarrow \tfrac{1}{2}H_2 \tag{21}$$

However, in the case of a weak acid, such as acetic acid (CAS 64-19-7), the additional step shown in Equation (22):

$$CH_3CO_2H \rightleftharpoons CH_3CO_2^- + H^+ \tag{22}$$

had to be taken into account because i_{ss} was controlled by the slower diffusion coefficient of undissociated acid rather than that of H^+. This is illustrated in Figure 17, where steady-state limiting currents are compared for the strong acid $HClO_4$ and weak acids acetic and ascorbic acid (CAS 50-81-7).[75]

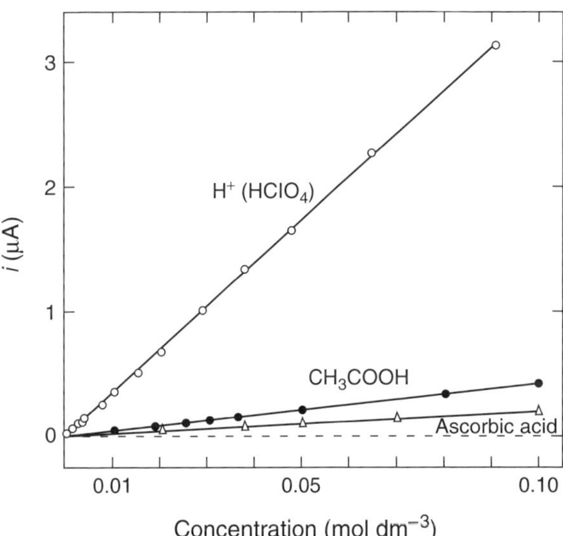

Figure 17 Mass transport-limited NPV currents for proton reduction in strong ($HClO_4$) and weak (acetic, ascorbic) acids as a function of concentration. (Reproduced by permission of The American Chemical Society from M. Ciszkowska, Z. Stojek, S.E. Morris, J.G. Osteryoung, 'Steady-state Voltammetry of Strong and Weak Acids With and Without Supporting Electrolyte', *Anal. Chem.*, **64**, 2372–2377 (1992).)

4.1.2 Chemistry Coupled to Electron Transfer

Pulse voltammetry has proven useful for study of chemical reactions coupled to electron transfer.[34,38,39] Some of the common reaction schemes encountered in practice include slow heterogeneous kinetics, preceding chemical reaction (prior to electron transfer) and following chemical reaction. Chemical conversions are commonly denoted C and electron transfers E in the electrochemical literature. Therefore, an ECE mechanism, for example, involves an electron transfer followed by a chemical reaction whose product undergoes a second electron transfer.

In slow electron transfer, the only new parameter to consider is the rate of the heterogeneous process, which is characterized by k^0, the rate of the reaction at zero overpotential and the transfer coefficient, α. While slow electron transfer is not coupled to chemistry per se, it is conveniently treated along with the other deviations from ideal reversible behavior.

Chemical reactions can either precede or follow an electron transfer. The voltammetric response will depend on whether the initial and final chemical forms are redox active and the rates and reversibility of the chemical steps will depend on the timescale of the pulse method. For example, in the scheme shown in Equations (23) and (24):

$$Y \longrightarrow O \tag{23}$$

$$O + ne^- \longrightarrow R \tag{24}$$

where Y is electrochemically silent (CE mechanism), one may expect that the current obtained for the reduction of O to R (Equation 23) will depend on the rate at which Y converts to O compared with the experimental timescale. Similarly, for an electron transfer (Equation 24), followed by (EC mechanism) Equation (25):

$$R \longrightarrow Y \tag{25}$$

the current for oxidation of R back to O when viewed by RPV will depend on the rate of Equation (25) compared with the experimental timescale. More complicated situations, such as ECE (two electron transfers separated by a chemical conversion), disproportionation, etc., with varying degrees of reversibility are also possible and are accordingly complicated to analyze.

For references see page 10032

SWV responses have been defined for many of the basic mechanistic cases.[38] A well-known example of homogeneous chemistry coupled to redox reactions is electroreduction of anthracene (CAS 120-12-7) (An) in acetonitrile[65] (Equations 26–28):

$$\mathrm{An} \rightleftharpoons \mathrm{An}^{\overset{+}{\cdot}} + e^- \quad (26)$$

$$\mathrm{An}^{\overset{+}{\cdot}} + \mathrm{CH_3CN} \longrightarrow \mathrm{An{-}CH_3CN}^{\overset{+}{\cdot}} \quad (27)$$

$$\mathrm{An{-}CH_3CN}^{\overset{+}{\cdot}} \rightleftharpoons \mathrm{AnCH_3CN^+} + \mathrm{H^+} + e^- \quad (28)$$

The radical cation generated in Equation (26) can react with nucleophilic CH_3CN to form an oxidizable intermediate. Disproportionation of the intermediates (Equation 29):

$$\mathrm{An}^{\overset{+}{\cdot}} + \mathrm{An{-}CH_3CN}^{\overset{+}{\cdot}} \rightleftharpoons \mathrm{An} + \mathrm{AnCH_3CN^+} \quad (29)$$

which reforms 1 mol of starting material and produces 1 mol of final product could affect the pulse voltammetric results if these conversions proceed significantly on the voltammetric timescale. Figure 18 shows, however, that the above reaction sequence was adequately explained by an ECE mechanism, i.e. the sequence of Equations (26–28), and that Equation (29) only had a small effect on the outcome.

An example of the application of RPV to the evaluation of unstable intermediates is provided by the electrochemical reduction of 1-iodoalkanes at Hg in

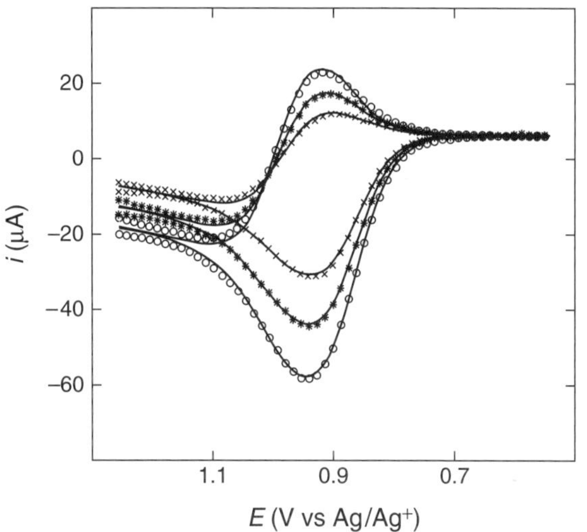

Figure 18 Forward and reverse square wave voltammograms with best-fit theoretical curves for an ECE mechanism. Square wave frequencies were (○) 3401, (∗) 2000 and (×) 1000 Hz, with $\Delta E_s = 10\,\mathrm{mV}$ and $E_{SW} = 50\,\mathrm{mV}$. (Reprinted from M.M. Murphy, Z. Stojek, J.J. O'Dea, J.G. Osteryoung, 'Pulse Voltammetry at Cylindrical Electrodes: Oxidation of Anthracene', *Electrochim. Acta*, **36**, 1475–1484 (1991), Copyright 1991, with permission from Elsevier Science.)

nonaqueous solvents[68] (Equations 30 and 31):

$$\mathrm{RI} + \mathrm{Hg} + e^- \longrightarrow \mathrm{RHg}^{\bullet}_{(ads)} + \mathrm{I}^- \quad (30)$$

$$\mathrm{RI} + 2e^- \longrightarrow \mathrm{R}^- + \mathrm{I}^- \quad (31)$$

Reduction of the iodoalkane (RI) at Hg by one electron generates an adsorbed radical (Equation 30), while two-electron reduction at more extreme potentials produces the anion, R^- and I^-. The radical can dimerize to form R_2 (Equation 32):

$$2\mathrm{RHg}^{\bullet}_{(ads)} \longrightarrow \mathrm{HgR_2} \quad (32)$$

and the anion R^- can react with a proton donor impurity (e.g. water) to form the alkane or with the original iodoalkane to form dimer and iodide (Equations 33 and 34):

$$\mathrm{R}^- + \mathrm{HD} \longrightarrow \mathrm{RH} + \mathrm{D}^- \quad (33)$$

$$\mathrm{R}^- + \mathrm{RI} \longrightarrow \mathrm{R_2} + \mathrm{I}^- \quad (34)$$

where HD is some proton donor. Figure 19 shows RPV of 2 mM 1-iododecane (CAS 2050-77-3) in pure propylene carbonate without supporting electrolyte at an Hg hemisphere microelectrode which was formed on a 12.5-μm radius Pt disk. Note how the use of a microelectrode in this highly resistive medium gives well-defined pulse voltammetric waves. This RPV experiment probed the reoxidation of products such as $RHg^{\bullet}$ and R^- generated at negative potentials. Two oxidation waves were found between ca. −0.5 and +1.0 V. The magnitude of the more negative wave decreased with increasing t_p, suggesting that the product formed during t_g at the negative limit was unstable on a timescale comparable to t_p. Furthermore, the ratio $i_{DC}/(i_{DC} - i_{RP})$ for the more negative wave increased with increasing RI concentration. It was suggested that this was caused by the effect of Equation (33), the rate of which would increase with increasing concentration of RI and subsequently attenuate i_{RP}. Additional information was obtained by examination of the effect of the negative potential limit at which the initial products were generated, shown in Figure 19. As this value was made less negative, the magnitude of both RPV waves decreased and the height of the second wave decreased at the expense of the first. This was consistent with increased rate of radical production by Equation (30) at less negative generation potentials and shows that the radical is actually fairly long lived in the pure solvent.

4.1.3 *Corrosion*

The study of corrosion processes is a broad field of electrochemical research and an extremely important one. Corrosion causes billions of dollars in losses per

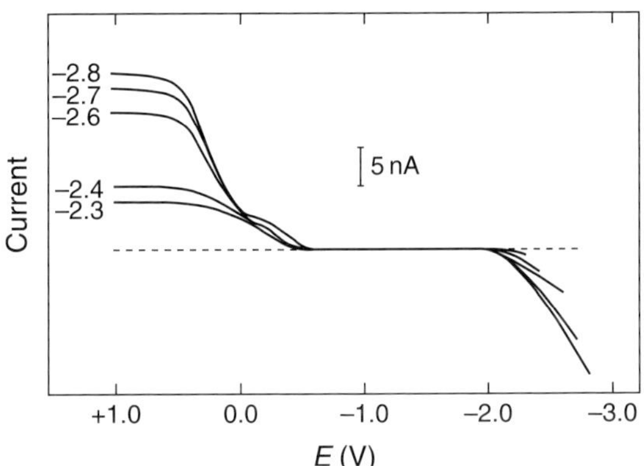

Figure 19 RPV of 2 mM 1-iododecane in propylene carbonate at various generation potentials, indicated at the left of each curve, at an Hg hemispherc electrode plated on to a 12.5-μm radius Pt disk. $t_p = 50$ ms, $t_g = 1$ s. (Reproduced by permission of The American Chemical Society from M. Ciszkowska, Z. Stojek, J. Osteryoung, 'Pulse Voltammetric Techniques at Microelectrodes in Pure Solvents', *Anal. Chem.*, **62**, 349–353 (1990).)

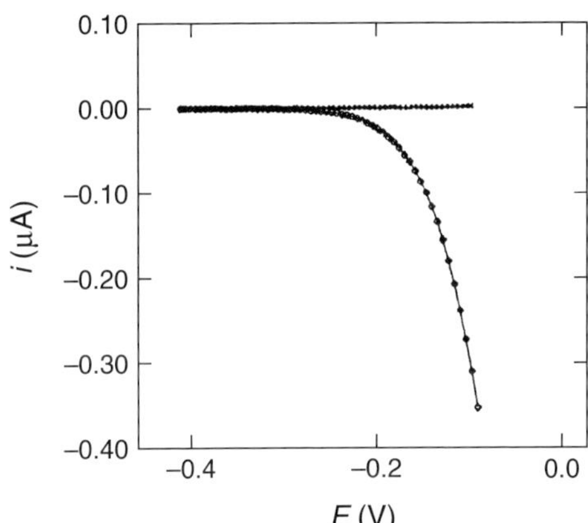

Figure 20 NPV for anodization of a 12.5-μm radius Cu electrode in 50% (v/v) ethylene glycol–water containing 1 M NaCl. t_p, 20 ms; t_w, 400 ms at −0.4 V. (Reprinted from K. Wikiel, J. Osteryoung, 'Pulse Voltammetric Techniques Applied to the Anodization of Copper Microelectrodes', *Electrochim. Acta*, **38**, 2291–2296 (1993), Copyright 1993, with permission from Elsevier Science.)

year in failure of equipment and civil infrastructure and is becoming an increasingly pressing problem for the military in the face of tight budgets for replacement of aging equipment such as military aircraft. Unfortunately, real corrosion phenomena are extremely complex and very difficult to simulate or reproduce in the laboratory.

Pulse voltammetry has been applied to the study of some extremely simple corrosion systems including anodic dissolution of copper (CAS 7440-50-8)[59,60] and silver (CAS 7440-22-4).[61] The virtue of the pulse voltammetric approach, in this case, is the ability to control precisely the potential applied to the corroding electrode and the timescale of potential application. Additionally, if metallic microelectrodes are used, as in the example below, a low current density allows high overpotentials to be investigated without significant iR_u distortion.

Anodizaton of a copper microelectrode (25-μm diameter, sealed in glass) was performed in 50% (v/v) ethylene glycol–water containing 1 M NaCl[60] (Equation 35):

$$Cu + 2Cl^- \rightleftharpoons CuCl_2^- + e^- \quad (35)$$

A typical NPV trace for this system is shown in Figure 20. The potential was held between pulses at −0.4 V and was progressively stepped (t_p, 20 ms) to more positive values to drive Equation (35). The flat line in Figure 20 is the current just prior to pulse application. Constancy of this value is a good measure of the degree to which initial boundary conditions of the experiment are being renewed from pulse to pulse. The use of a microelectrode allowed the analysis of the entire anodic dissolution process. The slopes of semilogarithmic plots may be related to the stoichiometry and formation constants of the products formed.[60] At low current densities (i.e at the foot of the wave), Equation (35) was operative and generated a soluble product. At more extreme values, neutral CuCl was formed, which precipitated on the electrode.

4.2 Analysis and Detection

Pulse voltammetric methods hold great promise for analytical applications in laboratory and field monitoring instruments. Of particular importance is the ability of the pulse techniques to discriminate against background processes to produce high S/N detection methods. Coupled to microelectrodes and other technologies, these methods are sometimes capable of the identification and quantification of multianalyte mixtures with minimum or no sample preparation. Under favorable circumstances, pulse voltammetry is capable of sensitivity and detection limits rivaling or surpassing those of other common analytical methodologies. A comparison of detection limits for some of these methods is presented in Table 2.

4.2.1 Pulse Voltammetry Coupled to Stripping Analysis for Metal Detection

A large fraction of papers published using pulse voltammetric detection methods deal with metal detection in water. This indicates not only the importance of environmental metal pollution, especially heavy metals such as Pb (CAS 7439-92-1) and Hg (CAS 7439-97-6), but also the ideal fit between these problems and electrochemical detection. A wide variety of soluble toxic metal ions are addressable by voltammetric methods,

For references see page 10032

Table 2 Comparison of detection limits for selected analytical methods

Technique	LLD (mol L^{-1})
DC voltammetry	10^{-6}
Visible spectrophotometry	10^{-6}
Atomic absorption spectroscopy	10^{-7}
Mass spectrometry	10^{-9}
Neutron activation analysis	10^{-10}
SWV	$10^{-8}-10^{-9}$
DPV	$10^{-8}-10^{-9}$
Stripping analysis with pulse voltammetry	$10^{-9}-10^{-10}$

including Pb(II) (CAS 10099-74-8), Hg(II) (CAS 7783-34-8), Cu(II) (CAS 10125-13-0), Zn(II) (CAS 10196-18-6), Cr(VI) (CAS 7440-47-3), Cd(II) (CAS 10022-68-1) and As(III) (CAS 7784-34-1).

Identification and quantification of these metals are possible with a sub-part per billion detection limit in some cases. A particularly noteworthy demonstration of the capability of pulse voltammetric detection was provided by Meyer et al.,[76] for the detection of Hg at the 5×10^{-14} M level, or ca. 0.01 ppt. This extremely low level of detection was achieved by ASV in conjunction with differential pulse anodic stripping voltammetry (DPASV) detection. The authors deposited Hg(II) from an acidic thiocyanate solution on to a GC working electrode. Unusually long deposition times, up to 40 min, were necessary to achieve the low detection limits reported. Nonetheless, this impressive demonstration of ultratrace metal detection shows how the very simple and relatively inexpensive electrochemical detection can rival or even surpass the performance of more complex and much more expensive instrumentation.

It is worthwhile digressing briefly to explain the ASV process[10] since it is commonly employed for electrochemical metal detection. ASV involves two steps. First, metal is plated by cathodic deposition on to the working electrode at a prescribed potential and for a fixed time. This step is a preconcentration in which metal ions are removed from a dilute solution and plated as elemental metal on to the working electrode. Since the working electrode is small compared with the volume of sample, the metal is concentrated by several orders of magnitude during this step. If a mercury electrode is used, the metal may dissolve in the electrode if it forms an amalgam. To a first approximation, the longer the deposition phase, the more metal is deposited at the electrode. Stirring often accompanies the preconcentration step to increase mass transfer of metal ion to the electrode surface. Following the preconcentration phase, metal is oxidatively removed from the electrode or "stripped" by application of a positive-going waveform. While the waveforms used to strip metal are identical with those described earlier, the difference between ASV and common pulse experiments is that the metal to be oxidized is located on or in the electrode, rather than in the solution at the electrode–solution interface. This has the important consequence that (ideally) the metal is quantitatively removed from the electrode during oxidative stripping, resulting in a symmetrical, peaked current–potential curve. This is true even if a nondifferential technique such as NPV is employed. True quantitative stripping of metal is usually achieved only with TFMEs (section 2.3.3). In the majority of recent applications either DPASV or square wave anodic stripping voltammetry (SWASV) waveforms are used, because of their favorable background discriminating abilities. The peak current or charge under the stripping peak can then be calibrated back to the original metal ion concentration in the sample. If the peaks are not overlapped on the potential axis, stripping voltammetry can be used to identify individual components of mixtures.[77,78] Detection limits in the single part per billion range are achievable, with care, in most cases.

CSV[11] is a related method which also has been used extensively in metal detection. This method is similar to ASV in its detection step, but metals (or organics) are adsorbed in their oxidized forms on an electrode surface, often with the help of a complexing agent. The adsorbed metal ion is then reductively stripped from the electrode in the detection step. CSV is typically employed for metal ions which are difficult to reduce directly to metal in aqueous media, including Cr(VI), Zn(II) and UO_2^{2+} (CAS 36478-76-9). CSV has also been employed widely in the analysis of organics, to be described in section 4.2.2.

Environmental trace metal analysis is becoming an increasingly important endeavor. Electrochemical methods provide a natural avenue to devices and instruments for field monitoring of metals. Stripping analysis by pulse voltammetry has played a large role in progress to move monitoring operations from the laboratory to the field.[79–85] Microelectrodes and microelectrode arrays prepared by photolithographic methods, coupled to DPASV and SWASV, offer the promise of fast, sensitive, selective and perhaps disposable devices compatible with field systems.[80,86–93]

Figure 21 illustrates the simultaneous detection of several metals in water using SWASV.[88] The working electrode in this case was an array of 19 Ir microdisks, each 10 µm in diameter, overcoated with electrodeposited Hg to form microhemispherical Hg electrodes. The metal ions Zn(II), Cd(II), Pb(II) and Cu(II) were codeposited on to the Hg–Ir array at −1.4 V vs Ag/AgCl for 4.33 min prior to the SWV scan. Figure 21 illustrates how multiple metals can be identified and quantified by SWASV. The

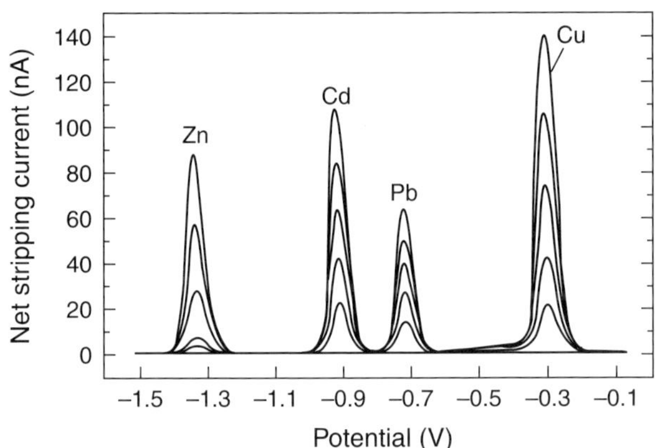

Figure 21 Simultaneous square wave anodic stripping voltammetric detection of Zn(II), Cd(II), Pb(II) and Cu(II) in 2 mM acetate buffer at 1, 2, 3, 4 and 5 ppb for each metal ion. $E_{SW} = 25\,mV$; $\Delta E_s = 5\,mV$; $f = 120\,Hz$; deposition potential $= -1.4\,V$ for 260 s. (Reprinted from G.T.A. Kovacs, C.W. Storment, S.P. Kounaves, 'Microfabricated Heavy Metal Ion Sensor', *Sens. Actuators B*, **23**, 41–47 (1995), Copyright 1995, with permission from Elsevier Science.)

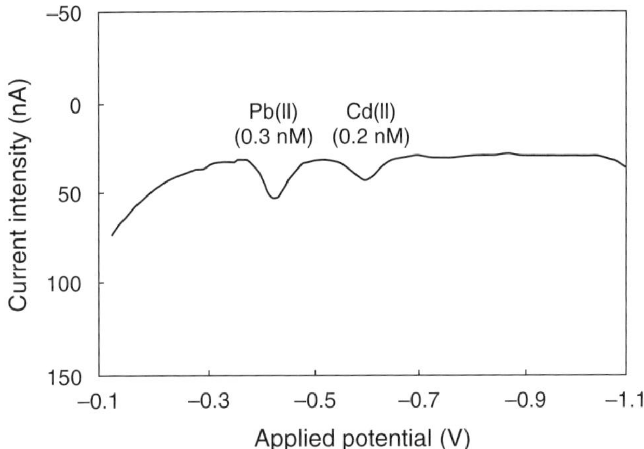

Figure 22 Square wave anodic stripping detection of Pb(II) and Cd(II) in raw river water at a gel-coated Hg-plated Ir microelectrode array composed of 12.2-µm diameter Hg hemispheres. Deposition potential, −1.1 V for 15 min; $E_{SW} = 25\,mV$; $\Delta E_s = 8\,mV$; $f = 50\,Hz$. (Reproduced by permission of The American Chemical Society from C. Belmont-Hébert, M.L. Tercier, J. Buffle, G.C. Fiaccabrino, N.F. de Rooij, M. Koudelka-Hep, 'Gel-integrated Microelectrode Arrays for Direct Voltammetric Measurements of Heavy Metals in Natural Waters and Other Complex Media', *Anal. Chem.*, **70**, 2949–2956 (1998).)

positions of the SWASV peaks on the potential axis identify the metal and the charge under the stripping peak, or equivalently the magnitude of the peak, quantifies the amount of each metal stripped, which is calibrated to the original sample concentration by use of appropriate calibration curves or tables. The extremely high S/N available by SWASV is clearly shown in Figure 21.

Figure 22 shows the use of SWASV for the detection of Pb(II) and Cd(II) in raw river water.[92] The electrode was a 5 × 20 array of 12.2-µm diameter Hg hemispheres on 5-µm diameter Ir disks, and the detection method was SWASV. The concentrations shown in the figure amount to 60 ppt Pb(II) and 22.4 ppt Cd(II). This demonstrates again that it is possible to use SWASV to detect trace components with high S/N at levels competitive with conventional instruments which are not commonly amenable to portability and field analysis. Table 3 summarizes some representative detection strategies for toxic metal ions in water using pulse voltammetry.

Table 3 Stripping analysis of selected metals with pulse voltammetric detection

Metal ion	Method	Application/conditions	LLD ($mol\,L^{-1}$)[a]	Working electrode	Ref.
Hg(II)	DPASV	Tap water acidified to pH 3.6	5×10^{-14}	Rotating GCE[b]	76
Hg(II)	SWASV	$HNO_3/NaNO_3$ media, pH 0–7	5×10^{-7}	Au microband/flow cell	93
Pb(II)	SWASV	Tap water Unmodified sample	5×10^{-9}	Hg UME array	90
Pb(II)	SWCSV	Seawater Acetate buffer, pH 5, xylenol orange	3×10^{-11}	HMDE	94
Pb(II)	DPASV	Acetate buffer, pH 4	2.4×10^{-8}	Hg UME array	89
Pb(II)	SWASV	0.1 M acetate buffer, pH 4	5×10^{-10}	Hg UME array	80
Cd(II)	DPASV	Acetate buffer, pH 4	$<1.8 \times 10^{-7}$	Hg UME array	89
Cd(II)	SWCSV	Seawater HEPES, pH 7.7, 8-hydroxyquinoline	4×10^{-10}	HMDE	95
As(III)	DPASV	1 M HCl	2.7×10^{-10}	Au disk	81
As(III)	SWCSV	Natural water Hydrazine sulfate, 2 M HCl	0.3×10^{-9}	HMDE	96
Cr(VI)	DPCSV	Groundwater Cupferron, PIPES buffer, pH 7	2×10^{-11}	HMDE	97

[a] ppb $= M \times 10^6 AW_{metal}$, where M is molarity ($mol\,L^{-1}$) and AW is atomic weight of metal ($g\,mol^{-1}$).
[b] GCE, glassy carbon electrode.

For references see page 10032

Table 4 AdSV of selected organics with pulse voltammetric detection

Compound	Method	Application/conditions	LLD ($mol\ L^{-1}$)[a]	Working electrode	Ref.
Uric acid (CAS 69-93-2), dopamine (CAS 62-31-7)	SWASV	Biomolecules, citrate–phosphate buffers	2×10^{-7} uric acid 2.7×10^{-9} dopamine	Clay–Nafion modified carbon UME	96
Terbinafine (CAS 91161-71-6)	SWCSV	Antimycotic, pH 6	1.7×10^{-10}	HMDE	97
Sunset Yellow	DPCSV	Synthetic colorants, phosphate–citrate buffer, KCl, pH 5.7	$0.03–0.2\ \mu g\ mL^{-1}$	HMDE	98
Ceftazidime	DPCSV	Pharmaceuticals, Britton–Robinson (BR) buffer, pH 4	5×10^{-9}	HMDE	99
Nitralin	SWCSV	Herbicides, pH >10	4.4×10^{-10}	HMDE	100
Doxazosin (CAS 77883-43-3)	DPCSV	Pharmaceuticals, BR buffer, pH 6.6	4×10^{-11}	Carbon paste	101
Atrazine	DPCSV	Herbicides, untreated groundwater	$5\ \mu g\ L^{-1}$	SMDE	102
Metamitron	DPCSV	Herbicides, BR buffer, pH 1.9	$<4 \times 10^{-9}$	Carbon paste	103
Naringin (CAS 10236-47-2)	DPCSV	Flavonoid, grapefruit juice	5.5×10^{-8}	HMDE	104
Methylamphetamine (CAS 51-57-0)	DPCSV	Stimulant, ethanol–water–0.25 M ammonium acetate	$0.125\ \mu g\ L^{-1}$	HMDE	105
Myoglobin (CAS 9008-45-1)	DPCSV	Biomolecules	$0.1\ \mu g\ mL^{-1}$	Chemically modified Ag electrode	106
Ofloxacin (CAS 82419-36-1)	DPCSV	Pharmaceuticals, BR buffer, pH 4–10	3×10^{-7}	DME	107

[a] detection limits in $mol\ L^{-1}$ unless indicated otherwise.

4.2.2 Analysis of Organic Compounds

Pulse voltammetric detection has been applied successfully to a wide range of organic compounds of biomedical and environmental interest. Various forms of nonelectrolytic preconcentration[11] coupled to CSV or ASV are applied to adsorb the analyte on an electrode, followed by SWV, DPV or similar detection. In some cases, selection of an appropriate potential, analogous to the deposition phase of ASV, is all that is necessary to facilitate adsorption of analyte. Hg electrodes have proven useful in potential-dependent adsorption of organics for preconcentration prior to CSV. In some cases a reagent may be added to effect adsorption of analyte, or the electrode itself can be modified to favor analyte adsorption. Detection limits in the $10^{-7}–10^{-9}$ M range are achievable.

A recent review of adsorptive stripping voltammetry (AdSV) in the trace analysis of organic molecules and biomolecules tabulates detection limits and conditions for approximately 80 biologically active compounds.[98] Many of these methods use pulse voltammetric detection because of their superior S/N and background discrimination. Additional details on specific analytical methods can be found in the reviews and monographs by Wang.[11,99] Table 4 shows a range of representative examples of CSV organic detection by pulse voltammetry.[98–111]

4.2.3 Chromatographic and Flow Injection Analysis Detectors

Chromatography and flow injection analysis (FIA) represent mature practical application areas in which pulse voltammetry has made contributions. Electrochemical detection in high-performance liquid chromatography (HPLC), for example, has been carried out successfully for many years and commercial units are available. The typical arrangement uses constant-potential amperometry as the detection scheme, in which the detector electrode is held at a potential where oxidizable or reducible components of a mixture are detected following separation. In principle, very rapid scan pulse methods, such as SWV, should provide additional potential resolution within the timescale of a chromatographic peak, although these methods have not been implemented in commercial devices.[112]

FIA is another relatively mature process analytical methodology in which samples are taken from a pot or

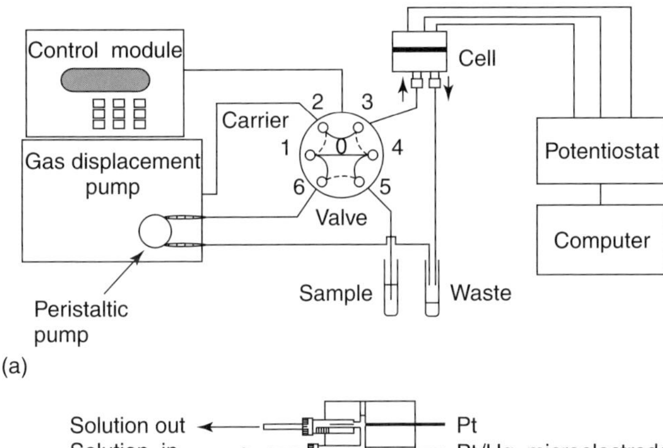

Figure 23 Schematic diagram of (a) a flow injection system and (b) a microelectrode flow cell for pulse voltammetric electrochemical detection in FIA. (Reproduced by permission of The American Chemical Society from F. Zhu, J.W. Ironstone, M.R. Ruegnitz, 'High Throughput Fast-scan Anodic Stripping Voltammetry in a Microflow System', *Anal. Chem.*, **69**, 728–733 (1997).)

pipe, injected into a flowing electrolyte and sent to a detector.(113) Pulse voltammetric applications have been reported for ASV of metals(114–117) and organics.(118,119) A typical experimental set-up for FIA with pulse electrochemical detection is shown in Figure 23.(116)

5 FUTURE PROSPECTS

Pulse voltammetry has been demonstrated to be useful and versatile in a variety of basic problems and applications. Further advances in instrumentation and techniques will no doubt occur in the future, in addition to many more convincing demonstrations of the utility of these methods for solving chemical problems. Mating sensitive pulse detection methods with other advanced technologies will, hopefully, speed the use of these methods in solving practical problems. Exciting recent uses and possibilities of pulse voltammetric detection in environmental analysis are capitalizing on the microfabrication methods of the semiconductor industry, for example, to produce arrays of electrodes which can do what single electrodes cannot. There is clearly sufficient interest and activity to generate many new innovations and applications. The key for future progress and wide use of pulse voltammetry will be to apply the techniques, which are now relatively mature, to new problems and new disciplines, particularly the difficult ones, i.e. real samples and complex matrices in chemistry, biology and the environment.

ACKNOWLEDGMENTS

Work performed in the authors' laboratories was supported by the US Department of Energy (M.T.C.) and by the Air Force Office of Scientific Research (R.A.O.).

ABBREVIATIONS AND ACRONYMS

AdSV	Adsorptive Stripping Voltammetry
ASV	Anodic Stripping Voltammetry
BAS	Bioanalytical Systems
CSV	Cathodic Stripping Voltammetry
CV	Cyclic Voltammetry
DC	Direct Current
DME	Dropping Mercury Electrode
DPASV	Differential Pulse Anodic Stripping Voltammetry
DPP	Differential Pulse Polarography
DPV	Differential Pulse Voltammetry
FIA	Flow Injection Analysis
GC	Glassy Carbon
GCE	Glassy Carbon Electrode
HMDE	Hanging Mercury Drop Electrode
HOPG	Highly Oriented Pyrolytic Graphite
HPLC	High-performance Liquid Chromatography
LLD	Lower Limit of Detection
NPP	Normal Pulse Polarography
NPV	Normal Pulse Voltammetry
PAR	EG&G Princeton Applied Research
RPP	Reverse Pulse Polarography
RPV	Reverse Pulse Voltammetry
RPW	Reverse Pulse Voltammetry with Waiting
SCE	Saturated Calomel Electrode
SMDE	Static Mercury Drop Electrode
S/N	Signal-to-noise Ratio
SWASV	Square Wave Anodic Stripping Voltammetry
SWV	Square Wave Voltammetry
TFME	Thin-film Mercury Electrode
TTL	Transistor-transistor Logic
UME	Ultramicroelectrode

RELATED ARTICLES

Biomolecules Analysis **(Volume 1)**
Voltammetry In Vivo for Chemical Analysis of the Living Brain • Voltammetry In Vivo for Chemical Analysis of the Nervous System

For references see page 10032

Clinical Chemistry **(Volume 2)**
Atomic Spectrometry in Clinical Chemistry • Electroanalysis and Biosensors in Clinical Chemistry • Electroanalytical Chemistry in Clinical Analysis

Environment: Water and Waste **(Volume 3)**
Heavy Metals Analysis in Seawater and Brines • Inorganic Environmental Analysis by Electrochemical Methods

Environment: Water and Waste cont'd **(Volume 4)**
Organic Analysis in Environmental Samples by Electrochemical Methods

Field-portable Instrumentation **(Volume 4)**
Electrochemical Sensors for Field Measurements of Gases and Vapors

Electroanalytical Methods **(Volume 11)**
Electroanalytical Methods: Introduction • Organic Electrochemical Mechanisms

Electronic Absorption and Luminescence **(Volume 12)**
Indirect Detection Methods in Capillary Electrophoresis

REFERENCES

1. A.J. Bard, L.R. Faulkner, 'Controlled Potential Microelectrode Techniques – Potential Step Methods', *Electrochemical Methods*, John Wiley and Sons, New York, Chapter 5, 136–212, 1980.
2. W.R. Heineman, P.T. Kissinger, 'Large Amplitude Controlled-potential Techniques', in *Laboratory Techniques in Electroanalytical Chemistry* 2nd edition, eds. P.T. Kissinger, W.R. Heineman, Marcel Dekker, New York, Chapter 3, 51–125, 1996.
3. D.T. Sawyer, J.L. Roberts, 'Controlled Potential Methods', *Experimental Electrochemistry for Chemists*, Wiley-Interscience, New York, Chapter 7, 329–395, 1974.
4. R.N. Adams, 'Mass Transfer to Stationary Electrodes in Quiet Solution', *Electrochemistry at Solid Electrodes*, Marcel Dekker, New York, Chapter 3, 43–66, 1969.
5. A.J. Bard, L.R. Faulkner, 'Introduction and Overview of Electrode Processes', *Electrochemical Methods*, John Wiley and Sons, New York, Chapter 1, 1–43, 1980.
6. A.J. Bard, L.R. Faulkner, 'Kinetics of Electrode Reactions', *Electrochemical Methods*, John Wiley and Sons, New York, Chapter 3, 86–118, 1980.
7. A.J. Bard, L.R. Faulkner, 'Double Layer Structure and Adsorbed Intermediates', *Electrochemical Methods*, John Wiley and Sons, New York, Chapter 12, 488–552, 1980.
8. E. Laviron, 'Voltammetric Methods for the Study of Absorbed Reactants', in *Electroanalytical Chemistry*, ed. A.J. Bard, Marcel Dekker, New York, 53–157, Vol. 12, 1982.
9. R.W. Murray (ed.), 'Molecular Design of Electrode Surfaces', *Techniques of Chemistry*, John Wiley and Sons, New York, Chapter 1, 1–48, Vol. 22, 1992.
10. J. Wang, *Stripping Analysis*, VCH, Deerfield Beach, FL, 1985.
11. J. Wang, 'Voltammetry Following Nonelectrolytic Preconcentration', in *Electroanalytical Chemistry*, ed. A.J. Bard, Marcel Dekker, New York, 1–88, Vol. 16, 1989.
12. J. Osteryoung, 'Pulse Voltammetry', *J. Chem. Educ.*, **60**, 296–298 (1983).
13. J. Osteryoung, R.A. Osteryoung, 'Analytical Pulse Voltammetry', in *Electrochemistry, Sensors and Analysis*, eds. M.R. Smyth, J.G. Vos, Elsevier, Amsterdam, 3–12, 1986.
14. J. Osteryoung, M. Schreiner, 'Recent Advances in Pulse Voltammetry', *CRC Crit. Rev. Anal. Chem.*, **19**, S1–S27 (1988).
15. R.A. Osteryoung, J.G. Osteryoung, 'Pulse Voltammetric Methods of Analysis', *Philos. Trans. R. Soc. London, Ser. A*, **302**, 315–326 (1981).
16. M.T. Carter, R.A. Osteryoung, 'Heterogeneous and Homogeneous Electron Transfer Reactions of Tetrathiafulvalene in Ambient Temperature Chloroaluminate Molten Salts', *J. Electrochem. Soc.*, **141**, 1713–1720 (1994).
17. J. Osteryoung, E. Kirowa-Eisner, 'Reverse Pulse Polarography', *Anal. Chem.*, **52**, 62–66 (1980).
18. T.R. Brumleve, J. Osteryoung, 'Spherical Diffusion and Shielding Effects in Reverse Pulse Voltammetry', *J. Phys. Chem.*, **86**, 1794–1801 (1982).
19. M. Wojciechowski, J. Osteryoung, 'Electrochemistry and Reverse Pulse Polarographic Determination of 1,2-Dibromo-2,4-dicyanobutane', *Anal. Chem.*, **57**, 927–933 (1985).
20. A. Brestovisky, E. Kirowa-Eisner, J. Osteryoung, 'Direct and Titrimetric Determination of Hydrogen Peroxide by Reverse Pulse Polarography', *Anal. Chem.*, **55**, 2063–2066 (1983).
21. C.-L. Ni, F.C. Anson, J.G. Osteryoung, 'A Pulse Polarographic Measurement of the Kinetics of Dissociation of the $[Ru^{III}(EDTA)]_2$ Dimer', *J. Electroanal. Chem.*, **202**, 101–109 (1986).
22. J. Osteryoung, D. Talmore, J. Hermolin, E. Kirowa-Eisner, 'Reverse Pulse Voltammetry. Application to Second Order Following Reactions', *J. Phys. Chem.*, **85**, 285–289 (1981).
23. E.P. Parry, R.A. Osteryoung, 'Evaluation of Analytical Pulse Polarography', *Anal. Chem.*, **37**, 1634–1637 (1965).
24. J.H. Christie, J. Osteryoung, R.A. Osteryoung, 'Instrumental Artifacts in Differential Pulse Polarography', *Anal. Chem.*, **45**, 210–215 (1973).
25. J.W. Dillard, J.A. Turner, R.A. Osteryoung, 'Digital Simulation of Differential Pulse Polarography with Incremental Time Change', *Anal. Chem.*, **49**, 1246–1250 (1977).

List of selected abbreviations appears in Volume 15

26. K. Aoki, J. Osteryoung, 'Theory of Differential Pulse Polarography at Expanding or Stationary Planar Electrodes for Quasi-reversible or Totally Irreversible Reactions', *J. Electroanal. Chem.*, **110**, 19–36 (1980).
27. D.J. Myers, J. Osteryoung, 'Amperometric Titrations Employing Differential Pulse Polarography', *Anal. Chem.*, **46**, 356–359 (1974).
28. D.J. Myers, R.A. Osteryoung, J. Osteryoung, 'Pulse Voltammetry at Rotated Electrodes', *Anal. Chem.*, **46**, 2089–2092 (1974).
29. J. Osteryoung, J.J. O'Dea, 'Square Wave Voltammetry', in *Electroanalytical Chemistry*, ed. A.J. Bard, Marcel Dekker, New York, 209–308, Vol. 14, 1986.
30. J. Osteryoung, R.A. Osteryoung, 'Square Wave Voltammetry', *Anal. Chem.*, **57**, 101A – 110A (1985).
31. E. Zachowski, M. Wojciechowski, J. Osteryoung, 'The Analytical Application of Square Wave Voltammetry', *Anal. Chim. Acta*, **183**, 47–57 (1986).
32. K. Aoki, K. Tokuda, H. Matsuda, J. Osteryoung, 'Reversible Square Wave Voltammograms: Independence of Electrode Geometry', *J. Electroanal. Chem.*, **207**, 25–39 (1986).
33. W. Go, J.J. O'Dea, J. Osteryoung, 'Square Wave Voltammetry for the Determination of Kinetic Parameters. The Reduction of Zinc(II) at Mercury Electrodes', *J. Electroanal. Chem.*, **255**, 21–44 (1988).
34. J.J. O'Dea, K. Wikiel, J. Osteryoung, 'Square Wave Voltammetry for ECE Mechanisms', *J. Phys. Chem.*, **94**, 3628–3636 (1990).
35. M. Wojciechowski, W. Go, J. Osteryoung, 'Square Wave Anodic Stripping Analysis in the Presence of Dissolved Oxygen', *Anal. Chem.*, **57**, 155–158 (1985).
36. K. Aoki, K. Maeda, J. Osteryoung, 'Characterization of Nernstian Square-wave Voltammograms', *J. Electroanal. Chem.*, **272**, 17–28 (1989).
37. M.J. Nuwer, J.J. O'Dea, J. Osteryoung, 'Analytical and Kinetic Investigations of Totally Irreversible Electron Transfer Reactions by Square-wave Voltammetry', *Anal. Chim. Acta*, **261**, 13–25 (1991).
38. J.J. O'Dea, J. Osteryoung, R.A. Osteryoung, 'Theory of Square-wave Voltammetry for Kinetic Systems', *Anal. Chem.*, **53**, 695–701 (1981).
39. J. Zeng, R.A. Osteryoung, 'Square-wave Voltammetry for a Pseudo-first-order Catalytic Process', *Anal. Chem.*, **58**, 2766–2771 (1986).
40. R. Samuelsson, J.J. O'Dea, J. Osteryoung, 'Rapid Scan Square Wave Voltammetric Detector for HPLC', *Anal. Chem.*, **52**, 2215–2216 (1980).
41. L. Mahoney, J. O'Dea, J. Osteryoung, 'Development and Characterization of an Automated Flow System for Voltammetric Analysis', *Anal. Chim. Acta*, **281**, 25–33 (1993).
42. Z. Galus, 'Mercury Electrodes', in *Laboratory Techniques in Analytical Chemistry*, 2nd edition, eds. W.R. Heineman, P.T. Kissinger, Marcel Dekker, New York, Chapter 9, 267–288, 1996.
43. A. Bond, *Modern Polarographic Methods of Analysis*, Marcel Dekker, New York, 1980.
44. L. Novotny, M. Heyrovsky, 'Renewable Mercury Electrodes – Versatile Research Tools in General Chemistry', *Croat. Chem. Acta*, **70**, 151–165 (1997).
45. J.A. Turner, J.H. Christie, M. Vukovic, R.A. Osteryoung, 'Square-wave Voltammetry at the Dropping Mercury Electrode: Experimental', *Anal. Chem.*, **49**, 1904–1908 (1977).
46. W.M. Peterson, 'Static Mercury Drop Electrode', *Am. Lab.*, **11**(12), 69–70 (1979).
47. J.A. Turner, U. Eisner, R.A. Osteryoung, 'Pulsed Voltammetric Stripping at the Thin-film Mercury Electrode', *Anal. Chim. Acta*, **90**, 25–34 (1977).
48. T.R. Copeland, J.H. Christie, R.A. Osteryoung, R.K. Skoerboe, 'Analytical Applications of Pulsed Voltammetric Stripping at Thin Mercury Electrodes', *Anal. Chem.*, **45**, 2171–2174 (1973).
49. K. Wikiel, J. Osteryoung, 'Square Wave Voltammetry at a Mercury Film Electrode: Experimental Results', *Anal. Chem.*, **61**, 2086–2092 (1989).
50. C. Wechter, J. Osteryoung, 'Square Wave and Linear Scan Anodic Stripping Voltammetry at Iridium Based Mercury Film Electrodes', *Anal. Chem.*, **61**, 2092–2097 (1989).
51. S. Kounaves, J.J. O'Dea, P. Chandrasekhar, J. Osteryoung, 'Square-wave Voltammetry at the Mercury Film Electrode: Theoretical Treatment', *Anal. Chem.*, **58**, 3199–3202 (1986).
52. S. Kounaves, J.J. O'Dea, P. Chandrasekhar, J. Osteryoung, 'Square Wave Anodic Stripping Voltammetry at the Mercury Film Electrode: Theoretical Treatment', *Anal. Chem.*, **59**, 386–389 (1987).
53. R.M. Wightman, D.O. Wipf, 'Voltammetry at Ultramicroelectrodes', in *Electroanalytical Chemistry*, ed. A.J. Bard, Marcel Dekker, New York, 267–353, Vol. 15, 1989.
54. D.K. Roe, 'Overcoming Solution Resistance with Stability and Grace in Potentiostatic Circuits', in *Laboratory Techniques in Analytical Chemistry*, 2nd edition, eds. W.R. Heineman, P.T. Kissinger, Marcel Dekker, New York, Chapter 7, 193–234, 1996.
55. D. Whelan, J.J. O'Dea, J. Osteryoung, K. Aoki, 'Square Wave Voltammetry at Small Disk Electrodes: Theory and Experiment', *J. Electroanal. Chem.*, **202**, 23–36 (1986).
56. J. O'Dea, M. Wojciechowski, J. Osteryoung, K. Aoki, 'Square Wave Voltammetry at Electrodes Having Small Dimensions', *Anal. Chem.*, **57**, 954–955 (1985).
57. L. Sinru, J. Osteryoung, J.J. O'Dea, R.A. Osteryoung, 'Normal and Reverse Pulse Voltammetry at Microdisk Electrodes', *Anal. Chem.*, **60**, 1135–1141 (1988).
58. L. Sinru, R.A. Osteryoung, 'Normal and Reverse Pulse Voltammetry for Poised Systems at Microdisk Electrodes', *Anal. Chem.*, **60**, 1845–1850 (1988).

59. K. Wikiel, M.M. dos Santos, J. Osteryoung, 'Determination of Stability Constants by Using Normal Pulse Voltammetry at Microelectrodes', *Electrochim. Acta*, **38**, 1555–1558 (1993).
60. K. Wikiel, J. Osteryoung, 'Pulse Voltammetric Techniques Applied to the Anodization of Copper Microelectrodes', *Electrochim. Acta*, **38**, 2291–2296 (1993).
61. M. Donten, J. Osteryoung, 'Pulse Techniques in Studies of Metal Dissolution: Anodization of Silver', *J. Electrochem. Soc.*, **13**, 82–88 (1991).
62. Z.J. Karpinski, R.A. Osteryoung, 'Short Time Pulse Voltammetric Studies of Fast Heterogeneous Electron Transfer Reactions', *J. Electroanal. Chem.*, **349**, 285–297 (1993).
63. Z.J. Karpinski, R.A. Osteryoung, 'Renewal of Boundary Conditions in Pulse Voltammetry at Microdisk Electrodes for Non-reversible Systems', *J. Electroanal. Chem.*, **307**, 47–62 (1991).
64. M.A.M. Nöel, J.J. O'Dea, R.A. Osteryoung, 'Short Time Pulse Voltammetry in Ambient Temperature Chloroaluminate Ionic Liquids', *J. Electrochem. Soc.*, **139**, 1231–1236 (1992).
65. M.M. Murphy, Z. Stojek, J.J. O'Dea, J.G. Osteryoung, 'Pulse Voltammetry at Cylindrical Electrodes: Oxidation of Anthracene', *Electrochim. Acta*, **36**, 1475–1484 (1991).
66. M.M. Murphy, J.J. O'Dea, J. Osteryoung, 'Pulse Voltammetry at Microcylinder Electrodes', *Anal. Chem.*, **63**, 2743–2750 (1991).
67. S.E. Morris, M. Ciszkowska, J.G. Osteryoung, 'Steady-state Voltammetry of Simple and Polyelectrolyte Strong Acids With and Without Supporting Electrolyte', *J. Phys. Chem.*, **97**, 19 453–19 457 (1993).
68. M. Ciszkowska, Z. Stojek, J. Osteryoung, 'Pulse Voltammetric Techniques at Microelectrodes in Pure Solvents', *Anal. Chem.*, **62**, 349–353 (1990).
69. J. Golas, Z. Galus, J. Osteryoung, 'Iridium-based Small Mercury Electrodes', *Anal. Chem.*, **59**, 389–392 (1987).
70. J. Golas, Z. Kowalski, 'Applications of Small Iridium-based Mercury Electrodes to High Concentrations of Depolarizers', *Anal. Chim. Acta*, **221**, 305–318 (1989).
71. C. Wechter, J. Osteryoung, 'Voltammetric Characterization of Small Platinum–Iridium-based Mercury Film Electrodes', *Anal. Chim. Acta*, **234**, 275–284 (1990).
72. J.L. Anderson, L.A. Coury, J. Leddy, 'Dynamic Electrochemistry: Methodology and Application', *Anal. Chem.*, **70**, 519R–589R (1998).
73. M. Ciszkowska, A. Jaworski, J.G. Osteryoung, 'Voltammetric Reduction of Hydrogen Ion in Solutions of Polyprotic Strong Acids With and Without Supporting Electrolyte', *J. Electroanal. Chem.*, **423**, 95–101 (1997).
74. M. Ciszkowska, Z. Stojek, J.G. Osteryoung, 'Voltammetric Reduction of Polyprotic Acids at the Platinum Microelectrode: Dependence on Supporting Electrolyte', *J. Electroanal. Chem.*, **398**, 49–56 (1995).
75. M. Ciszkowska, Z. Stojek, S.E. Morris, J.G. Osteryoung, 'Steady-state Voltammetry of Strong and Weak Acids With and Without Supporting Electrolyte', *Anal. Chem.*, **64**, 2372–2377 (1992).
76. S. Meyer, F. Scholz, R. Trittler, 'Determination of Inorganic Ionic Mercury Down to $5 \times 10^{-14}\,mol\,L^{-1}$ by Differential Pulse Anodic Stripping Voltammetry', *Fresenius' J. Anal. Chem.*, **356**, 247–252 (1996).
77. W. Jin, V.D. Nguyen, P. Valenta, H.W. Nürnberg, 'Simultaneous Determination of Seven Toxic Trace and/or Ultratrace Metals in Environmental Plants by Differential Pulse Voltammetry Without Change of Solution and Electrode', *Anal. Lett.*, **30**, 1235–1254 (1997).
78. C. Locatelli, 'Anodic and Cathodic Stripping Voltammetry in the Simultaneous Determination of Toxic Metals in Environmental Samples', *Electroanalysis*, **9**, 1014–1017 (1997).
79. J. Wang, B. Greene, 'Characteristics of a Flow Cell for the Determination of Arsenic(III) by Stripping Voltammetry', *J. Electroanal. Chem.*, **154**, 261–268 (1983).
80. J. Wang, J. Lu, B. Tian, C. Yarnitzky, 'Screen-printed Ultramicroelectrode Arrays for On-site Stripping Measurements of Trace Metals', *J. Electroanal. Chem.*, **361**, 77–83 (1993).
81. G. Forsberg, J.W. O'Laughlin, R.G. Megargle, S.R. Koirtyohann, 'Determination of Arsenic by Anodic Stripping Voltammetry and Differential Pulse Anodic Stripping Voltammetry', *Anal. Chem.*, **47**, 1586–1591 (1979).
82. A.M. Bond, W.A. Czerwinski, M. Llorente, 'Comparison of Direct Current, Derivative Direct Current, Pulse and Square Wave Voltammetry at Single Disc, Assembly and Composite Carbon Electrodes: Stripping Voltammetry at Thin Film Mercury Microelectrodes with Field-based Instrumentation', *Analyst*, **123**, 1333–1337 (1998).
83. R. Pongartz, K.G. Herrmann, 'Determination of Monomethylcadmium in the Environment by Differential Pulse Anodic Stripping Voltammetry', *Anal. Chem.*, **68**, 1262–1266 (1996).
84. D. Sancho, M. Vega, L. Debán, R. Pardo, G. González, 'Determination of Copper and Arsenic in Refined Beet Sugar by Stripping Voltammetry Without Sample Pretreatment', *Analyst*, **123**, 743–747 (1998).
85. M.-L. Tercier, N. Parthamaraty, J. Buffle, 'Reproducible, Reliable and Rugged Hg-plated Ir-based Microelectrode for In Situ Measurements in Natural Waters', *Electroanalysis*, **7**, 55–63 (1995).
86. S.P. Kounaves, W. Deng, P.R. Hallock, G.T.A. Kovacs, C.W. Storment, 'Iridium-based Ultramicroelectrode Array Fabricated by Microlithography', *Anal. Chem.*, **6**, 418–423 (1994).
87. B.J. Seddon, Y. Shao, N.H. Girault, 'Printed Microelectrode Array and Amperometric Sensor for Environmental Monitoring', *Electrochim. Acta*, **39**, 2377–2386 (1994).
88. G.T.A. Kovacs, C.W. Storment, S.P. Kounaves, 'Microfabricated Heavy Metal Ion Sensor', *Sens. Actuators B*, **23**, 41–47 (1995).

89. A. Uhlig, M. Paeschke, V. Schnakenberg, R. Hintsche, H.J. Diederich, F. Scholz, 'Chip-array Electrodes for Simultaneous Stripping of Trace Heavy Metals', *Sens. Actuators B*, **24–25**, 899–903 (1995).
90. R.J. Reay, A.F. Flannery, C.W. Storment, S.P. Kounaves, G.T.A. Kovacs, 'Microfabricated Electrochemical Analysis System for Heavy Metal Detection', *Sens. Actuators B*, **34**, 450–455 (1996).
91. M.A. Augelli, V.B. Nascimento, J.J. Pedrotti, I.G.R. Gutz, L. Angnes, 'Flow-through Cell Based on an Array of Gold Microelectrodes Obtained from Modified Integrated Circuit Chips', *Analyst*, **122**, 843–847 (1997).
92. C. Belmont-Hébert, M.L. Tercier, J. Buffle, G.C. Fiaccabrino, N.F. de Rooij, M. Koudelka-Hep, 'Gel-integrated Microelectrode Arrays for Direct Voltammetric Measurements of Heavy Metals in Natural Waters and Other Complex Media', *Anal. Chem.*, **70**, 1949–2956 (1998).
93. M.T. Carter, E.D. Cravens, 'Hybrid Electrochemical/Microfluidic Monitors for Trace Heavy Metals', *Proc. SPIE*, **3534**, 251–260 (1999).
94. Q.G. Wu, G.E. Batley, 'Determination of Sub-nanomolar Concentrations of Lead in Seawater by Adsorptive Stripping Voltammetry with Xylenol Orange', *Anal. Chim. Acta*, **309**, 95–101 (1995).
95. O. Abollino, M. Aceto, G. Sacchero, C. Sarzanini, E. Mentasti, 'Determination of Copper, Cadmium, Iron, Manganese, Nickel and Zinc in Antarctic Seawater. Comparison of Electrochemical and Spectroscopic Procedures', *Anal. Chim. Acta*, **305**, 200–206 (1995).
96. H. Li, R.B. Smart, 'Determination of Sub-nanomolar Concentration of Arsenic(III) in Natural Waters by Square Wave Cathodic Stripping Voltammetry', *Anal. Chim. Acta*, **325**, 25–32 (1996).
97. J. Wang, J. Lu, K. Olsen, 'Measurement of Ultratrace Levels of Chromium by Adsorptive Stripping Voltammetry in the Presence of Cupferron', *Analyst*, **117**, 1913–1917 (1992).
98. A.Z. Abuzuhri, W. Voelter, 'Applications of Adsorptive Stripping Voltammetry for the Trace Analysis of Metals, Pharmaceuticals and Biomolecules', *Fresenius' J. Anal. Chem.*, **360**, 1–9 (1998).
99. J. Wang, *Electroanalytical Techniques in Clinical Chemistry and Laboratory Medicine*, VCH, New York, 1988.
100. J.M. Zen, P.J. Chen, 'A Selective Voltammetric Method of Uric Acid and Dopamine Detection Using Clay-modified Electrodes', *Anal. Chem.*, **69**, 5087–5093 (1997).
101. A. Arranz, S.F. Debetono, J.M. Moreda, A. Cid, J.F. Arranz, 'Voltammetric Behaviour of the Antimycotic Terbinafine at the Hanging Mercury Drop Electrode', *Anal. Chim. Acta*, **351**, 97–103 (1997).
102. Y.N. Ni, J.L. Bai, L. Jin, 'Simultaneous Adsorptive Voltammetric Analysis of Mixed Colorants by Multivariate Calibration Approach', *Anal. Chim. Acta*, **329**, 65–72 (1996).
103. V.S. Ferreira, M.V.B. Zanoni, A.G. Fogg, 'Cathodic Stripping Voltammetric Determination of Ceftazidime in Urine at a Hanging Mercury Drop Electrode', *Microchem. J.*, **57**, 115–122 (1997).
104. A. Arranz, S.F. Debetono, J.M. Moreda, J.F. Arranz, 'Study of the Electro-adsorptive Behavior of the Herbicide Nitralin by Means of Voltammetric Techniques', *Talanta*, **45**, 417–424 (1997).
105. A. Arranz, S.F. Debetono, J.M. Moreda, A. Cid, J.F. Arranz, 'Cathodic Stripping Voltammetric Determination of Doxazosin in Urine and Pharmaceutical Tablets Using Carbon Paste Electrodes', *Analyst*, **122**, 849–854 (1997).
106. C.M.P. Vaz, S. Crestana, S.A.S. Machado, L.H. Mazo, L.A. Avaca, 'Electroanalytical Determination of the Herbicide Atrazine in Natural Waters', *Int. J. Environ. Anal. Chem.*, **62**, 65–76 (1996).
107. A. Arranz, S.F. Debetono, J.M Moreda, A. Cid, J.F. Arranz, 'Preconcentration and Voltammetric Determination of the Herbicide Metamitron with a Silica-modified Carbon Paste Electrode', Mikrochim. Acta, **127**, 273–279 (1997).
108. E. Reichart, D. Obendorf, 'Determination of Naringin in Grapefruit Juice by Cathodic Stripping Differential Pulse Voltammetry at the Hanging Mercury Drop Electrode', *Anal. Chim. Acta*, **360**, 179–187 (1997).
109. Y. Fushinuki, I. Taniguchi, 'Determination of Methylamphetamine in Urine by Differential Pulse Polarography', *Anal. Sci.*, **14**, 265–268 (1998).
110. Y.T. Long, J.J. Zhu, H.Y. Chen, 'Preconcentration and Voltammetric Determination of Trace Myoglobin at a 6-Mercaptopurine Modified Silver Electrode', *Fresenius, J. Anal. Chem.*, **360**, 614–617 (1998).
111. M. Rizk, F. Belal, F.A. Aly, N.M. Elenany, 'Differential Pulse Polarographic Determination of Ofloxacin in Pharmaceuticals and Biological Fluids', *Talanta*, **46**, 83–89 (1998).
112. J. O'Dea, J. Osteryoung, 'Rapid Scan Square Wave Voltammetric Detector for High Performance Liquid Chromatography', *Anal. Chem.*, **52**, 2215–2216 (1980).
113. E.D. Yalvac, R.A. Bredeweg, 'Status of Process Flow-injection Analysis and Current Trends', *Proc. Control Anal.*, **6**, 41–95 (1994).
114. C. Wechter, M. Sleszynski, J.J. O'Dea, J. Osteryoung, 'Anodic Stripping Voltammetry with Flow Injection Analysis', *Anal. Chim. Acta*, **175**, 45–53 (1985).
115. J. Alpízar, A. Cladera, V. Cerdá, E. Lastres, L. Garcia, M. Caduceus, 'Simultaneous Flow Injection Analysis of Cadmium and Lead with Differential Pulse Voltammetric Detection', *Anal. Chim. Acta*, **340**, 149–158 (1997).
116. F. Zhu, J.W. Ironstone, M.R. Ruegnitz, 'High Throughput Fast-scan Anodic Stripping Voltammetry in a Microflow System', *Anal. Chem.*, **69**, 728–733 (1997).
117. L.A. Mahoney, J. O'Dea, J.G. Osteryoung, 'Development and Characterization of an Automated Flow

System for Voltammetric Analysis', *Anal. Chim. Acta*, **281**, 25–33 (1993).

118. R. Samuelsson, J. Osteryoung, 'Determination of *N*-Nitrosamines by High Performance Liquid Chromatographic Separation with Voltammetric Detection', *Anal. Chim. Acta*, **123**, 97–105 (1981).

119. L.A. Mahoney, J. O'Dea, J.G. Osteryoung, 'Determination of Charge Transfer Parameters Using an Automated Flow System', *J. Electroanal. Chem.*, **366**, 81–92 (1994).

Scanning Tunneling Microscopy, In Situ, Electrochemical

Kingo Itaya
Tohoku University, Sendai, Japan

List of selected abbreviations appears in Volume 15

This article describes the in situ scanning tunneling microscopy (STM) operated at electrode–electrolyte interfaces. It is demonstrated that in situ STM makes it possible to monitor with atomic resolution a wide variety of electrode processes such as the adsorption of inorganic and organic species and the dissolution and deposition of metals and semiconductors. Owing to limitations on space, the focus is on selected topics and mainly on our own experimental results.

1 INTRODUCTION

Of the phenomena which occur at the interface between a solid and a liquid, common examples include the deposition and corrosion of metals, the charging and discharging of storage batteries, and the wet processing of semiconductor devices. Such processes, and many others of a similar nature, involve electrochemical (EC) oxidation–reduction reactions that take place at solid–electrolyte interfaces. Until recently, there had been few in situ methods available for the structural determination of an electrode surface, in solution, at the atomic level. Atomic level information had previously been acquired only via surface spectroscopic techniques under ultrahigh vacuum (UHV).[1,2]

However, since its invention by Binnig and Rohrer,[3] STM was immediately established as an invaluable and powerful surface analysis technique with atomic resolution in UHV. Belatedly, but assuredly, developments in STM operated at solid–liquid interfaces led to its valuation as arguably the premier technique for atomic-level surface structural investigations of chemical processes taking place at solid–liquid interfaces. It has been demonstrated that in situ STM makes it possible to monitor, under reaction conditions, a wide variety of electrode processes such as the adsorption of inorganic and organic species, the reconstruction of electrode surfaces, and the dissolution and deposition of metals and semiconductors. Several review articles on in situ STM and related techniques such as in situ atomic force microscopy (AFM) have been published.[4–7] The reviews by Gewirth and Niece[6] and the present author[7] are the most comprehensive in terms of results obtained on various substrates of metals and semiconductors.

This article describes the current status of in situ STM. Owing to limitations on space, the focus is on selected topics and mainly on our own experimental results. Experimental procedures are only briefly described, because detailed reviews on this aspect have already been published.[5]

2 EXPERIMENTAL ASPECTS

2.1 Principle of Scanning Tunneling Microscopy

In conventional STM operated in vacuum or air, a voltage (V) is applied between the substrate and tip electrodes as shown schematically in Figure 1. The tunneling current density (i_T) can be expressed by Equation (1) when a low-bias voltage is applied:[8]

$$i_T = \frac{e^2}{h^2}\frac{(2m\bar{\phi})^{1/2}}{\Delta s} V \exp(-A\bar{\phi}^{1/2}\Delta s) \qquad (1)$$

where e is the charge of an electron, h is Planck's constant, $\bar{\phi}$ is the mean barrier height and Δs is the distance between the two electrodes. A is defined by Equation (2):

$$A = 4\pi\frac{(2m)^{1/2}}{h} \qquad (2)$$

where m is the mass of an electron. As can be seen in Equation (1), the tunneling current is exponentially dependent on the width of the potential barrier (Δs) and the square root of the mean barrier height. This characteristic exponential dependence allows STM to achieve high resolution on the z-axis. For typical metals ($\bar{\phi} = 4{-}5\,\text{eV}$) a predicted change in the tunneling current (i_T) of one order of magnitude with a change $\Delta s = 0.1\,\text{nm}$ has been verified. If the tunneling current is kept constant to within 2%, then the tunneling gap (Δs) remains constant to within 0.001 nm.

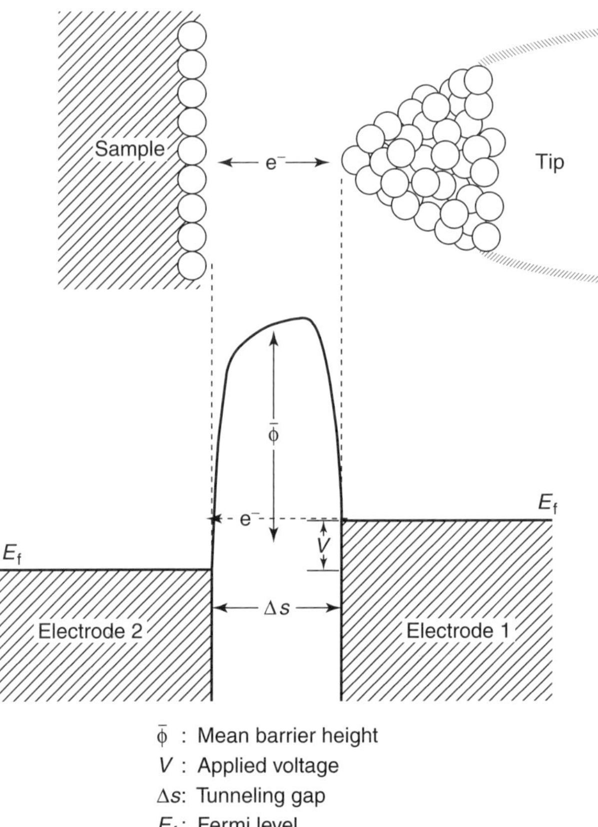

Figure 1 Tunneling barrier between two metal electrodes (substrate and tip).

2.2 Principle of In Situ Scanning Tunneling Microscopy

The review by Siegenthaler describes a detailed comparison between various types of electric circuits to control the electrode potentials of the tunneling tip and the substrate independently using the so-called bipotentiostat.[5] Figure 2 illustrates the apparatus for in situ STM with the four-electrode configuration and shows the EC cell. Using a bipotentiostat, the electrode potentials of the substrate [working electrode (WE) one (WE_1)] and the tunneling tip (WE_2) can be controlled independently with respect to a reference electrode (RE). The EC current (i_F) flowing through the substrate and the counter electrode (CE) can be monitored from the output of a current follower. The tunneling current (i_T) can be measured by the other amplifier. The potential difference between WE_1 and WE_2 is equivalent to the bias voltage (V) in Figure 1.

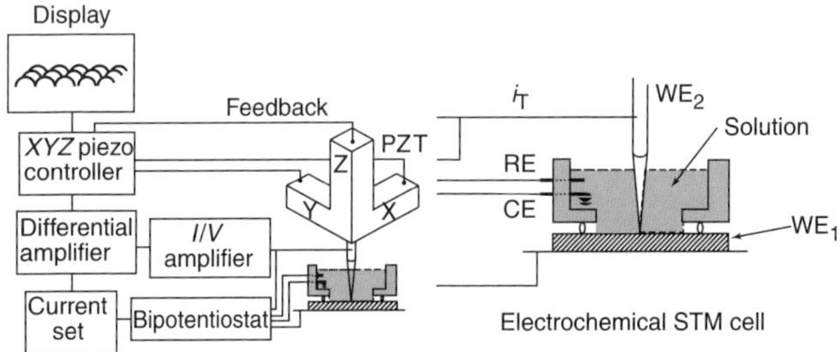

Figure 2 Apparatus for in situ STM with the four-electrode configuration.

For references see page 10066

The side wall of the tip must be isolated in order to reduce the background EC current flowing through the tip. Soft glass, organic polymers, and Apiezon wax have been used. Details of tip-coating methods have been described by Siegenthaler.[(5)]

2.3 Preparation of Well-defined Electrode Surfaces

As a fundamental basis for all STM studies, electrode–electrolyte interfaces must be prepared reproducibly, and methods must be established to observe these interfaces accurately. Well-defined single crystalline surfaces must be exposed to solution in order to understand surface structure–reactivity relationships on the atomic scale. It is still difficult to elucidate EC reactions on the atomic scale using polycrystalline electrodes. Efforts have succeeded in producing extremely well-defined, atomically flat surfaces of various electrodes made of noble metals, base metals, and semiconductors without either oxidation or contamination in solution.

2.3.1 Flame-annealing and Quenching Method

A unique and very convenient way to expose well-defined clean Pt to aqueous solution was proposed by Clavilier et al. in 1980, in which mechanically exposed single-crystal Pt was annealed in an oxygen flame and quenched in pure water.[(9)] They also established a method for preparing a single-crystal Pt electrode by melting a Pt wire in a flame. This technique was extended by Hamelin for Au,[(10)] by Motoo and Furuya for Ir,[(11)] and by us for Rh and Pd.[(12)]

Figure 3 shows typical examples of a cyclic voltammogram (CV) of the three low-indexed Pt surfaces in sulfuric acid solution obtained in our laboratory. The results show that the hydrogen adsorption–desorption reaction is a very structure-sensitive reaction on the Pt surfaces, as indicated by the different shapes and peak positions for the CVs of the different crystalline faces. Although Clavilier et al. quantitatively analyzed the binding sites of hydrogen using systematically prepared stepped surfaces,[(13)] more recent investigations using a CO replacement technique clearly indicated that the charges shown in Figure 3 include a significant contribution of the adsorption and desorption of sulfate/bisulfate.[(14)]

Nevertheless, direct evidence to support the existence of a well-defined surface in solution was presented by us in 1990 using in situ STM.[(15)] Figure 4(a) shows our first

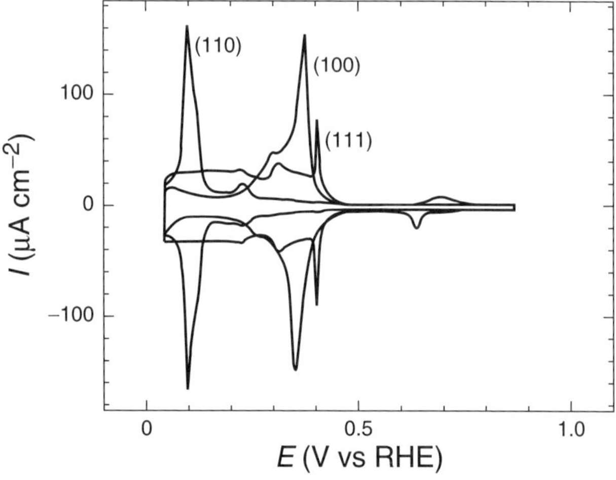

Figure 3 CVs of Pt(111), Pt(110), and Pt(100) in 0.5 M H_2SO_4. Scan rate: 50 mV s^{-1}. RHE, reversible hydrogen electrode.

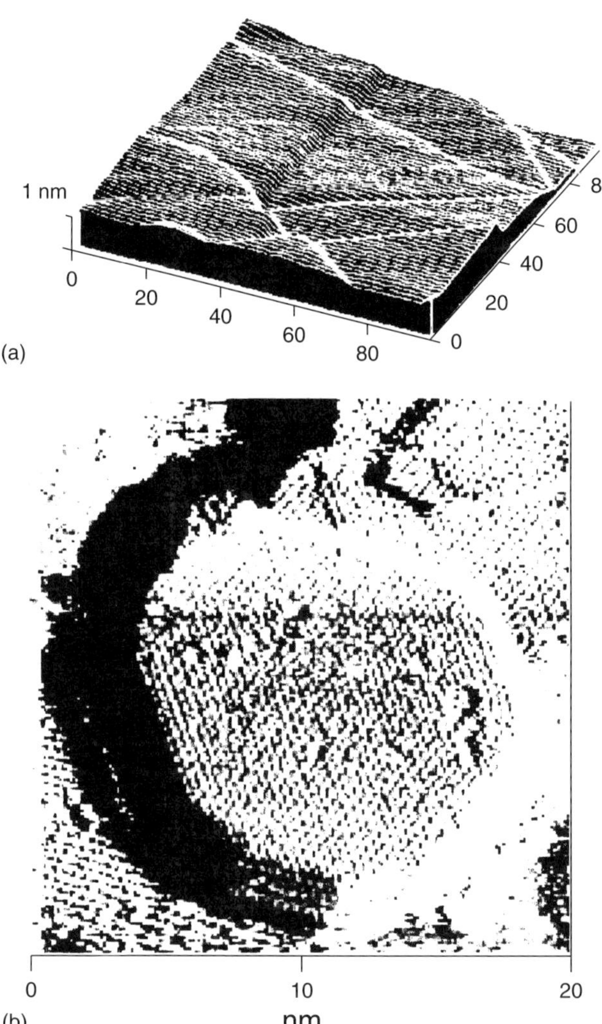

Figure 4 In situ STM images of flame-annealed Pt(111) in solution. (Figure 4(a) reprinted from K. Itaya, S. Sugawara, K. Sashikata, N. Furuya, 'In Situ Scanning Tunneling Microscopy of Platinum(III) Surface with the Observation of Monoatomic Steps', *J. Vac. Sci. Technol.*, **A8**, 515–519,[(15)] Copyright (1990) American Vacuum Society. Figure 4(b) reprinted from S. Tanaka, S.-L. Yau, K. Itaya, 'In-situ Scanning Tunneling Microscopy of Bromine Adlayers on Pt(111)', *J. Electroanal. Chem.*, **396**, 125–130,[(16)] Copyright (1995), with permission from Elsevier Science.)

STM image of a flame-annealed Pt(111) in sulfuric acid solution. The height of each step is ca. 0.23 nm, in accord with the monatomic step height of 0.238 nm on a Pt(111) surface. The monatomic steps observed on the surface are usually located on nearly parallel straight lines or form an angle of 60°, as expected for a surface with threefold symmetry. The terraces seem to be atomically flat. Later it was shown that the terrace was composed of Pt atoms forming a (1×1) structure as shown in Figure 4(b).[16] On the upper and lower terraces, the Pt$(111-1 \times 1)$ structure was clearly discerned at potentials near the hydrogen evolution reaction. The nearest-neighbor spacing and corrugation height were 0.28 and 0.03 nm, respectively.

Flame-annealed Au single crystals have been more frequently used for various studies including the potential-induced reconstruction investigated by several groups.[17] However, it must be emphasized that the flame-annealing method can be applied only to Au, Pt, Rh, Pd, Ir, and possibly Ag.

2.3.2 Ultrahigh Vacuum Electrochemical Methods

It is well established that clean surfaces are exposed in UHV by cycles of Ar-ion bombardment and high-temperature annealing. The surface structure and composition are usually determined by low-energy electron diffraction (LEED) and Auger electron spectroscopy (AES). By using an ultrahigh vacuum electrochemical (UHVEC) system, in which a chamber for EC measurements was interfaced to a UHV apparatus, well-defined substrates can be transferred into an EC cell in a purified Ar atmosphere. This experimental procedure was successfully applied to various metals such as Pt, Au, Pd, and Rh.[1,2] However, for some metals, such as Ni, the oxidation of the surface took place in the EC chamber before immersion of the electrode in electrolyte solutions owing to the presence of trace amounts of oxygen and water vapor.[18] The same difficulty was encountered with Cu electrodes.[19] It is clear that the problem of substrate oxidation of reactive metals occurring during the immersion and emersion processes is still unsolved in the UHVEC method.

2.3.3 Iodine–Carbon Monoxide Replacement Technique

It is important to note that iodine adlayers are known to protect highly sensitive surfaces of metal single crystals from oxidation and contamination in the ambient atmosphere, providing easy preparation and handling of well-defined surfaces during EC measurements.[1,2] The iodine–CO replacement is known to be a method for exposing well-defined and clean surfaces of such electrodes as Pt and Rh in solution.[20,21] The adsorbed iodine on these surfaces can be replaced by a CO adlayer. Clean surfaces are then exposed in solution by the EC oxidation of CO from the surface.

2.3.4 Electrochemical Etching Method

As described above, the flame-annealing and quenching method can only be applied to limited metals such as Pt, Au, Rh, Pd, and Ir and cannot be used for more industrially important less noble metals such as Ni, Co, Fe, and Cu, because they are heavily oxidized in the flame as well as in air. These metals are also difficult to transfer into electrolyte solutions without oxidation even by using UHVEC.

However, it was recently found that the anodic dissolution of various metals and semiconductors occurs only at the step edge under carefully adjusted EC conditions, resulting in atomically flat terrace-step structures. Although the etching method has not yet been well recognized as a promising method for exposing well-defined surfaces of various metals, we demonstrate in this article that layer-by-layer dissolution occurs on various semiconductors, such as Si, GaAs, and InP, resulting in the formation of atomically flat terrace-step structures. It was also found that the layer-by-layer dissolution occurs on various metals such as Ni, Ag, Co, Pd, and Cu.[7]

3 STRUCTURE OF SPECIFICALLY ADSORBED ANIONS

The adsorption of anions such as iodide, bromide, cyanide, and sulfate/hydrogensulfate on electrode surfaces is currently one of the most important subjects in electrochemistry. It is well known that various EC surface processes, such as the underpotential deposition (UPD) of hydrogen and metal ions are strongly affected by coadsorbed anions.[1,2] In particular, structures of the iodine adlayers on Pt, Rh, Pd, Au, and Ag surfaces have been extensively investigated using UHVEC techniques such as LEED.[1] For example, the commensurate $(\sqrt{3} \times \sqrt{3})R30°$, (3×3), and $(\sqrt{7} \times \sqrt{7})R19.1°$ adlattices were found to form on the well-defined Pt$(111-1 \times 1)$ surface, depending on the electrode potential and pH of the solution.[22] More recently, these structures were confirmed by STM in both air[23] and solution.[24] In contrast to Pt(111), only one phase of the commensurate $(\sqrt{3} \times \sqrt{3})R30°$ structure was observed on Pd(111) and Rh(111) surfaces with in situ STM.[25,26]

On the other hand, it has recently been recognized that iodine adlayer structures are more complicated on Au and Ag surfaces. Although several discrepancies about the iodine adlayer structure on Au(111) [I–Au(111)] are found in the literature,[27] surface X-ray scattering (SXS) studies carried out by Ocko et al. revealed

For references see page 10066

structural changes of I–Au(111) in KI solution.[28] They found an increasing degree of compression, the so-called electrocompression, of the iodine adlattice with increasing iodine coverage and electrode potential. Instead of commensurate structures found on Pt(111), Rh(111), and Pd(111) as described above, they proposed that the iodine adlayer should be characterized as two distinct series of incommensurate adlattices, a centered rectangular phase and a rotated hexagonal phase.[28] We have recently reported the structures of I–Au(111) in KI solution determined by both ex situ LEED and in situ STM,[27] which agree with Ocko et al.'s SXS results. Similar electrocompression was also found recently on Ag(111) using the LEED and in situ STM techniques.[29] Our results clearly demonstrate that the complementary use of LEED and in situ STM is a powerful technique for determining atomic structures of iodine adlayers on single-crystal electrodes.

3.1 Iodine Adlayers on Platinum(111)

The objective of this section is to describe the potential dependence of the structure of iodine on Pt(111). Hubbard et al. have extensively investigated the structure of the iodine adlayer formed on Pt(111) in aqueous iodide solutions using the UHVEC technique.[1,22] They demonstrated that the adlayer structures, mainly the commensurate (3×3) and $(\sqrt{7} \times \sqrt{7})R19.1°$, were formed on the well-defined Pt(111–1 × 1) surface in the double-layer potential range, depending on the electrode potential and pH of the solution. For example, in a solution containing 10 mM $KClO_4$ and 0.1 mM KI, adjusted to pH 4 with HI, they found the (3×3) and $(\sqrt{7} \times \sqrt{7})R19.1°$ structures in anodic and cathodic potential ranges, respectively [see Figure 4 in Hubbard et al.[22]].

Although previous STM studies revealed various atomic structures on Pt in air and in solution, no direct in situ STM investigation has been carried out for the structural change expected from the results reported by Hubbard et al. However, as described in our recent paper,[30] the structural transformation induced by changing the electrode potential did occur, but was surprisingly slow. Nearly perfect (3×3) and $(\sqrt{7} \times \sqrt{7})R19.1°$ adlayers could be prepared by immersion of the electrode in a solution containing iodide ions at anodic and cathodic potentials, which is consistent with the previous result.[22] However, in situ STM showed that both of the structures co-exist on the Pt(111) surface, when the clean Pt(111) electrode was immersed at potentials in the middle of the double-layer potential range. A typical example is shown in Figure 5(a), in which a $(\sqrt{7} \times \sqrt{7})$ domain can be clearly seen at the center of the image, surrounded by (3×3) domains.

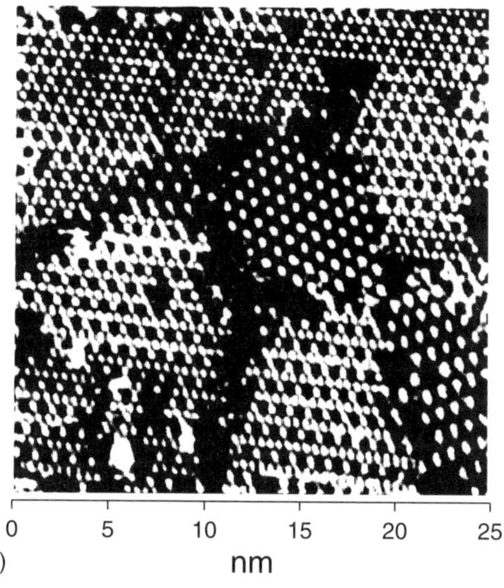

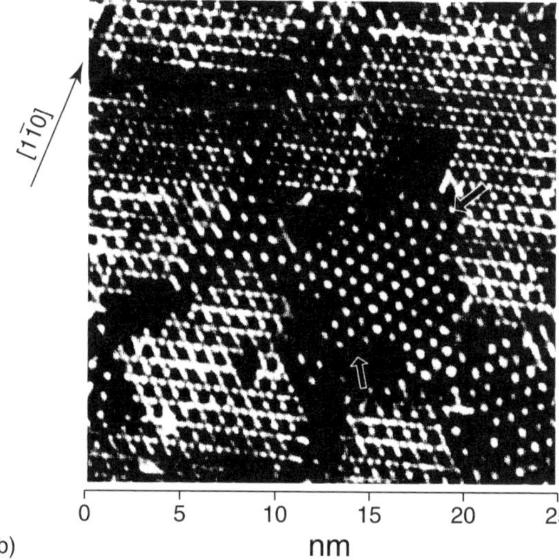

Figure 5 Adlayer structures of iodine on Pt(111): (a) at 0.45 V; (b) 10 min after the potential step from 0.45 to 0.15 V. (Reprinted with permission from J. Inukai, Y. Osawa, M. Wakisaka, K. Sashikata, Y.-G. Kim, K. Itaya, 'Underpotential Deposition of Copper on Iodine-Modified Pt(111): In Situ STM and Ex Situ LEED Studies', *J. Phys. Chem.*, **102**, 3498–3505,[30] Copyright (1998) American Chemical Society.)

The potential-dependent structural change was directly probed by a time-dependent in situ STM experiment. When the electrode potential was stepped to the cathodic potential limit, the $(\sqrt{7} \times \sqrt{7})$ structure was expected to appear upon consuming the (3×3) domains according to previous work.[22] When the electrode potential was stepped after the acquisition of the image shown in Figure 5(a), the image shown in Figure 5(b) was obtained after 10 min, indicating that the interconversion between the two structures was very slow. Only a few iodine atoms marked by arrows were incorporated into the $(\sqrt{7} \times \sqrt{7})$ domain.

The above result provides direct evidence that the surface diffusion of iodine atoms is very slow on Pt(111), suggesting that the iodine atoms are attached on Pt(111) through a strong chemical bond. The initially formed structure was almost insensitive to changes in the electrode potential under our experimental conditions.[30]

3.2 Iodine on Gold(111) and Silver(111)

Although the structures of the iodine adlayers are more complicated on Au(111) and Ag(111), in situ SXS studies by Ocko et al. revealed a series of I–Au(111) adlattices.[28] The adlattice constants varied continuously with the electrode potential in each of the two-dimensional phases designated by Ocko et al., the rectangular $c(p \times \sqrt{3})$ phase and the rotated hexagonal phase.[28]

We have extensively investigated the same system by the complementary use of in situ STM and ex situ LEED. Figure 6 shows a CV of a well-defined Au(111) in 1 mM KI. A pair of anodic–cathodic peaks associated with adsorption–desorption of iodine appear below 0 V vs Ag/AgI. The small pair of peaks at 0.5 V vs Ag/AgI is due to a phase transition between the two phases described in detail below. The transition between these two phases occurred at ca. 0.5 V, where a small reversible peak appears in the voltammogram in Figure 6. Adatoms were at the same height with atomic corrugation about 0.03–0.04 nm at potentials more negative than 0.4 V. The I adlattice obtained at −0.2 V possessed $(\sqrt{3} \times \sqrt{3})R30°$ symmetry with a characteristic interatomic distance of 0.50 nm. No distortion from the threefold symmetry was observed. A positive shift of the electrode potential

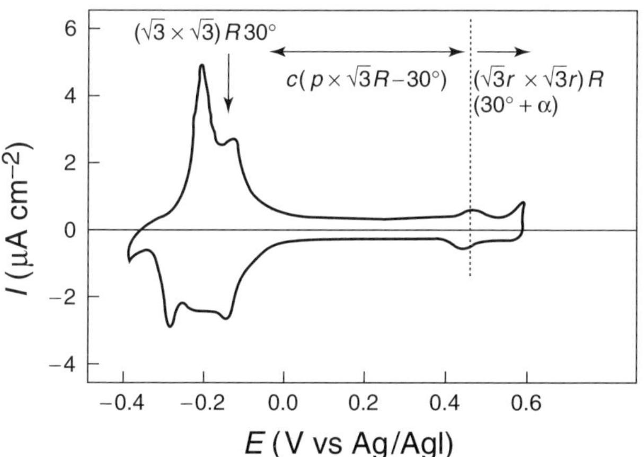

Figure 6 CV of Au(111) in 1 mM KI at a scan rate of 50 mV s^{-1}. (Reproduced from T. Yamada, N. Batina, K. Ogaki, S. Okubo, K. Itaya, 'Lateral Structure of Iodine Adlattices on Au(III) and Ag(III) Electrodes', in *Electrode Processes VI, Proceedings of the Sixth International Symposium*, eds. A. Wieckowski, K. Itaya, 1996, by permission of the Electrochemical Society, Inc.)

resulted in the formation of a more densely packed adlattice. The spacing of atoms in the $\sqrt{3}$ direction remained unchanged within the experimental error, regardless of the electrode potential between 0 and 0.4 V. Images obtained at more positive potentials contain more atoms in the direction perpendicular to the $\sqrt{3}$ direction, resulting in a uniaxial compression as described by Ocko et al. using SXS.

In the phase observed at potentials more positive than 0.5 V, the atomic adlattices of I possess a true sixfold symmetry. The nearest I–I distance was smaller than that of $(\sqrt{3} \times \sqrt{3})R30°$ (0.50 nm), and the entire lattice seemed to be rotated by several degrees with respect to the $(\sqrt{3} \times \sqrt{3})R30°$. This type of adlattice has been denoted as rot–hex, $(\sqrt{3}r \times \sqrt{3}r)R(30° + \alpha)$ by Ocko et al.[28] It is interesting that the adlattices are furthermore modulated with periodically arranged surface features. These features, namely groups of slightly elevated I atoms, are interpreted as Moiré patterns resulting from the mismatch between the adlattice and the lattice of Au(111).

The Moiré pattern can be analyzed by simulation to determine the adlattice constant.[31,32] The STM images were simulated by computer calculation based on a simple "hard-ball contact model". Various compression ratios (0–15% subtracted from the values for $(\sqrt{3} \times \sqrt{3})R30°$) and rotation angles [0–5° from $(\sqrt{3} \times \sqrt{3})R30°$] were tested. The rot–hex I–I distance obtained by this simulation varied from 0.45 to 0.43 nm and the rotation angle from 2 to 5°, the variations corresponding to the change of potential from 0.45 to 0.55 V.

We also carried out quantitative analysis of LEED patterns using a UHVEC system as described in our previous papers.[27,31,32] A $c(p \times \sqrt{3}R - 30°)$ structure was found in the potential range between −0.1 and 0.4 V, as shown in Figure 6. The variation of the adlattice constants is only along a particular direction of the Au atomic rows. As the largest value of p is equal to 3, at which the adlattice is identical with $(\sqrt{3} \times \sqrt{3})R30°$, a value of p smaller than 3 signifies compression of the $(\sqrt{3} \times \sqrt{3})R30°$ structure only in the direction of the Au(111) atom row. Hence the variation of p is referred to as uniaxial compression.

The lattice parameters determined by in situ STM and ex situ LEED are fairly consistent with those obtained by SXS, which is usually believed to give the most accurate values. Nevertheless, our results are consistent with those obtained by SXS, demonstrating that the complementary use of in situ STM and ex situ LEED is a powerful technique for characterizing the atomic structure of iodine on Au(111).

The same phase transition from the $c(p \times \sqrt{3} - R30°)$ to the rotated hexagonal structures was found and characterized by in situ STM and ex situ LEED for an iodine

For references see page 10066

adlayer on Ag(111) in an HI solution.[29] Our result is consistent with that obtained using SXS carried out by Ocko et al.[33] LEED patterns with six splitting spots were observed at potentials before the bulk formation of AgI, indicating that the rotated hexagonal structure was stable even in UHV.[29] The iodine adlayer on Ag(111) was complicated in an alkaline solution, showing several structures including square ($\sqrt{3} \times \sqrt{3}R - 30°$) and ($\sqrt{3} \times \sqrt{3})R30°$,[29] suggesting that there is a remarkable pH dependence of the structure of iodine on Ag(111).

3.3 Sulfate/Hydrogensulfate on Gold(111), Platinum(111), and Rhodium(111)

It has been demonstrated by several groups that in situ STM can be used to visualize adsorbed sulfate/hydrogensulfate species on Au(111), Pt(111), and Rh(111).

3.3.1 Gold(111)

An ordered structure with a ($\sqrt{3} \times \sqrt{7}$) symmetry was first observed for the adsorbed sulfate/hydrogensulfate on Au(111) in sulfuric acid by Magnussen et al.,[34] who proposed a model structure based on the assumption that the adsorbed species is hydrogensulfate, not sulfate, with a surface coverage of 0.4. More recently, Weaver et al. reported STM images with the same symmetry of ($\sqrt{3} \times \sqrt{7}$) on Au(111)[35] as that observed by Magnussen et al. However, they proposed a possibility of incorporation of hydronium cations in the ordered sulfate adlayer by taking into account the result that the surface coverage of sulfate on Au(111) determined by chronocoulometry and radiochemical assay is 0.2.[36] Note that this surface coverage is half of that for the structure proposed by Magnussen et al. as described above.

3.3.2 Platinum(111)

Stimming et al. found by in situ STM that adsorbed sulfate ions form the same adlayer structure as that found on Au(111).[37] Ordered domains with ($\sqrt{3} \times \sqrt{7}$) symmetry appeared in the potential range 0.5–0.7 V vs a RHE in 0.05 M H_2SO_4. As shown in Figure 3, only the (111) surface shows the characteristic butterfly peaks at potentials slightly more negative than 0.5 V. Their STM observations confirmed that the butterfly peaks are due to the adsorption and desorption of sulfate ions as indicated with the CO replacement technique by Clavilier et al. as described above.[14] STM images obtained on Pt(111) were interpreted in terms of the coadsorption of sulfate anions and water.

3.3.3 Rhodium(111)

High-resolution STM imaging conducted on atomically flat terraces at 0.5 V in H_2SO_4 readily discerned atomic features as shown in Figure 7(a) obtained near the step edges.[38] The image areas include three terraces with monatomic steps. It is clearly seen that parallel atomic rows in each domain are located in the directions forming angles of nearly 60° or 120°. It is also recognized that individual bright spots exist very near the monatomic step. This observation strongly indicates that the entire surface of Rh(111) is almost completely covered by adsorbed sulfate ions even very near the end of the terraces.

Figure 7(b) presents an STM image showing a more detailed internal structure acquired in an area where a single domain appeared on a wide terrace. It can be seen that there are two different parallel rows with a 30° rotation relative to the underlying Rh lattice. One appears as bright spots. The observed atomic distance in these bright rows along the A direction is equal to 0.46 nm. The average distance between neighboring bright rows is ca. 0.7–0.73 nm. The interatomic distance of 0.74–0.75 nm observed along the B direction in this particular STM image in Figure 7(b) is slightly larger than that of the $\sqrt{7}a_{Rh}$ (0.707 nm), probably owing to a small thermal drift during the acquisition of the image. However, it was ascertained that the distance along the B direction is very close to the $\sqrt{7}a_{Rh}$ based on the averaging of all atomic images obtained in this study. The angle between the directions marked by arrows A and B in Figure 7(b) is ca. 72°. The above results strongly indicate that the unit cell can be defined by the so-called ($\sqrt{3} \times \sqrt{7}$) structure. The rows along the direction marked by the arrow B are constituted of alternative bright and dark spots in the image shown in Figure 7(b). The dark spots appeared almost at the center between neighboring bright spots in the direction of B. Magnussen et al. have also found the position for the darker spots equidistant between two brighter spots along the B direction.[34]

The results described above are almost the same as those reported for Au(111) by Magnussen et al.[34] and other investigators[35] and those observed on Pt(111).[37] Magnussen et al. proposed a model structure with a unit cell, the so-called ($\sqrt{3} \times \sqrt{7}$), for the adlayer of hydrogensulfate (HSO_4^-) on Au(111). Both bright and dark spots were interpreted as hydrogensulfate ions adsorbed on Au(111). Therefore, the surface coverage of this proposed structure is 0.40.

According to the coverage value of ca. 0.2 obtained on Rh(111) by Zelenay and Wiekowski,[39] it is reasonable to expect that only the bright spots in the STM images observed on Rh(111) correspond to the adsorbed sulfate or hydrogensulfate. If sulfate or hydrogensulfate is assumed to be also trigonally coordinated on Rh(111),

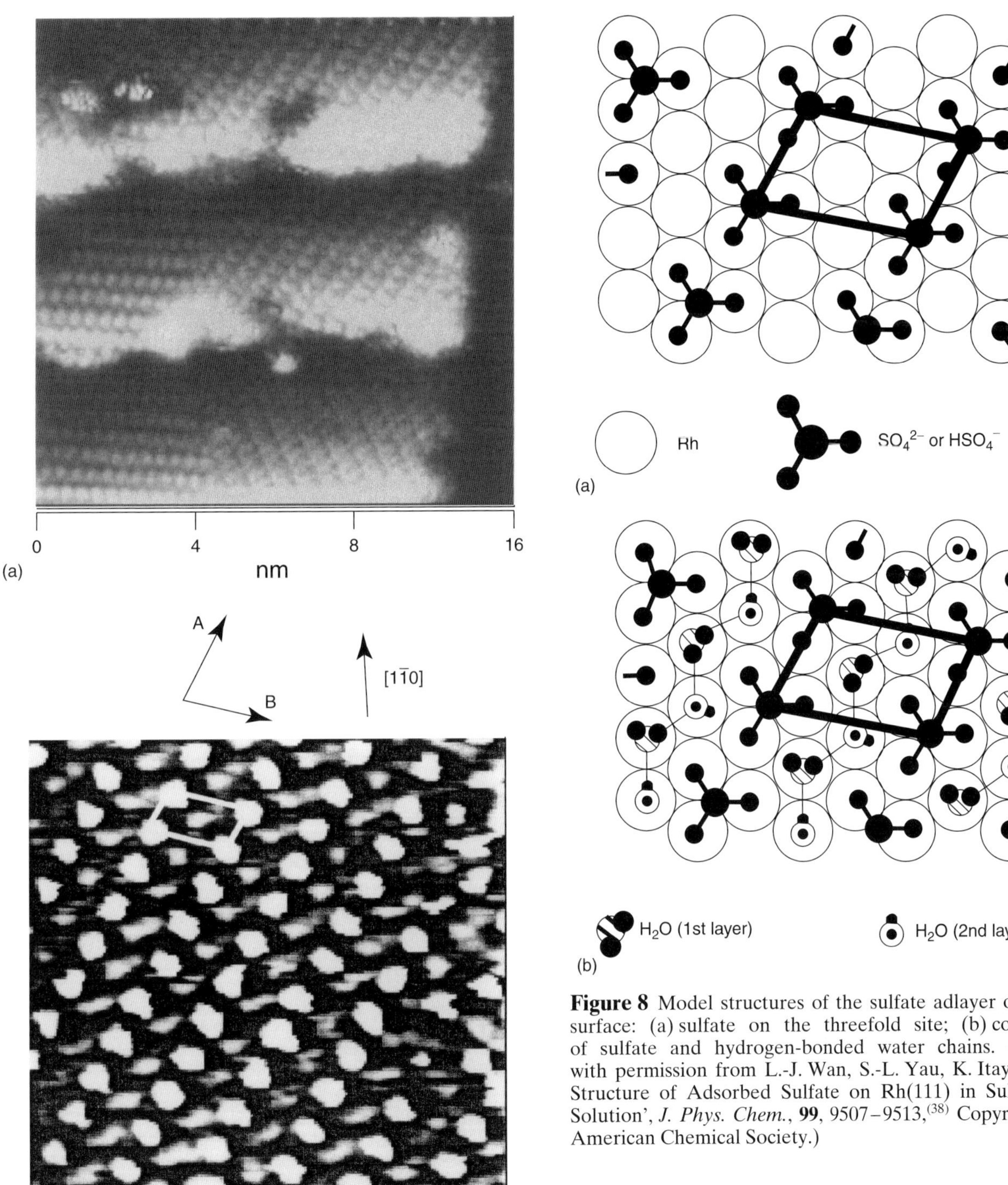

Figure 7 High-resolution STM images of sulfate adlayer on Rh(111) obtained in H_2SO_4. (Reprinted with permission from L.-J. Wan, S.-L. Yau, K. Itaya, 'Atomic Structure of Adsorbed Sulfate on Rh(111) in Sulfuric Acid Solution', *J. Phys. Chem.*, **99**, 9507–9513,[38] Copyright (1995) American Chemical Society.)

Figure 8 Model structures of the sulfate adlayer on Rh(111) surface: (a) sulfate on the threefold site; (b) coadsorption of sulfate and hydrogen-bonded water chains. (Reprinted with permission from L.-J. Wan, S.-L. Yau, K. Itaya, 'Atomic Structure of Adsorbed Sulfate on Rh(111) in Sulfuric Acid Solution', *J. Phys. Chem.*, **99**, 9507–9513,[38] Copyright (1995) American Chemical Society.)

a ball-model can be presented as shown in Figure 8(a), where the SO_4^{2-} (or HSO_4^-) is positioned at the threefold hollow sites. It can be seen in Figure 8(a) that the sulfate ions along the $\sqrt{3}$ direction form an almost close-packed row. On the other hand, an open space can be found between neighboring rows of the sulfates. As described above, Weaver et al. proposed a model where coadsorbed hydronium cations exist along the $\sqrt{3}$ direction between neighboring rows of sulfates.[35] The dark spots which appeared in the STM images

For references see page 10066

were assigned to be the coadsorbed hydronium cations. Such cation coadsorption was expected to minimize the coulombic repulsion between adjacent SO_4^{2-} on Au(111). Although the coadsorption of hydronium cations is thought to be a factor which explains the nonuniform interatomic distances of the $(\sqrt{3} \times \sqrt{7})$ structure, it is not clear why the adsorbed sulfates have the different spacings along the $\sqrt{3}$ and $\sqrt{7}$ directions.

In a previous paper, we proposed a new model to explain the nonuniform spacing in the unit cell of $(\sqrt{3} \times \sqrt{7})$.[38] It can be seen in Figure 8(a) that uncoordinated Rh atoms are arranged in a zig-zag form in the $\sqrt{3}$ direction between neighboring rows of the adsorbed sulfates. In the new model shown in Figure 8(b), hydrogen-bonded water chains are simply inserted along the $\sqrt{3}$ direction between neighboring rows of the sulfates. The model shown in Figure 8(b) includes the adsorbed sulfate/hydrogensulfate and hydrogen-bonded water chains formed along the $\sqrt{3}$ direction. The model shows only the first water bilayer. Water molecules in the first layer are bonded directly to Rh atoms at the top site via the oxygen lone pair. For the ice-like lattices on the fcc(111) surfaces, two hydrogen bonds form to oxygen lone pairs of two water molecules in the second layer. It is assumed, however, that only a hydrogen bond forms to an oxygen lone pair of a water molecule in the second layer as shown in Figure 8(b), although it is expected that the water in the second layer forms a hydrogen bond with an oxygen lone pair of sulfate, which is not drawn in Figure 8(b) for the sake of clarity. According to the model, it is possible that the dark spots which appeared in the STM image shown in Figure 7(b) arise from the water molecules in the second layer. Although the model presented here would seem to be equivalent or similar to the model proposed by Weaver et al.,[35] our model more confidently explains the feature of the nonuniform interatomic distances in the $(\sqrt{3} \times \sqrt{7})$ structure. It is also noteworthy that the hydrogen-bonded water chains are expected to form hydronium cations in acidic solutions. If the water molecule in the second layer is protonated to form hydronium cations, the model presented here is equivalent to the model proposed by Weaver et al.[35] We have calculated the heights of the adsorbed sulfate and the water molecule in the second layer for the model shown in Figure 8(b) using a hard-ball contact model, yielding values of 0.35 and 0.30 nm respectively. The outer sphere of the water molecule in the second layer is slightly lower than that of the sulfate. It is interesting to compare these values with the corrugation heights described above, although it is true from a theoretical point of view that corrugation heights observed in STM images should arise from electronic factors, such as wave functions of adsorbates rather than spatial structural factors. We believe that a similar coadsorption of water chains and sulfate might occur on Au(111) and also possibly on Pt(111).

The adlayer structure almost abruptly disappeared at potentials near the hydrogen evolution reaction, showing the Rh(111–1 × 1) structure, indicating that the adsorbed sulfate is completely desorbed at the potential of the hydrogen evolution.

Nevertheless, it is now clear that sulfate/hydrogensulfate form adlayers with the same structure and symmetry on at least three different substrates, Au(111), Rh(111), and Pt(111). It is noteworthy that the atomic diameters of Au, Rh, and Pt are 0.289, 0.268, and 0.278 nm, respectively, i.e. the diameter of Rh is the smallest. In general, structures of many adlayers depend strongly on the diameter of substrates. The appearance of the same $(\sqrt{3} \times \sqrt{7})$ structure on the substrates with different lattice parameters might suggest that the coadsorption of sulfate/hydrogensulfate and water illustrated in Figure 8(b) is flexible with respect to the change in lattice parameter of the substrate. Such flexibility may be due to the existence of water molecules between the sulfate/hydrogensulfate chains with a relatively weak hydrogen bonding.

3.4 Cyanide

In situ STM was recently employed by Weaver et al., for the first time, to examine the CN adlayer on Pt(111).[40] It was reported that six CN functional groups form a hexagonal ring with an additional CN in the center of the ring (see Figure 9b).

Our in situ STM observations revealed a new structure for the CN adlayer and the complexation of alkali metal cations such as Na^+ and K^+ with the CN adlayer on Pt(111).[41] An atomically resolved STM image acquired in an alkaline CN solution containing Na^+ cations is shown in Figure 9(a). Figure 9(a) shows an STM image of the hollow hexagonal pattern. This image (5 × 5 nm) was acquired in 0.1 mM NaCN + 0.1 M $NaClO_4$ (pH 9.5) at 0.6 V with a bias voltage of −50 mV and a tunneling current of 20 nA.

Under these experimental conditions, well-arranged hexagonal rings, aligned in a direction 30° rotated from the close-packed directions of Pt(111) lattice, are observed. The 0.95 ± 0.02 nm distance between the nearest-neighbor hexagonal rings, as measured from their centers, is roughly twice as large as the $\sqrt{3}$ lattice spacing of the Pt (0.2778 nm). This ordered atomic feature can be characterized as the $(2\sqrt{3} \times 2\sqrt{3})R30°$ structure. However, Weaver et al. reported an STM image [see Figures 3 and 4 in Stuhlmann et al.[40]] which was interpreted as the $(2\sqrt{3} \times 2\sqrt{3})R30°$–7CN structure with cyanides bound in symmetric top sites surrounded by hexagonal rings of near-top CN. It is important to see in Figure 9(a) that the center spots are now essentially

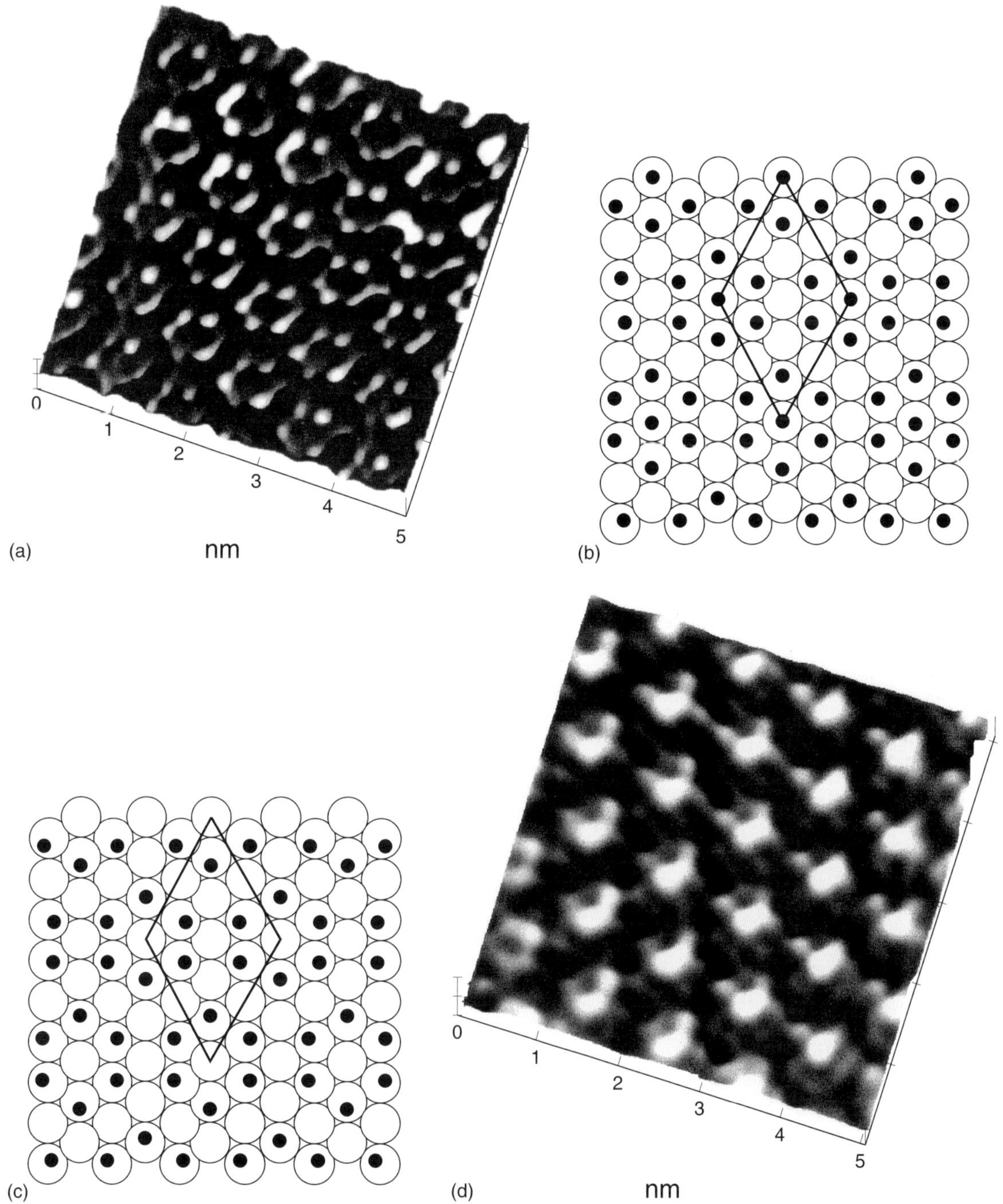

Figure 9 (a) High-resolution STM image of the hollow CN hexagonal arrangement. Two ball models of the $(2\sqrt{3} \times 2\sqrt{3})R30°$ structure are shown in (b) and (c). (d) STM image in the solution containing K^+ ions. (Reprinted with permission from Y.-G. Kim, S.-L. Yau, K. Itaya, 'Direct Observation of Alkali Cations on Cyanide-modified Pt(111) by Scanning Tunneling Microscopy', *J. Am. Chem. Soc.*, **118**, 393–400,[41] Copyright (1996) American Chemical Society.)

invisible in the image, strongly suggesting that there is no cyanide in the center of the hexagonal ring.

Based on the results described above, two model structures are presented in Figures 9(b) and (c) to outline atomic arrangements of the adsorbed CN on Pt(111). The model structure shown in Figure 9(b) is essentially a replica of the model proposed by Weaver et al.[40] in which the center spot was attributed to the adsorbed CN. However, the image shown in Figure 9(a) suggests that the adsorbed CN is not located in the center of the six-membered ring. Figure 9(c) is a new model where the adsorbed CN in the center is removed.

For references see page 10066

It is interesting that the six-membered ring is similar in structure to crown ethers. Crown ethers are known to complex effectively with alkali metal cations. The configuration of the CN adlayer is such that the C is bound to the Pt electrode with the N facing the solution side. Each nitrogen atom contains a lone pair of electrons which is expected to act as a binding site similar to the oxygen atoms in crown ethers.

It was surprising to find that the bright spots appear in the center of the six-membered ring in a solution containing K^+ as shown in Figure 9(d)[41] It was found that K^+ cations are more strongly bound than Na^+ in the center of the CN ring, because bright spots due to the coordinated Na^+ cations were only sparsely observed in the center of the hexagonal ring of CN. Our STM result for the coordinated K^+ described above is probably the first case to describe an outer Helmholtz layer, because almost all previous STM studies have elucidated the adlayer structure directly attached on the electrode surface.

4 UNDERPOTENTIAL DEPOSITION

The EC adsorption of hydrogen and metals on a foreign metal substrate taking place in a potential region positive to the thermodynamically reversible potential is called the UPD. Particularly the UPD of hydrogen on single-crystal Pt electrodes was intensively and systematically investigated by Clavilier et al.[13] The UPD of a metal, M, on a different metal substrate, Ms, is expected to occur at potentials more positive than the reversible potential for the bulk deposition of M when an interaction between M and Ms is greater than that of Ms–Ms. The UPD process is important in EC reactions, such as metal deposition, as the initial step of a series of reactions and also because of electrocatalytic effects induced by adatoms formed by the process. Although a large number of UPD systems have been investigated by using conventional EC techniques, such as CV, to evaluate the thermodynamics and kinetics, the structural information of UPD layers was first obtained mainly by Hubbard et al. using the UHVEC technique.[1,2] With a considerable amount of previous knowledge available on the UPD phenomenon itself, in situ STM was applied, for the first time, to determine the structure of the adlayer of Cu on Au(111) in sulfuric acid solutions with atomic resolution.

4.1 Copper Underpotential Deposition on Gold(111)

Magnussen et al. reported the first atomic image of a Cu adlayer on Au(111) in sulfuric acid solution.[42] They found a $(\sqrt{3} \times \sqrt{3})R30^\circ$ structure after the first UPD peak which transformed into a second phase of (5×5) structures. However, the appearance of the (5×5) structure was confirmed to be due to chloride contamination,[43] suggesting that the UPD process is extremely sensitive to co-adsorbates.

Figures 10(a) and (b) show CVs of an Au(111) electrode obtained in pure 0.05 M H_2SO_4 solution in the presence of 1 mM $CuSO_4$.[44] The main oxidation peak at 1.25 V and the cathodic peak at 0.82 V are due to

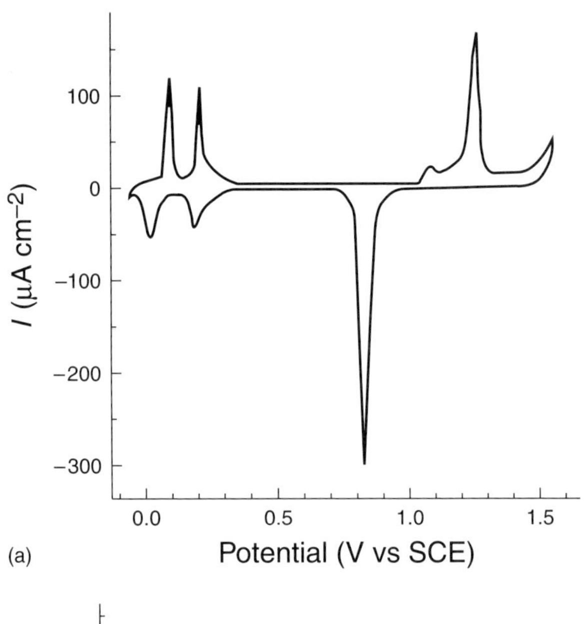

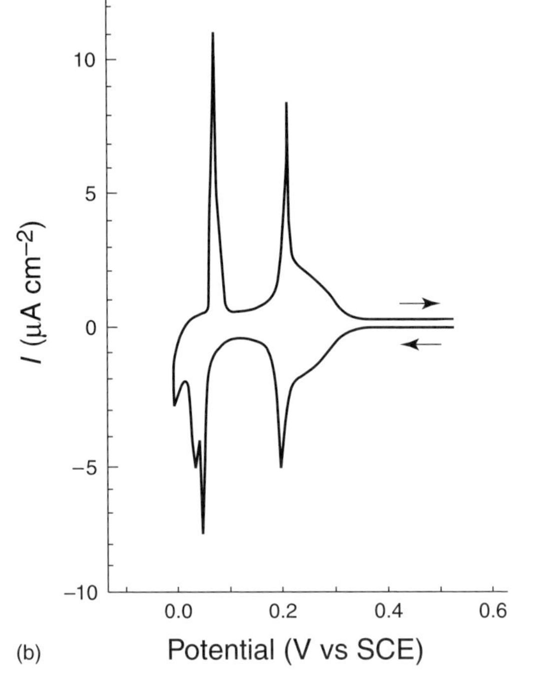

Figure 10 CVs for a Au(111) electrode in 0.05 M H_2SO_4 + 1 mM $CuSO_4$. Scan rate: (a) 20 and (b) 1 mV s^{-1}. (Reprinted from T. Hachiya, H. Honbo, K. Itaya, 'Detailed Underpotential Deposition of Copper on Gold(111) in Aqueous Solutions', *J. Electroanal. Chem.*, **315**, 275–291,[44] Copyright (1991), with permission from Elsevier Science.)

the oxidation of the surface of Au(111) and the reduction of the oxide layers, respectively. On the other hand, two different waves for the UPD of Cu are clearly observed in the potential region between 0.35 and 0 V vs a saturated calomel electrode (SCE) before the beginning of the bulk deposition. Figure 10(b) shows a detailed CV for the UPD observed at a scan rate (v) of $1\,\mathrm{mV\,s^{-1}}$. Two distinctly different processes can be seen clearly in the CV shown in Figure 10(b). The peak current is proportional to the scan rate only up to $5\,\mathrm{mV\,s^{-1}}$, and then is approximately proportional to $v^{1/2}$, suggesting that the UPD of Cu on Au(111) is a surprisingly slow process.

Figure 11(a) shows a high-resolution STM image obtained in 0.05 M H_2SO_4 + 1 mM $CuSO_4$ solution.[44] Although the wide terrace of the Au(111) surface was almost completely covered by the Cu adlayer with the $(\sqrt{3} \times \sqrt{3})R30°$ structure, several types of phase boundary can be seen in Figure 11(a). Figure 11(b) shows a model structure of the phase boundary marked by arrow (a), in which two $(\sqrt{3} \times \sqrt{3})R30°$ domains are shifted by a half position along the direction indicated by the arrow. In the model structure shown in Figure 11(b), it was assumed that the solid circles represent Cu atoms. The same $(\sqrt{3} \times \sqrt{3})R30°$ structure was also found by in situ AFM.[45]

However, the coulometric curve obtained simultaneously with CV shown in Figure 10(b) showed that the ratio of the charges consumed during the first and second UPD processes was roughly 2 : 1, suggesting that the surface coverage of Cu was about 2/3 after the first UPD peak.[44] According to the model structure shown in Figure 11(b), the surface coverage must be 1/3 because of the $(\sqrt{3} \times \sqrt{3})R30°$ structure.

This discrepancy was carefully investigated by Shi and Lipkowski using a chronocoulometric technique. They measured the Gibbs excess of Cu adatoms and that of coadsorbed sulfate (SO_4^{2-}) as a function of the electrode potential, and concluded that Cu adatoms are packed to form a honeycomb $(\sqrt{3} \times \sqrt{3})R30°$ structure with the center of each honeycomb cell occupied by a sulfate ion.[46]

Finally, Toney et al. examined the above system using an SXS technique,[47] and concluded that the Cu atoms form a honeycomb lattice and are adsorbed on threefold hollow sites with sulfate ions located at the honeycomb centers. They also concluded that three oxygens of each sulfate ion bond to Cu atoms. According to all of the results described above, the most reliable model structure can be presented as shown in Figures 12(a) and (b), in which top and perspective views, respectively, are given. This model structure is essentially the same as that shown by Toney et al.[48] According to the model structure, the corrugation observed by in situ STM and AFM must be considered to be due to the coadsorbed sulfate ions, and not the Cu atoms. Note that Blum et al. studied a statistical mechanical model for the UPD of Cu.[48] Their proposed structure is similar to that shown in Figures 12(a) and (b).[48] The study of the UPD of Cu on Au(111) was a very important lesson for understanding limitations and strengths of various in situ techniques. It is clear that STM and AFM cannot distinguish chemical species.

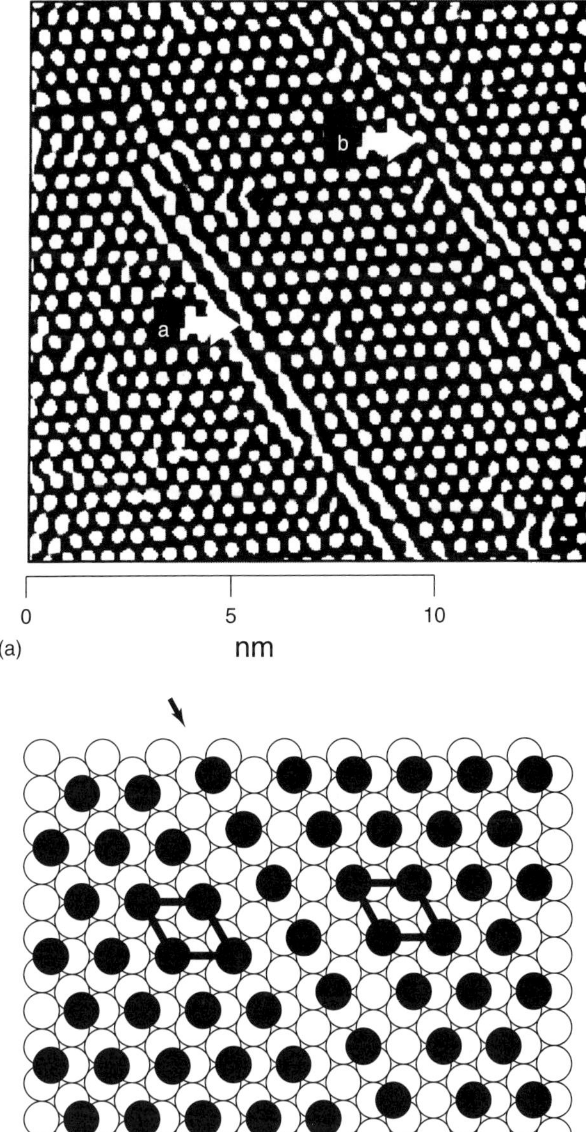

Figure 11 (a) High-resolution STM image and (b) a model structure of a copper adlattice on Au(111). (Reprinted from T. Hachiya, H. Honbo, K. Itaya, 'Detailed Underpotential Deposition of Copper on Gold(111) in Aqueous Solutions', *J. Electroanal. Chem.*, **315**, 275–291,[44] Copyright (1991), with permission from Elsevier Science.)

Review articles are useful for finding the literature on UPD of various metals on different substrates.[6,7] Here

For references see page 10066

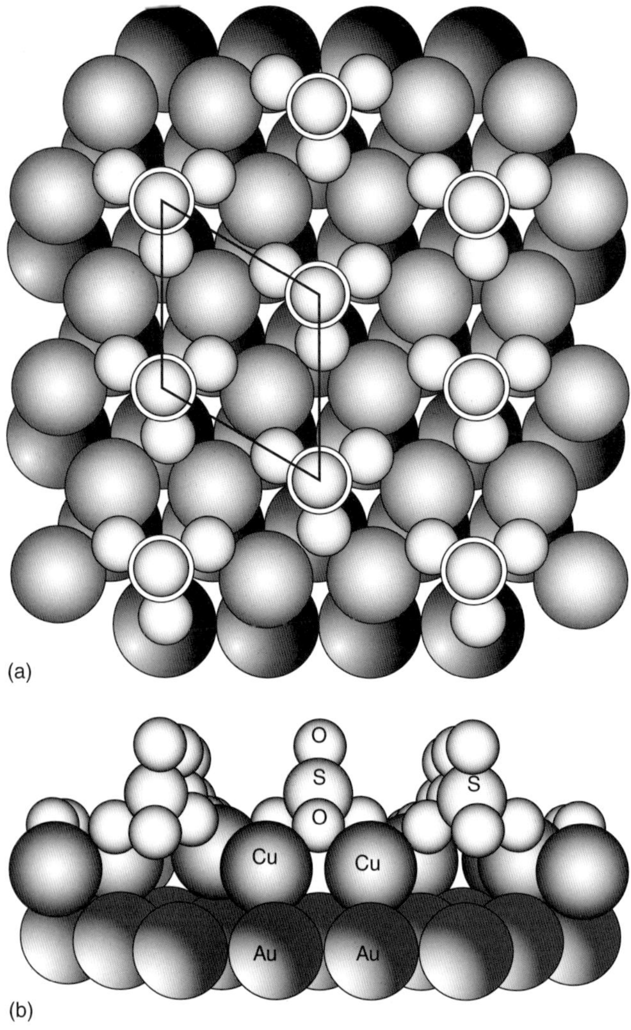

Figure 12 (a) Top and (b) perspective view of the coadsorbed Cu^{2+} and SO_4^{2-} adlayer on Au(111).

we only describe the UPD of Cu on Au(111), which is the most extensively studied system.

5 ADSORPTION OF ORGANIC MOLECULES ON MODIFIED ELECTRODES

STM has also made it possible to determine directly orientations, packing arrangements, and even internal structures of organic molecules adsorbed both on surfaces in UHV and at solid–liquid interfaces. For example, individual molecules and distinguishable molecular shapes of benzene on Pt(111), coadsorbed benzene and CO on Rh(111), naphthalene on Pt(111), and copper phthalocyanine on Cu(100) have successfully been resolved with STM under UHV conditions.(49) These results have stimulated a large number of STM studies of ordered molecular adlayers in UHV and air and at solid–liquid interfaces.

A variety of experimental procedures have been reported for the preparation of ordered molecular adlayers on well-defined substrates including single crystals of metals and layered materials, such as highly ordered pyrolytic graphite (HOPG) and MoS_2.(50) Alkanethiols have been intensively investigated on metals such as Au, because the –SH group is known to be chemically attached to the Au surface through the formation of a covalent bond between S and Au atoms, producing densely packed adlayers. On the other hand, it is well known that simple physical adsorption can also provide ordered adlayers of molecules, such as liquid crystals and *n*-alkanes on inert substrates such as HOPG and MoS_2.(50)

From the EC point of view, the adsorption of organic molecules at electrode–electrolyte interfaces can be considered as one of the most promising approaches not only for the preparation of ordered adlayers, but also for elucidating the role of properties of adsorbed molecules and the nature of electrode–electrolyte interfaces.(1,51) In spite of a large number of reports describing observations by STM and related techniques, such as AFM of adsorbed organic molecules in UHV, air, and organic liquids, only a few in situ STM studies have been carried out for organic molecules adsorbed at electrode–electrolyte interfaces under EC conditions. Recently, pioneering studies have shown high-resolution images of molecules such as DNA bases (adenine, guanine, and cytosine) adsorbed on HOPG and Au(111) in electrolyte solutions.(6,7) In another study, xanthine and its oxidized form(52) and porphyrins(53) were found to form ordered adlayers on HOPG. Further, the order–disorder transition in a monolayer of 2,2′-bipyridine on Au(111) was reported as a function of electrode potential.(54)

Although a number of successful in situ experiments using STM and AFM have been performed to determine the structure of organic adlayers, HOPG, similar layered crystals, and Au electrodes have almost exclusively been used as the substrate. Therefore, the role of the interaction between organic molecules on one hand and substrates on the other in ordering processes is not yet fully understood. We have long been interested in finding a more appropriate substrate to investigate the adsorption of organic molecules. Recently, we disclosed a novel property of iodine-modified electrodes for the adsorption of organic molecules.(55–59) It was found, for the first time, that a water-soluble porphyrine, 5,10,15,20-tetrakis(*N*-methylpyridinium-4-yl)-21*H*,23*H*-porphyrine (TMPyP), formed highly ordered molecular arrays via self-ordering on the iodine-modified Au(111) electrode in $HClO_4$ solution.(55) As described in the preceding section, the iodine adlayers are thought to protect metal surfaces from oxidation and contamination in the ambient atmosphere, providing an easy method of preparation of well-defined surfaces.[1,2] However, it should now be recognized that the I–Au(111) electrode is one of the most promising substrates for the investigation of organic molecules in solution. Indeed, we discovered this electrode, with great generality, to be a suitable one on which to form highly ordered adlayers not only of TMPyP but also various other molecules such as crystal violet.[56] It was also demonstrated in our papers that various iodine-modified metal electrodes such as Ag(111)[58] and Pt(100)[59] can also be employed as a substrate on which to investigate the adsorption of organic molecules. In this section, we briefly summarize in situ STM of organic molecules adsorbed on iodine-modified electrodes.

5.1 5,10,15,20-Tetrakis(*N*-methylpyridinium-4-yl)-21*H*,23*H*-porphyrine on Iodine–Gold(111)

The experimental procedure was fairly simple. A well-defined Au(111) surface prepared by the flame-annealing and quenching method was immersed in 1 mM KI solution for several minutes and then thoroughly rinsed with 0.1 M $HClO_4$ solution. The iodine-modified electrode, thus prepared, was installed in an EC cell containing pure $HClO_4$ solution for in situ STM measurements. Under potential control, the iodine adlayer structures can be determined by in situ STM as described above. After achieving an atomic resolution, a dilute solution of TMPyP was injected into the $HClO_4$ solution. After the addition of TMPyP, STM images for the iodine adlayer usually became unclear within the first 5–10 min, because of the adsorption of TMPyP, and then ordered adlayers became visible, extending over atomically flat terraces.[55]

Molecular orientations, packing arrangements and even internal molecular structures of the adsorbed TMPyP molecules can be seen in high-resolution STM images. Two typical STM images are presented in Figures 13(a) and (b). These high-resolution images directly demonstrate that the flat-lying TMPyP molecule can be recognized in the images as a square with four additional bright spots. The shape of the observed features in the image clearly corresponds to the chemical structure of the TMPyP molecule. The characteristic four bright spots located at the four corners of a square correspond to the pyridinium units of TMPyP. The center to center distance between the bright spots was found to be 1.3 nm measured diagonally, which is nearly equal to the distance between two diagonally located pyridinium units. In addition to the internal structure, the STM image shown in Figure 13(a) reveals details of the symmetry and the packing arrangement. It can be seen that there are three different molecular rows marked by arrows I, II, and III. In the row marked I, all TMPyP molecules show an identical orientation with an intermolecular distance of ca. 1.8 nm. An alternative orientation can be seen along the row II in which every second molecule shows the same orientation. On the other hand, the rotation angle

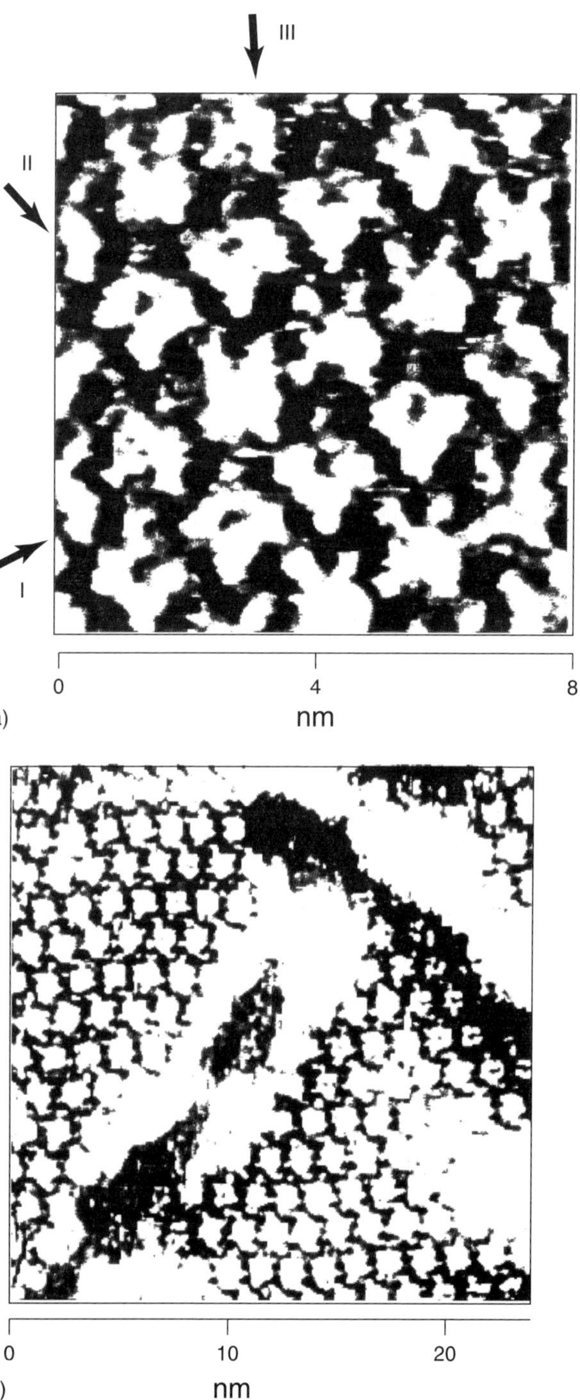

Figure 13 High-resolution STM images of TMPyP on I-modified Au(111) in $HClO_4$. (Reprinted from M. Batina, M. Kunitake, K. Itaya, 'Highly Ordered Molecular Arrays Formed on Iodine-modified Au(111) in Solution: In Situ STM Imaging', *J. Electroanal. Chem.*, **405**, 245–250,[56] Copyright (1996), with permission from Elsevier Science.)

of ca. 45° can be recognized between two neighboring molecules in row III.

The well-ordered TMPyP arrays were found to extend over atomically flat terraces as shown in Figure 13(b), which was obtained in areas involving monatomic step edges. It is clearly seen that the same molecular feature as that seen in Figure 13(a) extends over the lower and upper terraces. The ordered structure could be seen even on the relatively narrow terrace in the upper part of the image. It is also surprising to find that individual TMPyP molecules exist very near the monatomic step. The result shown in Figure 13(b) indicates that the entire surface of the I–Au(111) is almost completely covered by ordered TMPyP molecules even near the end of the terraces. Figures 14(a) and (b) illustrate a structural model showing a top view and a side view of the ordered TMPyP adlayer on the iodine on Au(111), respectively. We have also investigated the adsorption of TMPyP on a well-defined Au(111) in the absence of iodine adlayer.(57) After achieving an atomic resolution for the Au(111–1 × 1) structure in 0.1 M $HClO_4$, a dilute solution of TMPyP was injected into the solution in a manner similar to that described above. Although the TMPyP molecules adsorbed directly on Au(111) could be seen by STM, the adsorbed molecules did not form ordered adlayers. Disordered adlayers formed on bare Au(111) suggest that strong interactions including chemical bonds between the Au substrate and the organic molecules prevented self-ordering processes from occurring, which must involve surface diffusion of the adsorbed molecules. The surface diffusion of the molecules adsorbed on bare Au was found to be very slow. Relatively weak van der Waals-type interactions between the hydrophobic iodine adlayer and the organic molecules could be the key factor promoting self-ordering processes on the I–Au(111) substrate.

It should particularly be emphasized that the iodine layer plays a crucial role in the formation of highly ordered TMPyP arrays. The relatively weak van der Waals-type interaction on the iodine adlayer seems to be a key factor in the formation of ordered molecular arrays of such large molecules. However, the relationship between the TMPyP and iodine adlayer structures is not fully understood, as described above,(57,58) because the iodine adlayer structures on Au(111) and Ag(111) are complicated by a potential-dependent compression in the adlayers.(27,29)

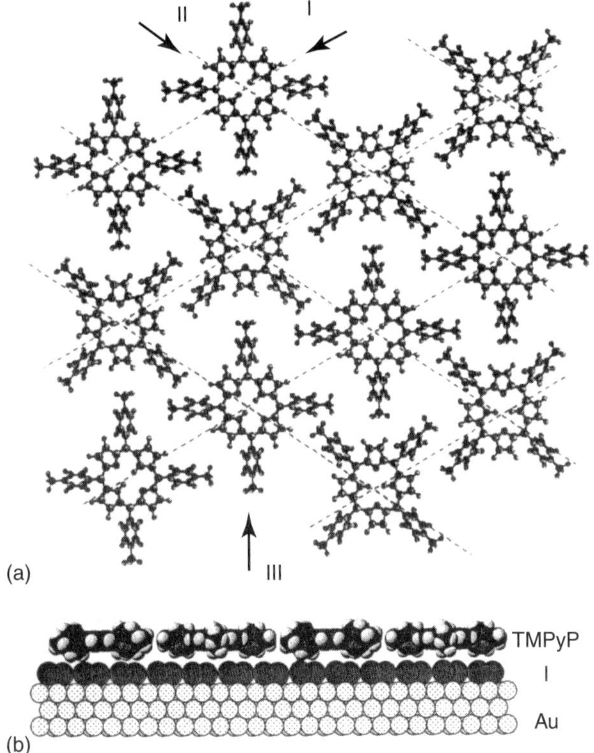

Figure 14 Illustrative depictions of the structure of TMPyP adlayer on I–Au(111): (a) top view; (b) side view. (Reprinted with permission from M. Kunitake, U. Akiba, N. Batina, K. Itaya, 'Structures and Dynamic Formation Processes of Porphyrine Adlayers on Iodine-modified Au(111) in Solution: In Situ STM Study', *Langmuir*, **13**, 1607–1615,(57) Copyright (1997), American Chemical Society.)

5.2 Other Molecules on Iodine–Gold(111)

Here, we briefly describe further evidence that the I–Au(111) electrode can be employed as an ideal substrate for in situ STM imaging of various adsorbed organic molecules in solution. Organic substances investigated were water-soluble cationic molecules purposely selected based on their characteristic shapes: triangular and linear. Hexamethylpararosaniline (crystal violet) and 4,4′-bis(*N*-methylpyridinium)-*p*-phenylenedivinylene (PPV) were also found to form highly ordered molecular arrays on top of the iodine monolayer adsorbed on Au(111).(56)

In situ STM with near-atomic resolution revealed their orientation, packing arrangement, and internal structure of each molecule. A typical high-resolution STM image of the molecular arrays of crystal violet (Figure 15b) is shown in Figure 15(a). It is also surprising to see that the STM image shows a distinctly characteristic, propeller-shaped feature for each molecule with highly ordered arrays. Each molecule has three benzene rings located at the apexes of a triangle with an equal distance from the central carbon atom. Three bright spots seem to correspond to the benzene rings. The center of each spot is located at a distance of ca. 0.35 nm from the center of the triangle. An additional spot can also be seen at the position of the central carbon atom of crystal violet. According to the STM image, it is clear that all molecules are oriented in the same

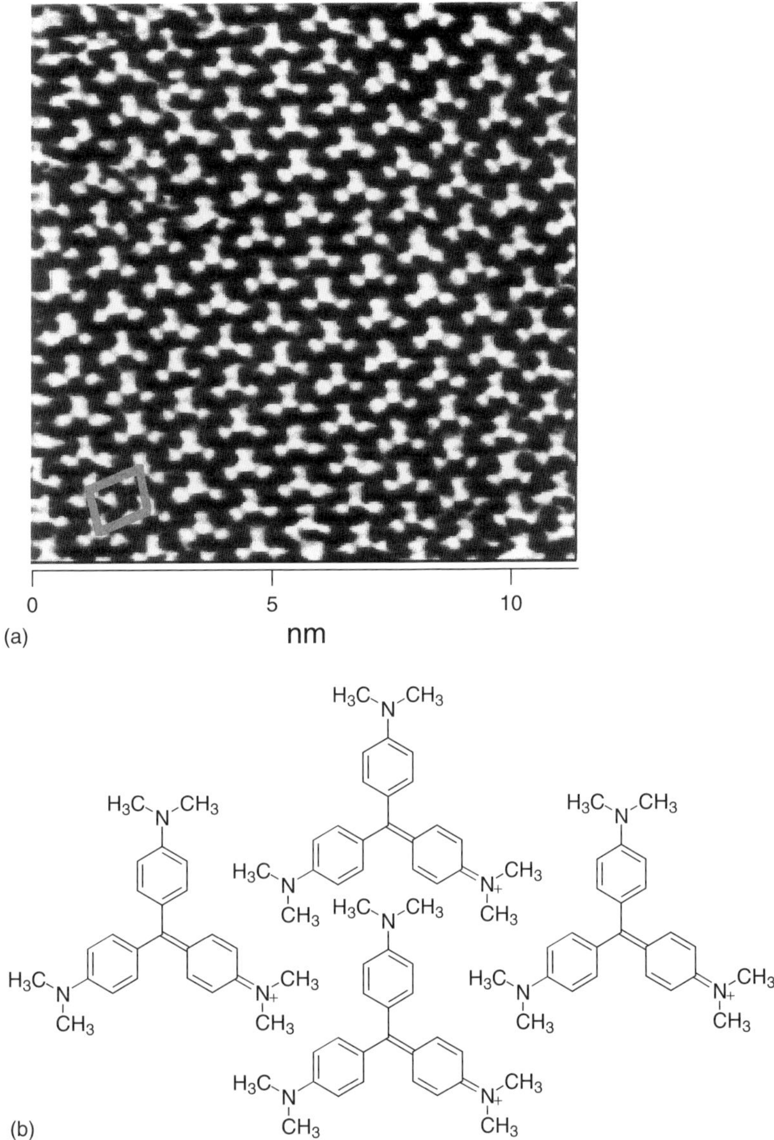

Figure 15 High-resolution STM image of crystal violet on I–Au(111). (Reprinted from M. Batina, M. Kunitake, K. Itaya, 'Highly Ordered Molecular Arrays Formed on Iodine-modified Au(111) in Solution: In Situ Scanning Tunneling Microscopy Imaging', *J. Electroanal. Chem.*, **405**, 245–250,[56] Copyright (1996), with permission from Elsevier Science.)

direction. The unit cell shown in Figure 15(a) can be characterized by the lattice parameters, $a = 0.9$ nm, $b = 1.1$ nm, and the angle of ca. $75°$, indicating that the crystal violet adlayer was slightly deformed from threefold symmetry.

The third compound investigated is the highly symmetric cationic PPV molecule with two terminal pyridinium rings connected with the straight phenylenedivinylene core. The image shown in Figure 16 is a typical STM image of the ordered PPV adlayers formed on the I–Au(111) surface. The individual flat-lying PPV molecule can be seen as a linearly aligned feature consisting of three bright spots that can be attributed to the three aromatic rings in PPV. Three bright spots aligned in a straight line suggest that the PPV molecules adsorb on the I–Au(111) surface with a straight configuration. PPV molecules are expected to form straight or bent configurations, depending on the relative orientation of the two trans CH=CH double bonds. The STM image shown in Figure 16(a) indicates that the two trans CH=CH double bonds are located on opposite sides, forming the straight configuration in the adlayer shown in Figure 16. It is also clearly seen that the tightly packed arrangement forms long striped domains. In each domain, all molecules show the same orientation as indicated by the model in Figure 16(b). The width of each domain along the molecular axis was found to be

For references see page 10066

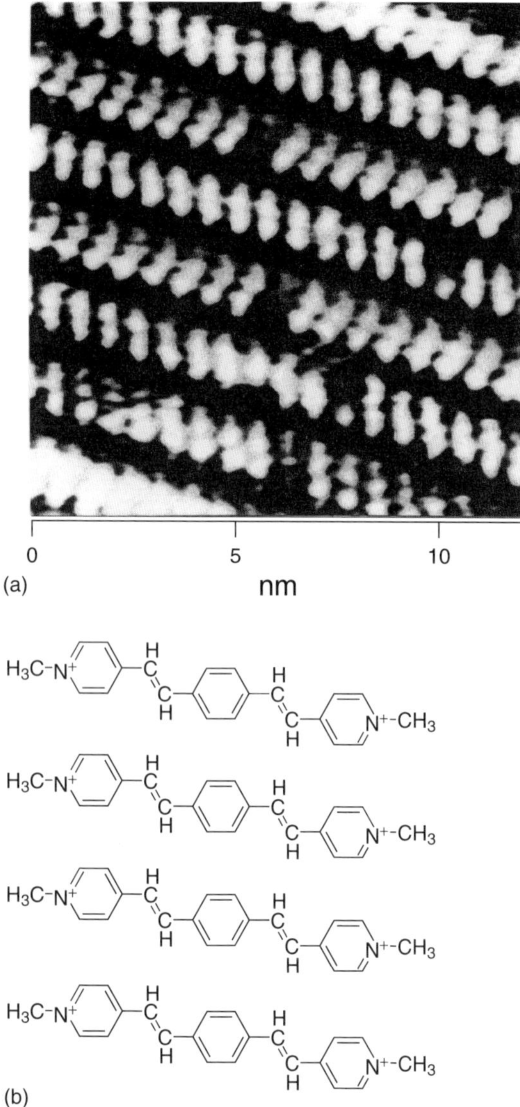

Figure 16 High-resolution STM image of PPV on I–Au(111). (Reprinted from M. Batina, M. Kunitake, K. Itaya, 'Highly Ordered Molecular Arrays Formed on Iodine-modified Au(111) in Solution: In Situ STM Imaging', *J. Electroanal. Chem.*, **405**, 245–250,[56] Copyright (1996), with permission from Elsevier Science.)

ca. 2.1 nm from the STM image, which corresponds to the total molecular length of PPV. It is also interesting that a zig-zag arrangement appears alternately in these striped domains.

In this section, we have presented further evidence that the I–Au(111) electrode can be employed as an ideal substrate for in situ STM imaging of adsorbed organic molecules in solution. The organic substances investigated were all water-soluble cationic molecules with characteristic shapes: triangular, linear, and square. Molecules of crystal violet, PPV, and TMPyP were all found to form highly ordered arrays on top of the iodine monolayer adsorbed on Au(111). The novel approach, using the iodine monolayer as an intermediate layer for the adsorption and formation of molecular arrays, has great potential for investigations of many organic molecules, including more complexed molecules and native biological materials.

In evaluating the structural relationship between iodine and TMPyP adlayers, we also investigated the structure of TMPyP on I–Pt(100).[59] It was found that adsorbed TMPyP molecules formed a highly ordered adlayer with a side-by-side configuration. The adlayer structure of TMPyP on I–Pt(100) is controlled by the interaction between the iodine adlayer and TMPyP.[59]

6 ADSORPTION OF AROMATIC MOLECULES ON CLEAN BARE ELECTRODES

The adsorption of organic molecules on bare electrode surfaces in electrolyte solutions under potential control has long been investigated for elucidating the role of the structures and properties of adsorbed molecules in EC reactions.[1,2,51] Although conventional EC and optical techniques, such as infrared (IR), Raman, and second harmonic generation (SHG) spectroscopy, have been extensively applied to the investigation of the molecular adsorption at electrode surfaces in solution,[51] they usually can provide only averaged information on the molecular orientation and packing within an adlayer. Ex situ techniques such as LEED and AES, using the UHVEC technique, have also been extensively employed for generating understanding of the relationship between the adsorbed molecules and the atomic structure of the electrode surfaces.[1,2] More recently, in situ STM has been well recognized as an important in situ method for structural investigation of adsorbed chemical species on well-defined electrode surfaces in electrolyte solution with atomic resolution.

Although small inorganic species, such as halide, sulfate, cyanide, and thiocyanate, adsorbed on the metal electrode surface can be visualized relatively easily by in situ STM as described already, high-resolution STM images have rarely been reported for organic molecules adsorbed on bare metal surfaces.

In the previous section, it was demonstrated that highly ordered molecular arrays of porphyrine, crystal violet, and a linear aromatic molecule, PPV, were easily formed on iodine-modified Au(111) rather than on bare Au, and they were visualized in solution by in situ STM with near-atomic resolution, revealing packing arrangements and even internal molecular structures. Such an extraordinarily high resolution achieved in solution strongly encouraged us to investigate the adsorption of relatively small organic molecules such as benzene directly attached to the electrode surface in order to understand electrocatalytic activities of noble metals such as Pt and Rh.

On the other hand, many reports have describe the investigation of the adsorption in UHV of aromatics, such as benzene and its derivatives on Pt, Rh, Ni, Ir, Ru, and Pd, which were performed by using various surface-sensitive techniques such as LEED, AES, and electron energy-loss spectroscopy (EELS). The purpose of those investigations was to evaluate gas-phase catalytic reactions, such as hydrogenation, dehydrogenation, and dehydrocyclization.(60) In UHV the (3×3) superlattice of benzene and CO coadsorbed on Rh(111) revealed a well-ordered array of ring-like features associated with adsorbed benzene molecules, whereas CO did not appear in STM images.(61) A more detailed STM study has been reported by Somorjai et al.(62)

Nevertheless, we described, for the first time, the adlayer structures of benzene adsorbed on Rh(111) and Pt(111) in HF solutions.(63) High-resolution STM images allowed us to determine the packing arrangement and even the internal structure of each benzene molecule in solution.

6.1 Benzene on Rhodium(111) in Hydrofluoric Acid

Figures 17(a) and (b) show CVs of Rh(111) and Pt(111) electrodes in the absence and presence of benzene in 0.01 M HF, respectively. In the absence of benzene, the CV obtained on the well-defined Rh(111) and Pt(111) exhibited several highly reversible characteristic peaks. It was noted that the heights and widths of these characteristic peaks depended on the quality of the surface of Rh(111) prepared by the flame-annealing and quenching method.

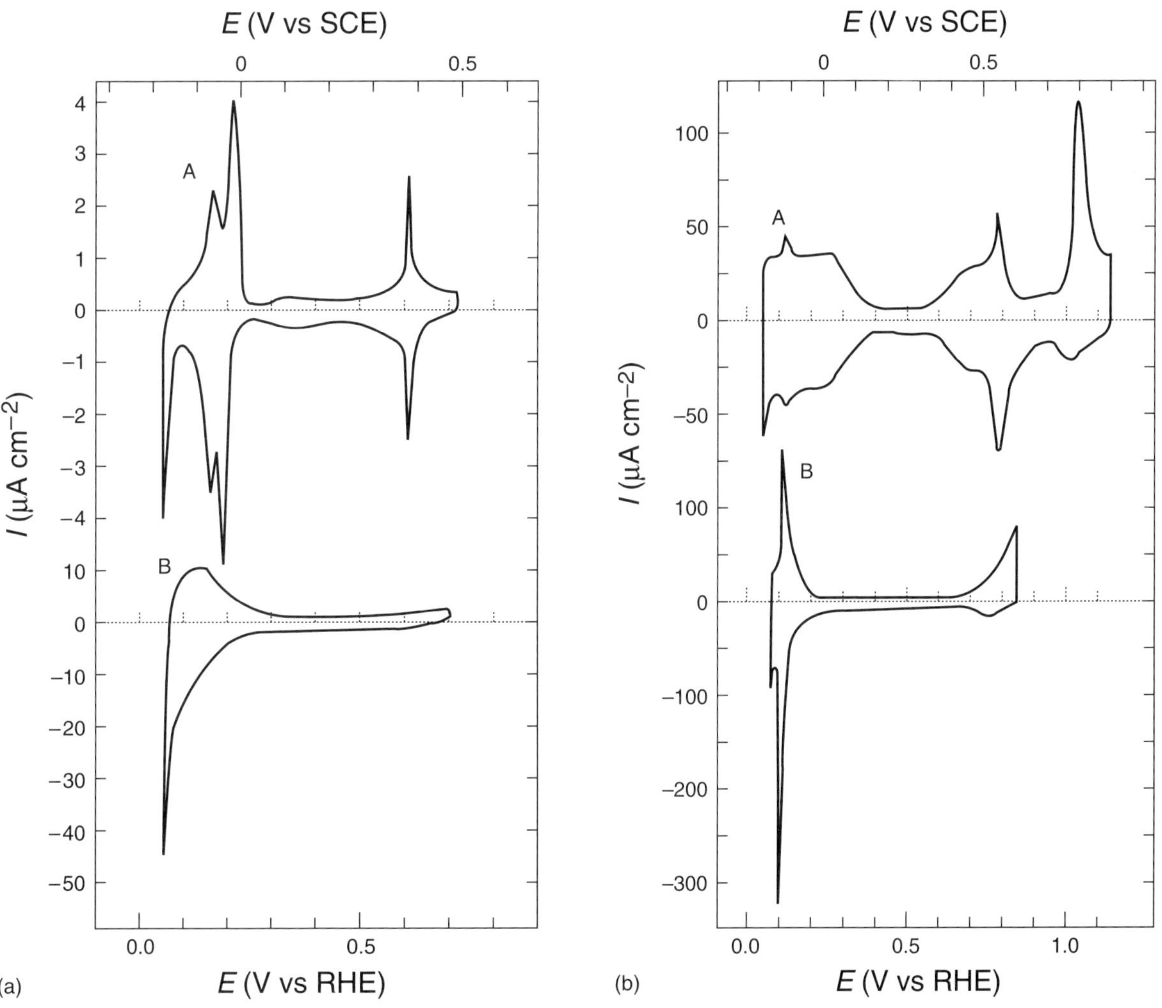

Figure 17 CV of (a) Rh(111) and (b) Pt(111), (A) without and (B) with 1 mM benzene. (Reprinted with permission from S.-L. Yau, Y.-G. Kim, K. Itaya, 'In Situ Scanning Tunneling Microscopy of Benzene Adsorbed on Rh(111) and Pt(111) in HF Solution', *J. Am. Chem. Soc.*, **118**, 7795–7803,(63) Copyright (1996), American Chemical Society.)

For references see page 10066

After the Rh(111) electrode had been subjected to the CV measurement in the pure HF solution, the electrode was transferred into 0.01 M HF solution containing ca. 1 mM benzene. The CV indicated a featureless double-layer region between 0.3 and 0.7 V, as shown in Figure 17(a). The cathodic current commencing at about 0.3 V was considered to be due to simultaneously occurring processes, such as the desorption of adsorbed benzene, the adsorption of hydrogen, and the irreversible hydrogenation of benzene to cyclohexane, according to previous studies using differential electrochemical mass spectrometry (DEMS).[64] A similar featureless CV was also obtained with a benzene-dosed Rh(111) electrode in pure 0.01 M HF. The Rh(111) electrode was immersed in 0.01 M HF containing 1 mM benzene for 1 min at the open-circuit potential (OCP) and then transferred to the pure HF solution. These results strongly suggest that benzene is chemisorbed and remains on the surface of Rh(111), at least in the potential range between 0.3 and 0.7 V.

A well-defined terrace-step structure was easily observed on the well-prepared Rh(111) face. The atomic image of Rh(111–1 × 1) was routinely discerned on the terrace in the pure HF solution. The almost perfectly aligned hexagonal structure can be seen with an inter-atomic distance of 0.27 nm, indicating that the structure of the Rh(111) surface is (1 × 1). Identical atomic images were consistently observed in the potential range between 0.1 and 0.75 V. No additional species were found in STM images at potentials corresponding to the butterfly peaks or the hydrogen adsorption and desorption peaks.[63]

After achieving the atomic resolution, a small amount of 1 mM benzene solution was directly added to the STM cell at 0.45 V. The average concentration of benzene in 0.01 M HF was 10 μM. Immediately after the injection of benzene, completely different patterns appeared in STM images. Figure 18(a) shows an example of the STM images acquired at 0.45 V. It is evident that the atomically flat terraces are now covered by ordered benzene adlayers. An averaged domain size was about 10 × 10 nm. The adsorbed benzene molecules appear to form a square adlattice in each domain. Furthermore, the molecular rows in a given domain cross each other, forming boundaries at an angle of either 60° or 120°.

More details of the orientation of benzene in the adlayer are revealed by the higher-resolution STM image shown in Figure 18(b). The acquisition of the STM image was performed specifically under conditions with minimal thermal drift in the x and y directions in order to determine the unit cell of the adlayer as accurately as possible. It is seen in Figure 18(b) that the molecular rows along the direction of arrows A and B cross each other at 90°, and they are always parallel with the close-packed and $\sqrt{3}$ directions of the Rh(111) substrate, respectively. The intermolecular distances along these directions are not equal to each other and were found to be, on average, 0.8 and 0.9 nm, respectively. Based on the orientation of molecular rows and the intermolecular distances, we concluded that the benzene adlayer was composed of rectangular unit cells, namely $c(2\sqrt{3} \times 3)rect$ ($\theta = 0.17$), as shown in Figure 18(b). The known lattice spaces of $2\sqrt{3}$ and 3 on Rh(111) (0.268 nm) correspond to 0.93 and 0.80 nm, respectively, which are consistent with our experimental values.

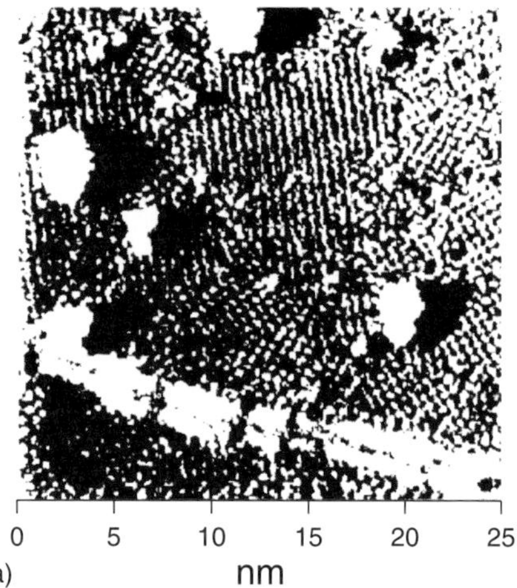

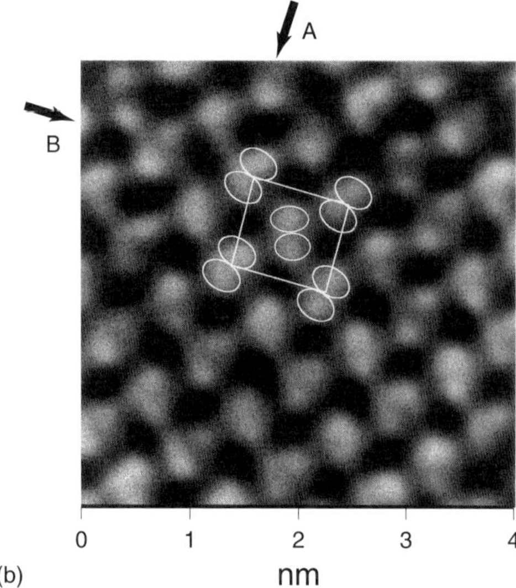

Figure 18 High-resolution STM images of the $c(2\sqrt{3} \times 3)rect$ benzene adlayer in HF. (Reprinted with permission from S.-L. Yau, Y.-G. Kim, K. Itaya, 'In Situ Scanning Tunneling Microscopy of Benzene Adsorbed on Rh(111) and Pt(111) in HF Solution', *J. Am. Chem. Soc.*, **118**, 7795–7803,[63] Copyright (1996), American Chemical Society.)

Surprisingly, the STM image allowed us to determine the internal structure and micro-orientation of each benzene molecule adsorbed on Rh(111). It is clear that each spot is split into two bright spots, forming a characteristic dumbbell shape for each benzene molecule. The STM discerned a 0.01 nm corrugation between the valley and ridge of each benzene molecule. It can also be seen in Figure 18(b) that the orientation of dumbbell shaped benzene is not the same for all molecules, but depends on their positions. The dumbbell-shape of the central benzene molecule in the unit cell shown in Figure 18(b) is clearly rotated by 60° with respect to the molecules located on the four corners of the unit cell. The molecules on the corners of the unit cell appeared with an identical feature, suggesting that they are situated on an identical binding site. It is also seen that the orientation of these dumbbells is always rotated by 30° with respect to the direction of close-packed rows [arrow A in Figure 18(b)] of the Rh(111) substrate. The STM image shown in Figure 18(b) provides more detailed information on the orientation of molecule in the unit cell as discussed below.

The $c(2\sqrt{3} \times 3)rect$ structure described above was consistently observed in the potential range between 0.4 and 0.7 V without additional structural transitions. On the other hand, it was found that the adlayer structure changed at negative potentials. A negative potential step from 0.45 to 0.35 V induced a reconstruction in the benzene adlayer from $c(2\sqrt{3} \times 3)rect$ symmetry to an ordered hexagonal pattern. The electrode potential of 0.35 V is near the onset potential of the cathodic current as shown in Figure 17(a).

Figures 19(a) and (b) show a set of STM images acquired in almost the same area in order to reveal the dynamic process of phase transition. It is clearly seen in Figure 19(a) that a new domain appeared with the hexagonal array of benzene on the upper-right corner marked by solid lines, while the $c(2\sqrt{3} \times 3)rect$ structure remained as the main phase. A further cathodic step to 0.25 V resulted in a predominantly hexagonal phase, while eliminating the $c(2\sqrt{3} \times 3)rect$ domains as shown in Figure 19(b). Such a long-range ordered hexagonal pattern could be seen over almost the entire area of the terrace at 0.25 V. All benzene molecules exhibited the same corrugation height of 0.07 nm, similar to that in the $c(2\sqrt{3} \times 3)rect$ structure.

To reveal the internal molecular structure in the hexagonal phase, STM images were acquired under particularly carefully adjusted experimental conditions with minimal thermal drift. Figure 20(a) shows one of the highest resolution images acquired on the terrace shown in Figure 19(b). Compared with the crystal orientation, $[1\bar{1}0]$, determined by the Rh(111–1 × 1) atomic image, it can be seen that all benzene molecules are almost perfectly aligned along three close-packed directions of Rh(111). The molecular rows cross each other at an angle of either 60° or 120° within experimental error (±2°). The intermolecular distance along these rows was found to be 0.8 nm, which corresponds to three times the lattice parameter of Rh(111). Therefore, we conclude that the hexagonal structure is (3×3)-C_6H_6 ($\theta = 0.11$), as shown by the unit cell superimposed in Figure 20(a). Moreover, a careful examination of the image reveals that each benzene molecule appears as a set of three spots with similar intensities. It can also be seen that a clear dip exists in the center of each triangle with three lobes. These features can be more clearly seen in the height-shaded surface plot obtained by applying a mild 2D Fourier transform filter method as shown in Figure 20(b). The spacing between the two lobes in each molecule was found to be about 0.3 nm. In addition, a

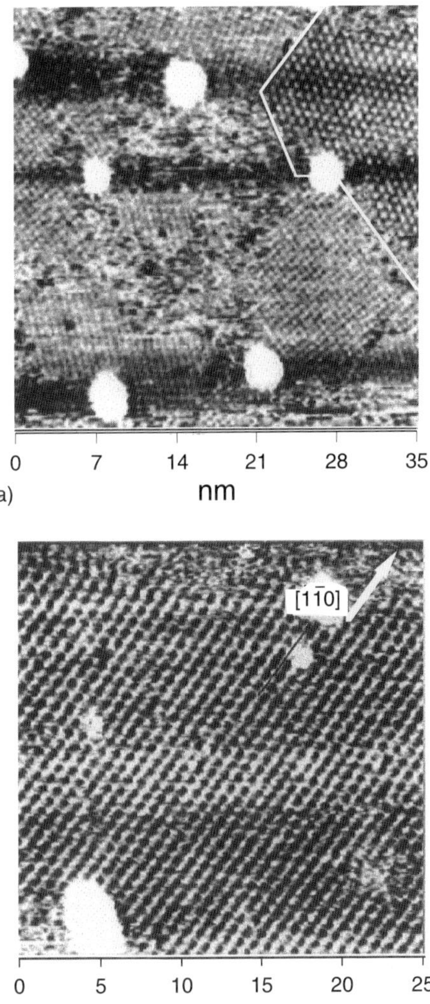

Figure 19 STM images of (a) domain boundaries of $c(2\sqrt{3} \times 3)$ *rect* and (3×3) benzene adlayers and (b) the pure (3×3) structure on Rh(111). (Reprinted with permission from S.-L. Yau, Y.-G. Kim, K. Itaya, 'In Situ Scanning Tunneling Microscopy of Benzene Adsorbed on Rh(111) and Pt(111) in HF Solution', *J. Am. Chem. Soc.*, **118**, 7795–7803,[63] Copyright (1996), American Chemical Society.)

For references see page 10066

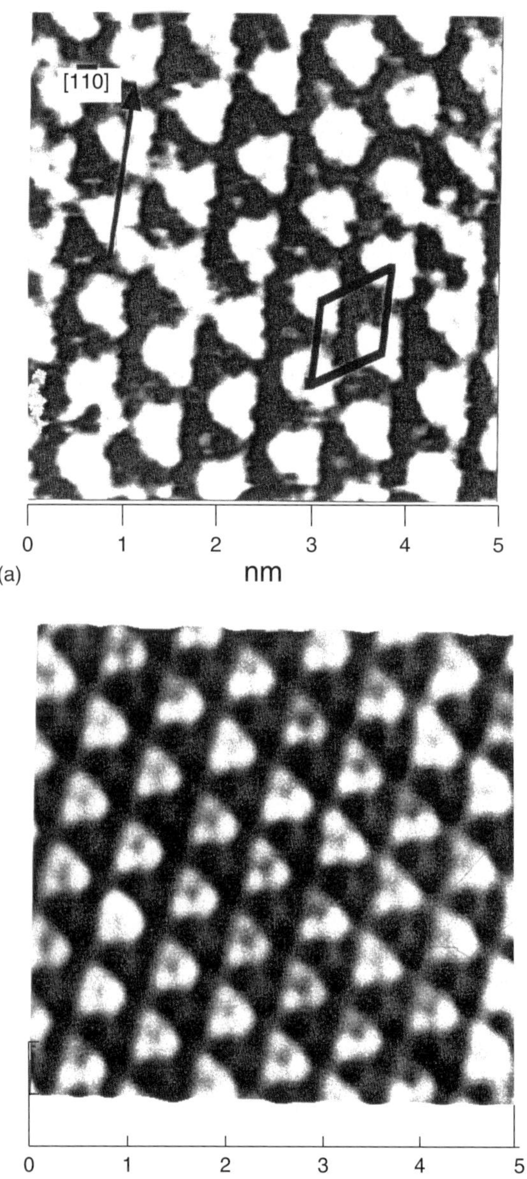

Figure 20 High-resolution STM images of the (3 × 3) structure on Rh(111). (a) Top view; (b) height-shaded plot. (Reprinted with permission from S.-L. Yau, Y.-G. Kim, K. Itaya, 'In Situ Scanning Tunneling Microscopy of Benzene Adsorbed on Rh(111) and Pt(111) in HF Solution', *J. Am. Chem. Soc.*, **118**, 7795–7803,[63] Copyright (1996), American Chemical Society.)

weaker additional spot with a small corrugation of about 0.02 nm can be seen in the unit cell. It is important to note that all these features of the STM image for the (3 × 3) adlayer observed at 0.25 V are essentially identical with those found for the coadsorbed benzene and CO adlayer on Rh(111) in UHV reported by Ohtani et al.[61] They also found the weaker spot, which was attributed to the coadsorbed CO or artifacts caused by asymmetric tips.[61]

When the electrode potential was stepped further in the negative direction, the ordered (3 × 3) domain became islands with the same internal structure, suggesting that the desorption of benzene occurred preferentially at edges of the islands of ordered (3 × 3) domains.[63] Eventually, all adsorbed benzene molecules were desorbed from the surface at 0.1 V owing partially to the hydrogen adsorption and partially to the hydrogenation as expected from the result obtained by DEMS,[64] and the Rh(111–1 × 1) structure was consistently discerned at 0.1 V. The structural changes described above were reversible. When the electrode potential was stepped back to the positive region, the (3 × 3) and $c(2\sqrt{3} \times 3)$*rect* phases returned at the potentials described above.

The structures and registries of chemisorbed benzene on Rh(111) have been thoroughly scrutinized by the surface-sensitive techniques, such as LEED, EELS, and angle-resolved ultraviolet photoemission spectroscopy (ARUPS), in UHV.[60] These previous studies revealed various structures for benzene, including well-known structures such as $c(2\sqrt{3} \times 4)$*rect* and (3 × 3), depending on whether CO was present unintentionally or intentionally in the UHV chambers. Although it has been repeatedly demonstrated that the adlayer structures of benzene on Rh and Pt were greatly affected by the presence of CO in the adlayer, the structure of the pure benzene adlayer is not yet fully understood. Neuber et al. have reported that a completely new structure with a $(\sqrt{19} \times \sqrt{19})R23.4°$ symmetry appeared for the pure benzene adsorption on Rh(111) under cleaner UHV conditions in the absence of CO, and the previously known structures of $c(2\sqrt{3} \times 4)$*rect* and (3 × 3) were found to appear upon admission of CO.[65]

However, it was found in our study[63] that the anodic peak due to the oxidation of CO was hardly detectable in CV even after a prolonged STM experiment for several hours in an air-saturated HF solution. We strongly believe that the adlayer structures found in HF solution described above did not result from contamination with CO.

It is extremely important to recognize that in the previous study of the adsorbed benzene on Rh(111) in UHV,[65] one of the structures of the pure benzene adlayer was attributed to the $c(2\sqrt{3} \times 3)$*rect* structure, which was found in solution. This result strongly suggests that the existence of water molecule on top of the benzene adlayer plays a minor role in determining the structure of benzene.

Figure 21(a) shows a proposed model for the $c(2\sqrt{3} \times 3)$*rect* structure. All of the adsorbed benzene molecules are assumed to be located on the twofold bridging sites. The benzene molecule at the center of the unit cell also occupies a twofold site, but it is rotated by 60° from the orientation of the molecules at the corners. Weiss and Eigler reported three distinct types of STM images for isolated benzene molecules located at threefold hollow,

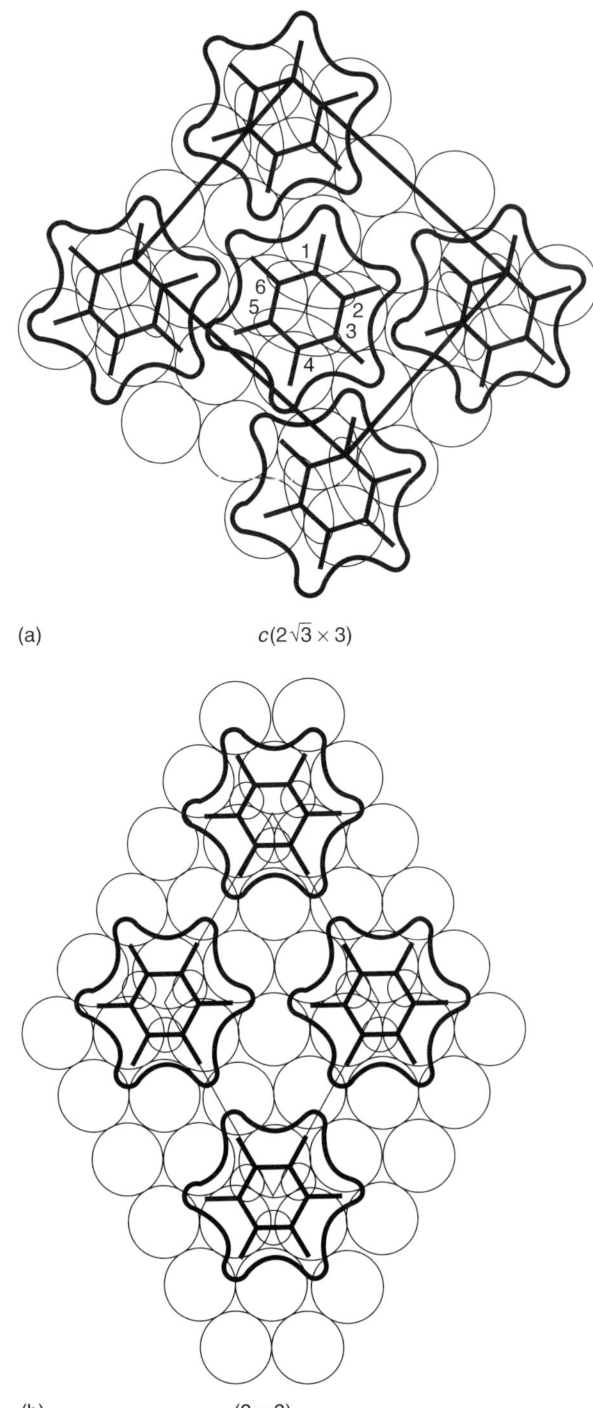

Figure 21 Space models of (a) the $c(2\sqrt{3} \times 3)rect$ and (b) the (3×3) structures. (Reprinted with permission from S.-L. Yau, Y.-G. Kim, K. Itaya, 'In Situ Scanning Tunneling Microscopy of Benzene Adsorbed on Rh(111) and Pt(111) in HF Solution', *J. Am. Chem. Soc.*, **118**, 7795–7803,[63] Copyright (1996), American Chemical Society.)

atop, and bridge sites on Pt(111) at 4 K.[66] They assigned the single bump elongated perpendicularly to the bridge to the bridge-bonded benzene. In Figure 18(b) each benzene molecule is seen with the dumbbell shape on Rh(111) and elongated perpendicularly to the bridge. It is clear that the direction of each elongated dumbbell is always rotated by approximately 30° with respect to that of the corresponding atomic row of Rh(111).

These detailed features can be explained by the model structure shown in Figure 21(a), where two lobes marked by the circles are assumed to be localized to near carbon atoms (1, 2, 6 and 3, 4, 5) bonded across the Rh atoms.

The STM image obtained at 0.25 V shown in Figure 20 can be explained by the structural model with the (3×3) symmetry illustrated in Figure 21(b). Although the structure proposed here is basically the same as that proposed previously, based on LEED, EELS,[60] and STM[61] studies in UHV, for the adlayer of coadsorbed benzene and CO on Rh(111), two CO molecules thought to be located at the threefold hollow sites in the unit cell are omitted in Figure 21(b). Each benzene molecule is assumed to bond at the threefold hollow site. The coadsorption of CO was unlikely to take place in the solution under the present conditions, because no oxidation peak was observed, as described above. Instead of CO, water molecules or hydronium cations might be coadsorbed near the uncoordinated threefold hollow sites to stabilize the (3×3) structure, their function being similar to that of the coadsorbed CO. The weak small spots seen in Figure 20 might be due to such coadsorbed water molecules or hydronium cations.

6.2 Other Molecules

The in situ STM imaging of benzene adlayers on Pt(111) was carried out in the same manner as that on Rh(111).[63] It was found in our study that benzene adlayers on Pt(111) mostly appeared as less ordered phases than those on Rh(111). The intermolecular distances and directions of molecular rows indicate that the structure of the benzene adlayer at 0.35 V is $c(2\sqrt{3} \times 3)rect$, the same as that found on Rh(111).

More interestingly, in situ STM revealed the reconstruction of the benzene adlayer that occurs upon the cathodic potential step from 0.35 to 0.25 V. On the basis of detailed results, we proposed that the benzene adlayer has a $(\sqrt{21} \times \sqrt{21})R10.9°$ structure ($\theta = 0.14$).[63]

After the atomic resolution was achieved with Rh(111), a saturated naphthalene solution was added to the STM cell at 0.3 V.[67] A high-resolution image acquired in an ordered domain is shown in Figures 22(a) and (b). It is clearly seen that the molecular rows parallel the ⟨110⟩ direction of the substrate indicated by the arrows in Figure 22(a). More importantly, the STM image allowed us to determine the internal structure and orientation of each naphthalene molecule. The elongated features

For references see page 10066

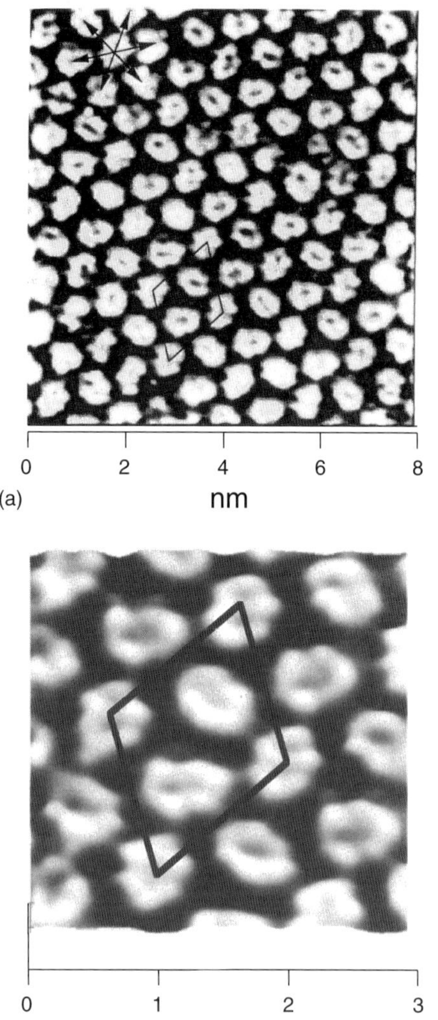

Figure 22 High-resolution STM images of naphthalene on Rh(111). (a) Top view; (b) perspective view. (Reprinted with permission from S.-L. Yau, Y.-G. Kim, K. Itaya, 'High-resolution Imaging of Aromatic Molecules Adsorbed on Rh(111) and Pt(111) in Hydrofluoric Acid Solution: In Situ STM Study', *J.Phys. Chem. B*, **101**, 3547–3553,[67] Copyright (1997) American Chemical Society.)

along the longer molecular axis (C_2) were discerned for each molecule. In addition, the images of some molecules clearly show a two-ring structure expected from the molecular model. It can be seen that naphthalene molecules are perfectly aligned with a regular micro-orientation along the molecular rows.

A typical arrangement of naphthalene molecules on Pt(111) was found to be a full monolayer of flat-lying naphthalene molecules. Although the overall appearance of the naphthalene adlayer on Pt(111) on the large scale was similar to that on Rh(111), close inspection revealed that the adlayer included many randomly oriented molecules. Because the molecular rows were nearly parallel to the close-packed Pt atomic rows, and because the intermolecular distance was three times that of the Pt substrate, the structure roughly fitted a (3×3) symmetry.

Periodic rotation of the molecules of naphthalene by 60° is seen within each molecular row with every third molecule being in the same orientation. A further magnified view in height-shaded mode is shown in Figure 22(b), in which the two-ring structure can be more clearly seen. The nearest-neighbor distance of 0.82 nm on average is equivalent to three times the Rh lattice parameter of 0.268 nm.

According to the results described above, the unit cell can be defined as a $(3\sqrt{3} \times 3\sqrt{3})R30°$ symmetry as shown in Figure 22(b).[67] It is now clear that all naphthalene molecules align their C_2 axes along the close-packed directions of Rh substrate. The molecules aligned along the ⟨112⟩ direction, which is the so-called $\sqrt{3}$ direction, have the same orientation. The spacing between two adjacent molecules along the $\sqrt{3}$ direction is measured to be 1.4 nm, which is three times the $\sqrt{3}$ spacing. In our model structure,[67] two carbon atoms at the 9- and 10-positions are assumed to be attached directly to a Rh atom. It is noteworthy that this structure is identical with that previously proposed from LEED results by UHV workers.[68] If one recalls the identical results for benzene adsorbed on Rh(111) in UHV and in HF solution as described above, the results obtained with naphthalene further support the predominant adsorbate–substrate interaction for hydrophobic molecules and the minor role of water molecules.

1,2- and 1,4-naphthoquinone can be considered as representative derivatives of naphthalene in order to understand the effect of functional groups on the molecular organization.[67] Generally, these quinones, similarly to naphthalene, formed ordered adlayers on Rh(111) and mostly disordered adlayers with the flat-lying orientation on Pt(111). The high-resolution STM images revealed the details of internal molecular structures. It was demonstrated that 1,2-naphthoquinone formed a well-ordered adlayer with the structure $(3\sqrt{3} \times 3\sqrt{3})R30°$, which was identical with that found for naphthalene. More interestingly, it was found that an additional bright spot, unseen in the image for naphthalene, is seen at the 2-position of each 1,2-naphthoquinone molecule.[67] This spot exhibits a ca. 0.03 nm higher corrugation with respect to the naphthalene ring, which is likely to be due to the oxygen at the 2-position. The oxygen at the 1-position was not clearly seen in the STM images.

An STM image of anthracene on Rh(111) unambiguously disclosed the internal molecular structure of anthracene.[67] It was also found that molecules of anthraquinone and 1,4,9,10-anthracenetetrol adsorbed in a manner similar to those of anthracene.

To elucidate the effect of molecular structure on the packing arrangement, biphenyl was further investigated on Rh(111).[69] In contrast to the planar structure of naphthalene and anthracene, the two aromatic rings of biphenyl are slightly off the coplanar configuration because of the restriction of hydrogen atoms at the 2,2′- and 6,6′-positions. However, it was found that biphenyl formed disordered adlayers on Rh(111) in HF.[69] Although STM images revealed the internal molecular structure of biphenyl with clear identification of two rings in each molecule, the nonplanar configuration was not clearly seen. We expected that two rings of biphenyl behave like two benzene molecules, and they prefer to be attached on the bridge sites as benzene molecules do.[69] The appearance of the disordered adlayer of biphenyl suggests that biphenyl is more strongly attached on Rh(111) than naphthalene.

It is of special interest to distinguish structures among a series of benzene derivatives such as phthalic acid, terephthalic acid, and hydroquinone.

7 ELECTROCHEMICAL DISSOLUTION PROCESSES OF SEMICONDUCTORS

The preparation of clean and stable semiconductor surfaces is the first step in the manufacture of semiconductor devices. The drive towards submicron technology for ultralarge-scale integrated circuits has focused special attention on wet chemical processes, since high-temperature-based procedures often lead to adverse effects that arise from new and difficult-to-control reaction channels; the pursuit of nanometer-scale technology has likewise necessitated the development of surface characterization methods that permit atomic-scale resolution. Since commercial integrated circuits are still based exclusively on silicon, the wet-chemical processing of Si single-crystal surfaces has been and continues to be widely investigated; much of the interest centers around the nature of the hydrogen termination of the Si surface atoms.

The first work to demonstrate that an Si(111) surface etched in aqueous NH_4F is ideally terminated with Si monohydride was based upon IR spectroscopic studies of the Si–H vibrational modes.[70] This original work stimulated further investigations, likewise based upon IR and other spectroscopic techniques, on Si(100) and Si(110) substrates. STM in vacuum directly established the fact that the NH_4F-etched Si(111) surface was atomically well defined with an ideal H-terminated Si(111) : H–(1 × 1) structure.[71,72] While these studies provided critical information on the post-etched Si surfaces, it was clear that in situ investigations had to be undertaken if the chemical etching process is to be understood at the atomic level.

In response to this need, in situ STM and AFM were adopted for semiconductor-etching studies. Our first in situ STM observation[73] of the Si(111) : H–(1 × 1) atomic structure in a noncorrosive solution (aqueous H_2SO_4) spurred investigation of the etching of Si(111) in corrosive solutions, such as aqueous NH_4F,[74–76] and NaOH.[77–79]

It was soon determined that the etching of Si(111) was potential dependent. At potentials markedly negative of the open-circuited potential (OCP), the etching rate decreased and the dissolution proceeded via a step-selective layer-by-layer mechanism.[74–78] At potentials near or more positive than the OCP, the etching rate increased and pit corrosion occurred on the terraces, which resulted in atomically roughened surfaces.[77,78] The potential dependence of the Si-etching process is a technologically relevant issue. Industrial wet-etching processes are usually performed without potential control, hence the surface chemical reactions that characterize the simple dipping of Si wafers in an etching bath are expected to bear a strong resemblance to what is observed from EC etching at the OCP. The ability to control the applied potential during the etching process may serve as an additional critical factor in the preparation of atomically well-defined semiconductor surfaces.

We have also demonstrated that in situ STM can be employed to monitor atomic-scale features of the etching process.[74–76] For example, we discovered that, in general, multiple H-terminated Si atoms at the kink and step sites were eroded more rapidly than the monohydride-capped atoms.

In this section, we first describe detailed etching processes of Si(111) in NH_4F solutions. Atomic images of GaAs and InP surfaces are also briefly discussed.

7.1 Silicon(111)

The STM imaging of the n-Si(111) electrode was performed immediately after the electrode was etched. After the Si electrode had been immersed in the NH_4F solution, the electrode potential of Si was immediately set to −1.1 V. Note that the potential effect on the etching rate of Si was reported in NaOH solutions.[77,78] It was found that both chemical and EC etching mechanisms operate at the OCP, leading to a higher etching rate.[78,79] We also found that applying a cathodic potential of −1.1 V drastically reduced the corrosion rate of Si in NH_4F solution.[74–76]

Figures 23(a) and (b) show a crystallographic orientation and a ball-and-stick model, respectively, of an H-terminated Si(111) with $[11\bar{2}]$ and $[\bar{1}\bar{1}2]$ oriented steps. One of the most interesting features of an Si(111) surface is the existence of two structurally different steps where the Si atoms have monohydride and dihydride configurations. They can be exemplified by the steps in the $[11\bar{2}]$ and

For references see page 10066

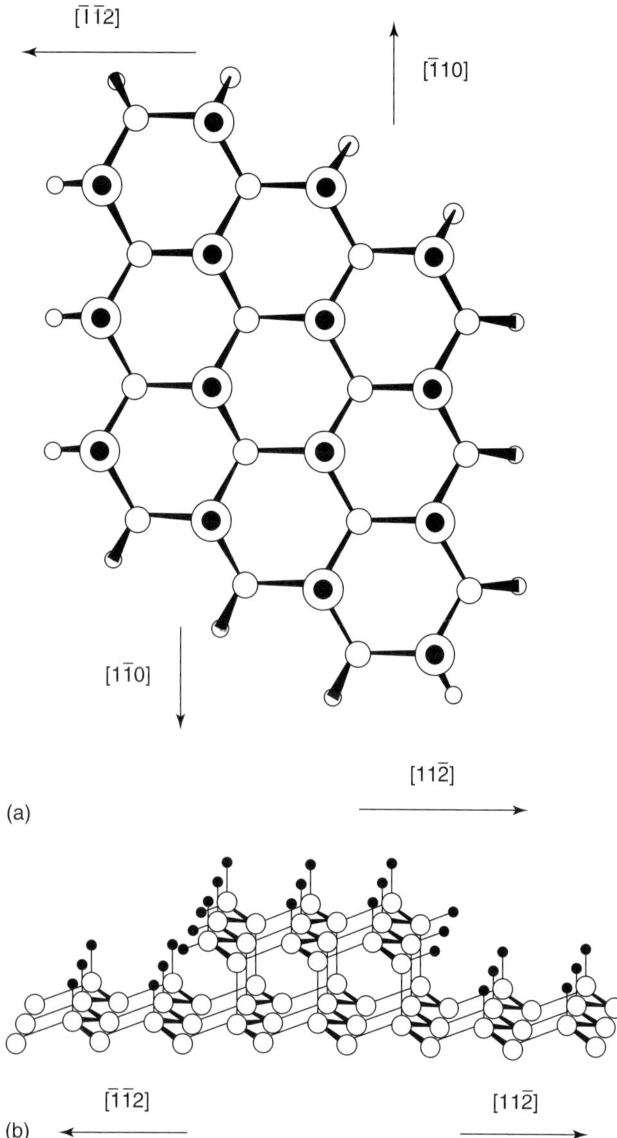

Figure 23 (a) Top view and (b) side view of the Si(111) surface. (Reprinted with permission from K. Kaji, S.-L. Yau, K. Itaya, 'Atomic Scale Etching Processes of n-Si(111) in NH_4F Solutions: In Situ Scanning Tunneling Microscopy', *J. Appl. Phys.*, **78**, 5727–5733,[75] (1995).)

$[\bar{1}\bar{1}2]$ directions, respectively (Figure 23b). Ex situ STM results have shown that the dihydride-bound Si atoms are more reactive than those of the monohydride ones in weakly alkaline HF solutions, resulting in the appearance of the most stable $[11\bar{2}]$ steps.[72] Those experiments were conducted with Si(111) samples tilted towards $[11\bar{2}]$ and $[\bar{1}\bar{1}2]$.[72] Furthermore, there are two possible dihydride structures for the Si atoms at the $[\bar{1}\bar{1}2]$ step. The dihydride axis is either perpendicular or parallel to the (111) plane. It is believed that there is a strong repulsive interaction among the horizontal dihydride structures, so that the perpendicular dihydride, as depicted in Figure 23(b), is in fact more stable than the horizontal one, as confirmed by an IR spectroscopic study.[80] The horizontal dihydride[72] might be too reactive to exist on an Si(111) surface in the presence of etching species of H_2O and F^-.

We focused our attention on the evaluation of the reactivity difference between the microscopically different steps on Si(111).[75] We were able to use in situ STM to locate some areas which contain both types of steps so that the reactivity of these steps can be simultaneously examined under identical conditions. This approach is evidently more advantageous than that of the previous study which used differently tilted Si(111) substrates in separated experiments.[72] As the initial surface feature of Si controlled the subsequent etching process, we first recorded an STM image at −1.1 V to show the initial surface morphology as shown in Figure 24(a), followed by stepping the electrode potential to a less negative value of −1.04 V to accelerate the erosion of Si. A series of STM images shown in Figures 24(b–f) was acquired successively with a time interval of 13 s.[75]

Figure 24(a) shows well-defined double-layer steps of 0.32 nm in height and terraces extending more than 25 nm on the (111) surface. The Si(111) was etched in 11 M NH_4F for 3 min at room temperature. The relative heights of terraces are reflected by their brightness in the STM image. The internal atomic structure of terrace (marked T) was readily discerned by a high-resolution STM scan. A well-ordered hexagonal pattern with an interatomic spacing of 0.38 nm was in good agreement with the ideal Si(111):H−1×1 structure.[73] Consequently, the treatment in 40% NH_4F yielded a long-range ordered monohydride-terminated Si(111) surface with no discernable vacancy defect in the hexagonal network. The step orientations, as defined by their outward normals, are shown in Figure 24(a). It is important to note that both the mono- and dihydride steps were probed by the in situ STM imaging at the same time. The shape of the terraces shown in Figure 24(a) was determined by the morphology of the step ledges, i.e. the monohydride steps are mostly straight, in strong contrast to the typical zig-zag pattern for the dihydride ones.

The small islands (3 nm in diameter, probably impurities) at the upper edge of the STM image were used as a guide against themal drift during the STM measurement. Their unchanged locations indicate low thermal drift. The time-dependent STM results presented below demonstrate the important role played by the atomic structure at the steps in controlling the etching rate.

Figures 24(b–f) present the time-dependent etching process of Si(111) after acquiring the image of Figure 24(a). During the first 13 s etching of the Si from Figures 24(a) to (b), the width of the upper portion of the terrace T marked by D decreased from 16 to 8 nm,

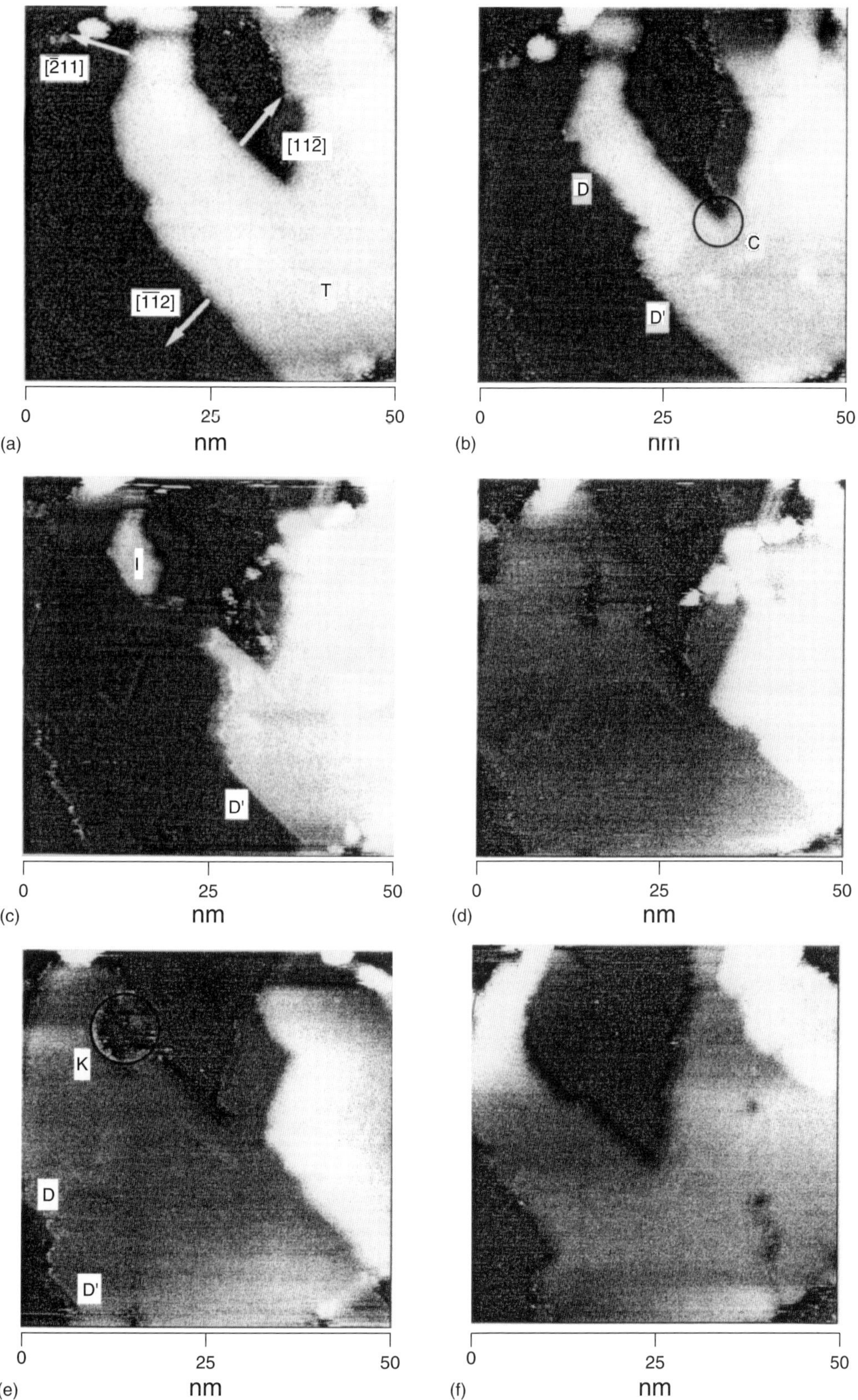

Figure 24 Successively recorded STM images for the etching process of Si(111) in NH_4F. The images were acquired at time intervals of 12.8 s. (Reprinted with permission from K. Kaji, S.-L. Yau, K. Itaya, 'Atomic Scale Etching Processes of n-Si(111) in NH_4F Solutions: In Situ Scanning Tunneling Microscopy', *J. Appl. Phys.*, **78**, 5727–5733,[75] (1995).)

For references see page 10066

while the lower portion marked by D′ retracted from 18 to 12.5 nm. The relatively faster erosion of the upper half of the $[\bar{1}\bar{1}2]$ dihydride step is thought to be due to the higher kink density within the zig-zag pattern step ledge. On the other hand, the monohydride step in the direction of $[11\bar{2}]$ seems to be unchanged from Figures 24(a) to (b). In particular, the steps within the circle marked C in Figure 24(b) remained still in both images, indicating that the ideal monohydride $[11\bar{2}]$ and $[\bar{2}11]$ step in the absence of kink sites is inactive under the present etching conditions. The dihydride-terminated steps continued to retract rapidly to dissolve the terrace T, leaving a small isolated island I in Figure 24(c). However, it was surprising to find that the $[\bar{1}\bar{1}2]$ step with the dihydride configuration marked by D′ in Figure 24(c) was essentially unchanged from Figures 24(b) to (c). The well-defined step ledge, marked D′ in Figure 24(c), suggests that an ideal dihydride step without kinks is also stable.

After the complete removal of the island I in Figure 24(c), a small, bilayer deep (0.32 nm) pit evolved in Figure 24(d). This newly formed depression is apparently a real pit, not an imaging artifact, because it expanded and coalesced with an adjacent $[11\bar{2}]$ step, as shown in Figure 24(e). This coalescence introduced many kink sites (marked K) in Figure 24(e) at the almost inactive $[11\bar{2}]$ step. A close examination of Figures 24(e) and (f) reveals the very important fact that the etching rate of the $[11\bar{2}]$ monohydride step is now increased by the introduction of the kinks into the nearly ideal monohydride step. The $[11\bar{2}]$ step line retracted by ca. 3 nm within 13 s, which corresponds to an etching rate of 14 nm min^{-1}. This is a significant increase from a negligible level for the ideal monohydride step as observed in Figures 24(a) and (b).

The aforementioned results provide compelling evidence that the difference in the chemical reactivity of the monohydrogen-terminated and dihydrogen-capped Si surface atoms profoundly influences the Si-etching process. We extended our investigations of the EC etching of Si(110) and Si(100).[74,76]

7.2 Other Semiconductors

It is becoming more urgent for workers in semiconductor technology to understand wet-chemical etching processes, particularly those of Si, GaAs, and InP with atomic resolution.

We acquired the first atomic STM images of GaAs surfaces in 0.05 M H_2SO_4 solution.[81–83] The results clearly demonstrate that the ideal GaAs(001−1 × 1) and (111−1 × 1) structures exist in a pure H_2SO_4 solution in a cathodic potential region. The samples were n-type Si-doped GaAs(001), (111)A and (111)B wafers, grown by the horizontal Bridgman method. The GaAs(001) and (111)B samples were treated in 1 M HCl for 10 min at room temperature. The GaAs(111)A sample was etched with an etching solution [H_2SO_4–H_2O_2–H_2O (1 : 8 : 1 by volume)]. The etching rate of the (111)A surface has been reported to be the lowest among all low-index planes in this mixed solution. After the etching, the solution was completely replaced with 0.05 M H_2SO_4. The replacement of the etching solution was carried out repeatedly to exclude HCl in the solution. It is important that the GaAs surface should always be kept submerged in the solution in order to protect it from oxidation and contamination in the ambient atmosphere.

Figure 25(a) shows a typical surface topography of a chemically etched GaAs(001) surface acquired in an area of 50 × 50 nm. It can be clearly seen that the surface of the (001) exhibits a well-defined step-terrace structure extending over a large area. Wider terraces are seen to extend over 30 nm. The relative brightness of the terraces in the STM image reflects their heights on the surface, i.e. the surface ascends from left to right. The rather uniform appearance of the terraces strongly suggests that the (001) surface has a structure that is well defined on an atomic scale. It is also clear that the steps intersect each other to form an angle of 90°, as expected for a surface with fourfold symmetry. These steps appearing as straight lines were confirmed as double-layer steps on the (001) surface based on the observed height of 0.28 nm obtained by a cross-section analysis. This unique height of 0.28 nm for the steps indicates straightforwardly that the (001) surface prepared by etching in HCl must be either Ga- or As-terminated. According to the crystallographic orientation of the GaAs(001) electrode, these steps were found to be parallel to either [110] or $[\bar{1}10]$ directions.

Figure 25(b) shows our first atomic STM image of an atomically flat terrace on a GaAs(001) surface. It is clear that the ideal square arrangement expected for the (001) surface with fourfold symmetry is discerned by in situ STM. The observed nearest interatomic distances in the [110] and $[\bar{1}10]$ directions were found to be 0.4 ± 0.02 nm. The atomic image shown in Figure 25(b) clearly demonstrates that the ideal, nonreconstructed GaAs(001−1 × 1) structure is exposed in H_2SO_4 solution under the cathodic polarization. Note that the ideal (1 × 1) structure seemed to be extended over the entire region of the terrace, because pits or even single atomic defects were rarely observed.

The (111)A surface etched in the mixed solution containing H_2O_2 was also found to have an atomically flat terrace-step structure in H_2SO_4 solution as shown in Figure 26(a). All steps observed were double-layer steps with a height of 0.33 nm. The steps in local areas were straight and parallel to the close-packed atomic row direction of (111) surface. Figure 26(b) shows a typical

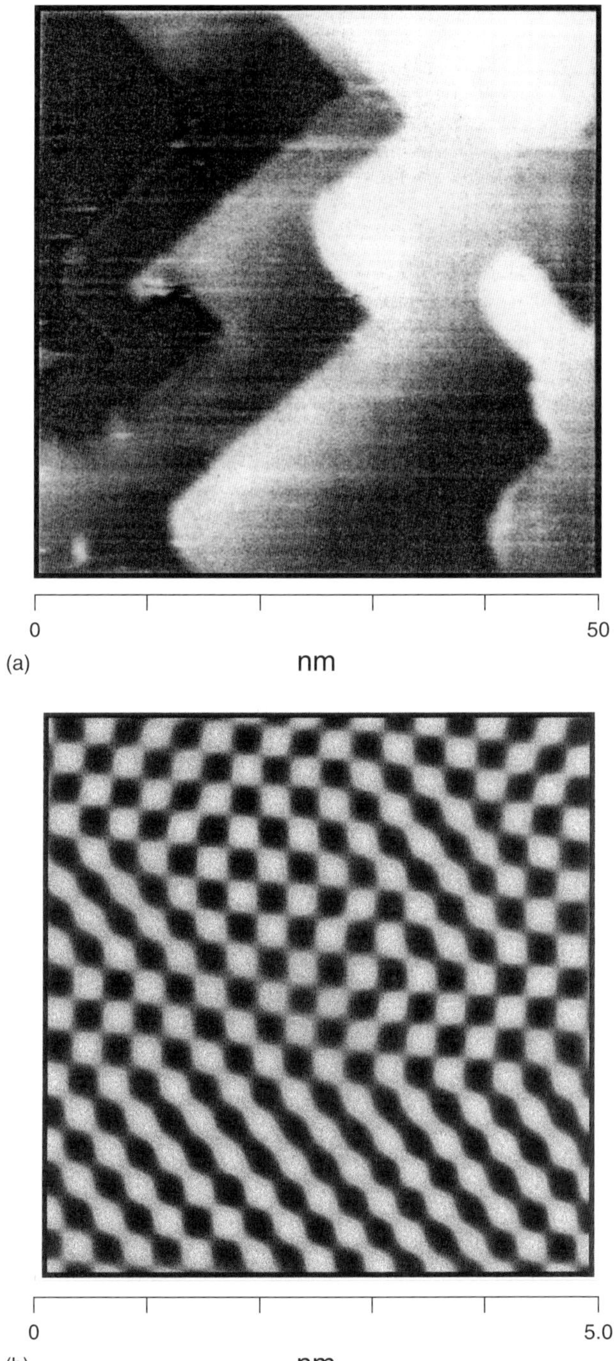

Figure 25 (a) Topographic and (b) high-resolution STM images of n-GaAs(001) in H_2SO_4. (Reprinted with permission from H. Yao, S.-L. Yau, K. Itaya, 'In Situ Scanning Tunneling Microscopy of GaAs(001), (111)A, and (111)B Surfaces in Sulfuric Acid Solution', *Appl. Phys. Lett.*, **68**, 1473–1475[82] (1996).)

atomic STM image, revealing an interatomic distance of 0.4 nm with an almost perfect hexagonal arrangement. This result clearly demonstrates that the ideal

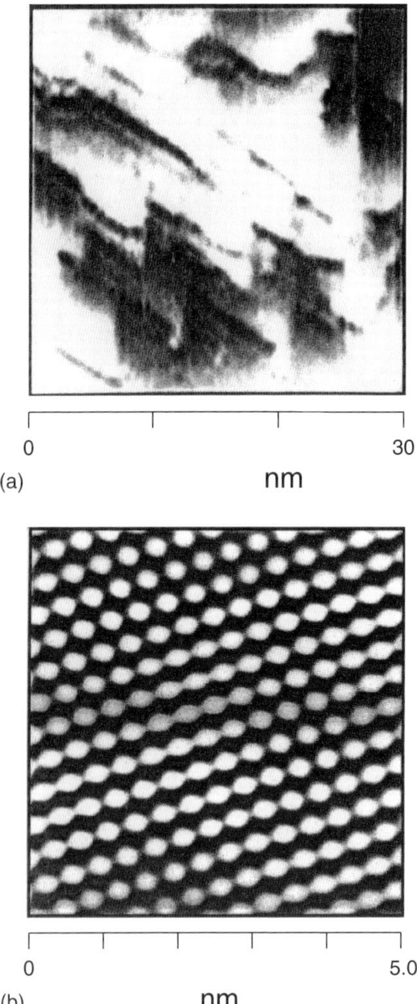

Figure 26 STM images of n-GaAs(111)A in 0.05 M H_2SO_4. (Reprinted with permission from H. Yao, S.-L. Yau, K. Itaya, 'In Situ Scanning Tunneling Microscopy of GaAs(001), (111)A, and (111)B Surfaces in Sulfuric Acid Solution', *Appl. Phys. Lett.*, **68**, 1473–1475[82] (1996).)

GaAs(111)A–(1×1) structure, as shown in Figure 27(c), is exposed in H_2SO_4 solution. It is reasonably expected that the uppermost layer on the (111)A surface consists of Ga atoms.

Finally, it is noteworthy that the GaAs(111)B surface prepared by etching in 1 M HCl also showed an atomically flat terrace-step structure in H_2SO_4 with a step height identical with that observed on the (111)A surface. Although the average terrace width was typically 5–10 nm, obviously smaller than that found on the A surface, an atomic STM image revealed a hexagonal arrangement of As atoms with an interatomic distance of 0.4 nm. The above results indicate that the GaAs(111)B surface has also the ideal (1×1) structure shown in Figure 27(d).[82]

For references see page 10066

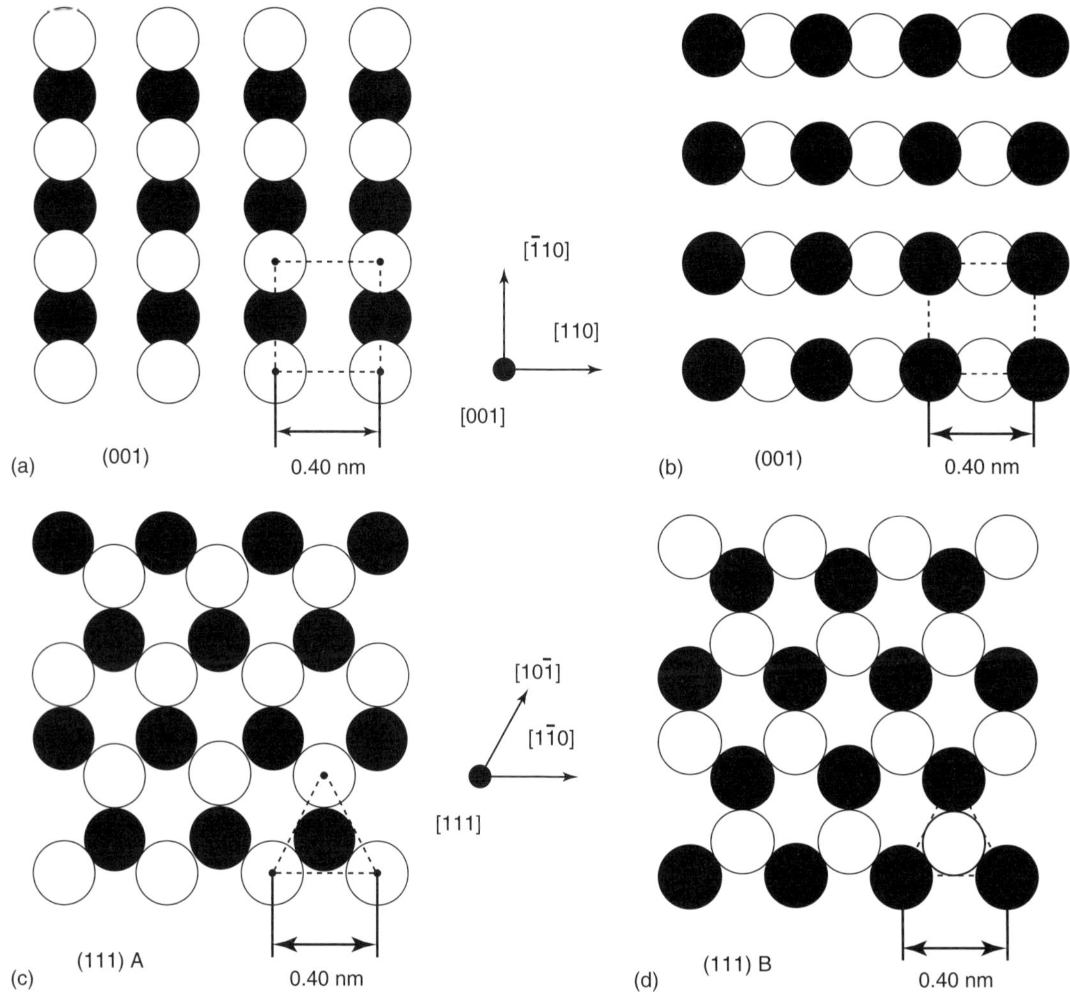

Figure 27 Top views of ball models for ideal (a) Ga-terminated GaAs(001), (b) As-terminated GaAs(001), (c) GaAs(111)A, and (d) GaAs(111)B. (Reprinted with permission from H. Yao, S.-L. Yau, K. Itaya, 'In Situ Scanning Tunneling Microscopy of GaAs(001), (111)A, and (111)B Surfaces in Sulfuric Acid Solution', *Appl. Phys. Lett.*, **68**, 1473–1475[82] (1996).)

In summary, it was demonstrated that the well-defined GaAs(001), (111)A, and (111)B surfaces can be prepared by chemical etching in solutions. Atomically flat terrace-step structures were consistently observed by in situ STM on all three surfaces in H_2SO_4 solution under potential control. Furthermore, we successfully obtained the first atomic STM images which showed that the ideal GaAs(001–1×1), GaAs(111)A–(1×1), and GaAs(111)B–(1×1) structures are exposed and persist in H_2SO_4 solution.

In spite of the fact that InP is a very important material for both optoelectronic and electronic device applications, there have been a fewer STM studies on InP than on GaAs. Previous STM studies of InP were mostly carried out in UHV.

We have recently shown the first atomically resolved STM images of InP(001), (111)A, and (111)B surfaces in an H_2SO_4 solution under cathodic potential control, which effectively protected the surfaces from oxidation.[84] These images clearly demonstrate that well-defined InP surfaces can be prepared by chemical etching in HCl solution. In STM images, individual atoms were relatively clearly observed on the atomically flat terraces of the (111) surface with a corrugation height of ca. 0.02 nm. Monolayer steps were found to be exactly parallel to the close-packed atomic row direction of the (111) surface. An atomically resolved STM image of the (111)A surface revealed a perfect hexagonal arrangement of In atoms with an interatomic distance of 0.42 nm, as expected for the ideal InP (111)A surface. These results clearly demonstrate that the ideal InP(111)A–(1×1) structure is exposed in the H_2SO_4 solution after the chemical etching. This (1×1) structure seemed to extend over the entire region of the terrace; even atomic defects were rarely observed.

On the other hand, the chemically etched (111)B surface also possessed a well-defined structure in H_2SO_4 solution. An atomically flat terrace-step structure was observed with a step height identical with that on the A surface, although the average terrace width was typically 5–10 nm, which was smaller than that on the A surface. The appearance of wider terraces on the (111)A surface is reasonable, because the etching rate of (111)A surface in HCl solution is the slowest of all low-index planes.

An almost ideal square arrangement expected for the (001) surface was clearly discerned by in situ STM. The observed interatomic distances in the [110] and [$1\bar{1}0$] directions were found to be equal to 0.42 nm. Atomic images clearly demonstrate that the ideal InP($001-1 \times 1$) structure exists in H_2SO_4 solution under cathodic potential control. We also presume the InP($001-1 \times 1$), (111)A–(1×1) and (111)B–(1×1) surfaces to be terminated by hydrogen, at least under cathodic polarization.

In summary, it has been demonstrated that well-defined InP(001), (111)A, and (111)B surfaces can be prepared by chemical etching in HCl solution.[84] In situ STM revealed atomically flat terrace-step structures on each surface in H_2SO_4 solution under proper potential control. Furthermore, we successfully obtained the first atomic resolution STM images of InP surfaces, which showed that the ideal InP($001-1 \times 1$), (111)A–(1×1) and (111)B–(1×1) structures are exposed and persist in H_2SO_4 solution under EC conditions.

8 CONCLUSION

The methods for exposing well-defined electrode surfaces in solution were reviewed. The flame-annealing and quenching method can be applied to Au, Pt, Rh, Pd, and Ir single-crystal electrodes. The UHVEC method can be used for Pt, Au, Pd and Rh. For some metals, such as Ni and Cu, surface oxidation takes place in the EC chamber before immersion of the electrode in electrolyte solutions. It was demonstrated that the EC etching method produced atomically flat terrace-step structures of semiconductors under carefully adjusted EC conditions. The method of anodic dissolution is expected to become an important in situ technique for exposing well-defined surfaces of various semiconductors and metals.[7]

The structures of specifically adsorbed iodine on Pt(111) and Au(111) and briefly on Ag(111) were discussed, demonstrating that complementary use of in situ STM and ex situ LEED is a powerful combination to characterize the atomic structure of adsorbed iodine. The adsorption of sulfate/hydrogensulfate adsorbed on Au(111), Pt(111), and Rh(111) was described, with emphasis on the fact that the same ($\sqrt{3} \times \sqrt{7}$) structure is formed on these three substrates. Our model indicates that hydrogen-bonded water chains are inserted along the $\sqrt{3}$ direction between neighboring rows of the adsorbed sulfates. The detailed structural analysis of the CN adlayer on Pt(111) was carried out by in situ STM, revealing that six CN groups form a hexagonal ring without an additional CN at the center of the ring. The complexation of K^+ with the CN adlayer was also discussed.

The UPD of Cu on Au(111) in H_2SO_4 was discussed in depth. In situ SXS indicated that the Cu atoms form a honeycomb lattice and are adsorbed on threefold hollow sites with sulfate ions located at the honeycomb centers. According to the model structure, the corrugation observed by in situ STM and AFM should be ascribed to the coadsorbed sulfate ions, and not the Cu atoms.

It was shown that, in general, the iodine-modified electrodes are suitable for producing highly ordered adlayers of various organic molecules. TMPyP forms highly ordered adlayers on I–Au(111), I–Ag(111), and I–Pt(100).

The adlayer structure of benzene on Rh(111) and Pt(111) was also described in detail; it was found to be dependent on the electrode potential. The (3×3) structure found on Rh(111) in the cathodic potential range is almost identical with that found in UHV for the coadsorbed benzene and CO. The molecular shapes of naphthalene and anthracene could be clearly discerned by in situ STM.

The EC etching processes of Si(111), Si(110), and Si(100) were discussed in relation to the atomic structures of the step-edges. It was shown that the chemical etching produces well-defined GaAs and InP single-crystal electrodes. The atomic structures of these electrodes in solution could be clearly seen.

This article clearly demonstrates that STM allows us not only to determine interfacial structures but also to follow EC reactions. It is certain that in situ STM will continue to be the premier technique in the study of the relationship between the reactivity and the structure of electrode surfaces.

ACKNOWLEDGMENTS

This work was supported by the Exploratory Research for Advanced Technology (ERATO)–Itaya Electrochemiscopy Project organized by Japan Science and Technology Corporation (JST) and partially by the Ministry of Education, Science, Sports and Culture, Japan, with a Grant-in-Aid for Science Research on the Priority Area of "Electrochemistry of Ordered Interfaces".

For references see page 10066

ABBREVIATIONS AND ACRONYMS

AES	Auger Electron Spectroscopy
AFM	Atomic Force Microscopy
ARUPS	Angle-resolved Ultraviolet Photoemission Spectroscopy
CE	Counter Electrode
CV	Cyclic Voltammogram
DEMS	Differential Electrochemical Mass Spectrometry
EC	Electrochemical
EELS	Electron Energy-loss Spectroscopy
HOPG	Highly Ordered Pyrolytic Graphite
IR	Infrared
LEED	Low-energy Electron Diffraction
OCP	Open-circuit Potential
PPV	4,4′-bis(*N*-methylpyridinium)-*p*-phenylenedivinylene
RE	Reference Electrode
RHE	Reversible Hydrogen Electrode
SCE	Saturated Calomel Electrode
SHG	Second Harmonic Generation
STM	Scanning Tunneling Microscopy
SXS	Surface X-ray Scattering
TMPyP	5,10,15,20-tetrakis(*N*-methylpyridinium-4-yl)-21*H*,23*H*-porphyrine
UHV	Ultrahigh Vacuum
UHVEC	Ultrahigh Vacuum Electrochemical
UPD	Underpotential Deposition
WE	Working Electrode

RELATED ARTICLES

Surfaces **(Volume 10)**
Surfaces: Introduction • Proximal Probe Techniques • Scanning Probe Microscopy, Industrial Applications of • Scanning Tunneling Microscopy/Spectroscopy in Analysis of Surfaces

Electroanalytical Methods **(Volume 11)**
Electroanalytical Methods: Introduction • Infrared Spectroelectrochemistry • Self-assembled Monolayers on Electrodes • Surface Analysis for Electrochemistry: Ultrahigh Vacuum Techniques • Ultraviolet/Visible Spectroelectrochemistry • X-ray Methods for the Study of Electrode Interaction

REFERENCES

1. A.T. Hubbard, 'Electrochemistry at Well-characterized Surfaces', *Chem. Rev.*, **88**, 633–656 (1988).

List of selected abbreviations appears in Volume 15

2. M.P. Soriaga, 'Ultra-high Vacuum Techniques in the Study of Single-crystal Electrode Surfaces', *Prog. Surf. Sci.*, **39**, 325–443 (1992).
3. H.-J. Guntherodt, R. Wiesendanger (eds.), *Scanning Tunneling Microscopy I*, Springer-Verlag, Berlin, 1991.
4. P.A. Christensen, 'Electrochemical Aspects of STM and Related Techniques', *Chem. Soc. Rev.*, **21**, 197–208 (1992).
5. H. Siegenthaler, 'STM in Electrochemistry', in *Scanning Tunneling Microscopy II*, eds. R. Wiesendanger, H.-J. Guntherodt, Springer-Verlag, Berlin, 7–49, 1992.
6. A.A. Gewirth, B.K. Niece, 'Electrochemical Applications of In Situ Scanning Probe Microscopy', *Chem. Rev.*, **97**, 1129–1162 (1997).
7. K. Itaya, 'In Situ Scanning Tunneling Microscopy in Electrolyte Solutions', *Prog. Surf. Sci.*, **58**, 121–247 (1998).
8. J.G. Simmons, 'Electric Tunnel Effect between Dissimilar Electrodes Separated by a Thin Insulating Film', *J. Appl. Phys.*, **34**, 2581–2590 (1963).
9. J. Clavilier, R. Faure, G. Guinet, R. Durand, 'Preparation of Monocrystalline Pt Microelectrodes and Electrochemical Study of the Plane Surfaces Cut in the Direction of the {111} and {110}', *J. Electroanal. Chem.*, **107**, 205–209 (1980).
10. A. Hamelin, 'Double-layer Properties at sp and sd Metal Single-crystal Electrodes', in *Modern Aspects of Electrochemistry*, eds. B.E. Conway, R.E. White, J.O'M. Bockris, Plenum Press, New York, 1–101, Vol. 16, 1985.
11. S. Motoo, N. Furuya, 'Hydrogen and Oxygen Adsorption on Ir(111), (100) and (110) Planes', *J. Electroanal. Chem.*, **167**, 309–315 (1984).
12. K. Sashikata, N. Furuya, K. Itaya, 'In Situ Electrochemical Scanning Tunneling Microscopy of Single-crystal Surfaces of Pt(111), Rh(111), and Pd(111) in Aqueous Sulfuric Acid Solution', *J. Vac. Sci. Technol.*, **B9**, 457–464 (1991).
13. J. Clavilier, A. Rodes, K. El. Achi, M.A. Zamakhchari, 'Electrochemistry at Platinum Single Crystal Surfaces in Acidic Media: Hydrogen and Oxygen Adsorption', *J. Chim. Phys.*, **88**, 1291–1337 (1991).
14. J.M. Feliu, J.M. Orts, R. Gomez, A. Aldaz, J. Clavilier, 'New Information on the Unusual Adsorption States of Pt(111) in Sulfuric Acid Solutions from Potentiostatic Adsorbate Replacement by CO', *J. Electroanal. Chem.*, **372**, 265–268 (1994).
15. K. Itaya, S. Sugawara, K. Sashikata, N. Furuya, 'In Situ Scanning Tunneling Microscopy of Platinum(111) surface with the Observation of Monatomic Steps', *J. Vac. Sci. Technol.*, **A8**, 515–519 (1990).
16. S. Tanaka, S.-L. Yau, K. Itaya, 'In-situ Scanning Tunneling Microscopy of Bromine Adlayers on Pt(111)', *J. Electroanal. Chem.*, **396**, 125–130 (1995).
17. D.M. Kolb, 'Reconstruction Phenomena at Metal–Electrolyte Interfaces', *Prog. Surf. Sci.*, **51**, 109–173 (1996).

18. K. Wang, G.S. Chottiner, D.A. Scherson, 'Electrochemistry of Nickel(111) in Alkaline Electrolytes', *J. Phys. Chem.*, **97**, 10 108–10 111 (1993).
19. J.L. Stickney, C.B. Ehlers, B.W. Gregory, 'Adsorption of Gaseous and Aqueous HCl on the Low-index Planes of Copper', *Langmuir*, **4**, 1368–1373 (1988).
20. M. Hourani, A. Wieckowski, 'Electrochemistry of the Ordered Rh(111) Electrode: Surface Preparation and Voltammetry in $HClO_4$ Electrolyte', *J. Electroanal. Chem.*, **227**, 259–264 (1987).
21. C.M. Vitus, S.-C. Chang, B.C. Schardt, M.J. Weaver, 'In Situ Scanning Tunneling Microscopy as a Probe of Adsorbate-induced Reconstruction at Ordered Monocrystalline Electrodes: CO on Pt(100)', *J. Phys. Chem.*, **95**, 7559–7563 (1991).
22. F. Lu, G.N. Salaita, H. Baltruschat, A.T. Hubbard, 'Adlattice Structure and Hydrophobicity of Pt(111) in Aqueous Potassium Iodide Solutions: Influence of pH and Electrode Potential', *J. Electroanal. Chem.*, **222**, 305–320 (1987).
23. B.C. Schardt, S.-L. Yau, F. Rinaldi, 'Atomic Resolution Imaging of Adsorbates on Metal Surfaces in Air: Iodine Adsorption on Pt(111)', *Science*, **243**, 1050–1053 (1989).
24. N. Shinotsuka, K. Sashikata, K. Itaya, 'In Situ Scanning Tunneling Microscopy of Underpotential Deposition of Ag on Pt(111) $(\sqrt{7} \times \sqrt{7})R19.1^\circ$–I', *Surf. Sci.*, **335**, 75–82 (1995).
25. K. Sashikata, Y. Matsui, K. Itaya, M.P. Soriaga, 'Adsorbed-iodine-catalyzed Dissolution of Pd Single-crystal Electrodes: Studies by Electrochemical Scanning Tunneling Microscopy', *J. Phys. Chem.*, **100**, 20 027–20 034 (1996).
26. L.-J. Wan, S.-L. Yau, G.M. Swain, K. Itaya, 'In-situ Scanning Tunneling Microscopy of Well-ordered Rh(111) Electrodes', *J. Electroanal. Chem.*, **381**, 105–111 (1995).
27. T. Yamada, N. Batina, K. Itaya, 'Structure of Electrochemically Deposited Iodine Adlayer on Au(111) Studied by Ultrahigh-vacuum Instrumentation and In Situ STM', *J. Phys. Chem.*, **99**, 8817–8823 (1995).
28. B.M. Ocko, G.M. Watson, J. Wang, 'Structure and Electrocompression of Electrodeposited Iodine Monolayers on Au(111)', *J. Phys. Chem.*, **98**, 897–906 (1994).
29. T. Yamada, K. Ogaki, S. Okubo, K. Itaya, 'Continuous Variation of Iodine Adlattices on Ag(111) Electrodes: In Situ STM and Ex Situ LEED Studies', *Surf. Sci.*, **369**, 321–335 (1996).
30. J. Inukai, Y. Osawa, M. Wakisaka, K. Sashikata, Y.-G. Kim, K. Itaya, 'Underpotential Deposition of Copper on Iodine-Modified Pt(111): In Situ STM and Ex Situ LEED Studies', *J. Phys. Chem.*, **102**, 3498–3505 (1998).
31. N. Batina, T. Yamada, K. Itaya, 'Atomic Level Characterization of the Iodine-modified Au(111) Electrode Surface in Perchloric Acid Solution by In-situ STM and Ex-situ LEED', *Langmuir*, **11**, 4568–4576 (1995).
32. T. Yamada, N. Batina, K. Ogaki, S. Okubo, K. Itaya, 'Lateral Structure of Iodine Adlattices on Au(111) and Ag(111) Electrodes', in *Electrode Processes VI, Proceedings of the Sixth International Symposium*, eds. A. Wieckowski, K. Itaya, Electrochemical Society, Pennington, NJ, 43–57, 1996.
33. B.M. Ocko, O.M. Magnussen, J.X. Wang, R.R. Adzic, Th. Wandlowski, 'The Structure and Phase Behavior of Electrodeposited Halides on Single-crystal Metal Surfaces', *Physica B*, **221**, 238–244 (1996).
34. O.M. Magnussen, J. Hagebock, J. Hotlos, R.J. Behm, 'In Situ Scanning Tunneling Microscopy Observations of a Disorder–Order Phase Transition in Hydrogensulfate Adlayers on Au(111)', *Faraday Discuss. Chem. Soc.*, **94**, 329–338 (1992).
35. G.J. Edens, X. Gao, M.J. Weaver, 'The Adsorption of Sulfate on Gold(111) in Acidic Aqueous Media: Adlayer Structural Inferences from Infrared Spectroscopy and Scanning Tunneling Microscopy', *J. Electroanal. Chem.*, **375**, 357–366 (1994).
36. Z. Shi, J. Lipkowski, M. Gamboa, P. Zelenay, A. Wieckowski, 'Investigations of SO_4^{2-} Adsorption at the Au(111) Electrode by Chronocoulometry and Radiochemistry', *J. Electroanal. Chem.*, **366**, 317–326 (1994).
37. A.M. Funtikov, U. Stimming, R. Vogel, 'Anion Adsorption from Sulfuric Acid Solutions on Pt(111) Single Crystal Electrodes', *J. Electroanal. Chem.*, **428**, 147–153 (1997).
38. L.-J. Wan, S.-L. Yau, K. Itaya, 'Atomic Structure of Adsorbed Sulfate on Rh(111) in Sulfuric Acid Solution', *J. Phys. Chem.*, **99**, 9507–9513 (1995).
39. P. Zelenay, A. Wieckowski, 'Radiochemical Assay of Adsorption at Single Crystal/Solution Interfaces' *J. Electrochem. Soc.*, **139**, 2552–2558 (1992).
40. C. Stuhlmann, I. Villegas, M.J. Weaver, 'Scanning Tunneling Microscopy and Infrared Spectroscopy as Combined In Situ Probes of Electrochemical Adlayer Structure. Cyanide on Pt(111)', *Chem. Phys. Lett.*, **219**, 319–324 (1994).
41. Y.-G. Kim, S.-L. Yau, K. Itaya, 'Direct Observation of Alkali Cations on Cyanide-modified Pt(111) by Scanning Tunneling Microscopy', *J. Am. Chem. Soc.*, **118**, 393–400 (1996).
42. O.M. Magnussen, J. Hotlos, R.J. Nichols, D.M. Kolb, R.J. Behm, 'Atomic Structure of Cu Adlayers on Au(100) and Au(111) Electrodes Observed by In Situ Scanning Tunneling Microscopy', *Phys. Rev. Lett.*, **64**, 2929–2932 (1990).
43. O.M. Magnussen, J. Hotlos, G. Beitel, D.M. Kolb, R.J. Behm, 'Atomic Structure of Ordered Copper Adlayers on Single-crystalline Gold Electrodes', *J. Vac. Sci. Technol.*, **B9**, 969–975 (1991).
44. T. Hachiya, H. Honbo, K. Itaya, 'Detailed Underpotential Deposition of Copper on Gold(111) in Aqueous Solutions', *J. Electroanal. Chem.*, **315**, 275–291 (1991).
45. S. Manne, P.K. Hansma, J. Massie, V.B. Elings, A.A. Gewirth, 'Atomic-resolution Electrochemistry with the

Atomic Force Microscopy: Copper Deposition on Gold', *Science*, **251**, 183–186 (1991).

46. Z. Shi, J. Lipkowski, 'Coadsorption of Cu^{2+} and SO_4^{2-} at the Au(111) Electrode', *J. Electroanal. Chem.*, **365**, 303–309 (1994).
47. M.F. Toney, J.N. Howard, J. Richer, G.L. Borges, J.G. Gordon, O.R. Melroy, D. Yee, L. B. Sorensen, 'Electrochemical Deposition of Copper on a Gold Electrode in Sulfuric Acid: Resolution of the Interfacial Structure', *Phys. Rev. Lett.*, **75**, 4472–4475 (1995).
48. D.A. Huckaby, L. Blum, 'A Model for Sequential First-order Phase Transitions Occurring in the Underpotential Deposition of Metals', *J. Electroanal. Chem.*, **315**, 255–261 (1991).
49. S. Chiang, 'Scanning Tunneling Microscopy Imaging of Small Adsorbed Molecules on Metal Surfaces in an Ultrahigh Vacuum Environment', *Chem. Rev.*, **97**, 1083–1096 (1997).
50. A. Ikai, 'STM and AFM of Bio/Organic Molecules and Structures', *Surf. Sci. Rep.*, **26**, 261–332 (1996).
51. J. Lipkowski, P.N. Ross (eds.), *Adsorption of Molecules at Metal Electrodes*, VCH, New York, 1992.
52. N.J. Tao, Z. Shi, 'Real-time STM/AFM Study of Electron Transfer Reactions of an Organic Molecule: Xanthine at the Graphite–Water Interfaces', *Surf. Sci. Lett.*, **321**, L149–L156 (1994).
53. N.J. Tao, G. Cardenas, F. Cunha, Z. Shi, 'In Situ STM and AFM Study of Protoporphyrin and Iron(III) and Zinc(II) Protoporphyrins Adsorbed on Graphite in Aqueous Solutions', *Langmuir*, **11**, 4445–4448 (1995).
54. F. Cunha, N.J. Tao, 'Surface Charge Induced Order–Disorder Transition in an Organic Monolayer', *Phys. Rev. Lett.*, **75**, 2376–2379 (1995).
55. M. Kunitake, N. Batina, K. Itaya, 'Self-organized Porphyrine Array on Iodine-modified Au(111) in Electrolyte Solutions: In Situ Scanning Tunneling Microscopy Study', *Langmuir*, **11**, 2337–2340 (1995).
56. M. Batina, M. Kunitake, K. Itaya, 'Highly Ordered Molecular Arrays Formed on Iodine-modified Au(111) in Solution: In Situ STM Imaging', *J. Electroanal. Chem.*, **405**, 245–250 (1996).
57. M. Kunitake, U. Akiba, N. Batina, K. Itaya, 'Structures and Dynamic Formation Processes of Porphyrine Adlayers on Iodine-modified Au(111) in Solution: In Situ STM Study', *Langmuir*, **13**, 1607–1615 (1997).
58. K. Ogaki, N. Batina, M. Kunitake, K. Itaya, 'In Situ Scanning Tunneling Microscopy of Ordering Processes of Adsorbed Porphyrine on Iodine-modified Ag(111)', *J. Phys. Chem.*, **100**, 7185–7190 (1996).
59. K. Sashikata, T. Sugata, M. Sugimasa, K. Itaya, 'In Situ Scanning Tunneling Microscopy Observation of a Porphyrine Adlayer on an Iodine-modified Pt(100) Electrode', *Langmuir*, **14**, 2896–2902 (1998).
60. G.A. Somorjai, *Introduction to Surface Chemistry and Catalysis I*, J. Wiley & Sons, New York, 1994.
61. H. Ohtani, R.J. Wilson, S. Chiang, C.M. Mate, 'Scanning Tunneling Microscopy Observations of Benzene Molecules on the Rh($111-3\times3$)(C_6H_6+2CO) Surface', *Phys. Rev. Lett.*, **60**, 2398–2401 (1988).
62. H.A. Yoon, M. Salmeron, G.A. Somorjai, 'Scanning Tunneling Microscopy (STM) Study of Benzene and its Coadsorption with Carbon Monoxide on Rh(111)', *Surf. Sci.*, **373**, 300–306 (1997).
63. S.-L. Yau, Y.-G. Kim, K. Itaya, 'In Situ Scanning Tunneling Microscopy of Benzene Adsorbed on Rh(111) and Pt(111) in HF Solution', *J. Am. Chem. Soc.*, **118**, 7795–7803 (1996).
64. H. Baltruschat, U. Schmiemann, 'The Adsorption of Unsaturated Organic Species at Single Crystal Electrodes Studied by Differential Electrochemical Mass Spectrometry', *Ber. Bunsenges. Phys. Chem.*, **97**, 452–460 (1993).
65. M. Neuber, F. Schneider, C. Zubragel, M. Neumann, 'A Dense and CO Free Benzene Structure on Rh(111)', *J. Phys. Chem.*, **99**, 9160–9168 (1995).
66. P.S. Weiss, D.M. Eigler, 'Site Dependence of the Apparent Shape of a Molecule in Scanning Tunneling Microscope Images: Benzene on Pt{111}', *Phys. Rev. Lett.*, **71**, 3139–3142 (1993).
67. S.-L. Yau, Y.-G. Kim, K. Itaya, 'High-resolution Imaging of Aromatic Molecules Adsorbed on Rh(111) and Pt(111) in Hydrofluoric Acid Solution: In Situ STM Study', *J. Phys. Chem. B*, **101**, 3547–3553 (1997).
68. R.F. Lin, R.J. Koestner, M.A. Van Hove, G.A. Somorjai, 'The Adsorption of Benzene and Naphthalene on the Rh(111) Surface: A LEED, AES and TDS Study', *Surf. Sci.*, **134**, 161–183 (1983).
69. S.-L. Yau, K. Itaya, 'Structures and Dynamic Processes of Molecular Adlayers on Rh(111) and Pt(111) in HF Solution: Naphthalene and Biphenyl', *Colloids Surf. A: Physicochem. Eng. Aspects*, **134**, 21–30 (1998).
70. G.S. Higashi, Y.J. Chabal, G.W. Trucks, K. Raghavachari, 'Ideal Hydrogen Termination of the Si(111) Surface', *Appl. Phys. Lett.*, **56**, 656–658 (1990).
71. R.S. Becker, G.S. Higashi, Y.J. Chabal, A.J. Becker, 'Atomic Scale Conversion of Clean Si(111):$H-1\times1$ to Si(111)-2×1 by Electron-stimulated Desorption', *Phys. Rev. Lett.*, **65**, 1917–1920 (1990).
72. H.E. Hessel, A. Feltz, M. Reiter, U. Memmert, R.J. Behm, 'Step-flow Mechanism versus Pit Corrosion: Scanning-tunneling Microscopy Observations on Wet Etching of Si(111) by HF Solutions', *Chem. Phys. Lett.*, **186**, 275–280 (1991).
73. K. Itaya, R. Sugawara, Y. Morita, H. Tokumoto, 'Atomic Resolution Images of H-terminated Si(111) Surfaces in Aqueous Solutions', *Appl. Phys. Lett.*, **60**, 2534–2536 (1992).
74. S.-L. Yau, K. Kaji, K. Itaya, 'Electrochemical Etching of Si(001) in NH_4F Solutions: Initial Stage and {111} Microfacet Formation', *Appl. Phys. Lett.*, **66**, 766–768 (1995).

75. K. Kaji, S.-L. Yau, K. Itaya, 'Atomic Scale Etching Processes of n-Si(111) in NH_4F Solutions: In Situ Scanning Tunneling Microscopy', *J. Appl. Phys.*, **78**, 5727–5733 (1995).
76. J.H. Ye, K. Kaji, K. Itaya, 'Atomic-scale Elucidation of the Anisotropic Etching of (110) n-Si in Aqueous NH_4F: Studies by In Situ Scanning Tunneling Microscopy', *J. Electrochem. Soc.*, **143**, 4012–4019 (1996).
77. P. Allongue, V. Kieling, H. Gerischer, 'Etching Mechanism and Atomic Structure of H-Si(111) Surfaces Prepared in NH_4F', *Electrochimica Acta*, **40**, 1353–1360 (1995).
78. P. Allongue, V. Costa-Kieling, H. Gerischer, 'Etching of Silicon in NaOH Solutions: I. In Situ Scanning Tunneling Microscopic Investigation of n-Si (111)', *J. Electrochem. Soc.*, **140**, 1009–1018 (1993).
79. P. Allongue, V. Costa-Kieling, H. Gerischer, 'Etching of Silicon in NaOH Solutions: II. Electrochemical Studies of n-Si(111) and (100) and Mechanism of the Dissolution', *J. Electrochem. Soc.*, **140**, 1018–1026 (1993).
80. P. Jakob, Y.J. Chabal, K. Kuhnke, S.B. Christman, 'Monohydride Structures on Chemically Prepared Silicon Surfaces', *Surf. Sci.*, **302**, 49–56 (1994).
81. H. Yao, S.-L. Yau, K. Itaya, 'Atomic Resolution Images of GaAs(111)A Surfaces in Sulfuric Acid Solution', *Surf. Sci.*, **335**, 166–170 (1995).
82. H. Yao, S.-L. Yau, K. Itaya, 'In Situ Scanning Tunneling Microscopy of GaAs(001), (111)A, and (111)B Surfaces in Sulfuric Acid Solution', *Appl. Phys. Lett.*, **68**, 1473–1475 (1996).
83. K. Uosaki, S. Ye, N. Sekine, 'Study of the Electronic Structure of GaAs(100) Single Crystal Electrode/Electrolyte Interfaces by Electrochemical Tunneling Spectroscopy', *Bull. Chem. Soc. Jpn.*, **69**, 275–288 (1996).
84. H. Yau, K. Itaya, 'Atomically Resolved Scanning Tunneling Microscopy Images of InP(001), (111)A, and (111)B Surfaces in Sulfuric Acid Solution', *J. Electrochem. Soc.*, **145**, 3090–3094 (1998).

Selective Electrode Coatings for Electroanalysis

Lee J. Klein and Dennis G. Peters
Indiana University, Bloomington, USA

Tremendous strides have taken place in recent years in our ability to modify electrodes chemically to enhance their selectivity and sensitivity for electroanalyses. Four different categories of selective electrode coatings have been chosen for discussion in this article: self-assembled monolayers (SAMs); polymer coatings; Langmuir–Blodgett films; and immobilized enzymes. Each of these chemical modifications is discussed in terms of the necessary materials, the methods of preparation and characterization, and some examples of analytical applications.

1 INTRODUCTION

In the realm of electroanalysis, one of the greatest challenges is to design and fabricate an electrode that ideally will respond both selectively and sensitively to a particular chemical species (e.g. a simple inorganic cation or anion, a metal–ion complex, or an organic or biological material). Over the past 30 years or so, tremendous advances have been made in our ability to tailor or modify chemically the surfaces of electrodes to enhance their selectivity and sensitivity for a variety of analyses.[1] In a sense, these advances constitute a form of molecular engineering, in which one knowledgeably and purposefully redesigns or alters the surface of an electrode to try to define and control how that electrode responds to one targeted species or to a family of species. In this short article, we attempt to introduce and describe some of these advances.

For references see page 10084

Four different categories of selective electrode coatings have been chosen for discussion. First, we begin with SAMs on electrodes; a SAM is an organized array of molecules strongly adsorbed or chemically bonded to an electrode surface, and such a monolayer can form spontaneously when a carefully prepared electrode surface is brought into contact with an appropriate species that can interact with the surface. Second, we consider polymer-coated electrodes; most simply, one can create a polymer-coated electrode (a) by electropolymerizing a selected (and perhaps functionalized) monomer directly onto the surface of an electrode or (b) by coating or depositing a preformed (and functionalized) polymer onto an electrode surface. Third, we describe the preparation of Langmuir–Blodgett films on electrodes; a Langmuir–Blodgett film is an ordered monolayer of amphiphilic molecules formed at a liquid/gas interface, and this film is subsequently transferred mechanically to the surface of an electrode. Fourth, we discuss the immobilization of enzymes onto electrodes, which can be accomplished in a variety of ways, including electrostatic self-adsorption and incorporation into SAMs, polymer films, or Langmuir–Blodgett films.

Recent progress in this area of research and development has been so great that we have been forced to be selective in choosing the topics, examples, and references included in this article. Thus, we can offer only a brief glimpse into this highly interesting subject. Obviously, by referring to the cited reviews and original papers, and by seeking out other publications by the major contributors to this field, the interested reader can find more information.

2 SELF-ASSEMBLED MONOLAYERS ON ELECTRODES

When an electrode (substrate) is immersed into a solvent in which an appropriate surfactant is dissolved, the molecules of surfactant can interact (react) spontaneously at the surface of the electrode to give a SAM. From the viewpoint of electroanalysis, an excellent example of such a self-assembly process is seen in the spontaneous formation of a SAM consisting of a functionalized alkanethiol on the surface of a gold electrode. For purposes of illustration, let us assume that we have a solution of $(C_5H_5)Fe(C_5H_4)CONH(CH_2)_6SH$ in ethanol. Typical of many surfactants, $(C_5H_5)Fe(C_5H_4)CONH(CH_2)_6SH$ consists of three moieties: (a) the surface-active head group, –SH, which tethers the molecule to the gold surface via the formation of an S–Au bond; (b) the alkyl group, $-(CH_2)_6-$, which is essentially a spacer that separates the head group from the end group; and (c) the end group itself, $(C_5H_5)Fe(C_5H_4)CONH-$, which is farthest from the surface of the gold electrode and which imparts desirable characteristics to the SAM on gold (such as selectivity toward a particular species in a solution to be analyzed). Figure 1 depicts the spontaneous process by which this functionalized alkanethiol, originally dissolved in ethanol, forms a SAM on gold.

Although details of the mechanism of adsorption of an alkanethiol are still unclear, the most important step in the formation of a SAM on gold is chemisorption of the deprotonated (thiolate) form, $(C_5H_5)Fe(C_5H_4)CONH(CH_2)_6S^-$, onto the gold surface.[2,3] Then, as more and more of the thiolate form becomes attached (adsorbed) to the gold surface, the assembly of a closely packed (ordered) SAM takes place.

2.1 Types of Materials Used

Since the pioneering work of Zisman,[4] who investigated the wetting behavior of solutions of surfactants, several classes of compounds have been used to prepare SAMs. First, long-chain carboxylic (*n*-alkanoic) acids can be adsorbed onto a variety of substrates, including aluminum oxide,[5–7] silver,[8,9] and iron oxide.[10] In addition, diacetylenic carboxylic acids, such as $CH_3(CH_2)_8C{\equiv}C{-}C{\equiv}C(CH_2)_8COOH$, have been used to prepare SAMs on glass, quartz, and silicon;[11] although such substrates cannot serve as electrodes, they could be rendered conductive by being coated with thin layers of indium-doped tin oxide (ITO) prior to the formation of the SAM. Second, alkylsilane derivatives can be bonded onto hydroxylated surfaces; for example, compounds such as $RSiX_3$, R_2SiX_2, and R_3SiX (where R is an alkyl chain,

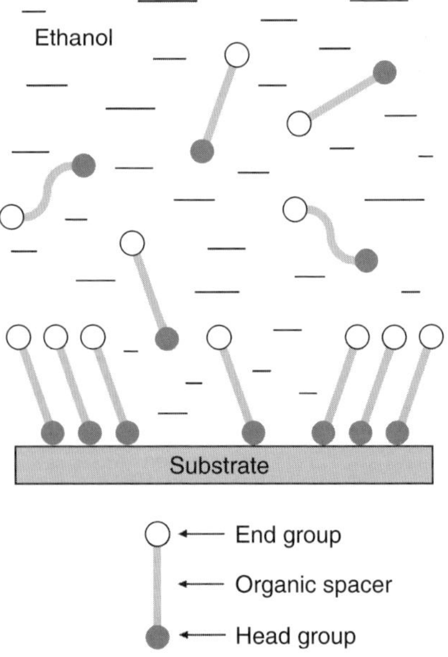

Figure 1 Spontaneous formation of a SAM of functionalized alkanethiol onto a gold surface.

which may or may not be functionalized with a suitable electroactive end group, and X is a chloro or alkoxy moiety) can be attached to glass,[12] silicon dioxide,[13] or aluminum oxide,[14] via the formation of stable Si–O bonds. Once again, the strategy of coating the substrate with a thin layer of ITO (which is also a hydroxylated surface) can provide a conductive electrode onto which the SAM can form. When a trichlorosilane ($RSiCl_3$) is used to prepare a SAM on one of these hydroxylated substrates, it is possible to obtain a cross-linked network, as illustrated in Figure 2.

Third, as described earlier, functionalized organosulfur compounds such as alkanethiols, dialkyl sulfides, and dialkyl disulfides can form SAMs on gold,[15–18] silver,[19] copper,[20] nickel,[21] mercury,[22] and platinum[23] surfaces. Gold is much preferred, however, because it has the least tendency to form a surface oxide and is, therefore, the easiest substrate to prepare and handle.

2.2 Methods of Preparation

Many of the articles cited in the preceding section contain specific procedures employed to prepare SAMs; the reader is encouraged to consult these sources. However, for illustrative purposes, it is useful to examine some key steps of a general protocol for the formation of a SAM consisting of a functionalized alkanethiol on gold. First, the particular gold substrate must be selected; one may use either a polycrystalline (rough) gold surface or an evaporated (smooth) gold surface. Second, the gold surface is usually cleaned (freed of adventitious impurities) by being heated, immersed in solutions of strong oxidants, exposed to an argon or oxygen plasma, or placed in a suitable electrolytic cell whereupon the potential of the gold electrode is cycled; often a combination of these techniques is used. Third, the desired functionalized alkanethiol (which must often be synthesized) is dissolved in an appropriate organic solvent; typically, the concentration of the alkanethiol is 10^{-3} M. Fourth, the gold substrate is immersed into the alkanethiol solution for times ranging from minutes to 12 h or longer at room temperature to promote the self-assembly process. Finally, after being rinsed, the monolayer-covered gold electrode may be annealed thermally to cause the monolayer to rearrange into a more ordered structure.

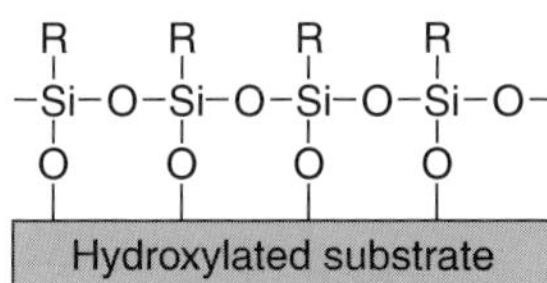

Figure 2 Cross-linked network formed by reaction of a trichlorosilane ($RSiCl_3$) with a hydroxylated substrate.

2.3 Methods of Characterization

Several electroanalytical techniques can be used to study and characterize SAMs on electrodes. Cyclic voltammetry is the most popular method, but differential pulse voltammetry, chronoamperometry, and hydrodynamic voltammetry deserve mention.

Ulman[24] and Finklea[25] have reviewed the literature on the various nonelectrochemical techniques that can be employed to characterize SAMs. Although space limitations prevent any detailed discussion of these methods, it is useful at least to list some of these techniques, most of which are completely described elsewhere in this encyclopedia. Many workers have utilized wetting contact angle to elucidate the structure of SAMs. Both infrared and Raman spectroscopy as well as ellipsometry and surface plasmon resonance techniques have seen extensive use. High-vacuum surface methods such as X-ray photoelectron spectroscopy (XPS), Auger electron spectroscopy (AES), high-resolution electron energy loss spectroscopy (HREELS), and low-energy electron diffraction spectroscopy (LEEDS) have been employed. Scanning tunneling microscopy (STM) and atomic force microscopy (AFM) can provide images of alkanethiol-based SAMs on gold substrates. Some investigators have used the electrochemical quartz crystal microbalance (EQCM) to probe the kinetics of self-assembly of alkanethiols onto gold surfaces.

2.4 Analytical Applications

Despite enormous interest in the preparation and properties of SAMs, especially those involving functionalized alkanethiols on gold electrodes, the list of practical analytical applications is still somewhat limited. A review by Mandler and Turyan[26] summarizes the literature through mid-1995 on the use of SAMs in electroanalytical chemistry. Citing a few examples of analytical applications will help to convey the excitement and promise of this area of research.

Steinberg and Rubinstein[27] found that a mixed SAM, composed of 2,2′-thiobisethyl acetoacetate and *n*-octadecyl mercaptan, on a gold substrate can be used to determine Cu^{2+} in the presence of Fe^{3+}, and that replacing *n*-octadecyl mercaptan with *n*-octadecyltrichlorosilane afforded an electrode system for the detection of trace levels of Cu^{2+} (10^{-7} M), Pb^{2+} (10^{-5} M), and Zn^{2+} (10^{-9} M). A novel microsensor for pH measurements has been developed that is based on the coadsorption of ferrocenyl and quinone thiols onto a gold microelectrode; the redox potential for the ferrocenyl moiety is independent of pH, but the redox potential for the quinone moiety is pH dependent, so the difference in these redox potentials is related to the pH of the sample.[28]

For references see page 10084

Turyan and Mandler[29] have devised a highly sensitive and selective method for the determination of Cd^{2+}; a detection limit as low as 4×10^{-12} M was obtained. These workers prepared SAMs, consisting of *w*-mercaptocarboxylic acids ($HS(CH_2)_nCOOH$, where $n = 2$, 3, 6, and 11) adsorbed onto both thin mercury film and gold electrodes. When placed into a buffered solution containing Cd^{2+}, the modified electrode acts to preconcentrate Cd^{2+}, possibly through formation of a complex involving the interaction of Cd^{2+} with both the sulfur and carboxylate moieties of the SAM. After the preconcentration step, the complexed cadmium(II) can be determined by means of voltammetry.

Another approach for gaining selectivity for electroanalysis involves the incorporation of so-called "molecular gates" in an otherwise blocking monolayer.[30] Two mercaptans, 1-hexadecanethiol and 4-hydroxythiophenol, are coadsorbed onto a gold surface to form the SAM; the first compound is a passivating (blocking) substance, whereas the second species acts as an electron-transfer site (the "molecular gate"). These "molecular gates" allow $[Ru(NH_3)_6]^{3+}$, but not $[Fe(CN)_6]^{3-}$, to undergo a reversible one-electron reduction.

Chlorpromazine in human urine has been measured by means of an amperometric flow detector coated with unsubstituted *n*-alkanethiols on gold.[31] By varying the chain length of the alkanethiol, one can tailor the permselective properties of the SAM so that other constituents of the urine are prevented from reaching the gold surface. In the same study, it was shown that such a SAM-coated detector can distinguish dopamine (a neurotransmitter) from ascorbate; this approach overcomes the difficult problem of detecting dopamine in cerebral fluid, which contains relatively high concentrations of ascorbate.

A method to determine glucose relies on the mediated oxidation of the reduced form of glucose oxidase (GOx(red)), by dicationic (ferricenylmethyl)dimethyloctadecyl ammonium ion ($C_{18}Fc^{2+}$), which is incorporated into a self-assembled bilayer.[32] Preparation of the bilayer begins with the spontaneous reaction of octadecyltrichlorosilane with a film of microporous aluminum oxide to form a cross-linked SAM. After gold is evaporated onto the back of the SAM–Al_2O_3 assembly, (ferrocenylmethyl)dimethyloctadecyl ammonium ion ($C_{18}Fc^{+}$) is allowed to interleave with the silanized surface to form a bilayer, and glucose oxidase (GOx) is immobilized on the bilayer. In the presence of glucose, the following reactions (Equations 1–3) occur

$$\text{GOx} + \text{glucose} \longrightarrow \text{GOx(red)} + \text{gluconolactone} \qquad (1)$$

$$\text{GOx(red)} + 2C_{18}Fc^{2+} \longrightarrow \text{GOx} + 2C_{18}Fc^{+} + 2H^{+} \qquad (2)$$

$$C_{18}Fc^{+} \rightleftharpoons C_{18}Fc^{2+} + e^{-} \qquad (3)$$

List of selected abbreviations appears in Volume 15

where the last process gives rise to a steady-state voltammetric current that is related to the original concentration of glucose.

3 POLYMER-COATED ELECTRODES

Although electrodes coated with a variety of polymer films have been prepared, there are essentially three major types of polymers to consider here: electronically conducting organic polymers, polymers containing electroactive metal-ion sites, and ion-exchange polymers.

3.1 Organic Conducting Polymers

3.1.1 Types of Materials Used

Two of the best examples of organic conducting polymers are polypyrrole **(1)** and polythiophene **(2)**. Other conducting polymers include polyacetylene, polyaniline, poly-*p*-phenylene, polyazulene, polyindole, and polyfuran. Moreover, it is possible to prepare copolymers consisting of two different monomers (such as a mixture of thiophene and pyrrole[33,34]), and even more exotic starting materials (such as 1-thienyl-2-cyano-2-phenylethylene[35,36]) can be used. Information about the properties of conducting polymers on electrodes can be found in two reviews.[37,38]

(1) **(2)**

3.1.2 Methods of Preparation

There are several important requirements for the successful electrochemical preparation of a polymer film on an electrode: the parent monomer must be readily oxidizable (reducible) at an electrode in the chosen solvent–electrolyte; the radical cation (anion) arising from oxidation (reduction) of the monomer must react rapidly with other monomeric species to grow the polymer; the polymer must have a relatively low solubility so that it will deposit on the electrode; and the polymer must have an oxidation (reduction) potential less (more) positive than that of the monomer if the polymer is to have a high conductivity. A common procedure[39] for the formation of a film of polypyrrole on a platinum substrate involves the controlled-potential electrolysis of a solution of 50 mM pyrrole in acetonitrile containing 0.10 M tetraethylammonium tetrafluoroborate. Films are grown slowly by proper choice of the potential, the thickness of a film (typically 20–80 nm) is controlled by the amount of charge passed, and oxygen is excluded from the system. Polypyrrole films can also be prepared by means of constant-current electrolysis, and polypyrrole can be deposited onto other substrates including carbon, gold, palladium, and ITO.

Electrochemical techniques for the preparation of some of the other polymeric films mentioned above are similar to those used to obtain polypyrrole coatings. In addition, one can prepare some conducting polymers by casting a soluble precursor polymer onto a chosen substrate and by heating the system to form the desired material.(37) Alternatively, if the conducting polymer can be synthesized by a convenient chemical procedure and if the polymer is soluble in a volatile solvent, a solution of the polymer can be applied to an electrode surface by means of dip- or spin-coating, and the solvent can be evaporated to produce the polymer-coated electrode.

3.1.3 Methods of Characterization

Most of the methods mentioned in section 2.3 can be utilized to characterize polymer films on electrodes. Scanning electron microscopy (SEM) and transmission electron microscopy (TEM) are other useful approaches. A rich source of both electrochemical and nonelectrochemical techniques for the study of conducting polymers is a review by Doblhofer and Rajeshwar.(40)

One of the most important properties of a conducting polymer film is its conductivity, which can be measured by means of the classic four-probe method.(41) Conductivity of a conjugated polymer film arises from partial oxidation (p-doping) or partial reduction (n-doping). In electrochemical p-doping – which all of the polymers mentioned so far undergo – electrons are withdrawn from the polymer, and the polymer becomes positively charged. Then anions (usually from the supporting electrolyte) diffuse into the polymer to provide charge compensation; thus, the choice of supporting electrolyte can affect the nature of the conducting polymer. In combination, all of these processes profoundly influence the formation, structure, and conductivity of organic polymers; these issues are discussed in two previously cited reviews.(37,38)

3.1.4 Analytical Applications

Several kinds of analytical applications have been developed that involve the use of an electrode coated with a film of conducting organic polymer. Examples in the present section have been chosen to exclude situations where the polymer film also contains either trapped transition-metal sites (see section 3.2) or an immobilized enzyme (see section 5).

Shiu et al.(42) have reported that, when pyrrole is oxidatively polymerized onto a glassy carbon electrode in the presence of either bathophenanthroline disulfonate or bathocuproine disulfonate, these electroinactive anionic complexing agents are incorporated into the polymer. When Cu^{+} or Cu^{2+} is extracted (via formation of a complex) from a sample solution into the polymer, a cyclic voltammetric peak current is seen that can be used to determine these cations; the electrode responds better to Cu^{+} than to Cu^{2+}, and the detection limit for Cu^{+} is approximately 2×10^{-6} M. In their paper, the authors review earlier efforts to develop polymer-coated electrodes for the determination of other metal cations.

Because p-doping of a polymer is associated with incorporation of anions, this phenomenon can be utilized to fabricate polymer-coated electrodes that respond to anions in a sample solution. This principle has been employed for the development of a polypyrrole-coated platinum detector for the flow injection analysis (FIA) of phosphate, carbonate, and acetate;(43) the electrode response to phosphate and carbonate is linear over the concentration range from 10^{-5} to 10^{-3} M. Similarly, a polyaniline-coated glassy carbon detector for the FIA of iodide, bromide, thiocyanate, and thiosulfate has been devised, with detection limits of 1, 5, 10, and 10 mg L^{-1}, respectively.(44) A poly(3-methylthiophene)-coated graphite electrode has been constructed for the potentiometric determination of iodide, chloride, bromide, and sulfite;(45) for the measurement of iodide, the linear dynamic range of the electrode is 1×10^{-7} to 5×10^{-1} M, and the detection limit for iodide is 1×10^{-8} M.

A platinum electrode coated with polypyrrole with incorporated anti-human serum albumin (serving as a charge-compensating anion) has been found to respond linearly in alternating current (AC) voltammetric experiments to human serum albumin at concentrations ranging from 10 to 50 µg mL^{-1}.(46) Polypyrrole on glassy carbon, as well as polypyrrole on mercury-filmed glassy carbon, provides a sensor for the amperometric determination of dichloramine separated from chloramine in potable water by means of high-performance liquid chromatography (HPLC); a detection limit of 4×10^{-6} M dichloramine is possible by use of the polypyrrole–mercury film–glassy carbon electrode.(47)

Slater et al.(48) have devised sensors consisting of platinum microband-electrode arrays coated with polypyrrole (or an alkylated polypyrrole). When such a detector is exposed to the vapors of an alcohol, molecules of the alcohol permeate the polymer, causing the polypyrrole film to swell and to undergo a change in conductivity. Time-resolved measurement of the change in conductivity of the polypyrrole sensors was used to separate and classify methanol, ethanol, and propanol vapors.

3.2 Metal Ion-containing Polymers

Several strategies can be employed to incorporate electroactive transition-metal sites into polymer films on

For references see page 10084

electrodes; these kinds of polymer films are often referred to as *redox polymers*, although this terminology can also embrace polymer films bearing electroactive organic groups (e.g. naphthalene, anthracene, or viologen). First, it is possible to polymerize onto the surface of an electrode a monomer in which the transition-metal site is *covalently bonded* to the polymerizable moiety; a classic example is vinylferrocene, which gives rise to poly(vinylferrocene) **(3)**.[49] Second, the monomer to be polymerized can consist of a transition-metal complex *coordinately linked* to the polymerizable moiety; an example of such a monomer is $[Ru(bpy)_2(vbpy)]^{2+}$ (where vbpy denotes 4-methyl-4′-vinyl-2,2′-bipyridine and bpy denotes 2,2′-bipyridine) which forms poly$[Ru(bpy)_2(vbpy)]^{2+}$ **(4)**.[50] Third, a transition-metal complex can be *coordinately linked* to a polymer after the polymer has been deposited onto an electrode; an example is the interaction of $[Ru(NH_3)_5OH_2]^{2+}$ with a pyridine moiety of a film of poly(4-vinylpyridine) (PVP) **(5)**.[51]

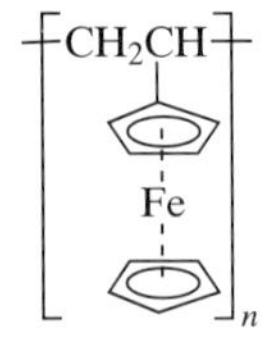

(3) 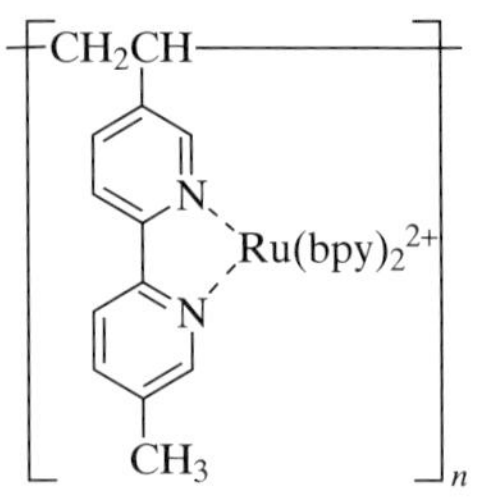 **(4)** 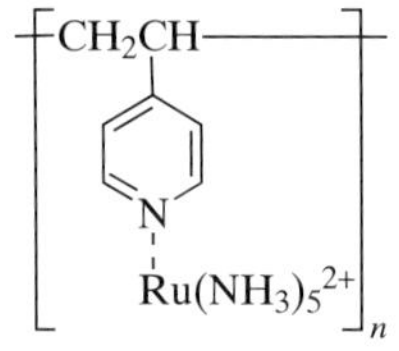 **(5)**

Fourth, an ion-exchange polymer can be coated onto an electrode, and the fixed charged sites (cationic or anionic) *electrostatically interact* with a metal complex; an example is the attachment of $[Ru(bpy)_3]^{2+}$ to sulfonate groups of Nafion®, a perfluorinated polysulfonated ion-exchange polymer. Only the first three kinds of redox polymers will be discussed further in this section; films of ion-exchange polymers on electrodes are described in section 3.3.

Before leaving this discussion, we should mention a few other ways to immobilize electroactive transition-metal sites onto electrode surfaces, although these approaches do not strictly involve polymer-coated electrodes. One method[52] entails adsorption of a molecule with an extended π-electron system (e.g. 1-(9-phenanthrene)-2-(4-pyridine)ethene) onto a suitable electrode (e.g. pyrolytic graphite or glassy carbon); then an appropriately substituted transition-metal species (e.g. $[Ru(NH_3)_5OH_2]^{2+}$) is coordinately linked to the adsorbate. In another approach, a functionalized silane (e.g. 4-[β-(trichlorosilyl)ethyl]pyridine) is attached to an electrode possessing surface hydroxy groups (e.g. Pt–OH), after which a transition-metal complex (e.g. $[Ru(bpy)_2(DME)Cl]^+$, where DME denotes dimethoxyethyl) is bonded to the functionalized (pyridyl) end of the silane.[53] In addition, one can functionalize the surface groups on edge planes of pyrolytic graphite by treating the graphite with a suitable reagent (e.g. thionyl chloride or cyanuric chloride); subsequent reaction with tetra(aminophenyl)porphyrin, followed by metallation of the porphyrin, completes the process.[54]

Many aspects of redox polymers bearing electroactive transition-metal sites, as well as other features of chemically modified electrodes, have been reviewed by Abruña.[55]

3.2.1 *Types of Materials Used*

Three examples of polymerizable species with covalently bonded transition-metal sites are (a) the aforementioned vinylferrocene (which can be polymerized onto platinum), (b) *N*-(2-ferrocenylethyl)pyrrole **(6)** (which can be copolymerized with pyrrole to produce a ferrocene-containing polypyrrole film on platinum),[56] and (c) *N*,*N*′-bis-[3-(pyrrol-1-yl)propyl]ferrocene-1,1′-dicarboxamide **(7)** (which can be polymerized onto glassy carbon).[57]

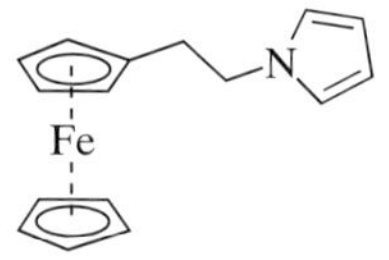

(6) 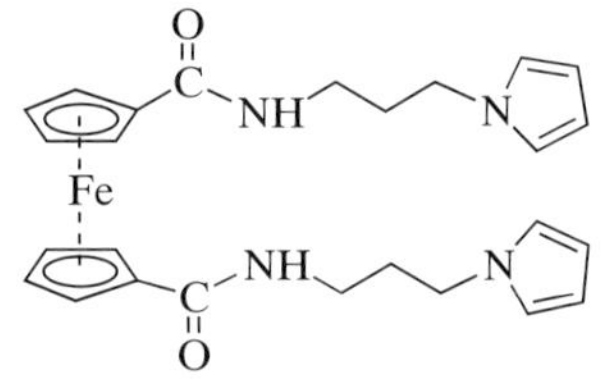 **(7)**

We now turn our attention to electrodes coated with polymer films bearing coordinately linked redox-active transition-metal sites; the transition metal can be (a) part of the original molecule (monomer) from which the polymer film is created or (b) introduced after the polymer film has been deposited onto the surface of the electrode.

Included in the first subclass are monomers having metal-containing ligands bound to pyrrole.[58] One example is $[Ru(bpy)L_2]^{2+}$, where L is 4-(2-pyrrol-1-ylethyl)-4′-methyl-2,2′-bipyridine, which can be anodically polymerized onto a platinum electrode; when either ferrocene or decamethyl ferrocene permeates the polymer film, there can be direct oxidization at the underlying platinum or catalytic oxidation by electrogenerated ruthenium(III) sites.[59] A second example is a nickel(II) pyridyltritolylporphyrin linked to *N*-(3-propyl)pyrrole; the resulting compound **(8)** can be oxidatively polymerized onto platinum, and the resulting polymer film exhibits both reversible oxidation and reduction processes.[60]

There has been much additional research dealing with polymer films that are formed from monomers

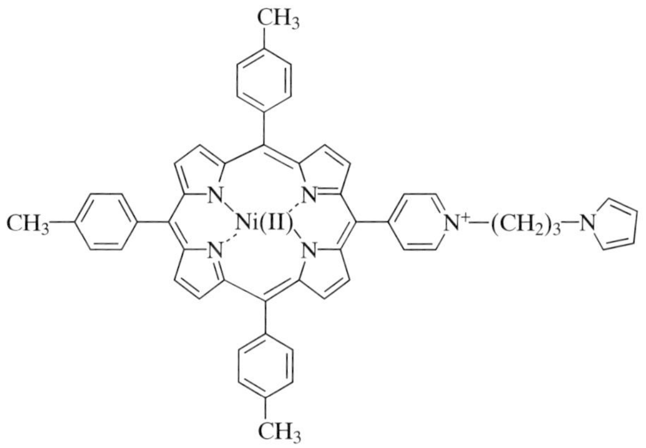

(8)

with already incorporated transition-metal sites. Earlier we mentioned work[(50)] in which the monomer $[Ru(bpy)_2(vbpy)]^{2+}$ can be reductively polymerized to form poly$[Ru(bpy)_2(vbpy)]^{2+}$ on platinum, glassy carbon, tin oxide, and titanium oxide electrodes; in the same study, related monomers containing ruthenium or iron were polymerized, and it was shown that pairs of such monomers can yield site-mixed copolymers as well as spatially segregated two-layer films. Glassy carbon electrodes dip-coated with aquobis(bpy)(poly-4-vinylpyridine)ruthenium(II) have been used by Samuels and Meyer[(61)] as a way to generate immobilized ruthenium(IV) sites for the catalytic oxidation of 2-propanol. Anodic polymerization of cobalt- and nickel-containing tetrakis(*o*-, *m*-, and *p*-aminophenyl)porphyrin, tetrakis(*p*-(dimethylamino)phenyl)porphyrin, tetrakis(*p*-hydroxyphenyl)porphyrin, and tetrakis(*p*-*N*-pyrrolylphenyl)porphyrin onto platinum has been described;[(62)] a film of electropolymerized cobalt tetrakis(*o*-aminophenyl)porphyrin catalyzes the reduction of oxygen. Electroreductive polymerization of $[Fe(vbpy)_3]^{2+}$ and $[Ru(4,4'\text{-dimethyl-bpy})_2(vbpy)]^{2+}$ has been accomplished at a platinum cathode; emphasis was placed on the structure of the polymers and their mechanism of formation.[(63)] When a copolymer consisting of 1-vinyl-2-pyrrolidone and *meso*-[tri(phenyl)mono (*p*-methacrylamidophenyl)]porphine is treated with iron(II), the resulting solid can be dissolved in tetrahydrofuran (THF), and the solution can be applied to the surface of a graphite electrode; evaporation of the solvent gives a coated electrode with redox-active iron sites that catalyze the reduction of oxygen.[(64)] Treatment of poly-*N*-vinylimidazole (pnvi) with $[Ru(bpy)_2Cl_2]$ gives several metallopolymers, such as $[Ru(bpy)_2(pnvi)_2]^{2+}$ and $[Ru(bpy)_2(pnvi)Cl]^{+}$, which can be coated onto carbon electrodes.[(65)]

For the second subclass, for which an electroactive transition-metal species is introduced after the polymer is deposited onto an electrode, we will cite just two representative examples. Oyama and Anson[(51)] dip-coated pyrolytic graphite with films of PVP or polyacrylonitrile (PAN), and demonstrated that ruthenium(III)(edta)OH_2 (where edta is ethylenediaminetetraacetate) loses H_2O in coordinating to a pendant pyridine group of PVP and that $[Ru(NH_3)_5OH_2]^{2+}$ loses H_2O in binding to a $-CN$ moiety of PAN; in addition, the interaction of $[Ru(NH_3)_5OH_2]^{2+}$ with a film consisting of a mixture of PVP and PAN was examined. Platinum electrodes coated with *p*-chlorosulfonated polystyrene have been treated with polypyridyl complexes of iron, ruthenium, and osmium as well as a ferrocene derivative, a nickel macrocycle, and a nickel porphyrin;[(66)] these inorganic species possess unbound amine, hydroxy, and carboxylate moieties that can react with the *p*-chlorosulfonated polystyrene to anchor the redox-active transition-metal complex to the polymer.

As a final group of metal-containing species that can be used to modify an electrode surface, metal phthalocyanines **(9)** and metal salens **(10)** deserve some mention.

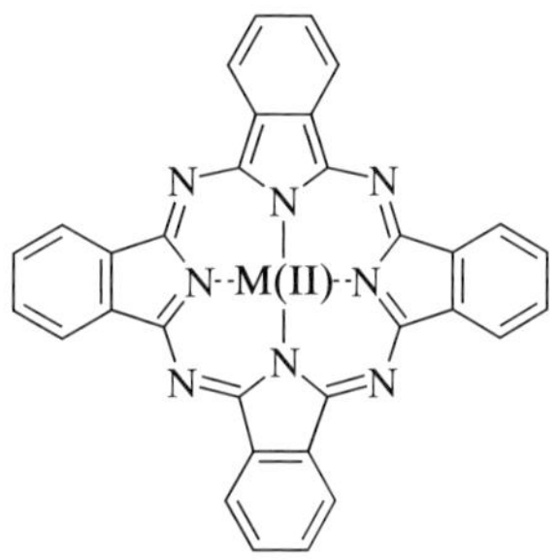

(9) 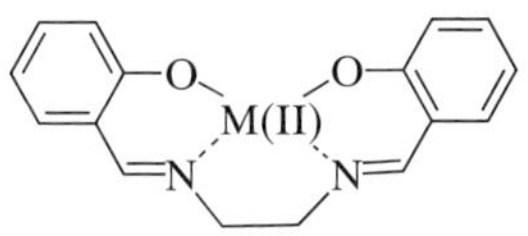 **(10)**

Over the years, metal phthalocyanines have received much attention as catalysts for the reduction of oxygen. Films of polypyrrole and polyaniline, each intercalated with tetrasulfonated iron phthalocyanine, have been electrodeposited onto gold; these chemically modified electrodes appear to catalyze the reduction of oxygen to water.[(67)] In addition, the reduction of oxygen has been examined at pyrolytic graphite electrodes spin-coated with a number of different metal phthalocyanines,[(68)] and at highly oriented pyrolytic graphite electrodes with adsorbed layers of iron and cobalt crown phthalocyanines.[(69)] Furthermore, oxidative electropolymerization of cobalt and nickel tetraaminophthalocyanines onto glassy carbon electrodes has been accomplished.[(70)] Metal salen complexes are distinguished by the fact that the ligand (a Schiff base) which coordinates the metal cation can be oxidatively polymerized onto a variety of electrodes; thus, the resulting polymers are conductive (due to their

For references see page 10084

extended π-conjugated systems) as well as redox-active (due to the presence of a transition metal). A number of investigators have studied the formation and properties of anodically polymerized films of metal salens.(71–74)

3.2.2 *Methods of Preparation*

As the preceding section has revealed, metal ion-containing polymers on electrodes have been prepared in many ways.(55) Anodic or cathodic electropolymerization of a metal-containing monomer onto an appropriate electrode is the most common and straightforward procedure, and many of the references cited above provide details.(49,50,56–60,62,63,70–74) Usually, it is most convenient to carry out the direct electropolymerization by means of cyclic voltammetry, because the thickness of the polymer film can be controlled by the number of consecutive scans and by the scan rate; an excellent example of this technique is provided by Figure 3, which depicts the oxidative polymerization of copper(II) salen onto a platinum electrode.(72)

Other methods of preparing metal-containing polymer films on electrodes include metallation of an already electrodeposited polymer film;(51) adsorption of an organic species, followed by metallation;(52) attachment of a complexing ligand via silanization, followed by metallation;(53) functionalization of edge-plane sites of graphite, followed by reaction with a complexing ligand and introduction of a metal ion;(54) dip-coating of an electrode with a metallated polymer;(61,64,65) covalent linkage of a functionalized metal-containing species to a functionalized polymer film;(66) intercalation of a polymer with a metal-containing compound;(67) spin-coating of an electrode with a metal-containing species;(68) and adsorption of a metallated compound.(69)

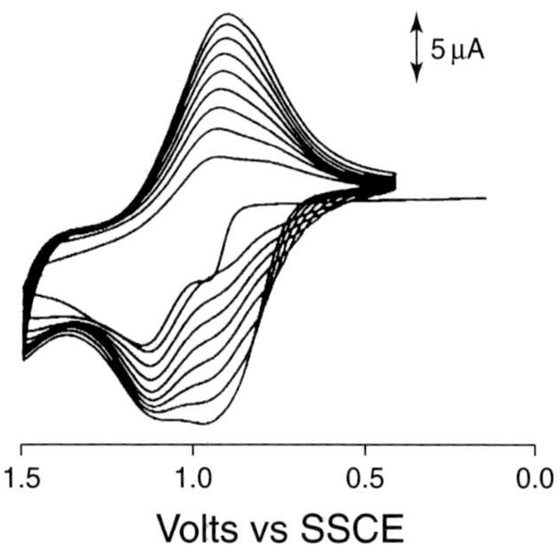

Figure 3 Consecutive cyclic voltammograms showing the oxidative polymerization of copper(II) salen in acetonitrile containing 0.10 M tetra-*n*-butylammonium hexafluorophosphate at a platinum disk electrode and at a scan rate of $100\,\mathrm{mV\,s^{-1}}$. (Reproduced by permission from Hofercamp and Goldsby.(72) Copyright 1989, American Chemical Society.)

3.2.3 *Methods of Characterization*

Virtually all of the electrochemical and spectroscopic techniques referred to in section 3.1.3 are suitable for the characterization of metal-containing polymer films on electrodes. Many of the references listed in the preceding section contain specific information about the methods employed to characterize the chemically modified electrodes that are the subjects of those studies. Most importantly, since the metal-ion sites within a polymer coating are ultimately what is essential to its redox activity, electroanalytical techniques are especially informative.

3.2.4 *Analytical Applications*

Most studies of metal-containing polymers on electrodes have focused on (a) understanding charge-transport (electron-transfer) mechanisms within such films, (b) investigating catalytic processes that take place within and at the surface of these films, or (c) characterizing the electrochemistry of the immobilized metal center itself. Practical analytical applications remain an area ripe for further development. We can, however, summarize several analytical uses for electrodes coated with metal-containing polymers.

Earlier, we indicated that poly-*N*-vinylimidazole reacts with $[Ru(bpy)_2Cl_2]$ to give ruthenium-containing polymers.(65) When coated onto a rotating disk electrode, these polymers provide a sensor that responds linearly to either $[Fe(CN)_6]^{4-}$ or $[Fe(H_2O)_6]^{2+}$ at concentrations as low as 5×10^{-6} M.(75) A glassy carbon electrode coated with polymerized nickel(II) phthalocyanine has been used for amperometric determination of phenolic antioxidants – *tert*-butylhydroxytoluene (BHT), *tert*-butylhydroxyanisole (BHA), propyl gallate (PG), and *tert*-butylhydroquinone (TBHQ) – after separation by HPLC;(76) detection limits for BHA, PG, and BHT are 0.11, 0.15, and 0.60 ppm, respectively. A sensor has been developed that utilizes a glassy carbon electrode dip-coated with $[Ru(bpy)_2(PVP)_{10}Cl]Cl$ for the flow injection measurement of nitrite;(77) this procedure has been applied to the determination of nitrate in fertilizer, after the nitrate in the sample is first converted to nitrite with the aid of a copper–cadmium reductor.

A carbon electrode bearing a film of anodically polymerized tetraaminophthalocyanatocobalt(II) responds potentiometrically in a Nernstian fashion to sulfide ion (3×10^{-6} to 10^{-3} M) and to mercaptoethanol (6×10^{-6} to 10^{-3} M).(78) Amperometric determination of BHA (an antioxidant) has been accomplished with a poly[tetraaminophthalocyanatonickel(II)]-modified glassy carbon electrode;[79] a linear response was observed for the concentration range from 5.6×10^{-8} to 5.6×10^{-5} M, and the detection limit was 1.5×10^{-8} M. Glassy carbon, coated with electropolymerized tetraaminophthalocyanatonickel(II), is a sensor for the determination of dopamine at concentrations from 2×10^{-7} to 2×10^{-5} M, with a detection limit of 9×10^{-8} M;[80] one can eliminate the interference due to ascorbic acid by coating the modified electrode with a layer of Nafion®.

3.3 Ion-exchange Polymers

As mentioned previously, Nafion® **(11)** is a perfluorinated, polysulfonated cation-exchange polymer.

$$-(CF_2CF_2)_x-(CFCF_2)_y-$$
$$O-CF_2-CFCF_3-O-CF_2CF_2SO_3H \text{ (side chain on CF)}$$

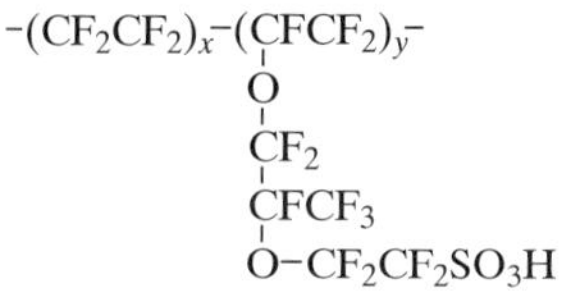

(11)

One can purchase Nafion®, in its protonated form, as a dilute solution of the polymer in alcohol–water. A measured small volume of this solution can be applied to the surface of a glassy carbon electrode; then, after the solvent has evaporated, a film of the polymer adheres to the electrode. This coated electrode can be soaked in an appropriate electrolyte solution (e.g. dilute sulfuric acid), and can then be inserted into an electrochemical cell containing a solution of some cationic metal complex (e.g. $[Ru(bpy)_3]^{2+}$) along with the same supporting electrolyte. Quickly, the $[Ru(bpy)_3]^{2+}$ becomes incorporated into the polymer film via a cation-exchange process (two protons at two adjacent $-SO_3^-$ sites are replaced by one $[Ru(bpy)_3]^{2+}$ moiety) until a steady state is attained, whereupon a variety of electroanalytical measurements can be performed.[81] One can employ an alternative approach, which involves preloading the Nafion® with $[Ru(bpy)_3]^{2+}$ (or some other electroactive species), by adding the latter to the alcohol–water solution of the polymer before it is applied to the electrode surface. This is the essence of many procedures for the preparation and use of ion-exchange polymer films on electrodes.

3.3.1 Types of Materials Used

Both cationic and anionic ion-exchange polymer films have been coated onto electrodes. Despite the overwhelming popularity of Nafion®, several other types of cation-exchange polymers have been employed. Majda and Faulkner[82] used platinum electrodes spin-coated with the sodium salt of poly(styrene sulfonate) to investigate the electrochemistry of incorporated $[Ru(bpy)_3]^{2+}$. When a pyrolytic carbon electrode filmed with deprotonated polyacrylic acid is placed into a solution of $[Ru(NH_3)_6]^{3+}$, the $-COO^-$ sites can electrostatically bind the positively charged metal complex.[83] Another approach taken by Vining and Meyer[84] involved coating glassy carbon electrodes with *p*-chlorosulfonated polystyrene; then the polymer-filmed electrode was exposed to an aqueous medium of pH 9 in order to convert the $-SO_2Cl$ sites to $-SO_3^-$ sites which can bind catalytically active, cationic ruthenium and osmium oxo-bridged dimers. In addition, poly(vinyl sulfate) has been proposed as a coating that can bind cations, including $[Ru(NH_3)_6]^{3+}$, $[Ru(bpy)_3]^{2+}$, $[Co(bpy)_3]^{3+}$, and $[Co(phen)_3]^{3+}$.[85] A mixture of a cystine-derivatized pyrrole and a tetraethylammonium-substituted pyrrole can be oxidatively copolymerized onto platinum or glassy carbon electrodes; chemical treatment of the polymer-coated electrode with dithiothreitol converts the polymer to its cysteine-SH form which tightly binds $[Fe_4S_4]^{2+}$ centers to give an electrode–polymer assembly with electroactive, cysteinyl-ligated, ferredoxin-like units.[86] Carbon electrodes coated with Eastman Kodak AQ 55 (a polysulfonated block copolymer) incorporate and strongly retain electroactive countercations (e.g. $[Ru(1,10\text{-phenanthroline})_3]^{2+}$ and $[Ru(9,10\text{-phenanthroline-5,6-dione})_3]^{2+}$) in aqueous and acetonitrile solutions.[87]

Electrodes coated with anion-exchange polymers that can electrostatically bind anionic metal complexes have received considerable attention. Oyama and Anson[88] demonstrated that graphite electrodes filmed with protonated PVP can strongly bind multiply charged anions such as $[Fe(CN)_6]^{3-}$ and $[IrCl_6]^{2-}$. Methyl- or benzyl-quaternized PVP provides a polyelectrolyte coating for graphite that is superior to PVP, because the density of cationic sites available for anion binding is independent of pH.[85] Protonated poly(L-lysine), coated onto a freshly cleaved graphite electrode, has been shown to interact electrostatically with both $[Fe(III)(edta)]^-$ and $[Fe(II)(edta)]^{2-}$, and a model to explain charge propagation through the film has been proposed.[89] A platinum electrode coated with poly[5,15-bis(2-aminophenyl)porphyrin], with and without metallation by cobalt(II), responds potentiometrically at pH 5.5 to a variety of anions, including iodide, bromide, fluoride, thiocyanate, salicylate, nitrate, and acetate.[90] An interesting new derivative of polyacetylene with pendant cationic moieties – namely, poly(dihexyldipropargyl ammonium bromide) – has been synthesized and has been coated (as a mixed film with Nafion®) onto an ITO electrode; however, its binding of anionic transition-metal complexes has not yet been explored.[91]

For references see page 10084

3.3.2 Methods of Preparation

By far the most common methods to prepare electrodes with films of ion-exchange polymers are the techniques of dip-coating, drop-coating, and spin-coating. These procedures are literally self-explanatory, but details can be found in the various references cited in the preceding section. Brumlik et al.[92] have described a new approach for the fabrication of sulfonated fluorochlorocarbon ionomer films on surfaces of stainless steel and microporous alumina, which entails the radiofrequency plasma polymerization of trifluorochloroethylene (TFCE) and trifluoromethane sulfonic acid (TFMSA); this technique offers the advantage of providing anion-exchange polymers that are not commercially available, and so it deserves attention by electroanalytical chemists.

3.3.3 Methods of Characterization

As for other previously described kinds of selective coatings on electrodes, the same methods of characterization (including all of the electroanalytical techniques) outlined in sections 2.3 and 3.1.3 can be employed for surfaces coated with films of ion-exchange polymers. In addition, some of the references cited in section 3.1.1 provide valuable insights. In particular: film thicknesses have been measured;[82,87] diffuse-reflectance Fourier transform infrared (FTIR) spectroscopy has been used to characterize a polymer;[86,92] the quantities of electrostatically bound transition-metal species have been determined spectrophotometrically through the use of optically transparent, polymer-coated electrodes;[89] the electrical conductivities of polymer films, along with their absorption, fluorescence, electron paramagnetic resonance (EPR), and X-ray diffraction spectra, have been determined;[91,92] and SEM and XPS of polymer films have proven to be useful.[92]

3.3.4 Analytical Applications

Electrodes coated with ion-exchange polymers, especially Nafion®, have been employed for many different kinds of electrochemical studies, including electroanalysis. In some applications the purpose of a Nafion® film has been to provide or to improve selectivity toward one or more specific analytes, whereas in other instances a Nafion® film has served to protect some other coating on the surface of an electrode.

To alleviate problems associated with an adventitious surfactant (alkaline phosphatase) during the adsorptive stripping voltammetric determination of copper(II) at the 10 nM concentration level, Economou and Fielden[93] used Nafion® to coat a mercury-filmed glassy carbon electrode. In a method for the measurement of copper(II) in natural waters,[94] a carbon electrode, first coated with a complexing agent (1-phenyl-3-methyl-4-octanoylpyrazol-5-one), was subsequently modified with a Nafion® film; the procedure is successful for the determination of nanomolar to micromolar concentrations of copper(II), with a relative standard deviation of 1–6%. Lead(II) has been measured by means of square-wave voltammetry with a glassy carbon electrode modified with a mercury film coated with Nafion® that contains 2,2′-bipyridine;[95] for a 5-min preconcentration step, the anodic stripping method gives a calibration curve for lead(II) that is linear from approximately 2×10^{-8} to 2×10^{-6} M, with a detection limit of 2×10^{-9} M.

A Nafion®-coated glassy carbon electrode, treated with ruthenium(III), can be employed as an amperometric sensor for the determination of hydrazine, dimethylhydrazine, and phenylhydrazine in drinking and river waters; the method provides a linear response for hydrazine in the range from 1.8×10^{-7} to 4.6×10^{-4} M, with a detection limit of 1.0×10^{-7} M, and similar analytical results can be achieved with the other hydrazines.[96] A method for the amperometric determination of nitric oxide has been devised that utilizes an ultramicroelectrode chemically modified with a film of poly[nickel(II) salen] and Nafion®; the electrode responds linearly to in vivo nitric oxide at concentrations from 1.0×10^{-8} to 4.0×10^{-6} M in biological systems.[97] A Nafion®-modified carbon paste electrode has been developed for the square-wave voltammetric determination of Doxazosin (an antihypertensive substance), after the compound is adsorptively preconcentrated onto the electrode;[98] the detection limit is 2.3×10^{-11} M, which is better (by five orders of magnitude) than a spectrophotometric method.

Zen and Wang[99] have employed a graphite electrode spin-coated with Nafion® which is (a) impregnated with different ratios of Pb^{2+} and Ru^{3+}, (b) treated with potassium hydroxide to form $Pb(OH)_2$ and $Ru(OH)_3$, and (c) conditioned in the presence of oxygen at 53 °C; this electrode responds linearly to dissolved oxygen at concentrations ranging from 1 to 10 ppm in both acidic and alkaline environments. Finally, nanoelectrode ensembles filmed with Eastman Kodak AQ 55 (a polysulfonated block copolymer) have been studied by Ugo et al.[100] In separate experiments, (ferrocenylmethyl)trimethylammonium cation and $[Ru(NH_3)_6]^{3+}$ were preconcentrated (partitioned) into these modified electrodes, and the cyclic voltammetric response for each system was examined. For the first cation, linear calibration curves for concentrations ranging from 20 nM to 50 μM were obtained (with a detection limit of 5 nM), and similar results were achieved for $[Ru(NH_3)_6]^{3+}$ (with a detection limit of 1 nM). This approach to highly sensitive analysis merits extension to systems containing other cations.

4 LANGMUIR–BLODGETT FILMS ON ELECTRODES

For many types of electrode coatings, the dual nature of surfactants plays a key role. In the language of Langmuir–Blodgett films, the surfactant molecule is termed an "amphiphile" and is mechanically forced into an organized monolayer within a liquid/gas interface. Typically, one accomplishes this by first preparing a dilute solution (~1%) of pure amphiphile(s) in a solvent of high vapor pressure (e.g. chloroform). Aliquots of this solution are then allowed to evaporate over a liquid (the subphase) in which the amphiphiles as a whole are insoluble. Usually, the subphase consists of purified water of controlled pH and ionic strength, but mercury, glycerol, hydrocarbons, and other liquids have been employed as well.[101] After the volatile solvent evaporates, the amphiphiles are in a state known as a two-dimensional gas in which they are widely and randomly distributed along the surface of the subphase. With aqueous subphases, the amphiphiles are oriented with their hydrophilic regions immersed, forcing hydrophobic side chains up and out of the water. As shown in Figure 4, the two-dimensional gas and subphase are contained in a specially designed (commercially available) trough with a movable barrier traversing the interface.

At this stage, the barrier is then moved along the trough to force the amphiphiles into successively more ordered states, while the surface pressure and area per molecule are recorded. A pressure–area isotherm results which frequently displays marked phase transitions. A theoretical isotherm for stearic acid on 0.01 M hydrochloric acid, depicted in Figure 5, illustrates the various phases of film formation.

If the area of a solid-phase film is decreased further by the movable barrier, a sharp decline in surface pressure results, indicating buckling of the monolayer and the formation of multilayer structures. For the purposes of electrode coatings, such films are to be avoided since they are not reproducible, and multilayer coatings on electrodes can easily be achieved through the repeated application of monolayers.

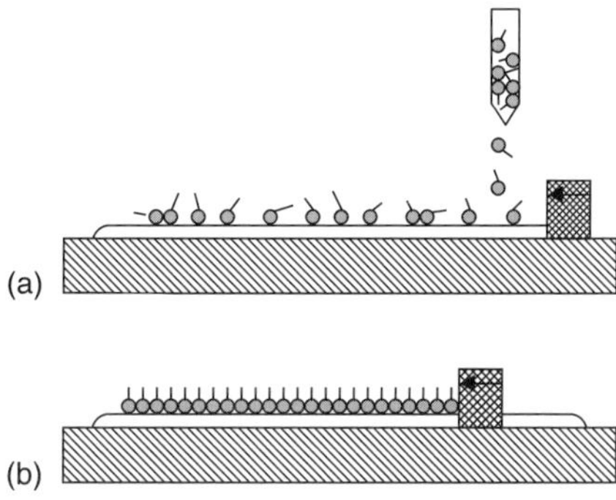

Figure 4 (a) Formation of the two-dimensional gas and (b) compression into an ordered monolayer. (Reproduced by permission from Ulman.[107] Copyright 1991, Academic Press.)

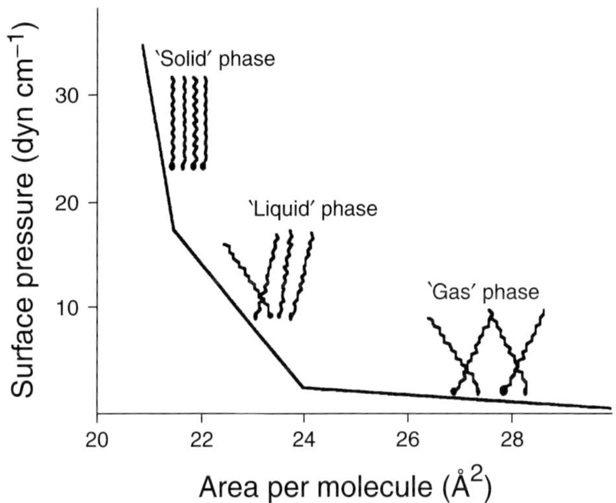

Figure 5 Theoretical isotherm for stearic acid on 0.01 M hydrochloric acid. (Reproduced by permission from Ulman.[107] Copyright 1991, Academic Press.)

4.1 Types of Materials Used

A wide variety of molecular structures can be incorporated into Langmuir–Blodgett films. DeArmond and Fried[102] and Goldenberg[103] have summarized much of the work that has been done with electroactive amphiphiles (which usually must be synthesized). Space limitations preclude a detailed discussion of these species, but some noteworthy examples include derivatized fullerenes, tetrathiafulvalenes (TTFs), quinones, and various metal-containing complexes such as porphyrins, phthalocyanines, cyclams, bipyridine complexes, and metallocenes. In addition, several polymerizable amphiphiles have been employed in electrode films. With these species, polymerization may be initiated either before or after transfer of the monolayer to the electrode, a feature which is unique to the Langmuir–Blodgett method.

In some applications, the amphiphile need not be electroactive, in which case a very large array of molecules may be employed for a Langmuir–Blodgett film. Hann[101] has provided detailed coverage of many of these species: long-chain (C_{16} or higher) fatty acids and their various derivatives comprise the largest and most well studied class. For biological applications, phospholipids and sterols are most commonly employed.[104]

4.2 Methods of Preparation

Once an ordered monolayer has been formed, it must be transferred onto an electrode surface; the method employed for this transfer depends on the shape and

For references see page 10084

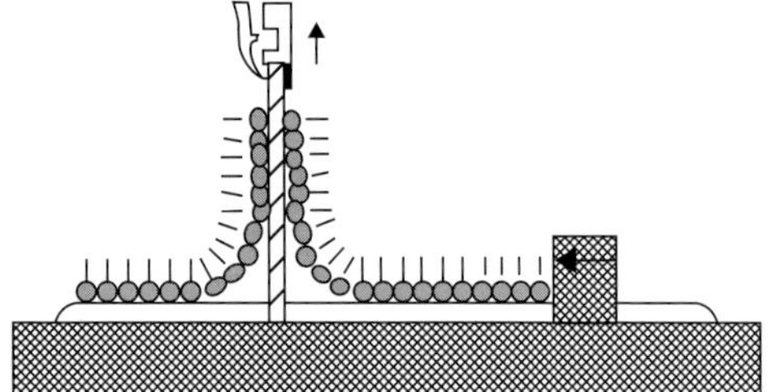

Figure 6 Transfer of a monolayer to a planar electrode by slow emersion. (Reproduced by permission from Ulman.[107] Copyright 1991, Academic Press.)

wetting characteristics of the electrode surface. For planar electrodes, transfer may be accomplished in two ways. First, but less well understood, is the method involving vertical immersion and emersion of the electrode through the interface. Figure 6 illustrates how this method was recently applied to the coating of optically transparent ITO electrodes.[105,106]

In the first stage of the process, the ITO electrode (a hydrophilic substrate) penetrates an ordered monolayer on water from above. Formation of an upward-curving meniscus prevents the amphiphiles from adsorbing as the electrode is immersed. Slow emersion of the electrode from the subphase then takes place ($5\,mm\,min^{-1}$), as the surface pressure is maintained constant by the movable barrier. If the substrate is sufficiently smooth, water will drain away and evaporate from the surface of the electrode as an even monolayer is transferred. Reimmersion of such an electrode into the interface can then be used to produce bilayers and multilayers whose structures closely resemble those of biological cell membranes.[104] One can make a preliminary evaluation of the quality and quantity of film transferred by simply comparing the area of the electrode to the change in area of the trough as dictated by the barrier (the transfer ratio).[107]

Alternatively, or if the electrode is not conveniently submerged, a horizontal-lifting method of transfer may be employed. Figure 7 illustrates how this procedure (Schaefer's method) has been applied to the coating of freshly cleaned (hydrophobic) glassy carbon electrodes.[108,109]

One coats the electrode by simply lowering it onto a monolayer parallel with the surface. Following reversed-phase adsorption of the film, the electrode is either withdrawn with a monolayer (or bilayer)[102] intact on its surface or held at the interface as excess monolayer is swept away. In the former case, recompression of the interface by means of the movable barrier generates a new ordered monolayer.

To perform electrochemical examination of the film, it is necessary to transfer the electrode to a different solution. In some cases, however, withdrawal of the electrode from the subphase is unnecessary, and the

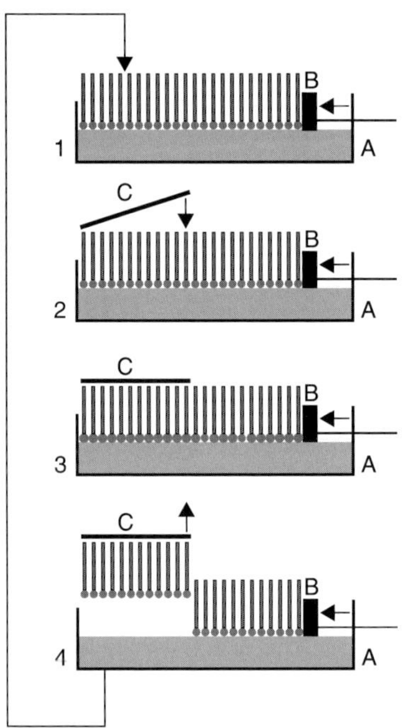

Figure 7 Transfer of a monolayer to an electrode by the horizontal-lifting technique (Schaefer's method): (A) trough; (B) movable barrier; (C) substrate. (Reproduced by permission from Ulman.[107] Copyright 1991, Academic Press.)

electrochemistry can be done in situ. In these experiments, counter- and reference electrodes are pre-positioned in a subphase containing a suitable electrolyte. So-called "horizontal-touch" techniques have been successfully employed with both glassy carbon[110] and ITO[111] electrodes.

Horizontal transfers avoid the surface flow of the monolayer inherent with vertical methods, and thus can preserve the original structure of the monolayer to a larger degree. Furthermore, surface-pressure requirements are much less stringent, making the technique applicable to a wider variety of amphiphiles.[111]

Nelson and Benton[112] described a vertical-type procedure for the preparation of a spherical hanging mercury drop electrode (HMDE) modified with a monolayer of naturally occurring phospholipids. In this case, film deposition occurs during the immersion step, with the long alkyl chains of the lipids next to the mercury.

As a consequence of their formation and structure, Langmuir–Blodgett films exhibit several clear advantages over other electrode coatings in terms of flexibility, ease of application, and sensitivity. First, one can easily make mixed films in any proportion by combining the appropriate ratio of one amphiphile with another in the evaporated solvent. As long as the amphiphiles are miscible with one another, the resulting films will have a large degree of homogeneity with no dependence on competing chemical reactions (as with mixed SAMs and copolymers). Second, the Langmuir–Blodgett technique allows one to prepare both highly ordered and multiply layered coatings on electrodes; the number of discrete layers can range from one to as many as 50. Third, the monolayer approach is very efficient, consuming only tiny quantities of synthetic amphiphiles,[111] while exposing the vast majority of these molecules to electroanalysis. Expensive amphiphiles can be further conserved by being mixed with well behaved long-chain carboxylic acids.[106] Casting and bulk polymerization methods tend to be far more wasteful. Fourth, several manufacturers offer fully automated, programmable-trough instruments, enabling one to prepare films with little operator involvement. Some of these are even equipped with multiple troughs, allowing one to prepare alternate-layered electrodes.

For some Langmuir–Blodgett films, a possible disadvantage lies in the fact that they are inherently less stable than other electrode coatings. Inclusion of solvent during deposition of a Langmuir–Blodgett film can lead to its post-transfer instability; evaporation of solvent trapped in irregularities or imperfections on the surface can cause collapse of the monolayer.[101] Another problem is that forming the film and transferring it to the surface of an electrode are both very time-consuming processes.

4.3 Methods of Characterization

An excellent summary of the nonelectrochemical techniques employed to characterize Langmuir–Blodgett and other organic thin films has been written by Ulman.[113] One of the most powerful techniques that can be applied in situ involves optical second harmonic generation (SHG). This method has been reviewed by Corn[114] and by Richmond,[115] and features submonolayer sensitivities, instantaneous response time, the ability to discriminate between surface and bulk species, and access to both film thickness and orientation. Although Langmuir–Blodgett films adhere relatively weakly to electrode surfaces, many are still amenable to all standard surface analyses requiring high vacuum. One recent example included the secondary ion mass spectrometric analysis of cadmium arachidate films on various metal substrates, with little or no sample preparation being required.[116]

4.4 Analytical Applications

Although electroanalytical applications for Langmuir–Blodgett films are largely unrealized, it is clear that electrochemical research in this area has only just begun. Many reports have focused either on the dynamic transport of electroactive species across a bilayer membrane or on the enzymatic generation of electroactive species at the surface of a modified electrode. Results from these systems are included in section 5 below.

Nelson[117] prepared phosphatidylserine-coated mercury electrodes (HMDEs) to study the reduction of Tl^{+}, Cd^{2+}, Cu^{2+}, Pb^{2+}, and Eu^{3+} in aqueous systems. It was found that the organization of the film is potential dependent and that the reduction potentials of the various metal ions depend strongly on the extent of film protonation (pH) and the presence of other metal species. Nelson et al.[118] observed that a variety of common aqueous pollutants, including polynuclear aromatic hydrocarbons (PAHs), polychlorinated biphenyls (PCBs) and various pesticides, can be concentrated from contaminated water samples into dioleoylphospholipid monolayers on mercury; for PAHs, these investigators established that the detection limit is $0.4\,\mu g\,L^{-1}$, with a relative standard deviation of 7%.

In a unique application of the Langmuir–Blodgett technique, Fujihira and Poosittisak[109] employed a glassy carbon electrode filmed with an electroactive bipyridinium amphiphile and coated with platinum. This two-layer electrode catalyzed the reduction of oxygen when used for rotating-disk voltammetry, a fact which simultaneously attests to the versatility and stability of Langmuir–Blodgett films.

Finally, Miller and Bard[111] designed a special trough to accommodate the filming of a small horizontally oriented ITO electrode. They demonstrated picomole sensitivities of this electrode toward alkylated $[Ru(bpy)_3]^{2+}$ species, noting that the current efficiency for the Langmuir–Blodgett method far exceeds that attainable by traditional casting methods with the same complexes.

5 IMMOBILIZED ENZYMES ON ELECTRODES

Chemical selectivity remains one of the great challenges facing analytical electrochemists today. In this regard, electrodes modified with enzymes (bioelectrochemical sensors) continue to hold the most promise for progress and are presently one of the most active areas of research in electrochemistry.[119–122] Over 3300 references (in English) from 1987 to mid-1999 were retrieved on this subject from the Institute of Scientific Information database alone.[123] These reports cover a host of enzymes (from multiple sources) and immobilization techniques for virtually every type of electrode. Common themes include (a) the effect of various deposition processes on the activity and stability of the enzyme(s), (b) the influence of any mediators and/or required cofactors on electronic communication between the enzyme and the electrode, (c) the sensitivity and selectivity of the modified electrode for a given substrate (especially within a complex matrix), (d) the effect of polymers, membranes,

For references see page 10084

and various other supports on the performance of the electrode system, and (e) the overall stability and applicability of the modified electrode system. With this in mind, our goal is merely to provide the reader with an entry point into the literature of enzyme-modified electrodes along with only the most basic of overviews.

5.1 Types of Materials and Methods of Preparation

Although enzymes typically do not exhibit reversible redox behavior at bare electrodes, they frequently do when immobilized in polymers, membranes, cross-linked networks, and a host of other media. Reasons for this behavior have been discussed by several authors.(124,125) In addition, it is usually necessary to employ "mediators" (small, organic, electron-transfer partners) to establish electronic communication between the electrode and the immobilized enzyme. Figure 8 illustrates a general scheme that applies to an electrode modified with an oxidase.

Usually, the link between electrode and enzyme is achieved with a combination of a synthetic mediator and an enzyme cofactor (e.g. reduced forms of nicotinamide adenine dinucleotide or flavin adenine dinucleotide). Enzyme-modified electrodes of all kinds are prepared from aqueous starting materials, but are not restricted to use in aqueous media.(126) A brief overview of some common materials and methods for the construction of enzyme-modified electrodes follows. Although this list is far from complete, recent examples and reviews therein should serve as an adequate starting point.

5.1.1 Enzymes and Mediators

Glucose sensors based on GOx derived from *Aspergillus niger* form the cornerstone of enzyme research in electrochemistry. Reports featuring a new bioelectrochemical sensor design typically employ GOx as a benchmark for the performance of the electrode – usually one notes the effect of ascorbate and uric acid since these are problematic interferents found in serum samples. Additional commonly employed enzymes include other oxidases (especially horseradish peroxidase),(127) dehydrogenases,(128) cytochromes,(125) and myoglobins.(125) For systems with synthetic mediators, derivatives of ferrocene appear to be the most common, but tetracyanoquinodimethane (TCNQ), TTF, ferricyanide, and ruthenium and osmium complexes are also frequently employed. Enzymes and the mediators used with them have been extensively reviewed.(119–122,124,126–130)

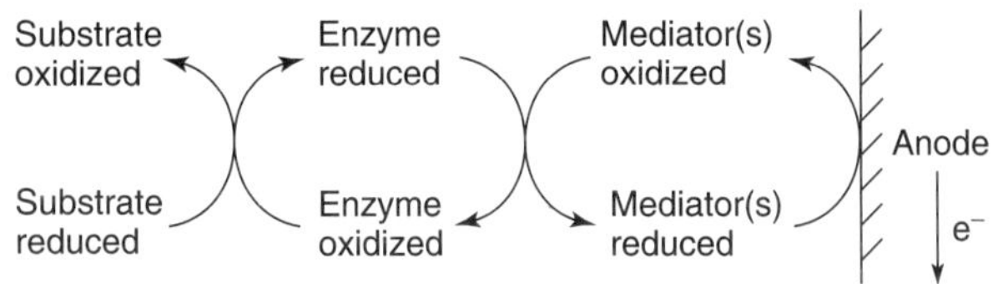

Figure 8 General format of an electrode modified with an oxidase and mediator(s).

List of selected abbreviations appears in Volume 15

5.1.2 Polymers for Electrodeposition

Electroactive polymers provide the experimenter with a rapid and simple method for the immobilization of enzymes onto an electrode surface. Pyrroles and aromatic amines (especially diamines) are the most commonly employed monomers and may be polymerized onto an electrode from an aqueous solution containing the enzyme of interest. This results in the entrapment of enzymes (and any other added reagents) within the bulk of the polymer film, which can also serve as part of the transduction mechanism. For example, the conductivity of polypyrrole is pH sensitive,(131) so it is the medium of choice for enzyme–substrate interactions which proceed with a net change in pH. Applications of electroactive polymers (and incorporated mediators) to enzyme-modified electrodes have been reviewed.(129,130)

5.1.3 Chemical Cross-linking

Glutaraldehyde can be employed for the chemical cross-linking of enzymes in solution. Typically, a mixture of enzyme(s), mediator(s), and frequently a stabilizer (such as bovine serum albumin (BSA)) is treated with glutaraldehyde and then cast onto an electrode surface. A recent paper by Sarkar et al.(132) illustrates the use of glutaraldehyde in an amino acid sensor based on screen-printed carbon electrodes. Cross-linking and polymeric supports are often used either in tandem or sequentially to enhance the stability and selectivity of an electrode. Thus, Moore et al.(133) utilized a mixture of glutaraldehyde, poly(vinyl alcohol), and GOx in a voltammetric sensor for acetaminophen and norepinephrine. Nishizawa et al.(131) prepared layered amperometric sensors for penicillin by coating interdigitated platinum arrays with polypyrrole followed by cross-linking of penicillinase with BSA.

5.1.4 Covalent Attachment

When a glassy carbon electrode is treated with dilute nitric acid and dichromate at positive potentials (+2.2 V versus saturated calomel electrode (SCE)), carboxylic acid functionalities are rapidly formed along the surface. Carbodiimide reagents may then be employed in a two-step synthesis which results in the covalent attachment of an enzyme to the surface (via an amide linkage). Bourdillon et al.(134) have provided an example pertaining to GOx.

5.1.5 Immobilization in Self-assembled Monolayers

Willner and Ricklin[135] recently prepared an amperometric biosensor based on an enzyme covalently attached to both an electron-transfer mediator (pyrroloquinolinequinone (PQQ)) and a monolayer of cysteamine on gold. They demonstrated that PQQ in this configuration is capable of oxidizing the reduced form of nicotinamide adenine dinucleotide in solution, while being oxidized at the electrode, thus linking the electrode to the activity of the enzyme. Willner et al.[136] have summarized other methods pertaining to enzymes on SAMs.

5.1.6 Langmuir–Blodgett Films

Biotinylated phospholipids are readily assembled into monolayer Langmuir–Blodgett films. These films can then be treated with streptavidin-modified enzymes prior to transfer onto an electrode surface. Rehak et al.[137] have employed this technology to prepare a xanthine-selective electrode. Applications of Langmuir–Blodgett films for other biosensors have been reviewed by Osa and Anzai[138] and by Aizawa.[139]

5.1.7 Direct Adsorption

Certain carbon electrode surfaces (such as pyrolytic graphite edge planes) have been shown to adsorb enzymes from solution electrostatically. In some cases, this results in the direct communication of an enzyme with the electrode (allowing for potential-controlled enzyme activity). Using this approach, Hirst et al.[140] demonstrated a potential-dependent fumarate–succinate equilibrium with succinate dehydrogenase.

5.1.8 Lipid and Surfactant Membranes (Bilayers and Multilayers)

Rusling[125] and Nikolelis et al.[141] have summarized the application of casting techniques to the study of membrane-bound proteins (especially cytochrome $P450_{cam}$ and myoglobin) on various electrode surfaces. Typically, phospholipids (or other synthetic surfactants) are initially cast onto the electrode surface from a volatile organic solvent (e.g. chloroform). As the solvent evaporates, the surfactant molecules spontaneously form multiple bilayer membranes on the surface. This modified electrode surface is then allowed to equilibrate (~20 min) with an aqueous solution of an enzyme, resulting in the incorporation of the protein into the bilayers. Alternatively, an aqueous vesicular dispersion of pure surfactant(s) is combined with a solution of the enzyme, and the mixture is cast onto the surface. As the water evaporates, vesicles in the mixture collapse, again resulting in the formation of multiple bilayers containing an enzyme. Lipid bilayers provide the enzyme with a more native environment, and so preserve the original structure and activity to a larger degree.

5.1.9 Zeolites and Clays

Recently, it has been demonstrated that certain zeolites and clays (immobilized onto the surfaces of electrodes) can act as effective hosts for both enzymes and electron-transfer mediators.[142] These "reagentless" sensors feature simple construction and enhanced strength, but are essentially restricted to neutral aqueous solutions. Responses to interferents may be diminished with further modifications of the zeolite surface.

5.2 Methods of Characterization

Primary concerns for any enzyme-modified electrode are for the activity and stability of the enzyme. To this end FIA (described elsewhere in this encyclopedia) is ordinarily employed. Response of the electrode to multiple injections over a period of minutes to weeks can indicate whether denaturing or leaching of enzyme or mediator occurs. FIA also affords rapid determination of the limit of detection and linear range, while revealing the response time of the electrode. In addition, known inhibitors and interferents can be introduced into a solution of substrate to show that the enzyme behaves normally and that the initial response can be recovered.

Several standard surface analytical techniques have also been applied to enzyme-modified electrodes. Recent investigations have included XPS,[143] SEM,[144] FTIR spectroscopy,[145] and STM.[146]

5.3 Analytical Applications

Immobilized enzymes impart unprecedented chemical selectivity to virtually any electrode and, in principle, allow for repeated use of the catalysts in continuous processes without contaminating the products. As a result, enzyme-modified electrodes are finding application in the food industry,[147,148] in clinical pharmacology (especially for glucose, lactate, urea, xanthine, glutamine, and various neurotransmitters),[149–151] and in environmental chemistry for the detection of pesticides[152,153] and pollutants.[154]

ABBREVIATIONS AND ACRONYMS

AC	Alternating Current
AES	Auger Electron Spectroscopy
AFM	Atomic Force Microscopy
BHA	*tert*-Butylhydroxyanisole

For references see page 10084

BHT	*tert*-Butylhydroxytoluene
BSA	Bovine Serum Albumin
EPR	Electron Paramagnetic Resonance
EQCM	Electrochemical Quartz Crystal Microbalance
FIA	Flow Injection Analysis
FTIR	Fourier Transform Infrared
GOx	Glucose Oxidase
HMDE	Hanging Mercury Drop Electrode
HPLC	High-performance Liquid Chromatography
HREELS	High-resolution Electron Energy Loss Spectroscopy
ITO	Indium-doped Tin Oxide
LEEDS	Low-energy Electron Diffraction Spectroscopy
PAHs	Polynuclear Aromatic Hydrocarbons
PAN	Polyacrylonitrile
PCBs	Polychlorinated Biphenyls
PG	Propyl Gallate
PQQ	Pyrroloquinolinequinone
PVP	Poly(4-vinylpyridine)
SAM	Self-assembled Monolayer
SCE	Saturated Calomel Electrode
SEM	Scanning Electron Microscopy
SHG	Second Harmonic Generation
STM	Scanning Tunneling Microscopy
TBHQ	*tert*-Butylhydroquinone
TCNQ	Tetracyanoquinodimethane
TEM	Transmission Electron Microscopy
TFCE	Trifluorochloroethylene
TFMSA	Trifluoromethane Sulfonic Acid
THF	Tetrahydrofuran
TTF	Tetrathiafulvalene
XPS	X-ray Photoelectron Spectroscopy

RELATED ARTICLES

Biomolecules Analysis **(Volume 1)**
Voltammetry In Vivo for Chemical Analysis of the Living Brain • Voltammetry In Vivo for Chemical Analysis of the Nervous System

Clinical Chemistry **(Volume 2)**
Biosensor Design and Fabrication • Electroanalysis and Biosensors in Clinical Chemistry • Electroanalytical Chemistry in Clinical Analysis • Glucose Measurement

Environment: Water and Waste **(Volume 4)**
Ion-selective Electrodes in Environmental Analysis

List of selected abbreviations appears in Volume 15

Field-portable Instrumentation **(Volume 4)**
Electrochemical Sensors for Field Measurements of Gases and Vapors

Electroanalytical Methods **(Volume 11)**
Electroanalytical Methods: Introduction • Ion-selective Electrodes: Fundamentals • Self-assembled Monolayers on Electrodes • Surface Analysis for Electrochemistry: Ultrahigh Vacuum Techniques

REFERENCES

1. R.W. Murray (ed.), *Techniques of Chemistry*, Wiley-Interscience, New York, Vol. 22, 1992.
2. C.J. Zhong, M.D. Porter, 'Evidence for Carbon–Sulfur Bond Cleavage in Spontaneously Adsorbed Organosulfide-based Monolayers at Gold', *J. Am. Chem. Soc.*, **116**(25), 11 616–11 617 (1994).
3. C.J. Zhong, M.D. Porter, 'Designing Interfaces at the Molecular Level', *Anal. Chem.*, **67**(23), 709A–715A (1995).
4. W.A. Zisman, 'Relation of the Equilibrium Contact Angle to Liquid and Solid Constitution', *Adv. Chem. Ser.*, **43**, 1–51 (1964).
5. D.L. Allara, R.G. Nuzzo, 'Spontaneously Organized Molecular Assemblies. 1. Formation, Dynamics, and Physical Properties of *n*-Alkanoic Acids Adsorbed from Solution on an Oxidized Aluminum Surface', *Langmuir*, **1**(1), 45–52 (1985).
6. D.L. Allara, R.G. Nuzzo, 'Spontaneously Organized Molecular Assemblies. 2. Quantitative Infrared Spectroscopic Determination of Equilibrium Structures of Solution-adsorbed *n*-Alkanoic Acids on an Oxidized Aluminum Surface', *Langmuir*, **1**(1), 52–66 (1985).
7. H. Ogawa, T. Chihara, K. Taya, 'Selective Monomethyl Esterification of Dicarboxylic Acids by Use of Monocarboxylate Chemisorption on Alumina', *J. Am. Chem. Soc.*, **107**(5), 1365–1369 (1985).
8. N.E. Schlotter, M.D. Porter, T.B. Bright, D.L. Allara, 'Formation and Structure of a Spontaneously Adsorbed Monolayer of Arachidic Acid on Silver', *Chem. Phys. Lett.*, **132**(1), 93–98 (1986).
9. D.L. Allara, S.V. Atre, C.A. Elliger, R.G. Snyder, 'Formation of a Crystalline Monolayer of Folded Molecules by Solution Self-assembly of α,ω-Alkanedioic Acids on Silver', *J. Am. Chem. Soc.*, **113**(5), 1852–1854 (1991).
10. Q. Liu, Z. Xu, 'Self-assembled Monolayer Coatings on Nanosized Magnetic Particles Using 16-Mercaptohexadecanoic Acid', *Langmuir*, **11**(12), 4617–4622 (1995).
11. D.Y. Huang, Y.T. Tao, 'Self-assembled Monolayer: Behavior of Diacetylenic Amphiphiles', *Bull. Inst. Chem., Acad. Sin.*, **33**, 73–80 (1986).
12. C.A. Goss, D.H. Charych, M. Majda, 'Application of (3-Mercaptopropyl)trimethoxysilane as a Molecular

Adhesive in the Fabrication of Vapor-deposited Gold Electrodes on Glass Substrates', *Anal. Chem.*, **63**(1), 85–88 (1991).

13. R. Maoz, J. Sagiv, 'Penetration-controlled Reactions in Organized Monolayer Assemblies. 1. Aqueous Permanganate Interaction with Monolayer and Multilayer Films of Long-chain Surfactants', *Langmuir*, **3**(6), 1034–1044 (1987).
14. C.J. Miller, C.A. Widrig, D.H. Charych, M. Majda, 'Microporous Aluminum Oxide Films at Electrodes. 4. Lateral Charge Transport in Self-organized Bilayer Assemblies', *J. Phys. Chem.*, **92**(7), 1928–1936 (1988).
15. R.G. Nuzzo, D.L. Allara, 'Adsorption of Bifunctional Organic Disulfides on Gold Surfaces', *J. Am. Chem. Soc.*, **105**(13), 4481–4483 (1983).
16. R.G. Nuzzo, F.A. Fusco, D.L. Allara, 'Spontaneously Organized Molecular Assemblies. 3. Preparation and Properties of Solution Adsorbed Monolayers of Organic Disulfides on Gold Surfaces', *J. Am. Chem. Soc.*, **109**(8), 2358–2368 (1987).
17. E.B. Troughton, C.D. Bain, G.M. Whitesides, R.G. Nuzzo, D.L. Allara, M.D. Porter, 'Monolayer Films Prepared by the Spontaneous Self-assembly of Symmetrical and Unsymmetrical Dialkyl Sulfides from Solution onto Gold Substrates: Structure, Properties, and Reactivity of Constituent Functional Groups', *Langmuir*, **4**(2), 365–385 (1988).
18. G.M. Whitesides, P.E. Laibinis, 'Wet Chemical Approaches to the Characterization of Organic Surfaces: Self-assembled Monolayers, Wetting, and the Physical–Organic Chemistry of the Solid–Liquid Interface', *Langmuir*, **6**(1), 87–96 (1990).
19. J.N. Richardson, S.R. Peck, L.S. Curtin, L.M. Tender, R.H. Terrill, M.T. Carter, R.W. Murray, G.K. Rowe, S.E. Creager, 'Electron-transfer Kinetics of Self-assembled Ferrocene Octanethiol Monolayers on Gold and Silver Electrodes from 115 to 170 K', *J. Phys. Chem.*, **99**(2), 766–772 (1995).
20. Y. Feng, W.K. Teo, K.S. Siow, Z. Gao, K.L. Tan, A.K. Hsieh, 'Corrosion Protection of Copper by a Self-assembled Monolayer of Alkanethiol', *J. Electrochem. Soc.*, **144**(1), 55–64 (1997).
21. Z. Mekhalif, J. Riga, J.J. Pireaux, J. Delhalle, 'Self-assembled Monolayers of *n*-Dodecanethiol on Electrochemically Modified Polycrystalline Nickel Surfaces', *Langmuir*, **13**(8), 2285–2290 (1997).
22. K. Slowinski, R.V. Chamberlain, R. Bilewicz, M. Majda, 'Evidence for Inefficient Chain-to-chain Coupling in Electron Tunneling Through Liquid Alkanethiol Monolayer Films on Mercury', *J. Am. Chem. Soc.*, **118**(19), 4709–4710 (1996).
23. M.A. Hines, J.A. Todd, P. Guyot-Sionnest, 'Conformation of Alkanethiols on Au, Ag(111), and Pt(111) Electrodes: a Vibrational Spectroscopy Study', *Langmuir*, **11**(2), 493–497 (1995).
24. A. Ulman, *Ultrathin Organic Films*, Academic Press, San Diego, CA, 1–99, 1991.
25. H.O. Finklea, 'Electrochemistry of Organized Monolayers of Thiols and Related Molecules on Electrodes', in *Electroanalytical Chemistry*, eds. A.J. Bard, I. Rubinstein, Dekker, New York, 109–335, Vol. 19, 1996.
26. D. Mandler, I. Turyan, 'Applications of Self-assembled Monolayers in Electroanalytical Chemistry', *Electroanalysis*, **8**(3), 207–213 (1996).
27. S. Steinberg, I. Rubinstein, 'Ion-selective Monolayer Membranes Based upon Self-assembling Tetradentate Ligand Monolayers on Gold Electrodes. 3. Application as Selective Ion Sensors', *Langmuir*, **8**(4), 1183–1187 (1992).
28. J.J. Hickman, D. Ofer, P.E. Laibinis, G.M. Whitesides, M.S. Wrighton, 'Molecular Self-assembly of Two-terminal, Voltammetric Microsensors with Internal References', *Science*, **252**(5006), 688–691 (1991).
29. I. Turyan, D. Mandler, 'Self-assembled Monolayers in Electroanalytical Chemistry: Application of ω-Mercaptocarboxylic Acid Monolayers for Electrochemical Determination of Ultralow Levels of Cadmium(II)', *Anal. Chem.*, **66**(1), 58–63 (1994).
30. O. Chailapakul, R.M. Crooks, 'Synthesis and Characterization of Simple Self-assembling Nanoporous Monolayer Assemblies: a New Strategy for Molecular Recognition', *Langmuir*, **9**(4), 884–888 (1993).
31. J. Wang, H. Wu, L. Angnes, 'On-line Monitoring of Hydrophobic Compounds at Self-assembled Monolayer Modified Amperometric Flow Detectors', *Anal. Chem.*, **65**(14), 1893–1896 (1993).
32. C. Bourdillon, M. Majda, 'Microporous Aluminum Oxide Films at Electrodes. 7. Mediation of the Catalytic Activity of Glucose Oxidase via Lateral Diffusion of a Ferrocene Amphiphile in a Bilayer Assembly', *J. Am. Chem. Soc.*, **112**(5), 1795–1799 (1990).
33. O. Inganäs, B. Liedberg, W. Chang-Ru, 'A New Route to Polythiophene and Copolymers of Thiophene and Pyrrole', *Synth. Met.*, **11**(4), 239–249 (1985).
34. J.P. Ferraris, D.J. Guerrero, 'Recent Advances in Heteroaromatic Copolymers', in *Handbook of Conducting Polymers*, 2nd edition, eds. T.A. Skotheim, R.L. Elsenbaumer, J.R. Reynolds, Dekker, New York, 259–276, 1998.
35. J. Roncali, 'Synthetic Principles for Bandgap Control in Linear π-Conjugated Systems', *Chem. Rev.*, **97**(1), 173–205 (1997).
36. J. Rault-Berthelot, C. Rozé, M.M. Granger, E. Raoult, 'Anodic Oxidation of Various Arylene–Cyanovinylenes Made of Alternating Fluorenyl, Thienyl and/or Phenyl Units', *J. Electroanal. Chem.*, **466**(2), 144–154 (1999).
37. G.P. Evans, 'The Electrochemistry of Conducting Polymers', in *Advances in Electrochemical Science and Engineering*, eds. H. Gerischer, C.W. Tobias, VCH, Weinheim, 1–74, Vol. 1, 1990.

38. G. Inzelt, 'Mechanism of Charge Transport in Polymer-modified Electrodes', in *Electroanalytical Chemistry*, ed. A.J. Bard, Dekker, New York, 89–241, Vol. 18, 1994.
39. A.F. Diaz, J.I. Castillo, J.A. Logan, W.Y. Lee, 'Electrochemistry of Conducting Polypyrrole Films', *J. Electroanal. Chem.*, **129**(1), 115–132 (1981).
40. K. Doblhofer, K. Rajeshwar, 'Electrochemistry of Conducting Polymers', in *Handbook of Conducting Polymers*, 2nd edition, eds. T.A. Skotheim, R.L. Elsenbaumer, J.R. Reynolds, Dekker, New York, 531–588, 1998.
41. L.B. Coleman, 'Technique for Conductivity Measurements on Single Crystals of Organic Materials', *Rev. Sci. Instrum.*, **46**(8), 1125–1126 (1975).
42. K.K. Shiu, S.K. Pang, H.K. Cheung, 'Electroanalysis of Metal Species at Polypyrrole-modified Electrodes', *J. Electroanal. Chem.*, **367**(1), 115–122 (1994).
43. Y. Ikariyama, W.R. Heineman, 'Polypyrrole Electrode as a Detector for Electroinactive Anions by Flow Injection Analysis', *Anal. Chem.*, **58**(8), 1803–1806 (1986).
44. E. Wang, A. Liu, 'Polyaniline Chemically Modified Electrode for Detection of Anions in Flow-injection Analysis and Ion Chromatography', *Anal. Chim. Acta*, **252**(1), 53–57 (1991).
45. A. Galal, Z. Wang, A.E. Karagözler, H. Zimmer, H.B. Mark, Jr, P.L. Bishop, 'A Potentiometric Iodide (and Other) Ion Sensor Based on a Conducting Polymer Film Electrode. Part II. Effect of Electrode Conditioning and Regeneration Techniques', *Anal. Chim. Acta*, **299**(2), 145–163 (1994).
46. R. John, M. Spencer, G.G. Wallace, M.R. Smyth, 'Development of a Polypyrrole-based Human Serum Albumin Sensor', *Anal. Chim. Acta*, **249**(2), 381–385 (1991).
47. Y. Lin, G.G. Wallace, 'Development of a Polymer-based Electrode for Sensitive Detection of Dichloramine', *Anal. Chim. Acta*, **263**(1), 71–75 (1992).
48. J.M. Slater, J. Paynter, E.J. Watt, 'Multi-layer Conducting Polymer Gas Sensor Arrays for Olfactory Sensing', *Analyst*, **118**(4), 379–384 (1993).
49. P.J. Peerce, A.J. Bard, 'Polymer Films on Electrodes. Part II. Film Structure and Mechanism of Electron Transfer with Electrodeposited Poly(vinylferrocene)', *J. Electroanal. Chem.*, **112**(1), 97–115 (1980).
50. H.D. Abruña, P. Denisevich, M. Umaña, T.J. Meyer, R.W. Murray, 'Rectifying Interfaces Using Two-layer Films of Electrochemically Polymerized Vinylpyridine and Vinylbipyridine Complexes of Ruthenium and Iron on Electrodes', *J. Am. Chem. Soc.*, **103**(1), 1–5 (1981).
51. N. Oyama, F.C. Anson, 'Polymeric Ligands as Anchoring Groups for the Attachment of Metal Complexes to Graphite Electrode Surfaces', *J. Am. Chem. Soc.*, **101**(13), 3450–3456 (1979).
52. A.P. Brown, F.C. Anson, 'Molecular Anchors for the Attachment of Metal Complexes to Graphite Electrode Surfaces', *J. Electroanal. Chem.*, **83**(1), 203–206 (1977).
53. H.D. Abruña, T.J. Meyer, R.W. Murray, 'Chemical and Electrochemical Properties of 2,2′-Bipyridyl Complexes of Ruthenium Covalently Bound to Platinum Oxide Electrodes', *Inorg. Chem.*, **18**(11), 3233–3240 (1979).
54. R.D. Rocklin, R.W. Murray, 'Chemically Modified Carbon Electrodes. Part XVII. Metallation of Immobilized Tetra(aminophenyl)porphyrin with Manganese, Iron, Cobalt, Nickel, Copper, and Zinc, and Electrochemistry of Diprotonated Tetraphenylporphyrin', *J. Electroanal. Chem.*, **100**(1), 271–282 (1979).
55. H.D. Abruña, 'Coordination Chemistry in Two Dimensions: Chemically Modified Electrodes', *Coord. Chem. Rev.*, **86**(1), 135–189 (1988).
56. A. Haimerl, A. Merz, 'Ferrocene-modified Polypyrrole Films by Electrochemical Copolymerization', *Angew. Chem. Int. Ed. Engl.*, **25**(2), 180–181 (1986).
57. J.G. Eaves, R. Mirrazaei, D. Parker, H.S. Munro, 'Redox-active Films of Ferrocene Covalently Attached to Polypyrrole', *J. Chem. Soc., Perkin Trans. 2*, (4), 373–376 (1989).
58. A. Deronzier, J.C. Moutet, 'Functionalized Polypyrroles. New Molecular Materials for Electrocatalysis and Related Applications', *Acc. Chem. Res.*, **22**(7), 249–255 (1989).
59. S. Cosnier, A. Deronzier, J.C. Moutet, 'Oxidative Electropolymerization of Polypyridinyl Complexes of Ruthenium(II)-containing Pyrrole Groups', *J. Electroanal. Chem.*, **193**(1), 193–204 (1985).
60. A. Deronzier, J.M. Latour, 'A Poly(pyrrole – Nickel(II)-pyridiniumporphyrin)-modified Electrode', *J. Electroanal. Chem.*, **224**(1), 295–301 (1987).
61. G.J. Samuels, T.J. Meyer, 'An Electrode-supported Oxidation Catalyst Based on Ruthenium(IV). pH "Encapsulation" in a Polymer Film', *J. Am. Chem. Soc.*, **103**(2), 307–312 (1981).
62. A. Bettelheim, B.A. White, S.A. Raybuck, R.W. Murray, 'Electrochemical Polymerization of Amino-, Pyrrole-, and Hydroxy-substituted Tetraphenylporphyrins', *Inorg. Chem.*, **26**(7), 1009–1017 (1987).
63. C.M. Elliott, C.J. Baldy, L.M. Nuwaysir, C.L. Wilkins, 'Electrochemical Polymerization of 4-Methyl-4′-vinyl-2,2′-bipyridine-containing Metal Complexes: Polymer Structure and Mechanism of Formation', *Inorg. Chem.*, **29**(3), 389–392 (1990).
64. G.X. Wan, K. Shigehara, E. Tsuchida, F.C. Anson, 'Virtues of a Copolymer Containing Pyrrolidone and Iron Porphyrin Groups in the Catalysis of the Reduction of Dioxygen at Graphite Electrodes', *J. Electroanal. Chem.*, **179**(1), 239–250 (1984).
65. S.M. Geraty, J.G. Vos, 'Synthesis, Characterization, and Photochemical Properties of a Series of Ruthenium Containing Metallopolymers Based on Poly-*N*-vinylimidazole', *J. Chem. Soc., Dalton Trans.*, (12), 3073–3078 (1987).
66. C.D. Ellis, T.J. Meyer, 'Incorporation of Redox Couples into *p*-Chlorosulfonated Polystyrene Coated Electrodes

by Chemical Binding', *Inorg. Chem.*, **23**(12), 1748–1756 (1984).
67. A. El Hourch, S. Belcadi, P. Moisy, P. Crouigneau, J.M. Léger, C. Lamy, 'Electrocatalytic Reduction of Oxygen at Iron Phthalocyanine Modified Polymer Electrodes', *J. Electroanal. Chem.*, **339**(1), 1–12 (1992).
68. J. Zagal, M. Páez, A.A. Tanaka, J.R. dos Santos, Jr, C.A. Linkous, 'Electrocatalytic Activity of Metal Phthalocyanines for Oxygen Reduction', *J. Electroanal. Chem.*, **339**(1), 13–30 (1992).
69. N. Kobayashi, P. Janda, A.B.P. Lever, 'Cathodic Reduction of Oxygen and Hydrogen Peroxide at Cobalt and Iron Crowned Phthalocyanines Adsorbed on Highly Oriented Pyrolytic Graphite Electrodes', *Inorg. Chem.*, **31**(25), 5172–5177 (1992).
70. H. Li, T.F. Guarr, 'Formation of Electronically Conductive Thin Films of Metal Phthalocyanines via Electropolymerization', *J. Chem. Soc., Chem. Commun.*, (13), 832–834 (1989).
71. C.P. Horwitz, R.W. Murray, 'Oxidative Electropolymerization of Metal Schiff-base Complexes', *Mol. Cryst. Liq. Cryst.*, **160**, 389–404 (1988).
72. L.A. Hofercamp, K.A. Goldsby, 'Surface-modified Electrodes Based on Nickel(II) and Copper(II) Bis(salicylaldimine) Complexes', *Chem. Mater.*, **1**(3), 348–352 (1989).
73. P. Audebert, P. Capdevielle, M. Maumy, 'Redox and Conducting Polymers Based on Salen-type Metal Units: Electrochemical Study and Some Characteristics', *New J. Chem.*, **16**(6), 697–703 (1992).
74. C.E. Dahm, D.G. Peters, J. Simonet, 'Electrochemical and Spectroscopic Characterization of Anodically Formed Nickel Salen Polymer Films on Glassy Carbon, Platinum, and Optically Transparent Tin Oxide Electrodes in Acetonitrile Containing Tetramethylammonium Tetrafluoroborate', *J. Electroanal. Chem.*, **410**(2), 163–171 (1996).
75. S.M. Geraty, D.W.M. Arrigan, J.G. Vos, 'The Use of Electrodes Coated with Ruthenium-containing Polymers as Sensors for Iron(II)', *Anal. Chem. Symp. Ser.*, **25**, 303–308 (1986).
76. M.A. Ruiz, E. Garcia-Moreno, C. Barbas, J.M. Pingarrón, 'Determination of Phenolic Antioxidants by HPLC with Amperometric Detection at a Nickel Phthalocyanine Polymer Modified Electrode', *Electroanalysis*, **11**(7), 470–474 (1999).
77. A.P. Doherty, M.A. Stanley, D. Leech, J.G. Vos, 'Oxidative Detection of Nitrite at an Electrocatalytic [Ru(bpy)$_2$poly-(4-vinylpyridine)$_{10}$Cl]Cl Electrochemical Sensor Applied for the Flow Injection Determination of Nitrate Using a Cu/Cd Reductor Column', *Anal. Chim. Acta*, **319**(1), 111–120 (1996).
78. Y.H. Tse, P. Janda, H. Lam, A.B.P. Lever, 'Electrode with Electropolymerized Tetraaminophthalocyanatocobalt(II) for Detection of Sulfide Ion', *Anal. Chem.*, **67**(5), 981–985 (1995).
79. M.A. Ruiz, M.G. Blázquez, J.M. Pingarrón, 'Electrocatalytic and Flow-injection Determination of the Antioxidant *tert*-Butylhydroxyanisole at a Nickel Phthalocyanine Polymer Modified Electrode', *Anal. Chim. Acta*, **305**(1), 49–56 (1995).
80. T.F. Kang, G.L. Shen, R.Q. Yu, 'Voltammetric Behavior of Dopamine at Nickel Phthalocyanine Polymer Modified Electrodes and Analytical Applications', *Anal. Chim. Acta*, **354**(1), 343–349 (1997).
81. C.R. Martin, I. Rubinstein, A.J. Bard, 'Polymer Films on Electrodes. 9. Electron and Mass Transfer in Nafion® Films Containing $Ru(bpy)_3^{2+}$', *J. Am. Chem. Soc.*, **104**(18), 4817–4824 (1982).
82. M. Majda, L.R. Faulkner, 'Electrochemical Behavior of Tris(2,2′-Bipyridine)ruthenium Complexes in Films of Poly(styrenesulfonate) on Electrodes', *J. Electroanal. Chem.*, **169**(1), 77–95 (1984).
83. N. Oyama, F.C. Anson, 'Electrostatic Binding of Metal Complexes to Electrode Surfaces Coated with Highly Charged Polymeric Films', *J. Electrochem. Soc.*, **127**(1), 249–250 (1980).
84. W.J. Vining, T.J. Meyer, 'Redox Properties of the Water Oxidation Catalyst $(bpy)_2(H_2O)RuORu(H_2O)(bpy)_2^{4+}$ in Thin Polymeric Films. Electrocatalytic Oxidation of Cl^- to Cl_2', *Inorg. Chem.*, **25**(12), 2023–2033 (1986).
85. N. Oyama, T. Shimomura, K. Shigehara, F.C. Anson, 'Electrochemical Responses of Multiply Charged Transition Metal Complexes Bound Electrostatically to Graphite Electrode Surfaces Coated with Polyelectrolytes', *J. Electroanal. Chem.*, **112**(2), 271–280 (1980).
86. C.J. Pickett, K.S. Ryder, J.C. Moutet, 'Synthesis and Anodic Polymerization of an L-Cystine Derivatized Pyrrole; Copolymerization with a Tetraalkylammonium Pyrrole Allows Reduction of the Cystinyl Film to a Cysteinyl State that Binds Electroactive $\{Fe_4S_4\}^{2+}$ Centres', *J. Chem. Soc., Chem. Commun.*, (9), 694–697 (1992).
87. F. Nguyen, F.C. Anson, 'Electrochemical Examination of Electrostatic Attraction of Cations and the Ejection of Anions by Polyanionic Electrode Coatings Stable in Non-aqueous Solvents', *Electrochim. Acta*, **44**(2), 239–245 (1998).
88. N. Oyama, F.C. Anson, 'Electrostatic Binding of Metal Complexes to Electrode Surfaces with Highly Charged Polymeric Films', *J. Electrochem. Soc.*, **127**(1), 247–248 (1980).
89. F.C. Anson, J.M. Savéant, K. Shigehara, 'New Model for the Interior of Polyelectrolyte Coatings on Electrode Surfaces. Mechanisms of Charge Transport Through Protonated Poly(L-lysine) Films Containing $Fe^{III}(edta)^-$ and $Fe^{II}(edta)^{2-}$ as Counterions', *J. Am. Chem. Soc.*, **105**(5), 1096–1106 (1983).
90. R. Volf, T.V. Shishkanova, P. Matejka, M. Hamplová, V. Král, 'Potentiometric Anion Response of Poly[5, 15-bis(2-aminophenyl)porphyrin] Electropolymerized Electrodes', *Anal. Chim. Acta*, **381**(2), 197–205 (1999).

91. K. Yoshino, K. Yoshimoto, S. Morita, T. Kawai, S.H. Kim, K.L. Kang, S.K. Choi, 'Electrical and Electrochemical Properties of Polyacetylene Derivatives with Pendant Cationic Groups', *Synth. Met.*, **69**(1), 81–82 (1995).
92. C.J. Brumlik, A. Parthasarathy, W.J. Chen, C.R. Martin, 'Plasma Polymerization of Sulfonated Fluorochlorocarbon Ionomer Films', *J. Electrochem. Soc.*, **141**(9), 2273–2279 (1994).
93. A. Economou, P.R. Fielden, 'Adsorptive Stripping Voltammetry on Mercury Film Electrodes in the Presence of Surfactants', *Analyst*, **118**(11), 1399–1404 (1993).
94. J. Labuda, M. Vanìcková, E. Uhlemann, W. Mickler, 'Applicability of Chemically Modified Electrodes for Determination of Copper Species in Natural Waters', *Anal. Chim. Acta*, **284**(3), 517–523 (1994).
95. J.M. Zen, S.Y. Huang, 'Square-wave Voltammetric Determination of Lead(II) with a Nafion® 2,2′-Bipyridyl Mercury Film Electrode', *Anal. Chim. Acta*, **296**(1), 77–86 (1994).
96. I.G. Casella, M.R. Guascito, A.M. Salvi, E. Desimoni, 'Catalytic Oxidation and Flow Detection of Hydrazine Compounds at a Nafion®/Ruthenium(III) Chemically Modified Electrode', *Anal. Chim. Acta*, **354**(1), 333–341 (1997).
97. L.Q. Mao, Y. Tian, G.Y. Shi, H.Y. Liu, L.T. Jin, K. Yamamoto, S. Tao, J.Y. Jin, 'A New Ultramicrosensor for Nitric Oxide Based on Electropolymerized Films of Nickel Salen', *Anal. Lett.*, **31**(12), 1991–2007 (1998).
98. S.F. de Betono, A.A. Garcìa, J.F.A. Valentin, 'UV Spectrophotometry and Square-wave Voltammetry at Nafion®-modified Carbon Paste Electrode for the Determination of Doxazosin in Urine and Formulations', *J. Pharm. Biomed. Anal.*, **20**(4), 621–630 (1999).
99. J.M. Zen, C.B. Wang, 'Determination of Dissolved Oxygen by Catalytic Reduction on a Nafion®/Ruthenium Oxide Pyrochlore Chemically Modified Electrode', *J. Electroanal. Chem.*, **368**(1), 251–256 (1994).
100. P. Ugo, L.M. Moretto, S. Bellomi, V.P. Menon, C.R. Martin, 'Ion-exchange Voltammetry at Polymer Film-coated Nanoelectrode Ensembles', *Anal. Chem.*, **68**(23), 4160–4165 (1996).
101. R.A. Hann, 'Molecular Structure and Monolayer Properties', in *Langmuir–Blodgett Films*, ed. G. Roberts, Plenum Press, New York, 17–92, 1990.
102. M.K. DeArmond, G.A. Fried, 'Langmuir–Blodgett Films of Transition Metal Complexes', in *Progress in Inorganic Chemistry*, ed. K.D. Karlin, Wiley, New York, 97–142, 1997.
103. L.M. Goldenberg, 'Electrochemical Properties of Langmuir-Blodgett Films', *J. Electroanal. Chem.*, **379**(1), 3–19 (1994).
104. R.M. Swart, 'Monolayers and Multilayers of Biomolecules', in *Langmuir–Blodgett Films*, ed. G. Roberts, Plenum Press, New York, 273–316, 1990.
105. A.J. Fernández, M.T. Martìn, J.J. Ruiz, E. Muñoz, L. Camacho, 'Electrochemical Behavior of LB Films Containing a Mixture of Viologen and a Phospholipid', *J. Phys. Chem. B*, **102**(35), 6799–6803 (1998).
106. L.M. Goldenberg, C. Pearson, M.R. Bryce, M.C. Petty, 'Preparation and Characterization of Conductive Langmuir–Blodgett Films of a Tetrabutylammonium-Ni(dmit)$_2$ Complex', *J. Mater. Chem.*, **6**(5), 699–704 (1996).
107. A. Ulman, *Ultrathin Organic Films*, Academic Press, San Diego, CA, 124–125, 1991.
108. S. Nagase, M. Kataoka, R. Naganawa, R. Komatsu, K. Odashima, Y. Umezawa, 'Voltammetric Anion Responsive Sensors Based on Modulation of Ion Permeability Through Langmuir–Blodgett Films Containing Synthetic Anion Receptors', *Anal. Chem.*, **62**(13), 1252–1259 (1990).
109. M. Fujihira, S. Poosittisak, 'Precise Control of Amount of Electrodeposited Platinum by Using the Langmuir–Blodgett Film and its Application to Electrocatalysis of Molecular Oxygen Reduction', *Chem. Lett.*, (2), 251–252 (1986).
110. M. Fujihira, T. Araki, 'In Situ Electrochemical Measurements of the Redox-active Monolayer by the Horizontal Touching Method Under a Controlled Surface Pressure', *Chem. Lett.*, (6), 921–926 (1986).
111. C.J. Miller, A.J. Bard, 'Horizontal Touch Voltammetric Analysis: Determination of Insoluble Electroactive Species in Films at the Air/Water Interface', *Anal. Chem.*, **63**(17), 1707–1714 (1991).
112. A. Nelson, A. Benton, 'Phospholipid Monolayers at the Mercury/Water Interface', *J. Electroanal. Chem.*, **202**(1), 253–270 (1986).
113. A. Ulman, *Characterization of Organic Thin Films*, Butterworth-Heinemann, Stoneham, MA, 1995.
114. R.M. Corn, 'Optical Second Harmonic Generation Studies of Adsorption, Orientation, and Order at the Electrochemical Interface', *Anal. Chem.*, **63**(2), 285A–295A (1991).
115. G.L. Richmond, 'Optical Second Harmonic Generation as an In Situ Probe of Electrochemical Interfaces', in *Electroanalytical Chemistry*, ed. A.J. Bard, Dekker, New York, 87–180, Vol. 17, 1991.
116. N. Ogura, Y. Ichinohe, S. Yoshida, T. Watanabe, T. Hoshi, K. Endo, M. Kudo, 'Substrate Dependence of Secondary Ion Intensities from Langmuir–Blodgett Films Investigated by TOF-SIMS', *J. Surf. Anal.*, **5**(2), 318–321 (1999).
117. A. Nelson, 'Voltammetry of Tl^{I}, Cd^{II}, Cu^{II}, Pb^{II}, and Eu^{III} at Phosphatidylserine-coated Mercury Electrodes', *J. Chem. Soc., Faraday Trans.*, **89**(16), 3081–3090 (1993).
118. A. Nelson, N. Auffret, J. Readman, 'Initial Applications of Phospholipid-coated Mercury Electrodes to the Determination of Polynuclear Aromatic Hydrocarbons and Other Organic Micropollutants in Aqueous Systems', *Anal. Chim. Acta*, **207**(1), 47–57 (1988).

119. J. Wang, 'Electroanalysis and Biosensors', *Anal. Chem.*, **71**(12), 328R–332R (1999).
120. J. Janata, M. Josowicz, P. Vanýsek, D.M. DeVaney, 'Chemical Sensors', *Anal. Chem.*, **70**(12), 179R–208R (1998).
121. J.L. Anderson, L.A. Coury, Jr, J. Leddy, 'Dynamic Electrochemistry: Methodology and Application', *Anal. Chem.*, **70**(12), 519R–589R (1998).
122. M. Aizawa, 'New Trends in Bioelectrochemistry', in *New Challenges in Organic Electrochemistry*, ed. T. Osa, Gordon and Breach Science Publishers, Amsterdam, 339–356, 1998.
123. The Institute for Scientific Information databases are site-configured and can be accessed at http://www.webofscience.com/.
124. P.N. Bartlett, P. Tebbutt, R.G. Whitaker, 'Kinetic Aspects of the Use of Modified Electrodes and Mediators in Bioelectrochemistry', *Prog. React. Kinet.*, **16**(2), 55–155 (1991).
125. J.F. Rusling, 'Enzyme Bioelectrochemistry in Cast Biomembrane-like Films', *Acc. Chem. Res.*, **31**(6), 363–369 (1998).
126. L. Stancik, 'Amperometric Biosensors for Non-aqueous Media', *Chem. Listy*, **91**(1), 30–37 (1997).
127. T. Ruzgas, E. Csöregi, J. Emnéus, L. Gorton, G. Marko-Varga, 'Peroxidase-modified Electrodes: Fundamentals and Application', *Anal. Chim. Acta*, **330**(2), 123–138 (1996).
128. I. Katakis, E. Dominguez, 'Catalytic Electrooxidation of NADH for Dehydrogenase Amperometric Biosensors', *Mikrochim. Acta*, **126**(1), 11–32 (1997).
129. P.N. Bartlett, J. Cooper, 'Applications of Electroactive Polymers in Biochemistry and Bioelectronics', in *Electroactive Polymer Chemistry*, ed. M.E.G. Lyons, Plenum Press, New York, 233–267, Vol. 2, 1996.
130. T.W. Lewis, G.G. Wallace, M.R. Smyth, 'Electrofunctional Polymers: their Role in the Development of New Analytical Systems', *Analyst*, **124**(3), 213–219 (1999).
131. M. Nishizawa, T. Matsue, I. Uchida, 'Penicillin Sensor Based on a Microarray Electrode Coated with pH-responsive Polypyrrole', *Anal. Chem.*, **64**(12), 2642–2644 (1992).
132. P. Sarkar, I.E. Tothill, S.J. Setford, A.P.F. Turner, 'Screen-printed Amperometric Biosensors for the Rapid Determination of L- and D-Amino Acids', *Analyst*, **124**(6), 865–870 (1999).
133. T.J. Moore, G.G. Nam, L.C. Pipes, L.A. Coury, Jr, 'Chemically Amplified Voltammetric Enzyme Electrodes for Oxidizable Pharmaceuticals', *Anal. Chem.*, **66**(19), 3158–3163 (1994).
134. C. Bourdillon, J.P. Bourgeois, D. Thomas, 'Covalent Linkage of Glucose Oxidase on Modified Glassy Carbon Electrodes. Kinetic Phenomena', *J. Am. Chem. Soc.*, **102**(12), 4231–4235 (1980).
135. I. Willner, A. Ricklin, 'Electrical Communication between Electrodes and $NAD(P)^+$-dependent Enzymes Using Pyrroloquinolinequinone–Enzyme Electrodes in a Self-assembled Monolayer Configuration: Design of a New Class of Amperometric Biosensors', *Anal. Chem.*, **66**(9), 1535–1539 (1994).
136. I. Willner, E. Katz, B. Willner, 'Electrical Contact of Redox Enzyme Layers with Electrodes: Routes to Amperometric Biosensors', *Electroanalysis*, **9**(13), 965–977 (1999).
137. M. Rehak, M. Snejdarkova, M. Otto, 'Application of Biotin–Streptavidin Technology in Developing a Xanthine Biosensor Based on a Self-assembled Phospholipid Membrane', *Biosens. Bioelectron.*, **9**(4), 337–341 (1994).
138. T. Osa, J. Anzai, 'Preparation of Biosensors by Immobilizing Enzymes on Surface of Langmuir–Blodgett Membrane', *Hyomen*, **30**(12), 985–990 (1992).
139. M. Aizawa, 'LB (Langmuir–Blodgett) Film Technology for Biosensors', *Yukagaku*, **39**(3), 166–170 (1990).
140. J. Hirst, A. Sucheta, B.A.C. Ackrell, F.A. Armstrong, 'Electrocatalytic Voltammetry of Succinate Dehydrogenase: Direct Quantification of the Catalytic Properties of a Complex Electron-transport Enzyme', *J. Am. Chem. Soc.*, **118**(21), 5031–5038 (1996).
141. D.P. Nikolelis, T. Hianik, U.J. Krull, 'Biosensors Based on Thin Lipid Films and Liposomes', *Electroanalysis*, **11**(1), 7–15 (1999).
142. B. Liu, F. Yan, J. Kong, J. Deng, 'A Reagentless Amperometric Biosensor Based on the Co-immobilization of Horseradish Peroxidase and Methylene Green in a Modified Zeolite Matrix', *Anal. Chim. Acta*, **386**(1), 31–39 (1999).
143. A. Griffith, A. Glidle, G. Beamson, J.M. Cooper, 'Determination of the Biomolecular Composition of an Enzyme-polymer Biosensor', *J. Phys. Chem. B*, **101**(11), 2092–2100 (1997).
144. P.C. Pandey, S. Upadhyay, H.C. Pathak, 'A New Glucose Biosensor Based on Sandwich Configuration of Organically Modified Sol-gel Glass', *Electroanalysis*, **11**(1), 59–64 (1999).
145. B. Wang, B. Li, Z. Wang, G. Xu, Q. Wang, S. Dong, 'Sol-gel Thin-film Immobilized Soybean Peroxidase Biosensor for the Amperometric Determination of Hydrogen Peroxide in Acid Medium', *Anal. Chem.*, **71**(10), 1935–1939 (1999).
146. Q. Chi, J. Zhang, S. Dong, E. Wang, 'Direct Electrochemistry and Surface Characterization of Glucose Oxidase Adsorbed on Anodized Carbon Electrodes', *Electrochim. Acta*, **39**(16), 2431–2438 (1994).
147. W.A.C. Somers, W. Van Hartingsveldt, E.C.A. Stigter, J.P. Van Der Lugt, 'Bio-electrochemistry: a Powerful Tool for the Production of Ingredients in the Food Industry', *Agro Food Ind. Hi-Tech*, **8**(2), 32–35 (1997).
148. A. Maines, D. Ashworth, P. Vadgama, 'Enzyme Electrodes for Food Analysis', *Food Technol. Biotechnol.*, **34**(1), 31–42 (1996).

149. J. Wang, 'Amperometric Biosensors for Clinical and Therapeutic Drug Monitoring: a Review', *J. Pharm. Biomed. Anal.*, **19**(1), 47–53 (1999).
150. L.V. Sigolaeva, A.V. Eremenko, A. Makower, G.F. Makhaeva, V.V. Malygin, I.N. Kurochkin, 'A New Approach for Determination of Neuropathy Target Esterase Activity', *Chem.-Biol. Interact.*, **120**, 559–565 (1999).
151. Y. Hu, K.M. Mitchell, F.N. Albahadily, E.K. Michaelis, G.S. Wilson, 'Direct Measurement of Glutamate Release in the Brain Using a Dual Enzyme-based Electrochemical Sensor', *Brain Res.*, **659**(1), 117–125 (1994).
152. P. Skladal, 'Biosensors Based on Cholinesterase for Detection of Pesticides', *Food Technol. Biotechnol.*, **34**(1), 43–49 (1996).
153. T. Noguer, B. Leca, G. Jeanty, J.L. Marty, 'Biosensors Based on Enzyme Inhibition: Detection of Organophosphorus and Carbamate Insecticides and Dithiocarbamate Fungicides', *Field Anal. Chem. Tech.*, **3**(3), 171–178 (1999).
154. W. Del Carlo, M. Mascini, 'Immunoassay for Polychlorinated Biphenyls (PCB) Using Screen Printed Electrodes', *Field Anal. Chem. Tech.*, **3**(3), 179–184 (1999).

Self-assembled Monolayers on Electrodes

Harry O. Finklea
West Virginia University, Morgantown, USA

List of selected abbreviations appears in Volume 15

Alkanethiols and related molecules spontaneously adsorb from solution or vapor phase onto oxide-free metals, especially gold, to form close-packed oriented monolayers. The ease and flexibility of the self-assembly process provides a convenient method for altering the properties of the metal as an electrode. The close-packed hydrocarbon layer blocks access of most solution species to the electrode surface. Consequently, interfacial capacitances are markedly reduced and most electron-transfer reactions are strongly inhibited. The monolayers also provide a foundation for attachment of additional molecules or molecular layers to electrodes. Self-assembled monolayers (SAMs) on electrodes have been used in voltammetric methods to improve the analytical sensitivity and to impart greater selectivity towards specific analytes. Redox molecules have been anchored at fixed and controllable distances from the electrode surface, thereby permitting the study of electron-transfer kinetics over long distances and at large driving forces. The ability of SAMs to inhibit metal corrosion and to promote better adhesion of electroactive polymers has been demonstrated.

1 INTRODUCTION

1.1 Definition of Self-assembled Monolayer

An organized SAM is a single layer of molecules on a substrate in which the molecules exhibit a high

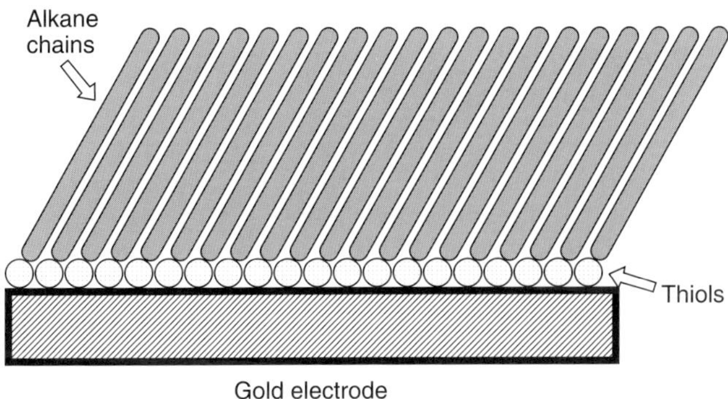

Figure 1 An organized monolayer on a substrate (electrode). The monolayer can be deposited by the LB method or by self-assembly. (Reprinted from Finklea,[1] p. 110, by courtesy of Marcel Dekker, Inc.)

degree of orientation, molecular order and packing (Figure 1). There are two common methods for depositing a monolayer. In the Langmuir–Blodgett (LB) method, amphiphilic molecules are spread on the air–water interface, compressed laterally, and transferred to the substrate either by dipping the substrate through the interface or by touching it to the interface.[2] In the self-assembly method, the monolayer spontaneously forms upon exposure of the substrate to a solution or vapor containing the molecules. Successful self-assembly requires a relatively strong bond between the substrate and an atom or moiety in the molecule, and an additional lateral interaction between molecules in the monolayer. The strength of the head group–substrate bonds, the lateral interactions and the density of packing result in sufficient stability that the monolayer resists removal by a solvent rinse. Unlike the popular LB monolayers, SAMs remained relatively obscure until the 1980s when several researchers discovered that long-chain thiols and disulfides spontaneously formed remarkably well-packed and stable monolayers on gold substrates.[3–9] Because SAMs provide a facile means of defining the chemical composition and structure of a surface (see below), they have become the focus of intensive investigation. Potential technological applications can be found in areas such as wetting, lubrication, adhesion, corrosion, biocompatibility, catalysis, chemical sensing and nanoscale lithography.

1.2 Types of Self-assembled Monolayers on Electrodes

SAMs have been formed on every common electrode material with the possible exception of carbon. The common electrode metals (gold, silver, platinum and mercury) have all been coated with SAMs containing sulfur compounds, especially thiols, disulfides, and sulfides. The sulfur compounds most commonly contain pendant alkane chains of varying lengths. Less common are SAMs based on adsorption of isonitriles.[10,11] Chlorosilanes with long alkyl chains self-assemble on doped metal oxides such as SnO_2,[12] silicon with a thin oxide coating[13] and even gold without any surface oxide.[14,15] Generation of alkane radicals near an oxide-free silicon surface results in the formation of a densely packed SAM in which a methylene carbon is directly bonded to a silicon surface atom.[16,17]

1.3 Advantages of Self-assembled Monolayers Based on Thiols and Related Molecules

The first advantage is the ease with which SAMs are formed when gold and other metals are exposed to thiols and related molecules (in subsequent discussion, the word thiols will also imply disulfides and sulfides). A monolayer is deposited on the metal in a matter of seconds to minutes. The self-assembly method does not require anaerobic or anhydrous conditions; nor does it require a vacuum. Self-assembly is relatively insensitive to the choice of solvent. While organic-free metal surfaces are desirable, the high affinity of the sulfur for the metal enables the assembling layer to displace more weakly adsorbed impurities. Curvature or accessibility of the metal surface is not a factor; substrates can range from macroscopic to submicroscopic, and from smooth to highly porous.

A second advantage arises from the affinity of the sulfur for the metal and the strength of the bond formed. SAMs survive prolonged exposure to vacuum. It is possible to have a wide range of functional groups in the adsorbing molecule without disrupting the self-assembly process or destabilizing the SAM. Considering just the family of w-substituted alkanethiols, the terminal substituent can be an alkane (linear, branched, perfluorinated, perdeuterated), alkene, alkyne, aromatic, halide, ether, alcohol, aldehyde, carboxylic acid, amide, ester, amine or nitrile.[18–20] The "body" of the molecule can contain, for example, heteroatoms,[21–24] aromatic groups,[25–27] conjugated unsaturated links[28–30] and other rigid rod structures,[31–33] sulfones[24,34] and amides.[35,36] If the SAM is uniform in composition and densely packed, then a single functional group is exposed on the external surface. This property permits the exploration of the effect of surface composition on such surface-sensitive properties as wetting, friction and adhesion.[20,37–40] The diversity of sulfur-based SAMs, their autophobic behavior and the slow exchange with solution molecules allows the preparation of micropatterned surfaces in which the SAM is laterally heterogeneous.[41]

Mixed SAMs (Figure 2) can be prepared by depositing different molecules simultaneously or sequentially. When both molecules are present in the deposition solution,

For references see page 10105

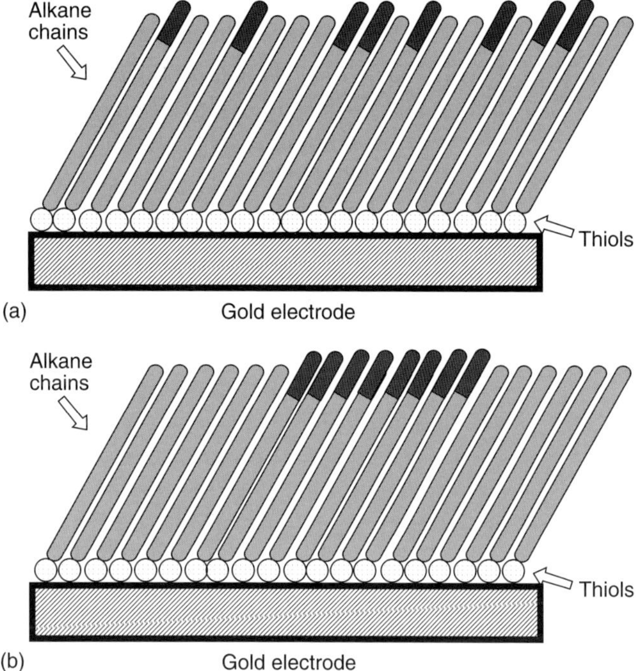

Figure 2 Two mixed SAMs exhibiting homogeneous mixing (a) and phase separation (b). Differences in chain length and terminal groups are shown.

the mole fraction of each molecule in the SAM can be controlled via the mole ratio of the two molecules in the solution, the identity and temperature of the solvent and the time of the deposition.[42] The distribution of the two molecules may vary from intimately mixed to completely phase-separated. It is possible to deposit SAMs with a 1 : 1 mole ratio of two chains and terminal groups with presumably perfect mixing via asymmetric sulfides or disulfides or thiols with two pendant chains.[43] In sequential deposition, a second thiol is incorporated into an existing SAM by prolonged immersion of the substrate in the second deposition solution.

In electrochemistry, the utility of thiol-based SAMs arises from their ability to survive the electrochemical experiment. The sulfur atoms resist oxidation, reduction and desorption. SAMs on electrodes are stable over a wide range of potentials and electrolyte compositions (especially aqueous electrolytes). They afford a means of controlling the electrode/electrolyte interfacial properties and the accessibility of the electrode surface to solution molecules. SAMs also provide a means of attaching to the electrode a diverse set of structures ranging from modified monolayers to multilayers. Specific applications include the development of more selective and sensitive electrochemical sensors (especially biosensors), control of faradaic reaction mechanisms, and a better understanding of the factors controlling electron transfer over long distances and under large driving forces.

1.4 Scope and Previous Reviews

The scope of this article will be primarily limited to SAMs composed of thiols and related molecules on metal electrodes. Since there is already a vast literature on the characterization of SAMs on metals, and since the findings therein are relevant to the behavior of SAMs on electrodes, a brief overview of nonelectrochemical characterization of SAMs follows the section on SAM preparation. Postdeposition modification reactions are summarized before discussion of the electrochemical characterization of the SAMs on electrodes. The final section covers a range of electrochemical applications of SAMs.

A previous review of SAMs on electrodes provides a more in-depth treatment of many topics in this article.[1] Two books by Ulman survey the field of monolayers prepared by both LB and self-assembly methods.[2,44] Reviews by Nuzzo cover both the use of SAMs as model surfaces and their technological applications.[45,46] Other discussions of applications of SAMs have been provided by Zhong and Porter,[47] Whitesides and Laibinis,[37] and Wink et al.[48]

2 PREPARATION OF SELF-ASSEMBLED MONOLAYERS ON ELECTRODES

2.1 Substrate Choice and Preparation

Gold is the most popular substrate for thiol SAMs. Owing to its noble character, gold substrates can be handled in air without the formation of an oxide surface layer, and can survive harsh chemical treatments which remove organic contaminants. Electrodes may be composed of bulk gold, either polycrystalline or single crystal, or thin films deposited on various substrates. To some extent, the choice of substrate is dictated by the application of the SAM. For the most highly oriented crystalline SAM, a single crystalline and atomically flat substrate is desired. When the SAM's function is to block access of molecules to the gold, a minimum density of pinholes is sought. While the origin of pinholes in SAMs is still unclear, impurities such as inorganic oxides are a likely source. Hence, the purity of the gold surface appears to be more important than its smoothness.[49] Both requirements are met by evaporated or sputtered gold films (typically 50–200 nm thick) on glass, silicon or cleaved mica substrates. To improve the adhesion of the gold film to the oxide surface, a thin (1–5 nm) layer of chromium or titanium is often deposited first. There is evidence that defects in the underlayer of Cr or Ti can be a source of pinholes in the final SAM.[50] If the deposited film is annealed at high temperatures, then the gold surface is largely composed of Au(111) domains and is frequently labeled as a Au(111) substrate. A comparison of grain size and crystallinity of thermally evaporated and sputtered gold on glass, silicon and cleaved mica substrates has been provided by Golan et al.[51] Some subtle effects of the substrate identity and the gold deposition protocol on the behavior of the subsequently deposited SAMs have been noted.[52–54] Procedures are available for the preparation of ultraflat[55,56] or optically transparent gold substrates.[57,58]

Cleaning and etching steps are often part of the deposition protocol. For bulk gold, heating the gold in a gas/air flame results in a hydrophilic surface, indicating that all organics have been removed. Electrochemical cycling into the oxide formation region in dilute acid functions as both a cleaning and an annealing process; the resulting voltammogram provides an assay of the cleanliness and crystallinity of the gold and its true surface area.[1] Etching the bulk gold in dilute aqua regia removes polishing damage and improves the blocking properties of the SAM.[59] Evaporated or sputtered gold films are frequently immersed into the thiol deposition solution as soon as they are removed from the gold deposition unit. Despite the fact that the gold surfaces are hydrophobic and therefore contaminated, SAM deposition proceeds unimpeded. Alternatively, organic contaminants can be removed via exposure of the gold to a powerful oxidant. Popular oxidants include "piranha" solution (a 1 : 3 mixture of 30% hydrogen peroxide and concentrated sulfuric acid at ca. 100 °C) (*Caution: this mixture reacts violently with organic material and has been known to explode when stored in closed containers!*), an oxygen plasma and ozone generated by UV (ultraviolet) light. These treatments leave a surface oxide which can become trapped under a self-assembling thiol monolayer, and which can affect the properties of the SAM deleteriously.[60,61] An ethanol rinse rapidly removes the gold oxide.

The other coinage metals, silver and copper, and platinum have been used as substrates for thiol SAMs. For freshly evaporated silver or copper thin films, surface oxides exist prior to SAM deposition. The self-assembly process appears to remove the oxide layer from silver but not from copper, yielding lower quality SAMs on the latter substrate.[62–64] The oxide layer can be removed by an acid soak[65] or electrochemical reduction.[66] A recent innovation has been to form a single atomic layer of copper or silver on a gold surface by underpotential deposition (UPD) and to assemble the monolayer on the UPD layer.[67–70] There is growing evidence that the bond between the sulfur and the UPD atom is stronger than the bond between the sulfur and gold atoms, resulting in a more stable SAM. It is even possible to form a UPD Cu layer after deposition of a short-chain alkanethiol SAM.[71]

Liquid mercury avoids the issues of surface crystallinity, roughness and morphology, but is easily oxidized in the presence of thiols. It is possible to prepare homogenous and densely packed SAMs of alkanethiols on mercury via solution deposition,[72–75] but not by vapor deposition.[76]

Other substrates for thiol SAMs include nickel,[77,78] indium tin-oxide,[79] indium phosphide,[80] and a Tl-Ba-Cu-O high-temperature superconductor.[81]

2.2 Solution Adsorption

Immersion of the substrate into a homogenous solution of the self-assembling molecule at room temperature followed by rinsing is the most common approach for depositing the SAM. Usually any solvent capable of dissolving the molecule is suitable. Ethanol is the most popular solvent.[19] The thiol concentration can be varied from micromolar levels to that of the neat thiol liquid. Very low concentrations of thiols are favored by those seeking large crystalline domains of alkanethiols via slow self-assembly,[82] but are not a guarantee of SAM quality.[83] There is evidence that the solvent can be trapped in the SAM and not removed by the subsequent rinsing step.[19,84] Insoluble thiols can be dispersed into water with the aid of surfactants[85,86] or cyclodextrins[87] and subsequently assembled on gold. Considerable effort has been made to ascertain the time needed to form a well-organized layer. For millimolar or higher concentrations of thiols, a disordered monolayer is deposited in a few seconds. There is a much slower transformation over a period of hours to days into a highly oriented and densely packed monolayer.[19,50] Scanning tunneling microscopy (STM) studies (see section 3) suggest that the gold is slightly etched during self-assembly; gold has been detected in the deposition solution.[88,89] To remove kinetically trapped disordered states, the SAM is sometimes annealed by soaking it in hot deposition solution,[90] exposing it to warm temperatures in a gaseous ambient,[91] or subjecting the electrode to a cyclic voltammogram (CV) and repeated immersions in the deposition solution.[92]

When two thiols are being co-deposited, the solvent, time and temperature all affect the mole fraction of each component in the SAM.[42,64,93,94] The less soluble thiol is preferentially deposited. High concentrations of the thiol, long adsorption times and high temperatures encourage the SAM composition to approach equilibrium with the solution; the result is an abrupt transition of the SAM composition from one component to the other as a function of the mole ratio of the two components in solution.[42]

For references see page 10105

2.3 Other Deposition Methods

Deposition of the SAM from an electrolyte with the substrate under potential control ensures that the metal is in the reduced state. A thiol SAM desorbs at very negative potentials in alkaline electrolytes (section 5.2). Consequently, a SAM can be formed in an alkaline electrolyte with dissolved alkanethiols by slowly shifting the electrode potential in the positive direction from the desorption potential.[95–97] When the thiol is sufficiently volatile, SAM deposition can be performed from the vapor phase, both in a vacuum[98] and at ambient pressures.[99] The resulting SAMs appear to be virtually identical to those obtained by solution deposition. Majda et al. have demonstrated that monolayers very similar to those obtained by self-assembly can be obtained by spreading a mixture of alkanethiol and long-chain alkanol on a LB trough, applying compression and transferring the monolayer to a gold substrate.[100–102] There are numerous approaches to producing micropatterned SAMs, but the most common is to press a patterned polysiloxane stamp inked with a thiol onto a gold substrate and then to fill in the gaps with solution deposition of a second thiol.[41]

3 NONELECTROCHEMICAL CHARACTERIZATION OF SELF-ASSEMBLED MONOLAYERS

SAMs have been subjected to virtually every surface analytical method known. Among the more frequently applied tools are wetting contact angle, ellipsometry, surface plasmon resonance (SPR) spectroscopy, surface IR (infrared) spectroscopy, Raman spectroscopy, X-ray absorption spectroscopy, X-ray photoelectron spectroscopy (XPS) and Auger electron spectroscopy (AES), temperature programmed desorption (TPD), scanning electron microscopy (SEM), surface ionization mass spectroscopy (SIMS) and laser desorption mass spectroscopy (LDMS), STM and atomic force microscopy (AFM), assorted diffraction (X-ray, electron, atom) and piezoelectric methods. The following discussion is a summary of observations from all of these methods.

The mechanism of self-assembly appears to follow two stages. Initially, alkanethiols adsorb horizontally onto the metal substrate. Subsequently, the thiols lift up to form the vertically oriented layer. The thickness of the SAM is linearly related to the chain length of the thiol. On single-crystal surfaces, SAMs generally form highly crystalline lattices which are commensurate with the metal atom lattice. Chain "melting" to a more liquid-like structure has been reported in several instances, but other observations indicate only a slow increase in the degree of chain disorder as the temperature is raised. The exact binding sites and nature of bonding between the sulfur and the surface metal atoms is still not clear. The Au–S bond strength is about 40 kcal mol^{-1}.[18] On the most common metal substrate, Au(111), alkanethiols form a $(\sqrt{3} \times \sqrt{3})R30^\circ$ hexagonal lattice with an average spacing of 5.0 Å (1 Å = 10^{-10} m) between alkane chains. The alkane chains are in the *trans*-conformation with very few gauche defects. The average tilt of the alkane chain with respect to the surface normal is typically less than or equal to 30°. The tilt angle is controlled by the headgroup spacing combined with a minimization of free volume in the alkane chain domain. A detectable $C(4 \times 2)$ unit cell of four thiols within the lattice arises from variations in the twist of the alkane chains. In highly ordered SAMs, the orientation of the terminal group can depend on the number of methylenes separating the terminal group from the thiol (the "odd–even" effect). Domains of SAMs with nanometer dimensions are separated by grain boundaries corresponding to changes in chain tilt and/or registry with the substrate surface atoms. Even when the surface of the substrate is not ordered or when a large group in the chain perturbs the packing, the exposed alkane chains often adopt a hexagonal close-packed structure. There is considerable evidence that thiols lose the hydrogen to form a surface thiolate. Disulfides cleave to form the identical thiolate. Surface metal atoms should carry a corresponding positive charge, but spectroscopic evidence for the oxidized metal layer has not been found. An X-ray diffraction study[103] suggests that the thiolates exist as a dimer on a Au(111) surface, but there has been no other supporting evidence for this hypothesis. The degree of molecular order is very similar in SAMs prepared from thiols and from disulfides, but noticeably lower in SAMs prepared from sulfides. A controversial proposal by Zhong and Porter suggests that the C–S bond can also cleave during the self-assembly process of sulfides.[104] STM images reveal the presence of "pits" after SAM deposition from either solution or vapor. These pits have been identified as missing single atom layers in flat terraces. The pits contain attached thiolates and are not the origin of pinholes discussed in section 5.3. Postdeposition thermal annealing causes migration of the surface gold atoms, and promotes Ostwald ripening of the "pits". The SAMs are air stable unless they are exposed to ozone, in which case the thiolates are rapidly oxidized to sulfonates.[66]

Most of the characterization methods indicate that mixed SAMs composed of molecules with similar chain lengths are homogeneously mixed rather than phase separated. Only in high-resolution STM images is there evidence of imperfect mixing.[105,106] When the mixed SAM contains molecules of greatly differing chain length, phase separation into microscopic domains is detectable. The longer-chain component exhibits a greater degree of disorder since it is not stabilized by packing over the entire surface.

Slight increases in disorder of the outer part of the SAM can be detected when the SAM is brought into contact with a liquid, depending on the liquid and the terminal group of the SAM. SAMs with hydrophilic surface groups, particularly an ethylene oxide oligomer, are very resistant to adsorption of proteins from solution.[22,57] Thiols with terminal acid or base groups exhibit marked changes in the ionization behavior when they are incorporated into SAMs. The ionization reaction which forms a charged species is less favored thermodynamically. Thus, alkyl carboxy groups (−COOH), which in homogeneous solution have a pK_a of about five, do not start to ionize significantly in SAMs until the pH of the contacting solution exceeds six. Complete ionization of the carboxy groups requires a much higher pH.

4 POSTDEPOSITION MODIFICATION OF SELF-ASSEMBLED MONOLAYERS

Because a diverse variety of moieties can be attached to the thiol molecule without inhibiting SAM formation, there is an abundance of chemical transformations available to tailor the surface composition and properties. For both bond cleavage and bond formation on a SAM, both steric factors and electrostatic attraction/repulsion strongly affect the reactivity of the terminal groups. A topic not covered here is the physical transformation of SAMs via lithographic methods into patterned arrays.

Thiol exchange is accomplished by immersing the SAM in a new thiol deposition solution. The kinetics of exchange indicate the presence of two populations of molecules in the SAM, one of which is rapidly exchanged. In general, the second population is not fully replaced by the new thiol even after prolonged exposure (days) to the second deposition solution.

Ester groups near the exposed surface of a SAM can be hydrolyzed by small nucleophiles,[107,108] but not by an enzyme.[109]

Internal polymerizations of SAMs include conversion of diacetylenes to a conjugated ene–yne structure,[110,111] cross-linking of a terminal vinyl group,[112] hydrolysis and cross-linking of a trimethoxysilane,[113] dehydration of a boronic acid to a borate glass,[114] and electrochemical polymerization of a terminal pyrrole.[115,116] The polymerized SAMs exhibit increased stability towards thermal desorption and greater resistance to exchange with solution thiols.

Numerous examples exist of reactions designed to couple a new moiety covalently to the surface of a SAM. Among the more common covalent transformations are the formation of amides and esters via acid chlorides or anhydrides or with the aid of carbodiimide coupling agents. Proteins, enzymes and other biomolecules are attached via reaction of exposed amines to SAMs with terminal COOH. Strategies for more site-specific binding of biomolecules take advantage the biotin–streptavidin reaction,[117] the high affinity of a dihistidine tag for Ni^{2+} complexed with nitriloacetic acid[118,119] and the reconstitution of flavin or nicotinamide adenine dinucleotides (NAD) with their enzymes.[120,121] Pendant DNA oligomers on a SAM can be hybridized with their complementary oligomers with remarkable specificity.[122,123]

SAMs with ionized terminal groups electrostatically attract ions of the opposite charge. For polyions (polymers with charge groups, charged planar inorganic oxide fragments and proteins), the binding is sufficiently strong that the attached layer resists removal by a water rinse. Because a bound polyion reverses the surface charge on an ionized SAM, it is possible to build up multiple layers of polyions with sequential immersion of the substrate in aqueous solutions of each component.[124,125] Multilayers can be also constructed on SAMs via sequential deposition of a metal cation and then a difunctional molecule.[126,127]

A second highly oriented layer of surfactant molecules can be deposited on the top of a SAM via LB transfer. The orientation of the outer layer depends on whether the SAM surface is hydrophobic[128] or hydrophilic.[129]

The importance of cell membrane phospholipid bilayers in controlling ion transport and signalling has led to a considerable effort to create analogous structures on or near solid substrates. Phospholipids spontaneously form a second layer on top of a methyl-terminated SAM from either a suspension of phospholipid vesicles,[130] a suspension of phospholipid and detergent micelles[131] or from the evaporation of a phospholipid solution.[132] The resulting structure is called a hybrid bilayer. To avoid the obvious disadvantage of a bilayer with one side pinned to the substrate, suspended bilayers have been formed by creating a SAM with a remote pendant phospholipid[133–135] or cholesterol[136] and assembling the bilayer around the remote seed molecules.

5 ELECTROCHEMICAL CHARACTERIZATION OF SELF-ASSEMBLED MONOLAYERS ON ELECTRODES

5.1 Double-layer Structure and Capacitance

Electrode/electrolyte interfaces exhibit a capacitance whose magnitude reflects the distribution of ions on the

For references see page 10105

solution side of the interface. The electrolyte double layer is composed of the Helmholtz layer, a layer of ions and solvent in physical contact with the electrode, and the diffuse layer, a layer of ions near the electrode whose concentration deviates from bulk concentrations. In relatively concentrated electrolytes, the capacitance of the Helmholtz layer dominates the interfacial capacitance. For most metals, typical Helmholtz capacitances range from 10–100 $\mu F\,cm^{-2}$, and are dependent on potential. When a long-chain alkanethiol self-assembles on a metal electrode, the Helmholtz layer changes from a mixture of ions and solvent with a high dielectric constant to an ion-free hydrocarbon layer with a low dielectric constant. Consequently, the interfacial capacitance is dramatically reduced and becomes virtually independent of potential.

Interfacial capacitances are often measured via the charging current in a CV. More detailed studies of capacitance behavior are obtained by AC (alternating current) impedance spectroscopy or AC voltammetry. A simple parallel-plate capacitor model predicts that the reciprocal capacitance increases linearly with the thickness of the dielectric layer. Plots of C^{-1} versus n (number of CH_2 groups in the alkane chain) are indeed linear for the longer chain lengths. Shorter chain-length SAMs appear to be permeable to some electrolyte ions and exhibit greater capacitances than expected (Figure 3).[6,85,137,138] Extraction of the SAM dielectric constant requires an assumption about the tilt of the alkane chains. Observed values fall in the range of 2.3–2.6. The dielectric constants are consistent with a close-packed layer of alkane chains with essentially no penetration by electrolyte solvent or ions. Mathematical models have been developed to predict the capacitances of a SAM-coated electrode when the surface charge of the SAM changes because of either an acid–base reaction or a redox reaction.[139–141] Deviations from the capacitances of simple alkanethiol SAMs are useful in monitoring phenomena such as poorly packed SAMs, increased solvent and/or ion penetration and variations in the tilt of the alkane chains. Capacitance measurements are also helpful in monitoring the self-assembly process in situ and in confirming the formation of a hybrid bilayer on a SAM.

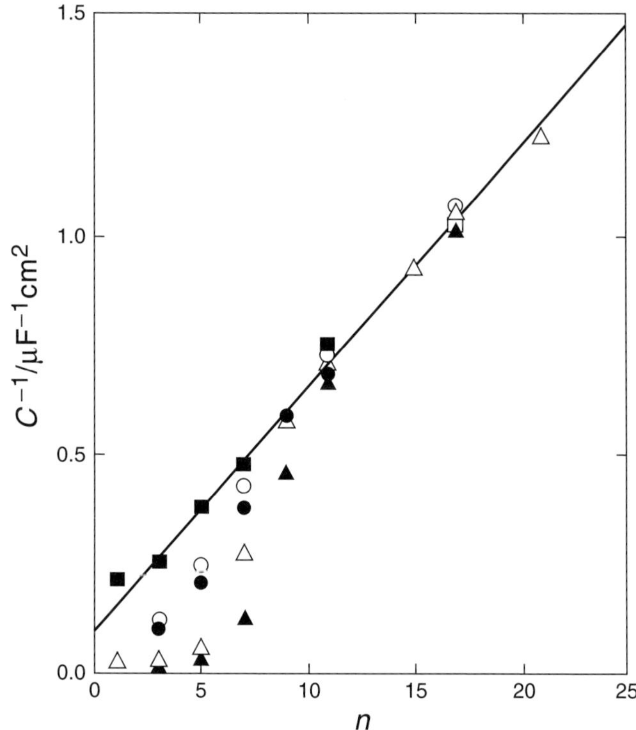

Figure 3 Reciprocal capacitance of alkanethiol SAMs versus chain length n. The symbols represent capacitances obtained from CVs. Filled symbols indicate 10 $mV\,s^{-1}$ and empty symbols indicate 100 $mV\,s^{-1}$ scan rate. △, ▲ 1 M KCl; ○, ● 1 M $HClO_4$; □, ■ 1 M NaF. (Reprinted with permission from Porter et al.,[6] Copyright 1987, American Chemical Society.)

5.2 Electrochemical Stripping and Deposition of Self-assembled Monolayers

SAMs from thiols, disulfides and sulfides resist desorption over a wide potential range, but at very negative potentials and in strongly alkaline electrolytes, they are desorbed quantitatively.[26,52,72,96,97,137,142] On single crystal surfaces or on mercury, a typical CV exhibits a rather sharp cathodic peak (or sometimes two peaks), corresponding to stripping of the thiols, and a somewhat broader anodic peak corresponding to readsorption of the thiol (Figure 4). The half reaction is thus written as Equation (1):

$$\mathbf{M}|S(CH_2)_nX + \mathbf{n}e^- \rightleftharpoons \mathbf{M}| + {}^-S(CH_2)_nX \quad (1)$$

where **M** is the electrode metal. The desorption proceeds either from a few nucleation centers or nearly homogeneously across the electrode, depending on the chain length of the thiol and the applied potential.[143]

A key observation is that the area under the desorption peak, corrected for charging current and true surface area, is independent of the chain length of the alkanethiol. The area, converted to moles of electrons per cm^2, is 7.8×10^{-10} for Au(111), 7.7×10^{-10} for Ag(111), and 9.8×10^{-10} for Hg. The theoretical coverages of thiols (in moles cm^{-2}) for a $(\sqrt{3} \times \sqrt{3})R30°$ lattice are 7.7×10^{-10} for Au(111) and 7.6×10^{-10} for Ag(111). The data support the conclusion that $\mathbf{n} = 1$ in Equation (1). Schneider and Buttry argue that the correction for charging current is underestimated, so that $\mathbf{n}$ is smaller than 1.[144] Even with this caveat, the area of the cathodic stripping peak is a useful measure of thiol coverage as a function of SAM composition[26,104] and electrochemical history.[145] For example, a racemic deposition mixture

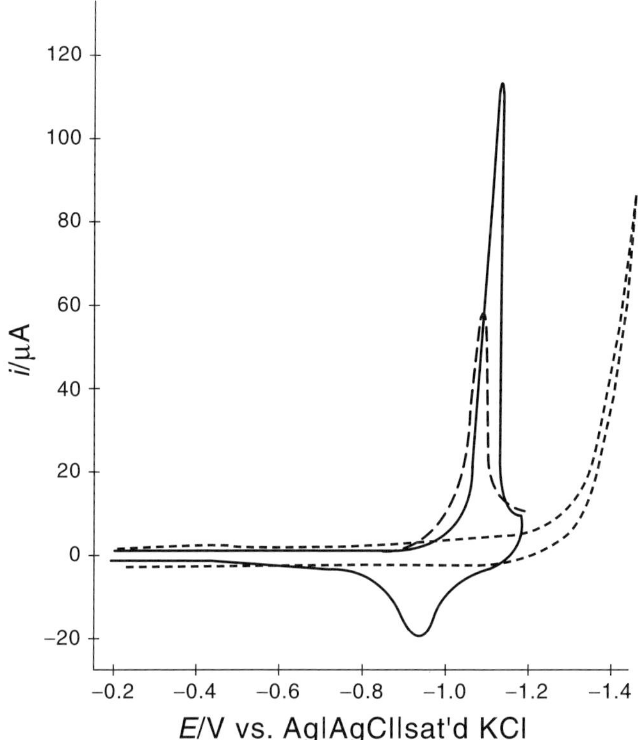

Figure 4 Electrochemical stripping and redeposition of a dodecanethiol SAM on Au(111) in 0.5 M KOH. The solid and dashed lines are the first and second scans, respectively. The dotted line is a CV of bare Au(111). (Reprinted with permission from Walczak et al.,[142] Copyright 1991, American Chemical Society.)

of an optically active thiol yields a higher coverage on mercury than do deposition solutions containing either pure enantiomer.[146]

The potentials of the cathodic stripping peaks supply information about the strength of the metal–sulfur bond and the interchain interaction, the accessibility of the metal–sulfur bond to cations from the solution, and the presence of any intermediate or weakly adsorbed states.[96] Gold substrates with a high density of steps connecting Au(111) terraces exhibit two stripping waves, suggesting that thiols are more strongly bonded to step sites.[52] The presence of two stripping peaks for a SAM formed from an asymmetrical sulfide is part of the evidence for cleavage of the C–S bond in the sulfide.[104] In some mixed SAMs, distinct stripping peaks can be assigned to each component and that component thiol selectively desorbed; both observations support the existence of phase-separated domains.[147,148]

The desorbed thiols tend to remain near the electrode and are readily readsorbed when the electrode potential is shifted to a more positive value. It is possible to create partial monolayers by controlling the electrode potential prior to and during emersion.[95]

5.3 Blocking Behavior

Alkanethiol SAMs suppress faradaic processes such as electrode oxidation and the exchange of electrons between the electrode and solution redox couples. This blocking property is attributed to the densely packed structure of the hydrocarbon chains which impede the approach of solution ions and molecules to the electrode surface. Possible applications of blocking SAMs can be found in the areas of corrosion prevention, nanoscale lithography and selective electrodes. However, closer examination of SAM-coated electrodes reveals the presence of pinholes (bare metal sites) and other defects, which permit a close approach of solution species. It is important to understand the nature, size and distribution of the pinholes and other defects before the applications mentioned above are achievable.

Of all the methods for probing SAM structures, voltammetry is the most sensitive tool for detection of pinholes and defects in a SAM. The extraordinary sensitivity arises from the ability to detect currents corresponding to oxidation or reduction of a fraction of a monolayer, and from the high rates of mass transfer of redox couples to small "hot spots" on an otherwise blocked electrode. For the subsequent discussion, we define the area fraction of pinholes as $1 - \Theta$, where Θ is the fractional coverage relative to a complete monolayer. Gold oxidation is suppressed by the thiol SAM, except at pinholes. Consequently, a measurement of $1 - \Theta$ can be obtained from the charge needed to reduce the oxide.[7] Cyclic voltammetry of solution redox couples at electrodes with imperfect SAMs exhibits current peaks or small current plateaus at low overpotentials (near the formal potential $E^{0'}$) (Figure 5). These current peaks or plateaus greatly exceed the extremely small tunneling currents found on electrodes with perfect SAMs (section 6.5). The degree of attenuation of the current by the SAM at low overpotentials relative to the currents at a bare electrode is often used as a qualitative assessment of pinholes and defects in the SAM.

Quantitative measurements of $1 - \Theta$ require a model. One model treats the pinholes as an array of microelectrodes in an insulating plane. Given the somewhat unrealistic assumptions of uniform size and uniform spacing for the microelectrodes, CVs can be fitted with calculated voltammograms (the symbols in Figure 5) to obtain the pinhole parameters (the pinhole radius R_a, the radius of the blocked area around the pinhole R_0, and $1 - \Theta$).[149] In alternating current impedance spectroscopy (ACIS), the electrode is biased at the Nernst potential (at or close to $E^{0'}$) in an electrolyte containing both the oxidized and the reduced forms of a redox couple and the impedance of the system is measured as a function of frequency. The same model is used to interpret

For references see page 10105

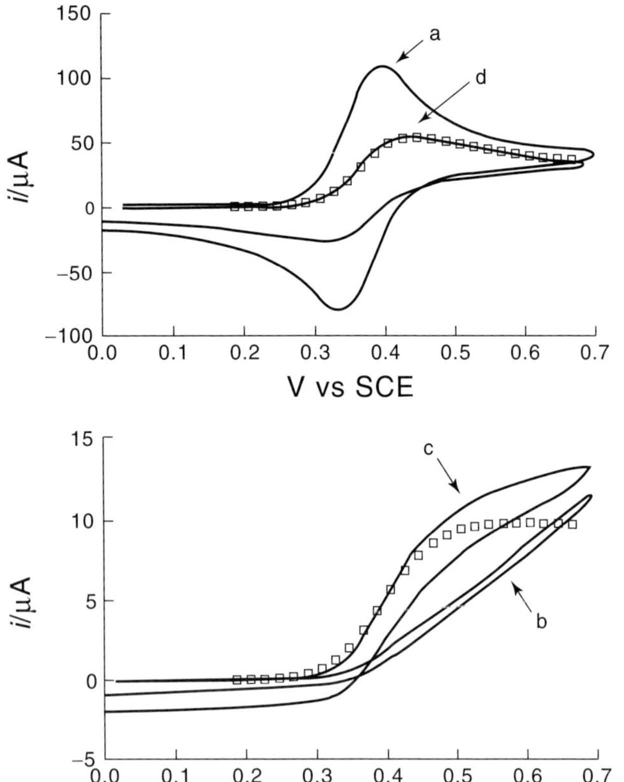

Figure 5 Electrochemical blocking of a solution redox couple by an octadecanethiol SAM. Ferrocenylmethyl-trimethylammonium in 0.5 M H_2SO_4, polycrystalline gold, $0.1\,V\,s^{-1}$ scan rate. Curve (a) is the reversible CV obtained on bare gold, curves (b), (c) and (d) are obtained on the same electrode with different monolayers. The symbols are theoretical fits to a microarray electrode model. (Reprinted with permission from Finklea et al.,[149] Copyright 1993, American Chemical Society.)

the data.[149] Typical values obtained are 10^{-2} to 10^{-4} for $1 - \Theta$ and micrometer to submicrometer dimensions for R_a. Deviations between the impedance behavior predicted by the model and the observed impedance behavior suggest that the pinholes are not uniformly distributed in size and in spacing.

Alternatives to the electrochemical approach for pinhole detection and mapping include deposition of metals[150] or etching the substrate metal in the presence of the SAM and then scanning the surface with STM, AFM or SEM.[50,151] These experiments suggest that the pinholes are of atomic dimensions, subnanometer rather than micrometers. Scanning electrochemical microscopy (SECM), a method which maps electroactive "hot spots" on an electrode, indicates that pinholes are less than 0.5 μm in diameter.[152] It is likely that the apparent submicrometers to micrometers dimensions of pinholes obtained by voltammetric methods arise from areas with a high density of subnanometer pinholes.

Unfortunately, no procedure has been developed which reliably yields pinhole-free SAMs. A survey of the literature indicates that the least defective SAMs are obtained most consistently on freshly deposited thin films of metal or on mercury. For bulk solid metal electrodes, surfaces should be annealed, cleaned of organic contaminants, and etched to expose fresh metal and to remove inorganic oxides, a potential source of pinholes.[153] In general, flexible chains yield better blocking properties than rigid ones. Postdeposition procedures that reduce the defectiveness of a SAM include electrochemical polymerization of phenol[154] and polymerization of a diacetylene SAM.[110] However, polymerization of a terminal vinyl group has the opposite effect.[112]

Nonaqueous electrolytes tend to increase the permeability of SAMs, presumably because of increased solvent interactions with the pendant chains. Some redox couples (e.g. ferrocenemethanol, benzoquinone) are not blocked by even pinhole-free SAMs in aqueous electrolytes by virtue of their hydrophobicity. Electrostatic attraction or repulsion between surface moieties on the SAM and solution redox couples have a powerful effect on the blocking behavior of SAMs. These last two observations form the basis of the creation of selective electrodes (section 6.2).

5.4 Attached Redox Centers

SAMs afford a highly flexible and convenient method for attaching redox centers to electrodes. The coverage of the redox center can be varied without lowering the packing density of the SAM, thereby controlling the lateral spacing between redox centers (Figure 6). The spacing between the electrode and the redox center is determined by the chain length; close packing prevents motion of the redox center towards the electrode. It is possible to control the local environment near the

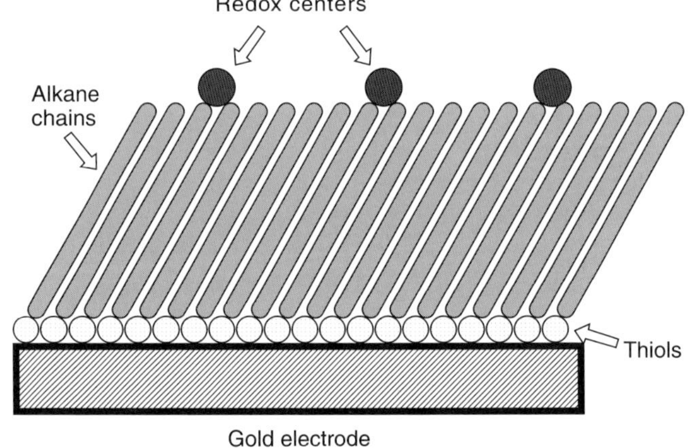

Figure 6 An electroactive SAM containing both redox and diluent thiols. (Reprinted from Finklea,[1] p. 240, by courtesy of Marcel Dekker, Inc.)

redox center by either partially burying the redox center in the hydrocarbon domain, or by varying the terminal group of diluent thiols. In particular, proteins are less likely to suffer denaturing structural changes when they are attached to a hydrophilic SAM surface compared to their interaction with a bare metal. The redox centers serve as a sensitive and nondestructive probe of the structure of SAMs (especially in mixed SAMs), their dynamics (thiol exchange) and their reactivity (surface attachment and hydrolysis reactions). For electron-transfer kinetic studies, electroactive SAMs are much less sensitive to pinholes and other defects than the blocking SAMs. The redox centers provide a means of catalyzing electron transfer to kinetically slow species (see section 6.4). A wide range of redox centers have been incorporated into SAMs, including ferrocenes, ruthenium and osmium complexes, viologens, porphyrins, heme proteins, (hydro)quinones, azobenzenes and fullerenes. Several strategies exist for the formation of multilayers of redox centers.

There are two main approaches to creating an electroactive SAM. The redox center may be prepared with a pendant thiol prior to deposition, or it may be coupled to the SAM via electrostatic binding, or amide or ester bond formation after deposition. In either approach, there are a number of issues that need to be addressed: the coverage of the redox center, the formal potential, the shape of the CV compared to that of an ideal voltammogram and causes for deviations from the ideal shape.

A CV of an electroactive SAM usually exhibits matching anodic and cathodic current peaks (Figure 7). The coverage of the redox center Γ on the electrode surface is obtained by current integration after extrapolating and subtracting the charging current baseline underneath either the anodic or the cathodic current peak, Equation (2).

$$\Gamma = \frac{Q}{\mathbf{n}FA} \tag{2}$$

Q is the peak area in coulombs, $\mathbf{n}$ is the number of electrons transferred in the redox center half reaction, F is Faraday's constant, and A is the electrode area. The peak area and coverage are independent of the scan rate. Coverages down to $10^{-12}\,\text{mol cm}^{-2}$ can be measured.

The CV is reversible if its shape and the peak positions do not change with scan rate. The peak current i_p is proportional to the scan rate v, Equation (3)

$$i_p = \frac{\mathbf{n}^2F^2A\Gamma v}{4RT} \tag{3}$$

For an ideally shaped CV, the current obeys Equation (4):

$$i = \frac{4i_p e^{\theta}}{(1+e^{\theta})^2} \tag{4}$$

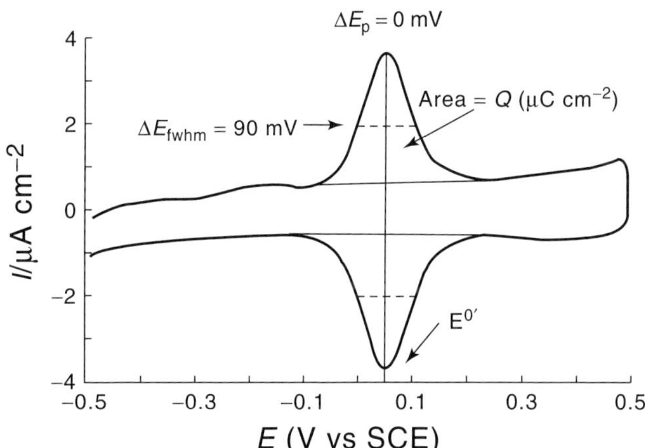

Figure 7 Reversible CV of an electroactive SAM. The redox thiol is $HS(CH_2)_{11}CONHCH_2pyRu(NH_3)_5{}^{2+/3+}$ and the diluent thiol is $HS(CH_2)_{11}COOH$. The electrolyte is 1 M Na_2SO_4 adjusted to pH 4. A linear charging current baseline is extrapolated under the faradaic waves to obtain the peak areas. (Reprinted from Finklea,[1] p. 241, by courtesy of Marcel Dekker, Inc.)

where θ is given by Equation (5)

$$\theta = \frac{\mathbf{n}F(E - E^{0'})}{RT} \tag{5}$$

The anodic and cathodic peak potentials are identical ($\Delta E_p = 0$) and are equal to $E^{0'}$. The peak half-width ΔE_{fwhm} is $3.53RT/\mathbf{n}F$, or $90.6/\mathbf{n}$ mV at 25 °C.

Departures from the ideal CV diagnostics are common and reveal molecular detail about the redox center and/or the SAM. A nonzero peak splitting indicates possible intermolecular interactions between the redox centers or a change in SAM structure with respect to the oxidation state of the redox centers (e.g. precipitation of the redox center with a counter ion). ΔE_{fwhm} can be either greater or smaller than the theoretical value. Large values of ΔE_{fwhm} may be caused by a spread of formal potentials (possibly indicating a disorganized SAM structure with a range of local environments about the redox centers), intermolecular interactions between redox centers or double-layer effects. In the latter case, the change in charge density of the SAM surface causes a variation in the local electrostatic potential.[139,155] The distortion of the CV is reduced when the coverage of redox centers is low and/or ion-pairing reduces the change in charge density with oxidation state. Sharp CVs with small ΔE_{fwhm} values are obtained if the redox centers tend to form a solid-like phase; this occurs only in the most densely packed electroactive SAMs.[156]

The formal potential is also informative. Generally, attached redox couples on the external surface of the SAM exhibit $E^{0'}$ values within 100 mV of the $E^{0'}$ of a solution analog, suggesting that the local solvation and dielectric constant are similar to that found at a bare electrode. For high coverages of redox centers with at least one charged oxidation state, $E^{0'}$ shifts at a Nernstian rate (60/**n** mV) with each decade change in concentration of the oppositely charged counter ion in the electrolyte.[157] This shift is due to the migration of the counter ion in or out of the SAM to provide charge compensation during the oxidation or reduction. The ion migration is also detected as mass changes for SAMs on quartz crystal microbalances (QCMs).[158,159] Formal potentials shift markedly as the redox center becomes progressively more buried in the hydrocarbon domain of the SAM.[160] The direction of the shift is consistent with the destabilization of the more highly charged oxidation state. Ion pairing is invoked to explain shifts in $E^{0'}$ with respect to the identity of the electrolyte counter ions.[160]

A brief overview of the behavior of some of the more common redox centers will be given; see Table 1 for characteristic parameters.

Ferrocene derivatives constitute the largest class of attached redox centers. Maximum coverages in SAMs are consistent with a close-packed layer of ferrocenes. The formal potential can be shifted over a 0.5 V range, depending on the presence of electron-donating or electron-withdrawing substitutents. Thiol exchange reactions in SAMs have been monitored by ferrocene thiols with distinguishable formal potentials. The reduced ferrocene is chemically stable, but the oxidized ferricenium is subject to nucleophilic attack by hydroxide with concomitant loss of coverage, so acidic electrolytes are preferred. The peak shape of 100% ferrocene thiol SAMs is usually distorted by double-layer effects or the formation of a solid-like phase. However, carefully prepared mixed SAMs containing a low density of ferrocene thiols exhibit ideal CV shapes, suggesting that the redox centers reside on the external surface of the SAM.[161]

The $pyRu(NH_3)_5{}^{2+/3+}$ (py = pyridine) redox center has been thoroughly studied.[92] The reduced coverage relative to that of ferrocene SAMs reflects the electrostatic repulsion between cationic redox centers and the need to incorporate counter ions in the SAM. Like ferrocene, the Ru redox centers are unstable in the oxidized state in alkaline electrolytes. Unlike the ferrocenes, nearly ideal CV peak shapes are obtained at all coverages in aqueous electrolytes. In nonaqueous electrolytes, the CV peak shapes become broader and more distorted; the change is reversible upon reimmersion in aqueous electrolytes.[162]

CVs of attached metalloporphyrins are generally poorly defined unless the porphyrin ring is incorporated into a protein or protein fragment. Cytochrome *c* has been studied extensively. This protein can be attached via amide formation or by electrostatic binding.[163] It is believed that the electrostatic binding orients the heme-containing cleft towards the SAM surface. The observation that ΔE_{fwhm} is consistently larger than 100 mV has been interpreted in terms of a Gaussian spread of formal potentials, suggesting a heterogeneity in the orientation of the proteins.[164,165] Microperoxidase-11 is an oligopeptide containing a heme group and several carboxy groups which are usually attached to a cystamine SAM.[166] This protein fragment is a useful electrocatalyst.[167]

Viologens (4,4′-bipyridiniums) exhibit two distinct one-electron waves at negative potentials. In SAMs, the viologens yield high coverages and nonideal CV shapes due in part to a tendency of the radical cation to dimerize and the hydrophobicity of the two reduction products. The CVs are sensitive to the identity and concentration of the counter ions.[159] Quinones and azobenzenes generally exhibit high and stable coverages, very slow electron-transfer kinetics and very broad peaks, due in part to the dense packing achieved by these neutral organic redox centers. The formal potentials show the expected Nernstian shift with pH for a $2H^+$, $2e^-$ half reaction.[168,169] In very densely packed SAMs, azobenzenes are electrochemically silent, presumably because of strong inhibition of ion ingress and the significant structural change during reduction.[170]

Table 1 Properties of common redox centers attached to SAMs

Redox center	Max. Γ[a]	$E^{0'}$ vs SCE	Comments on peak shapes	Refs.
Ferrocenes	4–5	+0.2 to +0.7 V	Ranges from narrow to ideal to broad	79, 158, 161
$pyRu(NH_3)_5$	1–2	0 V	Nearly ideal at all coverages	92, 162
Cytochrome *c*	0.2	0 V	Slightly broader than ideal	163–165
Microperoxidase-11	2	−0.4 V	Broad CVs	166, 167
Viologens	4	−0.3 to −0.5 V −0.8 to −1.0 V	Two waves, broad, distorted by dimerization reaction	159
Quinones	3–5	pH-sensitive	Very broad	168
Azobenzenes	3–5	pH-sensitive	Very broad, sensitive to packing	53, 169, 170

[a] Coverage × 10^{-10} mol cm^{-2}.

6 ELECTROCHEMICAL APPLICATIONS OF SELF-ASSEMBLED MONOLAYERS ON ELECTRODES

The following discussion focuses on applications of SAMs for the enhancement of signal-to-noise in voltammetry, the development of selective electrodes, electrocatalysis, the study of electron-transfer kinetics over long distances and at large driving forces and the control of corrosion and adhesion at electrodes.

6.1 Microarray Electrodes

Microarray electrodes offer advantages for both analytical and kinetic studies because the electrode capacitance is greatly reduced and mass transfer is greatly enhanced relative to surface faradaic processes. The resulting improved signal-to-noise on a SAM-coated electrode has been demonstrated for the voltammetry of $Ru(bpy)_3{}^{2+/3+}$ (bpy = 2, 2′-bipyridine) in the presence of gold oxidation and oxide reduction.[(171)] However, for practical applications, it is necessary to control the density and size of the pinholes in the SAM. Mixed SAMs composed of a long-chain alkane thiol and a short thiol (3-mercaptoproprionic acid,[(148)] 4-aminothiophenol[(172)]) form phase-separated domains of the short thiol with dimensions on the order of $10^4\,nm^2$ under the right conditions. Reproducible microarray behavior has been obtained for mixed SAMs of hexadecanethiol and 4-hydroxythiophenol.[(153,173)] It is probable that the "microelectrodes" are phase-separated domains of 4-hydroxythiophenol. At low microelectrode density, the mixed SAM shows a preferential response to neutral redox species over charged ones, suggesting the presence of adsorption or permselectivity through the hydrophobic SAM. A similar microarray electrode has been fabricated by LB deposition of a mixture of octadecanethiol plus octadecanol containing a small percentage of ubiquinone.[(102,174)] In this system, the density of "molecular gates" is readily controlled. At ubiquinone coverages less than $10^{-15}\,mol\,cm^{-2}$, the current per ubiquinone becomes independent of the ubiquinone coverage. This is evidence for electron transfer at isolated "molecular gates". Like the preceding system, the microarray electrode exhibits greater response for hydrophobic redox couples than for hydrophilic ones.

6.2 Selective Permeation

For SAMs that are less than perfectly blocking, selectivity arises from the inherent affinity of hydrophobic molecules for the alkane domain. This property has been applied in the development of a liquid chromatography (LC) electrochemical detector which is selective for neutral organic analytes over ionic compounds.[(175)] Even more dramatic are electrostatic effects. The current is greatly enhanced at an ionic SAM for redox species of the opposite charge and strongly suppressed for a redox species of the same charge.[(176–178)] For example, cationic dopamine can be detected in the presence of 100-fold higher concentration of anionic ascorbic acid at a gold electrode coated with w-mercaptoalkanoic acid.[(179)] Interestingly, a highly charged and hydrophilic SAM containing w-mercaptodecanesulfonic acid strongly blocks organic redox couples, regardless of their charge.[(180)]

By extension, a bilayer of molecules on or near the electrode surface should provide even greater selectivity for analytes if ion transport agents are incorporated into the bilayer.[(133,181)] A phospholipid bilayer can be formed near but not on an electrode coated with a SAM whose constituent molecules possess a long hydrophilic body and a hydrophobic phytanyl tail.[(135)] The hydrophobic tails serve as nucleation sites for the spontaneous assembly of a bilayer. When valinomycin is incorporated into the bilayer, the conductivity of the assembly exhibits a marked increase in the presence of K^+, and a smaller response in the presence of other cations.

A hybrid bilayer containing an alkanethiol and a phospholipid is strongly blocking towards ferricyanide until melittin is added to the solution.[(130)] Melittin, a cationic oligopeptide isolated from bee venom, is believed to aggregate in bilayers and to create pores that are much larger than most ions. Thus, selectivity should be minimal, except possibly for electrostatic effects due to the positive charge of melittin.

6.3 Preconcentration and Selective Binding

Preconcentration implies that the analyte partitions preferentially into the SAM or to its external surface, so that the analytical signal is enhanced. Preconcentration without designed binding sites depends on hydrophobic/hydrophilic or electrostatic forces. Thus, SAMs with a terminal COOH group preconcentrate Cd^{2+}, while SAMs with a terminal pyridinium bind chromate, allowing detection of these ions down to 10^{-10} M with cathodic stripping voltammetry.[(182,183)] Cationic surfactants are quantified by the current obtained for the reduction of ferricyanide at an octadecanethiol-coated electrode; adsorption of the cationic surfactant on the SAM enhances the concentration of ferricyanide and hence the current.[(184)]

Designed binding sites in the SAM can greatly enhance the selectivity of the electrode (Table 2). Cyclodextrins, either with a pendant thiol or attached to an existing SAM, bind electroactive analytes like ferrocene[(185)] or quinone.[(186)] Nonelectroactive *trans*-azo dyes, which also bind to cyclodextrins, are detected either by competitive inhibition[(187)] or by attachment of an electroactive viologen to the dye.[(188)] The formal potential of a

For references see page 10105

Table 2 SAMs with designed binding sites and the targeted analytes

Binding site	Analyte(s)	Refs.
Cyclodextrin	Ferrocenes	185, 187
Cyclodextrin	Quinone, azo dyes	186
Cyclodextrin	Azo dyes	188
Cyclic bis(bipyridinium)	Indole, catechol	189
bis(Acetylacetone)	4-Coordinate metal ions	190, 191, 192
Steroids	Cyclic polyols	193

cyclic bis(bipyridinium) molecule in a SAM shifts in the negative direction in the presence of indole or catechol.[189] A bis(acetylacetone)sulfide forms SAMs that bind Cu^{2+}, Pb^{2+} and Zn^{2+} selectively in the presence of Fe^{2+} or Fe^{3+}.[190–192] The first three ions readily form four-coordinate complexes, while iron cations prefers a six-coordinate geometry. The bound Cu^{2+} or Pb^{2+} act as nucleation sites for the electrodeposition of bulk copper or lead. Consequently, direct measurements are possible for Cu down to 10^{-7} M and Pb down to 10^{-5} M, and an indirect measurement of Zn can be made down to 10^{-9} M. A mixed SAM containing octadecanethiol and a flat oriented steroid partially blocks ferricyanide reduction.[193] The reduction current is greatly attenuated when cyclic polyols with equatorial OH groups are present in the solution. The polyols are thought to form ice-like structures in the cavities defined by the steroid molecules.

Incorporation of enzymes into SAM structures affords highly selective electrodes. Direct electron transfer between the enzyme and the electrode is relatively rare; most biosensors use a freely diffusing or attached redox couple to catalyze electron transfer to the enzyme (see section 6.4). Examples of the former approach include electrodes selective for the reduction of fumarate,[194] the oxidation of hydrogen peroxide,[167] and the oxidation of lactate.[195]

A particularly elegant approach to a biosensor is based on a suspended bilayer near a gold surface (Figure 8).[134] The bilayer contains gramicidin fragments in the inner layer which are tethered to the gold surface, and hence are fixed. Gramicidin fragments in the outer layer contain a biotin tag which is connected via streptavidin to a biotin-tagged antibody fragment. In the absence of the analyte, the outer-layer gramicidin molecules diffuse laterally and intermittently form ionically conducting dimers with the inner layer gramicidin; the conductivity of the bilayer, measured by AC impedance, is relatively high. In the presence of the analyte, the outer-layer gramicidins become locked to fixed antibody fragments and thus can no longer form conducting dimers. The signal is the rate of decrease in the conductance of the bilayer. This sensor is potentially competitive with enzyme-linked immunosorbent assay in terms of flexibility, selectivity and sensitivity, and does not require the washings or reagent additions of the latter method.

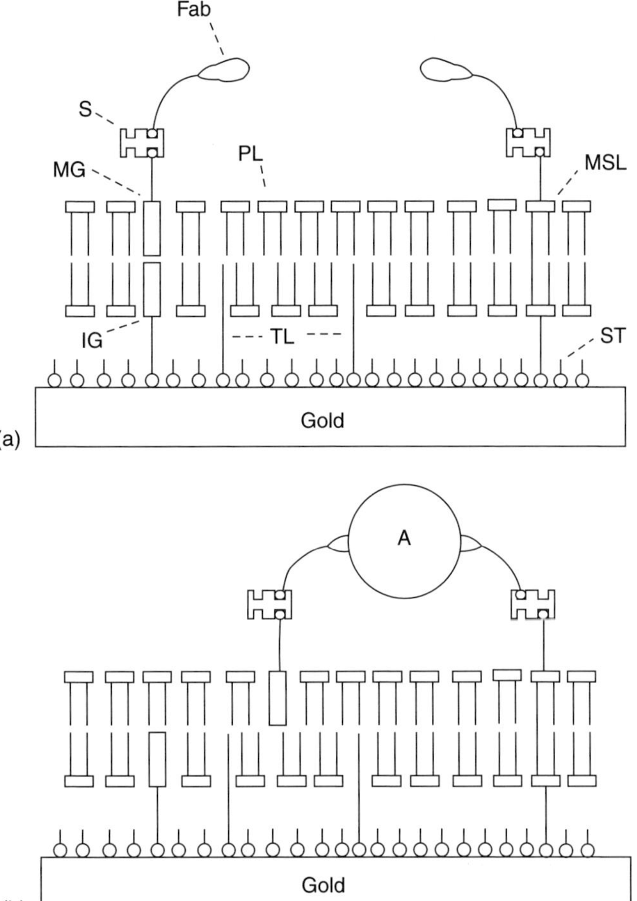

Figure 8 Biosensor based on a suspended lipid bilayer.[134] A = analyte; Fab = biotinylated antibody fragment; IG = immobilized gramacidin; MG = mobile gramacidin with pendant biotin; MSL = membrane-spanning lipid with pendant biotin; PL = phospholipid; S = streptavidin; ST = spacer thiol; TL = tethered lipid. The tethered lipid anchors the phospholipid bilayer to the gold electrode; a short spacer thiol (e.g. mercaptoacetic acid) helps to define the ionic reservoir under the lipid bilayer. The immobilized gramacidins are attached to the electrode via a thiol. Biotinylated antibody fragments are attached via streptavidin to both membrane-spanning lipids with fixed positions and gramacidins which diffuse freely across the external surface. When mobile and immobilized gramacidins coincide, ions can pass across the bilayer and conductivity is high (a). In the presence of the analyte, the Fab's are cross-linked, and the attached gramacidins are no longer able to diffuse freely. Ion channels cannot form, and the conductivity is low (b).

6.4 Electrocatalysis

In electrocatalysis (also known as mediated electron transfer), a redox couple, either attached or freely diffusing, transfers electrons between the electrode and target redox couple. The target redox couple is thermodynamically capable of being oxidized or reduced at the electrode, but the electron-transfer kinetics are extremely slow. Kinetically slow systems include multielectron redox centers and enzymes whose redox centers are inaccessible to the electrode. Because electron transfer between the electrocatalyst and the target molecule is usually thermodynamically downhill, a characteristic feature of voltammograms is catalytic current in one direction only (anodic or cathodic).

SAMs serve two roles in electrocatalysis. First, they provide a flexible means of anchoring the electrocatalyst and/or the enzyme. Second, SAMs block access of the target redox couple to the electrode, a useful feature for fundamental studies. For example, SAMs with pendant ferrocenes block reduction of Fe^{3+} at the $E^{0'}$ of the $Fe^{2+/3+}$ redox couple, but catalyze the reduction of Fe^{3+} at potentials where the attached ferriceniums are reduced.(196) It is possible to measure the rate constant for electron exchange between the catalyst and the target.(197) A SAM containing a remarkable quinone that exhibits photoisomerization between an electroactive and a nonelectroactive form is the basis of a light-switchable electrocatalytic electrode.(198)

Numerous examples exist of electrocatalysis of multi-electron redox systems. The two-electron reduction of O_2 to H_2O_2 is catalyzed by attached cobalt porphyrins,(199,200) while the four-electron reduction to water is catalyzed by an attached tetrone (diquinone), albeit at potentials well negative of the thermodynamic potential.(201) NADH oxidation is catalyzed by attached catechols,(202,203) NO reduction is catalyzed by $Cr(terpyridine)_2{}^{3+}$,(204) H_2O_2 reduction is catalyzed by iron porphyrin or heme protein,(166,167) and both ascorbate and hydrazine oxidation are catalyzed by immobilized ferricyanide (possibly as a Prussian blue-like complex).(205)

Likewise, enzymes have been immobilized on SAMs and their catalytic reaction driven by electrocatalysis. If the electrocatalyst is freely diffusing in the electrolyte, then the SAM must be made permeable to the electrocatalyst. A mixed SAM containing octadecanethiol and dibenzyldisulfide is the base for a hybrid bilayer containing pyruvate oxidase; the benzylthiolate components enhance access of the electrocatalyst ferrocenemethanol to the electrode.(206) Alternately, the electrocatalyst and the enzyme are both attached to the SAM. Glucose oxidase can be attached to a mixed SAM of 2-aminoethanethiol and a long chain ferrocenethiol.(207) More complex structures containing multiple layers of enzymes and electron catalysts have been constructed with the purpose of improving the sensitivity of the sensor or making the sensor light-switchable; in each case, the base coat is a SAM of cystamine.(120,121,208–210)

6.5 Long-range Electron Transfer

The close-packed structure of the alkanethiol SAM provides a useful spacer for anchoring redox centers at fixed and controllable distances from the electrode surface (Figure 6). Mixed SAMs allow adequate spacing between redox centers without loss of packing density. The redox centers can also be freely diffusing if the SAM is completely free of pinholes and other defects (section 5.3), but that condition is difficult to prove.(85,211,212) The alkane chain is an insulator, and as a consequence, electron transfer must proceed by a tunneling mechanism. The rate of electron transfer is diminished to the point that the rate constants can easily be measured over a wide range of overpotentials ($\eta = E - E^{0'}$). Both of these facts have been widely exploited by researchers interested in long-range electron transfer and in fundamental theories of electron transfer. The prevailing theory is the Marcus DOS (density of states) model.(213,214) This model focuses on the overlap of donor and acceptor energy levels in the electrode and in the redox center. It includes an important parameter in Marcus theory called the reorganization energy (λ), which is the energy needed to change the redox-center structure and its solvation sphere between the equilibrium states for the two oxidation states.

A number of electrochemical methods provide access to the rate of electron transfer: cyclic voltammetry,(215) ACIS,(216) square-wave voltammetry,(217) and chronoamperometry.(213,218) These electrochemical methods are useful for rate constants in the range 10^{-2} to $10^5\,s^{-1}$. Measurements of rate constants outside this range require spectroscopic(219) or temperature-jump methods.(220)

Three types of experimental results are of interest. A plot of $\ln k^0$ (k^0 is the standard rate constant at $\eta = 0$) versus the number of repeating units in the tether (i.e. the number of CH_2 groups in the alkane chain) is usually linear, indicating that tunneling rates decay exponentially with distance (Figure 9).(213,220–224) The slope of the plot yields the tunneling parameter β. For a single monolayer, a plot of $\ln k$ versus η (Tafel plot) is nearly linear, but with some curvature (Figure 10). Fitting the curvature to the Marcus DOS model yields λ, the reorganization energy. The reorganization energy can also be obtained from a plot of $\ln k^0$ versus T^{-1} (Arrhenius plot).

Detailed experimental studies have been performed on SAMs with attached ferrocenes, $pyRu(NH_3)_5$, and cytochrome *c*.(1) For alkanethiols on gold, plots of $\ln k^0$ versus the number of CH_2 groups are linear. Values for the tunneling parameter β are 1.0–1.1 per CH_2. The same β is obtained for blocking SAMs and various redox couples in solution.(212) Assuming a 30° tilt for an all-*trans* alkyl chain, β is 1.0–1.1 Å^{-1}. Theoretical calculations are in agreement with this value.(225,226) The tunneling mechanism invokes electronic coupling

For references see page 10105

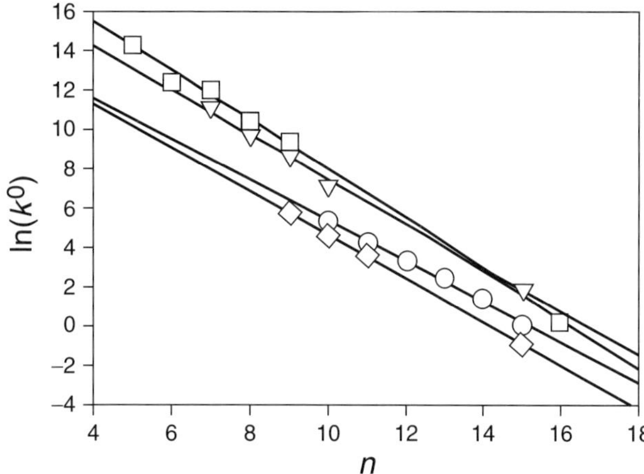

Figure 9 Semilog plots of standard rate constant k^0 versus n in alkanethiol SAM with an attached redox center. ○: $HS(CH_2)_nCONHCH_2pyRu(NH_3)_5^{2+}$ ($\beta = 1.0/CH_2$);[221] ▽: $HS(CH_2)_nNHCOFc$ (Fc = ferrocene) ($\beta = 1.1/CH_2$);[222] □: $HS(CH_2)_nOOCFc$ ($\beta = 1.3/CH_2$);[213,220] ◇: cytochrome c electrostatically adsorbed on $HS(CH_2)_nCOOH$ ($\beta = 1.1/CH_2$).[223,224]

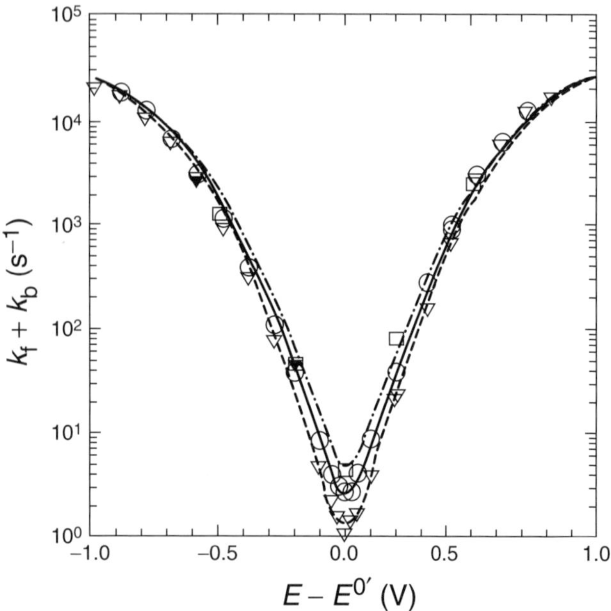

Figure 10 Tafel plots for a mixed SAM containing $HS(CH_2)_{16}OOCFc$ (Fc = ferrocene) and $HS(CH_2)_{15}CH_3$ in 1 M $HClO_4$ at three temperatures: 1 °C (▽), 25 °C (○), and 47 °C (□). The solid lines are predicted from Marcus DOS theory for a standard rate constant of 1.25 s^{-1} at 25 °C and a reorganization energy of 0.85 eV. (Reprinted with permission from Chidsey, Science, **251**, 919–922,[213] Copyright 1991 by the American Association for the Advancement of Science.)

of the metal and redox-center orbitals via the highest filled and lowest unoccupied molecular orbitals of the connecting alkane chain. From this model, a substantial lowering of β is expected if a fully conjugated spacer is used in the SAM. As predicted, a SAM composed of repeating phenyl–alkynyl spacer units yields a β of 0.6 Å^{-1}.[29] Insertion of a heteroatom, a double bond or a triple bond into the repeating CH_2 spacer unit in an alkane chain reduces the electronic coupling.[227] Several experiments suggest that electronic coupling through alkane chains not attached to or in contact with the redox center can contribute significantly to the total electronic coupling.[75,221]

In general, reorganization energies obtained from Tafel plots and Arrhenius plots are in agreement. The λ values are also in surprisingly good agreement with a simple theoretical equation based on the energy of a charged sphere in a dielectric continuum.[162,228] Table 3 lists selected λ values obtained in aqueous electrolytes.

6.6 Corrosion Control

The most popular metal substrate, gold, is hardly in need of corrosion protection in everyday use. However, protection from chemical etchants is a useful property in the preparation of patterned gold arrays. Two etchants for gold that are effectively blocked by long-chain alkanethiol SAMs are an oxygen-saturated alkaline cyanide bath and a solution of ferricyanide combined with thiocyanate or thiosulfate.[234,235] Various gold microelectrode geometries have been fabricated by this means.[236]

Corrosion protection of copper by SAMs has received more attention, principally by Aramaki et al.[237–241] An alkanethiol SAM affords modest protection against corrosion in aqueous electrolytes. Better corrosion protection is obtained by first depositing a SAM of 11-hydroxyundecanethiol, and then reacting the exposed OH groups with octadecyltrichlorosilane. The 5-nm bilayer inhibits both O_2 reduction and copper oxidation in solution and provides excellent protection against indoor atmospheric corrosion for nearly a year.[231] An XPS study of copper corrosion shows that both the copper and the sulfur are oxidized and that the rate of corrosion in air decreases dramatically with a small increase in SAM thickness.[242] Pit corrosion is clearly evident when a SAM-coated Cu(100) surface is exposed to dilute HCl

Table 3 Reorganization energies of selected redox centers in water

Redox center	λ (eV)	Refs.
Ferrocene	0.8–1.0	29, 213, 220, 229
Ferri/ferrocyanide	1.1	228
$M(bpy)_3$ (M = Fe, Ru, Os)	0.6	228
$pyRu(NH_3)_5$	0.8–1.2	221, 230, 231
Cytochrome c	0.6	232, 233

solution, suggesting that chemical attack occurs at defects in the SAM.[(65)]

6.7 Adhesion Control

Several organic compounds (aniline, pyrrole, thiophene) undergo polymerization reactions when oxidized to form conducting polymers. When the monomers are oxidized at an electrode, the polymers precipitate onto the electrode to form a coating. The conducting polymer coatings have potential applications in the areas of electrochromism, charge storage and organic semiconductors. It is desirable to create smooth and adherent coatings on the electrodes.

SAMs containing the monomer greatly improve the adhesion, smoothness and density of the conducting polymer coatings. The SAM appears to encourage the formation of many nucleation sites for growth of the polymer. Thus, a poly(aniline) film deposited on a gold electrode coated with a *p*-aminothiophenol SAM is denser and more uniform and exhibits a more uniform discharge than does a poly(aniline film) deposited on a bare gold electrode.[(243,244)] In a phase-separated mixed SAM of *p*-aminothiophenol and octadecanethiol, poly(aniline) is selectively deposited on the aminothiophenol domains.[(172,245)] A mechanism for initiation and growth of poly(aniline) on the SAM has been proposed.[(246)] SAMs of *w*-(*N*-pyrrole)alkanethiol greatly improve the adhesion[(247,248)] and smoothness of deposited poly(pyrrole).[(249)] A linear mass change versus charge obtained on a QCM indicates that the poly(pyrrole) is depositing layer by layer, rather than by growth of clumps.[(250)]

ABBREVIATIONS AND ACRONYMS

AC	Alternating Current
ACIS	Alternating Current Impedance Spectroscopy
AES	Auger Electron Spectroscopy
AFM	Atomic Force Microscopy
CV	Cyclic Voltammogram
DOS	Density of States
IR	Infrared
LB	Langmuir–Blodgett
LC	Liquid Chromatography
LDMS	Laser Desorption Mass Spectroscopy
NAD	Nicotinamide Adenine Dinucleotides
QCM	Quartz Crystal Microbalances
SAM	Self-assembled Monolayer
SECM	Scanning Electrochemical Microscopy
SEM	Scanning Electron Microscopy
SIMS	Surface Ionization Mass Spectroscopy
SPR	Surface Plasmon Resonance
STM	Scanning Tunneling Microscopy
TPD	Temperature Programmed Desorption
UPD	Underpotential Deposition
UV	Ultraviolet
XPS	X-ray Photoelectron Spectroscopy

RELATED ARTICLES

Electroanalytical Methods **(Volume 11)**
Electroanalytical Methods: Introduction • Selective Electrode Coatings for Electroanalysis • Surface Analysis for Electrochemistry: Ultrahigh Vacuum Techniques • Ultrafast Electrochemical Techniques

REFERENCES

1. H.O. Finklea, 'Electrochemistry of Organized Monolayers of Thiols and Related Molecules on Electrodes', in *Electroanalytical Chemistry*, eds. A.J. Bard, I. Rubinstein, Marcel Dekker, New York, 109–335, Vol. 19, 1996.
2. A. Ulman, *An Introduction to Ultrathin Organic Films: from Langmuir–Blodgett to Self-assembly*, Academic Press, San Diego, 1991.
3. D.L. Allara, R.G. Nuzzo, 'Modifications of Properties of Metals', U.S. Patent Application 389/775 (June 18), 1982.
4. R.G. Nuzzo, D.L. Allara, 'Adsorption of Bifunctional Organic Disulfides on Gold Surfaces', *J. Am. Chem. Soc.*, **105**, 4481–4483 (1983).
5. R.G. Nuzzo, F.A. Fusco, D.L. Allara, 'Spontaneously Organized Molecular Assemblies. 3. Preparation and Properties of Solution Adsorbed Monolayers of Organic Disulfides on Gold Surfaces', *J. Am. Chem. Soc.*, **109**, 2358–2368 (1987).
6. M.D. Porter, T.B. Bright, D. Allara, C.E.D. Chidsey, 'Spontaneously Organized Molecular Assemblies. 4. Structural Characterization of *n*-Alkyl Thiol Monolayers on Gold by Optical Ellipsometry, Infrared Spectroscopy, and Electrochemistry', *J. Am. Chem. Soc.*, **109**, 3559–3568 (1987).
7. H.O. Finklea, S. Avery, M. Lynch, T. Furtsch, 'Blocking Oriented Monolayers of Alkyl Mercaptans on Gold Electrodes', *Langmuir*, **3**, 409–413 (1987).
8. E. Sabatani, I. Rubinstein, R. Maoz, J. Sagiv, 'Organized Self-assembling Monolayers on Electrodes. Part I. Octadecyl Derivatives on Gold', *J. Electroanal. Chem.*, **219**, 365–371 (1987).
9. E.B. Troughton, C.D. Bain, G.M. Whitesides, R.G. Nuzzo, D.L. Allara, M.D. Porter, 'Monolayer Films Prepared by the Spontaneous Self-assembly of Symmetrical and Unsymmetrical Dialkyl Sulfides from Solution onto Gold Substrates: Structure, Properties, and Reactivity of Constituent Functional Groups', *Langmuir*, **4**, 365–385 (1988).

10. U.B. Steiner, W.R. Caseri, U.W. Suter, 'Adsorption of Alkanenitriles and Alkanedinitriles on Gold and Copper', *Langmuir*, **8**, 2771–2777 (1992).
11. J.J. Hickman, C. Zou, D. Ofer, P.D. Harvey, M.S. Wrighton, P.E. Laibinis, C.D. Bain, G.M. Whitesides, 'Combining Spontaneous Molecular Assembly With Microfabrication to Pattern Surfaces: Selective Binding of Isonitriles to Platinum Microwires and Characterization by Electrochemistry and Surface Spectroscopy', *J. Am. Chem. Soc.*, **111**, 7271–7272 (1989).
12. I. Tabushi, K. Kurihara, K. Naka, K. Yamamura, H. Hatakeyama, 'Supramolecular Sensor Based on SnO_2 Electrode Modified With Octadecasilyl Monolayer Having Molecular Binding Sites', *Tetrahedron Lett.*, **28**, 4299–4302 (1987).
13. Z.H. Jin, D.V. Vezenov, Y.W. Lee, J.E. Zull, C.N. Sukenik, R.F. Savinell, 'Alternating Current Impedance Characterization of the Structure of Alkylsiloxane Self-assembled Monolayers on Silicon', *Langmuir*, **10**, 2662–2671 (1994).
14. H.O. Finklea, L.R. Robinson, A. Blackburn, B. Richter, D. Allara, T. Bright, 'Formation of an Organized Monolayer by Solution Adsorption of Octadecyltrichlorosilane on Gold: Electrochemical Properties and Structural Characterization', *Langmuir*, **2**, 239–244 (1986).
15. D.L. Allara, A.N. Parikh, F. Rondelez, 'Evidence for a Unique Chain Organization in Long Chain Silane Monolayers Deposited on Two Widely Different Substrates', *Langmuir*, **11**, 2357–2360 (1995).
16. M.R. Linford, C.E.D. Chidsey, 'Alkyl Monolayers Covalently Bonded to Silicon Surfaces', *J. Am. Chem. Soc.*, **115**, 12631–12632 (1993).
17. M.R. Linford, P. Fenter, P.M. Eisenberger, C.E.D. Chidsey, 'Alkyl Monolayers on Silicon Prepared from 1-Alkenes and Hydrogen-terminated Silicon', *J. Am. Chem. Soc.*, **117**, 3145–3155 (1995).
18. R.G. Nuzzo, L.H. Dubois, D.L. Allara, 'Fundamental Studies of Microscopic Wetting on Organic Surfaces. 1. Formation and Structural Characterization of a Self-consistent Series of Polyfunctional Organic Monolayers', *J. Am. Chem. Soc.*, **112**, 558–569 (1990).
19. C.D. Bain, E.B. Troughton, Y.-T. Tao, J. Evall, G.M. Whitesides, R.G. Nuzzo, 'Formation of Monolayer Films by the Spontaneous Assembly of Organic Thiols from Solution onto Gold', *J. Am. Chem. Soc.*, **111**, 321–335 (1989).
20. C.D. Bain, G.M. Whitesides, 'Modeling Organic Surfaces With Self-assembled Monolayers', *Angew. Chem. Int. Ed. Engl.*, **28**, 506–512 (1989).
21. C.D. Bain, G.M. Whitesides, 'Depth Sensitivity of Wetting: Monolayers of w-Mercapto Ethers on Gold', *J. Am. Chem. Soc.*, **110**, 5897–5898 (1988).
22. C. Pale-Grosdemange, E.S. Simon, K.L. Prime, G.M. Whitesides, 'Formation of Self-assembled Monolayers by Chemisorption of Derivatives of Oligo(ethylene glycol) of Structure $HS(CH_2)_{11}(OCH_2CH_2)_mOH$ on Gold', *J. Am. Chem. Soc.*, **113**, 12–20 (1991).
23. P.E. Laibinis, C.D. Bain, R.G. Nuzzo, G.M. Whitesides, 'Structure and Wetting Properties of w-Alkoxy-n-alkanethiolate Monolayers on Gold and Silver', *J. Phys. Chem.*, **99**, 7663–7676 (1995).
24. S.D. Evans, E. Urankar, A. Ulman, N. Ferris, 'Self-assembled Monolayers of Alkanethiols Containing a Polar Aromatic Group: Effects of the Dipole Position on Molecular Packing, Orientation, and Surface Wetting Properties', *J. Am. Chem. Soc.*, **113**, 4121–4131 (1991).
25. S.-C. Chang, I. Chao, Y.-T. Tao, 'Structures of Self-assembled Monolayers of Aromatic-derivatized Thiols on Evaporated Gold and Silver Surfaces: Implication on Packing Mechanism', *J. Am. Chem. Soc.*, **116**, 6792–6805 (1994).
26. Y.-T. Tao, C.-C. Wu, J.-Y. Eu, W.-L. Lin, K.-C. Wu, C. Chen, 'Structure Evolution of Aromatic-derivatized Thiol Monolayers on Evaporated Gold', *Langmuir*, **13**, 4018–4023 (1997).
27. E. Sabatani, J. Cohen-Boulakia, M. Bruening, I. Rubinstein, 'Thioaromatic Monolayers on Gold: A New Family of Self-assembling Monolayers', *Langmuir*, **9**, 2974–2981 (1993).
28. T. Kim, R.M. Crooks, M. Tsen, L. Sun, 'Polymeric Self-assembled Monolayers. 2. Synthesis and Characterization of Self-assembled Polydiacetylene Mono- and Multilayers', *J. Am. Chem. Soc.*, **117**, 3963–3967 (1995).
29. S.B. Sachs, S.P. Dudek, R.P. Hsung, L.A. Sita, J.F. Smalley, M.D. Newton, S.W. Feldberg, C.E.D. Chidsey, 'Rates of Interfacial Electron Transfer Through π-Conjugated Spacers', *J. Am. Chem. Soc.*, **119**, 10563–10564 (1997).
30. R.W. Zehner, L.R. Sita, 'Electrochemical Evaluation and Enhancement via Heterogeneous Exchange of the Passivating Properties and Stability of Self-assembled Monolayers Derived from the Rigid Rod Arenethiols', *Langmuir*, **13**, 2973–2979 (1997).
31. J.M. Tour, L.J. Jones II, D.L. Pearson, J.J.S. Lamba, T.P. Burgin, G.M. Whitesides, D.L. Allara, A.N. Parikh, S.V. Atre, 'Self-assembled Monolayers and Multilayers of Conjugated Thiols, α, w-Dithiols, and Thioacetyl-containing Adsorbates. Understanding Attachments Between Potential Molecular Wires and Gold Surfaces', *J. Am. Chem. Soc.*, **117**, 9529–9534 (1995).
32. B. Liedberg, Z. Yang, I. Engquist, M. Wirde, U. Gelius, G. Gotz, P. Bauerle, R.-M. Rummel, Ch. Ziegler, W. Gopel, 'Self-assembly of Alpha-functionalized Terthiophenes on Gold', *J. Phys. Chem. B*, **101**, 5951–5962 (1997).
33. Y.S. Obeng, M.E. Laing, A.C. Freidli, H.C. Yang, D.N. Wang, E.W. Thulstrup, A.J. Bard, J. Michl, 'Self-assembled Monolayers of Parent and Derivatized [n]Staffane-3,3($n-1$)-dithiols on Polycrystalline Gold Electrodes', *J. Am. Chem. Soc.*, **114**, 9943–9952 (1992).

34. S.D. Evans, K.E. Goppert-Bearducci, E. Urankar, L.J. Gerenser, A. Ulman, 'Monolayers Having Large In-plane Dipole Moments: Characterization of Sulfone-containing Self-assembled Monolayers of Alkanethiols on Gold by Fourier Transform Infrared Spectroscopy, X-ray Photoelectron Spectroscopy, and Wetting', *Langmuir*, **7**, 2700–2709 (1991).
35. S.-W. Tam-Chang, H.A. Biebuyck, G.M. Whitesides, 'Self-assembled Monolayers on Gold Generated from Alkanethiols With the Structure $RNHCOCH_2SH$', *Langmuir*, **11**, 4371–4382 (1995).
36. R.S. Clegg, J.E. Hutchison, 'Hydrogen-bonding, Self-assembled Monolayers: Ordered Molecular Films for Study of Through-peptide Electron Transfer', *Langmuir*, **12**, 5239–5243 (1996).
37. G.M. Whitesides, P.E. Laibinis, 'Wet Chemical Approaches to the Characterization of Organic Surfaces: Self-assembled Monolayers, Wetting, and the Physical–Organic Chemistry of the Solid–Liquid Interface', *Langmuir*, **6**, 87–96 (1990).
38. M.T. McDermott, J.-B.D. Green, M.D. Porter, 'Scanning Force Microscopic Exploration of the Lubrication Capabilities of *n*-Alkanethiolate Monolayers Chemisorbed at Gold: Structural Basis of Microscopic Friction and Wear', *Langmuir*, **13**, 2504–2510 (1997).
39. R.C. Thomas, J.E. Houston, R.M. Crooks, T. Kim, T.A. Michalske, 'Probing Adhesion Forces at the Molecular Scale', *J. Am. Chem. Soc.*, **117**, 3830–3834 (1995).
40. H.I. Kim, T. Koini, T.R. Lee, S.S. Perry, 'Systematic Studies of the Frictional Properties of Fluorinated Monolayers With Atomic Force Microscopy: Comparison of CF_3- and CH_3-terminated Films', *Langmuir*, **13**, 7192–7196 (1997).
41. G.P. Lopez, H.A. Biebuyck, G.M. Whitesides, 'Scanning Electron Microscopy Can Form Images of Patterns in Self-assembled Monolayers', *Langmuir*, **9**, 1513–1516 (1993).
42. J.P. Folkers, P.E. Laibinis, G.M. Whitesides, J. Deutch, 'Phase Behavior of Two-component Self-assembled Monolayers of Alkanethiolates on Gold', *J. Phys. Chem.*, **98**, 563–571 (1994).
43. V. Chechik, H. Schonherr, G.J. Vancso, C.J.M. Stirling, 'Self-assembled Monolayers of Branched Thiols and Disulfides on Gold: Surface Coverage, Order and Chain Orientation', *Langmuir*, **14**, 3003–3010 (1998).
44. A. Ulman, *Characterization of Organic Thin Films: Materials Characterization Series–Surfaces, Interfaces, Thin Films*, Butterworth-Heinemann, Stoneham, MA, 1995.
45. L.H. Dubois, R.G. Nuzzo, 'Synthesis, Structure, and Properties of Model Organic Surfaces', *Annu. Rev. Phys. Chem.*, **43**, 437–463 (1992).
46. A.R. Bishop, R.G. Nuzzo, 'Self-assembled Monolayers: Recent Developments and Applications', *Curr. Opin. Colloid Interface Sci.*, **1**, 127–136 (1996).
47. C.-J. Zhong, M.D. Porter, 'Designing Interfaces at the Molecular Level', *Anal. Chem.*, **67**, 709A–715A (1995).
48. Th. Wink, S.J. van Suilen, A. Bult, W.P. van Bennekom, 'Self-assembled Monolayers for Biosensors', *Analyst*, **122**, 43R–50R (1997).
49. L.-H. Guo, J.S. Facci, G. McLendon, R. Mosher, 'Effect of Gold Topography and Surface Pretreatment on the Self-assembly of Alkanethiol Monolayers', *Langmuir*, **10**, 4588–4593 (1994).
50. X.-M. Zhao, J.L. Wilbur, G.M. Whitesides, 'Using Two-stage Chemical Amplification To Determine the Density of Defects in Self-assembled Monolayers of Alkanethiolates on Gold', *Langmuir*, **12**, 3257–3264 (1996).
51. Y. Golan, L. Margulis, I. Rubinstein, 'Vacuum-deposited Gold Films. I. Factors Affecting Film Morphology', *Surf. Sci.*, **264**, 312–326 (1992).
52. M.M. Walczak, C.A. Alves, B.D. Lamp, M.D. Porter, 'Electrochemical and X-ray Photoelectron Spectroscopic Evidence for Differences in the Binding Sites of Alkanethiolate Monolayers Chemisorbed on Gold', *J. Electroanal. Chem.*, **396**, 103–114 (1995).
53. W.B. Caldwell, K. Chen, B.R. Herr, C.A. Mirkin, J.C. Hulteen, R.P. Van Duyne, 'Self-assembled Monolayers of Ferrocenylazobenzenes on Au(111)/Mica Films: Surface-enhanced Raman Scattering Response vs Surface Morphology', *Langmuir*, **10**, 4109–4115 (1994).
54. V.K. Gupta, N.L. Abbott, 'Uniform Anchoring of Nematic Liquid Crystals on Self-assembled Monolayers Formed from Alkanethiols on Obliquely Deposited Films of Gold', *Langmuir*, **12**, 2587–2593 (1996).
55. D. Stamou, D. Gourdon, M. Liley, N.A. Burnham, A. Kulik, H. Vogel, C. Duschl, 'Uniformly Flat Gold Surfaces: Imaging the Domain Structure of Organic Monolayers Using Scanning Force Microscopy', *Langmuir*, **13**, 2425–2428 (1997).
56. P. Wagner, M. Hegner, H.-J. Gůntherodt, G. Semenza, 'Formation and in situ Modification of Monolayers Chemisorbed on Ultraflat Template-stripped Gold Surfaces', *Langmuir*, **11**, 3867–3875 (1995).
57. P.A. DiMilla, J.P. Folkers, H.A. Biebuyck, R. Harter, G.P. Lopez, G.M. Whitesides, 'Wetting and Protein Adsorption of Self-assembled Monolayers of Alkanethiolates Supported on Transparent Films of Gold', *J. Am. Chem. Soc.*, **116**, 2225–2226 (1994).
58. K.A. Peterlinz, R. Georgiadis, 'In situ Kinetics of Self-assembly by Surface Plasmon Resonance Spectroscopy', *Langmuir*, **12**, 4731–4740 (1996).
59. S.E. Creager, L.A. Hockett, G.K. Rowe, 'Consequences of Microscopic Surface Roughness for Molecular Self-assembly', *Langmuir*, **8**, 854–861 (1992).
60. H. Ron, I. Rubinstein, 'Alkanethiol Monolayers on Preoxidized Gold. Encapsulation of Gold Oxide Under an Organic Monolayer', *Langmuir*, **10**, 4566–4573 (1994).
61. H. Ron, S. Matlis, I. Rubinstein, 'Self-assembled Monolayers on Oxidized Metals. 2. Gold Surface Oxidative

Pretreatment, Monolayer Properties, and Depression Formation', *Langmuir*, **14**, 1116–1121 (1998).
62. P.E. Laibinis, C.D. Bain, G.M. Whitesides, 'Attenuation of Photoelectrons in Monolayers of *n*-Alkanethiols Adsorbed on Copper, Silver, and Gold', *J. Phys. Chem.*, **95**, 7017–7021 (1991).
63. P.E. Laibinis, G.M. Whitesides, D.L. Allara, Y.-T. Tao, A.N. Parikh, R.G. Nuzzo, 'Comparison of the Structure and Wetting Properties of Self-assembled Monolayers of *n*-Alkanethiols on the Coinage Metal Surfaces, Cu, Ag, Au', *J. Am. Chem. Soc.*, **113**, 7152–7167 (1991).
64. P.E. Laibinis, G.M. Whitesides, '*w*-Terminated Alkanethiolate Monolayers on Surfaces of Copper, Silver, and Gold Have Similar Wettabilities', *J. Am. Chem. Soc.*, **114**, 1990–1995 (1992).
65. J. Scherer, M.R. Vogt, O.M. Magnussen, R.J. Behm, 'Corrosion of Alkanethiol-covered Cu(100) Surfaces in Hydrochloric Acid Solution by in situ Scanning Tunneling Microscopy', *Langmuir*, **13**, 7045–7051 (1997).
66. M.H. Schoenfisch, J.E. Pemberton, 'Air Stability of Alkanethiol Self-assembled Monolayers on Silver and Gold Surfaces', *J. Am. Chem. Soc.*, **120**, 4502–4513 (1998).
67. G.K. Jennings, P.E. Laibinis, 'Underpotential Deposited Metal Layers of Silver Provide Enhanced Stability to Self-assembled Alkanethiol Monolayers on Gold', *Langmuir*, **12**, 6173–6175 (1996).
68. G.K. Jennings, P.E. Laibinis, 'Self-assembled *n*-Alkanethiolate Monolayers on Underpotentially Deposited Adlayers of Silver and Copper on Gold', *J. Am. Chem. Soc.*, **119**, 5208–5214 (1997).
69. J.D. Burgess, F.M. Hawkridge, 'Octadecyl Mercaptan Sub-monolayers on Silver Electrodeposited on Gold Quartz Crystal Microbalance Electrodes', *Langmuir*, **13**, 3781–3786 (1997).
70. F.P. Zamborini, J.K. Campbell, R.M. Crooks, 'Spectroscopic, Voltammetric, and Electrochemical Scanning Tunneling Microscopic Study of Underpotentially Deposited Cu Corrosion and Passivation With Self-assembled Organomercaptan Monolayers', *Langmuir*, **14**, 640–647 (1998).
71. M. Nishizawa, T. Sunagawa, H. Yoneyama, 'Underpotential Deposition of Copper on Gold Electrodes Through Self-assembled Monolayers of Propanethiol', *Langmuir*, **13**, 5215–5217 (1997).
72. A. Demoz, D.J. Harrison, 'Characterization of Extremely Low Defect Density Hexadecanethiol Monolayers on Hg Surfaces', *Langmuir*, **9**, 1046–1050 (1993).
73. N. Muskal, I. Turyan, D. Mandler, 'Thiol Self-assembled Monolayers on Mercury Surfaces', *J. Electroanal. Chem.*, **409**, 131–136 (1996).
74. O.F. Magnussen, B.M. Ocko, M. Deutsch, M.J. Regan, P.S. Pershan, D. Abernathy, G. Grubel, J.-F. Legrand, 'Self-assembly of Organic Films on a Liquid Metal', *Nature*, **384**, 250–252 (1996).
75. K. Slowinski, R.V. Chamberlain, C.J. Miller, M. Majda, 'Through-bond and Chain-to-chain Coupling. Two Pathways in Electron Tunneling Through Liquid Alkanethiol Monolayers on Mercury Electrodes', *J. Am. Chem. Soc.*, **119**, 11910–11919 (1997).
76. C. Bruckner-Lea, J. Janata, J. Conroy, A. Pungor, K. Caldwell, 'Scanning Tunneling Microscopy on a Mercury Sessile Drop', *Langmuir*, **9**, 3612–3617 (1993).
77. Z. Mekhalif, J.-J. Pireaux, J. Delhalle, 'Self-assembled Monolayers of *n*-Dodecanethiol on Electrochemically Modified Polycrystalline Nickel Surfaces', *Langmuir*, **13**, 2285–2290 (1997).
78. A.D. Vogt, T. Han, T.P. Beebe, 'Adsorption of 11-Mercaptoundecanoic Acid in Ni(111) and Its Interaction With Probe Molecules', *Langmuir*, **13**, 3397–3403 (1997).
79. T. Kondo, M. Takechi, Y. Sato, K. Uosaki, 'Adsorption Behavior of Functionalized Ferrocenylalkane Thiols and Disulfide on Au and ITO and Electrochemical Properties of Modified Electrodes: Effects of Acyl and Alkyl Groups Attached to the Ferrocene Ring', *J. Electroanal. Chem.*, **381**, 203–209 (1995).
80. Y. Gu, Z. Lin, R.A. Butera, V.S. Smentkowski, D.H. Waldeck, 'Preparation of Self-assembled Monolayers on InP', *Langmuir*, **11**, 1849–1851 (1995).
81. K. Chen, F. Xu, C.A. Mirkin, R.-K. Lo, K.S. Nanjundaswamy, J.-P. Zhou, J.T. McDevitt, 'Do Alkanethiols Adsorb onto the Surfaces of Tl-Ba-Cu-O-based High-temperature Superconductors? The Critical Role of H_2O Content in the Adsorption Process', *Langmuir*, **12**, 2622–2624 (1996).
82. P. Fenter, A. Eberhardt, K.S. Liang, P. Eisenberger, 'Epitaxy and Chainlength Dependent Strain in Self-assembled Monolayers', *J. Chem. Phys.*, **106**, 1600–1608 (1997).
83. N. Camillone III, T.Y.B. Leung, G. Scoles, 'A Low Energy Helium Atom Diffraction Study of Decanethiol Self-assembled on Au(111)', *Surf. Sci.*, **373**, 333–349 (1997).
84. A. Kudelski, P. Krysinski, 'Solvent Trapping During the Self-assembly of Octadecanethiol Monolayer on Roughened Gold Electrodes from Surface-enhanced Raman Scattering Studies', *J. Electroanal. Chem.*, **443**, 5–7 (1998).
85. C. Miller, P. Cuendet, M. Gratzel, 'Adsorbed *w*-Hydroxy Thiol Monolayers on Gold Electrodes: Evidence for Electron Tunneling to Redox Species in Solution', *J. Phys. Chem.*, **95**, 877–886 (1991).
86. J. Liu, A.E. Kaifer, 'Preparation of Self-assembled Monolayers from Micellar Solutions', *Israel J. Chem.*, **37**, 235–239 (1997).
87. J. Yan, S. Dong, 'Formation of Surface Inclusion Complexes Between Cyclodextrins and *n*-Alkanethiols and Their Self-assembled Behaviors on Gold', *Langmuir*, **13**, 3251–3255 (1997).

88. J.A.M. Sondag-Huethorst, C. Schonenberger, L.G.J. Fokkink, 'Formation of Holes in Alkanethiol Monolayers on Gold', *J. Phys. Chem.*, **98**, 6826–6834 (1994).
89. K. Edinger, A. Golzhauser, K. Demota, Ch. Woll, M. Grunze, 'Formation of Self-assembled Monolayers of *n*-Alkanethiols on Gold: A Scanning Tunneling Microscopy Study on the Modification of Substrate Morphology', *Langmuir*, **9**, 4–8 (1993).
90. E. Delamarche, B. Michel, Ch. Gerber, D. Anselmetti, H.-J. Guntherodt, H. Wolf, H. Ringsdorf, 'Real-space Observation of Nanoscale Molecular Domains in Self-assembled Monolayers', *Langmuir*, **10**, 2869–2871 (1994).
91. N. Camillone, C.E.D. Chidsey, P. Eisenberger, P. Fenter, J. Li, K.S. Liang, G.Y. Liu, G. Scoles, 'Structural Defects in Self-assembled Organic Monolayers via Combined Atomic Beam and X-ray Diffraction', *J. Chem. Phys.*, **99**, 744–747 (1993).
92. H.O. Finklea, D.D. Hanshew, 'Preparation and Reversible Behavior of Organized Thiol Monolayers With Attached Pentaminepyridineruthenium Redox Centers', *J. Electroanal. Chem.*, **347**, 327–340 (1993).
93. C.D. Bain, J. Evall, G.M. Whitesides, 'Formation of Monolayers by the Coadsorption of Thiols on Gold: Variation in the Head Group, Tail Group, and Solvent', *J. Am. Chem. Soc.*, **111**, 7155–7164 (1989).
94. C.D. Bain, G.M. Whitesides, 'Formation of Monolayers by the Coadsorption of Thiols on Gold: Variation in the Length of the Alkyl Chain', *J. Am. Chem. Soc.*, **111**, 7164–7175 (1989).
95. D.E. Weisshaar, B.D. Lamp, M.D. Porter, 'Thermodynamically Controlled Electrochemical Formation of Thiolate Monolayers at Gold: Characterization and Comparison to Self-assembled Analogs', *J. Am. Chem. Soc.*, **114**, 5860–5862 (1992).
96. D.W. Hatchett, R.H. Uibel, K.J. Stevenson, J.M. Harris, H.S. White, 'Electrochemical Measurement of the Free Energy of Adsorption of *n*-Alkanethiolates on Ag(111)', *J. Am. Chem. Soc.*, **120**, 1062–1069 (1998).
97. K.J. Stevenson, M. Mitchell, H.S. White, 'Oxidative Adsorption of *n*-Alkanethiolates at Mercury. Dependence of Adsorption Free Energy on Chain length', *J. Phys. Chem. B*, **102**, 1235–1240 (1998).
98. G.E. Poirier, E.D. Pylant, 'The Self-assembly Mechanism of Alkane Thiols on Au(111)', *Science*, **272**, 1145–1148 (1996).
99. O. Chailapakul, L. Sun, C. Xu, R.M. Crooks, 'Interactions Between Organized, Surface-confined Monolayers and Vapor-phase Probe Molecules. 7. Comparison of Self-assembling *n*-Alkanethiol Monolayers Deposited on Gold from Liquid and Vapor Phases', *J. Am. Chem. Soc.*, **115**, 12459–12467 (1993).
100. R. Bilewicz, M. Majda, 'Monomolecular Langmuir–Blodgett Films at Electrodes. Formation of Passivating Monolayers Incorporating Electroactive Reagents', *Langmuir*, **7**, 2794–2802 (1991).
101. P. Krysinski, R.V. Chamberlain, M. Majda, 'Partial Electron Transfer in Octadecanethiol Binding to Gold', *Langmuir*, **10**, 4286–4294 (1994).
102. R. Bilewicz, T. Sawaguchi, R.V. Chamberlain II, M. Majda, 'Monomolecular Langmuir–Blodgett Films at Electrodes. Electrochemistry at Single Molecule "Gate Sites"' *Langmuir*, **11**, 2256–2266 (1995).
103. P. Fenter, A. Eberhardt, P. Eisenberger, 'Self-assembly of *n*-Alkyl Thiols as Disulfides on Au(111)', *Science*, **266**, 1216–1218 (1994).
104. C.-J. Zhong, M.D. Porter, 'Evidence for Carbon–Sulfur Cleavage in Spontaneously Adsorbed Organosulfide-based Monolayers at Gold', *J. Am. Chem. Soc.*, **116**, 11616–11617 (1994).
105. S.J. Stranick, A.N. Parikh, Y.-T. Tao, D.L. Allara, P.S. Weiss, 'Phase Separation of Mixed-composition Self-assembled Monolayers into Nanometer Scale Molecular Domains', *J. Phys. Chem.*, **98**, 7636–7646 (1994).
106. S.J. Stranick, S.V. Atre, A.N. Parikh, M.C. Wood, D.L. Allara, N. Winograd, P.S. Weiss, 'Nanometer-scale Phase Separation in Mixed Composition Self-assembled Monolayers', *Nanotechnology*, **7**, 438–442 (1996).
107. H. Van Ryswyk, E.D. Turtle, R. Watson-Clark, T.A. Tanzer, T.K. Herman, P.Y. Chong, P.J. Waller, A.L. Taurog, C.E. Wagner, 'Reactivity of Ester Linkages and Pentaammineruthenium(III) at the Monolayer Assembly/Solution Interface', *Langmuir*, **12**, 6143–6150 (1996).
108. V. Chechik, C.J.M. Stirling, 'Reactivity in Monolayers versus Bulk Media: Intra- and Intermolecular Aminolysis of Esters', *Langmuir*, **13**, 6354–6356 (1997).
109. P. Neogi, S. Neogi, C.J.M. Stirling, 'Reactivity of Carboxy Esters in Gold–Thiol Monolayers', *J. Chem. Soc., Chem. Commun.*, 1134–1136 (1993).
110. T. Kim, Q. Ye, L. Sun, K.C. Chan, R.M. Crooks, 'Polymeric Self-assembled Monolayers: 5. Synthesis and Characterization of *w*-Functionalized, Self-assembled Diacetylenic and Polydiacetylenic Monolayers', *Langmuir*, **12**, 6065–6073 (1996).
111. T. Kim, K.C. Chan, R.M. Crooks, 'Polymeric Self-assembled Monolayers. 4. Chemical, Electrochemical, and Thermal Stability of *w*-Functionalized, Self-assembled Diacetylenic and Polydiacetylenic Monolayers', *J. Am. Chem. Soc.*, **119**, 189–193 (1997).
112. J.S. Peanasky, R.L. McCarley, 'Surface-confined Monomers on Electrode Surfaces. 4. Electrochemical and Spectroscopic Characterization of Undec-10-ene-1-thiol Self-assembled Monolayers on Au', *Langmuir*, **14**, 113–123 (1998).
113. W.R. Thompson, M. Cai, M. Ho, J.E. Pemberton, 'Hydrolysis and Condensation of Self-assembled Monolayers of (3-Mercaptopropyl)trimethoxysilane on Ag and Au Surfaces', *Langmuir*, **13**, 2291–2302 (1997).
114. R.I. Carey, J.P. Folkers, G.M. Whitesides, 'Self-assembled Monolayers Containing *w*-Mercaptoalkylboronic Acids Adsorbed onto Gold Form a Highly Cross-linked,

Thermally Stable Borate Glass Surface', *Langmuir*, **10**, 2228–2234 (1994).

115. R.J. Willicut, R.L. McCarley, 'Electrochemical Polymerization of Pyrrole-containing Self-assembled Alkanethiol Monolayers on Au', *J. Am. Chem. Soc.*, **116**, 10823–10824 (1994).
116. R.J. Willicut, R.L. McCarley, 'Surface-confined Monomers on Electrode Surfaces. Part 3. Electrochemical Reactions and Scanning Probe Microscopy Investigations of *w*-(*N*-Pyrrolyl)alkanethiol Self-assembled Monolayers on Gold', *Anal. Chim. Acta*, **307**, 269–276 (1995).
117. J. Spinke, M. Liley, F.-J. Schmitt, H.-J. Guider, L. Angermaier, W. Knoll, 'Molecular Recognition at Self-assembled Monolayers: Optimization of Surface Functionalization', *J. Chem. Phys.*, **99**, 7012–7019 (1993).
118. G.B. Sigal, C. Bamdad, A. Barberis, J. Strominger, G.M. Whitesides, 'A Self-assembled Monolayer for the Binding and Study of Histidine-tagged Proteins by Surface Plasmon Resonance', *Anal. Chem.*, **68**, 490–497 (1996).
119. M. Liley, T.A. Keller, C. Duschl, H. Vogel, 'Direct Observation of Self-assembled Monolayers, Ion Complexation, and Protein Conformation at the Gold/Water Interface: An FTIR Spectroscopic Approach', *Langmuir*, **13**, 4190–4192 (1997).
120. E. Katz, V. Heleg-Shabtai, B. Willner, I. Willner, A.F. Buckmann, 'Electrical Contact of Redox Enzymes With Electrodes: Novel Approaches for Amperometric Biosensors', *Bioelectrochem. Bioenerg.*, **42**, 95–104 (1997).
121. A. Bardea, E. Katz, A.F. Buckmann, I. Willner, 'NAD^+-dependent Enzyme Electrodes: Electrical Contact of Cofactor-dependent Enzymes and Electrodes', *J. Am. Chem. Soc.*, **119**, 9114–9119 (1997).
122. K.A. Peterlinz, R.M. Georgiadis, T.M. Herne, M.H. Tarlov, 'Observation of Hybridization and Dehybridization of Thiol-tethered DNA Using Two-color Surface Plasmon Resonance Spectroscopy', *J. Am. Chem. Soc.*, **119**, 3401–3402 (1997).
123. J.J. Storhoff, R. Elghanian, R.C. Mucic, C.A. Mirkin, R.L. Letsinger, 'One-pot Colorimetric Differentiation of Polynucleotides With Single Base Imperfections Using Gold Nanoparticle Probes', *J. Am. Chem. Soc.*, **120**, 1959–1964 (1998).
124. F. Caruso, K. Niikura, D.N. Furlong, Y. Okahata, 'Ultrathin Multilayer Polyelectrolyte Films on Gold: Construction and Thickness Determination', *Langmuir*, **13**, 3422–3426 (1997).
125. F. Caruso, K. Niikura, D.N. Furlong, Y. Okahata, 'Assembly of Alternating Polyelectrolyte and Protein Multilayer Films for Immunosensing', *Langmuir*, **13**, 3427–3433 (1997).
126. T.L. Freeman, S.D. Evans, A. Ulman, 'XPS Studies of Self-assembled Multilayer Films', *Langmuir*, **11**, 4411–4417 (1995).
127. H. Lee, L.J. Kepley, H.-G. Hong, S. Akhter, T.E. Mallouk, 'Adsorption of Ordered Zirconium Phosphonate Multilayer Films on Silicon and Gold Surfaces', *J. Phys. Chem.*, **92**, 2597–2601 (1988).
128. L.-H. Guo, J.S. Facci, G. McLendon, 'Distance Dependence of Electron Transfer Rates in Bilayers of a Ferrocene L–B Monolayer and a Self-assembled Monolayer on Gold', *J. Phys. Chem.*, **99**, 8458–8461 (1995).
129. P. Sanassy, S.D. Evans, 'Mixed Alkanethiol Monolayers on Gold Surfaces: Substrates for Langmuir–Blodgett Film Deposition', *Langmuir*, **9**, 1024–1027 (1993).
130. A.L. Plant, 'Self-assembled Phospholipid/Alkanethiol Biomimetic Bilayers on Gold', *Langmuir*, **9**, 2764–2767 (1993).
131. S. Terrettaz, T. Stora, C. Duschl, H. Vogel, 'Protein Binding to Supported Lipid Membranes: Investigation of the Cholera Toxin–Ganglioside Interaction by Simultaneous Impedance Spectroscopy and Surface Plasmon Resonance', *Langmuir*, **9**, 1361–1369 (1993).
132. L. Ding, J. Li, S. Dong, E. Wang, 'Supported Phospholipid Membranes: Comparison Among Different Deposition Methods for a Phospholipid Monolayer', *J. Electroanal. Chem.*, **416**, 105–112 (1996).
133. N. Bunjes, E.K. Schmidt, A. Jonczyk, F. Rippmann, D. Beyer, H. Ringsdorf, P. Graber, W. Knoll, R. Naumann, 'Thiopeptide-supported Lipid Layers on Solid Substrates', *Langmuir*, **13**, 6188–6194 (1997).
134. B.A. Cornell, V.L.B. Braach-Maksvytis, L.G. King, P.D. J. Osman, B. Ragues, L. Wieczorek, R.J. Pace, 'A Biosensor That Uses Ion-channel Switches', *Nature*, **387**, 580–583 (1997).
135. B. Raguse, V. Braach-Maksvystis, B.A. Cornell, L.G. King, P.D.J. Osman, R.J. Pace, L. Weiczorek, 'Tethered Lipid Bilayer Membranes: Formation and Ionic Reservoir Characterization', *Langmuir*, **14**, 648–659 (1998).
136. L.M. Williams, S.D. Evans, T.M. Flynn, A. Marsh, P.F. Knowles, R.J. Bushby, N. Boden, 'Kinetics of the Unrolling of Small Unilamellar Phospholipid Vesicles onto Self-assembled Monolayers', *Langmuir*, **13**, 751–757 (1997).
137. C.A. Widrig, C. Chung, M.D. Porter, 'The Electrochemical Desorption of *n*-Alkanethiol Monolayers from Polycrystalline Au and Ag Electrodes', *J. Electroanal. Chem.*, **310**, 335–359 (1991).
138. K. Slowinski, R.V. Chamberlain, R. Bilewicz, M. Majda, 'Evidence for Inefficient Chain-to-chain Coupling in Electron Tunneling Through Liquid Alkane Thiol Monolayer Films on Mercury', *J. Am. Chem. Soc.*, **118**, 4709–4710 (1996).
139. C.P. Smith, H.S. White, 'Theory of the Interfacial Potential Distribution and Reversible Voltammetric Response of Electrodes Coated With Electroactive Molecular Films', *Anal. Chem.*, **64**, 2398–2405 (1992).
140. C.P. Smith, H.S. White, 'Voltammetry of Molecular Films Containing Acid/Base Groups', *Langmuir*, **9**, 1–3 (1993).

141. W.R. Fawcett, M. Fedurco, Z. Kovacova, 'Double Layer Effects at Molecular Films Containing Acid/Base Groups', *Langmuir*, **10**, 2403–2408 (1994).
142. M.M. Walczak, D.D. Popenoe, R.S. Deinhammer, B.D. Lamp, C. Chung, M.D. Porter, 'Reductive Desorption of Alkanethiolate Monolayers at Gold: A Measure of Surface Coverage', *Langmuir*, **7**, 2687–2693 (1991).
143. D.-F. Yang, M. Morin, 'Chronoamperometric Study of the Reductive Desorption of Alkanethiol Self-assembled Monolayers', *J. Electroanal. Chem.*, **441**, 173–181 (1998).
144. T.W. Schneider, D.A. Buttry, 'Electrochemical Quartz Microbalance Studies of Adsorption and Desorption of Self-assembled Monolayers of Alkyl Thiols on Gold', *J. Am. Chem. Soc.*, **115**, 12391–12397 (1993).
145. W.R. Everett, I. Fritsch-Faules, 'Factors That Influence the Stability of Self-assembled Organothiols on Gold under Electrochemical Conditions', *Anal. Chim. Acta*, **307**, 253–268 (1995).
146. N. Muskal, I. Turyan, A. Shurky, D. Mandler, 'Chiral Self-assembled Monolayers', *J. Am. Chem. Soc.*, **117**, 1147–1148 (1995).
147. S. Imabayashi, D. Hobara, T. Kakiuchi, W. Knoll, 'Selective Replacement of Adsorbed Alkanethiols in Phase-separated Binary Self-assembled Monolayers by Electrochemical Partial Desorption', *Langmuir*, **13**, 4502–4504 (1997).
148. D. Hobara, M. Ota, S. Imabayashi, K. Niki, T. Kakiuchi, 'Phase Separation of Binary Self-assembled Thiol Monolayers Composed of 1-Hexadecanethiol and 3-Mercaptopropionic Acid on Au(111) Studied by Scanning Tunneling Microscopy and Cyclic Voltammetry', *J. Electroanal. Chem.*, **444**, 113–119 (1998).
149. H.O. Finklea, D.A. Snider, J. Fedyk, E. Sabatani, Y. Gafni, I. Rubinstein, 'Characterization of Octadecanethiol-coated Gold Electrodes as Microarray Electrodes by Cyclic Voltammetry and ac Impedance Spectroscopy', *Langmuir*, **9**, 3660–3667 (1993).
150. L. Sun, R.M. Crooks, 'Imaging of Defects Contained Within *n*-Alkylthiol Monolayers by Combination of Underpotential Deposition and Scanning Tunneling Microscopy: Kinetics of Self-assembly', *J. Electrochem. Soc.*, **138**, L23–L25 (1991).
151. L. Sun, R.M. Crooks, 'Indirect Visualization of Defect Structures Contained Within Self-assembled Organomercaptan Monolayers: Combined Use of Electrochemistry and Scanning Tunneling Microscopy', *Langmuir*, **9**, 1951–1954 (1993).
152. F. Forouzan, A.J. Bard, M.V. Mirkin, 'Voltammetric and Scanning Electrochemical Microscopic Studies of the Adsorption Kinetics and Self-assembly of *n*-Alkanethiol Monolayers on Gold', *Israel J. Chem.*, **37**, 155–163 (1997).
153. O. Chailapakul, R.M. Crooks, 'Interactions Between Organized, Surface-confined Monolayers and Liquid-phase Probe Molecules. 4. Synthesis and Characterization of Nanoporous Molecular Assemblies: Mechanism of Probe Penetration', *Langmuir*, **11**, 1329–1340 (1995).
154. H.O. Finklea, D.A. Snider, J. Fedyk, 'Passivation of Pinholes in Octadecanethiol Monolayers on Gold Electrodes by Electrochemical Polymerization of Phenol', *Langmuir*, **6**, 371–376 (1990).
155. R. Andreu, J.J. Calvente, W.R. Fawcett, M. Molero, 'Role of Ion Pairing in Double-layer Effects at Self-assembled Monolayers Containing a Simple Redox Couple', *J. Phys. Chem. B*, **101**, 2884–2894 (1997).
156. Y. Sato, K. Uosaki, 'Self-assembled Monolayer as a Mediator and Barrier for Electron Transfer Between Electrodes and Redox Species in Solution', in *Redox Mechanisms and Interfacial Properties of Molecules of Biological Importance*, eds. F.A. Schultz, I. Taniguchi, The Electrochemical Society, Pennington, NJ, 299–311, 1993.
157. J. Redepenning, H.M. Tunison, H.O. Finklea, 'Influence of Donnan Potentials on Apparent Formal Potentials Measured for Organized Thiol Monolayers With Attached Pentaminepyridineruthenium Redox Centers', *Langmuir*, **9**, 1404–1407 (1993).
158. K. Shimazu, I. Yagi, Y. Sato, K. Uosaki, 'In Situ and Dynamic Monitoring of the Self-assembling and Redox Processes of a Ferrocenylundecanethiol Monolayer by Electrochemical Quartz Crystal Microbalance', *Langmuir*, **8**, 1385–1387 (1992).
159. H.C. De Long, D.A. Buttry, 'Environmental Effects on Redox Potentials of Viologen Groups Embedded in Electroactive Self-assembled Monolayers', *Langmuir*, **8**, 2491–2496 (1992).
160. G.K. Rowe, S.E. Creager, 'Redox and Ion-pairing Thermodynamics in Self-assembled Monolayers' *Langmuir*, **7**, 2307–2312 (1991).
161. C.E.D. Chidsey, C.R. Bertozzi, T.M. Putvinski, A.M. Mujsce, 'Coadsorption of Ferrocene-terminated and Unsubstituted Alkane Thiols on Gold: Electroactive Self-assembled Monolayers', *J. Am. Chem. Soc.*, **112**, 4301–4306 (1990).
162. M.S. Ravenscroft, H.O. Finklea, 'Electron-transfer Kinetics to Attached Redox Centers on Gold Electrodes in Nonaqueous Electrolytes', *J. Phys. Chem.*, **98**, 3843–3850 (1994).
163. M.J. Tarlov, E.F. Bowden, 'Electron-transfer Reaction of Cytochrome *c* Adsorbed on Carboxylic Acid Terminated Alkanethiol Monolayer Electrodes' *J. Am. Chem. Soc.*, **113**, 1847–1849 (1991).
164. R.A. Clark, E.F. Bowden, 'Voltammetric Peak Broadening for Cytochrome *c*/Alkanethiolate Monolayer Structures: Dispersion of Formal Potentials', *Langmuir*, **13**, 559–565 (1997).
165. T.M. Nahir, E.F. Bowden, 'The Distribution of Standard Rate Constants for Electron Transfer Between Thiol-modified Gold Electrodes and Adsorbed Cytochrome *c*', *J. Electroanal. Chem.*, **410**, 9–13 (1996).

166. T. Lötzbeyer, W. Schuhmann, E. Katz, J. Falter, H.-L. Schmidt, 'Direct Electron Transfer Between the Covalently Immobilized Enzyme Microperoxidase MP-11 and a Cystamine-modified Gold Electrode', *J. Electroanal. Chem.*, **377**, 291–294 (1994).
167. T. Lötzbeyer, W. Schuhmann, H.-L. Schmidt, 'Minizymes. A New Strategy for the Development of Reagentless Amperometric Biosensors Based on Direct Electron-transfer Processes', *Bioelectrochem. Bioenerg.*, **42**, 1–6 (1997).
168. J.J. Hickman, D. Ofer, P.E. Laibinis, G.M. Whitesides, M.S. Wrighton, 'Molecular Self-assembly of Two-terminal Voltammetric Microsensors With Internal References', *Science*, **252**, 688–691 (1991).
169. H.-Z. Yu, Y.-Q. Wang, J.-Z. Cheng, J.-W. Zhao, S.-M. Cai, H. Inokuchi, A. Fujishima, Z.-F. Liu, 'Electrochemical Behavior of Azobenzene Self-assembled Monolayers on Gold', *Langmuir*, **12**, 2843–2848 (1996).
170. W.B. Caldwell, D.J. Campbell, K. Chen, B.R. Herr, C.A. Mirkin, A. Malik, M.K. Durbin, P. Dutta, K.G. Huang, 'A Highly Ordered Self-assembled Monolayer Film of an Azobenzenealkanethiol on Au(111): Electrochemical Properties and Structural Characterization by Synchrotron In-plane X-ray Diffraction, Atomic Force Microscopy, and Surface-enhanced Raman Spectroscopy', *J. Am. Chem. Soc.*, **117**, 6071–6082 (1995).
171. E. Sabatani, I. Rubinstein, 'Organized Self-assembling Monolayers on Electrodes. 2. Monolayer-based Ultramicroelectrodes for the Study of Very Rapid Electrode Kinetics', *J. Phys. Chem.*, **91**, 6663–6669 (1987).
172. W.A. Hayes, H. Kim, X. Yue, S.S. Perry, C. Shannon, 'Nanometer-scale Patterning of Surfaces Using Self-assembly Chemistry. 2. Preparation, Characterization, and Electrochemical Behavior of Two-component Organothiol Monolayers on Gold Surfaces', *Langmuir*, **13**, 2511–2518 (1997).
173. O. Chailapakul, R.M. Crooks, 'Synthesis and Characterization of Simple Self-assembling, Nanoporous Monolayer Assemblies: A New Strategy for Molecular Recognition', *Langmuir*, **9**, 884–888 (1993).
174. R. Bilewicz, M. Majda, 'Bifunctional Monomolecular Langmuir–Blodgett Films at Electrodes. Electrochemistry at Single Molecule Gate Sites', *J. Am. Chem. Soc.*, **113**, 5464–5466 (1991).
175. J. Wang, H. Wu, L. Angnes, 'On Line Monitoring of Hydrophobic Compounds at Self-assembled Monolayer Modified Amperometric Flow Detectors', *Anal. Chem.*, **65**, 1893–1896 (1993).
176. K. Takehara, H. Takemura, Y. Ide, 'Electrochemical Studies of the Terminally Substituted Alkanethiol Monolayers Formed on a Gold Electrode: Effects of the Terminal Group on the Redox Responses of $Fe(CN)_6^{3-}$, $Ru(NH_3)_6^{3+}$, and Ferrocenedimethanol', *Electrochim. Acta*, **39**, 817–822 (1994).
177. K. Doblhofer, J. Figura, J.-H. Fuhrhop, 'Stability and Electrochemical Behavior of "Self-assembled" Adsorbates With Terminal Ionic Groups', *Langmuir*, **8**, 1811–1816 (1992).
178. A. Doron, E. Katz, G. Tao, I. Willner, 'Photochemically, Chemically, and pH-controlled Electrochemistry at Functionalized Spiropyran Monolayer Electrodes', *Langmuir*, **13**, 1783–1790 (1997).
179. F. Malem, D. Mandler, 'Self-assembled Monolayers in Electroanalytical Chemistry: Application of w-Mercapto Carboxylic Acid Monolayers for the Electrochemical Detection of Dopamine in the Presence of a High Concentration of Ascorbic Acid', *Anal. Chem.*, **65**, 37–41 (1993).
180. I. Turyan, D. Mandler, 'Characterization and Electroanalytical Application of Omega-mercaptoalkanesulfonic Acid Monolayers on Gold', *Israel J. Chem.*, **37**, 225–233 (1997).
181. J. Li, L. Ding, E. Wang, S. Dong, 'The Ion Selectivity of Monensin Incorporated Phospholipid/Alkane Thiol Bilayers', *J. Electroanal. Chem.*, **414**, 17–21 (1996).
182. I. Turyan, D. Mandler, 'Self-assembled Monolayers in Electroanalytical Chemistry: Application of w-Mercaptocarboxylic Acid Monolayers for Electrochemical Determination of Ultralow Levels of Cadmium(II)', *Anal. Chem.*, **66**, 58–63 (1994).
183. I. Turyan, D. Mandler, 'Selective Determination of Cr(VI) by a Self-assembled Monolayer-based Electrode', *Anal. Chem.*, **69**, 894–897 (1997).
184. M. Gerlache, Z. Senturk, G. Quarin, J.-M. Kauffmann, 'Self-assembled Monolayer Gold Electrode for Surfactant Analysis', *J. Solid State Electrochem.*, **1**, 155–160 (1997).
185. M.T. Rojas, R. Königer, J.F. Stoddart, A.E. Kaifer, 'Supported Monolayers Containing Preformed Binding Sites. Synthesis and Interfacial Binding Properties of a Thiolated β-Cyclodextrin Derivative', *J. Am. Chem. Soc.*, **117**, 336–343 (1995).
186. Y. Maeda, T. Fukuda, H. Yamamoto, H. Kitano, 'Regio- and Stereoselective Complexation by a Self-assembled Monolayer of Thiolated Cyclodextrin on a Gold Electrode', *Langmuir*, **13**, 4187–4189 (1997).
187. A.E. Kaifer, 'Functionalized Self-assembled Monolayers Containing Preformed Binding Sites', *Israel J. Chem.*, **36**, 389–397 (1996).
188. M. Lahav, K.T. Ranjit, E. Katz, I. Willner, 'Photostimulated Interactions of Bipyridinium-Azobenzene With a beta-Aminocyclodextrin Monolayer-functionalized Electrode: An Optoelectronic Assembly for the Amperometric Transduction of Recorded Optical Signals', *Israel J. Chem.*, **37**, 185–195 (1997).
189. M.T. Rojas, A.E. Kaifer, 'Molecular Recognition at the Electrode–Solution Interface. Design, Self-assembly, and Interfacial Binding Properties of a Molecular Sensor', *J. Am. Chem. Soc.*, **117**, 5883–5884 (1995).
190. I. Rubinstein, S. Steinberg, Y. Tor, A. Shanzer, J. Sagiv, 'Ionic Recognition and Selective Response in

Self-assembling Monolayer Membranes on Electrodes', *Nature*, **332**, 426–429 (1988).

191. S. Steinberg, Y. Tor, E. Sabatani, I. Rubinstein, 'Ion-selective Monolayer Membranes Based on Self-assembling Tetradentate Ligand Monolayers on Gold Electrodes. 2. Effect of Applied Potential on Ion Binding', *J. Am. Chem. Soc.*, **113**, 5176–5182 (1991).
192. S. Steinberg, I. Rubinstein, 'Ion-selective Monolayer Membranes Based upon Self-assembling Tetradentate Ligand Monolayers on Gold Electrodes. 3. Application as Selective Ion Sensors', *Langmuir*, **8**, 1183–1187 (1992).
193. J.-H. Fuhrhop, T. Bedurke, M. Gnade, J. Schneider, K. Doblhofer, 'Hydrophobic Gaps of Steroid Size in a Surface Monolayer Collect 1,2-*trans*-Cyclohexanediol and Glucose from Bulk Water', *Langmuir*, **13**, 455–459 (1997).
194. K.T. Kinnear, H.G. Monbouquette, 'Direct Electron Transfer to *Escherichia coli* Fumarate Reductase in Self-assembled Alkanethiol Monolayers on Gold Electrodes', *Langmuir*, **9**, 2255–2257 (1993).
195. D.D. Schlereth, R.P.H. Kooyman, 'Self-assembled Monolayers With Biospecific Affinity for NAD(H)-dependent Dehydrogenases: Characterization by Surface Plasmon Resonance Combined With Electrochemistry "in situ"', *J. Electroanal. Chem.*, **444**, 231–240 (1998).
196. Y. Sato, H. Itoigawa, K. Uosaki, 'Unidirectional Electron Transfer at Self-assembled Monolayers of 11-Ferrocenyl-1-undecanethiol on Gold', *Bull. Chem. Soc. Jpn*, **66**, 1032–1037 (1993).
197. H.-G. Hong, 'Kinetic Behavior of a Rotating Gold Disk Electrode Modified With a Self-assembled Ferrocenecarboxamidoyl Monolayer Based on Zirconium Phosphonate', *Electrochim. Acta*, **42**, 2319–2326 (1997).
198. A. Doron, M. Portnoy, M. Lion-Dagan, E. Katz, I. Willner, 'Amperometric Transduction and Amplification of Optical Signals Recorded by a Phenoxynaphthacenequinone Monolayer Electrode: Photochemical and pH-gated Electron Transfer', *J. Am. Chem. Soc.*, **118**, 8937–8944 (1996).
199. J.E. Hutchison, T.A. Postlethwaite, C. Chen, K.W. Hathcock, R.S. Ingram, W. Ou, R.W. Linton, R.W. Murray, D.A. Tyvoll, L.L. Chang, J.P. Collman, 'Electrocatalytic Activity of an Immobilized Cofacial Diporphyrin Depends on the Electrode Material', *Langmuir*, **13**, 2143–2148 (1997).
200. J. Zak, H. Yuan, M. Ho, L.K. Woo, M.D. Porter, 'Thiol-derivatized Metalloporphyrins: Monomolecular Films for the Electrocatalytic Reduction of Dioxygen at Gold Electrodes', *Langmuir*, **9**, 2772–2774 (1993).
201. E. Katz, H.-L. Schmidt, 'Gold Electrode Modification With a Monolayer of 1,4,5,8-Naphthalenetetrone as a Four-electron Transfer Mediator: Electrochemical Reduction of Dioxygen to Water', *J. Electroanal. Chem.*, **368**, 87–94 (1994).
202. E. Lorenzo, L. Sanchez, F. Pariente, J. Tirado, H.D. Abruna, 'Thermodynamics and Kinetics of Adsorption and Electrocatalysis of NADH Oxidation With a Self-assembling Quinone Derivative', *Anal. Chim. Acta*, **309**, 79–88 (1995).
203. M. Kunitake, K. Akiyoshi, K. Kawatana, N. Nakashima, O. Manabe, 'Transmembrane Electron Transfer Through a New Hydrophobic Mediator Incorporated in Monolayer Assemblies on Metal Electrodes', *J. Electroanal. Chem.*, **292**, 277–280 (1990).
204. M. Maskus, H.D. Abruna, 'Syntheses and Characterization of Redox-active Metal Complexes Sequentially Self-assembled onto Gold Electrodes via a New Thiol-terpyridine Ligand', *Langmuir*, **12**, 4455–4462 (1996).
205. D.N. Upadhyay, V. Yegnaraman, G.P. Rao, 'Hexacyanoferrate Modification of Gold Electrode Through Monolayer Approach', *Langmuir*, **12**, 4249–4252 (1996).
206. O. Pierrat, C. Bourdillon, J. Moiroux, J.-M. Laval, 'Enzymatic Electrocatalysis Studies of *Escherichia coli* Pyruvate Oxidase, Incorporated into a Biomimetic Supported Bilayer', *Langmuir*, **14**, 1692–1696 (1998).
207. S. Rubin, J.T. Chow, J.P. Ferraris, T.A. Zawodzinski, 'Electrical Communication Between Components of Self-assembled Mixed Monolayers', *Langmuir*, **12**, 363–370 (1996).
208. V. Heleg-Shabtai, E. Katz, I. Willner, 'Assembly of Microperoxidase-11 and Co(II)-Protoporphyrin IX Reconstituted Myoglobin Monolayers on Au-electrodes: Integrated Bioelectrocatalytic Interfaces', *J. Am. Chem. Soc.*, **119**, 8121–8122 (1997).
209. V. Heleg-Shabtai, E. Katz, S. Levi, I. Willner, 'Microperoxidase-11 Functionalized Electrodes: An Active Monolayer Interface for the Electrocatalyzed Reduction of Co(II)-Protoporphyrin IX Reconstituted Myoglobin and for the Generation of Integrated Protein Electrodes for Bioelectrocatalyzed Hydrogenation of Acetylenes', *J. Chem. Soc., Perkin Trans.*, **2**, 2645 (1997).
210. I. Willner, B. Willner, 'Photoswitchable Biomaterials as Grounds for Optobioelectronic Devices', *Bioelectrochem. Bioenerg.*, **42**, 43–57 (1997).
211. C. Miller, M. Gratzel, 'Electrochemistry at w-Hydroxy Thiol Coated Electrodes. 2. Measurement of the Density of Electronic States Distributions for Several Outer-sphere Redox Couples', *J. Phys. Chem.*, **95**, 5225–5233 (1991).
212. A.M. Becka, C.J. Miller, 'Electrochemistry at w-Hydroxy Thiol Coated Electrodes. 3. Voltage Independence of the Electron Tunneling Barrier and Measurements of Redox Kinetics at Large Overpotentials', *J. Phys. Chem.*, **96**, 2657–2668 (1992).
213. C.E.D. Chidsey, 'Free Energy and Temperature Dependence of Electron Transfer at the Metal–Electrolyte Interface', *Science*, **251**, 919–922 (1991).
214. C.J. Miller, 'Heterogeneous Electron-transfer Kinetics at Metallic Electrodes', in *Physical Electrochemistry:*

Principles, Methods, and Applications, ed. I. Rubinstein, Marcel Dekker, New York, 27–79, 1995.

215. E. Laviron, 'General Expression of the Linear Potential Sweep Voltammogram in the Case of Diffusionless Electrochemical Systems', *J. Electroanal. Chem.*, **101**, 19–28 (1979).
216. E. Laviron, 'A.C. Polarography and Faradaic Impedance of Strongly Adsorbed Electroactive Species. Part I. Theoretical and Experimental Study of a Quasi-reversible Reaction in the Case of a Langmuir Isotherm', *J. Electroanal. Chem.*, **97**, 135–149 (1979).
217. J.H. Reeves, S. Song, E.F. Bowden, 'The Application of Square wave Voltammetry to Strongly Adsorbed Quasireversible Redox Molecules', *Anal. Chem.*, **65**, 683–688 (1992).
218. T.M. Nahir, R.A. Clark, E.F. Bowden, 'Linear Sweep Voltammetry of Irreversible Electron Transfer in Surface-confined Species Using Marcus Theory', *Anal. Chem.*, **66**, 2595–2598 (1994).
219. S. Ye, A. Yashiro, Y. Sato, K. Uosaki, 'Electrochemical in situ FT-IRRAS Studics of a Self-assembled Monolayer of 2-(11-Mercaptoundecyl)hydroquinone', *J. Chem. Soc., Faraday Trans.*, **92**, 3813–3821 (1996).
220. J.F. Smalley, S.W. Feldberg, C.E.D. Chidsey, M.R. Linford, M.R. Newton, Y.-P. Liu, 'The Kinetics of Electron Transfer Through Ferrocene-terminated Alkanethiol Monolayers on Gold', *J. Phys. Chem.*, **99**, 13141–13149 (1995).
221. H.O. Finklea, L. Liu, M.S. Ravenscroft, S. Punturi, 'Multiple Electron Tunneling Paths across Self-assembled Monolayers of Alkane Thiols With Attached Ruthenium(II/III) Redox Centers', *J. Phys. Chem.*, **100**, 18852–18858 (1996).
222. K. Weber, L. Hockett, S. Creager, 'Long-range Electronic Coupling Between Ferrocene and Gold in Alkanethiolate-based Monolayers on Electrodes', *J. Phys. Chem. B*, **101**, 8286–8291 (1997).
223. Z.Q. Feng, S. Imabayashi, T. Kakiuchi, K. Niki, 'Long Range Electron-transfer Reaction Rates to Cytochrome *c* Across Long- and Short-chain Alkanethiol Self-assembled Monolayers: Electroreflectance Studies', *J. Chem. Soc., Faraday Trans.*, **93**, 1367–1370 (1997).
224. S. Song, R.A. Clark, E.F. Bowden, M.J. Tarlov, 'Characterization of Cytochrome *c*/Alkanethiolate Structures Prepared by Self-assembly on Gold', *J. Phys. Chem.*, **97**, 6564–6572 (1993).
225. C. Liang, M.D. Newton, 'Ab Initio Studies of Electron Transfer. 2. Pathway Analysis for Homologous Organic Spacers', *J. Phys. Chem.*, **97**, 3199–3211 (1993).
226. C.-P. Hsu, R.A. Marcus, 'A Sequential Formula for Electronic Coupling in Long Range Bridge-assisted Electron Transfer. Formulation of Theory and Application to Alkane Thiol Monolayers', *J. Chem. Phys.*, **106**, 584–598 (1997).
227. J. Cheng, G. Saghi-Szabo, J.A. Tossell, C.J. Miller, 'Modulation of Electronic Coupling Through Self-assembled Monolayers via Internal Chemical Modification', *J. Am. Chem. Soc.*, **118**, 680–684 (1996).
228. S. Terrettaz, A.M. Becka, M.J. Traub, J.C. Fettinger, C.J. Miller, 'w-Hydroxythiol Monolayers at Au Electrodes. 5. Insulated Voltammetric Studies of Cyano/Bipyridyl Iron Complexes', *J. Phys. Chem.*, **99**, 11216–11224 (1995).
229. K. Weber, S.E. Creager, 'Voltammetry of Redox-active Groups Irreversibly Adsorbed onto Electrodes. Treatment Using the Marcus Relation Between Rate and Overpotential', *Anal. Chem.*, **66**, 3164–3172 (1994).
230. H.O. Finklea, D.D. Hanshew, 'Electron-transfer Kinetics in Organized Thiol Monolayers With Attached Penta-ammine(pyridine)ruthenium Redox Centers', *J. Am. Chem. Soc.*, **114**, 3173–3181 (1992).
231. H.O. Finklea, M.S. Ravenscroft, D.A. Snider, 'Electrolyte and Temperature Effects on Long Range Electron Transfer Across Self-assembled Monolayers', *Langmuir*, **9**, 223–227 (1993).
232. S. Terrettaz, J. Cheng, C.J. Miller, 'Kinetic Parameters for Cytochrome *c* via Insulated Electrode Voltammetry', *J. Am. Chem. Soc.*, **118**, 7857–7858 (1996).
233. J. Cheng, S. Terrettaz, J.I. Blankman, C.J. Miller, B. Dangi, R.D. Guiles, 'Electrochemical Comparison of Heme Proteins by Insulated Electrode Voltammetry', *Israel J. Chem.*, **37**, 259–266 (1997).
234. A. Kumar, H.A. Biebuyck, N.L. Abbott, G.M. Whitesides, 'The Use of Self-assembled Monolayers and a Selective Etch to Generate Patterned Gold Features', *J. Am. Chem. Soc.*, **114**, 9188–9189 (1992).
235. A. Kumar, H.A. Biebuyck, G.M. Whitesides, 'Patterning Self-assembled Monolayers: Applications in Materials Science', *Langmuir*, **10**, 1498–1511 (1994).
236. N.L. Abbott, D.R. Rolison, G.M. Whitesides, 'Combining Micromachining and Molecular Self-assembly to Fabricate Microelectrodes', *Langmuir*, **10**, 2672–2682 (1994).
237. Y. Yamamoto, H. Nishihara, K. Aramaki, 'Self-assembled Layers of Alkanethiols on Copper for Protection Against Corrosion', *J. Electrochem. Soc.*, **140**, 436–443 (1993).
238. M. Itoh, H. Nishihara, K. Aramaki, 'A Chemical Modification of Alkanethiol Self-assembled Monolayers With Alkyltrichlorosilanes for the Protection of Copper Against Corrosion', *J. Electrochem. Soc.*, **141**, 2018–2023 (1994).
239. M. Ishibashi, M. Itoh, H. Nishihara, K. Aramaki, 'Permeability of Alkane Thiol Self-assembled Monolayers Adsorbed on Copper Electrodes to Molecular Oxygen Dissolved in 0.5 M Na_2SO_4 Solution', *Electrochim. Acta*, **41**, 241–248 (1996).
240. R. Haneda, H. Nishihara, K. Aramaki, 'Chemical Modification of an Alkanethiol Self-assembled Layer to Prevent Corrosion of Copper', *J. Electrochem. Soc.*, **144**, 1215–1221 (1997).

241. R. Haneda, K. Aramaki, 'Protection of Copper Corrosion by an Ultrathin Two-dimensional Polymer Film of Alkanethiol Monolayer', *J. Electrochem. Soc.*, **145**, 1856–1861 (1998).
242. P.E. Laibinis, G.M. Whitesides, 'Self-assembled Monolayers of *n*-Alkanethiolates on Copper are Barrier Films That Protect the Metal against Oxidation by Air', *J. Am. Chem. Soc.*, **114**, 9022–9028 (1992).
243. I. Rubinstein, J. Rishpon, E. Sabatani, A. Redondo, S. Gottesfeld, 'Morphology Control in Electrochemically Grown Conducting Polymer Films. 1. Precoating the Metal Substrate With an Organic Monolayer', *J. Am. Chem. Soc.*, **112**, 6135–6136 (1990).
244. E. Sabatani, Y. Gafni, I. Rubinstein, 'Morphology Control in Electrochemically Grown Conducting Polymer Films. 3. A Comparative Study of Polyaniline Films on Bare Gold and on Gold Pretreated With *p*-Aminothiophenol', *J. Phys. Chem.*, **99**, 12305–12311 (1995).
245. W.A. Hayes, C. Shannon, 'Nanometer-scale Patterning of Surfaces Using Self-assembly Chemistry. 3. Template-directed Growth of Polymer Nanostructures on Organothiol Self-assembly Mixed Monolayers', *Langmuir*, **14**, 1099–1102 (1998).
246. J. Lukkari, K. Kleemola, M. Meretoja, T. Ollonqvist, J. Kankare, 'Electrochemical Post-self-assembly Transformation of 4-Aminothiophenol Monolayers on Gold Electrodes', *Langmuir*, **14**, 1705–1715 (1998).
247. R.J. Willicut, R.L. McCarley, 'Surface-confined Monomers on Electrode Surfaces. 1. Electrochemical and Microscopic Characterization of *w*-(*N*-Pyrrolyl)alkanethiol Self-assembled Monolayers on Au', *Langmuir*, **11**, 296–301 (1995).
248. C.N. Sayre, D.M. Collard, 'Self-assembled Monolayers of Pyrrole-containing Alkanethiols on Gold', *Langmuir*, **11**, 302–306 (1995).
249. D.B. Wurm, S.T. Brittain, Y.-T. Kim, 'Organic Monolayers as Nucleation Sites for Epitaxial Growth. 1. Electrochemical Polymerization of *N*-Alkylpyrrole', *Langmuir*, **12**, 3756–3758 (1996).
250. D.B. Wurm, K. Zong, Y.-H. Kim, Y.-T. Kim, M. Shin, I.C. Jeon, 'Electrochemical Quartz Crystal Microbalance Studies of Polypyrrole Growth on Self-assembled Nucleation Sites', *J. Electrochem. Soc.*, **145**, 1483–1488 (1998).

Surface Analysis for Electrochemistry: Ultrahigh Vacuum Techniques

Manuel P. Soriaga
Texas A&M University, Texas, USA

The study of processes that transpire at heterogeneous interfaces is an exceedingly difficult proposition. No single experimental technique can ever hope to unravel all the nuances of heterogeneous reactions; hence, in surface science, the use of multiple complementary methods is not uncommon. Ultrahigh vacuum electrochemistry (UHV/EC) is a term ascribed to the approach that rests upon the integration of classical electrochemical methods with surface-sensitive analytical techniques; this strategy parallels that successfully implemented in the study of gas–solid heterogeneous catalysis. The unique surface sensitivity of the techniques adopted emanates from the

For references see page 10139

use of particles (e.g. ions or electrons) that serve to interrogate the outermost layer(s) of the electrode. This surface sensitivity is tempered by the requirement that the analysis be performed in an environment (outside the electrochemical cell) that does not impede the mean-free paths of the probe particles. Since its inception in the early 1970s, more than a thousand UHV/EC-based studies have been published; most of the work involved polycrystalline materials and focused on the elemental composition at the electrode surface. While its importance in the study of polycrystalline surfaces cannot be trivialized, the greater value of UHV/EC appears to be in its ability to help resolve fundamental issues that intertwine interfacial structure and composition with electrochemical reactivity. It is in this context that the present review is written.

A complete mechanism of an electrochemical reaction must incorporate all the physical and chemical interactions that arise between an electrified surface and its environment. The extent of such interactions depends upon several factors such as solvent, supporting electrolyte, electrode potential, reactant concentration, electrode material and surface crystallographic orientation. The traditional approach is based upon a thermodynamic treatment of the interface and its response to external perturbations. Interpretation of the results relies on phenomenological models of the interface. Although a thermodynamic treatment cannot be ignored, the need for an atomic-level view has long been realized. One approach[1–12] *towards the establishment of an atomic-level description parallels that successfully implemented in the study of gas–solid heterogeneous catalysis; it rests upon the integration of classical electrochemical methods with surface-sensitive analytical techniques. The analytical methods that exhibit surface sensitivity are based upon the mass-selection and/or energy-discrimination of electrons, ions, atoms or molecules scattered from solid surfaces. These particles have shallow escape depths; hence the information they bear is characteristic of the near-surface layers. Their short mean-free paths, however, necessitate a high vacuum (*$<10^{-6}$ *torr) environment. The application of such surface techniques to electrochemistry requires that the analysis be performed outside the electrochemical cell. The possibility of structural and compositional changes that accompany the removal of the electrode from solution is the major concern in UHV/EC studies. Although a myriad of surface analytical techniques is currently available, those actually employed in UHV/EC have been limited to low-energy electron diffraction (LEED), Auger electron spectroscopy (AES), X-ray photoelectron spectroscopy (XPS), high-resolution electron energy loss spectroscopy (HREELS), reflection high-energy electron diffraction (RHEED), work-function changes, and thermal desorption mass spectrometry (TDMS).*

List of selected abbreviations appears in Volume 15

While most vacuum-based analytical methods do not require single-crystal surfaces, the use of uniform (monocrystalline) surfaces is a necessary aspect for fundamental studies. The low-index crystallographic faces [(100), (110) and (111)] have been widely used because of their low free energies, high symmetries, and relative stabilities. In addition, it may be possible to reconstruct the overall behavior of polycrystalline electrodes from the individual properties of the low-index planes.[1–4] *A handful of procedures for the preparation and preservation of well-defined single-crystal surfaces have been described.*[5–8] *The verification or identification of initial, intermediate, and final interfacial structures and compositions is an essential ingredient in electrochemical surface science.*

1 SURFACE CHARACTERIZATION TECHNIQUES

1.1 Surface Spectroscopy with Low-energy Electrons

The main difficulty in surface analysis lies in the exceedingly low population of atoms at the interface (10^{15} atoms cm^{-2}) relative to that in the bulk (10^{23} atoms cm^{-3}). Experiments intended to examine the physical and chemical properties of surfaces must employ methods that interact only with the interfacial layers; the majority of such techniques take advantage of the surface sensitivity of *low-energy* electrons. The surface sensitivity arises because the mean-free-path of an electron through a solid is dependent upon its kinetic energy. As shown in the "universal curve" reproduced in Figure 1, the electron mean-free-path falls to a minimum (0.4 to 2 nm) when the kinetic energy is between 10 and 500 eV. That is, information derived from experimental techniques based upon low-energy-electron incidence onto and/or emergence from surfaces will exclusively be from the topmost surface layers.

A solid surface subjected to a beam of electrons of incident (primary) energy E_p gives rise to the appearance of backscattered and emitted electrons; a plot of the number of electrons $N(E)$ as a function of energy E of these electrons is shown in Figure 2.[9–15] This spectrum can be divided into four regions according to the origin of the scattered electrons: (i) true secondary electrons, created as a result of inelastic interactions between the incident and bound electrons, give rise to the prominent broad band at the lower end of the spectrum; (ii) Auger electrons emitted and primary electrons scattered due to interactions with electronic states in the solid account for the small peaks in the medium-energy range of the spectrum; (iii) primary electrons scattered upon interactions with the vibrational states of the surface yield

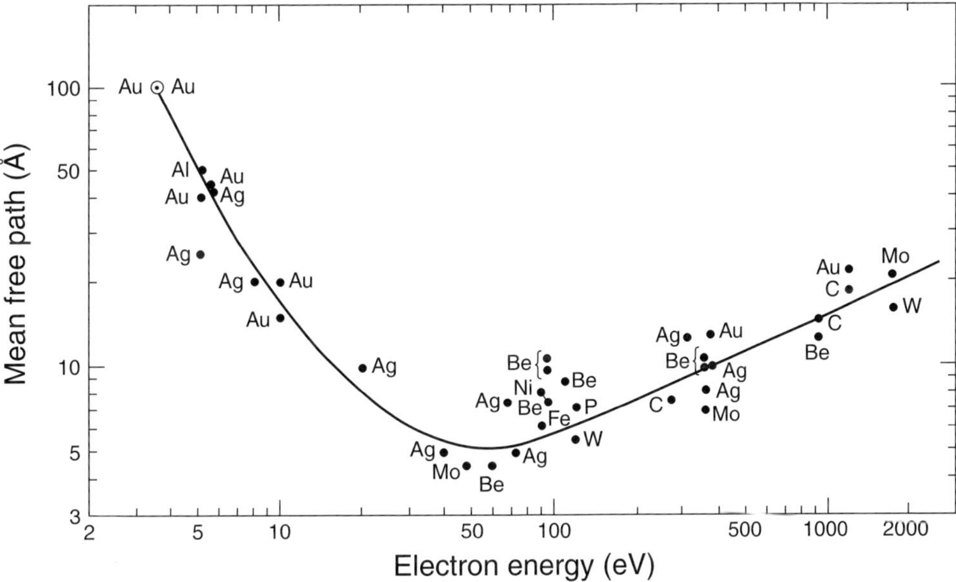

Figure 1 "Universal curve": electron mean free path as a function of electron kinetic energy. (Reproduced by permission from Somorjai.[9])

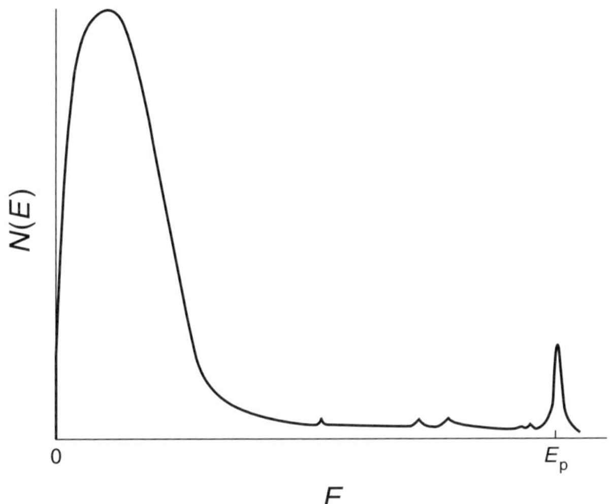

Figure 2 Experimental number [$N(E)$] of scattered electrons of energy E as a function of electron energy. (Reproduced by permission from Somorjai.[9])

peaks close to the elastic peak; (iv) primary electrons scattered elastically, which comprise only a few percent of the total incident electrons, appear at E_p. The spectral regions (ii) to (iv) have been exploited in modern surface analysis.

1.2 Low-energy Electron Diffraction

In this method,[9–15] the surface is irradiated with a monoenergetic beam of electrons and the elastically backscattered electrons are collected onto a phosphor screen. The virtue of LEED as a surface structural technique is a result of the low kinetic energies used (50 to 500 eV): (i) the electron mean free path is at a minimum; (ii) the de Broglie wavelengths, $\lambda_e \approx (150/E_e)^{1/2}$ (where E_e is in electronvolts and λ_e is in angstroms) correspond to crystal lattice dimensions; and (iii) electron backscattering is strong which minimizes incident electron fluxes at, and subsequent scattering from, nonsurface layers. In LEED, the presence (or absence) of diffraction patterns on the fluorescent screen is a consequence of the order (or disorder) of the atomic arrangements at the surface.

The locations of the diffraction spots define the reciprocal lattice of the real surface. The real-space surface structure can be reconstructed from the reciprocal lattice vectors.[9–15] The *coherence width* of electron beam sources in LEED is typically 10 nm. That is, sharp diffraction features appear only if well-ordered domains are *at least* 10 nm × 10 nm in size; diffraction from smaller domains leads to beam broadening. The analysis of LEED data based solely upon the geometry of the diffraction spots provides information on the periodicity of the electron scatterers on the surface. In some favorable instances, other information such as adsorbate coverages or point group symmetries can also be inferred. However, the actual location of the atoms within the surface lattice cannot be determined without an analysis of the intensities of the diffracted beams. Surface crystallography by LEED relies upon a comparison of the measured diffraction intensities with those calculated for model structures. LEED simulations are difficult because of multiple electron scattering.[9–15]

For references see page 10139

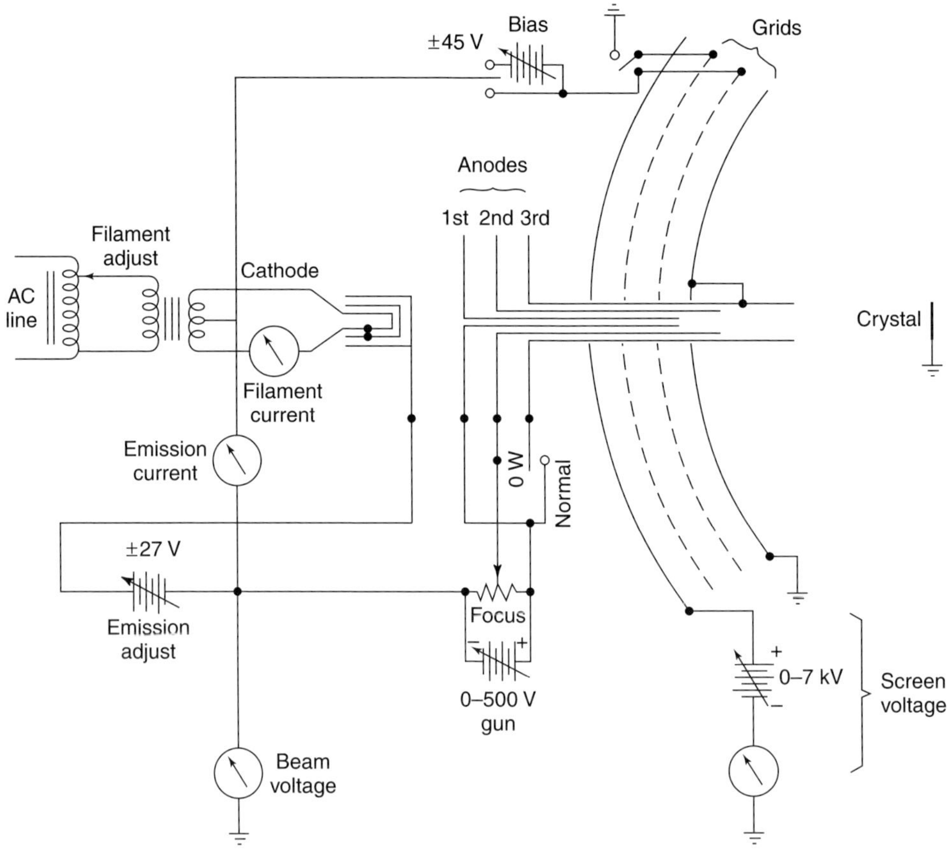

Figure 3 Schematic diagram of a LEED apparatus. (Reproduced by permission from Adamson.[17])

Two schemes are employed for the notation of interfacial adlattice structures. The matrix notation, which is applicable to any system, is based upon the relationship between the real-space lattice vectors of the *adsorbate* mesh (two-dimensional lattice) and the *substrate* mesh. For example, if the adsorbate unit cell vectors $\mathbf{a}'$ and $\mathbf{b}'$ are related to those of the substrate mesh, $\mathbf{a}$ and $\mathbf{b}$, according to Equations (1) and (2):

$$\mathbf{a}' = m_{11}\mathbf{a} + m_{12}\mathbf{b} \tag{1}$$

$$\mathbf{b}' = m_{21}\mathbf{a} + m_{22}\mathbf{b} \tag{2}$$

Then the matrix $\mathbf{M}$ defined by the coefficients m_{ij}, $\mathbf{M} = \begin{pmatrix} m_{11} & m_{12} \\ m_{21} & m_{22} \end{pmatrix}$, denotes the real-space surface structure. The other method, known as the Wood notation,[16] is more widely used but is applicable only if the angle between $\mathbf{a}'$ and $\mathbf{b}'$ is the same as that between $\mathbf{a}$ and $\mathbf{b}$. The adlattice structure is labeled using the general form $(n \times m)R\phi^\circ$ or $c(n \times m)R\phi^\circ$, where c designates a centered unit cell, $R\phi^\circ$ the angle of rotation of the adsorbate unit cell relative to the substrate unit mesh, and n and m are scale factors relating the adsorbate and substrate unit cell vectors $|\mathbf{a}'| = n|\mathbf{a}|$ and $|\mathbf{b}'| = m|\mathbf{b}|$.

A schematic diagram of a typical LEED instrument is shown in Figure 3.[17] The LEED "optics" consists of a phosphor-coated hemispherical screen at the center of which is a normal-incidence, electrostatically focused electron gun. In front are three concentric grids; the outer grid is held at ground potential, while the inner two are maintained at a voltage just below that of the electron gun in order to reject *inelastically* backscattered electrons. The elastically diffracted electrons which pass through the suppressor grids are accelerated onto the fluorescent screen by a 5-kV potential applied to the screen. For quantitative intensity measurements, additional provisions (e.g. Faraday cup, spot photometer, or digital camera) will be necessary.

A typical LEED pattern, that for an iodine-coated Pd(111) surface, is shown in Figure 4 along with the suggested real-space adlattice structure, $\mathrm{Pd}(111) - (\sqrt{3} \times \sqrt{3})R30^\circ - \mathrm{I}$.

1.3 Reflection High-energy Electron Diffraction

RHEED[2,12–15] represents an alternative to LEED. The principal difference between the two techniques is that, whereas low-energy electrons are utilized in LEED,

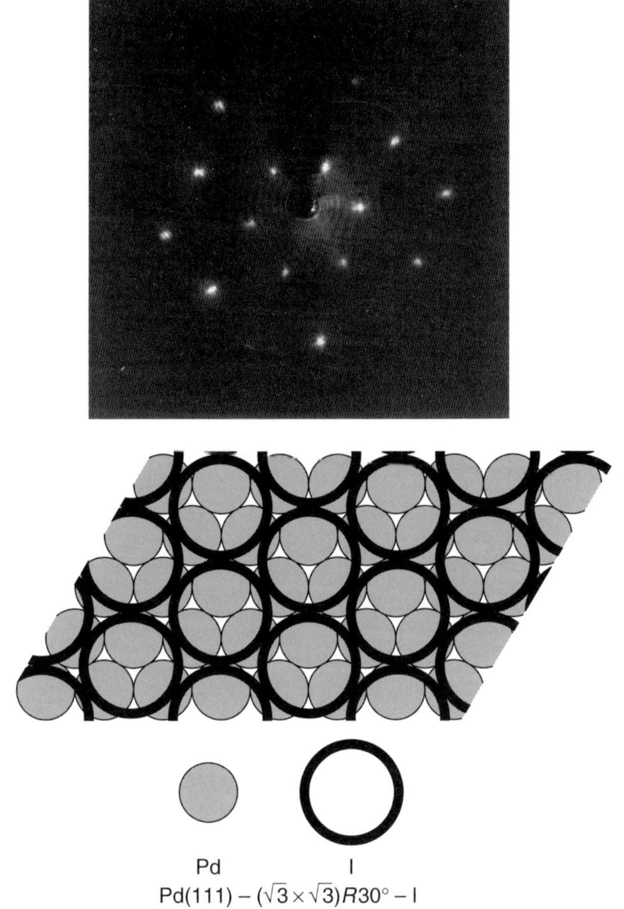

Figure 4 LEED pattern for a Pd(111) electrode coated with iodine from dilute aqueous NaI; also shown is the postulated adlattice structure.

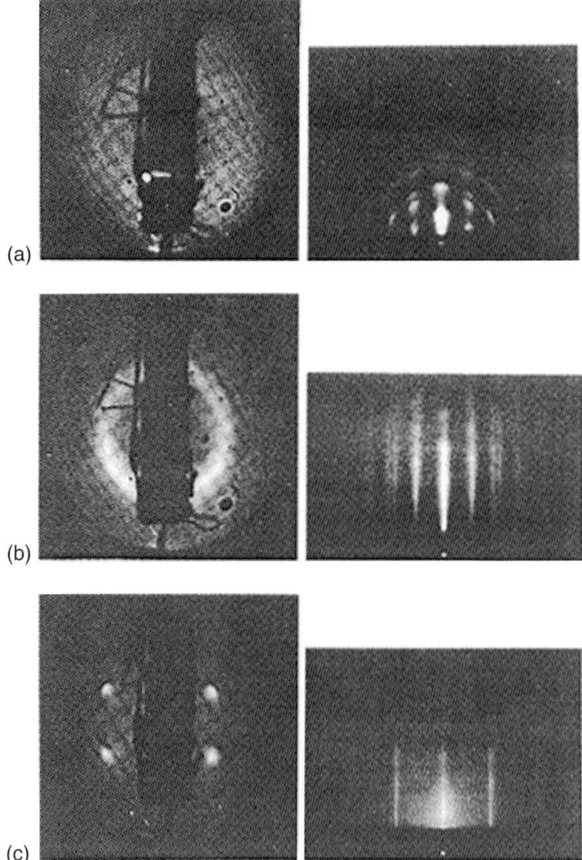

Figure 5 LEED and RHEED patterns for Au vapor deposited onto (a) glass at room temperature, (b) glass at 400 °C, and (c) mica at 360 °C. (Reproduced by permission of Zei et al.[18])

high-energy (30 to 100 keV) electrons are employed in RHEED. At such high energies, the mean-free-paths of the incident electrons are long (10 to 100 nm) and elastic scattering is predominantly in the forward direction. Hence, to afford the required surface sensitivity, RHEED experiments are performed at small angles ($<5°$) of incidence and diffraction. Energy filtering is not a requirement in RHEED because of the large energy difference between the elastically and inelastically scattered electrons; post-acceleration is likewise unnecessary as the primary electrons are sufficiently energetic to produce fluorescence on the phosphor screen.

Figure 5 shows LEED and RHEED patterns of gold films evaporated on glass and on mica;[18] the film deposited on glass (at room temperature) is rough but that on mica (at elevated temperatures) is smooth and well-ordered. The ordered-surface diffraction is manifested in LEED by distinct spots and by sharp streaks in RHEED. This difference is due to the fact that, in reciprocal space, the surface layer is represented by perpendicularly-oriented *rods* that pass through the points in the reciprocal net. In normal-incidence LEED, only a section *perpendicular* through these reciprocal lattice rods is displayed, leading to the spot pattern. In grazing-incidence RHEED, a section *parallel* through the rods is displayed, resulting in a streak pattern. For the same reason, RHEED is unable to detect changes in periodicity along the plane of incidence. Hence, if the full two-dimensional periodicity is to be established, the sample needs to be rotated.

RHEED is most useful in studies related to the structure and morphology of thin films and surface coatings. It is possible to continuously monitor film formation under deposition conditions since the front of the sample is unimpeded by either the electron source or analyzer. In view of its low-scattering-angle geometry, RHEED is sensitive to surface asperities.

1.4 Auger Electron Spectroscopy

AES is one of the more widely used techniques for surface elemental analysis.[9–15] In the Auger process, illustrated

For references see page 10139

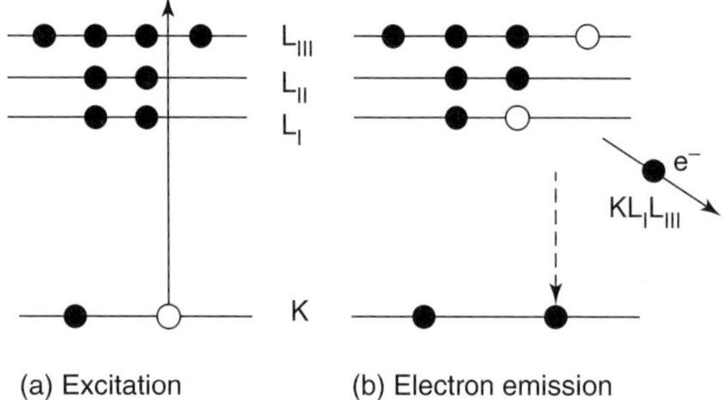

Figure 6 Schematic diagram of the Auger emission process: (a) excitation; (b) electron emission. The core (K) level electron is that which is ejected in XPS, while the Auger electron is the ejected electron designated as KL_IL_{III}.

schematically in Figure 6, a core (K) level electron is emitted when a beam of electrons, typically with energies between 2 and 10 keV, is impinged onto the sample surface. In the decay process, an electron in an upper (L_I) level falls into the vacant core level and another electron in a different upper (L_{III}) level is ejected; the latter is the Auger electron. This particular Auger process is labeled KL_IL_{III} in order to specify the energy levels involved. The kinetic energy of the Auger electron is dependent upon the binding energies of the K, L_I, and L_{III} electrons but not on E_p, as shown in Equation (3):

$$E_{KL_IL_{III}} = E_K - E_{L_I} - E_{L_{III}} - e\varphi_{sp} \tag{3}$$

where e is the electronic charge and φ_{sp} the spectrometer work function. The exact application of Equation (3) must realize that the energy difference is actually between singly ionized and doubly ionized states. Nevertheless, $E_{KL_IL_{III}}$, as obtained from empirical spectra, is characteristic of a given atom, which affords AES its element-specificity. It should be noted that, although the incident electrons are of high energies, AES is still a surface-sensitive method because the *emitted* Auger electrons are of much lower energies that fall within the minimum of the "universal curve" (Figure 1).

An inherent difficulty in AES is that the Auger emission peaks are of very low intensities superimposed on a large (secondary emission) background. The usual approach to negate the background is a combination of electron-energy analysis with suitable modulation techniques. In multi-technique instrumentation that includes LEED, a retarding field analyzer (RFA) is most affordable since it makes use of the LEED optics. The sensitivity and resolution of an RFA, however, leave much to be desired. A cylindrical mirror analyzer (CMA), Figure 7, is thus more widely used. Energy analysis with a CMA is achieved by a negative ramp voltage applied to the outer cylinder while the inner cylinder is held at ground potential. Only electrons of the appropriate energy

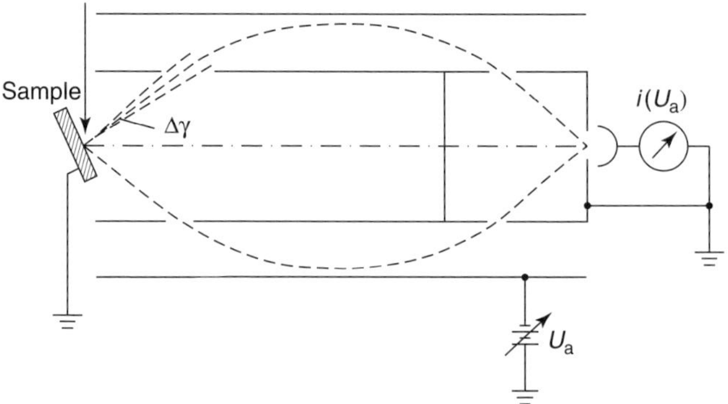

Figure 7 Schematic cross-section of a CMA. U_a is the potential applied between the two coaxial cylindrical electrodes, $\Delta\gamma$ is the angular spread of the electrons at the entrance slit, and $i(U_a)$ is the current at the exit aperture. (Reproduced by permission from Ertl and Kuppers.[11])

can pass unhindered through the CMA and into the detector, which is usually a channel electron multiplier. The pass energy of the CMA is modulated and then synchronously demodulated with a lock-in amplifier; the resultant spectrum is a derivative spectrum, $dN(E)/dE$, devoid of the large background. Newer instruments employ software-based modulation and filtering. Higher resolution can be achieved via a double-pass CMA or by the use of a cylindrical hemisphere analyzer (CHA), which is a double-focusing analyzer (see below).

A typical Auger spectrum is shown in Figure 8 for a Pd(100) surface exposed *serially* to dilute aqueous solutions of NaCl, NaBr and NaI. For quantitative and/or molecular compositional analysis, the derivative spectrum is difficult to process. An alternative approach, the collection of nonderivative Auger spectra, involves pulse counting electronics or direct current measurements; spectra thus generated can be deconvoluted by a fast Fourier transform algorithm[19] to obtain information on chemical shifts and lineshapes. Changes in Auger lineshapes reflect modifications in the valence band density of states.[20,21]

The use of derivative Auger spectra for the determination of adsorbate surface coverages has been the subject of numerous studies.[22] One method, the first to be used in surface electrochemical studies, makes use of Equation (4):[23–25]

$$\Gamma_a = \frac{I_a}{I_p\phi_c G_a} \tag{4}$$

where Γ_a is the absolute packing density (mol cm^{-2}) of the adsorbate, I_a the Auger current for the adsorbate, I_p the primary beam current, ϕ_c the measured collection efficiency of the Auger spectrometer, and G_a is the calculated Auger electron yield factor.[26] I_a is obtained using Equation (5) by double integration of the adsorbate

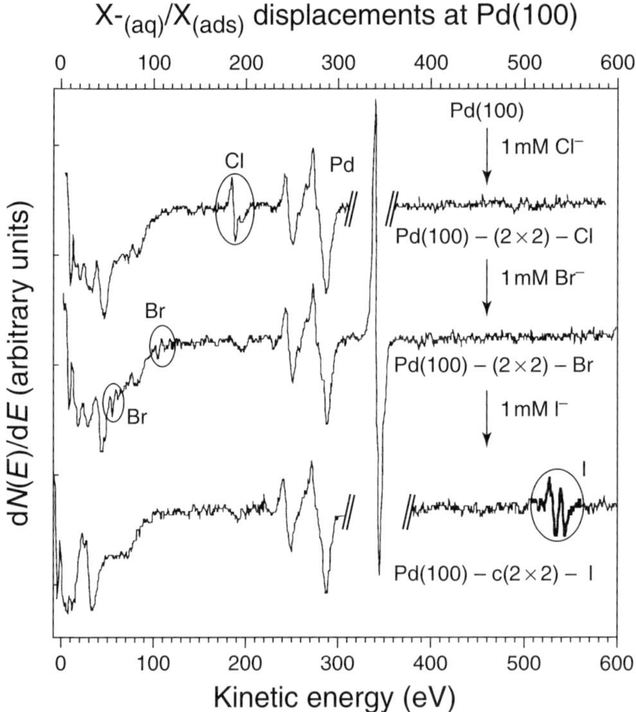

Figure 8 Auger electron spectra for a Pd(100) electrode after sequential exposure to 1 mM NaCl, NaBr, and NaI.

second-harmonic amplitude A_2 corrected for the clean-surface signal A_{2c}:[23]

$$I_a = \frac{4}{k^2}\int_0^{E_p}\int_0^{E}(A_2 - \phi_b A_{2c})\,dE'\,dE \qquad (5)$$

where ϕ_b is the observed attenuation of the substrate signal by the adsorbed species, and k is the modulation amplitude. For simple adsorbates for which well-defined adlayers are available for calibration purposes, Equation (4) can be expressed in purely empirical terms by Equation (6):[24,25]

$$\Gamma_a = \left(\frac{I_a}{I_M^0}\right)\frac{1}{B_a} \qquad (6)$$

where I_M^0 is the Auger signal for the clean substrate, and B_a is a calibration factor.

1.5 X-ray Photoelectron Spectroscopy

This technique, originally referred to as electron spectroscopy for chemical analysis (ESCA),[27] is the other widely used method for surface compositional analysis. In XPS, the solid surface is irradiated with X-rays which results in the ejection of a core-level electron. The kinetic energy E_{Kin} of the emitted photoelectron is given by Equation (7):

$$E_{Kin} = h\nu - E_B - e\varphi_{sp} \qquad (7)$$

where $h\nu$ is the energy of the incident X-ray photon and E_B, within the framework of Koopman's theorem,[11,12] is the binding energy of the core-level electron. For studies with metals, it is convenient to reference E_B with respect to the Fermi level; the latter is determined from the onset of electron emission at the highest kinetic energy.

The XPS source consists of an anode material which, upon bombardment by high-energy electrons, emits X-rays. The emitted radiation can be rendered monochromatic either by Bragg diffraction or by the use of the characteristic emission lines of the anode; for Mg and Al anodes, these lines are 1253.6 eV (Mg Kα) and 1486.6 eV (Al Kα), respectively. It is important to note that, for $E_B < 700$ eV, the E_{Kin} of the ejected photoelectron will *not* fall within the minimum of the universal curve. In such a case, the surface sensitivity of XPS becomes minimal. This can be remedied either by the use of near-grazing incidence or by the detection of electrons emitted at small angles with respect to the surface plane.

To afford the high resolution required for meaningful XPS studies, energy analysis is usually based upon a CHA (Figure 9). A potential difference U_k is applied across the inner and outer hemispheres of radii R_1 and R_2, respectively. Electrons of energy eV_e are focused at

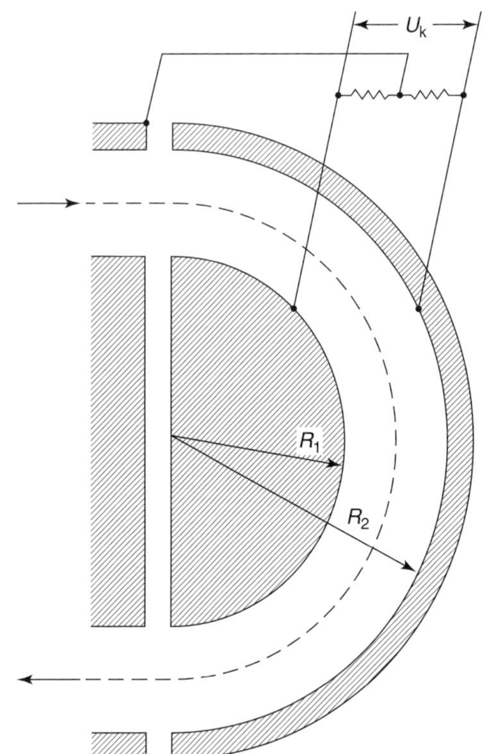

Figure 9 Schematic cross-section of the concentric hemisphere analyzer. (Reproduced by permission from Ertl and Kuppers.[11])

For references see page 10139

the exit slit only if Equation (8) is satisfied:

$$U_k = V_e\left(\frac{R_2}{R_1} - \frac{R_1}{R_2}\right) \tag{8}$$

The CHA is double focusing since it focuses in two planes. The resolution of a CHA can be improved significantly by electron pre-retardation via either an RFA or a retarding lens system. XPS has also been performed with a double-pass CMA. Detection is typically done with a channel electron multiplier. Due to inherently weak intensities, signal averaging and other data processing routines are always employed.

Qualitative elemental analysis of sample surfaces relies upon the comparison of measured E_B values with those for reference materials. In this regard, it is critical to note that XPS of an insulating layer (e.g. a thick oxide film) will result in a net loss of electrons and, hence, an excess of positive charge at the interface; such positive-charge excess leads to shifts in the experimental E_B values. This problem can be effectively countered by irradiation of the surface, at low currents, with low-energy electrons.

Quantitative analysis is based on the fact that the ionization cross-section of a core electron is essentially independent of the valence state of the element. Hence, the intensity will always be proportional to the number of atoms within the detected volume. For quantitative purposes, the area under the background-corrected peak is taken as the intensity. The intensity is a complicated function of several parameters some of which can be eliminated by the use of a reference state analyzed under identical conditions as the sample. If the spectrometer has a small aperture and the surface is uniformly irradiated, the equation for the intensity can be simplified to Equation (9):[11,28]

$$I_A = \sigma_A D L_A J_0 N_A \lambda_M G_1 \cos\theta_1 \tag{9}$$

where I_A is the integrated peak intensity for an element A, σ_A the photoionization cross-section, D the spectrometer detection efficiency, L_A the angular asymmetry of the emitted intensity with respect to the angle between the incidence and detection directions, J_0 the flux of primary photons, N_A the density of atoms A, λ_M the escape depth of the photoelectron, G_1 the spectrometer transmission, and θ_1 is the angle between the surface normal and the detection direction.

XPS is complementary to AES. The ionization cross-section for an Auger process decreases with E_B; the latter, in turn, increases with atomic number Z. Hence, AES is most sensitive for $Z < 45$ elements. The one distinct advantage that XPS offers is in the determination of oxidation states. This is possible because E_B of the core-level electrons is influenced by changes in the chemical environment. In principle, identical information can be obtained from Auger peak energy shifts and lineshapes; in practice, however, the task is nontrivial.

XPS spectra of a smooth polycrystalline Ir foil electrode surface before and after pretreatment with iodine, are shown in Figure 10; the peaks at 62.1 and 65.2 eV represent surface iridium oxide.

1.6 High-resolution Electron Energy Loss Spectroscopy

Almost all of the incident electrons impinged at a solid surface undergo inelastic events that cause them to be backscattered at energies lower than the primary energy E_p. If E_l is the energy lost to the surface, peaks would appear in the energy distribution spectrum (Figure 2) at energies $\Delta E = E_p - E_l$. Such peaks, commonly referred to as electron energy loss peaks, are of several types according to the origin of the energy loss. For vibrational excitations, the energy losses are small since

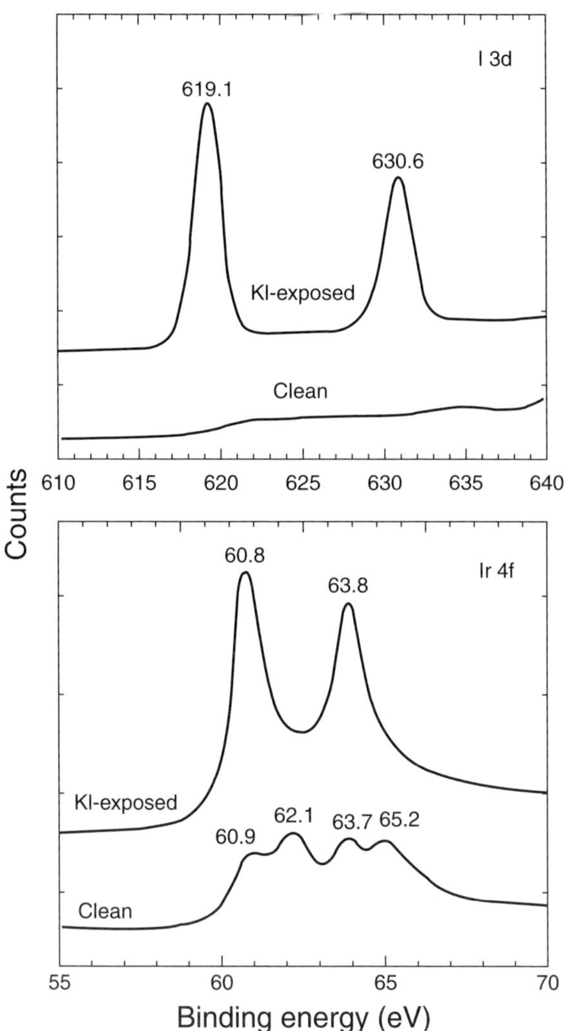

Figure 10 X-ray photoelectron spectra for a thermally annealed polycrystalline Ir foil electrode before and after pretreatment with iodine.

$E_{vib} < 4000\,cm^{-1} < 0.5\,eV$. Hence, the loss peaks due to vibrational interactions lie close to the elastic peak and can be observed only if the energy loss measurements are done at high resolution.

There are two mechanisms that give rise to vibrational HREELS:[9,11,12,29,30] dipole scattering and impact scattering. In dipole scattering, the incident electron interacts with the oscillating electric dipole moment induced by the vibration of surface species. Such interactions occur at long range and can be described either classically or quantum mechanically. Two important selection rules apply for surface dipole scattering: (i) only vibrations whose dynamic dipole moments perpendicular to the surface are nonzero contribute to HREELS spectra. This selection rule is the same as that for infrared reflection–absorption spectroscopy (IRAS);[31–34] (ii) the intensity distribution with respect to scattering angle is sharply peaked in the specular direction; that is, dipole-scattering loss peaks disappear when the backscattered electrons are collected at nonspecular angles.

Impact scattering, which can only be treated quantum mechanically, involves exceedingly short-range interactions between the incident electron and the oscillator at the surface. The surface dipole selection rules do not apply to impact scattering. Theoretical considerations have predicted, and experimental studies have confirmed, the following properties of this type of scattering mechanism:[29] (i) impact scattering vanishes in the specular direction; (ii) impact scattering is more likely to prevail at higher energies; (iii) strong dipole scatterers are weak impact scatterers; conversely, weak dipole scatterers are strong impact scatterers.

HREELS is an extremely sensitive technique. The limit of detection for strong dipole scatterers such as CO can be as low as 0.0001 monolayer; for weak scatterers such as hydrogen, the limit is 0.01 monolayer. In comparison, IRAS for chemisorbed CO, a strong absorber, is restricted to coverages above 0.1 monolayer. HREELS studies of non-CO organic molecules adsorbed at atomically smooth electrode surfaces are abundant; similar experiments using IRAS are meager. The energy accessible to HREELS ranges from $100\,cm^{-1}$ to $4000\,cm^{-1}$; IRAS detectors are not useful below $600\,cm^{-1}$. On the other hand, IRAS has higher resolution and can be utilized for experiments under electrochemical conditions.[33]

Figure 11 shows a schematic diagram of an HREELS spectrometer.[35] The energy of incident electrons can be varied from 1 to 10 eV. To afford high resolution, energy monochromation and analysis are done either with a CMA, cylindrical deflector, or spherical deflector analyzers in combination with retarding field optics. Off-specular collection of the backscattered electrons is afforded by rotation of either the sample or the analyzer. Due to extremely low signals (10^{-10} A), continuous dynode electron multiplier detectors are widely used.

An example HREEL spectrum, that of 2,5-dihydroxybenzenesulfonate chemisorbed on a Pd(100) surface, is shown in Figure 12.

1.7 Work Function Measurements

The work function is the energy required to remove a Fermi-level electron from the bulk to the vacuum just outside the surface.[36,37] In the bulk, the electron has

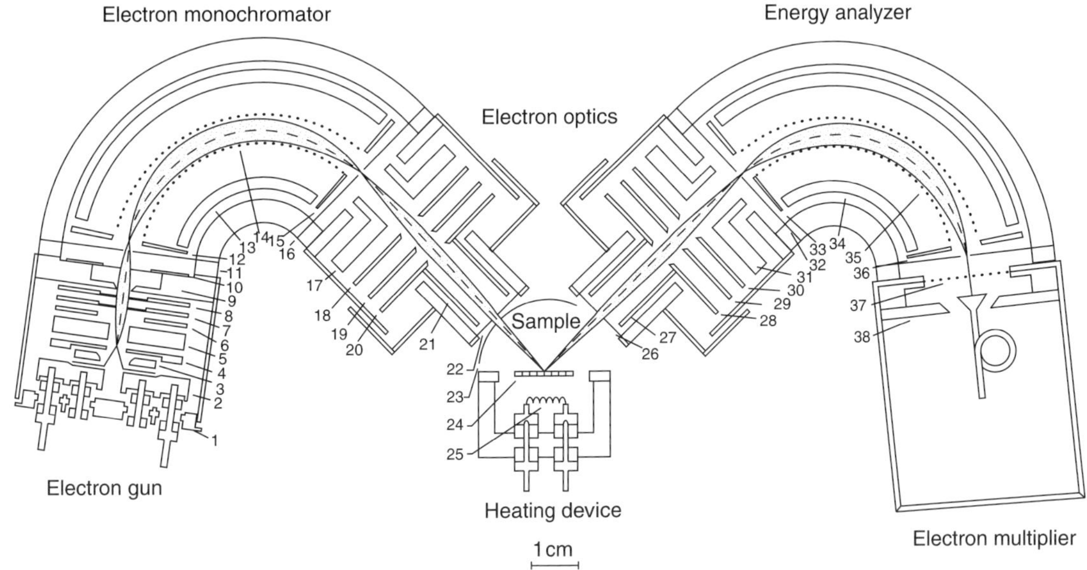

Figure 11 Schematic drawing of a tandem cylindrical deflector spectrometer used for HREELS. (Reproduced by permission from Froitzheim.[35])

For references see page 10139

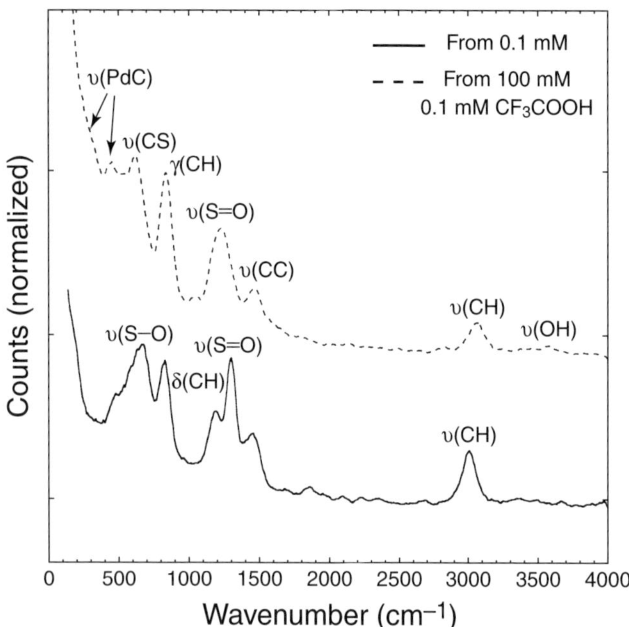

Figure 12 High-resolution electron energy loss spectra of the potassium salt of hydroquinone sulfonate chemisorbed onto Pd(100) from two different concentrations.

an electrochemical potential $\bar{\mu}_e$ which is equal to the Fermi energy. Once the electron is removed from the crystal and is at rest, its electrochemical potential is simply the electrostatic potential energy $-e\Phi_0$, where Φ_0 is the electrostatic potential just outside the surface (where the electron does not feel its image charge). The work function is thus defined by Equation (10):

$$\phi = -e\Phi_0 - \bar{\mu}_e = -e(\Phi_0 - \Phi_i) - \mu_e \qquad (10)$$

It should be noted that the Fermi energy is not a property only of the bulk, because it contains the electrostatic potential Φ_i inside the metal, which is determined by the surface dipole layers.

In surface adsorption studies, only the surface dipole part $e(\Phi_0 - \Phi_i)$ changes, and consequently, the change in work function is equal to the change in this quantity; absolute work functions are less important. For a clean metal surface, the exponential decay of the wave function into the vacuum (electron "overspill") creates the surface double layer. The dipole has the negative end outward, and is dependent on the surface crystallography.

Work function measurements lead to useful correlations between $\Delta\phi$ and the nature of adsorption. For example, the adsorption of an electronegative atom such as oxygen changes the surface dipole and renders the outside of the crystal more negative, thereby increasing the work function. Conversely, an electropositive adsorbate such as Cs increases the work function. For atomic adsorbates, the sign of $\Delta\phi$ is therefore correlated with the direction of charge transfer: positive $\Delta\phi$ is associated with adsorbate-to-substrate charge transfer and negative $\Delta\phi$ with substrate-to-adsorbate charge transfer. In the case of molecular adsorbates, it must be realized that the dipole moment of the molecule is a more important contributor to $\Delta\phi$ than the surface chemical bond.

Experimental measurements of ϕ and $\Delta\phi$ have been based upon the diode method, field emission, contact potential difference, and the photoelectric effect; the latter two are more common. The photoelectric method, which measures absolute values of the work function, is based upon the determination of the threshold energy $h\nu_0$ for photoelectron ejection; the work function is then calculated from the equation $h\nu_0 = e\phi$. The contact potential difference method, which monitors changes in work function, depends on the measurement of the potential difference between two plates in electrical contact. If one of the plates is used as a reference of constant work function, $\Delta\phi$ at the other plate is manifested as a change in the contact potential. The most common way of measuring the change in contact potential uses a vibrating tip close to the surface as the reference plate (the Kelvin probe method). In its balanced condition there is an electric-field-free region between the sample and the tip, and no induced alternating current flows in the circuit. Adsorption leads to a momentary field and, hence, an alternating current. Electronic feedback is used to adjust the potential on the sample until the field-free condition again applies.

1.8 Temperature Programmed Desorption

Thermal desorption techniques[9–15] exploit the fact that species adsorbed on a surface will desorb at a rate which increases with temperature; the temperature dependence of the desorption rate yields data on desorption energies which, for most cases, yield information on adsorption binding energy states. Thermal desorption can also be used to obtain surface coverages and, in combination with mass spectrometry, determine the identity of species desorbed from the surface.

Temperature-induced desorption methods can be classified according to whether the rise in temperature is fast (flash desorption) or gradual (temperature programmed desorption (TPD)). In flash desorption, the desorption rate is much greater than the rate at which desorbed gas is pumped out of the system; in this case, the desorption of a given binding state is marked by a plateau in the pressure–temperature curve. In TPD, the slow heating (desorption) rate allows the evolved gases to be pumped out; as a result, the desorption of a particular binding state appears as a peak instead of a plateau in the desorption curve. The most common method of extracting the activation energy for desorption (E_d) from the peak temperature (T_p) involves the use of the so-called Redhead equations.[38] Other approaches exist that, while more rigorous, are also more difficult.[39] The desorption activation energy is related to the heat of adsorption ΔH_{ads} as shown in Equation (11):

$$E_d = E_a + \Delta H_{ads} \quad (11)$$

where E_a is the activation energy for adsorption. Several instances can be found in which the adsorption is nonactivated; for such cases, E_d is equal to ΔH_{ads}.

The use of a mass spectrometer to monitor species emitted from the surface simplifies TPD measurements. A comparatively inexpensive means for mass detection is afforded by a quadrupole mass analyzer, a schematic diagram of which is shown in Figure 13.[40] This mass filter consists of four parallel rod-shaped electrodes arranged at the apices of a diamond. The existence of an appropriately varying electrical field between the pairs of opposite electrodes will cause all ions to impact on the rods during transit except those of a particular mass-to-charge ratio m/z.

In TDMS, the substrate is positioned as close as possible to the mass spectrometer; provision must be made to minimize temperature gradients at the sample surface. The temperature is monitored by a thermocouple wire placed in direct contact with the crystal. Problems associated with degassing from parts of the sample manipulator close to the crystal can be solved by masking the mass spectrometer with a small aperture such that only line-of-sight detection is possible. Microprocessor control allows multiplexed data acquisition.

A thermal desorption mass spectrum is shown in Figure 14 for CO desorbed from a Pd(111)–c(4 × 2)–CO adlattice.

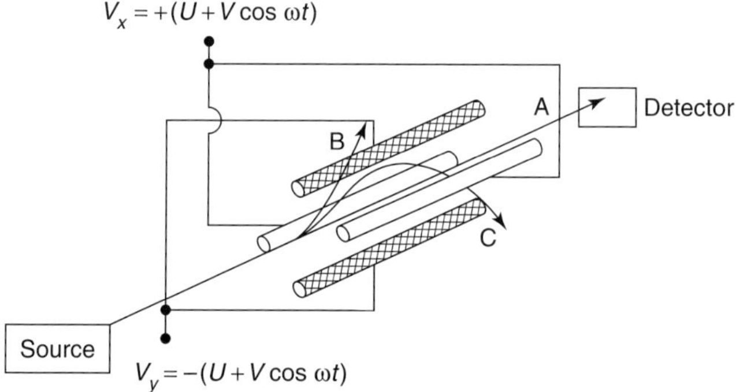

Figure 13 Schematic diagram of a quadrupole mass filter. Ions B and C have incorrect m/z ratio and are thrown against the rods; A has the proper m/z ratio and is transmitted through the rods onto the detector. (Reproduced by permission from Strobel and Heineman.[40])

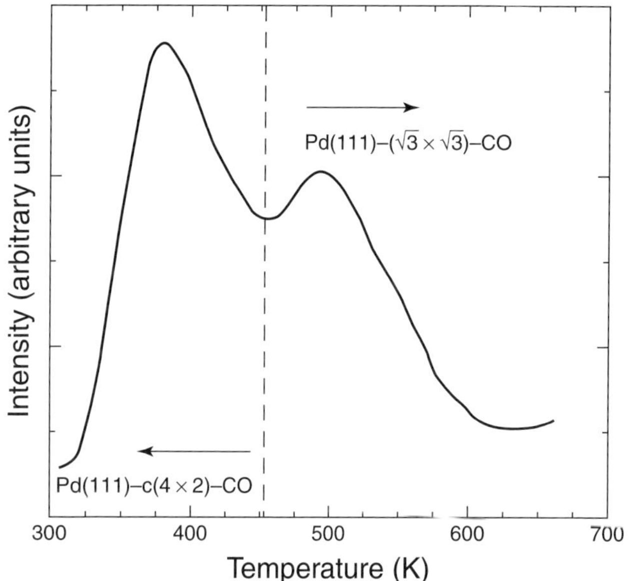

Figure 14 Thermal desorption mass spectrum of Pd(111)–c(4 × 2)–CO.

1.9 Instrument Designs

The single most critical step in UHV/EC experiments is the transfer of the electrode between the electrochemical cell (at ambient pressures) and the surface analysis chamber (under ultrahigh vacuum (UHV)). Ideally, the transfer is not accompanied by changes in surface structure and/or composition. The simplest approach, transfer through air is applicable only if the surface is inert (such as an oxide film) or is covered with a protective film (of solvent or electrolyte)[3] that can be removed by evacuation inside the UHV chamber. An alternative procedure that would allow electrode transfer in a controlled environment makes use of an inert-atmosphere glovebox attached to the surface-analysis instrument; electrochemical experiments are performed inside the box. However, regardless of the purity of the inert gas, the environment is not free of contaminants since the walls of the glovebox are replete with impurities that are slowly desorbed. The transfer-through-air and drybox methods have been used extensively in XPS and AES studies with oxided or chemically modified polycrystalline surfaces;[41,42] neither has been adopted for single-crystal work.

The approach employed most successfully in UHV/EC studies of single-crystal electrode surfaces involves the fabrication of a multi-technique surface-analysis apparatus to which an electrochemistry chamber is physically appended. The entire assembly is constructed of stainless steel and can be baked to about 200 °C in UHV to attain ultra-clean conditions. UHV is maintained by a combination of a titanium sublimation pump and either an

For references see page 10139

ion or a turbomolecular pump. Transfer of the electrode between the analysis and electrochemistry chambers is accomplished by a sample manipulator-translator. In some instruments, the crystal remains attached to the same sample holder as it is moved between the two compartments; in other systems, the crystal is actually transferred between two different manipulators. A gate valve isolates the electrochemistry compartment, whenever necessary, from the rest of the system. It is preferable to keep the electrochemistry chamber under UHV when not in use in order to preserve its cleanliness. The electrochemical cell itself is located inside a bellows-enclosed compartment separated from the electrochemistry chamber by another gate valve; the cell is inserted only after the electrochemistry chamber is brought to ambient pressures with ultra-high purity inert gas. Based upon these considerations, various types of UHV/EC instruments have been constructed;[1,5,43–55] three of these are shown in Figures 15–17[1,52,55] for illustrative purposes.

In the UHV/EC system depicted in Figure 17,[55] an isolable, differentially pumped antechamber is situated between the UHV and electrochemistry compartments. The main function of this antechamber is to minimize the influx of solvent and/or electrolyte vapor into the surface analysis compartment. In this context, it is important to mention that the pressure in the electrochemistry chamber is usually an order of magnitude higher than in the UHV chamber; mass spectrometric analysis of the residual gas has revealed that the pressure difference arises primarily from higher amounts of water in the electrochemistry compartment. Because water is only weakly surface-active, it is generally not of major concern in UHV/EC studies. However, in the presence of comparatively high quantities of water, impurity species may be dislodged from the walls of the chamber and onto the sample surface. Similar "knock-off" effects can arise when the chamber is backfilled with high purity inert gas. Hence, it is critical to maximize the cleanliness of the electrochemistry chamber and its associated manifold; this can be accomplished by frequent bakeout and continuous evacuation of the electrochemistry chamber when it is not in use.

It is also important to ensure that the backfill gas is of the highest purity to minimize surface contamination by trace-level impurities; argon of at least 99.99% purity is usually employed. Background contamination is metal specific; for example, Cu is more sensitive to residual O_2 while Pt is more susceptible to carbonaceous impurities. Hence, depending upon the nature of the investigation, it may be necessary to pass the high-purity inert gas through molecular scavengers (e.g. a Ti sponge heated to 900 °C) for still further purification. It must also be realized that electrode-surface contamination can also result from trace-level impurities in the electrolyte solution. Such impurities can originate from the solvent, electrolyte, glassware, and/or the inert gas employed for solution deaeration. The level of solution-based impurities can be minimized by the use of highly purified chemical reagents and gases; in the case of aqueous solutions, the utilization

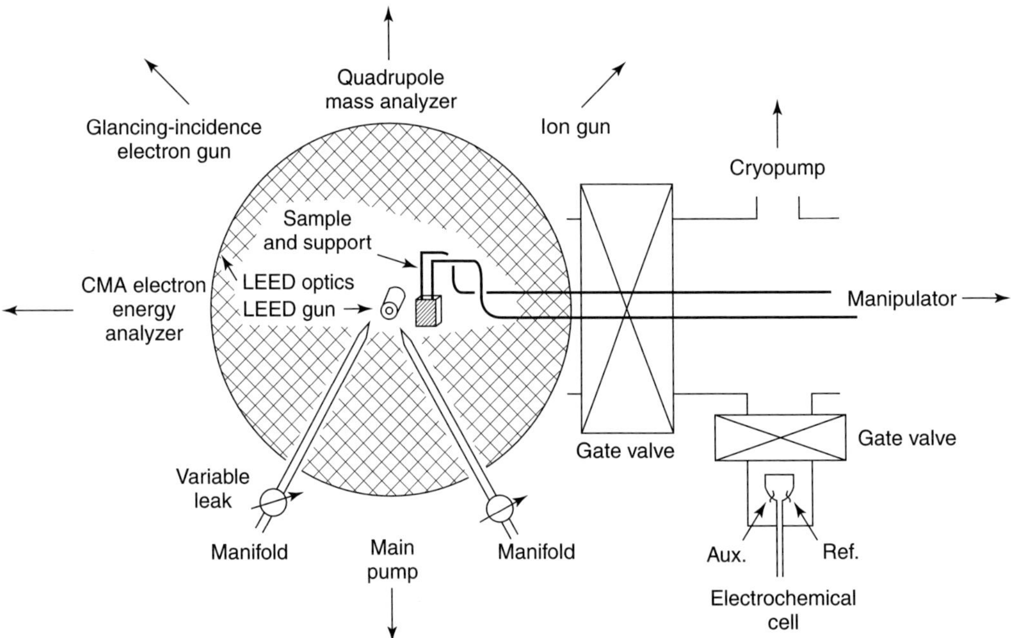

Figure 15 Schematic drawing of an experimental arrangement for UHV/EC studies. In this design, the crystal remains attached to only one sample manipulator. (Reproduced by permission from Hubbard et al.[1])

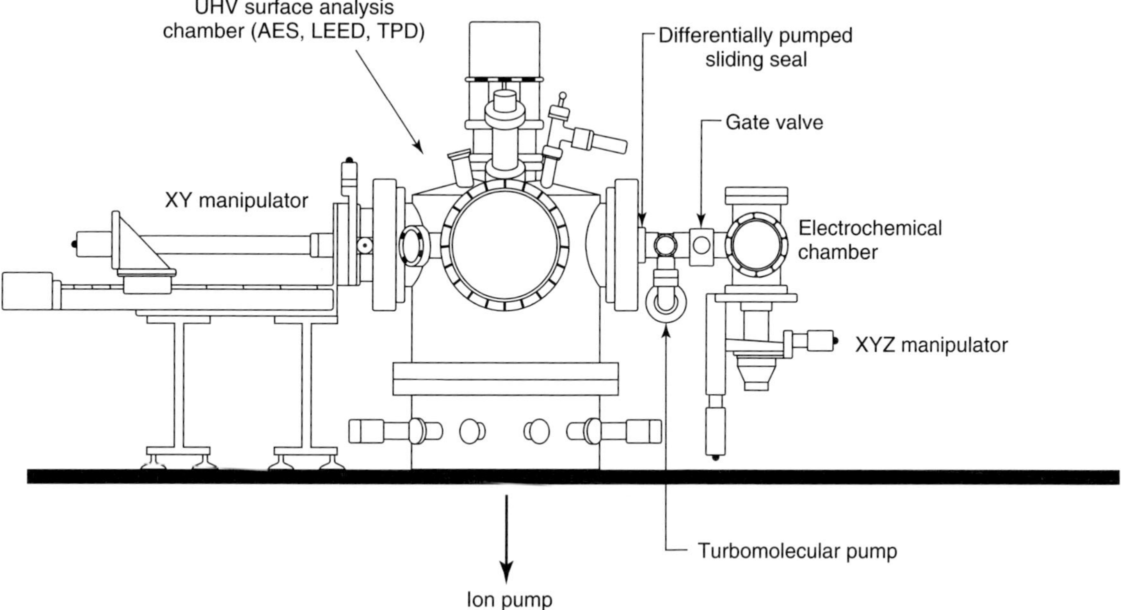

Figure 16 Schematic drawing of an experimental arrangement for UHV/EC studies. In this design, the electrode is transferred between different manipulators. (Reproduced by permission from Leung et al.[55])

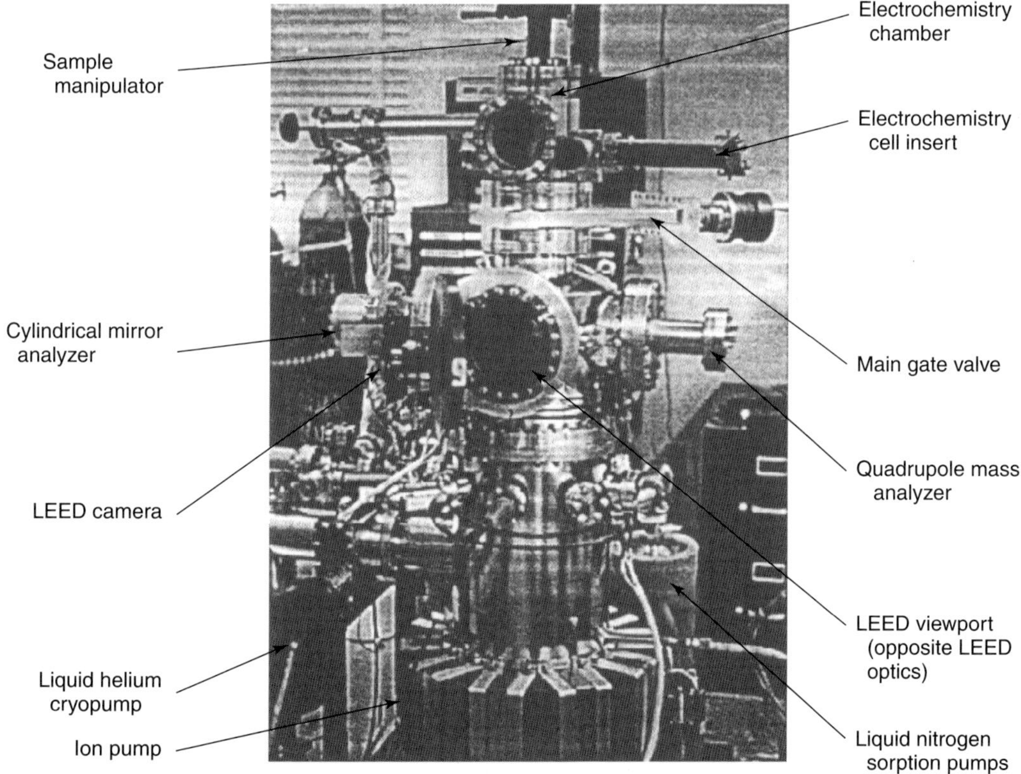

Figure 17 Photograph of a UHV/EC apparatus. In this design, the sample electrode remains attached to only one sample manipulator.

For references see page 10139

of pyrolytically triply distilled water is recommended,[56] although the use of water eluted through (commercial) multiple-filtration stages is now an acceptable alternative.

A typical UHV/EC experiment following instrument "bakeout" (at which point the base pressure should be less than 5×10^{-10} torr) would include the following steps. After surface preparation and characterization, the electrode is transferred into the electrochemistry chamber (cf. Figure 17) which, by closure of the main gate valve, is isolated from the UHV system and then backfilled with high-purity inert gas. The external gate valve is opened and the cell is inserted into the electrochemistry chamber. After completion of the electrochemical experiments, the cell is retracted, the external gate valve closed, and the chamber evacuated by a liquid helium cryogenic or turbomolecular pump to less than 10^{-6} torr. At this point, the main gate valve can be opened to complete the evacuation of the electrochemistry chamber and permit the transfer of the electrode into the analysis compartment. Pumpdown from ambient pressure to 10^{-8} torr vacuum is usually achieved in less than 15 min.

Electrochemistry experiments have been performed with cells in either the standard or thin-layer arrangement. The latter significantly reduces the level of surface contamination from solution-borne impurities. If the entire electrode is to be immersed in solution, all faces of the single-crystal should be oriented identically in order to obtain voltammetry characteristic of only one crystallographic face. As an alternative, the electrode can be positioned on top of the electrochemistry cell in such a way that only one crystal face is exposed to solution (Figure 18).[47] Such a configuration, however, often results in the adherence of a droplet of electrolyte on the crystal surface when the electrode is withdrawn from the solution; this problem does not arise if the electrode is withdrawn slowly ($1\,\mathrm{mm\,s^{-1}}$) in the vertical position, as the droplet would then form at the bottom of the crystal.[47]

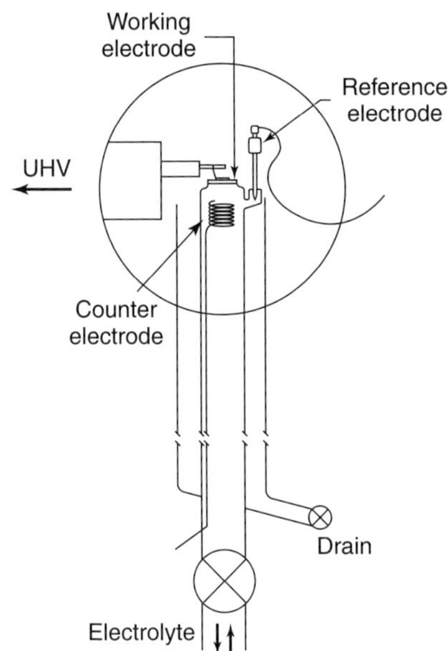

Figure 18 Schematic view of an electrochemical cell for use with a single-crystal disk electrode. (Reproduced by permission from Leung et al.[55])

2 FUNDAMENTAL ASPECTS

2.1 The Emersion Process

It is important to determine the changes at the electrode–electrolyte interface when the electrode is *emersed* (removed at a given potential) from the electrolyte solution. In the ideal process, the emersed electrode retains an interfacial layer identical in composition and structure to that when the electrode was still in solution. The electrode–solution interface, or the electrochemical double layer, is a structured assembly of solvent, electrolyte, and reactant. In the traditional view,[57–60] this ensemble, nominally 1 nm in thickness, is subdivided into an inner (compact) layer that consists of field-oriented adsorbed solvent molecules and *specifically* adsorbed anions, and an outer layer composed of solvated cations. The locus of the centers of the adsorbed anions delineates the so-called inner Helmholtz plane (IHP), whereas the line of centers of the nearest solvated cations defines the outer Helmholtz plane (OHP). Charge transfer reactions are thought to occur at this outer (reaction) plane. The solvated ions interact with the charge metal only through long-range electrostatic forces and, because of thermal agitation in the solution, are distributed in a three-dimensional region that extends from the OHP into the bulk of the solution. This region is identified as the *diffuse layer*, and its thickness is a function of electrolyte concentration; it is less than 30 nm for concentrations greater than 10^{-2} M.

Clearly, the electrode-withdrawal process involves a delicate balance with respect to the thickness of the emersion layer: it must be sufficiently thick to incorporate the intact electrochemical double layer but it should also be thin enough to exclude *residual* (bulk) electrolyte. Numerous studies have helped establish the fact that the electrochemical double layer can, under appropriate electrolyte concentrations, be retained intact when the electrode is withdrawn from solution under potential control. The optimum concentration depends upon whether emersion is hydrophobic or hydrophilic.[61–64] For the latter type, the concentration must not be much *higher* than 10^{-3} M if inclusion of bulk electrolyte is to be circumvented. For hydrophobic emersion, the concentration must not be much *lower* than 10^{-3} M if double-layer discharge is to be avoided. In cases where the mode of emersion is not known, an electrolyte concentration of 10^{-3} M appears to be a logical choice.

Investigations of hydrophobic emersion based upon electrode resistance measurements,[64] electroreflectance spectroscopy,[61,62,65] XPS,[66] and work function changes[36,37,61,63] have been able to: (i) demonstrate the existence of an emersed double layer; (ii) determine its stability; and (iii) monitor changes in its structure and composition brought about by the emersion process. The evidence has been compelling that the structure and composition of the double layer in the emersed phase are very similar, if not identical, to those in the solution state; that is, only a little or no double-layer discharge occurs upon emersion. Later studies focused on the effect of the emersion process on the structure of adsorbed molecular species;[67–71] the data obtained demonstrate that the structure and orientation of molecular adsorbates are essentially unperturbed by the emersion process.

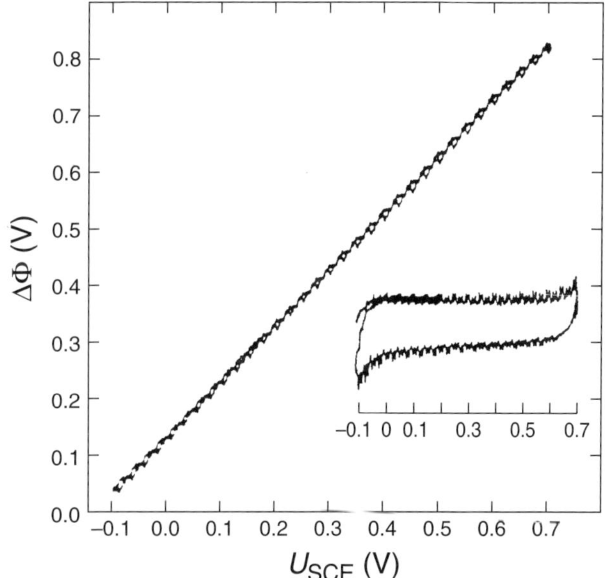

Figure 19 A plot of the work function of a polycrystalline Au emersed from 0.1 M $HClO_4$ into UHV as a function of the emersion potential. The work function of the clean metal was 5.2 eV. The lower and upper lines, respectively, represent the solution inner potential if the absolute normal hydrogen electrode (NHE) potential is 4.45 or 4.85 V. (Reproduced by permission from Hansen and Hansen.[37])

2.2 Perturbations Caused by Evacuation and Surface Analysis

Another important issue in UHV/EC centers around the perturbations caused by the evacuation and subsequent surface analytical processes. Alterations in the surface electronic structure can be studied by work function change measurements; representative results are shown in Figure 19, in which a plot of the work function of a polycrystalline Au emersed from 0.1 M $HClO_4$ into UHV as a function of the emersion potential is presented.[37,71] It can be seen here that the work function tracks the applied potential over a wide range, even into the oxide formation region. This, and other sets of pertinent data, demonstrates that the electronic properties of the electrochemical double layer are unaffected by emersion into either the ambient or UHV.

An expected effect of evacuation are changes in composition due to UHV-induced desorption; the extent will depend upon the heats of vaporization ΔH_{vap} or sublimation ΔH_{sub} of the *unbound* materials (excess water, unadsorbed gases, liquids, and sublimable solids) entrapped within the emersed layer. Water retained as part of the hydration sphere can survive the evacuation process if the hydration enthalpies ΔH_{hyd} are substantial.[72,73] Strongly chemisorbed species, such as iodine at the noble-metal electrodes, are expected to form stable well-ordered adlattices in solution[74–76] that would not reconstruct in vacuum. Similarly, the coverages and structures of irreversibly adsorbed molecules are expected to withstand the evacuation process. One example is provided by 3-pyridylhydroquinone (Py-H_2Q) which, on Pt(111), is chemisorbed through the N-heteroatom.[77] In such a mode of surface attachment, the diphenol group is pendant and able to undergo reversible quinone/hydroquinone redox reaction. It has been shown that the redox current–potential curves for chemisorbed Py-H_2Q, before and after a 1-h exposure to UHV, are identical (Figure 20).[77]

Perturbations can also arise from the surface analysis itself. For example, implicit in TPD is the requirement for complete desorption; hence, TPD and TDMS are inherently destructive techniques. On the other hand, analytical methods based upon electron and optical spectroscopies are not intended to damage the surface layer. Unless exceedingly high photon fluxes are used, optical methods are nondeleterious when compared with particle-based techniques. Several surface processes are known to be stimulated by electron impact. Examples are binding-site conversions, dissociative chemisorption, and particle desorption.[12] Such processes take place even at minimal sample heating (low electron power densities), which indicate that surface thermal effects are insignificant. It is now accepted that electron-stimulated reactions occur mainly via electronic excitations. These excitations can lead to bond dissociation and form the basis of the surface spectroscopic technique known as electron stimulated desorption ion angular distribution (ESDIAD).[78]

Pendant functional groups not directly bonded to the substrate surface, such as the diphenol moiety in Py-H_2Q,

For references see page 10139

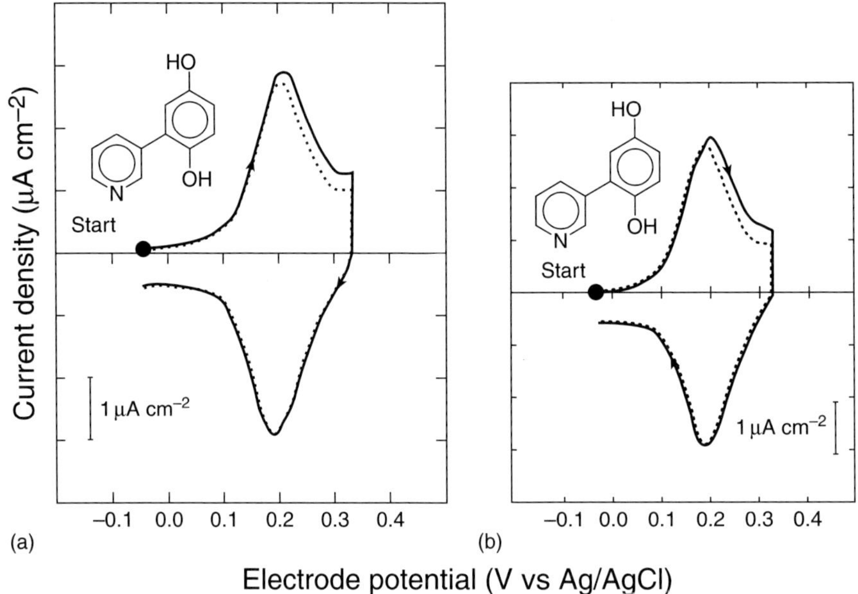

Figure 20 Cyclic voltammetry of Py-H_2Q at Pt(111). (a) Solid curve: first scan; dotted curve: after 1 h in UHV. (b) Solid curve: first scan; dotted curve: second scan. (Reproduced by permission from Stern et al.[77])

are most prone to electron-stimulated desorption. In other instances, electron irradiation can induce surface displacement reactions that involve species present as residual gas in the analysis chamber. It has been reported, for example, that prolonged electron-beam irradiation during LEED and/or AES in the presence of residual water vapor caused a $Ba(CN)_2$ adlayer to undergo a reaction that formed gaseous HCN and solid $Ba(OH)_2$.[72,73]

Even if changes in the emersed double layer are induced by the surface analysis, it must be realized that such alterations would be deleterious only if the post-analysis layers are to be used for further experiments. In those instances when additional experiments have to be performed, it is a simple matter to regenerate the surface to exactly the point just prior the surface analysis. If desired, beam damage can be assessed by repeated analysis over a period of time followed by extrapolation of the data to *zero* time.

3 REPRESENTATIVE STUDIES

UHV/EC investigations with single-crystal electrode surfaces can be broadly classified into three groups. The first is focused on the structure and constitution of the electrochemical double layer as functions of electrode potential and solution composition. The second centers on electrodeposition and dissolution processes; included in this category are extensive studies on hydrogen and oxygen adsorption/desorption. The third deals with the interfacial structure and reactivity of chemisorbed molecules.

3.1 Electrochemical Double Layer

Two general strategies have been adopted in UHV/EC studies of the electrical double layer. One, strictly a model approach, involves the synthesis of the double layer in UHV by sequential cryogenic adsorption of its constituents;[79–85] the temperature must be maintained below 160 K at all times in order to prevent the evaporation of unbound solvent. The other approach is based upon the structural and compositional analysis of the emersed layer; since surface characterization is done at ambient temperatures, excess water in the diffuse layer is pumped away.

The viability of the cryogenic coadsorption approach[79–82] was first demonstrated by comparison of work-function changes $\Delta\varphi$ for UHV-synthesized and actual Ag(110)–X–H_2O layers, where X denotes Cl^- or Br^-. The results showed that, as long as *coadsorbed* water is present in the cryogenic layer, good agreement exists between the ex situ and in situ results. The satisfactory agreement indicates that, at least under zero diffuse-layer-charge conditions, information from the UHV simulation work may have relevance to actual electrochemical systems. The requirement of solvation implies that the electronic properties of the unsolvated Ag(110)–Br interface are not identical to those of the fully solvated Ag(110)–Br–H_2O layer. Other more complex interfacial systems have also been modeled via cryogenic coadsorption. For example, UHV-synthesized H_2O–HF–CO coadsorbed layers were studied at Pt(111) and Rh(111) surfaces by HREELS, LEED, TPD and XPS.[83,84] In such work, the "control" of electrode potential was based upon the amount of coadsorbed H_2. The cryogenic coadsorption approach offers two main advantages: (i) the control of interfacial parameters far more precisely than can be achieved in solution; and (ii) the detailed characterization of fully solvated species by a host of surface-sensitive spectroscopic methods. The approach, however, provides little beyond structural models for the interfacial layers; its usefulness as a surface analytical technique is unclear.

The direct approach to the study of the electrochemical double layer involves the characterization of the electrolyte layer retained at the emersed surface. This approach is applicable only to cases in which the compact layer consists of materials that remain on the surface even after evacuation to UHV. For example, only limited information can be expected when emersion is from aqueous HF since both H_2O and HF will be completely desorbed, at ambient temperatures, in vacuum. It is also implied that the molecules pumped away are inconsequential in the formation and preservation of the electrochemical double layer. This is not an unreasonable premise since the double layer composed of chemisorbed species is governed by strong chemical bonds, and should only be minimally perturbed by physisorbed species.

UHV/EC studies of anions specifically adsorbed from aqueous solutions have been carried out at well-defined Pt(111),[86–92] Pt(100),[93,94] stepped Pt(s)[6(111) × (111)],[95] Cu(111),[96] Ag(111),[97] Au(111),[98] and Pd(111)[99] electrodes. The anions studied include the halides,[86–90,93–99] SH^-,[89,99] CN^-,[91,92] SCN^-,[100] and SO_4^{2-}.[96] All of these anions yield surface coverages and well-ordered structures that depend upon the solution pH and the applied potential.

3.2 Underpotential Electrodeposition and Anodic Dissolution

The deposition of *monolayer* quantities of one metal onto another generally occurs at potentials positive of that for *bulk* deposition because of preferential interactions between the substrate and the foreign-metal electrodeposit. The underpotential deposition (UPD) process is strongly influenced by the structure and composition of the substrate; hence, UPD research is an area in which UHV/EC methods have been widely adopted. The published literature on UHV/EC studies of UPD can be categorized according to whether the experiments were used to correlate substrate structure with the electrodeposition voltammograms, or to determine the interfacial properties of the adatom-modified substrate. Investigations devoted to structure–voltammetry correlations help establish reference states against which new experiments can be calibrated; those focused on post-deposition characterization yield important information concerning the electrocatalytic selectivity of the mixed-metal interfaces.

The first applications of LEED and AES in electrochemistry involved the correlation of the surface crystallographic orientation with the underpotential hydrogen deposition at Pt electrodes.[1,101–105] Those studies were motivated by earlier work with polycrystalline Pt electrodes whose cyclic voltammograms showed two hydrogen deposition peaks; the peaks were identified simply as weakly and strongly bound states of adsorbed hydrogen. Subsequent studies with single-crystalline but *uncharacterized* Pt(111), Pt(110), and Pt(100) electrodes associated the weakly bound hydrogen peak with the Pt(111) electrode, and the strongly bound hydrogen with the Pt(100) electrode. These studies, though not definitive enough due to the existence of atomically disordered surfaces because of multiple oxidation–reduction cycles, provided the impetus for further UHV/EC studies.

Later work based upon flame-annealed single-crystal surfaces[3] led to the discovery of new voltammetric features for Pt(111) in the form of highly reversible pseudocapacitance peaks at potentials *well positive* of the usual hydrogen deposition peaks.[106–108] UHV/EC studies that employed apparatus equipped with improved UHV-to-EC transfer technology were able to reproduce the new voltammetric results. Extensive follow-up work then ensued that clarified several aspects of this surface-sensitive reaction.[101,107,108]

While studies of hydrogen chemisorption at Pt single-crystal electrode surfaces have been extensive, investigations on the formation of underpotential states of oxygen have not been as numerous. Electrochemical experiments have been performed only with Pt(100) and Pt(111) electrodes.[109–111] Gas-phase and solution-state reactions with oxygenous species have been carried out at stainless-steel single crystals.[112–114] The occurrence of place-exchange during anodic film formation has been studied via LEED spot-profile analysis.[111] This irreversible place-exchange reaction accounts for the observation that the electrode surface loses its single crystallinity even after only minimal anodic oxidation.

The literature on monolayer metal deposits is extensive. Most of the work pertains to the geometric, electronic, and catalytic properties of foreign metals *vapor deposited in UHV* onto single-crystal substrates; a compilation of the adlattice structures of such metal adlattices has been published.[115] Studies of foreign metal monolayers *deposited electrochemically* have been primarily with

For references see page 10139

polycrystalline substrates. The first UHV-based investigation of electrodeposited admetals employed XPS to determine the core-level shifts of submonolayer Cu and Ag on *polycrystalline* Pt.[43,116] Quite a few investigations have been reported of admetal deposits on single-crystal surfaces but only a minority of these involve UHV/EC technology.

The first UHV/EC work on electrodeposition at well-defined electrode surfaces involved Ag at an iodine-coated Pt(111) electrode.[117,118] The iodine pretreatment was done in UHV to form a protective Pt(111)($\sqrt{7} \times \sqrt{7}$)R19.1°–I adlattice before immersion into a solution containing dilute Ag^+ in 1 M $HClO_4$. Subsequent studies included Ag electrodeposition on I-coated Pt(100)[119] and stepped Pt(s)[6(111) × (111)],[95] Cu on I-pretreated Pt(111),[120] and Pb on I-covered Pt(111).[121] Sn[122] and Pb[123] deposition onto *iodine-free* Pt(111) in Br^- or Cl^- solutions[124] has also been studied. Although the Pt substrate was not pretreated with I, the presence of halide ions in the plating solution led to specific adsorption of anions prior to the deposition process.

Electrodeposition from solutions free of surface-active anions has been studied. These investigations, carried out in ClO_4^- or F^- electrolyte, included UPD of Cu on Pt(111),[125–127] Tl, Pb, Bi, and Cu on Ag(111).[128] Invariably, the underpotentially deposited films showed unique adlattice geometries that were dependent upon the substrate orientation and the admetal coverage.

The atomic layer epitaxy (ALE) approach to deposition of a compound film, based upon the alternate layer-by-layer deposition of the elements of the compound, has been adopted in the electrochemical synthesis of compound semiconductors. This electrochemical analog, referred to as electrochemical atomic layer epitaxy (ECALE).[129,130] takes advantage of the fact that only monolayer quantities are produced by UPD. For Example, the UPD-based epitaxial growth of CdTe on Au(111) has been monitored by LEED and AES.[129]

While UHV/EC-based investigations on electrodeposition abound, little has been published on the reverse reaction, anodic dissolution; most of the studies reported have made use of electrochemical scanning tunneling microscopy (ECSTM).[131] One system, that of the adsorbate-catalyzed corrosion of Pd, has been extensively studied using both UHV/EC and ECSTM.[132,133] In the absence of a layer of chemisorbed iodine, anodic oxidation of Pd in sulfuric acid yields only a passivating metal oxide film; when iodine is chemisorbed on the surface, dissolution of Pd occurs without alteration of the structure or composition of the iodine adlattice (Figure 20).[132] It has been found that, at low dissolution rates, the reaction proceeds at the step-edges (Figure 21) at much higher currents, I–Pd place-exchange occurs that leads to the formation of pits (Figure 22).[133]

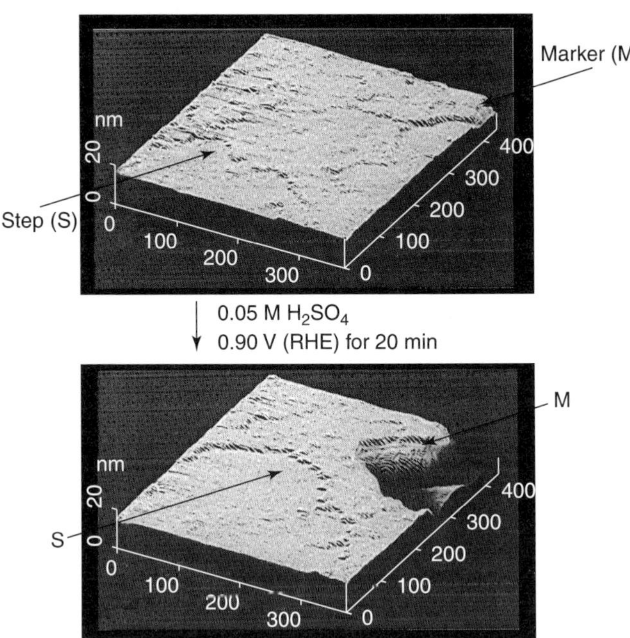

Figure 21 Wide-area ECSTM images of a Pd(111)–($\sqrt{3} \times \sqrt{3}$)R30°–I facet on a single-crystal bead at low-current adsorbate-catalyzed dissolution. The potential of the tip was held at 0.9 V; the tunneling current was 1 nA.

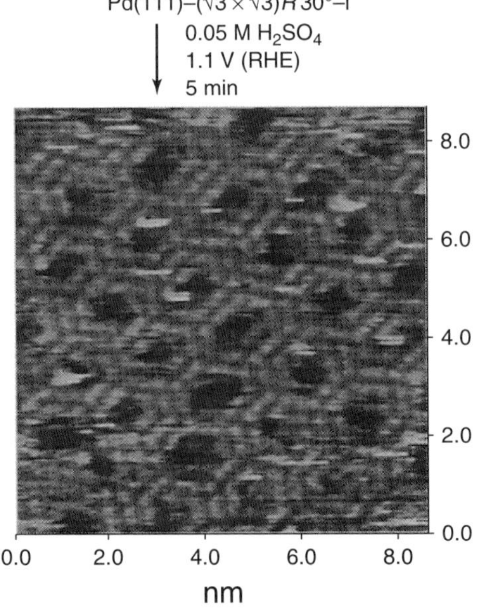

Figure 22 Atomic-resolution ECSTM image of a Pd(111)–($\sqrt{3} \times \sqrt{3}$)R30°–I facet on a single-crystal bead after medium-current anodic dissolution.

3.3 Molecular Adsorption[134]

The capability to prepare single-crystal surfaces by thermal treatment at ambient pressures[3] has fostered the proliferation of non-UHV studies of the adsorption of molecules at monocrystalline electrodes. The detail of information obtained from such in situ work, however, falls short of that provided by UHV/EC experiments. As one example, although in situ IRAS has provided much information about the structure-sensitivity of the chemisorption and anodic oxidation of CO, its sensitivity is too low to permit meaningful investigations with other molecules as simple as ethylene.

3.4 Solvent–Electrode Interactions

The nature of the interactions between the solvent and the electrode surface has significant ramifications in electrochemical surface science. For instance, the use of strongly surface-active solvents would severely repress electrocatalytic processes that rely on a direct interaction between the reactant and the metal surface. The bonding of water to metal surfaces is an important issue in aqueous electrochemistry. In models suggested to explain the potential dependence of double-layer capacity, the existence has been postulated of monomeric and clustered water molecules, both of which are able to adopt two opposite dipolar orientations.[135,136] The studies of water adsorption on single-crystal electrodes are all based upon vapor deposition in UHV, usually at cryogenic temperatures since water is not adsorbed on clean metal surfaces at ambient temperatures. Of significant interest to electrochemistry is the observation that, on Ni, Pt, Ag, Cu, and Pd, water is dissociatively chemisorbed if the surface contains *submonolayer* coverages of oxygen.[137] The reaction is thought to occur by hydrogen abstraction. This reaction is metal-specific since at other noble metals such as Ru(001), adsorbed oxygen is inactive towards water dissociation.

Nonaqueous solvents commonly used in electrochemistry include acetonitrile, dimethylformamide, *p*-dioxane, sulfolane, dimethylsulfoxide, pyridine, acetic acid, propylene carbonate, liquid ammonia, and dichloromethane.[138] Work on such materials can be categorized according to how the electrode is allowed to interact with the nonaqueous solvent: by vapor dosing in vacuum, or by exposure to aqueous solutions containing small quantities of nonaqueous-solvent material. Studies under the latter category are more abundant, but those are more accurately classified under electrode–solute, not electrode–solvent, interactions. Except for one case, all UHV-based studies with nonaqueous-solvent compounds were carried out purely in the context of gas–solid interactions.[139] The intent of the one exception[140] was to use the reactions between the solvent vapor and the metal surface as models for the electrochemical analogs; for better simulation of solution conditions, vapor dosing was up to 0.3 torr, approaching the vapor pressures of the liquid solvents. UHV/EC work in which the electrode is immersed in pure nonaqueous solvent has not been pursued.

3.5 Group IB Electrodes

Most organic compounds are only weakly adsorbed on Cu, Ag, and Au electrode surfaces; hence, unless the adsorbate itself is a solid or when adsorption is carried out at cryogenic temperatures, meaningful UHV/EC experiments with the coinage metals are limited. One study, which took advantage of the strong interaction of the -SH functional group with the coinage metals, used HREELS, LEED, AES, and voltammetry to determine the influence of the location of the N heteroatom on the adsorption properties of the isomers 2-mercaptopyridine and 4-mercaptopyridine at Ag(111) in aqueous HF.[141] The subject compounds were thought to undergo isomerization upon oxidative adsorption through the -SH moiety.

3.6 Group VIII Electrodes

The abundance of studies of organic molecular adsorption at electrode surfaces involves the platinum metals. This is not surprising since these metals are well known for their electrocatalytic activities and an immense body of work has already been amassed for these materials in their polycrystalline states.[142–145] Surface electrochemical studies of metal-organic compounds at single-crystal electrodes can be broadly classified according to whether the work was done with CO (and related small molecules) or with more complex molecules. The former are more numerous, although a vast majority of the studies have been carried out without UHV-based surface characterization. Work with well-defined surfaces have been limited to LEED of CO adlattices on Pt(111)[146] and Pd(111),[147,148] and HREELS, LEED, TPD and XPS of mixed H_2O–HF–CO layers generated in UHV by cryogenic adsorption at Pt(111) and Rh(111) surfaces.[83,84]

An impressive amount of detailed information on a wide variety of complex organic compounds chemisorbed at well-defined Pt(111) and Pt(100) electrode surfaces has been furnished by LEED, AES, TPD, and HREELS.[134] Electrocatalytic reactivity studies which accompanied these investigations were limited to anodic oxidation reactions; only correlations between the mode of adsorbate bonding and the extent of anodic oxidation were attempted.

3.7 Carbon Monoxide

Much of what is known about the structure and reactivity of CO chemisorbed at single-crystal electrodes, and their dependencies on surface crystallographic orientation,

For references see page 10139

electrode potential, and adsorbate coverage are based almost entirely upon in situ IRAS measurements.[134] Two UHV/EC studies on CO are noteworthy. One made use of a well-defined Pt(111) surface and sought to correlate anodic peak potentials with observed LEED structures.[146] The other, based upon LEED, AES, TPD, voltammetry, and coulometry, examined the chemisorption of CO at well-defined and anodically disordered Pd(111).[147] In this study, it was shown that CO adsorption from solution onto a UHV-prepared Pd(111) surface yielded an ordered adlattice, Pd(111)–c(4 × 2)–CO, in which the CO molecules occupy two-fold hollow sites. Two CO-to-CO_2 anodic oxidation peaks were observed: the first peak is partial oxidation of the c(4 × 2) layer followed by an *adlattice reconstruction* to form a Pd(111)($\sqrt{3} \times \sqrt{3}$)$R30°$–CO structure; the second peak is due to complete oxidative desorption of the Pd(111)($\sqrt{3} \times \sqrt{3}$)$R30°$–CO adlayer to yield a clean and well-ordered Pd(111) single-crystal surface.[148] At the surface disordered by extensive anodic oxidation, chemisorption of CO occurred spontaneously but no ordered CO adlayers were produced; on the disordered surface, the CO molecules were thought to reside on atop sites.

3.8 Other Organic Compounds

UHV/EC investigations with organic compounds more complex than CO have been focused on the nature of the adsorbate–substrate chemical interactions as a function of interfacial parameters, and on the influence of the mode of attachment on the reversible and/or catalytic electrochemistry of the adsorbed species. For example, the differences between gas-phase and solution-state chemisorption and catalytic hydrogenation of ethylene have been documented:[149–152] variations in the structures of ethylene chemisorbed at the solid–solution and gas–solid interfaces lead to different reaction pathways. In solution, ethylene chemisorption occurs molecularly through its π-electron system, whereas chemisorption in UHV is accompanied by molecular rearrangements to form a surface ethylidyne species. In electrocatalytic hydrogenation, ethylene is reduced on the Pt surface by adsorbed H atoms; in gas-phase hydrogenation, H atoms must be transferred from the Pt surface through a layer of irreversibly adsorbed ethylidyne to ethylene adsorbed on top of the ethylidyne layer. Further work with alkenes[153–157] has been centered on the effects of hydrocarbon chain length and the presence of weakly surface-active substituents such as carboxylates[158] and alcohols.[159] These studies showed that: (i) the primary mode of surface coordination of terminal alkenes, alkenols, and alkenoic acids is through the olefinic double bond; and (ii) the pendant alkyl chain is always extended outward on top of the propylene moiety. From coulometric measurements, it was concluded that anodic oxidation of the chemisorbed alkenes is limited to the anchor group, unless the pendant moieties are too close in close proximity to the electrode surface.

Early studies with smooth polycrystalline Pt indicated that aromatic compounds such as 1,4-dihydroxybenzene are chemisorbed in discrete, nonrandom orientations that depend upon various interfacial parameters.[160–162] Subsequent UHV/EC experiments with well-defined Pt(111) electrodes[163,164] supported the earlier findings, although the conditions at which the multiple orientational transitions occur are different for the polycrystalline and single-crystal electrodes. The electrocatalytic oxidation of multiply oriented aromatic molecules has been shown to be strongly dependent on their *initial* adsorbed orientations.

Sulfur-containing compounds investigated[165,166] include thiophenol, pentafluorothiophenol, 2,3,5,6-tetrafluorothiophenol, 2,3,4,5-tetrafluorothiophenol, 2,5-dihydroxythiophenol, 2,5-dihydroxy-4-methylbenzyl mercaptan, and benzyl mercaptan; chemisorption of these compounds occurs oxidatively through the sulfur group with loss of the sulfhydryl hydrogen. The tethered diphenolic moieties in the adsorbed dihydroxythiophenols show reversible quinone/diphenol redox chemistry. The S-heterocyclic compounds studied were thiophene, bithiophene, and their carboxylate and methyl derivatives. Experimental evidence indicates that these compounds are bound exclusively through the S heteroatom, although the chemisorption process may be accompanied by self-desulfurization. The electropolymerization of 3-methylthiophene at clean Pt(111) and monomer-treated Pt(111) pretreated has been studied, and the properties of the two types of polymer film were compared. In terms of the HREELS spectra, two major differences were noted which were attributed to changes in the physical nature of the polymer film.

The chemisorption of pyridine, bipyridine, multinitrogen heteroaromatic compounds, and their derivatives[165–168] has been examined as a function of isomerism and substituents. Pyridine forms a well-ordered layer of admolecules chemisorbed through the N heteroatom in a tilted vertical orientation. The derivatives are coordinated similarly unless the ring nitrogen is sterically hindered such as in 2,6-dimethylpyridine where chemisorption is in the flat orientation. Pyrazine, pyrimidine, and pyridazine are chemisorbed through only *one* nitrogen heteroatom in a tilted-vertical orientation. For the derivatives, adsorption occurs through the least hindered ring nitrogen. Carboxylate substituents located in positions *ortho* or *meta* to the nitrogen heteroatom interact with the Pt(111) surface at positive potentials. The chemisorbed layers were disordered as indicated by the absence of LEED patterns and were observed to be electrochemically unreactive. The adsorption behavior of the bipyridyls was found to be sensitive to steric hindrance at the positions *ortho* to the nitrogen heteroatom. Studies on the mode of chemisorption at well-defined Pt(111) of L-dopa, L-tyrosine, L-cysteine, L-phenylalanine, alanine, and dopamine have been reported.[167,168]

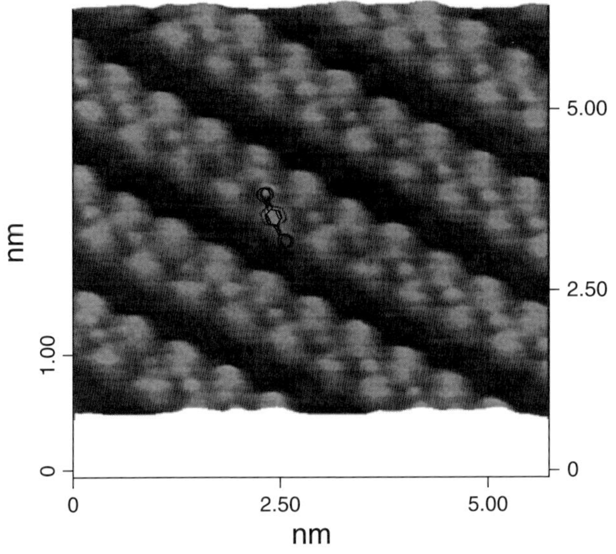

Figure 23 High-resolution ECSTM image of Pd(111) immersed in a 0.01 mM aqueous solution of hydroquinone in 0.05 M H_2SO_4.

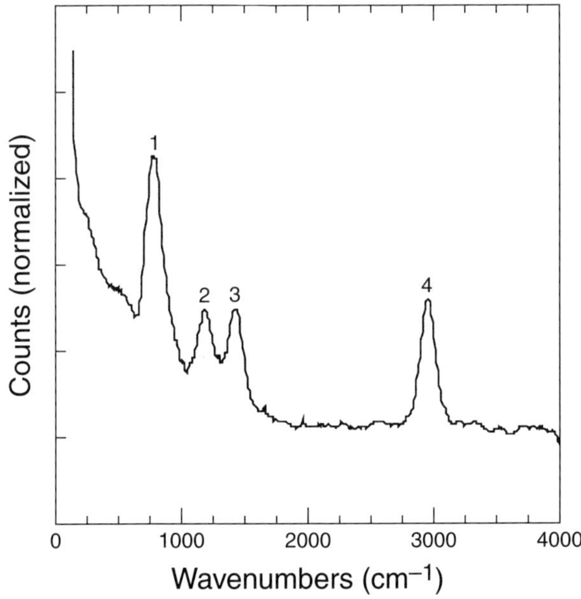

Figure 24 HREEL spectrum of a Pd(111) surface after emersion from a 0.1 mM aqueous solution of benzoquinone in 1 mM tetrafluoroacetic acid. Peaks 2, 3 and 4 correspond to *in-plane* C–H bending, C–C and C–H stretch modes, respectively; peak 1 is due to an *out-of-plane* C–H bending mode.[170]

The chemisorption of aromatic molecules on well-defined Pd(*hkl*) surfaces has recently been studied by a combination of UHV/EC and ECSTM.[169,170] Figure 23 shows a high-resolution STM image of a Pd(111) electrode immersed in a dilute solution of 1,4-dihydroxybenzene; Figure 24 shows the corresponding HREEL spectrum. Analysis of both sets of data indicate that hydroquinone is oxidatively chemisorbed as benzoquinone and, although bound in the flat orientation, is slightly tilted.[170]

4 EPILOGUE

The aim of modern electrochemical surface science is the design of superior electrode materials not only for the characterization but also for the control of important electron-transfer reactions. This goal can best be accomplished by the establishment of fundamental correlations between the structure, composition and chemical reactivity at the electrode–solution interface. As should be evident from the above review and the cited literature, the UHV/EC approach provides an arsenal of surface-sensitive techniques that help in the establishment of such fundamental correlations.

The UHV/EC approach offers three essential functions: (i) it can be used to tailor surfaces of well-defined structure and composition for the performance of specific electrochemical tasks; (ii) it can be employed to determine (or model via cryogenic deposition of the constituent species) the structure and composition of the compact layer; (iii) it can be exploited to interrogate the composition and lateral structure of nonspecifically absorbed materials in the diffuse layer. These tasks will always be paramount in fundamental electron-transfer investigations, whether those are directed towards biotechnology, electroanalysis, fuel-cell technology, or materials research. That UHV/EC-based results have been supported and complemented by in situ experiments essentially assures the continued utilization of this approach in future studies centered around the electrode–electrolyte interface.

ACKNOWLEDGMENTS

The author wishes to thank the National Science Foundation and the Robert A. Welch Foundation for their support of work cited in this review.

ABBREVIATIONS AND ACRONYMS

AES	Auger Electron Spectroscopy
ALE	Atomic Layer Epitaxy

For references see page 10139

CHA	Cylindrical Hemisphere Analyzer
CMA	Cylindrical Mirror Analyzer
ECALE	Electrochemical Atomic Layer Epitaxy
ECSTM	Electrochemical Scanning Tunneling Microscopy
ESCA	Electron Spectroscopy for Chemical Analysis
ESDIAD	Electron Stimulated Desorption Ion Angular Distribution
HREELS	High-resolution Electron Energy Loss Spectroscopy
IHP	Inner Helmholtz Plane
IRAS	Infrared Reflection–Absorption Spectroscopy
LEED	Low-energy Electron Diffraction
OHP	Outer Helmholtz Plane
Py-H_2Q	3-Pyridylhydroquinone
RFA	Retarding Field Analyzer
RHEED	Reflection High-energy Electron Diffraction
TDMS	Thermal Desorption Mass Spectrometry
TPD	Temperature Programmed Desorption
UHV	Ultrahigh Vacuum
UHV/EC	Ultrahigh Vacuum Electrochemistry
UPD	Underpotential Deposition
XPS	X-ray Photoelectron Spectroscopy

RELATED ARTICLES

Surfaces **(Volume 10)**
Auger Electron Spectroscopy in Analysis of Surfaces • Electron Energy Loss Spectroscopy in Analysis of Surfaces • X-ray Photoelectron Spectroscopy in Analysis of Surfaces

APPENDIX 1

Ultrahigh Vacuum Electrochemistry Bibliography

Presented in this Appendix is a near-exhaustive bibliography of UHV/EC-based investigations of single-crystal electrodes published since 1991. A compilation of earlier work has been published previously (G.J. Cali, J.R. McBride, M.P. Soriaga, *Prog. Surf. Sci.*, **39**, 422 (1992)). Owing to space limitations, no listing is provided for studies with polycrystalline electrodes which are more numerous. Dr Y.-G. Kim and Dr J.E. Soto assisted in the compilation of this bibliography.

A.T. Hubbard, J.Y. Gui, 'Molecular Chemistry at Well-defined Surfaces', *J. Chim. Phys.*, **88**, 1547 (1991).

S.C. Chang, M.J. Weaver, 'In-situ IR Spectroscopy at Single-crystal Electrodes: An Emerging Link Between EC and UHV Surface Science', *J. Phys. Chem.*, **95**, 5391 (1991).

J.Y. Gui, D.A. Stern, C.H. Lin, A.T. Hubbard, 'Potential-dependent Surface Chemistry of Hydroxypyridines Adsorbed at Pt(111) Studied By EELS, LEED, AES, and Electrochemistry', *Langmuir*, **7**, 3183 (1991).

C. Shannon, D.G. Frank, A.T. Hubbard, 'Electrode-reactions of Well-characterized Adsorbed Molecules', *Annu. Rev. Phys. Chem.*, **42**, 393 (1991).

J.Y. Gui, D.A. Stern, F. Lu, A.T. Hubbard, 'Surface-chemistry of 5-Membered Heteroaromatics at Pt(111) Studied by EELS, LEED, AES and Electrochemistry: Furan, Pyrrole and Thiophene', *J. Electroanal. Chem.*, **305**, 37 (1991).

D. Aberdam, R. Durand, R. Faure, 'Study of UPD on Metal Single-crystals by Surface Techniques', *J. Chim. Phys.*, **88**, 1519 (1991).

J.Y. Gui, D.A. Stern, D.G. Frank, A.T. Hubbard, 'Adsorption and Surface Structural Chemistry of Thiophenol, Benzyl Mercaptan, and Alkyl Mercaptans: Comparative-studies at Ag(111) and Pt(111): AES, EELS, LEED, and EC', *Langmuir*, **7**, 955 (1991).

N. Kizhakevariam, E.M. Stuve, R. Dohloelze, 'Coadsorption of Water and Chlorine on Ag(110): Evidence for Adsorbate-induced Hydrophilicity', *J. Chem. Phys.*, **94**, 670 (1991).

J.K. Sass, D. Lackey, J. Schott, 'Electrochemical Double-layer Simulations by Halogen, Alkali and Hydrogen Coadsorption with Water on Metal Surfaces', *Surf. Sci.*, **247**, 239 (1991).

T. Solomun, 'The Role of the Electrolyte Anion in Anodic Dissolution of Pd(100)', *J. Electroanal. Chem.*, **302**, 31–46 (1991).

A.T. Hubbard, J.Y. Gui, 'Molecular Chemistry at Well-defined Surfaces', *J. Chim. Phys.*, **88**, 1547 (1991).

X.P. Gao, A. Hamelin, M.J. Weaver, 'Atomic Relaxation at Ordered Electrode Surfaces Probed by STM: Au(111) in Aqueous-solution Compared with UHV Environments', *J. Chem. Phys.*, **95**, 6993 (1991).

K.R. Zavadil, D. Ingersoll, J.W. Rogers, 'Electrochemical and Surface Spectroscopic Investigation of Cu(111) Films on Ru(0001)', *J. Electroanal. Chem.*, **318**, 223 (1991).

I. Villegas, J.L. Stickney, 'GaAs Deposition on Au(100) and Au(110): A LEED, AES, and STM Study', *J. Vac. Sci. Technol. A*, **10**, 3032 (1992).

P.N. Ross, A.T. D'Agostino, 'The Effect of Surface Reconstruction on the Capacitance of Au(100) Surfaces', *Electrochim. Acta*, **37**, 615 (1992).

J.R. McBride, J.A. Schimpf, M.P. Soriaga, 'In-situ Chemisorption-induced Reordering of Disordered Pd(100) Surfaces', *J. Am. Chem. Soc.*, **114**, 10 950 (1992).

M.P. Soriaga, 'UHV Techniques in the Study of Single-crystal Electrode Surfaces', *Prog. Surf. Sci.*, **39**, 325 (1992).

N. Kizhakevariam, E.M. Stuve, 'Coadsorption of Water and Hydrogen on Pt(100): Formation of Adsorbed Hydronium Ions', *Surf. Sci.*, **275**, 223 (1992).

E.Y. Cao, P. Gao, J.Y. Gui, A.T. Hubbard, 'Adsorption and Electrochemistry of SCN^- at Ag(111) and Pt(111): AES,

CV, HREELS and LEED', *J. Electroanal. Chem.*, **339**, 311 (1992).

G.N. Salaita, A.T. Hubbard, 'Surface Characterization of Molecules at Pt(111) by LEED, Auger, HREELS and EC', *Catal. Today*, **12**, 465 (1992).

D.C. Zapien, J.Y. Gui, D.A. Stern, A.T. Hubbard, 'Surface Electrochemistry and Molecular Orientation: Studies of Pyridyl Hydroquinones at Pt(111)', *J. Electroanal. Chem.*, **330**, 469 (1992).

E.M. Stuve, N. Kizhakevariam, 'Chemistry and Physics of the Liquid–Solid Interface: A Surface Science Perspective', *Vac. Sci. Technol. A*, **11**, 2217 (1993).

R.A. Bradley, R. Georgiadis, S.D. Kevan, 'Nonlinear-optical Spectroscopy of the Ag(111) Surface in an Electrolyte and in Vacuum', *J. Chem. Phys.*, **99**, 5535 (1993).

K. Tanaka, 'UHV-EC: Electrochemical Preparation of Specific Metal Surfaces', *Denki Kagaku*, **61**, 392 (1993).

S.G. Sun, D.F. Yang, S.J. Wu, 'Electrochemical, AES and LEED of Au(210) in the Presence of Adsorbed Pyridine', *J. Electroanal. Chem.*, **349**, 211 (1993).

W.H. Hung, J. Schwartz, S.L. Bernasek, 'Adsorption of H_2O on Oxidized Fe(100) Surfaces', *Surf. Sci.*, **294**, 21 (1993).

J.A. Schimpf, J.R. McBride, M.P. Soriaga, 'Adsorbate-catalyzed Layer-by-layer Metal Dissolution in Halide-free Solutions', *J. Phys. Chem.*, **97**, 10 518 (1993).

P. Len, T. Solomun, 'Underpotential Deposition of Cu on Pd(100): An Electron Spectroscopy Study', *J. Electroanal. Chem.*, **353**, 131 (1993).

D.G. Frank, O.M.R. Chyan, T. Golden, 'I Monolayer Structures at Pt(111) by Means of Angular-distribution Auger Microscopy', *J. Phys. Chem.*, **97**, 5444 (1993).

J.F. Rodriguez, D.L. Taylor, H.D. Abruna, 'Concentration-dependence of the UPD of Ag on Pt(111): Electrochemical and UHV Studies', *Electrochim. Acta*, **38**, 235 (1993).

R.L. Borup, D.E. Sauer, E.M. Stuve, 'Electrodeposited Pb on Pt(111). Surface Redox Behavior of Lead and Dynamic Emersion', *Surf. Sci.*, **293**, 10 (1993).

R.L. Borup, D.E. Sauer, E.M. Stuve, 'Electrodeposited Pb on Pt(111). Vacuum Characterization and Thermal Desorption of the Emersed Adlayer', *Surf. Sci.*, **293**, 27 (1993).

E.Y. Cao, D.A. Stern, J.Y. Gui, A.T. Hubbard, 'Studies of Ru(001) Electrodes in Aqueous Electrolytes', *J. Electroanal. Chem.*, **354**, 71 (1993).

A. Krasnopoler, E.M. Stuve, 'Stabilization of Water by Fluorine on Ag(110)', *Surf. Sci.*, **303**, 355 (1994).

J.A. Schimpf, J.B. Abreu, M.P. Soriaga, 'In-situ Reordering of Extensively Disordered Pd(100) Surfaces', *Electrochim. Acta*, **39**, 2445 (1994).

J.A. Schimpf, J.B. Abreu, A. Carrasquillo, M.P. Soriaga, 'Layer-by-layer Metal Dissolution in Inert Electrolyte: Pd(100)–c(2 × 2)–I', *Surf. Sci.*, **314**, L909 (1994).

J.A. Schimpf, J.B. Abreu, M.P. Soriaga, 'Electrochemical Regeneration of Clean and Ordered Pd(100) Surfaces By Iodine Adsorption–Desorption', *J. Electroanal. Chem.*, **364**, 247 (1994).

N. Kizhakevariam, X.D. Jiang, M.J. Weaver, 'IR Spectroscopy of Model Electrochemical Interfaces in UHV', *J. Chem. Phys.*, **100**, 6750 (1994).

R.L. Borup, D.E. Sauer, E.M. Stuve, 'Electrochemical and Vacuum Behavior of CO and Pb Coadsorbed on Pt(111)', *J. Electroanal. Chem.*, **374**, 235 (1994).

R.L. Borup, D.E. Sauer, E.M. Stuve, 'Electrochemical and Vacuum Coadsorption of CO and Pb on Pt(111)', *J. Vac. Sci. Technol. A*, **12**, 1886 (1994).

A.T. Hubbard, 'Surface Electrochemistry of Metals', *Heterogen. Chem. Rev.*, **1**, 3 (1994).

M.P. Soriaga, 'Electrochemical Surface Science: Applications of UHV Techniques', *Kikan Kagaku Sosetsu*, **26**, 76 (1995).

E.M. Stuve, A. Krasnopoler, D.E. Sauer, 'Relating the In-situ, Ex-situ, and Non-situ Environments in Surface Electrochemistry', *Surf. Sci.*, **335**, 177 (1995).

A. Krasnopoler, E.M. Stuve, 'Evidence of Specific and Nonspecific Adsorption of Perchlorate (ClO_4) on Silver(110)', *J. Vac. Sci. Technol. A*, **13**, 1681 (1995).

A. Krasnopoler, A.L. Johnson, E.M. Stuve, 'Hydrogen Bonding and Molecular Orientation in Water–Fluorine Adlayers on Ag(110)', *Surf. Sci.*, **328**, 186 (1995).

P. Mrozek, Y.E. Sung, A. Wieckowski, 'Ag Deposition on Au(111)', *Surf. Sci.*, **335**, 44 (1995).

F. Bensebaa, T.H. Ellis, 'Water at Surfaces: What Can We Learn from Vibrational Spectroscopy?', *Prog. Surf. Sci.*, **50**, 173 (1995).

Y. Hori, H. Wakebe, T. Tsukamoto, 'Adsorption of CO Accompanied with Simultaneous Charge-transfer on Cu Single-crystals', *Surf. Sci.*, **335**, 258 (1995).

G.M. Brisard, E. Zenati, H.A. Gasteiger, 'UPD of Pb on Cu(111)', *Langmuir*, **11**, 2221 (1995).

M.P. Soriaga, J.A. Schimpf, A. Carrasquillo, 'Electrochemistry of the I-on-Pd Single-crystal Interface: Studies by UHV-EC and In-situ STM', *Surf. Sci.*, **335**, 273 (1995).

J.A. Schimpf, A. Carrasquillo, M.P. Soriaga, 'Electrochemical Digital Etching in Inert Electrolyte', *Electrochim. Acta*, **40**, 1203 (1995).

J.B. Abreu, R.J. Barriga, W. Temesghen, 'I-Catalyzed Dissolution and Reordering of Ion-bombarded Pd(111)', *J. Electroanal. Chem.*, **381**, 239 (1995).

R. Dalbeck, W. Vielstich, 'Ex-situ Analysis of Sulfate Adsorption on Pt via Thermal Desorption Spectroscopy in UHV', *Electrochim. Acta*, **40**, 2687 (1995).

I. Villegas, R. Gomez, M.J. Weaver, 'NO as a Probe Adsorbate for Linking Pt(111) Electrochemical and Model UHV Interfaces', *J. Phys. Chem.*, **99**, 14 832 (1995).

N. Kizhakevariam, I. Villegas, M.J. Weaver, 'Model Electrochemical Interfaces in UHV: Solvent-induced Surface-potential Profiles on Pt(111) from Work-function Measurements and Infrared Stark Effects', *Surf. Sci.*, **336**, 37 (1995).

N.M. Markovic, H.A. Gasteiger, C.A. Lucas, 'The Effect of Cl^- on the UPD of Cu on Pt(111)', *Surf. Sci.*, **335**, 91 (1995).

U.W. Hamm, V. Lazarescu, D.M. Kolb, 'Adsorption of Pyrazine on Au(111) and Ag(111): An XPS Study', *J. Chem. Soc. Faraday Trans.*, **T92**, 3785 (1996).

For references see page 10139

A. Krasnopoler, N. Kizhakevariam, E.M. Stuve, 'Hydrogen Bonding and Surface Interactions in Protic Solvents: Coadsorption of NH_3 and HF with Water on Ag(110)', *J. Chem. Soc. Faraday Trans.*, **T92**, 2445 (1996).

M.P. Soriaga, R.J. Barriga, M.R. Marrero, J.E. Sotó, B. Bravo-Cisneros, M.A. del Valle, 'Tecnicas de Ultravacio En La Electroquimica De Superficie', *Quim. Actualid. Futuro.*, **5**, 69 (1996).

J.B. Abreu, J.-J. Jeng, M.P. Soriaga, J.F. Garst, T.L. Wade, J.L. Stickney, 'The Solid–Liquid Interface in Grignard Reagent Formation', in *Sixth International Symposium on Electrode Processes*, eds. K. Itaya, A. Wieckowski, The Electrochemical Society, Pennington, NJ, 233–242, 1996.

I. Villegas, M.J. Weaver, 'IR Spectroscopy of Model Electrochemical Interfaces in UHV: Evidence for Coupled Cation–Anion Hydration in the Pt(111)/K^+,Cl^- System', *J. Phys. Chem.*, **100**, 19502 (1996).

I. Villegas, M.J. Weaver, 'IR Spectroscopy of Model Electrochemical Interfaces in UHV: Interfacial Cation Solvation by NH_3 on Pt(111)', *Surf. Sci.*, **367**, 162 (1996).

W.F. Temesghen, J.E. Soto, M.P. Soriaga, 'Coordination Chemistry Pd(110) Single-crystal Surfaces', in *Sixth International Symposium on Electrode Processes*, eds. K. Itaya, A. Wieckowski, The Electrochemical Society, Pennington, NJ, 58–67, 1996.

T.E. Lister, L.P. Colletti, J.L. Stickney, 'Electrochemical Formation of CdSe Monolayers on the Low-index Planes of Au', *Israel J. Chem.*, **37**, 287 (1997).

I.S.M. Thurgate, T. Naumovski, P. Hale, 'Surface Reconstructions on Au in Contact with Solution', *Surf. Rev. Lett.*, **4**, 1375 (1997).

G.M. Brisard, E. Zenati, H.A. Gasteiger, 'UPD of Pb on Cu(100) in the Presence of Cl^-: LEED, AES, and EC Studies', *Langmuir*, **13**, 2390 (1997).

G.M. Brisard, E. Zenati, H.A. Gasteiger, 'LEED, AES and EC Studies of the UPD of Pb on Cu(100) and Cu(111)', *ACS Symp. Ser.*, **656**, 142 (1997).

A. Carrasquillo, J.-J. Jeng, R.J. Barriga, 'Ligand Substitution and Competitive Coordination of Halides at Well-defined Pd(100) and Pd(111)', *Inorg. Chim. Acta*, **255**, 249 (1997).

R.G. Jones, M. Kadodwala, 'Br Adsorption on Cu(111)', *Surf. Sci.*, **370**, L219 (1997).

R.L. Borup, D.E. Sauer, E.M. Stuve, 'Electrolyte Interactions with Vapor- and Solution-dosed CO on Pt(111)', *Surf. Sci.*, **374**, 142 (1997).

D.E. Sauer, R.L. Borup, E.M. Stuve, 'Electrocatalysis of HCOOH and CO with Probe Adlayers of C and Ethylidyne on Pt(111)', *ACS Symp. Ser.*, **656**, 283 (1997).

I. Villegas, M.J. Weaver, 'Modeling Electrochemical Interfaces in UHV: Molecular Roles of Solvation in Double-layer Phenomena', *J. Phys. Chem. B*, **101**, 10166 (1997).

I. Villegas, M.J. Weaver, 'IR Spectroscopy of Model Electrochemical Interfaces in UHV: Acetone Chemisorption on Pt(111)', *J. Electoanal. Chem.*, **426**, 55 (1997).

List of selected abbreviations appears in Volume 15

A. Carrasquillo, J.-J. Jeng, M.P. Soriaga, 'The Interfacial Chemistry of Pd(100) Single-crystal Electrodes in Aqueous Cl^- Solutions', *Proc. Electrochem. Soc.*, **17**, 263 (1997).

M.P. Soriaga, K. Itaya, J.L. Stickney, 'STM and UHV-EC: Complementary, Noncompeting Techniques', in *Electrochemical Nanotechnology*, eds. W. Lorenz, W. Plieth, Springer-Verlag, Berlin, 267–276, 1998.

E.A. Lafferty, Y.-G. Kim, M.P. Soriaga, 'Selective and Quantitative Removal of Pd Films from Pt Substrates by Adsorbed-I-catalyzed Anodic Stripping', *Electrochim. Acta*, **44**, 1031 (1998).

J.B. Abreu, A.A. Ashley-Facey, J.E. Soto, M.P. Soriaga, 'The Interfacial Chemistry of the Grignard Reaction', *J. Coll. Interf. Sci.*, **206**, 247 (1998).

Y.-G. Kim, M.P. Soriaga, 'Evidence of I–Pd Place Exchange in the I-catalyzed Dissolution of Pd(111)', *J. Phys. Chem. B*, **102**, 6188 (1998).

G. Pirug, H.P. Bonzel, 'UHV Simulation of the Electrochemical Double Layer: Adsorption of $HClO_4/H_2O$ on Au(111)', *Surf. Sci.*, **405**, 87 (1998).

J. Inukai, Y. Osawa, K. Itaya, 'Adlayer Structures of Chlorine, Bromine, and Iodine on Cu(111) Electrode in Solution: STM and LEED Studies', *J. Phys. Chem. B*, **102**, 10034 (1998).

C. Stuhlmann, Z. Park, C. Bach, 'An Ex-situ Study of Cd UPD on Cu(111)', *Electrochim. Acta*, **44**, 993 (1998).

A. Wieckowski, A. Kolics, J.C. Polkinghorne, 'Radioactive Labeling, EC, and UHV Study of Sulfate Adsorption on Type 316 Stainless Steel', *Corrosion*, **54**, 800 (1998).

C. Tang, S. Zou, M.W. Severson, 'Coverage-dependent IR Spectroscopy of CO on Ir(111) in Aqueous Solution: Comparison Between EC and UHV Environments', *J. Phys. Chem. B*, **102**, 8796 (1998).

R. Gomez, J.M. Feliu, A. Aldaz, 'Double-layer Charge-corrected Voltammetry for Assaying CO Coverages: Comparisons Between EC and UHV Environments', *Surf. Sci.*, **410**, 48 (1998).

L.F. Li, D. Totir, Y. Gofer, 'The Electrochemistry of Ni in a Li-based Solid Polymer Electrolyte in UHV', *Electrochim. Acta*, **44**, 949 (1998).

L.F. Li, D. Totir, G.S. Chottiner, 'Electrochemical Reactivity of CO and S Adsorbed on Ni(111) and Ni(110) in a Li-based Solid Polymer Electrolyte in UHV', *J. Phys. Chem. B*, **102**, 8013 (1998).

M.J. Weaver, 'Potentials of Zero Charge for Pt(111)-aqueous Interfaces: A Combined Assessment from EC and UHV Measurements', *Langmuir*, **14**, 3932 (1998).

D.Y. Zemlyanov, E. Savinova, A. Scheybal, 'XPS Observation of OH Groups Incorporated in an Ag(111) Electrode', *Surf. Sci.*, **418**, 441 (1998).

T.D. Jarvi, T.H. Madden, E.M. Stuve, 'UHV and EC Behavior of Vapor-deposited Ru on Pt(111)', *Electrochem. Solid St.*, **2**, 224 (1999).

Y.-G. Kim, M.P. Soriaga, 'The Interfacial Chemistry of Pd Electrodes', in *Interfacial Electrochemistry*, ed. A. Wieckowski, Marcel Dekker, New York, 249–268, 1999.

J.E. Soto, Y.-G. Kim, M.P. Soriaga, 'UHV-EC and EC-STM Studies of Molecular Chemisorption at Well-defined Pd(hkl) Surfaces', *Electrochem. Commun.*, **1**, 135 (1999).

REFERENCES

1. A.T. Hubbard, R.P. Ishikawa, J. Katekaru, *J. Electroanal. Chem.*, **86**, 271 (1978).
2. G.G. Will, *J. Electrochem. Soc.*, **112**, 451 (1965).
3. J. Clavilier, *J. Electroanal. Chem.*, **107**, 205 (1980).
4. R. Adzic, in *Modern Aspects of Electrochemistry XXI*, eds. R.E. White, J.O'M. Bockris, B.E. Conway, Plenum, New York, 1990.
5. M.P. Soriaga, *Prog. Surf. Sci.*, **39**, 325 (1992).
6. M.P. Soriaga, D.A. Harrington, J.L. Stickney, A. Wieckowski, in *Modern Aspects of Electrochemistry XXVII*, eds. B.E. Conway, R.E. White, J.O'M. Bockris, Plenum, New York, 1996.
7. I.M. Villegas, M.J. Weaver, *J. Phys. Chem. B*, **101**, 10 166 (1997).
8. M.P. Soriaga, R.J. Barriga, M.R. Marrero, J.E. Sotó, B. Bravo-Cisneros, M.A. del Valle, *Quim. Actualid. Futuro.*, **5**, 69 (1996).
9. G.A. Somorjai, *Chemistry in Two Dimensions: Surfaces*, Cornell University, Ithaca, 1981.
10. T.H. Rhodin, G. Ertl (eds.), *The Nature of the Surface Chemical Bond*, North-Holland Publishing, New York, 1979.
11. G. Ertl, J. Kuppers, *Low Energy Electrons and Surface Chemistry*, VCH Publishers, New York, 1985.
12. D.P. Woodruff, T.A. Delchar, *Modern Techniques of Surface Science*, Cambridge University Press, New York, 1986.
13. A. Zangwill, *Physics at Surfaces*, Cambridge University Press, New York, 1988.
14. M.A. Van Hove, S.W. Wang, D.F. Ogletree, G.A. Somorjai, *Adv. Quantum Chem.*, **20**, 1 (1989).
15. G.A. Somorjai, *Introduction to Surface Chemistry and Catalysis*, Wiley, New York, 1994.
16. E.A. Wood, *Crystal Orientation Manual*, Columbia University Press, New York, 1963.
17. A.W. Adamson, *Physical Chemistry of Surfaces*, Wiley, New York, 1990.
18. M.S. Zei, Y. Nakai, G. Lehmpfuhl, D.M. Kolb, *J. Electroanal. Chem.*, **150**, 201 (1983).
19. K.W. Nebesny, N.R. Armstrong, *J. Electron Spectrosc.*, **37**, 355 (1986).
20. J.M. Morabito, *Surf. Sci.*, **49**, 318 (1975).
21. W.A. Coghlan, R.E. Clausing, *At. Data*, **5**, 317 (1973).
22. J. Houston, *Surf. Sci.*, **38**, 283 (1973).
23. J.A. Schoeffel, A.T. Hubbard, *Anal. Chem.*, **49**, 2330 (1977).
24. B.C. Schardt, J.L. Stickney, D.A. Stern, D.G. Frank, J.Y. Katekaru, S.D. Rosasco, G.N. Salaita, M.P. Soriaga, A.T. Hubbard, *Inorg. Chem.*, **24**, 1419 (1985).
25. G.N. Salaita, A.T. Hubbard, in *Molecular Design of Electrode Surfaces*, ed. R.W. Murray, Wiley, New York, 1989.
26. M. Gryzinski, *Phys. Rev.*, **A138**, 305 (1965).
27. K. Siegbahn, *ESCA, Atomic, Molecular and Solid State Structure Studies by Means of Electron Spectroscopy*, Almqvist-Wiksell Boktryckeri, Sweden, 1967.
28. M.P. Seah, in *Practical Surface Analysis by Auger and X-ray Photoelectron Spectroscopy*, eds. D. Briggs, M.P. Seah, Wiley, New York, 1983.
29. H. Ibach, D.A. Mills, *Electron Energy Loss Spectroscopy*, Academic Press, New York, 1982.
30. N.R. Avery, in *Vibrational Spectroscopy of Molecules on Surfaces*, eds. J.T. Yates, Jr, T.E. Madey, Plenum, New York, 1987.
31. F.M. Hoffman, *Surf. Sci. Rep.*, **3**, 2 (1983).
32. H.A. Pearce, N. Sheppard, *Surf. Sci.*, **59**, 205 (1976).
33. A. Bewick, B.S. Pons, in *Advances in Infrared and Raman Spectroscopy*, eds. R.J.H. Hester, R.E. Clark, Wiley-Hayden, London, 1985.
34. H.G. Tompkins, in *Methods of Surface Analysis*, ed. A. Czanderna, Elsevier, New York, 1975.
35. H. Froitzheim, in *Topics in Current Physics. IV*, ed. H. Ibach, Springer-Verlag, New York, 1977.
36. W.N. Hansen, *J. Electroanal. Chem.*, **150**, 133 (1983).
37. W.N. Hansen, G.J. Hansen, in *Electrochemical Surface Science*, ed. M.P. Soriaga, American Chemical Society, Washington, DC, 1988.
38. P.A. Redhead, *Vacuum*, **12**, 203 (1962).
39. C.N. Chittenden, E.D. Pylant, A.L. Schwaner, J.M. White, in *Surface Imaging and Visualization*, ed. A.T. Hubbard, CRC Press, Boca Raton, 1995.
40. H.A. Strobel, W.R. Heineman, *Chemical Instrumentation*, Wiley, New York, 1989.
41. J. Augustynski, L. Balsenc, in *Modern Aspects of Electrochemistry XIV*, eds. J.O'M. Bockris, B.E. Conway, Plenum, New York, 1979.
42. P.M.A. Sherwood, in *Contemporary Topics in Analytical and Clinical Chemistry*, eds. D.M. Hercules, G.M. Hieftje, L.R. Snyder, M.E. Evenson, Plenum, New York, 1982.
43. J.S. Hammond, N. Winograd, *J. Electroanal. Chem.*, **78**, 55 (1977).
44. A.T. Hubbard, *Acc. Chem. Res.*, **13**, 177 (1980).
45. P.N. Ross, in *Chemistry and Physics of Solid Surfaces*, eds. R. Vanselow, R. Howe, Springer-Verlag, New York, 1982.
46. E. Yeager, A. Homa, B.D. Cahan, D. Scherson, *J. Vac. Sci. Technol.*, **20**, 628 (1982).
47. D.M. Kolb, *Z. Phys. Chem. Neue Folge*, **154**, 179 (1987).
48. A.T. Hubbard, M.P. Soriaga, J.L. Stickney, in *New Directions in Chemical Analysis*, ed. B. Shapiro, Texas A&M University Press, College Station, 1985.
49. A.T. Hubbard, *Chem. Rev.*, **88**, 633 (1988).

List of selected abbreviations appears in Volume 15

50. M.E. Hanson, E. Yeager, in *Electrochemical Surface Science*, ed. M.P. Soriaga, American Chemical Society, Washington, DC, 1988.
51. J.L. Stickney, C.B. Ehlers, B.W. Gregory, in *Electrochemical Surface Science*, ed. M.P. Soriaga, American Chemical Society, Washington, DC, 1988.
52. T. Mebrahtu, J.F. Rodriguez, M.E. Bothwell, I.F. Cheng, D.R. Lawson, J.R. McBride, C.R. Martin, M.P. Soriaga, *J. Electroanal. Chem.*, **267**, 351 (1989).
53. T. Solomun, W. Richtering, H. Gerischer, *Ber. Bunsen-Ges. Phys. Chem.*, **91**, 412 (1987).
54. K.R. Zavadil, N.R. Armstrong, *J. Electrochem. Soc.*, **137**, 2371 (1990).
55. L.-W.H. Leung, T.W. Gregg, D.W. Goodman, *Rev. Sci. Instrum.*, **62**, 1857 (1991).
56. B.E. Conway, H. Angerstein-Kozlowska, W.B.A. Sharp, E.E. Criddle, *Anal. Chem.*, **45**, 1331 (1973).
57. A.J. Bard, H.D. Abruña, C.E. Chidsey, L.R. Faulkner, S. Feldberg, K. Itaya, O. Melroy, R.W. Murray, M.D. Porter, M.P. Soriaga, H.S. White, *J. Phys. Chem.*, **97**, 7147 (1993).
58. A.J. Bard, L.R. Faulkner, *Electrochemical Methods*, John Wiley, New York, 1980.
59. M.P. Soriaga (ed.), *Electrochemical Surface Science*, American Chemical Society, Washington, DC, 1988.
60. E. Yeager, J. Kuta, in *Physical Chemistry, An Advanced Treatise. IXA*, eds. H. Eyring, D. Henderson, W. Jost, Academic Press, New York, 1970.
61. W. Hansen, D.M. Kolb, *J. Electroanal. Chem.*, **100**, 493 (1979).
62. D. Rath, D.M. Kolb, *Surf. Sci.*, **109**, 641 (1981).
63. G.J. Hansen, W.N. Hansen, *J. Electroanal. Chem.*, **150**, 193 (1983).
64. W.N. Hansen, *Surf. Sci.*, **101**, 109 (1980).
65. J.D.E. McIntyre, *Adv. Electrochem. Electrochem. Engr.*, **9**, 61 (1973).
66. W. Hansen, D.M. Kolb, D. Rath, R. Wille, *J. Electroanal. Chem.*, **110**, 369 (1980).
67. O. Hofman, K. Doblhofer, H. Gerischer, *J. Electroanal. Chem.*, **161**, 337 (1984).
68. G.J. Hansen, W.N. Hansen, *Ber. Bunsen-Ges. Phys. Chem.*, **91**, 317 (1987).
69. J.E. Pemberton, R.L. Sobocinksi, M.A. Bryant, *J. Am. Chem. Soc.*, **112**, 6177 (1990).
70. J.E. Pemberton, R.L. Sobocinksi, *J. Electroanal. Chem.*, **318**, 157 (1991).
71. E.R. Koetz, H. Neff, R.H. Muller, *J. Electroanal. Chem.*, **215**, 331 (1986).
72. S.D. Rosasco, J.L. Stickney, G.N. Salaita, D.G. Frank, J.Y. Katekaru, B.C. Schardt, M.P. Soriaga, D.A. Stern, A.T. Hubbard, *J. Electroanal. Chem.*, **188**, 95 (1985).
73. D.G. Frank, J.Y. Katekaru, S.D. Rosasco, G.N. Salaita, B.C. Schardt, M.P. Soriaga, J.L. Stickney, A.T. Hubbard, *Langmuir*, **1**, 587 (1985).
74. J.F. Rodriguez, T. Mebrahtu, M.P. Soriaga, *J. Electroanal. Chem.*, **264**, 291 (1979).
75. T.E. Felter, A.T. Hubbard, *J. Electroanal. Chem.*, **100**, 473 (1979).
76. B.C. Schardt, S.L. Yau, F. Rinaldi, *Science*, **243**, 1050 (1989).
77. D.A. Stern, L. Laguren-Davidson, F. Lu, C.H. Lin, D.G. Frank, G.N. Salaita, N. Walton, J.Y. Gui, D.C. Zapien, A.T. Hubbard, *J. Am. Chem. Soc.*, **111**, 877 (1989).
78. T.E. Madey, F.P. Netzer, *Surf. Sci.*, **117**, 549 (1980).
79. J.K. Sass, K. Kretzschmar, S. Holloway, *Vacuum*, **31**, 483 (1981).
80. J.K. Sass, *Vacuum*, **33**, 741 (1983).
81. J.K. Sass, K. Bange, in *Electrochemical Surface Science*, ed. M.P. Soriaga, American Chemical Society, Washington, DC, 1988.
82. J.K. Sass, K. Bange, R. Dohl, E. Plitz, R. Unwin, *Ber. Bunsenges, Phys. Chem.*, **88**, 354 (1984).
83. F.T. Wagner, S.J. Schmieg, T.E. Moylan, *Surf. Sci.*, **195**, 403 (1988).
84. F.T. Wagner, T.E. Moylan, in *Electrochemical Surface Science*, ed. M.P. Soriaga, American Chemical Society, Washington, DC, 1988.
85. E.M. Stuve, A. Krasnopoler, D.E. Sauer, *Surf. Sci.*, **335**, 177 (1995).
86. D.A. Stern, H. Baltruschat, M. Martinez, J.L. Stickney, D. Song, S.K. Lewis, D.G. Frank, A.T. Hubbard, *J. Electroanal. Chem.*, **217**, 101 (1987).
87. G.N. Salaita, D.A. Stern, F. Lu, H. Baltruschat, B.C. Schardt, J.L. Stickney, M.P. Soriaga, D.G. Frank, A.T. Hubbard, *Langmuir*, **2**, 828 (1986).
88. F. Lu, G.N. Salaita, H. Baltruschat, A.T. Hubbard, *J. Electroanal. Chem.*, **222**, 305 (1987).
89. N. Batina, J.W. McCargar, L. Laguren-Davidson, C.-H. Lin, A.T. Hubbard, *Langmuir*, **5**, 123 (1989).
90. J.L. Stickney, S.D. Rosasco, G.N. Salaita, A.T. Hubbard, *Langmuir*, **1**, 66 (1985).
91. B.C. Schardt, J.L. Stickney, D.A. Stern, D.G. Frank, J.Y. Katekaru, S.D. Rosasco, G.N. Salaita, M.P. Soriaga, A.T. Hubbard, *Inorg. Chem.*, **24**, 1419 (1985).
92. S.D. Rosasco, J.L. Stickney, G.N. Salaita, D.G. Frank, J.Y. Katekaru, B.C. Schardt, M.P. Soriaga, D.A. Stern, A.T. Hubbard, *J. Electroanal. Chem.*, **188**, 95 (1985).
93. G.A. Garwood, A.T. Hubbard, *Surf. Sci.*, **112**, 281 (1982).
94. G.A. Garwood, A.T. Hubbard, *Surf. Sci.*, **121**, 396 (1982).
95. T. Solomun, A. Wieckowski, S.D. Rosasco, A.T. Hubbard, *Surf. Sci.*, **147**, 241 (1984).
96. C.B. Ehlers, J.L. Stickney, *Surf. Sci.*, **239**, 85 (1990).
97. G.N. Salaita, F. Lu, L. Laguren-Davidson, A.T. Hubbard, *J. Electroanal. Chem.*, **229**, 1 (1987).
98. B.G. Bravo, S.L. Michelhaugh, M.P. Soriaga, I. Villegas, D.W. Suggs, J.L. Stickney, *J. Phys. Chem.*, **95**, 5245 (1991).
99. T. Mebrahtu, M.E. Bothwell, J.E. Harris, G.J. Cali, M.P. Soriaga, *J. Electroanal. Chem.*, **300**, 487 (1991).

100. D.G. Frank, J.Y. Katekaru, S.D. Rosasco, G.N. Salaita, B.C. Schardt, M.P. Soriaga, J.L. Stickney, A.T. Hubbard, *Langmuir*, **1**, 587 (1985).
101. P.N. Ross, in *Electrochemical Surface Science*, ed. M.P. Soriaga, American Chemical Society, Washington, DC, 1988.
102. W.E. O'Grady, M.Y.C. Woo, P.L. Hagans, E. Yeager, *J. Vac. Sci. Technol.*, **14**, 365 (1977).
103. K. Yamamoto, D.M. Kolb, R. Kotz, G. Lempfuhl, *J. Electroanal. Chem.*, **96**, 233 (1979).
104. P.N. Ross, *J. Electroanal. Chem.*, **126**, 67 (1979).
105. A.S. Homa, E. Yeager, B.D. Cahan, *Electroanal. Chem.*, **150**, 181 (1983).
106. J. Clavilier, A. Rodes, K. El Achi, M.A. Zamakhchari, *J. Chim. Phys.*, **88**, 1291 (1991).
107. P.N. Ross, *J. Chim. Phys.*, **88**, 1353 (1991).
108. G. Jerkiewicz, B.E. Conway, *J. Chim. Phys.*, **88**, 1291 (1991).
109. F.T. Wagner, P.N. Ross, *Surf. Sci.*, **160**, 305 (1985).
110. J. Horkans, B.D. Cahan, E. Yeager, *Surf. Sci.*, **37**, 559 (1973).
111. F.T. Wagner, P.N. Ross, *J. Electroanal. Chem.*, **150**, 141 (1985).
112. J.B. Lumsden, G.A. Garwood, A.T. Hubbard, *Surf. Sci.*, **121**, 1524 (1982).
113. D.A. Harrington, A. Wieckowski, S.D. Rosasco, B.C. Schardt, G.N. Salaita, J.B. Lumsden, A.T. Hubbard, *Corrosion Sci.*, **25**, 849 (1985).
114. D.A. Harrington, A. Wieckowski, S.D. Rosasco, B.C. Schardt, G.N. Salaita, A.T. Hubbard, *Langmuir*, **1**, 232 (1985).
115. J.P. Biberian, G.A. Somorjai, *J. Vac. Sci. Technol.*, **16**, 2073 (1979).
116. J.S. Hammond, N. Winograd, *J. Electrochem. Soc.*, **124**, 826 (1977).
117. A.T. Hubbard, J.L. Stickney, S.D. Rosasco, M.P. Soriaga, D. Song, *J. Electroanal. Chem.*, **150**, 165 (1983).
118. J.L. Stickney, S.D. Rosasco, M.P. Soriaga, D. Song, A.T. Hubbard, *Surf. Sci.*, **130**, 326 (1983).
119. J.L. Stickney, S.D. Rosasco, B.C. Schardt, A.T. Hubbard, *J. Phys. Chem.*, **88**, 251 (1984).
120. J.L. Stickney, S.D. Rosasco, A.T. Hubbard, *J. Electrochem. Soc.*, **131**, 260 (1984).
121. J.L. Stickney, D.A. Stern, B.C. Schardt, D.C. Zapien, A. Wieckowski, A.T. Hubbard, *J. Electroanal. Chem.*, **213**, 293 (1986).
122. J.L. Stickney, B.C. Schardt, D.A. Stern, A. Wieckowski, A.T. Hubbard, *J. Electrochem. Soc.*, **133**, 648 (1986).
123. B.C. Schardt, J.L. Stickney, D.A. Stern, A. Wieckowski, D.C. Zapien, A.T. Hubbard, *Langmuir*, **3**, 239 (1987).
124. B.C. Schardt, J.L. Stickney, D.A. Stern, A. Wieckowski, D.C. Zapien, A.T. Hubbard, *Surf. Sci.*, **175**, 520 (1986).
125. P.C. Andricacos, P.N. Ross, *J. Electroanal. Chem.*, **167**, 301 (1984).
126. D. Aberdam, R. Durand, R. Faure, F. El-Omar, *Surf. Sci.*, **162**, 782 (1985).
127. L.-W.H. Leung, T.W. Gregg, D.W. Goodman, *Langmuir*, **7**, 3205 (1991).
128. L. Laguren-Davidson, F. Lu, G.N. Salaita, A.T. Hubbard, *Langmuir*, **4**, 224 (1988).
129. B.W. Gregory, J.L. Stickney, *J. Electroanal. Chem.*, **300**, 543 (1991).
130. D.W. Suggs, J.L. Stickney, *J. Phys. Chem.*, **95**, 10 056 (1991).
131. K. Itaya, *Prog. Surf. Sci.*, **58**, 121 (1998).
132. J.A. Schimpf, J.R. McBride, M.P. Soriaga, *J. Phys. Chem.*, **97**, 10 518 (1993).
133. Y.-G. Kim, M.P. Soriaga, *J. Phys. Chem. B*, **102**, 6188 (1998).
134. M.P. Soriaga, in *Frontiers of Electrochemistry II*, eds. P.N. Ross, J. Lipkowski, VCH, New York, 1992.
135. B.B. Damaskin, A.N. Frumkin, *Electrochim. Acta*, **19**, 173 (1974).
136. R. Parsons, *J. Electroanal. Chem.*, **59**, 229 (1975).
137. P.A. Thiel, T.E. Madey, *Surf. Sci. Rep.*, **7**, 211 (1987).
138. C.K. Mann, K.K. Barnes, *Electrochemical Reactions in Nonaqueous Solvents*, Marcel Dekker, New York, 1970.
139. H. Ohtani, C.T. Kao, M.A. Van Hove, G.A. Somorjai, *Prog. Surf. Sci.*, **23**, 155 (1987).
140. G.A. Garwood, Jr, A.T. Hubbard, *Surf. Sci.*, **118**, 233 (1982).
141. J.Y. Gui, F. Lu, D.A. Stern, A.T. Hubbard, *J. Electroanal. Chem.*, **292**, 245 (1990).
142. B.B. Damaskin, O.A. Petrii, V.V. Batrakov, *Adsorption of Organic Compounds on Electrodes*, Plenum, New York, 1971.
143. R.D. Snell, A.G. Keenan, *Chem. Soc. Rev.*, **8**, 259 (1979).
144. J.L. Stickney, M.P. Soriaga, A.T. Hubbard, S.E. Anderson, *J. Electroanal. Chem.*, **125**, 73 (1981).
145. S.L. Michelhaugh, C. Bhardwaj, G.J. Cali, B.G. Bravo, M.E. Bothwell, G.M. Berry, M.P. Soriaga, *Corrosion*, **47**, 322 (1991).
146. D. Zurawski, M. Wasberg, A. Wieckowski, *J. Phys. Chem.*, **94**, 2076 (1990).
147. G.M. Berry, M.E. Bothwell, S.L. Michelhaugh, J.R. McBride, M.P. Soriaga, *J. Chim. Phys.*, **88**, 1591 (1991).
148. G.M. Berry, J.R. McBride, J.A. Schimpf, M.P. Soriaga, *J. Electroanal. Chem.*, **353**, 281 (1993).
149. A. Wieckowski, S.D. Rosasco, G.N. Salaita, A.T. Hubbard, B.E. Bent, F. Zaera, D. Godbey, G.A. Somorjai, *J. Am. Chem. Soc.*, **107**, 5910 (1985).
150. A.T. Hubbard, M.A. Young, J.A. Schoeffel, *J. Electroanal. Chem.*, **114**, 273 (1980).
151. F. Zaera, G.A. Somorjai, *J. Am. Chem. Soc.*, **106**, 2288 (1984).
152. M. Hourani, A. Wieckowski, *Langmuir*, **6**, 379 (1990).
153. J.Y. Gui, D.A. Stern, D.C. Zapien, G.N. Salaita, F. Lu, C.H. Lin, B.E. Kahn, A.T. Hubbard, *J. Electroanal. Chem.*, **252**, 169 (1988).
154. B.E. Kahn, S.A. Chaffins, J.Y. Gui, F. Lu, D.A. Stern, A.T. Hubbard, *Chem. Phys.*, **141**, 21 (1990).

155. G.N. Salaita, C. Lin, P. Gao, A.T. Hubbard, *Arabian J. Sci. Engr.*, **15**, 319 (1990).
156. J.Y. Gui, L. Laguren-Davidson, C.H. Lin, F. Lu, G.N. Salaita, D.A. Stern, B.E. Kahn, A.T. Hubbard, *Langmuir*, **5**, 819 (1989).
157. N. Batina, D.C. Zapien, F. Lu, C.H. Lin, J.W. McCargar, B.E. Kahn, J.Y. Gui, D.G. Frank, G.N. Salaita, D.A. Stern, A.T. Hubbard, *Electrochim. Acta*, **34**, 1031 (1989).
158. N. Batina, S.A. Chaffins, J.Y. Gui, F. Lu, J.W. McCargar, J.W. Rovang, D.A. Stern, A.T. Hubbard, *J. Electroanal. Chem.*, **284**, 81 (1990).
159. N. Batina, J.W. McCargar, C.H. Lin, G.N. Salaita, B.E. Kahn, A.T. Hubbard, *Electroanal.*, **1**, 213 (1989).
160. M.P. Soriaga, E. Binamira-Soriaga, A.T. Hubbard, J.B. Benziger, K.W.P. Pang, *Inorg. Chem.*, **24**, 65 (1985).
161. M.P. Soriaga, J.H. White, V.K.F. Chia, D. Song, P.O. Arrhenius, A.T. Hubbard, *Inorg. Chem.*, **24**, 73 (1985).
162. M.P. Soriaga, J.L. Stickney, A.T. Hubbard, *J. Mol. Catal.*, **21**, 211 (1983).
163. N. Batina, B.E. Kahn, J.Y. Gui, F. Lu, J.W. McCargar, H.B. Mark, C.H. Lin, B.N. Salaita, H. Zimmer, D.A. Stern, A.T. Hubbard, *Langmuir*, **5**, 588 (1989).
164. D.A. Stern, L. Laguren-Davidson, F. Lu, C.H. Lin, D.G. Frank, G.N. Salaita, N. Walton, J.Y. Gui, D.C. Zapien, A.T. Hubbard, *J. Am. Chem. Soc.*, **111**, 877 (1989).
165. S.A. Chaffins, J.Y. Gui, B.E. Kahn, C.H. Lin, F. Lu, G.N. Salaita, D.A. Stern, D.C. Zapien, A.T. Hubbard, C.M. Elliott, *Langmuir*, **6**, 951 (1990).
166. S.A. Chaffins, J.Y. Gui, C.H. Lin, F. Lu, G.N. Salaita, D.A. Stern, A.T. Hubbard, *Langmuir*, **6**, 1273 (1990).
167. A.T. Hubbard, D.G. Frank, D.A. Stern, M.J. Tarlov, N. Batina, N. Walton, E. Wellner, J.W. McCargar, in *Redox Chemistry and Interfacial Behavior of Biological Molecules*, eds. G. Dryhurst, R. Niki, Plenum, New York, 1988.
168. D.A. Stern, N. Walton, J.W. McCargar, G.N. Salaita, L. Laguren-Davidson, F. Lu, C.H. Lin, J.Y. Gui, N. Batina, D.G. Frank, A.T. Hubbard, *Langmuir*, **4**, 711 (1988).
169. M.P. Soriaga, Y.-G. Kim, in *Interfacial Electrochemistry*, ed. A. Wieckowski, Marcel Dekker, New York, 1999.
170. J.E. Soto, Y.-G. Kim, M.P. Soriaga, *Electrochem. Commun.*, **1**, 135 (1999).

Ultrafast Electrochemical Techniques

Robert J. Forster
Dublin City University, Dublin, Ireland

List of selected abbreviations appears in Volume 15

Ultrafast electrochemical techniques provide information about the kinetics and thermodynamics of redox processes that occur at submillisecond or even nanosecond timescales. This short timescale is achieved either by making very rapid transient measurements or by using ultrasmall probes to achieve very high rates of diffusion under steady-state conditions. Microelectrodes (i.e. electrodes with critical dimensions in the micrometer range) play pivotal roles in both approaches. Electrochemistry has several advantages over spectroscopy in that it provides direct information about electron transfer (E) and coupled chemical (C) reactions. Ultrafast electrochemical techniques now allow it to do so at times as short as 10 ns. In transient measurements, decreasing the lower accessible timescale depends critically on fabricating ultramicroelectrodes that continue to respond ideally as their critical dimension (e.g. the radius of a microdisc) decreases. It is now possible to assemble microelectrodes that respond to changes in applied potential within less than a few nanoseconds. In steady-state approaches, ultrasmall probes are required to make short timescale measurements and various approaches that yield nanodes (i.e. electrodes of nanometer dimension) have been proposed. However, beyond the need for smaller probes and faster instrumentation, the continued development of new theory describing electron transfer is essential, where the dimensions of the zone that is depleted of reactant because of a Faradaic reaction and the electrochemical double layer become comparable.

1 PRINCIPLES AND OBJECTIVES OF ULTRAFAST ELECTROCHEMICAL TECHNIQUES

1.1 Introduction

Irrespective of the approach taken, from the detection of highly reactive intermediates to the tantalizing possibility of directly probing the energetics and dynamics of facile electron transfers, ultrafast electrochemical techniques are revolutionizing investigations into redox processes. In doing so, they promise to revolutionize the type, quality and range of information available to test and develop new theories of electron transfer. Moreover, these high-speed approaches will allow new devices to be developed, such as sensors that exploit differences in reaction dynamics rather than energetics to achieve a selective response.

1.2 The Need for Speed

Many significant electrochemical events, such as electron and proton transfers, ligand exchanges, isomerizations, and ejection of leaving groups, occur on the low microsecond and nanosecond time domains. To achieve a meaningful insight into these redox processes, it must be possible to measure rate constants under a wide range of experimental conditions, such as driving force, temperature and so on. However, conventional electrochemical methods cannot fulfil this role since they are restricted to millisecond, or longer, timescales. Thus, while modern laser-based spectroscopy has provided a powerful new insight into chemical processes that occur at picosecond and even femtosecond timescales, it is only recently that electrochemists have meaningfully probed redox processes occurring on the submicrosecond timescale.

The objective of this article is not to provide an extensive catalogue of recent studies conducted in this area. Rather, the intention is to describe the theory and practice of ultrafast electrochemical techniques, to convey a sense of the current state of the art in the area, and to look at likely future developments.

Much of our discussion will focus on microelectrodes; the ability routinely to fabricate electrodes with radii smaller than one hundredth of the thickness of a human hair has profoundly changed the way electrochemistry is undertaken. The impact of microelectrodes in redefining the spatial and time limits of electrochemistry began in the late 1960s with Fleischmann's pioneering work.[1] Since the 1970s, researchers have used these ultrasmall probes to study the kinetics and mechanism of fast heterogeneous electron transfers and homogeneous chemical reactions, to extend the range of electroanalytical measurements, to map redox activity topographically using scanning electrochemical microscopy (SECM), and to provide platforms for the fabrication of sensors for both in vitro and in vivo applications.[2] The areas of microelectrodes are more than one million times smaller than conventional electrodes, allowing electrochemical measurements to be performed under unusual conditions and in nonconventional media. For example, it is now possible to monitor the concentrations of radicals in organic solvents that live for only a few nanoseconds. Moreover, measurements can be performed with a high degree of spatial resolution, e.g. the detection of neurotransmitter release from single brain cells.

2 DIRECT ELECTROCHEMICAL METHODS

2.1 Microelectrode Fabrication and Characterization

Microelectrodes are miniature electrodes where the critical electrode dimension is less than about 10 µm yet remains much greater than the thickness of the electrical double layer, which is typically 10–100 Å ($1\ \text{Å} = 10^{-10}$ m). The physicochemical properties of microelectrodes are distinctly different from the properties observed for macroelectrodes and from those predicted for electrodes whose dimensions approach the molecular level. Although microelectrodes have been used since the 1940s, e.g. to measure oxygen concentrations within tissue,[3] it is only since the 1980s that the availability of microscopic gold and platinum wires has allowed them to be widely used.

Figure 1 illustrates the five common microelectrode geometries. The microdisc is the most popular geometry and is employed in approximately 50% of all investigations. Other common geometries include cylinders (20%) and arrays (20%), with the remaining 10% comprising bands and rings and less frequently spheres, hemispheres, and more unusual assemblies. The most popular materials include platinum, carbon fibers, and gold, although mercury, iridium, nickel, silver, and superconducting ceramics have also been used.[2] Microdisc electrodes predominate because of their ease

For references see page 10168

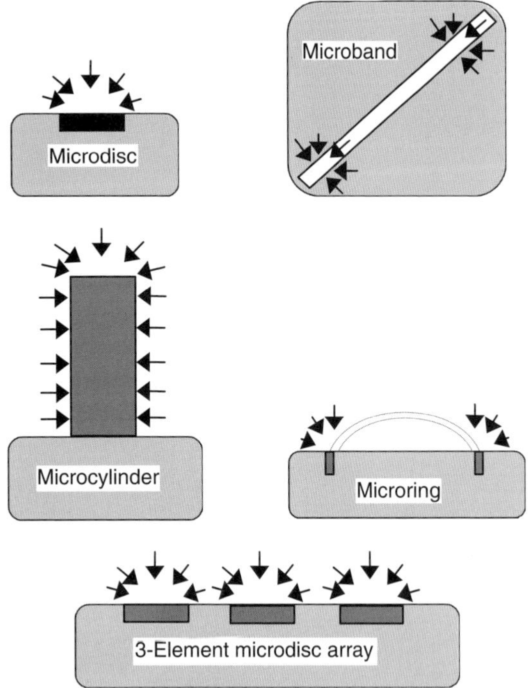

Figure 1 The most common microelectrode geometries and their associated diffusion fields.

of construction and because the sensing surface of the electrode can be mechanically polished. Microelectrodes in the form of discs, cylinders, and bands are commonly fabricated by sealing a fine wire or foil into a nonconducting electrode body such as glass. Microlithographic techniques are perhaps the best method of producing well-defined microelectrode arrays. Other array fabrication methods include immobilizing large numbers of metal wires within a nonconducting support and electrodeposition of mercury or platinum within the pores of a polymer membrane. Spherical and hemispherical microelectrodes are typically formed by electrodepositing mercury onto platinum or iridium microdiscs.

2.2 Cell Time Constants

Every electrochemical measurement has a lower timescale limit that is imposed by the RC cell time constant, i.e. the product of the solution resistance, R, and the double layer capacitance, C, of the working electrode. Meaningful electrochemical data can only be extracted at timescales that are longer than the cell time constant, typically five to ten times the RC time constant.[(4)] Therefore, an important objective when seeking to make high-speed transient measurements is to minimize the cell time constant.

As discussed in detail by Newman[(5)] and described in Equation (1), the solution resistance for a disc-shaped ultramicroelectrode is inversely proportional to the electrode radius,

$$R = \frac{1}{4\kappa r} \tag{1}$$

where κ is the conductivity of the solution and r is the radius of the microdisc. Equation (1) shows that R increases as the electrode radius decreases. Thus, changes in the cell resistance with decreasing electrode radius do not have the desired effect of reducing the RC cell time constant.

Altering the potential that is applied to an electrode causes the charge on the metal side of the interface to change and some reorganization of the ions and solvent dipoles in the double layer on the solution side of the interface will occur. This process causes electrons to flow into or out of the surface giving rise to the charging or capacitive response. The double layer capacitance for a disc-shaped ultramicroelectrode is proportional to the area of the electrode surface and is given by Equation (2),

$$C = \pi r^2 C_o \tag{2}$$

where C_o is the specific double layer capacitance of the electrode. Thus, shrinking the size of the electrode causes the interfacial capacitance to decrease with decreasing r^2.

The existence of the double layer capacitance at the working electrode complicates electrochemical measurements at short timescales. Figure 2 is an equivalent circuit of an electrochemical cell, where Z_F is the Faradaic impedance corresponding to the electrochemical reaction. In seeking to make transient measurements, the electrochemical cell must respond to the applied potential waveform much more rapidly than the process one is seeking to measure. However, the potential across a capacitor cannot be changed instantaneously and the double layer capacitance must be charged through the solution resistance in order to change the potential across the Faradaic impedance.

The time constant for this charging process is given by Equation (3):

$$RC = \frac{\pi r C_o}{4\kappa} \tag{3}$$

and is typically hundreds of microseconds for a conventional millimeter-size electrode, placing a lower limit on the useful timescale of the order of several milliseconds.

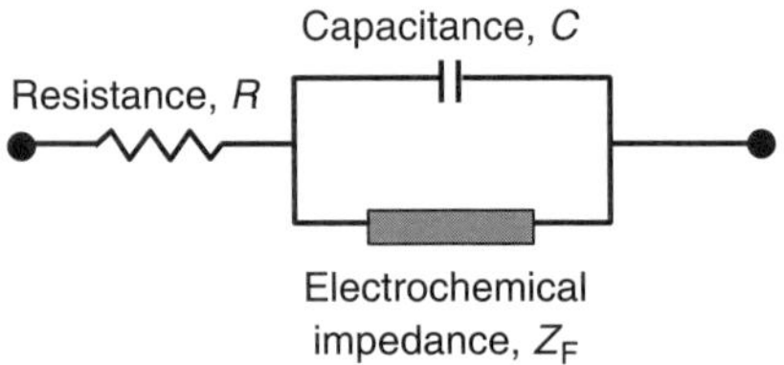

Figure 2 Equivalent circuit for an electrochemical cell and associated electrochemical process.

The use of ultramicroelectrodes with critical dimensions in the micrometer and even nanometer range has opened new possibilities for fast kinetic studies because of the greatly diminished capacitance of these ultrasmall probes.

For example, Figure 3 shows how the *RC* cell time constant as measured in 0.1 M HCl depends on the radius of platinum microdisc electrodes. In these experiments, the double layer charging process was monitored using chronoamperometry conducted at microsecond timescales following a potential step within the double layer region of the voltammogram. As the electrode radius decreases from 25 to 1 μm, the cell time constant decreases linearly from approximately 2 μs to 80 ns. The slope of the best fit line is consistent with Equation (3) where the double layer capacitance is about 40 μF cm^{-2}. Moreover, the intercept is approximately 4.3 ns indicating that the stray capacitance (see below) of these microelectrodes is very small. In conclusion, it is apparent that by using ultramicroelectrodes with radii of microns, cell time constants of tens of nanoseconds can be achieved.

The preceding analysis indicates that the overall cell resistance influences the cell time constant. Therefore, decreasing the resistance of the solution, through which the Faradaic and charging currents must flow, will decrease the cell time constant. Figure 4 illustrates the decrease in cell time constant that is observed for a 5 μm radius platinum microdisc as the solution conductivity is systematically varied by changing the supporting electrolyte concentration from 0.05 to 2.0 M. As predicted by Equation (3), this figure shows that a linear response is obtained, and the specific double layer capacitance is estimated as 41.7 ± 2.1 μF cm^{-2} over this range of supporting electrolyte concentrations. Significantly, Figures 3 and 4 indicate that microdisc electrodes can be manufactured that respond ideally to changes in the applied potential at timescales as short as 30 ns. Experiments performed in more conducting solutions, e.g. highly concentrated acids, indicate that *RC* cell time constants as short as 5 ns can be achieved.

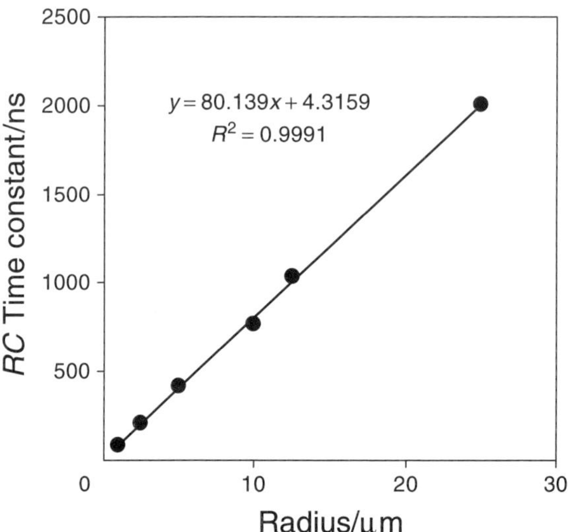

Figure 3 Relationship between the *RC* cell time constant and the radius of platinum microdiscs where the supporting electrolyte is 0.1-M HCl. Cell time constants were measured using chronoamperometry conducted on a microsecond to submicrosecond timescale by stepping the potential from 0.200 to 0.250 V vs Ag/AgCl.

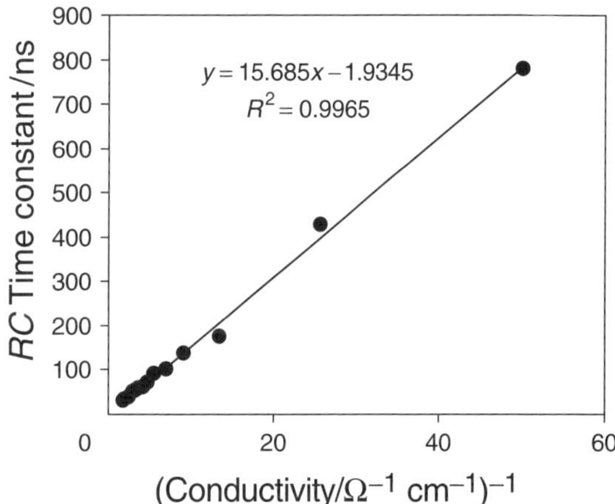

Figure 4 Relationship between the *RC* cell time constant of a 5 μm radius platinum microdisc and the reciprocal of the solution conductivity. Cell time constants were measured using chronoamperometry conducted on a microsecond to submicrosecond timescale by stepping the potential from 0.200 to 0.250 V vs Ag/AgCl.

However, there are a number of practical problems associated with the design and fabrication of microelectrodes that cause micrometer-sized electrodes to have *RC* time constants that greatly exceed those predicted by Equation (3).

An important cause of nonideal responses is stray capacitance within the electrochemical system that may arise from the electrode itself, the leads, or electrical connections. Stray capacitance will increase the cell time constant as described by Equation (4).

$$RC = \frac{1}{4\kappa r}(\pi r^2 C_0 + C_{\text{Stray}}) \tag{4}$$

where C_{Stray} is the stray capacitance. Although it depends on the microelectrode design and the experimental setup, this stray capacitance is typically between a few picofarads and several tens of picofarads. The cell time constant observed in these circumstances depends strongly on the relative magnitudes of the double layer and stray capacitances. At a normal size electrode, the stray capacitance is negligible compared to the double layer capacitance and therefore does not significantly affect the observed cell time constant. However, when the magnitude of the double layer capacitance is reduced

For references see page 10168

by shrinking the size of the electrode to micrometer and submicrometer dimensions, the stray and double layer capacitances can become comparable. For example, taking a typical value of $40\,\mu F\,cm^{-2}$ as the specific double layer capacitance for a platinum electrode in contact with 1 M aqueous electrolyte solution, the double layer capacitance of a 1 mm radius disc will be approximately 1 µF. This value is significantly larger than the picofarads stray capacitance found in a typical electrochemical experiment. However, for a 1 µm radius microdisc, the interfacial capacitance will decrease by six orders of magnitude to approximately 1 picofarad. Therefore, stray capacitance of even a few picofarads will cause the observed *RC* time constant to increase significantly beyond the minimum value dictated by double layer charging alone. This increased cell time constant will cause the transient response characteristics of the electrodes to become nonideal. Thus, an important objective in seeking to implement ultrafast transient techniques is to minimize the stray capacitance.

There are two major sources of stray capacitance. First, the capacitance of the cell leads and capacitive coupling between leads. Second, the microelectrode itself. By using high-quality cable of minimum length, e.g. by mounting the current-to-voltage converter directly over the electrochemical cell, and by avoiding the use of switches as far as possible, stray capacitance from the electrochemical system can be minimized. However, the importance of good microelectrode fabrication and design should not be overlooked. For example, if there is a small imperfection in the seal between the insulator and the electrode material then solution leakage will cause the *RC* cell time constant to increase massively and the Faradaic response may become obscured by charging/discharging processes. Moreover, as shown by Wightman's group[(6)] and Faulkner et al.,[(7)] using silver epoxy or mercury to make the electrical connection between the microwire and a larger hook-up wire can cause the *RC* cell time constant to increase dramatically. This increase arises because the electronically conducting mercury/glass insulator/ionically conducting solution junctions cause significant stray capacitance. It is important to note that these effects may only become apparent in high-frequency measurements.

2.3 Ohmic Effects

When Faradaic and charging currents flow through a solution, they generate a potential that acts to weaken the applied potential by an amount iR, where i is the total current. This is an undesirable process that leads to distorted voltammetric responses. It is important to note that, as described by Equation (1), microelectrodes exhibit higher resistances than macroelectrodes because of their smaller size. However, the currents observed at microelectrodes are typically six orders of magnitude smaller than those observed at macroelectrodes. These small currents often completely eliminate ohmic drop effects even when working in organic solvents. For example, the steady-state current observed at a 5 µm radius microdisc is approximately 2 nA for a 1.0 mM solution of ferrocene. Taking a reasonable value of $0.01\,\Omega^{-1}\,cm^{-1}$ as the specific conductivity, then Equation (1) indicates that the resistance will be of the order of 50 000 Ω. This analysis suggests that the iR drop in this organic solvent is a negligible 0.09 mV. In contrast, for a conventional macroelectrode the iR drop would be of the order of 5–10 mV. Under these circumstances, distorted current responses and shifted peak potentials would be observed in cyclic voltammetry.

It is useful at this point to investigate the effect of experimental timescale on the iR drop observed at microelectrodes. The following section discusses the way that the diffusion field at microelectrodes depends on the characteristic time of the experiment. However, in general, at short times, the dominant mass transport mechanism is planar diffusion and the microelectrode behaves like a macroelectrode. Therefore, at short times the current i decreases with decreasing electrode area (r^2). Since the resistance increases with decreasing electrode radius rather than electrode area, the product iR decreases with decreasing electrode radius in short timescale experiments. Thus, beyond the reduced iR drop because of low currents, decreasing the electrode radius from say 1 mm to 10 µm decreases the ohmic iR drop observed at short times by a factor of one hundred. In contrast, at long experimental timescales the Faradaic current depends directly on the radius making the product iR independent of the electrode radius. Thus, while the low currents observed at microelectrodes reduce ohmic effects for all experimental timescales, using the transient rather than the long timescale response offers even better performance.

2.4 Mass Transport

Oxidation or reduction of a redox-active species at an electrode surface generates a concentration gradient between the interface and the bulk solution. This redox process requires electron transfer across the electrode/solution interface. The rate at which electron transfer takes place across the interface is described by the heterogeneous electron transfer rate constant, k. If this rate constant is large, then diffusional mass transport will control the current observed. Our objective is to describe how these diffusion fields evolve in time. The experiment of interest involves stepping the potential from an initial value where no electrode reaction occurs, to one where electrolysis proceeds at a diffusion controlled rate. Consider the case of a spherical electrode of radius r_s placed in a solution that contains only supporting electrolyte and a redox-active species of concentration C. The concentration gradient at the electrode surface is obtained by solving Fick's second law in spherical coordinates, Equation (5)

$$\frac{\partial C(r,t)}{\partial t} = D\left[\frac{\partial^2 C(r,t)}{\partial r^2} + \frac{2}{r}\frac{\partial C(r,t)}{\partial r}\right] \tag{5}$$

The boundary conditions for the potential step experiments described above are shown in Equation (6)

$$\lim_{r\to\infty} C(r,t) = C^{\infty}$$

$$C(r,0) = C^{\infty}$$

$$\text{and} \qquad C(r,t) = 0 \quad \text{for } t > 0 \tag{6}$$

where r is the distance from the center of the sphere, D is the diffusion coefficient for the redox active species, and C is the concentration as a function of distance r and time t.

Equation (5) can be solved using Laplace transform techniques to give the time evolution of the current, $i(t)$, subject to the boundary conditions described. Equation (7)

$$i(t) = \frac{nFADC^{\infty}}{r_s} + \frac{nFAD^{1/2}C^{\infty}}{\pi^{1/2}t^{1/2}} \tag{7}$$

is obtained where n is the number of electrons transferred in the redox reaction, F is Faraday's constant, and A is the geometric electrode area.

Equation (7) shows that the current response following a potential step contains both time-independent and time-dependent terms. The differences in the electrochemical responses observed at macroscopic and microscopic electrodes arise because of the relative importance of these terms at conventional electrochemical timescales. It is possible to distinguish two limiting regimes depending on whether the experimental timescale is short or long.

2.4.1 Short Times

At sufficiently short times, the thickness of the diffusion layer that is depleted of reactant is much smaller than the electrode radius and the spherical electrode appears to be planar to a molecule at the edge of this diffusion layer. Under these conditions, the electrode behaves like a macroelectrode and mass transport is dominated by linear diffusion to the electrode surface as illustrated in Figure 5(a). At these short times, the $t^{-1/2}$ dependence of the second term in Equation (7) makes it significantly larger than the first term and the current response induced by the potential step initially decays in time according the Cottrell equation, Equation (8)

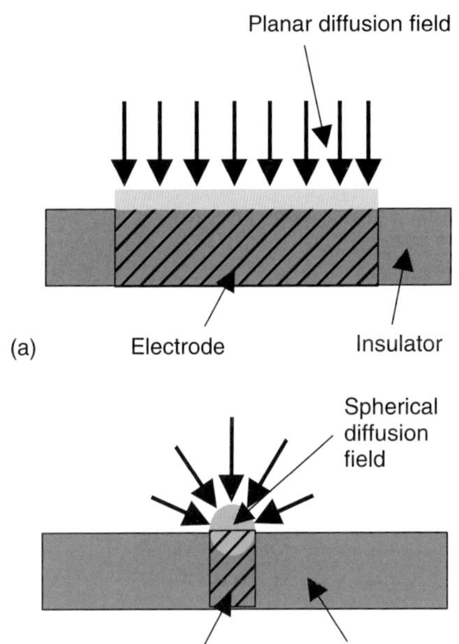

Figure 5 Diffusion fields observed at microelectrodes. (a) Linear diffusion observed at short times. (b) Radial (convergent) diffusion observed at long times.

$$i(t) = \frac{nFAD^{1/2}C^{\infty}}{\pi^{1/2}t^{1/2}} \tag{8}$$

Figure 6 shows the relationship that exists between the range of useable scan rates and electrode radius subject to the condition that ohmic drop is negligible and that the dominant mass transport regime is linear diffusion.[8]

2.4.2 Long Times

At long times, the transient contribution given by the second term of Equation (7) has decayed to the point where its contribution to the overall current is negligible. At these long times, the spherical character of the electrode becomes important and the mass transport process is dominated by radial (spherical) diffusion as illustrated in Figure 5(b).

The current attains a time-independent steady-state value given by Equation (9)

$$i_{ss} = \frac{nFADC^{\infty}}{r_s} \tag{9}$$

The steady-state response arises because the electrolysis rate is equal to the rate at which molecules diffuse to the electrode surface.

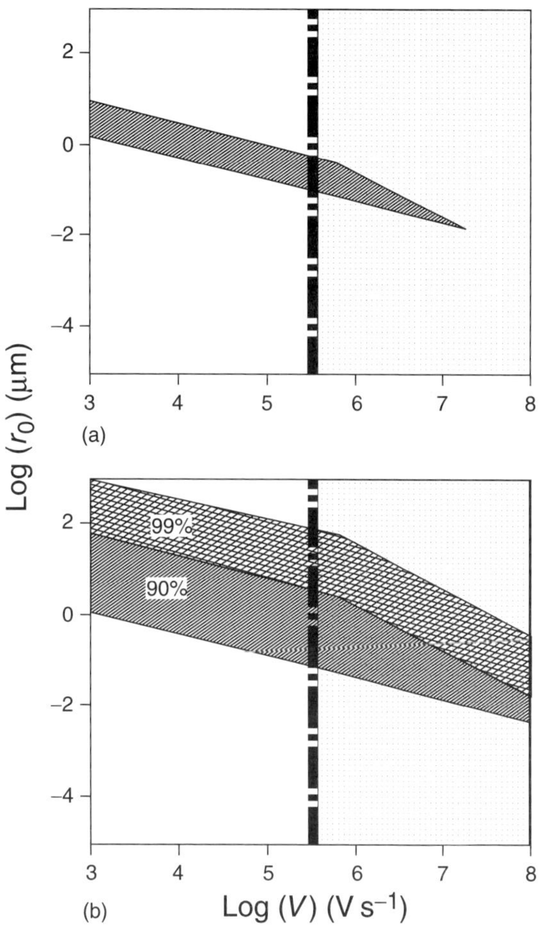

Figure 6 Theoretical limitations to ultrafast cyclic voltammetry. The shaded (or cross-hatched) area between the slanted lines represents the radius that a microdisc must have if the ohmic drop is to be less than 15 mV and distortions due to nonplanar diffusion account for less than 10% of the peak current. (a) Without *iR* drop compensation by positive feedback; (b) with 90% and 99% ohmic drop compensation. The dotted areas in (a) and (b) represent the regions where transport within the double layer affects the voltammetric response. Limits are indicative and correspond approximately to a 5-mM anthracene solution in acetonitrile, with 0.3-M tetrafluoroborate as supporting electrolyte. (Reprinted from C. Amatore, *Electrochemistry at Microelectrodes*, ed. I. Rubenstein, Chapter 4, 1995 by courtesy of Marcel Dekker, Inc.)

Since *short* and *long* times are relative terms, it is useful to determine the times over which transient and steady-state behaviors will predominate and how this time regime is affected by the electrode radius.

This objective can be achieved by considering the ratio of the transient to steady-state current contributions (Equations (8) and (9), respectively). This analysis gives a dimensionless parameter $(\pi Dt)^{1/2}/r_s$ that can be used to calculate a lower time limit at which the steady-state contribution will dominate the total current to a specified extent. For example, the time required for the steady-state current contribution, i_{ss}, to be ten times larger than the transient component, i_t can be calculated. Taking a typical value of D as $1 \times 10^{-5}\,cm^2\,s^{-1}$ for an aqueous solution, then for an electrode of radius 5 mm, the experimental timescale must be longer than 80 s. Therefore, steady state is not observed for macroelectrodes at the tens of millivolts per second timescale typical of conventional cyclic voltammetry experiments. However, reducing the electrode radius by a factor of a thousand to 5 μm, means that a steady-state response can be observed for times longer than 80 μs. Since the steady-state current becomes more dominant with increasing time, steady-state responses are easily observed for microelectrodes in electrochemical experiments run at conventional timescales. Figure 7(a) shows the sigmoidal-shaped responses that characterize steady-state mass transfer in slow scan rate cyclic voltammetry. In contrast, at short experimental timescales (high scan rates) peaked responses (Figure 7b) similar to those observed at conventional macroelectrodes are seen.

The preceding analysis considered a spherical electrode because its surface is uniformly accessible and a simple closed-form solution to the diffusion equation exists.[9] The microdisc is the most widely used geometry, but derivation of rigorous expressions describing their experimental response is complicated because the surface is not uniformly accessible. For discs, electrolysis at the

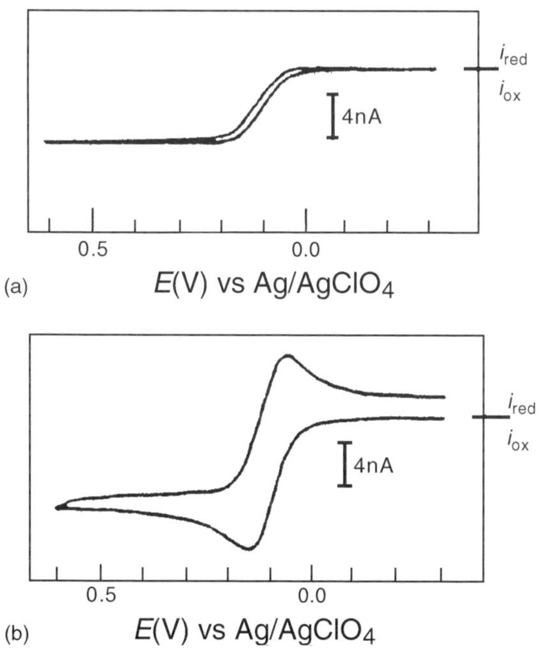

Figure 7 Effect of scan rate on the cyclic voltammetry of 1.0 mM ferrocene at a 6.5-μm gold microdisc where the supporting electrolyte is 0.1-M tetrabutyl ammonium perchlorate in acetonitrile. (a) Scan rate is $0.1\,V\,s^{-1}$; (b) scan rate is $10\,V\,s^{-1}$. (Reprinted with permission from J.O. Howell and R.M. Wightman, *Anal. Chem.*, **56**, 524. Copyright (1984) American Chemical Society.)

outer circumference of the disc diminishes the flux of the electroactive material to the center of the electrode. However, microdisc and microring geometries share the advantage of spherical microelectrodes in that quasi-spherical diffusion fields are established in relatively short periods of time. The steady-state current for a disc is given by Equation (10)

$$i_{ss} = 4nFDCr, \tag{10}$$

where r is the radius of the disc.

Observing a steady-state response depends on all the electrode dimensions being small, not just the radius, and is therefore not achieved for every geometry at the timescales considered above. For example, band electrodes whose thickness is in the micrometer range, but whose length is several millimeters, do not exhibit true steady-state responses. However, a high analyte flux to the ends of the band often makes it possible to observe a pseudo-steady-state condition in a practical sense.

Radial diffusion gives very high rates of mass transport to the electrode surface with a mass transport coefficient of the order of D/r. Therefore, even at rotation rates of 10^4 rpm, convective transport to a rotating macroelectrode is smaller than diffusion to a 1 μm microdisc. The high flux at a microelectrode means that a reverse wave under steady-state conditions is not observed (Figure 5a), because the electrolysis product leaves the diffusion layer at an enhanced rate.

2.5 Instrumental Challenges

Taking typical values of 1 mM and $1 \times 10^{-5}\,\mathrm{cm^2\,s^{-1}}$ for the concentration and diffusion coefficient, respectively, the steady-state current is of the order of 2 nA for a 5 μm radius microdisc. As described by Equation (10), the magnitude of this current will decrease with decreasing electrode radius and picoamp currents will be observed for a 10 nm radius electrode. These are certainly small currents but can be measured with relatively simple circuits based on operational amplifiers. This success can only be achieved because the measurement is performed at long timescales.

In contrast, trying to measure the currents generated at microelectrodes at short timescales poses real technological challenges. The current observed at short timescales is typically in the microamp range. For example, when the potential is stepped to a value where electrolysis proceeds at a diffusion-controlled rate, the Faradaic current observed 1 μs after the application of the potential step will be approximately 135 nA for a 5 μm radius microdisc,. i.e. almost two orders of magnitude larger than the steady-state value. However, there is a difficulty in using operational amplifiers to measure these short timescale currents. For a given operational amplifier, the product of the gain and frequency bandwidth is a constant. Therefore, if a large amplification of a small current is required then the time resolution will be adversely affected and it will become increasingly more difficult to probe fast electron transfer dynamics accurately. The development of faster electronic components, notably for the capture, processing, and streaming of video images, will certainly continue to decrease the lower accessible limit. However, even today, measuring submicroamp currents with nanosecond time resolution remains a significant technological challenge.

Another approach to solving this amplification problem is to use microelectrode arrays. Arrays are ensembles of microelectrodes that may consist of regularly or irregularly spaced assemblies of identical electrodes, ensembles of electrodes with identical shape but uneven dimensions, or disordered arrangements of irregularly shaped electrodes. In seeking to make ultrafast measurements, the advantage of using a microelectrode array is that it provides a nonelectronic approach to current amplification. Often the individual elements are not independently addressable and all of the electrodes operate at the same applied potential. Under these conditions, the total response of the array depends on the relative size of the individual electrodes and the thickness of the diffusion layer that develops around each element during electrolysis. When seeking to make ultrafast measurements, the individual diffusion fields need to act independently of one another, i.e. the separation of the electrodes must be much larger than the diffusion layer thickness throughout the experiment. If the diffusion layers that develop at the individual electrodes merge, then a planar diffusion layer is created that extends over the entire surface of the array. Under these conditions, the array behaves like a large electrode whose surface area is given by the sum of the electrochemically active and inactive areas. This condition does not yield any benefit in terms of probing fast electron transfer dynamics.

Thus, arrays used for dynamic measurements typically consist of N widely separated and noninteracting discs that will provide N times the current observed at a single electrode. In terms of the RC cell time constant, such a microelectrode array can be considered as a circuit of N parallel combinations of each single microelectrode. Thus, the equivalent resistance and capacitance are R/N and $N \times C$, respectively. The product, RC, of the array remains unchanged from that found at a single element of the assembly. Microelectrode arrays appear to offer the same advantages as a single microelectrode, but it is not necessary to construct such a high-gain current-to-voltage converter, making measurements at shorter timescales possible. However, in the short timescale limit (nonoverlapping diffusion fields), the

For references see page 10168

gain achieved is necessarily finite and fixed by the total number of microelectrodes in the array. In this sense, the amplification achieved is less flexible and of lower quality than is possible using electronic means. Moreover, fabrication of microelectrode arrays with electrode radii smaller than ten micrometers, in which each electrode exhibits its theoretical *RC* time constant, remains technologically challenging.

2.6 Transient Techniques

In transient electrochemical measurements involving a solution-phase redox couple, one seeks to create a competition between the reaction of interest, electron transfer at the electrode surface or coupled homogeneous steps, and diffusion of the species to and from the electrode surface.[10] In the following sections, consideration is given to how high diffusion rates that allow fast reactions to be investigated are obtained by controlling the experimental timescale directly, i.e. by applying a shorter and shorter time perturbation to the system.

In order to extract useful kinetic information from transient electrochemical responses, the Faradaic and charging currents must be separated. This requirement presents a number of challenges when attempting to perform experiments at very short timescales. For example, in fast scan rate cyclic voltammetry, the charging current increases proportionally to the scan rate while the Faradaic current for diffusive species is proportional to the square root of the scan rate. These dependencies cause the ratio of the Faradaic to charging current ratio to *decrease* with *increasing* scan rate. In some sense, this process corresponds to a decreasing signal-to-noise ratio since the magnitude of the charging current places a lower limit on the detectable Faradaic current. One strategy for dealing with capacitive charging in high frequency experiments is to measure the double layer charging current in blank electrolyte and then to subtract this signal from the total current observed in the presence of the analyte of interest. However, the usefulness of this approach is limited since at high scan rates the magnitudes of the charging and Faradaic currents are often similar.

While microelectrodes can significantly decrease the deleterious effects of ohmic drop, the large currents observed in short-timescale transient experiments can result in significant *iR* drop and distorted voltammograms. Determination of heterogeneous electron transfer rate constants often relies on measuring the scan rate dependence of the peak-to-peak separation, ΔE_P. Ohmic effects represent a serious problem, not only because they will cause a significant ΔE_P to be observed even when heterogeneous electron transfer is fast, but also because the magnitude of the ohmic effect depends on the experimental timescale. Several strategies have been used to decrease ohmic effects. First, the Faradaic information is extracted by means of convolution of the voltammograms with the diffusion operator $(\pi t)^{-1/2}$. Second, the experimental voltammogram is simulated using a model that incorporates ohmic and capacitive factors. Third, corrections for ohmic drop are made on-line, using positive feedback circuitry. Savéant et al.[11] have developed high-speed potentiostats capable of providing positive feedback to compensate on-line for ohmic drop at scan rates up to $5 \times 10^5\ \mathrm{V\,s^{-1}}$.

In contrast to cyclic voltammetry, for chronoamperometric experiments the decay rates for double layer charging and Faradaic processes are different, allowing the two processes to be separated on a kinetic basis. Ideally, the double charging current undergoes a single exponential in time while the Faradaic current for a diffusive species decreases with the square root of time. Therefore, at short times the current will be dominated by the charging current while at long times the diffusion controlled Faradaic current will be relatively more significant.

2.6.1 *Heterogeneous Electron Transfer Dynamics*

As discussed above, at short times the diffusion layer thickness is much smaller than the microelectrode radius and the dominant mass transport mechanism is planar diffusion. As observed for macroelectrodes at conventional millivolt per second scan rates, the Faradaic current increases in proportion to the electrode area. Under these conditions, the classical theories, e.g. that of Nicholson and Shain,[12] can be used to extract kinetic parameters from the scan rate dependence of the separation between the anodic and cathodic peak potentials. The peak-to-peak separation increases with increasing scan rate thus improving the accuracy of the rate constant determination. However, as discussed above, stray capacitance, ohmic drop, double layer charging currents, and difficulties posed by bandwidth limitations of the instrument must be recognized. These effects can be recognized by performing measurements across the whole dynamic range available, i.e. from kilovolts per second to megavolts per second.

As shown in Table 1, the heterogeneous electron transfer dynamics of a diverse range of organic and inorganic species have been investigated using transient techniques. However, cyclic voltammetry of the anthracene/anthracene anion radical has been used extensively as a reference system for characterizing new electrodes or instruments. This focus arises because the aromatic nature of the molecule leads to very small changes in the bond lengths and angles, i.e. the Marcus inner sphere reorganization energy is small,[13] causing the heterogeneous electron transfer rate constant to be large.

Table 1 Rate constants for heterogeneous electron transfer as determined using transient methods

Analyte	Electrode[a]; Element, Dimension (μm)	k^o (s^{-1})	Refs.
Anthracene	Au, 6.5	3.46 ± 0.55	14
	Au, 5	3.3	16
	Au, 5	2.6	17
	Au, 3.0 and 8.5	3–4.8	18, 19
9-Bromoanthracene	Au, 5	2.9 ± 0.3	31
	Au, 6	10	20
Anthraquinone	Pt cylinder, $r = 25.4\,\mu m$, $\ell < 0.25$ cm	1.78 ± 0.35	14
	Pt, 5	1.5	17
	Au, 6.5	2.4	21
Benzoquinone	Au, 6.5	0.39 ± 0.1	14
	Pt, 5	0.14	17
Ferrocene	Pt, 5	1.1	17
	Au, 5	3.1 ± 1.1	22
	Pt, 5–25	1.4–3.6	21
	Au, 5	5.5	15
	Pt, 10		23
Ferrocyanide	C cylinder, $r = 15\,\mu m$, $\ell = 500\,\mu m$	0.0114 ± 0.0022	24
	Pt, 10, 50	0.42 ± 0.03	25
	C, 5	0.06 ± 0.05	26
	Pt, 1–30	0.64–0.79	27
	Pt, 20	0.2	28
9-Fluorenone	Pt, 6	3	29
10-Methylacridan cation radical	Au, Pt, 8.5, 10	3.2 ± 0.5	30
$[Ru(bpy)_3]^{2+}$	Hg, 5.5	0.45	22
	Au, 5	2.5	22

[a] Dimension given is the radius of a microdisc electrode unless otherwise stated.

Howell and Wightman[(14)] have studied this reaction using cyclic voltammetry at scan rates up to 10^5 V s^{-1} and good agreement between the predictions of the Nicholson and Shain theory and experiment was found after correcting for the nonspherical nature of the microdisc used.

A useful strategy in trying to extend the upper limit of measurable electron transfer rate constants is to perform measurements at lower temperatures. This strategy is successful because even for heterogeneous electron transfers with negligible inner sphere reorganization energies, activation barriers of the order of 20–25 kJ mol^{-1} are expected. Therefore, considerably slower rates of heterogeneous electron transfer ought to be observed even by decreasing the temperature of the electrochemical cell by a few tens of degrees. Measurements of this type are facilitated greatly by microelectrodes, since solvents, such as alcohols or nitriles that remain liquid over a wide temperature range, can be used without catastrophic ohmic effects. For example, Weaver and co-workers investigated the ferrocene, *o*-nitrotoluene, and nitromesitylene systems in acetonitrile, propionitrile, and butyronitrile at a gold microdisc using scan rates up to 10^4 V s^{-1} between 200 and 300 K.[(15)] The experimental voltammograms were interpreted with the aid of simulated responses that accounted for the activation enthalpy, temperature-dependent diffusion coefficients, and double layer capacitance. For ferrocene, the standard heterogeneous electron transfer rate constants ranged from 0.083 cm s^{-1} at 198 K to approximately 5.5 cm s^{-1} at 298 K yielding an activation enthalpy of 20 kJ mol^{-1}.

There is little doubt that temperature-dependent studies offer a useful new insight into the electrochemical behavior of fast redox couples especially by providing activation enthalpies, entropies, and reaction free energies that allow electronic coupling terms to be evaluated. However, as exemplified by the careful work of Evans et al.[(32)] and Safford and Weaver,[(15)] attention must be paid to two important processes. First, the rate of diffusional mass transport will decrease with decreasing temperature. This process will make the voltammetric response less sensitive to the electrode kinetics. Second, lower temperatures will cause the solution resistance to increase, which can cause an enhanced peak-to-peak separation to be observed that could be incorrectly interpreted as slow heterogeneous electron transfer.

For references see page 10168

Irrespective of whether steady-state or transient approaches are used, the electrode surface activation, pretreatment, roughness, degree of sample adsorption, and electrode material as well as the nature and concentration of the supporting electrolyte, sample concentration, and solvent type and purity, can have a profound effect on the magnitude of the heterogeneous electron transfer rate observed.

Despite the many elegant investigations that have been conducted into heterogeneous electron transfer dynamics using ultrafast electrochemical techniques, the magnitude of the diffusion-controlled current at short times ultimately places a lower limit on the accessible timescale. For diffusive species, the thickness of the diffusion layer, δ, is defined as $\delta = (\pi Dt)^{1/2}$ and is therefore proportional to the square root of the polarization time, t. The diffusion layer thickness can be estimated to be approximately 50 Å if the diffusion coefficient is $1 \times 10^{-5}\,\mathrm{cm^2\,s^{-1}}$ and the polarization time is 10 ns. Given a typical bulk concentration of the electroactive species of 1 mM, this analysis reveals that only 10 000 molecules would be oxidized or reduced at a 1 µm radius microdisc under these conditions. The average current for this experiment is only 170 nA, which is too small to be detected with low-nanosecond time resolution.

Therefore, in order to probe the dynamics and energetics of ultrafast heterogeneous electron transfer dynamics this diffusion limitation must be eliminated. One successful approach to achieving this objective is to use self-assembled or spontaneously adsorbed monolayers. When immobilized on an electrode surface the electroactive species no longer needs to diffuse to the electrode to undergo electron transfer. Moreover, the electroactive species is preconcentrated on the electrode surface. For example, in the situation considered above, there will be approximately 1.7×10^{-20} mol of electroactive material within the diffusion layer. Given that the area of a 1 µm disc is approximately $3.1 \times 10^{-8}\,\mathrm{cm^2}$, this translates into an "equivalent surface coverage" of about $5.4 \times 10^{-13}\,\mathrm{mol\,cm^{-2}}$. In contrast, the surface coverage, Γ, observed for dense monolayers of adsorbates is typically more than two orders of magnitude larger with coverages of the order of $10^{-10}\,\mathrm{mol\,cm^{-2}}$ being observed. This higher concentration gives rise to much larger currents that are easier to detect at short timescales.

For example, Faulkner et al. have investigated the dynamics and energetics of electron transfer across mercury/anthraquinone monolayer interfaces.(33) The surface coverage of an anthraquinone-2,6-disulfonic acid (2,6-AQDS) monolayer formed from a solution containing micromolar concentrations of the quinone was approximately $1.6 \times 10^{-10}\,\mathrm{mol\,cm^{-2}}$. Therefore, in transient electrochemical experiments performed on nanosecond timescales, the average Faradaic current would be expected to be of the order of 50 µA. It is technologically feasible to detect currents of this magnitude with circuitry based on existing wide bandwidth operational amplifiers.

As exemplified by the work of Chidsey,(34) Acevedo and Abruña,(35) Forster and Faulkner,(36) and Finklea and Hanshew,(37) electroactive adsorbed monolayers have been developed that exhibit close to ideal reversible electrochemical behavior under a wide variety of experimental conditions of timescale, temperature, solvent and electrolyte. In order for a monolayer to be an attractive model system for understanding those factors that control the rate of heterogeneous electron transfer, it should have a number of properties. The nature and physical location of the electroactive center within the molecule should be well defined. The electron transfer should be mechanistically simple with the electrochemical responses being close to ideally reversible and stable over prolonged periods. The molecules should adsorb strongly onto electrode surfaces in both oxidized and reduced forms. It should be possible to control the surface coverage by controlling the concentration of the adsorbate in solution (reversible physisorption), by controlling the time that the pristine electrode is exposed to the deposition solution, by using electroinactive diluents, or by controlled desorption from a high-coverage monolayer. The layers should be stable in a wide variety of solvents, especially water, to allow the effect of solvent parameters, e.g. relaxation rate and dielectric constant, on simple electron transfer processes to be investigated.

In these respects, monolayers containing ferrocene or ruthenium hexamine are attractive model systems for studying heterogeneous electron transfer across electrode/monolayer interfaces. However, in many of these studies the bridging ligand linking the electroactive group to the electrode surface has been carefully chosen with regard to its length and electronic structure so that electron transfer occurs on a millisecond or longer timescale. Therefore, ultrafast electrochemical techniques are not required to probe the details of electron transfer. In contrast, species that are directly adsorbed onto the electrode surface (e.g. anthraquinones) or where the bridging ligand is shorter (e.g. 4,4′-dipyridyl-type linkers) can undergo electron transfer on the submicrosecond timescale. For example, Faulkner et al.(33) found that 2,6-AQDS monolayers can be reversibly converted to the hydroquinone form in a two-electron, two-proton mechanism and virtually ideal cyclic voltammetry and chronoamperometry are observed. The standard heterogeneous electron transfer rate constant is in the range of 10^4 to $10^5\,\mathrm{s^{-1}}$ depending on the driving force (overpotential, $\eta \equiv E_{\mathrm{app}} - E^{o'}$) for the reaction and the proton concentration.

Acevedo and Abruña[35] formed spontaneously adsorbed monolayers of the complex $[Os(bpy)_2 Cl (dipy)]^+$, where dipy is 4,4′-trimethylenedipyridine, and investigated their dynamic and energetic properties. The voltammetric responses observed for the $Os^{2+/3+}$ redox reaction at scan rates over $8000\,V\,s^{-1}$ were close to the behavior expected for an ideal reversible one-electron transfer reaction involving a surface-confined species. They used Laviron's formalism[38] describing the variation of the peak potential with scan rate to estimate the rate of heterogeneous electron transfer as $2 \times 10^5\,s^{-1}$. The formation and structure of these monolayers on Pt(111) surfaces was investigated using scanning tunneling microscopy (STM) and electrochemical scanning tunneling microscopy (ECSTM). Molecularly resolved images were obtained that demonstrated that the monolayer exists as a tightly packed two-dimensional crystal on the electrode surface. Abruña also demonstrated that the exchange dynamics associated with the desorption and displacement of these adsorbates are controlled by the rate of desorption via a dissociative mechanism and that the applied potential does not significantly affect the free energy of adsorption.

The ideality of the electrochemical response exhibited by these osmium-containing monolayers makes them attractive model systems for probing the effect of solvent, electrolyte, and temperature on the electron transfer dynamics. Faulkner and Forster[36] investigated these effects for $[Os(bpy)_2\ Cl\ (pNp)]^+$, where pNp is 4,4′-dipyridyl, 1,2-bis(4-pyridyl)ethane, or dipy, monolayers on platinum microelectrodes. The rate of heterogeneous electron transfer was measured using chronoamperometry conducted on a low microsecond timescale.

For the $Os^{2+/3+}$ redox reaction, heterogeneous electron transfer is a rapid first-order process characterized by a single unimolecular rate constant (k/s^{-1}). Tafel plots of the dependence of $\ln k$ on the electrochemical driving force (overpotential) show curvature, indicating that the transfer coefficient is potential-dependent. For sufficiently large overpotentials, k tends to become independent of the free energy driving force, which is consistent with Marcus theory. The response is asymmetric with respect to overpotential, with the slope for the oxidation process tending towards zero more rapidly than that for the reduction process. This response was modeled as a tunneling process between electronic manifolds on the two sides of the interface. Temperature-resolved measurements of k and the formal potential were made from −5 to +40 °C to provide enthalpies and entropies of activation, respectively. The corresponding free energies of activation ranged from $12.3\,kJ\,mol^{-1}$ in acetonitrile to $6.4\,kJ\,mol^{-1}$ in chloroform. There is weak coupling between the electronic manifolds on the two sides of the electrochemical interface. Surprisingly however, k depends linearly on the longitudinal relaxation rate of the solvent. This behavior is not predicted on the basis of contemporary theories of electron transfer. These osmium- and related ruthenium-containing monolayers have been used to investigate interfacial field effects on reductive chloride elimination and as model systems for investigating how Faradaic responses that occur at identical formal potentials can be separated on a kinetic basis.[39]

More recently, Forster has probed the dynamics of both metal-centered oxidation and ligand-based reduction processes to address how electronic states of the bridging ligand contribute to the electron tunneling pathway.[40] This work indicates that the close proximity of the redox potentials of the bridge and the remote bipyridyl ligands leads to stronger electronic coupling between the electrode and bipyridyl ligands than between the electrode and the osmium center. Forster and O'Kelly also investigated monolayers that incorporated a ligand capable of undergoing protonation/deprotonation reactions within the coordination shell of the metal center.[41] This study showed that the rate of heterogeneous electron transfer across the metal/monolayer interface depends on the pH of the contacting electrolyte. From a molecular electronics perspective, pH-induced *conformational gating* of this type offers the possibility of developing pH-triggered electrical switches.

There are few literature reports addressing the issue of separating charging and Faradaic currents in chronoamperometric experiments involving surface-confined reactants. Separating these contributions is complicated by the fact that both are expected to exhibit single exponential decays in time. Measuring the capacitive response in blank supporting electrolyte is not a useful strategy since the double-layer capacitance is altered by adsorption. Moreover, the interfacial capacitance depends on the redox composition of the monolayer. Much of the work discussed above relies on a kinetic separation of these currents, i.e. a microelectrode with an RC cell time constant at least five times smaller than the redox process under investigation is used causing the short and long timescale currents to be dominated by double-layer charging and the Faradaic reactions, respectively. However, approaches based on digital simulation and Laplace transform methods have been developed.[42]

2.6.2 *Homogeneous Chemical Kinetics*

There are two important advantages in using micro- rather than macroelectrodes for probing the dynamics of homogeneous chemical reactions. First, because of their relative immunity to ohmic effects, microelectrodes have greatly extended the range of useful media, e.g. low-dielectric-constant solvents, solids, and solutions with no deliberately added electrolyte. This capability

For references see page 10168

often makes it feasible to compare electrochemical rate constants directly with those obtained using other techniques, notably transient absorption or luminescence spectroscopy. Second, very high diffusion rates and high-quality data can be obtained at short experimental timescales. These decreases in the lower accessible time limit have important implications for probing the dynamics of rapid heterogeneous electron transfer and homogeneous chemical reactions alike. For example, bimolecular reactions in solution cannot proceed faster than the rate at which molecules come into close contact. Thus, bimolecular rate constants cannot exceed the diffusion-limited rate constant that is of the order of $10^9-10^{10}\,M^{-1}\,s^{-1}$ in most organic solvents. Since the characteristic time of cyclic voltammetry is RT/Fv, where v is the scan rate, experiments performed at $mV\,s^{-1}$ scan rates allow kinetic information, such as lifetimes that are close to the diffusion limit, to be obtained.[43]

In the past, when the rate constant for the following reaction was large, the voltammetric response was irreversible because the intermediate underwent a following chemical reaction before the direction of the potential scan was reversed. With the development of microelectrodes and high-scan-rate voltammetry, it is now possible to detect the fast-decaying intermediate. In many circumstances, it is even possible to observe reversible electrochemical responses since the experimental timescale can be made shorter than the lifetime of the electrogenerated reactant. For example, as illustrated in Figure 8, Howell and Wightman[44] have shown that the irreversible response observed for the oxidation of anthracene at slow scan rates becomes fully reversible at a scan rate of $10^4\,V\,s^{-1}$. This behavior is opposite to that expected when heterogeneous electron transfer is slow and suggests that the cation radical undergoes a following chemical reaction. The ability to make the voltammetric response reversible means that the formal potentials of highly reactive species can be measured accurately. However, it is important that ohmic drop is negligible, or is made so using on-line current interrupt or positive feedback approaches, otherwise the experimental responses have to be modeled if true thermodynamic information is to be obtained.

Beyond this thermodynamic information, transient measurements can be used to probe the mechanism of the following chemical reaction. For example, the oxidation of anthracene has been reported widely as proceeding through an ECE mechanism, where E denotes an electron transfer step and C a chemical reaction. However, Osteryoung et al.[45] report that closer agreement is found between staircase, square wave, normal pulse, and reverse pulse responses and a DISP1 mechanism, where DISP denotes a disproportionation (DISP) reaction. First-order rate constants range from $310-750\,s^{-1}$. However, for true

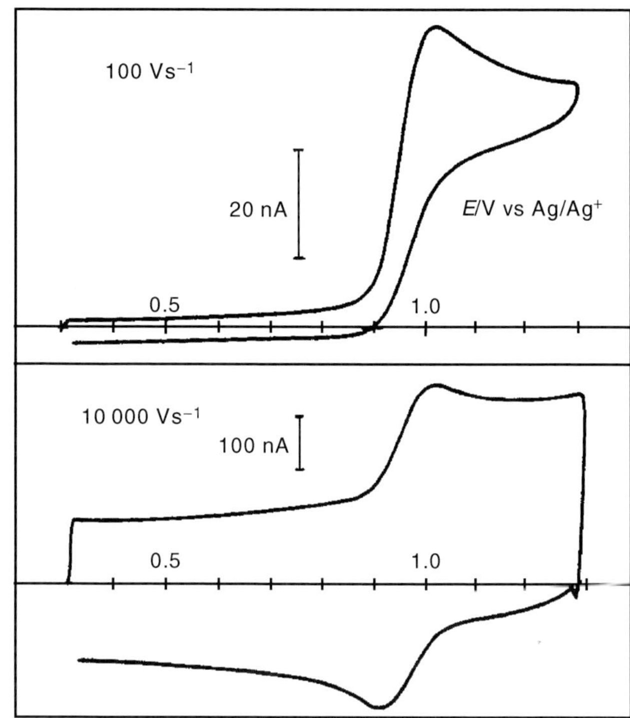

Figure 8 Effect of scan rate on the voltammetric response observed at a 5-µm platinum microdisc for the oxidation of 2.36-mM anthracene in dimethylformamide (DMF) containing 0.6-M tetraethyl ammonium perchlorate. (Reprinted with permission from J.O. Howell and R.M. Wightman, *J. Phys. Chem.*, **88**, 3915. Copyright (1984) American Chemical Society.)

first-order kinetics, the same rate constant ought to be obtained irrespective of the time at which the transient response is analyzed. In the case of anthracene oxidation, smaller rate constants are observed at long times. This behavior may arise because of low concentrations of reactive impurities in the sample that compete with the main reaction at long times.

As indicated in Table 2, the kinetics of many following chemical reactions have been investigated using transient methods at microelectrodes including the oxidation of polyalkylbenzenes, anthracene, aromatic hydrocarbons, and ascorbic acid; the reduction of quinones, aryl halides, chloroquinoline, acetophenones, and butylpyridinium derivatives; and the initial stages of electropolymerization of the conducting polymer polypyrrole. The shortest lifetime attainable is currently in the tens of nanoseconds range, e.g. the lifetime for reductive dimerization of the 2,6-diphenylpyrillium cation in acetonitrile has been determined to be 50 ns.[43]

The ability of individual solvents to stabilize transition states can have a profound effect on the observed rate constants. For example, Wipf and Wightman[52] studied the cleavage reaction involved in the reduction of 9-chloroanthracene in low dielectric solvents including

Table 2 Rate constants for homogeneous chemical reactions as determined using transient methods

Analyte	Electrode[a]; element, dimension (μm)	k	Refs.
Anthracene oxidation	Pt, 5	$7.6 \pm 1.0 \times 10^3\ s^{-1}$	44, 46
Ascorbic acid oxidation	Hg, 5	$1.4 \times 10^3\ s^{-1}$	47
3-Bromoacetophenone reduction	Au, 5	$6.3 \times 10^4\ s^{-1}$	31
	Au, 5	$8 \times 10^3\ s^{-1}$	20
Acetophenone oxidation	Au, 5	$6.3 \times 10^4\ s^{-1}$	31
Chlorpromazine + dopamine	Pt, 25	$10^8\ M^{-1}\ s^{-1}$	48
N,*N*-dimethylaniline oxidation	Pt, 12.5	$6.3 \times 10^5\ M^{-1}\ s^{-1}$	49
$Fe(CO)_5 + CO$	Au, 100	$6 \pm 2 \times 10^6\ M^{-1}\ s^{-1}$	50
Methylbenzenes	Pt, 5	$10^9\ M^{-1}\ s^{-1}$	51

[a] Dimension given is the radius of a microdisc electrode unless otherwise stated.

tetrahydrofuran, dichloromethane, and chlorobenzene. These measurements would simply be impossible using conventional macroelectrodes. The transition state in this reduction involves transfer of the negative charge of the anion radical to the chlorine atom. The localized negative charge can be stabilized by the solvent thus lowering the energy of the transition state and yielding a more rapid cleavage. Measurements of this type not only provide fundamental mechanistic information but also play an important role in seeking to optimize industrial processes.

Bond formation is also amenable to investigation using ultrafast transient electrochemical techniques. For example, dimerization reactions triggered by electron transfer have been investigated using both cyclic voltammetry and potential-step methods. Andrieux, Hapiot, and Savéant[31] have made seminal contributions in this area. For example, using chronoamperometry conducted on a microsecond timescale, these authors showed that reduction of 1-methyl-4-*tert*-butylpyridinium cations in DMF leads to radical–radical coupling at a rate of $6.4 \pm 0.4 \times 10^6\ M^{-1}\ s^{-1}$. Their method of analysis involves stepping the potential to a value that is sufficiently negative of the formal potential for reduction to proceed at a diffusion-controlled rate. At a transition time, τ, the current is measured and the potential stepped back to its original value. The current is measured again at 2τ. The is essentially a generator–collector experiment and, as shown in Equation (11), the ratio of the currents obtained at τ and 2τ are normalized relative to the values that would be obtained in the absence of a following chemical reaction

$$R = \frac{I(2\tau)/I(\tau)}{[I(2\tau)/I(\tau)]_{\text{dif}}} \tag{11}$$

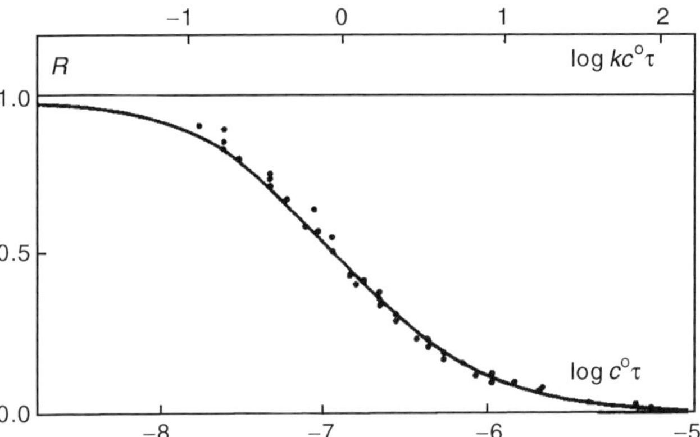

Figure 9 Variation of R with τ in double potential step chronoamperometry of 1-methyl-4-*tert*-butylpyridinium cation at concentrations of 2.5, 5, and 10 mM at 2.5- and 8.5-μm radius gold disc electrodes. (Reprinted with permission from C.P. Andrieux, P. Hapiot, J.-M. Savéant, *J. Phys. Chem.*, **92**, 5992. Copyright (1988) American Chemical Society.)

Figure 9 shows how R depends on the characteristic time of the experiment, τ, and the concentration of 1-methyl-4-*tert*-butylpyridinium cation. By combining cyclic voltammetric and potential-step techniques, Savéant has studied more complicated chemical systems such as the methylacridan (AH)/10-methylacridinium (A^+) couple in acetonitrile, which is an analogue of the coenzyme NADH/NAD^+ couple.[30] The formal potentials and the heterogeneous electron transfer rates, as well as the reaction rates of the following chemical reactions, were determined. These investigations revealed that oxidation of AH, as well as reduction of A^+, were kinetically controlled by following homogeneous chemical steps, rather than by the initial electron transfer, which appears to be quite fast in both cases.

For references see page 10168

2.7 Steady-state Electrochemistry

In the preceding sections, we have focused on the use of transient methods to address short timescales in electrochemistry. However, in an electrochemical measurement involving a solution-phase reactant, the appropriate timescale is not necessarily the actual duration of the experiment. Typically, the timescale is imposed by the time taken for diffusion, t_{dif}, of the analyte of interest from the bulk solution to the electrode surface or vice versa. In seeking to measure electrochemical kinetics, one seeks to match this diffusion time to the half-life of the chemical process (homogeneous reaction dynamics) or of the interfacial electron transfer process (heterogeneous electron transfer dynamics). In transient methods, t_{dif} is directly related to the duration of the experiment. However, under steady-state conditions, t_{dif} is not related to the actual duration of the experiment since the diffusion-layer thickness depends only on the radius. Under these conditions, t_{dif} is approximately equal to r^2/D. This property has a profound impact on the size of microelectrodes required to make measurements at steady state, e.g. given a typical diffusion coefficient of $10^{-5}\,cm^2\,s^{-1}$, microelectrodes with radii of less than 30 nm are required to address submicrosecond timescales. Thus, it is clear that probing ultrafast electrochemical processes demands the fabrication of vanishingly small electrodes. However, although specialized, this is not an impossible task and well-characterized band electrodes with widths of 2 nm have been fabricated by sealing a deposited metal film between insulators.

In contrast to transient techniques, a steady-state response is unaffected by charging currents, is insensitive to low levels of reactant adsorption, and requires less complex instrumentation. The choice between transient or steady-state approaches centers on the need to produce a relatively larger, but ideally responding, electrode for transient measurements, e.g. a 25-µm radius microelectrode will exhibit a submicrosecond RC time constant in high concentrations of supporting electrolyte, compared to the need to fabricate nanometer or even angstrom dimension electrodes for use in a steady-state approach. However, the ability of transient reversal techniques, such as double step chronoamperometry or cyclic voltammetry, to create and subsequently to detect electrochemical intermediates, represents a significant advantage when attempting to characterize reactive intermediates or elucidate reaction mechanisms.

Perhaps the most convenient way to measure the steady-state current is to apply a potential sweep to the microelectrode. The appropriate timescale can be determined as outlined in section 2.4 and this time can then be converted into a scan rate by noting that the characteristic time of cyclic voltammetry is $RT/F\nu$. It is important to stress that the magnitude of the steady-state current is intrinsic to the size of the microelectrode. It is independent of the technique (e.g. slow-scan voltammetry or long-timescale potential step) and of the direction of the potential scan or step. Moreover, at steady state all concentrations in the vicinity of the electrode surface are unchanging over time.

Figure 10 shows that while reversible, quasireversible, and irreversible systems all exhibit sigmoidal-shaped current responses and the same limiting currents in slow-scan voltammetric experiments, the experimental formal potential shifts from its thermodynamic value. This shift arises from competition between kinetic and diffusion control, with less steep curves being obtained for quasi-reversible and irreversible systems. Quasi-reversibility is observed if the kinetic distance of the reaction, D/k^o, where k^o is the standard heterogeneous electron transfer rate constant, sufficiently exceeds the dimension of the microdisc (e.g. $D/k^o = 10^4\,r$). Under these conditions, the difference between the operational and thermodynamic formal potentials can be used to determine k^o without any interference from the charging current. Reversible responses are observed when D/k^o is much smaller than the radius of the microdisc.

This analysis places upper limits on the magnitude of k^o that can be determined using a particular radius microdisc under steady-state conditions. For example, for a 1-µm radius microdisc, the upper limit on the measurable k^o is approximately $10^{-5}\,cm\,s^{-1}$. Using microdiscs of 50 nm, which are about the smallest microdiscs currently routinely available, the upper limit can be extended to about $2\times10^{-4}\,cm\,s^{-1}$. However, in practice this upper limit cannot be extended indefinitely by shrinking the

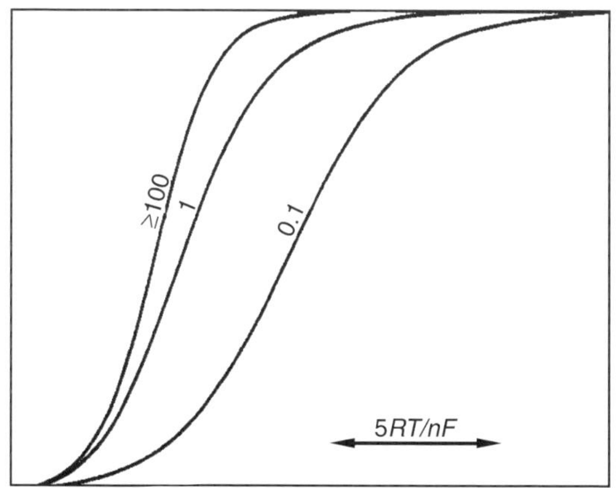

Figure 10 Shapes of the reversible, quasi-reversible, and irreversible steady-state voltammograms. The value of $k^o r/D$ is given on each curve. (Reproduced with kind permission of Kluwer Academic Publishers, Dordrecht, The Netherlands from K.B. Oldham *Microelectrodes: Theory and Applications, Steady-state Voltammetry*, eds. M.I. Montenegro, M.A. Queirós, J.L. Daschbach, NATO ASI Series E, Vol. 197, Figure 6, 1991.)

electrode. Imperfect seals, often caused by mismatching of the thermal expansion coefficients of the glass shroud and the metallic microwire, often enhance non-Faradaic currents.

Another important consideration when trying to fabricate increasingly smaller microelectrodes is the formation of a microcavity around the electrode tip. Creating a slightly recessed electrode (e.g. because a gold microwire is used that is much softer than the insulating glass shroud), may not significantly compromise the response observed for a 10 μm radius microdisc. However, such a cavity can become very important if the electrode size is reduced to nanometer dimensions, causing distortions that are attributed to kinetic effects to become significant. Moreover, the use of an ion-exchange material as an insulator may lead to a double layer being set up on the shroud, generating an electrical potential that would compromise the measurement of the interfacial kinetics. However, these effects are likely to be important only when the Debye length of the electrolyte (typically 10–100 Å) is larger than the electrode radius. A more important effect arises from surface contamination by impurities found within the analyte, solvent, or supporting electrolyte. It is important therefore to maintain high levels of cell cleanliness and to confirm the pristine state of the electrode surface by measuring the kinetics of a reference system before and after a new analyte is investigated.

2.7.1 Heterogeneous Electron Transfer Dynamics

Understanding the elementary steps involved in heterogeneous electron transfer across a metal/solution interface is of considerable technological importance in areas ranging from battery performance to corrosion inhibition.(53) In trying to use steady-state methods to determine kinetic parameters for fast reactions, experimental conditions are usually chosen so that there is mixed control by kinetics and diffusion. Aoki et al.,(54) Fleischmann et al.,(55) and Oldham and Zoski(56) have addressed this difficult issue and provided several equivalent approaches for analyzing the experimental current–voltage curves. More recently, Mirkin and Bard(57) developed a new approach to determining kinetic parameters for a simple quasi-reversible electron transfer reaction. In principle, only the one-quarter ($E_{1/4}$), one-half ($E_{1/2}$), and three-quarters ($E_{3/4}$) potentials from a single steady-state voltammogram are required, although more precise values of k^o and α, the transfer coefficient, can be obtained by fitting the full voltammogram. Moreover, the analysis is independent of the electrode area and the concentration of the electroactive species that improves the reliability of the analysis.

The parameters describing the kinetics of heterogeneous electron transfer for several solution-phase redox-active molecules have been reported and a representative sample is given in Table 3. As expected, these data show that the rate of electron transfer across the electrode/solution interface is significantly influenced by the identity of the redox couple. These variations reflect differences in the reaction adiabaticity and activation barriers that exist for the individual systems. However, even for a single species, a considerable range of k^o values are observed, e.g. for the archetypal reversible couple ferrocene, reported k^o values range from 0.09 to 220 cm s^{-1}. There are numerous reasons for these discrepancies, including the electrode preparation regime, the purity of the ferrocene, solvent and electrolyte, and the method used to analyze the data. Moreover, for high ferrocene concentrations the electrode becomes pacified by an insoluble layer. This blocking of the electrode surface impedes diffusion and prevents the current observed from increasing proportionally with increasing ferrocene concentration. Thus, even for the standard electrochemical test system considerable care must be taken over all aspects of the experiment, chemicals, electrode preparation, experimental setup, as well as data collection and analysis if accurate kinetic parameters are to be obtained.

2.7.2 Homogeneous Chemical Kinetics

A second important application of steady-state measurements is in studies of chemical reactivity. Steady-state measurements using electrodes of different radii can provide a powerful insight into the kinetics of homogeneous reactions where the limiting current density depends on the magnitude of the homogeneous rate constant. Hence, as described in Table 4, coupled chemical (C) and electron transfer (E) reactions (e.g. EC mechanisms), catalytic follow-up processes, as well as reactions involving DISP have been characterized. It is important to note that reactions, such as chemical reactions that follow electron transfer mechanisms, cannot be investigated in the same way since the current density is not influenced by the following chemical reaction. In these circumstances, the homogeneous reaction does not affect the height or shape of the reversible steady-state voltammogram. However, as indicated by Figure 11, the position of the wave on the potential axis depends on the homogeneous reaction rate, and kinetic information can be obtained by probing how $E_{1/2}$ depends on the electrode radius.

The technique is best applied to reactions whose rate constants are of the same magnitude as D/r^2. Therefore, given that it is now feasible to fabricate microelectrodes with submicrometer critical dimensions, the dynamics of first-order chemical reactions with rate constants between 10^2 and 10^4 s^{-1} can be investigated.

For references see page 10168

Table 3 Rate constants for heterogeneous electron transfer determined using steady-state methods

Analyte	Electrode[a]; element, dimension (μm)	k^o (cm s^{-1})	Ref.
Anthracene	Au ring $\Delta r = 0.09$ μm $r = 5$ μm	3.33 ± 0.05	58
(C_6H_6) Cr $(CO)_3^+$	Pt, 25	≥ 0.3	59
$(CpCOOCH_3)_2C_o^+$	Pt, 23 Å to 4.7 μm	130 ± 70	60
Cytochrome *c*	C, 6.3	>0.4	61
9,10-Diphenylanthracene	Au ring $\Delta r = 90$ μm $r = 5$ mm	5.7 ± 0.1	58
Ferrocene	Au ring $\Delta r = 0.09$ μm $r = 5$ μm	0.09 ± 0.005	58
	C, 6	2.3 ± 0.8	62
	Pt, 0.3–25	≥ 6	63
	Pt, 1	>2	64
	Pt, 16 Å to 2.6 μm	220 ± 120	60
$Fe(OEP)(N\text{-}MeIM)_2^+$	Pt, 1–25	0.4	65
	Pt, 0.5–12.5	0.38	66
	Pt, 1–25	0.35	65
$Fe(TPP)(HIM)_2^+$	Pt, 1–25	0.5	65
$Fe(TPP)py_2^+$	Pt, 1–25	0.6	65
$MV^{2+(F)}$	Pt, 22 Å to 0.21 μm	170 ± 90	60
Naphthalene	Au ring $\Delta r = 0.2$ μm $r = 20.5$ μm	0.88 ± 0.02	58
Oxygen	Pt, 12.7–250	0.63 ± 0.05	67
$Ru(NH_3)_6^{3+}$	Au, 5	0.076	68
	Pt, 1.3–4.6	0.26 ± 0.13	60
	Pt, 11 Å to 11.1 μm	79 ± 44	60
$[Ru(bpy)_3]^{2+}$	Pt, 11.2		69
Tetracyanoethylene	Au ring $\Delta r = 0.2$ μm $r = 20.5$ μm	0.15 ± 0.01	58
Tetracyanoquinodiethane	Au ring $\Delta r = 0.2$ μm $r = 20.5$ μm	0.23 ± 0.01	58
Zn(TPP)	Pt, 1 to 25 μm	>1	65

[a] Dimension given is the radius of a microdisc electrode unless otherwise stated. OEP, octa-ethyl-porphyrin; MeIM, methyl imidazole; TPP, tetra-phenyl porphyrin; HIM, imidazole; MV, methyl viologen.

Table 4 Rate constants for homogeneous chemical reactions determined using steady-state methods

Analyte	Electrode[a]; element, dimension (μm)	k	Refs.
Anthracene oxidation	Pt, 0.3–62.5	190 ± 50 s^{-1}	70
Ascorbic acid oxidation at a Prussian blue film	Pt, 2.5 to 25	1.3×10^5 M^{-1} s^{-1}	71
Ferrocyanide oxidation in the presence of ascorbic acid	Pt band pair, gap = 2–12	27 ± 4 M^{-1} s^{-1}	72
9,10-Diphenylanthracene + 4,4-dibromodiphenyl	C, 6–9	3.9 ± 0.6 M^{-1} s^{-1}	73
Anion radicals + alkyl halides	Pt, 0.25	9×10^{-4} to 1.7×10^4	74, 75
$[Fe(CN)_6]^{3-}$ + aminopyridine	Pt, 0.3–25	$3.0 \pm 0.6 \times 10^3$ M^{-1} s^{-1}	76
	Pt, 2.5–432	1.8×10^3 M^{-1} s^{-1}	77
	Pt band pair, gap = 2–12	$8 \pm 1 \times 10^2$ M^{-1} s^{-1}	72
Hexamethylbenzene oxidation	Pt, 0.3–25	720 ± 100 s^{-1}	70
H_2 evolution on Pt from acetic acid solution	Pt, 0.3–25	4.1×10^{10} M^{-1} s^{-1}	76
1-Napthylamine oxidation	Pt, 0.5–12.5	4.1×10^3 s^{-1}	78
Triphenylamine oxidation	Pt, 0.3–20	$> 3 \times 10^4$ M^{-1} s^{-1}	70
Thioselenanthrene		8.87 ± 1.1 s^{-1}	79
Dibenzo-1,2-diselenine		20.7 ± 2.8 s^{-1}	80

[a] Dimension given is the radius of a microdisc electrode unless otherwise stated.

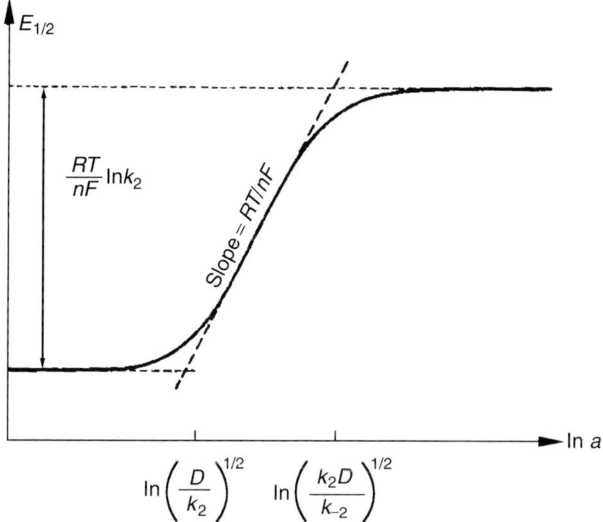

Figure 11 Dependence of $E_{1/2}$ for an electron transfer reaction on the logarithm of the electrode radius. (Reprinted from K.B. Oldham, *J. Electroanal. Chem.*, **313**, 3. Copyright (1991) with the permission of Elsevier Science.)

3 INDIRECT METHODS

As discussed above, with direct electrochemical measurements a competition is created between the reaction of interest and diffusional mass transport either by controlling the time over which the system is electrochemically perturbed (transient methods) or by controlling the electrode size (steady-state methods). In both approaches, diffusion provides a time reference for the measurement. In contrast, for indirect methods, the reaction of interest competes with another chemical reaction whose rate constant is known from independent measurements. Thus, this second reaction acts as a kinetic reference instead of the diffusion to or from the electrode that applies in the direct methods.

Competitive preparative-scale electrolysis is an approach that can provide a limited insight into the relative rates of reactions. Here, the distribution of products between two competitive reactions is measured, one of which is the reaction of interest and the other is known from the independent sources and acts as reference. For example, Savéant et al.[(81)] have investigated the competition that takes place between electron and hydrogen atom transfer pathways for the reduction of aryl halides in H-atom donor solvents such as acetonitrile and dimethylsulfoxide (DMSO). Mathematical models describing how the product distributions depend on the reaction mechanism and rate constant have been determined. Not only do these analyses provide important fundamental information about those factors that control chemical reactivity, they also allow important industrial processes to be optimized, e.g. reduction of CO_2 in media of low proton mobility.

3.1 Redox Catalysis

Redox catalysis can be viewed as the ultimate limit of small probe electrochemistry in which the electrode is decreased to the size of a single molecule. Figure 12 illustrates how the dynamics of an electron transfer reaction can be probed using a mediator in a redox catalysis scheme. Only the case in which the dynamics of a reduction reaction are being probed is considered.

In this approach, a substrate A that would normally be reduced at an electrode surface is now reduced by a mediator Q, i.e. the mediator Q replaces the electrode surface and reduces the substrate A. If the reduced form of the redox couple of interest undergoes a following chemical reaction, then its decay can be monitored using this approach. Figure 13 illustrates the differences between redox catalysis and direct electrochemistry.

While it provides great insight into a wide variety of reaction mechanisms, the useful time domain of this technique is somewhat unusual in light of the discussion on direct methods. Redox catalysis is only useful for monitoring species with lifetimes less than approximately 10 µs. As discussed previously, the rate of diffusion increases with decreasing electrode size. In redox catalysis, the electrode has been reduced to the dimensions of a single molecule allowing extremely low timescales to be accessed. In the case where the driving force is large (i.e. large values of $E^{o}_{PQ} - E^{o}_{AB}$), k_+ will have a value close to the diffusion limit. Taking a maximum value of 50 mM for $[P]_0$, it is apparent that the shortest measurable lifetime, $k\,[P]_0$, is of the order of 50 ns.

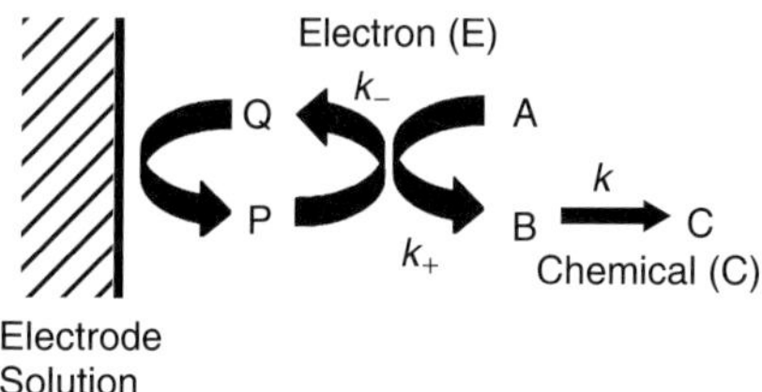

Figure 12 Dynamics of an EC reaction.

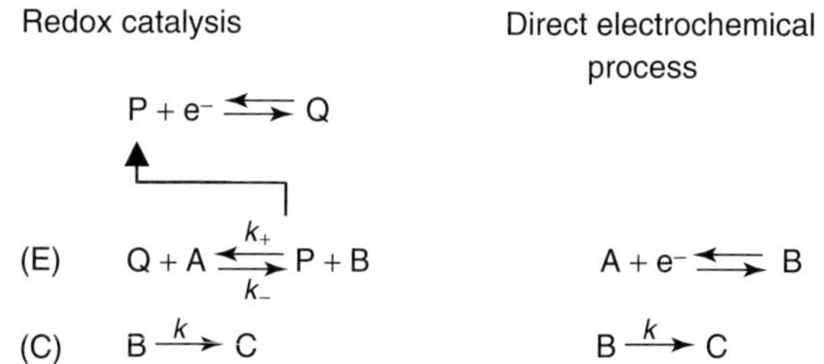

Figure 13 Comparison of redox catalysis and direct electrochemistry.

For references see page 10168

In redox catalysis, the current flowing through the electrode is monitored. The mediating P/Q couple is selected to be electrochemically fast and reversible as well as having a standard potential located positive of the substrate (A) reduction wave. Thus, in cyclic voltammetry experiments performed on the mediator alone, reversible on-electron waves are observed even at high scan rates because the rate of heterogeneous electron transfer is fast. However, when the analyte of interest is added, the voltammetry for the mediator becomes irreversible and the reduction current increases. These effects provide kinetic information about the electron transfer and coupled chemical reactions. As illustrated in Figure 13, the rate-determining step is either the forward electron transfer or the homogeneous chemical reaction with the electron transfer step acting as a pre-equilibrium. It is the competition that exists between the backward electron transfer step and the chemical reaction that dictates which process will be rate determining.

The work of Utley et al. on *ortho*-quinodimethanes (*o*-QDM) provides a good illustration of how redox catalysis affects the voltammetric response.[82] As illustrated in Figure 14, co-electrolysis of 1,2-bis(bromomethylarenes) and dienophiles (e.g. maleic anhydrides or quinones), gives the Diels-Alder adduct expected for reaction of the *o*-QDM derived from the dihalide. It is known that direct reduction of the dihalides gives *o*-QDMs but in the absence of a dienophile they polymerize. However, Utley performed co-electrolysis experiments at a potential several hundred millivolts more positive than the reduction potential of the dihalide. Cyclic voltammetry of this system is illustrated in Figure 15 which shows that a chemically reversible reduction is observed for the mediator (dienophile) in the absence of the dihalide. Moreover, this reversible response is observed at slow scan rates, indicating that the radical-anion is stable over a significant lifetime. However, in the presence of the analyte of interest, the radical-anion can transfer an electron to the added dihalide, which rapidly undergoes a following chemical reaction, i.e. it loses bromide anion. As shown in Figure 14, at long times an *o*-QDM is formed, which can react relatively slowly with the regenerated mediator. On the cyclic voltammetry timescale the regenerated mediator is rapidly reduced, and it is this process that is responsible for the loss of reversibility and the more than doubling of the peak current seen in Figure 15. This increased current is known as the catalytic current.

The effect of kinetics on the diffusion processes of P, Q and A is given by Equation (12)

$$\frac{d[P]}{dt} = -\frac{d[Q]}{dt} = \frac{-d[A]}{dt} = k_+k\frac{[A][Q]}{k_-[P]+k} \qquad (12)$$

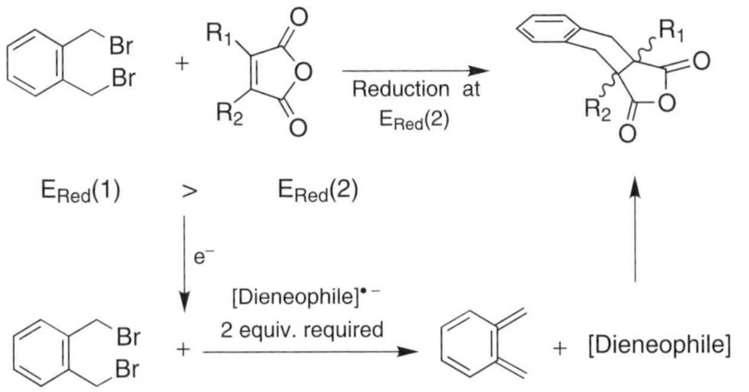

Figure 14 Redox catalyzed reaction scheme for 1,2-bis(bromomethylarenes) and functionalized maleic anhydride.

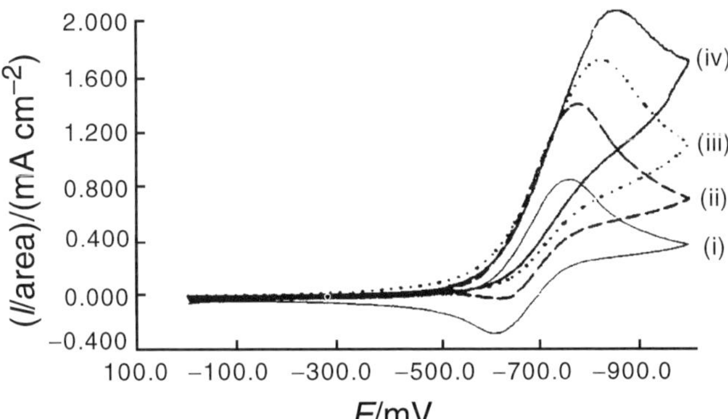

Figure 15 Characteristic cyclic voltammetry of a redox catalyzed system at a mercury electrode where the supporting electrolyte is 0.1-M tetraethyl ammonium perchlorate in DMF. The scan rate in all cases is $0.3\,V\,s^{-1}$. (i) Anhydride only, (ii), (iii), and (iv) contain 0.5, 1.0 and 2.0 equiv. of dibromide, respectively. (Reprinted from E. Oguntoye, S. Szunerits, J.H.P. Utley and P.B. Wyatt, *Tetrahedron*, **52**, 7771. Copyright (1996) with permission of Elsevier Science.)

Kinetic control by the forward electron transfer step, Equation (13)

$$\frac{d[P]}{dt} = -\frac{d[Q]}{dt} = \frac{-d[A]}{dt} = k_+[A][Q] \qquad (13)$$

is achieved when $k_-[P]_0 \ll k$. In contrast, when $k_-[P]_0 \gg k$, the kinetics of the homogeneous chemical reaction are rate determining, Equation (14)

$$\frac{d[P]}{dt} = -\frac{d[Q]}{dt} = \frac{-d[A]}{dt} = \frac{k_+k[A][Q]}{k_-[P]} \qquad (14)$$

This analysis reveals that the rate-determining step depends on the concentration of the mediator, P, and can be switched from electron transfer to homogeneous chemical reaction by changing $[P]_0$. Solutions to these systems of differential equations have been obtained for a wide variety of electrochemical techniques, including cyclic voltammetry and chronoamperometry, allowing rate data to be obtained for unstable intermediates having lifetimes down to the low nanosecond time range.

The technique has certain limitations, e.g. if the chemical reaction is so fast that the kinetic control is always by the forward electron transfer step then it is not possible to characterize the homogeneous chemical reaction kinetically. However, under these conditions it is often still possible to measure the formal potential and heterogeneous electron transfer rate constant for the A/B couple if experiments are conducted using a range of mediators with different formal potentials. Savéant et al.[81] have elucidated the relationship between the electron transfer rate constant and the formal potential of the PQ couple where the homogeneous chemical reaction is so fast that it occurs within the molecular diffusion layer of the mediator. Moreover, this group has extended the theory to allow more complex reaction schemes, including bimolecular following reactions, than the electron transfer reaction described above to be characterized. These advances are significant since simple electron transfer reaction schemes are rarely found in real experimental systems.

3.2 Indirect Laser-induced Temperature Jump Method

There is a long and distinguished record of using temperature jump to elucidate the kinetics of homogeneous chemical phenomena, e.g. Sutin et al.[83] used a laser-induced temperature jump method to achieve rise times of tens of nanoseconds. Moreover, approaches have been developed to allow the temperature of the electrode/solution interface to be rapidly changed. However, electrical heating methods typically allow the interfacial temperature to be perturbed only at the millisecond or longer timescale. The use of lasers has radically changed this field allowing much more rapid temperature jumps to be achieved allowing fast electron transfer processes to be investigated. These laser-based approaches stem from early work in which laser flashes impinged on the electrode/solution interface causing electrons to be photoemitted. For example, Barker et al.[84] studied the effects of irradiation by a 20 ns laser pulse ($\lambda = 694.3$ nm) on mercury and suggested that laser-induced temperature changes might be the cause of some of the observed potential changes since the laser energy was insufficient to cause photoemission at the applied potentials investigated. This "front side" irradiation approach without stimulated photoemission was also used by Benderskii et al.[85] to examine double-layer phenomena at mercury electrodes with nanosecond time resolution. Coufal and Hefferle[86] have also demonstrated the ability of laser-induced temperature steps to resolve nonelectrochemical events at subnanosecond timescales.

An extremely important variation on these approaches has been developed by Feldberg and co-workers.[87] Their technique, the indirect laser-induced temperature jump method (ILIT) involves indirect irradiation of the electrode, i.e. the laser pulse impinges on the "back face" or nonelectrolyte electrode surface. In this method, some of the incident photons are absorbed by the thin foil or film electrode and thermalise virtually instantaneously causing a rapid rise in the interfacial temperature. The time dependence of the thermal perturbation is determined by the shape of the laser pulse and by the thickness, thermal diffusion constant, and heat capacity of the electrode material, also the heat capacities and thermal diffusion constants of the two adjacent media (electrolyte at the front face, and dielectric surrounding the rear of the electrode). Typically, this interfacial temperature jump destroys the equilibrium between the double-layer potential and the redox potential. This process changes the open-circuit potential as the system attempts to re-establish equilibrium under the new conditions. In contrast to classical charge-injection methods, thermal perturbation does not cause a large iR spike, making data interpretation more straightforward. Moreover, the ILIT offers the possibility of extracting information about thermal as well as electrochemical properties of electrodes and adsorbed layers.

When compared with direct irradiation of the electrode/solution interface, Feldberg's indirect approach has several advantages. Photoemission of electrons into the solution is not possible; the optical density of the solution is irrelevant and photolysis and/or direct thermal heating of the solution is avoided; the form of the interfacial temperature perturbation approximates a step function more closely than that associated with direct irradiation; deconvolution of the data is facilitated and the rates of the rise and fall of the interfacial temperature can be adjusted (but not independently) by changing the thickness of the electrode.

There are three major processes that can cause the open-circuit potential to change in response to the temperature jump. First, there is a junction potential between the (hot) electrode and the (cold) contact wire. Second, a change occurs in the potential across the electrode double layer which can be caused by a change in the capacitance or dipole reorientation (equivalent to a change in the potential of zero charge); by charge transfer (electron transfer between the electrode and a redox system) located at the outer Helmholtz plane (solution-phase reactant) or inner Helmholtz plane (adsorbed reactant), or ion transfer between the inner and outer Helmholtz planes. Third, a Soret potential arises from the temperature gradient between the interfacial region and the bulk electrolyte. Therefore, this approach can be used to determine kinetic and thermodynamic parameters of potential-dependent double-layer restructuring, heterogeneous

For references see page 10168

electron transfer to the solution phase and adsorbed electroactive species and ion transfer from the inner to outer Helmholtz planes, all at nanosecond or shorter timescales. The theoretical analysis for each of these situations yields an analytic equation provided that the temperature changes are small (controllable through the laser power) and the charge-transfer resistance is constant.

As an example, determination of the heterogeneous electron transfer rate constant for ferro/ferri cyanide and the ways in which ILIT can provide a powerful insight into adsorption phenomena are reviewed here. As illustrated in Figure 16, the typical ILIT experiment uses a Nd:YAG laser ($\lambda = 1.06\,\mu m$, full width at half height maximum of 8 ns) that is capable of resolving relaxations as short as 10 ns. The pulse from this laser passes through a neutral density filter that allows the incident power to be controlled before being directed by a mirror onto the back surface of the working electrode. As shown in Figure 17, the cell contains two cylinders of $25\,\mu m$ Pt foil approximately 1 cm wide. Onc cylinder acts as a pseudoreference electrode against which the change in the open-circuit potential is measured while the other is used as an auxiliary electrode for a three-electrode potentiostat.

Figure 18, illustrates examples of the ILIT responses obtained for a $1.07\,\mu m$ gold film electrode in contact with 1.0 M KF before and after the addition of 4×10^{-5} M of $K_3[Fe(CN)_6]$ and $K_4[Fe(CN)_6]$. In both cases the electrode potential is poised at the formal potential of the redox couple. The magnitude of the ILIT response is of the order of a few millivolts and is typical of these experiments. In the absence of any dissolved electroactive species, the open-circuit potential changes abruptly within a few nanoseconds and then relaxes over several hundred nanoseconds. Both before and after addition of the electroactive species, simple thermal responses are obtained, i.e. the change in the open-circuit potential precisely follows

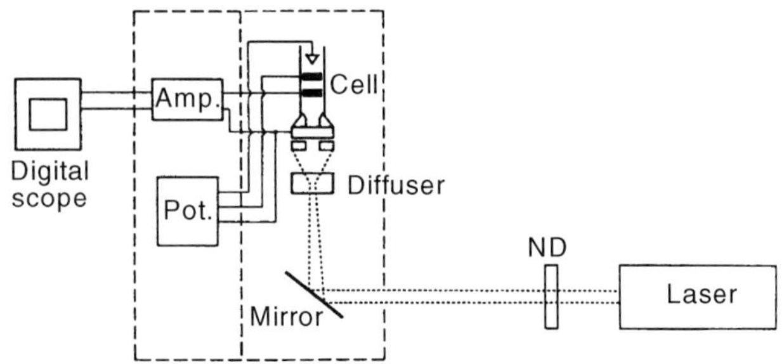

Figure 16 Block diagram of the ILIT apparatus. ND, neutral density filter; Amp., amplifier; Pot., potentiostat. The broken line denotes the Faraday cage. (Reprinted from J.F. Smalley, L. Geng, S.W. Feldberg, L.C. Rogers, J. Leddy, *J. Electroanal. Chem.*, **356**, 181. Copyright (1993) with the permission of Elsevier Science.)

Figure 17 Diagram of the ILIT cell. (Reprinted from J.F. Smalley, L. Geng, S.W. Feldberg, L.C. Rogers, J. Leddy, *J. Electroanal. Chem.*, **356**, 181. Copyright (1993) with the permission of Elsevier Science.)

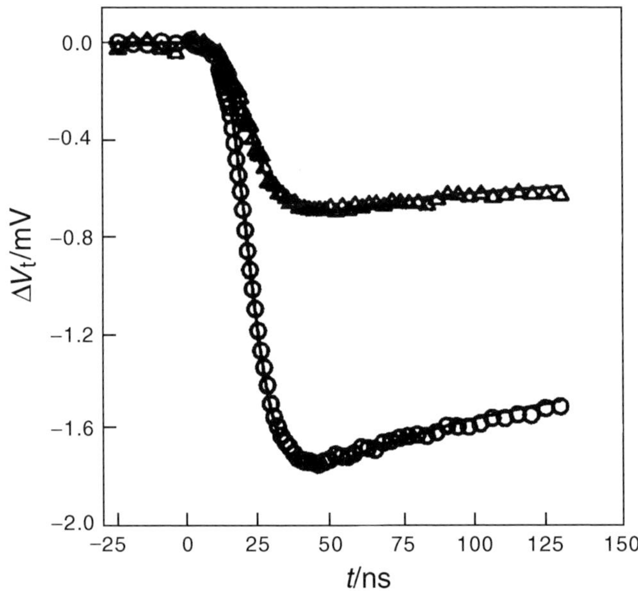

Figure 18 The ILIT responses of a gold electrode in 1.0-M KF at 0.2 V vs saturated calomel electrode (SCE) before (Δ) and after ($\circ$) addition of 4×10^{-5} M each of $K_3[Fe(CN)_6]$ and $K_4[Fe(CN)_6]$ at $E = E^{\circ} = 0.214$ V vs SCE. The solid curves represent theoretical fits. (Reprinted from J.F. Smalley, L. Geng, S.W. Feldberg, L.C. Rogers, J. Leddy, *J. Electroanal. Chem.*, **356**, 181. Copyright (1993) with the permission of Elsevier Science.)

the change in the interfacial temperature. Moreover, the time dependence of these responses follow, almost exactly, that predicted by theory. Analyzing the ILIT response in the presence of the electroactive species provides a value of $1.64 \times 10^{-2}\,\mathrm{cm\,s^{-1}}$ for the standard heterogeneous electron transfer rate constant which is in reasonable agreement with previous results obtained using traditional electrochemical techniques.

Implicit in the analysis is the assumption that the change in the interfacial temperature does not cause the rate constant to change significantly. For the case of the $Fe(CN)_6^{3-/4-}$ couple, the activation energy is of the order of $13\,\mathrm{kJ\,mol^{-1}}$ so that changes in the electrode temperature of up to 4 K will result in k^o being constant to within 10%. However, for reactions featuring large activation energies, e.g. those with large inner or outer sphere reorganization energies, temperature-induced changes in k^o may be significant.

Although the open-circuit potential becomes constant with time in all cases, suggesting that equilibrium is achieved, the magnitude of the ILIT response depends on the concentration of the ferrocyanide and ferricyanide. While effects such as adsorption of an impurity, electron transfer between the electrode, and solution-phase reactant cannot be excluded, this observation is considered to arise because of $Fe(CN)_6^{3-/4-}$ adsorption onto the electrode surface. Adsorption will modify the structure of the electrochemical double layer thus changing the thermal response. The concentration-dependent ΔV_t can be used as a measure of the surface coverage of the adsorbate and the concentration-dependent surface coverage modeled using the Frumkin adsorption isotherm. This analysis reveals that the $Fe(CN)_6^{3-/4-}$ layer is strongly bound, $K_{ads} = 4.7 \times 10^9\,\mathrm{M^{-1}}$. That positive values of the interaction parameter, g, are observed suggests either that there are strong repulsive interactions between the adsorbates, or that the potential difference between the inner and outer Helmholtz planes changes with changing surface coverage.

A very important example of the ultrafast capabilities of this method is in the determination of heterogeneous electron transfer rate constants for electroactive self-assembled monolayers. As discussed in the section dealing with fast transient methods, when a short bridge is used to tether the redox active species to the electrode surface, the rate of heterogeneous electron transfer becomes immeasurably fast even for chronoamperometry at microelectrodes. The ILIT was used to probe the kinetics of electron transfer between a gold electrode and a self-assembled monolayer formed from $CH_3(CH_2)_{n-1}SH$ and $(\eta^5{-}C_5H_5)Fe(\eta^5{-}C_5H_4)CO_2(CH_2)_nSH$ (i.e. a diluted ferrocene alkane thiol monolayer), where n is the number of methylene groups in the bridge and $5 \le n \le 9$.[88] In these experiments, as well as recording the time-dependent change in the open-circuit potential, the size

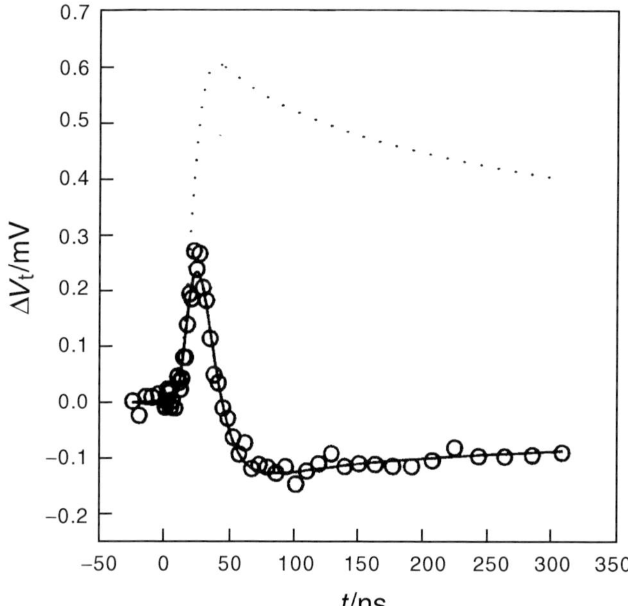

Figure 19 ILIT response obtained for a gold electrode coated with a mixed monolayer formed from $CH_3(CH_2)_4SH$ and $(\eta^5{-}C_5H_5)Fe(\eta^5{-}C_5H_4)CO_2(CH_2)_5SH$ with $N_T = 1.3 \times 10^{-10}$ mol, $T = 31.0\,^\circ$C, thickness of the electrode $= 0.96 \times 10^{-4}$ cm. Open circles are the experimental data. $E_{app} = 0.475$ V vs standard SCE, solid line is the theoretical fit. The dotted line represents the response which would be observed if there were no relaxation of the ILIT signal caused by electron transfer between the electrode and the redox couple. (Reprinted with permission from J.F. Smalley, S.W. Feldberg, C.E.D. Chidsey, M.R. Linford, M.D. Newton, Y.-P. Liu, *J. Phys. Chem.*, **99**, 13141. Copyright (1995) American Chemical Society.)

of the temperature jump at the electrode/electrolyte interface was monitored using a pressure transducer.

Figure 19 shows the ILIT response obtained close to the formal potential of the ferrocene/ferricenium redox reaction for a gold film electrode modified with a ferrocene alkane thiol monolayer in which the bridge contains five methylene groups. Fitting these data using theory developed for heterogeneous electron transfer involving a surface confined species provides an estimate of k as $7.8 \times 10^7\,\mathrm{s^{-1}}$. This value is comparable with the upper limit of that possible using state-of-the-art transient chronoamperometry. By controlling the applied potential, the potential dependence of the heterogeneous electron transfer rate constant can be determined. For this system, this dependence appears to be adequately described by the Butler–Volmer formalism of electrode kinetics.

Beyond potential-dependent rate constants, the approach was extended to measure activation (reorganization) energies by using a thermostated cell. Figure 20 shows that the electron transfer rate constant depends on the number of methylene groups in the bridge. At $25\,^\circ$C, the standard electron transfer rate constants

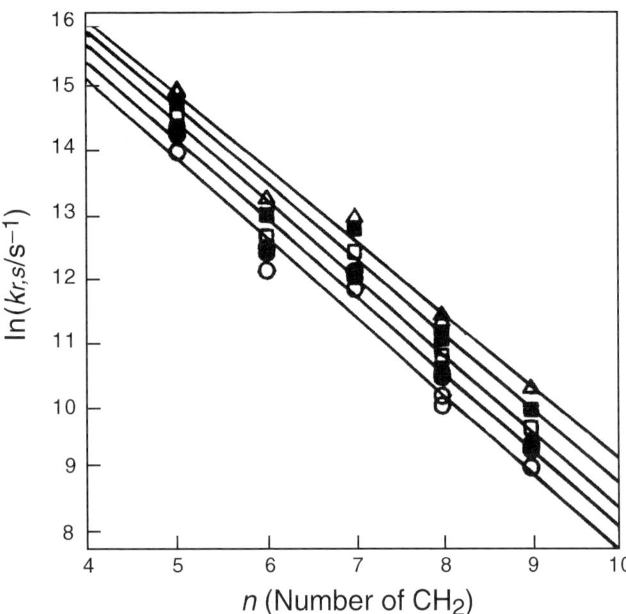

Figure 20 Plots of $k_{\Gamma,s}$ versus n at five different temperatures. Open circles, $T = 14.9\,°C$; filled circles, $T = 25\,°C$; open squares, $T = 35.0\,°C$; filled squares, $T = 44.8\,°C$; open triangles, $T = 55.2\,°C$. (Reprinted with permission from J.F. Smalley, S.W. Feldberg, C.E.D. Chidsey, M.R. Linford, M.D. Newton, Y.-P. Liu, *J. Phys. Chem.*, **99**, 13 141. Copyright (1995) American Chemical Society.)

vary according to $k_{\Gamma,s,n=0}\exp[-\beta_n n]$ where $k_{\Gamma,s,n=0}$ is the extrapolated rate constant for the electron transfer at $n = 0$ and is equal to $6 \times 10^8\,s^{-1}$ and β_n is 1.21 ± 0.05. Moreover, for each of the monolayers investigated, linear Arrhenius plots are observed. The slopes of these plots have been used to determine the reorganization energy, λ, assuming that λ is the sole contributor to the activation energy for electron transfer. The reorganization energies range from $0.70 \pm 0.04\,eV$ for $n = 5$ to 0.91 ± 0.05 for $n = 9$.

3.3 Scanning Electrochemical Microscopy

Bard et al. have developed an important technique, SECM,[89] in which a small electrode is precisely positioned close to a surface in an electrochemical cell arrangement to obtain information about the surface topography and reactions that occur in the solution space between tip and sample. As described in Equation (10), at long timescales a steady-state potential-independent current is observed for the oxidation or reduction of an electroactive species at a microelectrode. Figure 21 shows how the current to this microelectrode tip is perturbed by the presence of a substrate near the tip. These effects form the basis of the SECM technique. When the tip is brought near a nonconducting surface, the current will be smaller than the steady-state value of Equation (10)

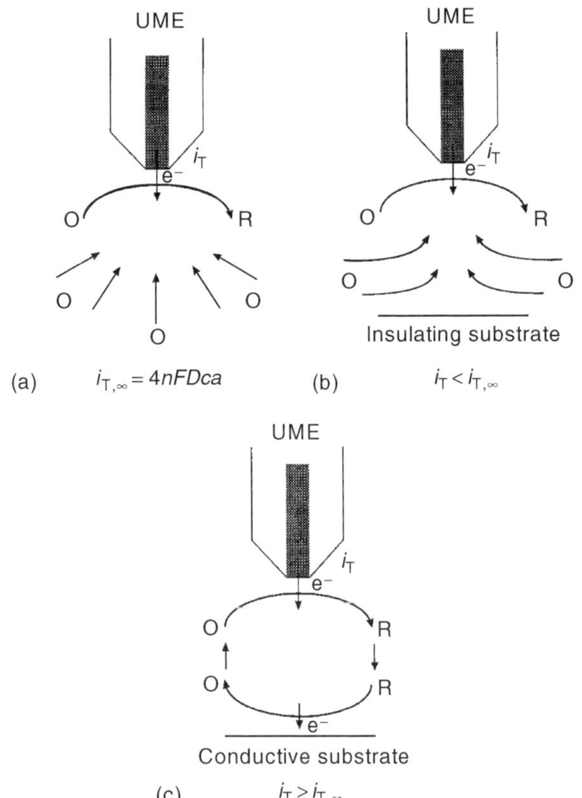

Figure 21 Basic principles of SECM. (a) With the microelectrode far from the substrate, diffusion of O leads to a steady-state current, $i_{T,\infty}$. (b) With the microelectrode near an insulating substrate, hindered diffusion of O leads to $i_T < i_{T,\infty}$. (c) With the microelectrode near a conducting surface, positive feedback of O to the tip leads to $i_T > i_{T,\infty}$. (Reproduced with the permission of the Royal Society of Chemistry from M. Arca, A.J. Bard, B.R. Horrocks, T.C. Richards, D.A. Treichel, *Analyst*, **119**, 719 (1994).)

because the surface blocks diffusion of the redox active species to the tip. The closer the tip is to the surface, the smaller the current, until when the tip-to-surface spacing, d, becomes zero, the tip current will approach zero. The tip current, i_T, as a function of separation from the surface can be approximately described by Equation (15)

$$\frac{i_T}{i_{T,\infty}} = \left[0.292 + \frac{1.515}{L} + 0.655\exp\left(\frac{-2.4035}{L}\right)\right]^{-1} \quad (15)$$

where the normalized tip–substrate separation, L, is equal to d/r. Thus, the presence of a nonreactive substrate near the tip can always be recognized by the condition $i_T < i_{T,\infty}$.

A different behavior is observed when the tip is near a surface where tip-generated product can be re-oxidized or re-reduced, e.g. when the substrate is a conductive material at a sufficiently positive or negative potential. Under these circumstances, there is a flux of the analyte to the tip from both the substrate and the bulk solution. Owing to this positive feedback from the substrate, $i_T > i_{T,\infty}$. When the redox reaction occurring at the surface proceeds at a diffusion-controlled rate, Equation (16) applies

$$\frac{i_T}{i_{T,\infty}} = 0.68 + \frac{0.738}{L} + 0.3315 \exp\left(\frac{-1.0672}{L}\right) \quad (16)$$

Figure 22 shows approach curves, i.e. plots of $i_T/i_{T,\infty}$ versus L for the cases where the microelectrode tip approaches conducting and nonconducting surfaces.

An important attribute of the technique is that if the electrode radius is known, the Faradaic current can be used to determine the tip–surface separation without having prior knowledge of the bulk concentration or diffusion coefficients.

3.3.1 Heterogeneous Electron Transfer Dynamics

SECM has recently emerged as a useful technique for measuring the rate constants of fast heterogeneous electron transfer reactions. To achieve this objective, the SECM apparatus is used to form a twin-electrode thin-layer cell between the microelectrode tip and a conducting substrate. During voltammetric scans, this configuration induces high rates of mass transfer between the substrate and tip electrodes because large concentration gradients exist between the two closely spaced electrodes. The overall current observed can be limited by either the rate of mass transfer to the electrode or the intrinsic heterogeneous electron transfer rate. Therefore, the increase in mass transfer of the thin layer cell allows faster rates of electron transfer to be measured in the SECM arrangement than can be accomplished with the same electrode in bulk solution.

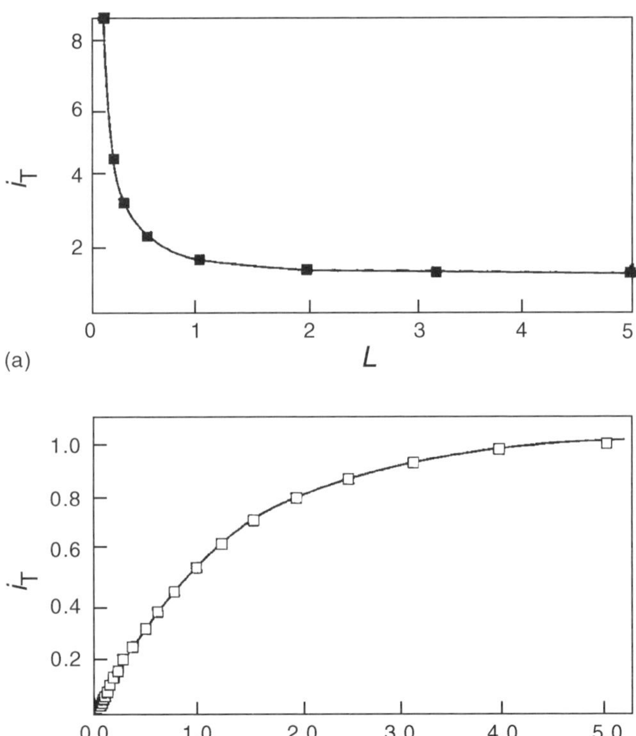

Figure 22 Diffusion controlled steady-state tip current as a function of tip–substrate separation. $i_T = i_T/i_{T,\infty}$, $L = d/r$. (a) Substrate is a conductor, (b) conductor is an insulator. The solid lines are theoretical fits. (Reproduced with the permission of the Royal Society of Chemistry from M. Arca, A.J. Bard, B.R. Horrocks, T.C. Richards, D.A. Treichel, *Analyst*, **119**, 719 (1994).)

Most electrochemical responses are sensitive to the heterogeneous electron transfer rate whenever the rate of mass transfer is of the same order of magnitude or greater than the standard heterogeneous electron transfer rate constant. The advantage of SECM can be seen by comparing mass-transfer coefficients for two different electrode configurations. For a microdisc in bulk solution, the steady-state mass-transfer coefficient is approximately D/r. For thin-layer cells, the mass transfer coefficient is approximately D/d. However, while the smallest well-characterized microdisc electrodes are of the order of 500 nm, stable layers as thin as 40 nm have been obtained for a 1-μm diameter Pt disc electrode opposed by a mercury substrate. Clearly, the rate of mass transfer can be enhanced by more than an order of magnitude in going from a microelectrode in bulk solution to the thin-layer configuration in SECM. Achieving a 40 nm substrate to tip separation allows rate constants of the order of several centimeters per second to be measured under steady-state conditions.

Figure 23 shows experimental data and theoretical calculations for the oxidation of ferrocene system taken at different tip–substrate separations.[90] This figure shows that the wave broadens as the tip–substrate separation decreases and the limiting current increases owing to enhanced feedback of the analyte by reaction at the conducting substrate. More significantly however, the waves broaden because the mass-transfer rate increases to the point where is becomes comparable to, or greater than, the electron transfer rate. It is this increased sensitivity of the voltammetric response to k that allows large rate constants to be measured under steady-state conditions. The measured k^o for ferrocene was $3.7 \pm 0.6\,\mathrm{cm\,s^{-1}}$, confirming the rapid nature of heterogeneous electron transfer in this system.

3.3.2 Homogeneous Chemical Kinetics

SECM can also be used to measure the rate of homogeneous chemical reactions that follow oxidation or reduction by monitoring the effect of the reaction

For references see page 10168

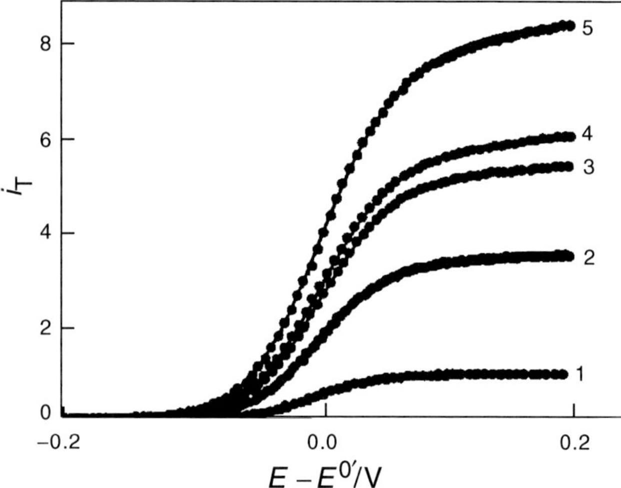

Figure 23 Experimental data and theoretical fits for the oxidation of ferrocene at a 2-μm-diameter platinum tip opposed by a 1.5-mm-diameter platinum substrate electrode. Voltammograms 1–5 show the transition from data recorded in bulk solution (curve 1) to those recorded with the tip electrode 100 nm from the substrate. Currents are normalized such that $i_T = i_T/i_{T,\infty}$. (Reproduced with the permission of the Royal Society of Chemistry from M. Arca, A.J. Bard, B.R. Horrocks, T.C. Richards, D.A. Treichel, *Analyst*, **119**, 719 (1994).)

of interest on the tip current. For example, in the case of a reversible redox reaction, when d is small and the substrate area is large in comparison to the microelectrode tip, essentially all of the material that is electrogenerated at the tip is collected by the substrate. However, a following chemical reaction will consume some of the tip product, decreasing the collection efficiency. In this sense, the tip and substrate electrodes act as a generator–collector assembly in an analogous way to the ring and disc electrodes in a rotating ring-disc electrode. Theory that describes both first- and second-order homogeneous chemical reactions following electron transfer has been developed.[89] This theory is most accessible in the form of working curves describing how the currents observed at both tip and substrate depend on their mutual separation.

A typical reaction scheme which can be studied by SECM is shown by Equation (17)

$$\begin{aligned} &\text{tip:} && R \longrightarrow O + ne^- \\ &\text{gap:} && 2O \longrightarrow \text{products} \\ &\text{substrate:} && O + ne^- \longrightarrow R \end{aligned} \tag{17}$$

Rate constants can be determined if the tip and substrate are positioned sufficiently close so that a detectable amount of R can cross the gap before reacting. The time required for a species to cross the gap is approximately d^2/D. Given that the half-life of the electrogenerated intermediate undergoing a second-order following reaction is $1/kC$, where C is the bulk concentration, rate constants as large as $10^8\,M^{-1}\,s^{-1}$ can be measured using separations of the order of 100 nm. For unimolecular reactions, first-order rate constants as large as $10^5\,s^{-1}$ are measurable.

Bard et al.[91] used the SECM technique to investigate the reductive coupling of dimethyl fumarate as well as fumaronitrile in DMF. For fumaronitrile, the rate constant was shown to be approximately constant as the analyte concentration was changed over two orders of magnitude. This consistency indicates that second-order kinetics are being followed for the following chemical reaction of the fumaronitrile anion radical. The measured rate constant is large, $2 \times 10^5\,M^{-1}\,s^{-1}$, demonstrating the ability of SECM to provide high-quality kinetic information about fast homogeneous chemical reactions. In contrast, the other alkene, dimethyl fumarate, appears to be less activated and undergoes a much slower chemical reaction when reduced and a rate constant of $170\,M^{-1}\,s^{-1}$ was determined using both tip feedback and substrate current measurements. Good agreement of experiment with theory was seen for both modes of SECM operation.

Despite the rapid nature of these processes, Bard and his group have extended the use of SECM to the study of even faster reactions, e.g. the dimerization of 4-nitrophenolate,[92] which has been shown to undergo a fast irreversible dimerization following oxidation. The calculated rate constant, approximately $8 \times 10^7\,M^{-1}\,s^{-1}$, agrees well with that measured by fast-scan cyclic voltammetry.

4 FUTURE DIRECTIONS

4.1 Analytical Applications of Fast Electrochemistry

When attempting to determine the concentration of a particular redox active species in a complex mixture, the response of the target analyte is typically separated from redox active interferences on the basis of different formal potentials. In fact, there have been relatively few reports using differences in electrochemical reactivity (i.e. electrode kinetics) to determine the concentration of a target analyte by separating its voltammetric response from that of an interferent on the basis of different time constants for the two reactions. Since the width of the electrochemical response for any species is a sizeable fraction of the potential scale, relying on the potential axis alone to generate a selective response provides only a very limited ability to resolve an analyte's response from that of an interfering species. If the time axis as well as the potential axis can be used, separating the analyte's response from that of an interfering species becomes considerably more likely. This time-resolved approach ought to benefit significantly from the dramatic expansion in the range of timescales that can be resolved and exploited in electrochemistry with the advent of high-speed instrumentation and microelectrodes.

It is widely recognized that double-layer charging can represent a significant obstacle to achieving very low limits of detection in electroanalysis. Often, background subtraction is achieved by measuring the voltammetric response in the absence of the electroactive species of interest. However, differences in rates at which charging and Faradaic processes occur can be exploited to time gate the response so that an enhanced Faradaic to capacitive current ratio is obtained.

4.2 Ultimate Limits on Ultrafast Electrochemical Techniques

Decreasing the timescale of t_{dif} of an electrochemical experiment involving a solution-phase reactant can be achieved either by shortening the experimental timescale (faster voltammetric scan rates or shorter potential steps), or by shrinking the critical dimension of the microelectrode in steady-state measurements. Both of these effects cause the thickness of the diffusion layer to shrink and it will eventually become less than that of the electrochemical double layer. This effect has profound consequences for migration of the analyte within the electric field of the double layer.

In conventional electrochemical cells containing a dilute (≤ 0.1 M) supporting electrolyte, the electrical field that exists because of the electrode–solution potential difference decays exponentially from the electrode surface out into solution over a distance of a few tens of angstroms. Under these conditions, the effects of diffusion or migration within the double layer are negligible. In classical electrochemical theories, it is assumed that the concentrations of charged species within the interfacial region are described by a Boltzmann distribution imposed by the electrical potential of the double layer. Further out in solution, diffusion is the mechanism of mass transport of molecules and ions. The Frumkin correction allows the true electrode concentrations to be related to those existing at the end of the electrochemical double layer. While these concentrations do not represent those actually found at the electrode surface, they are usually used as boundary conditions in solving diffusion problems.

The difficulty with the Frumkin approximation arises when dealing with ultrafast electrochemical techniques since the sizes of the double and diffusion layers become comparable. Therefore, using the concentrations found at the end of the interfacial region as a boundary condition ceases to be appropriate. For example, given a typical diffusion coefficient of $10^{-5}\,cm^2\,s^{-1}$, the diffusion layer thickness will be of the order of 100 Å for a cyclic voltammetry experiment in which the scan rate is $300\,000\,V\,s^{-1}$ or for a 10 nm electrode in a steady-state measurement. This diffusion-layer thickness is comparable to the typical double-layer thickness. The difficulty arises because the voltammetric response will no longer be controlled by diffusion alone but migration within the double layer will influence the observed response.

Recently, two models describing this mixed control situation have been developed for transient and steady-state methods. Both of these models predict that the voltammetric waves appear to be kinetically controlled by heterogeneous electron transfer when analyte migration within the double layer becomes important. However, the predicted behavior contrasts with that predicted by conventional theories of electrode kinetics. In the Butler–Volmer formalism of electrode kinetics, the heterogeneous electron transfer rate constant is expected to increase without limit as the overpotential (electrochemical driving force) is increased. However, coupling between diffusion and migration in the interfacial region causes the apparent heterogeneous electron transfer rate constant eventually to become potential independent, corresponding to the maximum rate, not of electron transfer, but of the reactant crossing the double layer. Further decreases in the critical dimension of ultramicroelectrodes used in steady-state measurements, and the timescale of transient electrochemical techniques, will allow further testing of these theories to be performed.

In order to achieve faster perturbation of the chemical system in transient measurements, it will be necessary to generate pulse signals with rise and fall times on the picosecond timescale if deconvolution of the instrumental and electrode responses is to be avoided. With the development of ever faster chips (e.g. even personal computers have cycle times on the order of nanoseconds), programmable pulse generators with this performance level are becoming more widely available commercially. Another key issue is the transduction of the fast current response. The transient current response for electrodes with subnanosecond *RC* time constants will be on the nano- and even picoamp scale. One strategy for detecting these very small currents with high speed is to insert a second stage of amplification following fast current-to-voltage conversion. Amplifiers of this type that operate at gigahertz frequencies are also available, although their cost is high. Ultrahigh-speed transient digitizers with gigahertz bandwidth and gigahertz sampling rate are also available. However, the total cost of assembling such a

For references see page 10168

system with low to subnanosecond time resolution would be of the order of $175 000.

5 CONCLUSIONS

This article has attempted to convey the significant new insights and opportunities that ultrafast electrochemical techniques have provided. This revolution in electrochemistry has greatly extended the range of conditions of solvent, temperature, and timescale under which it is now possible to obtain information about redox processes directly. In particular, today microelectrodes allow experiments (e.g. voltammetry in oil or concrete) to be performed that would simply have been impossible a few years ago. This advance has not only revolutionized the field internally, it has broadened the impact of electrochemistry into new dimensions of space and time, e.g. microsecond monitoring of neurotransmitter release with single-cell spatial resolution.

ACKNOWLEDGMENTS

The Irish Science and Technology Agency, Enterprise Ireland, that has funded our research in this area over several years is gratefully acknowledged. Sincere thanks go to Professor Larry R. Faulkner of the University of Austin at Texas for a wonderful introduction to this area of study.

ABBREVIATIONS AND ACRONYMS

dipy	4,4′-Trimethylenedipyridine
DISP	Disproportionation
DMF	Dimethylformamide
DMSO	Dimethylsulfoxide
ECSTM	Electrochemical Scanning Tunneling Microscopy
HIM	Imidazole
ILIT	Indirect Laser-induced Temperature Jump Method
MeIM	Methyl Imidazole
MV	Methyl Viologen
OEP	Octa-ethyl-porphyrin
o-QDM	*ortho*-Quinodimethanes
SCE	Saturated Calomel Electrode
SECM	Scanning Electrochemical Microscopy
STM	Scanning Tunneling Microscopy
TPP	Tetra-phenyl Porphyrin
2,6-AQDS	Anthraquinone-2,6-disulfonic acid

List of selected abbreviations appears in Volume 15

RELATED ARTICLES

Biomolecules Analysis **(Volume 1)**
Voltammetry In Vivo for Chemical Analysis of the Living Brain

Electroanalytical Methods **(Volume 11)**
Electroanalytical Methods: Introduction • Organic Electrochemical Mechanisms • Pulse Voltammetry

REFERENCES

1. M. Fleischmann, S. Pons, D.R. Rolison, P. Schmidt (eds.), *Ultramicroelectrodes*, Datatech Science, Morgantown, NC, 1987.
2. R.J. Forster, 'Microelectrodes: New Dimensions in Electrochemistry', *Chem. Soc. Rev.*, 289–297 (1994).
3. P.W. Davies, F. Brink, 'Microelectrodes for Measuring Local Oxygen Tension in Animal Tissues', *Rev. Sci. Instrum.*, **13**, 524–533 (1942).
4. A.J. Bard, L.R. Faulkner, *Electrochemical Methods: Fundamentals and Applications*, J. Wiley & Sons, New York, 1980.
5. J. Newman, 'Resistance for Flow of Current to a Disk', *J. Electrochem. Soc.*, **113**, 501–502 (1968).
6. D.O. Wipf, A.C. Michael, R.M. Wightman, 'Microdisk Electrodes. Part II Fast Scan Cyclic Voltammetry with Very Small Electrodes', *J. Electroanal. Chem.*, **269**(1), 15–27 (1989).
7. L.R. Faulkner, M.R. Walsh, C. Xu, 'New Instrumental Approaches to Fast Electro-chemistry at Ultramicroelectrodes', in *Contemporary Electroanalytical Chemistry*, ed. A. Ivaska, Plenum Press, New York, 5–14, 1990.
8. W.J. Bowyer, E.E. Engelman, D.H. Evans, 'Kinetics by Cyclic Voltammetry at Low Temperatures Using Microelectrodes', *J. Electroanal. Chem.*, **262**(1–2), 67–83 (1989).
9. C. Amatore, B. Fosset, 'Equivalence Between Microelectrodes of Different Shapes: Between Myth and Reality', *Anal. Chem.*, **68**(24), 4377–4388 (1996).
10. M.I. Montenegro, 'Applications of Microelectrodes in Kinetics', *Research in Chemical Kinetics*, eds. R.G. Compton, G. Hancock, Elsevier, The Netherlands, 1, Vol. 2, 1994.
11. D. Garreau, P. Hapiot, J.-M. Savéant, 'Instrumentation for Fast Voltammetry at Ultramicroelectrodes. Stability and Bandpass Limitations', *J. Electroanal. Chem.*, **272**(1–2), 1–16 (1989).
12. R.S. Nicholson, I. Shain, 'Theory of Stationary Electrode Polarography. Single Scan and Cyclic Methods Applied to Reversible, Irreversible and Kinetic Systems', *Anal. Chem.*, **36**(4), 706–723 (1964).
13. R.A. Marcus, 'On the Theory of Electron-transfer Reactions. VI. Unified Treatment for Homogeneous and

Electrode Kinetics', *J. Chem. Phys.*, **43**(2), 679–701 (1965).

14. J.O. Howell, R.M. Wightman, 'Ultra-fast Voltammetry and Voltammetry in Highly Resistive Solutions with Microvoltammetric Electrodes', *Anal. Chem.*, **56**(3), 524–529 (1984).
15. L.K. Safford, M.J. Weaver, 'The Evaluation of Rate Constants for Rapid Electrode-reactions Using Microelectrode Voltammetry–Virtues of Measurements at Lower Temperatures', *J. Electroanal. Chem.*, **331**(1–2), 857–876 (1992).
16. C.P. Andrieux, D. Garreau, P. Hapiot, J.Pinson, J.M. Savéant, 'Fast Sweep Cyclic Voltammetry at Ultramicroelectrodes. Evaluation of the Method for Fast Electron Transfer Kinetic Measurements', *J. Electroanal. Chem.*, **243**, 321–335 (1988).
17. M.I. Montenegro, D. Pletcher, 'The Determination of the Kinetics of Electron Transfer Using Fast Sweep Cyclic Voltammetry at Microdisk Electrodes', *J. Electroanal. Chem.*, **200**, 371–374 (1986).
18. C. Amatore, C. Lefrou, F. Pfluger, 'On-line Compensation of Ohmic Drop in Submicrosecond Time Resolved Cyclic Voltammetry at Ultramicroelectrodes', *J. Electroanal. Chem.*, **270**, 43–59 (1989).
19. C. Amatore, C. Lefrou, 'New Concept for a Potentiostat for Online Ohmic Drop Compensation in Cyclic Voltammetry Above 300 kV s^{-1}', *J. Electroanal. Chem.*, **324**(1–2), 33–58 (1992).
20. D.O. Wipf, R.M. Wightman, 'Rapid Cleavage Reactions of Haloaromatic Radical Anions Measured with Fast Scan Cyclic Voltammetry', *J. Phys. Chem.*, **93**, 4286–4291 (1989).
21. I. Lavagnini, P. Pastore, F. Magno, 'Application of Cyclic Voltammograms Under Mixed Spherical Semiinfinite Linear Diffusion at Microdisk Electrodes for Measurement of Fast Electrode-kinetics', *J. Electroanal. Chem.*, **333**(1–2), 1–10 (1992).
22. D.O. Wipf, E.W. Kristensen, M.R. Deakin, R.M. Wightman, 'Fast Scan Cyclic Voltammetry as a Method to Measure Rapid Heterogeneous Electron Transfer Kinetics', *Anal. Chem.*, **60**, 306–310 (1988).
23. H.F. Zhou, N.Y. Gu, S.J. Dong, 'Studies of Ferrocene Derivative Diffusion and Heterogeneous Kinetics in Polymer Electrolyte by Using Microelectrode Voltammetry', *J. Electroanal. Chem.*, **441**(1–2), 153–160 (1998).
24. A. Neudeck, J. Dittrich, 'The Determination of Diffusion Coefficients and Rate Constants from the Dependence of the Peak Separation and Peak Current on the Scan Rate of Cyclic Voltammograms at Micro-cylindrical Electrodes', *J. Electroanal. Chem.*, **313**(1–2), 37–59 (1991).
25. W. Huang, R. McCreery, 'Electron-transfer Kinetics of $[Fe(CN)_6]^{3-/4-}$ on Laser-activated and CN^- Modified Pt Electrodes', *J. Electroanal. Chem.*, **326**(1–2), 1–12 (1992).
26. N. Oyama, T. Ohsaka, N. Yamamoto, J. Matsui, O. Hatozaki, 'Determination of the Heterogeneous Electron Transfer Rate Constants for the Redox Couples Octacyanomolybdate $[Mo(CN)_8]^{4-/3-}$, Octacyanotungstate $[W(CN)_8]^{4-/3-}$, Hexacyanoferrate $[Fe(CN)_6]^{4-/3-}$, Hexacyanoosmate $[Os(CN)_6]^{4-/3-}$, Hexachloroiridate $[Ir(Cl)_6]^{3-/2-}$, Using Fast Sweep Cyclic Voltammetry at Carbon Fiber Electrodes', *J. Electroanal. Chem.*, **265**(1), 297–304 (1989).
27. C. Beriet, D. Pletcher, 'A Further Microelectrode Study of the Influence of Electrolyte Concentration on the Kinetics of Redox Couples', *J. Electroanal. Chem.*, **375**(1–2), 213–218 (1994).
28. K. Winkler, 'The Kinetics of Electron-transfer in $[Fe(CN)_6]^{4-/3-}$ Redox System on Platinum Standard-size and Ultramicroelectrodes', *J. Electroanal. Chem.*, **388**(1–2), 151–159 (1995).
29. D.O. Wipf, M.R. Wightman, 'Submicrosecond Measurements with Cyclic Voltammetry', *Anal. Chem.*, **60**, 2460–2464 (1988).
30. P. Hapiot, J. Moiroux, J.-M. Savéant, 'Electrochemistry of NADH–NAD^+ Analogs – A Detailed Mechanistic Kinetic and Thermodynamic Analysis of the 10-Methylacridan–10-Methylacridinium Couple in Acetonitrile', *J. Am. Chem. Soc.*, **112**(4), 1337–1343 (1990).
31. C.P. Andrieux, P. Hapiot, J.-M. Savéant, 'Fast Chemical Steps Coupled with Outer-sphere Electron Transfers: Application of Fast Scan Voltammetry at Ultramicroelectrodes to the Cleavage of Aromatic Halide Anion Radicals in the Microsecond Lifetime Range', *J. Phys. Chem.*, **92**(21), 5987–5992 (1988).
32. W.J. Bowyer, E.E. Engelman, D.H. Evans, 'Kinetic Studies by Cyclic Voltammetry at Low Temperatures Using Ultramicroelectrodes', *J. Electroanal. Chem.*, **262**(1–2), 67–82 (1989).
33. P. He, R.M. Crooks, L.R. Faulkner, 'Adsorption and Electrode-reactions of Disulfonated Anthrquinones at Mercury-electrodes', *J. Phys. Chem.*, **94**(3), 1135–1141 (1990).
34. C.E.D. Chidsey, 'Free-energy and Temperature-dependence of Electron-transfer at the Metal–Electrolyte Interface', *Science*, **251**(4996), 919–922 (1991).
35. D. Acevedo, H.D. Abruña, 'Electron-transfer Study and Solvent Effects on the Formal Potential of a Redox-active Self-assembling Monolayer', *J. Phys. Chem.*, **95**(23), 9590–9594 (1991).
36. R.J. Forster, L.R. Faulkner, 'Electrochemistry of Spontaneously Adsorbed Monolayers–Equilibrium Properties and Fundamental Electron-transfer Characteristics', *J. Am. Chem. Soc.*, **116**(20), 5444–5452 (1994).
37. H.O. Finklea, D.D. Hanshew, 'Electrochemistry of Spontaneously Adsorbed Monolayers – Effects of Solvent, Potential, and Temperature on Electron-transfer', *J. Am. Chem. Soc.*, **114**(20), 5453–5461 (1992).
38. E. Laviron, 'General Expression of the Linear Potential Sweep Voltammogram in the Case of Diffusionless Electrochemical Systems', *J. Electroanal. Chem.*, **101**(1), 19–28 (1979).

39. R.J. Forster, L.R. Faulkner, 'Interfacial Field Effects on Reductive Chloride Elimination from Spontaneously Adsorbed Monolayers', *Langmuir*, **11**(3), 1014–1023 (1995).
40. R.J. Forster, 'Heterogeneous Kinetics of Metal and Ligand Based Redox Reactions within Adsorbed Monolayers', *Inorg. Chem.*, **35**, 3394–3401 (1996).
41. J.P. O'Kelly, R.J. Forster, 'Potential Dependent Adsorption of Anthraquinone-2,7-disulfonate on Mercury', *Analyst*, **123**(10), 1987–1993 (1998).
42. A.A. Pilla, *Computers in Chemistry and Instrumentation*, Vol. 2, eds. J.S. Mattson, H.B. Mark, H.C. MacDonald, Marcel Dekker, New York, 1972.
43. C.A. Amatore, A. Jutand, F. Pflüger, 'Nanosecond Time-resolved Cyclic Voltammetry Direct Observation of Electrogenerated Intermediates with Biomolecular Diffusion Controlled Decays Using Scan Rates in the Megavolt per Second Range', *J. Electroanal. Chem.*, **218**(1–2), 361–365 (1987).
44. J.O. Howell, R.M. Wightman, 'Ultrafast Voltammetry of Anthracene and 9,10-Diphenylanthracene', *J. Phys. Chem.*, **88**(18), 3915–3922 (1989).
45. M. Murphy, Z. Stojek, J. O'Dea, J. Osteryoung, 'Pulse Voltammetry at Cylindrical Electrodes: Oxidation of Anthracene', *Electrochim. Acta.*, **36**(9), 1475–1484 (1991).
46. S. Okazaki, N. Oyama, S. Nomura, 'Correlation Between the Redox Potentials of 9-Substituted Anthracenes and the Results of PM3 Calculation', *Electroanalysis*, **9**(16), 1242–1246 (1997).
47. K.R. Wehmeyer, R.M. Wightman, 'Cyclic Voltammetry and Anodic Stripping Voltammetry with Mercury Microelectrodes', *Anal. Chem.*, **57**, 1989–1993 (1985).
48. R.S. Robinson, R.L. McCreery, 'Submicrosecond Spectroelectrochemistry Applied to Chlorpromazine Cation Radical Charge Transfer Reactions', *J. Electroanal. Chem.*, **182**(1), 61–72 (1985).
49. H. Yang, D.O. Wipf, A.J. Bard, 'Application of Rapid Scan Cyclic Voltammetry to a Study of the Oxidation and Dimerization of *N,N*-Dimethylaniline in Acetonitrile', *J. Electroanal. Chem.*, **331**(1–2), 913–924 (1992).
50. C. Amatore, J.N. Verpeaux, P.J. Krusic, 'Electrochemical Reduction of Iron Pentacarbonyl Revisited', *Organometallics*, **7**(11), 2426–2428 (1988).
51. C.J. Schlesener, C. Amatore, J.K. Kochi, 'Marcus Theory in Organic Chemistry. Mechanisms of Electron and Proton Transfer for Aromatics and Their Cation Radicals', *J. Phys. Chem.*, **90**, 3747–3756 (1986).
52. D.O. Wipf, R.M. Wightman, 'Voltammetry with Microvoltammetric Electrodes in Resistive Solvents Under Linear Diffusion Conditions', *Anal. Chem.*, **62**(2), 98–102 (1990).
53. A.J. Bard, H.D. Abruña, C.E.D. Chidsey, L.R. Faulkner, S.W. Feldberg, K. Itaya, M. Majda, O. Melroy, R.W. Murray, M.D. Porter, M.P. Soriaga, H.S. White, 'The Electrode–Electrolyte Interface – A Status-report', *J. Phys. Chem.*, **97**(28), 7147–7173 (1993).
54. K. Aoki, K. Tokuda, H. Matsuda, 'Simulation in the Theory of Step Experiments at Microelectrodes', *J. Electroanal. Chem.*, **235**(1–2), 87–89 (1987).
55. M. Fleischmann, S. Bandyopadhyay, S. Pons, 'The Behavior of Microring Electrodes', *J. Phys. Chem.*, **89**(25), 5537–5541 (1985).
56. K.B. Oldham, C.G. Zoski, 'Simulation of Voltammetric Steady States at Hemispherical and Disk Microelectrodes', *J. Electroanal. Chem.*, **256**(1), 11–21 (1988).
57. M.V. Mirkin, A.J. Bard, 'Simple Analysis of Quasi-reversible Steady-state Voltammograms', *Anal. Chem.*, **64**(19), 2293–2302 (1992).
58. A. Russell, K. Repka, T. Dibble, J. Ghoroghchian, J. Smith, M. Fleischmann, C. Pitt, S. Pons, 'Determination of Electrochemical Heterogeneous Electron Transfer Reaction Rates from Steady-state Measurements at Ultramicroelectrodes', *Anal. Chem.*, **58**(14), 2961–2964 (1986).
59. C.G. Zoski, D.A. Sweigart, N.J. Stone, P.H. Rieger, E. Mocellin, T.F. Mann, D.R. Mann, D.K. Gosser, M.M. Doeff, A.M. Bond, 'An Electrochemical Study of the Substitution and Decomposition Reactions of (Arene) Tricarbonyl Chromium Radical Cations', *J. Am. Chem. Soc.*, **110**(7), 2109–2116 (1988).
60. R.M. Penner, M.J. Heben, T.L. Longin, N.S. Lewis, 'Fabrication and Use of Nanometer-sized Electrodes in Electrochemistry', *Science*, **250**(4984), 1118–1121 (1990).
61. F.N. Büchi, A.M. Bond, 'Interpretation of the Electrochemistry of Cytochrome-*c* at Macro-sized and Micro-sized Carbon Electrodes Using a Microscopic Model Based on a Partially Blocked Surface', *J. Electroanal. Chem.*, **314**(1–2), 191–206 (1991).
62. A. Owlia, J.F. Rusling, 'Non-linear Regression Analysis of Steady-state Voltammograms Obtained at Microelectrodes', *Electroanalysis*, **1**(2), 141–149 (1989).
63. A.M. Bond, T.L.E. Henderson, D.R. Mann, T.F. Mann, W. Thormann, C.G. Zoski, 'Fast Electron Transfer Rate for the Oxidation of Ferrocene in Acetonitrile or Dichloromethane at Platinum Disk Ultra-microelectrodes', *Anal. Chem.*, **60**(18), 1878–1882 (1988).
64. J. Daschbach, D. Blackwood, J.W. Pons, S. Pons, 'Electrochemistry of Ferrocene in Acetonitrile. Evidence for Irreversible Kinetic due to Passive Film Formation', *J. Electroanal. Chem.*, **237**(2), 269–275 (1987).
65. Y. Zhang, C.D. Baer, C. Camaioni-Neto, P. O'Brien, D.A. Sweigart, 'Steady-state Voltammetry with Microelectrodes – Determination of Heterogeneous Charge-transfer Rate Constants for Metalloporphyrin Complexes', *Inorg. Chem.*, **30**(8), 1682–1685 (1991).
66. K.B. Oldham, C.G. Zoski, A.M. Bond, D.A. Sweigart, 'Measurement of Ultrafast Electrode Kinetics via Steady-state Voltammograms at Microdisk Electrodes', *J. Electroanal. Chem.*, **248**(2), 467–473 (1988).

67. B.R. Scharifker, P. Zelenay, J. O'M Bockris, 'The Kinetics of Oxygen Reduction in Molten Phosphoric Acid at High Temperatures', *J. Electrochem. Soc.*, **134**(11), 2714–2725 (1987).
68. A.J. Bard, M.V. Mirkin, P.R. Unwin, D.O. Wipf, 'Scanning Electrochemical Microscopy.12. Theory and Experiment of the Feedback Mode with Finite Heterogeneous Electron-transfer Kinetics and Arbitrary Substrate Size', *J. Phys. Chem.*, **96**(4), 1861–1868 (1992).
69. C. Amatore, S.C. Paulson, H.B. White, 'Successive Electron-transfers in Low Ionic Strength Solutions. Migration Flux Coupling by Homogeneous Electron Transfer Reactions', *J. Electroanal. Chem.*, **439**(1), 173–182 (1997).
70. M. Fleischmann, F. Lassere, J. Robinson, 'The Application of Microelectrodes to the Study of Homogeneous Processes Coupled to Electrode Reactions. Part II ECE and DISP1 Reactions', *J. Electroanal. Chem.*, **177**(1–2), 115–127 (1984).
71. S. Dong, G. Che, 'Electrocatalytic Oxidation of Ascorbic-acid at a Prussian Blue Film Modified Microdisk Electrode', *J. Electroanal. Chem.*, **315**(1–2), 191–199 (1991).
72. T.V. Shea, A.J. Bard, 'Digital Simulation of Homogeneous Chemical Reactions Coupled to Heterogeneous Electron Transfer and Applications at Platinum/Mica/Platinum Ultramicroband Electrodes', *Anal. Chem.*, **59**, 2101–2111 (1987).
73. C.L. Miaw, J.F. Rusling, A. Owlia, 'Simulation of 2-electron Homogeneous Electrocatalysis for Steady-state Voltammetry at Hemispherical Microelectrodes', *Anal. Chem.*, **62**(3), 268–273 (1990).
74. S.U. Pedersen, K. Daasbjerg, 'Ultramicroelectrodes for Electrochemical Monitoring of Homogeneous Reactions', *Acta Chem. Scand.*, **43**, 301–303 (1989).
75. K. Daasbjerg, S.U. Pedersen, H. Lund, 'Electrochemical Measurements of Rate Constants for the Electron Transfer Reaction to Sterically Hindered Alkyl Halides', *Acta Chem. Scand.*, **43**, 876–881 (1989).
76. M. Fleischmann, F. Lassere, J. Robinson, D. Swan, 'The Application of Microelectrodes to the Study of Homogeneous Processes Coupled to Electrode Reactions. Part II ECE and DISP1 Reactions', *J. Electroanal. Chem.*, **177**(1–2), 97–115 (1984).
77. Dong, G. Che, 'Application of Ultramicroelectrodes in Studies of Homogeneous Catalytic Reactions. 2. A Theory of Quasi-1st and 2nd-order Homogeneous Catalytic Reactions', *Electrochim. Acta*, **37**(15), 2695–2699 (1992).
78. S. Daniele, P. Ugo, G.A. Mazzocchin, 'The Use of Microelectrodes for Studying the Processes Involved in 1-naphthylamine Oxidation in Dimethyl Sulfoxide', *J. Electroanal. Chem.*, **267**(1–2), 129–140 (1989).
79. R. Muller, L. Lamberts, M. Evers, 'The Electrochemical Oxidation of Thioselenanthrene in Acetonitrile at Conventional Electrodes and Microelectrodes', *J. Electroanal. Chem.*, **417**(1–2), 35–43 (1996).
80. R. Muller, L. Lamberts, M. Evers, 'The Electrochemical Oxidation of Dibenzo(C,E)-1,2-Diselenine to its Cation Radical. A Voltammetric Study in Acetonitrile at Conventional Electrodes and Microelectrodes', *J. Electroanal. Chem.*, **401**(1–2), 183–189 (1996).
81. C.P. Andrieux, P. Hapiot, J.-M. Savéant, 'Fast Kinetics by means of Direct and Indirect Electrochemical Techniques', *Chem. Rev.*, **90**(5), 723–738 (1990).
82. E. Oguntoye, S. Szunertis, J.H.P. Utley, P.B. Wyatt, 'Electro-organic Reactions. 46. Diels–Alder Trapping of *O*-Quinodimethane Generated by Redox-mediated Cathodic Reduction of α, α'-Dibromo-*O*-xylene in the Presence of Hindered Dienophiles', *Tetrahedron*, **52**(22), 7771–7778 (1996).
83. D.H. Turner, G.W. Flynn, N. Sutin, J.V. Beitz, 'Laser Raman Temperature-jump Study of the Kinetics of the Triiodide Equilibrium. Relaxation Times in the 10^{-8}–10^{-7} Second Range', *J. Am. Chem. Soc.*, **94**(5), 1554–1562 (1972).
84. G.C. Barker, A.W. Gardner, G. Bottura, 'Laser Induced Potential Changes at a Mercury Electrode', *J. Electroanal. Chem.*, **45**(1), 21–30 (1973).
85. V.A. Benderskii, G.I. Velichko, 'Temperature Jump in Electric Double-layer Study. Part I. Method of Measurements', *J. Electroanal. Chem.*, **140**, 1–22 (1982).
86. H. Coufal, P. Hefferle, 'Photothermal Analysis of Thin Films', *Springer Ser. Chem. Phys*, **39**, 497–501 (1984).
87. J.F. Smalley, C.V. Krishnan, M. Goldman, S.W. Feldberg, I. Ruzic, 'Laser Induced Temperature Jump Coulostatics for the Investigation of Heterogeneous Rate Processes. Theory and Application', *J. Electroanal. Chem.*, **248**(2), 255–283 (1988).
88. J.F. Smalley, S.W. Feldberg, C.E.D. Chidsey, M.R. Linford, M.D. Newton, Y.-P. Liu, 'The Kinetics of Electron Transfer Through Ferrocene-terminated Alkane Thiol Monolayers on Gold', *J. Phys. Chem.*, **99**(35), 13 141–13 149 (1995).
89. A.J. Bard, F.-R.F. Fan, M.V. Mirkin, *Electroanalytical Chemistry*, Vol. 18, ed. A.J. Bard, Marcel Dekker, New York, 243, 1993.
90. M.V. Mirkin, L.O.S. Bulhões, A.J. Bard, 'Determination of the Kinetic-parameters for the Electroreduction of C_{60} by Scanning Electrochemical Microscopy and Fast Scan Cyclic Voltammetry', *J. Am. Chem. Soc.*, **115**(1), 201–204 (1993).
91. F. Zhou, P.R. Unwin, A.J. Bard, *J. Phys. Chem.*, **96**, 4917 (1992).
92. D.A. Treichel, M.V. Mirkin, A.J. Bard, 'Scanning Electrochemical Microscopy. 27. Application of a Simplified Treatment of an Irreversible Homogeneous Reaction Following Electron-transfer to the Oxidative Dimerization of 4-Nitrophenolate in Acetonitrile', *J. Phys. Chem.*, **98**(22), 5751–5757 (1994).

Ultraviolet/Visible Spectroelectrochemistry

Daniel A. Scherson, Yuriy V. Tolmachev, and Ionel C. Stefan
Case Western Reserve University, Cleveland, USA

This article reviews the field of solution-phase ultraviolet/visible (UV/VIS) absorption spectroelectrochemistry and its application to problems in thermodynamics, kinetics, and mass transport, with emphasis on both experimental and theoretical aspects. Examples are provided for transmission in thin layer and semi-infinite media, using optically transparent electrodes, and external and internal reflection in quiescent media. Also illustrated is the coupling of UV/VIS spectroscopy to systems under forced convection, such as rotating disk and channel electrodes, under steady state and transient conditions. Attention is also focused on spatially resolved spectroelectrochemistry for the imaging of diffusion and reaction layers. The last section is devoted to the more complex and particularly powerful modulation techniques involving diffraction, refraction and absorption in stagnant media and in the presence of convective flow. Factors that limit the sensitivity and spatial and temporal resolution of UV/VIS absorption spectroelectrochemistry, as well as its future prospects in analytical chemistry are briefly discussed.

List of selected abbreviations appears in Volume 15

1 INTRODUCTION

Light in the UV/VIS range, i.e. 190–700 nm, can promote excitation of electronic states in atomic and molecular species in gas and solution phases, and also in condensed matter, including crystalline solids and pure liquids. As such, it represents a valuable probe of the structure and properties of materials.(1–3) UV/VIS spectroelectrochemistry refers collectively to a wide array of techniques that employ radiation in this frequency domain for the study of the optical properties of electrodes and electrolyte solutions, particularly those induced by changes in the applied potential across electrode–solution interfaces.(4–8) Implicit in this definition is the fact that measurements are performed in situ, i.e. with the electrode immersed in the solution under potential control. Particularly amenable to UV/VIS spectroelectrochemical investigation are electrode processes that generate soluble species that absorb radiation in this spectral region, elicit spatial variations in the index of refraction, or modify the optical properties of interfaces. The dependence of the molar absorptivity on the wavelength of the probing beam, which is often specific to each and every chromophore, provides an added dimension for the identification and monitoring of species involved in electrochemical reactions as compared to simple current–potential relationships.

This article reviews the field of solution phase UV/VIS absorption spectroelectrochemistry and its application to problems in kinetics, mechanisms and mass transport. Within this rather narrow scope, it excludes equally important areas that rely on the analysis of light reflected from the electrode surface, such as ellipsometry(9,10) and electroreflectance,(11,12) as well as simple refraction, notably probe beam deflection(13) and interferometry,(14) for the study of interfacial and bulk phenomena. As the vast majority of heterogeneous electron transfer reactions involve generation or depletion of species in solution, diffusion plays an important role in controlling the current that flows across the interface and, thus, the time evolution of concentration profiles.(15) Hence, it is of key importance to highlight at the outset fundamental aspects of the two most relevant physical phenomena, namely, the interaction of light and matter, and mass transport of species in condensed media, to gain a full appreciation of the various factors associated with the design and interpretation of spectroelectrochemical experiments.

2 THEORETICAL CONSIDERATIONS

2.1 Optics

2.1.1 Light Propagation through Single-phase Materials

The optical properties of an isotropic nonmagnetic media are defined in terms of its complex refractive index $\hat{n} = n - ik$, or its complex dielectric function, $\hat{\varepsilon} = \varepsilon' - i\varepsilon''$, where the index of refraction n and the extinction coefficient k are both functions of the wavelength λ, and $\varepsilon' = (n^2 - k^2)$ and $\varepsilon'' = 2nk$.[16,17] Modifications in the composition of an otherwise homogeneous solution, such as those derived from electrode processes, will often bring about variations in the local magnitudes of n and k, so that the intensity, as well as the direction of light propagating through the media may be altered. These effects can be exploited to obtain time- and space-resolved concentration profiles, allowing important aspects of electrochemical systems to be investigated, including identification of products of redox reactions, measurements of diffusion coefficients, elucidation of reaction mechanisms, and determination of the rates of both heterogeneous and homogeneous electron transfer reactions involving electrogenerated species.

Correlations between n and k, and the concentration of a single species in solution are usually linear, a factor that simplifies considerably the analysis of experimental data. Furthermore, if ionic migration in an electric field is sufficiently minimized, for example by using an excess of supporting electrolyte, mass transport under either quiescent or laminar flow conditions can be approximated by relatively simple differential equations, allowing solutions to be expressed in terms of common analytic functions. It is precisely the interaction of light with such spatially and temporally varying concentration fields that constitutes the basis for a rigorous mathematical treatment of spectroelectrochemical experiments that rely on absorption, refraction, and diffraction of light. In view of the nature of the material being reviewed, it seems appropriate to focus general attention on light absorption, and defer discussion of refraction and diffraction to those few specialized sections where methods based on these phenomena are introduced.

2.1.2 Absorption

The instantaneous attenuation in the intensity I of a collimated light beam propagating along an axis y, $-\mathrm{d}I$, through a media of thickness $\mathrm{d}y$, containing, without loss of generality, a single absorbing species, is most often proportional to the intensity of the light incident on the infinitesimal volume, $\mathrm{d}x\,\mathrm{d}y\,\mathrm{d}z$, $I(x, y, z, t)$, to the local concentration $c(x, y, z, t)$ and to the molar absorptivity of that species $\varepsilon(\lambda)$:

$$-\mathrm{d}I(x, y, z, t) = I(x, y, z, t)\kappa(\lambda)c(x, y, z, t)\,\mathrm{d}y \qquad (1)$$

In Equation (1) $\kappa(\lambda) = k\varepsilon(\lambda)$, $k = 2.303$, and λ is the wavelength of light at which the measurements are being carried out. The intensity of the light emerging from a cell of length d along y, $I(x, d, z, t)$, is given by Equation (2),

$$I(x, d, z, t) = I(x, 0, z)\exp\left\{-\int_0^d \kappa(\lambda)c(x, y, z, t)\,\mathrm{d}y\right\} \qquad (2)$$

where, for simplicity, the intensity of the beam incident on the cell, $I(x, 0, z)$ has been assumed independent of time. Equation (2) can be integrated over the entire cross-sectional area of the beam xz, to yield Equation (3)

$$\int_x\int_z I(x, d, z, t)\,\mathrm{d}x\,\mathrm{d}z = \int_x\int_z I(x, 0, z)\exp\left\{-\int_0^d \kappa(\lambda) \times c(x, y, z, t)\,\mathrm{d}y\right\}\mathrm{d}x\,\mathrm{d}z \qquad (3)$$

If absorption is weak,[18] as is assumed throughout this article, the exponential function can be approximated by the first two terms in its Taylor's series expansion. On this basis, the instantaneous absorbance $A(t)$, i.e. the log (base 10) of the ratio of the light intensity incident on, and that emerging from the cell, is expressed by Equation (4):

$$A(t) = \frac{\int_x\int_z I(x, 0, z)\left\{\varepsilon(\lambda)\int_0^d c(x, y, z, t)\,\mathrm{d}y\right\}\mathrm{d}x\,\mathrm{d}z}{\int_x\int_z I(x, 0, z)\,\mathrm{d}x\,\mathrm{d}z} \qquad (4)$$

In cases in which the concentration is only a function of y and t, $A(t)$ reduces to the much simpler Equation (5):

$$A(t) = \varepsilon\int_0^d c(y, t)\,\mathrm{d}y \qquad (5)$$

where the explicit dependence of ε on λ will, hereafter, be omitted. This expression specifies that the intensity of the light emerging from the cell is proportional to the integral of the concentration profile along y, and would be applicable for measurements in which a beam of light is incident normal to a semitransparent, planar electrode placed in the xz plane (Figure 1a). Alternatively, the beam could propagate through the solution and reflect off the surface of a highly polished electrode at an angle strictly smaller than 90° (Figure 1b). This geometry will increase the effective optical path compared to the transparent

For references see page 10222

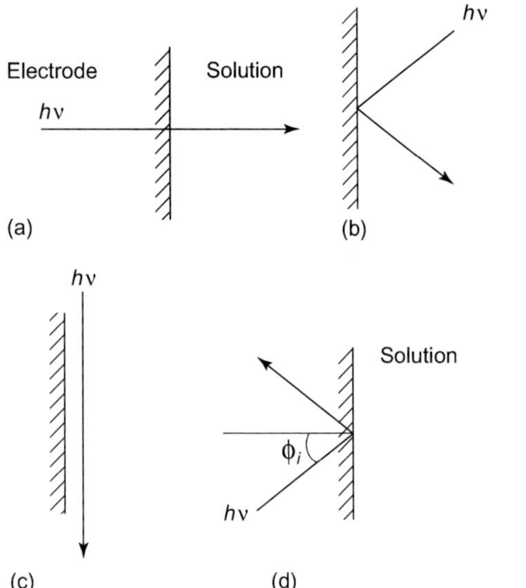

Figure 1 Experimental arrangements for UV/VIS spectroelectrochemical experiments: (a) optically transparent electrode transmission; (b) external reflection; (c) parallel geometry; (d) internal reflection.

electrode, which will appear as a correction factor to Equation (5) (see section 4.4.1).

A far more powerful strategy involves the use of a well-collimated beam of monochromatic light of finite width propagating parallel to and very close to the surface of a planar electrode (Figure 1c). This can then be imaged onto a photodiode array detector, for example, to yield spatially resolved profiles of absorbing species.

It becomes evident from these arguments, that a quantitative analysis of dynamic spectroelectrochemical experiments involving solution-phase chromophores requires time dependent concentration profiles to be calculated. Such problems are inextricably linked to finding solutions of the equations that govern the mass transport of species in solution, subject to the appropriate boundary conditions, especially those prescribed at the electrode surface.[15]

2.1.3 *Internal Reflection*

Consider a beam of light propagating in the xy plane in phase 1 with index of refraction n_1 impinging onto the plane xz that separates phase 1 from another phase 2 with different index of refraction n_2, at an angle of incidence ϕ_i with respect to the normal to the interface (Figure 1d). Classical optics predicts that ϕ_i is related to the angle the transmitted light will make with respect to the same plane, ϕ_t, through the familiar Snell's law (Equation 6),

$$\sin\phi_t = \frac{\sin\phi_i}{n_{12}} \tag{6}$$

where $n_{12} = n_2/n_1$. If $n_2 < n_1$ and $\sin\phi_i > n_{12}$, ϕ_t becomes imaginary leading physically to a condition known as total internal reflection.[16] Based on purely trigonometric arguments, Equation (6) can be rewritten as

$$\cos\phi_t = \pm i\sqrt{\frac{\sin^2\phi_i}{n_{12}^2} - 1} \tag{7}$$

Based on Equations (6) and (7), the amplitude of the electric field, E, may be shown to be proportional to Equation (8):

$$e^{-i\omega(t-((x\sin\phi_i)/(n_{12}v_2)))}e^{-((\omega y)/(v_2))\left(\sqrt{(\sin^2\phi_i)/(n_{12}^2)-1}\right)} \tag{8}$$

where $\omega = 2\pi v_j/\lambda_j$ and v_j and λ_j are the velocity and wavelength of light in medium j, respectively. This expression represents a wave travelling along the x-axis within the interfacial plane, for which its intensity decreases exponentially along the y-axis, i.e. normal to the plane. The distance at which the magnitude of the electric field E decreases to 1/e of its value at the precise interface is known as the depth of penetration δ (Equation 9):

$$\frac{E_\delta}{E_o} = \frac{1}{e} = e^{-((\omega\delta)/(v_2))\left(\sqrt{(\sin^2\phi_i)/(n_{12}^2)-1}\right)} \tag{9}$$

Hence, in terms of $\omega = 2\pi v_j/\lambda_j$,

$$\delta = \frac{\lambda_2}{2\pi\sqrt{\dfrac{\sin^2\phi_i}{n_{12}^2} - 1}} = \frac{\lambda_1}{2\pi\sqrt{\sin^2\phi_i - n_{12}^2}} \tag{10}$$

where $\lambda_1 = n_{12}\lambda_2$ in Equation (10).

This evanescent wave can be used to probe species present at, and in the near vicinity of the interface, as has been widely popularized in the infrared spectral region under the acronym of attenuated total reflection (ATR).[19] It thus follows from Equation (10) that λ and δ are of the same order of magnitude, i.e. a few hundred nanometers in the UV/VIS region.

2.2 Mass Transport

2.2.1 *Quiescent Solutions*

In the absence of complications derived from ionic migration in an electric field, natural convection, and homogeneous chemical reactions, mass transport in stagnant solutions is governed to a good degree of approximation by Fick's second law,[20]

$$\frac{\partial c}{\partial t} = D\nabla^2 c \tag{11}$$

where D is the diffusion coefficient of the species in question. The symbol ∇^2 in Equation (11) represents the

Laplacian, which in cartesian coordinates is given by Equation (12):

$$\nabla^2 = \frac{\partial^2}{\partial x^2} + \frac{\partial^2}{\partial y^2} + \frac{\partial^2}{\partial z^2} \quad (12)$$

For measurements performed over relatively short periods of time, as is often the case, the changes in concentration of reactants and products are confined to a small volume of solution close to the electrode surface. Under these conditions, the composition of the media far away from the electrode may be assumed to remain unaltered during the entire data acquisition period, and thus the actual size of the container may be regarded as being arbitrarily large.

The theoretical treatment of a vast number of UV/VIS spectroelectrochemical experiments involving planar, cylindrical and spherical electrodes reduces to finding solutions of Equation (11) in one dimension under the initial and boundary conditions specified in the left column of Table 1, where C_R is the concentration of an electroactive reactant R, and the superscript 'o' represents its bulk value. As will be illustrated for a few simple cases in this article, the Laplace transform method[21] affords a particularly powerful tool for solving problems of this type.[15,22]

It is convenient from a computational viewpoint to define a dimensionless concentration, i.e.

$$\theta_R(y,t) = \frac{C_R^o - C_R}{C_R^o} \quad (13)$$

Substitution of Equation (13) into Equation (11) renders a similar differential equation subject to initial and boundary conditions shown in the center column in Table 1, which do not depend explicitly on the value of C_R^o.

Subsequent application of the Laplace transform with respect to t reduces the problem to solving a simple linear differential equation, as prescribed in the third column in that same table, for which a solution can be easily obtained to yield

$$\bar{\theta}_R(y,s) = \bar{\theta}_R(0,s)\mathrm{e}^{-(s/D_R)^{1/2}y} \quad (14)$$

In Equation (14) the bar above indicates that the function has been Laplace transformed, and $\bar{\theta}_R(0,s)$ is the concentration of the reactant at the boundary for $t > 0$ in Laplace space. As will be shown, the quantitative analysis of a variety of electrochemical techniques requires for $\bar{\theta}_R(0,s)$ to be explicitly specified.

Many of the systems to be described in this article involve simple reactions of the Equation (15) type:

$$\mathrm{R} \longrightarrow \mathrm{P} \quad (15)$$

where, unless otherwise stated, both R and P are solution phase species. A relationship between the concentration profiles of R and P can be obtained from conservation of mass at the surface (Equation 16)

$$D_R \left.\frac{\partial C_R(y,t)}{\partial y}\right|_{y=0} = -D_P \left.\frac{\partial C_P(y,t)}{\partial y}\right|_{y=0} \quad (16)$$

The terms on the right and left of this equation represent the fluxes of P and R through the electrode, respectively, bearing opposite signs, as required.

2.2.1.1 Chronocoulometry Consider a planar electrode embedded in the xz plane immersed in a quiescent, homogeneous solution containing an electrochemically active species R, and assume a potential step is applied to the electrode of a magnitude large enough to reduce C_R to zero at the surface. This experimental protocol constitutes the basis of chronocoulometry, an electrochemical technique in which the current generated following the potential step is monitored as a function of time. The time evolution of the concentration profile for R in this case may be obtained by setting $\theta_R(0,t) = 1$. Inserting its Laplace transform, $\bar{\theta}_R(0,s) = 1/s$, in Equation (14) gives Equation (17),

$$\bar{\theta}_R(y,s) = \frac{1}{s}\mathrm{e}^{-(s/D_R)^{1/2}y} \quad (17)$$

which yields Equation (18) upon Laplace inversion:

$$C_R = C_R^o \mathrm{erf}\left(\frac{y}{2\sqrt{D_R t}}\right) \quad (18)$$

where erf(z) is the error function of argument z.[23]

Table 1 Differential equations and boundary conditions for mass transport of species in quiescent media in real and Laplace spaces

Real space		Laplace space
$\frac{\partial C_R(y,t)}{\partial t} = D_R \frac{\partial^2 C_R(y,t)}{\partial y^2}$	$\frac{\partial \theta_R(y,t)}{\partial t} = D_R \frac{\partial^2 \theta_R(y,t)}{\partial y^2}$	$D_R \frac{\mathrm{d}^2\bar{\theta}_R(y,s)}{\mathrm{d}y^2} = s\bar{\theta}_R(y,s)$
$C_R(y,0) = C_R^o$	$\theta_R(y,0) = 0$	$\bar{\theta}_R(y,s) = 0$
$C_R(\infty,t) = C_R^o$	$\theta_R(\infty,t) = 0$	$\bar{\theta}_R(\infty,s) = 0$

For references see page 10222

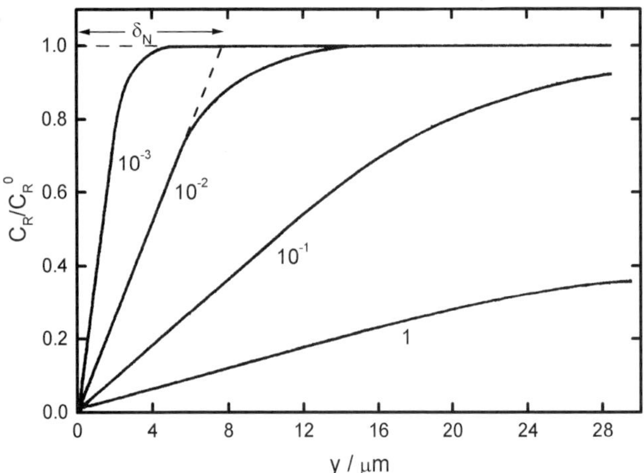

Figure 2 Time evolution of the concentration profile for a reactant R following application of a potential step to a planar electrode based on Equation (18) for $D_R = 10^{-5}\,\mathrm{cm^2\,s^{-1}}$. Each of the curves was calculated for the specified times in units of seconds.[73]

A series of C_R/C_R^o versus y plots show that for fixed values of t, the concentration near the surface is proportional to the distance from the electrode (Figure 2). Linear extrapolation to the bulk concentration, i.e. $C_R/C_R^o = 1$, as indicated in that figure defines an imaginary region known as the Nernst diffusion layer. This simple diffusional theory predicts that, as time elapses, the thickness of this layer δ_N extends progressively into the bulk solution. In practice, however, variations in concentration give rise to differences in the density of the media, and thus to the onset natural convection, a phenomenon not considered by Fick's second law, leading ultimately to the formation of a rather static boundary layer.[24]

If the product of the reaction P is not present at the beginning of the experiment, its time-varying concentration profile can be calculated in a very similar way by defining an appropriate dimensionless concentration (Equation 19),

$$\theta_P(y, t) = -\frac{C_P}{C_R^o} \tag{19}$$

rendering a problem identical to that specified for R in Table 1. In Laplace space this gives Equation (20),

$$\bar{\theta}_P(y, s) = \bar{\theta}_P(0, s)\mathrm{e}^{-(s/D_P)^{1/2}y} \tag{20}$$

where $\bar{\theta}_P(0, s)$ is the dimensionless concentration of P at the electrode surface. The latter can be obtained from Equation (16) in Laplace space, as the additional boundary condition, which in terms of dimensionless concentrations reads

$$\bar{\theta}_P(0, s) = -\bar{\theta}_R(0, s)\left(\frac{D_R}{D_P}\right)^{1/2} \tag{21}$$

Insertion of Equation (21) into Equation (20) and replacing $\bar{\theta}_R(0, s)$ by $1/s$ yields Equation (22),

$$\bar{\theta}_P(y, s) = -\frac{1}{s}\left(\frac{D_R}{D_P}\right)^{1/2} \mathrm{e}^{-(s/D_P)^{1/2}y} \tag{22}$$

which upon subsequent Laplace inversion gives the time-dependent concentration profile of P for a chronocoulometric-type experiment (Equation 23):

$$C_P(y, t) = \left(\frac{D_R}{D_P}\right)^{1/2} C_R^o \operatorname{erfc}\left(\frac{y}{2\sqrt{D_P t}}\right) \tag{23}$$

where $\operatorname{erfc}(z) = 1 - \operatorname{erf}(z)$ is the complementary error function of argument z.[23]

An explicit expression for the flux of P at the surface can be obtained from Equation (20) as Equation (24),

$$-D_P \left.\frac{\partial\bar{\theta}_P(y, s)}{\partial y}\right|_{x=0} = \sqrt{D_P s}\,\bar{\theta}_P(0, s) \tag{24}$$

a quantity proportional to the charge that flows across the electrode per unit time due to a heterogeneous electron transfer reaction, or faradaic current.

The total amount of material produced by the electrode reaction up to time t is then proportional to the integral of the flux with respect to t, which in Laplace space is equivalent to simply dividing Equation (24) by s, to yield a quantity denoted as $Q(s)$.

$$Q(s) = \sqrt{\frac{D_P}{s}}\,\bar{\theta}_P(0, s) \tag{25}$$

Equation (25) is also obtained by integrating the dimensionless concentration profile along the entire y-axis, a quantity proportional to the absorbance (see Equation 5), provided P is the only absorbing species in the media (Equation 26):

$$\int_0^\infty \bar{\theta}_P(y, s)\,\mathrm{d}y = \int_0^\infty \bar{\theta}_P(0, s)\mathrm{e}^{-(s/D_R)^{1/2}y}\,\mathrm{d}y = \sqrt{\frac{D_P}{s}}\,\bar{\theta}_P(0, s) \tag{26}$$

Such an identity should not be surprising, as the total amount of solution-phase material produced by the electrode, is indeed directly proportional to that calculated based on the flux at the interface. However, no such rigorous relationship exists involving the *measured* charge, as in many cases the current may contain nonfaradaic contributions derived primarily from interfacial double layer capacitive effects.

2.2.1.2 Coupled Chemical Reactions Many electrochemical processes of fundamental and technological interest involve generation of species that are either intrinsically unstable, i.e. undergo spontaneous decomposition, or are capable of reacting with other solution-phase species. The latter include redox-mediated catalysis in which reduction or oxidation of the reactant R at an electrode produces P (Equation 27), which in turn reacts with Q via homogeneous electron transfer to yield a desired product Z, regenerating R (Equation 28):

$$R \longrightarrow P \tag{27}$$

$$P + Q \xrightarrow{k} Z + R \tag{28}$$

A mathematical analysis of this reaction sequence requires that the appropriate diffusion equations for R and P are solved simultaneously. Under conditions in which there is great excess of Q in solution, the homogeneous electron transfer reaction may be assumed to be pseudo first order in P. Furthermore, if both Q and Z are electrochemically inactive in the potential range in which R is either oxidized or reduced, the system of differential equations governing the overall process may be shown to be given by Equations (29) and (30),

$$\frac{\partial C_R}{\partial t} = D_R \frac{\partial^2 C_R}{\partial y^2} + \beta^2 C_P \tag{29}$$

$$\frac{\partial C_P}{\partial t} = D_P \frac{\partial^2 C_P}{\partial y^2} - \beta^2 C_P \tag{30}$$

where $\beta^2 = kC_Q$ is a pseudo-first-order rate constant. A solution to the concentration step problem for P in Laplace space, assuming $D_R = D_P = D$, may be obtained by the same techniques introduced above,[25] yielding Equation (31):

$$\bar{\theta}_P(y, s) = e^{-([s+\beta^2]/D)^{1/2} y} \tag{31}$$

where $\bar{\theta}_P(y, s) = \overline{C}_P / C_R^0$ and $\overline{C}_P$ is the Laplace transform of C_P.

2.2.2 *Diffusion in the Presence of Convective Flow*

The relationship that governs convective diffusion of solution-phase species that do not participate in any homogeneous phase reactions may be written in general as Equation (32):

$$\frac{\partial c}{\partial t} = D\nabla^2 c - \mathbf{v}\nabla c \tag{32}$$

where $\mathbf{v}$ is a vector that represents the fluid velocity. Considerable mathematical simplifications to this rather formidable problem can be obtained provided that two conditions are fulfilled:[26]

1. The thickness of the diffusion boundary layer must be small compared to the thickness of the hydrodynamic boundary layer, i.e. that region of the solution in which velocity gradients occur, so that the velocity components along the relevant axes can be approximated by the first terms of their power series expansion along an axis normal to the electrode surface.
2. Convection, rather than diffusion, is the predominant mode of mass transport along the axis parallel to the electrode surface in the direction of fluid flow.

Under the same conditions specified for diffusion in quiescent media (section 2.2.1), the convective diffusion equations for a rotating disk (and ring-disk), channel-type, and tube-type geometries under steady state are given by Equations (33) in Table 2.[27] Solutions for transient convective diffusion for rotating disk and channel-type electrodes are discussed in section 4.

The dimensionless variables defined in Equations (34–36) enable a single equation to be obtained that is applicable to a reactant (and under certain conditions to a product[27]) for all three electrode arrangements (see Equation 37), subject to the boundary conditions specified in Equations (38) and (39) in that same table. By taking the Laplace transform of Equations (37) and (39) with respect to X, denoted as L_X, and using Equation (38), yields Equations (40) and (41):

$$s\bar{\theta}(s, Y) = \frac{1}{Y} \frac{\partial^2 \bar{\theta}(s, Y)}{\partial Y^2} \tag{40}$$

$$\lim_{Y\to\infty} \bar{\theta}(s, Y) = 0 \tag{41}$$

where $L_X\{\theta(X, Y)\} = \bar{\theta}(s, Y)$

The general solution to this problem may be written as Equation (42),

$$\bar{\theta}(s, Y) = A_1(s)\mathrm{Ai}(s^{1/3} Y) \tag{42}$$

where Ai(z) is the Airy function of argument z.[23]

At $Y = 0$ Equation (43) holds

$$\bar{\theta}(s, 0) = \frac{A_1(s)}{3^{1/3}\Gamma(2/3)} \tag{43}$$

Hence, $A_1(s) = 3^{2/3}\Gamma(2/3)\bar{\theta}(s, 0)$, and therefore,

$$\bar{\theta}(s, Y) = 3^{2/3}\Gamma\left(\frac{2}{3}\right)\bar{\theta}(s, 0)\mathrm{Ai}(s^{1/3} Y) \tag{44}$$

From Equation (44), and recalling Equation (45),

$$\left.\frac{\partial \mathrm{Ai}(s^{1/3} Y)}{\partial Y}\right|_{Y=0} = -\frac{3^{-1/3}}{\Gamma(1/3)} s^{1/3} \tag{45}$$

the flux at the surface is given by

$$\left.\frac{\partial \bar{\theta}(s, Y)}{\partial Y}\right|_{Y=0} = -3^{1/3}\frac{\Gamma(2/3)}{\Gamma(1/3)}\bar{\theta}(s, 0)s^{1/3} \tag{46}$$

For references see page 10222

Table 2 Steady-state convective diffusion equations, dimensionless variables, boundary conditions and universal dimensionless differential equation for rotating disk, channel and tube electrodes

Rotating disk	Channel	Tube	
	Steady-state convective diffusion equation		
$ay\Omega\left(\frac{\Omega}{\nu}\right)^{1/2}\left\{r\frac{\partial C_i}{\partial r} - y\frac{\partial C_i}{\partial y}\right\} = D_i\frac{\partial^2 C_i}{\partial y^2}$	$2v_o\frac{y}{h}\frac{\partial C_i}{\partial x} = D_i\frac{\partial^2 C_i}{\partial y^2}$	$2v_o\left(1-\frac{r}{R}\right)\frac{\partial C_i}{\partial x} = D_i\frac{\partial^2 C_i}{\partial r^2}$	(33)
	Dimensionless variables		
$X = \left(\frac{r}{r_0}\right)^3 = \rho^3$	$X = \frac{x}{l}$	$X = \frac{x}{l}$	(34)
$Y_i = \left(\frac{r}{r_0}\right)\left(\frac{3a\nu}{D_i}\right)\left(\frac{\Omega}{\nu}\right)^{1/2} y$	$Y_i = \left(\frac{2v_o h^2}{D_i l}\right)\frac{y}{h}$	$Y_i = \left(\frac{2v_o R^2}{D_i l}\right)\left(1-\frac{r}{R}\right)$	(35)
i = R or P	$\theta_R = \frac{C_R^o - C_R}{C_R^o}$	$\theta_P = K\frac{C_P}{C_R^o}$	(36)
	Universal equation		
	$Y\frac{\partial\theta}{\partial X} = \frac{\partial^2\theta}{\partial Y^2}$		(37)
	Boundary conditions		
	$\lim_{X\to\infty}\theta(X,Y) = 0$		(38)
	$\lim_{Y\to\infty}\theta(X,Y) = 0$		(39)

Table 3 Fundamental relationships for concentrations and fluxes at the surface in Laplace and real spaces

Laplace space	Real space
$\bar{\theta}(s,Y) = 3^{2/3}\Gamma(2/3)\mathrm{Ai}(s^{1/3}Y)\bar{\theta}(s,0)$	$\theta(X,Y) = \frac{Y}{3^{2/3}\Gamma(1/3)}\int_0^X \theta(x,0)(X-x)^{-4/3}\exp\left\{-\frac{Y^3}{9(X-x)}\right\}dx$
$\bar{\theta}(s,Y) = -3^{1/3}\Gamma(1/3)\frac{\mathrm{Ai}(s^{1/3}Y)}{s^{1/3}}\left.\frac{\partial\bar{\theta}}{\partial Y}\right\|_{Y=0}$	$\theta(X,Y) = \frac{-1}{3^{1/3}\Gamma(2/3)}\int_0^X \left.\frac{\partial\theta}{\partial Y}\right\|_{Y=0}(X-x)^{-2/3}\exp\left\{-\frac{Y^3}{9(X-x)}\right\}dx$
$\bar{\theta}(s,0) = -3^{-1/3}\frac{\Gamma(1/3)}{\Gamma(2/3)}s^{-1/3}\left.\frac{\partial\bar{\theta}}{\partial Y}\right\|_{Y=0}$	$\theta(X,0) = \frac{-1}{3^{1/3}\Gamma(2/3)}\int_0^X \left.\frac{\partial\theta}{\partial Y}\right\|_{Y=0}(X-x)^{-2/3}dx$
$\left.\frac{\partial\bar{\theta}}{\partial Y}\right\|_{Y=0} = -3^{1/3}\frac{\Gamma(2/3)}{\Gamma(1/3)}s^{1/3}\bar{\theta}(s,0)$	$\left.\frac{\partial\theta}{\partial Y}\right\|_{Y=0} = -\frac{3^{1/3}}{\Gamma(1/3)}\int_0^X (X-x)^{-1/3}d\theta(x,0)$

Equation (46) can be rearranged as Equation (47)

$$\bar{\theta}(s,0) = -3^{-1/3}\frac{\Gamma(1/3)}{\Gamma(2/3)}\left.\frac{\partial\bar{\theta}(s,Y)}{\partial Y}\right|_{Y=0} s^{-1/3} \quad (47)$$

and then replaced into Equation (44) to yield Equation (48):

$$\bar{\theta}(s,Y) = -3^{1/3}\Gamma(1/3)\left.\frac{\partial\bar{\theta}(s,Y)}{\partial Y}\right|_{Y=0} s^{-1/3}\mathrm{Ai}(s^{1/3}Y) \quad (48)$$

Equations (44) and (46–48) and their corresponding inverse Laplace transforms (see Table 3),[28] provide useful relationships between fluxes and concentrations, which can be used to solve a variety of complex problems of electrochemical and spectroelectrochemical interest.

3 EXPERIMENTAL CONSIDERATIONS

Spectroelectrochemical experiments in the UV/VIS range may be broadly classified as transmission, external reflection, and internal reflection (see Figure 1). This brief section summarizes general aspects of these measurements with particular emphasis on factors that limit their capabilities in terms of spatial and temporal resolution, as well as detection sensitivity. More specialized issues are discussed in the applications section.

With the exception of the thin layer technique in the static mode, for which cells can be readily adapted to fit in the sample compartment of commercial spectrophotometers, implementation of all other UV/VIS spectroelectrochemical methodologies requires the custom assembly of optical components to bring radiation into the cell and, after interaction with the electrolyte solution, redirect the beam to the detection stage.

The spatial resolution of profile images is governed primarily by diffraction limits imposed by the optics, which set bounds to how small a volume can be reliably imaged, about 2–4 µm for this type of application, although in some cases the size of individual pixels or diodes in arrays may become the limiting factor. Beam collimation, an important aspect of this type of experiment, can be most readily achieved with lasers. However, wavelength tunability may at times be required and, in the absence of expensive instrumentation such as tunable solid-state and dye lasers, conventional optical components can be utilized to modify beams from arc or filament lamps into the desired shape. Even when collimation is optimum, a beam traveling parallel and too close to the electrode surface will undergo diffraction. Although analysis of the pattern generated is closely related to the profile itself, it has been difficult to obtain reliable quantitative spatial information using this approach. Advantage has been taken, however, of the spatial specificity of the diffracted beam to probe a small volume of solution, within less than 10 µm from an electrode surface especially for transient type measurements.(29)

Technical advances in electronics have rendered electrochemistry (more precisely, the ability to establish a specific working electrode potential over a short period of time) as the factor that limits time-resolution, rather than data acquisition. Conventional potentiostats suffer from serious limitations, in that the powers required to achieve potential control are higher than the compliance limits of regular operational amplifiers. For example, if a 0.5 V step is applied to an electrode with a double layer capacity of 1 µF, over a period of 1 µs, the potentiostat must supply a current of 0.5 A. If, in addition, the resistance due to the electrolyte is of about 0.5 kΩ, the voltage required to drive the system would be 250 V, or a peak power of 125 W.

A means of overcoming some of these restrictions, implemented by Winograd,(30) involves the use of a capacitor to inject charge into the double layer at very fast rates, coupled to a potentiostat, to establish potential control at a later time. Further reductions in the time constant of the cell τ_{cell}, defined as the product of the cell resistance and the cell capacitance, can be achieved using microelectrodes. Their size leads to smaller currents and, thus, reduced power requirements. This strategy allows use of a two-electrode cell arrangement with a comparatively much larger counter electrode, so that its potential may not be greatly affected following the potential step. Using this approach, charge can be delivered to the working electrode over tens of nanoseconds. It is indeed serendipitous that UV/VIS spectroelectrochemical methods have played a key role in verifying that potentials can indeed be controlled over such short periods of time.(31)

Detection sensitivity is determined by the response of the photomultiplier, photodiode or charge-coupled device (CCD), including their noise characteristics. Most often, signal averaging or modulation schemes are required to extract signals for systems involving species displaying small molar absorptivities at low concentrations, or cells with thin pathlengths. Although double-beam instruments have been used for UV/VIS spectroelectrochemical studies,(12) it is usually not possible to employ a reference beam to normalize absorption from the cell. Under such conditions, special efforts must be made to stabilize the single beam by using, for instance, a split beam in conjunction with its own photomultiplier to control the voltage at the primary photomultiplier.

A powerful experimental approach to enhancing sensitivity via signal averaging involves the use of spectrophotometers capable of acquiring spectra over a wide wavelength region in times of the order of milliseconds. Some of these instruments are based on mechanical motion, such as the vibrating grating/mirror developed by Harrick, the ingenious ScanDisk of OLIS's subtractive monochromator, or on the use of CCD such as that described by Park.(32,33) Wavelength modulation provides yet another viable means of extracting very weak spectral signals.(34) This technique relies on the application of a periodic perturbation to the wavelength by, for example, mechanically vibrating a grating or a slit.(35) Hence, using a conventional lock-in amplifier it becomes possible to measure a signal proportional to the derivative of the response, which can then be integrated to generate, within an additive constant, a more conventional spectrum. Another type of signal averaging involves single wavelength measurements in which the same experiment is performed repetitively. Implementation of this methodology relies on returning the system to its original state following each acquisition over as short a time as possible. Such conditions are most readily achieved by probing volumes of solution very close to the electrode surface using, for example, internal reflection or diffraction, which can be "filled" and "dumped" by potential control.

Because of differences in the acquisition times and other considerations, it is difficult to assess absolute sensitivities of the various techniques developed to date. Nevertheless, a very useful semiquantitative comparison of these methods has been provided by McCreery et al.(29)

4 APPLICATIONS

This section presents experimental and theoretical aspects of the techniques that constitute the foundations of

For references see page 10222

modern UV/VIS solution-phase spectroelectrochemistry. The intent has not been to review exhaustively the literature in this area, but rather to provide illustrative examples that may be used as a basis for the design and execution of experiments and interpretation of data.

The material selected has been organized in increasing order of complexity, which coincides, by and large, with their chronological development. In this spirit, the first subsection introduces the now classical thin-layer cell technique in the static mode, a method that provides primarily thermodynamic and kinetic information of strictly solution-phase reactions, followed by dynamic measurements in quiescent media involving optically transparent electrodes. Attention is then focused on the monitoring of solution composition changes in the immediate vicinity of the electrode by internal reflection, and of the entire diffusion layer by the more powerful and versatile specular reflection mode. This latter approach enhances the detection sensitivity by increasing the effective optical path allowing, in addition, use of a much wider variety of electrode materials. Illustrations of reflection techniques are provided for a number of electrode geometries, including microelectrodes, under stagnant and forced convection conditions. Described in the subsequent section are imaging techniques to obtain detailed dynamic information regarding evolution of concentration profiles using mostly multidiode array detectors. Owing to its own mathematical nuances, steady and transient UV/VIS spectroelectrochemical methods under forced convection were grouped in a separate subsection. Finally, the last subsection addresses methods that rely on potential modulation for the study of kinetics and mass transport phenomena associated with electron-transfer reactions.

Because of their unique spectroscopic and electrochemical properties, a few systems have been used as models for the development of most of these techniques. The most useful absorption features and the corresponding diffusion coefficients of some of the most representative species involved are compiled in Table 4.

Table 4 Molar absorptivities at the specified wavelengths and diffusion coefficients for selected chromophores commonly used in spectroelectrochemical experiments

Chromophore	$\varepsilon(M^{-1}\,cm^{-1}) \times 10^{-3}(\lambda(nm))$	$D(cm^2\,s^{-1}) \times 10^6$	Ref.
CPZ cation	12 (520)	4.1[a]	36
Orthodianisidine cation	22.7 (515)	4.4[a]	37
Ferricyanide	1.0 (425)	6.32	38
MB	56 (633)	4.1	39
TAA cation[b]	11 ± 0.1(633)	1.25	40
Methyl promazine	5.46 (600)	2.83	41
Methyl viologen cation	9.7 (632.8)	8.6[a]	42

[a] Values correspond to *D* of the parent neutral species.
[b] TAA, tris(4-methoxyphenyl)amine; CPZ, chlorpromazine; MB, methylene blue.

4.1 Transmission Thin-layer Spectroelectrochemistry

The simplest type of UV/VIS spectroelectrochemical experiment, both from experimental and theoretical viewpoints, involves measurements of fully homogeneous solutions that contain a single species that absorbs light in this spectral region, either a redox-active reactant or a product generated directly or indirectly by an electrochemical process. Such conditions can be achieved by using the electrode as a source or sink of electrons to convert reactants into products to the desired extent, rendering a solution for which the concentration of all species is uniform within the thin layer cell and in equilibrium with the applied potential. On this basis, the absorbance may be interpreted in the same fashion as in conventional nonelectrochemical spectroscopic measurements.

Owing primarily to their ease of construction, high versatility and compatibility with commercial spectroscopic instrumentation, thin-layer cells have become a common tool for thermodynamic and also for certain kinetic studies of redox-active materials containing UV/VIS chromophores.[5,7] In particular, the lack of spatial dependence allows conventional kinetic analyses for determination of rate constants and reaction orders and elucidation of mechanisms.

The archetypal UV/VIS thin-layer spectroelectrochemical cell is comprised of two flat plates made out of a material transparent in this wavelength range, such as fused silica or quartz, placed parallel to and at a small distance from each other (10–200 μm) forming a small volume cavity.[8] Either a thin metal, or a conductive oxide layer, such as In-doped Sn oxide, deposited on one of the plates, or a thin self-standing open metal grid interposed between the plates is used as the optically transparent working electrode. Structural integrity and isolation from the ambient atmosphere are achieved by means of adhesives, gaskets or mechanical compression. Depending on the design, the cell can be filled by suction or by capillary action through an opening at the bottom in contact with a bulk solution reservoir, where the reference and auxiliary electrodes are placed (Figure 3). Various types of optically transparent thin-layer electrode (OTTLE) cells have been designed to fit in the sample compartment of conventional spectrophotometers enabling transmission spectroelectrochemical data to be acquired normal to the electrode in a rather straightforward fashion.[7]

Vast improvements in the spectral sensitivity of thin layer cells can be obtained by placing the beam parallel as opposed to normal to the electrode surface, thereby

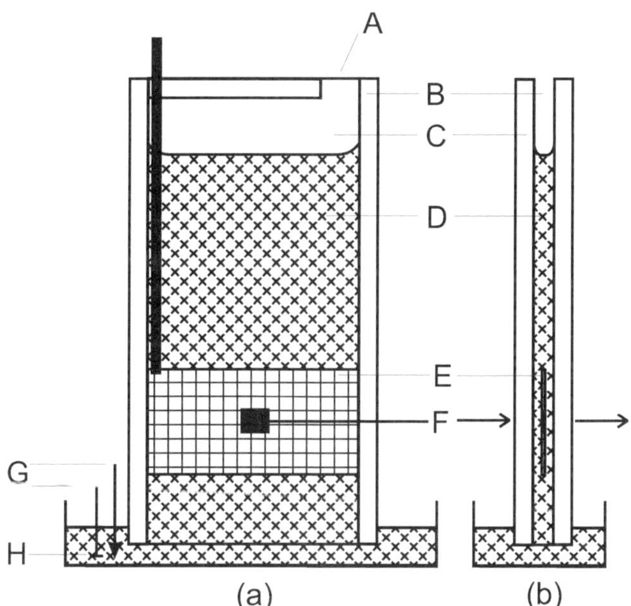

Figure 3 Optically transparent thin-layer electrochemical cell: (a) front view; (b) side view. A = point of solution suction; B = Teflon tape spacers; C = microscope slides; D = solution; E = gold minigrid electrode; F = optical path of spectrometer; G = reference and auxiliary electrodes; H = solution exit.[5]

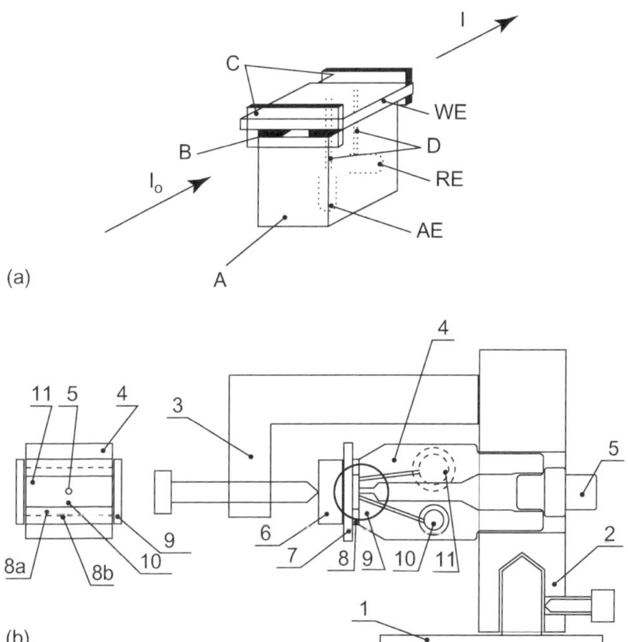

Figure 4 (a) View of the LOPTLC: A = cell body; B = thickness spacers; C = quartz windows; D = solution contact channels; WE = working electrode; RE = reference electrode; AE = auxiliary electrode. (b) Diagrams of the LOPTLC and the interior of the cell cavity: 1 = base plate; 2 = cell mount; 3 = thumbscrew and mount; 4 = cell; 5 = auxiliary electrode chamber; 6 = pressure plate; 7 = working electrode; 8a = Teflon thickness spacer; 8b = silicone rubber seal; 9 = quartz window; 10 = reference electrode chamber; 11 = solution inlet chamber.

increasing the optical pathlength, a configuration that also allows the use of nontransparent electrodes. At least two versatile long optical pathlength thin-layer cells (LOPTLCs) have been described in the literature using either conventional windows[43] (Figure 4) or arrays of optical fibers[44] to bring the beam in and out of the cell.

Among its many advantages, transmission thin layer spectroelectrochemistry enables expedient and quantitative conversion of reactant into products, which in many cases may be difficult to synthesize, handle or stabilize over long periods of time. In addition, the small area of contact between the solution in the thin layer and that in the bulk reservoir increases the time required to achieve chemical equilibration. In other words, the composition of the solution within the layer following polarization remains virtually unaltered after the electrode is open circuited for periods of time on the order of minutes. As the electrolyte solutions are at all times homogeneous, the absorbance in Equation (5) reduces to the most familiar form of Beer's law, allowing the spectral properties of electrogenerated species to be determined.

However, the thin character of the cell brings about large ohmic resistances. Consequently, the time required for a uniform potential to be established within the cell can be relatively long, thereby restricting the range of accessible homogeneous rate constants that can be measured with this technique. In fact, about 30 s are often required for complete electrolysis of material confined within the OTTLE cell.

Despite these shortcomings, transmission UV/VIS spectroelectrochemistry has become a well-established tool in inorganic chemistry and biologically oriented areas for the characterization of redox-active species under equilibrium conditions, or for reactions that proceed via fast electrochemical steps followed by a relatively slower chemical step. Examples of some of these applications are given below.

4.1.1 Thermodynamics

4.1.1.1 Determination of the Formal Reduction Potential $E^{o\prime}$ *and the Number of Electrons Transferred* n *for Redox Reactions Involving Stable Solution-phase Species*

Consider a generalized reversible redox couple (Equation 49)

$$R + ne^- \rightleftharpoons P \tag{49}$$

where both R and P represent stable solution phase species, for which the kinetics of electron transfer for

For references see page 10222

both the forward and reverse process are relatively fast. Under these conditions, the ratio of the concentrations of R and P at the electrode–solution interface will be determined by the applied potential E_{appl}, via the Nernst equation. If the electrode is polarized at a fixed value of E_{appl} for a sufficiently long time, so that C_R and C_P become uniform throughout the region probed by the beam, an explicit relationship can be obtained between E_{appl} and the relative concentrations of R and P (Equation 50):

$$E_{appl} = E^{o\prime} + \left(\frac{RT}{nF}\right)\ln\left(\frac{C_R}{C_P}\right) \qquad (50)$$

where $E^{o\prime}$ is the formal reduction potential of the redox couple.

If it is further assumed that at least one of these species displays one or more clearly defined features in the UV/VIS region, it becomes possible to determine C_R/C_P as a function of E_{appl} directly from the spectroscopic data, i.e.

$$\frac{C_R}{C_P} = \frac{A_i - A_P}{A_R - A_i} \qquad (51)$$

In Equation (51) A_P and A_R are the absorbances due to P and R, respectively, at judiciously selected wavelengths, measured at potentials sufficiently positive or negative so that only P or R are present in the solution, and A_i is the corresponding absorbance at some intermediate E_{appl} value. Combination of Equations (50) and (51) yields Equation (52)

$$E_{appl} = E^{o\prime} + \left(\frac{RT}{nF}\right)\ln\left(\frac{A_i - A_P}{A_R - A_i}\right) \qquad (52)$$

from which $E^{o\prime}$ and n can be determined, respectively, from the intercept and slope of E_{appl} versus $(A_i - A_P)/(A_R - A_i)$, or Nernstian plots. The subsections to follow will show examples of application of these principles.

Reversible Systems. Figure 5 shows in situ UV/VIS spectra of a 2.0 mM $K_3[Fe(CN)_6]$ aqueous solution in 1 M KCl at different potentials recorded in a thin layer cell.(45) As indicated, $[Fe(CN)_6]^{3-}$ displays two rather prominent bands centered at 213 and 420 nm, which can be used to construct Nernstian plots, such as those shown in the insert to Figure 5. Analysis of these data yielded values of $E^{o\prime} = 0.256 \pm 0.001$ V versus Ag/AgCl and $n = 0.965 \pm 0.0015$, in fairly good agreement with results obtained by other means.

Quasireversible Systems. The virtual diffusional decoupling between the solution contained within the thin-layer cell and that in the bulk reservoir allows spectroelectrochemical experiments to be conducted over periods of time long enough to investigate slow

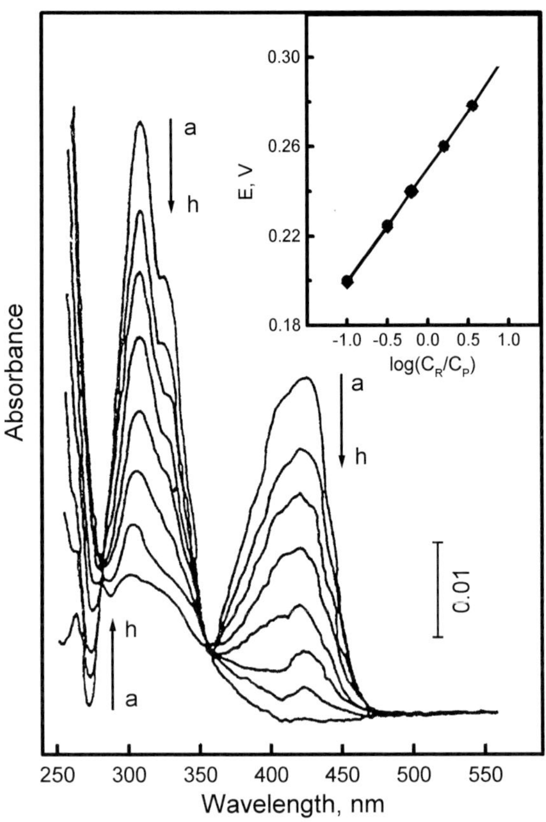

Figure 5 UV/VIS absorption spectra of 2.0 mM $K_3Fe(CN)_6$ in aqueous 1 M KCl at different potentials in the range +0.5 (a) and 0.0 (h). The inset shows a plot of E_{appl} versus $\log(C_R/C_P)$ at (•) $\lambda = 312$ nm and (♦) $\lambda = 420$ nm obtained in a thin-layer cell.(45)

redox reactions. For example, vitamin B_{12}, cyano-$[Co^{III}]$cobalamin, exhibits no cyclic voltammetric reduction features at potentials positive to the observed two-electron reduction wave, (Figure 6), regardless of scan rate.(46) Clear evidence that this reaction proceeds via two consecutive one-electron transfer steps was obtained by Mark et al.(46,47) in buffered solutions (pH = 6.86), using an Hg–Au minigrid working electrode in an OTTLE cell. For these measurements, the potential was stepped to different values in the range $-1.15 < E < -0.1$ V versus SCE (saturated calomel electrode) and the UV/VIS spectrum recorded after equilibrium had been established. Two distinct absorption peaks were observed for potentials in the range $-0.58 < E < -0.75$ V at 311 and 361 nm (Figure 7a), and $-0.77 < E < -0.95$ V versus SCE (Figure 7b) at 368 nm. Nernstian plots of the absorbance measured at these wavelengths yielded for the two one-electron transfer processes values of $E_{1/2} = -0.655$ V and $E_{1/2} = -0.88$ V versus SCE.

Mediated Electron Transfer. The rates of heterogeneous electron transfer of large redox-active molecules,

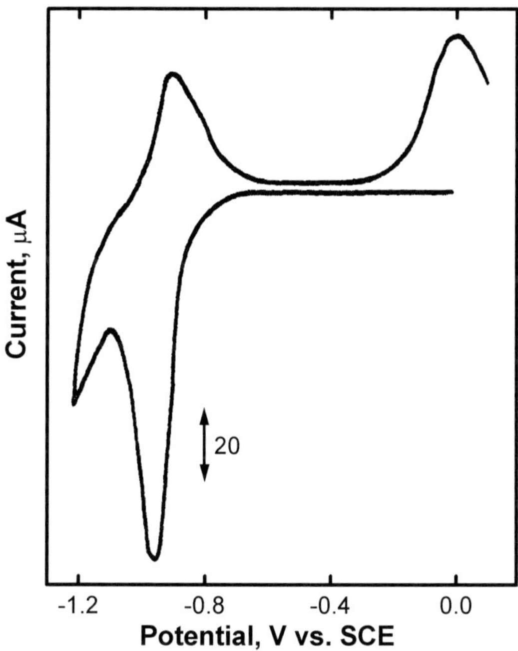

Figure 6 Cyclic voltammogram of 1 mM vitamin B_{12}, Britton Robinson buffer pH = 6.86, 0.5 M Na_2SO_4 obtained with an Hg–Au minigrid in a thin-layer cell. Scan rate $2\,mV\,s^{-1}$.[46]

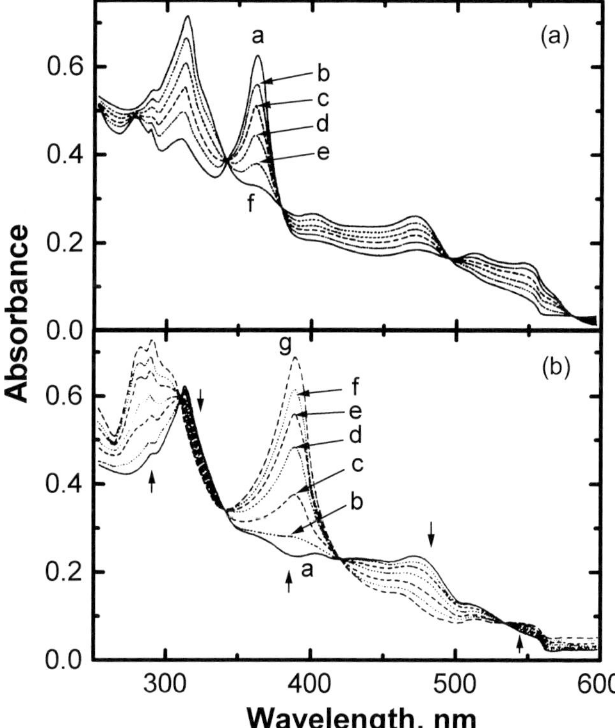

Figure 7 UV/VIS spectra obtained in a thin-layer cell for (a) the one-electrode reduction of vitamin B_{12} under the same conditions specified in the caption to Figure 6. Applied potentials: a = −0.550; b = −0.630; c = −0.660; d = −0.690; e = −0.720; f = −0.770 V versus SCE and (b) the second one-electron reduction. Applied potentials: a = −0.770; b = −0.820; c = −0.860; d = −0.880; e = −0.900; f = −0.920; g = −1.000 V versus SCE.[46]

such as proteins, are very small, due at least in part to the insulating character of the chemical environment surrounding the redox site. A common strategy to overcome this difficulty, shown schematically in Scheme 1, is to employ an electron-transfer mediator M, either in solution or bound to the electrode surface, capable of exchanging electrons with such larger molecules in solution at much higher rates.

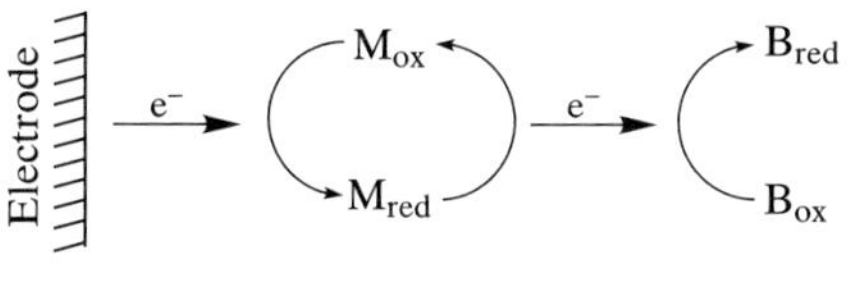

Electrode $\xrightarrow{e^-}$ M_{ox} / M_{red} $\xrightarrow{e^-}$ B_{red} / B_{ox}

Scheme 1

This approach enables spectra of redox-active proteins or other large molecules to be recorded as a function of E_{appl}, allowing values of $E^{o\prime}$ and n to be determined by the same method described above. A classic example of this principle is provided by cytochrome c, a species that exhibits no clear voltammetric features on Au electrodes. However, large enhancements in the electron transfer rates can be obtained by using 2,6-dichlorophenolindophenol (DCIP) as a mediator, as shown in Scheme 2.

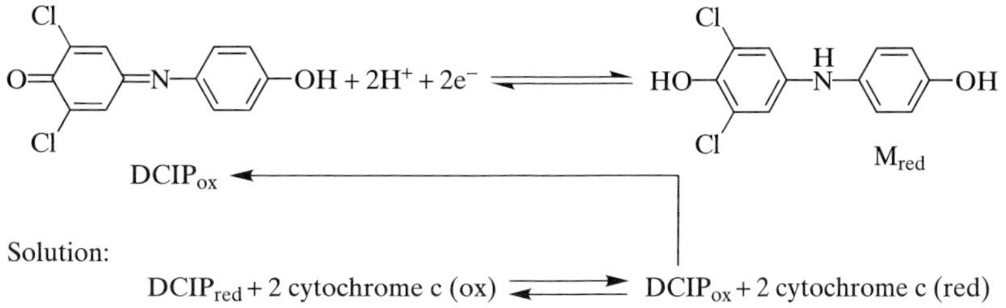

Electrode:

[Cl, Cl-substituted quinone imine]=N–C₆H₄–OH + $2H^+ + 2e^-$ ⇌ HO–[Cl, Cl-substituted ring]–NH–C₆H₄–OH (M_{red})

$DCIP_{ox}$

Solution:

$DCIP_{red}$ + 2 cytochrome c (ox) ⇌ $DCIP_{ox}$ + 2 cytochrome c (red)

Scheme 2

For references see page 10222

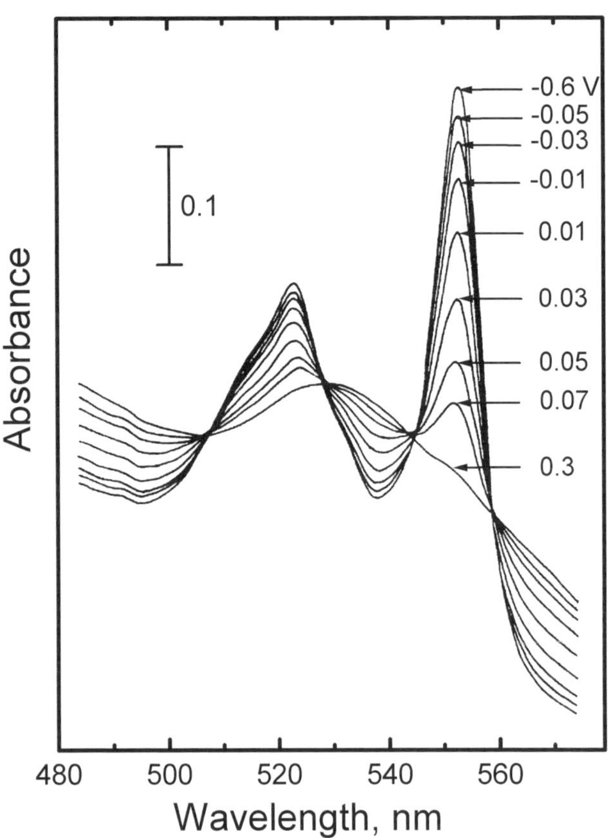

Figure 8 UV/VIS spectra of cytochrome c in 0.5 phosphate buffer (pH = 7.0), obtained in a thin-layer cell in the presence of 2,6-dichlorophenol–indophenol mediator–titrant, for a series of applied potentials versus SCE.[48]

This mediated process was monitored spectroelectrochemically in phosphate buffer (pH = 7), using a Au minigrid working electrode in an OTTLE cell, at potentials in the range −0.600 (fully reduced cytochrome c) to 0.300 V (fully oxidized cytochrome c) (Figure 8).[48] Nernstian plots for absorbance measured at 520 and 550 nm, the characteristic absorption maxima for cytochrome c(oxidized) and cytochrome c(reduced), yielded values of $E^{o\prime} = +0.264$ V versus SHE and $n = 1$.

4.1.2 Coupled Chemical Reactions

Electrochemistry provides an expedient means of preparing solutions of a given composition by controlling the charge that flows through the interface over a certain period of time. It thus becomes possible to use this method to monitor rates of some chemical reactions involving electrogenerated species under various initial conditions as a function of time. Application of this simple methodology assumes that the total time of charge injection is long enough for the concentration of all species to be uniform throughout the thin-layer cell before any significant bulk reaction occurs.

This approach has been exploited for studies of certain inorganic and biological processes that proceed via complex mechanisms, including chemical reactions that precede or follow an electron-transfer step (Scheme 3), for which one of the species involved absorbs light in the UV/VIS range.

$$\mathrm{R} + ne^- \rightleftharpoons \mathrm{P}$$
$$\mathrm{P} \xrightarrow{k} \mathrm{Q}$$

Scheme 3

Three different techniques will be featured in this section to illustrate the principles involved.

4.1.2.1 Single Potential Step Perhaps the simplest transient spectroelectrochemical experiment of the type discussed above involves monitoring the absorbance of one or both of the redox species as a function of time following a potential step. An excellent illustration of this technique is provided by the work of Owens and Dryhurst,[49] who studied the electrochemical oxidation of 5,6-diaminouracil in aqueous buffered solutions pH = 4–6, for which the product, the corresponding diimine, undergoes hydrolysis, as depicted in Scheme 4. This system may be regarded as a model for a variety of biologically fundamental redox processes, including the enzymatic oxidation of purines.

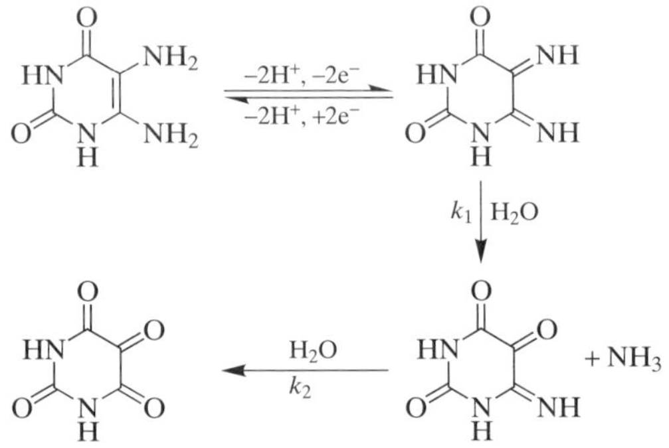

Scheme 4

UV/VIS spectra obtained using an OTTLE cell incorporating a Au minigrid electrode, recorded at different time intervals after the potential step was applied, revealed a broad band centered at 320 nm (not shown here) attributed to the diimine intermediate. If it is assumed that the decomposition of diimine follows first-order kinetics, the time dependence of its decay can be related to the change in absorbance via Equation (53):

$$\ln(A_t - A_\infty) = -kt + \ln(A_0 - A_\infty) \qquad (53)$$

where A_0, A_t and A_∞ are the absorbances at times 0, t and at the end of the experiment. A linear relationship

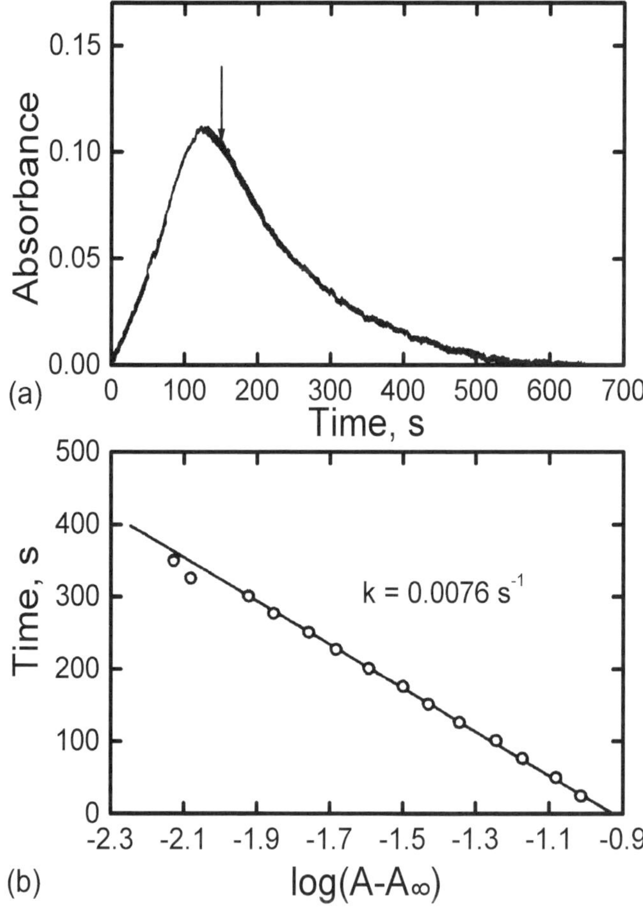

Figure 9 (a) Absorbance versus time recorded at $\lambda = 320$ nm during the electrooxidation of 10 mM 5,6-diaminouracil in McIlvaine buffer (pH = 5) at 0.35 V in a thin-layer cell with an Au minigrid electrode. The arrow indicates the time at which 5,6-diaminouracil was fully electrolyzed. (b) Plot of t versus $\log(A - A_\infty)$ obtained from the data in (a) by setting $t = 0$, at the point specified by the arrow.[49]

was observed between t and $\ln(A_t - A_\infty)$ recorded at $\lambda = 320$ nm (Figure 9), from which the rate constant k_1 was calculated, yielding a value of $0.04\,s^{-1}$ at pH = 4.

4.1.2.2 Double Potential Step A slight variation of the method just described involves application of a second potential step, after allowing a period of time t_R to elapse following the initial potential step. Assume, for example, that a certain absorbing species produced during the initial step, is then partially consumed via a first-order chemical reaction as specified in Scheme 3 above. If a potential step is applied in precisely the opposite direction at t_R, so as to convert all remaining P back to R, the difference in the change in absorbance between the first, ΔA_f (proportional to the amount of P generated), and the second step, ΔA_b (proportional to the amount of P remaining in solution), provides a measure of the amount of product decomposed during t_R. Based on information collected for various values of t_R, the rate of chemical decomposition of the product can be determined from Equation (54):

$$\ln \frac{\Delta A_f}{\Delta A_b} = kt_R \quad (54)$$

This method was employed to determine k_f for benzidine rearrangement[50] in an aqueous acidic electrolyte, a process that proceeds via the electrochemical–chemical (EC) mechanism shown in Scheme 5, using an Au minigrid electrode in an OTTLE cell.

For these experiments, azobenzene was first completely reduced to hydrazobenzene during the first potential step from 0.0 to −0.6 V versus SCE, the potential was then maintained at −0.6 for a reaction time t_R, and the remaining hydrazobenzene reoxidized to azobenzene during the second potential step to 0.3 V. All spectroscopic data were acquired at $\lambda = 325$ nm, a value at which only azobenzene displays significant absorption. Typical optical response curves are shown in Figure 10, and the plot based on Equation (54) is given in Figure 11.

The advantage of the double-potential step versus single-potential step method is that either P or R can be monitored spectroscopically. However, the results obtained with this approach are less precise than for

Electrode:

Azobenzene $+ 2e^- + 2H^+ \rightleftharpoons$ Hydrazobenzene

Solution:

Hydrazobenzene $\xrightarrow[H^+]{k}$ Diphenyline + Benzidine

Scheme 5

For references see page 10222

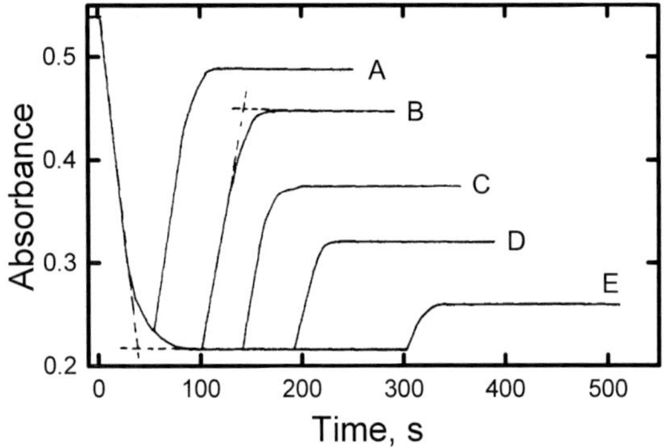

Figure 10 Absorbance versus time curves obtained at $\lambda = 325$ nm for a double-potential-step experiment performed in a thin-layer cell in 1 mM azobenzene, 0.100 M HCl, 0.150 M KCl, 44% ethanol aqueous solution. First step from +0.0 to −0.6 V and second step to +0.3 V versus SCE; $t_R = 50$ (A), 100 (B), 150 (C), 200 (D), and 300 s (E).[50]

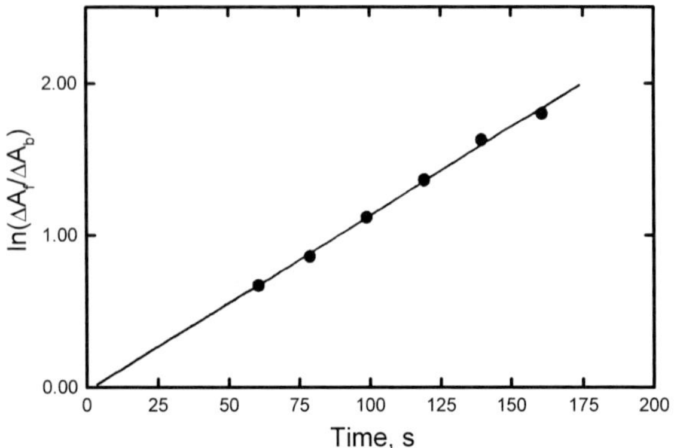

Figure 11 Kinetic plot, $\ln(\Delta A_f/\Delta A_b)$ versus time based on the absorbance versus time data for the double-potential-step experiment described in the caption to Figure 10.[50]

single potential step, as t_R has to be calculated graphically (dotted lines in Figure 10).

4.1.2.3 Single Potential Step Open Circuit Relaxation In some cases, species P in Scheme 3 can yield upon subsequent chemical reaction an electroactive species Q, which might be the original electroactive reactant R itself, a factor that can complicate analysis of spectroscopic data of the type introduced in previous sections. If k in Scheme 3 is sufficiently small, it becomes possible to monitor the reaction progress by bringing the electrochemical reaction to completion and then opening the electrical circuit enabling P to undergo further chemical transformations. This technique, known as single potential step open circuit relaxation, has been used to elucidate the mechanism of oxidation of the psychoactive drug CPZ.[51] This material can be electrochemically oxidized in strongly acidic media via two sequential one-electron steps to produce first, a deep red cation radical chlorpromazine cation radical ($CPZ^{+\bullet}$) (see Table 4), and, at more negative potentials, a colorless sulfoxide-type derivative (CPZO). However, the cation radical undergoes disproportionation according to the reaction sequence shown in Scheme 6.

$$\text{Electrode: CPZ} \longrightarrow CPZ^{+\bullet} + e^-$$

$$\text{Solution: } 2\,CPZ^{+\bullet} \xrightarrow[k]{H_2O} \text{CPZ} + \text{CPZO}$$

Scheme 6

For this process, Q in Scheme 3, happens to be identical to R. Evidence in support of this mechanism was obtained by spectroelectrochemical experiments using an Au minigrid working electrode in an OTTLE cell in a strongly acidic solution. Figure 12 shows a series of spectra recorded for a potential step from 0.250 V to 0.620 mV versus SCE. Curves a and b in this figure represent spectra collected before (CPZ) and after the step allowing 2.8 min for equilibrium to be reached, whereas all other spectra were recorded after the circuit was opened. The peaks at 262 and 525 nm, characteristic of $CPZ^{+\bullet}$, decreased during open circuit, and new features appeared at 292 and 337 nm, attributed to the final reaction product CPZO.

Plots of $1/C$ versus t for absorbance measured at 525 and 337 nm were linear, and therefore consistent with a

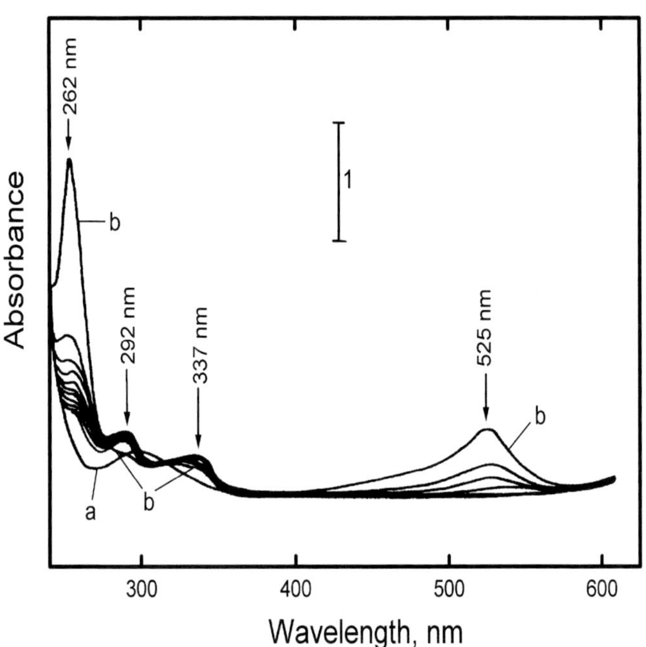

Figure 12 UV/VIS spectra recorded during a thin-layer chronoabsorptometry experiment with open-circuit relaxation in a 4.8 mM CPZ in 3 M H_2SO_4: (a) spectrum of CPZ at 250 mV; (b) spectrum of CPZ at open circuit. Spectra were acquired every 2.8 min.[51]

second order mechanism for $CPZ^{+\bullet}$ decomposition, with slopes of $8.6\,M^{-1}s^{-1}$ and $4.4\,M^{-1}s^{-1}$, i.e. a ratio of 2 : 1, in accordance with the reaction stoichiometry.

4.1.3 Slow Scan Rate Cyclic Voltammetry

The small amount of electroactive material contained within a thin-layer cell allows changes in composition induced by the applied potential to occur within rather short times, of the order of tens of seconds. Under these conditions it becomes possible to follow optically electrochemical processes involving absorbing species, by scanning the electrode potential linearly between two prescribed limits, while monitoring the current, a technique known as cyclic voltammetry. Analysis of these measurements can be greatly simplified by scanning the potential at slow enough rates to achieve at every instant quasi-equilibrium conditions throughout the cell.

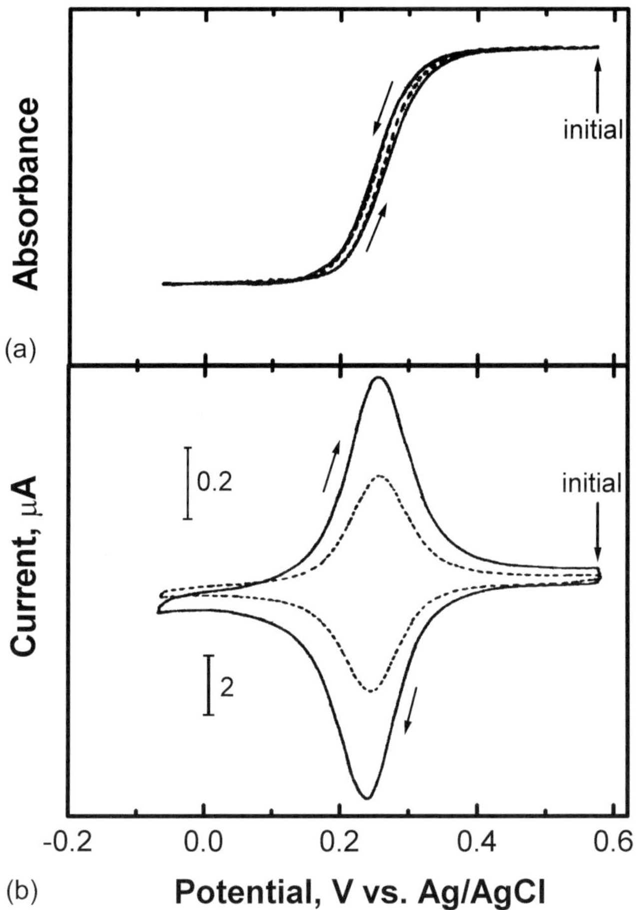

Figure 13 Curve (a) is the absorption versus time plot recorded during the cyclic voltammogram shown by curve (b) at $\lambda = 420\,nm$ using a glassy carbon electrode in a 0.4 mM ferricyanide in 0.5 M KCl solution at linear scan rates of (——) $2\,mV\,s^{-1}$ and (- - -) $1\,mV\,s^{-1}$.[43]

Figure 13 shows simultaneous absorbance and current data obtained on a glassy carbon electrode in the long pathlength thin layer cell (LPTLC) shown in Figure 4 in a solution of 0.4 mM ferricyanide in 0.5 M KCl at $\lambda = 420\,nm$.[43] For this LPTLC, the absorbance change, following full reduction or subsequent oxidation, is about two orders of magnitude larger than those observed for OTTLE cells described earlier.

4.1.4 Molecular Adsorption on Solid Electrodes

Advantage can be taken of the much enhanced sensitivity of LPTLC spectroelectrochemical cells and the possibility of using flat opaque electrodes to measure minute changes in concentration ΔC, such as those derived from adsorption of UV/VIS chromophores on electrode surfaces, by monitoring the corresponding decrease in the absorbance ΔA. As may be inferred from simple arguments, these two quantities are related to the width of the optical cell w and $\varepsilon(\lambda)$ (Equation 55):

$$\Delta C = \frac{\Delta A}{w\varepsilon(\lambda)} \qquad (55)$$

On this basis, material lost from the solution phase, as measured by spectroscopic means, may be assumed to be adsorbed on the electrode surface, and the coverage Γ may be calculated from Equation (56),

$$\Gamma = \frac{V\Delta C}{a} \qquad (56)$$

where V is the volume of the cell, and a is the area of the electrode. The ability to measure Γ accurately is thus determined by how small an absorbance change the instrument is capable of detecting. Saturation coverages of small size organic molecules on common metals are about $10^{-10}\,mol\,cm^{-2}$; hence, assuming a cell volume of 10 μL, the change in concentration associated with formation of a full monolayer would be 10^{-5} M. Minimum detectable values of ΔA are of the order of 0.002 at a signal-to-noise ratio of 2; therefore, for $\varepsilon(\lambda)$ of $1000\,M^{-1}cm^{-1}$, $w = 1$ and $a = 1\,cm^2$, ΔC as small as 2×10^{-6} M, and, thus, smaller that those corresponding to $\Gamma = 1$ could be observed.

Measurements involving a series of heterocyclic aromatics were performed by Kuwana et al. with the same LPTLC spectroelectrochemical cell shown in Figure 4 incorporating a Pt electrode.[52] Figure 14(a) shows absorbance versus time plots obtained after injecting a solution 32.9 μM *trans*-1,2-bis-(4-pyridine)ethylene (Pyc=cPy) in buffered (pH = 7) 0.1 M $NaClO_4$ (first curve) monitored at $\lambda = 299\,nm$ with the Pt electrode polarized at −0.1 V versus Ag/AgCl. The two other curves were obtained after a second and third injection without removing Pyc=cPy adsorbed on the Pt surface during the

For references see page 10222

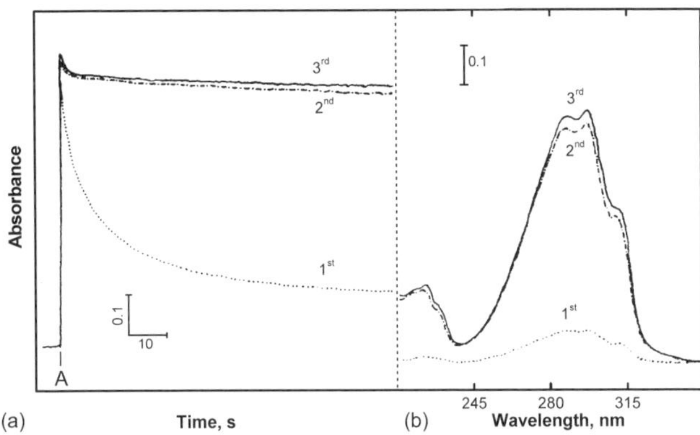

Figure 14 (a) Plots of absorbance versus time recorded at $\lambda = 299$ nm after the first, second, and third fillings of a thin-layer cell with a solution 32.9 µM PyC-CPy (*trans*-1,2-bis(4-pyridine)-ethylene) in 0.1 M $NaClO_4$ in 0.02 M phosphate buffer (pH = 7) using a Pt electrode polarized at −0.1 V versus Ag/AgCl immediately after each filling. Experiments began at point A. (b) Absorbance spectra acquired at $E = -0.1$ V after the corresponding absorbance versus time response has reached a steady state value for the 1st (· · ·), 2nd (-·-·-) and 3rd (——) fillings.[52]

first injection. Figure 14(b) shows the corresponding spectra of the solution phase recorded after the absorbance had reached steady state following each injection. The very small differences observed between the second and third injections indicate that the surface had attained saturation, i.e. $\Gamma = 1$.

These data may be used to determine molecular packing densities and from these to infer the mode of bonding of species to the surface as thoroughly discussed by Soriaga and Hubbard.[53] Proposed orientations of four of the species investigated by Kuwana et al. are shown in Scheme 7.[52]

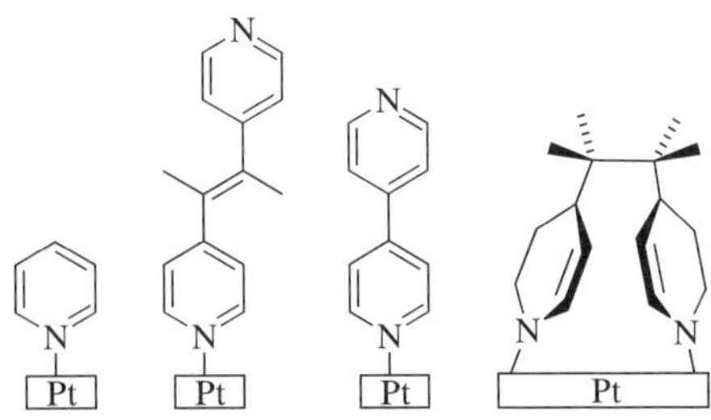

Scheme 7

4.2 Transmission Spectroelectrochemistry in Semi-infinite Media

Methods based on the dynamic monitoring of absorbance changes induced by electrochemical perturbations can provide kinetic information regarding rates of homogeneous electron transfer rates involving electrogenerated species, as well as mass transport phenomena, not accessible by the static techniques described in the previous section. For simplicity, consider an OTE immersed in a solution containing a single, nonabsorbing solution-phase species R capable of undergoing heterogeneous electron transfer to yield an absorbing solution phase product P (Equation 57):

$$\mathrm{R} + ne^- \longrightarrow \mathrm{P} \qquad (57)$$

If C_P is independent of x and z, the instantaneous absorbance $A(t)$ in Equation (4) reduces to Equation (58),

$$A(t) = \varepsilon \int_0^d C_\mathrm{P}(y, t)\,\mathrm{d}y \qquad (58)$$

where d is the length of the optical cell along the axis of propagation of the light. Provided the measurements are made over relatively shorts periods of time, so that the diffusion layer thickness is always smaller than d, the media may be regarded as semi-infinite and $A(t)$ may be rewritten as Equation (59),

$$A(t) = \varepsilon \int_0^\infty C_\mathrm{P}(y, t)\,\mathrm{d}y \qquad (59)$$

or, in Laplace space (Equation 60), as

$$\overline{A}(s) = \varepsilon \int_0^\infty \overline{C}_\mathrm{P}(y, s)\,\mathrm{d}y \qquad (60)$$

If it is further assumed the overall measurement time is short so that no complications arise from natural convection, an explicit form for $\overline{C}_\mathrm{P}(y, s)$ for a potential step experiment of the type described in section 2.2.2 may be obtained from Equation (22):

$$\overline{C}_\mathrm{P}(y, s) = -\frac{C_\mathrm{R}^\mathrm{o}}{s}\left(\frac{D_\mathrm{R}}{D_\mathrm{P}}\right)^{1/2} \mathrm{e}^{-(s/D_\mathrm{P})^{1/2} y} \qquad (61)$$

Equation (61) can be substituted into Equation (60) to yield Equation (62),

$$\overline{A}(s) = \frac{\varepsilon C_\mathrm{R}^\mathrm{o} D_\mathrm{R}^{1/2}}{s^{3/2}} \qquad (62)$$

and inverse Laplace transformed to afford an expression for the time dependence of the absorbance due to P for such diffusion-controlled process (Equation 63):

$$A(t) = \varepsilon C_\mathrm{R}^\mathrm{o}\left(\frac{4D_\mathrm{R} t}{\pi}\right)^{1/2} \qquad (63)$$

Based on the analysis of such chronoabsorptometric-type experiment, it becomes possible to determine D_R if ε is known or, conversely, ε if D_R is known.

A rather versatile cell design reported by Kuwana et al.[4] suitable for this type of measurements incorporates an OTE in a glass sandwich-type cell (Figure 15), in

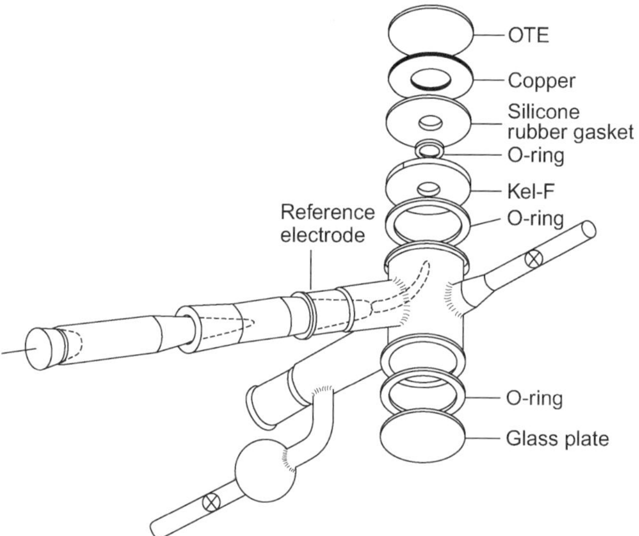

Figure 15 Sandwich cell for transmission UV/VIS spectroelectrochemical experiments.[4]

which the light travels along the axis of the cylinder. The small bent tube, known as a Luggin capillary, allows electrolytic contact between the main and reference electrode compartments.

Potential step chronoabsorptometry at an optically transparent tin oxide electrode has been used[54] to study the oxidation of $[Fe(CN)_6]^{4-}$ to $[Fe(CN)_6]^{3-}$ in aqueous electrolytes by monitoring $A(t)$ at $\lambda = 420\,nm$ (Table 4). In accordance with theoretical predictions, plots of $A(t)$ versus $t^{1/2}$ (Figure 16) were found to be linear, yielding values of D for $[Fe(CN)_6]^{4-}$ determined from the slope, of $6.5 \times 10^{-6}\,cm^2\,s^{-1}$, in reasonable agreement with those obtained by other methods.

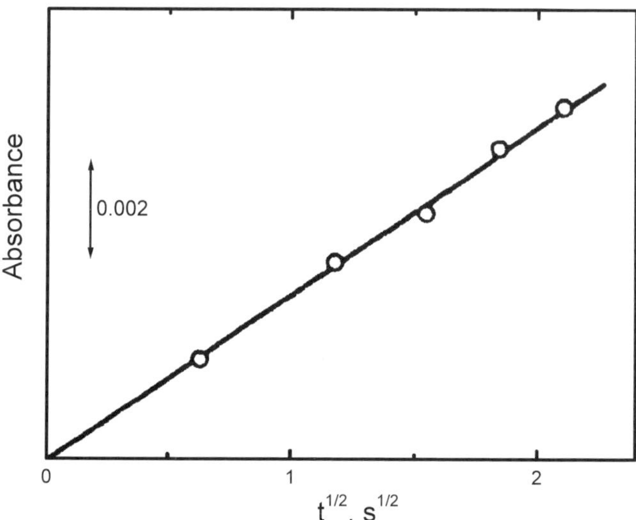

Figure 16 Plot of A versus $t^{1/2}$ for the chronoamperometric oxidation of $[Fe(CN)6]^{4-}$ at a tin oxide optically transparent electrode.[54]

One of the advantages of using large-volume cells, as opposed to thin-layer cells, is the decrease in τ_{cell}, which leads to improvements in transient response. However, the relatively low spectral sensitivity due to the very short optical paths associated with OTE, the restricted number of electrode materials with suitable electronic and optical characteristics, and the high resistance associated with thin films, severely limit the applicability of the technique described in this section. As explored in detail later below, some of these difficulties can be overcome by implementation of dynamic methods in both internal and external reflection modes.

4.3 Internal Reflection

As briefly discussed in section 2.1.3, light traveling through a media of high index of refraction, impinging on a planar interface with a media of smaller index of refraction at a sufficiently large angle, will penetrate slightly into the adjacent phase and interact with species present therein. For spectroelectrochemical applications, the surface of the internal reflection element (IRE) is coated with a thin film of an electronically conducting material, such as a metal or a highly doped semiconductor, which serves as the optically transparent working electrode. This technique was pioneered by Kuwana, Winograd et al. in the 1970s.[4]

The volume of solution probed by the evanescent wave is of the same order of magnitude as the wavelength of light, about 100–300 nm in the UV/VIS range, and therefore the size of the cell becomes immaterial. Under such conditions, the time response of the cell is limited not by the solution resistance, but by the thin character of the electrode, as was the case with the OTE-based techniques described in the previous sections.

Owing to the very short optical path and the relatively low concentrations of chromophores often used, the extent to which the exponentially decaying evanescent wave in the solution is attenuated is very small, and therefore its spatial dependence may be regarded as unperturbed by the presence of absorbing material therein. Under these conditions, the instantaneous absorbance is given by Equation (64),

$$A(t) = \varepsilon N_{eff} \int_0^{\infty} C_P(y, t) \exp\left(-\frac{y}{\delta}\right) dy \qquad (64)$$

where δ is the penetration depth (see Equation 10), and N_{eff} is a sensitivity factor that depends on the geometry and electrode material.

Insertion of Equation (22) into Equation (64) in Laplace space and subsequent inversion, yields Equation (65) for a chronoabsorptometric experiment:

$$A(t) = \varepsilon\delta N_{\text{eff}} C_{\text{R}}^{\text{o}} \left(\frac{D_{\text{R}}}{D_{\text{P}}}\right)^{1/2} [1 - \exp(\beta^2 t)\text{erfc}(\beta t^{1/2})] \quad (65)$$

where $\beta = D_{\text{R}}^{1/2}/\delta$.

For $\delta \sim 100\,\text{nm}$, $A(t)$ reaches steady state in about 1 ms, which may be regarded as the time needed to fill the cell. If the potential is then returned to its original value by applying a reverse step, the cell can be dumped over as short a time, making it possible to perform numerous *identical* experiments that can then be co-added and averaged to improve signal detection.

A multiple reflection internal reflection spectroscopy (IRS) cell designed by Kuwana and Winograd[55,56] is shown in Figure 17. In this optical arrangement two prisms are used to guide the beam into and out of the IRE, allowing five internal reflections to probe the solution phase of the interface, thereby increasing the effective pathlength.

Sensitivities better than one part in 10^5 with time resolutions in the microsecond range have been obtained for systems in which $N_{\text{eff}} \gg 1$, and long averaging times. This is illustrated in Figure 18 for MV^{++} reduction, for which the time resolution, using charge injection to achieve fast potential control, was of the order of 4 μs.[57] Advantage was taken of the extraordinary capabilities of this measurement scheme for kinetics studies involving fast processes,[58] such as the second-order catalytic mechanism (Scheme 8) for the reaction of cyanide with electrogenerated tri-*p*-anisylamine cation radical ($TAA^{+\bullet}$) in acetonitrile.

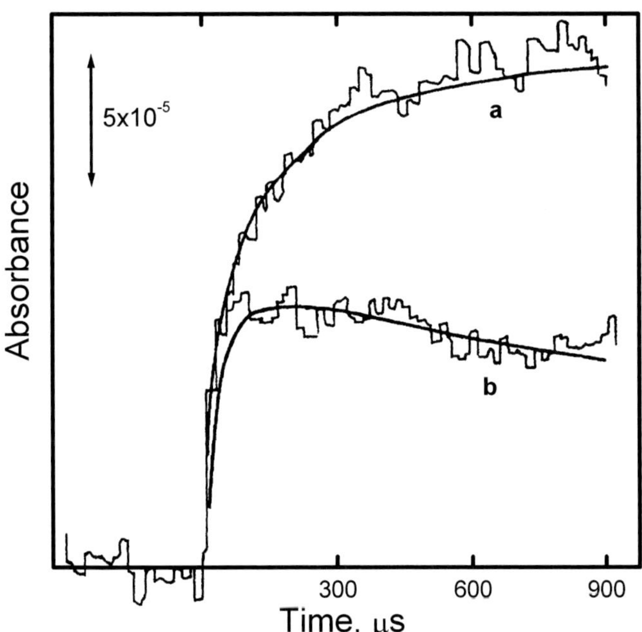

Figure 18 Absorbance versus time curves for the reduction of methyl viologen dication (MV^{++}) reduction, monitored at $\lambda = 605\,\text{nm}$ in the internal reflection mode. Curve (a) corresponds to the diffusion-limited reduction of MV^{2+} to its cation radical. Curve (b) is for the reduction to MV^0. Solid lines are theoretical transients; noisy curves are experimental.[57]

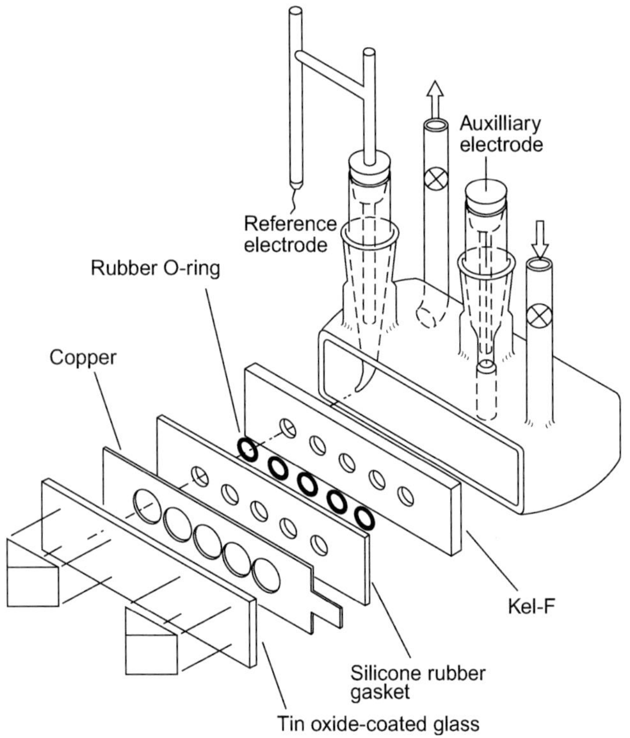

Figure 17 Cell for UV/VIS IRS measurements.[59]

$$TAA \rightleftharpoons TAA^{+\bullet} + e^-$$
$$TAA^{+\bullet} + CN^- \xrightarrow{k_1} TAA + CN^\bullet$$
$$2\,CN^\bullet \xrightarrow{k_2} (CN)_2$$

Scheme 8

Plots of absorbance versus time following a potential step recorded at $\lambda = 715\,\text{nm}$, a wavelength at which $TAA^{+\bullet}$ exhibits a characteristic absorption band, were recorded both in the presence and in the absence of tetraethylammonium cyanide (TEACN; Figure 19). Analysis of these data yielded values for the k_1 in Scheme 8 of $2.7 \times 10^5\,\text{M}^{-1}\,\text{s}^{-1}$, in agreement with those obtained by other means.

IRS also lends itself to the study of homogeneous electron-exchange reactions involving electrogenerated species for values of k in Scheme 8 larger than $10^6\,\text{M}^{-1}\,\text{s}^{-1}$. Under such conditions, the intermediate is formed in close

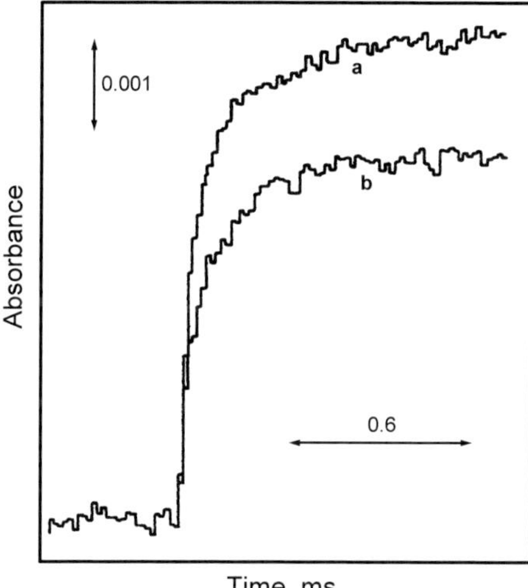

Figure 19 Absorbance versus time curves obtained at $\lambda = 715$ nm in the internal reflection mode following a potential step for (a) 0.5 mM TAA and (b) 0.5 mM TAA with 15 mM TEACN added, in 0.3 M tetraethylammonium perchlorate (TEAP) in acrylonitrile (AN) using an optically transparent Pt electrode for 800 repetitions. $N_{\text{eff}} = 15$, $\sqrt{D}/\delta = 115$.[58]

proximity to the electrode surface and can, therefore, be detected by the evanescent wave.

From a general perspective, quantitative analysis of experimental data can be pursued using digital simulation techniques to predict, for a proposed mechanism, time-dependent concentration profiles for all species involved,[59] as illustrated for an electrochemical–electrochemical–chemical (EEC)-type process in Figure 20. Numerical values for the various parameters involved can then be calculated based on best fits to the experimental curves. In fact, for the reduction of MV^{++} mentioned above, which has been found to follow the mechanism in Scheme 6, a fit to the data obtained by monitoring $A(t)$ at $\lambda = 605$ nm, associated with the radical cation intermediate $MV^{+\bullet}$, yielded values of $k_f = 3 \times 10^9\ M^{-1}\ s^{-1}$.[57]

More recently, Heineman et al.[60–63] have implemented the use of IRS to develop spectroelectrochemical sensors with high selectivity and sensitivity. For this application, the IRE is covered first with a thin conducting layer as before, and then with a film displaying molecular specificity or selective film, as shown in Figure 21. This device allows three independent sensing techniques to be employed simultaneously, i.e. electrochemistry, spectroscopy, and selective partitioning, thereby enhancing the degree of specificity for chemical detection. As a means of illustration, Figure 22 shows results obtained with simultaneous optical (panels a and b) and electrochemical techniques (panel c) for the detection of ferrocyanide partitioned into a poly(dimethyldiallyl ammonium chloride)-SiO_2 composite sol-gel film from an aqueous solutions at two different concentrations.

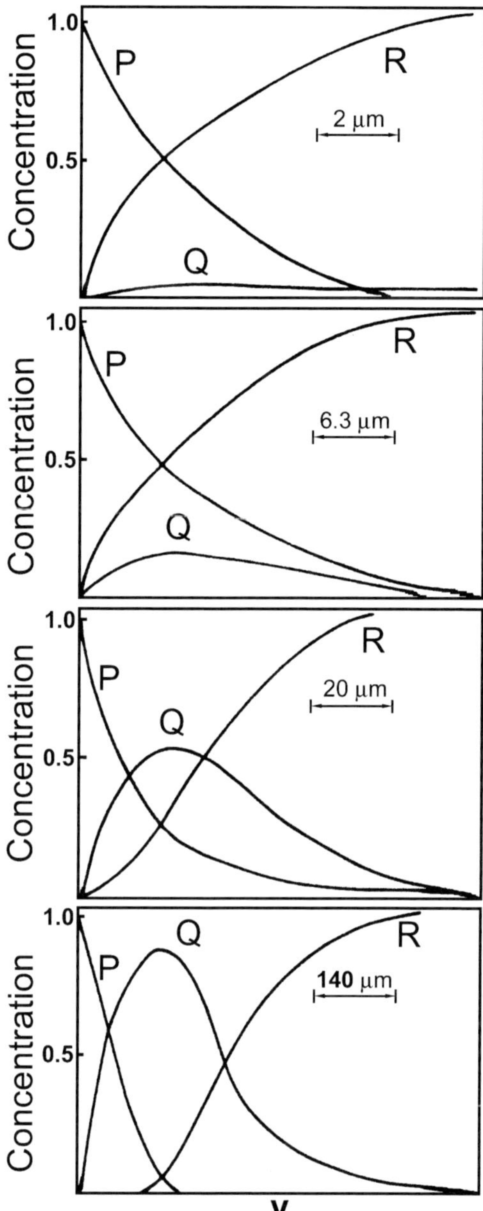

Figure 20 Concentration profiles obtained by digital simulation for an EEC mechanism with $k_f C = 10^2\ s^{-1}$. The different curves are drawn for $t = 1$, 10, 100, and 500 ms, from (a) to (d).[59]

4.4 External Reflection

The rather short effective pathlengths associated with techniques based on OTEs limits their applicability to systems involving strong chromophores, or long times for sufficient absorbing material to accumulate and be detected. Although gains in sensitivity can indeed be

For references see page 10222

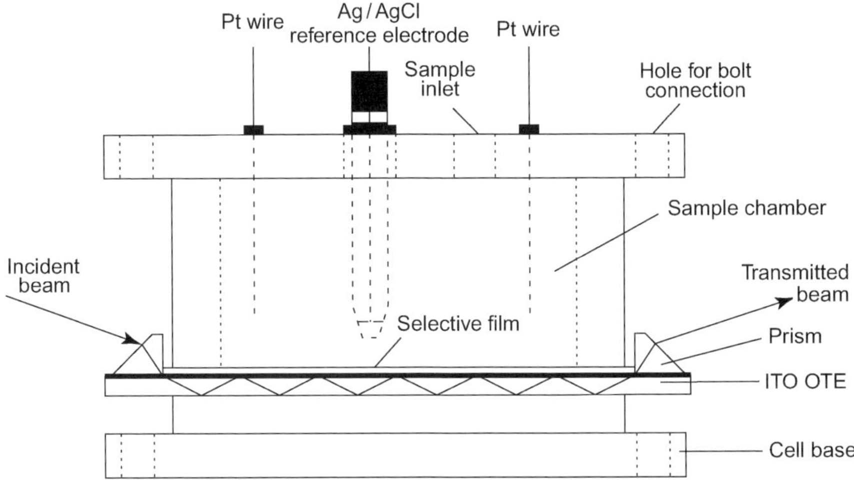

Figure 21 Simplified diagram of an internal reflection-based spectroelectrochemical sensor cell. ITO, indium-tin oxide.[60]

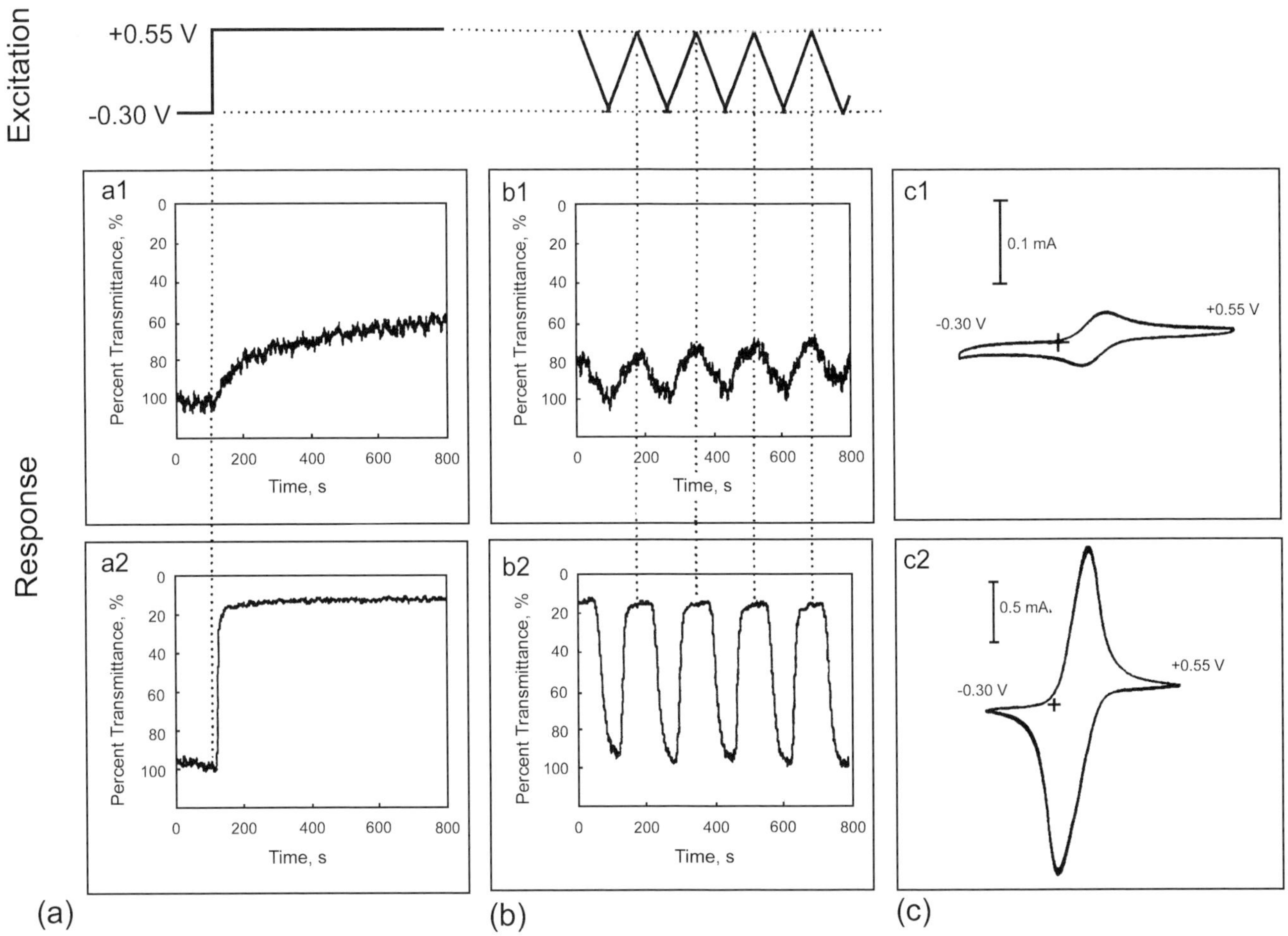

Figure 22 Optical response at $\lambda = 420\,\text{nm}$ of the sensor in Figure 20 to (a) potential step and (b) cyclic voltammetry for two concentrations of $Fe(CN)_6^{4-}$. Panel 1, 0.025 mM; panel 2, 2.5 mM. Panels (c) show cyclic voltammograms recorded in those solutions recorded at a scan rate of $10\,\text{mV}\,\text{s}^{-1}$. Potentials are referred to an Ag/AgCl reference electrode.[60]

achieved by IRS using extensive signal averaging, the region probed by the evanescent wave is very close to the electrode surface. Hence, it is not possible to monitor changes in composition at times *longer* than a few milliseconds. In addition, IRS is very sensitive to wavelength and refractive index effects on the pathlength, compromising the analysis of spectral data, particularly when N_{eff} is large.

Many of these problems can be minimized by experiments in which light propagating in the solution is reflected from, as opposed to transmitted through, the electrode. This approach allows transient studies to be performed under either quiescent or convective flow conditions, involving weakly absorbing and short-lived species. Such improvements are derived primarily, but not exclusively, from increases in the optical pathlength, achieved for example by employing grazing incidence or multiple reflection, better current distribution, and from the possibility of using microelectrodes.

Implementation of external reflection geometries often requires custom assembly of components, including a light source, either a laser or a conventional arc lamp, a monochromator, a detector, and other optical elements, as prescribed by the specific application. A few optical arrangements designed for this type of spectroelectrochemical experiments are depicted in Figure 23.[31] Considerable care has been taken to develop electrochemical cells displaying optimum optical and electrochemical response. For example, Figure 24[39,64] shows two cells reported in the literature for (a) external reflection at constant angle of incidence and (b) for multiple reflection experiments. One of the most interesting applications of external reflection, reported by McCreery et al.,[31]

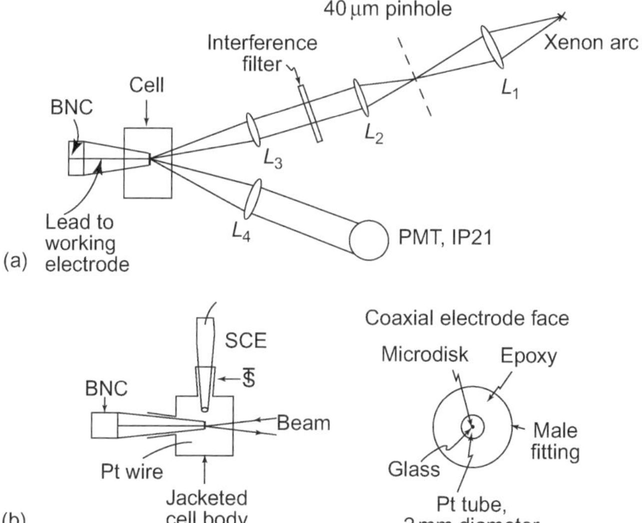

Figure 23 (a) Optical arrangement and (b) schematic diagram of a cell for spectroelectrochemistry at a microdisk electrode.[31]

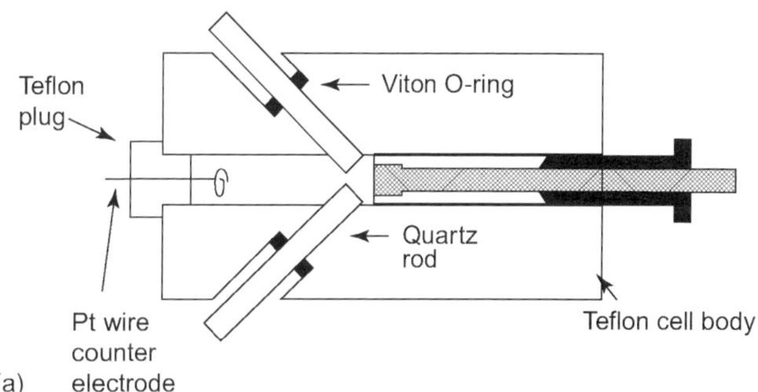

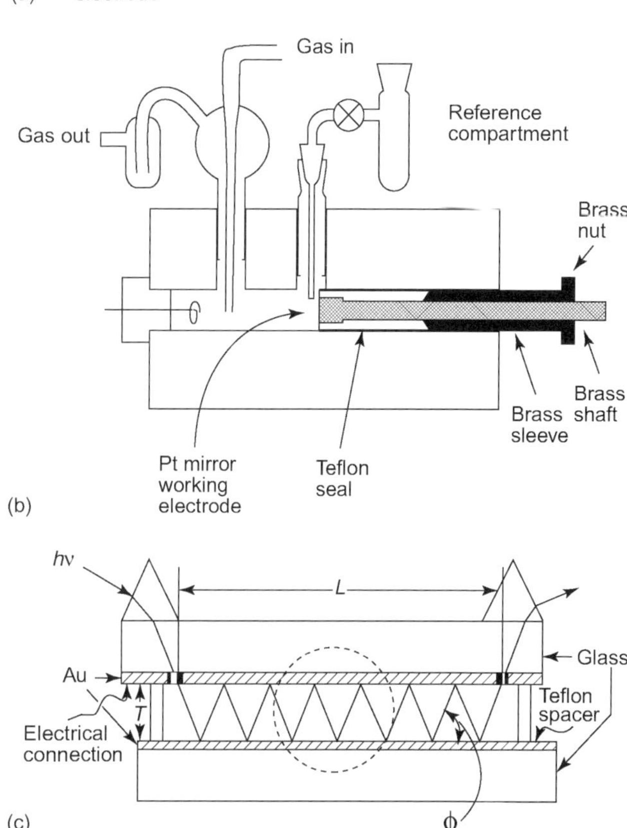

Figure 24 (a) Spectroelectrochemical cell for chronoabsorptometric studies: (a) top view, (b) side view,[64] (c) cross-section of a multiple specular reflection spectroelectrochemical cell.[39]

exploits the advantages of the small currents associated with micro-electrodes to eliminate the need for a potentiostat and use instead a capacitor-based injection circuit to charge the double layer, coupled to a pulse generator. As shown by these workers, and discussed later in this chapter, such a strategy enables a fixed electrode potential to be established following a step, within times shorter than 35 ns, making it possible to follow the time evolution of the absorbance from times as short as 0.15 μs.

Metals, even when optically polished are not perfect reflectors in the UV/VIS region; in addition, their reflectance is often a function of the state of charge and coverage of adsorbed species on the surface.[12] However,

For references see page 10222

it is possible to either minimize, or account for these effects, allowing quantitative information to be acquired regarding the identity and reactivity of electrogenerated solution phase species produced at conventional solid electrodes.

4.4.1 Planar Electrodes

The solution-phase absorbance measured with light reflected from a planar electrode at an angle of incidence ϕ (Figure 24a[64]) is $2/\cos\phi$ larger than that observed, under otherwise identical conditions, using a beam normal to an OTE in the standard transmission configuration. The factor of 2 accounts for the beam crossing the diffusion layer twice, whereas the divisor of $\cos\phi$ reflects the increase in pathlength for a single crossing of the layer. This enhancement factor, denoted as η, arises purely from geometric considerations, assuming the electrode surface to be perfectly reflecting, and that the incident and reflected beams traverse a diffusion layer uniform along an axis normal to the surface, i.e. possible edge effects are ignored. Based on these considerations, η will range from 2, for normal incidence, up to about 400 for ϕ slightly less than 89.7°. Comparisons between theoretical and experimental data, as well as quantitative calculations of edge effects, have been provided in the literature.[38,42,65]

The higher absorbances achieved at glancing incidence enable detection of short-lived electrogenerated chromophores with increased sensitivity, making it possible to monitor, for example, species with small ε, or strongly absorbing materials at lower concentrations. In addition, the higher electrical conductivity of massive, compared to thin film electrodes, allows measurements to be performed with better time resolution. Some of these aspects are discussed in more detail later in this section.

4.4.1.1 Identification of a Solution-phase Corrosion Product Polarization of metals at sufficiently high potentials may result in the formation of species derived from the electrode displaying finite solubility in the electrolyte solution. An illustrative example of the use of UV/VIS spectroscopy in the external reflectance mode for the identification of a metal corrosion product was reported by Kolb et al.[66] for the case of Ru in acid solutions. For these experiments, a Harrick rapid scan spectrophotomer was employed to collect spectra by reflecting light from an optically smooth Ru film electrode sputtered on glass as a function of the applied potential. Figure 25 shows the normalized reflectivity of Ru polarized at 1.17 V versus SCE in 0.5 M H_2SO_4 using the reflectance spectrum of pristine Ru at −0.1 V as a reference. Based on a comparison of the features observed with those of the genuine material (shown by the inset in the figure), the corrosion product could be identified as RuO_4.

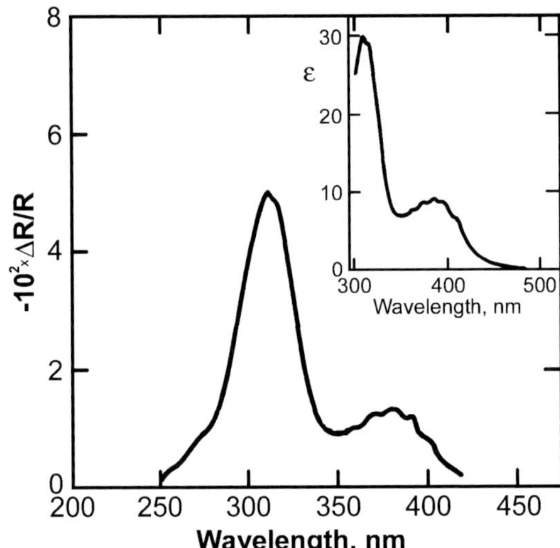

Figure 25 Background-corrected differential reflectance spectra for an Ru electrode in 0.5 M H_2SO_4. (· · ·) Result obtained following a potential step from −0.1 to +1.17 V versus SCE. The inset shows the molar extinction coefficient ε versus λ for RuO_4.[66]

4.4.1.2 Chronoabsorptometry

Glancing Incidence – Systems Involving a Single Absorbing Product. Some of the advantages associated with external reflection compared to transmission through an OTE may be illustrated using potential step chronoabsorptometry (see section 4.1) as an example. Assuming that the only optically absorbing species at the wavelength selected for the measurements (λ) is the product P of the electrochemical reaction, and that the heterogeneous electron transfer process is under strict diffusion control, $A(t)$ is given by Equation (63) corrected for the difference in optical path, i.e. $\eta = 2/\cos\phi$, namely

$$A(t) = \varepsilon_P C_R^o \frac{2}{\cos\phi}\left(\frac{4D_R t}{\pi}\right)^{1/2} \quad (66)$$

Equation (66) has been verified in a number of laboratories for simple electrochemical processes. In particular, Figure 26(a)[42] shows absorbance versus time and Figure 26(b) the corresponding absorbance versus $t^{1/2}$ plots calculated from these data for a buffered solution (pH = 7) of 0.59 mM MV^{++}. The potential in this case was stepped from −0.3 to −0.8 V versus SCE using an Au electrode with an He–Ne laser beam ($\lambda = 632.8$ nm) incident on the sample at $\phi = 88.52°$. At this energy $MV^{+\bullet}$ displays significant absorption. The straight line in the figure was calculated based on values of ε_P and D_R reported in the literature (see Table 4). It is evident from these results that the agreement between theory and experiment is excellent for times longer than 10 ms. The discrepancies found at

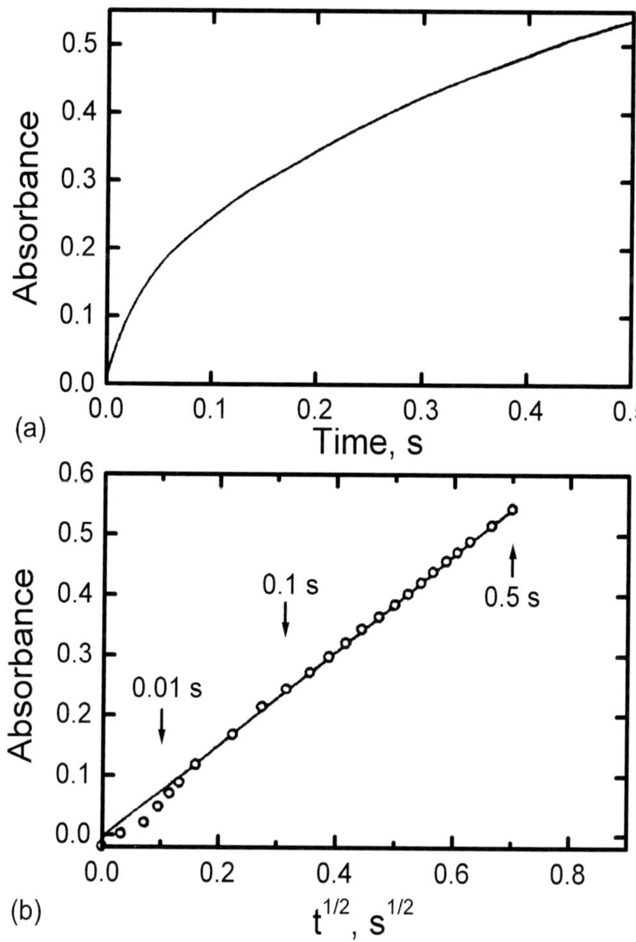

Figure 26 (a) Absorption versus time and (b) absorption versus $t^{1/2}$ plots obtained in a solution 0.59 mM methyl viologen in 0.3 M phosphate buffer (pH = 7.0) for a potential step from −0.3 to −0.8 V versus SCE. The data represent the average of three runs.[42]

shorter times were attributed to a variety of factors, a thorough discussion of which may be found in the original literature.[42] Good quantitative agreement (<2%) was also observed between the theoretical and experimental values of η for ϕ in the range 88–89°, which corresponds to η values of about 50–100, for electrolysis times of the order of 50 ms. However, deviations were detected at shorter times and smaller ϕ. This technique has been successfully extended to the entire UV/VIS range by using a continuum light source and a well-collimated beam, augmenting significantly its overall versatility.

Glancing Incidence – Systems Involving an Absorbing Reactant and an Absorbing Product. In the case of systems in which both R and P in Equation (57) absorb at the same wavelength λ, Equation (66) must be modified to Equation (67):

$$A(t) = \Delta\varepsilon(\lambda)C_R^o \frac{2}{\cos\phi}\left(\frac{4D_R t}{\pi}\right)^{1/2} \tag{67}$$

where $\Delta\varepsilon(\lambda) = \varepsilon_P(\lambda) - \varepsilon_R(\lambda)$. Assume that after a certain period of time τ, following application of the step, the electrode potential is stepped to a new value at which the reverse reaction occurs under diffusion control. Equation (68) presents this in mathematical terms (see section 2.2.1),

$$C_R(0,t) = S_\tau \text{ where } S_\tau = \begin{cases} 0, & t \le \tau \\ 1, & t > \tau \end{cases}, \theta_R(0,t) = 1 - S_\tau,$$

$$\text{and } \bar{\theta}_R(0,s) = \frac{1}{s} - \frac{e^{-\tau s}}{s} \tag{68}$$

and, hence,

$$Q(s) = \left(\frac{D_R}{s}\right)^{1/2}\bar{\theta}_R(0,s) = \frac{D_R^{1/2}}{s^{3/2}} - D_R^{1/2}\frac{e^{\tau s}}{s^{3/2}} \tag{69}$$

After inversion of Equation (69) and substitution in Equation (25) the transient absorbance is given by Equation (70):

$$A(t) = \Delta\varepsilon(\lambda_o)C_R^o \frac{2}{\cos\phi}\left(\frac{4D_R}{\pi}\right)^{1/2} \times \left[t^{1/2} - S_\tau(\tau - t)^{1/2}\right] \tag{70}$$

which reduces to Equation (54) for $t \le \tau$, and, for the second step only Equation (71) holds:

$$A(t > \tau) = A(\tau) - \Delta\varepsilon(\lambda_o)C_R^o \frac{2}{\cos\phi}\left(\frac{4D_R}{\pi}\right)^{1/2} \times \left[\tau^{1/2} - t^{1/2} + (\tau - t)^{1/2}\right] \tag{71}$$

An interesting application of this formalism was described by Jones and Hinman,[64] who examined the redox properties of iron tetraphenylporphyrin chloride (FeTPPCl) in a nonaqueous solution, using the cell shown in Figure 24(a). Figure 27 shows a plot of ΔA versus t obtained for a 10.6 µM FeTPPCl solution in 0.1 M tetra-*n*-butyl ammonium perchlorate (TBAP) in CH_2Cl_2 for an experiment in which the potential was stepped from 0.9 V to 1.3 V versus SCE for 1 min, and then stepped back to 0.9 V. A clear indication that the system behaves in an ideal way was obtained from the linear character of the plots of ΔA versus $t^{1/2}$, and ΔA versus $\tau^{1/2} - t^{1/2} + (\tau - t)^{1/2}$ (Figure 28), which may be used as a criterion to ascertain that, within the timescale of the measurements, the product of the reaction P is stable. As the spectrum of R is often known, measurements of this type at different values of λ, allow the spectrum of the other species to be determined, yielding in this case

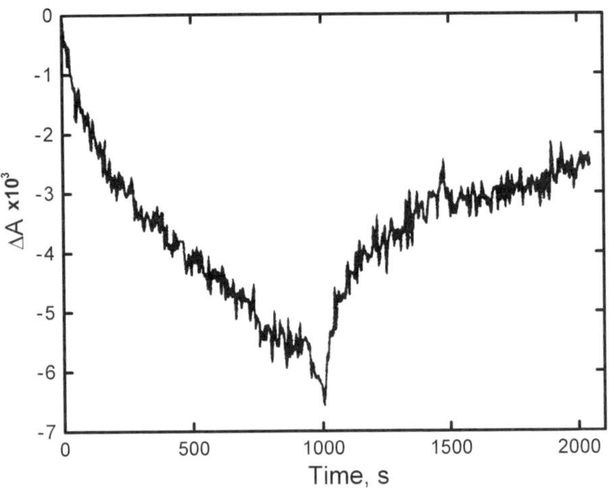

Figure 27 Plot of ΔA versus t obtained at $\lambda = 410\,\mathrm{nm}$ in a $10\,\mu\mathrm{M}$ FeTPPCl solution in CH_2Cl_2 containing 0.1 M TBAP during a double potential step chronoabsorptometric experiment. The data represent an average of 64 measurements.(64)

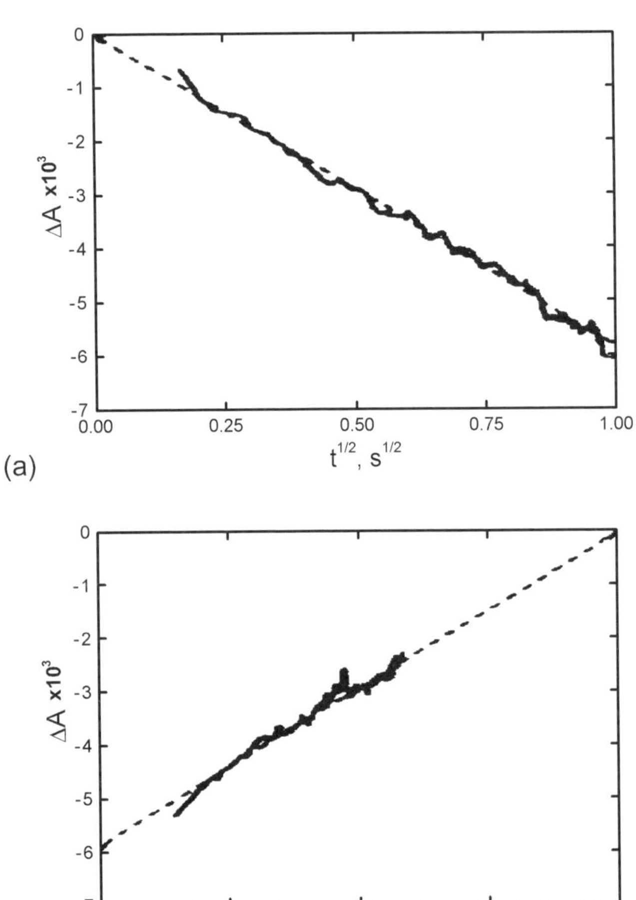

Figure 28 Plots of (a) ΔA versus $t^{1/2}$ and (b) ΔA versus $\varphi = \tau^{1/2} - t^{1/2} + (\tau - t)^{1/2}$ obtained from the data in Figure 27. The broken lines represent a linear least-squares fit.(64)

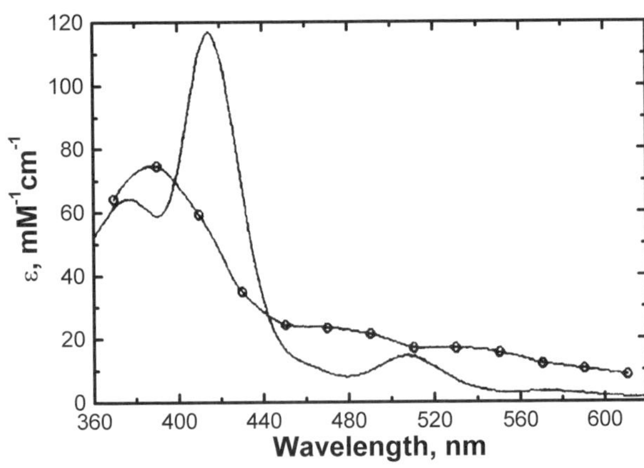

Figure 29 UV/VIS spectrum of FeTPPCl (——) and its one-electron oxidation product (○) calculated from chronoabsorptometric data.(64)

excellent agreement with that of the oxidized form of FeTPPCl obtained by other means (Figure 29).

Multiple Reflection. A different approach(39) to increasing the pathlength is to introduce a well-collimated beam into a cell consisting of two working electrodes placed parallel, and at close distance from each other, as shown schematically in Figure 24. Under these conditions and assuming P is the only absorbing species, $A(t)$ for this multiple reflection configuration is given to a good degree of approximation by

$$A(t) = \varepsilon_{\mathrm{P}} C_{\mathrm{R}}^{\mathrm{o}} \frac{2w}{T \sin\phi} \left(\frac{4D_{\mathrm{R}}t}{\pi}\right)^{1/2} \tag{72}$$

where w and T are the cell length and cell thickness, respectively. According to Equation (72) the further enhancement in absorbance derived from multiple, compared to single, reflection amounts to a factor $L/(T\tan\phi)$. For systems in which the only absorbing species at λ is the reactant R, the decrease in absorbance after a potential step will also be linear in $t^{1/2}$, namely,

$$A(t) = \frac{\varepsilon_{\mathrm{R}} C_{\mathrm{R}}^{\mathrm{o}} L}{\sin\phi} - \frac{2L}{T \sin\phi} \varepsilon_{\mathrm{R}} C_{\mathrm{R}}^{\mathrm{o}} \left(\frac{4D_{\mathrm{R}}t}{\pi}\right)^{1/2} \tag{73}$$

The leading term in Equation (73) represents the initial absorbance of the solution, a factor which is independent of T, and $L/\sin\phi$ represents the *total* pathlength.

Shown in Figure 30 are A versus $t^{1/2}$ plots for the reduction of MB on Au in 0.1 M KCl solutions buffered (pH = 7) obtained with the cell shown in Figure 24(c), using an He–Ne laser, following a potential step as a function of MB concentration. Excellent linearity was observed within the timescale of the measurements, except for the most concentrated solution, for which the electrogenerated radical undergoes precipitation.

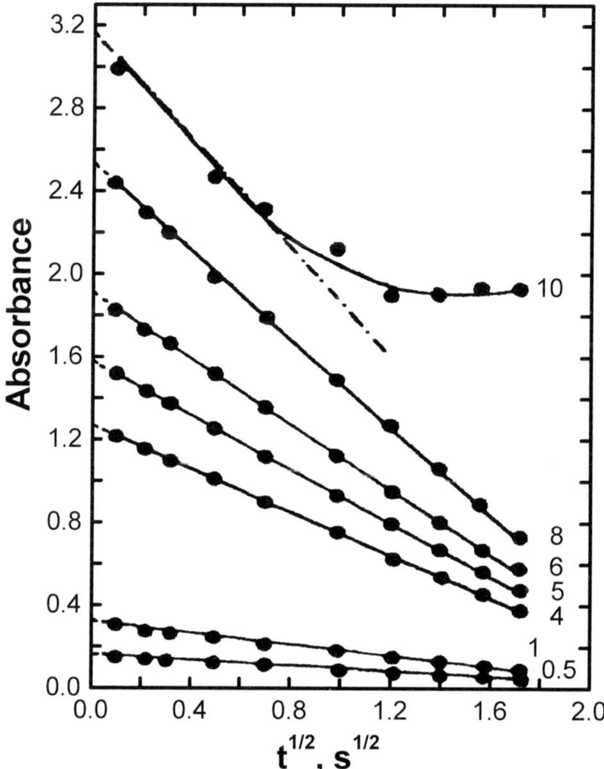

Figure 30 Plots of A versus $t^{1/2}$ for the reduction of MB at concentrations in the range 0.5 to 10 μM.[39]

Furthermore, the magnitude of A, based on ε for MB at 632.8 nm (see Table 4), the geometry of the cell and the angle of incidence, was in agreement with that found experimentally with $\phi < 45°$ for about 330 reflections.

4.4.1.3 Determination of Rate Constants of Homogeneous Electron Transfer Reactions Involving Electrogenerated Species Consider a reaction involving exclusively solution-phase species, A, $A^{+\bullet}$, B and B′, in which A is reduced at the electrode surface to yield $A^{+\bullet}$, which then reacts with B to yield B′ regenerating A (Equation 74):

$$2A^{+\bullet} + B \longrightarrow 2A + B' \tag{74}$$

Two simple reaction mechanisms consistent with this stoichiometry may be envisaged as Equation (75) with (76), and Equation (75) with (77):

$$A^{+\bullet} + B \xrightarrow{k_2} B^{+\bullet} + A \tag{75}$$

$$A^{+\bullet} + B^{+\bullet} \xrightarrow{\text{fast}} B' + A \tag{76}$$

$$2B^{+\bullet} \xrightarrow{\text{fast}} B + B' \tag{77}$$

Assume further, that the rate-determining step is first order both in $A^{+\bullet}$ and B, such as a collision between these species, and that the equilibrium constant for the overall reaction is large so that all reactions following the slow step will proceed to completion. Hence, if Equation (75) is slower than (76) and (77), then Equation (78) holds:

$$\frac{d[A^{+\bullet}]}{dt} = -2k_2[A^{+\bullet}][B] \tag{78}$$

If an electrode immersed in a homogeneous solution containing A and B at known concentrations is stepped to a potential sufficiently positive to form the corresponding radical cation $A^{+\bullet}$ under strict diffusion control diffusing into the solution, then $A^{+\bullet}$ will encounter and react with B diffusing toward the electrode in an electron transfer-type process to yield $B^{+\bullet}$. As the amount of B at, and near the surface is expected to be small, Equation (75) will proceed a distance away from the electrode. Analytic solutions of this rather complex problem do not exist in general; however, it is possible to generate concentration profiles of each of the species involved by digital simulation techniques, provided that all the relevant parameters are specified and, from these, to predict the absorbance associated with the chromophores.

The power of UV/VIS spectroelectrochemistry as a tool to determine rate constants of reactions between an electrogenerated radical with a species in solution, i.e. k_2 in Equations (76) and (77), was demonstrated by McCreery et al.,[67] who measured the rates of $CPZ^{+\bullet}$ with a catecholamine, using, among other techniques, grazing incidence external reflection at a Pt electrode.

Figure 31 shows the results of numerical simulations for the reaction sequence shown above, as $C_{CPZ^{+\bullet}}$ observed in the presence and in the absence of catecholamine, denoted as DA, as a function of distance, expressed in terms of units of $(Dt)^{1/2}$, for four different values of kt assuming equal bulk concentrations for both CPZ and DA. For simplicity, the diffusion coefficients of all species involved were assumed to be the same. Figure 31(a) corresponds to a situation in which $k_2 = 0$, i.e. no subsequent reaction involving $CPZ^{+\bullet}$. For k_2 finite and $t \neq 0$, the concentrations of both $C_{CPZ^{+\bullet}}$ and C_{DA} at a fixed distance from the electrode decrease, and therefore the extent of overlap between the two profiles of the two species, also diminishes.

Based on arguments set forward in previous sections, the absorbance of a chromophore is proportional to the integral of the concentration profile, enabling comparisons to be made with experimental measurements. Figure 32 shows normalized absorbances, i.e. those obtained in the presence and in the absence of equimolar amounts of DA, monitored at wavelengths at which only DOQ, the oxidation product of DA ($\lambda = 400$ nm, upper curve), and $CPZ^{+\bullet}$ ($\lambda = 525$ nm, all

For references see page 10222

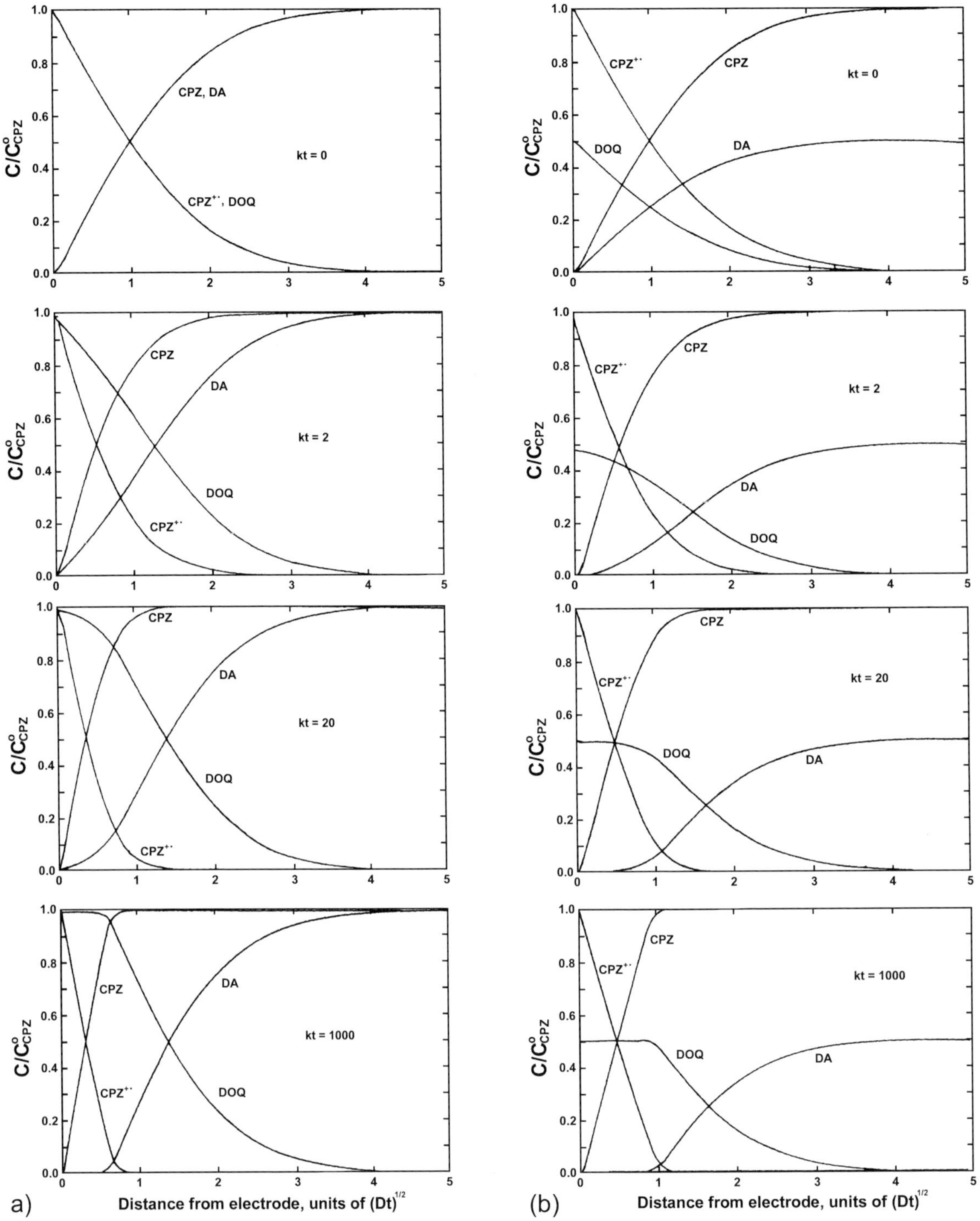

Figure 31 Simulated concentration versus distance profiles for the EE′C reaction sequence at various values of β and kt: (a) $C^{o}_{CPZ} = C^{o}_{DA}$ ($\beta = 1$); (b) $C^{o}_{CPZ} = 2C^{o}_{DA}$ ($\beta = 2.0$). In all cases the ordinate represents fractional concentration relative to the C^{o}_{CPZ} value.[67] DOQ, dopamine quinone.

other curves) absorb, respectively. Also included in these plots as solid lines are best-fit numerical simulations for $\beta = C_{CPZ}/C_{DA}$, of 0.5, 1, 2, and 5. 5, yielding k_2 values of $2.12 \pm 0.25\,M^{-1}\,s^{-1}$.

4.4.2 Microelectrodes

Planar electrodes of dimensions larger than about 0.1 mm suffer from relatively slow diffusional relaxation; hence, long times are required to recover the initial conditions

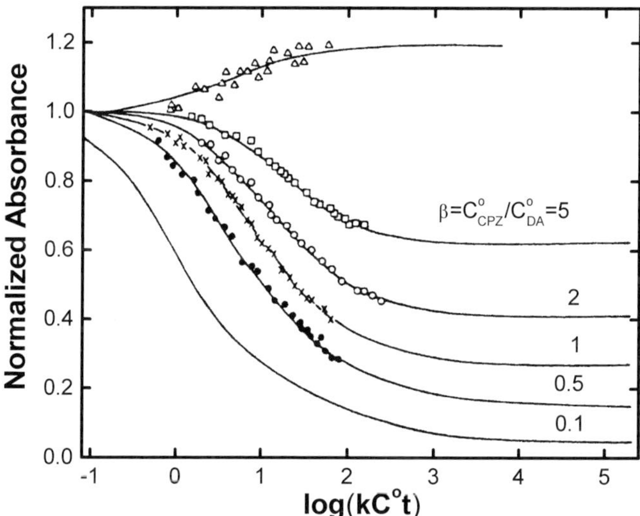

Figure 32 Normalized absorbance versus $\log(kC^{b}t)$ for various values of β. The solid curves are theoretical plot for DOQ monitored at $\lambda = 400\,\text{nm}$ (top curve) or $CPZ^{+\bullet}$ monitored at $\lambda = 525\,\text{nm}$ (remaining curves). Points are best fits to experimental data.[67]

and, thus, acquire sufficient measurements over reasonable times for efficient signal averaging. Considerable enhancements in mass transport in quiescent solutions can be realized by using electrodes of micrometer dimensions, including disks and cylinders. Despite their size, the absorbance observed for such microelectrodes in the external reflection mode is larger than that using an OTE for similar electrolysis times. Because of their small area, both capacitive and faradaic currents are small, placing lower demands on the potentiostat power, increased time averaging, higher duty cycle, low cell-time constants, more precise solution resistance and, most importantly, better conditions for studies of fast reactions.

4.4.2.1 Microcylinders An analytic expression for the time-dependent concentration profile of a species P in a stagnant solution, produced at the circumference of a circular electrode of radius r_0, via oxidation or reduction of R under chronocoulometric conditions, can be obtained by solving the appropriate Fick's law using the Laplace transform method. The solution to this problem may be shown to be given by Equation (79):[37]

where $J_0(\zeta)$, and $Y_0(\zeta)$ are Bessel functions of order zero of the first and second kind of argument ζ, respectively.[23] Figure 33 shows a plot of $A_{\text{cyl}}/A_{\text{pl}}$ where A_{cyl} and A_{pl} represent the absorbances observed with a cylinder by integration of Equation (79) along r, and with a planar electrode (see Equation 64), respectively, as a function of the dimensionless variable $(D_{\text{R}}t)^{1/2}/r_0$. As would be expected, $A_{\text{cyl}}/A_{\text{pl}}$ is close to unity for $\delta_{\text{N}} < r_0$ (see Figure 33), but deviates significantly at longer times for fixed r_0.

Figure 34 shows average absorbance versus $t^{1/2}$ plots in the range 0.1–100 ms following a potential step from 0.3 to 1.1 V versus SCE in solutions of 3.5 mM OD in 1 M H_2SO_4 using a Pt microcylinder with $r_0 = 12.5\,\mu\text{m}$. The open circles in this figure were obtained based on the theory introduced above without adjustable parameters. Similar data over a much shorter timescale, down to 4 µs, could be recorded using a Pt microcylinder with $r_0 = 5\,\mu\text{m}$ in 2.2 mM OD in 1 M H_2SO_4 after 2700 averages, under otherwise identical conditions. In this latter case, deviations of only up to 5% could be observed at 1 ms. Both of these experiments involved the use of a laser focused onto the cylinder wall as the radiation source, and optical fibers to collect the reflected light in a two-electrode configuration (Figure 35). The primary factor that limits a quantitative analysis of data acquired at longer times relates to thermal effects induced by heat dissipation within the electrode. Such phenomena restrict the range of validity of the model to times at which δ_{N} is less than twice the electrode radius, i.e.

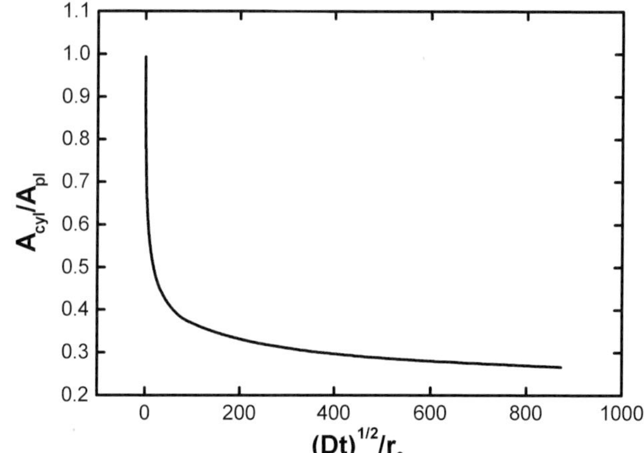

Figure 33 Ratio of absorbance for a cylindrical (cyl) and a planar (pl) electrode as a function of the dimensionless variable $(D_{\text{R}}t)^{1/2}/r_0$.[37]

$$C_{\text{P}}(r,t) = C_{\text{R}}^{\text{o}}\left[1 + \frac{2}{\pi}\int_0^{\infty} \frac{J_0\left(\dfrac{r}{r_0}\xi\right)Y_0(\xi) - Y_0\left(\dfrac{r}{r_0}\xi\right)J_0(\xi)}{J_0^2(\xi) + Y_0^2(\xi)}\left(\frac{\exp[(-D_{\text{R}}t/r_0^2)\xi^2]}{\xi}\right)\mathrm{d}\xi\right] \qquad (79)$$

For references see page 10222

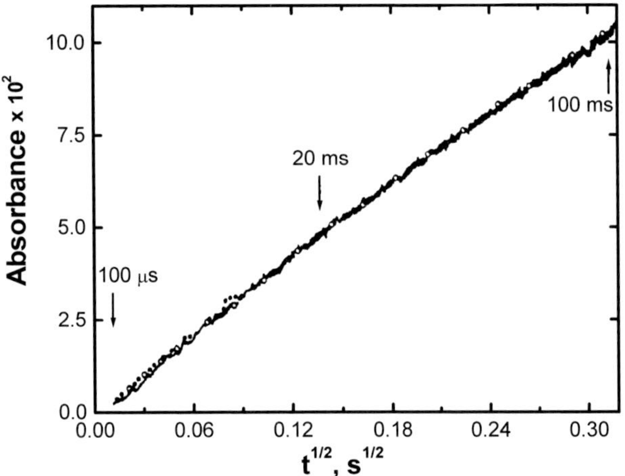

Figure 34 Absorbance versus $t^{1/2}$ for a 12.5 µm radius platinum electrode in 1 M H_2SO_4 containing 3.5 mM OD (*o*-dianisidine): open circles = theoretically predicted response; solid line = experimental response for a conventional three-electrode arrangement, average of 100 runs; solid circles = response for a two-electrode configuration, average of 60 runs. E_{app} was stepped from 0.3 to 1.1 V versus SCE.[37]

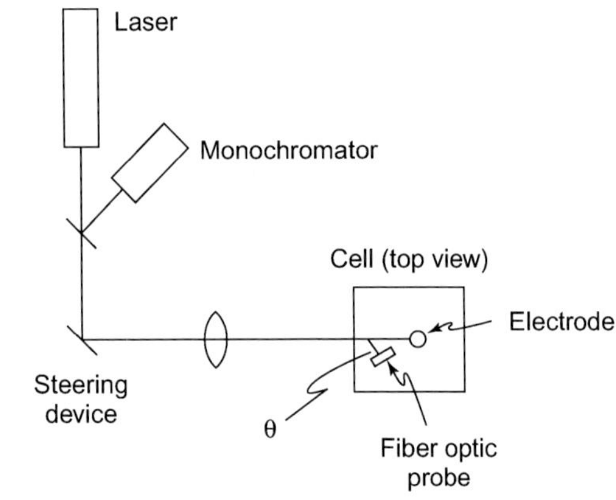

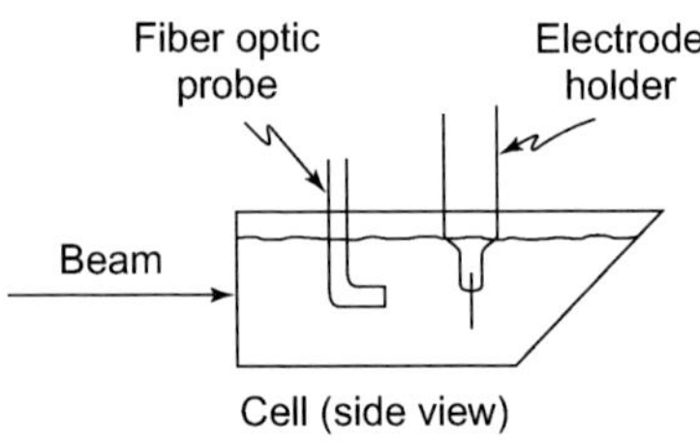

Figure 35 Experimental arrangement for UV/VIS spectroelectrochemical measurements involving a microcylinder electrode.[37]

$(Dt)^{1/2}/r_0 < 2$. As discussed by these authors, both the IR drop, and the time constant of the cell increase as r_0 decreases; hence, smaller-diameter electrodes will not allow access to measurements for times shorter than a few microseconds. Nevertheless, it is possible to achieve submicrosecond resolution by employing a microdisk electrode, as described in the next subsection.

4.4.2.2 Microdisk Electrodes Unlike the case of cylindrical electrodes, the solution resistance R_{sol} for a microdisk is directly, as opposed to inversely, proportional to r_0. As for a typical microdisk electrode, the contribution of the capacity to τ_{cell} is very small compared to that ascribed to R_{sol}, decreasing r_0 should lead to a better transient response. Indeed, chronoabsorptometric experiments involving the use of a coaxial Au microdisk electrode with $r_0 = 30$ µm in a two-electrode cell configuration, allowed for absorbances associated with electrogenerated $MV^{+\bullet}$ to be monitored in the millimeter range with an Xe arc lamp down to 150 ns following extensive signal averaging (Figure 36).[31] This improved methodology enables measurements of second-order homogeneous rate constants of reactions involving electrochemically generated strong chromophores (k_2 in Equation 75), of the order of $10^8\,M^{-1}\,s^{-1}$, as illustrated by the reaction of CPZ with DA at physiological pH. It should be emphasized that these values exceed those achievable with more conventional stopped flow experiments.

Shown in Figure 37 are absorbance versus time plots for generation of $CPZ^{+\bullet}$ both in the absence and in the presence of DA. The solid lines in the lower curve represent best digital simulation fits to the average of 1120 runs, for which k_2 for the reduction of the radical

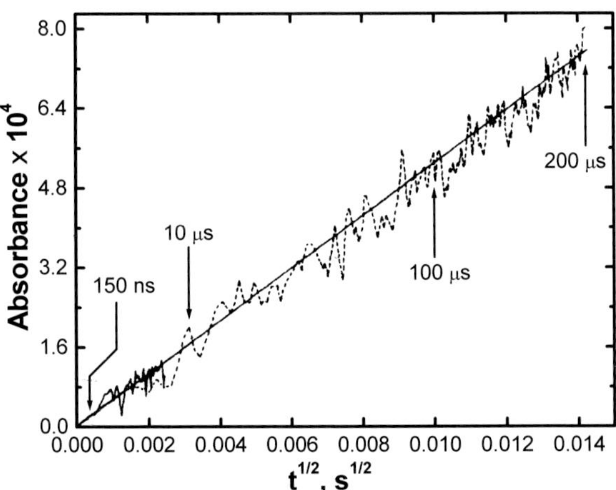

Figure 36 Absorbance versus time curves for the generation of $MV^{+\bullet}$ from 5.6 mM $MVCl_2$ in 2 M phosphate buffer, (pH = 7.0) recorded using a coaxial Au microdisk working electrode (30 µm radius) with an He–Ne laser at $\lambda = 632.8$ nm. The dashed line is an average of 60 000 runs covering 1–200 µs. The solid noisy line is a shorter transient at higher time resolution. The straight line is a least squares fit to all of the data, and has the same slope as 5 ms runs.[31]

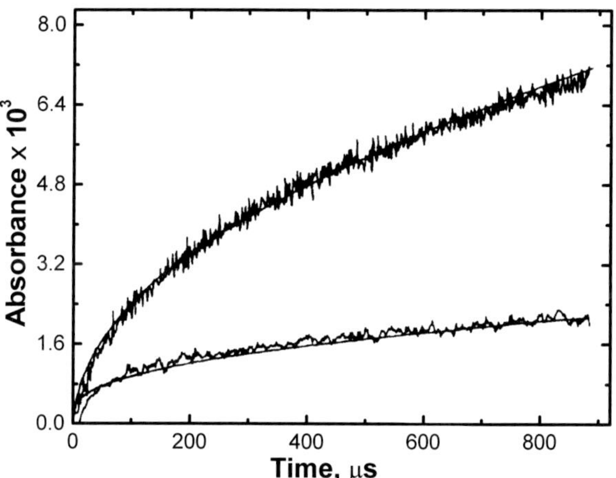

Figure 37 Absorbance versus time curves for the generation of $CPZ^{+\bullet}$ in the presence (lower curve) and absence (upper curve) of equimolar DA in 40% $MeOH/H_2O$, containing 0.2 M dimethylarsenate buffer (pH = 6.8) plus KCl for a total ionic strength of 1.8 M. The noisy lines are averages of 1120 runs. The upper smooth curve was calculated for $k = 0$, the lower curve for $k = 6.2 \times 10^7\,M^{-1}\,s^{-1}$ for the reduction of the radical by dopamine.(31)

by DA is $6.2 \times 10^7\,M^{-1}\,s^{-1}$ (lower curve), whereas that in the upper curves assumes $k_2 = 0$.

4.5 Spatially Resolved Spectroelectrochemistry

Transmission and external reflection spectroelectrochemical methods afford information about integrated profiles of the species being monitored. Additional insight into mechanistic aspects of homogeneous-phase redox reactions and transport phenomena can be obtained from the concentration profiles themselves. Not surprisingly, considerable effort has been devoted to monitor their temporal and spatial evolution using both interferometry(14) and transmission spectroscopy. This section summarizes the most salient features of the latter of these two methodologies and discusses with various illustrations its scope of applicability.

The problems associated with the lack of selectivity and rather weak sensitivity of solution-phase imaging techniques based on interferometry so far reported can be circumvented to a significant extent by using absorption, as opposed to refraction, as the physical phenomenon being monitored. Although restricted to systems involving one or more chromophores, absorption-imaging methods have provided a wealth of high spatial resolution information regarding the time evolution of diffusion profiles induced by heterogeneous electron transfer reactions at electrode–solution interfaces approaching the limits imposed by optical diffraction. Because of the rather low concentrations of reactants often involved in spectroelectrochemical experiments, distortions in

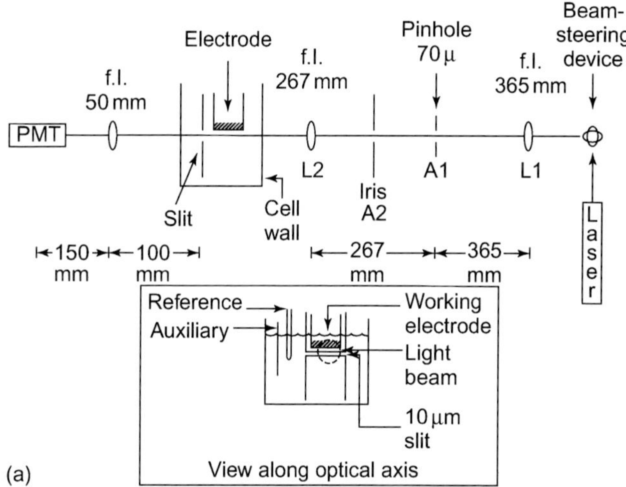

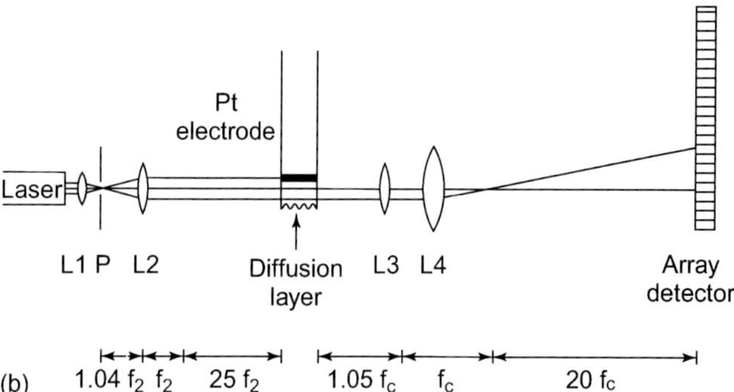

Figure 38 Diagrams of experimental arrangements for profile imaging using (a) a scanning slit(68) and (b) a multidiode array.(40)

the spatially resolved profiles measured by absorption, induced by changes in the index of refraction, can be neglected.

Two types of absorption-based profile imaging methods(40,68) have been described in the literature, involving in both cases weakly focused light propagating parallel to the electrode surface. The first, depicted in Figure 38(a),(68) employs a moving mechanical slit aligned along an axis normal to the electrode, and a photomultiplier to measure the extent of light attenuation through small volumes of solution. The second approach(40,69) (Figure 38b), relies on a photodiode array detector to capture the magnified image of the beam produced either by a laser or a lamp, after crossing the diffusion boundary layer, as a function of time. Analysis of the projection of an expanded image onto such an array enables, within diffraction-limited resolution, construction of time-resolved and spatially resolved maps of light intensities, which can then be related to a physical variable of the system being examined, such as concentration, as a function of distance normal to the probing beam.

For references see page 10222

4.5.1 Imaging of Diffusion Layers

4.5.1.1 Potential Step

Planar Electrodes. Figure 39 shows a series of $C_{TAA^{+\bullet}}/C^{o}_{TAA}$ versus y (the distance from the electrode surface) plots obtained for single and double potential step experiments in 2.56 and 4.06 mM TAA in acetonitrile solutions, respectively.(40) As shown in Table 4, the species $TAA^{+\bullet}$ exhibits a large ε at $\lambda = 632.8$ nm. Each of the solid curves in (a) and (b) were acquired with an He–Ne laser at different times after application of a potential step from 0.0 to +0.8 V versus SCE, whereas those in (c) and (d) were recorded at various times after returning the potential to 0 V, following polarization at +0.8 for 2 s (see caption). Excellent agreement was found between the results of these experiments using values of ε and D obtained independently, based on solutions of Fick's law (see Equation 23). Deviations observed at times longer than 8 s in (a) and (b), are attributed to the onset of natural convection.

Modifications to the optical set-up(70) in Figure 40, to include an Xe arc lamp instead of a laser, makes it possible to gain access to a much wider spectral region and thereby monitor profiles of a more general class of chromophores. Using such an approach, direct verification of theoretical predictions for times up to 10 s could be obtained for the potential step oxidation of ferrocyanide ion $[Fe(CN)_6]^{4-}$ to the corresponding ferric species.

Overall, analysis of these type of data allows for D for both R and P to be determined without specific knowledge of electrode areas, concentration, molar absorptivity, absorbance, number of electrons transferred or current, as would be required by chronoamperometry and chronoabsorptometry.

Microelectrodes. Imaging techniques have also been implemented for the analysis of cylindrical diffusion fields around thin wires.(71) In this case, a parallel beam of light of dimensions larger than the diameter of the wire is directed normal to the cylinder axis, as

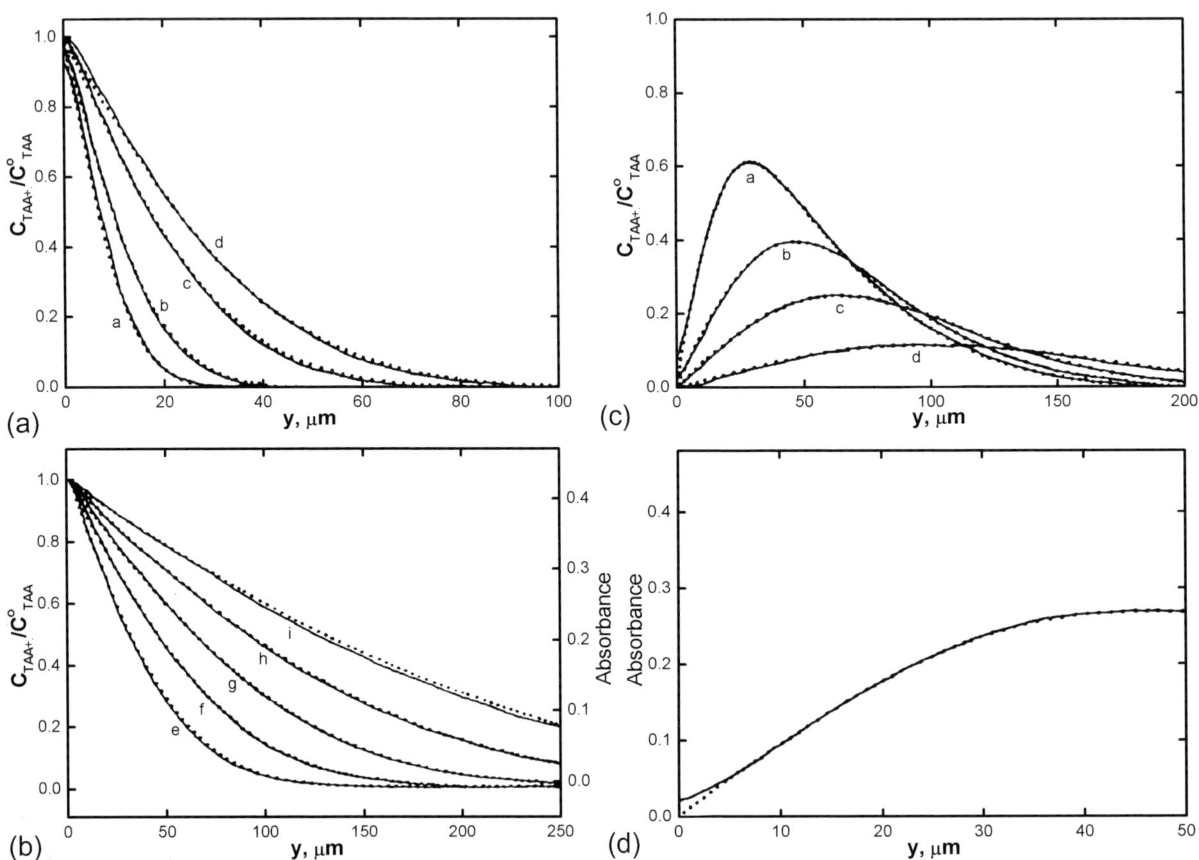

Figure 39 (a and b) Plots of $C_{TAA^{+\bullet}}$ relative to C^{o}_{TAA} as a function of y after various times following application of a single potential step from 0.0 to 0.8 V versus SCE: (a) 0.05 s, (b) 0.1 s, (c) 0.3 s, (d) 0.5 s, (e) 1.0 s, (f) 2.0 s, (g) 4.0 s, (h) 8.0 s, (i) 16 s. Solid lines are experimental profiles: points were calculated from Fick's law and Beer's law with pathlength = 0.015 cm and $\varepsilon = 11\,000\ M^{-1}\ cm^{-1}$; $C^{o}_{TAA} = 2.56$ mM. (c) As for (a) following application of a subsequent step after 2 s at 0.8 V, back to 0.0 V versus SCE: (a) 0.1 s, (b) 0.4 s, (c) 1.0 s, (d) 3 s. (d) Expanded profile of curve b in (c) above; $C^{o}_{TAA} = 4.06$ mM.(40)

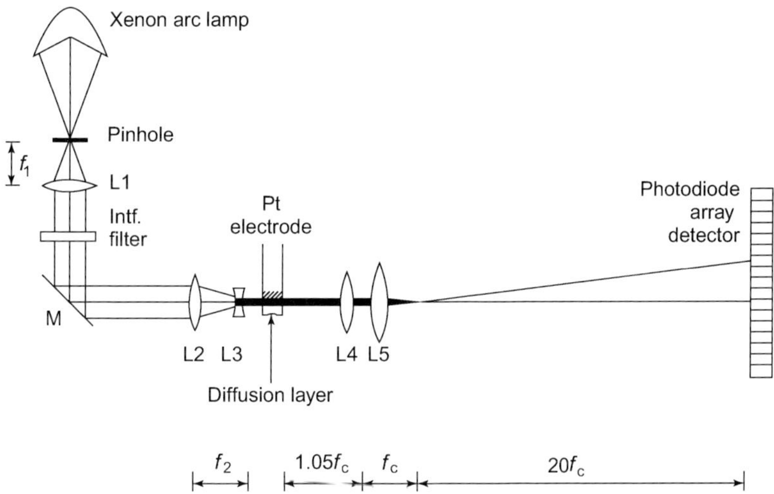

Figure 40 Optical arrangement for profile imaging using an arc lamp and a photodiode array.[70]

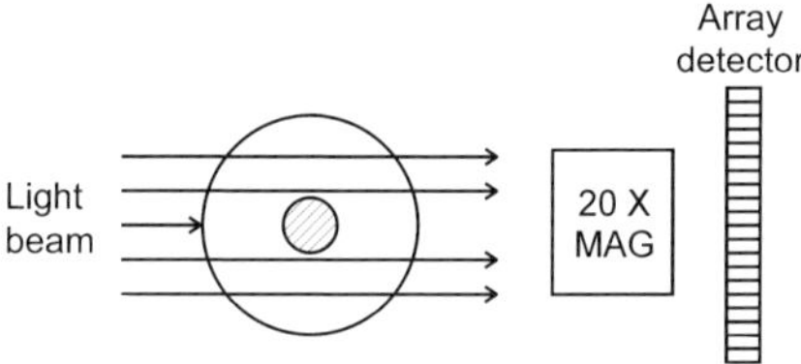

Figure 41 Orientation of microwire (cross-hatched) diffusion field, and detector. Magnification was accomplished with a two-element simple magnifier.[71]

shown in Figure 41, and then magnified before striking the array detector. Concentration profiles based on information obtained from lateral absorbances can be obtained by a mathematical technique known as Abel inversion, a procedure readily amenable to computer implementation. Experimental results in agreement with theoretical predictions have been obtained for microwires of radii 6–25 µm and electrolysis times ranging from 50 ms to several seconds. Good agreement between theory and experiments were obtained for $C(r)$ for various times following the potential step, as shown in Figure 42.

4.5.1.2 Cyclic Voltammetry The same experimental approach enables images of concentration profiles to be obtained for planar electrodes during cyclic voltammetry.[40] For example, Figure 43(a) shows profiles recorded during the forward and Figure 43(b) reverse sweeps for the $TAA/TAA^{+\bullet}$ redox system in acetonitrile. Also depicted in the insert to Figure 43(a) is a plot of absorbance versus potential obtained with a photomultiplier tube (PMT), instead of a diode array, through a slit (see Figure 38) illuminating the region $0.0 \pm 0.2\,\mu m$, i.e. very close to the electrode surface, under the same conditions. For scan rates up to ca. $80\,mV\,s^{-1}$, these

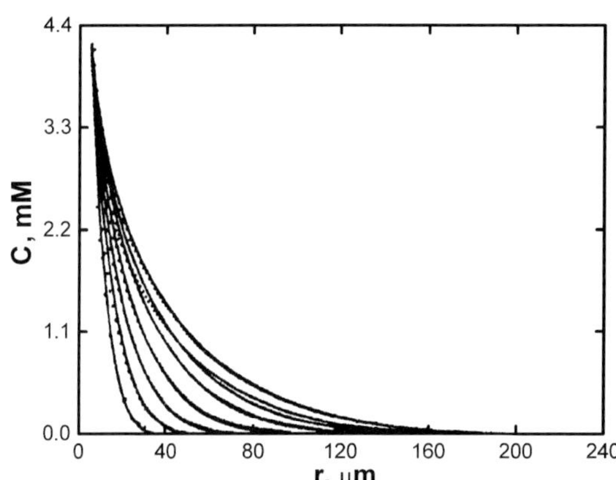

Figure 42 Radial concentration profiles at a 6 µm radius microwire, determined by Abel inversion. Dotted curves are experimental; solid lines are simulated $C(r)$. From the bottom, electrolysis times are 0.2, 0.5, 1, 2, 4, 6, and 10 s; $C^{o}_{CPZ} = 4.2\,mM$ in 1 M $HClO_4$, 40% MeOH.[71]

curves displayed virtually no hysteresis. At the half-wave potential, i.e. $E_{1/2} = 0.54\,V$ versus SCE, the surface concentrations of TAA and $TAA^{+\bullet}$ are equal. Under these conditions, the surface absorbance should then be half that of the maximum absorbance for full TAA oxidation, which can be measured at the most positive potentials (see Figure 43). Furthermore, the overall magnitude and shapes of these curves are consistent with the predicted Nernstian behavior for fast heterogeneous electron transfer.

Although the concentration profiles of $TAA^{+\bullet}$ recorded during the forward and reverse scans show common values for the surface concentration, large differences are observed away from the electrode (see Figure 43). This is not surprising as $TAA^{+\bullet}$ accumulates

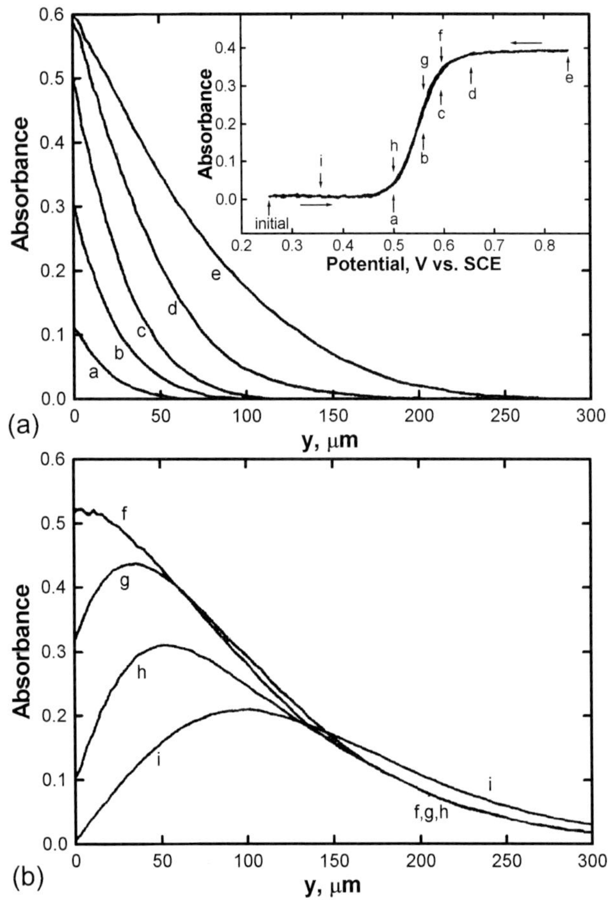

Figure 43 (a) Absorbance versus y, for the forward scan. (b) Same as (a) for the reverse scan during a cyclic voltammogram recorded at a scan rate of $80\,\text{mV}\,\text{s}^{-1}$ in solution, $C^{\circ}_{\text{TAA}} = 3.6\,\text{mM}$ (see text for details). The inset shows the potentials at which the diode array was triggered to obtain the absorbance profiles.[40]

during the forward scan, diffusing away from the electrode.

4.5.2 *Imaging of Reaction Layers*

The ability of measuring time-, and space-resolved profiles can be advantageous for the evaluation of rate constants of homogeneous electron transfer reactions involving electrogenerated species and the diagnosis of reaction mechanisms.

4.5.2.1 Determination of Rate Constants Scheme 9 represents a process in which a species S, produced by the oxidation of R, reacts with another electroactive species P to regenerate R and the oxidized form of P, denoted as Q. The equilibrium constant of the overall reaction is given by Equation (80):

$$\frac{[\text{R}]^{n_2}[\text{Q}]^{n_1}}{[\text{S}]^{n_2}[\text{P}]^{n_1}} = K_{\text{eq}} \qquad (80)$$

$$\text{R} \longrightarrow \text{S} + n_1\text{e}^-$$
$$\text{P} \longrightarrow \text{Q} + n_2\text{e}^-$$
$$n_2\text{S} + n_1\text{P} \Longleftrightarrow n_2\text{R} + n_1\text{Q}$$

Scheme 9

The magnitude of K_{eq} can be calculated directly from the values of the formal reduction potentials of the two redox couples involved. As demonstrated by theoretical calculations, the concentration profiles of these species are very sensitive to the reaction parameters, including stoichiometry (i.e. n_2/n_1), $C^{\circ}_{\text{P}}/C^{\circ}_{\text{R}}$, K_{eq} and the rate constant, often expressed in terms of a dimensionless quantity $\gamma = k_1 C^{\circ}_{\text{R}} t$.

An example of this type of mechanism[41,72] is provided by the oxidation of methoxypromazine (MPZ) by the electrochemically generated $\text{CPZ}^{+\bullet}$ (=S) at potentials at which $\text{MPZ}^{+\bullet}$ (=Q) is also produced as shown in Scheme 10. Equation (81) gives the equilibrium constant for the overall reaction:

$$K_{\text{eq}} = \frac{k_1}{k_{-1}} = \exp\left[\left(\frac{nF}{RT}\right)(E_1 - E_2)\right] = 159 \qquad (81)$$

$$\text{CPZ} \rightleftharpoons \text{CPZ}^{+\bullet} + \text{e}^- \qquad E_1 = +0.68\ \text{V versus SCE}$$
$$\text{MPZ} \rightleftharpoons \text{MPZ}^{+\bullet} + \text{e}^- \qquad E_2 = +0.55\ \text{V versus SCE}$$
$$\text{CPZ}^{+\bullet} + \text{MPZ} \underset{k_{-1}}{\overset{k_1}{\rightleftharpoons}} \text{CPZ} + \text{MPZ}^{+\bullet}$$

Scheme 10

For these studies, the He–Ne laser was replaced by an Xe arc lamp, thereby allowing the beam to be tuned to the specific absorption bands of the chromophores, i.e. $\text{CPZ}^{+\bullet}$ and $\text{MPZ}^{+\bullet}$. The contributions due to $\text{MPZ}^{+\bullet}$ to the absorbance at $\lambda = 520\,\text{nm}$, the absorption maxima of $\text{CPZ}^{+\bullet}$, can be determined from the results obtained at 600 nm, a wavelength at which $\text{CPZ}^{+\bullet}$, CPZ, and MPZ do not absorb. A comparison between experimental results (solid lines) and best fit simulations (dotted lines) in the form of $C(y,t)/C^{\circ}_{\text{CZP}}$ versus y, where y is the distance normal from the electrode surface, for $\lambda = 600\,\text{nm}$ and 520 nm, are given in Figure 44(a) and (b), respectively, for $t = 1$, 3 and 6 s, assuming $K_{\text{eq}} = 159$, $n_1 = n_2$, $\gamma = k_1 C^{\circ}_{\text{R}} t = \infty$. The latter condition simply implies that the reaction rate is very large, i.e. no kinetic information can be inferred from this analysis.

Equally large rate constants were found for the cross-reaction between triflupromazine hydrochloride (TPZ) and hydroquinone (H_2Q) for which $n_2/n_1 = 2$. As for K_{eq}, absolute values of n_1 and n_2, can be obtained using conventional electrochemical techniques, or spatially resolved spectroelectrochemistry for the individual redox couples.

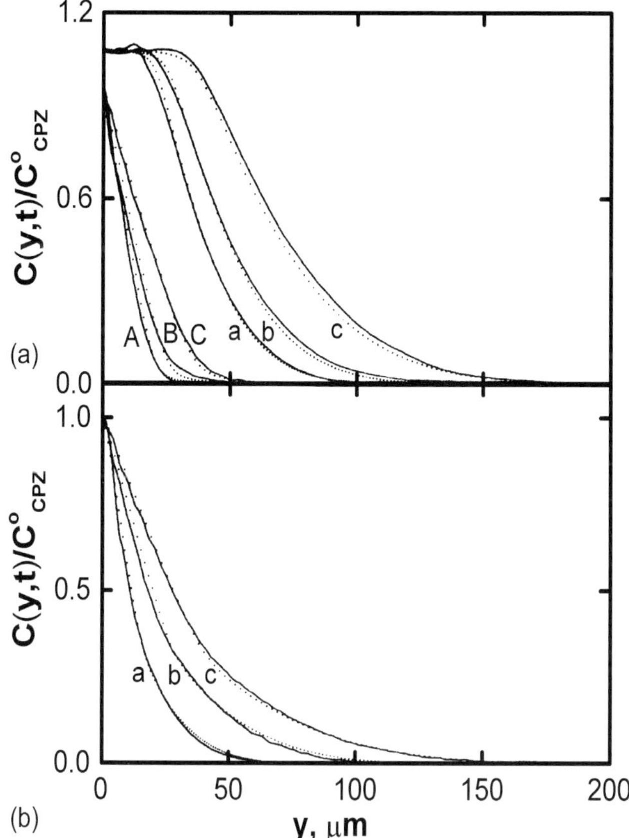

Figure 44 (a) Experimental (——) and simulated (· · ·) concentration profiles for the $CPZ^{+\bullet}$/MPZ cross-reaction: curves A–C are $CPZ^{+\bullet}$, curves a–c are $MPZ^{\bullet+}$. Simulations are for k_1, $\gamma = \infty$, $K_{eq} = 159$. Curves: (A, a) $t = 2$ s; (B, b) $t = 3$ s; (C, c) $t = 6$ s. Experimental profiles determined at $\lambda = 600$ nm for $MPZ^{+\bullet}$, and at $\lambda = 520$ nm for $CPZ^{+\bullet}$, corrected for absorption due to $MPZ^{\bullet+}$. (b) Experimental (——) and simulated (· · ·) absorbance profiles determined at $\lambda = 520$ nm for the CPZ/$CPZ^{+\bullet}$ cross-reaction. γ, $k_1 = \infty$, $K_{eq} = 159$. Curves: (a) 1 s; (b) 3 s; (c) 6 s.[41]

Careful inspection of simulated profiles for the reaction mechanism in Scheme 10 reveals for large values of K_{eq} and infinitely large values of γ, a rather linear region extending from the electrode, at which the dimensionless concentration of S under the operating diffusion control conditions is unity, into the solution. Extrapolation of this profile onto the abscissa may be used to define a dimensionless reaction layer thickness, denoted δ_r, a parameter that depends on n_2/n_1, and the relative bulk concentrations of P and R.

4.5.2.2 Mechanistic Diagnosis Methods based on the measurement of integrated profiles by absorption techniques yielded data consistent with at least two proposed mechanisms for the oxidation of DA by $CPZ^{+\bullet}$, already given in Equations (75–77). In other words, it was not possible to discern whether $DA^{+\bullet}$ reacts with $CPZ^{+\bullet}$ or undergoes dismutation (see Equations 76 and 77) to yield the specified products. As evidenced by numerical simulations,[36] however, the spatial and temporal profiles of $CPZ^{+\bullet}$ do indeed show differences for these two mechanisms and, therefore, direct observation could help identify the preferred pathway.

Comparisons between experimental and theoretical data[36] provided unambiguous proof that Equation (75) with (77) is not operative, thus favoring stepwise electron transfer (Equations 75 and 76) as being the most likely mechanism. Moreover, analysis of the data obtained with a wire electrode revealed that under very acidic conditions the reaction between the neutral species CPZ and DOQ cannot be neglected, i.e. the rate constant is indeed significant. Overall, there appears to be no other method available to date capable of providing this level of detail regarding such a complex reaction.

4.6 Ultraviolet/Visible Spectroelectrochemistry in the Presence of Convective Flow

The specificity of spectroscopic techniques, together with the well-defined mass transport characteristics of the rotating disk electrode (RDE), the rotating ring-disk electrode (RRDE), the channel electrode and, to a lesser extent tube electrodes,[73] provide excellent experimental tools for the quantitative analysis of complex electrode processes. This section illustrates with representative examples, theoretical and experimental aspects of the coupling of UV/VIS spectroscopy and forced convection systems, both under steady-state and transient conditions.

4.6.1 Steady-state Measurements

4.6.1.1 Near Normal Incidence Ultraviolet/Visible Reflection–Absorption Spectroscopy at Rotating Disk Electrodes Optical access to the diffusion boundary layer of an RDE under near-normal incidence conditions requires an experimental arrangement of the type depicted in Figure 45.[74] The beam enters through a window parallel to the electrode surface, travels along the axis of rotation of the disk, reflects off the electrode surface and, after emerging from the cell through the same window, is directed toward the detector. Care must be exercised to set the distance between the electrode surface and the window (Figure 45) sufficiently long to preserve the laminar fluid flow characteristics undisturbed. The amount of product generated during the experiments will be assumed to be sufficiently small so that the bulk solution composition remains unaltered during data acquisition. Such requirements will be closely fulfilled provided the currents are small, the experiments short and the volume of the solutions large. For small incidence angles, such as those used in practice, the

For references see page 10222

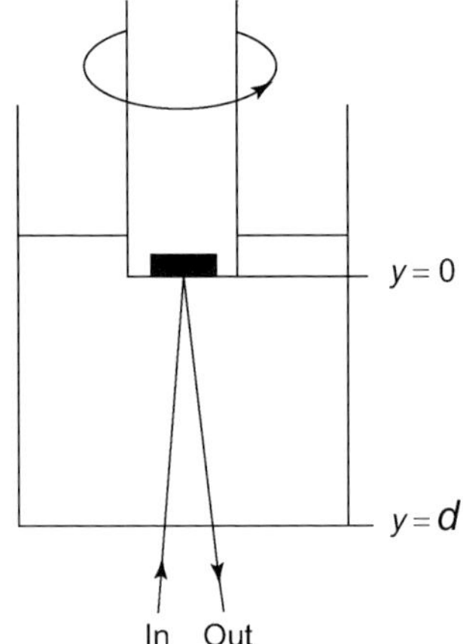

Figure 45 Diagram of the experimental arrangement for reflection–absorption UV/VIS spectroscopy at a RDE at near normal incidence angle. The labels 'In' and 'Out' refer to the incoming and outgoing beams, respectively. Other components have been omitted for clarity.(74)

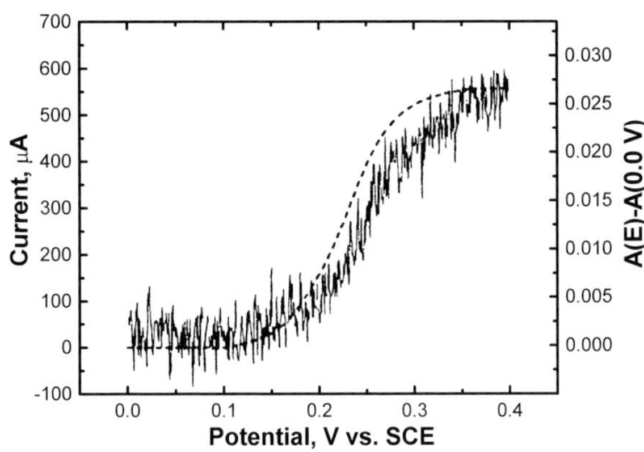

Figure 46 Combined plot of the current (dashed line) and $A(E) - A(0.0\,\text{V})$, as defined in Equation (81), as a function of the applied potential for the oxidation of 0.01 M $[Fe(CN)_6]^{4-}$ in 0.5 M K_2SO_4 aqueous solutions. These curves were recorded at $\omega = 530$ rpm and $\lambda = 420$ nm. The optical and electrochemical data were recorded at scan rates of 100 and $2\,\text{mV s}^{-1}$, respectively.(74)

errors introduced by assuming strict normal incidence are relatively small and therefore will be neglected.

Simple Redox Reactions. Steady-state concentration profiles for a stable product P generated at a RDE along the rotation axis of the disk can be obtained by solving Equation (33) in Table 2, assuming variations in the radial coordinate can be ignored, to yield Equation (82):

$$C_P(\zeta) = \frac{C_P^s}{\Gamma(4/3)} \int_{\zeta_P}^{\infty} \exp(-u^3)\,du \tag{82}$$

On this basis, the absorbance due to P is given by Equation (83):(74)

$$A(E) = 1.62936\varepsilon\nu^{1/6} D_R^{2/3} D_P^{-1/3} \omega^{-1/2} \frac{1 - C_R^s(E)}{C_R^o} \tag{83}$$

where the superscript 's' refers to the concentration at the surface, and the potential E in this case is positive (or negative) enough for oxidation (or reduction) to ensue. Shown in Figure 46 is a comparison of plots of the current I versus E (dashed line, left ordinate) and A versus E (solid line, right ordinate) for an Au RDE in a 0.01 M $Fe(CN)_6{}^{4-}$ solution in aqueous 0.5 M K_2SO_4 recorded at $\omega = 530$ rpm(74) using the arrangement shown in Figure 45. For these experiments, the optical response was monitored while scanning E at relatively high rates of $100\,\text{mV s}^{-1}$ whereas the current was recorded at $2\,\text{mV s}^{-1}$. At high scan rates, the concentration profiles cannot achieve steady state; hence, the optical signal is smaller than that expected from the current. This effect is particularly pronounced for potentials at which the current is below its diffusion-limited value i_{lim}, i.e. $C_R^s = 0$. Nevertheless, as predicted by theory, a plot of absorbance at the diffusion-limited current as a function of $\omega^{-1/2}$ was found to be linear, yielding values for ε of $[Fe(CN)_6]^{3-}$ within about 1% of those reported in the literature.

Determination of Faradaic Efficiencies of Complex Processes. Electrosynthesis affords an expedient route for the industrial production of a variety of high-valued chemicals. In addition to cell design considerations and electrode stability, the *overall* efficiency of these processes is often controlled by the kinetics of heterogeneous electron transfer, as well as by the rates of preceding and subsequent chemical reactions. Of common occurrence are situations in which materials, other than the desired product, are generated during operation, thereby decreasing the *specific* faradaic efficiencies, i.e. the fraction of the total current (or charge) involved in the generation of a given product. Quantitative aspects of this phenomenon may be assessed by determining product distributions using conventional analytical techniques external to the reactor or cell. A far more desirable strategy is to acquire such information in situ or on line, i.e. by probing directly the solution adjacent to the electrode surface under well-defined conditions of mass transfer, especially when dealing with rather unstable products.

Within the Levich formalism, the flux j_P of a nonadsorbing product P at the surface of a RDE, may be expressed in terms of the partial current density due to the formation of P (Equation 84):

$$D_P \left(\frac{dC_P}{dy}\right)_0 = \frac{-j_P}{n_P F} \tag{84}$$

where n_P is the number of electrons required to form one molecule of P. This parameter is defined as negative or positive depending on whether the process is a reduction or an oxidation, respectively. As is customary, the sign of the current is positive for an oxidation and negative for a reduction.

In terms of the dimensionless variable $\zeta_P = y/\delta_P$, where $\delta_P = 1.805\, D_P^{1/3} \nu^{1/6} \omega^{-1/2}$ is the thickness of the diffusion boundary layer for P, Equation (84) can be rewritten as Equation (85):

$$\left(\frac{dC_P}{d\zeta_P}\right)_0 = \frac{-j_P \delta_P}{n_P F D_P} \tag{85}$$

The solution for the profile of a stable product P, i.e. displaying no decomposition, along an axis normal to the electrode surface at steady state may be expressed in terms of the flux at the surface (Equation 86):[26]

$$C_P(\zeta_P) = -\left(\frac{dC_P}{d\zeta_P}\right)_0 \int_{\varsigma_P}^{\infty} \exp(-u^3)\, du = \frac{-j_P \delta_P}{n_P F D_P} \int_{\varsigma_P}^{\infty} \exp(-u^3)\, du \tag{86}$$

For normal incidence reflection absorption experiments at an RDE, and assuming P is the only optically absorbing species in the media, the absorbance A is given by

$$A = 2\varepsilon_P \int_0^{\infty} C_P(y)\, dy = 2\varepsilon_P \delta_P \int_0^{\infty} C_P(\varsigma_P)\, d\varsigma_P = 2\varepsilon_P \frac{\delta_P^2}{D_P} \frac{j_P}{n_P F} \int_0^{\infty} \int_{\varsigma_P}^{\infty} \exp(-u^3)\, du\, d\varsigma_P \tag{87}$$

The value of the double integral in the right-hand side of Equation (87) is $\Gamma(2/3)/3$, where $\Gamma(2/3)$ is the gamma function of argument 2/3.[23] Hence, upon rearrangement, Equation (87) may be written as

$$\frac{A\omega}{j_P} = 2.9408 \frac{\varepsilon_P}{n_P F} \left(\frac{\nu}{D_P}\right)^{1/3} \tag{88}$$

As the right-hand side of Equation (88) involves parameters intrinsic either to the product (ε_P, n_P, D_P) or to the solution (ν), the quantity $A\omega/j_P$ is constant. In fact, the constancy of this ratio for different E, ω, and C_R may be regarded as proof that the process indeed satisfies the requirements of the model. Note that in the derivation of Equation (88), no restrictions were imposed on the nature of the step-limiting the values of j_P, such as diffusion, heterogeneous or homogeneous reactions, or a combination thereof.

The reduction of bisulfite in mildly acidic solutions generates dithionite $S_2O_4^{2-}$, a material that displays a high ε band centered at $\lambda = 316$ nm, i.e. within a spectral region that does not interfere with other species in the media.

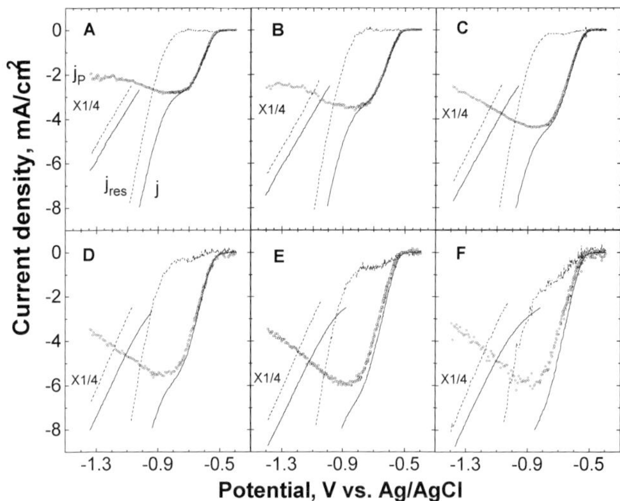

Figure 47 Current density j (solid line), partial current density for dithionite generation j_P (points), and difference between these two currents j_{res} (dotted line), obtained with an Au disk (area 0.452 cm^2) of an Au–Au RRDE assembly in 10 mM Na_2SO_3 in 0.50 M phosphate buffer solution (pH = 5.25), at different rotation rates: A = 200, B = 400, C = 900, D = 1600, E = 2500, and F = 3600 rpm. Values of j_P were calculated from Equation (86) based on the absorbance measured in the reflection absorption mode at near-normal incidence at $\lambda = 316$ nm.[92]

Plots of j versus E, known as polarization curves, where j is the total current density, were recorded with an Au RDE in solutions of 10 mM Na_2SO_3 in 0.50 M NaH_2PO_4 adjusted to pH 5.25 with NaOH using the arrangement shown in Figure 45. At all rotation rates examined, the currents at the inflection (solid lines in Figure 47), were significantly smaller than the diffusion-limited current densities (j_{lim}) for the two-electron oxidation of sulfite in the same electrolyte at 1.00 V, e.g. $j_{lim} = 19.1$ mA cm^{-2} at 900 rpm.

Plots of A at 316 nm versus E recorded during acquisition of I versus E curves in Figure 47, were used to construct j_P versus E plots using Equation (86), and are given in scattered form in the same figure. For these calculations, the constant $A\omega/j_P$ was determined based on ε of $S_2O_4^{2-}$, i.e. $7.3 \pm 0.3 \times 10^6$ cm^2 mol^{-1}, and the diffusion coefficient of $S_2O_4^{2-}$, i.e. $9.1 \pm 0.03 \times 10^{-6}$ cm^2 s^{-1}, respectively, $\nu = 0.010$ cm^2 s^{-1}, the kinematic viscosity of water, and $n = 2$, yielding a value of 9.9 ± 0.6 rpm cm^2 mA^{-1}. Also given in this figure are plots of j_{res} versus E, where j_{res} is the contribution to the total current j due to processes other than bisulfite reduction to dithionite, $j_{res} = j - j_P$.

In the range $-0.60 > E > -0.75$ V, j and j_P were found to virtually coincide at j_{res} about 0. This observation provides convincing evidence that j and j_P are not only proportional, but also that the *specific* faradaic efficiency (j_P/j) for dithionite generation under these conditions is

For references see page 10222

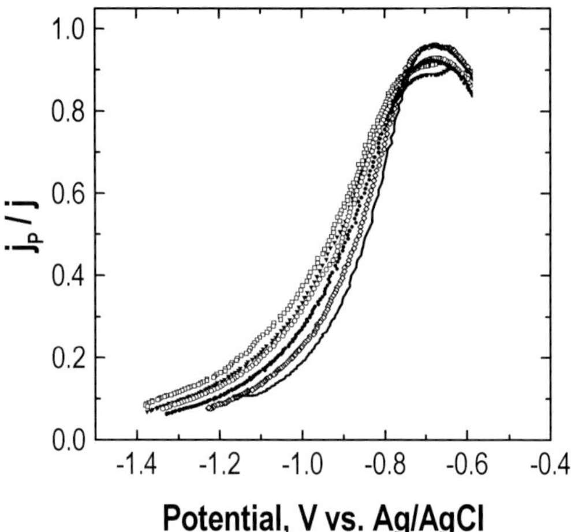

Figure 48 Faradaic efficiency for dithionite generation, i.e. j_P/j, as a function of the applied potential based on the data shown in Figure 47 for six different rotation rates: solid line = 200, diamonds = 400, solid circles = 900, open circles = 1600, solid triangles = 2500, and open squares = 3600 rpm.[92]

within experimental error about 100% (Figure 48). The decrease in the *specific* faradaic efficiency at more positive potentials is probably caused by contributions to the total current due to the reduction of adventitious oxygen in the solution. Furthermore, j_P increased with ω over the entire potential range, indicating that the reaction proceeds under partial mass transport control. However, despite the increase in j as the potential was made more negative, j_P reached a maximum at about -0.80 V (j_P^{max}) for all ω, and decreased steadily thereafter. In addition, plots of j_P^{max} versus $\omega^{1/2}$ and $1/j_P^{max}$ versus $1/\omega^{1/2}$ yielded straight lines with nonzero intercepts, and j_{res} was found to be independent of ω.

4.6.1.2 Rotating Disk Electrode with a Concentric Transparent Ring An interesting forced convection system introduced by Debrodt and Heusler,[75] involves a conventional RRDE assembly in which the metal ring is replaced by a material transparent in the UV/VIS region. This device makes it possible to detect optically absorbing species generated at the disk electrode (Figure 49). Although certain theoretical aspects of this RRDE electrode have been analyzed by digital simulation techniques,[76] analytical solutions can be obtained for simple reactions via the unified formalism given in Tables 2 and 3.[77] Specifically, the dimensionless concentration and flux of a reactant, or a product along the surface of an RDE are constant, and given by $\theta^D = \{1+\beta^{-1}\}^{-1}$, and $(\partial\theta/\partial\zeta)|_{\zeta=0} = -3\theta^D/\Gamma(1/3)$, respectively, where $\beta = [\Gamma(1/3)/3]k[3/(a\nu D_R^2)]^{1/3}(\nu/\Omega)^{1/2}$.

Based on the second equation for real space in Table 3, the concentration along an axis normal to the surface of an RDE beyond the electrode edge ($1 < \rho, \zeta$) can be expressed as Equation (89),

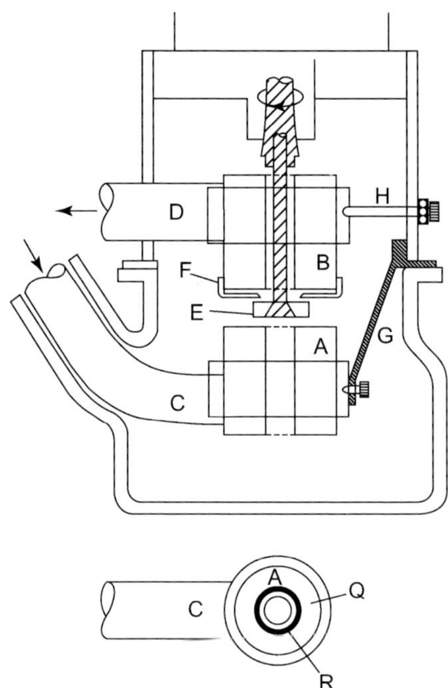

Figure 49 Diagram of a transparent ring RRDE.[75] A, circular light inlet; Q, quartz glass cover; C, optical fiber cable; R, ring; B, D, receivers built similarly; E, rotating disk with quartz ring and electrode; F, aperture; G, H, adjusting devices.

$$\Theta = \frac{\theta}{\theta^D} = 3\left[\frac{\Gamma(1/3)}{\Gamma(2/3)}\right]^{-1}\int_0^1 \exp\left[\frac{-\varsigma^3}{(1-\rho/\rho')}\right] \times \left[1-\left(\frac{\rho}{\rho'}\right)^3\right]^{-2/3}\rho'\,d\rho' \quad (89)$$

which at $\zeta = 0$ reduces to Equation (90):

$$\Theta(\rho > 1, \varsigma = 0) = \frac{3}{4} + \sqrt{\frac{3}{4}}\pi \ln\left[\frac{1+g^3}{(1+g)^3}\right] - \frac{3}{2}\pi \arctan\left[(2g-1)\sqrt{3}\right] \quad (90)$$

where $g = (\rho^3 - 1)^{1/3}$. Concentration profiles along ζ for three values of ρ are given in Figure 50.

The integrated profile along ζ, defined as $I(\rho)$ may be written as Equation (91):

$$I(\rho) = \int_0^\infty \Theta(\rho,\varsigma)\,d\varsigma = \left[\frac{\Gamma(1/3)}{\Gamma(2/3)}\right]^{-1}\int_0^1 \rho'(\rho^3-\rho'^3)^{-2/3} \times \int_0^\infty \exp\left\{\frac{-\varsigma^3}{1-(\rho'/\rho)^3}\right\} d\varsigma\,d\rho' \quad (91)$$

A plot of $I(\rho)$ versus ρ is shown in Figure 51(a).

An explicit expression for the total amount of material probed by the light beam, defined by a cylindrical shell

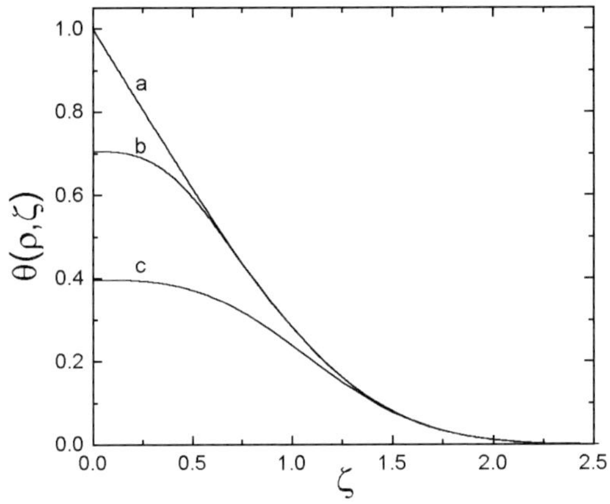

Figure 50 Plot of $\theta(\rho, \zeta)$ versus ζ for three values of ρ, i.e. curve a = 1, b = 65.5/64.5, and c = 74.5/64.5.[77]

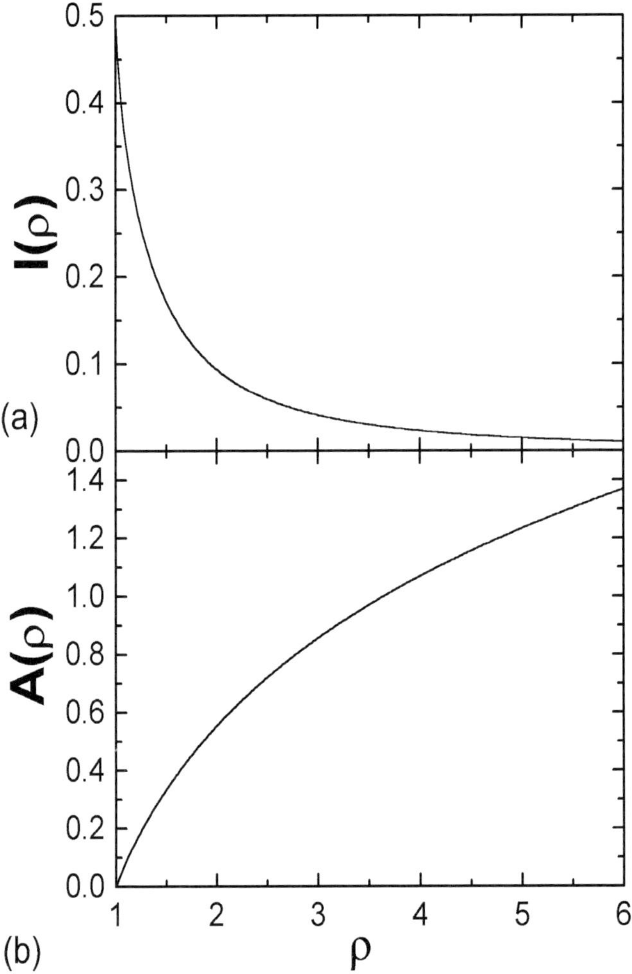

Figure 51 Plots of (a) $I(\rho)$ versus ρ and (b) $A(\rho)$ versus ρ for a transparent ring RRDE.[77]

with inner, and outer dimensionless radii, $\rho_1 = r_1/r_0$ and $\rho_2 = r_2/r_0$, respectively, can be shown to be given by Equation (92):

$$N_{vol} = N_o\theta^D \left\{ 2\int_{\rho_1}^{\rho_2} I(\rho;\beta)\rho\,d\rho \right\} = N_o\theta^D[A(\rho_2) - A(\rho_1)] \quad (92)$$

where $A(\rho)$ is given by Equation (93),

$$A(\rho) = 2\int_0^{\rho} I(\rho)\rho\,d\rho \quad (93)$$

and N_o is the same as for the RDE.[77] A plot of $A(\rho)$ versus ρ is shown in Figure 51(b).

4.6.1.3 Channel Electrodes

Profile Imaging Downstream from the Electrode Surface. The composition of a solution downstream from a channel-type electrode depends, among other factors, on the rates of heterogeneous electron transfer and homogeneous reactions involving electrogenerated species, and their mass transport characteristics. Information regarding various aspects of these processes can be obtained from the analysis of spectroscopic measurements in which a beam oriented normal to the electrode surface is used to monitor the absorbance of the solution along an axis parallel to the fluid flow. A cell arrangement suitable for this type of measurements is shown in Figure 52.[78]

Simple Electrode Reaction. In the case of uniform surface concentration, the integrated concentrations profile of the absorbing product P, denoted as I, at a distance X from the downstream edge of the electrode ($X > 1$) along an axis normal to its surface may be shown to be given by Equation (94):

$$I(X > 1;\infty) = \int_0^{\infty} \theta_P(X, Y)\,dY = \frac{1}{3^{1/3}}\Gamma\left(\frac{2}{3}\right) \times \int_0^1 x^{-1/3}(X - x)^{-1/3}\,dx \quad (94)$$

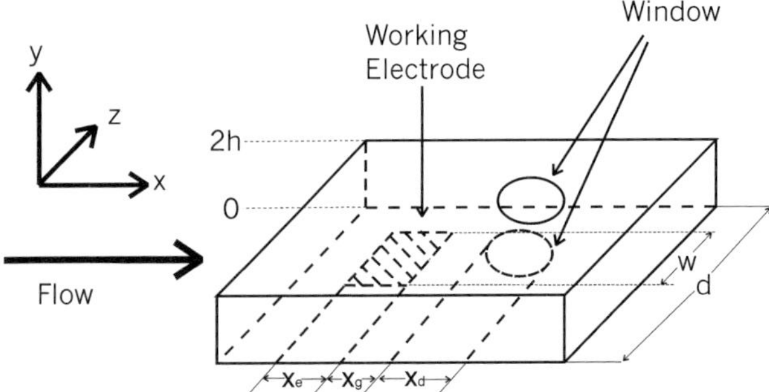

Figure 52 Diagram of a channel cell for spectroelectrochemical studies.[78]

For references see page 10222

where $\theta_P = (D_P/D_R)^{2/3}(C_P/C_R^o)$. This requirement is fulfilled when the reaction is reversible and, thus, C_R and C_P are prescribed by the Nernst equation, or under diffusion limited conditions.

Figure 53 shows plots of current and absorbance versus E obtained in a solution 0.01 M $K_4Fe(CN)_6$ in 0.25 M K_2SO_4 recorded simultaneously using a potential scan rate of $1\,mV\,s^{-1}$ and a flow rate of $1.74\,mL\,min^{-1}$,[79] where the optical monitoring was performed at 1.1 cm from the downstream edge of the electrode. The ordinates of the two curves in this figure are scaled based on the values of A at i_{lim}.

Experiments performed under diffusion-limited conditions[80] showed that the absorbance as a function of X for $X > 1$ (solid squares in Figure 54), which may be regarded as an image of the integrated concentration profile, was in quantitative agreement with that predicted

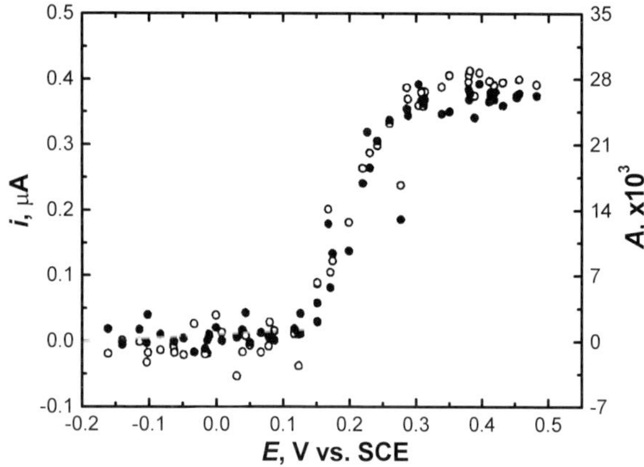

Figure 53 Combined plots of current (left ordinate, solid circles) and absorbance (right ordinate, open circles) versus E for a solution 0.01 M $K_4Fe(CN)_6$ in 0.25 M K_2SO_4 solution. Data were collected simultaneously as E was scanned at $1\,mV\,s^{-1}$ at a flow rate of $1.74\,mL\,min^{-1}$. The ordinates of the two curves have been scaled using an appropriate conversion factor.[79]

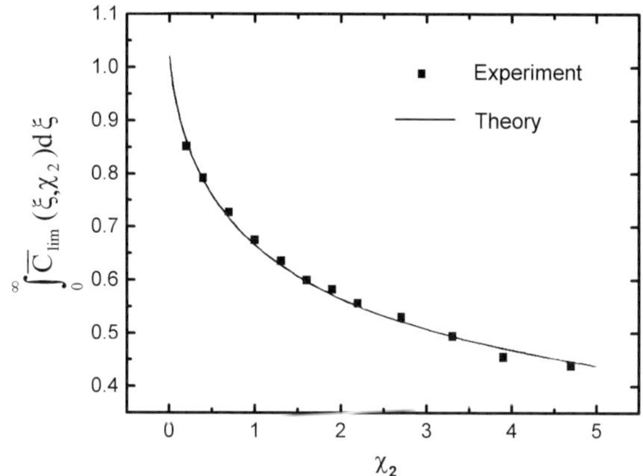

Figure 54 Plot of the steady-state dimensionless integrated concentration profile of $[Fe(CN)_6]^{3-}$, generated at the surface of an Au electrode by the oxidation of 0.01 M $[Fe(CN)_6]^{4-}$ in 0.25 M K_2SO_4, as a function of the dimensionless distance from the downstream edge of the electrode $\chi_2 = x_2/l$. All measurements were performed at $\lambda = 420\,nm$ along the center axis of the channel in the direction of the flow (solid squares) under diffusion-limited conditions for the oxidation of $[Fe(CN)_6]^{4-}$. The solid curve represents theoretical results (the original paper gives full details).[80]

Table 5 Differential equations and boundary conditions for an EC-type mechanism at a channel-type electrode cell (note that $K = kx_e/6U$)

$D\frac{\partial^2 C_R}{\partial y^2} = U_x\frac{\partial C_R}{\partial x}$ $D\frac{\partial^2 C_P}{\partial y^2} - kC_P = U_x\frac{\partial C_P}{\partial x}$			$w\frac{\partial^2 \overline{C}_R}{\partial Y^2} = (Y - Y^2)\frac{\partial \overline{C}_R}{\partial X}$ $w\frac{\partial^2 \overline{C}_P}{\partial Y^2} = (Y - Y^2)\frac{\partial \overline{C}_P}{\partial X} + K\overline{C}_P$		
$x = 0$	$y \geq 0$	$C_R = C_R^o$ $C_P = 0$	$X = 0$	$Y \geq 0$	$\overline{C}_R = 1$ $\overline{C}_P = 0$
$0 < x \leq x_e$	$y = 0$	$C_R = C_R^s$ $C_P = C_P^s$	$0 < X \leq 1$	$Y = 0$	$\overline{C}_R = \frac{C_R^s}{C_R^o}$ $\overline{C}_P = \frac{C_P^s}{C_R^o}$
$x > x_e$	$y = 0$	$\frac{\partial C_R}{\partial y} = 0$ $\frac{\partial C_P}{\partial y} = 0$	$X > 1$	$Y = 0$	$\frac{\partial \overline{C}_R}{\partial Y} = 0$ $\frac{\partial \overline{C}_P}{\partial Y} = 0$
$x > 0$	$y = 2h$	$\frac{\partial C_R}{\partial y} = 0$ $\frac{\partial C_P}{\partial y} = 0$	$X > 0$	$Y = 1$	$\frac{\partial \overline{C}_R}{\partial Y} = 0$ $\frac{\partial \overline{C}_P}{\partial Y} = 0$

by theory (solid line in the same figure), lending strong support to the validity of the underlying assumptions of the model.

Determination of the Rate Constant of an Electrochemical–Chemical Mechanism. Consider a simple EC mechanism of the type shown in Scheme 2, where k is the rate constant of the homogeneous chemical reaction. Within the framework of approximations specified in section 2.2.2, and assuming $D_P = D_R = D$, the steady state convective diffusion equations for R and P are given in the right-hand panel of Table 5, where the fluid velocity U_x is given by

$$U_x = \frac{3}{2}U\left[1 - \frac{(h-y)^2}{h^2}\right] \quad (95)$$

In Equation (95) U is the mean flow velocity, and h is the half-height of the channel.

The appropriate boundary conditions, summarized in the lower section of the same panel, specify that the concentration of the two species is constant over the entire surface of the electrode, which implies that the electrochemical process is infinitely fast. The last entry in that table means that no reactions occur at the optically transparent surface directly opposite to the electrode along the entire length of the channel. This problem, cast in terms of the dimensionless variables $Y = y/2h, X = x/x_e$, $\overline{C}_R = C_R^s/C_R^o$, and $\overline{C}_P = C_P^s/C_R^o$ is given on the right-hand panel in the table.

If P is the only absorbing species at the selected λ, plots of absorbance versus $-\ln w$, where $w = (Dx_e)/(24Uh^2)$, $K = (kx_e)/(6U)$, determined by digital simulations yielded for different X and $w/K = 0.01$, showed clearly defined maxima for each of the curves (Figure 55a).[81] Maxima were also observed for a fixed value of $X = 1.75$ and different values of w/K, as shown in Figure 55(b).

Based on these results, an empirical relationship was obtained between the flow rate and the position along the x-axis at the absorption maxima, V_m and x, respectively (Equation 96),

$$\log V_m = \log\left(\frac{2}{3}h^2 x_e b\right) + 1.50\log k_1 - 0.50\log D + \log\left(-4.45 + 4.72\frac{x}{x_e}\right) \quad (96)$$

where, as before, x_e is the length of the electrode, and b the channel width. Hence, it becomes possible to extract values for k_1 by determining experimentally such maxima using the required physical dimensions of the cell and the electrode, as well as the diffusion coefficient. From a practical viewpoint it is more convenient to fix x and measure the relative absorbance for various values of V_m.

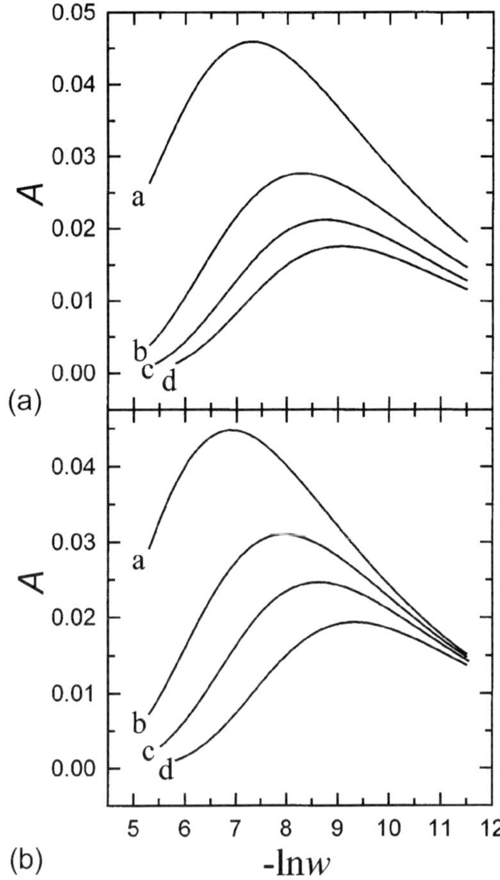

Figure 55 (a) Absorbance versus $-\ln w$ at $w/K = 0.01$ for different X: curve a = 1.25, b = 1.75, c = 2.25, d = 2.75. (b) Absorbance versus $-\ln w$ at $X = 1.75$ for different w/K: curve a = 0.025, b = 0.0125, c = 0.008, d = 0.005.[81]

Figure 56 shows a plot of f/q, a quantity proportional to the normalized absorbance ratio at two values of V_m, corrected for contributions due to other absorbing species in the media, as a function of V_m, for the oxidation of p-aminophenol (PAP) to yield p-benzoquinoneimine (BQI). This latter species undergoes hydrolysis to form benzoquinone (BQ).[82] The solid line in this figure represents the best fit to the experimental points. Based on the maximum in this curve, the pseudo-first-order rate constant k for the irreversible hydrolysis of BQI can be evaluated using Equation (96), yielding values of about $0.33\,s^{-1}$.

4.6.2 Chronoamperometry under Forced Convection

4.6.2.1 Rotating Disk Electrodes Consider a chronoamperometric experiment in which the potential of an RDE immersed in a solution containing a nonabsorbing

For references see page 10222

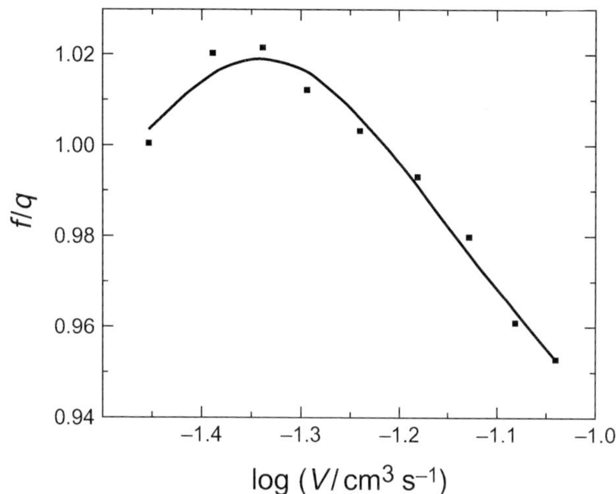

Figure 56 Plot of f/q versus $\log V$ for $X = 2.5$. Solid squares are experimental data and the solid line is a fit to obtain the maximum of f/q.[82]

electroactive species R is stepped from a value E_i, at which no reaction occurs, to a potential sufficiently positive (or negative) E_f for R to undergo oxidation (or reduction) generating an optically absorbing product, P. According to Beer's law, the instantaneous change in absorbance A along y can be written as

$$A(t) - A(t=0) = 2\varepsilon_P \int_0^\infty C_P(y,t)\,dy \qquad (97)$$

In Equation (97), $C_P(y,t)$ is the transient concentration profile of P, which can be determined based on the solution of the transient convective diffusion equation subject to the appropriate initial and boundary conditions. Solutions to this problem, for a reactant, have been reported for short times by Krylov and Babak,[83] in terms of parabolic cylinder functions, $D_k(z)$, and for long times by Nicancioglu and Newman in terms of (numerically evaluated) eigenvectors and eigenfunctions

Table 6 Transient convective diffusion to a RDE

Governing equation

$$D\frac{\partial^2 C_R}{\partial y^2} - v_y\frac{\partial C_R}{\partial y} - \frac{\partial C_R}{\partial t} = 0$$

Boundary conditions for concentration step problem

$$C_R(0,t) = 0$$
$$C_R(\infty,t) = C_o$$
$$C_R(y,0) = C_o$$

Short-time solution	Long-time solution
Dimensionless variables	
$c = \dfrac{C_R}{C_o}, \tau = (DA^2)^{1/3}t, x = \left(\dfrac{A}{D}\right)^{1/3} y; A = a\left(\dfrac{\omega^3}{\nu}\right)^{1/2};$ $z = x(2\tau)^{-1/2}$	$\Theta = \dfrac{C_o - C_R}{C_o}, \theta = \omega\left(\dfrac{D}{\nu}\right)^{1/3}\left(\dfrac{a}{3}\right)^{2/3} t; \varsigma = \left(\dfrac{a\nu}{3D}\right)^{1/3}\left(\dfrac{\omega}{\nu}\right)^{1/2} y$
Convective diffusion equation	
$\dfrac{\partial^2 c}{\partial x^2} + x^2\dfrac{\partial c}{\partial x} - \dfrac{\partial c}{\partial \tau} = 0$	$\dfrac{\partial^2 \Theta}{\partial \zeta^2} + 3\zeta^2\dfrac{\partial \Theta}{\partial \zeta} - \dfrac{\partial \Theta}{\partial \theta} = 0$
Boundary conditions	
$c(0,\tau) = 0$; $c(\infty,\tau) = 1$; $c(z,0) = 1$	$\Theta(0,\theta) = 1$; $\Theta(\infty,\theta) = 0$; $\Theta(\zeta,0) = 0$
Dimensionless profile	
$c(z,\tau) = \sum_{n=0}^{\infty} \tau^{3n/2} G_n(z)$	$\Theta(\zeta,\theta) = \Theta^{ss} - \Theta^{t}$
$G_n(z) = e^{-z^2/4} \sum_{k=-3n-1}^{3n-1} {}^{**}b_k^{(n)} D_k(z)$	$\Theta^{ss} = \dfrac{1}{\Gamma(4/3)}\int_\zeta^\infty e^{-x^3}\,dx, \quad \Theta^{t} = \sum_{n=0}^{\infty} B_n Z_n(\zeta) e^{\lambda_n \theta}$
	$B_n = \dfrac{\int_0^\infty \Theta^{ss} e^{\zeta^3} Z_n(\zeta)\,d\zeta}{\int_0^\infty e^{\zeta^3} Z_n^2(\zeta)\,d\zeta}$
Dimensionless flux	
$\dfrac{j(\tau)}{j(\infty)} = \dfrac{a^{1/3}}{0.62(\pi\tau)^{1/2}}\left[1 + \sum_{n=1}^{\infty}\left(-\pi \sum_{k=-3n-1}^{3n-1} {}^{**}\dfrac{2^{k/2} b_k^{(n)}}{\Gamma(-k/2)}\right)\tau^{3n/2}\right]$	$\dfrac{j(\theta)}{j(\infty)} = -\dfrac{3a^{1/3}}{0.62}\left(\left.\dfrac{\partial \Theta^{ss}}{\partial \zeta}\right\|_{\zeta=0} - \left.\dfrac{\partial \Theta^{\tau}}{\partial \zeta}\right\|_{\zeta=0}\right)$

of the associated Sturm-Liouville system.[84] A summary of the two mathematical formalisms, which include the appropriate dimensionless variables and the dimensionless profiles, are given in Table 6. Explicit expressions for the coefficients $b_k(n)$ may be found in the original literature.[83] Exact concentration profiles based on these two solutions spanning a wide range of dimensionless times are shown in Figure 57(a). These were used to calculate the corresponding transient integral profiles involved in the analysis of the spectroscopic data, assuming $D_R = D_P$ (Figure 57b).

Figure 58 shows curves of $[A(t) - A(t = 0)]$ versus t for measurements performed in 0.01 M $K_4Fe(CN)_6$ in 0.5 M K_2SO_4 aqueous solutions using an Au RDE in the arrangement shown in Figure 45,[85] with the spectrophotometer set at $\lambda = 420$ nm (see Table 4), where the ε value of $Fe(CN)_6{}^{4-}$ is very small. Reflection absorption data were acquired at various rotation rates following a potential step from 0.0 to 0.4 V versus Ag/AgCl, which is positive enough for the oxidation of the

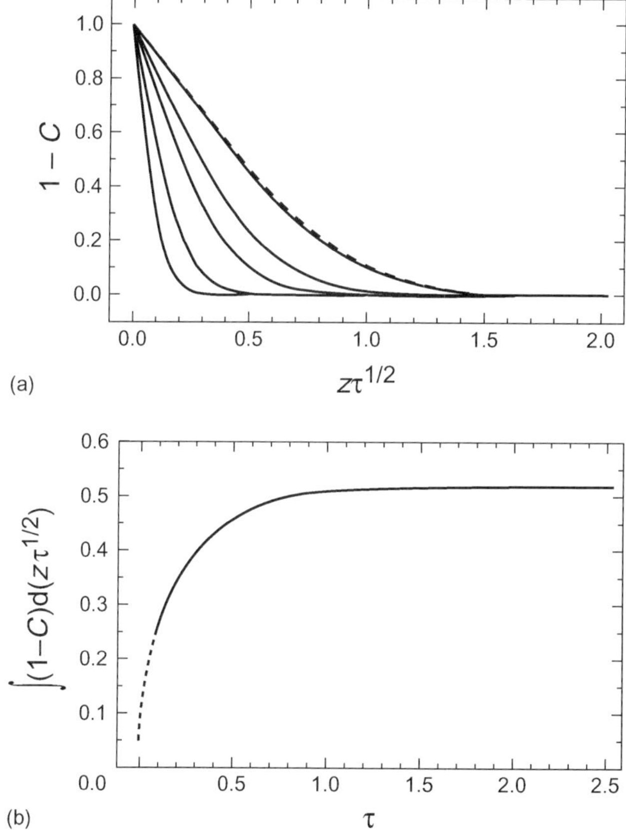

Figure 57 (a) Exact concentration profiles obtained using the short- and long-time solutions, for (curves from left to right) $\tau = 0.01$, 0.03, 0.1, 0.2, 1.0, and the steady state (dashed line). (b) Theoretical integral concentration profiles versus time using the data shown in (a); the dashed and the solid lines were obtained from the short-time and long-time solutions.[85]

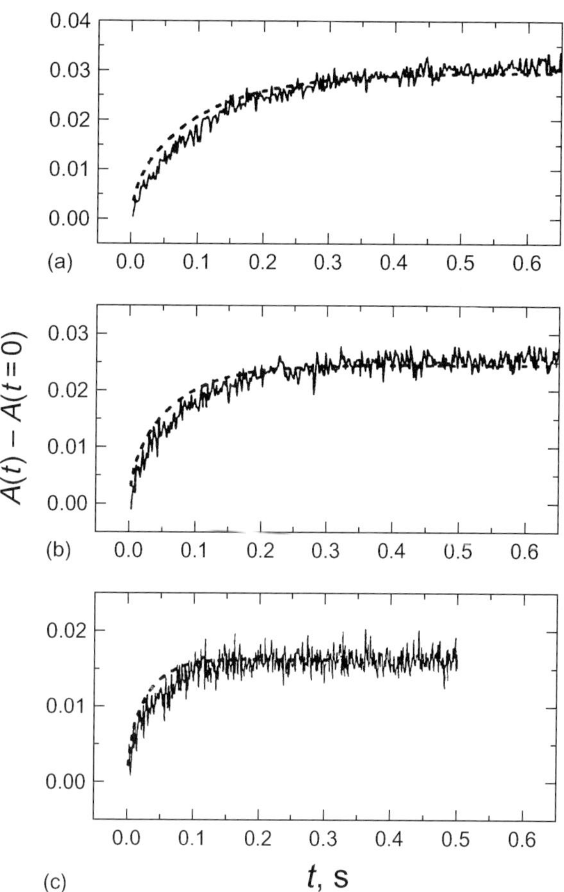

Figure 58 Experimental (difference) absorbance $[A(t) - A(t = 0)]$ versus t: (a) $\omega = 400$ rpm, (b) $\omega = 650$ rpm, (c) $\omega = 1400$ rpm; the dashed lines are the corresponding theoretical curves.[85]

ferrous species to occur under complete mass transport control. The results obtained for each cycle, consisting of a forward and backward step, were co-added using a signal averager. Also shown in this figure as dotted lines are the theoretical results based on values of ε for $[Fe(CN)_6]^{3-}$ at 420 nm $= 1.14 \times 10^6$ mol^{-1} cm^2, $D = 7.3 \times 10^{-6}$ cm^2 s^{-1}, $\nu = 0.01$ cm^2 s^{-1}, and the specified rotation rate ω (radians per second), using the appropriate conversion between dimensionless and actual integral profiles (Equation 98):

$$\int (1 - C)\,\mathrm{d}(z\tau^{1/2}) = a^{1/3}\left(\frac{\omega}{2}\right)^{1/2} D^{-1/3}\nu^{-1/6} \int c\,\mathrm{d}y \tag{98}$$

where $a = 0.51024$. As can be seen, the agreement between theory and experiment for the three rotation rates examined is very good.

4.6.2.2 Channel Electrodes The transient convective diffusion equation governing mass transport of a solution

For references see page 10222

phase species C in a channel is given by Equation (99):

$$\frac{\partial C}{\partial t} = D\frac{\partial^2 C}{\partial y^2} - v_x\frac{\partial C}{\partial x} \tag{99}$$

where $v_x = (3V_f/4hd)[1 - (h-y)^2/h^2]$ and other terms have their usual meanings.

Consider a chronopotentiometric experiment in which a potential of large enough magnitude is applied to an electrode in the channel to reduce instantaneously C_R to zero at the surface, assuming, that $C_P = 0$ before the potential step is applied. The initial and boundary conditions for R and P may thus be summarized as Equations (100–103):

$$\text{for } t < 0: \quad y = 0, 0 < x < (x_e + x_g + x_d), \quad C_R = C_R^o, C_P = 0 \tag{100}$$

$$\text{for } t > 0: \quad 0 < x < x_e; C_R = 0, \quad D_R\left.\frac{\partial C_R(y,t)}{\partial y}\right|_{y=0} = -D_P\left.\frac{\partial C_P(y,t)}{\partial y}\right|_{y=0} \tag{101}$$

$$\text{for all } t: \quad y = 0, x_e < x < (x_e + x_g + x_d); \quad D_R\left.\frac{\partial C_R(y,t)}{\partial y}\right|_{y=0} = D_P\left.\frac{\partial C_P(y,t)}{\partial y}\right|_{y=0} = 0 \tag{102}$$

$$y = 2h, 0 < x < (x_e + x_g + x_d): \quad D_R\left.\frac{\partial C_R(y,t)}{\partial y}\right|_{y=0} = D_P\left.\frac{\partial C_P(y,t)}{\partial y}\right|_{y=0} = 0 \tag{103}$$

where x_e is the width of the electrode and x_g and x_d are the distances from the downstream edge of the electrode to the area of the solution being monitored optically along an axis normal to the plane of the electrode, as shown in Figure 52.

The time-dependent concentration profiles for all species involved can be obtained by digital simulation techniques, and the values of the parameters then adjusted until a best fit to the experimental data is found. Figure 59(a) shows values of the normalized absorbance for the one-electron oxidation of ferrocyanide, and Figure 59(b) is the reduction of ferricyanide via chronopotentiometric techniques acquired at wavelengths at which the absorbance of each of the species is maximum.[78] The solid lines in these figures are calculated best-fit curves to the experimental data. Using the value of D_R determined from i_{lim} (Equation 104),

$$i_{lim} = 0.925nFwC_R^o\left(\frac{V_f D_R^2 x_e^2}{h^2 d}\right)^{1/3} \tag{104}$$

for different flow rates and the physical dimensions of the channel, the ratio of the diffusion coefficients of the two species was found to be in excellent agreement with data reported in the literature, i.e. $D_{ferro}/D_{ferri} = 1.17 \pm 0.05$.

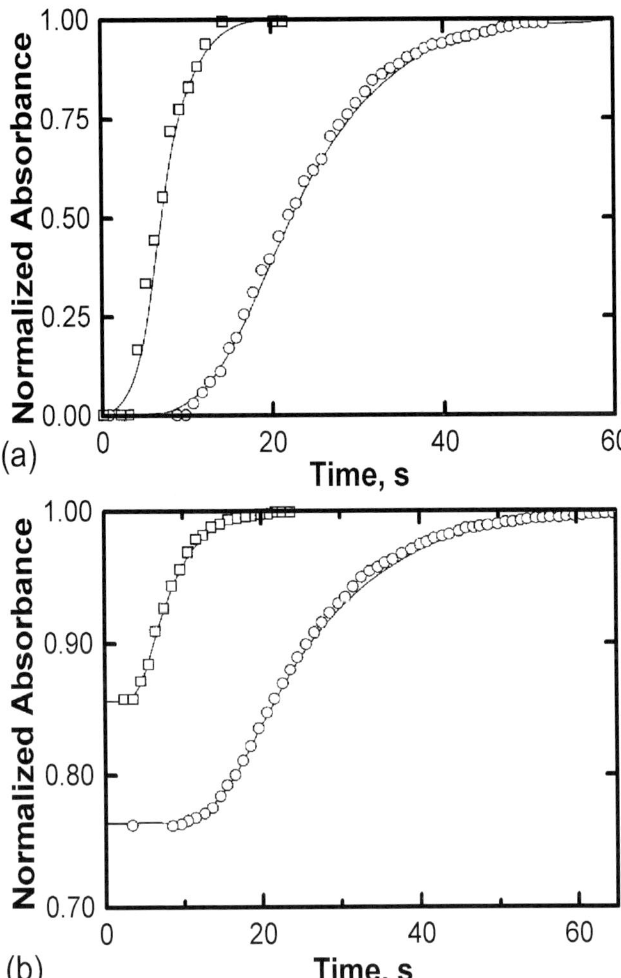

Figure 59 (a) Transient absorbance at $\lambda = 420$ nm for the one-electron oxidation of $Fe(CN)_6^{4-}$ and (b) $\lambda = 218$ nm for the one-electron reduction of $Fe(CN)_6^{3-}$ measured using the channel cell with cell depths of 0.0805 cm and 0.0792 cm, respectively, following a potential step at the working electrode. Solution flow rates: □ = 1.44×10^{-3} cm^3 s^{-1}; (○) 7.51×10^{-3} cm^3 s^{-1}. Solid lines represent theoretical curves.[78]

Application of this method revealed very small differences between the diffusing coefficients of electrogenerated radical species and the neutral precursors in nonaqueous solutions, which suggests that the interactions of the two types of species with the solvent are indeed quite similar.

4.7 Modulation Techniques

Application of an external periodic potential to an electrical system, such as an electrochemical cell, will elicit corresponding periodic variations in the current. Analysis of the amplitudes and phases of the response as a function of the perturbation frequency constitutes the basis of a general method for the study of the dynamic behavior of electronic circuitry known as impedance spectroscopy. The mathematical treatment of modulation techniques can be simplified considerably by employing periodic signals of a small enough magnitude so that the behavior of the system may be represented in terms of linear equations. This approach allows contributions to the observed signals derived from other sources, such as noise, to be minimized or eliminated, by extracting that component of the response that oscillates at the same frequency as the exciting perturbation. Also amenable to experimental determination are periodic changes in the concentration profiles of species consumed or generated by faradaic processes brought about by the modulating potential, which can be monitored optically by relying either on absorption of UV/VIS chromophores, or on beam-bending effects derived from local changes in the index of refraction along an axis normal to the beam direction.

4.7.1 General Considerations

Consider an experimental arrangement in which light propagating through the solution, reflects at normal incidence from a flat electrode, placed parallel to and at a distance d from a planar transparent cell window. It will be assumed, for simplicity, that the redox process involves consumption of a nonabsorbing species R, and generation of a single absorbing product P, where both R and P are solution phase species, and that only R is present at the beginning of the experiments. If the attenuation in light intensity induced by other components in the path of the beam, including reflection from the electrode, can be ignored, the intensity I of the light exiting the cell is given by (Equation 105),

$$I = I_o \exp\left[-2\varepsilon k \int_0^d c(y,t)\,\mathrm{d}y\right] \qquad (105)$$

where I_o is the light intensity entering the cell. Attention is focused hereafter on solutions of one-dimensional mass transport equations of the form

$$c(y,t) = \bar{c}(y,t) + \tilde{c}(y,t) \qquad (106)$$

where in Equation (106) $\bar{c}(y,t)$ is a slowly varying function of time, or steady state, and $\tilde{c}(y,t)$ is the oscillatory contribution. Introducing Equation (106) into Equation (105), a normalized absorbance may be defined as follows (Equation 107):

$$\frac{I}{\bar{I}} = \exp\left[-2\varepsilon k \int_0^d \tilde{c}(y,t)\,\mathrm{d}y\right] \qquad (107)$$

Provided that the changes in concentration induced by the periodic perturbation are small, the argument in the exponential can be approximated by the first two terms in the Taylor series expansion, yielding Equation (108) after rearrangement:

$$\frac{\bar{I} - I}{\bar{I}} = \frac{I_{\mathrm{ac}}}{I_{\mathrm{dc}}} = 2\varepsilon k \int_0^d \tilde{c}(y,t)\,\mathrm{d}y \qquad (108)$$

If it is further assumed that the changes in concentration are confined to a volume of solution very close to the electrode surface, d may be regarded as arbitrarily large, and therefore the upper limit of integration may be set as infinity, leading to much simpler mathematical expressions.

From an experimental viewpoint, the ratio $I_{\mathrm{ac}}/I_{\mathrm{dc}}$, can be measured very accurately, such as by using a lock-in amplifier. A quantitative analysis of the results obtained using this modulation technique requires the integral on the right-hand side of Equation (108) to be evaluated, a problem that is equivalent to finding solutions for the mass transport equations subject to the appropriate boundary conditions.

Of particular interest here is derivation of expressions for the oscillatory component of the response of the form

$$\tilde{c}(y,t) = A^*\Psi(y)\exp(j\omega t) \qquad (109)$$

where A^* in Equation (109) is the maximum amplitude of the sinusoidally varying concentration, and $\Psi(y)$ is a function of distance.

4.7.2 Quiescent Solutions

Substitution of Equation (109) into Fick's second law leads to a differential equation in y in total derivatives, which can be easily solved analytically to yield Equation (110):

$$\Psi(y) = \exp\left[-\left(\frac{j\omega}{D}\right)^{1/2} y\right] \qquad (110)$$

Hence, $I_{\mathrm{ac}}/I_{\mathrm{dc}}$, also denoted as $I(\omega t)$, may be written as Equation (111):

$$\begin{aligned}\frac{I_{\mathrm{ac}}}{I_{\mathrm{dc}}} = I(\omega t) &= 2\varepsilon k \int_0^\infty A^* \exp\left[-\left(\frac{j\omega}{D}\right)^{1/2} y\right] \exp(j\omega t)\,\mathrm{d}y \\ &= 2\varepsilon k A^* \left(\frac{D}{j\omega}\right)^{1/2} \exp(j\omega t) \\ &= 2\varepsilon k A^* \left(\frac{D}{\omega}\right)^{1/2} \exp\left[j\left(\omega t - \frac{\pi}{4}\right)\right]\end{aligned} \qquad (111)$$

that is, the optical response lags the modulated potential by $\pi/4$.

For references see page 10222

In the case of very fast heterogeneous electron transfer reactions, the applied potential prescribes the ratio of concentrations of reactants and products through the Nernst equation. For perturbations of very small amplitude, this equation can be linearized, to yield Equation (112) that relates the amplitudes of the applied potential E^* and those of the concentrations of R and P, A_R^* and A_P^*, respectively:

$$E = \overline{E} + \widetilde{E} = E_{dc} + E^* \exp(j\omega t) \tag{112}$$

where E_{dc} and E^* are given by Equation (113):

$$E_{dc} = E^{o\prime} + \frac{RT}{nF} \ln \frac{\bar{c}_R(0,t)}{\bar{c}_P(0,t)};$$

$$\text{and} \quad E^* = \frac{RT}{nF}\left(\frac{A_R^*}{\bar{c}_R(0,t)} - \frac{A_P^*}{\bar{c}_P(0,t)}\right) \tag{113}$$

Because of conservation of mass at the electrode surface (see Equation 16) $A_P^* = -A_R^* \xi$, where $\xi = (D_R/D_P)^{1/2}$; hence, it becomes possible to express either one of the A_i^* coefficients in terms of E^*, to yield Equation (114)

$$E^* = \frac{4RTA_R^*}{nFc_R^o} \cosh^2\left(\frac{a}{2}\right) \tag{114}$$

where a is given by Equation (115):

$$a = \frac{nF}{RT}(E_{dc} - E_{1/2}); \text{ and } E_{1/2} = E^{o\prime} + \frac{RT}{nF} \ln \left(\frac{D_P}{D_R}\right)^{1/2} \tag{115}$$

The current associated with the faradaic process is related to the flux of material at the electrode surface via Equation (116),

$$\begin{aligned} i &= -nFAD_R \left.\frac{\partial c_R}{\partial y}\right|_{y=0} = -nFAA_P^*(j\omega D_P)^{1/2} \exp(j\omega t) \\ &= -nFAA_P^*(\omega D_P)^{1/2} \exp\left[j\left(\omega t + \frac{\pi}{4}\right)\right] \end{aligned} \tag{116}$$

where A is the area of the electrode. Hence, from Equations (111) and (116), the current and the optical response are 90° out of phase with respect to one another.

4.7.2.1 Alternating Current Solution-phase Reflectance Spectroscopy Based on Equations (105) and (109), the ratio of the magnitudes of the current and reflectance signal induced by the modulated applied potential is given by Equation (117),

$$\frac{|i(\omega t)|}{|I(\omega t)|} = \frac{nFA\omega}{2\varepsilon k} \tag{117}$$

and, therefore, for a fixed perturbation frequency, the electrical and optical responses are proportional to each other. In the case of solution-phase redox couples involving at least one UV/VIS chromophore, these considerations afford a basis for implementing a spectroscopic analog of a highly sensitive electrochemical technique known as alternating current (AC) voltammetry. This method relies on the superposition of an AC signal of small amplitude onto a slow linear potential scan, while monitoring the amplitude of the current response. Although restricted to optically absorbing species, this spectroscopic approach offers a definite advantage over current measurements in that the response is not affected by contributions to the electrical signal derived from other sources, particularly double layer capacity effects.

By way of illustration,[86] Figure 60 displays a plot of $I(\omega t)$ versus E_{dc} for the oxidation of an unbuffered solution 10 mM $K_4Fe(CN)_6$ in aqueous 1 M KCl obtained at a frequency of 43.9 Hz at a scan rate of $2\,mV\,s^{-1}$, for which both for the forward and reverse scans, as indicated by the arrows, were found to superimpose. The position of the peak at 0.219 V versus SCE agrees very well with the reversible half-wave potential for the redox couple. Moreover, the width of the peak at half-height was 94 mV, in harmony with that predicted by theory, i.e. 90 mV. This overall behavior is characteristic of the AC response for a reversible one-electron system.

The direct relationship between the optical and electrical response can also be valuable for the determination of rate constants by examining the dependence of the phase angle ϕ on the perturbation frequency. In particular, in the absence of effects due to IR drop in the solution, ϕ may be shown to be related to the rate of a simple heterogeneous

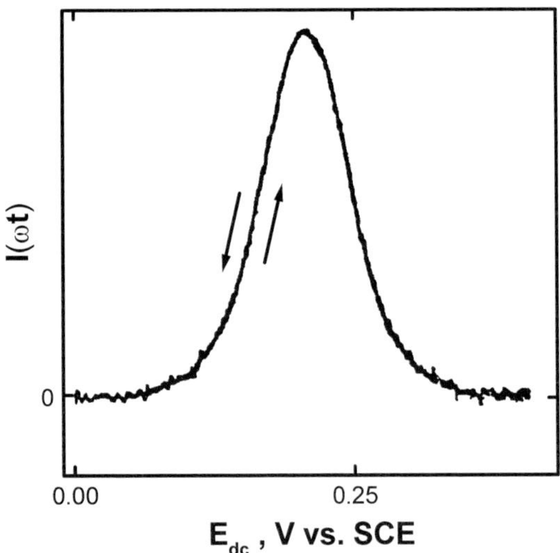

Figure 60 AC reflectance spectra for the oxidation of 10 mM $K_4Fe(CN)_6$ in 1 M KCl, at $\lambda = 420\,nm$, $\omega/2\pi = 43.9\,Hz$, $\nu = 2\,mV\,s^{-1}$.[86]

electron transfer reaction k_s (Equation 118):

$$\cot\phi = 1 + \frac{(\omega D/2)^{1/2}}{k_s} \quad (118)$$

In fact, a plot of the iR-corrected $\cot\phi$ versus $\omega^{1/2}$ (Figure 61) for the ferro–ferricyanide couple obtained from potential modulated spectroelectrochemical measurements was found to be linear, yielding values for k_s around $0.096 \pm 0.008\,\mathrm{cm\,s^{-1}}$ in agreement with data obtained by more conventional means.[86]

4.7.2.2 Diffractive Alternating Current Modulation Light striking the edge of a solid object will be scattered producing a diffraction pattern on a plane normal to the beam direction, including regions in the shadow of the object, with much of the intensity of the diffracted beam originating from areas very close to the edge. Furthermore, the diffracted light will be attenuated by a UV/VIS absorber present very near the edge, and the extent of attenuation will be proportional to the concentration of the absorber therein, as predicted by Beer's law. Assuming the concentration of product in the area sampled by the diffracted beam, i.e. at and near the electrode edge C_P^s is uniform, the attenuation of the beam may be written as Equation (119):

$$A = \log\left(\frac{I_o}{I_t}\right) = \varepsilon w C_P^s \quad (119)$$

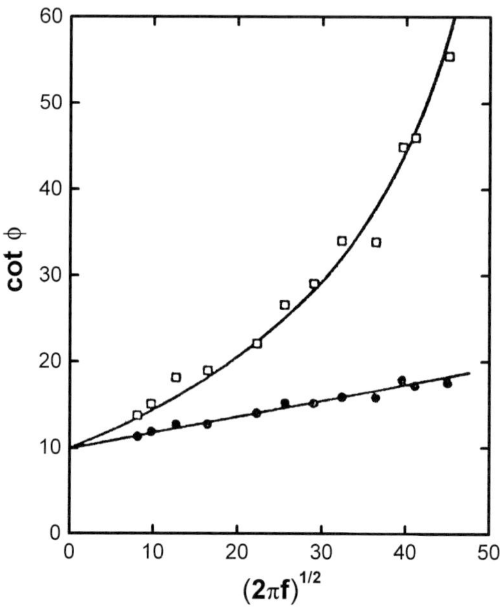

Figure 61 Plots of $\cot\phi$ versus $\omega^{1/2}$ for the oxidation of 10 mM $K_4Fe(CN)_6$ in 1 M KCl. Squares represent $\cot\varphi_S$, circles represent calculated $\cot\phi$; $E_{dc} = 0.219$ V versus SCE, $\Delta E = 18$ mV.[86]

where w is the electrode length along the optical axis. Furthermore, I_o and I_t are the intensities of the diffracted beam at a specific angle of observation measured before P is generated, and after C_P^s reaches a limiting value, respectively. This situation is somewhat equivalent to that of a conventional spectroelectrochemical cell of a pathlength equal to the length of the electrode along the direction of beam propagation, which can be electrochemically filled or emptied with absorber over a very short period of time.

If A is small, the normalized difference in light intensities $I_o - I_t$, denoted as $\Delta I_{diff}/I_o$, may be shown to be given by Equation (120):

$$\frac{\Delta I_{diff}}{I_o} = k\varepsilon b\xi C_P^o \quad (120)$$

where $C_R^o\xi = C_P$, C_R^o is the bulk concentration of R, and $\xi = (D_R/D_P)^{1/2}$. On this basis, it becomes possible to monitor the modulated diffracted light intensity, and, thus, the concentration of electrogenerated product P in a region of the solution very close to electrode surface. More specifically, the amplitude of the modulated diffracted light intensity I_{ac}^{diff} for AC voltammetry is related to the amplitude of the applied voltage by Equation (121):

$$\frac{I_{ac}^{diff}}{I_o} = \frac{k\varepsilon bn\xi FC_R^o E^*}{4RT\cosh^2(a/2)} \quad (121)$$

As before, this latter expression is valid provided the heterogeneous electron transfer reaction is infinitely fast.

Experiments involving TAA in TEAP/acetonitrile and square wave modulation of relatively large amplitude, i.e. 0.5 V about $E^{o\prime}$, yielded values of I_{ac}^{diff} lower than those predicted theoretically.[29,87] However, the output of the lock-in amplifier using a modulation frequency of 10 Hz was found to be proportional to the concentration of TAA in the range from 3×10^{-7} to 1×10^{-4} M. Furthermore, judicious choice of experimental conditions, including higher modulation frequencies, lower angles and phase adjustments, made it possible to increase the detection sensitivity to less than 2×10^{-8} M. Also in qualitative agreement with the expected response, were the results of spectroelectrochemical AC voltammetry experiments, which yielded superimposable curves for forward and backward scans for slow scan rates, about $2\,\mathrm{mV\,s^{-1}}$, and E^* value in the range 10–50 mV (Figure 62).

4.7.2.3 Transmission Alternating Current Voltammetry Application of a sinusoidally varying potential to a planar electrode induces a modulation in the concentration of an optically absorbing redox product. Changes along an axis parallel to the electrode surface can be monitored spectroscopically using a CCD or a photomultiplier in

For references see page 10222

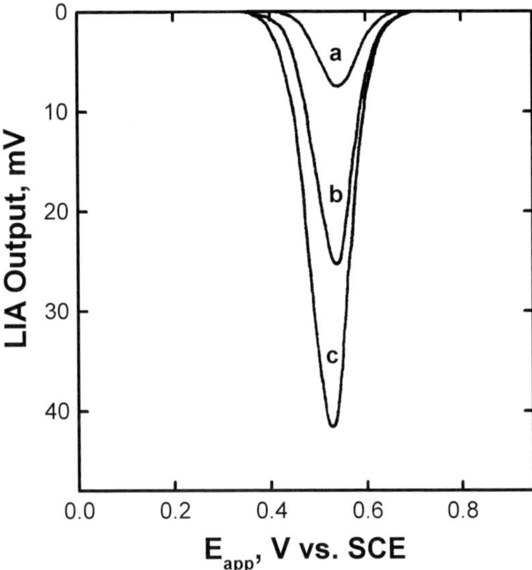

Figure 62 Lock-in amplifier output versus direct current (DC) potential for small-amplitude square-wave AC potential superimposed on a slow DC scan ($2\,\mathrm{mV\,s^{-1}}$): modulation frequency 10 Hz; phase shift −20°; time constant 0.3 s; $C^{\circ}_{\mathrm{TAA}} = 0.257\,\mathrm{mM}$; curve a, $\Delta E = 10\,\mathrm{mV}$; b, $\Delta E = 30\,\mathrm{mV}$; c, $\Delta E = 50\,\mathrm{mV}$. Reverse scans were superimposed on forward scans within the pen width. Peaks occur at +0.52 V versus SCE.[87]

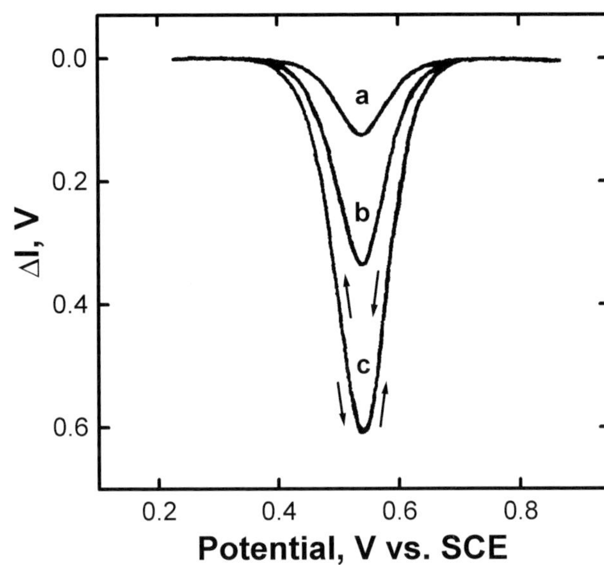

Figure 63 Optically monitored AC voltammograms for TAA solutions. Square-wave perturbations of varying height were superimposed on a slow DC ramp ($2\,\mathrm{mV\,s^{-1}}$). Frequency = 5.0 Hz, phase = −20°. Curve a, $\Delta E = 10\,\mathrm{mV}$; b, $\Delta E = 30\,\mathrm{mV}$; c, $\Delta E = 60\,\mathrm{mV}$. $y = 3\,\mu\mathrm{m}$, $C^{\circ}_{\mathrm{TAA}} = 0.38\,\mathrm{mM}$.[87]

conjunction with a narrow slit to illuminate a small volume of solution. The normalized optically modulated signal may thus be expressed as Equation (122):

$$\frac{I_{\mathrm{ac}}}{I_{\mathrm{dc}}} = \varepsilon kw\tilde{c}(y,t) = \varepsilon kwA^{*}\exp\left[-\left(\frac{j\omega}{D}\right)^{1/2}y\right]\exp(j\omega t)$$
$$= \varepsilon kwA^{*}\exp\left[-\left(\frac{\omega}{2D}\right)^{1/2}y\right] \times \exp\left\{j\left[\omega t + \left(\frac{\omega}{2D}\right)^{1/2}y\right]\right\} \quad (122)$$

where w is the length of the electrode along the direction of light propagation, and y is the distance normal to the electrode surface. Although results have been reported for a CCD, the most interesting data have been collected with the slit arrangement.[87] In particular, Figure 63 shows results for the optically monitored AC spectroelectrochemical voltammogram of TAA obtained for three different scan rates at a distance of 3 µm from the surface, which are similar to those obtained using diffraction as described in the previous section. The expression above indicates that the phase is a function not only of the frequency of the perturbation, but also of the specific distance from the electrode surface.

4.7.2.4 Alternating Current Modulated Solution-phase Refraction According to Fermat's principle, a beam travelling along an axis z, in a medium in which the index of refraction n varies normal to the direction of propagation, say y, will bend toward regions having larger n, and that the extent of bending will be proportional to $\partial c/\partial y$. Assume, more specifically, that z and y are parallel and perpendicular to the electrode surface. A modulation in the concentration profile of solution phase species induced by the applied potential will modify the index of refraction along y, and consequently the extent of beam bending. Although refraction is insensitive to the nature of the species that causes the change in the index of refraction, it is useful from a didactic viewpoint to regard these effects as caused by a single species, and that the changes in n are linear in the concentration of that species, i.e. $\partial n/\partial c = N$. This situation can be realized experimentally for systems in which a single species is consumed or generated at the electrode surface. On this basis, and neglecting the effects due to the much slower nonoscillatory transient contribution, the oscillatory degree of bending $\tilde{\Psi}$ will be given by Equation (123),

$$\tilde{\Psi} = \frac{w}{n}N\frac{\partial\tilde{c}}{\partial y} \quad (123)$$

where w is the length of the electrode along z. An explicit expression for the derivative can be obtained from Equations (109) and (110), yielding

$$\frac{\partial\tilde{c}}{\partial y} = A^{*}\left\{-\left(\frac{j\omega}{D}\right)^{1/2}\right\}\exp\left[-\left(\frac{j\omega}{D}\right)^{1/2}y\right]\exp(j\omega t)$$

$$= -A^* \left(\frac{\omega}{D}\right)^{1/2} \exp\left[-\left(\frac{\omega}{2D}\right)^{1/2} y\right]$$
$$\times \exp\left[j\left\{\omega t + y\left(\frac{\omega}{2D}\right)^{1/2} + \frac{\pi}{4}\right\}\right] \quad (124)$$

Based on Equations (109) and (124) above, this analysis predicts that $\tilde{c}$ and $\partial\tilde{c}/\partial y$ should be 45° out of phase with respect to one another. In fact, large differences in the magnitude of the lock-in amplifier output as a function of phase angle have been reported[87] for potential modulation experiments in solutions containing 0.66 mM TAA and 10 mM BQ, which generate, respectively, an absorbing, and a nonabsorbing product upon oxidation and reduction (Figure 64). This phenomenon makes it possible to cnhance the response of the absorbing versus the nonabsorbing product, by simply adjusting the phase angle. This is illustrated in Figure 64(b–d) in the form of amplitude of the measured light intensity at the image plane $\Delta I_{\mathrm{image}}$, as a function of applied DC potential (E_{dc}) for three different phase angles. Figure 64(a) is the magnitude of the AC current as a function of the applied DC potential, which reflect the much higher concentration of BQ compared to TAA.

4.7.3 Warburg Impedance

The dynamic properties of electronic circuits are customarily defined in terms of a frequency-dependent function known as the impedance, Z (Equation 125):

$$Z = \frac{\mathrm{d}E}{\mathrm{d}i} \quad (125)$$

where E is the potential and i the current. It is then possible to define a diffusional impedance, also known as the Warburg impedance, Z_{w} as the ratio of the time derivatives of the voltage and the current.

4.7.3.1 Quiescent Solutions The Warburg impedance for quiescent solutions can be obtained from Equations (112) and (116), namely Equation (126),

$$Z_{\mathrm{w}} = \frac{\sigma\sqrt{2}}{(j\omega)^{1/2}} \quad (126)$$

where σ is defined by Equation (127):

$$\sigma = \frac{RT}{n^2F^2A\sqrt{2}}\left(\frac{1}{D_{\mathrm{R}}^{1/2}\bar{c}_{\mathrm{R}}(0,t)} + \frac{1}{D_{\mathrm{P}}^{1/2}\bar{c}_{\mathrm{P}}(0,t)}\right) \quad (127)$$

For a reversible reaction, Z_{w} reduces to Equation (128):

$$Z_{\mathrm{w}} = \frac{4RT}{n^2F^2A(j\omega D_{\mathrm{o}})^{1/2}c_{\mathrm{o}}^*}\cosh^2\left(\frac{a}{2}\right) \quad (128)$$

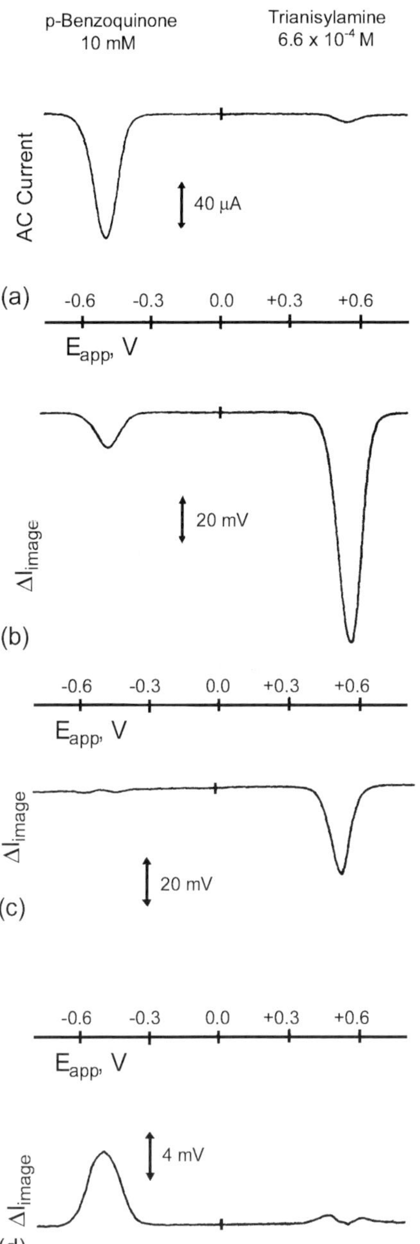

Figure 64 The AC current and AC voltammetric response for a mixture of TAA (0.66 mM) and BQ (0.01 M) for several phase angles. $\Delta E = 70\,\mathrm{mV}$, scan rate $= 2\,\mathrm{mV\,s^{-1}}$, $y = 3\,\mu\mathrm{m}$: (a) AC current on a 3.2 mm diameter Pt electrode (phase angle = 0°); (b) optical monitoring at a phase of −20°; (c) at −100°; (d) at −130°.[87]

Based on the results presented in the previous section, the Warburg impedance is in phase with the optical response (see Equation 111).

4.7.3.2 Forced Convection Expressions for the Warburg impedance of an RDE can be obtained[88] by taking the Laplace transform of the concentration step problem

For references see page 10222

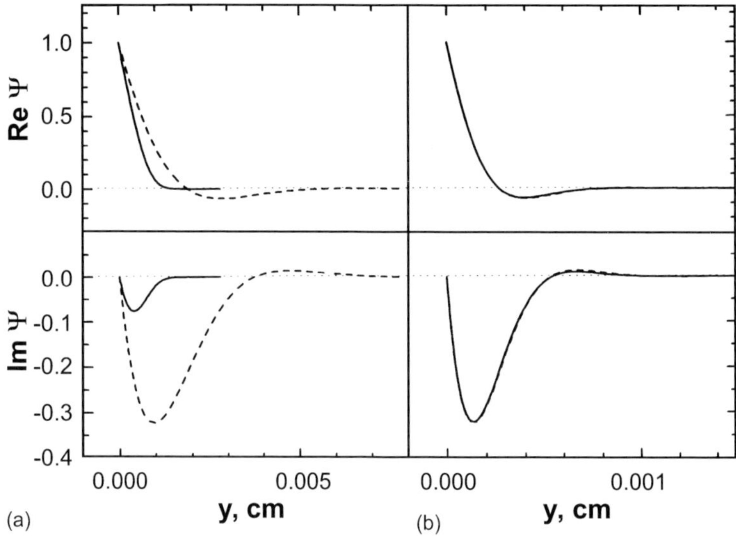

Figure 65 Plots of Re ψ and Im ψ for (a) $K = 1.2$ and (b) $K = 60$ as functions of the real axial distance for $\Omega = 0$ (dashed line) and $\Omega = 377\,\text{rad s}^{-1}$ (3600 rpm; solid line).[88]

as specified in section 2.2.2, and regarding the complex parameter s as $j\omega$.[89] Plots of the real and imaginary parts of the integral of the concentration profiles as a function of the dimensionless frequency K are shown in Figure 65. These data can also be represented by plotting the real versus the imaginary part of the function in question and specifying the values of the frequency directly on the curve (Figure 66), also known as Cole–Cole or Nyquist diagrams. It is significant to note that for $K > 40$, the angle reaches $-45°$, which implies that at high enough frequencies, the response of the electrochemical system in the presence of forced convection approaches that in a quiescent media.

4.7.4 Determination of Molar Absorptivities of Electrogenerated Species

Provided that $D_R = D_P$, rearrangement of Equation (102) yields Equation (129) for $\varepsilon(k)$ in terms of the integral of the sinusoidal profile:

$$\varepsilon(\lambda) = \frac{I_{ac}}{I_{dc}}\left[2k\int_0^{\infty}\tilde{c}(y,t)\,dy\right]^{-1} \tag{129}$$

Setting $I_{ac} = I^* \exp[j(K\theta + \phi)]$, where ϕ is the phase, this equation can be rearranged to Equation (130),

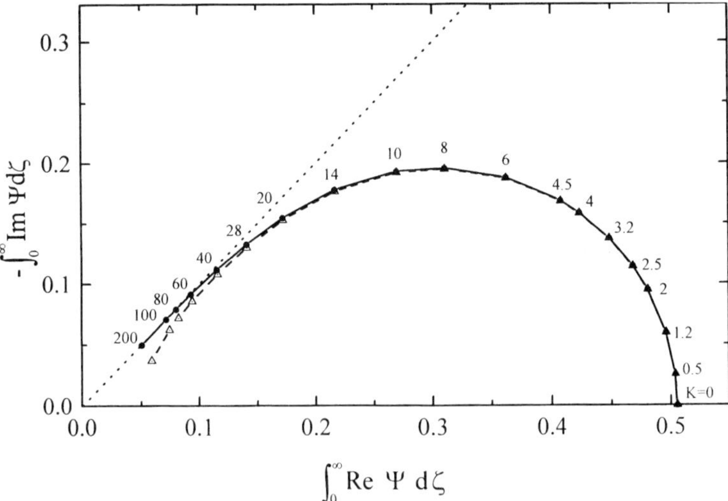

Figure 66 Plots of the real and imaginary components of the integral of the dimensionless concentration profile in the Cole–Cole representation: • = exact and Δ = long-time solutions.[88]

$$\varepsilon(\lambda) = \frac{I^*}{I_{dc}}\left(2A^*k\gamma\int_0^\infty \Psi\,d\varsigma\right)^{-1} \qquad (130)$$

where A^* is the amplitude of the time-varying concentration, and γ the factor that converts y into the dimensionless variable ζ. As $\varepsilon(\lambda)$ is real, the phase may be expressed as Equation (131):[90]

$$\phi = \arctan\left\{\frac{\int_0^\infty \mathrm{Im}\Psi\,d\varsigma}{\int_0^\infty \mathrm{Re}\Psi\,d\varsigma}\right\} \qquad (131)$$

Plots of $\left|\int_0^\infty \Psi\,d\varsigma\right|$ versus K, and ϕ versus K for an RDE are shown Figure 67.

The most striking verification of the validity of this theory, is provided by the excellent agreement between $\varepsilon(\lambda)$ obtained for ferricyanide from potential modulation spectroelectrochemical measurements (Figure 68) at an RDE in $K_4Fe(CN)_6$/0.5 M K_2SO_4 aqueous solutions (see scattered plots), and from conventional transmission experiments given by the continuous curve in the same figure.[91]

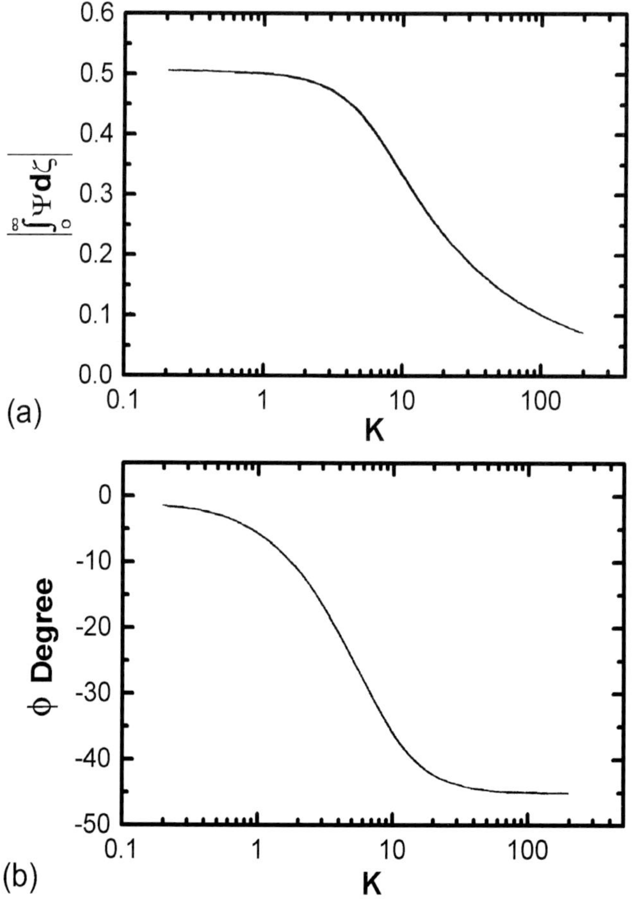

Figure 67 (a) Plot of $\left|\int_0^\infty \Psi\,d\zeta\right|$ as a function of K, based on the exact solutions. (b) Plot of the phase ϕ, as a function of the dimensionless frequency K.[91]

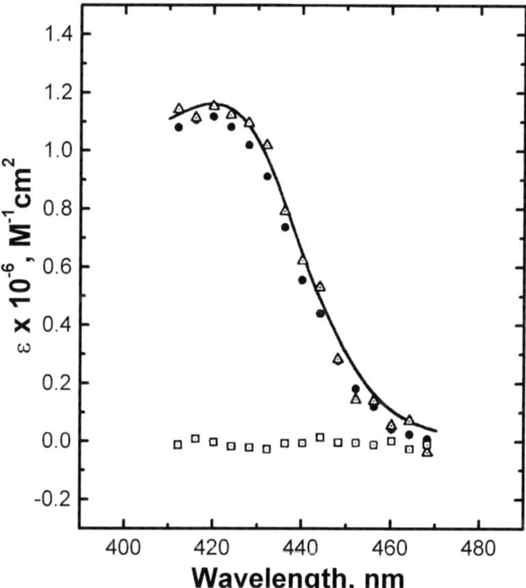

Figure 68 Comparison between $\varepsilon(\lambda)$ versus λ data obtained by conventional transmission spectroscopy (solid line) and that derived from the magnitude of I^*/I_{dc} determined from potential modulation experiments and the value of $\left|\int_0^\infty \Psi\,d\zeta\right|$ determined by numerical means for $c_o = 10$ mM (triangles) and $c_o = 3$ mM (solid circles). The open squares were obtained in measurements in which the potential was modulated about 0.0 V.[88]

CONCLUDING REMARKS

The quantitative analysis of UV/VIS spectroelectrochemical techniques relies on well-known optical and mass transport principles. Although analytic solutions for transient problems appear to be available only for the simplest of systems, digital simulations involving an inexpensive desk computer provide a means of diagnosing mechanisms and extracting kinetic parameters of interest from experimental data. Furthermore, the advent of computer-controlled instrumentation including data acquisition makes it possible to collect large amounts of information so as to achieve reliable statistics. Intrinsic theoretical and practical limitations imposed by the very nature of these measurements, sets limits on the space-, and time-resolution of absorption-based UV/VIS spectroelectrochemical techniques. Without diminishing its importance and future impact in scientific and technological fields, it seems reasonable to conclude that this specialized subfield of analytical chemistry has reached a solid level of maturity. On this basis, new developments may not be expected in instrumental principles, but in the applications area, especially sensor miniaturization.

For references see page 10222

It may be envisaged that new generations of inexpensive, fully tunable lasers and widespread implementation of chemometrics, will lead to versatile and expedient means for quantitative detection of weak chromophores and thus join the array of existing analytical techniques. Also to be expected over the next few years, are applications of UV/VIS imaging analysis to the study of current distribution in electrochemical cells, which are expected to have an impact on the design of electrochemical reactors for electrosynthesis. It may be interesting to note that in contrast to the absorption-based techniques featured in this article, the analysis of in situ UV/VIS ellipsometric and electroreflectance measurements, continues to pose formidable problems in terms of a quantitative theoretical interpretation. It is hoped that the newly discovered links between electrochemistry and surface science will prompt scientists in both fields to find novel approaches for gaining new insights into the fundamental basis of this phenomenon and thus fully exploit its exquisite surface specificity and sensitivity, which greatly surpasses our theoretical understanding.

ACKNOWLEDGMENTS

Support for this work was provided by the National Science Foundation, and the National Institutes of Health. The authors are deeply indebted to Prof. Richard L. McCreery from Ohio State University for his insightful comments during the preparation of this manuscript.

ABBREVIATIONS AND ACRONYMS

AC	Alternating Current
AN	Acrylonitrile
ATR	Attenuated Total Reflection
BQ	Benzoquinone
BQI	*p*-Benzoquinoneimine
CCD	Charge-coupled Device
CPZ	Chlorpromazine
$CPZ^{+\bullet}$	Chlorpromazine Cation Radical
DC	Direct Current
DCIP	2,6-Dichlorophenolindophenol
DOQ	Dopamine Quinone
EC	Electrochemical–Chemical
EEC	Electrochemical–Electrochemical–Chemical
FeTPPCl	Iron Tetraphenylporphyrin Chloride
H_2Q	Hydroquinone
IRE	Internal Reflection Element
IRS	Internal Reflection Spectroscopy
LOPTLC	Long Optical Pathlength Thin-layer Cell
LPTLC	Long Pathlength Thin Layer Cell
MB	Methylene Blue
MPZ	Methoxypromazine
MV^{++}	Methyl Viologen Dication
OD	*o*-Dianisidine
OTTLE	Optically Transparent Thin-layer Electrode
PAP	*p*-Aminophenol
PMT	Photomultiplier Tube
Pyc=cPy	1,2-Bis-(4-pyridine)ethylene
RDE	Rotating Disk Electrode
RRDE	Rotating Ring-disk Electrode
SCE	Saturated Calomel Electrode
$TAA^{+\bullet}$	Tri-*p*-anisylamine Cation Radical
TBAP	Tetra-*n*-butyl Ammonium Perchlorate
TEACN	Tetraethylammonium Cyanide
TEAP	Tetraethylammonium Perchlorate
TPZ	Triflupromazine Hydrochloride
UV/VIS	Ultraviolet/Visible

List of selected abbreviations appears in Volume 15

RELATED ARTICLE

Electroanalytical Methods **(Volume 11)**
Electroanalytical Methods: Introduction

REFERENCES

1. F. Wooten, *Optical Properties of Solids*, Academic Press, New York, 1972.
2. A.V. Sokolov, *Optical Properties of Metals*, American Elsevier Publishing Company, Inc., New York, 1967.
3. J.I. Steinfeld, *Molecules and Radiation*, The MIT Press, Cambridge, MA, 1985.
4. T. Kuwana, N. Winograd, 'Spectroelectrochemistry at Optically Transparent Electrodes. 1. Electrodes Under Semi-infinite Diffusion Conditions', in *Electroanalytical Chemistry*, ed. A.J. Bard, Marcel Dekker, New York, 1–78, Vol. 7, 1974.
5. W.R. Heineman, F.M. Hawkridge, H.N. Blount, 'Spectroelectrochemistry at Optically Transparent Electrodes. II. Electrodes Under Thin-layer and Semi-infinite Diffusion Conditions and Indirect Coulometric Titrations', in *Electroanalytical Chemistry*, ed. A.J. Bard, Marcel Dekker, New York, 1–113, Vol. 13, 1984.
6. R.L. McCreery, 'Spectroelectrochemistry', in *Physical Methods of Chemistry*, eds. B.W. Rossiter, S.F. Hamilton, John Wiley & Sons, New York, 591–661, Vol. II, 1986.
7. J. Niu, S. Dong, 'Transmission Spectroelectrochemistry', *Reviews in Analytical Chemistry*, **XV1**, 1 (1996).
8. W.R. Heineman, 'Spectroelectrochemistry', *Anal. Chem.*, **50**, 390A (1978).

9. R.H. Muller, 'Principles of Ellipsometry', in *Advances in Electrochemistry and Electrochemical Engineering*, ed. R.H. Muller, J. Wiley & Sons, New York, 167, Vol. 9, 1973.
10. S. Gottesfeld, 'Ellipsometry', in *Physical Electrochemistry: Principles, Methods and Applications*, ed. I. Rubinstein, Marcel Dekker, New York, 1995.
11. J.D.E. McIntyre, 'Specular Reflection Spectroscopy of the Electrode Solution Interphase', in *Advances in Electrochemistry and Electrochemical Engineering*, ed. R.H. Muller, J. Wiley & Sons, New York, 61, Vol. 9, 1973.
12. D.M. Kolb, 'UV Visible Reflectance Spectroscopy', in *Spectroelectrochemistry Theory and Practice*, ed. R.J. Gale, Plenum Press, New York, 87, 1988.
13. C. Barbero, M.C. Miras, R. Kotz, 'Electrochemical Mass Transport Studied by Probe Beam Deflection: Potential Step Experiments', *Electrochim. Acta*, **37**, 429 (1992).
14. R.H. Muller, 'Double Beam Inteferometry for Electrochemical Studies', in *Advances in Electrochemistry and Electrochemical Engineering*, ed. R.H. Muller, J. Wiley & Sons, New York, 281, Vol. 9, 1973.
15. A.J. Bard, L.R. Faulkner, *Electrochemical Methods, Fundamentals and Applications*, Wiley, New York, 1980.
16. M. Born, E. Wolf, *Principles of Optics*, 6th edition, Pergamon Press, Oxford, 1987.
17. M.V. Klein, T.E. Furtak, *Optics*, 2nd edition, Wiley, New York, 1986.
18. J.D. Brewster, J.L. Anderson, 'On Absorbance Measurements in Spatially Inhomogeneous Fields', *Appl. Spectrosc.*, **43**, 710–714 (1989).
19. N.J. Harrick, *Internal Reflection Spectroscopy*, Interscience Publishers, New York, 1967.
20. H.S. Carslaw, J.C. Jaeger, *Conduction of Heat in Solids*, 2nd edition, Clarendon Press, Oxford, 1967.
21. R.V. Churchill, R. Vance, *Operational Mathematics*, 2nd edition, McGraw-Hill, New York, 1958.
22. D.D. Macdonald, *Transient Techniques in Electrochemistry*, Plenum Press, New York, 1977.
23. M. Abramowitz, I.A. Stegun, *Handbook of Mathematical Functions with Formulas, Graphs, and Mathematical Tables*, U.S. Govt. Print. Off., Washington, 1964.
24. R.B. Bird, W.E. Stewart, E.N. Lightfoot, *Transport Phenomena*, Wiley, New York, 1960.
25. N. Winograd, H. Blount, T. Kuwana, 'Spectroelectrochemical Measurement of Chemical Reaction Rates First-order Catalytic Processes', *J. Phys. Chem.*, **73**, 3456 (1969).
26. V.G. Levich, *Physicochemical Hydrodynamics*, Prentice-Hall Inc., Englewood Cliffs, 1962.
27. Y.V. Tolmachev, Z. Wang, D.A. Scherson, 'In Situ Spectroscopy in the Presence of Convective Flow Under Steady-state Conditions: A Unified Mathematical Formalism', *J. Electrochem. Soc.*, **143**, 3539 (1996).
28. A. Erdelyi, *Table of Integral Transforms*, McGraw-Hill, New York, Vol. 1, 1954.
29. C.-C. Jan, B.K. Lavine, R.L. McCreery, 'High-sensitivity Spectroelectrochemistry Based on Electrochemical Modulation of an Absorbing Analyte', *Anal. Chem.*, **57**, 752–758 (1985).
30. J.E. Davis, N. Winograd, 'Application of Coulostatic Charge Injection. Techniques to Improve Potentiostat Risetimes', *Anal. Chem.*, **44**, 2152 (1972).
31. R.S. Robinson, R.L. McCreery, 'Submicrosecond Spectroelectrochemistry Applied to Chlorpromazine Cation Radical Charge Transfer Reactions', *J. Electroanal. Chem.*, **182**, 61–72 (1985).
32. C.H. Pyun, S.M. Park, 'Construction of a Microcomputer Controlled Near Normal Incidence Reflectance Spectroelectrochemical System and Its Performance Evaluation', *Anal. Chem.*, **58**, 251 (1986).
33. C. Zhang, S.M. Park, 'Simple Technique for Constructing Thin-layer Electrochemical-cells', *Anal. Chem.*, **60**, 1639 (1988).
34. Y.R. Shen, 'Optical Wavelength-modulation Spectroscopy', *Surface Sci.*, **37**, 522–539 (1973).
35. S. Kim, D.A. Scherson, 'In-situ UV-visible Reflection Absorption Wavelength Modulation Spectroscopy of Species Irreversible Adsorbed on Electrode Surfaces', *Anal. Chem.*, **64**, 3091 (1992).
36. A. Deputy, H.-P. Wu, R.L. McCreery, 'Spatially Resolved Spectroelectrochemical Examination of the Oxidation of Dopamine by Chlorpromazine Cation Radical', *J. Phys. Chem.*, **94**, 3620–3624 (1990).
37. R.S. Robinson, C.W. McCurdy, R.L. McCreery, 'Microsecond Spectroelectrochemistry by External Reflection from Cylindrical Microelectrodes', *Anal. Chem.*, **54**, 2356–2361 (1982).
38. J.P. Skully, R.L. McCreery, 'Glancing Incidence External Reflection Spectroelectrochemistry with a Continuum Source', *Anal. Chem.*, **52**, 1885–1889 (1980).
39. C.E. Baumgartner, G.T. Marks, D.A. Alkens, H.H. Richtol, 'Spectroelectrochemical Monitoring of Electrode Reactions by Multiple Specular Reflection Spectroscopy', *Anal. Chem.*, **52**, 167–270 (1980).
40. C.-C. Jan, R.L. McCreery, 'High-resolution Spatially Resolved Visible Absorption Spectrometry of the Electrochemical Diffusion Layer', *Anal. Chem.*, **58**, 2771–2777 (1986).
41. A.L. Deputy, R.L. McCreery, 'Spatially Resolved Spectroelectrochemistry for Examining an Electrochemically Initiated Homogeneous Electron Transfer Reaction', *J. Electroanal. Chem.*, **257**, 57–70 (1988).
42. R.L. McCreery, R. Pruiksma, R. Fagan, 'Optical Monitoring of Electrogenerated Species Via Specular Reflection at Glancing Incidence', *Anal. Chem.*, **51**, 749 (1979).
43. J. Zak, M.D. Porter, T. Kuwana, 'Thin-layer Electrochemical Cell for Long Optical Path Length Observation of Solution Species', *Anal. Chem.*, **55**, 2219–2222 (1983).
44. J.D. Brewster, J.L. Anderson, 'Fiber Optic Thin-layer Spectroelectrochemistry with Long Optical Path', *Anal. Chem.*, **54**, 2560–2566 (1982).

45. J. Niu, S. Dong, 'An Integrated Calcium Fluoride Crystal Thin-layer Cell and its Application to Identification of Electrochemical Reduction Product of Bilirubin', *Electrochim. Acta*, **40**, 823–828 (1995).
46. K.A. Rubinson, E. Itabashi, H.B.J. Mark, 'Electrochemical Oxidation and Reduction of Methylcobalamin and Coenzyme B12', *J. Am. Chem. Soc.*, (1982).
47. T.M. Kenyhercz, T.P. DeAngelis, B.J. Norris, W.R. Heineman, J. Mark, B. Harry, 'Thin Layer Spectroelectrochemical Study of Vitamin B12 and Related Cobalamin Compounds in Aqueous Media', *J. Am. Chem. Soc.*, **98**, 2469–2477 (1975).
48. W.C. Anderson, B.H. Halsall, W.R. Heineman, 'A Small-volume Thin-layer Spectroelectrochemical Cell for the Study of Biological Components', *Anal. Biochem.*, **93**, 366–372 (1979).
49. J.L. Owens, G. Dryhurst, 'Electrochemical Oxidation of 5,6-Diaminouracil an Investigation by Thin-layer Spectroelectrochemistry', *J. Electroanal. Chem.*, **80**, 171 (1977).
50. E. Blubaugh, M.A. Yacynych, W.R. Heineman, 'Thin-layer Spectroelectrochemistry for Monitoring Kinetics of Electrogenerated Species', *Anal. Chem.*, **51**, 561 (1979).
51. T.B. Jarbawi, W.R. Heineman, 'Thin-layer Spectroelectrochemistry of a Regenerative Mechanism', *J. Electroanal. Chem.*, **132**, 323–328 (1982).
52. Y.P. Gui, T. Kuwana, 'Long Optical Path Length Thin-layer Spectroelectrochemistry. Quantitation and Potential Dependence of Electroinactive Species Adsorbed on Platinum', *J. Electroanal. Chem.*, **222**, 321–330 (1987).
53. M.P. Soriaga, A.T. Hubbard, 'Determination of the Orientation of Aromatic-molecules Adsorbed on Platinum-electrode – The Effect of Solute Concentration', *J. Am. Chem. Soc.*, **104**, 3937 (1982).
54. J.W. Strojek, T. Kuwana, 'Electrochemical-spectroscopy Using Tin Oxide-coated Optically Transparent Electrodes', *J. Electroanal. Chem.*, 741 (1968).
55. N. Winograd, T. Kuwana, 'Characteristics of the Electrode-solution Interface Under Faradaic and Non-faradaic Conditions as Observed by Internal Reflection Spectroscopy', *J. Electroanal. Chem.*, **23**, 333 (1969).
56. N. Winograd, T. Kuwana, 'High Sensitivity Internal Reflection Spectroelectrochemistry for Direct Monitoring of Diffusing Species Using Signal Averaging', *Anal. Chem.*, **43**, 252 (1971).
57. N. Winograd, T. Kuwana, 'Evaluation of Fast Homogeneous Electron-exchange Reaction Rates Using Electrochemistry and Reflection Spectroscopy', *J. Am. Chem. Soc.*, **92**, 224 (1970).
58. H.N. Blount, N. Winograd, T. Kuwana, 'Spectroelectrochemical Measurements of Second-order Catalytic Reaction Rates Using Signal Averaging', *J. Phys. Chem.*, **74**, 3231 (1970).
59. N. Winograd, T. Kuwana, 'Homogeneous Electron-transfer Reactions Studied by Internal Reflection Spectroelectrochemistry', *J. Am. Chem. Soc.*, **93**, 4343 (1971).
60. Y. Shi, A.F. Slaterbeck, C.J. Seliskar, W.R. Heineman, 'Spectroelectrochemical Sensing Based on Multimode Selectivity Simultaneously Achievable in a Single Device. 1. Demonstration of Concept with Ferricyanide', *Anal. Chem.*, **69**, 3679–3686 (1997).
61. Y. Shi, C.J. Seliskar, W.R. Heineman, 'Spectroelectrochemical Sensing Based on Multimode Selectivity Simultaneously Achievable in a Single Device. 2. Demonstration of Selectivity in the Presence of Direct Interferences', *Anal. Chem.*, **69**, 4819–4827 (1997).
62. A.F. Slaterbeck, T.H. Ridgway, C.J. Seliskar, W.R. Heineman, 'Spectroelectrochemical Sensing Based on Multimode Selectivity Simultaneously Achievable in a Single Device. 3. Effect of Signal Averaging on Limit of Detection', *Anal. Chem.*, **71**, 1196–1203 (1999).
63. L. Gao, C.J. Seliskar, W.R. Heineman, 'Spectroelectrochemical Sensing Based on Multimode Selectivity Simultaneously Achievable in a Single Device 4. Sensing with Poly(vinyl alcohol) – Polyelectrolyte Blend Modified Optically Transparent Electrodes', *Anal. Chem.*, **71**, 4061–4068 (1999).
64. D.H. Jones, S. Hinman, 'Determination of the Visible Spectra of Electrode Reaction Products in Strongly Absorbing Media by Diffusion-controlled Chronoabsorptometry', *Can. J. Chem.*, **74**, 1403 (1996).
65. E. Ahlberg, D.P. Parker, V.D. Parker, 'Sensitivity Enhancement During Electrochemical Specular Reflectance Measurements', *Acta Chem. Scand.*, **B33**, 760–762 (1979).
66. R. Kotz, S. Stucki, D. Scherson, D.M. Kolb, 'In-situ Identification of RuO_4 as the Corrosion Product During Oxygen Evolution on Ruthenium in Acid Media', *J. Electroanal. Chem.*, **172**, 211–219 (1984).
67. J.S. Mayausky, R.L. McCreery, 'Spectroelectrochemical Examination of Chloripromazine Cation Radical Reactions with Mono- and Bifunctional Nucleophiles', *J. Electroanal. Chem.*, **145**, 117–126 (1983).
68. R. Pruiksma, R.L. McCreery, 'Observation of Electrochemical Concentration Profiles by Absorption Spectroelectrochemistry', *Anal. Chem.*, **51**, 2253 (1979).
69. C.-C. Jan, R.L. McCreery, F.T. Gamble, 'Diffusion Layer Imaging: Spatial Resolution of Electrochemical Concentration Profiles', *Anal. Chem.*, **57**, 1763–1765 (1985).
70. H.-P. Wu, R.L. McCreery, 'Spatially Resolved Absorption Spectroelectrochemistry', *J. Electrochem. Soc.*, **136**, 1375 (1989).
71. H.-P. Wu, R.L. McCreery, 'Observation of Concentration Profiles at Cylindrical Microelectrodes by a Combination of Spatially Resolved Absorption Spectroscopy and the Abel Inversion', *Anal. Chem.*, **61**, 2347–2352 (1989).
72. A.L. Deputy, R.L. McCreery, 'Spatially Resolved Absorption Examination of the Redox Catalysis Mechanism: Equilibrium and Near Equilibrium Cases', *J. Electroanal Chem.*, **285**, 1–9 (1989).
73. C.M.A. Brett, M. Oliveira Brett, *Electrochemistry Principles Methods and Applications*, Oxford University Press, Oxford, 1993.

74. M. Zhao, D.A. Scherson, 'UV-Visible Reflection Absorption Spectroscopy in the Presence of Convective Flow', *Anal. Chem.*, **64**, 3964 (1992).
75. H. Debrodt, K.E. Heusler, 'Eine rotierende Scheibenelektrode mit optisch durchlassigem Ring zur spektrophotometrischen Untersuchung von Produkten elektrochemischer Reaktionen', *Ber. Bunsen-Ges. Phys. Chem.*, **81**, 1172 (1977).
76. R. Dorr, E.W. Grabner, 'Digital Simulation of a Rotating Disk Electrode with Optically Transparent Ring', *Ber. Bunsen-Ges. Phys. Chem.*, **82**, 164 (1978).
77. Y.V. Tolmachev, Z. Wang, D.A. Scherson, 'In-situ Spectroscopy in the Presence of Convective Flow Under Steady State Conditions: A Unified Mathematical Approach', *J. Electrochem. Soc.*, **143**, 3539 (1996).
78. R.L. Wang, K.Y. Tam, R.G. Compton, 'Applications of the Channel Flow Cell for UV Visible Spectroelectrochemical Studies. Part 3. Do Radical Cations and Anions have Similar Diffusion Coefficients to their Neutral Parent Molecules?', *J. Electroanal. Chem.*, **434**, 105 (1997).
79. Z. Wang, M. Zhao, D. Scherson, 'Channel Flow Cell for UV/Visible Spectroelectrochemistry', *Anal. Chem.*, **66**, 4560 (1994).
80. Z. Wang, D. Scherson, 'In Situ UV-visible Spectroscopic Imaging of Laminar Flow in a Channel-type Electrochemical Cell', *J. Electrochem. Soc.*, **142**, 4225 (1995).
81. S. Ma, Y. Wu, Z. Wang, 'Spectroelectrochemistry for a Coupled Chemical Reaction in the Channel Cell. Part I. Theoretical Simulation of an EC Reaction', *J. Electroanal. Chem.*, **464**, 176 (1999).
82. Z. Wang, X. Li, Y. Wu, Y. Tang, S. Ma, 'Spectroelectrochemistry for a Coupled Chemical Reaction in the Channel Cell. Part II. Kinetics of Hydrolysis and the Absorption Spectrum of *p*-Benzoquinoneimine', *J. Electroanal. Chem.*, **464**, 181 (1999).
83. V.S. Krylov, V.N. Babak, 'Nonsteady-state Diffusion to the Surface of a Rotating Disk', *Soviet Electrochem.*, **7**, 626 (1971).
84. K. Nicancioglu, J.S. Newman, 'Transient Convective Diffusion to a Disk Electrode', *J. Electroanal. Chem.*, **50**, 23 (1974).
85. M. Zhao, D.A. Scherson, 'Transient Convective Diffusion to a Rotating Disk Electrode as Monitored by Near-normal Incidence Reflection Absorption Ultraviolet-visible Spectroscopy', *J. Electrochem. Soc.*, **140**, 729 (1993).
86. A.S. Hinman, J.F. McAleer, S. Pons, 'Spectroscopic Determination of the AC Voltammetric Response', *J. Electroanal Chem.*, **154**, 45–56 (1983).
87. C.-C. Jan, R.L. McCreery, 'Spectroelectrochemical Determination of Trace Concentrations by Diffusion Layer Imaging', *J. Electroanal. Chem.*, **220**, 41 (1987).
88. M. Zhao, D.A. Scherson, 'Theoretical Aspects of Potential Modulation Normal Incidence Reflection Absorption UV-visible Spectroscopy Under Forced Convection', *J. Electrochem. Soc.*, **140**, 1671–1676 (1993).
89. D. Scherson, J. Newman, 'The Warburg Impedance in the Presence of Convective Flow', *J. Electrochem. Soc.*, **127**, 110 (1980).
90. M. Zhao, D.A. Scherson, 'Theoretical Aspects of Potential Modulation Reflection Absorption UV-visible Spectroscopy Under Forced Convection', *J. Electrochem. Soc.*, **140**, 1671 (1993).
91. M. Zhao, D.A. Scherson, 'Potential Modulation Normal Incidence Reflection Absorption UV-visible Spectroscopy Under Forced Convection', *J. Electrochem. Soc.*, **140**, 2877–2879 (1993).
92. Y.V. Tolmachev, Z. Wang, Y. Hu, I.T. Bae, D.A. Scherson, 'In-situ Spectroscopic Determination of Faradaic Efficiencies in Systems with Forced Convection Under Steady State: Electroreduction of Bisulfite to Dithionite on Gold in an Aqueous Electrolyte', *Anal. Chem.*, **70**, 1149–1155 (1998).

X-ray Methods for the Study of Electrode Interaction

Enrique Herrero
Universidad de Alicante, Alicante, Spain

For references see page 10249

X-ray methods provide an excellent tool for the determination of the structure and composition of the electrode/solution interphase. The possibility of carrying out in situ experiments, because of the low absorbability of the X-rays by the solution, allows direct correlation to be made between the structural changes observed in the sample and the potential applied. Depending on the interaction of the X-rays with matter, these techniques can be classified in two main groups: X-ray absorption techniques and X-ray scattering techniques. Within the absorption techniques, extended X-ray absorption fine structure (EXAFS) and X-ray absorption near edge structure (XANES) will be considered. EXAFS allows determination of the short-range order of the sample, i.e. it provides information about the nature and distances of the closest neighbors to the absorbing atom. On the other hand, identification of the oxidation state of the absorbing atom can be done with XANES. Grazing incident X-ray diffraction (GIXD) and crystal truncation rod (CTR) measurements belong to the scattering techniques. The information obtained from these measurements can be considered complementary since GIXD allows precise determination of the long-range order structure of the interphase whereas CTR analysis gives information about the out-of-plane structure of the interphase. The last technique covered, X-ray standing waves (XSW), is not a pure scattering technique, but rather one that combines X-ray scattering and X-ray interference. With this last technique, the exact position of a foreign atom in the interphase can be determined.

1 INTRODUCTION

Understanding the electrode/electrolyte interphase is one of the major areas of study in electrochemistry.[1] The structure and properties of the interphase and the interactions both between the electrode and electrolyte, and within the different species in the electrolyte, affect and control the behavior of electrochemical systems. The complex nature of the electrochemical interphase requires the use of different techniques, since a single technique is able to render only a partial picture of the system. The first available techniques, pure electrochemical ones, provided macroscopic information about the reactions occurring at the interphase, and the thermodynamics and kinetics of these processes. It is clear that a complete picture of the interphase cannot be achieved with the information collected by these techniques, since no information about the interactions between the electrode and electrolyte and the interphase structure is obtained.

The existence of a condensed phase in contact with the electrode caused a major problem in using the techniques employed in other fields of surface chemistry, since it constituted a barrier that limited the penetration of the probes. The first approach to studying the surface structure of an electrode was the ex situ approach, using the techniques for surface characterization that the ultrahigh vacuum (UHV) environment provided. Although UHV techniques, such as low-energy electron diffraction (LEED) and Auger electron spectroscopy (AES), are powerful surface characterization techniques, the removal of the electrolyte and the loss of potential control can lead to changes in the structure and composition in the interphase. This drawback has limited the application of UHV techniques to the characterization of very stable layers, normally metal layers, on the electrode surface.

In recent years, the advent of in situ spectroscopic scanning probe microscopy (SPM) and X-ray techniques for electrochemical systems have started to fill in the gap in information about the interphase. The spectroscopic techniques, such as in situ Fourier transform infrared spectroscopy (FTIRS) and related techniques,[2,3] enhanced Raman spectroscopy,[4–7] and nonlinear optical spectroscopy,[8,9] give information about the chemical composition and the orientation of the different species on the interphase. The SPM techniques[10] (scanning tunneling microscopy (STM) and atomic force microscopy (AFM)) provide information about the surface structure on the atomic level, when ordered structures are found. However, interpretation of the images obtained with these techniques is not always straightforward, since the maxima appearing in the image cannot be unequivocally identified with a single species in the system.

Details about the surface structure can also be obtained by X-ray techniques. When compared with SPM techniques, X-ray techniques present several advantages:

- Distances and angles calculated from X-ray measurements have a higher accuracy than those calculated from SPM techniques. The typical error in the distance measurements with STM or AFM is ca. 10%, whereas errors below 2% can be routinely obtained with X-ray techniques. This leads to a better distinction between different possible structures and the detection of small shifts in the positions of the atoms with different experimental conditions.

- Penetration of the X-rays in matter allows collection of data about the surface structure of the electrode, which can also be affected by the experimental conditions. SPM techniques provide information about the outermost layer (only in extreme conditions can the underlying substrate be imaged).
- They can also be used to determine the out-of-plane structure (the structure of the interphase perpendicular to the electrode surface), the short-range order in nonordered interphases, and the chemical nature of the species in the interphase.

The main disadvantage of the X-ray techniques is the requirement of a source bright enough to be surface sensitive. Although some studies have been done using laboratory X-ray sources, their brightness does not allow data about the structure of the interphase to be obtained without very long acquisition times, and a more powerful source has to be used: a synchrotron.

X-ray techniques can be divided in two main families, depending on the interaction of the X-rays with matter: absorption techniques and scattering techniques. Within the absorption techniques,(11–14) we can consider the surface extended X-ray absorption fine structure (SEXAFS), which provides information about the short-range order structure of the substrate, and XANES, which allows identification of the oxidation state of the atoms. The main scattering techniques(11–13,15–19) are GIXD that allows a precise determination of the long-range order structure of the interphase and CTR analysis, a reflectivity technique that gives information about the out-of-plane structure of the interphase. The XSW technique is not a pure scattering technique, but rather one that combines X-ray scattering and X-ray interference. However, in the present article, it has been included as a scattering technique. With this technique, the atomic positioning of foreign atoms in the interphase can be determined.

1.1 X-ray Sources

X-rays were discovered by Röntgen in 1895 using an evacuated tube with two metal plates (anode and cathode). When a voltage is applied between anode and cathode, the residual gas in the tube is ionized and the cations generated collide with the cathode, which releases electrons under this bombardment. These electrons are accelerated by the electric field to the anode and collide with it, generating the X-rays in the sudden deceleration of the electrons. If all the energy of the impinging electrons is converted to X-ray photon energy, the wavelength of the resulting X-rays will be given by the Duane–Hunt law (Equation 1):

$$\lambda_{\min} = \frac{hc}{eV} = \frac{12\,398\,\text{Å}\,\text{V}^{-1}}{V} \tag{1}$$

where h is Planck's constant, c is the velocity of light, e is the electron charge and V the applied voltage between anode and cathode. ($1\text{Å} = 10^{-10}\,\text{m}$.) However, not all the kinetic energy of the electrons is transformed into X-ray photon energy. Part of the incident energy can be lost in multiple collisions and inelastic scattering, resulting in a continuous "bremsstrahlung" spectrum (white radiation). Therefore, the wavelength obtained according to Equation (1) ($\lambda_{\min}$) represents the minimum wavelength emitted by the tube.

When the accelerating voltage reaches a specific value (which depends on the target material), the electrons are able to knock out a core electron from the atoms of the target material, producing a core hole. If an electron from the outer shells descends to fill this hole, the emitted energy depends on the energy of the two levels involved. This results in a sharp spike in the emission spectrum (Figure 1). These characteristic lines are well defined and have an intensity that can be up to three orders of magnitude higher than that of the bremsstrahlung radiation. Thus, conventional X-ray tubes can be used as sources of both a continuous X-ray spectrum and discrete characteristic lines.

The main disadvantage of these X-ray tubes is the low brightness of the generated X-rays. Rotating anode sources, which take advantage of bent crystal optics, are able to generate 10^2–10^3 times higher intensity than a conventional tube. However, the application of these sources to the study of the solid/liquid interphase requires

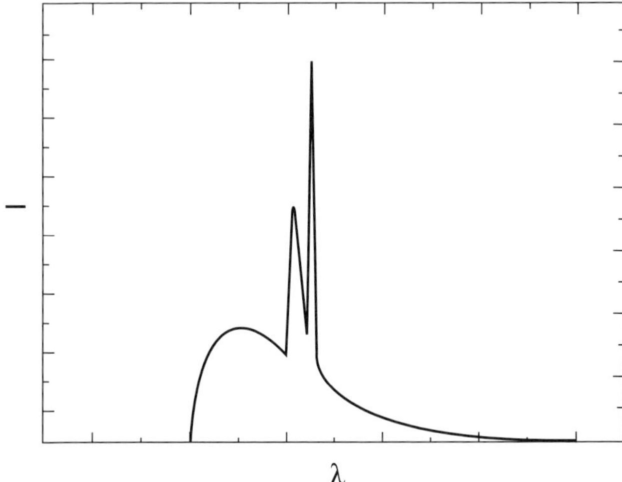

Figure 1 X-ray spectrum with the characteristic lines emitted by an X-ray tube.

For references see page 10249

long acquisition times. For that reason, most of the X-ray studies of the electrochemical interphase have been carried out using synchrotron radiation, which provides brightness at least 10^3 times higher than that obtained with a rotating anode.

1.2 Synchrotron Radiation

The basis of synchrotron radiation is the electromagnetic radiation emitted by a charged particle under acceleration. When a charged particle (in this case, an electron or positron) describes a circular trajectory as a result of a centripetal acceleration, it emits radiation perpendicular to the trajectory. For electrons traveling at low velocity, the radiation is emitted isotropically in all the directions perpendicular to the acceleration, as shown in Figure 2(a). However, at relativistic velocities the radiation is focused into a cone in the plane of orbit of the electrons and in the forward direction (Figure 2b). The divergence angle (θ_v) is approximately inversely proportional to the wavelength and given by Equation (2):

$$\theta_v \propto \frac{1}{\lambda}, \quad \lambda = \frac{E}{m_0c^2} \tag{2}$$

where m_0 is the rest mass of the electron.

The radiation emitted by relativistic electrons has a very small divergence (around 0.2 milliradian) which results in a high brightness (defined as the number of photons per unit band width, per unit solid angle, per unit area source) of the radiation. Another important quality of the radiation is that it is almost 100% polarized in the plane of the orbit.

The spectrum of the radiation emitted by a relativistic electron in a circular orbit is continuous down to a critical wavelength, which in turn depends on the energy of the electrons and the radius of the orbit that it follows. The high brightness of continuous radiation allows a specific wavelength to be selected for the experiment without being limited to the characteristic lines of a tube generator. Moreover, the intensity of synchrotron radiation is at least 10^3 times higher than that obtained by a rotating anode tube.

A synchrotron (Figure 3) is an approximately circular vacuum chamber with bending magnets for directing the beam, quadrupole magnets for focusing the beam and radiofrequency cavities for accelerating the electron beam. A synchrotron working properly requires that the energy of the beam be at least 10 MeV. Therefore, a linear accelerator (LINAC) with radiofrequency cavities injects the electron beam continuously in the ring. This classical synchrotron has some disadvantages:[20]

- low stability of the beam;
- the intensity and the critical energy of the radiation changes very quickly depending on the accelerating cycle;

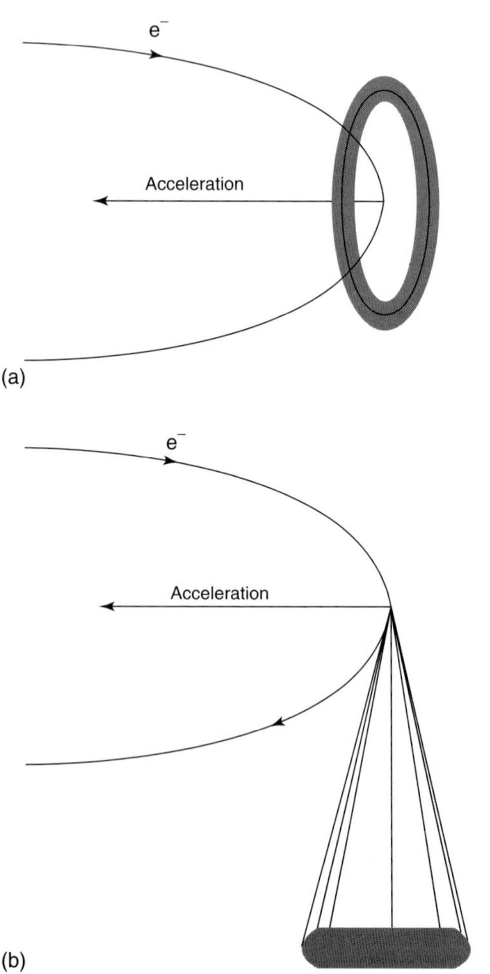

Figure 2 Radiation emitted by an orbiting electron at (a) low and (b) relativistic velocities.

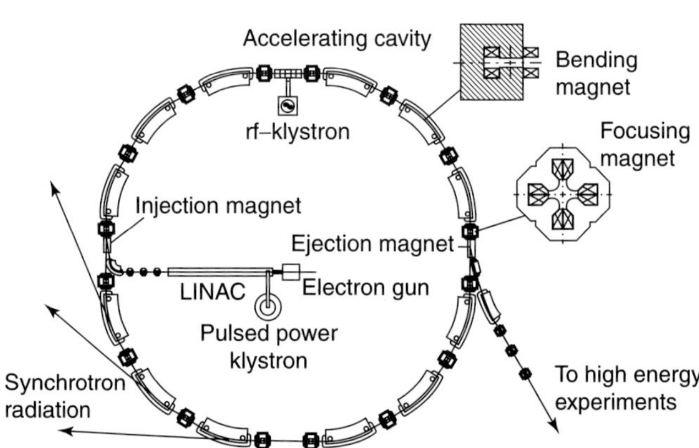

Figure 3 Sketch of a synchrotron. (Reproduced from K. Wilke, 'Synchrotron Radiation Sources', *Rep. Prog. Phys.*, **54**, 1005–1068. © 1991 by permission of the Institute of Physics and the author.)

- low beam currents and therefore low intensity;
- the beam is not focused to the extremely small dimensions required by some experiments.

For those reasons, synchrotrons have been replaced by storage rings, which in design are similar to synchrotrons. In these storage rings, the accelerated beam is injected at a given energy, and the machine keeps the beam circulating in the ring for several hours without change in energy. Therefore, the spectrum and the intensity are extremely stable. A radiofrequency cavity is also needed to compensate losses of energy due to the emission of radiation. The electrons circulating in the ring are grouped in bunches, conferring on the beam a time structure which can be used for several experiments, especially in kinetic studies.

In the storage ring, the X-rays are generated in the insertion devices: wigglers, undulators and wavelength sifters. In general, they consist of magnetic structures with alternating poles. As the critical wavelength is proportional to the bending radius, these devices force the electron beam to move with very short radius of curvature. A detailed discussion of the advantage of each insertion device can be found in Wilke.[(20)]

2 X-RAY ABSORPTION TECHNIQUES

2.1 Interaction of X-rays with Matter

X-rays can interact in two different ways with matter: the X-rays are either scattered or absorbed by the matter. In the first case, the electromagnetic wave of the X-ray beam makes the electrons of matter oscillate, so that they emit radiation in all directions. When no momentum is transferred to the electrons (elastic scattering), the scattered beam has the same wavelength as the incident beam. On the other hand, when momentum is transferred to loosely bound electrons (Compton effect), the scattered beam has a higher wavelength than the incident beam. The scattering phenomenon will be discussed extensively in section 3, therefore in this section we will concentrate on the absorption phenomena.

When monochromatic X-rays travel through matter, part of the incident beam is absorbed by it. The intensity of transmitted beam (I) follows the Beer–Lambert Law (Equation 3):

$$I = I_0 e^{-\mu x} \tag{3}$$

where μ is the linear absorption coefficient (which depends on the wavelength), I_0 the intensity of the incident beam and x the thickness of the sample. Sometimes, Equation (3) is expressed in terms of the mass absorption coefficient μ/ρ, which is obtained by dividing the linear absorption coefficient by the density (ρ). Therefore (Equation 4):

$$I = I_0 e^{-\left(\frac{\mu}{\rho}\right)\rho x} \tag{4}$$

The mass absorption coefficient for an element is almost independent of the physical state and approximately additive. This way, the mass absorption coefficient for a compound, mixture or alloy can be expressed as Equation (5):

$$\frac{\mu}{\rho} = \sum_i w_i \left(\frac{\mu}{\rho}\right)_i \tag{5}$$

where w_i is the mass fraction of the element i and $(\mu/\rho)_i$ its mass absorption coefficient. Except for the discontinuities for the absorption edges (see below), the mass absorption coefficient approximately follows the relationship (Equation 6):

$$\frac{\mu}{\rho} = CZ^4\lambda^3 \tag{6}$$

where C is a constant and Z the atomic number of the element. For practical uses, values of the mass absorption coefficients are tabulated in Ibers and Hamilton.[(21)]

If the energy of the incident beam is increased, it can reach a value where a core electron from an atom in the sample can be ejected, originating an abrupt increase in the absorption coefficient, which is called an absorption edge. There is an absorption edge for each shell and subshell in the excited atom, i.e. there is a K absorption edge, three L edges, five M edges etc. (Figure 4). As expected, the energy of the absorption edge increases in the order M < L < K. After the edge, the diminution in the absorption mass coefficient can also be described by Equation (6), but the constant is different to that obtained at energies below the edge.

Once a hole in the core levels has been created, the excited atom can relax in several ways. First, an electron of the outer shells can descend to fill in the hole, emitting radiation. The energy of this radiation depends on the energy difference between the two levels involved in the electronic transition. This phenomenon is known as fluorescence. The allowed transitions between two electronic levels are restricted to changes in the quantum number of $\Delta l = \pm 1$, and $\Delta m = \pm 1, 0$. The intensity of the fluorescence signal generated depends on the probability of the transition between the two electronic levels.

The relaxation of the atom can also give rise to the emission of Auger electrons or secondary electrons. In the Auger process, one electron descends to occupy the hole in the core levels, and the energy emitted is used to eject an electron from another level. The secondary electrons are emitted when the Auger electrons or fluorescence

For references see page 10249

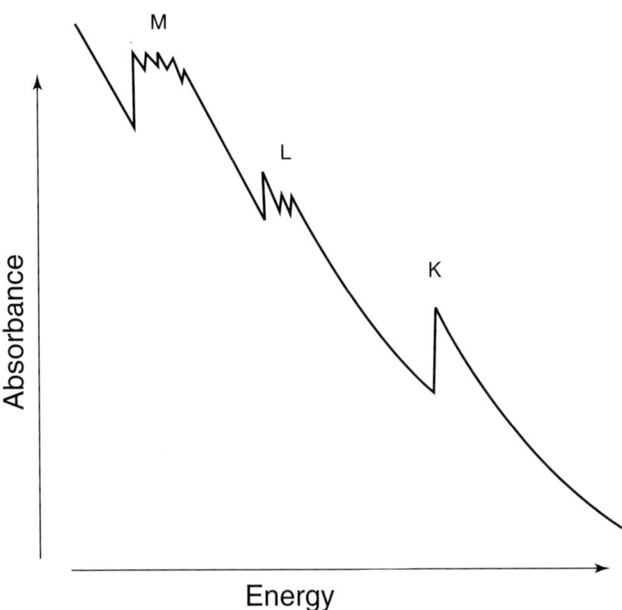

Figure 4 Absorbance spectrum of an atom as a function of the energy of the incident beam.

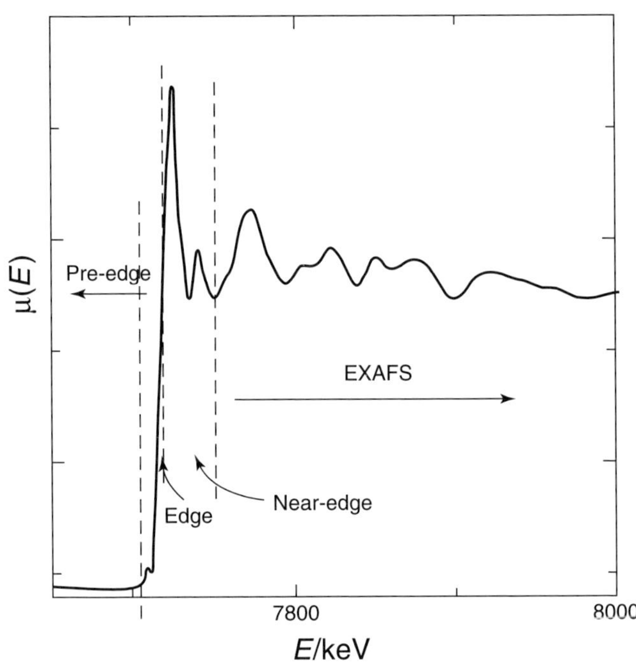

Figure 5 Absorbance spectrum of a CoO foil divided into different regions.

radiation interacts with the absorbing atom, ejecting one electron from the outer shells.

When the absorption (or fluorescence) spectrum of a substance is recorded, some oscillations are observed after the edge, superimposed on the decay in absorbance that should approximately follow Equation (6). Figure 5 shows the typical absorption spectrum of the K edge of a CoO foil. Although the presence of oscillations (EXAFS) was first discovered by Kronig in 1931,[22,23] the phenomenon was not fully understood until the 1970s, when Stern, Lytle and Sayers developed their theory.[24] From the beginning, it was known that the oscillations contained structural information, since the absorption coefficient for monoatomic diluted gases, for which no interactions between the atoms are expected, showed no oscillations. However, it was not clear whether the oscillations were due to long- or short-range order interactions. The theories developed in the 1970s confirmed that the nature of the oscillations were short-range order interactions.[24–29]

The oscillation in the EXAFS region is the result of an interference phenomenon. The electron ejected from the core level can be treated as a wave because of wave–particle duality. In the presence of another atom in the near neighborhood, this wave can be backscattered and interfere with the outgoing wave (Figure 6). From the interference nature of the EXAFS phenomenon, it is clear that the frequency of the oscillations depends on the distance between the absorbing atom and the backscatterer, and the amplitude is function of the nature and number of backscatterers. Therefore, a detailed

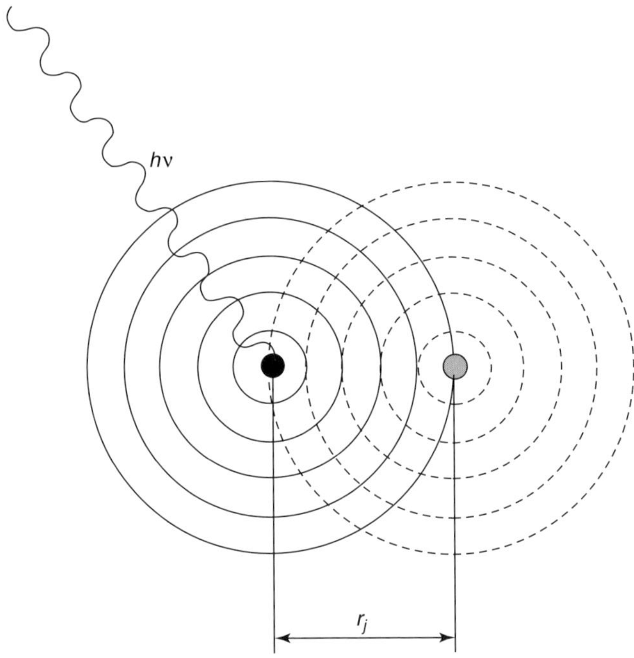

Figure 6 Scheme of the backscattering phenomenon that originates the EXAFS spectra. r_j is the distance between the absorbing and backscattering atoms.

analysis of the EXAFS zone can provide information about the short-range order of the sample.

The theories for describing EXAFS are based on the single-electron single-scattering formalism.[24–29] This assumes that the photoelectron has an energy high enough to suffer only one backscattering event. This condition is normally fulfilled for energies 40 eV above the edge. However, the absorption spectrum of a substance has other regions with important features (Figure 5). Near or below the edge, some absorption peaks may appear, caused by the excitation of core electrons to bond states. The edge position gives information about the effective charge of the absorbing atom, and can be used in the assignment of oxidation states of the atom. In the near-edge region, the photoelectron has a very small momentum and multiple backscattering events take place. All these regions are normally termed XANES. Although the structural information that can be obtained from this region can be very rich, the complex nature of the interaction makes the development of a complete mathematical model difficult. The present status of XANES theory has been reviewed in Bianconi[30] and Stöhr.[31]

2.2 Extended X-ray Absorption Fine Structure and X-ray Absorption Near Edge Structure Theory

As already mentioned, EXAFS refers to the region in the absorption spectrum between 40 and 1000 eV above the edge. In order to obtain the structural information contained in the spectrum, the absorption coefficient as a function of the energy has to be known ($\mu(E)$). The first step towards obtaining the structural information is to isolate the oscillatory part of the spectrum from the rest. This process is carried out by subtracting the absorption coefficient obtained for isolated atoms, $\mu_0(E)$, from the experimental absorption coefficient $\mu(E)$. $\mu_0(E)$ could be obtained for a diluted monoatomic gas, since the interactions between the different atoms are very small and the atoms can be considered as isolated. The result is normalized to $\mu_0(E)$ to obtain the EXAFS function in the energy space ($\chi(E)$) on a per atom basis (Equation 7):

$$\chi(E) = \frac{\mu(E) - \mu_0(E)}{\mu_0(E)} \tag{7}$$

The dual behavior of the emitted electron allows its wavelength to be calculated from Equation (8):

$$\lambda = \frac{2\pi}{k} \tag{8}$$

where k is the wave vector, calculated according to Equation (9):

$$k = \sqrt{\frac{2m}{\hbar^2}(E - E_0)} \tag{9}$$

Here $\hbar$ is $h/2\pi$, m is the mass of the electron and E_0 the energy of the absorption edge. When E is in eV and k in Å^{-1}, Equation (9) is transformed to Equation (10):

$$k = \sqrt{0.2625(E - E_0)} \tag{10}$$

The structural information of the EXAFS can only be extracted after transformation of the EXAFS function to the wave vector space ($\chi(k)$) using Equation (9).

As each shell of atoms surrounding the absorber will give an interference event, the EXAFS spectrum will involve the summation over all these interference events (Equation 11):

$$\chi(k) = \sum_j A_j(k) \sin\left(2kr_j + \phi_{ij}(k)\right) \tag{11}$$

where the subscripts j and i refer to the jth shell of neighboring atoms and the absorbing atom, respectively, $A_j(k)$ is the amplitude term, $\sin(2kr_j + \phi_{ij}(k))$ the oscillatory term, r_j the distance between the atoms in the jth shell to the absorbing atom and $\phi_{ij}(k)$ the total phase shift experienced by the photoelectron. We will discuss each term separately.

2.2.1 *Oscillatory Term:* $\sin\left(2kr_j + \phi_{ij}(k)\right)$

This term contains the information about the distance between the backscattering and absorbing atoms. As seen in Figure 6, the wave travels twice the distance between the absorbing and backscattering atom. Therefore, in the final state, when it returns to the absorbing atom, it has experienced a phase shift of $2kr_j$. However, the backscattering phenomenon is mainly due to the repulsion between the photoelectron and the electronic cloud of the backscattering atom and the effective distance traveled by the photoelectron wave is shorter than $2kr_j$. An additional term, $\phi(k)$, is required to account for this effect.

2.2.2 *The Amplitude Term:* $A_j(k)$

This term, which contains information about the nature and number of the atoms in the jth shell, can be written as Equation (12):

$$A_j(k) = \frac{N_j}{kr_j^2} S_i(k) F_j(k) \mathrm{e}^{-2\sigma_j^2 k^2} \mathrm{e}^{-2r_j/\lambda_j} \tag{12}$$

where N_j is the number of atoms in the jth shell, $F_j(k)$ the backscattering amplitude for a single atom, $S_i(k)$ an amplitude reduction associated to many-body effects, σ_j^2 the Debye–Waller factor and $\lambda_j(k)$ the mean free path of the electron.

As can be seen, the maximum amplitude, given by $N_jF_j(k)$ is reduced by a series of factors. The first one, $1/(kr_j^2)$, has a double diminution effect. First, the EXAFS function will decrease rapidly as the distance

For references see page 10249

between the absorbing and the backscattering atom increases. This means that the contributions of the farther backscattering atoms to the total EXAFS function will be smaller than those arising from the closer atoms, restricting the EXAFS phenomenon to the closest shells and justifying the short-range nature of the EXAFS. The second damping factor is related to k, implying a diminution of the oscillation amplitude as the energy increases. This is one of the factors that restricts the presence of oscillations to approximately 1000 eV above the absorption edge. Another important factor in the decrease of the oscillations with increasing energy is the Debye–Waller factor (see below).

An additional damping factor is related to many-body effects ($S_i(k)$). As the subscript indicates, this term is associated with the absorbing atom and includes losses in the photoelectron energy due to electron shake-up (excitation of other electrons in the absorber) and shake-off (ionization of the electrons of the outer shells).

The reduction in the amplitude due to the thermal vibration and static disorder is contained in $e^{-2\sigma_j^2 k^2}$. When the backscattering event takes place, not all the backscattering atoms in the shell are exactly at the same distance. On the contrary, they may appear displaced from their "equilibrium" positions with a mean standard deviation of σ_j^2. The presence of a negative exponential means that the effect of the Debye–Waller factor is more intense at high-energy values (high k values). The EXAFS spectra of samples with a higher Debye–Waller factor (because of vibrations or static disorder) will exhibit oscillations that disappear rapidly above the edge.

The last damping term corresponds to the inelastic losses in the photoelectron energy (e^{-2r_j/λ_j}). The photoelectron that suffers an inelastic loss is not able to contribute to the interference phenomenon. This term is also responsible for the short-range limitation of the EXAFS phenomenon, since the probability of an inelastic loss increases with the distance between the adsorbing and backscattering atoms.

As already mentioned, there is not a complete theory that describes the XANES region of the spectra.[30,31] The information contained in this region is mainly related to the absorbing atom. The position of the edge is related to the electronic density of the absorbing atom, in a similar way to X-ray photoelectron spectroscopy (XPS). Other features of the region, such as the so-called "white line" (a sharp peak appearing after the edge) is associated with transitions to unoccupied states. These transitions are also affected by the symmetry of the adsorbing atom.

2.3 Data Analysis

In order to obtain the structural information contained in the EXAFS spectrum, the data have to be treated and manipulated. The reliability of the data obtained from the EXAFS spectrum depends on the correctness of the data manipulation. A very detailed data analysis can be found in Teo.[32] It normally comprises the following steps:

1. background removal and normalization;
2. conversion to the k space;
3. k weighting;
4. Fourier transforming and filtering;
5. fitting for the phase and amplitude of the filtered spectrum.

We will discuss now in detail every point in the data analysis.

2.3.1 Background Removal and Normalization

Prior to background subtraction, the measured data has to be converted to absorption coefficient. When the measured property is the transmitted intensity, the total linear absorption coefficient can be calculated from Equation (13):

$$\mu(E)x = \ln \frac{I_0}{I} \tag{13}$$

Another possibility is to measure the fluorescence intensity (F). As the probability of emitting a photoelectron is directly proportional to the probability of absorption, the

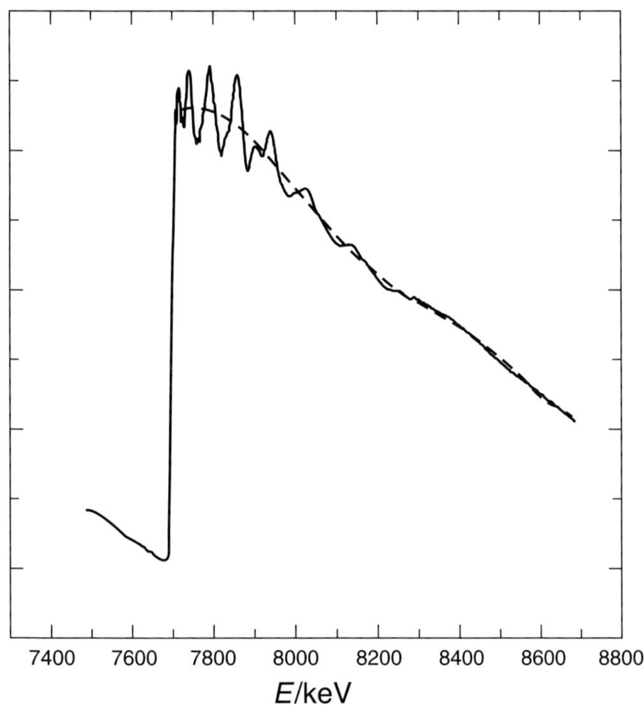

Figure 7 EXAFS spectrum of Co. The dashed line represents the fitted spline to calculate $[\mu(E) - \mu_0(E)]$.

absorption coefficient is given by Equation (14):

$$\mu(E)x = \frac{F}{I_0} \tag{14}$$

In order to calculate the EXAFS function ($\chi(E)$) according to Equation (7), the value of $\mu_0(E)$ has to be known. Normally $\mu_0(E)$ values are not available and thus other strategies have to be used to calculate $\chi(E)$. Moreover, the spectrum also contains additional background elements, such as the spectrometer baseline, beam harmonics, elastic scattering, etc. The strategy normally followed to calculate $\mu_0(E)$ and to remove all the additional background is to fit a polynomial spline to the EXAFS part of spectrum. Figure 7 shows the spline obtained for the Co spectrum.

As mentioned above, the spline obtained is not only $\mu_0(E)$ but also contains the additional background, thus it is not possible to use it as normalization factor. One of the most used strategies is to divide $[\mu(E) - \mu_0(E)]$ by the edge jump ($\mu_{E'}$). Figure 8(a) shows the EXAFS function obtained for a Co film after background subtraction and normalization.

2.3.2 *Conversion to k Space*

The conversion to k space according to Equation (9) requires the determination of the edge threshold (E_0). However, the edge position cannot be identified with any feature of the spectrum. Several points can be chosen as the edge threshold, i.e. the onset of the jump, the inflexion point, the mid-point in the edge jump, the edge peak, etc. Despite such an unclear definition of edge threshold, the exact position of the edge is not very important, provided that it is chosen similarly for related compounds. The final analysis of the EXAFS spectrum normally uses comparison with model compounds or theoretical models, in which the value of E_0 is treated as an adjustable parameter.

2.3.3 *k Weighting*

The oscillations in the EXAFS function are attenuated at high k values by a series of factors. The amplitude of the oscillation has $1/k$ factor and normally the backscattering amplitude $[F_j(k)]$ also depends on $1/k^2$. Thus, the oscillations appearing at low energies (low

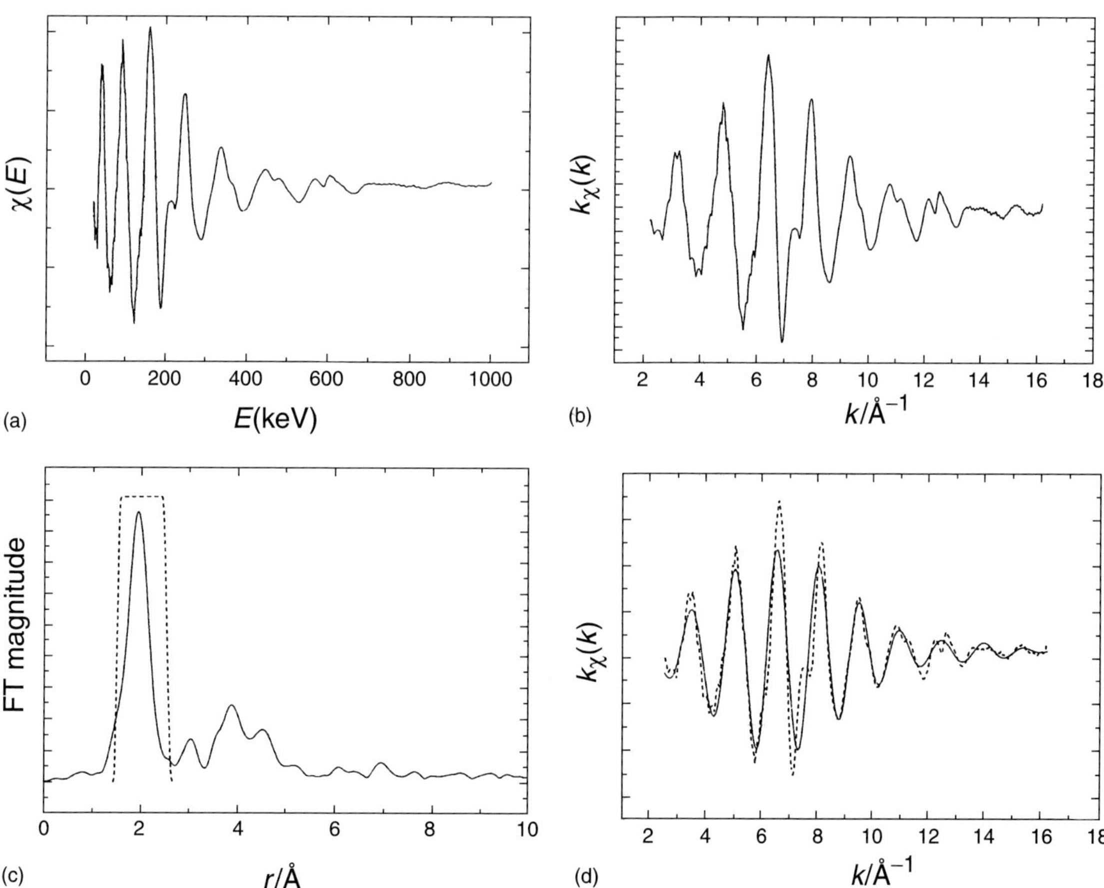

Figure 8 Four steps in the data analysis of an EXAFS spectrum: (a) EXAFS spectrum after background removal and normalization, (b) k^1 weighting, (c) Fourier transform and (d) EXAFS function after Fourier filtering. The dashed line in (c) represents the window used for the Fourier filtering and in (d) represents the original k weighted function (panel (b)).

For references see page 10249

wave vector) dominate over those at high energies. On the other hand, the interatomic distances depend only on the frequency, not on the amplitude of the oscillations. To prevent any effect of the amplitude on the distance determination, the EXAFS function is normally weighted by k^1, k^2 or k^3. k^1 factor cancels the $1/k$ factor in the amplitude functions and k^3 cancels both the $1/k$ and the $1/k^2$ factors. k^3 is normally preferred for light absorbing atoms and k^1 for heavy absorbing atoms. When comparing the k weighted function (Figure 8b) to the original EXAFS function (Figure 8a), it can be seen that the oscillations on the k weighted function shows a higher relative amplitude for high energies.

2.3.4 *Fourier Transform and Filtering*

Fourier transform allows the EXAFS function to be translated from the k space to the frequency (or distance) space (Equation 15):

$$\phi(r) = \frac{1}{\sqrt{2\pi}} \int_{k_{\min}}^{k_{\max}} \mathrm{w}(k)k^n\chi(k)\mathrm{e}^{2ikt}\,\mathrm{d}k \qquad (15)$$

where $\mathrm{w}(k)$ is a window function which selects the k range to be transformed. Fourier transform yields a function that resembles a radial distribution function, in which a number of peaks appear at given distances (Figure 8c). The advantage of Fourier transform is that the transform of a summation of functions is the summation of the Fourier transform of the individual functions. Thus every peak appearing in the radial distribution function can be assigned to a given shell (unless the distances between two shells are small enough to prevent peak resolution in the frequency space). The distance at which each peak appears does not correspond exactly to the distance between the absorbing atom and the shell since these distances have to be corrected for the total phase shift ($\phi_{ij}(k)$) (see section 2.2). Fourier back transforming the desired shell or shells using a window as shown in Figure 8(c) (dashed line) allows filtering the $k^n\chi(k)$ function and eliminating the undesired contributions (noise, high frequencies...) (Figure 8d).

2.3.5 *Fitting for Phase and Amplitude*

The last part of the data analysis is obtaining the structural parameters. As shown in section 2.2, the EXAFS function depends on several parameters: r_j, N_j, σ_j and λ_j. In order to obtain these parameters from the spectrum, the total phase shift ($\phi_{ij}(k)$) and backscattering amplitude ($\mathrm{F}_j(k)$) have to be known. One of the usual approaches to calculating both functions is to use the spectrum of model compounds, from which the different parameters are known. Another possibility is to generate these functions from theoretical models, such as the one proposed by Teo and Lee[33] or from the program code FEFF.[34–37]

The use of $\phi_{ij}(k)$ and $\mathrm{F}_j(k)$ obtained theoretically or from model compounds implies that they are similar to those found in our spectrum, and this is normally termed as phase and amplitude transferability. The phase transferability is normally regarded as very good, since the phase shift depends on the core electrons of the absorbing and backscattering atoms. Therefore, it is almost independent of the chemical state of both atoms and only depends on their nature. This allows very accurate determination of the interatomic distances, with an estimated error of ±0.01 Å. On the other hand, amplitude transferability is considered worse than the phase transferability, since it depends on several factors. Except when very similar model compounds are used, the determination of N_j is not better than ±20%.

The strategy for obtaining the different structural parameters is a nonlineal least-squares fitting with the following function (Equation 16) to the k^n-weighted filtered function:

$$\mathrm{Y}(k) = \sum_j \frac{N_j}{r_j^2}\mathrm{F}_j(k_j)k_j^{n-1}\mathrm{e}^{-2\sigma_j^2k_j^2}\mathrm{e}^{-2\frac{r_j}{\lambda_j(k_j)}} \times \sin\left(2k_jr_j + \phi_{ij}(k_j)\right) \qquad (16)$$

where $\mathrm{F}_j(k_j)$ and $\phi_{ij}(k_j)$ functions are obtained from model compounds or theoretical models. Here the value of E_0 is allowed to vary for every shell, giving a different k_j value for the shell according to Equation (17):

$$k_j = \sqrt{k^2 - 0.2625(E_{0_j} - E_0)} \qquad (17)$$

2.4 Experimental Aspects

2.4.1 *Detection*

When the X-ray beam interacts with matter, four different processes occur: absorption, fluorescence, Auger electron emission and secondary electron emission.[38] As they are proportional to the absorption coefficient, any of these processes can be use to record an EXAFS spectrum. However, the solvent in the electrochemical cell readily absorbs the Auger or secondary electrons. Therefore, the use of Auger or secondary electrons in in situ electrochemical studies is completely precluded.

The mode of detection is generally dictated by the concentration of the species of interest. In electrochemical environments, we are normally interested in the species in the interphase area, which represent a very small amount of the total number of species in the whole electrochemical system. For that reason, the preferred method of detection is fluorescence using a solid-state detector, which has a high energy resolution and sensitivity. The main drawback of the solid-state detector is the long counting times required to record a spectrum. Total reflection geometry can be used to increase sensitivity.(11,12) The absorption detection strategy is normally restricted to very porous materials, which have a very high surface-to-volume ratio.

In recording an EXAFS spectrum of a surface species (SEXAFS), it is also important to consider the polarization geometry. Synchrotron radiation is highly polarized in the orbit plane. Because of selection rules, a backscattering event will take place only when the position vector of the backscattering atom has a component in the polarizing plane. The number of neighbors obtained from the spectrum (N_{eff}) is related to the real number of neighbors (N) for a K edge according to Equation (18):

$$N_{\mathrm{eff}} = 3\sum_{i=1}^{N} \cos^2 \alpha_i \qquad (18)$$

where α_i is the angle between the position vector of the i-esim atom and the polarization vector. In-plane (polarization vector parallel to the surface) and out-of-plane (polarization vector perpendicular to the surface) polarization geometries have been used to characterize the exact position of the atoms of an adsorbed layer with respect to the surface atoms.(39)

2.4.2 *Time Resolved Extended X-ray Absorption Fine Structure*

In a normal EXAFS experiment, the energy of the incident beam is scanned with the use of a monochromator. The time required to cover the full EXAFS region by the monochromator prevents the use of this technique in kinetic studies. The alternative for time resolved studies is the use of dispersive arrangements.(40–44) In this case, the use of bend optics permits a range of energies to be focused on the sample (normally 600 eV). A photodiode array is employed for the simultaneous detection of the full range and the whole spectrum can be recorded in a millisecond range.

2.4.3 *Electrochemical Cell*

The design of an electrochemical cell depends on the electrode material and the type of research. There are two main categories, classified by the type of electrode surface: smooth or rough. Figure 9 depicts the basic design used in all the cells employed for smooth electrodes (generally single crystal electrodes).(45–47) The cell body has two ports for electrolyte exchange, two ports for the counter and reference electrodes and an X-ray transparent polymer film which serves as a window. The polymer film has two positions, inflated and deflated.

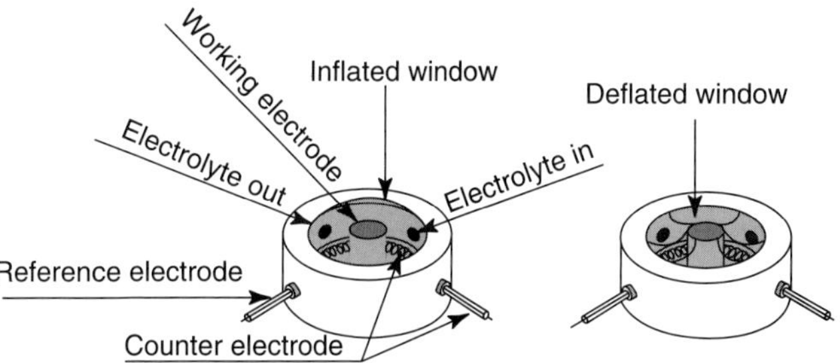

Figure 9 Typical cell design for an X-ray cell used with smooth electrodes, with the polymer window in the inflated and deflated positions.

The deflated position, in which the cell attains a thin layer configuration, is used during the data acquisition. This configuration is required to minimize the scattering due to the solvent. When the electrolyte or potential is changed, the polymer is set in the inflated position, since the equilibrium conditions are reached faster.

A typical cell used in the studies of rough electrodes is depicted in Figure 10.(48) In this cell, the working electrode is set perpendicular to the beam. Unlike the previous cell, this design allows working in absorption or fluorescence mode. The cell is also in thin layer configuration.

A different approach has been used in the study of passive films.(49,50) In this case, the electrodes were prepared by sputtering thin films of the electrode material (iron) onto Mylar™ film, in which a thin film of gold or tantalum had previously been sputtered to improve the electrical conductivity. This film is then mounted in the electrochemical cell. The EXAFS and XANES spectra are taken in fluorescence mode, through the back of the electrode (the polymer film). The advantage of this configuration is that the X-ray does not travel through the electrolyte solution, and the electrochemical cell is not restricted to a thin layer configuration. The main disadvantage is that the EXAFS and XANES spectra contain information not only of the electrode surface but also from the bulk electrode material. For this reason, the studies are restricted to thin films.

A complete review of the different cell designs can be found in Sharpe et al.(14)

3 X-RAY SCATTERING TECHNIQUES

As mentioned in section 2.1, the X-rays can be absorbed or scattered by matter. In the latter case, the different waves scattered from different atoms can interfere and maxima in the scattered intensity can be obtained at given incident angles. This is normally known as diffraction. However, this is not the only use of the scattering

For references see page 10249

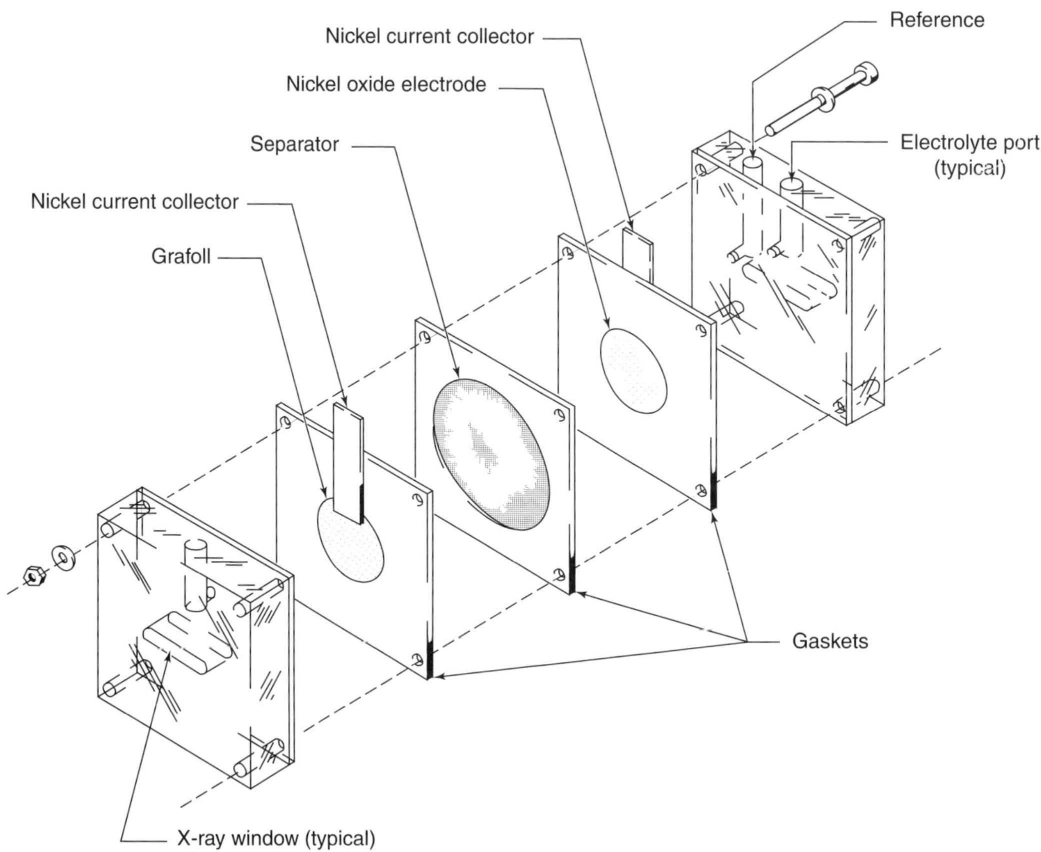

Figure 10 EXAFS cell used in the study of nickel oxide electrodes. (Reprinted with permission from J. McBreen, W.E. O'Grady, K.I. Pandya, R.W. Hoffman, D.E. Sayers, *Langmuir*, **3**, 428–433 (1987). © American Chemical Society.)

measurements. Additional information can be obtained by studying the scattered intensity in the regions between the diffraction peaks, normally known as CTR and Bragg rods. In the following text, the diffraction and the CTR and Bragg rod measurements will be treated in two different sections. However, the separation is just made for clarity, since both diffraction and rod studies are different ways of measuring the same property: the X-ray scattering.

3.1 X-ray Diffraction

When elastic scattering occurs, the different waves scattered from different atoms can interfere. As the path length is different for each wave, they can interfere constructively or destructively, depending on the incident and scattering angles. In crystalline materials, which have the atoms in fixed positions, the interference phenomenon will give rise to very characteristic diffraction patterns.

A crystalline material is composed of a lattice of identical structural units. Considering a material that has only one atom per unit cell, the position of each atom in the crystal can be found according to Equation (19):

$$\mathbf{R}_{xyx} = x\mathbf{a} + y\mathbf{b} + z\mathbf{c} \tag{19}$$

where **a**, **b** and **c** are the lattice vectors and x, y and z integers. In the Bragg approach, the atoms are considered occupying a set of parallel planes with a d spacing. A maximum in the scattered radiation will take place when the reflections from successive planes interfere constructively (Figure 11). The difference in the path length for the two waves is $2d\sin\theta$. Thus, a constructive interference will occur when this difference be equal to an integral number of wavelengths (Bragg's law) (Equation 20):

$$n\lambda = 2d\sin\theta \tag{20}$$

In X-ray diffraction, it is convenient to define the scattering vector $\mathbf{Q} = \mathbf{K_0} - \mathbf{K_H}$, as the difference between the incident ($\mathbf{K_0}$) and scattered ($\mathbf{K_H}$) wave vector. In elastic scattering, $\mathbf{K_0}$ and $\mathbf{K_H}$ have the same magnitude ($2\pi/\lambda$). Therefore, **Q** is given by Equation (21):

$$\mathbf{Q} = \frac{4\pi}{\lambda}\sin\theta \tag{21}$$

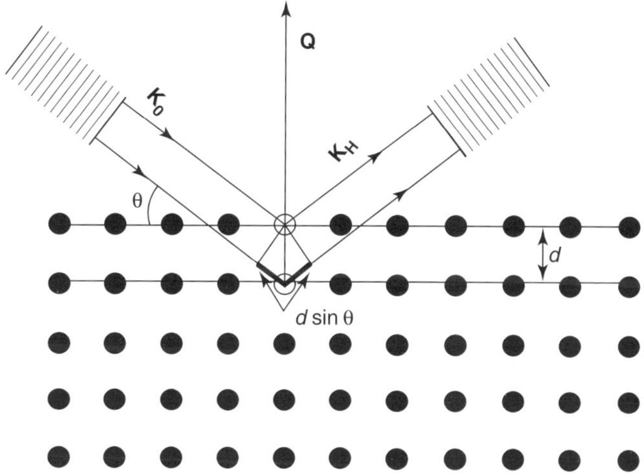

Figure 11 Bragg diffraction from lattice planes with a d spacing.

which leads to an alternate expression (Equation 22) for Bragg's law using Equation (20):

$$\mathbf{Q} = n\frac{2\pi}{d} \tag{22}$$

In order to study the X-ray diffraction is useful to employ the Von Laue approach. In the first step it is convenient to define $\mathbf{Q}$ as a linear combination of the primitive reciprocal lattice vectors, $\mathbf{a}^*$, $\mathbf{b}^*$ and $\mathbf{c}^*$ (Equation 23):

$$\mathbf{Q} = Q_a\mathbf{a}^* + Q_b\mathbf{b}^* + Q_c\mathbf{c}^* \tag{23}$$

$\mathbf{a}^*$, $\mathbf{b}^*$ and $\mathbf{c}^*$ are defined by Equation (24):

$$\mathbf{a}^* = 2\pi\frac{\mathbf{b}\times\mathbf{c}}{V} \quad \mathbf{b}^* = 2\pi\frac{\mathbf{c}\times\mathbf{a}}{V} \quad \mathbf{c}^* = 2\pi\frac{\mathbf{a}\times\mathbf{b}}{V} \tag{24}$$

where V is the unit cell volume given by $(\mathbf{a}\cdot\mathbf{b}\times\mathbf{c})$.

Within the kinematic approximation (also known as the Born approximation), which assumes a single scattering event for the photon, the scattering amplitude (A) can be calculated from Equation (25):

$$A = \frac{\mathrm{e}^2}{mc^2R}\int \mathrm{e}^{iQr}\rho(\mathbf{r})\,\mathrm{d}^3\mathbf{r} \tag{25}$$

where e^2/mc^2R is the scattering from a single electron neglecting the polarization factor $(1 + 1/2\cos^2\theta)$, R is the distance between the sample and detector and $\rho(\mathbf{r})$ is the electron density. The total electron density is obtained by summing the electron density for each atom in the crystal (Equation 26), giving a scattered amplitude of:

$$\begin{aligned} A &= \frac{\mathrm{e}^2}{mc^2R}\left[\int \mathrm{e}^{iQr}\rho(\mathbf{r})\,\mathrm{d}^3r\right]\left[\sum_{xyz}\mathrm{e}^{iQR_{xyz}}\right] \\ &= A_e f(\mathbf{Q})\mathrm{F}(\mathbf{Q}) \end{aligned} \tag{26}$$

where the atomic form factor ($f(\mathbf{Q})$) is the Fourier transform of the atomic electron density ($\rho(\mathbf{r})$) and the interference function ($\mathrm{F}(\mathbf{Q})$) is the summation term. Thus (Equation 27), the scattering intensity(I) is equal to:

$$I = |A|^2 = A_e^2|f(\mathbf{Q})|^2|\mathrm{F}(\mathbf{Q})|^2\mathrm{e}^{-2\sigma^2|\mathbf{Q}|^2} \tag{27}$$

where the Debye–Waller factor has been included to account for thermal vibrations. Since in Equation (27), the $|\mathrm{F}(\mathbf{Q})|^2$ changes rapidly with $\mathbf{Q}$, the position of the diffraction spots will be given by the maxima in $|\mathrm{F}(\mathbf{Q})|^2$. Using Equation (23) the interference function can be written as Equation (28):

$$\begin{aligned} |\mathrm{F}(\mathbf{Q})|^2 &= |\mathrm{F}(Q_a, Q_b, Q_c)|^2 \\ &= \left|\sum_{x=1}^{N_a}\sum_{y=1}^{N_b}\sum_{z=1}^{N_c}\mathrm{e}^{2\pi i(xQ_a+yQ_b+zQ_c)}\right|^2 \\ &= \frac{\sin^2\pi N_aQ_a}{\sin^2\pi Q_a}\frac{\sin^2\pi N_bQ_b}{\sin^2\pi Q_b}\frac{\sin^2\pi N_cQ_c}{\sin^2\pi Q_c} \end{aligned} \tag{28}$$

where N_a, N_b and N_c are the total number of unit cells in the $\mathbf{a}$, $\mathbf{b}$ and $\mathbf{c}$ directions, respectively. This function maximizes (Equation 29) when:

$$Q_a = h \quad Q_b = k \quad Q_c = l \tag{29}$$

with h, k and l integers. The position of the diffraction (Bragg) peak is normally termed as (h, k, l). Since N_a, N_b and N_c tend to infinity, Equation (28) can be transformed to Equation (30):

$$\begin{aligned} |\mathrm{F}(\mathbf{Q})|^2 &\longrightarrow (N_aN_bN_c)\delta(Q_a - h)\delta(Q_b - k)\delta(Q_c - l) \\ &= N\delta(\mathbf{Q} - \mathbf{G}_{hkl}) \end{aligned} \tag{30}$$

where δ is the Dirac function, N the total number of atoms in the crystal and $\mathbf{G}_{hkl}$ is defined as $h\mathbf{a}^* + k\mathbf{b}^* + k\mathbf{c}^*$. The intensity of the Bragg peak is therefore (Equation 31):

$$I_{hkl} = |A|^2 = A_e^2|f(\mathbf{Q})|^2(N_aN_bN_c)^2\mathrm{e}^{-2\sigma^2|\mathbf{Q}|^2} \tag{31}$$

For ideal crystals, the diffraction pattern consists of perfect δ-functions, with infinite value for a Bragg diffraction peak and zero value for the rest of the points of the reciprocal space. However, real crystals present imperfections, defects, grain boundaries, etc. A real crystal can be considered as a set of domains (a lattice of atoms perfectly oriented), in which each domain can have an orientation that differs from the next, giving different diffraction patterns. This results in a broadening of the diffraction peaks. Defining $\Delta\mathbf{Q}$ as the full width at half-maximum (fwhm) of the diffraction peak, the domain size (L) can be estimated from Equation (32):

$$L \approx \frac{2\pi}{\Delta\mathbf{Q}} \tag{32}$$

For references see page 10249

When the unit cell has several atoms, the atomic form factor ($f(\mathbf{Q})$) is replaced by the structure factor defined by Equation (33):

$$S(\mathbf{Q}) = \sum_{j=1}^{n} f_j(\mathbf{Q})e^{i\mathbf{Q}r_j} \quad (33)$$

where n represents the total number of atoms in the unit cell and the subscript j the jth atom in the cell.

When considering diffraction from monolayers and surfaces, they can be treated as a two-dimensional lattice structure, that is, a crystal that has $N_c = 1$. Substituting that value in Equation (28) provides the new diffraction conditions (Equation 34):

$$|\mathrm{F}(\mathbf{Q})|^2 = \frac{\sin^2 \pi N_a Q_a}{\sin^2 \pi Q_a} \frac{\sin^2 \pi N_b Q_b}{\sin^2 \pi Q_b} \quad (34)$$

with maxima at $Q_a = h$ and $Q_b = k$. No condition is limiting the value of Q_c, so the maxima are obtained regardless of its value. This results in the transformation of the diffraction spots into rods with $Q_a = h$ and $Q_b = k$. Such rods are normally termed Bragg rods. This way, the diffracted intensity of the rod will be (Equation 35):

$$I_{hk} = |A|^2 = A_e^2 |f(\mathbf{Q})|^2 (N_a N_b)^2 e^{-2\sigma^2|\mathbf{Q}|^2} \quad (35)$$

This equation can be used to calculate the different intensity of the diffraction spots from an adsorbed single layer on a surface. As the diffraction from monolayers and surfaces is independent of the value of l, surface diffraction spots are normally termed as (h, k).

3.2 Crystal Truncation Rods and Bragg Rods

The termination of a crystal in a surface or interface will give to the intensity profile some characteristics of the Bragg rods. Thus, the scattered intensity is not zero in the regions connecting two Bragg diffraction spots along directions normal to the surface, giving rise to the CTRs. For a given diffraction peak at (h, k, l), a CTR will occur at (h, k, Q_c). The intensity of such rods can be used to probe the structure of the interphase normal to the electrode surface. Normally, the CTR is measured as reflectivity (defined as the ratio of the integrated reflected intensity to the integrated incident intensity).[51–55] The integrated intensity is preferred in CTR studies since it is independent of the crystal imperfections and instrumental resolution. Thus, integrating Equation (27) for a given value of h, k over the angles accepted by the detector, Equation (36) is obtained:

$$R = T^4(Q_z) \frac{4\pi^2 r_0^2}{\Gamma^2 |\mathbf{k_0}|^2 \sin^2 \alpha} |f(Q_z)|^2 e^{-2\sigma^2 Q_z^2} \left| \sum_{n=0}^{\infty} e^{iQ_z dn} \right| e^{-\frac{Q_{abs}}{Q_z}} \quad (36)$$

where $T(Q_z)$ is the Fresnel factor and has been added to account for enhanced surface scattering process, r_0 the Thomson radius (e^2/mc^2), Q_z the projection of $\mathbf{Q}$ in the direction normal to the surface, Γ the area per atom in each layer, d the layer spacing, and α the incident angle. The term e^{-Q_{abs}/Q_z} accounts for the losses in the intensity due to the absorption of the window material and the thin layer of solvent.[52] This equation has been obtained for equal incident and exit angles. For specular CTR, i.e. when $\alpha = \theta$ or when $h = k = 0$, Equation (36) is converted to Equation (37):

$$R = T^4(Q_z) \frac{16\pi^2 r_0^2}{\Gamma^2 Q_z^2} |f(Q_z)|^2 e^{-2\sigma^2 Q_z^2} \left| \sum_{n=0}^{\infty} e^{iQ_z dn} \right| e^{-\frac{Q_{abs}}{Q_z}} \quad (37)$$

Solving the summation term of Equation (36) gives Equation (38):

$$\left| \sum_{n=0}^{\infty} e^{iQ_z dn} \right| = \frac{1}{2\sin^2(\pi Q_z)} \quad (38)$$

Therefore, for an ideally truncated crystal, in which the topmost layer has the same lattice structure and d spacing as the bulk crystal, the CTR should show a dependence on $[\sin^2(\pi Q_z)]^{-1}$. However, crystals are not ideally truncated in most cases. The topmost layer of the crystal may be reconstructed, the d spacing may be longer or shorter than in the bulk crystal, resulting in a truncation rod that does not follow the dependence on

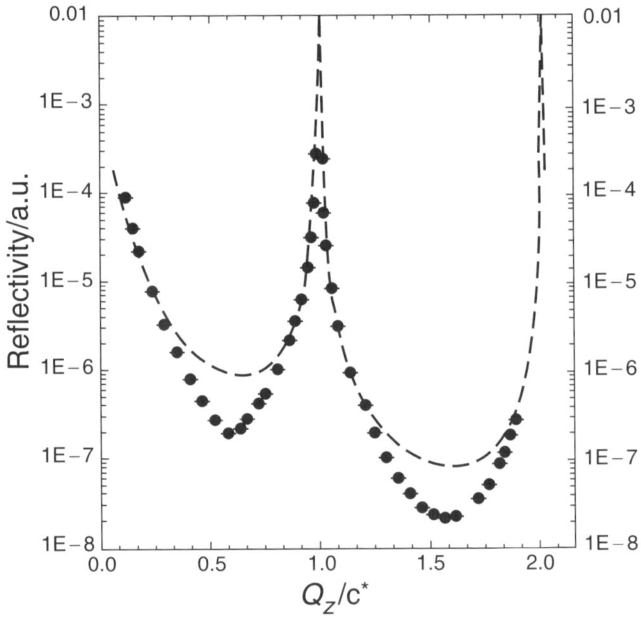

Figure 12 CTR measurement of a Au(111) electrode in 0.10 M H_2SO_4 containing 1.0 mM Br^- + 1.0 mM Cu^{2+} measured at E = +0.70 V (vs Ag/AgCl) (●). The dashed line is the calculated CTR curve from an ideally truncated Au(111) crystal.

$[\sin^2(\pi Q_z)]^{-1}$. Moreover, the presence of overlayers on the surface greatly affects the reflectivity of the truncation rod, creating asymmetries in the CTR profile and valleys that have a lower intensity than those expected for an ideally truncated surface (Figure 12).

The presence of overlayers or reconstruction phenomena has a clear effect on the CTR profile. For those cases, Equation (36) has to be modified to account for the presence of such "imperfections". For instance, for a surface in which the topmost layer can be reconstructed and t different overlayers are present, the reflectivity can be calculated from Equation (39):

$$R = T^4(Q_z)\frac{4\pi^2 r_0^2}{\Gamma^2|\mathbf{k_0}|^2 \sin^2\alpha}|S(Q_z)|^2 e^{-\frac{Q_{\mathrm{abs}}}{Q_z}} \tag{39}$$

where $S(Q_z)$ contains the sum over the atomic layers represented by Equation (40):

$$\begin{aligned} S(Q_z) = &\sum_{n=1}^{\infty}\left(f_{\mathrm{bulk}}(Q_z)\rho_{\mathrm{bulk}} e^{-Q_z^2\sigma_{\mathrm{bulk}}^2} e^{iQ_z n d_{\mathrm{bulk}}}\right) \\ &+ f_s(Q_z)\rho_s e^{-Q_z^2\sigma_s^2} e^{-iQ_z(d_s - d_{\mathrm{bulk}})} \\ &+ \sum_{m=1}^{t}\left(f_m(Q_z)\rho_m e^{-Q_z^2\sigma_m^2} e^{-iQ_z\left(d_s - d_{\mathrm{bulk}} + \sum_{j=1}^{m} d_j\right)}\right) \end{aligned} \tag{40}$$

Each layer is defined by a Debye–Waller factor (σ), a layer distance from the layer below (d) and a relative atomic density (ρ) defined as the ratio of the number of atoms in the nth layer to the number of atoms in a layer of the bulk material (obviously $\rho_{\mathrm{bulk}} = 1$). The first summation term corresponds to the different layers in the bulk material, the second term to the surface layer and the last one to the overlayers deposited on the surface. In Equation (40), only the topmost layer of the crystal is allowed to reconstruct, expand or contract with respect to the layer structure of the bulk material. Fitting this equation to the experimentally obtained CTR profile will allow determination of the different layer parameters.

CTR and diffraction measurements can be considered as complementary techniques. Whereas diffraction gives the in-plane structure, CTR provides information about the out-of-plane structure, allowing determination of the interlayer distances and of the nature of the species in the layer. The main disadvantage of the CTR measurements is that for complicated systems, the number of adjustable parameters is very high and several sets of parameters can be found that fit the experimental data. In such cases, the CTR measurements only allow some of the possible models for the interphase, proposed according to the diffraction measurements, to be discarded. Moreover, the error in the coverage values given by the CTR measurements is higher than that obtained from diffraction. On the other hand, CTR analysis is not restricted to the presence of ordered layers, unlike diffraction. Information on the coverage and the d spacing can be obtained for disordered overlayers or surfaces.

For ideal flat overlayers and surfaces, the intensity of the Bragg rods is independent of Q_z. However, real overlayers and surfaces can have atoms with different d spacing or can be constituted of two different species. In such cases the intensity (or reflectivity) is not constant and depends on Q_z. For such cases, the analysis of the Bragg rods is similar to that used for the CTR measurements, but with a structure factor containing only the terms related to the overlayers (or the surface). A different term has to be added for every different species appearing at a given d distance from the surface.

3.3 X-ray Standing Waves

Two different phenomena are combined in this technique: diffraction from a perfect crystal and interference. The interference between the incident and diffracted beams creates the standing waves that can be used to probe the position of the foreign atoms in the crystal, or adsorbed on the surface.

When the diffraction theory was considered, the intensity of the diffraction beam was calculated by using the kinematic approach. However, when perfect crystals are to be considered, this theory does not describe the intensity adequately and a dynamic approach has to be used.

In the vicinity of a Bragg diffraction peak, the incident and diffracted beams interfere to generate standing waves. These waves extend up to 1000 Å from the interface in both directions (towards the bulk and the interphase), allowing probing of the interphase and the bulk material. These waves have nodal and antinodal planes that are parallel to the diffracting planes with a nodal wavelength equal to the d spacing of the diffracting planes. As the angle of incidence is advanced through the Bragg reflection, the relative phase between the incident and diffracted beam changes by $-\pi$. Owing to this change, the nodal plane moves $d/2$ in the direction perpendicular to the surface (Figure 13).

If the energy of the incident beam is set slightly above the absorption edges of the atom we want to study in the interphase, the fluorescence yield of this species will exhibit a modulation as the angle is scanned through the Bragg peak. Analysis of such modulation will serve to determine the distance of the species relative to the surface. However, only one measurement will not be enough to determine the distance r, since two atoms separated by a distance equal to d would give the same modulation. Therefore, several Bragg diffractions or specular reflections have to be measured in order to unequivocally assign a distance.

For references see page 10249

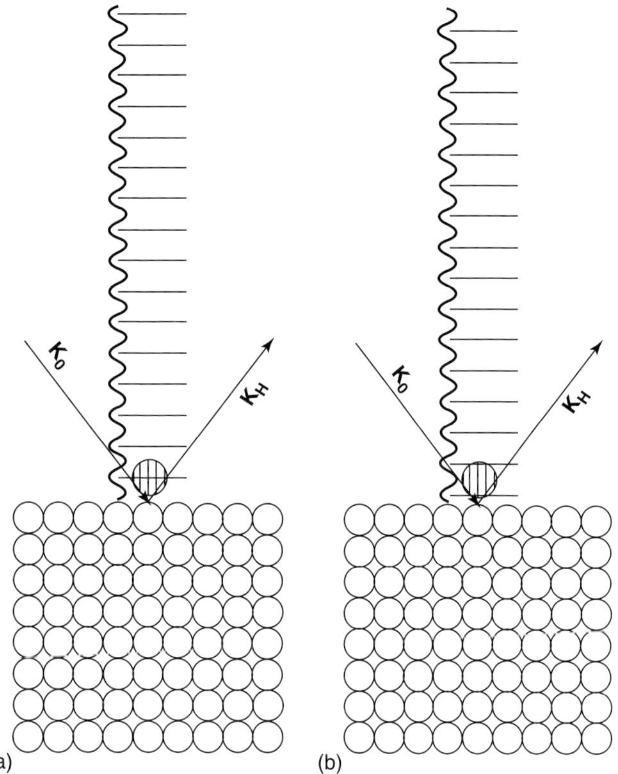

Figure 13 Schematic representation of the standing wave generated by the interference of the incident and the scattered beam before a Bragg diffraction peak (a) and after it (b).

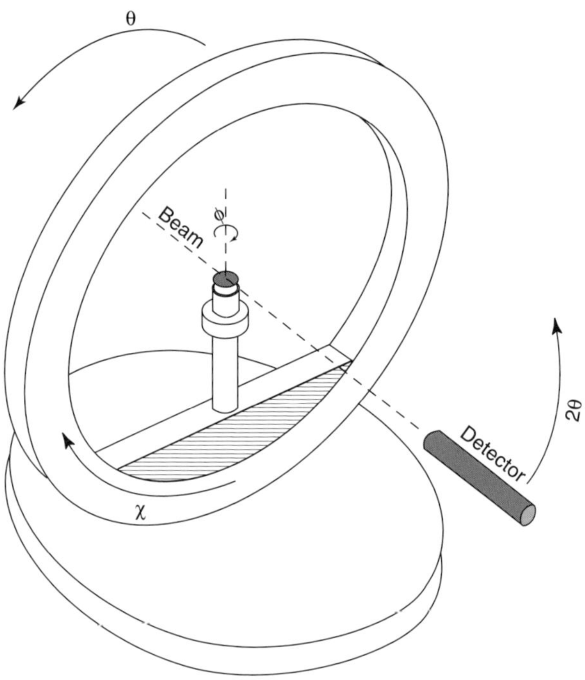

A complete review on the XSW analysis can be found in Zegenhagen.(56)

In some cases, layered synthetic microstructures (LSM) has been used instead of perfect crystals. LSM are periodic structures consisting in alternating layers of low and high electron density materials. These materials have several advantages:

- They can be manufactured with a great variety of d spacings and materials, allowing the study of the diffuse layer.
- The reflection curves compare well with the dynamic diffraction theory.
- They have a rather large energy band pass.(11,12)

3.4 Experimental Aspects

3.4.1 Four Circle Diffractometer

A four circle diffractometer (Figure 14) is used in order to search all the positions in the reciprocal space in scattering experiments. The sample has three rotational degrees of freedom: θ, χ and ϕ and the detector one, 2θ. Obviously, there is an indeterminacy in the position in the reciprocal space, since only three rotational degrees are required to define a position. The degeneracy is solved using an additional constrain in the angle position. For normal operational conditions the 2θ angle is set to be equal to twice the value of θ. This is equivalent to have incident and scattered angles equal. CTR equations were derived in such conditions.

Figure 14 Sketch of a four-circle diffractometer.

3.4.2 Electrochemical Cell

The cell used in scattering measurements is essentially the same as used in the EXAFS experiments for smooth electrodes (Figure 9). This cell presents the drawback of the thin layer configuration, which makes obtaining kinetic data very difficult. To solve that problem, different cells have been devised and used that do not require a thin layer configuration using a transmission geometry (Figure 15).(57–59) In these cells, a column of solution surrounded by a polymer film is maintained over the electrode. The counter and reference electrode are then positioned over the working electrode, allowing a much better current distribution over the working electrode. However, the thick column of electrolyte on top of the electrode surface increases the X-ray absorption by the electrolyte and requires the use of the brightest synchrotron sources.

3.4.3 Grazing Incident X-ray Diffraction Geometry

The determination of the in-plane structure of a surface requires the use of surface diffraction techniques. In

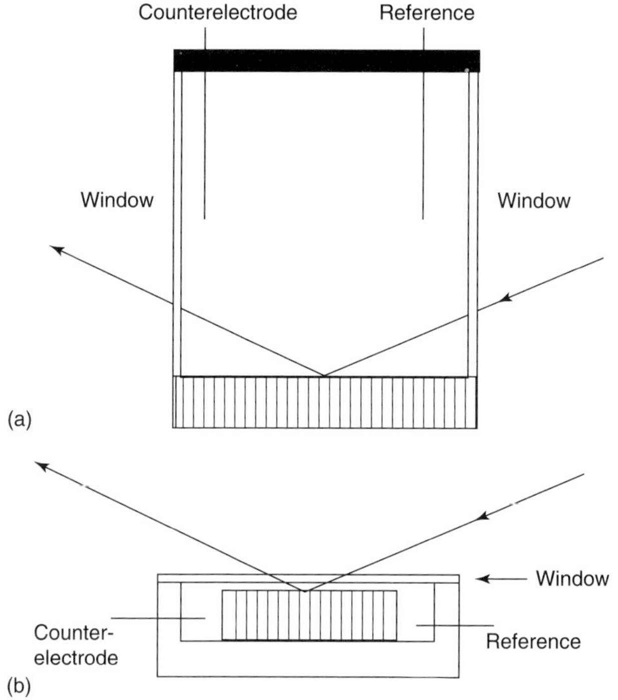

Figure 15 Scheme of the electrochemical cell used in X-ray scattering measurements using (a) transmission geometry and (b) reflection geometry.

order to limit the penetration of the X-rays into matter, the incident angle is kept small. With small incident angles, the scattering vector lies predominantly in the surface plane, probing the in-plane structure of the surface or interphase. Grazing incident geometry has several advantages over other possible geometries, such as an increase in the resolution of the surface peaks and a reduction in the background scattering from the bulk crystal.

3.4.4 Diffraction from Overlayers and Surfaces

The case of a crystal ideally terminated is seldom found in nature. The electronic deficiency caused in the topmost layer because of truncation favors the appearance of surface relaxation and reconstruction phenomena. In the simplest case, surface relaxation, the d spacing of the surface layer is altered from that found in the bulk material (maintaining the same lattice in the surface layer as in the bulk layers). Using grazing angle geometry (low Q_c), diffraction peaks for the surface can be found at the positions where the CTR are found. As explained in section 3.1, these peaks will correspond to positions where h and k are integers. The surface relaxation can be measured from CTR studies.

When there is a surface reconstruction process, the lattice of the surface layer will be different from that found in the rest of the crystal layers. For that reason, additional diffraction peaks will appear which correspond to the diffraction from the surface. These new diffraction peaks will appear at positions where h and k are not integers (known as noninteger peaks). For each additional noninteger peak, a new Bragg rod will be found, since the surface can be considered as an additional layer. As before, CTR measurements will give the d spacing of the surface layer, and will serve to confirm the diffraction structure proposed according to the diffraction measurements.

The presence of overlayers is similar to that found with reconstructed surfaces. The overlayer will give additional diffraction peaks, depending on its lattice. For each lattice peak, Bragg rods will appear. A nice example of the application of Bragg rods to the study of overlayers is found in the work by Toney et al. studying copper underpotential deposition (UPD) on Au(111) electrodes. The analysis of the intensity of Bragg rods served to discriminate the real structure of the Cu UPD from the other structures proposed (see section 4.2.2).[60] As in the previous cases, CTR measurements will give additional information of the surface and overlayer structure.

4 APPLICATION OF X-RAY METHODS TO ELECTROCHEMISTRY

4.1 Surface Structure of Metal Electrodes

A logical first step in the interphase studies deals with the surface characterization, especially when single crystal electrodes are used. For single crystal electrodes three different phenomena may take place when the electrode potential is changed: surface reconstruction, expansion or contraction of the topmost layer and surface roughening. For surface reconstruction studies, the best technique is GIXD, since it allows an unequivocal characterization. For reconstructed surfaces, is logical to expect an interlayer distance different to that found in bulk material, especially in the cases where the atomic surface density is different to that observed in the bulk layer. CTR studies can be used to measure the interlayer distance, jointly with the atomic density of the surface layer. CTR studies can also be used to measure surface roughness.

4.1.1 Au(111) Electrodes

Unlike most of the (111) planes of the face-centered cubic (fcc) crystals, the Au(111) surface reconstructs. The studies of this electrode surface can be used as a model of the information that can be obtained from the system using X-ray techniques. The first point to determine is the nature of the reconstruction and its stability range.

For references see page 10249

In UHV environments, the reconstruction leads to the formation of a $(23 \times \sqrt{3})$ structure (also called herringbone reconstruction) with an increase in the atomic surface density of 4.4%. In electrochemical environments, there is an additional variable, the electrode potential, and the reconstruction can be sensitive to changes of potential. In this particular case, the reconstruction is found to be potential dependent.[61–63] Below a critical potential, which is always more negative than the potential of zero change (p.z.c.), the structure found in three different electrolytes (0.01 M NaF, NaCl and NaBr) is the same as that found in the UHV environments: the $(23 \times \sqrt{3})$ surface structure (Figure 16). Above this potential, the reconstruction is partially lifted to give a $(p \times \sqrt{3})$ structure with $23 < p < 30$. At potentials positive to the p.z.c., the reconstruction disappears and the (1×1) surface structure is found. The phase transition between the $(23 \times \sqrt{3})$ and the (1×1) structure is reversible, that is, at a given potential the surface structure obtained is the same, irrespective of the initial conditions.

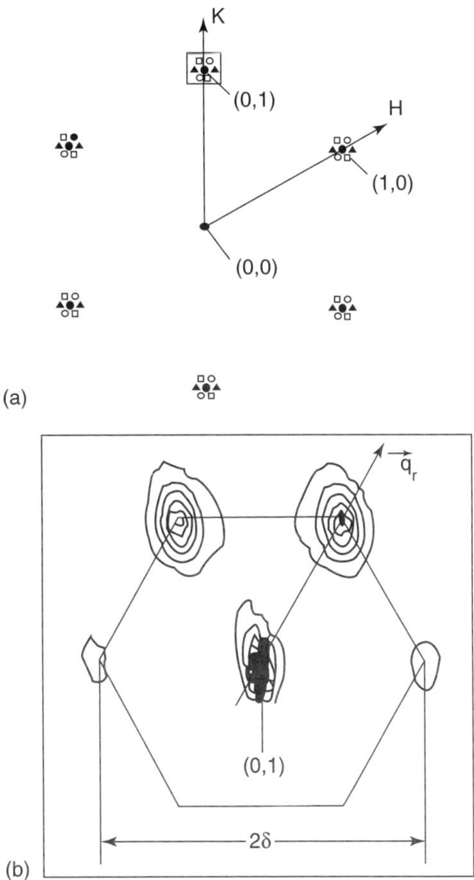

Figure 16 (a) In-plane diffraction pattern of the $(23 \times \sqrt{3})$ reconstruction in hexagonal coordinates. The solid circles are the periodicities from the underlying bulk substrate. (b) X-ray scattering equal-intensity contours in the vicinity of the (0,1) reflection in 0.01 M NaCl at −0.3 V vs Ag/AgCl. (Reproduced with permission from J. Wang, B.M. Ocko, A.J. Davenport, H.S. Isaacs, *Phys. Rev. B*, **46**, 10 321–10 338 (1992). © American Physical Society.)

The kinetics of the phase transition can be obtained following the time evolution of the peak intensity of the reconstructed surface. It was shown that the transition to the (1×1) phase is much faster than the formation of the reconstructed phase. Dependence in the rate of the phase transition with the anion present in the electrolyte was also observed. Chloride and bromide anions accelerate the transition,[63] whereas the presence of pyridine, 2,2′-bypyridine and uracil slows down the phase transition.[64]

Finally, the expansion or compression of the topmost layer was studied by CTR measurements. The presence of a reconstructed surface (which has a higher atomic density than the bulk layer) leads to an expansion of the interlayer distance for the reconstructed surface, i.e. the d spacing has increased 3.3% with respect to that found in the bulk material. No change in the distances was observed for the (1×1) structure.

4.1.2 Au(100) Electrodes

The general behavior of the Au(100) electrode is similar to that described for the Au(111) electrode.[65] At negative potential with respect to the p.z.c., the surface layer exhibits an hexagonal reconstructed surface, whereas at positive potentials, the (1×1) structure is found. The hexagonal reconstruction has an atomic surface density 22% greater than that found in the bulk Au(100) layer. This reconstructed phase is the same that was found in UHV environments.[66,67] When the reconstruction is lifted at positive potentials these excess of atoms is segregated to an additional layer. This new layer has an atomic surface density that corresponds to 22% of the bulk layer. In perchloric acid, a partial reconstruction is observed at potentials negative to 0.4 V vs a standard calomel electrode (SCE).[57]

4.1.3 Pt(111) and Pt(100) Electrodes

No evidence of reconstruction has been found for these surfaces in electrochemical environments.[68–71] In these surfaces, the effect of the anion and hydrogen adsorption in the relaxation/contraction of the topmost layer and the effect of oxide formation has been studied.

4.2 Adsorbed Layers on Metal Surfaces

One of the most studied phenomena in surface electrochemistry in recent years is the UPD of metals. The UPD process is mainly controlled by the interaction between the depositing metal and the electrode surface. However, the anion present in the electrolyte also plays an important role in this process since the interaction of the anions with the electrode surface and with the depositing metal affects the whole UPD process. Therefore, a logical first step when studying UPD processes is to understand the process of anion adsorption on the bare surface.

4.2.1 Anion Adsorption on Single Crystal Electrodes

Studies of anion adsorption on single crystal electrodes have focused mainly on halide adsorption. These studies highlight the influence of the surface structure and nature on the electrode. A clear example of that is the adsorption of iodine on Au(111), Ag(111) and Pt(111) electrodes. For Au(111) and Ag(111) electrodes, iodine adsorption exhibits very complex behavior. The first structure observed is a $(\sqrt{3} \times \sqrt{3})$ R30° (only present in Ag(111) electrodes).[72] As the potential is made more positive, the structure changes to a uniaxial incommensurate structure, $c(p \times \sqrt{3})$, where p decreases with the potential.[72,73] Finally, the structure changes to a rotated incommensurate hexagonal phase. On the other hand, Pt(111) electrodes exhibit completely different behavior. Only two commensurate different phases are observed: (3×3) and a (7×7)R 21.8°.[72] Moreover, the I–I distance obtained is smaller in Pt that in Au or Ag, clearly the result of a stronger Pt–I interaction. The Au–I distance obtained by CTR measurements is in the range of 2.28 to 2.45 Å, depending on the iodine surface structure.[73] XSW studies performed with LSM materials showed changes in the diffuse layer of iodide.[15,74] At potentials for which the iodine adlayer was not saturated, the diffuse adlayer showed an accumulation of iodide associated to the adsorbed iodine. Upon saturation of the iodine adsorbed layer, there is a depletion of iodine in the diffuse layer.

Bromide adsorption on Au(111) only shows a rotated hexagonal structure that compresses when the potential is scanned in the positive direction,[72,75,76] similar to what is observed for chloride adsorption.[75] The structure of bromide on a Pt(111) electrode is also a hexagonal structure that compresses, but in this case the structure is aligned with the (1,0) surface direction.[77] Commensurate structures are found only on Ag(111) electrodes with a (7×7)R 21.8° structure.[72] Ordered structures and the phase transition between them are also observed on Au(100) and Ag(100) electrodes.[78–80]

4.2.2 Copper Underpotential Deposition on Gold Single Crystal Electrodes

The cyclic voltammogram of Cu UPD on Au(111) shows two well-defined and sharp pairs of peaks, corresponding to two different adsorption–desorption processes. The pair of peaks (deposition/stripping) at high underpotentials has a broad shoulder, whereas the deposition peak at low underpotentials splits in two on high-quality Au(111) single crystals.[81,82] For these reasons, two different surface structures have to be expected after each adsorption peak.

The first in situ measurements on this system were done by SEXAFS, which show the existence of a Cu-(1×1) structure on the electrode surface after the second deposition peak.[83,84] SEXAFS measurements provided an accurate determination of the Cu–Cu distance in this structure (2.92 ± 0.03 Å,[84] 2.89 ± 0.03 Å[85]). This value is the same as the Au–Au distance in the (111) plane (2.88 Å), implying that a Cu-(1×1) adlayer is commensurate with the Au(111) substrate.[84,85] In this structure, Cu adatoms probably sit on threefold hollow sites, as derived from the polarization dependence of the SEXAFS[84] or by estimating the number of gold nearest neighbors.[85] SEXAFS spectra also indicate the existence of oxygen (likely arising) from (bi)sulfate on the Cu adlayer in which Cu presents a +1 oxidation state, as suggested by XANES.[83,85]

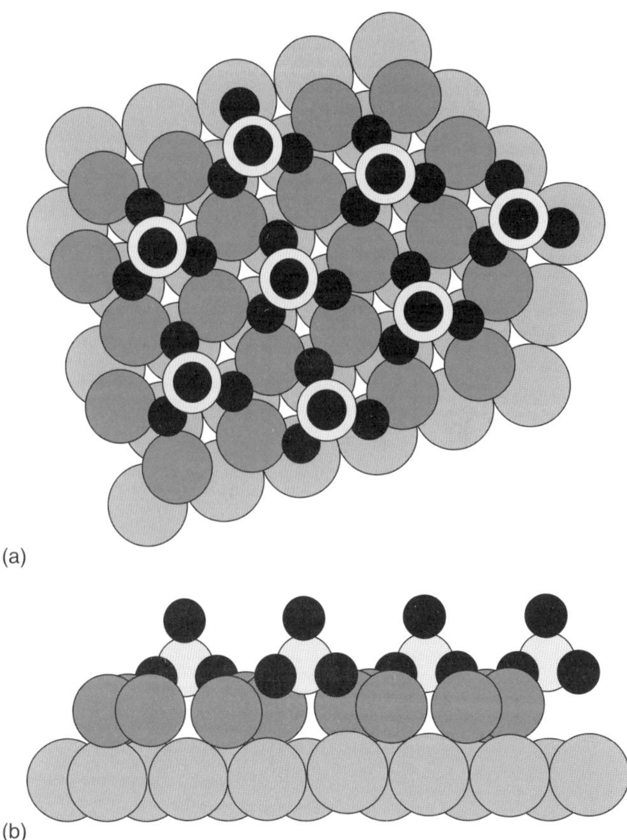

Figure 17 Surface structure of the $(\sqrt{3} \times \sqrt{3})$ honeycomb structure found for Cu UPD on Au(111) electrodes in sulfuric acid media. (a) Top view (b) side view. (Reproduced with permission from M.F. Toney, J.N. Howard, J. Richer, G.L. Borges, J.G. Gordon, O.R. Melroy, *Phys. Rev. Lett.*, **75**, 4772–4775 (1995). © American Physical Society.)

For references see page 10249

The determination of the structure after the first UPD peak was more difficult and was, for some time, controversial. The first SPM[82,86–89] images showed a $(\sqrt{3} \times \sqrt{3})R30^\circ$ structure in which the maxima were interpreted as Cu adatoms. Based on this structure the total Cu coverage was 0.33. The $\Theta_{Cu} = 0.33$ was in clear disagreement with the results from chronocoulometry and quartz crystal microbalance (QCM), which indicated that the Cu coverage in the adlayer at these potentials was 0.67 and the (bi)sulfate coverage value was 0.33.[90–95] Also theoretical calculations predicted the formation of a honeycomb $(\sqrt{3} \times \sqrt{3})R30^\circ$ structure,[96] whose coverages were in agreement with those found with QCM,[94,95] and in UHV.[97,98] This apparent contradiction was reconciled by Toney et al. using X-ray CTR measurements.[60] They found that the Cu adlayer, at intermediate coverage values, had a honeycomb $(\sqrt{3} \times \sqrt{3})R30^\circ$ structure in which $\Theta_{Cu} = 0.67$ and with (bi)sulfate anions occupying the centers of the honeycomb ($\Theta_{anion} = 0.33$) in agreement with the coverage values found with chronocoulometry and QCM (Figure 17). This meant that the scanning probe techniques (STM, AFM) were not imaging the Cu adatoms but the (bi)sulfate anions that occupied the centers of the honeycomb and protruded well above the Cu plane.[60] Recent SEXAFS measurements also support the honeycomb model (Figure 18).[39] As was the case for the Cu-(1×1) structure, XANES clearly suggested that the oxidation state of the Cu adatoms after the first UPD peak is approximately +1,[99] again probably as a result of a polar bond or surface dipoles. For this structure, it has been proposed that the Cu adatoms occupy on-top positions.[100]

The voltammogram of Cu UPD in presence of chloride shows two pairs of peaks, as in the case of sulfuric acid, but the pair at high underpotentials is reversible and appears at potentials that are more positive than in sulfuric acid. The second pair of peaks almost overlaps with bulk Cu deposition. After the first peak, a bilayer structure was proposed after SEXAFS measurements, in which chloride is adsorbed on the copper ($\Theta_{Cu} = 0.62$) and the copper atoms are in registry with the top layer of chloride ions.[39,101] The distances in the proposed

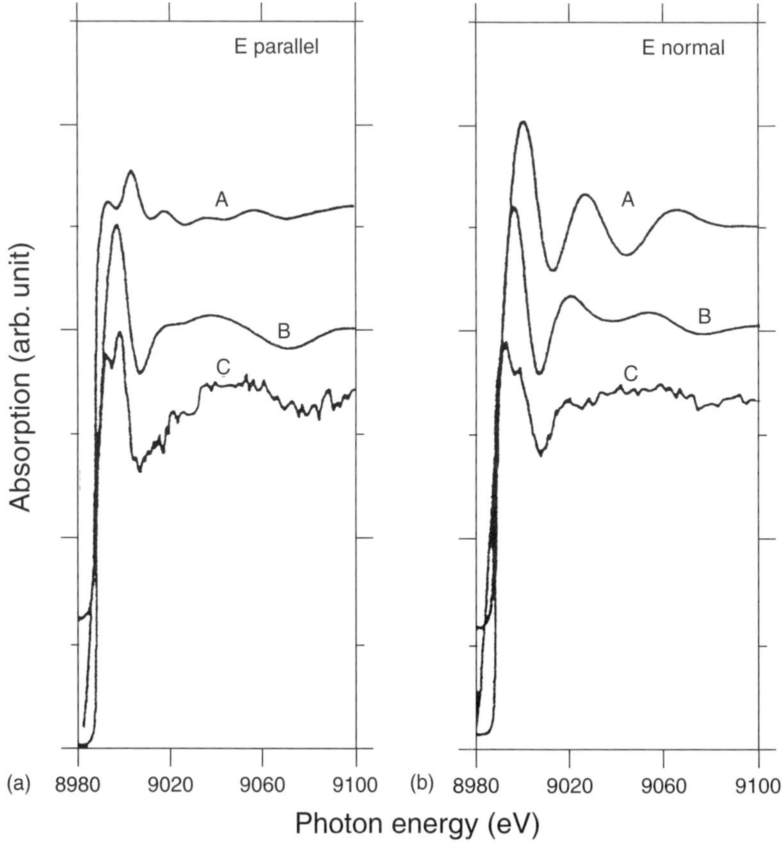

Figure 18 Comparison of (A) and (B) calculated FEFF6.0 and (C) experimental Cu XAFS spectra for the Cu-$(\sqrt{3} \times \sqrt{3})$ honeycomb structure on Au(111) electrodes. The sulfate has one oxygen atom pointing to the surface in model (a) and three oxygen atoms in model (b). (Reproduced from S. Wu, J. Lipkowski, T. Tyliszczak, A.P. Hitchcock, 'Effect of Anion Adsorption on Early Stages of Copper Electrocrystallization at Au(111) Surface', *Prog. Surf. Sci.*, **50**, 227–236. © 1995 with permission from Elsevier Science.)

structure are quite similar to those found in solid CuCl.[102]

In the presence of bromide, the behavior is very similar to that found in the presence of chloride, with a predicted bilayer structure.[92,103,104] Recent X-ray scattering studies revealed the presence of an ordered hexagonal bromide adlayer at the onset of copper deposition. This adlayer undergoes a phase transition to form a (4×4) commensurate structure ($\Theta_{Br} = 0.56$) at the peak at 0.32 V vs Ag/AgCl. The bromide adlayer remains stable until bulk copper deposition.[104] Copper is deposited in-between the gold surface and the bromide layer.[104] As a general trend for all Cu–halide adlayers, the structure of these adlayers is governed by the halide–halide and the Cu–halide interactions in contrast to the Cu–(bi)sulfate adlayers in which the Cu–Au interaction is dominant. However, all interactions contribute to the UPD processes.

On Au(100) electrodes in sulfuric acid media, Cu UPD occurs in a relatively broad peak.[105] STM studies showed that the surface structure after the deposition peak is a pseudomorphic (1×1) structure in which the Cu atoms occupy the fourfold hollow site.[105] XSW measurements also indicated that the Cu adatoms occupy the fourfold hollow site.[106] However, SEXAFS measurements suggested a different picture of the adlayer at full coverage.[99] In the model proposed according to the experimental data (which is incompatible with STM and AFM images), the Cu–Cu distance is shorter than the Au–Au distance in the unreconstructed Au(100) surface and the Cu adatoms occupy the a-top sites.[99,107] This would imply that the topmost gold layer is rearranged or reconstructed. As on Au(111) electrodes, the Cu oxidation state is close to +1.[99,107]

4.2.3 Lead Underpotential Deposition on Au(111) Electrodes

Lead and gold have very different atomic sizes (lead is ca. 20% larger than gold), which favors the formation of incommensurate adlayers. In fact, at potentials negative of the main UPD peaks, a hexagonal incommensurate structure has been found using X-ray diffraction techniques.[108,109] This structure was also observed by SPM.[110–113] In this structure, the lead–lead distances are compressed 0.7% with respect to bulk lead.[108,109] The lead adlayer is rotated with respect to the Au(111) plane with an angle which varies between 2.5° and 0° depending on the applied potential (Figure 19),[109] which is consistent with the change in lead–gold distance over the same potential region observed with another X-ray diffraction technique.[114]

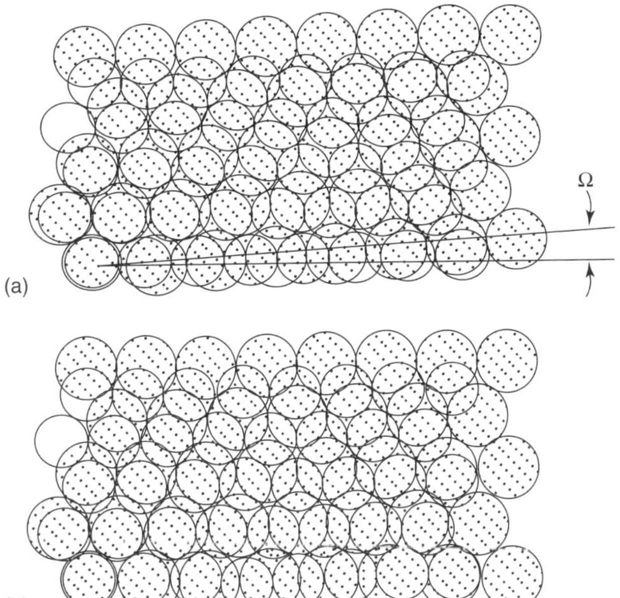

Figure 19 Surface structure of the Pb layer on Au(111) electrodes at (a) E < 0.13 V and (b) E > 0.16 V vs Ag/AgCl in 0.1 M $HClO_4$. (Reprinted with permission from M.F. Toney, J.G. Gordon, M.G. Sammant, G.L. Borges, O.R. Melroy, D. Yee, L.B. Sorensen, *J. Phys. Chem.*, **99**, 4733–4744 (1995).)

4.2.4 Mercury Underpotential Deposition on Au(111) Electrodes

The behavior of the mercury UPD on Au(111) electrodes resembles in some cases that of copper UPD on the same electrodes. The voltammetric profile of mercury UPD on Au(111) electrodes in 0.1 M H_2SO_4 shows several pairs of peaks. The voltammogram shows a main deposition/stripping peak (centered at around +0.90 V vs Ag/AgCl) that splits into two. The initial stages of mercury UPD appear to trigger an order–disorder transition in the (bi)sulfate adlayer that gives rise to the first pair of peaks.[115] After the second pair of peaks an ordered surface structure is found. STM measurements have shown the existence of two different ordered structures after this peak:[116]

$$\begin{pmatrix} 2 & 0 \\ 3 & 2/3 \end{pmatrix} \quad \text{and} \quad \begin{pmatrix} 1 & \bar{1} \\ 4 & 4 \end{pmatrix}$$

The first structure was later observed by in situ X-ray diffraction and identified as a $(\sqrt{3} \times \sqrt{19})$ surface structure.[117] CTR measurements indicated that the adlayer is probably constituted by Hg_2^{2+} cations and (bi)sulfate anions.[117] Based on the atomic distances derived from the CTR measurements, an adlayer structure similar (although distorted) to the honeycomb structure observed for copper UPD in the same medium

For references see page 10249

was proposed.[117] At +0.82 V vs Ag/AgCl, the ordered structure disappears.[117]

Apart from the surface processes of mercury UPD, a process controlled by diffusion also appears in the voltammogram at potentials around +0.54 V vs Ag/AgCl. This redox reaction corresponds to the oxidation–reduction of mercury species in solution according to Equation (41):

$$2Hg^{2+} + 2e^- \longleftrightarrow Hg_2{}^{2+} \quad E^\circ = +0.698\,\text{V vs Ag/AgCl} \tag{41}$$

Coinciding with this diffusion-controlled process other additional ordered structures have been found. AFM measurements have identified a hexagonal structure at these potentials.[118,119] X-ray diffraction studies identified two different hexagonal adlayers in this region: one at potentials between +0.63 and +0.68 V (vs Ag/AgCl) with a Hg–Hg distance of 3.84 Å, and another at potentials below +0.63 V (vs Ag/AgCl) with a Hg–Hg distance of 3.33 Å.[120] The first hexagonal adlayer appears to be metastable and evolves, with time, to give the second adlayer.[120]

The effect of anions is also evident in the case of mercury UPD. Unlike copper UPD, the voltammetric profile for mercury UPD on Au(111) in the presence of chloride resembles that observed in sulfuric acid media alone.[115] This would indicate that in both chloride and sulfuric acid media the UPD process is governed by the mercury–substrate interactions. After the main UPD peaks, CTR measurements showed that a Hg_2Cl_2 bilayer is formed,[121] in which mercury is bonded to the gold surface and chloride is deposited on top of the mercury adatoms. In acetic acid media, hexagonal structures have been observed with X-ray diffraction after the main UPD peak.[121]

4.2.5 Copper Underpotential Deposition on Pt(111) Electrodes

EXAFS studies have shown that the copper layer in sulfuric acid media is not completely discharged on the surface at potentials after the UPD peak.[122,123] The charge of the copper species on the Pt(111) surface is close to +1 and not zero as expected based on complete charge transfer, while the platinum surface retains some negative charge, as in the case of Au(111) electrodes.[83,85] At full coverage, EXAFS results show that the copper layer has a close-packed structure with a copper–copper distance of 2.77 ± 0.03 Å. This structure corresponds to the copper residing in the threefold hollow sites of the Pt(111) surface.[122] GIXD and CTR measurement have determined that the UPD process of copper on Pt(111) electrodes is very similar to the one taking place on Au(111) electrodes.[124] The first step in the UPD process is the formation of the honeycomb structure, followed by the formation of a full copper monolayer with (bi)sulfate anions adsorbed on top of the copper layer.

Bromide and chloride anions also show a dramatic effect upon copper UPD onto platinum single crystals. In both media, the deposition and stripping again take place in two distinct steps on Pt(111),[125–128] in which copper retains a partial charge.[129] In chloride media, GIXD measurements showed that the adlayer presents a Bragg rod at the (0.765, 0, L) position after the first UPD peak, which corresponds to the formation of an incommensurate hexagonal CuCl adlayer (Figure 20).[125,126] After the second UPD peak, a commensurate (1 × 1) copper adlayer forms on the electrode surface, which, in turn, is covered by a disordered layer of chloride anions.[127] In this adlayer, SEXAFS findings also indicated that in the presence of chloride, the copper–copper bond distance of the deposited layer was close to that of bulk

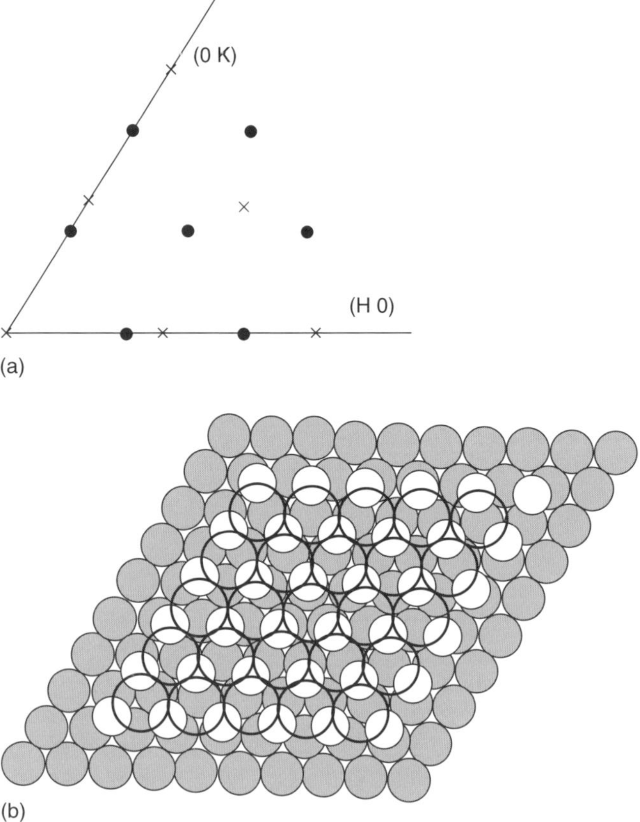

Figure 20 (a) Schematic representation of the in-plane reciprocal space in the potential region between the two main peaks in the Cu UPD in presence of chloride anions. (b) Surface structure of the adlayer. Pt atoms: shadowed circles; Cu atoms: white circles and Cl atoms: open circles. (Reproduced from N.M. Markovic, H.A. Gasteiger, C.A. Lucas, I.M. Tidswell, P.N. Ross, 'The Effect of Chloride on the Underpotential Deposition of Copper on Pt(111): AES, LEED, RRDE and X-ray Scattering Studies', *Surf. Sci.*, **335**, 91–100. © 1995 with permission from Elsevier Science.)

copper, unlike that of copper deposited in the absence of chloride.[130] The kinetics of the phase transitions between the incommensurate and the (1 × 1) structures were followed by GIXD.[131,132] These studies revealed that the transition from the (1 × 1) structure to the incommensurate structure occurs through a two-step mechanism. In the first step, the copper is desorbed which gives rise to the current response. On a much longer timescale, the reorganization of the layer takes place to give an ordered CuCl bilayer.

In bromide medium, an incommensurate hexagonal structure aligned along the (1,0) surface direction is observed by GIXD after the first UPD peak, corresponding to a CuBr layer with an interatomic distance of 3.74 Å.[125,126,133] At potentials negative of the second UPD peak, it has been proposed that copper forms a (1 × 1) structure, which is, in turn, covered by a disordered bromide layer.

4.2.6 Lead Underpotential Deposition on Ag(111) Electrodes

Attempts to elucidate the structure of lead adsorbed on Ag(111) have been done using surface X-ray scattering techniques.[108,109,134–137] Deposition from either perchloric acid or acetate media shows the presence of peaks in the azimuthal scans (varying ϕ angle) at $\phi = \pm 4.5°$.[109,134,135] This marks the presence of two equivalent domains of an incommensurate lead overlayer that are each ~4.5° from the Ag[011] direction as confirmed in the STM measurements.[138–140] GIXD measurements have also shown that lead deposited onto Ag(111) at full monolayer coverage undergoes a compression of 1.4% relative to bulk lead.[134–136] The compression increases linearly with applied potential until the onset of bulk deposition where the compression is 2.8%. X-ray studies done employing surface differential diffraction have found that the distances measured for the lead layer were between 3.00 ± 0.05 and 3.2 ± 0.1 Å.[137] These distances can be ascribed to lead atoms being adsorbed between the a-top and the bridge sites.

In situ EXAFS has been performed in acetate media.[141] These studies have found that the deposited lead is in a zero-valent state and that the lead layer is deposited incommensurate to the silver substrate. There is also scattering observed from an oxygen atom, which is most likely coming from the water or acetate. This scattering implies that these molecules are adsorbed on the deposited lead layer.

4.2.7 Thallium Underpotential Deposition on Silver Single Crystal Electrodes

GIXD studies of thallium UPD on Ag(111) have shown that after complete deposition of one monolayer, there is an incommensurate, hexagonal structure on the surface.[109,142,143] This structure is compressed relative to bulk thallium by 1.4–3.0% and rotated from the Ag[$01\bar{1}$] direction by $\Omega = 4–5°$, depending on potential.[109,142,143] Vapor deposited thallium in UHV presents the same structure, suggesting that the solvent molecules do not affect the deposition structure.[143] Upon completion of deposition of the second monolayer, the thallium forms a bilayer that has also an incommensurate hexagonal structure.[109,142,143] The second layer is commensurate with the layer beneath it and the newly deposited thallium atoms sit in the threefold hollow sites of the bottom layer.[143] In this structure, the compression is 1.0% relative to bulk thallium and the rotation is 3.9°.[142,143]

Thallium deposition onto Ag(100) has also been studied by surface X-ray scattering.[144] A disordered phase is formed after the first voltammetric peak, whereas an ordered monolayer is obtained after the second voltammetric peak. This layer has a $c(p \times 2)$ close-packed structure which compresses uniaxially (p decreasing from 1.185 to 1.168) with decreasing potential.[144] With deposition of the second layer, it has been found that the first layer expands slightly and both layers form a $c(1.2 \times 2)$ bilayer.

4.3 Fuel Cell Electrodes

Most fuel cells use platinum as an electrode material (especially for the hydrogen fuel cell and the direct methanol fuel cell). The high cost of platinum makes the dispersion of the metal on a supporting material such as carbon advisable in practical applications. On some occasions, platinum has been alloyed in order to increase its catalytic activity towards the desired reaction.

The dispersion of platinum creates small particles that can have different electrocatalytic properties than the bulk material. The first studies carried out confirmed that the platinum oxide formed at positive potentials has a short-range structure similar to α-PtO_2.[145] Alloying the platinum resulted in a diminution in the d-band occupancy with respect to the platinum particles and a contraction in the Pt–Pt distances.[146] All the alloys studied showed an increase in the catalytic activity for the oxygen reduction except the PtMn/C alloy. A volcano type correlation was found between the Pt–Pt distance and the d band occupancy with the electrocatalytic activity for oxygen reduction. In contrast to the oxygen reduction, the electrocatalytic oxidation of hydrogen shows no significant differences between Pt and the Pt alloys, indicating that the structural parameters and d band occupancies have no effect on the reaction.[147] For Pt particles, structural changes are observed upon hydrogen adsorption, whereas the alloy particles exhibit no changes.[147,148] These changes in the Pt–Pt distance

For references see page 10249

and in the Pt coordination number have been attributed to a reversible reconstruction that takes place upon hydrogen adsorption.[148,149]

The effect of the Pt particle size has been investigated in relation to methanol oxidation. The EXAFS and XANES results show that the reduced activity towards methanol oxidation is the consequence of a stronger CO and OH adsorption in the small particles.[149] The strong OH adsorption also results in a lower activity towards oxygen reduction. The effect of particle size has also been studied for Cu and Pb UPD on carbon-supported platinum.[150–152]

Pt–Ru particles exhibit an increase in the catalytic activity towards methanol oxidation due mainly to a bifunctional mechanism in which the Ru atoms facilitated the removal of the CO species. This effect has been confirmed by XANES, which shows no change in the oxidation state of the Ru atoms when CO is present in the solution for potentials at which the oxidation of ruthenium takes place in absence of CO.[153] The presence of Ru in the alloy diminishes the Pt–Pt distances and diminished the *d* band occupancy, a clear indication that the Ru has also an electronic effect on the platinum particles.[154] The formation of RuOH species in the alloy takes place at 0.24 V vs SCE.[154]

4.4 Battery and Oxide Electrodes

Oxidation of $Ni(OH)_2$ has been extensively studied by EXAFS and XANES.[155–161] The uncharged material (β-$Ni(OH)_2$) shows changes in the (0001) plane of the $Ni(OH)_2$ with charging–discharging cycles.[157] The charge material corresponds to a nickel oxide in an oxidation state of ca. +3.5, determined by using the edge position and the distances and coordination numbers of the oxygen atoms.[159] The structure of the charged material showed two different Ni–O interactions at 1.88 and 2.07 Å and with coordination numbers of 4.1 and 2.2 respectively, and two Ni–Ni interactions at 2.82 and 3.13 Å with coordination numbers of 4.7 and 1.0 respectively. The longer Ni–O interactions were interpreted as Ni–OH distances and the shorter were attributed to Ni–O in tetravalent nickel compounds. The same oxidation state has been obtained by simulating the near edge part of the spectrum with the FEFF6 code program.[160] Oxidation of α-$Ni(OH)_2$ also gives nickel oxides with an oxidation state of 3.5.[161]

Additional studies have been conducted with V_2O_5 electrodes,[162] CuO_2 electrodes,[163] PbO_2[164] and $Li_xMn_2O_4$.[165]

4.5 Corrosion

The nature of the passive film formed on iron electrodes has been investigated with in situ XANES in borate buffer solution. Comparison with model compounds indicates that the structure of the oxide passive film is γ-Fe_2O_3/Fe_3O_4.[50] Diffraction studies of the passive film have been used to discriminate between the two possible structures.[166] Calculating the theoretical structure factors for different structures and comparing them with those obtained from diffraction measurements, allowed the identification of the structure of the passive film. The film structure is based on the structure of Fe_3O_4, but with cation vacancies on the tetrahedral sites and octahedral sites and cations occupying the interstitial octahedral sites. The oxidation state of iron in the passive film is mainly Fe^{3+}, with a Fe^{2+} proportion that oscillates between 4 and 20% depending on the applied potential.[167] When the passive film was obtained at potentials below −0.6 V vs mercury sulfate electrode (MSE), dissolution is observed during the first stages of the film formation.[167] The cathodic dissolution of the passive films was studied using Fe_2O_3 and Fe_3O_4 films.[168] The reduction of the Fe_2O_3 film takes place in a two-step mechanism. In the first step, the Fe_2O_3 is reduced to Fe_3O_4 with partial dissolution. In the second step, the formed Fe_3O_4 film is reduced to Fe^{2+}. The dissolution of the passive film depends on the solubility of the Fe^{2+} species. Thus, in basic media no dissolution is observed, owing to the insolubility of Fe(II) species. In borate buffer, dissolution is only observed during step 2. The anions also affect the dissolution rate of the passive film. During anodic dissolution, chloride anions accelerate the dissolution when compared to sulfate anions, as a result of the complexation of Fe cations by chloride anions.[169] Cathodic dissolution does not show any dependence on the anions. The effects of the presence of Cr and other additives to iron have also been studied.[170–172] These studies showed the presence of Cr(IV) at positive potentials.

The addition of chromium confers corrosion resistance to passive films. For that reason, it has been used extensively in aluminum alloys. The XANES studies indicate the possibility of oxidizing the initial Cr(III) to Cr(VI).[49,173,174] Depending on the experimental conditions, the Cr(VI) species are stable on the passive film and can be reduced to the initial oxidation state. The Cr(VI) species accumulate on the outer region of the passive film. Owing to the toxicity of chromium, cerium oxides have been used as a possible substitute in passive films. XANES studies indicate that cerium is in an oxidation state of +3.[175]

4.6 Other Examples

Several studies have been conducted on polymer electrodes modified with several salts[176–181] and modified electrodes.[182–185] These studies focus mainly in the determination by EXAFS of the short-range structure around the metallic atom in the polymer or modified electrodes.

ACKNOWLEDGMENTS

The author is indebted to Dr J.M. Feliu for critical review of the manuscript. This work is partially supported by the Ministry of Education and Culture (Spain), grant no. PB96-0409.

ABBREVIATIONS AND ACRONYMS

AES	Auger Electron Spectroscopy
AFM	Atomic Force Microscopy
CTR	Crystal Truncation Rod
EXAFS	Extended X-ray Absorption Fine Structure
fcc	Face-centered Cubic
FTIRS	Fourier Transform Infrared Spectroscopy
fwhm	Full Width at Half-maximum
GIXD	Grazing Incident X-ray Diffraction
LEED	Low-energy Electron Diffraction
LSM	Layered Synthetic Microstructures
MSE	Mercury Sulfate Electrode
p.z.c.	Potential of Zero Change
QCM	Quartz Crystal Microbalance
SCE	Standard Calomel Electrode
SEXAFS	Surface Extended X-ray Absorption Fine Structure
SPM	Scanning Probe Microscopy
STM	Scanning Tunneling Microscopy
UHV	Ultrahigh Vacuum
UPD	Underpotential Deposition
XANES	X-ray Absorption Near Edge Structure
XPS	X-ray Photoelectron Spectroscopy
XSW	X-ray Standing Waves

RELATED ARTICLES

Electroanalytical Methods **(Volume 11)**
Electroanalytical Methods: Introduction • Infrared Spectroelectrochemistry • Microbalance, Electrochemical Quartz Crystal • Scanning Tunneling Microscopy, In Situ, Electrochemical • Surface Analysis for Electrochemistry: Ultrahigh Vacuum Techniques

X-ray Spectrometry **(Volume 15)**
X-ray Techniques: Overview • Absorption Techniques in X-ray Spectrometry • Structure Determination, X-ray Diffraction for

REFERENCES

1. J.O'M. Bockris, B.E. Conway, E. Yeager, *Comprehensive Treatise of Electrochemistry*, Plenum Press, New York, Vol. 1, 1980.
2. S. Pons, 'The Use of Fourier Transform Infrared Spectroscopy for In Situ Recording of Species in the Electrode–Electrolyte Interphase', *J. Electroanal. Chem.*, **150**, 495–504 (1983).
3. A. Bewick, 'In Situ Infrared Spectroscopy of the Electrode/Electrolyte Interphase', *J. Electroanal. Chem.*, **150**, 481–483 (1983).
4. M. Fleischmann, P.J. Hendra, A.J. McQuillan, 'Raman Spectra of Pyridine Adsorbed at a Silver Electrode', *Chem. Phys. Lett.*, **26**, 163–166 (1974).
5. M. Fleischmann, P.J. Hendra, A.J. McQuillan, R.L. Paul, E.S. Reid, 'Raman Spectroscopy at the Electrode–Electrolyte Interface', *J. Raman. Spectrosc.*, **4**, 269–274 (1976).
6. M. Fleischmann, P.J. Hendra, A.J. McQuillan, A. James, 'Raman Spectra from Electrode Surfaces', *J. Chem. Soc. Chem. Commun.*, 80–81 (1973).
7. D.J. Jeanmarie, R.P. Van Duyne, 'Surface Raman Spectrum Electrochemistry. Part I Heterocyclic, Aromatic and Aliphatic Amines Adsorbed on the Anodized Silver Electrode', *J. Electroanal. Chem.*, **84**, 1–20 (1977).
8. G. Richmond, 'Investigations of Electrochemical Interfaces by Nonlinear Optical Methods', in *Electrochemical Interfaces*, ed. H.D. Abruña, VCH Publishers, New York, 267–331, 1991.
9. G. Richmond, G.M. Robinson, V.L. Shannon, 'Second Harmonic Generation Studies of Interfacial Structure and Dynamics', *Prog. Surf. Sci.*, **28**, 1–20 (1988).
10. A.A. Gewirth, B.K. Niece, 'Electrochemical Applications of In Situ Scanning Probe Microscopy', *Chem. Rev.*, **97**, 1129–1162 (1997).
11. H.D. Abruña, 'X-rays as Probes of Electrochemical Interfaces', in *Modern Aspects of Electrochemistry*, eds. J. O'M. Bockris, R.E. White, B.E. Conway, Plenum Press, New York, 265–326, Vol. 20, 1989.
12. H.D. Abruña, 'Probing Electrochemical Interfaces with X-rays', *Adv. Chem. Phys.*, **77**, 255–335 (1990).
13. J. McBreen, 'In Situ Synchrotron Techniques in Electrochemistry', in *Physical Electrochemistry*, ed. I. Rubistein, Marcel Dekker, New York, 339–391, 1994.
14. L.E. Sharpe, W.R. Heineman, R.C. Elder, 'EXAFS Spectroelectrochemistry', *Chem. Rev.*, **90**, 705–722 (1990).
15. H.D. Abruña, G.M. Bommarito, D. Acevedo, 'The Study of Solid/Liquid Interfaces with X-ray Standing Waves', *Science*, **250**, 69–74 (1990).

16. M.F. Toney, O.R. Melroy, 'Surface X-ray Scattering', in *Electrochemical Interfaces: Modern Techniques for In Situ Interface Characterization*, ed. H.D. Abruña, VCH Verlag Chemical, Berlin, 55–129, 1991.
17. B.M. Ocko, 'X-ray Reflectivity and Surface Roughness', in *Spectroscopic and Diffraction Techniques in Interfacial Electrochemistry*, eds. C. Gutiérrez, C. Melendres, Kluwer Academic Publishers, Dordrecht, 343–359, 1990.
18. J. Robinson, 'X-ray Diffraction at the Electrode–Solution Interface', in *Spectroscopic and Diffraction Techniques in Interfacial Electrochemistry*, eds. C. Gutiérrez, C. Melendres, Kluwer Academic Publishers, Dordrecht, 313–341, 1990.
19. H. You, Z. Nagy, 'Application of X-ray Scattering Techniques for the Study of Electrochemical Interphases', *Current Topics in Electrochemistry*, **2**, 21–43 (1993).
20. K. Wilke, 'Synchrotron Radiation Sources', *Rep. Prog. Phys.*, **54**, 1005–1068 (1991).
21. J.A. Ibers, W.C. Hamilton, *International Tables for X-ray Crystallography*, Kynoch, Birmingham, 1974.
22. R. de L. Kronig, 'Theory of Fine Structure in the X ray Absorption Spectra', *Z. Phys.*, **70**, 317–323 (1931).
23. R. de L. Kronig, 'Theory of Fine Structure in the X-ray Absorption Spectra-II', *Z. Phys.*, **75**, 191–210 (1932).
24. D.E. Sayers, E.A. Stern, F.W. Lytle, 'New Technique for Investigating Non-crystalline Structures: Fourier Analysis of the Extended X-ray Absorption Fine Structure', *Phys. Rev. Lett.*, **27**, 1204–1207 (1971).
25. E.A. Stern, 'Theory of the Extended X-ray Absorption Fine Structure', *Phys. Rev. B*, **10**, 3027–3037 (1974).
26. E.A. Stern, D.E. Sayers, F.W. Lytle, 'Extended X-ray Absorption-fine-structure Technique. II. Determination of Physical Parameters', *Phys. Rev. B*, **11**, 4836–4846 (1975).
27. C.A. Ashley, S. Doniach, 'Theory of Extended X-ray Absorption Edge Fine Structure Spectroscopy', *Phys. Rev. B*, **11**, 1279–1290 (1975).
28. P.A. Lee, J.B. Pendry, 'Theory of the Extended X-ray Absorption Fine Structure', *Phys. Rev. B*, **11**, 2795–2811 (1975).
29. P.A. Lee, G. Beni, 'New Method for the Calculation of Atomic Phase Shifts: Application to Extended X-ray Absorption Fine Structure (EXAFS) in Molecules and Crystals', *Phys. Rev. B*, **15**, 2862–2883 (1977).
30. A. Bianconi, 'XANES Spectroscopy', in *X-ray Absorption: Principles, Applications, Techniques of EXAFS, SEXAFS and XANES*, John Wiley & Sons, New York, 573–625, 1988.
31. J. Stöhr, 'NEXAFS Spectroscopy', in *Springer Series in Surface Science*, eds. G. Ertl, R. Gomer, D.L. Mills, H.K.V. Lotsch, Springer, Berlin, Vol. 25, 1992.
32. B.K. Teo, *EXAFS: Basic Principles and Data Analysis*, Springer-Verlag, Berlin, 1986.
33. B.K. Teo, P.A. Lee, 'Ab Initio Calculation of Amplitude and Phase Functions for Extended X-ray Absorption Fine Structure Spectroscopy', *J. Am. Chem. Soc.*, **101**, 2815–2832 (1979).
34. J.J. Rehr, J. Mustre de Leon, S.I. Zabinsky, R.C. Albers, 'Theoretical X-ray Absorption Fine Structure Standards', *J. Am. Chem. Soc.*, **113**, 5135–5140 (1991).
35. J. Mustre de Leon, J.J. Rehr, S.I. Zabinsky, R.C. Albers, 'Ab Initio Curved-wave X-ray-absorption Fine Structure', *Phys. Rev. B*, **44**, 4146–4156 (1991).
36. J.J. Rehr, S.I. Zabinsky, A. Ankudinov, R.C. Albers, 'Atomic-XAFS and XANES', *Physica B.*, **209**, 23–26 (1995).
37. S.I. Zabinsky, J.J. Rehrm, A. Ankudinov, R.C. Albers, M.J. Ella, 'Multiple Scattering Calculations of X-ray Absorption Spectra', *Phys. Rev. B*, **52**, 2995–3009 (1995).
38. P.A. Lee, P.H. Citrin, P. Eisenberg, B.M. Kinkaid, 'Extended X-ray Absorption Fine Structure – Its Strength and Limitations as Structural Tool', *Rev. Mod. Phys.*, **53**, 769–806 (1981).
39. S. Wu, J. Lipkowski, T. Tyliszczak, A.P. Hitchcock, 'Effect of Anion Adsorption on Early Stages of Copper Electrocrystallization at Au(111) Surface', *Prog. Surf. Sci.*, **50**, 227–236 (1995).
40. E. Dartyge, L. Depautex, J.M. Dubisson, A. Fontaine, A. Jucha, P. Leboucher, G. Tourillon, 'X-ray Absorption in Dispersive Mode: a New Spectrometer and a Data Acquisition System for Fast Kinetics', *Nucl. Instrum. Methods*, **A246**, 452–460 (1986).
41. H. Tolentino, E. Dartyge, A. Fontaine, G. Tourillon, 'X-ray Absorption Spectroscopy in the Dispersive Mode with Synchrotron Radiation: Optical Considerations', *J. Appl. Crys.*, **21**, 15–21 (1988).
42. F. D'Acapito, F. Boschenerini, A. Marcelli, S. Mobilio, 'Dispersive EXAFS Apparatus at Frascati', *Rev. Sci. Instrum.*, **63**, 899–901 (1992).
43. P.G. Allen, S.D. Conradson, J.E. Penner-Hahn, 'A 4-point Crystal Bender for Dispersive X-ray Diffraction', *J. Appl. Cryst.*, **26**, 172–179 (1993).
44. A.J. Dent, M.P. Wells, R.C. Farrow, C.A. Ramsdale, G.E. Derbyshire, G.N. Greaves, J.W. Couves, J.M. Thomas, 'Combined Energy Dispersive EXAFS and X-ray Diffraction', *Rev. Sci. Instrum.*, **63**, 902–905 (1992).
45. M.G. Samant, M.F. Toney, G.L. Borges, L. Blum, O.R. Melroy, 'Grazing Incident X-ray Diffraction of Lead Monolayers at a Silver(111) and a Gold(111) Electrode/Electrolyte Interface', *J. Phys. Chem.*, **92**, 220–225 (1988).
46. M.G. Samant, G.L. Borges, J.G. Gordon, O.R. Melroy, L. Blum, 'In Situ Surface Extended X-ray Absorption Fine Structure Spectroscopy of a Lead Monolayer at a Silver(111) Electrode/Electrolyte Interface', *J. Am. Chem. Soc.*, **109**, 5970–5974 (1987).
47. M.J. Albarrelli, J.H. White, M.G. Bommarito, M. McMillan, H.D. Abruña, 'In Situ EXAFS at Chemically Modified Electrodes', *J. Electroanal. Chem.*, **248**, 77–86 (1988).

48. J. McBreen, W.E. O'Grady, K.I. Pandya, R.W. Hoffman, D.E. Sayers, 'EXAFS Study of the Nickel Hydroxide Electrode', *Langmuir*, **3**, 428–433 (1987).
49. A.J. Davenport, H.S. Isaacs, G.S. Frankel, A.G. Schrott, C.V. Jahnes, M.A. Russak, 'In Situ X-ray Absorption Study of Chromium Valency Changes in Passive Oxides on Sputtered AlCr Thin Films Under Electrochemical Control', *J. Electrochem. Soc.*, **138**, 337–338 (1991).
50. A.J. Davenport, M. Sansone, 'High Resolution In Situ XANES Investigation of the Nature of the Passive Film on Iron in a pH 8.4 Borate Buffer', *J. Electrochem. Soc.*, **142**, 725–730 (1995).
51. A.R. Sandy, S.G.J. Mochrie, D.M. Zhener, K.G. Huang, D. Gibbs, 'Structure and Phases of the Au(111) Surface: X-ray Scattering Measurements', *Phys. Rev. B*, **43**, 4667–4687 (1991).
52. B.M. Ocko, J. Wang, A. Davenport, H. Isaacs, 'In Situ X-ray Reflectivity and Diffraction Studies of the Au(100) Reconstruction in an Electrochemical Cell', *Phys. Rev. Lett.*, **65**, 1466–1469 (1990).
53. I.K. Robinson, 'Crystal Truncation Rods and Surface Roughness', *Phys. Rev. B*, **33**, 3830–3836 (1986).
54. I.K. Robinson, 'Surface Crystallography', in *Handbook on Synchrotron Radiation*, eds. G. Brown, D.E. Moncton, Elsevier Science, Amsterdam, 221, Vol. 3, 1991.
55. D. Gibbs, B.M. Ocko, D.M. Zehner, S.G.J. Mochrie, 'Absolute X-ray Reflectivity Study of the Au(100) Surface', *Phys. Rev. B*, **38**, 7303–7310 (1988).
56. J. Zegenhagen, 'Surface Structure Determination with X-ray Standing Waves', *Surf. Sci. Reports*, **18**, 199–271 (1993).
57. K.M. Robinson, W.E. O'Grady, 'X-ray Diffraction and Electrochemical Study on the Oxidation of Flame-annealed Au(100) Single-crystal Surfaces', *J. Electroanal. Chem.*, **384**, 139–144 (1995).
58. H. You, C.A. Melendres, Z. Nagy, V.A. Maroni, W. Yun, R.M. Yonco, 'X-ray-reflectivity Study of the Copper–Water Interface in Transmission Geometry Under In Situ Electrochemical Control', *Phys. Rev. B*, **45**, 11 288–11 289 (1992).
59. F. Brossard, V.H. Etgens, A. Tadjeddine, 'In Situ X-ray Diffraction Using a New Electrochemical Cell Optimised for Third Generation Synchrotron Light Sources', *Nucl. Instrum. Meth. Phys. Res. B*, **129**, 419–422 (1997).
60. M.F. Toney, J.N. Howard, J. Richer, G.L. Borges, J.G. Gordon, O.R. Melroy, 'Electrochemical Deposition of Copper on a Gold Electrode in Sulfuric Acid – Resolution of the Interfacial Structure', *Phys. Rev. Lett.*, **75**, 4772–4775 (1995).
61. J. Wang, A.J. Davenport, H.S. Isaacs, B.M. Ocko, 'Surface Charge Induced Ordering of the Au(111) Surface', *Science*, **255**, 1416–1419 (1992).
62. B.M. Ocko, A. Gibaud, J. Wang, 'Surface X-ray Diffraction Study of the Au(111) Electrode in 0.01 M NaCl: Electrochemical Induced Surface Reconstruction', *J. Vac. Sci. Technol. A*, **10**, 3019–3025 (1992).
63. J. Wang, B.M. Ocko, A.J. Davenport, H.S. Isaacs, 'In Situ X-ray -Diffraction and -Reflectivity Studies of the Au(111)/Electrolyte Interface: Reconstruction and Anion Adsorption', *Phys. Rev. B*, **46**, 10 321–10 338 (1992).
64. S. Wu, J. Lipkowski, O.M. Magnussen, B.M. Ocko, Th. Wandlowski, 'The Driving Force for the $(p \times \sqrt{3}) \leftrightarrow (1 \times 1)$ Phase Transition of Au(111) in the Presence of Organic Adsorption: A Combined Chronocoulometric and Surface X-ray Scattering Study', *J. Electroanal. Chem.*, **446**, 67–77 (1998).
65. B.M. Ocko, J. Wang, A. Davenport, H.S. Isaacs, 'In Situ X-ray Reflectivity and Diffraction Studies of the Au(100) Reconstruction in an Electrochemical Cell', *Phys. Rev. Lett.*, **65**, 1466–1469 (1990).
66. D. Gibbs, B.M. Ocko, D.M. Zehner, S.G.J. Mochrie, 'Absolute X-ray Reflectivity Study of the Au(100) Surface', *Phys. Rev. B*, **38**, 7303–7310 (1988).
67. S.G.J. Mochrie, D.M. Zehner, B.M. Ocko, D. Gibbs, 'Structure and Phases of the Au(001) Surface: X-ray Scattering Measurements', *Phys. Rev. Lett.*, **64**, 2925–2928 (1990).
68. I.M. Tidswell, N.M. Markovic, P.N. Ross, 'Potential Dependent Surface Relaxation of the Pt(001)/Electrolyte Interface', *Phys. Rev. Lett.*, **71**, 1601–1604 (1993).
69. I.M. Tidswell, N.M. Markovic, P.N. Ross, 'Potential Dependent Surface Structure of the Pt(111)/Electrolyte Interface', *J. Electroanal. Chem.*, **376**, 119–126 (1994).
70. H. You, D.J. Zurawski, Z. Nagy, R.M. Yonco, 'In Situ X-ray Reflectivity Study of Incipient Oxidation of Pt(111) Surface in Electrolyte Solutions', *J. Chem. Phys*, **100**, 4699–4702 (1994).
71. H. You, Z. Nagy, 'Oxidation–Reduction-induced Roughening of Platinum(111) Surface', *Physica B*, **198**, 187–194 (1994).
72. B.M. Ocko, O.M. Magnussen, J.X. Wang, R.R. Adzic, Th. Wandlowski, 'The Structure and Phase Behavior of Electrodeposited Halides on Single-crystal Metal Surfaces', *Physica B*, **211**, 238–244 (1996).
73. B.M. Ocko, G.M. Watson, J. Wang, 'Structure and Electrocompression of Electrodeposited Iodine Monolayers on Au(111)', *J. Phys. Chem.*, **98**, 897–906 (1994).
74. G.M. Bommarito, J.H. White, H.D. Abruña, 'Electrodeposition of Iodide on Platinum: Packing Density and Potential-dependent Distributional Changes Observed In Situ with X-ray Standing Waves', *J. Phys. Chem.*, **94**, 8280–8288 (1990).
75. O.M. Magnussen, B.M. Ocko, R.R. Adzic, J.X. Wang, 'X-ray Diffraction Studies of Ordered Chloride and Bromide Monolayers at the Au(111)–Solution Interface', *Phys. Rev. B*, **51**, 5510–5513 (1995).
76. O.M. Magnussen, B.M. Ocko, J.X. Wang, R.R. Adzic, 'In Situ X-ray Diffraction and STM Studies of Bromide Adsorption on Au(111) Electrodes', *J. Phys. Chem.*, **100**, 5500–5508 (1996).

77. C.A. Lucas, N.M. Markovic, P.N. Ross, 'Observation of an Ordered Bromide Monolayer at the Pt(111)–Solution Interface by In Situ Surface X-ray Scattering', *Surf. Sci.*, **340**, L954–L959 (1995).
78. B.M. Ocko, O.M. Magnussen, J.X. Wang, Th. Wandlowski, 'One-dimensional Commensurate–Incommensurate Transition: Bromide on the Au(100) Electrode', *Phys. Rev. B*, **53**, R7654–7657 (1996).
79. Th. Wandlowski, J.X. Wang, O.M. Magnussen, B.M. Ocko, 'Structural and Kinetic Aspects of Bromide Adsorption on Au(100)', *J. Phys. Chem.*, **100**, 10 277–10 287 (1996).
80. B.M. Ocko, J.X. Wang, Th. Wandlowski, 'Bromide Adsorption on Ag(001): A Potential Induced Two Dimensionalising Order–Disorder Transition', *Phys. Rev. Lett*, **79**, 1511–1514 (1997).
81. M.H. Hölze, V. Zwing, D.M. Kolb, 'The Influence of Steps on the Deposition of Cu onto Au(111)', *Electrochim. Acta*, **40**, 1237–1247 (1995).
82. T. Hachiya, H. Honbo, K. Itaya, 'Detailed Underpotential Deposition of Copper on Gold(111) in Aqueous-solutions', *J. Electroanal. Chem.*, **315**, 275–291 (1991).
83. L. Blum, H.D. Abruña, J. White, J.G. Gordon II, G.L. Borges, M.G. Samant, O.R. Melroy, 'Study of the Underpotentially Deposited Copper on Gold by Fluorescence Detected Surface EXAFS', *J. Chem. Phys.*, **82**, 6732–6738 (1986).
84. O.R. Melroy, G.L. Borges, L. Blum, H.D. Abruña, M.J. Albarelli, M.G. Samant, M. McMillan, J.H. White, J.G. Gordon II, 'In-plane Structure of Underpotentially Deposited Copper on Gold(111) Determined by Surface EXAFS', *Langmuir*, **4**, 728–732 (1988).
85. A. Tadjeddine, G. Tourillon, D. Guay, 'Structural and Electronic Characterization of Underpotentially Deposited Copper on Gold Single-crystal Probed by In Situ X-ray Absorption-spectroscopy', *Electrochim. Acta*, **36**, 1859–1862 (1991).
86. O.M. Magnussen, J. Hotlos, R.J. Nichols, D.M. Kolb, R.J. Behm, 'Atomic-structure of Cu Adlayers on Au(100) and Au(111) Electrodes Observed by In Situ Scanning Tunneling Microscopy', *Phys. Rev. Lett.*, **64**, 2929–2932 (1990).
87. N. Batina, T. Will, D.M. Kolb, 'Study of the Initial-stages of Copper Deposition by In Situ Scanning-tunneling-microscopy', *Faraday Discuss.*, **94**, 93–100 (1992).
88. M.P. Green, K.P. Hanson, 'Copper Adlayer Formation on Au(111) from Sulfuric-acid Electrolyte', *J. Vac. Sci. Technol. A*, **10**, 3012–3018 (1992).
89. S. Manne, P.K. Hansma, J. Massie, V.B. Elings, A.A. Gewirth, 'Atomic-resolution Electrochemistry with the Atomic Force Microscope – Copper Deposition on Gold', *Science*, **251**, 183–186 (1991).
90. Z. Shi, J. Lipkowski, 'Investigations of SO_4^{2-} Adsorption at the Au(111) Electrode in the Presence of Underpotentially Deposited Copper Adatoms', *J. Electroanal. Chem.*, **364**, 289–294 (1994).
91. Z. Shi, J. Lipkowski, 'Coadsorption of Cu^{2+} and SO_4^{2-} at the Au(111) Electrode', *J. Electroanal. Chem.*, **365**, 303–309 (1994).
92. Z. Shi, S. Wu, J. Lipkowski, 'Coadsorption of Metal Atoms and Anions – Cu UPD in the Presence of SO_4^{2-}, Cl^- and Br^-', *J. Electrochim. Acta*, **40**, 9–15 (1995).
93. J.G. Gordon, O.R. Melroy, M.F. Toney, 'Structure of Metal Electrolyte Interfaces – Copper on Gold(111), Water on Silver(111)', *Electrochim. Acta*, **40**, 3–8 (1995).
94. G.L. Borges, K.K. Kanazawa, J.G. Gordon II, K. Ashley, J. Richer, 'An In Situ Electrochemical Quartz-crystal Microbalance Study of the Underpotential Deposition of Copper on Au(111) Electrodes', *J. Electroanal. Chem.*, **364**, 281–284 (1994).
95. M. Watanabe, H. Uchida, M. Miura, N. Ikeda, 'Electrochemical Quartz-crystal Microbalance Study of Copper Ad-atoms on Highly Ordered Au(111) Electrodes in Sulfuric Acid', *J. Electroanal. Chem.*, **384**, 191–195 (1995).
96. L. Blum, D.A. Huckaby, M. Legault, 'Phase Transitions at Electrode Interfaces', *Electrochim. Acta*, **41**, 2207–2227 (1996).
97. Y. Nakay, M.S. Zei, D.M. Kolb, G. Lehmpfuhl, 'A LEED and RHEED Investigation of Cu on Au(111) in the Underpotential Region', *Ber. Bunsenges. Phys. Chem.*, **88**, 340–345 (1984).
98. M.S. Zei, G. Qiao, G. Lehmpfuhl, D.M. Kolb, 'The Influence of Anions on the Structure of Underpotentially Deposited Cu on Au(111): A LEED, RHEED, and AES Study', *Ber. Bunsenges. Phys. Chem.*, **91**, 349–353 (1987).
99. A. Tadjeddine, A. Lahrichi, G. Tourillon, 'In Situ X-ray-absorption Near-edge Structure – A Probe of the Oxidation-state of Underpotentially Adsorbed Metals', *J. Electroanal. Chem.*, **360**, 261–270 (1993).
100. E.D. Chabala, J. Cairns, T. Rayment, 'In Situ Real Time Study of an Electrode Process by Differential X-ray Diffraction. Part 2. Cu Underpotential Deposition on Au(111)', *J. Electroanal. Chem.*, **412**, 77–84 (1996).
101. S. Wu, Z. Shi, J. Lipkowski, A.P. Hitchcock, T. Tyliszczak, 'Early Stages of Copper Electrocrystallization: Electrochemical and In Situ X-ray Absorption Fine Structure Studies of Coadsorption of Copper and Chloride on the Au(111) Electrode Surface', *J. Phys. Chem. B*, **101**, 10 310–10 322 (1997).
102. R.G.W. Wyckoff, *Crystal Structures*, Interscience, New York, Vol. 1, 1960.
103. E. Herrero, S. Glazier, J.L. Buller, H.D. Abruña, 'X-ray and Electrochemical Studies of Cu UPD on Single Crystal Electrodes in the Presence of Bromide: Comparison Between Au(111) and Pt(111) Electrodes', *J. Electroanal. Chem.*, **461**, 121–130 (1999).
104. E. Herrero, S. Glazier, H.D. Abruña, 'X-ray and Electrochemical Studies of Cu UPD on Au(111) Single Crystal Electrodes in the Presence of Bromide', *J. Phys. Chem. B*, **102**, 9825–9833 (1998).

105. F.A. Möller, O.M. Magnussen, R.J. Behm, 'In Situ STM Studies of Cu Underpotential Deposition on Au(100) in the Presence of Sulfate and Chloride Anions', *Phys. Rev. B*, **51**, 2484–2490 (1995).
106. H.D. Abruña, T. Gog, G. Mateerlik, W. Uelhoff, 'X-ray Standing-wave Study of Copper Underpotentially Deposited on Au(100)', *J. Electroanal. Chem.*, **360**, 315–323 (1993).
107. G. Tourillon, D. Guay, A. Tadjeddine, 'In-plane Structural and Electronic Characteristics of Underpotentially Deposited Copper on Gold-(100) Probed by In Situ X-ray Absorption-spectroscopy', *J. Electroanal. Chem.*, **289**, 263–278 (1990).
108. M.G. Samant, O.R. Melroy, M.F. Toney, G.L. Borges, L. Blum, 'Grazing-incidence X-ray-diffraction of Lead Monolayers at a Silver(111) and Gold(111) Electrode/Electrolyte Interface', *J. Phys. Chem.*, **92**, 220–225 (1988).
109. M.F. Toney, J.G. Gordon, M.G. Sammant, G.L. Borges, O.R. Melroy, D. Yee, L.B. Sorensen, 'In Situ Atomic-structure of Underpotentially Deposited Monolayers of Pb and Tl on Au(111) and Ag(111) – A Surface X-ray-scattering Study', *J. Phys. Chem.*, **99**, 4733–4744 (1995).
110. M.P. Green, K.J. Hanson, R. Carr, I. Lindau, 'STM Observations of the Underpotential Deposition and Stripping of Pb on Au(111) Underpotential Sweep Conditions', *J. Electrochem. Soc.*, **137**, 3493–3498 (1990).
111. M.P. Green, K.J. Hanson, 'Alloy Formation in an Electrodeposited Monolayer', *Surf. Sci.*, **259**, L743–L749 (1991).
112. N.J. Tao, J. Pan, Y. Li, P.I. Oden, J.A. Derose, S.M. Lindsay, 'Initial-stage of Underpotential Deposition of Pb on Reconstructed and Unreconstructed Au(111)', *Surf. Sci.*, **271**, L338–L344 (1992).
113. C.-H. Chen, N. Washburn, A.A. Gewirth, 'In Situ Atomic-force Microscope Study of Pb Underpotential Deposition on Au(111) – Structural Properties of the Catalytically Active Phase', *J. Phys. Chem.*, **97**, 9754–9760 (1993).
114. E.D. Chabala, B.H. Harji, T. Rayment, M.D. Archer, 'In Situ X-ray-diffraction Study of Underpotential Deposition at the Gold(111) Surface', *Langmuir*, **8**, 2028–2033 (1992).
115. E. Herrero, H.D. Abruña, 'Underpotential Deposition of Mercury on Au(111): Electrochemical Studies and Comparison with Structural Investigations', *Langmuir*, **13**, 4446–4453 (1997).
116. J. Inukai, S. Sugita, K. Itaya, 'Underpotential Deposition of Mercury on Au(111) Investigated by In Situ Scanning-tunneling-microscopy', *J. Electroanal. Chem.*, **403**, 159–168 (1996).
117. J. Li, H.D. Abruña, 'Coadsorption of Sulfate/Bisulfate Anions with Hg Cations During Hg Underpotential Deposition on Au(111): An In Situ X-ray Diffraction Study', *J. Phys. Chem. B*, **101**, 244–252 (1997).
118. C.-H. Chen, A.A. Gewirth, 'AFM Study of the Structure of Underpotentially Deposited Ag and Hg on Au(111)', *Ultramicroscopy*, **42**, 437–444 (1992).
119. C.-H. Chen, A.A. Gewirth, 'In Situ Observation of Monolayer Structures of Underpotentially Deposited Hg on Au(111) with the Atomic Force Microscope', *Phys. Rev. Lett.*, **68**, 1571–1574 (1992).
120. J. Li, H.D. Abruña, 'Phases of Underpotentially Deposited Hg on Au(111): and In Situ Surface X-ray Diffraction Study', *J. Phys. Chem. B*, **101**, 2907–2916 (1997).
121. J. Li, E. Herrero, H.D. Abruña, 'The Effects of Anions in the Underpotential Deposition of Hg on Au(111): An Electrochemical and In Situ X-ray Diffraction Study', *Colloids Surf. A*, **134**, 113–131 (1998).
122. H.S. Yee, H.D. Abruña, 'In Situ X-ray Studies of the Underpotential Deposition of Copper on Platinum(111)', *J. Phys. Chem.*, **97**, 6278–6288 (1993).
123. H.S. Yee, H.D. Abruña, 'Ab-initio XAFS Calculations and In Situ XAFS Measurements of Copper Underpotential Deposition on Pt(111) – A Comparative Study', *J. Phys. Chem.*, **98**, 6552–6558 (1994).
124. C.A. Lucas, N.M. Markovic, P.N. Ross, 'Electrochemical Deposition of Copper onto Pt(111) in the Presence of (Bi)sulfate Anions', *Phys. Rev. B*, **56**, 3651–3654 (1997).
125. C.A. Lucas, N.M. Markovic, I.M. Tidswell, P.N. Ross, 'In Situ X-ray Scattering Study of the Pt(111)–Solution Interface – Ordered Anion Structures and Their Influence on Copper Underpotential Deposition', *Physica B*, **221**, 245–250 (1996).
126. I.M. Tidswell, C.A. Lucas, N.M. Markovic, P.N. Ross, 'Surface Structure Determination Using Anomalous X-ray Scattering – Underpotential Deposition of Copper on Pt(111)', *Phys. Rev. B*, **51**, 10 205–10 208 (1995).
127. N.M. Markovic, H.A. Gasteiger, C.A. Lucas, I.M. Tidswell, P.N. Ross, 'The Effect of Chloride on the Underpotential Deposition of Copper on Pt(111): AES, LEED, RRDE and X-ray Scattering Studies', *Surf. Sci.*, **335**, 91–100 (1995).
128. R. Gómez, H.S. Yee, G.M. Bommarito, J.M. Feliu, H.D. Abruña, 'Anion Effects and the Mechanism of Cu UPD on Pt(111) – X-ray and Electrochemical Studies', *Surf. Sci.*, **335**, 101–109 (1995).
129. J.H. White, H.D. Abruña, 'Coadsorption of Copper with Anions on Platinum(111) – The Role of Surface Redox Chemistry in Determining the Stability of a Metal Monolayer', *J. Phys. Chem.*, **94**, 894–900 (1990).
130. H.S. Yee, H.D. Abruña, 'In Situ X-ray-absorption Spectroscopy Studies of Copper Underpotentially Deposited in the Absence and Presence of Chloride on Platinum(111)', *Langmuir*, **9**, 2460–2469 (1993).
131. A.C. Finnefrock, L.J. Buller, K.L. Ringland, J.D. Brock, H.D. Abruña, 'Time-resolved Surface X-ray Scattering Study of Surface Ordering of Electrodeposited Layers', *J. Am. Chem. Soc.*, **119**, 11 703–11 704 (1997).

132. H.D. Abruña, J.M. Feliu, J.D. Brock, L.J. Buller, E. Herrero, J. Li, R. Gómez, A. Finnefrock, 'Anion and Electrode Surface Structure Effects on the Deposition of Metal Monolayers: Electrochemical and Time-resolved Surface Diffraction Studies', *Electrochim. Acta*, **43**, 2899–2909 (1998).
133. N.M. Markovic, C.A. Lucas, H.A. Gasteiger, P.N. Ross, 'The Structure of Adsorbed Bromide Concurrent with the Underpotential Deposition (UPD) of Cu on Pt(111)', *Surf. Sci.*, **372**, 239–254 (1997).
134. M.G. Samant, M.F. Toney, G.L. Borges, L. Blum, O.R. Melroy, 'In Situ Grazing-incidence X-ray-diffraction Study of Electrochemically Deposited Pb Monolayers on Ag(111)', *Surf. Sci.*, **193**, L29–L36 (1988).
135. O.R. Melroy, M.F. Toney, G.L. Borges, M.G. Samant, J.B. Kortright, P.N. Ross, L. Blum, 'Two-dimensional Compressibility of Electrochemically Adsorbed Lead on Silver(111)', *Phys. Rev. B*, **38**, 10 962–10 965 (1988).
136. O.R. Melroy, M.F. Toney, G.L. Borges, M.G. Samant, J.B. Kortright, P.N. Ross, L. Blum, 'An In Situ Grazing-incidence X-ray Scattering Study of the Initial-stages of Electrochemical Growth of Lead on Silver(111)', *J. Electroanal. Chem.*, **258**, 403–414 (1989).
137. E.D. Chabala, T. Rayment, 'In Situ Surface Differential Diffraction Study of Metal Monolayer Formation by Underpotential Deposition on Silver(111) Oriented Surfaces', *Langmuir*, **10**, 4324–4329 (1994).
138. W. Obretenov, U. Schmidt, W.J. Lorenz, G. Staikov, E. Budevski, D. Carnal, U. Müller, H. Siegenthaler, E. Schmidt, 'Underpotential Deposition and Electrocrystallization of Metals – An In Situ Scanning-tunneling-microscopy Study with Lateral Atomic-resolution', *Faraday Discuss.*, **94**, 107–116 (1992).
139. W.J. Lorenz, L.M. Gassa, U. Schmidt, W. Obretenov, G. Staikov, V. Bostanov, E. Budevski, 'STM Studies in Underpotential Overpotential Metal-deposition', *Electrochim. Acta*, **37**, 2173–2178 (1992).
140. U. Müller, D. Carnal, H. Siegenthaler, E. Schmidt, W.J. Lorenz, W. Obretenov, U. Schmidt, G. Staikov, E. Budevski, 'Superstructures of Pb Monolayers Electrochemically Deposited on Ag(111)', *Phys. Rev. B*, **46**, 12 899–12 901 (1992).
141. M.G. Samant, G.L. Borges, J.G. Gordon, O.R. Melroy, L. Blum, 'In Situ Surface Extended X-ray Absorption Fine Structure Spectroscopy of a Lead Monolayer at a Silver(111) Electrode/Electrolyte Interface', *J. Am. Chem. Soc.*, **109**, 5970–5974 (1987).
142. M.F. Toney, J.G. Gordon, M.G. Samant, G.L. Borges, O.R. Melroy, L.-S. Kau, D.G. Wiesler, D. Yee, L.B. Sorensen, 'Surface X-ray-scattering Measurements of the Substrate-induced Spatial Modulation of an Incommensurate Adsorbed Monolayer', *Phys. Rev. B*, **42**, 5594–5603 (1990).
143. M.F. Toney, J.G. Gordon, M.G. Samant, G.L. Borges, O.R. Melroy, D. Yee, L.B. Sorensen, 'Underpotentially Deposited Thallium on Silver(111) by In Situ Surface X-ray-scattering', *Phys. Rev. B*, **45**, 9362–9374 (1992).
144. J.X. Wang, R.R. Adzic, O.M. Magnussen, B.M. Ocko, 'Structural Evolution During Electrocrystallization – Deposition of Tl on Ag(100) from Monolayer to Bilayer and to Bulk Crystallites', *Surf. Sci.*, **344**, 111–121 (1995).
145. M.E. Herron, S.E. Doyle, S. Pizzini, K.J. Roberts, J. Robinson, G. Hards, F.C. Walsh, 'In Situ Studies of a Dispersed Platinum on Carbon Electrode Using X-ray Absorption Spectroscopy', *J. Electroanal. Chem.*, **324**, 243–258 (1992).
146. S. Mukerjee, S. Srinivasan, M.P. Soriaga, J. McBreen, 'Role of Structural and Electronic Properties of Pt and Pt Alloys on Electrocatalysis of Oxygen Reduction: An In Situ XANES and EXAFS Investigation', *J. Electrochem. Soc.*, **142**, 1409–1422 (1995).
147. S. Mukerjee, J. McBreen, 'Hydrogen Electrocatalysis by Carbon Supported Pt and Pt Alloys: An In Situ X-ray Absorption Study', *J. Electrochem. Soc.*, **143**, 2285–2294 (1996).
148. P.G. Allen, S.D. Conradson, M.S. Wilson, S. Gottesfeld, I.D. Raistrick, J. Valerio, M. Lovato, 'In Situ Structural Characterization of a Platinum Electrocatalyst by Dispersive X-ray Absorption Spectroscopy', *Electrochim. Acta*, **39**, 2415–2418 (1994).
149. S. Mukerjee, J. McBreen, 'Effect of Particle Size on the Electrocatalysis by Carbon-supported Pt Electrocatalyst: An In Situ XAS Investigation', *J. Electroanal. Chem.*, **448**, 163–171 (1998).
150. J. McBreen, W.E. O'Grady, G. Tourillon, E. Dartyge, A. Fontaine, 'XANES Study of Underpotential Deposited Copper on Carbon Supported Platinum', *J. Electroanal. Chem.*, **307**, 229–240 (1991).
151. J. McBreen, 'EXAFS Studies of Adsorbed Copper on Carbon Supported Platinum', *J. Electroanal. Chem.*, **357**, 373–386 (1993).
152. J. McBreen, M. Sansone, 'In Situ X-ray Absorption Spectroscopic Study of Adsorbed Pb on Carbon-supported Pt', *J. Electroanal. Chem.*, **373**, 227–233 (1994).
153. D. Aberdam, R. Durand, R. Faure, F. Gloaguen, L.J. Hazemann, E. Herrero, A. Kabbabi, O. Ulrich, 'X-ray Absorption Near Edge Structure Study of the Electro-oxidation of Co on $Pt_{50}Ru_{50}$ Nanoparticles', *J. Electroanal. Chem.*, **398**, 43–47 (1995).
154. J. McBreen, S. Mukerjee, 'In Situ X-ray Absorption Studies of a Pt–Ru Electrocatalyst', *J. Electrochem. Soc.*, **142**, 3399–3404 (1995).
155. J. McBreen, W.E. O'Grady, K.I. Pandya, R.W. Hoffman, D.E. Sayers, 'EXAFS Study of the Nickel Hydroxide Electrode', *Langmuir*, **3**, 428–433 (1987).
156. J. McBreen, W.E. O'Grady, G. Tourillon, E. Dartyge, A. Fontaine, K.I. Pandya, 'In Situ Time Resolved X-ray Absorption Near Edge Structure of the Nickel Oxide Electrode', *J. Phys. Chem.*, **93**, 6308–6311 (1989).

157. K.I. Pandya, R.W. Hoffman, J. McBreen, W.E. O'Grady, 'In Situ X-ray Absorption Spectroscopy Studies of Nickel Oxide Electrodes', *J. Electrochem. Soc.*, **137**, 383–388 (1990).
158. K.I. Pandya, W.E. O'Grady, D.A. Corrigan, J. McBreen, R.W. Hoffman, 'Extended X-ray Absorption Fine Structure Investigations of Nickel Hydroxides', *J. Phys. Chem.*, **94**, 21–26 (1990).
159. W.E. O'Grady, K.I. Pandya, K.E. Swider, D.A. Corrigan, 'In Situ X-ray Absorption Near Edge Structure Evidence for Quadrivalent Nickel in Nickel Battery Electrodes', *J. Electrochem. Soc.*, **143**, 1613–1616 (1996).
160. X. Quian, H. Sambe, D.E. Ramaker, K.I. Pandya, W.E. O'Grady, 'Quantitative Interpretation of the K-edge NEXAFS Data for Various Nickel Hydroxides and the Charged Nickel Electrode', *J. Phys. Chem. B*, **101**, 9441–9446 (1997).
161. T.W. Capehart, D.A. Corrigan, R.S. Conell, K.I. Pandya, R.W. Hoffman, 'In Situ Extended X-ray Absorption Fine Structure Spectroscopy of a Thin Film Nickel Hydroxide Electrode', *Appl. Phys. Lett.*, **58**, 865–867 (1991).
162. C. Cartier, A. Tranchant, M. Verdaguer, R. Messina, H. Dexpert, 'X-ray Diffraction and X-ray Absorption Studies of the Structural Modifications Induced by Electrochemical Lithium Intercalation into V_2O_5', *Electrochim. Acta*, **35**, 889–898 (1990).
163. P. Druska, H.H. Strehblow, 'In Situ Examination of Electrochemically Formed Cu_2O by EXAFS in Transmission', *J. Electroanal. Chem.*, **335**, 55–65 (1992).
164. M.E. Herron, D. Pletcher, F.C. Walsh, 'A Combined Electrochemical and In Situ X-ray Diffraction Study of the Cycling of Well-defined Lead Dioxide Layers on Platinum', *J. Electroanal. Chem.*, **332**, 183–197 (1992).
165. S. Mukerjee, T.R. Thurston, N.M. Jisrawi, X.Q. Yang, J. McBreen, M.L. Daroux, X.K. Xing, 'Structural Evolution of $Li_xMn_2O_4$ in Lithium-ion Battery Cells Measured In Situ Using Synchrotron X-ray Diffraction Techniques', *J. Electrochem. Soc.*, **145**, 466–472 (1998).
166. M.F. Toney, A.J. Davenport, L.J. Oblonsky, M.P. Ryan, C.M. Vitus, 'Atomic Structure of the Passive Oxide Film Formed on Iron', *Phys. Rev. Lett.*, **79**, 4282–4285 (1997).
167. L.J. Oblonsky, A.J. Davenport, M.P. Ryan, H.S. Isaacs, R.C. Newman, 'In Situ X-ray Absorption Near Edge Structure Study of the Potential Dependence of the Formation of the Passive Film on Iron Borate Buffer', *J. Electrochem. Soc.*, **144**, 2398–2404 (1997).
168. P. Schmuki, S. Virtanen, A.J. Davenport, C. Vitus, 'In Situ X-ray Absorption Near Edge Spectroscopic Study of the Cathodic Reduction of Artificial Iron Oxide Passive Films', *J. Electrochem. Soc.*, **143**, 574–582 (1996).
169. S. Virtanen, P. Schmuki, A.J. Davenport, C. Vitus, 'Dissolution of Thin Iron Oxide Films Used as Models for Iron Passive Films Studied by In Situ X-ray Absorption Near-edge Spectroscopy', *J. Electrochem. Soc.*, **144**, 198–204 (1997).
170. A.J. Davenport, M. Sansone, J.A. Bardwell, A.J. Aldykiewicz, M. Taube, C. Vitus, 'In Situ Multielement XANES Study of Formation and Reduction of the Oxide Film in Stainless Steel', *J. Electrochem. Soc.*, **141**, L6–L8 (1994).
171. J.A. Bardwell, G.I. Sproule, B. Macdougall, M.J. Graham, A.J. Davenport, H.S. Isaacs, 'In Situ XANES Detection of Cr(VI) in the Passive Film on Fe–^{26}Cr', *J. Electrochem. Soc.*, **139**, 371–373 (1992).
172. P. Schmuki, S. Virtanen, H.S. Isaacs, M.P. Ryan, A.J. Davenport, H. Böhni, T. Stenberg, 'Electrochemical Behavior of Cr_2O_3/Fe_2O_3 Artificial Passive Films Studied by In Situ XANES', *J. Electrochem. Soc.*, **143**, 791–801 (1998).
173. G.S. Frankel, A.J. Davenport, H.S. Isaacs, A.G. Schrott, C.V. Jahnes, M.A. Russak, 'X-ray Absorption Study of Electrochemically Grown Oxide Films an Al–Cr Sputtered Alloys: I. Ex situ Studies', *J. Electrochem. Soc.*, **139**, 1812–1820 (1992).
174. G.S. Frankel, A.G. Schrott, A.J. Davenport, H.S. Isaacs, C.V. Jahnes, M.A. Russak, 'X-ray Absorption Study of Electrochemically Grown Oxide Films an Al–Cr Sputtered Alloys: II. In Situ Studies', *J. Electrochem. Soc.*, **141**, 83–90 (1994).
175. A.J. Davenport, H.S. Isaacs, M.W. Kendig, 'X-ray Absorption Study of Cerium in the Passive Film on Aluminum', *J. Electrochem. Soc.*, **136**, 1837–1838 (1989).
176. D.A. Smith, M.J. Heeg, W.R. Heineman, R.C. Elder, 'Direct Determination of Fe–C Bond Lengths in Iron(II) and Iron(III) Cyanide Solutions Using EXAFS Spectroelectrochemistry', *J. Am. Chem. Soc.*, **106**, 3053–3054 (1984).
177. R.C. Elder, C.E. Lunte, A.F.M.N. Rahaman, J.R. Kirchoff, H.D. Dewald, W.R. Heineman, 'In Situ Observation of Copper Redox in a Polymer Modified Electrode Using EXAFS Spectroelectrochemistry', *J. Electroanal. Chem.*, **240**, 361–364 (1988).
178. J. McBreen, I.-C. Lin, 'X-ray Studies of PEO–$ZnBr_2$ Complexes', *J. Electrochem. Soc.*, **139**, 960–966 (1992).
179. J. McBreen, X.Q. Yang, H.S. Lee, Y. Okamoto, 'X-ray Absorption Studies of Mixed Salt Polymer Electrolytes', *J. Electrochem. Soc.*, **142**, 348–354 (1995).
180. J. McBreen, X.Q. Yang, H.S. Lee, Y. Okamoto, 'EXAFS Studies of Mixed Salt $CoBr_2$/LiBr–PEO Complexes', *Electrochim. Acta*, **40**, 2115–2118 (1995).
181. M. Aziz, R.J. Latham, R.G. Linford, W.S. Schlindwein, 'X-ray Spectroscopy of Polymer Electrolytes', *Electrochim. Acta*, **40**, 2119–2122 (1995).
182. M.J. Albarelli, J.H. White, M.G. Bommarito, M. McMillan, H.D. Abruña, 'In Situ Surface EXAFS of Chemically Modified Electrodes', *J. Electroanal. Chem.*, **248**, 77–86 (1988).
183. G. Tourillon, E. Dartyge, H. Dexpert, A. Fontaine, A. Jucha, P. Lagarde, D.E. Sayers, 'Electrochemical Inclusion of Metallic Clusters in Organic Conducting

Polymers: An In Situ Dispersive X-ray Absorption Study', *J. Electroanal. Chem.*, **178**, 357–366 (1984).

184. G. Tourillon, E. Dartyge, H. Dexpert, A. Fontaine, A. Jucha, P. Lagarde, D.E. Sayers, 'Electrochemical Inclusion of Metallic Clusters in Organic Conducting Polymers: An In Situ Dispersive X-ray Absorption Study', *Surf. Sci.*, **156**, 536–547 (1985).

185. G. Tourillon, H. Dexpert, P. Lagarde, 'Charge and Size Displacement Effects on the Conductivity of Poly-3-methylthiophene: An EXAFS Study', *J. Electrochem. Soc.*, **134**, 327–331 (1987).

Group 1 (I)	Group 2 (II)	Group 3	Group 4	Group 5	Group 6	Group 7	Group 8	Group 9
1 HYDROGEN **H** **1.008** 1								
3 LITHIUM **Li** **6.941** 2.1	4 BERYLLIUM **Be** **9.012** 2.2							
11 SODIUM **Na** **22.99** 2.8.1	12 MAGNESIUM **Mg** **24.31** 2.8.2							
19 POTASSIUM **K** **39.10** 2.8.8.1	20 CALCIUM **Ca** **40.08** 2.8.8.2	21 SCANDIUM **Sc** **44.96** 2.8.9.2	22 TITANIUM **Ti** **47.87** 2.8.10.2	23 VANADIUM **V** **50.94** 2.8.11.2	24 CHROMIUM **Cr** **52.00** 2.8.13.1	25 MANGANESE **Mn** **54.94** 2.8.13.2	26 IRON **Fe** **55.85** 2.8.14.2	27 COBALT **Co** **58.93** 2.8.15.2
37 RUBIDIUM **Rb** **85.47** 2.8.18.8.1	38 STRONTIUM **Sr** **87.62** 2.8.18.8.2	39 YTTRIUM **Y** **88.91** 2.8.18.9.2	40 ZIRCONIUM **Zr** **91.22** 2.8.18.10.2	41 NIOBIUM **Nb** **92.91** 2.8.18.12.1	42 MOLYBDENUM **Mo** **95.94** 2.8.18.13.1	43 TECHNETIUM **Tc** **[98.91]** 2.8.18.13.2	44 RUTHENIUM **Ru** **101.1** 2.8.18.15.1	45 RHODIU **Rh** **102.9** 2.8.18.16.
55 CAESIUM **Cs** **132.9** 2.8.18.18.8.1	56 BARIUM **Ba** **137.3** 2.8.18.18.8.2	57−71 LANTHANIDES	72 HAFNIUM **Hf** **178.5** 2.8.18.32.10.2	73 TANTALUM **Ta** **180.9** 2.8.18.32.11.2	74 TUNGSTEN **W** **183.8** 2.8.18.32.12.2	75 RHENIUM **Re** **186.2** 2.8.18.32.13.2	76 OSMIUM **Os** **190.2** 2.8.18.32.14.2	77 IRIDIU **Ir** **192.2** 2.8.18.32.15
87 FRANCIUM **Fr** **[223.0]** 2.8.18.32.18.8.1	88 RADIUM **Ra** **[226.0]** 2.8.18.32.18.8.2	89−103 ACTINIDES	104 RUTHERFORDIUM **Rf** **[261.1]**	105 DUBNIUM **Db** **[262.1]**	106 SEABORGIUM **Sg** **[263.1]**	107 BOHRIUM **Bh** **[264.1]**	108 HASSIUM **Hs** **[265.1]**	109 MEITNERI **Mt** **[268]**

LANTHANIDES

57 LANTHANUM **La** **138.9** 2.8.18.18.9.2	58 CERIUM **Ce** **140.1** 2.8.18.19.9.2	59 PRASEODYMIUM **Pr** **140.9** 2.8.18.21.8.2	60 NEODYMIUM **Nd** **144.2** 2.8.18.22.8.2	61 PROMETHIUM **Pm** **[146.9]** 2.8.18.23.8.2	62 SAMARIU **Sm** **150.4** 2.8.18.24.8

ACTINIDES

89 ACTINIUM **Ac** **[227.0]** 2.8.18.32.18.9.2	90 THORIUM **Th** **232.0** 2.8.18.32.18.10.2	91 PROTACTINIUM **Pa** **231.0** 2.8.18.32.20.9.2	92 URANIUM **U** **238.0** 2.8.18.32.21.9.2	93 NEPTUNIUM **Np** **[237.0]** 2.8.18.32.22.9.2	94 PLUTONIU **Pu** **[239.1]** 2.8.18.32.24

Key

ATOMIC NUMBER
NAME
SYMBOL
ATOMIC WEIGHT
ELECTRONIC CONFIGURATION